5th International Conference on New Frontiers in Physics 2016

EPJ Web of Conferences Volume 164 (2017)

Crete, Greece
6 - 14 July 2016

Part 1 of 2

Editors:

Larissa Bravina
Yiota Foka
Sonja Kabana

ISBN: 978-1-5108-5347-8

Printed from e-media with permission by:

Curran Associates, Inc.
57 Morehouse Lane
Red Hook, NY 12571

Some format issues inherent in the e-media version may also appear in this print version.

Printed by Curran Associates, Inc. (2018)

For additional information, please contact EDP Sciences – Web of Conferences at the address below.

EDP Sciences – Web of Conferences
17, Avenue du Hoggar
Parc d'Activité de Courtabœuf
BP 112
F-91944 Les Ulis Cedex A
France

Phone: +33 (0) 1 69 18 75 75
Fax: +33 (0) 1 69 28 84 91

contact-edps@webofconferences.org

Additional copies of this publication are available from:

Curran Associates, Inc.
57 Morehouse Lane
Red Hook, NY 12571 USA
Phone: 845-758-0400
Fax: 845-758-2633
Email: curran@proceedings.com
Web: www.proceedings.com

TABLE OF CONTENTS

PART 1

PART 2

5th International Conference on New Frontiers in Physics
6 – 14 July 2016, Kolymbari, Crete, Greece

Conference webpage: https://indico.cern.ch/event/icnfp2016/

International Advisory Committee

Halina Abramowicz (Tel Aviv University, Israel)
Charalampos Anastasiou (ETH Zurich, Switzerland)
Ignatios Antoniades (Ecole Polytechnique, Paris, France)
Marco Genovese (INRIM, Torino, Italy)
Rohini Godbole (Indian Institute of Science, Bangalore, India)
Elias Kiritsis (APC and University of Crete, Greece)
Shahn Majid (Queen Mary, University of London, UK)
Angela V. Olinto (University of Chicago, USA)
Valery Rubakov (INR, Moscow, Russia)
Valerio Scarani (National University of Singapore, Singapore)
Eduard Shuryak (Stony Brook University, USA)
Emmanuel Tsesmelis (CERN, Switzerland and University of Oxford, UK)
Urs Wiedemann (CERN, Switzerland)

Local Organising Committee

Laszlo Pal Csernai (University of Bergen, Norway)
Roar Emaus (University of Oslo, Norway)
Maxime Gouzevich (University of Lyon, France)
Nikolay Gulitskiy (Saint Petersburg State University, Russia)
Oleksii Ivanytskyi (National Academy of Sciences of Ukraine, Kyiv, Ukraine)
Thanos Kalamaris (Institute for Nuclear and Particle Physics "Demokritos", Athens, Greece)
Nikos Kallithrakas (Technical University of Crete, Chania, Greece)
Elias Kiritsis (UoC, Heraklion, Greece and U. Paris 7, Paris, France)
Vladimir Kovalenko (Saint Petersburg State University, Russia)
Christine Kourkoumelis (University of Athens, Athens, Greece)
Emanuela Larentzakis (OAC, Kolymbari, Greece)
Anton Makarov (Saint Petersburg State University, Russia)
Lisa Paspalaki (Institute for Nuclear and Particle Physics "Demokritos", Athens,Greece)
Violetta Sagun (Bogolyubov Institute for Theoretical Physics, Kyiv, Ukraine)
Maria Stefaniak (WUT, Warsaw, Poland)
Nan Su (FIAS, University of Frankfurt, Germany)
Giorgos Vasileiadis (University of Athens, Greece)

Conference Chairs

Larissa Bravina (University of Oslo, Norway)
Yiota Foka (GSI, Germany)
Sonia Kabana (University of Nantes and Subatech, France)

Organising Institutes and Sponsors

CERN - European Organization for Nuclear Research
EMMI - ExtreMe Matter Institute
GSI - Helmholtzzentrum fur Schwerionenforschung
FAIR - Facility for Antiproton and Ion Research
University of Oslo
Subatech
University of Nantes
Warsaw University of Technology
University of Crete
Academia Europaea
University of Bergen
University of Lyon
University of Frankfurt

5th International Conference on New Frontiers in Physics

New Frontiers in Physics
ICNFP 2016

6 - 14 July 2016, Kolymbari, Crete, Greece

http://indico.cern.ch/event/icnfp2016

International Advisory Committee

Halina Abramowicz (Tel Aviv University, Israel)
Charalampos Anastasiou (ETH Zurich, Switzerland)
Ignatios Antoniadis (Ecole Polytechnique, Paris, France)
Marco Genovese (INRIM, Torino, Italy)
Rohini Godbole (Indian Institute of Science, Bangalore, India)
Elias Kiritsis (APC and University of Crete, Greece)
Shahn Majid (Queen Mary, University of London, UK)
Angela V. Olinto (University of Chicago, USA)
Valery Rubakov (INR, Moscow, Russia)
Valerio Scarani (National University of Singapore, Singapore)
Eduard Shuryak (Stony Brook University, USA)
Emmanuel Tsesmelis (CERN, Switzerland and University of Oxford, UK)
Urs Wiedemann (CERN, Switzerland)

Organising Committee

Maria Boura (University of Athens, Greece)
Laszlo Csernai (University of Bergen, Norway)
Roar Emaus (University of Oslo)
Bengt Henrik Brusheim Johansson (University of Oslo)
Thanos Kalamaris (Institut for Nuclear and Particle Physics "Demokritos", Athens, Greece)
Nikos Kalithrakas (TU, Chania, Greece)
Elias Kiritsis (UoC, Heraklion, Greece and U. Paris 7, Paris, France)
Vladimir Kovalenko (Saint Petersburg State University, Russia)
Emanuela Larentzakis (OAC, Kolymbari, Greece)
Daniela Radulescu, conference secretary (University of Frankfurt, Germany)
Lisa Paspalaki (Institut for Nuclear and Particle Physics "Demokritos", Athens, Greece)
Maria Stefaniak (WUT, Warsaw, Poland)
Giorgos Vasileiadis (University of Athens, Greece)

Conference Chairs

Larissa Bravina (UiO, Oslo, Norway)
Yiota Foka (GSI, Darmstadt, Germany)
Sonia Kabana (Univ. of Nantes & Subatech, France)

Main topics of the Conference

High Energy Particle Physics
Heavy Ion Physics
Superconductivity, Critical Phenomena
Quantum Physics, Quantum Entanglement
Quantum Field Theory, String theory
Cosmology, Astrophysics, Gravity
Mathematical Physics

Image by ancient.eu

Saint Petersburg State University www.spbu.ru
Technical University of Crete
HELLENIC REPUBLIC National and Kapodistrian University of Athens
GOETHE UNIVERSITÄT FRANKFURT AM MAIN

DOI: 10.1051/epjconf/201716400001

Preface

5th International Conference on New Frontiers in Physics 2016

The conference series "New Frontiers in Physics" aims to promote scientific exchange and development of novel ideas in science with particular emphasis on interdisciplinarity. The main topics of the conference are high energy particle and nuclear physics, heavy ion physics, quantum optics and quantum entanglement, cosmology, astrophysics, gravity, mathematical physics.

The 5th International Conference on New Frontiers in Physics (ICNFP 2016) attracted more than 200 participants. At the opening of the Conference Dr Konstantinos Zormpas, the General Director of the Academy which hosted the conference, welcomed the participants. During the conference 7 lectures, 85 plenary and 91 parallel talks, 3 mini-workshops with 24 talks, two public talks (T. Tomaras and V. Mukhanov), altogether, 209 oral presentations and 20 posters, were given by 233 participants. A QCD session dedicated to V. Gribov's confinement scenario and G. Altarelli Memorial Session have been organized. Also, special session took place on 11th July 2016.

A science exhibition and outreach day has been organized in the Sailing Club of Chania on 8th July 2016 by Dr. Marge Bardeen and Prof. Ch. Kourkoumeli with music and a public talk on “Gravitational waves: the confirmation which came 100 years after their prediction” by Prof. Th. Tomaras in Greek in the evening. In addition, there was a Public talk in English entitled “Gravitational waves, Cosmic Primordial Radiation and the Origin of the Universe” by Vyacheslav Mukhanov, an Opera Gala (Kalliopi Petrou (soprano), Alessia Toffanin (piano)), two classical concerts by Ruben Muradyan (piano), Svetlana (Violin), Vladimir(Cello) and Mikhail Nor (Tenor), as well as excursions and a conference dinner.

We would like to thank all the conference participants as well as the members of the International Advisory Committee, the Local Organizing Committee, the Organizing Institutes and Sponsors, the Technical University of Chania, the University of Crete at Herakleion, Christina Kourkoumeli from University of Athens, the "Go-lab European Outreach project" and the Chania Sailing Club for the organization of public events and outreach, and especially the hosting Orthodox Academy of Crete for the extremely warm atmosphere of the conference. We would also like to thank Vladimir Kovalenko from Saint Petersburg State University and the staff of “European Physical Journal Web of Conferences” for the help with publishing the proceedings.

Larissa Bravina, Yiota Foka and Sonia Kabana,

Editors

DOI: 10.1051/epjconf/201716400001

edp sciences EPJ Web of Conferences

Statement of Peer review

In submitting conference proceedings to EPJ Web of Conferences, I certify to the Publisher that I adhere to the **Policy on Publishing Integrity** of the journal in order to safeguard good scientific practice in publishing.

1. All articles have been subjected to peer review administered by the proceedings editors.

2. Reviews have been conducted by expert referees, who have been requested to provide unbiased and constructive comments aimed, whenever possible, at improving the work.

3. Proceedings editors have taken all reasonable steps to ensure the quality of the materials they publish and their decision to accept or reject a paper for publication has been based only on the merits of the work and the relevance to the journal.

Title, date and place of the conference

International Conference on New Frontiers in Physics, 6-14 July 2016
OAC, Kolymbari, Crete, Greece

Proceedings editor(s):

Larissa Bravina, Yiota Foka, Sonja Kabana

Date and editor's signature

November 10th, 2017

EPJ Web of Conferences **164**, 01001 (2017) DOI: 10.1051/epjconf/201716401001
ICNFP 2016

A framework for unified Dirac gauginos

Karim Benakli[1,a]

[1] *Sorbonne Universités, UPMC Univ Paris 06, UMR 7589, LPTHE, F-75005, Paris, France*
CNRS, UMR 7589, LPTHE, F-75005, Paris, France

Abstract. We identify the Minimal Dirac Gaugino Supersymmetric Standard Model (MDGSSM) as the minimal field content with Dirac gauginos allowing unification of gauge coupling. We stress that its parameter space describes also other most popular models as the MSSM, NMSSM and MRSSM. We discuss the generation of trilinear couplings in models of gauge mediation that has been overlooked in the past. We study the different source of Higgs mixings and constraints from the ρ parameter. Finally, we provide new experimental limits on the masses of the scalar octets.

1 Introduction

The possibility of extending the MSSM to allow Dirac gaugino masses has been subject to many studies (see for example [1–23]. Multiple realisations have been proposed that differ by the fate of R-symmetry, the presence or absence of additional states and interactions (for a short introduction see for example [14]). Here we argue that the field content of the MDGSSM is the minimal content of extra states to describe al these models that automatically preserve unification of gauge couplings while allowing the new couplings to the Higgs to enhance naturalness and allow the boundary conditions to be unified at a high energy scale.

The additional states compare to the MSSM are:

- A singlet, a triplet and an octet supermultiplets **S**, **T** and **O**, respectively, introduced to give the gauginos Dirac masses.
- There are extra vector-like charged states, subsequently called "fake leptons" [16] as they carry the same quantum numbers as the Standard Model leptons. They were introduced in order to restore the automatic gauge coupling unification that was spoiled by the addition of the adjoint representations of the Standard Model gauge group.

One of the important novelty of the Dirac gaugino models with respect to the NMSSM is its mixing with the Standard Model Higgs. This mixing is present at tree level due to the supersymmetric operator describing the Dirac gaugino bino mass leads to a modification of the $U(1)_Y$ D-term as

$$D_1 = D_Y^{(0)} \to D_1 = -2m_{1D}S_R + D_Y^{(0)} \qquad \text{with} \qquad D_Y^{(0)} = -g' \sum_j Y_j \varphi_j^* \varphi_j \tag{1.1}$$

[a] e-mail: kbenakli@lpthe.jussieu.fr

where S_R is the real part of S and φ_j a scalar field with charge Y_j under $U(1)_Y$. Upon elimination of the auxilliary fields, this implies an interaction of the form:

$$g' m_{1D} S_R (|H_u^0|^2 - |H_d^0|^2). \tag{1.2}$$

thereof a tree-level induced mixing. However, as this can typically be compensated by the presence in the superpotential of a term of the form:

$$W \supset \lambda_S S H_u H_d. \tag{1.3}$$

A precise evaluation of this mixing at the tree and one-loop level needs to be carried out carefully.

The parameter space is constrained by the requirements of stability of the vacuum avoiding existence of directions in the phase space of the model taking the fields expectation values to charge- and colour- breaking vacua. This is important due to trilinear terms, some have not been explicitly discussed in the existing literature while they are expected to be generically present in the model. This is the case for example of soft terms mixing three adjoint scalars that we will show that they are generated in models of gauge mediation. It is also constrained if one needs to keep couplings small enough if we want to preserve perturbativity up to the GUT scale. This restriction can be of course relaxed if one allows for Landau poles below the GUT scale. However, as one of the virtues of the MSSM was to predict perturbative unification of gauge couplings, and was one of the motivations for introducing the MDGSSM, we shall place emphasis on regions of the parameter space which respect this condition.

2 The MDGSSM field content

The MDGSSM field content can be seen as the minimal set providing the MSSM gauginos a Dirac mass while preserving two-loop unification and perturbativity of gauge couplings. We summarised it in Table 1. In addition to the chiral multiplets transforming under the adjoint representations of the gauge groups, it includes new fields charged under the lepton number global symmetry. They consist of extra Higgs-like doublets [1] $\mathbf{R}_u, \mathbf{R}_d$ as well as two pairs of vector-like right-handed electron superfields $\mathbf{E'}_{1,2}$ in $(\mathbf{1},\mathbf{1})_1$ and $\tilde{\mathbf{E}}'_{1,2}$ in $(\mathbf{1},\mathbf{1})_{-1}$. Such states are compatible with an $(SU(3))^3$ Grand Unification gauge group. This is the minimal set which enables a "natural" unification (unification without mass thresholds tuning) similar to the MSSM. The adjoint chiral multiplets contain new complex adjoint scalars, S, T and O:

$$\begin{aligned}
S &= \frac{S_R + iS_I}{\sqrt{2}} \\
T &= \frac{1}{2\sqrt{2}} \begin{pmatrix} T_R + iT_I & \sqrt{2}(T_{+R} + iT_{+I}) \\ \sqrt{2}(T_{-R} + iT_{-I}) & -(T_R + iT_I) \end{pmatrix} \\
O^{(a)} &= \frac{O_R^{(a)} + iO_I^{(a)}}{\sqrt{2}}
\end{aligned} \tag{2.1}$$

where the $S_R, O_R^{(a)}, T_R, T_{-R}, T_{+R}$ and the $S_I, O_I^{(a)}, T_I, T_{-I}, T_{+I}$ are real scalars and pseudo-scalars, respectively.

[1]The hypercharges are opposite with respect to the Higgs doublet in the MSSM to match the MRSSM notation for the same fields.

Names		Spin 0	Spin 1/2	Spin 1	$(SU(3), SU(2), U(1)_Y)$	R-charge
Quarks	$\mathbf{Q}$	$\tilde{Q} = (\tilde{u}_L, \tilde{d}_L)$	(u_L, d_L)		$(\mathbf{3}, \mathbf{2}, 1/6)$	1
	$\mathbf{U^c}$	$\tilde{U}^c_L$	U^c_L		$(\bar{\mathbf{3}}, \mathbf{1}, -2/3)$	0
(×3 families)	$\mathbf{D^c}$	$\tilde{D}^c_L$	D^c_L		$(\bar{\mathbf{3}}, \mathbf{1}, 1/3)$	0
Leptons	$\mathbf{L}$	$(\tilde{\nu}_{eL}, \tilde{e}_L)$	(ν_{eL}, e_L)		$(\mathbf{1}, \mathbf{2}, -1/2)$	1
(×3 families)	$\mathbf{E^c}$	$\tilde{E}^c$	E^c		$(\mathbf{1}, \mathbf{1}, 1)$	0
Higgs	$\mathbf{H_u}$	(H_u^+, H_u^0)	$(\tilde{H}_u^+, \tilde{H}_u^0)$		$(\mathbf{1}, \mathbf{2}, 1/2)$	1
	$\mathbf{H_d}$	(H_d^0, H_d^-)	$(\tilde{H}_d^0, \tilde{H}_d^-)$		$(\mathbf{1}, \mathbf{2}, -1/2)$	1
Gluons	$\mathbf{W}_{3\alpha}$		$\lambda_{3\alpha}$ $[\equiv \tilde{g}_\alpha]$	g	$(\mathbf{8}, \mathbf{1}, 0)$	1
W	$\mathbf{W}_{2\alpha}$		$\lambda_{2\alpha}$ $[\equiv \tilde{W}^\pm, \tilde{W}^0]$	$W^\pm, W^0$	$(\mathbf{1}, \mathbf{3}, 0)$	1
B	$\mathbf{W}_{1\alpha}$		$\lambda_{1\alpha}$ $[\equiv \tilde{B}]$	B	$(\mathbf{1}, \mathbf{1}, 0)$	1
DG-octet	$\mathbf{O}$	O	χ_g $[\equiv \tilde{g}']$		$(\mathbf{8}, \mathbf{1}, 0)$	0
DG-triplet	$\mathbf{T}$	$\{T^0, T^\pm\}$	$\{\chi_T^0, \chi_T^\pm\}$ $[\equiv \{\tilde{W}'^\pm, \tilde{W}'^0\}]$		$(\mathbf{1}, \mathbf{3}, 0)$	0
DG-singlet	$\mathbf{S}$	S	χ_S $[\equiv \tilde{B}']$		$(\mathbf{1}, \mathbf{1}, 0)$	0
Higgs-like	$\mathbf{R_u}$	R_u	$\tilde{R}_u$		$(\mathbf{1}, \mathbf{2}, -1/2)$	1
leptons	$\mathbf{R_d}$	R_d	$\tilde{R}_d$		$(\mathbf{1}, \mathbf{2}, 1/2)$	1
Fake	$\hat{\mathbf{E}}(\times 2)$	$\hat{E}$	$\tilde{\hat{E}}$		$(\mathbf{1}, \mathbf{1}, 1)$	0
electrons	$\hat{\mathbf{E}}'(\times 2)$	$\hat{E}'$	$\tilde{\hat{E}}'$		$(\mathbf{1}, \mathbf{1}, -1)$	2

Table 1: Chiral and gauge multiplet fields in the model.

The superpotential for these fields can be written as

$$W = W_{Yukawa} + W_{DG} + W_{RV} \tag{2.2}$$

where W_{Yukawa} contains the usual MSSM Yukawas part

$$W_{Yukawa} = Y_u^{ij} \mathbf{U^c}_i \mathbf{Q}_j \mathbf{H}_u - Y_d^{ij} \mathbf{D^c}_i \mathbf{Q}_j \mathbf{H}_d - Y_e^{ij} \mathbf{E^c}_i \mathbf{L}_j \mathbf{H_d} \tag{2.3}$$

W_{DG} contains the R-symmetric (according to the choice of R-charges in table 1) contributions of the non-MSSM fields

$$\begin{aligned} W_{DG} =&(\mu + \lambda_S \mathbf{S})\mathbf{H_d H_u} + \sqrt{2}\lambda_T \mathbf{H_d T H_u} \\ &(\mu_R + \lambda_{SR}\mathbf{S})\mathbf{R_u R_d} + 2\lambda_{TR}\mathbf{R_u T R_d} \\ &+ (\mu_{\hat{E}\,ij} + \lambda_{S\hat{E}^c\,ij}\mathbf{S})\hat{\mathbf{E}}_i\hat{\mathbf{E}}'_j + \lambda_{S\,Eij}\mathbf{S}\mathbf{E}^c{}_i\hat{\mathbf{E}}'_j \\ &+ \lambda_{S\,LRi}\mathbf{S}\mathbf{L}_i\mathbf{R}_d + 2\lambda_{TLRi}\mathbf{L}_i\mathbf{T}\mathbf{R_d} - Y_{\hat{E}i}\mathbf{R_u H_d}\hat{\mathbf{E}}_i \\ &- Y^{ij}_{LFV}\mathbf{L_i}\cdot\mathbf{H_d}\hat{\mathbf{E}}_j - Y^j_{EFV}\mathbf{R_u H_d}\mathbf{E}^c{}_j\,, \end{aligned} \tag{2.4}$$

while W_{RV} gathers the R-symmetry violating terms

$$\begin{aligned} W_{RV} =& L\mathbf{S} + \frac{\hat{M}_1}{2}\mathbf{S}^2 + \frac{\kappa}{3}\mathbf{S}^3 + \hat{M}_2\mathrm{tr}(\mathbf{TT}) + \hat{M}_3\mathrm{tr}(\mathbf{OO}) \\ &+ \lambda_{ST}\mathbf{S}\mathrm{tr}(\mathbf{TT}) + \lambda_{SO}\mathbf{S}\mathrm{tr}(\mathbf{OO}) + \frac{\kappa_O}{3}\mathrm{tr}(\mathbf{OOO}) - Y_{\hat{E}'i}\mathbf{R_d H_u}\hat{\mathbf{E}}'_i \\ \underset{\mathrm{R-symmetry}}{\longrightarrow}&\ 0 \end{aligned} \tag{2.5}$$

For the soft SUSY-breaking terms, from the MSSM we retain only the bilinear terms – i.e. conventional mass-squared terms and the B_μ term. All the scalar trilinear and Majorana gaugino mass terms violate R-symmetry; while for B_μ we suppose that, since R-symmetry is a chiral symmetry, we are breaking R-symmetry in the Higgs sector – and in fact it is only in combination with the superpotential terms $\mu, \lambda_S, \lambda_T$ that the R-symmetry is violated. Hence in principle we can have an entirely R-preserving supersymmetry-breaking sector.

The soft SUSY breaking terms beyond those of the MSSM consist of[2]:

- Dirac gaugino masses:

$$W_{\mathrm{supersoft}} = \int d^2\theta\,\sqrt{2}\theta^\alpha\Big[m_{D1}\mathbf{S}W_{Y\alpha} + 2m_{D2}\mathrm{tr}(\mathbf{T}W_{2\alpha}) + 2m_{D3}\mathrm{tr}(\mathbf{O}W_{3\alpha})\Big]. \tag{2.6}$$

- soft terms associated with the adjoint scalars

$$\begin{aligned} -\Delta\mathcal{L}^{\mathrm{scalar\ soft}}_{\mathrm{adjoints}} =& m_S^2|S|^2 + \frac{1}{2}B_S(S^2 + h.c.) + 2m_T^2\mathrm{tr}(T^\dagger T) + (B_T\mathrm{tr}(TT) + h.c.) \\ &+ 2m_O^2\mathrm{tr}(O^\dagger O) + (B_O\mathrm{tr}(OO) + h.c.) \\ &+ [T_S S H_u\cdot H_d + 2T_T H_d\cdot T H_u + \frac{1}{3}\kappa A_\kappa S^3 + t_S S + h.c.] \\ &+ [T_{SO}S\,\mathrm{tr}(O^2) + T_{ST}S\,\mathrm{tr}(T^2) + \frac{1}{3}T_O\mathrm{tr}(O^3) + h.c.] \end{aligned} \tag{2.7}$$

  The terms on the last line have generally been neglected, but will play an important role in this work.

- soft terms involving the new vector-like leptons:

$$\begin{aligned} -\Delta\mathcal{L}^{\mathrm{scalar\ soft}}_{\mathrm{vector-like}} =& m^2_{R_u}|R_u|^2 + m^2_{R_d}|R_d|^2 + [B_R R_d R_u + h.c.] \\ &+ \hat{E}_i(m^2_{\hat{E}})^i_j\hat{E}^j + \hat{E}'^i(m^2_{\hat{E}'})^j_i\hat{E}'_j + [B^{ij}_{\hat{E}}\hat{E}_i\hat{E}'_j + h.c.] \\ &+ [T^{ij}_{SE}S\,\hat{E}_i\hat{E}'_j + T_{SR}S\,R_dR_u + h.c.]\,. \end{aligned} \tag{2.8}$$

[2]We suppress gauge indices while retaining generation indices and denote the complex conjugation of fields by upper versus lower indices.

 DOI: 10.1051/epjconf/201716401001

Let us highlight that in an R-symmetry conserving model, one cannot simultaneously have the trilinears T_{SE} (respectively T_{SR}) from (2.8) and the superpotential couplings λ_{SE} (respectively λ_{SR}) from (2.4) as each term requires a different R-charge for the fields $\hat{E}$ and $\hat{E}'$ (respectively R_u and R_d) to be R-invariant.

We consider now the scalar mass matrix. We use the notation

$$\begin{aligned} \tilde{m}_S^2 &= \tilde{m}_{SR}^2 + \lambda_S^2 \frac{v^2}{2} \\ \tilde{m}_T^2 &= \tilde{m}_{TR}^2 + \lambda_T^2 \frac{v^2}{2}\,, \end{aligned} \tag{2.9}$$

where the effective masses for the real parts of S and T read:

$$\tilde{m}_{SR}^2 = m_S^2 + 4m_{1D}^2 + B_S\,, \qquad \tilde{m}_{TR}^2 = m_T^2 + 4m_{2D}^2 + B_T\,. \tag{2.10}$$

Then, at tree level the scalar mass matrix in the basis $\{h, H, S_R, T_R^0\}$ is [15]:

$$\begin{pmatrix} M_Z^2 + \Delta_h s_{2\beta}^2 & \Delta_h s_{2\beta} c_{2\beta} & \Delta_{hS} & \Delta_{hT} \\ \Delta_h s_{2\beta} c_{2\beta} & M_A^2 - \Delta_h s_{2\beta}^2 & \Delta_{HS} & \Delta_{HT} \\ \Delta_{hS} & \Delta_{HS} & \tilde{m}_S^2 & \lambda_S \lambda_T \frac{v^2}{2} \\ \Delta_{hT} & \Delta_{HT} & \lambda_S \lambda_T \frac{v^2}{2} & \tilde{m}_T^2 \end{pmatrix} \tag{2.11}$$

where we have defined:

$$\Delta_h = \frac{v^2}{2}(\lambda_S^2 + \lambda_T^2) - M_Z^2 \tag{2.12}$$

which vanishes when λ_S and λ_T take their $N = 2$ values,

$$\Delta_{hS} = -2\frac{v_S}{v}\tilde{m}_{SR}^2, \qquad \Delta_{hT} = -2\frac{v_T}{v}\tilde{m}_{TR}^2 \tag{2.13}$$

and

$$\Delta_{HS} = g' m_{1D} v s_{2\beta}, \qquad \Delta_{HT} = -g m_{2D} v s_{2\beta}\,. \tag{2.14}$$

This matrix is diagonalised by the mixing matrix S_{ij}. Of particular interest will be S_{11} which measures if the lightest scalar eigenstate is Standard Model Higgs like, and S_{13} which measures the proportion of the scalar singlet S_R in this lightest eigenstate.

3 Higgs mixing

The singlet S is that it will in general mix with the Higgs eigenstates. This will lead to tree-level decays of S into tops, W, Z or Higgs which are experimentally constrained. The minimisation condition of v_S on the off-diagonal element Δ_{hS} of the scalar mass matrix given in (2.13) leads to (see [13])

$$\begin{aligned} \Delta_{hS} &= v[v_S \lambda_S^2 - g' m_{1D} c_{2\beta} + \sqrt{2}\lambda_S \mu + \lambda_S \lambda_T v_T] \\ &= v[\sqrt{2}\lambda_S \tilde{\mu} - g' m_{1D} c_{2\beta}]\,, \end{aligned} \tag{3.1}$$

where we used the effective mass parameter

$$\tilde{\mu} = \mu + \frac{1}{\sqrt{2}}(\lambda_S\, v_S + \lambda_T\, v_T)\,. \tag{3.2}$$

From this basic analytical calculation, we see that we can minimise the tree-level mixing by choosing:

$$\lambda_S \sim \frac{g_Y m_{1D} c_{2\beta}}{\sqrt{2}\tilde{\mu}} . \tag{3.3}$$

In general, this relation will be modified at one-loop, but the property that one value of λ_S is favoured will remain.

Also, the v.e.v. of T gives a contribution to the W boson mass and the electroweak precision data give bounds on it. The induced correction to the ρ-parameter:

$$\rho \equiv \frac{M_W^2}{c_{\theta_W}^2 M_Z^2} = 1 + \Delta\rho , \tag{3.4}$$

with $\Delta\rho$ is given analytically at tree-level by ([13])

$$\Delta\rho \sim \frac{4v_T^2}{v^2} , \tag{3.5}$$

where v is the usual Standard Model Higgs v.e.v.. In order to be below the experimental constraints, we need $\Delta\rho \lesssim (4.2 \pm 2.7) \times 10^{-4}$, ([13] – see also [24, 25] –). At tree level, we have

$$v_T \simeq \frac{v^2}{2\tilde{m}_{TR}^2} \left[-g m_{2D} c_{2\beta} - \sqrt{2}\tilde{\mu}\lambda_T \right] , \tag{3.6}$$

with $\tilde{m}_{TR}^2 = m_T^2 + 4m_{2D}^2 + B_T$, therefore, small $\Delta\rho$ require large triplet Dirac and soft masses. This requirement can often be at odd with naturalness which prefers smaller triplet masses. Indeed, radiative corrections induced by the adjoint triplet scalars to $m_{H_{u,d}}^2$ are [13]:

$$\delta m_{H_{u,d}}^2 \supset -\frac{1}{16\pi^2}(2\lambda_T^2 m_T^2)\log\left\{\frac{\Lambda}{\text{TeV}}\right\} , \tag{3.7}$$

with Λ the UV cut-off, $m_{H_{u,d}}^2, m_T^2$ the squared masses for Higgses and scalar triplet T, and λ_T the coupling defined in (2.4). For Λ at the Planck scale, requiring a fine-tuning $\Delta_T = \delta m_H^2/m_H^2$ better than 10% finally gives us

$$m_T \lesssim \frac{1}{\lambda_T} 450\ \text{GeV} . \tag{3.8}$$

4 Trilinear couplings

Previous studies of Dirac gaugino models have generally neglected the phenomenology of adjoint self-coupling terms, with an exception being a superpotential term $\frac{\kappa}{3}S^3$ used in [13] to generate μ/B_μ as in the NMSSM, and a recent brief discussion in [26]. In the case of superpotential terms such as λ_{SO} these can be neglected when considering an R-symmetric visible sector; however, trilinear soft couplings such as T_{SO}, T_O (see (2.7)) are always allowed. It is therefore interesting to consider what values we expect from models of supersymmetry-breaking mediation.

Starting with a spurion analysis where supersymmetry is broken by either a D-term D or F-term F, then if the mediating dynamics is at a scale M the terms in our effective Lagrangian should be given by powers of $\frac{D}{M}, \frac{F}{M}, \frac{D}{M^2}, \frac{F}{M^2}$ with appropriate factors of couplings and $\kappa_l \equiv 1/16\pi^2$. Furthermore, quartic and higher-order couplings – which are "hard" SUSY-breaking parameters – are always generated, but do not lead to quadratic divergences because they appear suppressed by powers of the scale M which is the cutoff of our effective theory. Important in this work are the quartics such as $\mathcal{L} \supset \frac{\lambda_{4S}}{24} S^4$ which must have size $\lambda_{4S} \sim \kappa_l^p \left(\frac{D}{M^2}\right)^q$ for some integer p, q (or similarly for F-terms with *even* q)); taking $p = 1, q = 1$ for a D-term we naively have a quadratic divergence in the scalar mass proportional to λ_{4S} but this yields $\Delta m_S^2 \sim \kappa_l \lambda_{4S} M^2 \sim \kappa_l^2 D \ll M^2$, while for $q = 2$ we have $\kappa_l^2 \frac{D^2}{M^2}$. In fact, this tells us that the case $q = 1$ is special because it implies a much larger correction at one loop than the direct mass, and could therefore destabilise the calculation. We shall return to this below.

As a first observation, if the mediation is by gravity, then M should be identified with the Planck scale (unless there is significant sequestering) and we should only consider the leading order terms. We would therefore require the quantum gravity theory to give us the terms T_{SO}, T_O at leading order $D/M, F/M$ and the quartics must, by the above reasoning, be negligible.

On the other hand, in the case of low-scale supersymmetry breaking – where it was argued in [21] that this requires Dirac gauginos – $\sqrt{F} \sim \sqrt{D} \sim M \sim$ TeV, and we generate all terms at a similar order, which would include T_{SO}, T_O. However, the phenomenology is significantly changed by the presence of higher-dimensional operators and the goldstino couplings [27] and, since it is difficult to reconcile with perturbative unification, we shall not discuss this further here.

Finally, for gauge mediation M could be as small as $\sqrt{F}$ or $\sqrt{D}$ but there is no a priori upper limit on M until we choose a particular quantum gravity embedding. The Dirac gaugino masses are expected to be generated at one loop and be of order $\kappa_l \frac{D}{M}$ or $\kappa_l \frac{F^2}{M^3}$. For F-term breaking the standard gauge-mediation soft mass-squareds for the squarks/sleptons are of order $\kappa_l^2 \frac{F^2}{M^2}$, while in D-term breaking they may be suppressed. Therefore if we imagine that $\kappa_l \frac{D}{M} \sim$ TeV, then for terms $\kappa_l \frac{D^2}{M^3}$ to be significant we would need $D \sim M^2$ and furthermore $M \sim 100$ TeV.

4.1 Adjoint couplings in gauge mediation

One of the most interesting issues in the construction of gauge mediation models with Dirac gaugino masses has been that of the adjoint scalar masses: in the simplest realisation, only a B-type mass-squared $\mathcal{L} \supset -\frac{1}{2} B_\Sigma \Sigma^2$ is generated at leading order in D/M^2, and not a conventional mass-squared $\mathcal{L} \supset -m_\Sigma^2 |\Sigma|^2$. This happens for one pair of vector-like messengers $Q, \tilde{Q}$ having charges under a hidden $U(1)$ of $+1, -1$, where the $U(1)$ obtains a D-term. This was noticed from the earliest models [3, 5] with the original proposed solution being to add a supersymmetric mass for the adjoint – which would also violate the R-symmetry and generate Majorana masses for the gauginos, with a see-saw effect. However, an alternative solution was found to be to introduce additional messenger states with non-diagonal couplings to either the adjoints (in the D-term case) [10, 12] or an F-term spurion [8, 10, 12]; in the D-term case this requires the couplings to violate the $U(1)$-charges. In [12] examples were given where the ratio of B-type to conventional masses is arbitrary. The general ansatz was to couple the adjoint to messenger fields $Q_i, \tilde{Q}_j$ and to possible F-term spurions X via superpotential couplings

$$W \supset M Q_i \tilde{Q}_i + \lambda_{i\bar{j}} Q_i \Sigma \tilde{Q}_j + \mu_{i\bar{j}} X Q_i \tilde{Q}_j \tag{4.1}$$

and D-terms via charges $e_i, \tilde{e}_i$ which we can write as a matrix $e_{i\bar{j}}(Q_i Q_j^* - \tilde{Q}_i \tilde{Q}_j^*)$.

More in the spirit of the earlier works, the issue was rephrased in the language of effective operators in [26, 28, 29], where it was claimed that the explanation for the absence of conventional mass-squared terms for the adjoints at leading order is that the operator responsible for the generation of a leading-order mass-squared term should be

$$\mathcal{L} \supset \int d^4\theta \frac{1}{M^2} [\psi^\dagger e^{qV} \psi + \tilde{\psi}^\dagger e^{-qV} \tilde{\psi}] \Sigma^\dagger \Sigma, \tag{4.2}$$

where $\psi, \tilde{\psi}$ are a pair of fields charged under the hidden $U(1)$ with charges $\pm q$ which obtain v.e.v.s (and thus generate a contribution to the hidden D-term). The above operator is generated by including terms in the superpotential that mix the messengers $Q, \tilde{Q}$ with other pairs of fields $N, \tilde{N}$ which are neutral (or at least have different charges) under the hidden $U(1)$, so that the v.e.v.s of $\psi, \tilde{\psi}$ generate messenger mixing terms. This is clearly nearly equivalent to the above ansatz, and can be written in the form

$$W \supset M_{ij} Q_i \tilde{Q}_j + \lambda_i Q_i \Sigma \tilde{Q}_i \tag{4.3}$$

where we now write the mass terms as violating the $U(1)$ charges instead.

If we start with the case of no couplings/mass mixing terms that violate the $U(1)$ D-term charges, we shall first give a simple proof that the conventional mass term $|\Sigma|^2$ vanishes at leading order for any number of messengers, and then look at higher-order terms. Considering first the visible gauge group to be $U(1)$, we have the effective potential contribution from the messenger scalars (since the fermion potential is independent of D):

$$\begin{aligned} V = &\int \frac{d^d q}{(2\pi)^d} \mathrm{tr} \log(q^2 + \mathcal{M}_Q^2 + De) + \mathrm{tr} \log(q^2 + \mathcal{M}_{\tilde{Q}}^2 - De) \\ \equiv &V_+ + V_- . \end{aligned} \tag{4.4}$$

Here we have $\mathcal{M}_Q^2 = (M+\lambda\Sigma)(M^\dagger+\lambda^\dagger\overline{\Sigma})$, $\mathcal{M}_{\tilde{Q}}^2 = (M^\dagger+\lambda^\dagger\overline{\Sigma})(M+\lambda\Sigma)(M^\dagger+\lambda^\dagger\overline{\Sigma})$ are the supersymmetric mass-squared matrices. Now, if we take the couplings to preserve the $U(1)$ charges then we can write the

$$\begin{aligned} V_+ =& De \int \frac{d^d q}{(2\pi)^d} \mathrm{tr}\left(\frac{1}{q^2 + \mathcal{M}_Q^2}\right) - \frac{1}{2} D^2 e^2 \int \frac{d^d q}{(2\pi)^d} \mathrm{tr}\left(\frac{1}{q^2 + \mathcal{M}_Q^2}\right)^2 + O(D^3) \\ \to 16\pi^2 V =& D^2 e^2 \mathrm{tr}\left(\log \mathcal{M}_Q^2/\mu^2\right) + O(D^4) \end{aligned} \tag{4.5}$$

since the eigenvalues of $\mathcal{M}_Q^2$ and $\mathcal{M}_{\tilde{Q}}^2$ are equal. Next, by taking the derivative with respect to Σ we find only a holomorphic function of Σ:

$$\begin{aligned} 16\pi^2 \frac{\partial V}{\partial \Sigma} =& D^2 e^2 \mathrm{tr}\left([M + \lambda\Sigma]^{-1}\lambda\right) + O(D^4) \\ \to V =& \frac{D^2 e^2}{16\pi^2}\left[\mathrm{tr}\left(\log MM^\dagger/\mu^2\right) + \tilde{V}(\Sigma) + \overline{\tilde{V}}(\overline{\Sigma})\right] + O(D^4). \end{aligned} \tag{4.6}$$

As an example, consider the simple model of a single messenger where the matrices become numbers; then we have

$$\tilde{V}(\Sigma) = -\sum_{n=1}^{\infty} \frac{1}{n}\left(\frac{-\lambda\Sigma}{M}\right)^n . \tag{4.7}$$

This potential manifestly has trilinear and quartic couplings, although at order $\frac{D^2}{M^3}, \frac{D^2}{M^4}$ respectively. Indeed, if we continue with the ansatz (4.1) then it is easy to see that there are no terms of linear order in D, because $\mathcal{M}_Q^2 = (M + \lambda\Sigma)(M + \lambda^\dagger\Sigma^\dagger) = \mathcal{M}_{\tilde{Q}}^2$ and

$$V = D \int \frac{d^d q}{(2\pi)^d} \mathrm{tr}\left\{\left([q^2 + \mathcal{M}_Q^2]^{-1} - [q^2 + \mathcal{M}_{\tilde{Q}}^2]^{-1}\right)e\right\} + O(D^2). \tag{4.8}$$

Hence to have large cubic interactions we should start from ansatz (4.3). In this way, in order to have an interesting phenomenology we require either $D \sim M^2$ with both at a low scale, or we require (as proposed in [28]) that

$$B_\Sigma < m_\Sigma^2 \sim aD + b\frac{D^2}{M^2} \tag{4.9}$$

with some cancellation between the two terms so that we can have $m_\Sigma \sim T_{SO}$. Note that once we take this ansatz with non-zero mixing between the messengers and $[M, e] \neq 0$ we typically generate trilinear terms in the potential – but also tadpoles. The issue of tadpoles is then easily circumvented by embedding the coupling of the singlet adjoint S to the $SU(3)$ and $SU(2)$ adjoints into the generator $T^Y = \frac{1}{\sqrt{60}}\mathrm{diag}(2, 2, 2, -3, -3)$. This then also means that the couplings of the singlet adjoint S are related to those of T and O; for example, for T_{SO}, if we have calculated the coupling for $U(1)$ messengers as being $\mathcal{L} \supset \frac{1}{6}T_\Sigma \Sigma^3$, then we have

$$\begin{aligned} T_{SO}\delta^{ab} =& T_\Sigma \mathrm{tr}(T^Y T_3^a T_3^b) \\ =& \frac{1}{\sqrt{15}} T_\Sigma \delta^{ab} \end{aligned} \tag{4.10}$$

where T_3^a, T_3^b are $SU(3)$ generators. However, exploring sets of messengers which give these desired properties with sufficiently large trilinear couplings and exploring the vacuum stability of the total system would be very interesting, but is beyond the scope of this work.

5 Octets production

Consider the case when either the scalar or pseudoscalar colour octets are lighter than a TeV. Even though such light scalars should be copiously produced in pairs at both 8 and 13 TeV, as shown in figure 1, their decays are loop suppressed and this inhibits single production.

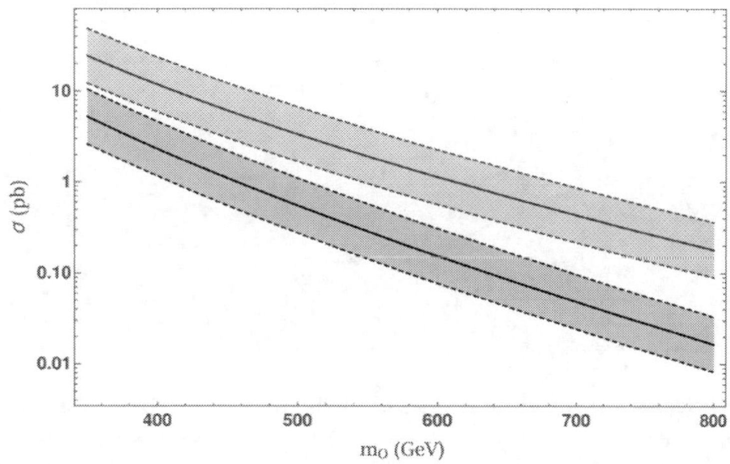

Figure 1: Pair production cross-section of octets at tree-level, at 8 TeV (blue, lower curves) and 13 TeV (red, upper curves). The bands indicate a variation of a factor of 2 each way relative to the values obtained in `MadGraph`.

Since current limits place all squarks above about 800 GeV, then, as discussed in [30], the octets decay only to gluons and quarks – in particular almost entirely top quarks. This means that the possible signatures are four jets, dijet/ditop searches, and four tops. Up until relatively recently the constraints on them were rather weak, with dijets providing no constraint, and a mild constraint from ditops [31]. However, now the four top channel is particularly important: [32] placed a limit of 32 fb at 8 TeV, and [33] found 370 fb for Standard-Model-like kinematics, or 140 fb with and EFT pointlike interaction, at 13 TeV.

We consider the cross-section times branching ratio approach, taking the most conservative values of twice the tree-level cross-section (i.e. a K-factor of 2) and a limit at 13 TeV of 140 fb. To compute the branching ratio into four tops, we require the widths into gluons and tops. Expressions given in [30] used *complex* octets, which is not appropriate for our case where the necessarily large gluino mass causes a large splitting. Instead we use the expressions presented in [27].

The first important observation is that the pseudoscalar octet *does not couple* to gluons, and so pair production of pseudoscalar octets yields only four-top events, and by our above criteria excludes pseudoscalars below about 880 GeV *by the* 13 *TeV data*. These are therefore less interesting for our analysis.

On the other hand, the scalar octet couples to squarks via its D-term coupling, and so couples to gluons. Since it couples to *all* coloured squarks, this can potentially be large. However, to be very conservative, we show production times branching ratio of four-tops via scalar octets in figure 2 at 8 and 13 TeV with the limits shown using a K-factor of 2, as we vary the octet mass and for three different values of the Dirac gluino mass, where the first two generations of squarks are decoupled (i.e. heavy and degenerate). To produce these, we take left-handed stops and sbottoms of 1200 GeV, right-handed stops of 800 GeV, and decoupled right-handed sbottoms (at 4 TeV). We neglect all squark mixing (which is a good approximation in this model). Since the couplings involve a cancellation between left- and right-handed squarks, this is very conservative: if we took heavier left-handed squarks, we would enhance the gluon rate relative to the top rate (because it has a contribution from sbottoms as well as stops) weakening the bounds.

We conclude that for 2.5 GeV gluinos, the octet scalars must be heavier than 500 GeV; but for 3 TeV gluinos there is no constraint.

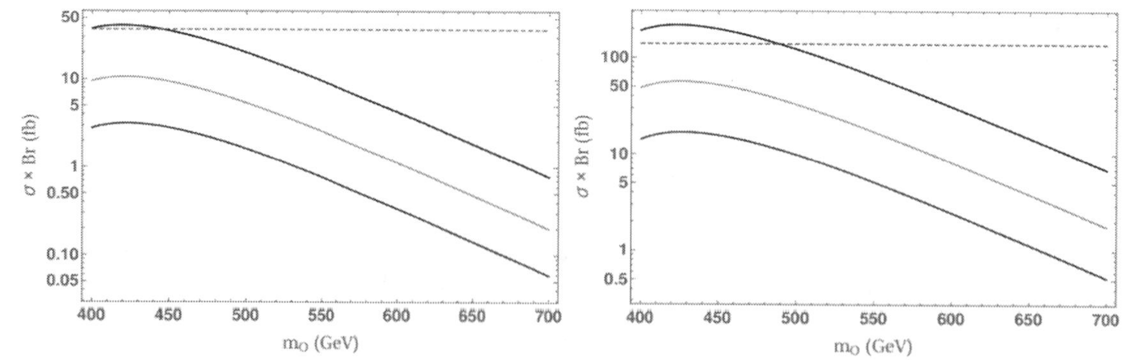

Figure 2: Four-top production times branching ratio from scalar colour octets as a function of the octet mass, for gluino masses of 2.5 TeV (upper curve, blue), 3 TeV (middle curve, orange) and 3.5 TeV (lower curve, purple). The experimental limit is shown as the dashed red horizontal line. The left plot is computed for $\sqrt{s} = 8$ TeV, and right is for $\sqrt{s} = 13$ TeV.

Acknowledgments

This work is supported by the European Research Council (ERC) under the Advanced Grant Higgs@LHC (ERC-2012-ADG20120216-321133), the support of the Agence Nationale de Recherche under grant ANR-15-CE31-0002 "HiggsAutomator", the Labex "Institut Lagrange de Paris" (ANR-11-IDEX-0004-02, ANR-10-LABX-63).

References

[1] P. Fayet, *Phys.Lett.* **B78** 417 (1978)
[2] L. J. Hall and L. Randall, Nucl. Phys. **B352** 289 (1991)
[3] P. J. Fox, A. E. Nelson, and N. Weiner, JHEP **0208** 035 (2002)
[4] A. E. Nelson, N. Rius, V. Sanz, and M. Unsal, JHEP **08** 039 (2002)
[5] I. Antoniadis, K. Benakli, A. Delgado, and M. Quiros, Adv. Stud. Theor. Phys. **2** 645 (2008)
[6] I. Antoniadis, K. Benakli, A. Delgado, M. Quiros and M. Tuckmantel, Nucl. Phys. B **744** 156 (2006)
[7] G. D. Kribs, E. Poppitz, and N. Weiner, Phys. Rev. **D78** 055010 (2008)
[8] S. D. L. Amigo, A. E. Blechman, P. J. Fox, and E. Poppitz, JHEP **01** 018 (2009)
[9] G. Belanger, K. Benakli, M. Goodsell, C. Moura, and A. Pukhov, JCAP **0908** 027 (2009)
[10] K. Benakli and M. D. Goodsell, Nucl. Phys. **B816** 185 (2009)
[11] K. Benakli and M. D. Goodsell, Nucl. Phys. **B830** 315 (2010)
[12] K. Benakli and M. D. Goodsell, Nucl. Phys. **B840** 1 (2010)
[13] K. Benakli, M. D. Goodsell, and A.-K. Maier, Nucl. Phys. **B851** 445 (2011)
[14] K. Benakli, Fortsch.Phys. **59** 1079 (2011)
[15] K. Benakli, M. D. Goodsell, and F. Staub, JHEP **06** 073 (2013)
[16] K. Benakli, L. Darmé, M. D. Goodsell, and P. Slavich, JHEP **05** 113 (2014)
[17] K. Benakli, M. Goodsell, F. Staub, and W. Porod, Phys. Rev. **D90** no. 4, 045017 (2014)
[18] E. Dudas, M. Goodsell, L. Heurtier, and P. Tziveloglou, Nucl. Phys. **B884** 632 (2014)
[19] K. Benakli, L. Darmé and M. D. Goodsell, JHEP **1511** 100 (2015)
[20] S. Abel and M. Goodsell, JHEP **06** 064 (2011)
[21] T. Gherghetta and A. Pomarol, JHEP **12** 069 (2011)
[22] S. P. Martin, Phys. Rev. **D92** no. 3, 035004 (2015)
[23] R. Davies, J. March-Russell, and M. McCullough, JHEP **1104** 108 (2011)
[24] E. Bertuzzo, C. Frugiuele, T. Gregoire, and E. Ponton, JHEP **04** 089 (2015)
[25] H. Beauchesne and T. Gregoire, JHEP **05** 051 (2014)
[26] L. M. Carpenter and J. Goodman, JHEP **07** 107 (2015)
[27] M. D. Goodsell and P. Tziveloglou, Nucl. Phys. **B889** 650 (2014)
[28] C. Csaki, J. Goodman, R. Pavesi, and Y. Shirman, Phys. Rev. **D89** no. 5, 055005 (2014)
[29] L. M. Carpenter, JHEP **09** 102 (2012)
[30] S. Y. Choi, M. Drees, J. Kalinowski, J. M. Kim, E. Popenda, and P. M. Zerwas, Phys. Lett. **B672** 246 (2009)
[31] **CMS** Collaboration, S. Chatrchyan *et al.*, Phys. Rev. Lett. **111** no. 21, 211804 (2013), [Erratum: Phys. Rev. Lett.**112**, no.11, 119903 (2014)].
[32] **CMS** Collaboration, V. Khachatryan *et al.*, JHEP **11** 154 (2014)
[33] T. A. collaboration, "Search for production of vector-like top quark pairs and of four top quarks in the lepton-plus-jets final state in pp collisions at $\sqrt{s} = 13$ TeV with the ATLAS detector,".

EPJ Web of Conferences **164**, 01002 (2017) DOI: 10.1051/epjconf/201716401002
ICNFP 2016

A Model for Soft Interactions based on the CGC/Saturation Approach and the BFKL Pomeron

Errol Gotsman[1,a]

[1] *School of Physics and Astronomy, Tel Aviv University, Ramat Aviv, Israel*

Abstract. We propose a consistent approach to describe soft interactions at LHC and higher energies based on the BFKL Pomeron and CGC/saturation.We assume that the same Pomeron describes both "soft" and "hard" interactions. Having eight parameters we obtain a good description of the elastic and inclusive proton-proton data at energies W ≤ 0.546 TeV.

1 Introduction

Strong interactions at high energy are considered a typical example of processes that occur at long distances, where the unknown mechanism which confines quarks and gluons, plays a confusing role in undermining all efforts to treat these process theoretically. Consequently, the description of these soft processes are usually made using high energy phenomenology, based on Pomeron-Reggeon calculus [1].

To ameliorate this difficulty, the Tel Aviv group (Gotsman, Levin and Maor) has suggested and formalized an approach, which attempts to describe soft interactions by adopting well established aspects of QCD in the saturation region [2], and extending these to the region of soft interactions (low Q^2). Our basic assumption, is that the BFKL Pomeron [3], at high energies, describes both hard and soft scattering.

The processes that our model has been successfully applied are proton-proton scattering over the energy range 546 GeV ≤ W ≤ 57 TeV: include total, elastic and diffractive cross sections as well as the forward slope [4]: inclusive production [5]:long range correlations [6]: and the survival probability of central exclusive production [7].

To illustrate our approach, I will limit myself and only discuss our results, and compare these to the relevent experimental data for the proton-proton total, elastic and differential cross sections, as well as inclusive production.

2 Background

1. In the Regge limit of pQCD, when $s \gg \Lambda_{hard}$, as the energy increases the parton density becomes more dense, and the scattering amplitude A(s,t) grows.

2. As long as densities are not too high, growth is described by BFKL evolution equation.

[a] e-mail: gotsman@post.tau.ac.il

3. Density becomes higher as A(s,t) $\rightarrow$ 1, and one enters a regime called "saturation", where the BFKL evolution fails.

4. Non linearities lead to "saturation" + "unitarization" of A(s,t).

5. Balitsky-Kovchegov equation [8] is the simplest and most accurate way to describe the saturation regime of QCD. It is non-linear and resums QCD fan diagrams in the LLA.

3 Phenomenological Input

One of the most well established principles of high energy scattering is that the resulting cross section should not violate the Froissart bound (F.B.).

There is a very simple criterion due to Heisenberg (quoted by Kovner and Wiedemann [9]) relating the impact parameter dependence of the profile of the matter density in any target, and the violation of the F.B.

(i) If the matter density decays exponentially at the periphery i.e. $\rho(b) \propto e^{-mb}$, then for the production of the lightest particle of mass m, ($E \sim \frac{s}{m}$), the target energy density is $E\rho(b)$. Scattering can only take place for $b \leq b_{max}$. Since

$$Ee^{-mb_{max}} = m \Rightarrow b_{max} \leq \frac{1}{m} ln(\frac{s}{m^2}) \Rightarrow \sigma_{tot} \leq b^2_{max} \leq \frac{c}{m^2} ln^2(\frac{s}{m^2})$$

this satisfies the F.B.

(ii) Conversely, if the density distribution in the target is power-like, i.e. $\rho(b) \propto b^{-\lambda}$. Hence,

$$Ee^{mb_{max}} = m \Rightarrow b^{-\lambda}_{max} \sim (\frac{s}{m^2})^{1/2} \Rightarrow \sigma_{tot} \leq b^2_{max} \sim (\frac{s}{m^2})^{\frac{2}{\lambda}}.$$

i.e. $\sigma_{tot} \propto$ a power of s, which leads to a violation of the F.B.

In building our model, we need to overcome a deficiency i.e. the fact that the BFKL Pomeron does not lead to shrinkage of the diffractive peak, and has no slope for the Pomeron trajectory. This can be cured by introducing a non-perturbative correction at large impact parameter, which also assures satisfying the Froissart-Martin bound for σ_{tot}.

Consequently, in our model we fix the large b behaviour by assuming that the saturation momentum Q_s^2, has the following form:

$Q_s^2(b, Y) = Q_{0s}^2(b, Y_0)e^{\lambda(Y-Y_0)}$ and $Q_{0s}^2(b, Y_0) = (m^2)^{(1-\frac{1}{\bar{\gamma}})}[S(b, m)]^{\frac{1}{\bar{\gamma}}}$; $S(b, m) = \frac{m^2}{2\pi}e^{-mb}$ and $\bar{\gamma} = 0.63 = 1 - \gamma_{cr}$.

The parameter $\lambda = \bar{\alpha}_S\chi(\gamma_{cr})/(1 - \gamma_{cr})$, in leading order of perturbative QCD ($\lambda = 0.2$ to 0.3).

The parameter m which is introduced to describe the large b behaviour, and determines the typical sizes of dipoles inside the hadron.

Our model also includes two additional scales m_1 and m_2, which describe two typical sizes in the proton wave function. One can associated these with: (i) the distance between the constituent quarks; size of the proton $R_p \approx \frac{1}{m_1}$. and (ii) m_2 can be associated with the size of the constituent quark; $R_q \approx \frac{1}{m_2}$.

Altinoluk et al [10] have proved the equivalence of the CGC/saturation approach and the BFKL Pomeron calculus for a wide range of rapidities , $Y \leq \frac{2}{\Delta_{\rm BFKL}} \ln\left(\frac{1}{\Delta^2_{\rm BFKL}}\right)$.

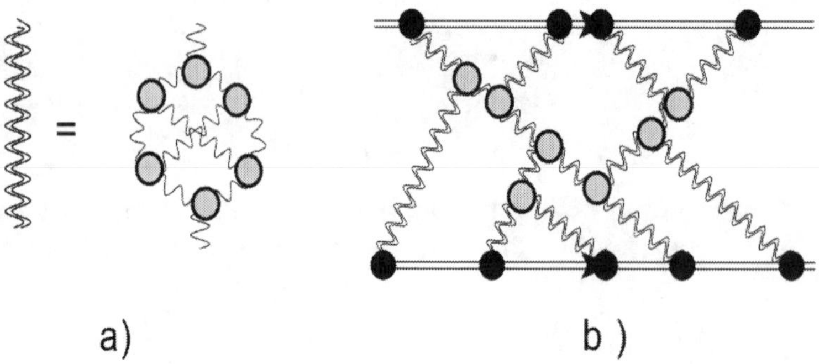

Figure 1. a) Dressed Pomeron in MPSI approximation and b) Sum of net diagrams. Wavy lines describe BFKL Pomerons. The grey blobs stand for triple Pomeron vertices, while black blobs show the hadron-Pomeron vertex g(b).

3.1 Dressed Pomeron in MPSI approximation

For a description of the process see Fig.1.

Since the typical rapidity is $O(Y - Y_i) \approx \frac{1}{\Delta_{BFKL}}$, only large Pomeron loops with rapidity $O(Y)$ contribute at high energies → can sum such loops using MPSI approximation.

For the BFKL Pomeron $\lambda = 4.88\bar{\alpha}_s$ while $\Delta_{BFKL} = 4ln2\bar{\alpha}_s \approx 0.2$.

The resulting Green function of the Dressed Pomeron is given by:

$$G_{P}^{\rm dressed}(Y - Y_0, r, R, b) =$$
$$a^2\left\{1 - \exp(-T(Y - Y_0, r, R, b))\right\} + 2a(1-a)\frac{T(Y - Y_0, r, R, b)}{1 + T(Y - Y_0, r, R, b)}$$
$$+ (1-a)^2\left\{1 - \exp\left(\frac{1}{T(Y - Y_0, r, R, b)}\right)\frac{1}{T(Y - Y_0, r, R, b)}\Gamma\left(0, \frac{1}{T(Y - Y_0, r, R, b)}\right)\right\}$$

The BFKL amplitude in the vicinity of the saturation scale is denoted by

$$T(Y - Y_0, r, R, b) = \frac{\bar{\alpha}_S^2}{4\pi} G_P(z \to 0) = \phi_0\left(r^2 Q_s^2(R, Y, b)\right)^{1-\gamma_{cr}} = \phi_0 S(b)\, e^{\lambda(1-\gamma_{cr})Y}$$

$G_P(Y, r, R)$ denotes the BFKL Pomeron's Green function.

z = $ln(r^2 Q_s^2(b, Y))$, a = 0.65, the critical anamolous dimension is given by $\gamma_{cr} \approx 0.37$.

3.2 Parameters of the Model

We used two inputs: $r = R$ and $Q_s^2 = \left(1/(m^2R^2)\right) S(b) \exp(\lambda Y)$.

The phenomenological profile function $S(b) = \frac{m^2}{2\pi} e^{-mb}$ with normalization $\int d^2b\, S(b) = 1$. We need to introduce four constants: g_i and m_i ($i = 1, 2$), to describe the vertices of the hadron-Pomeron interaction

$$g_i(b) = g_i S_P(b) \text{ with } S_P(b) = \frac{m_i^3 b}{4\pi} K_1(m_i b)\,;\; S_P(b) \xrightarrow{\text{Fourier image}} \left(\frac{m_i^2}{q^2 + m_i^2}\right)^2$$

The opacity is given by:

$$\Omega_{i,k}(Y;b) = \int d^2b' \frac{g_i(\vec{b}')\, g_k(\vec{b}-\vec{b}')\, \bar{G}_{\boldsymbol{P}}^{\text{dressed}}(Y)}{1 + 1.29\, \bar{G}_{\boldsymbol{P}}^{\text{dressed}}(Y)\left[g_i(\vec{b}') + g_k(\vec{b}-\vec{b}')\right]},$$

where $\bar{G}_{\boldsymbol{P}}^{\text{dressed}}(Y) = \int d^2b''\, G_{\boldsymbol{P}}^{\text{dressed}}(Y;b'')$.
The factor 1.29 originates from estimates of the $3\boldsymbol{P}$ vertex in the CGC/saturation approach.

3.3 Basic formalism for the two channel model

Following Good-Walker [11] the observed physical hadronic and diffractive states are written

$$\psi_h = \alpha\Psi_1 + \beta\Psi_2; \qquad \psi_D = -\beta\Psi_1 + \alpha\Psi_2; \qquad \text{where} \qquad \alpha^2 + \beta^2 = 1.$$

Functions ψ_1 and ψ_2 form a complete set of orthogonal functions $\{\psi_i\}$ which diagonalize the interaction matrix $\mathbf{T}$

$$A_{i,k}^{i'k'} = <\psi_i\,\psi_k|\mathbf{T}|\psi_{i'}\,\psi_{k'}> = A_{i,k}\,\delta_{i,i'}\,\delta_{k,k'}.$$

The unitarity constraints can be written as

$$2\,\text{Im}\,A_{i,k}(s,b) = |A_{i,k}(s,b)|^2 + G_{i,k}^{in}(s,b)$$

At high energies a simple solution to this equation is

$$A_{i,k}(s,b) = i\left(1 - \exp\left(-\frac{\Omega_{i,k}(s,b)}{2}\right)\right)$$

$$G_{i,k}^{in}(s,b) = 1 - \exp\left(-\Omega_{i,k}(s,b)\right).$$

$G_{i,k}^{in}(s,b)$ denotes the contribution of all non-diffractive inelastic processes.

3.4 Physical observables for elastic and low mass diffraction

elastic amplitude :	$a_{el}(s,) = i\left(\alpha^4 A_{1,1} + 2\alpha^2\beta^2 A_{1,2} + \beta^4 A_{2,2}\right);$
elastic observables :	$\sigma_{tot} = 2\int d^2b\, a_{el}(s,b); \quad \sigma_{el} = \int d^2b\, |a_{el}(s,b)|^2;$
optical theorem :	$2\,\text{Im}A_{i,k}(s,t=0) = 2\int d^2b\,\text{Im}A_{i,k}(s,b) = \sigma_{el} + \sigma_{in} = \sigma_{tot};$
single diffraction :	$\sigma_{sd}^{GW} = \int d^2b\left(\alpha\beta\{-\alpha^2 A_{1,1} + (\alpha^2-\beta^2)A_{1,2} + \beta^2 A_{2,2}\}\right)^2;$
double diffraction :	$\sigma_{dd}^{GW} = \int d^2b\,\alpha^4\beta^4\{A_{1,1} - 2A_{1,2} + A_{2,2}\}^2.$

'GW' denotes the Good -Walker component, that is responsible for diffraction in the small mass region.
The parameters of our best fit are given in Table 1, while our results and comparison with the relevant experimental data are displayed in Fig.2.

 DOI: 10.1051/epjconf/201716401002

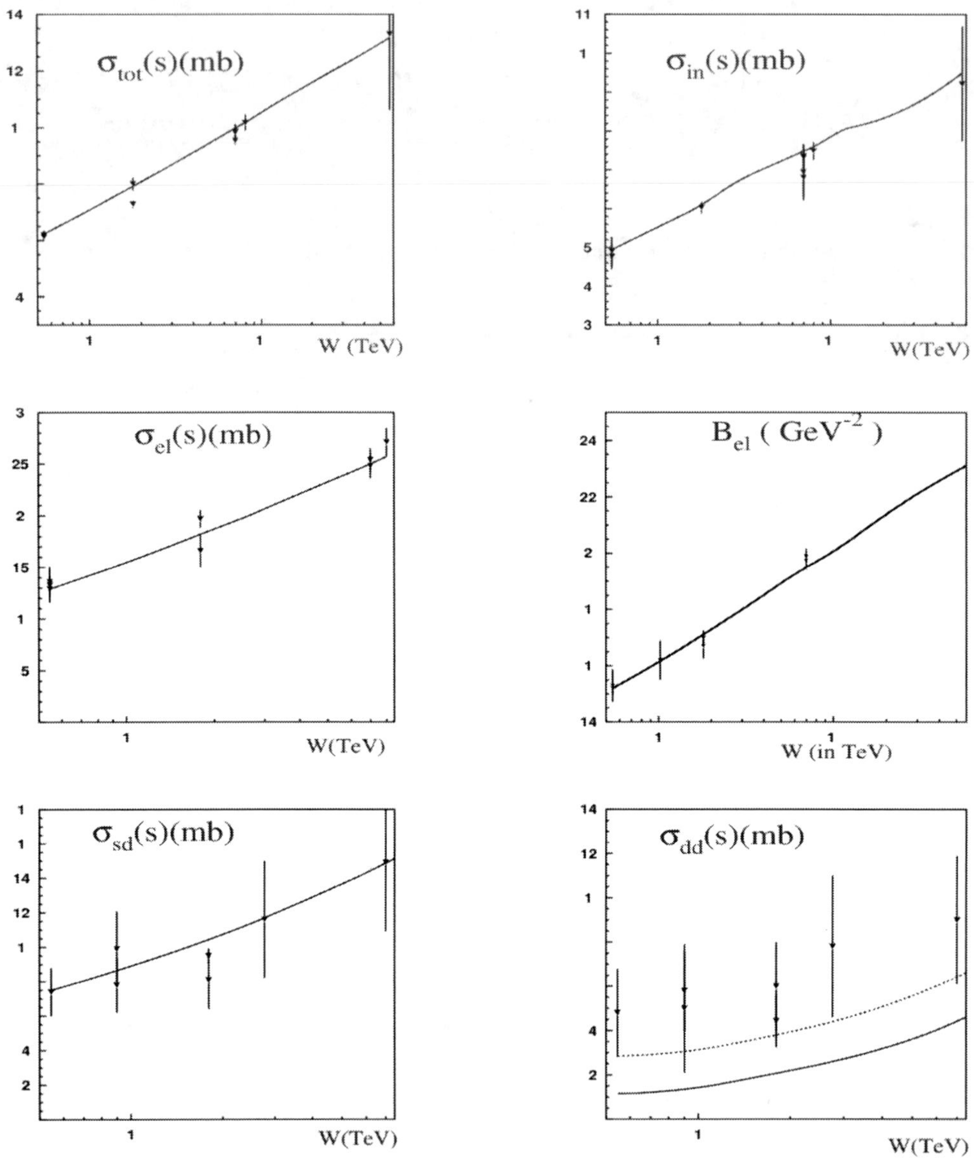

Figure 2. Comparison of model results with experimental data: the energy behaviour of the total (Fig.2-a), inelastic (Fig.2-b), elastic cross section (Fig.2-c), as well as the elastic slope B_{el} (Fig.2-d), and single diffraction (Fig.2-e) and double diffraction (Fig.2-f) cross sections. The solid lines show our fit. The data has been taken from Ref.[12] for energies less than the LHC energy. At the LHC energy for total and elastic cross sections we use [13] data and for the single and double diffraction cross sections are taken from ref.[14]. The dotted line in Fig.2-f) which is obtained assuming factorization is discussed in [4].

Table 1. Table 1. Fitted parameters of the model

model	λ	ϕ_0	g_1 (GeV^{-1})	g_2 (GeV^{-1})	$m(GeV)$	$m_1(GeV)$	$m_2(GeV)$	β
2 channel	0.38	0.0019	110.2	11.2	5.25	0.92	1.9	0.58

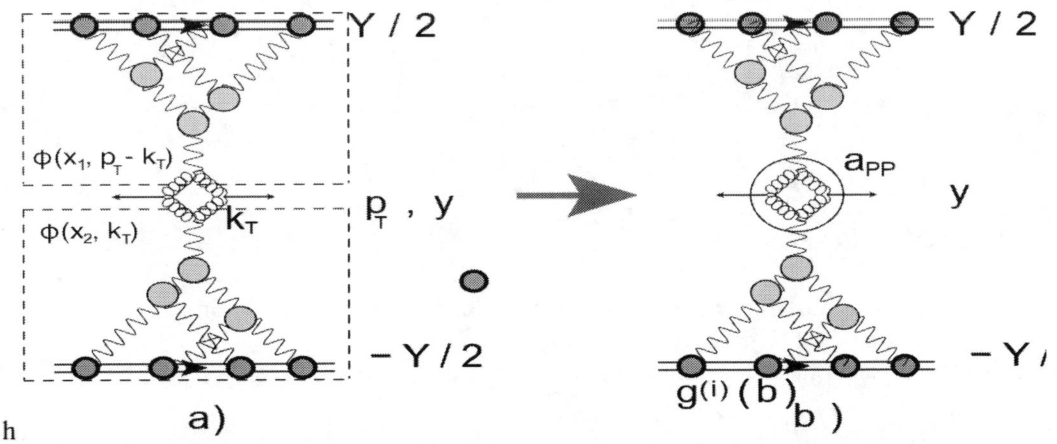

Figure 3. The graphical representation of Eqn. (1) (Fig-3a). Wavy lines denote the BFKL Pomerons, while the helical lines illustrate the gluons. In Fig-3b the Mueller diagram for inclusive production is shown.

4 Application of the CGC/saturation approach to Inclusive Production

See [5] for details.

Inclusive production occurs in two stages:
First stage: Production of a mini-jet with typical transverse momentum Q_s: Q_s (saturation scale)$\gg$ soft scale.
Second stage: Decay of minijets into hadrons, which is treated phenomenologically.

For mini-jet production we use the k_T factorization formula:

$$\frac{d\sigma}{dy\, d^2 p_T} = \frac{2\pi \bar{\alpha}_S}{p_T^2} \int d^2 k_T\ \phi_G^{h_1}\left(x_1; \vec{k}_T\right) \phi_G^{h_2}\left(x_2; \vec{p}_T - \vec{k}_T\right) \tag{1}$$

where $\phi_G^{h_i}$ denotes the probability to find a gluon that carries the fraction x_i of energy with $k_\perp$ transverse momentum. and $\bar{\alpha}_S = \alpha_s N_c/\pi$, with the number of colours equal to N_c.

Where $\frac{1}{2}Y+y = \ln(1/x_1)$ and $\frac{1}{2}Y-y = \ln(1/x_2)$. $\phi_G^{h_i}$ is the solution of the Balitsky-Kovchegov(BK) non-linear evolution equation, and can be viewed as the sum of 'fan' diagrams of the BFKL Pomeron interactions, shown in Fig-3.

For the sake of simplicity all other indices in $\phi\,(x_1, p_T - k_T)$ and $\phi\,(x_2, k_T)$ are omitted.

Eqn.(1) can be rewritten as a Mueller diagram (Fig-3b), and the inclusive cross section is given by:

$$\begin{aligned}\frac{d\sigma}{dy} &= \int d^2p_T \frac{d\sigma}{dy\,d^2p_T} = a_{PP}\ \ln(W/W_0)\left\{\alpha^4\, In^{(1)}\left(\frac{1}{2}Y+y\right) In^{(1)}\left(\frac{1}{2}Y-y\right)\right.\\ &+ \alpha^2\beta^2\left(In^{(1)}\left(\frac{1}{2}Y+y\right) In^{(2)}\left(\frac{1}{2}Y-y\right) + In^{(2)}\left(\frac{1}{2}Y+y\right) In^{(1)}\left(\frac{1}{2}Y-y\right)\right)\\ &+ \left.\beta^4\, In^{(2)}\left(\frac{1}{2}Y+y\right) In^{(2)}\left(\frac{1}{2}Y-y\right)\right\}\end{aligned}$$

$$In^{(i)}(y) = \int d^2b\ N^{BK}\left(g^{(i)}\,S\,(m_i,b)\ \tilde{G}_P\,(y)\right)$$

where $\tilde{G}_P\,(y) = \phi_0 \exp(\lambda\,(1-\gamma_{cr})\,y)$

and N^{BK} is an approximation to the numerical solution of the BK equation.

The mass of mini jet is given by $m^2_{jet} = 2m_{soft}p_T$.

Since the typical transverse momentum is equal to the saturation scale, we have

$$\frac{m^2_{jet}}{p^2_T} = \frac{2\,m_{soft}}{Q_s\,(W)} = r^2_0\left(\frac{W}{W_0}\right)^{-\frac{1}{2}\lambda}$$

Values of parameters have been extracted from the diffractive and elastic data.

The only free parameters are a_{PP} and r^2_0.

Our curves are calculated for $a_{PP} = 0.21$ and $r^2_0 = 8$, which have been determined from the experimental data.

4.1 Our Model Results for Inclusive Production

The results of our fit and the relevent experimental data are displayed in Fig (4) and Fig.(5).

5 Conclusions

Our constructed model based on the BFKL Pomeron and the CGC/saturation approach, successfully describes data in the Regge region, for high energy hadron scattering.

We do not require that the soft Pomeron appear as a Regge pole.

We suggest a procedure where the matching with long distance physics (where confinement of quarks and gluons is essential) can be reached within the CGC/saturation approach.

We have also shown that our formalism for soft (long distance) interactions, successfully describes inclusive production.

References

[1] P.D.B. Collins, *An introduction to Regge theory and high energy physics*(Cambridge University Press, 1977)

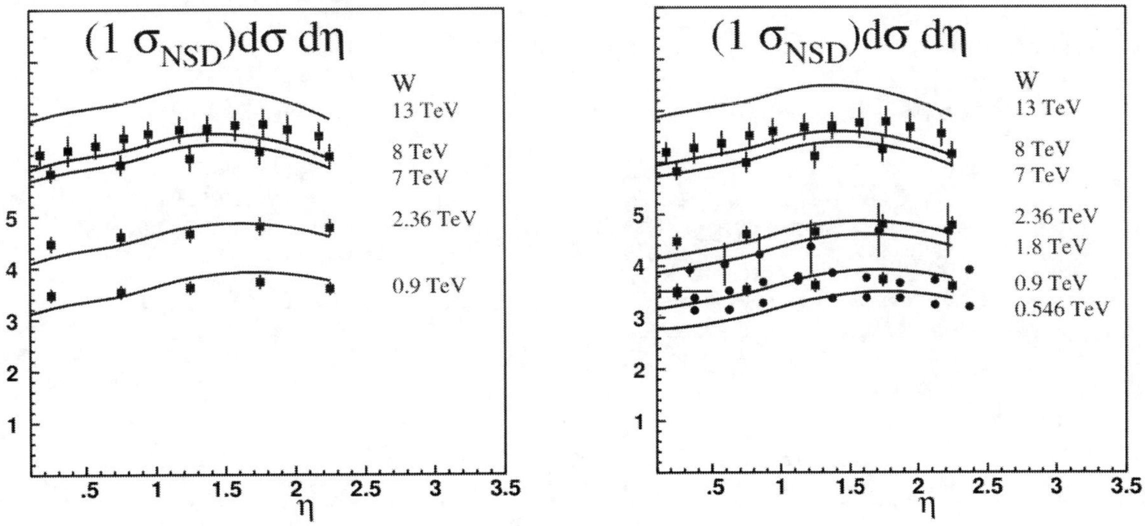

(a) (b)

Figure 4. The single inclusive density $(1/\sigma_{NSD}) d\sigma/d\eta)$ versus energy. The data were taken from ALICE [18],CMS [19], ATLAS [22] and from PDG [17]. The description of the CMS data is plotted in in fig.(2-a) while fig.(2-b) presents the comparison with all inclusive spectra with W ≥ 0.546 TeV.

[2] Yuri V. Kovchegov and Eugene Levin, *Quantum Chromodynamics at High Energies*(Cambridge Monographs on Particle Physics, Nuclear Physics and Cosmology, Cambridge University Press, 2012).
[3] E.A. Kuraev, L. N. Lipatov and F. S. Fadin, Sov.Phys.JETP **45**, 199 (1977); Ya. Ya. Balitsky and L. N. Lipatov, Sov.J.Nucl.Phys. **28**, 22 (1978).
[4] E. Gotsman, E. Levin and U. Maor, Eur.Phys.J. **C75**, 179 (2015).
[5] E. Gotsman, E. Levin and U. Maor, Phys.Lett. **B746**, 154 (2015).
[6] E. Gotsman, E. Levin and U. Maor, Eur.Phys.J. **C75**, 518 (2015).
[7] E. Gotsman, E. Levin and U. Maor, Eur.Phys.J. **C76**, 177 (2016).
[8] I. Balitsky, Phys.Rev. **D60** 014020 (1999); Y.V. Kovchegov, Phys. Rev **D60**, 034008 (1999).
[9] A.Kovner and U. Wiedemann, Phys. Letts. **B551** 311 (2003).
[10] T. Altinoluk et al JHEP, **1404**, 075 (2014).
[11] M.L. Good and W.D. Walker, Phys.Rev. **120**, 1857 (1960).
[12] J. Beringer et al [particle Data Group], Phys. Rev. **D86**, 010001 (2012).
[13] F. Ferro [Totem collaboration] AIP Conf.Proc. **1350**, 172 (2011); G.Antchev et. al. [TOTEM collaboration], Europhys. Lett.**96**, 21002 (2011),Europhys. Lett. **95**, 41001 (2011).Nucl.Phys. **899**, 527 (2015).
[14] M. G. Poghosyan, J.Phys.**G38**, 124044 (2011), [ALICE Collaboration], arXiv:0911.5430[hep-ex].
[15] G. Aad et al. [ATLAS Collaboration], Nature Commun. **2**, 463 (2011); Phys. Lettrs. **B754**, 214 (2016).

DOI: 10.1051/epjconf/201716401002

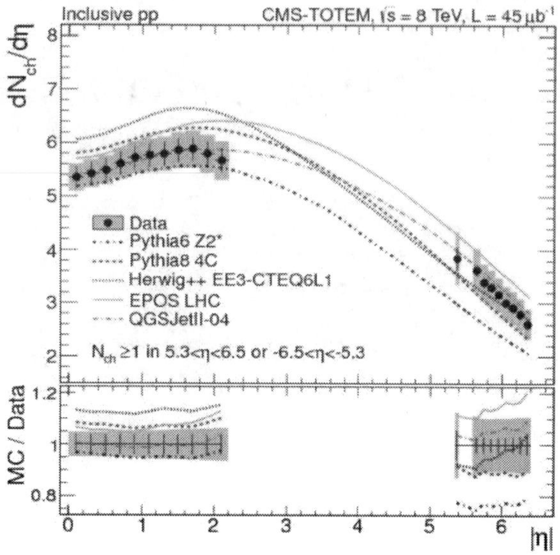

(a)

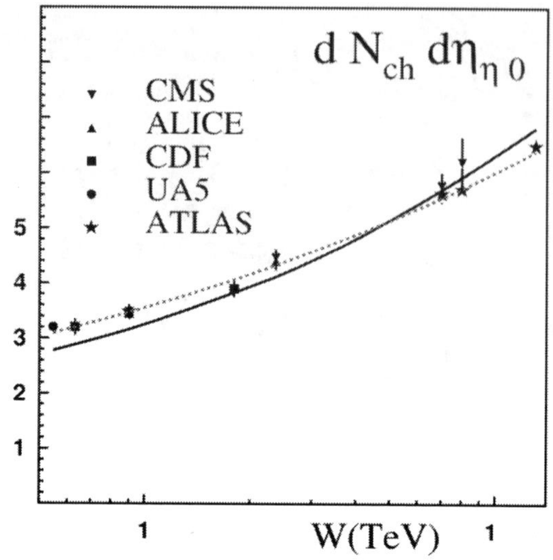

(b)

Figure 5. The comparison of the inclusive production at $W = 8\,TeV$ with the Monte Carlo models is shown in fig (a). The figure is taken from [CMSIN and TOTEMIN] In fig.(b) $dN_{ch}/d\eta$ at $\eta = 0$ versus energy W is displayed. Our estimates are shown by the solid line. The dotted line corresponds to fit: $0.725\,(W/W_0)^{0.23}$ with $W_0 = 1\,GeV$. The data are taken from [17],[18],[19],[20],[21],[22].

[16] CMS Collaboration,Phys. Letts. **B749**, 242, (2015).

[17] J. Beringer, et al. (Particle Data Group), Phys.Rev. **D86**,010001 (2012).

[18] K. Aamodt et al.[ALICE Collaboration], Eur. Phys. J.**C68**, 89 (2010); [arXiv:1004.3034 [hep-ex]]; ALICE Collaboration, Eur. Phys. J.**C65**, 65 (2010) 111 [arXiv:0911.5430 [hep-ex]].

[19] S. Chatrchyan et al. [CMS and TOTEM Collaborations], Eur. Phys. J. **C74**, 3053 (2014), [arXiv:1405.0722 [hep-ex]]; V. Khachatryan et al. [CMS Collaboration], Phys. Rev. Lett. **105**, 022002 (2010); [arXiv:1005.3299 [hep-ex]]. V. Khachatryan et al.[CMS Collaboration], JHEP **1002**, 041 (2010) [arXiv:1002.0621 [hep-ex].

[20] T. A. Altonen, et al [CDF Collaboration] Phys. Rev. **D79**, 112005 (2009).

[21] G.J. Alner, et al [UA5 Collaboration] Z.Phys. **C33** , 1 (1986).

[22] [ATLAS Collaboration] arXiv:1003.3124 [hep-ex].

DOI: 10.1051/epjconf/201716401003

A new measurement of the Cosmic-Ray Electron Spectrum with Fermi-LAT

Francesca R. Spada[1,a] on behalf of the Fermi-LAT Collaboration

[1] *Istituto Nazionale di Fisica Nucleare sez. di Pisa*
Largo B. Pontecorvo, 56127 Pisa, Italy

Abstract. Cosmic-ray electrons provide a probe of the origin and propagation mechanisms of cosmic rays in local galactic environment.
The Large Area Telescope (LAT), on board the Fermi mission, is a pair-conversion space telescope for high-energy electromagnetic radiation, operating in low Earth orbit since June 2008. The LAT has collected the largest high-energy cosmic-ray electron sample to date, with more than 10k events above 1 TeV. The event selection and reconstruction capability exploits the performance of the recently improved event-level analysis developed by the Fermi-LAT Collaboration.
We present a new measurement of the inclusive spectrum of cosmic-ray electrons and positrons performed on seven years of data, in an energy range from 7 GeV up to 2 TeV.

1 Introduction

While propagating throughout the Galaxy, high-energy electrons and positron (CRE) rapidly lose energy by inverse Compton scattering on the interstellar radiation field and by synchrotron emission on the Galactic magnetic field. As a consequence, the CRE diffusion distance is of the order of 100 pc at 1 TeV, much shorter than the radial scale of the Galaxy.

The shape of the CRE spectrum from about 100 GeV up to few TeV can provide evidence of the presence of local CRE sources. These can be of astrophysical, like Supernova Remnants (SNR) or Pulsar Wind Nebulae (PWN), or of exotic nature, as the elusive dark matter.

Lower-energy CRE are affected by energy-dependent diffusive losses, convection in the interstellar medium, and perhaps reacceleration by second-order Fermi processes during transport from their sources. Since all these processes can affect the CRE spectrum after its injection by the sources, the observed spectrum is sensitive to the environment, determining how CRE propagate through the Galaxy.

A recent measurement by AMS-02 [1] has shown that the CRE spectrum can be fit with a single power law up to 1 TeV, with an index of -3.170 ± 0.008. Measurements by H.E.S.S. [2, 3] give an indication of a cutoff at 2 TeV. These results can be jointly interpreted with the presence of local CRE sources with a spectral cutoff at about this energy.

[a] e-mail: francesca.spada@cern.ch

Figure 1: The Fermi satellite. The obsevatory carries two instruments on-board: the γ-ray Burst Monitor (GBM) [4] and the Large Area Telescope (LAT). The GBM, sensitive in the energy range between 8 keV and 40 MeV, observes the full unocculted sky with rough directional capabilities – at the level of one to a few degrees – for the study of transient sources, in particular Gamma Ray Bursts (GRBs). The LAT is designed to detect γ-rays of energy greater than 20 MeV, where pair-conversion is the dominant interaction process, and is also naturally suitable to detect electrons and positrons.

2 The Fermi mission

The Fermi γ-ray Space Telescope is a NASA mission designed to survey the sky in the broad energy range from 20 MeV to more than 300 GeV, with the additional capability of studying transient phenomena at lower energies and of detecting charged species, notably leptons, from GeV to TeV energies.

The Fermi satellite, shown in figure 1, was launched from Kennedy Space Center, Cape Canaveral, FL, on board a Delta II 7920-H rocket on July 11, 2008, and is in nominal science operation since August 4, 2008, orbiting at an altitude of 565 km with a 25.6° inclination. The large field of view of the telescope (2.5 sr) provides the ability to observe 20% of the sky at any time which, in the nominal scanning mode of operation, translates in a full sky survey every three hours.

Onboard Fermi, the Large Area Telescope (LAT) [5], exploiting state-of-the-art, high-energy physics detector techniques, is designed to detect γ-rays of energy greater than 20 MeV, where pair-conversion is the dominant interaction process.

2.1 The LAT instrument

The LAT is a 4×4 array of identical towers, each made by a tracker-converter (TRK) and a calorimeter module (CAL). A segmented anti-coincidence detector (ACD) covers the tracker array, and a programmable trigger and data acquisition system completes the instrument.

Each TRK module features 16 tungsten layers, inducing the conversion of γ-rays into e^+/e^- pairs, and 18 $x - y$ pairs of single-sided silicon strip detector planes, for a total of 1.5 radiation lengths of material on-axis. The silicon sensor technology allows precise tracking with no detector-induced dead time and no use of consumables, and the capability to self-trigger.

The calorimeter, described in more detail in section 2.2, provides an intrinsically three-dimensional image of the shower development, which is crucial for both the energy reconstruction,

Figure 2: Exploded view of a calorimeter module with the hodoscopic 12 × 8 array of CsI blocks in the carbon fibre structure. Also shown are the readout electronics and the aluminum side panels providing shield against electromagnetic interference.

especially at high-energy where a significant part of the shower can leak out of the back of the instrument, and for background rejection.

The ACD, a set of plastic scintillators surrounding the tracker, serves as a veto against the overwhelming background due to charged cosmic rays. In order to limit the "self-veto" effect due to the backsplash of secondaries from high-energy particles hitting the calorimeter, it is segmented in 89 tiles providing spatial information that can be correlated with the signal from the tracker and the calorimeter. The overall power budget of the LAT is 650 W.

Since electrons and positrons, as well as photons, develop electromagnetic showers in the CAL, and are detected in the TRK silicon modules, the LAT is naturally suitable to measure CRE direction and energy, up to the TeV scale. An on-board filter reduces the data rate to be downliked to Earth from about 10 kHz to 300–500 Hz. The so called *gamma filter* is specifically designed to reject charged particles, but it accepts all events with a raw deposited energy in the CAL greater than 20 GeV, so it does not prevent CRE to be accepted in the data sample. However, extending the spectrum measurement below this energy requires the use of an unbiased sample of all trigger types, that actually exists and is prescaled on-board by a factor 250. A dedicated analysis is needed, described in detail in section 3.3, and a unified approach is used wherever possible, with ad-hoc strategies when necessary.

2.2 The electromagnetic calorimeter

Figure 2 shows an exploded view of a calorimeter module, consistsing of 96 CsI(Tl) scintillation crystals with PIN photodiode readouts, arranged in a hodoscopic configuration for a total depth of ~8.6 radiation lengths on axis. The eight layers contained in a module, each formed of 12 crystals, are rotated by 90° with respect to the preceding and the following one, thus forming an $x - y$ array.

Each of the 1526 crystals forming the CAL provides a measurement of both the amount of energy released, and the three-dimensional location of the energy deposition, thus allowing the imaging of

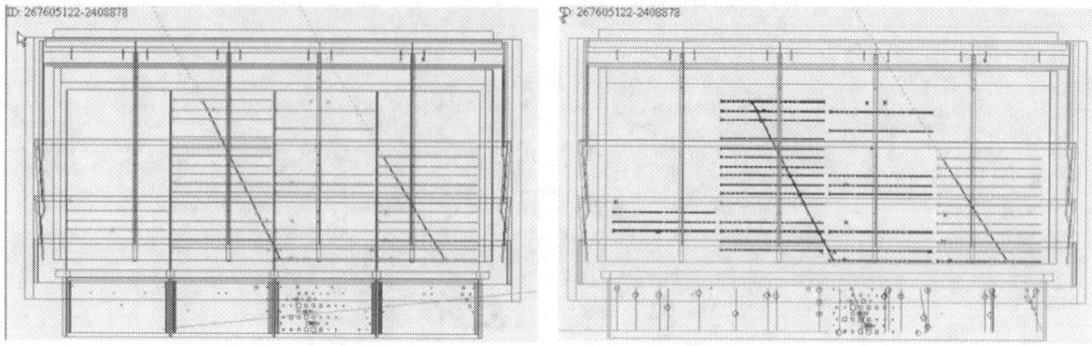

Figure 3: Effect of the CAL clustering stage on the event reconstruction, in a typical event in which ghost activity appears alongside with the signal produced by a real gamma-ray crossing the detector. Left panel: a single particle in the LAT is assumed, resulting in a flipped shower axis (green line at the bottom). Right panel: clustering the activity in the CAL, the gamma-ray is isolated and the shower axis correctly reconstructed (red line).

the shower development. Two dual-photodiode assemblies, measuring the scintillation light produced in the crystal and containing a large-area photodiode for small energy depositions, and a small-area photodiode to measure large energy releases, are located at each end. The readout is performed by a custom analog amplifier ASIC, that minimized the power consumption and allows to cover the large dynamic range of the input signal.

Nonetheless, extending the CRE energy measurement beyond 1 TeV with the LAT is challenging because, at such high energy, only 35% of the shower is typically contained in the CAL and a significant fraction of the crystals along the shower axis are saturated, since crystal saturation occurs for CRE above 600 GeV. This makes both the energy reconstruction and the hadronic background rejection extremely difficult.

2.3 Pass 8 event-level analysis

The event selection developed for the LAT before launch has been periodically updated over the years to reflect the constantly improving knowledge of the detector and the environment in which it operates. Pass 8, a radical update of the entire event-level analysis released to the public in 2015, represents the most recent iteration [6]. The revision involved every aspect of the data processing, from the simulation of the detector to the event reconstruction and the background rejection. The improvements include (but are not limited to) a significant reduction in background contamination, an increased effective area, a better point-spread function, a better understanding of the systematic uncertainties and an extension of the energy reach for the photon analysis below 100 MeV and above a few hundred GeV. The calorimeter reconstruction has been extensively revisited at all levels – the crystal simulation, the energy and direction reconstruction, the background-rejection algorithms.

Extremely relevant to the CRE spectrum measurement is the fact that in pass 8 a clustering stage in the calorimeter was introduced, aimed at identifying and isolating residual *ghost signals* away from the gamma-ray shower, that can introduce substantial errors in the measurement of the energy, centroid and direction of the shower. This allowed to recover the loss in effective area coming from events misclassified as background, due to the fact that the matching between the tracker and the

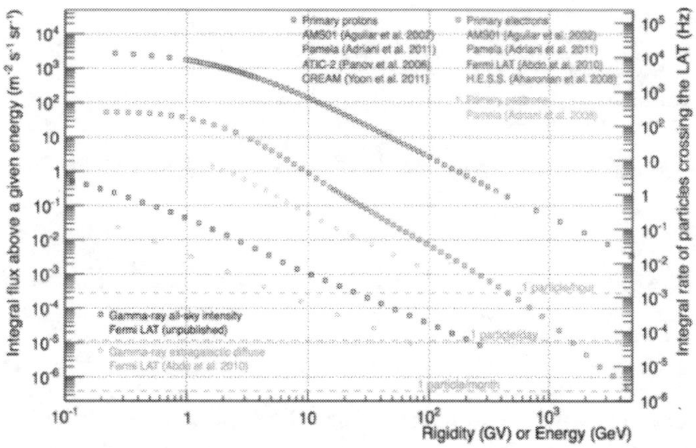

Figure 4: Integral fluxes for a few of the most abundant particle species at low-Earth orbit. Blue: primary protons, that represent the most important background source for this analysis; red: primary electrons and yellow: primary positrons, i.e. (summed) the signal.

calorimeter is one of the main inputs to the background rejection. Figure 3 shows the effect of the CAL clustering stage on the event reconstruction: the signal produced by a gamma-ray crossing the detector is decoupled from the ghost activity far from the real particle.

Above a few GeV, the particle energy is determined from a three-dimensional profile fit to the calorimeter layer energies. This approach is optimal up to 1 TeV, where the average energy release per crystal at the shower maximum starts exceeding the dynamic range of the readout electronics and saturation becomes an issue, as anticipated in section 2.2. In order to overcome this limitation, the profile fit has been extensively reworked taking into account individual crystal energies at each layer. Concerning the contributions from saturated channels, Monte Carlo simulations indicate that their energy deposits can be recovered to some extent by using the information from the nearby (non-saturated) crystals, thus extending the energy measurement range up to ~3 TeV.

3 The analysis

Figure 4 shows the integral fluxes for a few particle species in cosmic rays at low-Earth orbit. It is evident that to extract our signal, consisting of electrons and positrons reaching the detector from all directions, we must be able to get rid of a much more abundant background, mainly due to protons, since $n_p/n_e \sim 500$ for E > 100 GeV.

Thanks to the new pass 8 event analysis, with its improved algorithms for track and shower reconstruction, as well as powerful multivariate methods for background suppression, we can achieve a level of background contamination smaller than 25% and an energy resolution better than 20% up to 2 TeV.

The LAT on-board *gamma filter* is designed to reject charged particles, but it accepts all events with a deposited energy in the CAL larger than 20 GeV, thus preserving high-energy charged particles from being lost from the data sample. However, extending this analysis below 20 GeV requires the use of the unbiased sample of all trigger types, that is prescaled on-board by a factor 250.

As we did in the previous LAT measurement of the CRE spectrum [7], we performed two independent analyses: a High-Energy (HE) analysis above 42 GeV, and a Low-Energy (LE) analysis between

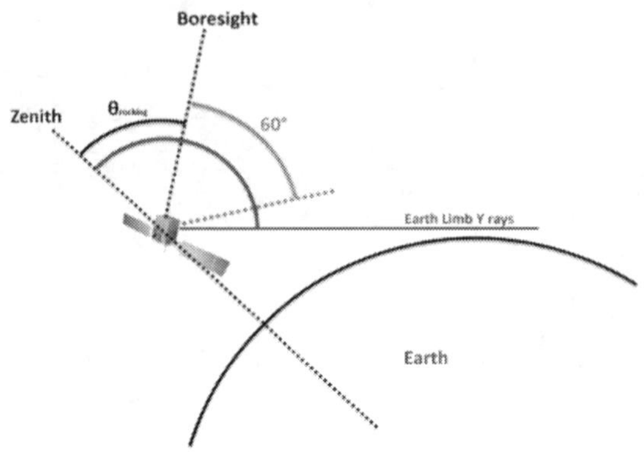

Figure 5: At the average Fermi orbital heigth, Earth limb photons reach the detector at a 113° angle from the Earth zenith direction. In standard survey mode, i.e. with a LAT rocking angle of 51°, getting rid of the background due to limb photons requires rejecting particles coming at an angle greater than 60° from the boresight direction.

7 and 70 GeV. We chose to use a unified approach wherever possible, with ad-hoc strategies when needed. Both analyses, however, require a high-quality starting sample, a first background rejection step, and a final selection to reduce as much as possible the residual proton contamination.

3.1 Event selection

The starting sample consists of LAT pass 8 data collected in standard survey mode, i.e. when the rocking angle of the LAT from the zenith is less than 51°, for an overall live time of 4.68 years.

Earth limb photons appear as a sharp peak located at 113° from the Earth zenith direction. The rocking angle selection, with an additional one on events coming within 60° from the LAT boresight, eliminate a potential contamination from this source (see figure 5).

Quality cuts consist in requiring events with a well reconstructed track, and whose path length through the CAL is larger than 8 radiation lengths. Additionally, events with a χ^2/n.d.f. > 60 for the shower profile fit are rejected.

In order to remove α particles and heavier ions, we add requirements on the path-length corrected signal in the ACD and on the TKR time over threshold: both variables are in fact sensitive to the ionization signal, that is proportional to Z^2, thus providing charge-deposition information. Applying a cut in the plane spanned by these two variables around the region corresponding to $Z = 1$ reduces the contamination to less than a few per mill with respect to protons. The residual contribution from celestial photons also is reduced to less than 1% of the CRE flux.

For the LE analysis, an additional selection is applied that requires more than 2 GeV deposited in the CAL and depends on the geomagnetic cutoff rigidity.

We are now left with a proton background only, that requires a dedicated strategy. The chosen approach is based on the multivariate analysis toolkit TMVA [8] making use of boosted decision trees (BDT), that were trained on Monte Carlo-simulated datasets of electrons and positrons (the signal) and protons (the background).

To account for rapid changes with the energy of the typical event topology in the detector, we define 8 energy bins in $\log_{10}$(E/GeV) equally spaced between 1.5 ad 3.5 (corresponding to 31.6 GeV

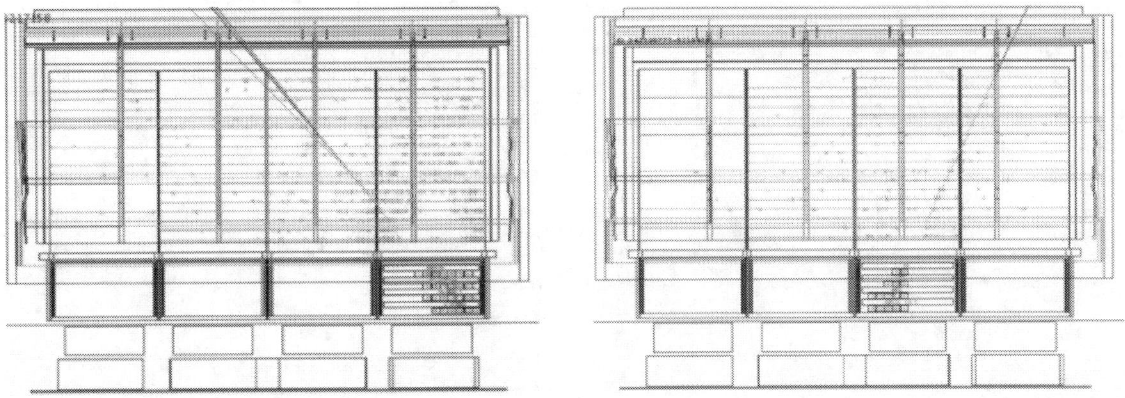

Figure 6: Different incident particles determine different event topologies in the LAT. Left: 1 TeV proton (background). Right: 1 TeV electron (signal).

and 3.16 TeV) for the HE, plus 1 energy bin for the LE, and separately train a BDT in each bin. All BDTs are built from the same 19 observables, chosen according to good Monte Carlo-data agreement and high separation efficiency, that are related to the event topology in TKR and CAL. Five concern the position of the shower in the detector, while the others describe the shower topology in the TKR and CAL – as e.g. the shower transverse size that, as evident from figure 6, looks quite different in the case of signal (right) and background (left), thus providing a high discriminating power.

3.2 High energy

The validity of the BDT approach strongly relies on the assumption that Monte Carlo well reproduces the distributions of the variables in the data. For a few of the original Monte Carlo outputs this was not the case above a few hundreds of GeV, and this discrepancy had a significant impact on the effectiveness of the BDT method. In order to mitigate these discrepancies, we derived additive individual

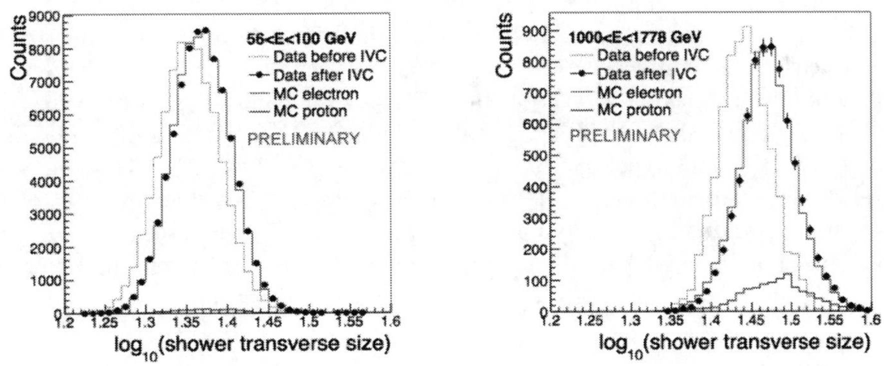

Figure 7: The shower transverse size before (dotted lines) and after (black circles) IVC correction for events between 56 and 100 GeV (left) and between 1 and 1.78 TeV (right). The red histograms correspond to the electron simulation, and the blue ones to the residual background.

 DOI: 10.1051/epjconf/201716401003

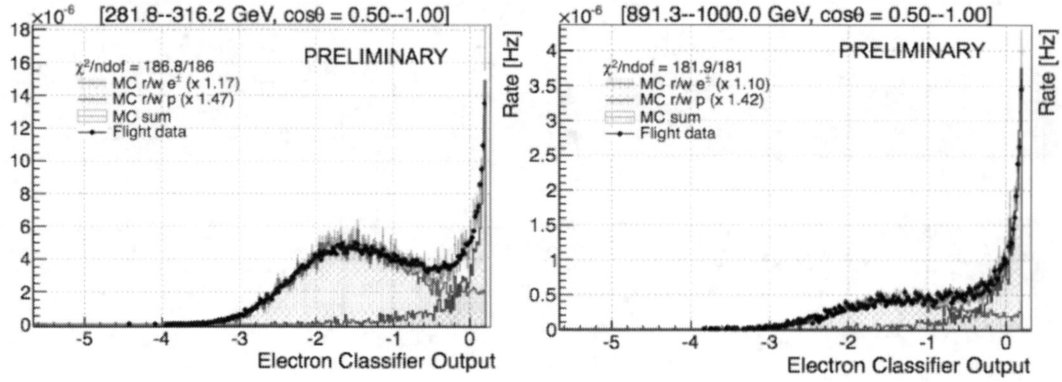

Figure 8: Template fit results for the HE analysis in two different energy bins. The Monte Carlo prediction (grey line) is the sum of the signal (red) and background (blue) templates, scaled by the normalization factors given by the template fit and reported in brackets in the caption. The thus rescaled simulation well reproduces the data distribution (black dots).

variable corrections (IVC) for the data as a function of energy and incidence angle of the particle. The calibration is applied to the data, which is equivalent to applying it to the simulation, but has the advantage that there is no need to retrain the BDTs. The IVCs provide a significant improvement in the agreement for all the variables used as input to the BDTs, and the residual differences are taken into account in the estimate the systematic uncertainties. Figure 7 shows, as an example, the IVC impact on the data-simulation agreement for the shower transverse size in two different energy bins.

Each BDT provides an output variable p_{BDT} assuming values between -1 (candidate protons) and 1 (candidate electrons), whose distribution for electrons peaks very sharply at 1. It is then convenient to build an estimator as

$$P_{CRE} = \log_{10}(1 - p_{BDT}) \tag{1}$$

For each energy bin, we then construct the Monte Carlo electron and proton distributions of P_{CRE} and use them as templates. The sum of the two templates is fit to the data by varying their normalizations, and the renormalized proton template is used to estimate the number of background events in the data integrating the P_{CRE} distribution up to a chosen cut value. The so obtained number of background events is subtracted to the number of events in the data to obtain the number of CRE events. The cut on P_{CRE} is energy-dependent and is such that the efficiency for electrons goes from 90% at 30 GeV to 80% at 300 GeV, and then down to 50% at 2 TeV and above. This choice is driven by the increasing difficulty of the background rejection with energy, particularly above 300 GeV. The analysis stops when the residual contamination reaches 20%, which occurs at 2 TeV. One example of the result of the template fit, for two different energy bins, is shown in figure 8, where the Monte Carlo prediction is the sum of the signal and background templates scaled by the normalization factors given by the template fit to the data, and clearly well reproduces the data distribution.

3.3 Low energy

Due to the on-board gamma filter, extending the analysis below 20 GeV requires the use of the unbiased sample of all trigger types, that is prescaled by a factor 250. The LE selection is based on the

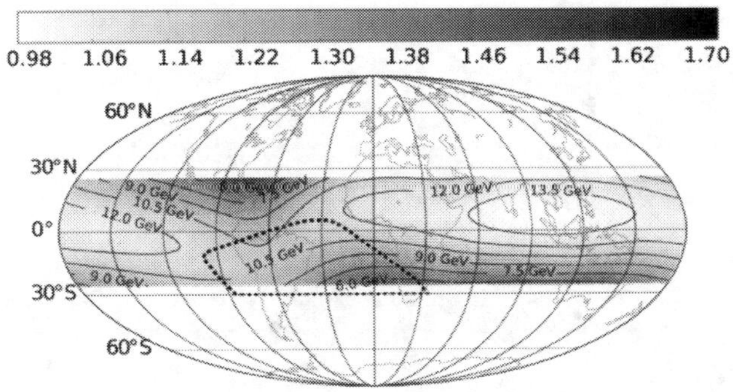

Figure 9: Map of the Earth showing McIlwain-L values (greyscale) in the region spanned by the LAT orbit. Higher McIlwain-L values (i.e., darker regions) correspond to lower rigidity cutoff values (contour lines).

same multivariate analysis approach as used for the HE selection, and a dedicated classification tree is trained in the low-energy region. Contrary to the HE case, the background Monte Carlo sample used for the training is a realistic simulation of cosmic rays in low Earth orbit and of Earth limb photons [9]. Since the LE analys stops at 70 GeV, no IVC correction, relevant at high energy, is applied to the data. The residual background contamination is estimated by the same template fitting technique used in the HE analysis.

Below ~20 GeV, the CRE flux reaching the LAT is strongly influenced by the shielding effect of the Earth magnetic field: at any given geomagnetic position, Galactic charged particles can reach the detector only if they are above a certain rigidity R_C. The dependence of this *rigidity cutoff* on the coordinates can be conveniently parametrized by the McIlwain-L parameter [10], that describes the set of geomagnetic field lines crossing the magnetic equator at a distance of L Earth radii. Geographic coordinates that are magnetically equivalent from the standpoint of incoming charged particles thus share the same McIlwain-L, which is then an optimal parameter for describing cutoff rigidities. Measuring the CRE spectrum at a given energy E requires selecting data collected in a McIlwain-L region in which $R_C < E$.

A map of the Earth in McIlwain-L is shown in figure 9. The LAT orbit spans an interval $L = \{0.98 - 1.73\}$, corresponding to vertical rigidity cutoff values from ~6 GeV to ~14 GeV. The L range is thus divided in 15 bins, and the corresponding R_C values are determined from a fit of the count spectrum. This provides an empirical relation $R_C = f(L)$, that can then be used to find, for each energy bin of the CRE spectrum, the McIlwain-L value $L_{\rm min}$ corresponding to the lower boundary of the energy bin itself, and select data collected in regions with $L > L_{\rm min}$.

However, the cutoff is smooth, and even above the nominal R_C there is still a non-negligible fraction of electrons and positrons which are prevented from reaching the detector by the magnetic shadow of the Earth. The fraction of undetected CRE is estimated with the help of the particle trajectory tracing code developed by Smart and Shea [10] and the 2010 model of the Earth's magnetic field from the International Geomagnetic Reference Field (IGRF) [11]. The tracer computes the trajectory of test particles – generated according to a power law with an index of −3.2 using the abundance ratios measured by AMS-02 for electrons and positrons [12] and passing the standard analysis event selection – starting from the spacecraft, that is each time assigned a realistic orbital position with a probability based on real data. Test particles with a trajectory reaching 20 Earth radii correspond to CRE that could actually be observed by the instrument, while trajectories intersecting the Earth's atmosphere

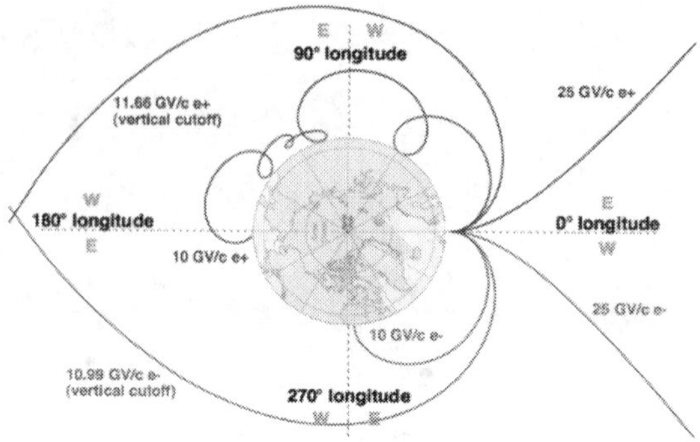

Figure 10: Trajectories of test particles in the Earth magnetic field computed by the tracing code. Red lines represent electrons, blue ones represent positrons. At a given orbital position, particles behave depending on their energy. Test particles with a trajectory reaching 20 Earth radii correspond to CRE that could actually be observed by the instrument, while trajectories intersecting the Earth's atmosphere correspond to lost particles.

correspond to lost particles, as explained in figure 10. The observed flux is then corrected accordingly, with the derived correction factor increasing from +10% at 7 GeV to +40% at 15 GeV, decreasing down to +5% at 30 GeV.

3.4 Systematics

The control of the absolute energy scale is key to the measurement of the CRE spectrum. A self-consistent check comes from a comparison of the CRE cutoff measured in LAT data with the one predicted by the tracer as described in section 3.3. We find an average ratio $R_C^{\mathrm{data}}/R_C^{\mathrm{tracer}} = 1.033 \pm 0.004(\mathrm{stat}) \pm 0.020(\mathrm{syst})$. The calibration of the CAL CsI crystals ensures a linearity better than 1% up to the saturation level. Assuming that the light yield is linear for electron showers up to 2 TeV, we rescale the event energies in data by −3.3% in both the LE and HE analyses (here the IVC correction is applied afterwards). The in-flight absolute energy scale measurement gives a systematic uncertainty on the energy scale at 10 GeV of 2%. We estimate the uncertainty at 1 TeV to be 6%.

For the HE analysis, we considered three additional sources of systematic uncertainty, as shown in figure 11:

A *Acceptance*
The main uncertainty is due to the estimation of the effective area. We vary the cut on P_{CRE} in a range corresponding to ±20% in signal efficiency, obtaining a flux variation of ~2% up to 1 TeV, increasing to 9% at 2 TeV;

B *IVC correction*
We move the correction factors, derived from the difference between the peak position of the data and Monte Carlo distributions for each BDT input variable, by its $1-\sigma$ uncertainty, getting a signal variation from 2% at 40 GeV to 12% at 2 TeV;

C *Residual fraction of background events*
To take into account the uncertainty on the GEANT4 prediction of the fraction of protons mim-

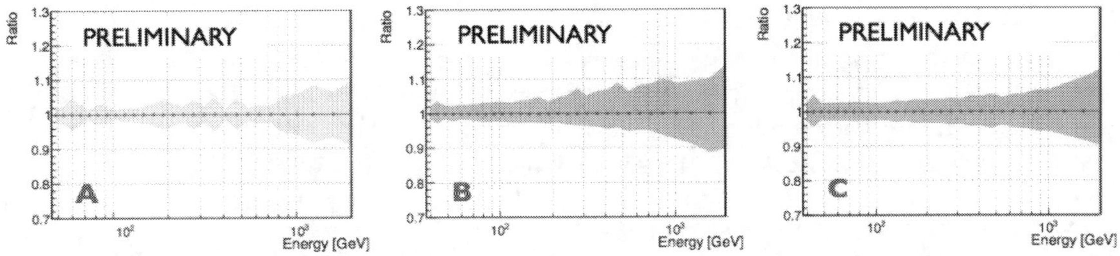

Figure 11: Main sources of systematic error in the CRE measurement, represented with energy as a band including the ratio of the values obtained by varying analysis parameters with the central ones.

icking electromagnetic showers, we assume a 20% uncertainty on the number of background events, getting a signal variation < 2% up to 1 TeV that increases to 6% at 2 TeV.

4 Conclusions

The CRE inclusive spectrum is shown in figure 12. The shaded band represents the systematics and does not include the one on the energy scale. Between 50 GeV and 2 TeV, the spectrum is compatible with a single power law with a spectral index -3.07 ± 0.02. Below 100 GeV, the new LAT measurement differs from the previous one [7] by 10-30%. A large part of this difference is due to the fact that, in the previous analysis, the simulation that was used did not take into account the presence of *ghosts* described in section 2.3. Taking into account the uncertainties on the energy scale, this measurement is in agreement with the AMS and HESS ones.

References

[1] M. Aguilar *et al.* [AMS Collaboration], Phys. Rev. Lett. **113** (2014) 221102
[2] F. Aharonian *et al.* [H.E.S.S. Collaboration], Phys. Rev. Lett. **101** (2008) 261104

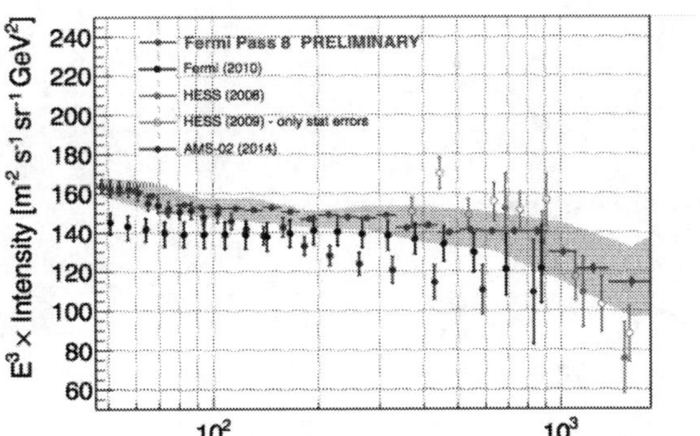

Figure 12: The CRE inclusive spectrum. The shaded band represents the systematics and does not include the one on the energy scale.

DOI: 10.1051/epjconf/201716401003

[3] F. Aharonian *et al.* [H.E.S.S. Collaboration], Astron. Astrophys. **508** (2009) 561
[4] C. Meegan *et al.*, AIP Conf. Proc. **921** (2007) 13
[5] W. B. Atwood *et al.* [Fermi-LAT Collaboration], Astrophys. J. **697** 1071 (2009)
[6] W. Atwood *et al.* [Fermi-LAT Collaboration], arXiv:1303.3514 [astro-ph.IM]
[7] M. Ackermann *et al.* [Fermi-LAT Collaboration], Phys. Rev. D **82** (2010) 092004
[8] A. Hocker *et al.*, PoS ACAT (2007) 040 [physics/0703039 [PHYSICS]]
[9] T. Mizuno, T. Kamae, G. Godfrey, T. Handa, D. J. Thompson, D. Lauben, Y. Fukazawa and M. Ozaki, Astrophys. J. **614** (2004) 1113
[10] D. F. Smart and M. A. Shea, Advances in Space Research **36** (2005) 2012
[11] C. C. Finlay *et al.*, Geophys. J. Int. **183** (2010) 12161230
[12] L. Accardo *et al.* [AMS Collaboration], Phys. Rev. Lett. **113** (2014) 121101

EPJ Web of Conferences **164**, 01004 (2017) DOI: 10.1051/epjconf/201716401004
ICNFP 2016

A project to measure quantum spin correlations of relativistic electron pairs in Møller scattering

Jacek Ciborowski[1,a], Paweł Caban[2], Michał Drągowski[1], Joachim Enders[3], Yuliya Fritzsche[3], Artem Poliszczuk[1], Jakub Rembieliński[2], and Marta Włodarczyk[1]

[1] *University of Warsaw, Faculty of Physics, Pasteura 5, 02-093 Warsaw, Poland*
[2] *University of Łódź, Faculty of Physics and Applied Informatics, Pomorska 149/153, 90-236 Łódź, Poland*
[3] *Technische Universität Darmstadt, Institut für Kernphysik, Schlossgartenstraße 2, 64289 Darmstadt, Germany*

Abstract. A project to measure quantum spin correlations for relativistic particles with mass is presented. It aims at testing spin observables in relativistic quantum mechanics. The measurement will be performed for pairs of entangled electrons originating from Møller scattering. A double Mott polarimeter, allowing for simultaneous measurement of the spin projections of both electrons, was designed for this purpose.

1 Introduction

1.1 Review of spin correlations

At the basis of the modern quantum mechanics and the quantum information theory lies the famous paper of Einstein, Podolsky and Rosen (EPR) [1] published in 1935. In this paper the completeness of quantum mechanics was put to doubt from the point of view of the local realism. The operational criterion of the local realism was given in 1964 by J. S. Bell [2]. It has a form of inequalities [2, 3] which should be fulfilled for every local realistic theory. The Bell-type inequalities involve the correlation functions, defined as: $C(A, B) = \sum_{\alpha\beta} \alpha\beta P_{\alpha\beta}$, where α and β are the possible outcomes of measurements of some arbitrary observables A and B, performed by two distant observers, Alice and Bob, respectively, and $P_{\alpha\beta}$ are the corresponding joint probabilities. The most common choices for A and B are the spin projections along given directions or polarisations. It has been shown experimentally that real quantum systems violate the Bell-type inequalities, both for photons [4] and massive particles [5].

A large majority of correlation experiments was performed with use of photons or nonrelativistic particles with mass. Since the pioneering paper by Czachor [6] one can notice a raise of interest in relativistic aspects of the quantum information theory, expressed in the most important papers on that subject [6, 9–13], including the papers of our group [7, 8, 14–19]. However, the theoretical analysis of spin correlations in relativistic regime given in these papers is full of ambiguities and interpretational difficulties. They arise mostly due to the fact of an apparent conflict between nonlocality of the relativistic quantum mechanics and the Lorentz symmetry. Rembieliński *et al.* exhaustively discussed the

[a] e-mail: cib@fuw.edu.pl

DOI: 10.1051/epjconf/201716401004

relativistic EPR correlation function for a bipartite system of spin-1/2 fermions in the framework of the relativistic quantum mechanics [17] and the quantum field theory [18]. The function was calculated in both pseudoscalar and vector state, which are the relativistic counterparts of the non-relativistic singlet and triplet states, respectively. The further analysis of the relativistic correlation functions for the first time showed their unexpected behaviour [7, 8]. Namely, for some fixed configurations it was found that the correlation functions are non-monotonic functions of particle momenta and may have local extrema. The degree of the violation of the Bell-type inequalities also depends on the particle momenta and in some cases is a non-monotonic function of the momentum. The correlation functions and the degree of violation of the Bell-type inequalities are different for different choices of the relativistic spin projection operator, but their qualitative behaviour is similar. The observed effect was also found for many other systems. An analysis of spin correlations in bipartite systems of spin-1/2 and spin-1 particles was performed [20]. The correlation functions were calculated for states which are not Lorentz scalars, for two different choices of relativistic spin projection operators. Again, in both cases there existed configurations for which the functions had local extrema. Local extrema were found in the correlation function in a hybrid system of a photon and a spin-1/2 fermion [21]. Correlations of relativistic vector bosons assuming that their linear polarisations (determined by helicity, a quantity which in contrast to spin is uniquely defined in relativistic quantum mechanics) is measured, were discussed by Caban [14]. The results confirm the earlier observation that the correlation function can, for some fixed configurations, have local extrema. Moreover, it was shown that for certain configurations the degree of the inequality violation increases with the particle momenta, reaching its maximal value in the ultrarelativistic limit, in contrast to suggestions of some authors [6, 9, 22]. All the above mentioned results strongly suggest that the existence of the local extrema, independent of the choice of the relativistic spin projection operator, is an immanent feature of relativistic correlations. It is also clear that the behaviour of the relativistic correlation functions may be different than in the non-relativistic case.

As far as we know, up to date there have been only three experiments testing Bell-type inequalities by means of massive relativistic spin one-half particles: the Lamehi-Rachti–Mittig (LRM) experiment [23] performed about thirty years ago in CEN-Saclay and two recent experiments: the first one given at the Kernfysisch Versneller Instituut (KVI, Holland) by S. Hamieh *et al.* [24] and the second one performed by H. Sakai *et al.* [25] in RIKEN Accelerator Research Facility (Japan). In all three experiments the proton-proton spin correlations were measured. LRM team tested Bell-type inequalities with use of the low energy (13.5 MeV) proton beam which corresponds to the proton velocity $v \sim 0.17c$. For the other hand, in KVI experiment, the spin correlations of proton pairs in a $^1S_0$ intermediate state, obtained from $^{12}C(d, ^2He)^{12}B$ nuclear charge-exchange reaction, were measured for protons with the kinetic energy $\sim$ 86 MeV ($v \sim 0.4c$). Finally, in the RIKEN experiment the proton pair was created in the $^1H(d, ^2He)n$ charge-exchange reaction with the proton energy $\sim$ 135 MeV ($v \sim 0.5c$). All these experiments were in agreement with the non-relativistic quantum mechanics predictions. The particles were too slow to give a significant difference between predictions of relativistic and non-relativistic quantum mechanics. From our estimation it follows that in a conclusive experiment the kinetic energy of the particles used should be at least of order of the particles rest mass.

In view of the above, we have undertaken a task of measuring the Einstein-Podolsky-Rosen (EPR) spin correlations for electron pairs originating from the Møller scattering in the relativistic regime, which is already achieved for MeV-energy electrons. This is the first project addressing the issue of quantum spin correlations of relativistic massive fermions.

DOI: 10.1051/epjconf/201716401004

1.2 Spin correlations in Møller scattering – theory

Given two correlated spin-1/2 fermions, one can measure their spin projections separately on given directions and the results are either +1/2 or -1/2 for each of them. Let $P_{s1\,s2}$, denote the probability that the fermions had spin projections $s1$ and $s2$, respectively, where $s1$ and $s2$ are either + or - (along the positive or negative direction of the axis). The correlation function can then be expressed as: $C = P_{++} + P_{--} - P_{-+} - P_{+-}$. The probabilities $P_{s1\,s2}$ are the primary subject of our measurement using the method of Mott polarimetry (below). Calculations regarding the correlation function and the corresponding probabilities, in the context of the present project, have already been completed [26] and also reported at this Conference [27]. As an example, the dependence of P_{++}, P_{--}, P_{-+} and P_{+-} on beam energy is shown in fig. 1. The correlation function and the probabilities depend on the beam energy and polarisation as well as on the final state electron configurations, including a dependence on the directions along which spin projections are measured, $\vec{a}$ and $\vec{b}$.

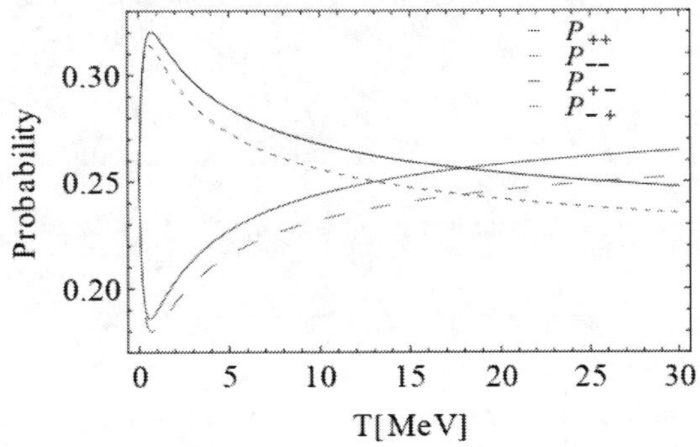

Figure 1. Dependence of the probabilities on beam energy (3 MeV chosen for this measurement). Plot made for a symmetric scattering and vectors $\vec{a}$ and $\vec{b}$ in the scattering plane.

2 Experiment

2.1 Preliminaries

We are planning to carry out the measurement using a 3 MeV polarised electron beam available at the S-DALINAC accelerator in the Institut für Kernphysik, Technische Universität Darmstadt. Beam electrons will scatter on atomic electrons in a 25μm-thick Be target (as well as on Be nuclei which constitutes the main source of background). This particular choice of beam energy allows to collect sufficient statistics while the Møller electrons are still highly relativistic, approx. 1.5 MeV each, in a symmetric scattering configuration, i.e. at $\theta = 16.75°$. The corresponding differential cross section for this process amounts to approx. 100 mb/sr. Longitudinal, transverse and intermediate beam polarisations are available in the range of values from 0.35 to 0.85.

2.2 Mott scattering

Spin projections on predefined directions will be measured by means of the Mott polarimetry. The differential cross section for scattering of a polarised electron off a nucleus is given by the following formula:

$$\frac{d\sigma}{d\Omega} = \left(\frac{d\sigma}{d\Omega}\right)_0 \left[1 + S(\theta)\vec{P}\vec{n}\right], \tag{1}$$

where $(d\sigma/d\Omega)_0$ corresponds to the scattering of unpolarised electrons, $S(\theta)$ is the Sherman function (depending on the electron scattering angle, θ) and $\vec{P}\vec{n}$ is the electron spin projection on a direction given by a unit vector $\vec{n}$, perpendicular to the scattering plane. The Sherman function, calculated for a single nucleus (zero-thickness target), in terms of the electron energy and scattering angle (contour plot), is shown in fig. 2. The higher value of the Sherman function, the higher sensitivity of scattering

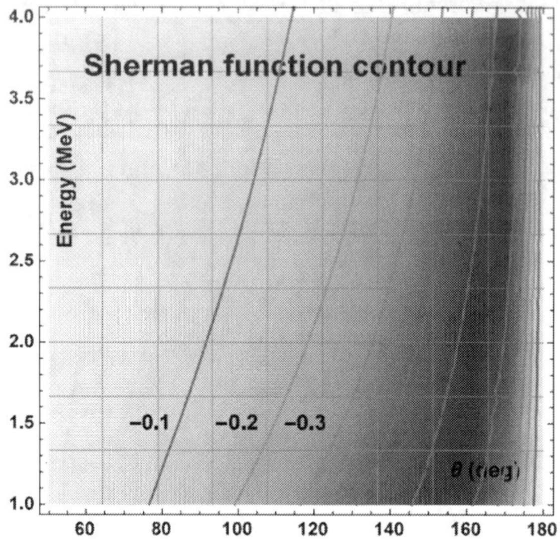

Figure 2. Sherman function contour: energy (MeV) vs. scattering angle θ (degrees), in the MeV-energy range.

to the electron polarisation. For the Møller electrons of about 1.5 MeV energy, the lowest acceptable value of $S \approx -0.2$ implies scattering at about 120°. In practice, Mott targets must have finite thickesses, d, and in consequence the corresponding effective Sherman function take smaller values. The effective Sherman function can be approximately parametrised as: $S(d) = S_0/(1 + \alpha d)$, where S_0 is the function for a single nucleus (zero-thickness target) and α – a parameter (the so called dilution factor). The optimal target thickness equals $1/\alpha$ for this parametrisation and can be derived by maximising the figure-of-merit, given by: $S^2(d)\,d$. The effective Sherman function drops to a half of its theoretical value, S_0, for a target of the optimal thickness. Since the procedure of deriving probabilities $P_{s1\ s2}$ requires knowledge of the effective Sherman function in the conditions of the experiment, it has to be measured or at least reliably simulated, which is also a part of dedicated activities in this project [28].

2.3 Double Mott polarimetry

The idea of measuring spin correlations of electron pairs described in this article, boils down to measuring four numbers, as described below. The sequence of Møller – Mott scatterings is illustrated on fig. 3 which depicts forward Mott scatterings for clarity while the real measurement will be done for backward scattering (angles about 120°). Assume two hypothetical planes, each perpendicular to the respective Møller electron and a ring drawn at each plane with the centre at the incident electron track. Two oppositely located areas (ring sections) are marked black at each ring, corresponding to scattering to the left or to the right off the centre, marked by letters L and R for each of the two rings. Events of interest are those in which each electron hits either the respective area L or R. This way four coincidences can be defined and the corresponding numbers of events counted, marked: L_1L_2, L_1R_2, R_1L_2 and R_1R_2. The centres of the two areas and the entry point at the Mott target for each Møller electron define a plane to which the perpendicular direction is that at which the spin projection of each electron is measured. These directions are marked by vectors $\vec{a}$ and $\vec{b}$, corresponding to $\vec{n}$ in (1) .

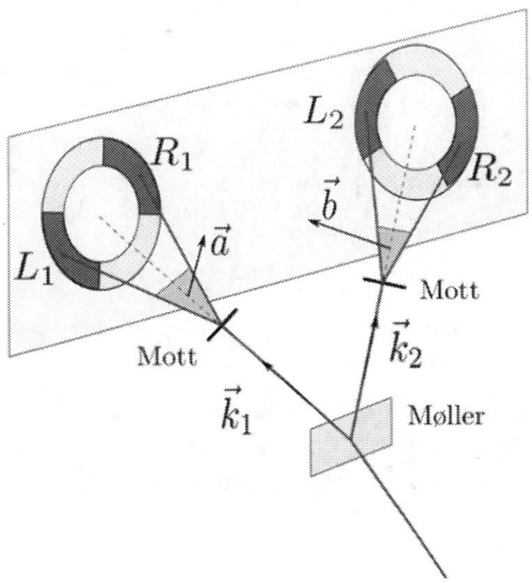

Figure 3. Illustration of the method of measuring coincidences L_1L_2, L_1R_2, R_1L_2 and R_1R_2, needed for deriving probabilities of spin projections, P_{++}, P_{--}, P_{-+}, P_{+-} (see text).

2.4 Derivation of results

The probabilities of spin projections P_{++}, P_{--}, P_{-+}, P_{+-} can be determined from four numbers of coincidences L_1L_2 etc. as follows. First the probabilities of the above coincidences are defined, P_{LL}, P_{RR}, P_{LR} and P_{RL}, by normalising the numbers to the total. Then, following Sakai *et al.* [25], one can compute the probabilities from the following formula:

$$\begin{pmatrix} P_{++} \\ P_{--} \\ P_{+-} \\ P_{-+} \end{pmatrix} = \frac{1}{\bar{S}_1 \bar{S}_2} \begin{pmatrix} c_{++} & c_{--} & -c_{+-} & -c_{-+} \\ c_{--} & c_{++} & -c_{-+} & -c_{+-} \\ -c_{+-} & -c_{-+} & c_{++} & c_{--} \\ -c_{-+} & -c_{+-} & c_{--} & c_{++} \end{pmatrix} \begin{pmatrix} P_{LL} \\ P_{RR} \\ P_{LR} \\ P_{RL} \end{pmatrix}, \quad (2)$$

where $P_{LL} = N_{LL}/N_0$ etc., N_{LL}, N_{RR}, N_{LR} and N_{RL} denote the number of given coincidences and $N_0 = N_{LL} + N_{RR} + N_{LR} + N_{RL}$. The coefficients c_{ij} are defined as: $c_{++} = (1 + \bar{S}_1)(1 + \bar{S}_2)/4$, $c_{--} = (1 - \bar{S}_1)(1 - \bar{S}_2)/4$ etc., and $\bar{S}_1$, $\bar{S}_2$ are the effective values of the Sherman function for the Mott scattering into the selected ring section (fig. 3) for the first and the second Møller electron, respectively. It can be seen from (2) that precise knowledge of the Sherman function in terms of the electron scattering angle and energy (in the range of this measurement) is necessary for determining accurate values of probabilities, given the measured numbers of coincidences.

DOI: 10.1051/epjconf/201716401004

2.5 The double polarimeter

The prototype of the double polarimeter is shown in fig. 4. It is built of the following parts:

- The main pipe, which will be a part of the accelerator vacuum system, attached with a flange (top left) and closed with a beam dump at the bottom right end; 3 MeV electrons will enter from the side of a flange
- Two pipes (Møller legs) welded into the main pipe at $\theta = 16.75^\circ$ which will select two electrons after a symmetric scattering of the beam on atomic electrons in a 20 μm Be target, which will be placed at the crossing of the leg axes
- Two Mott polarimeters, each consisting of a pair of pipes (Mott arms) welded at 120° into each Møller leg (backscattering). Each Mott arm will be equipped with a 10 μm Au Mott target. The downstream part of the Møller leg will have a larger diameter to reduce backscattering background and will be closed with an Møller electron dump
- Each Mott arm will be equipped with a scintillation counter
- All above enumerated pipes will be equipped with collimators
- Directions for measuring spin projections, $\vec{a}$ and $\vec{b}$, can be selected by appropriate orientation of the arms of the Mott polarimeters.

The angular acceptance for a 3 MeV beam is planned to be approx. $3x10^{-4}$ sr for the Møller electrons and $7x10^{-3}$ sr for the Mott electrons. The event rate is expected to be approximately 10^{-16} per beam electron. The prototype is currently under construction and testing.

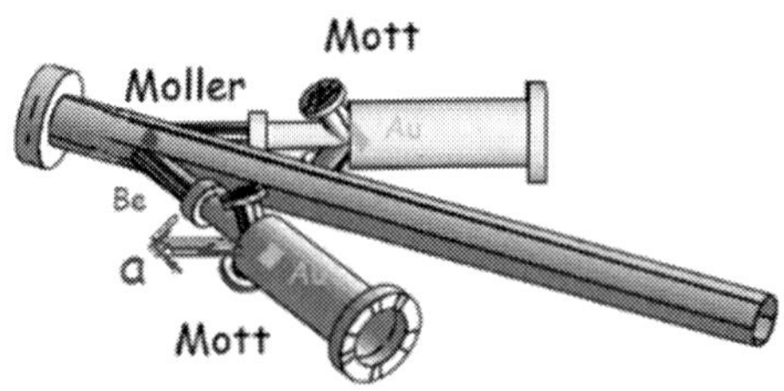

Figure 4. A view of the double polarimeter. The electron beam enters from the upper left corner and scatters on the Møller (Be) target. Two final state electrons enter the pipes at which two Mott polarimeters are installed. Each electron scatters off the Mott Au target in the respective polarimeter and is recorded in the scintillation counter, attached through a flange to each polarimeter arm (not shown). Selecting a relative orientation of the vectors $\vec{a}$ and $\vec{b}$ (in a limited range) is done by appropriate screwing of the polarimeter and the Møller pipe flanges (one of the vectors, $\vec{a}$, is marked on the figure).

Simulations of the response of the double polarimeter using the Geant4 [29] package is in progress. Electrons detected in scintillation counters are predominantly the beam electrons (3 MeV) and Møller electrons (about 1.5 MeV). A significant fraction of electrons emerging from the Be target loose some energy by undergoing scatterings on inner walls of the pipes. Since the polarimeter has no tracking, it is of key importance to have a correct modelling and understanding of its performance. Preliminary results also indicate a low-energy photon background entering the scintillators. It is expected that the requirement of a coincidence in two counters (as explained above) will eliminate most of this background as well as beam scattering events. The response of the scintillation counters is simulated in parallel using both Geant4, as well as the FLUKA [30] and PHOTRACK [31] packages, the latter dedicated to light collection in the scintillator–lightguide assembly.

2.6 The beam

The measurement is planned at the S-DALINAC electron accelerator, located in the Institut für Kernphysik of the Technische Univeristät Darmstadt, Germany. It offers a 3 MeV polarised electron beam with longitudinal and transverse polarisation ranging from 0.35 to 0.85. The beam intensity reaches approx. 10^{14} electrons/s which would yield of order 10^2 useful events per day.

3 Summary and conclusions

In summary, the present project constitutes a first attempt to measure spin correlations of relativistic electron pairs (final state of the Møller scattering). The main aim of the measurement is to draw conclusions regarding spin observables in relativistic Quantum Mechanics. A prototype of a dedicated double Mott polarimeter is currently under construction in parallel to partial laboratory tests using a beta-electron source. Data taking, depending on the S-DALINAC availability, is forseen in 2018.

This work was partially supported from the funds of the National Science Centre, contract/decision DEC-2012/06/M/ST2/00430 and as a postdoc project (M.W.) contract/decision DEC-2013/08/S/ST2/00551.

References

[1] A. Einstein, B. Podolsky, and N. Rosen, Phys. Rev. **47**, 777 (1935)
[2] J. S. Bell, Physics, **1**, 195 (1964)
[3] J. F. Clauser *et al.*, Phys. Rev. Lett. **23**, 880 (1969)
[4] A. Aspect, J. Dalibard, and G. Roger, Phys. Rev. Lett. **49**, 1804 (1982)
[5] M. A. Rowe *et al.*, Nature **409**, 791 (2001)
[6] M. Czachor, Phys. Rev. **A55**, 72 (1997)
[7] P. Caban, J. Rembieliński and M. Włodarczyk, Phys. Rev. **A77**, 012103 (2008)
[8] P. Caban, J. Rembieliński and M. Włodarczyk, Phys. Rev. **A79**, 014102 (2009)
[9] H. Terashima and M. Ueda, Int. J. Quantum Information, **1**, 93 (2003)
[10] A. Peres, P. F. Scudo and D. R. Terno, Phys. Rev. Lett. **88** 230402, (2002)
[11] M. Czachor and M. Wilczewski, Phys. Rev. A68, 010302(R) (2003)
[12] A. Peres and D. R. Terno, Rev. Mod. Phys. **76**, 93 (2004)
[13] P. M. Alsing and G. J. Milburn, Quantum Information and Computation, **2**, 487 (2002)
[14] P. Caban, Phys. Rev. **A77**, 062101 (2008)
[15] P. Caban and J. Rembieliński, Phys. Rev. **A59**, 4187 (1999)
[16] J. Rembieliński and K. A. Smoliński, Phys. Rev. **A66**, 052114 (2002)
[17] P. Caban and J. Rembieliński, Phys. Rev. **A72**, 012103 (2005)
[18] P. Caban and J. Rembieliński, Phys. Rev. **A74**, 042103 (2006)
[19] P. Caban, Phys. Rev. **A76**, 052102 (2007)
[20] P. Caban, J. Rembieliński and M. Włodarczyk, Phys. Rev. **A83**, 034101 (2010)

[21] P. Caban, J. Rembieliński, P. Witas and M. Włodarczyk, Phys. Rev. **A83**, 032115 (2011)

[22] D. Ahn *et al.*, Phys. Rev. **A67**, 012103 (2003)

[23] M. Lamehi-Rachti and W. Mittig, Phys. Rev. **D14**, 2543 (1976)

[24] S. Hamieh *et al.*, J. Phys. G: Nucl. Part. Phys. **30**, 481 (2004)

[25] H. Sakai *et al.*, Phys. Rev. Lett. **97**, 150405 (2006)

[26] P. Caban, J. Rembieliński and M. Włodarczyk, Phys. Rev. **A88**, 032116 (2013)

[27] M. Wlodarczyk *et al.*, this conference

[28] M. Drągowski *et al.*, accepted to Nucl. Inst. Meth. in Phys. Res. B (2016)

[29] S. Agostinelli *et al.*, Nucl. Inst. Meth. in Phys. Res. A **506**, 250 (2003)

[30] A. Ferrari, P.R. Sala, A. Fasso and J. Ranft, FLUKA: A multi-particle transport code (Program version 2005), No. INFN-TC-05-11 (2005)

[31] N. Ghal-Eh, M.C. Scott, R. Koohi-Fayegh and M.F. Rahimi, Nucl. Inst. Meth. in Phys. Res. A **516**, 116 (2004)

 DOI: 10.1051/epjconf/201716401005

Basic features of proton-proton interactions at ultra-relativistic energies and RFT-based quark-gluon string model

E. Zabrodin[1,2,3,a], L. Bravina[2,3], and J. Bleibel[4,5]

[1] *Skobeltsyn Institute of Nuclear Physics, Moscow State University, RU-119991 Moscow, Russia*
[2] *Department of Physics, University of Oslo, PB 1048 Blindern, N-0316 Oslo, Norway*
[3] *National Research Nuclear University "MEPhI" (Moscow Engineering Physics Institute), Kashirskoe highway 31, Moscow, RU-115409, Russia*
[4] *Max-Planck-Institut für Intelligente Systeme, Heisenbergstr. 3, D-70569 Stuttgart, Germany*
[5] *Institut für Augewandte Physik, Universität Tübingen, Auf der Morgenstelle 10, D-72076 Tübingen, Germany*

Abstract. Proton-proton collisions at energies from $\sqrt{s}$ = 200 GeV up to $\sqrt{s}$ = 14 TeV are studied within the microscopic quark-gluon string model. The model is based on Gribov's Reggeon Field Theory accomplished by string phenomenology. Comparison with experimental data shows that QGSM describes well particle yields, rapidity- and transverse momentum spectra, rise of mean $\langle p_T \rangle$ and forward-backward multiplicity correlations. The latter arise in QGSM because of the addition of various processes with different mean multiplicities. The model also indicates fulfillment of extended longitudinal scaling and violation of Koba-Nielsen-Olesen scaling at LHC. The origin of both features is traced to short-range particle correlations in the strings. Predictions are made for $\sqrt{s}$ = 14 TeV.

1 Introduction

Despite the significant progress achieved in the last years, the theory of multiparticle production in elementary proton-proton collisions at ultrarelativistic energies is not completed yet. We all know, of course, that the processes with large momentum transfer Q^2 are well described by the perturbative chromodynamics (pQCD). Unfortunately, even at very high energies the main contribution to multiparticle production in hadronic interactions comes from the processes with small momentum transfer. This means that the running coupling constant $\alpha_s(Q^2)$ is not small and, therefore, the perturbative series expansion is not very helpful. Other techniques, based on non-perturbative methods, should be utilized. The quark-gluon string model (QGSM) [1] and similar to it dual parton model (DPM) [2] is one of the possible approaches to solution of this very interesting and difficult problem. Both models are based on the Reggeon Field theory (RFT) [3]. Basic principles of the QGSM are sketched below.

[a] e-mail: eugen.zabrodin@fys.uio.no

2 The QGSM model

The quark-gluon string model, formulated about 35 years ago [1], employs the $1/N$ series expansion of the amplitude of a process in QCD, where N is either the number of colors [4] or the number of flavors [5]. This method is also called *topological expansion*, because of emergence of diagrams of various topologies. Although it is not possible to assign weights for the diagrams within the QCD, there is one-to-one mapping between the diagrams in $1/N$-expansion and the processes with exchange of Regge singularities in the t-channel. For instance, exchange of quantum numbers via Reggeons corresponds to *planar* diagrams, whereas the *cylinder* diagrams are represented by the reactions without the quantum number exchange. The latter proceed via the exchange of Pomerons. Therefore, the perturbative Reggeon Field Theory (RFT) [3] is directly linked to quantum chromodynamics.

The Monte Carlo version of the QGSM [6] employs statistical weights, hadron structure functions and leading quark fragmentation functions obtained from the Regge approach in [1] to choose subprocesses of string production, to compute mass and momentum of strings and to simulate string decays, respectively. The hadron-hadron collision part of the model includes single and double diffraction subprocesses, antibaryon-baryon annihilation and elastic scattering. The hadron inelastic interaction cross section $\sigma_{inel}(s) = \sigma_{tot}(s) - \sigma_{el}(s)$ is split further into the cross section for single diffractive interactions $\sigma_{SD}(s)$ and the cross section for non-diffractive reactions $\sigma_{ND}(s)$, similar to analysis of experimental data. By means of the Abramovskii-Gribov-Kancheli (AGK) cutting rules [7] the inelastic non-diffractive interaction cross section $\sigma_{ND}(s)$ can be expressed via the sum of the cross sections for the production of $n = 1, 2, \ldots$ pairs of quark-gluon strings, or cut Pomerons, and the cross section of double diffractive process

$$\sigma_{ND}(s) = \sum_{n=1}^{\infty} \sigma_n(s) + \sigma_{DD}(s) \, . \tag{1}$$

To find $\sigma_n(s)$ one can utilize the quasi-eikonal model [8, 9] which states that

$$\sigma_{tot}(s) = \sum_{n=0}^{\infty} \sigma_n(s) = \sigma_P \, f\left(\frac{z}{2}\right) , \tag{2}$$

$$\sigma_n(s) = \frac{\sigma_P}{nz}\left(1 - \exp(-z) \sum_{k=0}^{n-1} \frac{z^k}{k!}\right) , \; k \geq 1 \tag{3}$$

$$\sigma_0 = \sigma_P \left(f\left(\frac{z}{2}\right) - f(z) \right) , \tag{4}$$

$$f(z) = \sum_{\nu=1}^{\infty} \frac{(-z)^{\nu-1}}{\nu \nu!} , \tag{5}$$

Here

$$\sigma_P = 8\pi\gamma_P \exp(\Delta\xi) , \tag{6}$$

$$z = \frac{2C\gamma_P}{(R_P^2 + \alpha'_P \xi)} \exp(\Delta\xi) . \tag{7}$$

The cross section σ_0 corresponds to diffraction contribution. The parameters γ_P and R_P are Pomeron-nucleon vertex parameters, quantity $\Delta \equiv \alpha_P(0) - 1$, and $\alpha_P(0)$ and α'_P is the intercept and the slope of the Pomeron trajectory, respectively. The quantity C takes into account the deviation from the pure eikonal approximation ($C = 1$) due to intermediate inelastic diffractive states, $\xi = \ln(s/s_0)$ and s_0 is a scale parameter. Note that at very high energies the enhanced diagrams [10], describing the multiple

Table 1. Parameters of both soft and hard Pomeron used in the present calculations.

Parameter	Soft Pomeron	Hard Pomeron
$\alpha_P(0)$	1.15615	1.3217
α'_P	0.25	0
R_P	2.0	2.4
γ_P	1.27475	0.021

gluon production from Pomeron vertices, can violate [11] the AGK cutting rules. These diagrams, however, are not implemented in the MC version of QGSM employed for present calculations.

At ultrarelativistic energies the model includes the new processes which represent the hard (multi)Pomeron exchanges. - The concept of soft and (semi)hard Pomerons is attributed to a major part of Monte Carlo models based on RFT, see [2, 12–14]. - Parameters of the soft and hard Pomerons used in present calculations are listed in Table 2. We see that the intercept of hard Pomeron trajectory is significantly larger than that of the soft Pomeron, whereas its slope is zero.

The diagrams describing subprocesses in ultrarelativistic pp collisions are displayed in Fig. 1. Here diagram (a) shows the soft multi-Pomeron exchange, diagram (b) represents both hard and soft Pomeron exchange, diagrams (c) and (d) show the single diffractive (SD) process with small (c) and large (d) mass excitation, and diagrams (e)-(g) display double diffractive (DD) processes with large (e) and small (f) mass excitation, and central diffraction (g). The cross sections of SD and DD processes in pp collisions were calculated within the QGSM in [15]. In our MC version, we utilize the following parametrization of these results proposed in [16]

$$\sigma_{SD}(s) = 0.68\left(1 + \frac{36}{s}\right)\ln(0.2\,s) + 0.6\,, \tag{8}$$

$$\sigma_{DD}(s) = 0.27\ \ln s + 1.65\,. \tag{9}$$

Cross sections of both processes rise as $\sigma_D \propto \ln s$, which satisfies the Froissart limit $\sigma_{tot} \propto (\ln s)^2$.

Cross sections σ_{tot}, σ_{el}, σ_{SD} and σ_{DD} obtained by QGSM are compared in Fig. 2 with the predictions of other RFT-based models [14, 17–20]. At Tevatron energy $\sqrt{s} = 1.8$ TeV all models agree within 5% accuracy limit for all cross sections except of the double-diffraction one, whereas at $\sqrt{s} = 14$ TeV the model predictions start to deviate. Still, results for σ_{tot} and σ_{SD} are close to each other. Two versions of the Durham model, KMR-1 and KMR-2, predict excess of about 50% of double-diffraction cross section compared to other models. QGSM results are quite close to the calculations of the GLMM model [17]. QGSJET model [14], which also employs the scheme with soft and hard Pomerons, yields larger total and elastic cross sections at $\sqrt{s} = 14$ TeV. Further discussion of the differences and similarities between the models can be found in [18].

It is easy to see that the main contribution to particle production in pp collisions at ultrarelativistic energies comes from the multi-Pomeron exchanges. The amplitude of the n-Pomeron exchange is suppressed by factor $\left(\frac{1}{N}\right)^{2n}$, but it grows proportionally to $\left(\frac{s}{s_0}\right)^{n\Delta}$. The last term dominates over the suppression factor at LHC energies.

Finally, the Field-Feynman algorithm [21] is employed to describe the fragmentation of strings into hadrons. Here the hadron emission proceeds from both ends of the string with equal probabilities under imposed conditions of the preservation of the quark numbers and the energy-momentum conser-

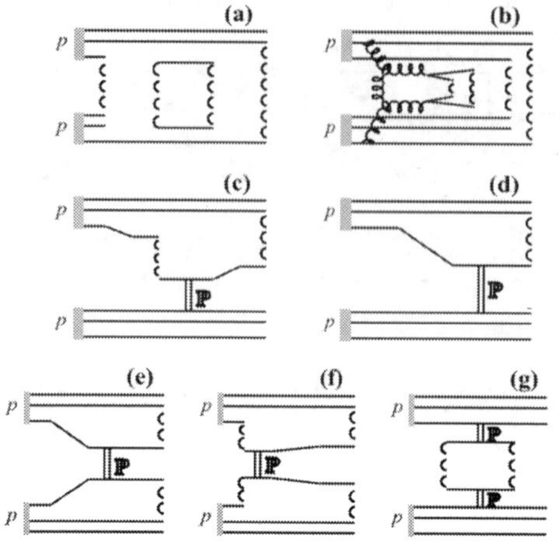

Figure 1. Diagrams taken into account in QGSM in the modeling of pp interactions at ultrarelativistic energies: (a) multi-Pomeron exchange, (b) (semi)hard gluon-gluon interaction and soft Pomeron exchange, (c)-(d) single diffraction with high-mass and low-mass excitation, (e)-(f) double diffraction with low-mass and high-mass excitation, (g) central diffraction.

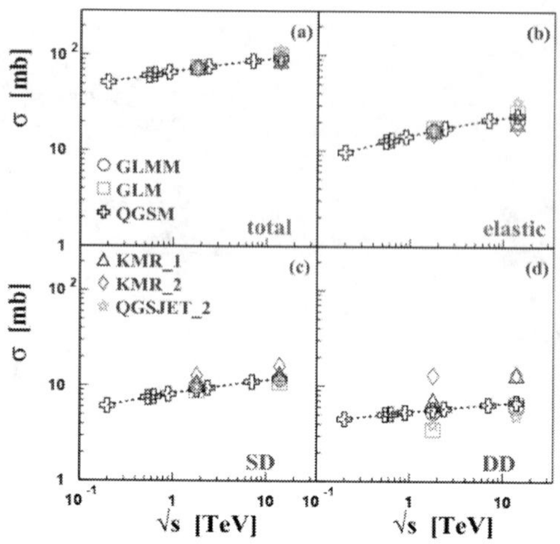

Figure 2. Model predictions for (a) total, (b) elastic, (c) single-diffractive and (d) double-diffractive cross sections as functions of $\sqrt{s}$. The models are as follows: GLMM (circles) [17], GLM (squares) [18], QGSM (crosses), KMR-1 (triangles) [19], KMR-2 (diamonds) [20], and QGSJET-2 (stars) [14]. Dashed lines between the QGSM points are drawn to guide the eye.

vation. The transverse momenta of the produced quarks and diquarks obey the power-law probability

$$f(p_T^2)\,dp_T^2 \;=\; \frac{3\,D\,b_2(s)}{\pi}\,(1+Dp_T^2)^{-4}\,dp_T^2\,, \tag{10}$$

$$b_2(s) \;=\; 0.325+0.016\,\ln s\,,\quad D=0.34\,(\mathrm{GeV}/c)^{-2} \tag{11}$$

QGSM was successfully extended to the description of hadron-nucleus and nucleus-nucleus collisions at energies spanning from several GeV up to energies available at RHIC and LHC, see, e.g., [22–30]. Our present study, however, deals merely with proton-proton collisions in the energy range 200 GeV $\le \sqrt{s} \le$ 14 TeV.

3 Results

3.1 Transverse momentum and rapidity distributions

Here results for inelastic and non-single diffractive pp collisions are presented. Figure 3 displays dependence of the invariant cross section $E\dfrac{d^3\sigma}{dp^3}$ of charged hadrons on their transverse momentum p_T in NSD events. Calculations are performed for c.m. energies varying from 200 GeV to 14 TeV. Experimental data used for the comparison are taken from [31–43]. One can see that the combination of soft and hard processes in the model describes fairly well the p_T-spectra in the interval $0 \le p_T \le$ 10 GeV/c for all energies. The slopes of the distributions become less steep with rising collision

DOI: 10.1051/epjconf/201716401005

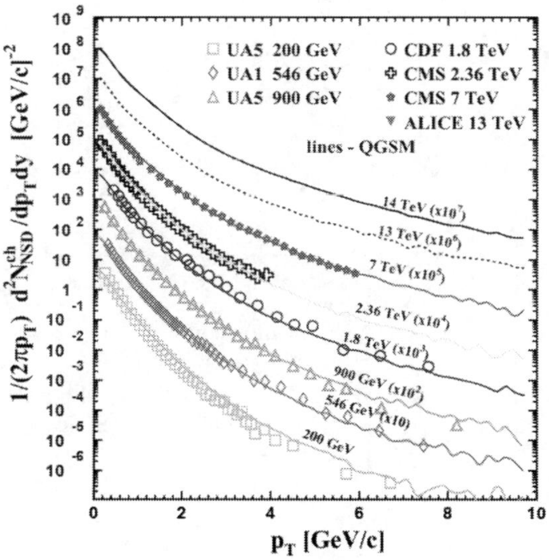

Figure 3. Transverse momentum distributions of the invariant cross section of charged particles in NSD pp collisions obtained in QGSM at $|y| \leq 2.5$ for energies from 200 GeV up to 14 TeV. Data are taken from [31–33, 39, 41].

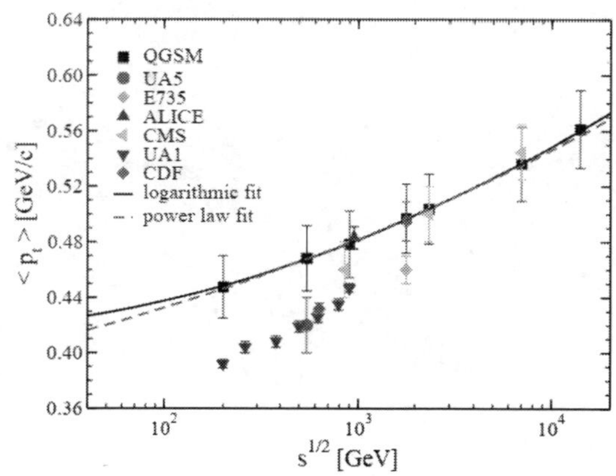

Figure 4. $\langle p_T \rangle$ vs. $\sqrt{s}$. Squares indicate the QGSM results; other symbols show the experimental data from [31–34, 37, 41]. Solid and dashed lines present fit to Eq. (12) and Eq. (13), respectively.

energy, i.e. the spectra become harder because of the increasing contribution of the hard processes. This leads to increase of the average transverse momentum of charged hadrons shown in Fig. 4. Here the model results are plotted onto the experimental data, obtained in pp and $\bar{p}p$ interactions at $\sqrt{s} \leq 7$ TeV. The LHC data show clearly a faster rise of the mean p_T compared to linear extrapolation $\langle p_T \rangle = a + b \ln s$ of data below 900 GeV. Two parametrizations of $\langle p_T \rangle$ were proposed. The first one is purely phenomenological quadratic dependence on $\ln s$, and the second one is a power-law dependence on $E = \sqrt{s}/2$ obtained within the color glass condensate (CGC) approach [44, 45]. By fitting the QGSM results, we got the following parameters for both approximations

$$\langle p_{\rm T} \rangle = 0.417 - 0.0035 \ln s + 0.00059 \ln^2 s\,, \quad (12)$$

$$\langle p_{\rm T} \rangle = 0.243 + 0.12 {\rm E}^{0.1107}\,. \quad (13)$$

Both curves are shown in Fig. 4 also. We see that the curves almost coincide within the whole investigated energy range, and so it is impossible to discriminate between the two parametrizations. Note, however, that Eq. (13) contains one less parameter compared to Eq. (12) because the exponent 0.1107 is not a free parameter but equals to one half of the exponent of the power-law fit to c.m. energy dependence of magnitudes of $dN^{charge}/d\eta$ distributions at midrapidity [44, 45].

To check this correspondence, the $dN/d\eta$ distributions of charged hadrons in inelastic and NSD events are displayed in Fig. 5(a) and Fig. 5(b), respectively. Comparison with experimental data shows that QGSM correctly reproduces the data in both central and fragmentation region. The charged particle density at midrapidity increases with rising $\sqrt{s}$, i.e., no Feynman scaling [46] is observed at LHC. The model indicates further rise of the central charged particle density at $\sqrt{s} = 14$ TeV of about 20% compared to that at $\sqrt{s} = 7$ TeV; the values at midrapidity are $dN^{inel}/d\eta = 5.8$ and $dN^{NSD}/d\eta = 6.7$, respectively. It is interesting to compare these results with the predictions in Ref. [47] made by

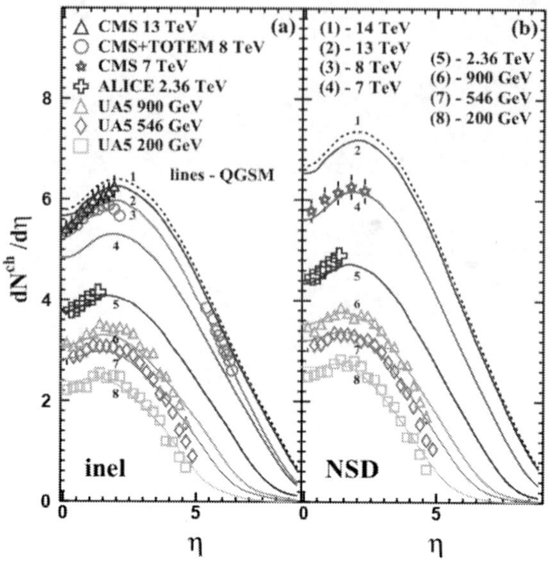

Figure 5. Pseudorapidity spectra of charged hadrons in (a) inelastic and (b) NSD pp collisions at energies from 200 GeV up to 14 TeV. Lines show the QGSM calculations, symbols denote the data taken from [31, 33, 36, 39, 41, 43].

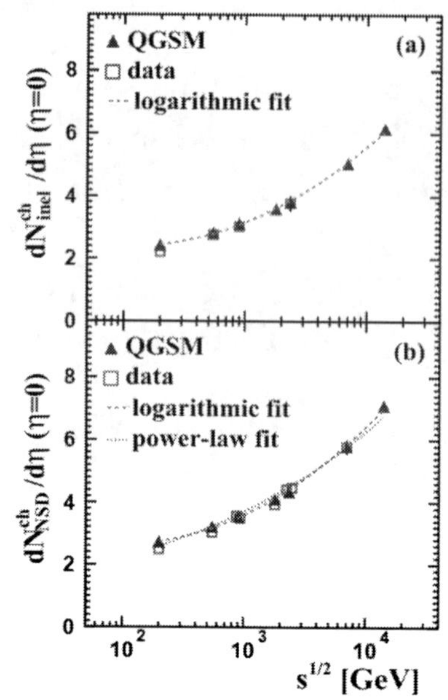

Figure 6. The density of charged particles at midrapidity as a function of $\sqrt{s}$ for (a) inelastic and (b) NSD pp collisions. Dashed lines show the results of the fit to quadratic dependence on $\ln s$, whereas dotted line (b) – to the power-law dependence $\sim \sqrt{s}^{\lambda}$.

extrapolation of the data at lower energies. The extrapolation method relies on the proportionality to $\ln s$ of both the height and the width of $dN/d\eta$-distributions, and the energy independence of the slopes of η-spectra. This provides $dN^{inel}/d\eta(\eta = 0) = 4.6 \pm 0.4$ and $dN^{NSD}/d\eta(\eta = 0) = 5.25 \pm 0.7$ [47]. Both results are, obviously, significantly below the LHC experimental data.

The increase of midrapidity density as function of $\sqrt{s}$ is presented in Fig. 6 for both inelastic and non-single diffractive interactions. One can see clearly the deviation from the linear rise, proportional to $\ln s$, which is valid for energies below 2 TeV. Similarly to p_T spectra, $dN/d\eta$ distributions calculated by QGSM were also fitted to the quadratic logarithmic dependence and to the power-law dependence, respectively, in the energy interval 200 GeV $\leq \sqrt{s} \leq$ 14 TeV. The results are as follows:

$$\left.\frac{dN_{inel}}{d\eta}\right|_{\eta=0}(s) = 4.36 - 0.507 \ln s + 0.03 \ln^2 s\,, \tag{14}$$

$$\left.\frac{dN_{NSD}}{d\eta}\right|_{\eta=0}(s) = 5.015 - 0.60 \ln s + 0.036 \ln^2 s\,, \tag{15}$$

$$\left.\frac{dN_{NSD}}{d\eta}\right|_{\eta=0}(s) = 0.77\, E^{0.22}\,. \tag{16}$$

Both fitting curves are plotted onto the spectra in Fig. 6. Again, there is no visible difference between the two parametrizations. Note that the exponent $\lambda = 0.22$ is indeed twice larger than the corresponding exponent of the power-law fit to the mean-p_T data.

DOI: 10.1051/epjconf/201716401005

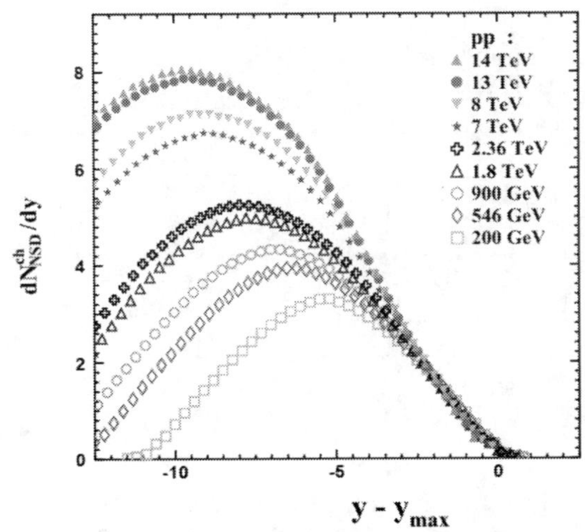

Figure 7. The distributions $\frac{1}{\sigma_{NSD}}\frac{d\sigma_{NSD}}{dy}$ of charged hadrons as functions of rapidity difference $y - y_{max}$ obtained in QGSM for pp collisions at all energies in question.

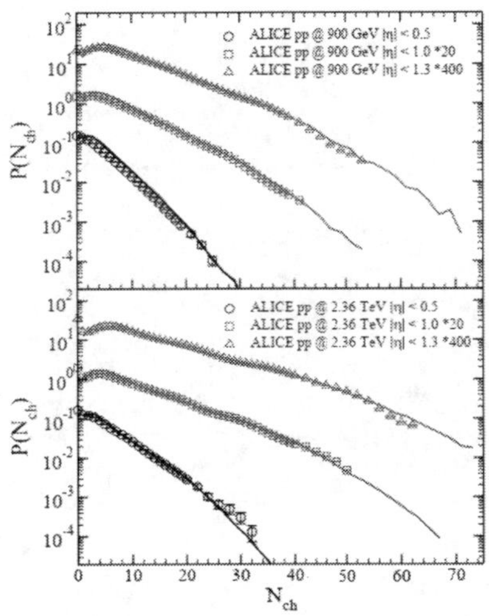

Figure 8. Charged particle multiplicity distributions in $|\eta| < 0.5$, $|\eta| < 1.0$ and $|\eta| < 1.3$ intervals, obtained in QGSM for pp collisions at (a) $\sqrt{s} = 900$ GeV and at (b) $\sqrt{s} = 2.36$ TeV. Open symbols denote the ALICE data [36].

3.2 Extended longitudinal scaling and KNO scaling

Extended longitudinal scaling (ELS) was observed first in Ref. [48] in the fragmentation region of dN/dy (or $dN/d\eta$) spectra. Here the particle rapidity spectra have the same slope independent on the c.m. energy $\sqrt{s}$. To demonstrate this distinctly, the dN/dy distributions at different energies are shifted to the beam rapidity, y_{max}. The ELS was found to hold to 10% in a broad energy range 30.8 GeV $\leq \sqrt{s} \leq$ 900 GeV [31]. Some time ago, in Ref. [49] the violation of the ELS in pp collisions was predicted at energies available at LHC. To obtain the dN/dy particle spectrum at LHC the authors of [49] have fitted the rapidity spectra of hadrons at lower energies to the Gaussian, extracted the widths σ_G, and extrapolated the obtained $\sigma_G(\sqrt{s})$ dependence to the LHC energy range. The predicted dN/dy spectrum appeared to be quite narrow, thus indicating violation of the ELS.

QGSM does not support this prediction. We plot in Fig. 7 dN^{ch}_{NSD}/dy spectra as functions of $y - y_{max}$ in pp interactions at energies from $\sqrt{s}$ = 200 GeV to 14 TeV. It is easy to see that in the fragmentation region all curves sit on the top of each other. The fulfillment of the extended longitudinal scaling in the model is due to short-range rapidity correlations of hadrons in a string [16]. The correlation function of any two particles has exponential dependence

$$C(y_1, y_2) \propto \exp\left[-\lambda(y_1 - y_2)\right] , \tag{17}$$

therefore, particles are uncorrelated provided $\Delta y = y_1 - y_2 \gg 1$. Then, the i-th particle inclusive cross section for the process $1 + 2 \longrightarrow i + X$ is

$$f_i = \frac{d^2\sigma(y_1 - y_i, y_i - y_2, (p_T^{(i)})^2)}{dy_i d^2(p_T^{(i)})^2} \tag{18}$$

DOI: 10.1051/epjconf/201716401005

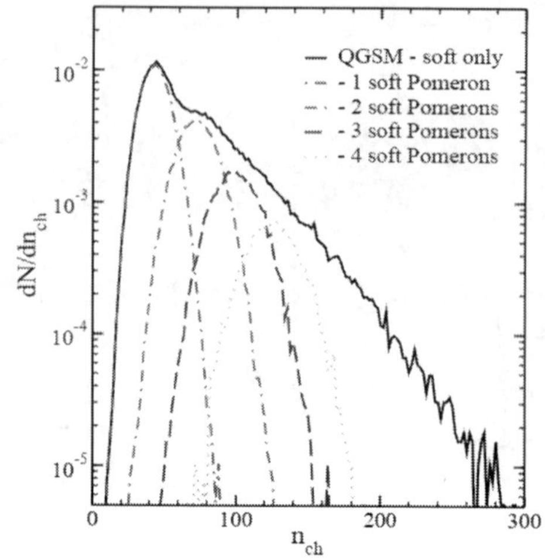

Figure 9. QGSM calculations of charged particle multiplicity distribution for processes going via the exchange of soft Pomerons in pp collisions at $\sqrt{s}$ = 14 TeV. Partial contributions of the 1P, 2P, 3P, and 4P processes are shown by dash-dotted, double-dash–dotted, dashed, and dotted lines, respectively.

If particle i is produced in the fragmentation region of particle 1, this means that $y_1 - y_i \approx 1$ and $y_i - y_2 \approx y_1 - y_2 \gg 1$, i.e., f_i depends only on two variables, $y_1 - y_i$ and $(p_T^{(i)})^2$. Recalling that $\ln x_F^{(i)} \propto y_i - y_1$, the condition of the Feynman scaling for the particle inclusive density $n_i = f_i/\sigma_{inel}$ is recovered:

$$n_i = \phi(x_F^{(i)}, (p_T^{(i)})^2)\,. \tag{19}$$

Another interesting form of the scaling dependence is the so-called Koba-Nielsen-Olesen (KNO) scaling suggested in Ref. [50]. According to it, the multiplicity distribution $P_n(s) = \sigma_n(s)/\sigma_{tot}(s)$, where $\sigma_n(s)$ is the partial cross section of the process with n particles in the final state, depends on the c.m. energy $\sqrt{s}$ only through the normalized multiplicity $n/\langle n\rangle$, i.e.,

$$P_n(s) = \frac{1}{\langle n\rangle}\Psi\left(\frac{n}{\langle n\rangle}\right)\,. \tag{20}$$

Here $\langle n\rangle$ is the averaged multiplicity and $\Psi(n/\langle n\rangle)$ is an energy independent function. The KNO scaling was observed in pp and $\bar{p}p$ collisions up to energies available at ISR accelerator, $\sqrt{s} \leq$ 62 GeV. Its violation at higher energies was predicted by the RFT specialists, see, e.g., [1, 7, 51]. The mechanism of the KNO violation within the Reggeon Field Theory is as follows.

The multiplicity of the produced particles in ultrarelativistic hadronic interactions is determined mainly by the number of cut Pomerons. Each cut Pomeron produces two strings, and the short range correlations of particles within the strings provide a Poisson multiplicity distribution of secondaries. At not very high energies or in the central rapidity window $|y| \leq 1$, the contributions of the processes with different number of cut Pomerons strongly overlap. Therefore, the KNO scaling approximately holds. With rising c.m. energy, the number of strings and the energy deposited into a single string increases. As a result, we have an enhancement of the high multiplicity events. The particle distribution starts to deviate from the Poissonian, and violation of the KNO scaling takes place.

The QGSM calculations of the charged particle multiplicity distributions in three midrapidity intervals for pp interactions at 900 GeV and 2.76 TeV are compared in Fig. 8 with the experimental data

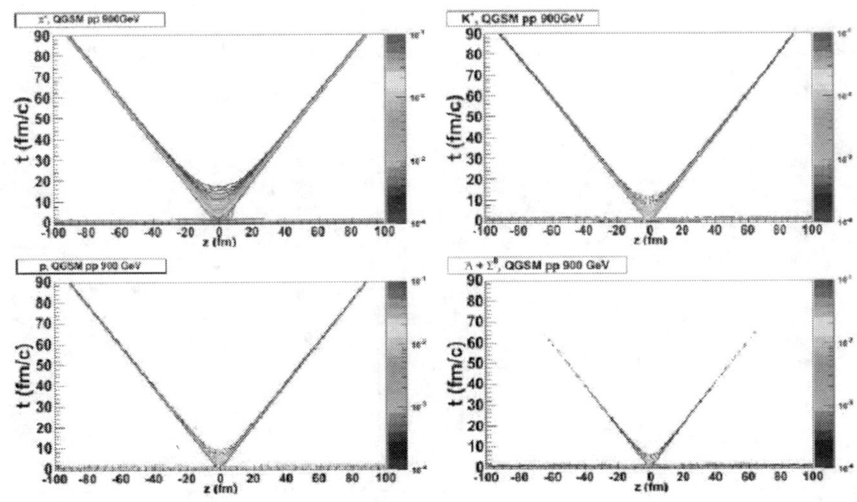

Figure 10. $d^2N/dzdt$ distribution of positive pions (upper left), positive kaons (upper right), protons (bottom left) and $\Lambda + \Sigma^0$ (bottom right) in QGSM for pp collisions at $\sqrt{s} = 900$ GeV.

of the ALICE Collaboration [36]. The model results agree well with the data. Investigation of these spectra in terms of the variable $n/\langle n \rangle$, see Ref. [52], has confirmed the violation of the KNO scaling in pp collisions at LHC energies. Namely, (i) formation of the more pronounced peak at not very high multiplicities, and (ii) rise of the high multiplicity tail of the multiplicity distribution have been observed. Figure 9 reveals the underlying mechanisms of these phenomena. For the sake of simplicity, we consider here events with only soft Pomerons. The more Pomerons in the event, the larger the high multiplicity tail of its spectrum. The average multiplicities of the multi-Pomeron distributions do not coincide anymore, and the local peak corresponding to the single-Pomeron exchange becomes apparent.

3.3 Particle freeze-out

In this section we present the distributions of coordinates and momenta for four abundant particle species, namely π^+, K^+, p and $\Lambda + \Sigma^0$. Recall that the distributions are averaged over the whole ensemble of events. Therefore, the fluctuations on the event-by-event basis are disregarded. Apparently, particles can be produced in inelastic and elastic collisions, and in decays of resonances, but here we will not distinguish between the last interaction points.

The phase-space distribution of the particles on the mass shell depends on seven independent variables: coordinates $\vec{x}$, momenta $\vec{p}$, and time t. In both hadron-hadron interactions and central collisions of symmetric nuclei, there are only ten different coordinate pairs [26]. Two of them which give a nice visual information about the evolution of the system are shown in Figs. 10 and 11. Figure 10 displays the longitudinal position of particles versus the creation time, $\frac{d^2N}{dzdt}$. Calculations are done for pp collisions at $\sqrt{s} = 900$ GeV. One can see that a large number of secondary hadrons freezes out within

 DOI: 10.1051/epjconf/201716401005

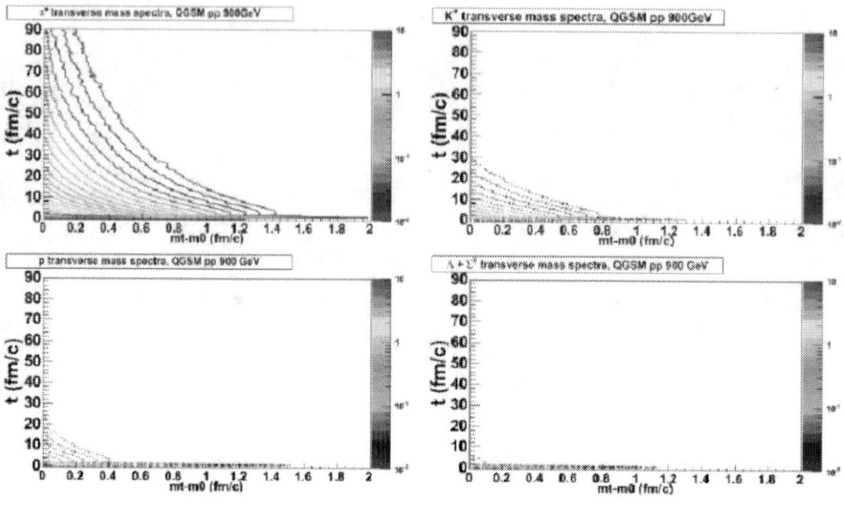

Figure 11. The same as Fig. 10 but for $d^2N/m_T dm_T dt$ distribution.

the first fermi-second. Particles demonstrate clear mass hierarchy: heavy particles freeze out earlier compared to the light ones, as heavy strings become less prevalent. Production of charged pions takes place up to $t \approx 18$ fm/c in the central zone of the collision, whereas for charged kaons, protons and Λ's these times reduce to 12 fm/c, 10 fm/c, and 7 fm/c, respectively.

The transverse mass distribution, $\dfrac{d^2N}{m_T dm_T dt}$, is shown in Fig. 11. It appears that particles produced at the very beginning of the collision carry larger transverse mass and, therefore, larger transverse momentum. This finding is in line with the results of femtoscopy analysis that hadrons with the largest p_T indicate the smallest emission zone. A second peak seen in just a few fm/c after the initial collision comes from the decays of short-lived resonances. The continuous production of hadrons from the resonance decays takes place up to quite long times. The transverse mass of the created hadrons drops almost exponentially with increasing time.

3.4 Forward-backward multiplicity correlations

Forward-backward multiplicity correlations between charged hadrons emitted in forward and backward hemispheres were first observed in [53]. The strength of the correlations is defined as

$$b = \frac{\langle (n_F - \langle n_F \rangle)(n_B - \langle n_B \rangle) \rangle}{\left[\langle (n_F - \langle n_F \rangle)^2 \rangle \langle (n_B - \langle n_B \rangle)^2 \rangle \right]^{1/2}} , \tag{21}$$

where n_B and n_F represent multiplicities of charged particles in backward and forward hemispheres, respectively. Figure 12 depicts the dependence of the mean charged-particle multiplicity in the backward hemisphere $\langle n_B \rangle$, measured in the range $-4 \leq \eta \leq 0$, on the multiplicity in the forward hemisphere n_F for the symmetric range $0 \leq \eta \leq 4$ at four c.m. energies, $\sqrt{s}$ = 200 GeV, 546 GeV,

 DOI: 10.1051/epjconf/201716401005

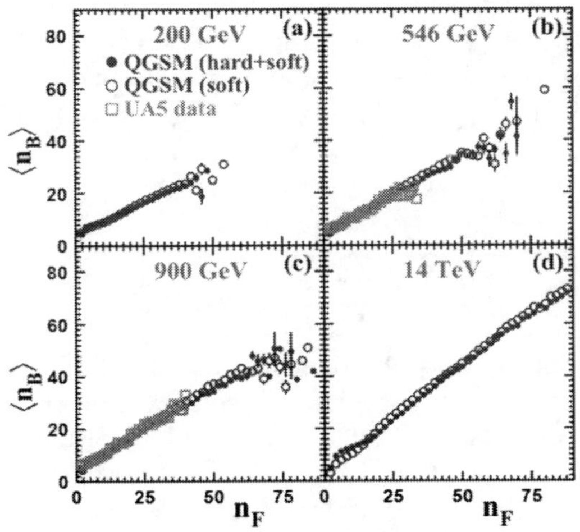

Figure 12. Backward-forward multiplicity correlations $\langle n_B(n_F)\rangle$ for $0 \le |\eta| \le 4$ in NSD pp interactions at $\sqrt{s}$ = 200 GeV (a), 546 GeV (b), 900 GeV (c) and 14 TeV (d). Open circles denote contributions of soft processes, full symbols are for all processes. Data are from [31].

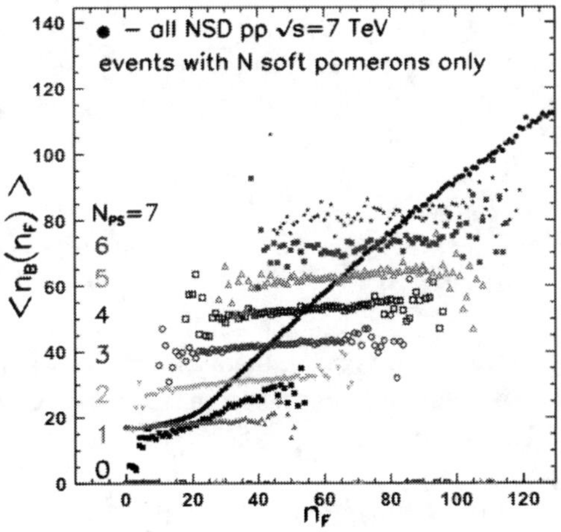

Figure 13. Backward-forward multiplicity correlations $\langle n_B(n_F)\rangle$ in QGSM calculations of NSD pp interactions at $\sqrt{s}$ = 7 TeV proceeding via the soft Pomeron exchanges. Also shown are correlations for the collisions with fixed amount of soft Pomerons (from zero to seven).

900 GeV, and 14 TeV. Comparison with the experimental data at $\sqrt{s}$ = 546 GeV and 900 GeV shows a good agreement between the model results and the data. The dependence $\langle n_B(n_F)\rangle$ looks pretty linear at all energies,

$$\langle n_B(n_F)\rangle = a + b\, n_F\ , \qquad (22)$$

revealing a sort of long-range correlations. The correlation strength is characterized by the slope b, which increases with rising energy of the collisions. As shown in Fig. 12, the correlations between the $\langle n_B\rangle$ and n_F are mainly determined by the soft processes.

The linear profile of the forward-backward correlations can be reproduced by the mixture of several emitting sources. Moreover, the multiplicity of the secondary hadrons should be described by the Poisson distribution, and all sources should possess different average multiplicities [31], see also [54, 55] and references therein. Let us sketch briefly how this picture emerges in the QGSM.

It is worth noting that the model successfully reproduces the forward-backward multiplicity correlations in pp and $\bar{p}p$ collisions not only at energies above hundred GeV, but also at energies about few GeV in the c.m. frame [56]. In this energy range the variety of subprocesses describing the $\bar{p}p$ interactions is more copious compared to that describing the pp collisions, because of, e.g., annihilation processes, planar diagrams, and so forth. As a result, the correlation strength is stronger in $\bar{p}p$ interactions. At ultrarelativistic energies the contribution of these diagrams to the total cross section rapidly drops to zero. The difference between the b^{pp}_{corr} and $b^{\bar{p}p}_{corr}$ disappears, but the slopes of the distributions become steeper at higher energies. It happens because of the increase of the variety of multi-Pomeron processes with energy. If, however, we are picking up just one particular subprocess with the fixed amount of Pomeron exchanges, the correlation slope is essentially zero, as seen in Fig. 13. In this figure the forward-backward correlations in NSD pp collisions at $\sqrt{s}$ = 7 TeV are

DOI: 10.1051/epjconf/201716401005

shown for the subprocesses with N soft Pomerons, where N varies from zero to seven. Despite the significant event-by-event multiplicity fluctuations, the slopes of the partial distributions are absent. But when we are combining the subevents with different mean multiplicities together, a nice picture of positive correlations $\langle n_B(n_F)\rangle$ arises. This circumstance explains also the nonzero slope of the distribution for events with just one soft Pomeron. Here we are mixing up several diagrams corresponding to the double diffraction displayed in Fig. 1. Detailed study of the forward-backward multiplicity correlations within the QGSM and comparison with the available experimental data can be found in [55].

4 Conclusions

The Monte Carlo version of the quark-gluon string model is applied for the description of proton-proton collisions at energies from $\sqrt{s} = 200$ GeV to 14 TeV. The model is based on Gribov's Reggeon Field Theory and employs theoretical estimates for the total cross sections, as well as for the cross sections of single diffraction and double diffraction processes. Parameters of soft and hard Pomerons are tuned to provide the best fit to the data at LHC energies. It is shown that QGSM reproduces well the bulk observables, such as yields, rapidity- and transverse momentum spectra of charged particles at all energies in question.

Several scaling hypothesis have been examined.The model indicates fulfillment of the extended longitudinal scaling accompanied by the violation of both Feynman scaling and KNO scaling. The origin of these phenomena is traced to short range momentum correlations of particles in strings and copious production of hadrons in multi-Pomeron processes. The latter leads to the formation of distinct peak in the multiplicity distribution at lower multiplicities and rise of the high-multiplicity tail of the distribution.

Although the main part of the most abundant particle species is produced at the very beginning of the collision, there is no sharp freeze-out of hadrons in pp interactions. Hadron emissivity rapidly drops with time t. However, the production of pions in the central zone of pp collisions at $\sqrt{s} = 900$ GeV, for instance, takes place up to $t \sim 18$ fm/c. For other hadron species this time is shortened, thus revealing the mass hierarchy: the heavier the particle, the faster the freeze-out proceeds. Hadrons with large transverse momentum are produced at the first fm/c, whereas soft hadrons are emitted during the whole time of evolution. Secondary peak in the $d^2N/m_T dm_T dt$ distribution appears because of decays of short-lived resonances.

Finally, the model was employed to study forward-backward multiplicity correlations. Note that no extra tuning of the model parameters was performed. Because of the rich set of various subprocesses at partonic level, QGSM reproduces nicely almost linear positive FB correlations in pp and $\bar{p}p$ interactions in a very broad energy range. We show that for each particular subprocess the FB multiplicity correlations are absent despite of the significant multiplicity fluctuations on event-by-event basis. But when we add up all these events with different average multiplicities, the picture of strong positive FB correlations emerges.

Acknowledgements

We would like to thank B. Kopeliovich, V. Khose, E. Levin, L. Lipatov, J. Nyiri, and V. Fadin for the fruitful discussions and valuable comments. J.B. thanks the German Research Foundation (DFG) for the financial support through the Project BL 1286/2-1. L.B. acknowledges financial support of the Alexander von Humboldt Foundation.

References

[1] A.B. Kaidalov, Phys. Lett. B **116**, 459 (1982);
A.B. Kaidalov and K.A. Ter-Martirosyan, Phys. Lett. B **117**, 247 (1982).
[2] A. Capella, U. Sukhatme, C.-I. Tan, and J. Tran Thanh Van, Phys. Rep. **236**, 225 (1994).
[3] V.N. Gribov, Sov. Phys. JETP **26**, 414 (1968).
[4] G. t'Hooft, Nucl. Phys. B **72**, 461 (1974).
[5] G. Veneziano, Phys. Lett. B **52**, 220 (1974).
[6] N.S. Amelin and L.V. Bravina, Sov. J. Nucl. Phys. **51**, 133 (1990) .
[7] V. Abramovskii, V. Gribov, O. Kancheli, Sov. J. Nucl. Phys. **18**, 308 (1974).
[8] A. B. Kaidalov, Surveys in High Energy Phys. **13**, 265 (1999).
[9] M. Baker, K. A. Ter-Martirosyan, Phys. Rep. **28C**, 1 (1976).
[10] O.V. Kancheli, JETP Lett. **11**, 267 (1970);
A.H. Mueller, Phys. Rev. D **2**, 2963 (1970).
[11] Y. V. Kovchegov and K. Tuchin, Phys. Rev. D **65**, 074026 (2002);
E. Levin and A. Prygarin, Phys. Rev. C **78**, 065202 (2008).
[12] K. Werner, F.-M. Liu, and T. Pierog, Phys. Rev. C **74**, 044902 (2006).
[13] R. Engel, J. Ranft, and S. Roesler, Phys. Rev. D **52**, 1459 (1995).
[14] S. Ostapchenko, Nucl. Phys., Proc. Suppl. **151**, 143 (2006); Phys. Rev. D **83**, 014018 (2011).
[15] A.B. Kaidalov and M.G. Poghosyan, Eur. Phys. J. C **67**, 397 (2010).
[16] J. Bleibel, L.iV. Bravina, and E.E. Zabrodin, Phys. Rev. D **93**, 114012 (2016).
[17] E. Gotsman, E. Levin, U. Maor, and J. S. Miller, Eur. Phys. J. C **57**, 689 (2008).
[18] E. Gotsman, E. Levin, and U. Maor, Eur. Phys. J. C **71**, 1553 (2011).
[19] M. G. Ryskin, A. D. Martin, and V. A. Khoze, Eur. Phys. J. C **54**, 199 (2008).
[20] M. G. Ryskin, A. D. Martin, V. A. Khoze, and A. G. Shuvaev, J. Phys. G **36**, 093001 (2009).
[21] R. D. Field, R. P. Feynman, Nucl. Phys. B **136**, 1 (1978).
[22] N.S. Amelin, E.F. Staubo, and L.P. Csernai, Phys. Rev. D **46**, 4873 (1992).
[23] N.S. Amelin *et al.*, Phys. Rev. C **47**, 2299 (1993).
[24] L. Bravina, L.P. Csernai, P. Levai, D. Strottman, Phys. Rev. C **50**, 2161 (1994).
[25] L.V. Bravina, Phys. Lett. B **344**, 49 (1995).
[26] L.V. Bravina *et al.*, Phys. Rev. C **60**, 044905 (1999).
[27] E.E. Zabrodin *et al.*, Phys. Lett. B **508**, 184 (2001).
[28] E.E. Zabrodin *et al.*, Phys. Rev. C **63**, 034902 (2001).
[29] L.V. Bravina *et al.*, Phys. Rev. C **78**, 014907 (2008).
[30] G. Burau *et al.*, Phys. Rev. C **71**, 054905 (2005).
[31] G. J. Alner *et al.* (UA5 Collab.), Phys. Rep. **154**, 247 (1987).
[32] G. Arnison *et al.* (UA1 Collab.), Phys. Lett. B **118**, 167 (1982);
C. Albajar *et al.* (UA1 Collab.), Nucl. Phys. B **335**, 261 (1990).
[33] F. Abe *et al.* (CDF Collab.), Phys. Rev. Lett. **61**, 1819 (1988); Phys. Rev. D **41**, R2330 (1990).
[34] T. Alexopoulos *et al.* (E735 Collab.), Phys. Rev. D **48**, 984 (1993).
[35] ALICE Collaboration (K. Aamodt *et al.*), Eur. Phys. J. C **68**, 89 (2010).
[36] ALICE Collaboration (K. Aamodt *et al.*), Eur. Phys. J. C **68**, 345 (2010).
[37] ALICE Collaboration (K. Aamodt *et al.*), Phys. Lett. B **693**, 53 (2010).
[38] ALICE Collaboration (K. Aamodt *et al.*), Report No. ALICE-PUBLIC-2013-001.
[39] ALICE Collaboration (J. Adam *et al.*), Phys. Lett. B **753**, 319 (2016).

[40] CMS Collaboration (K. Khachatryan *et al.*), J. High Energy Phys. 02, 041 (2010).
[41] CMS Collaboration (K. Khachatryan *et al.*), Phys. Rev. Lett. **105**, 022002 (2010).
[42] CMS Collaboration (K. Khachatryan *et al.*), Phys. Lett. B **751**, 143 (2015).
[43] CMS and TOTEM Collaborations (S. Chatrchyan *et al.*), Eur. Phys. J. C **74**, 3053 (2014).
[44] L. McLerran and M. Praszalowicz, Acta Phys. Polon. B **41**, 1917 (2010).
[45] E. Levin and A.H. Rezaeian, Phys. Rev. D **82**, 014022 (2010).
[46] R. Feynman, Phys. Rev. Lett. **23**, 1415 (1969);
R. Feynman, *Photon-hadron interactions* (Benjamin, NY, 1972).
[47] W. Busza, J. Phys. G **35**, 044040 (2008).
[48] G.J. Alner *et al.* (UA5 Collab.), Z. Phys. C **33**, 1 (1986).
[49] J. Cleymans, J. Strümpfer, and L. Turko, Phys. Rev. C **78**, 017901 (2008).
[50] Z. Koba, H.B. Nielsen, and P. Olesen, Nucl. Phys. B **40**, 317 (1972).
[51] E.M. Levin and M.G. Ryskin, Yad. Fiz. **19**, 389 (1974).
[52] V. Zaccolo *et al.* (ALICE Collab.), Nucl. Phys. A **956**, 529 (2016).
[53] C.J. Alner *et al.* (UA5 Collab.), Nucl. Phys. B **291**, 445 (1987).
[54] V.V. Vechernin, *these Proceedings*.
[55] L. Bravina, J. Bleibel, R. Emaus, E. Zabrodin, *these Proceedings*.
[56] L.V. Bravina *et al.*, Sov. J. Nucl. Phys. **50**, 245 (1989).

EPJ Web of Conferences **164**, 01006 (2017) DOI: 10.1051/epjconf/201716401006
ICNFP 2016

BEH Boson properties from CMS results

Paolo Checchia[1,a] on behalf of CMS Collaboration
[1] *INFN sezione di Padova, Padova, Italy.*

Abstract. The most relevant results on Higgs sector from CMS are presented. The status of the measurements of the Higgs Boson properties after the complete analysis of Run I dataset and an overview of the results obtained with the limited luminosity delivered at 13 TeV in 2015 are given. Implications of the results and future perspectives are also briefly discussed.

1 Introduction

The discovery of the Brout-Engler-Higgs (BEH) boson [1–5], at CERN by the ATLAS[6] and CMS[7] experiments has been one of the most relevant results in particle physics. After several decades of searches for the BEH (hereafter simply H) particle mainly at LEP, Tevatron and LHC, the observation of a new state in the mass region still not excluded, is a step forward in the comprehension of the mechanism capable to generate the mass of elementary particles. This mechanism, called "BEH" or "Higgs" mechanism, is a crucial part of the standard model (SM) of particle physics and the discovery of the H particle with a mass compatible with the SM expectations is an additional success of the model. However the discovered particle could be the lightest state of a more complex Higgs sector as foreseen by several extensions of the SM, including supersymmetric models [8]. To establish the actual nature of the particle it is necessary to provide precision measurements of its properties, and this is one of the goals of the LHC experiments in the near future.

2 Status of searches before the discovery

The legacy of LEP and SLC SM precision measurements [9], and of LEP, Tevatron and LHC direct searches for the H boson is contained in Fig. 1. In the SM, several physical observables can be predicted by calculating the effect of radiative corrections that depend on the Higgs boson mass. The $\Delta\chi^2$ as obtained from the SM fit to the precision measurements is superimposed to the regions excluded by direct searches. It is evident that the fit prefers a low value for the Higgs boson mass. In addition, only a very narrow window in the mass range remained not excluded by direct searches.

3 Expected properties of the SM Higgs boson

Within the standard model all Higgs boson properties depend only on its mass.

 DOI: 10.1051/epjconf/201716401006

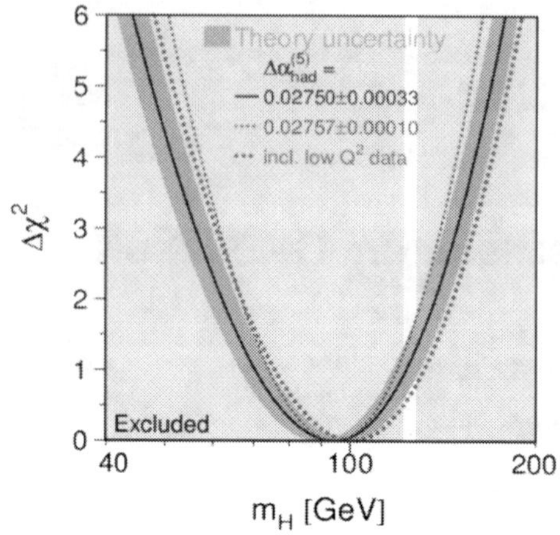

Figure 1. The status of the $\Delta\chi^2$ as obtained from the SM fit to the precision measurements superimposed to the direct searches for the SM Higgs boson. The blue band represents an estimate of the theoretical error. This figure is taken from Ref. [9].

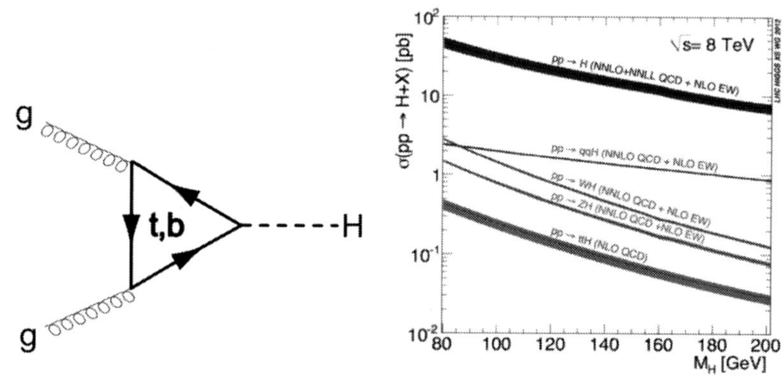

Figure 2. Gluon-gluon fusion (left) and production cross sections (right).

At a hadron collider the SM Higgs boson production is largely dominated by the two gluons collision (Fig. 2). However, other production mechanisms such as vector boson fusion (VBF) and associated production (VH, where $V = W^\pm$ or Z, or $t\bar{t}H$) can play a relevant role since the presence of other energetic particles in the final state can be useful in reducing the background (Fig. 3).

In the low mass region where the Higgs particle cannot decay to on-shell gauge bosons, the total decay with (Γ_H) is expected to be very narrow. In that region the decay is dominated by b-quark pairs but several channels have a branching ratio with a reasonable possibility to be measured (Fig 4).

[a]e-mail: paolo.checchia@pd.infn.it

DOI: 10.1051/epjconf/201716401006

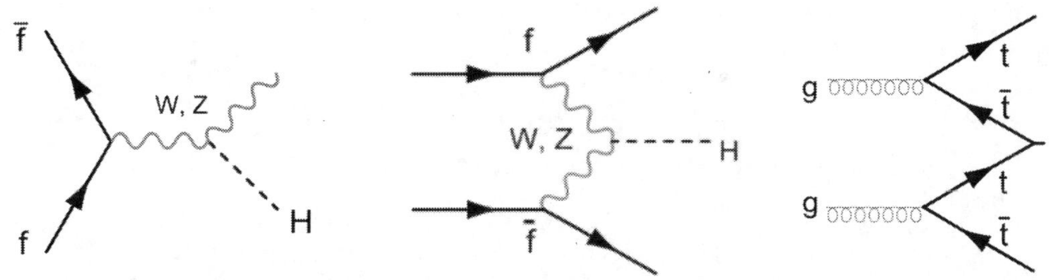

Figure 3. Higgs-strahlung or VH (left), vector boson fusion or VBF (centre) and top-fusion (right) production processes.

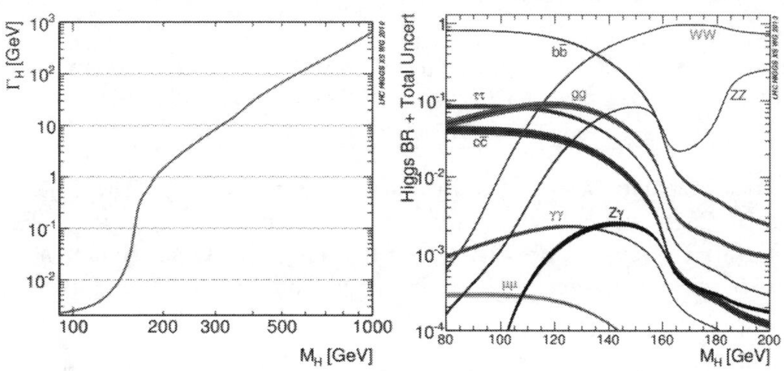

Figure 4. Total decay width (left) and decay branching fractions to the most relevant final states (right) as a function of the SM Higgs boson mass.

4 The Higgs boson properties after LHC Run 1

4.1 The mass

The measurement of the Higgs boson mass (m_H), necessary to quantify the SM predictions on the particle properties, relies on the same final states relevant for the discovery: the $H \to ZZ^* \to 4\ell$ ($\ell = e, \mu$) and $H \to \gamma\gamma$ since for both channels the detector resolution allows to obtain a narrow peak on the reconstructed invariant mass.

In the case of the four lepton final state [10], both lepton flavours are well measured and in addition the $Z \to 2\ell\gamma^* \to 4\ell$ decay can be used to cross check the calibration (Fig. 5). This channel offers additional advantages since the background is small and it is also possible to enhance the signal significance by using discriminant variables based on the event kinematics. The measured mass is $m_H = 125.6 \pm 0.4$ (stat) ± 0.2 (syst) GeV. The statistical error dominates and the systematic uncertainty is mainly due to energy-momentum scale.

Also the diphoton decay channel [11] produces a narrow peak, given the excellent CMS electromagnetic calorimeter, but it has to deal with a large background. Therefore, to improve the analysis, the data are subdivided in categories with different invariant mass resolutions and different signal-to-background ratios. It is then possible to visualize the signal by summing all the categories with a weight proportional to signal/signal plus background ratio: $w = S/(S + B)$. The mass measurement is

$m_H = 124.70 \pm 0.31$ (stat) ± 0.15 (syst) GeV. Also in this case the error is dominated by the statistical contribution, with a dominant systematic uncertainty due to the photon energy scale.

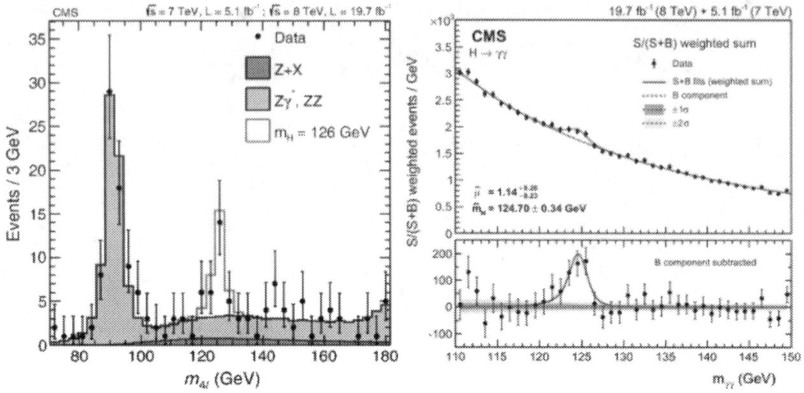

Figure 5. Invariant mass distribution of four lepton final state [10](left). Weighted invariant mass distribution of diphoton final state (see text) together with the background-subtracted distribution (right).

The two measurements are compatible within $\sim$ 1.6 σ. They have been combined giving the final CMS determination of the Higgs boson mass after RUN 1: $m_H = 125.02^{+0.26}_{-0.27}$ (stat) $^{+0.14}_{-0.15}$ (syst) GeV [12]. Furthermore, the CMS results have been combined with the ATLAS mass measurements of the Higgs boson obtaining [13]: $m_H = 125.09 \pm 0.21$ (stat) ± 0.11 (syst) GeV.

4.2 Spin-parity

The new particle with a mass of $\sim$ 125 GeV is a Higgs boson if it can be proven to be a scalar state. It is then necessary to show that the hypothesis $J^P = 0^+$ is the most probable. The direct decay into the diphoton final state excludes the spin 1 hypothesis. Other hypotheses are compared to the SM 0^+ hypothesis by considering the kinematic properties of several final state decays. In particular, in the $H \to ZZ^* \to 4\ell$ final state, the distributions of the decay angles Φ, θ_1 and θ_2 (Fig. 6) bring information on the parity of the initial state.

A signal plus background likelihood function $\mathcal{L}_{\mathcal{J}^P}$, for a given J^P hypothesis, is defined by combining all the discriminant observables. The test statistic to compare the J^P hypothesis with the SM (0^+) hypothesis is $q = -2ln(\mathcal{L}_{J^P}/\mathcal{L}_{0^+})$. The expected q distributions for 0^- hypothesis and SM Higgs boson are shown in the right panel of Fig. 6. Defining the cumulative probability $P_{J^P} = P(q > q_{obs}|J^P + bkg)$, from the ratio $\mathrm{CL}_s = P_{0^-}/P_{0^+}$ the pseudoscalar hypothesis is excluded at 99.9% C.L..

Other spin-parity hypotheses have been considered but they are all disfavoured and hence the discovered particle is very likely a scalar state.

4.3 WW channel

For a Higgs boson with a mass of about 125 GeV, the SM foresees that the $H \to WW$ decay channel has the second largest branching ratio. However, only the final states with double $W \to \ell\nu$ decay can be easily distinguished from background, given the presence of several relevant contributions. This decay chain has two neutrinos in the final state and hence it is not possible to fully reconstruct the invariant mass. Therefore, the signal must be detected as an excess of events over the expected

 DOI: 10.1051/epjconf/201716401006

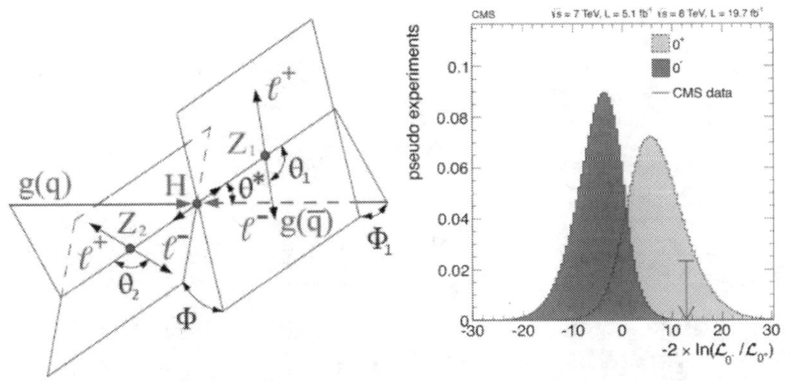

Figure 6. Illustration of the production and decay of a particle H. The angles θ^*, Φ_1 and Φ are given in the H rest frame, and θ_1, θ_2 in the Z_1, Z_2 rest frames, respectively (left). Distributions of the test statistics $q = -2ln(\mathcal{L}_{J^P}/\mathcal{L}_{0^+})$. The distribution of the SM Higgs boson (0^+) is shown in the yellow histogram while the distribution for the pseudoscalar hypothesis is represented by the blue histogram. The observed value is indicated by the red arrow (right).

background in regions of properly defined kinematic quantities. The azimuthal angle $\Delta\phi_{\ell\ell}$ between the leptons, their mass $m_{\ell\ell}$ and the transverse mass m_T are used as discriminant kinematic variables [14]. The last is defined as $m_T = \sqrt{2p_T^{\ell\ell}E_T^{miss}(1 - \cos\Delta\phi_{\ell\ell,E_T^{miss}})}$ and its distribution is shown in Fig.7. To optimise the signal/background ratio, the events are grouped in categories based on the lepton flavour and on the number of jets in the final state. There is a clear excess of events with respect to the background only hypothesis. From a $(m_T, m_{\ell\ell})$ template fit the observed excess is quantified in 4.3 σ, while the expected signal significance for a SM Higgs boson with m_H = 125.6 GeV is 5.8 σ .

4.4 Fermionic channels

Since in the SM the Yukawa coupling is proportional to the fermion mass, the $H \rightarrow \tau\tau$ decay has a large branching ratio compared to the other leptons. On the other hand, in this channel the mass reconstruction is problematic given the presence of neutrinos from the τ decay in the final state. The background from $Z \rightarrow \tau\tau$ can be separated on the basis of the difference in mass only once a dedicated mass reconstruction is implemented. After LHC RUN 1 CMS observes a significance of 3.2 σ with respect to the background only hypothesis to be compared with an expected significance of 3.7 σ.

The inclusive $H \rightarrow b\bar{b}$ production, despite its high rate, is difficult to detect given the huge QCD $b\bar{b}$ background. For this reason CMS limits the analysis to the associate production (VH) [15] and to VBF production [16]. For the former, the presence of additional high p_T objects from gauge boson decay, allows to reduce the background and also to implement an efficient trigger for the events. An excess of events over the background-only hypothesis has been observed, corresponding to a significance of 2.1 standard deviations (2.1 expected). This result, combined with the $H \rightarrow \tau\tau$ channel, gives an overall significance of 3.8 σ [17] that constitutes an evidence that the particle at 125 GeV is coupled to fermions as predicted for the Higgs boson in SM. Concerning the VBF production, CMS observes a deviation with respect to the background only hypothesis with a significance of 2.2 σ, larger than expectations from SM predictions (0.8 σ).

In SM, the coupling to the top quark is the largest Yukawa coupling of the Higgs boson to a fermion. Therefore, in absence of direct decay, the $t\bar{t}H$ production is the experimental tool to measure the coupling between top quark and Higgs boson (Fig. 3). To obtain the best sensitivity, the events are

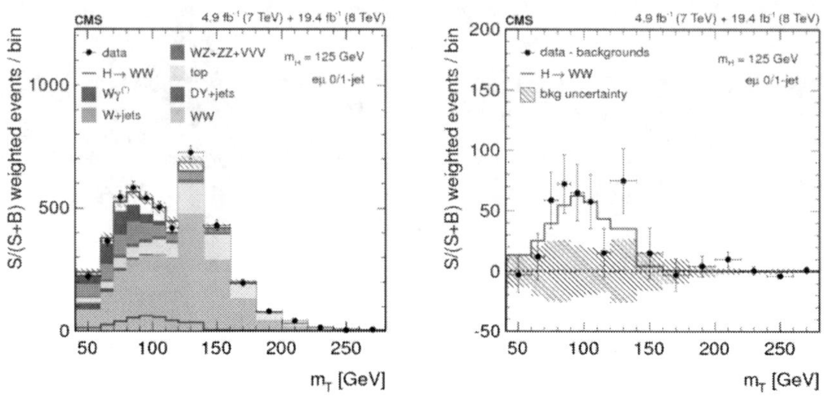

Figure 7. Distributions of m_T in the $e\mu$ final state. The observed data (black points) are superimposed to the simulated events for signal+background and signal alone (left). The background-subtracted data with the fitted signal component (right).

divided in several categories depending on the top and Higgs boson decay. After RUN 1, CMS observes [18] a deviation with respect to the background only hypothesis which is higher than expected: 3.4 σ to be compared to an expectation of 1.2 σ. The deviation with respect to the background plus SM $t\bar{t}H$ signal is about two standard deviations. A new method based on matrix elements [19], limited to $H \to b\bar{b}$ decay, does not produce evidence of a signal, within the analysed statistics (19.5 fb^{-1}), but it represents a promising strategy for future precise measurements of the top Yukawa coupling.

4.5 Couplings

In order to understand more on the nature of the discovered particle, a direct comparison with the SM expectations is convenient. Given the vacuum expectation value (VEV) $v = (\sqrt{2}G_F)^{-1/2} \approx 246$ GeV, the SM couplings to vector bosons are $g_{HVV} = \frac{2m_V^2}{v}$ while the Yukawa couplings to fermions are proportional to the fermion masses: $g_{Hff} = \frac{m_f}{v}$. It is then useful to introduce the parameters given in Table 1 [20] to compare deviations from the SM predictions. When the coupling is not direct as in gluon fusion production or in diphoton decay, the coupling modifiers are calculated in terms of the direct parameters: $\kappa_g = \kappa_g(\kappa_b, \kappa_t, m_H)$, $\kappa_\gamma = \kappa_\gamma(\kappa_b, \kappa_t, \kappa_W, m_H)$.

The parameters obtained by fitting all RUN 1 data are shown in Fig. 8 [12]. The combined signal strength is $\mu = 1.0 \pm 0.14$ in perfect agreement with SM expectations for $m_H = 125$ GeV. The compatibility with the SM is evident for most of the results, with a few exceptions as for the top production mentioned above. In addition, the so-called "custodial symmetry" [21] has been tested, that foresees equal coupling modifiers for gauge bosons. The combined analysis of all channels gives $\lambda_{WZ} = 0.91^{+0.14}_{-0.12}$ and then the assumption $\lambda_{WZ} = 1$ is reasonable. As a consequence, further tests based on the hypothesis $\kappa_W = \kappa_Z = \kappa_V$ are exploited. In particular, with additional assumptions, including the hypothesis of equal coupling modifiers for fermions (κ_f), a test of coupling to fermions and vector

Table 1. Definition of some parameters to compare deviations from SM predictions.

Name	meaning	symbol
Signal strength	$\frac{\sigma(BR)}{\sigma^{SM}(BR^{SM})}$	μ
coupling modifiers	$\frac{g_{HXX}}{g_{HXX}^{SM}}$	κ_X
total width modifier	$\frac{\Gamma_{tot}}{\Gamma_{tot}^{SM}}$	κ_H^2
ratio of coupling modifiers	$\frac{\kappa_X}{\kappa_Y}$	λ_{XY}

bosons is performed. In Fig. 9 the Confidence Level contours of the fit are shown in the κ_f, κ_V plane. The best fit results are $\kappa_V = 1.01 \pm 0.07, \kappa_f = 0.89^{+0.14}_{-0.13}$, fully compatible with the SM expectations.

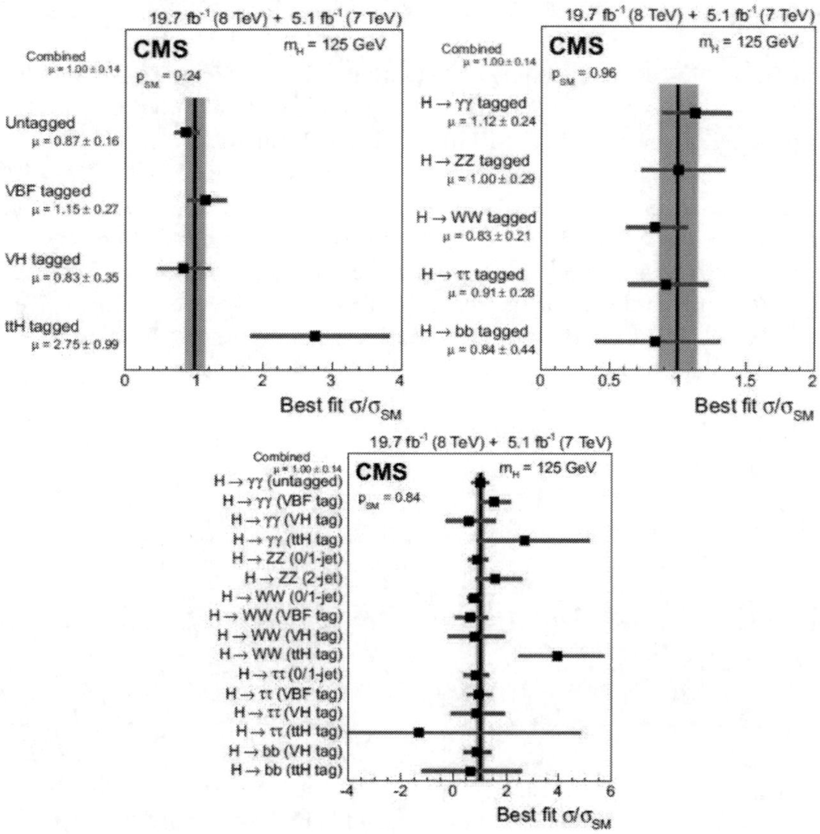

Figure 8. Signal strength for different production (top left), decay modes (to right) and combinations grouped by decay modes and tags of particular production mechanisms (bottom). The horizontal bars indicate the total uncertainty including both the statistical and systematic errors. The vertical green band shows the overall uncertainty on the signal strength.

DOI: 10.1051/epjconf/201716401006

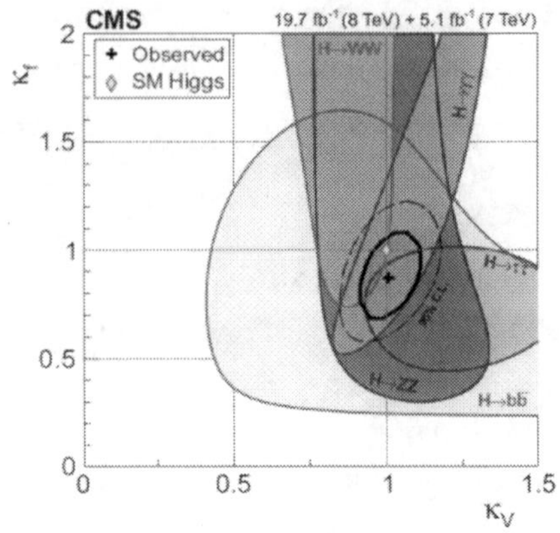

Figure 9. Contours of the 68% C.L. regions for individual channels and for the combination (full line). The cross corresponds to the best fit and the dashed line represents the 95% C.L. contour. The SM expectation is indicated by the diamond.

5 First results at 13 TeV

In this section the first results obtained by CMS with the data collected in 2015, at the center-of-mass energy of 13 TeV, are presented. The dataset corresponds to a total luminosity ranging from 2.3 fb^{-1} to 2.8 fb^{-1} depending on the final state. All results are preliminary.

The presence of the Higgs boson peak has been searched in the $H \rightarrow ZZ^{*} \rightarrow 4\ell$ [22] and in the $H \rightarrow \gamma\gamma$ [23] final states. In the former, there is an indication of the signal with a significance of 2.5 σ corresponding to $\mu = 0.82^{+0.57}_{-0.43}$. The energy dependence of the fiducial cross section, considering $\sqrt{s} = 7$, 8 and 13 TeV data is compatible, within the large uncertainty, with the SM expectation as shown in Fig. 10. For the diphoton final state analysis [23], the observed (expected) significance with respect to the background only hypothesis is 1.7 σ (2.7 σ) corresponding to $\mu = 0.69^{+0.47}_{-0.42}$. The weighted sum of the contributions of all categories is shown in Fig. 11.

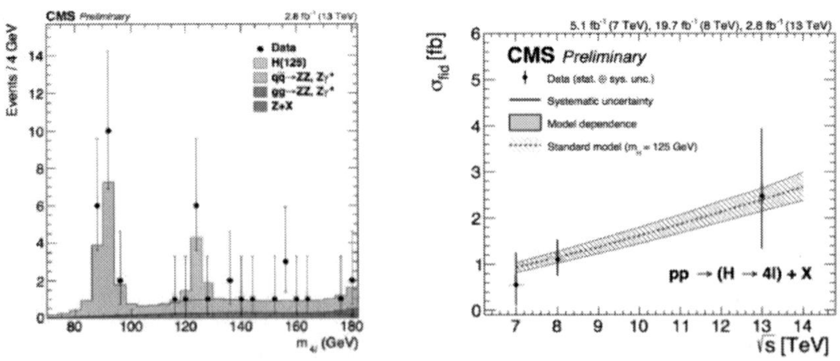

Figure 10. Invariant mass distribution of four lepton final state at 13 TeV center-of-mass energy (left). The measured fiducial cross-section as a function of the center-of-mass energy (right).

Several analyses of production/decay channels successfully finalised using the RUN 1 dataset have been repeated with the first data at higher energy. For the $H \rightarrow WW$ decay, only the $e\mu$ final state has been considered [24], at a preliminary stage, in order to reduce the large Drell-Yan background

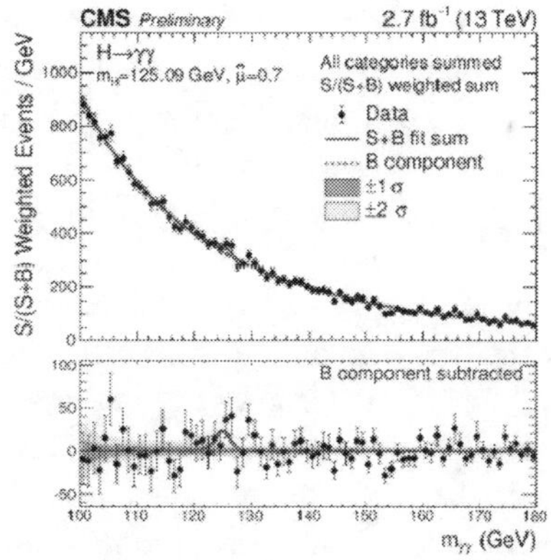

Figure 11. Weighed invariant mass distribution of diphoton final state at 13 TeV center-of-mass energy.

present in the same-flavour dilepton events. Following the same strategy as in RUN 1 analysis, the observed (expected) significance with respect to the background only hypothesis is 0.7 σ (2.0 σ). No evidence of signal has been found in the 2015 dataset for the VBF production of $H \to b\bar{b}$ decay [25]. However, this negative result has been combined with 8 TeV data result obtaining a global significance of 1.2 σ corresponding to a signal strength $\mu = 1.3^{+1.2}_{-1.1}$ for $m_H = 125$ GeV.

Given the particular importance of $t\bar{t}H$ production, several searches have been accomplished for (i) (di)leptonic top decay associated to $H \to b\bar{b}$ [26], (ii) top tagged $H \to \gamma\gamma$ decay [23], (iii) multilepton final states [27]. Within the available dataset, none of the searches has found a signal evidence and the combined signal strength is $\mu = 0.15^{+0.95}_{-0.81}$ [28].

6 Implications

The results obtained by CMS can be summarised as follows [29]:

- the 125 GeV particle is in the range preferred by fits to SM precision measurements;
- it is a scalar so it is a BEH-like boson;
- according to SM, at that mass several final states are accessible;
- it is coupled to bosons and fermions;
- the measured production/decay properties are compatible with the SM expectations at that mass;
- no other states from additional Higgs doublets have been found despite several dedicated searches have been performed.

The above does not prove that the discovered particle is the SM Higgs boson since lightest states of more complex Higgs sectors may have similar properties. In addition it is worth to notice that the value $m_H = 125$ GeV is within the SUSY-accepted range. More precise measurements are necessary to clarify the situation.

7 Future perspectives

In a short time scale, the 2016 LHC run is proceeding well and the CMS is taking data with a good efficiency. Furthermore, the production cross sections at $\sqrt{s} = 13$ TeV increase with respect to 8 TeV

 DOI: 10.1051/epjconf/201716401006

by a factor ~ 2.2 for gluon fusion and by a factor ~ 3.9 for $t\bar{t}H$ production. These conditions, at the end of 2016, would allow to obtain a relevant improvement on the measured values, with statistical uncertainties reduced by more than a factor two.

In a middle time scale the perspectives of the LHC program have to be considered. Following the plans, the machine should deliver 300 ÷ 350 fb^{-1} at $\sqrt{s}$ = 13 TeV or 14 TeV. The expected CMS precision on the Higgs couplings [30], with an integrated luminosity of 300 fb^{-1} at $\sqrt{s}$ = 14 TeV, are 6-8% for $\kappa_\gamma, \kappa_g, \kappa_W, \kappa_Z, \kappa_\tau$, about 13% for κ_b and 15% for κ_t. As shown in the left panel of Fig. 12, in an optimistic scenario with a reduction of a factor two of the theoretical uncertainties which constitutes the dominant part of systematic errors, the expected uncertainties on the Higgs couplings would range from about 5% for $\kappa_\gamma, \kappa_g, \kappa_W, \kappa_Z, \kappa_\tau$, to 7 – 10% for κ_b, κ_t. The increase in center-of-mass energy and luminosity gives an opportunity to observe heavy H states as foreseen by MSSM or to extend the excluded region in $m_A - tan\beta$ plane. If the discovered particle is the lightest CP-even state h, values of m_A smaller than 230-260 GeV will be excluded for all $tan\beta$ values [31].

In a longer time scale the High Luminosity LHC program would give the ultimate opportunity for CMS to contribute to the Higgs sector. In the right panel of Fig. 12, the estimated precision on the coupling modifiers for an assumed luminosity of 3000 fb^{-1} is shown. The contribution from theoretical errors is more relevant in this case. A high luminosity program gives also an unique chance to observe rare channels as the decay $H \rightarrow \mu\mu$, or the Higgs boson pair production which is related to the Higgs self-coupling λ.

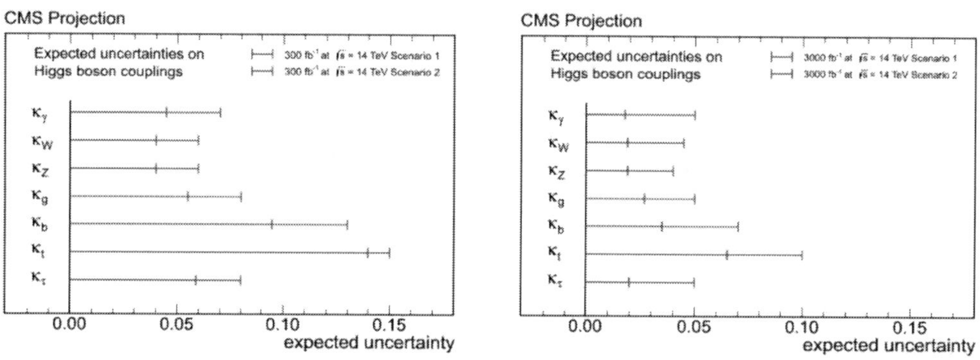

Figure 12. Estimated precision on the measurements for coupling modifiers for a SM-like Higgs boson at a center-of-mass energy of 14 TeV and a dataset with integrated luminosity of 300 (left) and 3000 fb^{-1} (right). Two scenarios for systematic uncertainties are considered (see text for details).

8 Conclusions

The discovery of the BEH particle has been not only one of the most important results in particle physics but also one of the most relevant scientific achievements. The discovery requires several further actions that will engage the HEP community for several years to try to give an answer to the following questions [32]:

- is the new particle really the SM BEH Boson?
- is it an elementary particle? or...
- is it a composite particle?

- is it natural?
- are there other Higgs fields?
- is it really responsible for the mass of all elementary particles?

and also

- is it at the origin of the matter-antimatter asymmetry?
- does it have a role in the inflationary expansion of the Universe?

To obtain a lot of more precise measurements, as expected in the long LHC and High Luminosity LHC programs, is the unique way to answer to the previous questions.

References

[1] F. Englert and R. Brout, Phys. Rev. Lett. **13**, 321 (1964), doi:10.1103/PhysRevLett.13.321.
[2] P. W. Higgs, Phys. Lett. **12**, 132 (1964), doi:10.1016/0031-9163(64)91136-9.
[3] P. W. Higgs, Phys. Rev. Lett. **13**, 508 (1964), doi:10.1103/PhysRevLett.13.508.
[4] G. S. Guralnik, C. R. Hagen and T. W. B. Kibble, Phys. Rev. Lett. **13**, 585 (1964), doi:10.1103/PhysRevLett.13.585.
[5] P. W. Higgs, Phys. Rev. **145**, 1156 (1966), doi:10.1103/PhysRev.145.1156.
[6] G. Aad et al., Phys. Lett. B **716**, 1 (2012), doi:10.1016/j.physletb.2012.08.020.
[7] S. Chatrchyan et al., Phys. Lett. B **716**, 30 (2012), doi:10.1016/j.physletb.2012.08.021.
[8] G.C. Branco et al., Phys. Rep. **516**, 1 (2012), doi:10.1016/j.physrep.2012.02.002.
[9] The ALEPH, DELPHI, L3, OPAL Collaborations, the LEP Electroweak Working Group, Phys. Rep. **532**, 119 (2013), doi:10.1016/j.physrep.2013.07.004.
[10] S. Chatrchyan et al., Phys. Rev. D **89**, 092007 (2014), doi:10.1103/PhysRevD.89.092007.
[11] V. Khachatryan et al., Eur. Phys. J. C (2014) 74: 3076, doi:10.1140/epjc/s10052-014-3076-z
[12] V. Khachatryan et al., Eur. Phys. J. C (2015) 75:212, doi:10.1140/epjc/s10052-015-3351-7
[13] G. Aad et al., Phys. Rev. Lett. **114**, 191803, doi:10.1103/PhysRevLett.114.191803
[14] S. Chatrchyan et al., J. High Energ. Phys. **01** 096 (2014). doi:10.1007/JHEP01(2014)096
[15] S. Chatrchyan et al., Phys. Rev. D **89**, 012003 (2014), doi:10.1103/PhysRevD.89.012003.
[16] V. Khachatryan et al.,Phys. Rev. D **92**, 032008 (2015), doi:10.1103/PhysRevD.92.032008.
[17] S. Chatrchyan et al., Nature Phys. **10**, 557 (2014), doi:10.1038/nphys3005.
[18] V. Khachatryan et al., J. High Energ. Phys. **09** 087 (2014). doi:10.1007/JHEP09(2014)087
[19] V. Khachatryan et al., Eur. Phys. J. C (2015) 75:251 doi:10.1140/epjc/s10052-015-3454-1
[20] S. Heinemeyer et al., Handbook of LHC Higgs Cross Sections: 3. Higgs Properties: Report of the LHC Higgs Cross Section Working Group. arXiv:1307.1347. CERN-2013-004 https://twiki.cern.ch/twiki/bin/view/LHCPhysics/CrossSections .
[21] M.J.G. Veltman, Nucl. Phys. B 123, 89 (1977). doi:10.1016/0550-3213(77)90342-X
[22] CMS Collaboration: CMS-PAS-HIG-15-004.
[23] CMS Collaboration: CMS-PAS-HIG-15-005.
[24] CMS Collaboration: CMS-PAS-HIG-15-003.
[25] CMS Collaboration: CMS-PAS-HIG-16-003.
[26] CMS Collaboration: CMS-PAS-HIG-16-004.
[27] CMS Collaboration: CMS-PAS-HIG-15-008.
[28] https://twiki.cern.ch/twiki/bin/view/CMSPublic/TTHCombMoriond2016

DOI: 10.1051/epjconf/201716401006

[29] P. Checchia, Int.J.Mod.Phys. A30 (2015) 07, 1530003, doi: 10.1142/S0217751X15300033
[30] CMS Collaboration: CMS NOTE-13-002,arXiv:1307.7135.
V. Khachatryan et al.,CERN-LHCC-2015-10.
[31] I. M. Lewis, arXiv:1308.1742.
[32] INFN, CSN1: Proposal for a long term strategy for accelerator based experiments, Phys. Ser. 60 (2015) pp. 1-291.

COMPASS results on proton longitudinal spin structure function g_1 and quark fragmentation functions

Fabienne Kunne[1]

[1] IRFU, CEA Saclay, Université Paris-Saclay, France

On behalf of the COMPASS Collaboration

Abstract. We present a few recent highlights from the COMPASS experiment at CERN, related to the nucleon spin and structure: the proton spin structure function $g_1^p(x)$ measured using 200 GeV polarized muons, together with a QCD fit of g_1 world data; pion and kaon multiplicities in DIS in view of constraining quark fragmentation functions.
The statistical precision on g_1^p is improved by a factor of ~2 at low x. A re-evaluation of the Bjorken sum rule based on COMPASS data alone confirms its validation within 9% accuracy. An NLO QCD fit to g_1 world data leads to a value of the quark spin contribution to the nucleon spin $\Delta\Sigma$, between 0.26 and 0.36 at $Q^2 = 3$ (GeV/c)^2, the size of the uncertainty being driven by the unknown shape of the polarized parton distributions.
The results on pion and kaon multiplicities produced in DIS of muons off an isoscalar target constitute an impressive data set of more than 400 points in π and 400 in K, covering a large kinematical domain in (x, z, Q^2), which will be used in future NLO QCD fits to extract quark fragmentation functions into π and K. A LO fit is performed to extract the favoured and unfavoured quark fragmentation functions into pions.

1 Measurements of g_1^p proton spin structure function at 200GeV

The 1/2 nucleon spin can be decomposed along the longitudinal projection as follows

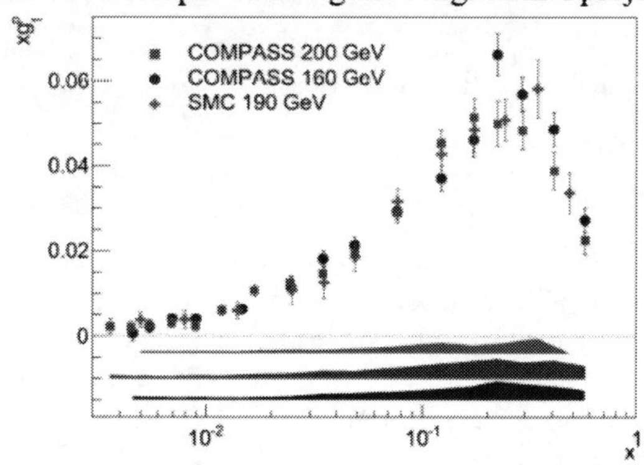

Fig. 1. COMPASS and SMC data for the proton longitudinal spin structure function g_1^p.

1/2 = ½ ΔΣ + ΔG + L_z, where ΔΣ represents the quark spin contribution, ΔG the gluon spin one and L_z the contribution from the orbital angular momenta of quarks and gluons. We present here COMPASS results on the proton longitudinal spin structure function $g_1^p(x)$ [1]. The measurements were performed using 200 GeV polarized muons from the CERN SPS (tertiary beam), scattered off a longitudinally polarized NH_3 target [2]. Results [1] are shown in Fig.1 together with previous results taken at 160 GeV [3] and earlier results from SMC [4]. The new data cover smaller x values, down to 0.0036. They improve the statistical precision of previous COMPASS data by a factor of about 2.

2 Test of the Bjorken Sum Rule

The Bjorken sum rule is a fundamental QCD prediction relating the difference of first moment of the proton and the neutron spin structure functions to the ratio of the axial-vector to vector weak coupling constants g_A/g_V assuming SU(2) flavour symmetry. It can be tested by combining the COMPASS proton [1] and deuteron [5] results to calculate the non-singlet spin structure function $g_1^{NS}(x) = g_1^p(x) - g_1^n(x)$, and compare its first moment to the prediction.

Preliminary results for $g_1^{NS}(x)$ are shown in Fig.2 *left*, and the corresponding truncated integral values (integrated from the minimum $x=x_{min}$ to 1), are shown versus x_{min} in Fig.2 *right*. The full integral corresponds to a value of g_A/g_V = 1.220 ± 0.053(stat) ± 0.095(syst). Comparing this result [1] to the value obtained from neutron decay measurements, 1.2701 ± 0.0025 [6], we can say that the Bjorken sum rule is validated within 9% accuracy. The dominant uncertainty comes from systematic uncertainties on beam and target polarisation and target dilution factor (fraction of polarisable material in the target). Note that the data cover 94% of the contribution to the integral, a remarkable feature compared to other evaluations.

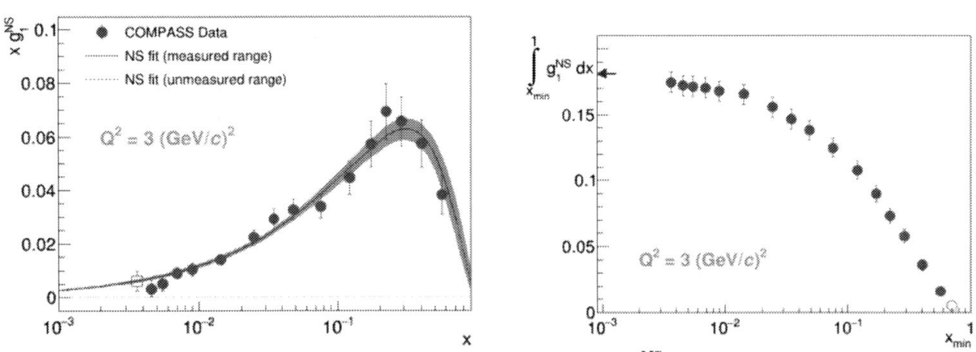

Fig. 2. *Left:* The nucleon non-singlet spin structure function g_1^{NS} vs x, measured from COMPASS p and d data. *Right:* truncated integral of g_1^{NS} from x_{min} to 1, vs x_{min}. The arrow shows the value extrapolated to x_{min}=0.

3 NLO QCD fit of g_1 world data

The COMPASS g_1^p and g_1^d measurements are shown in Fig. 3 as a function of Q^2 for various x values, together with other world data. The COMPASS coverage is very significant in the low x and large Q^2 regions. To extract quark and gluon helicity distributions, we have performed a global NLO QCD fit to all these and existing neutron data. A functional form is assumed at an initial Q_0^2 scale for the singlet, the triplet and the octet SU(3) flavour combinations of the quark distributions as well as for the gluon distribution. The first moments of the non

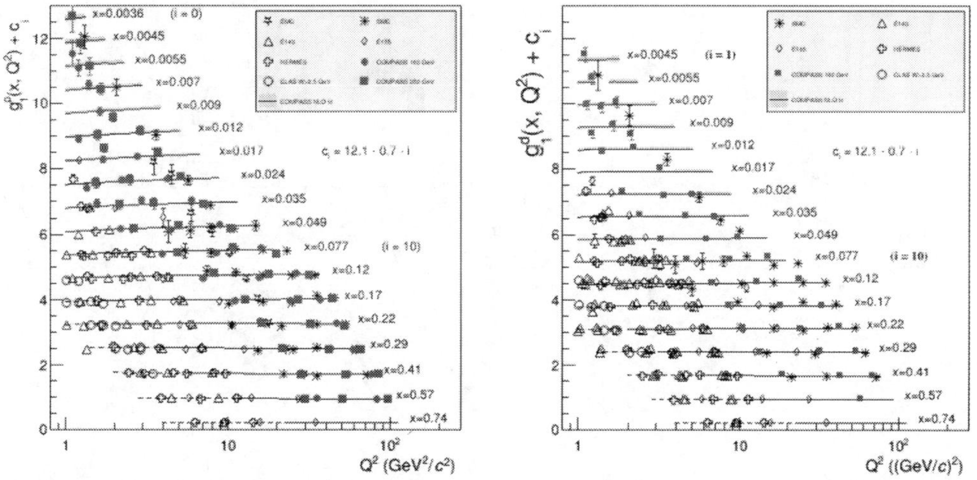

Fig. 3. World data on g_1 proton *(left)* and deuteron *(right)* vs Q^2, for various x. COMPASS data are in red. The dotted lines represent the solutions of a COMPASS global NLO QCD fit (see text)..

singlet distributions are fixed to the values of the axial-vector weak coupling constants measured in neutron and hyperon β-decay. The x-Q^2 evolution is made in the MS factorisation scheme (program #1 taken from [7]). Depending on the choice of the functional forms at the initial Q_0^2 scale, two classes of possible solutions are found. They correspond to a positive and negative ΔG value and values in between are allowed. The value of $\Delta\Sigma$, correlated to ΔG, ranges between 0.26 and 0.36 at the same scale. The dominant uncertainty in the extraction comes from the choice of the functional forms chosen at the initial Q_0^2 scale. Results for the two solutions are shown in Fig.4 at $Q^2 = 3$ (GeV/c)^2, with on top the quark singlet helicity $\Delta q^S(x)$ (the integral of which gives $\Delta\Sigma$) and the gluon helicity $\Delta G(x)$ distributions. In Fig.4 bottom, the quark helicity distributions are shown separately for the u, d and s flavours. The solid and dashed lines correspond to positive and negative ΔG solutions of the NLO QCD fit. The dark and light bands correspond respectively to the propagation of the statistical error from the data and to the combined statistical and systematic uncertainties. The latter largely dominates and comes mainly from the choice of functional forms.

 DOI: 10.1051/epjconf/201716401007

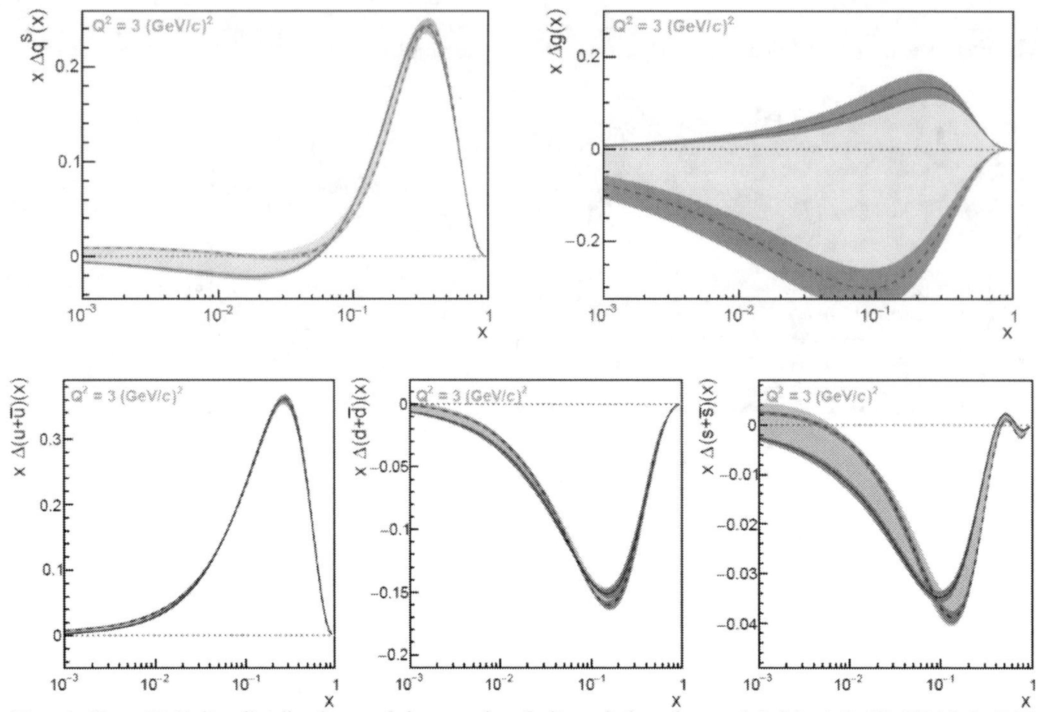

Fig. 4. *Top:* Helicity distributions of the quarks *(left)* and the gluons (*right*) at 3 (GeV/c)2 in MS scheme [1]. *Bottom:* Helicity distributions per flavour. The solid and dashed lines correspond to the solutions with positive and negative ΔG. Dark bands correspond to propagation of statistical errors, and the light bands to the combined statistical and systematic uncertainties (dominating and mainly attributed to choice of functional forms).

4 Measurement of π and K multiplicities in semi inclusive DIS

Parton fragmentation functions (FFs), like parton distribution functions (PDFs), are non-perturbative objects. They describe the hadronization of partons. The FFs are process independent and are used to describe the probability that a quark of flavor q fragments into a hadron of type *h* (D_q^h). They are needed e.g. for the extraction of flavor dependent quark helicity distributions Δq(*x*) [8,9] from semi-inclusive polarized deep inelastic scattering. In particular, to access the strange quark polarization Δs from polarized SIDIS, the strange quark FF into kaon D_s^K is needed and it constitutes by far the largest contribution to the uncertainty in the extraction. Data sensitive to FFs exist from e^+e^- and p p(bar) reactions, but they are insufficient for a good flavor separation; they also lie at relatively too high Q^2 values. This is why it was decided to measure charged hadron multiplicities in SIDIS with a high precision and a fine binning in several variables.
At leading order (LO), the hadron multiplicities $M(x,Q^2,z)$, defined as the mean number of hadrons *h* produced in a semi inclusive deep inelastic scattering event $\mu\, p \rightarrow \mu\, h\, X$, are simply related to PDFs, $q(x, Q^2)$ (for flavour *q*), and quark FFs, $D_q^h(z,Q^2)$:

$$\frac{\mathrm{d}M^{\mathrm{h}}(x,z,Q^2)}{\mathrm{d}z} = \frac{\mathrm{d}^3\sigma^{\mathrm{h}}(x,z,Q^2)/\mathrm{d}x\mathrm{d}Q^2\mathrm{d}z}{\mathrm{d}^2\sigma^{\mathrm{DIS}}(x,Q^2)/\mathrm{d}x\mathrm{d}Q^2} \tag{1}$$

$$\frac{\mathrm{d}^2\sigma^{\mathrm{DIS}}}{\mathrm{d}x\mathrm{d}Q^2} = C(x,Q^2)\sum_{\mathrm{q}} e_{\mathrm{q}}^2 q(x,Q^2), \qquad \frac{\mathrm{d}^3\sigma^{\mathrm{h}}}{\mathrm{d}x\mathrm{d}Q^2\mathrm{d}z} = C(x,Q^2)\sum_{\mathrm{q}} e_{\mathrm{q}}^2 q(x,Q^2) D_{\mathrm{q}}^{\mathrm{h}}(z,Q^2) \tag{2}$$

 DOI: 10.1051/epjconf/201716401007

where x is the Bjorken variable, *i.e.* the fraction of momentum carried by the struck quark, Q^2 is the momentum transfer and z is the energy fraction taken by the hadron. C depends on kinematics and e_q is the quark fractional charge. Note that apart from the Q^2 dependence, PDFs depend upon x, while FFs depend upon z. This will be useful in the disentanglement of FFs from PDFs, while only the product of both is measured.

The π [10] and K multiplicities [11] were measured using COMPASS SIDIS data taken in 2006 with an isoscalar target (^{6}LiD). They were corrected by the value of the global acceptance of the apparatus estimated in each (x, y, z) bin, where y is the lepton energy fraction carried by the virtual photon. The acceptance of the spectrometer was evaluated from a Monte Carlo simulation, comparing the number of generated and reconstructed events in the whole chain of analysis. For this study, the y variable is chosen instead of the Q^2 one, since Q^2 is too much correlated to x. For the π and K identification a RICH detector is used. Additional corrections including RICH unfolding and radiative corrections were applied. As an illustration, final multiplicities for positive charge kaons are shown in Fig.5 in the three dimensional binning. Similar data were obtained for negative charge kaons, as well as for positive and negative charge pions.

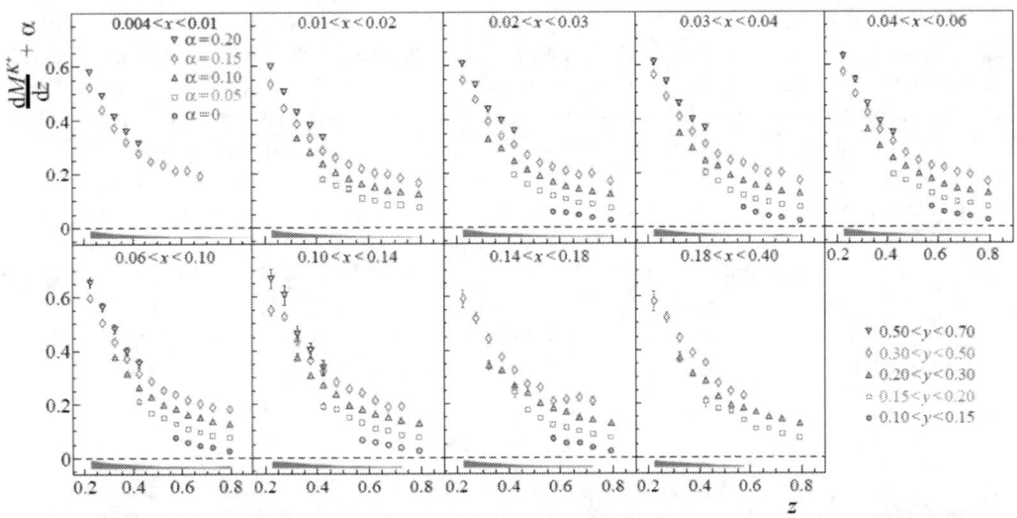

Fig. 5. K$^+$ multiplicities [11] shown vs z for nine x ranges and 5 y ranges (for better visibility shifted vertically by a constant α). Statistical errors are small and hardly visible. The bands correspond to the total systematic uncertainty for the range 0.3 < y < 0.5.

The sum of positively and negatively charged hadron multiplicities integrated over z is of special interest. For kaon multiplicities the sum is used by HERMES [12] to extract at LO pQCD the product of the strange quark distribution and the fragmentation function of strange quarks into kaons. For pions, the sum allows one to verify the applicability of the LO pQCD formalism in the COMPASS kinematic domain. In Fig. 5 COMPASS results are compared to HERMES ones [12] for pions on the left and kaons on the right. Results differ significantly in both cases. However they were taken in different kinematic ranges, COMPASS data covering higher energies than HERMES. NLO QCD analyses of the data should help understanding the discrepancy.

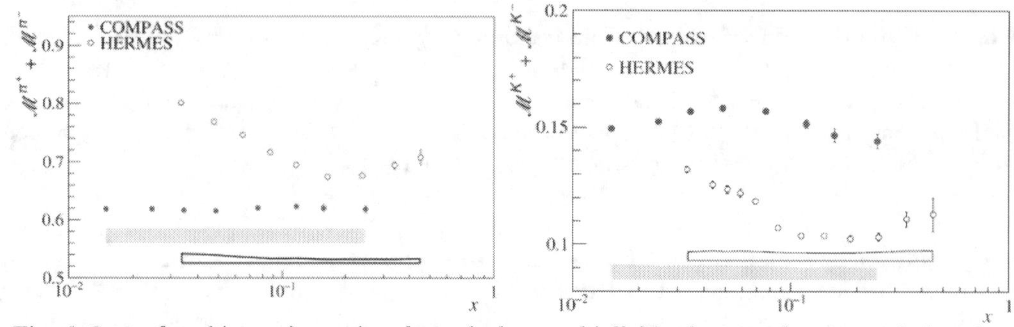

Fig. 6. Sum of positive and negative charge hadron multiplicities integrated over z and plotted vs x. COMPASS data (full points) are compared to HERMES ones (open points). The bands show the total systematic uncertainties.

5 Extraction of quark fragmentation functions into pions from a LO fit to pion multiplicities

Starting from the equation (1) given at LO, FFs can be extracted once PDFs are taken from the literature. In each (x,y,z) bin, two equations are given by the π^+ and π^- multiplicity measured in that bin. We can extract two FFs, namely the favored and unfavored quark FFs into pions. They correspond to pion valance and sea quarks respectively, and are the two independent pion FFs left after applying isospin and charge symmetry relations between FFs, and assuming in addition that the strange quark FF is equal to the unfavored one. In order to use simultaneously all data measured at various Q^2, a LO fit is performed. For this, functional forms are chosen for the shape of the two FFs at an initial Q_0^2 value, and the DGLAP equations are used for the evolution to the measured Q^2 of each data point. Results for the favored and unfavored quark FF into pions obtained from the fit of the pion multiplicities are shown in Fig.6. The statistical error band was obtained with a replica method, varying the data points within their error. Our result [10] is in fair agreement with fits realized at NLO on similar SIDIS data [13]. They differ, as expected from fits including only $e^+ e^-$ data [7].

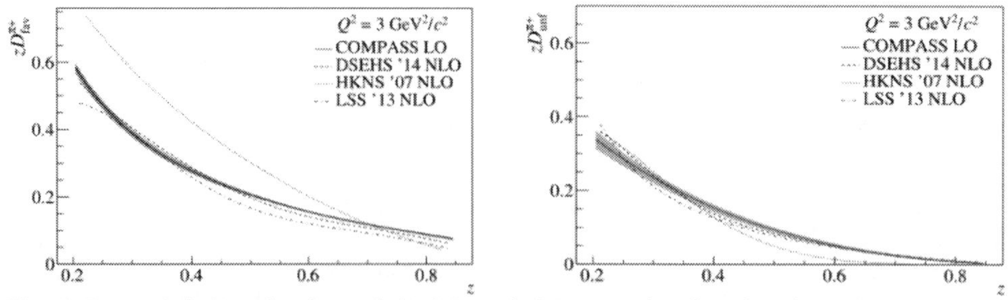

Fig. 6. Favored *(left)* and unfavored *(right)* quark fragmentation functions into pions extracted from a LO fit to COMPASS π multiplicities [10].

6 Conclusion

Recent COMPASS results are show on the spin structure function $g_1^p(x)$ measured using a 200 GeV polarized muon beam. They improve the statistical precision by a factor of ~2 at low x. A re-evaluation of the Bjorken sum rule based on COMPASS data alone confirms its validation within 9% accuracy. The NLO QCD fit was performed to g_1 proton, deuteron and neutron world data. It leads to a value of the quark spin contribution to the nucleon spin $\Delta\Sigma$, between 0.26 and 0.36 at $Q^2 = 3$ $(GeV/c)^2$. The rather large uncertainty comes essentially from the unknown shape of the polarized parton distributions. The Bjorken sum rule was verified within 9% accuracy using COMPASS data alone, which cover 94% of the contribution to the integral, a remarkable fact compared to other measurements.
Precise results on pion and kaon multiplicities produced in DIS of muons off an isoscalar target were presented. They constitute an impressive data set covering a large kinematical domain, which will be used in future global NLO QCD fits to extract quark fragmentation functions into π and K. A first LO fit was performed to extract the favoured and unfavored quark fragmentation functions into pions. Results agree with previous NLO fits which included a preliminary subset of the present data. The sum of positively and negatively charged hadron multiplicities integrated over z is found in significant disagreement with HERMES data, both for pions and kaons.
The results presented here were previously discussed in a preliminary form [15, 16]. More data on the spin structure function of the deuteron will soon be published by COMPASS, together with an update on the verification of the Bjorken sum rule [17].

References

1. COMPASS Collab., C.Adolph et al., PLB **753**,18 (2016)
2. COMPASS Collab., P.Abbon el al., NIM A **517**, 455 (2007).
3. COMPASS Collab., M. G. Alekseev et al., PLB **690**, 466 (2010)
4. SMC Collab., B. Adeva et al., Phys. Rev. D 58, 11201 (1998)
5. COMPASS Collab., V. Alexakhin et al., PLB **647**, 8 (2007)
6. J. Beringer et al.,, Particle Data Group, Phys. Rev. D86, 010001 (2012)
7. M. Hirai, S. Kumano, T.-H. Nagal and K. Sudoh, Phys. Rev. D **75**,094009 (2007)
8. COMPASS Collab., M. G. Alekseev et al., Phys. Lett. B **693** (2010) 227
9. COMPASS Collab., M. G. Alekseev et al., Phys. Lett. B **680** (2009) 217
10. COMPASS Collab., C.Adolph et al., arXiv 1604.02695, Phys. Lett. B to be published
11. COMPASS Collab., C.Adolph et al., arXiv 1608.06760, subm. to Phys. Lett. B
12. HERMES Collab., A. Airapetian et al., Phys. Rev.D **89** (2014) 097101
13. D. de Florian, et al., Phys. Rev. D **91** (2015) 014035.
14. E. Leader et al., arXiv 1312.5200, Proceedings of DSPIN-13, Dubna, Russia (2013)
15. F. Kunne, Proceedings of SPIN2014, Int. Journal of Modern Physics A: Conf. Series Vol. **40** (2016) 1660017
16. F. Kunne, Proceedings of SPIN2014, Int. Journal of Modern Physics A: Conf. Series Vol. **40** (2016) 1660027
17. M.Wilfert, Contribution to SPIN2016 "Final COMPASS results on the spin dependent structure functions g_1^d and g_1^p"

Diphoton searches (CMS)

Milena Quittnat[1,*] on behalf of the CMS collaboration

[1] *ETH Zurich*

Abstract. The search for high mass resonances decaying into two photons is well motivated by many physics scenarios beyond the standard model. This note summarizes the results of this search in proton-proton collision with a center-of-mass energy of $\sqrt{s}$ = 13 TeV and an integrated luminosity of 3.3 fb^{-1} by the CMS experiment. It presents the interpretation of the results under spin-0 and spin-2 hypotheses with a relative width up to 5.6×10^{-2} in a mass region of 500 – 4500 GeV. The results of the $\sqrt{s}$ = 13 TeV analysis are combined statistically with previous searches performed by the CMS collaboration, employing a center-of-mass energy of $\sqrt{s}$ = 8 TeV and an integrated luminosity of 19.7 fb^{-1}.

1 Introduction

Several extensions of the standard model (SM) of particle physics predict the production of a resonance decaying into a high mass photon pair. The Landau-Yang theorem postulates a spin of 0 or an integer value of higher than 2 for these resonances [1]. The analysis described focuses on two benchmark models:

- the production of a spin-0 resonance via gluon-gluon fusion.
- an extra dimensional model, the Randall-Sundrum model, predicting a spin-2 resonance.

The so called two-Higgs-doublet model (2HDM) predicts the existence of five additional scalar or pseudo-scalar particles in the Higgs sector [2]. The lightest neutral Higgs particle could be identified with the 125 GeV Higgs boson which was discovered by the CMS and ATLAS experiments in 2012 [3, 4]. In the alignment limit, one of the additional neutral Higgs particles with mass m_S could decay into two photons at a significant rate and be detected in the CMS or ATLAS detector [5].
Additional space-like dimensions provide an explanation for the observed difference between the Planck and electroweak scales by introducing a gravitational field which also propagates through at least one extra dimension. The metric of the Randall-Sundrum (RS) graviton model [6] is that one of a five-dimensional anti-de-Sitter space with one warped extra-dimension. The RS-model postulates two brane worlds. The SM fields exist only on the 'electroweak brane' whereas gravity is also allowed

*e-mail: milenaq@phys.ethz.ch

to propagate to the second 'gravity brane'. The mediator of the gravitational force, the graviton, might then be observed as a spin-2 resonance with mass m_G in the ATLAS or CMS detector and would decay, among others, into diphotons. In 2015, the ATLAS and CMS collaborations could explore a new energy range using proton-proton collisions at a center-of-mass energy of $\sqrt{s} = 13\,\text{TeV}$. Both collaborations presented results on searches for diphoton resonances in the mass ranges of 200 GeV – 2 TeV (500 GeV – 5 TeV) for spin-0 (spin-2) hypotheses at ATLAS [7] and 500 GeV – 4.5 TeV at CMS [8]. This note summarizes the results from the CMS collaboration at $\sqrt{s} = 13\,\text{TeV}$ with an integrated luminosity of 3.3 fb^{-1} and reports further on the statistical combination of this search with similar searches by the CMS collaboration performed at $\sqrt{s} = 8\,\text{TeV}$. The note is based on the results reported in Ref. [9].

2 Diphoton reconstruction with the CMS detector

A detailed description of the CMS apparatus, together with the relevant kinematic variables, can be found in Ref. [10]. The CMS detector is designed to include the central tracking and calorimetric system within its solenoid, which provides a magnetic field of B = 3.8 T. The muon detection system surrounds the solenoid and is placed within the steel skeleton of the detector. The central 'barrel' part of the detector is complemented by a forward region at each side, the so-called 'endcaps'. The reconstruction of photon candidates rely mainly one the electromagnetic calorimeter (ECAL) and to some extent on the tracker system. The electromagnetic calorimeter is built from lead-tungstate crystals providing an excellent intrinsic diphoton energy resolution of approximately 0.5 % for masses above 100 GeV. Photon candidates are formed by clustering spatially correlated energy deposits in ECAL. They are reconstructed by the particle-flow event algorithm [11, 12], applying an optimized combination of the information of various detector parts. The energy of the photon candidates is evaluated by a multivariate regression technique. For a detailed description of the applied calibrations, corrections and techniques, Ref. [13] may be consulted.

3 Event selection and reconstruction

The full list of the used simulation samples of this analysis can be found in Ref. [9]. The 13 TeV datset with an integrated luminosity $\mathcal{L}$ of 3.3 fb^{-1} is recorded with two different magnetic field configurations: 2.7 fb^{-1} with B = 3.8 T and 0.6 fb^{-1} with B = 0 T. The diphoton final state has a clean experimental signature and two well isolated high momentum photon candidates. The trigger selection thus requires at least two photon candidates of transverse momentum above 60(40) GeV for B = 3.8 T (B = 0 T) and is fully efficient for the search range of masses m_X greater than 0.5 TeV [9]. For the final event selection, photon candidates are required to have a transverse momentum of $p_T > 75\,\text{GeV}$ and a pseudorapidity of $|\eta_{SC}| < 2.5$ as well as $|\eta_{SC}| < 1.44$ for at least one of them. To maximize the sensitivity of the analysis, the selected events are separated into two categories:

- EBEB: both photons are centrally detected, in the barrel (EB) region of ECAL with $|\eta_{SC}| < 1.44$.
- EBEE: one photon is centrally detected, the other one in the forward region, the endcaps (EE) of ECAL with $1.56 < |\eta_{SC}| < 2.5$.

The invariant mass $m_{\gamma\gamma}$ of the diphoton pair is required to be above 230 GeV (320 GeV) for EBEB (EBEE) events. The diphoton mass resolution depends also on the vertex identification. If the difference between the reconstructed vertex and the true vertex is smaller than 1 cm, the effect on the mass resolution is negligible. For the B = 3.8 T dataset, a boosted decision tree is used for the vertex identification and the interaction vertex is correctly assigned for about 90 % of the signal events [9]. Since this approach relies on the p_T of the tracks, a different identification method has to be used for B = 0 T and subsequently the vertex with the highest track multiplication is chosen. The chosen vertex algorithm improves the expected mass resolution of a possible resonance by approximately 20% in comparison to the average vertex position as depicted in Figure 1. The probability for the correct vertex assignment is about 60 %.

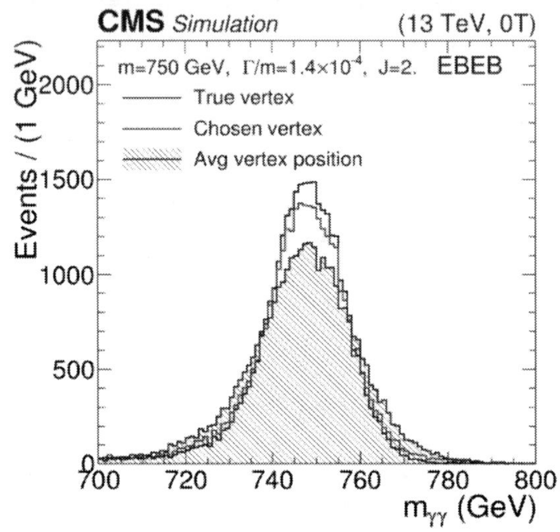

Figure 1. Expected reconstructed invariant mass distribution for a narrow width diphoton resonance at a mass of $m = 750$ GeV with B = 0 T for different primary vertex identifications, when both photons are centrally detected (EBEB).

For the final event selection, the photon candidates are required to satisfy a set of identification criteria. The respective variables and thresholds depend on the value of the magnetic field and are detailed in Ref. [9]. The per-photon identification efficiency is above 90 (85)% in the barrel (endcaps) for B = 3.8 T. In the B = 0 T dataset, the per-photon identification efficiency is above 85 (70)% for prompt isolated photon candidates in the barrel (endcaps), mainly due to a less efficient electron-veto. The overall signal selection efficiency as a function of m_X is shown in Figure 2 and varies between 0.5 – 0.7 (0.4 – 0.5) for B = 3.8 T (B = 0 T), depending on the signal hypothesis. The kinematic acceptance for the RS graviton resonances is approximately 20% lower than for scalar resonances for masses below 1 TeV due to the different angular distribution of the decay products. The two acceptances become similar for $m_X > 3$ TeV. There is nearly no difference in the expected sensitivity for the spin-0 or spin-2 hypothesis.

A data-driven tag and probe method is employed to verify the photon selection efficiency using $Z \to e^+e^-$ events. The electrons have to satisify the selection for the high momentum photon candidates with an inverted electron veto. The photon selection efficiency in data and the ratio to the Monte-Carlo (MC) prediction as a function of the transverse momentum p_T is shown in Figure 3. The scale factor is compatible with 1 in almost all p_T-bins and flat within a few percents. A 8 (16) % un-

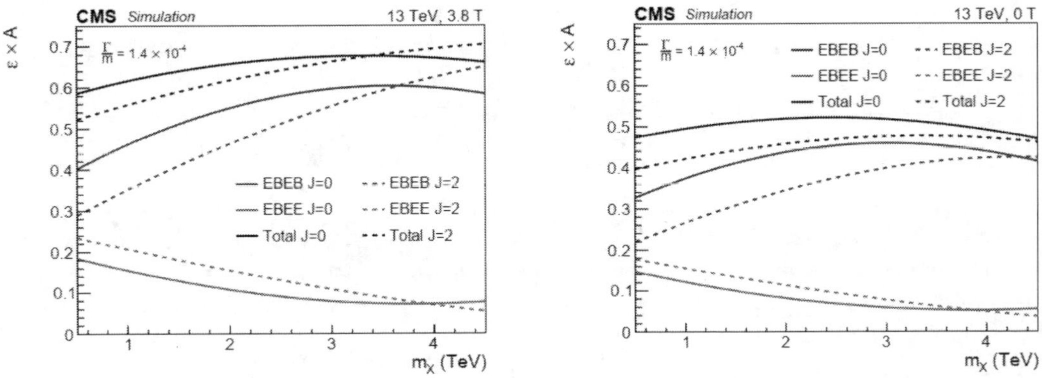

Figure 2. Fraction of events selected by the analysis categories for a possible signal of 500 GeV $< m_X <$ 4.5 TeV and $\Gamma_X/m_X = 1.4\times 10^{-4}$. The prediction for both, scalar and RS graviton resonances, are shown, on the left for the B = 3.8 T sample and on the right for the B = 0 T one.

certainty on the selection efficiency is propagated to the signal normalization at B = 3.8 T (B = 0 T).

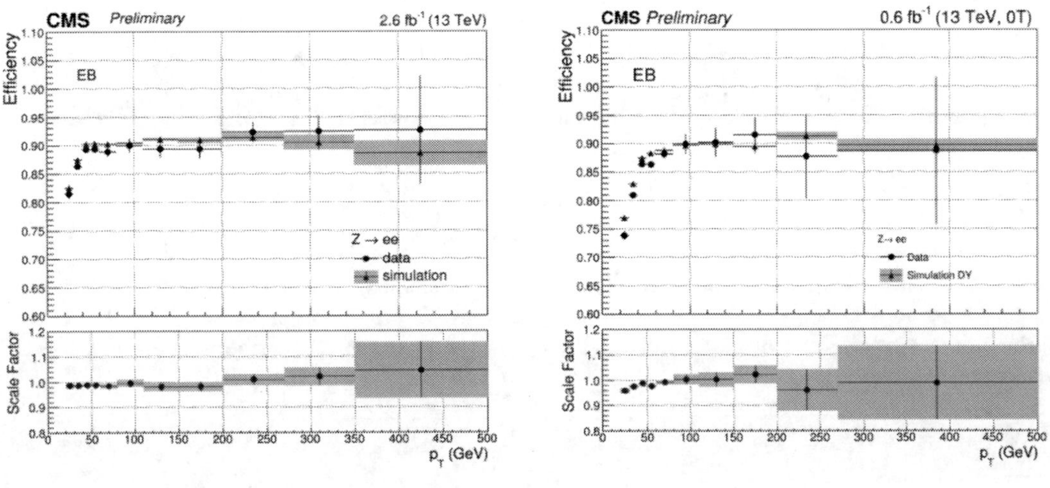

Figure 3. Photon selection efficiency measured with B = 3.8 T (left) and B = 0 T (right) using the tag and probe method (all cuts are applied except for electron rejection) for the EB category. Bottom: data over Monte-Carlo scale factors.

3.1 Energy corrections

The reconstruction and energy assignment of photon candidates is generally derived for lower energy photons than the ones entering this analysis. Thus, the energy scale and energy resolution might need to be corrected for higher values. As the energy determination is only dependent on ECAL, well-known electron objects from $Z \rightarrow e^+e^-$ events can be used instead of photon objects. Residual differences in the observed and predicted distributions of a $Z \rightarrow e^+e^-$ sample are thus determined by a tag and probe method with a procedure described in Ref. [13]. The energy scale corrections are about 0.5 (1.5) % for photon candidate compatibles at B = 3.8 T (B = 0 T). The additional Gaussian smearing for the MC prediction is similar for both values of the magnetic field and of the order of 0.8 % and 1.5 % (2 – 2.5 %) for photon candidates which are centrally (in the forward region) detected. The energy scale correction factors measured for the B = 0 T dataset are found to be about 1 % higher than the B = 3.8 T factors. The comparison with the MC prediction after all energy corrections are applied is reported in Figure 4. The variation of the corrections is studied with $Z \rightarrow e^+e^-$ events as a function of the transverse momentum p_T up to 150 (100 GeV) for EB (EE). The variation is smaller than 0.5 (0.7) % for EB (EE) in both datasets.

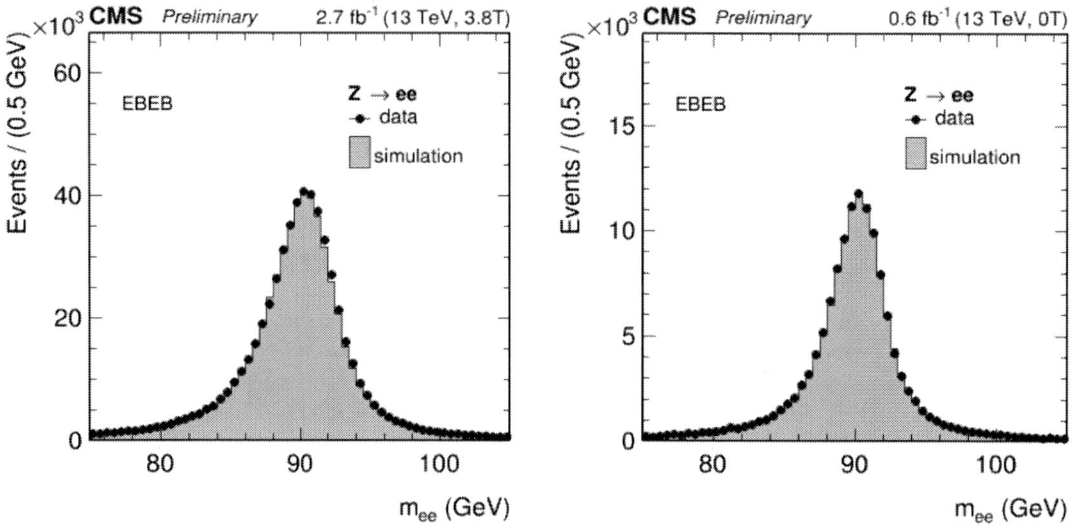

Figure 4. Comparison between the predicted and observed invariant mass distribution of electron pairs obtained after the application of energy scale and resolution corrections. Distributions are shown for events where both electrons are reconstructed for B = 3.8 T (left) and B = 0 T (right) in the EBEB category. The simulation predictions are scaled to match the number of events observed in data.

4 Diphoton mass spectrum

The diphoton spectrum of the events entering the analysis is shown in Figure 5 for all four categories ((B = 3.8 T, B = 0 T)×(EBEB,EBEE)) .

The individual background components of the diphoton sample at B = 3.8 T are estimated by a data-driven method. This evaluation measures that about 90 (80) % of the background events in the EBEB (EBEE) category originate from the irreducible $\gamma\gamma$-SM background processes. This estimate is compared to a precise next-to-next-to leading-order (NNLO) prediction of the $\gamma\gamma$-SM background processes. It is obtained by calculating the so-called 'k-factor', the ratio of the theoretical calculation of

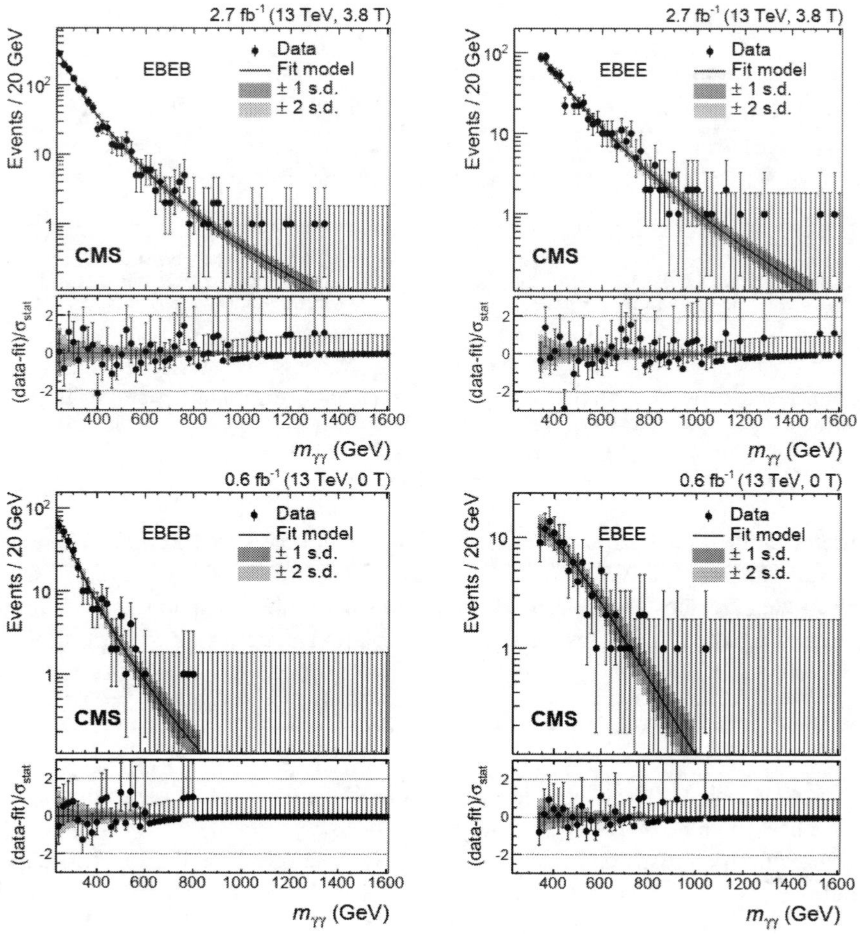

Figure 5. The observed invariant mass spectra are shown for the EBEB (left) and EBEE category (right). The top (bottom) row shows the results for the B = 3.8 T (B = 0 T) dataset. The results of the parametric fits to the data are depicted with their uncertainties.

the 2gNNLO [14] simulation program with the leading-order SHERPA simulation at gen-level. The leading-order $\gamma\gamma$-prediction, a SHERPA event generator sample emulated with the detector response, is then reweighed with this k-factor. The comparison of the $\gamma\gamma$ spectrum with the NNLO MC prediction is reported in Figure 6. The predictions from perturbative QCD match the observations within their uncertainties.

5 Statistical analysis

A simultaneous unbinned likelihood fit to $m_{\gamma\gamma}$ evaluates the diphoton distributions in all four analysis categories ((B = 3.8 T, B = 0 T)×(EBEB,EBEE)) and for each signal hypothesis S:

$$L(\mu, \theta) = \Pi_{i=1}^{N_{events}}[\mu \cdot S(m_i|\theta_S)] + B(m_i|\theta_B)] \cdot Poisson(N_{events}|N_B + \mu \cdot N_S)$$

DOI: 10.1051/epjconf/201716401008

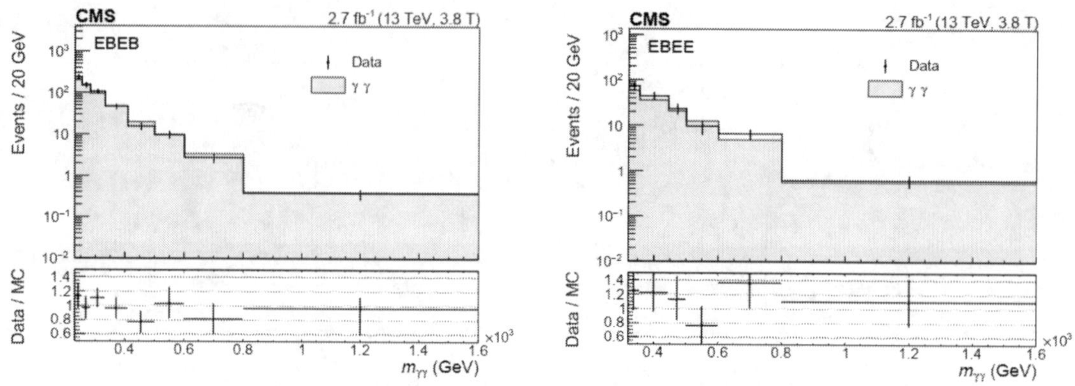

Figure 6. Comparison between the predicted and measured $\gamma\gamma$ background spectrum. Events in the EBEB (EBEE) category are shown on the left (right).

where B denotes the SM background model. The signal and background model depend on the mass m_i of event i and their respective nuisance parameters θ, which model the systematic uncertainties. The parameter μ denotes the signal strength of the considered model. The interpretation is performed with a composite statistical hypothesis test. The test statistics are based on a profile likelihood ratio of the form:

$$q(\mu) = -log\lambda(\mu) := -log\frac{L(\mu \cdot S + B|\hat{\hat{\theta}}_{\mu})}{L(\hat{\mu} \cdot S + B|\hat{\theta})}$$

relying on the likelihood L. The $\hat{x}$ notation denotes the best fit value of parameter x, the notation $\hat{\hat{x}}_y$ symbolizes the best fit value of x, conditionally on y.
The shape of the diphoton distribution as a function of $m_{\gamma\gamma}$ models the background expectation using the functional form:

$$g(m_{\gamma\gamma}) = m_{\gamma\gamma}{}^{a+b\cdot log(m_{\gamma\gamma})}$$

The coefficients are obtained through the maximum likelihood fit to data. Independent coefficients are used for each analysis category and are considered as unconstrained nuisance parameters. The uncertainty on the accuracy of the background determination is assessed and accounted for in the hypothesis test [9].
The signal distribution as a function of $m_{\gamma\gamma}$ is described as the convolution of the intrinsic shape of the resonance and the ECAL detector response [9, 19]. The signal hypotheses considered herein are simulated at leading order with the PYTHIA [20] event generator, using NNPDF2.3 [21] parton distribution functions. Three different values of relative width are used as benchmark models:

- Γ_X is much smaller than the detector resolution: $\Gamma_X/m_X = 1.4 \times 10^{-4}$.
- Γ_X is of the same size as the detector resolution: $\Gamma_X/m_X = 1.4 \times 10^{-2}$.
- Γ_X much larger than the detector resolution: $\Gamma_X/m_X = 5.6 \times 10^{-2}$.

For the RS graviton model, the relative widths correspond to the dimensionless coupling parameter

$$\tilde{k} = \sqrt{\frac{\Gamma_X/m_X}{1.4}} = 0.01, 0.1, \text{ and } 0.2$$

 DOI: 10.1051/epjconf/201716401008

as described in Ref. [22]. In order to determine the signal normalization, the efficiency of the final event selection is combined with the kinematic acceptance as shown in Figure 2. In this analysis, the statistical uncertainties dominate over the systematic ones. In addition to the above mentioned systematics on the background model, the selection efficiency and the photon energy scale, the luminosity uncertainty and the uncertainty on the parton distribution function enter the hypothesis test.

For the interpretation of the results, frequentist statistics and asymptotic formulas [15] are used throughout and their validity is tested [9]. Thus, upper limits on the resonant diphoton production rate under different signal hypotheses are evaluated with the modified frequentist method, also known as CL_s [16, 17]. The agreement of the observed result with the background-only hypothesis is estimated by the background-only p−value, the 'local p-value' p_0 [18].

6 Results of the search at $\sqrt{s} = 13$ TeV

For the $\sqrt{s} = 13$ TeV dataset, the search range for a possible resonance with mass m_X covers the region $500\,\text{GeV} < m_X < 4.5\,\text{TeV}$. The median expected and observed 95% C.L. exclusion limits on the production of scalar and RS graviton resonances are reported in Fig. 7 for the narrow and the largest width hypotheses. The complete list of results for the exlusion limits as well as for the background-only p−value can be found in Ref. [9]. In Figure 8, the results for a relative width of $\Gamma_X/m_X = 1.4 \times 10^{-2}$ are shown on the left. This signal hypothesis shows the largest excess in data among all considered widths hypotheses at a mass of $m_X = 760\,\text{GeV}$. The background-only p-value for the RS graviton model is depicted on the right side of that figure and shows a local significance of 2.9 standard deviations. The separate B = 3.8 T and B = 0 T contributions point out that the B = 0 T category adds approximately 20 % to the observed excess due to its lower luminosity. The probability of observing at least one excess more significant than that in the full search range and among all considered models of this analysis, is the so-called "look-elsewhere-effect" [23]. It is derived from a sampling distribution of max(p_0) of an ensemble of background-only pseudo datasets and found to be less than one standard deviation.

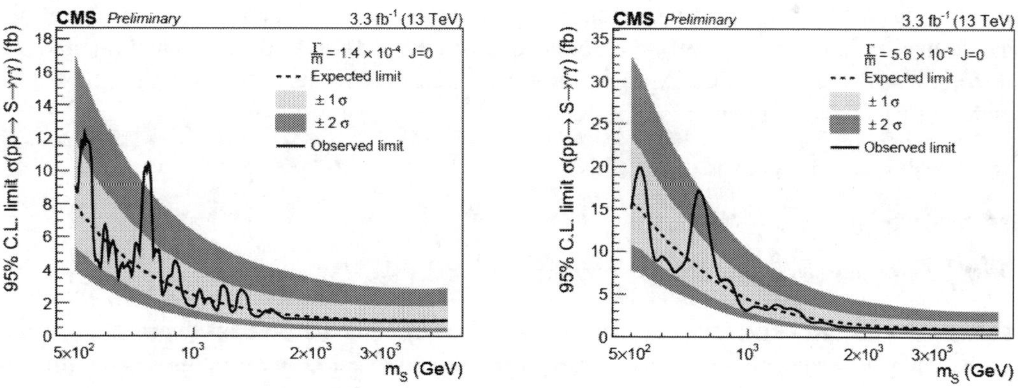

Figure 7. Expected and observed 95% C.L. exclusion limits in the range $500\,\text{GeV} < m_X < 4.5\,\text{TeV}$ for $\Gamma_X/m_X = 1.4 \times 10^{-4}$ (left) and 5.6×10^{-2} (right) for the scalar (J=0) resonance hypothesis.

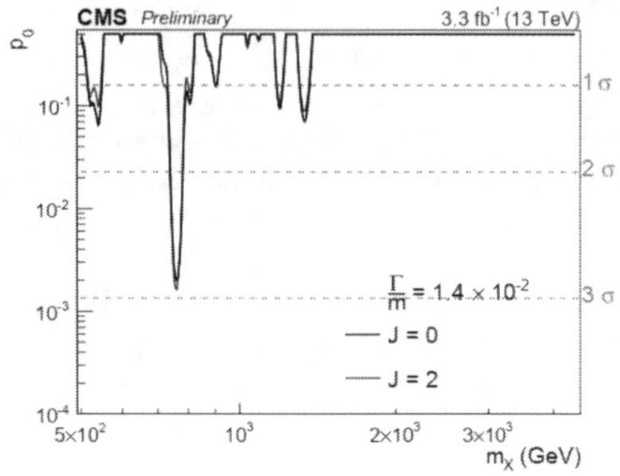

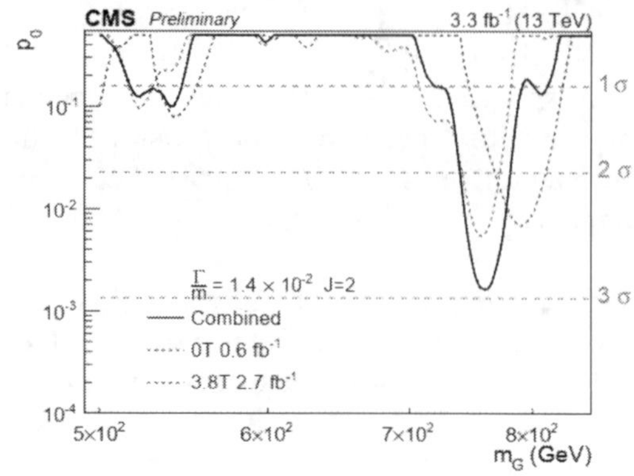

Figure 8. Observed background only p-values obtained on the 13 TeV,dataset. For $\Gamma_X/m_X = 1.4 \times 10^{-2}$, the range 500 GeV $< m_X <$ 4.5 TeV (850 GeV) is shown on the left (right). Results corresponding to both spin hypotheses are shown on the left. The contributions of the B = 3.8 T and B = 0 T datasets are shown separately on the right for the spin-2 hypothesis. Due to the different integrated luminosity, the weight of the B = 0 T dataset at $m_X = 760$ GeV is approximately 20 % of the B = 3.8 T one in the combined result.

7 Results of the combined analysis of $\sqrt{s} = 8$ TeV and 13 TeV datasets

The results of the $\sqrt{s} = 13$ TeV dataset are combined statistically with results obtained by the CMS collaboration at a center-of-mass energy of $\sqrt{s} = 8$ TeV employing an integrated luminosity of 19.7 fb^{-1}. Two searches for resonant diphoton production were performed with the 8 TeV dataset:

- an analysis searching for spin-0 and spin-2 resonances in the mass range 150 – 850 GeV [24].
- a second analysis focusing on spin-2 resonances with a search range of 500 GeV – 3 TeV [25].

The two event samples partially overlap, thus at each m_X, the analysis with the most stringent median expected exclusion limit on resonant diphoton production is taken for the final combination. For resonance masses below $m_X <$ 850 GeV, the results of Ref. [24] enter the combination, and those of Ref. [25] otherwise. The ratio of the signal production cross sections at 8 TeV and 13 TeV has been calculated [9]. It is approximately

- 0.29 (0.27) for m_G (m_S) = 500 GeV,
- 0.24 (0.22) for m_G (m_S) = 750 GeV,
- 0.04 (0.03) for masses in the TeV range (m_G or $m_S \approx$ 3 TeV).

For further details, Ref. [9] may be consulted. The combined result is obtained by a simultaneous fit to the $m_{\gamma\gamma}$ spectra in all event categories, assuming a common signal strength modifier for all categories. The median expected and observed 95% CL$_s$exclusion limits on the equivalent 13 TeV diphoton resonance production cross section, $\sigma_{G,S}^{13\,\text{TeV}} \cdot \mathcal{B}_{\gamma\gamma}$, for the combined analysis are reported in Figure 9 for all studied signal hypothesis of this analysis. The comparison with Figure 7 shows an improvement of the exclusion limits by 20 – 40 % for masses below 1.5 TeV. For masses above 1.5 TeV, the exclusion limits are mainly driven by the sensitivity of the 13 TeV analysis and the benefits of the combination decrease. Randall-Sundrum gravitons with masses below approximately 1.6, 3.3, and 3.8 TeV are excluded at a 95 % confidence level for $\tilde{\kappa} = 0.01, 0.1,$ and 0.2, respectively, corresponding to a relative widths of $\Gamma_X/m_X = 1.4 \times 10^{-4}, 1.4 \times 10^{-2}$ and 5.6×10^{-2}. Similar exclusion limits are found for the scalar signal hypotheses. The result of the local significance for a scalar resonance under the narrow width hypothesis is depicted in Figure 10. This model shows the largest excess in the combination among all considered signal hypotheses with a local significance of roughly 3.4 standard deviations at a mass of $m_S = 750\,\text{GeV}$. Further results can be found in Ref. [9]. Taking into account the effect of searching for a diphoton resonance among all six signal hypotheses, the global significance is estimated to be approximately 1.6 standard deviations [9].

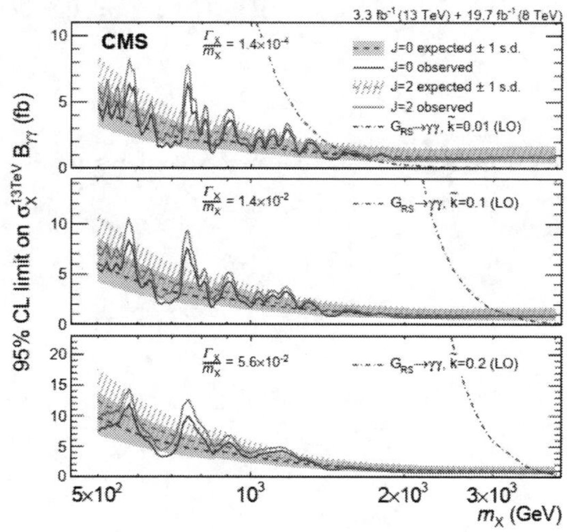

Figure 9. Expected and observed 95% C.L. exclusion limits in the range $500\,\text{GeV} < m_X < 3\,\text{TeV}$ for both spin hypothesis with a relative width of $\Gamma_X/m_X = 1.4\times10^{-4}$ (top), $\Gamma_X/m_X = 1.4\times10^{-2}$ (middle) and $\Gamma_X/m_X = 5.6\times10^{-2}$ (bottom).

8 Summary and outlook

This note summarizes the search for high mass resonances decaying into the diphoton final state by the CMS experiment [9]. The analysis employs a dataset of $3.3\,\text{fb}^{-1}$ and 19.7 $\text{fb}^{-1}$of proton-proton collisions at a center-of-mass energy of $\sqrt{s} = 13$ TeV and $\sqrt{s} = 8$ TeV, respectively. The search sets 95% CL_sexclusion limits on the resonant diphoton production of scalar resonances (J = 0) and RS gravitons (J = 2) in the mass range of $500\,\text{GeV} < m_X < 4.5$ TeV for relative width up to $\Gamma_X/m_X = 5.6 \times 10^{-2}$, using leading-order simulation for the signal processes. Randall-Sundrum gravitons with masses below approximately 1.6, 3.3, and 3.8 TeV are excluded at a 95 % confidence level for $\tilde{\kappa} = 0.01, 0.1,$ and 0.2, respectively, corresponding to a relative widths of $\Gamma_X/m_X = 1.4 \times 10^{-4}, 1.4 \times 10^{-2}$ and 5.6×10^{-2}. A modest excess of events over the background-only hypothesis is observed for m_X =750 GeV. The corresponding local significance under the narrow-width hypothesis ($\Gamma_X/m_X = 1.4 \times 10^{-4}$) for a spin-0 particle is approximately 3.4 standard deviations. Taking into

DOI: 10.1051/epjconf/201716401008

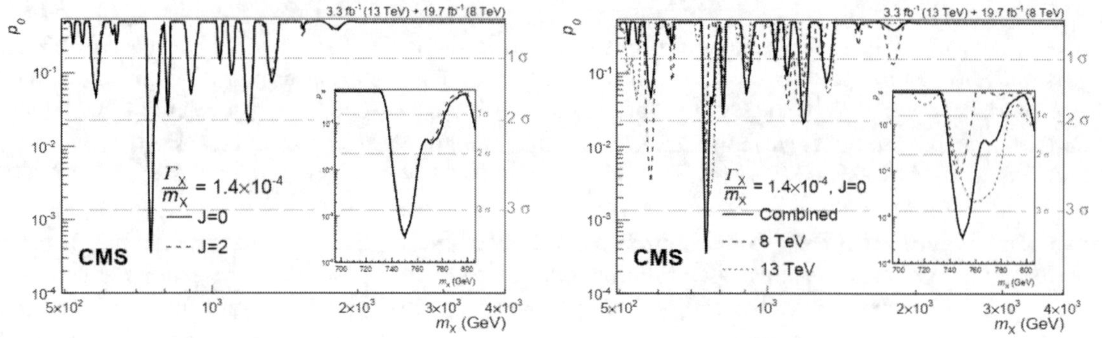

Figure 10. Left: observed background-only p-values obtained on the 8 and 13 TeV combination for both spin hypothesis and a relative width of $\Gamma_X/m_X = 1.4\times10^{-4}$. Right: observed background-only p-values for the spin-0 hypothesis with a relative width of $\Gamma_X/m_X = 1.4\times10^{-4}$. The contributions of the 8 and 13 TeV datasets are shown separately.

account all signal hypotheses considered in the analysis, the global significance of this excess is estimated to 1.6 standard deviations.

References

[1] L.D. Landau, Dokl. Akad. Nauk SSSR **60**, 207 (1948)
[2] N. Craig, J. Galloway, S. Thomas (2013), 1305.2424
[3] G. Aad et al. (ATLAS), Phys. Lett. **B716**, 1 (2012), 1207.7214
[4] S. Chatrchyan et al. (CMS), Phys. Lett. **B716**, 30 (2012), 1207.7235
[5] P.S.B. Dev, A. Pilaftsis (2014), 1408.3405
[6] L. Randall, R. Sundrum, Phys. Rev. Lett. **83**, 3370 (1999), hep-ph/9905221
[7] Journal of High Energy Physics **2016**, 1 (2016)
[8] CMS Collaboration (CMS), CMS Physics Analysis Summary CMS-PAS-EXO-15-004 (2015), https://cds.cern.ch/record/2114808
[9] V. Khachatryan et al. (CMS Collaboration), Phys. Rev. Lett. **117**, 051802 (2016)
[10] S. Chatrchyan et al. (CMS), JINST **3**, S08004 (2008)
[11] CMS Collaboration (CMS), CMS Physics Analysis Summary CMS-PAS-PFT-09-001 (2009), http://cdsweb.cern.ch/record/1194487
[12] CMS Collaboration (CMS), CMS Physics Analysis Summary CMS-PAS-PFT-10-001 (2010), http://cdsweb.cern.ch/record/1247373
[13] V. Khachatryan et al. (CMS), JINST **10**, P08010 (2015), 1502.02702
[14] S. Catani, L. Cieri, D. de Florian, G. Ferrera, M. Grazzini, Phys.Rev.Lett. **108**, 072001 (2012), 1110.2375
[15] G. Cowan, K. Cranmer, E. Gross, O. Vitells, Eur.Phys.J. **C71**, 1554 (2011), 1007.1727
[16] T. Junk, Nucl. Instrum. Meth. A **434**, 435 (1999), hep-ex/9902006
[17] A. Read, Tech. Rep. CERN-OPEN-2000-005, CERN (2000), http://cdsweb.cern.ch/record/451614
[18] ATLAS and CMS Collaborations, LHC Higgs Combination Group, ATL-PHYS-PUB/CMS NOTE 2011-11, 2011/005 (2011), http://cdsweb.cern.ch/record/1379837
[19] M. Baak, S. Gadatsch, R. Harrington, W. Verkerke, Nucl. Instrum. Meth. A **771**, 39 (2015), 1410.7388
[20] T. Sjöstrand, S. Ask, J.R. Christiansen, R. Corke, N. Desai, P. Ilten, S. Mrenna, S. Prestel, C.O. Rasmussen, P.Z. Skands, Comput. Phys. Commun. **191**, 159 (2015), 1410.3012
[21] R.D. Ball et al., Nucl. Phys. **B867**, 244 (2013), 1207.1303
[22] H. Davoudiasl, J.L. Hewett, T.G. Rizzo, Phys. Rev. Lett. **84**, 2080 (2000), hep-ph/9909255
[23] E. Gross, O. Vitells, Eur. Phys. J. **C70**, 525 (2010), 1005.1891
[24] V. Khachatryan et al. (CMS), Phys. Lett. **B750**, 494 (2015), 1506.02301
[25] CMS Physics Analysis Summary CMS-PAS-EXO-12-045, CERN (2015), http://cds.cern.ch/record/2017806

EPJ Web of Conferences **164**, 01009 (2017) DOI: 10.1051/epjconf/201716401009
ICNFP 2016

Exploring phases of dense QCD with compact stars

Armen Sedrakian[1,a]

[1] *Institute for Theoretical Physics, J. W. Goethe-University, D-60438 Frankfurt-Main, Germany*

Abstract. I review a number of recent developments in the physics of compact stars containing deconfined quark matter, including (a) their cooling with possible phase transition from a fully gapped to a gapless phase of QCD at low temperatures and large isospin; (b) the transport coefficients of the 2SC phase and the role played by the Aharonov-Bohm interactions between flux-tubes and unpaired fermions; (c) rapidly rotating compact stars and spin-down and spin-up induced phase transition between hadronic and QCD matter as well as between different phases of QCD.

1 Introduction

A unique way to discern the phase structure of dense quantum chromodynamics (QCD) is offered by the phenomenology of compact stars. Their global parameters, such as the mass, the radius, the moment of inertia etc. depend sensitively on the equation of state (i.e. pressure vs density relation at zero temperature; the many-body methods to derive an equation of state appropriate for compact stars are reviewed, for example, in Refs. [1–3]). The recent measurements of pulsars masses in binary orbits with white dwarfs inferred masses in the range 1.8 - 2.0 solar masses (hereafter $M_{\odot}$) [4–6]. These observations suggest that the equation of state of densest regions of compact stars must be moderately stiff. There are observational indications that the pulsar radii are not very large, possibly in the range of $11 \leq R \leq 14$ km, which suggest that at lower densities the equation of state cannot be too stiff. The mass bounds quoted above were important in constraining various models of equations of state of dense matter in recent years.

The information gained from the global parameters of compact stars is however limited, as the equation of state may be insensitive to the detailed structure of QCD phases, which is dictated by the form of the quasiparticle spectrum in the vicinity of the Fermi surface. An obvious example is the pairing, which introduces a gap in the spectrum of quasiparticles and leads to the phenomenon of superconductivity of quarks [7–10]. The cooling and transport in compact stars is sensitive to such details and most of this review is dedicated to the understanding of these properties of various phases of dense QCD.

This review is structured as follows. Section 2 discusses the cooling of compact stars with quark cores, in particular the differences in the neutrino emission from phases with and without fully gapped Fermi spheres. Section 3 reviews the recent computations of the transport coefficients of two-flavor cold quark matter where two of the colors are paired (the so-called 2SC phase). The possible role of color-magnetic flux tubes is also discussed. Finally, in Sec. 4 we discuss the possibility of observing a

[a] e-mail: sedrakian@th.physik.uni-frankfurt.de

phase transition between the hadronic and superconducting quark matter as well as phase transitions between two different phases of dense QCD in spinning up and down compact stars. Some final remarks are given in Sec. 5. An earlier account of the topics listed above can be found in Ref. [11]. This review does not provide a comprehensive coverage of the subjects listed above and is focused on the research of the author.

2 Cooling of compact stars

Currently the deconfinement of baryons into quark is not well understood, consequently the densities at which such a phase transition can take in compact stars is uncertain. The scale at which we may expect such a transition is set by the characteristic size of a baryon, which should be ≤ 0.84 fm - the measured charge radius of the proton. Once the interparticle distance becomes of the order of this scale the baryons will lose their identity, as their wave-functions will start to overlap. Furthermore, the phase diagram of quark matter at not very high densities is not precisely known, because at the densities relevant to compact stars the perturbative QCD is not valid and non-perturbative methods (such as lattice QCD) face technical obstacles.

Cooling of compact stars can provide indirect information on the properties of quark phases if the density of deconfinement is indeed reached in some heavier stars. Such stars are known as *hybrid compact star*. In this Section we summarize the results of the studies of cooling of hybrid compact stars carried out in Refs. [12–14]. As part of this program, the rapid cooling of the compact star in Cassiopea A (hereafter Cas A) was modelled as a phase transition within the QCD phase diagram [13]. The latest updated data [15] that covers the 10 year period from 2003 to 2013 was fitted in Ref. [14]. The unprecedented fast cooling of the compact star in Cas A, in fact, requires fast transient cooling model for this object consistent with the estimate of Ref. [15], which indicates a decline in the temperature of the star $2.9 \pm 0.9\%$ over 10 years of observation. However, this data cannot be interpreted unambiguously for a number of reasons, see the discussion in Refs. [15, 16].

Before turning to the QCD based description of cooling of compact stars and the Cas A case, it should be mentioned that a number of alternative models describe theoretically the Cas A cooling behaviour on the basis of physics unrelated to QCD degrees of freedom. Nucleonic stellar models (i.e., those containing neutrons, protons, and electrons) attribute the rapid cooling to the onset of Cooper pair-breaking process in the neutron superfluid component [17–22]. An alternative nucleonic model of Ref. [23] uses rates of modified Urca and bremsstrahlung processes that are enhanced compared to rates used in the minimalistic models quoted above by several orders of magnitude due to a softening of pionic modes, as initially discussed in Ref. [24]. A further model requires a fine-tuned onset of Urca process in rotating compact stars [25].

We consider models of hybrid compact stars which contain a color superconducting quark core surrounded by a nucleonic envelope. These phases are separated by a sharp interface, see Ref. [26]. Stars with such hybrid structure naturally correspond to the most massive members of the sequence of stellar equilibria modelled with an equation of state which contains a phase transition from nucleonic to quark matter. These massive members must be heavy enough to account for the inferred masses of pulsars [4–6].

Non-superconducting quarks cool matter via the direct Urca processes $d \to u + e + \bar{\nu}$ and $u + e \to d + \nu$ at a rate which is much larger than for the ordinary nucleonic matter with low proton fraction [12, 27–30]. However, cold quark matter is a (color) superconductor [7–10]. Because of β-equilibrium and non-zero strange quark mass, the Fermi surfaces of up and down quarks are shifted by amount which is of the order of the electron chemical potential or strange quark mass. This implies that new superconducting phases will emerge, in which the Cooper pairing differs from the

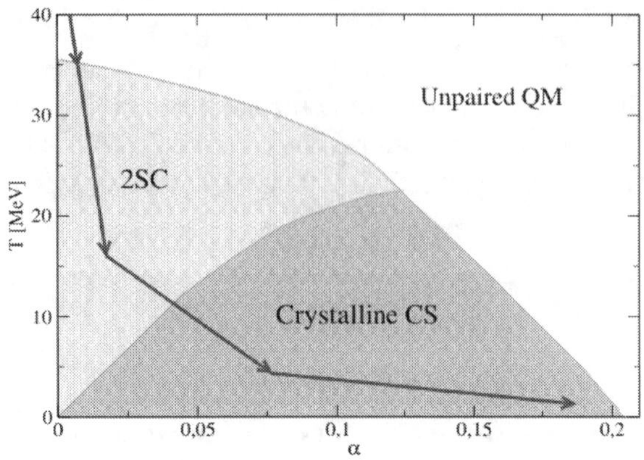

Figure 1. The 2SC and FF phases in the phase diagram of two-flavor quark matter spanned by the isospin asymmetry $\alpha = (n_d - n_u)/(n_u + n_d)$, where n_d and n_u are the number densities of d and u flavors of quarks, and temperature. The path that a compact star traverses during the cooling process is shown by arrows.

ordinary Bardeen-Cooper-Schrieffer pairing. Examples are the gapless two-flavor phases [31, 32] or the the crystalline color-superconducting phase [33, 34]. The latter phase has a number of realizations, because there are many ways to break the spatial symmetries by allowing Cooper pairs to carrying finite momentum. Ref. [14] assumed the so-called Fulde-Ferrell phase (hereafter FF phase), which postulates a single plane wave modulation of the condensate. As far as the cooling is concerned , the general properties of crystalline phases - the existence of gapless excitations [33–36] - is present already in the FF model.

A compact star born in a supernova explosion is hot and matter is nearly isospin symmetrical. In the subsequent minutes to hours the matter cools rapidly and the isospin asymmetry increases to the values characteristic for neutron stars. For temperatures $T \leq T_c$, where T_c is the critical temperature of phase transition, and not too high isospin asymmetries one finds the homogeneous 2SC phase in which the rotational/translational symmetries are unbroken, see Fig. 1. For low temperature and large asymmetries one finds that the ground state corresponds to the FF state.

The consequences of the transition from the symmetrical two-flavor BCS phase to the crystalline phase at some temperature T^* for the cooling of hybrid stars with application to the Cas A case were studied in Ref. [14]. Fits were performed to obtain the value of this temperature. Because the cross-flavor pairing occurs between the green and red colored quarks and the blue quarks do not participate in this type of the pairing a second parameter - the gap for blue-colored quarks Δ_b - was introduced in the color $\mathbf{6}_S$ and flavor $\mathbf{3}_S$ channel. Note that this same-flavor and same-color pairing is unaffected by the flavor asymmetry [37]. An addition parameter w is introduced to account for the finite time-scale of the phase transition, see Ref. [14] for details. Illustrative results are shown in Fig. 2, where we display the dependence of the (redshifted) surface temperature on time for a stellar model with mass $M/M_\odot = 1.93$. The parameter T^* is varied as indicated in the figure whereas the parameters w and Δ_b are held fixed.

For small T^* the quark core has no influence on the cooling because the neutrino emission from the red-green and blue condensates is suppressed by the pairing gaps in these channels. In the opposite case of high values of $T^* \sim 0.2$ MeV the transition to the FF phase occurs early in time and fast neutrino emission via the direct Urca process cools the star rapidly below the value measured for Cas A. After fine tuning of the phase-transition temperature to the value $T^* = 0.009385$ MeV the temperature of Cas A can be reproduced accurately (Fig. 2 left panel). As seen from the right panel of Fig. 2 a further tuning of the parameter T^* allows us to fit the details of the data with good accuracy.

Thus, we conclude that the rapid cooling of the compact star in Cas A can be accounted for via the phase transition from the 2SC to FF-type phase. It is interesting that the phase transition takes place *within the phase diagram of QCD* and is the consequence of the ordering of various

 DOI: 10.1051/epjconf/201716401009

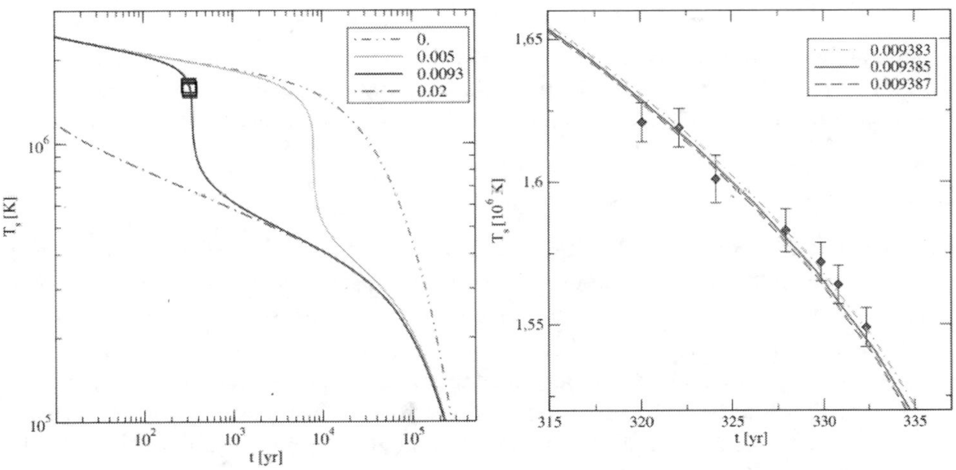

Figure 2. Dependence of the surface temperature of a CSs given in K on time in year. The lines correspond to different phase transition temperature T^* for fixed values of the width $w = 0.2$ and the blue quark pairing gap $\Delta_b = 0.15$ (both in MeV). The labels correspond to values of T^* in MeV. The Cas A data is shown by squares.

superconducting phases in the temperature, density, and isospin spaces. Such ordering was observed in numerous studies of other *population imbalance systems*, such as the nucleonic superfluids under isospin asymmetry or strong magnetic fields [38, 39].

In addition to the study of Cas A behaviour, Ref. [14] studied the dependence of the cooling curves on the variations of the mass of the compact star model. It was found that the heavy stars that are close to the maximal mass will cool to lower temperatures much faster; therefore some refitting of the parameters will be required to adjust these models to the case of Cas A. It was also found that low mass stars $M \sim 1.1\, M_\odot$ remain warm over longer time scales and are thus hotter than their heavy counterparts.

Some issues remain to be studied, for example the role of strange quarks in the quark phases. These may influence the cooling in case they are not involved in the pairing pattern. However, at asymptotically high density, when the strange quark mass vanishes, the color-flavor-locked (CFL) phase in which the pairing is "perfect", in the sense that all colors and flavors are involved in the condensate, is the ground state of matter [40]. Because the CFL phase has very low heat capacity as well as neutrino emissivity strong modifications are expected only in the case where imperfect CFL phases arise, such as the gapless CFL phase [10]. In any event, the low number of strange quarks can render the process on strange quarks subdominant.

3 Transport phenomena

The treatment of transport phenomena in dense QCD phases of neutron stars requires the knowledge of excitation spectra of quarks near their Fermi-surface. The pairing patterns in the color and flavor spaces are essential in determining the content of excitations, especially those that are ungapped. At large densities the dominant CFL phase is a superfluid, whose excitation spectrum contains only phonons in analogy liquid ^{4}He, see Refs. [41–43]. As the density is lowered to the densities relevant to compact stars the precise from of pairing becomes uncertain. The most robust phase at this density is the the two-flavor color-superconducting (2SC) phase, which pairs up and down quarks in a color antitriplet state. The blue color is either unpaired or undergoes single-flavor and color pairing with very small gaps of the order of keV [7]. In addition to the bulk fermions, the 2SC phase can contain color-magnetic flux tubes which will carry partially the magnetic field through the color superconductor [44]. The flux tubes can effectively scatter bulk fermions and, therefore, contribute to the transport and relaxation in a certain parameter range, which we will discuss below.

3.1 Bulk fermions

The bulk transport coefficients of the 2SC phase were computed in Ref. [45] using the Boltzmann equation in the transport relaxation approximation. The matrix elements of the scattering processes were computed to the leading order in the electromagnetic and strong coupling (α and α_s). The screening of these interactions were taken in the Hard-Thermal-Loop approximation. In addition it was assumed that the matter is strongly degenerate with temperature $T \ll \mu_q$, where μ_q is the chemical potential of quarks. An important feature of the interactions is the mixing of the gauge fields into a linear combination of electromagnetism and gluonic fields, called $\tilde{Q}$ and X. The first field is unscreened in the 2SC phase and penetrates unimpeded in the bulk; the second field is Meissner screened over the distances of the order of the penetration depth.

The ungapped fermions of the 2SC phase are the blue up (u) and down (d) quarks and electrons. Ref. [45] found the following key features of the transport in the 2SC phase: the thermal conductivity and shear viscosity are dominated by the blue-d quarks for sufficiently low temperatures. At higher temperatures the electrons take over the dominant role. This transition for thermal conductivity occurs at $T/\mu_q \sim \alpha/7.7 \sim 10^{-3}$, so most of the temperature range of interest for neutron stars is in the blue-d dominated regime with the scaling $\kappa_{bd} = 0.00617/(T/\mu_q)$. For the shear viscosity the crossover from blue-d to electron domination occurs at $T/\mu_q \sim 10^{-5}$, so electrons become dominant at relatively low temperature $T \sim 10\,\mathrm{keV}$; the shear viscosity scales as $\eta_e \simeq 0.00231(\mu_q^4/5\pi^2)\alpha^{-5/3}T^{-1}(T/\mu_q)^{-2}$. On the other hand the electrical conductivity of the rotated $\tilde{Q}$ charge is completely dominated by the electron, because the blue-d quarks are $\tilde{Q}$ neutral, whereas blue-u quarks are much less abundant; the electrical conductivity of electrons is given by $\sigma_e = 0.0433(\mu_q^2 e^2/\pi^2)\alpha^{-5/3}T^{-1}(T/\mu_q)^{-2/3}$; see Ref. [45] for the expressions of less dominant contributions. It was also shown that $SU(2)_{rg}$ gluons do not contribute to the transport coefficients in 2SC quark matter significantly.

3.2 Color-magnetic flux-tubes

The magnetic field will penetrate the 2SC phase in form of quantized color-magnetic flux tubes because for realistic values of parameters the 2SC phase is a type-II X-charge superconductor [44]. The corresponding Ginzburg-Landau parameter is given by

$$\kappa_{2SC} \approx 11\frac{\Delta}{\mu_q} \geq \frac{1}{\sqrt{2}}. \tag{1}$$

where Δ is the gap in the red-green channel. The critical field for formation of flux tubes is very high, $H_{c1} \sim 10^{17}$ G compared to typical field of a neutron stars, but as the 2SC phase nucleates, domains of superconducting matter will form, which will be Meissner screened. This will leave the normal regions with intense magnetic fields above the average value. Once these regions eventually undergo superconducting phase transition the magnetic field intensity within them will be larger than H_{c1} with X flux trapped in the form of flux tubes. The detailes of the configurations and density of the X-tubes is not known and depends, among several things, on the dynamics of the phase transition. However, their average density is expected to be of the order of that of the ordinary flux tubes in the hadronic core by simple flux conservation.

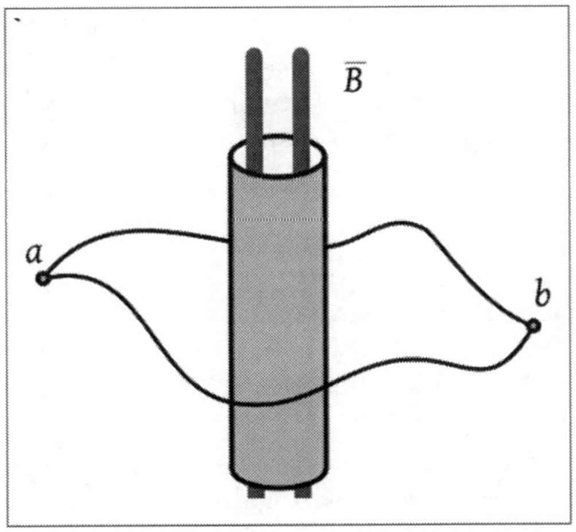

Figure 3. Illustration of the Aharonov-Bohm effect: an electron or blue u-quark acquires a phase shift as it passes from point a to point b around a color-magnetic flux tube (shown as the cylindrical slab).

The Aharonov-Bohm scattering cross-section of gapless fermions with the X flux tubes was calculated in terms of the cross-section given by

$$\frac{d\sigma}{d\vartheta} = \frac{\sin^2(\pi\tilde{\beta})}{2\pi k \sin^2(\vartheta/2)}, \tag{2}$$

where k is the momentum in the plane perpendicular to the flux-tube, ϑ is the scattering angle, $\tilde{\beta} = q_p/q_c$, q_p is the charge of the scattering particle and q_c is the charge of the condensate whose winding by a phase of 2π characterizes the flux-tube. The values of $\tilde{\beta}$ are given by

$$\tilde{\beta} = \mathrm{diag}\Big(\frac{1}{2} + \frac{3}{2}\sin^2\varphi, \frac{1}{2} - \frac{3}{2}\sin^2\varphi, \frac{1}{2} - \frac{3}{2}\sin^2\varphi, \frac{1}{2} + \frac{3}{2}\sin^2\varphi, -1 + 3\sin^2\varphi, -1, -3\sin^2\varphi\Big) \tag{3}$$

in the basis $\psi = (ru, gd, rd, gu, bu, bd, e^-)$, where rgb refer to red, green, and blue colors, e^- refers to electron, and the mixing angle is defined as $\cos\phi = \sqrt{3}g/\sqrt{e^2 + 3g^2}$, where e and g are the electromagnetic and strong coupling constants.

The relaxation rate for particles of species i scattering off flux tubes of area density n_v is then given by [44]

$$\tau_{if}^{-1} = \frac{n_v}{p_{Fi}} \sin^2(\pi\tilde{\beta}_i)\,. \tag{4}$$

where n_v is the number density of color-magnetic flux tubes, p_{Fi} is the Fermi momentum of fermion i. This result corresponds to the standard form for relaxation rate known for classical gases $\tau^{-1} = cn_v\sigma$, where $c = 1$ is the speed of the particles, n_v is the density of scattering centers, and $\sigma \propto \sin^2(\pi\tilde{\beta})/p_F$ is the cross section of Aharonov-Bohm scattering. Eq. (4) sets the scale for the thermal relaxation of the blue quarks and electrons in the 2SC phase. Note however that the blue-d quark, has zero Aharonov-Bohm interaction with the flux-tubes because $\tilde{\beta} = 0$. The other two, the electron and blue-u quark, have the same Aharanov-Bohm factors, see Eq. (3), but different Fermi momenta.

Ref. [44] argued that because the magnetic field in hadronic envelope is below the lower critical field H_{c1} for X-flux-tubes into 2SC quark matter, the trapped flux-tubes will be pulled out of the quark core. The force pulling the flux-tubes is balanced by the mutual friction force on the moving flux-tube due to its Aharonov-Bohm interaction with electrons and blue-u quarks. Apart from the rearrangement of the magnetic field, which is characterized by time-scales comparable to pulsar life-times, the flux expulsion has other physical significance: it could be a significant source of heating of the lower base

DOI: 10.1051/epjconf/201716401009

of the nuclear mantle, and therefore the inner core (which is roughly isothermal after 10^3 yr). Internal heating is known to affect the thermal evolution of compact star in the later stages of their thermal evolution (photo cooling era).

Consider now the combined effect of scattering of a fermion off other fermions and flux-tubes. If there are several scattering channels then the net relaxation time for fermion species i is given by the

$$\tau_i^{-1} = \tau_{i,v}^{-1} + \sum_j \tau_{ij}^{-1} \tag{5}$$

where $\tau_{i,v}^{-1}$ and τ_{ij}^{-1} are the fermion-flux-tubes and fermion-fermion relaxation rates, respectively. In a first approximation one can assume that the fermion-fermion and fermion–flux-tubes scattering do not interfere. Then, it is clear that the fermion–flux-tubes scattering shortens the mean-free-path of the partciles and, therefore, suppresses the transport coefficients.

With the picture above, some conclusions can be drawn for the individual transport coefficients [45]. Because the dominant contribution to the thermal conductivity is from blue-d quarks, which have zero Aharanov-Bohm scattering with vortices, the scattering off the flux-tubes has no significant effect on the thermal conductivity. The electrical conductivity and shear viscosity are dominated by electrons where the interactions are dominated by the $\tilde{Q}$ interaction.

The electron-flux-tubes relaxation rates for electrical conductivity and shear viscosity are approximately the same, as these involve the momentum relaxation up to a constant of order 1. Clearly the electron-flux-tubes sattering becomes important, when its relaxation time is comparable to the fermion-fermion relaxation time. According to the estimates of Ref. [45] flux-tubes can lower the magnitude of transport coefficients for realistic values of the external magnetic field.

4 Spin-up and spin-down induced phase transitions

The studies of compact stars with quark cores in a wide range of rotation periods from the Keplerian limit down to the static limit is of great interest in the interpretation of the phenomena related to the physics of pulsars. Compact stars that are born rapidly rotating evolve adiabatically by losing energy to gravitational waves and electromagnetic (to lowest order, dipole) radiation. Such evolution can be approximated by a sequence of configurations with constant rest mass and varying spin. The massive rapidly rotating configurations may not have a stable static limit. Consequently, the spin-down evolution can lead to the formation of a black hole when the support through centrifugal force cannot balance the pull of gravity anymore. The configurations which do not have static counterparts are known as *supramassive configurations* [46]. The time-reversed counterpart of the spin-down scenario is also realized in nature. According to the formation scenario of the millisecond pulsars, these are spun-up to millisecond frequencies by accretion from a companion. In this case the evolution involves some mass transfer via accretion to the compact star, therefore the rest mass of the star is not a constant anymore. However, the changes in the rest mass during the accretion are not significant and do not change the qualitative picture. Therefore, the changes in the structure of the compact stars as the star spin increases with time can be considered, in a first approximation, as "time reversed" evolution, which does not require additional physical input, such as the mass accretion rate from the companion.

How the spin evolution of hybrid stars, i. e., stars with quark-matter cores surrounded by a nuclear envelope, may differ from the evolution of purely nucleonic stars? The most dramatic effect is the spin-down or spin-up induced phase transition(s) between the hadronic and quark matter phases. The reason is that rapidly rotating stars are *less compressed* and are effectively more "low-density" objects than their static counterparts. Changes in the spin of the star induce changes in its density profile,

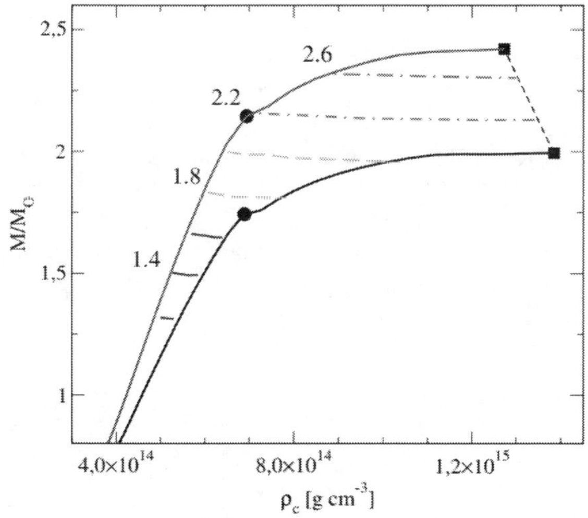

Figure 4. Dependence of masses of hybrid compact stars on the central density. The lower and upper solid lines corresponds to non-rotating and maximally rotating stars. The dashed and dash-dotted lines show stars with a fixed rest masses, as labeled in the plot, with the mass step between two sequential lines equal 0.2. Among these the dashed-dotted lines correspond to the class of the stars which do not have static counterparts. The squares show the maximal mass stars, which are connected by the line at which the stability region terminates. The stars located to the right of dots are hybrid stars that contain quark matter. All mass units are given in units of the solar mass $M_\odot$.

which in turn can lead to density-induced phase transitions from quark to hadronic matter and back. The physics of spin-up and -down induced phase transitions in *non-superconducting* quark matter was first discussed discussed by Glendenning, Pei and Weber in Ref. [47] and subsequently elaborated in many studies, see e. g., Refs. [48–52].

Rotating compact stars containing color superconducting cores were considered initially in Ref. [53]. The color superconducting matter was taken to be in the crystalline phase. The case where the quark matter pairs either in the 2SC or the CFL phases was subsequently studied in Ref. [54] on the basis of an equation of state constructed to reproduce the nuclear and quark matter phenomenologies as well as the astrophysical constraint on the value of the maximum mass of a compact star $M/M_\odot \geq 2$. Below, our discussion follows the study of Ref. [54], where the equation of state of a hybrid star was constructed assuming a sharp interface between the phases. This is the case when the surface tension of the interface is not too small. The hadronic matter description is based on relativistic density functional of nuclear matter with density-dependent couplings, see Refs. [55, 56] for the parameterization and further details. The quark matter equation of state is based on the Nambu–Jona-Lasinio model which includes in addition to common interactions also repulsive vector interactions as given in Ref. [26]. The mass vs central density relations for hybrid stars are shown in Fig. 4 in the static and Keplerian limits. These limits can be connected by constant *rest mass* lines (near horizontal lines in Fig. 4), which correspond to the scenarios of spin-up or spin-down of the stars without mass transfer. There are three classes of constant rest mass sequences in the case of hybrid stars:

1. The well-known *supramassive* compact stars which are stable only due to the (uniform) rotation (dashed-dotted lines in Fig. 4) . They collapse to a black hole as they spin down below certain critical angular velocity; the instability sets-in in the region to the right of the line connecting the maximal masses of non-rotating and maximally fast rotating compact stars in Fig. 4.

2. The second class can be defined as *transitional* compact stars (dashed lines in Fig. 4). These are purely nucleonic in the limit of maximally fast rotation and are hybrid in the static limit. This implies that there is a phase transition from nucleonic phase to the quark matter phase *at the center* of the star as the star spins down or up. Note that phase transition take also place within the supramassive sequences which contain quark matter already in the limit of maximal rotation. Because the central density and the density profiles of these stars change with their

DOI: 10.1051/epjconf/201716401009

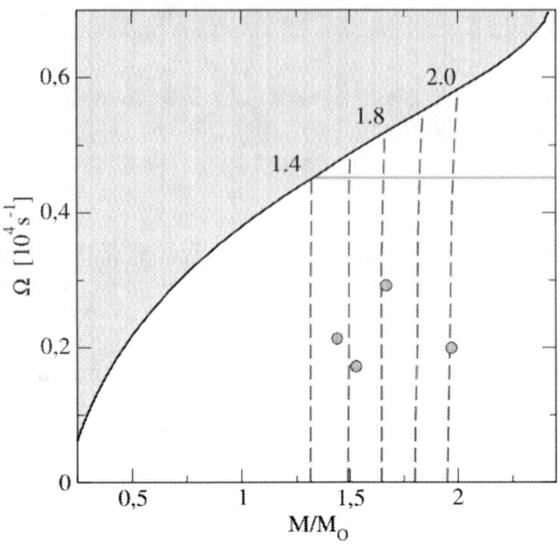

Figure 5. Frequency vs mass diagram for compact stars. Only the most massive object J1614-2230 is massive enough to contain quark matter. The solid line indicates the Keplerian frequency as a function of mass; the area above the line is excluded for uniformly rotating stars.

spin there is a continuous conversion of quark matter to nucleonic matter and back depending on whether the star spins down or up.

3. The third class of stars do not contain quark matter in either limit, i.e., these are purely nucleonic as the central densities are not sufficiently large for quark matter to form (solid lines in Fig. 4). The masses of these stars are below $1.7 M_{\odot}$ for the equation of state used in our example.

Figure 5 displays the frequency vs mass diagram for compact stars, with dots corresponding to objects for which the mass and the spin frequency are measured, see Ref. [54] for details. The shaded region is inaccessible to our models in the case of uniform rotation. The horizontal line shows the spin frequency of the fastest rotating pulsar J1748-2446, for which the mass is unknown. For present equation

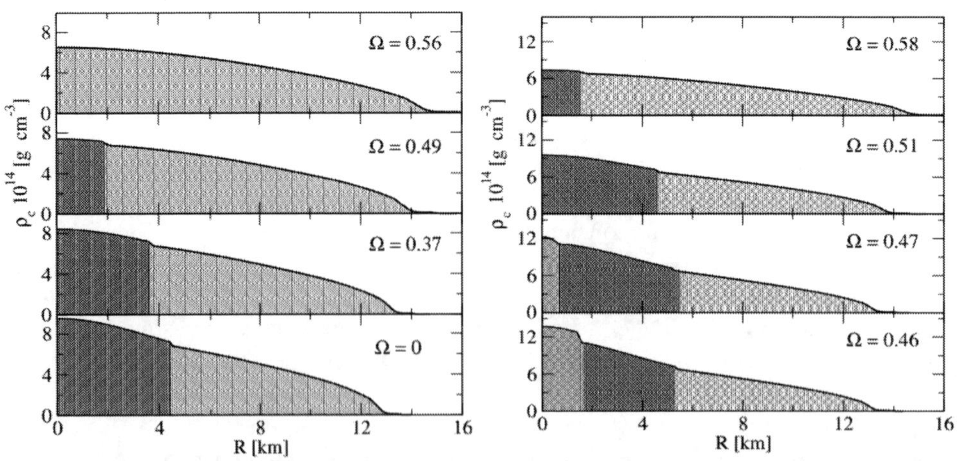

Figure 6. Equatorial density profiles of a rotating compact stars with rest masses 2.2 $M_{\odot}$ (left panel) and 2.4 $M_{\odot}$ (right panel). The low-density region corresponds to nucleonic matter (blue), which is followed by the 2SC phase (red) and CFL phase (green). The frequencies are given in units of 10^4 s^{-1}.

of state this object's mass needs to be above $1.4M_{\odot}$, which illustrates the possibility of constraining the equations of state of dense matter by future measurements of the spin of sub-millisecond pulsars.

Figure 6 shows the dependence of the equatorial density profile of 2.2 $M_{\odot}$ (left panel) and 2.4 $M_{\odot}$ (right panel) rest-mass stars on the internal radius for four spin frequencies in the range $0 \leq \Omega \leq 5.6\times10^3$ Hz. The upper panels in both cases show the stars rotating at the Keplerian limit. As expected, the region occupied by the 2SC phase becomes larger as the star's spin is reduced. Eventually, the CFL phase may form at the center of the most massive stars. Because the phase transition from nucleonic phase to the 2SC phase and from the 2SC to the CFL phase is first order, both are accompanied by jumps in the density, as seen in Fig. 6

5 Final remarks

The exploration of the dense QCD via the astrophysics of compact stars is an actively pursued subject. We expect substantial observational progress in the studies of compact stars in the upcoming years, including further measurements of the masses of pulsars, the radii of neutron stars, and eventually the gravitational wave signals from the mergers of two neutron stars or neutron stars and black holes. This motivates strongly further studies of dense matter in all its facets, including the physics that goes beyond the equation of state. Better understanding of the transport coefficients of dense QCD, weak interactions in dense matter, and dynamics of phase transitions would largely improve our modelling of astrophysical phenomena that are relevant for the current observational programs.

Acknowledgments

I would like to thank my collaborators M. Alford, L. Bonanno, G. Colucci, X.-G. Huang, H. Nishimura, and D. Rischke, for their contribution and insight into the research described above. This work was supported by the Deutsche Forschungsgemeinschaft (Grant No. SE 1836/3-2), by the Helmholtz International Center for FAIR, and by the NewCompStar COST Action MP1304.

References

[1] F. Weber, *Pulsars as astrophysical laboratories for nuclear and particle physics*, IOP, Bristol (1999).
[2] A. Sedrakian, Prog. Part. Nucl. Phys. **58**, 168 (2007).
[3] J. M. Lattimer and M. Prakash, Phys. Rep. **621**, 127 (2016).
[4] P.B. Demorest, T. Pennucci, S.M. Ransom, *et al.*, Nature **467**, 1081 (2010).
[5] J. Antoniadis, P.C.C. Freire, N. Wex, *et al.*, Science **340**, 448 (2013).
[6] E.D. Barr, P.C.C. Freire, M. Kramer, *et al.*, `arXiv:1611.03658` (2016).
[7] D. Bailin, A. Love, Phys. Rep. **107**, 325 (1984).
[8] K. Rajagopal, F. Wilczek, `arXiv:hep-ph/0011333`.
[9] M. Alford, Annual Rev. of Nucl. and Part. Science **51**, 131 (2001).
[10] M.G. Alford, A. Schmitt, K. Rajagopal, T. Schäfer, Rev. Mod. Phys. **80**, 1455 (2008).
[11] A. Sedrakian, PoS (Confinment X), 251 (2013), `arXiv:1301.2675`.
[12] D. Hess, A. Sedrakian, Phys. Rev. D **84**, 063015 (2011).
[13] A. Sedrakian, A&A **555**, L10 (2013).
[14] A. Sedrakian, Eur. Phys. J. A **52**, 44 (2016).
[15] K.G. Elshamouty, C.O. Heinke, G.R. Sivakoff, *et al.*, ApJ **777**, 22 (2013).

[16] B. Posselt, G.G. Pavlov, V. . Suleimanov, O. Kargaltsev, ApJ **779**, 186 (2013).
[17] D. Page, M. Prakash, J.M. Lattimer, A.W. Steiner, Phys. Rev. Lett. **106**, 081101 (2011).
[18] P.S. Shternin, D.G. Yakovlev, C.O. Heinke, *et al.*, MNRAS **412**, L108 (2011).
[19] P.S. Shternin, D.G. Yakovlev, MNRAS **446**, 3621 (2015).
[20] W.C.G. Ho, K.G. Elshamouty, C.O. Heinke, A.Y. Potekhin, Phys. Rev. C **91**, 015806 (2015).
[21] W.G. Newton, K. Murphy, J. Hooker, B.A. Li, ApJ Lett. **779**, L4 (2013).
[22] L.B. Leinson, Phys. Lett. B **741**, 87 (2015).
[23] D. Blaschke, H. Grigorian, D.N. Voskresensky, F. Weber, Phys. Rev. C **85**, 022802 (2012).
[24] C. Schaab, D. Voskresensky, A.D. Sedrakian, F. Weber, M.K. Weigel, A&A **321**, 591 (1997).
[25] R. Negreiros, S. Schramm, F. Weber, Phys. Lett. B **718**, 1176 (2013).
[26] L. Bonanno, A. Sedrakian, A&A **539**, A16 (2012).
[27] D. Page, M. Prakash, J.M. Lattimer, A.W. Steiner, Phys. Rev. Lett. **85**, 2048 (2000).
[28] M. Alford, P. Jotwani, C. Kouvaris, J. Kundu, K. Rajagopal, Phys. Rev. D **71**, 114011 (2005).
[29] R. Anglani, G. Nardulli, M. Ruggieri, M. Mannarelli, Phys. Rev. D **74**, 074005 (2006).
[30] H. Grigorian, D. Blaschke, D. Voskresensky, Phys. Rev. C **71**, 045801 (2005).
[31] I. Shovkovy, M. Huang, Phys. Lett. B **564**, 205 (2003).
[32] H. Müther, A. Sedrakian, Phys. Rev. D **67**, 085024 (2003).
[33] M. Alford, J.A. Bowers, K. Rajagopal, Phys. Rev. D **63**, 074016 (2001).
[34] R. Anglani, R. Casalbuoni, M. Ciminale, *et al.*, Rev. Mod. Phys. **86**, 509 (2014).
[35] A. Sedrakian, D.H. Rischke, Phys. Rev. D **80**, 074022 (2009).
[36] X.G. Huang, A. Sedrakian, Phys. Rev. D **82**, 045029 (2010).
[37] M.G. Alford, J.A. Bowers, J.M. Cheyne, G.A. Cowan, Phys. Rev. D **67**, 054018 (2003).
[38] M. Stein, A. Sedrakian, X.G. Huang, J.W. Clark, Phys. Rev. C **93**, 015802 (2016).
[39] M. Stein, A. Sedrakian, X.G. Huang, J.W. Clark, Phys. Rev. C **90**, 065804 (2014).
[40] M. Alford, K. Rajagopal, F. Wilczek, Nucl. Phys. B **537**, 443 (1999).
[41] I.A. Shovkovy, P.J. Ellis, Phys. Rev. C **66**, 015802 (2002).
[42] C. Manuel, F.J. Llanes-Estrada, J. Cosmo. Astropart. Phys. **8**, 001 (2007).
[43] C. Manuel, A. Dobado, F.J. Llanes-Estrada, JHEP **9**, 076 (2005).
[44] M.G. Alford, A. Sedrakian, J. of Phys. G **37**, 075202 (2010).
[45] M.G. Alford, H. Nishimura, A. Sedrakian, Phys. Rev. C **90**, 055205 (2014).
[46] G.B. Cook, S.L. Shapiro, S.A. Teukolsky, ApJ **422**, 227 (1994).
[47] N.K. Glendenning, S. Pei, F. Weber, Phys. Rev. Lett. **79**, 1603 (1997).
[48] N.K. Glendenning, F. Weber, ApJ Lett. **559**, L119 (2001).
[49] H. Grigorian, D. Blaschke, G. Poghosyan, Nucl. Phys. A **715**, 831 (2003).
[50] J.L. Zdunik, M. Bejger, P. Haensel, E. Gourgoulhon, A&A **450**, 747 (2006).
[51] E.B. Abdikamalov, H. Dimmelmeier, L. Rezzolla, J.C. Miller, MNRAS **392**, 52 (2009).
[52] F. Weber, M. Orsaria, R. Negreiros, ArXiv e-prints (2013), 1307.1103.
[53] N.D. Ippolito, M. Ruggieri, D.H. Rischke, *et al.*, Phys. Rev. D **77**, 023004 (2008).
[54] N.S. Ayvazyan, G. Colucci, D.H. Rischke, A. Sedrakian, A&A **559**, A118 (2013).
[55] G.A. Lalazissis, T. Nikšić, D. Vretenar, P. Ring, Phys. Rev. C **71**, 024312 (2005).
[56] G. Colucci, A. Sedrakian, Phys. Rev. C **87**, 055806 (2013).

EPJ Web of Conferences **164**, 01010 (2017) DOI: 10.1051/epjconf/201716401010
ICNFP 2016

Far-from-equilibrium energy flow and entanglement entropy *

Eugenio Megías[1,2,a]

[1] *Max-Planck-Institut für Physik (Werner-Heisenberg-Institut), Föhringer Ring 6, D-80805, Munich, Germany*
[2] *Departamento de Física Teórica, Universidad del País Vasco UPV/EHU, Apartado 644, 48080 Bilbao, Spain*

Abstract. The time evolution of the energy transport triggered in a strongly coupled system by a temperature gradient is holographically related to the evolution of an asymptotically AdS black brane. We study the far-from-equilibrium properties of such a system by using the AdS/CFT correspondence. In particular, we describe the appearance of a steady state, and study the information flow by computing the time evolution of the holographic entanglement entropy. Some universal properties of the quenching process are presented.

1 Introduction

The study of out-of-equilibrium dynamics is a notoriously challenging problem. The hydrodynamic approach has some limitations: it can be applied only for systems close to equilibrium, as it is based on the assumption that the mean free path (time) of particles is much shorter than the characteristic size (time scale) of the system. However, many physical systems are affected by far-from-equilibrium dynamics. Some examples are the the initial stages of the Quark-Gluon plasma thermalization [1], quenches in some condensed matter systems and fluctuations in the fractional Hall effect [2]. These kinds of studies are very difficult with the current techniques, and nowadays the results are limited to simple models. An interesting and tractable class of non-equilibrium configurations are the steady state flows which emerge between two heat baths [3–7]. The energy flow between these two systems is characterized by a time-independent configuration, but it does not correspond to equilibrium.

AdS/CFT is an excellent tool to study the thermalization process of theories with a Conformal Field Theory (CFT) fixed point in the ultraviolet, as it is then mapped to black hole formation in asymptotically AdS space [8]. A strong motivation to apply the AdS/CFT correspondence to far-from-equilibrium dynamics is that it might help to establish some universal properties of these systems. In particular, it could give some insight into the universal quenching properties of the systems. Apart from the energy flow, it is very interesting to study the information flow between two systems which are initially isolated. On this respect, some important quantity is the entanglement entropy which measures how the information get exchanged between two systems. Holographically it can be measured as a generalization of the Bekenstein-Hawking entropy formula [9–11]. Another quantity derived from the entanglement entropy is the mutual information, which measures the amount of information that can be obtained from one of the systems by looking at the other system.

*Plenary talk given by E. Megías at the 5th International Conference on New Frontiers in Physics (ICNFP 2016), 6-14 July 2016, Kolymbari, Crete, Greece.
[a] e-mail: emegias@mppmu.mpg.de

Figure 1. Two isolated systems initially at equilibrium are put in contact at $t = 0$. A spatially homogeneous non-equilibrium steady state develops at late times, and it carries an energy current $J_E = \langle T^{tx} \rangle_s$.

In this work we will study, within the AdS/CFT correspondence, the formation and time evolution of the steady state between two asymptotic equilibrium systems at different temperatures. In a second step we will address the time evolution of the entanglement entropy, and study some universal properties of the quenching process.

2 Energy flow in strongly coupled systems

It was shown in [3] the existence of a homogeneous steady state in a class of (1 + 1)-dim CFTs. A universal formula for the heat flow and the generalization to higher dimensions was derived in [4, 5]. We will study in this section the steady state formation in these systems in any number of dimensions.

2.1 Steady state formation

Let us consider two thermal reservoirs in $(d + 1)$-dim, each of them initially at equilibrium but at different temperatures, T_L and T_R. The initial energy density reads

$$\varepsilon(x, t = 0) = d a_d \left[T_L^{d+1} \Theta(-x) + T_R^{d+1} \Theta(x) \right] , \tag{1}$$

where a_d depends on the number of degrees of freedom in the CFT. After bringing the two systems into thermal contact at $t = 0$, a spatially homogeneous steady state develops, carrying a heat flow J_E which transfers energy from the hottest to the coldest system. Such a physical situation is presented in Fig. 1. The steady state configuration in the CFT can be described by the Lorentz-boosted stress tensor

$$\langle T^{\mu\nu} \rangle = a_d T^{d+1} \left(\eta^{\mu\nu} + (d+1) u^\mu u^\nu \right) , \tag{2}$$

where $\eta^{\mu\nu} = \text{diag}(-1, 1, \cdots, 1)$ is the CFT metric, and $u^\mu = (\cosh\theta, \sinh\theta, 0, \cdots, 0)$ is the fluid velocity with θ a boost parameter in the x-direction [5]. By using the conversation of energy and momentum and traceless of the stress tensor in the CFT,

$$\partial_\mu \langle T^{\mu\nu} \rangle = 0 \,, \qquad \langle T^\mu_\mu \rangle = 0 \,, \tag{3}$$

it has been obtained solutions consisting of "*shockwaves*" emanating from the interface [3–5, 7, 12], of the form

$$\langle T^{tt} \rangle = \mathcal{G}(x - c_s t) + \mathcal{G}(x + c_s t) \,, \qquad \langle T^{tx} \rangle = \mathcal{G}(x - c_s t) - \mathcal{G}(x + c_s t) \,, \tag{4}$$

where $\mathcal{G}(z)$ is some function which depends on the initial conditions. In this solution the intermediate steady state is bounded by two *shocks* propagating outwards at speed c_s. In the rest of this section we will study the shockwaves solutions in AdS/CFT in any number of dimensions. In Sec. 3 we will particularize the system to $d = 1$, and study the information flow in this case.

2.2 Holographic model

We will present a holographic model to study the system described above at strong coupling . Other out-of-equilibrium stationary configurations in holography can be found in e.g. [13–17] and references therein. Let us consider the Einstein-Hilbert action in $(d+2)$-dim given by

$$S = \frac{1}{16\pi G}\int d^{d+2}x\,\sqrt{-g}\,\{R + 2\Lambda\}\,, \tag{5}$$

where $\Lambda = -d(d+1)/2$ is a negative cosmological constant. The equations of motion write

$$R_{MN} - \frac{1}{2}g_{MN}R + g_{MN}\Lambda = 0\,, \qquad M, N = 1, \cdots, d+2\,. \tag{6}$$

As mentioned above, the energy transport is characterized by a Lorentz-boosted thermal distribution, Eq. (2). Its gravity dual is a boosted black brane with metric [5]

$$ds^2 = g_{MN}dx^M dx^N = \frac{L^2}{z^2}\left[\frac{dz^2}{f(z)} - f(z)\,(\cosh\theta\,dt - \sinh\theta\,dx)^2 + (\cosh\theta\,dx - \sinh\theta\,dt)^2 + dx_\perp^2\right], \tag{7}$$

where $x_\perp$ parameterizes the transverse spatial coordinates. L is the AdS radius, $f(z) = 1 - \left(\frac{z}{z_h}\right)^{d+1}$ and $z_h = \frac{d+1}{4\pi T}$. It is straightforward to check that Eq. (7) is a solution of the equations of motion Eq. (6) as long as the black hole horizon z_h and the boost parameter θ are constant and uniform, i.e. $z_h \neq z_h(t,x)$ and $\theta \neq \theta(t,x)$. While this solution is adequate to study the spatially homogeneous steady state regime corresponding to $t \gg |x|$, we expect the existence of more general (t,x)-dependent solutions which characterize the formation process of the steady state. In the following we will focus on these solutions.

2.3 Linearized solution

A convenient method to find a solution of Eq. (6) valid at any point of the space-time consists in the linearization of the problem. Let us consider a solution of the form [7]

$$z_h(t,x) = z_{h(0)} + \epsilon\, z_{h(1)}(t,x) + \cdots\,, \qquad \theta(t,x) = \theta_{(0)} + \epsilon\,\theta_{(1)}(t,x) + \cdots\,, \tag{8}$$

which means that all the space-time dependence is treated as a perturbation around the background $z_{h(0)}$ and $\theta_{(0)}$, which we keep constant and uniform. Then the equations of motion reduce to the following two equations

$$\partial_t^2 z_{h(1)}(t,x) - c_s^2\partial_x^2 z_{h(1)}(t,x) = 0\,, \qquad \partial_t z_{h(1)}(t,x) - c_s^2 z_{h(0)}\partial_x\theta_{(1)}(t,x) = 0\,, \tag{9}$$

where $c_s^2 = 1/d$ is the squared speed of sound. The solution of these equations was obtained in [7] under appropriate boundary conditions, in particular

- $T_{\rm ini}(x) \equiv T(t=0,x)\,,$
- $\langle T^{tx}(t=0,x)\rangle = 0 \quad$ for $\quad x \in (-\infty, +\infty)\,.$

The first boundary condition corresponds to the freedom in the initial profile of the temperature $T_{\rm ini}(x)$. Only when this profile is non-uniform, the system has a nontrivial time evolution. The second boundary condition states that there is no energy flow at $t = 0$, as the two subsystems are initially isolated.

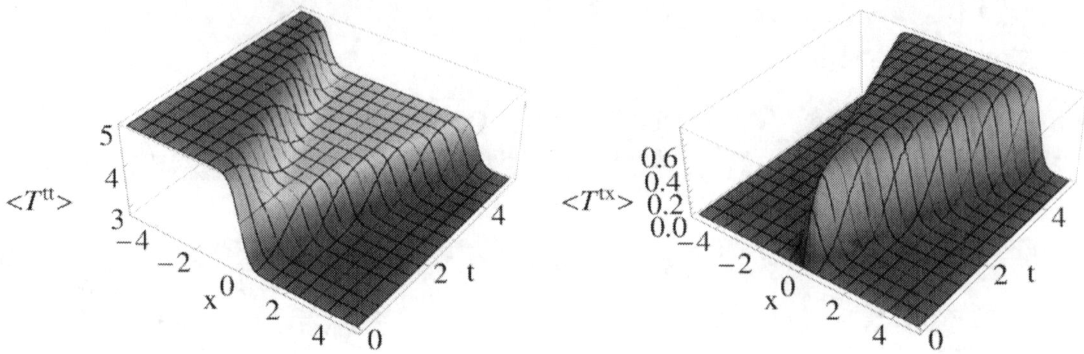

Figure 2. *Energy density and energy current computed with the linearized solution given by Eqs. (10) and (11) respectively. It is used the initial profile* $T_{ini}(x) = \frac{T_R+T_L}{2} + \frac{T_R-T_L}{2}\tanh(\alpha x)$*, with* $T_L = 1.2$*,* $T_R = 1$ *and* $\alpha = 3$*. We have considered* $d = 2$ *and set* $G = 1$*.*

Finally, the linearized solution leads to the following t-dependent energy density and energy flow [7]

$$\langle T^{tt}(t,x)\rangle = \frac{d}{8G}\frac{1}{z_{h(0)}^d}\left[-\frac{d}{2\pi z_{h(0)}} + \epsilon\left(T_{\text{ini}}(x-c_st) + T_{\text{ini}}(x+c_st)\right)\right], \tag{10}$$

$$\langle T^{tx}(t,x)\rangle = \epsilon\frac{1}{c_s}\frac{1}{8G}\frac{1}{z_{h(0)}^d}\left[T_{\text{ini}}(x-c_st) - T_{\text{ini}}(x+c_st)\right], \tag{11}$$

where $z_{h(0)} = \frac{d+1}{2\pi(T_L+T_R)}$. It is easy to see that corrections of order $O(\epsilon)$ are always proportional to factors $\propto \frac{T_L-T_R}{T_L+T_R}$. This illustrates the fact that the ϵ-expansion in Eq. (8) is equivalent to a small gradient expansion, i.e. $\left|\frac{T_L-T_R}{T_L+T_R}\right| \ll 1$, and ultimately to linearized hydrodynamics. This solution leads to the existence of "*shockwaves*" propagating at speed c_s, i.e. it is of the form of Eq. (4) and it fulfills Eq. (3). [1] We display in Fig. 2 the numerical result of Eqs. (10)-(11). The formation of the steady state and the propagation of the shockwaves is properly described in the regime of small difference of temperatures.

2.4 Exact solution for d = 1

To the best of our knowledge, no exact shockwave solution has been found for the holographic model of Eq. (5) in $d \geq 2$, so that in these cases one has to resort to numerics [6] or to linearization methods like the one presented above. However, an exact solution of the model with $d = 1$ was found in [5]. This is possible due to the absence of propagating gravitational degrees of freedom. We will present here this solution for completeness. In Fefferman-Graham coordinates it writes

$$ds^2 = \frac{L^2}{z^2}\left(dz^2 + \tilde{g}_{\mu\nu}dx^\mu dx^\nu\right), \tag{12}$$

[1] See [4, 12] for an alternative derivation of this solution by using hydrodynamic considerations.

where

$$\tilde{g}_{tt} = -\left(1 - \frac{z^2}{L^2}(F(x-t) + F(x+t))\right)^2 + \left(\frac{z^2}{L^2}(F(x-t) - F(x+t))\right)^2, \quad (13)$$

$$\tilde{g}_{tx} = -2\frac{z^2}{L^2}(F(x-t) - F(x+t)), \quad (14)$$

$$\tilde{g}_{xx} = \left(1 + \frac{z^2}{L^2}(F(x-t) + F(x+t))\right)^2 - \left(\frac{z^2}{L^2}(F(x-t) - F(x+t))\right)^2. \quad (15)$$

Note that this solution fulfills the UV boundary condition $\tilde{g}_{\mu\nu}(z=0) = \eta_{\mu\nu}$. The vacuum expectation value of the energy-momentum tensor writes

$$\langle T^{tt}\rangle = \langle T^{xx}\rangle = \frac{c}{6\pi L^2}(F(x-t) + F(x+t)), \quad (16)$$

$$\langle T^{tx}\rangle = \frac{c}{6\pi L^2}(F(x-t) - F(x+t)), \quad (17)$$

where c is the central charge of the CFT, and it fulfills the initial condition $\langle T^{tx}(t=0,x)\rangle = 0$. For numerical computations we will consider the profile

$$F(v) = \frac{\pi^2 L^2}{4}\Big((T_L^2 + T_R^2) + (T_R^2 - T_L^2)\tanh(\alpha v)\Big). \quad (18)$$

This function tends to a stepwise function $F(v) \to \frac{\pi^2 L^2}{2}\left(T_L^2\Theta(-v) + T_R^2\Theta(v)\right)$ in the limit $\alpha \to \infty$. We will use this solution in the next section.

3 Information flow

In the previous section we have studied the problem of energy flow between two initially isolated systems. The question then arises: how does information get exchanged between these two systems? Our next goal is to answer this question, and to this end we will study the information flow in the system.

3.1 Entanglement entropy

On the field theory side the *entanglement entropy* is defined in the following way. Let us divide the system into two subsystems A and B. Then the entanglement entropy of the subsystem A is defined as the von Neumann entropy of the reduced density matrix obtained by tracing out the degrees of freedom of the subsystem B. On the gravity side, the entanglement entropy can be computed as a generalization of the Bekenstein-Hawking entropy formula [9–11]. It follows from the area law relation

$$\mathcal{S}_A = \frac{\text{Area}(\gamma_A)}{4G}, \quad (19)$$

where the manifold γ_A is the d-dimensional minimal surface in AdS_{d+2} whose boundary is given by ∂A, and G is the $(d+2)$-dimensional Newton constant. The computation of the holographic entanglement entropy requires the study of minimal surfaces, and this demands the solution of geodesic equations.

In the following we will restrict to the case $d = 1$ for simplicity, with coordinates (t, x, z). We will study minimal surfaces whose boundary at $z = 0$ is in $x = x_L$ and $x = x_R$, and consider space-like intervals with $t(x_L) = t(x_R)$. The minimal surface compatible with these boundary conditions corresponds to a geodesic line in AdS_3, and it follows from a solution of the geodesic equations, which read

$$\frac{d^2x^P}{ds^2} + \Gamma^P_{MN}\frac{dx^M}{ds}\frac{dx^N}{ds} = 0\,, \qquad P = t, x, z\,. \tag{20}$$

The induced metric on the minimal surface reads

$$h_{ab} = \frac{\partial x^M}{\partial x^a}\frac{\partial x^N}{\partial x^b} g_{MN} = h_{ss}\,, \tag{21}$$

where s is the coordinate of the surface. The entanglement entropy then follows from the area of the manifold γ_A, which can be computed from the induced metric as

$$\mathcal{S}_A = \frac{1}{4G}\int_{s(x_L)}^{s(x_R)} ds\, \mathcal{L}\,, \qquad \text{with} \qquad \mathcal{L} = \sqrt{h_{ss}}\,. \tag{22}$$

At this point it is convenient to consider s as an affine parameter, so that

$$\frac{\partial x^M}{\partial s}\frac{\partial x^N}{\partial s} g_{MN} = 1\,, \tag{23}$$

and the entanglement entropy of Eq. (22) reduces to the trivial integration $\mathcal{S}_A = \frac{1}{4G}\int_{s(x_L)}^{s(x_R)} ds$. The solution of the geodesic equations leads to the behavior $z \sim e^{-|s|}$ in the regime $s \to \pm\infty$, and then one concludes that the entanglement entropy is divergent. In the present case the divergence behaves as $\text{Area}(\gamma_A^{div}) \sim -2L \log z_{uv} + \cdots$, and a renormalization scheme is required. We will use a minimal subtraction scheme, so that the renormalized entanglement entropy is defined as

$$\mathcal{S}_A^{ren} = \frac{1}{4G}\left(\text{Area}(\gamma_A) - \text{Area}(\gamma_A^{div})\right) \qquad \text{with} \qquad \text{Area}(\gamma_A^{div}) = -2L \log z_{uv}\,. \tag{24}$$

In the rest of the manuscript we will compute renormalized entropies according to this formula, and we will omit the superscript *ren*.

3.2 Geodesic equations

The geodesic equations of Eq. (20) consist of three coupled differential equations of second order, whose solution can be expressed in the parametric form

$$t = t(s)\,, \qquad x = x(s)\,, \qquad z = z(s)\,. \tag{25}$$

These equations can be solved by imposing six boundary conditions, which are:

$$\begin{cases} t(s_L) = t(s_R) = t_0 \\ x(s_L) = x_L\,, \qquad x(s_R) = x_R \\ z(s_L) = z(s_R) = z_{uv} \end{cases} \tag{26}$$

We will use the shooting method for the numerical computation of the geodesic equations: we shoot the solution from $s = 0$ with given values of $\{t(0), x(0), z(0)\}$ and $\{t'(0), x'(0), z'(0)\}$, and then find the values of these initial conditions that lead to the desired boundary values at $s \to s_{L,R}$. [2]

[2]There are in the literature other numerical methods for the solution of this two-point boundary value problem. An example are the relaxation methods, in which the solution is determined by starting with an initial guess and improving it iteratively, see e.g. [16].

DOI: 10.1051/epjconf/201716401010

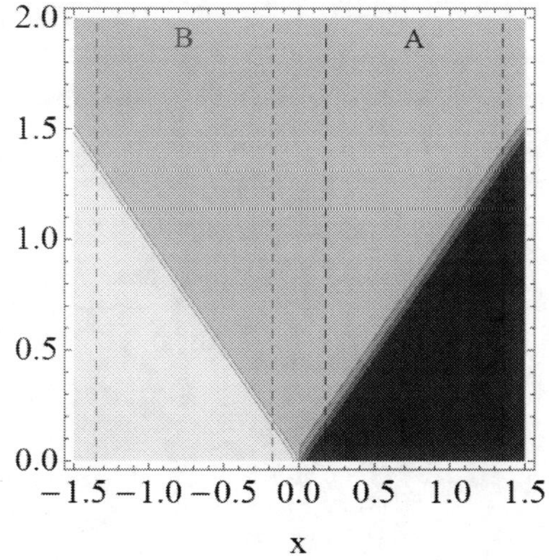

Figure 3. *Contour plot of energy density* $\langle T^{tt}(t,x)\rangle$ *with the model in* $d = 1$, *cf. Sec. 2.4. Dashed lines are the time evolution of the extrema of the intervals* A *and* B, *in the positive and negative semiplane respectively. We consider the intervals* $x^A \in [0.175, 1.35]$ *(blue) and* $x^B \in [-1.35, -0.175]$ *(red), temperatures* $T_L = 0.2$, $T_R = 0.195$ *and* $\alpha = 25$.

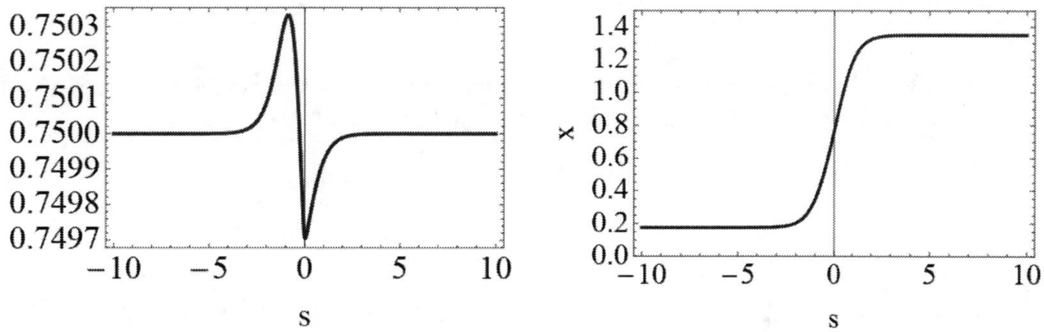

Figure 4. *Parametric dependence of the geodesic as a function of the affine parameter s. We show* $t = t(s)$ *(left) and* $x = x(s)$ *(right). We have considered the interval* $x^A \in [0.175, 1.35]$ *as shown in Fig. 3, and* $t_0 = 0.75$, *cf. Eq. (26).*

We have introduced a cutoff $z_{uv} \ll 1$ to regularize the problem, and this induces also a cutoff in the affine parameter, i.e. $s_L \sim -|\log z_{uv}|$ and $s_R \sim |\log z_{uv}|$. In the following we will consider space-like intervals A and B as shown in Fig. 3. It is displayed in Figs. 4 and 5 a typical solution of the geodesic equations, which fulfills the boundary conditions of Eq. (26). Once the geodesics are obtained, the next step is to compute the area of these curves and then the entanglement entropy from Eq. (24). In the rest of this section we will present some results for the time evolution of the entanglement entropy in the system of Sec. 2.

3.3 Entanglement entropy and universal time evolution

For the moment we will consider a single interval $x \in [x_L, x_R]$, that we denote by A, placed in the positive semiplane, i.e. $x_{L,R} > 0$. Let us study the time evolution of the entanglement entropy $\mathcal{S}_A$ during the process of quenching. We are considering the model with $d = 1$, so that the shockwaves are at $t = |x|$. This means that the shockwaves touch the two ends of the interval at times $t = |x_L|$ and $t = |x_R|$, see Fig. 3. We will denote these values as t_L and t_R respectively. If one assumes the limit

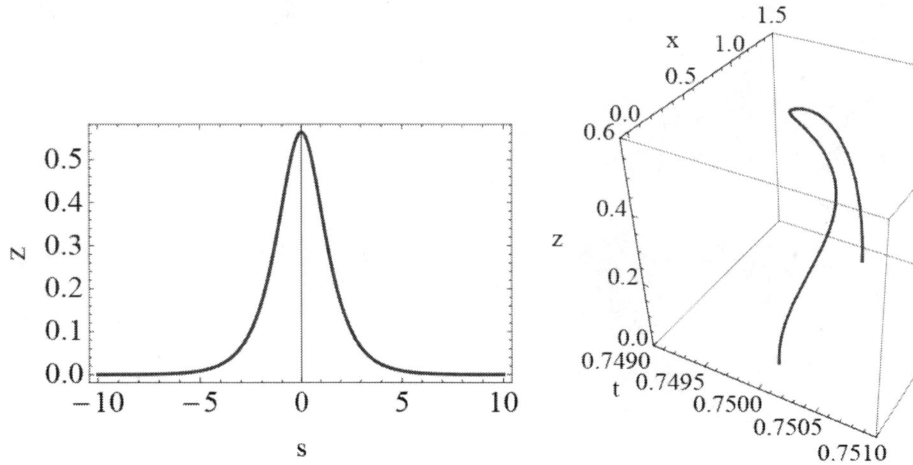

Figure 5. *Parametric dependence of the geodesic. (Left) We show $z = z(s)$. (Right) Geodesic in the space (t, x, z). See Fig. 4 for further details.*

$\alpha \to \infty$ in Eq. (18), the entanglement entropy turns out to be constant in the regimes $0 \le t \le t_L$ and $t_R \le t$, and there is a non trivial time evolution only in the interval $t_L \le t \le t_R$, i.e.

$$\mathcal{S}_A(t) = \begin{cases} \mathcal{S}_A(t=0) & 0 \le t \le t_L \\ \mathcal{S}_A(t) & t_L \le t \le t_R \\ \mathcal{S}_A(t=\infty) & t_R \le t \end{cases} . \tag{27}$$

We display in Fig. 6 (left) the time evolution of the entanglement entropy of interval A of Fig. 3, from a numerical computation of the geodesic equations. Let us focus on the regime $t_L \le t \le t_R$. It is convenient to define the normalized entanglement entropy $f_A(\rho)$ as

$$f_A(\rho) \equiv \frac{\mathcal{S}_A(t) - \mathcal{S}_A(t=0)}{\mathcal{S}_A(t=\infty) - \mathcal{S}_A(t=0)} \qquad \text{with} \qquad \rho \equiv (t - t_0)/\Delta t \,, \tag{28}$$

where $t_0 = t_L$ and $\Delta t = \ell = |x_R - x_L|$. This corresponds to the function $\mathcal{S}_A(t)$ normalized to the interval [0,1] in both horizontal and vertical axes. It is clear from Eqs. (27) and (28) that $f_A(\rho)$ has the extreme values $f_A(0) = 0$ and $f_A(1) = 1$. We have computed numerically the entanglement entropy $\mathcal{S}_A(t)$ in a wide variety of configurations with different temperatures T_L, T_R and lengths ℓ, and find that the behavior of $f_A(\rho)$ in every case is

$$f_A(\rho) \simeq 3\rho^2 - 2\rho^3 \,, \qquad 0 \le \rho \le 1 \,. \tag{29}$$

This function fits extremely well the numerical results of the entanglement entropies up to moderate gradients, i.e. $|T_L - T_R| < T_L + T_R$. This is illustrated in Fig. 6 (right) for a particular case. The result of Eq. (29) is independent of the values of the parameters T_L, T_R and ℓ, and so it implies the existence of an '*almost*' universal time-evolution of entanglement entropy in the theory with $d = 1$. [3]

The analysis presented above applies also to intervals in the negative semiplane. We show in Fig. 6 (left) the entanglement entropy of interval B of Fig. 3. Note that both functions, $\mathcal{S}_A(t)$ and $\mathcal{S}_B(t)$, tend to the same value when the intervals reach the steady state regime.

[3]The deviations from universality are stronger for bigger temperature differences.

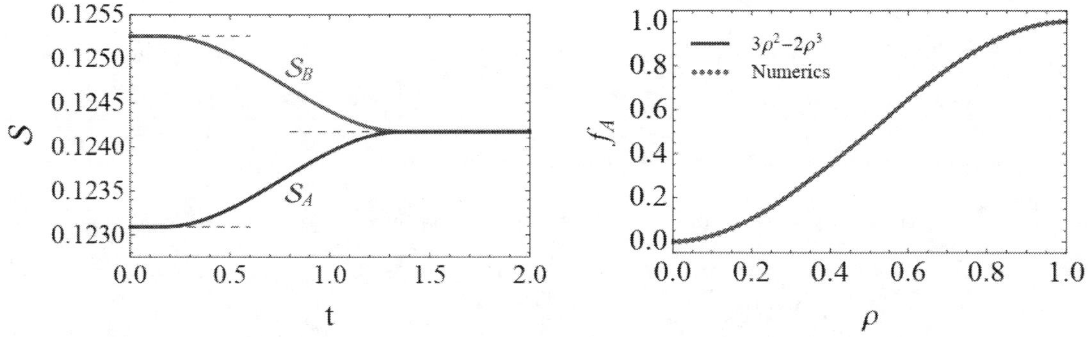

Figure 6. *(Left) Entanglement entropies of intervals A and B as a function of time, see Fig. 3. The (dashed) horizontal lines correspond to the results by using the analytical formulas, Eqs. (31) and (32). We have set G = 1 and L = 1. (Right) Entanglement entropy $\mathcal{S}_A$ as a function of time, normalized to* [0, 1] *in both horizontal and vertical axes, cf. Eq. (28). The dots correspond to the numerical result with the interval A in Fig. 3, while the continuous line is the universal behavior $f_A(\rho) = 3\rho^2 - 2\rho^3$.*

3.4 Conservation of entanglement entropy

Let us consider the two extreme regimes $t = 0$ and $t \to \infty$. It is possible to obtain analytical results for the entanglement entropies in these cases for the model with $d = 1$ presented in Sec. 2.4. When $t = 0$ the metric corresponds to a stepwise black hole, i.e.

$$ds^2 = ds_L^2\Theta(-x) + ds_R^2\Theta(x)\,. \tag{30}$$

If we place the interval in just one semiplane, i.e. $x_{L,R} > 0$ (or $x_{L,R} < 0$), the entanglement entropy corresponds to the one for a stationary black hole at temperature T, which reads

$$\mathcal{S}(T,\ell;t=0) = \frac{L}{2G}\log\left(\frac{1}{\pi T}\sinh(\pi\ell T)\right), \qquad \ell := |x_R - x_L|\,. \tag{31}$$

In this equation $T = T_L$ (or T_R) when $x_{L,R} < 0$ (or $x_{L,R} > 0$). In the other extreme, $t \to \infty$, the system is in the steady state regime, and the entanglement entropy is the one for a boosted black hole, that reads (see [18] for a derivation) [4]

$$\mathcal{S}(T_L,T_R,\ell;t=\infty) = \frac{L}{4G}\log\left(\frac{1}{\pi^2 T_L T_R}\sinh(\pi\ell T_L)\sinh(\pi\ell T_R)\right). \tag{32}$$

These analytical results, Eqs. (31) and (32), correspond to $\mathcal{S}_A(t=0)$ and $\mathcal{S}_A(t=\infty)$ in Eq. (27), respectively. From these formulas one can easily obtain the following property:

$$\mathcal{S}_A(t=0) + \mathcal{S}_B(t=0) = \mathcal{S}_A(t=\infty) + \mathcal{S}_B(t=\infty)\,, \tag{33}$$

where we have considered intervals A with $x_{L,R}^A > 0$, and B with $x_{L,R}^B < 0$, and lengths $\ell = \ell_A = \ell_B$. This property is non trivial, as in the lhs of Eq. (33) there is the contribution of stationary black

[4]Note that Eq. (32) is valid when $t \geq \max(|x_L|, |x_R|)$ if the initial profile $F(v)$ in Eq. (18) is a stepwise function, i.e. in the limit $\alpha \to \infty$. When $F(v)$ is a smooth function, the rhs of Eq. (32) corresponds to the asymptotic value of the entanglement entropy at very late times, i.e. for $t \gg \max(|x_L|, |x_R|)$.

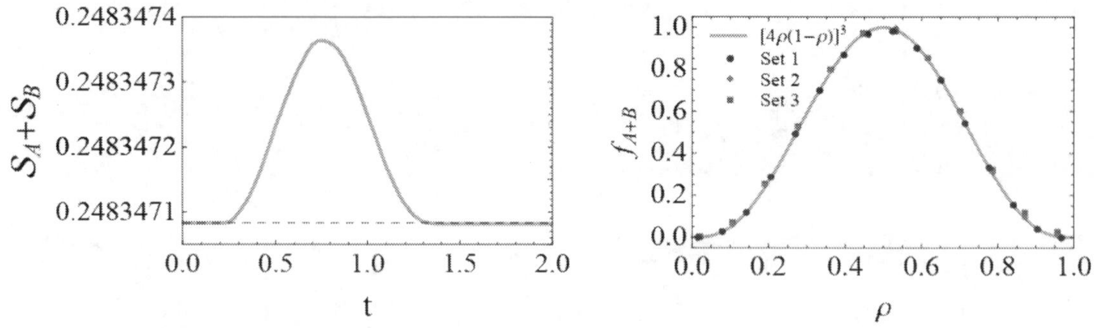

Figure 7. *(Left) Entanglement entropy $\mathcal{S}_A + \mathcal{S}_B$ as a function of time, see Fig. 3. The (dashed) horizontal line corresponds to the result by using the analytical formulas, Eqs. (31) and (32). We have set G = 1 and L = 1. (Right) Entanglement entropy $\mathcal{S}_{A+B}$ as a function of time, normalized to $[0, 1]$ in both horizontal and vertical axes, cf. Eq. (34). The dots correspond to the numerical result with intervals A and B, placed symmetrically with respect to $x = 0$ as shown in Fig. 3, in different configurations: Set 1 is $(T_L = 0.2, T_R = 0.195, \ell_A = \ell_B = 1.175)$, Set 2 is $(T_L = 0.2, T_R = 0.175, \ell_A = \ell_B = 1.175)$ and Set 3 is $(T_L = 0.2, T_R = 0.175, \ell_A = \ell_B = 1.475)$. The continuous line is the universal behavior $f_{A+B}(\rho) = [4\rho(1-\rho)]^3$.*

hole solutions at temperatures T_L and T_R, while in the rhs there is a boosted black hole and the corresponding energy flow contributes as well to the entanglement entropy. This relation is very interesting as it implies the *'conservation'* of entanglement entropies between $t = 0$ and $t = \infty$.

It is displayed in Fig. 7 (left) the time evolution of $\mathcal{S}_{A+B} \equiv \mathcal{S}_A + \mathcal{S}_B$. We can see that our numerics confirms the conservation rule of Eq. (33). In the next subsection we will study this system in the quenching regime, i.e. $t_L \leq t \leq t_R$ in Eq. (27), and characterize some violations of the entanglement entropy conservation in this case.

3.5 Non-universal effects in time evolution

As it is shown in Fig. 7 (left), we find from our numerics that $\mathcal{S}_{A+B}(t) \neq$ cte in the quenching regime. This implies that the entanglement entropy is not exactly conserved at intermediate times. A straightforward computation shows that these non-conservation effects are only possible if there are non-universal contributions in Eq. (29), otherwise this equation would predict $\mathcal{S}_{A+B}(t) =$ cte.

In the following we will restrict to intervals A and B with the same length and placed symmetrically with respect to $x = 0$, i.e. $\ell_A = \ell_B$ and $x^A_{L,R} = -x^B_{R,L}$. While the function $\mathcal{S}_{A+B}(t)$ has the same value at $t = 0$ and $t = \infty$ (see Eq. (33)), we find from our numerics that it has a maximum at $t_{\max} = (t_L + t_R)/2$. In order to characterize the time evolution of $\mathcal{S}_{A+B}(t)$, let us define the normalized entanglement entropy

$$f_{A+B}(\rho) \equiv \frac{\mathcal{S}_{A+B}(t) - \mathcal{S}_{A+B}(t=0)}{\mathcal{S}_{A+B}(t_{\max}) - \mathcal{S}_{A+B}(t=0)} \qquad \text{with} \qquad \rho \equiv (t - t_0)/\Delta t\,, \tag{34}$$

where t_0 and Δt are defined as in Eq. (28). Finally, from a numerical computation of $f_{A+B}(\rho)$ in a wide variety of intervals, we find that its behavior is universal and it is given by

$$f_{A+B}(\rho) \simeq [4\rho(1-\rho)]^3\,, \qquad 0 \leq \rho \leq 1\,. \tag{35}$$

This is illustrated in Fig. 7 (right) for several configurations. From a combination of the results in Eqs. (29) and (35), we conclude that the normalized entanglement entropy defined in Eq. (28) can be written as

$$f_A(\rho) = 3\rho^2 - 2\rho^3 + \Delta_A(\rho), \quad \text{with} \quad \Delta_A(\rho) \simeq C(T_L, T_R, \ell) \cdot [4\rho(1-\rho)]^3 . \tag{36}$$

The factor $C(T_L, T_R, \ell)$ has a non-universal dependence on the parameters of the interval, so that $\Delta_A(\rho)$ is a non-universal contribution to $f_A(\rho)$. Note, however, that $C(T_L, T_R, \ell)$ does not affect the universal behavior of $f_{A+B}(\rho)$, see Eq. (35). Some remarks deserve to be mentioned: on the one hand, $\Delta_A(\rho)$ is a correction of order $O(\rho^3)$, so that it does not jeopardize the behavior $S_A(t) \sim t^2$ which seems to be general in a wide variety of systems. On the other hand, the effect of $\Delta_A(\rho)$ is extremely small in the configurations we have studied numerically. [5]

4 Conclusions and outlook

In this work we have studied a holographic model for far-from-equilibrium dynamics that allows to describe the time dependent properties of energy flow and information flow of two thermal reservoirs initially isolated. Of relevance in this system is the appearance of a universal steady state, described by a boosted black brane. By using a linearization method, we have described the formation of the steady state and the propagation of shockwaves for any space-time dimension. A relevant observable that provides physical insight into the evolution of the system is the entanglement entropy, which measures the information flow between two subsystems. By using the exact solution for $d = 1$ provided in [5], we have studied numerically the time evolution of the entanglement entropy, and characterized some universal properties of the quenching process.

There remain some open questions. It would be interesting to study the entanglement entropy at higher dimensions, i.e. for $d \geq 2$. This can be easily done by using the linearized background presented in this work. An analysis beyond the linear response regime, i.e. for $0 < T_R/T_L < 1$, would demand a full numerical solution of the equations of motion for the background, see e.g. [6, 16, 18–20]. A quantity of interest related to the entanglement entropy is the *mutual information*, which measures which information of subsystem A is contained in subsystem B. An advantage of this quantity is that it is finite, so that it does not need to be regularized. The Ryu-Takayanagi formula for entanglement entropy satisfies some inequalities that set special restrictions, and have profound implications for the structure of holographic quantum states and their correlations. Apart from the monogamy of mutual information and strong subadditivity, other inequalities involving a large number of subsystems have been proved in the static case, see e.g. [21]. It would be interesting the study the time evolution of the mutual information, and to check the validity of the entanglement entropy inequalities in time dependent systems. These and other issues will be addressed in a forthcoming publication [18].

Acknowledgements

I would like to thank J. Erdmenger, D. Fernández, M. Flory, A.K. Straub and P. Witkowski for valuable discussions. This work has been supported by Plan Nacional de Altas Energías Spanish MINECO grant FPA2015-64041-C2-1-P, and by the Spanish Consolider Ingenio 2010 Programme CPAN (CSD2007-00042). I thank the

[5] One can see from Fig. 7 (left) that in this case the peak in $S_{A+B}(t_{\max})$ is a correction of order $O(10^{-6})$ with respect to $S_{A+B}(0)$, so that the order of magnitude of the non-universal contribution in Eq. (36) is

$$C(T_L, T_R, \ell) \simeq \Delta_A\left(\rho = \frac{1}{2}\right) \simeq \frac{1}{2}\frac{S_{A+B}(t_{\max}) - S_{A+B}(0)}{S_A(\infty) - S_A(0)} \sim O(10^{-4}). \tag{37}$$

Instituto de Física Teórica UAM/CSIC, Madrid, Spain, for their hospitality during the completion of the final stages of this work. The research of E.M. is supported by the European Union under a Marie Curie Intra-European fellowship (FP7-PEOPLE-2013-IEF) with project number PIEF-GA-2013-623006, and by the Universidad del País Vasco UPV/EHU, Bilbao, Spain, as a Visiting Professor.

References

[1] T. Ishii, E. Kiritsis, C. Rosen, JHEP **08**, 008 (2015)
[2] A. Polkovnikov, K. Sengupta, A. Silva, M. Vengalattore, Rev. Mod. Phys. **83**, 863 (2011)
[3] D. Bernard, B. Doyon, J. Phys. **A45**, 362001 (2012)
[4] H.C. Chang, A. Karch, A. Yarom, J. Stat. Mech. **1406**, P06018 (2014)
[5] M.J. Bhaseen, B. Doyon, A. Lucas, K. Schalm, Nature Physics **11**, 509-514 (2015)
[6] I. Amado, A. Yarom, JHEP **10**, 015 (2015)
[7] E. Megias, PoS **EPS-HEP2015**, 366 (2015), 1510.04219
[8] E. Witten, Adv. Theor. Math. Phys. **2**, 505 (1998)
[9] S. Ryu, T. Takayanagi, Phys. Rev. Lett. **96**, 181602 (2006)
[10] S. Ryu, T. Takayanagi, JHEP **08**, 045 (2006)
[11] V.E. Hubeny, M. Rangamani, T. Takayanagi, JHEP **07**, 062 (2007)
[12] J. Smoller, B. Temple, Commun. Math. Phys. **156**, 67 (1993)
[13] S. Khlebnikov, M. Kruczenski, G. Michalogiorgakis, JHEP **07**, 097 (2011)
[14] S. Fischetti, D. Marolf, J.E. Santos, Class. Quant. Grav. **30**, 075001 (2013)
[15] R. Emparan, M. Martinez, JHEP **09**, 068 (2013)
[16] C. Ecker, D. Grumiller, S.A. Stricker, JHEP **07**, 146 (2015)
[17] C.P. Herzog, M. Spillane, A. Yarom, JHEP **08**, 120 (2016)
[18] J. Erdmenger, D. Fernandez, M. Flory, E. Megias, A.K. Straub, P. Witkowski, work in progress (2016)
[19] J. Abajo-Arrastia, J. Aparicio, E. Lopez, JHEP **11**, 149 (2010)
[20] P.M. Chesler, L.G. Yaffe, JHEP **07**, 086 (2014)
[21] N. Bao, S. Nezami, H. Ooguri, B. Stoica, J. Sully, M. Walter, JHEP **09**, 130 (2015)

EPJ Web of Conferences **164**, 01011 (2017)
ICNFP 2016
DOI: 10.1051/epjconf/201716401011

Galaxy Formation from the Primordial Black Holes

Masahiro Morikawa[a]

[1] *Department of Physics, Ochanomizu University, 2-1-1 Otsuka, Bunkyo, Tokyo 112-8610, Japan.*

Abstract. Supermassive black hole (SMBH) of size $M_{BH} = 10^{6-10} M_{\odot}$ is common in the Universe and it defines the center of the galaxy. A galaxy and the SMBH are generally thought to have co-evolved. However, the SMBH cannot evolve so fast as commonly observed even at redshift $z > 6$. Therefore, we explore a natural hypothesis that the SMBH has been already formed mature at $z \gtrsim 10$ before stars and galaxies. The SMBH forms energetic jets and outflows which trigger massive star formation in the ambient gas. They eventually construct globular clusters and classical bulge as well as the body of elliptical galaxies. We propose simple models which implement these processes. We point out that the globular clusters and classical bulges have a common origin but are in different phases. The same is true for the elliptical and spiral galaxies. Physics behind these phase division is the runaway star formation process with strong feedback to SMBH. This is similar to the forest-fire model that displays self-organized criticality.

1 Introduction

Almost all the galaxy harbors a supermassive black hole (SMBH) of mass $M_{BH} = 10^{6-10} M_{\odot}$ in its center[1], where $M_{\odot} \approx 2 \times 10^{30}$kg is the solar mass. The masses of these SMBH are observed to have firm correlations with the basic components of the galaxy. For example in the case of classical bulge (CB), its mass M_{CB} has the relation $M_{BH} \approx 10^{-3} M_{CB}$[3][1]. In the case of globular clusters (GC), the number of them N_{GC} in a galaxy has the relation $M_{BH} \approx 10^{5.5} N_{GC} M_{\odot}$[4]. Therefore it might be natural to think that the galaxy and its SMBH have co-evolved interacting with each other.

However, the standard coagulation of the stellar size black holes to form SMBH within a limited time scale turns out to be very difficult [5]. This difficulty is enhanced by the recent observations [6][7] that report many mature SMBH exist at around $z \approx 6$ (only 7.4×10^{8}years after Big Bang). Various attempts to form these SMBH through the co-evolution with galaxies seem to be unnatural [8], without assuming seed black holes of huge mass $10^{6} M_{\odot}$. Moreover there seems to be an evidence that the above black hole/bulge mass relation $M_{BH} \approx 10^{-3} M_{CB}$ has already been established or even the coefficient increases toward the past[9]. All of these facts strongly suggest that the SMBH are primordial.

[a] hiro@phys.ocha.ac.jp

[1] We do not consider the similar structure called pseudo-bulge which is more like disks of spiral galaxies[2].

Therefore in this paper, we start from the hypothesis that mature SMBH have been formed at around $z \approx 10$ although the mass may somewhat increase by the later accretion. Then the formation of all the structures, *i.e.* the stars, globular clusters, and bulges, should be attributed to the SMBH. These structures will quickly feed back to the SMBH. Thus the galaxy-SMBH co-evolution would have taken place in the early stage and several correlations above [3][4] would have already been established.

The early formation of SMBH at around $z > 10$ has been studied in the model that the Bose-Einstein condensation of the self-interacting boson fields forms the dark energy [10–13][2]. In this model, the unstable uniform condensation of the field can collapse to form SMBH everywhere in the Universe. This is possible because the coherent condensation does not have velocity dispersion which prevents the collapse. Furthermore in this model, the non-condensate component of the boson gas can contribute as the thermal dark matter around the SMBH.

The gravitational potential of this dark matter attracts baryon gas, and the SMBH will form a strong jet beyond the present size of the galaxy. This is plausible from the many observations at present although the jet ejection mechanism is not at all clear so far[15]. This jet will compress the surrounding gas and trigger massive star formation there[16][17]. These first stars will build globular clusters (GC), the classical bulges (CB) in spiral galaxies (SG), and the main body of the elliptical galaxies (EG). This physics is closely related with the mechanism in [18] which displays the correlation between the SMBH mass M_{BH} and the velocity dispersion σ of the stars in CB. This scenario is briefly explained in section II.

According to the above scenario, the stars of GC and CB in the spiral galaxies,would have the same origin and are indistinguishable with each other. On the other hand, their clustering features are different; GC are extended in halo and CB forms the core of the galaxy. This divide may come from the two distinguished flows of gas in the primordial galaxy. We clarify these flows and the origin of distinct clustering features in section III.

Furthermore an apparent similarity of CB (in SG) and EG suggests that the EG and SG have the common origin. The fact that SG is smaller than EG, in average, suggests that the strength of the jet from SMBH or the mass of SMBH are thought to be the major factor to distinguish SG and EG. We will find much interesting parameter which clearly distinguish SG and EG in our model described in section IV.

We try to find the simplest model extracting the most relevant physics from very complicated galaxy formation processes. Limitations and prospects of our approach are described in the final section V.

2 Early formation of supermassive black holes

We briefly examine a possible origin of the primordial SMBH which was formed first before any other components of the galaxy. It is clear that the coagulation of the stellar size black holes to form SMBH takes too long time, more than the dynamical relaxation time scale $\tau_{rel} = \sigma^3/(G^2 m \rho \ln N)$, which turns out to be 2×10^{14}years, where σ, m, ρ, N, G are the velocity dispersion, mass, mass density, number of the black holes, and the gravitational constant, respectively. To bring this time scale within the cosmic age 1.38×10^{10}years, we need seed black holes of mass $10^5 M_\odot$ to start with. Furthermore dark matter gas is hopeless to collapse into SMBH because of its velocity dispersion and the angular momentum. The

[2]The existence of the mature SMBH before the recombination is impossible; accretion effects on the CMB significantly distort its spectrum beyond the observational limit[14].

only possibility will be the collapse of dark energy [10, 11]. This is possible if the dark energy is the Bose-Einstein condensation of fields (mass m) with an attractive self-interaction($\lambda < 0$). The condensate is characterized by the classical scalar field $\Psi(t,r)$ in the metric, assuming spherical symmetry,

$$ds^2 = \alpha^2 \mathrm{d}t^2 - a^2 \mathrm{d}r^2 - r^2 \mathrm{d}\theta^2 - r^2 \sin\theta^2 \mathrm{d}\phi^2 \tag{1}$$

and obeys the equation of motion,

$$\begin{aligned}
&r a \Psi' \dot{\Psi} - 2\dot{a} = 0, \\
&-2a'\alpha^2 - 2a\alpha\alpha' + r a \alpha^2 \Psi'^2 + r a^3 \dot{\Psi}^2 = 0, \\
&\tfrac{2a(r\alpha' + \alpha) - 2ra'\alpha}{r^2 a^3 \alpha} - 2m^2\Psi^2 + \lambda\Psi^4 - \tfrac{2}{r^2} = 0, \\
&r\left(\begin{array}{c} \dot{a}a^2\alpha\dot{\Psi} + a'\alpha^3\Psi' \\ + \quad a^3\left(2\alpha^3\Psi\left(m^2 - \lambda\Psi^2\right) - \dot{\alpha}\dot{\Psi} + \alpha\ddot{\Psi}\right) \end{array} \right) \\
&= a\alpha^2\left((r\alpha' + 2\alpha)\Psi' + r\alpha\Psi''\right).
\end{aligned} \tag{2}$$

This set of equations easily forms black hole even if mass-less non-interacting case, which was often used to analyze critical behavior in the black hole formation process[19][20]. It was concluded that the resultant black hole mass shows scaling properties. On the other hand, more realistic bound for the black hole formation comes from the quantum fluctuations. However, it turns out that even the quantum fluctuations cannot prevent the collapse of the condensation to black hole if the mass of the boson field exceeds the Kaup limiting mass [21]

$$M_{kaup} = 0.633\frac{\hbar c}{Gm} \approx \frac{m_{pl}^2}{m}, \tag{3}$$

where m_{pl} is the Planck mass. If the boson mass is the order of the present dark energy $0.01eV$, then the limiting mass becomes1.7×10^{22}kg, almost the mass of the planet Pluto.

After the adiabatic collapse of the dark energy, some portion of the condensation becomes a black hole [19] and some other portion will melt to form thermal boson gas around the black hole[10]. This latter melting process of the condensation depends on many complex conditions and has not yet been clarified. However it would be natural to suppose that it settles down to the thermal equilibrium of mass density $\rho(r) = \rho_0(r_0/r)^2$ where r is the distance from the central SMBH and ρ_0, r_0 are constants. Then this thermal uncondensed gas behaves as dark matter [10, 11] since it yields the commonly observed flat rotation curve.

Suppose that the SMBH thus formed at $z \approx 10$ is already surrounded by the thermal gas of dark matter with well developed gravitational potential. Then baryons are attracted within the free fall time scale about 5×10^7 years. At the same time the SMBH would yield energetic jets, which will trigger massive star formation along it through the ram pressure of the bow shock[16][17]. However the effect of the jet for star formation is not yet fully understood as well. If the jet conveys the momentum to the ambient baryon gas, then the pressure compresses the gas to promote the star formation. On the other hand, if the jet conveys the energy to the gas, then the heat makes the gas expand to prevent the star formation. We assume the former, the jet promotes the star formation. This assumption is partially supported by the following argument.

All the above arguments are for a steady jet with fixed direction. If the jet changes its direction rapidly less than the free fall time scale, as we will argue in the next section, then the trajectory envelope of the jet will form a superposed shock wave shells. A similar argument appears in [18] which displays the formation of the correlation between the mass M_{BH} of the SMBH and the velocity dispersion σ in galactic bulge. If the typical outflow radiation balances in momentum with the ambient gas of size R, then we have $GM_b M_{tot}/R^2 = L_{Edd}/c$, where $M_b = fM_{tot}$ is the baryon mass of the galaxy, *i.e.* the fraction f of the total mass M_{tot}. $L_{Edd} = 4\pi GM_{BH}c/\kappa$ is the Eddington limiting luminosity where κ is the electron scattering opacity. Then using the virial equilibrium relation $\sigma^2 = GM_{tot}/R$, we have [18],

$$M_{BH} = \frac{f\kappa\sigma^4}{4\pi G^2} \propto \sigma^4 \tag{4}$$

This successfully describes the observations without any free parameters[3]. On the other hand if we supposed energy balance, we have an extra factor (σ/c) on the right hand side of Eq.(4), and $M_{BH} \propto \sigma^5$. This predicts too small mass of SMBH and conflicts with observations. Thus the observed $M_{BH} - \sigma$ relation is consistent with the momentum balance, which suggests the jet/outflow-induced star formation.

3 Separation into globular clusters and ellipsoids

Now we examine how SMBH triggers star formation and makes basic components of galaxy. The first stars, as well as their direct descendants formed after the first supernova explosions, are thought to be the main ingredients of the galaxy components: classical bulge (CB), globular clusters (GC) and the elliptical galaxies (EG). These components must be formed at the same time and same mechanism since observations indicate that all of them are composed from very old population-II stars[3]. Then how these separations into components are processed?

We first consider the separation into GC and ellipsoids (*i.e.* the CB and the main body of EG). We consider that there had been two different kinds of gas velocity fields. One is caused by the local gravitational potential, and the other caused by the global cosmic turbulent flow[22][4]. The former velocity is given by the virial equilibrium and becomes a constant $v_{in} \equiv (4\pi G r_0^2 \rho_0)^{1/2}$. This is implied by the previous isothermal distribution of dark matter density $\rho = \rho_0 (r_0/r)^2$. The latter is the scaling velocity $v_{out} \equiv (\epsilon r)^{1/3}$, where ϵ is a constant. This is derived by the scaling property of the cosmic turbulence[22]. This cosmic turbulence is naturally expected from the equation for the self-gravitating fluid, whose Fourier transform

$$\frac{dv^{\alpha}_{\vec{k}}}{dt} = -ik_{\beta} \sum_{\vec{p}+\vec{q}=\vec{k}} \left(\begin{array}{c} \left(\delta_{\alpha\gamma} - \frac{k_{\alpha}k_{\gamma}}{k^2}\right) v^{\beta}_{\vec{p}} v^{\gamma}_{\vec{q}} - \nu k^2 v^{\alpha}_{\vec{k}} \\ + i\frac{k^{\alpha}}{k^2} 4\pi G \delta_{\vec{k}} \end{array} \right) \tag{5}$$

is similar to the Smoluchowski coagulation equation. The scaling relation derived from this equation is applied to various observations such as the scale dependent mass density, L/M ratio, magnetic field distributions, etc. They all consistently point the value $\epsilon \approx 0.3 cm^2/sec^3$[22]. The former virial velocity dominates for $r < r_*$ and the latter turbulent velocity dominates for $r > r_*$, where $r_* \approx 8.6 kpc$ if we assume $v_{in} = 200 km/sec$, as shown in Fig. 1.

[3]Some GC are reported to have several newer generation stars which might be triggered by merger for example. Here we concentrate on the main oldest generation of GC.

[4]This turbulence might be caused by the collapse of the condensation when SMBH are formed everywhere in the Universe or triggered by the tidal torque effect.

 DOI: 10.1051/epjconf/201716401011

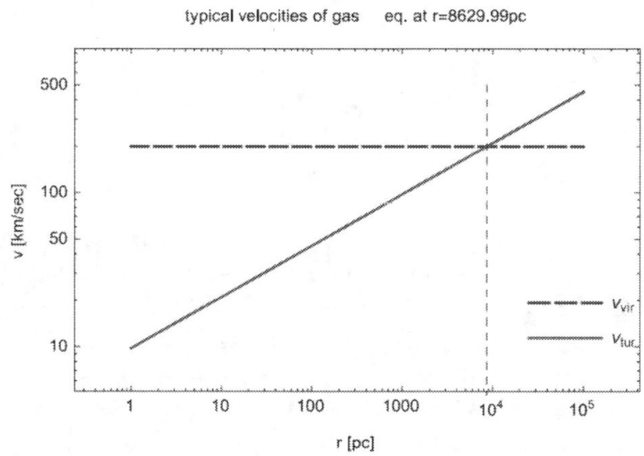

Figure 1. Velocity dispersion of ambient gas as a function of the distance r from the central SMBH. There are two distinct velocity fields: The constant virial velocity field $v_{in} = 200km/sec$ (broken blue, dominates inward) and the increasing turbulent velocity field $v_{out} \equiv (\epsilon r)^{1/3}$(solid red, dominates outward). They are equal with each other at $r_* \approx 8.6kpc$. A star formed at $r < r_*$falls down toward the SMBH and forms a bulge, while the star at $r > r_*$ stays far from the SMBH and forms globular clusters.

The baryon gas is exerted the ram pressure from the jet and forms stars. The stars in the near region $r < r_*$ falls down toward the center of the potential loosing the pressure balance and suffered friction. Eventually they will form a big cluster that we call classical bulge (CB). Supernova explosions and subsequent star formation may continue. On the other hand the stars formed from the gas in the far region $r > r_*$ have higher speed beyond the virial equilibrium and have less probability to fall. Therefore these stars remain far from the center possibly forming small clusters that we call globular clusters (GC).

There are some specific features of the above scenario which may be significant in comparison with observations: (a) All the stars in GC and CB have the same age and chemical components because they are formed by the the same gas triggered by the common jet at the same time. (b) Each GC has almost no local angular momentum. This is because each GC is formed in the finite region where the relative gas speed is small and coherent according to the relation $v_{out} \equiv (\epsilon r)^{1/3}$ (Fig.1). Moreover from the same reason, once the stars are formed, they experience the cold collapse to form tight-binding cluster which has the very similar profile with each other. (c) GC are loosely bounded to the galaxy since they tend to have higher velocities v_{out} than the virial equilibrium v_{in}. Therefore there may be significant number of stray GC in between galaxies. This point should be considered in wider viewpoint including the dwarf spheroidal galaxies[23]. They may be all the same family. These properties should be checked quantitatively. In summary, CB and GC are the same species but separated into two phases by two distinct velocity fields.

4 Separation into spiral and elliptical galaxies

Next we consider the separation into the spiral (SG) and elliptical (EG) galaxies in our scenario of mature SMBH. Apart from the disk, the classical bulge (CB) in SG, and EG

DOI: 10.1051/epjconf/201716401011

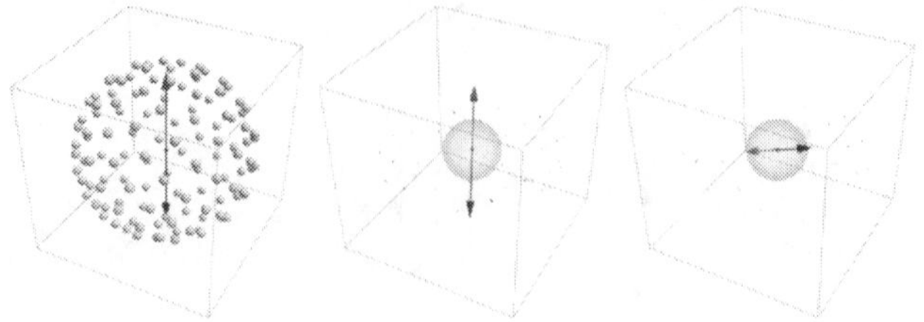

Figure 2. The galaxy formation history in case (a), high accretion rate $\lambda = 1$. These are the snapshots of the gas distribution (small balls), jet (arrows), and the bulge (central pink ball) derived by the numerical calculations of Eq.6. The time flows from left to right ($t = 0,\ 150,\ 300$). The gas was isotropically distributed at $t = 0$. Accreted objects keep exerting torque on SMBH and the jet wildly changes its direction rapidly. This rampaging jet triggers star formation in the whole solid angle. All the gas is exhausted and an elliptical galaxy is left. The parameters are $\mu = 1,\ n = 10,\ \kappa = 0.01,\ N = 256$.

are composed from old population-II stars as globular clusters (GC). Therefore it would be natural to speculate that CB and EG are the same species but in different phases.

In order to demonstrate this process of separation, we introduce a simple model for the massive star formation by jets. Suppose the energetic jet from SMBH hits the ambient gas that is isotropically distributed around the SMBH. We assumed that the jet triggers the gas to form stars. Then those stars formed in the near region $r < r_*$ fall down toward the center. Some of them will give torque on the jet from SMBH through the perturbation on the accretion disk. Then the jet from SMBH changes its direction due to this torque. In the new direction of the jet, there will be plenty of fresh gas ready to form stars. Thus the jet will trigger new star formation in this rich gas environment. This yields further torque on the SMBH. This feedback dynamics will be simply represented by the following model:

$$\begin{aligned} \dot{g}_i(t) &= -\mu \left| [\overrightarrow{J}(t)]_n \cdot \overrightarrow{s_i} \right|^{\alpha} |\overrightarrow{J}(t)| g_i(t), \\ \dot{\vec{J}}(t) &= -\lambda [\overrightarrow{J}(t) \times \textstyle\sum_i \overrightarrow{s_i} \dot{g}_i(t)]_n \sum_i \dot{g}_i(t) - \kappa \overrightarrow{J}, \end{aligned} \quad (6)$$

where the vector $\vec{J}(t)$ represents the jet (direction=axes of the jet, amplitude=strength of the jet). The scalar $g_i(t)$ represents the amount of gas in the i-th direction $\overrightarrow{s_i}$, where $\overrightarrow{\{s_i\}}_{1\leqq i\leqq N}$ covers the whole solid angle. The parameter α simply controls the beam widths and the symbol $[*]_n$ represents the unit vector with the same direction $*$. The directions $\overrightarrow{\{s_i\}}_{1\leqq i\leqq N}$ are designed based on the Fibonacci-Himawari coordinate system using the golden angle $\beta = 137.5°...$ so that it covers the whole solid angle with uniform density: $\overrightarrow{s_i} = (s\sin\gamma,\ s\cos\gamma,\ c)$ with $c = 1 - 2\,(i/i_{total})\,,\ s = \sqrt{1-c^2},\ \gamma = \beta i$. The first line of Eq.(6) represents the gas reduction process by the star formation triggered by the jet. This is proportional to the star formation rate μ, the strength of the jet toward the direction i with the jet collimation parameter α, and the amount of gas at that direction $g_i(t)$. The second line represents the time change rate of the jet. This is proportional to the feeding efficiency λ of the formed star to the SMBH, and the torque exerted by the falling stars just formed. A natural fade-out term parameterized by κ is added.

There are two typical cases in this model(Figs.2, 3). (a) The jet is active and rapidly changing its direction until finally all the ambient gas is exhausted to form stars (Fig. 2). (b) The jet is active until some intermediate time when the final jet direction is fixed. There are

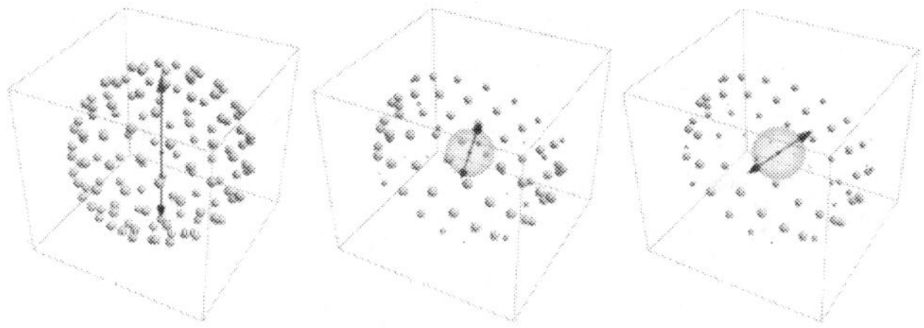

Figure 3. The galaxy formation history in case (b), low accretion rate $\lambda = 0.1$. The same as Fig.2 but with low accretion rate. The jet is inactive and its direction does not change much. Therefore the gas is left in a torus form almost perpendicular to the final jet direction. The gas will eventually settle to form a disk and a spiral galaxy is left.

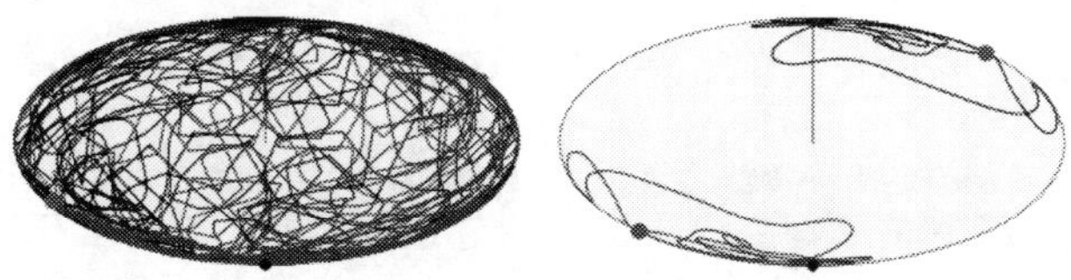

Figure 4. Trajectories of the moving jet-pair direction in the Mollweide projection of the whole solid angles; horizontal and vertical axes respectively represent the longitude and latitude. Small black and big red point pairs represent the initial and final jet directions, respectively. (left) The jet trajectory for the case with large accretion rate ($\lambda = 1$). The jet runs violently in the whole solid angle. This is the case of elliptical galaxy. (right) The same but small accretion rate ($\lambda = 0.055$). The jet runs gentle and eventually the direction is settled. This is the case of spiral galaxy.

finite remaining gas that failed to form stars and distributed almost perpendicular direction to the jet (Fig. 3)[5]. In the former case (a), the resultant structure has almost no gas but has a big star cluster at the center and small clusters GC around it, both composed from the stars of the same age. This is the typical elliptical galaxy. On the other hand in the latter case (b), the structure is the central star cluster bulge and GC, which are surrounded by the remaining gas. This gas would eventually form disk structure, in which new stars are going to be formed spontaneously. This is the typical spiral galaxy. Thus the main parts of the elliptical and spiral galaxies born at the same time by the same mechanism. Only the degree of jet activity separates them into the two phases SG and EG. The jet direction trajectories in both cases are compared in Fig.4.

The distinction of the two phases seems to be sharp if parameterized by the accretion rate λ, but not by the other parameters, as shown in Fig.5. If λ is large, then the torque exerted from the falling stars strongly changes the jet direction so that more star formation takes place in the fresh ambient gas. This further exert strong torque and this runaway continues until all the gas is exhausted. On the other hand if λ is small, then the falling stars exert only weak torque to change the jet direction. Then the jet cannot hit the sufficient amount of fresh gas to yield torque and eventually the jet direction is settled, leaving the torus shape

[5]If the settled jet-pair happens to point toward the gas remaining regions, then the small scale star formation activity may be still triggered which further induce density wave arm structure emanating from there.

DOI: 10.1051/epjconf/201716401011

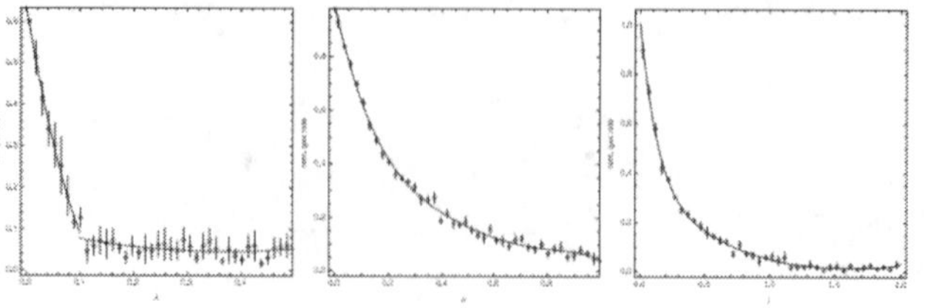

Figure 5. The fraction of the remaining gas that didn't form stars at $t = 300$. Each parameter is changed from the standard set $\lambda = 1$, $\mu = 1$, $j = 1$. (left) The case the accretion rate λ is changed. The remaining gas fraction has a sharp bend only in this case. (center) The case the star formation rate μ is changed. (right) The initial jet strength j is changed. If the jet activity does not exist at all ($j = 0$), it may yield a bulge-less spiral galaxy. The other parameters are the same as in Fig. 2.

gas distribution around the central bulge. The positive feedback eventually stops, and no runaway takes place.

This galaxy formation process has similar physics to the forest-fire model in complex systems[24, 25]. This model is composed of many trees on a two-dimensional lattice. Each tree on a lattice site has a finite probability p to ignite spontaneously. If ignite, the fire spreads to the neighboring trees with some probability. On the site of the burnt tree, a new tree has a chance to grow with some probability. The main feature is that the fire-fire correlation length $\xi(p)$, depends on p, determines the asymptotic two distinct states: (a) The fire dies out if $\xi(p)$ exceed the system size L. (b) The fire is sustained if $\xi(p) < L$. Percolation caused by the positive feed back or runaway process is the common feature in our galaxy model and the forest-fire model although the former is deterministic, as Eq.(6), and the latter is probabilistic.

The parameter of accretion rate λ may particularly be important to divide the galaxies into elliptical and spiral (Fig.5 left). The special value $\lambda_* \approx 0.1$ divides the EG ($\lambda > \lambda_*$) and SG($\lambda < \lambda_*$). The marginal case $\lambda \approx \lambda_*$ may correspond to the S0 galaxies. If the system has larger angular momentum in total, then the falling stars have less chance to fall near SMBH and therefore yield smaller accretion rate λ. Then this system tends to become a spiral galaxy. Moreover in this spiral galaxy case, the central cluster bulge is generally smaller than the elliptical galaxy provided the initial amount of ambient gas were the same with each other.

Other parameters are also important to divide the phases and yield variety of galaxies. The star formation rate μ would be higher in the denser environment of gas such as in the bottom of the gravitational potential. Therefore EG have much chance to be formed in the cluster and SG in the field.

Any relic of the above star-forming jet activity may be observed in the very early galaxies at distance. Most of the early stage spiral galaxies are generally thought to have very irregular shape as the rampaging jet which randomly forms stars and the remaining gas is not in ordered distribution. Thus, early irregular galaxies evolves into SG later. However this irregularity of early galaxies may be contaminated by the frequent collisions of galaxies in the dense environment in the past.

The jet activity may leave its trace in the faint structures in a spiral galaxy. The final stage of the jet would be gentle and the jet will generally have small precession. This jet may form a double-cone shape region in which the stars are massively formed. Subsequent supernova explosions of those stars may leave high energy electrons and protons there captured and stored in the magnetic fields. This relic ionized region will be observed in the early galaxies. However this relic structure may be contaminated with the relatively recent jet activity which often observed[26].

5 conclusions and prospects

We proposed a scenario that the primordial supermassive black hole (SMBH) made the galaxy. Energetic jets from the SMBH give ram pressure to the ambient gas to trigger the massive star formation. We focused on the following two aspects of this scenario.

The first aspect is the comparison of a globular cluster (GC) and a classical bulge (CB). According to our scenario, they are the same species but in distinct phases. We proposed that the difference comes from two kinds of velocity fields of the gas around the SMBH: (a) the virial velocity field associated with the dark matter distribution that dominates inside of the galaxy, and (b) the cosmic turbulent field, which obeys the Kolmogorov scaling, that dominates outside. Some portion of the star formed by the jet in (a) falls into the center and yields CB. On the other hand the star formed in (b) stays in halo and yields GC. Thus the two species CB and GC are formed simultaneously from the same gas, but in distinct velocity fields.

The second aspect is the comparison of the spiral galaxies (SG) and the elliptical galaxies (EG). According to our scenario, CB (in SG) and EG are the same species but in distinct phases. We proposed that the difference comes from the amount of jet activity. We introduced a simple model that describes this discrimination. According to this, (a) large feeding efficiency λ makes strong torque on SMBH that causes runaway flipping of the jet direction in the whole solid angle. Finally all the gas is exhausted and an EG is left. On the other hand (b) small λ makes weak torque and the jet direction change mildly without runaway. This makes smaller cluster in the center (CB) and the leftover gas will eventually form a disk, leaving a SG. Thus the existence/absence of the runaway separates the same species into two phases. This process has similar physics to the forest-fire model in complex systems.

Thus a SMBH defines the center of a galaxy and various star clusters in the galaxy (GC, CB, EG) are nurtured by the SMBH through the energetic jet emanating from the SMBH. According to this scenario, the hypothesis of the population-III stars that formed spontaneously may not be necessary.

Our simple analysis may be a useful supplement to the solid simulations of galaxy formation including all physical processes. We ignored many relevant dynamics such as major/minor mergers, detail of the star formation process, back reaction to the jet, jet collimation,... We would like to check whether the present model does include the most relevant process among them and refine our model through further study.

The author would like to thank Hideaki Mouri, Tohru Tashiro and all the laboratory members for fruitful discussions.

References

[1] J. Kormendy and D. Richstone, Annu. Rev. Astron. Astrophys. **33**, 581 (1995).
[2] J. Kormendy, arXiv:1504.03330v1 (2015).
[3] K. Gultekin et al. ApJ **698**, 198 (2009).

[4] G. Harris and W. Harris, MNRAS **410**, 2347H (2011).
[5] M. Rees, Annu. Rev. Astron. Astrophys. **22**, 471 (1984).
[6] P. Marziani and J. Sulentic, New Astron. Rev. **56**, 49 (2012).
[7] X. Wu et al., Nature **518,** 512 (2015).
[8] M. Latif et al., arxiv1404.5773v2 (2014).
[9] A. Schulze & L. Wisotzki, MNRAS **438**, 3422 (2014).
[10] M. Nishiyama, M. Morita, and M. Morikawa, arXiv:astro-ph/0403571 (2004).
[11] T. Fukuyama and M. Morikawa, Prog. Theor. Phys. **115,** 1047 (2006).
[12] T. Fukuyama, T. Tatekawa and M. Morikawa, JCAP**06**, 033 (2008).
[13] T. Fukuyama and M. Morikawa, Phys. Rev. **D80**, 063520 (2009).
[14] B.J. Carr, K. Kohri, Y. Sendouda, J. Yokoyama, Phys.Rev.**D81,** 104019 (2010).
[15] *The Formation and Disruption of Black Hole Jets*, ed. by I. Contopoulos, D. Gabuzda, and N. Kylafis, Astrophysics and Space Science Library **414** (2015).
[16] A. Reines and A. Deller, ApJ **750,** L24R (2012).
[17] C. Liu, Z. Gan, and F. Xie, Research in Astronomy and Astrophysics **13**, 899 (2013).
[18] A. King, ApJ **59**, L27 (2003).
[19] M. Choptuik, Phys. Rev. Lett. **70,** 9 (1993)
[20] C. Gundlach, Living Rev. Relativity **10**, 5 (2007).
[21] D. Kaup, Phys. Rev. **172** 1331 (1968).
[22] A. Nakamichi and M. Morikawa, JCAP**01,** 011 (2010).
[23] S. van den Bergh, Monthly Notices of the Royal Astronomical Society **385**, L20 (2007).
[24] C. Henley, Bull. Am. Phys. Soc. **34,** 838 (1989).
[25] P. Bak, and C. Tang, Phys. Lett. **A 147**, 297 (1990).
[26] G. Dobler et al., ApJ **717,** 825 (2010).

EPJ Web of Conferences **164**, 01012 (2017) DOI: 10.1051/epjconf/201716401012
ICNFP 2016

Haloscope searches for dark matter axions at the Center for Axion and Precision Physics Research

Eleni Petrakou[1,a] for CAPP/IBS

[1] *Center for Axion and Precision Physics Research, Institute for Basic Science (IBS), Daejeon 34141, Republic of Korea*

Abstract. The Center for Axion and Precision Physics Research (CAPP) was founded in 2013, with the ambition of shedding light on the strong CP problem and the proposed existence of axions. Much of CAPP's effort focuses on the direct detection of dark matter candidate axions with a series of local haloscope experiments, which endeavour to expand dramatically the coverage on the "invisible axion" mass range. The first two of them plan experimental runs during this year, tapping into ultra-low cryogenics and toroidal cavity geometries. The overall programme builds on cutting-edge technology, including developments in superconducting films, SQUID amplifiers and novel magnets.
This article presents the planned advancements and the status of the programme, while it can also be considered a pedagogical introduction to haloscope experiments.

1 Motivation for axion searches

The axion was introduced in 1978 as a consequence of the Peccei-Quinn $U(1)$ quasisymmetry[1], a popular proposed solution to the so-called strong CP problem. The strong CP problem refers to the fact that the strong force seems to conserve the CP symmetry – although the correctness of QCD predictions requires the presence of an additional, CP- and P-violating, term of gluon interactions, unless one quark mass is zero. Adding to the puzzle, nucleons have not so far been observed to possess electric dipole moments, whose presence would imply CP- and P-violation in QCD.
The introduction of the, spontaneously broken, Peccei-Quinn symmetry results in the presence of a new pseudoscalar field, known as axion[2], with properties suitable for a dark matter constituent[3]. (For a comprehensive overview of the Peccei-Quinn theory and the axion properties, see for example [4].)

The Center for Axion and Precision Physics Research (CAPP)[5] was founded in 2013 by the Institute for Basic Science in Daejeon, South Korea, with the goal of covering both sides of the coin of the strong CP problem: Definite measurements of the hadronic electric dipole moments and direct axion detection. Its direct detection searches, whose status is presented here, comprise of a series of experiments with the aim of excluding a significant portion of the axionic cold dark matter models' parameter space over the next years – or, if there is an axion, finding it.

[a]e-mail: petrakou@ibs.re.kr

2 Principles of haloscopes

The detection principle of haloscope experiments was first published by P. Sikivie[6] in 1983, turning the thus-far considered "invisible" candidate into a detectable entity. The principle relies on the "reverse Primakoff effect": In the presence of a magnetic field, the axion can mix with the virtual photon at loop-level, due to the axion's couplings to the strong sector. The end result is a real photon emitted at a frequency equal to the converted axion's mass. As axions are predicted to be abundant if they form the galactic dark matter, the observable will consist of a continuous electromagnetic signal at the microwave frequency range.
Specifically, the coupling of the axion to the electromagnetic field is described by

$$L_{a\gamma\gamma} = \frac{\alpha}{\pi}\frac{g_\gamma}{f_a}\, a\, \vec{E}\cdot\vec{B},$$

where a denotes the axion field, α is the fine structure constant, f_a is the axion decay constant, and g_γ is a model-dependent coefficient of order 1. The coupling is often expressed as $g_{a\gamma\gamma} \equiv \frac{\alpha}{\pi}\frac{g_\gamma}{f_a}$, and forms one of the parameters of the axion searches.

Since the introduction of this principle, a number of experiments have been set up to look for axion conversions, employing strong magnetic fields. They can be broadly divided in helioscopes, which point towards the Sun for streams of solar axions, and haloscopes, which look for ambient axions from the Milky Way's dark matter halo.
For realistic values of the theory, the power of the electromagnetic signal is expected to be eminently weak, with typical values lying around the range of 10^{-21}W [7]. In order to achieve sensitivity to this level of signal, haloscopes routinely operate in cryogenic temperatures for reduced noise and make use of radiofrequency cavities ("RF cavities"). The advantage of the latter is the enhancement of the signal by formation of standing waves at the cavity's resonant frequencies. The relevant expression for the resulting signal power, inserting realistic theoretical and experimental values, is:

$$P = 1.5\cdot 10^{-24} W \left(\frac{g_{a\gamma\gamma}}{10^{15} GeV}\right)^2 \left(\frac{C}{0.69}\right)\left(\frac{B}{8T}\right)^2 \left(\frac{V}{5l}\right)\left(\frac{\rho_a}{300\, MeV/cm^3}\right)\left(\frac{20\mu eV}{m_a}\right)\left(\frac{Q}{50\cdot 10^3}\right),$$

where ρ_a and m_a are the local density and mass of the axion, C is the form factor (an expression of the overlapping of the resonant mode's electric field with the external magnetic field), B is the magnetic field, and V and Q are the volume and loaded quality factor of the cavity.
The operation of RF cavities involves the appropriate choice of available electromagnetic modes, and a tuning mechanism for adjusting the resonant frequencies and so covering a range of the possible axion masses.
For illustration, Fig.1 shows an aluminum RF cavity attached to the inside of an open dilution refrigerator at CAPP's "RF Lab".

3 Haloscope advancement at CAPP

As seen from the last equation, the experimental settings which affect the sensitivity of a haloscope are the surrounding magnetic field, the RF cavity's volume and quality factor, and the form factor. In addition to these, the signal-to-noise ratio improves with reduced environment and device noise temperatures.
CAPP prepares advancements in each of these areas with an intense in-house R&D programme, while at the same time tapping into both local expertise and international collaboration.

DOI: 10.1051/epjconf/201716401012

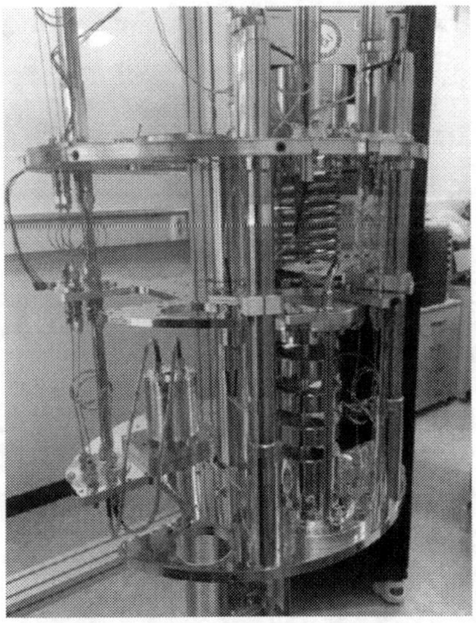

Figure 1: An aluminum RF cavity (bottom left, close to image center) attached to the inside of an open dilution refrigerator and linked to the read-out electronics.

Cryogenic operation
The haloscopes are designed to be installed inside dilution refrigerators for target operation at temperatures of a few dozen mK, and in any case well below 100mK. As of summer 2016 two BlueFors LD400 refrigerators are commissioned. In total, an array of six wet and dry dilution refrigerators will be operated in a dedicated low-vibration hall, at CAPP's Munji Campus lab space.

Low-noise amplifiers
HEMT amplifiers, the read-out devices of choice for weak signal, are being taken over by SQUID-based amplifiers. SQUID devices are commercially available for certain microwave frequencies, while dedicated development is carried out for the 2 – 10GHz range in collaboration with the Korea Research Institute for Standards and Science *(KRISS)*.

High field superconducting magnets
High-field superconducting magnets of different strengths and bores will accommodate the various haloscopes. The first of these, to be commissioned in 2017, will provide 18T. Stronger field values will rely critically on novel high-T_C cables, developed partially by the local company SuNAM. Based on this technology, the main milestone in the series of magnets will be the delivery of a 25T magnet of 10cm diameter in two years' time, as the first result of a joint project with Brookhaven National Laboratory (BNL)[8].

Superconducting cavity
RF cavities with superconducting surfaces will decrease dramatically the power losses. While typical values for the quality factor of a copper or aluminum cavity lie at $10^4 - 10^5$, a fully superconducting interior would rise by several orders of magnitude (although an increase much above 10^6 would not be beneficial to the experiments, as they would be limited by the quality factor of the axion signal). The obvious issue is maintaining superconductivity in the surrounding magnetic field. As the walls

of a cylindrical cavity are parallel to the field, deposition of a thin superconducting film is expected to bring about the desired improvement. For the endcaps, a novel solution will be attempted, with etching of appropriate hole patterns allowing the magnetic flux to penetrate the surface without destroying the superconducting state of the electrons. The development of this "vortex pinning" technique is carried out in collaboration with the Korea Advanced Institute of Science and Technology *(KAIST)*.

Phase-matched cavities

As the "transverse magnetic" resonant modes chiefly used for the axion searches oscillate between the cylindrical walls, probing higher regions of the mass range requires cavities of increasingly smaller diameter. Given this restriction, one solution for increasing the effective cavity volume is the simultaneous read-out of several frequency-matched cavities of the same dimensions; as the axion oscillation length is expected to be of the order of meters, the signal from them will be coherent. A dedicated project is under way for this technique[9] which has not been previously exploited in haloscope searches.

Toroidal RF cavities

An increase in the value of the form factor can be obtained by the use of toroidal, instead of cylindrical, geometry. In addition, toroid cavities offer the advantages of reduced fringe magnetic fields and avoidance of electromagnetic mode overlapping[10].

4 Haloscope programme and status

Cylindrical geometry

The first of the haloscope experiments employing the "classical" cylindrical geometry is CULTASK[11] *(CAPP Ultra-Low Temperature Axion Search in Korea)*. Using a copper cavity of 9cm diameter whose lowest mode corresponds to about 2GHz, it is expected to probe the 10μeV axion mass vicinity and act as a precursor to the rest of the experiments. As of summer 2016 the installation, engineering runs and commissioning of the electronics chain and data acquisition system are completed, and data of reasonable sensitivity are expected within the year.
Fig.2 shows the design of CULTASK with the cryogenic refrigerator, and the schematic of the system including the electronics and DAQ chain.

Over the next few years stronger results are foreseen from the delivery of progressively stronger magnets and the ongoing R&D. The commissioning of the main experiment is expected in 2018 when its key components, the 18T magnet and the custom SQUID amplifier, become available.
The projected sensitivity of the experiments is shown in Fig.3, in terms of the axion's mass and $g_{a\gamma\gamma}$ coupling. The estimations correspond to a signal-to-noise ratio of 5 and 100s scanning time per frequency value, for conventional quality factor cavities and estimated amplifier noise around 1K. With the addition of each further individual development, exploration of the cold dark matter parameter space will deepen, and coverage up to its higher end is expected within less than one decade.

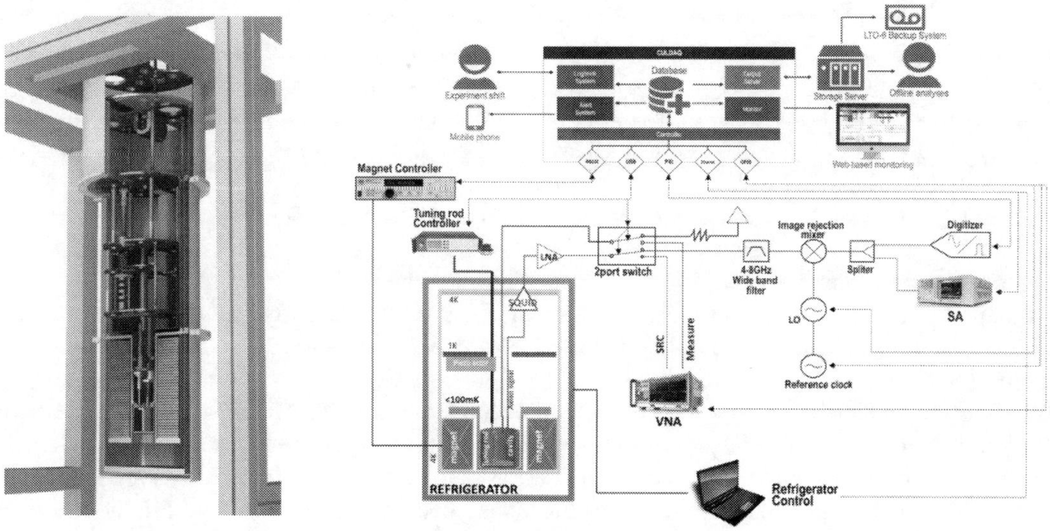

Figure 2: Left: Design of the CULTASK experiment with the BlueFors LD400 dilution regrifrator. Right: Schematic of the CULTASK electronics and data-taking system.

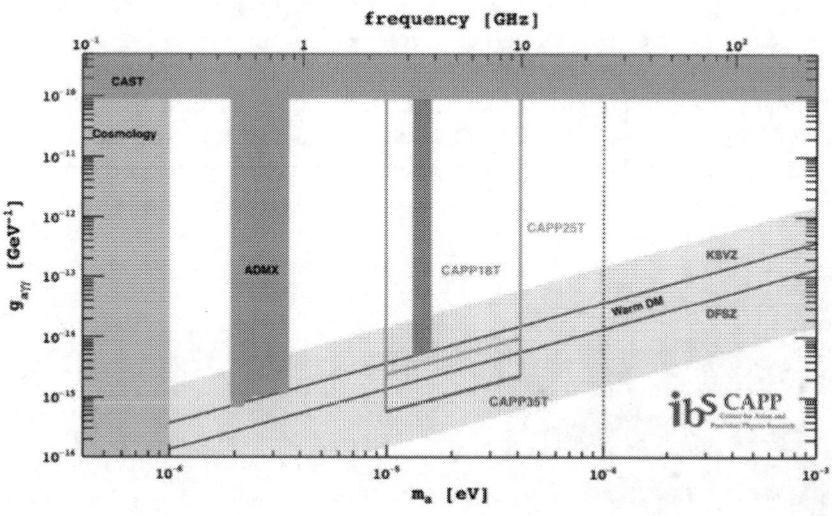

Figure 3: Estimated sensitivity of CAPP haloscopes for projected magnetic fields. The blue area to the left of the dashed line corresponds to the theoretically favorable axionic cold dark matter parameters.

Toroidal geometry
The lower end of the axion mass range can be covered effectively by toroid cavities and magnets. A small copper toroid haloscope, the *Toy CAPPuccino*[10], is commissioned for data taking at room temperature during 2016 (Fig.4). With a minor/major radius of 2/4cm, it is a precursor to the operation of an 80lt cavity at 12T and cryogenic temperatures. The latter will probe the mass spectrum around 1.3 – 1.8GHz, with sensitivity in the $g_{a\gamma\gamma}$ values relevant for cold dark matter.
An international project led by CAPP is under way for, eventually, a cavity of 50/200cm radius operating at 5T, which will probe the 190 – 850MHz mass range.

Rectangular geometry
In addition to the local haloscopes, a rectangular cavity has recently been included in the CAST helioscope[12] at CERN, forming the *CAST-CAPP/IBS Experiment*[13], and is currently in commissioning phase.
CAST is the leading helioscope experiment since 2003, using a 9T LHC magnet to induce the reverse Primakoff conversion. The RF cavity, housed in one of the magnet bores, forms a new addition to its existing detection techniques. The projected axion mass coverage will be 20 – 24μeV and, although corresponding to a small part of the mass range, analyzed data are expected within the next couple of years.

5 Summary

A series of direct axion detection searches at the Center for Axion and Precision Physics Research will form part of answering two of the most prominent questions in particle physics: The strong CP problem and the composition of dark matter. The experiments, some in commissioning phase and several under R&D, will employ cutting-edge technologies improving on all aspects of haloscope operations.
These proceedings presented an introduction to the principles of haloscopes, the CAPP R&D plans and the status of the programme. The Center's goal over the next few years is to cover the major part of the mass and coupling values of dark matter axions, either placing strong bounds on the theoretical expectations or, if the axion exists, discovering it.

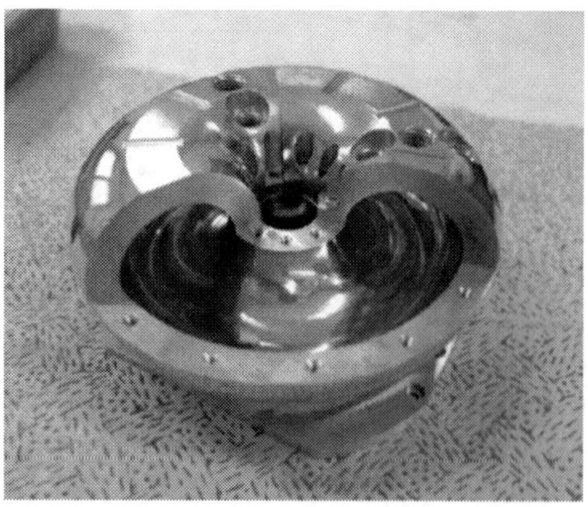

Figure 4. View of the cavity used in the Toy CAPPuccino experiment.

6 Acknowledgments

This work was supported by grant IBS-R017-D1-2016-a00 of the Republic of Korea.

References

[1] R. D. Peccei and H. Quinn, Phys. Rev. Lett. **38**, 1440 (1977)
[2] S. Weinberg, Phys. Rev. Lett. **40**, 223 (1978)
F. Wilczek, Phys. Rev. Lett. **40**, 279 (1978)
[3] J. E. Kim, Phys. Rev. Lett. **43**, 103 (1979)
M. A. Shifman, A. I. Vainshtein, and V. I. Zakharov, Nucl. Phys. B **166**, 493 (1980)
A. R. Zhitnitskii, Sov. J. Nucl. Phys. **31**, 260 (1980)
M. Dine, W. Fischler, and M. Srednicki, Phys. Lett. B **140**, 199 (1981)
[4] R. Bradley et al, Rev of Mod. Phys. **75**, 777 (2003)
[5] http://capp.ibs.re.kr/html/capp_en/
[6] P. Sikivie, Phys. Rev. Lett. **51**, 1415 (1983)
[7] W. U. Wuensch et al, Phys. Rev. D **40**, 3153 (1989)
[8] R. Gupta et al, IEEE Trans. Appl. Supercond. **26** (2016)
[9] S. W. Youn, 12th Patras Workshop:
https://indico.desy.de/getFile.py/access?contribId=46&resId=0&materialId=slides&confId=13889
[10] B. R. Ko, 12th Patras Workshop:
https://arxiv.org/abs/1609.03752
[11] W. H. Chung, 12th Patras Workshop:
https://indico.desy.de/getFile.py/access?contribId=50&resId=0&materialId=slides&confId=13889
[12] http://cast.web.cern.ch/CAST/
[13] L. Micelli, 12th Patras Workshop:
https://indico.desy.de/getFile.py/access?contribId=53&resId=0&materialId=slides&confId=13889

DOI: 10.1051/epjconf/201716401013

Highlights on signals from Dark Matter particles

R. Bernabei[1,2,a], P. Belli[1,2], A. Di Marco[1,2], F. Montecchia[2b], A. d'Angelo[3,4], A. Incicchitti[3,4], F. Cappella[5], V. Caracciolo[5], R. Cerulli[5], C.J. Dai[6], H.L. He[6], H.H. Kuang[6], X.H. Ma[6], X.D. Sheng[6], R.G. Wang[6], and Z.P. Ye[6c]

[1]*Dip. di Fisica, Università di Roma "Tor Vergata", I-00133 Rome, Italy*
[2]*INFN sez. Roma "Tor Vergata", I-00133 Rome, Italy*
[3]*Dip. di Fisica, Università di Roma "La Sapienza", I-00185 Rome, Italy*
[4]*INFN sez. Roma, I-00185 Rome, Italy*
[5]*INFN, Laboratori Nazionali del Gran Sasso, I-67100 Assergi (AQ), Italy*
[6]*Key Laboratory of Particle Astrophysics, IHEP, 100049 Beijing, P. R. China*

Abstract. Many experimental observations and theoretical arguments have pointed out that a large fraction of the Universe is composed by Dark Matter particles. Many possibilities are open on the nature and interaction types of such relic particles. In particular, this paper summarizes the main results obtained by exploiting the model independent Dark Matter annual modulation signature for the presence of Dark Matter particles in the galactic halo by DAMA experiment.

1 Introduction

Experimental observations and theoretical arguments have pointed out that most of the matter in the Universe has a non baryonic nature and is in form of Dark Matter (DM) particles. Many candidates – different in nature and with different and various interaction types – as DM particles of the Universe have been proposed within theories beyond the Standard Model of particle physics.

Depending on the DM candidate, the interaction processes can be various. Moreover, many other experimental and theoretical uncertainties exist and must be properly considered in a suitable interpretation and comparison among experiments of direct detection of DM particles.

Large efforts are dedicated all over the world to investigate the DM with different strategies and techniques that can give complementary information. However, let us briefly comment few items. Firstly, the DM indirect search – that is the study of possible products either of decay or of annihilation of DM particles in the galactic halo or in celestial body – is performed as by-product of experiments located underground, under-water, under-ice, or in space. The interpretation of such a study is strongly dependent on the chosen assumptions for the modeling of the background and is restricted to some DM candidates with peculiar features. Therefore, all that shows the intrinsic uncertainties of the DM indirect searches to unambiguously assess presence of DM in the galactic halo. On the other hand,

[a]rita.bernabei@roma2.infn.it
[b]also Dip. di Ingegneria Civile e Ingegneria Informatica, Università di Roma "Tor Vergata", I-00133 Rome, Italy
[c]also University of Jinggangshan, Jiangxi, China

experiments at accelerators may prove – when they can state a solid model independent result – the existence of some possible DM candidate particles, but they could never credit by themselves that a certain particle is a/the only solution for DM particle(s). Moreover, DM candidate particles and scenarios (even e.g. in the case of the neutralino candidate) exist which cannot be investigated at accelerators.

In order to pursue a widely sensitive direct detection of DM particles in the galactic halo, a model independent approach, a ultra-low-background suitable target material, a very large exposure and the full control of running conditions are strictly necessary.

In conclusion, suitable experiments offering a model independent signature for the presence of DM particles in the galactic halo are mandatory.

2 DM model independent signature

To obtain a reliable signature for the presence of DM particles in the galactic halo, it is necessary to exploit a suitable model independent signature. With the present technology, one feasible and able to test a large range of cross sections and of DM particle halo densities, is the so-called DM annual modulation signature [1]. The annual modulation of the signal rate originates from the Earth revolution around the Sun. In fact, as a consequence of its annual revolution around the Sun, which is moving in the Galaxy traveling with respect to the Local Standard of Rest towards the star Vega near the constellation of Hercules, the Earth should be crossed by a larger flux of DM particles around ~2 June (when the Earth orbital velocity is summed to the one of the solar system with respect to the Galaxy) and by a smaller one around ~2 December (when the two velocities are subtracted). Thus, this signature has a different origin and peculiarities than effects correlated with seasons (consider the expected value of the phase as well as the other requirements listed below). This DM annual modulation signature is very distinctive since the effect induced by DM particles must simultaneously satisfy all the following requirements: (1) the rate must contain a component modulated according to a cosine function; (2) with one year period; (3) with a phase that peaks roughly around ~ 2nd June; (4) this modulation must be present only in a well-defined low energy range, where DM particles can induce signals; (5) it must be present only in those events where just a single detector, among all the available ones in the used set-up, actually "fires" (*single-hit* events), since the probability that DM particles experience multiple interactions is negligible; (6) the modulation amplitude in the region of maximal sensitivity has to be $\lesssim 7\%$ in case of usually adopted halo distributions, but it may be significantly larger in case of some particular scenarios such as e.g. those in Ref. [2, 3]. This signature is model independent and might be mimicked only by systematic effects or side reactions able to simultaneously satisfy all the requirements given above; no one is available. At present status of technology it is the only DM model independent signature available in direct DM investigation that can be effectively exploited.

3 DAMA DM annual modulation results with highly radiopure NaI(Tl)

The DM annual modulation signature has been exploited with large exposure – using highly radiopure NaI(Tl) as target material – by the former DAMA/NaI ($\simeq$ 100 kg sensitive mass) experiment [4–13], and by the currently running DAMA/LIBRA ($\simeq$ 250 kg sensitive mass) [14–28], within the DAMA project. The DAMA project is dedicated to the development and use of low background scintillators for underground physics.

In particular, the experimental observable in DAMA experiments is the modulated component of the signal in NaI(Tl) target and not the constant part of it, as done in the other approaches aforementioned.

DOI: 10.1051/epjconf/201716401013

The full description of the DAMA/LIBRA set-up and performances during the phase1 and phase2 (presently running) and other related arguments have been discussed in details in Refs. [14–17, 19–21, 26] and references therein. Here we just remind that the sensitive part of this set-up is made of 25 highly radiopure NaI(Tl) crystal scintillators (5-rows by 5-columns matrix) having 9.70 kg mass each one. In each detector two 10 cm long UV light guides (made of Suprasil B quartz) act also as optical windows on the two end faces of the crystal, and are coupled to two low background photomultipliers (PMTs) working in coincidence at single photoelectron level. The low background 9265-B53/FL and 9302-A/FL PMTs, developed by EMI-Electron Tubes with dedicated R&Ds, were used in the phase1; for details see Ref. [8, 10, 14] and references therein. The detectors are housed in a sealed low-radioactive copper box installed in the center of a low-radioactive Cu/Pb/Cd-foils/polyethylene/paraffin shield; moreover, about 1 m concrete (made from the Gran Sasso rock material) almost fully surrounds (mostly outside the barrack) this passive shield, acting as a further neutron moderator. A threefold-levels sealing system prevents the detectors to be in contact with the environmental air of the underground laboratory [14]. The light response of the detectors during phase1 typically ranges from 5.5 to 7.5 photoelectrons/keV, depending on the detector. The hardware threshold of each PMT is at single photoelectron, while a software energy threshold of 2 keV electron equivalent (hereafter keV) is used [8, 14]. Energy calibration with X-rays/γ sources are regularly carried out in the same running condition down to few keV [14]; in particular, double coincidences due to internal X-rays from ^{40}K (which is at ppt levels in the crystals) provide (when summing the data over long periods) a calibration point at 3.2 keV close to the software energy threshold (for details see Ref. [14]). The radiopurity, the procedures and details are discussed in Ref. [14–17, 21] and references therein.

The data of DAMA/LIBRA–phase1 correspond to 1.04 ton $\times$ yr collected in 7 annual cycles; when including also the data of the DAMA/NaI experiment the total exposure is 1.33 ton $\times$ yr collected in 14 annual cycles. In order to investigate the presence of an annual modulation with proper features in the data, many analyses have been carried out. All these analyses point out the presence of an annual modulation satisfying all the requirements of the signature [15–17, 21]. In Fig. 1, as example, it is plotted the time behaviour of the experimental residual rate of the *single-hit* scintillation events for DAMA/LIBRA–phase1 in the (2–6) keV energy interval. When fitting the *single-hit* residual rate of DAMA/LIBRA–phase1 together with the DAMA/NaI ones, with the function: $A\cos\omega(t-t_0)$, considering a period $T = \frac{2\pi}{\omega} = 1$ yr and a phase $t_0 = 152.5$ day (June 2nd) as expected by the DM annual modulation signature, the following modulation amplitude is obtained: $A = (0.0110 \pm 0.0012)$ cpd/kg/keV, corresponding to 9.2 σ C.L..

When the period, and the phase are kept free in the fitting procedure, the modulation amplitude is (0.0112 ± 0.0012) cpd/kg/keV (9.3 σ C.L.), the period $T = (0.998 \pm 0.002)$ year and the phase $t_0 = (144 \pm 7)$ day, values well in agreement with expectations for a DM annual modulation signal. In particular, the phase is consistent with about June 2nd and is fully consistent with the value independently determined by Maximum Likelihood analysis [17].

For completeness, we recall that a slight energy dependence of the phase could be expected in case of possible contributions of non-thermalized DM components to the galactic halo, such as e.g. the SagDEG stream [12, 29, 30] and the caustics [31]. For more details see Ref. [17].

The modulation amplitudes singularly calculated for each annual cycle of DAMA/NaI and DAMA/LIBRA–phase1 are compatible among them and are normally fluctuating around their best fit values [15–17]. In particular, for the (2–6) keV energy interval the χ^2 is 10.8 over 13 *d.o.f.* corresponding to an upper tail probability of 63%, while the *run test* yields a lower tail probabilities of 23%. This analysis confirms that the data collected in all the annual cycles with DAMA/NaI and DAMA/LIBRA–phase1 are statistically compatible and can be considered together.

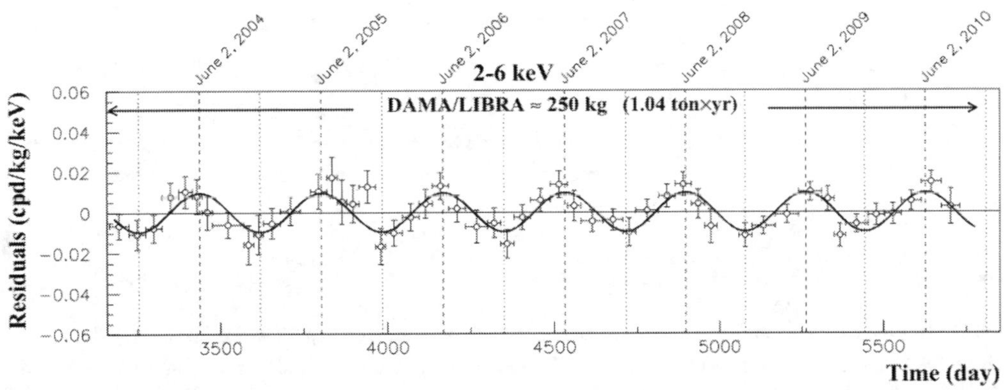

Figure 1. Experimental residual rate of the *single-hit* scintillation events measured by DAMA/LIBRA–phase1 in the (2–6) keV energy interval as a function of the time. The superimposed curve is the cosinusoidal function behaviour $A\cos\omega(t-t_0)$ with a period $T = \frac{2\pi}{\omega} = 1$ yr, a phase $t_0 = 152.5$ day (June 2nd) and modulation amplitude, A, equal to the central values obtained by best fit on the data points of the entire DAMA/LIBRA–phase1. The dashed vertical lines correspond to the maximum expected for the DM signal (June 2nd), while the dotted vertical lines correspond to the minimum.

The DAMA/LIBRA–phase1 *single-hit* residuals of Fig. 1 and those of DAMA/NaI have also been investigated by a Fourier analysis. The data analysis procedure has been described in details in Ref. [21]. A clear peak corresponding to a period of 1 year is evident for the (2–6) keV energy interval; the same analysis in the (6–14) keV energy region shows only aliasing peaks instead. Neither other structure at different frequencies has been observed (see also Ref. [21]).

Absence of any other significant background modulation in the energy spectrum has been verified in energy regions not of interest for DM; e.g. the measured rate integrated above 90 keV, R_{90}, as a function of the time has been analysed [17]. Similar result is obtained in other energy intervals. It is worth noting that the obtained results account of whatever kind of background and, in addition, no background process able to mimic the DM annual modulation signature (that is able to simultaneously satisfy all the peculiarities of the signature and to account for the measured modulation amplitude) is available (see also discussions e.g. in Ref. [14–17, 20, 21, 25, 32–38]).

A further relevant investigation in the DAMA/LIBRA–phase1 data has been performed by applying the same hardware and software procedures, used to acquire and to analyse the *single-hit* residual rate, to the *multiple-hit* one. In fact, since the probability that a DM particle interacts in more than one detector is negligible, a DM signal can be present just in the *single-hit* residual rate. Thus, the comparison of the results of the *single-hit* events with those of the *multiple-hit* ones corresponds practically to compare between them the cases of DM particles beam-on and beam-off. This procedure also allows an additional test of the background behaviour in the same energy interval where the positive effect is observed. In particular, in Fig. 2 the residual rates of the *single-hit* events measured over the DAMA/LIBRA–phase1 annual cycles are reported, as collected in a single cycle, together with the residual rates of the *multiple-hit* events, in the (2–6) keV energy interval. While, as already observed, a clear modulation, satisfying all the peculiarities of the DM annual modulation signature, is present in the *single-hit* events, the fitted modulation amplitude for the *multiple-hit* residual rate is well compatible with zero: $-(0.0005 \pm 0.0004)$ cpd/kg/keV in the energy region (2–6) keV. Thus, again evidence of annual modulation with the features required by the DM annual modulation signature is present in the *single-hit* residuals (events class to which the DM particle induced events belong), while it is absent in the *multiple-hit* residual rate (event class to which only background events belong). Similar

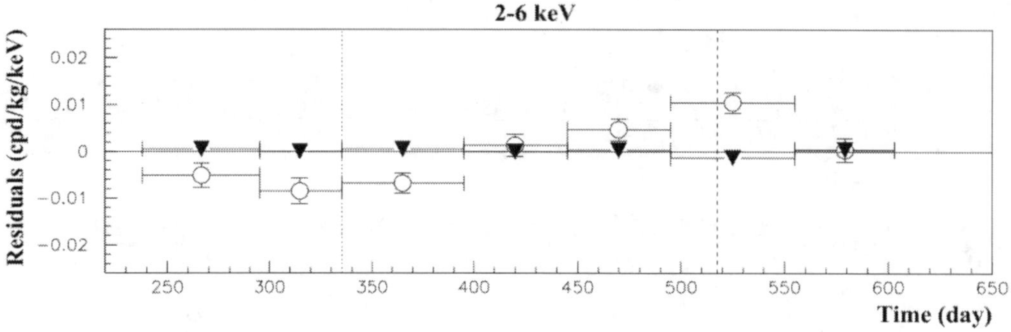

Figure 2. Experimental residual rates of DAMA/LIBRA–phase1 *single-hit* events (open circles), class of events to which DM events belong, and for *multiple-hit* events (filled triangles), class of events to which DM events do not belong. They have been obtained by considering for each class of events the data as collected in a single annual cycle and by using in both cases the same identical hardware and the same identical software procedures. The initial time of the figure is taken on August 7th. The experimental points present the errors as vertical bars and the associated time bin width as horizontal bars. Analogous results were obtained for the DAMA/NaI data [11].

results were also obtained for the last two annual cycles of the DAMA/NaI experiment [11]. Since the same identical hardware and the same identical software procedures have been used to analyse the two classes of events, the obtained result offers an additional strong support for the presence of a DM particle component in the galactic halo.

The annual modulation present at low energy can also be pointed out by depicting – as a function of the energy – the modulation amplitude, $S_{m,k}$, obtained by maximum likelihood method considering T =1 yr and t_0 = 152.5 day. For such purpose the likelihood function of the *single-hit* experimental data in the k–th energy bin is defined as: $\mathbf{L_k} = \mathbf{\Pi}_{ij} e^{-\mu_{ijk}} \frac{\mu_{ijk}^{N_{ijk}}}{N_{ijk}!}$, where N_{ijk} is the number of events collected in the i-th time interval (hereafter 1 day), by the j-th detector and in the k-th energy bin. N_{ijk} follows a Poisson's distribution with expectation value $\mu_{ijk} = \left[b_{jk} + S_{ik}\right] M_j \Delta t_i \Delta E \epsilon_{jk}$. The b_{jk} are the background contributions, M_j is the mass of the j–th detector, Δt_i is the detector running time during the i-th time interval, ΔE is the chosen energy bin, ϵ_{jk} is the overall efficiency. Moreover, the signal can be written as $S_{ik} = S_{0,k} + S_{m,k} \cdot \cos \omega(t_i - t_0)$, where $S_{0,k}$ is the constant part of the signal and $S_{m,k}$ is the modulation amplitude. The usual procedure is to minimize the function $y_k = -2ln(\mathbf{L_k}) - const$ for each energy bin; the free parameters of the fit are the $(b_{jk} + S_{0,k})$ contributions and the $S_{m,k}$ parameter. Hereafter, the index k is omitted for simplicity.

In Fig. 3 the obtained S_m are shown in each considered energy bin (there ΔE = 0.5 keV) when the data of DAMA/NaI and DAMA/LIBRA–phase1 are considered. It can be inferred that positive signal is present in the (2–6) keV energy interval, while S_m values compatible with zero are present just above. In fact, the S_m values in the (6–20) keV energy interval have random fluctuations around zero with χ^2 equal to 35.8 for 28 degrees of freedom (upper tail probability of 15%). All this confirms the previous analyses.

As described in Ref. [15–17, 21], the observed annual modulation effect is well distributed in all the 25 detectors at 95% C.L.

Among further additional tests, the analysis of the modulation amplitudes as a function of the energy separately for the nine inner detectors and the remaining external ones has been carried out for the entire DAMA/LIBRA–phase1. The obtained values are fully in agreement; in fact, the hypothesis that the two sets of modulation amplitudes as a function of the energy belong to same distribution

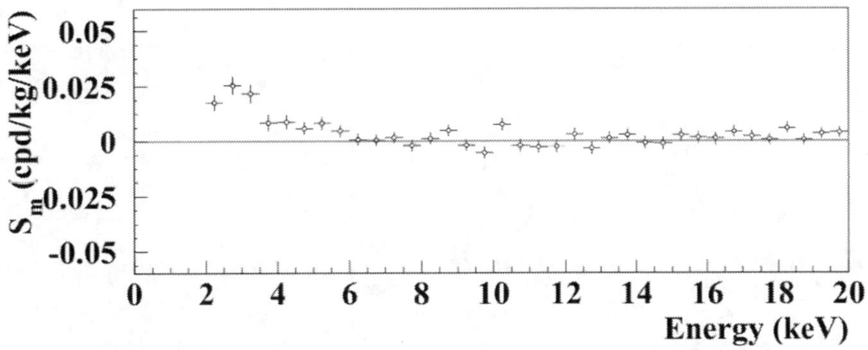

Figure 3. Energy distribution of the S_m variable for the total cumulative exposure 1.33 ton×yr. The energy bin is 0.5 keV. A clear modulation is present in the lowest energy region, while S_m values compatible with zero are present just above. In fact, the S_m values in the (6–20) keV energy interval have random fluctuations around zero with χ^2 equal to 35.8 for 28 degrees of freedom (upper tail probability of 15%).

has been verified by χ^2 test, obtaining: $\chi^2/d.o.f.$ = 3.9/4 and 8.9/8 for the energy intervals (2–4) and (2–6) keV, respectively (ΔE = 0.5 keV). This shows that the effect is also well shared between inner and outer detectors.

Let us, finally, release the assumption of a phase t_0 = 152.5 day in the procedure to evaluate the modulation amplitudes. In this case the signal can be written as:

$$\begin{aligned} S_{ik} &= S_{0,k} + S_{m,k} \cos\omega(t_i - t_0) + Z_{m,k} \sin\omega(t_i - t_0) \\ &= S_{0,k} + Y_{m,k} \cos\omega(t_i - t^*). \end{aligned} \tag{1}$$

For signals induced by DM particles one should expect: i) $Z_{m,k} \sim 0$ (because of the orthogonality between the cosine and the sine functions); ii) $S_{m,k} \simeq Y_{m,k}$; iii) $t^* \simeq t_0 = 152.5$ day. In fact, these conditions hold for most of the dark halo models; however, as mentioned above, slight differences can be expected in case of possible contributions from non-thermalized DM components, such as e.g. the SagDEG stream [12, 29, 30] and the caustics [31].

Considering cumulatively the data of DAMA/NaI and DAMA/LIBRA–phase1 the obtained 2σ contours in the plane (S_m, Z_m) for the (2–6) keV and (6–14) keV energy intervals are shown in Fig. 4–*left* while in Fig. 4–*right* the obtained 2σ contours in the plane (Y_m, t^*) are depicted.

Finally, setting S_m in eq. (1) to zero, the Z_m values as function of the energy have also been determined by using the same procedure. The values of Z_m are well compatible with zero, as expected [15–17].

No modulation has been found in any possible source of systematics or side reactions; thus, cautious upper limits on possible contributions to the DAMA/LIBRA–phase1 measured modulation amplitude have been obtained (see Refs. [9–11, 15–17, 20, 26]). It is worth noting that they do not quantitatively account for the measured modulation amplitudes, and also are not able to simultaneously satisfy all the many requirements of the signature. Similar analyses have also been performed for the DAMA/NaI data [10, 11].

Sometimes naive statements were put forward as the fact that in nature several phenomena may show some kind of periodicity. It is worth noting that the point is whether they might mimic the annual modulation signature in DAMA/NaI and in DAMA/LIBRA, i.e. whether they might be not only quantitatively able to account for the observed modulation amplitude but also able to contemporaneously satisfy all the requirements of the DM annual modulation signature. The same is for side reactions too. This has already been deeply investigated and discussed in DAMA literature.

 DOI: 10.1051/epjconf/201716401013

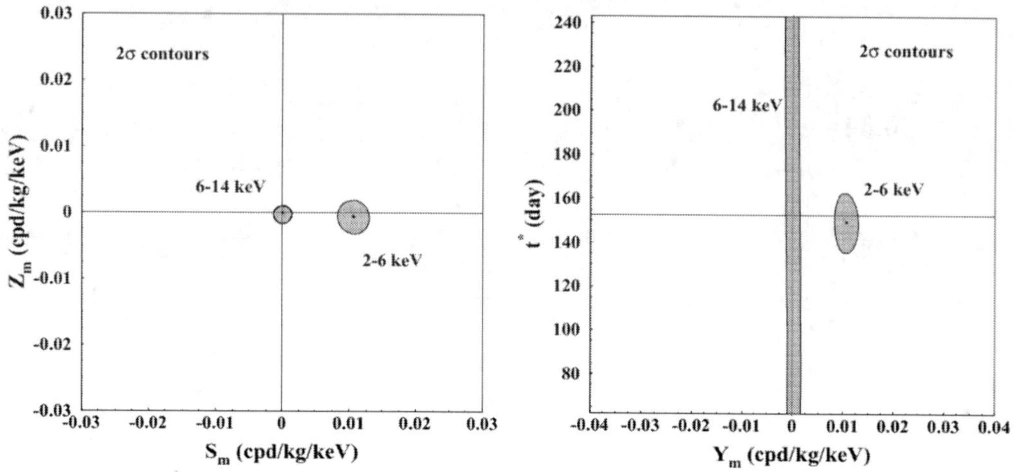

Figure 4. 2σ contours in the plane (S_m, Z_m) (*left*) and in the plane (Y_m, t^*) (*right*) for the (2–6) keV and (6–14) keV energy intervals. The contours have been obtained by the maximum likelihood method, considering the cumulative exposure of DAMA/NaI and DAMA/LIBRA–phase1. A modulation amplitude is present in the lower energy intervals and the phase agrees with that expected for DM induced signals. See text.

Table 1. Summary of the contributions to the total neutron flux at LNGS; the value, $\Phi^{(n)}_{0,k}$, the relative modulation amplitude, η_k, and the phase, t_k, of each component is reported. It is also reported the counting rate, $R_{0,k}$, in DAMA/LIBRA for *single-hit* events, in the $(2-6)$ keV energy region induced by neutrons, muons and solar neutrinos, detailed for each component. The modulation amplitudes, A_k, are reported as well, while the last column shows the relative contribution to the annual modulation amplitude observed by DAMA/LIBRA, $S_m^{exp} \simeq 0.0112$ cpd/kg/keV [17]. For details see Ref. [26] and references therein.

	Source	$\Phi^{(n)}_{0,k}$ (neutrons cm^{-2} s^{-1})	η_k	t_k	$R_{0,k}$ (cpd/kg/keV)	$A_k = R_{0,k}\eta_k$ (cpd/kg/keV)	A_k/S_m^{exp}
SLOW neutrons	thermal n ($10^{-2} - 10^{-1}$ eV)	1.08×10^{-6}	$\simeq 0$ however $\ll 0.1$	–	$< 8 \times 10^{-6}$	$\ll 8 \times 10^{-7}$	$\ll 7 \times 10^{-5}$
	epithermal n (eV-keV)	2×10^{-6}	$\simeq 0$ however $\ll 0.1$	–	$< 3 \times 10^{-3}$	$\ll 3 \times 10^{-4}$	$\ll 0.03$
FAST neutrons	fission, $(\alpha, n) \to$ n (1-10 MeV)	$\simeq 0.9 \times 10^{-7}$	$\simeq 0$ however $\ll 0.1$	–	$< 6 \times 10^{-4}$	$\ll 6 \times 10^{-5}$	$\ll 5 \times 10^{-3}$
	$\mu \to$ n from rock (> 10 MeV)	$\simeq 3 \times 10^{-9}$	0.0129	end of June	$\ll 7 \times 10^{-4}$	$\ll 9 \times 10^{-6}$	$\ll 8 \times 10^{-4}$
	$\mu \to$ n from Pb shield (> 10 MeV)	$\simeq 6 \times 10^{-9}$	0.0129	end of June	$\ll 1.4 \times 10^{-3}$	$\ll 2 \times 10^{-5}$	$\ll 1.6 \times 10^{-3}$
	$\nu \to$ n (few MeV)	$\simeq 3 \times 10^{-10}$	0.03342*	Jan. 4th*	$\ll 7 \times 10^{-5}$	$\ll 2 \times 10^{-6}$	$\ll 2 \times 10^{-4}$
	direct μ	$\Phi^{(\mu)}_0 \simeq 20\,\mu$ m^{-2}d^{-1}	0.0129	end of June	$\simeq 10^{-7}$	$\simeq 10^{-9}$	$\simeq 10^{-7}$
	direct ν	$\Phi^{(\nu)}_0 \simeq 6 \times 10^{10}\,\nu$ cm^{-2}s^{-1}	0.03342*	Jan. 4th*	$\simeq 10^{-5}$	3×10^{-7}	3×10^{-5}

* The annual modulation of solar neutrino is due to the different Sun-Earth distance along the year; so the relative modulation amplitude is twice the eccentricity of the Earth orbit and the phase is given by the perihelion.

In particular, in Refs. [20, 26] a quantitative evaluation why the neutrons, the muons and the solar neutrinos cannot give any significant contribution to the DAMA annual modulation results and cannot mimic this signature is outlined. Table 1 summarizes the safety upper limits on the contributions to the observed modulation amplitude due to the total neutron flux at LNGS, either from (α, n) reactions, from fissions and from muons' and solar-neutrinos' interactions in the rocks and in the lead around the experimental set-up; the direct contributions of muons and solar neutrinos are reported there too.

In any case no systematic effects or side reactions able to account for the whole observed modulation amplitude and to simultaneously satisfy all the requirements of the exploited DM signature have been found. A detailed discussion about all the related arguments can be found in Refs. [9–11, 14–17, 20, 21, 25, 26].

Analyses on the presence of possible diurnal effects in the DAMA/LIBRA–phase1 data [25] and on the so called "Earth Shadow Effect" [27] will be summarized in the following. Finally, the annual modulation result has been also interpreted in terms of Asymmetric Mirror DM [28], as briefly shown later.

4 Implications and comparisons

The long-standing annual-modulation evidence measured in DAMA experiments is model-independent, i.e. in particular independent on theoretical interpretations of the identity of DM and specifics of its interactions. It can be related to a variety of interaction mechanisms of DM particles with the detector materials and is compatible with a wide set of scenarios regarding the nature of the DM candidate and related astrophysical, nuclear and particle physics. For example, some of the scenarios available in literature and the different parameters are discussed in Refs. [4–7, 10–13, 15, 21] and references therein, and recently e.g. in Refs. [28, 39] . Further large literature is available on the topics (see for example in Ref. [21]) and many possibilities are open.

It is worth noting that no other experiment exists, whose result can be directly compared in a model-independent way with those by DAMA/NaI and DAMA/LIBRA. Some activities claim model-dependent exclusion under many largely arbitrary assumptions (see for example discussions in Ref. [10, 11, 15, 40–42]). Moreover, often some critical points exist in their experimental aspects, as mentioned above, and the existing experimental and theoretical uncertainties are generally not considered in their presented single model dependent result; moreover, implications of the DAMA results are often presented in incorrect/partial/unupdated way. Both the accounting of the existing uncertainties and the existence of alternative scenarios (see literature) allow one to note that model dependent results by indirect and direct experiments actually are not in conflict with the DAMA model independent result.

The model independent annual modulation effect observed by the DAMA experiments has been investigated in terms of many DM candidates. Here we just recall the case of a mirror-type dark matter candidates in some scenarios [28].

In the framework of asymmetric mirror matter, the DM originates from hidden (or shadow) gauge sectors which have particles and interaction content similar to that of ordinary particles. It is assumed that the mirror parity is spontaneously broken and the electroweak symmetry breaking scale v' in the mirror sector is much larger than that in the Standard Model, v = 174 GeV. In this case, the mirror world becomes a heavier and deformed copy of our world, with mirror particle masses scaled in different ways with respect to the masses of the ordinary particles. Then, in this scenario dark matter would exist in the form of mirror hydrogen composed of mirror proton and electron, with mass of about 5 GeV which is a rather interesting mass range for dark matter particles.

The data analysis in the Mirror DM model framework allows the determination of the $\sqrt{f}\epsilon$ parameter (where f is the fraction of DM in the Galaxy in form of mirror atoms and ϵ is the coupling constant). In the analysis several uncertainties on the astrophysical, particle physics and nuclear physics models have been taken into account in the calculation. The obtained values of the $\sqrt{f}\epsilon$ parameter in the case of mirror hydrogen atom ranges between 7.7×10^{-10} to 1.1×10^{-7}; they are well compatible with cosmological bounds [28].

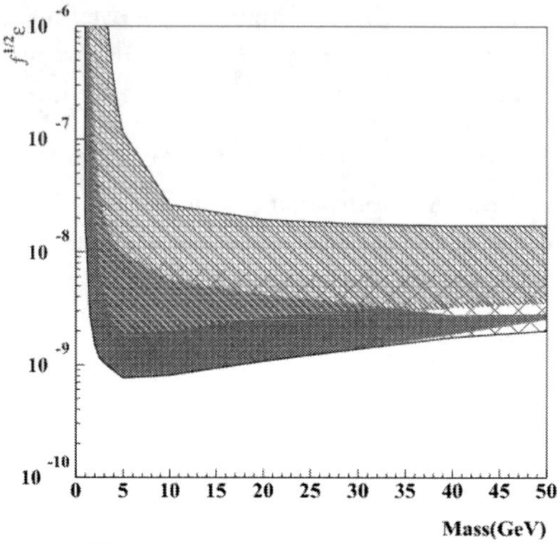

Figure 5. Allowed regions for the $\sqrt{f}\epsilon$ parameter as function of mirror hydrogen mass, obtained by marginalizing all the models for each considered scenario. The allowed intervals identify the $\sqrt{f}\epsilon$ values corresponding to C.L. larger than 5σ from the *null hypothesis*, that is $\sqrt{f}\epsilon = 0$. The allowed regions corresponding to five different scenarios are depicted in different hatching; the black line is the overall boundary; for details see Ref. [28].

In addition, releasing the assumption $M_{A'} \simeq 5m_p$, the allowed regions for the $\sqrt{f}\epsilon$ parameter as function of $M_{A'}$, mirror hydrogen mass, obtained by marginalizing all the models for each considered scenario, are shown in Fig. 5.

5 Diurnal modulation

The results obtained by investigating the presence of possible diurnal variation in the low-energy *single-hit* scintillation events collected by DAMA/LIBRA–phase1 (1.04 ton × year exposure) have been analysed in terms of a DM second order model-independent effect due to the Earth diurnal rotation around its axis [25]. In particular, the data were analysed using the sidereal time referred to Greenwich, often called GMST.

This daily modulation of the rate on the sidereal time, expected when taking into account the contribution of the Earth rotation velocity, has several requirements as the DM annual modulation effect does. The interest in this signature is that the ratio R_{dy} of this diurnal modulation amplitude over the annual modulation amplitude is a model independent constant at given latitude; considering the LNGS latitude one has $R_{dy} = \frac{S_d}{S_m} \simeq 0.016$.

Taking into account R_{dy} and the DM annual modulation effect pointed out by DAMA/LIBRA–phase1 for *single-hit* events in the low energy region, it is possible to derive the diurnal modulation amplitude expected for the same data. In particular, when considering the (2–6) keV energy interval, the observed annual modulation amplitude in DAMA/LIBRA–phase1 is: (0.0097 ± 0.0013) cpd/kg/keV [17] and the expected value of the diurnal modulation amplitude is $\simeq 1.5\times 10^{-4}$ cpd/kg/keV.

Fig. 6 shows the time and energy behaviour of the experimental residual rates of *single-hit* events both as a function of solar (*left*) and of sidereal (*right*) time, in the (2–6) keV interval. The used time bin is 1 (either solar or sidereal) hour.

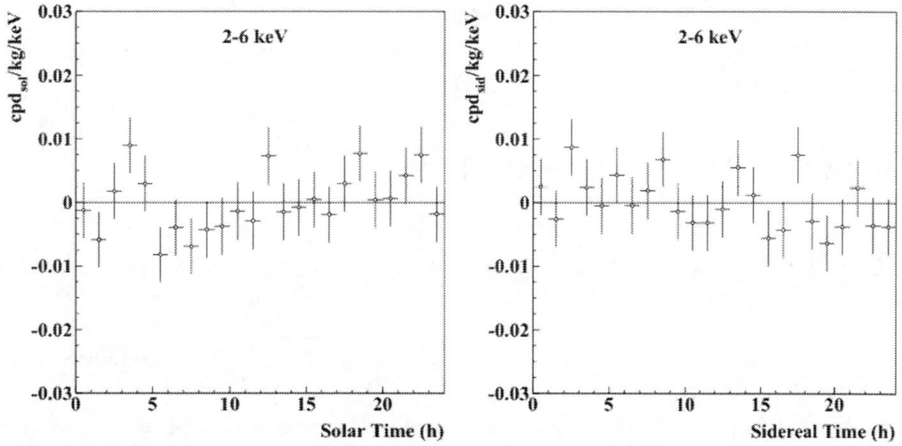

Figure 6. Experimental model-independent diurnal residual rate of the *single-hit* scintillation events, measured by DAMA/LIBRA–phase1 in the (2–6) keV energy interval as a function of the hour of the solar (*left*) and sidereal (*right*) day. The experimental points present the errors as vertical bars and the associated time bin width (1 hour) as horizontal bars. The cumulative exposure is 1.04 ton × yr. See Ref. [25] for details.

The null hypothesis (absence of residual rate diurnal variation) has been tested by a χ^2 test and run test [25]. Thus, the presence of any significant diurnal variation and of time structures can be excluded at the reached level of sensitivity (see Fig. 6).

In order to compare the experimental data with the DM diurnal effect due to the Earth rotation around its axis, the sidereal diurnal modulation amplitude of the (2–6) keV energy interval is taken into account: $A_d^{exp} = -(1.0 \pm 1.3) \times 10^{-3}$ cpd/kg/keV. Following the Feldman-Cousins procedure an upper limit can be obtained for the measured diurnal modulation amplitude: $A_d^{exp} < 1.2 \times 10^{-3}$ cpd/kg/keV (90% C.L.); thus, the effect of DM diurnal modulation (expected amplitude $\simeq 1.5\times 10^{-4}$ cpd/kg/keV) is out the present sensitivity [25].

In conclusion, at that level of sensitivity of DAMA/LIBRA–phase1 the presence of a significant diurnal variation and of diurnal time structures in the data can be excluded for both the cases of solar and sidereal time. In particular, the sidereal diurnal modulation amplitude expected – because of the Earth diurnal motion – on the basis of the DAMA DM annual modulation results cannot be investigated at the present sensitivity; DAMA/LIBRA–phase2, presently running, with a lower software energy threshold [19] can also offer the possibility to increase sensitivity to such an effect.

6 Daily effect on the sidereal time due to the shadow of the Earth

The results obtained in the investigation of possible diurnal effects for low-energy *single-hit* scintillation events of DAMA/LIBRA–phase1 have been analysed in terms of Earth Shadow Effect, a model-dependent effect that is expected for DM candidates inducing only nuclear recoils and having high cross-section (σ_n) with ordinary matter [27].

In fact a diurnal variation of the low energy rate could be expected for these specific candidates, because of the different thickness of the shield due to the Earth during the sidereal day, eclipsing the wind of DM particles. The induced effect should be a daily variation of their velocity distribution, and therefore of the signal rate measured deep underground. However, this effect is very small and would be appreciable only in case of high cross-section spin independent coupled candidates. Such candidates must constitute a little fraction (ξ) of the Galactic dark halo in order to fulfil the positive

DAMA result on annual modulation. By the fact, this analysis decouples ξ from σ_n. Considering the measured DM annual modulation effect and the absence – at the present level of sensitivity – of diurnal effects, the analysis selects allowed regions in the three-dimensional space: ξ, σ_n and DM particle mass in some model scenarios; for details see Ref. [27].

7 Conclusions and Perspectives

The cumulative exposure with ultra low background NaI(Tl) target by the former DAMA/NaI and DAMA/LIBRA–phase1 is 1.33 ton × yr (orders of magnitude larger than those available in the field) giving a model-independent positive evidence at 9.3 σ C.L. for the presence of DM candidates in the galactic halo with full sensitivity to many kinds of astrophysical, nuclear and particle physics scenarios. Other rare processes have also been searched for by DAMA/LIBRA-phase1 (see for details Refs. [22–24]) and by DAMA/NaI [43].

After the phase1, an important upgrade has been performed when all the PMTs have been replaced with new ones having higher Quantum Efficiency (QE). In this new configuration a software energy threshold below 2 keV has been reached [19]. DAMA/LIBRA is thus in its phase2, and after optimization periods it is continuously running with higher sensitivity.

The main goals of DAMA/LIBRA–phase2 are: (1) to increase the experimental sensitivity thanks to the lower software energy threshold of the experiment; (2) to improve the corollary investigation on the nature of the DM particle and related astrophysical, nuclear and particle physics arguments; (3) to investigate other signal features; (4) to investigate rare processes other than DM with high sensitivity.

Future improvements to increase the sensitivity of the set-up can be considered by using high QE and ultra-low background PMTs directly coupled to the NaI(Tl) crystals. In this way a further large improvement in the light collection and a further lowering of the software energy threshold would be obtained.

Finally, for completeness, we also mention that low background $ZnWO_4$ crystal scintillators have recently been proposed within the DAMA collaboration for the study of the directionality of DM candidates inducing just nuclear recoils. The features and performances of such anisotropic scintillators are very promising [44] and are under exploration.

References

[1] A.K. Drukier, K. Freese, and D.N. Spergel, *Phys. Rev. D* **33**, 3495 (1986); K. Freese, J. A. Frieman and A. Gould, *Phys. Rev. D* **37**, 3388 (1988).
[2] D. Tucker-Smith and N. Weiner, *Phys. Rev. D* **64**, 043502 (2001).
[3] K. Freese et al., *Phys. Rev. D* **71**, 043516 (2005); K. Freese et al., *Phys. Rev. Lett.* **92**, 11301 (2004).
[4] R. Bernabei et al., *Int. J. Mod. Phys. A* **22**, 3155 (2007).
[5] R. Bernabei et al., *Mod. Phys. Lett. A* **23**, 2125 (2008).
[6] R. Bernabei et al., *Phys. Rev. D* **77**, 023506 (2008).
[7] R. Bernabei et al., *Int. J. Mod. Phys. A* **21**, 1445 (2006).
[8] R. Bernabei et al., *Il Nuovo Cim. A* **112**, 545 (1999).
[9] R. Bernabei et al., *Eur. Phys. J. C* **18**, 283 (2000).
[10] R. Bernabei el al., *La Rivista del Nuovo Cimento* **26** n.1, 1-73 (2003).
[11] R. Bernabei et al., *Int. J. Mod. Phys. D* **13**, 2127 (2004).
[12] R. Bernabei et al., *Eur. Phys. J. C* **47**, 263 (2006).

[13] R. Bernabei et al., *Eur. Phys. J. C* **53**, 205 (2008).
[14] R. Bernabei et al., *Nucl. Instr. and Meth. A* **592**, 297 (2008).
[15] R. Bernabei et al., *Eur. Phys. J. C* **56**, 333 (2008).
[16] R. Bernabei et al., *Eur. Phys. J. C* **67**, 39 (2010).
[17] R. Bernabei et al., *Eur. Phys. J. C* **73**, 2648 (2013).
[18] P. Belli et al., *Phys. Rev. D* **84**, 055014 (2011).
[19] R. Bernabei et al., *J. of Instr.* **7**, P03009 (2012).
[20] R. Bernabei et al., *Eur. Phys. J. C* **72**, 2064 (2012).
[21] R. Bernabei et al., *Int. J. of Mod. Phys. A* **28**, 1330022 (2013).
[22] R. Bernabei et al., *Eur. Phys. J. C* **62**, 327 (2009).
[23] R. Bernabei et al., *Eur. Phys. J. C* **72**, 1920 (2012).
[24] R. Bernabei et al., *Eur. Phys. J. A* **49**, 64 (2013).
[25] R. Bernabei et al., *Eur. Phys. J. C* **74**, 2827 (2014).
[26] R. Bernabei et al., *Eur. Phys. J. C* **74**, 3196 (2014).
[27] R. Bernabei et al., *Eur. Phys. J. C* **75**, 239 (2015).
[28] A. Addazi et al, *Eur. Phys. J. C* **75**, 400 (2015).
[29] K. Freese et al., *Phys. Rev. D* **71**, 043516 (2005); *New Astr. Rev.* **49**, 193 (2005); astro-ph/0310334; astro-ph/0309279.
[30] G. Gelmini, P. Gondolo, *Phys. Rev. D* **64**, 023504 (2001).
[31] F.S. Ling, P. Sikivie and S. Wick, *Phys. Rev. D* **70**, 123503 (2004).
[32] R. Bernabei et al., *AIP Conf. Proceed.* **1223**, 50 (2010) [arXiv:0912.0660].
[33] R. Bernabei et al., *J. Phys.: Conf. Series* **203**, 012040 (2010) [arXiv:0912.4200]; *http://taup2009.lngs.infn.it/slides/jul3/nozzoli.pdf*, talk given by F. Nozzoli.
[34] R. Bernabei et al., in the volume *Frontier Objects in Astrophysics and Particle Physics*, ed. S.I.F. (Vulcano, 2010), p. 157 [arXiv:1007.0595].
[35] R. Bernabei et al., *Can. J. Phys.* **89**, 11 (2011).
[36] R. Bernabei et al., *Physics Procedia* **37**, 1095 (2012).
[37] R. Bernabei et al., arXiv:1210.6199.
[38] R. Bernabei et al., arXiv:1211.6346.
[39] S. Scopel et al., arXiv:1512.08577, to appear in the Proceed. of the 14th Marcel Grossmann Meeting.
[40] R. Bernabei et al., in *Liquid Noble gases for Dark Matter searches: a synoptic survey* (Exorma Ed., Roma, ISBN 978-88-95688-12-1, 2009), pp. 1–53; arXiv:0806.0011v2.
[41] R. Bernabei et al., *Int. J. Mod. Phys. A* **30**, 1530053 (2015), 73 pages.
[42] J.I. Collar and D.N. McKinsey, arXiv:1005.0838; arXiv:1005.3723; J.I. Collar, arXiv:1006.2031; arXiv:1010.5187; arXiv:1103.3481; arXiv:1106.0653; arXiv:1106.3559.
[43] R. Bernabei et al., *Phys. Lett. B* **408** 439 (1997); P. Belli et al., *Phys. Lett. B* **460** 236 (1999); R. Bernabei et al., *Phys. Rev. Lett.* **83** 4918 (1999); P. Belli et al., *Phys. Rev. C* **60** 065501 (1999); R. Bernabei et al., *Il Nuovo Cimento A* **112** 1541 (1999); R. Bernabei et al., *Phys. Lett. B* **515** 6 (2001); F. Cappella et al., *Eur. Phys. J.-direct C* **14** 1 (2002); R. Bernabei et al., *Eur. Phys. J. A* **23** 7 (2005); R. Bernabei et al., *Eur. Phys. J. A* **24** 51 (2005); R. Bernabei et al., *Astrop. Phys.* **4** (1995).
[44] F. Cappella et al., *Eur. Phys. J. C* **73**, 2276 (2013).

DOI: 10.1051/epjconf/201716401014

Holography for Heavy Ions Collisions at LHC and NICA

Irina Aref'eva[1,a]

[1] *Steklov Mathematical Institute, Russian Academy of Science*

Abstract. This is a contribution for the Proceedings of 5th International Conference on New Frontiers in Physics (ICNFP 2016), held at Crete, 6-14 July 2016. Our goal is to obtain phenomenologically reliable insights for the physics of the quark-gluon plasma (QGP) from the holography. I briefly review how in the holographical setup one can describe the QGP formation in heavy ion collisions and how to get quantitatively the main characteristics of the QGP formation – the total multiplicity and the thermalization time. To fit the experimental form of dependence of total multiplicity on energy, obtained at LHC, we have to deal with a special anisotropic holographic model, related with the Lifshitz-type background.
Our conjecture is that this Lifshitz-type background with non-zero chemical potential can be used to describe future data expected from NICA. In particular, we present the results of calculations the holographic confinement/deconfinement phase transition in the (μ, T) (chemical potential, temperature) plane in this anizotropic background and show the dependence of the transition line on the orientation of the quark pair. This dependence leads to a non-sharp character of physical confinement/deconfinement phase in the (μ, T)-plane. We use the bottom-up soft wall approach incorporating quark confinement deforming factor and vector field providing the non-zero chemical potential. In this model we also estimate the holographic photon production.

1 Introduction

Quark Gluon Plasma (QGP) produced in heavy ion collisions (HIC) at RHIC and LHC is a new form of matter formed from quarks and gluons at hight temperature [1, 2]. It is a strong coupling fluid [3], or an "operator boiling fluid" [4]. This makes perturbative methods inapplicable to study properties of QGP and especially it formation. The lattice QCD, the powerful tool of non-perturbative study, is formulated in the Euclidean spacetime and cannot be applied to QGP formation, since this is time depending phenomena (cf.[5]). This gives a strong motivation to study the process of QGP formation in HIC through the gauge/string duality. The gauge/string duality (or AdS/CFT correspondence) is directly applicable only for a very special 4-dimensional theory, namely, $\mathcal{N}$=4 SUSY Yang Mills theory [6]. This theory is conformal invariant on the quantum level and the conformal invariance is the main request for the AdS/CFT application. Of course, the real QCD is not conformal invariant, however, as we know from lattice calculations, the high-temperature QCD is. More precise, the deviation from conformality decreases as the temperature increases, see Fig.1.A. In Fig.1.B the QCD phase diagram is presented. The phase diagram of QCD is not well known either experimentally or

[a]e-mail: arefeva@mi.ras.ru

 DOI: 10.1051/epjconf/201716401014

theoretically. A commonly conjectured form of the phase diagram, temperature T vs quark chemical potential μ, is shown in Fig.1.B. The chemical potential μ is a measure of the imbalance between quarks and antiquarks in the system. The phase transition is not sharp and it is supposed to be the 1-st order.

Ordinary nuclear matter in this diagram is at $\mu = 310$ MeV and T close to zero. If we increase the quark density, i.e. increase μ, keeping the temperature low, we go into a phase of more and more compressed nuclear matter neutron stars. Above the (blue on the on-line version of the paper) smeared line there is a transition to the quark-gluon plasma. At ultra-high densities one expects to find the phase of color-superconducting quark matter. In ultra-relativistic heavy ion collisions one studies this matter in the regime of extreme energy density. In Fig.1.B. the typical values of μ and T in heavy-ion collisions (RHIC and LHC) are shown by a filled (cyan) region near the T-axis. The regions expected to be available at NICA (Nuclotron-based Ion Collider fAcility) [7] and FAIR (Facility for Antiproton and Ion Research) are indicated by arrows.

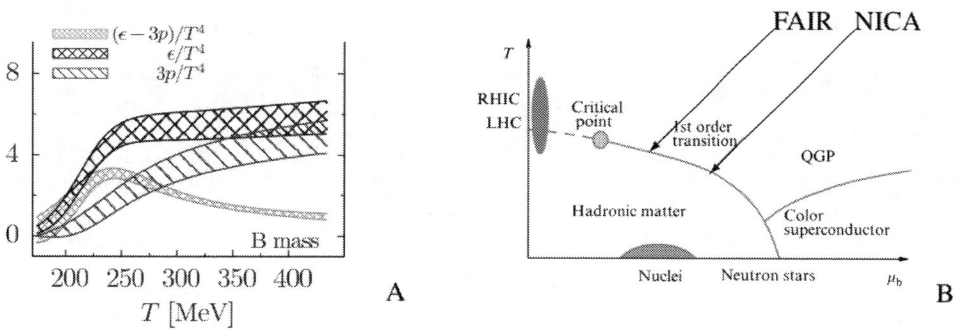

Figure 1. A. The trace anomaly, energy density and pressure for two flavors of twisted mass Wilson fermions at $m_\pi = 360$ MeV. The plot from [8]. B.The QCD phase diagram

The exact dual description of the real QCD is unknown, but holographic QCD models that fit perturbative (two loops β-function) and lattice QCD results (in particular, the quark confinement potential) have been proposed [9, 10]. Using these models several static properties of QGP have been reproduced [11].

The description of the QGP formation in HIC is a difficult subject, since it supposes to study a complicated real time phenomena – thermalization. We also do not know much from experiments about the details of the QGP formation in HIC, one can just estimates the time of QGP formation as well as the total multiplicity (there are arguments that the main part of particles is produced during the QGP formation) [12, 13]. The QGP formation has been the subject of the active studies within holographic approach in last years (see [14–16] and refs therein). Initially this problem has been considered in AdS background [17–23] and the total multiplicity within this approach has been estimated as

$$\mathcal{M}_{AdS} \sim s^{0.33}, \tag{1}$$

For the improved holographic background the estimation was [24]

$$\mathcal{M}_{IHQCD} \sim s^{0.22}(1 + \log \text{ corrections}), \tag{2}$$

DOI: 10.1051/epjconf/201716401014

The experimental multiplicity dependence on energy [12, 13] is

$$\mathcal{M}_{LHC} \sim s^{0.155}, \tag{3}$$

and it has been shown in [25] that the dependence (3) requires an unstable background.

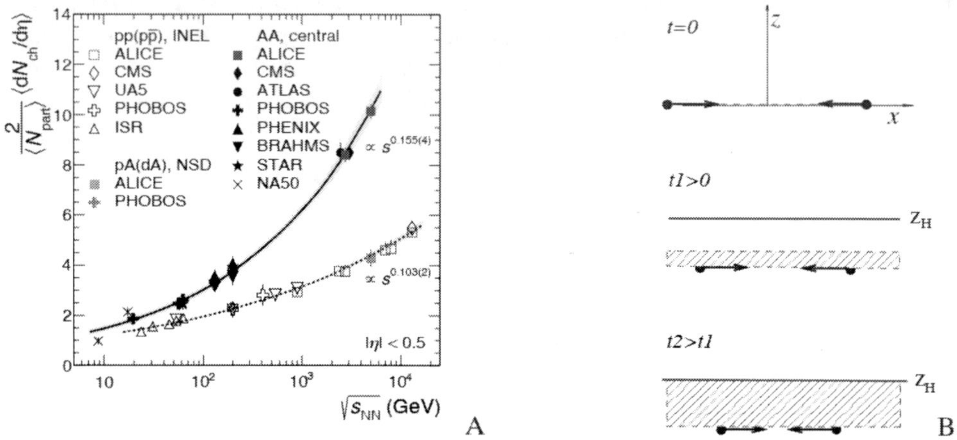

Figure 2. A. The plot of the total number of charged particles versus energy $\sqrt{s}$ for protons and $\sqrt{s_{NN}}$ per pair of nucleons for ions collisions. Plot from ALICE collaboration [13]. B. The schematic cartoon of the black hole formation in the bulk after collision of particles on the boundary.

In [26] it has been shown that the model that reproduces the Cornell potential also gives a correct energy dependence of multiplicities if we assume that the multiplicity is related with the dual entropy produced during a limited time period. However in this consideration there is a limitation on the possible energy of colliding shock walls[26]. Since in this consideration we have used a more or less general isotropic background reproducing AdS at UV and confinement at IR, we can think that the assumption about the isotropic background prevent to reproduce (3) at hight energy.

In [30] we have considered a special *anisotropic* backgrounds, parametrized by a paranmeter ν and have got

$$\mathcal{M}_\nu \sim s^{1/(\nu+2)}, \tag{4}$$

and therefore to get an estimation (3) we take $\nu = 4.45$. Note, that $\nu = 4$ gives $\mathcal{M}_{\nu=4} \sim s^{0.167}$, that is rather closed to (3).. $\nu = 1$ corresponds to the AdS case.

Let us note, that the idea to use anisotropic background in the holographic approach to QGP is not new. There were several proposals to deal with anisotropic backgrounds, see for example [35] and refs therein. Up to couple years ego, it has been believed that the matter was in the pre-equilibrium period up to 1 fm/c and then the QGP appears and it is isotropic. However now there is a believe that the QGP is created after a very short time after collision, $\tau_{therm} \sim 0.1 fm/c$, and it is anisotropic ("anisotropic" means a spatially anisotropy) for a short time τ after collision, $0 < \tau_{therm} < \tau < \tau_{iso}$, and the time of locally isotropization is about $\tau_{iso} \sim 2fm/c$ [36].

2 Holographic Thermalization

2.1 Thermalization

Suppose we deal with a correlator of a time depending quantity $O(t)$, $< O(t) >$. If at large t the system exhibits the behaviour

$$< O(t) > \underset{t\to\infty}{\to} << O >>, \tag{5}$$

where

$$<< O >>= \mathrm{tr}\left(e^{-\beta H} O\right) \tag{6}$$

one says that the system goes to the stationary state and this state is a thermal state, or in other words the system thermalizes. The simples thermalization process can be described by interaction with thermostat. For one oscillator interacting with thermostat one can show the thermalization explicitly.

2.2 Thermalalization as a black hole creation in AdS

According the holographic scenario the thermalization process in d-dimensional space-time can be understood as horizon formation in a d+1 dimensional gravitational theory. To initiate the process of BH formation one has to perturb the initial AdS metric. According to AdS/CFT correspondence this deformation should be related with the deformation of the energy-momentum tensor of the boundary theory. So the idea is to make some perturbation of AdS metric that near the boundary mimics the heavy ions collisions and see what happens, see Fig.2.B.

One can consider several deformations of AdS metric:

- a deformation by adding colliding gravitational shock waves.
- we can drop of a shell of matter with vanishing rest mass ("null dust"),
- we can study the toy 3-dim model with colliding ultra relativistic particles

It seems natural to think about an ultrarelativistic nucleus as a shock wave in 4-dimensional space-time. One can also assume that this shock has a profile function. According AdS CFT correspondence to this energy-momentum tensor corresponds the metric of a shock wave in AdS_5 of the following form

$$ds^2 = L^2 \frac{-du\,dv + dx_\perp^2 + \phi(x_\perp, z, L)\delta(u)\,du^2 + dz^2}{z^2}, \tag{7}$$

here u, v are light-cone variables, $u = x_0 + x_\parallel$, $v = x_0 - x_\parallel$, $\phi(x_\perp, z, L)$ is the profile function, L is the scale parameter. For $\phi(x_\perp, z, L) = 0$ the metric (7) is nothing but the Poincare metric for AdS_5. For two colliding nuclei corresponds the following metric in AdS_5

$$ds^2 = L^2 \frac{-du\,dv + dx_\perp^2 + \phi_1(x_\perp, z, L)\delta(u)\,du^2 + \phi_2(x_\perp, z, L)\delta(v)\,dv^2 + dz^2}{z^2}. \tag{8}$$

It is interesting to note that gaussian regularization of the point shock wave profile function $\phi(x_\perp, 0)$ exactly corresponds to the Woods-Saxon profile for the nuclear density. This correspondence give us L equal to 4.3 fermi for gold and L equal to 4.4 fermi for lead.

It is natural to ask the question: can we guaranty the black hole formation under collisions of these shock waves. If the answer is "yes", then I'll take the entropy of this formed black hole as a multiplicity of particles production. This idea has been explored in several papers [17–24]. The idea to relate multiplicities with the thermodynamic characteristic of the media produced as a result of a collision of particles comes back to Landau and Fermi [27–29].

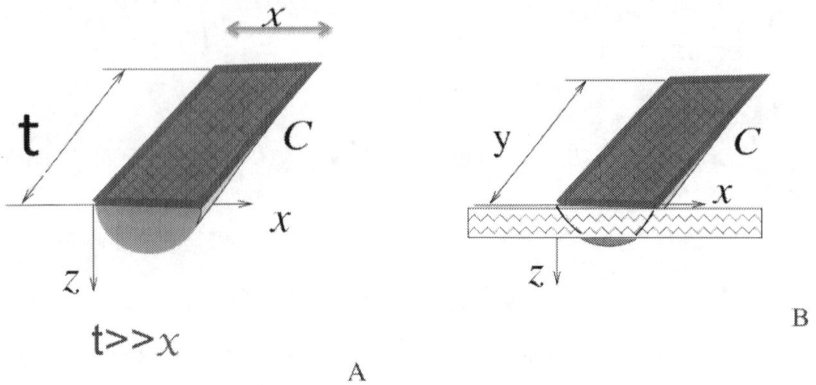

Figure 3. A. Wilson loop in the static background. B. Wilson loop in the Vaidya background

2.3 Thermalization time

Thermalization time depends on the physical quantities that expect to be thermalized. This issue has been discuss in details in our talk [37] and is not discussed here, see also [38–40] for estimations of thermalization time of different variables in the Lifshitz-type background.

3 Physics in the Lifshitz-type background

3.1 Lifshitz-type background

In [30] we have studied collisions of shock waves in the anisotropic background

$$ds^2 = 2\pi\alpha'\left(\frac{-dt^2 + dx^2}{z^2} + \frac{dy_1^2 + dy_2^2}{z^{2/\nu}} + \frac{dz^2}{z^2}\right), \tag{9}$$

where ν is the critical exponent.

After collision the black hole is produced and the stationary black hole metric in this background is [31]

$$ds^2 = z^{-2}\left(-f(z)dt^2 + dx^2\right) + z^{-2/\nu}(dy_1^2 + dy_2^2) + \frac{dz^2}{z^2 f(z)}, \tag{10}$$

where the blackening function is

$$f = 1 - mz^{2/\nu+2}. \tag{11}$$

Motivated by [32, 33] we deform this background as [1]

$$ds^2 = \frac{b(z)}{z^2}\left(\left(-f(z)dt^2 + dx^2\right) + z^{2-2/\nu}(dy_1^2 + dy_2^2) + \frac{dz^2}{f(z)}\right), \tag{12}$$

where the blackening function takes into account a non-zero chemical potential

$$f \;=\; 1 - \left(\frac{1}{z_h^{2/\nu+2}} + q^2 z_h^2\right) z^{2/\nu+2} + q^2 z^{2/\nu+4} \tag{13}$$

[1]The problem of finding of shock waves in the background (12) and estimation of the trapped surface in this background has not solved yet, compare with [22, 34]

and $b(z)$ is a factor that provides the quark confinement in the isotropic case [32]

$$b(z) = \exp(\frac{cz^2}{2}) \tag{14}$$

and the zero-component of the vector field is given by

$$A_0(z) = i\left(\mu - \frac{4\sqrt{3}\,q}{c}\left(1 - e^{-cz^2/4}\right)\right). \tag{15}$$

According the AdS/CFT dictionary, the boundary value $A_0(0)$ is related to the quark chemical potential:

$$A_0(0) = i\mu \tag{16}$$

The vanishing of $A_0(z)$ at the horizon, $A_0(z_h) = 0$, sets a relation between the chemical potential and the charge of the black hole:

$$\mu = \frac{\sqrt{3}\,q}{c}\left(1 - e^{-cz_h^2/4}\right) \tag{17}$$

The temperature T is defined by the relation

$$T = \frac{1}{4\pi}\left|\frac{df}{dz}\right|_{z=z_h} = \frac{1}{2\pi}\frac{\frac{1}{\nu}+1}{z_h}\left(1 - \frac{Q_\nu^2}{2}\right), \tag{18}$$

where $Q_\nu = \sqrt{\frac{2}{\frac{1}{\nu}+1}}\,q z_h^{\frac{1}{\nu}+2}$. We can also present the dependence of T on the chemical potential

$$T = \frac{\left(\frac{1}{\nu}+1\right)}{2\pi z_h}\left(1 - \frac{c^2\mu^2 z_h^{\frac{2}{\nu}+4}}{3\left(\frac{1}{\nu}+1\right)\left(1 - e^{-\frac{cz_h^2}{2}}\right)^2}\right) \tag{19}$$

see Fig.4.

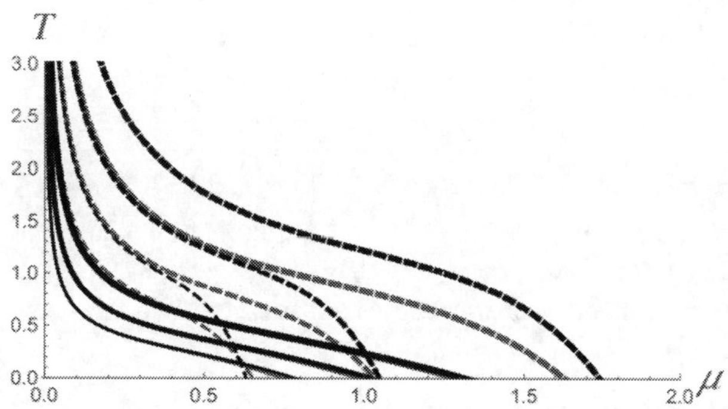

Figure 4. Relation between T and μ given by formula (18) and (17) for isotropic (solid lines) and anisotropic case with parameter $\nu = 4$ (dashed lines) for different values of the charge q: $q = 2$ (solid/dashed thick line), $q = 1$ (solid/dashed line) and $q = 0.5$ (thin solid/dashed line) and different confinement factors: $c = 0.2$ (blue lines) and $c = 0.3$ (green lines)

3.2 Temporal Wilson loop in the charged quark confinement background

3.2.1 Energy between quarks located along x-direction in the quark confinement background (12)

The potential of the interquark interaction along x-direction can be extracted from the rectangular time-like Wilson loop of size $T \times X$, i.e. the loop in which one side is infinite along the time direction, and the other is along the x direction

$$W(T, X) = \langle \mathrm{Tr}_F \, e^{i \oint_{T\times X} dx_\mu A_\mu} \rangle \sim e^{-V(X)T}, \tag{20}$$

F means the fundamental representation. Following the holographic approach [41–43] the expectation value of the Wilson loop in the fundamental representation calculated on the gravity side reads as:

$$W[C] = e^{-S_{(xt)}}, \tag{21}$$

where $C = T \times X$ in a contour on the boundary, S_{xt} is the minimal Nambu-Goto action of the string hanging from the contour C in the bulk

$$S_{xt} = \frac{1}{2\pi\alpha'} \int d\sigma^1 d\sigma^2 \sqrt{-\det(h_{\alpha\beta})}, \tag{22}$$

where $h_{\alpha\beta}$ is the induced metric of the world-sheet $h_{\alpha\beta} = g_{MN}\partial_\alpha X^M \partial_\beta X^N$, $\alpha, \beta = 0, 1$, g_{MN} is the background metric, $M, N = 0, \ldots, 4$, $X^M = X^M(\sigma^0, \sigma^1)$ specify the string worldsheet with the worldsheet coordinates σ^1, σ^2. We parametrize the worldsheet as $X^0 \equiv t = \sigma^0$ and $X^1 \equiv x = \sigma^1$ and consider the static configuration $X^4 \equiv z = z(x)$. The action $S_{(xt)}$ is:

$$S_{xt} = \frac{T}{2\pi\alpha'} \int \frac{b(z)}{z^2} \sqrt{f(z) + z'^2}\, dx. \tag{23}$$

where $'$ means $\frac{d}{dx}$. If we take $z' = 0$ in (23) we get the "potential"

$$V_x(z) = \frac{b(z)}{z^2} \sqrt{f(z)} \tag{24}$$

We consider the symmetrical parametrization $z(\pm\ell) = 0$ and $z(0) = z_*$, $z'(0) = 0$. The distance between the two endpoints of the string can be represented as

$$L_x = 2 \int_\infty^{z_*} \frac{dz}{z'} = 2 \int_0^{z_*} \frac{dz}{\sqrt{f(z)} \sqrt{\left(\frac{V_x^2(z)}{V_x^2(z_*)} - 1\right)}} \tag{25}$$

We consider the case when $z_* < z_{h_{i0}}$, where $z_{h_{i0}}$ is the smallest of the horizons, i.e. if $f(z_{h_i}) = 0$, then $z_{h_{i0}} < z_{h_i}$. Therefore $f(z) > 0$ for $0 < z < z_*$.

We get the energy of the string we subtract the mass of the two free quark [35, 42, 43]

$$\pi\alpha' E_x = \int_0^{z_*} \frac{dz}{z^2} \left[\frac{b(z)V(z)}{\sqrt{V^2(z) - V^2(z_*)}} - 1 \right] - \frac{1}{z_*} + m^{\frac{\nu}{2\nu+2}} e^{c\, m^{-\frac{2\nu}{2\nu+2}}} - \sqrt{\pi c}\, \mathrm{erfi}\left(\sqrt{c}\, m^{-\frac{\nu}{2\nu+2}}\right). \tag{26}$$

To guaranty that the expression under the second square root in (25) is positivity we have to integrate along the curve where the potential $V_x(z)$ is a decreasing function. Depending on the parameters c, z_h, ν and q the function $V_x(z)$ can have two extremal points, one minimum and one maximum, or have no extremal point at all at the interval $0 < z < z_{h_{i0}}$, see Fig. 5 and Fig. 6.

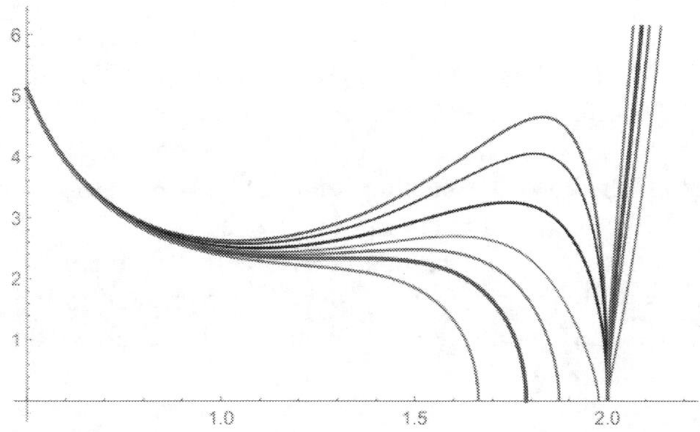

Figure 5. The potential $V(z) = V(z, c, q, \nu, z_h)$ given by (24) with $b(z)$ given by (14) with $c = 2$ and $f(z) = f(z, q, \nu, z_h)$ given by (13) with $\nu = 1$ and $z_h = 2$. The red line shows $V(z)$ for $q = 0$. We see that the corresponding potential has minimum and maximum. The blue lines show $V(z)$ for $q = 0.1$ and $q = 0.15$. The corresponding potential have minimum and maximum. The green lines show $V(z)$ for $q = 0.18$ and $q = 0.195$. The corresponding potential have minimum and maximum and the second zero of the corresponding function $f(z) = f(z, q, 1, z_h)$ are on the left of $z_h = 2$. The magenta line show $V(z)$ for $q = 0.21$. The corresponding potential has no local extremum and the second zero of $f(z)$ is at $z_{h_2} = 1.784$. For $q = 0.21$ (the brown line) the corresponding potential has no minimum and maximum on $0 < z < z_{h_2}$ and the potential is the decreasing function at $z < z_{h_2} = 1.662$

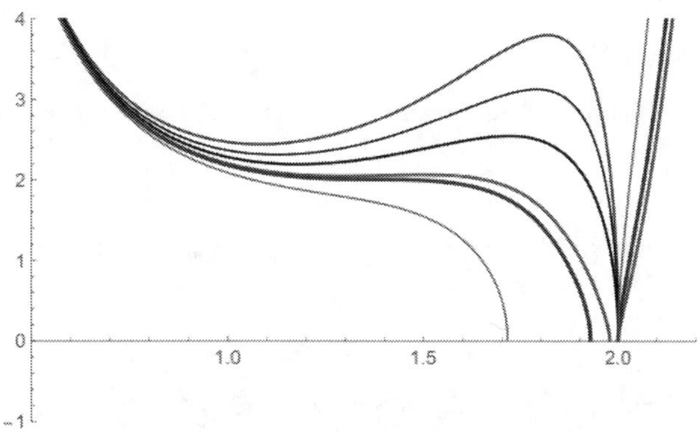

Figure 6. The potential $V_x(z) = V(z, c, q, \nu, z_h)$ given by (24) with $b(z)$ given by (14) with $c = 2$ and $f(z) = f(z, q, \nu, z_h)$ given by (13) with $\nu = 4$ and $z_h = 2$. The red line shows $V(z)$ for $q = 0$. We see that the corresponding potential has minimum and maximum. The blue lines show $V(z)$ for $q = 0.15, 0.2$. The corresponding potential have minimum and maximum. The green line shows $V(z)$ for $q = 0.238$. The second zeros of the corresponding function $f(z)$ at $z = 1.97$. The magenta line shows $V(z)$ for $q = 0.245$. The corresponding potential has no minimum and maximum and the second zero of the corresponding function $f(z) = f(z, q, 1, z_h)$ is at $z_{h_2} = 1.893$. The brown line shows $V(z)$ for $q = 0.28$ and the corresponding potential has no minimum and maximum on $0 < z < z_{h_2}$, where $z_{h_2} = 1.71$ is the second zero of the corresponding functions $f(z)$. The corresponding potential (the brown lines) are the decreasing functions. The critical charge $q_{cr}|_{c=2, z_h=2} = 0.245$

DOI: 10.1051/epjconf/201716401014

Therefore, as for isotropic cases [44–50] [2], we have two cases:

- There is no extremal point in the interval $0 < z < z_{h_{i_0}}$. This case corresponds to the deconfinement phase and dependence of ℓ on z_* has the form as presented in Fig.7.a). We see that there are two branches indicated by blue and red color, the same ℓ can be obtained by two different z_*, except $z_* = z_{*0}$ where ℓ reaches its maximal value ℓ_0, i.e. $\ell = \ell(z_*)$ monotonically increases from $\ell(0) = 0$ to $\ell(z_{*_0}) = \ell_0$ (the first branch shown by the blue color) and $\ell = \ell(z_*)$ monotonically decreases from $\ell(z_{*_0}) = \ell_0$ to $\ell(z_{h_0}) = 0$ (the second branch). In Fig.7.a) $z_{h_0} = z_h$. The plot in Fig.7.b) shows the values of the energy between quarks located along x-direction as function of z_* for two branches. The corresponding values of the energy as functions ℓ are presented in Fig.7.c). As shown in this plot, the values of energy corresponding to the second branch are large than those for the first one.

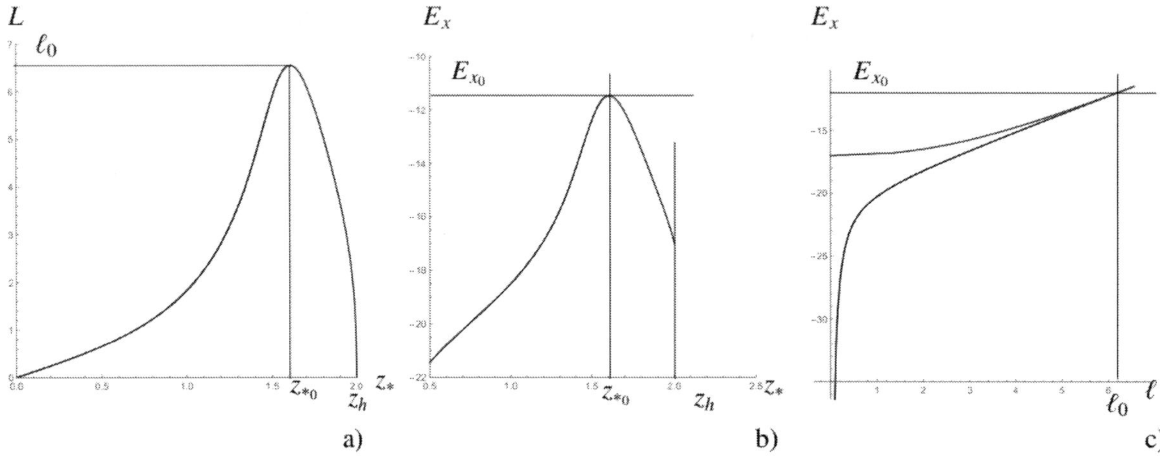

Figure 7. Above the critical point (deconfinment): a) $L = L(z_*)$, b) $E_x = E_x(z_*)$ and c)$E_x = E_x(\ell)$. Here we put $\alpha' = 1/\pi$

- There are two extremal points in the interval $0 < z < z_{h_{i_0}}$, $z_{min} = z_0$ and $z_{max} = z_0'$ and the potential is decreasing function only on the intervals

$$0 < z < z_{min} \quad \text{and} \quad z_{max} < z < z_{h_{i_0}} \tag{27}$$

we can guarantee that $V(z)/V(z_*) < 1$ only in the region (27).

This case corresponds to the confinement phase and the dependence of L on z_* is presented in Fig.8.a). Suppose that $V(z)$ has a local minimum at $z = z_0$. In this case the potential $V_x(z)$ is a decreasing function from $z = 0$ up to $z = z_0$. The first branch $L = L(z)$ is shown by the blue line in the plot in Fig.8.a). and $L(z) \to \infty$ when $z \to z_0 - 0$. For $z_0 < z < z_0'$ gets $L = L(z)$ an imaginary part and at $z = z_0'$ starts the second branch. For the second branch $E_2 < E(\ell) < E_2'$.

In Fig.9 we present the same quantities as inFig.7 and Fig.8 at the same parameters except that in Fig.7 and Fig.8 $\nu = 1$ (isotropic case) and in Fig.9 $\nu = 4$ (anisotropic case, longitudinal orientation of the Wilson loop). The confinement solution is realized on the first branch discussed above. Let us estimate the string tension for this configuration. Suppose that

$$V'(z)|_{z=z_0} = 0, \quad V''(z_0) > 0 \tag{28}$$

[2]Note, that the isotropic case with the confinement factor has been studied at zero chemical potential in [44, 45] and with nonzero chemical potential in [46–50]

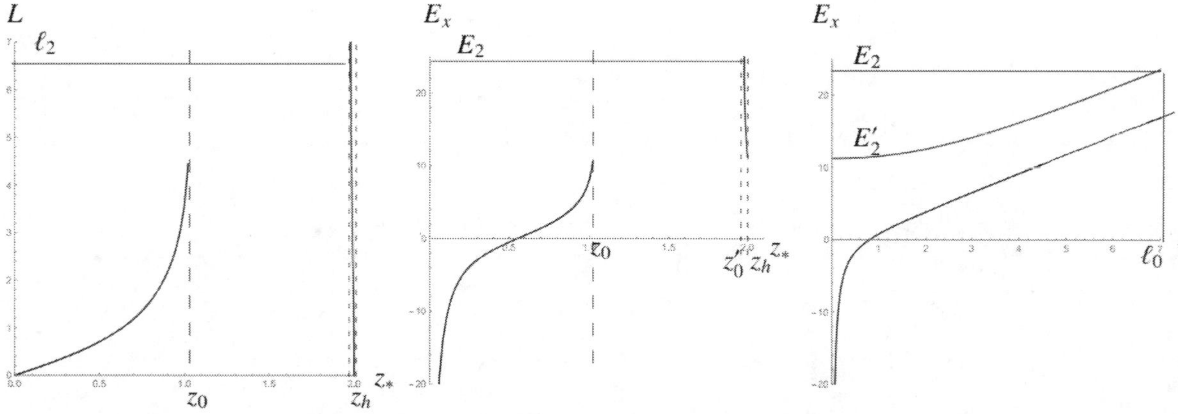

a) b) c)

Figure 8. Below the critical point (confinment): a) $L = L(z_*)$, b) $E_x = E_x(z_*)$ (not that in this case we make a different subtraction) and c) $E_x = E_x(\ell)$

Near $z = z_0$ we have $V''(z_0) > 0$ and

$$\sqrt{\frac{V(z)}{V(z_0)}^2 - 1} = \sqrt{\frac{V''(z_0)}{V(z_0)}}(z - z_0) + O(z - z_0)^2 \tag{29}$$

$$L_1 \underset{z_*\sim z_0}{\sim} -2\sqrt{\frac{V(z_0)}{f(z_0)V''(z_0)}}\ln(z_0 - z_*) \tag{30}$$

We can also calculate the energy near $z_* = z_0$ using (26). We have

$$S_{xt} \underset{z_*\sim z_0}{\sim} -\frac{T}{2\pi\alpha'}\frac{b(z_0)}{z_0^2}\frac{1}{\sqrt{\frac{V''(z_0)}{V(z_0)}}}\ln(z_0 - z_*) \tag{31}$$

Therefore we get

$$S_{xt} \underset{L\to\infty}{\sim} \frac{T}{\pi\alpha'}\frac{L_1}{2}\frac{b(z_0)\sqrt{f(z_0)}}{z_0^2}$$

and

$$\sigma_x = \frac{V_x(z_0)}{2\pi\alpha'} \tag{32}$$

3.2.2 Energy between quarks located along y-direction in the quark confinement background (12)

The Nambu-Goto action S_{yt} of the string hanging from the contour C, being the rectangular with sites along the time direction and one of transversal directions, in the bulk is

$$S_{yt} = \frac{Ti}{2\pi\alpha'}\int\frac{b(z)}{z^2}\sqrt{z^{2-2/\nu}f(z) + z'^2}\,dx. \tag{33}$$

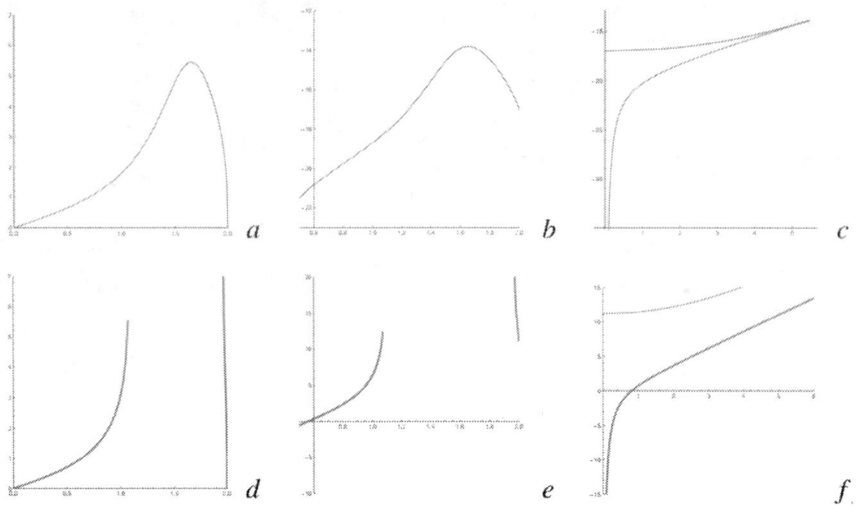

Figure 9. The same quantities as in Fig.8 and 8 at the same values of parameters, except that now we deal with the anisotropic case, $\nu = 4$. a),b) and c) above the critical point (deconfinment): a) $L = L(z_*)$, b) $W = W(z_*)$ and c) $W = W(L)$. d), e) and f) above the critical point (deconfinment): d) $L = L(z_*)$, e) $W = W(z_*)$ (note that in this case we make a different subtraction) and f) $W = W(L)$.

The corresponding "potential" is, see Fig.10,

$$V_y(z) = \frac{b(z)\sqrt{f(z)}}{z^{1/\nu+1}} \tag{34}$$

and for L_y we have

$$L_y = 2\int_0^{z_*} \frac{z^{1/\nu-1}\,dz}{\sqrt{f(z)}\sqrt{\frac{V^2(z)}{V^2(z_*)}-1}}. \tag{35}$$

To get for the energy of the string between two quarks stretched in the y-direction we perform the subtraction of the two quark mass that corresponds the subtraction of the action on the sheet starting at the boundary $z = 0$ and ending at the horizon

$$\pi\alpha' E_y = \int_0^{z_*} \frac{dz}{z^2}\left[\frac{b(z)V(z)}{\sqrt{V^2(z)-V^2(z_*)}} - 1\right] - \frac{1}{z_*} + m^{\frac{\nu}{2\nu+2}} e^{c\,m^{-\frac{2\nu}{2\nu+2}}} - \sqrt{\pi c}\,\mathrm{erfi}\left(\sqrt{c}\,m^{-\frac{\nu}{2\nu+2}}\right). \tag{36}$$

Now, suppose that $V_y(z)$ has a minimum at $z = z_0$, we can estimate the contributions near this minimum, we have

$$\pi\alpha' E_y \underset{z_*\sim z_0}{\sim} \frac{\ln(z_0 - z_*)}{z_0^2}\frac{b(z_0)}{\sqrt{\frac{V''(z_0)}{V(z_0)}}} \tag{37}$$

$$L_y \underset{z_*\sim z_0}{\sim} \frac{z_0^{1/\nu-1}\ln(z_0 - z_*)}{\sqrt{f(z_0)}}\frac{1}{\sqrt{\frac{V''(z_0)}{V(z_0)}}} \tag{38}$$

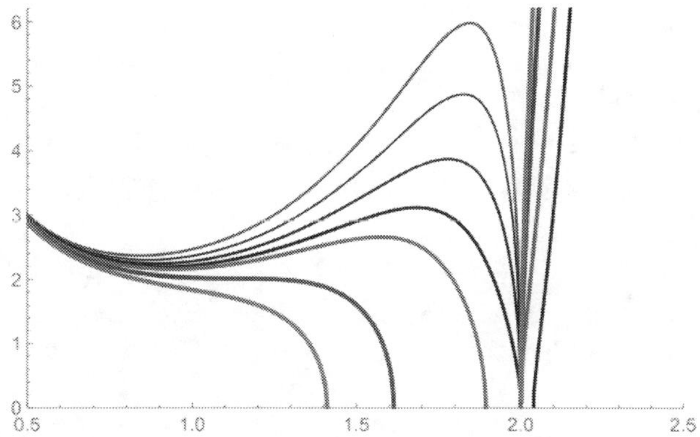

Figure 10. The potential $V_y(z) = V(z, c, q, \nu, z_h)$ given by (34) with $b(z)$ given by (14) with $c = 2$ and $f(z) = f(z, q, \nu, z_h)$ given by (13) with $\nu = 4$ and $z_h = 2$. The red line shows $V_y(z)$ for $q = 0$. We see that the corresponding potential has minimum and maximum. The blue lines show $V_y(z)$ for $q = 0.15, 0.2$ and $q = 0.23$. The corresponding potentials have minimum and maximum. The green line shows $V_y(z)$ for $q = 0.25$ and second zeros of the corresponding function $f(z)$ is at $z_{h_2} = 1.89$. The magenta line show $V(z)$ for $q = 0.3$. The corresponding potential has no minimum and maximum and the second zero of the corresponding function $f(z) = f(z, q, 1, z_h)$ is at $z_{h_2} = 1.61$. The brown line shows $V_y(z)$ for $q = 0.35$ and the corresponding potential have no minimum and maximum on $0 < z < z_{h_2}$, where $z_{h_2} = 1.40$ is the second zero of the corresponding functions $f(z)$. The critical charge $q_{cr}|_{c=2,z_h=2} = 0.3$

Therefore,

$$\frac{\pi\alpha' S_{yt}}{T} = L_y V(z_0) \tag{39}$$

and

$$\sigma_y = \frac{V_y(z_0)}{2\pi\alpha'}, \tag{40}$$

i.e. we get the answer similar to (32).

In Fig.11 we present the same quantities as inFig.7 and Fig.8 at the same parameters except that in Fig.7 and Fig.8 $\nu = 1$ (isotropic case) and in Fig.11 $\nu = 4$ (anisotropic case, transversal orientation of the Wilson loop). We see that for the transversal orientation at the same parameters as for the isotropic case we can get different solutions: for the isotropic case, as well as for the anisotropic longitudinal case, $c = 1.2$ corresponds to the deconfined phase, meanwhile this value of c in the anisotropic transversal case case corresponds to the confined phase. This observation indicates that the phase diagram should be essential depends on the orientation in the case of the anisotropic background.

3.3 Holographic anisotropic QCD phase diagrams

In Fig.12 we present the phase diagram for isotropic and anisotropic E_x and E_y cases. Holographic isotropic QCD phase diagram has been studied previously in bottom-up approaches [46–48]. We see that for zero chemical potential the deconfinement occurs for the low temperature in the anisotropic case, and for zero temperature the deconfinement phase survives for the large chemical potential in the anisotropic case as well.

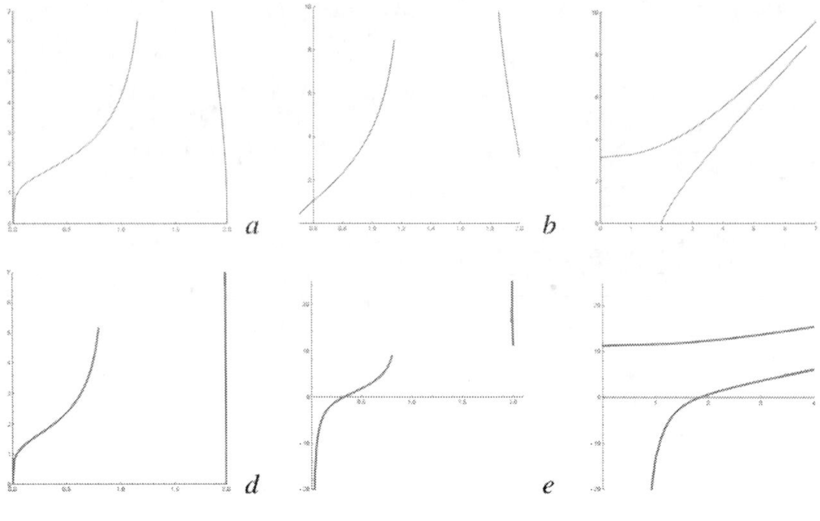

Figure 11. The same quantities at the same parameters as in Fig.9 except that now we deal with the transversal orientation of the Wilson loop, $\nu = 4$. We see that both values of c, $c = 1.2$ and $c = 2$, correspond to the confinement phase: a) and d) $L = L(z_s)$, b) and e) $W = W(z_s)$ (note that in this case we make a different subtraction), c) and f) $W = W(L)$.

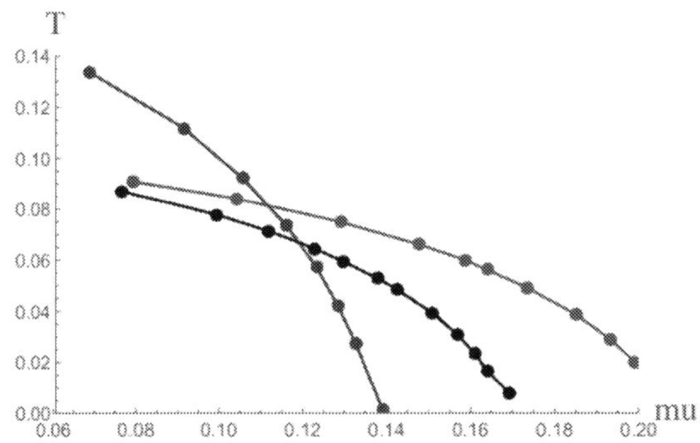

Figure 12. The deconfinement transition line in the (μ, T) plane. Phase transitions lines dividing the plane in two regions, a hadron phase near the origin, and a deconfined phase beyond the curve. The red line(points) corresponds to the isotropic case, the blue line (points) corresponds to the E_x in the anisotropic case with $\nu = 4$, the green to E_y in the same anisotropic case. The dots represent numerical data points, while the solid interpolating lines are included "by hands"

Fig.12 shows the dependence of the transition line on the orientation. Since quarks can be arbitrary oriented in respect to the collision line, this dependence on the orientation leads to a broadening of the line separating confinement and deconfinement phases in the (μ, T)-plane. Decreasing of anisotropy decreases the broadening of the phase transition boundary.

Let us remind that in isotropic case the confinement/deconfinment diagram has been studied in lattice QCD. Experimentally the phase boundary between hadronic matter and the quark - gluon plasma in relativistic heavy ion collisions is probing using the HRM [51, 52]

4 Direct photons and electric conductivity

The thermal-photon production from the QGP plays an essential role, since photons after they are produced in HIC almost do not interact with the QGP and, therefore, they give us the local information in heavy ion collisions. The photon-emission rate is related to the retarded correlator of currents in momentum space [53]

$$G^R_{\mu\nu}(k) = i\int d^4(x-y)e^{ik\cdot(x-y)}\theta(x^0)\langle[J^a_\mu(x), J^b_\nu(0)]\rangle, \tag{41}$$

so that in the thermal equilibrium it is given by the light-like retarded correlator as

$$d\Gamma = -\frac{d^3k}{(2\pi)^3}\frac{e^2 n_b(|\mathbf{k}|)}{|\mathbf{k}|}\mathrm{Im}\left[\mathrm{tr}\left(\eta^{\mu\nu}G^{ab\,R}_{\mu\nu}\right)\right]_{k^0=|\mathbf{k}|}, \tag{42}$$

where Γ denotes the number of photons emitted per unit time per unit volume and $n_b(|\mathbf{k}|)$ denotes the thermal distribution function for bosons.

The imaginary part of the retarded correlator is also related with the spectral function

$$\chi_{\mu\nu}(k) = -2\Im\, G^R_{\mu\nu}(k) \tag{43}$$

and due to the Kubo formula with the conductivity tensor

$$\sigma_{\mu\nu}(k) = -\frac{G^R_{\mu\nu}(k)}{iw} \tag{44}$$

The spectral function can be evaluated by holography using the flow method related with the membrane paradigm [54] . Below we shortly remind this prescription, see also [55, 56]. One starts from the Maxwell action in the given background

$$S_M = -\mathcal{N}\int d^5x\sqrt{-g}\frac{V_{dil}(\phi)}{4}F^{MN}F_{MN}, \tag{45}$$

where it is assumed also the presence of the dilaton field ϕ, and denote by $\mathcal{N}$ the normalization constant including the 5-dimensional gravitational constant G_5 etc.

The gauge field $A_\mu(t,\mathbf{x},r)$ it is taken to be

$$A_\mu(t,\mathbf{x},z) = \int\frac{d^4k}{(2\pi)^4}e^{-i\omega t+i\mathbf{kx}}\mathcal{A}_\mu(z,\omega,k), \quad \mathcal{A}_\mu(z,\omega,k) = A_\mu(z,\omega,k)a_\mu(\omega,k) \tag{46}$$

$A_\mu(t,\mathbf{x},z)$ satisfies the Maxwell equation, and $A_\mu(r,\omega,k)$ satisfies the boundary conditions at the boundary

$$\lim_{z\to 0}A_\mu(z,\omega,k) = 1 \tag{47}$$

and the infalling boundary conditions at the horizon,

$$\lim_{z\to z_h}A_\mu(z,\omega,k) = 0 \tag{48}$$

One can consider for simplicity the case $k = k_x$, $k_{y_1} = k_{y_2} = 0$.

Let us parametrize the metric as

$$ds^2 = \frac{b^2(z)}{z^2}\left(-f(z)dt^2 + dx^2 + p(z)(dy_1^2 + dy_2^2) + \frac{dz^2}{f(z)}\right), \tag{49}$$

In our model $p(z) = z^{2-2/\nu}$. The Maxwell action (with the dilaton) has the form

$$\int d^4xdzV(z)\left(-\frac{F_{01}^2+F_{02}^2}{fp} - \frac{F_{03}^2}{f} + f\left(\frac{F_{1r}^2+F_{2r}^2}{p} + F_{3r}^2\right) - F_{0r}^2 + \frac{F_{12}^2}{p^2} + \frac{F_{13}^2+F_{23}^2}{p}\right) \tag{50}$$

here

$$V(z) = V_{dil}(\phi(z))\frac{b(z)}{z} \tag{51}$$

and

$$F_{MN} = \partial_M A_N - \partial_N A_M \tag{52}$$

Assume that the fluctuating vector field depends only on x_0, x_3, r-coordinates, one can write the E.O.M. in term of E_L and $E_\perp$

$$E_L'' + \left(\frac{f'}{f}\frac{w^2}{w^2-fk^2} + \frac{V'}{V}\right)E_L' + \frac{w^2-fk^2}{f^2}E_L = 0 \tag{53}$$

$$E_{\perp,i}'' + \left(-\frac{p'}{p} + \frac{V'}{V} + \frac{f'}{f}\right)E_{\perp,i}' + \frac{w^2-k^2}{f}E_{\perp,i} = 0, \quad i = 1,2 \tag{54}$$

here $E_L = kA_0 + wA_3$ and $E_{\perp,i} = wA_{\perp,i}$, $i = 1,2$.

The boundary terms come from that terms the total action (50) which have the z-derivative and we get

$$S_{boundary} = S_{boundary,1} + S_{boundary,2} \tag{55}$$

where $S_{boundary,1}$ and $S_{boundary,2}$ are

$$S_{boundary,1} = \int d^4x\frac{f}{w^2-k^2f}E_LE_L' \tag{56}$$

$$S_{boundary,2} = \int d^4xV\frac{f}{pw^2}E_\perp E_\perp', \quad E_\perp E_\perp' \equiv \sum_{i=1,2} E_{\perp,i}E_{\perp,i}' \tag{57}$$

In the analogy with the isotropic case we introduce

$$\zeta_\perp = -\frac{Vf}{pw}\frac{\partial_r E_i^\perp}{E_i^\perp}, \tag{58}$$

$$\zeta_L = -\frac{Vf}{\omega}\frac{\partial_r E_L}{E_L}, \tag{59}$$

which satisfy the following equations

$$\zeta_\perp' - \frac{w}{f}\left[\frac{p}{V}\zeta_\perp^2 + \frac{V}{p}\left(1 - f\frac{k^2}{w^2}\right)\right] = 0. \tag{60}$$

$$\zeta_L' + \frac{k^2f'}{w^2-fk^2}\zeta_L - \frac{w}{f}\left[\frac{\zeta_L^2}{V} + V\left(1 - f\frac{k^2}{w^2}\right)\right] = 0 \tag{61}$$

From these eqs follows that if $\zeta_\perp' \neq \infty$ and $\zeta_L' \neq \infty$ then on the horizon should be

$$\left.\frac{p}{V}\zeta_\perp^2 + \frac{V}{p}\right|_{z=z_h} = 0 \Rightarrow \zeta_\perp = i\frac{V(z_h)}{p} \tag{62}$$

$$\left.\frac{1}{V}\zeta_L^2 + V\right|_{z=z_h} = 0 \Rightarrow \zeta_L = iV(z_h) \tag{63}$$

Since for $w = 0$ the derivative $\zeta'_\perp = 0$ and we have

$$\zeta_\perp|_{z=0} = \zeta_\perp|_{z=z_h} = i\frac{V}{p}|_{z=z_h} \tag{64}$$

$1/z_h$ and it dependence on the temperature is read from (18) and approximately

$$\frac{1}{z_h} \approx \frac{2\pi\nu}{1+\nu}\frac{T}{1-(\frac{\nu}{\nu+1})^{\frac{2+3\nu}{\nu}}q^2\left(\frac{1}{2\pi T}\right)^{\frac{2+4\nu}{\nu}}} \tag{65}$$

In what follows we use this approximation. Therefore, the dependence of the electric conductivity on the temperature and the chemical potential is given by

$$|\zeta_\perp(T,\nu,q)| \approx \left(\frac{2\pi\nu}{1+\nu}\right)^{3-2/\nu}\frac{T^{3-2/\nu}}{\left(1-(\frac{\nu+1!!}{\nu!!})^{\frac{2+3\nu}{\nu}}q^2\left(\frac{1}{2\pi T}\right)^{\frac{2+4\nu}{\nu}}\right)^{3-2/\nu}} \tag{66}$$

Form this formula we see, that increasing anisotropy we increase the electric conductivity at hight temperatures, see Fig.13. We also see that there is a critical temperature $T_0 = T_0(\nu, q)$ such that for $T < T_0$ the conductivity decreases as we increase ν.

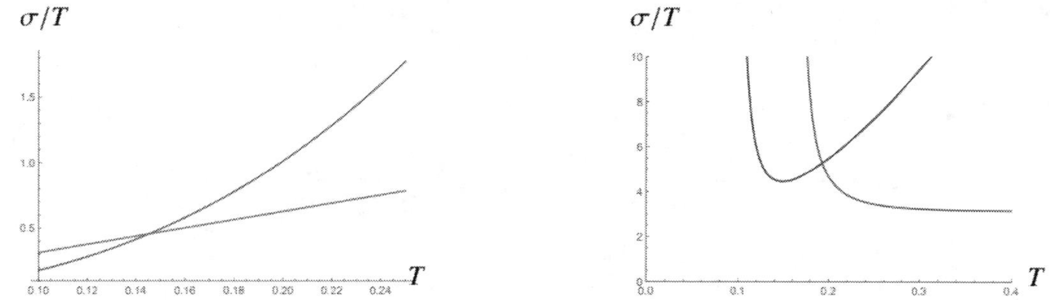

Figure 13. The electric conductivity for $\nu = 1$ and $\nu = 4$ and different values of q: a) $q = 0$, b) $q = 0.2$

Here we have to note that equation (66) is valued only for $T > T_{app}$, see Fig.14. In Fig.14 and in Fig.15 we plot the electric conductivity as function of T for $\nu = 1$ and $\nu = 4$ for different values of q and the constant c specifying the factor b. We see that the confining factor $b(z)$ does not change the qualitative picture too much.

Note, that in the end of this sector few comments are in order. Almost all theoretical predictions underestimate the direct-photon spectra. Several attempts [58, 59], including the effects from strong magnetic fields [60], have been undertaken to fit theoretical predictions to LHC experimental data. As it is stressed in [56] direct photons calculations [53, 56, 61, 62] from holography have to be complemented by the medium evolution and the photons production from other phases. This also concerns to our consideration. Here we have just presented the preliminary estimation of the role of the chemical potential, the Lifshitz type anisotropy and the confining factor on photon emission rate. To give predictions for direct photons that can be observed at NICA one has to perform the study similar to [56]. We have seen that at hight temperature the anisotropy of the Lifshitz type increases the photon production and the chemical potential also increases it, meanwhile for temperature less then the critical one, they act to different direction.

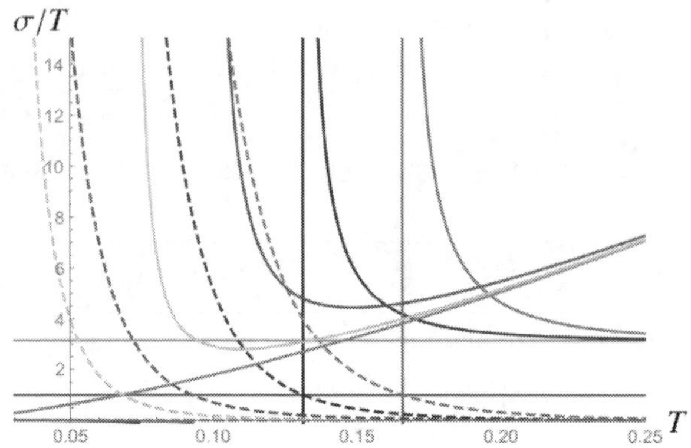

Figure 14. a) The electric conductivity for $\nu = 1$ and $\nu = 4$ and different values of q. The cases $\nu = 1$ and $q = 0, 0.1, 0.2$ are shown by gray, blue and green lines and the cases $\nu = 4$ and $q = 0, 0.1, 0.2$ are shown by brown, darker cyan and darker green. Dashed lines show validity of the approximation: only in the region where the dashed lines are below the red line we can use this approximation.

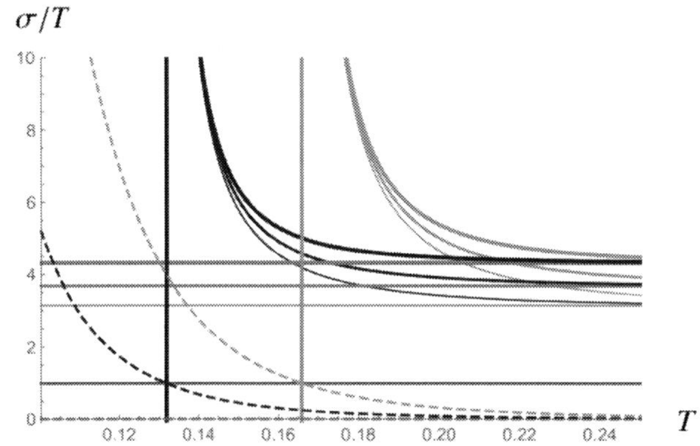

Figure 15. The electric conductivity for $\nu = 1$, different values of q and the constant c specifying the factor b. The gray solid lines show the electric conductivity for $\nu = 1$ and different values of $q = 0, 0.1, 0.2$ (thin, middle and thick lines). The blue and green solid lines show the electric conductivity for $\nu = 1$ and different values of $q = 0, 0.1, 0.2$ (thin, middle and thick lines) and $c = 1$ and $c = 1.2$, respectively. The solid lines are the same as in a) and the dashed lines show validity of the approximation: only in the regions where the dashed lines are below the red line we can use our approximation.

5 Conclusion

As a conclusion let me summarize. Motivated by the fact that to fit the experimental form of dependence of total multiplicity on energy we have to deal with a special anisotropic holographic model, related with the Lifshitz-like background, we have estimated the holographic confinement/deconfinement phase transition in the (μ, T) plane in this anizotropic background. We have

DOI: 10.1051/epjconf/201716401014

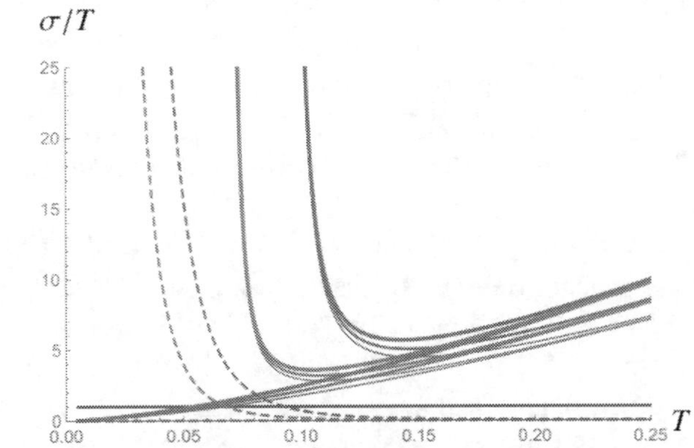

Figure 16. The electric conductivity for $\nu = 4$, different values of q and the constant c specifying the factor b. The brown solid lines show the electric conductivity for $\nu = 4$ and different values of $q = 0, 0.1, 0.2$ (thin, middle and thick lines). The darker cyan and green solid lines show the electric conductivity for $\nu = 4$ and different values of $q = 0, 0.1, 0.2$ (thin, middle and thick lines) and $c = 1$ and $c = 1.2$, respectively. The dashed lines show validity of the approximation: only in the regions where the dashed lines are below the red line we can use our approximation.

found the dependence of the transition line on the orientation of the quark pair. This dependence leads to a non-sharp character of physical confinement/deconfinement phase in the (μ, T)-plane. This calculation seems relevant in the context of the future project NICA and FAIR.

We are also going to estimate non-zero magnetic effects in the the Lifshitz-like background, in particular, by analogy with [63] one can estimate the crossover temperature in a magnetic field. As in isotropic case [64] it is interesting to find Debye screening mass near deconfinement in the anisotropic background.

Acknowledgments

I would like to thank the Organizers of ICNFP 2016 for the kind invitation and warm hospitality.

References

[1] J. Adams et al. (STAR Collab.) *Nucl. Phys. A* **757** 102 (2005)
[2] K. Aamodt et al. (ALICE Collab.) *Phys. Rev. Lett.* **105** 252301 (2010)
[3] E. Shuryak *Nucl. Phys. A* **750** 64 (2005)
[4] E.L. Feinberg, *Phys. Usp.* **168** 697 (1998)
[5] K. G. Wilson at al, Phys. Rev. D **49**, 6720 (1994)
[6] O. Aharony et al. *Phys. Rept.* **323** 183 (2000)
[7] V. Kekelidze et al, Nucl.Phys.A,956 (2016) 846
[8] A. Bazavov, arXiv:1505.05543 [hep-lat].
[9] U. Gursoy, E. Kiritsis, L. Mazzanti and F. Nitti, *JHEP* **05**, 033 (2009);
[10] O. DeWolfe, S. S. Gubser and C. Rosen, Phys. Rev. D **84**, 126014 (2011)

[11] J. Casalderrey-Solana, H. Liu, D. Mateos, K. Rajagopal, U. A. Wiedemann, *Gauge/String Duality, Hot QCD and Heavy Ion Collisions*, Cambridge University Press 2014
[12] G. Aad *et al.* [ATLAS Collaboration], Phys. Lett. B **710**, 363 (2012)
[13] J. Adam *et al.* [ALICE Collaboration], Phys. Rev. Lett. **116**, no. 22, 222302 (2016)
[14] I. Ya. Aref'eva, *Phys. Usp.* **57** (2014) 527.
[15] O. DeWolfe, S. S. Gubser, C. Rosen and D. Teaney, *Prog. Part. Nucl. Phys.* **75**, 86 (2014).
[16] W. van der Schee, arXiv:1407.1849.
[17] S. S. Gubser, S. S. Pufu and A. Yarom, Phys. Rev. D **78** (2008) 066014; JHEP **0911** (2009) 050
[18] L. Alvarez-Gaume et al., JHEP **0902** (2009) 009
[19] S. Lin and E. Shuryak, Phys. Rev. D **79** (2009) 124015
[20] J. L. Albacete, Y. V. Kovchegov and A. Taliotis, JHEP **0905** (2009) 060
[21] I. Y. Aref'eva, A. A. Bagrov and E. A. Guseva, JHEP **0912**, 009 (2009)
[22] I. Y. Aref'eva, A. A. Bagrov and L. V. Joukovskaya, JHEP **1003** , 002, (2010)
[23] Y. V. Kovchegov and S. Lin, JHEP **1003** (2010) 057; Y. V. Kovchegov, Prog. Theor. Phys. Suppl. **187** (2011) 96
[24] E. Kiritsis and A. Taliotis, JHEP **1204**, 065 (2012)
[25] I. Ya. Aref'eva, E. O. Pozdeeva and T. O. Pozdeeva, Teor. Mat. Fiz. **180**, 35 (2014).
[26] D. S. Ageev and I. Ya. Aref'eva, J. Exp. Theor. Phys. **120**, no. 3, 436 (2015)
[27] L.D. Landau, *Izvestya AN. Ser. phys.* **17** 51 (1953)
[28] Fermi E *Progr. Theoret. Phys.* **5** 570 (1950)
[29] I.Ya. Pomeranchuk, *DAN, SSSR* **78** 884 1951
[30] I. Ya. Aref'eva and A. A. Golubtsova, *JHEP* **1504** (2015) (011).
[31] I. Y. Aref'eva, A. A. Golubtsova and E. Gourgoulhon, JHEP **1609**, 142 (2016)
[32] O. Andreev and V. I. Zakharov, Phys. Rev. D **74**, 025023 (2006)
[33] A. Chamblin, R. Emparan, C. V. Johnson and R. C. Myers, Phys. Rev. D **60** 064018 (1999)
[34] I. Ya. Arefeva, A. A. Bagrov and E. O. Pozdeeva, JHEP **1205**, 117 (2012)
[35] D. Giataganas, *PoS Corfu* **2012**, 122 (2013)
[36] M. Strickland, *Pramana* **84** 671 (2015).
[37] I. Aref'eva, EPJ Web Conf. **125**, 01007 (2016).
[38] I. Ya. Aref'eva, *Theor. Math. Phys.*, 184:3 (2015), 1239-1255
[39] D. S. Ageev, I. Y. Aref'eva, A. A. Golubtsova and E. Gourgoulhon, arXiv:1606.03995.
[40] D. Ageev, EPJ Web Conf. **125**, 04007 (2016).
[41] J. M. Maldacena, Phys. Rev. Lett. **80**, 4859 (1998);
[42] S. J. Rey, S. Theisen and J. T. Yee, Nucl. Phys. B **527**, 171 (1998)
[43] A. Brandhuber, N. Itzhaki, J. Sonnenschein and S. Yankielowicz, Phys. Lett. B **434**, 36 (1998)
[44] O. Andreev and V. I. Zakharov, JHEP **0704**, 100 (2007);
[45] M. Mia, K. Dasgupta, C. Gale and S. Jeon, arXiv:1006.0055 [hep-th].
[46] P. Colangelo, F. Giannuzzi and S. Nicotri, Phys. Rev. D **83**, 035015 (2011).
[47] R. G. Cai, S. He and D. Li, JHEP **1203**, 033 (2012)
[48] Y. Yang and P. H. Yuan, JHEP **1512**, 161 (2015)
[49] C. Ewerz, O. Kaczmarek and A. Samberg, arXiv:1605.07181 [hep-th].
[50] Z. Fang, S. He and D. Li,
[51] P. Braun-Munzinger and J. Stachel, Nucl. Phys. A **606**, 320 (1996)
[52] A. Andronic, P. Braun-Munzinger, K. Redlich and J. Stachel, arXiv:1611.01347 [nucl-th].

[53] S. Caron-Huot, P. Kovtun, G. D. Moore, A. Starinets and L. G. Yaffe, JHEP **0612**, 015 (2006)
[54] N. Iqbal and H. Liu, Phys. Rev. D **79**, 025023 (2009)
[55] S. I. Finazzo and R. Rougemont, Phys. Rev. D **93**, no. 3, 034017 (2016)
[56] I. Iatrakis, E. Kiritsis, C. Shen and D. L. Yang, arXiv:1609.07208 [hep-ph].
[57] I. Iatrakis, E. Kiritsis, C. Shen and D. L. Yang, arXiv:1611.04848 [hep-ph].
[58] L. McLerran and B. Schenke, Nucl. Phys. A **946**, 158 (2016)
[59] J. F. Paquet, C. Shen, G. S. Denicol, M. Luzum, B. Schenke, S. Jeon and C. Gale, Phys. Rev. C **93**, no. 4, 044906 (2016)
[60] B. Muller, S. Y. Wu and D. L. Yang, Phys. Rev. D **89**, 026013 (2014)
[61] K. A. Mamo, JHEP **1308**, 083 (2013)
[62] S. Y. Wu and D. L. Yang, JHEP **1308**, 032 (2013)
[63] S. I. Finazzo and J. Noronha, Phys. Rev. D **90**, 115028 (2014)
[64] R. Rougemont, R. Critelli and J. Noronha, Phys. Rev. D **93**, 045013 (2016)

Initial conditions in chaotic inflation with modified gravity

Dmitry Gorbunov[1,a]

[1] *Institute for Nuclear Research of Russian Academy of Sciences, 117312 Moscow, Russia*

Abstract. Evolution in chaotic inflationary models begins at a high energy density with semi-classical initial conditions presumably consistent with a universal quantum nature of all the fundamental interactions: each quantum contributes the same amount to the energy density. There is an upper limit on this amount inherent in the approach, ensuring that all the quanta are inside the weak-coupling domain. This issue is investigated in realistic models of modified gravity, R2- and Higgs-inflations, with emphasize on the specific change of the initial conditions with metric frame, while all the quanta still contribute equal parts. This illustrates that inflationary models with flat potential may be natural.

1 Introduction

The Standard Model of particle physics (SM) and General Relativity (GR) fail to describe the cosmology. The major phenomenological problems here are numerous dark matter phenomena and the matter-antimatter asymmetry of the Universe. There are also several puzzling coincidences between cosmological parameters, among them the widely recognized is an order-of-magnitude equality of baryon, dark matter and dark energy contributions to the present energy density of the Universe. The vanilla ΛCDM cosmological model predicts too large number of dwarf galaxies and has other issues probably related to evolution of the matter perturbations at small scales.

There is a separate class of problems called *the problem of initial conditions for the Hot Big Bang Theory.* Since the homogeneous Universe may form spherical or hyperbolic 3-dimensional manifold, its *flatness* at present is mysterious. Both at radiation and at matter dominated stages of the Universe expansion the size of causally connected region, *particle horizon*, grows faster than a physical spatial size does, so the Universe must be highly inhomogeneous at large scales, that is quite contrary to all observations. The Universe expansion is adiabatic, keeps the *entropy* conserved in a comoving volume, hence the entropy density in the very early Universe was enormously high $s \ggg M_{\rm Pl}^3$, which is weird. Both radiation and matter dominated stages imply *singular energy density* at the very beginning of expansion, $\rho \propto 1/t^2$. If SM is a remnant of some Grand Unified Theory, the corresponding phase transition(s) at the Hot stage generally populates the Universe with *topological defects*; later some of these relics, in particular monopoles and domain walls, become dominant and overclose the Universe.

The initial condition problems can be elegantly solved by introducing an (almost exponentially) fast expansion preceding the Hot stage. Then the curvature (if any) grows exponentially making the Universe exponentially flat. An originally causally connected region, whatever tiny, becomes exponentially large and today we actually observe only its minuscule part. Any heavy stable relics

[a]e-mail: gorby@ms2.inr.ac.ru

(like monopoles) are at exponentially large distances one from each other, so the late Hot Universe is safe. There are no initial singularities in the exponentially expanding Universe.

This stage, called *inflation*, can be naturally organized by a slowly evolving homogeneous scalar field ϕ, *inflaton*, see Fig. 1. One of the most popular idea here is chaotic inflation [1].

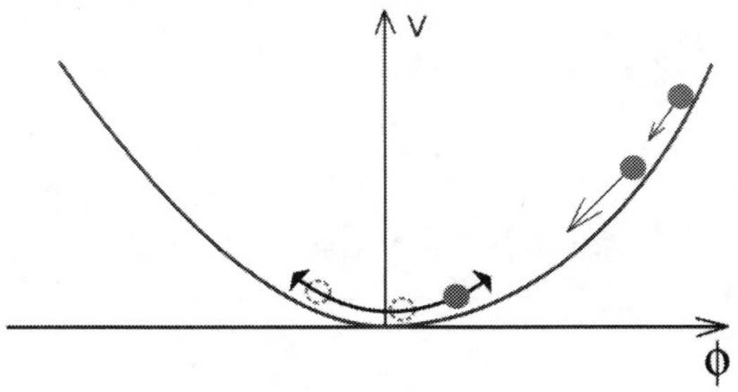

Figure 1. In a model of chaotic inflation with large field [1] the inflaton slowly rolls along the potential slope mimicking almost constant energy density, so the Universe exhibits (almost) exponential expansion. The slow rolling stops at some point and the inflaton starts to oscillate, so the Universe enters a stage of decelerating expansion. The hot stage starts after reheating, when the inflaton transfers its energy density into ultrarelativistic (SM) particles.

2 Chaotic inflation and unlikeness of models with flat potential

In the chaotic inflation field ϕ rolls down along the potential $V(\phi)$ starting from large, generally transplanckian values, $\phi > M_{\rm Pl}$, see Fig. 1. In the Robertson–Walker metric, $ds^2 = dt^2 - a^2(t)\,\mathbf{dx}^2$, with $a(t)$ being the scale factor, $\phi(t,\mathbf{x})$ solves equation

$$\ddot{\phi} - \frac{\Delta\phi}{a^2} + 3\mathcal{H}\dot{\phi} + V'(\phi) = 0\,. \tag{1}$$

The Hubble parameter $\mathcal{H} \equiv \dot{a}/a$ obeys the Friedmann equation

$$\mathcal{H}^2 = \frac{8\pi}{3M_{\rm Pl}^2}\left(\frac{1}{2}\dot{\phi}^2 + \frac{1}{2a^2}(\nabla\phi)^2 + V(\phi)\right) \tag{2}$$

and plays a role of dynamical friction in Eq. (1). Actually, if the field evolves very slowly, then its energy density is dominated by the scalar potential $V(\phi)$ and the Hubble parameter remains almost constant, see Eq. (2), ensuring the almost exponential expansion, $a(t) \propto \exp(\mathcal{H}t)$. Then the gradient terms in Eqs. (1), (2), being suppressed by the growing scale factor, get nullified and the Hubble friction governs the field homogeneous slow rolling towards the origin of scalar potential, $\phi = \phi(t) \to 0$. The kinetic terms in Eqs. (1), (2) are negligible if the so-called slow roll parameters are small,

$$\frac{M_{\rm P}^2}{2}\left(\frac{V'}{V}\right)^2 \equiv \epsilon \ll 1\,, \quad M_{\rm P}^2\frac{V''}{V} \equiv \eta \ll 1\,. \tag{3}$$

where we introduce the reduced Planck mass as $M_{\rm P} \equiv M_{\rm Pl}/\sqrt{8\pi}$. One can check that the latter conditions are satisfied for any power-law scalar potential, $V(\phi) \propto \phi^\alpha$, if the scalar field takes a transplanckian value, $\phi_i > M_{\rm Pl}$. Then the field goes down slowly and, once $\epsilon \simeq 1$ or $\eta \simeq 1$, the inflation terminates.

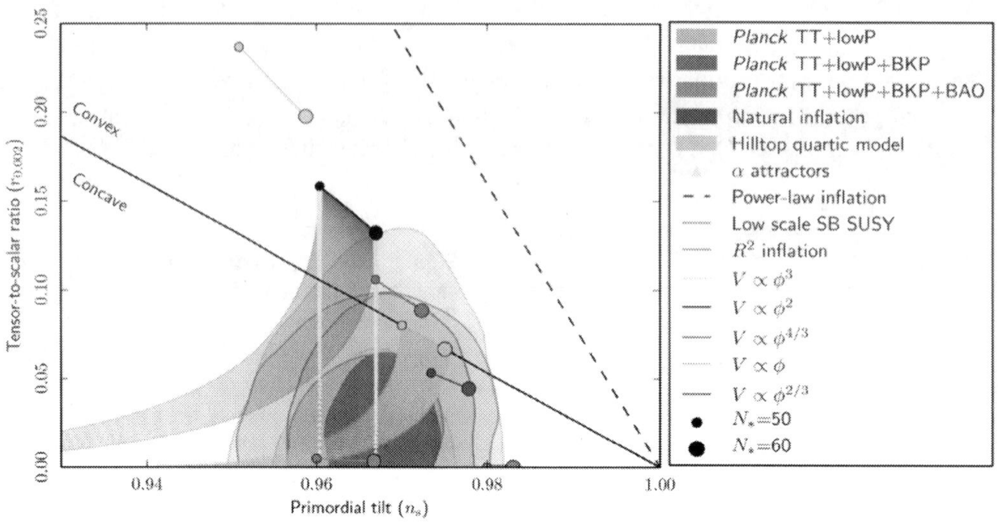

Figure 2. Cosmological constraints [4] on inflationary models. The colored regions are allowed at 95% CL, there are predictions of various cosmological models.

The chaotic initial conditions for the inflaton field imply [1, 2] that the scalar field fluctuates in any Planck-size region, and all the contributions to the energy density –kinetic term, gradient term and potential term– are (on average) of the same order, which is Planckian, i.e.

$$\frac{1}{2}\dot{\phi}^2 \sim \frac{1}{2}(\partial_i\phi)^2 \sim V(\phi) \sim M_{\rm Pl}^4 \,. \tag{4}$$

Then if in a particular Planck-size patch the potential term becomes somewhat dominant by chance, as a result of the fluctuations, the dynamics of the scalar field (1) rapidly approaches the slow roll and the chaotic inflation begins in this spatial region.

Inflation itself is almost untestable: the only exception is the prediction of exponentially flat Universe, that is in agreement with observations at present. However, it is a general prediction, inherent in the very paradigm of inflation. To recognize a particular inflationary model we need something else. Quite remarkably, the inflation provides with an appropriate feature. Namely, inflaton inhomogeneities $\delta\phi(t, \mathbf{x})$ (linear perturbations about the homogeneous inflationary solution) originated from the inflaton quantum fluctuations can seed the matter inhomogeneities (dubbed scalar perturbations) in the late Universe. Their subsequent evolution initiates large scale structure formation and explain the anisotropy of cosmic microwave background (CMB) radiation $\delta T/T \sim 10^{-4}$.

The scalar perturbation spectrum is almost flat. Both its amplitude and small tilt depend on the inflationary model parameters, and a combined analysis of sensitive to matter inhomogeneities cosmological observables allows one to discriminate between various inflationary models, for details see e.g. Ref. [3]. Figure 2 shows that the present data tend to favor the large field inflation models with *concave* scalar potential, e.g. the power-law potential $V(\phi) \propto \phi^\alpha$ with $\alpha < 1$. Among them are theoretically well motivated models with asymptotically flat potentials: R^2-inflation [5], Higgs-inflation [6], moduli inflation in supersymmetric extensions of the SM, etc. To explain the CMB anisotropy

$\delta T/T \sim 10^{-4}$ the inflaton plateau must be at about

$$V_0 \simeq 10^{-12}\, M_{\rm Pl}^4\,. \tag{5}$$

At first glance [7], however, these models seem to be unlikely in the framework of the chaotic inflation. Indeed, at the Planck epoch, when kinetic and gradient terms take the Planckian values according to conditions (4), the potential (being flat) is fixed by the cosmological observations at much lower level (5),

$$\frac{1}{2}\dot{\phi}^2 \sim \frac{1}{2}(\partial_i\phi)^2 \sim M_{\rm Pl}^4 \gg V(\phi)\,, \tag{6}$$

so that slow roll never happens. In such a model the chaotic inflation requires a beginning with all terms of the same order that is much smaller than the Planckian value, which seems unnatural. Below we illustrate the problem with two prominent examples and explain its solution by correctly imposing the chaotic initial conditions, for more details see Refs. [8, 9].

3 An example: R^2-inflation

Let us consider the first well-developed inflationary model [5] based on modified gravity with action

$$S^{JF} = -\frac{M_P^2}{2}\int\sqrt{-g}\,d^4x\left(R - \frac{R^2}{6\mu^2}\right), \tag{7}$$

which after the convenient Weyl transformation (from the Jordan frame to the Einstein frame),

$$g_{\mu\nu} \to \tilde{g}_{\mu\nu} = \Omega^2\, g_{\mu\nu}\,, \qquad \Omega^2 \equiv \exp\left(\sqrt{\frac{2}{3}}\frac{\phi}{M_P}\right),$$

casts into the standard Einstein–Hilbert term and terms describing the scalar degree of freedom, dubbed *scalaron*, emerging in the gravity sector,

$$S^{EF} = \int\sqrt{-\tilde{g}}\,d^4x\left[-\frac{M_P^2}{2}\tilde{R} + \frac{1}{2}\tilde{g}^{\mu\nu}\partial_\mu\phi\partial_\nu\phi - \frac{3\mu^2 M_P^2}{4}\left(1 - \frac{1}{\Omega^2(\phi)}\right)^2\right].$$

The model has an exponentially flat potential (see left panel of Fig. 3), which approaches (5) to be consistent with the observed CMB anisotropy. Naively, within the chaotic paradigm, the model suffers from unlikeness problem (6).

This is not true in fact. Indeed, in the Jordan frame the chaotic initial conditions look like $|R^{JF}| \sim M_{\rm Pl}^2$, and matching to cosmological data infers

$$\Omega^2 = 1 - \frac{R^{JF}}{3\mu^2} \sim 10^{12}\,. \tag{8}$$

The scalar curvature in the Jordan frame is related to that in the Einstein frame and scalaron as follows,

$$-R^{JF} = -\Omega^2 R^{EF} + \Omega^2\frac{8\pi}{M_{\rm Pl}^2}g_{\mu\nu}^{EF}\partial^\mu\phi\partial^\nu\phi - \Omega^2\frac{4\sqrt{3\pi}}{M_{\rm Pl}}g_{\mu\nu}^{EF}\partial^\mu\partial^\nu\phi\,.$$

The first and second terms in the r.h.s. above refer to the pure gravity and scalaron contributions to the energy density, respectively. Hence they must be of the same order within the paradigm of chaotic inflation. The third term happens to be of the same order provided the equations of motion.

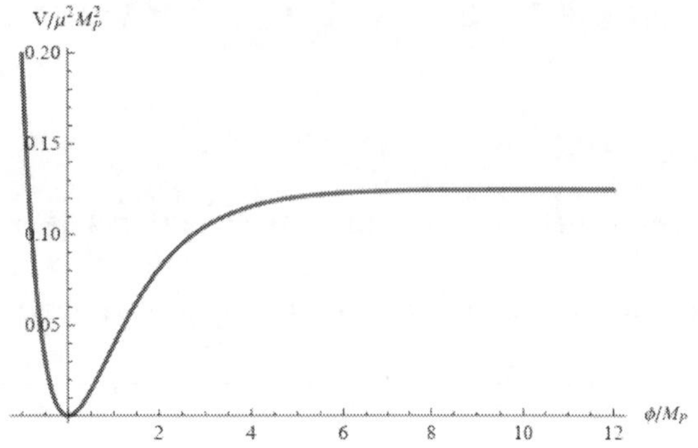

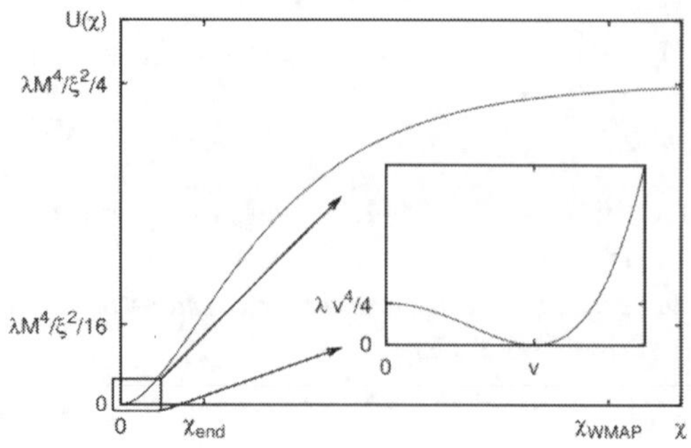

Figure 3. Examples of a flat potential in R^2-inflation (left panel) and Higgs-inflation (right panel).

Therefore one has

$$\left|R^{JF}\right| \Omega^{-2} \sim \left|R^{EF}\right| \sim \dot{\phi}^2 M_{\rm Pl}^2 \sim (\partial_i \phi)^2 M_{\rm Pl}^2 \,, \tag{9}$$

which means that all the terms in the Einstein frame are actually of the same order, consistent with the requirement of successful inflation,

$$\frac{1}{2}\dot{\phi}^2 \sim \frac{1}{2}(\partial_i\phi)^2 \sim V(\phi) \sim 10^{-12} \times M_{\rm Pl}^4 \,. \tag{10}$$

There is no unlikeness problem in the R^2-inflation.

Moreover, the very idea of starting R^2-inflation with Einstein-frame initial conditions (6) is ridiculous, since in the Jordan frame, due to the Weyl transformation with scale-factor (8), the curvature then much exceeds the Planckian values, see eq. (9). One simply cannot treat semiclassically the quanta naturally adopted to formulate the initial conditions in the chaotic approach. On the contrary, one can start with all the quanta at the upper border of the weak coupling domain, which in the Jordan frame with lagrangian (7) implies large curvature $R^{JF} \gg \mu^2$. Then the border under discussion moves from the Planck scale to

$$\Lambda^{JF} \simeq \frac{M_{\rm P}}{\sqrt{6}\mu} \sqrt{R^{JF}} \,,$$

and the contribution of dominating in (7) R^2-term to the energy density fluctuations,

$$E^{JF4} \sim \frac{M_{\rm P}^2}{\mu^2} R^{JF2} \,,$$

is well inside the weak coupling domain, limited from above by the new strong coupling scale Λ^{JF}, as $E^{JF4} \ll \Lambda^{JF4}$. Note, that the corresponding to E^{JF} energy scale in the Einstein frame *does not depend* on the value E^{JF},

$$E^{EF} \sim \Omega^{-1} E^{JF} \sim \frac{M_{\rm P}}{\Lambda^{JF}} E^{JF} \sim \sqrt{\mu M_{\rm P}} \,, \tag{11}$$

where in the Weyl scale-factor (8) we express the curvature R^{JF} through the strong coupling scale Λ^{JF}. One concludes that in the Einstein frame all the terms contributing to energy density (10) are of the same order E^{EF^4}, cf. (11), and inside the weak-coupling domain.

4 One more example: Higgs inflation

To cook the successful inflation with the SM Higgs weak doublet H one must introduce its non-minimal coupling to gravity [6],

$$S = \int d^4x\,\sqrt{-g}\left(-\frac{M_P^2}{2}R - \xi H^\dagger HR + \mathcal{L}_{SM}\right)$$

or written in the unitary gauge $H^T = \left(0\,,(h+v)/\sqrt{2}\right)$ (neglecting $v = 246\,\mathrm{GeV}$)

$$S = \int d^4x\,\sqrt{-g}\left(-\frac{M_P^2+\xi\,h^2}{2}R + \frac{(\partial_\mu h)^2}{2} - \frac{\lambda\,h^4}{4} + \mathcal{L}_{SM}\right). \tag{12}$$

Performing the Weyl transformation

$$g_{\mu\nu} = \Omega^{-2}\tilde{g}_{\mu\nu}\,, \qquad \Omega^2 = 1 + \frac{\xi\,h^2}{M_P^2}$$

one finds in the Einstein frame for the canonically normalized field χ

$$\frac{d\chi}{dh} = \frac{M_P\sqrt{M_P^2+(6\xi+1)\xi\,h^2}}{M_P^2+\xi\,h^2}$$

the scalar potential (see right panel of Fig. 3)

$$U(\chi) = \frac{\lambda M_P^4\,h^4(\chi)}{4(M_P^2+\xi\,h^2(\chi))^2} = \frac{\lambda M_P^4}{4\xi^2}\left(1-\exp\left(-\frac{\sqrt{2}\chi}{\sqrt{3}M_P}\right)\right)^2$$

at relevant for inflation values $h \gg M_P/\sqrt{\xi}$. It looks exactly as the scalaron potential in the previous example of R^2-inflation,[1] see Fig. 3, and the correct height of the potential plateau (5) is achieved with [6]

$$\frac{\xi}{\sqrt{\lambda}} \simeq 5\times10^4\,, \tag{13}$$

assuming the Higgs self-coupling is positive at the inflationary scale. Since the two potentials coincide one naively expects [7] here the same unlikeness problem.

Actually, there is no unlikeness problem here. Similarly to the case of R^2-inflation, one starts with the chaotic initial conditions implemented in the Jordan frame,

$$\Omega^2\,M_{\rm Pl}^2\,R^{JF} \sim \xi\dot{h}^2 \sim \xi(\partial_i h)^2 \sim \lambda h^4 \sim M_{\rm Pl}^4\,. \tag{14}$$

Here we account for the fact that the nonminimal gravity coupling to the Higgs field effectively changes the Higgs kinetic and gradient terms, and chose the dominant parts. Consequently,

$$R^{JF} \sim \Omega^{-2}\,M_{\rm Pl}^2,$$

and since the cosmological observations fix the potential in the Einstein frame at the value (5), hence the Weyl scale-factor is

$$\Omega^2 \sim 10^6\,. \tag{15}$$

[1] Predictions of these two models for the scalar and tensor perturbation spectra differ only slightly, see Refs. [10, 11], because of different post-inflationary history which culminates in different reheating temperatures [12, 13].

Then in the Einstein frame one arrives at the same picture: all the terms are of the same order needed for the successful inflation, namely

$$\frac{1}{2}\dot{\phi}^2 \sim \frac{1}{2}(\partial_i\phi)^2 \sim \Omega^{-4} M_{\rm Pl}^4 \sim 10^{-12} \times M_{\rm Pl}^4 \,. \tag{16}$$

There is no unlikeness problem in the Higgs-inflation either.

Analogously to the case of R^2-inflation one can raise the question of the strong-coupling scale in this model and its relation to the values associated with the chaotic initial conditions. Even for small fields the strong-coupling scale in the model (12) is well below the Planck scale [14], namely it is at $M_P\sqrt{\lambda}/\xi \sim 10^{14}$ GeV given the much exceeding unity value of the non-minimal coupling (13). At large homogeneous fields the non-minimal coupling effectively changes the Planck mass, pushing the scale of strong gravity up to

$$\Lambda^{JF} \simeq \sqrt{8\pi\xi}\, h \sim \Omega M_{\rm Pl} \,,$$

and the semiclassical treatment of the gravitational sector is applicable only at energy densities below ${\Lambda^{JF}}^4$. The chaotic initial conditions (14) obey this constraint, moreover, the corresponding energy scale $\sim h$ is right at the upper border of the weak-coupling domain, which is limited by the scalar, rather than gravity, dynamics, see Ref. [14] for details. The Weyl transformation with scale-factor (15) leads to the chaotic initial conditions in the Einstein frame (16), which implies the energy scale $\sim \Omega^{-1} h \sim M_{\rm Pl}/\sqrt{\xi}$ right at the upper border of the weak-coupling domain in this frame [14]. As in case of the R^2-inflation this scale *does not depend* on the value h of strong coupling scale in the Jordan frame. Again, the initial conditions like (6) imposed in the Einstein frame would imply working outside the weak-coupling regime, which is not treatable.

5 Conclusion: initial conditions in an arbitrary frame

We find that the *chaotic paradigm* works perfectly well for both R^2-inflation and Higgs-inflation: they do not suffer from the unlikeness problem. The presented analysis suggests a simple prescription for implementing the chaotic initial conditions in any model as follows. Start in any frame with all contributions to the energy density being of the same order (say, $\sim M_{\rm Pl}^4$ or any other value). Then in any other frame the corresponding contributions are always of the same (though another) order (say, $\sim 10^{-12} \times M_{\rm Pl}^4$ in the EF). The natural energy scale of the corresponding contributions is argued to be the strong coupling scale in a given frame. This statement is consistent with the Weyl transformation, which matches the energy scales of different frames.

The work is supported by the RSF grant 14-12-01430.

References

[1] Linde A.D., Phys. Lett. B **129**, 177 (1983).
[2] Linde A.D., Rept. Prog. Phys. **47**, 925 (1984).
[3] Gorbunov D.S. and Rubakov V.A., *Introduction to the theory of the early universe: Cosmological perturbations and inflationary theory*, (Hackensack, USA: World Scientific, 2011), 489.
[4] Ade P.A.R. et al. [Planck Collaboration], Astron. Astrophys. **594**, A20 (2016).
[5] Starobinsky A.A., Phys. Lett. B **91**, 99 (1980).
[6] Bezrukov F.L. and Shaposhnikov M.E., Phys. Lett. B **659**, 703 (2008).
[7] Ijjas A., Steinhardt P.J. and Loeb A., Phys. Lett. B **723**, 261 (2013).

[8] Gorbunov D.S. and Panin A.G., Phys. Lett. B **743**, 79 (2015).
[9] Gorbunov D. and A. Panin A., "Strong-coupling scale and frame-dependence of the initial conditions for chaotic inflation in models with modified (coupling to) gravity," arXiv:1607.05268 [gr-qc].
[10] Bezrukov F.L. and Gorbunov D.S., Phys. Lett. B **713**, 365 (2012).
[11] Gorbunov D. and Tokareva A., JCAP **1312**, 021 (2013).
[12] Gorbunov D.S. and Panin A.G., Phys. Lett. B **700**, 157 (2011).
[13] Bezrukov F., Gorbunov D. and Shaposhnikov M., JCAP **0906**, 029 (2009).
[14] Bezrukov F.L., Magnin A. and Shaposhnikov M., Phys. Lett. B **675** 88 (2009).

EPJ Web of Conferences **164**, 01016 (2017) DOI: 10.1051/epjconf/201716401016
ICNFP 2016

LHCb detector and trigger performance in Run II

Dordei Francesca[1,a], on behalf of the LHCb Collaboration

[1] *European Organization for Nuclear Research (CERN), Geneva, Switzerland*

Abstract. The LHCb detector is a forward spectrometer at the LHC, designed to perform high precision studies of b- and c- hadrons. In Run II of the LHC, a new scheme for the software trigger at LHCb allows splitting the triggering of events into two stages, giving room to perform the alignment and calibration in real time. In the novel detector alignment and calibration strategy for Run II, data collected at the start of the fill are processed in a few minutes and used to update the alignment, while the calibration constants are evaluated for each run. This allows identical constants to be used in the online and offline reconstruction, thus improving the correlation between triggered and offline selected events. The required computing time constraints are met thanks to a new dedicated framework using the multi-core farm infrastructure for the trigger. The larger timing budget, available in the trigger, allows to perform the same track reconstruction online and offline. This enables LHCb to achieve the best reconstruction performance already in the trigger, and allows physics analyses to be performed directly on the data produced by the trigger reconstruction. The novel real-time processing strategy at LHCb is discussed from both the technical and operational point of view. The overall performance of the LHCb detector on the data of Run II is presented as well.

1 Introduction

The LHCb experiment is dedicated to flavour physics studies at the Large Hadron Collider (LHC) at CERN. In particular, it is designed for precision measurements of *CP* violation and rare decays of beauty (b) and charm (c) hadrons, in order to indirectly search for new physics effects beyond the Standard Model. The LHCb detector [1][2] has been designed to accomplish these delicate measurements. It is a single-arm forward spectrometer, shown in figure 1, covering a pseudorapidity, η, range unique among the LHC detectors, $2 < \eta < 5$. The detector includes a high-precision tracking system, which provides a measurement of momentum of charged particles, with a relative uncertainty that varies from 0.5% at low momentum to 1.0% at 200 GeV/c. The minimum distance of a track to a primary vertex, the impact parameter, is measured with a resolution of $(15 + 29/p_T)$ μm, where p_T is the component of the momentum transverse to the beam in GeV/c. The tracking system consists of a silicon-strip vertex detector (VELO) surrounding the proton-proton (pp) interaction region [3], a large-area silicon-strip detector (TT) located upstream of a dipole magnet with a bending power of about 4 Tm, and three T stations of silicon-strip detectors, Inner Tracker (IT), and straw drift tubes, Outer Tracker (OT) [4], placed downstream of the magnet. After travelling about 10 m in the LHCb

[a] e-mail: francesca.dordei@cern.ch

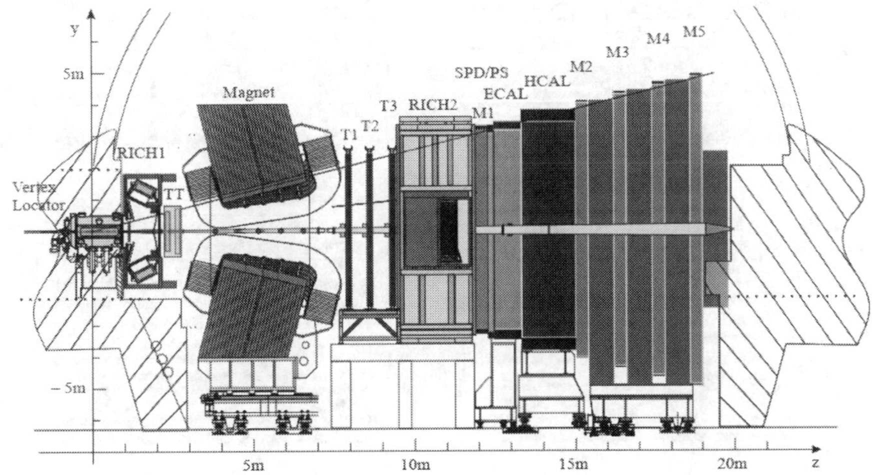

Figure 1. Schematic view of the LHCb detector in the $y-z$ plane. The right-handed coordinate system used has the z axis along the beam direction pointing from the interaction point into the experiment, the y axis vertically upwards and the x axis pointing towards the outside of the LHC ring. Figure taken from [1].

detector, from the VELO to the end of the three T stations, a particle has traversed about 60% of a radiation length and 20% of an absorption length.

Different types of charged hadrons are distinguished using information from two ring-imaging Cherenkov detectors (RICHs) [5]. Photons, electrons and hadrons are identified by a calorimeter system consisting of scintillating-pad and preshower detectors (SPD/PS), an electromagnetic calorimeter (ECAL) and a hadronic calorimeter (HCAL). Muons are identified by a system composed of alternating layers of iron and multiwire proportional chambers (M1-M5) [6]. In Run I, this system allowed to obtain a kaon identification efficiency larger than 95%, together with a pion misidentification fraction of $\sim$10% over the full momentum range and a muon identification efficiency larger than 97% [2].

The nominal LHC rate, including empty bunches, is 40 MHz. This rate is too large to allow data to be written to storage. The main goal of the trigger is to reduce this rate to $O(10)$ kHz, keeping the most relevant pp interactions for subsequent physics analysis. In particular, LHCb is mainly interested in heavy flavour physics and the signatures of these decays are the presence of high transverse momentum (p_T) tracks, high transverse energy (E_T) in the calorimeters and displaced vertices, because b and c hadrons fly on average over few centimetres before they decay. Moreover, the presence of displaced vertices implies that some of the decay products have a large displacement from the pp interaction vertex. The trigger [7] consists of a hardware stage (L0), based on information from the calorimeter and muon systems, followed by a software stage (HLT), split into two steps, which applies a full event reconstruction. The trigger system as a crucial role in LHCb since for precision measurements the main gain comes from the availability of more data. Thus, increasing the speed of the system and improving the quality of the data coming from the trigger was one of the main priority in view of the Run II of the LHC.

In Run I (2010-2012) of the LHC, the LHCb experiment has collected a total luminosity of 3 fb^{-1} of pp collisions at 7 and 8 TeV centre-of-mass energy, as well as 5 TeV centre-of-mass proton-ion data. A very successful physics program has been carried out, leading to over 300 publications, thanks to the excellent performance of the detector and the LHC, covering a wide range of physics

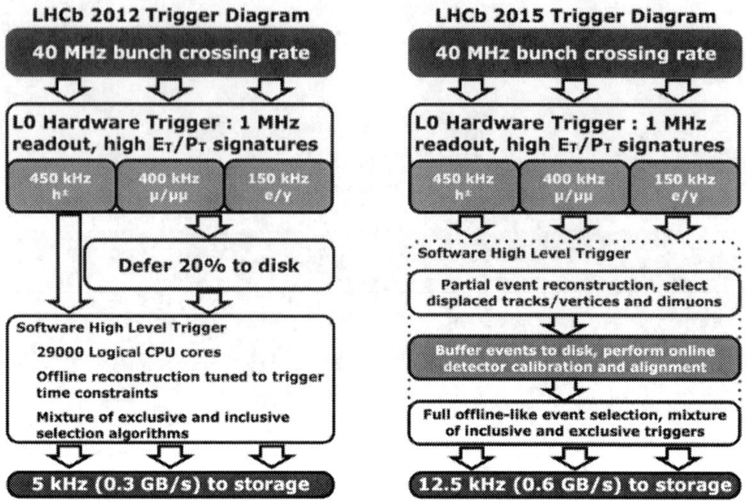

Figure 2. LHCb trigger scheme in Run I (left) and in Run II (right). The rate of collisions, including empty bunch crossings, is 40 MHz.

topics. After the Long Shutdown 1 (LS1), in 2015 LHC started the Run II, which will continue until 2018, allowing to collect a total luminosity of ~5 fb^{-1} pp collision data. The centre-of-mass energy delivered by the LHC was increased to 13 TeV, leading to a ~60% increase in the $b\bar{b}$ and $c\bar{c}$ cross-sections. In these new conditions, LHCb will continue pursuing its core physics program with increased statistics, trying to keep (and increase) the variety of studied topics. During LS1, a large amount of work has been spent in consolidation, maintenance and repair of the subdetectors and of the general infrastructure. The main challenge has been to keep (and even improve) the performance achieved during Run I despite the increase in centre-of-mass energy. The key point to achieve this delicate point was a completely revisited trigger strategy that allows to get the same performance obtained offline already at the trigger level.

2 The trigger system

In Run I the visible rate, i.e. the rate of collisions, was 15 MHz while the output rate of events saved on disk was 5 kHz, see figure 2 on the left. Almost all events accepted by the HLT were sent to permanent offline storage containing all raw information from the detector. The online event reconstruction, i.e. the one performed in the trigger, was simpler than the one used offline, i.e. the one performed on the LHC Computing Grid to recreate particles in the event from the raw data. This was done in order to fulfil the tight time constraints imposed in the trigger. This implied the usage of a preliminary version of the alignment and calibration and a faster, but less performing, track reconstruction and particle identification determination. The final detector calibration and alignment parameters were obtained offline on triggered data and applied afterwards during scheduled campaigns where the data were centrally re-processed. During these campaigns, all the events were re-reconstructed offline using a different version of the reconstruction software that allowed to achieve the best performance regardless of the timing required. Clearly, this strategy costs a lot of computing resources, since the reconstruction software was run twice and could also cause a loss of imperfectly reconstructed data in the trigger due to the usage of preliminary calibration and alignment.

In Run II the visible rate is almost the double compared to Run I. The larger cross-sections in Run II, especially in the case of charm hadrons, demand a more efficient trigger strategy, since the offline processing resources won't go up as much as the signal rate. To achieve this, a completely new trigger scheme, see figure 2 on the right, has been devised that allows to have the same offline-data quality already at the trigger level. Clearly, this is a difficult challenge due to the time constraints imposed by the trigger. The major change to the trigger strategy that allowed to achieve this goal is that the processing of the second step of the software trigger (HLT2) has been completely deferred, in order to optimise the usage of the event filter farm (EFF) where the software trigger is run. The EFF is a computing cluster consisting of 1800 server nodes, with a combined storage space of 5.2 PB in 2015 and $\sim$ 10 PB in 2016. This can accommodate up to two weeks of LHCb data taking in nominal conditions. Since LHC does not spend 100% of its time in stable running due to e.g. planned technical stops, machine development phases and time between data taking fills needed for the ramping of the LHC dipole magnets, it is possible to use the time in which the farm is not busy to process the HLT2 data. In order to achieve this, the farm nodes were equipped with local storage space permitting to buffer the first level of the HLT (HLT1) output into the nodes and to process it during the LHC downtime. This strategy allows to have more time to process a single event and gives the opportunity to calibrate and align the subdetectors in real time using data from HLT1. Here, real time is defined as the interval between the collisions in the detector and the moment the data are sent to permanent storage. This task, however, presents an unprecedented challenge as it requires to align more than 1700 detector components and compute almost 2000 calibration constants on the fly; in Sect. 3 it will be explained the strategy used to align and calibrate the detector in such a small time window. This continuous calibration and alignment, which only updates the constants when necessary, provides offline-quality variables, such as particle identification (PID), at the trigger level; this fact, combined with the doubling of the trigger farm capacity and the improved track reconstruction, allows to have the same online reconstruction as the offline one and to increase the output rate to 12.5 kHz. Indeed, a lot of work has been put into optimizing the code and algorithms used in the trigger and the reconstruction:

- Faster implementations of large areas of the code thanks to vector instructions.
- New algorithms have been put in place to improve the performance and/or reduce the execution time with respect to their Run I counterparts.
- A faster reconstruction has been achieved thanks to a linear extrapolation of tracks from the VELO to the TT as an intermediate step before the so-called forward tracking.
- The primary vertex reconstruction has been updated and optimized, with the same algorithm now used online and offline.

This optimized trigger strategy has several advantages. First, it minimises the difference between the online and offline performance, allowing to use a more effective trigger selection that can take advantage of the hadron identification information. For example charm physics is limited by trigger output rate constraints; using hadron identification in the trigger allows to have a higher selection efficiency and purity for the doubly Cabibbo suppressed modes and, at the same time, satisfy the output rate constraints by pre-scaling the more abundant Cabibbo favourite modes. Second, it ensures the stability of the alignment quality and hence of the physics performance.

3 Alignment and calibration in Run II

In order to achieve the same offline performance already in the trigger one key point is to perform the spatial alignment and calibration of the detector in real time. Both the alignment and calibration are

DOI: 10.1051/epjconf/201716401016

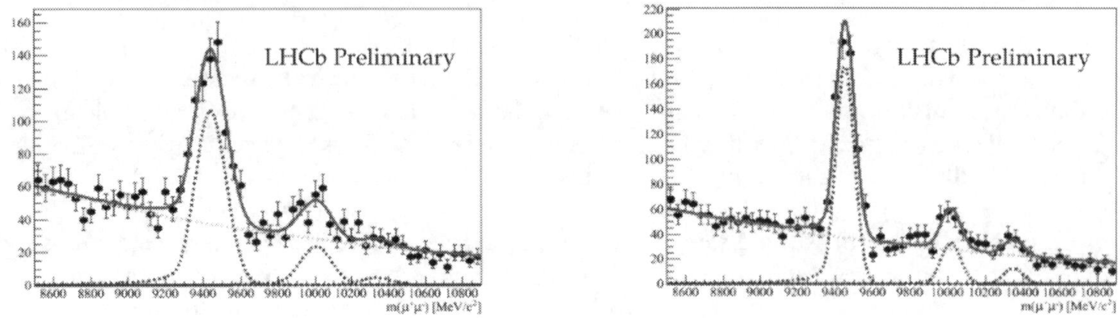

Figure 3. Invariant mass distribution for $\Upsilon \to \mu^+\mu^-$. The mass resolution is 92 MeV/c^2 with the first alignment (left) and is enhanced to 49 MeV/c^2 with the improved alignment (right).

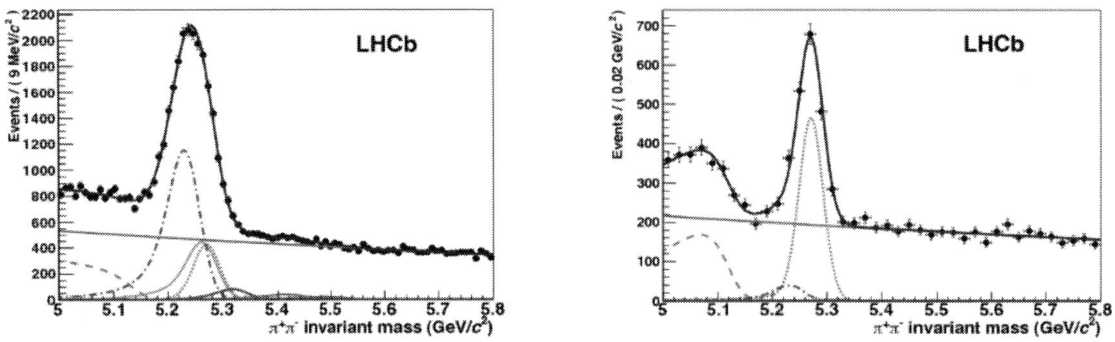

Figure 4. Invariant mass distribution for $B^0 \to h^+h^-$ decays [9] in the LHCb data before the use of the RICH information (left), and after applying RICH particle identification (right). The signal under study is the decay $B^0 \to \pi^+\pi^-$, represented by the turquoise dotted line. The contributions from different b-hadron decay modes ($B^0 \to K\pi$ red dashed-dotted line, $B^0 \to$ 3-bodies orange dashed-dashed line, $B_s \to KK$ yellow line, $B_s \to K\pi$ brown line, $\Lambda_b \to pK$ purple line, $\Lambda_b \to p\pi$ green line), are eliminated by positive identification of pions, kaons and protons. Only the signal and two background contributions remain visible after applying hadron identification requirements. The grey solid line is the combinatorial background.

crucial to achieve the best physics performance. In particular, the correct alignment of the VELO is needed to identify secondary vertices from the decay of particles with b or c quarks while a misalignment of the all tracking system would degrade the mass resolution. As an example, the improvement of the alignment significantly increases the Υ mass resolution from 92 MeV/c^2 with the first alignment to 49 MeV/c^2 with the improved alignment, as is shown in figure 3. Similarly, good alignment and calibration of the ring-imaging Cherenkov detectors are necessary in order to use hadron identification criteria in exclusive selections to suppress background. Figure 4 shows the effect in the $B^0 \to h^+h^-$ mass spectrum of hadron identification criteria: the ratio between the signal, $B^0 \to \pi^+\pi^-$, and the combinatorial background increases by approximately a factor 2 and the ratio between the signal and the favoured $B^0 \to K\pi$ increases by a factor 35.

It is then clear that a more effective selection and a higher signal purity of studied channels can be achieved mainly thanks to a real-time alignment and calibration of the detector. This allows to

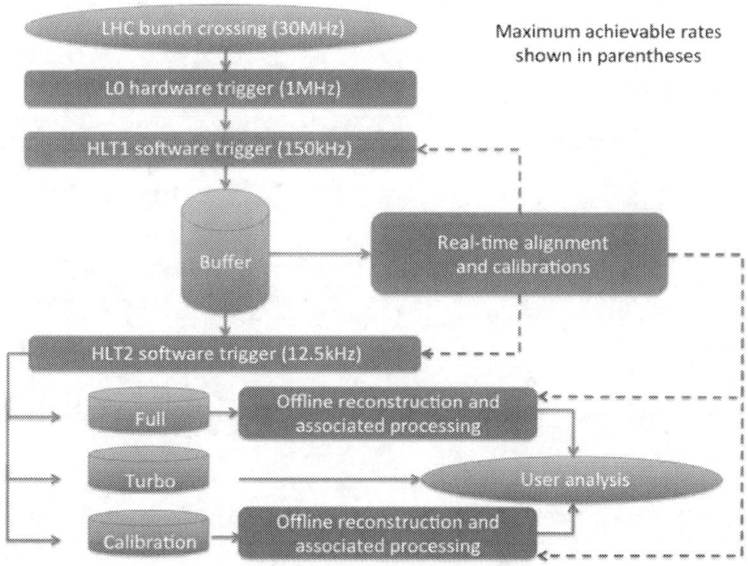

Figure 5. Schematic diagram showing the overall data processing model in Run-II, where the blue solid line represents data flow, and the red dashed line the propagation of calibration and alignment constants.

better select the channels interesting to pursue LHCb's physics program without increasing the overall trigger rate.

The automatic calibration and alignment is performed in the trigger farm within a few minutes once the necessary data samples are collected. Dedicated samples from HLT1 are taken as input for each system, for instance $D^0 \to K^+\pi^-$ candidates for the Tracker alignment and $J/\psi \to \mu^+\mu^-$ candidates for the alignment of the Muon system. When the new alignment and calibration are available, and a significant variation is observed, a change of run is triggered. The new constants are updated for the new run to be used online by the two stages of the software trigger, and offline, for every further reconstruction and selection. The events collected in the previous run, during the automatic alignment and calibration processing, are still reconstructed in HLT2 and offline with the previous constants for consistency with HLT1. This flow is schematically shown in Figure 5.

The spatial alignment of the LHCb detector consists in the alignment of tracking system, namely the VELO, the Tracker (TT, IT e OT) and the Muon system and of the RICH mirrors. Here only the main features will be highlighted, more information can be found in [8]. Concerning the calibration, at LHCb the Ring imaging Cherenkov detectors and the Outer Tracker must be calibrated in real time in order to get the relation connecting the measured quantity to the physical quantity that one wants to achieve. Again, only the main features will be highlighted, more information can be found in [8].

Alignment of the Tracking system

The alignment of the LHCb tracking detector uses information from optical and mechanical surveys and from reconstructed charged particle trajectories. Although the final alignment precision is obtained with reconstructed tracks, a precise survey is essential both as a starting point for the track-based alignment and to constrain degrees of freedom to which fitted track trajectories are insensitive. The alignment is achieved through minimisation of the residuals of a Kalman fit [10] to a set of well

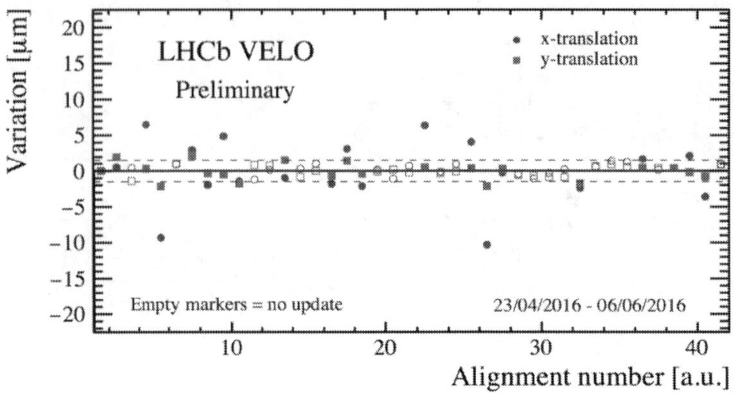

Figure 6. Stability of the alignment of the VELO halves during Run II fills. See the text for more explanations.

reconstructed tracks from HLT1. The alignment of the VELO is performed first, followed by the corresponding alignment of the tracking stations. This order is chosen due to the nature of the VELO, which centres itself around the collision point. On average the VELO alignment constants change significantly every 2-3 fills, while tracker constants change only every several fills. The muon system alignment is run only in monitoring mode, and the constants are updated only after hardware intervention. As an example, in figure 6 the stability of the alignment of the VELO halves during the first data taking period of Run II is shown. Each point is obtained running the online alignment procedure and shows the difference between the initial alignment constants (the ones used in the previous fill) and the new ones computed by the alignment. The alignment is updated (filled dots) only if the variation is significant (in the plot the red or the blue points must be outside the horizontal lines at ± 2 μm). However, since there are more degrees of freedom for which the VELO is aligned for than those shown in figure 6, the filled dots do not necessarily corresponds to points outside the horizontal lines.

RICH mirror alignment

Both RICH detectors have two sets of mirrors where photons are reflected before to be deflected out of the LHCb acceptance onto the photon detection plane. Misalignment of the RICH detectors causes the circular rings on the detection plane to become distorted. Due to the fact that the projected position of the track is not centered with respect to the Cherenkov ring, the Cherenkov opening angle varies as a function of the azimuthal angle. This distortion requires an individual correction for each mirror, that is calculated using a set of well reconstructed tracks from HLT1.

RICH calibration

The refractive index of the gas radiators used in the RICH varies as a function of the ambient temperature and pressure. Using a set of well reconstructed tracks originating from a particle of known mass the expected Cherenkov angle can be calculated using accurate momentum measurements provided by the tracking stations. By comparing the measured Cherenkov angles with the expected angles it is possible to extract scale factors to calibrate the refractive index of the radiator. Similarly, the anode images produced by the Hybrid Photon Detectors (HPDs) that detect the Cherenkov photons have to be calibrated, since they are affected by magnetic and electric fields.

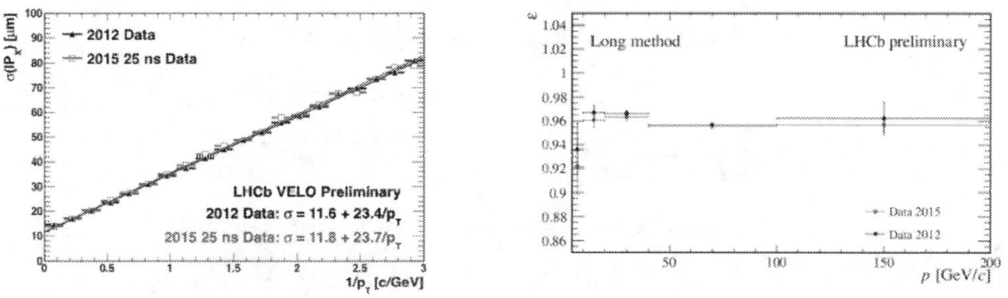

Figure 7. Left: Resolution of the x component of the impact parameter in 2012 50 ns data (black) and 2015 25 ns data (red). Right: Track reconstruction efficiency for long tracks for 2012 50 ns (black) and 2015 25 ns (red) data.

Outer Tracker calibration

The outer tracker is composed by straw tubes in which the drift time in the gas is measured. This measurement could be different from the time estimated from the distance of the track to the wire since it can be affected by differences between the true collision time and the LHCb clock. On top of this effect, a delay in the electronic readout also causes a time difference that is characteristic for each module. However this contribution is small compared to the global offset and can be calibrated offline. The drift time residuals of a sample of well reconstructed tracks are used to provide the global drift time offset correction.

4 Trigger and detector performance in Run II

Thanks to the fact that the alignment and calibration are performed in real time and thanks to all the improvements in the reconstruction sequence, it has been possible to achieve the same offline performance of Run I already in the data coming from the trigger. As an example, in Figure 7 (left) it is shown a comparison of the resolution of the x component of the impact parameter, defined as the distance in x between a track and the associated primary proton-proton vertex (PV), between 2012 data (black) and 2015 data (red), with a bunch spacing of 50 ns and 25 ns, respectively. In this figure, only events with one reconstructed PV are used, where the PV fit is reperformed excluding each track in turn. The resulting PV is required to have more than 25 tracks to minimise the contribution from PV resolution. It is possible to see that the same performance as Run I is obtained. Similarly, in Figure 7 (right) it is shown a comparison of the track reconstruction efficiency in Run I (black) and in Run II (red), with a bunch spacing of 50 ns and 25 ns, respectively. Here, only so called long tracks are considered, which are reconstructed using the information from the whole tracking system in LHCb. Again, despite the more harsh environment due to the greater centre-of-mass energy in Run II, the same performance is achieved. Moreover, this completely revisited trigger strategy and the excellent performance obtained allows to achieve better trigger efficiencies. Thanks to the usage, e.g., of more powerful particle identification criteria in the trigger, the signal to background ratio have been increased. As an example, the efficiency to select $B^0 \to D^0\pi^+$ events in the HLT2 was around 75% in Run I while it is greater than 90% in Run II.

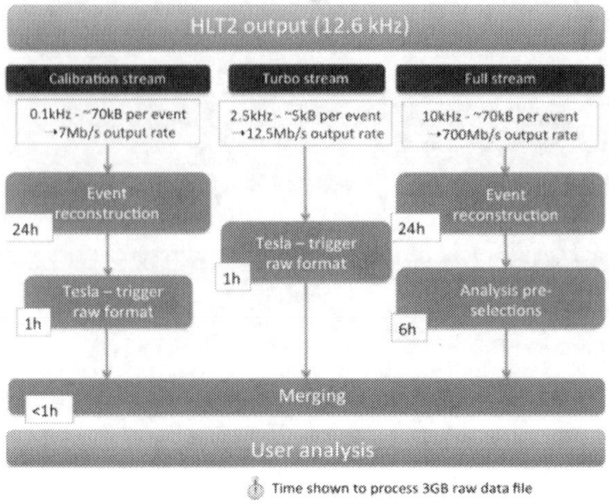

Figure 8. Turbo data processing versus the traditional approach (Full stream). The time taken for each step in hours is provided for a 3 GB raw data file.

5 Turbo and Turbo++

Finally, as in Run II the last level of the trigger performs the same reconstruction as the one performed offline, it is possible to run physics analyses, with the same offline performance, directly on the output of the trigger using the special stream of data known as Turbo stream [11]. The Turbo stream is designed to store part of the trigger bandwidth (around 20%) in a special format that allows to bypass the offline reconstruction, saving CPU and storage resources. The schematic data flow of the Turbo stream compared to the traditional data flow (represented by the Full stream) is shown in Figure 8. This approach has the advantage that, for some events, it will be possible to save only the information on the signal candidate tracks (~15 kB per event) instead of all the electronic signals recorded from the detector (~70 kB per event). This decrease of almost one order of magnitude of the event size allows a higher selection rate, for example in the charm analyses that during Run I used trigger lines which were pre-scaled due to output rate requirements. It is obvious that the real-time alignment and calibration becomes essential when the raw event is not saved and it was an essential achievement to have the Turbo stream functioning efficiently. For standard Run-II conditions ~20% of the HLT2 selected events will be sent to the Turbo stream at a cost of less than 2% of the output bandwidth. In order to perform physics analyses with the online reconstruction, decay candidates must appear in the same format as expected by the output of the traditional processing, such that the existing analysis infrastructure can be used. This is the purpose of the Tesla application [11]. The use of the Turbo stream in 2015 proved to be successful. The first two published physics measurements from the LHCb experiment based on data collected in the 2015 run were based on the Turbo stream [12][13]. Moreover, since 2016, it is also possible to store in the Turbo stream not only the candidates that pass the trigger selection but also the full event, allowing to use in the trigger selection higher level variables. This enabled even more analyses to profit of the Turbo stream. In 2016, out of the 420 HLT2 trigger lines developed, around 150 use the Turbo stream. Using this stream it has been possible to successfully select key channels and to monitor the data distributions already few days after the beginning of the data taking in 2016, as shown in Figure 9, where the invariant mass distribution of $B^+ \to J/\psi(\mu^+\mu^-)K^+$ events is shown.

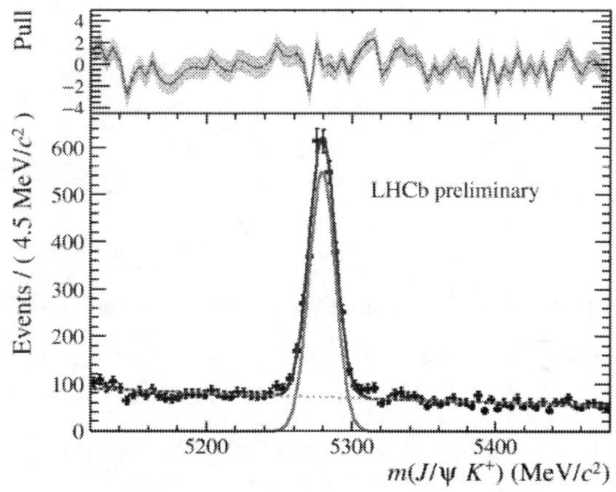

Figure 9. Invariant mass distribution of $B^+ \to J/\psi(\mu^+\mu^-)K^+$ events obtained using the Turbo stream.

6 Conclusions

LHCb is the first High Energy Physics experiment with a full calibration, alignment and reconstruction done in real-time. The revisited trigger strategy allows to achieve Run I offline performances already in the trigger. Reaching the ultimate precision of the LHCb experiment already in real time as the data arrive has the power to transform the experimental approach to processing large quantities of data. In particular, this allowed the usage of the so-called Turbo stream to perform physics analyses directly on the trigger output, increasing the physics program that can be covered.

References

[1] A. A. Alves Jr., et al. (LHCb collaboration), JINST **3**, S08005 (2008)
[2] R. Aaij, et al. (LHCb collaboration), Int. J. Mod. Phys. **A30**, 1530022 (2015)
[3] R. Aaij, et al., JINST **9**, P09007 (2014)
[4] R. Arink, et al., JINST **9**, P01002 (2014)
[5] M. Adinolfi, et al., Eur. Phys. J. **C73**, 2431 (2013)
[6] A. A. Alves Jr., et al., JINST **8**, P10020 (2013)
[7] R. Aaij, et al., JINST **8**, P04022 (2013)
[8] G. Dujany and B. Storaci, J. Phys. Conf. Ser. **664**, no. 8 082010 (2015)
[9] R. Aaij, et al. (LHCb collaboration), JHEP **037**, 10 (2012)
[10] W. Hulsbergen, Nucl. Instrum. Meth. **A600**, 471 (2009)
[11] R. Aaij, et al., arXiv **1604.05596**, (2016)
[12] R. Aaij, et al. (LHCb collaboration), JHEP **10**, 17 (2015)
[13] R. Aaij, et al. (LHCb collaboration), JHEP **09**, 13 (2016)

New results from T2K

A. Longhin[1,a] on behalf of the T2K Collaboration

[1] *Istituto Nazionale di Fisica Nucleare, Sez. di Padova, via Marzolo 8, 35131 Padova, Italy.*

Abstract. The T2K experiment is a 295-km long-baseline neutrino experiment in Japan employing an off-axis muon neutrino beam with a ~ 0.6 GeV peak energy. The beam, produced from 30-GeV protons at the J-PARC complex on the Pacific coast, is directed to the Super-Kamiokande detector. T2K released the first long-baseline measurement of a nonzero value for the θ_{13} mixing parameter through the observation of electron neutrino appearance ($\nu_\mu \to \nu_e$) and produced the most precise measurement of θ_{23} through the observation of muon neutrino disappearance ($\nu_\mu \to \nu_\mu$). T2K data, in combination with reactor experiments, also excludes at 90% C.L. a significant region of the Dirac CP phase: $\delta_{CP} < -3.02(-1.87)$ and $\delta_{CP} > -0.49(-0.98)$ for the normal (inverted) hierarchy. A full joint appearance and disappearance fit including both neutrino (7×10^{20} protons on target, PoT) and anti-neutrino (4×10^{20} PoT) data and, for the first time, a constraint from water target data in the near detector, is presented yielding improved sensitivity on δ_{CP} and improved precision on $\sin^2 2\theta_{23}$ and the atmospheric mass splitting.

1 The T2K experiment

T2K is a long-baseline neutrino oscillation experiment located in Japan, exploiting 30 GeV protons from the Japan Proton Accelerator Research Center (J-PARC, in Tōkai, Ibaraki Prefecture) to create a muon neutrino beam. The proton beam is directed onto a graphite target and the resulting pions and kaons focussed by a system of three magnetic horns into a 96 m-long decay tunnel filled with helium gas. Depending on the direction of the 250 kA pulsed current, horns can select either positively charged particles, for a beam composed mainly of ν_μ (ν-mode), or negatively charged particles, for a beam composed mainly of $\bar{\nu}_\mu$ ($\bar{\nu}$-mode).

The neutrino beam is measured by two near detectors (INGRID, ND280) located 280 m from the target and a far detector, Super-Kamiokande (SK). The far detector and one of the near detectors (ND280) are placed 2.5° off-axis with respect to the protons' direction, which results in a narrow neutrino energy spectrum with a peak at about 0.6 GeV. The baseline between neutrino production and the far detector, 295 km, is chosen to correspond to the first minimum in the ν_μ survival probability at the peak energy.

T2K can measure neutrino oscillation in two channels: ν_μ disappearance ($\nu_\mu \to \nu_\mu$), which is dominated by the oscillation parameters $\sin^2 2\theta_{23}$ and Δm^2_{32} and ν_e appearance ($\nu_\mu \to \nu_e$), which is sensitive to $\sin^2 2\theta_{13}$, the octant of θ_{23} and through subleading terms to δ_{CP} and the mass hierarchy[1]

[a] e-mail: longhin@pd.infn.it

[1] "Normal" mass hierarchy (NH) indicates the $\Delta m^2_{32}, \Delta m^2_{31} > 0$ case (state m_3 having the largest mass) while the other case is denoted as "inverted" mass hierarchy (IH).

via matter effects [2]. Variations in δ_{CP} induce asymmetries in the probabilities for the CP-conjugate channels $\nu_\mu \to \nu_e$ and $\bar{\nu}_\mu \to \bar{\nu}_e$. Negative values of $\sin\delta_{CP}$ enhance $\nu_\mu \to \nu_e$ and suppresses $\bar{\nu}_\mu \to \bar{\nu}_e$ oscillations and vice versa.

T2K has been taking data since 2010. The beam power has been steadily increasing (see Fig. 1) and stable running at 420 kW beam power was achieved in 2016. From mid-2013 until spring 2016, the beam ran in $\bar{\nu}$-mode. The event rates in $\bar{\nu}$-mode are significantly lower than in the ν-mode beam, mainly due to the lower $\bar{\nu}$ cross section and secondly to charge asymmetries in hadroproduction. Sensitivity studies show that the optimal strategy to constrain δ_{CP} is achieved when collecting roughly equal amounts of ν-mode and $\bar{\nu}$-mode protons on target (PoT) [1]. Using both beam polarities T2K can test the PMNS framework and search in an optimal way for CP violation ($\sin\delta_{CP} \neq 0$) implying $P(\nu_\mu \to \nu_e) \neq P(\bar{\nu}_\mu \to \bar{\nu}_e)$ or CPT violation or non-standard matter effects by comparing $P(\nu_\mu \to \nu_\mu) = P(\bar{\nu}_\mu \to \bar{\nu}_\mu)$. The results considered in this report are based on a sample of 1.1×10^{21} PoT, 0.7 and 0.4 $\times 10^{21}$ in ν-mode and $\bar{\nu}$-mode respectively (the full Run 1–7b data set).

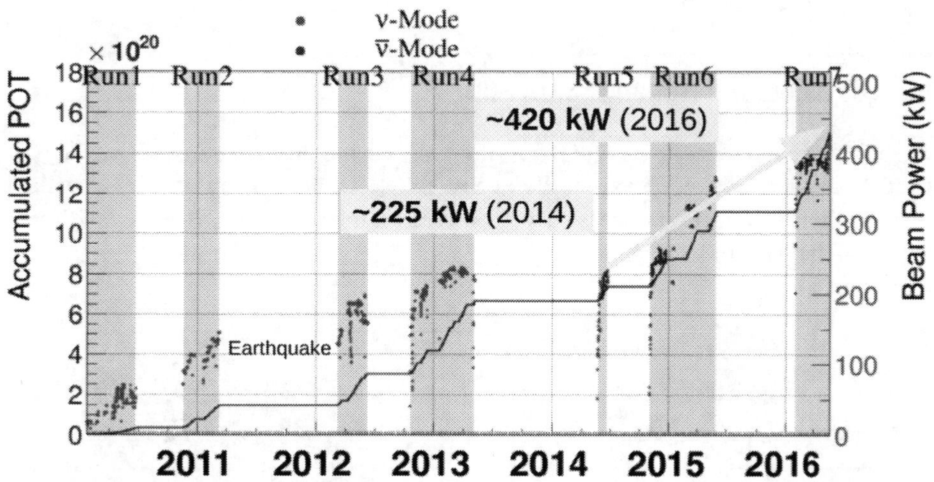

Figure 1. Time evolution of the integrated PoT (cumulative histogram, left side scale) and beam power on target (markers, right side scale).

1.1 The neutrino beam

A data-driven simulation is used to predict the neutrino flux from the T2K beam line. It is based on FLUKA (primary proton-target interactions) and the GEANT3 (propagation, reinteractions) Monte Carlo and incorporates data from the NA61/SHINE experiment [2–4], which has provided measurements of the hadron production cross sections using a thin carbon and a T2K replica target [5]. Beam monitor data (direction, profile and stability) as measured by the on-axis near detector INGRID [6], are also used as inputs to estimate the flux and its systematic uncertainty, currently at the $O(10\%)$ level [7]. The predicted fluxes in both ν and $\bar{\nu}$ modes at SK are shown in Fig. 2.

[2]Matter effects are moderate due to the 295 km baseline amounting at most to $\sim$ 1/3 of the effects ruled by δ_{CP}.

DOI: 10.1051/epjconf/201716401017

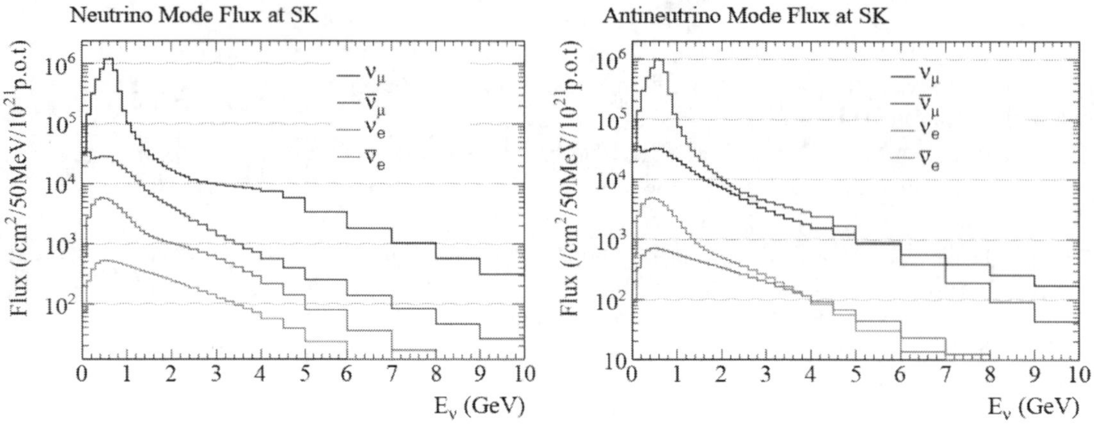

Figure 2. Predicted fluxes at the SK detector in the absence of oscillation effects for ν-mode (left) and $\bar{\nu}$-mode (right).

1.2 The near detectors: ND280 and INGRID

The INGRID [6] near detector (Fig. 3, left) is aligned with the proton beam axis while ND280 (Fig. 3, right) is at the same effective off-axis angle as SK. INGRID, is an array of iron/scintillator (scintillator/water) detectors arranged in a cross shape centered on the beam axis. The beam direction has been shown to be stable within 0.4 mrad.

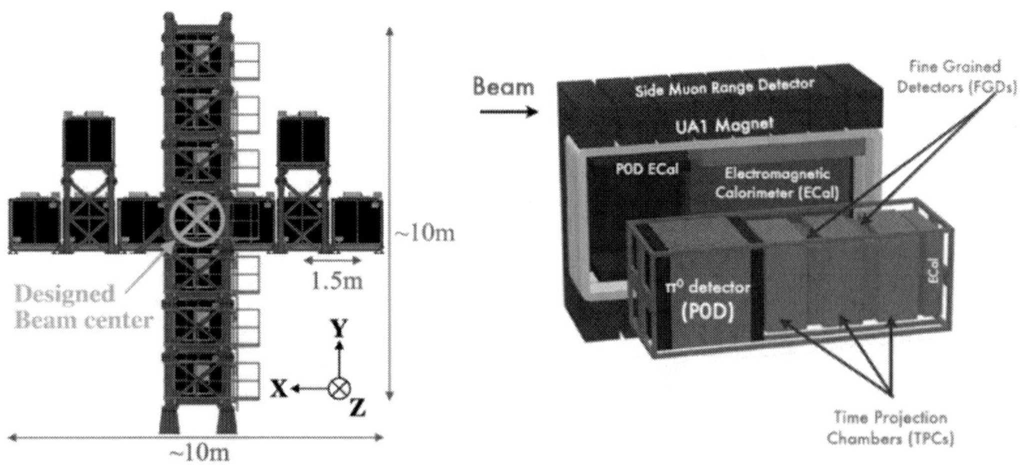

Figure 3. The T2K near detectors. Left: INGRID (on-axis). Right: ND280 (2.5°) off-axis.

ND280 consists of several sub-detectors installed inside the refurbished UA1/NOMAD magnet, providing a 0.2 T uniform magnetic field for charge identification. Three time projection chambers (TPC [8]) are interleaved with two fine-grained detectors (FGD [9]). The upstream FGD detector

(FGD1) is composed of 15 scintillators layers, while the downstream one (FGD2) contains 7+6 scintillator+water alternated layers. Energy loss in the TPC gas and track bending allow for good particle identification. ν_μ and $\bar{\nu}_\mu$ interactions are separated by tagging the primary $\mu^\pm$ and the neutrino energy can be measured (especially for charged-current quasi-elastic interactions, CCQE, where, to a good approximation, it only depends on lepton kinematics). ND280 data (in particular from TPCs and FGD 1, 2) are used directly in the oscillation fits to reduce the flux and cross-section uncertainties and provide a data-driven contraint to the expected number of interactions at SK.

1.3 The far detector: Super-Kamiokande

SK [10] is a 50 kt (22.5 kt fiducial) water Cherenkov detector located 1000 m underground in the Kamioka mine in the Japanese Alps (Gifu Prefecture). The water volume is divided into an outer detector which acts as a veto for entering particles (1885 outward-facing 8-inch photomultipliers - PMTs) and an inner detector lined with 11129, 20-inch PMTs for imaging Cherenkov rings. SK is capable of very good μ/e separation by the pattern of Cherenkov light cones from the charged lepton.

2 The oscillation analysis

The T2K oscillation results feature a joint appearance and disappearance fit using the ν-mode and $\bar{\nu}$-mode data. Data samples of charged current (CC) interactions are fit at ND280 to provide a tuned prediction of the unoscillated spectrum at the far detector and its associated uncertainty. This is then compared to the data at the far detector, where μ-like or e-like data samples are fit to estimate the oscillation parameters.

2.1 The near detector fit

The near detector fit takes inputs from the flux, cross section and detector models with associated uncertainties. The cross-section and flux uncertainties in the predicted rates at SK are reduced by fitting the parameter values in the underlying models. The differences in target materials in the near detector (carbon, oxygen) and far detector (primarily oxygen) is dealt by including large prior uncertainties as the differences in the interactions on different nuclei are not well understood. Separate parameters are used for Fermi momentum, binding energy, multinucleon event normalisation and CC coherent pion production nomalisation on oxygen. The flux model takes into account the correlations between the neutrino fluxes at the near and far detectors.

The ν-mode ν_μ-CC candidates originating in the FGD fiducial volumes are split into three samples depending on the number of tagged pions: CC0π, CC1π^+ and CC-other. In terms of interaction processes these samples are dominated by CCQE interactions ($\nu_\mu n \to \mu^- p$ or $\bar{\nu}_\mu p \to \mu^+ n$, the SK golden sample), CC resonant pion production and deep inelastic scattering, respectively [12]. For $\bar{\nu}$-mode data, both ν_μ and $\bar{\nu}_\mu$ CC interactions are selected and then divided into two subsamples, defined by the number of reconstructed tracks: $\bar{\nu}_\mu$ (ν_μ) CC-1-track, dominated by CCQE interactions and $\bar{\nu}_\mu$ (ν_μ) CC-N-tracks consisting of resonant production and deep inelastic scattering [13].

An important update in the near detector analysis is the use of interactions in the water modules of FGD2 which allow reducing the uncertainties related to the extrapolation across differing nuclear targets in the near and far detectors.

Using FGD1 and FGD2 interaction samples separately results in a total of six samples in ν-mode and eight samples in $\bar{\nu}$-mode. The binned distributions in p_μ and $\cos\theta_\mu$ for the 14 samples are fitted simultaneously to constrain parameters representing the systematic uncertainties in the flux and

DOI: 10.1051/epjconf/201716401017

	ν-mode		$\bar{\nu}$-mode	
Total	1-ring μ-like	1-ring e-like	1-ring μ-like	1-ring e-like
without constraint	12.0%	11.9%	12.5%	13.7%
with constraint	5.0%	5.4%	5.2%	6.2%

Table 1. Relative total systematic error on the number of expected neutrino interactions in SK (single-ring e-like and μ-like) before and after the near detector data constraint.

interaction models. As an example, the muon momentum distributions are shown in Fig. 4 for the CC-0π, CC-1π^+ and CC-other samples in ν-mode in FGD1 before (top row) and after (bottom row) the ND280 fit.

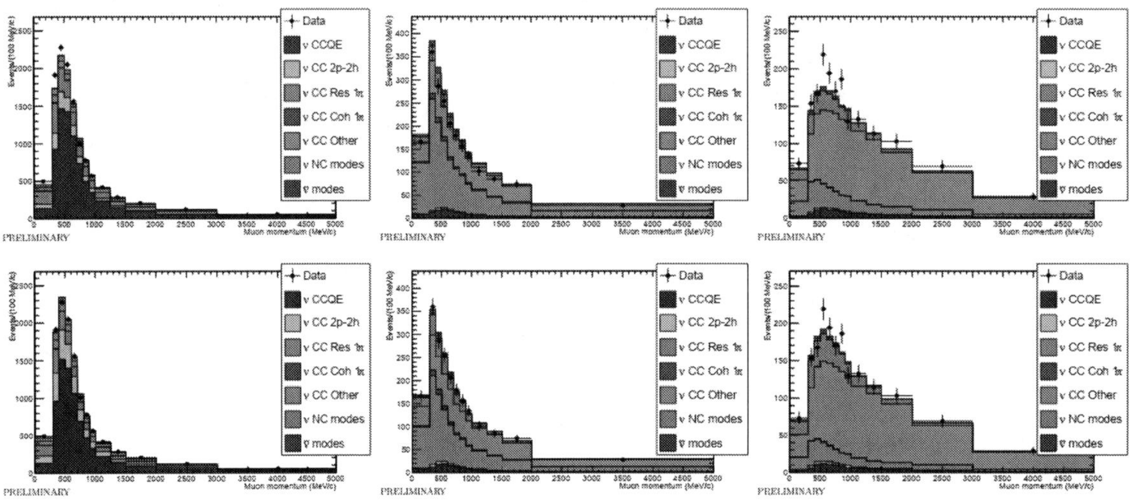

Figure 4. Example distributions of momentum for the muon candidate for the CC-0π (left column), CC1π (middle column) and CC other (right column) samples in the FDG1 fiducial volume. The upper row shows the distributions before the ND280 fit while the distributions after the fit are in the lower row (see Sect. 2.1).

The goodness of fit is 8.6%. The reduction of the uncertainty on the expected number of events at SK is substantial passing from 12–14% to 5–6%, as it can be seen in Tab. 1. The overall error is dominated by the cross-section parameters which are not constrained by the near detector (in particular the multinucleon event normalisation systematic).

2.2 Far detector fit

To enhance the purity of the selected sample in CCQE interactions candidates must have a single identified Cherenkov ring associated to the outgoing lepton. The neutrino energy (E_ν) for CC-QE interactions can be reconstructed from momentum and direction of the lepton assuming a two body interaction. Candidates must be on time with the accelerator, fully contained in the fiducial volume and must have $E_\nu < 1.25$ GeV.

The ν_μ and $\bar{\nu}_\mu$ samples are defined by a single ring muon-like pattern (1Rμ) with a visible energy (E_{vis}) above 200 MeV and having at most one delayed electron signature (from the primary μ decay).

The ν_e and $\bar{\nu}_e$ samples are defined by a single ring electron-like candidatee (1Re) with $E_{vis} > 100$ MeV and no decay electrons that may signal the presence of pions in the event. In addition the 1Re events are classified by a likelihood discriminator that separates out events containing a π^0. The number of events selected at each step in the ν_e selection chain are shown in the upper plots of Fig. 5 for ν and $\bar{\nu}$-modes both for data (bullets) and Monte Carlo (histogram). The distributions of the reconstructed neutrino energy for the final ν_e or $\bar{\nu}_e$ in both beam modes are shown in the bottom plots of the same figure.

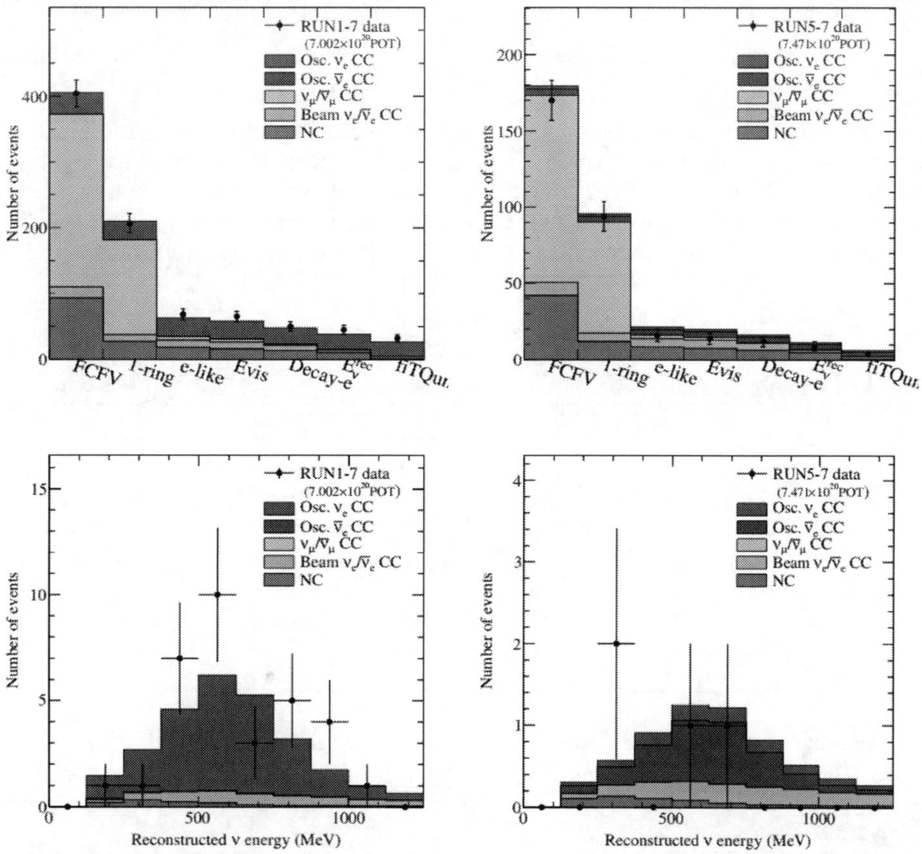

Figure 5. Reconstructed energy distributions for ν_e candidates in ν-mode running (left) and $\bar{\nu}_e$ candidates in $\bar{\nu}$-mode running (right). "Osc." stands for "Oscillated" and indicates the ν_e component arising from $\nu_\mu \rightarrow \nu_e$ transitions. The most relevant backgrounds are the intrinsic ν_e or $\bar{\nu}_e$ components in the beam and Neutral Current (NC) interactions due to π^0 production.

T2K observes a total of 32 ν_e candidate events in the far detector with a purity above 80% and 4 $\bar{\nu}_e$ candidates with ~ 50% purity. The reconstructed energy distributions of the 125 ν_μ candidates in ν-mode running are shown in the left plot of Fig. 6. The same distribution for the 66 $\bar{\nu}_\mu$ candidates in $\bar{\nu}$-mode running is shown by the right-side plot in the same figure.

Table 7 shows the measured and expected event rates under different assumptions for the $\delta_{\rm CP}$ phase: 0, $\pm\pi/2$, π. In ν-mode, the observed 1Re events are larger than the expected ν_e appearance

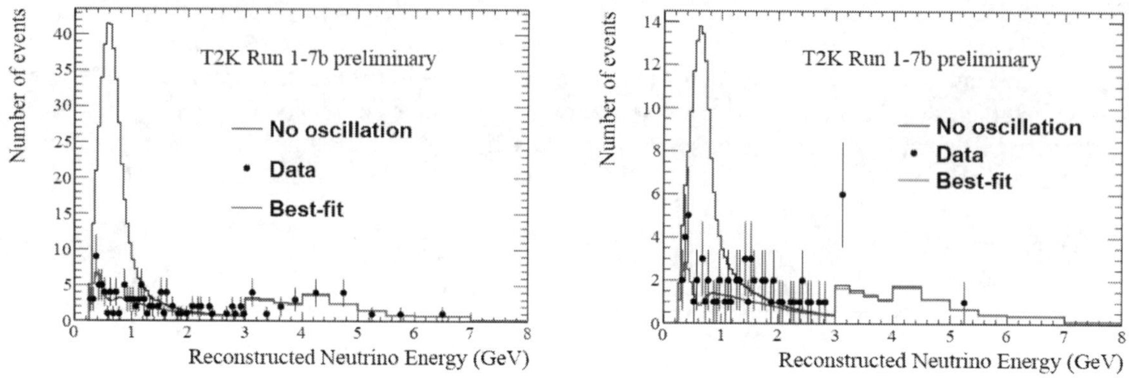

Figure 6. Reconstructed energy distributions for ν_μ candidates in ν-mode running (left) and $\bar{\nu}_\mu$ candidates in $\bar{\nu}$-mode running (right). Left: best-fit neutrino reconstructed energy spectra for the ν-mode μ-like sample corresponding to the Run 1–7b statistic (red). The distribution is obtained by fitting $\sin^2\theta_{23}$, δ_{CP} and mass hierarchy with reactor constraint and marginalising all the systematic parameters. The best-fit distributions are compared with the predicted unoscillated spectrum (blue) as well as the Run 1–7b data set (black dots). Right: anti-ν mode.

Normal hierarchy						
Beam mode	Sample	$\delta_{CP}=-\pi/2$	$\delta_{CP}=0$	$\delta_{CP}=+\pi/2$	$\delta_{CP}=\pi$	Observed
neutrino	μ-like	127.5	127.2	127.4	127.7	125
neutrino	e-like	26.9	22.7	18.4	22.7	32
antineutrino	μ-like	64.2	64.1	64.2	64.4	66
antineutrino	e-like	6.0	6.9	7.7	6.8	4
Inverted hierarchy						
Beam mode	Sample	$\delta_{CP}=-\pi/2$	$\delta_{CP}=0$	$\delta_{CP}=+\pi/2$	$\delta_{CP}=\pi$	Observed
neutrino	μ-like	126.9	127.1	126.8	126.6	125
neutrino	e-like	23.9	20.0	16.1	20.0	32
antineutrino	μ-like	63.8	64.0	63.8	63.7	66
antineutrino	e-like	6.5	7.4	8.4	7.4	4

Figure 7. Expected event rate for SK sample for each mass hierarchy and the following values of δ_{CP}. $-\pi/2$, 0, $+\pi/2$, π. The other oscillation parameters are fixed to $\sin^2\theta_{13} = 0.0217$, $\sin^2\theta_{23} = 0.528$, $\Delta m^2_{32}(\Delta m^2_{13}) = 2.509\times10^{-3}eV^2/c^4$, $\sin^2\theta_{12} = 0.846$, $\Delta m^2_{21} = 7.53\times10^{-5}eV^2/c^4$. The observed event rate during the Run 1–7b data taking is also shown.

events for any choice of δ_{CP}, the lowest difference happening for $\delta_{CP} = -\pi/2$ while in $\bar{\nu}$-mode the observation lies below the expectation also in this case favoring $\delta_{CP} = -\pi/2$ which is the value inducing the largest ν-$\bar{\nu}$ asymmetry. This behavior in the observed data is such that the resulting constraints on δ_{CP} are tighter than what is expected given the sensitivity of the measurement (see Fig. 9).

The result of the near detector fit is propagated to the far detector. This is used as a prior for the far detector fit, which includes additional uncertainties from the SK detector model. The electron and muon samples in the ν-mode and $\bar{\nu}$-mode are fitted simultaneously with nuisance parameters shared between the samples. The results in the δ_{CP}-$\sin^2 2\theta_{13}$ space are shown in Fig. 8. The θ_{13} determination is consistent with the reactor measurement and data show a preference for the maximally CP-violating value of $\delta_{CP} = -\pi/2$ which enhances $\nu_\mu \to \nu_e$ and suppresses $\bar{\nu}_\mu \to \bar{\nu}_e$.

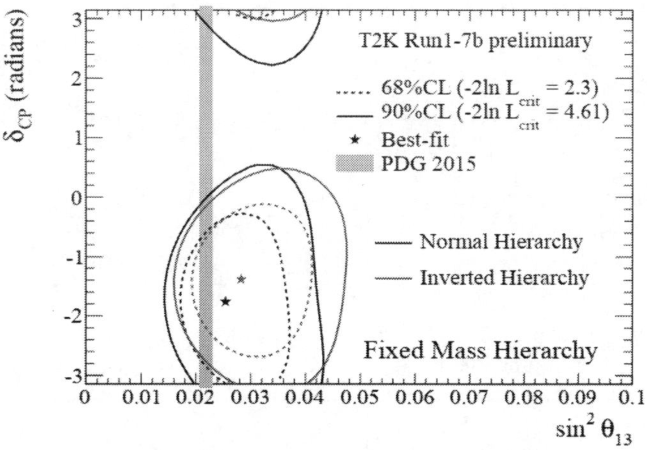

Figure 8. Measured $\Delta\chi^2$ distributions as a function of $\sin^2\theta_{13}$, δ_{CP} and mass hierarchy with the full Run 1–7b data set. Both normal (black solid line) and inverted (black dashed line) hierarchies are shown. The constant $\Delta\chi^2$ critical values in the gaussian approximation are shown. The mass hierarchy was fixed, i.e. the contours are not produced with common χ^2 minimum value. All the oscillation parameters which are not parameters of interest are marginalised. The reactor constraint is not applied.

Figure 9 (left) shows the observed $\Delta\chi^2$ as a function of δ_{CP} and mass hierarchy (black is NH, red IH), with the reactor constraint on $\sin^2 2\theta_{13}$ [14–16]. The 90% confidence levels are built with the Feldman-Cousins method, resulting in intervals of [-3.02, -0.49] for NH and [-1.87, -0.98] for IH. CP conservation ($\sin\delta_{CP} = 0$) is excluded then for both mass hierarchy hypotheses at 90% confidence level. Toy Monte Carlo studies indicate that if $\delta_{CP} = -\pi/2$ the probability to exclude $\delta_{CP} = 0$ or $\pm\pi$ with present data is about 20%. The right plot in Fig.9 shows the expected contraints achievable on average on δ_{CP} with this data sample. In these plots both NH and IH $\Delta\chi^2$ are with reference to the same global minimum which occurs in the NH hypothesis. All the oscillation parameters which are not parameters of interest are marginalised. The global values from PDG 2016 are used as uncertainties on $\sin^2\theta_{13}$, $\sin^2\theta_{12}$ and Δm^2_{21}.

The ν_μ and $\bar{\nu}_\mu$ selection shown in Fig. 6 and Tab. 7 show maximal disappearance occurring exactly in the highest statistics region of the energy spectrum of interacting neutrinos providing world-leading constraints in the $\sin^2\theta_{23}$ and Δm^2_{23} parameters as shown in left plot of Fig. 10. The T2K contour (black) is obtained using $\Delta\chi^2$ critical values in the gaussian approximation. All the oscillation parameters which are not parameters of interest are marginalised. The reactor constraint on θ_{13} is applied. The T2K contour is compared to the other experiment results (SK [18], MINOS+, NOνA [19] and IceCube DeepCore [20]). In the right plot of Fig. 10 it can be seen how T2K data tend to favour maximal disappearance ($\sin^2 2\theta_{23} = 1$) also when fitting ν_μ (red) and $\bar{\nu}_\mu$ (blue) data separately. The best fit for NH occurs at $\sin^2 2\theta_{23} = 0.532^{+0.044}_{-0.060}$ and $\Delta m^2_{23} = 2.545^{+0.084}_{-0.082}$ while at $\sin^2 2\theta_{23} = 0.534^{+0.041}_{-0.059}$ and $\Delta m^2_{23} = 2.510^{+0.082}_{-0.083}$ for IH.

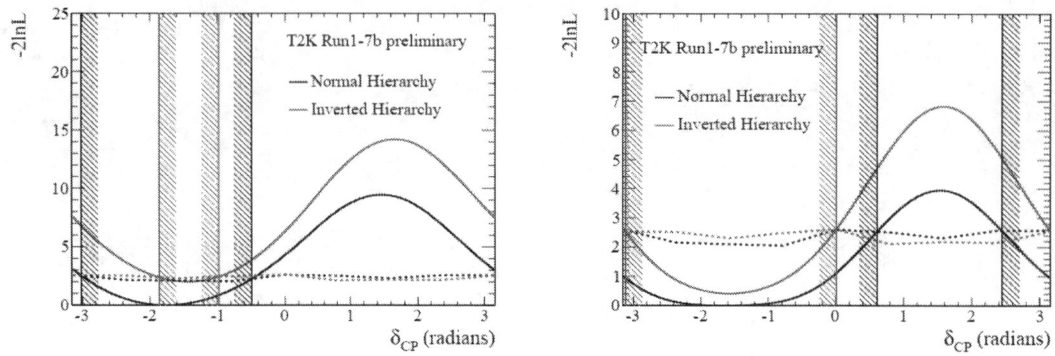

Mass Hierarchy	Allowed region at 90% CL (radians)
Normal	$-3.02 \leq \delta_{CP} \leq -0.49$
Inverted	$-1.87 \leq \delta_{CP} \leq -0.98$

Figure 9. Observed (left) and expected (right) 90% CL intervals on $\delta_{\rm CP}$. Left: measured allowed 90% CL intervals for $\Delta\chi^2$ distributions as a function of $\delta_{\rm CP}$ and mass hierarchy with the Run 1–7b data set. Both normal (black lines) and inverted (red lines) hierarchies are shown. Right: expected allowed 90% CL intervals assuming ($\delta_{\rm CP} = -1.601$, $\theta_{23} = 0.528$ and $\Delta m^2_{32} = 2.509 \times 10^{-3} eV^2/c^4$ (Δm^2_{13}).

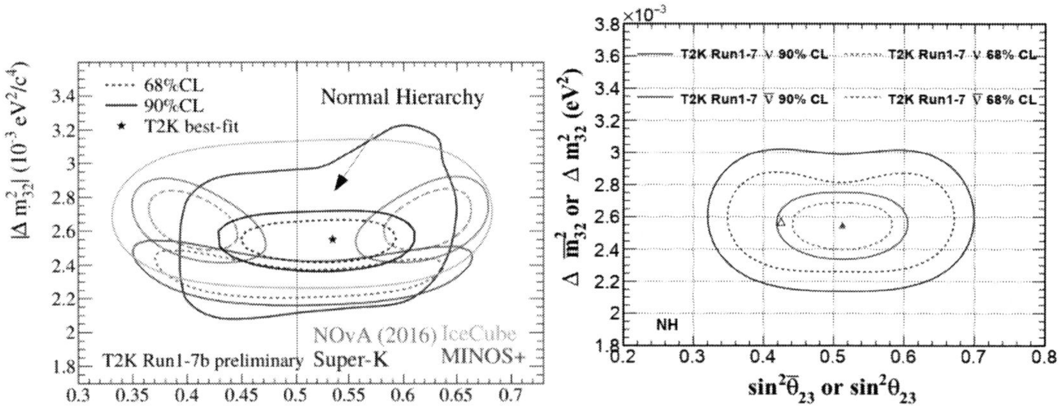

Figure 10. Left: Measured $\Delta\chi^2$ distributions as a function of $\sin^2\theta_{23}$, $\Delta m^2_{32}(\Delta m^2_{13})$ and mass hierarchy with the full Run 1–7b data set. The mass hierarchy is fixed to normal. Right: comparison of T2K ν_μ (red) and $\bar{\nu}_\mu$ (blue) data.

3 Summary and future prospects

The results of a joint analysis of ν and $\bar{\nu}$ oscillations at T2K with the ν_μ ($\bar{\nu}_\mu$) disappearance and $\nu_\mu(\bar{\nu}_\mu) \to \nu_e(\bar{\nu}_e)$ appearance channels, based on 7×10^{20} PoT in ν-mode and 4.0×10^{20} in $\bar{\nu}$-mode have been presented. Data show a preference for maximal θ_{23} mixing with no hints for a violation of the CPT theorem. In the appearance channel, we observe a tendency for large ν_e appearance and a low $\bar{\nu}_e$ appearance with respect to expectations. This favours the $\delta_{\rm CP} \sim -\pi/2$ solution, with a 90% confidence interval of $[-3.02, -0.49]$ in normal mass hierarchy hypothesis and $[-1.87, -0.98]$ in inverted mass hierarchy. The first combined neutrino and anti-neutrino oscillation analysis at T2K is well-advanced.

3.1 The T2K-II phase

T2K was originally approved to collect 7.8×10^{21} PoT by 2021. An extension of the experiment (T2K-II phase) has been proposed recently aiming at a three-fold increase in the baseline sample by 2026 (20×10^{21} PoT). This goal can be met by a staged upgrade program of the accelerator beamlines. The first-stage of the program has been approved to improve the J-PARC Main Ring power supplies and reduce the machine cycle from 2.48 s to 1.3 s, allowing the beam power to be raised from the current 420 kW to $\sim$ 800 kW. Further accelerator and secondary beamline upgrades will allow it to reach 1.3 MW. Figure 11 shows the expected future trends in terms of accelerator power and collected PoT. Further improvements will allow the magnetic horns to be operated with an increased current

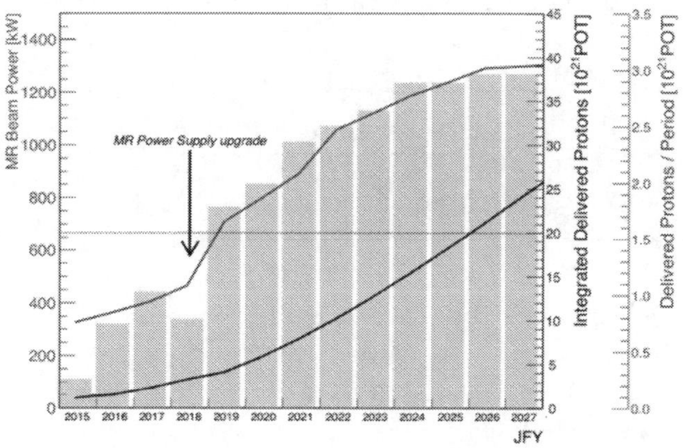

Figure 11. Projection of PoT (blue curve) and power on target (red curve) for the T2K-II program.

from 250 to 320 kA, an optimization of the far detector fiducial volume and of the samples used in the analysis. This could allow a 50% increase in the effective PoT.

The potential impact of the outlined program is quite significant: it may allow T2K-II to exclude CP conservation at more than 3 σ for any value of θ_{23} provided that the true value would turn out to be the currently favoured one i.e. $\delta_{CP} = -\pi/2$ (see Fig. 12).

References

[1] K. Abe et al. (T2K Coll.) PTEP 2015 (2015) no.4, 043C01
[2] N. Abgrall et al. (NA61/SHINE Coll.), Phys. Rev. C 85, 035210 (2012).
[3] N. Abgrall et al. (NA61/SHINE Coll.), Phys. Rev. C 84, 034604 (2011).
[4] N. Abgrall et al. (NA61/SHINE Coll.), Eur. Phys. J. C76, 84 (2016).
[5] N. Abgrall et al. (NA61/SHINE Coll.), Eur. Phys. J. C76 11, 617 (2016).
[6] K. Abe et al. (T2K Coll.), Nucl. Instrum. Methods A694, 211 (2012).
[7] K. Abe et al. (T2K Coll.), Phys.Rev. D87 (2013) no.1, 012001
[8] N. Abgrall et al. (T2K ND280 TPC Coll.), Nucl. Instrum. Methods A637, 25 (2011).

DOI: 10.1051/epjconf/201716401017

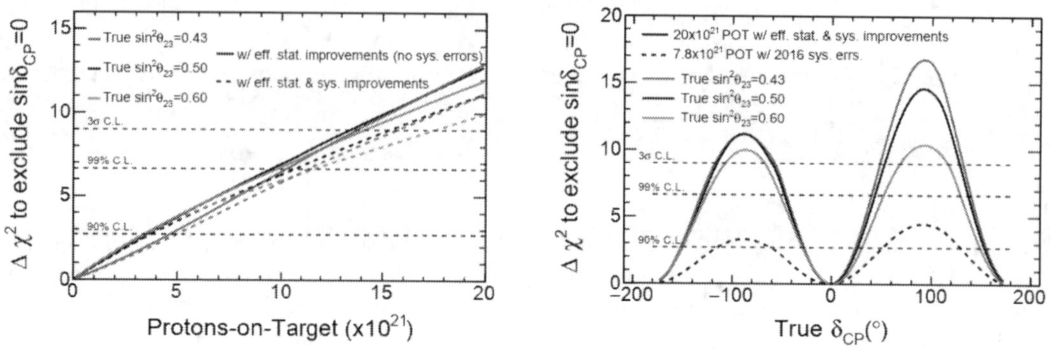

Figure 12. T2K-II physics reach. $\Delta\chi^2$ to exclude $\sin\delta_{CP} = 0$ as a function of PoT with a 50% effectively statistic improvement of T2K Phase 2 statistics and improved systematic errors. The $\delta_{CP} = -\frac{\pi}{2}$ and normal mass hierarchy are assumed to be true. Right: $\Delta\chi^2$ to exclude $\sin\delta_{CP} = 0$ as function of true δ_{CP}. Sensitivities at different values of $\sin^2\theta_{23}$ (0.43, 0.5, 0.63) are compared. Mass hierarchy is assumed to be known.

[9] P. Amaudruz et al. (T2K ND280 FGD Coll.), Nucl. Instrum. Methods A696, 1 (2012).
[10] K. Abe et al. Nucl. Instrum. Meth., A659, 106-135, 2011.
[11] K. Abe et al. (T2K Coll.), Nucl. Instrum. Methods A659, 106 (2011).
[12] K. Abe et al. (T2K Coll.), Phys. Rev. D91, 072010 (2015).
[13] K. Abe et al. (T2K Coll.), Phys. Rev. Lett. 116 (2016) no.18, 181801.
[14] F. P. An et al. Phys. Rev. Lett., 108, 171803, 2012.
[15] J. K. Ahn et al. Phys. Rev. Lett., 108, 191802, 2012.
[16] Y. Abe et al. Phys. Rev. Lett., 108, 131801, 2012.
[17] K. Abe et al. e-Print: 1609.04111.
[18] R. Wendell PoS ICRC2015 (2015) 1062.
[19] P. Adamson et al. (NOνA Coll.), Phys. Rev. D93 (2016) 051104, e-Print: 1601.05037.
[20] M. G. Aartsen et al. (IceCube Coll.), Phys.Rev. D91 (2015) 072004, e-Print: 1410.7227.

DOI: 10.1051/epjconf/201716401018

Novel scenario for production of heavy flavored mesons in heavy ion collisions

B.Z. Kopeliovich[1a], J. Nemchik[2,3], I.K. Potashnikova[1], and Ivan Schmidt[1]

[1] *Departamento de Física, Universidad Técnica Federico Santa María; and Centro Científico-Tecnológico de Valparaíso; Casilla 110-V, Valparaíso, Chile*
[2] *Czech Technical University in Prague, FNSPE, Břehová 7, 11519 Prague, Czech Republic*
[3] *Institute of Experimental Physics SAS, Watsonova 47, 04001 Kosice, Slovakia*

Abstract. The observed strong suppression of heavy flavored hadrons produced with high p_T, is caused by final state interactions with the created dense medium. Vacuum radiation of high-p_T heavy quarks ceases at a short time scale, as is confirmed by pQCD calculations and by LEP measurements of the fragmentation functions of heavy quarks. Production of a heavy flavored hadrons in a dense medium is considerably delayed due to prompt breakup of the hadrons by the medium. This causes a strong suppression of the heavy quark yield because of the specific shape of the fragmentation function. The parameter-free description is in a good accord with available data.

1 Introduction

In the popular scenario, explaining jet quenching, observed in heavy ion collisions, by induced energy loss in the hot medium created in the nuclear collision, a much weaker suppression, compared with light hadrons, was anticipated [1] for heavy flavors, caused by the dead-cone effect.

Later, however, measurements revealed similar magnitudes of suppression for heavy and light hadrons. Here we propose an alternative scenario for production of heavy flavored hadrons from a hot medium. The novel mechanism well explains data in a parameter-free way.

2 Hard parton collision

High-pt parton-parton scattering leads to formation of 4 cones of gluon radiation: (i) the color field of the colliding partons is shaken off in forward-backward directions; (ii) the scattered partons carrying no field up to transverse frequences $k < p_T$, are regenerating the lost components of their field, radiating gluons and forming two high-p_T jets. This process is illustrated in figure 1.

The coherence length/time of gluon radiation by a quark of mass m_q and energy E_q has the form,

$$l_c = \frac{2E_q x(1-x)}{k_T^2 + x^2 m_q^2}, \tag{1}$$

[a]e-mail: boris.kopeliovich@usm.cl

DOI: 10.1051/epjconf/201716401018

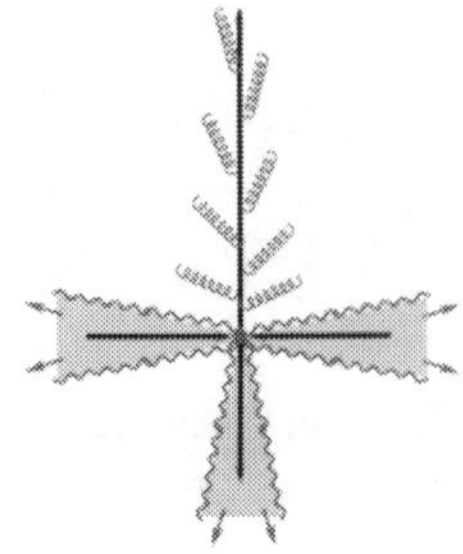

Figure 1. High-p_T parton scattering. The two forward-backward jets are formed by the shaken-off gluon fields. The high-p_T partons regenerate the lost field, radiating gluons.

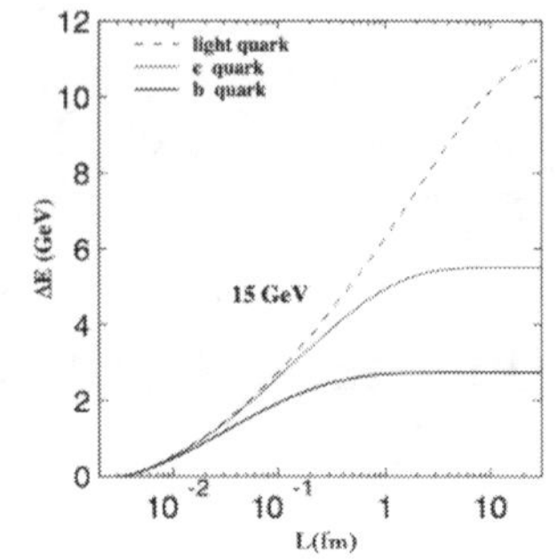

Figure 2. Radiational energy loss of light, c and b quarks having initial energy $E = \sqrt{p_T^2 + m_q^2} = 15\,\text{GeV}$, versus path length.

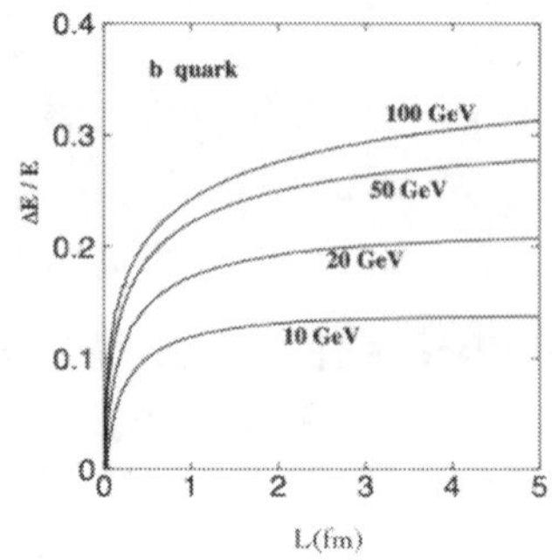

Figure 3. Fractional radiational energy loss by a high-p_T b-quark, produced with different initial energies.

where x is the fractional light-cone (LC) momentum of the radiated gluon. Apparently, first of all are radiated and regenerated gluons with small longitudinal and large transverse momenta.

The peculiar feature of these jets is closeness of their initial virtuality, imposed by p_T, and energy $E = \sqrt{p_T^2 + m_q^2}$. Therefore, increasing the jet energy one unavoidably intensifies radiation and dissipation of energy.

It is instructive to evaluate the amount of energy, radiated after the hard collision by the scattered parton over path length L [2],

$$\Delta E_{rad}(L) = E \int\limits_{\Lambda^2}^{p_T^2} dk^2 \int\limits_0^1 dx\, x\, \frac{dn_g}{dx\, dk^2} \Theta(L - l_c), \tag{2}$$

where the radiation spectrum reads,

$$\frac{dn_g}{dx\, dk^2} = \frac{2\alpha_s(k^2)}{3\pi x} \frac{k^2[1 + (1 - x)^2]}{[k^2 + x^2 m_q^2]^2} \tag{3}$$

The results of calculations for absolute and fractional radiated energy loss are presented in figures 2 and 3 respectively. We see that radiation of heavy quarks ceases shortly. Only a small fraction of the initial quark energy, $\Delta z = \Delta E_{rad}/E$, is radiated even after a long time interval. This is quite different from the hadronization pattern for light quarks, which keep radiating long time and lose most of the initial energy (see figure 2). Therefore, the final B, or D mesons carry almost the whole momentum of the jet. This expectation is confirmed by the direct measurements of the fragmentation functions in e^+e^- annihilation [3]. The example of the $b \to B$ fragmentation function depicted in figure 4 indeed shows that the distribution strongly peaks at $z \sim 0.8$. A similar behavior was observed for the $c \to D$ fragmentation function [4]. At the same time the fragmentation functions of light quarks to light mesons are well known to fall steadily and steeply from small z towards $z = 1$ [5]

3 Production length of heavy mesons

In what follows we mainly consider for concreteness B meson production. If a B-meson (or a $b\bar{q}$ dipole) is produced at the length l_p, the momentum of B equals to the momentum of the b-quark at

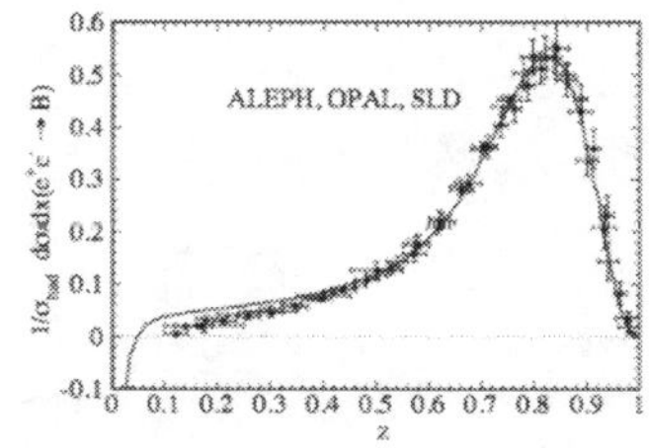

Figure 4. The b fragmentation function, from e^+e^- annihilation. The curve is the DGLAP fit [3].

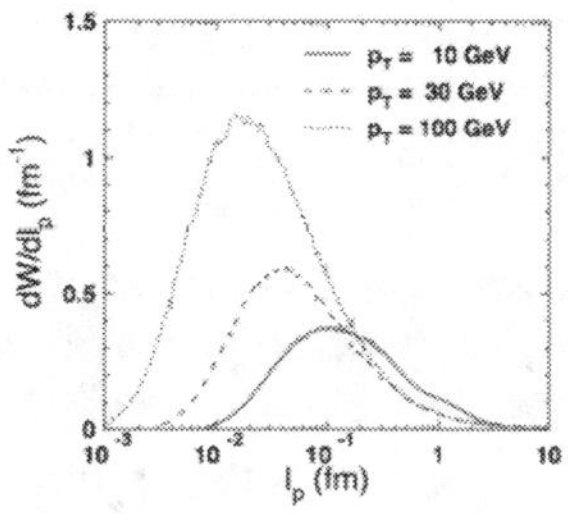

Figure 5. The l_p-distribution of B-mesons produced with different p_T in pp collisions.

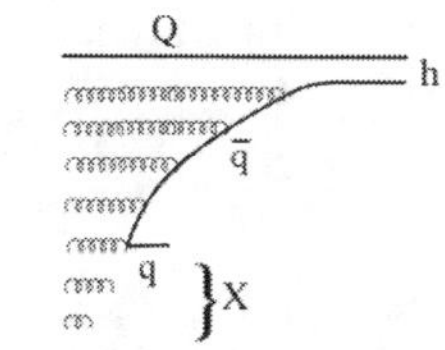

Figure 6. Redistribution of the energy inside a $Q\bar{q}$ dipole by gluon radiation by Q absorbed by $\bar{q}$.

this point. The fraction of the momentum carried by the light q is very small, $x \approx m_q/m_b$, i.e. is about 5%. We neglect this correction in what follows. Since the the radiational vacuum energy loss dE/dl is known, we can directly relate the production length distribution $W(l_p)$ to the $b \to B$ fragmentation function $D_{b/B}(z)$,

$$\frac{dW}{dl_p} = \left.\frac{\partial \Delta p_+^b / p_+^b}{\partial l}\right|_{l=l_p} D_{b/B}(z), \tag{4}$$

where the production length probability distribution and the fragmentation function are normalized to unity, $\int_0^\infty dl_p\, dW/dl_p = 1$ and $\int_0^1 dz\, D_{b/B}(z) = 1$ respectively; $z \equiv p_+^B/p_+^b = 1 - \Delta p_+^b(l_p)/p_+^b$; and

$$\Delta p_+^b(l_p) = \int\limits_0^{l_p} dl\, \frac{dp_+^b(l)}{dl}. \tag{5}$$

The rate of LC momentum loss is related to energy loss in accordance with $p_+^b = E + \sqrt{E^2 - m_b^2}$. Thus, knowledge of $D_{b/B}(z)$ and dE/dl gives a direct relation between z and l_p. As far as we are able to calculate $\Delta z(L)$, we can extract the production length of B-mesons directly from data for $D_{b/B}(z)$. The results for dW/dl_p are plotted vs l_p in figure 5. Remarkably, the mean value of l_p shrinks with rising p_T, like it happens for production of high-p_T light hadrons [6].

Concluding, the fragmentation of a b-quark in vacuum looks like radiational energy loss up to a point $l = l_p$, where it picks up a light $\bar{q}$ forming a colorless $b\bar{q}$ dipole, which performs a direct transition to B without loss the b-quark momentum.

4 In-medium fragmentation

While in vacuum production of B at $l = l_p$ is the end of the story, in a dense medium the B-meson (or $b\bar{q}$ dipole) can easily breakup interacting with the medium, and release the b-quark, which will continue hadronization and pick-up another light antiquark, and so on. Such recreations and breakups of B-mesons will be multiply repeated, until the final production of the detected B-meson, which will survive escaping from the medium. We should understand what happens with the b-quark, while it propagates either as a constituent of a $b\bar{q}$ dipole, or is released and is losing energy to hadronization.

After the color of the b-quark is neutralized by $\bar{q}$, the dipole propagates without radiation and loss of energy. However, if the b-quark did not finish regeneration of its color field, it keeps radiating inside the dipole. The only difference with the preceding radiation process by a single quark, is that the radiation inside the dipole is reabsorbed by accompanying $\bar{q}$, as is illustrated in figure 6. Thus, the $b\bar{q}$ dipole does not radiate, its energy remains constant, however the b-quark energy is redistributed inside the dipole, decelerating the b-quark and accelerating the $\bar{q}$.

As was emphasised above and and demonstrated in figures 2, 3, perturbative radiation of a heavy quark ceases shortly, within a distance of about 1 fm (for a typical p_T range). One might think, looking at figure 2 that on longer distances the quark propagates like a free particle with a constant energy. However, this would certainly contradict confinement. A popular model for the non-perturbative mechanism of energy loss is the string model with $dE_{str}/dl = -\kappa$, where the string tension in vacuum is $\kappa \approx 1\,\mathrm{GeV}/fm$.

While in vacuum a heavy flavored meson is produced on a very short length scale, $l_p \ll 1$ fm, in a hot medium strong absorption pushes the production point to the dilute medium surface. Therefore non-perturbative energy loss becomes important for such a long-lasting hadronization process, which is continuing throughout the whole area occupied by the medium. However, in a deconfined hot medium no string can be formed. Therefore the magnitude of the string tension, and even its very existence, depends on the medium temperature. We rely on the model [7, 8] based on the lattice simulations, of temperature dependence of the string tension, $\kappa(T) = \kappa(1 - T/T_c)^{1/3}$, where the critical temperature is fixed at $T_c = 200\,\mathrm{MeV}$.

Thus, the full rate of energy loss comes from both perturbative and nonperturbative mechanisms,

$$\frac{dE}{dl} = \frac{dE_{rad}}{dl} - \kappa(T). \tag{6}$$

After the b-quark has promptly radiated the whole spectrum of gluons and decreased its virtuality down to the soft scale, the string becomes the only source of energy loss. If the b quark pick-up a light $\bar{q}$ they are connected by a string, which is also decelerating the b-quark and accelerating the $\bar{q}$ with a rate given by the string tension. Such an exchange of energy between b abd $\bar{q}$ is similar to what we observed above for perturbative radiation. So we can conclude that the b-quark is constantly loosing energy with a rate, which does not depend of whether it propagates alone, or as a constituent of a $b\bar{q}$ dipole. The process will finalize only after the last recreation of a B-meson, which escapes from the medium without further breakups. Apparently, the final B-meson will have a reduced energy compared with a B-meson produced at $l = l_p$ in vacuum, where it has no possibility for further breakups. In other words, with the same starting momentum p_+^b the final momentum of the B coming out of a medium, will be smaller than in vacuum. This causes suppression because of steeply falling p_T distribution of the perturbatively produced b-quarks, and due to the steep fall-off of the b-quark fragmentation function at small z , as is shown in figure 4. Notice that above description of fragmentation and time-dependent energy loss holds for charm quarks as well.

5 Where does the suppression come from?

The cross section of production of a B-meson with momentum p_T in pp collision reads,

$$\frac{d\sigma(pp \to BX)}{d^2p_T} = \int d^2p_+^b \,\frac{d\sigma(pp \to QX)}{d^2p_+^b}\,\frac{1}{z} D_{b/B}(z), \tag{7}$$

where $p_+^b = p_T^b + \sqrt{(p_T^b)^2 + m_b^2}$ is the initial LC momentum of the b-quark;

$$z \equiv \frac{(p_T + \sqrt{p_T^2 + M_B^2})}{p_+^b} = 1 - \frac{\Delta p_+^b(l_p)}{p_+^b}. \tag{8}$$

 DOI: 10.1051/epjconf/201716401018

Similar relation holds for AA collisions, however l_p^{AA}, the production length of the final, last created colorless $Q\bar{q}$ dipole, is longer than in pp collisions, so LC momentum loss Δp_+^b is larger, and z_{AA} is smaller.

Besides, the cross section acquires a suppression factor $S(l_p^{AA})$, which is the survival probability of the $Q\bar{q}$ dipole, created at the point l_p^{AA}, to escape the medium without being broken-up, and to develop the hadronic wave function. Thus in AA collision Eq. (7) is modified as,

$$\frac{d\sigma(AA \to BX)}{d^2p_T} = \int d^2p_+^b \, \frac{d\sigma(pp \to QX)}{d^2p_+^b} \, \frac{1}{z_{AA}} D_{b/B}(z_{AA})\, S(l_p^{AA}), \tag{9}$$

with $z_{AA} = 1 - \Delta p_+^b(l_p^{AA})/p_+^b$.

The last factor $S(l_p^{AA})$ in (9) is an important player, making the hadronization processes in pp and AA different. Indeed, if this factor were unity, $S = 1$, then there would be no reason to delay production point to a longer distance z_{AA} compared with hadronization in vacuum, equation (200). However, absorption terminates the colorless $b\bar{q}$ dipoles produced "too early", so it pushes the production point to the diluted surface of the hot medium, making the production length long, $l_p^{AA} \gg l_p$, and $z_{AA} \ll z$, resulting in a strong suppression of the fragmentation function $D_{b/B}(z_{AA})$ according to figure 4.

Thus, the two last factors in (9) work in opposite directions, causing suppression of produced B or D mesons:

(i) The fragmentation function $D_{b/B}(z)$, peaking at large z (fig. 4), tends to reduce momentum loss $\Delta p_+^b(l_p^{AA})$, selecting shorter l_p^{AA}.

(ii) However, a shorter l_p^{AA} means a longer path length for further propagation of the colorless $Q\bar{q}$ dipole in the medium, increasing its chance to brake-up.

6 Attenuation of a $Q\bar{q}$ dipole in a hot medium

A $Q\bar{q}$ pair produced perturbatively with initially small separation, quickly expands. Indeed, the light quark in the $Q\bar{q}$-meson carries a tiny fraction of the momentum, $x \sim m_q/m_Q$. Therefore, even if the produced $b\bar{q}$ dipole has a small transverse separation, its size expands with a high speed, enhanced by $1/x$. The formation time of the B-meson wave function (in the medium rest frame) is very short,

$$t_f^B = \frac{\sqrt{p_T^2 + m_B^2}}{2m_B\omega}, \tag{10}$$

where $\omega = 300\,\text{MeV}$ is the oscillator frequency, which determines the splitting of the ground state and the first radial excitation. For instance at $p_T = 10\,\text{GeV}$ the B meson is formed on a distance $l_f^B = 0.8\,\text{fm}$.

Thus, not a small-size $r^2 \sim 1/p_T^2$ dipole, like for light hadron production [9], but a nearly formed large $Q\bar{q}$ dipole is propagating through the medium. It can be easily broken-up, so its mean free path is quite short. Indeed, the B-meson is nearly as big as a pion, $\langle r_{ch}^2\rangle_B = 0.378\,\text{fm}^2$ [10]. The mean free path of such a meson in a hot medium is very short, $\lambda_B \sim \left[\hat{q}\,\langle r_T^2\rangle\right]^{-1}$, where $\langle r_T^2\rangle = 8\langle r_{ch}^2\rangle/3$. The so called transport coefficient $\hat{q}$ is the rate of broadening of the quark transverse momentum in the medium, this is why it controls the dipole absorption cross section. For instance, at $\hat{q} = 1\,\text{GeV}^2/\,\text{fm}$ (compare with [9]) the mean free pass $\lambda_B = 0.04\,\text{fm}$, i.e. the b-quark propagates through the hot medium, frequently picking up and losing light antiquark comovers. Meanwhile the b-quark keeps losing energy with a rate, enhanced by medium-induced effects. Eventually the detected B-meson is formed and can survive in the dilute periphery of the medium.

6.1 The suppression factor $S(l_p^{AA})$

Thus, the finally detected B-meson is produced at $l = l_p^{AA}$. In the low energy limit the formation length (10) is very short and one can use the eikonal Glauber approximation,

$$S(l_p^{AA}) = \exp\left[-\frac{\langle r_B^2\rangle}{2}\int\limits_{l_p^{AA}}^{\infty} dl\,\hat{q}(l)\right]. \tag{11}$$

In contrast, in another extreme, the high-energy limit, the dipole size is "frozen" by Lorentz time dilation. Then,

$$S(l_p^{AA}) = \int d^2r\,dx\,|\Psi_B(r,x)|^2 \exp\left[-\frac{r^2}{2}\int\limits_{l_p^{AA}}^{\infty} dl\,\hat{q}(l)\right], \tag{12}$$

where $\Psi_B(r,x)$ is the LC wave function of the B-meson.

The general description interpolating between these two limits, is the path-integral technique [11], summing all paths of the Q and $\bar{q}$. The result has the form,

$$S(l_1,l_2) \propto \left|\int\limits_0^1 dx \int d^2r_1 d^2r_2\,\Psi_M^\dagger(r_2,x)\,G_{Q\bar{q}}(l_1,\vec{r}_1,x;l_2,\vec{r}_2,x)\,\Psi_{in}(r_1,x)\right|^2, \tag{13}$$

where in the case under consideration $l_1 = l_p^{AA}$, $l_2 \to \infty$. The initial distribution amplitude $\Psi_{in}(r_1,x)$ is taken in the Gaussian form with mean separation $\langle r_1^2\rangle = \langle r_B^2\rangle$. The Green function $G_{Q\bar{q}}(l_1,\vec{r}_1,x;l_2,\vec{r}_2,x)$ describing propagation of the dipole between longitudinal coordinates l_1, l_2 with initial and final separations $\vec{r}_1$ and $\vec{r}_2$ respectively, satisfies the 2-dimensional LC equation,

$$i\frac{d}{dl_2}G_{Q\bar{q}}(l_1,\vec{r}_1,x;l_2,\vec{r}_2,x) = \left[-\frac{\Delta_{r_2}}{2\,p_T\,x(1-x)} + V_{Q\bar{q}}(l_2,\vec{r}_2)\right]G_{Q\bar{q}}(l_1,\vec{r}_1;l_2,\vec{r}_2) \tag{14}$$

The imaginary part of the LC potential is responsible for absorption, $\mathrm{Im}\,V_{Q\bar{q}}(l,\vec{r}) = -\hat{q}(l)r^2/4$. The real part is the phenomenological Cornell-type potential, adjusted to reproduce the masses and decay constants for B and D mesons [10, 12].

7 Results

While the employed phenomenology allows a parameter-free description of data, one parameters is unavoidably present in such kind of analysis. This is the transport coefficient $\hat{q}$, which cannot be reliably predicted, in particular its coordinate and time dependence. We employ here the popular model [13],

$$\hat{q}(l,\vec{b},\vec{\tau}) = \frac{\hat{q}_0\,l_0}{l}\,\frac{n_{part}(\vec{b},\vec{\tau})}{n_{part}(0,0)}\,\Theta(l-l_0), \tag{15}$$

where $\vec{b}$ is the impact parameter of nuclear collision, $\vec{\tau}$ is the impact parameter of the hard parton-parton collision relative to the center of one of the nuclei, $n_{part}(\vec{b},\vec{\tau})$ is the number of participants, and $\hat{q}_0$ is the rate of broadening of a quark propagating in the maximal medium density produced at impact parameter $\tau = 0$ in central collisions ($b = 0$) at the time $t = t_0$ after the collision. The time interval after the hard collision is $t = l/v_{Q\bar{q}}$ where $v_{Q\bar{q}}$ is the speed of the $Q\bar{q}$ dipole. We fixed the medium equilibration time at $t_0 = 1$ fm.

In such a scheme $\hat{q}_0$ is the only fitted parameter, which, however, has been already determined from other hard processes in AA collisions. In particular, it the analysis [9] of data on quenching of light high-p_T hadrons it was found $\hat{q}_0 = 1.6\,\mathrm{GeV}^2/\,\mathrm{fm}$ at $\sqrt{s} = 200\,\mathrm{GeV}$ and $\hat{q}_0 = 2\,\mathrm{GeV}^2/\,\mathrm{fm}$ at $\sqrt{s} = 2760\,\mathrm{GeV}$.

Different sources of time-dependent medium-induced energy loss were added, including radiative and collisional mechanisms [14]. Medium-induced energy loss is much smaller than the vacuum one, and do not produce a dramatic effect. They are particularly small for heavy flavors.

The results of calculations for B-meson production are compared with data on indirect production of J/ψ, originating from B decays. Comparison is made vs p_T and centrality. In figure 7 the dashed curve is calculated with pure vacuum energy loss (radiative plus string) neglecting induced energy loss for $\hat{q}_0 = 2.5$. The upper and bottom solid curves are calculated with $\hat{q}_0 = 2.5$ and $3\,\mathrm{GeV}^2/\,\mathrm{fm}$ respectively, at $\sqrt{s} = 2.76\,\mathrm{TeV}$ in comparison with data [15]. Figure 8 demonstrate comparison

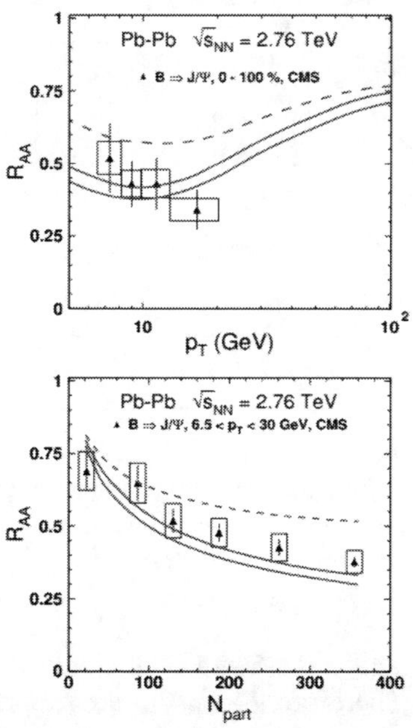

Figure 7. Comparison with CMS data for indirect J/ψ production [16] at $\sqrt{s} = 2.76\,\mathrm{TeV}$. The dashed curves are calculated neglecting induced energy loss. The upper and bottom solid curves are calculated with $\hat{q}_0 = 2.5$ and $3\,\mathrm{GeV}^2/\,\mathrm{fm}$ respectively.

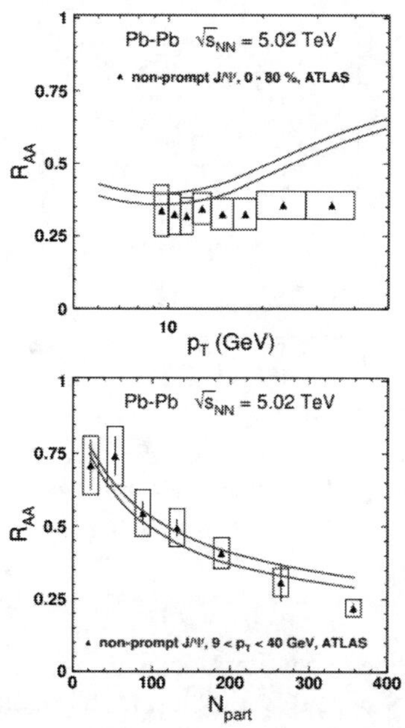

Figure 8. The same as in Fig. 7, but at $\sqrt{s} = 5.02\,\mathrm{TeV}$. The upper and bottom curves are calculated with $\hat{q}_0 = 2.6$ and $3.2\,\mathrm{GeV}^2/\,\mathrm{fm}$ respectively. Data are from [16].

with recent data from ATLAS [16]. The upper and bottom curves are calculated with $\hat{q}_0 = 2.6$ and $3.2\,\mathrm{GeV}^2/\,\mathrm{fm}$ respectively, at $\sqrt{s} = 5.02\,\mathrm{TeV}$. The lack of rise of R_{AA} at high p_T in data looks unusual, compared with the typical behavior of other hadrons, so we would restrain of claiming a serious disagreement.

DOI: 10.1051/epjconf/201716401018

Notice that the rise of $\hat{q}_0$ with energy is natural, but its value , should not depend on the process used to measure it. However, different analyses of different reactions rely on many simplifications and model-dependent assumptions. Therefore, would be naive to expect exact correspondence of $\hat{q}_0$ values extracted from data on different hard processes. The values found here are pretty close to those extracted in [9] from data on high-p_T production of light hadrons.

The approach developed here can also be applied to production of D-mesons. The results are compared with data in figures 9 and 10 vs p_T and centrality.

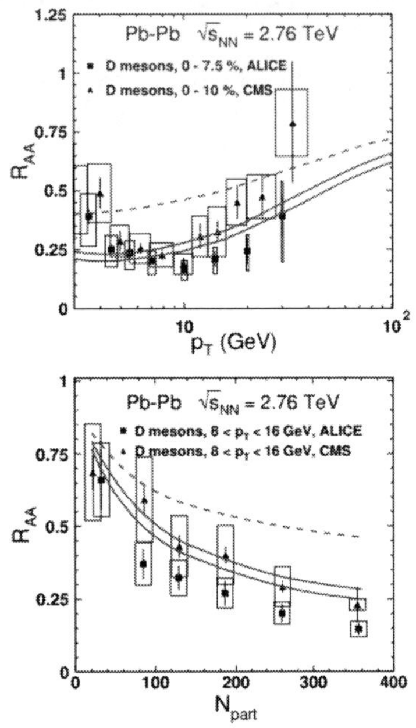

Figure 9. The same as in figure 7, but for D-mesons at $\sqrt{s}$ = 2.76 TeV. Calculations are done with the same values of $\hat{q}_0$ = 2.5 and 3 GeV^2/ fm. Data are from [17–19] and [20, 21].

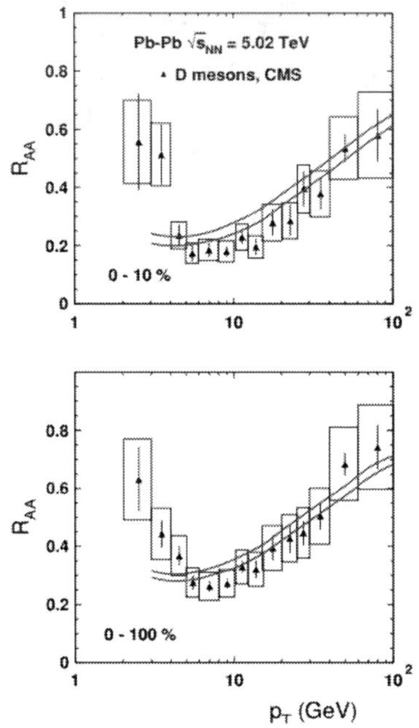

Figure 10. The same as in figure 9, but at $\sqrt{s}$ = 5.02 TeV for centralities 0-80% and minimum bias events. Data from CMS measurements [22].

Notice that c-quarks radiate in vacuum much more energy than b-quarks, while the effects of absorption of $c\bar{q}$ and $b\bar{q}$ dipoles in the medium are similar. Therefore, D-mesons are suppressed in AA collisions more than B-mesons.

8 Summary

Fragmentation of high-p_T heavy quarks expose nontrivial features.

- Heavy and light quarks produced in high-p_T partonic collisions radiate differently. Heavy quarks regenerate their stripped-off color field much faster than light ones, and radiate a significantly smaller fraction of the initial energy.
- This peculiar feature of heavy-quark jets leads to a specific shape of the fragmentation functions. Differently from light flavors, the heavy quark fragmentation functions strongly peak at large frac-

tional momentum z, i.e. the produced heavy-light meson, B or D, carry the main fraction of the jet momentum. This is a clear evidence of a short production time of heavy-light mesons.

- Contrary to the propagation of a small $\bar{q}q$ dipole, which survives in the medium due to color transparency, a $\bar{q}Q$ dipole promptly expands to a large size. Such a big dipole has no chance to survive intact in a hot medium. On the other hand, a breakup of such a dipole does not suppress directly the production rate of $\bar{q}Q$ colorless dipoles, but increase energy loss preceding the final production of a heavy flavored meson. This is different from the scenario of high-p_T production of light $\bar{q}q$ mesons [9].
- Data for production of high-p_T B and D mesons are explained in a parameter-free way. The extracted values of the transport coefficient agree with the results of previous analyses within unavoidable uncertainties, related to employed simplifications and model dependent assumptions, made in the calculations.
- We have disregarded so far the small initial state suppression of heavy flavors due to higher twist heavy dipole attenuation and leading twist shadowing [23]. Inclusion of these effect will lead to a small decrease of the values of $\hat{q}_0$ extracted from the analysis.

Acknowledgements

This work was supported in part by Fondecyt (Chile) grants 1130543, 1130549, 1140842, 1140377, by Proyecto Basal FB 0821 (Chile), and by CONICYT grant PIA ACT1406 (Chile) . J.N. work was partially supported by the grant 13-20841S of the Czech Science Foundation (GAČR), by the Grant MŠMT LG15001, by the Slovak Research and Development Agency APVV-0050-11 and by the Slovak Funding Agency, Grant 2/0020/14.

References

[1] Y. L. Dokshitzer and D. E. Kharzeev, Phys. Lett. B **519**, 199 (2001); [hep-ph/0106202].
[2] B. Z. Kopeliovich, I. K. Potashnikova and I. Schmidt, Phys. Rev. C **82**, 037901 (2010).
[3] B. A. Kniehl, G. Kramer, I. Schienbein and H. Spiesberger, Phys. Rev. D **77**, 014011 (2008); [arXiv:0705.4392 [hep-ph]].
[4] T. Kneesch, B. A. Kniehl, G. Kramer and I. Schienbein, Nucl. Phys. B **799**, 34 (2008); [arXiv:0712.0481 [hep-ph]].
[5] B. A. Kniehl, G. Kramer and B. Potter, Nucl. Phys. B **597**, 337 (2001); [hep-ph/0011155].
[6] B. Z. Kopeliovich, H. -J. Pirner, I. K. Potashnikova and I. Schmidt, Phys. Lett. B **662**, 117 (2008).
[7] H. Toki, S. Sasaki, H. Ichie and H. Suganuma, Austral. J. Phys. **50**, 199 (1997); [hep-ph/9602367].
[8] H. Ichie, H. Suganuma and H. Toki, Phys. Rev. D **54**, 3382 (1996); [hep-ph/9602412].
[9] B. Z. Kopeliovich, J. Nemchik, I. K. Potashnikova and I. Schmidt, Phys. Rev. C **86**, 054904 (2012); [arXiv:1208.4951 [hep-ph]].
[10] C. W. Hwang, Eur. Phys. J. C **23**, 585 (2002); [hep-ph/0112237].
[11] B. Z. Kopeliovich and B. G. Zakharov, Phys. Rev. D **44**, 3466 (1991).
[12] M. Z. Yang, Eur. Phys. J. C **72**, 1880 (2012); [arXiv:1104.3819 [hep-ph]].
[13] X. F. Chen, C. Greiner, E. Wang, X. N. Wang and Z. Xu, Phys. Rev. C **81**, 064908 (2010); [arXiv:1002.1165 [nucl-th]].
[14] R. Baier, Nucl. Phys. A **715**, 209 (2003); [hep-ph/0209038].
[15] S. Chatrchyan et al. (CMS Collaboration); JHEP **1205**, 063 (2012); arXiv:1201.5069 [nucl-ex]. CMS-PAS-HIN-12-014.

[16] ATLAS-CONF-2016-109 (2016).

[17] A. Dubla (for the ALICE Collaboration) "Measurements of heavy-flavour nuclear modification factor and elliptic flow in *Pb-Pb* collisions at $\sqrt{s_{NN}} = 2.76$ TeV with ALICE", Quark Matter 2015, Sep. 27 - Oct. 3, 2015, Kobe, Japan.

[18] J. Adam et al. (ALICE Collaboration); JHEP **03**, 081 (2016); arXiv:1509.06888 [nucl-ex], CERN-PH-EP-2015-25.

[19] J. Adam et al. (ALICE Collaboration); JHEP **1511**, 205 (2015) arXiv:1506.06604 [nucl-ex], CERN-PH-EP-2015-151.

[20] Jian Sun (for the CMS Collaboration); "Open Heavy Flavor Measurements in Heavy Ion Collisions with CMS", Quark Matter 2015, Sep. 27 - Oct. 3, 2015, Kobe, Japan.

[21] S. Chatrchyan at al. (CMS Collaboration); "Nuclear Modification Factor of prompt D_0 in *Pb-Pb* Collisions at $\sqrt{s_{NN}} = 2.76$ TeV" CMS-PAS-HIN-15-005.

[22] A.M. Sirunyan et al. (the CMS Collaboration) D^0 meson nuclear modification factor in *Pb-Pb* collisions at $\sqrt{s_{NN}} = 5.02$ TeV", CMS PAS HIN-16-001.

[23] B. Z. Kopeliovich and A. V. Tarasov, Nucl. Phys. A **710**, 180 (2002) [hep-ph/0205151].

EPJ Web of Conferences **164**, 01019 (2017)
ICNFP 2016
DOI: 10.1051/epjconf/201716401019

Overview of recent azimuthal correlation measurements from ALICE

You Zhou (for the ALICE Collaboration)[1,a]

[1] *Niels Bohr Institute, University of Copenhagen, Blegdamsvej 17, 2100 Copenhagen, Denmark*

Abstract. Azimuthal correlations are a powerful tool to probe the properties and the evolution of the collision system. In this proceedings, we will review the recent azimuthal correlation measurements from ALICE at the LHC. The comparison to other experimental measurements and various theoretical calculations will be discussed as well.

1 Introduction

One of the fundamental questions in the phenomenology of Quantum Chromo Dynamics (QCD) is what are the properties of matter at extreme densities and temperatures where quarks and gluons are in a new state of matter, the so-called Quark Gluon Plasma (QGP) [1, 2]. For a non-central collision between two nuclei of lead, the initial nuclear overlap region is spatially asymmetric with an "almond-like" shape. The translation of such anisotropies in the fluctuating initial conditions into anisotropies in momentum space would not occur if individual nucleon-nucleon collisions emitted independently of each other. Thus, the collective expansion of matter created in collisions of heavy-ions, refereed to phenomenon of "collective flow", proved to be one of the best probes to study the detailed properties of these unknown states of matter. Measurements of the collective expansion of the QGP, in particular the azimuthal anisotropies in the expansion, provide us with constraints on the fluctuating initial conditions and the properties of the system such as the equation of state (EoS) and transport coefficients like shear viscosity over entropy density ratio η/s. The characterization of an anisotropic "shape" in the final momentum space is typically performed using a Fourier decomposition of the azimuthal angle distribution of final state particles $P(\varphi)$ [3, 4]:

$$P(\varphi) = \frac{1}{2\pi} \sum_{n=-\infty}^{+\infty} \overrightarrow{V_n}\, e^{-in\varphi} \tag{1}$$

where φ is the azimuthal angle of emitted particles, $\overrightarrow{V_n}$ is the n-th order flow-vector defined as $\overrightarrow{V_n} = v_n\, e^{in\Psi_n}$, its magnitude v_n is the n-th order anisotropic flow harmonic and its orientation is symmetry plane (participant plane) angle Ψ_n. Alternatively, this anisotropy can be generally given by the joint probability density function ($p.d.f.$) in terms of v_n and Ψ_n as:

$$P(v_m, v_n, ..., \Psi_m, \Psi_n, ...) = \frac{1}{N_{event}} \frac{dN_{event}}{v_m v_n \cdots \mathrm{d}v_m\, \mathrm{d}v_n \cdots \mathrm{d}\Psi_m\, \mathrm{d}\Psi_n \cdots}. \tag{2}$$

[a] e-mail: you.zhou@cern.ch

Anisotropic flow was a major piece of the RHIC program [5–7], and its characterization in terms of hydrodynamics [8–12] was taken as one of the most important evidences of observation of the strongly-coupled quark-gluon plasma (sQGP) at RHIC. The precision measurements of anisotropic flow based on the huge data collected at the LHC experiments [13–16] and the successful descriptions by hydrodynamic calculations, demonstrate that the QGP created at the LHC continually behaves like a strongly coupled liquid with a very small η/s [17–22] which is close to a quantum limit $1/4\pi$ [23].

The anisotropic flow measurements are recently extended by ALICE Collaboration to the forward pseudorapidity region with its unique coverage ($-3.5 < \eta < 5$) [24], to identified particles with its powerful particle identifications [25]. In addition, the correlations between different order flow harmonics with newly proposed symmetric cumulants are investigated in Pb–Pb collisions at $\sqrt{s_{\rm NN}}$ = 2.76 TeV [26]. Furthermore, based on the data taken in recent LHC Run2 program, the anisotropic flow of charged particle are measured at the highest collision energy so far at $\sqrt{s_{\rm NN}}$ = 5.02 TeV Pb–Pb collisions [27]. All the measurements provide crucial information on the initial conditions and the dynamic of the system which were poorly known before. In this proceedings we will review recent ALICE measurements of anisotropic flow at the LHC.

2 Forward flow

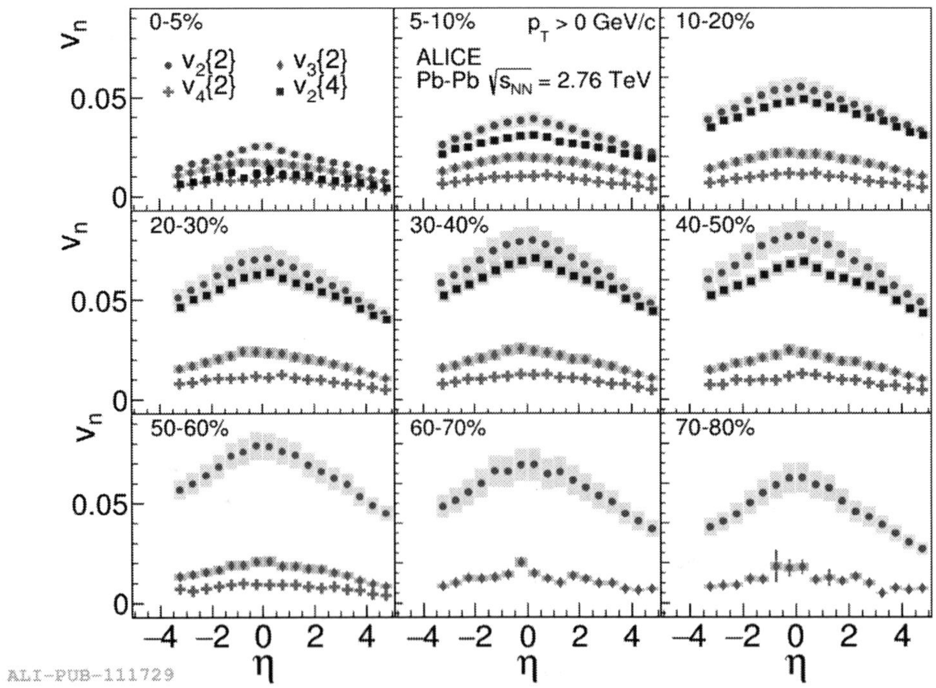

Figure 1. Pseudorapidity dependence of v_2, v_3 and v_4 in various centrality classes. Figure is taken from [24].

The pseudorapidity dependence of v_2 over a wide range ($-5.0 < \eta < 5.3$) and various of collision energies, and systems was investigated by PHOBOS Collaboration [7, 28–30]. It was reported that in the rest frame of one of the colliding nuclei ($\eta - y_{\rm beam}$), v_2 is independent of collision energy. Similar feature was observed in multiplicity density distributions [31]. They suggest that particle production is independent of the collision energy at forward rapidity, the fragmentation region. Such effect is known as extended longitudinal scaling. Not only v_2 but also v_3 and v_4 have been measured over the widest η-range at the LHC, in Pb–Pb collisions at $\sqrt{s_{_{\rm NN}}}$ = 2.76 TeV using the ALICE detector [24]. The pseudorapidity dependence of anisotropic flow in nine centrality classes is shown in Fig. 1. First of all, it is seen that v_2 measured with both 2- and 4-particle cumulants increases from 0-5% to 40-50% centrality classes, $v_2\{2\}$ decreases slightly in more peripheral collisions over the entire pseudorapidity range. Compared to v_2, the higher harmonic flow v_3 and v_4 show weaker centrality dependence. This might be due to the fact that initial state fluctuations play a prominent role for higher harmonics, as the centrality dependence of the corresponding eccentricities ε_n (with $n \geq 3$) are more modest relative to ε_2.

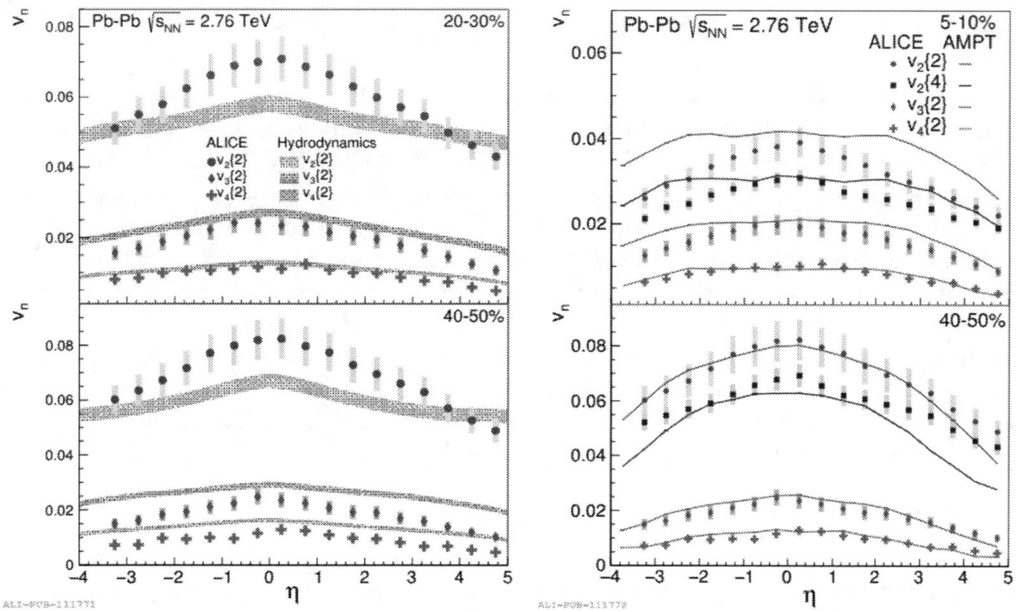

Figure 2. Pseudorapidity dependence of v_2, v_3 and v_4 in selected centrality class. Calculations from hydrodynamic (left) and AMPT (right) are presented for comparisons [32, 33]. Figure is taken from [24].

In order to understand the role of η/s and hadronic interactions in forward pseudorapidity region, the measurements are compared to viscous hydrodynamic calculations [32] whose input parameters e.g. temperature dependent shear viscosity over entry density ratio η/s(T) are tuned to fit the RHIC data, shown in Fig. 2 (left). Although it can nicely describe the PHOBOS measurements of $v_2(\eta)$ [7, 28, 29], it clearly fails to quantitatively describe the LHC data using the same parameterization. More specifically, for selected two centralities, hydrodynamic calculations underestimates the measured v_2 while they overestimate the measurements of v_3 and v_4. Further tuning of parameterization of η/s(T) and/or the modification of initial conditions are necessary to describe the data better. At the same time, A multiphase transport (AMPT) model [33] with a string melting scenario is used for comparisons. This is a hybrid transport model consists of four main stages: initial conditions, partonic interactions, hadronization, and hadronic rescattering. It could reproduce compatible $p_{\rm T}$ integrated v_2, v_3 and v_4 as data at the LHC [33]. Good agreements of anisotropic flow from AMPT and data are observed

 DOI: 10.1051/epjconf/201716401019

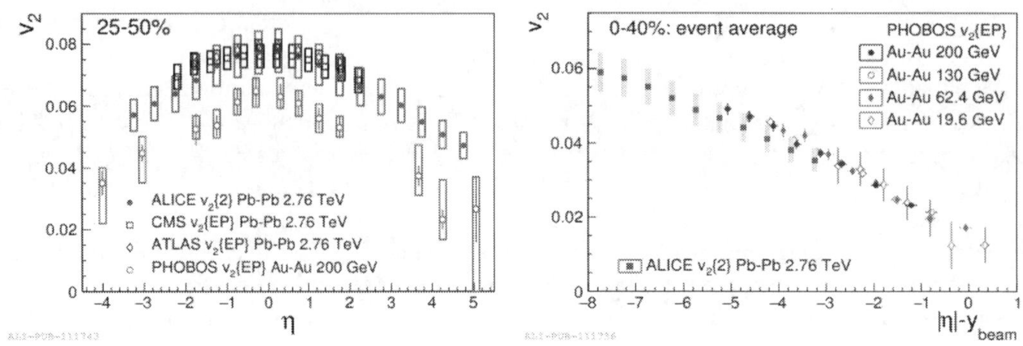

Figure 3. $v_2(\eta)$ measured at RHIC [7, 28, 29] and the LHC [34, 35] (left); observed v_2 in the rest frame of the projectiles for the event averaged 0-40 % centrality range (right). Figure is taken from [24].

in Fig. 2 (right) in 40-50% centrality class, except $v_2\{4\}$ at forward pseudorapidity. In more central collisions, AMPT overestimates v_2 and v_3 nevertheless agrees with v_4 measurements.

The pseudorapidity dependence of v_2 was measured both at RHIC and other LHC experiments before, comparisons to ALICE measurements are shown in Figure 3 (left) with a wider centrality range 25-50 %. It is found that for the overlapped pseudorapidity region, the recent $v_2(\eta)$ measurements by ALICE is consistent with published results from ATLAS [34] and CMS [35] Collaborations within the systematic uncertainties. All the results at the LHC are systematically above the previous $v_2(\eta)$ from PHOBOS at RHIC. As mentioned above, it was reported that in the rest frame of one of the colliding nuclei ($\eta - y_{beam}$), v_2 is energy independent. This is examined also at the LHC as shown in Fig. 3 (right). It is seen that the extended longitudinal scaling is found to hold in Au–Au collisions at 19.6 GeV up to 2.76 TeV in Pb–Pb collisions, similar conclusion was obtained by ATLAS and CMS with smaller pseudorapidity coverage [34, 35].

3 Anisotropic flow of identified particles

The measurements of anisotropic flow for charged particles provided a strong testing ground for hydrodynamical calculations that attempts to describe the dynamical evolution of the system created in heavy-ion collisions. In addition, the measurements for identified particles, together with the comparisons to various theoretical calculations, can help us further constrain initial conditions and dynamics of the collisions, i.e. the properties of the system (e.g. η/s) as well as probe the particle production mechanism (e.g. coalescence). Using the unique capability of particle identification of ALICE detector, not only the v_2 but also higher harmonic flow v_3, v_4 and v_5 are measured in Pb–Pb collisions at $\sqrt{s_{NN}}$ = 2.76 TeV [25, 37]. It is observed in Fig. 4 (top left) that the p_T-differential v_2 for charged pions, kaons and protons show an obvious mass dependence feature for $p_T < 2-3$ GeV/c, in all presented centrality classes. This mass ordering originates from the fact that radial flow creates a depletion in the particle spectrum at low p_T, which increases with increasing particle mass and transverse velocity. This depletion leads to heavier particles (e.g. protons) having smaller v_2 values compared to lighter ones (e.g. pions) at given values of p_T. A crossing of identified particles v_2 is observed around $p_T \approx 2.5$ GeV/c in the most central collisions and happens at lower p_T values in more peripheral collisions, agree with what reported before [37]. Similar to v_2, a clear mass ordering feature is seen in higher harmonic flow v_3, v_4 and v_5 of charged pions, kaons and protons for $p_T < 2-3$ GeV/c. This could be understood as the interplay between the radial flow and anisotropic flow. Furthermore,

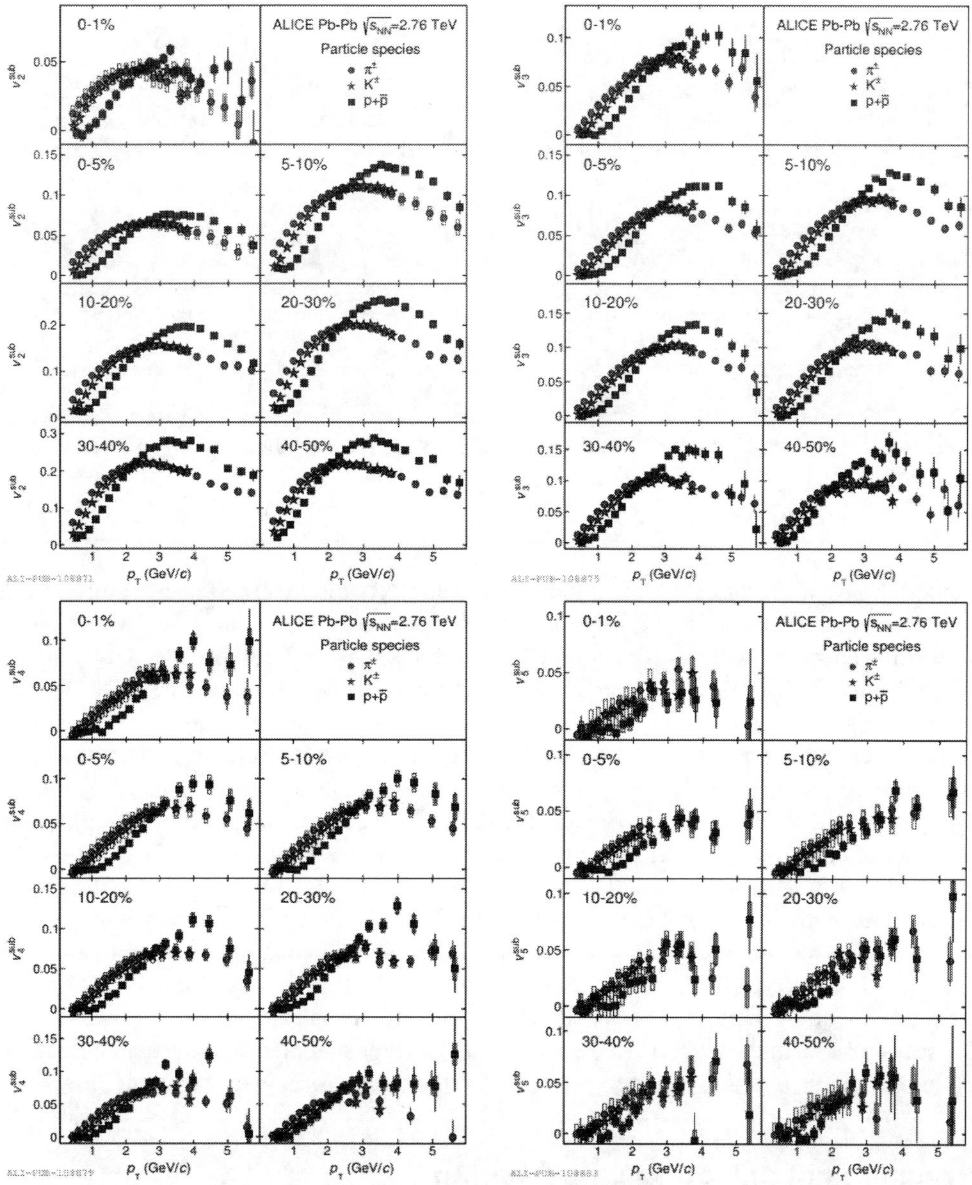

Figure 4. The p_T-differential v_2^{sub} (top left), v_3^{sub} (top right), v_4^{sub} (bottom left) and v_5^{sub} (bottom right) for identified particle species in various centrality classes in Pb–Pb collisions at $\sqrt{s_{NN}}$ = 2.76 TeV. Figure is taken from [25].

the crossing of baryon and meson v_n is also observed. The p_T region where this crossing happens depending on on the centrality and the order of flow harmonic n.

To better extract information on dynamic of system evolution, especially the information of freeze-out conditions, the anisotropic flow of identified particles are compared to hydrodynamic calculations based on iEBE-VISHNU [36]. This is an event-by-event based hybrid model, which uses AMPT as initial conditions and couples 2+1 dimension viscous hydrodynamic calculations to a hadronic

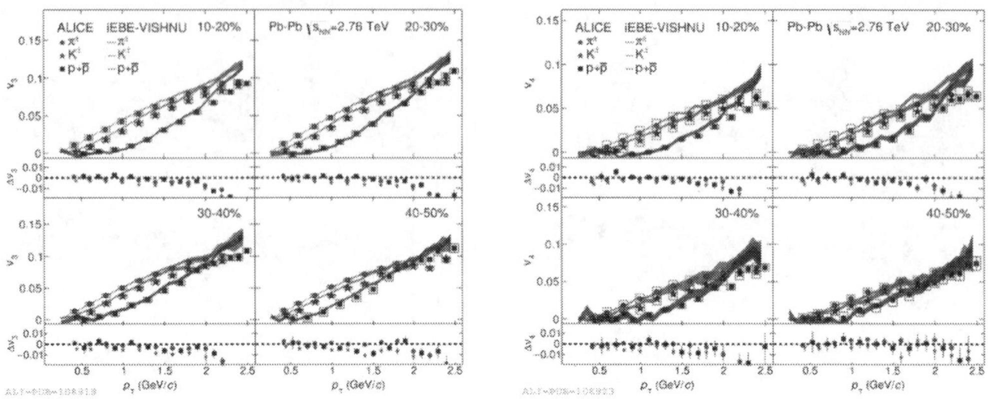

Figure 5. p_{T}-differential v_3^{sub} (left) and v_4^{sub} (right) for identified particle species in various centrality classes in Pb–Pb collisions at $\sqrt{s_{\mathrm{NN}}} = 2.76$ TeV. Comparisons to iEBE-VISHNU calculations [36] are presented. Figure is taken from [25].

cascade model UrQMD. Figure 5 shows that iEBE-VISHNU model could successfully produce the mass ordering feature of anisotropic flow for identified particles. More specifically, the model over-predicts v_2 in average by a 10 % for pion, 10-15 % for kaons and compatible with protons in the level of approximation of 10 %. The agreements with data seems become better in more peripheral collisions. For higher harmonic flow v_3 and v_4, the model describe the data with a reasonable accuracy, they are consistent within 5% for $p_{\mathrm{T}} < 2$ GeV/c in presented centrality classes.

It was observed at RHIC that the v_2 of baryons is larger than that of mesons at the intermediate p_{T} region (roughly $2 < p_{\mathrm{T}} < 4$ GeV/c) [38–40]. It was suggested that if both v_2 and p_{T} are scaled by the number of constituent quarks (n_q), an universe scaling was expected for the v_2/n_q as a function of p_{T}/n_q for different particle species. This so-called number of constituent quark scaling (NCQ) was observed at RHIC with exceptional of pion v_2 [41], but holds at an approximate level of ±20% for v_2 at the LHC [37]. Such scaling is explained by the fact that the quark coalescence might be the dominant particle production mechanism in the intermediate p_{T} region. Figure 6 presents the NCQ scaling for v_2, v_3, v_4 and v_5 for charged pions, kaons and protons in various centrality classes. It was observed that the NCQ scaling holds at the level of ±20 % within the current statistical and systematic uncertainties, not only for elliptic flow which reported before in [37], but also for higher harmonic flow.

4 Correlations of anisotropic flow harmonics

Recently, it is proposed that the correlations between different order flow harmonics have unique sensitivities to initial conditions and η/s of the produced matter in hydrodynamic calculations. This was firstly investigated by ATLAS Collaboration using so-called Event Shape Engineering (ESE) [42]. Later on, a new observable named Symmetric 2-harmonic 4-particle Cumulants, $SC(m,n)$, was proposed to quantitatively study the relationship between different orders of anisotropic flow harmonics v_m and v_n [43]. It is defined as:

$$\begin{aligned} SC(m,n) &= \langle\langle\cos(m\phi_1 + n\phi_2 - m\phi_3 - n\phi_4)\rangle\rangle - \langle\langle\cos(m\phi_1 - m\phi_2)\rangle\rangle\,\langle\langle\cos(n\phi_1 - n\phi_2)\rangle\rangle \\ &= \langle v_m^2\, v_n^2\rangle - \langle v_m^2\rangle\,\langle v_n^2\rangle, \end{aligned} \quad (3)$$

DOI: 10.1051/epjconf/201716401019

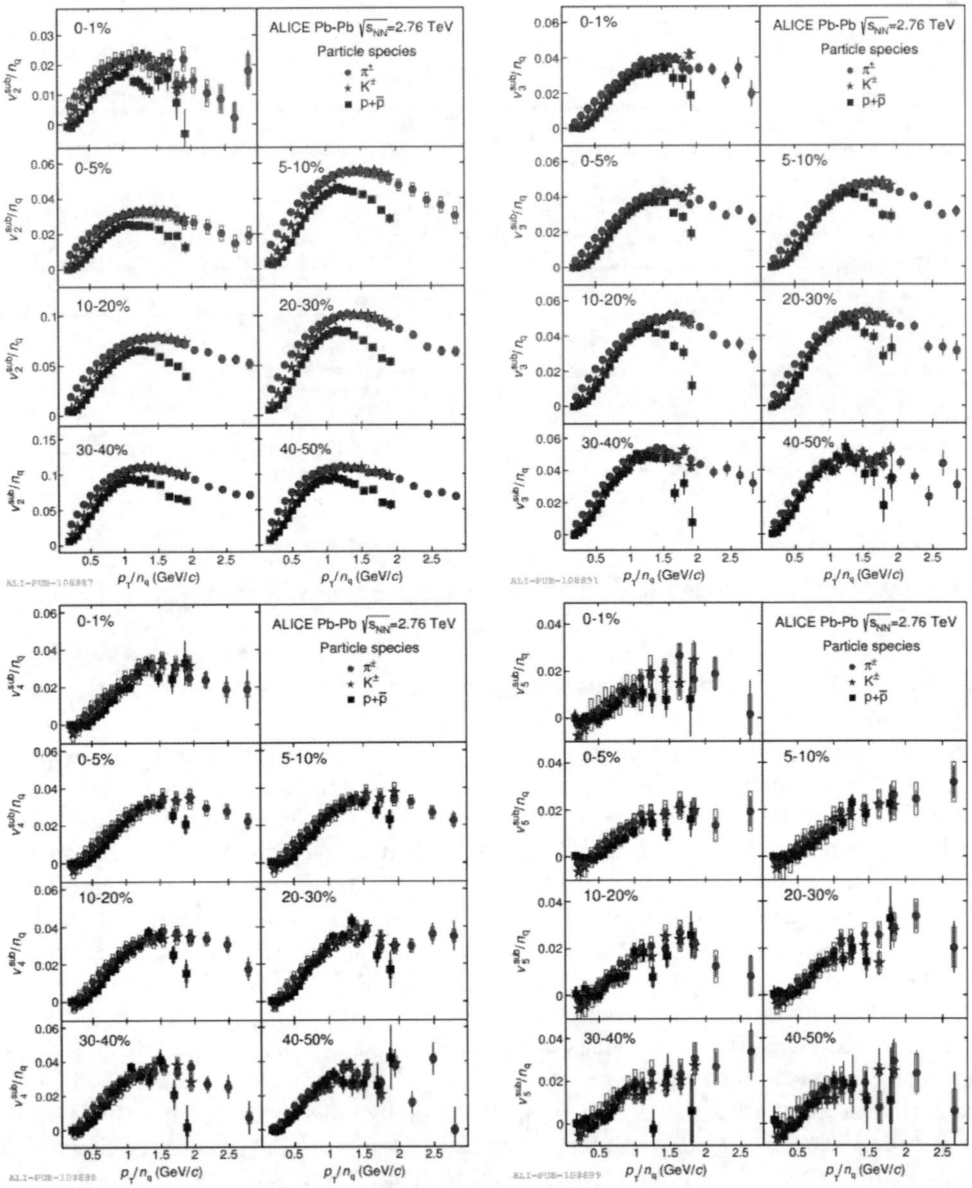

Figure 6. v_2^{sub}/n_q (top left), v_3^{sub}/n_q (top right), v_4^{sub}/n_q (bottom left) and v_5^{sub}/n_q (bottom right) as a function of p_{T}/n_q for identified particles in various centrality classes in Pb–Pb collisions at $\sqrt{s_{\mathrm{NN}}} = 2.76$ TeV. Figure is taken from [25].

with $n \neq m$. By construction, this observable should be insensitive to non-flow effects due to usage of 4-particle cumulant and it is independent on the correlations between various symmetry planes. Thus, it is non-zero only if there is correlation (or anti-correlation) between v_m and v_n. Previous AMPT model calculations showed that $SC(3, 2)$ is negative (anti-correlations between v_2 and v_3) while $SC(4, 2)$ is positive (correlations between v_2 and v_4). The observed (anti-)correlations are also sensitive to the transport properties of the created system, e.g. the partonic and hadronic interactions [43, 44].

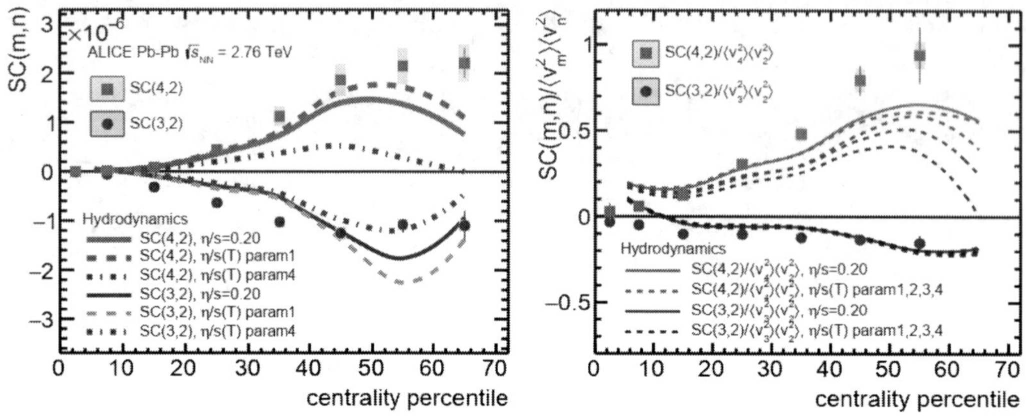

Figure 7. The centrality dependence of $SC(m,n)$ (left) and $NSC(m,n)$ (right) at $\sqrt{s_{\rm NN}}$ = 2.76 TeV Pb–Pb collisions, the hydrodynamic calculations from [45] are presented for comparisons.

The first ALICE measurements of centrality dependence of $SC(4,2)$ (red squares) and $SC(3,2)$ (blue circles) are presented in Fig. 7 (left). Positive values of $SC(4,2)$ and negative results of $SC(3,2)$ are observed for all centralities, as correctly predicted by AMPT [43]. This confirms a correlation between the event-by-event fluctuations of v_2 and v_4, and an anti-correlation between v_2 and v_3. The like-sign technique, which is a powerful approach to suppress non-flow effects [13], is applied. It is found that the difference between correlations for like-sign and all charged combinations are much smaller compared to the magnitudes of $SC(m,n)$ itself. Therefore, the non-zero values of $SC(m,n)$ presented here cannot be explained by non-flow effects solely. To further study the properties of QGP, we show in Fig. 7 (right) the comparison of $SC(m,n)$ measurements and hydrodynamic model calculations. This calculation is from the event-by-event perturbative-QCD+saturation+hydrodynamic ("EKRT") framework [45]. It incorporates both initial conditions and hydrodynamic evolution. It is observed that $SC(m,n)$ is very sensitive to the values of η/s in this model. Thus, the investigation of $SC(m,n)$ provides a new approach to constrain the input of η/s in hydrodynamic calculations, in addition to standard anisotropic flow studies. We also notice that although this hydrodynamic model reproduced the centrality dependence of v_n for $n \leq 2$ fairly well using various η/s(T) parametrization [45], its predictions of $SC(m,n)$ cannot quantitatively describe the data. There is no single centrality for which a given η/s(T) parameterization describes simultaneously $SC(4,2)$ and $SC(3,2)$. Therefore, it is believed that the new observable $SC(m,n)$ provides stronger constrains on the η/s and initial conditions than individual v_n measurement alone.

In order to further investigate the correlations between v_m and v_n without contributions from flow harmonics, it is proposed to divide $SC(m,n)$ by the products $\langle v_m^2\rangle\langle v_n^2\rangle$. This new observable is named $NSC(m,n)$. The measurements from ALICE are presented in Fig. 7 (right). Here 2-particle correlations $\langle v_m^2\rangle$ and $\langle v_n^2\rangle$ are obtained with a pseudorapidity gap of $|\Delta\eta| > 1.0$ to suppress short range non-flow effects. Similar to $SC(4,2)$, the normalized $SC(4,2)$ observable still exhibits a significant sensitivity to different η/s parameterizations and the initial conditions, which again provides a good opportunity to discriminate between various possibilities of the detailed setting of η/s(T) of the produced QGP and the initial conditions used in hydrodynamic calculations. However, $NSC(3,2)$ is independent of the setting of $\eta/s(\mathrm{T})$. In addition, it was demonstrated in [46] that $NSC(3,2)$ is compatible with its corresponding observable $NSC^{\varepsilon}(3,2)$ in the initial state. Thus, the $NSC^{v}(3,2)$ measurements could be taken as a crucial observable to constrain initial conditions without concern of transport properties of the system. Furthermore, none of existing theoretical calculations can reproduce the data, future developments of $SC(m,n)$ and $NSC(m,n)$ involved higher harmonics and more particle correlations [47], together with comparisons to further tuned hydrodynamic calculations, are crucial for deep understanding of correlations between different oder flow harmonics.

5 Anisotropic flow at $\sqrt{s_{\mathrm{NN}}} = 5.02$ TeV

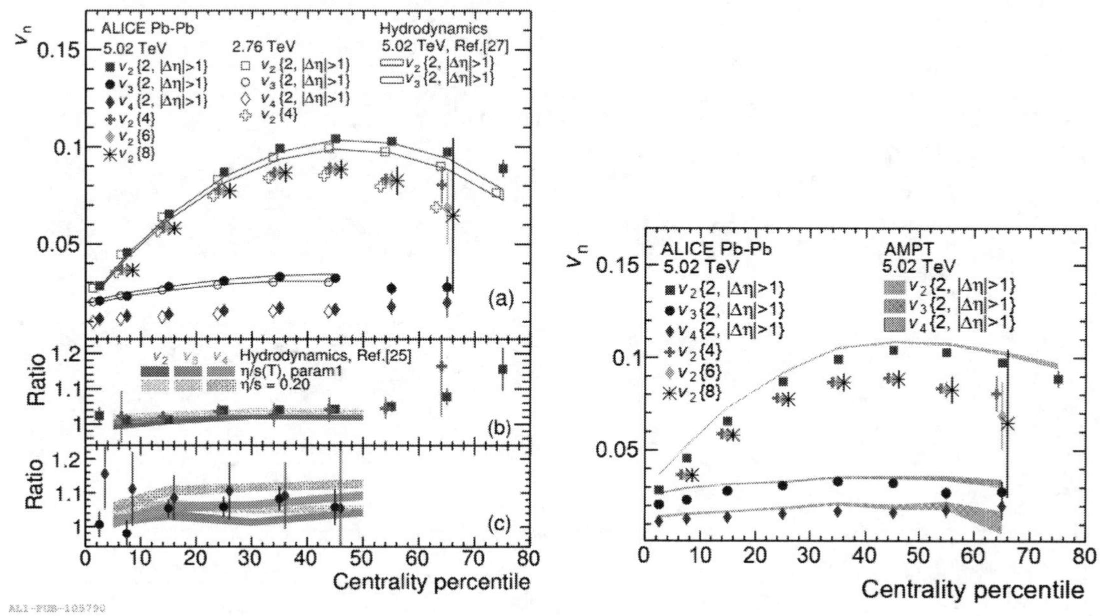

Figure 8. Anisotropic flow as a function of centrality. Hydrodynamic [48, 49] and AMPT [50] calculations are presented for comparisons.

Exploiting the data collected during November 2015 with Pb–Pb collisions at record breaking energies of $\sqrt{s_{\mathrm{NN}}} = 5.02$ TeV, ALICE measured for the first time v_2, v_3 and v_4 of charged particles at this energy [27]. Centrality dependence of anisotropic flow using two-particle correlations with pseudorapidity and multi-particle cumulants methods are shown in Fig. 8. It is observed that v_2 increases from central to peripheral collisions, saturates in the 40–50% centrality class with a maximum value about 0.10 for two-particle correlations and is around 0.09 for multi-particle cumulants. The multi-particle cumulants show similar centrality dependence and the results from four-, six- and eight-particle cumulants are consistent with each other. On the other hand, the significant differences between two- and multi-particle cumulants could be attribute to different sensitivities to elliptic flow fluctuations. The higher harmonic flow v_3 and v_4 are smaller and their centrality dependence is much weaker compared to v_2. In addition, anisotropic flow measurements are compared to hydrodynamic calculations which made a robust prediction without relying on a particular model for initial conditions and without precise knowledge of medium properties such as viscosity [48]. As shown in Fig. 8 left (a), the hydrodynamic calculation are compatible with the measured anisotropic flow v_n. Furthermore, the AMPT model based on a new set of input parameters only slightly overestimates the anisotropic flow measurements [50], as shown in Fig. 8 (right).

Figure 8 left (b) and (c) show the ratios of v_n measured at 5.02 TeV to 2.76 TeV. The increase of v_2 and v_3 is rather moderate from the two energies, with an increase of (3.0±0.6)%, (4.3±1.4)%, respectively. The increase is more significant for v_4, which is (10.2±3.8)%, over the centrality range 0–50% in Pb–Pb collisions. This increase of anisotropic flow is compatible with hydrodynamic predictions [49]. The data seems support a low value of η/s for the created matter in Pb–Pb collisions at $\sqrt{s_{\rm NN}}$ = 5.02 TeV, suggests that the η/s does not change significantly with respect to the one at lower energy.

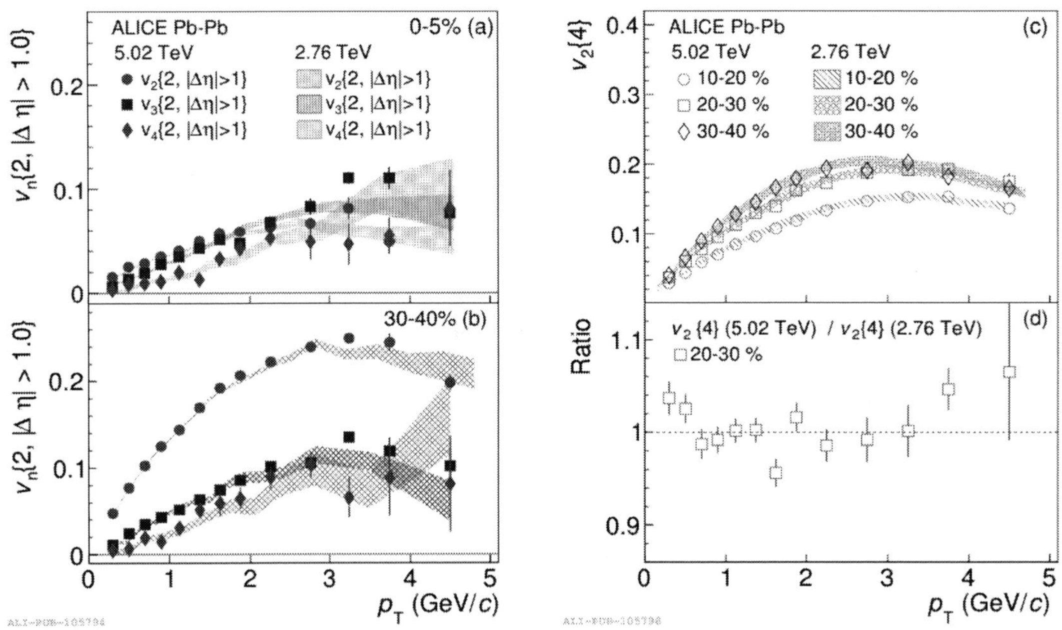

Figure 9. The p_T-differential flow measured in selected centrality classes. The same measurements from 2.76 TeV are presented for comparisons.

The transverse momentum (p_T) dependent anisotropic flow v_n ($n = 2, 3, 4$) are presented for selected centrality classes. The results are consistent with the previous $v_n(p_T)$ measurements from 2.76 TeV, as shown in Fig. 9. It suggests that the observed increase of p_T integrated flow results seen in Fig. 8 could be mainly attributed to stronger radial flow produced at 5.02 TeV. This stronger radial flow results in an increase of mean transverse momentum $\langle p_T \rangle$ from 2.76 to 5.02 TeV. Future comparisons of p_T-differential flow between experimental measurements and hydrodynamic calculations will provide important information to constrain further details of the theoretical calculations, e.g. determination of radial flow and freeze-out conditions.

Acknowledgments

This work is supported by the Danish Council for Independent Research, Natural Sciences, and the Danish National Research Foundation (Danmarks Grundforskningsfond).

References

[1] T.D. Lee, Phys. Rev. **D19**, 1802 (1979), [,174(1978)]
[2] E.V. Shuryak, Phys. Rept. **61**, 71 (1980)
[3] J.Y. Ollitrault, Phys. Rev. **D46**, 229 (1992)
[4] S. Voloshin, Y. Zhang, Z. Phys. **C70**, 665 (1996), hep-ph/9407282
[5] K.H. Ackermann et al. (STAR), Phys. Rev. Lett. **86**, 402 (2001), nucl-ex/0009011
[6] K. Adcox et al. (PHENIX), Phys. Rev. Lett. **89**, 212301 (2002), nucl-ex/0204005
[7] B.B. Back et al. (PHOBOS), Phys. Rev. Lett. **89**, 222301 (2002), nucl-ex/0205021
[8] P. Huovinen, P.F. Kolb, U.W. Heinz, P.V. Ruuskanen, S.A. Voloshin, Phys. Lett. **B503**, 58 (2001), hep-ph/0101136
[9] P.F. Kolb, P. Huovinen, U.W. Heinz, H. Heiselberg, Phys. Lett. **B500**, 232 (2001), hep-ph/0012137
[10] M. Luzum, P. Romatschke, Phys. Rev. **C78**, 034915 (2008), [Erratum: Phys. Rev.C79,039903(2009)], 0804.4015
[11] H. Song, U.W. Heinz, Phys. Rev. **C77**, 064901 (2008), 0712.3715
[12] H. Song, S.A. Bass, U. Heinz, T. Hirano, C. Shen, Phys. Rev. Lett. **106**, 192301 (2011), [Erratum: Phys. Rev. Lett.109,139904(2012)], 1011.2783
[13] K. Aamodt et al. (ALICE), Phys. Rev. Lett. **105**, 252302 (2010), 1011.3914
[14] K. Aamodt et al. (ALICE), Phys. Rev. Lett. **107**, 032301 (2011), 1105.3865
[15] G. Aad et al. (ATLAS), Phys. Lett. **B707**, 330 (2012), 1108.6018
[16] S. Chatrchyan et al. (CMS), Eur. Phys. J. **C72**, 2012 (2012), 1201.3158
[17] U. Heinz, R. Snellings, Ann. Rev. Nucl. Part. Sci. **63**, 123 (2013), 1301.2826
[18] M. Luzum, H. Petersen, J. Phys. **G41**, 063102 (2014), 1312.5503
[19] P. Huovinen, Int. J. Mod. Phys. **E22**, 1330029 (2013), 1311.1849
[20] E. Shuryak (2014), 1412.8393
[21] H. Song, Pramana **84**, 703 (2015), 1401.0079
[22] K. Dusling, W. Li, B. Schenke, Int. J. Mod. Phys. **E25**, 1630002 (2016), 1509.07939
[23] P. Kovtun, D.T. Son, A.O. Starinets, Phys. Rev. Lett. **94**, 111601 (2005), hep-th/0405231
[24] J. Adam et al. (ALICE) (2016), 1605.02035
[25] J. Adam et al. (ALICE), JHEP **09**, 164 (2016), 1606.06057
[26] J. Adam et al. (ALICE) (2016), 1604.07663
[27] J. Adam et al. (ALICE), Phys. Rev. Lett. **116**, 132302 (2016), 1602.01119
[28] B. Alver, G. Roland, Phys. Rev. **C81**, 054905 (2010), [Erratum: Phys. Rev.C82,039903(2010)], 1003.0194
[29] B.B. Back et al. (PHOBOS), Phys. Rev. Lett. **94**, 122303 (2005), nucl-ex/0406021
[30] B. Alver et al. (PHOBOS), Phys. Rev. Lett. **98**, 242302 (2007), nucl-ex/0610037
[31] I.G. Bearden et al. (BRAHMS), Phys. Rev. Lett. **88**, 202301 (2002), nucl-ex/0112001
[32] G. Denicol, A. Monnai, B. Schenke, Phys. Rev. Lett. **116**, 212301 (2016), 1512.01538
[33] J. Xu, C.M. Ko, Phys. Rev. **C83**, 034904 (2011), 1101.2231

[34] G. Aad et al. (ATLAS), Eur. Phys. J. **C74**, 2982 (2014), 1405.3936
[35] S. Chatrchyan et al. (CMS), Phys. Rev. **C87**, 014902 (2013), 1204.1409
[36] H.j. Xu, Z. Li, H. Song, Phys. Rev. **C93**, 064905 (2016), 1602.02029
[37] B.B. Abelev et al. (ALICE), JHEP **06**, 190 (2015), 1405.4632
[38] J. Adams et al. (STAR), Phys. Rev. Lett. **92**, 052302 (2004), nucl-ex/0306007
[39] B.I. Abelev et al. (STAR), Phys. Rev. **C75**, 054906 (2007), nucl-ex/0701010
[40] S.S. Adler et al. (PHENIX), Phys. Rev. Lett. **91**, 182301 (2003), nucl-ex/0305013
[41] A. Adare et al. (PHENIX), Phys. Rev. **C85**, 064914 (2012), 1203.2644
[42] G. Aad et al. (ATLAS), Phys. Rev. **C92**, 034903 (2015), 1504.01289
[43] A. Bilandzic, C.H. Christensen, K. Gulbrandsen, A. Hansen, Y. Zhou, Phys. Rev. **C89**, 064904 (2014), 1312.3572
[44] Y. Zhou, K. Xiao, Z. Feng, F. Liu, R. Snellings (2015), 1508.03306
[45] H. Niemi, K.J. Eskola, R. Paatelainen (2015), 1505.02677
[46] X. Zhu, Y. Zhou, H. Xu, H. Song (2016), 1608.05305
[47] Y. Zhou, Adv. High Energy Phys. **2016**, 9365637 (2016), 1607.05613
[48] J. Noronha-Hostler, M. Luzum, J.Y. Ollitrault, Phys. Rev. **C93**, 034912 (2016), 1511.06289
[49] H. Niemi, K.J. Eskola, R. Paatelainen, K. Tuominen, Phys. Rev. **C93**, 014912 (2016), 1511.04296
[50] Z. Feng, G.M. Huang, F. Liu (2016), 1606.02416

EPJ Web of Conferences **164**, 01020 (2017) DOI: 10.1051/epjconf/201716401020
ICNFP 2016

Overview of the CLIC detector and its physics potential

Rickard Ström[1a] on behalf of the CLICdp Collaboration

[1] *CERN, Geneva, Switzerland*

Abstract. The CLIC detector and physics study (CLICdp) is an international collaboration that investigates the physics potential of the Compact Linear Collider (CLIC). CLIC is a high-energy electron-positron collider under development, aiming for centre-of-mass energies from a few hundred GeV to 3 TeV. In addition to physics studies based on full Monte Carlo simulations of signal and background processes, CLICdp performs cutting-edge hardware R&D. In this contribution CLICdp will present recent results from physics prospect studies, emphasising Higgs studies. Additionally the new CLIC detector model and the recently updated CLIC baseline staging scenario will be presented.

1 Introduction

The Compact Linear Collider (CLIC) [1] is a proposal for a future electron-positron collider at CERN in the post HL-LHC era. It is the only mature option for an electron-positron multi-TeV accelerator. Figure 1 shows the CLIC project timeline including important milestones such as the decision at the update of the European Strategy for Particle Physics in 2019-2020, and first beams scheduled for 2035 following a decade of development and construction. Compared to hadron colliders CLIC has a superior sensitivity to electroweak states. This is due both to the cleaner experimental environment (including well-defined initial states) as well as the lower background from QCD processes.

CLIC is using a novel two-beam acceleration scheme going up to the full design energy of 3 TeV in 50 km, providing an instantaneous luminosity of $6{\times}10^{34}$ cm^{-2}s^{-1}. The accelerator is foreseen to be built in several stages, each operated for 5-7 years [2]. The staging is designed to make full use of the physics potential of CLIC with initial operation at lower centre-of-mass energy focusing on precision top and Higgs physics. The initial stage is at a collision energy of 380 GeV (500 fb^{-1}) with an additional dedicated run to measure the top quark mass using a threshold scan around 350 GeV (100 fb^{-1}). Subsequent stages at higher energies of 1.5 TeV (1.5 ab^{-1}) and 3 TeV (3 ab^{-1}) focus on exploring physics beyond the Standard Model (BSM) as well as studies dedicated to the top Yukawa coupling, direct probing of the Higgs self-coupling through double-Higgs production, rare decays, etc. Overall, the CLIC project offers a rich physics program for $\sim$20 years, with discovery potential to new physics, that can reach scales of up to several tens of TeV, through indirect searches with precision measurements.

CLICdp is a collaboration consisting of 28 institutes from 17 different countries. The work focuses on evaluating the physics prospects through simulation studies, detector optimisation, and R&D for a future CLIC detector.

[a] e-mail: lars.rickard.stroem@cern.ch

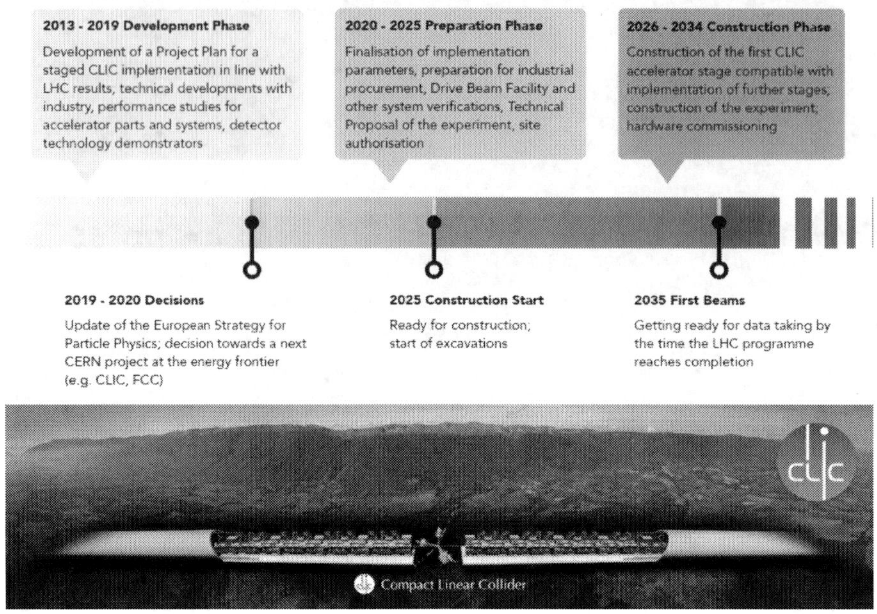

Figure 1. CLIC project timeline including important decisions and various planned phases for development, preparation, and construction. First beams are planned for 2035 and will be followed by 20 years of operation resulting in a rich physics program of both precision measurements as well as discovery. Credit: CLIC and CLICdp Collaboration.

In section 2 we will give a brief introduction to the accelerator complex and the challenging beam conditions at CLIC. In section 3 we will go through the detector requirements for a future CLIC experiment followed by an update of the latest CLIC detector model in section 4. The following sections go through the physics potential using illustrative example studies to map out the detector key capabilities. We conclude in section 8 with a short summary of these proceedings.

2 The CLIC accelerator

The CLIC accelerator [3] uses normal-conducting radio-frequency (RF) cavities to accelerate the main beams of electrons and positrons. In order to meet the energy and luminosity target within a few tens of kilometres many accelerating cavities (~150,000) are needed, each with a high acceleration field of $\sim 100\,\mathrm{MV\,m^{-1}}$. Since standard RF sources scale unfavourably to high frequency, the CLIC approach is to make use of a low-frequency high-current drive beam to generate high-frequency RF (12 GHz). The drive beam itself is accelerated using conventional klystrons, and its frequency is increased using a series of delay loops and combiner rings as indicated in figure 2. The drive beam is then decelerated through a series of Power Extraction and Transfer Structures (PETS) creating high RF power that can be used to accelerate the main beams to 1.5 TeV. This two-beam concept, where the drive beam is used to deliver RF power to all the cavities along the main beam, has been demonstrated at CERN in the dedicated CLIC Test Facility (CTF).

The large number of accelerating cavities and decelerating structures constitute a big challenge, not only in precision during production but also in terms of operation where the breakdown rate needs

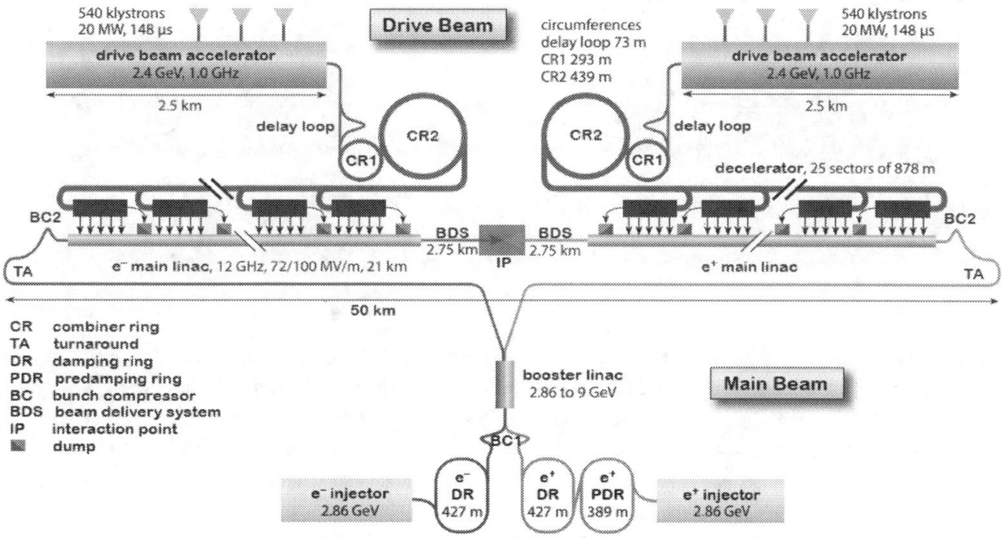

Figure 2. The CLIC accelerator at 3 TeV, including the drive beam complex at the top and the main beam at the bottom. The drive beam frequency is increased through a series of delay loops and combiner rings before reaching the desired bunch structure used to generate the high frequency RF power needed for the main beam acceleration. Credit: CLIC Collaboration.

to be kept at a manageable level. Further challenges for the CLIC accelerator complex come from the luminosity requirement, where a very small and dense beam, 40 nm x 1 nm in the transverse plane and $\sim 10^9$ particles per bunch, is needed to reach high enough statistics for rare processes at high energy. This puts constraints on both beam stability and alignment.

The beam and bunch structure of CLIC is rather distinct, with a bunch train repetition rate of 50 Hz. Each train consists of 312 bunches with a bunch spacing of 0.5 ns. Individual bunch crossings cannot be identified online, but the slow train repetition rate allows a readout of the complete detector during the 20 ms between bunch trains, making an offline selection possible (it is estimated that only about one hard e^+e^- physics interaction will be produced per bunch train crossing) [1].

The very small beam size and high bunch charge leads to large electromagnetic fields and interactions between colliding bunches. The resulting background of e^+e^- pairs and $\gamma\gamma \to$ hadrons, so-called beamstrahlung, constitute a large experimental background for CLIC, predominately in the forward regions. It is estimated that around 3 $\gamma\gamma \to$ hadrons interaction occurs for every bunch crossing at 3 TeV, depositing $\sim$20 TeV per bunch train in the calorimeters. The background can be reduced to a manageable level by applying combined p_T and timing cuts for the different subdetectors. Figure 3 shows an $e^+e^- \to H^+H^- \to t\bar{b}b\bar{t}$ event before (left) and after (right) such cuts have been applied. The cuts are designed to protect high p_T physics objects. Using hadron-collider-like jet clustering algorithms with beam-jets further reduces the impact of the background on physics measurements.

The chiral nature of the weak force offers enhancement or suppression of different physical processes by introducing polarisation of the main beams. For example WW-fusion Higgs production is greatly enhanced but the effect is less significant in other channels. In the baseline CLIC design a polarisation of 80% in the electron beam is foreseen. Positron polarisation is considered an option for future upgrades of the CLIC facility.

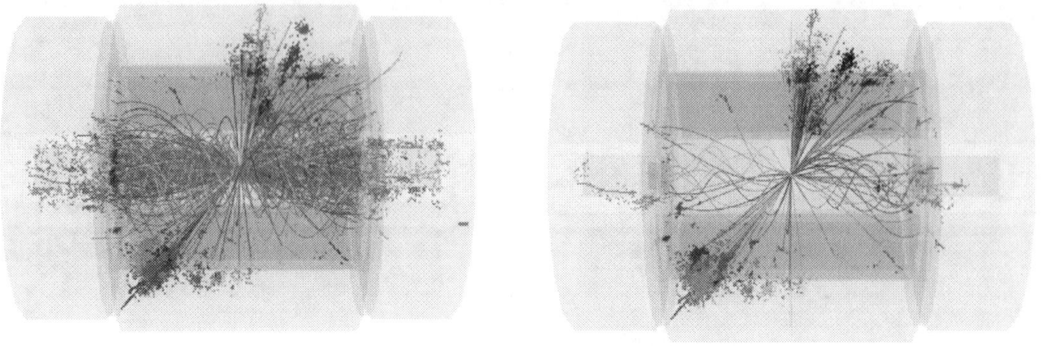

Figure 3. The beam and bunch structure at CLIC induces a challenging experimental background of beamstrahlung events. The background is reduced by combined cuts in the different subdetectors protecting the important high p_T physics objects. The left (right) hand side show an multi-jet $e^+e^- \rightarrow H^+H^- \rightarrow t\bar{b}b\bar{t}$ event at 3 TeV overlaid with $\gamma\gamma \rightarrow$ hadrons background before (after) such cuts have been applied. Figures taken from [1].

3 Detector requirements

The performance requirements for the CLIC detector are driven by the physics goals, requiring:

- Precise impact parameter resolution: $\sigma^2_{d_0} = (5\ \mu\text{m})^2 + (15\ \mu\text{m GeV})^2/p^2 \sin^3\theta$, needed for accurate vertex reconstruction to enable flavour-tagging with clean b-, c- and light-quark jet separation,
- Excellent track-momentum resolution: $\sigma_{p_T}/p_T^2 \lesssim 2 \cdot 10^{-5}\ \text{GeV}^{-1}$, needed for Higgs recoil analysis (see section 5), extraction of the position of kinematic edges (see section 7.1), etc. The impact of different resolutions is illustrated in the left of figure 4 [4], showing the reconstructed di-muon invariant mass for $H \rightarrow \mu^+\mu^-$ events.
- Jet-energy resolution for high-energy jets: at least 3.5% above 100 GeV, needed to distinguish e.g. W/Z/H di-jet decays. The effect of using various jet-mass resolutions is shown in the right of figure 4 [1] (a mass resolution of 2.5% corresponds to a jet-energy resolution of 3.5%),
- Overall need for precise timing to suppress background: translates to about 10 ns hit time-stamping resolution in the tracking region and about 1 ns accuracy for calorimeter hits (the resolution for a cluster of hits < 0.5 ns),
- Overall detector coverage (hermeticity), particularly in the forward region. This is needed for lepton identification and missing energy.

4 CLIC detector overview

The CLIC detector design is based on the requirements outlined in section 3 and is optimised for a wide range of physics benchmark analyses. The current detector model is illustrated in figure 5 with the main features described below.

The innermost part of the CLIC detector consists of a low-mass vertex detector, containing 3 cylindrical double layers in the barrel and 3 spiral double layers in the forward regions. Each layer corresponds to 0.2% X_0 (equivalent to about 200 μm of Si). Such a small radiation length is needed to reach the impact parameter resolution described in section 3. The detector design has roughly 2 billion pixels, each 25 μm square with a single point resolution of $\sim$ 3 μm. The vertex detector covers the area $\theta > 7^{\circ}$ and is designed to consume very little power since there is no material budget for liquid cooling. Cooling is achieved via an active air cooling strategy that induces a spiral airflow, and the power consumption is kept low via power-pulsing of the front-end electronics.

Moving outwards, the tracking system is complemented by a large silicon tracker with a radius of 1.5 m (4.6 m long) operating in a 4 T solenoid field. The inner (outer) tracker has 3 (3) barrel layers and 7 (4) forward disks. The material budget allowed is only 1-2% X_0 per layer. The sensors will contain either large pixels or short strips with a granularity of approximately 50 μm x (1 - 10) mm, and a single point resolution of $\sim$ 7 μm. A minimum number of hits is required for pattern recognition, which is why the tracking system is designed for at least 8 hits at $\theta > 8^{\circ}$ per event.

The Electromagnetic calorimeter consists of 40 layers of tungsten absorbers interleaved with silicon sensors of 300 micron thickness and with 5x5 mm^2 pads. The full depth of the electromagnetic calorimeter corresponds to 23 X_0. The hadronic calorimeter consists of 60 layers of steel absorbers interleaved with scintillator tiles of 30x30 mm^2. The total depth of the hadronic calorimeter is 7.5 λ_{int}. The whole calorimeter system is optimised for particle flow analysis (PFA) [5, 6].

The above mentioned detector components are all enclosed in a 4 T superconducting solenoid magnet with an inner radius of 3.4 m and a length of 8.3 m. The iron return yoke enclosing the magnet is instrumented with muon chambers than are used for muon identification. The detector also has a complex forward region with two important sub-detector systems: LumiCal (luminosity monitoring through Bhabha scattering) and BeamCal (beam monitoring and extended coverage for electromagnetic particles in the very forward region).

The physics studies presented in this proceeding were done using predecessors to the CLIC detector model presented above. These predecessors, CLIC_ILD and CLIC_SiD [1], are variants of the ILD and SiD [7] designs used at the International Linear Collider (ILC) but adapted for the CLIC environment. These two models differ from each other primarily in the tracking system where the former is using a time projection chamber (TPC), whereas the latter utilises a silicon tracker. All studies were performed using a detector simulation of the physics events of interest overlaid with $\gamma\gamma \rightarrow$ hadrons background.

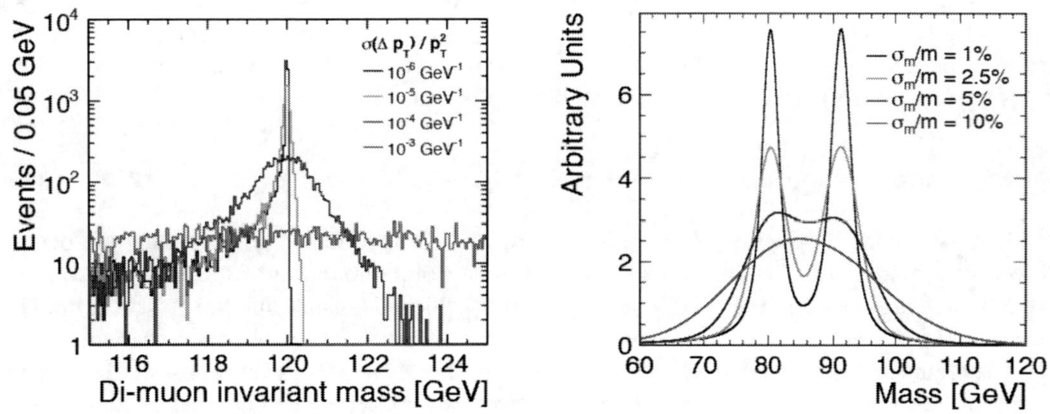

Figure 4. Left: Di-muon invariant mass distribution in a search for $H \rightarrow \mu^+\mu^-$ decays at $\sqrt{s}$ =3 TeV for different σ_{p_T}/p_T^2 resolutions. Figure taken from [4]. Right: The required W/Z jet mass separation is obtained for mass resolutions below 2.5%, corresponding to a jet-energy resolution of 3.5%. Figure taken from [1].

DOI: 10.1051/epjconf/201716401020

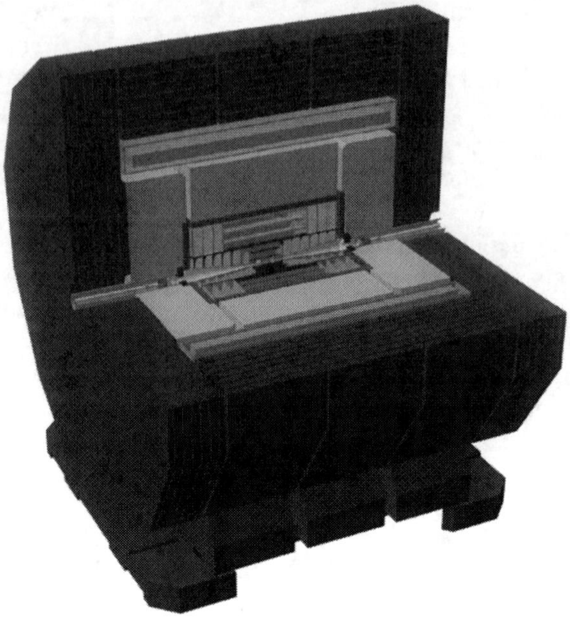

Figure 5. The CLIC detector model. Starting from the innermost part it consists of a low mass vertex detector, a large radius tracker detector, and a calorimeter system optimised for particle flow analysis. These parts are enclosed in a 4 T solenoid field. The outermost layer consists of the return yolk of the magnetic field and is instrumented with muon detectors. Credit: CLICdp Collaboration.

5 Higgs physics at CLIC

The clean collision environment of an electron-positron collider is ideal for precision physics studies of the Higgs boson, a prime target of CLIC operation at 380 GeV. CLIC is a Higgs factory covering several Higgs production processes as can be seen on the left hand side of figure 6 where the cross-sections of various Higgs processes are shown as a function of centre-of-mass energy of the collisions. At the initial energy stage roughly 70,000 Higgsstrahlung events are produced, and about 3 million Higgs bosons will be produced at the two higher energy stages.

5.1 Higgsstrahlung

The Higgsstrahlung production mode ($e^+e^- \rightarrow HZ$), illustrated in the leftmost diagram of figure 7, allows for a model-independent measurement of the absolute coupling of the Higgs boson to the Z boson, which is unique for lepton colliders [8]. This is achieved by studying the mass recoiling against the Z boson for events with two oppositely charged leptons that have an invariant mass consistent with m_Z. The right side of figure 6 shows the Z-recoil mass for $Z \rightarrow \mu^+\mu^-$ events at a studied benchmark energy of 350 GeV where the cross-section is dominant and close to maximal. The remainder of the event, i.e. the particles from the Higgs decay, is not considered in the event selection, giving a result that is independent of specific Higgs decay modes. Note that this also applies to potential invisible final states. This technique for determining the absolute HZZ coupling is possible due to the well-defined initial state and clean experimental conditions at a linear collider.

The uncertainty on the HZZ coupling is calculated to 2% using the leptonic Z decays. These decays give the cleanest signature, but the result can be improved significantly by adding the large statistics from the hadronic Z decays. The combined uncertainty on the HZZ coupling is about 0.8%. The analysis described above can also be used to put constraints on the Higgs branching ratio to invisible states: the estimated uncertainty is below 1% at 90% confidence level [9].

5.2 Higgs production at higher energies

At higher energies the Higgs production is dominated by vector boson fusion where either W or Z bosons fuse to produce a Higgs boson, see the middle and right diagram of figure 7. The dominant WW-fusion process gives a constraint on the H coupling to W and also gives the best sensitivity for the Higgs mass and Higgs total decay width. The Higgs mass can be extracted with a precision of 24 MeV (including electron polarisation) using a combination of both high energy stages (1.5 and 3 TeV). This precision compares favourably with the HL-LHC projection of 50 MeV per experiment [10].

Rarer Higgs decays, with branching ratios at the sub-% level, such as $H \to \mu^+\mu^-$, $H \to \gamma\gamma$, and $H \to Z\gamma$, also benefit from the higher statistics at high energies. Here we reach an uncertainty on the cross-section × branching ratio for the three processes mention above of 25% (with CLIC at 3 TeV), 15% (with CLIC at 1.4 TeV), and 42% (with CLIC at 1.4 TeV) respectively [9].

Once the Higgs coupling to the Z, g_{HZZ}, has been determined using the recoil mass technique, the Higgs coupling to W, g_{HWW}, can be determined by studying the cross-section × branching ratio fraction between Higgsstrahlung and WW-fusion events, e.g. using b-tagged events,

$$\frac{\sigma(e^+e^- \to ZH) \times BR(H \to b\bar{b})}{\sigma(e^+e^- \to \nu_e\bar{\nu}_e H) \times BR(H \to b\bar{b})} \propto \left(\frac{g_{HZZ}}{g_{HWW}}\right)^2 . \quad (1)$$

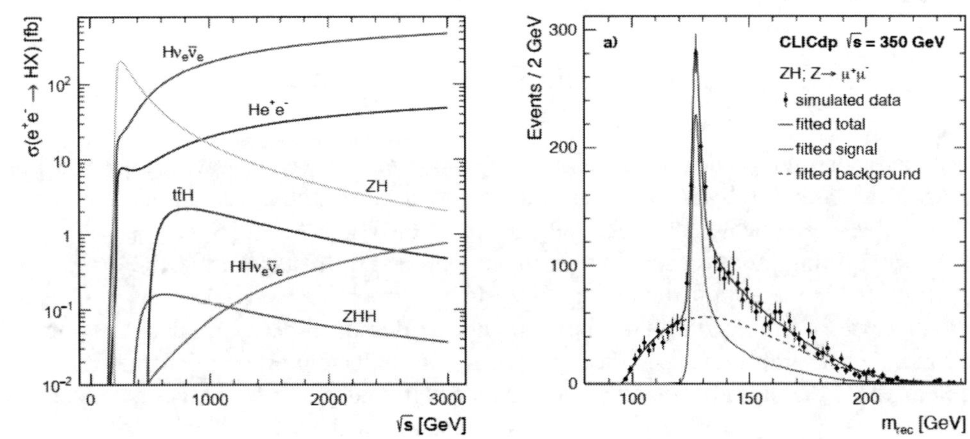

Figure 6. Left: Dominant Higgs production cross-sections at CLIC as a function of center-of-mass collision energy. The distributions shown correspond to unpolarised beams and do not include the effects of initial-state radiation (ISR) or beamstrahlung. The plot is produced using perturbative NLO calculation with the generator Whizard 1.95. Figure taken from [9]. Right: The Z-recoil mass for $Z \to \mu^+\mu^-$ events at a studied benchmark energy of 350 GeV. Figure taken from [9].

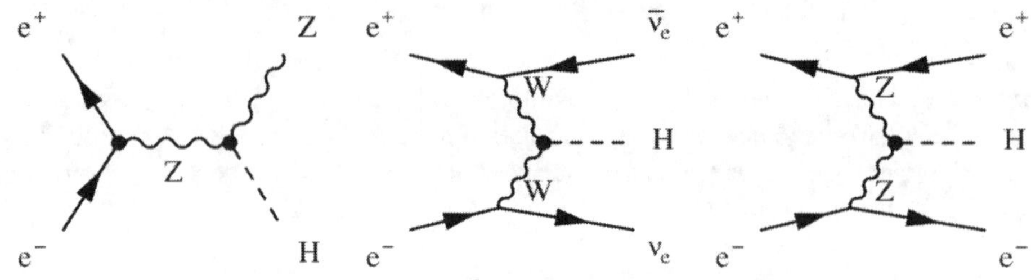

Figure 7. The three highest cross-section Higgs production processes at CLIC.

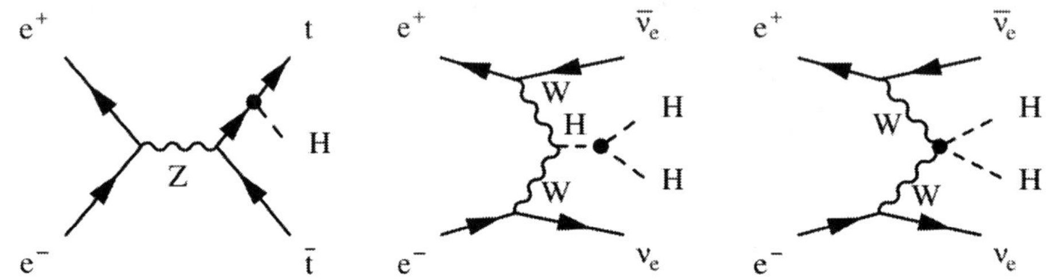

Figure 8. Higgs processes at CLIC involving the top Yukawa coupling g_{Htt}, the Higgs boson trilinear self-coupling λ and the quartic coupling g_{HHWW}.

In order to determine absolute measurements of other Higgs couplings, the Higgs total decay width needs to be derived from the data[1]. For example it can be determined by measuring the Higgs decay to WW^* in the WW-fusion process,

$$\sigma(\mathrm{H}\nu_e\overline{\nu}_e) \times \mathrm{BR}(\mathrm{H} \to \mathrm{WW}^*) \propto \frac{g^4_{\mathrm{HWW}}}{\Gamma_{\mathrm{H}}}, \tag{2}$$

given that the model-independent measurement of g_{HWW} was derived following equation 1 above. In practice however, both the Higgs couplings and the total decay width are determined through a global fit of all available experimental measurements involving the Higgs boson processes, see section 5.2.1.

Another interesting process accessible at the second and third stages of CLIC is the top Yukawa process illustrated in the left diagram of figure 8. In this process a Higgs boson is produced in association with a pair of top quarks. The channel with the dominant decay mode, $\mathrm{H} \to \mathrm{b}\bar{\mathrm{b}}$, has eight fermions in the final sample, including four b-jets. Naturally flavour tagging information is one of the crucial elements to distinguish the signal from the background. Therefore this process also provides an excellent detector benchmark test. The precision reached with CLIC using both the fully-hadronic and semi-leptonic final states is 4-5% [9] at 1.4 TeV. As a comparison, for HL-LHC the prediction is 7-10% [10] per experiment.

Double Higgs production $\mathrm{e}^+\mathrm{e}^- \to \mathrm{HH}\nu_e\overline{\nu}_e$, as illustrated in the middle and right diagrams of figure 8, is best reachable at the highest energy scale of 3 TeV. The high energy is crucial in order to

[1] The total Higgs decay width, Γ_{H}, is less than 5 MeV and hence too small to be measured directly with current experiments.

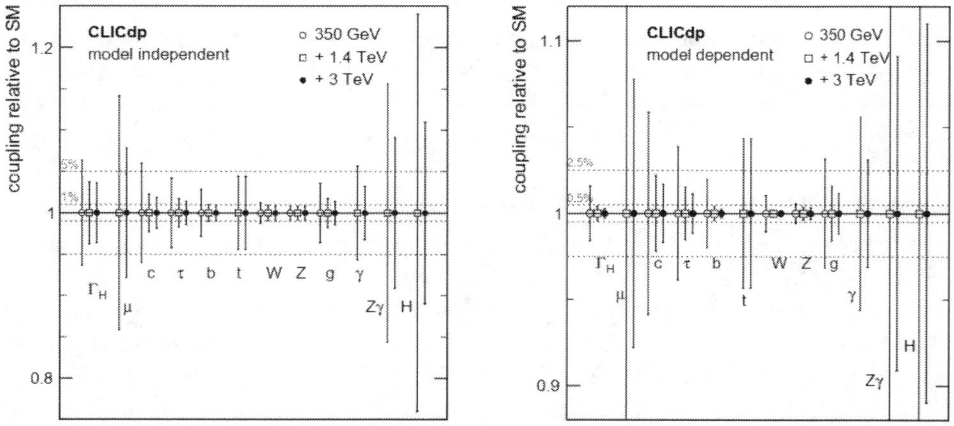

Figure 9. Illustration of the precision of the Higgs couplings and decay width of the studied three-stage CLIC programme. Figures taken from [9]. Left: Results using a model-independent fit. Right: Results using a model-dependent fit as described in 5.2.1.

profit from the higher cross-section and higher luminosity at this energy. This channel gives access to the Higgs tri-linear self-coupling λ, a direct probe of the Higgs potential, at the precision of about 10% for 1.4 TeV and 3 TeV operation combined [9]. This is considerably more precise than the predicted precision of ~ 30% for the HL-LHC.

5.2.1 Global Higgs fit

The global fit results of the full CLIC program, i.e. ~5-7 years of running at each stage, is presented in figure 9, where the model-independent (model-dependent) couplings are shown to the left (right). In general, model-independent couplings can be determined down to ~2% for most couplings while model-dependent couplings can be determined to ~0.1-1%[2]. The accuracy on the Higgs total decay width is ~3.6% (model-independent) [9], ~0.3% (model-dependent, derived) [9]. It should be noted that CLIC sensitivity is significantly better than HL-LHC for the couplings to c, b, W, Z, and g, the self-coupling and the Higgs total decay width. Further the couplings to μ, τ, t, γ, and $Z\gamma$ are at a similar level of sensitivity.

6 Top physics at CLIC

The top quark is of particular interest due to its strong coupling to the Higgs field; a precise measurement of the top quark mass is needed to verify the relation between the top quark and the SM gauge bosons. Further, precision top quark studies may be a prominent place where indications of potential new physics show up. The determination of the top quark mass and the top quark couplings to Z and γ with high precision are among the main focuses of the initial stage of CLIC. Rare decays such as $t \rightarrow cH$ and $t \rightarrow c\gamma$ can also be studied and here CLIC provides competitive limits.

[2]The fit is done in the same manner as for the LHC experiments, i.e. assuming no invisible Higgs decays, leading to a model-dependence.

As mentioned in section 1 the CLIC staging program includes a dedicated threshold scan around 350 GeV to measure the top quark mass. This is shown in the left of figure 10, where the $t\bar{t}$ production cross-section is shown as a function of centre-of-mass energy for 10 measurements of $10\,\text{fb}^{-1}$ each [11]. The position of the threshold is related to the top quark mass, and the slope of the rise in cross-section around the threshold reflects its natural width. From the normalisation we can extract a measurement of the strong coupling constant and the top Yukawa coupling. While the former is sensitive to QCD corrections close to the threshold where the $t\bar{t}$-system experiences QCD bound state effects, the latter is related to H loop corrections.

Extraction of the theoretically well-defined 1S top quark mass can be done with a statistical accuracy of about 30 MeV [11, 12]. At this level of precision we expect the experimental and theoretical systematic uncertainties to become relevant. The theoretical uncertainty includes the uncertainty in the NNNLO description of the threshold shape and the uncertainty in the conversion of the threshold mass to the minimal subtraction $\overline{\text{MS}}$ scheme [13]. A total uncertainty on the top quark mass of about 50 MeV [12] seems feasible, which is about one order of magnitude beyond the capabilities of HL-LHC.

The top gauge couplings can be determined through measurements of cross-sections and forward-backward asymmetries for different polarisations [14]. The general $t\bar{t}V$ vertex can be parameterised using either form factors or effective operators. The related parameters are substantially modified in many BSM models, e.g. in models with extra dimensions and new heavy gauge bosons. The right side of figure 10 shows a comparison of the uncertainty on the measured top quark form factors. The precision reached by CLIC (at the level of 1% or below) is shown in green, and foresees an order of magnitude better precision than HL-LHC shown in red [2]. Further, the fact that the top quark decays before it hadronizes allows access to its polarisation by measuring the angular distributions of the decay products. This is a sensitive probe for CP violation in the top sector.

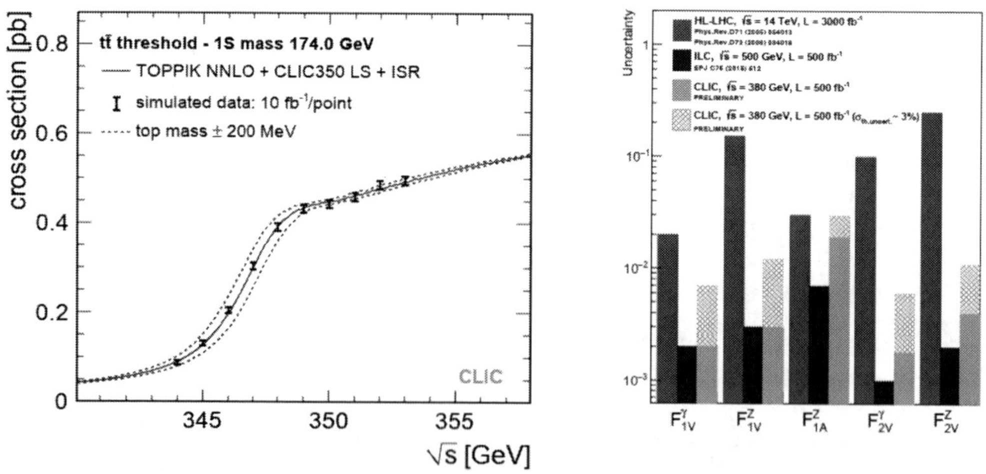

Figure 10. Left: Cross-section of $t\bar{t}$ production as function of centre-of-mass energy for 10 measurement points of $10\,\text{fb}^{-1}$ each around the top quark threshold mass. For illustration a ±200 MeV mass shift is indicated by dashed lines. Figure taken from [11]. Right: Comparisons of the uncertainty on the measured top quark form factors (considering parity conserving couplings). Figure taken from [2].

DOI: 10.1051/epjconf/201716401020

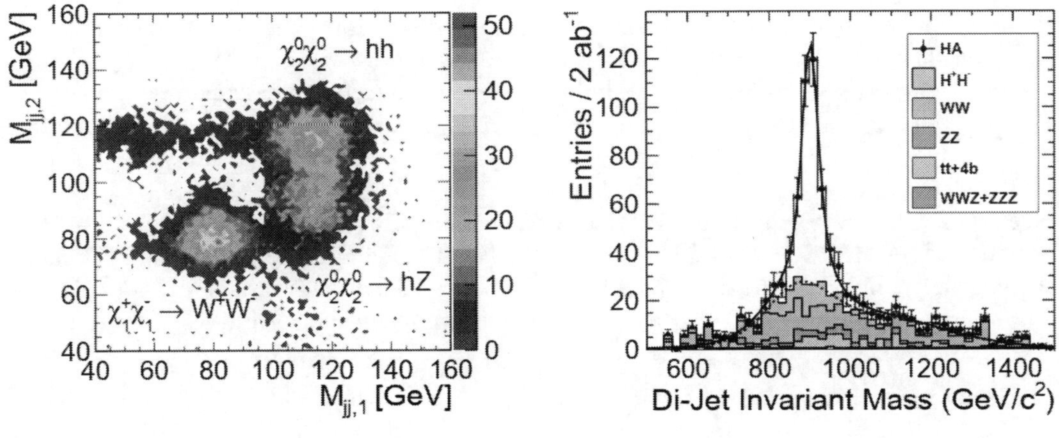

Figure 11. Left: The reconstructed di-jet masses from the decay product of directly produced charginos and neutralinos. Figure taken from [1]. Right: The reconstructed di-jet mass in a scenario with heavy Higgs bosons. Figure taken from [1].

At high-energy CLIC operation, an increased boost leads to better separation between the decay products of the two top quarks. A top tagging algorithm based on the method presented in [15] is under development and currently being re-optimised for electron-positron collisions. The focus is on further exploration of top gauge couplings. For example for effective D6 operators we see a significant improvement going to higher energies. Further, initial studies of the CKM matrix element v_{tb}, through the study of single top production $e^+\gamma \to t + b + \bar{\nu}$ look promising. This process has no background from $t\bar{t}$.

7 BSM physics at CLIC

In addition to the Higgs and top physics programs outlined in section 5 and 6, CLIC also has an excellent sensitivity to BSM physics through both direct and indirect measurements. Due to the clean collision environment CLIC is particularly suited to study non-colored TeV-scale particles such as for example sleptons and charginos in supersymmetry (SUSY). These might be hidden in the large QCD backgrounds at the LHC. To quantify the performance and give input to the detector optimisation effort, three different SUSY scenarios were defined and used as benchmark signal samples for various BSM analyses [1]. In general each analysis was able to measure the mass and other basic properties of the particles set out to find. A small selection of the BSM benchmark analyses are highlighted in section 7.1 and 7.2.

7.1 Direct searches for BSM physics

Direct production of new particles is possible up to the kinematic limit ($\sqrt{s}/2$ for particles produced in pairs). This includes studying e.g. mono-photon events allowing model-independent searches for dark matter. Below we show a few examples of analyses concerning the reconstruction and direct measurements of possible SUSY particles.

Slepton signatures are particularly clean with two high-momentum leptons and missing energy in the final state ($e^+e^- \rightarrow \tilde{\mu}_R^+\tilde{\mu}_R^- \rightarrow \mu^+\mu^-\tilde{\chi}_1^0\tilde{\chi}_1^0$). The slepton and gauginos masses are extracted from the position of the kinematic edges of the lepton energy distribution, the so-called endpoints. The precision on slepton masses is below 1 % for sleptons with a mass ~1 TeV.

Also hadronic events were studied, e.g. through the pair-production of charginos ($e^+e^- \rightarrow \tilde{\chi}_1^1\tilde{\chi}_1^1 \rightarrow \tilde{\chi}_1^0\tilde{\chi}_1^0W^+W^-$) and neutralinos (e.g. $e^+e^- \rightarrow \tilde{\chi}_2^0\tilde{\chi}_2^0 \rightarrow HH\tilde{\chi}_1^0\tilde{\chi}_1^0$), reconstructed into 4 jets and missing energy, see figure 11. The precision, $\Delta m/m$, on gaugino masses of a few hundred GeV is roughly 1-1.5%.

Scenarios with heavy Higgs bosons typically have complex final states, e.g. $e^+e^- \rightarrow H^0A \rightarrow b\bar{b}b\bar{b}$ and $e^+e^- \rightarrow H^+H^- \rightarrow t\bar{b}b\bar{t}$, requiring precise flavour tagging. The H_0, A_0, and $H^\pm$ states are almost degenerate in mass. The right side of figure 11 shows the di-jet invariant mass distribution for the $b\bar{b}b\bar{b}$ final states. Here CLIC reaches as mass precision $\Delta m/m = 0.3\%$ [1].

7.2 Indirect searches for BSM physics

Indirect searches are conducted by studying the variables and couplings in the SM with very high precision and allow discovery of BSM signals beyond the centre-of-mass energy of the collider. For example Z' and Higgs compositeness models can be probed up to scales of tens of TeV [10, 16].

Z' is a hypothetical gauge boson arising from extensions of the electroweak symmetry of the SM. This can be described e.g. in so-called Minimal anomaly-free Z' models (AFZ). These are accessible by precision studies of $e^+e^- \rightarrow \mu^+\mu^-$ including measurements of the cross-section and both the forward-backwards and left-right asymmetry. Model-dependence arise through mixing of the Z' and Z and allow for discovery up to 50 TeV [16]. In case a Z' is indeed found in LHC or HL-LHC data, CLIC can measure the effective couplings with high precision.

Another important analysis is the study of vector boson scattering (VBS). While the SM Higgs preserve the unitarity of scattering amplitudes in longitudinal VBS, anomalous couplings or additional resonances can be introduced in BSM models. At first glance, CLIC at 3 (1.5) TeV is roughly two (one) orders of magnitude more precise for anomalous couplings than LHC at 8 TeV.

8 Conclusions & Summary

CLIC has a well-established physics program that spans over several decades with a three-stage implementation. CLIC opens up the possibility of e^+e^- collisions with $\sqrt{s} \gg 1$ TeV, giving improved precision of many observables, access to rare Higgs decays, and discovery potential for BSM physics at the energy frontier with a reach in indirect search up to tens of TeV.

A model-independent measurement of the HZZ coupling can be extracted at the initial stage at 380 GeV. This stage also includes a dedicated threshold scan to precisely measure the top mass. At the two higher energy stages, 1.5 TeV and 3 TeV, statistics improve for the rare Higgs decays and we get direct access to e.g. the top Yukawa coupling and the Higgs self-coupling.

The feasibility of the CLIC program has been demonstrated through extensive simulation, prototyping, and accelerator and detector R&D. Results from the LHC provide an important input for the CLIC physics program, a strategy that can be adapted to potential LHC/ HL-LHC discoveries.

References

[1] L. Linssen et al., *CLIC Conceptual Design Report: Physics and Detectors at CLIC*, CERN-2012-003 (2012)

[2] M. J. Boland et al. [CLIC and CLICdp Collaborations], *Updated baseline for a staged Compact Linear Collider*, CERN-2016-004 (2016)
[3] M. Aicheler et al., *CLIC Conceptual Design Report: A Multi-TeV Linear Collider Based on CLIC Technology*, CERN-2012-007 (2012)
[4] C. Grefe, *Detector Optimization Studies and Light Higgs Decay into Muons at CLIC*, PhD Thesis, CERN-THESIS-2012-344, BONN-IR-2013-14 (2012)
[5] M. A. Thomson, *Particle Flow Calorimetry and the PandoraPFA Algorithm*, Nucl. Instrum. Meth. **A611** (2009)
[6] J. S. Marshall, A. Munnich and M. A. Thomson, *Performance of Particle Flow Calorimetry at CLIC*, Nucl. Instrum. Methods Phys. Res. A**700** (2012)
[7] H. Abramowicz et al., *The International Linear Collider Technical Design Report - Volume 4: Detectors*, ILC-REPORT-2013-040 (2013)
[8] M. A. Thomson, *Model-independent measurement of the $e^+e^- \rightarrow HZ$ cross section at a future e^+e^- linear collider using hadronic Z decays*, Eur. Phys. J. **C76** (2016)
[9] H. Abramowicz et al., *Higgs Physics at the CLIC Electron-Positron Linear Collider*, arXiv:1608.07538 (2016)
[10] S. Dawson et al., *Higgs Working Group Report of the Snowmass 2013 Community Planning Study*, Proceedings of the Community Summer Study 2013: Snowmass on the Mississippi, CNUM: C13-07-29.2, arXiv:1310.8361 (2013)
[11] K. Seidel, et al., *Top quark mass measurements at and above threshold at CLIC*, Eur. Phys. J., **C73** (2013)
[12] F. Simon, *A First Look at the Impact of NNNLO Theory Uncertainties on Top Mass Measurements at the ILC*, Proceedings for the International Workshop on Future Linear Colliders, arXiv:1603.04764 (2015)
[13] P. Marquard et al., *Quark Mass Relations to Four-Loop Order in Perturbative QCD*, Phys. Rev. Lett. **114** (2015)
[14] M. Amjad et al., *A precise characterisation of the top quark electro-weak vertices at the ILC*, Eur. Phys. J. **C75** (2015)
[15] D. E. Kaplan, et al., *Top Tagging: A Method for Identifying Boosted Hadronically Decaying Top Quarks*, Phys. Rev. Lett. **101** (2008)
[16] J.-J. Blaising and J. D. Wells, *Physics performances for Z' searches at 3 TeV and 1.5 TeV CLIC*, LCD-NOTE-2012-009, arXiv:1208.1148 (2012)

DOI: 10.1051/epjconf/201716401021

Performance of the ATLAS Detector in Run-2

Nicoletta Garelli[1,a] on behalf of the ATLAS Collaboration

[1] *SLAC National Accelerator Laboratory, 2575 Sand Hill Rd, Menlo Park, CA 94025, United States*

Abstract. The second LHC run has started in June 2015 with a p-p center-of-mass collision energy of 13 TeV and ATLAS has taken 3.2 fb^{-1} of data at the new energy in 2015. In this talk the improvements made to the ATLAS experiment during the two-year shutdown 2013/2014 will be discussed. Furthermore, reconstruction algorithms of tracks, e/γ, muons, taus, jets and flavour tagging have been improved for Run-2. Reconstruction efficiency, isolation performance, transverse momentum resolution and momentum scales are measured in various regions of the detector and in momentum intervals enlarged with respect to those measured in the Run-1. The performance of the upgraded detector and the new reconstruction algorithms measured using these data taken in 2015 will be discussed.

1 Introduction

The second three-year run of the Large Hadron Collider (LHC) marks a new era for high energy physics. More data are needed in order to assess the nature of the Higgs boson observed in 2012 with the ATLAS experiment, while the existence of dark matter and unnatural fine tuning in the Standard Model also motivate potential new physics near the electroweak energy scale. The discovery of the Higgs boson with mass $126.0 \pm 0.4(stat) \pm 0.4(sys)$ GeV is detailed here [1]; the searches have been performed in three channels: $H \to ZZ^{(*)} \to 4l$, $H \to \gamma\gamma$ and $H \to WW^{(*)} \to e\nu\mu\nu$.

The LHC took two years of shutdown from February 2013, referred to as Long Shutdown 1, for upgrading the machine to reach a center of mass energy of 13 TeV and peek luminosity up to $1.5 \cdot 10^{34}\,\mathrm{cm}^{-2}\mathrm{s}^{-1}$, beyond original design, operating with a bunch spacing of 25 ns. The experiments took advantage from this time to repair broken components, update their online systems and install new detector components to cope with the upgraded LHC and the expected pile-up (μ) of about 40.

2 The ATLAS Detector

A Toroidal LHC Apparatus (ATLAS) is a general purpose detector at the LHC installed at Point 1 of the accelerator ring [2]. It has a forward-backward symmetry which provides almost full solid angle coverage around the interaction point [1]. ATLAS is a big cylinder weighing 7000 t, 44 m long and 22 m high, composed of an inner tracker, a calorimeter system and a muon spectrometer. The Inner Detector

[a] e-mail: nicoletta.garelli@cern.ch

[1] ATLAS uses a right-handed coordinate system with its origin at the nominal interaction point (IP) in the center of the detector and the z-axis along the beam pipe. The x-axis points from the IP to the center of the LHC ring, and the y-axis points upwards. Cylindrical coordinates (r, ϕ) are used in the transverse plane, ϕ being the azimuthal angle around the z-axis. The pseudo-rapidity is defined in terms of the polar angle θ as $\eta = -ln(tan\frac{\theta}{2})$.

(ID) is surrounded by a superconducting solenoid providing a 2 T axial magnetic field. It consists of silicon pixels and the SemiConductor Tracker (SCT) built from silicon strips, with pseudo-rapidity coverage up to $|\eta| = 2.5$ and a gas filled straw tubes detector, the Transition Radiation Tracker (TRT), with pseudo-rapidity coverage up to $|\eta| = 2$. The calorimeter system surrounds the solenoid. Showers produced by photons, electrons and positrons are contained in the electromagnetic calorimeter (EM) which uses liquid argon as active medium. The hadronic showers are adsorbed by the Tile calorimeter in the barrel and more liquid argon calorimeter in the endcaps. The calorimeter detectors cover the pseudo-rapidity range of $|\eta| < 4.9$. The muon spectrometer is composed of precision measurement chambers and trigger chambers located within three super-conducting toroids arranged with an eight-fold azimuthal symmetry, covering $|\eta| < 2.7$. The toroidal magnetic field in the barrel region is 0.5 T, while in the end-caps 1 T.

2.1 Major ATLAS Activities during the LS1

ATLAS underwent a series of updates to cope with the new LHC working point for Run-2. The most significant ones are summarized below.

The pixel detector was extracted and equipped on surface with new services. A fourth additional layer, referred to as Insertable B-Layer (IBL) was added between a new beam pipe and the current inner pixel layer (B-layer) to improve the b-tagging efficiency and helping preserving tracking performance in face of higher luminosity effects. It is installed at 33 mm from the beam line and equipped with planar and 3D sensors [3].

During the Run-1, significant TRT gas leaks were observed, and many repaired during the LS1. For regions not fully repaired, it has been decided to switch to an argon based gas mixture instead of mitigate the unaffordable xenon loss.

New muon detector chambers were installed to close the coverage holes both in the feet region and in the shafts area where elevators are located to lift material from the ground level up to below the calorimeter systems.

A new off-detector readout complex was deployed for the Cathode Strip Chambers (CSC) to remove the previous limitations to sustain 100 kHz level-1 trigger rate. It is the first ATCA-based system of the ATLAS experiment and worked very reliably from the beginning of operations in 2015 [4].

All the accessible front-ends were repaired to improve the overall ATLAS efficiency reaching more than 96% of operational fraction across all detector systems prior to the start of the Run-2.

2.1.1 Consolidation Work during the YETS 2015

The Year-End-Technical-Stop (YETS) of 2015 lasted few months. The most significant ATLAS task during this shutdown was the repair of the two vacuum bellows in one of the endcap toroid cryostats. A damage was in fact discovered around the leads of the endcap toroid C, and the repair was carefully prepared and completed on time.

3 The ATLAS Performance

On 3 June 2015 the LHC delivered the first collisions at 13 TeV, after two months of beam commissioning. The rest of the year was characterized by a step-wise increase in the number of bunches that allowed to reach a peak luminosity of $5 \cdot 10^{33}\,cm^{-2}s^{-1}$, although still below the maximum value of Run-1 ($7.7 \cdot 10^{33}\,cm^{-2}s^{-1}$). The luminosity performance was excellent, thanks to the small beam losses and good emittance preservation through the beam cycle. In 2016 the peak luminosity exceeded the

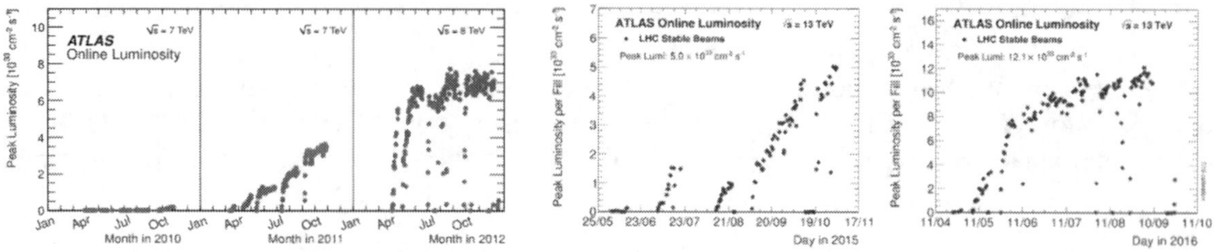

Figure 1: The peak instantaneous luminosity delivered to ATLAS per day versus time during stable beams for p-p collisions in 2010, 2011 and 2012 (*left*) [5], in 2015 (*center*), and in 2016 (*right*) [6].

nominal value of $1 \cdot 10^{34}\ \mathrm{cm^{-2}s^{-1}}$ by early summer, and reached up to $1.2 \cdot 10^{34}\ \mathrm{cm^{-2}s^{-1}}$ in September, as shown in figure 1.

The luminosity integrated by ATLAS over the course of 2015 p-p runs was above $4\,\mathrm{fb}^{-1}$ with an average pile-up of about 14, while between 16 April and 25 July of 2016 ATLAS collected over $18\,\mathrm{fb}^{-1}$ with an average pile-up of about 23, as reported in figure 2. The luminosity is determined from luminosity detectors which have been calibrated with the use of the van-der-Meer beam-separation method.

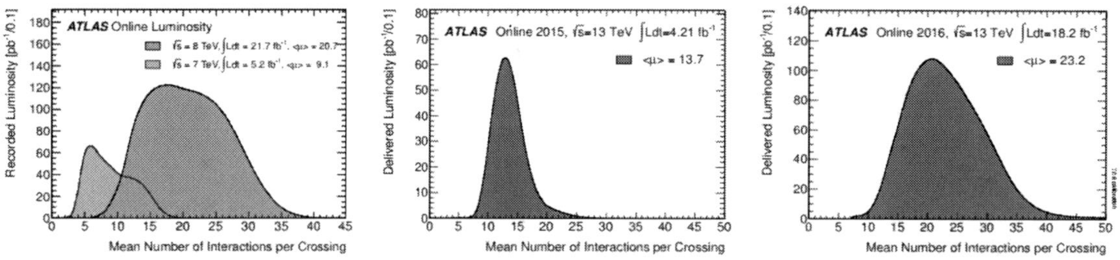

Figure 2: The luminosity-weighted distribution of the mean number of interactions per crossing for the 2011 and 2012 data (*left*), the 2015 (*center*) [5], and the 16 April to 25 July 2016 (*right*) p-p collision data [6]. All data delivered to ATLAS during stable beams is shown, and the integrated luminosity and the mean pile-up (μ) values are given in the plots.

The ratio between the luminosity recorded by ATLAS and the total delivered luminosity establishes the data taking efficiency. In 2012 it was 93.5%, while in 2015 and between April and September 2016 it was 92%.

The data quality efficiency instead takes into account the relative detector uptime and good data quality efficiency during stable beam pp collisions. The data quality efficiency in the 2015 25ns bunch spacing p-p runs was 87.1%. The two long LHC fills where IBL was turned off, as reported in section 3.1, significantly contributed to the lower efficiency with respect to the 95.5% value of Run-1.

In the following sections the status of the ATLAS detector, the performance of the reconstruction of the main physics objects and the efficiency measurements in 2015 are summarized.

3.1 IBL Commissioning

During the cosmic ray commissioning of the IBL in early 2015 a mechanical distortion of its support structure, the staves, was observed. The bowing of the staves depends linearly on the temperature with a gradient of about $10\,\mu$m/K twisting the staves in the r-ϕ direction. The impact of this effect is mitigated by a temperature control to 0.2 K level [7].

For two LHC fills in October 2015 the IBL was off and excluded by the ATLAS data taking due to the observation of a significant front-end low voltage current increase and drift in the calibration parameters (the threshold and the time over threshold). This effect is compatible with the front-end radiation damage which was initially somewhat surprising given the FE ASIC design was qualified for about 100 times the dose at that point. This was soon understood to be similar to several other observations of a transient effect related to typical designs also on 130νm technologies and the current is expected to decrease after about 1 Mrad dose [8].

After the observations in 2015, it was decided to run IBL at 15 °C in early 2016 instead of the nominal −10 °C, and at the decreased digital voltage of 1 V. After measurements in the laboratories, in the summer it was decided to reduce the temperature to 5 °C and increase the voltage back to 1.2 V. IBL accumulated so far about 9 Mrad and the observed current has significantly reduced from the maximum peak. Figure 3 compares the correction needed for the IBL distortion in the three different temperature settings. The effects are regularly corrected online before the reconstruction of the data. After applying the alignment correction, they do not impact the tracking performance significantly.

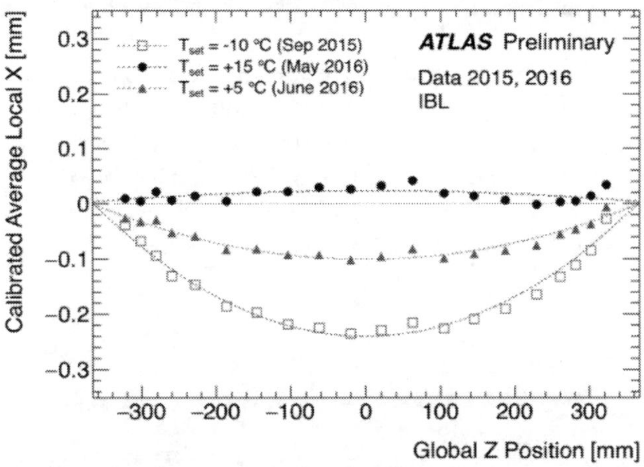

Figure 3. IBL local-x correction in the transverse plane averaged over all 14 IBL staves for 2015 data (*red open squares*), and for 2016 data at different operating temperatures (15 °C, *solid blue circles*; 5 °C, *solid green triangles*) [9].

The IBL is composed of twelve million pixel cells of $50 \times 250\,\mu\text{m}^2$ versus $50 \times 400\,\mu\text{m}^2$ for the original pixel detector. The smaller size in the z-direction results in an improved Z spatial resolution of $40\,\mu$m instead of $75\,\mu$m. Figure 4 shows the improved resolution of the impact parameter with respect to 2012. At low p_T both d_0 and z_0 resolution improve close to a factor of two because of the IBL addition so close to the interaction point and lighter stave material reducing the impact of multiple scattering. At high p_T, where the detector granularity dominates, the improved resolution is only in the longitudinal impact parameter due to the IBL pixel size.

3.2 ID Tracking Performance

Tracks above a certain p_T threshold are reconstructed offline with the full acceptance range of the ID, with multi-stage track identification algorithms. The inside-out algorithm reconstructs mostly

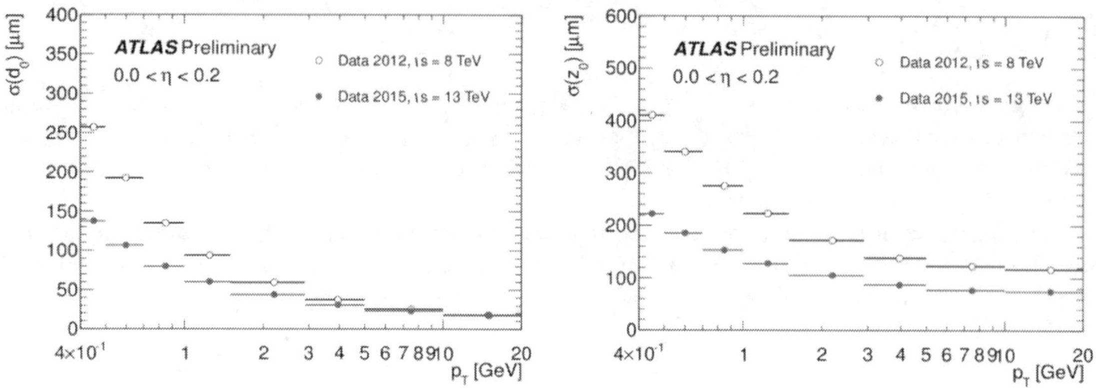

Figure 4: Comparison of the impact parameter resolution in Run-1 (black empty circles) and 2015 (red filled circles). [10]

particles produced in the primary p-p collisions. The tracks are seeded in the pixel detector (three seeds) and the hits from the neighboring SCT layers are added. The track candidates found in the silicon detectors are then extrapolated to include measurements in the TRT.

The outside-in algorithm starts from segments reconstructed in the TRT and extends them inwards by adding silicon hits not yet used by inside-out tracking. It reconstructs secondary charged tracks (*e.g.* photon conversions, material interactions).

Reconstructed tracks are selected by applying quality criteria for typical physics analyses. The *Loose* tracks have $|p_T| > 400\,\mathrm{MeV}$, $|\eta| < 2.5$, number of silicon hits ≥ 7, number of shared modules ≤ 1, number of silicon holes ≤ 2, number of pixel holes ≤ 1. The *Tight Primary* tracks, in addition to the *Loose* selection requirements have number of silicon hits ≥ 9 if $|\eta| \leq 1.65$, number of silicon hits ≥ 11 if $|\eta| \leq 1.65$, at least one hit on the two innermost pixel layers, and no pixel holes.

The tracking efficiency for primary tracks in 2015 for tracks above 5 GeV remained high, as shown in figure 5: 90% for Loose and 85% for Tight Primary selections [11].

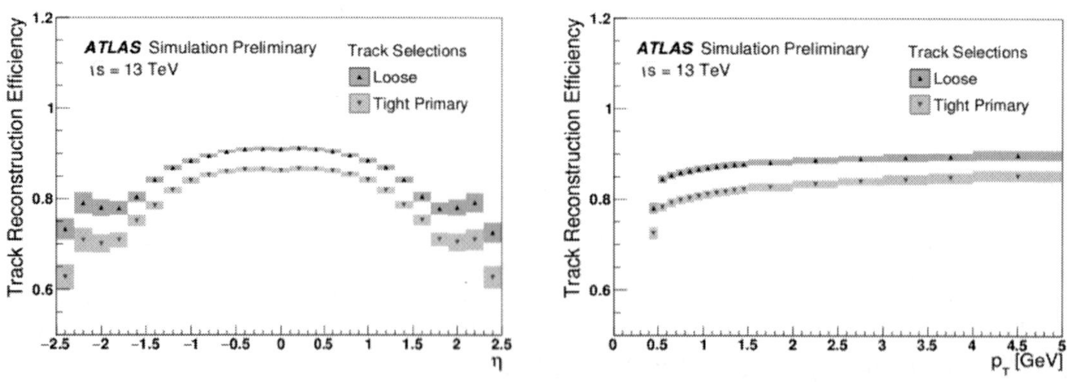

Figure 5: Track reconstruction efficiency in 2015 [11].

3.3 *b*-tagging Performance

The identification of b-quark jets plays a crucial role for the Run-2 measurements, both within the Standard Model and the New Physics scenarios. The addition of the IBL and the several enhancements to the tracking and the b-tagging algorithms significantly improve the performance, as illustrated in [12]. Figure 6 shows the comparison between the performance of the default b-tagging algorithm for Run-1 (MV1c) and the default one of Run-2 (MV2c20). Comparing the two algorithms, the light-flavour jet rejection is improved by a factor of about 4 and the c-jet rejection by a factor of between 1.5-2 for a 70% b-jet efficiency.

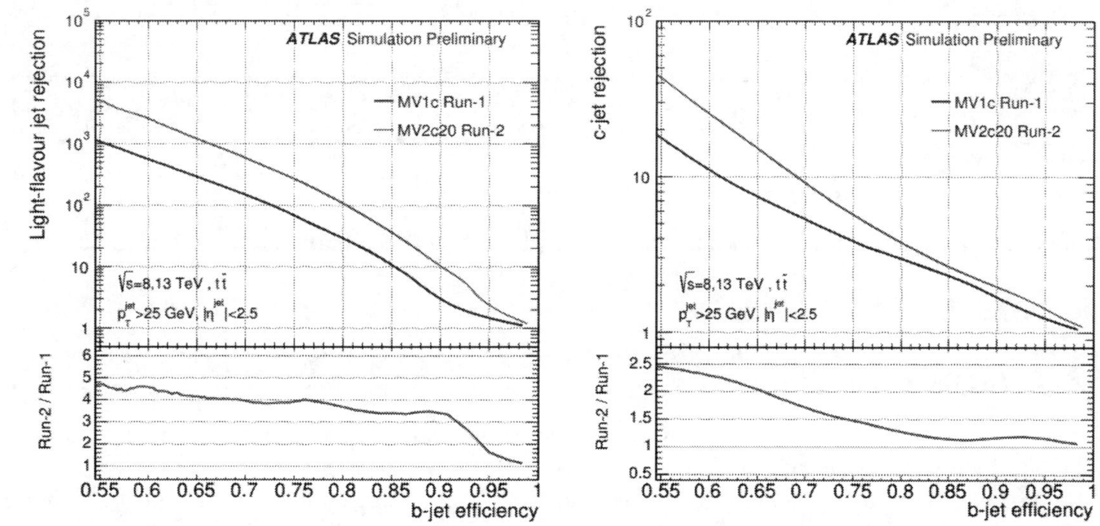

Figure 6: The light (*left*) and c-jet rejection (*right*) versus b-jet efficiency for the Run-1 detector and reconstruction software (blue) compared to algorithm using the Run-2 setup (red) [12].

3.4 Jet Reconstruction Performance

Nearly all the LHC physics arise from quark and gluon interactions, thus a detailed understanding of the quantum chromodinamics (QCD) is fundamental for analyzing new interesting phenomena and their background. Precision measurements of the production of jets are crucial tests of QCD.

The details of the ATLAS jet calibration in 2015 can be found here [13]. To cluster particles belonging to the same jet, the anti$-k_t$ algorithm with radius parameter $R = 0.4$ is used. The input is the positive-energy topological clusters (topoclusters) of the calorimeter cell energies. The uncertainty of the jet energy scale is the dominant experimental uncertainty and an estimate can be derived by in situ measurements. The first in situ results of the jet energy scale using the p-p 13 TeV dataset of 2015, approximately 3.3 fb^{-1} of data, show an agreement between data and Monte Carlo better than 2% up to 3 TeV, as seen in Figure 7. The analog signals of the EM calorimeter are sampled digitally once per bunch crossing over four bunch crossings and those measurements are converted to an energy measurement using constants calculated using dedicated calibration runs. During Run-1, five samples instead of four were used. Noise-levels defined for the calibration of the topoclustering algorithm in 2015 have changed with respect to Run-1 in order to better match noise measurements made in 2012 data.

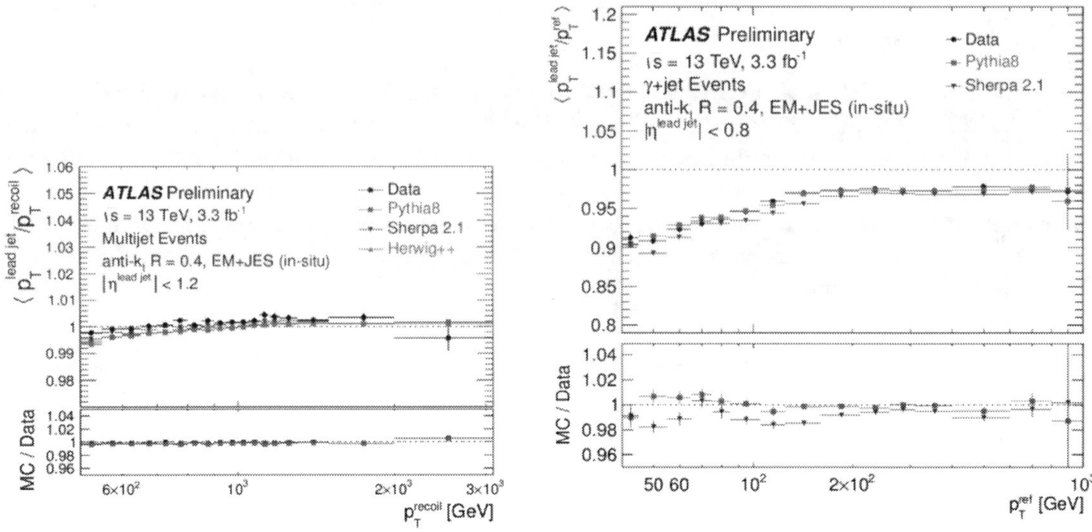

Figure 7: Multijet balance validation (*left*) and γ plus jet balance validation (*right*) in 13 TeV data and MC simulation [14].

3.5 e/γ Reconstruction Performance

Electrons and photons are excellent probes for interesting signatures in p-p collisions, including the Standard Model Higgs decay channel $H \to \gamma\gamma$.

The ATLAS electron efficiency measurements in 2015 are detailed in [15]. The electron identification is based on a likelihood combining the electromagnetic calorimeter showers, tracking, track-cluster matching and TRT particle-identification to separate isolated electrons. Several changes to the input variables used for the electron identification have been introduced in Run-2. In particular, the number of IBL hits is used for discriminating between electrons and converted photons, and new discriminating variables in the algorithms have been used due to the change in the TRT gas. The reconstruction efficiency is comparable to the one observed in 2012. The combined reconstruction and identification efficiency in $Z \to ee$ events are shown in Figure 8.

The identification of prompt photons relies on the high granularity of the calorimeter system. Three independent methods have been exploited to identify photons with the 2015 p-p collisions data. At low energy radiative photons from $Z \to ll\gamma$ are used; in the medium energy range shower shape properties of electrons from $Z \to ee$ decays are extracted and extrapolated to photons; while at high energy the efficiency on samples of reconstructed photons is directly measured after subtracting the hadronic background with a technique based on track isolation. The identification efficiency measurements are detailed in [16] and a comparison of the data driven measurements of the identification efficiency for unconverted photons is shown in Figure 8. The photon identification efficiency increases from 53−64% (47−61%) for unconverted (converted) photons at $E_T \sim 10$ GeV to 88−92% (96−98%) for $E_T \geq 100$ GeV.

3.6 Tau Reconstruction Performance

The tau leptons play an important role in the Higgs physics and new physics searches, and in the detector performance studies such as the transverse missing energy determination. The hadronically

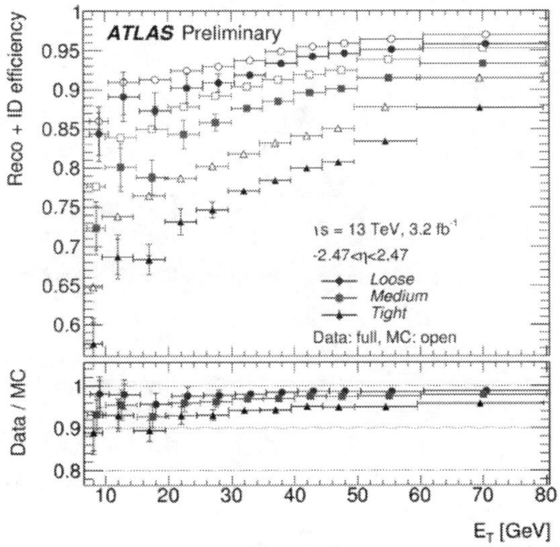

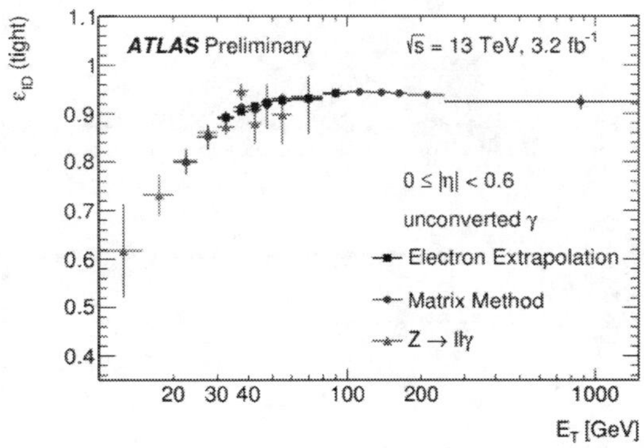

Figure 8: Combined electron reconstruction and identification efficiency in $Z \to ee$ events as a function of the transverse energy E_T (*left*) [15]. Comparison of the data-driven measurements of the identification efficiency for unconverted photons (*right*) [16].

decaying taus are narrow jets with low track multiplicity. They are identified using boosted decision tree to distinguish them from other jets and the performance is evaluated using $Z \to \tau\tau$ candidates.

A good agreement between data and simulation both in 2015 and 2016 is registered, as shown in Figure 9 and described in [17].

3.7 Muon System Performance

The ATLAS muon spectrometer performed very well during Run 2, with an alignment precision of about 50 μm for both barrel and endcap.

The reconstruction of the muons is crucial to some of the most important physics results published by ATLAS. This article [19] describes the performance of the ATLAS muon reconstruction measured using 3.2 fb^{-1} of data in 2015. The calibration sample consists of $Z \to \mu\mu$ decays and $J/\Psi \to \mu\mu$ decays. The dimuon invariant mass distributions are shown in figure 10. The muon reconstruction efficiency is measured to be close to 99% over most of the covered phase space ($|\eta| < 2.5$ and $5 < p_T < 100$ GeV). The isolation efficiency varies between 93% and 100% depending on the selection applied and on the momentum of the muon. The p_T resolution in data and simulation agree to better than 5% for most of the η range.

4 Conclusion

ATLAS underwent several upgrades during the two-year shutdown 2013/2014 referred to as LS1. The restart after the LS1 and the data taking through-out 2015 has been very successful. Despite the challenging conditions, the system stability and the data taking efficiency have quickly reached the Run-1 level. Detailed performance studies demonstrate good understanding of the 2015 data. The 2016 data taking started on 25 April, with an excellent LHC performance, an integrated luminosity in

DOI: 10.1051/epjconf/201716401021

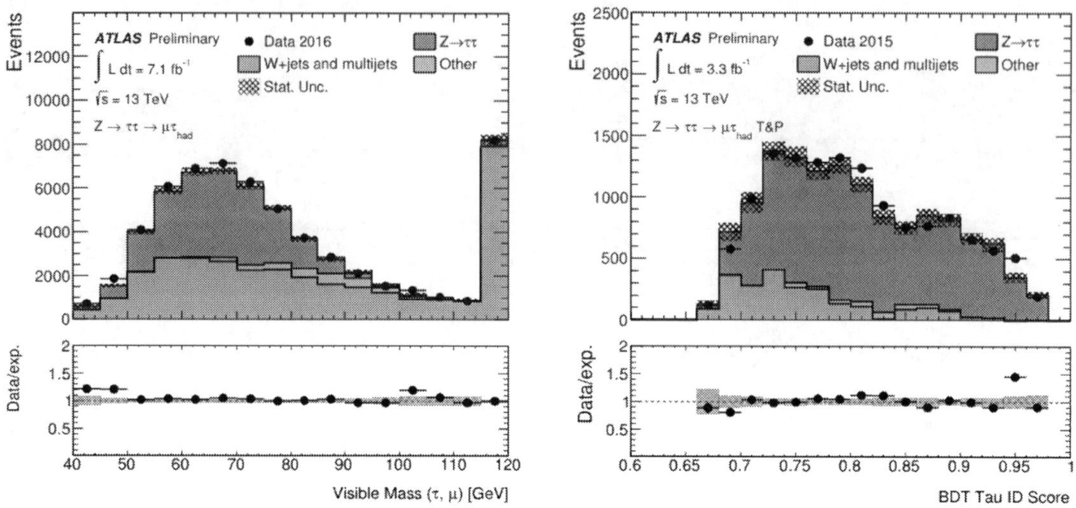

Figure 9: The visible mass reconstructed using isolated muons and offline tau candidates passing the offline loose identification requirement (*left*). The BDT tau identification score for offline tau candidates passing the offline medium identification requirement (*right*) [18].

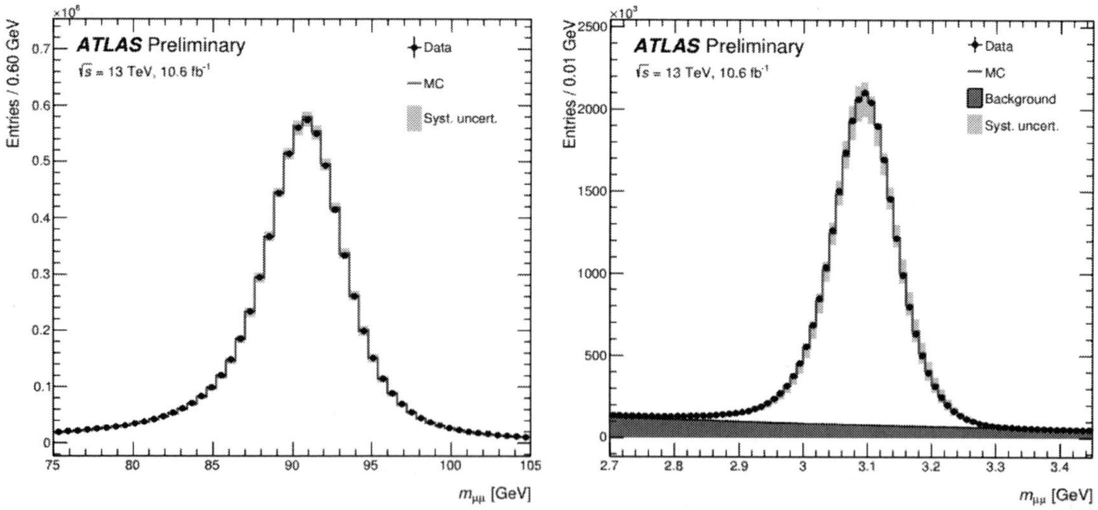

Figure 10: Dimuon invariant mass distribution of $Z \to \mu\mu$ events (*left*) and $J/\Psi \to \mu\mu$ events (*right*) reconstructed with combined muons. The points show the data. The continuous line corresponds to the simulation with the momentum corrections applied. Simulations are normalized to data [19].

early summer already higher than in 2015, and the instantaneous luminosity beyond the design level of $1 \cdot 10^{34}\,\text{cm}^{-2}\text{s}^{-1}$.

References

[1] The ATLAS Collaboration, Physics Letters B **Volume 716, Issue 1**, pages 1-29 (2012)
[2] The ATLAS Collaboration, JINST 3 **S08003**, (2008)
[3] Capeans, M. et al., *ATLAS Insertable B-Layer Technical Design Report* (2010)
[4] The ATLAS Collaboration, ATL-MUON-PROC-2015-009 (2015)
[5] https://twiki.cern.ch/twiki/bin/view/AtlasPublic/LuminosityPublicResults
[6] https://twiki.cern.ch/twiki/bin/view/AtlasPublic/LuminosityPublicResultsRun2
[7] The ATLAS Collaboration, ATL-INDET-PUB-2015-001 (2015)
[8] F. Faccio, G. Cervelli, IEEE Transactions on Nuclear Science **Volume 52, Issue 6**, pages 2413-2420 (2005)
[9] https://atlas.web.cern.ch/Atlas/GROUPS/PHYSICS/PLOTS/IDTR-2016-005
[10] https://atlas.web.cern.ch/Atlas/GROUPS/PHYSICS/PLOTS/IDTR-2015-007
[11] The ATLAS collaboration, ATL-PHYS-PUB-2015-051 (2015)
[12] The ATLAS collaboration, ATL-PHYS-PUB-2015-022 (2015)
[13] The ATLAS collaboration, ATL-PHYS-PUB-2015-015 (2015)
[14] http://atlas.web.cern.ch/Atlas/GROUPS/PHYSICS/PLOTS/JETM-2015-003/
[15] The ATLAS collaboration, ATL-CONF-2016-024 (2016)
[16] The ATLAS collaboration, ATL-PHYS-PUB-2016-014 (2016)
[17] The ATLAS collaboration, ATL-PHYS-PUB-2015-025 (2015)
[18] https://twiki.cern.ch/twiki/bin/view/AtlasPublic/TauPublicCollisionPlots#Performance_plots_for_LHCC_2015
[19] The ATLAS collaboration, Eur. Phys. J. **C76**, 292 (2016)

DOI: 10.1051/epjconf/201716401022

Precision measurements and searches with single and multiple gauge bosons with the ATLAS detector

E. Yu. Soldatov[1,a]
on behalf of the ATLAS Collaboration

[1]*National Research Nuclear University «Moscow Engineering Physics Institute», 115409, Kashirskoe shosse, 31, Moscow, Russia*

Abstract. The ATLAS collaboration has carried out a set of measurements at center-of-mass energies of 8 TeV and 13 TeV, which provide stringent tests of the electroweak sector of the Standard Model, specifically on single, di- and multi-boson production cross sections.
This paper gives a review of the latest results on these topics from ATLAS. It covers the inclusive production measurement of single W and Z/γ^* bosons and the measurement of the complete set of coefficients that describe the angular distributions of the decay leptons in Z events. Also it includes the review of the inclusive and differential cross sections measurements for the production of two heavy gauge bosons (WW, WZ, ZZ) in the leptonic channels, the production measurement of the Z boson in association with one or two photons and the search for the WZ production via vector-boson scattering. These measurements are compared to (N)NLO predictions of the Standard Model and provide model-independent constraints on the so-called «new physics», by setting the limits on the anomalous triple or quartic gauge-boson couplings.

1 Introduction

During several last years ATLAS experiment [1] has performed a lot of new Standard Model (SM) measurements. The range of these measurements has shown in figure 1. It starts from the production of single bosons and ends on tree-boson production and also on vector-boson fusion/scattering (VBF/VBS) processes. Data measurements become more precise, so we need next-to-next-to-leading-order (NNLO) theory predictions for comparisons.

Two main motivations of the SM measurements in ATLAS are: test the theory with high precision and find the signs of «new physics». From one hand, we are doing measurements of integrated, differential cross sections and different angular distributions. They help us to prove the validity of the Standard Model at the TeV scale, to perform a comparison of data with theory predictions of higher order QCD and QED effects, to probe the proton structure and to understand irreducible diboson backgrounds into Higgs and exotic analyses. From another hand, we are extrapolating self-coupling structure of gauge bosons. They

[a]e-mail: EYSoldatov@mephi.ru

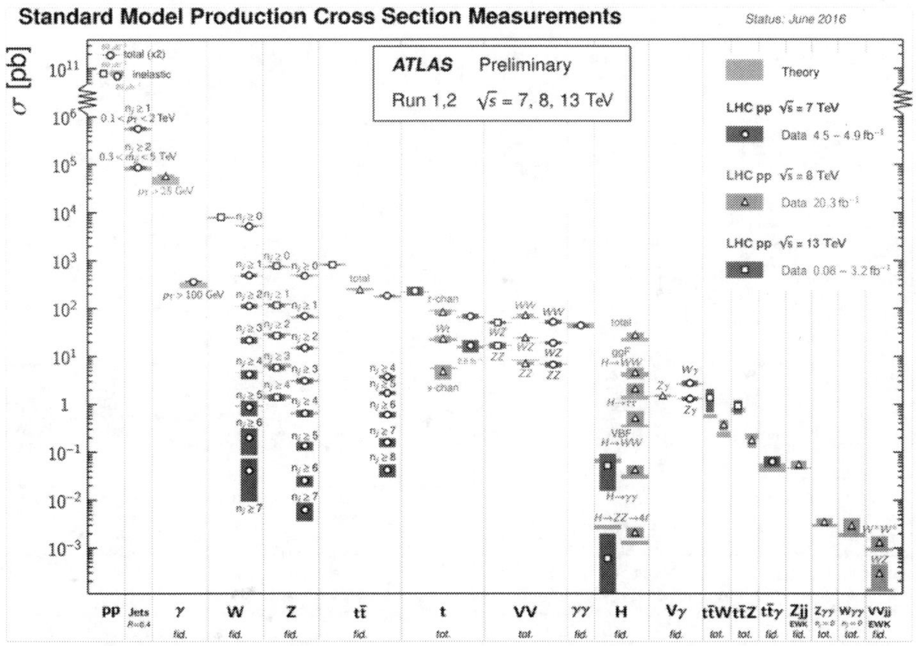

Figure 1. Summary of several Standard Model total and fiducial production cross-section measurements by ATLAS, corrected for leptonic branching fractions, compared to the corresponding theoretical expectations [2].

should improve our understanding of electroweak symmetry breaking mechanism and S-matrix unitarity treatment as well as intersect with determination of Higgs couplings and indicate «new physics», if anomalous triple/quartic gauge couplings (aTGC/aQGC) are present.

This paper contains description of the measurements of the inclusive Z and W boson productions, angular coefficients in Z boson events, as well as the multi-boson productions (WZ, ZZ, WW, $Z\gamma$, $Z\gamma\gamma$) by using data from the Run1 and first data from the Run2.

2 The inclusive W/Z production

The inclusive measurement of single W and Z boson production [3] uses only 81 pb^{-1} of 13 TeV Run2 data. Main backgrounds are comming from the multi-jet events and inclusive Z/W events. Fiducial region uses selections on leptons kinematics and on full or transverse invariant mass of the reconstructed objects. The cross sections are in a good agreement with the theoretical predictions of different PDF sets as shown in figure 2. Main uncertainties for the measurement are comming from the luminosity, JES and the dominant multi-jet background.

This study also gives the results on the cross-section ratios for W^+/W^- and W/Z. The agreement of the measured and predicted values is actually very good and has shown in figure 3. The main idea of this study is that the ratios will be sensitive to the PDFs since some of the uncertainties will be cancelled out in the ratios. The first ratio shows better agreement with CT14nnlo and MMHT14nnlo. For the second ratio, 2% precision of the measurement is not enough to distinguish predictions with different PDFs.

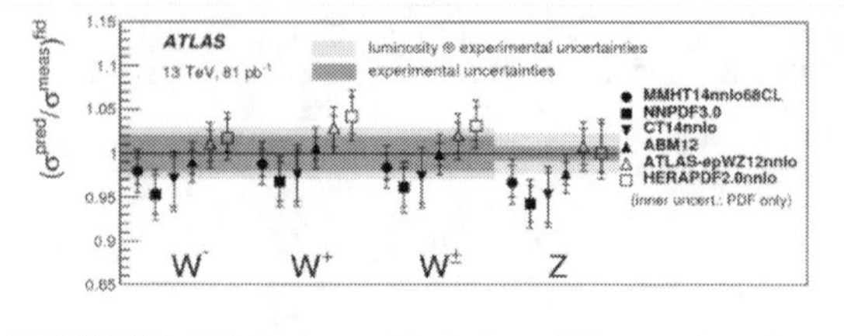

Figure 2. Results for the inclusive W and Z cross-section measurements and its comparison with theory [3].

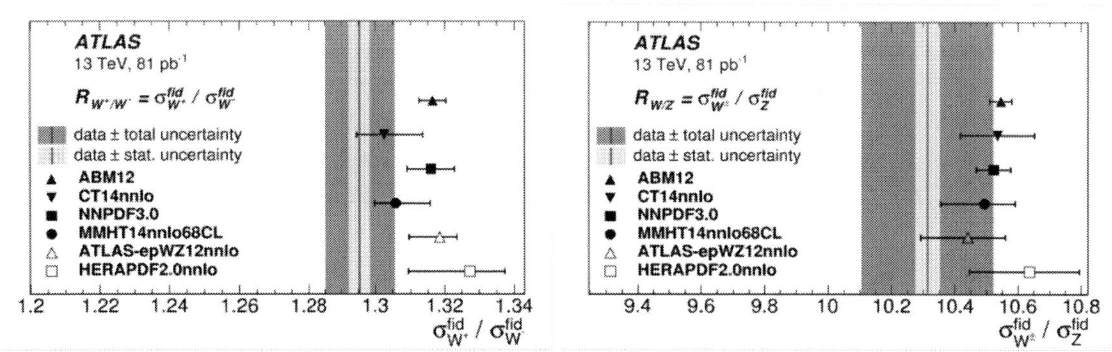

Figure 3. Cross-section ratios for W^+/W^- (left), W/Z (right) and comparison with theory predictions with different PDF sets [3].

Another single Z boson study is the measurement of the angular coefficients in Z boson events [4]. It was done by using the full 2012 dataset of 20.3 fb^{-1}. This study is about to measure the production dynamics through the spin-1 Z boson via spin correlation between initial and final state partons in the Collins Soper's reference frame. Coefficients can be expressed as a function of θ and ϕ.

The event selection contains some kinematical selections of leptons for central-central and central-forward channels as well as two-lepton invariant mass window requirement. Coefficients A_{0-7} are in a good agreement with theory. A_{0-2} are not in agreement (see figure 4) since they are very sensitive probe of the higher order QCD corrections.

3 Di- and multi-boson productions

The first considered study is the leptonic WZ measurement, which was done with full 8 TeV 2012 dataset [5].

The event selection includes kinematic selections for the leptons from Z and W and also contains requirements on the two-lepton invariant mass (for Z) and on the transverse mass build from lepton and E_T^{miss} (for W).

This study gave the measurement of the integrated and differential cross sections and also the measurement of the cross-section ratio of W^+Z/W^-Z (shown in figure 5). Moreover, this

 DOI: 10.1051/epjconf/201716401022

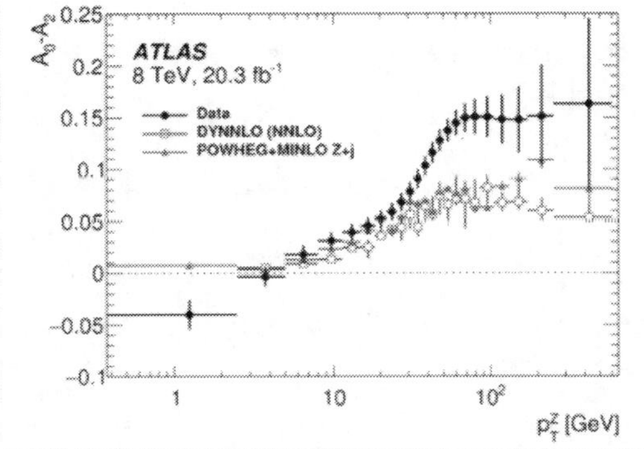

Figure 4. Distributions of the angular coefficients A_0–A_2 as a function of p_T^z [4]. The results from the y_Z-integrated measurements are compared to the DYNNLO and the POWHEG MINLO predictions.

study set the limits on aTGC and on aQGC by using the vector-boson scattering subset of the WZ events.

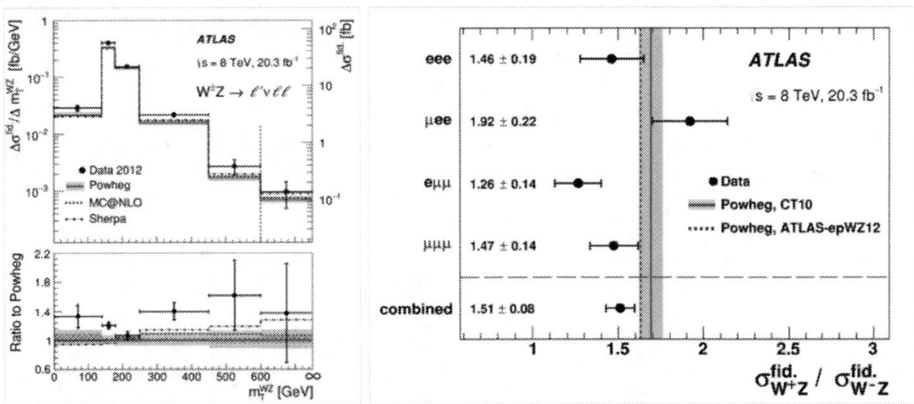

Figure 5. Comparison of the measured observables of WZ production with theoretical predictions [5]. The left figure shows distribution of differential cross section versus m_T^{WZ}. The lower plot shows the ratios of the data to theory predictions. The right figure shows cross-section ratio W^+Z/W^-Z for different decay channels of WZ and comparison with theory predictions.

The cross-section measurements show fair agreement with theory. It is mainly connected with the usage of NLO thoretical predictions. The NNLO predictions should improve the situation.

The uncertainty of the measurement is dominated by electron ID, luminosity, muon efficiency and degree of knowledge of mis-ID background. Ratios show also fair agreement inside 2σ. NNLO predictions can improve the picture.

The study of the same final state was also done by using the full 2015 13 TeV dataset of 3.2 fb^{-1} [6]. Selection is the same as for 8 TeV analysis. However, 13 TeV analysis was used

DOI: 10.1051/epjconf/201716401022

just for the measurement of the integrated cross section, the differential cross section versus jet multiplicity and the cross sections ratio (shown in figure 6).

The measured integrated cross section ($\sigma^{Observed}_{W\pm Z} = 50.6 \pm 2.6(stat) \pm 2.0(sys) \pm 0.9(th) \pm 1.2(lumi)$ pb) was compared with theoretical NNLO prediction by the MATRIX generator ($\sigma^{Expected}_{W\pm Z} = 48.2^{+1.1}_{-1.0}(scale)$ pb). Comparison shows very good agreement.

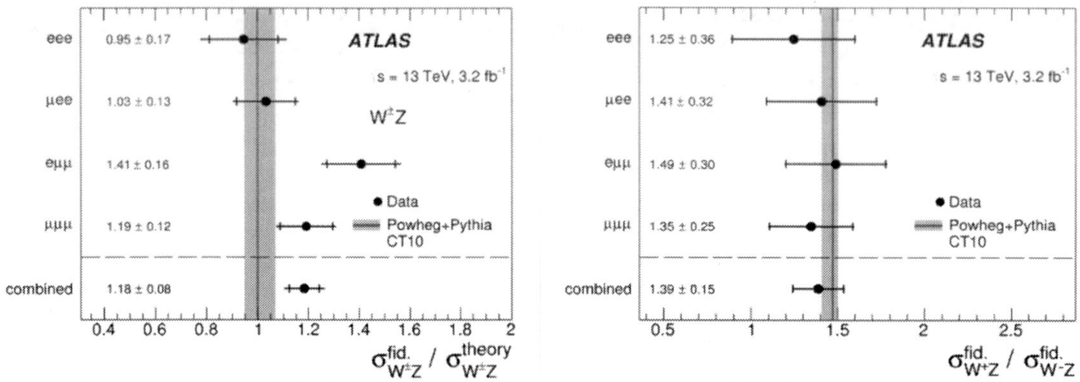

Figure 6. Ratio of the measured $W^{\pm}Z$ integrated cross sections in the fiducial phase space to the NLO SM prediction from the Powheg+Pythia in each of the four channels and for their combination (left). Measured ratios $\sigma^{fid}_{W+Z}/\sigma^{fid}_{W-Z}$ of W^+Z and W^-Z integrated cross sections in the fiducial phase space in each of the four channels and for their combination (right) [6].

Other predictions were made by using the NLO theory predictions and at the figure 7 one can see how different the NLO predictions are compared to the NNLO case for 13 TeV.

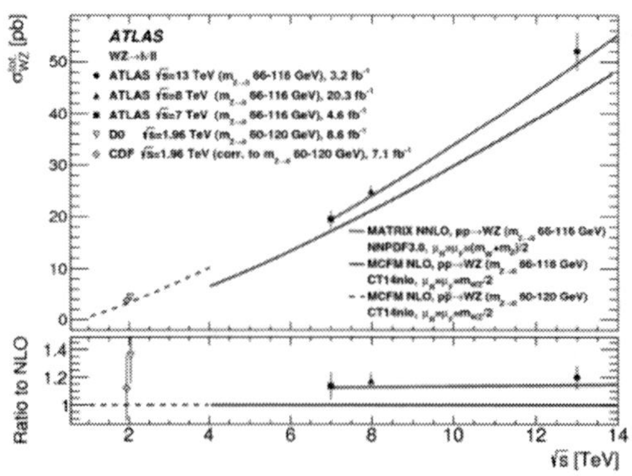

Figure 7. Comparison of the $W^{\pm}Z$ cross section measurements at various center-of-mass energies with NLO and NNLO Standard Model expectations [6]. The lower plot shows the ratios of the NNLO predictions to the NLO predictions.

Statistical uncertainties are quite large for this new 13 TeV data measurement, so the agreement with NLO theory predictions is good within these errors.

The next measurement was focused on ZZ production [7]. It was done using full 2015 13 TeV dataset. Only charged lepton (e/μ) decay channels of the Z bosons were used. All leptons were kinematically selected for the fiducial region. The requirement on the invariant mass of both lepton pairs was imposed. The measured integrated cross section is in a good agreement with NNLO SM prediction as shown at figure 8.

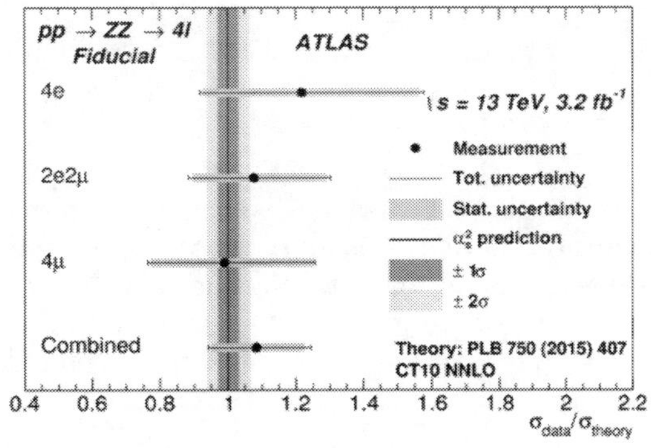

Figure 8. Comparison between measured fiducial cross sections for different channels of ZZ measurement and $O(\alpha_s^2)$ theory predictions [7].

The latest stady of leptonic WW production was done using full 2012 dataset of 8 TeV data [8]. For the fiducial region selection, transverse momenta requirement for the sum of 2 neutrinos and invariant mass of two same flavour leptons anti-window requirement ($|m_Z - m_{\ell\ell}| > 15$ GeV) were used. Moreover, kinematic selections for the leptons as well as the jet veto were applied.

This study gave the integrated and differential cross-section measurements as well as the setting the limits on aTGC. It is shown in figure 9, that measured integrated and differential cross sections are in agreement with the NNLO theory predictions within errors.

Another 8 TeV WW final state study, relying upon the same dataset, is the exclusive $\gamma\gamma \to WW$ production [9]. The fiducial event selection includes the kinematic selections for the leptons, as well as selections on transverse momentum of electron+muon pair and invariant mass of two leptons requirements. Also, it uses specially developed exclusivity selection (no extra tracks within $\Delta z = 1$ mm from z average of 2 selected leptons). This measurement shows good agreement between data ($\sigma^{Observed}_{\gamma\gamma\to WW} = 6.9 \pm 2.2(stat) \pm 1.4(sys)$ fb) and MC predictions ($\sigma^{Predicted}_{\gamma\gamma\to WW} = 4.4 \pm 0.3$ fb). Statistical uncertainty is dominant for this measurement.

ATLAS recently published a measurement of the $Z\gamma$ and $Z\gamma\gamma$ production [10]. This study uses full 8 TeV dataset. For the $Z\gamma$ there are three final states: $ee\gamma$, $\mu\mu\gamma$ and $\nu\nu\gamma$. Event selection contains kinematic requirements on leptons and photon. Charged lepton channels also have two lepton invariant mass requirement. Neutrino channel has a requirement on the missing transverse momentum and on the angles between the final state objects of this study.

The study provides the measurement of the integrated (shown in figure 10) and differential (shown in figure 11) cross sections and the setting the limits on aTGC. The measured cross

 DOI: 10.1051/epjconf/201716401022

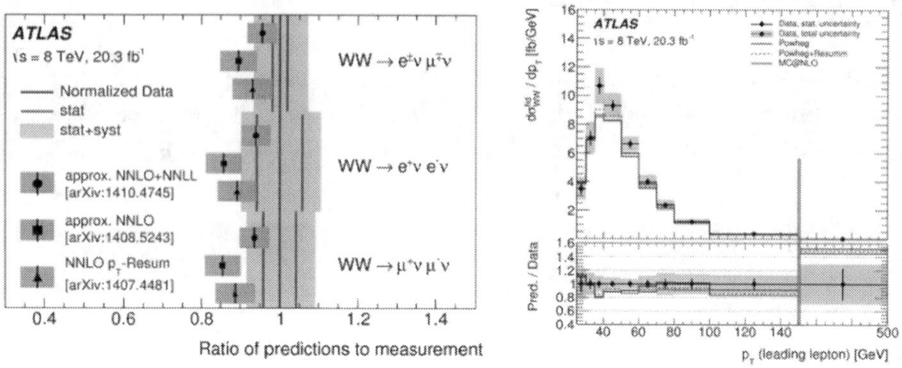

Figure 9. Comparison of the measured cross sections of WW production with various theoretical predictions [8]. The left figure shows the measured and predicted fiducial cross sections normalised to the respective measured values for all final states. The right figure shows the measured unfolded differential cross sections of WW production in the $e\mu$ final state for the transverse momentum of the leading lepton, p_T^{lead}. The lower plot shows the ratios of the predictions to data.

sections are in agreement with the NNLO theory predictions. The main uncertainties are comming from the lepton isolation, JES, luminosity, photon isolation and the backgrounds determination.

The second part of the analysis dedicated to the two photon case. Selection was slightly changed only for neutrino channel in order to use the diphoton trigger. This first ATLAS study of the $Z\gamma\gamma$ provides the measurement of integrated cross section and the setting the limits on aQGC. The measured cross sections for $ee\gamma\gamma$, $\mu\mu\gamma\gamma$ and $\nu\nu\gamma\gamma$ channels are in fair agreement with the NLO theory predictions (see figure 10). Combined $ee\gamma\gamma$ and $\mu\mu\gamma\gamma$ measurement has significance equal to 6σ.

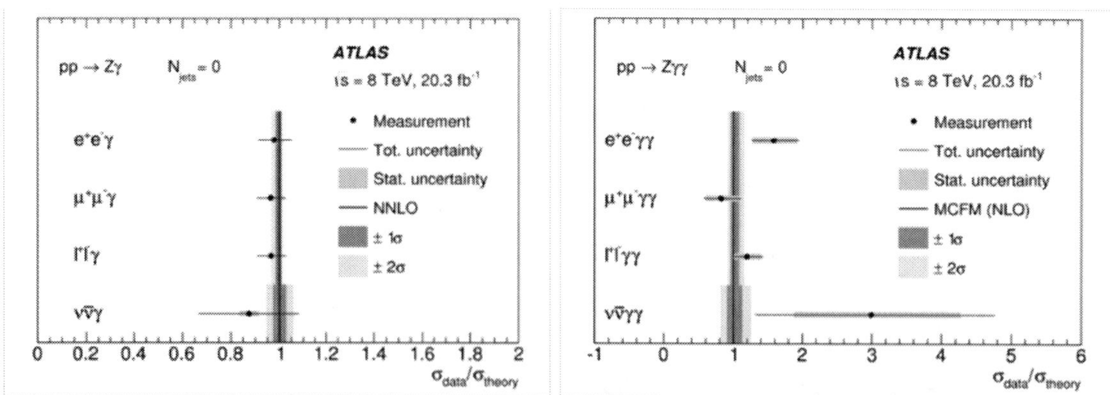

Figure 10. Comparison between the measured cross sections and the theory predictions for the $pp \rightarrow Z\gamma$ (left) and $pp \rightarrow Z\gamma\gamma$ (right) processes in the inclusive $N_{jets} = 0$ extended fiducial region [10].

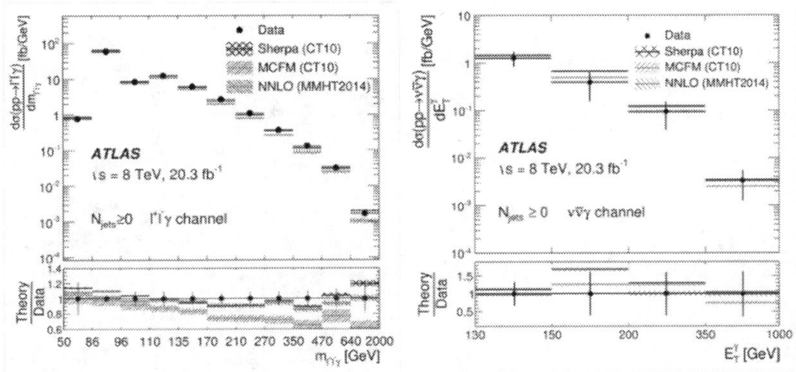

Figure 11. The measured and predicted differential cross sections as a function of E_T^γ for the $pp \to \ell\ell\gamma$ process (left) and for the $pp \to \nu\nu\gamma$ process (right) using inclusive ($N_{jet} \geq 0$) selection [10]. The lower plots show the ratios of the predictions to data (shaded bands).

4 Anomalous couplings study

Only several triple and quartic vertices are allowed in the Standard Model on the tree level, especially: $WW\gamma$, ZWW, $WWZZ$, $WWZ\gamma$ and $WW\gamma\gamma$. Effects of «new physics» can be observed in the terms of anomalous couplings. At the figure 12 one can see the effect of nonzero aGC coupling parameter for the cross section in the high energy region.

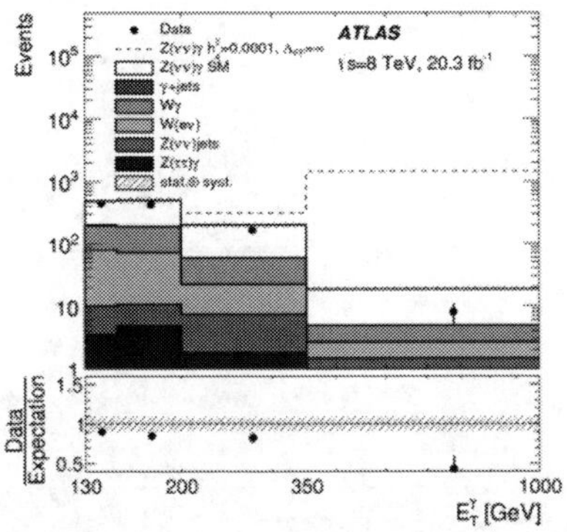

Figure 12. The photon transverse energy E_T^γ distribution from exclusive ($N_{jet} = 0$) $Z(\nu\nu)\gamma$ events [10]. The numbers of candidate events observed in data (points with error bars) are compared to the sum of the SM signal and the various backgrounds (black solid line) and also to the sum of the signal prediction with one nonzero aTGC parameter and the various backgrounds (red dashed line). The ratio of the numbers of candidate events observed in data to the sum of expected SM signal and backgrounds is also shown.

For the charged triple boson vertices ATLAS studies of WW and WZ set the most stringent limits on $\Delta\kappa_Z$, λ_Z and Δg_1^Z effective Lagrangian parameters (see figure 13).

For other two parameters ($\Delta\kappa_\gamma$ and λ_Z) of the charged vertices ATLAS results of the same order as the CMS and previous results.

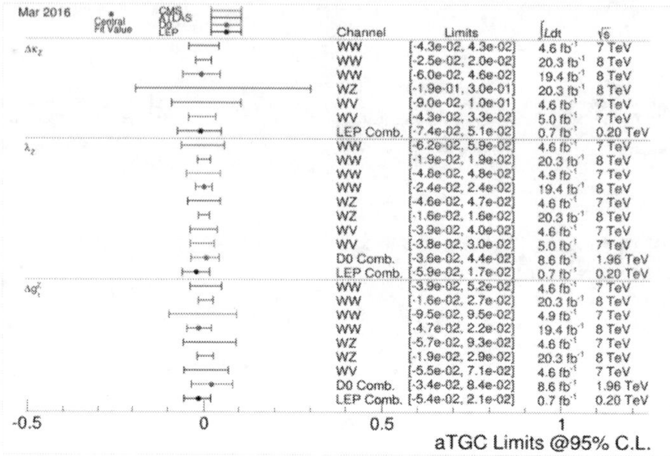

Figure 13. Comparisions of the observed limits for $\Delta\kappa_Z$, λ_Z and Δg_1^Z effective Lagrangian parameters of the WWZ vertex obtained by the various collider experiments [11].

Neutral triple vertices, such as $ZZ\gamma$, $Z\gamma\gamma$ and ZZZ, are forbidden in the SM. The ATLAS $Z\gamma$ measurement contributed to the limits for the neutral coupling parameters h_3^γ, h_3^Z, h_4^γ and h_4^Z. The latest results from ATLAS have shown at figure 14 and they are slightly better than the results from CMS.

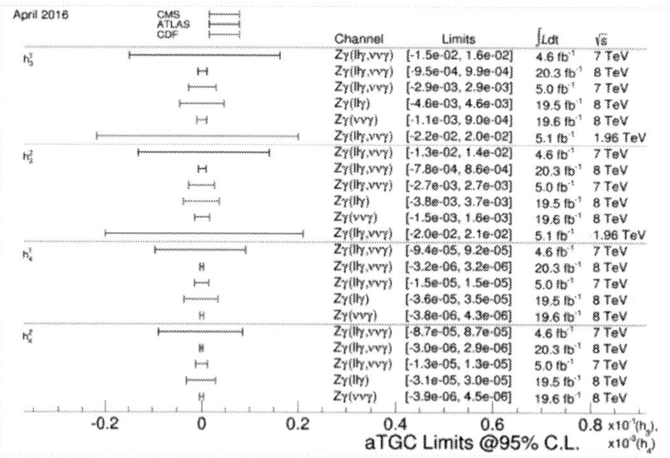

Figure 14. Comparisions of the observed limits for h_3^γ, h_3^Z, h_4^γ and h_4^Z effective Lagrangian parameters of the $ZZ\gamma$ and $Z\gamma\gamma$ vertices obtained by the various collider experiments [11].

The quartic gauge couplings use parameterization from the Effective field theory approach, which gives several families of parameters: «M-family», «T-family» and «S-family». Unfortunately, there are not so many results from ATLAS for «M-family» and «T-family». Only the $Z\gamma\gamma$ study obtained such results, however this channel is not so sensitive for these parameters. Thus CMS limits are slightly better for all considered parameters.

Limits on "S-family" parameters α_4 and α_5 were obtained from the VBS part of the WZ production measurement and shown at figure 15.

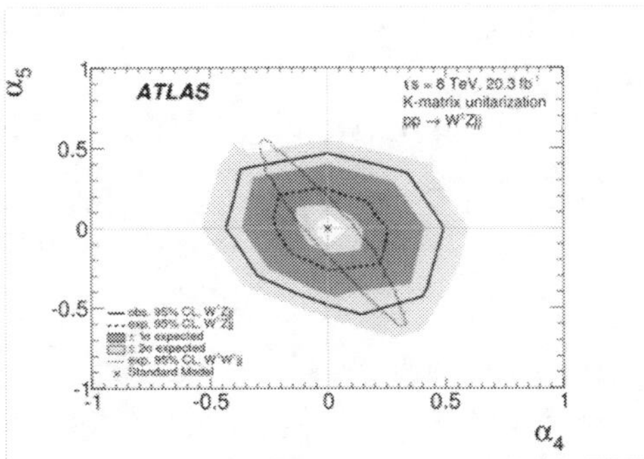

Figure 15. Expected and observed 95% CL limit contours on the aQGC parameters α_4 and α_5 from $W^{\pm}Z$ measurement [5]. The solid and dashed lines represent the observed and expected limits, respectively. The regions outside the black contours are excluded. The green and yellow bands correspond to the 1σ and 2σ uncertainty on the expected limit, respectively. The expected exclusion contour from an ATLAS analysis using $W^{\pm}W^{\pm}$jj events [12] is indicated by the dotted line.

5 Conclusions

The new Standard Model results from the ATLAS experiment were discussed.

The latest single and multi-boson measurements using the Run1 and the Run2 datasets are compatible with the SM expectations (NLO/NNLO).

Very strong aTGC and aQGC limits were set. No deviations from the SM were observed.

6 Acknowledgements

We acknowledge the support of MEPhI Academic Excellence Project (contract No. 02.a03.21.0005, 27.08.2013).

References

[1] ATLAS Collaboration, JINST **3**, S08003 (2008)
[2] https://twiki.cern.ch/twiki/bin/view/AtlasPublic/StandardModelPublicResults
[3] ATLAS Collaboration, Phys. Lett. B **759**, 601 (2016)
[4] ATLAS Collaboration, JHEP **1608**, 159 (2016)
[5] ATLAS Collaboration, Phys. Rev. D **93**, 092004 (2016)
[6] ATLAS Collaboration, Phys. Lett. B **762**, 1 (2016)
[7] ATLAS Collaboration, Phys. Rev. Lett. **116**, 101801 (2016)
[8] ATLAS Collaboration, JHEP **1609**, 029 (2016)
[9] ATLAS Collaboration, Phys.Rev. D **94**, 032011 (2016)
[10] ATLAS Collaboration, Phys. Rev. D **93**, 112002 (2016)
[11] https://twiki.cern.ch/twiki/bin/view/CMSPublic/PhysicsResultsSMPaTGC
[12] ATLAS Collaboration, Phys. Rev. Lett. **113**, 141803 (2014)

Quantum Mechanics, vacuum, particles, Gödel-Cohen incompleteness and the Universe

Luis Gonzalez-Mestres[1,a]

[1] *Cosmology Laboratory, John Naisbitt University, Belgrade and Paris Goce Delceva 8, 11070 Novi Beograd, Serbia*

Abstract. Are the standard laws of Physics really fundamental principles? Does the physical vacuum have a more primordial internal structure? Are quarks, leptons, gauge bosons... ultimate elementary objects? These three basic questions are actually closely related. If the deep vacuum structure and dynamics turn out to be less trivial than usually depicted, the conventional "elementary" particles will most likely be excitations of such a vacuum dynamics that remains by now unknown. We then expect relativity and quantum mechanics to be low-energy limits of a more fundamental dynamical pattern that generates them at a deeper level. It may even happen that vacuum drives the expansion of the Universe from its own inner dynamics. Inside such a vacuum structure, the speed of light would not be the critical speed for vacuum constituents and propagating signals. The natural scenario would be the superbradyon (superluminal preon) pattern we postulated in 1995, with a new critical speed c_s much larger than the speed of light c just as c is much larger than the speed of sound. Superbradyons are assumed to be the bradyons of a super-relativity associated to c_s (a Lorentz invariance with c_s as the critical speed). Similarly, the standard relativistic space-time with four real coordinates would not necessarily hold beyond low-energy and comparatively local distance scales. Instead, the spinorial space-time (SST) with two complex coordinates we introduced in 1996-97 may be the suitable one to describe the internal structure of vacuum and standard "elementary" particles and, simultaneously, Cosmology at very large distance scales. If the constituents of the preonic vacuum are superluminal, quantum entanglement appears as a natural property provided $c_s \gg c$. The value of c_s can even be possibly found experimentally by studying entanglement at large distances. It is not excluded that preonic constituents of vacuum can exist in our Universe as free particles ("free" superbradyons), in which case we expect them to be weakly coupled to standard matter. If a preonic vacuum is actually leading the basic dynamics of Particle Physics and Cosmology, and standard particles are vacuum excitations, the Gödel-Cohen incompleteness will apply to vacuum dynamics whereas the conventional laws of physics will actually be approximate and have error bars. We discuss here the possible role of the superbradyonic vacuum and of the SST in generating Quantum Mechanics, as well as the implications of such a dynamical origin of the conventional laws of Physics and possible evidences in experiments and observations. Black holes, gravitational waves, possible "free" superbradyons or preonic waves, unconventional vacuum radiation... are considered from this point of view paying particular attention to LIGO, VIRGO and CERN experiments.

This lecture is dedicated to the memory of John Bell

[a]e-mail: luis.gonzalez-mestres@cosmology.megatrend.edu.rs;personale-mail:lgmsci@yahoo.fr

1 Introduction

Is Quantum Mechanics an ultimate principle of Physics, or an approximate property of standard particles generated at a more fundamental level by a deeper dynamics?

Quantum Mechanics (QM) is usually assumed to be a basic principle of standard physics, and plays a central role in the description of conventional "elementary" particles. In particular, Quantum Field Theory (QFT) successfully describes many aspects of particle physics, and has generated the "standard model" of electro-weak interactions as well as Quantum Chromodynamics (QCD). No deviation from Quantum Mechanics has by now been demonstrated experimentally.

In spite of such evidences, the question of the foundations of Quantum Mechanics is far from being just an academic one. It has a direct interest for practical applications in electronics, computers and other industrial fields as well as in various domains of fundamental research.

1.1 The uncertainties about entanglement

Quantum entanglement plays a crucial role in the theory of quantum information, including quantum teleportation that is claimed to be successful up to distances of $\simeq$ 100 Km [1] and quantum key distribution (QKD) where a system able to operate over 307 Km of optical fibre has been described [2]. More recently, the results of measurement device independent QKD (MDIQKD) over 404 km of ultralow-loss optical fibre and 311 km of standard optical fibre have been reported [3].

Simultaneously, quantum telescopes [4, 5] and other uses of entanglement are being seriously considered. A satellite devoted to test the transfer of quantum information between space and Earth has also been launched [6, 7].

In all cases, it is hoped that Quantum Mechanics remains valid at the relevant distance scales. On this fundamental question, further work is required.

Understanding the deep origin of Quantum Mechanics can help to better determine its validity and possible modifications, and to develop all the above mentioned fields.

Standard quantum mechanics has often been considered as counter-intuitive, in spite of experimental evidence [8]. Entanglement appears as a mysterious phenomenon from the point of view of relativity, as no signal is expected to travel faster than light.

Direct experimental tests of quantum entanglement, involving a small time interval comptatible with space-like separation between two distant local events, have been performed up to distances of 1.3 Km, confirming the validity of QM theory at this scale. [9, 10]. This is the largest distance scale at which entanglement has been direcly checked in this way until now.

1.2 On alternatives to standard theories

Although there is no practical incompatibility between relativity and Quantum Mechanics for the description of standard matter and QFT is a successful approach, the combination of both theories has nontrivial implications such as excluding most local hidden variable approaches to QM [9, 10].

A major challenge for QFT is the elaboration of a quantum theory of gravity [11, 12]. In the standard approach, quantum gravity is usually expected to provide an ultimate theory for both particle physics and Big Bang cosmology. Such a point of view can, however, be questioned as new physics (preons...) may manifest itself at very high energy and very small time and distance scales [13, 14]. Quantum Mechanics and relativity are then in most cases expected to be of composite origin [15, 16].

As emphasized in [14, 15], and already considered in [17, 18], the most relevant preonic pattern to generate possible violations of standard relativity and Quantum Mechanics would be a nontrivial internal structure of the physical vacuum. Contrary to the original preon model [20] where preons were assumed to be direct constituents of the standard particles, the basic preon model dealt with in [17, 18] and in [14, 15] assumes vacuum to be made of preonic constituents and standard particles to be excitations of such a vacuum (see also [19]). Conventional relativity and Quantum Mechanics become then low-energy limits of a more sophisticated and fundamental dynamics.

The existence of a pre-Big Bang era would also be a natural consequence of this, yet unexplored, preonic vacuum structure [21, 22]. In such a scenario, the standard Big Bang must be replaced by a new cosmology at the early stages and cosmic inflation is no longer necessary [13, 23].

1.3 The possible role of space-time structure

The structure of space-time is also a crucial issue to understand the possible origin of Quantum Mechanics [14, 15]. In particular, nothing is known about the properties of space-time inside the physical vacuum. Just as the speed of sound is much smaller than the critical speed of the particles forming a solid, it seems natural to assume [17, 18] that the critical speed c_s of the preonic constituents of vacuum is much larger than that the speed of light c. Then, the propagation of superluminal signals in vacuum would provide a natural explanation for the observed quantum entanglement [14, 15] and, simultaneously, a possible large distance limit for the validity of Quantum Mechanics.

Furthermore, the space-time structure at this deeper level can also go beyond the standard real space-time coordinates and their conventional extensions. In [24, 25], we introduced a spinorial space-time (SST) where two complex coordinates replace the four real ones with the symmetry group SU(2) instead of the usual SO(3) for space rotations. Spin-1/2 particles become then actual representations of the space symmetry group, and it is tempting to consider a possible relation between the SU(2) complex space-time coordinates and the complex wave functions of Quantum Mechanics [14, 15].

The SST cosmic geometry locally predicts the existence of a privileged space direction (PSD) for each comoving observer as already noticed in [24, 25] and further discussed in [26, 27] as well as in more recent work [13, 23]. It must be emphasized, however, that this property does not by itself imply any anisotropy in the global expansion of the Universe. The existence of the PSD may have been confirmed by Planck data [28, 29]. But such a PSD is strictly a local phenomenon and the SST defines a globally uniform and isotropic cosmic geometry. Thus, a cosmic space-time geometry based on the SST appears to be consistent with Planck results and simultaneously with recent work confirming the isotropy of the expansion of the Universe [30].

1.4 A new approach to Quantum Mechanics

As discussed in [14, 15], [13, 23] and in previous papers, the SST geometry and a preonic vacuum structure can be reliably combined to build an alternative approach to Quantum Mechanics. And, subsequently, to the Big Bang cosmology based on conventional QM and relativity at scales where the validity of these principles has not really been tested .

In this lecture, we present un updated detailed discussion of such a pattern, including possible experimental and cosmological tests as well as the implications of Gödel-Cohen incompleteness [31].

2 Bell's theorems and experimental tests

What would be the implications of strict standard relativistic causality for the theoretical and phenomenological content, origin and implications of Quantum Mechanics?

The original theorems (1964, 1975) by John Stuart Bell [32, 33] have plaid a crucial role in the understanding of the grounds of QM and in the conception of possible experimental tests of its basic propeties. They still lead nowadays to interesting conceptual controversies [34, 35].

John Bell actually taught us what is, and what is not, Quantum Mechanics and how one can test the validity of standard QM. Bell's work and life are remembered in [36, 37].

2.1 Bell's work and its relevance

How to conceive experiments allowing to confirm the validity of standard quantum mechanics and to exclude interpretations based on local hidden variables? Bell's inequalities have clearly answered the theoretical side of this question and opened the way to successful experimental tests.

The first paper by John Bell (1964), *On The Einstein Podolsky Rosen Paradox*, refers to the 1935 article by Albert Einstein, Boris Podolsky and Nathan Rosen (EPR) *Can Quantum-Mechanical Description of Physical Reality Be Considered Complete?* [38]. The end of the EPR abstract explicitly states on the properties of quantum wave functions:

"One is thus led to conclude that the description of reality as given by a wave function is not complete".

In [32], Bell subsequently cites this point of wiew by Albert Einstein [39]:

"But on one supposition we should, in my opinion, hold fast: the real factual situation of the system S_2 is independent of what is done with the system S_1, which is spatially separated from the former".

To refute the arguments of [38, 39] Bell wrote in particular in 1964:

"In a theory in which parameters are added to quantum mechanics to determine the results of individual measurements, without changing the statistical predictions, there must be a mechanism whereby the setting of one measuring device can influence the reading of another instrument, however remote. Moreover, the signal involved must propagate instantaneously, so that such a theory could not be Lorentz invariant".

Later, in his book *Speakable and Unspeakable in Quantum Mechanics* [40], commenting on a hidden variable approach to Quantum Mechanics considered by Georges Lochak [41], John Bell wrote explicitly:

"... if his extension is local it will not agree with quantum mechanics, and if it agrees with quantum mechanics it will not be local".

This was the main result of John Bell's work that recent loophole-free experiments appear to have clearly confirmed more than 50 years after his first pioneering article. To reach such an experimental accomplishment, a remarkable long-term effort has been necessary.

2.2 Bell's inequalities

In what follows, to discuss the basic content of Bell's inequalities and the associated experiments, we refer to the CHSH formulation [42]. The CHSH approach to Bell's theorems is used in [9, 44] and also in [45] and in previous tests of Bell's inequalities. We also assume that the recent experiments presented as "loophole-free" have really this property.

In the CHSH formulation of Bell's result, the test of the possible existence of local hidden variables with standard relativity uses two boxes, A and B, able to receive a binary input (0 or 1) and produce a binary output (+1 or -1). We are interested in the expectation value $< x.y >_{(a,b)}$ where $x.y$ stands for the product of the output values (x for box A, y for box B) given the random inputs a for box A and b for box B. Then, the "local realism" discussed by John Bell leads to:

$$S \ = < x.y >_{(0,0)} \ + \ < x.y >_{(0,1)} \ + \ < x.y >_{(1,0)} \ - \ < x.y >_{(1,1)} \leq \ 2 \tag{1}$$

The value $S \ = \ 0$ is obtained if x and y take random values. If x and y are both systematically equal to 1 (or to -1), one gets $S \ = \ 2$. Values of S larger than 2 are obtained introducing correlations between x and y contrary to the locality assumption. Measuring experimentally the value of S and of similar quantities emerging from Bell's work is therefore a very important task.

Since the 1970s, experimentalists worked to test Bell inequalities [46] and the results confirmed their violation as expected from standard Quantum Mechanics. However, the existence of various possible loopholes was often claimed [47]. To face these criticisms, experiments explicitly presented as loophole-free have been performed in the recent years [43].

2.3 Loophole-free experiments

In particular, the result of a loophole-free Bell test involving a comparatively large distance scale (1.3 Km) has recently been reported [9, 10].

Using an experimental scheme with high-fidelity entanglement between distant electron spins, the measurements described in [9] get $S \ = \ 2.42 \ \pm \ 0.20$. The time window during which A and B are space-like separated and measurements can be performed is 4.27 μs . A second experiment with minor changes [10] has more recently yielded $S \ = \ 2.35 \ \pm \ 0.18$.

Simultaneously, loophole-free experiments using entangled photon pairs [44, 45] yield equally results against the hypothesis of local realism.

The results of [9, 10] and of [44, 45] appear to be clearly incompatible with most conventional local hidden variable interpretations of Quantum Mechanics based on standard relativity. However, the authors of [9] clearly emphasize that: i) their work *"rules out large classes of local realist theories"*; ii) *"no Bell experiment can exclude the infinite number of local realist theories"*. To these comments, [10] adds a conclusion claiming that the second data run has *"strengthened the rejection of a broad class of local realistic theories"*. A more recent test of Bell inequalities can be found in [48].

As remarkably foreseen and explicitly developed by John Bell, the measurement of S and of related parameters has been shown to be a direct, efficient way to test standard QM and exclude most local hidden variable approaches provided loopholes can be ruled out.

3 Beyond Bell's theorems

What can be the future of Quantum Mechanics after the successful loophole-free tests of the expected violation of Bell's inequalities?

As underlined in [9, 10], local hidden variables cannot be totally excluded by the recent and experimental tests of Bell's inequalities. Furthermore, Bell's theorems assumed special relativity to hold exactly and to be a universal law. There is obviously room for further studies, including possible new physics, without contradicting the successful use of the work of John Bell to check experimentally the specific properties of standard QM.

In particular, the confirmed violation of Bell's inequalities does not contradict the possibility that Quantum Mechanics be of composite origin, emerging from a more fundamental set of laws. Conventional QM is not incomplete as a physical law, but it can actually be an approximation holding with some limitations for standard particles at low energy [14, 15].

Bell's inequalities were obtained assuming locality and standard relativity, without considering any possible internal structure of the physical vacuum. Going beyond such a scenario appears today as a natural option [13, 23].

In particular, the time window of 4.27 μs during which A and B are space-like separated in [9] corresponds to the distance of 1.28 Km divided by the speed of light c. This allows to exclude a standard relativistic exchange between the two boxes. But if the physical vacuum is made of superluminal preons, as considered in [17, 18] and in subsequent papers, the propagation of superluminal signals in vacuum can generate a new kind of causal connection between A and B without contradicting the results of [9, 10] and of the other loophole-free experiments that have confirmed the violation of Bell's inequalities for conventional particles with standard relativity.

A natural question is then: why would the superluminal signal propagation inside the physical vacuum act on the internal structure and properties of standard particles in such a way that quantum entanglement can be generated? A logical answer is that in such an approach the conventional "elementary" particles are actually excitations of the physical vacuum. Then, in a sense, two entangled particles form a single vacuum excitation at the time scales of current experiments.

3.1 Quantization and complex wave functions

Interpreting standard particles as excitations of vacuum raises the basic questions of origin of quantization and of the generation of complex wave functions. Natural quantization and Fermi statistics for spinorial wave functions, leading subsequently to quantization and Bose statistics for composite bosons, was considered in [14, 15] and in [13, 23]. If standard particles are just solutions of the vacuum internal equations, Fermi statistics naturally emerges if such equations lead to spinorial wave functions as the solutions at the basic level, and if these solutions are unique around each point of space-time for each relevant set of quantum numbers. The standard exclusion principle would then just reflect the uniqueness of the solutions to vacuum equations.

As discussed in [14, 15] and in [13, 23], complex wave functions and Fermi quantization can automatically emerge from the fundamental equations of a vacuum with a SST. Space-time coordinates would then be complex and spinorial wave functions would naturally correspond to the structure of the spinorial space-time. Fermi statistics would just reflect the existence of a single local solution to the basic dynamical equations for each kind of spinorial excitations, leading to the observed quantization. Bosons can be obtained as composite objects made of two fermions, or as new solutions to vacuum equations subject to similar constraints. A scalar complex wave function can be the product of two complex spinors from which the scalar object is made.

In such an approach, as in standard Quantum Mechanics, quantum uncertainty would be a natural consequence of the wave nature of the particles generated as vacuum excitations. Precision experiments [49] do not seem to confirm the "measurement-disturbance relationship" suggested by Werner Heisenberg [50]. In 1927, Heisenberg wrote in particular [51]:

"At the instant of time when the position is determined, that is, at the instant when the photon is scattered by the electron, the electron undergoes a discontinuous change in momentum. This change is the greater the smaller the wavelength of the light employed, i.e., the more exact the determination of the position".

But reference [49] concludes, on the grounds of experimental results, that *"although correct for uncertainties in states, the form of Heisenberg's precision limit is incorrect if naively applied to measurement"*.

More generally, in connection with the possible origin of wave functions, the existence of complex soliton solutions with real energy to complexified classical equations (Korteweg - de Vries, Sine-Gordon...) must also be noticed [52].

4 A possible origin of Quantum Mechanics

In [14, 15], [53, 54], [13, 55] and in other papers, we explicitly considered the possible generation of Quantum Mechanics from a preonic vacuum with a SST space-time geometry. In [14], we illustrated this idea using cosmic spinorial coordinates (ξ , ξ_0) and writing a classical wave function for a spin-1/2 particle at rest around a space-time origin ξ_0 of the form:

$$\Psi_{sp}\,(\xi) = F(|\,\xi - \xi_0\,|^2)\,(\xi - \xi_0) \qquad (2)$$

together with:

$$\Psi^*_{sp}\,(\xi) = F(|\,\xi - \xi_0\,|^2)\,(\xi - \xi_0)^* \qquad (3)$$

where * stands for complex conjugate corresponding to opposite spin. The real function F is assumed to contain an appropriate cutoff in $|\,\xi - \xi_0\,|^2$.

It was emphasized in [14] that the wave functions Ψ_{sp} and Ψ^*_{sp} violate local standard causality, as they take nonzero values for present and future values of cosmic time around $|\,\xi_0\,|$ (see next Section). Space-like distances are thus replaced by direct spinorial distances on the SST. F tacitly defines a space-time distance scale λ_{SST} below which causality does no longer hold in the conventional sense used for the standard relativistic space-time.

As indicated before and suggested in previous papers, Quantum Mechanics and Fermi statistics can naturally be generated if Ψ_{sp} and Ψ^*_{sp} provide indeed a unique solution to the internal equations of a preonic physical vacuum for the relevant set of quantum numbers.

The expression $|\,\xi - \xi_0\,|$ actually corresponds to a local time scale that will be associated by dynamics to a relevant space scale. In any case, as the upper bounds on the size of the electron are still far from the Planck length, the dynamical structure of expressions like (2) and (3) can manifest itself well before Planck scale introducing new physics for standard "elementary" particles. The study of ultra-high energy cosmic rays (UHECR) can thus play a crucial role to search for deformations and violations of standard relativity and quantum mechanics generated by a new dynamics [56, 57]. Experiments like AUGER [58] can yield important results in this respect [59, 60].

4.1 SST and the propagation of extended objects

An additional mechanism can contribute to the inset of Quantum Mechanics, as discussed in [15, 53], [54, 55], and later in this lecture. For a given value of ξ_0 , the spinorial particle structure defined by equations like (2) and (3) is not only extended in space but also in time.

As a result, the time propagation of a spin-1/2 particle with an internal structure defined by (2) or (3) generates an overlap between wave functions corresponding to different values of ξ_0 and of the cosmic time $|\,\xi_0\,|$. Thus, a continuously propagating particle is not, strictly speaking, a solution of the vacuum equations that generate the static solutions (2) and (3). Continuous propagation of such an extended object can be forbidden for this reason.

A natural solution to avoid this difficulty would be a discrete, and basically random, propagation of particles at very small time scales. The situation will then look close to that generating the Feynman path integral of standard quantum mechanics at larger scales. Thus, QM will emerge as an approximate large distance and low energy description of a more involved dynamics.

4.2 Particle wave functions and the physical vacuum

Clearly, the hypothesis that Ψ_{sp} and Ψ^*_{sp} describe excitations of the inner vacuum dynamics is a natural one and can reasonably challenge the standard QFT approach where all the dynamics is carried by the "elementary" particles and the vacuum is basically empty except for the condensates of conventional particle fields related to spontaneous symmetry breaking and for the zero modes of boson fields.

Conventional QFT particles do not interact with any inner vacuum dynamics. But the established validity domain of such a quantum field theory is not larger than that covered by accelerator experiments. Thus, new physics involving a direct role of the physical vacuum may manifest itself at higher energy scales. The generation of Quantum Mechanics from vacuum dynamics would basically occur at very small distance scales, largely beyond the reach of LHC experiments even if some signatures at accelerators can be imagined as discussed in Section 10.

If fermions are generated through equations like (2) and (3) corresponding to unique solutions satisfying the requirements of vacuum internal dynamics (see also Subsection 3.1), and the harmonic-oscillator nature of standard boson fields allows for alternative descriptions concerning the presence of bosonic zero modes in vacuum [26, 61], possible solutions to the standard cosmological constant problem naturally emerge. Understanding the deep nature of vacuum as well as its actual relation with the dynamics of our Universe and of standard matter is then a crucial subject on which complementary material has been presented in [62].

4.3 Superluminal signals, quantum dynamics and Cosmology

Quantum entanglement is not the only phenomenon of quantum physics where superluminal signals can be present. In [61, 63] and in [26], we considered the possibility that the standard condensates and bosonic zero modes of QFT would actually not be permanently present in a superbradyonic vacuum. This can be possible if, as a conventional particle (vacuum excitation) gets close to the local region considered, the vacuum constituents react in order to temporarily generate the relevant bosonic components allowing the particle to propagate and interact.

In such a mechanism, a superluminal signal emitted by the approaching particle can play an important role reaching the vacuum structure in the relevant region and activating its dynamical reorganization in order to allow the particle to propagate and interact. This nonstandard dynamics would be difficult to observe in laboratory experiments because of the permanent presence of surrounding matter, but can play a crucial role at cosmic level.

In all cases, the generation of particles and Quantum Mechanics from the vacuum inner structure would open the way to new dynamical potentialities that are never considered in standard particle phyisics and QFT. If particles are actually vacuum excitations and Quantum Mechanics is generated by vacuum dynamics, a new cosmology should be built [13, 26]: instead of a Universe "made of particles", the relevant subject would be the expanding vacuum together with its excitations [64, 65].

If Quantum Mechanics is not a fundamental principle of Physics but results from the dynamics of the physical vacuum, a pre-Big Bang phase of the history of the Universe is obviously required [13, 26] beyond the "primeval quanta" postulated in 1931 by Georges Lemaître as *The Beginning of the World* [66]. Since then, Physics has evolved and new concepts have been developed.

5 The spinorial space-time

We briefly remind and update here part of the material presented in previous contributions to this conference (see also [53, 67]). The SST appears as the natural space-time geometry to describe a physical world where fermions exist with spinorial wave functions that should be actual representations of the space symmetry group. The implications of the SST geometry for Particle Physics and Cosmology turn out to be surprisingly strong and deserve careful attention.

In particular, the cosmic SST automatically yields the $H\,t = 1$ relation (H = ratio between relative velocities and distances at cosmic level, t = age of the Universe) in a purely geometric way, without requiring any explicit presence of conventional matter and of the associated dynamics. This property can substantially change the formulation of cosmological laws

The spinorial space-time introduced in [24, 25] is a basic SU(2) representation with two complex coordinates replacing the standard four real ones of the relativistic space-time. Its properties, with possible cosmological implications directly generated by its mathematical structure, have been further considered in subsequent papers including [26, 27], [68, 69] and [13, 23].

5.1 SST, cosmic time, age of the Universe

To a SU(2) spinor ξ describing the cosmic SST coordinates (two complex variables instead of the conventional four real ones) of a point of our space-time, it is possible to associate a positive SU(2) scalar $\mid \xi \mid$ defined by the relation $\mid \xi \mid^2 = \xi^\dagger \xi$ (the dagger stands for hermitic conjugate).

Then, a suitable definition of the cosmic time naturally equivalent to the age of the Universe can be $t = \mid \xi \mid$. The associated space is given by the S^3 hypersphere $\mid \xi \mid = t$ with an additional spinorial structure that does not exist in the standard space. Other definitions of the cosmic time t in terms of $\mid \xi \mid$ (f.i. $t = \mid \xi \mid^2$) can also be used, but they lead to similar cosmological results as long as a single-valued function of $\mid \xi \mid$ provides the definition of the cosmic time.

In this definition of the age of the Universe, the value $t = 0$ is reduced to a point. There can actually be some nontrivial structure around this value in the real history of the Universe, but the relevant values of its size in space and time would be very small and can be neglected in practice except for specific studies of the initial singularity.

5.2 SST, cosmic coordinates and expanding Universe

Using the definition $t = \mid \xi \mid$, and taking ξ_0 to be the observer position on the $\mid \xi \mid = t_0$ hypersphere, space translations can be defined inside this associated space hypersphere (the space at $t = \mid \xi_0 \mid$) and simultaneously on all the space hyperspheres of the SST by SU(2) transformations acting on the whole cosmic spinor space. One can then write $\xi = U\,\xi_0$ with:

$$U = exp\,(i/2\ t_0^{-1}\ \vec{\sigma}.\vec{\mathbf{x}}) \equiv U(\vec{\mathbf{x}}) \tag{4}$$

where $\vec{\sigma}$ is the vector formed by the usual Pauli matrices and the vector $\vec{\mathbf{x}}$ the spatial position (in time units, at the present stage) of ξ with respect to ξ_0 at constant cosmic time t_0 as measured on the S^3 hypersphere. Equation (4) tacitly implies that $\vec{\mathbf{x}}$ must be proportional to t_0 .

The origin of cosmic time, naturally associated to the beginning of the Universe, is given by the point $\xi = 0$ where, as just stressed, the initial space is contracted to a single point. One then gets an expanding universe where cosmological comoving frames correspond to straight lines in the SST starting from the time origin $\xi = 0$ and are transformed into eachother under the action of the cosmic

SU(2) group. Thus, the cosmic SST geometry naturally provides for each comoving observer a local privileged rest frame corresponding to the relevant straight line (ξ_0 = t times a constant spinor) in a way compatible with existing cosmological observations.

As already noticed in [24, 25], attempts to associate to the cosmic spinor ξ a space-like position vector with real cosmic space coordinates defined by $\vec{\mathbf{x}}_c = \xi^\dagger \vec{\sigma} \xi$ do not actually generate such spatial coordinates. Instead, one gets $\mid \xi \mid^2$ times a unit vector corresponding to a local privileged space direction (PSD) "parallel" (in the SST) to the cosmic spinor ξ. The direction in the S^3 hypersphere associated to the vector $\xi^\dagger \vec{\sigma} \xi$ corresponds to the set of points whose cosmic SST space-time position is equal to ξ times a complex phase.

Thus, to define the standard real space coordinates in the SST, a space origin ξ_0 at the cosmic time $t_0 = \mid \xi_0 \mid$ is required, as in equation (1). Such coordinates correspond to a local description of the S^3 hypersphere at cosmic time t_0 as viewed from the space origin ξ_0 .

The new space-time geometry introduced by the SST clearly leads to natural limitations for general relativity and standard cosmology. Rather than an intrinsic fundamental property of space and time defining a global space-time geometry, conventional relativity is expected to be a low-energy symmetry of standard matter similar to the effective Lorentz-like symmetry of the kinematics of low-momentum phonons or solitons in a condensed medium [17, 18]. In condensed matter, the speed of sound or the maximum speed of solitons plays the role of the critical speed. The speed of light c would then be the critical speed of a family of excitations of the physical vacuum (the standard particles) not directly associated to an intrinsic geometry of space-time in our Universe.

Space rotations around a fixed point ξ_0 in the SST at $t = t_0$ are given by SU(2) transformations acting on the spatial position vector $\vec{\mathbf{x}}$ defined by (1). A standard spatial rotation with respect to the local origin ξ_0 is now given by a SU(2) element $U(\vec{\mathbf{y}})$ turning $U(\vec{\mathbf{x}})$ into $U(\vec{\mathbf{y}})\ U(\vec{\mathbf{x}})\ U(\vec{\mathbf{y}})^\dagger$. The vector $\vec{\mathbf{y}}$, related to $U(\vec{\mathbf{y}})$ in a similar way to (1), provides the rotation axis and angle. If a spin-1/2 object is present at the position $\vec{\mathbf{x}}$ with an associated local spinor coordinate ξ_p describing its internal structure, then ξ_p transforms into $\xi'_p = U(\vec{\mathbf{y}})\ \xi_p$.

5.3 Some properties of the SST Universe

The cosmic SST just described automatically generates three basic cosmological phenomena in a purely geometric way [26, 27] and without any explicit presence of matter and energy:

i) The standard Lundmark-Lemaître-Hubble (LLH) relation between relative velocities and distances at cosmic scale, with a ratio H (velocity/distance) equal to the inverse of the age of the Universe ($H = t^{-1}$). As t is the radius of the spatial hypersphere, and comoving frames correspond to straight lines starting from the time origin, distances between comoving objects will be proportional to t and the (constant) associated speeds will be given by these distances divided by t. Thus, t^{-1} turns out to be the automatic geometric value of the LLH constant H in the SST.

ii) The existence of a privileged space direction (PSD) for each comoving observer, as already obtained in [24, 25] and further studied in [26, 27] and in subsequent papers. As previously stressed, the existence of such a local PSD does not imply any global anisotropy in the expansion of the Universe. The global expansion is totally isotropic on purely geometric grounds, and space remains a homogeneous hypersphere for all values of t .

iii) In the direct SST formulation, space translations form a (non-abelian) compact group, contrary to standard space-time. This allows to escape standard no-go theorems for symmetries.

More details, including a study of the cosmological implications of the unconventional properties of the cosmic SST, are given in [26, 27], [68, 69], [13, 23] and in related papers.

In all cases, a fundamental feature of the cosmic spinorial space-time geometry is that it is not dominated by standard matter or conventional dark energy, and that its structure is defined in a totally independent way suggesting a more primordial origin (vacuum dynamics?). In the SST without standard matter, space units are not required, as time provides an effective distance scale.

The possible connection between the cosmic SST structure and the ultimate dynamics of matter and vacuum is a crucial open question requiring further fundamental research (theory, experiment, observation, phenomenology). Preonic and and pre-Big Bang scenarios should be carefully considered in this respect [13, 15]. Furthermore, the definition of the direction of time with the SST geometry leads to a natural contradiction between local and cosmic levels.

5.4 SST and the space-time contradiction

What would be the dynamical implications of introducing a local SST geometry around a space-time origin ξ_0 different from the cosmic origin $\xi = 0$?

As discussed in [15] and previously in [53, 54], introducing a local SST (the above considered wave functions are just an example) automatically leads to a contradiction between the time and space directions defined at $\xi = \xi_0$ by the cosmic SST ($t = \mid \xi \mid$) and the structure generated by the local SST where time for ξ would be defined by $t_{local} = \mid \xi - \xi_0 \mid$.

Similarly, the internal time of the moving particle considered in (2) and (3) would not coincide with the t_{local} defined at a given ξ_0 when the particle propagates beyond ξ_0.

Quantum Mechanics can then be a direct consequence of such a local space-time contradiction. An extended spinorial object will be necessarily extended in time, and this property is expected to forbid continouous motion at scales close to the size of the particle. Thus, the dynamical role of the local SST can be directly at the origin of the Feynman path integral (see also subsection 4.1).

An essential question would then be that of the pre-Big Bang evolution leading to the generation of standard matter and Quantum Mechanics. If the internal vacuum dynamics is at the origin of the cosmic SST geometry, the emission of standard matter and radiation can just correspond to this expansion reflecting the fact that such a phenomenon is energetically favoured [64, 65].

6 Vacuum structure and dynamics

A nontrivial question is how to define precisely the physical vacuum. If vacuum has its own internal structure and dynamics, and if standard particles are actually vacuum excitations, how can theory and phenomenology distinguish between an object present "inside the vacuum" and an equivalent item "propagating in our Universe"? And how to define vacuum excitations?

Assuming that the physical vacuum is made of preons as already suggested in [17, 18], is it possible to unambiguously define a "matter Universe" lying on this preonic vacuum? Should one just take into account local vacuum excitations to describe this matter?

At this stage, a related question must be seriously considered: is the structure of vacuum the same in all the regions of the SST Universe corresponding to the hypersphere $\mid \xi \mid = t$ at a given t ? In particular, can some parts of the global SST Universe be populated by free preons able to reach our matter Universe and to propagate in it? And how does the vacuum structure vary between the different cosmic regions of the global SST?

In a solid, photons are present through the electromagnetic interaction. But external free photons can also reach a transparent solid and propagate in it. What is the situation for "free" preons traveling in the part of the Universe associated to conventional matter and to our physical vacuum? Just as matter is not homogeneous around us and several kinds of matter exist where light can propagate, there

DOI: 10.1051/epjconf/201716401023

is no reason to exclude the possibility that vacuum be inhomogeneous at the scale of the cosmic SST and that some preons can travel as free particles through different regions of the global SST Universe. The cosmic SST geometry seems to indicate the existence of a fundamental dynamics of the physical vacuum common to all the regions of the global Universe. But simultaneously, some differences can also exist and play a significant role, including possible local fluctuations of the expansion rate.

The study by all possible ways of the structure, dynamical properties and cosmic influence of the vacuum must be the highest priority in Particle Physics and Cosmology.

In the present situation, a practical choice can be to define the vacuum as the set of preons, including virtual ones, trapped in its internal structure. "Real" preons able to travel over long distances will then be dealt with as free particles.

The question of the local vacuum rest frame (VRF) must also be seriously considered [17, 70]. The VRF would be basically the local rest frame of the internal vacuum structure. A natural choice seems to be suggested by cosmological observations (isotropy of the cosmic microwave background...) [56, 71] and can in any case be a useful starting point.

The possibility that the internal dynamics of the physical vacuum is actually leading the expansion of the Universe and that standard matter just corresponds to the energy emitted in this energetically favoured process has also been considered in previous papers, including [64, 65]. Such a scenario would be radically different from standard cosmology where conventional matter (including the dark one) is assumed to lead the Universe evolution.

6.1 Superbradyons

Such as introduced in [17, 18] and further dealt with in subsequent papers including [14, 15], superbradyons are the preonic constituents of the physical vacuum and the bradyons (positive mass and energy, real momentum particles) of a super-relativity (a Lorentz-like symmetry with a critical speed $c_s \gg c$). They are not tachyons, but new particles that do not obey the standard Lorentz symmetry.

What can be the properties of "free" superbradyons in our matter Universe?

In [17, 18] and it subsequent work, it has been systematically assumed that at a speed larger than c (up to some possible small corrections), "free" superbradyons are not really stable and release "Cherenkov" radiation in the form of ordinary particles as long as such decays are kinematically allowed. A "free" superbradyon would then interact with standard matter and, in particular, undergo "Cherenkov" decays until it reaches a speed equal or close to c.

This is the strongest constraint (no specific refractive index, no associated speed) that can be introduced on superbradyon propagation in our Universe, even if it is simultaneously conjectured that superbradyons couple weakly to standard matter. It is also tacitly assumed that superbradyons have a specific identity as physical objects and cannot just disappear by decaying into ordinary matter.

Except for their coupling to standard matter, the "free" superbradyons just considered can in principle propagate in vacuum just like photons in a transparent crystal. They would correspond to a specific family of superbradyons, but not necessarily to all the kinds of superbadyons present in the vacuum structure, just as electrons are different from photons.

It must also be noticed that "free" superbradyons can equally increase their energy in some interactions with standard matter. Globally, their interaction with matter in our Universe will depend in particular on the matter density and distribution as well as on the nature of this matter, including dark matter of which superbradyons themselves can be a significant component [24, 63].

The question can also be raised of the possible emission of superbradyons by the physical vacuum, if the vacuum is leading the expansion of the Universe and such an expansion is energetically favoured. Cosmological equations should leave room for this kind of new physics.

If the physical vacuum has indeed a nontrivial, preonic, internal structure, the study if its physical and cosmological implications is the most fundamental research field in the relevant domains. In this context, the search and study of "free" superbradyons is a very fundamental subject.

Clearly, the physics of superbradyons is full of unknowns and open questions. A crucial one is: how to produce, detect and identify such objects? In this respect, the possibility that CERN beams locally excite the vacuum structure leading to superbradyon emission deserves serious consideration.

Similarly, the possible implications of a superbradyonic vacuum for electronics, informatics and associated fields must be carefully studied including the role of superluminal signals.

7 Standard QM and relativity as low-energy approximations

What can be the validity domain of the standard principles of Physics? What is really known with the required precision? It is a fact that 13 orders of magnitude separate the highest LHC proton energies from the Planck scale. Deviations from conventional relativity and Quantum Mechanics can appear above the LHC energies and become stronger as the energy scale increases [14]. Relativity and QM can then be low-energy approximations to a deeper description of matter [17, 72].

In practice, the degree of validity of QM and relativity for ultra-high energy (UHE) particles including the observed ultra-high energy cosmic rays (UHECR) remains by now unknown [56, 73, 74] and requires further experimental, phenomenological and theoretical work.

Present data and analyses do not allow yet to elucidate [75] if the observed fall of the UHECR spectrum [59, 60] is really a signature of the Greisen-Zatsepin-Kuzmin (GZK) cutoff [76, 77] or corresponds, for instance, to the maximum energies of particles emitted by astrophysical sources. New physics can also generate unconventional mechanisms able to fake the GZK cutoff [27, 73]. It therefore seems difficult from a theoretical point of view, in the present situation, to reliably interpret existing data for high-energy cosmic rays like those presented in [78, 79], [80, 81] and [82, 83]. In all cases, new generations of UHECR experiments are required.

Another serious unknown is the degree of validity and precision of models and algorithms used to describe the interactions involving UHECR [84], especially if new physics can be present and modify QFT. This situation naturally generates significant experimental uncertainties.

The properties of UHECR should be carefully studied as far as possible, including satellite experiments. A systematic search for all kinds of possible signatures of new physics is especially required, together with a study of possible connections with the basic physics involved in the early Universe dynamics. Basic ingredients of these searches must be systematic tests of Lorentz symmetry at UHE, as already suggested in 1996-97 [18, 24], together with UHE tests of all the fundamental principles of standard physics including QM [16, 27] in all possible ways.

Interactions and mixing with superbradyons can also be at the origin of deformations of relativity and QM. As already emphasized, even at low energy QM and relativity are expected to be incomplete descriptions of reality in the presence of a preonic vacuum structure [14, 15].

7.1 Possible deformations of relativity and QM

The subject of possible deformations of particle kinematics was already dealt with in [18, 72] and in subsequent work, in order to test Lorentz symmetry at very high energy.

Lorentz symmetry violation (LSV) at high energy was in particular described, assuming the existence of a local VRF, by the high-energy equation:

$$E \simeq p\,c + m^2\,c^3\,(2\,p)^{-1} - p\,c\,\alpha\,(p\,c\,E_a^{-1})^2/2 \qquad (5)$$

DOI: 10.1051/epjconf/201716401023

p is the momentum modulus assumed to be $\ll E_a c^{-1}$, m the mass of the particle, α a positive constant describing the strength of the deformation and E_a an effective fundamental energy scale.

In such a scenario, a possible negative deformation term accounting for LSV:

$$\Delta E \simeq - p\, c\, \alpha\, (p\ c\, E_a^{-1})^2/2 \tag{6}$$

would become larger than the mass term $m^2\, c^3\, (2\ p)^{-1}$ above the transition energy E_{trans}:

$$E_{trans} \simeq \alpha^{-1/4}\, (E_a\, m)^{1/2}\, c \tag{7}$$

In its simplest form, such a mechanism can possibly suppress the GZK cutoff as suggested in [18, 72]. But more sophisticated formulations can also be considered [27, 73], leading to various possible scenarios at ultra-high energy.

In all cases, the existence of a local VRF is an essential ingredient of the pattern considered. Otherwise, if there is no privileged reference frame, suitable symmetry transformations will remove the effects of apparent deformations. In a similar way to special relativity, Quantum Mechanics, energy-momentum conservation and other particle properties can also be deformed at high energy and become an incomplete low-energy limit of a more sophisticated dynamics.

Deformed Heisenberg algebras and scenarios involving a non-commutative space-time have been considered in [16, 85], in [86, 87] and in several other approaches including strings [88, 89]. In particular, q-deformations of the quantum algebra were studied in [85].

In [16, 56], we considered new commutation relations between momentum variables where the commutators between components associated to different directions would vanish in the zero-momentum limit and become significant at UHE. An explicit example of the deformations of commutation relations involving momentum operators was:

$$\Delta p_x\, \Delta p_y \gtrsim \Phi(p^2) \tag{8}$$

$$\Delta p_y\, \Delta p_z \gtrsim \Phi(p^2) \tag{9}$$

$$\Delta p_z\, \Delta p_x \gtrsim \Phi(p^2) \tag{10}$$

where x, y and z design three orthogonal space directions, and $\Phi(0) = 0$. Again, a local privileged reference frame is tacitly involved in such a formulation of Quantum Mechanics that can be associated to the explicit emergence of fundamentally new physics at energies far below Planck scale.

8 Possible signatures of a preonic vacuum

The direct detection of "free" superbradyons is not the only possible way to search for potential signatures of a preonic vacuum. In particular, black holes, gravitational waves and CERN experiments can provide examples of such a quest for new physics of preonic origin [64, 65].

The interaction with a preonic vacuum structure can modify the internal black hole dynamics. Not only the equations of standard physics would cease to hold below some critical distance scale and black hole dynamics would be modified, but a nontrivial vacuum structure may interact with the matter trapped in the black hole. Preons and other "exotic" objects released by vacuum inside the black hole or produced through interactions would potentially be able to escape the black hole region if, for instance (as in the case of superbradyons) : i) they can travel at a speed much larger than that of light ; ii) their coupling to gravitation is much smaller than that of standard matter.

Several kinds of unconventional mechanisms can be at work in such a scenario. An example is the possibility that the preonic vacuum structure absorbs matter trapped in the black hole and releases

 DOI: 10.1051/epjconf/201716401023

it outside the black hole. In this and other ways, a superbradyonic vacuum can potentially generate new forms of black hole radiation mechanisms involving higher energies than the thermal radiation considered by Stephen Hawking [90].

The recent observations ([91, 92] and [93, 94] reported by the collaborations LIGO [95] and VIRGO [96, 97] indicate the detection of gravitational waves generated by two binary black hole mergers (GW150914 and GW151226). A careful look to the published data analyses based on standard techniques allows to better explicit the potential role of the preonic vacuum and the modifications one can expect in such an unconventional scenario.

In [97], the LIGO and VIRGO collaborations present an updated analysis of GW150914 data. The abstract of the article refers to two black holes of approximately 35 solar masses each that orbited each other as close as $\sim$ 350 Km before merging to a single black hole. An attempt is presented to estimate the distance of such an event to earth and the energy radiated in the form of gravitational waves. Obviously, a preonic vacuum may have influenced the dynamics of the two black holes, as well as the merging process and the associated emission of waves.

In the introduction, the authors of [97] define the black hole as *"a region of space-time where the gravitational fiel is so intense that neither matter nor radiation can escape"*. As an initial approximation, they refer to the standard Schwarzschild radius $r_{Schwarz}$ through the expression:

$$r_{Schwarz} \; = \; 2\,G\,m\,c^{-2} \; = \; 2.95\;m\;M_{\odot}^{-1}\;Km \qquad (11)$$

where $M_{\odot}$ is the solar mass, m the mass of the black hole and G the usual Newton's gravitational constant. Clearly, such an expression cannot be applied to preonic vacuum constituents present inside the black hole. Futhermore, the presence of nonstandard objects is expected to modify black hole dynamics including formulae like (11) and its more precise versions that incorporate general-relativistic effects and are also considered in [97].

In Section 1 of [97], the authors write: *"During the period when gravitational wave frequency and amplitude are increasing orbital motion of two bodies is the only possible explanation"*. Again, alternatives to such an interpretation can be naturally generated by the interaction of the preonic vacuum with the melting black holes. The same objection can be raised when later, in Section 4, the frequencies reached by gravitational waves are used to reject families of black hole models.

Preonic waves must also be considered as a possible source of high-frequency gravitational waves, potentially modifying their internal structure and properties in a significant way.

Simultaneously, ANTARES [98] and IceCube [99] have found no high energy neutrino event that could be in coincidence with the GW150914 signal such as reported by LIGO and VIRGO [100]. Again, preonic constituents of vacuum may have replaced the emission of standard particles in the merging black hole dynamics. In this case, they can also modify the standard timing for gravitational wave emission and propagation.

Not only a preonic vacuum would modify the black hole dynamics described in [101, 102] or in [103, 104], but it can play a strong role in primordial black holes and, more globally, in a related pre-Big Bang scenario where standard physics would be progressively generated.

9 Gödel-Cohen incompleteness

An introduction to Gödel's incompleteness theorems can be found in [105]. Applying it to the situation described in the present lecture is a far from trivial task.

In [106, 107], the authors explicitly claim that *"for any consistent, recursive axiomatisation of mathematics, there exist specific Hamiltonians for which the presence or absence of a spectral gap is independent of the axioms"*. The same philosophy is developed in [108, 109].

But what can be the actual situation in our world if the version of Quantum Mechanics used for quantum calculations is not a consistent system of axioms but a set of approximations to the physical laws, and if the laws of Physics result from the preonic vacuum dynamics where standard particles are vacuum excitations and Quantum Mechanics is generated from such a deeper dynamics?

If Quantum Mechanics is in practice the expression of more complete laws resulting from the dynamics of a superbradyonic vacuum, is it possible to obtain a deformed, approximate version of the quantum theory for standard particles such that all relevant calculations of Quantum Physics escape the usual consequences of Gödel incompleteness?

In the presence of a preonic vacuum, no real law of Physics is expected to refer only to conventional particles. If this is the case in practical formulations, intrinsic error bars should actually be made explicit. These error bars must refer not only to standard matter itself but also to preonic objects unknown in terms of this matter. They should be present in the values of all the conventional physical quantities, including the most fundamental ones.

A new approach to the formulation of the laws of Physics is clearly required, searching in particular for a dynamical pattern where incompleteness would apply to the preonic vacuum dynamics but the approximate quantum calculations for standard matter would escape its implications using a well-suited version of Quantum Mechanics.

10 Further dynamical and cosmological considerations

As already discussed in previous papers [23, 26], "free" massive superbradyons with a speed close to c can form a new kind of cosmic dark matter in our Universe [27, 61]. An important open question is in such case how would two "free" superbradyons interact, and what can emerge from such a process.

It also remains unclear how well "free" superbradyons would obey the approximate quantum-mechanical laws that describe vacuum excitations. In all cases, a "free" massive superbradyon with a speed equal to c would have a rest energy $m\, c_s^2$ (m = inertial mass) much larger than its kinetic energy equal to $m\, c^2/2$. Even coupling very weakly to gravitation, such an extremely large rest energy can be expected to produce significant gravitational effects. In particular, "free" massive superbradyons would in principle be trapped by strong enough gravitational fields.

The possibility that signatures of the preonic vacuum are produced at LHC or spontaneously emitted by vacuum inside this accelerator in the presence of strong particle beams should also be considered. A possible signal can be an important Cherenkov-like emission of standard particles by an object with low momentum. In a recent experimental search, the CMS collaboration has displayed [110] some atypical multi-jet events that could potentially be associated to the production of black holes. The possibility that similar events actually involve a superbradyonic component should also be studied, together with the search for possible black hole candidates.

Issues related to Quantum Gravity (QG) and to the role of strings have also been discussed in previous contributions to this Conference (see, for instance, [19, 57]). Preonic and pre-Big Bang approaches, together with new space-time geometries like the SST, can radically change the situation with respect to standard theories. In practice, the notion of QG as it is usually formulated, and the Planck scale itself, would no longer make sense [26].

As strings are naturally composite objects [16, 57], new Regge-like and string-like structures are expected [14, 57] to be generated from a preonic vacuum with a spinoral space-time geometry, including descriptions of standard particles as excitations. The SST geometry can naturally lead to Regge trajectories with a spin spacing of 1/2 instead of 1.

Some of the topics discussed in this lecture are dealt with in more detail in two contributions to this Conference [62].

11 Conclusion and comments

The origin of Quantum Mechanics is a crucial issue, not only for fundamental Physics but also for several fields of high-technology applications (electronics, computers [113, 114], telecommunications [115, 116]...). If conventional particles are actually excitations of a preonic vacuum, standard laws like Quantum Mechanics and relativity are expected to be low-energy approximations to a more fundamental (preonic) dynamics. Then, Gödel-Cohen incompleteness will not apply in its usual form.

Research on Particle Physics and Cosmology is actually just beginning, and will require a very long period including new generations of experiments and observations.

If the expansion of our Universe has always been driven by the dynamics of a preonic vacuum, why did such an expansion have a previous phase different from the present one?

References

[1] See, for instance, H. Takesue et al., *Quantum teleportation over 100 km of fiber using highly efficient superconducting nanowire single-photon detectors*, *Optica* **2**, 832 (2015), and references therein.

[2] B. Korzh et al., *Provably secure and practical quantum key distribution over 307 km of optical fibre*, *Nature Photonics* **9**, 163 (2015), arXiv:1407.7427

[3] H.-L. Yin et al., *Measurement device independent quantum key distribution over 404 km optical fibre*, arXiv:1606.06821

[4] A. Kellerer, *Quantum Telescopes*, *Astronomy and Geophysics*, **55** (3): 3.28 (2014), arXiv:1403.6681

[5] A.R. Kurek et al., *Quantum Telescopes: feasibility and constraints*, *Optics Letters* **41** (6), 1094 (2016), arXiv:1508.04275

[6] See, for instance, on the site of the Austrian Academy of Sciences, *Fist Quantum Satellite Successfully Launched*, (press release).

[7] *Xinhua Insight: China launches first-ever quantum communication satellite*, http://news.xinhuanet.com/english/2016-08/16/c_135601064_2.htm

[8] P. Shadbolt, J.C.F. Mathews, A. Laingand J.L. O'Brien, *Testing foundations of quantum mechanics with photons*, *Nature Physics* **10**, 278 (2014) and arXiv:1501.03713

[9] B. Hensen et al., *Loophole-free Bell inequality violation using electron spins separated by 1.3 kilometres*, *Nature* **526**, 682 (2015), arXiv:1508.05949

[10] B. Hensen et al., *Loophole-free Bell test using electron spins in diamond: second experiment and additional analysis*, arXiv:1603.05705

[11] See, for instance, M. Van Raamsdonk, *Lectures on Gravity and Entanglement*, arXiv:1609.00026 , and references therein.

[12] J. M. Maldacena, *The large n limit of superconformal field theories and Supergravity*, hep-th/9711200

[13] See, for instance, L. Gonzalez-Mestres, *Big Bang, inflation, standard Physics... and the potentialities of new Physics and alternative cosmologies*, talk given at the 4th International Conference on New Frontiers in Physics, Kolymbari, Crete, Greece, 23-30 August 2015, *EPJ Web of Conferences* **126**, 0212 (2016), and references therein.

[14] See, for instance, L. Gonzalez-Mestres, *Tests and prospects of new physics at very high energy*, contribution the 3rd International Conference on New Frontiers in Physics (ICNFP 2014), Kolymbari, Crete, Greece, 28 July - 6 August 2014, *EPJ Web of Conferences* **95**, 05007 (2015), and references therein.

[15] See, for instance, L. Gonzalez-Mestres, *Spinorial space-time and the origin of Quantum Mechanics*, contribution to the 4th International Conference on New Frontiers in Physics, Kolymbari, Crete, Greece, 23-30 August 2015, *EPJ Web of Conferences* **126**, 05006 (2016), and references therein.

[16] L. Gonzalez-Mestres, *Preon models, relativity, quantum mechanics and cosmology (I)*, arXiv:0908.4070 , and references therein.

[17] L. Gonzalez-Mestres, *Properties of a possible class of particles able to travel faster than light*, Proceedings of the January 1995 Moriond Workshop, Ed. Frontières, arXiv:astro-ph/9505117

[18] L. Gonzalez-Mestres, *Vacuum Structure, Lorentz Symmetry and Superluminal Particles*, arXiv:physics/9704017

[19] L. Gonzalez-Mestres, *Pre-Big Bang, space-time structure, asymptotic Universe*, 2nd International Conference on New Frontiers in Physics, Kolymbari, Crete, Greece, August 28 - September 5, 2013, *EPJ Web of Conferences* **71**, 00063 (2014), references therein and Post Scriptum to the preprint hal-00983005.

[20] See, for instance, the 1979 Nobel lecture by Abdus Salam, and references therein.

[21] L. Gonzalez-Mestres, *Cosmological Implications of a Possible Class of Particles Able to Travel Faster than Light*, Proceedings of the TAUP 1995 Conference, *Nucl. Phys. Proc. Suppl.* **48** (1996), 131, arXiv:astro-ph/9601090

[22] L. Gonzalez-Mestres, *Pre-Big Bang, vacuum and noncyclic cosmologies*, 2011 Europhysics Conference on High Energy Physics, Grenoble, July 2011, *PoS EPS-HEP2011* 479, and references therein.

[23] L. Gonzalez-Mestres, *BICEP2, Planck, spinorial space-time, pre-Big Bang*, contribution the 3rd International Conference on New Frontiers in Physics, Kolymbari, Crete, Greece, August 23 - 30, 2014, *EPJ Web of Conferences* **95**, 03014 (2015), and references therein.

[24] L. Gonzalez-Mestres, *Physical and Cosmological Implications of a Possible Class of Particles Able to Travel Faster than Light*, contribution to the 28th International Conference on High Energy Physics, Warsaw 1996, arXiv:hep-ph/9610474, and references therein.

[25] L. Gonzalez-Mestres, *Space, Time and Superluminal Particles*, arXiv:physics/9702026

[26] L. Gonzalez-Mestres *Pre-Big Bang, fundamental Physics and noncyclic cosmologies*, presented at the International Conference on New Frontiers in Physics, ICFP 2012, Kolymbari, Crete, June 10-16 2012, *EPJ Web of Conferences* **70**, 00035 (2014), and references therein. Preprint version at mp_arc 13-18.

[27] L. Gonzalez-Mestres, *Cosmic rays and tests of fundamental principles*, CRIS 2010 Proceedings, *Nucl. Phys. B, Proc. Suppl.* **212-213** (2011), 26, and references therein. The *arXiv.org* version arXiv:1011.4889 includes a relevant Post Scriptum.

[28] Planck Collaboration, ESA site: http://www.cosmos.esa.int/web/planck

[29] Planck Collaboration, *Planck 2013 results. XXIII. Isotropy and statistics of the CMB*, *Astronomy and Astrophysics* **571**, A23 (2014), also available at arXiv:1303.5083

[30] D. Saadeh et al., *How isotropic is the Universe?*, *Phys. Rev. Lett.* **117**, 131302 (2016), arXiv:1605.07178 , and references therein.

[31] For an introduction to Gödel-Cohen incompleteness, see for instance J. Steinmetz, *An Intuitively Complete Analysis of Godel's Incompleteness*, arXiv:1512.03667

[32] J.S. Bell, *On The Einstein Podolsky Rosen Paradox*, *Physics* **1**, 195 (1964). Available at the address http://cds.cern.ch/record/111654/files/vol1p195-200_001.pdf

[33] J.S. Bell, *The Theory of Local Beables*, in *Speakable and Unspeakable in Quantum Mechanics* (Cambridge Univ. Press, Cambridge, 2004), 52. 1975 CERN preprint CERN preprint TH.2053

available at the address http://cds.cern.ch/record/980036/files/197508125.pdf

[34] See, for instance, H.M. Wiseman, *The Two Bell's Theorems of John Bell*, *J. Phys. A: Math. Theor.* **47** 424001 (2014), arXiv:1402.0351

[35] See, for instance, S. Boughn, *A Modest View of Bell's Theorem*, arXiv:1604.08529

[36] R.A. Bertlmann, *John Bell and the Nature of the Quantum World*, *J. Phys. A: Math. Theor.* **47**, 424007 (2014), arxiv:1411.5322

[37] R.A. Bertlmann, *Bell's Universe: A Personal Recollection*, in *Quantum [Un]Speakables II: Half a Century of Bell's Theorem*, eds. R. A. Bertlmann and A. Zeilinger, Springer 2016, arxiv:1605.08081

[38] A. Einstein, B. Podolsky and N. Rosen, *Phys. Rev.* 47, 777 (1935).

[39] In his 1964 article, John Bell cites: *Albert Einstein, Philosopher Scientist* (Edited by P.A. Schilp), Library of Living Philosophers, Evanston, Illinois (1949), p.85.

[40] J.S. Bell, *Speakable and Unspeakable in Quantum Mechanics*, Cambridge University Press, Cambridge, 2004.

[41] G. Lochak, *Has Bell's inequality a general meaning for hidden variable theories?*, *Foundations of Physics* **6**, 173 (1976).

[42] J.F. Clauser, M.A. Horne, A. Shimony, A. and R.A. Holt, *Proposed Experiment to Test Local Hidden-Variable Theories*, *Phys. Rev. Lett.* **23**, 880 (1969). Available at the address https://www.researchgate.net/publication/228109500

[43] See also A. Asin and L. Masanes, *Certified randomness in quantum physics*, *Nature* **540**, 213 (2016), and references therein.

[44] M. Giustina et al., *A significant-loophole-free test of Bell's theorem with entangled photons*, *Phys. Rev. Lett.* **115**, 250401 (2015), arXiv:1511.03190

[45] L.K. Shalm et al., *A strong loophole-free test of local realism*, *Phys. Rev. Lett.* **115**, 250402 (2015), arXiv:1511.03189

[46] See, for instance, A. Aspect, *Bell's Theorem : The Naive View of an Experimentalist*, in *Quantum [Un]speakables - From Bell to Quantum information*, edited by R. A. Bertlmann and A. Zeilinger, Springer (2002), arxiv:quant-ph/0402001 , and references therein.

[47] See, for instance, J-A. Larsson, *Loopholes in Bell Inequality Tests of Local Realism,Journal of Physics A* **47**, 424003 (2014), arXiv: 1407.0363 , and references therein.

[48] W. Rosenfeld et al., *Event-ready Bell-test using entangled atoms simultaneously closing detection and locality loopholes*, arXiv:1611.04604

[49] See, for instance, L.A. Rozema et al., *Violation of Heisenberg's Measurement-Disturbance Relationship by Weak Measurements*, *Phys. Rev. Lett.* **109**, 100404 (2012), arXiv:1208.0034

[50] W. Heisenberg, *Über den anschaulichen Inhalt der quantentheoretischen Kinematik und Mechanik*, *Zeitschfrift für Physik* **43**, 172 (1927).

[51] English translation from W. Heisenbeg, *The Physical Content of Quantum Kinematics and Mechanics*, in *Quantum Theory and Measurement*, Edited by J. A. Wheeler and W. H. Zurek, Princeton Univ. Press, 1984.

[52] See, for instance, J. Cen and A. Fring, *Complex solitons with real energies*, *Journal of Physics A: Math. and Theor.* **49** 365202 (2016], arxiv:1602.05465 , and references therein.

[53] L. Gonzalez-Mestres, *Quantum Mechanics and the Spinorial Space-Time*, mp_arc 15-86

[54] L. Gonzalez-Mestres, *Quantum Mechanics, preonic vacuum and space-time contradiction*, mp_arc 15-90

[55] L. Gonzalez-Mestres, *Quantum Mechanics, space-time, preons and entanglement*, mp_arc 15-92

[56] L. Gonzalez-Mestres, *High-energy cosmic rays and tests of basic principles of Physics*, presented at the International Conference on New Frontiers in Physics, ICFP 2012, Kolymbari, Crete, June 10-16 2012, *EPJ Web of Conferences* **70**, 00047 (2014), and references therein. Preprint version at mp_arc 13-19.

[57] See also L. Gonzalez-Mestres, *Spinorial Regge trajectories and Hagedorn-like temperatures*, presented at ICNFP 2015, EPJ Web of Conferences **126**, 05005 (2016).

[58] Pierre Auger Observatory, https://www.auger.org/

[59] The Pierre Auger Collaboration, *The Pierre Auger Observatory Upgrade - Preliminary Design Report*, arxiv:1604.03637 , and references therein.

[60] The Pierre Auger Collaboration, *Testing Hadronic Interactions at Ultrahigh Energies with Air Showers Measured by the Pierre Auger Observatory*, *Phys. Rev. Lett.* **117**, 192001 (2016), arxiv:1610.08509

[61] L. Gonzalez-Mestres, *Lorentz symmetry violation, dark matter and dark energy*, Invisible Universe International Conference, Paris June 29 - July 3, 2009. The *arXiv.org* version arXiv:0912.0725 includes a relevant Post Scriptum.

[62] L. Gonzalez-Mestres, *The present status of Cosmology and new approaches to particles and Cosmos* and *Value of H, spinorial space-time and Universe's expansion*, ICNFP 2016 Conference.

[63] L. Gonzalez-Mestres, *Superbradyons and some possible dark matter signatures*, arXiv:0905.4146

[64] L. Gonzalez-Mestres, *Can matter accelerate the expansion of the Universe? (I)* (April 26, 2016), Part of a contribution to the 5th International Conference on New Frontiers in Physics, Kolymbari, Crete, Greece, July 5 - 15, 2016. Available at mp_arc 16-33.

[65] L. Gonzalez-Mestres, *Cosmological implications of a preonic vacuum (I)* (August 2, 2016), Part of a contribution to the 5th International Conference on New Frontiers in Physics, Kolymbari, Crete, Greece, July 5 - 15, 2016. Available at mp_arc 16-62.

[66] G. Lemaître, *The Beginning of the World from the Point of View of Quantum Theory*, Letter to *Nature*, *Nature* **127**, 706, 9 May 1931.

[67] L. Gonzalez-Mestres, *Planck data, spinorial space-time and asymptotic Universe*, mp_arc 13-33, and references therein.

[68] L. Gonzalez-Mestres, *Spinorial space-time and privileged space direction (I)*, mp_arc 13-75, and references therein.

[69] L. Gonzalez-Mestres, *Spinorial space-time and Friedmann-like equations (I)*, mp_arc 13-80, and references therein.

[70] See, for instance, L. Gonzalez-Mestres, *Lorentz symmetry violation at Planck scale, cosmology and superluminal particles*, talk given at COSMO-97, Ambleside September 15-19 1997, Proceedings edited by L. Rozskowski, World Scientific 1998, arXiv:physics/9712056, references therein and subsequent papers.

[71] See, for instance, the Planck Collaboration, *Planck 2015 results. XVI. Isotropy and statistics of the CMB*, *Astronomy & Astrophysics* **594**, A16 (2016), arXiv:1506.07135, and references therein.

[72] L. Gonzalez-Mestres, *Absence of Greisen-Zatsepin-Kuzmin Cutoff and Stability of Unstable Particles at Very High Energy, as a Consequence of Lorentz Symmetry Violation*, Proceedings of the 25th International Cosmic Ray Conference, Potchefstroomse Universiteit 1997, Vol. **6**, p. 113. arXiv:physics/9705031.

[73] L. Gonzalez-Mestres, *Testing fundamental principles with high-energy cosmic rays*, 2011 Europhysics Conference on High Energy Physics, Grenoble, July 2011, *PoS EPS-HEP2011* 390, and references therein.

[74] L. Gonzalez-Mestres, *Ultra-high energy physics and standard basic principles*, contribution the 2nd International Conference on New Frontiers in Physics, Kolymbari, Crete, Greece, August 28 - September 5, 2013, *EPJ Web of Conferences* **71**, 00062 (2014). See also the Post Scriptum to the preprint version, mp_arc 14-31.

[75] A.Watson, *High-Energy Cosmic Rays and the Greisen-Zatsepin-Kuzmin Effect*, *Rept.Prog.Phys.* **77** (2014) 036901, arXiv:1310.0325.

[76] K. Greisen, *End to the Cosmic-Ray Spectrum?* *Phys.Rev.Lett.* **16** (1966), 748, http://physics.princeton.edu/ mcdonald/examples/EP/greisens_prl_16_748_66.pdf

[77] G.T. Zatsepin and V.A. Kuz'min, *Upper Limit on the Spectrum of Cosmic Rays*, *JETP Letters* **4**, 78

[78] The Telescope Array Collaboration, *Indications of Intermediate-Scale Anisotropy of Cosmic Rays with Energy Greater Than 57 EeV in the Northern Sky Measured with the Surface Detector of the Telescope Array Experiment*, *Astrophysical Journal***790**, L21 (2014), arXiv:1404.5890.

[79] The Telescope Array Collaboration, *The Energy Spectrum of Cosmic Rays above $10^{17.2}$ eV Measured by the Fluorescence Detectors of the Telescope Array Experiment in Seven Years*, *Astropart.Phys.* **80**, 131 (2016), arXiv:1511.07510, and references therein.

[80] The Pierre Auger Observatory, *Large scale distribution of ultra high energy cosmic rays detected at the Pierre Auger Observatory with zenith angles up to 80°*, arXiv:1411.6953, and references therein.

[81] The Pierre Auger Collaboration, *Multi-resolution anisotropy studies of ultrahigh-energy cosmic rays detected at the Pierre Auger Observatory*, arXiv:1611.06812, and references therein.

[82] R. Abbasi et al., *Report of the Working Group on the Composition of Ultra High Energy Cosmic Rays*, Proceedings of the UHECR workshop, Springdale, USA, 2014, arXiv:1503.07540, and references therein.

[83] Telescope Array and Pierre Auger Collaborations, *Pierre Auger Observatory and Telescope Array: Joint Contributions to the 34th International Cosmic Ray Conference (ICRC 2015)*, arXiv:1511.02103, and references therein.

[84] See, for instance, The Pierre Auger Collaboration, *Testing Hadronic Interactions at Ultrahigh Energies with Air Showers Measured by the Pierre Auger Observatory*, *Phys. Rev. Lett.* **117**, 192001 (2016), arXiv:1610.08509, and references therein.

[85] J. Wess, *q-Deformed Heisemberg Algebras*, Lectures given at the 38. Internationale Universitätswochen für Kern- und Teilchenphysik, Schladming (Austria), January 1999, arXiv:math-ph/9910013, and references therein.

[86] S. Majid and H. Ruegg, *Bicrossproduct structure of the Poincaré group and noncommutative geometry*, *Physics Letters B* **334**, 348-354 (1994), arXiv:hep-th/9405107arXiv:hep-th/9405107.

[87] A. Connes and J. Lott, *Particle models and noncommutative geometry*, Nucl. Phys. Proc. Suppl. **B 18**, 29 (1990), http://deepblue.lib.umich.edu/bitstream/handle/2027.42/29524/0000611.pdf

[88] N.E. Mavromatos and R.J. Szabo, *arXiv.org*, arXiv:hep-th/9811116

[89] N. Seiberg and E. Witten, *String theory and noncommutative geometry*, JHEP 09, 032 (1999), arXiv:hep-th/9908142.

[90] S. Hawking, *Black hole explosions?*, *Nature* **248**, 30 (1974).

[91] B.P. Abbott et al. (LIGO Scientific Collaboration and Virgo Collaboration), *Observation of Gravitational Waves from a Binary Black Hole Merger*, *Phys. Rev. Lett.* **116**, 061102 (2016).

[92] B.P. Abbott et al. (LIGO Scientific Collaboration and Virgo Collaboration), *Astrophysical Implications of the Binary Black-Hole Merger GW150914*, *The Astrophysical Journal Letters* **818**, L22 (2016).

[93] B.P. Abbott et al. (LIGO Scientific Collaboration and Virgo Collaboration), *Observation of Gravitational Waves from a 22-Solar-Mass Binary Black Hole Coalescence*, *Phys. Rev. Lett.* **116**, 241103 (2016).

[94] B.P. Abbott et al. (LIGO Scientific Collaboration and Virgo Collaboration), *Binary Black Hole Mergers in the first Advanced LIGO Observing Run*, arXiv:1606.04856

[95] The LIGO Scientific Collaboration, http://www.ligo.org/

[96] VIRGO Collaboration, http://www.virgo-gw.eu/

[97] The LIGO Scientific Collaboration, the Virgo Collaboration, *The basic physics of the binary black hole merger GW150914*, arXiv:1608.01940

[98] ANTARES Collaboration, http://antares.in2p3.fr/

[99] IceCube South Pole Neutrino Observatory, http://icecube.wisc.edu/

[100] ANTARES Collaboration, IceCube Collaboration, LIGO Scientific Collaboration, Virgo Collaboration, *High-energy Neutrino follow-up search of Gravitational Wave Event GW150914 with ANTARES and IceCube*, arXiv:1602.05411

[101] T. Jacobson, *Introductory Lectures on Black Hole Thermodynamics*, 1996 Lectures given at the University of Utrecht.

[102] G. 't Hooft, *Introduction to the Theory of Black Holes*, 2009 Lectures presented at the Utrecht University, ITP-UU-09/11, SPIN-09/11.

[103] L. Gualtieri and V. Ferrari, *Black Holes in general Relativity*, Università degli studi di Roma, 2011.

[104] Harvey Reall, Lecture notes on Black Holes (2016), University of Cambridge.

[105] Stanford Encyclopedia of Philosophy, *Gödel's Incompleteness Theorems*, https://plato.stanford.edu/entries/goedel-incompleteness/

[106] T. Cubitt, D. Perez-Garcia and M. Wolff, *Undecidability of the Spectral Gap*, *Nature* **528**, 207 (2015), arXiv: 1502.04135.

[107] T. Cubitt, D. Perez-Garcia and M. Wolff, *Undecidability of the Spectral Gap (full version)*, arXiv:1502.04573.

[108] J. Bausch et al., *Size-Driven Quantum Phase Transitions*, arXiv:1512.05687.

[109] G. De las Cuevas et al., *Fundamental limitations in the purifications of tensor networks*, *Journal of Mathematical Physics* **57**, 071902 (2016), arXiv:1512.05709.

[110] CMS Collaboration, *Search for black holes at* $\sqrt{s}$ = *13 TeV*, http://cms-results.web.cern.ch/cms-results/public-results/preliminary-results/EXO-15-007 . See also [111, 112]

[111] CMS Collaboration, *Search for resonances and quantum black holes using dijet mass spectra in proton-proton collisions at sqrt(s) = 8 TeV*, *Phys. Rev. D* **91**, 052009 (2015), arXiv:1501.04198

[112] CMS Collaboration, *Search for lepton flavour violating decays of heavy resonances and quantum black holes to an e-mu pair in proton-proton collisions at sqrt(s) = 8 TeV*, *The European Physical Journal C* **76**, 317 (2016), arXiv:1604.05239

[113] S.J. Devitt, *Performing Quantum Computing Experiments in the Cloud*, *Phys. Rev. A* **94**, 032329 (2016), arXiv:1605.05709

[114] S. Boixo et al., *Characterizing Quantum Supremacy in Near-Term Devices*, arXiv:1608.00263

[115] See also QiChao Sun et al., *Quantum teleportation with independent sources over an optical fibre network*, *Nature Photonics* **10**, 671 (2016), arXiv: 1602.07081 , and references therein.

[116] R. Valivarthi et al., *Quantum teleportation across a metropolitan fibre network*, *Nature Photonics* **10**, 676 (2016), arXiv:1605.08814

 DOI: 10.1051/epjconf/201716401024

Recent results of Daya Bay Experiment

Dongmei Xia[1,a], on behalf of the Daya Bay Collaboration

[1] *College of Power Engineering, Chongqing University, Chongqing 400044, China*

Abstract. The Daya Bay reactor neutrino experiment aimed to precisely measure the least known mixing angle θ_{13}. In March 2012, Daya Bay announced the non-zero value with more than 5σ. With more statistics, less background and better control of systematics, the Daya Bay experiment is continuously improving the precision of $\sin^2 2\theta_{13}$ as well as the effective neutrino mass squared differences. In this paper, I will report the recent oscillation results, which are the most precise measurement of the oscillation parameter θ_{13} and $|\Delta m^2_{ee}|$. With 1230 days of data, $\sin^2 2\theta_{13}$ is measured to be [8.41 ± 0.27(stat.) ± 0.19(syst.)]$\times 10^{-2}$, and $|\Delta m^2_{ee}|$ = [2.50±0.06(stat.)±0.06(syst.)]$\times 10^{-3}$eV^2 with χ^2/NDF = 232.6/263. An independent measurement with the inverse beta decay neutron captured on hydrogen will also be presented.

1 Introduction

Using the PMNS matrix U_{PMNS}, a neutrino with flavor α can be expressed as a combination of the mass eigenstates of neutrino.

$$\nu_\alpha = \sum_i U_{\alpha i}\nu_i \,, \tag{1}$$

where $\alpha = e, \mu, \tau$ for flavor eigenstates, and $i = 1, 2, 3$ for mass eigenstates. In the three-flavor model, U_{PMNS} is commonly parameterized with three mixing angles (θ_{12}, θ_{23}, and θ_{13}) and the CP-violating phase δ_{CP}. Of these mixing angles, θ_{13} is the least known. Solar and long baseline reactor experiments have measured θ_{12} [1, 2] and atmospheric and accelerator experiments measured θ_{23} [3, 4].

Previous short baseline reactor and accelerator neutrino experiments found no evidence of reactor antineutrino disappearance, but limited the value of θ_{13} to be less than 10 degree at 90% confidence level. The remain unknowns in neutrino oscillation include the mass hierarchy of neutrino and the CP phase. A non-zero value for θ_{13} makes it possible to measure the unknowns. And the magnitude of θ_{13} is the signpost for future experiments to determine the neutrino mass hierarchy and to search for neutrino CP-violation.

The reactor antineutrino survival probability can be expressed using

$$P_{\bar{\nu}_e \to \bar{\nu}_e}(L) = 1 - \cos^4\theta_{13}\sin^2 2\theta_{12}\sin^2\Delta_{21} - \sin^2 2\theta_{13}\left(\cos^2\theta_{12}\sin^2\Delta_{31} + \sin^2\theta_{12}\sin^2\Delta_{32}\right) \,, \tag{2}$$

where $\sin^2\Delta_{ee} \equiv \cos^2\theta_{12}\sin^2\Delta_{31} + \sin^2\theta_{12}\sin^2\Delta_{32}$. KamLAND experiment measured θ_{12} with an effective baseline of about 180 km [2]. For short baseline reactor experiments, Δ_{31} and Δ_{32} are indistinguishable, therefore, Eq. (2) can be approximately expressed with a single effective parameter $\sin^2\Delta_{ee}$.

[a] e-mail: xiadm@cqu.edu.cn

Previous searches of θ_{13} with reactor neutrinos were limited by uncertainty from reactors. A far versus near relative measurement can offset this uncertainty.

$$\frac{N_f}{N_n} = \left(\frac{N_{p,f}}{N_{p,n}}\right)\left(\frac{L_n}{L_f}\right)^2\left(\frac{\epsilon_f}{\epsilon_n}\right)\frac{P_{sur}(E_\nu, L_f)}{P_{sur}(E_\nu, L_n)}. \tag{3}$$

Of this method, sensitivity to neutrino oscillation depends on relative uncertainties between detectors, including the number of target protons N_p, the baseling L and the detecting efficiency. Three experiments were constructed based on this technique, Daya Bay [5], RENO [6], and Double CHOOZ [9]. In March 2012, Daya Bay reported the discovery of the non-zero θ_{13} [5]. and then be confirmed by the other two experiments [6, 9]. The large value of θ_{13} has allowed the measurement of Δm_{ee} from the oscillation of the neutrino energy spectrum.

2 The Daya Bay Experiment

Figure 1. Layout of the Daya Bay Experiment.

The experiment is located in the Daya Bay nuclear power station in ShenZhen, which is in the south of China, and about 55 km away from Hong Kong. Figure 1 shows the layout of the experiment. There are 6 reactor cores with a total power of 17.4 GW_{th} (giga walts), and three experimental halls (The Daya Bay near site, LingAo near site and the far site) with 8 operating detectors , the total target mass of the detectors is 160 ton. The coordinates of each detector and reactor can be found in [19]. Th e coordinates were detected using GPS, and the uncertainties of the coordinates were 18mm. Details of the experimental layout can be found in [10].

In the experimental hall, a water cerenkov detector and the resistive plate chambers are used to tag muons, two identical designed antineutrino detectors(AD) are mounted in each near experimental hall, and 4 identical ADs were mounted in the far hall [17].

Each AD is constructed as a nested three zone structure, as shown in Figure 2. The inner most is the 3-meter acrylic tank, containing 20 tons of gadolinium loaded liquid scintillator(GdLS), which is the antineutrino target. In the 4-meter acrylic tank, it contains 22 tons of normal liquid scintillator (LS), with a function of gamma catcher. The outermost is the buffer shielding. Three automatic calibration units (ACU-A, ACU-B, ACU-C) were mounted on the top of each AD for detector calibration. 192 PMTs are used to collect photons. Details of the antineutrino detector can refer to [17]

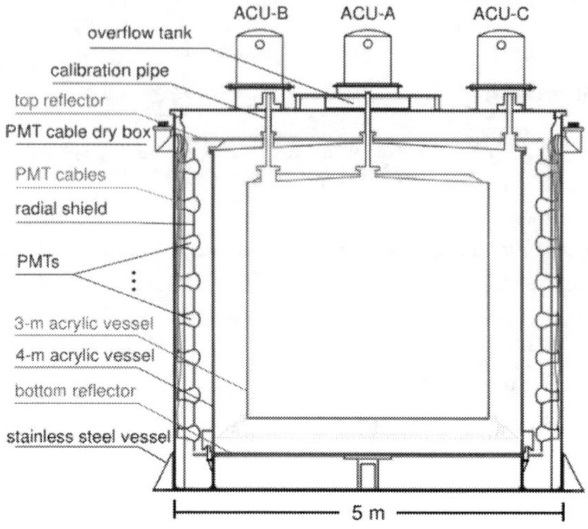

Figure 2. Scheme of the antineutrino detector.

The experiment started the data taking in Dec 2011, with the operating of 6 AD. The 8AD data taking period started in October 2012. Last year, we published the 621 days data result which combined the 6-AD and 8-AD period data [10], and the 1230 days data result will be published soon.

3 Energy Calibration

To do the energy calibration, the first step is to do the PMT gain calibration. There are two methods for PMT gain calibration, one is based on the single photoelectron from the PMT dark noise, and the other is based on the weekly deployment of LED. The second step is the energy calibration, both the calibration sources as well as the spallation neutrons were used for energy calibration.

The AD-to-AD relative energy scale was also measured using 13 different calibration sources, including both deployed and naturally occurred sources. The calibration sources include ^{68}Ge, ^{60}Co and AmC neutron sources; Neutrons from IBD and muon spallation captured both on H and Gd were used. Signals generated by natural radioactivity including α decays and γ rays from ^{40}K and ^{208}Tl were also compared between detectors. Figure 3 shows the comparison of the mean reconstructed energy between the ADs for these 13 various sources. Error bars in the figure are statistical only, and relative energy scale uncertainty for all calibration sources were less than 0.2%. The 8 MeV n-Gd capture γ peaks from Am-C sources were used to define the energy scale of each detector.

The energy model of Daya Bay experiment include the nonlinearity from liquid scintillator and the readout electronics. The energy model was built based on various γ peaks and continuous ^{12}B beta spectrum. This model was also validated using independent calibration data, include the Michel electron, the β+γ continuous spectra from Bi and Tl decays, and the Compton scattering electron data. Figure 4 shows the energy model, which the uncertainty band corresponds to the model consistent with the calibration data within 68% C.L. The prominent nonlinearity below 4 MeV was attributed to scintillator light yield (from ionization quenching and Cherenkov light production) and the charge response of the electronics. Gamma rays from both deployed and intrinsic sources as well as spallation ^{12}B decay determined the energy model, and provided an envelope of curves consistent with the data within a 68.3% C.L. (grey band). An independent estimate using the β+γ energy spectra from ^{212}Bi,

DOI: 10.1051/epjconf/201716401024

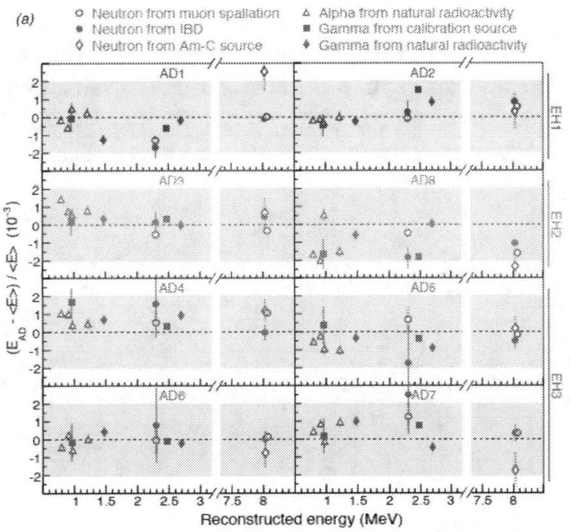

Figure 3. Comparison of the reconstructed energy between antineutrino detectors for a variety of calibration references. E_{AD} is the reconstructed energy determined using each AD, and $\langle E \rangle$ is the 8-detector average.

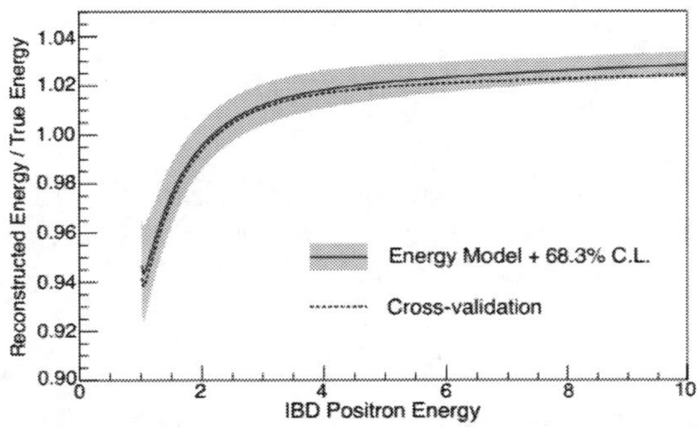

Figure 4. Estimated energy response of the detectors to positrons, including both kinetic and annihilation gamma energy (red solid curve).

^{214}Bi, ^{208}Tl, and the 53-MeV edge in the Michael electron spectrum gave a similar result (blue dashed line), with larger systematic uncertainties [10].

4 Event selection

Antineutrinos are detected via inverse beta decay (IBD), which provides two time-correlated signals of specific energies. The first, or prompt signal is produced by positron. The delayed signal is neutron, which can be captured on Gd or H. If it was captured on Gd, the capture time is about 30 us, and the excited Gd will release ~8MeV of γ rays. If it was captured on H, the capture time is about 200us, and releasing ~2.2MeV γ rays.

To select IBD events, the PMT flashers are removed which are instrumental backgrounds[11]. Then the coincidence cut is applied, with the prompt energy between 0.7 and 12 MeV, the delayed energy between 6 and 12MeV. The coincidence time cut was applied as more than 1us and less than 200us. The combined efficiency was estimated to be 80.6%. The absolute efficiencies and their correlated uncertainties were canceled when comparing the ratio of signals in the far versus near detectors. Therefore, only the uncorrelated uncertainties were considered. The uncorrelated uncertainty of detection efficiency was estimated to be 0.13% [10].

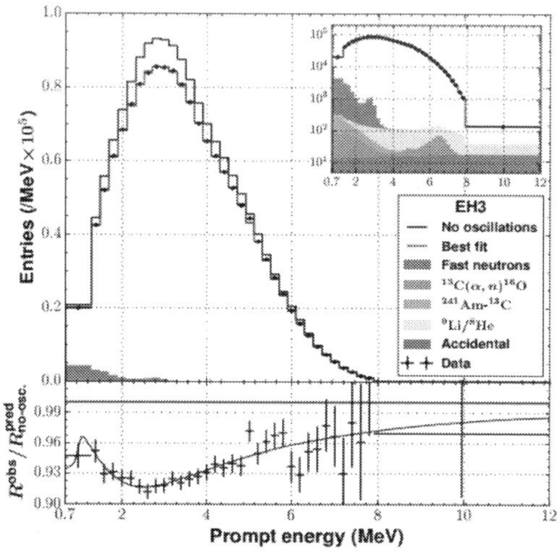

Figure 5. (upper) The observed reconstructed positron energy spectrum for EH3(black points), compared with the prediction assuming no oscillation (blue line) and the best three-flavor neutrino oscillation model (red line). (Lower) The ratio of the background-subtracted spectra to the prediction assuming no oscillation.

There are five main backgrounds including the accidentals, the fast neutrons, the $^9Li/^8He$, $^{13}C(\alpha,n)^{16}O$, and the background from the calibration source ^{241}Am-^{13}C. The uncertainty of these backgrounds were limited to $\leqslant 0.2\%$ for each experimental hall. In summary, the backgrounds were estimated to contribute 1.8% to the sample from EH1, 1.5% to EH2, and 2.0% to EH3. And the uncertainties of background were 0.2%, 0.15% and 0.2% for the three experimental hall, respectively. The upper panel of Figure 5 shows the positron energy spectra distribution in EH3, and compared the experimental data to prediction assuming no oscillation. The estimated background are barely visible as can be seen from the figure. The ratios of the data to predictions assuming no oscillation are shown in the lower panel of figure 5, to highlight the spectral distortion. The distortion of the energy spectra is consistent with oscillation, which allowed to measure Δm_{ee}.

With the 1230 days of data, over 2.8 million IBD candidates were obtained. This is double statistics of the result we published with 621 days data in 2015. Figure 6 shows the survival probability versus L/E for the three experimental hall using the 1230 days of data. The points are the ratios of the observed antineutrino spectra to expectation assuming no oscillation. The solid line is the expectation using the best estimates of the two parameters, $\sin^2 2\theta_{13}$ an $|\Delta_{ee}|$. Error bars are statistical only. $< E_\nu >$ was obtained using the estimated detector response, while L_{eff} was calculated by equating the actual

 DOI: 10.1051/epjconf/201716401024

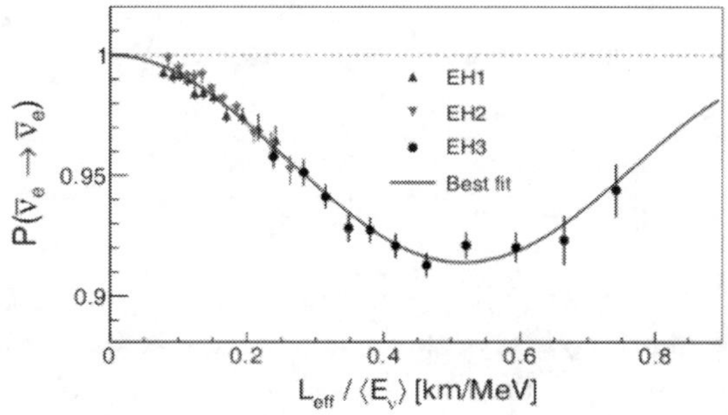

Figure 6. Electron antineutrino survival probability versus effective propagation distance L_{eff}, divided by the average antineutrino energy$< E_\nu >$.

flux to an effective antineutrino flux using a single baseline. And all the three experimental halls were consistent with the oscillation hypnosis.

5 Oscillation results

Figure 7 shows the 2-D confidence and the $\Delta\chi^2$ versus Δm_{ee} and $\sin^2 2\theta_{13}$. The upper panel shows the 1-D confidence of θ_{13} when profiling Δm_{ee}, and the right panel gives the 1-D confidence of Δm_{ee} when profiling θ_{13}. The best fit gives the central value of θ_{13} and Δm_{ee}.

$$\sin^2 2\theta_{13} = [8.41 \pm 0.27(stat.) \pm 0.19(syst.)] \times 10^{-2} , \quad (4)$$

$$|\Delta m^2_{ee}| = [2.50 \pm 0.06(stat.) \pm 0.06(syst.)] \times 10^{-3}\text{eV}^2 , \quad (5)$$

$$\chi^2/NDF = 232.6/263 . \quad (6)$$

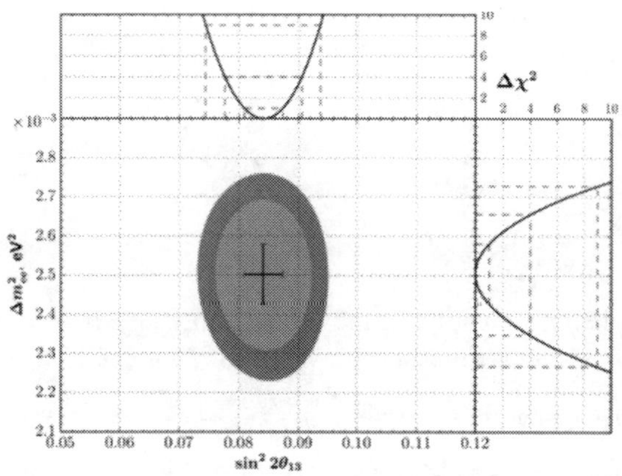

Figure 7. Confidence intervals for $\sin^2 2\theta_{13}$and $|\Delta m^2_{ee}|$ from comparison of the antineutrino rate and prompt positron spectra observed in the far versus near detectors.

With the 1230 days of data, Daya Bay experiment gives the most precise measurement of both θ_{13} and Δm_{ee}. And the result is consistent with the MeV-scale reactor experiments like RENO [7],

Double CHOOZ[8], and consistent with the GeV-scale accelerator experiments like T2K[13], MINOS [12] and NOvA[14], and consistent with the atmospheric experiments like supreK [15] and Icecube [16]. Δm^2_{32} is $2.45 \times 10^{-3} eV^2$ when consider normal hierarchy. While $-2.55 \times 10^{-3} eV^2$ when consider inverted mass hierarchy.

6 Independent measurement of θ_{13} using nH

An independent measurement of $\sin^2 2\theta_{13}$ has been measured via the detection of IBDs tagged by neutron capture on hydrogen (nH) by Daya Bay experiment. The background of nH analysis is higher, the neutron capture time (200ms) is longer, and the energy neutron captured on hydrogen is lower (2.2MeV). To meet these challenges, new analysis approaches have been developed. Based on the rate deficit observed at the far-site detectors for the 6-AD period, a value of $\sin^2 2\theta_{13} = 0.071 \pm 0.011$ is extracted. The result is shown in Figure 8, which is consistent with the result obtained in the nGd analysis. Details of the nH analysis can refer to [18]

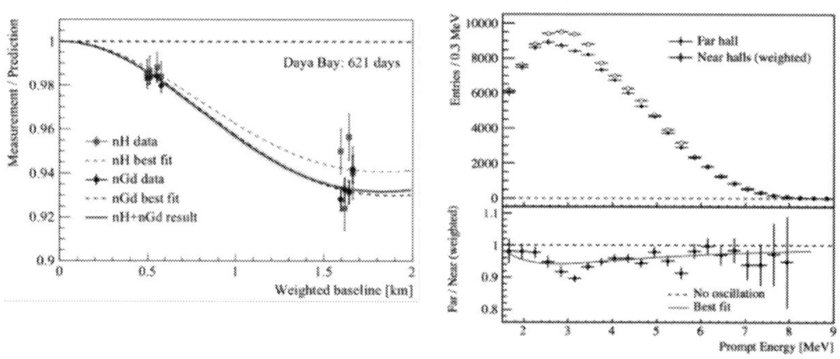

Figure 8. Left: Comparison of the nGd and nH analysis of the measurement over prediction assuming no oscillation versus weighted baseline, using 621 days of data. Right: The top panel shows the reconstructed prompt-energy spectrum of the far hall (solid blue points) and the expectation based on the measurements of the two near halls (empty black points). Spectra are background-subtracted. Error bars are purely statistical. The bottom panel shows the ratio of the Far/Near halls, and the curve representing the best value of $\sin^2 2\theta_{13} = 0.071 \pm 0.011$.

7 Summary

In this article, I present the latest results of the oscillation analysis from Daya Bay Experiment. With 1230 days of data, the most precise measurement of $\sin^2 2\theta_{13}$ and $|\Delta m^2_{ee}|$ are given

$$\begin{aligned} \sin^2 2\theta_{13} &= [8.41 \pm 0.33] \times 10^{-2} , \\ |\Delta m^2_{ee}| &= [2.50 \pm 0.08] \times 10^{-3} \mathrm{eV}^2 , \\ \Delta m^2_{32}(NH) &= [2.45 \pm 0.08] \times 10^{-3} \mathrm{eV}^2 , \\ \Delta m^2_{32}(IH) &= [-2.55 \pm 0.08] \times 10^{-3} \mathrm{eV}^2 . \end{aligned}$$

Independent measurement $\sin^2 2\theta_{13}$ using neutron captured on hydrogen with 631-day of data is also presented. $\sin^2 2\theta_{13} = 0.071 \pm 0.011$.

8 Acknowledgments

Part of this work is supported by the National Natural Science Foundation of China (Grant No. 11505018) and Chongqing Science and Technology plan project (Grant No. Cstc2015jcyj40031). Daya Bay is supported in part by the Ministry of Science and Technology of China, the U.S.Department of Energy, the Chinese Academy of Sciences, the CAS Center for Excellence in Particle Physics, the National Natural Science Foundation of China, the Guangdong provincial government, the Shenzhen municipal government, the China General Nuclear Power Group, Key Laboratory of Particle and Radiation Imaging (Tsinghua University), the Ministry of Education, Key Laboratory of Particle Physics and Particle Irradiation (Shandong University), the Ministry of Education, Shanghai Laboratory for Particle Physics and Cosmology, the Research Grants Council of the Hong Kong Special Administrative Region of China, the University Development Fund of The University of Hong Kong, the MOE program for Research of Excellence at National Taiwan University, National Chiao-Tung University, and NSC fund support from Taiwan, the U.S. National Science Foundation, the Alfred P. Sloan Foundation, the Ministry of Education, Youth, and Sports of the Czech Republic, the Joint Institute of Nuclear Research in Dubna, Russia, the CNFCRFBR joint research program, the National Commission of Scientific and Technological Research of Chile, and the Tsinghua University Initiative Scientific Research Program. We acknowledge Yellow River Engineering Consulting Co., Ltd., and China Railway 15th Bureau Group Co., Ltd., for building the underground laboratory. We are grateful for the ongoing cooperation from the China General Nuclear Power Group and China Light and Power Company

References

[1] Q. R. Ahmad et al. (SNO Collaboration), Phys. Rev. Lett. 89,2676 011301 (2002).
[2] S. Abe et al. (KamLAND Collaboration), Phys. Rev. Lett. 100,2678 221803 (2008).
[3] Y. Fukuda et al. (Super-Kamiokande Collaboration), Phys. Rev.2670 Lett. 81, 1562 (1998).
[4] M. H. Ahn et al. (K2K Collaboration), Phys. Rev. Lett. 90,2672 041801 (2003).
[5] F. P. An et al. (Daya Bay Collaboration), Phys. Rev. Lett. 108,171803 (2012).
[6] J. Ahn et al. (RENO Collaboration), Phys. Rev. Lett. 108,191802 (2012).
[7] J. H. Choi et al. (RENO), Phys. Rev. Lett. 116, 211801 (2016).
[8] M. Ishitsuka, "Double chooz," (2016), moriond 2016.
[9] Y. Abe (Double Chooz Collaboration), Phys. Rev. Lett. 108,131801 (2012).
[10] F.P.An et al. (Daya Bay Collaboration), Phys. Rev. Lett. 115, 111802 (2015).
[11] F.P.An et al. (Daya Bay Collaboration), Nucl. Instr. Meth. A 685, 78 (2012) .
[12] P. Adamson et al. (MINOS Collaboration), Phys. Rev. Lett. 110,171801(2013).
[13] K. Abe et al. (T2K), Phys. Rev. D91, 072010 (2015).
[14] P. Adamson et al. (NOvA), Phys. Rev. Lett. 116, 151806 (2016).
[15] R. Wendell (Super-Kamiokande), (2014), arXiv:1412.5234[hep-ex].
[16] M. G. Aartsen et al. (IceCube), Phys. Rev. D91, 072004 (2015).
[17] F. P. An et al. (Daya Bay Collaboration), Chin. Phys. C 37,011001 (2013).
[18] F. P. An et al. (Daya Bay Collaboration), Phys. Rev. D90, 071101(2014).
[19] F. P. An et al. (Daya Bay Collaboration), arXiv:1610.04802 [hep-ex].

Results from the OPERA experiment

A. Di Crescenzo[1,2,a] on behalf of the OPERA Collaboration

[1] *Dipartimento di Fisica dell'Università Federico II di Napoli*
[2] *INFN Sezione di Napoli*

Abstract. The OPERA neutrino experiment was designed to perform a unique ν_τ appearance measurement in the ν_μ CNGS beam to confirm the oscillation mechanism in the atmospheric sector $\nu_\mu \to \nu_\tau$. The detection of τ leptons produced in ν_τ CC interactions and of their decays is accomplished exploiting the high spatial resolution of nuclear emulsions. Five ν_τ candidate events have been detected in the full data sample from 2008-2012 CNGS runs, with an expected background of 0.25 events. The background only hypothesis is rejected with a significance larger than 5 σ. The analysis of the tau neutrino sample in the framework of the 3+1 neutrino model is also presented. Furthermore OPERA good capabilities in detecting electron neutrino interactions allow setting limits on the $\nu_\mu \to \nu_e$ oscillation channel.

1 The OPERA detector

The aim of the OPERA (Oscillation Project with Emulsion-tRacking Apparatus) experiment [1, 2] at the Gran Sasso underground Laboratory (LNGS) was to perform the first direct detection in the appearance mode of $\nu_\mu \to \nu_\tau$ oscillations. The direct appearance search was based on the detection of τ leptons produced in charged current interactions (CC) of ν_τ. The CERN neutrino beam is an almost pure ν_μ beam produced by protons accelerated in the CERN SPS and injected in the CNGS beam line. The OPERA experiment [3, 4] was installed in the LNGS Hall C.
OPERA was a hybrid apparatus with a modular structure. The detector concept was based on the Emulsion Cloud Chamber (ECC) technique, combined with real-time detection techniques (electronic detectors). The ECC basic unit in OPERA is a brick made of 56 lead plates, providing the required mass, interleaved with 57 nuclear emulsion films, providing the necessary spatial and angular resolution. The electronic detectors were used to trigger the neutrino interactions, to locate the brick in which the interaction took place, to identify muons and measure their momentum and charge. The analysis of nuclear emulsion films is performed in the OPERA scanning laboratories in Europe and Japan by fast automatic scanning systems developed to reconstruct with sub-micrometric accuracy the neutrino interaction events.
The detector shown in Fig. 1 is composed of two identical supermodules (SM1 and SM2), each consisting of a target section made of arrays of bricks coupled to a scintillator Target Tracker detector (TT) to trigger the read-out and localize neutrino interactions within the target. Each target section is followed by a muon spectrometer, namely a dipole magnet instrumented with resistive plate chambers

[a]e-mail: dicrescenzo@na.infn.it

(RPCs) and drift tubes detectors (PT), to measure the muon charge and momentum.

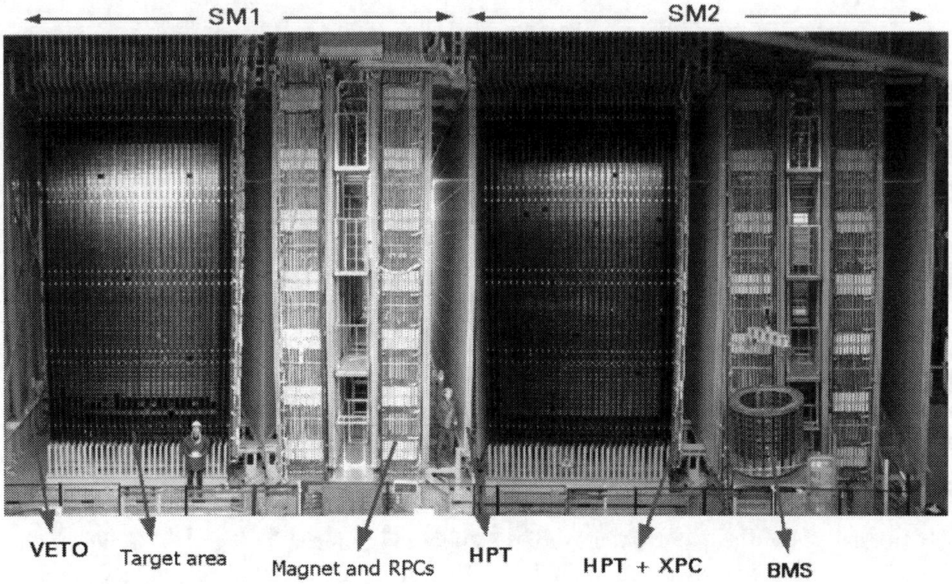

Figure 1. View of the OPERA detector. The upper horizontal lines indicate the position of the two identical supermodules (SM1 and SM2). Arrows show the position of a target area (ECC brick walls interleaved with planes of plastic scintillators), the VETO planes, drift tube detector (HPT) planes, the magnet and the RPC. A part of the Brick Manipulator System (BMS) is also visible.

2 Event reconstruction and analysis

The electronic detector data are used to identify the brick in which the neutrino interaction occurred. Events are classified as Charged Current like (1μ) if a track is tagged as a muon or if a minimum number of detector walls is traversed, or as Neutral Current like (0μ) in the complementary case. The bricks tagged as containing a neutrino interaction vertex were extracted from the detector, their emulsion films developed and sent to the scanning laboratories in Europe and Japan, where a search for τ decays is performed using custom automated optical microscopes.
The search for the neutrino interaction vertex starts from a set of tracks predicted using information from the electronic detectors. Confirmed tracks are followed upstream in the brick from the most downstream emulsion film to the possible interaction vertex. A volume of 1 cm^2 by 5 films upstream and 10 films downstream is scanned in order to fully explore the vertex area and possibly find other associated particle tracks. A procedure called *decay search* is then applied [5] in order to search for charged or neutral decay topologies, secondary interactions or gamma-ray conversions. If a secondary vertex is found, a full kinematical analysis is performed combining the measurements in the nuclear emulsion with data from the electronic detectors. The momentum of charged particles can be measured in emulsions by the Multiple Coulomb Scattering up to 6 GeV/c with resolution better than 22%

using the angular deviations [6]; it can be measured up to 12 GeV/c with a resolution better than 33% using position deviations. For muons crossing the spectrometers, the momentum is measured with a resolution better than 22% up to 30 GeV/c. The muon charge is also determined [7]. The hint of a decay topology is the observation of an impact parameter, defined as the minimum distance between the track and the reconstructed vertex, excluding low momentum tracks, greater than 10 μm.

3 Search for $\nu_\mu \rightarrow \nu_\tau$ oscillations

In the analysed sample, five ν_τ candidates were observed. The first ν_τ candidate, observed in the 2008-2009 data sample [8], has seven prongs at the primary vertex. One track exhibits a kink topology; the daughter track is identified as a hadron by its interaction. In addition, two γ-rays point to the secondary vertex. Their invariant mass is $120 \pm 20(stat.) \pm 35(syst.)$ MeV/c^2 , compatible with the π^0 mass. Its combination with the secondary hadron, assumed to be a π^-, gives an invariant mass of $640^{+125}_{-80}(stat.)^{+100}_{-90}(syst.)$ MeV/c^2. The decay mode is therefore compatible with $\tau \rightarrow \rho(770)\nu_\tau$, whose branching ratio is about 25%.
The second ν_τ candidate event, found in the 2011 data sample [9], is a 2-prong neutrino interaction with a 1.54 mm track (associated with τ lepton) and a longer track identified as a hadron being incompatible with a muon track because of its range. The τ lepton exhibits a 3-prong decay topology. The daughter particles are identified as hadrons on the basis of momentum-range consistency. The kinematical analysis of the event satisfies all the specified criteria for the $\tau \rightarrow 3h$ decay channel.
The third ν_τ candidate was observed in the 2012 1μ data sample [10]. The primary vertex is given by two tracks and a γ-ray. The first track is associated to a hadron from the momentum-range correlation. The other track from the τ lepton decaying into a muon. The muon momentum determined from the track reconstructed by the electronic detectors is 2.8 ± 0.2 GeV/c; its negative charge is assessed with 5.6σ significance. All the kinematical criteria for the selection of $\tau \rightarrow \mu$ decays are satisfied.
The fourth ν_τ event was observed in 2012 0μ data sample. The primary vertex is defined by four tracks and two γ-rays. One track exhibits a kink topology. The daughter track exits the target, stopping in the spectrometer after crossing three RPC planes. Momentum-range correlation allows to identify it as a hadron.
The fifth ν_τ candidate was found in 2012 0μ sample [11]. The primary vertex consists of the τ candidate track, since it exhibits a kink topology, and a charged particle track (p_1, see Fig. 2). In addition, four forward-going and two backward-going nuclear fragments are observed, all pointing to the primary vertex. The τ candidate decays, after a flight length of 960 ± 30 μm, into one charged particle identified as a hadron since a secondary interaction is reconstructed 22 lead plates downstream yielding four charged particles and a photon. Figure 2 shows the display of the event as reconstructed in the brick. The momentum of the charged particle producing the primary track p_1 is 1.0 ± 0.1 GeV/c. The particle is identified as a hadron from its interaction in the downstream brick. This, together with the null result from the search for large angle tracks, allows to rule out the presence of a muon at the primary vertex. The kinematical analysis of the event satisfies all the specified criteria for the $\tau \rightarrow h$ decay channel.
In the analysed data sample the expected number of background events is 0.04, 0.17, 0.004 and 0.03 in the $\tau \rightarrow h$, $\tau \rightarrow 3h$, $\tau \rightarrow \mu$ and $\tau \rightarrow e$ channels, respectively, mainly from charmed events with an undetected primary muon, hadronic re-interactions (for the hadronic decay channels) and large angle muon scattering (for the $\tau \rightarrow \mu$ channel). From the observation of the five candidate events and taking into account the signal-to-noise ratio for each decay channel, non-null observation can be assessed with 5.1 sigma significance.

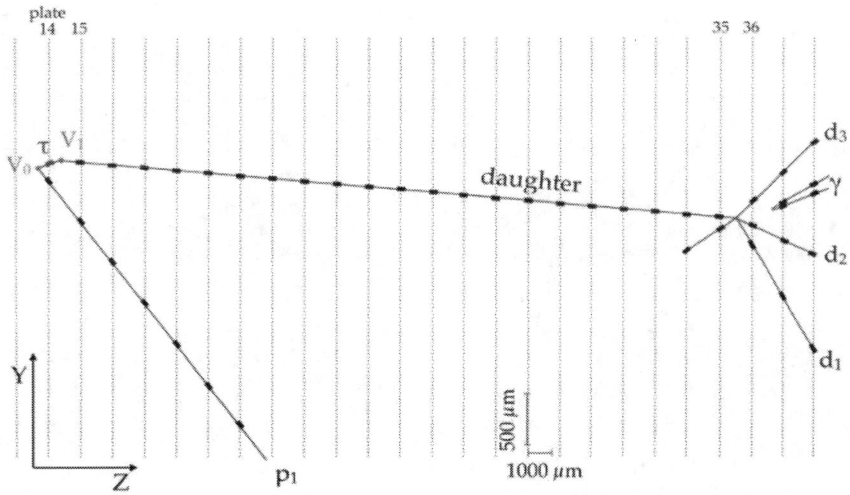

Figure 2. Event display of the fifth ν_τ candidate event in the horizontal projection longitudinal to the neutrino direction. The event primary and secondary vertices are indicated as V_0 and V_1, respectively.

4 Search for sterile neutrinos

The OPERA experiment can test existence of a fourth (sterile) neutrino state at the eV mass scale by looking for deviations from the standard three-neutrino model predictions. In presence of a fourth sterile neutrino with mass m_4, the oscillation probability is a function of the 4×4 mixing matrix U and of the three squared mass differences. In the framework of the 3+1 model, at values of $\Delta m^2_{14} > 1$ eV^2 , the measured 90% C.L. upper limit on the mixing term $\sin 2\theta_{\mu\tau} = 2 \mid U_{\mu 4} \| U_{\tau 4} \mid$ is 0.116, independently of the mass hierarchy of the three standard neutrinos. The OPERA experiment extends the exclusion limits on Δm^2_{14} in the $\nu_\mu \rightarrow \nu_\tau$ appearance channel down to values of 10^{-2} eV^2 at large mixing for $\sin^2 2\theta_{\mu_\tau} \geq 0.5$, as shown in Fig. 3.
A systematic search for ν_e events was applied to 505 0μ events located in 2008 and 2009 data sample [16]. The number of observed ν_e interactions (19 events) is compatible with that expected from the ν_e beam contamination (19.8 ± 2.8). In Fig. 4 the reconstructed energy distribution of the detected candidates is compared with the energy spectra of the ν_e beam contamination, the expected oscillated ν_e from the three-flavour oscillation and the background.
The observation is compatible with the non-oscillation hypothesis; an upper limit $\sin^2(2\theta_{13}) < 0.44$ is derived at the 90% C. L..
In the non-standard oscillation framework and in the one mass scale dominance approximation (2 neutrino mixing) the exclusion plot in Fig. 5 was derived. A complete analysis in the 3+1 scheme based on the statistics of the full data sample (factor of 2.5 increase) is ongoing.

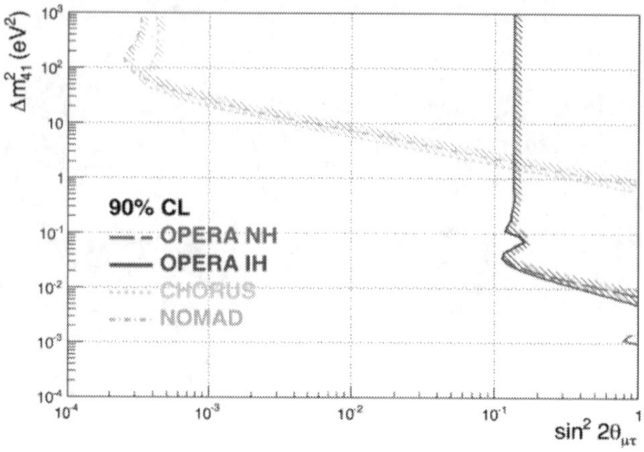

Figure 3. OPERA preliminary 90% C.L. exclusion limits in the Δm^2_{41} vs $\sin^2 2\theta_{\mu\tau}$ parameter space for the normal (NH, red) and inverted (IH, blue) hierarchy of the three standard neutrino masses. The exclusion plots by NOMAD [14] and CHORUS [15] are also shown. Shaded bands are drawn to indicate the excluded regions.

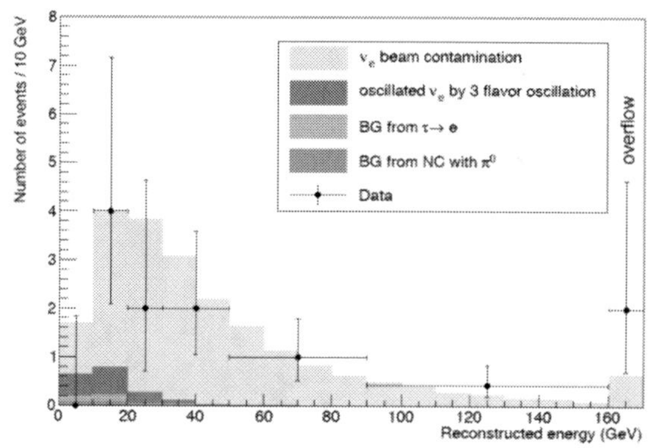

Figure 4. Energy distribution of the detected ν_e events comparing to the expectations of beam ν_e and oscillated ν_e.

References

[1] M. Guler et al. (OPERA Collab.), CERN-SPSC-97-24, 1997.
[2] M. Guler et al. (OPERA Collab.), CERN-SPSC-2001-25, 2001.
[3] R. Acquafredda et al. (OPERA Collab.), JINST **4** (2009) P04018.
[4] R. Acquafredda et al. (OPERA Collab.), New J. Phys. **8** (2006) 303.
[5] N. Agafanova et al. (OPERA Collab.), Eur. Phys. J C 89 (2014) 2986.
[6] N. Agafanova et al. (OPERA Collab.), New J. Phys. **14** (2012) 013026.
[7] N. Agafanova et al. (OPERA Collab.), New J. Phys. **13** (2011) 053051.
[8] N. Agafanova et al. (OPERA Collab.), Phys. Lett. B **691** (2010) 138.
[9] N. Agafanova et al. (OPERA Collab.), JHEP **11** (2013) 036.

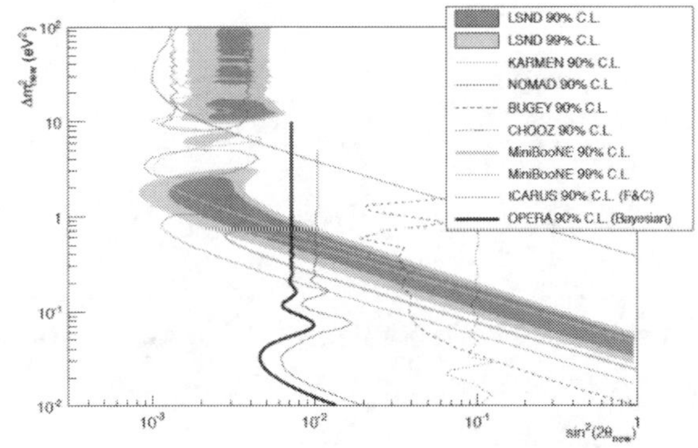

Figure 5. The exclusion plot for the parameters of the non-standard $\nu_\mu \rightarrow \nu_e$ oscillation, obtained from this analysis using the Bayesian method, is shown. Limits from other experiments are also shown.

[10] N. Agafanova et al. (OPERA Collab.), Phys. Rev. D **89** (2014) 051102.
[11] N. Agafanova et al. (OPERA Collab.), Prog. Theor. Exp. Phys. (2014) 101C01.
[12] A. Aguilar-Arevalo et al. (LSND Collaboration), Phys. Rev. D **64** (2001) 112007;
A. A. Aguilar-Arevalo et al. (MiniBooNE Collaboration), Phys. Rev. Lett. **110** (2013) 161801;
G. Mention et al., Phys. Rev. D **83** (2011) 073006;
Th. A. Muller et al., Phys. Rev. C **83** (2011) 054615;
C. Giunti and M. Laveder, Phys. Rev. C **83** (2011) 065504.
[13] N. Agafonova et al. (OPERA Collab.) JHEP **1506** (2015) 069.
[14] P. Astier et al. (NOMAD Collaboration), Nucl. Phys. **B 611** 3 (2001).
[15] E. Eskut et al. (CHORUS Collaboration), Nucl. Phys. **B 793** 326 (2008).
[16] N. Agafanova et al. OPERA Collab.) JHEP **07**(2013) 004.

DOI: 10.1051/epjconf/201716401026

Results from the OPERA experiment at the CNGS beam

N. Mauri[1,a] on behalf of the OPERA Collaboration

[1] *INFN - Sezione di Bologna, Viale Berti Pichat 6/2, I-40127 Bologna, Italy*

Abstract. The OPERA experiment at the Gran Sasso underground laboratory was designed to study $\nu_\mu \rightarrow \nu_\tau$ oscillations in appearance mode in the CNGS neutrino beam. Five ν_τ candidate events have been observed, allowing to assess the discovery of $\nu_\mu \rightarrow \nu_\tau$ transitions in the atmospheric sector with a significance of 5.1σ. In this paper the ν_τ data analysis will be discussed, with emphasis on the background constraints obtained using dedicated data-driven control samples. Results on the search for $\nu_\mu \rightarrow \nu_e$ oscillations, on the search for sterile neutrino mixing and on the atmospheric muon charge ratio will also be presented.

1 Introduction

Neutrino oscillations [1, 2] are an experimental evidence that updated the standard framework founding most particle physics theories. The results of solar, atmospheric, reactor and accelerator experiments established that flavour mixing occurs not only in the hadronic sector, but in the leptonic sector as well. This phenomenon was discovered in 1998 through the disappearance of atmospheric ν_μ. A deficit in the solar ν_e flux was already observed since 30 years at that time. The flavour disappearance is produced with an oscillatory pattern by the quantum-mechanical interference between neutrino mass eigenstates. The leptonic flavour transitions are parametrised by the PMNS unitary matrix, through which the weak interaction eigenstates are represented as linear combinations of the mass eigenstates, and by two distinct squared mass differences, Δm^2_{sol} and Δm^2_{atm}, determining the oscillation frequencies at the solar and at the atmospheric scale.

The observation of flavour appearance is the last step of the discovery phase of neutrino oscillations. While the solar ν_e beam cannot be exploited for direct appearance search, due to the very low energy spectrum (well below the μ mass threshold), the atmospheric and artificial ν_μ beams have been used for this challenging purpose in the last decade. In particular, the OPERA experiment in the CNGS beam was designed to unambiguously evidence the oscillations at the atmospheric scale through the direct ν_τ appearance.

2 The OPERA experiment

OPERA [3] is a long baseline neutrino experiment located in the underground Gran Sasso Laboratory (LNGS). The direct appearance search is based on the detection of τ leptons produced in Charged Current (CC) ν_τ interactions with a signal to noise ratio of $O(10)$. The observation of the short-lived τ

[a] e-mail: nicoletta.mauri@bo.infn.it

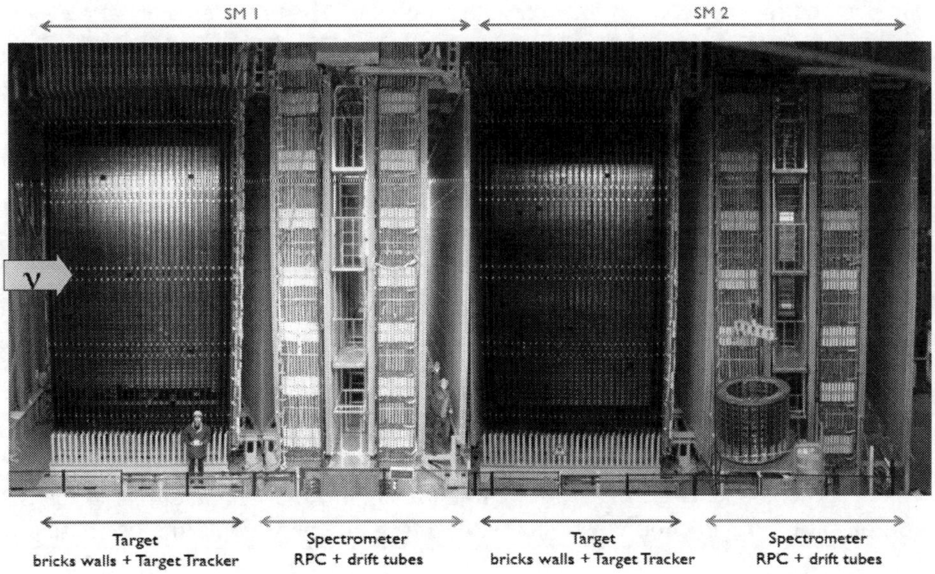

Figure 1. The OPERA detector in the Hall C at LNGS.

decay is a demanding experimental challenge, since it requires a very high tracking accuracy, $O(\mu\text{m})$, integrated in a large mass to maximise the number of ν_τ CC interactions. The OPERA experiment accomplishes these two requirements with a modular and hybrid apparatus composed by Emulsion Cloud Chambers (ECC) complemented with electronic detectors (Fig. 1). The emulsion technique allows to distinguish the three leptonic neutrino CC interations, thus OPERA has a unique capability to study ν_e, ν_μ and ν_τ on event-by-event basis. In addition to its main goal, the OPERA detector at LNGS is well suited to cosmic ray physics measurements, in particular to determine the atmospheric muon charge ratio at the highest observable energies.

2.1 The OPERA detector

The ECC basic unit in OPERA is the "brick", a sandwich made of 56 lead plates, 1 mm thick, interspaced with 57 nuclear emulsion films. The submicrometre spatial resolution of the nuclear emulsion allows a precise three-dimensional reconstruction of the neutrino vertex as well as of the decay vertex associated with short-lived particles, like the τ lepton. Moreover, each brick is a compact standalone detector measuring electromagnetic showers and charged particle momentum through multiple Coulomb scattering [4]. The overall OPERA target is composed of about 150000 bricks, easily removable for the analysis, for a total mass of 1.25 kton.

In each of the two target sections, the bricks are arranged in 29 vertical "walls", transverse to the beam direction, interleaved with Target Tracker walls (TT), planes of horizontal and vertical scintillator strips. The TT trigger the read-out and allow to localise the brick containing the neutrino interaction with an accuracy of $O(\text{cm})$. Tightly packed removable doublets of emulsion films called Changeable Sheets (CS) [5] are attached downstream of each brick in a separate envelope. They serve as interfaces between the TT and the brick, in order to confirm the brick selection and to provide a more accurate prediction of the track position. The target section is followed by a magnetic spectrometer. A dipolar iron magnet is instrumented with RPC and drift tube detectors (Precision Trackers, PT), in order to measure the muon charge and momentum. OPERA is made of two identical Super Modules (SM) each consisting of a target section and a muon spectrometer, as shown in Fig. 1.

2.2 The CNGS neutrino beam

OPERA was exposed from 2008 to 2012 to the long baseline CNGS (CERN Neutrinos to Gran Sasso) ν_μ beam, 730 km away from the source. The CNGS was a conventional neutrino beam produced by 400 GeV/c protons extracted from the CERN SPS and collided with a graphite target. The prompt ν_τ contamination was negligible, $O(10^{-6})$. The ν_e component was relatively small: in terms of CC interactions, the ν_e and $\bar{\nu}_e$ contaminations were together less than 1%.

The beam was designed and optimised for ν_τ appearance search. Differently from other neutrino beams designed to measure the disappearance signal, the mean energy of the CNGS was not tuned around the oscillation peak since at the atmospheric scale it is well below the τ production threshold (for $L = 730$ km, the oscillation maximum is at $E_\nu \sim 1.5$ GeV). The neutrino energy spectrum ($\langle E_{\nu_\mu} \rangle = 17$ GeV) was set up to maximise the number of ν_τ CC interactions at LNGS, i.e. the convolution of the CC interaction cross section (rising with energy) and the oscillation probability (higher at low energy).

At the end of the CNGS operation, after the 2012 Run, 17.97×10^{19} protons on target (pot) had been delivered and integrated by the OPERA detector, almost 80% of the nominal design value. The CNGS beam intensity and the total number of pot per year accumulated by OPERA are shown in Fig. 2, together with the corresponding OPERA target mass depletion.

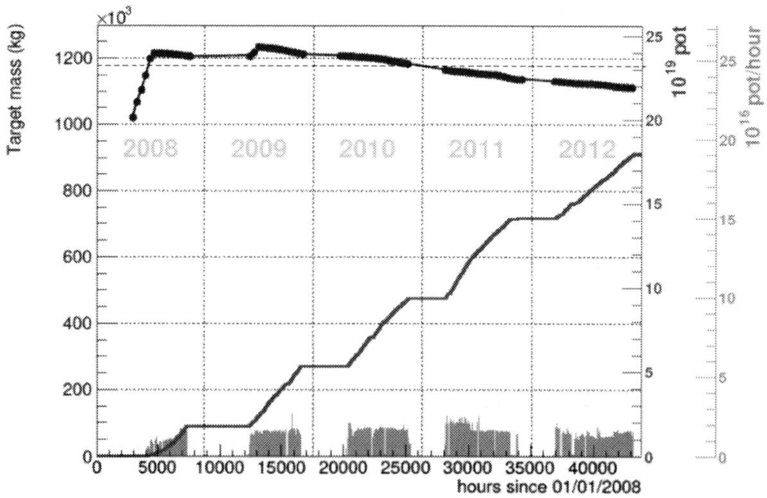

Figure 2. CNGS beam luminosity in pot and corresponding OPERA target mass evolution [12].

2.3 Data Processing

During the data taking from 2008 to 2012, the electronic detectors acquired both cosmic ray and CNGS beam events. The latter are selected through the timing coincidence with CERN and tagged as "on time" events. These are divided by an offline algorithm in "external" events, i.e. mostly muons produced by ν_μ CC interactions in rock, and "internal" events, i.e. interactions occurring inside the OPERA target. For oscillation studies, only internal events are analysed. A total of 19505 internal interactions have been registered in 5 years of data taking. They are further classified as CC-like (1μ) or as NC-like (0μ) [6].

A dedicated program reconstructs the tracks in the electronic detectors and builds a 3D probability map for bricks to contain the neutrino vertex. The brick with the highest probability is extracted from the target, the CS films are detached, developed and analysed with high-speed automatic optical microscopes [7], searching for tracks compatible with the TT prediction. In case this search is unsuccessful, the brick is equipped with a new CS doublet and inserted back into the target, while the next brick in the probability map is extracted. In case of positive CS result the brick is unpacked and not replaced. The target mass evolution is shown in Fig. 2 in parallel with the CNGS integrated luminosity. The emulsion films are developed and dispatched to one of the scanning laboratories in Europe or in Japan.

The tracks found in the CS doublet are extrapolated to the most downstream film of the brick and followed upstream using predictions from the scanning in each film, until they converge to the stopping point, i.e. they are not found in three consecutive films. Vertex confirmation is obtained by scanning a large volume of $\sim$2 cm^3 around the stopping point.

The final phase of the event analysis is the decay search procedure [8]. This procedure is applied to detect charged and neutral decay topologies, secondary interactions or photon conversions in the neighbourhood of the primary vertex. If a secondary vertex is found, a full kinematical analysis is performed extending the volume scanning and following the tracks also in the downstream bricks. This analysis integrates the complementary information provided by emulsions and electronic detectors, making use of the angles measured in the emulsion films, the momenta determined by multiple Coulomb scattering measured in the brick, the momenta measured by the magnetic spectrometers, and the total energy deposited in the instrumented target acting as a calorimeter [6]. The energy of photons and electrons is also estimated using calorimetric techniques [9].

3 Neutrino oscillation results

3.1 $\nu_\mu \rightarrow \nu_\tau$ appearance search

3.1.1 Data Sets

For the $\nu_\mu \rightarrow \nu_\tau$ appearance search the data sample consists of all the 0μ events and of the 1μ events with muon momentum below 15 GeV/c. The results here reported are related to events reconstructed in the first and in the second bricks in the probability map [10]. The numbers of fully analysed events for each year of data taking are summarised in Tab. 1.

Table 1. CNGS pot and corresponding neutrino interactions in the OPERA detector for the analysed data set.

	2008	2009	2010	2011	2012	Total
pot (10^{19})	1.74	3.53	4.09	4.75	3.86	17.97
0μ	148	250	209	223	149	979
1μ ($p_\mu <$ 15 GeV/c)	534	1019	814	749	590	3706
All events	682	1269	1023	972	739	4685

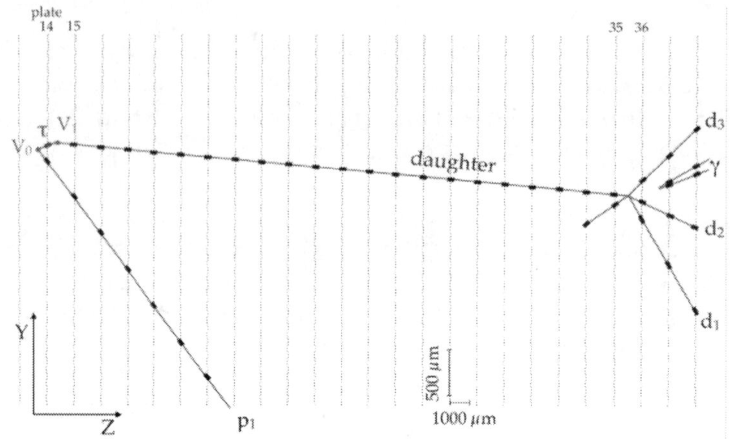

Figure 3. Display of the 5th ν_τ candidate event ($\tau \to 1h$) in the brick (side view). The primary interaction vertex is labeled as V_0, the secondary decay vertex is labeled as V_1 [10].

After the application of the decay search procedure and the kinematical analysis, 5 events fulfill all the topological and kinematical cuts required to be ν_τ candidates. In one of them, the τ lepton decays into a μ [11], one of them is in the $\tau \to 3h$ decay channel [12], and three are in $\tau \to 1h$ [13, 14]. The cuts and requirements for each decay channel are detailed in the corresponding references. Fig. 3 shows the display in the brick of the 5th ν_τ candidate event, a $\tau \to 1h$ decay.

3.1.2 Data-driven background estimation

The three main sources of background for the ν_τ appearance search are charmed particle decays, hadronic interactions and large-angle muon scattering (LAS). Charmed particle decays in which the muon produced at the primary vertex in ν_μ CC interactions is not identified contribute to the background for all the τ decay channels. Hadronic re-interactions contribute only to the $\tau \to 1h, 3h$ decay channels, while LAS can mimic the $\tau \to \mu$ decay.

The contributions of the above mentioned processes are validated using real data samples. The observed CNGS events with charm production represent an important data set to benchmark the τ detection efficiency, due to the similar mass and decay topologies. Fig. 4 shows the distributions of the flight length of the charm candidates, the angular separation between the charmed particle and the primary muon in the beam transverse plane (ϕ), the impact parameters of the secondary particles with respect to the primary vertex, and the primary muon momentum for data and Monte Carlo. The expected number of charm events and the shape of the distributions are in very good agreement. The uncertainty on the charm background has been estimated to be 20% [8].

The background associated to hadronic interactions has been checked using test-beam data. Nuclear fragments rates and angular distributions have been characterised up to high angle ($\tan\theta = 3$). Further constraints are given by the analysis of hadronic tracks produced in CNGS ν_μ CC interactions. The accuracy of the expected hadronic background is at the 30% level [10].

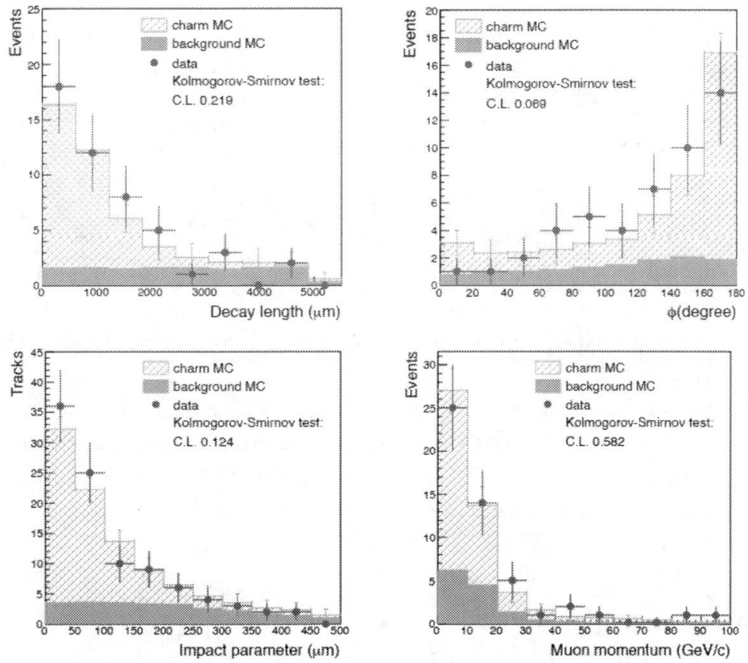

Figure 4. Shape comparison between observed ν_μ CC interactions with candidate charm decays and MC expectations. The expected background contribution is also shown (stacked histogram) [8]. Top Left: distribution of the decay length of the candidate charmed particles. Top Right: distribution of the angle between the candidate charmed particle and the primary muon in the ν transverse plane. Bottom Left: distribution of the impact parameters of the candidate charm daughter particles with respect to the neutrino interaction vertex. Bottom Right: distribution of muon momentum.

The LAS background contribution has been recently reevaluated and strongly reduced with respect to the value considered in the experiment proposal [15]. The LAS rate is calculated using refined Monte Carlo simulations which are benchmarked against experimental data.

The estimated background events for the analysed data set with the corresponding uncertainties are summarised in Table 2. The total expected background amounts to 0.25 ± 0.05 events. Details on the background expectation are given in [12].

3.1.3 Results

The expected numbers of ν_τ events for each decay channel are computed assuming $\Delta m^2_{23} = 2.44 \times 10^{-3}$ eV^2 [16] and maximal mixing, and are reported in Table 2. The total expected signal amounts to 2.64 ± 0.53 events.

The significance of the observation of 5 ν_τ events is evaluated testing the null hypothesis, i.e. computing the probability of a background fluctuation. In this counting analysis, two test statistics are used, one based on the Fisher's method, the other one based on the profile likelihood ratio. Both methods exclude the background-only hypothesis with a significance of 5.1 standard deviations [10]. The

Table 2. Expected signal and background events in the analysed data set [10].

Channel	Exp. Background				Exp. Signal	Observed
	Charm	Hadronic re-int	LAS	Total		
$\tau \to 1h$	0.017 ± 0.003	0.022 ± 0.006	–	0.04 ± 0.01	0.52 ± 0.10	3
$\tau \to 3h$	0.17 ± 0.03	0.003 ± 0.001	–	0.17 ± 0.03	0.73 ± 0.14	1
$\tau \to \mu$	0.004 ± 0.001	–	0.0002 ± 0.0001	0.004 ± 0.001	0.61 ± 0.12	1
$\tau \to e$	0.03 ± 0.01	–	–	0.03 ± 0.01	0.78 ± 0.16	0
Total	0.22 ± 0.04	0.02 ± 0.01	0.0002 ± 0.0001	0.25 ± 0.05	2.64 ± 0.53	5

observed number of ν_τ candidates is also compatible with the expectations from the PMNS neutrino oscillation framework.

For the CNGS baseline and energy range the rate of ν_τ CC interactions varies as $(\Delta m^2_{23})^2$. Thus, the number of observed signal candidates is used to measure the atmospheric squared mass difference in appearance mode for the first time. Using three different approaches (profile likelihood ratio, Feldman-Cousins method, Bayesian statistics) and assuming maximal mixing, the 90% C.L. interval for Δm^2_{23} is $[2.0, 5.0] \times 10^{-3}$ eV^2 [10].

3.2 Sterile neutrino mixing search

The search for sterile neutrino mixing can be explored by OPERA in several ways. Results have been published on the search for oscillations induced by the mixing with a sterile neutrino in the $\nu_\mu \to \nu_e$ [9] and $\nu_\mu \to \nu_\tau$ [17] channels in appearance mode.

The $\nu_\mu \to \nu_e$ oscillation channel is studied thanks to the excellent capabilities of nuclear emulsions to disentangle electrons from photon conversions in identified electromagnetic showers. Here the results published in Ref. [9] are presented, which concern the analysis of 2008-2009 data, corresponding to 5.3×10^{19} pot. The analysis of the complete data sample is ongoing and will be published in a near future.

The ν_e search is systematically applied to analysed 0μ events. A special procedure is used to select a sample of events with "shower hints", based on track multiplicity found in the CS. For the selected sample an additional scanning volume is required, from the last film of the vertex standard volume up to the most downstream film of the brick. If an electromagnetic shower is reconstructed and if the primary track initiating the shower is recognised as a single track on the vertex emulsion film, the event is classified as ν_e candidate event. Applying the above mentioned procedure, 19 ν_e candidate events are observed in the 2008-2009 data set. The expected number of ν_e CC interactions due to the intrinsic beam contamination (see Sect. 2.2) is $n_e = 19.8 \pm 2.8$ (syst.). This number is consistent with the observed ν_e candidates, showing no excess signal resulting from $\nu_\mu \to \nu_e$ oscillations. The ν_e energy spectrum is shown in Fig. 5, and compared with the expected energy spectra of the ν_e beam contamination, of the oscillated ν_e from the three-flavour model and of the background [9]. The observation is used to set limits on the oscillation parameters of the standard 3ν framework, with a limited sensitivity to the actual θ_{13} region [16]. An upper limit $\sin^2 2\theta_{13} < 0.44$ is derived at 90% C.L.

The results are used to constrain the effective mixing parameters with a sterile neutrino. The number of ν_e candidates is compared to the expectation from an approximated two-state model parametrised in terms of two effective parameters, Δm^2_{new} and θ_{new}. The approximation is valid assuming CP conservation, neglecting the standard oscillations, here treated as a background, and for large values of Δm^2_{new} (>0.1 eV^2). To optimise the sensitivity, an energy cut at 30 GeV is applied. The

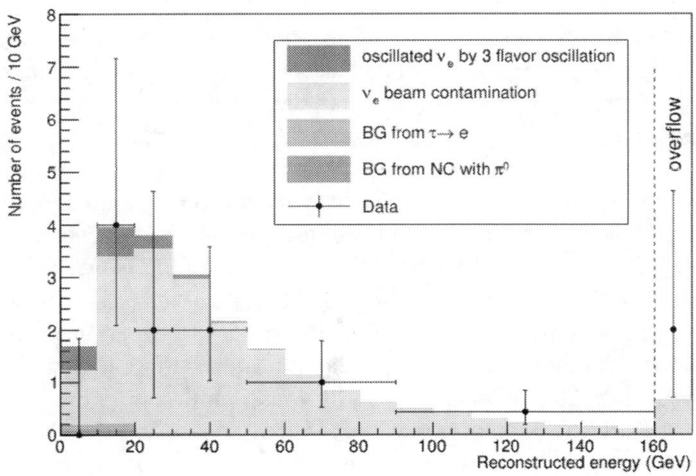

Figure 5. Energy spectrum of the observed ν_e candidate events (data points), compared with the expected energy spectra of the ν_e beam contamination, of the oscillated ν_e from the three-flavour model and of the background (histograms) [9].

number of observed events below 30 GeV is 6, to be compared to an expectation of 9.4 ± 1.3 (syst.) events. A 90% C.L. upper limit $\sin^2 2\theta_{new} < 7.2 \times 10^{-3}$ is thus derived for large Δm^2_{new} values. The final analysis on the complete data set will consider the so-called 3+1 model, providing results on the 4-state mixing parameters, Δm^2_{41} and $\sin^2 2\theta_{\mu e}$.

The search for sterile neutrino mixing is explored also in the $\nu_\mu \rightarrow \nu_\tau$ appearance channel. The results are presented in these Proceedings [19] and detailed in [17, 18]. The results on ν_τ appearance are interpreted in the context of the 3+1 neutrino model, deriving limits on the oscillations induced by a massive sterile neutrino. Exclusion regions are obtained in the (Δm^2_{41}, $\sin^2 2\theta_{\mu\tau}$) parameter space. The limits on Δm^2_{41} are extended up to 10^{-2} eV^2 for relatively large mixing, $\sin^2 2\theta_{\mu\tau} > 0.5$. At large values of Δm^2_{41}(> 1 eV^2), marginalising over the *CP*-violating phase, values of the effective mixing parameter $\sin^2 2\theta_{\mu\tau} > 0.119$ are excluded at 90% C.L.

4 Cosmic ray physics results

The OPERA detector located in the LNGS cavern is in a privileged location to study TeV scale cosmic rays, at an average depth of 1400 m of rock corresponding to 3800 m.w.e. The minimum surface muon energy is ~1 TeV (1.4 TeV averaged over all the directions and rock depths). OPERA is the deepest experiment able to measure charge-separated atmospheric muons.

The atmospheric muon charge ratio is a highly informative observable to shed light on the primary cosmic ray composition and interactions in the very forward region. Atmospheric muons carry information on their parents, mainly pions and kaons, produced in the collisions of primaries on atmospheric nuclei.

Cosmic ray induced events were acquired by OPERA together with CNGS neutrino events and easily classified through the timing coincidence with CERN (see Sect. 2.3). In five years of data taking, more than 3 million atmospheric muon events were recorded, among which about 110000 muon bundles, i.e. events with multiple muon tracks. Data were collected with both magnetic field polarities in order to minimise systematic errors due to misalignment.

OPERA has reported the final results on the atmospheric muon charge ratio R_μ using the complete statistics from 2008 up to 2012 [20]. The two data sets collected with opposite magnet polarities are combined reaching the most accurate measurement to date of R_μ in the TeV energy region. The muon charge ratio is computed separately for single muons, $R_\mu(n_\mu = 1) = 1.377 \pm 0.006\,(stat.)^{+0.007}_{-0.001}\,(syst.)$, and for muon bundles, $R_\mu(n_\mu > 1) = 1.098 \pm 0.023\,(stat.)^{+0.015}_{-0.013}\,(syst.)$. This is the first observation of a decrease in the charge ratio of high multiplicity events with respect to single muon events. The dilution effect in R_μ is expected since the multiple muon sample selects events generated by heavier primary cosmic rays and secondaries with a low value of Feynman-x, coming from the central region [20].

For single muons, R_μ is studied as a function of the vertical surface energy $E_\mu \cos\theta^*$, where θ^* is the zenith angle at the muon production point, fitting data to a parametrised model taking into account pions and kaons as muon parents [21]. The atmospheric muon charge ratio is measured in a wide $E_\mu \cos\theta^*$ range, from 500 GeV up to $\sim$ 10 TeV, and plotted in Fig. 6. With an average value $\langle E_\mu \cos\theta^* \rangle \simeq 2$ TeV, OPERA is the magnetised experiment measuring the charge ratio at the largest vertical surface energy. The results show an increase of R_μ as a function of $E_\mu \cos\theta^*$ compatible with a simple parametric model where the rise is due to the increasing kaon contribution to the muon flux. The fit of OPERA and L3+C data, shown in Fig. 6, yields the fractions of charged mesons decaying into positive muons $f_{\pi^+} = 0.5512 \pm 0.0014$ and $f_{K^+} = 0.705 \pm 0.014$. The prompt muon component does not significantly contribute to R_μ up to $E_\mu \cos\theta^* \lesssim 10$ TeV.

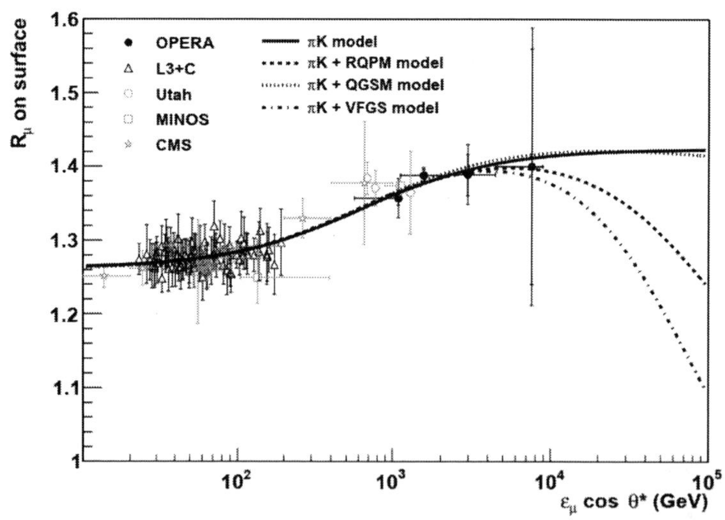

Figure 6. The atmospheric muon charge ratio measured by OPERA (*black points*) as a function of $E_\mu \cos\theta^*$ [20].

Taking into account the primary cosmic ray composition requires to disentangle the energy and zenith angle dependencies [21]. The fit in two dimensions $(E_\mu, \cos\theta^*)$ yields the proton excess in primary cosmic rays $\delta_0 = 0.61 \pm 0.02$ at primary energy $\langle E_N \rangle \approx 20$ TeV/nucleon, consistent with direct measurements of the primary composition [21], and the factor related to the associated kaon production, $Z_{pK^+} = 0.0086 \pm 0.0004$, here determined for the first time. No similar results on the associated kaon production cross section are available to date from accelerator experiments.

5 Conclusions

The OPERA experiment has observed neutrino oscillation at the atmospheric scale in appearance mode detecting 5 ν_τ candidate events. The discovery of ν_τ appearance is assessed with a significance of 5.1σ.

The results on $\nu_\mu \to \nu_\tau$ search, compatible with the standard 3ν model, have been used to constrain the parameter space of oscillations induced by a massive sterile neutrino. Limits on the sterile neutrino mixing have also been derived in the $\nu_\mu \to \nu_e$ appearance channel, given the number of observed ν_e interactions.

Moreover, the OPERA detector was exploited to measure the atmospheric muon charge ratio up to the largest surface muon energy to date, $O(10)$ TeV. The OPERA measurement of R_μ in the TeV energy range determined for the first time the spectrum weighted moment Z_{pK^+}, fundamental to accurately predict the atmospheric $\bar{\nu}/\nu$ ratio.

References

[1] B. Pontecorvo, Zh. Eksp. Teor. Fiz. **33** 549 (1957); B. Pontecorvo, Zh. Eksp. Teor. Fiz. **34** 247 (1958); B. Pontecorvo, Sov. Phys. JETP **26** 984 (1968)

[2] Z. Maki, M. Nakagawa and S. Sakata, Prog. Theor. Phys. **28** 870 (1962)

[3] R. Acquafredda *et al.* (OPERA Collaboration), JINST **4**, P04018 (2009)

[4] N. Agafonova *et al.* (OPERA Collaboration), New J. Phys. **14**, 013026 (2012)

[5] A. Anokhina *et al.* (OPERA Collaboration), JINST **3** P07005 (2008)

[6] N. Agafonova *et al.* (OPERA Collaboration), New J. Phys. **13**, 053051 (2011)

[7] N. Armenise *et al.*, Nucl. Instrum. and Methods A **551** 261 (2005);
M. De Serio *et al.*, Nucl. Instrum. and Methods A **554** 247 (2005);
L. Arrabito *et al.*, Nucl. Instrum. and Methods A **568** 578 (2006);
C. Bozza *et al.*, Nucl. Instrum. and Methods A **703** 204 (2013)

[8] N. Agafonova *et al.* (OPERA Collaboration), Eur. Phys. J. C **74**, 2986 (2014)

[9] N. Agafonova *et al.* (OPERA Collaboration), JHEP **07**, 004 (2013)

[10] N. Agafonova *et al.* (OPERA Collaboration), Phys. Rev. Lett. **115**, 121802 (2015)

[11] N. Agafonova *et al.* (OPERA Collaboration), Phys. Rev. D **89**, 051102 (2014)

[12] N. Agafonova *et al.* (OPERA Collaboration), JHEP **11**, 036 (2013), Erratum: ibidem **04**, 014 (2014).

[13] N. Agafonova *et al.* (OPERA Collaboration), Phys. Lett. B **691**, 138 (2010)

[14] N. Agafonova *et al.* (OPERA Collaboration), PTEP, 101C01 (2014)

[15] A. Longhin *et al.*, IEEE Trans. Nucl. Sc. **62**, 2216 (2015)

[16] K. A. Olive *et al.*, [Particle Data Group], Chin. Phys. C **38** 090001 (2014)

[17] N. Agafonova *et al.* (OPERA Collaboration), JHEP **06**, 069 (2015)

[18] M. Tenti, proceedings of the XVII Lomonosov Conference on Elementary Particle Physics, Moscow, 2015

[19] N. Mauri, "Search for sterile neutrino mixing in the $\nu_\mu \to \nu_\tau$ appearance channel with the OPERA detector", these proceedings, ICNFP 2015

[20] N. Agafonova *et al.* (OPERA Collaboration), Eur. Phys. J. C **74**, 2933 (2014)

[21] T. K. Gaisser, Astropart. Phys. **35**, 801 (2012)

Search for a high mass diphoton resonance using the ATLAS detector

Faïrouz Malek on behalf of the ATLAS Collaboration[1,a]

[1] *LPSC-Grenoble, CNRS-IN2P3, Université Grenoble-Alpes*

Abstract. High-mass states decaying into two photons are predicted in many extensions of the Standard Model (SM). The diphoton final state provides a clean experimental signature with good invariant mass resolution and moderate backgrounds. Searches for high-mass resonances decaying into two photons for a spin-0 or spin-2 state are presented. The latest ATLAS results using p-p collision data at 13 TeV and covering a large mass range are discussed.

1 Introduction

Searches for high-mass resonances decaying into two photons using CERN Large Hadron Collider (LHC) proton–proton collision data at $\sqrt{s}$= 13 TeV recorded in 2015 by the ATLAS detector and corresponding to an integrated luminosity of 3.2 fb^{-1} are described.

During the LHC Run-1 period, at $\sqrt{s}$= 7 TeV and 8 TeV, more than 20 fb^{-1} of data were recorded and searches for diphoton resonances have been reported by the ATLAS and CMS collaborations [1–5]. The analyses using the 2015 LHC pp data have also been published [6, 7].

The analyses were performed using two benchmark models: the diphoton decay resulting from a hypothetical particle of spin-0 or spin-2. This implies that diphotons have different kinematic properties depending on the model used. This means that one needs to apply two different selections, with looser kinematic selection requirements for the spin-2 resonance search. The photon identification criteria and the event pre-selection are common to both searches.

2 Search for a $\gamma\gamma$ resonance

2.1 The spin-2 case

The search for a spin-2 $\gamma\gamma$ resonance uses the Randall–Sundrum (RS) model [8] graviton as a benchmark. In this model, we look for the lightest Kaluza–Klein [9] spin-2 graviton excitation (G^*) with a dimensionless coupling $k/\overline{M}_{\rm Pl}$, where $\overline{M}_{\rm Pl} = M_{\rm Pl}/\sqrt{8\pi}$ is the reduced Planck scale and k the curvature scale of the extra dimension. The lightest graviton excitation is expected to be a fairly narrow resonance for $k/\overline{M}_{\rm Pl} < 0.3$ [10], with the width given by $1.44(k/\overline{M}_{\rm Pl})^2 m_{G^*}$, where m_{G^*} is the mass of the lightest graviton state. For $k/\overline{M}_{\rm Pl} = 0.1$, the natural width increases from 11 GeV at $m_{G^*} = 800$ GeV

[a] e-mail: fmalek@lpsc.in2p3.fr

to 30 GeV at $m_{G^*} = 2200$ GeV. The shape of the invariant mass distribution of the main background from the production of prompt photon pairs is estimated from theoretical computations, and the contribution from the reducible background of events where at least one jet is misidentified as a photon is added from data-driven estimates. The search is performed in the mass range above 500 GeV and in the $k/\overline{M}_{\rm Pl}$ range 0.01 to 0.3, searching for an excess over the estimated background diphoton invariant mass distribution. To model such an excess, the RS graviton resonance shape is convolved with the experimental resolution.

2.2 The spin-0 case

Spin-0 $\gamma\gamma$ resonances are predicted in theories with an extended Higgs sector [11–17]. The search for a spin-0 resonance uses a more restricted kinematic range for the photon selection. The background is estimated by fitting the diphoton invariant mass distribution to an analytical function, searching for an excess modelled by a spin-0 resonance convolved with the experimental resolution. The search is performed in the mass range 200–2000 GeV, and for width values up to 10% of the mass of the hypothesized particle.

3 ATLAS detector

The ATLAS detector [18] is a multi-purpose detector with a forward-backward symmetric cylindrical geometry.[1] At small radii, the inner detector (ID) is immersed in a 2 T magnetic field and is made up of fine-granularity pixel and microstrip detectors covering a pseudorapidity range $|\eta| < 2.5$. A gas-filled straw-tube transition radiation tracker (TRT) complements the silicon tracker at larger radii and also provides electron identification capabilities based on transition radiation. The electromagnetic (EM) calorimeter is a lead/liquid-argon sampling calorimeter with accordion geometry. The calorimeter is divided into a barrel section covering $|\eta| < 1.475$ and two end-cap sections covering $1.375 < |\eta| < 3.2$. Hadronic calorimetry in the region $|\eta| < 1.7$ uses steel absorbers and scintillator tiles as the active medium. Liquid-argon calorimetry with copper absorbers is used in the hadronic end-cap calorimeters, which cover the region $1.5 < |\eta| < 3.2$. A forward calorimeter using copper or tungsten absorbers with liquid argon completes the calorimeter coverage up to $|\eta| = 4.9$. The muon spectrometer measures the deflection of muon tracks within $|\eta| < 2.7$. The deflection is provided by a toroidal magnetic field from air-core superconducting magnets, with an integral of approximately 3 T·m and 6 T·m in the central and end-cap regions, respectively. The muon spectrometer is instrumented with trigger chambers covering $|\eta| < 2.4$. Events are selected using a first-level trigger implemented in custom electronics, which reduces the event rate to a design value of 100 kHz. Software algorithms with access to the full detector information are then used in the high-level trigger to yield a recorded event rate of about 1 kHz [19].

4 Photon selection

Photon and electron candidates are reconstructed from clusters of energy deposited in the electromagnetic calorimeter. Candidates without a matching track or reconstructed conversion vertex in the inner detector are classified as unconverted photons. Those with a matching reconstructed conversion

[1]The ATLAS experiment uses a right-handed coordinate system with its origin at the nominal interaction point (IP) in the centre of the detector and the z-axis along the beam pipe. The x-axis points from the IP to the centre of the LHC ring, and the y-axis points upward. Cylindrical coordinates (r, ϕ) are used in the transverse plane, ϕ being the azimuthal angle around the beam pipe. The pseudorapidity is defined in terms of the polar angle θ as $\eta = -\ln\tan(\theta/2)$.

vertex or a matching track, consistent with originating from a photon conversion, are classified as converted photons. Those matched to a track consistent with originating from an electron produced in the beam interaction region are kept as electrons.

Only photon candidates with $|\eta| < 2.37$ are considered, not including the transition region $1.37 < |\eta| < 1.52$ between the barrel and end-cap calorimeters.

The efficiency of the photon identification increases with the transverse energy of the photon E_{T} from 85% at 50 GeV to 95% at 200 GeV. For $E_{\mathrm{T}} > 50$ GeV, the uncertainty in the photon identification efficiency varies between $\pm 1\%$ and $\pm 5\%$ depending on η and E_{T} [20].

To further reject the background from jets misidentified as photons, the photon candidates are required to be isolated using both calorimeter and tracking detector information. The calorimeter isolation variable, $E_{\mathrm{T}}^{\mathrm{iso}}$, is defined as the sum of the E_{T} of energy clusters deposited in a cone of size $\Delta R = \sqrt{(\Delta\eta)^2 + (\Delta\phi)^2} = 0.4$ around the photon candidate, excluding an area of size $\Delta\eta \times \Delta\phi = 0.125 \times 0.175$ centred on the photon cluster; the expected photon energy deposit outside the excluded area is subtracted. The pile-up and underlying-event contribution to the calorimeter isolation variable is subtracted from the isolation energy event-by-event [21–23]. The selection requirement on the calorimeter isolation variable is defined by $E_{\mathrm{T}}^{\mathrm{iso}} < 0.022 E_{\mathrm{T}} + 2.45$ GeV. The track isolation variable ($p_{\mathrm{T}}^{\mathrm{iso}}$) is defined as the scalar sum of the transverse momenta of the tracks in a cone of $\Delta R = 0.2$ around the photon candidate. The tracks are required to have $p_{\mathrm{T}} > 1.0$ GeV and to be consistent with originating from the diphoton primary vertex. For converted photons, the one or two tracks associated with the photon conversion are excluded from the $p_{\mathrm{T}}^{\mathrm{iso}}$ computation. The requirement applied for the track isolation variable is $p_{\mathrm{T}}^{\mathrm{iso}} < 0.05 E_{\mathrm{T}}$.

5 Data and simulated event samples

Data were collected by the ATLAS detector in 2015 using pp collisions at a centre-of-mass energy of $\sqrt{s} = 13$ TeV. Events were recorded using a diphoton trigger with E_{T} thresholds of 35 GeV and 25 GeV for the leading and subleading photon candidates, respectively. The trigger has a signal efficiency close to 99% for events fulfilling the final event selections. After data-quality requirements, the data sample corresponds to an integrated luminosity of 3.2 fb^{-1}. The measurement of the integrated luminosity has an uncertainty of $\pm 5\%$ [24].

Simulated Monte Carlo (MC) samples are used to optimize the search strategy and to study background sources.The generated events are passed through a full detector simulation [25] based on Geant4 [26]. Pile-up from additional pp collisions in the same and neighbouring bunch crossings is simulated by overlaying each MC event with a variable number of simulated inelastic pp collisions generated using Pythia8 with the A2 tune [27]. The MC events are weighted to reproduce the distribution of the average number of interactions per bunch crossing observed in the data.

5.1 Spin-2 signal

Signal samples for the RS graviton model are generated using Pythia8 [28], version 8.186, with the NNPDF23LO [29] parton distribution functions (PDF) and the A14 [30] set of tuned parameters for the underlying event, for different choices of the graviton mass and the parameter $k/\overline{M}_{\mathrm{Pl}}$. A mass range from 500 GeV to 5000 GeV and $k/\overline{M}_{\mathrm{Pl}}$ values from 0.01 to 0.3 are chosen. Samples for any mass or $k/\overline{M}_{\mathrm{Pl}}$ value are obtained by reweighting an event sample generated with a uniform mass distribution using the Breit–Wigner and parton luminosity terms.

5.2 Spin-0 signal

The signal in the spin-0 particle search is simulated as if it were a SM Higgs boson produced in pp collisions via gluon fusion and decaying into two photons. MC samples are produced for different hypotheses of the spin-0 boson mass (m_X) in the range 200 GeV to 2000 GeV and of the decay width (Γ_X) up to 10% of m_X. Gluon fusion events are generated with Powheg-box [31, 32], version 2, interfaced with Pythia8 for the underlying event, parton showering and hadronization. To model signals with large decay widths, a function parameterizing the theoretical line-shape of the resonance is used. The Powheg-box implementation of a large-width spin-0 resonance with couplings like those of the Higgs boson in the SM is chosen for this function.

5.3 Background

Events containing two prompt photons, representing the largest irreducible background to the search, are simulated using the Sherpa [33] generator version 2.1.1. Matrix elements are calculated with up to two partons at leading order in QCD and merged with the Sherpa parton shower [34] using the ME+PS@LO prescription [35]. Samples of the photon+jet reducible background component are also generated using Sherpa.

6 Event selection

For the spin-2 resonance, the transverse energy of each photon is required to satisfy $E_T > 55$ GeV. For the spin-0 resonance , we apply tighter selections which were optimized on simulated background and signal samples. For a given value of $m_{\gamma\gamma}$, the transverse energy is required to be $E_T > 0.4m_{\gamma\gamma}$ for the photon with the highest E_T and $E_T > 0.3m_{\gamma\gamma}$ for the photon with the second-highest E_T.

The selected samples consist of events from diphoton production, followed by photon+jet production, with one jet misidentified as a photon, and dijet production with two jets misidentified as photons. Background sources from Drell–Yan, $W\gamma$ or $Z\gamma$ production, with either one or two isolated electrons misidentified as photons, are negligible.

Control regions are built from events failing the isolation requirement and/or some of the tight photon identification requirements, Ref. [36, 37]. They are used to estimate the relative contribution of the various sources of background directly from data. Figure 1 shows the decomposition of the selected data sample into the contributions from diphoton, photon+jet or jet+photon, and dijet events for both selections and the corresponding purities, defined as the ratio of diphoton events to the total number of events in the sample. The purity is $(94^{+3}_{-7})\%$ for the spin-2 selection and $(93^{+3}_{-8})\%$ for the spin-0 selection. The estimate of the uncertainties is sensitive to the small number of events in some of the control regions.

For the spin-2 resonance search, the acceptance for the benchmark RS graviton model varies from 66% at a mass of 500 GeV, to 91% at a mass of 5000 GeV. For the spin-0 resonance search, the acceptance ranges from 52% to 62% in the mass range from 200 GeV to 700 GeV for a particle similar to a Higgs boson produced by gluon fusion and is almost constant above 700 GeV.

7 Signal modelling

The invariant mass distribution of the diphoton pair for the signal is expected to peak near the assumed mass of the new resonance, with a spread given by the convolution of its intrinsic decay width with the experimental resolution. For both searches, the experimental resolution of the invariant mass is

DOI: 10.1051/epjconf/201716401027

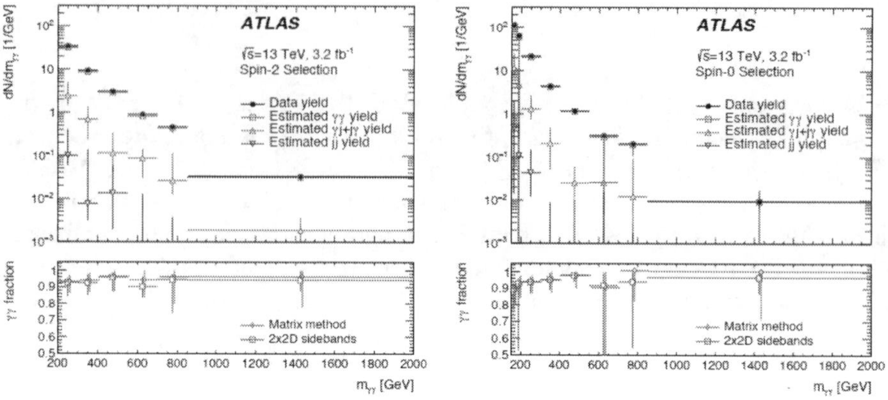

Figure 1. The diphoton invariant mass distributions (upper panels) of the data for the spin-2 and spin-0 selections and their decomposition into contributions from genuine diphoton, photon+jet plus jet+photon and jet+jet events. The bottom panels show the purity of diphoton events [6]. The total uncertainties are shown, including statistical and systematic components.

modelled with a double-sided Crystal Ball (DSCB) function. Interference effects between signal and background are ignored.

For the spin-2 resonance search, the signal mass distribution for any value of the mass and $k/\overline{M}_{\rm Pl}$ is obtained by a convolution of the intrinsic detector resolution, modelled by a DSCB function, with the predicted distribution of the mass line-shape at generator level.

For the spin-0 resonances with larger natural widths, the reconstructed line-shapes are well described by DSCB functions. The function effectively parameterizes the combined effects of the theoretical line-shape and the detector response for a signal with width values up to 10% of the resonance mass.

8 Background estimates

Two different methods are used to estimate the background contributions to the $m_{\gamma\gamma}$ distribution: MC extrapolation method and functional-form approach. In the spin-2 search, the shape of the invariant mass distribution of the main diphoton background is predicted using the next-to-leading (NLO) order in QCD DIPHOX [38] computation, version 1.3.2. The background from photon+jet and dijet production is added using control samples from the data. The second approach, more appropriate for the spin-0 search, is based on using a smooth functional form, with fully data-driven parameters to model the total background. In this approach, the mass distribution from data is fitted in the range above 150 GeV.

The two methods used for the background estimate were compared in the mass range where they are both used (500-5000 GeV for spin-2 and 200-2000 GeV for spin-0). For the MC extrapolation method, the statistical uncertainty is directly related to the total number of events in the data sample and thus is at the level of $\pm 1.5\%$ for the spin-2 resonance search selection. For the functional-form approach, the statistical uncertainty originating from the uncertainty in the determination of the parameters of the function from the fit to the data is larger, especially at high masses while the systematic uncertainty is given directly by the "spurious" signal uncertainty, which depends on the signal hypothesis being considered. This systematic uncertainty is lower than that of the MC extrapolation approach. For all masses, the total uncertainty in the background estimate is significantly smaller than the expected statistical fluctuations of the background yield in the different signal regions being considered. There is good agreement between the background predictions from the two methods.

9 Results

The diphoton mass distributions from the two event samples are fitted assuming the background-only or signal-plus-background hypotheses. The uncertainties related to the overall normalization of the signal yield only affect the limits on the production cross sections, while the uncertainty in the signal mass resolution affects both the compatibility with the background-only hypothesis and the limits on the production cross section. The signal mass resolution uncertainty ranges, depending on the mass, from 20 to 60% for spin-2 and from 30 to 60% for spin-0. The uncertainties in the background estimates affect both the compatibility with the background-only hypothesis and the cross-section limits. The relative systematic uncertainty in the background after the fit to the data is typically ±2% to ±4% at masses near 750 GeV depending on the selection and on the method used to compute the background.

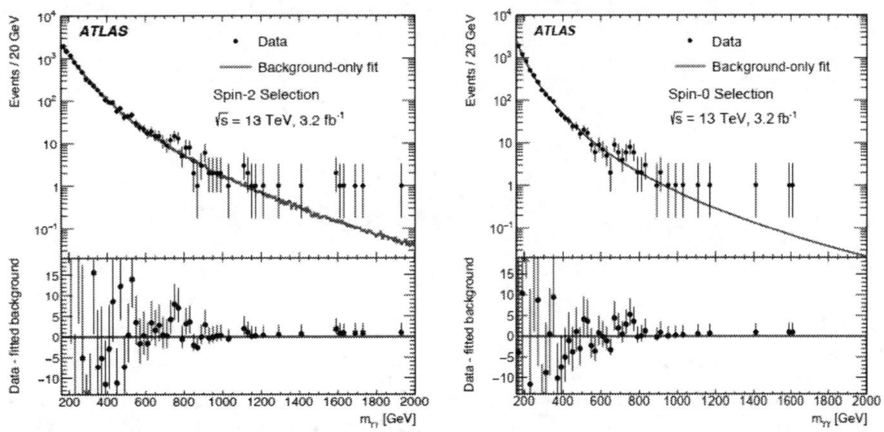

Figure 2. Distribution of the diphoton invariant mass for the selection used in the search for a spin-2 and a spin-0 resonance, with the best background-only fit [6]. The difference between the data and the fit is shown in the bottom panel for $m_{\gamma\gamma}$> 200 GeV. The arrow shown in the lower panel indicates a values outside the range with more than one standard deviation. There is no data event with $m_{\gamma\gamma}$> 2000 GeV.

Figure 2 shows the diphoton invariant mass distributions for the events selected in the spin-2 (spin-0) resonance search together with the best background-only fit using the MC extrapolation approach (the functional-form approach). For the spin-2 case, the largest deviation from the background-only hypothesis is observed near a mass of 750 GeV, for a $k/\overline{M}_{\rm Pl}$ value of 0.23, corresponding to a local excess of 3.8 standard deviations. The width associated with $k/\overline{M}_{\rm Pl}$ = 0.23 at m_{G^*} = 750 GeV is 57 GeV. The global significance evaluated using the search region of 500–2000 GeV in mass and 0.01–0.3 in $k/\overline{M}_{\rm Pl}$ is 2.1 standard deviations. For the spin-0 resonance search, the largest deviation is also observed near a mass of 750 GeV. It corresponds to a local excess over the background-only hypothesis with a significance of 3.9 standard deviations for a width of 45 GeV. The global significance evaluated using the search region of 200–2000 GeV in mass and 0%–10% in Γ_X/m_X is 2.1 standard deviations.

The compatibility of the excesses observed in the two analyses is assessed with a bootstrap statistical procedure, assuming a common signal. If the spin-2 (spin-0) signal is assumed, the two analyses are compatible within 0.5 (0.2) standard deviations.

10 Cross-section limits

Since for the spin-2 case a larger model dependence exists and the analysis is performed for a specific benchmark model of a spin-2 graviton, limits on the total cross section times branching ratio to two photons are given assuming specific model parameters. Figure 3 is shown for a $k/\overline{M}_{\rm Pl}$ value of 0.1. The predicted cross sections are computed at LO in QCD using PYTHIA8. The uncertainty band represents the PDF uncertainty estimated from the variations of the NNPDF23LO PDF set. Outside of the excess region, the observed limit on the cross section times branching ratio is of 1 fb for an RS graviton mass between 500 GeV and 5000 GeV for the $k/\overline{M}_{\rm Pl} = 0.1$ and the graviton mass is excluded to 3.2 TeV, a neat improvement from the 8 TeV reults (2.66 TeV).

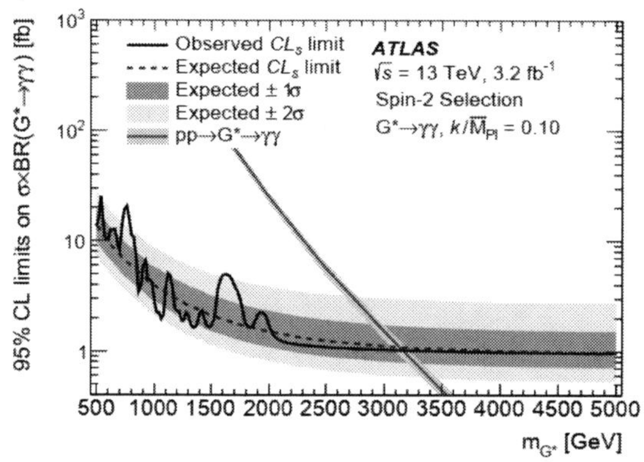

Figure 3. Upper limits on the production cross section of an RS graviton as a function of the assumed mass, for a $k/\overline{M}_{\rm Pl}$ value of 0.1 [6]. The predicted cross section times branching ratio to two photons for the RS graviton model, computed at LO, is also shown. The uncertainty in the cross-section values represent the PDF uncertainty.

For the spin-0 case, limits are interpreted in a nearly model-independent way in terms of the fiducial cross section, defined as the product of the cross section times the branching ratio to two photons within the fiducial acceptance defined. Figure 4 shows the limits on the signal fiducial cross section times branching ratio to two photons as a function of the hypothesized mass for various assumptions about the width. Except near 750 GeV, the observed limit is in agreement with the expected limit assuming the background-only hypothesis.

11 Conclusion

Search for high mass resonances decaying into two photons in the ATLAS experiment at the LHC are presented. The analyses are performed with pp collision data corresponding to an integrated luminosity of 3.2 fb^{-1} at a centre-of-mass energy of $\sqrt{s}$ = 13 TeV, recorded in 2015. The analyses

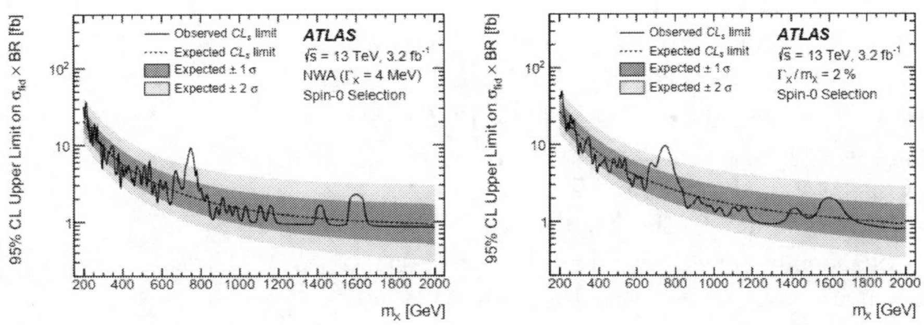

Figure 4. Upper limits on the fiducial cross section at $\sqrt{s}$ = 13 TeV of a spin-0 particle as a function of the assumed mass m_X, for different values of the decay width divided by the mass [6]. In (Left) picture a narrow-width signal, with Γ = 4 MeV, is assumed.

were optimized for two benchmark searches: spin-2 Randall–Sundrum graviton resonance with mass above 500 GeV and spin-0 resonance with mass above 200 GeV. The data are consistent with the background-only hypothesis and 95% CL limits are derived on the cross section for the production of the two benchmark resonances as a function of their masses and widths. The largest deviation from the background-only hypothesis is observed in a broad region near a mass of 750 GeV and with a width of about 50 GeV, with global significances of 2.1 standard deviations for both searches.

The intriguing resonance at 750 GeV decaying into photon pairs, which caused considerable interest from the 2015 data, has not reappeared in the much larger 2016 data set and thus appears to be a statistical fluctuation as shown in 2016 summer conferences [39].

References

[1] ATLAS Collaboration, New J. Phys. **15** (2013) 043007.
[2] CMS Collaboration, Phys. Rev. Lett. **108** (2012) 111801.
[3] ATLAS Collaboration, Phys. Rev. D **92** (2015) 032004.
[4] ATLAS Collaboration, Phys. Rev. Lett. **113** (2014) 171801.
[5] CMS Collaboration, Phys. Lett. B **750** (2015) 494.
[6] ATLAS Collaboration, JHEP **09** (2016) 001.
[7] CMS Collaboration, Phys. Rev. Lett. **117** (2016) 051802.
[8] L. Randall and R. Sundrum, Phys. Rev. Lett. **83** (1999) 3370.
[9] T. Appelquist, A. Chodos and P. Freund, *Modern Kaluza-Klein theories*, vol. **65**, Frontiers in Physics, Addison-Wesley, 1987.
[10] H. Davoudiasl, J. Hewett and T. Rizzo, Phys. Rev. Lett. **84** (2000) 2080.
[11] A. Hill and J.J. van der Bij, Phys. Rev. D **36** (1987) 3463.
[12] M. J. G. Veltman and F. J. Yndurain, Nucl. Phys. B **325** (1989) 1.
[13] T. Binoth and J.J. van der Bij, Z. Physics. C **75** (1997) 17.
[14] R. Schabinger and J. D. Wells, Phys. Rev. D **72** (2005) 093007.
[15] B. Patt and F. Wilczek, arXiv: hep-ph/0605188.
[16] G. M. Pruna and T. Robens, Phys. Rev. D **88** (2013) 115012.
[17] T. D. Lee, Phys. Rev. D **8** (1973) 1226.
[18] ATLAS Collaboration, JINST **3** (2008) S08003.
[19] ATLAS Collaboration, ATL-DAQ-PUB-2016-001, url: http://cds.cern.ch/record/2136007.
[20] ATLAS Collaboration, arXiv: 1606.01813 [hep-ex].
[21] ATLAS Collaboration, Phys. Rev. D **83** (2011) 052005.
[22] M. Cacciari, G. P. Salam and G. Soyez, JHEP **04** (2008) 005.
[23] M. Cacciari, G. P. Salam and S. Sapeta, JHEP **10** (2010) 065.
[24] ATLAS Collaboration, Eur. Phys. J. C **73** (2013) 2518.
[25] ATLAS Collaboration, Eur. Phys. J. C **70** (2010) 823.
[26] S. Agostinelli et al., Nucl. Instrum. Meth. A **506** (2003) 250.
[27] ATLAS Collaboration, ATL-PHYS-PUB-2012-003, url: http://cds.cern.ch/record/1474107.
[28] T. Sjöstrand, S. Mrenna and P. Z. Skands, Comput. Phys. Commun. **178** (2008) 852.
[29] S. Carrazza, S. Forte and R. Rojo, arXiv: 1311.5887 [hep-ph].
[30] ATLAS Collaboration, ATL-PHYS-PUB-2014-021, url: http://cds.cern.ch/record/1966419.
[31] S. Alioli et al., JHEP **06** (2010) 043.
[32] E. Bagnaschi et al., JHEP **02** (2012) 088.
[33] T. Gleisberg et al., JHEP **02** (2009) 007.
[34] S. Schumann and F. Krauss, JHEP **03** (2008) 038.
[35] S. Höche et al., JHEP **05** (2009) 053.
[36] ATLAS Collaboration, JHEP **01** (2013) 086.
[37] ATLAS Collaboration, Phys. Rev. D **85** (2012) 012003.
[38] T. Binoth et al., Eur. Phys. J. C **16** (2000) 311.
[39] ATLAS Collaboration, ATLAS-CONF-2016-059, url: http://cds.cern.ch/record/2206154.

EPJ Web of Conferences **164**, 01028 (2017) DOI: 10.1051/epjconf/201716401028
ICNFP 2016

Search for Supersymmetry at the LHC

Christian Autermann[1,a]
on behalf of the ATLAS and CMS collaborations

[1] *I. Phys. Inst. B, RWTH Aachen University, Germany*

Abstract. Searches for supersymmetry at the ATLAS and CMS experiments are discussed. The analyzed data were recorded at the Large Hadron Collider in 2015 at a center of mass energy of 13 TeV and correspond to an integrated luminosity of up to 3 fb^{-1} per experiment. No physics beyond the standard model has been observed. Different inclusive and specialized analysis strategies targeting various signal scenarios are introduced. The analysis sensitivities are summarized as exclusion contours for different simplified signal model spectra and compared to the previous 8 TeV results.

1 Introduction

After the long shutdown, the Large Hadron Collider (LHC) at CERN has delivered in 2015 the first proton-proton collision data at a center of mass energy of 13 TeV. The two experiments, ATLAS [1] and CMS [2], have currently analyzed the data corresponding to an integrated luminosity of up to 3 fb^{-1}. The collaborations focus strongly on searches for physics beyond the standard model, especially after the discovery of the Higgs boson [3, 4]. The sensitivity to signals of new physics scenarios already matches or exceeds the results obtained on the 8 TeV data set collected until 2012 corresponding to up to 20 fb^{-1}.

Supersymmetry (SUSY) [5] is a very popular theory for physics beyond the standard model, developed since the 1970s as an extension of the Poincaré symmetry. SUSY links each particle of the standard model (SM) with a supersymmetric partner with equal properties, except for the spin which differs by 1/2 and the mass. The supersymmetry fixes the hierarchy problem associated with the divergent first-loop order correction terms to the Higgs-mass parameter m_H^2. SUSY provides a dark matter candidate particle if the lightest supersymmetric particle (LSP) is neutral and stable, e.g. due to R-parity conservation. The unification of gauge couplings at a high energy scale becomes possible in supersymmetry, motivating believes in the existence of grand unified theories.

The standard model of particle physics is very successfully describing the fundamental particles and their interactions studied at the collider experiments. Excellent agreement between theory predictions and measurements at the ATLAS and the CMS experiments is observed over all studied production and decay channels. A particular example is the observation of the $B_s^0 \to \mu^+\mu^-$ decay and the measurement of its branching fraction [6]. This decay is rare in the SM because it is suppressed at loop- or box-diagram level to an order, at which also new particles such as predicted from supersymmetry may contribute and interfere. The measured branching fractions are consistent with standard

[a]e-mail: auterman@cern.ch

DOI: 10.1051/epjconf/201716401028

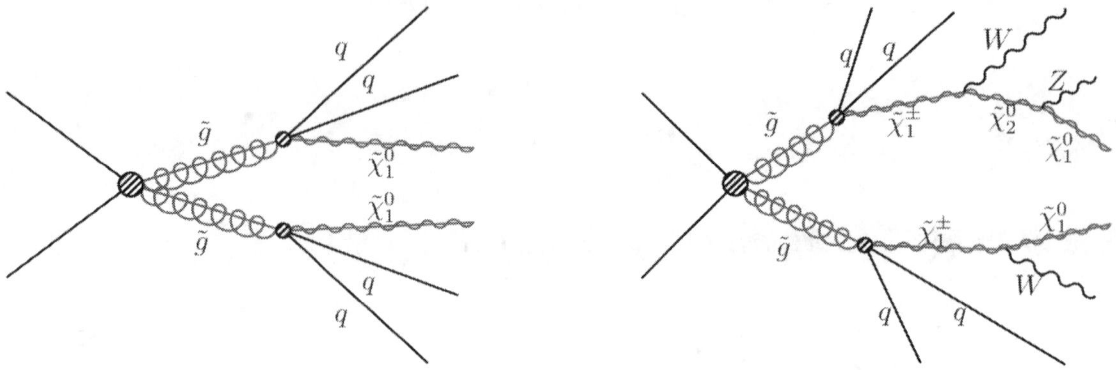

(a) $\tilde{g}\tilde{g}$, decay via off-shell squarks (b) $\tilde{g}\tilde{g}$, decay via off-shell squarks and heavy gauginos

Figure 1. Effective Feynman diagrams for strong production of gluinos leading to highly energetic jets (a) or to final states with large jet multiplicities (b), depending on the gluino and neutralino masses.

model expectations and can be used to constrain models of supersymmetry [7]. In order to interpret the analyis results generally simplified model spectra (SMS) [8, 9] are used, which allow to model individual decay topologies through effective on-shell theories, involving only the masses of typically two SUSY particles as free parameters.

The good understanding of SM physics requires dedicated searches for new physics. Different strategies for searches for supersymmetry are discussed, from very inclusive searches for strong production in hadronic final states to more specific signatures such as dilepton resonances and kinematic edges are discussed in the following.

2 Direct inclusive searches for strong production of SUSY

Strong production of supersymmetry and decay channels into fully hadronic final states, as shown in Fig. 1, typically have the largest production cross sections and branching fractions and offer therefore generally good signal sensitivities. As SUSY is a broken symmetry, the SUSY partner masses in the supermultiplets differ. Supersymmetric particle masses are typically heavy, due to existing direct limits and SM precision measurements. Events containing SUSY particles contain therefore more energy than the SM background and accumulate in the high energy tails of distributions like E_T^{miss}. Various other signal sensitive variables such as M_{T2} [10], α_T [11], or "Razor" [12, 13] are used to enrich the signal selection. The classic variable missing transverse energy E_T^{miss}, utilized e.g. in [14, 15], is defined as the absolute value of the vectorial sum of all reconstructed particle momenta, and is therefore sensitive to the amount of energy of particles that did not interact with the detector, such as neutrinos or the supersymmetry LSPs.

The precise estimation of the standard model background contributions to the tails of the signal sensitive variables is of crucial importance to achieve sensitive results. In the case of the fully hadronic final states this SM background is due to QCD multijet production, where the artificial E_T^{miss} is caused mostly by jet energy resolution effects, and by $t\bar{t}$, W+jets, and $Z \to \nu\bar{\nu}$ processes with real E_T^{miss} from hard neutrinos and no reconstructed leptons. Especially the QCD multijet processes are hard to simulate with Monte Carlo methods, as on one hand the various effects leading to E_T^{miss} are very rare and hard to model and on the other hand the background production cross section is very large requiring large statistics.

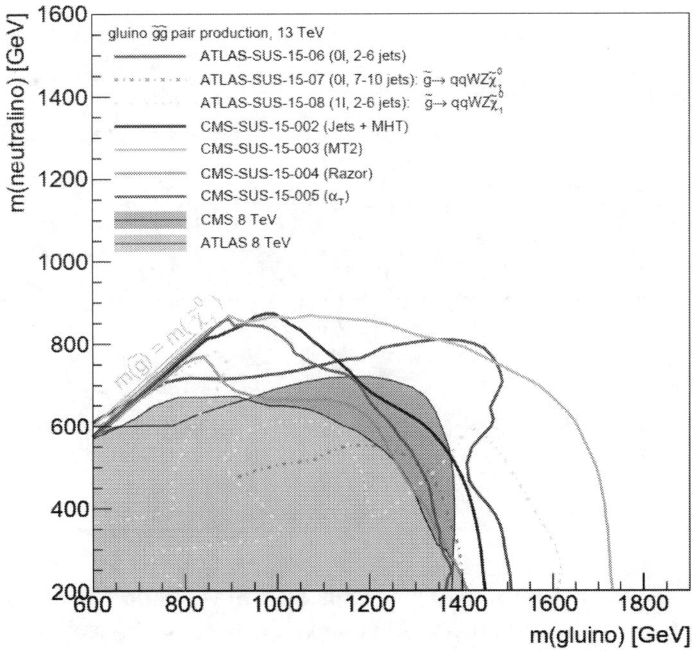

Figure 2. Resulting exclusion contours for inclusive searches in the fully-hadronic final state for simplified scenarios of gluino-pair production with either short decays via off-shell squarks $\tilde{g} \to qq\tilde{\chi}_1^0$ into highly energetic jets or longer decays $\tilde{g} \to qqZ\tilde{\chi}_1^0$ leading to higher jet multiplicities. ATLAS exclusion contours are taken from [14, 16, 17], CMS from taken from [10–12, 15], the gray shaded areas correspond to the envelope of all 8 TeV contours, taken from [18–20].

The estimation of the standard model background is often improved by the use of data sideband regions, where either the Monte Carlo simulation is corrected and validated, or events kinematically similar to the background in the signal region are extracted. The discussed results from the same experiment, the same signal scenario, and therefore a similar final state are not necessarily strongly correlated, because of the set of different signal-sensitive variables and the rich variety of data-driven background estimation strategies.

In Fig. 2 the exclusion contours with respect to the gluino and the lightest neutralino masses for gluino-pair production scenarios are shown for the different inclusive supersymmetry search analyses performed at ATLAS with data corresponding to an integrated luminosity of 3.2 fb^{-1} and at CMS with 2.2 fb^{-1} of data. The best sensitivity of the analyses performed on data collected at a center-of-mass energy of 8 TeV and about 20 fb^{-1} of data are shown as a gray area [20]. Two $\tilde{g}\tilde{g}$ SMS with different gluino decays leading to different final state particle topologies are shown. Results for long cascades, where the gluinos decay through charginos into the neutralino LSP lead to many jets and W/Z-bosons in the final state and are shown dashed. High jet multiplicities are targeted by [16], while the W/Z may lead to leptons which are analyzed by [17, 21]. Direct gluino decays $\tilde{g} \to qq\tilde{\chi}_1^0$ through squarks assumed to be off-shell lead to rather low multiplicities of energetic jets, significant E_T^{miss} due to the two neutralinos $\tilde{\chi}_1^0$ LSPs, and no leptons. This challenging final state requires analyses

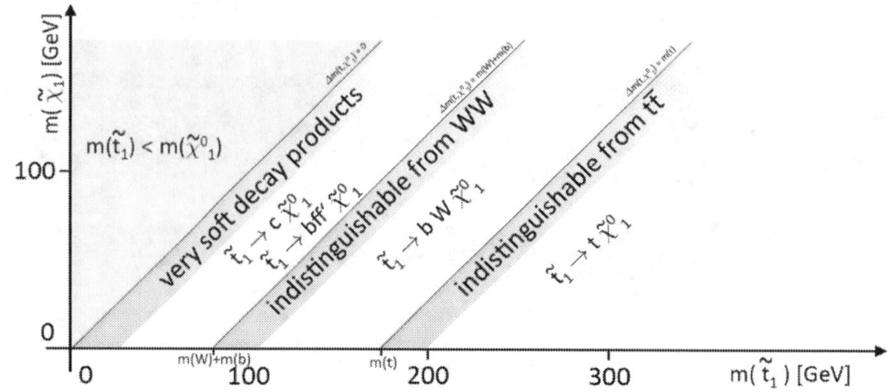

Figure 3. Phase-space region of the stop decay for $\tilde{t}_1\tilde{t}_1$ pair production in the m($\tilde{t}_1$) and m($\tilde{\chi}^0_1$) plane [20]. The decay topology and the kinematics of the decay products differ significantly, requiring a rich set of experimental analyses.

with sophisticated background estimation strategies [10–15]. While CMS typically uses multiple exclusive signal region combined into one likelihood for best signal sensitivity, ATLAS typically uses one dedicated single-channel region per signal scenario to allow for easier reinterpretations of the result. Further interpretations in more simplified scenarios and cross-section limits can be found under the quoted references.

3 Third generation squark searches

Naturalness arguments motivate searches for direct production of third generation squarks, i.e. top and bottom squarks. The third generation has strong couplings to the Higgs sector, therefore third generation squarks are required to remain light, at the order of the TeV scale, in order to not reintroduce a new little hierarchy problem. Pair production of top squarks may lead to different decay channels depending on the mass difference to the lightest neutralino Δm = m($\tilde{t}$) − m($\tilde{\chi}^0_1$) as illustrated in Fig. 3. Different analysis strategies are developed, to cover the stop and neutralino mass space with optimal sensitivity.

In the case of large mass differences Δm between the stop and the neutralino on-shell top-quarks are produced in the $\tilde{t}_1 \to t\tilde{\chi}^0_1$ decay, as shown in Fig. 4. The SM top transverse momentum p_T depends on Δm, so that the signal can be distinguished from SM top-pair production. For smaller Δm on-shell top quarks or on-shell W-bosons may no longer occur in the stop decays $\tilde{t}_1 \to bff'\tilde{\chi}^0_1$, leading to final states with only a small amount of visible energy. If the mass gap Δm is even smaller than the b-quark mass, only the flavor changing decay $\tilde{t}_1 \to c\tilde{\chi}^0_1$ into a c-jet of very low energy and $E_T^{\rm miss}$ is possible. In particular if the mass gap Δm is equal to the top-mass or to the W-boson mass it becomes very difficult to distinguish the potential $\tilde{t}_1\tilde{t}_1$ signal from standard model top-pair production or from SM W-pair production. Here, dedicated analyses using e.g. the spin-spin correlation or the SM top-pair cross-section are developed.

In the first 3 fb^{-1} of the 13 TeV data several searches, comparable to the inclusive analyses discussed before but using b-tagged jet information, have been performed in the all-hadronic final state [22, 23], and in the final state with one lepton [24, 25], or with two leptons [26]. These analyses

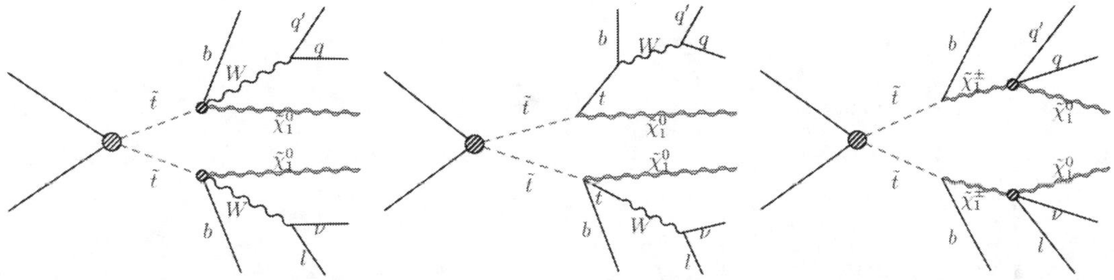

(a) $\tilde{t}\tilde{t} \to bbWW\tilde{\chi}_1^0\tilde{\chi}_1^0$ (b) $\tilde{t}\tilde{t} \to tt\tilde{\chi}_1^0\tilde{\chi}_1^0$ (c) $\tilde{t}\tilde{t} \to b\bar{b}\tilde{\chi}_1^{\pm}\tilde{\chi}_1^{\pm}$

Figure 4. Effective Feynman diagrams for direct stop production and the following stop three body decay into the neutralino, a b-quark and an on-shell W-boson (a), an on-shell top (b), or through trough a chargino $\tilde{t} \to b\tilde{\chi}_1^{\pm}$ followed by a three-body decay of the $\tilde{\chi}_1^{\pm}$ (c).

probe the phase-space where the standard model particles from the stop decay have significant energy and can be clearly separated from SM processes, as illustrated in Fig. 5. The lightest stop mass up to 800 GeV depending on m($\tilde{\chi}_1^0$) can be excluded for simplified $\tilde{t}_1\tilde{t}_1$-production scenarios.

At low Δm, the decay products of the off-shell top-quarks and W-bosons have only low energy and are hard to identify. Analyses [27] of final states with at least one soft lepton but large missing transverse momentum can probe top squark masses below 340 GeV for small mass gaps of Δm = 50 GeV. Mono-jet analyses [28] provide additional sensitivity. Here, the most energetic jet is most likely due to initial-state radiation, while the visible standard model particles of the stop quark decay do not pass the reconstruction thresholds. Previous analyses carried out on the 8 TeV data [20] shown as gray area in Fig. 5 made also use of c-tagging algorithms, that attempt to identify the low energetic c-jet of the flavor changing stop decay possible at low Δm.

4 Searches for gauge-mediated broken supersymmetry

In models of gauge-mediated supersymmetry breaking the lightest SUSY particle is the gravitino $\tilde{G}$, the spin-3/2 partner of the spin-2 graviton. The experimental signature of these models is determined by the nature of the next-to-lightest SUSY particle (NLSP). The results discussed in the following assume that the NLSP is the lightest neutralino $\tilde{\chi}_1^0$. The neutralino decays $\tilde{\chi}_1^0 \to B\tilde{G}$ into the gravitino LSP and a neutral standard model boson B, which can be either γ, Z^0, or H as illustrated in Fig. 6. The branching fraction depends on the neutralino mass and mixing, the $\tilde{\chi}_1^0$ lifetime depends on the gravitino mass. The m($\tilde{G}$) and thus the $\tilde{\chi}_1^0$ lifetime are assumed to be negligible.

Bino- and wino-like neutralino mixings may lead to up to two photons in the final state. The analyses [29, 30] carried out on the first 3 fb^{-1} of 13 TeV data already exceed the sensitivity of previous analyses [20] as shown in Fig. 7. Final states with photons arise in the standard model from QCD multijet like γ+jet production, initial-, and final state radiation of photons. Additionally, misidentified jets or electrons may be reconstructed as photons. The "fake photon" background from misidentified electrons can be modeled using a sample of isolated well identified electrons, that pass all signal selection cuts, except for the photon identification. This sample is then weighted according to the electron identification ϵ_e efficiency and the electron misidentification probability $f_{e\to\gamma}$. Because $f_{e\to\gamma}$ is about 50 times smaller than ϵ_e, the statistical precision of this data-driven approach is very good.

DOI: 10.1051/epjconf/201716401028

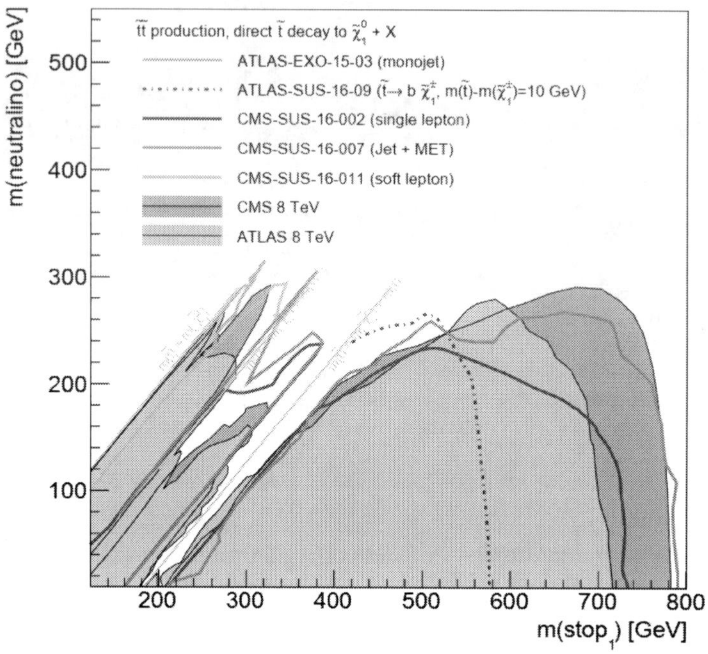

Figure 5. Selected ATLAS [28] and CMS [23, 25, 27] exclusion contours for top squark pair production $\tilde{t}_1\bar{\tilde{t}}_1$ directly decaying via on- or off-shell top-quarks and W-bosons $\tilde{t}_1 \rightarrow bff'\tilde{\chi}_1^0$ or results [26] for decays via charginos $\tilde{t}_1 \rightarrow b\tilde{\chi}_1^{\pm} \rightarrow bff'\tilde{\chi}_1^0$. The 8 TeV contour envelopes are from [18–20].

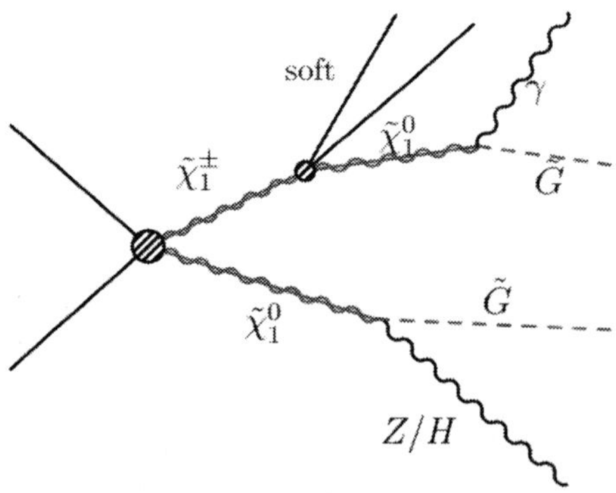

(a) $\tilde{g}\tilde{g}$, decay via off-shell squarks

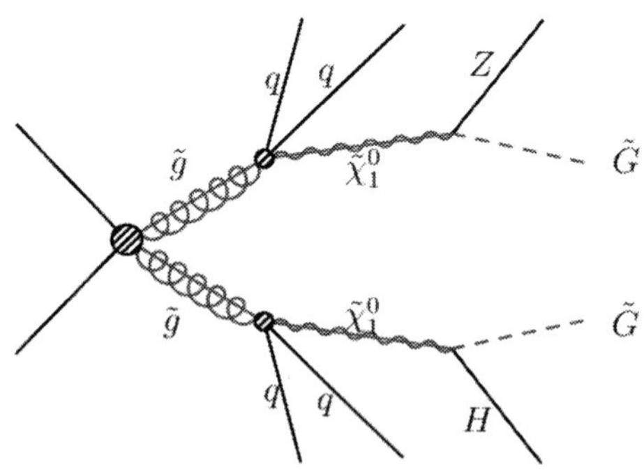

(b) $\tilde{g}\tilde{g}$, decay via off-shell squarks and heavy gauginos

Figure 6. Effective Feynman diagrams for SUSY models with gauge-mediated symmetry breaking for electroweak production (a) and strong production (b). The type of the standard model boson created together with the gravitino depends generally on the mixing of the neutralino mother particle.

 DOI: 10.1051/epjconf/201716401028

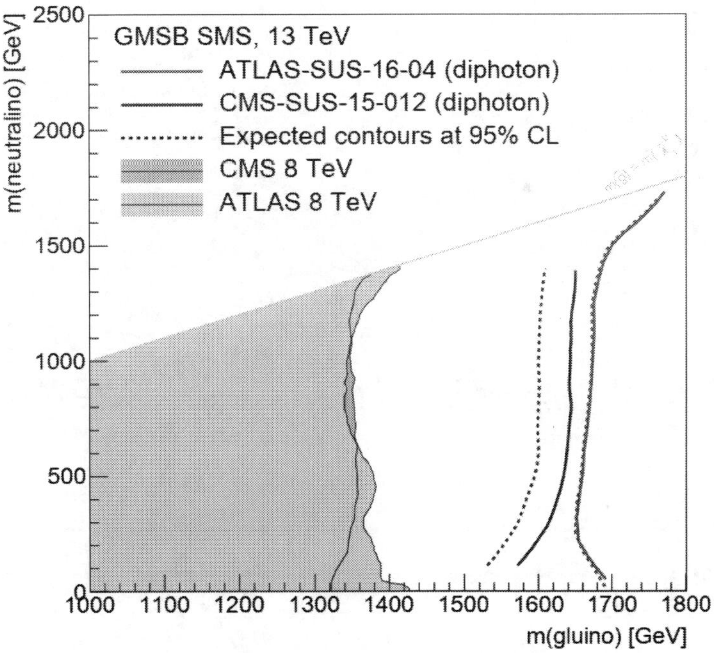

Figure 7. Resulting exclusion contours at 13 TeV [29, 30] compared to the results at 8 TeV [31, 32] for gluino-pair production $\tilde{g}\tilde{g}$ within gauge-mediated symmetry breaking SUSY scenarios with bino-like neutralinos and $\tilde{g} \to qq\tilde{\chi}_1^0 \to qq\gamma\tilde{G}$ leading to final states with diphotons, jets, and E_T^{miss}.

5 Dilepton mass-edge and resonance searches

Supersymmetry searches for dilepton resonances or mass-edges are typical examples for highly sensitive and optimized analyses for specific signal scenarios. The signal in the dilepton invariant mass m_{ll} spectrum has the form of a sharp peak or a triangular shape. Heavy neutral SUSY particles can decay into a light neutral particle such as the neutralino LSP via a slepton $\tilde{\chi}_2^0 \to l\tilde{l} \to ll\tilde{\chi}_1^0$ or a standard model boson $\tilde{\chi}_2^0 \to Z\tilde{\chi}_1^0 \to ll\tilde{\chi}_1^0$ as shown in Fig. 9(a). The slepton can be off- or on-shell, which influences slightly the curvature of the lower edge of the triangular shaped signal in the invariant dilepton mass spectrum. The position of the upper edge is determined by the mass difference of the two involved neutral SUSY particles $\Delta m = \text{m}(\tilde{\chi}_2^0) - \text{m}(\tilde{\chi}_1^0)$.

The prominent dilepton resonance or edge-like shape signal is an important feature of the analysis, but the excellent signal sensitivity is also due to the very reliable and precise estimation of the standard model backgrounds. The dilepton final state with E_T^{miss} and jets is dominated to a large fraction by standard model tt-production and similar backgrounds symmetric in lepton flavor. Lepton flavor-symmetric (SF) backgrounds contain as many $e^+e^- + \mu^+\mu^-$ events as there are $\mu^+e^- + \mu^-e^+$ events, while the signal does not contain opposite-flavor (OF) lepton events. The amount of flavor symmetric backgrounds in the SF signal region can therefore be estimated [33] from a OF control region. Differences in electron and muon efficiencies lead to small corrections that can be derived by explicitly measuring the relevant electron and muon efficiencies or by measuring the $R_{\text{SF/OF}}$ difference in data sidebands. The systematic uncertainty of this method, i.e. the uncertainty on $R_{\text{SF/OF}}$ is about 4%

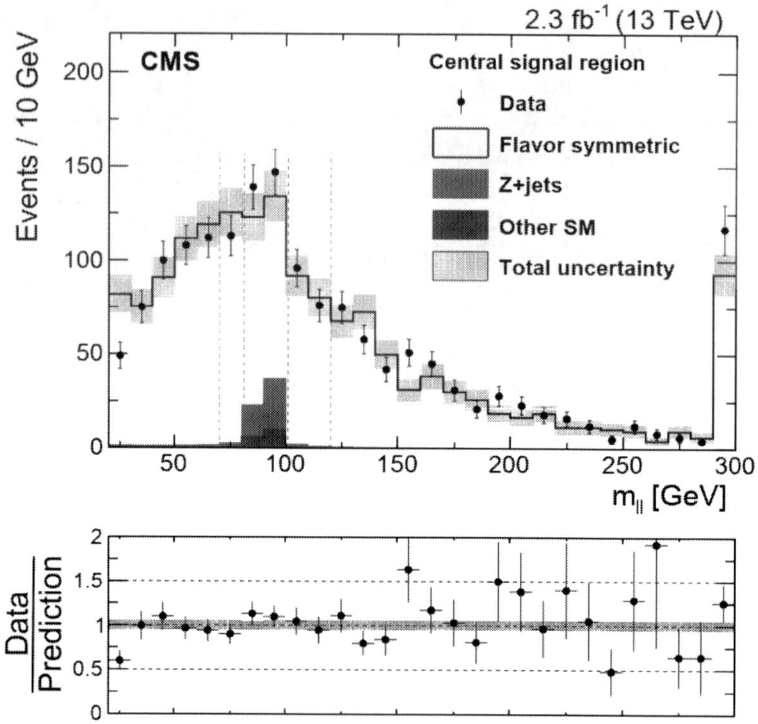

Figure 8. Invariant dimuon mass distribution for the data compared to the standard model background estimation [33] in the region, where the Run-I analysis [35] reported an excess using 20 fb^{-1} of 8 TeV data.

and of the same order as the statistical uncertainty of the OF control region, which is of the order of the irreducible statistical uncertainty of the data.

The precise determination of the subdominant $Z/\gamma^* \rightarrow l^+l^-$ background is relevant in particular for signal resonances on the Z-peak. Here [34], the similarity between γ-jet events used as data control sample with Z+jets events is exploited. In both processes the E_T^{miss} originates from similar mechanisms such as jet resolution and mismeasurements and not from genuine unobserved particles. The control sample is selected with the same kinematic cuts as applied on the data, and the events are reweighed to match the boson p_T distribution of the $Z\gamma^*$ events. Differences in resolution between photons and muons are taken into account. The E_T^{miss} of the events in the control sample is recalculated and is normalized to the data in a sideband region.

The analyses carried out in the 8 TeV data [35, 36] received some attention, because excesses of the order of $2.5 - 3.0\,\sigma$ were observed in the low invariant dilepton mass region at CMS, and on the Z-resonance at ATLAS, respectively. The ATLAS and CMS analyses have been repeated on the first set of data recorded at 13 TeV. ATLAS [34] has analyzed 3.2 fb^{-1} of data and finds good agreement of the data and background expectation in invariant dilepton mass regions below and above the Z-region, not confirming the excess observed by CMS in 8 TeV, but again finds an excess on the Z-peak corresponding to 2.2σ. CMS [33] has analyzed 2.2 fb^{-1} of 13 TeV data finding generally good agreement between data and standard model expectations as shown in Fig. 8, ruling out the possibility

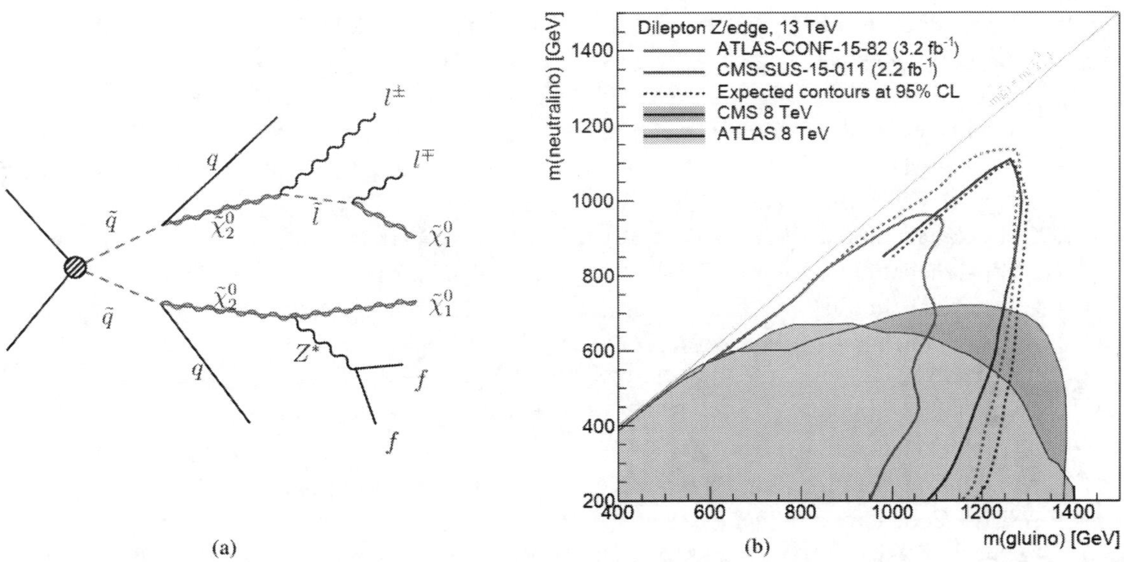

Figure 9. Effective Feynman diagram for sbottom pair production leading to a final state with dileptons (a). In the upper cascade a heavy neutralino $\tilde{\chi}_2^0$ decays via a slepton $\tilde{l}$ into a light neutralino $\tilde{\chi}_1^0$, the invariant mass of the dilepton system exhibits a characteristic triangular shaped form. In the lower cascade dileptons through the Z-resonance can be produced. The exclusion contours obtained at 13 TeV [33, 34] for a simplified scenario of gluino pair-production $\tilde{g}\tilde{g}$ with dilepton resonant decays $\tilde{g} \rightarrow qqZ\tilde{\chi}_1^0$ and negligible neutralino LSP masses $m(\tilde{\chi}_1^0) = 1$ GeV is compared to the 8 TeV contours [35, 36] in (b).

that the 8 TeV excesses were due to a SUSY signal. The resulting exclusion contours for a simplified scenario of gluino pair-production leading to resonant decays through Z-bosons are shown in Fig. 9(b) comparing the 8 TeV and 13 TeV results of ATLAS and CMS.

6 Conclusion

The ATLAS and CMS collaborations have searched in the first 3 fb^{-1} of data collected at 13 TeV for signs of supersymmetry. The observed data are in good agreement with the expectations from standard model processes. Limits on supersymmetric model parameters and masses in simplified model spectra have been derived that begin to exceed previous results obtained with data collected at lower center-of-mass energies. A wide range of signal scenarios, production and decay channels, and SUSY particle mass regions have been studied using sophisticated analysis methods in particular for the data-driven estimation of the SM backgrounds. The LHC continued to deliver about 40 fb^{-1} of 13 TeV data in 2016, that will be analyzed for the upcoming conferences. Natural supersymmetry is expected at the TeV scale, the data currently being collected will soon allow to challenge these scenarios.

References

[1] ATLAS Collaboration, JINST **3**, S08003 (2008)

[2] CMS Collaboration, JINST **3**, S08004 (2008)
[3] ATLAS Collaboration, Phys. Lett. B **716**, 1 (2013), hep-ex/1207.7214
[4] CMS Collaboration, Phys. Lett. B **716**, 30 (2012), hep-ex/1207.7235
[5] S.P. Martin, Adv. Ser. Direct. High Energy Phys. **18** (1997), hep-ph/9709.356
[6] LHCb, CMS Collaborations, Nature **522**, 68 (2015), hep-ex/1411.4413
[7] P. Bechtle et al. (2015), hep-ph/1508.05951
[8] N. Arkani-Hamed, P. Schuster, N. Toro, J. Thaler, L.T. Wang et al. (2007), hep-ph/0703.088
[9] J. Alwall, P. Schuster, N. Toro, Phys. Rev. D **79**, 075020 (2009), hep-ph/0810.3921
[10] CMS Collaboration, JHEP **10**, 006 (2016), hep-ex/1603.04053
[11] CMS Collaboration, Submitted to Eur. Phys. J. C (2016), hep-ex/1611.00338
[12] CMS Collaboration, Submitted to Phys. Rev. D (2016), hep-ex/1609.07658
[13] ATLAS Collaboration, Eur. Phys. J. **C73**, 2362 (2013), hep-ex/1212.6149
[14] ATLAS Collaboration, Eur. Phys. J. **C76**, 392 (2016), hep-ex/1605.03814
[15] CMS Collaboration, Phys. Lett. **B758**, 152 (2016), hep-ex/1602.06581
[16] ATLAS Collaboration, Phys. Lett. **B757**, 334 (2016), hep-ex/1602.06194
[17] ATLAS Collaboration, Eur. Phys. J. **C76**, 565 (2016), hep-ex/1605.04285
[18] ATLAS Collaboration, *Public Results* (2016), https://twiki.cern.ch/twiki/bin/view/AtlasPublic/SupersymmetryPublicResults
[19] CMS Collaboration, *Public Results* (2016), http://cms-results.web.cern.ch/cms-results/public-results/publications/
[20] C. Autermann, Prog. Part. Nucl. Phys. **90**, 125 (2016), hep-ex/1609.01686
[21] CMS Collaboration, subm. to Phys. Rev. D (2016), hep-ex/1609.09386
[22] CMS Collaboration, PAS SUS-16-001 (2016), https://cds.cern.ch/record/2140975
[23] CMS Collaboration, PAS SUS-16-007 (2016), https://cds.cern.ch/record/2141543
[24] ATLAS Collaboration, Phys. Rev. **D94**, 052009 (2016), 1606.03903
[25] CMS Collaboration, PAS SUS-16-002 (2016), https://cds.cern.ch/record/2139650
[26] ATLAS Collaboration, CONF-2016-009 (2016), https://cds.cern.ch/record/2139643
[27] CMS Collaboration, PAS SUS-16-011 (2016), https://cds.cern.ch/record/2161097
[28] ATLAS Collaboration, Phys. Rev. **D94**, 032005 (2016), hep-ex/1604.07773
[29] ATLAS Collaboration, Eur. Phys. J. **C76**, 517 (2016), 1606.09150
[30] CMS Collaboration, PAS SUS-15-012 (2016), https://cds.cern.ch/record/2143897
[31] ATLAS Collaboration, Phys. Rev. **D92**, 072001 (2015), 1507.05493
[32] CMS Collaboration, Phys. Rev. **D92**, 072006 (2015), 1507.02898
[33] CMS Collaboration, Accepted by JHEP (2016), hep-ex/1607.00915
[34] ATLAS Collaboration, CONF-2015-082 (2015), https://cds.cern.ch/record/2114854
[35] CMS Collaboration, JHEP **04**, 124 (2015), 1502.06031
[36] ATLAS Collaboration, Eur. Phys. J. **C75**, 318 (2015), [Erratum: Eur. Phys. J.C75,no.10,463(2015)], hep-ex/1503.03290

EPJ Web of Conferences **164**, 01029 (2017) DOI: 10.1051/epjconf/201716401029
ICNFP 2016

SHiP: a new facility with a dedicated detector to search for new neutral particles and studying tau neutrino properties

V.Shevchenko[1,2,a], on behalf of the SHiP Collaboration.

[1] *National Research Centre "Kurchatov Institute" ac.Kurchatova sq., 1, Moscow 123182 Russia*
[2] *Far Eastern Federal University, Sukhanova str. 8, Vladivostok 690950 Russia*

Abstract.
SHiP (Search for Hidden Particles) is a new general purpose fixed target facility, whose Technical Proposal has been recently reviewed by the CERN SPS Committee and by the CERN Research Board. The two boards recommended that the experiment proceeds further to a Comprehensive Design phase in the context of the new CERN Working group "Physics Beyond Colliders", aiming at presenting a CERN strategy for the European Strategy meeting of 2019. In the initial phase of SHiP, the 400 GeV proton beam extracted from the SPS will be dumped on a heavy target with the aim of integrating 2×10^{20} pot in 5 years. A dedicated detector, based on a long vacuum tank followed by a spectrometer and particle identification detectors, will allow probing a variety of models with light long-lived exotic particles and masses below $O(10)$ GeV/c^2. The main focus will be the physics of the so-called Hidden Portals, i.e. search for Dark Photons, Light scalars and pseudo-scalars, and Heavy Neutrinos. The sensitivity to Heavy Neutrinos will allow for the first time to probe, in the mass range between the kaon and the charm meson mass, a coupling range for which Baryogenesis and active neutrino masses could also be explained. Another dedicated detector will allow the study of neutrino cross-sections and angular distributions.

1 Introduction

There had been many expectations before LHC started, notably Higgs boson or strong interaction of vector bosons («guaranteed discovery») and possible New Physics phenomena in TeV ballpark (SUSY particles? extra dimensions? compositeness? etc). After five years the situation can be characterized as the Standard Model impressive triumph: the Higgs boson is discovered, many tiny and highly nontrivial SM effects in flavor physics and rare decays are confirmed, and nothing is found beyond SM. On the other hand, we still know (not from LHC) that there is physics beyond the SM: neutrino masses and oscillations; dark matter; baryon asymmetry of the Universe show that clearly. Besides those observational facts, there are many «why» and «how» in the SM itself: how is EW scale so smaller than UV scale? why hierarchy between SM scales? why are lefts doublets and rights singlets? why 3 generations? why CKM hierarchy? why CP? etc. Speaking in more theoretical prospective one can ask, is the SM a consistent field theory in the first place? Generally

[a] e-mail: Vladimir.Shevchenko@cern.ch

speaking, the answer to this question is negative. Existence of Landau pole, anomalies, naturalness issues precludes one to consider SM as fully satisfactory theoretical framework. On the other hand, it is also a fact of life that the things in the SM are arranged in such a tricky way, that (almost) all the above mentioned problems are either cured or seem not to be physically relevant. Nice example is anomaly cancellation between quark and lepton sectors, which would not be possible without delicate SM charge distribution between particles.

Let us remind that there are many various scales and small numbers in the SM with intricate relations between them. They include

- dimensionless gauge coupling constants: $\alpha_{em} = (137.035999679)^{-1}$; $\alpha_s(m_Z) = (8.503)^{-1}$
- Yukawa constants: $m_{u,d}/\Lambda = (0.5 \div 2)\%$; $\Lambda/m_b = (6 \div 24)\%$; $m_e/m_t = 3 \cdot 10^{-6}$
- QCD to EW scale ratio: $G_F m_p^2 = 10^{-5}$
- CKM matrix parameters: $\lambda = |V_{us}| = 0.22$

There are even more mysterious relations like Koide relation between lepton masses or Zeldovich-Bjorken relation between QCD and cosmological scales. Another instructive example is UV-regularization, corresponding to inclusion of some (artificial) information about theory behavior in ultraviolet region. In renormalizable theories all physics above Λ can be encoded just in a couple of numbers - coefficients in front of marginal operators like the famous fine structure constant $1/137$ in QED. Renormalizability, however, does not mean naturalness. Computing, for example, average of some local d-dimensional operator $\mathcal{T}[\phi]$ in a theory with UV cutoff Λ, one gets typically (up to, perhaps, logarithms):

$$\int \mathcal{D}\phi\, \mathcal{T}[\phi]\, e^{iS[\phi]} \sim c \cdot \Lambda^d + \text{finite part} \qquad (1)$$

The first term characterizes *ad hoc* assumptions about the integration measure/detector (for example, geometry of the lattice with the link size $a \sim 1/\Lambda$ used for computation), but not the genuine physics of this operator. Of course in many cases symmetries of the theory guarantee $c = 0$, but if not, we have to work out a way of disentangling "the physics of the detector" from "the physics of the physics" (which presumably is hidden in the finite part). Notable examples include vacuum energy density, Higgs boson mass and gluon condensate, where only in the latter case the answer is known (qualitatively, but not quantitatively) - quantum scale anomaly of Yang-Mills theory and dimensional transmutation take care of unphysical UV scale Λ to become physical scale Λ_{QCD}. In other cases of this sort, like cosmological constant problem or hierarchy problem a solution is yet to be found.

Perhaps the most important qualitative result of LHC is experimental observation that the observable Higgs mass corresponds to metastability of the SM vacuum (see Fig.1).

Incomplete list of researchers who contributed to this issue includes I.V. Krive, A.D.Linde, N.Cabibbo, L.Maiani, G.Parisi, R.Petronzio, M.Lindner, H.B.Nielsen, C.Froggatt, J. Elias-Miro, J. R. Espinosa, G. F. Giudice, G. Isidori, A. Riotto, A. Strumia, J. R. Espinosa, M. Quiros, G.Altarelli. This result gives a strong hint that there is no heavy New Physics degrees of freedom up to the Planck scale. It is worth reminding that historically the top and Higgs masses were predicted [1] from criticality to be 173 ± 5 GeV and 135 ± 9 GeV back to 1995.

If indeed no NP degrees of freedom can be found at energy frontier it is perhaps more useful to think not about NP that could solve theoretical problems of the SM, but about NP that could explain observed effects beyond SM. It is remarkable that such scenarios could have NP degrees of freedom as relatively light (a few GeV) but very weekly interacting SM singlets (see Fig.2). It is clear that there are two crucial requirements one has to obey for experimental search of such particles: huge initial intensity and large (to be more precise, very long) controlled detector volume. A project of experimental facility capable to do the job is described below.

2 Motivations for the SHiP Project

SHiP (Search for Hidden Particles) [3, 4] is a new general purpose fixed target facility proposed at the CERN SPS accelerator to search for particles predicted by Hidden Portals. These particles are expected to be predominantly accessible through the decays of heavy hadrons. The facility is therefore designed to maximise the production and detector acceptance of charm and beauty mesons, while providing the cleanest possible environment. The 400 GeV proton beam extracted from the SPS will be dumped on a high density target with the aim of accumulating 2×10^{20} protons on target during 5 years of operation. The charm production at SHiP exceeds any existing and planned facility. SHiP facility is supposed to be located in the North Area and shares TT20 transfer line and the slow extraction mode with other fixed target programmes.

A general layout is depicted on Fig.3. Water cooled target will dissipate 4×10^{13} protons at 7 sec which gives average power 355 kW (2.56 MW during 1 sec spill). Initial dose is estimated to be around 50 Sv/h. The target is segmented: high-Z hybrid solution is composed of molybdenum alloy strips and pure tungsten ones.

The main decay vessel (in vacuum option) has the size $10 \times 5 \times 60$ meters. Liquid or plastic scintillator options for veto system (with the readout by optical modules) are being considered.

A dedicated detector, based on a long vacuum tank followed by a spectrometer and particle identification detectors, will allow probing a variety of models with light long-lived exotic particles and masses below $O(10)$ GeV/c^2. Since the hidden sector particles originating from charm and beauty are produced with a significant transverse momentum with respect to the beam axis, the detector should be placed as close as possible to the target. A critical component of SHiP is the muon shield, which deflects the high flux of muons produced in the target, that would represent a very serious background for the particle searches, away from the detector. The detector is designed to fully reconstruct the exclusive decays of hidden particles and to reject the background down to below 0.1 events in the sample of 2×10^{20} protons on target. The spectrometer is designed to accurately reconstruct the decay vertex, mass and impact parameter of the decaying particle at the target. A set of calorimeters followed by muon chambers provide identification of electrons, photons, muons and charged hadrons. A dedicated timing detector measures the coincidence of the decay products, which allows the rejection of combinatorial backgrounds. The decay volume is surrounded by background taggers to tag neutrino and muon inelastic scattering in the surrounding structures, which may produce long-lived SM particles, such as K_L, that have similar topologies to the expected signals.

From theoretical point of view it is useful to discuss search for Hidden Particles in terms of portals. The total Lagrangian has the following form:

$$L = L_{SM} + L_{HS} + L_{Portal} \tag{2}$$

where Hidden Sector (HS) contains Dark Matter and can have very complicated structure, while L_{Portal} contains terms responsible for interaction between HS particles and some singlet operator built from SM fields. People analyse renormalizable portals (scalar, vector and neutrino) as well as higher dimensional ones (see references in [4]). Of particular interest is the so called νMSM model [5], with the Largangian

$$L = L_{SM} + \bar{N}_I i\partial_\mu \gamma^\mu N_I - F_{\alpha I}\Phi\bar{N}_I L_\alpha - \frac{M_I}{2}\bar{N}_I^c N_I + \text{h.c.} \tag{3}$$

The new particles - Heavy Neutral Leptons - are schematically depicted on Fig.4. Particle N_1 with mass in keV range plays a role of dark matter, while N_2 and N_3 with GeV-scale masses are responsible for the SM neutrino masses via see-saw mechanism and also for baryogenesis. These particles are among the objects to search for at the SHiP facility.

Let us estimate the number of Hidden Sector events. In general it is given by

$$n(HS) = N(pot) \times \chi(pp \to HS) \times \mathcal{P}_{vtx} \times \mathcal{A}_{tot}(HS \to \text{visible}) \tag{4}$$

where $N(pot) = 2 \times 10^{20}$ and $\mathcal{P}_{vtx}$ is probability that Heavy Neutral Lepton (HNL) of a given mass and couplings decays in the SHiP fiducial volume. One gets for the detector acceptance for all HNL final (visible) states

$$\mathcal{A}_{tot}(HS \to \text{visible}) = \sum_i \text{Br}(HNL \to i) \times \mathcal{A}(i) \tag{5}$$

with

$$\chi(pp \to HNL) = 2 \sum_{q=c,b} \chi(pp \to q\bar{q}) \times \text{Br}(q \to HNL) \times U^2 \tag{6}$$

where mixing matrix $U^2 = U_e^2 + U_\mu^2 + U_\tau^2$ and conservative numerical estimates give $\mathcal{P}_{vtx} \times \mathcal{A} \times$ Selection $\sim 10^{-6}$, with $\chi(pp \to c\bar{c}) = 1.7 \times 10^{-3}$, $\chi(pp \to b\bar{b}) = 1.6 \times 10^{-7}$ for 400 GeV protons on Molybdenum target. This shows crucial importance of initial multiplicity. Typical SHiP sensitivity plots from [3, 4] are depicted on Fig.5 and Fig.6.

The experimental facility is also ideally suited for studying interactions of tau neutrinos. It will therefore host an emulsion cloud chamber based on the OPERA detector concept, upstream of the hidden-particle decay volume, followed by a muon spectrometer. For the total number of neutrino one can estimate

$$N_{\nu_\tau+\bar{\nu}_\tau} = 4N_{pot}\frac{\sigma_{c\bar{c}}}{\sigma_{pN}} f_{D_s}\text{Br}(D_s \to \tau) = 2.85 \times 10^{-5} N_{pot} \tag{7}$$

The resulting expected multiplicity factors in SHiP in comparison with DONUT and CHARM experiments for tau-neutrino and Heavy Neutral Leptons are

$$\frac{N(\nu_\tau)@\,SHiP}{N(\nu_\tau)\,@\,DONUT} \sim 600 \;\text{ and }\; \frac{N(HNL)\,@\,SHiP}{N(HNL)\,@\,CHARM} \sim 10000 \tag{8}$$

3 Conclusions

It had been customary to argue, before LHC started, that hierarchy problem pointed to a few TeV scale as a reasonable one for New Physics (NP). Supersymmetric extension and other scenarios beyond the Standard Model had been intensively studied. In reality no these new particles have been discovered by ATLAS and CMS experiments (leaving aside the 126 GeV Higgs boson). Moreover, indirect searches of NP by LHCb and other precision experiments provided very high, hundreds-TeV bounds for NP degrees of freedom in some channels (without special assumptions about flavor alignment). On the other hand the experimentally known New Physics phenomena (neutrino oscillations, dark matter etc) give no indication to the intrinsic scales of the corresponding physics. From theoretical side, all that calls for reconsideration of the Standard Model as being the theory probably adequately describing physics up to the Planck scale. From experimental point of view, this can be interpreted as an argument in favor of NP (responsible, in particular, for neutrino oscillations and dark matter) manifested by relatively light, but very weekly interacting particles. The search of the latter should be done on intensity frontier rather than on energy frontier, and the proposed Search for Hidden Particles (SHiP) experiment at CERN's SPS, discussed in the present talk, is an example of this kind. Search for Hidden Particles (SHiP) experiment is complement to searches for New Physics at energy frontier at CERN. It is proposed to search for New Physics in the largely unexplored domain of new, very weakly interacting particles with masses $O(10)$ GeV/c^2. It provides unique opportunities for tau-neutrino studies. Sensitivity improves previous experiments by factor $O(10^4)$ for Hidden Sector and by $O(10^2)$ for neutrino physics. The impact of the discovery of new light hidden particles is hard to overestimate.

References

[1] C. D. Froggatt and H. B. Nielsen, *Standard model criticality prediction: Top mass 173 +- 5-GeV and Higgs mass 135 +- 9-GeV,* Phys. Lett. B **368** (1996) 96, hep-ph/9511371.

[2] D. Buttazzo, G. Degrassi, P. P. Giardino, G. F. Giudice, F. Sala, A. Salvio and A. Strumia, JHEP **1312** (2013) 089, arXiv:1307.3536.

[3] The SHiP Collaboration, *A facility to Search for Hidden Particles (SHIP) at the CERN SPS,* April 2015, arXiv:1504.04956.

[4] S. Alekhin *et al., A facility to Search for Hidden Particles at the CERN SPS: the SHiP physics case,* Rept. Prog. Phys. **79** (2016) no.12, 124201, arXiv:1504.04855 [hep-ph].

[5] T. Asaka, S. Blanchet and M. Shaposhnikov, *The nuMSM, dark matter and neutrino masses,* Phys. Lett. B **631** (2005) 151, hep-ph/0503065.

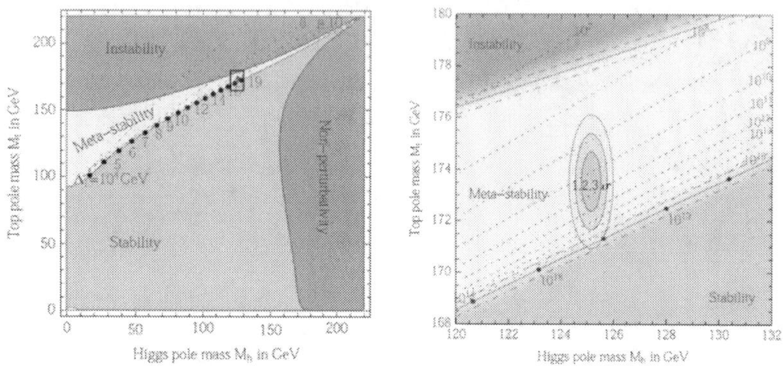

Figure 1. (Meta)-Stability of the SM vacuum (from [2]).

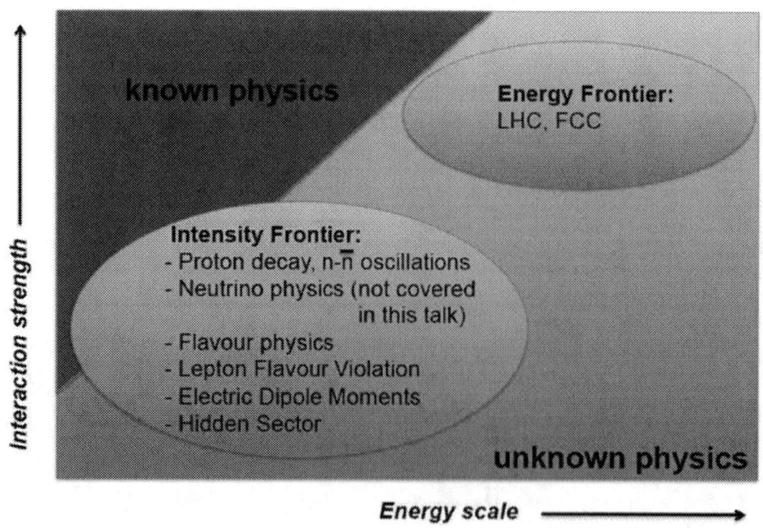

Figure 2. Energy Frontier vs Intensity Frontier

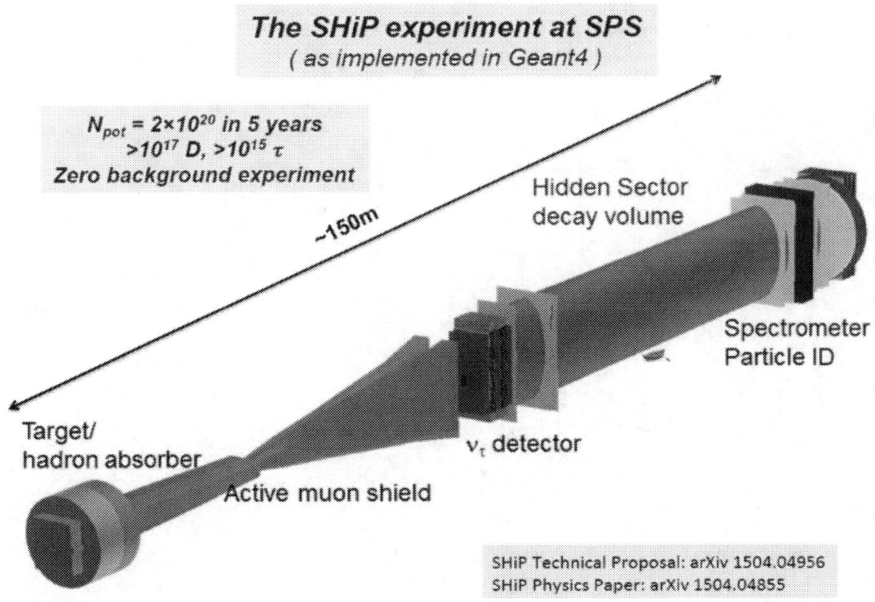

Figure 3. General Layout of SHiP experiment

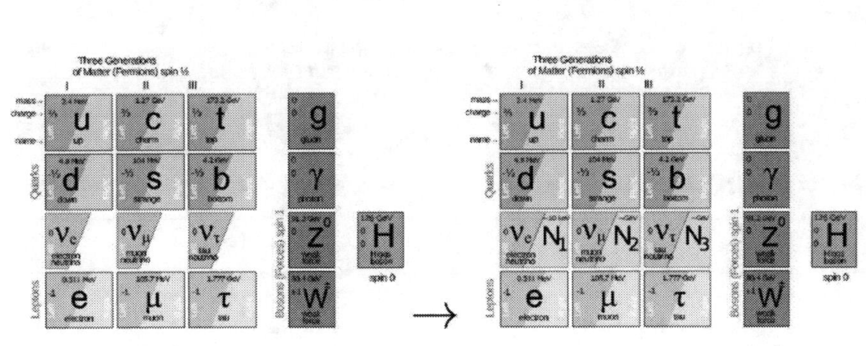

Figure 4. nu-Minimal Standard Model

DOI: 10.1051/epjconf/201716401029

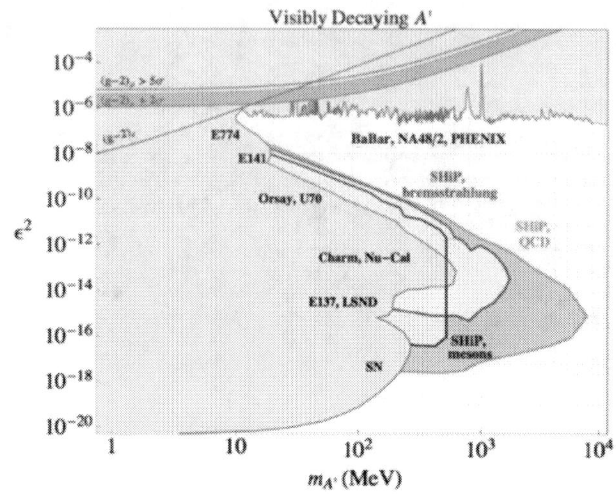

Figure 5. SHiP sensitivity to dark photons

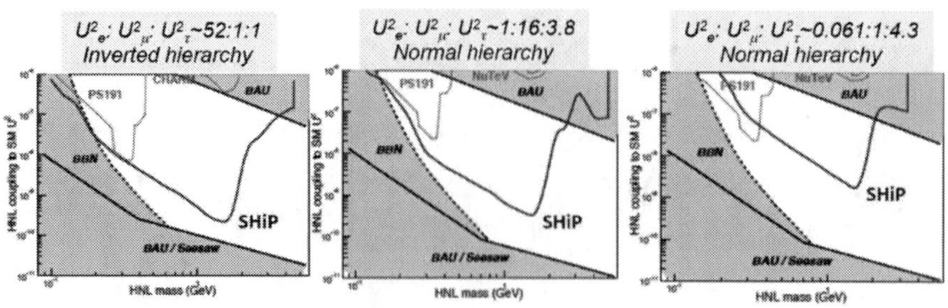

Figure 6. SHiP sensitivity to HNL

SUSY searches with the ATLAS detector

Riccardo Maria BIANCHI[1,a], on behalf of the ATLAS Collaboration

[1] *University of Pittsburgh, Department of Physics and Astronomy, Pittsburgh PA 15260, USA*

Abstract. Despite the absence of experimental evidence, weak scale supersymmetry remains one of the best motivated and studied Standard Model extensions. This talk summarises recent ATLAS results for searches for supersymmetric (SUSY) particles, with focus on those obtained using proton-proton collisions at a centre of mass energy of 13 TeV. Strong production in both R-Parity conserving and R-Parity violating SUSY scenarios are considered. The searches involved final states including jets, missing transverse momentum, light leptons, as well as long-lived particle signatures.

1 Introduction

Supersymmetry (SUSY) postulates the existence of a twin particle for each Standard Model particle. Masses, and nature, of the supersymmetric particles depend on different parameters, addressed in many different SUSY models.

Figure 1 shows an overview of all the SUSY searches run by the ATLAS Experiment [1], which amount to about 50 analyses. In the figure, the searches are divided in categories; from the top: analyses involving strong production of squarks and gluinos; strong production involving third generation squarks (both $\tilde{g}$-mediated and from direct production); electroweak production of non-coloured SUSY partners; long-lived particles and R-parity violation decays. The light-blue bars designate the analyses that have been updated in order to analyze the experimental data taken during the Run 2 phase of the LHC operation, at a centre-of-mass energy of 13 TeV. The green bars, instead, show the mass scale explored by the analyses using the data taken in the Run 1 phase of LHC at a centre-of-mass energy of 7 and 8 TeV. All public results from the ATLAS SUSY analyses can be found at [2].

This paper presents the latest results of four searches which have been recently updated to analyze the latest available Run 2 datasets at a centre-of-mass of 13 TeV. Three of them analyze a dataset of 3.2 fb^{-1}, while the fourth exploits the latest 14 fb^{-1}dataset.

2 Heavy long-lived charged R-hadrons

Gluino and squark production in proton-proton collisions can produce composite colorless states of gluinos and squarks, in addition to ordinary particles (gluon and quarks) from the Standard Model. These states are called *R-hadrons*.

At the LHC R-hadrons, if they exist, are expected to be produced as charged or neutral states, and to be heavy and slow, with a small β velocity. They also should have a ionization energy loss dE/dx

[a]e-mail: riccardo.maria.bianchi@cern.ch

DOI: 10.1051/epjconf/201716401030

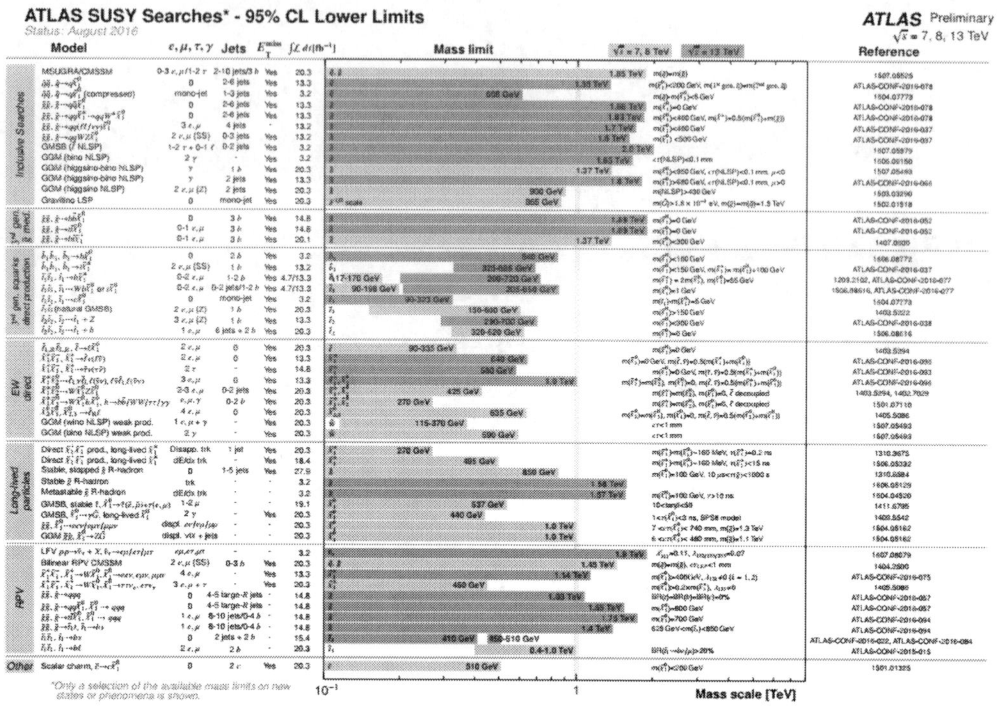

Figure 1: Mass reach of ATLAS searches for Supersymmetry [2]. Only a representative selection of the available results is shown.

larger than that of Standard Model particles. Moreover, they are expected to be able to change charge when interacting with the detector material [3, 4].

The search presented here [5] analyzes a dataset of 3.2 $fb^{-1}$and makes use of information from both the pixel-based inner tracking detector and the hadronic calorimeter, in a combined way. This search extends the analysis [6], which only uses Pixel information. In this search the information from the muon spectrometer has been omitted, in order to be sensitive to scenarios when the R-hadrons decay or turn neutral before arriving in the muon system. In the search, the dE/dx ratio and velocity measurements are used to infer the R-hadron mass.

2.1 Object selection

To tag a particle as an R-hadron candidate different criteria must be fullfilled: a very well-reconstructed isolated track, with at least 7 hits in the silicon-based tracking detector; at least 2 calorimeter clusters used to measure the dE/dx ratio; a transverse momentum of the particle larger than 50 GeV and the fullfilment of other requirements used to prevent misidentification.

Moreover, cosmic-rays are rejected by omitting a track when a similar, symmetrical track is observed in the detector. And particles from the decay of the Z boson to muons are rejected by omitting reconstructed tracks whose invariant mass is in the [81, 101] GeV window.

2.2 Event selection and background estimation

For the online event selection a trigger with an E_T^{miss} signature and a cut at 70 GeV has been used, whose signal efficiency is in the range 32%–50%, depending on the R-hadron mass. For the offline event selection, the requirements are a primary vertex reconstructed from at least two tracks with p_T > 400 MeV, and one R-hadron candidate. Then, for the final event selection, extra requirements are imposed: a p_T above 200 GeV; a R-hadron β velocity smaller than 0.75 and a product $\beta\gamma$ smaller than 1.35 or 1.15, according to the R-hadron mass.

The signal region is defined in the m_β-$m_{\beta\gamma}$ plane, as shown in figure 2.

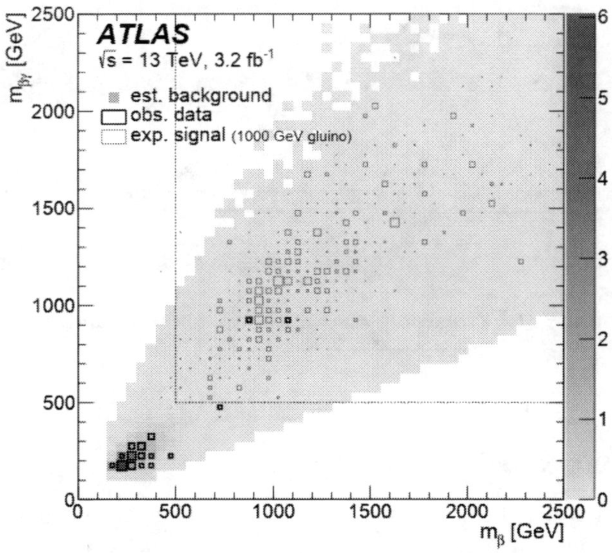

Figure 2: Data (bold boxes) and background estimates (colour fill) for m_β vs. $m_{\beta\gamma}$ for the gluino R-hadron search for the 1000 GeV mass hypothesis. The blue thin-line boxes illustrate the expected signal on top of the background. The signal region is shown by the two black dashed vertical/horizontal lines at 500 GeV [5].

The final signal selection efficiency is of 9%–15% for gluino and stop R-hadrons and 6%–8% for sbottom R-hadrons. The lower efficiency for bottom squarks is expected, as bottom R-hadrons tend to be neutral more often than those with up-type squarks [5].

The background is evaluated in a data-driven manner, randomly picking momentum, β and $\beta\gamma$ values sampled from probability distributions functions (PDF) determined from data; that yields mass distributions estimating the background.

2.3 Results

Two events with masses above 500 GeV pass the final event selection for the 1000 GeV mass hypothesis (figure 2), while only one of these events passes the final event selection for the 1600 GeV mass hypothesis. However no statistically significant excess is observed in the whole analized mass range, and therefore 95% CL exclusion limits are placed on the R-hadron production cross section, as shown in figure 3. The cross-section limits are translated into expected lower limits at 95% CL on R-hadron masses of 1655 GeV, 865 GeV and 945 GeV for the production of long-lived gluino, sbottom and stop R-hadrons, respectively. Corresponding observed lower mass limits at 95% CL for gluino, sbottom and stop are set, as well, at 1580 GeV for the gluino mass, at 805 GeV for the sbottom mass and at 890 GeV for the stop mass. The figure also shows the limits set by the previous search with the data collected at a centre-of-mass energy of 8 TeV[7], for comparison. The current search greatly improves the older limits.

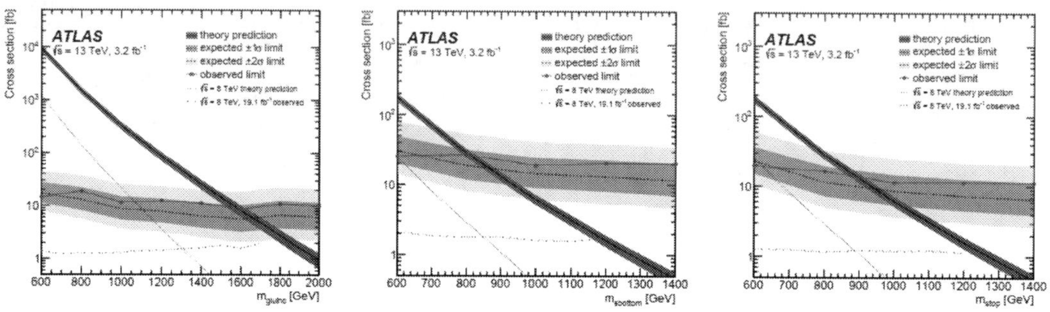

(a) gluino exclusion plot (b) sbottom exclusion plot (c) stop exclusion plot

Figure 3: Exclusion plots for gluino, sbottom and stop [5].

3 Two leptons, jets and $E_{\mathrm{T}}^{\mathrm{miss}}$ search

This section presents two searches for new phenomena in final states containing a same-flavour (SF) opposite-sign (OS) lepton pair (electrons or muons), jets, and large $E_{\mathrm{T}}^{\mathrm{miss}}$: the so-called "on-shell Z" and "edge" searches [8]. The "on-shell Z" channel requires a lepton pair whose invariant mass m_{ll} is consistent with the mass of the Z boson mass, while the "edge" search considers all SFOS lepton pairs. The requirement of two leptons in the final state suppresses large SM backgrounds from QCD multijet and $W+$jets production.

The diagrams of the processes considered for this search are shown in figure 4. SFOS leptons can come from the $\tilde{\chi}_2^0$ decay, which depend on the mass difference $\Delta m_X = m_{\tilde{\chi}_2^0} - m_{\tilde{\chi}_1^0}$, the chargino-neutralino mixing and on the presence of SUSY particles with masses $< m_{\tilde{\chi}_2^0}$ which can be produced in the decay.

The "on-shell Z" search targets the processes where $\Delta m_X > M_Z$, leading to a peak in the invariant-mass distribution near $m_{ll} \approx m_Z$. The "edge" search analyzes the processes where $\Delta m_X < M_Z$, which lead to a rising m_{ll} distribution truncated at a kinematic endpoint below the Z mass; depending on the mass difference m_X and the mass of the sleptons produced in the decay, other kinematic endpoints can appear as well. Three models are considered for the "on-shell Z" search, involving squark-pair and gluino-pair production; and two models are considered for the edge search, both involving the direct gluino-pair production, which differ by the $\tilde{\chi}_2^0$ decay mode.

SUSY signals are generated over a two-dimensional plane varying the gluino, squark and neutralino masses. All other supersymmetric particles are decoupled: they are considered having masses compatible with a much higher energy scale than that at the LHC, and thus not contributing in the analyzed processes.

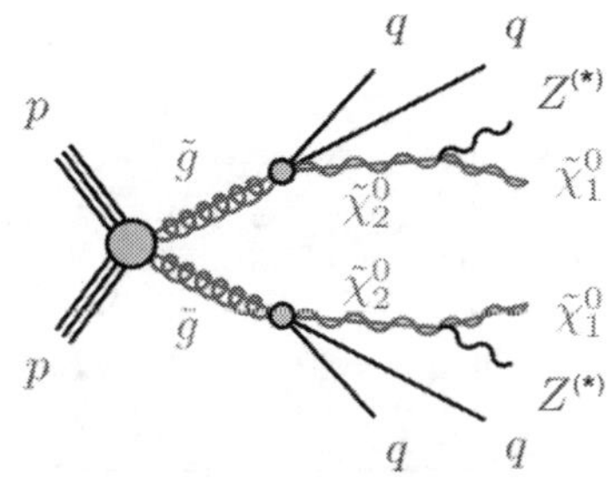

(a) "on-shell Z" channel diagram

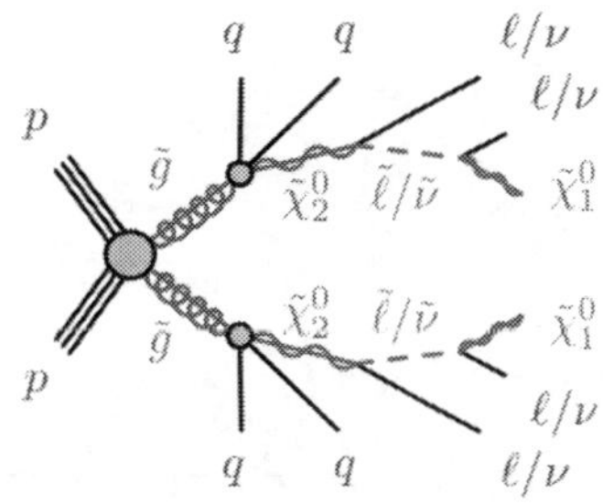

(b) "edge" channel diagram

Figure 4: The diagrams of the processes considered for the di-lepton "on-shell Z" (a) and "edge" (b) searches [8].

The search has been recently updated to exploit 14.7 fb^{-1}of LHC pp collision data at $\sqrt{s}$ = 13 TeV collected by the ATLAS detector in 2015 and 2016.

3.1 Event selection and background estimation

Data events are collected using a combination of single-lepton and di-lepton triggers. For both searches events are required to contain at least two signal leptons (electrons or muons) with $p_T >$ 25 GeV, which must pass at least one of the leptonic triggers. If more leptons are present, the selection process continues based on the two leading leptons (those with the highest p_T).

For each search signal regions (SRs) are defined to accept events from the SUSY signals, control regions (CRs) are built to be empty of SUSY events and to estimate the background events, and validation regions (VRs) are designed to help validating the SR and CR selections. Figure 5 shows all the most relevant regions defined for the two searches.

One signal region (SRZ) is defined for the "on-shell Z" search, requiring a higher p_T for the leading electrons ($p_T >$ 50 GeV), a large missing transverse energy ($E_T^{miss} >$ 225 GeV), two or more jets, two leptons whose invariant mass is in the range [81, 101] GeV, and a large $H_T >$ 600 GeV[1].

The "edge" search requires at least two leptons with $p_T >$ 25 GeV, and it explores the full m_{ll} spectrum, with the only exception for the region with $m_{ll} <$ 12 GeV, rejected to suppress Drell-Yann processes and low mass resonances. For this search three SRs are defined, according to the value of the gluino-neutralino mass difference $\Delta m_{\tilde{g}} = m_{\tilde{g}} - m_{\tilde{\chi}_1^0}$, called SR-low, SR-medium and SR-high. All the three regions require $E_T^{miss} >$ 200 GeV, while SR-medium and SR-high also require an H_T quantity of 400 GeV and 700 GeV respectively.

The SRs dominant background for this searches comes from "flavor-symmetric" (FS) processes[2]: $t\bar{t}$ (which dominates), WW, ZZ and $Z\tau\tau$. Those FS processes amount to the 60%–90% of the total background and are estimated with a data-driven approach, using control samples of $e\mu$ events. The Z/γ + jets contribution is small but can mimic the signal, so it is estimated with a data-driven method as well. Di-boson production (WZ and ZZ) gives a contribution of 5%–20% to the total background and are estimated with simulated Monte Carlo data and then validated with dedicated $3l$ (WZ) and $4l$

[1] H_T is the *total visible transverse energy*, generally defined as the scalar sum of the transverse momenta of the photons, leptons and jets in the event. For the "edge" search, leptons are not included in the H_T definition.

[2] In FS processes the di-leptonic ratio is $ee : \mu\mu : e\mu = 1 : 1 : 2$ because of the two leptons coming from independent semi-leptonic W decays.

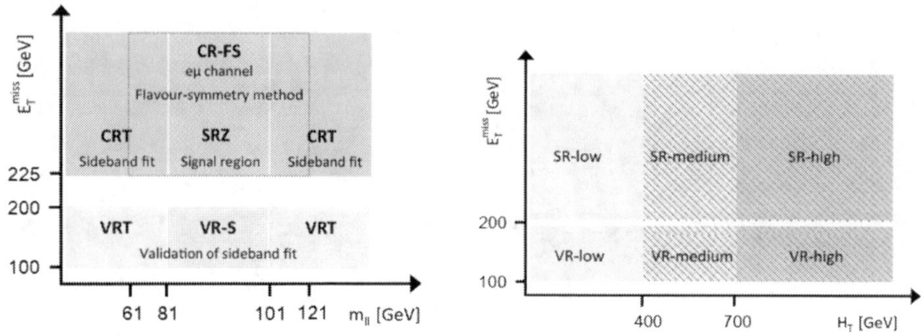

(a) "on-shell Z" channel diagram (b) "edge" channel diagram

Figure 5: Schematic diagrams of the signal, control and validation regions (SR, CR and VR respectively) for the "on-shell Z" (a) and "edge" (b) searches [8]. For the "on-shell Z" search the various regions are shown in the m_{ll}-$E_{\mathrm{T}}^{\mathrm{miss}}$ plane, whereas in the case of the "edge" search the signal and validation regions are depicted in the H_T–$E_{\mathrm{T}}^{\mathrm{miss}}$ plane.

(ZZ) validation regions (VRs). All other sources of background — like ttW, ttW, $ttWW$ — provide small contributions and are estimated with simulated data only.

3.2 Results

The observed yields and the expected background in the SRs for both searches are shown in figure 6. For the "on-shell Z" search (figure 6a), a total of 60 events are observed in data, with a predicted background of 53.5 ± 9.3 events; the significance[3] corresponds to 0.47σ, expressed in terms of standard deviations. For the "edge" search, since signal models may produce kinematic endpoints at any value of m_{ll}, a "sliding window" approach is used to define possible di-lepton mass windows. Events selected in the three signal regions are then further grouped into non-orthogonal windows based on the m_{ll} value, chosen because they are the most sensitive ones for at least one grid point in the signal model parameter space. The expected and observed yields in the 24 (overlapping) m_{ll} ranges of SR-low, SR-medium, and SR-high signal regions are shown in figure 6b, together with their significance.

No statistically significant excesses are observed, so exclusion limits are set. The results of the "on-shell Z" search are interpreted in a simplified model with gluino-gluino pair production, where the gluinos decay as $\tilde{g} \rightarrow q\bar{q}\tilde{\chi}_1^0$ and the $\tilde{\chi}_2^0$ decays as $\tilde{\chi}_2^0 \rightarrow Z\tilde{\chi}_1^0$, with the mass of $\tilde{\chi}_1^0$ set to 1 GeV. The "edge" search results are interpreted in two simplified models with gluino-gluino pair production, with gluinos decaying as $\tilde{g} \rightarrow q\bar{q}\tilde{\chi}_2^0$.

Figures 7a–7c show the exclusion plots for the "on-shell Z" search, for different signal grids; while figure 7d shows the exclusion plot for the "edge" search. In all figures the observed limits are shown by the solid red line, with the dotted red lines indicating the variation resulting from varying the signal cross-section within its uncertainty ($\pm 1\sigma_{theory}^{SUSY}$). The dashed blue line indicates the expected limits at 95% CL and the yellow band shows the 1σ variation of the expected limit as a consequence of the uncertainties in the background prediction and the experimental uncertainties in the signal ($\pm 1\sigma_{exp}$).

[3]*Significance* is the probability for the background to produce a fluctuation greater than or equal to that observed in data, when expressed in terms of the number of standard deviations σ.

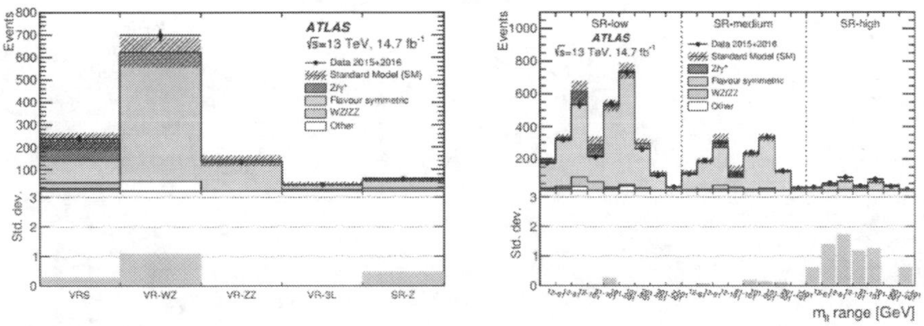

(a) "on-shell Z" search (b) "edge" search

Figure 6: The expected and observed yields in the signal regions for the "on-shell Z" (a) and "edge" (b) searches [8]. The data are compared to the sum of the expected backgrounds. The significance of the difference between the data and the expected background is shown in the bottom plot; for regions in which the data yield is less than expected, the significance is set to zero. The hashed bands include the statistical and systematic uncertainties in the background prediction.

The two searches set a lower limit on the $\tilde{g}$ mass $m_{\tilde{g}}$ at 1.7 TeV, and at 980 GeV on the $\tilde{q}$ mass $m_{\tilde{q}}$.

4 Two photons plus $E_{\mathrm{T}}^{\mathrm{miss}}$ search

In this search [9] the results are interpreted in the context of *general-gauge-mediation* (GGM) SUSY model, with the Gravitino $\tilde{G}$ as the Lightest Supersymmetric Particle (LSP) with a very low mass ($m_{\tilde{G}} \ll 1$ GeV). In this model-dependent search the mass of the gluino ($m_{\tilde{g}}$) and the mass of the neutralino ($m_{\tilde{\chi}_1^0}$) are considered as free parameters, with the only constraint of the mass of the neutralino to be smaller than the gluino mass ($m_{\tilde{\chi}_1^0} < m_{\tilde{g}}$). All other SUSY masses are considered decoupled. In the examined model, R-parity conservation is assumed, so sparticles are produced in pairs; and the *bino*-like neutralino $\tilde{\chi}_1^0$ is considered being the NLSP, which decays to $\tilde{G}$ plus Standard Model particles, with a high probability of photons ($\gamma + \tilde{G}$). In the end a long decay chain is expected, with two photons and missing transverse energy in the final states; the diagram is shown in figure 9a.

4.1 Event selection

For the online event selection a di-photon trigger is used, with a p_{T} threshold of 50 GeV. For the offline event selection a signal region is defined, requiring two isolated photons with p_{T} larger than 75 GeV, large missing transverse energy ($E_{\mathrm{T}}^{\mathrm{miss}} > 175$ GeV) and large effective mass ($m_{\mathrm{eff}} > 1500$ GeV), defined as the scalar sum of the transverse momenta of the photons, leptons and jets in the event, plus the missing transverse energy. Additional cleaning cuts are applied to reject events from beam, cosmic rays and detector noise. A single signal region for all gluino-neutralino mass points is used, since all the mass points give very similar results. The SR has been optimized on $E_{\mathrm{T}}^{\mathrm{miss}}$, m_{eff} and p_{γ} for two benchmark signal points, with high and low neutralino mass: $(m_{\tilde{g}}, m_{\tilde{\chi}_1^0}) = (1500, 1300)$ GeV and $(m_{\tilde{g}}, m_{\tilde{\chi}_1^0}) = (1500, 100)$ GeV.

 DOI: 10.1051/epjconf/201716401030

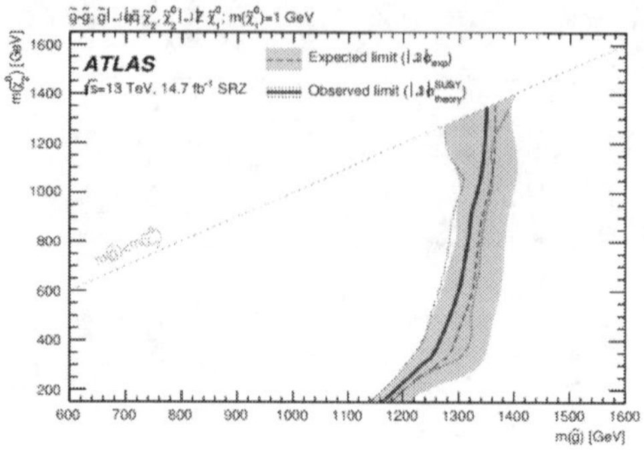

(a) "On-shell Z" ($\tilde{g}$-$\tilde{\chi}_2^0$)

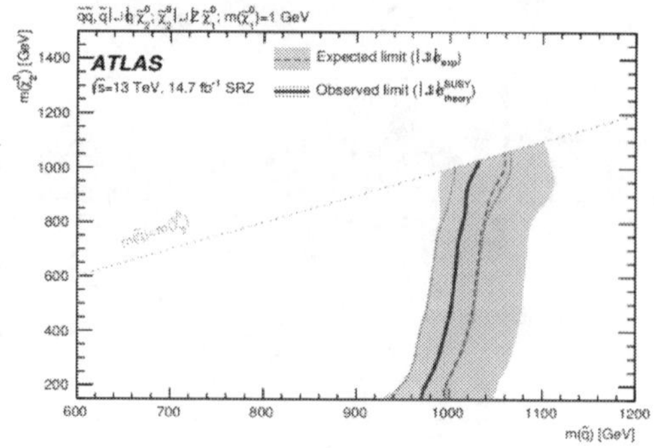

(b) "On-shell Z" ($\tilde{q}$-$\tilde{\chi}_2^0$)

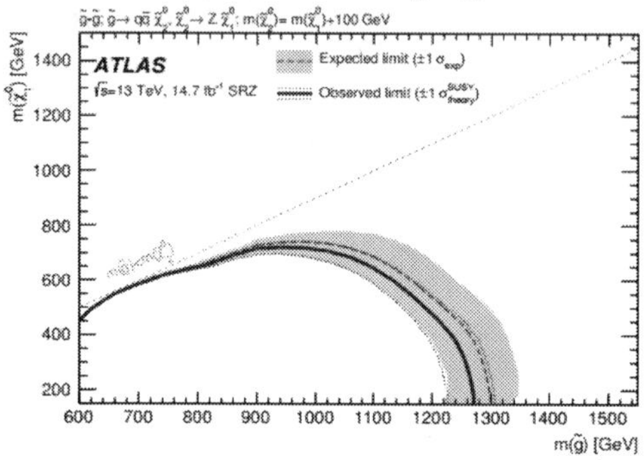

(c) "On-shell Z" ($\tilde{g}$-$\tilde{\chi}_1^0$)

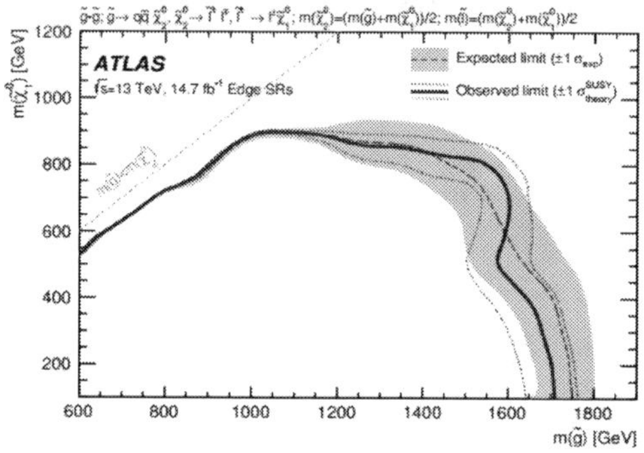

(d) "edge"

Figure 7: Exclusion plots: (a, b, c) "on-shell Z" search, for different signal grids; (d) "edge" search [8].

In figure 8 the $E_{\mathrm{T}}^{\mathrm{miss}}$ and m_{eff} distributions from data and background are shown, together with those from the two SUSY signal benchmark points. The yellow band represents the uncertainty in the data/SM ratio that arises from the statisically limited estimations of the SM background.

4.2 Background estimation

The main sources of background for this search are: QCD processes, both from real di-photon plus jets events and from jet-faking events, estimated with a data-driven method; the electron-faking events from W, Z and $t\bar{t}$ processes where an electron is misreconstructed as a photon, estimated with a data-driven method as well; irreducible background from $W\gamma\gamma$ and $Z\gamma\gamma$, estimated with simulation (the $W\gamma\gamma$, being an important contribution, is also normalized from a $l\gamma\gamma$ control region).

The total experimental uncertainty is estimated to be about 4.7% (the list of all sources can be found in [9]).

4.3 Results

No events are observed in the signal region, so lower limits on particle masses are set. Figure 9b shows the exclusion plot for the di-photon search, varying the value of the gluino and neutralino masses in

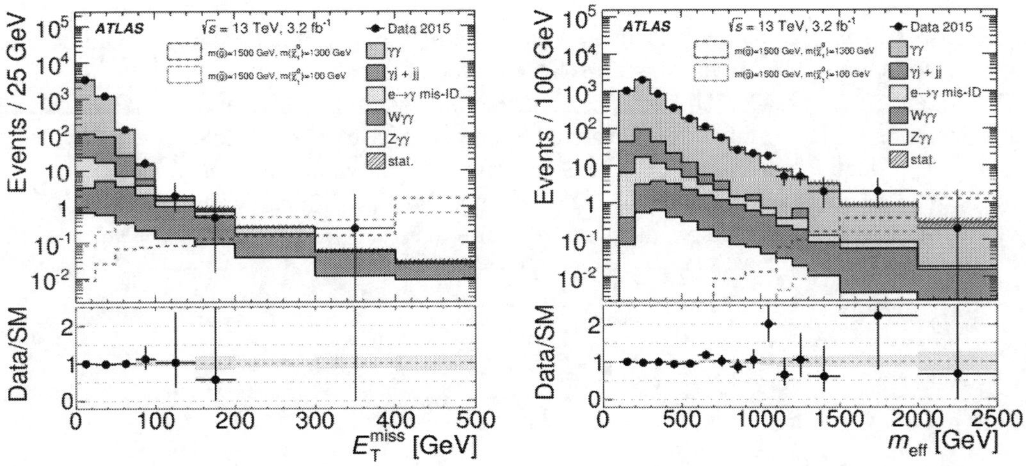

(a) (b)

Figure 8: Distributions of $E_{\mathrm{T}}^{\mathrm{miss}}$ (a) and m_{eff} (b) for the di-photon search [9].

the $m_{\tilde{g}}$-$m_{\tilde{\chi}_1^0}$ plane. The observed limits are shown for the nominal SUSY model cross section, as well as for a SUSY cross section increased and lowered by one standard deviation σ of the systematic uncertainty on the process cross-section. The expected limit is also shown, together with its $\pm 1\sigma$ range. The area previously excluded from ATLAS using 8 TeV data [10] is also shown, in grey.

The plot shows the observed lower limit on the gluino mass as roughly independent of the *bino*-like neutralino mass. A lower limit has been set on the mass of a GGM degenerate octet of gluino states at 1650 GeV, for a neutralino mass of 250 GeV, which extends the corresponding limit of 1340 GeV from the similar ATLAS search at 8 TeV.

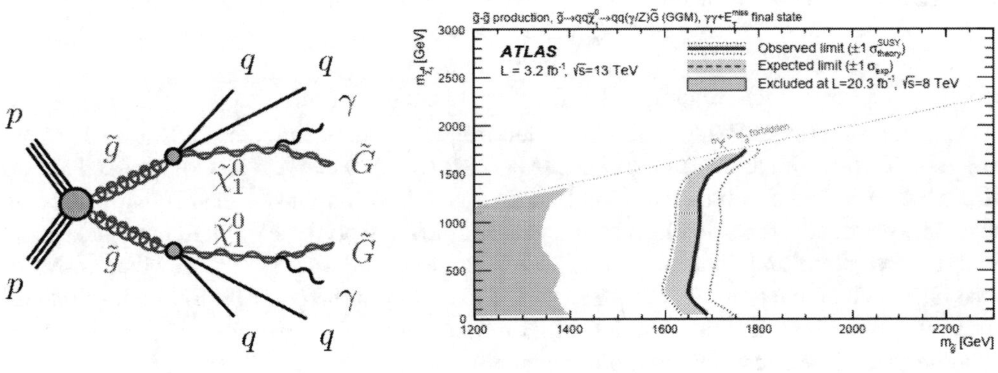

(a) (b)

Figure 9: (a) The process considered for the di-photon plus missing transverse energy search. (b) The exclusion plot for the di-photon search [9].

5 Taus plus jets plus $E_{\mathrm{T}}^{\mathrm{miss}}$ search

The ATLAS Collaboration recently updated the search for new physics in events with hadronically-decaying tau leptons, jets and missing transverse momentum [11].

Two channels are considered in this search: the one containing exactly one τ lepton, and the one with 2 or more τ. The results are interpreted in the context of two models: a simplified model of gluino-pair production with τ-rich decay cascade (figure 10a), where the masses of the gluino and the neutralino are taken as free parameters ; and a Gauge-Mediated-Supersymmetry-Breaking (GMSB) (figure 10b), where Λ and $\tan\beta$ are taken as free parameters.

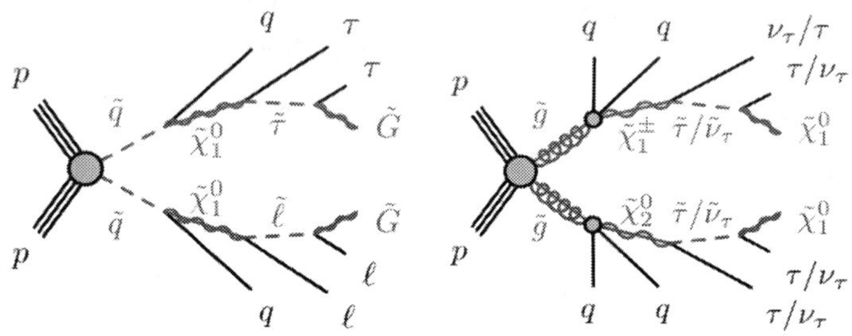

(a) (b)

Figure 10: The processes considered in the tau search for (a) the simplified gluino-pair production model and (b) for the GMSB model [11].

5.1 Event selection and background estimation

For the online event selection, an $E_{\mathrm{T}}^{\mathrm{miss}}$ trigger is used, with a threshold of $E_{\mathrm{T}}^{\mathrm{miss}} > 180$ GeV. For the offline event selection, many observables are used as discriminating variables: the transverse mass m_l^T, the total visible transverse energy H_T (defined as the sum of the transverse momentum of the taus plus the transverse momentum of the jets), the transverse missing energy $E_{\mathrm{T}}^{\mathrm{miss}}$, the effective mass m_{eff} (defined as the sum of H_T and $E_{\mathrm{T}}^{\mathrm{miss}}$), the *stransverse mass* $m_{T2}^{\tau\tau}$ [12, 13], the sum of transverse masses, the total number of jets and the total number of b-jets.

Three signal regions (SR) are chosen for the 1-τ channel, according to different mass splittings between the gluino and the LSP (Gravitino $\tilde{G}$ or Neutralino $\tilde{\chi}_1^0$): the Compressed SR (< 100 GeV), the Medium-Mass SR (500–900 GeV) and the High-Mass SR (> 1200 GeV). The Compressed SR targets topologies where a jet from initial-state radiation (ISR), and with high p_{T}, recoils against the gluino pair whose decay products receive a Lorentz boost. Both LSPs tend to be emitted opposite to the ISR jet in the transverse plane and, therefore, such events would have large $E_{\mathrm{T}}^{\mathrm{miss}}$. The Medium-Mass and High-Mass SRs include cuts on the transverse mass, to suppress the contribution of $W(\tau\nu)$+jets and of tau $t\bar{t}$ events. No SR is defined for the GMSB model, since the sensitivity of the 1-τ channel is expected to be much lower than that of the 2-τ channel.

For the 2-τ channel, two signal regions have been defined: the Compressed SR (< 900 GeV) and the High-Mass SR (> 1200 GeV). The Compressed SR features a cut on the *stransverse mass* $m_{T2}^{\tau\tau}$ to suppress SM background contributions, which exhibit a kinematic endpoint around the W and Z masses. A cut on the sum of transverse masses is also applied, to take advantage of the large $E_{\mathrm{T}}^{\mathrm{miss}}$ and the large τ and jet multiplicity expected for signal events. The High-Mass SR includes a requirement

DOI: 10.1051/epjconf/201716401030

on H_T, which is efficient for high-mass gluino signals. A signal region is also defined for the GMSB model, and targets more specifically events in which squark–antisquark production takes place, in the region $\Lambda \geq 80$ TeV, which has not been not excluded by Run 1 searches.

In both channels, the dominant background originates from SM processes involving the top quark or a massive vector boson and jets. Different control and validation Regions (CRs and VRs) are defined, to help estimating the background.

5.2 Results

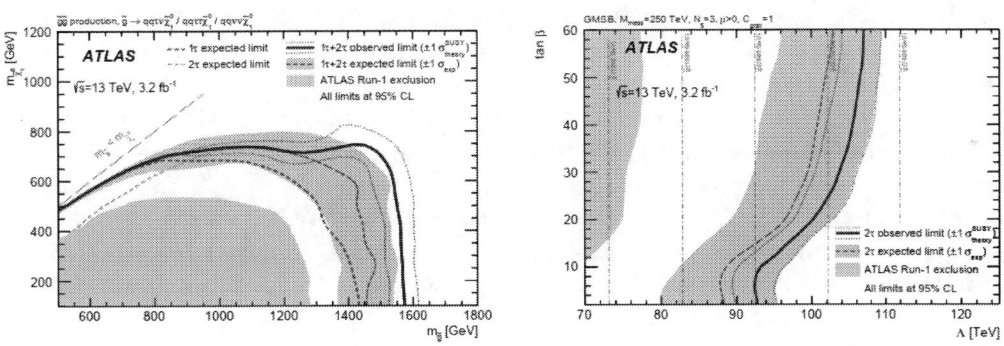

(a) (b)

Figure 11: Exclusion contours at 95% CL for the (a) simplified gluino-pair production model and (b) for the GMSB model [11].

No excesses are observed by this search, therefore exclusion limits are set. Figure 11 shows the exclusion plots for the two models. Figure 11a, in particular, shows the exclusion contours at 95% CL for the simplified model of gluino-pair production, in the gluino-neutralino mass plane, based on the results from both the 1-τ and the 2-τ channels, plus their combination. Figure 11b shows, instead, the exclusion plot for the GMSB model, in the Λ-tanβ plane. In both plots, the red solid line and the blue dashed line correspond to the observed and median expected limits, respectively: for the combination of the two channels in figure 11a and for the 2-τ channel in figure 11b. The yellow band shows the one standard-deviation range of expected limits around the median value. The effect of the uncertainty on the signal cross-section is shown as a red dotted line. The grey-shaded areas in the plots show the areas previously excluded by the searches at 8 TeV, for comparison.

In the context of the simplified gluino-pair production model, the analysis sets a lower limit on the mass of the gluino $m_{\tilde{g}}$ at 1.57 TeV, for a neutralino mass $m_{\tilde{\chi}_1^0}$ around 100 GeV; and a lower limit on the mass of the neutralino $m_{\tilde{\chi}_1^0}$ at 700 GeV for a gluino mass in the range 800 GeV$< m_{\tilde{g}} <$ 1.5 TeV, and at 750 GeV for a gluino mass of $m_{\tilde{g}} \approx$ 1.4 TeV. For the GMSB model, the analysis sets a lower limit on the gluino mass $m_{\tilde{g}}$ at 2.05 TeV for small tanβ values; and at 2.32 TeV for larger values of tanβ.

For both models, the new exclusion limits set analyzing 3.2 fb^{-1}of LHC collision data at $\sqrt{s}$ = 13 TeV significantly improve the previous limits obtained by ATLAS with 20.3 fb^{-1}of 8 TeV data [14, 15].

References

[1] ATLAS Collaboration, *The ATLAS Experiment at the CERN Large Hadron Collider*, JINST 3 (2008) S08003.

[2] ATLAS Collaboration, *ATLAS SUSY Results*, https:/twiki.cern.ch/twiki/bin/view/AtlasPublic/SupersymmetryPublicResults.

[3] A. C. Kraan, *Interactions of heavy stable hadronizing particles*, Eur. Phys. J. C 37 (2004) 91–104, arXiv:hep-ex/0404001.

[4] R. Mackeprang and D. Milstead, *An updated description of heavy-hadron interactions in GEANT-4*, Eur. Phys. J. C 66 (2010) 493–501, arXiv:0908.1868 [hep-ph].

[5] ATLAS Collaboration, *Search for heavy long-lived charged R-hadrons with the ATLAS detector in 3.2 fb-1 of proton–proton collision data at* $\sqrt{s}$ = 13 *TeV*, Physics Letters B (2016), pp. 647-665, doi: 10.1016/j.physletb.2016.07.042, arXiv: 1606.05129.

[6] ATLAS Collaboration, *Search for metastable heavy charged particles with large ionization energy loss in pp collisions at* $\sqrt{s}$ = 13 *TeV using the ATLAS experiment*, arxiv: 1604.04520.

[7] ATLAS Collaboration, *Searches for heavy long-lived charged particles with the ATLAS detector in proton-proton collisions at* $\sqrt{s} = 8$ *TeV*, JHEP 01 (2015) 068, arXiv:1411.6795 [hep-ex].

[8] ATLAS Collaboration, *Search for new phenomena in events containing a same-flavour opposite-sign dilepton pair, jets, and large missing transverse momentum in* $\sqrt{s} = 13$ *TeV pp collisions with the ATLAS detector*, arXiv: 1611.05791, submitted to Eur. Phys. J. C.

[9] ATLAS Collaboration, *Search for supersymmetry in a final state containing two photons and missing transverse momentum in* $\sqrt{s} = 13$ *TeV pp collisions at the LHC using the ATLAS detector*, arXiv: 1606.09150, accepted by Eur. Phys. J. C.

[10] ATLAS Collaboration, *Search for photonic signatures of gauge-mediated supersymmetry in 8 TeV pp collisions with the ATLAS detector*, Phys. Rev. D 92 (2015) 072001, arXiv:1507.05493 [hep-ex].

[11] Aaboud, M., Aad, G., Abbott, B. *et al.* (ATLAS Collaboration), *Search for squarks and gluinos in events with hadronically decaying tau leptons, jets and missing transverse momentum in proton–proton collisions at* $\sqrt{s} = 13$ *TeV recorded with the ATLAS detector*, Eur. Phys. J. C (2016) 76: 683, doi:10.1140/epjc/s10052-016-4481-2.

[12] C. G. Lester and D. J. Summers, *Measuring masses of semiinvisibly decaying particles pair produced at hadron colliders*, Phys. Lett. B463 (1999) 99–103, arXiv:hep-ph/9906349 [hep-ph].

[13] C. G. Lester and B. Nachman, *Bisection-based asymmetric MT 2 computation: a higher precision calculator than existing symmetric methods*, JHEP 03 (2015) 100, arXiv:1411.4312 [hep-ph].

[14] ATLAS Collaboration, *Search for supersymmetry in events with large missing transverse momentum, jets, and at least one tau lepton in 20* fb^{-1} *of* $\sqrt{s} = 8$ *TeV proton–proton collision data with the ATLAS detector*, JHEP 1409 (2014) 103, arXiv:1407.0603 [hep-ex].

[15] ATLAS Collaboration, *Summary of the searches for squarks and gluinos using* $\sqrt{s} = 8$ *TeVpp collisions with the ATLAS experiment at the LHC*, JHEP 1510 (2015) 054, arXiv:1507.05525 [hep-ex].

EPJ Web of Conferences **164**, 01031 (2017) DOI: 10.1051/epjconf/201716401031
ICNFP 2016

The next challenge for neutrinos: the mass ordering

Luca Stanco[1,a]

[1] *I.N.F.N. Padova, Via Marzolo, 8 I-35131 Padova, Italy*

Abstract. Neutrino physics is nowadays receiving more and more attention as a possible source of information for the long–standing investigation of new physics beyond the Standard Model. This is also supported by the recent change of perspectives in neutrino researches since the discovery period is almost over and we are entering the phase of precise measurements. Despite the limited statistics collected for some variables, the three–flavour neutrino framework seems well strengthening. However some relevant pieces of this framework are still missing. The amount of a possible CP violation phase and the mass ordering are among the most challenging and probably those that will be known in the near future. In this paper we will discuss these two correlated issues and a very recent new statistical method introduced to get reliable results on the mass ordering.

1 Introduction

The current scenario of the Standard Model (SM) of particle physics, being arguably stalled by the discovery of the Higgs boson, is *desperately* looking for new experimental inputs to provide a more comfortable theory. In parallel, experiments on neutrinos so far have been an outstanding source of novelty and unprecedented results. In the last two decades several results were obtained studying atmospheric, solar or reactor neutrinos, and more recently neutrino productions from accelerator–based beams. Almost all these results have contributed to strengthen the flavour–SM. Nevertheless, relevant parts like the leptonic CP phase and the neutrino masses are still missing, a critical ingredient being the still undetermined neutrino mass ordering. On top of that the possibility of lepton flavour violation (e.g. if neutrinos are Majorana particles) and the absolute masses of neutrinos are very open issues. Unfortunately these latter questions will probably take sometime to be answered, while the previous ones are matter of debates and experimental proposals.

The achievements of the last two decades brought out a coherent picture, namely the mixing of three neutrino flavour–states, ν_e, ν_μ and ν_τ, with three ν_1, ν_2 and ν_3 mass eigenstates. In the three–neutrino framework still an unknown parameter, highly correlated with the masses and the CP angle, δ_{CP}, is strongly pursed by current and proposed experiment, i.e. the neutrino mass ordering of the neutrino mass eigenstates. Specifically, it is still largely unconstrained the sign of $|\Delta m^2_{31}| = |m^2_3 - m^2_1|$, the difference of the squared masses of ν_3 and ν_1. The mass ordering (MO) is usually identified either as normal hierarchy (NH) when $\Delta m^2_{31} > 0$ and inverted hierarchy (IH) in the opposite case. Its importance is enormous to provide inputs for the next studies and experimental proposals, to finally

[a] e-mail: luca.stanco@pd.infn.it

clarify the needs or not for new projects, and to constraint analyses in other fields like cosmology and astrophysics.

From the recent review [1] a synthetic flow chart can be extracted (figure 1), where some key steps have been identified, the *anni mirabiles*. From the picture three specific years were fundamental to understand the neutrino paradigm: 1998, when the oscillations were first discovered by Super-Kamiokande [2], not forgetting the missing-observation of Chooz [3]; 2002, when the discovery of the solar oscillations by SNO [4], together with the oscillation pattern of reactors' antineutrino measured by KAMLAND [5], opened the way to the 3×3 mixing scenario; 2012, when θ_{13}, the third mixing angle, was finally measured by Daya-Bay, Reno and Double-Chooz reactor experiments [6] and T2K [7], an accelerator beam experiment. Next year might be the fourth key–year since NOvA provided first indications [8] on the mass ordering, and T2K recently released [9] exclusion of $\delta_{CP} = 0$ at 90% C.L.. The issue of the measurement on the mass ordering will be further discussed in this paper, while a robust measurement of $\delta_{CP} \neq 0$ could be soon obtained. Finally, from the year 2020 T2K should begin a campaign of measurements on δ_{CP} [10].

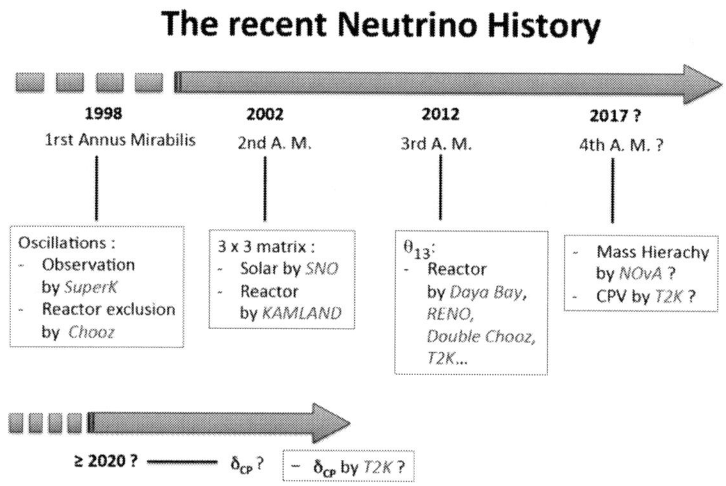

Figure 1. A synthetic view of the recent history of oscillation neutrino discoveries. Prospects for the next future are also drawn.

2 Neutrino mass ordering

The neutrino mass ordering indicates whether there are one or two highest standard-neutrino masses, as shown in figure 2 that illustrates the relative flavour sharing due to their mixing. Sensitivities to the mass ordering are given by the contemporary presence of two or more contributions in the neutrino propagator (see e.g. [11]). Specifically, as far as oscillations are concerned, neutrino mass hierarchy can be measured observing the interference of oscillations driven by $\pm\Delta m^2_{31}$ with oscillations driven by another quantity Q with known sign. For example the dependence on MO at the long-baseline experiments with neutrino beams rises up from the interference terms between the pure flavour oscillation $\nu_\mu \to \nu_e$, and the Q term that describes the resonant flavour transitions when neutrinos propagate in a medium with varying density. The high statistic atmospheric experiments like INO (India) [12], KM3/ORCA (Europe) [13] and PINGU (USA) [14] can access the MO based on the same feature. The same occurs for the collective effects in neutrino coming from supernovae explosions. Instead, the JUNO (China) experiment [15] foresees to establish the MO by looking at the interference between the solar and the atmospheric oscillations, i.e. the oscillations driven by $\Delta m^2_{21} = m^2_2 - m^2_1$ and Δm^2_{31}, respectively.

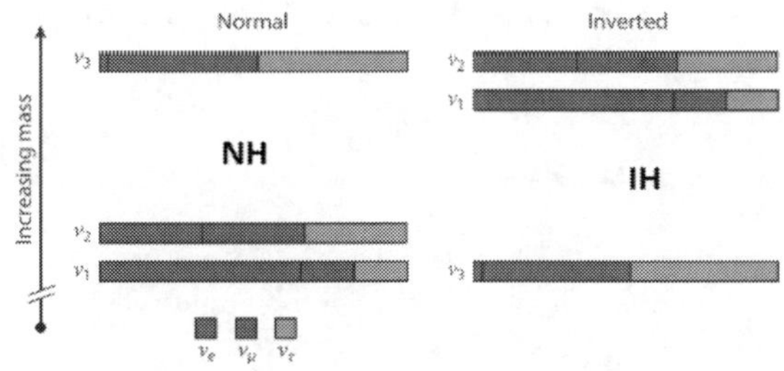

Figure 2. Illustration of the neutrino mass ordering for the two hypotheses, Normal Hierarchy (NH) and Inverted Hierarchy (IH). The flavour sharing due to the mixing is also drawn for each mass eigenstate.

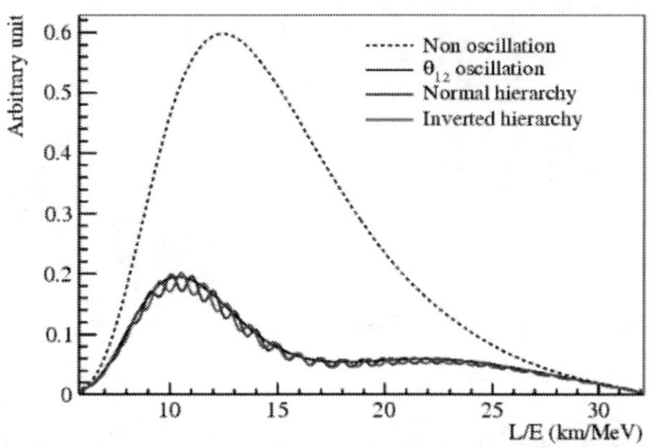

Figure 3. The relative shape difference of the reactor antineutrino flux for the two different neutrino mass ordering as function of L/E (baseline distance L over the neutrino energy E). The large discrepancy with the non-oscillation expectation is due to the solar oscillation, while the ripples are due to the atmospheric one. Taken from [16].

For example in figure 3, the interference between the solar and atmospheric oscillations brings up the ripple behaviour and its relative shift for the two options, NH and IH. The separation between the two patterns is manly matter of energy resolution. JUNO aims to achieve less than a $3\%/\sqrt{E(MeV)}$ uncertainty.

For the accelerator basis searches the NOvA experiment foresees some information be available after several years of running (figure 4). Adding measurements on δ_{CP} from few years of T2K exposure will allow to slightly increase the separation between the two options in different portions of δ_{CP} range (a nice review on the combined challenging for the current running experiments is given in [17]). However the expectation is not exciting, only a 3-sigma significance could be obtained and only in some favorable δ_{CP} regions.

 DOI: 10.1051/epjconf/201716401031

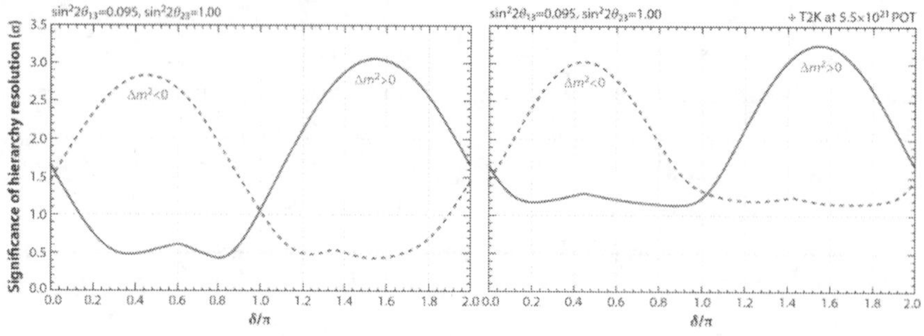

Figure 4. Prospect of the mass ordering disentanglement after several years (3 with neutrino and 3 with antineutrino) of NOvA data taking, as function of δ_{CP} (left). On the right the slight improvement in significance obtained by adding few years of T2K exposure is shown. Taken from [18].

As a matter of fact the perspectives for the determination of the neutrino mass ordering in the near future are rather poor, even less favorable than the prospects for the δ_{CP} measurement. The overall scenario is summarized in figure 5[1].

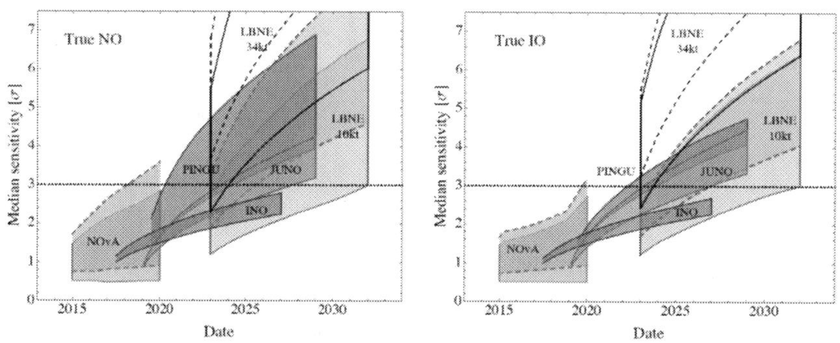

Figure 5. Perspectives for the neutrino mass ordering in the next future. The left (right) panel shows the median sensitivity in number of sigmas for rejecting the IO (NO) if the NO (IO) is true for different facilities as a function of the date. The width of the bands correspond to different true values of the CP phase δ for NOvA and LBNE (now DUNE [19]), different true values of θ_{23} between 40^0 and 50^0 for INO and PINGU, and energy resolution between $3\%/\sqrt{E(MeV)}$ and 3.5% for JUNO. For the long baseline experiments, the bands with solid (dashed) contours correspond to a true value for θ_{23} of 40^0 (50^0). NO (IO) stands for NH (IH). Taken from [20].

There are concerns for each of the facilities:

- NOvA would need some luckiness;
- PINGU (and ARCA/ORCA) are not yet fully funded;
- INO would not reach a good significance;

[1]The recent update given by A. Heijboer [21] at the NOW2016 conference does not really change the perspectives, but a shortening of the period needed to PINGU/ORCA to reach a high sensitivity: starting hopefully in 2020 only 4 years instead of 8 would be needed to get around 5 sigmas. See also the recent ARCA/ORCA Letter of Intent [13].

 DOI: 10.1051/epjconf/201716401031

- JUNO would cope with a very challenging energy resolution;
- DUNE should be able to get it in few years but it is not clear if and when it would start.

Thus it becomes highly worth to look for new techniques on analyses of the mass ordering in order to improve the significance level and to gain in time schedule. Following the discussion in [1] a change of perspective is first of all needed. Technically one should focus on the rejection of the wrong hierarchy and not on the observation of the true one. Therefore it is mandatory to introduce new test statistics that would allow to distinguish between NH and IH within that approach.

Moreover it is also time to work out a comprehensive handling of all the future measurements on MO. Reasons to alternatively obtain a robust result from only one experiment are due to the lack of confidence on the 3-neutrino framework and/or the cross-correlation of the systematic errors between different experiments. The first concern should be targeted with specific experiments and it should not attain the extraction of the oscillation parameters. The second concern about the systematic errors should not prevent to conservatively use one experiment as a guideline adding information from the other ones in a controlled way.

In the second part of this paper focus will be given on a very recent new statistical technique [22] that could be extensively applied to single or multiple measurements of the neutrino mass ordering.

3 Standard techniques for MO

Provided the current knowledge of the oscillation parameters [23], with uncertainties from few percents to more than 10%, all the methods developed so far for establishing whether MO is normal or inverted are based on the computation of the χ^2 difference with respect to the best fit solutions for the NH and IH cases [24]. These analyses make use of the test statistic

$$\Delta\chi^2_{min} = \chi^2_{min}(IH) - \chi^2_{min}(NH), \tag{1}$$

where the two minima are evaluated spanning the uncertainties of the three-neutrino oscillation parameters, namely Δm^2_{21}, $\pm\Delta m^2_{31}$, θ_{12}, θ_{23}, θ_{13} and δ_{CP}, θ_{ij} $(i, j = 1, 2, 3)$ being the mixing angles in the standard parameterization. The statistical significance in terms of σ's is computed as $\sqrt{\Delta\chi^2}$. The limits of such procedures are well known [25]. In particular, the significance corresponds only to the median expectation and does not consider the intrinsic statistical fluctuations. Thus errors of type I and II [27] should be taken into account when comparing the probability density functions of each χ^2_{min}, the corrected significance is lower and more sigma's have to be gained to reach a robust observation. Despite this and other caveats no alternative test statistic has been outlined so far.

Meanwhile first results from NOvA became available last year even if only with an exposure of 2.74×10^{20} p.o.t. [26]. The standard analysis via $\Delta\chi^2_{min}$ on mass ordering performed on the 2015 NOvA $\nu_\mu \to \nu_e$ appearance is reported in figure 6. It is confirmed that no real sensitivity is available yet on MO.

4 A new techniques for MO

The new technique reported in [22] is based on a new test statistic that properly weights the intrinsic statistical fluctuations of the data and extracts the significances of either NH or IH. First, the Poisson distributions for n_i observed events $f_{MO}(n_i; \mu_{MO}|\delta_{CP})$ are considered, where $\mu_{MO}(\delta_{CP})$ are the expected number of events as function of δ_{CP}, MO standing for IH or NH. For a specific n the left and right cumulative functions of f_{IH} and f_{NH} are computed and their ratios, q_{MO}, are evaluated. The ratios are similar to the CL$_s$ test statistic [28] used for the Higgs discovery. Since for the ν_e appearance at NOvA

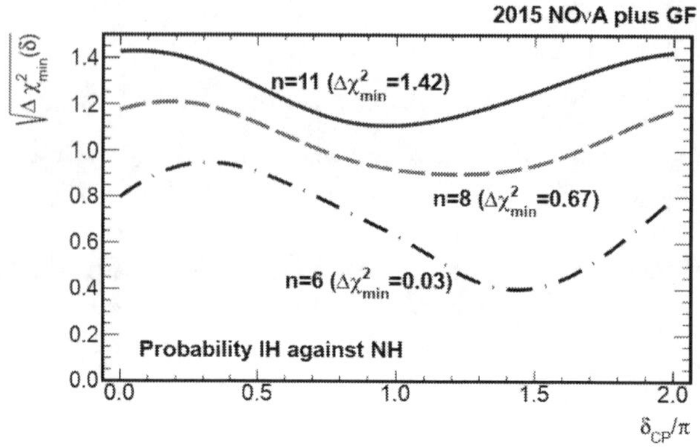

Figure 6. The standard technique based on $\Delta\chi^2_{min}$ applied to the 2015 NOvA results. The significance of IH exclusion as function of δ_{CP} is shown. The three curves correspond to the significances for different numbers of observed events (6, 8 or 11) by NOvA in the $\nu_\mu \to \nu_e$ channel, for a total exposure of 2.74×10^{20} p.o.t. Minimization over θ_{23} and θ_{13} and use of the best fit values from the global fit (GF) analysis [23] for the other parameters, are performed. The quoted $\Delta\chi^2_{min}$ correspond to $\chi^2_{min}(IH) - \chi^2_{min}(NH)$ computed in the whole range of δ_{CP}. Taken from [22].

the number of expected events as function of δ_{CP} is asymmetric towards IH and NH (less events are expected for IH than for NH), the CL_s are defined for either the IH or the NH case:

$$q_{IH}(n,\delta_{CP}) = \frac{\sum_{n_i^{IH}\geq n} f_{IH}(n_i^{IH};\mu_{IH}|\delta_{CP})}{\sum_{n_i^{NH}\geq n} f_{NH}(n_i^{NH};\mu_{NH}|\delta_{CP})}, \quad q_{NH}(n,\delta_{CP}) = \frac{\sum_{n_i^{NH}\leq n} f_{NH}(n_i^{NH};\mu_{NH}|\delta_{CP})}{\sum_{n_i^{IH}\leq n} f_{IH}(n_i^{IH};\mu_{IH}|\delta_{CP})}.$$

q_{IH} and q_{NH} are two discretized random variables comprised to the [0, 1] interval. As n goes to zero q_{IH} goes to one, while when n increases q_{IH} asymptotically tends to zero. q_{NH} behaves the other way around towards n. For illustration purpose the behaviours of f_{MO} and q_{MO} are shown in Fig. 7 for a typical case ($n = 8$).

The probability mass functions, $P(n)$, of each q_{MO} have been computed via toy Monte Carlo simulations based either on f_{IH} (test of IH against NH) or f_{NH} (test of NH against IH). They are further compared to the observed data n_D, the number of observed events either in real data or in Monte Carlo simulation, evaluating the p–value probabilities for n_D:[2]

$$p_{IH}(n_D,\delta_{CP}) = \sum_{q'_{IH}\leq q_{IH}(n_D)} P_{IH}(q'_{IH}|\delta_{CP}), \qquad p_{NH}(n_D,\delta_{CP}) = \sum_{q'_{NH}\leq q_{NH}(n_D)} P_{NH}(q'_{NH}|\delta_{CP}).$$

Finally the significance has been computed with the one–sided option since this corresponds to 0 sigma ($Z = 0$) for $p = 50\%$ that equalizes the IH and NH probabilities. Within that choice Z is defined as $Z = \Phi^{-1}(1-p)$, where Φ^{-1} is the quantile of the standard Gaussian and Z is the number of standard deviations.

With the new method an averaged increase of 0.5 σ with respect to the standard $\Delta\chi^2_{min}$ is obtained [22]. Worth to note that the increase is not constant but it depends on the discrimination threshold n_D and δ_{CP}: the gain of the new method in terms of the number of sigma's strongly raises with n_D and "favorable" regions of δ_{CP}. As demonstrated in the appendix of [29] the new method is

[2] It is worthwhile to note that the method does not really need the double definition of q_{IH} and q_{NH} since they are complementary. All the results can be obtained using only one expression, taking the p–value in the proper domain. The procedure described in the text is at ease of the reader.

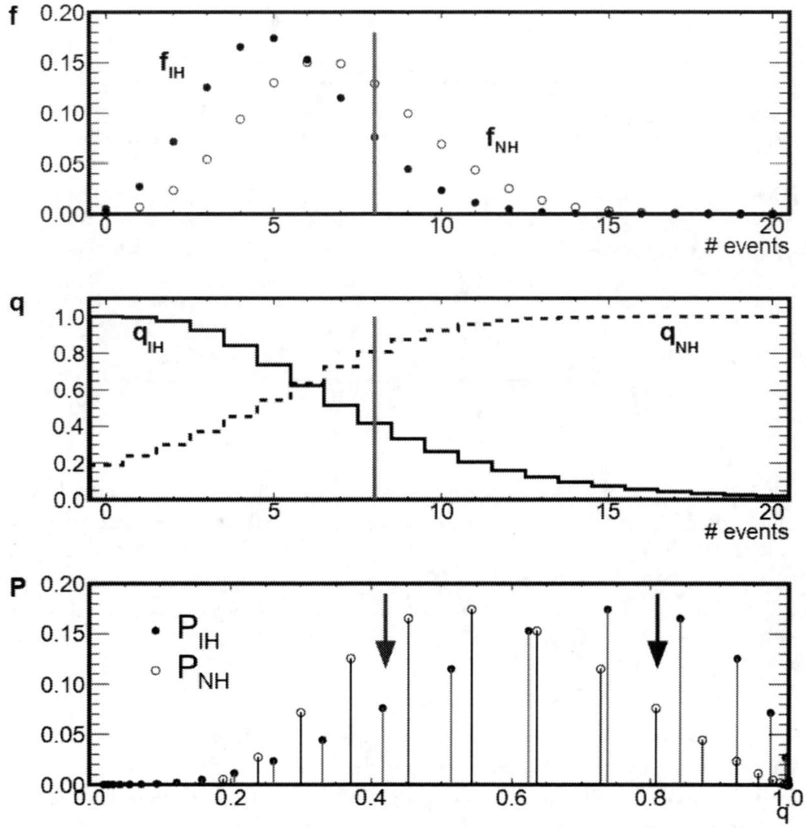

Figure 7. The behaviours of the distributions related to the new estimator q_{MO}. Top: the expected Poisson distributions (signal plus background) for IH (plain dots) and NH (void dots), for $\delta_{CP} = 3/2\,\pi$ and 2.74×10^{20} p.o.t., in the 2015 NOvA analysis. Middle: the corresponding values assumed by q_{IH} (plain line) and q_{NH} (dashed line). Bottom: the probability mass functions of q_{IH} (plain dots) and q_{NH} (void dots). The arrows indicate the thresholds used to compute q_{IH} and q_{NH} for $n = 8$. Taken from [22].

generally better than $\Delta\chi^2_{min}$ for many reasons: it deals with the full probability distributions, it profits of the intrinsic fluctuations of the data and, most relevant, it looks to the right question (to disprove one MO option). In fact the new q estimator focusses on the possibility to reject the wrong hierarchy, disregarding the other one. Therefore, once one option is selected (e.g. rejection of IH) it does not provide any evaluation on the other option (rejection of NH). Instead the $\Delta\chi^2_{min}$ method treats the two option in a symmetric way with the disadvantage of mixing up the information.

Results are much more promising when the data sample increases. For example if a factor 3 in exposure is applied to the 2015 NOvA analysis the rejection of IH can reach the 95% C.L. in almost the full range of δ_{CP}, even including the current uncertainties on θ_{23} and θ_{13} in a Bayesian way (figure 8). The new method works better when some statistical fluctuations with respect to the median expectations should be observed. The 95% C.L. exclusion is obtained for the case $3 \times 8 = 24$ events (3× 2015 NOvA conditions) that is slightly far from the median expectations (at $\delta_{CP} = 1.5\,\pi$, $3 \times 5.3 = 15.8$ and $3 \times 6.9 = 20.8$ are expected for IH and NH, respectively). Including the systematic errors, scaled from the 2015 NOvA analysis, the achievement is not spoiled [22], with about a decrease of 0.3-0.4 σ in significance.

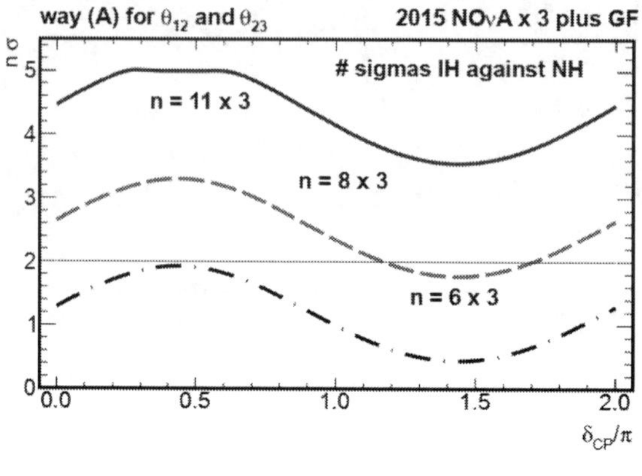

Figure 8. The significance from the the new method in terms of σ (one-sided option) is shown for a total exposure of 8.22×10^{20} p.o.t. (top), assuming the same efficiency for the signal and the same level of background rejection of 2015 NOvA analysis. The tested hypothesis is IH against NH. The different curves correspond to different possible observations of events, 18, 24 and 33 (signal plus background), as estimated by 2015 NOvA in the $\nu_\mu \to \nu_e$ channel. The uncertainties on θ_{23} and θ_{13} are treated as nuisances (way A). Taken from [22].

NOvA collaboration just released preliminary new results [30] obtained from a total exposure of 6.05×10^{20} p.o.t., a factor 2.2 with regard to the 2015 results. 33 events for $\nu_\mu \to \nu_e$ appearance production were found. However the background level was highly enhanced (factor 4.5) against an increase of the efficiency of a factor 2.5. By scaling these number to the 2015 analysis and exposure, the 33 events of 2016 corresponds to about 6 events in 2015. That is around the median expectation without an even moderate fluctuation. Any how, applying the new method [29], the gain in exposure from 2015 to 2016 is sufficient to obtain a first important result: the inverted hierarchy can be excluded at 95% C.L. in the δ_{CP} interval [0.10 π, 0.77 π] (Fig. 9).

5 Conclusions

The neutrino mass ordering (MO) is one of the most relevant questions to be answered in the near future. Despite the huge investments in running experiments (NOvA and T2K) and future projects (JUNO and ORCA/PINGU at first level) the real possibility to reach a robust result on MO is still too low, even in 10 years from now. Its strong correlation with the δ_{CP} phase contributes to puzzling the scenario. It is mandatory to explore new ways in statistical analysis, also motivated by the too simple methods applied so far, all based on the minimization of the χ^2. The new method introduced in [22] is quite promising. In the next few years, working out first the NOvA data sample, it would provide reliable results on the mass hierarchy, within the framework of the 3–flavour neutrinos. If the future NOvA results would be in line with its 2015 result where some fluctuation from the median expectations were observed, and assuming the three-neutrino oscillation paradigm be true, then the inverted hierarchy could be disproved at 95% C.L. in the full range of δ_{CP}. Meanwhile, with the preliminary 2016 NOvA data release, the new method allows to exclude IH at 95% C.L. in the δ_{CP} interval [0.10 π, 0.77 π]. Moreover, plans of the authors are to add T2K data and to extend their technique to the next projects, JUNO and ORCA/PINGU.

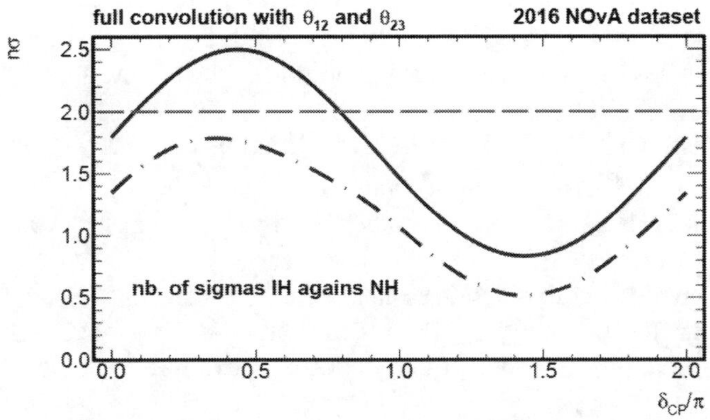

Figure 9. The exclusion of the inverted hierarchy as obtained by the new statistical method applied to the recent release of NOvA 2016 results. The uncertainties on θ_{23} and θ_{13} are treated as nuisances, i.e. with a full convolution of their probability densities. IH is rejected at 95% C.L. in the δ_{CP} interval [0.10 π, 0.77 π]. The standard analysis with $\Delta\chi^2_{min}$ is also shown as dashed-dot line. Taken from [29].

To conclude it is time to look forward and evaluate the consequences of a possible MO knowledge to the future neutrino programs:

- T2K would have an easier life for the first measurement of δ_{CP};
- JUNO would become relevant mainly for checking the consistency of the 3ν framework, since it will measure MO almost in vacuum conditions, without the δ_{CP} dependence;
- DUNE/HyperK may be forced to partially re-tune their goals.

Acknowledgements

I wish to acknowledge the kind invitation to present for the first time publicly our new results on the neutrino mass ordering. I am also delightful of the warm hospitality of the organizers of ICFNP 2016. I also acknowledge R. Brugnera and S. Dusini for the proof-reading of the manuscript and providing relevant criticisms.

References

[1] L. Stanco, Rev. in Phys. **1**, 90 (2016) [arXiv:1511.09409].
[2] Y. Fukuda et al. (Super-Kamiokande collaboration), Phys. Rev. Lett., **81**, 1562 (1998) [hep-ex/9807003].
[3] M. Apollonio et al. (CHOOZ collaboration) Phys. Lett. B **420**, 397 (1998) [hep-ex/9711002].
[4] Q. R. Ahmad et al. (SNO collaboration), Phys. Rev. Lett., **89**, 011301 (2002) [nucl-ex/0204008].
[5] K. Eguchi et al. (KamLAND collaboration), Phys. Rev. Lett., **90**, 021802 (2003) [hep-ex/0212021].
[6] F. P. An et al. (Daya Bay collaboration), Phys. Rev. Lett. **108**, 171803 (2012) [arXiv:1203.1669]; J. K. Ahn et el. (RENO collaboration), Phys. Rev. Lett. **108**, 191802 (2012) [arXiv:1204.0626]; Y. Abe et al. (Double Chooz collaboration), Phys. Rev. Lett. **108**, 131801 (2012) [arXiv:1112.6353].
[7] K. Abe et al., Phys. Rev. Lett. **107**, 041801 (2011) [arXiv:1106.2822].
[8] R. Patterson (NOvA collaboration), talk given at the Joint Experimental-Theoretical Physics Seminar, Fermilab, 6 August 2015.
[9] H. A. Tanaka (for the T2K collaboration), talk at Neutrino2016, London (UK), 4-9 July 2016.
[10] K. Abe et al. (T2K collaboration), arXiv:1607.08004.
[11] M. Blennow and A.Yu. Smirnov, Adv. High Energy Phys. **2013** 972485 (2013).
[12] S. Ahmed et al. (INO collaboration), arXiv:1505.07380.
[13] S. Adrián-Martinez et al. (KM3 collaboration), Jour. Phys. G **43**, 084001 (2016) [arXiv:1601.07459].
[14] M. G. Aartsen et al. (PINGU collaboration), Letter of Intent, arXiv:1401.2046.
[15] JUNO collaboration, JUNO Conceptual Design Report, arXiv:1508.07166 (2015).
[16] F. An et al. (Juno collaboration), Jour. Phys. G **43**, 030401 (2016) [arXiv:1507.05613].
[17] T. Nagaya and R.K.Plunkett, New J. Phys. **18**, 015009 (2016) [arXiv:1507.08134].
[18] R.B. Patterson, Ann. Rev. Nucl. Part. Sci. **65** 177 (2015) [arXiv:1506.07917].
[19] R. Acciarri et al. (DUNE collaboration), arXiv:1512.06148.
[20] M. Blennow et al., JHEP 03, 028 (2014)[arXiv:1311.1822].
[21] A. Heijboer, talk at NOW2016, Otranto (Italy), 4-11 September 2016.
[22] L. Stanco et al., arXiv:1606.09454v1.
[23] F. Capozzi et al., Nucl. Phys. B **908**, 2018 (2016) [arXiv:1601.07777].
[24] There are many analyses for the MO determination. Some of the most relevant are: X. Qian et al., Phy. Rev. D **86** 113011 (2012); M. Blennow et al., JHEP **03**, 028 (2014); M.C. Gonzalez-Garcia et al., JHEP **11**, 052 (2014).
[25] E. Ciuffoli et al., JHEP **1401** (2014) 095; M. Blennow, JHEP **01** (2014) 139.
[26] P. Adamson et al., Phys. Rev. Lett. **116**, 151806 (2016) [arXiv:1601.05022].
[27] K.A. Olive et al. (Particle Data Group), Review of Particle Physics, Chin. Phys. C **38** (2014) 090001.
[28] A.L. Read, J. Phys. G 28 (2002); G. Cowan et al., EPJC 71 1554 (2011).
[29] L. Stanco et al., arXiv:1606.09454v2.
[30] P. Vahle (for the NOvA collaboration), talk at Neutrino2016, London (UK), 4-9 July 2016.

The true quantum face of the "exponential" decay: Unstable systems in rest and in motion

K. Urbanowski[1,a]

[1] *University of Zielona Góra, Institute of Physics, ul. Prof. Z. Szafrana 4a, 65–516 Zielona Góra, Poland.*

Abstract. Results of theoretical studies and numerical calculations presented in the literature suggest that the survival probability $\mathcal{P}_0(t)$ has the exponential form starting from times much smaller than the lifetime τ up to times $t \gg \tau$, and that $\mathcal{P}_0(t)$ exhibits inverse power–law behavior at the late time region for times longer than the so–called crossover time $T \gg \tau$ (The crossover time T is the time when the late time deviations of $\mathcal{P}_0(t)$ from the exponential form begin to dominate). More detailed analysis of the problem shows that in fact the survival probability $\mathcal{P}_0(t)$ can not take the pure exponential form at any time interval including times smaller than the lifetime τ or of the order of τ and it has has an oscillating form. We also study the survival probability of moving relativistic unstable particles with definite momentum $\vec{p} \neq 0$. These studies show that late time deviations of the survival probability of these particles from the exponential–like form of the decay law, that is the transition times region between exponential–like and non-exponential form of the survival probability, should occur much earlier than it follows from the classical standard considerations.

1 Introduction

The studies of the process of decay of radioactive elements were initialized by the discovery of radioactivity in the nineteenth century. Experiments have shown that the number of radioactive elements in the sample, which emits radioactive radiation, decreases with the passage of time. These observations and assumption that the decay rate follows the laws of probability led Rutherford and Sody to the formulation of radioactive decay law as a function of time [1–3]. This radioactive decay law allows to determine the number $N(t)$ of atoms of the radioactive element at the instant t knowing the initial number $N_0 = N(0)$ of them at initial instant of time $t_0 = 0$ and has the exponential form: $N(t) = N_0 \exp[-\lambda t]$, where $\lambda > 0$ is a constant. Since then, the belief that the decay law has the exponential form has become common. The rise of Quantum Mechanics led to an understanding that the radioactive decay similarly to the process of emission of photons by excited atoms are time dependent quantum processes. Considering the excited atomic levels and applying the Shrödinger equation to describe the time evolution Weisskopf and Wigner found that to a good approximation the non–decay probability of the exited levels is a decreasing function of time having exponential form [4]. Further studies of the quantum decay process showed that basic principles of the quantum theory does not allow it to be described by an exponential decay law at very late times [5, 6] and at

[a]e-mail: K.Urbanowski@if.uz.zgora.pl;k.a.urbanowski@gmail.com

initial stage of the decay process (see [6] and references therein). Theoretical analysis shows that at late times the survival probability (i. e. the decay law) should tends to zero as $t \to \infty$ much more slowly than any exponential function of time and that as function of time it has the inverse power–like form at this regime of time [5, 6]. There was many unsuccessful attempts to verify experimentally predicted deviations from the exponential form of the decay law at late times regime (see eg. [7]). The first experimental evidence of deviations of the decay law from exponential form at long time regime was reported in [8]. The early times properties of the decay process lead to the so called Quantum Zeno Effect [9, 10], that is to slowing down sufficiently frequently observed decay process up to stop it down in the case of the continuously observed an unstable system. The experimental confirmation of this effect was reported, e.g. in [11] and recently in [12]. All these results of theoretical and experimental researches caused that there are rather widespread belief that a universal feature of the quantum decay process is the presence of three time regimes of the decay process: the early time (initial), exponential (or "canonical"), and late time having inverse–power law form [13]. This belief is reinforced by a numerous presentations in the literature decay curves obtained for quantum models of unstable systems. The typical form of such a decay curve one can find in Fig. (1). In this context, each experimental evidence of oscillating decay curve at times of the order of life times is considered as an anomaly caused by a new quantum effects or new interactions: As an example of such a situation one can recall the result of the GSI experiment, where an oscillating decay rate of the ionized isotopes ^{140}Pr and ^{142}Pm moving with relativistic velocity ($\gamma \simeq 1.43$, where γ is the Lorentz factor) was observed [14, 15]. The question arises, if indeed in the case of one component quantum unstable systems these oscillations of the decay process at the "exponential" regime are an anomaly, or perhaps universal feature of quantum decay processes.

In the light of the mentioned results of the GSI experiment the another question arises: Whether and how the oscillations depend on the motion of the unstable quantum system. To find an answer to this question we need to know how to describe the decay process of unstable quantum systems in motion. This is quite general problem problem encountered by researchers studying properties of moving unstable particles: Namely physicists studying the decay processes are often confronted with the problem of how to predict the form of the decay law of the particle moving in respect to the rest reference frame of the observer knowing the decay law of this particle decaying in its rest frame. From the standard, text book considerations one finds that if the decay law of the unstable particle in rest has the exponential form $\mathcal{P}_0(t) = \exp[-\frac{\Gamma_0 t}{\hbar}]$ then the decay law of the moving particle with momentum $p \neq 0$ is $\mathcal{P}_p(t) = \exp[-\frac{\Gamma_0 t}{\hbar\gamma}]$, where t denotes time, Γ_0 is the decay rate (time t and Γ_0 are measured in the rest reference frame of the particle) and γ is the relativistic Lorentz factor, $\gamma \equiv 1/\sqrt{1-\beta^2}$, $\beta = v/c$, $v = |\vec{v}|$ is the velocity of the particle, $\vec{v} = c\vec{p}/\sqrt{\vec{p}^2 + m_0^2 c^2}$ and m_0 – is the rest mass. This equality is the classical physics relation. It is almost common belief that this equality is valid also for any t in the case of quantum decay processes and does not depend on the model of the unstable particles considered. For the proper interpretation of many accelerator experiments with high energy unstable particles as well as of results of observations of astrophysical processes in which a huge numbers of elementary particles (including unstable one) are produced we should be sure that this belief is supported by theoretical analysis of quantum models of decay processes. The problem seems to be extremely important because from some theoretical studies it follows that in the case of quantum decay processes this relation is valid to a sufficient accuracy only for not more than a few lifetimes $\tau_0 = \hbar/\Gamma_0$ [16–19]. On the other hand all known tests of the relation $\mathcal{P}_p(t) = \exp[-\frac{\Gamma_0 t}{\hbar\gamma}]$ were performed for times of the order of τ_0 (see, eg. [20, 21]) and for times longer than a few lifetimes this relation was not tested till now. What is more it appears that this relation may not apply in the case of the mentioned famous result of the GSI experiment. So we can see that the problem requires a deeper analysis. In these considerations the basis of such an analysis will be the formalism developed in [16, 17] where within the quantum field theory the formula for the survival amplitude of moving particles was derived. We will follow the method used in [19] and we will analyze numerically properties survival probability for a model of the unstable particle based on the Breit–Wigner mass distribution and considered therein.

So, in the following it will be shown that that in the case of unstable systems in rest there is no time interval in which the survival probability (decay law) could be a decreasing function of time of the purely exponential form. We also show that even in the case of a single component unstable system the decay curve has an oscillatory form with a smaller or a large amplitude of oscillations depending on the model considered. Next it will also be shown that the relativistic treatment of the problem within the Stefanovich–Shirokov theory [16, 17] yields decay curves tending to zero as $t \to \infty$ much slower than one would expect using classical time dilation relation which confirms and generalizes some conclusions drawn in [19]. Our results shows that conclusions relating to the quantum decay processes of moving particles based on the use of the classical physics time dilation relation need not be universally valid.

2 Preliminaries

The main information about properties of quantum unstable systems is contained in their survival probability. Let the reference frame O_0 be the common inertial rest frame for the observer and for the unstable system. Then if one knows that the system in the rest frame is in the initial unstable state $|\phi\rangle \in \mathcal{H}$, ($\mathcal{H}$ is the Hilbert space of states of the considered system), which was prepared at the initial instant $t_0 = 0$, one can calculate its survival probability (the decay law), $\mathcal{P}_0(t)$, of the unstable state $|\phi\rangle$ decaying in vacuum, which equals

$$\mathcal{P}_0(t) = |a_0(t)|^2, \tag{1}$$

where $a_0(t)$ is the probability amplitude of finding the system at the time t in the rest frame O_0 in the initial unstable state $|\phi\rangle$,

$$a_0(t) = \langle\phi|\phi(t)\rangle. \tag{2}$$

and $|\phi(t)\rangle$ is the solution of the Schrödinger equation for the initial condition $|\phi(0)\rangle = |\phi\rangle$, which has the following form within the system units $\hbar = c = 1$ used in the next parts of this paper:

$$i\frac{\partial}{\partial t}|\phi(t)\rangle = H|\phi(t)\rangle, \tag{3}$$

where $|\phi\rangle, |\phi(t)\rangle \in \mathcal{H}$, and H is the total self–adjoint Hamiltonian for the system considered. It is important to remember that if $|\phi\rangle$ represents an unstable state then it cannot be an eigenvector for H: In such a case the eigenvalue equation $H|\phi\rangle = \epsilon_\phi|\phi\rangle$ has no solutions for $|\phi\rangle$ under considerations.

We have $|\phi(t)\rangle = U(t)|\phi\rangle$, where $U(t)$ is unitary evolution operator and $U(0) = \mathbb{I}$ is the unit operator. Therefore $a(t) \equiv \langle\phi|U(t)|\phi\rangle$. The one–parameter family of unitary operators $U(t)$ forms group: $U(t_1)\,U(t_2) = U(t_1 + t_2)$. The the total Hamiltonian H of the system is a generator of this group and operators H and $U(t)$ have common eigenfunctions.

The rest reference frame O_0 is defined using common solution of the eigenvalue problem for H and the momentum operator $\mathbf{P}$:

$$\mathbf{P}|\mu; p\rangle = \vec{p}|\mu; p\rangle, \tag{4}$$

and

$$H|\mu; p\rangle = E'(\mu, p)\,|\mu; p\rangle, \tag{5}$$

where $\mu \equiv E'(\mu, 0)$ and $\sigma_c(H)$ is the continuous part of the spectrum of the Hamiltonian H. Operators H and $\mathbf{P}$ act in the state space $\mathcal{H}$.

DOI: 10.1051/epjconf/201716401032

In the rest reference frame of the quantum unstable system O_0, when $\vec{p} = 0$, we have $|\mu;0\rangle = |\mu;p=0\rangle$,

$$\mathbf{P}|\mu;0\rangle = 0, \tag{6}$$

and

$$H|\mu;0\rangle = \mu\,|\mu;0\rangle, \quad \mu \in \sigma_c(H), \tag{7}$$

Eigenvectors $|\mu;0\rangle$ are normalized as usual:

$$\langle 0;\mu|\mu';0\rangle = \delta(\mu - \mu'). \tag{8}$$

Now we can model the unstable system in the rest system O_0 as the following wave–packet $|\phi_0\rangle \equiv |\phi_{\vec{p}=0}\rangle \stackrel{\text{def}}{=} |\phi\rangle$,

$$|\phi_0\rangle \equiv |\phi\rangle = \int_{\mu_0}^{\infty} c(\mu)\,|\mu;0\rangle\,d\mu, \tag{9}$$

where expansion coefficients $c(\mu)$ are functions of the mass parameter μ, that is of the rest mass μ. (Here μ_0 is the lower bound of the spectrum $\sigma_c(H)$ of H). We require the state $|\phi_0\rangle$ to be normalized: So it has to be $\int_{\mu_0}^{\infty}|c(\mu)|^2\,d\mu = 1$.

The expansion (9) and (7) allow one to find the amplitude $a_0(t)$ and to write

$$a_0(t) \equiv \int_{\mu_0}^{\infty} \omega(\mu)\,e^{-\,i\,\mu\,t}\,d\mu, \tag{10}$$

where $\omega(\mu) \equiv |c(\mu)|^2 > 0$.

So the amplitude $a_0(t)$, and thus the decay law $\mathcal{P}_0(t)$ of the unstable state $|\phi\rangle$, are completely determined by the density of the mass (energy) distribution $\omega(\mu)$ for the system in this state [22] (see also: [5, 6, 23–27]. From (10) and from the Riemann–Lebesque lemma it follows that $|a(t)| \to 0$ as $t \to \infty$. It is because from the normalization condition $a_0(0) \equiv \int_{Spec.(H)} \omega(\mu)\,dm = 1$ it follows that $\omega(\mu)$ is an absolutely integrable function. (Note that this approach is also applicable in Quantum Field Theory models [28, 29]).

Khalfin assuming that the spectrum of H must be bounded from below, $\mu_0 > -\infty$, and using the Paley–Wiener Theorem [30] proved that in the case of unstable states there must be $|a_0(t)| \geq A\,\exp\,[-b\,t^q]$, for $|t| \to \infty$, (see [5]). Here $A > 0$, $b > 0$ and $0 < q < 1$. Therefore the decay law $\mathcal{P}_0(t)$ of unstable states decaying in the vacuum, (1), can not be described by an exponential function of time t if time t is suitably long, $t \to \infty$, and that for these lengths of time $\mathcal{P}_0(t)$ tends to zero as $t \to \infty$ more slowly than any exponential function of t.

Note that the use of the Schrödinger equation (3) allows one to find that within the problem considered.

$$i\frac{\partial}{\partial t}\langle\phi|\phi(t)\rangle = \langle\phi|H|\phi(t)\rangle. \tag{11}$$

This relation leads to the conclusion that the amplitude $a_0(t)$ satisfies the following equation

$$i\frac{\partial a_0(t)}{\partial t} = h(t)\,a_0(t), \tag{12}$$

where

$$h(t) = \frac{\langle\phi|H|\phi(t)\rangle}{a_0(t)}, \tag{13}$$

and $h(t)$ is the effective Hamiltonian governing the time evolution in the subspace of unstable states $\mathcal{H}_{||} = P\mathcal{H}$, where $P = |\phi\rangle\langle\phi|$ (see [31] and also [32, 33] and references therein). The subspace

$\mathcal{H} \ominus \mathcal{H}_{\parallel} = \mathcal{H}_{\perp} \equiv Q\mathcal{H}$ is the subspace of decay products. Here $Q = \mathbb{I} - P$. There is the following equivalent formula for $h(t)$ [31–33]:

$$h(t) \equiv \frac{1}{a_0(t)} \frac{\partial a_0(t)}{\partial t}. \tag{14}$$

If $\langle\phi|H|\phi\rangle$ exists then using unitary evolution operator $U(t)$ and projection operators P and Q the relation (13) can be rewritten as follows

$$h(t) = \langle\phi|H|\phi\rangle + \frac{\langle\phi|HQ\,U(t)|\phi\rangle}{a_0(t)}. \tag{15}$$

One meets the effective Hamiltonian $h(t)$ when one starts with the Schrödinger equation for the total state space $\mathcal{H}$ and looks for the rigorous evolution equation for a distinguished subspace of states $\mathcal{H}_{\parallel} \subset \mathcal{H}$ [27, 31]. In general $h(t)$ is a complex function of time and in the case of $\mathcal{H}_{\parallel}$ of dimension two or more the effective Hamiltonian governing the time evolution in such a subspace it is a non–hermitian matrix $H_{\parallel}$ or non-hermitian operator. There is

$$h(t) = \mu_{\phi}(t) - \frac{i}{2}\gamma_{\phi}(t), \tag{16}$$

and

$$\mu_{\phi}(t) = \Re\,[h(t)], \qquad \gamma_{\phi}(t) = -2\,\Im\,[h(t)], \tag{17}$$

are the instantaneous mass (energy) $\mu_{\phi}(t)$ and the instantaneous decay rate, $\gamma_{\phi}(t)$ [31–33]. Here $\Re\,(z)$ and $\Im\,(z)$ denote the real and imaginary parts of z respectively. The relations (12), (14) and (17) are convenient when the density $\omega(\mu)$ is given and one wants to find the instantaneous mass $\mu_{\phi}(t)$ and decay rate $\gamma_{\phi}(t)$: Inserting $\omega(\mu)$ into (10) one obtains the amplitude $a_0(t)$ and then using (14) one finds the $h(t)$ and thus $\mu_{\phi}(t)$ and $\gamma_{\phi}(t)$.

In closing this Section we should pay attention to another problem: The state vector $|\phi\rangle$ of the form (9) corresponding to a quantum unstable system can not be an eigenvector of the Hamiltonian H, otherwise it would be that $\mathcal{P}_0(t) = |\langle\phi|\phi(t)\rangle|^2 = |\langle\phi|\exp[-itH]\phi\rangle|^2 \equiv 1$ for all times t. The fact that the vector $|\phi\rangle$ describing the unstable quantum system is not the eigenvector for H means that the mass (energy) of this object is not defined. Simply the mass can not take the exact constant value in this state $|\phi\rangle$. In such a case quantum systems are characterized by the mass (energy) distribution density $\omega(\mu)$ and the average mass $< m >= \int_{\mu_0}^{\infty} \mu\,\omega(\mu)\,d\mu$ or by the instantaneous mass (energy) $\mu_{\phi}(t)$ but not by the exact value of the mass.

3 The Breit–Wigner model

It turns out that the decay curves obtained in the very large class of models are very similar in form to the curves calculated for $\omega(\mu)$ having a Breit–Wigner form (see [23] and analysis in [6]). So to find the most typical properties of the decay curve it is sufficient to make the relevant calculations for $\omega(\mu)$ modeled by the the Breit–Wigner distribution of the energy density $\omega_{BW}(\mu)$. The typical form of the survival probability $\mathcal{P}_0(t)$ obtained in such a way is presented in Fig (1), where the calculations were made for $\omega(\mu) \equiv \omega_{BW}(\mu)$,

$$\omega_{BW}(\mu) = \frac{N}{2\pi}\,\Theta(\mu - \mu_0)\,\frac{\varGamma_0}{(\mu - m_0)^2 + (\frac{\varGamma_0}{2})^2}, \tag{18}$$

where N is a normalization constant.

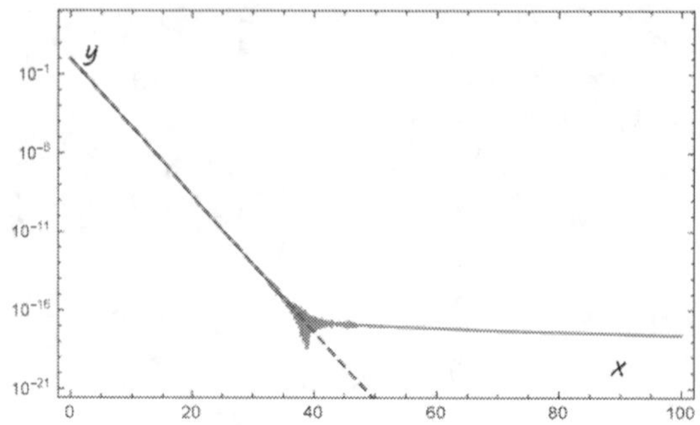

Figure 1. Decay curves obtained for $\omega_{BW}(E)$ given by Eq. (18). Axes: $x = t/\tau_0$ — time t is measured in lifetimes τ_0, y — survival probabilities on a logarithmic scale (The solid line: the decay curve $\mathcal{P}_0(t) = |a_0(t)|^2$; The dotted line: the canonical decay curve $\mathcal{P}_c(t) = |a_c(t)|^2$. The case $s_R = \frac{E_R}{\Gamma_0} = 1000$.

The standard canonical form of the survival amplitude $a_c(t)$, is given by the following relation,

$$a_c(t) = \exp\,[-i\frac{t}{\hbar}\,(m_0 - \frac{i}{2}\,\Gamma_0)]. \tag{19}$$

Γ_0 is the decay rate and $\frac{\hbar}{\Gamma_0} \equiv \frac{1}{\Gamma_0} = \tau_0$ is the lifetime within the assumed system of units $\hbar = c = 1$ (time t and Γ_0 are measured in the rest reference frame of the particle).

The case $\omega(\mu) = \omega_{BW}(\mu)$ is the typical case considered in numerous papers and used therein to model decay processes. Therefore it is very important to analyze real form of the decay curves obtained using $\omega(\mu) = \omega_{BW}(\mu)$ and this is why we consider this case in this paper. What is more inserting, substituting into $\omega_{BW}(\mu)$ into (10) allows one to find the analytical formula for the amplitude $a_0(t)$. The result is (see, eg. [32–34]),

$$a_0(t) = N\,e^{-\frac{i}{\hbar}(m_0 - i\frac{\Gamma_0}{2})t}\Big\{1 - \frac{i}{2\pi}\Big[e^{\frac{\Gamma_0 t}{\hbar}}\,E_1\Big(-\frac{i}{\hbar}(m_R + \frac{i}{2}\Gamma_0)t\Big) - E_1\Big(-\frac{i}{\hbar}(m_R - \frac{i}{2}\Gamma_0)t\Big)\Big]\Big\}, \tag{20}$$

where $E_1(x)$ denotes the integral–exponential function defined according to [35, 36] and $m_R = m_0 - \mu_0$.

It is convenient to consider the following function

$$\zeta(t) \stackrel{\text{def}}{=} \frac{a_0(t)}{a_c(t)}. \tag{21}$$

There is

$$|\zeta(t)|^2 = \frac{\mathcal{P}_0(t)}{\mathcal{P}_c(t)}, \tag{22}$$

where $\mathcal{P}_c(t) = |a_c(t)|^2$ is the canonical exponential form of the decay law. Analysis of properties of this function allows one to visualize all the more subtle differences between $\mathcal{P}_0(t)$ and $\mathcal{P}_c(t)$. For example, if one finds a time interval $[t_1, t_2]$ such that $\zeta(t) = const$ for $t \in [t_1, t_2]$ this will mean that the survival probability $\mathcal{P}(t)$ has purely exponential form in this time interval.

The function $\zeta(t)$ takes the following form in the case of the unstable system modeled by $\omega_{BW}(m)$:

$$\zeta(t) \equiv N\,\Big\{1 - \frac{i}{2\pi}\Big[e^{\frac{\Gamma_0 t}{\hbar}}\,E_1\Big(-\frac{i}{\hbar}(m_R + \frac{i}{2}\Gamma_0)t\Big) - E_1\Big(-\frac{i}{\hbar}(m_R - \frac{i}{2}\Gamma_0)t\Big)\Big]\Big\}. \tag{23}$$

This function was used to find numerically $|\zeta(t)|^2$ for $\omega(m) = \omega_{BW}(m)$. Results of numerical calculations are presented in Figs (2) and (3): It turns out that in the case considered the form of $|\zeta(t)|^2$ and $\mathcal{P}_0(t)$depend on the ratio $s_R \stackrel{\text{def}}{=} \frac{m_R}{\Gamma_0} \equiv \frac{m_0-\mu_0}{\Gamma_0}$.

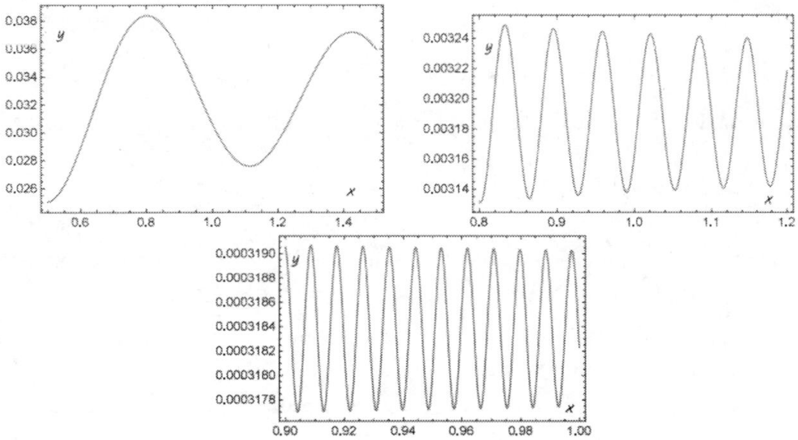

Figure 2. A comparison of decay curves obtained for $\omega_{BW}(\mu)$ given by Eq. (18) with canonical decay curves. Axes: $x = t/\tau_0$ — time t is measured in lifetimes τ_0, y — The function $f(t) = (|\zeta(t)|^2 - 1) = \frac{\mathcal{P}_0(t)}{\mathcal{P}_c(t)} - 1$, where $\zeta(t)$ is defined by the formula (21). The top panel: $s_R = 10$. The middle panel: $s_R = 100$. The lower panel: $s_R = 1000$.

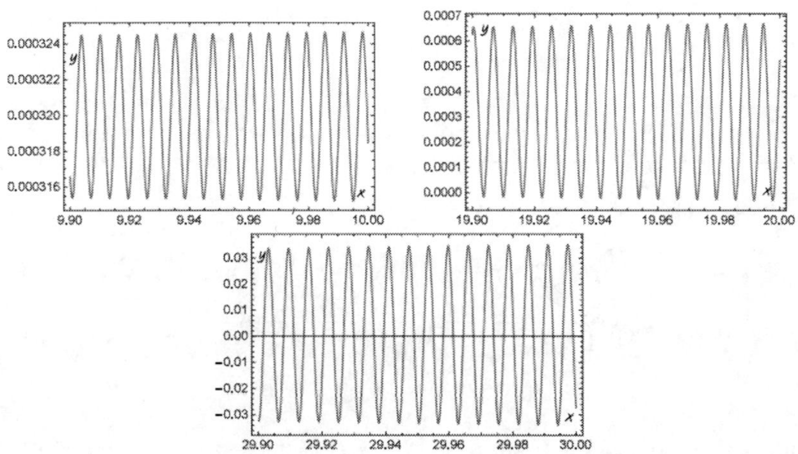

Figure 3. A comparison of decay curves obtained for $\omega_{BW}(\mu)$ given by Eq. (18) with canonical decay curves. Axes: $x = t/\tau_0$ — time t is measured in lifetimes τ_0, y — The function $f(t) = (|\zeta(t)|^2 - 1) = \frac{\mathcal{P}_0(t)}{\mathcal{P}_c(t)} - 1$, where $\zeta(t)$ is defined by the formula (21), $\mathcal{P}_0(t) = |a_0(t)|^2$, $\mathcal{P}_c(t) = |a_c(t)|^2$. The case $s_R = 1000$.

The derivative of $\zeta(t)$ given by (23) equals [32, 33]

$$\frac{\partial \zeta(t)}{\partial t} = i\,\frac{N}{2\pi}\,\frac{\Gamma_0}{\hbar}\,e^{\frac{\Gamma_0}{\hbar}t}\,E_1\Big(-\frac{i}{\hbar}(m_R + \frac{i}{2}\Gamma_0)t\Big). \tag{24}$$

From the properties of the integral–exponential function $E_1(x)$ it follows that the equation $\frac{\partial\zeta(t)}{\partial t} = 0$ can be satisfied at most for some isolated values of time t. So, from the formula (24) the conclusion follows: Within the model considered there is no time interval in which $\zeta(t) = const$, that is, there is no time interval in which the survival probability $\mathcal{P}_0(t)$ has a pure exponential form. This conclusion explains the results presented in Figs (2) and (3).

4 The general case of $\zeta = \frac{a_0(t)}{a_c(t)}$

Let us consider now the general case of $\zeta(t)$. From the definition (21) it follows that

$$\frac{\partial\zeta(t)}{\partial t} \equiv \frac{i}{\hbar}(m_0 - \frac{i}{2}\Gamma_0)\,\zeta(t) \;+\; e^{+i\frac{t}{\hbar}(m_0 - \frac{i}{2}\Gamma_0)}\,\frac{\partial a(t)}{\partial t} = \frac{i}{\hbar}(m_0 - \frac{i}{2}\Gamma_0)\zeta(t) \;-\; \frac{i}{\hbar}\,h(t)\,\zeta(t). \tag{25}$$

Let us use now the relation (15) and assume that $\langle\phi|H|\phi\rangle$ exists and there exists instants $0 < t_1 < t_2 < \infty$ of time t such that for any $t \in (t_1, t_2)$ there is $\zeta(t) = \zeta(t_1) = \zeta(t_2) = const \stackrel{\text{def}}{=} c_\phi \neq 0$. In this case there should be $\frac{\partial\zeta(t)}{\partial t} = 0$ for all $t \in (t_1, t_2)$. Taking into account that by definition $\zeta(t) \neq 0$ from (25) we conclude that it is possible only and only if

$$h(t) \;-\; (m_0 - \frac{i}{2}\Gamma_0) = 0, \quad \text{for } t_1 \leq t \leq t_2, \tag{26}$$

that is if and only if

$$h(t_1) = h(t) = h(t_2) = const \stackrel{\text{def}}{=} c_h \neq 0 \quad \text{for } t_1 \leq t \leq t_2. \tag{27}$$

Using (15) and the property $|\phi(t)\rangle = U(t)\,|\phi\rangle$ one concludes that the equality $h(t_1) = h(t) = c_h$ can take place if

$$\frac{\langle\phi|HQ\,U(t_1)|\phi\rangle}{a_0(t_1)} = \frac{\langle\phi|HQ\,U(t)|\phi\rangle}{a_0(t)}. \tag{28}$$

we can replace the operator $U(t)$ by the product $U(t_1)\,U(t-t_1) \equiv U(t)$. Taking into account the group properties of the one–parameter family of unitary operators $U(t)$ we can use in (28) $U(t_1)\,U(t-t_1) \equiv U(t)$ instead of $U(t)$. Next keeping in mind that $a_0(t) \neq 0$, $a_0(t_1) \neq 0$ and taking into account that $\lambda(t, t_1) \stackrel{\text{def}}{=} \frac{a_0(t)}{a_0(t_1)}$ is a complex function one can replace the relation (28) by the following one

$$\langle\phi|HQ\,U(t_1)\Big[\,\lambda(t, t_1)|\phi\rangle \;-\; U(t-t_1)|\phi\rangle\Big] = 0. \tag{29}$$

This condition can be satisfied in two cases: The first one is

$$U(t-t_1)|\phi\rangle \;-\; \lambda(t, t_1)|\phi\rangle \;=\; 0, \tag{30}$$

and the second occurs when $[\,\lambda(t, t_1)|\phi\rangle \;-\; U(t-t_1)|\phi\rangle] \neq 0$ and vectors $(\langle\phi|H)^+ = H|\phi\rangle$ and $Q\,U(t_1)\,[\lambda(t, t_1)|\phi\rangle \;-\; U(t-t_1)|\phi\rangle]$ are orthogonal to each other.

The first case means that $h(t_1) = h(t) = c_h = const$ which by (27) means that $\frac{\partial\zeta(t)}{\partial t} = 0$ if and only if the vector $|\phi\rangle$ representing an unstable state of the system is an eigenvector for the unitary evolution operator $U(t)$. As we noted earlier the evolution operator $U(t)$ and the total Hamiltonian H of the system have common eigenvectors. This means that $h(t_1) = h(t) = c_h = const$ and thus $\frac{\partial\zeta(t)}{\partial t} = 0$ for $t \in (t_1, t_2)$ if and only if the unstable state $|\phi\rangle$ of the system is an eigenvector for H, which is in contradiction with the property that the vector $|\phi\rangle$ representing the unstable state cannot be the eigenvector for the total Hamiltonian H.

The second case: From the definition of the projectors P and Q it follows that this case can be realized only if the vector $H|\phi\rangle$ is proportional to the vector $|\phi\rangle$: $H|\phi\rangle = \alpha_\phi|\phi\rangle$, that is similarly to the first case $\frac{\partial\zeta(t)}{\partial t} = 0$ if and only if the vector $|\phi\rangle$ representing the unstable state of the system considered is an eigenvector for the total Hamiltonian H, which is again in clear contradiction with the condition that the vector $|\phi\rangle$ representing the unstable state cannot be the eigenvector for the total Hamiltonian H.

Taking into account implications of the above to possible realizations of the relation (29) we conclude the supposition that such time interval $[t_1, t_2]$ can exist that $h(t_1) = h(t) = c_h = const$ for $t \in (t_1, t_2)$ and thus $\zeta(t) = const = \zeta(t_1) = \zeta(t_2)$ for $t \in (t_1, t_2)$ is false. So taking into account the definition of $\zeta(t)$ the following conclusion follows: Within the approach considered in this paper for any time interval $[t_1, t_2]$ the decay law can not be described by the exponential function of time. This conclusion is the general one. It does not depend on models of quantum unstable states and confirms the similar conclusion drawn earlier for the Breit–Wigner model.

The another important conclusion is that at any time interval $[t_1, t_2]$ the effective Hamiltonian $h(t)$ can not be constant. This means that at any time interval $[t_1, t_2]$ the instantaneous mass $\mu_\phi(t) = \Re\,[h(t)]$ and decay rate $\gamma_\phi(t) = -2\Im\,[h(t)]$ can not be constant in time:

$$\mu_\phi(t) \neq const., \qquad \gamma_\phi(t) \neq const. \tag{31}$$

In other words, as it follows from the above analysis the case $\mu_\phi(t) = const$ and $\gamma_\phi(t) = const$ can be realized only if the state $|\phi\rangle$ is an eigenvector for the total Hamiltonian H, that is if an only if there is no any decay od the state $|\phi\rangle$. Results of numerical calculations performed for $\omega(\mu) = \omega_{BW}(\mu)$ and presented in Fig (4) confirm the conclusion (31). In these Figures the function

$$\kappa(t) = \frac{\mu_\phi(t) - \mu_0}{m_0 - \mu_0}, \tag{32}$$

is presented and calculations were performed for $s_R = \frac{m_R}{\Gamma_0} = \frac{m_0 - \mu_0}{\Gamma_0} = 1000$. This function illustrates a typical form of time-varying $\mu_\phi(t)$.

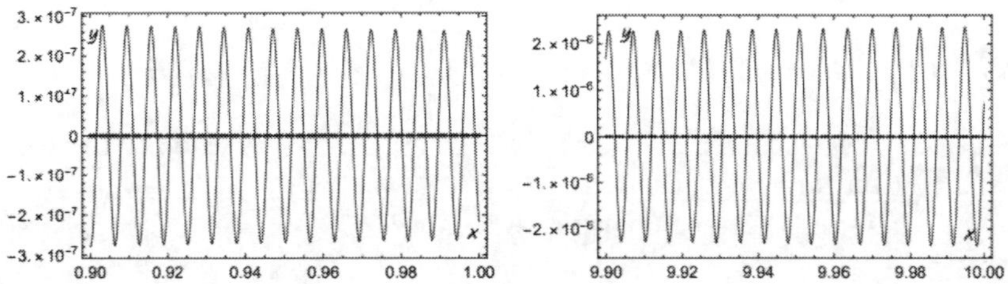

Figure 4. The instantaneous mass $m_\phi(t)$ as a function of time obtained for $\omega_{BW}(\mu)$. Axes: $y = \kappa(t) - 1$, where $\kappa(t)$ is defined by (32); $x = t/\tau_\phi$: Time is measured in lifetimes. The horizontal dashed line represents the value of $\mu_\phi(t) = m_0$

Summing up the oscillating decay curves of one component unstable system can not be considered as something extraordinary or as anomaly: It seems to be a universal feature of the decay process. The oscillatory modulation of decay curves takes place even in the quantum unstable system modeled by the Breit–Wigner distribution of the energy density. In general, the oscillatory modulation of the survival probability and thus the decay curves with model depending amplitude and oscillations

 DOI: 10.1051/epjconf/201716401032

period takes place even in the case of one component unstable systems. From results of the model calculations presented in Figs (2) and (3) it follows that at the initial stage of the "exponential" (or "canonical") decay regime the amplitude of these oscillations may be much less than the accuracy of detectors. Then with increasing time the amplitude of oscillations grows (see Fig. (3)), which increases the chances of observing them. This is a true quantum picture of the decay process at the so–called "exponential" regime of times.

5 Moving unstable systems with constant momentum

In the case of moving quantum unstable systems we need the probability amplitude $a_p(t) = \langle \phi_p | \phi_p(t) \rangle$, which defines the survival probability $\mathcal{P}_p(t) = |a_p(t)|^2$. There is $|\phi_p(t)\rangle \stackrel{\text{def}}{=} \exp[-itH]\,|\phi_p\rangle$ in $\hbar = c = 1$ units. So we need the vector $|\phi_p\rangle$ and eigenvalues $E'(\mu, p)$ solving Eq. (5). Vectors $|\phi\rangle, |\phi_p\rangle$ are elements of the same state space $\mathcal{H}$ connected with the coordinate rest system of the observer O: We are looking for the decay law of the moving particle measured by the observer O. If to assume for simplicity that $\mathbf{P} = (P_1, 0, 0)$ and that $\vec{v} = (v_1, 0, 0) \equiv (v, 0, 0)$ then there is $\vec{p} = (p, 0, 0)$ for the eigenvalues $\vec{p}$ of the momentum operator $\mathbf{P}$. Let $\Lambda_{p,\mu}$ be the Lorentz transformation from the reference frame O, where the momentum of the unstable particle considered is zero, $\vec{p} = 0$, into the frame O' where the momentum of this particle is $\vec{p} \equiv (p, 0, 0) \neq 0$ and $p \geq 0$, or, equivalently, where its velocity equals $\vec{v} = \vec{v}_{p,\mu} \equiv \frac{\vec{p}}{\mu\,\gamma_\mu}$, (where μ is the rest mass and $\gamma_\mu \equiv \sqrt{p^2 + (\mu)^2/\mu}$). Now using the unitary representation, $U(\Lambda_{p,\mu})$, of the transformation $\Lambda_{p,\mu}$, which acts in the Hilbert space $\mathcal{H}$ of states $|\phi\rangle \equiv |\phi; 0\rangle, |\phi_p\rangle \in \mathcal{H}$, one can show that the vector $U(\Lambda_{p,\mu})|\mu; 0\rangle$ is the common eigenvector for operators H and $\mathbf{P}$, that is that there is $|\mu; p\rangle \equiv U(\Lambda_{p,\mu})|\mu; 0\rangle$ (see, eg. [37]). Indeed, taking into account that operators H and $\mathbf{P}$ form a 4–vector $P_\nu = (P_0, \mathbf{P}) \equiv (P_0, P_1, 0, 0)$, and $P_0 \equiv H$, we have $U^{-1}(\Lambda_{p,\mu}) P_\nu U(\Lambda_{p,\mu}) = \Lambda_{p,\mu;\,\nu\lambda} P_\lambda$, where $\nu, \lambda = 0, 1, 2, 3$ (see, e.g., [37], Chap. 4). From this general transformation rule it follows that

$$U^{-1}(\Lambda_{p,\mu}) P_0 U(\Lambda_{p,\mu}) = \gamma_\mu\,(P_0 \,+\, v_\mu\,P_1) \equiv \gamma_\mu (H \,+\, v_\mu\,P_1). \tag{33}$$

Based on the relation (33), one can show that that vectors $U(\Lambda_{p,\mu})|\mu; 0\rangle$ are eigenvectors for the Hamiltonian H. There is

$$H\,U(\Lambda_{p,\mu})|\mu; 0\rangle = U(\Lambda_{p,\mu})\,U^{-1}(\Lambda_{p,\mu})\,H\,U(\Lambda_{p,\mu})|\mu; 0\rangle = \gamma_\mu\,U(\Lambda_{p,\mu})\,(H \,+\, v_\mu\,P_1)\,|\mu; 0\rangle. \tag{34}$$

The Lorentz factor γ_μ corresponds to the rest mass μ being the eigenvalue of the vector $|\mu; 0\rangle$. There are $\gamma_\mu \neq \gamma_{\mu'}$ and $v_\mu \neq v_{\mu'}$ for $\mu \neq \mu'$. From (4), (6) it follows that $P_1\,|\mu; 0\rangle = 0$ for $p = 0$, which means that using (7) the relation (34) can be rewritten as follows

$$H\,U(\Lambda_{p,\mu})|\mu; 0\rangle = \mu\gamma_\mu\,U(\Lambda_{p,\mu})|\mu; 0\rangle. \tag{35}$$

Taking into account the form of the γ_μ forced by the condition $p = const$ one concludes that in fact the eigenvalue found, $\mu\gamma_\mu$, equals $\mu\gamma_\mu \equiv \sqrt{p^2 + \mu^2}$. So, $E'(\mu, p) = \sqrt{p^2 + \mu^2}$. (Similarly one can show that vectors $U(\Lambda_{p,\mu})|\mu; 0\rangle$ are the eigenvectors of the momentum operator $\mathbf{P}$ for the eigenvalue $\mu\gamma_\mu\,v_\mu \equiv p$, that is that $U(\Lambda_{p,\mu})|\mu; 0\rangle \equiv |\mu; p\rangle$). Thus finally we come to desired results:

$$H|\mu; p\rangle = \sqrt{p^2 + \mu^2}\;|\mu; p\rangle \tag{36}$$

which replaces Eq. (5).

The moving quantum unstable particle ϕ with constant momentum, $\vec{p}$, can be modeled analogously as the quantum unstable system in the rest frame (when $\vec{p} = 0$) as the following wave–packet $|\phi_p\rangle$,

$$|\phi_p\rangle = \int_{\mu_0}^{\infty} c(\mu)\, |\mu; p\rangle\, dm, \tag{37}$$

where expansion coefficients $c(\mu)$ are functions of the mass parameter μ, that is of the rest mass μ, which is Lorentz invariant and therefore the scalar functions $c(\mu)$ of μ are also Lorentz invariant and are the same as in the rest reference frame O_0.

Now using (36) and the equation (37) we obtain the final, required relation for the amplitude $a_p(t)$ (see [16, 17, 38]),

$$a_p(t) = \int_{\mu_0}^{\infty} \omega(\mu)\, e^{-i\mu\gamma_\mu t}\, d\mu \equiv \int_{\mu_0}^{\infty} \omega(\mu)\, e^{-i\sqrt{p^2+\mu^2}\; t}\, d\mu. \tag{38}$$

This is the place when it should be explained why the Lorentz factor γ_μ is used in (38) (and earlier in relations (33) — (36) instead of $\gamma_\mu = \gamma$. In the rest reference frame the unstable quantum system is modeled as the wave packet given by the relation (9). Let us choose some eigenvalues, $\mu_1 < \mu_2 < \ldots < \mu_k < \ldots < \mu_n \in \sigma_c(H) = [\mu_0, \infty)$, of the spectrum $\sigma_c(H)$ of H. These eigenvalues are connected with corresponding eigenvectors $|\mu_k; 0\rangle$ of H as follows:

$$\begin{array}{ccccccc}
 & & O_0 & & O_p & & \\
\mu_1 & \leftrightarrow & |\mu_1;0\rangle & \overset{U(\Lambda_{p,\mu_1})}{\longmapsto} & |\mu_1;p\rangle & \leftrightarrow & p = \mu_1\gamma_{\mu_1}v_{\mu_1} \\
\mu_2 & \leftrightarrow & |\mu_2;0\rangle & \overset{U(\Lambda_{p,\mu_2})}{\longmapsto} & |\mu_2;p\rangle & \leftrightarrow & p = \mu_2\gamma_{\mu_2}v_{\mu_2} \\
\ldots & \ldots & \ldots & \ldots & \ldots & \ldots & \ldots \\
\mu_k & \leftrightarrow & |\mu_k;0\rangle & \overset{U(\Lambda_{p,\mu_k})}{\longmapsto} & |\mu_k;p\rangle & \leftrightarrow & p = \mu_k\gamma_{\mu_k}v_{\mu_k} \\
\ldots & \ldots & \ldots & \ldots & \ldots & \ldots & \ldots
\end{array} \tag{39}$$

As it is seen from the above analysis each vector $|\mu; 0\rangle$ numbered by $\mu \in \sigma_c(H)$ can be transformed correctly in the vector $|\mu; p\rangle$ connected with the reference frame O_p only if one takes into account that every point μ from the spectrum of H, considered as the rest mass, has the "own" Lorentz factor $\gamma = \sqrt{\mu^2 + p^2}/\mu \equiv \gamma_\mu$. (The reference frame O_p was defined by condition $\vec{p} = const \neq 0$, and $O_0 = O_{p=0}$). In other words it is impossible to realize the above transformations of vectors $|\mu; 0\rangle$ assigned to a reference frame O_0 to the reference frame O_p if $\vec{v} = const$.

The above derivation of the expression for $a_p(t)$ is similar to that of [19]. It is based on [37] and it is reproduced here for the convenience of readers. This is a shortened and slightly changed, simplified version of the considerations presented in [16] and mainly in [17] and more explanations and more details can be found therein and in [39, 40], where this formula was derived using the quantum field theory theory approach.

Inserting $\omega(\mu) \equiv \omega_{BW}(\mu)$ into (10) and into (38) one can find decay curves (survival probabilities) $\mathcal{P}_0(t)$ and $\mathcal{P}_p(t)$. Results of numerical calculations are presented in Fig (5), where calculations were performed for $\mu_0 = 0$, $E_0/\varGamma_0 \equiv m_0/\varGamma_0 = 1000$ and $cp/\varGamma_0 \equiv p/\varGamma_0 = 1000$. Values of these parameters correspond to $\gamma = \sqrt{2}$, which is very close to γ from the experiment performed by the GSI team [14, 15] and this is why such values of them were chosen in our considerations. According to the literature for laboratory systems a typical value of the ratio $m_0/\varGamma_0$ is $m_0/\varGamma_0 \geq O(10^3 - 10^6)$ (see eg. [41]) therefore the choice $m_0/\varGamma_0 = 1000$ seems to be reasonable minimum. Similarly to the case of quantum unstable systems in rest one can calculate the ratio $\mathcal{P}_p(t)/\mathcal{P}_c(t/\gamma)$ in the case of moving particles. Results of numerical calculations of this ratio are presented in Figures (6) and (7), and

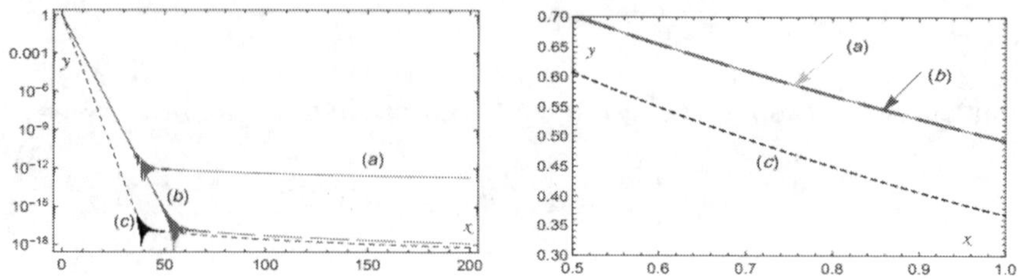

Figure 5. Decay curves obtained for $\omega_{BW}(m)$ given by Eq. (18). Axes: $x = t/\tau_0$ — time t is measured in lifetimes τ_0, y — survival probabilities (The left panel: the logarithmic scales, (a) the decay curve $\mathcal{P}_p(t)$, (b) the decay curve $\mathcal{P}_0(t/\gamma)$, (c) the decay curve $\mathcal{P}_0(t)$; The right panel: $(a) - \mathcal{P}_p(t)$, $(b) - \mathcal{P}_0(t/\gamma)$, $(c) - \mathcal{P}_0(t)$).

calculations were performed for $\omega(\mu) = \omega_{BW}(\mu)$ and for $\mu_0 = 0$, $m_0/\Gamma_0 = 1000$, $cp/\Gamma_0 \equiv p/\Gamma_0 = 1000$ and $\gamma = \sqrt{2}$.

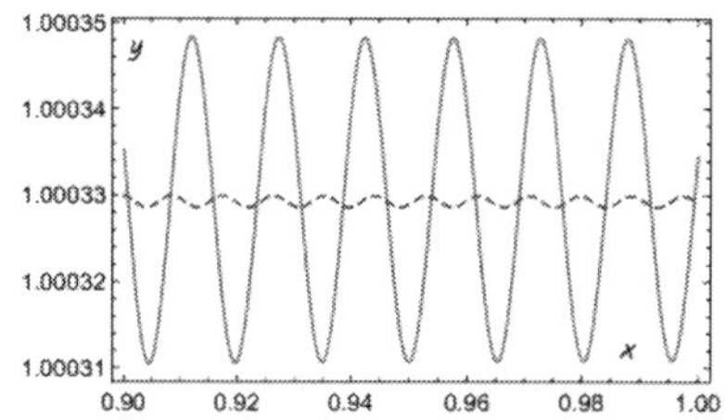

Figure 6. Axes: $x = t/\tau_0$ — time t is measured in lifetimes τ_0, y — Ratio of probabilities — Solid line: $\mathcal{P}_p(t)/\mathcal{P}_c(t/\gamma)$; Dashed line $\mathcal{P}_0(t/\gamma)/\mathcal{P}_c(t/\gamma)$.

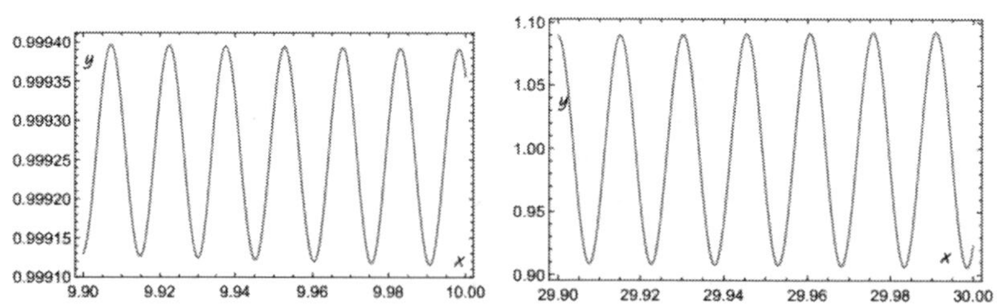

Figure 7. Axes: $x = t/\tau_0$ — time t is measured in lifetimes τ_0, y — Ratio of probabilities: $\mathcal{P}_p(t)/\mathcal{P}_c(t/\gamma)$.

6 Summary

- The mass of the system in the unstable state $|\phi\rangle$ is not defined: It can not take the exact value. Unstable system can be characterized by the mass distribution $\omega(\mu)$, the average mass $< m >= \int_{\mu_0}^{\infty} \mu\, \omega(\mu) d\mu$ and by instantaneous mass (energy) $\mu_\phi(t)$ but not by the mass.

- There is no any time interval in which the survival probability (decay) law could be a decreasing function of time of the purely exponential form: Even in the case of the Breit–Wigner model in so–called "exponential regime" the decay curves are oscillatory modulated with smaller or large amplitude of oscillations depending on the parameters of the model.
- At any time interval the instantaneous mass $\mu_\phi(t)$ and instantaneous decay rate $\gamma_\phi(t)$ can not be constant in time.
- In the case of moving relativistic quantum unstable system moving with constant momentum $\vec{p}$, when unstable systems are modeled by the Brei–Wigner mass distribution $\omega(\mu)$, only at times of the order of lifetime τ_ϕ it can be $\mathcal{P}_p(t) \simeq \mathcal{P}_0(t/\gamma)$ to a better or worse approximation. At times longer than a few lifetimes the decay process of moving particles observed by an observer in his rest system is much slower that it follows from the classical physics relation $\mathcal{P}_p(t) \stackrel{?}{=} \exp{[-\frac{t}{\gamma}\,\Gamma_0]}$.
- In the case of moving relativistic quantum unstable system moving with constant momentum $\vec{p}$ decay curves are also oscillatory modulated but the amplitude of these oscillations is higher than in the case of unstable systems in rest.
- There is a need to test the decay law of moving relativistic unstable system for times much longer than the life time

Acknowledgments:
The work was supported by the Polish NCN grant No DEC-2013/09/B/ST2/03455.

References

[1] E. Ruthheford, Philosophical Magazine, **XLIX**, 1, 161, (1900).
[2] E. Rutherford, F. Soddy, Philosophical Magazine, **IV**, 370 – 96 and 569 – 85 (1902).
[3] E. Rutherford, *Radioactive substances and their radiations*, Cambridge Unversity Press 1913.
[4] V. F. Weisskopf, E. T. Wigner, Z. Phys., **63**, 54, (1930); **65**, 18, (1930).
[5] L. A. Khalfin, Zh. Eksp. Teor. Fiz., **33**, p.p. 1371 – 1382, (1957) [in Russian]; [Sov. Phys. — JETP **6**, (1958), 1053].
[6] L. Fonda, G. C. Ghirardii and A. Rimini, Rep. on Prog. in Phys., **41**, p.p. 587 – 631, (1978).
[7] E. B. Norman, *et al.*, Phys. Rev. Lett., **60**, 2246 (1988).
[8] C. Rothe, S. I. Hintschich and A. P. Monkman, Phys. Rev. Lett., **96**, 163601 (2006).
[9] B. Misra, E. C. G. Sudarshan, Journal of Mathematical Physics, **18**, 745, (1977).
[10] M. C. Fischer, *et al.*, Phys. Rev. Lett., **87**, 040402, (2001).
[11] W. M. Itano, *et al.*, Phys. Rev., **A 41**, 2295, (1990).
[12] Y. S. Patil, *et al.*, Phys. Rev. Lett., **115**, 140402 (2015).
[13] M. Peshkin, A. Volya and V. Zelevinsky, Europhysics Letters, **107**, 40001, (2014).
[14] Yu. A. Litvinov, *et al.*, Phys. Lett., **B 664**, 162, (2008).
[15] P. Kienle, *et al.*, Phys. Lett., **B 726**, 638, (2013).
[16] E. V. Stefanovitch, International Journal of Theoretical Physics, **35**, p.p. 2539 – 2554, (1996).
[17] M. Shirokov, International Journal of Theoretical Physics, **43**, p.p. 1541 – 1553, (2004).
[18] P. Exner, Phys. Rev., **D 28**, p.p. 2621 – 2627, (1983).
[19] K. Urbanowski, Physics Letters, **B 737**, 346 – 351, (2014).
[20] J. Bailye, *et al*, Nature, **268** (5618), p.p. 301 – 305, (1977).
[21] David H. Frisch, James H. Smith, American Journal of Physics, **31**, p.p. 342 – 355, (1963).

[22] N. S. Krylov, V. A. Fock, Zh. Eksp. Teor. Fiz., **17**, p.p. 93 – 107, (1947) [in Russian]. V. A. Fock, *Fundamentals of Quantum mechanics*, Mir Publishers, Moscow 1978.

[23] N. G. Kelkar, M. Nowakowski, J. Phys. A: Math. Theor., **43**, 385308 (2010), 9 pp.

[24] J. Martorell, J. G. Muga, and D. W.L. Sprung, *Lect. Notes Phys.*, **789**, 239 – 275, (2009) (Springer–Verlag, Berlin, Heidelberg 2009).

[25] E. Torrontegui, J. G. Muga, J. Martorell, and D. W. L. Sprung, Advances in Quantum Chemistry, **60**, 485 – 535, (2010).

[26] Gaston Garcia-Calderon, Roberto Romo and Jorge Villavicencio, Physical Review B **76**, 035340 (2007), 9 pp.

[27] F. Giraldi, Eur. Phys. J. D., **69**: 5, (2015), 8 pp.

[28] F. Giacosa, Found. of Phys., **42**, 1262 (2012).

[29] M. L. Goldberger, K. M. Watson, *Collision theory*, Wiley 1964.

[30] R. E. A. C. Paley, N. Wiener, *Fourier transforms in the complex domain*, American Mathematical Society, New York, 1934.

[31] K. Urbanowski, Phys. Rev., **A 50**, 2847 – 2853, (1994).

[32] K. Urbanowski, Eur. Phys. J., **C 58**, 151, (2008).

[33] K. Urbanowski, Cent. Eur. J. Phys., **7**, 696 – 703, (2009).

[34] K. M. Sluis, E. A. Gislason, Phys. Rev. **A 43**, (1991), 4581.

[35] *NIST Handbook of Mathematical Functions*, Natl. Inst.. Stand. and Technology, Eds.: F. W. J. Olver, D. W. Lozier, R. F. Boisvert and Ch. W. Clark,(Cambridge, 2010).

[36] *Handbook of Mathematical Functions With Formulas, Graphs, and Mathematical Tables*, Eds.: Milton Abramowitz and Irene A. Stegun, New York: Dover Publications, Inc. (1964).

[37] W. M. Gibson, B. R. Polard, *Symmetry principles in elementary particle physics*, Cambridge, 1976.

[38] L. A. Khalfin, Quantum theory of unstable particles and relativity, PDMI PREPRINT–6/1997 (St. Petersburg Department of Steklov Mathematical Institute, St. Petersburg, Russia, 1997).

[39] E. V. Stefanovich, Relativistic quantum dynamics, arXiv: 0504062 [physics]. E. V. Stefanovich, Violations of Eistein's time dilation formula in particle decays, arXiv: 0603043 [physics]. B. T. Shields, *et al*, Phys. Rev., **A 82**, 052116 (2010).

[40] M. I. Shirokov, Concepts of Physics, **3**, 193 – 205 (2006); arXiv: quant–ph/0508087v.2.; Physics of Particles and Nuclei Letters, **6**, 14 – 17, (2009).

[41] L. M. Krauss, J. Dent, Phys. Rev. Lett., **100**, 171301 (2008), 4 pp.

Vladimir Sudakov and double-logarithmic asymptotics of amplitudes in QED, QCD and gravity

L. N. Lipatov[1,2,a]

[1] *Petersburg Nuclear Physics Institute, Gatchina, Orlova Roshcha, 188300, St.Petersburg, Russia*
[2] *Saint Petersburg State University, St. Petersburg, Russia*

Abstract.
We review the Sudakov results on the double logarithmic asymptotics of the electron form-factor which were based on his parametrization of the virtual particle momenta in the Feynman diagrams. The high energy amplitudes for various QED and QCD processes in the double-logarithmic approximation are obtained by using the Bethe-Salpeter approach and the evolution equations. The ultraviolet divergency of the graviton Regge trajectory allows to derive the infrared evolution equation for the graviton-graviton scattering amplitude with a double-logarithmic accuracy. The asymptotic behavior of this amplitude depends essentially on the rank N of the super-symmetry.

1 Introduction

The investigation of scattering amplitudes at high energies $\sqrt{s}$ in QED was initiated by Vladimir Vasil'evich Sudakov 60 years ago in his paper [1], devoted to the asymptotic behavior of the vertex function in the double logarithmic approximation (DLA) valid in the region

$$\alpha \ln^2 \frac{s}{m_e} \sim 1\,,\ \alpha = \frac{e^2}{4\pi} \ll 1\,,\ s = (p_e + p_{\overline{e}})^2\,, \tag{1}$$

where p_e and $p_{\overline{e}}$ are momenta of initial electron and positron. Really V. Sudakov considered the case of the high energy scattering of the virtual electron off a charged center. In the case of the real electron there are infrared divergencies which can be removed by the introduction of the photon mass λ. These divergencies are canceled in the inclusive cross-section for the inelastic scattering with the emission of an arbitrary number of photons [2]. In one loop approximation the vertex function $\gamma^{(1)}$ for the e^+e^- annihilation in a photon with the large virtuality s can be calculated with the Sudakov parametrization of the momentum k (see Figure)

$$k = \alpha\, p'_e + \beta\, p'_{\overline{e}} + k_\perp\,,\ p_e = p'_e + \frac{m_e^2}{s'} p'_{\overline{e}}\,,\ p_{\overline{e}} = p'_{\overline{e}} + \frac{m_e^2}{s'} p_e\,,\ k_\perp p'_e = k_\perp p'_{\overline{e}} = 0\,,$$

$$d^4k = \frac{|s'|}{2}\, d\alpha\, d\beta\, d^2k_\perp\,, |\alpha| \ll 1\,,\ |\beta| \ll 1\,,\ s' = 2p'_e p'_{\overline{e}} \approx s\,,\ -k_\perp^2 \ll \sqrt{s}\,.$$

[a]The investigation is supported by the grants 11.38.223.2015 of SPSU and 14-2200281 of RFBI.
e-mail: lipatov@thd.pnpi.spb.ru

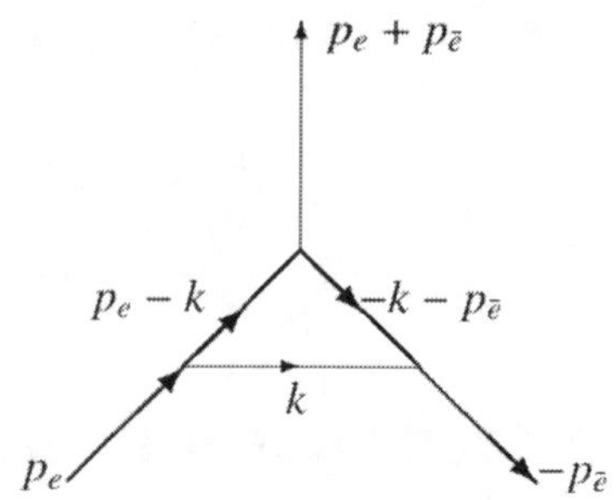

According to V. Sudakov the vertex function $\gamma^{(1)}$ after its integration over $k_\perp$ can be written as follows

$$\frac{\gamma^{(1)}}{\gamma_{Born}} = -\frac{e^2}{8\pi^2}\int_0^1\int_0^1 \frac{d\alpha\, d\beta\ \theta(s\alpha\beta - \lambda^2)}{\left(\alpha + \frac{m_e^2}{s}\beta\right)\left(\beta + \frac{m_e^2}{s}\alpha\right)} = -\frac{e^2 \ln\frac{s}{m_e^2}}{8\pi^2}\left(\frac{\ln\frac{s}{m_e^2}}{2} + \ln\frac{m_e^2}{\lambda^2}\right).$$

One can derive the evolution equation for γ in the infrared cut-off λ in DLA

$$\frac{\partial\gamma}{\partial \ln m_e^2/\lambda^2} = -\frac{e^2}{8\pi^2}\ln\frac{s}{m_e^2}\,\gamma\,.$$

Its solution allows to calculate the Sudakov vertex function in the DL approximation

$$\gamma = \gamma_{Born}\exp\left(-\frac{e^2\ln\frac{s}{m_e^2}}{8\pi^2}\left(\frac{\ln\frac{s}{m_e^2}}{2} + \ln\frac{m_e^2}{\lambda^2}\right)\right). \tag{2}$$

The probability for the emission of n additional soft photons obeys the Poisson distribution

$$w_n = e^{-a}\frac{a^n}{n!}\,,\quad a = 2\frac{e^2\ln\frac{s}{m_e^2}}{8\pi^2}\left(\frac{\ln\frac{s}{m_e^2}}{2} + \ln\frac{m_e^2}{\lambda^2}\right), \tag{3}$$

but one should take into account the energy conservation for the total frequency ω of the emitted photons. It turns out, that the QED radiative corrections to the production of narrow resonances are large. For example, the double-logarithmic result for the cross-section of the inclusive Z-boson production in the e^+e^- collisions at $\sqrt{s} - M_Z \ll M_Z$ has the form (cf. [3])

$$\sigma_{e^+e^-\to Z+photons} = \int_0^{M_Z}\sigma(W-\omega)\frac{d\omega}{\omega}\frac{4\alpha}{\pi}\ln\frac{W}{m_e}\left(\frac{W}{\omega}\right)^{-\frac{4\alpha}{\pi}\ln\frac{W}{m_e}},\quad W = \sqrt{s}\,, \tag{4}$$

where $\sigma(W)$ is the well known Born cross-section of the Z-boson production in the e^+e^- collisions

$$\sigma(W) = \frac{3\pi}{M_Z^2}\frac{\Gamma_{e\bar{e}}\Gamma_{tot}}{(W - M_Z)^2 + \frac{\Gamma_{tot}^2}{4}}. \tag{5}$$

Note, that $\sigma_{e^+e^-\to Z+photons}$ does not depend on λ and behaves as $\sim 1/(W - M_Z)$ above the resonance.

2 Backward Compton scattering and electron reggeization

A.A. Abrikosov in ref. [2] and in subsequent papers of 1956 investigated the backward γe scattering at high energies and fixed momentum transfers (k and k' are initial and final momenta of the photon)

$$s = (k+p)^2 \gg u = (k'-p)^2 \sim m_e^2\,.$$

He obtained the simple Sudakov-type result for the corresponding amplitude in DLA

$$A_{DL}(s,u) = A_B \exp\left(-\frac{\alpha}{2\pi} \ln\frac{s}{m_e^2} \ln\frac{|u|}{\lambda^2}\right).$$

Ten years later M. Gell-Mann with collaborators discovered, that $A(s,u)$ for fixed $u \sim \lambda^2$ has the form

$$A(s,u) = A_B \left(\frac{s}{m_e^2}\right)^{\omega(u)}, \quad u = -|q|_\perp^2,$$

where the Regge trajectory $\omega(\omega)$ in one loop is given below

$$\omega(u) = \frac{\alpha}{2\pi}(\hat{q} - m_e) \int \frac{d^2k_\perp}{\pi} \frac{1}{k_\perp^2 - \lambda^2} \frac{\hat{q}_\perp - \hat{k}_\perp + m_e}{(q-k)_\perp^2 - m_e^2}.$$

V. G. Gorshkov, V. N. Gribov and G. V. Frolov performed similar calculations in the real case $\lambda \to 0$ and found, that the integration region $k_\perp^2 \ll q^2$ over the virtual photon momenta gives an universal contribution and produces a divergent factor in the amplitude. The universality of this factorization for small $k_\perp$ was proven by V.N. Gribov with the use of the dispersion relations also for the case of the QED corrections to the strong interactions in the Regge kinematics. Below we shall apply the Gribov arguments for the derivation of the non-linear evolution equations for the amplitudes in DLA.

3 Charge particle scattering and annihilation in QED

The forward scattering of charged particles in QED is related to the Pomeranchuk singularity responsible for the high energy behavior of total cross-sections which does not contain double logarithms.

But one can consider the process $e^+e^- \to \mu^+\mu^-$ in the kinematics when μ^- flies along e^-

$$s = (p_e + p_{\bar{e}})^2 \gg t = (p_e - p_\mu)^2 \sim m^2.$$

It turns out [4], that the amplitude for this process is factorized in DLA

$$A_{DL}(s) = A_{Born}\, f(s), \quad f(s) = f(s,p^2)_{p^2=m^2}$$

and the main contribution to it goes from ladder diagrams satisfying the Bethe-Salpeter equation [4]

$$f(s,p^2) = 1 + \frac{\alpha}{2\pi} \int_{m^2}^{s} \frac{ds'}{s'} \int_{\max(m^2, p^2\frac{s'}{s})}^{s'} \frac{dp'^2}{p'^2} f(s',p'^2).$$

The solution of this equation allows to find $f(s)$ on the mass shell $p^2 = m^2$ in DLA [4]

$$f(s) = \int_{a-i\infty}^{a+i\infty} \frac{dj}{2\pi i} \left(\frac{s}{m^2}\right)^j \frac{2}{j + \sqrt{j^2 - \frac{2\alpha}{\pi}}} = I_0\left(\sqrt{\frac{\alpha}{\pi}} \ln\frac{s}{m^2}\right).$$

For the e^+e^- backward scattering in the Regge kinematics

$$s = (p_1 + p_2)^2 \gg u = (p_1 - p_2')^2 \sim m_e^2$$

the amplitude is also factorized

$$A_{DL}(s) = A_{Born}\, f(s), \quad f(s) = f(s,p^2)_{p^2=m^2}.$$

In this case apart from the ladder diagrams a large contribution appears also from the Sudakov-type photon insertions. The corresponding Bethe-Salpeter equation is more complicated but it can be solved explicitly, which allows one to obtain the corresponding amplitude on the mass shell [5]

$$f(s) = 4\int_{a-i\infty}^{a+i\infty} \frac{d\,l}{2\pi i}\, e^{l\rho}\, \frac{d}{d\,l} \ln D_{-\frac{1}{4}}(l)\,,\ \rho = \sqrt{\frac{2\alpha}{\pi}}\ln\frac{s}{m^2}\,.$$

Here we omitted the Sudakov factor from soft photons with $k_\perp \ll m$ and used the expression

$$D_p(x) = \frac{e^{-\frac{x^2}{4}}}{\Gamma(-p)}\int_0^\infty \frac{dt}{t^{1+p}}\, e^{-xt-\frac{t^2}{2}}\,.$$

for the parabolic cylinder function.

4 Infrared evolution equations

A simple approach for the derivation of scattering amplitudes in QED and QCD at DLA was suggested in the paper [6]. It is based on the introduction of the infrared cut-off μ at the transverse momenta of virtual particles in the Feynman diagrams

$$|k_\perp|^2 > \mu^2$$

and the subsequent construction of the evolution equation in μ for the amplitudes on the mass shell. This method is similar to the renormalization group approach in the Sudakov interpretation [7].

For example, for the e^+e^- forward scattering the μ^2-evolution equation can be written as follows

$$f(s,\mu^2) = 1 + e^2\int_{-\infty}^{\infty}\frac{ds\alpha}{s\alpha}\frac{ds\beta}{s\beta}\int_\mu \frac{d^2k_\perp}{i(2\pi)^4}\, f(s\alpha,|k_\perp|^2)\,\frac{|k_\perp|^2}{(s\alpha\beta-|k_\perp|^2)^2}\, f(s\beta,|k_\perp|^2)\,.$$

With the transition to the j-representation it can be reduced to the algebraic relation

$$f(s,\mu^2) = \int_{a-i\infty}^{a+i\infty}\frac{dj}{\pi i}\left(\left(-\frac{s}{\mu^2}\right)^j + \left(\frac{s}{\mu^2}\right)^j\right)\frac{f_j}{j}\,,\quad f_j = 1 + \frac{\alpha}{2\pi}\frac{f_j^2}{j^2}\,,$$

which leads to the above result for $f(s)$ expressed in terms of the Bessel function $I_0(x)$.

In an analogous way the evolution equation for the backward e^+e^- scattering amplitude is reduced to the differential equation which can be easily solved in terms of the parabolic cylinder function [6]

$$f_j = 1 + \frac{2\alpha}{\pi}\frac{d}{dj}\frac{f_j}{j} - \frac{\alpha}{2\pi}\frac{f_j^2}{j^2}\,,\quad \frac{f_j}{j} = 4\sqrt{\frac{\pi}{2\alpha}\frac{d}{dl}}\ln\left(e^{l^2/4}D_{-1/4}(l)\right)\,.$$

In this equation the non-linear term corresponds to the ladder contribution and the term with the derivative describes the soft Sudakov photons attached to the external legs according to the Gribov factorization theorem discussed above.

In QCD the large logarithms of Q^2 in the electron-proton scattering are summed with the help of the DGLAP equation [8]. At $x \to 0$ this equation for a non-singlet parton distribution

$$\frac{d}{d\ln Q^2}\, n_q(Q^2,x) = \frac{n_c^2-1}{2N_c^2}\frac{\alpha_c(Q^2)}{2\pi}\int_x^1\frac{d\beta}{\beta}\, n_q(Q^2,\beta)$$

can be modified by a simple deformation of the integration region

$$n_q(Q^2,x) = \delta(x-1) + \frac{n_c^2-1}{2N_c^2}\frac{\alpha_c}{2\pi}\int_x^1 \frac{d\beta}{\beta}\int_{m^2}^{Q^2\frac{\beta}{x}} \frac{dk^2}{k^2}\, n_q(k^2,\beta)\,.$$

Its solution corresponding to the non-singlet quark distribution in DLA is given below

$$n_q(Q^2,x) = \int_{a-i\infty}^{a+i\infty} \frac{dj}{2\pi}\left(\frac{1}{x}\right)^j \frac{j\gamma(j)}{\frac{N_c^2-1}{2N_c}\frac{\alpha}{2\pi}}\left(\frac{Q^2}{m^2}\right)^{\gamma(j)}, \quad j\gamma(j) = \frac{N_c^2-1}{2N_c}\frac{\alpha}{2\pi} + \gamma^2(j)\,.$$

In a similar way one can modify and solve the equations at $x \to 0$ for the singlet quark and gluon distributions in DLA [9].

5 Graviton Regge trajectory and scattering amplitudes in super-gravity

The graviton Regge trajectory in LLA was calculated many years ago [10]

$$\omega(q^2) = \frac{\alpha}{\pi}\int \frac{q^2\, d^2k}{k^2(q-k)^2} f(k,q)\,, \quad \alpha = \frac{\kappa^2}{8\pi^2}\,,$$

where $\gamma = \frac{\kappa^2}{8\pi}$ is the Newton constant and

$$f(k,q) = (k,q-k)^2\left(\frac{1}{k^2}+\frac{1}{(q-k)^2}\right) - q^2 + \frac{N}{2}(k,q-k)\,.$$

Here N is the number of different gravitinos appearing in the corresponding action

$$S_{3/2} = -\frac{1}{2}\epsilon^{\mu\nu\rho\sigma}\int d^4x \sum_{r=1}^{N} \bar{\psi}^r_\mu \gamma_5\gamma_\nu\partial_\rho\psi^r_\sigma .$$

The graviton Regge trajectory contains infrared and ultraviolet divergencies

$$\omega(q^2) = -\alpha|q|^2\left(\ln\frac{|q|^2}{\lambda^2} + \frac{N-4}{2}\ln\frac{|\Lambda|^2}{|q|^2}\right).$$

The ultraviolet divergency is fictive, because in one loop the gravity is a renormalizable theory. The cut-off Λ^2 should be substituted by s, which leads to an appearance of double-logarithmic terms in the corresponding scattering amplitude. The summation of all such terms is performed below [10].

It is convenient to use the Mellin representation of the graviton-graviton scattering amplitude

$$A(s,t) = A_{Born}\, s^{-\alpha|q|^2\ln\frac{|q|^2}{\lambda^2}}\,\Phi(\xi)\,, \quad \Phi(\xi) = \int_{a-i\infty}^{a+i\infty}\frac{d\omega}{2\pi i\,\omega}\left(\frac{s}{|q|^2}\right)^{\omega} f_\omega\,.$$

The partial wave f_ω satisfies the infrared evolution equation similar to the QCD case [11]

$$f_\omega = 1 + b\frac{d}{d\omega}\frac{f_\omega}{\omega} - b\,\frac{N-6}{2}\frac{f_\omega^2}{\omega^2}\,, \quad b = \alpha|q|^2\,, \quad \alpha = \frac{\kappa^2}{8\pi^2}\,, \quad \xi = \alpha\,|q|^2\ln^2\frac{s}{|q|^2}\,.$$

Its solution in the perturbation theory is

$$\Phi(\xi) = 1 - \frac{N-4}{2}\frac{\xi}{2} + \frac{(N-4)(N-3)}{2}\frac{\xi^2}{4!} - \frac{N-4}{8}(5N^2-26N+36)\frac{\xi^3}{6!} + \ldots\,.$$

The analytic solution of the equation can be constructed in terms of the parabolic cylinder function

$$\frac{f_\omega^{(N)}}{\omega} = \frac{2}{6-N}\frac{1}{\sqrt{b}}\frac{d}{dx}\ln d^{(N)}(x)\,,\ d^{(N)}(x) = e^{\frac{x^2}{4}}D_{\frac{6-N}{2}}(x)\,,\ x = \frac{\omega}{\sqrt{b}}\,.$$

One can verify [11], that the quantity $A(s,t)/A_{Born}$ grows with energy for $N < 4$ and tends to zero at $s \to \infty$ for $N > 4$.

References

[1] V. V. Sudakov, *Sov. Phys. JETP* **30** 87 (1956).
[2] A. A. Abrikosov, *Sov. Phys. JETP* **30** 96 (1956).
[3] Ya. I. Azimov, V. A. Khoze, L. N. Lipatov, *JETP Letters* **21** 378 (1975).
[4] V. G. Gorshkov, V. N. Gribov, G. V. Frolov, L. N. Lipatov, *Phys. Lett.* **22** 671 (1966).
[5] V. G. Gorshkov, V. N. Gribov, L. N. Lipatov, G. V. Frolov, *Yad. Fiz.* **6** 361 (1967).
[6] R. Kirschner, L. N. Lipatov, *Nucl. Phys.* **123** 122 (1983).
[7] V. V. Sudakov, *Sov. Phys. JETP* **31** 729 (1956).
[8] V. N. Gribov, L. N. Lipatov, *Sov. J. Nucl. Phys.* **18** (1972) 438, 675;
L. N. Lipatov, *Sov. J. Nucl. Phys.* **20** (1975) 93;
G. Altarelli and G. Parisi, *Nucl. Phys.* **B126** (1977) 298;
Yu. L. Dokshitzer, *Sov. Phys. JETP* **46** (1977) 641.
[9] B. L. Ioffe, V. S. Fadin, L. N. Lipatov, *Quantum Chromodynamics, Perturbative and non-perturbative aspects*, Cambridge University Press, (2010).
[10] L. N. Lipatov, Phys. Lett. **116**, 411 (1982).
[11] J. Bartels, L. N. Lipatov, A. Sabio Vera, J. High Energy Phys. **07** 056 (2014).

EPJ Web of Conferences **164**, 01034 (2017)
ICNFP 2016

DOI: 10.1051/epjconf/201716401034

What could we learn about high energy particle physics from cosmological observations at largest spatial scales ?

Dmitry Gorbunov[1,a]

[1] *Institute for Nuclear Research of Russian Academy of Sciences, 117312 Moscow, Russia*

Abstract. The very well known example of cosmology testing particle physics is the number of relativistic particles (photons and three active neutrinos within the Standard Model) at primordial nucleosynthesis. These days the earliest moment we can hope to probe with present cosmological data is the early time inflation. The particle physics conditions there and now are different because of different energy scales and different values of the scalar fields, that usually prohibits a reliable connection between the particle physics parameters at the two interesting epochs. The physics at the highest energy scales may be probed with observations at the largest spatial scales (just somewhat smaller than the size of the visible Universe). However, we are not (yet) ready to make the tests realistic, because of lack of a self-consistent theoretical description of the presently favorite cosmological models to be valid right after inflation.

1 Introduction

The Universe is expanding, which is a most natural conclusion from observations of the Doppler redshift of the spectra emitted by distant astrophysical sources. The redshift z is defined through the ratio of wavelengths of absorbed and emitted photons, respectively,

$$\frac{\lambda_{\text{abs.}}}{\lambda_{\text{em.}}} \equiv 1 + z\,. \tag{1}$$

The larger the redshift is, the larger distance has been covered by a photon and hence the earlier it has been emitted. The host of observed galaxies mostly refer to redshift $z \lesssim 4$, but there are exceptions with higher redshifts upto $z \sim 10$.

It is convenient to introduce a time-dependent scale factor $a(t)$, so that all the cosmological physical distances, like distances between galaxy clusters, grow proportionally to the scale factor, $L \propto a(t)$, while the Universe expands. The scale factor is not physical by itself: one can always rescale everything, e.g. replace meters with feet. What matters is its time-dependence, and so the physical quantity describing the expansion is the Hubble parameter,

$$H(t) \equiv \frac{\dot{a}(t)}{a(t)}\,, \tag{2}$$

[a] e-mail: gorby@ms2.inr.ac.ru

the bigger its value, the higher the Universe expansion rate is. At small redshifts, $z \ll 1$, from eqs. (1), (2) one finds the Hubble law

$$H(t_0)r = z\,, \tag{3}$$

which states that the distance to the object r is proportional to its redshift z. The factor entering this relation, dubbed the Hubble constant, $H(t_0) \equiv H_0$, is actually the value of the Hubble parameter at present. Studying the distance-redshift relation (3) enables one to measure the present expansion rate H_0.

The Universe is homogeneous and isotropic, that we deduced from the analysis of distribution of the galaxies and galaxy clusters in the visible part of the Universe. An enormous progress here happened in the last decade thanks to spectacular performance of the astronomical telescopes deeply surveying the sky, especially Sloan Digital Sky Survey project. There were several claims of space-anisotropy, which check failed. Admittedly, there are large voids, the hugest one is particularly interesting, but it is the only void of that scale, and may be just a fluctuation.

The Universe is occupied by "thermal" photons: cosmic microwave background (CMB) radiation with spectrum identical to that of the black body with temperature $T_0 = 2.726\,\mathrm{K}$. Both shape and normalization of the spectrum fit, the number density of the relic photons is $n_\gamma = 411\,\mathrm{cm}^{-3}$. The spectrum has been measured over the three decades in frequency (logarithmic scale) with accuracy in temperature better than a per mille.

The widely accepted general conclusion is that our Universe is homogeneous, isotropic, hot and expanding. With expanding Euclidean manifold as a space part the interval between events gets modified as

$$\Delta s^2 = c^2\,\Delta t^2 - a^2(t)\,\Delta \mathbf{x}^2\,.$$

If we can measure how the photons propagate, we can extract $a(t)$ and thus understand the Universe expansion law. The next step is to connect it with fundamental physics through a bridge between General Relativity and particle physics. The bridge is called the Friedman equation (the first of the Einstein equations),

$$\left(\frac{\dot{a}}{a}\right)^2 = H^2\,(t) = \frac{8\pi}{3}\,G\rho^{\mathrm{energy}}_{\mathrm{density}}(t)\,, \tag{4}$$

and it relates the expansion rate $H(t)$ with the total energy density of the Universe $\rho^{\mathrm{energy}}_{\mathrm{density}}(t)$. The higher the density is, the faster the Universe expands.

The Hubble law (3) allows for estimating the present Hubble parameter H_0, and hence the energy density of the Universe from the Friedman equation (4). To understand the content of the present Universe, one must measure the evolution of the Hubble parameter in time. The latter allows one to discriminate between different contributions because their different dependence on the scale factor: matter, radiation and curvature behave as

$$\rho_{\mathrm{matter}} \propto \frac{1}{a^3(t)}\,, \qquad \rho_{\mathrm{radiation}} \propto \frac{1}{a^4(t)}\,, \qquad \rho_{\mathrm{curvature}} \propto \frac{1}{a^2(t)}\,,$$

respectively. Presently dominating dark energy exhibits no noticeable (so far) dependence on the scale factor,

$$\rho_{\mathrm{dark\ energy}} \approx \mathrm{const}\,.$$

Since the Friedman equation (4) is symmetric with respect to time reversal, one concludes, that in the past the matter density was higher, our Universe was "hotter", and was filled with electromagnetic plasma. Actually, following the Friedman equation (4) and based on the cosmological observations we can confidently track the history of our Universe back to the epoch when the plasma temperature was as high as 2-3 MeV. At that epoch active neutrinos decouple from the primordial plasma populated by electrons, positrons, photons, protons and neutrons. As a collisionless component, neutrinos, however, proceed to contribute to the energy density and pressure and hence influence the Universe expansion. The timetable of the late Universe, with major events like primordial nucleosynthesis, the transition from radiation domination to matter domination (RD/MD equality epoch) and recombination, is determined by the Universe expansion rate in the corresponding epochs. Carefully measuring the chemical composition of the detached cold gas clouds we extract the light chemical element abundances and by comparing them with the nucleosynthesis predictions we determine the Universe expansion rate at that epoch. Then the Friedman equation (4) gives the amount of relativistic species at that time, in this way we are sure the Standard Model (SM) neutrinos have been there. The Universe population at RD/MD and recombination we know thanks to the cosmological observations of CMB and matter distribution (galaxies, galaxy clusters) in the late Universe. Likewise, observations of quasars, supernovae type Ia, and galaxy clusters formation are used to investigate reionization and deceleration-acceleration transition.

Cosmology definitely asks for new physics, most probably new particle physics. However, it gives no any hints of the corresponding energy scales. Instead of that, cosmology indicates the time-scales associated with corresponding new physics. A physical process usually becomes of importance in the expanding Universe when its rate Γ are of about the Universe expansion rate,

$$\Gamma \sim H \propto \frac{T^2}{M_{\rm Pl}}, \tag{5}$$

where the Planck mass is related to the gravitational constant as $M_{\rm Pl} \equiv 1/\sqrt{G_N}$. From the analysis of expanding history of the early Universe one concludes that the baryogenesis must happened some time before the temperature of primordial plasma dropped down to 1 MeV, as the matter-antimatter asymmetry has generated by the nucleosynthesis epoch. The dark matter must be produced well before the epoch of RD/MD equality, $T \sim 1$ eV. The dark energy must be present by $T \simeq 5$ K. Rates Γ of the corresponding new physics processes must exceed the expansion rates H at the epochs with plasma of the corresponding temperatures T in accordance with eq. (5). In this way cosmological observations constrain the new physics scales. The earliest epoch which we know about directly, from analysis of the observational data (chemical element abundances), is the pre-nucleosyntesis epoch of $T \sim 1$ MeV. Was the Universe hotter we know not, at least for sure.

2 Large scale structures, matter inhomogeneities and inflation

However, there are observations which most probably advertise the presence of other new physics operating when the Universe was much younger, and its energy density significantly exceeding that of the pre-nucleosynthesis epoch. These observations concern the inhomogeneous Universe: CMB anisotropy and Large Scale Structure. Both are originated from the matter inhomogeneities in the early Universe. Their seeds are initial scalar perturbations of matter density with amplitude of $\delta\rho/\rho \sim \delta T/T \sim 10^{-4}$. Combined analysis of the cosmological data revealed the properties of these perturbations. They are adiabatic, so that while the total matter energy density fluctuates, the relative concentration of all species remains intact,

$$\delta\left(\frac{n_B}{s}\right) = \delta\left(\frac{n_{DM}}{s}\right) = \delta\left(\frac{n_L}{s}\right) = 0\,.$$

The initial matter inhomogeneities are Gaussian, so all the correlation functions in the Fourier space are described by only the two-point function, which behaves as

$$\langle\frac{\delta\rho}{\rho}(\mathbf{k})\frac{\delta\rho}{\rho}(\mathbf{k}')\rangle \propto \left(\frac{\delta\rho}{\rho}(\mathbf{k})\right)^2 \times \delta(\mathbf{k}+\mathbf{k}')\,.$$

 DOI: 10.1051/epjconf/201716401034

The scalar perturbations exhibit a very flat power spectrum,

$$\langle\left(\frac{\delta\rho}{\rho}(\mathbf{x})\right)^2\rangle = \int_o^\infty \frac{\mathbf{dk}}{\mathbf{k}}\mathcal{P}_S(\mathbf{k}), \qquad \mathcal{P}_S(\mathbf{k}) \approx \text{const}, \tag{6}$$

and the combined analysis of cosmological data on CMB and large scale structure favors [1]

$$\mathcal{P}_S \equiv A_S \times \left(\frac{k}{k_*}\right)^{n_s-1}, \qquad A_S \approx 2.5\times 10^{-9}, \qquad n_S \approx 0.97. \tag{7}$$

There are no viable sources of the initial scalar perturbations fueling during the standard Hot Universe expansion. Most probably they emerge earlier, at the inflationary stage of almost exponential expansion specially designed to explain while the Universe is flat, homogeneous and isotropic. The exponential expansion is provided by a specific dynamics of a scalar field dubbed inflaton. At the inflationary stage it evolves very slowly, mimicking the system of constant energy density. It makes the scale factor grow exponentially, as follows from the Friedman equation (4). Quantum fluctuations of this field, as well as spin-2 quantum fluctuations of metric —so-called tensor perturbations or gravity waves —evolve very peculiarly. The amplitude of a perturbation of wave-length $\lambda \propto 1/a$ decreases inversely proportional to the scale factor until the moment, when the 3-momentum $q = 2\pi/\lambda$ becomes of about the Hubble parameter, $q \approx H$, so-called horizon crossing. From this moment on, the wavelength proceeds to stretch but in the regime $\lambda \gg 1/H$ (the mode is called superhorizon) the amplitude remains frozen at the value $\delta\varphi_\lambda = H/(2\pi)$. The process is illustrated in Fig. 1. For the

Figure 1. At inflationary stage, the amplitude of subhorizon modes, $q \gg H$, *left panel*, decreases as $\delta\varphi_\lambda \propto q$, while the mode wave-length stretches as $\lambda = 2\pi/q \propto a$; meanwhile the amplitude of superhorizon modes, $q \ll H$, *right panel*, remains constant at the value of Hubble parameter taken at the moment of the horizon crossing, $\delta\varphi_\lambda = H_{infl}(q = H)/2\pi$.

superhorizon mode the proper quantum amplitude must become exponentially small $\delta\varphi_\lambda^{quantum} \propto 1/a$ with exponentially growing scale factor. On the contrary, the amplitude remains constant, and hence becomes populated by exponentially large number of quanta, i.e. *becomes classical.* This miraculous process results in production of classical spatial perturbations of inflaton field and metric.

Later, at the post-inflationary stage, reheating, and hot stages of the Universe evolution, the wave-length continues to grow proportionally to the scale factor, while the horizon $l_H \propto 1/H$ grows slower and hence at some moment the mode crosses the horizon once again becoming subhorizon, see Fig. 2. Then the amplitude starts to evolve obeying the linear perturbation equations written above the cosmological background determined by the homogeneous component of the primordial plasma. Later, after RD/MD equality, the dark matter perturbations start to dominate and finally decouple from the Hubble flow. They make the local gravitational potentials which form the proto-halos of future galaxies and galaxy clusters the ordinary matter (baryons) inhabited after recombination.

The inflaton field perturbations $\delta\phi(x)$ give birth to the inhomogeneities of the matter distribution $\delta\rho(x)$, while the metric perturbations give rise to the primordial gravitational waves $h_{\mu\nu}(x)$. The contrasts of the matter distribution $\delta\rho/\rho$ and amplitudes of the gravitational waves h are related to the

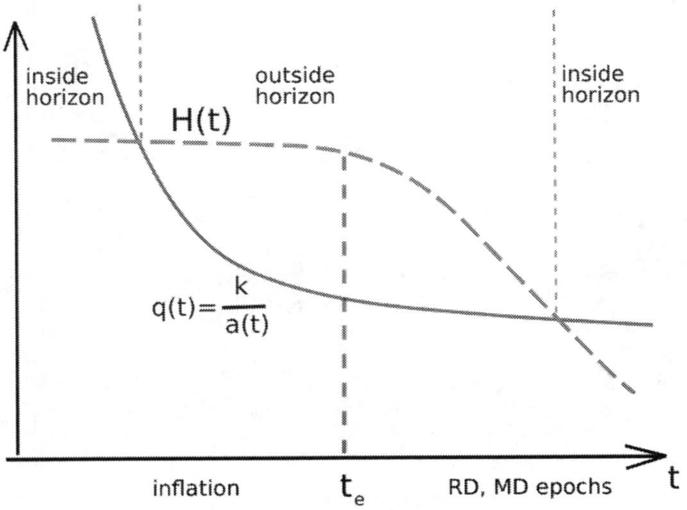

Figure 2. Evolution of the physical momentum $q = k/a(t)$ of the mode of conformal momentum k and the Hubble parameter $H(t)$ in the realistic Universe. Originated from the quantum fluctuation at the inflationary stage, the mode decreases exponentially at that time until the horizon crossing, $H = q$. Then, while outside the horizon it remains constant, and starts to evolve later, only after the horizon entrance at the hot stage of the Universe expansion.

Hubble parameter, inflaton field and its potential at inflation as follows (for details see e.g. [2])

$$\delta\phi \to \frac{\delta\rho}{\rho} \sim \frac{H^2}{\dot\phi} \propto \frac{V^{3/2}}{V'}, \qquad h \sim \frac{H}{M_{Pl}} \propto V^{1/2}. \tag{8}$$

From the absence of any hints of the tensor modes one places an upper limit on the Hubble parameter at inflation, $H_{\rm inf}$, and hence at the energy scale of inflation at the level of

$$V^{1/4} \lesssim 10^{16}\,\text{GeV}.$$

Therefore, *the new physics,* we can hope to probe with cosmological observables sensitive to the inflationary dynamics, *dwells below this scale.*

One of the consequences of the absence of the relic gravity waves and the corresponding upper limit on the inflationary expansion rate is a severe constraint on the electroweak Higgs sector. In the SM the quantum corrections to the effective Higgs potential turn the Higgs self-coupling to negative values at large Higgs fields, and the SM Higgs potential gains another, global minimum in the (super)planckian region, see right panel of Fig. 3. At the inflationary stage all the fields fluctuate at $\delta\phi \sim H/2\pi$, including the SM Higgs. The cosmological evolution definitely must not end up with the SM Higgs trapped in the wrong planckian vacuum. *This requirement either places an upper limit on* H_{inf} and hence on the amplitude of the relic gravity waves one may expect to observe, *or asks for the new physics at the lower energy scale* which changes the Higgs potential in a way to make the inflationary dynamics safe from falling into the wrong vacuum (e.g. by getting rid of this vacuum). The limit depends on the exact values of the Higgs boson mass m_h and the top-quark mass m_t as shown on left panel of Fig. 3. This is one insight into the new high energy physics we gain from consideration of the early-time cosmology.

3 Inflationary models with concave potential and perturbative quantum corrections

Detailed measurements of matter power spectrum parameters allows one to extract the information about the primordial spectrum of scalar perturbations, parameterised as shown by (7). These observ-

DOI: 10.1051/epjconf/201716401034

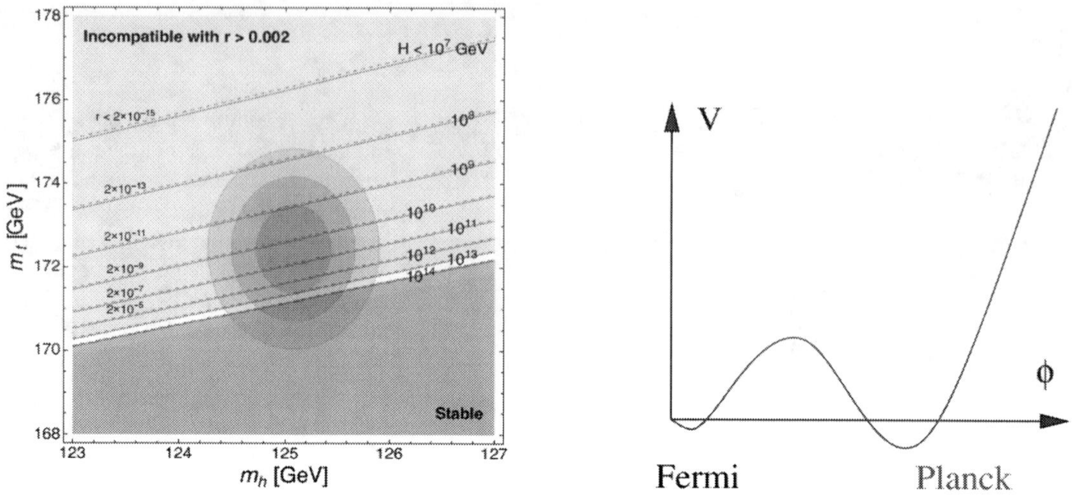

Figure 3. *Left panel:* Upper limits on the Hubble parameter at inflation and on the tensor-to-scalar ratio r (9) depending on (m_h, m_t) [3]. *Right panel:* a sketch of the SM effective Higgs potential corresponding to the presently measured central values of (m_h, m_t).

ables are related to the inflaton potential,

$$A_S \to \frac{V^{3/2}}{V'}, \qquad n_S \to \frac{V''}{V}, \left(\frac{V'}{V}\right)^2, \qquad r \equiv \frac{A_T}{A_S} \to \left(\frac{V'}{V}\right)^2, \tag{9}$$

where A_T refers to the magnitude of the tensor power spectrum defined analogously to the scalar case (6), (7). The experimental data are normalized to a particular value of k_* in (7), which refers to the mode exiting the horizon some $N_e \sim 50$ e-foldings before the end of inflation (that means, after the exit the scale factor grows by factor e^{N_e} till the inflation gets terminated). The exact number depends on the post-inflationary history of the Universe. If the inflaton sector is fully determined, and the inflaton coupling to the SM fields, responsible for the subsequent reheating of the Universe, is defined too, then N_e is fixed either, as it is obvious from Fig. 2.

So far, no evidences for the gravitational waves have been found, but the amplitude and the tilt of the scalar perturbations are measured with high accuracy, the results are presented in Fig. 4. Note that the most natural and simple models of scalar field with power-law potential, $V(\phi) \propto \phi^\alpha$, the quadratic, $V(\phi) \propto \phi^2$, and quartic ones, $V(\phi) \propto \phi^4$, [7] are excluded (or the former is at least strongly disfavored) by the present cosmological data. The strong upper limit on the gravity waves separates as viable the models with very flat potentials, as clearly seen from the expression for the tensor-to-scalar ratio,

$$r = \frac{A_T}{A_S} \propto \frac{\dot{\phi}^2}{H^2 M_{Pl}^2} \propto \left(\frac{V'}{V}\right)^2 \ll 1,$$

the latter constraint is inherent in the slow-roll approximation guaranteeing the almost exponential expansion during inflation.

Moreover, as a general tendency one can observes with a closer look at Fig. 4, that the presently available data point at the large field inflation models with *concave scalar potentials* (check for $V(\phi) \propto \phi^\alpha$ with $\alpha < 1$) as certainly more preferable. This finding seems good for robustness of the inflation

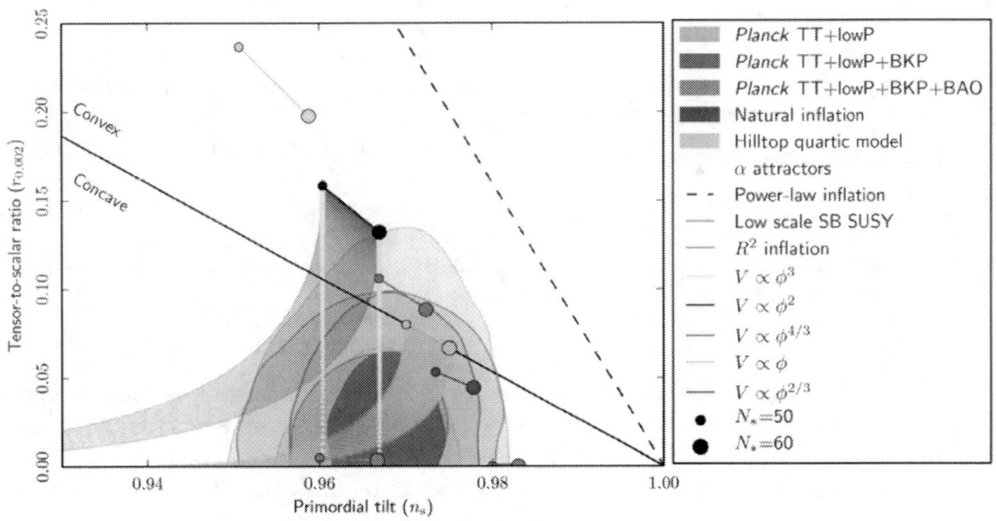

Figure 4. Present cosmological constraints [1] on (r, n_s) and predictions of a set of popular inflationary models. The colored regions are allowed at 65% and 95% CL, the model predictions depend on the number of e-folding N_e remained after the horizon crosssing.

predictions of the cosmological parameters, but seems bad for any other possible low-energy check of the inflaton sector, including the reheating dynamics, which in turn limit the accuracy of the inflation predictions of cosmological parameters.

The problem can be illustrated with an example of exponentially flat potential, typical for many well-developed cosmologically viable models, including very popular R^2-inflation [4], Higgs-inflation [5] and α-attractors [6]. In the former two cases the effective scalar potential at the inflationary stage coincide,

$$V(\phi) = \frac{3\mu^2 M_P^2}{4}\left[1 - \exp\left(-\sqrt{2/3}\,\phi/M_P\right)\right]^2, \tag{10}$$

while in the latter case the numerical factor in the exponent is an arbitrary coefficient (proportional to α !). Here we introduced the reduced Planck mass $M_P^2 \equiv M_{Pl}^2/8\pi$ and dimension-of-mass parameter μ. All the models in the asymptotic of large fields exponentially approach the constant value $V(\phi \to \infty)$ which defines the amplitude of the tensor perturbations by eq. (8). In both R^2-inflation and Higgs-inflation there is only one model parameter, represented here as μ. And it is fixed by matching to the measurements of the CMB anisotropy $\delta T/T \sim 10^{-4}$ as

$$\mu \approx 10^{13}\,\text{GeV}.$$

In case of α-attractors there is some freedom after the matching to CMB because there are two model parameters.

The inflaton potential (10) is plotted in Fig. 5. In Higgs-inflation and R^2-inflation the post-inflationary stage and reheating are fixed by the SM interactions and gravity, respectively, so that both the expansion law and the reheating temperatures are known [8, 9]. Therefore, for the mode k_* used as a normalization in the present cosmological data analyses we know the number of e-foldings N_e remained from its horizon exit till the end of inflation, and hence the exact value of the field ϕ

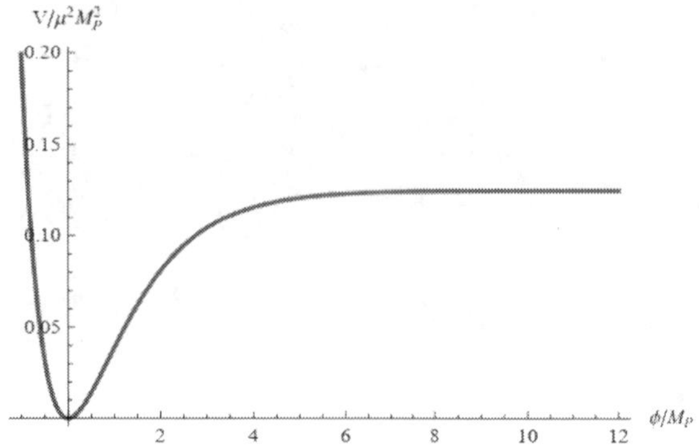

Figure 5. An example of the concave-type flat inflaton potential shared by R^2-, Higgs- and α-attractor inflations. At inflationary stage the inflaton starts with superplanckian fields and slowly rolls along the potential slope mimicking almost constant energy density, so the Universe exhibits (almost) exponential expansion. The slow rolling stops at some point with inflaton field of a subplanckian value and the inflaton starts to oscillate, so the Universe enters a stage of decelerating expansion.

where we probe the scalar potential with the cosmological data, see Figs. 2, 5. Being exponentially flat at this region, the potentials are "exponentially insensitive" to possible quantum corrections [10], protected by the approximate shift-symmetry, $\phi \to \phi + \text{const}$. It proves the robustness of the inflationary predictions of the cosmological parameters, such as amplitudes and tilts of the scalar and tensor spectra.

However, the scalar potential is obviously non-renormalizable, and there are no symmetries to protect its form at the quantum level. Naturally, close to the origin, the potentials in all the three models take their unique renormalizable polynomial form, saturated by either quadratic or quartic terms. The quantum corrections to these couplings are under control. The corresponding coupling terms may be naively extracted from the original large-field potential (10) and thus related to the small-field dynamics important for cosmology (e.g. reheating) or even direct tests of the inflaton sector like in Ref. [11]. This procedure, unfortunately, is not justified at all, given the fact that to relate the small-field polynomial behavior and large-field exponential behavior of the scalar potential one must cross the intermediate region where neither the renormalizability nor the shift symmetry prevents the potential from getting unknown and hence untamable quantum corrections.

This observations prevents from matching the small-field and large-field parameters of the model. Hence the power of the model predictions are rather limited. In particular, in the Higgs-inflation one cannot predict the cosmological parameters from the measurement of the Higgs boson mass. In R^2-model, the mass term μ in (10), which value is fixed from the CMB anisotropy measurements, may not be the same at reheating. Note, that since reheating temperature enters logarithmically to the number of efoldings N_e, it introduces the corresponding uncertainty into the model predictions of the cosmological parameters. Naturally it is small, however in some cases it may be crucial. For instance, the difference between the predictions of cosmological parameters obtained for Higgs-inflation and R^2-inflation is quite small and mostly attributed to the difference in reheating temperatures [12, 13]. The uncertainty associated with nonrenormalizability of the models seems enough to prevent from precise distinguishing between the two.

To cure the situation one must either invoke some symmetry to protect the form of the inflaton potential or to learn how to properly account for quantum corrections from the gravity sector.

4 Summary

The present cosmological observations confirm that the Universe was as hot as 2-3 MeV and most probably (much) hotter. The presence of dark matter and the baryon asymmetry of the Universe certainly imply new physics active in the very early Universe, but the cosmology does not point at particular energy scales. Inflation must proceed keeping save the electroweak vacuum, the requirement which can suggest the new physics coupled to the SM Higgs sector. At the moment (and probably in the near future), it is difficult to prove, because a new collider is needed to pin down m_t and m_h and thus settle the matter of where is the large-field minimum of the SM Higgs potential. Finally, matter inhomogeneities of the largest sizes can allow us to probe physics at the highest energies (inflation). However, (most) cosmologically favorable models are nonrenormalizable, which prevents any direct connection between small- and large-field model parameters. We are not (yet) ready to test properly the high energy physics with large-scale cosmological observations, a breakthrough in our understanding of quantum gravity and large-field dynamics is needed to open this possibility.

The work is supported by the RSF grant 14-12-01430.

References

[1] Ade P.A.R. et al. [Planck Collaboration], Astron. Astrophys. **594**, A20 (2016).
[2] Gorbunov D.S. and Rubakov V.A., *Introduction to the theory of the early universe: Cosmological perturbations and inflationary theory*, (Hackensack, USA: World Scientific, 2011), 489.
[3] W. E. East, J. Kearney, B. Shakya, H. Yoo and K. M. Zurek, [arXiv:1607.00381 [hep-ph]].
[4] Starobinsky A.A., Phys. Lett. B **91**, 99 (1980).
[5] Bezrukov F.L. and Shaposhnikov M.E., Phys. Lett. B **659**, 703 (2008).
[6] R. Kallosh, A. Linde and D. Roest, Phys. Rev. Lett. **112** (2014) no.1, 011303 doi:10.1103/PhysRevLett.112.011303 [arXiv:1310.3950 [hep-th]].
[7] Linde A.D., Phys. Lett. B **129**, 177 (1983).
[8] Bezrukov F., Gorbunov D. and Shaposhnikov M., JCAP **0906**, 029 (2009).
[9] Gorbunov D.S. and Panin A.G., Phys. Lett. B **700**, 157 (2011).
[10] Bezrukov F.L., Magnin A. and Shaposhnikov M., Phys. Lett. B **675** 88 (2009).
[11] F. Bezrukov and D. Gorbunov, JHEP **1005** (2010) 010 doi:10.1007/JHEP05(2010)010 [arXiv:0912.0390 [hep-ph]].
[12] Bezrukov F.L. and Gorbunov D.S., Phys. Lett. B **713**, 365 (2012).
[13] Gorbunov D. and Tokareva A., JCAP **1312**, 021 (2013).

EPJ Web of Conferences **164**, 02001 (2017)
ICNFP 2016
DOI: 10.1051/epjconf/201716402001

Historical and personal recollections of Guido Altarelli

Giorgio Parisi[1,2,3,a]

[1] *Dipartimento di Fisica, Università degli studi di Roma La Sapienza,*
[2] *Nanotec-CNR, UOS Rome,*
[3] *INFN-Sezione di Roma 1, Piazzale A. Moro 2, 00185, Rome*

Abstract. In this paper I will present a short scientific biography of Guido Altarelli, briefly describing some of his most important seminal works. I will analyze in great details the paper of the q^2 evolution of the effective quark distribution: I will put this paper in a historical perspective, describing our theoretical understanding at that time and the reasons why the paper was so successful.

1 A short scientific biography

Guido Altarelli, a great theoretical physicists, who spent most of its scientific career in Rome and at CERN, was born in Rome on 12 July 1941 and passed away on 30 September 2015 in his 74th year.

He received three very prestigious awards: the Julius Wess Award (2011), the J. J. Sakurai Prize of the APS (2012) and the High Energy and Particle Physics Prize of the EPS (2015). These three very celebrated distinctions only partially reflect the richness of his contributions: his deep and lasting influence is ubiquitous in the field of high energy physics.

Guido graduated in Physics from Rome University in 1963, the supervisor being Raul Gatto (who was in Florence at that time). In 1964 he went to Florence to join the very lively and large group of young researchers (*gattini*, i.e. small cats) who were working together under the careful and inspiring supervision of Gatto[1]. In 1968-1970 he spent two years in New York (at New York University and at Rockefeller University). In 1970 he became professor of theoretical physics in Rome. In this period he spent long periods in other institutions, e.g. the ENS in Paris and the Boston University.

In 1987 Guido became Senior Staff Physicist at the Theory Division of CERN (where remained up to the age of 65 years in 2006), and he was the Theory Division Leader in 2000-04. In 1992 he moved to the newly founded University Rome 3, where he remained up to formal retirement, dividing his time between teaching (in Rome) and research (mostly at CERN). He continued his research activity up to his last months.

As usually in Italy, his first research work was his Laurea thesis: it was done in collaboration with his friend Franco Buccella on the process $e^+ + e^- \to e^+ + e^- + \gamma$. Gatto suggested them to perform this computation in preparation for future experiments at the ADONE colliding beams of Frascati; it was part of a wide spectrum program lead by Nicola Cabibbo and Gatto, aimed to compute the cross sections for every possible process that could be observed [1] in $e^+ + e^-$ colliding beam experiments.

[a] e-mail: Giorgio.Parisi@Roma1.Infn.It
[1] In Italian "gatto" means "cat".

Guido Altarelli and Franco Buccella had to evaluate the cross section for the process $e^+ + e^- \to e^+ + e^- + \gamma$, paying attention to the one-loop infrared divergences in the process $e^+ + e^- \to e^+ + e^-$. The formulae they wrote in their first attempt were correct. Unfortunately, for very small transfer momentum of the γ the computer was nearly computing 0/0 and, due to rounding effects in floating point arithmetics, the output was extraordinarily large. This first attempt produced an extremely large cross section that left everybody puzzled: however they were fast enough to understand the reason of the strange result: they finally wrote a successful thesis [2].

In Florence he worked with Gatto and the other gattini (among them Marco Ademollo, Buccella, Luciano Maiani and Giuliano Preparata). Most of the topics they investigated were fashionable at that time: e.g. phenomenological applications (mainly to to hadron spectroscopy) of various symmetries ($SU(6)$, $U(6)\otimes U(6)$ and $U(12)$)[3–5]; he also started to be interested in Regge poles [6]. He continued to work on the theoretical evaluations of the cross section of different reactions in electron-positron colliding beams.

In New York, he worked mainly with Hector Rubinstein on Regge pole predictions coming from the resummation of ladder diagrams, on duality and on bootstrap [9]. At that time the dominant school was asserting that the mystery of strong interaction could be understood using only general self-consistency properties of the S matrix (S matrix bootstrap). A quantum field theory approach was considered pointless: I remember that in some circles a too deep knowledge of quantum field theory was considered to be a handicap. This phenomenological program culminated in the duality and in the superconvergence sum rules. This line of research reached a definitive success with the Veneziano model [7] (and a few month later with the Virasoro model [8]). It is amusing to note that these two models were the starting point of the most sophisticated existing field theory, i.e. string theory; they correspond respectively to the open and to the closed string models.

After his arrival in Rome, his scientific interests moved firstly toward the parton model and later concentrated on what we now call the Standard Theory. This change was in agreement with the strong tradition of the Rome school (led by Cabibbo who became a full professor at the Rome University in 1966 at the age of 31), of using field theory tools to understand particle physics: it was the alternative viewpoint respect to the S matrix bootstrap.

Cabibbo was deeply interested in weak interaction physics. The theory of weak interaction is based on local currents, Lagrangians: one needs the whole the apparatus of local quantum field theory and Cabibbo's interest in field theory was a necessity. He created a strong school of younger collaborators and students and the Rome school gave a crucial contribution to the construction of the standard model.

Many of the Roman works of Guido were written with Cabibbo, Mario Greco, Keith Ellis, Maiani, Guido Martinelli, Giorgio Parisi and Roberto Petronzio. Among the most important contributions are the papers on the octet enhancement of non-leptonic weak interactions in asymptotically free gauge theories [10, 11]: for the first time it was shown that the newly born QCD was able to contribute to solving some of the old mysteries of weak interactions (in this case the dominance of the $\Delta I = 1/2$ strangeness changing processes over the much weaker $\Delta I = 3/2$ processes).

Other important contributions of this period are:

- His paper [12] on the equations for the q^2 dependence of the proton structure functions where effective parton model distributions were introduced and the equation for the q^2 evolutions were written (the so called DGLAP equations).
- The first computation of the next to the leading order (NLO) corrections to the naïve parton model [13]: in this paper he discovered that there are quite large one-loop NLO corrections to the standard parton prediction for the $\mu^+ + \mu^-$ production in hadronic corrections: there is a correction factor of order 1 at not extremely large values of the invariant mass of the muon pairs.

- The first computation of the transverse momentum distribution of the $\mu^+ + \mu^-$ pairs produced in hadronic collisions [14]. This was an important check of QCD: one could see directly the effects of hard scattering of the partons that producing very large momentum pairs. The process also provided an independent way to measure the running strong coupling constant.
- A careful analysis of the production of vector mesons at colliders where theoretical QCD based considerations mix with more phenomenological considerations [15]. This paper was crucial to construct a well-founded model for the production of vector mesons as a function of total energy, transverse momentum, and rapidity.

In the 70's Guido and I have been working side by side: also when I was not actually working together with him I was quite aware of his works. This went on up to the beginning of the 80's. At that time I became mostly interested in problems in statistical mechanics and my only residual activity in high energy physics was lattice QCD. Guido, especially after going to CERN, started to explore new territories in high energy physics and our roads unfortunately separated.

His interaction with CERN's environment strongly reflected on his choice of research subjects. He gave crucial contributions to the understanding of the physics of the colliders: the planning of experiments in CERN was strongly influenced by his theoretical analysis. His work has also been crucial in putting an order into the huge mass of experimental data and theories (sometimes one contradictory with the other) and in identifying the crucial experiments or theoretical analysis that had to be done to remove the contradictions.

The number results that he obtained in this period is so abundant that I will only make an arbitrary selection, strongly biased by my personal tastes:

- In the 80's he got interested in predictions for future CERN experiments, e.g. for the production of jets, heavy-vector mesons [16] and other exotic objects like supersymmetric particles [17] and the Higgs meson [18]. Of course, the production of the Higgs meson is strongly dependent on his mass, so it was crucial to get the most accurate bounds on his mass from the comparison of the experimental data with the standard model [19].
- He has also written seminal papers on the decay of heavy quarks (e.g on inclusive beauty decays and on the spectator model [20]). In this context he continued the analysis of the consequences of QCD on weak interactions, computing the two loops contributions [21]: it was the first NLO computation for the strong corrections to weak interactions.
- He became to be deeply interested in polarized proton structure function [22], where he discovered, together with Graham Ross [23], the crucial interplay between the gluon anomaly and polarization effects.
- Mainly with Riccardo Barbieri, he constructed a model-independent analysis of the electroweak data [24–27]. This approach was quite new and the methodology of this study became a standard for future analysis.

With the new millennium, Guido starts to study new subjects. He is mostly interested in the attempt of deciphering the information contained in the huge mass of experimental data, especially in the newly obtained mass matrices of neutrino. He looks for fingerprints of symmetries; his old expertise in symmetry groups (from Florence times) has certainly been very useful to him.

- He studies the detailed predictions of realistic SU(5) unification (also with supersymmetry) both for proton decay [28] and for neutrino mixing [29].
- He is fascinated by the elegance of the tribimaximal neutrino mixing: many of his papers [30] (mostly with Ferruccio Feruglio) are dedicated to the search for the origins of this baffling symme-

try. Although nowadays we know that neutrino mass matrix is not tribimaximal, it is still possible that deviations from a tribimaximalily come from a perturbative expansion in some small parameter.

- He is still interested to the evolutions of parton distribution, especially in the small x region, an important experimental region, where many data are available. The theoretical analysis is quite complex, however, his careful analysis shows that the resummation of the leading logarithms in x gives a precise prediction that is in very good agreement with the experiments [31].

2 The effective q^2 dependent parton distribution functions and their evolution equations

Here I will present some historical recollections on the paper *Asymptotic freedom in parton language* that we wrote when Guido and myself were both in Paris (Guido was in the ENS in the center of Paris and I were in the IHES at Bures-sur-Yvette). I will concentrate my attention on this paper also because of it the paper of Guido I know better; I believe that I am qualified to describe the motivations and the perspectives. I will not discuss other people's important works [32, 33] in the same direction that we were not aware of at that time.

2.1 Field theories preliminaries

Generally speaking, the relevance of a paper can be understood only in the context of the physics that was known at the time that the paper is written. One should be very carefully: if we miss the historical context when reading a paper after a long time, we risk to consider trivial the main advances of that paper and to consider very relevant something that we quite trivial. The historical reconstruction of the knowledge and of the understanding at the time when the paper was written is crucial to avoid misinterpretations.

At the time of the paper, very few things were known in strongly interacting quantum field. About 10 years before Murray Gell-Mann wrote [34] in a very influential paper: *We use the method of abstraction from a Lagrangian field theory model. In other words, we construct a mathematical theory of the strongly interacting particles, which may or may not have anything to do with reality, find suitable algebraic relations that hold in the model, postulate their validity and then throw away the model. We compare this process to a method sometimes employed in French cuisine: a piece of pheasant meat is cooked between two slices of veal, which are then discarded.*

As remarked by David Gross *until 1973 it was not thought proper to use field theories without apologies* [35]. Feynman himself refrained to speak of field theory and invented partons. In 1977 the situation was much better; however, people were cautious with field theory: they were very careful in making assumptions that were not proved at least at all orders in perturbation theory and nearly nothing was proved.

One of the most solid results was Wilson [36] short distance operator product expansion that states that in the limit of small x the product of two operators can be written as

$$A(x)B(0) \rightarrow_{x\rightarrow 0} \sum_C C(0)|x|^{-d_A-d_B+d_C} , \qquad (1)$$

where the sum runs over all the local operators C, including the identity. The leading terms come from the operators C with the lowest dimensions. The dimensions of the operators were the canonical one in free field theory and this simple case everything was clear.

Strong interactions are not a free field theory. In the case of a non-free theory, we have to use the renormalization group. The dimensions of the operators could be different from the canonical ones in an interacting theory. In a seminal paper [37] Symanzik studied the case of a ϕ^4 theory: he obtained

$$\phi(x)\phi(0) \to_{x\to 0} \alpha|x|^{-2d_\phi}\mathbf{1} + \beta|x|^{-2d_\phi+d_{\phi^2}}\phi^2(0)\,, \tag{2}$$

where **1** is the identity operator.

In this case, the Wilson expansion was proved, and the dimensions were not the naive one, provided that the renormalization group had a non-trivial fixed point: the dimensions contained a non-trivial coupling constant dependent part, the called anomalous dimensions. The renormalization group with a running coupling constant has to be used in order to compute these anomalous dimensions.

In order to explain the importance of Symanzik's paper, I have to show its relevance of the short distance operator expansion on deep inelastic scattering. The total cross section for the process "virtual gamma+ proton" (i.e. deep inelastic scattering) can be written in terms of the appropriate Fourier transform of the function:

$$G(x^2, x_0) = \langle p|J(x)J(0)|p\rangle\,, \tag{3}$$

where $|p\rangle$ denotes the one proton state and we have suppressed the Lorentz indices of the currents. Kinematical considerations imply that the region relevant for deep inelastic scattering is $x^2 \approx 0$, i.e. the light cone [38].

At the end of the sixties, it was shown experimentally that in inelastic electron-proton scattering in the region of large momentum transfer, the data were compatible with a simple scaling of the structure functions (the so-called Bjorken scaling). The deep roots of Bjorken scaling for deep inelastic scattering have been soon identified. Bjorken scaling was shown to be a consequence of the naive light cone expansion [38] of Brandt and Preparata (1969)

$$J(x)J(0) \to_{x^2\to 0} \frac{O(x_0, 0)}{x^2}\,, \tag{4}$$

where $O(x_0, 0)$ is a bilocal operator.

This expansion has the same form of the one valid in a free field theory. Indeed in free field theories, the naive light cone expansion can be derived by the Wilson expansion by doing a Taylor expansion at $x_0 = 0$. An infinite number of terms in the Taylor expansion have to be considered, each term corresponding to a different moment of the experimental structure function. In this way, the Wilson short distance operator product becomes deeply related to the experimentally observed approximate Bjorken scaling.

However, the crucial question was what happens for an interacting theory? Christ, Hasslacher and Muller (1972) computed the coupling dependent anomalous dimensions of the operators relevant for deep inelastic scattering (the so-called twist-two operators) for a Yang-Mills theory [39].

If we neglect the dependence of the running coupling constant (α) on the momenta, when we consider only non-singlet contribution (the gluon contribution is a singlet) they obtained something like

$$M_n(q^2) \equiv \int_0^1 dx x^{n-1} F(x, q^2)\,; \quad M_n(q^2) = C_n \exp(\gamma_n(\alpha)\log(q^2))\,.$$

Here $F(x, q^2)$ is the structure function of the proton, that enters in the computation of deep inelastic scattering.

If we specialize the Christ-Hasslacher-Muller formulae to QCD, we get a simple result always in the case of the non-singlet contribution. Taking care of the running coupling constant $\alpha(q^2)$, we get the final formulae:

$$\frac{\partial M_n(q^2)}{\partial \log(q^2)} = \gamma_n(\alpha(q^2)), \tag{5}$$

where the linear term in $\alpha(q^2)$ of is obtained from formulae of [39].

If we use the standard properties of the Mellin trasform and of the inverse Mellin we finally get

$$\frac{\partial F(x,q^2)}{\partial \log(q^2)} = \int_x^1 \frac{dy}{y} F(x,q^2) P_{q,q}(x/y, \alpha(q^2)), \tag{6}$$

where the quark fragmentation function $P_{q,q}(z, \alpha(q^2))$ is the inverse Mellin transform of the anomalous dimensions $\gamma_n(\alpha(q^2))$. Using the first order in α for $\gamma_n(\alpha)$ computed in [39], we finally get for the non singlect contribution:

$$P_{q,q}(z) = \frac{8}{3}\frac{\alpha(q^2)}{4\pi}\left(\frac{1+z^2}{(1-z)_+} + \frac{3}{2}\delta(z-1)\right) \tag{7}$$

When we wrote our work (spring '77), all that was very well known. In '73 I obtained the previous formula for the valence quarks [40]. Later on, similar formulae were written for the gluons and sea-quarks contribution using inverse Mellin transform [41, 42].

2.2 The parton model

In theories where the interaction could be neglected in the large momentum transfer region, deep inelastic structure functions measure the number of charged partons (quarks in QCD) carrying a fraction of momentum x in the infinite momentum frame. This observation leads to Bjorken scaling in the parton model. Guido was familiar with the details of the parton model construction and its interpretation in $p = \infty$ frame and he wrote an influential paper on it [43].

However, the Bjorken scaling was not exact and the effect of the strong interactions in the large momentum region was weak, but it was not vanishing. At that time was a big question how to deal with the parton model concepts in QCD [44, 45]. Many phenomenological arguments were simply formulated using the parton language and the QCD results were obtained via operator product expansion. Most of the relations coming from the parton model where out of reach of the operator product expansion.

However, there was a different case where the physical interpretation was quite clear: quantum electrodynamics, where electromagnetic radiative corrections produce perturbative effects that are not negligible, i.e. they are not simply proportional to α with a prefactor of order 1. In QED one often finds that the net effect of radiative corrections is proportional to $\alpha \log(E/m_e)$ or $(\alpha \log(E/m_e))^k$. Various techniques can be used. The most famous approach that leads to a simple and fast computation of these effects is the *Equivalent photon approximation in quantum electrodynamics* (due to Weizsaker-Williams.).

Nicola Cabibbo was strongly interested in computing these effects for applications to the experimental analysis of events collected in e^+e^- collisions. Cabibbo and Rocca in 1974 wrote [46] formulae like:

$$P_{e\to e\gamma}(\eta) = \frac{\alpha}{\pi}\frac{1+(1-\eta)^2}{\eta}\log(E/m_e); \quad P_{\gamma\to e^+e^-}(\epsilon) = \frac{\alpha}{2\pi}(1+(1-2\epsilon)^2)\log(E/m_e) \tag{8}$$

where η is the fraction of longitudinal momentum carried by the photon and ϵ is the fraction of longitudinal momentum carried by the electron.

These probabilities can be combined. They computed the probability of finding inside a γ a triplet γ, e^+, e^-: it is proportional to $P_{\gamma\to e^+e^-}P_{e\to e\gamma}\log(E/m_e)^2$. Nicola and I [2] used these ideas [47] to estimate the cross section for the process $e^+e^- \to 2e^+2e^-$ that was a candidate to be an important background in e^+e^- colliding beam experiments.

2.3 The paper

This paper stemmed from a proposal from Guido, i.e. to make previously-obtained results on scale violations clearer and more exploitable. It was written while both authors were in Paris, and Guido liked to remark that it is the most cited *French* paper in the field of high energy physics.

The motivations of the paper were clearly stated in the introduction. When speaking of the standard operator product approach to scaling violations in QCD, we wrote: *In spite of the relative simplicity of the final results, their derivation, although theoretically rigorous, is somewhat abstract and formal, being formulated in the language of renormalization group equations for the coefficient functions of the local operators which appear in the light cone expansion for the product of two currents.*

We also added: *In this paper, we show that a alternative derivation of all results of current interest for the* Q^2 *behavior of deep inelastic structure functions is possible. In this approach all stages of the calculation refer to parton concepts and offer a very illuminating physical interpretation of the scaling violations. In our opinion the present approach, although less general, is remarkably simpler than the usual one since all relevant results can be derived in a direct way from the basic vertices of QCD, with no loop calculations being involved (the only exception is the lowest order expression for the running coupling constant which we do not rederive). This method can be described as an appropriate generalization of the equivalent photon approximation in quantum electrodynamics (Weizsaker-Williams .... Cabibbo-Rocca).*

In spring '77 Guido and I were often discussing QCD scaling violations. Guido suggested that it would be pedagogically useful to derive all the equations for scaling violations using the same techniques of Cabibbo-Rocca; no loops were involved: only the evaluation of the vertices in the infinite momentum frame. As I remarked before, Guido was an expert on parton model in the infinite momentum frame and the Cabibbo Rocca computation was already done in the infinite momentum. Most of the formulae we needed were written there: only the gluon splitting into two gluons function was missing. However, in our paper, we rederived all the spitting functions.

The final master [3] equations are well known:

$$\frac{dq^i(x,t)}{dt} = \frac{\alpha(t)}{2\pi}\int_x^1 \frac{dy}{y}\left[q^i(y,t)P_{q,q}\left(\frac{x}{y}\right) + G(y,t)P_{q,G}\left(\frac{x}{y}\right)\right],$$

$$\frac{dG^i(x,t)}{dt} = \frac{\alpha(t)}{2\pi}\int_x^1 \frac{dy}{y}\left[\sum_{i=1}^{2f} q^i(y,t)P_{G,q}\left(\frac{x}{y}\right) + G(y,t)P_{G,G}\left(\frac{x}{y}\right)\right], \tag{9}$$

where $t = \log(q^2)$ and the index i runs over quarks and antiquarks of all flavour.

We eventually obtained the expression for all the splitting functions ($P_{q,q}, P_{q,G}, P_{G,q}, P_{G,G}$). Our computations were much simpler that the original computations based on the one loop corrections to the vertices of the operators entering in the Wilson product expansion of two currents (a là Christ-Hasslacher-Muller). The computations were particular transparent and simple when we extended it to the case of polarized partons as an using the helicity formulation in the infinite momentum frame.

[2]This computation was never published: Cabibbo and Rocca wrote [46] that I did such a computation and I wrote that the computation was done by Cabibbo [42]. I remember that I had many discussions with Nicola on these generalized Weizsaker-Williams relations, but I forgot what exactly happened in this case.

[3]We used the wording *master equations* because they were classical probabilistic equations derived in a quantum setting.

The paper was a well-done cocktail of renormalization group results, parton model, and perturbation theory in infinite momentum frame: easy to drink and to swallow. The paper was very successfully, also because it was very clearly written: it was written by Guido, not by myself. It was really pedagogic: it contained all the logical steps and the final receipt was quite easy to follow.

2.4 Aftermath

I am convinced that the most important result of the paper was *not* the construction of a practical way to compute scaling violations in deep inelastic scattering. Indeed this task was already accomplished using the Mellin transformation and related techniques: these techniques were slightly more cumbersome but they were easy to use. Moreover, the equations (9) were well known, at least for the valence quarks.

The important point in our paper was to shift the focus from Wilson operator expansion to the effective number of partons that was dependent on the resolution, i.e. q^2. It was more than a computation: it was a *change in the language* we use. The appropriate choice of the language is one of the most important scientific tools.

For example, the Drell-Yan process (i.e. $pp \to l^+l^- + \cdots$) could be not studied by a Wilson operator expansion: the Brandt-Preparata analysis did not work in this case. Similar conclusions are also true for the jet production in hadronic collisions and for the computation of the cross section for large transverse momenta that were related to the hard scattering of partons.

However, it was now possible to study these process using the new language: we have to factorize the amplitude for the process into a part containing the effective parton distribution at the relevant energy and into a part containing the hard scattering that could be treated in perturbation theory in the running coupling constant. Moreover, in presence of energy dependent effective parton distributions the parton model predictions were ambiguous at order α_{QCD} and next to the leading order corrections (NLO) have to be computed in order to refine the predictions.

The solution of all these problems was at hand after our paper. This opportunity was immediately taken by Guido. All the following papers are published in '78 just after our '77 paper:

- *Leptoproduction and Drell-Yan processes beyond the leading approximation in chromodynamics* [13]. This is the first NLO computation we can find in the literature: it was simply done by computing the first corrections in α_{QCD} to the process $q + \bar{q} \to l^+l^- + \cdots$ and comparing them to the first order for deep inelastic scattering. In the experimental region, the corrections were quite surprisingly of order 1, because of a factor π^2 that was present only at positive q^2 and not at negative q^2.
- *Transverse momentum of jets in electroproduction from quantum chromodynamics* [48]. The large transverse momentum distribution of jets in deep inelastic scattering was analytically computed and successfully compared to the experiments. These large transverse momentum events have the same origin of the hard scattering processes that induce scaling violations in deep inelastic scattering.
- *Transverse momentum in Drell-Yan processes* [14]. There are two papers on this subjects that are nearly simultaneous to the previously mentioned papers. Here it was shown that the transverse momentum distribution of large mass muon pairs produced in proton-nucleus collisions can be quantitatively explained the combination of two ingredients: the hard scattering of partons (gluon and quarks) and the intrinsic transverse distribution of constituent partons inside the hadrons.
- *Processes involving fragmentation functions beyond the leading order in QCD* [49]. Here the whole analysis is based on energy dependent effective parton distributions and in the comparison between the naïve parton model and the one loop results.

In all these four problems the matching of the experimental data with the theoretical predictions allowed and independent measure of the QCD running coupling constant. At that times it was possible to achieve this goal in the first three cases. The compatibility of these independent determinations of α_{QCD} with the value coming from deep inelastic scattering (and from ψ decay) was instrumental in convincing physicists that QCD was the correct theory.

3 We miss him very much

His scientific success was inseparable from his human qualities. Guido aimed to get a deep understanding of the world, hence his great passion for history, especially that of the many countries he was traveling through.

Perhaps his most characteristic traits were his great kindness and intellectual honesty, coupled with a rather ironic view of himself and life in general. His great inquisitiveness, the enjoyment he derived from learning new things and putting the pieces of a puzzle together, allowed him to make great summaries of topical subjects, which allowed us to take stock of the current state of a field of research, indicating new directions to take.

He liked clear, precise formulations, which could be understood by all. He was not a reclusive or selfish scientist, interested only in what personal prestige might be gained from his research. Guido was also a researcher who enjoyed to work with others within the large community of high-energy particle physicists.

Many of Guido's works, from the most famous to the lesser known, were conceived in a spirit not only of research but also in a spirit of service to the community. It is difficult to think which would be the status of the field without his *seminal* contributions.

We all miss him very much, not only as an invaluable scientist but also as a dear friend, always ready to help.

References

[1] N. Cabibbo and R. Gatto. Phys. Rev. **124**, 1577 (1961).
[2] G. Altarelli and F. Buccella. Nuovo Cimento **34**, 1337 (1964) .
[3] G. Altarelli, F. Buccella, and R. Gatto. Phys. Lett. **14**, 70 (1965).
[4] M. Ademollo, G. Altarelli, and R. Gatto, Phys. Rev. Lett. **14**, 420 (1965).
[5] G. Altarelli, R. Gatto, L. Maiani, and G. Preparata, Phys. Rev. Lett. **16**, 918 (1966).
[6] G. Altarelli, R. Gatto, and L. Maiani, Nuovo Cimento A **58**, 259 (1968).
[7] G. Veneziano, Nuovo Cimento A **57**, 190 (1968).
[8] M. A. Virasoro, Phys. Rev. **177**, 2309 (1969).
[9] G. Altarelli and H. R. Rubinstein, Phys. Rev. **187**, 2111 (1969); Phys. Rev. **183**, 1469 (1969).
[10] G. Altarelli and L. Maiani, Phys. Lett. B **52**, 351 (1974).
[11] G. Altarelli, N. Cabibbo, and L. Maiani, Nucl. Phys. B **88**, 285 (1975).
[12] G. Altarelli and G. Parisi, Nucl. Phys. B **126**, 298 (1977).
[13] G. Altarelli, R. K. Ellis, and G. Martinelli, Nucl. Phys. B **157**, 461 (1979).
[14] G. Altarelli., G. Parisi, and R. Petronzio, Phys. Lett. B **76**, 351 (1978); Phys. Lett. B **76**, 356 (1978).
[15] G. Altarelli, R. K. Ellis, M. Greco and G. Martinelli, Nucl. Phys. B **246**, 12 (1984).
[16] G. Altarelli, M. Diemoz, G. Martinelli, and P. Nason, Nucl. Phys. B **308**, 724 (1988).
[17] G. Altarelli, G. Martinelli, B. Mele, B., and R. Rückl, Nucl. Phys. B **262**, 204 (1985).

[18] G. Altarelli, B. Mele, and F. Pitolli, Nucl. Phys. B **287**, 205 (1987).
[19] G. Altarelli and G. Isidori. Phys. Lett. B **337**, 141 (1994).
[20] G. Altarelli and S. Petrarca, Phys. Lett. B **261**, 303 (1991).
[21] G. Altarelli, G. Curci, G. Martinelli and S. Petrarca, Nucl. Phys. B **187** (1981) 461.
[22] G. Altarelli, R. D. Ball, S. Forte, and G. Ridolfi, Nucl. Phys. B **496**, 337 (1997).
[23] G. Altarelli, and G. G. Ross, Phys. Lett. B, **212**, 391 (1988).
[24] G. Altarelli and R. Barbieri, Phys. Lett. B **253**, 161 (1991).
[25] G. Altarelli, R. Barbieri, and S. Jadach, Nucl. Phys. B **369**, 3 (1992).
[26] G. Altarelli, R. Barbieri, and F. Caravaglios, Nucl. Phys. B **405**, 3 (1993).
[27] G. Altarelli, R. Barbieri, and F. Caravaglios, Phys. Lett. B **314**, 357 (1993).
[28] G. Altarelli and F. Feruglio, Phys. Lett. B **511**, 257, (2001).
[29] G. Altarelli, F. Feruglio, and I. Masina, J. High Ener. Phys. **11**, 040 (2000).
[30] G. Altarelli and F. Feruglio, Nucl. Phys. B **741**, 215 (2006); Nucl. Phys. B **720**, 64 (2005); Rev. Mod. Phys. **82**, 2701 (2010).
[31] G. Altarelli, R. D. Ball, and S. Forte, Nucl. Phys. B, **674**, 459 (2003).
[32] V.N. Gribov and L.N. Lipatov, Sov. J. Nucl. Phys. **15**, 438 (1972).
[33] Yu.L. Dokshitzer, Sov. Phys. JETP **46**, 641 (1977).
[34] M. Gell-Mann, Physics **1**, 63 (1964).
[35] D. J. Gross in *History of Original Ideas and Basic Discoveries in Particle Physics*, ed. by H. B. Newman and T. Ypsilantis, Springer (2012).
[36] K. G. Wilson, Phys. Rev. **179**, 1499 (1969).
[37] K. Symanzik, Comm. Math. Phys. **23**, 49 (1971).
[38] R. A. Brandt and G. Preparata, Nucl. Phys. B **27**, 541 (1971).
[39] N. Christ, B. Hasslacher and A. H. Mueller, Phys. Rev. D **6**, 3543 (1972).
[40] G. Parisi, Phys. Lett. B **50**, 367 (1974).
[41] G. Parisi and R. Petronzio, Phys. Lett. B **62**, 331 (1976).
[42] G. Parisi, in *1976 Moriond Proceeding: Weak interactions and neutrino physics*, ed. by J. Tran Thanh Van (1976); reprinted in G. Parisi, *Field Theory, Disorder and Simulations* (World Scientific 1992).
[43] G. Altarelli, N. Cabibbo, L. Maiani and R. Petronzio, Nucl. Phys. B **69**, 531 (1974).
[44] A. M. Polyakov, Sov. Phys. JEPT **32**, 296 (1971).
[45] J. Kogut and L. Susskind, Phys. Rev. D **9**, 3391 (1974).
[46] N. Cabibbo and M. Rocca, *The ρ Bremsstrahlung*, Ref TH. 1872-CERN (1974).
[47] N. Cabibbo and G. Parisi, unpublished (1975).
[48] G. Altarelli and G. Martinelli, Phys. Lett. B **76**, 89 (1978).
[49] G. Altarelli, R. K. Ellis, G. Martinelli and S.-Y. Pi, Nucl. Phys. B **16**, 301 (1979).

EPJ Web of Conferences **164**, 02002 (2017)
ICNFP 2016
DOI: 10.1051/epjconf/201716402002

Effective actions for high energy scattering in QCD and in gravity

L. N. Lipatov[1,2,a]

[1] *Petersburg Nuclear Physics Institute, Gatchina, Orlova Roshcha, 188300, St.Petersburg, Russia*
[2] *Saint Petersburg State University, St. Petersburg, Russia*

Abstract.
The scattering amplitudes in QCD and gravity at high energies are described in terms of reggeized gluons and gravitons, respectively. In particular, the BFKL Pomeron in $N = 4$ SUSY is dual to the reggeized graviton living in the 10-dimensional anti-de-Sitter space. The effective actions for the reggeized gluons and gravitons are local in their rapidities. The Euler-Lagrange equations for these effective theories are constructed and their solutions are used for calculations of corresponding Reggeon vertices and trajectories.

1 Introduction

The hadron scattering amplitudes $A(s,t)$ at high energies $\sqrt{s} = \sqrt{(p_1+p_2)^2}$ and fixed momentum transfers $q = \sqrt{-t}$ can be expressed in terms of the t-channel partial waves $f_j(t)$ having the Regge poles $f_j(t) \sim \frac{1}{j-j(t)}$. The Pomeron is a special Regge pole responsible for a slow growth of total cross sections σ_t and for the fulfilment of the Pomeranchuk theorem on the equality of the particle-particle and particle-anti-particle cross sections at large energies. The exchange of two or more reggeons generates more complicated Mandelstam singularities of $f_j(t)$ in the j-plane. To take into account all possible Pomeron interactions V.N. Gribov constructed the Reggeon calculus based on the 2+1 quantum field theory of a complex scalar field [1].

The scattering amplitudes in QCD with the gluon quantum numbers in the t-channel have the Regge form in the so-called leading logarithmic approximation (LLA) [2]. Here the Pomeron is a colorless composite state of the reggeized gluons. Its wave function in QCD and in other field models with the gauge group $SU(N_c)$ satisfies the BFKL equation [2]. It is remarkable, that the equations for the composite states of several reggeized gluons with the color singlet and octet quantum numbers are integrable at large N_c [3, 4]. Thus, it looks natural to reformulate QCD and other gauge models in terms of the effective degrees of freedom corresponding to the reggeized gluons. Indeed, the gluon Regge trajectory and various reggeon couplings in upper orders of perturbation theory can be calculated from the effective action presented in Ref. [5, 6]. In the $N = 4$ SUSY the Pomeron is dual to the reggeized graviton propagating in the anti-de-Sitter space due to AdS/CFT correspondence [7–9]. The effective action for the high energy scattering in the Einstein gravity was derived in ref. [10]. Below we consider the actions for reggeized gluons and gravitons and the corresponding effective Euler-Lagrange equations.

[a]The investigation is supported by the grants 11.38.223.2015 of SPSU and 14-2200281 of RFBI.
e-mail: lipatov@thd.pnpi.spb.ru

DOI: 10.1051/epjconf/201716402002

2 Effective theory for high energy processes in QCD

It is natural to construct the effective theory in QCD for a cluster of quarks, gluons and reggeized gluons having their rapidities y in the interval η around its central value y_0. Apart from the anti-hermitian matrix $N_c \times N_c$ corresponding to the gluon field $v^a_\mu(x)$ we introduce also the fields $A_\pm = p^\mu_{1,2} A_\mu = A_0 \pm A_3$ describing the production and annihilation of the reggeized gluons

$$v_\mu(x) = -iT^a v^a_\mu(x)\,,\;\; A_\pm(x) = -iT^a A^a_\pm(x)\,,\;\; [T^a, T^b] = i f^{abc} T_c\,. \tag{1}$$

The matrices T^a are generators of the gauge group $SU(N_c)$ in the fundamental representation.

Contrary to $v_\mu(x)$ the fields $A^\mp = A_\pm$ are invariant under the local gauge transformations

$$\delta v_\mu = \frac{1}{g}\,[D_\mu, \chi(x)]\,,\;\; D_\mu = \partial_\mu + g v_\mu\,,\;\; \delta A_\pm = 0 \tag{2}$$

providing that parameters $\chi(x)$ vanish at $x \to \infty$. Under the global SU_n rotations the Reggeon fields $A^\pm$ are transformed as v_μ. They satisfy also the kinematical constraints

$$\partial_- A^- = \partial_+ A^+ = 0\,,\;\; \partial_\pm = \frac{\partial}{\partial x^0} \pm \frac{\partial}{\partial x^3} \tag{3}$$

corresponding to the fact, that the Sudakov components α_i, β_i of the cluster momenta k_i are strongly ordered $\beta_i \gg \beta_{i+1}$, $\alpha_i \ll \alpha_{i+1}$ in the multi-Regge kinematics.

The effective action for a cluster of real and virtual particles with their rapidities belonging to the small rapidity interval η has the form [5]

$$S_{eff} = \int d^4x \left(L_{QCD} + Tr\,\partial_\mu A^+ \partial_\mu A^-\right) + S_{ind}\,,\;\; S_{ind} = Tr \int d^4x (V_+ \partial^2_\mu A^+ + V_- \partial^2_\mu A^-)\,, \tag{4}$$

where L_{QCD} is the usual QCD lagrangian. The anti-hermitian matrices $V_\pm$ in the induced action S_{ind} are expressed in terms of eikonal amplitudes for a massless particle scattered off the external gluon field v_μ [5]

$$V_\pm = \frac{1}{g}\,\partial_\pm \frac{1}{D_\pm}\,\overleftarrow{\partial}_\pm \equiv -\frac{\partial_\pm}{g} + v_\pm - v_\pm \frac{g}{\partial_\pm} v_\pm + v_\pm \frac{g}{\partial_\pm} v_\pm \frac{g}{\partial_\pm} v_\pm + \ldots\,, \tag{5}$$

where by definition the derivative $\overleftarrow{\partial}$ acts on functions situated to the left from it. The contribution $-\partial_\pm/g$ is negligible in S_{int}, because the fields $A^\pm$ do not depend on $x^\pm$. It is natural to define the action of the integral operator $1/\partial_\pm$ in the symmetric form

$$\frac{1}{\partial_\pm} f(x^\pm) = \frac{1}{2}\left(\frac{1}{\partial_0} \mp \frac{1}{\partial_3}\right) f(x^\pm) = \frac{1}{4}\left(\int_{-\infty}^{x^\pm} dx'^\pm f(x'^\pm) - \int_{x^\pm}^{\infty} dx'^\pm f(x'^\pm)\right). \tag{6}$$

The gauge invariance of the action is a consequence of the gauge symmetry of $V_\pm$ valid after its integration due to the vanishing of the parameter $\chi(x)$ at $x \to \infty$.

The matrix $V_\pm$ can be written in one of two factorized forms

$$V_\pm = -\frac{\partial_\pm}{g}\, O(x^\pm) = O^+(x^\pm)\,\frac{\overleftarrow{\partial}_\pm}{g}\,,\;\; O(x^\pm) \equiv -\frac{1}{D_\pm}\,\overleftarrow{\partial}_\pm\,,\;\; O^+(x^\pm) = \partial_\pm \frac{1}{D_\pm}\,. \tag{7}$$

Here the poles $1/D_\pm$ appear from propagators of the massless particles with different rapidities emitting gluons inside the given rapidity interval η. Their virtualities $\sim k_\pm$ are large. Hence it looks

DOI: 10.1051/epjconf/201716402002

natural to define the operators $1/\partial_\pm$ with the above principal value prescription but in all terms for the perturbative expansion of $1/D_\pm$. It would lead to the anti-hermicity of $V_\pm$.

Integrating over $x^\pm$ we can present S_{ind} in terms of asymptotic values of functions $O(x^\pm)$

$$S_{ind} = -\frac{Tr}{g}\int d^2x_\perp \left(\int_{-\infty}^{\infty} dx^-\, O|_{x^+=-\infty}^{x^+=\infty}\, \partial_\sigma^2 A^+ + \int_{-\infty}^{\infty} dx^+ O|_{x^-=-\infty}^{x^-=\infty} \partial_\sigma^2 A^- \right). \tag{8}$$

It is real due to the anti-hermicity of $O|_{x^\pm=-\infty}^{x^\pm=\infty}$.

The product of the operators $O^+(x^\pm)$ and $O(x^\pm)$ does not depend on $x^\pm$, but generally $O(x^\pm)$ is not an unitary operator. It is related to our use of the principal value prescription $\theta(|k^\pm|-\epsilon)/k^\pm$ for propagators $1/\partial_\pm$. Note, that ϵ can be chosen here in the form $e^{-\eta}$ where η is a low limit for the relative rapidity of neighboring clusters described by the effective action.

One can use the following representation for $1/D_\pm$ in terms of modified P-exponents

$$\frac{1}{D_\pm} = P\left(\frac{e^{-\frac{g}{4}\int_{-\infty}^{x_\pm} d\tilde{x}^\pm v_\pm}}{e^{-\frac{g}{4}\int_{x_\pm}^{\infty} d\tilde{x}^\pm v_\pm}}\right)\frac{1}{\partial_\pm}\,\bar{P}\left(\frac{e^{-\frac{g}{4}\int_{x_\pm}^{\infty} d\tilde{x}^\pm v_\pm}}{e^{-\frac{g}{4}\int_{-\infty}^{x_\pm} d\tilde{x}^\pm v_\pm}}\right). \tag{9}$$

It allows to write $O(x^\pm)$ in the form

$$O(x^\pm) = P\left(\frac{e^{-\frac{g}{4}\int_{-\infty}^{x_\pm} d\tilde{x}^\pm v_\pm}}{e^{-\frac{g}{4}\int_{x_\pm}^{\infty} d\tilde{x}^\pm v_\pm}}\right)\frac{1}{2}\left(Pe^{\frac{g}{4}\int_{-\infty}^{\infty} d\tilde{x}^\pm v_\pm} + \bar{P}e^{-\frac{g}{4}\int_{-\infty}^{\infty} d\tilde{x}^\pm v_\pm}\right). \tag{10}$$

The first factor here is an unitary matrix and second one is an hermitian operator which does not depend on x. We obtain a simple representation for the difference of $O(x^\pm)$ at $x^\pm = \pm\infty$

$$O|_{x^\pm=-\infty}^{x^\pm=\infty} = \frac{1}{2}\left(Pe^{-\frac{g}{4}\int_{-\infty}^{\infty} d\tilde{x}^\pm v_\pm} - \bar{P}e^{\frac{g}{4}\int_{-\infty}^{\infty} d\tilde{x}^\pm v_\pm}\right)\left(Pe^{\frac{g}{4}\int_{-\infty}^{\infty} d\tilde{x}^\pm v_\pm} + \bar{P}e^{-\frac{g}{4}\int_{-\infty}^{\infty} d\tilde{x}^\pm v_\pm}\right), \tag{11}$$

which leads to an expression for S_{ind} having the reality property.

Effective vertices for the reggeized gluon with the indices ± 1, color index c and momentum q, transformed to $r+1$ gluons with the polarization indices $\mu_0, \mu_1, ..., \mu_r$, color indices $a_0, a_1, ..., a_r$ and momenta $k_0, k_1, ..., k_r$, can be written in the form [5, 6]

$$\Delta^{\mu_0\mu_1...\mu_r\pm}_{a_0a_1...a_rc} = -q^2\prod_{i=0}^{r}\delta^{\mu_i}_{\pm}\Delta_{a_0a_1...a_r c}(k_0^\pm, k_1^\pm, ...k_r^\pm)\,,\;\; \Delta_{a_0c} = \delta_{a_0c}\,,\;\; \sum_{i=0}^{r} k_r^\pm = q^\pm = 0\,. \tag{12}$$

Here $\Delta_{a_0...a_r}$ has the Bose symmetry and satisfies the recurrent relation (the Ward identity) [5]

$$\Delta_{a_0a_1...a_r}(k_0^\pm, k_1^\pm, ...k_r^\pm) = \frac{1}{k_r^\pm}\sum_{t=0}^{r-1} if_{aa_ta_r}\,\Delta_{a_0...a_{t-1}aa_{t+1}...a_r}(k_0^\pm, k_1^\pm, ...k_{t-1}^\pm, k_t^\pm + k_r^\pm, k_{t+1}^\pm, ...k_r^\pm)\,,$$

where f_{abc} is the structure constant of the gauge group. Thus, these vertices for the principal value prescription do not depend on the color representations of the matrices $v_\pm(x)$ in S_{ind}.

3 Euler-Lagrange equation for the effective action in QCD

One can obtain the following expression for the variation of S_{ind} over $v_\pm$

$$\delta S_{ind} = -Tr\int d^4x\left(\delta v_+\, O(x^+)\,\partial_\mu^2 A^+ O^+(x^+) + \delta v_-\, O(x^-)\,\partial_\mu^2 A^- O^+(x^-)\right). \tag{13}$$

It allows to derive the Euler-Lagrange equations for the effective action in a pure glue-dynamics

$$[D_\mu, G^{\mu\nu}]^\perp = 0\,, \ [D_\mu, G^{\mu\pm}] = j^\pm_{ind}\,, \ j^\pm_{ind} = O(x^\pm)(\partial^2_\sigma A^\pm)O^+(x^\pm)\,. \tag{14}$$

Due to the relations $D_\pm\, O(x^\pm) = 0$ and $\partial_\pm A^\pm = 0$ the currents $j^\pm_{ind}$ are conserved

$$[D_\pm, j^\pm_{ind}] = 0\,. \tag{15}$$

In the quasi-elastic kinematics, where $A^+ = 0$, in the gauge $v'_- = 0$, we have

$$j'^{-}_{ind} = \lambda\, \partial^2_\sigma A^- \lambda^+\,, \ \lambda = \frac{1}{2}\left(P e^{\frac{g}{4}\int_{-\infty}^{\infty} d\tilde{x}^- v'_-} + \bar{P} e^{-\frac{g}{4}\int_{-\infty}^{\infty} d\tilde{x}^- v'_-}\right) = 1\,. \tag{16}$$

In this case there is an explicit solution of the Euler-Lagrange equations

$$\widetilde{v}^\sigma = \delta^\sigma_- A^-. \tag{17}$$

Thus, the reggeon field A^- at $A^+ = 0$ has a physical interpretation of the classical solution being a superposition of the shock waves $\delta(x^\mp - x_0^\mp)\ln|x^\perp - x_0^\perp|^2$.

Inserting the solution of the Euler-Lagrange equation for a general kinematics in S_{eff}, one can obtain a generating function for the reggeized gluon vertices in the tree approximation [5, 6]. Note, that even for the quasi-elastic kinematics the Euler-Lagrange equations have other solutions corresponding to more complicated initial and final conditions. The asymptotic behavior of these solutions is fixed at $t = -\infty$ and $t = \infty$ in terms of arbitrary functions $v_1(\vec{x})$ and $v_2(\vec{x})$, respectively. The effective action calculated on the general solution depending on $A^\pm$ and $v_{1,2}(\vec{x})$ gives a possibility to find a generating functional in a tree approximation for all possible gluon and quark transitions on an arbitrary number of reggeized gluons [5]. Furthermore, by calculating the functional integral over the quantum fluctuations δv_μ around the classical solutions we can find various effective vertices with loop corrections [11–13].

4 Effective theory for the high energy gravity

Due to the AdS/CFT correspondence the $N = 4$ super-symmetric gauge theory is equivalent to the 10-dimensional super-string model living on the 10-dimensional anti-de-Sitter space [7–9]. Here the BFKL Pomeron is dual to the reggeized graviton (see, for example, [14]). Note, that the graviton Regge trajectory and its various couplings in a leading order are known [15].

The high energy effective action in the Einstein gravity is constructed for a cluster of gravitons and reggeized gravitons having the rapidities in an interval around their central value [10]. Apart from the usual Einstein-Hilbert action and a kinetic term for the reggeon fields $A^{\pm\pm}$

$$S = -\frac{1}{2\kappa^2}\int d^4x\left(\sqrt{-g}\,R + \partial_\sigma A^{++}\partial_\sigma A^{--}\right) + S_{ind}\,, \tag{18}$$

it contains the additional contribution S_{ind}

$$S_{ind} = -\frac{1}{2\kappa^2}\int d^4x\left(\frac{j_{++}}{2}\,\partial^2_\sigma A^{++} + \frac{j_{--}}{2}\,\partial^2_\sigma A^{--}\right), \tag{19}$$

where the induced currents $j_{\pm\pm}$ are functionals of the metric tensor $g^{\mu\nu}$.

The fields $A^{\pm\pm}$ describe the production and annihilation of reggeized gravitons in the t-channel. They are invariant under the general coordinate transformations reduced to the Poincare subgroup at large distances and satisfy the kinematical constraints

$$\partial_+ A^{++} = \partial_- A^{--} = 0\,, \tag{20}$$

corresponding to the strong ordering of the Sudakov components for momenta of produced clusters in the multi-Regge kinematics. The graviton fields $h_{\mu\nu}$ are introduced as fluctuations of $g_{\mu\nu}$ around the Minkowsky metric tensor $\eta_{\mu\nu}$ having the diagonal structure $(1,-1,-1,-1)$.

The functional form of the current $j_{\pm\pm} \equiv \partial_\pm j^\mp$ is fixed by the general covariance of the action. It turns out, that the function $\omega^\mp = j^\mp + 2x^\mp$ satisfies the Hamilton-Jacobi equation [10]

$$g^{\rho\sigma}\,(\partial_\rho \omega^\mp)(\partial_\sigma \omega^\mp) = 0\,. \tag{21}$$

Physically this equation describes the light-front wave moving in an arbitrary gravitational field. On the other hand, the quantity $x'^\mp = \omega^\mp/2$ can be considered as a light-cone component of the coordinate transformation $x'(x)$ to the system with the global light-cone time $x'^\mp$, corresponding to the equality $g'^{\mp\mp} = 0$. The Hamilton-Jacobi equation does not fix the global light-cone time system completely. It is naturally to impose on $g'^{\rho\sigma}$ more restrictive constraints

$$g'^{\rho\mp} = \eta^{\rho\mp}\,,\ \eta^{\mp\mp} = \eta_\perp^{\rho\mp} = 0\,,\ \eta^{\pm\mp} = 1\,. \tag{22}$$

In this inertial system we have in particular $g'^{\rho\mp}\frac{\partial}{\partial x'^\rho} = \frac{\partial}{\partial x'^\pm}$.

Note, that the induced action can be expressed only in terms of the functions $\omega^\mp$ at $x^\pm = \infty$

$$S_{ind} = \frac{1}{2\kappa^2}\int d^2x^\perp \left(\int_{-\infty}^{\infty}\frac{dx^-}{4}\,\omega^-\,\partial^2_{\sigma\perp}A^{++}\,|^{x^+=\infty}_{x^+=-\infty} + \int_{-\infty}^{\infty}\frac{dx^+}{4}\,\omega^+\,\partial^2_{\sigma\perp}A^{--}|^{x^-=\infty}_{x^-=-\infty}\right). \tag{23}$$

5 Classical equations for the effective gravity

The variations $\delta g^{\mu\nu}$ and $\delta\omega^\mp$ are not independent due to the Hamilton-Jacobi equation for $\omega^\mp$

$$\partial_\mu\omega^\mp\,\partial_\nu\omega^\mp\,\delta_\tau g^{\mu\nu} + 2\,g^{\rho\sigma}\,\partial_\rho\omega^\mp\,\delta_\tau\partial_\sigma\omega^\mp = 0 \tag{24}$$

and expressed through the corresponding infinitesimal shifts in the proper time τ

$$\delta_\tau g^{\mu\nu} = d\tau\,(\partial_\sigma g^{\mu\nu})\,\frac{d}{d\tau}x^\sigma\,,\ \ \delta_\tau\partial_\sigma\omega^\mp = d\tau\frac{d}{d\tau}\,\partial_\sigma\omega^\mp\,. \tag{25}$$

They can be calculated with the help of the Hamilton equations

$$\frac{d}{d\tau}x^\sigma = g^{\sigma\rho}\,\partial_\rho\omega^\mp\,,\ \ \frac{d}{d\tau}\,\partial_\sigma\omega^\mp = -\frac{1}{2}\,(\partial_\sigma g^{\mu\nu})\,\partial_\mu\omega^\mp\partial_\nu\omega^\mp\,. \tag{26}$$

Because $x'^\mp = \omega^\mp/2$ are light-cone components of $x'(x)$ in systems with $g'^{\mp\mp} = 0$ we have

$$2g^{\rho\sigma}\,\partial_\rho\omega^\mp\,\delta_\tau\partial_\sigma\omega^\mp = -2d\tau\,g'^{\chi\mp}\,\frac{\partial g^{\mu\nu}}{\partial x'^\chi}\,\partial_\mu\omega^\mp\partial_\nu\omega^\mp = -2d\tau\,\frac{\partial g^{\mu\nu}}{\partial x'^\pm}\,\partial_\mu\omega^\mp\partial_\nu\omega^\mp = 4\delta_\tau\frac{\partial\omega^\mp}{\partial x'^\pm}\,, \tag{27}$$

where the property $g'^{\chi\mp} = \eta^{\chi\mp}$ of the global light-cone time inertial systems was used. By integrating these equalities over the coordinates x' it is possible to verify the relations

$$\int d^4x'\,4\delta_\tau\frac{\partial\omega^\mp}{\partial x'^\pm}\,\partial^2_{\sigma\perp}\,A^{\pm\pm}(x') = \int d^4x\,\sqrt{-g}\,2g^{\rho\sigma}\,\partial_\rho\omega^\mp\,\delta_\tau(\partial_\sigma\omega^\mp)\,\partial^2_{\sigma\perp}\,A^{\pm\pm}(x')\,,$$

where we performed a transition to the integration over the coordinates x with the general metric tensor $g^{\mu\nu}$ by introducing the additional factor $\sqrt{-g}$ for the covariance of this expression.

The left hand side of last equality after its integration over $x'^{\pm}$ enters in the variation of the induced action

$$\delta S_{ind} = \frac{1}{2\kappa^2}\int d^2x^{\perp}\left(\int_{-\infty}^{\infty}\frac{dx^-}{4}\delta\omega^-\,\partial^2_{\sigma\perp}A^{++}|_{x^+=-\infty}^{x^+=\infty} + \int_{-\infty}^{\infty}\frac{dx^+}{4}\delta\omega^+\,\partial^2_{\sigma\perp}A^{--}|_{x^-=-\infty}^{x^-=\infty}\right). \tag{28}$$

Therefore, taking into consideration again the above relation between $\delta_\tau g^{\mu\nu}$ and $\delta_\tau\omega^{\mp}$ on particle trajectories, one can present the variation of S_{ind} over $g^{\mu\nu}$ in a simple covariant form

$$\delta S_{ind} = -\frac{1}{2\kappa^2}\int d^4x\,\sqrt{-g}\,\delta g^{\mu\nu}\,\frac{1}{4}\left(\partial_\mu\omega^-\,\partial_\nu\omega^-\,\partial^2_{\sigma\perp}A^{++}(x') + \partial_\mu\omega^+\,\partial_\nu\omega^+\partial^2_{\sigma\perp}\,A^{--}(x')\right). \tag{29}$$

We remind, that the factor $\sqrt{-g}$ is obtained as a result of the transformation from a special coordinate system with the global light-cone time $x'^{\mp}$ to a general system with coordinates x.

The above expression for δS_{ind} allows one to calculate easily the induced energy-momentum tensor $\theta_{\mu\nu}$ in the Euler-Lagrange equations

$$R_{\mu\nu} - \frac{1}{2}g_{\mu\nu}R = -\theta_{\mu\nu}\,,\ \theta_{\mu\nu} = \partial_\mu x'^-\partial_\nu x'^-\partial^2_\sigma A^{++}(x') + \partial_\mu x'^+\partial_\nu x'^+\partial^2_\sigma A^{--}(x'), \tag{30}$$

where $x'^{\pm} = \omega^{\pm}/2$ are coordinates in the corresponding global light-cone systems. Below this result is obtained independently from considerations related to the general covariance of the Euler-Lagrange equations.

To begin with, in a quasi-elastic kinematics, where $A^{++} = 0$, it is convenient to work in the inertial system with the metric tensor obeying the constraints $g'^{\rho+} = \eta^{\rho+}$ with the global light-cone time x'^+, because here the energy-momentum tensor is $T_{\mu\nu} \sim \delta^+_\mu\delta^+_\nu$ and there is a simple classical solution

$$g'^{\rho\sigma} = \eta^{\rho\sigma} + \delta^\rho_-\delta^\sigma_-A^{--}(x'). \tag{31}$$

It is a superposition of the plane-wave solutions of Aichelburg and Sexl with the gravitation centers situated at $x^+ = z^+$, $x^\perp = z^\perp$ and distributed with the weight function $\partial^2_{\tau\perp}A^{--}(z^+, z^\perp)$. The coordinate transformation $x' = x'(x)$ to the light-cone time system satisfies the equations

$$g^{\mu\nu}\frac{\partial x'^\rho}{\partial x^\mu}\frac{\partial x'^+}{\partial x^\nu} = \eta^{\rho+} \tag{32}$$

and the tensors $\theta_{\mu\nu}$ and $T_{\rho\sigma}$ in these two systems are related as follow

$$\theta_{\mu\nu}\frac{\partial x^\mu}{\partial x'^\rho}\frac{\partial x^\nu}{\partial x'^\sigma} = T_{\rho\sigma}\,. \tag{33}$$

It means, that the covariant energy-momentum tensor for the quasi-elastic kinematics is

$$\theta_{\mu\nu} = \partial_\mu x'^+\partial_\nu x'^+\partial^2_\sigma A^{--}(x')\,, \tag{34}$$

where partial derivatives $\partial_\mu x'^+$ are found from the solution of the Hamilton-Jacobi equation

$$g^{\mu\nu}\,\partial_\mu x'^+\partial_\nu x'^+ = 0\,. \tag{35}$$

The covariant energy-momentum tensor for a general kinematics where $A^{\pm\pm} \neq 0$ is conserved

$$D_\mu\,\theta^{\mu\nu} = 0 \tag{36}$$

due to the kinematical constraints $\partial_\pm A^{\pm\pm} = 0$. As a consequence of the Hamilton-Jacobi equation for $x'^{\pm}$ the tensor $\theta_{\mu\nu}$ is traceless

$$g^{\mu\nu}\,\theta_{\mu\nu} = 0\,. \tag{37}$$

Thus, we constructed the generally covariant Euler-Lagrange equations for the effective action in the high energy gravity. In a quasi-elastic kinematics with $A^{++} = 0$ one of their classical solutions at the coordinate system with the global light-cone time x'^{+} has a simple form $g'^{--} = A^{--}$. In a general quasi-multi-Regge kinematics there are other solutions parameterized by the functions with positive and negative frequencies at $t \to -\infty$ and $t \to \infty$, respectively. The effective action calculated on these solutions allows to construct a generating functional for the effective multi-graviton scattering amplitudes with an arbitrary number of reggeized gravitons in a tree approximation and to reproduce independently the known results for the graviton Regge trajectory and effective vertices [15]. The integration over fluctuations around the classical solutions gives a possibility to calculate loop corrections to reggeized graviton interactions and to reggeon vertices. Note, that using the one-loop Regge trajectory obtained in Refs [10, 15], the graviton scattering amplitude in the double-logarithmic approximation was constructed in [16]. A possible generalization of our approach to the super-gravity in the 10-dimensional AdS space will be interesting for the construction of the Gribov Pomeron calculus in N=4 SUSY.

References

[1] V. N. Gribov, *Sov. Phys. JETP* **26**, 414 (1968).
[2] L. N. Lipatov, *Sov. J. Nucl. Phys.* **23**, 338 (1976);
V. S. Fadin, E. A. Kuraev, L. N. Lipatov, *Phys. Lett.* **B 60**, 50 (1975);
E. A. Kuraev, L. N. Lipatov, V. S. Fadin, Sov. Phys. JETP **44**, 443 (1976);
Ya. Ya. Balitsky, L. N. Lipatov, *Sov. J. Nucl. Phys.* **28**, 822 (1978).
[3] L. N. Lipatov *High energy asymptotics of multi-colour QCD and exactly solvable lattice models*, hep-th/9311037, unpublished.
[4] L. N. Lipatov, *J. Phys.* **A 42**:304020 (2009).
[5] L. N. Lipatov, *Nucl. Phys.* **B 452**, 369 (1995); *Phys. Rept.* **286**, 131 (1997).
[6] E. N. Antonov, L. N. Lipatov, E. A. Kuraev, I. O. Cherednikov, *Nucl. Phys.* **B 721**, 111 (2005).
[7] J. M. Maldacena, *Adv. Theor. Math. Phys.* **2**, 231 (1998).
[8] S. S. Gubser, I. R. Klebanov, A. M. Polyakov, *Phys. Lett.* **B 428**, 105 (1998).
[9] E. Witten, *Adv. Theor. Math. Phys.* **2**, 253 (1998).
[10] L. N. Lipatov, Phys. Part. Nucl. **44** 391 (2013), arXiv: 1105.31277 [hep-ph].
[11] V. S. Fadin, L. N. Lipatov, *Phys. Lett.* **B 429**, 127 (1998);
M. Ciafaloni and G. Camici, *Phys. Lett.* **B 430**, 349 (1998).
[12] A. V. Kotikov, L. N. Lipatov, *Nucl. Phys.* **B 582**, 19 (2000).
[13] A. V. Kotikov, L. N. Lipatov, *Nucl. Phys.* **B 661**, 19 (2003).
[14] A. V. Kotikov, L. N. Lipatov, Nucl. Phys. **B 874**, 889 (2013).
[15] L. N. Lipatov, Phys. Lett. **116**, 411 (1982).
[16] J. Bartels, L. N. Lipatov, A. Sabio Vera, J. High Energy Phys. **07** 056 (2014).

EPJ Web of Conferences **164**, 02003 (2017) DOI: 10.1051/epjconf/201716402003
ICNFP 2016

Aspects of string phenomenology in particle physics and cosmology

I. Antoniadis[1,2,a]

[1] *LPTHE, UMR CNRS 7589 Sorbonne Universités, UPMC Paris 6, 75005 Paris, France*
[2] *Albert Einstein Center, Institute for Theoretical Physics, Bern University, Sidlerstrasse 5, CH-3012 Bern*

Abstract. I discuss possible connections between several scales in particle physics and cosmology, such the the electroweak, inflation, dark energy and Planck scales. In particular, I discuss the physics of extra dimensions and low scale gravity that are motivated from the problem of mass hierarchy, providing an alternative to low energy supersymmetry. I describe their realization in type I string theory with D-branes and I present the main experimental predictions in particle accelerators and their implications in cosmology. I also show that low-mass-scale string compactifications, with a generic D-brane configuration that realizes the Standard Model by open strings, can explain the relatively broad peak in the diphoton invariant mass spectrum at 750 GeV recently reported by the ATLAS and CMS collaborations.

1 Introduction

If String Theory is a fundamental theory of Nature and not just a tool for studying systems with strongly coupled dynamics, it should be able to describe at the same time particle physics and cosmology, which are phenomena that involve very different scales from the microscopic four-dimensional (4d) quantum gravity length of 10^{-33} cm to large macroscopic distances of the size of the observable Universe $\sim 10^{28}$ cm spanned a region of about 60 orders of magnitude. In particular, besides the 4d Planck mass, there are three very different scales with very different physics corresponding to the electroweak, dark energy and inflation. These scales might be related via the scale of the underlying fundamental theory, such as string theory, or they might be independent in the sense that their origin could be based on different and independent dynamics. An example of the former constrained and more predictive possibility is provided by TeV strings with a fundamental scale at low energies due for instance to large extra dimensions transverse to a four-dimensional braneworld forming our Universe [1]. In this case, the 4d Planck mass is emergent from the fundamental string scale and inflation should also happen around the same scale [2].

The appropriate framework of low string scale is type I string theory with D-branes. Gauge and gravitational interactions have different origin. The latter are described by closed strings, while the former emerge as excitations of open strings with endpoints confined on D-branes [3]. This leads to a braneworld description of our universe, which should be localized on a hypersurface, i.e. a membrane extended in p spatial dimensions, called p-brane (see Fig. 1). Closed strings propagate in all nine

[a] e-mail: ignatios.antoniadis@upmc.fr

dimensions of string theory: in those extended along the p-brane, called parallel, as well as in the transverse ones. On the contrary, open strings are attached on the p-brane. Obviously, our p-brane

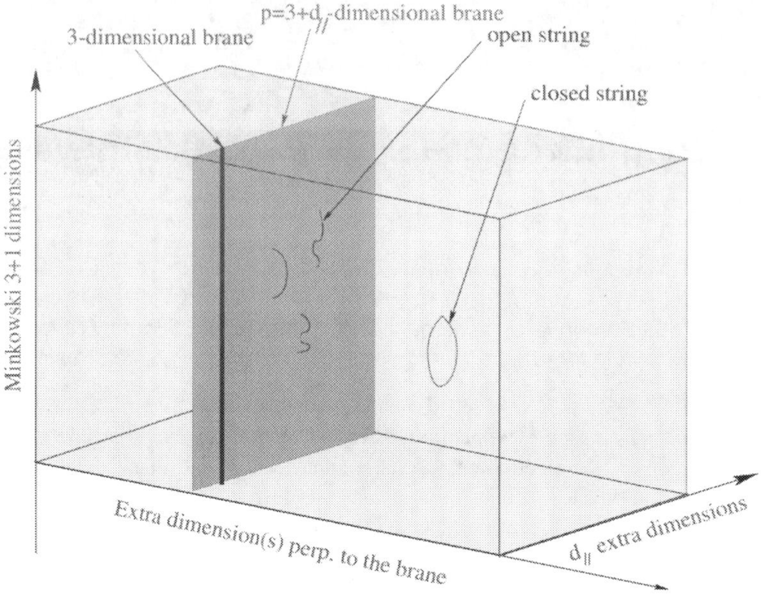

Figure 1. In the type I string framework, our Universe contains, besides the three known spatial dimensions (denoted by a single blue line), some extra dimensions ($d_{\parallel} = p - 3$) parallel to our world p-brane (green plane) where endpoints of open strings are confined, as well as some transverse dimensions (yellow space) where only gravity described by closed strings can propagate.

world must have at least the three known dimensions of space. But it may contain more: the extra $d_{\parallel} = p - 3$ parallel dimensions must have a finite size, in order to be unobservable at present energies, and can be as large as $\text{TeV}^{-1} \sim 10^{-18}$ m [4]. On the other hand, transverse dimensions interact with us only gravitationally and experimental bounds are much weaker: their size should be less than about 0.1 mm [5].

2 Framework of low scale strings

In type I string theory, the different origin of gauge and gravitational interactions implies that the relation between the Planck and string scales is not linear as in the heterotic string. The requirement that string theory should be weakly coupled, constrain the size of all parallel dimensions to be of order of the string length, while transverse dimensions remain unrestricted. Assuming an isotropic transverse space of $n = 9 - p$ compact dimensions of common radius $R_{\perp}$, one finds:

$$M_P^2 = \frac{1}{g^4} M_s^{2+n} R_{\perp}^n \,, \qquad g_s \simeq g^2 \,. \tag{1}$$

where g_s is the string coupling. It follows that the type I string scale can be chosen hierarchically smaller than the Planck mass at the expense of introducing extra large transverse dimensions felt only by gravity, while keeping the string coupling small [1]. The weakness of 4d gravity compared to gauge interactions (ratio M_W/M_P) is then attributed to the largeness of the transverse space $R_\perp$ compared to the string length $l_s = M_s^{-1}$.

An important property of these models is that gravity becomes effectively $(4+n)$-dimensional with a strength comparable to those of gauge interactions at the string scale. The first relation of Eq. (1) can be understood as a consequence of the $(4+n)$-dimensional Gauss law for gravity, with

$$M_*^{(4+n)} = M_s^{2+n}/g^4 \tag{2}$$

the effective scale of gravity in $4+n$ dimensions. Taking $M_s \simeq 1$ TeV, one finds a size for the extra dimensions $R_\perp$ varying from 10^8 km, .1 mm, down to a Fermi for $n = 1, 2$, or 6 large dimensions, respectively. This shows that while $n = 1$ is excluded, $n \geq 2$ is allowed by present experimental bounds on gravitational forces [5, 6]. Thus, in these models, gravity appears to us very weak at macroscopic scales because its intensity is spread in the "hidden" extra dimensions. At distances shorter than $R_\perp$, it should deviate from Newton's law, which may be possible to explore in laboratory experiments (see Fig. 2).

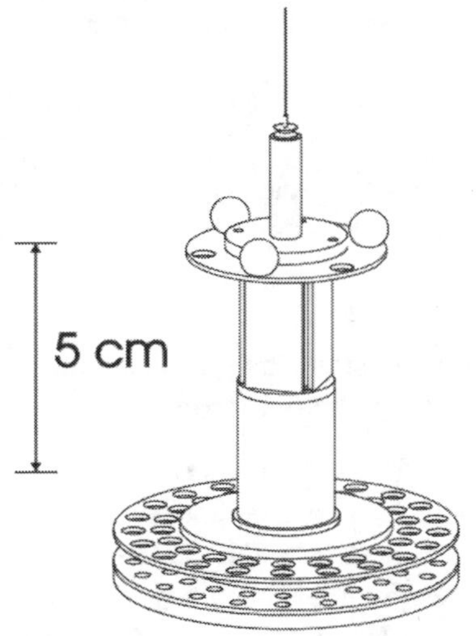

Figure 2. Torsion pendulum that tested Newton's law at 55 μm.

2.1 Experimental implications in accelerators

We now turn to the experimental predictions of TeV scale strings. Their main implications in particle accelerators are of four types, in correspondence with the four different sectors that are generally present:

1. New compactified parallel dimensions; In this case $RM_s \gtrsim 1$, and the associated compactification scale $R_\parallel^{-1}$ would be the first scale of new physics that should be found increasing the beam energy [4, 7]. The main consequence is the existence of KK excitations for all Standard Model particles that propagate along the extra parallel dimensions. These can be produced on-shell at LHC as new resonances [8] (see Fig. 3).

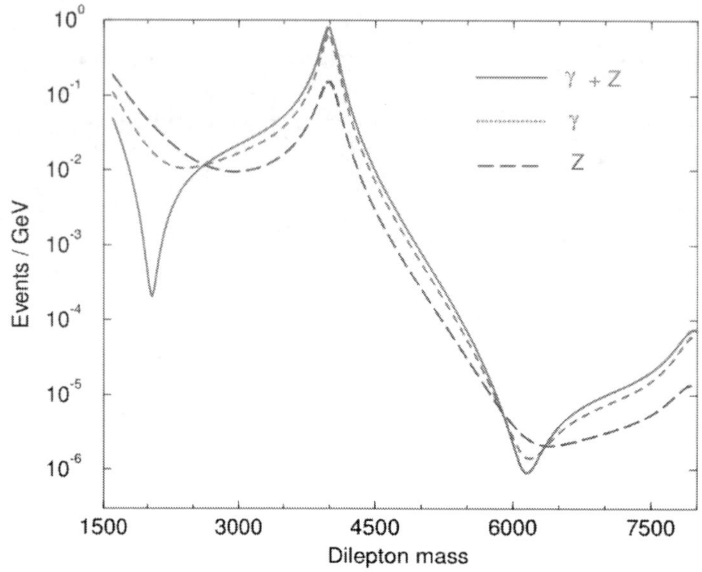

Figure 3. Production of the first KK modes of the photon and of the Z boson at LHC, decaying to electron-positron pairs. The number of expected events is plotted as a function of the energy of the pair in GeV.

2. New extra large transverse dimensions and low scale quantum gravity,. The main experimental signal is gravitational radiation in the bulk from any physical process on the world-brane [9].

3. Genuine string and quantum gravity effects. Direct production of string resonances in hadron colliders leads generically to a universal deviation from Standard Model in jet distribution [10]. In particular, the first Regge excitation of the gluon has spin 2 and a width an order of magnitude lower than the string scale, leading to a characteristic peak in dijet production; similarly, the first excitations of quarks have spin 3/2. The dijet (left) and γ + jet (right) cross-sections are shown in Fig. 4 for LHC energies, while Fig 5 shows the Signal-to-Noise ratio of the lowest massive Regge excitations for a 100 TeV future hadron collider [11].

4. Extra $U(1)$'s arising generically in D-brane models as part of unitary gauge group factors. They obtain in general masses due to four- or higher-dimensional anomalies, via the so-called Green-Schwarz anomaly cancellation mechanism involving axionic fields from the closed string sector. The resulting masses are therefore suppressed by a loop factor compared to the string scale. From the low energy point of view, they gauge global symmetries of the Standard Model (SM), such as the baryon and lepton number. An important property of the anomaly cancellation mechanism is that the anomalous $U(1)$ gauge bosons acquire masses leaving behind the corresponding global symmetries unbroken in perturbation theory. Thus, this is a way to guarantee proton stability (from unbroken baryon number) and avoid large Majorana neutrino masses (from unbroken lepton number) due to dimension-5 operators involving two higgses and two leptons that are suppressed only by the TeV string scale. Such extra $U(1)$s have interesting properties and distinct experimental signatures [12–15].

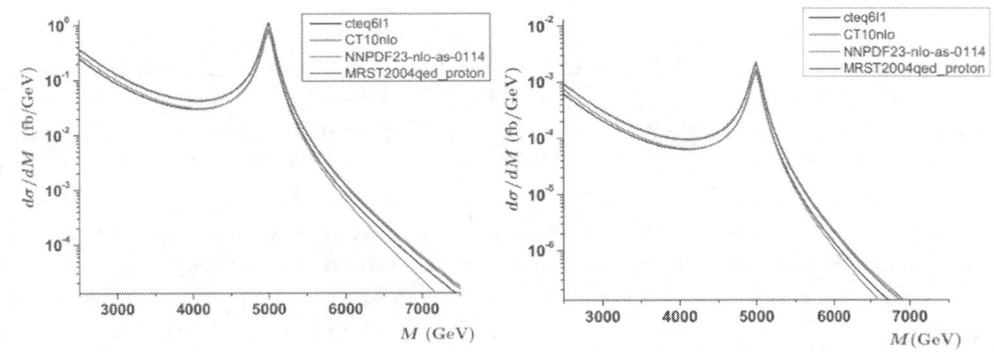

Figure 4. Production of the first Regge excitations at LHC in the dijet (left) and γ + jet (right) channels, for M_s = 5 TeV. The cross-section is plotted as a function of the invariant mass M.

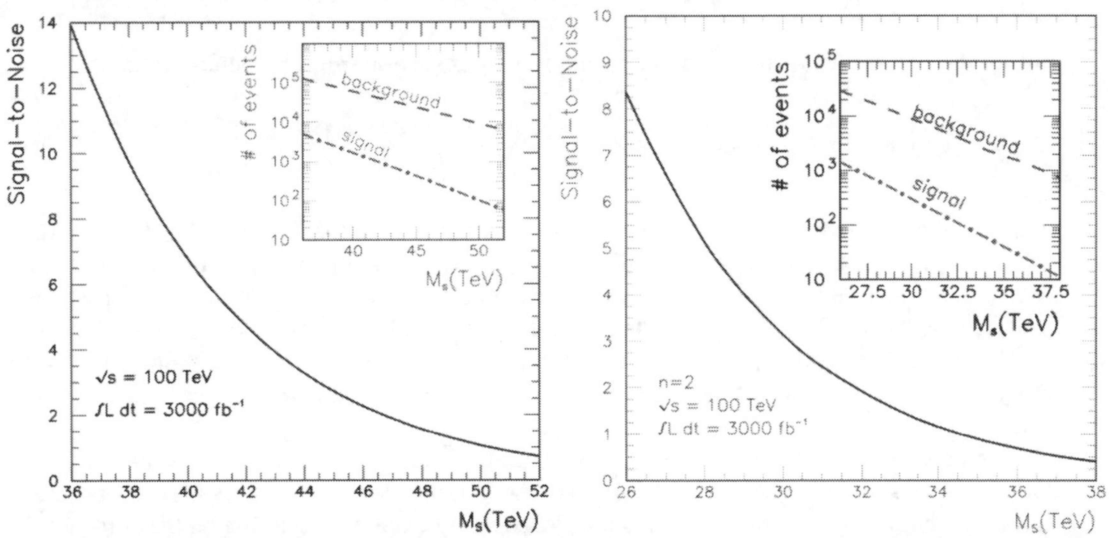

Figure 5. Dijet signal-to-noise ratio of the lowest n = 1 (left) and n = 2 (right) Regge excitations for a future 100 TeV hadron collider.

5. Concerning possible micro-black hole production, note that a string size black hole has a horizon radius $r_H \sim 1$ in string units, while the Newton's constant behaves as $G_N \sim g_s^2$. It follows that the mass of a d-dimensional black hole is [16]: $M_{\rm BH} \sim r_H^{d/2-1}/G_N \simeq 1/g_s^2$. Using the value of the SM gauge couplings $g_s \simeq g^2 \sim 0.1$, one finds that the energy threshold $M_{\rm BH}$ of micro-black hole production is about four orders of magnitude higher than the string scale, implying that one would produce 10^4 string states before reaching $M_{\rm BH}$.

On the other hand, there exist interesting implications in non accelerator table-top experiments due to the exchange of gravitons or other possible states living in the bulk.

DOI: 10.1051/epjconf/201716402003

3 750 GeV diphotons from closed string states

One can show that low-mass-scale string compactifications, with a generic D-brane configuration that realizes the standard model by open strings, can explain the peak in the diphoton invariant mass spectrum at 750 GeV recently reported by the ATLAS and CMS collaborations.

Indeed, the ATLAS [17] and CMS [18] collaborations reported excesses of events over expectations from standard model (SM) processes in the diphoton mass distribution around 750 GeV, using (respectively) 3.2 fb^{-1} and 3.3 fb^{-1} of data recorded at a center-of-mass energy $\sqrt{s} = 13$ TeV, targeting spin-0 and spin-2 resonances. This could be interpreted as decays of a new massive particle φ. The ATLAS Collaboration, reported for spin-0 the largest deviation from the background-only hypothesis for $M_\varphi \sim 750$ GeV and $\Gamma_{\text{total}} \approx 45$ GeV. The local significance iq 3.9σ, while the global significance is around 2σ. For the spin-2 resonance, both the local and global significances are somewhat smaller: 3.6σ and 1.8σ, respectively. The CMS analysis gives the largest excess for $M_\varphi = 760$ GeV and $\Gamma_{\text{total}} \approx 11$ GeV and has a local significance of 2.8σ for spin-0 and 2.9σ spin-2 hypothesis. After taking into account the effect of searching for several signal hypotheses, the significance of the excess is reduced to $< 1\sigma$. The CMS Collaboration also reported a combined search on data collected at $\sqrt{s} = 13$ TeV and $\sqrt{s} = 8$ TeV. For the combined analysis, the largest excess is observed at $M_\varphi = 750$ GeV and $\Gamma_{\text{total}} = 0.1$ GeV. The local significance is $\approx 3.4\sigma$ and the global significance 1.6σ.

Fitting the LHC13 data with a resonance yields a cross section times branching ratio of

$$\sigma_{\text{LHC13}}(pp \to \varphi + \text{anything}) \times \mathcal{B}(\varphi \to \gamma\gamma) \approx \begin{cases} (10 \pm 3)\ \text{fb} & \text{ATLAS} \\ (6 \pm 3)\ \text{fb} & \text{CMS} \end{cases}, \tag{3}$$

at 1σ [19]. On the other hand, no diphoton resonances were seen in the data at $\sqrt{s} = 8$ TeV, although both ATLAS [20] and CMS [21] data show a mild upward fluctuation at invariant mass of 750 GeV. The lack of an excess at $\sqrt{s} = 8$ TeV allows a quite precise limit to be placed on the corresponding cross section at $\sqrt{s} = 13$ TeV. The most stringent limit comes from the CMS search $\sigma_{\text{LHC8}}(pp \to \varphi + \text{anything}) \times \mathcal{B}(\varphi \to \gamma\gamma) < 2.00$ fb at 95% CL [21]. This implies that if the diphoton cross section grows by less than about a factor of 3 or 3.5 the LHC8 data are incompatible with the LHC13 data at 95% CL.

More recently, we proposed a model [22] to explain the data in which the resonance production mechanism is calculable in string based dynamics, with large extra dimensions [1]. In our proposal the observed diphoton excess originates from a closed string excitation φ living on the compact space of generic intersecting D-brane models that realize the SM chiral matter contents and gauge symmetry [23, 24]. Namely we consider extensions of the SM based on D-brane string compactifications with large extra dimensions [1]. The basic unit of gauge invariance for D-brane constructions is a $U(1)$ field, so that a stack of N identical D-branes eventually generates a $U(N)$ theory with the associated $U(N)$ gauge group; for $N = 2$, the gauge group can be $Sp(1) \cong SU(2)$ rather than $U(2)$. In the presence of many D-brane types, the gauge group becomes a product form $\prod U(N_i)$, where N_i reflects the number of D-branes in each stack. In the perturbative regime, gauge interactions emerge as excitations of open strings ending on D-branes, with gauge bosons due to strings attached to stacks of D-branes and chiral matter due to strings stretching between intersecting D-branes.

Our explanation of the peak in the diphoton invariant mass spectrum is independent of the structure of the D-brane model. Nevertheless, to motivate the discussion we adopt a minimal model containing 4 stacks of D-branes. The basic setting of the gauge theory is given by $U(3)_C \times Sp(1)_L \times U(1)_L \times U(1)_{I_R}$ [15, 25, 26]. In the bosonic sector the open strings terminating on the QCD stack contain, in addition to the $SU(3)_C$ octet of gluons g^a_μ, an extra $U(1)$ boson C_μ, most simply the manifestation

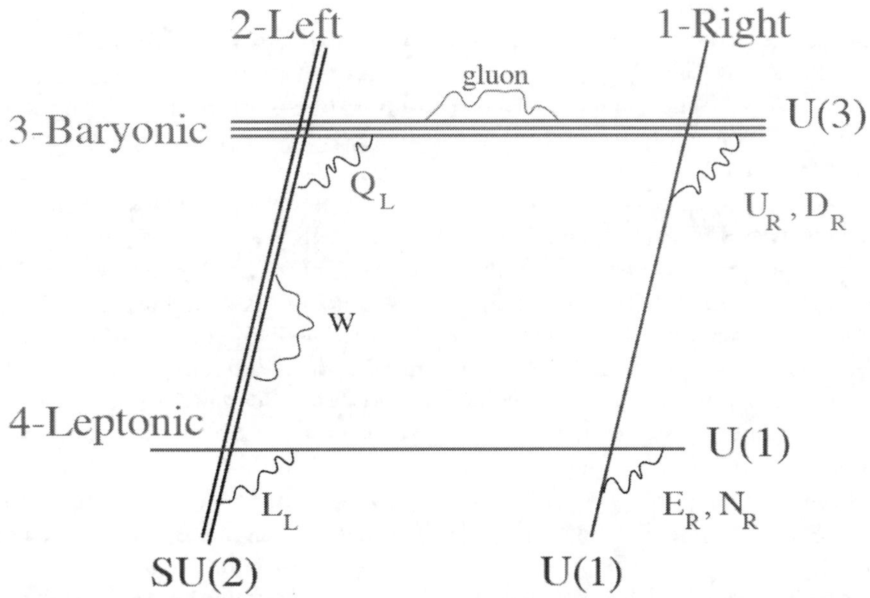

Figure 6. Pictorial representation of the $U(3)_B \times SU(2)_L \times U(1)_L \times U(1)_{I_R}$ D-brane model.

Table 1. Chiral spectrum of SM fields in the 4 stack D-brane model. We have added the right handed neutrino stretching between the lepton brane and the right brane.

Fields	Sector	Representation	Q_B	Q_L	Q_{I_R}	Q_Y
U_R	$3 \leftrightharpoons 1^*$	$(3,1)$	1	0	1	$\frac{2}{3}$
D_R	$3 \leftrightharpoons 1$	$(3,1)$	1	0	−1	$-\frac{1}{3}$
L_L	$4 \leftrightharpoons 2$	$(1,2)$	0	1	0	$-\frac{1}{2}$
E_R	$4 \leftrightharpoons 1$	$(1,1)$	0	1	−1	−1
Q_L	$3 \leftrightharpoons 2$	$(3,2)$	1	0	0	$\frac{1}{6}$
N_R	$4 \leftrightharpoons 1^*$	$(1,1)$	0	1	1	0
H	$2 \leftrightharpoons 1$	$(1,2)$	0	0	1	$\frac{1}{2}$

of a gauged baryon number. The $Sp(1)_L \simeq SU(2)$ stack is a terminus for the electroweak gauge bosons W^a_μ. The $U(1)_Y$ boson Y_μ that gauges the usual electroweak hypercharge symmetry is a linear combination of C_μ, and the $U(1)$ bosons B_μ and X_μ terminating on the separate $U(1)_L$ and $U(1)_{I_R}$ branes. A schematic representation of the D-brane construct is shown in Fig. 6. The general properties of the chiral spectrum are summarized in Table 1.

One can check by inspection that the hypercharge,

$$Q_Y = \frac{1}{2} Q_{I_R} + \frac{1}{6} Q_B - \frac{1}{2} Q_L, \tag{4}$$

 DOI: 10.1051/epjconf/201716402003

is anomaly free. However, the Q_B (gauged baryon number) is not anomaly free and we expect this anomaly to be canceled via a Green-Schwarz mechanism involving the exchange of twisted Ramond-Ramond (RR) closed string states [27–31]. There is an explicit mass term in the Lagrangian for the new gauge field $-\frac{1}{2}M'^2 Y'_\mu Y'^\mu$ whose scale comes from the compactification scheme. The scalar that gets eaten up to give the longitudinal polarization of the Y' is a closed string field and there is no extra Higgs particle [32]. In addition to the intermediate RR field, which is absorbed by the Y' in the anomaly cancellation, there is a closed string mode φ which couples to the anomaly free combination of the hypercharge (4). It can be either a scalar field from the Neveu-Schwarz sector that is complexified with the RR state absorbed by Y', or another RR pseudo-scalar (axion) coupled to $F\tilde{F}$.

There are two properties of the scalar φ that are necessary for explaining the 750 GeV signal. It should be a special closed string state with dilaton-like or axion-like coupling to F^2 (respectively to $F\tilde{F}$) of the electromagnetic field, but *may be decoupled* from G^2 of color $SU(3)$. The couplings of closed string states to gauge fields do indeed distinguish between different D-brane stacks, depending on the localization properties of D-branes with respect to φ in the compact dimensions. More specifically, it is quite natural to assume that φ is a closed string mode that is associated to the wrapped cycles of the (lepton) $U(1)_L$ and (right isospin) $U(1)_{I_R}$ stack of D-branes, however is not or only weakly attached to the wrapped cycle of (left) $Sp(1)_L$ or the color $SU(3)$ stack of D-branes. In this way, we may avoid unwanted dijet signals. Actually, within a selection of string based explanations of the resonance [33–42] our proposal is uniquely exemplified by the *possible suppression* of dijet topologies in the final state.[1] By choice, we may also allow a coupling φ to G^2. This is possible by modifying the localization properties of D-branes with respect to φ in the internal space.

Since the string mass scale is now known to be larger than $M_s \approx 7$ TeV [49], the mass $M_\varphi \approx 750$ GeV must be suppressed with respect to the string scale by some anomalous loop corrections. Because φ is a twisted closed string localized at an orbifold singularity, its coupling to $\gamma\gamma$ should be suppressed by M_s^{-1}, provided the bulk is large [14]. With this in mind, we parametrize the coupling of φ to the photon by the following vertex

$$\frac{c_{\gamma\gamma}}{M_s}\varphi F^2 . \tag{5}$$

To remain in the perturbative range, we also require $c_{\gamma\gamma} \lesssim 2\pi$. The partial decay width of φ to diphotons then follows as

$$\Gamma_{\gamma\gamma} = \frac{c_{\gamma\gamma}^2}{4\pi}\frac{M_\varphi^3}{M_s^2} . \tag{6}$$

Let us first assume the diphoton signal is produced via photon-photon fusion with φ as the resonance state. Following [50], herein we include the elastic-elastic processes (already considered in [22]) as well as elastic-inelastic and inelastic-inelastic contributions. The elastic production is suppressed with respect to inelastic by about an order of magnitude. However, elastic photoproduction events result in forward and backward protons which can be detected by forward detectors installed by ATLAS [51] and CMS [52]. Therefore, the detection of two unbroken protons in the final state, with M_{pp} paired to $M_{\gamma\gamma}$, may be a promising way to reduce the background in the future [53, 54].

The total photo-production cross section at LHC13 is [50]

$$\sigma_{\rm LHC13}(\gamma\gamma \to \varphi \to \gamma\gamma) = 4.1\ {\rm pb}\left(\frac{\Gamma_{\rm total}}{45\ {\rm GeV}}\right)\mathcal{B}^2(\varphi \to \gamma\gamma) , \tag{7}$$

[1] A related stringy explanation in which φ can be produced through photon fusion has been put forward in [43]. Alternative axion and dilaton models have been discussed in [44–48].

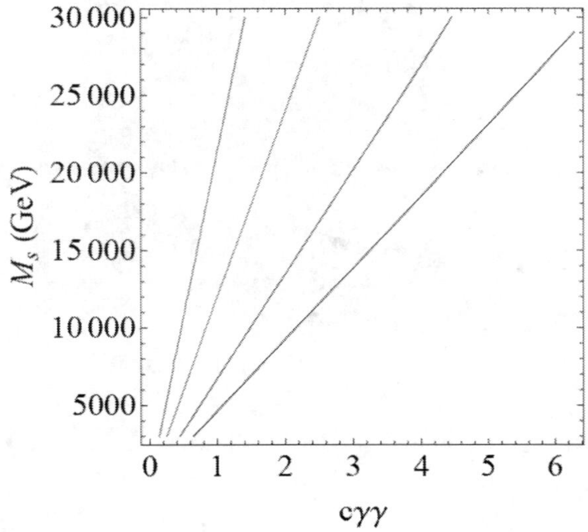

Figure 7. Best fit contours of diphoton cross section $\sigma_{\rm LHC13} \sim 5$ fb produced via photon fusion $\gamma\gamma \to \varphi \to \gamma\gamma$. The four lines (blue, red, yellow, green) are for $\Gamma_{\rm total} = 45, 10, 1.0, 0.1$ GeV, respectively.

where

$$\mathcal{B}(\varphi \to \gamma\gamma) = \frac{2.3 \times 10^6 c_{\gamma\gamma}^2}{\pi} \left(\frac{M_s}{\rm GeV}\right)^{-2} \left(\frac{\Gamma_{\rm total}}{45 \text{ GeV}}\right)^{-1} . \tag{8}$$

Substituting (8) into (7), and demanding (7) reproduces the diphoton signal (3) we obtain an equation connecting $c_{\gamma\gamma}$ with M_s for given $\Gamma_{\rm total}$. In Fig. 7 we show the best fit contours for $\sigma_{\rm LHC13} \sim 5$ fb and total widths $\Gamma_{\rm total} = 45, 10, 1, 0.1$ GeV. We assume that for a broad resonance the missing fraction of the decay width arises from the coupling of φ to some fermion bulk fields. These hidden fermions could make a contribution to the dark matter content of the universe [55]. We conclude that for both the narrow and the broad resonance hypotheses there is an allowed region of the parameter space which is consistent with the experimental lower bound of $M_s \simeq 7$ TeV [49] and reproduces the LHC13 signal. For the broad resonance hypothesis, $7 \lesssim M_s/\text{TeV} \lesssim 30$.

The total photo-production cross section at LHC8 is [50]

$$\sigma_{\rm LHC8}(\gamma\gamma \to \varphi \to \gamma\gamma) = 1.4 \text{ pb} \left(\frac{\Gamma_{\rm total}}{45 \text{ GeV}}\right) \mathcal{B}^2(\varphi \to \gamma\gamma) , \tag{9}$$

showing consistency with the 95% CL upper limit [21].

The assumed coupling of φ to the hypercharge field strength yields additional decay channels in the visible sector, namely $\varphi \to \gamma Z$ and $\varphi \to ZZ$, with

$$\frac{\Gamma_{\gamma Z}}{\Gamma_{\gamma\gamma}} = 2 \tan^2 \theta_W \approx 0.6 \quad \text{and} \quad \frac{\Gamma_{ZZ}}{\Gamma_{\gamma\gamma}} = \tan^4 \theta_W \approx 0.08 . \tag{10}$$

This prediction is in agreement with the recent upper limit reported in by the ATLAS Collaboration from searches in the γZ channel [56].

We now turn to discuss the production via gluon fusion. We parametrize the coupling of φ to the gluon by the following vertex

$$\frac{c_{gg}}{M_s} \varphi G^2 . \tag{11}$$

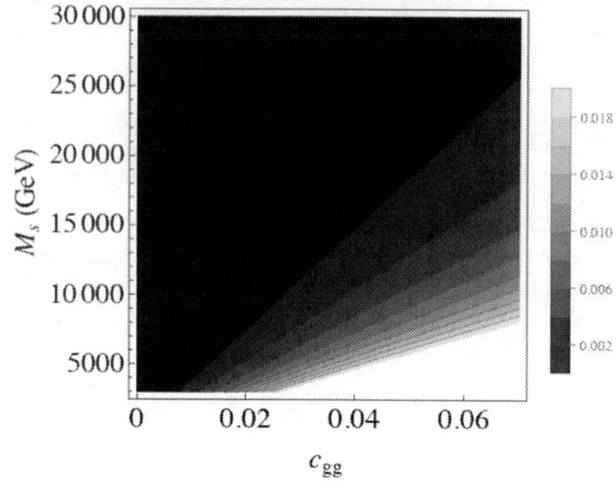

Figure 8. Contours of constant partial width Γ_{gg}. The color encoded scales are in GeV.

The partial decay width of φ to dijets is

$$\Gamma_{gg} = 8 \frac{c_{gg}^2}{4\pi} \frac{M_\varphi^3}{M_s^2} \,. \tag{12}$$

In Fig. 8 we show a scan of the parameter space (c_{gg}, M_s) for constant values of Γ_{gg} as obtained from (12).

In the narrow width approximation the cross section for diphoton production via gluon fusion is given by [53]

$$\sigma_{\rm LHC13}(gg \to \varphi \to \gamma\gamma) = 5.8 \times 10^3 \text{ pb } c_{gg}^2 \left(\frac{M_s}{\rm TeV}\right)^{-2} \mathcal{B}(\varphi \to \gamma\gamma) \tag{13}$$

and

$$\sigma_{\rm LHC8}(gg \to \varphi \to \gamma\gamma) = 1.2 \times 10^3 \text{ pb } c_{gg}^2 \left(\frac{M_s}{\rm TeV}\right)^{-2} \mathcal{B}(\varphi \to \gamma\gamma) \,. \tag{14}$$

Substituting (3) and (8) into (13) we arrive at the targeting constraint equation connecting c_{gg}, $c_{\gamma\gamma}$, and M_s, for given $\Gamma_{\rm total}$.

As of today the upper limit on dijet production at $M_{jj} = 750$ GeV is dominated by LHC8 data, $\sigma_{\rm LHC8}(pp \to jj) < 2.5$ pb at 95%CL [57]. The cross section for dijet production is

$$\sigma_{\rm LHC8}(gg \to \varphi \to gg) = 7.6 \times 10^3 \text{ pb } c_{gg}^4 \left(\frac{M_s}{\rm TeV}\right)^{-4} \left(\frac{\Gamma_{\rm total}}{45 \text{ GeV}}\right) . \tag{15}$$

Imposing the dijet constraint, $\sigma_{\rm LHC8}(pp \to jj) < 2.5$ pb, on (15) we obtain the excluded region of the $(c_{gg}, c_{\gamma\gamma})$ plane. The allowed region of the $(c_{gg}, c_{\gamma\gamma})$ parameter space, which explains the observed diphoton excess at LHC13 and is consistent with LHC8 data, is shown in Fig. 9 for illustrative values of the string scale, $M_s = 7$ TeV and 12 TeV. From (13) and (14) the LHC13/LHC8 luminosity ratio is found to be

$$\frac{\mathcal{L}_{gg}(\sqrt{s} = 13 \text{ TeV})}{\mathcal{L}_{gg}(\sqrt{s} = 8 \text{ TeV})} = 4.7. \tag{16}$$

Therefore, the production of φ by gluon fusion is consistent with the lack of a diphoton excess in LHC8 data. As we have seen, imposing the LHC8 dijet limit [57] further constrains the $(c_{gg}, c_{\gamma\gamma})$

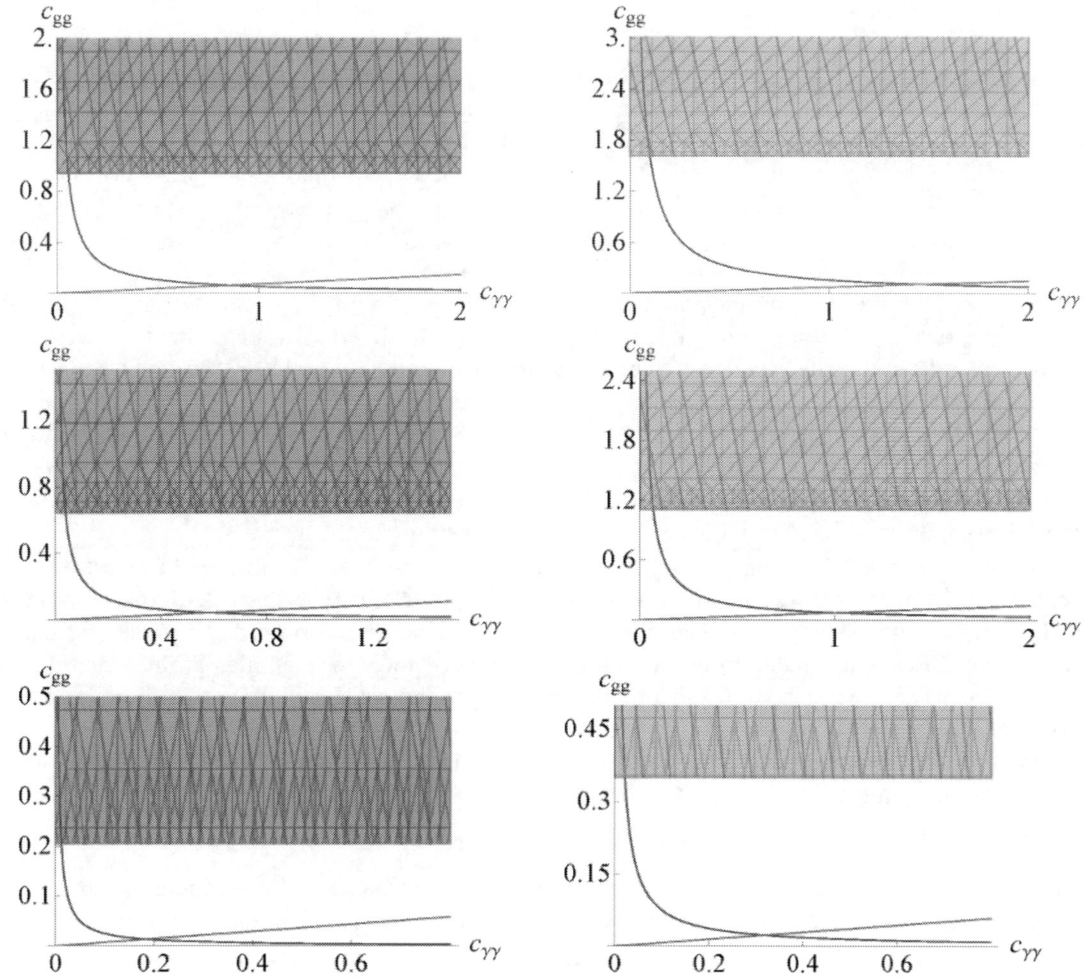

Figure 9. Allowed parameter space on the c_{gg} vs $c_{\gamma\gamma}$ plane for $M_s = 7$ TeV (left) and $M_s = 12$ TeV. The blue curves indicate the best fit of diphoton cross section ($\sigma_{\rm LHC13} = 5$ fb) for $M_s = 7$ TeV (left) and $M_s = 12$ TeV (right). The (red and orange) shaded regions are excluded at 95% CL by null results on dijet searches. The slanted (brown) curve determines the transition between c_{gg} and $c_{\gamma\gamma}$ dominance at production. Gluon fusion dominates above this curve. Results for three possible total widths are exhibited in the vertical columns, $\Gamma_{\rm total} = 45, 10, 0.1$ GeV, downwards.

parameter space. However, it remains the case that for $7 \lesssim M_s/\text{TeV} \lesssim 30$ there is always some allowed region of the $(c_{gg}, c_{\gamma\gamma})$ plane.

4 Effective Planck mass and the inflation scale

Low scale gravity with large extra dimensions is actually a particular case of a more general framework, where the UV cutoff is lower than the Planck scale due to the existence of a large number of particle species coupled to gravity [58]. Indeed, it was shown that the effective UV cutoff M_{UV} is given by

$$M_{UV}^2 = M_P^2/N\,, \tag{17}$$

where the counting of independent species N takes into account all particles which are not broad resonances, having a width less than their mass. For instance, in braneworld gravity with n large extra dimensions of average size R, the particle species are the Kaluza-Klein (KK) excitations of the graviton (and other possible bulk modes), whose number at a given energy scale E_* is given by

$$N \simeq R^n E_*^n\,. \tag{18}$$

Here, we work out the consequences of this scale dependence of the strength of gravity for inferring various quantities during inflation [2], which we take to be driven by a single field for economy of discussion and because the data doesn't compel us to consider otherwise [59]. As is to be expected, all dimensionless observables such as the amplitude and spectral properties of the perturbations are unaffected by the changing strength of gravity at inflationary energies. However, when one tries to *infer* an absolute energy scale for inflation, one finds that it is undetermined commensurate with (17) up to the unknown spectrum of universally coupled species between laboratory scales and the inflationary scale, the details of which we elaborate upon in the following.

According to the inflationary paradigm, the primordial perturbations observed in the CMB were created at horizon crossing during the quasi de Sitter (dS) phase of early accelerated expansion sourced by the inflaton field. Therefore all quantities that enter calculations of primordial correlation functions (which we subsequently relate to observables in the CMB) refer to quantities at the scale at which inflation occurred. We denote all quantities measured at the scale of inflation with a starred subscript. The dominant contribution to the temperature anisotropies comes from adiabatic perturbations [2] sourced by the comoving curvature perturbation $\mathcal{R}$, defined as the conformal factor of the 3-metric h_{ij} in comoving gauge:

$$h_{ij}(t,x) = a^2(t)e^{2\mathcal{R}(t,x)}\hat{h}_{ij};\quad \hat{h}_{ij} := \exp[\gamma_{ij}] \tag{19}$$

with $\partial_i\gamma_{ij} = \gamma_{ii} = 0$ defining transverse traceless graviton perturbations. The temperature anisotropies are characterized by the dimensionless power spectrum for $\mathcal{R}$, whose amplitude is given by

$$\mathcal{P}_{\mathcal{R}} := \frac{H_*^2}{8\pi^2 M_*^2\epsilon_*} = \mathcal{A}\times 10^{-10}, \tag{20}$$

where $\epsilon_* := -\dot{H}_*/H_*^2$, H_* being the Hubble factor during inflation. Given that $\mathcal{R}$ is conserved on super-horizon scales (in the absence of entropy perturbations), this immediately relates to the amplitude of the late time CMB anisotropies, which fixes $\mathcal{A} \sim 22.15$ [59]. The tensor anisotropies are characterized by the tensor power spectrum

$$\mathcal{P}_\gamma := 2\frac{H_*^2}{\pi^2 M_*^2}, \tag{21}$$

Taking the ratio of the above with (20), we find the tensor to scalar ratio

$$r_* := \frac{\mathcal{P}_\gamma}{\mathcal{P}_{\mathcal{R}}} = 16\epsilon_*. \tag{22}$$

[2] In what follows, we assume that all of the extra species have sufficiently suppressed couplings to the inflaton during inflation (e.g. either through derivative couplings or as Planck suppressed interactions) so that isocurvature perturbations are not significantly generated. This is trivially true for hidden sector fields.

Therefore any determination of r_*, either through direct measurements of the stochastic background of primordial gravitational waves or through their secondary effects on the polarization of the CMB [60–62] allows us in principle to fix the scale of inflation:

$$H_* = M_* \left(\frac{\pi^2 \mathcal{A} r_*}{2 \cdot 10^{10}} \right)^{1/2} := \Upsilon = 1.05 \sqrt{r_*} \times 10^{-4}. \tag{23}$$

We see that any measurements of r_* determines the scale of inflation *up to our ignorance of the effective strength of gravity at the scale* H_*, given by $M_* \sim \frac{M_P}{\sqrt{N}}$, where N is the effective number of all universally coupled species up to the scale H_*– whether they exist in the visible sector or in any hidden sector. Note that as one lowers the scale of strong gravity, the maximum reheating temperature T_i is necessarily lowered as well, since it cannot be higher than the inflation scale. Conservatively, T_i cannot be too far below the TeV scale without spoiling the standard scenarios of big bang cosmology– in particular, mechanisms for Leptogenesis and Baryogenesis which can occur no lower than the electroweak scale [63].

An example of extra species is KK graviton modes in theories of low scale gravity and large extra dimensions. In this case $N_* = N_{\rm TT}/2$, where $N_{\rm TT}$ counts the two polarizations of the massless graviton plus the number of contributing TT Kaluza-Klein (KK) resonances with masses below the momentum transfer in question. On a Minkowski background for example, in the presence of large (but stabilized) extra dimensions, this is nothing more than the usual manner in which extra dimensions open up[3]. An important qualification to the discussion initiated in [2] was raised by the authors of [64], who pointed out that on a dS background $N_{\rm TT}$ could not receive any contributions from KK gravitons since the former must respect the Higuchi bound [65] whereby massive gravitons are gapped as $m^2_{\rm KK} > 2H^2$. Therefore, it was reasoned that KK gravitons could not enhance any processes involving graviton exchange at inflationary energies. However, even in this very special limit (where an enhanced symmetry results in important qualitative differences compared to quasi dS backgrounds), KK gravitons with masses $m^2_{\rm KK} \gg 2H^2$ can still contribute non-trivially to the observed tensor power. This occurs without modifications to the spectral properties of the curvature perturbations, rendering the inferred energy scale of inflation effectively uncertain from any potential observation of primordial B-modes. This is because of the well known fact that KK modes of the graviton can couple much more strongly to 4-dimensional (4D) localized sources than the zero mode in the presence of warped extra dimensions [66]. Indeed, in an example studied in [67], one finds that individual KK modes of the graviton couple with the characteristic effective strength

$$M_* \sim M_{\rm pl}\, e^{-3z_c/2} \ll M_{\rm pl} \tag{24}$$

whenever $z_c \gg 1$, where e^{z_c} parametrizes the warping. If this were the context in which inflation actually occurs, attempts to infer an absolute scale for inflation from any putative observation of primordial B-modes in the CMB *at a fixed scale* would be rendered uncertain up to the factor

$$N^{\rm eff}_{\rm TT} = 2\left(1 + e^{3(z_c - N_{\surd})}\right) \tag{25}$$

where $N_{\surd}$ is the number of e-folds since horizon exit of the pivot scale at which we fix the tensor amplitude.

[3]We remind the reader that the moduli mass that stabilizes any extra dimensions is independent of the expectation value of the moduli (i.e. its size). It is the latter that determines the mass splitting of the KK excitations.

References

[1] I. Antoniadis, N. Arkani-Hamed, S. Dimopoulos and G. R. Dvali, Phys. Lett. B **436** (1998) 257 [arXiv:hep-ph/9804398]

[2] I. Antoniadis and S. P. Patil, Eur. Phys. J. **C75** (2015) 182 [arXiv:1410.8845 [hep-th]].

[3] C. Angelantonj and A. Sagnotti, *Phys. Rept.* **371** (2002) 1 [Erratum-ibid. **376** (2003) 339] [arXiv:hep-th/0204089].

[4] I. Antoniadis, *Phys. Lett.* B **246** (1990) 377.

[5] D. J. Kapner, T. S. Cook, E. G. Adelberger, J. H. Gundlach, B. R. Heckel, C. D. Hoyle and H. E. Swanson, Phys. Rev. Lett. **98** (2007) 021101.

[6] J. C. Long and J. C. Price, *Comptes Rendus Physique* **4** (2003) 337; R. S. Decca, D. Lopez, H. B. Chan, E. Fischbach, D. E. Krause and C. R. Jamell, *Phys. Rev. Lett.* **94** (2005) 240401; R. S. Decca et al., arXiv:0706.3283 [hep-ph]; S. J. Smullin, A. A. Geraci, D. M. Weld, J. Chiaverini, S. Holmes and A. Kapitulnik, arXiv:hep-ph/0508204; H. Abele, S. Haeßler and A. Westphal, in 271th WE-Heraeus-Seminar, Bad Honnef (2002).

[7] I. Antoniadis and K. Benakli, *Phys. Lett.* B **326** (1994) 69.

[8] I. Antoniadis, K. Benakli and M. Quirós, *Phys. Lett.* B **331** (1994) 313 and *Phys. Lett.* B **460** (1999) 176; P. Nath, Y. Yamada and M. Yamaguchi, *Phys. Lett.* B **466** (1999) 100 T. G. Rizzo and J. D. Wells, *Phys. Rev.* D **61** (2000) 016007; T. G. Rizzo, *Phys. Rev.* D **61** (2000) 055005; A. De Rujula, A. Donini, M. B. Gavela and S. Rigolin, *Phys. Lett.* B **482** (2000) 195.

[9] G.F. Giudice, R. Rattazzi and J.D. Wells, *Nucl. Phys.* B **544** (1999) 3; E.A. Mirabelli, M. Perelstein and M.E. Peskin, *Phys. Rev. Lett.* **82** (1999) 2236; T. Han, J.D. Lykken and R. Zhang, *Phys. Rev.* D **59** (1999) 105006; K. Cheung, W.-Y. Keung, *Phys. Rev.* D **60** (1999) 112003; C. Balázs *et al.*, *Phys. Rev. Lett.* **83** (1999) 2112; J.L. Hewett, *Phys. Rev. Lett.* **82** (1999) 4765.

[10] L.A. Anchordoqui, H. Goldberg, D. Lust, S. Nawata, S. Stieberger and T.R. Taylor, Phys. Rev. Lett. **101** (2008) 241803 [arXiv:0808.0497 [hep-ph]].

[11] L. A. Anchordoqui, I. Antoniadis, D. C. Dai, W. Z. Feng, H. Goldberg, X. Huang, D. Lust and D. Stojkovic *et al.*, Phys. Rev. D **90** (2014) 6, 066013 [arXiv:1407.8120 [hep-ph]].

[12] I. Antoniadis, E. Kiritsis and T. N. Tomaras, Phys. Lett. B **486**, 186 (2000) [arXiv:hep-ph/0004214]; I. Antoniadis, E. Kiritsis, J. Rizos and T. N. Tomaras, Nucl. Phys. B **660**, 81 (2003) [arXiv:hep-th/0210263].

[13] G. Shiu and S.-H. H. Tye, *Phys. Rev.* D **58** (1998) 106007; Z. Kakushadze and S.-H. H. Tye, *Nucl. Phys.* B **548** (1999) 180; L. E. Ibáñez, C. Muñoz and S. Rigolin, *Nucl. Phys.* B **553** (1999) 43.

[14] I. Antoniadis, E. Kiritsis and J. Rizos, Nucl. Phys. B **637**, 92 (2002) [arXiv:hep-th/0204153];

[15] L. A. Anchordoqui, I. Antoniadis, H. Goldberg, X. Huang, D. Lüst and T. R. Taylor, Phys. Rev. D **85**, 086003 (2012) [arXiv:1107.4309 [hep-ph]].

[16] G.T. Horowitz and J. Polchinski, Phys. Rev. D **55** (1997) 6189.

[17] ATLAS Collaboration, ATLAS-CONF-2015-081; M. Delmastro [on behalf of the ATLAS Collaboration], 51st Rencontres de Moriond (Electroweak session) 17 May 2016, La Thuile (Italy).

[18] CMS Collaboration, CMS-PAS-EXO-15-004; P. Musella [on behalf of the CMS Collaboration], 51st Rencontres de Moriond (Electroweak session) 17 May 2016, La Thuile (Italy); CMS Collaboration, CMS-PAS-EXO-16-018.

[19] R. Franceschini *et al.*, JHEP **1603** (2016) 144 [arXiv:1512.04933 [hep-ph]].

[20] G. Aad *et al.* [ATLAS Collaboration], Phys. Rev. D **92**, 032004 (2015) [arXiv:1504.05511 [hep-ex]].

[21] V. Khachatryan *et al.* [CMS Collaboration], Phys. Lett. B **750**, 494 (2015) [arXiv:1506.02301 [hep-ex]].

[22] L. A. Anchordoqui, I. Antoniadis, H. Goldberg, X. Huang, D. Lüst and T. R. Taylor, Phys. Lett. B **755**, 312 (2016) [arXiv:1512.08502 [hep-ph]]; arXiv:1603.08294 [hep-ph].

[23] R. Blumenhagen, M. Cvetic, P. Langacker and G. Shiu, Ann. Rev. Nucl. Part. Sci. **55**, 71 (2005) [hep-th/0502005].

[24] R. Blumenhagen, B. Kors, D. Lüst and S. Stieberger, Phys. Rept. **445**, 1 (2007) [hep-th/0610327].

[25] D. Cremades, L. E. Ibanez and F. Marchesano, JHEP **0307**, 038 (2003) [hep-th/0302105].

[26] L. A. Anchordoqui, I. Antoniadis, H. Goldberg, X. Huang, D. Lüst and T. R. Taylor, Phys. Lett. B **749**, 484 (2015) [arXiv:1507.05299 [hep-ph]].

[27] M. B. Green and J. H. Schwarz, Phys. Lett. B **149**, 117 (1984).

[28] E. Witten, Phys. Lett. B **149**, 351 (1984).

[29] M. Dine, N. Seiberg, X. G. Wen and E. Witten, Nucl. Phys. B **289**, 319 (1987).

[30] W. Lerche, B. E. W. Nilsson, A. N. Schellekens and N. P. Warner, Nucl. Phys. B **299**, 91 (1988).

[31] L. E. Ibanez and F. Quevedo, JHEP **9910**, 001 (1999) [hep-ph/9908305].

[32] D. M. Ghilencea, L. E. Ibanez, N. Irges and F. Quevedo, JHEP **0208**, 016 (2002) [hep-ph/0205083].

[33] J. J. Heckman, Nucl. Phys. B **906** (2016) 231 [arXiv:1512.06773 [hep-ph]].

[34] M. Cvetic, J. Halverson and P. Langacker, arXiv:1512.07622 [hep-ph].

[35] L. E. Ibanez and V. Martin-Lozano, arXiv:1512.08777 [hep-ph].

[36] E. Palti, Nucl. Phys. B **907** (2016) 597 [arXiv:1601.00285 [hep-ph]].

[37] A. Karozas, S. F. King, G. K. Leontaris and A. K. Meadowcroft, Phys. Lett. B **757** (2016) 73 [arXiv:1601.00640 [hep-ph]].

[38] A. E. Faraggi and J. Rizos, Eur. Phys. J. C **76** (2016) no.3, 170 [arXiv:1601.03604 [hep-ph]].

[39] P. Anastasopoulos and M. Bianchi, arXiv:1601.07584 [hep-th].

[40] M. Cvetic, J. Halverson and P. Langacker, arXiv:1602.06257 [hep-ph].

[41] T. Li, J. A. Maxin, V. E. Mayes and D. V. Nanopoulos, arXiv:1602.09099 [hep-ph].

[42] G. K. Leontaris and Q. Shafi, arXiv:1603.06962 [hep-ph].

[43] S. Abel and V. V. Khoze, JHEP **1605** (2016) 063 [arXiv:1601.07167 [hep-ph]].

[44] T. Higaki, K. S. Jeong, N. Kitajima and F. Takahashi, Phys. Lett. B **755**, 13 (2016) [arXiv:1512.05295 [hep-ph]].

[45] E. Megias, O. Pujolas and M. Quiros, JHEP **1605** (2016) 137 [arXiv:1512.06106 [hep-ph]].

[46] I. Ben-Dayan and R. Brustein, arXiv:1601.07564 [hep-ph].

[47] N. D. Barrie, A. Kobakhidze, M. Talia and L. Wu, Phys. Lett. B **755**, 343 (2016) [arXiv:1602.00475 [hep-ph]].

[48] L. Aparicio, A. Azatov, E. Hardy and A. Romanino, JHEP **1605** (2016) 077 [arXiv:1602.00949 [hep-ph]].

[49] V. Khachatryan *et al.* [CMS Collaboration], Phys. Rev. Lett. **116** (2016) no.7, 071801 [arXiv:1512.01224 [hep-ex]].

[50] L. A. Harland-Lang, V. A. Khoze and M. G. Ryskin, JHEP **1603** (2016) 182 [arXiv:1601.07187 [hep-ph]].

[51] ATLAS Collaboration, Technical Report CERN- LHCC-2011-012. LHCC-I-020, CERN, Geneva, Nov. 2011.

[52] M Albrow *et al.* [CMS and TOTEM Collaborations], Technical Report CERN-LHCC-2014-021. TOTEM-TDR-003. CMS-TDR-13, CERN, Geneva, Sep. 2014.

[53] C. Csaki, J. Hubisz, S. Lombardo and J. Terning, arXiv:1601.00638 [hep-ph].

[54] A. D. Martin and M. G. Ryskin, J. Phys. G **43**, no. 4, 04LT02 (2016) [arXiv:1601.07774 [hep-ph]].

[55] K. R. Dienes and B. Thomas, Phys. Rev. D **85**, 083523 (2012) [arXiv:1106.4546 [hep-ph]]; Phys. Rev. D **85**, 083524 (2012) [arXiv:1107.0721 [hep-ph]].

[56] ATLAS Collaboration, ATLAS-CONF-2016-010.

[57] G. Aad *et al.* [ATLAS Collaboration], Phys. Rev. D **91**, 052007 (2015) [arXiv:1407.1376 [hep-ex]].

[58] G. Dvali, arXiv:0706.2050 [hep-th]; Int. J. Mod. Phys. A **25** (2010) 602 [arXiv:0806.3801 [hep-th]]; G. Dvali and M. Redi, Phys. Rev. D **77** (2008) 045027 [arXiv:0710.4344 [hep-th]]; R. Brustein, G. Dvali and G. Veneziano, JHEP **0910** (2009) 085 [arXiv:0907.5516 [hep-th]]; G. Dvali and C. Gomez, Phys. Lett. B **674** (2009) 303.

[59] P. A. R. Ade *et al.* [Planck Collaboration], Astron. Astrophys. (2014) [arXiv:1303.5076 [astro-ph.CO]]; arXiv:1303.5082 [astro-ph.CO].

[60] A. G. Polnarev, Sov. Astron. **29**, 607 (1985).

[61] M. Kamionkowski, A. Kosowsky and A. Stebbins, Phys. Rev. Lett. **78**, 2058 (1997) [astro-ph/9609132].

[62] U. Seljak and M. Zaldarriaga, Phys. Rev. Lett. **78**, 2054 (1997) [astro-ph/9609169].

[63] M. Trodden, Rev. Mod. Phys. **71**, 1463 (1999) [hep-ph/9803479]; S. Davidson, E. Nardi and Y. Nir, Phys. Rept. **466**, 105 (2008) [arXiv:0802.2962 [hep-ph]].

[64] M. Kleban, M. Mirbabayi and M. Porrati, JCAP **1601**, no. 01, 017 (2016) [arXiv:1508.01527 [hep-th]].

[65] A. Higuchi, Nucl. Phys. B **282**, 397 (1987).

[66] L. Randall and R. Sundrum, Phys. Rev. Lett. **83**, 3370 (1999) [hep-ph/9905221].

[67] I. Antoniadis and S. P. Patil, *in preparation.*

Hunting for Dark Particles with Gravitational Waves

Gian F. Giudice[1,a]

[1] *CERN, Theoretical Physics Department, Geneva, Switzerland*

Abstract. The LIGO observation of gravitational waves from a binary black hole merger has begun a new era in fundamental physics. If new dark sector particles, be they bosons or fermions, can coalesce into exotic compact objects (ECOs) of astronomical size, then the first evidence for such objects, and their underlying microphysical description, may arise in gravitational wave observations. In this work we study how the macroscopic properties of ECOs are related to their microscopic properties, such as dark particle mass and couplings. We then demonstrate the smoking gun exotic signatures that would provide observational evidence for ECOs, and hence new particles, in terrestrial gravitational wave observatories. Finally, we discuss how gravitational waves can test a core concept in general relativity: Hawking's area theorem.

1 Introduction

The LIGO detection of GW150914 [1] – the first observed event of gravitational wave (GW) emission from binary black-hole merging – has ushered in a new era in the exploration of our universe. Not only does this result open new prospects for the study of the astrophysical properties of objects in extreme density conditions, but also allows for new ways of testing fundamental physics. Indeed, soon after the LIGO announcement, it was suggested that GW observations can be used to probe violations of the equivalence principle and Shapiro delay [2–5], modifications of gravity [4, 6–8], the presence of event horizons [9], the quantum structure of black holes [10], the hypothesis that dark matter is in the form of primordial black holes [11–13], the propagation of gravity waves [4, 14–21], phase transitions in the early universe [22, 23], neutrino properties from associated neutrino emission [24, 25], and axion clouds around black holes [26].

Here we want to investigate another way in which GW detection can contribute to fundamental science: the search for new physics in the form of exotic compact objects (ECO). Dark matter is a primary motivation to expect that new forms of stable matter must exist and be abundant in our universe. Moreover, quite independently of dark matter considerations, many new-physics theories predict the existence of stable or long-lived light particles with feeble interactions with ordinary matter. Because of their elusive nature, here we will generally refer to them as dark particles. It is conceivable that dark particles could coalesce into macroscopic objects. Although at present we are not able to predict the mass distribution of such exotic objects, it is possible that they could reach astronomical sizes. In this case GW detection from the merging of two ECOs or the collapse of an ECO into a stellar black hole could be used as a new and formidable way to probe the existence of dark particles and test their

[a] e-mail: Gian.Giudice@cern.ch

DOI: 10.1051/epjconf/201716402004

properties. Actually, in some cases, this could be the only way in which dark particles are discovered, because their extremely feeble interactions make them completely invisible to collider searches or other particle physics experiments.

This presentation is based on ref. [27], where an extended discussion and a complete list of references can be found.

2 Signatures of Exotic Compact Objects

2.1 LIGO Sensitivity on ECO parameters

Let us consider two point masses M_1 and M_2 orbiting each other under the force of gravity. This is equivalent to a single body with reduced mass $\mu \equiv M_1 M_2/M_{\rm tot}$ (where $M_{\rm tot} = M_1 + M_2$) moving in an external gravitational potential. The equivalent body moves in an elliptic orbit with major semi-axis l, describing the separation of the two bodies.

The orbital period P is related to $M_{\rm tot}$ and l by Kepler's third law

$$P^2 = \frac{4\pi^2 l^3}{M_{\rm tot}} \,. \tag{1}$$

The frequency f of GW emission is twice the orbital frequency $\nu = 1/P$, and thus is given by

$$f = \sqrt{\frac{M_{\rm tot}}{\pi^2 l^3}} \,. \tag{2}$$

For the system of two BHs the innermost stable circular orbit (ISCO) is defined by

$$R_{BH}^{\rm ISCO} \equiv 6 M_{\rm tot} \,. \tag{3}$$

The ISCO determines the end of the inspiral phase and the beginning of the merger phase, thus this radius characterises the typical frequency $f_{BH}^{\rm ISCO}$ expected in a binary merger. For a BH-BH merger, we find

$$f_{BH}^{\rm ISCO} = \frac{1}{6^{3/2}\pi M_{\rm tot}} \qquad \text{(for BH)} \,. \tag{4}$$

The maximum frequency at the end of inspiral, obtained from numerical simulations of the waveform of BH mergers, is given by $f = f_{BH}^{\rm ISCO}(1+\Delta)$, where the correction term Δ is computed in post-Newtonian approximation and is a function of the mass ratio M_1/M_2 and of a single combination of the two BH spins. Its analytical expression can be found in [28]. The correction term Δ vanishes in the limit of negligible BH spin and large mass ratio. For non-rotating BHs, it reaches a maximum value $\Delta = 1.1$ for $M_1/M_2 = 3.56$, while $\Delta = 0.7$ for equal BH masses. For spinning BHs, in the parameter range for which the expansion given in [28] is valid, the maximum value reached is $\Delta = 3.1$. Thus, for discussion purposes, it is adequate to take the expression in eq. (4) as the frequency determining the end of inspiral. The correction term Δ is fully included in our numerical analysis.

By analogy with eq. (3), for ECOs we may define a typical GW frequency determined by the ISCO radius $R_{ECO}^{\rm ISCO} \equiv 3\, M_{\rm tot}/C$, where $C \equiv M/R$ is the compactness of the ECO, assumed to be the same for the two merging bodies. This ansatz is corroborated by numerical analyses of NS orbits [29].

Consequently, we may define the typical frequency $f_{ECO}^{\rm ISCO}$ in the merging of two ECOs, using eq. (2) for $l = 3M_{\rm tot}/C$:

$$f_{ECO}^{\rm ISCO} = \frac{C^{3/2}}{3^{3/2}\pi M_{\rm tot}} \qquad \text{(for ECO)} \,. \tag{5}$$

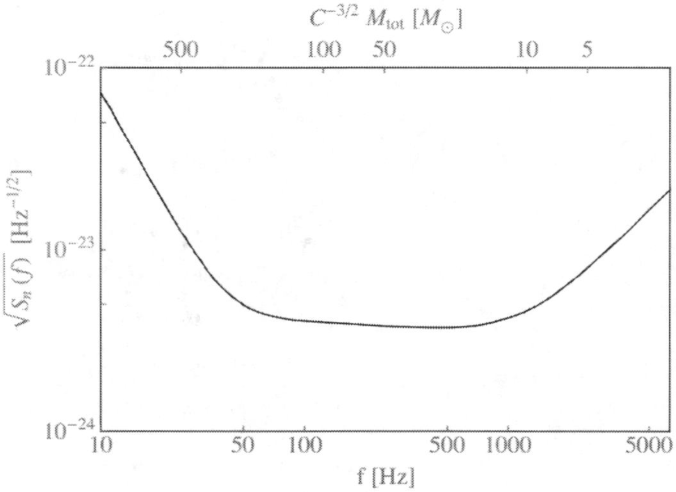

Figure 1. *LIGO noise power spectral density. For a given characteristic GW frequency for a binary merger, on the upper axes we show the corresponding combination of ECO compactness $C = M/R$ and total mass $M_{\rm tot}$.*

Note that eq. (5) becomes equal to eq. (3), when we approach the Schwarzschild BH compactness $C_{BH} = 1/2$.

To observe the merger and ringdown phase it is crucial that the characteristic frequency $f^{\rm ISCO}$ falls within the LIGO sensitivity range. LIGO sensitivity can be described in terms of the so-called noise power spectral density [30]. We plot this function in fig. 1. The most sensitive interval, where the signal-to-noise is maximised, is $f = [50 - 1000]$ Hz. On the upper axes of fig. 1 we also show the sensitivity in terms of C and $M_{\rm tot}$ by employing eq. (5). Thus we see that the signal-to-noise ratio is maximised for objects in the range

$$\left(\frac{C}{0.2}\right)^{3/2} 1.1\, M_\odot \lesssim M_{\rm tot} \lesssim \left(\frac{C}{0.5}\right)^{3/2} 88\, M_\odot\,. \tag{6}$$

For a GW signal with strain $h(t)$ the expectation value of the signal-to-noise ratio is

$$\rho^2 = 4 \int_0^\infty \frac{|\tilde{h}(f)|^2}{S_n(f)} df\,, \tag{7}$$

where $\tilde{h}(f)$ is the Fourier transform of the signal. As we are interested in the region of best sensitivity, in eq. (7) we have taken the case of optimal orientation (face-on, overhead source). Averaging over sky position, inclination, and polarisation, one finds that $\langle\rho^2\rangle$ is equal to $4/25$ times the expression in eq. (7). A criterion for detectability is the requirement $\rho \geqslant 8$ [31]. Assuming that most of the signal-to-noise ratio is accumulated during the inspiral phase, we can use the quadrupole approximation truncated at the Newtonian order [32]

$$\tilde{h}(f) \approx \frac{\sqrt{5/24}}{\pi^{2/3} D_L} M_c^{5/6} f^{-7/6}\,, \tag{8}$$

where

$$M_c = \frac{(M_1 M_2)^{3/5}}{(M_1 + M_2)^{1/5}}\,, \tag{9}$$

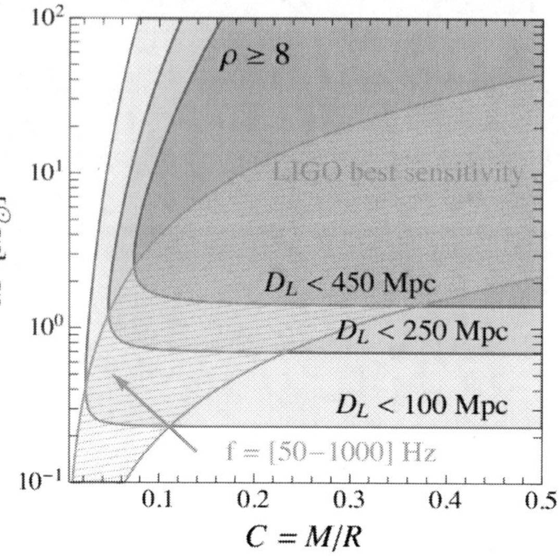

Figure 2. *The LIGO best sensitivity range in ECO mass M and compactness C, for equal mass binary GW events. The yellow band corresponds to the GW frequency range $f = [50 - 1000]$ Hz, the green regions to a signal-to-noise ratio for an inspiral event occurring within the luminosity distance D_L, taking $\rho \geqslant 8$ as a criterion for detection.*

is the so-called chirp mass of the system. It is important to remark that GW detectors do not measure the source-frame chirp mass, but the redshifted mass $\widetilde{M}_c$, related to the source-frame mass by the relation $\widetilde{M}_c = (1+z)M_c$, where z is the redshift of the merging system with respect to the detector. The amplitude of a GW is proportional to the inverse of the luminosity distance $D_L = (1+z)D_c$, where the comoving distance is

$$D_c = \frac{c}{H_0}\int_0^z \frac{dt}{E(t)}, \qquad E(z) = \sqrt{\Omega_M(1+z)^3 + \Omega_k(1+z)^2 + \Omega_\Lambda}\,, \tag{10}$$

with $t_H \equiv 1/H_0 = 13.969$ Gyr the Hubble time. For a given luminosity distance, and assuming standard cosmology (that corresponds to $\Omega_k = 0$, $\Omega_\Lambda = 0.7$, and $\Omega_M = 0.3$), it is possible to extract the redshift z, and convert the detector-frame chirp mass into the source frame chirp mass.

We can use the signal-to-noise ratio criterion $\rho \geqslant 8$ to estimate the sensitivity range of the LIGO detector in terms of mass and compactness. For illustration, we take the limit of equal masses $M_1 = M_2 = M$. Since we are limiting the analysis to the inspiral phase, eq. (8) is valid in the frequency range $f < f_{\rm ISCO}$ and we cutoff the integral in eq. (7) at $f = f_{\rm ISCO}$. We show our result in fig. 2. The yellow band shows the LIGO frequency sensitivity range, given in eq. (6). The green regions show the signal intensity sensitivity range (corresponding to $\rho \geqslant 8$), for a single GW event occurring within a distance D_L. We choose three particular values of the maximum luminosity distance, namely $D_L = 100, 250, 450$ Mpc. Note that the corresponding redshift is very small, ranging from $z = 0.023$ for $D_L = 100$ Mpc to $z = 0.1$ for $D_L = 450$ Mpc. We therefore neglect the redshift effects in fig. 2. The overlap between the yellow and green regions identifies the LIGO best sensitivity in ECO mass M and compactness C.

2.2 Features in a Binary Mass Census

In sect. 2.1 we focused on ECO signatures that can be extracted from the waveform of a single GW event. Here we discuss the physics information that can be extracted from statistical distributions of GW events, which will be available once a significant number of mergers have been observed.

We also remark that, even if ECO signatures were identified in single waveforms, distributions of events will be helpful to understand the ECO mass function and possibly gain knowledge about their formation. To explore this possibility we must first understand the expected distribution of known compact objects, as this will form the known 'background' distribution.

2.2.1 On the mass distribution of compact merging binaries in the local Universe

The mass distribution of compact object binaries can be effectively organised in a plane of the chirp mass M_c and the symmetric mass ratio η. While the chirp mass is defined in eq. (9), the symmetric mass ratio is related to the primary and secondary binary constituent masses M_1, M_2 and it corresponds to the ratio between reduced and total mass.

When the population of mergers is projected onto the $M_c - \eta$ plane, binaries involving neutron stars and black holes (in the combinations NS-NS, NS-BH, and BH-BH) generally populate different regions. The mass distribution of these 'conventional' compact object binaries depends on the assumed model of stellar binary evolution. In turn, models of stellar binary evolution depend on a number of parameters that are largely unconstrained or poorly known. In the following we briefly mention some of the most relevant sources of uncertainty and features of these models, while we refer the reader to [27] for a more detailed discussion and for the relevant references.

Details of the 'common envelope (CE) phase'. The formation of a compact binary object is preceded by an evolutionary phase in which the binary system evolves enclosed into a gas envelope originated from the interactions between the donor and the companion star. The most important physical quantity describing the CE phase is the binding energy of the envelope, which in turn depends on the characteristic mass and radius of the binary system. If the binding energy is too high the stars will merge during the CE phase thus preventing the formation of a compact binary object. If the binding energy is too low the decrease in orbital separation during the CE phase is too slow, and the resulting compact binary object will never merge within an Hubble time. Furthermore, the formation of a CE strongly depends on the properties of the donor star. This is particularly true if the donor star is situated in the Hertzsprung gap (HG) of the Hertzsprung-Russell diagram; in this case, two alternative choices are commonly adopted: either HG stars are CE donors or they are not.

Unknown maximum mass of NS and minimal mass of BH. All observed NS have masses below 2 $M_\odot$, and theoretical modelling suggests that this is likely to be an upper bound. On the other hand, stellar BH, which are expected to be generated by the gravitational collapse of massive stars, have a mass distribution known to start from about 5 $M_\odot$ and to extend to tens of solar masses. In other words, observations (mostly based on X-ray measurements) favour the existence of a *mass gap* between the distribution of NS and stellar BH, although it cannot be excluded that this is only the result of some observational bias.

The presence of this mass gap may seem surprising, given that the progenitor stellar mass distribution is known to be smooth. Hence, the distribution of the ensuing compact remnants might be expected to be smooth as well. However, it was shown that the presence of this mass gap can be linked to the evolutionary mechanism of dying stars. For sufficiently fast growth of the instabilities causing the stellar death, the evolution leads to two distinct outcomes: either a violent explosion that ejects most of the mass from the star, leaving a NS remnant; or a failed supernova, in which all the

DOI: 10.1051/epjconf/201716402004

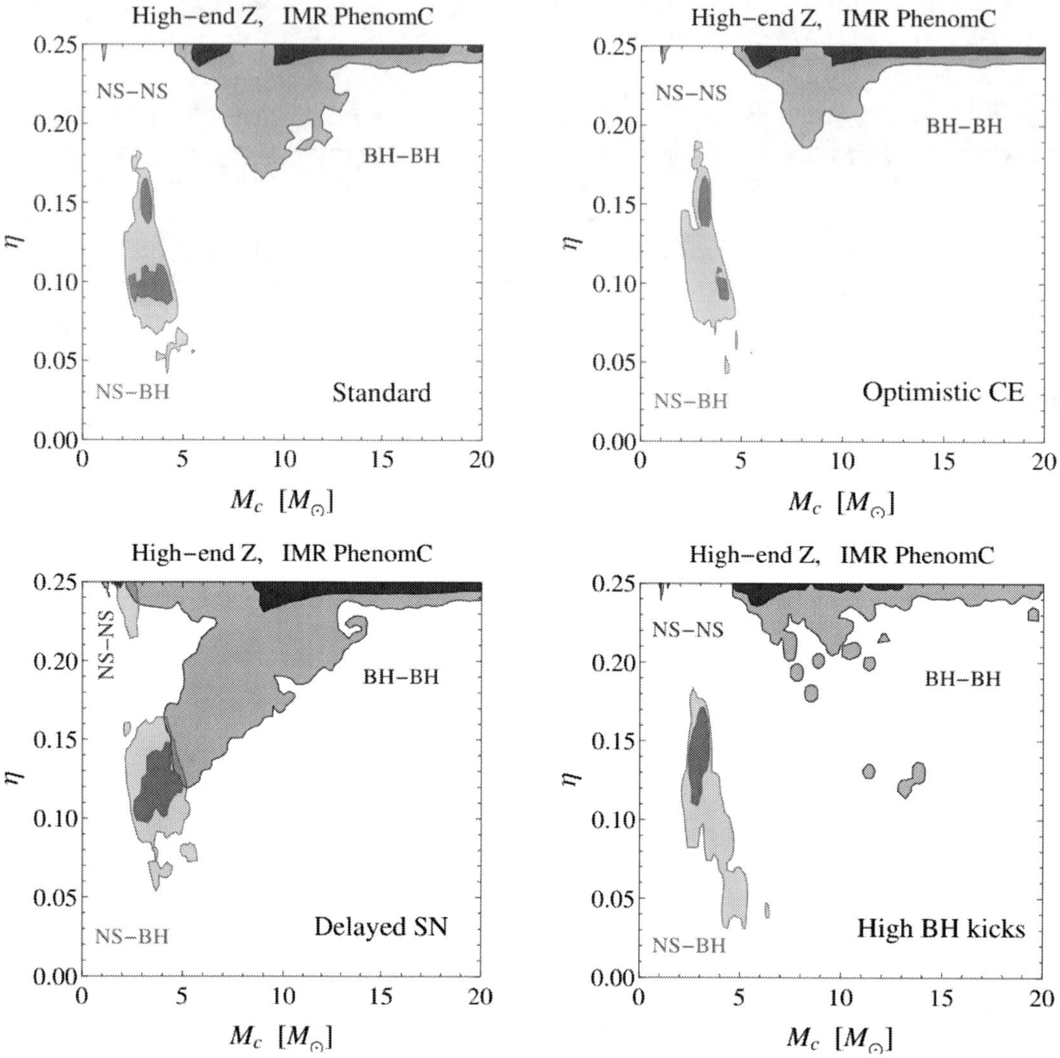

Figure 3. *Binary mass distribution – in the plane chirp mass vs. symmetric mass ratio* (M_c, η) *– for four stellar evolution models. For each distribution the contours of different colour shading represent the* 68% *and* 99% *CL regions.*

stellar mass collapses in a BH. A mass gap is dynamically explained, as long as the growth timescale is sufficiently short (less than about 200 ms).

Physics of the explosions that form compact objects. The physics describing star evolution, supernova explosions and the subsequent formation of compact objects (either NS or BH) is plagued by many uncertainties. *i)* The evolution of a main sequence star toward its eventual supernova stage is affected by mass loss via stellar winds. *ii)* Supernova explosions may be modelled by means of different mechanism, depending on the type of instability that eventually lead to the ejection of in-falling matter. *iii)* Finally, the formation of a NS or BH from supernova explosions may be accompanied by a natal kick.

Metallicity Z. The information about the metallicity – which in general describes the fraction of mass that is not in Hydrogen or Helium – is of fundamental importance to model the chemical evolution of the local Universe.

The impact of these uncertainties on the expected distribution of binary merger primaries may be estimated using simulations. The resulting distributions in the $M_c - \eta$ plane for these models are shown in fig. 3. Clearly, variations in the underlying model do lead to differences in the expected distributions. However, a qualitative feature that persists across all of these scenarios is that NS-NS, NS-BH, and BH-BH mergers populate reasonably distinct regions. It is then tempting to conclude that, if future GW observations fell outside the regions in fig. 3 covered by 'conventional' mergers, we could infer evidence for an exotic population of compact objects. However, to assess this possibility it is necessary to consider more carefully the GW observables.

A GW observatory will observe the GW emitted in a binary merger. Physical properties of the binary system are then extracted by comparing the observed waveform with waveforms predicted within general relativity. Thus M_c and η are not themselves observables, but they may be determined from this comparison. M_c is relatively universal and may be extracted from the inspiral phase of the wave. However, an independent extraction of η requires consideration of the merger phase and is thus only accessible by comparison with waveforms generated numerically. If a detected waveform is studied by comparing with expected NS-NS, NS-BH, and BH-BH waveforms, then only best-fit parameters populating these regions in the $M_c - \eta$ plane will be found. If the waveform turns out sufficiently exotic that none of these waveforms give a good fit, then discovery would come from single-event analysis. Thus, it seems unlikely that evidence for ECOs would first arise by analyses of distributions in the $M_c - \eta$ plane.

However, the chirp mass may be extracted from the inspiral phase and thus does not require a detailed comparison between the observed wave and the numerical GR waveforms generated for specific scenarios. Thus, once many merger events have been observed, it may be possible to search for a new distribution of ECOs by considering the chirp mass alone, without relying too heavily on specific expected waveforms. In fig. 4 we show the four distributions of fig. 3 projected onto the chirp mass axes alone. To make a concrete connection with observations, we superimpose the expected LIGO sensitivity as a function of the chirp mass. At each value of red-shifted chirp mass – under the simplifying assumption $M_1 = M_2$ – we computed the maximum luminosity distance at which the signal-to-noise ratio in eq. (7) reaches the detectability threshold $\rho = 8$. The corresponding redshift, evaluated according to eq. (10), allows us to reconstruct the maximum volume V inside which the analysed binary could be seen. The expected upper limit on the merger rate density, after an operating time T, is $\mathcal{R} = 1/(\epsilon VT)$, where ϵ is an experimental efficiency which we take to be 30%.

Note that this result, although strictly valid only for the case of equal-mass BH-BH mergers, should give a correct order-of-magnitude estimate of the expected LIGO sensitivity.

Clearly, there are large variations in the expected distributions based on the underlying model. However, in three of the four choices an identifiable chirp mass gap appears between the NS-NS and BH-BH distributions. This mass gap does not arise in the model with delayed SN explosion, as also clear from the corresponding mass distribution in fig. 3. However, when projected onto the chirp mass axes, the merger rate density of BH-BH binaries populating the mass gap turns out to be extremely low, order of magnitudes below the LIGO sensitivity. Finally, we note that the BH-BH binary merger rate expected from the scenario with high natal BH kicks is smaller than for the other models and appears to be in some tension with the observation of GW150914, given the LIGO sensitivity. Thus, the high natal BH kick scenario already appears disfavoured. These simulations thus seem to suggest that a mass gap is likely for binary mergers of known objects.

DOI: 10.1051/epjconf/201716402004

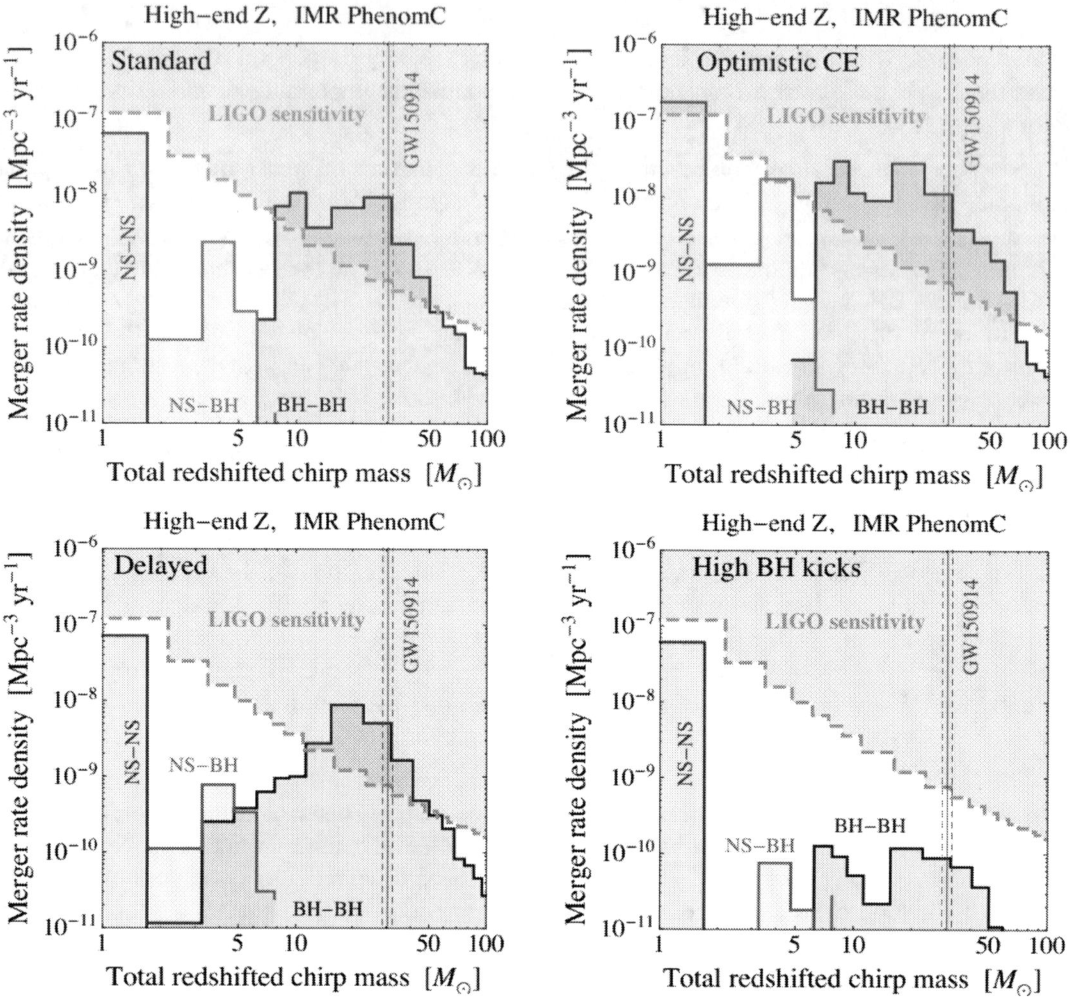

Figure 4. *Merger rate density as a function of the redshifted chirp mass for four stellar evolution models. In each case, we show the distributions for NS-NS, NS-BH and BH-BH event rates. The vertical red line corresponds to the event GW150914, with measured detector-frame chirp mass* $M_c = 30.5^{+1.7}_{-1.8}\ M_\odot$ *and source redshift* $z = 0.093^{+0.028}_{-0.036}$*. Also shown is the LIGO detection sensitivity, obtained with the estimate described in the text.*

There is no reason to expect ECOs to have a mass distribution that exhibits a mass gap in precisely the same location. Thus, events filling the gap in the chirp mass distribution would point towards a population of ECOs. This is especially true if the waveforms appear to be of the NS-NS or NS-BH type, because massive ECOs with compactness $C < 1/2$ could resemble NS, although with anomalously large masses.

To make a quantifiable assessment of this possibility would require not only a better understanding of the systematic uncertainties in the expected distribution of known objects, but also concrete predictions for the distribution of ECOs. As this distribution will depend in detail on the formation mechanism it is highly model dependent and would require treatment on a case-by-case basis for each particle physics model. As this is beyond the scope of this work we will simply speculate that the comparison of the observed distribution of chirp masses with that expected from theoretical models can give some information about a possible population of ECOs. Unfortunately, the large astrophysical uncertainties in the modelling of the 'background' are a severe limiting factor for the discovery of a 'signal'.

3 Test of the Area Theorem

Hawking's area theorem [34] states that the sum of the horizon areas of a system of BHs never decreases. This remarkable result follows directly from general relativity and the null energy condition. However, once we use the thermodynamical identification of a BH temperature (inversely proportional to mass, $T = 1/8\pi M$ [35, 36]) and BH entropy (proportional to area, $S = A/4$ [37–39]), Hawking's theorem acquires a new interpretation [40–42]. The statement about the ever increasing area can be viewed as a generalised second law of thermodynamics for the ever increasing entropy.

This thermodynamical interpretation of the area theorem has a deep physical significance that goes beyond classical general relativity. Indeed, the area theorem is apparently violated by the quantum process of Hawking's radiation. During BH evaporation, conservation of energy implies that the BH mass must decrease, hence the BH area must decrease as well. This contradiction with the area theorem is resolved in the context of the generalised second law of thermodynamics. Once we account for the entropy of the emitted matter and radiation, and not only the entropy associated with the BH area, the second law of thermodynamics is respected by BH evaporation. All this to say that any observational test of the area theorem is a way of probing the fundamental principles of general relativity and statistical mechanics.

It is remarkable that GW detection can indeed provide a test of the area theorem. The horizon area of a Kerr BH of mass M and angular momentum J is given by

$$A = 8\pi M^2\left(1 + \sqrt{1-a^2}\right) \qquad a \equiv \frac{J}{M^2}\,. \tag{11}$$

For the process of two BH with horizon areas $A_{1,2}$ merging into a final BH of area A_f, Hawking's theorem predicts

$$A_f > A_1 + A_2 \qquad \text{(area theorem)}\,. \tag{12}$$

The area theorem does not only provide a test of fundamental principles, but can also probe the existence of ECOs. In order to understand how ECOs could give an apparent violation of the area theorem, if erroneously taken for BHs, it is useful to clarify the physical content of the area theorem when applied to merger events. eq. (12) can be written as a lower bound on the final BH mass

$$M_f > \sqrt{M_1^2\, s_1 + M_2^2\, s_2}\,, \qquad s_{1,2} \equiv \frac{1+\sqrt{1-a_{1,2}^2}}{1+\sqrt{1-a_f^2}}\,. \tag{13}$$

The absence of naked singularities (aka cosmic censorship conjecture [43]) requires $0 \le a \le 1$, which implies $s_{1,2} > 1/2$. Using this condition in eq. (13), we find

$$M_f > \sqrt{\frac{M_1^2 + M_2^2}{2}}\,. \tag{14}$$

In words, *Hawking's theorem implies a simple lower bound on the mass M_f of the BH resulting from the merging of two BHs with masses $M_{1,2}$*.

Another way of reading eq. (12) is as an inequality on the efficiency in GW emission, $\epsilon \equiv (M_1 + M_2 - M_f)/(M_1 + M_2)$,

$$\epsilon < 1 - \frac{\sqrt{s_1 + r^2\, s_2}}{1 + r}, \qquad r \equiv \frac{M_2}{M_1} \tag{15}$$

The upper bound on ϵ is maximised for maximally-rotating initial BHs and non-rotating final BH ($s_{1,2} = 1/2$), giving $\epsilon < 1 - \sqrt{(1 + r^2)/2}/(1 + r)$. In turn, this is optimised for $r = 1$, giving $\epsilon < 1/2$. In the special case of slowly-rotating initial BHs ($1 \leq s_1 = s_2 \leq 2$), the maximum upper bound is $\epsilon < 1 - \sqrt{1 + r^2}/(1 + r)$, which is optimised for $r = 1$ giving $\epsilon < 1 - 1/\sqrt{2}$.

To summarise, *the area theorem prescribes a maximum efficiency in GW emission for the merging process of two BHs*. An apparent violation of the theorem could occur if the merging objects are not really BHs and emit other forms of energy, which goes undetected and exceeds the bound in eq. (15). In other words, the apparent violation of the area theorem would occur in the same way as for Hawking's BH evaporation. Emission of non-gravitational radiation leads to entropy production, left unaccounted for by the area theorem. As a result, the area theorem is not satisfied, although the generalised second law of thermodynamics holds valid.

As a concrete example of observational test of the area theorem, let us consider the detection of GW150914, for which the extracted parameters are $M_1 = 36^{+5}_{-4}\, M_\odot$ and $a_1 = 0.3^{+0.5}_{-0.3}$ for the primary, $M_2 = 29 \pm 4\, M_\odot$ and $a_2 = 0.5^{+0.5}_{-0.4}$ for the secondary, and $M_f = 62 \pm 4\, M_\odot$ and $a_f = 0.67^{+0.05}_{-0.07}$ for the final BH [33]. This gives

$$A_1 + A_2 = (2.2 \pm 0.2) \times 10^5\ \text{km}^2\,, \qquad A_f = (3.7 \pm 0.4) \times 10^5\ \text{km}^2\,, \tag{16}$$

where we have taken into account the error correlation among the input parameters using the information in [33]. The result in eq. (16) shows that the observation of GW150914 has verified eq. (12), hence Hawking's area theorem, well within errors. This is also visualised in fig. 5, where we show how the LIGO result falls within the region of plane $A_1 + A_2$ vs A_f allowed by the area theorem. In the same plane we also show the region excluded by the condition on conservation of energy $M_f < M_1 + M_2$, which corresponds to $A_f < (\sqrt{A_1/s_1} + \sqrt{A_2/s_2})^2$. Recalling that $s_{1,2} > 1/2$, we obtain

$$A_f < 4\,(A_1 + A_2) \qquad \text{(energy conservation)}\,. \tag{17}$$

It should be stressed that the experimental determination of A_f in eq. (16) is not independent of $A_1 + A_2$. Once one assumes that the merging binaries are BH and fixes the initial parameters ($M_{1,2}$, $a_{1,2}$, the relative angles between the spin vectors, the angular velocity, and orbit eccentricity), the values of M_f and a_f are uniquely determined by the general relativistic calculation. A genuine test of the area theorem would require an independent extraction of the initial and final BH parameters. This is not impossible as the inspiral and ringdown phases are mostly sensitive to initial and final BH parameters, respectively.

Future GW observations of BH mergers will further test the area theorem. What if these observations do not agree with the theorem's expectations? Unless one is ready to abandon the basic principles of general relativity, measurements of apparent violations of the area theorem can be used to argue for the existence of ECOs. Depending on their nature, ECOs could emit significant fractions of their energy in non-gravitational forms, in practice exceeding the bound in eq. (15). This can happen if they emit intense electromagnetic radiation in collimated narrow beams not pointing towards the line of sight, or if they radiate particles (such as dark photons or light axion-like particles) that remain invisible to us.

The possibility of testing the area theorem has a theoretical interest independent of ECOs, since its validity is not guaranteed in various extension of general relativity (GR). The key observation for

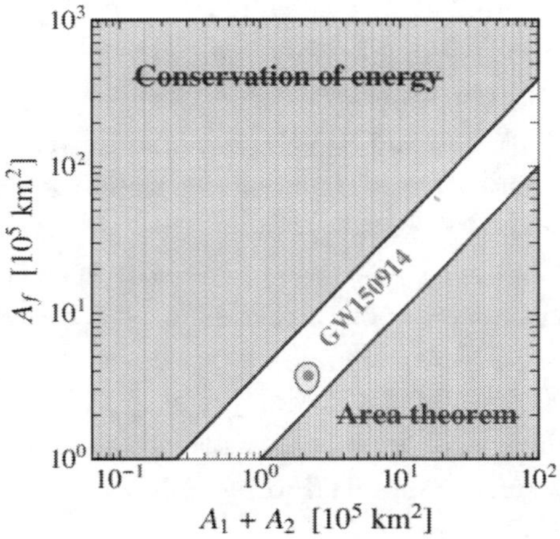

Figure 5. *In the plane of the horizon areas of initial ($A_1 + A_2$) and final (A_f) BH binary mergers, the dashed regions indicate where Hawking's area theorem is violated [$A_f < A_1 + A_2$] and where energy conservation does not hold [$A_f > 4(A_1 + A_2)$]. The result from GW150914 corresponds to the green area (at 90% CL) and lies within the allowed region (white band). The correlation between the extraction of A_f and $A_1 + A_2$ is not taken into account.*

proving the theorem is that in GR the null energy condition – that is $T_{ab}k^a k^b \geqslant 0$ for any null vectors k^a – implies, as a direct consequence of Einstein equations ($R_{ab} - g_{ab}R/2 = 8\pi G_N T_{ab}$), the null convergence condition $R_{ab}k^a k^b \geqslant 0$. In turn, the null convergence condition ensures, by means of the Raychaudhuri equation describing geodesics congruence, the validity of the area theorem. The proof is not valid in extended gravity theories, when the field equations differ from those of GR. This is the case of the Brans-Dicke theory (and, more generally, scalar-tensor gravity) [44, 45], in which Einstein equations are modified by new terms beyond the energy-momentum tensor involving the scalar field.

As a consequence, the null energy condition does not imply anymore the null convergence condition, thus invalidating Hawking's argument. Even if static asymptotic BH must recover GR solutions (the no-hair theorem), during the dynamical stages of the merging process the validity of the area theorem is not guaranteed. In terms of energy balance, this corresponds to the emission – in addition to tensor GW perturbations – of the extra scalar degrees of freedom.

In conclusion, the search for violations of the area theorem can be used as a tool to confirm potential ECO discoveries and to derive information on undetected forms of radiation. Furthermore, testing the area theorem is an interesting task independently of searches for ECOs, as it offers a probe of fundamental principles.

References

[1] B. P. Abbott *et al.* [LIGO Scientific and Virgo Collaborations], Phys. Rev. Lett. **116** (2016) 6, 061102.

[2] E. O. Kahya and S. Desai, arXiv:1602.04779 [gr-qc].

[3] X. F. Wu *et al.*, arXiv:1602.01566 [astro-ph.HE].

[4] N. Yunes, K. Yagi and F. Pretorius, arXiv:1603.08955 [gr-qc].

[5] M. Liu, Z. Zhao, X. You, J. Lu and L. Xu, arXiv:1604.06668 [gr-qc].

[6] R. Konoplya and A. Zhidenko, arXiv:1602.04738 [gr-qc].

[7] J. W. Moffat, arXiv:1603.05225 [gr-qc].
[8] J. Vainio and I. Vilja, arXiv:1603.09551 [astro-ph.CO].
[9] V. Cardoso, E. Franzin and P. Pani, Phys. Rev. Lett. **116**, 171101 (2016).
[10] S. B. Giddings, arXiv:1602.03622 [gr-qc].
[11] M. Sasaki, T. Suyama, T. Tanaka and S. Yokoyama, arXiv:1603.08338 [astro-ph.CO].
[12] S. Bird, I. Cholis, J. B. Muñoz, Y. Ali-Haïmoud, M. Kamionkowski, E. D. Kovetz, A. Raccanelli and A. G. Riess, arXiv:1603.00464 [astro-ph.CO].
[13] S. Clesse and J. García-Bellido, arXiv:1603.05234 [astro-ph.CO].
[14] T. E. Collett and D. Bacon, arXiv:1602.05882 [astro-ph.HE].
[15] P. Bicudo, arXiv:1602.04337 [gr-qc].
[16] D. Blas, M. M. Ivanov, I. Sawicki and S. Sibiryakov, arXiv:1602.04188 [gr-qc].
[17] E. Calabrese, N. Battaglia and D. N. Spergel, arXiv:1602.03883 [gr-qc].
[18] J. García-Bellido, S. Nesseris and M. Trashorras, arXiv:1603.05616 [astro-ph.CO].
[19] M. Schreck, arXiv:1603.07452 [gr-qc].
[20] M. Arzano and G. Calcagni, arXiv:1604.00541 [gr-qc].
[21] V. Branchina and M. De Domenico, arXiv:1604.08530 [gr-qc].
[22] P. S. B. Dev and A. Mazumdar, arXiv:1602.04203 [hep-ph].
[23] J. Jaeckel, V. V. Khoze and M. Spannowsky, arXiv:1602.03901 [hep-ph].
[24] K. Langaeble, A. Meroni and F. Sannino, arXiv:1603.00230 [hep-ph].
[25] L. Lehner, S. L. Liebling, C. Palenzuela, O. L. Caballero, E. O'Connor, M. Anderson and D. Neilsen, arXiv:1603.00501 [gr-qc].
[26] A. Arvanitaki, M. Baryakhtar, S. Dimopoulos, S. Dubovsky and R. Lasenby, arXiv:1604.03958 [hep-ph].
[27] G. F. Giudice, M. McCullough and A. Urbano, JCAP **1610** (2016) no.10, 001.
[28] P. Ajith *et al.*, Phys. Rev. Lett. **106** (2011) 241101.
[29] D. Lai and A. G. Wiseman, Phys. Rev. D **54** (1996) 3958.
[30] P. Ajith, Phys. Rev. D **84**, 084037 (2011).
[31] M. Dominik *et al.*, Astrophys. J. **806**, no. 2, 263 (2015).
[32] S. Khan, S. Husa, M. Hannam, F. Ohme, M. Pürrer, X. Jimènez Forteza and A. Bohè, Phys. Rev. D **93**, no. 4, 044007 (2016).
[33] B. P. Abbott *et al.* [LIGO Scientific and Virgo Collaborations], arXiv:1602.03840 [gr-qc].
[34] S. W. Hawking, Phys. Rev. Lett. **26** (1971) 1344.
[35] S. W. Hawking, Nature **248** (1974) 30.
[36] S. W. Hawking, Commun. Math. Phys. **43** (1975) 199 [Commun. Math. Phys. **46** (1976) 206].
[37] J. D. Bekenstein, Lett. Nuovo Cim. **4** (1972) 737.
[38] J. D. Bekenstein, Phys. Rev. D **7** (1973) 2333.
[39] J. D. Bekenstein, Phys. Rev. D **9** (1974) 3292.
[40] T. Jacobson, G. Kang and R. C. Myers, Phys. Rev. D **49** (1994) 6587.
[41] J. D. Bekenstein, gr-qc/9409015.
[42] T. Jacobson, G. Kang and R. C. Myers, Phys. Rev. D **52** (1995) 3518.
[43] R. Penrose, Riv. Nuovo Cim. **1** (1969) 252 [Gen. Rel. Grav. **34** (2002) 1141].
[44] G. Kang, Phys. Rev. D **54**, 7483 (1996).
[45] V. Faraoni, Entropy **12**, 1246 (2010).

EPJ Web of Conferences **164**, 02005 (2017) DOI: 10.1051/epjconf/201716402005
ICNFP 2016

QCD evolution equations at small x: Pomerons, Odderons and more

Jamal Jalilian-Marian[1,2,a]

[1]*Natural Sciences Department, Baruch College, 17 Lexington Ave., New York, NY 10010, USA*
[2]*CUNY Graduate Center, 365 Fifth Ave., New York, NY 10016, USA*

Abstract. We give an overview of QCD at small x, and the equation that governs evolution of observables with energy (or x), knows as the JIMWLK equation. We show that it reduces, in the appropriate limits, to all the known perturbative evolution equations at small x.

1 Introduction

Quantum ChromoDynamics (QCD) is an extremely rich and at the same time complicated theory of strong interactions between quarks and gluons, the colored fundamental constituents of matter. Due to its complexity, often one is forced to resort to approximations in order to perform theoretical calculations of physical observables. One such approximation is perturbation theory where one expands the expressions for the observables in a power series in the coupling constant. Due to the non-Abelian nature of QCD and asymptotic freedom, the running coupling constant α_s becomes small at high momentum transfer p_t. Perturbative QCD (pQCD) is a systematic approach to particle production in QCD at high p_t. The justification for applying pQCD to scattering and production of strongly interacting particles lies on the proof and use of collinear factorization theorems which guarantee a clean separation of short and long distance physics effects. In the collinear factorization formalism, particle production cross sections are written as a convolution of parton distribution functions xf, non-perturbative yet universal quantities describing the momentum distribution of partons in a nucleon, of the incoming hadrons/nuclei with the hard scattering matrix elements $d\sigma^{parton}$ computable in pQCD to any order in the strong coupling α_s. This can symbolically written as

$$d\,\sigma^{hadron} \sim xf_1 \otimes xf_2 \otimes d\,\sigma^{parton} \otimes zD \tag{1}$$

where fragmentation function zD describes hadronization; how a colored parton (quark or gluon) turns into a colorless hadron. The fact that these parton distributions are universal (process independent) means they can be measured in one process, say Deep Inelastic Scattering (DIS) of electrons on a nucleon, and used in the above formula to predict the cross section for particle production in other processes. This universality of parton distribution functions is responsible for the predictive power of pQCD. Experimentally observed spectrum of produced particle in various experiments are in excellent

[a]e-mail: jamal.jalilian-marian@baruch.cuny.edu

DOI: 10.1051/epjconf/201716402005

agreement with predictions of QCD based on collinear factorization over many orders of transverse momenta.

The parton distribution functions xf which are an essential ingredient in a collinear factorization theorem depend on and evolve with scale Q^2 as described by the DGLAP evolutions equations (see [1] for a review). This "evolution" is due to radiation of extra partons as allowed by the available kinematic phase space in Q^2. As quarks can radiate gluons and gluons radiate other gluons (and quark anti-quark pairs), the DGLAP equation is a coupled differential-integral equation which can symbolically written as

$$\frac{d}{d\log Q^2}\begin{pmatrix} xq \\ xG \end{pmatrix} = \begin{pmatrix} P_{qq} & P_{qg} \\ P_{gq} & P_{gg} \end{pmatrix} \otimes \begin{pmatrix} xq \\ xG \end{pmatrix} \tag{2}$$

where $\otimes$ stands for integration over momentum fraction x and $P_{qq}, \cdots$ are the so-called splitting functions describing the probability for a parton to radiate another one and are calculated in pQCD to any order in the coupling constant. As this is a differential equation one needs initial conditions for the quark and gluon distribution functions xq, xg at some initial scale Q_0^2 (and at all x), typically taken to be of order of $1\, GeV^2$, after which one can solve the DGLAP equation (2) to obtain the distribution functions $xq(x, Q^2), xg(x, Q^2)$ which can then be used in the production cross section as described in the collinear factorization approach.

DGLAP evolution equation predicts a sharp rise in the (sea) quark and gluon distribution functions as the Q^2 or $1/x$ increases. There are more and more partons in the wave unction of a hadron as Q^2 increases, however each parton has an effective size of $1/Q^2$ so that partons which are radiated are occupying smaller and smaller areas. The number of prtons also increases when one increases $1/x$ as can be seen by noting that $P_{gg} \to \frac{1}{x}$ as $x \to 0$. However here there is a profound difference; at fixed Q^2, as one decreases x (the fraction of the hadron energy carried by a parton) more and more partons are radiated into the same effective area and will start to "overlap". This would lead to an interaction among these partons which are treated as quasi-free in the DGLAP formalism and lead to its breakdown. This untamed growth of parton distribution function would also lead to an unbounded growth of perturbative ross section and an eventual violation of Froissart (unitary) bound. Therefore one expects new QCD dynamics to become manifest when probing a hadronic/nuclear wave function small x.

This idea that gluons at small x will start to overlap and that the DGLAP equation would need to be modified was first pointed out by Gribov, Levin and Ryskin (GLR) (see [2] for reviews and comprehensive references) who pointed out that there should be a saturation of gluon density at small enough x such that the growth of perturbative cross sections are tamed. Based on this idea Mueller and Qiu calculated the Feynman diagrams leading to gluon saturation at small x and derived a new evolution equation for gluon distribution function which can be written symbolically as

$$\frac{d}{d\log Q^2}\, xG(x, Q^2) \sim xG(x, Q^2) - xG^2(x, Q^2) \tag{3}$$

The second term is due to "recombination" of gluons and comes with a negative sign; it will lead to a slowdown and eventual saturation of the gluon distribution function. Later McLerran and Venugopalan [3, 4] proposed an effective action approach to gluon saturation, they introduced a new scale, the saturation momentum Q_s and argued that this scale can be much larger than the genuinely non-perturbative scale in QCD, Λ_{QCD} at small x and/or for large nuclei. Therefore, they reasoned that one could apply weak coupling techniques to address gluon saturation even though one deals with a many-body system of interacting gluons. They realized that the relevant time scales are very different for small and large x degrees of freedom. Utilizing this they introduced the notion of color charge density

ρ^a to represent the color charges at large x which in turn generate a classical color field describing the small x degrees of freedom. As one is interested in the small x limit, the coupling between the color charges ρ^a and the classical field is taken to be eikonal.

The McLerran-Venugopalan (MV) effective action can be written as

$$S \equiv -\frac{1}{4}\int d^4x\, F^{a\,2}_{\mu\nu}(x) + \frac{i}{N_c}\int d^2x dx^-\, Tr\rho(x^-,x_t)W_{-\infty,\infty}[A^-](x^-,x_t) \\ +i\int d^2x\, dx^- F[\rho(x_t,x^-)] \tag{4}$$

where

$$W_{-\infty,\infty}[A^-](x^-,x_t) \equiv Pexp\Big[-ig\int_{-\infty}^{\infty} dx^+ A^-_a(x^+,x^-,x_t)\,T_a\Big] \tag{5}$$

and

$$\exp\Big[i\Big\{i\int d^2x dx^- F[\rho(x_t,x^-)]\Big\}\Big] \tag{6}$$

can be thought as the statistical weight of a given color charge configuration. In the original MV model, F was taken to be a Gaussian, given by

$$\int dx^-\; F[\rho(x_t,x^-)] = \frac{1}{2\mu^2}\rho^2(x_t) \tag{7}$$

In the MV effective action approach one first solves the classical equation of motion for a given configuration of color current $J^\mu_a(x) = \delta^{\mu\,+}\,\delta(x^-)\,\rho_a$

$$D_\mu\, F^{\mu\,\nu}_a = J^\nu_a \tag{8}$$

Given $A^\mu[\rho]$, the solution of the classical equation of motion (8) for a given color charge configuration one then averages over all color charges using the statistical weight given in (6) in order to compute a physical observable which typically involves correlations of the gluon field (or Wilson lines involving the gluon fields).

The procedure described above is valid at the classical level and needs to be augmented with quantum loop effects. In the kinematics of interest (high energy) this corresponds to a resummation of loop diagrams which are divergent in the limit $x \longrightarrow 0$. This can be physically understood as including more of the small x degrees of freedom into our definition of color charges at large x which leads to a renormalization of the color charge ρ, or more accurately, a renormalization of the weight functional describing the distribution of color charges ρ. This resummation of small x effects can be done at the effective action level, and consequently for the x evolution of any correlation function computed using this effective action. The resulting equation is known as the JIMWLK evolution equation [5] which is written as

$$\frac{\partial}{d\log\, 1/x} < O >_x = < H\, O >_x \tag{9}$$

where the Hamiltonian H is

$$H = -\frac{1}{2(2\pi)^3}\int d^2x_t\, d^2y_t\, d^2z_t\, M_{xyz}\Big(1 + U^\dagger_x U_y - U^\dagger_x U_z - U^\dagger_z U_y\Big)^{ab}\frac{\delta}{\delta\alpha^a_x}\frac{\delta}{\delta\alpha^b_y}, \tag{10}$$

 DOI: 10.1051/epjconf/201716402005

Here x_t, y_t, z_t are two-dimensional vectors on the transverse plane and a and b are color indices. The dipole kernel M_{xyz} is defined as

$$M_{xyz} \equiv \frac{(x_t - y_t)^2}{(x_t - z_t)^2(z_t - y_t)^2} \tag{11}$$

The Wilson line U_z is in the adjoint representation and is given by

$$U_z \equiv e^{-ig\,\alpha^a(z_t)\,t^a}, \tag{12}$$

where $\alpha^a(z_t)$ is the classical solution. Given the JIMWLK equation (9) one can derive evolution equations for any correlation function in the theory. The most ubiquitous one is the so called dipole cross section, defined as the normalized trace of Wilson lines (in the fundamental representation). It is

$$\frac{d}{dy}\langle \mathrm{tr} V_r^\dagger V_{\bar{r}}\rangle = -\frac{N_c\,\alpha_s}{2\pi^2}\int d^2z\,\frac{(r-\bar{r})^2}{(r-z)^2(\bar{r}-z)^2}\left\langle \mathrm{tr} V_r^\dagger V_{\bar{r}} - \frac{1}{N_c}\mathrm{tr} V_r^\dagger V_z\,\mathrm{tr}\, V_{\bar{r}}\, V_z^\dagger\right\rangle . \tag{13}$$

with $y \equiv \log 1/x$. This equation reduced to the BK equation [6] after making a Gaussian approximation after which it can be written as

$$\frac{d}{dy}S(r-s) = -\frac{N_c\,\alpha_s}{2\pi^2}\int d^2z\,\frac{(r-s)^2}{(r-z)^2(s-z)^2}\left[S(r-s) - S(r-z)\,S(z-s)\right] \tag{14}$$

where the S matrix is defined as

$$\langle S(r-\bar{r})\rangle \equiv \frac{1}{N_c}\,\langle \mathrm{tr} V_r\, V_{\bar{r}}^\dagger\rangle \tag{15}$$

BK equation is the simplest non-linear equation which contains the dynamics of gluon saturation. Expanded in powers of the gluon field $g\,A^\mu$ and keeping the first non-trivial term it reduces to the famous BFKL equation [7] for the evolution of gluonic two-point function. The BFKL equation sums the ladder diagrams where the rapidity (x) variable is ordered along the ladder. This is the so-called (perturbative or BFKL) pomeron which corresponds to the state of two Reggeized gluons exchanged in the t channel. Using the JIMWLK equation (9) one can also derive a evolution equation for the state of 3 or more Reggeized gluons where state of 3 Reggeized gluons in t channel is given its own name and is usually referred to as a odderon. For exchange of 4 or more Reggeized gluons one can show that defining the T matrix for the $2n$-point function

$$T^{(2n)}(x_1, x_2, \cdots x_{2n}) = 1 - \frac{1}{N_c}Tr\left(V_{x_1}V_{x_2}^\dagger\, V_{x_3}V_{x_4}^\dagger \cdots V_{x_{2n-1}}V_{x_{2n}}^\dagger\right) \tag{16}$$

one gets (in the linear region where all Wilson lines are expanded) [8]

$$\begin{aligned}
\frac{d}{dy}T^{(2n)}(l_1, l_2, \cdots l_{2n-1}, l_{2n}) = &-\frac{\bar{\alpha}}{2\pi}\sum_{j=1}^{2n}\int d^2p_t\left[\frac{l_j^2}{p_t^2\,[p_t^2 + (p_t - l_j)^2]}\right]T^{(2n)}_{(\prod_{k=1}^{2n} l_k)}\\
&+\frac{\bar{\alpha}}{4\pi}\int d^2p_t\left[\frac{l_1^2}{p_t^2\,(p_t+l_1)^2} + \frac{l_{2n}^2}{p_t^2\,(p_t-l_{2n})^2} - \frac{(l_1+l_{2n})^2}{(p_t+l_1)^2\,(p_t-l_{2n})^2}\right]T^{(2n)}_{(l_1+p_t)\,(\prod_{k=2}^{2n-1} l_k)\,(l_{2n}-p_t)}\\
&+\frac{\bar{\alpha}}{4\pi}\sum_{j=2}^{2n}\int d^2p_t\left[\frac{l_{j-1}^2}{p_t^2(p_t+l_{j-1})^2} + \frac{l_j^2}{p_t^2(p_t-l_j)^2} - \frac{(l_{j-1}+l_j)^2}{(p_t+l_{j-1})^2(p_t-l_j)^2}\right]T^{(2n)}_{(\prod_{k=1}^{j-2} l_k)(l_{j-1}+p_t)(l_j-p_t)(\prod_{k=j+1}^{2n} l_k)}
\end{aligned} \tag{17}$$

Eq. (17) describes evolution of a state of $2n$ Reggeized gluons exchanged in the t channel and is identical to the BJKP equation [9]. We *emphasize* that the JIMWLK equation for the $2n$-point function in general includes non-linear terms; here we have ignored the non-linearities of the JIMWLK equation to demonstrate that it includes the BJKP equation as its linear limit (in the region where the gluon field strength is not large). The full non-linear evolution equation for $2n$-point function can be written as [8]

$$H\,\hat{T}^{(2n)} = \frac{\bar{\alpha}}{4\pi}\int d^2 z_t \Big[\, H_1\,\hat{T}^{(2n)} + H_2\,\hat{T}^{(2n)} - H_3\,\hat{T}^{(2n)} - H_4\,\hat{T}^{(2n)} \Big] \tag{18}$$

where $H_{1,2,3,4}$ act on $\hat{T}^{(2n)}$ as

$$\begin{aligned} H_1\,\hat{S}^{(2n)} = &-\sum_{i=1}^{n}\,[M_{x_{2i-1}x_{2i}z}]\,\hat{S}^{(2n)}_{(\prod_{k=1}^{2n} x_k)} \\ &+ \sum_{j=2\,;\,j>i}^{n}\,\sum_{i=1}^{n-1}\,[\,M_{x_{2i-1}x_{2j-1}z} - M_{x_{2i-1}x_{2j}z} - M_{x_{2i}x_{2j-1}z} + M_{x_{2i}x_{2j}z}\,]\,\hat{S}^{(2j-2i)}_{(\prod_{k=2i}^{2j-1} x_k)}\hat{S}^{(2n-2j+2i)}_{(\prod_{k=1}^{2i-1} x_k)(\prod_{k=2j}^{2n} x_k)} \end{aligned}$$

and

$$\begin{aligned} H_2\,\hat{S}^{(2n)} \;=\; &-\sum_{i=1}^{n}\,[\,M_{x_{2i-1}x_{2i}z}\,]\,\hat{S}^{(2)}_{(x_{2i-1}x_{2i})}\hat{S}^{(2n-2)}_{(\prod_{k=1}^{2i-2} x_k)(\prod_{k=2i+1}^{2n} x_k)} \\ &+ \sum_{j=2\,;\,j>i}^{n}\,\sum_{i=1}^{n-1}[\,M_{x_{2i-1}x_{2j-1}z}\,]\,\hat{S}^{(2j-2i)}_{(\prod_{k=2i-1}^{2j-2} x_k)}\hat{S}^{(2n-2j+2i)}_{(\prod_{k=1}^{2i-2} x_k)(\prod_{k=2j-1}^{2n} x_k)} \\ &- \sum_{j=2\,;\,j>i}^{n}\,\sum_{i=1}^{n-1}[\,M_{x_{2i-1}x_{2j}z}\,]\,\hat{S}^{(2j-2i+2)}_{(\prod_{k=2i-1}^{2j} x_k)}\hat{S}^{(2n-2j+2i-2)}_{(\prod_{k=1}^{2i-2} x_k)(\prod_{k=2j+1}^{2n} x_k)} \\ &- \sum_{j=2\,;\,j>i}^{n}\,\sum_{i=1}^{n-1}[\,M_{x_{2i}x_{2j-1}z}\,]\,\hat{S}^{(2j-2i-2)}_{(\prod_{k=2i+1}^{2j-2} x_k)}\hat{S}^{(2n-2j+2i+2)}_{(\prod_{k=1}^{2i} x_k)(\prod_{k=2j-1}^{2n} x_k)} \\ &+ \sum_{j=2\,;\,j>i}^{n}\,\sum_{i=1}^{n-1}[\,M_{x_{2i}x_{2j}z}\,]\,\hat{S}^{(2j-2i)}_{(\prod_{k=2i+1}^{2j} x_k)}\hat{S}^{(2n-2j+2i)}_{(\prod_{k=1}^{2i} x_k)(\prod_{k=2j+1}^{2n} x_k)} \end{aligned}$$

and

$$
\begin{aligned}
H_3\,\hat{S}^{(2n)} \;=\;& -\sum_{i=1}^{n} \left[M_{x_{2i-1}x_{2i}z} \right] \frac{1}{2}\hat{S}^{(2)}_{z\,x_{2i-1}} \hat{S}^{(2n)}_{(\prod_{k=1}^{2i-2} x_k)\, z\, (\prod_{k=2i}^{2n} x_k)} \\
& -\sum_{i=1}^{n} \left[M_{x_{2i-1}x_{2i}z} \right] \frac{1}{2}\hat{S}^{(2)}_{z\,x_{2i}} \hat{S}^{(2n)}_{(\prod_{k=1}^{2i-1} x_k)\, z\, (\prod_{k=2i+1}^{2n} x_k)} \\
& + \sum_{j=2\,;\,j>i}^{n} \sum_{i=1}^{n-1} \left[M_{x_{2i-1}x_{2j-1}z} - M_{x_{2i-1}x_{2j}z} \right] \frac{1}{2}\hat{S}^{(2j-2i+2)}_{z\left(\prod_{k=2i-1}^{2j-1} x_k\right)} \hat{S}^{(2n-2j+2i)}_{(\prod_{k=1}^{2i-2} x_k)\, z\left(\prod_{k=2j}^{2n} x_k\right)} \\
& + \sum_{j=2\,;\,j>i}^{n} \sum_{i=1}^{n-1} \left[- M_{x_{2i}x_{2j-1}z} + M_{x_{2i}x_{2j}z} \right] \frac{1}{2}\hat{S}^{(2j-2i)}_{z\left(\prod_{k=2i+1}^{2j-1} x_k\right)} \hat{S}^{(2n-2j+2i+2)}_{(\prod_{k=1}^{2i} x_k)\, z\left(\prod_{k=2j}^{2n} x_k\right)} \\
& + \sum_{j=2\,;\,j>i}^{n} \sum_{i=1}^{n-1} \left[M_{x_{2j-1}x_{2i-1}z} - M_{x_{2j-1}x_{2i}z} \right] \frac{1}{2}\hat{S}^{(2j-2i)}_{z\left(\prod_{k=2i}^{2j-2} x_k\right)} \hat{S}^{(2n-2j+2i+2)}_{(\prod_{k=1}^{2i-1} x_k)\, z\left(\prod_{k=2j-1}^{2n} x_k\right)} \\
& + \sum_{j=2\,;\,j>i}^{n} \sum_{i=1}^{n-1} \left[- M_{x_{2j}x_{2i-1}z} + M_{x_{2j}x_{2i}z} \right] \frac{1}{2}\hat{S}^{(2j-2i+2)}_{z\left(\prod_{k=2i}^{2j} x_k\right)} \hat{S}^{(2n-2j+2i)}_{(\prod_{k=1}^{2i-1} x_k)\, z\left(\prod_{k=2j+1}^{2n} x_k\right)}
\end{aligned}
$$

and

$$
\begin{aligned}
H_4\,\hat{S}^{(2n)} \;=\;& -\sum_{i=1}^{n} \left[M_{x_{2i-1}x_{2i}z} \right] \frac{1}{2}\hat{S}^{(2)}_{z\,x_{2i}} \hat{S}^{(2n)}_{(\prod_{k=1}^{2i-1} x_k)\, z\, (\prod_{k=2i+1}^{2n} x_k)} \\
& -\sum_{i=1}^{n} \left[M_{x_{2i-1}x_{2i}z} \right] \frac{1}{2}\hat{S}^{(2)}_{z\,x_{2i-1}} \hat{S}^{(2n)}_{(\prod_{k=1}^{2i-2} x_k)\, z\, (\prod_{k=2i}^{2n} x_k)} \\
& + \sum_{j=2\,;\,j>i}^{n} \sum_{i=1}^{n-1} \left[M_{x_{2i-1}x_{2j-1}z} - M_{x_{2i}x_{2j-1}z} \right] \frac{1}{2}\hat{S}^{(2j-2i)}_{z\left(\prod_{k=2i}^{2j-2} x_k\right)} \hat{S}^{(2n-2j+2i+2)}_{(\prod_{k=1}^{2i-1} x_k)\, z\left(\prod_{k=2j-1}^{2n} x_k\right)} \\
& + \sum_{j=2\,;\,j>i}^{n} \sum_{i=1}^{n-1} \left[- M_{x_{2i-1}x_{2j}z} + M_{x_{2i}x_{2j}z} \right] \frac{1}{2}\hat{S}^{(2j-2i+2)}_{z\left(\prod_{k=2i}^{2j} x_k\right)} \hat{S}^{(2n-2j+2i)}_{(\prod_{k=1}^{2i-1} x_k)\, z\left(\prod_{k=2j+1}^{2n} x_k\right)} \\
& + \sum_{j=2\,;\,j>i}^{n} \sum_{i=1}^{n-1} \left[M_{x_{2j-1}x_{2i-1}z} - M_{x_{2j}x_{2i-1}z} \right] \frac{1}{2}\hat{S}^{(2j-2i+2)}_{z\left(\prod_{k=2i-1}^{2j-1} x_k\right)} \hat{S}^{(2n-2j+2i)}_{(\prod_{k=1}^{2i-2} x_k)\, z\left(\prod_{k=2j}^{2n} x_k\right)} \\
& + \sum_{j=2\,;\,j>i}^{n} \sum_{i=1}^{n-1} \left[- M_{x_{2j-1}x_{2i}z} + M_{x_{2j}x_{2i}z} \right] \frac{1}{2}\hat{S}^{(2j-2i)}_{z\left(\prod_{k=2i+1}^{2j-1} x_k\right)} \hat{S}^{(2n-2j+2i+2)}_{(\prod_{k=1}^{2i} x_k)\, z\left(\prod_{k=2j}^{2n} x_k\right)}
\end{aligned}
$$

It is not possible, in principle, to solve these equations analytically. Therefore either one has to make further approximations or resort to numerical techniques. The most common approximation being the Gaussian truncation which allows one to write all higher point functions in terms of the two point function S which satisfies the BK equation,

$$S^{(2n)} \simeq S^{n} \tag{19}$$

where $S^{(2n)} = 1 - T^{(2n)}$ and $T^{(2n)}$ is defined in (16). The dipole cross section $S(r_t = x_t - y_t)$ goes to zero as $r_t \to 0$ (color transparency) and approaches 1 as $r_t \to \infty$ (black disk limit). The dipole cross section is the ubiquitous quantity appearing in the DIS structure functions F_2, F_L and in single inclusive particle production in the forward rapidity region of hadron-hadron (nucleus) collisions [10]. Higher point functions such as the Quadrupole [11], normalized trace of 4 Wilson lines, on the other hand appear in less inclusive quantities such as di-hadron production in DIS and in proton-nucleus collisions. The evolution of Quadrupole has been studied numerically and has been shown to grow with the same power of energy (x) and exhibit geometric scaling as the dipole cross section does. The small x evolution equations are not closed form equations; evolution of a $2n$-point function involves higher point functions [12]; however it can be shown that with leading N_c accuracy they involve only dipoles and quadrupoles. All other higher point functions are suppressed bu powers of N_c.

In summary, the effective action approach to QCD at small x has enabled one to study the dynamics of QCD at small x systematically and in a theoretically controlled way. Using this effective action a non-linear evolution equation known as the JIMWLK equation has been derived which describes the x (or equivalently energy or rapidity) dependence of $2n$-point functions of Wilson lines which are the ingredients of particle production cross sections in high energy hadronic collisions. It has been shown that the JIMWLK equation for $2n$-point function of Wilson lines reduces to the BFKL and BJKP equations in the linear (dilute) regime. While there has been recent progress in computing higher order corrections to the JIMWLK equation, yet the ultimate goal, to include the physics of DGLAP evolution, remains to be realized.

References

[1] H. L. Lai, J. Botts, J. Huston, J. G. Morfin, J. F. Owens, J. w. Qiu, W. K. Tung and H. Weerts, Phys. Rev. D **51**, 4763 (1995) doi:10.1103/PhysRevD.51.4763.

[2] H. Weigert, Prog. Part. Nucl. Phys. **55**, 461 (2005); F. Gelis, E. Iancu, J. Jalilian-Marian, R. Venugopalan, Ann. Rev. Nucl. Part. Sci. **60**, 463-489 (2010). [arXiv:1002.0333 [hep-ph]]; J. Jalilian-Marian, Y. V. Kovchegov, Prog. Part. Nucl. Phys. **56**, 104-231 (2006); J. L. Albacete and C. Marquet, Prog. Part. Nucl. Phys. **76**, 1 (2014).

[3] L. D. McLerran and R. Venugopalan, Phys. Rev. D **49**, 2233 (1994), Phys. Rev. D **49**, 3352 (1994); J. Jalilian-Marian, A. Kovner, L. D. McLerran and H. Weigert, Phys. Rev. D **55**, 5414 (1997).

[4] S. Jeon and R. Venugopalan, Phys. Rev. D **71**, 125003 (2005); A. Dumitru, J. Jalilian-Marian, E. Petreska, Phys. Rev. **D84**, 014018 (2011); J. Jalilian-Marian, S. Jeon and R. Venugopalan, Phys. Rev. D **63**, 036004 (2001); A. Ayala, J. Jalilian-Marian, L. D. McLerran and R. Venugopalan, Phys. Rev. D **52**, 2935 (1995) doi:10.1103/PhysRevD.52.2935.

[5] J. Jalilian-Marian, A. Kovner, L. D. McLerran and H. Weigert, Phys. Rev. D **55**, 5414 (1997); J. Jalilian-Marian, A. Kovner, A. Leonidov and H. Weigert, Nucl. Phys. B **504**, 415 (1997), Phys. Rev. D **59**, 014014 (1999), Phys. Rev. D **59**, 014015 (1999), Phys. Rev. D **59**, 034007 (1999), A. Kovner, J. G. Milhano and H. Weigert, Phys. Rev. D **62**, 114005 (2000); A. Kovner and J. G. Milhano, Phys. Rev. D **61**, 014012 (2000); E. Iancu, A. Leonidov and L. D. McLerran, Nucl. Phys. A **692**, 583 (2001), Phys. Lett. B **510**, 133 (2001); E. Ferreiro, E. Iancu, A. Leonidov and L. McLerran, Nucl. Phys. A **703**, 489 (2002).

[6] I. Balitsky, Nucl. Phys. B **463**, 99 (1996); Y. V. Kovchegov, Phys. Rev. D **60**, 034008 (1999); Phys. Rev. D **61**, 074018 (2000).

[7] L. N. Lipatov, Sov. J. Nucl. Phys. **23**, 338 (1976) [Yad. Fiz. **23**, 642 (1976)]; E. A. Kuraev, L. N. Lipatov and V. S. Fadin, Sov. Phys. JETP **45**, 199 (1977) [Zh. Eksp. Teor. Fiz. **72**, 377 (1977)]; I. I. Balitsky and L. N. Lipatov, Sov. J. Nucl. Phys. **28**, 822 (1978) [Yad. Fiz. **28**, 1597 (1978)].

[8] A. Ayala, E. R. Cazaroto, L. A. Hernández, J. Jalilian-Marian and M. E. Tejeda-Yeomans, Phys. Rev. D **90**, no. 7, 074037 (2014) doi:10.1103/PhysRevD.90.074037; J. Jalilian-Marian, Phys. Rev. D **85**, 014037 (2012) doi:10.1103/PhysRevD.85.014037.

[9] J. Bartels, Nucl. Phys. B **175**, 365 (1980); T. Jaroszewicz, Acta Phys. Polon. B **11**, 965 (1980); J. Kwiecinski and M. Praszalowicz, Phys. Lett. B **94**, 413 (1980); Z. Chen, A. H. Mueller, Nucl. Phys. **B451**, 579-604 (1995).

[10] A. Dumitru and J. Jalilian-Marian, Phys. Rev. Lett. **89**, 022301 (2002); D. Kharzeev, Y. V. Kovchegov and K. Tuchin, Phys. Rev. D **68**, 094013 (2003); J. L. Albacete, N. Armesto, A. Kovner, C. A. Salgado and U. A. Wiedemann, Phys. Rev. Lett. **92**, 082001 (2004); J. Jalilian-Marian, Y. Nara and R. Venugopalan, Phys. Lett. B **577**, 54 (2003) doi:10.1016/j.physletb.2003.09.097; J. Jalilian-Marian, J. Phys. G **30**, S751 (2004), Nucl. Phys. A **748**, 664 (2005), J. Jalilian-Marian, Nucl. Phys. A **753**, 307 (2005) doi:10.1016/j.nuclphysa.2005.02.156, Phys. Rev. C **70**, 027902 (2004), Nucl. Phys. A **739**, 319 (2004); A. Dumitru, A. Hayashigaki and J. Jalilian-Marian, Nucl. Phys. A **765**, 464 (2006), Nucl. Phys. A **770**, 57 (2006); E. M. Henley and J. Jalilian-Marian, Phys. Rev. D **73**, 094004 (2006) doi:10.1103/PhysRevD.73.094004;
F. Gelis and J. Jalilian-Marian, Phys. Rev. D **76**, 074015 (2007), Phys. Rev. D **67**, 074019 (2003) doi:10.1103/PhysRevD.67.074019, Phys. Rev. D **66**, 094014 (2002) doi:10.1103/PhysRevD.66.094014, Phys. Rev. D **66**, 014021 (2002) doi:10.1103/PhysRevD.66.014021;
D. Boer, A. Utermann and E. Wessels, Phys. Rev. D **77**, 054014 (2008); E. R. Cazaroto, V. P. Gonçalves, F. S. Navarra, Nucl. Phys. A **872**, 196 (2011); J. Jalilian-Marian and A. H. Rezaeian, Phys. Rev. D **85**, 014017 (2012), J. Jalilian-Marian and A. H. Rezaeian, Phys. Rev. D **86**, 034016 (2012); A. H. Rezaeian, Phys. Rev. D **86**, 094016 (2012).

[11] J. Jalilian-Marian and Y. V. Kovchegov, Phys. Rev. D **70**, 114017 (2004) [Erratum-ibid. D **71**, 079901 (2005)]; A. Kovner and M. Lublinsky, JHEP **0611**, 083 (2006), Phys. Rev. D **84**, 094011 (2011), Phys. Rev. D **83**, 034017 (2011); J. Jalilian-Marian, Nucl. Phys. A **770**, 210 (2006); C. Marquet, Nucl. Phys. A **796**, 41 (2007); J. L. Albacete and C. Marquet, Phys. Rev. Lett. **105**, 162301 (2010).

[12] Y. V. Kovchegov, J. Kuokkanen, K. Rummukainen and H. Weigert, Nucl. Phys. A **823**, 47 (2009); C. Marquet and H. Weigert, Nucl. Phys. A **843**, 68 (2010); A. Dumitru and J. Jalilian-Marian, Phys. Rev. D **82**, 074023 (2010), Phys. Rev. D **81**, 094015 (2010); A. Dumitru, J. Jalilian-Marian, T. Lappi, B. Schenke and R. Venugopalan, Phys. Lett. B **706**, 219 (2011); A. Dumitru, K. Dusling, F. Gelis, J. Jalilian-Marian, T. Lappi and R. Venugopalan, Phys. Lett. B **697**, 21 (2011) doi:10.1016/j.physletb.2011.01.024; A. Ayala, M. Hentschinski, J. Jalilian-Marian and M. E. Tejeda-Yeomans, Phys. Lett. B **761**, 229 (2016) doi:10.1016/j.physletb.2016.08.035.

Preface to Special session on QCD dedicated to V. Gribov's confinement scenario

Julia Nyiri[1,a]

[1] *Wigner Research Centre for Physics, Budapest, Hungary*

V.N. Gribov would have been 85 years old in 2015. There were many events on this occasion, among others a Gribov-85 Memorial Workshop at the Landau Institute. Physicists from different parts of Russia and the world came to Chernogolovka, and those who could not come sent their contributions to the proceedings to pay tribute to a great physicist who was one of the creators of theoretical high energy physics. (The proceedings: Gribov-85 Memorial Volume, Exploring Quantum Field Theory is already published at World Scientific.)

On his 85th birthday a seminar was organized at the St. Petersburg Nuclear Physics Institute in Gatchina, where he was the leader of the Theory Department for many years. It was a quite unexpected event, and, as it turned out when I already arrived there, all the talks were just about remembering Volodya. I am his wife, so for me emotionally it was not so easy. On the other hand, it is very important that his memory, his ideas and works are kept alive.

My son told me a couple of years ago that everybody can be characterized by one word, if you want to underline the most important quality. I asked him what he would say about Volodya. The answer was: a good human being.

I remember a sentence Gribov told me long-long ago: "life is a chain of accidents". And, indeed: in 1958 I was admitted to the physical faculty of the university in Budapest. After that, I saw in a newspaper that there is a possibility to study abroad - at that time that meant the Soviet Union or Eastern Germany. I decided to try to go to the Soviet Union. I got the permission, and landed at the Leningrad University. From the fourth year at the university students from foreign countries (there were just a few) got the possibility to participate at seminars of the Theory Department in Gatchina, so I went there once a week. It was very interesting, I learned there a lot. And I saw there a young man, who, listening to the talks, many times jumped to the blackboard, and explained, what is so, what not, and why - I had the feeling that I would never dare to say anything in front of him. The man was V. Gribov, the leader of the Theory Department.

After finishing the university, I began to work at the Central Research Institute for Physics in Budapest - it is still my working place. In July 1977 an EPS conference was organized in Budapest, and we could invite Gribov - for quite a long time he was not allowed to travel abroad. And when we met in Budapest, it became immediately clear that we have to be together.

We got married in 1980, and I occured in a fantastic new world of physics. I remember all moments of our life. The pleasure when I was cooking something in the kitchen, and Volodya was standing behind me, telling in details what he is thinking about, what problem he tries to solve. For him there existed just one world, and any new question to what he got an answer had to fit in this picture.

[a]e-mail: nyiri.julia@wigner.mta.hu

I will never forget the incredible feeling when I looked at Volodya, who lay many hours on his couch with eyes closed. He did not sleep, and one could see he was thinking.

By the way, explaining something, he was always smoking. I was, of course, a bit worried, but I understood that he is used to it, and without smoking it is difficult for him to concentrate on his work.

Volodya told me the story how and when he began to smoke. He was born and grew up in Leningrad (now St. Petersburg). When the city was blockaded by the German army in WWII, he and his mother were evacuated to Siberia. Their life was not in a direct danger, but they had very little to eat. Children rolled up dry leaves to smoke and assuage their hunger. This must have been around 1942, when Volodya was 12 years old.

To return to physics: it was an extraordinary experience intellectually and emotionally to see, how Volodya was working on different subjects, and how he was always thinking. Long ago he found a consistent description of high energy hadron interaction (Gribov reggeon field theory). He predicted the existence of the Higgs-boson, and, together with Pontecorvo, the neutrino oscillations. What a pity that he could not see their discovery.

I am glad that the 5th International Conference on New Frontiers in Physics includes a Gribov session. Today the day begins with Gribov's theory of confinement, which he finished almost 19 years ago, in the hospital, before he died. It was published later with the help of Yu. Dokshitzer. (Also books in which his lectures on Quantum Electrodynamics, The Theory of Complex Angular Momenta, and Strong Interactions of Hadrons at High Energies are collected.)

Volodya had the habit not to publish his results immediately, he was checking again and again his calculations and the conclusions he came to. So he left many almost ready papers, and very precise calculations (sometimes more than 100 pages). With my colleagues we collected them, brought them to the institute. We are still in the process of scanning the papers, and our director, Peter Levai, has the plan to put them on a special Gribov website. And you can always find these papers in Budapest.

A hidden classical symmetry of QCD

L. Ya. Glozman[1,*]

[1] *Institute of Physics, University of Graz, A-8010 Graz, Austria*

Abstract. The classical part of the QCD partition function (the integrand) has, ignoring irrelevant exact zero modes of the Dirac operator, a local $SU(2N_F) \supset SU(N_F)_L \times SU(N_F)_R \times U(1)_A$ symmetry which is absent at the Lagrangian level. This symmetry is broken anomalously and spontaneously. Effects of spontaneous breaking of chiral symmetry are contained in the near-zero modes of the Dirac operator. If physics of anomaly is also encoded in the same near-zero modes, then their truncation on the lattice should recover a hidden classical $SU(2N_F)$ symmetry in correlators and spectra. This naturally explains observation on the lattice of a large degeneracy of hadrons, that is higher than the $SU(N_F)_L \times SU(N_F)_R \times U(1)_A$ chiral symmetry, upon elimination by hands of the lowest-lying modes of the Dirac operator. We also discuss an implication of this symmetry for the high temperature QCD.

1 Introduction

The QCD Lagrangian in Minkowski space-time has in the chiral limit the chiral symmetry:

$$U(N_F)_L \times U(N_F)_R = SU(N_F)_L \times SU(N_F)_R \times U(1)_A \times U(1)_V. \tag{1}$$

In the following we will always drop the $U(1)_V$ symmetry that is irrelevant to our subject. The $U(1)_A$ symmetry is invariance of the Lagrangian upon the axial flavor-neutral transformation

$$\Psi(x) \rightarrow e^{i\alpha\gamma_5}\Psi(x); \quad \bar{\Psi}(x) \rightarrow \bar{\Psi}(x)e^{i\alpha\gamma_5}. \tag{2}$$

The $SU(N_F)_L \times SU(N_F)_R$ symmetry implies invariance under a pure flavor rotation $SU(N_F)$ (for $N_F = 2$ it is the isospin symmetry) as well as invariance under the axial flavor transformation

$$\Psi(x) \rightarrow e^{i\gamma_5 \frac{\vec{\lambda}\cdot\vec{\alpha}}{2}}\Psi(x); \quad \bar{\Psi}(x) \rightarrow \bar{\Psi}(x)e^{i\gamma_5 \frac{\vec{\lambda}\cdot\vec{\alpha}}{2}}, \tag{3}$$

where $\vec{\lambda}$ are $SU(N_F)$ generators.

The axial $U(1)_A$ symmetry is broken anomalously, which is due to a noninvariance of the integration measure in the functional integral under a local $U(1)_A$ transformation [1]. The $SU(N_F)_A$ "symmetry"[1] (3) is broken spontaneously, because the ground state of the theory, the vacuum, is not invariant under the transformation (3). The latter noninvariance is encoded in the nonzero quark condensate of the vacuum, $< 0|\bar{\Psi}(x)\Psi(x)|0 > \neq 0$. The quark condensate of the vacuum can be expressed

*e-mail: leonid.glozman@uni-graz.at

[1] These transformations do not form a closed subgroup of the chiral group.

through a density of the near-zero modes $\lambda \to 0$ of the Euclidean Dirac operator (the Banks-Casher relation [2])

$$\lim_{m \to 0} < 0|\bar{\Psi}(x)\Psi(x)|0 >= -\pi\rho(0) . \quad (4)$$

The idea of the $N_F = 2$ lattice studies [3, 4] was to understand what would happen with $J = 1, 0$ mesons upon truncation of the near-zero modes of the Dirac operator. One expects that after truncation correlators of operators that are connected with each other through the $SU(2)_A$ transformation would become identical. If hadrons survive this truncation, then masses of chiral partners should be equal. It has turned out that a very clean exponential decay of correlators was detected in all $J = 1$ mesons. This implies that confined bound states survive the truncation. In the $J = 0$ mesons, while all $J = 0$ correlators become identical, the ground states disappear , because there is no exponential decay of the corresponding correlators: The near-zero modes are crucially important for existence of the (pseudo) Goldstone bosons, which is not surprising.

It has also turned out that the truncation restores in hadrons not only the $SU(2)_L \times SU(2)_R$ symmetry but also the $U(1)_A$ symmetry. One then concludes that the same lowest-lying modes are responsible for both $SU(2)_L \times SU(2)_R$ and $U(1)_A$ breakings which is consistent with the instanton-induced mechanism of both breakings [7–10].

However, a larger degeneracy that includes all possible chiral multiplets of the $J = 1$ mesons was detected, which was completely unexpected. This surprising degeneracy implies a symmetry that is higher than the $SU(2)_L \times SU(2)_R \times U(1)_A$. This not yet known symmetry has been reconstructed in refs. [11, 12] and turned out to be

$$SU(2N_F) \supset SU(N_F)_L \times SU(N_F)_R \times U(1)_A. \quad (5)$$

Transformations of this group include both the flavor rotations of the left- and right-handed quarks as well as transformations that mix the left- and right-handed components. This symmetry has been confirmed in lattice simulations with the $J = 2$ mesons [5] and in baryons [6].

While such an enlarged symmetry, that is larger than the symmetry of the QCD Lagrangian, was observed in the lattice experiment, its origin was mysterious. The light on this issue has been shed in ref. [13]. It has been shown that the classical part of the QCD partition function has, excluding irrelevant exact zero modes, a $SU(2N_F)$ local symmetry. This symmetry is not a symmetry of the Euclidean QCD Lagrangian because the irrelevant exact zero modes of the Euclidean Dirac operator break it. Consequently, we refer it as a hidden classical symmetry. This hidden classical symmetry is broken by the anomaly to $SU(N_F)_L \times SU(N_F)_R$. Spontaneous breaking of chiral symmetry reduces it to $SU(N_F)_V$. Effects of spontaneous and anomalous breakings are encoded in the near-zero modes of the Dirac operator. Consequently, truncation on the lattice of the lowest-lying modes restores the hidden $SU(2N_F)$ symmetry, which naturally explains degeneracies observed in lattice experiment.

Now we will present some details.

2 What can be apriori expected from the low-mode truncation?

A truncation of the low-lying modes of the Dirac operator means the following. The Euclidean Lagrangian with N_F degenerate quarks in a given gauge background is:

$$\mathcal{L} = \Psi^\dagger(x)(\gamma_\mu D_\mu + m)\Psi(x), \quad (6)$$

where

DOI: 10.1051/epjconf/201716403002

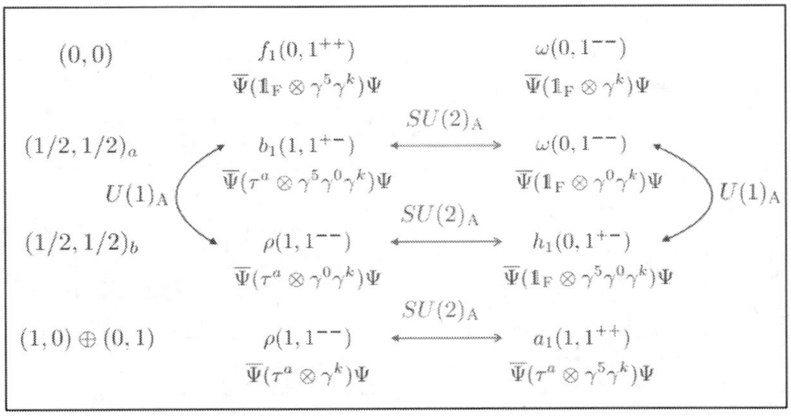

Figure 1. $SU(2)_L \times SU(2)_R$ and $U(1)_A$ classification of the $J = 1$ meson operators.

$$D_\mu = \partial_\mu + ig\frac{t^a}{2}A^a_\mu. \tag{7}$$

The hermitian Dirac operator, $i\gamma_\mu D_\mu$, has in a volume V a discrete spectrum with real eigenvalues λ_n:

$$i\gamma_\mu D_\mu \Psi_n(x) = \lambda_n \Psi_n(x). \tag{8}$$

We subtract from the full quark propagator S_{Full} the lowest k eigenmodes of the Dirac operator

$$S(x, y) = S_{Full}(x, y) - \sum_{n=1}^{k} \frac{1}{\lambda_n + im} \Psi_n(x)\Psi_n^\dagger(y). \tag{9}$$

At the same time we keep the gauge configurations intact. Given these truncated quark propagators we apply standard procedures to extract hadron spectra using the variational approach. Then we study dependence of hadron masses on the truncation number k. We perform $N_F = 2$ dynamical lattice calculations with the overlap Dirac operator with the gauge configurations generated by the JLQCD collaboration, for details see refs. [3–6].

When the near-zero modes, that are responsible for spontaneous chiral symmetry breaking are subtracted we can expect the $SU(2)_L \times SU(2)_R$ symmetry in correlators. If in addition the hadron states survive this surgery, then we can expect a mass degeneracy of chiral partners. For the $J = 1$ mesons the chiral partners are linked by the red arrows on Fig. 1.

The $U(1)_A$ transformation connects operators that are linked by the blue arrows on Fig. 1. Consequently, if the whole chiral symmetry of QCD $SU(2)_L \times SU(2)_R \times U(1)_A$ is restored, then we can expect a degeneracy of four mesons from the $(1/2, 1/2)_a$ and $(1/2, 1/2)_b$ representations on the one hand, and on the other hand a degeneracy of the ρ and a_1 mesons from the $(1, 0) + (0, 1)$ chiral representation. Note that in the chirally symmetric world there two independent orthogonal ρ-mesons that belong to two different chiral representations. In the real world with chiral symmetry breaking these two chiral representations are mixed in the meson wave function and two different ρ operators couple to one and the same ρ-meson.

The $SU(2)_L \times SU(2)_R \times U(1)_A$ symmetry does not connect, however, the four mesons from the $(1/2, 1/2)_a$ and $(1/2, 1/2)_b$ representations with the other mesons in Fig. 1. Consequently, given only

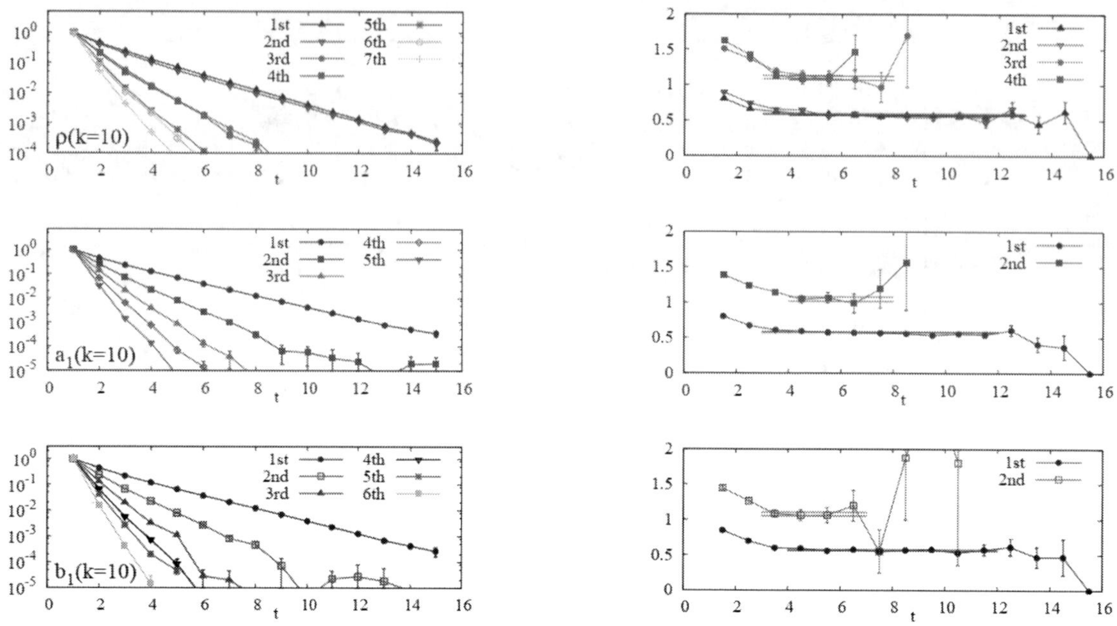

Figure 2. The eigenvalues of the cross-correlation matrix and effective mass plateaus for the isovector $J = 1$ mesons with $k = 10$

the $SU(2)_L \times SU(2)_R \times U(1)_A$ chiral symmetry we cannot expect a degeneracy larger than it is shown by arrows on Fig. 1.

3 Results

We do not show here our results for the real world ($k = 0$), since they are typical for all lattice studies and the experimental meson spectra are reasonably reproduced. The results for the eigenvalues of the cross-correlation matrix and effective mass plateau at $k = 10$ (10 lowest Dirac eigenmodes have been removed) for the isovector mesons are shown in Fig. 2.

A very clean exponential decay of the correlators is obvious, which means that there are physical states. It is much cleaner than in the untruncated (real) world. The reason for this is intuitively clear: After truncation there are no pion fluctuations in the system. We can conclude that mesons (which are bound states now) survive the truncation.

Note a double degeneracy of the rho-meson eigenvalues, which is absent in the untruncated world. This double degeneracy can be obtained only if both rho-operators from Fig. 1 are used in the cross-correlation matrix. This double degeneracy tells that there are two independent orthogonal degenerate ρ-mesons. If we put two lowest rho-eigenvalues, the lowest a_1 and the lowest b_1 eigenvalues on the same plot, then we will see that they all are identical (the same is true with the higher eigenvalues, but the result is less precise). This means that there is a symmetry in the system that is higher than $SU(2)_L \times SU(2)_R \times U(1)_A$.

Evolution of meson masses is shown in Fig. 3. We clearly see a larger degeneracy than the chiral $SU(2)_L \times SU(2)_R \times U(1)_A$ symmetry of the QCD Lagrangian. What does it mean?! The same results persist for the $J = 2$ mesons and baryons.

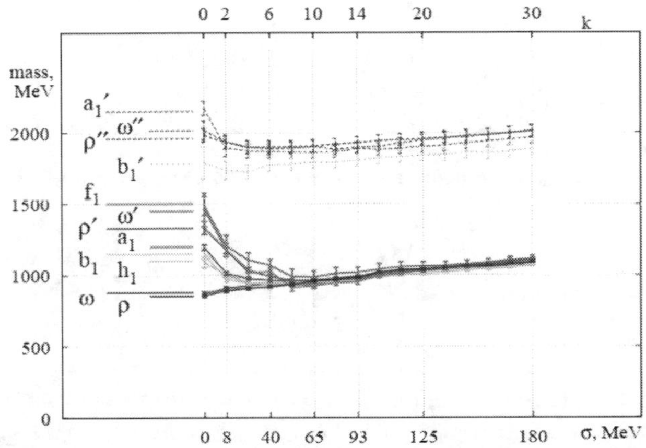

Figure 3. $J = 1$ meson mass evolution as a function of the truncation number k. σ shows energy gap in the Dirac spectrum.

4 $SU(4)$ symmetry of the meson spectra

Given this degeneracy first we need to understand what symmetry group does it correspond. This unexpected new symmetry has been reconstructed in ref. [11].

Given the standard spin, parity, etc. quantum numbers we can construct explicitly basis vectors for all irreducible representations of the chiral group shown in Fig. 1.

(i) (0,0):

$$|(0,0);\pm;J\rangle = \frac{1}{\sqrt{2}}|\bar{R}R \pm \bar{L}L\rangle_J. \quad (10)$$

(ii) $(1/2,1/2)_a$ **and** $(1/2,1/2)_b$**:**

$$|(1/2,1/2)_a;+;I=0;J\rangle = \frac{1}{\sqrt{2}}|\bar{R}L + \bar{L}R\rangle_J, \quad (11)$$

$$|(1/2,1/2)_a;-;I=1;J\rangle = \frac{1}{\sqrt{2}}|\bar{R}\vec{\tau}L - \bar{L}\vec{\tau}R\rangle_J, \quad (12)$$

$$|(1/2,1/2)_b;-;I=0;J\rangle = \frac{1}{\sqrt{2}}|\bar{R}L - \bar{L}R\rangle_J, \quad (13)$$

$$|(1/2,1/2)_b;+;I=1;J\rangle = \frac{1}{\sqrt{2}}|\bar{R}\vec{\tau}L + \bar{L}\vec{\tau}R\rangle_J. \quad (14)$$

(iii) (0,1)⊕(1,0):

$$|(0,1)+(1,0);\pm;J\rangle = \frac{1}{\sqrt{2}}|\bar{R}\vec{\tau}R \pm \bar{L}\vec{\tau}L\rangle_J, \quad (15)$$

Now we need to to find a minimal group that contains $SU(2)_L \times SU(2)_R \times U(1)_A$ as a subgroup and that combines all these vectors into one irreducible representation. These new symmetry transformations must connect all these basis vectors. The latter requirement can be achieved if these new symmetry transformations mix the left- and right-handed quarks. Consequently, the required symmetry group must contain as a subgroup the $SU(2)_{CS}$ *chiralspin* rotations that act on the following doublets:

$$\mathrm{U} = \begin{pmatrix} u_L \\ u_R \end{pmatrix}, \qquad \mathrm{D} = \begin{pmatrix} d_L \\ d_R \end{pmatrix}. \tag{16}$$

A three-dimensional imaginary space where these rotations are performed is called the *chiralspin* space. The chiralspin rotations mix the right- and left-handed components of the fermion fields. It is similar to the well familiar isospin space: Rotations in the isospin space mix particles with different electric charges.

If we combine the $SU(2)_{CS}$ and the isospin $SU(2)$ group into one larger group one arrives at the $SU(4)$ group with the fundamental vector

$$\Psi = \begin{pmatrix} u_{\mathrm{L}} \\ u_{\mathrm{R}} \\ d_{\mathrm{L}} \\ d_{\mathrm{R}} \end{pmatrix}. \tag{17}$$

The dim=15 irreducible representation of this group connects all $(0,0), (1/2,1/2)_a, (1/2,1/2)_b, (1,0)+(0,1)$ vectors. One of the $(0,0)$ basis vectors, namely $|(0,0); -; J=1\rangle >$ is a singlet of $SU(4)$.

We can construct an explicit realization of the $SU(2)_{CS}$ and $SU(4)$ algebra that acts on Dirac spinors [12]. Then the $SU(2)_{CS}$ chiralspin rotations are generated through

$$\Sigma = \{\gamma^0, i\gamma^5\gamma^0, -\gamma^5\}, \qquad [\Sigma^i, \Sigma^j] = 2i\epsilon^{ijk}\Sigma^k .$$

The Dirac spinor transforms under a global or local $SU(2)_{CS}$ transformation as

$$\Psi \to \Psi' = e^{i\varepsilon \cdot \Sigma/2}\Psi . \tag{18}$$

The $SU(4)$ group contains at the same time $SU(2)_L \times SU(2)_R$ and $SU(2)_{CS} \supset U(1)_A$ and has the following set of generators:

$$\{(\tau^a \otimes 1_D), (1_F \otimes \Sigma^i), (\tau^a \otimes \Sigma^i)\} .$$

The global and local $SU(4)$ transformations of the Dirac spinor are defined through

$$\Psi \to \Psi' = e^{i\epsilon \cdot \mathrm{T}/2}\Psi . \tag{19}$$

The $SU(2)_{CS}$ and $SU(4)$ transformations of all operators from Fig. 1 are shown in Fig. 4.

5 Zero modes and hidden classical symmetries of Euclidean QCD

The $SU(4)$ symmetry is obtained in lattice simulations upon subtraction of the near-zero modes of the Dirac operator. It implies that this symmetry should be encoded in the Euclidean QCD. Obviously the Lagrangian (6) does not have this symmetry. This is because the Dirac operator does not commute with the $SU(2)_{CS}$ transformations.[2] Then we should recover at which level this symmetry is hidden in Euclidean QCD. More explicitly, we have to find a part of the Euclidean QCD formalism that breaks this symmetry.

Consider the zero modes of the Dirac equation,

$$\gamma_\mu D_\mu \Psi_0(x) = 0. \tag{20}$$

Given standard antiperiodic boundary conditions for the quark field along the time direction, the zero modes are solutions of the Dirac equation with the gauge configurations of a nonzero global topological charge. The difference of numbers of the left-handed and right-handed zero modes is according to the Atiyah-Singer theorem fixed by the global topological charge Q of the gauge configuration:

$$n_L - n_R = Q. \tag{21}$$

Some $SU(2)_{CS}$ transformations rotate the right-handed spinor into the left-handed one and vice versa. Consequently, the zero modes explicitly violate the $SU(2)_{CS}$ and $SU(2N_F)$ symmetries: The zero modes introduce an

[2]In Euclidean space we have to substitute the γ_0 matrix through the γ_4 matrix.

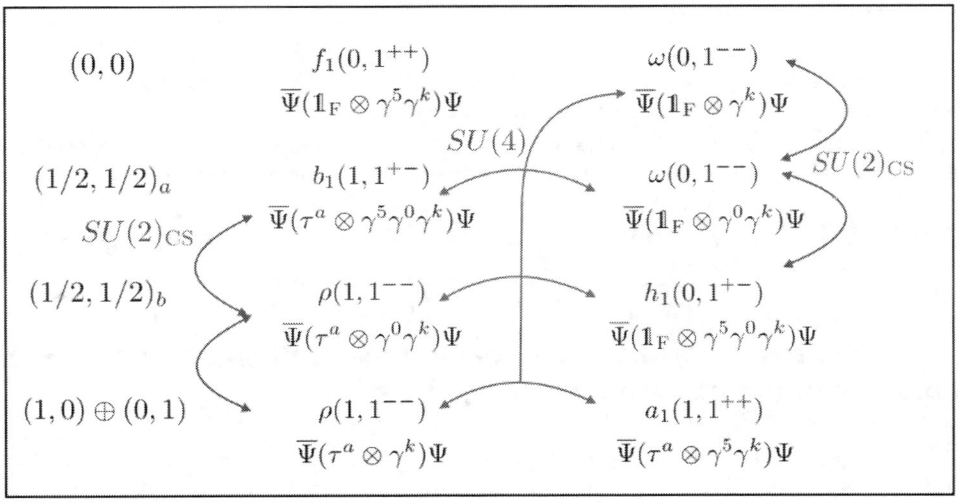

Figure 4. The green arrows connect operators that belong to the $SU(2)_{CS}$ triplets. The f_1 and a_1 operators are the $SU(2)_{CS}$ singlets. The purple arrows show the $SU(4)$ 15-plet. The f_1 operator is a singlet of $SU(4)$.

asymmetry between the left- and right-handed degrees of freedom and break the $SU(2)_{CS}$ invariance. The latter is possible only if there is no asymmetry between the left and the right.

It is well understood, however, that the exact zero modes are completely irrelevant since their contributions to the Green functions and observables vanish in the thermodynamical limit $V \to \infty$ [14–16]. Consequently, in the finite volume calculations we can subtract the irrelevant exact zero modes.

We can expand fields $\Psi(x)$ and $\Psi^\dagger(x)$ in the Lagrangian over a complete and orthonormal set $\Psi_n(x)$ of the eigenvalue problem (8):

$$\Psi(x) = \sum_n c_n \Psi_n(x), \qquad \Psi^\dagger(x) = \sum_k \bar{c}_k \Psi_k^\dagger(x), \tag{22}$$

where $\bar{c}_k, c_n$ are Grassmannian numbers. Then the fermionic part of the QCD partition function takes the following form

$$Z = \int \prod_{k,n} d\bar{c}_k dc_n e^{\sum_{k,n} \int d^4x \bar{c}_k c_n (\lambda_n + im) \Psi_k^\dagger(x) \Psi_n(x)}. \tag{23}$$

Now we can directly read-off symmetry properties of the classical part of the partition function, i.e. of the integrand. This functional contains only a superposition of terms $\Psi_k^\dagger(x)\Psi_n(x)$. It is precisely $SU(2)_{CS}$ and $SU(2N_F)$ symmetric, because

$$(U\Psi_k(x))^\dagger U\Psi_n(x) = \Psi_k^\dagger(x)\Psi_n(x), \tag{24}$$

where U is any local or global transformation from the groups $SU(2)_{CS}$ and $SU(2N_F)$, $U^\dagger = U^{-1}$. The exact zero modes, for which the equation (24) does not hold, have been subtracted from the partition function. We conclude that classically the Euclidean QCD without the irrelevant exact zero mode contributions is invariant with respect to both global and local $SU(2)_{CS}$ and $SU(2N_F)$ transformations.

The term "hidden classical $SU(2N_F)$ symmetry" should be correctly understood. It is not a global symmetry of the Lagrangian and consequently there are no respective conserved Noether currents. However, it is a real local symmetry of the classical part of the QCD partition function ignoring irrelevant exact zero modes.

How is this hidden classical symmetry broken? The integration measure in (23) is not invariant under a local $U(1)_A$ transformation [1]. Consequently, the $U(1)_A$ anomaly breaks the classical $U(1)_A$ symmetry. Since

 DOI: 10.1051/epjconf/201716403002

the $U(1)_A$ is a subgroup of $SU(2)_{CS}$, the anomaly breaks either the $SU(2)_{CS}$ symmetry. Hence the classical $SU(2N_F) \supset SU(N_F)_L \times SU(N_F)_R \times U(1)_A$ symmetry is broken by anomaly to $SU(N_F)_L \times SU(N_F)_R$.

The quark condensate in Minkowski space breaks all $U(1)_A$, $SU(2)_{CS}$, $SU(N_F)_L \times SU(N_F)_R$ and $SU(2N_F)$ symmetries to the vector flavor symmetry $SU(N_F)_V$. Hence, the new hidden classical $SU(2)_{CS}$ and $SU(2N_F)$ symmetries are broken both by the condensate and anomalously.

Spontaneous chiral symmetry breaking is encoded in the near-zero modes of the Dirac operator, as it follows from the Banks-Casher relation. If anomaly is also encoded in the near-zero modes, as suggested e.g. by the instanton mechanism of both breakings, then removal on lattice of the near-zero modes should restore not only chiral $SU(N_F)_L \times SU(N_F)_R$ and $U(1)_A$ symmetries, but also a larger $SU(2N_F)$ symmetry, which naturally explains lattice observations reviewed in previous sections.

6 What symmetries should one expect in mesons and baryons upon truncation of the near-zero modes?

From the results presented in Fig. 3 it is clearly seen that the degeneracy pattern is larger than $SU(4)$, because the $SU(4)$ singlet (f_1) and the $SU(4)$ 15-plet mesons ($\rho, \rho', a_1, b_1, h_1, \omega, \omega'$) are also degenerate. This implies that actually some higher symmetry is observed that includes the $SU(4)$ as a subgroup [12]. It was found that no higher symmetry exists, that would connect *local* quark bilinears from the 15-plet and singlet of $SU(4)$ within the same irreducible representation [17].

This challenging problem has been solved in ref. [13]. Hadron spectra are extracted from the correlation functions calculated with the gauge-invariant source operators. At each time slice "t" a meson correlator contains minimum the lowest Fock $\bar{q}q$ component with a quark and an antiquark located at different space points **x** and **y**. Both q and $\bar{q}$ interact with the same gauge configuration. Then all arguments of the previous section apply independently for q and $\bar{q}$. Since the $SU(2N_F)$ invariance is local, we can perform $SU(2N_F)$ rotations at points **x** and **y** with different rotation parameters. It is then clear that the meson correlation function with the $\bar{q}q$ valence content has a *bilocal* $SU(2N_F) \times SU(2N_F)$ symmetry. A symmetry of higher Fock components is obviously larger, but the whole correlator has a symmetry of the lowest $\bar{q}q$ component. Obviously, averaging over gauge configurations does not change this symmetry property.

The same argument applies to baryons and in this case we expect a trilocal $SU(2N_F) \times SU(2N_F) \times SU(2N_F)$ symmetry.

One of the irreducible representations of the $SU(4) \times SU(4)$ is 16-dimensional and is a direct sum of the 15-plet and singlet of $SU(4)$. Hence a direct prediction of this bilocal symmetry is a degeneracy of the $SU(4)$-singlet and of the $SU(4)$ 15-plet, in agreement with the lattice observations. This symmetry is bilocal and cannot be represented by the local composite operators which is consistent with conclusions of Ref. [17].

7 A short summary of our findings.

Our main findings can be summarized as follows.

1. The classical part of the partition function (the integrand), excluding irrelevant exact zero mode contributions, has $SU(2)_{CS}$ and $SU(2N_F)$ local symmetries. Since these symmetries are not symmetries of the QCD Lagrangian we refer them as hidden classical symmetries of QCD. There are no respective conserved Noether currents. These symmetries are broken at the quantum level by the axial anomaly and by the quark condensate. The physics of chiral symmetry spontaneous breaking and of anomaly is contained in the near-zero modes of the Dirac operator. Their truncation on the lattice should restore not only the $SU(N_f)_L \times SU(N_F)_R \times U(1)_A$ chiral symmetry but actually higher hidden classical symmetries $SU(2)_{CS}$ and $SU(2N_F)$.

2. We have shown that elimination of the near-zero modes leads to $SU(2N_F) \times SU(2N_F)$ and $SU(2N_F) \times SU(2N_F) \times SU(2N_F)$ symmetries in mesons and baryons.

3. The bilocal $SU(4) \times SU(4)$ symmetry explains a degeneracy of the $SU(4)$ singlet f_1 correlator with the $SU(4)$ 15-plet $\rho, \rho', \omega, \omega', h_1, a_1, b_1$ correlators.

DOI: 10.1051/epjconf/201716403002

8 Implications

It is natural to expect many different implications of the hidden classical symmetry. Here we will mention a most dramatic one [18].

At high temperature the quark condensate of the vacuum vanishes. There are lattice indications that above the critical temperature the $U(1)_A$ symmetry is restored and a gap opens in the Dirac spectrum [19, 20].[3] Then it follows that the $SU(2)_{CS}$ and $SU(2N_F)$ symmetries are manifest in Euclidean correlation functions and observables. Such symmetries cannot be obtained in terms of deconfined quarks and gluons in Minkowski space, where we live. Hence at high temperatures QCD is also in the confining regime and elementary objects are color singlet $SU(4)$ symmetric "hadrons". "Hadrons" with such a symmetry can be directly constructed in Minkowski space [23].

We acknowledge a partial support from the Austrian Science Fund (FWF) through the grant P26627-N27.

References

[1] K. Fujikawa, Phys. Rev. D **21** 2848 (1980). Erratum: [Phys. Rev. D **22** 1499 (1980)].
[2] T. Banks and A. Casher, Nucl. Phys. B **169** 103 (1980).
[3] M. Denissenya, L. Y. Glozman and C. B. Lang, Phys. Rev. D **89** 077502 (2014).
[4] M. Denissenya, L. Y. Glozman and C. B. Lang, Phys. Rev. D **91** 034505 (2015).
[5] M. Denissenya, L. Y. Glozman and M. Pak, Phys. Rev. D **91** 114512 (2015).
[6] M. Denissenya, L. Y. Glozman and M. Pak, Phys. Rev. D **92** 074508 (2015).
[7] G. 't Hooft, Phys. Rept. **142** 357 (1986).
[8] E. V. Shuryak, Nucl. Phys. B **203** 93 (1982).
[9] D. Diakonov and V. Y. Petrov, Nucl. Phys. B **272** 457 (1986).
[10] T. Schäfer and E. V. Shuryak, Rev. Mod. Phys. **70** 323 (1998).
[11] L. Y. Glozman, Eur. Phys. J. A **51** 27 (2015).
[12] L. Y. Glozman and M. Pak, Phys. Rev. D **92** 016001 (2015).
[13] L. Y. Glozman, arXiv:1511.05857 [hep-ph].
[14] H. Leutwyler and A. V. Smilga, Phys. Rev. D **46** 5607 (1992).
[15] R. Brower, S. Chandrasekharan, J. W. Negele and U. J. Wiese, Phys. Lett. B **560** 64 (2003).
[16] S. Aoki, H. Fukaya and Y. Taniguchi, Phys. Rev. D **86** 114512 (2012).
[17] T. D. Cohen, Phys. Rev. D **93** 034508 (2016).
[18] L. Y. Glozman, arXiv:1512.06703 [hep-ph].
[19] G. Cossu, S. Aoki, H. Fukaya, S. Hashimoto, T. Kaneko, H. Matsufuru and J. I. Noaki, Phys. Rev. D **87**, no. 11, 114514 (2013) [Phys. Rev. D **88**, no. 1, 019901 (2013)].
[20] A. Tomiya, G. Cossu, H. Fukaya, S. Hashimoto and J. Noaki, PoS LATTICE **2014**, 211 (2015) [arXiv:1412.7306 [hep-lat]]; G. Cossu *et al.* [JLQCD Collaboration], arXiv:1511.05691 [hep-lat];
[21] S. Sharma, V. Dick, F. Karsch, E. Laermann and S. Mukherjee, arXiv:1602.02197 [hep-lat].
[22] S. Aoki [JLQCD Collaboration], PoS(CD15) 045 (2015) [arXiv:1603.00997 [hep-lat]].
[23] M. Shifman, Phys. Rev. D **93**, no. 7, 074035 (2016).

[3] See also the opposite statement [21] and its critique [22].

DOI: 10.1051/epjconf/201716403003

Callan-Symanzik equations for infrared Yang-Mills theory

Axel Weber[1,a] and Pietro Dall'Olio[1]

[1] *Instituto de Física y Matemáticas, Universidad Michoacana de San Nicolás de Hidalgo, Edificio C-3, Ciudad Universitaria, A. Postal 2-82, 58040 Morelia, Michoacán, Mexico*

Abstract. Dyson-Schwinger equations have been successful in determining the correlation functions in Yang-Mills theory in the Landau gauge, in the infrared regime. We argue that similar results can be obtained, in a technically simpler way, with Callan-Symanzik renormalization group equations. We present generalizations of the infrared safe renormalization scheme proposed by Tissier and Wschebor in 2011, and show how the renormalization scheme dependence can be used to improve the matching to the existing lattice data for the gluon and ghost propagators.

Historically, the first semi-analytical method that has made it possible to access the deep infrared (IR) regime of Landau gauge Yang-Mills theory, have been Dyson-Schwinger equations. The first solutions of these equations that have been found show a scaling behavior of the propagators in the IR [1]. Several years later, another type of solutions of the same equations was discovered [2], with a massive gluon propagator and a finitely enhanced ghost propagator in the IR (decoupling solutions). To date, Dyson-Schwinger equations are still the main tool for the (semi-)analytical exploration of the IR regime of Yang-Mills theory.

In a parallel development that actually initiated much earlier with Gribov's observation of the existence of gauge copies in his famous 1978 paper [3] and was later worked out in great detail by Zwanziger [4], the theoretical foundations of IR Yang-Mills theory in the Landau gauge were laid. While this so-called Gribov-Zwanziger scenario seemed to confirm the existence of the scaling solutions, a more recent refinement [5] favors the decoupling type of solutions. Finally, the results of simulations of Yang-Mills theory in the Landau gauge on huge lattices [6] have been interpreted by most workers in the field as confirming the realization of the decoupling type of solutions in three and four space-time dimensions.

In this contribution, we will present a different technique for the exploration of the IR regime of Yang-Mills theory. We intend to reproduce the results of lattice simulations in the Landau gauge that restrict the gauge field configurations to the (first) Gribov region, but make no attempt to reach the fundamental modular region. The restriction to the Gribov region implies the breaking of BRST invariance in the continuum formulation [4]. The most important consequence of the broken BRST invariance for the IR regime is the appearance of a mass term for the gluon field [5]. Indeed, the addition of a gluonic mass term to the Yang-Mills action has been shown to reproduce all the solutions (two types of scaling solutions and the decoupling solution) found before with the help of Dyson-Schwinger equations when solving the Callan-Symanzik equations for this theory in the IR regime in an epsilon expansion around the upper critical space-time dimension [7]. This analysis also shows that

[a] e-mail: axel@ifm.umich.mx

the decoupling solution is the one that is physically realized since it is the only one that corresponds to an IR attractive fixed point (for dimensions above two).

For the rest of this contribution, we will focus on the decoupling solution. Tissier and Wschebor [8] have shown that straightforward one-loop perturbation theory applied to Yang-Mills theory with a gluon mass term yields results for the gluon and ghost propagators that reproduce surprisingly well the propagators found on the lattice in three and four space-time dimensions in the IR regime. It is apparent, however, that a renormalization group improvement is necessary for a quantitative description of the ultraviolet (UV) regime in four space-time dimensions, and also for a quantitatively good fit to the propagators in the IR regime in three dimensions. In renormalization group language, what is required is a precise description of the crossover between the UV attractive Gaussian fixed point and the IR attractive (high-temperature) fixed point. It turns out that an action that contains all the terms that are relevant or marginal with respect to one or the other fixed point, is sufficient to describe the complete crossover, see Eq. (1) below.

Two comments are in order before we move on: first, the BRST invariance of Yang-Mills theory is essential for the standard construction of the Hilbert space of the theory, the proof of unitarity and also the Kugo-Ojima confinement criterion [9]. It is not clear at present how these constructions could be adapted to the case of broken BRST symmetry which we have assumed here from the outset. In fact, adding a gluon mass term to the Yang-Mills action gives rise to a Curci-Ferrari model [10] which is perturbatively renormalizable, but is commonly considered not to be unitary [11]. We should like to point out that the statement of non-unitarity refers to the standard construction of the Hilbert space, and it has actually not been shown that no Hilbert space construction exists that would lead to a unitary theory. We refer, in this context, to the recent efforts in Refs. [12, 13] for the Curci-Ferrari model and the (refined) Gribov-Zwanziger action. In this contribution, we will not be concerned with this fundamental and important issue, but rather focus our attention on calculating the n-point functions of the theory in the most efficient and accurate way.

The second comment concerns the gluon propagator in the presence of a mass term: it has been shown in numerous occasions that a gluon propagator of the decoupling type violates spectral positivity (just as in the more obvious case of the scaling type), and it thus does not describe the propagation of a physical particle, contrary to the case of mass generation through the Higgs mechanism. Also, gobal color symmetry remains an exact symmetry of the theory even though BRST invariance is broken.

We shall now set up renormalization schemes for the theory that allow for a consistent formulation of Callan-Symanzik equations, i.e., that avoid the appearance of a Landau pole. Let us remark that we use the term "Callan-Symanzik equations" to refer to the equations that quantify the change of the renormalized correlation functions with respect to variations of the renormalization scale (not with respect to variations of the mass parameter), in accord with the use of the term in modern textbooks on quantum field theory.

We begin with the Euclidean action of Yang-Mills theory in Landau gauge, including, in addition, a mass term for the gluon field (we then arrive at a Curci-Ferrari model). It is convenient to introduce the Nakanishi-Lautrup auxiliary field B^a to implement the gauge condition. Then the action is, in four space-time dimensions,

$$S = \int d^4x \left(\frac{1}{4} F^a_{\mu\nu} F^a_{\mu\nu} + \frac{1}{2} A^a_\mu m^2 A^a_\mu + \partial_\mu \bar{c}^a D^{ab}_\mu c^b + i B^a \partial_\mu A^a_\mu \right) . \tag{1}$$

Note that there are (at this point) *two* parameters in the theory, the gluon mass parameter m and the coupling constant g that appears in the covariant derivative D^{ab}_μ and in the field strength tensor $F^a_{\mu\nu}$. The mass parameter m can be thought of as representing the Gribov parameter that is introduced in order to implement the restriction to the Gribov region [3]. The Gribov parameter, and as a consequence the mass parameter, is in principle fixed through the horizon condition [4] in terms of the unique scale of the theory, Λ_{QCD} (equivalently, in terms of the value of g at an arbitrary scale). Since the exact relation between m and Λ_{QCD} is not known, at least not beyond the lowest perturbative order, we treat m here as an independent parameter that has to be fixed by comparison with the "experiment" (lattice simulations), just like the mass of a physical particle in a theory without confinement is set equal to its physical value.

We emphasize that the breaking of BRST invariance and the appearance of a mass term for the gluon field should not alter the well-known perturbative behavior of Yang-Mills theory in the UV regime. In general, fine-tuning conditions will have to be implemented in the UV limit to make sure that the usual perturbative results and the BRST symmetry are recovered in this limit. These conditions express the fact that we are describing a gauge theory within a wider class of effective models.

The first renormalization scheme we shall consider is defined through the conditions (in momentum space)

$$\Gamma^T_{AA}(p^2 = \mu^2) = \mu^2 + m^2 , \tag{2}$$

$$\Gamma^L_{AA}(p^2 = \mu^2) = m^2 , \tag{3}$$

$$\Gamma_{c\bar{c}}(p^2 = \mu^2) = \mu^2 , \tag{4}$$

imposed on the renormalized proper two-point functions. These conditions determine the field renormalization constants Z_A and Z_c and the mass parameter m at the renormalization scale μ. The symbols Γ^T_{AA} and Γ^L_{AA} refer to the transverse and longitudinal part, respectively, of the proper gluonic two-point function. We have suppressed the delta functions for momentum conservation and the trivial color factor δ^{ab} resulting from global color symmetry in all two-point functions. Note that there is no trace of the gauge fixing function left in the longitudinal part because of our use of the auxiliary field B^a in the action (1). Contrary to the proper two-point function, the gluon propagator is exactly transverse (as it should be in the Landau gauge).

The normalization conditions (2)–(4) were first proposed by Tissier and Wschebor in Ref. [14], where they were written in a different, but equivalent form. For the definition of the renormalized coupling constant, Tissier and Wschebor use the renormalized ghost-gluon vertex in the Taylor limit where the ghost momentum vanishes. In this limit, the quantum corrections to the vertex vanish exactly [15] which makes the calculation of the beta function in this renormalization scheme particularly simple.

For a first assessment of the renormalization scheme dependence we propose to consider the ghost-gluon vertex at the symmetry point for the definition of the renormalized coupling constant, instead of using the Taylor limit. The calculation of the beta function in this scheme involves an explicit evaluation of the one-loop corrections to the ghost-gluon vertex (at the symmetry point). In Fig. 1 we compare the running of the strong fine structure constant $\alpha(\mu) = g^2(\mu)/4\pi$ and the square of the mass parameter $m^2(\mu)$ as functions of the renormalization scale, obtained from the integration of the corresponding beta functions, in the two renormalization schemes. For the integration of the differential equations, we have been careful to use initial conditions (at $\mu = 3$ GeV) in the two schemes that correspond to the same bare theory, i.e., that are physically equivalent. Although the qualitative behavior of the curves is the same in both schemes, a considerable quantitative dependence on the renormalization scheme is seen.

We should like to clarify that at any fixed order of perturbation theory, the results for all correlation functions have to be exactly the same in any renormalization scheme (except for possible

 DOI: 10.1051/epjconf/201716403003

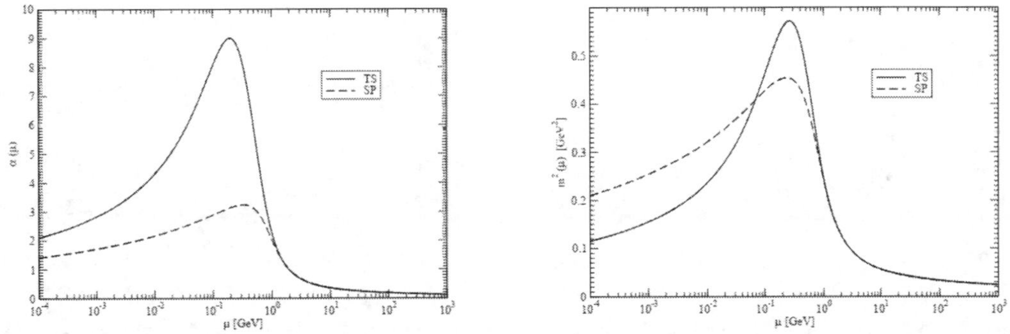

Figure 1. Running fine structure constant (to the left) and running mass parameter (to the right) in the Taylor scheme (TS) vs. the symmetry point scheme (SP).

constant rescalings of the fields, and assuming that physically equivalent values of the parameters g and m are chosen in all schemes). However, the results of applying Callan-Symanzik equations to different renormalization schemes can be different since they sum up parts of higher-loop diagrams by "extrapolating" the information obtained from the correlation functions to a given loop order in different ways.

Notice from Fig. 1 that the running coupling constant (or, rather, the fine structure constant) tends logarithmically to zero in the large-momentum limit as it should (asymptotic freedom). The same is true for the mass parameter. However, in the beta functions it is really the dimensionless combination m^2/μ^2 that matters, so the UV limit of a massless and BRST-invariant theory is actually reached much faster than logarithmically.

Interestingly, both the coupling constant and the mass parameter also tend logarithmically to zero in the IR limit of zero momentum, in four space-time dimensions. In the case of the coupling constant, this means that the IR fixed point is trivial. The triviality of the fixed point explains the success of straightforward perturbation theory in the IR regime (in four dimensions) [8]. On the other hand, the fact that the mass parameter tends to zero does not mean that one recovers a massless theory in the deep IR: the scale dependence of the gluon field renormalization constant leads to a finite limit of the gluon propagator at zero momentum. The trivial IR fixed point is thus essentially different from the trivial UV fixed point.

Let us now compare the results for the gluon and ghost propagators obtained from the integration of the Callan-Symanzik equations in the two different renormalization schemes with the data of lattice simulations. To this end, we have to determine those values of the parameters g and m at some (arbitrary) scale that lead to the best possible fits to the lattice data. The actual values of these parameters in the two renormalization schemes will generally not be physicall equivalent. Our strategy is to fix the two parameters in such a way that we obtain the best possible fit to the ghost propagator and its dressing function, a procedure that has turned out to determine the values of these parameters to a fair precision. Using the same values of the parameters in the calculation of the gluon propagator and comparing the result for the gluon propagator and its dressing function to the corresponding lattice data then allows to assess the degree to which the renormalization scheme is able to reproduce the lattice results (the dressing functions are the ratios of the full propagators to the tree-level propagators, the latter taken to be massless in the case of the gluon propagator).

DOI: 10.1051/epjconf/201716403003

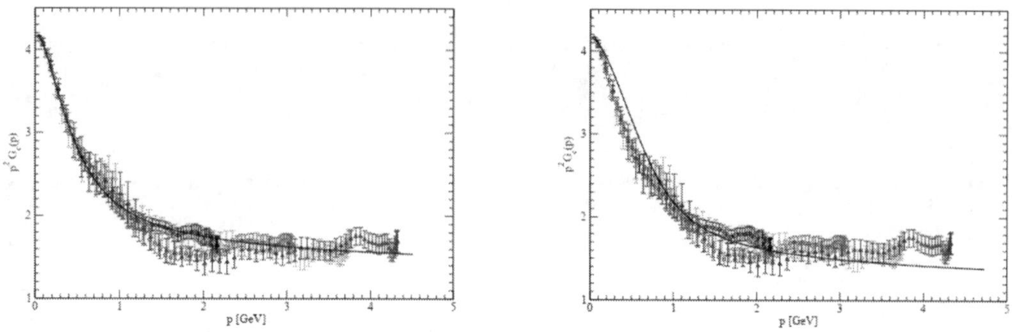

Figure 2. The ghost dressing function $F_c(p^2) = p^2 G_c(p^2)$, where $G_c(p^2) = 1/\Gamma_{c\bar{c}}(p^2)$ is the ghost propagator, with the definition of the renormalized coupling constant in the Taylor scheme (or Tissier-Wschebor's original scheme, to the left) and in the symmetry point scheme (to the right), compared to the lattice results of Ref. [16]. The different colors of the lattice data visible in the online version correspond to different directions on the lattice.

There are two circumstances that complicate the fits, or the determination of the free parameters. One is the fact that the lattice propagators are not normalized in any specific way, hence, for the comparison to the propagators that are obtained from the integration of the Callan-Symanzik equations, they are still to be multiplied with arbitrary overall constants (in this sense, there are two more parameters to be fitted). The other problem is the comparatively lower precision of the lattice data in the UV regime, in particular, the breaking of (continuous) rotational invariance on the lattice which becomes visible in this regime, cf. the plots of the ghost and gluon dressing functions below. In addition, one has to keep in mind that our results are renormalization group improvements of perturbative one-loop expressions, while it is well-known that, for a reliable fit to the UV data, renormalization group improved two-loop results are needed.

In Figs. 2–4 we show the best fits to the lattice data we have obtained in the original Tissier-Wschebor scheme with the definition of the renormalized coupling constant in the Taylor limit of the ghost-gluon vertex, and in the scheme that uses the symmetry point of the vertex for the definition of the coupling constant. The results obtained from the integration of the Callan-Symanzik equations for the ghost dressing function in the Tissier-Wschebor (or Taylor) scheme in Fig. 2 (to the left) are slightly different from the ones obtained by Peláez, Tissier and Wschebor in Ref. [18]. The difference lies in their fitting of the renormalization group improved result for the ghost dressing function to the lattice data for momenta in the directions $(0, 0, 1, 1)$ and $(1, 1, 1, 1)$ (red and blue data points in the online version), in the momentum regime where the breaking of rotational invariance on the lattice is visible, while we have found better overall fits by adjusting our curves to the data for momentum directions $(0, 0, 0, 1)$ and $(0, 1, 1, 1)$. Considering the data in the latter directions, our fit in Fig. 2 for the Tissier-Wschebor scheme is perfect for momenta up to 3.7 GeV, in the sense that the renormalization group improved curve stays within the statistical error bars for all data points.

In contrast, in the renormalization scheme that uses the ghost-gluon vertex at the symmetry point for the definition of the renormalized coupling constant, no such perfect fit to the ghost dressing function (or the ghost propagator, for that matter) can be achieved. Furthermore, the fit to the gluon propagator and the gluon dressing function in Fig. 4 is clearly poorer than the corresponding fit in Tissier-Wschebor's original (or Taylor) scheme. It is, hence, the latter choice that leads to better results for the gluon and ghost propagators. Notice how the plots of the gluon propagator give a

DOI: 10.1051/epjconf/201716403003

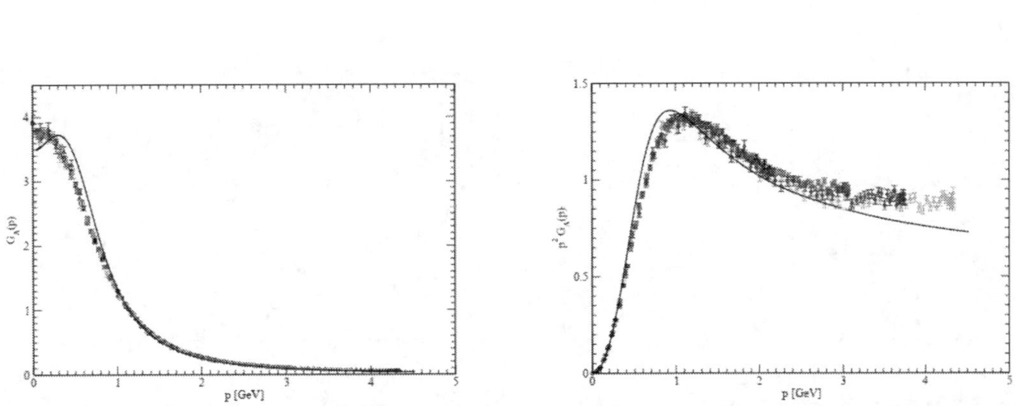

Figure 3. The gluon propagator $G_A(p^2) = 1/\Gamma^T_{AA}(p^2)$ (to the left) and the gluon dressing function $F_A(p^2) = p^2 G_A(p^2)$ (to the right) in the original Tissier-Wschebor or Taylor scheme, compared to the lattice results of Ref. [17].

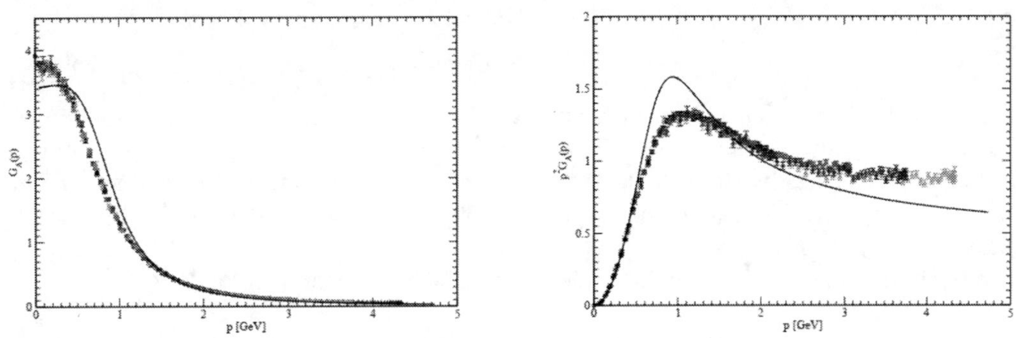

Figure 4. Gluon propagator (to the left) and gluon dressing function (to the right) in the symmetry point scheme, compared to the lattice results.

clearer picture of the quality of the fits in the extreme IR regime, while the plots of the gluon dressing function emphasize the mid- and large-momentum regimes.

We have also considered a different class of renormalization schemes that use the p^2-derivatives of the proper two-point functions in order to fix the field renormalization constants. Note that Z_A is determined in the Tissier-Wschebor scheme from the difference of Eqs. (2) and (3),

$$\left[\Gamma^T_{AA}(p^2) - \Gamma^L_{AA}(p^2)\right]\Big|_{p^2=\mu^2} = \mu^2 \,. \tag{5}$$

In the derivative schemes, we replace the latter condition with

$$\frac{\partial}{\partial p^2}\left[\Gamma^T_{AA}(p^2) - \zeta\,\Gamma^L_{AA}(p^2)\right]\Big|_{p^2=\mu^2} = 1 \,, \tag{6}$$

where we have, furthermore, introduced an *a priori* arbitrary parameter ζ, taking advantage of the fact that the "classical" longitudinal part of the proper two-point function, i.e., the tree-level contribution

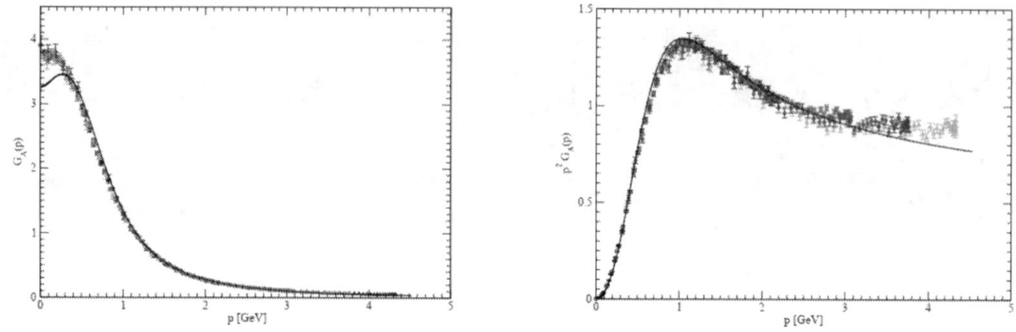

Figure 5. The gluon propagator $G_A(p^2) = 1/\Gamma^T_{AA}(p^2)$ (to the left) and the gluon dressing function $F_A(p^2) = p^2 G_A(p^2)$ (to the right) in the $\zeta = 1$ derivative scheme, compared to the lattice results of Ref. [17].

to the longitudinal part of the proper gluonic two-point function that is derived from Eq. (1), is p^2-independent. We keep the normalization condition (3) for the mass, but replace condition (4) with

$$\frac{\partial}{\partial p^2}\Gamma_{c\bar{c}}(p^2)\bigg|_{p^2=\mu^2} = 1\,, \tag{7}$$

in analogy with Eq. (6). Given our previous experience with the two "natural" definitions of the renormalized coupling constant, we will use the Taylor limit of the ghost-gluon vertex for its definition in the derivative schemes [we have actually checked that its definition from the symmetry point of the vertex leads to poorer fits to the gluon and ghost propagators, just as in the case of the normalization conditions (2)–(4)].

The IR limit of the beta function in the derivative schemes is (for N colors)

$$\beta_g = \mu\frac{d}{d\mu}g(\mu) = \frac{3\zeta-1}{12}\frac{Ng^2}{(4\pi)^2}g\,. \tag{8}$$

In order to avoid a Landau pole in the integration of the renormalization group equations, one needs the beta function to be positive in the IR, hence $\zeta > 1/3$ (in particular, for $\zeta = 0$ a Landau pole is inevitable). The "critical" value $\zeta = 1/3$ still leads to a finite (non-zero) value of $g(\mu)$ [and also of $m^2(\mu)$] in the limit $\mu \to 0$.

We have plotted our best fits to the gluon propagator and dressing function for the derivative schemes with $\zeta = 1$ and $\zeta = 1/3$ in Figs. 5 and 6, employing the same strategy for the fits as before. The corresponding fits to the ghost dressing function are practically identical to the one for the original Tissier-Wschebor scheme in Fig. 2 (to the left), perfect for momenta up to 3.7 GeV in the sense explained above. We conclude that the renormalization group improved results in the derivative schemes, in particular for the critical case $\zeta = 1/3$, fit the lattice data better than in the original Tissier-Wschebor scheme. Our fits should also be compared to more recent results of the solution of the Dyson-Schwinger equations or the in many aspects similar functional renormalization group equations, see Ref. [19].

Note that there is a decrease of the gluon propagator towards the extreme IR in the renormalization group improved curves, for all renormalization schemes considered. This decrease is not seen in the lattice data, however, a slight decrease is not excluded by the data, either. The decrease gets more

DOI: 10.1051/epjconf/201716403003

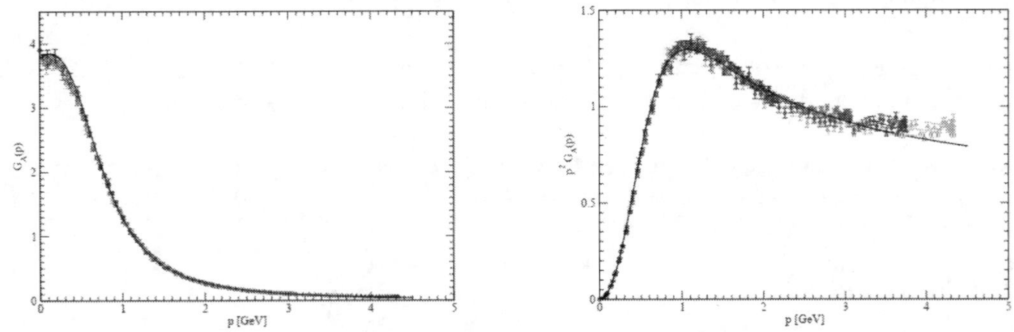

Figure 6. Gluon propagator (to the left) and gluon dressing function (to the right) in the critical $\zeta = 1/3$ derivative scheme, compared to the lattice results.

pronounced in the derivative schemes with increasing value of the parameter ζ, which is why the critical value $\zeta = 1/3$, the smallest possible, gives the best fit to the lattice data.

In conclusion, we have demonstrated that a renormalization group improvement of the perturbative one-loop results is sufficient to successfully reproduce the lattice data for the gluon and ghost propagators in Landau gauge Yang-Mills theory over the whole momentum range from the far UV to the deep IR, even better in a quantitative sense than through the application of Dyson-Schwinger equations (or functional renormalization group equations) [19], if one allows for a non-zero gluon mass parameter in the "classical" action. The usual massless Yang-Mills theory and its well-known behavior including the BRST symmetry is recovered in the UV regime, while the running mass parameter avoids the appearance of a Landau pole. In the Callan-Symanzik approach, only two parameters, the values of the coupling constant and the mass parameter at a given renormalization scale, have to be adjusted to achieve a good fit to the lattice results. The renormalization scheme dependence of the results can be used to improve the matching to the lattice data. Within the class of renormalization schemes we have considered, the best fits are obtained in the derivative scheme with the critical value $1/3$ of the ζ parameter [see Eq. (6)].

One of the important advantages of the Callan-Symanzik approach, as compared to the use of Dyson-Schwinger equations, or functional renormalization group equations for that matter, is the possibility of systematically improving the results by going to higher loop orders, without introducing any additional parameters. The calculation of the flow functions is analytical (the evaluation of some of the appearing Feynman diagrams may require numerical integration, but at least the IR and UV limits can usually be extracted analytically), and the renormalization group equations constitute a set of coupled ordinary first-order differential equations which can be handled numerically without difficulty.

The application of the Callan-Symanzik equations is not limited to the gluon and ghost propagators, but can be extended to the vertex functions. The advantages of the approach are even more evident in this case: there is no need to guess at the most important structures or to construct *ansätze* for the vertex functions, everything is in principle determined (to the order considered) by the corresponding Feynman diagrams once the renormalization scheme and the two free parameters are fixed, compare, e.g., Ref. [18] to Ref. [20]. As an example, the analytical evaluation in the IR limit of the one-loop corrections to the ghost-gluon and the three-gluon vertices at the respective symmetry points (with the tensor structures of the tree-level vertices, which give the dominant contributions in the UV as well as in the IR limit) shows that the ghost-gluon vertex function tends towards a non-zero finite constant, while the three-gluon vertex function becomes proportional to $\ln(p^2/\Lambda_1^2)$ for small momenta (at the symmetry point) $p_1^2 = p_2^2 = p_3^2 = p^2$, and Λ_1 is a new IR scale where the three-gluon vertex function changes sign. This sign change ("zero crossing") has been obtained both in the Callan-Symanzik and the Dyson-Schwinger approaches [18, 20], and it has been observed in lattice simulations in three space-time dimensions, while the data in four dimensions are still inconclusive [21]. The new scale Λ_1 is obtained as a function of the two free parameters when the Callan-Symanzik equations are integrated completely (the value of the coupling constant at a given renormalization scale may be expressed through $\Lambda_{\rm QCD}$).

Acknowledgements

The authors would like to thank Attilio Cucchieri and Tereza Mendes for allowing them to use their lattice data. A.W. wishes to thank the organizers of the conference, and in particular Hugo Reinhardt, for the invitation, and Dan Zwanziger for a brief discussion. Support by Conacyt project CB-2013/222812 and CIC-UMSNH is gratefully acknowledged.

References

[1] L. von Smekal, A. Hauck and R. Alkofer, Phys. Rev. Lett. **79**, 3591 (1997); C. Fischer and R. Alkofer, Phys. Lett. B **536**, 177 (2002)

[2] A.C. Aguilar and A.A. Natale, JHEP**0408**, 057 (2004); P. Boucaud, T. Brüntjen, J.P. Leroy *et al.*, JHEP**0606**, 001 (2006); M. Frasca, Phys. Lett. B **670**, 73 (2008)

[3] V.N. Gribov, Nucl. Phys. B **139**, 1 (1978)

[4] D. Zwanziger, Nucl. Phys. B **323**, 513 (1989); Nucl. Phys. B **399**, 477 (1993)

[5] D. Dudal, J.A. Gracey, S.P. Sorella *et al.*, Phys. Rev. D **78**, 065047 (2008)

[6] I.L. Bogolubsky, E.M. Ilgenfritz, M. Müller-Preussker and A. Sternbeck, Proc. Sci. LAT2007, 290 (2007); A. Cucchieri and T. Mendes, Proc. Sci. LAT2007, 297 (2007); A. Sternbeck, L. von Smekal, D.B. Leinweber and A.G. Williams, Proc. Sci. LAT2007, 340 (2007)

[7] A. Weber, Phys. Rev. D **85**, 125006 (2012); A. Weber, P. Dall'Olio and F. Astorga, Int. J. Mod. Phys. E **25**, 1642002 (2016)

[8] M. Tissier and N. Wschebor, Phys. Rev. D **82**, 101701 (2010)

[9] T. Kugo and I. Ojima, Prog. Theor. Phys. Suppl. **66**, 1 (1979)

[10] G. Curci and R. Ferrari, Nuovo Cim. A **32**, 151 (1976)

[11] G. Curci and R. Ferrari, Nuovo Cim. A **35**, 1 (1976); I. Ojima, Z. Phys. C **13**, 173 (1982); J. de Boer, K. Skenderis, P. van Nieuwenhuizen and A. Waldron, Phys. Lett. B **367**, 175 (1996)

[12] K.-I. Kondo, K. Suzuki, H. Fukamachi *et al.*, Phys. Rev. D **87**, 025017 (2013); M. Schaden and D. Zwanziger, Phys. Rev. D **92**, 025001 (2015)

[13] D. Dudal and S.P. Sorella, Phys. Rev. D **86**, 045005 (2012); M.A.L. Capri, D. Dudal, D. Fiorentini *et al.*, Phys. Rev. D **94**, 025035 (2016)

[14] M. Tissier and N. Wschebor, Phys. Rev. D **84**, 045018 (2011)

[15] J.C. Taylor, Nucl. Phys. B **33**, 436 (1971)

[16] A. Cucchieri and T. Mendes, Phys. Rev. D **78**, 094503 (2008)

[17] A. Cucchieri and T. Mendes, Phys. Rev. Lett. **100**, 241601 (2008)

[18] M. Peláez, M. Tissier and N. Wschebor, Phys. Rev. D **88**, 125003 (2013)

[19] C.S. Fischer, A. Maas and J.M. Pawlowski, Ann. Phys. **324**, 2408 (2009); M.Q. Huber and L. von Smekal, JHEP**1304**, 149 (2013)
[20] A.L. Blum, M.Q. Huber, M. Mitter and L. von Smekal, Phys. Rev. D **89**, 061703 (2014)
[21] A. Cucchieri, A. Maas and T. Mendes, Phys. Rev. D **77**, 094510 (2008)

Hamiltonian approach to QCD in Coulomb gauge: Gribov's confinement scenario at work*

H. Reinhardt[1,a], G. Burgio[1], D. Campagnari[1], M. Quandt[1], P. Vastag[1], H. Vogt[1], and E. Ebadati[1]

[1] *Universität Tübingen*
Institut für Theoretische Physik
Auf der Morgenstelle 14
D-72076 Tübingen
Germany

Abstract. I will review essential features of the Hamiltonian approach to QCD in Coulomb gauge showing that Gribov's confinement scenario is realized in this gauge. For this purpose I will discuss in detail the emergence of the horizon condition and the Coulomb string tension. I will show that both are induced by center vortex gauge field configurations, which establish the connection between Gribov's confinement scenario and the center vortex picture of confinement. I will then extend the Hamiltonian approach to QCD in Coulomb gauge to finite temperatures, first by the usual grand canonical ensemble and second by the compactification of a spatial dimension. I will present results for the pressure, energy density and interaction measure as well as for the Polyakov loop.

1 Introduction

V. N. Gribov's investigation of the problem of gauge fixing in Yang–Mills theory has led to a novel picture of confinement. In this scenario confinement comes about due to the field configurations near the Gribov horizon, which give rise to an infrared diverging ghost form factor. This approach was worked out in more detail and developed further by D. Zwanziger and is now referred to as the Gribov–Zwanziger confinement mechanism [1, 2]. In my talk, I will show that this picture is realized in the Hamiltonian approach to Yang–Mills theory in Coulomb gauge. Gribov already pointed out that in Coulomb gauge the use of the Hamiltonian approach may be advantageous over the usual functional integral approach and this, indeed, turns out to be the case. The variational approach to the Hamilton formulation of QCD in Coulomb gauge developed in refs. [3, 4] has proven to be much more efficient than the formulation of QCD in Coulomb gauge by means of the functional integral approach [5]. In my talk I will concentrate on a few aspects of the Hamiltonian approach to QCD in Coulomb gauge. I will discuss in detail the emergence of the horizon condition which is the core of the Gribov–Zwanziger confinement scenario. Furthermore, I will show that the Coulomb string tension is not related to the temporal but to the spatial Wilsonian string tension. I will extend then

*Talk given by H. Reinhardt at "5th International Conference on New Frontiers in Physics", 6-14 July 2016, Kolymbari, Greece.

[a] e-mail: hugo.reinhardt@uni-tuebingen.de

the Hamiltonian approach to finite temperatures, first by means of the grand canonical ensemble and second by compactification of a spatial dimension.

2 Hamiltonian approach to Yang–Mills theory in Coulomb gauge at $T = 0$

After canonical quantization and implementation of Gauß's law in Weyl and Coulomb gauge, the gauge fixed Yang–Mills Hamiltonian reads [6]

$$H = \frac{1}{2}\int \mathrm{d}^3x \left(J^{-1}[A]\boldsymbol{\Pi}^a(\boldsymbol{x}) \cdot J[A]\boldsymbol{\Pi}^a(\boldsymbol{x}) + \boldsymbol{B}^a(\boldsymbol{x}) \cdot \boldsymbol{B}^a(\boldsymbol{x})\right) + H_C \,. \tag{1}$$

Here

$$B_k^a(\boldsymbol{x}) = \varepsilon_{klm}\left(\partial_l A_m^a(\boldsymbol{x}) - \frac{g}{2} f^{abc} A_l^b(\boldsymbol{x}) A_m^c(\boldsymbol{x})\right) \tag{2}$$

is the non-Abelian magnetic field, $\Pi_k^a(\boldsymbol{x}) = \delta/(\mathrm{i}\delta A_k^a(\boldsymbol{x}))$ is the momentum operator, which represents the color electric field and

$$J[A] = \mathrm{Det}(-\hat{\boldsymbol{D}} \cdot \boldsymbol{\partial}) \,, \qquad \hat{D}_k^{ab}(\boldsymbol{x}) = \delta^{ab}\partial_k^x + g\hat{A}_k^{ab}(\boldsymbol{x}) \tag{3}$$

is the Faddeev–Popov determinant, where $\hat{A}_k^{ab} = f^{acb} A_k^c$ is the gauge field in the adjoint representation. Furthermore,

$$H_C = \frac{1}{2}\int \mathrm{d}^3x \int \mathrm{d}^3y \, J[A]^{-1}\rho^a(\boldsymbol{x}) J[A] \left[(-\hat{\boldsymbol{D}} \cdot \boldsymbol{\partial})^{-1}(-\boldsymbol{\partial}^2)(-\hat{\boldsymbol{D}} \cdot \boldsymbol{\partial})^{-1}\right]^{ab}(\boldsymbol{x}, \boldsymbol{y})\rho^b(\boldsymbol{y}) \tag{4}$$

is the so-called Coulomb term, which arises from the kinetic energy of the longitudinal part of the momentum operator after resolving Gauß's law. Here,

$$\rho^a(\boldsymbol{x}) = -f^{abc}\boldsymbol{A}^b(\boldsymbol{x}) \cdot \boldsymbol{\Pi}^c(\boldsymbol{x}) + \rho_m^a(\boldsymbol{x}) \tag{5}$$

is the color charge density, where the first term on the right-hand side is the contribution of the gauge field itself while the last term, ρ_m, refers to the matter fields. Due to the implementation of Coulomb gauge the scalar product between wave functionals

$$\langle \phi | \ldots | \psi \rangle = \int \mathcal{D}A \, J[A] \phi^*[A] \ldots \psi[A] \tag{6}$$

is defined by the functional integral over transversal gauge fields with the Faddeev–Popov determinant (3) in the integration measure representing the Jacobian of the change of variables from "Cartesian" to "curvilinear" variables in Coulomb gauge. With the gauge fixed Hamiltonian (1) one has to solve the stationary Schrödinger equation $H\psi[A] = E\psi[A]$ for the vacuum wave functional $\psi[A]$. Once $\psi[A]$ is known, all observables and correlation functions can, in principle, be calculated. This has been attempted by means of the variational principle using Gaussian type ansätze for the vacuum wave functional [7, 8]. However, the first attempts did not properly include the Faddeev–Popov determinant, which turns out to be crucial in order to describe the confinement properties of the theory. Below, I am discussing the variational approach developed in refs. [3, 4], which differs from previous attempts by the ansatz for the vacuum wave functional, the treatment of the Faddeev–Popov determinant and, equally important, in the renormalization, see ref. [9] for further details.

Figure 1: Dyson–Schwinger equation for the ghost propagator.

2.1 Variational solution of the Schrödinger equation

The ansatz for the vacuum wave functional is inspired by the quantum mechanics of a particle in a spherically symmetric potential for which the ground state wave functional reads $\psi(r) = u(r)/r$ where the radial wave functional $u(r)$ satisfies a usual one-dimensional Schrödinger equation and r represents (the square root of the radial part of) the Jacobian of the transformation from the Cartesian to spherical coordinates. Our ansatz for the vacuum wave functional is given by

$$\psi[A] = \frac{1}{\sqrt{J[A]}} \exp\left[-\frac{1}{2}\int \mathrm{d}^3x \int \mathrm{d}^3y \, A_k^a(\boldsymbol{x})\omega(\boldsymbol{x},\boldsymbol{y})A_k^a(\boldsymbol{y})\right] . \tag{7}$$

The inclusion of the pre-exponential factor has the advantage that it eliminates the Faddeev–Popov determinant from the integration measure in the scalar product (6). Furthermore, for the wave function (7) the gluon propagator is given up to a factor of $\frac{1}{2}$ by the inverse of the variational kernel $\omega(\boldsymbol{x},\boldsymbol{y})$. It turns out that in the Yang–Mills sector the Coulomb term H_C (4) can be ignored. Calculating the expectation value of the remaining parts of the Yang–Mills Hamiltonian (1) with the wave functional (7) up to two loops, the minimization of the energy density with respect to $\omega(\boldsymbol{x},\boldsymbol{y})$ yields the following gap equation in momentum space[1]

$$\omega^2(k) = \boldsymbol{k}^2 + \chi^2(k) + c\,, \tag{8}$$

where c is a finite renormalization constant resulting from the tadpole and

$$\chi_{kl}^{ab}(\boldsymbol{x},\boldsymbol{y}) = -\frac{1}{2}\left\langle\psi\left|\frac{\delta^2 \ln J[A]}{\delta A_k^a(\boldsymbol{x})\delta A_l^b(\boldsymbol{y})}\right|\psi\right\rangle \tag{9}$$

represents the ghost loop. This can be expressed in terms of the ghost propagator

$$G(\boldsymbol{x},\boldsymbol{y}) = \langle\psi|(-\hat{\boldsymbol{D}}\cdot\boldsymbol{\partial})^{-1}(\boldsymbol{x},\boldsymbol{y})|\psi\rangle\,, \tag{10}$$

which is evaluated with the vacuum wave functional (7) in an approximate way resulting in a Dyson–Schwinger equation for the form factor

$$d(\boldsymbol{k}) = g\boldsymbol{k}^2 G(\boldsymbol{k}) \tag{11}$$

of the ghost propagator which is diagrammatically illustrated in fig. 1. This equation has to be solved together with the gap equation (8). The numerical solutions are shown in fig. 2.

[1] Due to translational and rotational invariance, kernels such as $\omega(\boldsymbol{x},\boldsymbol{y})$ can be Fourier transformed as

$$\omega(\boldsymbol{x},\boldsymbol{y}) = \int \frac{d^3k}{(2\pi)^3}\, e^{i\boldsymbol{k}\cdot(\boldsymbol{x}-\boldsymbol{y})}\,\omega(k)\,,$$

where the new kernel in momentum space depends on $k = |\boldsymbol{k}|$ only. For simplicity, we will use the same symbol for the kernel in position and momentum space and go back and forth between both representations with impunity.

 DOI: 10.1051/epjconf/201716403004

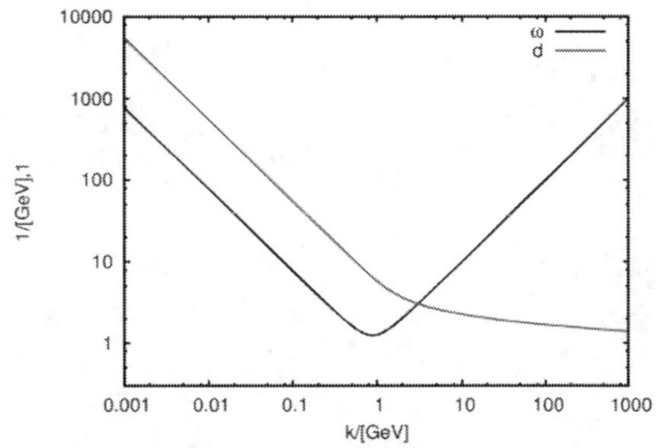

Figure 2: Numerical solution of the coupled gap equation for ω (8) and Dyson–Schwinger equation for the ghost form factor d for the renormalization constant $c = 0$ [10].

2.2 Physical implications of the ghost form factor

The ghost form factor expresses the deviation of Yang–Mills theory from QED. Dyson–Schwinger equations are functional differential equations and their solutions are uniquely determined after providing appropriate boundary conditions. In the present case the so-called horizon condition

$$d^{-1}(0) = 0 \tag{12}$$

is assumed, which is the key point in Gribov's confinement scenario.

Alternatively to the variational approach, one can indirectly determine the vacuum wave functional by solving the functional renormalization group flow equations for the various propagators and functions of the Hamiltonian approach. Restricting the flow equations to those for the ghost and gluon propagators, one finds for the ghost form factor the result shown in fig. 3. Starting with a constant ghost form factor in the ultraviolet, the ghost form factor develops an infrared singularity as the momentum cutoff of the flow equation tends to zero. This is nicely seen in fig. 3 (b), which shows a cut through fig. 3 (a) at fixed renormalization group scale k.

Let us also mention that the horizon condition (12) need not be assumed in the case of $D = 2 + 1$ dimensions, where it is a necessary consequence of the coupled equations for the ghost and gluon propagators obtained with the variational principle. Finally, the horizon condition (12) is also seen in the lattice data for the ghost form factor shown in fig. 4.

Coulomb gauge is called a physical gauge since in QED the remaining transversal components are the gauge invariant degrees of freedom. This is not the case for Yang–Mills theory. However, Coulomb gauge can be viewed as a physical gauge also in the case of Yang–Mills theory in the sense that the inverse ghost form factor in Coulomb gauge represents the dielectric function of the Yang–Mills vacuum [13]

$$\epsilon(k) = d^{-1}(k)\,. \tag{13}$$

The horizon condition (12) guarantees that this function vanishes in the infrared, $\epsilon(k = 0) = 0$. This implies that the Yang–Mills vacuum is a perfect color dielectricum, i.e. a dual superconductor. In this way the Hamiltonian approach in Coulomb gauge relates Gribov's confinement scenario to the dual Meißner effect, a confinement mechanism realized through the condensation of magnetic monopoles and proposed by Mandelstam and 't Hooft [14, 15]. The dielectric function obtained here as inverse

DOI: 10.1051/epjconf/201716403004

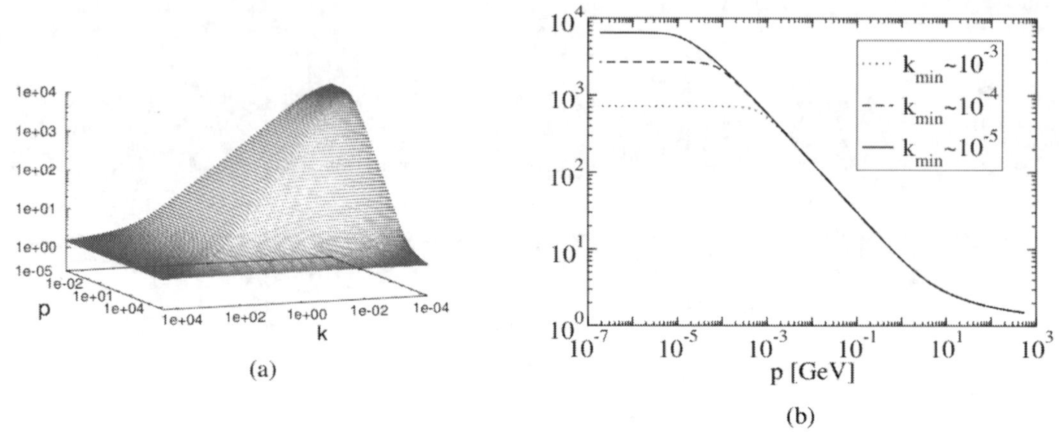

Figure 3: (a) The ghost form factor obtained in ref. [11] from the solution of the renormalization group flow equations. Here p represents the momentum variable of the ghost form factor while k is the infrared momentum cutoff of the flow equations. (b) Cuts through subfigure (a) at various values of the momentum scale k of the flow equations.

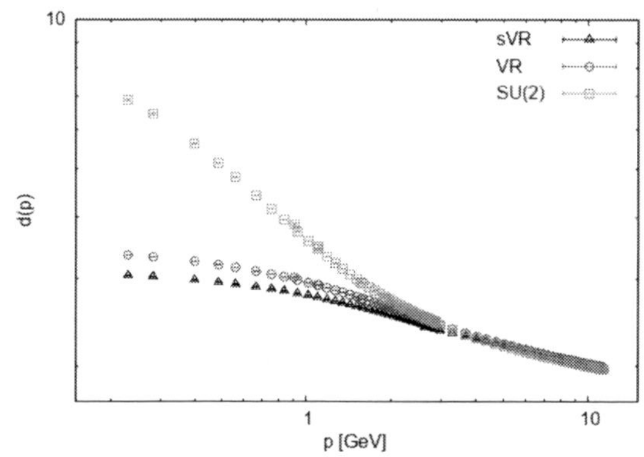

Figure 4: The ghost form factor in Coulomb gauge calculated on the lattice in ref. [12] (green symbols). The red and black symbols show the results obtained for the ghost form factor when all center vortices or the spatial center vortices are removed from the ensemble of gauge field configurations, see text.

ghost form factor is also in accord with the phenomenological bag model picture of hadrons. Inside the hadron, i.e. at small distance, the dielectric function is that of a normal vacuum while outside the physical hadrons the vanishing of the dielectric constant implies the absence of free charges by Gauß's law.

One may now ask, what field configurations induce the horizon condition and thus confinement? Given the relation of Gribov's confinement scenario to the dual superconductor, we expect magnetic monopoles to play a substantial role. Lattice calculations carried out in the so-called indirect maximum center gauge, which contains the maximum Abelian gauge in an intermediate step, show that magnetic monopoles are tied to center vortices. The latter are string-like gauge field configurations

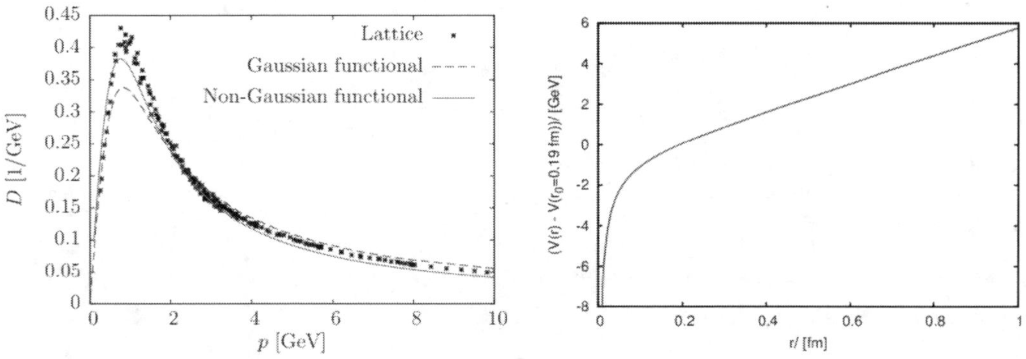

(a) (b)

Figure 5: (a) The static gluon propagator in Coulomb gauge calculated on the lattice for SU(2) gauge theory (crosses). The dashed and the full curves show the result of the variational calculation using, respectively, a Gaussian and non-Gaussian ansatz for the vacuum wave functional. (b) The non-Abelian Coulomb potential (16) obtained in the variational approach [10].

for which the Wilson loop equals a non-trivial center element of the gauge group, provided the loop has non-trivially linking with the center vortex string. When the center vortex content of the gauge field configurations is removed, which can be done on the lattice, one finds that the Wilsonian string tension and thus confinement disappears [16]. Figure 4 shows the ghost loop obtained on the lattice when the center vortices are removed from the ensemble of gauge field configurations [12]. The ghost loop becomes infrared flat and the horizon condition is lost. This shows that center vortices induce the horizon condition which is the corner stone of Gribov's confinement scenario. This also shows that Gribov's confinement scenario is tied to the center vortex picture of confinement. This is in accord with the observation that center vortices and magnetic monopoles are located on the Gribov horizon of Coulomb gauge [17].

2.3 Comparison with lattice calculation

Let us now compare the results of the variational solution with lattice calculations [18]. Figure 5 (a) shows the gluon propagator of Coulomb gauge obtained in SU(2) gauge theory. It is remarkable that the lattice data can be nicely fitted by Gribov's formula [1]

$$\omega(p) = \sqrt{p^2 + \frac{M^4}{p^2}} \tag{14}$$

where M is the so-called Gribov mass. The variational calculations reproduce the infrared behavior of the lattice propagator perfectly and are also in reasonably agreement with the lattice data in the ultraviolet. However, in the mid-momentum regime some strength is missing in the variational calculation. This missing strength is the result of the Gaussian type ansatz for the vacuum wave functional. In ref. [19], the ansatz for the vacuum wave functional was extended to include also cubic and quartic terms of the gauge field in the exponent of the vacuum wave functional

$$\psi[A] \sim \exp[-S[A]]\,, \tag{15a}$$

$$S[A] = \frac{1}{2}\int A\omega A + \frac{1}{3!}\int \gamma^{(3)}AAA + \frac{1}{4!}\int \gamma^{(4)}AAAA \tag{15b}$$

and one finds the full curve in fig. 5 (a), which gives a much better agreement with the lattice data in the mid-momentum regime.

2.4 The Coulomb string tension

So far we have ignored the Coulomb Hamiltonian H_C (4) which is reasonable in the Yang–Mills sector. However, this term becomes extremely important in the quark sector. Let us consider the piece of H_C quadratic in the color density of the matter fields ρ_m. From this term the Faddeev–Popov determinant drops out and its Yang–Mills vacuum expectation value provides the static potential for the matter fields

$$V_C(|\boldsymbol{x}-\boldsymbol{y}|) = g^2\left\langle\psi\middle|\left[(-\hat{\boldsymbol{D}}\cdot\boldsymbol{\partial})^{-1}(-\boldsymbol{\partial}^2)(-\hat{\boldsymbol{D}}\cdot\boldsymbol{\partial})^{-1}\right](\boldsymbol{x},\boldsymbol{y})\middle|\psi\right\rangle, \tag{16}$$

which is referred to as the non-Abelian Coulomb potential. The Coulomb potential resulting from the variational approach is shown in fig. 5 (b). At small distance it behaves like an ordinary Coulomb potential but rises linearly at large distance with a coefficient given by the so-called Coulomb string tension σ_C, which can be shown to represent an upper bound to the Wilsonian string tension σ_W [20]. On the lattice one finds a value of $\sigma_C = 2\ldots4\,\sigma_W$ [12, 17, 21]. At finite temperatures the Wilsonian string tension σ_W splits into a spatial and a temporal one. Above the deconfinement phase transition, the temporal string tension σ_t vanishes while the spatial string tension σ_s even slightly increases with the temperature. On the lattice one finds that the temporal and spatial string tension are induced by the spatial and temporal, respectively, center vortex content of the gauge field. When one removes the spatial center vortices from the ensemble of gauge fields, the temporal string tension is untouched while the spatial string tension disappears. Figure 6 (a) shows the non-Abelian Coulomb potential calculated on the lattice. When one removes the spatial center vortices this Coulomb potential becomes infrared flat, i.e. the Coulomb string tension σ_C disappears. This shows that the Coulomb string tension is tied to the spatial (Wilsonian) string tension and not to the temporal one, because both are sensitive to the same underlying degrees of freedom (the spatial center vortices). Since the spatial string tension increases with the temperature above the critical temperature we should expect the same behavior for the Coulomb string tension. This, indeed, is observed on the lattice as can be seen from fig. 6 (b), where the quantity $p^4 V_C(p)$ is shown for various temperatures. As can be seen from this figure, the Coulomb string tension

$$\sigma_C = \lim_{p\to0} p^4 V_C(p) \tag{17}$$

remains more or less constant as the temperature is increased from 0 to 1.5 times the critical temperature, and then increases strongly when further increasing the temperature to 3 times the critical one. This is in accordance with the temperature behavior of the spatial string tension found in lattice calculations [22].
The variational approach to Yang–Mills theory developed in refs. [3, 4] has also been extended to full QCD, see refs. [23–25]. Due to lack of time I will not consider the quark sector within my talk.

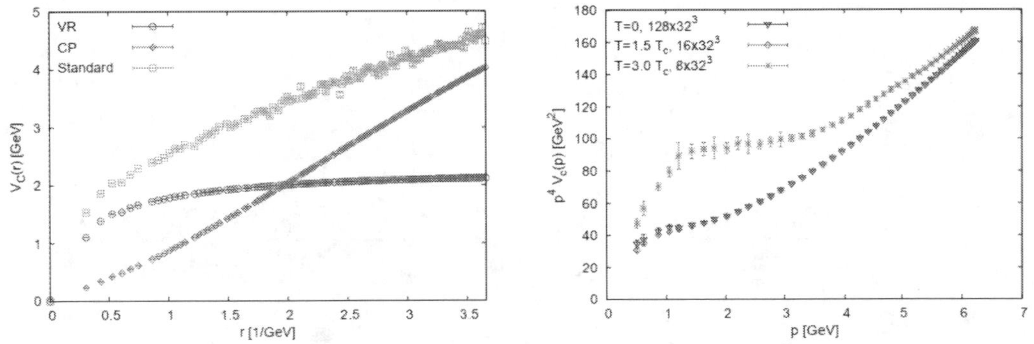

(a) (b)

Figure 6: (a) Non-Abelian Coulomb potential V_C in coordinate space calculated on the lattice (green symbols). Red and violet symbols show the results obtained by vortex removal and center projection, respectively. (b) Measurements for the quantity $p^4 V_C(p)$ as function of the momentum p for various temperatures [12].

3 Hamiltonian approach to Yang–Mills theory at finite temperatures

3.1 Grand canonical ensemble

In refs. [26, 27], the variational approach to Yang–Mills theory in Coulomb gauge was extended to finite temperatures T by making a quasi-particle ansatz for the density matrix of the grand canonical ensemble with zero chemical potential

$$\tilde{D} = \exp\left(-\tilde{H}/T\right) , \tag{18}$$

where $\tilde{H}$ denotes the quasi-gluon (single-particle) Hamiltonian whose quasi-particle energies were determined by minimizing the free energy

$$F = \langle H \rangle_T - TS \to \min . \tag{19}$$

Here,

$$\langle \ldots \rangle_T = \frac{\operatorname{tr}\left(\tilde{D} \ldots\right)}{\operatorname{tr} \tilde{D}} \tag{20}$$

denotes the thermal expectation value of the grand canonical ensemble (i.e. the trace extends over the whole gluonic Fock space) and

$$S = -\operatorname{tr}\left(\frac{\tilde{D}}{\operatorname{tr}\tilde{D}} \ln \frac{\tilde{D}}{\operatorname{tr}\tilde{D}}\right) = -\left\langle \ln \frac{\tilde{D}}{\operatorname{tr}\tilde{D}} \right\rangle_T \tag{21}$$

is the usual entropy defined with the density matrix (18). The variational principle (19) has been solved in ref. [27] analogously to the one at zero temperature. Recently this approach has been used to calculate the pressure, energy density and interaction strength of SU(2) gluon dynamics [28]. The results are shown in fig. 7. The agreement with lattice data is not very impressive but one should keep in mind that these are microscopic self-consistent calculations.

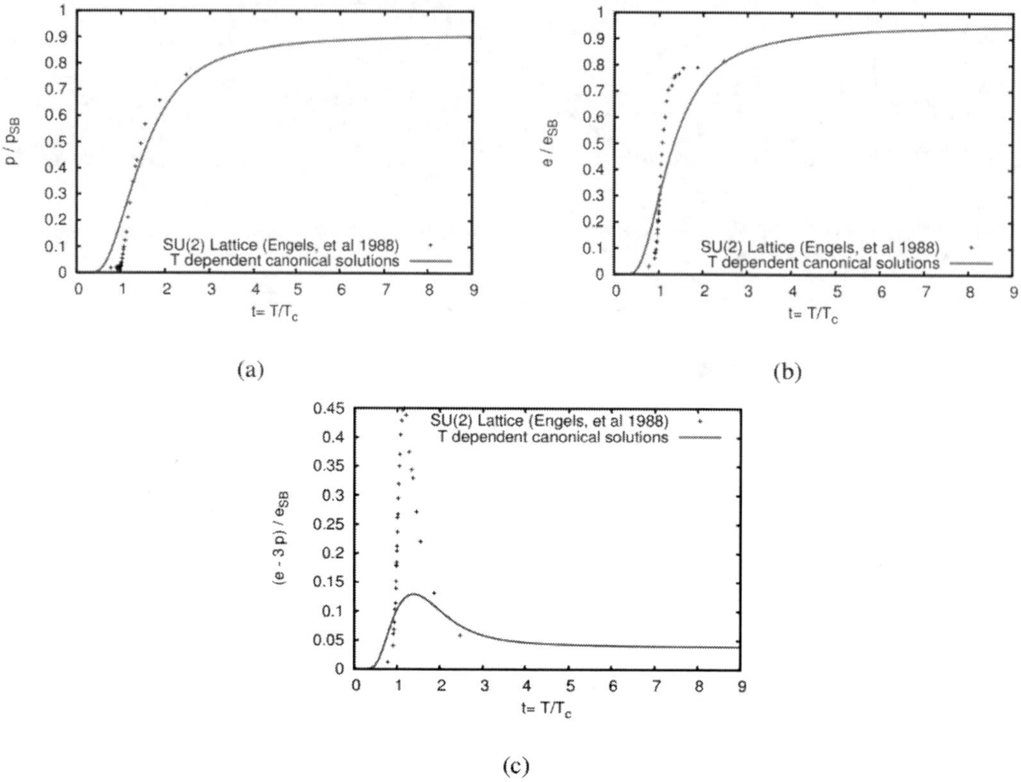

Figure 7: The (a) pressure, (b) energy density and (c) interaction strength of SU(2) gluon dynamics calculated in the finite temperature variational approach to Yang–Mills theory in Coulomb gauge developed in refs. [26, 27].

In ref. [29], gluon dynamics has been studied in a quasi-particle picture using Gribov's formula (14) for the gluon quasi-particle energy. Thereby the Gribov mass was adjusted to reproduce the high temperature tail of the interaction strength yielding a Gribov mass of $M = 705$ MeV which is substantially smaller than the value $M = 880$ MeV found on the lattice [18]. It is therefore not surprising that the model calculation of ref. [29] did not properly describe the interaction strength in the low temperature region. If one wants to reproduce pressure, energy density or interaction strength of Yang–Mills theory within the quasi-particle model used in ref. [29], a temperature dependent Gribov mass is required. This is seen from the lattice calculation of the gluon propagator $D(p)$ at finite temperature presented in fig. 8, where the quantity $D(p)/p$ is shown. Assuming Gribov's formula (14) for the gluon energy at all temperatures, this quantity reduces in the zero momentum limit to $1/(2M^2)$. As can be seen from fig. 8, the Gribov mass M does not change much as one increases the temperature from 0 to $1.5T_C$ but increases substantially when the temperature is further increased to 3 times the critical one.

3.2 Finite temperatures by compactification of a spatial dimension

There is a more efficient way to treat Yang–Mills theory at finite temperature within the Hamiltonian approach. The motivation comes from the Polyakov loop

$$P[A_0](\boldsymbol{x}) = \frac{1}{d_r} \mathrm{tr} P \exp\left[\mathrm{i} \int_0^L \mathrm{d}x^4 \, A_0(\boldsymbol{x}, x^4) \right] , \tag{22}$$

where d_r denotes the dimension of the representation of the gauge group and L is the inverse temperature. [2] The expectation value of the Polyakov loop $\langle P[A_0] \rangle$ is an order parameter for confinement [31]. In the center symmetric confined phase this quantity vanishes while it approaches unity in the deconfined phase. This quantity cannot be calculated straightforwardly in the Hamiltonian approach due to the unrestricted time interval and the use of the Weyl gauge $A_0 = 0$. One can, however, exploit $O(4)$ invariance and interchange the Euclidean time axis with a spatial axis. The temporal (anti-)periodic boundary conditions to the fields become then spatial boundary conditions while the new (Euclidean) time axis has infinite extent, as is required for the Hamiltonian approach. As the result, the partition function is entirely given by the ground state calculated on the spatial manifold $\mathbb{R}^2 \times S^1(L)$ where L is again the inverse temperature. The whole thermodynamics is then encoded in the vacuum calculated on the partially compactified spatial manifold $\mathbb{R}^2 \times S^1(L)$. This approach has been worked out in ref. [32] and was used in refs. [33, 34] to calculate the Polyakov loop within the Hamiltonian approach.

In refs. [35, 36] it was shown that instead of using the expectation value of the Polyakov loop as order parameter of confinement one can also use the Polyakov loop of the expectation value $P[\langle A_0 \rangle]$ or the expectation value of the gauge field $\langle A_0(\boldsymbol{x}) \rangle$ whereas the gauge field has to be in the Polyakov gauge, i.e. it has to be independent of the (Euclidean) time and diagonal in color space. Using this result it turns out that the most efficient way to obtain the Polyakov loop is to calculate the effective potential of a temporal background field a_0 fulfilling the Polyakov gauge and calculate the Polyakov loop from the minimum of this effective potential, say $\bar{a}_0$. The quantity $P[\bar{a}_0]$ can then be used as alternative to $\langle P[A_0] \rangle$. This was done within the Hamiltonian approach in refs. [33, 34] using the finite temperature formulation of quantum field theory developed in ref. [32]. The color diagonal constant background field $\boldsymbol{a} = \boldsymbol{a}_3 t_3$ (t denotes the generator of the color group in the fundamental representation) has to be directed along the compactified spatial axis. Using the zero temperature gluon and ghost propagator as input one finds within this approach for SU(2) the effective potential $e[a]$ shown in fig. 9 (a). The potential shows a second order phase transition. Using the Gribov mass of $M = 880\,\mathrm{MeV}$ for fixing the scale one finds from the self-consistent (zero temperature) solution a critical temperature of $T_C = 269\,\mathrm{MeV}$.

Since the SU(3) algebra consists of three SU(2) algebrae characterized by the three non-zero positive roots

$$\boldsymbol{\sigma} = (1, 0) , \quad \left(\frac{1}{2}, \frac{1}{2}\sqrt{3} \right) , \quad \left(\frac{1}{2}, -\frac{1}{2}\sqrt{3} \right) \tag{23}$$

the effective potential for SU(3) is given by the following sum of SU(2) potentials

$$e_{\mathrm{SU}(3)}[a] = \sum_{\sigma > 0} e_{\mathrm{SU}(2)(\boldsymbol{\sigma})}[a] , \tag{24}$$

[2] Recall that in the Euclidean functional integral approach, finite temperatures are introduced by compactifying the time axis.

 DOI: 10.1051/epjconf/201716403004

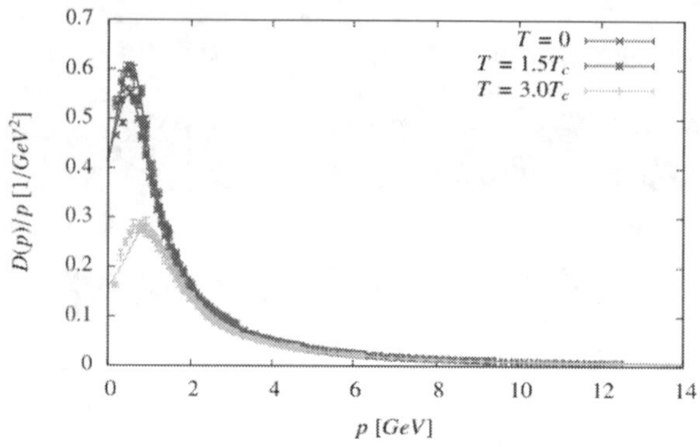

Figure 8: $D(p)/p$ for $T = 0,\ 1.5T_C$ and $3T_C$ [30].

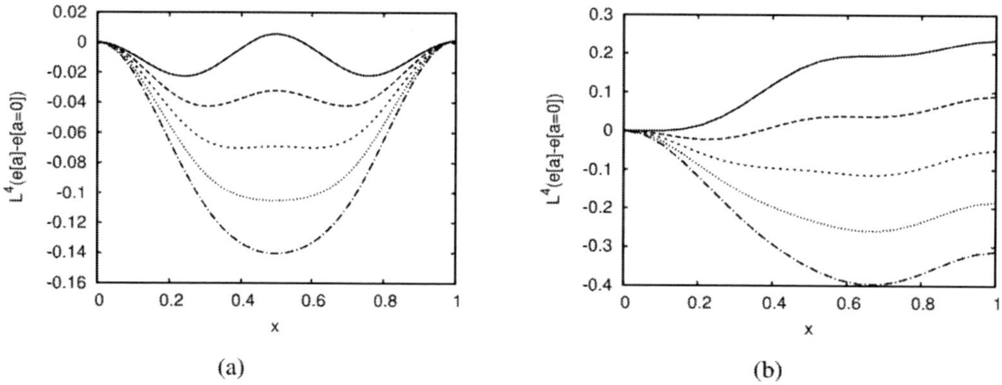

Figure 9: (a) The effective potential of the Polyakov loop (more precisely of the background field along the compactified spatial dimension) for the case of SU(2) for various temperatures (increasing from bottom to top) near the deconfinement phase transition; $x = \frac{a_3 L}{2\pi}$. (b) The $a_8 = 0$ section of the effective potential of the Polyakov loop for the gauge group SU(3) for various temperatures increasing from bottom to top, see also fig. 10.

where the sum runs over all positive non-zero roots given in eq. (23). SU(3) has two generators of the Cartan algebra (t_3, t_8). Therefore the effective potential of the Polyakov loop depends on two background fields a_3 and a_8 and is shown in fig. 10 for two temperatures, one below and one above the critical one. As one observes in these figures both for $T < T_C$ and $T > T_C$ an absolute minimum occurs for $a_8 = 0$. Cutting the effective potential along the $a_8 = 0$ axis one finds the temperature dependent potential shown in fig. 9 (b), which exhibits a first order phase transition. With a Gribov mass of $M = 880$ MeV as input one finds from the self-consistent (zero-temperature) solution a critical temperature of $T_C = 283$ MeV.

Finally, if one includes fermions the deconfinement phase transition is turned into a crossover for both SU(2) and SU(3) [37].

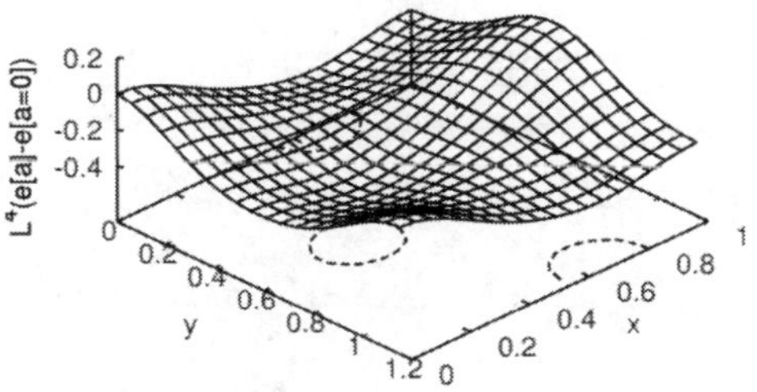

(a)

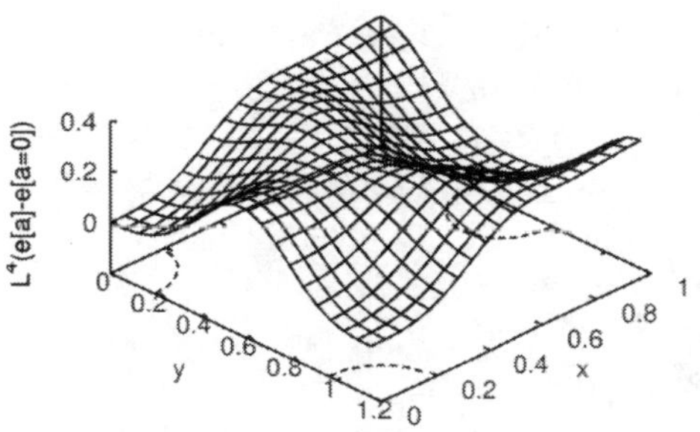

(b)

Figure 10: The effective potential of the Polyakov loop as function of the gauge fields along the Cartan generators a_3 and a_8 (a) for a temperature below the critical one and (b) for a temperature above the critical one; $x = \frac{a_3 L}{2\pi}$, $y = \frac{a_8 L}{2\pi}$.

4 Summary and Conclusions

In my talk I have summarized basic features of the Hamiltonian approach to Yang–Mills theory in Coulomb gauge. I have shown how the Gribov–Zwanziger confinement scenario is realized in this approach, and I have also established the connection with two alternative pictures of confinement, namely, the condensation of magnetic monopoles (dual Meißner effect) and the center vortex picture. In particular, I have shown shown that the Coulomb string tension is not related to the temporal but rather to the spatial Wilsonian string tension and hence has to increase with the temperature above the deconfinement phase transition. I have then extended the Hamiltonian approach to finite temperatures in two ways, first by means of the standard grand canonical ensemble and second by compactifying one spatial dimension. The latter formulation is advantageous since it does not require any assumption for the density operator of the grand canonical ensemble; instead the finite temperature theory is fully encoded in the vacuum state calculated on the partially compactified spatial manifold $\mathbb{R}^2 \times S^1(L)$. Within this approach the effective potential of the Polyakov loop was calculated and a second and first, respectively, order phase transition was found for the gauge group SU(2) and SU(3). These phase transitions turn into a crossover when the quarks are included. All these features are also corroborated by lattice calculations.

Acknowledgement

This work was supported by DFG under contract no. DFG-Re856/9-2.

References

[1] V. N. Gribov, Nucl. Phys. **B139**, 1 (1978).
[2] D. Zwanziger, Nucl. Phys. **B518**, 237 (1998).
[3] C. Feuchter and H. Reinhardt, Phys. Rev. D **70**, 105021 (2004); arXiv:hep-th/0402106.
[4] H. Reinhardt and C. Feuchter, Phys. Rev. D **71**, 105002 (2005); arXiv:hep-th/0408237.
[5] P. Watson and H. Reinhardt, Phys. Rev. D **85**, 025014 (2012); arXiv:1111.6078.

[6] N. H. Christ and T. D. Lee, Phys. Rev. D **22**, 939 (1980).
[7] D. Schütte, Phys. Rev. D **31**, 810 (1985).
[8] A. P. Szczepaniak and E. S. Swanson, Phys. Rev. D **65**, 025012 (2001); arXiv:hep-ph/0107078.
[9] J. Greensite, H. Matevosyan, Š. Olejník, M. Quandt, H. Reinhardt and A. P. Szczepaniak, Phys. Rev. D **83**, 114509 (2011); arXiv:1102.3941.
[10] D. Epple, H. Reinhardt and W. Schleifenbaum, Phys. Rev. D **75**, 045011 (2007); arXiv:hep-th/0612241.
[11] M. Leder, J. M. Pawlowski, H. Reinhardt and A. Weber, Phys. Rev. D **83**, 025010 (2011); arXiv:1006.5710.
[12] G. Burgio, M. Quandt, H. Reinhardt and H. Vogt, Phys. Rev. D **92**, 034518 (2015); arXiv:1503.09064.
[13] H. Reinhardt, Phys. Rev. Lett. **101**, 061602 (2008); arXiv:0803.0504.
[14] S. Mandelstam, Phys. Rep. **C23**, 245 (1976).
[15] G. 't Hooft, Phys. Scr. **25**, 133 (1982).
[16] P. de Forcrand and M. D'Elia, Phys. Rev. Lett. **82**, 4582 (1999); arXiv:hep-lat/9901020.
[17] J. Greensite, Š. Olejník and D. Zwanziger, Phys. Rev. D **69**, 074506 (2004); arXiv:hep-lat/0401003.
[18] G. Burgio, M. Quandt, H. Reinhardt, Phys. Rev. Lett. **102**, 032002 (2009); arXiv:0807.3291.
[19] D. R. Campagnari and H. Reinhardt, Phys. Rev. D **82**, 105021 (2010); arXiv:1009.4599.
[20] D. Zwanziger, Nucl. Phys. **B485**, 185 (1997); arXiv:hep-th/9603203.
[21] A. Voigt, E.-M. Ilgenfritz, M. Müller-Preussker, A. Sternbeck, Phys. Rev. D **78**, 014501 (2008); arXiv:0803.2307.
[22] G. Bali et al., Phys. Rev. Lett. **71**, 3059 (1993); arXiv:hep-lat/9306024.
[23] M. Pak and H. Reinhardt, Phys. Lett. B **707**, 566 (2012); arXiv:1107.5263.
[24] M. Pak and H. Reinhardt, Phys. Rev. D **88**, 125021 (2013); arXiv:1310.1797.
[25] P. Vastag and H. Reinhardt, Phys. Rev. D **93**, 065003 (2016); arXiv:1512.06733.
[26] H. Reinhardt, D. R. Campagnari and A. P. Szczepaniak, Phys. Rev. D **84**, 045006 (2011); arXiv:1107.3389.
[27] J. Heffner, H. Reinhardt and D. R. Campagnari, Phys. Rev. D **85**, 125029 (2012); arXiv:1206.3936.
[28] J. Heffner and H. Reinhardt, to be published
[29] D. Zwanziger, Phys. Rev. Lett. **94**, 182301 (2005); arXiv:hep-ph/0407103.
[30] H. Vogt, G. Burgio, M. Quandt and H. Reinhardt Proc. Sci. LATTICE2013, 363 (2014); arXiv:1311.5707.
[31] B. Svetitsky, Phys. Rep. **132**, 1 (1986).
[32] H. Reinhardt, arXiv:1604.06273.
[33] H. Reinhardt and J. Heffner, Phys. Lett. B **718**, 2 (2012); arXiv:1210.1742.
[34] H. Reinhardt and J. Heffner, Phys. Rev. D **88**, 045024 (2013); arXiv:1304.2980.
[35] J. Braun, H. Gies and J. M. Pawlowski, Phys. Lett. B **684**, 262 (2010); arXiv:0708.2413.
[36] F. Marhauser and J. M. Pawlowski, arXiv:0812.1144.
[37] J. Heffner, H. Reinhardt and P. Vastag, to be published.

EPJ Web of Conferences **164**, 03005 (2017) DOI: 10.1051/epjconf/201716403005
ICNFP 2016

Mass generation in Yang-Mills theories*

J. Papavassiliou[1,a], A. C. Aguilar[2], D. Binosi[3], and C. T. Figueiredo[2]

[1]*Department of Theoretical Physics and IFIC, University of Valencia and CSIC, E-46100, Valencia, Spain*
[2]*University of Campinas - UNICAMP, Institute of Physics "Gleb Wataghin", 13083-859 Campinas, SP, Brazil*
[3]*European Centre for Theoretical Studies in Nuclear Physics and Related Areas (ECT*) and Fondazione Bruno Kessler, Villa Tambosi, Strada delle Tabarelle 286, I-38123 Villazzano (TN) Italy*

Abstract. In this talk we review recent progress on our understanding of the nonperturbative phenomenon of mass generation in non-Abelian gauge theories, and the way it manifests itself at the level of the gluon propagator, thus establishing a close contact with a variety of results obtained in large-volume lattice simulations. The key observation is that, due to an exact cancellation operating at the level of the Schwinger-Dyson equations, the gluon propagator remains rigorously massless, provided that the fully-dressed vertices of the theory do not contain massless poles. The inclusion of such poles activates the well-known Schwinger mechanism, which permits the evasion of the aforementioned cancellation, and accounts for the observed infrared finiteness of the gluon propagator both in the Landau gauge and away from it.

1 Introduction

As has been well-established during the past decade by means of high-quality lattice simulations [1–4] and a variety of nonperturbative approaches in the continuum [5–7], the gluon propagator, $\Delta(q^2)$, reaches a finite value in the deep infrared, both in the Landau gauge as well as considerably away from it [8–10]. These results have induced a profound change in our understanding of the infrared sector of Yang-Mills theories, affecting our description of fundamental QCD phenomena such as confinement, chiral symmetry breaking, and bound-state formation [11].

A particular set of physical concepts and formal techniques for casting such a picture in a well-defined theoretical structure has been developed in a series of articles [5, 12, 13] based on the Schwinger-Dyson equations (SDEs) [14] derived within the powerful framework obtained from the fusion of the pinch technique (PT) [15–20] and the background field method (BFM) [21, 22].

In this presentation we outline the key elements of a particular mechanism that accounts for the observed finiteness of the gluon propagator, while preserving intact the fundamental symmetries of the theory [23].

*Talk given by J. Papavassiliou at the "5th International Conference on New Frontiers in Physics", 6-14 July 2016, Kolymbari, Greece.

[a]e-mail: Joannis.Papavassiliou@uv.es

2 The PT-BFM framework

The PT-BFM formalism provides a manifestly BRST preserving truncation scheme for the gluon propagator SDE [12, 13]. Within this framework one writes the gauge field, A^a_μ, as the sum of a background, B^a_μ, and a quantum, Q^a_μ, part. This introduces mixed Green's functions, describing combinations of B and Q fields. For instance, there are three types of gluon propagators: (*i*) the conventional propagator, with two Q-type gluons (Q^2), denoted by $\Delta^{ab}_{\mu\nu}(q)$, (*ii*) the background-quantum propagator, with one Q- and one B-type gluon (QB or BQ), denoted by $\widetilde{\Delta}^{ab}_{\mu\nu}(q)$, and (*iii*) the background propagator, with two B-type gluons, denoted by $\widehat{\Delta}^{ab}_{\mu\nu}(q)$.

The Q^2 gluon propagator, $\Delta^{ab}_{\mu\nu}(q) = \delta^{ab}\Delta_{\mu\nu}(q)$, is given by

$$\Delta_{\mu\nu}(q) = -i\left[\Delta(q^2)P_{\mu\nu}(q) + \xi\frac{q_\mu q_\nu}{q^4}\right], \qquad P_{\mu\nu}(q) = g_{\mu\nu} - \frac{q_\mu q_\nu}{q^2}. \tag{1}$$

where the function $\Delta(q^2)$ is related to the gluon self-energy $\Pi_{\mu\nu}(q) = P_{\mu\nu}(q)\Pi(q^2)$ through

$$\Delta^{-1}(q^2) = q^2 + i\Pi(q^2). \tag{2}$$

Similar expressions hold for the QB and BB propagators. For example, in the former case one has $\widetilde{\Delta}^{ab}_{\mu\nu}(q)$, with $\Delta(q^2) \to \widetilde{\Delta}(q^2)$ and $\Pi(q^2) \to \widetilde{\Pi}(q^2)$. Additionally, $\Delta(q^2)$ relates to $\widetilde{\Delta}(q^2)$ and $\widehat{\Delta}(q^2)$ by the so called "background-quantum identities" [24–26] which establishes that

$$\Delta(q^2) = [1 + G(q^2)]\widetilde{\Delta}(q^2), \qquad \widetilde{\Delta}(q^2) = [1 + G(q^2)]\widehat{\Delta}(q^2), \tag{3}$$

where $G(q^2)$ is an auxiliary function typical of the PT-BFM scheme [13]. In the ensuing analysis we will concentrate on the QB self-energy, $\widetilde{\Pi}(q^2)$, given by the sum of diagrams given in Figs. 1, 2 and 3.

3 Ward Identities

3.1 From Takahashi to Ward identities

The vertices appearing in the definition of the self-energy, $\widetilde{\Pi}(q^2)$, will be denoted by $\widetilde{\Gamma}_{\mu\alpha\beta}$ (BQ^2), $\widetilde{\Gamma}_\alpha$ ($B\bar{c}c$) and $\widetilde{\Gamma}^{abc}_{\mu\alpha\beta\gamma}$ (BQ^3). When contracted with the momentum carried by the B gluon entering the diagrams of the Figs. 1, 2 and 3, such vertices satisfy the following Takahashi identities

$$\begin{aligned}
q^\mu\widetilde{\Gamma}_{\mu\alpha\beta}(q,r,p) &= i\Delta^{-1}_{\alpha\beta}(r) - i\Delta^{-1}_{\alpha\beta}(p),\\
q^\mu\widetilde{\Gamma}_\mu(q,r,-p) &= D^{-1}(q+r) - D^{-1}(r),\\
q^\mu\widetilde{\Gamma}^{mnrs}_{\mu\alpha\beta\gamma}(q,r,p,t) &= f^{mse}f^{ern}\Gamma_{\alpha\beta\gamma}(r,p,q+t) + f^{mne}f^{esr}\Gamma_{\beta\gamma\alpha}(p,t,q+r)\\
&\quad + f^{mre}f^{ens}\Gamma_{\gamma\alpha\beta}(t,r,q+p),
\end{aligned} \tag{4}$$

where $D(q^2)$ represents the fully dressed ghost propagator, and the vertices $\Gamma_{\alpha\beta\gamma}$ on the r.h.s. of the last equation are the conventional three-gluon vertices (Q^3).

Here we are interested in the behaviour of the gluon propagator at the origin, $\Delta(0)$ [or equivalently $\widetilde{\Delta}(0)$ - see Eq. (3)]. Therefore, the relevant Ward identities (WI) are those obtained by taking the limit of the corresponding identity of Eq. (4) when the background gluon momentum q is taken to vanish.

After carrying out a Taylor expansion around $q = 0$ on both sides of the first two identities of Eq. (4) while assuming the absence of $1/q^2$ poles in the form factors of the vertices, we obtain the following WIs for BQ^2 e $B\bar{c}c$,

$$\widetilde{\Gamma}_{\mu\alpha\beta}(0,-p,p) = i\frac{\partial\Delta^{-1}_{\alpha\beta}(p)}{\partial p^{\mu}}; \qquad \widetilde{\Gamma}_{\mu\alpha\beta}(0,r,-r) = -i\frac{\partial\Delta^{-1}_{\alpha\beta}(r)}{\partial r^{\mu}}; \tag{5}$$

$$\widetilde{\Gamma}_{\mu}(0,-p,p) = i\frac{\partial D^{-1}(p^2)}{\partial p^{\mu}}; \qquad \widetilde{\Gamma}_{\mu}(0,r,-r) = -i\frac{\partial D^{-1}(r^2)}{\partial r^{\mu}}. \tag{6}$$

The corresponding WI for the BQ^3 vertex is somewhat more complicated and reads

$$\widetilde{\Gamma}^{mnrs}_{\mu\alpha\beta\gamma}(0,r,p,-r-p) = \left(f^{mne}f^{esr}\frac{\partial}{\partial r^{\mu}} + f^{mre}f^{ens}\frac{\partial}{\partial p^{\mu}}\right)\Gamma_{\alpha\beta\gamma}(r,p,-r-p),$$
$$\widetilde{\Gamma}^{mnrs}_{\mu\alpha\beta\gamma}(0,-r,-p,r+p) = -\left(f^{mne}f^{esr}\frac{\partial}{\partial r^{\mu}} + f^{mre}f^{ens}\frac{\partial}{\partial p^{\mu}}\right)\Gamma_{\alpha\beta\gamma}(-r,-p,r+p). \tag{7}$$

3.2 Ward identities and vertex form factors

The WIs just derived are bound to affect the form factors appearing in the tensorial decomposition of the corresponding vertices. Consider for example the simplest case of the gluon-ghost vertex, $\widetilde{\Gamma}^{\mu}$ ($B\bar{c}c$), whose the most general Lorentz decomposition is given by

$$\widetilde{\Gamma}^{\mu}(q,r,p) = \widetilde{\mathcal{A}}_1(q^2,r^2,p^2)q^{\mu} + \widetilde{\mathcal{A}}_2(q^2,r^2,p^2)r^{\mu}, \tag{8}$$

with q denoting the momentum carried by the background gluon. In agreement with the assumptions made when deriving the WIs of Eqs. (5), (6) and (7), we consider that the form factors $\widetilde{\mathcal{A}}_1$ and $\widetilde{\mathcal{A}}_2$ do not contain (kinematic or dynamical) poles in q^2; therefore, when $q = 0$, Eqs. (6) and (8) imply that

$$\widetilde{\mathcal{A}}_2(r^2) = -2i\frac{\partial D^{-1}(r^2)}{\partial r^2}. \tag{9}$$

Next, we consider the three-gluon vertex $\widetilde{\Gamma}^{\mu\alpha\beta}$, whose tensorial decomposition may be written as

$$\widetilde{\Gamma}^{\mu\alpha\beta}(q,r,p) = \sum_{i=1}^{14}\widetilde{A}_i(q^2,r^2,q\cdot r)\,b_i^{\mu\alpha\beta}, \tag{10}$$

where the basis $b^i_{\mu\alpha\beta}$ is chosen to be

$$\begin{aligned}
&b_1^{\mu\alpha\beta} = q^{\mu}g^{\alpha\beta}; && b_2^{\mu\alpha\beta} = q^{\mu}q^{\alpha}q^{\beta}; && b_3^{\mu\alpha\beta} = q^{\mu}q^{\alpha}r^{\beta}; && b_4^{\mu\alpha\beta} = q^{\mu}r^{\alpha}q^{\beta}; && b_5^{\mu\alpha\beta} = q^{\mu}r^{\alpha}r^{\beta},\\
&b_6^{\mu\alpha\beta} = r^{\mu}g^{\alpha\beta}; && b_7^{\mu\alpha\beta} = r^{\mu}q^{\alpha}q^{\beta}; && b_8^{\mu\alpha\beta} = r^{\mu}q^{\alpha}r^{\beta}; && b_9^{\mu\alpha\beta} = r^{\mu}r^{\alpha}q^{\beta}; && b_{10}^{\mu\alpha\beta} = r^{\mu}r^{\alpha}r^{\beta},\\
&b_{11}^{\mu\alpha\beta} = q^{\alpha}g^{\beta\mu}; && b_{12}^{\mu\alpha\beta} = q^{\beta}g^{\alpha\mu}; && b_{13}^{\mu\alpha\beta} = r^{\alpha}g^{\beta\mu}; && b_{14}^{\mu\alpha\beta} = r^{\beta}g^{\alpha\mu}.
\end{aligned} \tag{11}$$

As we assume that the form factors $\widetilde{A}_i$ do not contain poles in q^2, only q-independent tensors, *i.e.*, b_6, b_{10}, b_{13} and b_{14}, will survive in the limit $q \to 0$, therefore

$$\widetilde{\Gamma}^{\mu\alpha\beta}(0,r,-r) = \widetilde{A}_6(r^2)r^{\mu}g^{\alpha\beta} + \widetilde{A}_{10}(r^2)r^{\mu}r^{\alpha}r^{\beta} + \widetilde{A}_{13}r^{\alpha}g^{\beta\mu} + \widetilde{A}_{14}r^{\beta}g^{\alpha\mu}. \tag{12}$$

Then, using the identity (5) together with the inverse of the gluon propagator

$$-i\Delta^{-1}_{\mu\nu}(q) = \Delta^{-1}(q^2)P_{\mu\nu}(q) + \xi^{-1}q_\mu q_\nu, \tag{13}$$

and matching the corresponding tensorial structures, lead us to the relations

$$\widetilde{A}_6(r^2) = 2\frac{\partial\Delta^{-1}(r^2)}{\partial r^2} \qquad \widetilde{A}_{10}(r^2) = -2\frac{\partial}{\partial r^2}\left(\frac{\Delta^{-1}(r^2)}{r^2}\right) \qquad \widetilde{A}_{13}(r^2) = \widetilde{A}_{14}(r^2) = \frac{1}{\xi} - \frac{\Delta^{-1}(r^2)}{r^2}. \tag{14}$$

The analogous procedure involving the four-gluon vertex, $\widetilde{\Gamma}^{mnrs}_{\mu\alpha\beta\gamma}(q,r,p,t)$, would be particularly cumbersome in virtue of the large number of tensorial structures involved in its Lorentz decomposition. Luckily however, this decomposition is not needed in the computation of $\Delta(0)$, which is all we are interested in.

4 Seagull Identity

A second crucial ingredient in this analysis is the so-called seagull identity [27]. To derive it, let us consider the class of vector functions

$$\mathcal{F}_\mu(k) = f(k^2)\, k_\mu, \tag{15}$$

with $f(k^2)$ an arbitrary scalar function. Since $\mathcal{F}_\mu(-k) = -\mathcal{F}_\mu(k)$, we have that

$$\int_k \mathcal{F}_\mu(k) = 0; \quad \text{where} \quad \int_k \equiv \frac{\mu^\epsilon}{(2\pi)^d}\int \mathrm{d}^d k, \tag{16}$$

with $d = 4 - \epsilon$ and μ the 't Hooft mass.

If we now assume that the function $f(k^2)$ vanishes rapidly enough when $k^2 \to \infty$, so that the integral (in spherical coordinates, with $y = k^2$)

$$\int_k f(k^2) = \frac{1}{(4\pi)^{\frac{d}{2}}\Gamma(\frac{d}{2})}\int_0^\infty dy y^{\frac{d}{2}-1} f(y), \tag{17}$$

converges for all positive values d below a certain value d^*. Due to the translational invariance of dimensional regularization, we may shift the argument of the function defined in the Eq. (15) by an arbitrary momentum q without affecting the result of Eq. (16). Then, carrying out a Taylor expansion around $q = 0$, we obtain

$$\begin{aligned}\mathcal{F}_\mu(q+k) &= \mathcal{F}_\mu(k) + q^\nu\left\{\frac{\partial}{\partial q^\nu}\mathcal{F}_\mu(q+k)\right\}_{q=0} + O(q^2)\\ &= \mathcal{F}_\mu(k) + q^\nu\frac{\partial\mathcal{F}_\mu(k)}{\partial k^\nu} + O(q^2),\end{aligned} \tag{18}$$

and, after integrating both sides of the expansion and using Eq. (16), we find

$$q^\nu\int_k \frac{\partial\mathcal{F}_\mu(k)}{\partial k^\nu} = 0. \tag{19}$$

Since the integral above has two free Lorentz indices and no momentum scale, it must be proportional to the metric tensor $g_{\mu\nu}$. Moreover, given that the momentum q is arbitrary, we find that Eq. (19) implies the seagull identity [23],

$$\int_k \frac{\partial\mathcal{F}_\mu(k)}{\partial k^\mu} = 0. \tag{20}$$

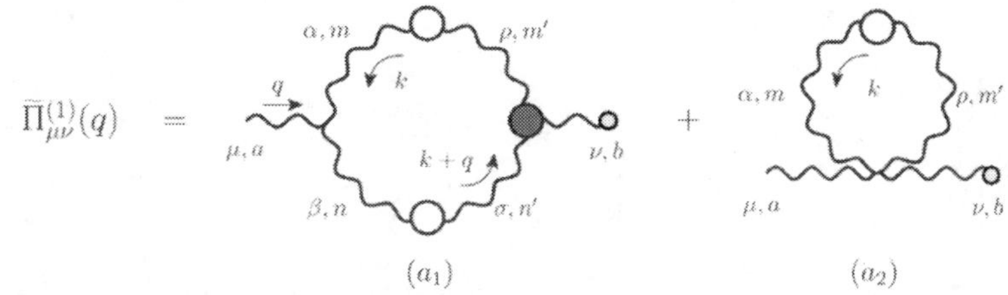

Figure 1. The gluonic one-loop dressed diagrams contributing to the SDE of the QB gluon self-energy, $\widetilde{\Pi}_{\mu\nu}(q^2)$. The white circles indicate fully dressed gluon propagators and the red circle indicates a fully dressed three gluon vertex, $\widetilde{\Gamma}_{\nu\sigma\rho}$ (BQ^2).

5 The zero momentum gluon propagator

The WIs derived in Sect. 3.1 may be used to cast the zero momentum gluon self-energy $\widetilde{\Pi}(0)$ in a form suitable for the application of the seagull identity given by Eq. (20). To do so, consider first the contribution given by the gluonic one-loop dressed diagrams (a_1) and (a_2) of Fig. 1. In particular, at $q = 0$ we have

$$d\,\widetilde{\Pi}^{(1)}(0) = a_1(0) + a_2(0), \tag{21}$$

with

$$a_1(0) = \frac{1}{2}g^2 C_A \int_k \Gamma^{(0)}_{\mu\alpha\beta}(0,k,-k)\Delta^{\alpha\rho}(k)\Delta^{\beta\sigma}(k)\widetilde{\Gamma}^{\mu}_{\sigma\rho}(0,k,-k)\,, \tag{22}$$

$$a_2(0) = -ig^2 C_A(d-1)\int_k \Delta^{\alpha}_{\alpha}(k), \tag{23}$$

where $C_A = N$ represents the Casimir eigenvalue of the adjoint representation for SU(N), while $\Gamma^{(0)}_{\mu\alpha\beta}$ is the (conventional) tree-level three-gluon vertex, which is given by

$$\Gamma^{(0)}_{\mu\alpha\beta}(0,k,-k) = 2k_\mu g_{\alpha\beta} - k_\beta g_{\alpha\mu} - k_\alpha g_{\beta\mu}\,. \tag{24}$$

Then, using Eq. (5), we derive the relation

$$\Delta^{\alpha\rho}(k)\Delta^{\beta\sigma}(k)\widetilde{\Gamma}^{\mu}_{\sigma\rho}(0,k,-k) = -i\frac{\partial}{\partial k^\mu}\Delta^{\alpha\beta}(k), \tag{25}$$

so that we can rewrite Eq. (22) in the form

$$a_1(0) = -\frac{i}{2}g^2 C_A\left\{\int_k \frac{\partial}{\partial k^\mu}[\Gamma^{(0)}_{\mu\alpha\beta}(0,k,-k)\Delta^{\alpha\beta}(k)] - \int_k \Delta^{\alpha\beta}(k)\frac{\partial}{\partial k^\mu}\Gamma^{(0)}_{\mu\alpha\beta}(0,k,-k)\right\}. \tag{26}$$

Finally, observing

$$\frac{\partial}{\partial k^\mu}\Gamma^{(0)}_{\mu\alpha\beta}(0,k,-k) = 2(d-1)g_{\alpha\beta}, \tag{27}$$

we note that the second term of Eq. (26) cancels against the expression for $a_2(0)$, leaving us with the final result

$$d\widetilde{\Pi}^{(1)}(0) = -g^2 C_A(d-1)\int_k \frac{\partial \mathcal{F}^{(1)}_\mu(k)}{\partial k^\mu}; \qquad \mathcal{F}^{(1)}_\mu(k) = \Delta(k^2)k_\mu. \tag{28}$$

Consequently, by virtue of the seagull identity, $\widetilde{\Pi}^{(1)}(0) = 0$.

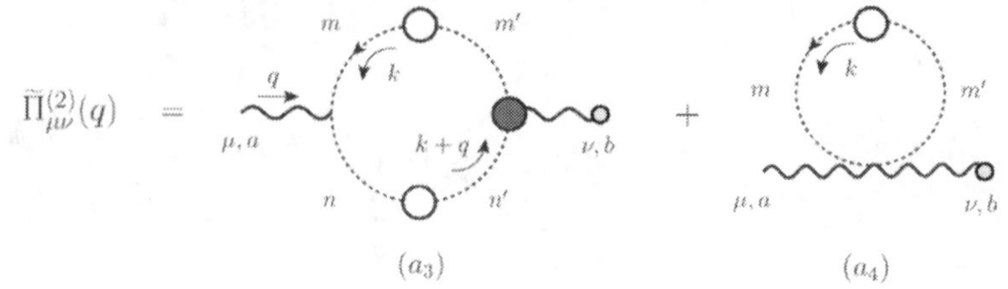

Figure 2. One-loop dressed ghost diagrams contributing to the SDE of the QB gluon self-energy, $\widetilde{\Pi}_{\mu\nu}(q^2)$. White circles indicate fully dressed ghost propagators and the red circle indicates a fully dressed gluon-ghost vertex, $\widetilde{\Gamma}_\mu\ (B\bar{c}c)$.

Next, consider the ghost loop diagrams of Fig. 2. At $q = 0$, we have

$$a_3(0) = -g^2 C_A \int_k k_\mu D^2(k^2)\widetilde{\Gamma}^\mu(0,k,-k), \tag{29}$$

$$a_4(0) = dg^2 C_A \int_k D(k^2), \tag{30}$$

and from Eq. (6), it is possible to obtain

$$a_3(0) = g^2 C_A \left\{ \int_k \frac{\partial}{\partial k^\mu}[\Gamma^{(0)}_\mu D(k^2)] - d\int_k D(k^2) \right\}. \tag{31}$$

Thus, the second term in Eq. (31) cancels against $a_4(0)$ and we are left with

$$d\widetilde{\Pi}^{(2)}(0) = g^2 C_A \int_k \frac{\partial \mathcal{F}^{(2)}_\mu(k)}{\partial k^\mu}; \qquad \mathcal{F}^{(2)}_\mu(k) = D(k^2)k_\mu\,, \tag{32}$$

so that the seagull identity implies that also $\widetilde{\Pi}^{(2)}(0) = 0$

Finally, the gluon two-loop dressed of Fig. 3 at $q = 0$ yield

$$a_5^{ab}(0) = \frac{i}{6} g^4 \Gamma^{(0)amnr}_{\mu\alpha\beta\gamma} \int_k \int_\ell \Delta^{\gamma\tau}(k)\Delta^{\beta\sigma}(\ell)\Delta^{\alpha\rho}(k+\ell)\widetilde{\Gamma}^{brnm}_{\mu\tau\sigma\rho}(0,k,\ell,-k-\ell),$$

$$a_6(0) = N_{\mu\alpha\beta\gamma} \int_k \Delta^{\gamma\tau}(k)\Delta^{\lambda\delta}(k) Y^{\alpha\beta}_\delta(k)\widetilde{\Gamma}^\mu_{\tau\lambda}(0,k,-k)\,, \tag{33}$$

where we have defined

$$N_{\mu\alpha\beta\gamma} = \frac{3}{4} i g^4 C_A^2 (g_{\mu\alpha}g_{\gamma\beta} - g_{\mu\beta}g_{\gamma\alpha})\,, \tag{34}$$

and

$$Y^{\alpha\beta}_\delta(k) = \int_\ell \Delta^{\alpha\rho}(\ell)\Delta^{\beta\sigma}(k+\ell)\Gamma_{\sigma\rho\delta}(-k-\ell,\ell,k)\,, \tag{35}$$

which is proportional to the subdiagram nested inside (a_6), and was studied in Ref. [28]. Employing the WI of Eq. (7), one obtains [23],

$$d\widetilde{\Pi}^{(3)}(0) = -i(d-1)g^4 C_A^2 \int_k \frac{\partial \mathcal{F}^{(3)}_\mu(k)}{\partial k^\mu}; \qquad \mathcal{F}^{(3)}_\mu(k) = \Delta(k^2)Y(k^2)k_\mu. \tag{36}$$

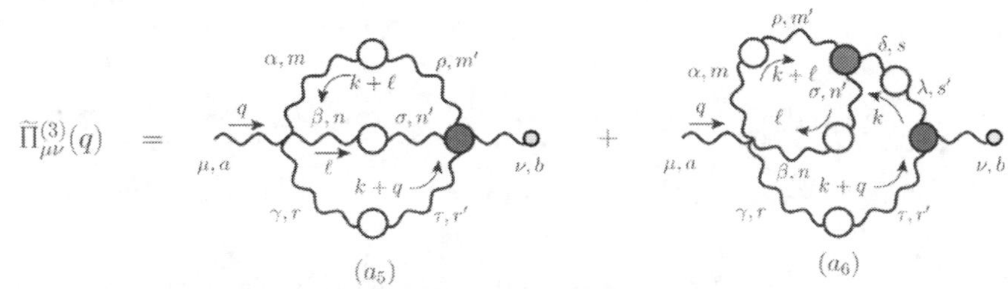

Figure 3. The gluonic two-loop dressed diagrams contributing to the SDE of the QB gluon self-energy, $\widetilde{\Pi}_{\mu\nu}(q^2)$. White circles indicate the fully gluon dressed propagators and the red circle in (a_5) indicates the fully dressed four gluon vertex, $\widetilde{\Gamma}^{abc}_{\mu\alpha\beta\gamma}\,(BQ^3)$ while in (a_6) is the three gluon vertex, $\widetilde{\Gamma}_{\mu\alpha\beta}\,(BQ^2)$.

The results of Eqs. (28), (32) and (36) establish that, in the absence of massless poles in the form factors of the fundamental vertices,

$$\widetilde{\Pi}^{(i)}(0) = 0, \quad i = 1,2,3; \quad \Rightarrow \quad \widetilde{\Pi}(0) = \sum_{i=1}^{3} \widetilde{\Pi}^{(i)}(0) = 0, \tag{37}$$

and since $\widetilde{\Delta}^{-1}(q^2) = q^2 + i\widetilde{\Pi}(q^2)$, we are lead to the conclusion that $\widetilde{\Delta}^{-1}(0) = 0$. The conventional gluon propagator is then obtained by using Eq. (3),

$$\Delta^{-1}(0) = \frac{\widetilde{\Delta}^{-1}(0)}{1 + G(0)}. \tag{38}$$

Now, if we introduce the additional assumption that $1 + G(0)$ is finite for every ξ, we conclude that $\Delta^{-1}(0) = 0$, *i.e., the gluon is massless in the absence of poles.*

We emphasize that the proof elaborated in this section is valid for any value of the gauge-fixing parameter ξ within the class of linear covariant gauges (R_ξ). Additionally, in this class of gauges, the $G(q^2)$ function is related to the ghost dressing function $F(q^2)$ as [26]

$$F^{-1}(q^2) = 1 + G(q^2) + L(q^2) + \xi K(q^2), \tag{39}$$

where K represents an auxiliary function related to the antighost coupling with a certain anti-BRST source necessary for formulating the theory in the BFM formalism. In the Landau gauge ($\xi = 0$), we know that $L(0) = 0$, while the ghost dressing function is non-zero at $q^2 = 0$ [29]; thus, we obtain $F^{-1}(0) = 1 + G(0)$ which ensures the finiteness of $G(0)$. However, for $\xi \neq 0$ the situation is more complex. Recent analytical studies [9, 10] indicate that the gluon propagator still saturates in the IR in this case, a result that has been corroborated by lattice simulations [30]. This points towards the finiteness of $1 + G(0)$ even in the case $\xi \neq 0$.

6 Vertices with massless poles

Evading the result established in the previous section requires relaxing the underlying assumption of the absence of massless poles. Thus, in the following we are going to assume that some of the vertices form factors contain such poles. Thus they will be divided in two parts, which will be indicated by a the corresponding superscript: "**p**" (for "pole" parts) or "**np**" (for "no-pole" parts):

$$\widetilde{\Gamma}_{\mu\alpha\beta}(q,r,p) = \widetilde{\Gamma}^{\mathbf{np}}_{\mu\alpha\beta}(q,r,p) + \widetilde{\Gamma}^{\mathbf{p}}_{\mu\alpha\beta}(q,r,p); \qquad \widetilde{\Gamma}_{\mu}(q,r,p) = \widetilde{\Gamma}^{\mathbf{np}}_{\mu}(q,r,p) + \widetilde{\Gamma}^{\mathbf{p}}_{\mu}(q,r,p),$$
$$\widetilde{\Gamma}^{mnrs}_{\mu\alpha\beta\gamma}(q,r,p,t) = \widetilde{\Gamma}^{\mathbf{np},mnrs}_{\mu\alpha\beta\gamma}(q,r,p,t) + \widetilde{\Gamma}^{\mathbf{p},mnrs}_{\mu\alpha\beta\gamma}(q,r,p,t). \tag{40}$$

Since the pole parts should act like a "dynamical Nambu-Goldstone bosons" and decouple from physical observables, they must be longitudinally-coupled; therefore we can write

$$\widetilde{\Gamma}^{\mathbf{p}}_{\mu\alpha\beta}(q,r,p) = \frac{q_\mu}{q^2}\widetilde{C}_{\alpha\beta}(q,r,p); \qquad \widetilde{\Gamma}^{\mathbf{p}}_{\mu}(q,r,p) = \frac{q_\mu}{q^2}\widetilde{C}(q,r,p),$$
$$\widetilde{\Gamma}^{\mathbf{p},mnrs}_{\mu\alpha\beta\gamma}(q,r,p,t) = \frac{q_\mu}{q^2}\widetilde{C}^{mnrs}_{\alpha\beta\gamma}(q,r,p,t). \tag{41}$$

In order to keep the BRST invariance intact, one requires that the Takahashi identities maintain their exact form in the presence of these poles. Consequently, Eq. (4) is now written as

$$q^\mu\widetilde{\Gamma}^{\mathbf{np}}_{\mu\alpha\beta}(q,r,p) + \widetilde{C}_{\alpha\beta}(q,r,p) = i\Delta^{-1}_{\alpha\beta}(r) - i\Delta^{-1}_{\alpha\beta}(p),$$
$$q^\mu\widetilde{\Gamma}^{\mathbf{np}}_{\mu}(q,r,p) + \widetilde{C}(q,r,p) = iD^{-1}(r^2) - iD^{-1}(p^2),$$
$$q^\mu\widetilde{\Gamma}^{\mathbf{np},mnrs}_{\mu\alpha\beta\gamma}(q,r,p,t) + \widetilde{C}^{mnrs}_{\alpha\beta\gamma}(q,r,p,t) = f^{mse}f^{ern}\Gamma_{\alpha\beta\gamma}(r,p,q+t) + f^{mne}f^{esr}\Gamma_{\beta\gamma\alpha}(p,t,q+r)$$
$$+ f^{mre}f^{ens}\Gamma_{\gamma\alpha\beta}(t,r,q+p). \tag{42}$$

Additionally, we derive the equivalent of Eqs. (5), (6) and (7) by carrying out a Taylor expansion of both sides of Eqs. (42) around $q = 0$. Clearly, we see that the zeroth order terms vanish in all cases,

$$\widetilde{C}_{\alpha\beta}(0,r,-r) = 0; \qquad \widetilde{C}(0,-r,r) = 0; \qquad \widetilde{C}^{mnrs}_{\alpha\beta\gamma}(0,r,p,-p-r) = 0, \tag{43}$$

while the terms linear in q^μ yield, for the BQ^2 vertex

$$\widetilde{\Gamma}^{\mathbf{np}}_{\mu\alpha\beta}(0,r,-r) = -i\frac{\partial\Delta^{-1}_{\alpha\beta}(r)}{\partial r^\mu} - \left\{\frac{\partial}{\partial q^\mu}\widetilde{C}_{\alpha\beta}(q,r,-r-q)\right\}_{q=0};$$
$$\widetilde{\Gamma}^{\mathbf{np}}_{\mu\alpha\beta}(0,p,-p) = i\frac{\partial\Delta^{-1}_{\alpha\beta}(p)}{\partial p^\mu} - \left\{\frac{\partial}{\partial q^\mu}\widetilde{C}_{\alpha\beta}(q,-p-q,p)\right\}_{q=0}. \tag{44}$$

Similarly, for the $B\bar{c}c$ vertex, we have

$$\widetilde{\Gamma}^{\mathbf{np}}_{\mu}(0,r,-r) = -i\frac{\partial D^{-1}(r^2)}{\partial r^\mu} - \left\{\frac{\partial}{\partial q^\mu}\widetilde{C}(q,r,-r-q)\right\}_{q=0},$$
$$\widetilde{\Gamma}^{\mathbf{np}}_{\mu}(0,p,-p) = i\frac{\partial D^{-1}(p^2)}{\partial p^\mu} - \left\{\frac{\partial}{\partial q^\mu}\widetilde{C}(q,-p-q,p)\right\}_{q=0}, \tag{45}$$

and for the BQ^3 vertex,

$$\widetilde{\Gamma}^{\mathbf{np},mnrs}_{\mu\alpha\beta\gamma}(0,r,p,-r-p) = \left(f^{mne}f^{esr}\frac{\partial}{\partial r^\mu} + f^{mre}f^{ens}\frac{\partial}{\partial p^\mu}\right)\Gamma_{\alpha\beta\gamma}(r,p,-r-p)$$
$$- \left\{\frac{\partial}{\partial q^\mu}\widetilde{C}^{mnrs}_{\alpha\beta\gamma}(q,r,p,-q-r-p)\right\}_{q=0},$$
$$\widetilde{\Gamma}^{\mathbf{np},mnrs}_{\mu\alpha\beta\gamma}(0,-r,-p,r+p) = -\left(f^{mne}f^{esr}\frac{\partial}{\partial r^\mu} + f^{mre}f^{ens}\frac{\partial}{\partial p^\mu}\right)\Gamma_{\alpha\beta\gamma}(-r,-p,r+p)$$
$$+ \left\{\frac{\partial}{\partial q^\mu}\widetilde{C}^{mnrs}_{\alpha\beta\gamma}(-q,-r,-p,q+r+p)\right\}_{q=0}. \tag{46}$$

Now, we can repeat the calculations of Sect. 5, using the new WIs of Eqs. (44), (45) and (46), which have additional terms containing the derivatives of the $\widetilde{C}$ functions. Consequently, the terms already present in Eqs. (5), (6) and (7) will trigger again the seagull identity, vanishing as before, whereas the contributions originated from the $\widetilde{C}$ functions will survive. As a result, one obtains

$$d\widetilde{\Pi}^{(1)}(0) = -\frac{1}{2}g^2 C_A \int_k \Gamma^{(0)}_{\mu\alpha\beta}(0,k,-k)\Delta^{\alpha\rho}(k)\Delta^{\beta\sigma}(k)\left\{\frac{\partial}{\partial q^\mu}\widetilde{C}_{\sigma\rho}(q,-k-q,k)\right\}_{q=0}, \tag{47}$$

$$d\widetilde{\Pi}^{(2)}(0) = -g^2 C_A \int_k \Gamma^{(0)}_{\mu}(0,k,-k)D^2(k^2)\left\{\frac{\partial}{\partial q^\mu}\widetilde{C}(q,-k-q,k)\right\}_{q=0}, \tag{48}$$

and

$$\begin{aligned} d\widetilde{\Pi}^{(3)}(0)\delta^{ab} &= \frac{1}{6}g^4\Gamma^{(0)amnr}_{\mu\alpha\beta\gamma}\int_k\int_\ell \Delta^{\alpha\rho}(k+\ell)\Delta^{\beta\sigma}(\ell)\Delta^{\gamma\tau}(k)\left\{\frac{\partial}{\partial q^\mu}\widetilde{C}^{brnm}_{\tau\sigma\rho}(q,-k-q,-\ell,k+\ell)\right\}_{q=0} \\ &+ iN_{\mu\alpha\beta\gamma}\delta^{ab}\int_k Y^{\alpha\beta}_{\delta}(k)\Delta^{\gamma\tau}(k)\Delta^{\lambda\delta}(k)\left\{\frac{\partial}{\partial q^\mu}\widetilde{C}_{\tau\lambda}(q,-k-q,k)\right\}_{q=0}. \end{aligned} \tag{49}$$

Therefore, the presence of $1/q^2$ poles allows for a non-vanishing value of the gluon self-energy at $q=0$, so that we can have $\widetilde{\Delta}^{-1}(0) \neq 0$.

Let us end this section by observing that the extended WIs of Eqs. (44), (45) and (46) yield interesting results for the vertices form factors. To begin with, notice that the condition of longitudinal coupling implies that only some of the vertex form factors may contain massless poles; specifically one has

$$\begin{aligned} &\widetilde{\mathcal{A}}_1 = \widetilde{\mathcal{A}}_1^{\mathbf{np}} + \widetilde{\mathcal{A}}_1^{\mathbf{p}}; && \widetilde{\mathcal{A}}_2 = \widetilde{\mathcal{A}}_2^{\mathbf{np}}, \\ &\widetilde{A}_i = \widetilde{A}_i^{\mathbf{np}} + \widetilde{A}_i^{\mathbf{p}}, \quad i=1,...,5; && \widetilde{A}_i = \widetilde{A}_i^{\mathbf{np}}, \quad i=6,...,14. \end{aligned} \tag{50}$$

Then, using the relation

$$\left\{\frac{\partial}{\partial q^\mu}\widetilde{C}(q,r,p)\right\}_{q=0} = 2r_\mu\widetilde{C}'(r^2), \tag{51}$$

we see that Eq. (9) gets modified to

$$\widetilde{\mathcal{A}}_2^{\mathbf{np}}(r^2) = -2\left[i\frac{\partial}{\partial r^2}D^{-1}(r^2) + \widetilde{C}'(r^2)\right], \tag{52}$$

Similarly, assuming that out of the five possible tensorial structures of $\widetilde{C}_{\sigma\rho}$ only the one proportional to $g_{\alpha\beta}$ develops a pole in q^2, one has

$$\left\{\frac{\partial}{\partial q^\mu}\widetilde{C}_{\alpha\beta}(q,r,p)\right\}_{q=0} = 2r_\mu g_{\alpha\beta}\widetilde{C}_1'(r^2). \tag{53}$$

so that the relation (14) for the A_6 form factor gets modified into

$$\widetilde{A}_6^{\mathbf{np}}(r^2) = 2\left[\frac{\partial}{\partial r^2}\Delta^{-1}(r^2) - \widetilde{C}_1'(r^2)\right]. \tag{54}$$

DOI: 10.1051/epjconf/201716403005

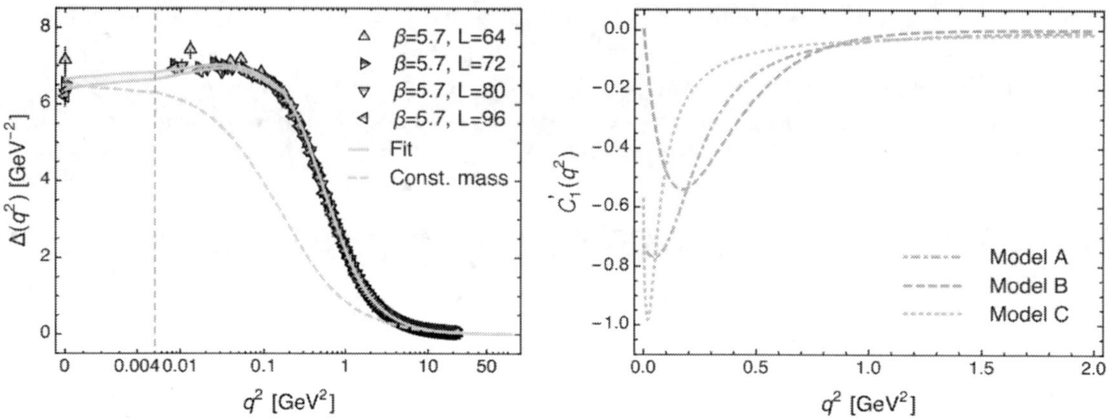

Figure 4. Left panel: The quenched SU(3) lattice data for the gluon propagator, $\Delta(q^2)$, renormalized at $\mu = 4.3$ GeV (symbols) and its corresponding fit (continuous line). Right panel: The functions $\widetilde{C}_1'(y)$ corresponding to: Model A with $a = -9.21\,\text{GeV}^{-4}$, $b = 0.94\,\text{GeV}^{-2}$, $c = -0.47$, Model B with $a = -25.13\,\text{GeV}^{-2}$, $b = 0.17\,\text{GeV}^2$, and Model C with $a = -12.79\,\text{GeV}^{-2}$, $b = 3.59\,\text{GeV}^{-1}$, $c = -0.60$.

7 Numerical analysis

In order to obtain some quantitative understanding of the equations derived, we finally carry out a numerical analysis under some simplifying hypotheses. Specifically, we assume that the contributions coming from the poles of $\widetilde{\Gamma}^{\mathbf{p}}_{\mu}(q,r,p)$ and $\widetilde{\Gamma}^{\mathbf{p},mnrs}_{\mu\alpha\beta\gamma}(q,r,p,t)$ are suppressed with respect to that of $\widetilde{\Gamma}^{\mathbf{p}}_{\mu\alpha\beta}(q,r,p)$, so that they can be neglected. Additionally, we consider only the $\widetilde{C}_1$ component of $\widetilde{C}_{\alpha\beta}$, that is only the term proportional to the metric tensor.

In this way, Eq. (47) results in

$$d\widetilde{\Pi}^{(1)}(0) = 2g^2 C_A \int_k (d-1)k^2\Delta^2(k^2)\widetilde{C}_1'(k^2), \tag{55}$$

which yields (in Euclidean space and spherical coordinates with $k^2 = y$)

$$\Delta^{-1}(0) = -\frac{3C_A\alpha_s}{8\pi}F(0)\int_0^{\infty} \mathrm{d}y\, y^2\Delta^2(y)\widetilde{C}_1'(y). \tag{56}$$

The precise form of the function $\widetilde{C}_1'(y)$ is undetermined at this level (we only know that it must be negative as $\Delta^{-1}(0)$ and $F(0)$ are positive). Thus, in order to proceed with our analysis we will consider three possible models describing its functional form, inspired by the solutions obtained from the Bethe-Salpeter equations governing the formation of massless poles [31, 32]:

$$\widetilde{C}_1'(y) = \begin{cases} 1/(ay^2 + by + c), & \text{Model A} \\ ay\exp(-y/b), & \text{Model B} \\ 1/(ay + b\sqrt{y} + c), & \text{Model C} \end{cases} \tag{57}$$

with a, b and c being suitable parameters which will be adjusted in order to satisfy Eq. (56).

Solutions of Eq. (56) are finally obtained when a realistic model for the gluon propagator is used. To this purpose we use the results for the gluon propagator from SU(3) quenched lattice simulations in the Landau gauge [3]. On the left panel of Fig. 4, we present a fit for the propagator (continuous

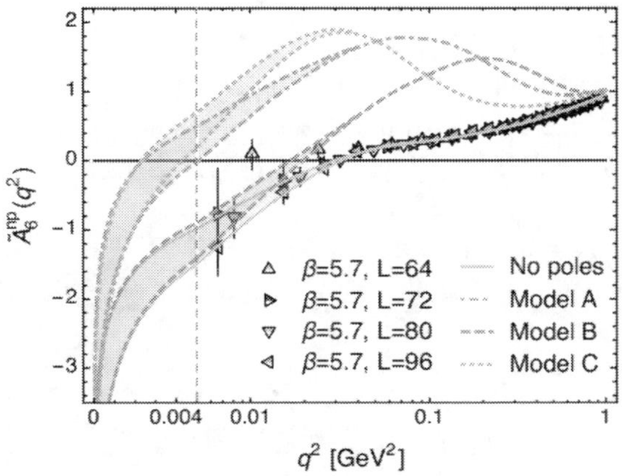

Figure 5. The form factor $\widetilde{A}_6^{\mathbf{np}}$. In the absence of massless poles, this quantity is proportional to the derivative of the inverse propagator (here evaluated directly from the lattice data). When massless poles are generated, the presence of the function $\widetilde{C}_1'$ modifies this dependence according to Eq. (54), and a positive maximum appears in the region $q^2 < 1$ GeV^2, whose height and exact location depend on the details of the model.

line), along with the lattice data renormalized at 4.3 GeV. On the right panel of the same figure we plot instead the typical solutions obtained corresponding to each of the model considered.

We conclude by presenting a short qualitative description of how the study of the form factor $\widetilde{A}_6^{\mathbf{np}}(q^2)$ can confirm or invalidate the need for longitudinally-coupled massless poles. Recall that, unlike the gluon, the ghost remains massless: its propagator behaves as $1/p^2$ in the IR multiplied by a dressing function that saturates at a finite nonvanishing value. This implies that [33]: (*i*) it must display a maximum in the deep IR (see again the left panel of Fig. 4), and (*ii*) the derivative of its inverse will display a logarithmic divergence and a zero crossing:

$$\frac{\partial}{\partial q^2}\Delta^{-1}(q^2) \underset{q\to 0}{\sim} \log q^2. \tag{58}$$

Therefore, if the finiteness of the gluon propagator was not due to the massless poles, Eq. (14) implies that the form factor $\widetilde{A}_6^{\mathbf{np}} \equiv \widetilde{A}_6$ would display the same IR behavior as that of Eq. (58). However, the presence of massless poles will change the shape of $\widetilde{A}_6$ [see Eq. (54)], forcing it to have a positive maximum, as can be appreciated in Fig. 5.

8 Conclusions

A unified framework for the self-consistent treatment of the IR finiteness of the gluon propagator at the level of the SDEs of the theory has been formulated within the PT-BFM framework. Particular attention has been dedicated to the extensive cancellations induced by the WIs of the theory, leading to the necessity of introducing massless poles in order to achieve the desired effect in a self-consistent way. The analysis has been carried out for a general value of the gauge-fixing parameter, reverting to the Landau gauge only in order to simplify the numerical analysis.

Particularly relevant is the observation that the presence of these poles is bound to affect not only the two-point but also the three-point sector of the theory. Therefore, determining special form factors of, *e.g.*, the (background) three-gluon vertex by means of lattice techniques, has the potential to confirm or discard massless poles as the IR mechanism underlying the dynamical generation of a gluon mass.

References

[1] A. Cucchieri, T. Mendes, PoS **LAT2007**, 297 (2007).
[2] A. Cucchieri, T. Mendes, Phys. Rev. Lett. **100**, 241601 (2008).
[3] I. Bogolubsky, E. Ilgenfritz, M. Muller-Preussker, A. Sternbeck, Phys. Lett. **B676**, 69 (2009).
[4] O. Oliveira and P. J. Silva, Phys. Rev. D **86**, 114513 (2012).
[5] A.C. Aguilar, J. Papavassiliou, JHEP **12**, 012 (2006).
[6] A.C. Aguilar, D. Binosi, J. Papavassiliou, Phys. Rev. **D78**, 025010 (2008).
[7] D. Dudal, J.A. Gracey, S.P. Sorella, N. Vandersickel, H. Verschelde, Phys. Rev. **D78**, 065047 (2008).
[8] A. Ayala, A. Bashir, D. Binosi, M. Cristoforetti, J. Rodriguez-Quintero, Phys. Rev. **D86**, 074512 (2012).
[9] A. C. Aguilar, D. Binosi and J. Papavassiliou, Phys. Rev. D **91**, no. 8, 085014 (2015).
[10] M.Q. Huber, Phys. Rev. **D91**, 085018 (2015).
[11] A. C. Aguilar, D. Binosi and J. Papavassiliou, Front. Phys. (Beijing) **11**, no. 2, 111203 (2016).
[12] D. Binosi, J. Papavassiliou, Phys. Rev. **D77**, 061702 (2008).
[13] D. Binosi, J. Papavassiliou, JHEP **0811**, 063 (2008).
[14] C.D. Roberts, A.G. Williams, Prog. Part. Nucl. Phys. **33**, 477 (1994).
[15] J.M. Cornwall, Phys. Rev. **D26**, 1453 (1982)
[16] J.M. Cornwall, J. Papavassiliou, Phys. Rev. **D40**, 3474 (1989).
[17] A. Pilaftsis, Nucl. Phys. **B487**, 467 (1997).
[18] D. Binosi, J. Papavassiliou, Phys. Rev. **D66**, 111901(R) (2002).
[19] D. Binosi, J. Papavassiliou, J.Phys.G **G30**, 203 (2004).
[20] D. Binosi, J. Papavassiliou, Phys. Rept. **479**, 1 (2009).
[21] L.F. Abbott, Nucl. Phys. **B185**, 189 (1981).
[22] L.F. Abbott, Acta Phys. Polon. **B13**, 33 (1982).
[23] A.C. Aguilar, D. Binosi, C.T. Figueiredo, J. Papavassiliou, Phys. Rev. **D94**, 045002 (2016).
[24] P.A. Grassi, T. Hurth, M. Steinhauser, Annals Phys. **288**, 197 (2001).
[25] D. Binosi, J. Papavassiliou, Phys. Rev. **D66**, 025024 (2002).
[26] D. Binosi, A. Quadri, Phys. Rev. **D88**, 085036 (2013).
[27] A.C. Aguilar, J. Papavassiliou, Phys. Rev. **D81**, 034003 (2010).
[28] D. Binosi, D. Ibañez, J. Papavassiliou, Phys. Rev. **D86**, 085033 (2012).
[29] A. Aguilar, D. Binosi, J. Papavassiliou, JHEP **0911**, 066 (2009).
[30] P. Bicudo, D. Binosi, N. Cardoso, O. Oliveira, P.J. Silva, Phys. Rev. **D92**, 114514 (2015).
[31] A. Aguilar, D. Ibanez, V. Mathieu, J. Papavassiliou, Phys. Rev. **D85**, 014018 (2012).
[32] A.C. Aguilar, D. Binosi, J. Papavassiliou (2016), 1611.02096.
[33] A.C. Aguilar, D. Binosi, D. Ibañez, J. Papavassiliou, Phys. Rev. **D89**, 085008 (2014).

EPJ Web of Conferences **164**, 03006 (2017) DOI: 10.1051/epjconf/201716403006
ICNFP 2016

Precision RENORM Tensor-Pomeron Cross Sections

Konstantin Goulianos[1,a]

[1] *The Rockefeller University, 1230 York Avenue, New York, NY 10065–6399, USA*

Abstract. Precision predictions of soft and hard diffraction, elastic, and total proton-proton cross sections, based on a Regge-theory inspired tensor-Pomeron implementation of the RENORM model, are compared to TOTEM, ATLAS, and CMS results at the LHC.

1 Introduction

In DIS-2015 (Spring 2015) we summarized [1] the pre-LHC predictions of the total, elastic and total-inelastic, as well as the single diffractive (SD) and double diffractive (DD) components of the proton-proton cross section at high energies, based on the RENORM / MBR model [2]. We compared the measurements of the SD and DD cross sections from the Tevatron and the LHC with the predictions of the model and found excellent agreement. Good agreement was also observed between the model predictions and the total, elastic, and total inelastic cross sections obtained at the Tevatron at $\sqrt{s}$ = 1.8 TeV, and at the LHC at $\sqrt{s}$ = 7 and 8 TeV.

The confirmation of the predictions of all the above cross sections at the Tevatron and the LHC up to $\sqrt{s}$ = 8 TeV prompted a calculation/extrapolation to $\sqrt{s}$ = 13 TeV, the nominal foreseen colliding-beam energy at the LHC for Summer 2015. For $\sigma_{\rm tot}$, $\sigma_{\rm el}$ and $\sigma_{\rm inel}$, we predicted 108 mb, 32 mb, and 77 mb, respectively, with uncertainties of $\sim$ 11% in all cases, mainly due to the uncertainty in the energy-squared scale parameter s_0 of the model.

In Summer 2015 we updated the value of s_0 to a more precise one based on a tensor glueball interpretation of the Axial Field Spectrometer (AFS) exclusive charged di-pion data [3–5]. This change in RENORM / NBR decreases the uncertainties in the predictions of the total, elastic, and total-inelastic cross sections to less than 2% from Tevatron to LHC energies, with little or no effect on the mean values. The predictions were compared with measurements by ATLAS at $\sqrt{s}$ = 7 TeV and by TOTEM at $\sqrt{s}$ = 7 and 8 TeV at Moriond QCD in March 2016 [6] and found to be in good agreement.

This is an update of [6] that includes verbatim certain sections and describes new measurements of the total inelastic cross section at $\sqrt{s}$ = 13 TeV by ATLAS and CMS at DIS-2016 (April 2016) by Trzebinski ATLAS) and Van Haevermaet (CMS) [7]. There are disagreements among the measurements themselves and the RENORM predictions at the $\sim$ 2 sigma level. In the present paper, we discuss these findings in an attempt to understand the origin of the observed discrepancies.

2 RENORM cross sections

The total, elastic, and total-inelastic cross sections in the RENORM / NBR model depend on the value of the energy-squared scale parameter, s_0. Quoting verbatim from Ref. [1],

[a] e-mail: dino@rockefeller.edu

"The total cross section ($\sigma_{\rm tot}$) is expressed as [8]

$$\sigma_{\rm tot}^{p^{\pm}p} = 16.79 s^{0.104} + 60.81 s^{-0.32} \mp 31.68 s^{-0.54} \quad \text{for } \sqrt{s} \leq 1.8 \text{ TeV}, \tag{1}$$

$$\sigma_{\rm tot}^{p^{\pm}p} = \sigma_{\rm tot}^{\rm CDF} + \frac{\pi}{s_0}\left[\left(\ln\frac{s}{s_F}\right)^2 - \left(\ln\frac{s^{\rm CDF}}{s_F}\right)^2\right] \quad \text{for } \sqrt{s} \geq 1.8 \text{ TeV}, \tag{2}$$

where s_0 and s_F are the energy and (Pomeron flux) saturation scales, $s_0 = 3.7 \pm 1.5$ GeV^2 and $\sqrt{s_F} = 22$ GeV, respectively. For $\sqrt{s} \leq 1.8$ TeV, where there are Reggeon contributions, we use the global fit expression [9], while for $\sqrt{s} \geq 1.8$ TeV, where Reggeon contributions are negligible, we employ the Froissart-Martin formula [10–12]. The two expressions are smoothly matched at $\sqrt{s} \approx 1.8$ TeV. The $\sigma_{\rm el}$ for $\sqrt{s} \leq 1.8$ TeV is obtained from the global fit [9], while for $1.8 < \sqrt{s} \leq 50$ TeV we use an extrapolation of the global-fit ratio of $\sigma_{\rm el}/\sigma_{\rm tot}$, which is slowly varying with $\sqrt{s}$, multiplied by σ_{tot}. The total non-diffractive cross section is given by $\sigma_{\rm ND} = (\sigma_{\rm tot} - \sigma_{\rm el}) - (2\sigma_{\rm SD} + \sigma_{\rm DD} + \sigma_{\rm CD})$."

2.1 Tensor-Pomeron predictions

The partial wave analysis of the AFS exclusive $\pi^{\pm}$ data [5], performed in terms of a fit with a model with S-wave and D-wave amplitudes as a function of the di-pion mass up to 2.3 GeV, leads to the results presented in Fig. 1.

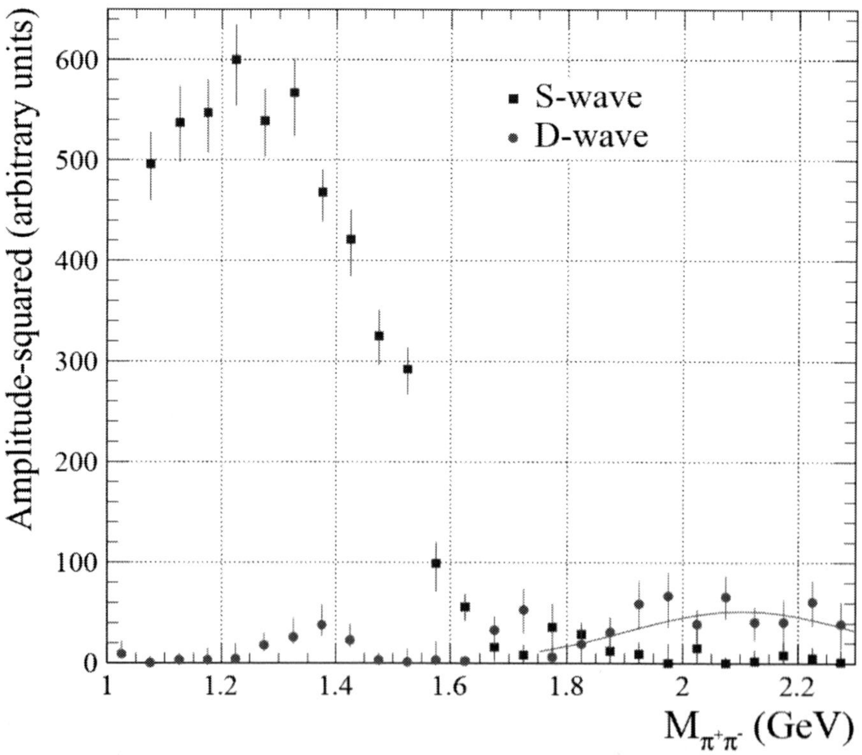

Figure 1. Extraction of tensor-Pomeron parameters from a Gaussian fit to the exclusive $\pi^{\pm}$ Axial Field Spectrometer data. The fit yields a mean mass value $\langle M_{\pi^+\pi^-}\rangle = 2.10$ GeV of width $\Delta = \pm 0.68$ GeV.

The D-wave dominates at masses above $\sim$ 2 GeV, and according to the presumed interpretation in Ref. [5] it corresponds to a spin-2 tensor glueball of mass $M_{\rm tgb}$. A Gaussian fit to this enhancement

 DOI: 10.1051/epjconf/201716403006

yields $M_{tgb} = 2.10 \pm 0.68$ GeV. Identifying M^2_{tgb} with the saturated glueball-like enhancement of the MBR-model parameter s_0 (see Eq. 2) yields $s_0 = 4.42 \pm 0.34$ GeV^2. Using this value in Eq. 2 we predicted for σ_{tot}, $\sigma_{el\pm6.6\ (lumi)}$, and σ_{inel} at 13 TeV cross sections of 103.7 ± 1.9 mb, 30.2 ± 0.8 mb, and 73.5 ± 1.3 mb, respectively. The ATLAS- and TOTEM-measured cross sections at $\sqrt{s} = 7$ and 8 TeV [13–15] are shown in Table 1 along with the NBR predictions.

Table 1. The total, elastic, and total inelastic MBR predictions (in mb) at $\sqrt{s} = 7$, 8 and 13 TeV compared to measurements at the LHC by TOTEM and ATLAS.

$\sqrt{s}$ (TeV)	MBR Experiment	σ_{tot} (mb)	σ_{el} (mb)	σ_{inel} (mb)
7	MBR	95.4 ± 1.2	26.4 ± 0.3	69.0 ± 1.0
	TOTEM	98.3 ± 2.9	24.8 ± 1.3	73.7 ± 1.7
	TOTEM_Lum_ind	98.0 ± 2.5	25.2 ± 1	72.9 ± 1.5
	ATLAS	95.35 ± 1.36	24.00 ± 0.60	71.34 ± 0.90
8	MBR	97.1 ± 1.4	27.2 ± 0.4	69.9 ± 1.0
	TOTEM	101.7 ± 2.9	27.1 ± 1.4	74.7 ± 1.7
13	MBR	103.7 ± 1.9	30.2 ± 0.8	73.5 ± 1.3
	ATLAS	xxx	xxx	73.1 ± 0.9(exp)±3.8(extr)±6.6(lumi)
	CMS	xxx	xxx	71.3 ± 0.6(exp)±2.7(extr)±2.1(lumi)

The ATLAS and CMS cross-section entries are imported from Refs. [7] (a) and (b), respectively. Below, we list all the ATLAS results presented in [7] (a).

1. Elastic cross section:
   σ_{el}^{ALFA}(7 TeV) = 24.00 ± 0.19 (stat) ± 0.57 (syst) mb
2. Nuclear slope:
   b^{ALFA}(7 TeV) = 19.73 ± 0.14 (stat) ± 0.26 (syst) GeV^{-2}
3. Total cross section
   σ_{tot}^{ALFA}(7 TeV = 95.35 ± 0.38 (stat) ± 1.25 (exp) ± 0.37 (extr) mb
4. Inelastic cross section:

- σ_{inel}^{ALFA}(7 TeV) = 71.34 ± 0.36 (stat) ± 0.83 (syst) mb
- σ_{inel}^{MBTS}(7 TeV) = 69.4 ± 2.4 (exp) ± 6.9 (extr) mb
- σ_{inel}^{MBTS}(13 TeV) = 73.1 ± 0.9 (exp) ± 3.8 (extr) ± 6.6 (lumi) mb

Shown in Table 1 is also a recent measurement of the total inelastic cross section by ATLAS at $\sqrt{s} = 13$ TeV [16], $\sigma_{inel} = 73.1 \pm 0.9$ (exp) ± 3.8 (extr) ± 6.6 (lumi) mb, which, apart from the extrapolation and luminosity uncertainties, is in excellent agreement with MBR. All the other measurements are also in good agreement with the predictions. The tensor-Pomeron-based prediction of σ_{inel} at $\sqrt{s} = 13$ TeV agrees with the ATLAS measurement at the 1% level.

It should be emphasized that the tensor-Pomeron hypothesis predicts directly only the total cross section. As discussed above, the elastic cross section for $\sqrt{s} \leq 1.8$ TeV is obtained from the global fit [9], while for $1.8 < \sqrt{s} \leq 13$ TeV we use an extrapolation of the global-fit ratio of σ_{el}/σ_{tot}, which is slowly varying with $\sqrt{s}$, multiplied by σ_{tot}.

The total inelastic cross section is calculated as the difference between the total and the elastic cross sections. Thus, a measured lower σ_{el} would result in a higher σ_{inel}. As seen in Table 1, the MBR σ_{el} is larger than the TOTEM and CMS measurements by $\sim$ 2 mb at $\sqrt{s}$=7 TeV, which would imply a higher MBR prediction for σ_{inel} at 13 TeV by $\sim$ 2 mb as well. This interplay between σ_{el} and σ_{inel} should be carefully taken into consideration as more measurements of both σ_{el} and σ_{tot} at $\sqrt{s}$ = 8 TeV and $\sqrt{s}$ = 13 TeV with reduced luminosity and extrapolation uncertainties become available.

3 Summary and conclusions

We present the predictions of the total, elastic scattering, and total-inelastic proton-proton cross sections at $\sqrt{s}$ = 7, 8 and 13 TeV of the MBR model, based on a Regge-theory inspired tensor-Pomeron implementation of the RENORM model for hadronic diffraction, and compare them with experimental results by the TOTEM, ATLAS, and CMS Collaborations. All measured cross sections are in good agreement within the experimental uncertainties of the data and the theoretical uncertainties of the model, reaching down to the $\sim$1% level.

We emphasize that the tensor-Pomeron model predicts directly only to the total cross section, while the predicted σ_{el} is calculated by an extrapolation of the global-fit [9] ratio of σ_{el}/σ_{tot} multiplied by the tensor-Pomeron σ_{tot} and therefore carries the larger uncertainties of this ratio. Future measurements of both σ_{el} and σ_{tot} at $\sqrt{s}$ = 8 TeV and $\sqrt{s}$ = 13 TeV with reduced luminosity and extrapolation uncertainties are needed to clarify the current experimental status.

Acknowledgments

Warmly acknowledged are Dr. Michael Albrow for providing a copy of the thesis of Peter C. Cesil, where the tensor-Pomeron data and analysis are discussed, and Dr. Robert Ciesielski for extracting the tensor-Pomeron parameters from the data and for many invaluable discussions.

References

[1] Konstantin Goulianos, *Review of RENORM Diffractive Predictions for LHC up to 8 TeV and Extension to 13 TeV*, PoS(DIS 2015)073.

[2] K. Goulianos, *Hadronic diffraction: where do we stand?* [arXiv:hep-ph/0407035]; K. Goulianos, *Diffraction in QCD* [arXiv:hep-ph/0203141].

[3] M.G. Albrow, T.D. Coughlin, J.R. Forshaw, Prog. Part. Nucl. Phys. **65**, 149 (2010) [DOI:10.1016/j.ppnp.2010.06.001 [arXiv:1006.1289].

[4] M.G. Albrow, private communication, discussions about the Axial Field Spectrometer results on exclusive $\pi^+\pi^-$.

[5] Peter Charles Cesil, *A Search for Centrally Produced Glueballs in Proton-Proton Interactions*, PhD thesis, University of Cambridge, September 1984.

[6] K. Goulianos, *RENORM tensor-Pomeron diffractive predictions*, in *51st Rencontres de Moriond on QCD and High Energy Interactions*, 19-26 Mar 2016. La Thuile, Italy C16-03-19.1 [arXiv:1606.01302v2 [hep-ph] 3 Jun 2016].

[7] Presented at *24th International Workshop on Deep-Inelastic Scattering and Related Subjects (DIS 2016)*, Hamburg, Germany, April 11-25, 2016 [https://dis2016.desy.de/]:
(a) Maciej Trzebinzki (ATLAS Collaboration), *Measurements of Total, Elastic, Inelastic and Diffractive Cross Sections with the ATLAS Detector*;
(b) Hans Van Haevermaet (CMS Collaboration), *Measurement of the inelastic cross section at 13 TeV*.

[8] R. Ciesielski and K. Goulianos, *MBR Monte Carlo Simulation in PYTHIA-8* [arXiv:1205.1446].

[9] R. J. M. Covolan, J. Montanha and K. Goulianos, Phys. Lett. B **389**, 176 (1996).

[10] M. Froissart, Phys. Rev. **3**, 123 (1961).

[11] A. Martin, Nuovo Cimento **42**, 930 (1966).

[12] A. Martin, Phys. Rev. D **80**, 065013 (2009).

[13] Marek Tasevsky (ATLAS Collaboration), *Diffraction and forward physics of the ATLAS experiment from Run I*, Diffraction 2014, AIP Conf. Proc. 1654 (2015) 040002, (2015-04-10) [DOI:10.1063/1.4915967].

[14] G. Antchev et al. (TOTEM Collaboration), Eur. Phys. Lett. **101** (2013) 21002, Eur. Phys. Lett. **101** (2013) 21004, Eur. Phys. Lett. **101** (2013) 21003.

[15] G. Antchev et al. (TOTEM Collaboration), Phys. Rev. Lett. **111**, 012001 (2013).

[16] D. Schaefer (CERN), *Recent soft QCD results from the LHC*, in *2016 Aspen Winter Conference in Particle Physics: Particle Physics on the Verge of Another Discovery?*, 11-16 January 2016, ACP, Aspen, CO.

Some Developments in Gribov's Approach to QCD

Patrick Cooper[1,a] and Daniel Zwanziger[2,3,b]

[1] *Physics Department, Duquesne University, 600 Forbes Ave, Pittsburgh, PA 15282, USA*
[2] *Physics Department, New York University, 4 Washington Place, New York, NY 10003, USA*
[3] *Speaker*

Abstract. We review several developments in the formulation of QCD provided by the GZ action. These include the GZ-action at finite temperature, the relation of the horizon condition and the Kugo-Ojima confinement criterion, the relation of the horizon condition and the dual-Meisssner effect, the alternative derivation of the GZ action provided by the Maggiore-Schaden shift, and the spontaneous breaking of BRST symmetry. We conclude with a proposal for the definition of physical states in the presence of BRST breaking.

1 Introduction

This article is dedicated to the memory of Vladimir Gribov. In it we review several developments that emerged from his paper of 1977 [1] on the quantization of gauge fields. In it he explained how the non-perturbative nature of non-abelian gauge theory, manifested in what are now called "Gribov copies." leads to a long-range, potentially confining force.

2 No Confinement without Coulomb Confinement

Gribov's insight into the mechanism of confinement is substantiated by the theorem [2],

$$V_{\rm coul}(R) \geq V_{\rm wilson}(R) \tag{1}$$

for $R \to \infty$, where the color-Coulomb potential is the temporal gluon propagator in Colomb gauge,

$$D_{00}(R, T) = V_{\rm coul}(R)\, \delta(T) + {\rm non-instantaneous.} \tag{2}$$

When the Wilson potential is linearly rising, the color-Coulomb potential is linear or super-linear.

3 All horizons are one horizon

Gribov proposed that the functional integral with respect to A should extend only over what is now called the Gribov region [1]. It is the region where all eigenvalues of the Faddeev-Popov operator

$$M^{ac}(gA) = -D_i^{ac}(gA)\partial_i - \partial_i^2 \delta^{ac} - f^{abc} gA_i \partial_i, \tag{3}$$

[a] e-mail: cooperp@duq.edu
[b] e-mail: dz2@nyu.edu

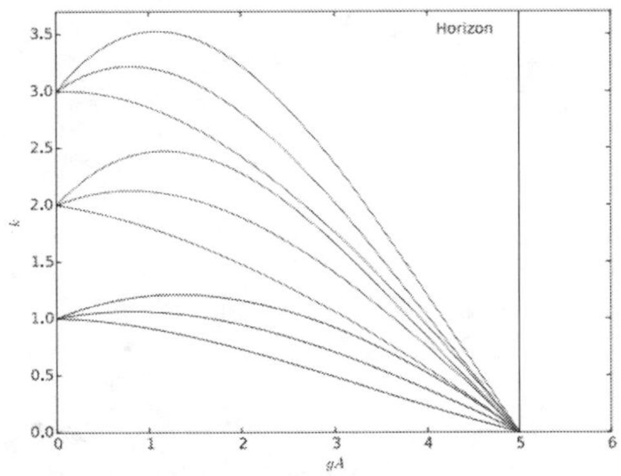

Figure 1. Cartoon of the eigenvalues of the Fadeev-Popov operator.

are positive. It was subsequently found that the eigenvalues are given by [3]

$$\lambda(\vec{k}; gA) = \vec{k}^2 \left(1 - \frac{H(gA)}{(N^2-1)dV}\right) + \text{higher} - \text{order in } \vec{k}, \tag{4}$$

where[1]

$$H(gA) = \int d^d x d^d y D^a_\mu(x) D^b_\mu(y) (M^{-1})^{ab}(x, y; A) \tag{5}$$

is called the "horizon function." The eigenvalues $\lambda(\vec{k}; gA)$ are continuous functionals of gA. At $gA = 0$ the eigenvalues are given by $\lambda(\vec{k}; 0) = \vec{k}^2$, where $\vec{k}_i = 2\pi n_i/L$, and $V = L^d$ is the quantization volume, and the n_i are integers. Here $\vec{k}$ serves as a label for the eigenvalue $\lambda(\vec{k}; gA)$ that is continuously connected to $\lambda(\vec{k}; 0) = \vec{k}^2$. In the limit of large volume L^d, the spectrum approaches a continuum, and at small $\vec{k}$ the terms of higher order in $\vec{k}$ become negligible. There is a common factor of $\vec{k}^2$ in (4), so in the limit of large quantization volume an infinite number of eigenvalues pass through 0 together and become negative together, as illustrated in Fig. 1. We call this phenomenon "all horizons are one horizon." This is illustrated in Fig. 2. Gribov regions are sometimes represented as in the left side of Fig. 2, with distinct horizons for different eigenvalues.[2] This is correct at finite volume L^d. At large volume, an infinite number of horizons coincide, as in the right side of Fig. 2.

4 Finite Temperature

So far we have been discussing zero temperature. At finite temperature T, a different picture emerges. In the Euclidean formulation, finite temperature corresponds to finite period $1/T$ in the thermal direction, with discrete Matsubara frequencies $2\pi nT$. In contrast, at large spatial volume the spectrum of the Faddeev-Popov operator for each n approaches a continuous spectrum, as illustrated in Fig. 3. As a result the horizon condition takes the form [6]

$$\left\langle \int d^D x d^D y D^{ab}_i D^{ac}_i(y) (M^{-1})^{bc}(x, y) \right\rangle = Vd(N^2 - 1), \tag{6}$$

[1]For reviews see [4, 5]

[2]Each zero of a every eigenvalue defines a Gribov horizon.

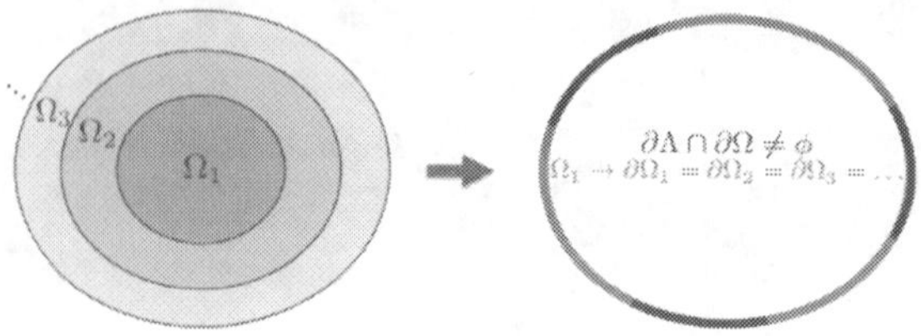

Figure 2. All horizons are one horizon

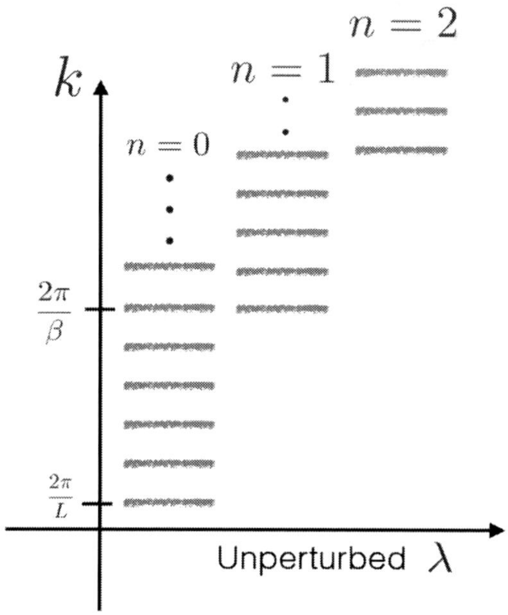

Figure 3. A sketch of the spectrum of the unperturbed Faddeev-Popov operator in Landau gauge $M(0) = -\partial_\mu^2$ at finite temperature. The index n labels the finite Matsubara frequencies. When the perturbation is turned on, $M(0) \to M(gA)$, the eigenvalues corresponding to the zero-Matsubara frequency cross the Gribov horizon first.

where $M = -D_\mu \partial_\mu$, $D = d + 1$, $\mu = 1, ...D$, and $i = 1, ...d$.

5 Horizon Function and Non-local Action

We return to zero temperature. The Euclidean action is given by,

$$S = S_{\rm FP} + \gamma H - \gamma \int d^D x\, (N^2 - 1) D, \tag{7}$$

where $S_{\rm FP}$ is the Faddeev-Popov action, and the horizon function H is given in (5).

The horizon function cuts off the functional integral at the Gribov horizon, as one sees from the eigenfunction expansion

$$(M^{-1})^{ab} = \sum_n \frac{\psi_n^a(x)\psi_n^b(y)}{\lambda_n(gA)}. \tag{8}$$

6 Horizon Condition and Kugo-Ojima Confinement Condition

The Gribov parameter is fixed by the horizon condition,

$$\langle H \rangle = (N^2 - 1)d \int d^d x. \tag{9}$$

It is a remarkable fact that the horizon condition and the famous Kugo-Ojima confinement condition are the same statement [7, 8]

$$-i \int d^d x \, \left\langle (D_\mu c)^a(x)(D_\mu \bar{c})^a(0) \right\rangle = (N^2 - 1)d. \tag{10}$$

This may indicate that color confinement is assured in this theory, although the precise hypotheses of the Kugo-Ojima theorem are not satisfied in this approach.

7 Horizon Condition and Dual Meissner effect

It is a remarkable fact that the horizon condition is equivalent to the statement that the QCD vacuum is a perfect color-electric superconductor, which is the dual Meissner effect [9],

$$G(\vec{k}) = \frac{d(\vec{k})}{\vec{k}^2} = \frac{1}{\epsilon(\vec{k})\vec{k}^2} \tag{11}$$

$$d^{-1}(\vec{k} = 0) = 0 \Longleftrightarrow \epsilon(\vec{k} = 0), \tag{12}$$

where $G(\vec{k})$ is the ghost propagator and $\epsilon(\vec{k})$ is the dielectric constant.

8 Auxiliary ghosts

Just as the Faddeev-Popov determinant is localized by introducing ghosts,

$$\det M = \int dc d\bar{c} \exp\left(- \int d^d x \, \bar{c} M c \right), \tag{13}$$

likewise, the horizon function in the action may be localized by introducing "auxiliary" ghosts [3],

$$\exp(-\gamma H) = \int d\varphi d\bar{\varphi} d\omega d\bar{\omega} \exp\left(- \int d^d x \, \left[\bar{\varphi} M \varphi - \bar{\omega} M \omega + \gamma^{1/2} D \cdot (\varphi - \bar{\varphi}) \right] \right). \tag{14}$$

9 Local Action and Physical Degrees of Freedom in Coulomb Gauge

In terms of the additional fields, the local action is given by [10],

$$\begin{aligned} S &= \int d^{d+1}x\, \{i\tau_i D_0 A_i + (1/2)[\tau_i^2 + (\partial_i\lambda)^2 + (1/4)F_{ij}^2 + i\partial_i\lambda D_i A_0 - \partial_i\bar{c} D_i c + \partial_i\bar{\varphi}_j \cdot D_i\varphi_j \\ &\quad - \partial_i\bar{\omega}_j \cdot (D_i\omega_j + D_i c \times \varphi_j) + \gamma^{1/2}\mathrm{Tr}[D_j(\varphi - \bar{\varphi})_j - D_j c \times \bar{\omega}_j] - \gamma(N^2-1)d\}, \quad (15) \end{aligned}$$

Here we have introduced the canonically conjugate color-electric field π, and decomposed it into its transverse and longitudinal parts

$$\pi_i = \tau_i + \partial_i\lambda, \tag{16}$$

where τ_i is the momentum conjugate to A_i, and both are transverse,

$$\partial_i\tau_i = \partial_i A_i = 0. \tag{17}$$

Only the first term,

$$i\tau_i D_0 A_i = i\tau_i\partial_0 A_i + i\tau_i A_0 \times A_i, \tag{18}$$

contains a time derivative. It acts on the two transverse degree of of freedom of the gluon which are the would-be physical degrees of freedom. All the remaining terms impose constraints.

10 Maggiore-Schaden Shift

It is another remarkable fact that the same action may be derived by a completely different method developed by Maggiore and Schaden, without reference to the Gribov horizon [11, 12].

Start with the usual Faddeev-Popov fields and action, on which the BRST operator s acts according to

$$\begin{aligned} sA_\mu &= D_\mu c & sc &= -(g/2)(c \times c) \\ s\hat{\bar{c}} &= i\hat{b} & s\hat{b} &= 0 \\ s\pi_i &= g\pi_i \times c, & &, \quad (19) \end{aligned}$$

Introduce auxiliary quartets of bose and fermi ghosts whose determinants cancel when they are integrated out, on which s acts according to

$$\begin{aligned} s\phi_B &= \omega_B & s\omega_B &= 0 \\ s\bar{\omega}_B &= \bar{\phi}_B & s\bar{\phi}_B &= 0. \end{aligned} \tag{20}$$

The resulting action is BRST-invariant by construction,

$$\mathcal{L} = \mathcal{L}^{\mathrm{YM}} + \mathcal{L}^{\mathrm{gf}} = \mathcal{L}^{\mathrm{YM}} + s\Psi \qquad s\mathcal{L} = 0, \tag{21}$$

where $\mathcal{L}^{\mathrm{FP}} = (1/4)F_{\mu\nu}^2$ and

$$\Psi \equiv \partial_i\hat{\bar{c}} \cdot A_i + \partial_i\bar{\omega}_j \cdot D_i\phi_j \tag{22}$$

$$s\Psi = i\partial_i\hat{b} \cdot A_i - \partial_i\hat{\bar{c}}_j \cdot D_i c + \partial_i\bar{\phi}_j \cdot D_i\phi_j - \partial_i\bar{\omega}_j \cdot (D_i\omega_j D_i c \times \phi_j). \tag{23}$$

Now make the change of variable

$$\begin{aligned} \phi_{jb}^a(x) &= \varphi_{jb}^a(x) - \gamma^{1/2}x_j\delta_b^a \\ \bar{\phi}_{jb}^a(x) &= \bar{\varphi}_{jb}^a(x) + \gamma^{1/2}x_j\delta_b^a \\ \hat{\bar{c}}^a(x) &= \bar{c}^a(x) - \gamma^{1/2}x_j f^{abc}\bar{\omega}_{jc}^b(x) \\ \hat{b}^a(x) &= b^a(x) + i\gamma^{1/2}x_j f^{abc}\bar{\varphi}_{jc}^b(x). \end{aligned} \tag{24}$$

 DOI: 10.1051/epjconf/201716403007

This procedure, by a completely different derivation, yields precisely the same x-independent Lagrangian density that was derived from a cut-off at the Gribov horizon. The resulting lagrangian density remains BRST-invariant, $s\mathcal{L} = 0$, but the *form* of the BRST operator s acting on the new fields is changed. (For an alternative approach and further references, see [13].)

11 Spontaneous breaking of BRST

The BRST operator acts on all the shifted fields exactly as it does on the corresponding unshifted fields, except for

$$s\bar{\omega}^a_{jb} = \bar{\varphi}^a_{jb} + \gamma^{1/2} x_j \delta^a_b. \tag{25}$$

We have regained an s-invariant, local, x-independent lagrangian density, but the BRST symmetry is spontaneously broken

$$\langle\{Q_B, \bar{\omega}^a_{ib}\}\rangle = \langle s\bar{\omega}^a_{ib}\rangle = \gamma^{1/2} x_i \delta^a_b. \tag{26}$$

When BRST is unbroken, physical states are characterized as being invariant under the action of the BRST operator, $Q_B \Phi_{\rm phys} = 0$ (and we mod out the states of norm zero). The last equation implies that the vacuum state is not BRST-invariant,[3]

$$\langle\{Q_B, \bar{\omega}^a_{ib}\}\rangle = (Q_B\Phi_0, \bar{\omega}^a_{ib}\Phi_0) + (\Phi_0, \bar{\omega}^a_{ib} Q_b \Phi_0) = \gamma^{1/2} x_i \delta^a_b \neq 0, \tag{27}$$

so we require another characterization of physical states.

12 Physical states in Faddeev-Popov Theory

For purposes of orientation, let us review quantization of Faddeev-Popov theory. The Faddeev-Popov determinant is non-local. It is localized at the cost of introducing the Faddeev-Popov ghost which creates unphysical states. There is also a benefit however, because the local action $\mathcal{L} = \mathcal{L}^{\rm YM} + s\Psi$, has a new symmetry, the BRST-symmetry, with $s\mathcal{L} = 0$, that is not present in the non-local theory. This symmetry allows us to characterize physical observables: they are the operators G that are invariant under the BRST symmetry,

$$\mathcal{W}_{\rm phys} \equiv \{G : sG = 0\}. \tag{28}$$

Physical states are obtained by applying physical operators to the vacuum state.

13 Physical states in the GZ theory

Localization of the cut-off at the Gribov horizon yields

$$\mathcal{L} = \mathcal{L}^{\rm YM} + s\chi = \mathcal{L}^{\rm YM} + \mathcal{L}_{\rm aux.gh}. \tag{29}$$

where $\mathcal{L}_{\rm aux.gh}$ now depends not only on the Faddeev-Popov ghosts, but also on the auxiliary ghosts. This action possesses many new symmetries, with generators Q_Y that leave ordinary physical observables invariant,

$$[Q_Y, \mathcal{L}] = [Q_Y, F^2] = [Q_Y, \bar{\psi}\psi] = 0. \tag{30}$$

A classification of the these symmetries may be found in [14, 15]. We call them "phantom symmetries" because they act on these physical operators like the zero-operator, but they act non-trivially on

[3] This implies that physical states cannot be defined as the cohomology of s.

the auxiliary ghost fields. It is natural to define physical observables as the class of operators that are invariant under all the new phantom symmetries [14],

$$\mathcal{W}_{\rm phys} \equiv \{G : sG = 0; [Q_Y, G] = 0\} \text{ for all } Q_Y. \tag{31}$$

We offer as a *conjecture* that BRST symmetry is not broken by s-exact operators in $\mathcal{W}_{\rm phys}$,

$$\langle sY \rangle = 0 \text{ for all } sY \in \mathcal{W}_{\rm phys}. \tag{32}$$

In other words, BRST symmetry is preserved *where it is needed*, namely in the physical subspace.

Explicit calculation of several examples reveals that BRST symmetry-breaking apparently afflicts the unphysical sector, but may be unbroken where it is needed, namely in cases of physical interest. For example, the BRST-exact part of the conserved energy-momentum tensor has vanishing expectation-value,

$$T_{\mu\nu} = T^{\rm YM}_{\mu\nu} + s\Xi_{\mu\nu} \tag{33}$$

$$\langle s\Xi_{\mu\nu} \rangle = 0. \tag{34}$$

As before physical states are obtained by applying physical operators to the vacuum state.

14 Conclusion

We have a local, renormalizable quantum field theory with the following interesting properties:
- It provides a cut-off at the Gribov horizon.
- The Kugo-Ojima color confinement condition is satisfied.
- The vacuum is a perfect dielectric.
- There is an alternate derivation by the Maggiore-Schaden shift that provides a BRST-invariant Lagangian.
- BRST-symmetry is spontaneously broken, but perhaps only in the unphysical sector.
- In Coulomb gauge, the color-Coulomb potential is linear or super-linear when the Wilson potential is linearly rising.

References

[1] V.N. Gribov, Quantization of non-Abelian Gauge Theories, Nucl. Phys. B **139**, 1 (1978).

[2] Daniel Zwanziger, No Confinement without Coulomb Confinement, Phys. Rev. Letts, **90,** 102001 (2003) and arXiv:hep-lat/0209105.

[3] Daniel Zwanziger, Local and Renormalizable Action from the Gribov Horizon, Nucl. Phys. B**323**, 513 (1989).

[4] R.F. Sobreira, S.P. Sorella, Introduction to the Gribov Ambiguities In Euclidean Yang-Mills Theories, arXiv:hep-th/0504095.

[5] N. Vandersickel, D. Zwanziger, The Gribov problem and QCD dynamics, Phys. Rept.**520**, 175 (2012) and arXiv:1202.1491 [hep-th].

[6] Patrick Cooper and Daniel Zwanziger, Local QCD Action at Finite Temperature, Phys. Rev. **D 93,** 105026 (2016) and arXiv:1512.08858

[7] T. Kugo, I. Ojima, Prog. Theor. Phys. Suppl.**66**, 1 (1979) ;Erratum Prog. Theor. Phys.**71**, 1121 (1984).

[8] T. Kugo, The Universal renormalization factors Z(1) / Z(3) and color confinement condition in nonAbelian gauge theory, arXiv:hep-th/9511033.

[9] H. Reinhardt, Phys. Rev. Lett. 101, 061602 (2008), 08030504.

[10] Patrick Cooper and Daniel Zwanziger, Spectral decomposition of the ghost propagator and a necessary condition for confinement, Phys. Rev. **D 93,** 105024 (2016) and arXiv:1512.05725.

[11] N. Maggiore, M. Schaden, Landau gauge within the Gribov horizon, Phys. Rev. D**50**, 6616 (1994).

[12] M. Schaden, habilitation thesis on *Kovariante Eichfixierung und Dynamische Brechung der BRS Symmetrie von Yang-Mills Eichtheorien*, Technische Universität München, Germany (1996).

[13] M. A. L. Capri, D. Fiorentini, M. S. Guimaraes, B. W. Mintz, L. F. Palhares, S. P. Sorella, Phys. Rev. D 94, 065009 (2016); arXiv:1606.06601 [hep-th].

[14] Martin Schaden and Daniel Zwanziger, Phys. Rev. **D92,** 025001 (2015), and arXiv:1412.4823 [hep-ph].

[15] Martin Schaden and Daniel Zwanziger, Living with spontaneously broken BRST symmetry. II. Poincaré invariance, Phys. Rev. **D92,** 025002 (2015) and arXiv:1501.05974 [hep-th].

The Gribov mode in hot QCD

Nan Su[1,a]

[1] *Frankfurt Institute for Advanced Studies, 60438 Frankfurt am Main, Germany*

Abstract. In this proceedings, I summarize recent findings of a novel massless mode, dubbed as "Gribov mode", generated by the (chromo)magnetic scale g^2T in hot QCD. The Gribov mode is a genuine non-Abelian mode inducing effects such as positivity violation.

1 Introduction

Gauge fixing is a mathematical procedure for removing redundant degrees of freedom in the field variables of gauge theories. A popular gauge-fixing procedure was invented by Faddeev and Popov [1] that leads to satisfactory results for Abelian gauge theories. However in the seminal work of Ref. [2], Gribov pointed out that there are still residue redundant degrees of freedom – namely the Gribov copies – after the Faddeev-Popov procedure. The Gribov copies reside in the IR of the non-Abelian gauge fields, and they are intimately related to the confinement of color charges. Later on, Zwanziger generalized Gribov's semi-classical approach to all orders that gave birth to the Gribov-Zwanziger action [3]. The Gribov-Zwanziger scenario has stimulated flourishing developments in the study of color confinement (see Refs. [4, 5] for reviews).

Since the last decade, there has been an increasing effort in generalizing the Gribov-Zwanziger scenario to finite temperature stimulated by the Linde problem that invalids conventional thermal perturbation theory at the (chromo)magnetic scale g^2T [6, 7]. The non-perturbative nature of the magnetic scale is intimately relathard-thermal-looped to the confining property of the dimensionally reduced Yang-Mills theory at high temperature. This suggests the need of incorporating a confinement mechanism in perturbative expansions even when dealing with the deconfined quark-gluon plasma phase. The Gribov-Zwanziger action provides an ideal framework for this purpose. It regulates the IR behavior of QCD by fixing the Gribov copies that remain after applying the Faddeev-Popov procedure. The Gribov-Zwanziger action is renormalizable, therefore it provides a systematic framework for perturbative calculations (i.e., $g \ll 1$) incorporating confinement effects. The gluon propagator in general covariant gauge reads

$$D^{\mu\nu}(P) = \left[\delta^{\mu\nu} - (1-\xi)\frac{P^\mu P^\nu}{P^2}\right]\frac{P^2}{P^4 + \gamma_G^4}\,, \tag{1}$$

where ξ is the gauge parameter. The Gribov parameter γ_G is solved self-consistently from a gap equation that is defined to infinite loop orders. The Gribov-Zwanziger gluon propagator is IR suppressed,

[a] e-mail: nansu@fias.uni-frankfurt.de

manifesting confinement effects, and it is a significant improvement over the one from the Faddeev-Popov quantization which forms the basis of conventional perturbative calculations. The gap equation at one-loop order can be solved analytically at asymptotically high temperatures and gives [8]

$$\gamma_G = \frac{D-1}{D}\frac{N_c}{4\sqrt{2}\pi}g^2T\,, \tag{2}$$

where D is the space-time dimensions and N_c is the number of colors. Eq. (2) provides a fundamental IR cutoff at the magnetic scale for the finite-temperature Gribov-Zwanziger action. In this way, the magnetic scale is intrinsically embedded in the Gribov-Zwanziger action.

2 Results and discussions

Self-energies of quarks and gluons are important measures for the collective behavior of the quark-gluon plasma, since thermal masses, dispersion relations, and spectral functions of collective excitations are derived from them. The Euclidean one-loop quark self-energy reads

$$\Sigma(P) = (ig)^2 C_F \sum\!\!\!\!\!\!\!\int_{\{K\}} \gamma^\mu S(K)\gamma^\nu D^{\mu\nu}(P-K)\,, \tag{3}$$

where $S(P)$ is the quark propagator and $D^{\mu\nu}(P)$ is the gluon propagator taken from Eq. (1). It is worth noting that there have been similar studies for the quark self-energy with non-perturbative gluons at finite density [9, 10] and in strong magnetic fields [11].

At $g \ll 1$ (i.e., high temperatures), we may apply the hard-thermal-loop systematics [13] in analyzing Eq. (3). As a result, the gauge-invariant contribution to Eq. (3) reads [12]

$$\begin{aligned}\Sigma(P) \;\simeq\; & -(ig)^2 C_F \sum_{\pm}\int_0^\infty \frac{dk}{2\pi^2}k^2 \int\frac{d\Omega}{4\pi}\frac{\tilde{n}_\pm(k,\gamma_G)}{4E^0_\pm} \\ & \times\left[\frac{i\gamma_0+\hat{\boldsymbol{k}}\cdot\boldsymbol{\gamma}}{iP_0+k-E^0_\pm+\frac{\boldsymbol{p}\cdot\boldsymbol{k}}{E^0_\pm}}+\frac{i\gamma_0-\hat{\boldsymbol{k}}\cdot\boldsymbol{\gamma}}{iP_0-k+E^0_\pm-\frac{\boldsymbol{p}\cdot\boldsymbol{k}}{E^0_\pm}}\right],\end{aligned} \tag{4}$$

where $\hat{\boldsymbol{k}} = \boldsymbol{k}/k$ with $k = |\boldsymbol{k}|$, $E^0_\pm = \sqrt{k^2 \pm i\gamma_G^2}$, $\tilde{n}_\pm(k,\gamma_G) \equiv n_B(\sqrt{k^2\pm i\gamma_G^2}) + n_F(k)$ with n_B and n_F the Bose-Einstein and Fermi-Dirac distributions, and $\int d\Omega = \int_0^{2\pi} d\phi \int_0^\pi d\cos\theta$.

The quark thermal mass incorporating effects from g^2T reads

$$m_q^2(\gamma_G) = \frac{g^2C_F}{4\pi^2}\sum_\pm \int_0^\infty dk\frac{k^2\tilde{n}_\pm(k,\gamma_G)}{E^0_\pm}\,. \tag{5}$$

It reduces to the conventional hard-thermal-loop one, $m_q^2(0) = C_F g^2T^2/8$, when setting $\gamma_G = 0$. $m_q^2(\gamma_G)$ receives negative contributions from γ_G, which is a manifestation of anti-screening effects generated by g^2T (see Fig. 1 in Ref. [12] for details). This is a profound signal of the build-up of long-range correlations in the system.

The dispersion relation is obtained by analytically continuing the self-energy (4) to Minkowski space and then solving the poles in the corresponding quark propagator $iS^{-1}(P) = \not{P} - \Sigma(P) = 0$. In contrast to the conventional hard-thermal-loop case, there are three poles in the propagator (see Fig. 2 in Ref. [12] for details). Firstly, the screened quasi-particle excitations are recovered,

$$\omega = \omega_+(p;\gamma_G)\,, \qquad \omega = \omega_-(p;\gamma_G)\,, \tag{6}$$

which are the so-called particle ω_+ and plasmino ω_- modes, with $\omega_\pm(0;\gamma_G) = m_q(\gamma_G)$ as expected. Both $\omega_\pm/m_q(\gamma_G)$ and their residues $Z_\pm$ are g-independent in the studied range. This property is identical to the conventional hard-thermal-loop case, which provides a non-trivial consistency check of the setup. Furthermore, there exists a novel excitation named *Gribov pole* as in Ref. [12],

$$\omega = \omega_G(p;\gamma_G)\,. \tag{7}$$

It describes *massless* fermionic excitations in the quark-gluon plasma with dispersion relation $\omega = v_s p$ at small momenta, with $v_s \approx 1/\sqrt{3}$ (speed of sound) independent of g for the studied range. The Gribov mode "grows" in the (ω, p)-plane while increasing the magnetic scale, and this effectively introduces a new *magnetic scaling* behavior to the non-Abelian plasma. At larger momenta than the permitted ones for each coupling, we hit branch cuts and Landau damping consequently takes place. The Gribov pole goes along with a residue $Z_G(p) < 0$ that induces *positivity violation* in the corresponding spectral functions in the region of space-like momenta. These novel features are direct manifestations of long-range confinement effects surviving at finite T in the non-Abelian plasma. The results reflect common features of Gribov-like approaches [2, 3, 14], though the calculation was done via the Gribov-Zwanziger action. It is tempting to explore the impact of the setup to heavy-ion phenomenology [15–18]. It is also interesting to understand whether there are any relations between the Gribov mode and the QCD transition.

Acknowledgments

This proceedings is based on the work in collaboration with Konrad Tywoniuk. I acknowledge the support from the GSI Helmholtzzentrum für Schwerionenforschung and Horst Stöcker.

References

[1] L. D. Faddeev and V. N. Popov, Phys. Lett. B **25**, 29 (1967).
[2] V. N. Gribov, Nucl. Phys. B **139**, 1 (1978).
[3] D. Zwanziger, Nucl. Phys. B **323**, 513 (1989).
[4] Y. L. Dokshitzer and D. E. Kharzeev, Ann. Rev. Nucl. Part. Sci. **54**, 487 (2004).
[5] N. Vandersickel and D. Zwanziger, Phys. Rept. **520** (2012) 175.
[6] A. D. Linde, Phys. Lett. B **96**, 289 (1980).
[7] D. J. Gross, R. D. Pisarski and L. G. Yaffe, Rev. Mod. Phys. **53**, 43 (1981).
[8] K. Fukushima and N. Su, Phys. Rev. D **88**, 076008 (2013).
[9] T. Kojo, Y. Hidaka, L. McLerran and R. D. Pisarski, Nucl. Phys. A **843**, 37 (2010).
[10] T. Kojo, Y. Hidaka, K. Fukushima, L. D. McLerran and R. D. Pisarski, Nucl. Phys. A **875**, 94 (2012).
[11] T. Kojo and N. Su, Phys. Lett. B **720**, 192 (2013).
[12] N. Su and K. Tywoniuk, Phys. Rev. Lett. **114**, 161601 (2015).
[13] E. Braaten and R. D. Pisarski, *Nucl. Phys. B* **337**, 569 (1990).
[14] D. Dudal, J. A. Gracey, S. P. Sorella, N. Vandersickel and H. Verschelde, Phys. Rev. D **78**, 065047 (2008).
[15] W. Florkowski, R. Ryblewski, N. Su and K. Tywoniuk, Acta Phys. Polon. B **47**, 1833 (2016).
[16] A. Bandyopadhyay, N. Haque, M. G. Mustafa and M. Strickland, Phys. Rev. D **93**, 065004 (2016).
[17] W. Florkowski, R. Ryblewski, N. Su and K. Tywoniuk, Phys. Rev. C **94**, 044904 (2016).
[18] V. Begun, W. Florkowski and R. Ryblewski, arXiv:1602.08308 [nucl-th].

Physics reach of MoEDAL at LHC: magnetic monopoles, supersymmetry and beyond

Nick E. Mavromatos[1,a] and Vasiliki A. Mitsou[2,b]
On behalf of the MoEDAL Collaboration

[1] *Theoretical Particle Physics and Cosmology Group, Department of Physics, King's College London, Strand, London WC2R 2LS, UK*
[2] *Instituto de Física Corpuscular (IFIC), CSIC – Universitat de València, C/ Catedrático José Beltrán 2, E-46980 Paterna (Valencia), Spain*

Abstract. MoEDAL is a pioneering experiment designed to search for highly ionising messengers of new physics such as magnetic monopoles or massive (pseudo-)stable charged particles, that are predicted to exist in a plethora of models beyond the Standard Model. Its ground-breaking physics program defines a number of scenarios that yield potentially revolutionary insights into such foundational questions as, are there extra dimensions or new symmetries, what is the mechanism for the generation of mass, does magnetic charge exist, what is the nature of dark matter, and, how did the big-bang develop at the earliest times. MoEDAL's purpose is to meet such far-reaching challenges at the frontier of the field. The physics reach of the existing MoEDAL detector is discussed, giving emphasis on searches for magnetic monopoles, supersymmetric (semi)stable partners, doubly charged Higgs bosons, and exotic structures such as black-hole remnants in models with large extra spatial dimensions and D-matter in some brane theories.

1 Introduction

MoEDAL (Monopole and Exotics Detector at the LHC) [1–3], the 7th experiment at the Large Hadron Collider (LHC) [4], was approved by the CERN Research Board in 2010. It is designed to search for manifestations of new physics through highly-ionising particles in a manner complementary to ATLAS and CMS [5]. The most important motivation for the MoEDAL experiment is to pursue the quest for magnetic monopoles and dyons at LHC energies. Nonetheless the experiment is also designed to search for any massive, stable or long-lived, slow-moving particles [6] with single or multiple electric charges arising in many scenarios of physics beyond the Standard Model (SM). A selection of the physics goals and their relevance to the MoEDAL experiment are described here and elsewhere [7]. For an extended and detailed account of the MoEDAL discovery potential, the reader is referred to the recently published *MoEDAL Physics Review* [8].

The structure of this paper is as follows. Section 2 provides a brief description of the MoEDAL detector. The physics reach of MoEDAL as far as magnetic monopoles and monopolia is discussed

[a]e-mail: nikolaos.mavromatos@kcl.ac.uk
[b]e-mail: vasiliki.mitsou@ific.uv.es

in Section 3, whilst Section 4 is dedicated to supersymmetric models predicting massive (meta)stable states. Scenarios with doubly-charged Higgs bosons and their observability in MoEDAL are highlighted in Section 5. Highly-ionising exotic structures in models with extra spatial dimensions, namely microscopic black holes and D-matter, relevant to MoEDAL are briefly mentioned in Sections 6 and Sections 7, respectively. The paper concludes with a summary and an outlook in Section 8.

2 The MoEDAL detector

The MoEDAL detector [2] is deployed around the intersection region at Point 8 of the LHC in the LHCb experiment Vertex Locator (VELO) [9] cavern. A three-dimensional depiction of the MoEDAL experiment is presented in Fig. 1. It is a unique and largely passive LHC detector comprised of four sub-detector systems.

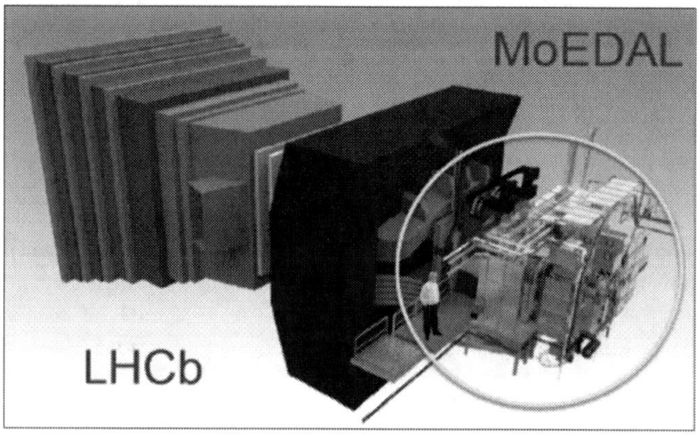

Figure 1. A three-dimensional schematic view of the MoEDAL detector (on the right) around the LHCb VELO region at Point 8 of the LHC.

The main sub-detector system is made of a large array of CR39®, Makrofol® and Lexan® nuclear track detector (NTD) stacks surrounding the intersection area. The passage of a highly-ionising particle through the plastic detector is marked by an invisible damage zone along the trajectory. The damage zone is revealed as a cone-shaped etch-pit when the plastic detector is etched using a hot sodium hydroxide solution. Then the sheets of plastics are scanned looking for aligned etch pits in multiple sheets. The MoEDAL NTDs have a threshold of $Z/\beta \sim 5$, where Z is the charge and $\beta = v/c$ the velocity of the incident particle. In proton-proton collision running, the only source of known particles that are highly ionising enough to leave a track in MoEDAL NTDs are spallation products with range that is typically much less than the thickness of one sheet of the NTD stack. In that case the ionising signature will be that of a very low-energy electrically-charged *stopped* particle. This signature is distinct to that of a *penetrating* electrically or magnetically charged particle that will usually traverse every sheet in a MoEDAL NTD stack, accurately demarcating a track that points back to the collision point. During the heavy-ion running one might expect a background from high ionising fragments, which however are produced in the far forward direction and do not fall into the acceptance of the MoEDAL detector.

A unique feature of the MoEDAL detector is the use of paramagnetic magnetic monopole trappers (MMTs) to capture electrically- and magnetically-charged highly-ionising particles. The aluminium

DOI: 10.1051/epjconf/201716404001

absorbers of MMTs will be subject to an analysis looking for magnetically-charged particles at a remote magnetometer facility [10, 11]. The search for the decays of long-lived electrically charged particles that are stopped in the trapping detectors will subsequently be carried out at a remote underground facility.

The only non-passive MoEDAL sub-detector system comprises an array of several TimePix pixel devices that form a real-time radiation monitoring system dedicated to the monitoring of highly-ionising background sources in the MoEDAL cavern.

3 Magnetic monopoles

The MoEDAL detector is designed to fully exploit the energy-loss mechanisms of magnetically charged particles [12–16] in order to optimise its potential to discover these messengers of new physics. There are various theoretical scenarios in which magnetic charge would be produced at the LHC [8]: (light) 't Hooft-Polyakov monopoles [14, 17], electroweak monopoles [16] and monopolium [13, 18–20]. Magnetic monopoles that carry a non-zero magnetic charge and dyons possessing both magnetic and electric charge are among the most fascinating hypothetical particles. Even though there is no generally acknowledged empirical evidence for their existence, there are strong theoretical reasons to believe that they do exist, and they are predicted by many theories including grand unified theories and superstring theory [21, 22].

The theoretical motivation behind the introduction of magnetic monopoles is the symmetrisation of the Maxwell's equations and the explanation of the charge quantisation [12]. Dirac showed that the mere existence of a monopole in the universe could offer an explanation of the discrete nature of the electric charge, leading to the Dirac Quantisation Condition (DQC),

$$\alpha\, g = \frac{N}{2} e, \quad N = 1, 2, ..., \tag{1}$$

where e is the electron charge, $\alpha = \frac{e^2}{4\pi\hbar c\varepsilon_0} = \frac{1}{137}$ is the fine structure constant (at zero energy, as appropriate to the fact that the quantization condition of Dirac pertains to long (infrared) distances from the centre of the monopole), ε_0 is the vacuum permitivity, and g is the monopole magnetic charge. In Dirac's formulation, magnetic monopoles are assumed to exist as point-like particles and quantum mechanical consistency conditions lead to Eq. (1), establishing the value of their magnetic charge. Although monopoles symmetrise Maxwell's equations in form, there is a numerical asymmetry arising from the DQC, namely that the basic magnetic charge is much larger than the smallest electric charge. A magnetic monopole with a single Dirac charge ($g_{\rm D}$) has an equivalent electric charge of $\beta(137e/2)$. Thus for a relativistic monopole the energy loss is around 4,700 times (68.5^2) that of a minimum-ionising electrically-charged particle. The monopole mass remains a free parameter of the theory.

3.1 Electroweak monopoles

The original monopole solution of 't Hooft and Polyakov [14] pertain to a simple compact (from the point of view of its topology in internal space) gauge group, such as $SU(2)$ or the more phenomenologically realistic Grand Unified Theory (GUT) group $SU(5)$. An important ingredient of the monopole in such models is played by the scalar (spontaneous symmetry breaking Higgs-like) sector of the model. The monopole is an extended object, with complicated structure, with a spherical core of radius R_c. inside which the scalar fields acquires a zero vacuum expectation value, $\langle\vec{\phi}\rangle = 0$, while outside the fields approach their non-trivial v.e.v. $\langle\vec{\phi}\rangle = \eta \neq 0$ asymptotically (where $\vec{\phi}$ denotes the appropriate field multiplet).

The mass of such monopoles are finite. In the Georgi-Glashow model of spontaneous symmetry breaking analysed by 't Hooft [14], based on the $SU(2)$ gauge group (with a Higgs triplet), the mass of the monopole may be of order of the electroweak scale. In GUT gauge group cases, the monopole mass is of order of the GUT scale, and thus outside of the LHC reach. The quantisation condition for the magnetic charge of the monopole in such theories arises because of the non trivial homotopy groups of the pertinent gauge group. In the case, e.g., of the $SU(2)$ group in the Georgi-Glashow model, the internal-space sphere S^2 winds around a configuration space sphere a number of times, defining different topological sectors, characterised by discrete monopole charges.

From this point of view, the Standard Model, with a Higgs doublet of fields and a gauge group $SU(2)\otimes U_Y(1)$, cannot support monopole solutions because its gauge group is not a simple one. However, Cho and Maison [16] argued that the presence of $U_Y(1)$ is important in providing CP^1 structures by considering the quotient group $SU(2) \otimes U_Y(1)/U_{em}(1)$, where $U_{em}(1)$ is the (unbroken) group of electromagnetism, thus admitting monopole solutions as a result of the fact that $\pi_2(CP^1) = n \in Z$, with the standard Higgs doublet as a CP^1 field. This monopole solution unfortunately had an energy functional which was singular at the origin $r \to 0$. Because of this singularity, the monopole mass could not be determined without embedding the theory into a larger group. It could also be possible that although such solutions are topologically allowed in the Standard Model, nevertheless because of this infinite energy are strongly coupled and bound into finite energy monopole-antimonopole pairs ("monopolia") which are not highly ionising. Such states, although not relevant directly for MoeDAL physics, may be constrained by means of other indirect methods at colliders, which we shall not discuss here.

Cho and collaborators have recently suggested [23] that a finite energy Cho-Maison monopole, but with magnetic charge twice the elementary Dirac charge $2g_D \equiv 137\,e$, can be a solution of a modified Standard Model Langrangian with a non-trivial vacuum permitivity (gauge invariant) function of the Higgs field modulus $\phi^\dagger\phi$ in front of the kinetic term of the hypercharge $U_Y(1)$ gauge field:

$$\mathcal{L} \ni -\frac{1}{4}\varepsilon(\phi^\dagger\phi)\,\mathcal{G}_{\mu\nu}\,\mathcal{G}^{\mu\nu}\,, \tag{2}$$

where $\mathcal{G}_{\mu\nu}$ denotes the $U_Y(1)$ gauge potential field strength, and the function $\varepsilon(\phi^\dagger\phi)$ is such that far away from the monopole centre approaches one, i.e. $\varepsilon(\phi^\dagger\phi) \to 1, r \to \infty$, while as $r \to 0$ vanishes in such a way that the energy of the monopole is finite. The term (2) ensures that there is an effective running $U_Y(1)$ coupling $\overline{g}' = g'/\varepsilon$, with g' the Standard Model hypercharge coupling. Simple functions $\epsilon(\phi^\dagger\phi)$ with such properties are [23]

$$\varepsilon(\phi^\dagger\,\phi) = \left|\frac{\phi^\dagger\,\phi}{\eta^2}\right|^{N/2}, \quad N \geq 4 + 2\sqrt{3} \simeq 7.46\,, \tag{3}$$

with η the (real) v.e.v. of ϕ in the symmetry broken phase. Unfortunately the LHC phenomenology of such simple models excludes them, in the sense that they can produce unacceptably high signals of $H \to \gamma\gamma$, where H is the Higgs field, defined as $\phi = \eta + H$ in the symmetry broken phase, i.e. at LHC energies and hence for distances far away from the monopole core. Expanding in such a case ε in powers of $H \ll \eta$, one obtains from (2) dimension-six operators in an effective field theory framework [24]. Applying then the phenomenological analysis of Ref. [25], one observes that acceptable LHC phenomenology of the modified standard model can be obtained for more complicated $\varepsilon(\phi^\dagger\phi)$ functions [24], which consist of algebraic combinations of various powers of the Higgs modulus yielding acceptably suppressed rates of $H \to \gamma\gamma$. Such constructions can yield monopole masses at least of order 5 TeV [24], which makes the case interesting for MoeDAL [1]. Unfortunately at present there is no microscopic justification of such modifications, nevertheless from a phenomenological viewpoint such models can be falsified at MoEDAL or other LHC experiments, given that they can in principle allow for the production of highly ionising monopole/antimonopole pairs from proton-proton collisions. The extended nature of the respective monopole solutions, however, most probably imply strong suppression factors in the corresponding cross sections, which make such processes rare.

3.2 Global $O(3)$ monopoles

Global monopoles have been proposed [27] as space time (cosmological) defects allowing for the spontaneous breaking of internal global $SO(3)$ symmetries in non gauged Georgi Glashow models. The appropriate scalar sector of such models consists of triplets of spin zero fields χ^a, $a = 1,2,3$, with an appropriate symmetry breaking potential that allows (some of) the fields to acquire v.e.v. η. There are global (non gauge) monopole solutions in such a model (and its variants), whose stability is still under debate [28], but which may have interesting physics results of relevance to collider physics, given that the associated monopole mass is proportional to η, $\mathcal{M} \sim \eta$ and the latter can be at TeV scale (if the scalar fields represent somehow new physics fields at TeV scales). The monopole has a core of radius $R_{\rm core} \sim 1/\eta$ and the probability of its production in pair with its antimonopole, can be estimated as [27, 29]

$$\mathcal{P} \propto e^{-4R_{\rm core}/\lambda_{\rm Compt}} \propto e^{-{\rm const}'/\lambda}, \tag{4}$$

where $\lambda_{\rm Compt} = 1/\mathcal{M}$ is the Compton wavelength, and λ the self-interaction coupling of the scalar $SO(3)$ symmetry breaking sector of the model.

For large monopoles $R_{\rm core} > \lambda_{\rm Compt}$, and therefore weak self-interaction couplings $\lambda < 1$, a semi-classical situation is reached where the form factor has an exponential suppression and the event is rare. However, there may be models of phenomenological interest in which the coupling $\lambda \gg 1$, is strong. In such cases $\mathcal{P}$ is expected to be large, and thus of relevance to collider (including LHC) phenomenology (although strong coupling can complicate analytical calculations). It must be noted, however, that once a monopole/antimonopole pair is produced, energy losses to the massless Goldstone fields associated with the breaking of the global $O(3)$ symmetry, at a rate of order η^2 [27] are probably expected, which must be taken into account when considering the relevant phenomenology, as they may lead to the monopole stopping inside the LHC detectors, for instance.

The global monopoles of Ref. [27] are not highly ionising particles, as they carry no magnetic charge. However, as was argued in [27], their gravitational effects far away from the monopole centre (i.e. at radial distances $r \to \infty$) are significant, in the sense that the spacetime is not the Minkowski one but has a deficit angle $\eta^2/M_{\rm P}^2$, with $M_{\rm P}$ is the Planck mass. Although minute, such an effect affects [30] the forward scattering amplitude of standard model particles that propagate in such backgrounds, leading to ring-like angular regions, where the scattering amplitude is very large. The size of such ring-like regions is determined by the ratio of the global monopole mass to the Planck mass. Such peculiar scattering patterns of ordinary standard model particles may indicate indirectly the presence of a neutral global monopole in the detector, despite that fact that the defect is not highly ionising. In this sense, such structures are of interest also to MoEDAL, whose detectors can be sensitive to such phenomena.

Recently [31] a variant of the global monopole model, including axion fields (with a given "axion charge") and a real electromagnetic field, which couples only gravitationally to the scalar $SO(3)$

[1]Self-gravitating monopole solutions have been considered in this context in Ref. [26], with the conclusion, as expected, that gravity has the tendency to reduce the monopole mass. However the order of the mass remains the same as in the flat spacetime case above.

symmetry breaking sector, has been considered, with the result that the axions are capable of inducing electromagnetic monopole solutions with a real magnetic charge of order of the "axion charge". Such solutions can of course be highly ionising and thus of direct relevance to MoeDAL, provided their mass is of TeV scale. In such a case both the high ionisation and the peculiar effects [30] of the monopole background on the scattering of ordinary particles on them, mentioned in the previous paragraph, are in operation.

There are several other models that we shall not list here and which contain some form of monopole or dyon solutions, with masses that could be of order of a few TeV, thus producible in principle at LHC. All such models are extensions of the Standard Model and thus contain new fields in their spectra, that may lead to several signatures of new physics at colliders, beyond the high ionisation induced by the monopoles. Lacking a concrete theoretical framework, therefore, for magnetic monopoles, an experimental, model independent (as much as possible) search for them, is a pressing need, which can provide the theoretical quest with important guidance, especially if a discovery is made! Fortunately the structure of LHC allow for such searches, and this is the topic of our talk, which will be outlined in the following sections.

3.3 Searches for light monopoles in MoEDAL

The high ionisation of slow-moving magnetic monopoles and dyons, implies quite characteristic trajectories when such particles interact with the MoEDAL NTDs, which can be revealed during the etching process [2, 8]. In addition, the high magnetic charge of a monopole (which is expected to be at least one Dirac charge $g_D = 68.5e$ (*cf.* Eq. (1)) implies a strong magnetic dipole moment, which in turn may result in a strong binding of the monopole with the $^{27}_{13}$Al nuclei of the aluminium MoEDAL MMTs. In such a case, the presence of a monopole trapped in an aluminium bar of an MMT would de detected through the existence of a persistent current, defined as the difference between the currents in the SQUID of a magnetometer before and after the passage of the bar through the sensing coil.

In the context of MoEDAL searches, 160 kg of prototype MoEDAL trapping detector samples were exposed to 8-TeV proton-proton collisions at the LHC, for an integrated luminosity of 0.75 fb^{-1} during Run I. No magnetic charge exceeding $0.5g_{\rm D}$ was detected in any of the exposed samples, allowing limits to be placed on monopole production in the mass range 100 GeV$\leq m \leq$ 3500 GeV [32]. Model-independent cross-section limits have been presented in fiducial regions of monopole energy and direction for $1g_{\rm D} \leq |g| \leq 6g_{\rm D}$, and model-dependent cross-section limits are obtained for Drell-Yan (DY) pair production of spin-1/2 and spin-0 monopoles for $1g_{\rm D} \leq |g| \leq 4g_{\rm D}$ (*cf.* Fig. 2). Under the assumption of Drell-Yan cross sections, mass limits are derived for $|g| = 2g_{\rm D}$ and $|g| = 3g_{\rm D}$ for the first time at the LHC, surpassing the previous result from ATLAS Collaboration [33], which placed limits only for monopoles with magnetic charge $|g| = 1g_{\rm D}$. Caution, however, should be exerted here in the sense that the non-perturbative nature of the large magnetic Dirac charge of the monopole invalidate any perturbative treatment based on Drell-Yan calculations of the pertinent cross sections and hence any result based on the latter is only indicative, due to the lack of any other concrete theoretical treatment. New results with MMTs exposed at 13-TeV pp LHC collisions have been released recently [34] extending further the limits in terms of monopole mass and charge up to $5g_{\rm D}$ in DY processes. A comparison of the limits on monopole production cross sections set by other colliders with those set by MoeDAL is given in Fig. 3 [22]. For more general limits including searches in cosmic radiation, see Ref. [35].

A possible explanation for this lack of experimental confirmation of monopoles is Dirac's proposal [12, 13, 18] that monopoles are not seen freely because they form a bound state called *monopolium* [19, 20, 36] being confined by strong magnetic forces. Monopolium is a neutral state, hence it is difficult to detect directly at a collider detector, although its decay into two photons would give a

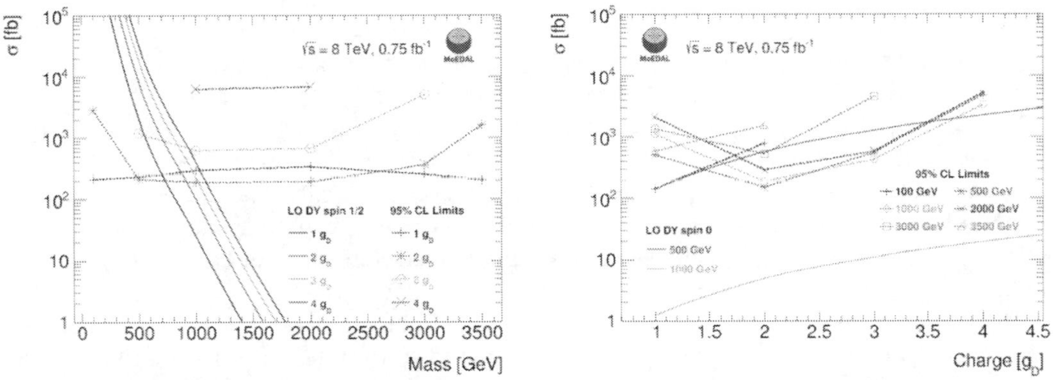

Figure 2. Cross-section upper limits at 95% confidence level for DY monopole production as a function of mass for spin-1/2 (left) and as a funchion of magnetic charge for spin-0 monopoles (right). The various line styles correspond to different monopole charges (left) or masses (right). The solid lines are DY cross-section calculations at leading order. From Ref [32].

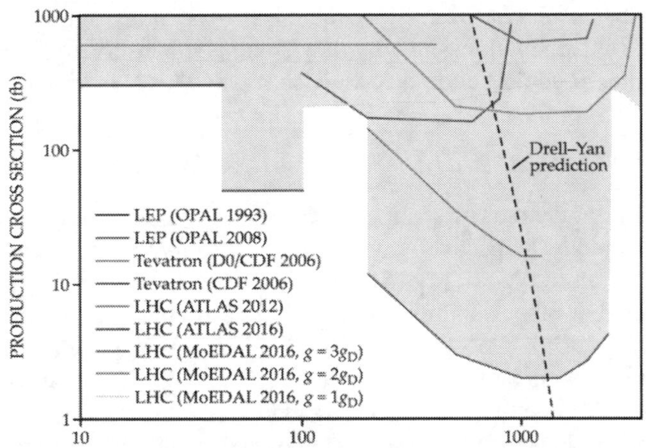

Figure 3. Upper bounds on the monopole production cross sections in various colliders, including MoEDAL. From Ref. [22].

rather clear signal for the ATLAS and CMS detectors [37, 38], which however would not be visible in the MoEDAL detector. Nevertheless according to a novel proposal [39], the LHC radiation detector systems can be used to turn the LHC itself into a new physics search machine by detecting final-state protons $pp \to pXp$ exiting the LHC beam vacuum chamber at locations determined by their fractional momentum losses. Such technique would be appealing for detecting monopolia. Furthermore the monopolium might break up in the medium of MoEDAL into highly-ionising dyons, which subsequently can be detected in MoEDAL [8]. Moreover its decay via photon emission would produce a peculiar trajectory in the medium, should the decaying states are also magnetic multipoles [8].

 DOI: 10.1051/epjconf/201716404001

4 Electrically-charged long-lived particles in supersymmetry

Supersymmetry (SUSY) is an extension of the Standard Model which assigns to each SM field a superpartner field with a spin differing by a half unit. SUSY provides elegant solutions to several open issues in the SM, such as the hierarchy problem, the identity of dark matter, and the grand unification. SUSY scenarios propose a number of massive slowly moving ($\beta \lesssim 0.5$) electrically charged particles. If they are sufficiently long-lived to travel a distance of at least $O(1\text{m})$ before decaying and their $Z/\beta \gtrsim 0.5$, then they will be detected in the MoEDAL NTDs. No highly-charged particles are expected in such a theory, but there are several scenarios in which supersymmetry may yield massive, long-lived particles that could have electric charges ± 1, potentially detectable in MoEDAL if they are produced with low velocities.

The lightest supersymmetric particle (LSP) is stable in models where R parity is conserved. The LSP should have no strong or electromagnetic interactions, for otherwise it would bind to conventional matter and be detectable in anomalous heavy nuclei [40]. Possible weakly-interacting neutral candidates in the Minimal Supersymmetric Standard Model (MSSM) include the sneutrino, which has been excluded by LEP and direct searches, the lightest neutralino $\tilde{\chi}_1^0$ (a mixture of spartners of the Z, H and γ) and the gravitino $\tilde{G}$.

4.1 Supersymmetric scenarios with R-parity violation

Several scenarios featuring metastable charged sparticles might be detectable in MoEDAL. One such scenario is that R parity *may not be exact*, since there is no exact local symmetry associated with either L or B, and hence no fundamental reason why they should be conserved. One could consider various ways in which L and/or B could be violated in such a way that R is violated, as represented by the following superpotential terms:

$$W_{RV} = \lambda''_{ijk}\bar{U}_i\bar{D}_j\bar{D}_k + \lambda'_{ijk}L_iQ_j\bar{D}_k + \lambda_{ijk}L_iL_j\bar{E}_k + \mu_iL_iH, \tag{5}$$

where $Q_i, \bar{U}_i, \bar{D}_i, L_i$ and $\bar{E}_i$ denote chiral superfields corresponding to quark doublets, antiquarks, lepton doublets and antileptons, respectively, with i, j, k generation indices. The simultaneous presence of terms of the first and third type in Eq. (5), namely λ and λ'', is severely restricted by lower limits on the proton lifetime, but other combinations are less restricted. The trilinear couplings in Eq. (5) generate sparticle decays such as $\tilde{q} \to \bar{q}\bar{q}$ or $q\ell$, or $\tilde{\ell} \to \ell\ell$, whereas the bilinear couplings in Eq. (5) generate Higgs-slepton mixing and thereby also $\tilde{q} \to q\ell$ and $\tilde{\ell} \to \ell\ell$ decays [41]. For a nominal sparticle mass $\sim$ 1 TeV, the lifetime for such decays would exceed a few nanoseconds for $\lambda, \lambda', \lambda'' < 10^{-8}$.

If R parity is broken, the LSP would be unstable, and might be charged and/or coloured. In the former case, it might be detectable directly at the LHC as a massive slowly-moving charged particle. In the latter case, the LSP would bind with light quarks and/or gluons to make colour-singlet states, the so-called *R-hadrons*, and any charged state could again be detectable as a massive slowly-moving charged particle. If $\lambda \neq 0$, the prospective experimental signature would be similar to a stau next-to-lightest sparticle (NLSP) case to be discussed later. On the other hand, if λ' or $\lambda'' \neq 0$, the prospective experimental signature would be similar to a stop NLSP case, yielding the possibility of charge-changing interactions while passing through matter. This could yield a metastable charged particle, created whilst passing through the material surrounding the intersection point, that would be detected by MoEDAL.

4.2 Metastable lepton NLSP in the CMSSM with a neutralino LSP

However, even if R parity *is* exact, the NLSP may be long lived. This would occur, for example, if the LSP is the gravitino, or if the mass difference between the NLSP and the neutralino LSP is small, offering more scenarios for long-lived charged sparticles. In *neutralino dark matter* scenarios based on the constrained MSSM (CMSSM), for instance, the most natural candidate for the NLSP is the lighter stau slepton $\tilde{\tau}_1$ [42], which could be long lived if $m_{\tilde{\tau}_1} - m_{\tilde{\chi}_1^0}$ is small. There are several regions of the CMSSM parameter space that are compatible with the constraints imposed by unsuccessful searches for sparticles at the LHC, as well as the discovered Higgs boson mass. These include a strip in the focus-point region where the relic density of the LSP is brought down into the range allowed by cosmology because of its relatively large Higgsino component, a region where the relic density is controlled by rapid annihilation through direct-channel heavy Higgs resonances, and a strip where the relic LSP density is reduced by coannihilations with near-degenerate staus and other sleptons. It was found in a global analysis that the two latter possibilities are favoured [43].

In the coannihilation region of the CMSSM, the lighter $\tilde{\tau}_1$ is expected to be the lightest slepton [42], and the $\tilde{\tau}_1 - \tilde{\chi}_1^0$ mass difference may well be smaller than m_τ: indeed, this is required at large LSP masses. In this case, the dominant stau decays for $m_{\tilde{\tau}_1} - m_{\tilde{\chi}_1^0} > 160$ MeV are expected to be into three particles: $\tilde{\chi}_1^0 \nu \pi$ or $\tilde{\chi}_1^0 \nu \rho$. If $m_{\tilde{\tau}_1} - m_{\tilde{\chi}_1^0} < 1.2$ GeV, the $\tilde{\tau}_1$ lifetime is calculated to be so long, in excess of ~ 100 ns, that it is likely to escape the detector before decaying, and hence would be detectable as a massive, slowly-moving charged particle [44, 45]. The relevance of such scenarios while considering cosmological constraints is demonstrated in Fig. 4. Even is lepton-flavor violating couplings are allowed, the long lifetime of the staus remains [45].

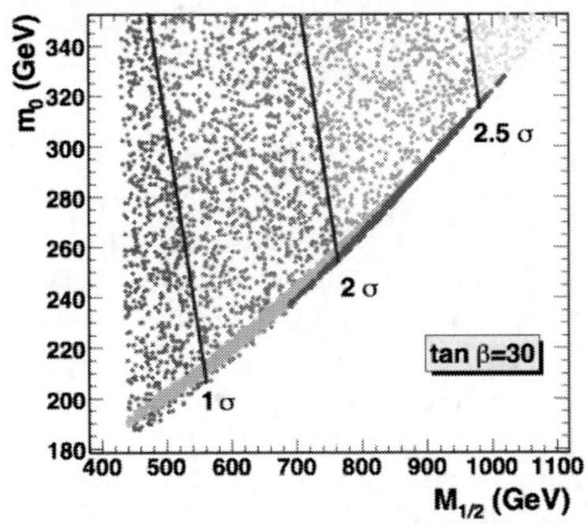

Figure 4. Allowed parameter regions in $M_{1/2} - m_0$ plane fixing $A_0 = 600$ GeV and $\tan\beta = 30$. The red (dark) narrow band is consistent region with dark matter abundance and $\delta m < m_\tau$ and the yellow (light) narrow band is that with $\delta m > m_\tau$. The green regions are inconsistent with the dark matter abundance, and in the white (excluded) area the LSP is the stau. The favoured regions of the muon anomalous magnetic moment at 1σ, 2σ and 2.5σ confidence level are indicated by solid lines. From Ref. [45].

4.3 Metastable sleptons in gravitino LSP scenarios

On the other hand, in *gravitino dark matter* scenarios with more general options for the pattern of supersymmetry breaking, other options appear quite naturally, including the lighter selectron or smuon, or a sneutrino [46], or the lighter stop squark $\tilde{t}_1$ [47]. If the gravitino $\tilde{G}$ is the LSP, the decay rate of a slepton NLSP is given by

$$\Gamma(\tilde{\ell} \to \tilde{G}\ell) = \frac{1}{48\pi M_*^2} \frac{m_{\tilde{\ell}}^5}{M_{\tilde{G}}^2} \left[1 - \frac{M_{\tilde{G}}^2}{m_{\tilde{\ell}}^2}\right]^4, \qquad (6)$$

where M_* is the Planck scale. Since M_* is much larger than the electroweak scale, the NLSP lifetime is naturally very long.

Gravitino (or axino) LSP with a long-lived charged stau may arise in gauge mediation and minimal supergravity models [48]. Large part of the parameter space potentially attractive for long-lived slepton searches with MoEDAL are compatible with cosmological constraints on the dark-matter abundance in superweakly interacting massive particle scenarios [49].

4.4 Long-lived gluinos in split supersymmetry

The above discussion has been in the context of the CMSSM and similar scenarios where all the supersymmetric partners of Standard Model particles have masses in the TeV range. Another scenario is "split supersymmetry", in which the supersymmetric partners of quarks and leptons are very heavy, of a scale m_s, whilst the supersymmetric partners of SM bosons are relatively light [50]. In such a case, the gluino could have a mass in the TeV range and hence be accessible to the LHC, but would have a very long lifetime:

$$\tau \approx 8\left(\frac{m_s}{10^9\ \text{GeV}}\right)^4\left(\frac{1\ \text{TeV}}{m_{\tilde{g}}}\right)^5\ \text{s}. \tag{7}$$

Long-lived gluinos would form long-lived gluino R-hadrons including gluino-gluon *(gluinoball)* combinations, gluino-$q\bar{q}$ *(mesino)* combinations and gluino-qqq *(baryino)* combinations. The heavier gluino hadrons would be expected to decay into the lightest species, which would be metastable, with a lifetime given by Eq. (7), and it is possible that this metastable gluino hadron could be charged.

In the same way as stop hadrons, gluino hadrons may flip charge through conventional strong interactions as they pass through matter, and it is possible that one may pass through most of a conventional LHC tracking detector undetected in a neutral state before converting into a metastable charged state that could be detected by MoEDAL.

4.5 Experimental considerations

There are several considerations supporting the complementary aspects of MoEDAL w.r.t. ATLAS and CMS when discussing the observability of (meta-)stable massive electrically-charged particles. Most of them stem from MoEDAL being "time-agnostic" due to the passive nature of its detectors. Therefore signal from very slowly moving particles will not be lost due to arriving in several consecutive bunch crossings. ATLAS and CMS, on the other hand, perform triggered-based analyses relying either on triggering on accompanying "objects", e.g. missing transverse energy, or by developing and deploying specialised triggers. In both cases the efficiency may lower and in the former the probed parameter space may be reduced. MoEDAL is mainly limited by the geometrical acceptance of the detectors, especially the MMTs, and by the requirement of passing the Z/β threshold of NTDs. In general ATLAS and CMS have demonstrated to cover high-velocities $\beta \gtrsim 0.2$, while MoEDAL is sensitive to lower ones $\beta \lesssim 0.2$.

When discussing the detection of particles stopped *(trapped)* in material that they may decay later, different possibilities are explored. CMS and ATLAS look in empty bunch crossings for decays of trapped particles into jets. MoEDAL MMTs may be monitored in a underground/basement laboratory for tracks arising from such decays. The background in the latter case, coming from cosmic rays, should be easier to control and assess. The probed lifetimes should be larger due to the unlimited monitoring time.

5 Doubly-charged Higgs bosons

Doubly-charged particles appear in many theoretical scenarios beyond the SM. For example, doubly-charged scalar states, usually termed doubly-charged Higgs fields, appear in left-right symmetric models [51–53] and in see-saw models for neutrino masses with Higgs triplets. A number of models encompasses additional symmetries and extend the SM Higgs sector by introducing doubly-charged Higgs bosons. A representative example of such a model is the L-R Symmetric Model (LRSM) [51–53], proposed to amend the fact that the weak-interaction couplings are strictly left handed by extending the gauge group of the SM so as to include a right-handed sector. The simplest realisation is an LRSM [51, 52] that postulates a right-handed version of the weak interaction, whose gauge symmetry is spontaneously broken at high mass scale, leading to the parity-violating SM. This model naturally accommodates recent data on neutrino oscillations and the existence of small neutrino masses. The model generally requires Higgs triplets containing doubly-charged Higgs bosons ($H^{\pm\pm}$) Δ_R^{++} and Δ_L^{++}, which could be light in the minimal supersymmetric left-right model [54].

Single production of a doubly-charged Higgs boson at the LHC proceeds via vector boson fusion, or through the fusion of a singly-charged Higgs boson with either a $W^{\pm}$ or another singly-charged Higgs boson. The amplitudes of the $W_L W_L$ and $W_R W_R$ vector boson fusion processes are proportional to $v_{L,R}$, the vacuum expectation values of the neutral members of the scalar triplets of the LRSM. For the case of Δ_R^{++} production, the vector boson fusion process dominates. Pair production of doubly-charged Higgs bosons is also possible via a Drell-Yan process, with γ, Z or Z_R exchanged in the s-channel, but at a high kinematic price since substantial energy is required to produce two heavy particles. In the case of Δ_L^{++}, double production may nevertheless be the only possibility if v_L is very small or vanishing.

The decay of a doubly-charged Higgs boson can proceed via several channels. The dilepton signature leads to the (experimentally clean) final state $q\bar{q} \to \Delta_L^{++}\Delta_L^{--} \to 4\ell$. However as long as the triplet vacuum expectation value, v_Δ, is much larger than 10^{-4} GeV, the doubly-charged Higgs decay predominantly into a pair of same-sign W bosons. For very small Yukawa couplings $H_{\ell\ell} \lesssim 10^{-8}$, the doubly-charged Higgs boson can be quasi-stable [55]. In Fig. 5, the partial decay width of the doubly charged Higgs boson into a W boson pair is shown as a function of its mass. For $v_\Delta \gg 10^{-4}$ GeV, this partial width is roughly equal to the total width of the doubly charged Higgs boson. In the case of long lifetimes, slowly moving pseudo-stable Higgs bosons could be detected in the MoEDAL NTDs. For example with CR39, one could detect doubly-charged Higgs particles moving with speeds less than around $\beta \simeq 0.4$.

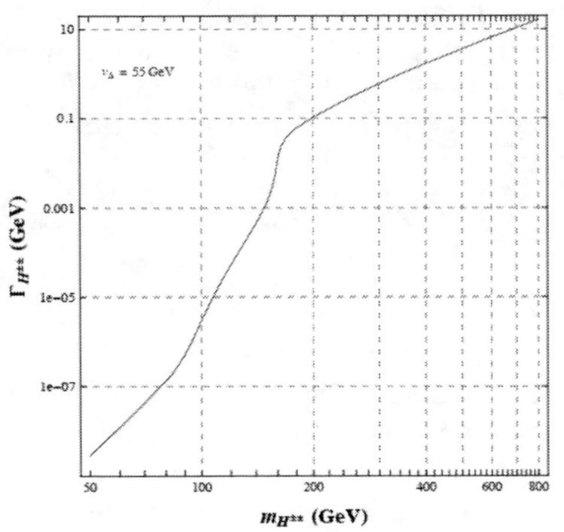

Figure 5. Partial decay width of $H^{\pm\pm} \to W^{\pm}W^{\pm}$ as a function of $m_{H^{\pm\pm}}$ for $v_\Delta = 55$ GeV. From Ref. [55].

DOI: 10.1051/epjconf/201716404001

6 Black hole remnants in large extra dimensions

Over the last decades, models based on compactified extra spatial dimensions (ED) have been proposed in order to explain the large gap between the electroweak (EW) and the Planck scale of $M_{\rm EW}/M_{\rm PL} \approx 10^{-17}$. The four main scenarios relevant for searches at LHC the Arkani-Hamed-Dimopoulos-Dvali (ADD) model of large extra dimensions [56], the Randall-Sundrum (RS) model of warped extra dimensions [57], TeV^{-1}-sized extra dimensions [58], and the Universal Extra Dimensions (UED) model [59].

The existence of extra spatial dimensions [56, 57] and a sufficiently small fundamental scale of gravity open up the possibility that microscopic black holes be produced and detected [60–65] at the LHC. Once produced, the black holes will undergo an evaporation process categorised in three stages [60, 61]: the *balding phase*, the actual *evaporation phase*, and finally the Planck phase. It is generally assumed that the black hole will decay completely to some last few SM particles. However, another intriguing possibility is that the remaining energy is carried away by a stable remnant.

The prospect of microscopic black hole production at the LHC within the framework of models with large extra dimensions has been studied in Ref. [56]. Black holes produced at the LHC are expected to decay with an average multiplicity of $\sim 10-25$ into SM particles, most of which will be charged, though the details of the multiplicity distribution depend on the number of extra dimensions [66]. After the black holes have evaporated off enough energy to reach the remnant mass, some will have accumulated a net electric charge. According to purely statistical considerations, the probability for being left with highly-charged black hole remnants drops fast with the deviation from the average. The largest fraction of the black holes should have charges ± 1 or zero, although a smaller but non-negligible fraction would be multiply charged.

The fraction of charged black-hole remnants has been estimated [66, 67] using the `PYTHIA` event generator [68] and the `CHARYBDIS` program [65]. It was assumed that the effective temperature of the black hole drops towards zero for a finite remnant mass, M_R. The value of M_R does not noticeably affect the investigated charge distribution, as it results from the very general statistical distribution of the charge of the emitted particles.

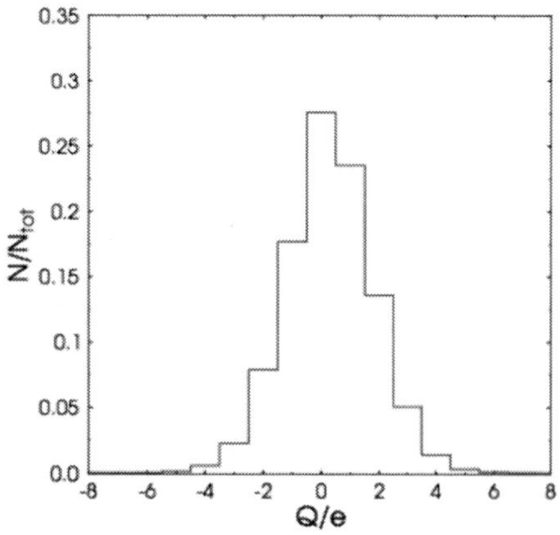

Figure 6. The distribution of black-hole remnant charges in proton-proton interactions at $\sqrt{s} = 14$ TeV calculated with the `PYTHIA` event generator [68] and the `CHARYBDIS` program [65]. From Ref. [67].

Thus, independent of the underlying quantum-gravitational assumption leading to the remnant formation, it was found that about 30% of the remnants are neutral, whereas $\sim 40\%$ would be singly-charged black holes, and the remaining $\sim$30% of remnants would be multiply-charged. The distribution of the remnant charges obtained is shown in Fig. 6. The black hole remnants considered here are heavy, with masses of a TeV or more. A significant fraction of the black-hole remnants produced

DOI: 10.1051/epjconf/201716404001

would have a Z/β of greater than five, high enough to register in the CR39 NTDs forming the LT-NTD sub-detector of MoEDAL.

7 D-matter

Some versions of string theory include higher-dimensional "domain-wall"-like membrane *(brane)* structures in space-time, called *D-branes*. In some cases the bulk is "punctured" by lower-dimensional D-brane defects, which are either point-like or have their longitudinal dimensions compactified [69]. From a low-energy observer's perspective, such structures would effectively appear to be point-like *D-particles*. The latter have dynamical degrees of freedom, thus they can be treated as quantum excitations above the vacuum [69, 70] collectively referred to as *D-matter*. D-matter states are non-perturbative stringy objects with masses of order $m_D \sim M_s/g_s$, where g_s is the string coupling, typically of order one so that the observed gauge and gravitational couplings is reproduced. Hence, the D-matter states could be light enough to be phenomenologically relevant at the LHC.

Depending on their type, D-branes could carry integral or torsion (discrete) charges with the lightest D-particle (LDP) being stable. Therefore the LDPs are possible candidates for cold dark matter [70]. D-particles are solitonic non-perturbative objects in the string/brane theory. As discussed in the relevant literature [70], there are similarities and differences between D-particles and magnetic monopoles with non-trivial cosmological implications [69, 71–73]. An important difference is that they could have *perturbative* couplings, with no magnetic charge in general. Nonetheless, in the context of brane-inspired gauge theories, brane states with magnetic charges can be constructed, which would manifest themselves in MoEDAL in a manner similar to magnetic monopoles.

Non-magnetically-charged D-matter, on the other hand, could be produced at colliders and also produce interesting signals of direct relevance to the MoEDAL experiment. For instance, excited states of D-matter ($D^\star$) can be electrically-charged. For typical string couplings of phenomenological relevance, the first few massive levels may be accessible to the LHC. Depending on the details of the microscopic model considered, and the way the SM is embedded, such massive charged states can be relatively long-lived, and could likewise be detectable with MoEDAL. D-matter/antimatter pairs can be produced [71, 72] by the decay of intermediate off-shell Z-bosons, as shown in Fig. 7.

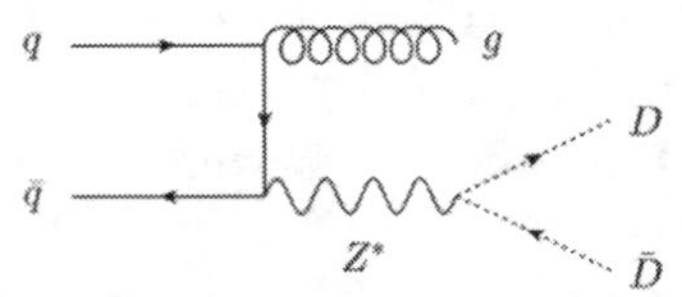

Figure 7. An example of parton-level diagrams for production of D-particles by $q\bar{q}$ collisions in a generic D-matter low-energy model [72].

8 Summary and outlook

MoEDAL is going to extend considerably the LHC reach in the search for (meta)stable highly ionising particles. The latter are predicted in a variety of theoretical models and include: magnetic monopoles, SUSY stable (or, rather, long-lived) spartners, quirks, strangelets, Q-balls, fractionally-charged massive particles, etc [8]. Such particles can be light enough to be producible at the LHC energies (see e.g. Q-balls in the context of some SUSY or brane models [74]). In the talk we have described searches for only a subset of those particles, due to lack of space. Specifically, we discussed monopoles, partners in some SUSY models, doubly charged Higgs bosons, black hole remnants in models with large extra spatial dimensions, as well as some (more exotic) scenarios on D-matter which characterises some brane models.

The MoEDAL design is optimised to probe precisely all such long lived states, unlike the other LHC experiments [5]. Furthermore it combines different detector technologies: plastic nuclear track detectors (NTDs), trapping volumes and pixel sensors [2]. The first physics results, pertaining to magnetic monopole trapping detectors, obtained with LHC Run I data, have been published [32] and the corresponding analysis at 13 TeV has been released recently [34]. The MoEDAL Collaboration is preparing new analyses with more Run II data, with other detectors (NTDs) and with a large variety of interpretations involving not only magnetic but also electric charges.

Acknowledgements

The authors are grateful to the ICNFP2016 organisers for scheduling the "Mini-Workshop on MoEDAL" during the Conference and for the kind invitation to present these talks in it. The work of NEM is supported in part by the UK Science and Technology Facilities Council (STFC) via the grants ST/L000326/1 and ST/P000258/1. VAM acknowledges support by the Spanish Ministry of Economy and Competitiveness (MINECO) under the project FPA2015-65652-C4-1-R, by the Generalitat Valenciana through the project PROMETEO II/2013-017 and by the Spanish National Research Council (CSIC) under the CT Incorporation Program 201650I002.

References

[1] For general information on the MoEDAL experiment, see: http://moedal.web.cern.ch/

[2] MoEDAL Collaboration, Technical Design Report of the MoEDAL Experiment *CERN Preprint* CERN-LHC-2009-006, MoEDAL-TDR-1.1 (2009), and references therein.

[3] J. L. Pinfold, "The search for the magnetic monopole and other highly ionizing particles at the LHC," these proceedings.

[4] L. Evans and P. Bryant, JINST **3**, S08001 (2008).

[5] A. De Roeck, A. Katre, P. Mermod, D. Milstead and T. Sloan, Eur. Phys. J. C **72**, 1985 (2012).

[6] M. Fairbairn, A. C. Kraan, D. A. Milstead, T. Sjostrand, P. Z. Skands and T. Sloan, Phys. Rept. **438**, 1 (2007); S. Burdin, M. Fairbairn, P. Mermod, D. Milstead, J. Pinfold, T. Sloan and W. Taylor, Phys. Rept. **582**, 1 (2015).

[7] V. A. Mitsou [MoEDAL Collaboration], EPJ Web Conf. **95** (2015) 04042.

[8] B. Acharya *et al.* [MoEDAL Collaboration], Int. J. Mod. Phys. A **29**, 1430050 (2014).

[9] A. A. Alves, Jr. *et al.* [LHCb Collaboration], JINST **3**, S08005 (2008).

[10] M. D. Joergensen, A. De Roeck, H.-P. Hachler, A. Hirt, A. Katre, P. Mermod, D. Milstead and T. Sloan, "Searching for magnetic monopoles trapped in accelerator material at the Large Hadron Collider," arXiv:1206.6793 [physics.ins-det] (2012).

[11] A. De Roeck, H. P. Hächler, A. M. Hirt, M.-D. Joergensen, A. Katre, P. Mermod, D. Milstead and T. Sloan, Eur. Phys. J. C **72**, 2212 (2012).

[12] P. A. M. Dirac, Proc. Roy. Soc. Lond. A **133**, 60 (1931).

[13] P. A. M. Dirac, Phys. Rev. **74**, 817 (1948).

[14] G. 't Hooft, Nucl. Phys. B **79**, 276 (1974); A. M. Polyakov, JETP Lett. **20**, 194 (1974) [Pisma Zh. Eksp. Teor. Fiz. **20**, 430 (1974)].

[15] B. Julia and A. Zee, Phys. Rev. D **11**, 2227 (1975); Y. Nambu, Nucl. Phys. B **130**, 505 (1977); E. Witten, Phys. Lett. **86B**, 283 (1979); G. Lazarides, M. Magg and Q. Shafi, Phys. Lett. **97B**, 87 (1980); R. d. Sorkin, Phys. Rev. Lett. **51**, 87 (1983); D. J. Gross and M. J. Perry, Nucl. Phys. B **226**, 29 (1983); J. S. Schwinger, Science **165**, 757 (1969); J. Preskill, Ann. Rev. Nucl. Part. Sci. **34**, 461 (1984); A. Achucarro and T. Vachaspati, Phys. Rept. **327**, 347 (2000) [Phys. Rept. **327**, 427 (2000)]; T. W. Kephart, C. A. Lee and Q. Shafi, JHEP **0701**, 088 (2007); D. G. Pak, P. M. Zhang and L. P. Zou, Int. J. Mod. Phys. A **30**, no. 27, 1550164 (2015); A. Rajantie, JHEP **0601**, 088 (2006).

[16] Y. M. Cho and D. Maison, Phys. Lett. B **391**, 360 (1997); W. S. Bae and Y. M. Cho, J. Korean Phys. Soc. **46**, 791 (2005).

[17] V. Vento and V. S. Mantovani, "On the magnetic monopole mass," arXiv:1306.4213 [hep-ph] (2013).

[18] Y. B. Zeldovich and M. Y. Khlopov, Phys. Lett. **79B**, 239 (1978).

[19] C. T. Hill, Nucl. Phys. B **224**, 469 (1983).

[20] V. K. Dubrovich, Grav. Cosmol. Suppl. **8N1**, 122 (2002).

[21] A. Rajantie, Contemp. Phys. **53**, 195 (2012).

[22] A. Rajantie, Phys. Today **69**, no. 10, 40 (2016).

[23] Y. M. Cho, K. Kimm and J. H. Yoon, Mod. Phys. Lett. A **31**, no. 09, 1650053 (2016).

[24] J. Ellis, N. E. Mavromatos and T. You, Phys. Lett. B **756**, 29 (2016).

[25] J. Ellis, V. Sanz and T. You, JHEP **1407**, 036 (2014).

[26] Y. M. Cho, K. Kimm and J. H. Yoon, Phys. Lett. B **761**, 203 (2016).

[27] M. Barriola and A. Vilenkin, Phys. Rev. Lett. **63**, 341 (1989).

[28] A. S. Goldhaber, Phys. Rev. Lett. **63**, 2158 (1989); S. H. Rhie and D. P. Bennett, Phys. Rev. Lett. **67**, 1173 (1991); L. Perivolaropoulos, Nucl. Phys. B **375**, 665 (1992); G. W. Gibbons, M. E. Ortiz, F. Ruiz Ruiz and T. M. Samols, Nucl. Phys. B **385**, 127 (1992); M. Hindmarsh, Nucl. Phys. B **392**, 461 (1993); G. Arreaga, I. Cho and J. Guven, Phys. Rev. D **62**, 043520 (2000); A. Achucarro and J. Urrestilla, Phys. Rev. Lett. **85**, 3091 (2000); R. Gregory and C. Santos, Class. Quant. Grav. **20**, 21 (2003); S. B. Gudnason and J. Evslin, Phys. Rev. D **92**, no. 4, 045044 (2015); A. Achucarro, B. Hartmann and J. Urrestilla, JHEP **0507**, 006 (2005).

[29] A. K. Drukier and S. Nussinov, Phys. Rev. Lett. **49**, 102 (1982).

[30] P. O. Mazur and J. Papavassiliou, Phys. Rev. D **44**, 1317 (1991).

[31] N. E. Mavromatos and S. Sarkar, "Magnetic Monopoles from Global Monopoles in the presence of Kalb-Ramond Torsion," arXiv:1607.01315 [hep-th] (2016).

[32] B. Acharya *et al.* [MoEDAL Collaboration], JHEP **1608**, 067 (2016).

[33] G. Aad *et al.* [ATLAS Collaboration], Phys. Rev. Lett. **109**, 261803 (2012); Phys. Rev. D **93**, no. 5, 052009 (2016).

[34] B. Acharya *et al.* [MoEDAL Collaboration], "Search for magnetic monopoles with the MoEDAL forward trapping detector in 13 TeV proton-proton collisions at the LHC," arXiv:1611.06817 [hep-ex] (2016).

[35] L. Patrizii and M. Spurio, Ann. Rev. Nucl. Part. Sci. **65**, 279 (2015).

[36] L. N. Epele, H. Fanchiotti, C. A. Garcia Canal and V. Vento, Eur. Phys. J. C **56**, 87 (2008); L. N. Epele, H. Fanchiotti, C. A. G. Canal and V. Vento, Eur. Phys. J. C **62**, 587 (2009).

[37] L. N. Epele, H. Fanchiotti, C. A. G. Canal, V. A. Mitsou and V. Vento, Eur. Phys. J. Plus **127**, 60 (2012).

[38] L. N. Epele, H. Fanchiotti, C. A. G. Canal, V. A. Mitsou and V. Vento, arXiv:1607.05592 [hep-ph] (2016).

[39] M. Kalliokoski, J. W. Lm̃sä, M. Mieskolainen and R. Orava, "Turning the LHC Ring into a New Physics Search Machine," arXiv:1604.05778 [hep-ex] (2016).

[40] H. Goldberg, Phys. Rev. Lett. **50**, 1419 (1983) Erratum: [Phys. Rev. Lett. **103**, 099905 (2009)]; J. R. Ellis, J. S. Hagelin, D. V. Nanopoulos, K. A. Olive and M. Srednicki, Nucl. Phys. B **238**, 453 (1984).

[41] V. A. Mitsou, PoS PLANCK **2015** (2015) 085 [arXiv:1510.02660 [hep-ph]].

[42] J. R. Ellis, K. A. Olive, Y. Santoso and V. C. Spanos, Phys. Lett. B **565**, 176 (2003).

[43] O. Buchmueller *et al.*, Eur. Phys. J. C **72**, 2243 (2012); Eur. Phys. J. C **74**, no. 6, 2922 (2014).

[44] T. Jittoh, J. Sato, T. Shimomura and M. Yamanaka, Phys. Rev. D **73**, 055009 (2006) Erratum: [Phys. Rev. D **87**, no. 1, 019901 (2013)].

[45] S. Kaneko, J. Sato, T. Shimomura, O. Vives and M. Yamanaka, Phys. Rev. D **78**, no. 11, 116013 (2008) Erratum: [Phys. Rev. D **87**, no. 3, 039904 (2013)].

[46] J. R. Ellis, K. A. Olive and Y. Santoso, JHEP **0810**, 005 (2008).

[47] J. L. Diaz-Cruz, J. R. Ellis, K. A. Olive and Y. Santoso, JHEP **0705**, 003 (2007).

[48] K. Hamaguchi, M. M. Nojiri and A. de Roeck, JHEP **0703**, 046 (2007).

[49] J. L. Feng, S. Iwamoto, Y. Shadmi and S. Tarem, JHEP **1512**, 166 (2015).

[50] N. Arkani-Hamed and S. Dimopoulos, JHEP **0506**, 073 (2005); G. F. Giudice and A. Romanino, Nucl. Phys. B **699**, 65 (2004) Erratum: [Nucl. Phys. B **706**, 487 (2005)].

[51] J. C. Pati and A. Salam, Phys. Rev. D **10**, 275 (1974) Erratum: [Phys. Rev. D **11**, 703 (1975)].

[52] R. N. Mohapatra and J. C. Pati, Phys. Rev. D **11**, 566 (1975).

[53] G. Senjanovic and R. N. Mohapatra, Phys. Rev. D **12**, 1502 (1975); R. N. Mohapatra and G. Senjanovic, Phys. Rev. D **23**, 165 (1981).

[54] C. S. Aulakh, A. Melfo and G. Senjanovic, Phys. Rev. D **57**, 4174 (1998); Z. Chacko and R. N. Mohapatra, Phys. Rev. D **58**, 015003 (1998); C. S. Aulakh, K. Benakli and G. Senjanovic, Phys. Rev. Lett. **79**, 2188 (1997).

[55] C. W. Chiang, T. Nomura and K. Tsumura, Phys. Rev. D **85**, 095023 (2012).

[56] N. Arkani-Hamed, S. Dimopoulos and G. R. Dvali, Phys. Lett. B **429**, 263 (1998); Phys. Rev. D **59**, 086004 (1999); I. Antoniadis, N. Arkani-Hamed, S. Dimopoulos and G. R. Dvali, Phys. Lett. B **436**, 257 (1998).

[57] L. Randall and R. Sundrum, Phys. Rev. Lett. **83**, 4690 (1999).

[58] I. Antoniadis, Phys. Lett. B **246**, 377 (1990); I. Antoniadis and K. Benakli, Phys. Lett. B **326**, 69 (1994); I. Antoniadis, K. Benakli and M. Quiros, Phys. Lett. B **331**, 313 (1994).

[59] T. Appelquist, H. C. Cheng and B. A. Dobrescu, Phys. Rev. D **64**, 035002 (2001).

[60] S. B. Giddings and S. D. Thomas, Phys. Rev. D **65**, 056010 (2002).

[61] W. Fischler, "A Model for high-energy scattering in quantum gravity," arXiv:hep-th/9906038 (1999).

[62] P. C. Argyres, S. Dimopoulos and J. March-Russell, Phys. Lett. B **441**, 96 (1998); S. Dimopoulos and G. L. Landsberg, Phys. Rev. Lett. **87**, 161602 (2001); G. L. Alberghi, R. Casadio and A. Tronconi, J. Phys. G **34**, 767 (2007); M. Cavaglia, R. Godang, L. Cremaldi and D. Summers, Comput. Phys. Commun. **177**, 506 (2007); D. C. Dai, G. Starkman, D. Stojkovic, C. Issever, E. Rizvi and J. Tseng, Phys. Rev. D **77**, 076007 (2008).

[63] R. Casadio and B. Harms, Int. J. Mod. Phys. A **17**, 4635 (2002).

[64] M. Cavaglia, Int. J. Mod. Phys. A **18**, 1843 (2003); P. Kanti, Int. J. Mod. Phys. A **19**, 4899 (2004).

[65] C. M. Harris, P. Richardson and B. R. Webber, JHEP **0308**, 033 (2003).

[66] B. Koch, M. Bleicher and H. Stoecker, J. Phys. G **34**, S535 (2007).

[67] S. Hossenfelder, B. Koch and M. Bleicher, "Trapping black hole remnants," hep-ph/0507140 (2005).

[68] T. Sjostrand, S. Mrenna and P. Z. Skands, JHEP **0605**, 026 (2006).

[69] J. R. Ellis, N. E. Mavromatos and D. V. Nanopoulos, Gen. Rel. Grav. **32**, 943 (2000); J. R. Ellis, N. E. Mavromatos and D. V. Nanopoulos, Phys. Lett. B **665**, 412 (2008); J. R. Ellis, N. E. Mavromatos and M. Westmuckett, Phys. Rev. D **70**, 044036 (2004); *ibid.* **71**, 106006 (2005).

[70] G. Shiu and L. T. Wang, Phys. Rev. D **69**, 126007 (2004).

[71] N. E. Mavromatos, S. Sarkar and A. Vergou, Phys. Lett. B **696**, 300 (2011).

[72] N. E. Mavromatos, V. A. Mitsou, S. Sarkar and A. Vergou, Eur. Phys. J. C **72**, 1956 (2012).

[73] See, e.g., E. Witten, "Comments on string theory," arXiv:hep-th/0212247 (2002), and references therein.

[74] S. Kasuya and M. Kawasaki, Phys. Rev. Lett. **85**, 2677 (2000); S. Kasuya, M. Kawasaki and T. T. Yanagida, PTEP **2015**, no. 5, 053B02 (2015); S. Abel and A. Kehagias, JHEP **1511**, 096 (2015).

 DOI: 10.1051/epjconf/201716405001

High-mass Higgs searches at ATLAS and CMS

Scott Snyder[1,a], on behalf of the ATLAS and CMS collaborations

[1] *PO Box 5000, Upton NY, 11973, USA*

Abstract. Several theories beyond the Standard Model, such as the two-Higgs doublet model, predict the existence of high-mass neutral and charged Higgs particles. The ATLAS and CMS experiments have completed an extensive program of searches for such particles using the $\sim$ 3 fb^{-1} of $\sqrt{s}$ = 13 TeV data recorded during 2015. This note summarizes a selection of these results.

1 Introduction

Since the first observations of the Higgs boson [1, 2], all measurements of its properties have been very consistent with expectations from the Standard Model (SM). There are, however, reasons to suspect that the known electroweak sector of the SM may not be complete, including the unnaturalness of the observed Higgs boson mass and the possibility that new particles could constitute part of the inferred "dark matter" component of the universe. Therefore, both the ATLAS and CMS experiments have been searching for additional heavy Higgs-like scalars. The searches reported here mostly use the $\sim$ 3 fb^{-1} of $\sqrt{s}$ = 13 TeV data recorded by each experiment during 2015.

Some searches reported here are interpreted in the context of a simple extension to the SM, the two-Higgs doublet model (2HDM) [3]. In the 2HDM, an additional Higgs doublet is introduced, yielding five physical Higgs bosons: two CP-even scalars h, H; one CP-odd scalar A; and two charged scalars $H^{\pm}$. (In the following, the observed 125 GeV resonance is denoted h; H is a putative heavier resonance.) Besides the particle masses (m_h, m_H, m_A, $m_{H^{\pm}}$), model parameters include the ratio of the vacuum expectation values of the two doublets $\tan\beta$; the mixing angle between h and H α; and the potential parameter mixing the two doublets m_{12}^2. There are many ways in which the doublets can be coupled to other particles. The most studied classes of models are "Type-I," in which all quarks couple to only one doublet, and "Type-II," in which up-type quarks couple to one doublet and down-type quarks to the other. The minimal supersymmetric standard model (MSSM) is a subset of 2HDM, and results are often reported in the context of a set of defined benchmark models [4, 5].

2 Higgs boson decays to ZZ

This section covers decays of a high-mass Higgs boson to two Z bosons, at least one of which decays into a lepton pair. Other diboson decay modes, including $Z\gamma$, are covered in [6].

The $H \to ZZ \to 4\ell$ channel is notable for having a very low background, though the branching ratio is also small. The high-mass analysis for both experiments is based on the analyses for

[a] e-mail: snyder@bnl.gov

m_h = 125 GeV. Selected events must have two same-flavor, opposite-sign lepton pairs. The CMS analysis [7] requires one pair with $40 < m_{\ell\ell} < 120$ GeV and the other with $12 < m_{\ell\ell} < 120$ GeV. Backgrounds are further reduced by requirements on lepton momenta, angular separation, and on track isolation and impact parameter. The resulting background is quite small and consists mostly of ZZ production, which is estimated from Monte Carlo (MC) simulation. Results are shown in Figure 1, including exclusion limits in $\sigma\times$BR for the case of a narrow resonance over the range 200–1000 GeV, and also limits for several larger widths as well as an interpretation in terms of a Type-I 2HDM. The corresponding ATLAS analysis is described in [8].

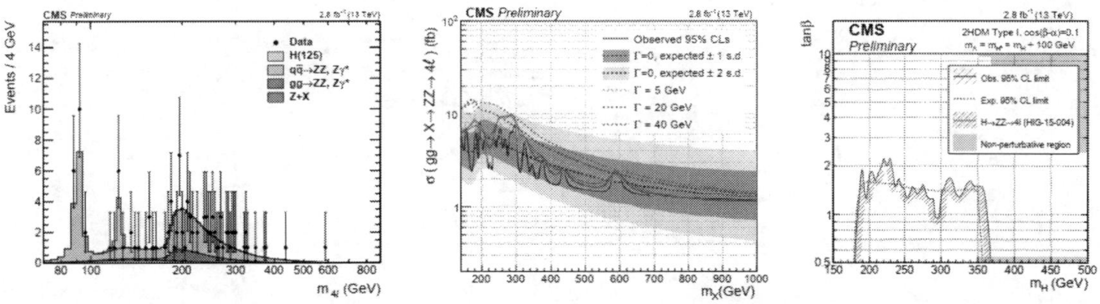

Figure 1. Selected results of the CMS $H \to ZZ \to 4\ell$ search [7]. Left: distribution of four-lepton invariant mass $m_{4\ell}$. Center: 95% CL exclusion in $\sigma \times$ BR as a function of $m_{4\ell}$ for the narrow-width approximation and several larger widths. Right: 95% CL exclusions in $\tan\beta/m_H$ for a Type-I 2HDM.

The ATLAS search in the $H \to ZZ \to \ell\ell\nu\nu$ channel [8] requires a $Z \to \ell\ell$ decay, $E_{\mathrm{T}}^{\mathrm{miss}} > 120$ GeV not pointing along a jet, and no b-tags. The discriminating variable is $m_T^{ZZ} = m_T(\ell\ell, E_{\mathrm{T}}^{\mathrm{miss}})$. There are additional requirements on p_T^Z, angular separations, and event transverse momentum balance. The major backgrounds are SM diboson production, which is estimated from MC simulations with the WZ background normalized to data, and Z + jets, which is estimated from data. Selected results are shown in Figure 2. The corresponding CMS analysis is described in [9].

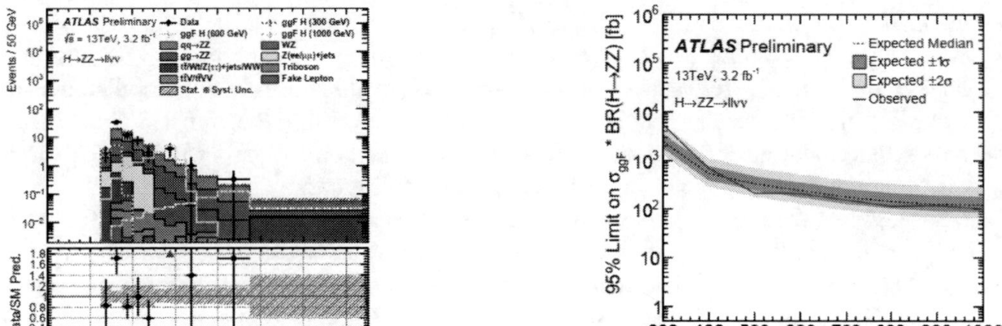

Figure 2. Selected results of the ATLAS $H \to ZZ \to \ell\ell\nu\nu$ search [8]. Left: m_T^{ZZ} distribution. Right: 95% $\sigma\times$BR limit for a narrow-width Higgs boson produced via gluon-gluon fusion.

In the $H \to ZZ \to \ell\ell qq$ channel, both Z bosons are reconstructed; however, there is a large background from Z + jets. For large m_H, the decay products become boosted, and the jets from the $Z \to qq$ tend to merge into a single jet. Therefore, the ATLAS $H \to ZZ \to \ell\ell qq$ search [10] is divided into "merged" and "resolved" analyses. Besides the $Z \to \ell\ell$ decay, the merged analysis requires the presence of a large-R jet with $p_T > 200$ GeV that is consistent with originating from Z boson decay, as well as $p_T(\ell\ell) > 0.3m_{\ell\ell}$. The resolved analysis requires two jets with $70 < m_{jj} < 105$ GeV and $\sqrt{p_T^2(\ell\ell) + p_T^2(jj)} > 0.5m_{\ell\ell jj}$. This analysis is further divided into tagged (with 2 b-tagged jets) and untagged categories. The dominant backgrounds are Z + jets and $t\bar{t}$, both of which are evaluated from data. Selected results are shown in Figure 3.

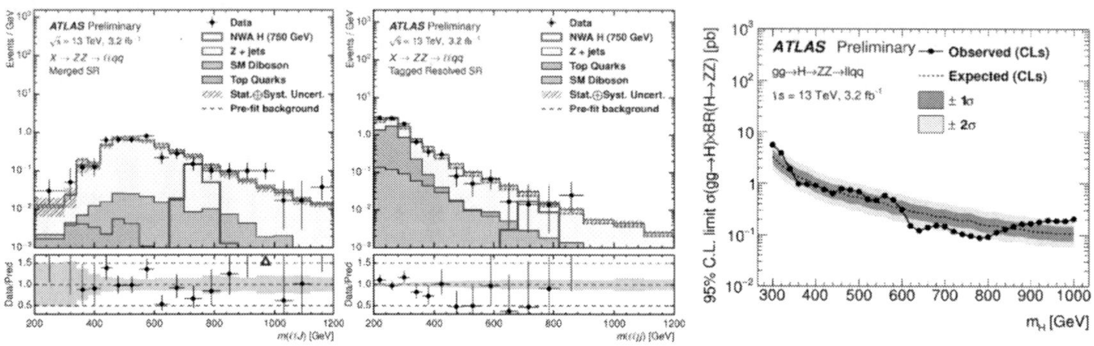

Figure 3. Selected results of the ATLAS $H \to ZZ \to \ell\ell qq$ search [10]. Left, middle: Observed $m_{\ell\ell qq}$ distribution compared to background expectations for the (left) merged analysis and (middle) tagged category of the resolved analysis. Right: 95% CL $\sigma \times$ BR limit for the production of a heavy scalar with narrow width.

3 Higgs boson decays to other Higgs particles

This section covers searches involving the CP-odd scalar A. Searches for a heavy scalar decaying into two SM Higgs bosons are covered in [11, 12].

CMS has searched for the process $H \to ZA \to \ell\ell bb$ [13]. The event selection requires a $Z \to \ell\ell$ decay (with $75 < m_{\ell\ell} < 105$ GeV) and two b-tagged jets. The analysis proceeds by choosing a set of hypothesized m_H/m_A pairs. For each specific pair, $\mathcal{H}_A^H$, a signal region $\mathcal{S}_A^H$ is defined in the $m_{bb}/m_{\ell\ell bb}$ plane as a rectangle centered on $\mathcal{H}_A^H$ with sizes given by three times the experimental resolution. The dominant backgrounds are Z + jets and $t\bar{t}$; the background is evaluated by fitting simulated samples to the control region defined by the area of the $m_{bb}/m_{\ell\ell bb}$ plane outside the signal region. (The $m_{\ell\ell}$ invariant mass requirement is also relaxed to 60–120 GeV in the control region.) The 95% CL limit on $\sigma \times$ BR is found for each $\mathcal{H}_A^H$; selected results are shown in Figure 4.

ATLAS has also searched for the process $A \to Zh$, with $h \to bb$ and either $Z \to \ell\ell$ or $Z \to \nu\nu$ [14]. Events must have at least one b-tagged jet, and are categorized as having either one or two tagged jets. If $p_T^Z < 500$ GeV (where p_T^Z is either E_T^{miss} or $m_{\ell\ell}$), then the analysis requires two jets with $110 < m_{jj} < 140$ GeV ("low-p_T^Z"); otherwise, it requires a single large-R jet consistent with Z boson decay with $75 < m_J < 145$ GeV ("high-p_T^Z"). If the event has two leptons, then there must be a $Z \to \ell\ell$ decay with $70 < m_{\ell\ell} < 110$ GeV (widened to 55–125 GeV for the high-p_T^Z case with muons), and in the low-p_T^Z case $E_T^{\text{miss}}/\sqrt{H_T} < 3.5\sqrt{\text{GeV}}$ (to suppress $t\bar{t}$ background). The final discriminant in this case is the overall invariant mass $m(Zh)$. If the event has no leptons, then the analysis requires

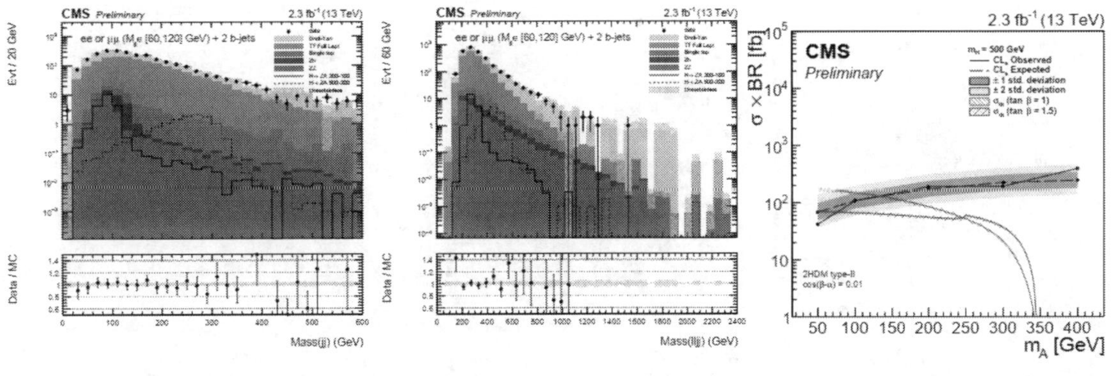

Figure 4. Selected results from the CMS $H \to ZA \to \ell\ell bb$ search [13]. Left/middle: $m_{bb}/m_{\ell\ell bb}$ distributions. Right: Resulting $\sigma \times$ BR limit as a function of m_A for m_H = 500 GeV. Expectations for a Type-II 2HDM model are also shown for two different values of $\tan\beta$.

$E_{\mathrm{T}}^{\mathrm{miss}} > 150$ GeV and, for the low-p_T^Z category, that the scalar p_T sum of the highest three p_T jets be > 150 GeV (or > 120 GeV if there are only two jets). Additional requirements are made on the track-based missing momentum and on the angular separation between objects to reduce backgrounds due to mismeasured $E_{\mathrm{T}}^{\mathrm{miss}}$. The final discriminant for this case is the transverse mass $m_T(Zh)$. The dominant backgrounds are Z + jets and $t\bar{t}$ and are studied by defining background-enriched control samples. For the Z + jets background, the m_{jj} (or m_J) requirement is inverted. A two-lepton low-p_T^Z $t\bar{t}$ control sample is also defined by requiring that the two leptons have different flavor with $m_{e\mu} > 40$ GeV. Backgrounds are estimated from MC simulations, constrained by including the control regions in the final fit. Selected results are shown in Figure 5, including $\sigma \times$BR limits as a function of m_A and limits in the $\tan\beta/\cos(\beta - \alpha)$ plane in the context of the MSSM.

4 Higgs boson decays to fermions

As mentioned earlier, models with two Higgs doublets include charged Higgs bosons. For $m_{H^\pm} > m_t$, production is dominantly in association with a top quark, $pp \to t(b)H^\pm$. Decay is mostly $H^\pm \to tb$, but $H^\pm \to \tau\nu$ is significant for large $\tan\beta$. Both experiments have reported searches for a charged Higgs boson, and also for a heavy H or A scalar decaying to a pair of tau leptons.

CMS has searched for $H^\pm \to tb$ and $H^\pm \to \tau\nu$ produced via $pp \to t(b)H^\pm$ using 8 TeV data from 2012 [15]. The hadronic channel is also sensitive to $t\bar{t} \to bH^\pm\bar{b}H^\pm$ and $t\bar{t} \to bH^\pm\bar{b}W$. This search is performed in four separate channels.

The first channel is τ_{had} + jets, which is sensitive to the $\tau\nu$ decay mode. This requires a hadronic tau decay, at least three jets, at least one b-tagged jet, $E_{\mathrm{T}}^{\mathrm{miss}} > 60$ GeV, and no leptons. Additional requirements are made on the azimuthal separations between objects. The discriminating variable used is $m_T(\tau_{\mathrm{had}}, E_{\mathrm{T}}^{\mathrm{miss}})$. The dominant backgrounds include SM diboson processes, W/Z + jets, and top quark decays into tau leptons; these are evaluated by replacing the muon in μ + jets data with a simulated tau lepton. The QCD multijet background is estimated by measuring the rate for a jet to be misidentified as a tau lepton and applying that to multijet data.

The second channel is $\ell\ell$, which is sensitive to both decay modes. This channel requires two opposite-charge leptons not consistent with Z boson decay, at least two jets, $E_{\mathrm{T}}^{\mathrm{miss}} > 40$ GeV, and at least two b-tagged jets. The final discriminating variable is the number of b-tagged jets. The dominant

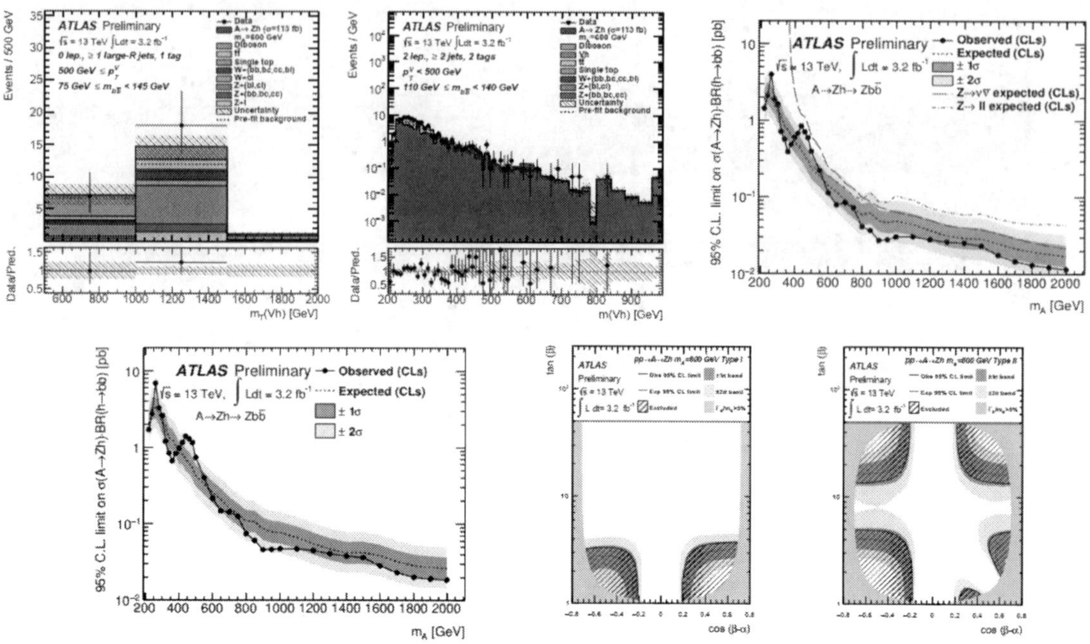

Figure 5. Selected results from ATLAS searches involving a CP-odd scalar A [14]. Top left: $m_T(Zh)$ for the 0-lepton, 1-tag, high-p_T^Z category. Top middle: $m(Zh)$ for the 2-lepton, 2-tag, low-p_T^Z category. Top right: 95% $\sigma \times$ BR limit for $A \to Zh \to Zb\bar{b}$ as a function of m_A, for production via the gluon-gluon fusion process. Bottom left: Same, for production with associated b quarks. Bottom middle: Interpretation of the $A \to Zh \to Zb\bar{b}$ search in terms of a Type-I 2HDM. Bottom right: Same for Type-II.

background is $t\bar{t}$ decay with two leptons; backgrounds are evaluated from MC simulations. The third channel is the same except that instead of two leptons, a muon and a tau lepton are required, and only one b-tagged jet is required. Backgrounds are again evaluated from MC, convoluted with the probability for a jet to be misidentified as a tau lepton, which is measured from data.

The fourth channel is ℓ + jets, which is sensitive to the tb decay mode. Events must have one lepton, $E_T^{\text{miss}} > 20$ GeV, at least two jets, and at least one b-tagged jet. Events are categorized based on whether they have one or two b-tagged jets. Background shapes are taken from MC simulations and normalized by fitting to data in both the signal and several control regions.

Selected results from this search are shown in Figure 6.

ATLAS has also searched for $H^\pm \to \tau\nu$ using 13 TeV data [16]. (The ATLAS $H^\pm \to tb$ search using 8 TeV data [17] is not covered here.) To select the final state $(t \to jjb)b\tau_{\text{had}}\nu$, events must contain a hadronically-decaying tau lepton, no other leptons, at least three jets with at least one b-tagged jet, and $E_T^{\text{miss}} > 150$ GeV. The final discriminant is the transverse mass $m_T(\tau_{\text{had}}, E_T^{\text{miss}})$, which must be > 50 GeV. The dominant backgrounds are those with a real τ_{had}, including $W \to \tau\nu$ and $t\bar{t}$; these are estimated with MC simulation and validated with data. Backgrounds with jets misidentified as tau leptons are estimated using data, while backgrounds with other leptons misidentified as tau leptons are small and are also estimated with MC simulation. Selected results are shown in Figure 7.

Both CMS and ATLAS have also searched for a heavy scalar decaying into two tau leptons. This is particularly interesting in the context of the MSSM, where the couplings of Higgs bosons

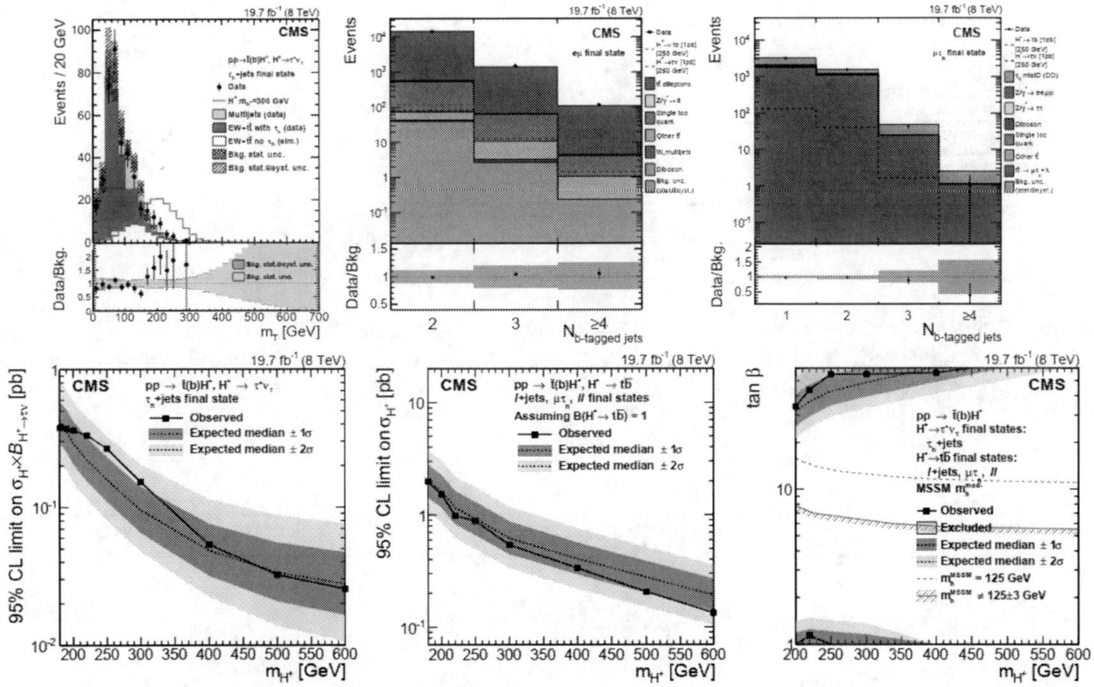

Figure 6. Selected results from the CMS charged Higgs boson search [15]. Top: observed data compared to MC simulations in the (left) $\tau_{\rm had}$ + jets, (middle) $e\mu$, and (right) $\mu\tau_{\rm had}$ channels. Bottom: 95% $\sigma \times$ BR limit as a function of $m_{H^\pm}$ for the (left) $\tau_{\rm had}$ + jets channel and (middle) the other three channels. Bottom right: 95% CL exclusion in $\tan\beta/m_{H^\pm}$ for the MSSM benchmark model $m_h^{\rm mod-}$.

to tau leptons may be enhanced for large $\tan\beta$. The CMS search [18] is divided into four channels: $\tau_e\tau_{\rm had}$, $\tau_\mu\tau_{\rm had}$, $\tau_e\tau_\mu$, and $\tau_{\rm had}\tau_{\rm had}$. The two tau lepton candidates must have opposite sign and be well-separated in ΔR ($>$ 0.3 for $\tau_e\tau_\mu$ and $>$ 0.5 for other channels). In the $\tau_e\tau_{\rm had}$ and $\tau_\mu\tau_{\rm had}$ channels, there must be no opposite-sign ee or $\mu\mu$ pairs and also $m_T(\ell, E_{\rm T}^{\rm miss}) < 40$ GeV (e) or < 30 GeV (μ). An additional topological requirement is applied in the $\tau_e\tau_\mu$ channel to reject $t\bar{t}$ events, and events are then categorized as having either 0 or ≥ 1 b-tagged jets. The final discriminant is the $\tau\tau$ transverse mass $m_{T,\tau\tau}$, computed using the likelihood technique described in [19]. The dominant $Z \to \tau\tau$ background is evaluated using MC simulations and cross-checked with $Z \to \mu\mu$ data. Other backgrounds include $t\bar{t}$, also evaluated using MC and checked against data, and W + jets and multijets, which are evaluated using same-sign $\tau_{\rm had}$ candidates and high-m_T control regions. Selected results are shown in Figure 8.

The ATLAS $H/A \to \tau\tau$ search [20] is divided into two channels. The first requires exactly one electron or muon along with a $\tau_{\rm had}$ candidate. Additional requirements include $\Delta\phi(\tau_{\rm had}, \ell) > 2.4$, $m_T(\ell, E_{\rm T}^{\rm miss}) < 40$ GeV, and for electrons, $m(e, \tau_{\rm had})$ must be outside the range 80–110 GeV, to reduce background from $Z \to ee$. The second channel requires two $\tau_{\rm had}$ candidates with opposite charge, leading/subleading $p_T > 110/55$ GeV, and $\Delta\phi(\tau_{\rm had}, \tau_{\rm had}) > 2.7$. For both channels, events are categorized as either b-veto (no b-tagged jets) or b-tagged (at least one b-tagged jet). The final discriminant is $(m_T^{\rm tot})^2 = m_T^2(E_{\rm T}^{\rm miss}, \tau_1) + m_T^2(E_{\rm T}^{\rm miss}, \tau_2) + m_T^2(\tau_1, \tau_2)$. The dominant background for the $\tau_\ell\tau_{\rm had}$ channel is $Z \to \tau\tau$, while that for the $\tau_{\rm had}\tau_{\rm had}$ channel is multijets. Backgrounds containing real $\tau_{\rm had}$ are estimated from MC simulations. Those with misidentified $\tau_{\rm had}$ are estimated using a "fake-factor" method

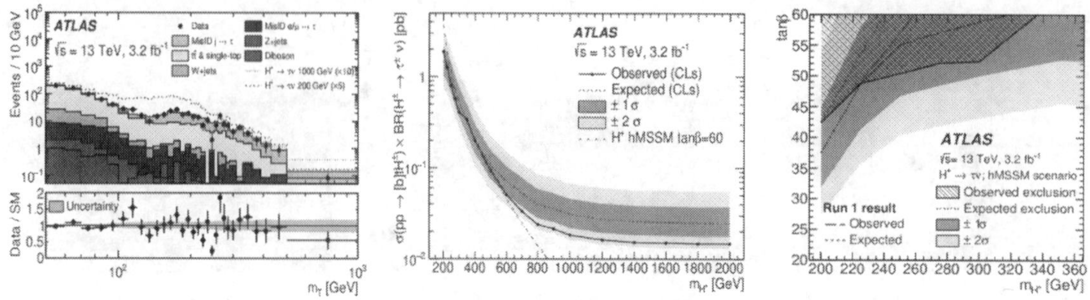

Figure 7. Selected results from the ATLAS charged Higgs boson search [16]. Left: m_T distribution after event selection. Middle: 95% $\sigma \times$ BR limit for charged Higgs boson production. Right: 95% exclusion limits in $\tan\beta/m_{H^\pm}$ for the hMSSM scenario.

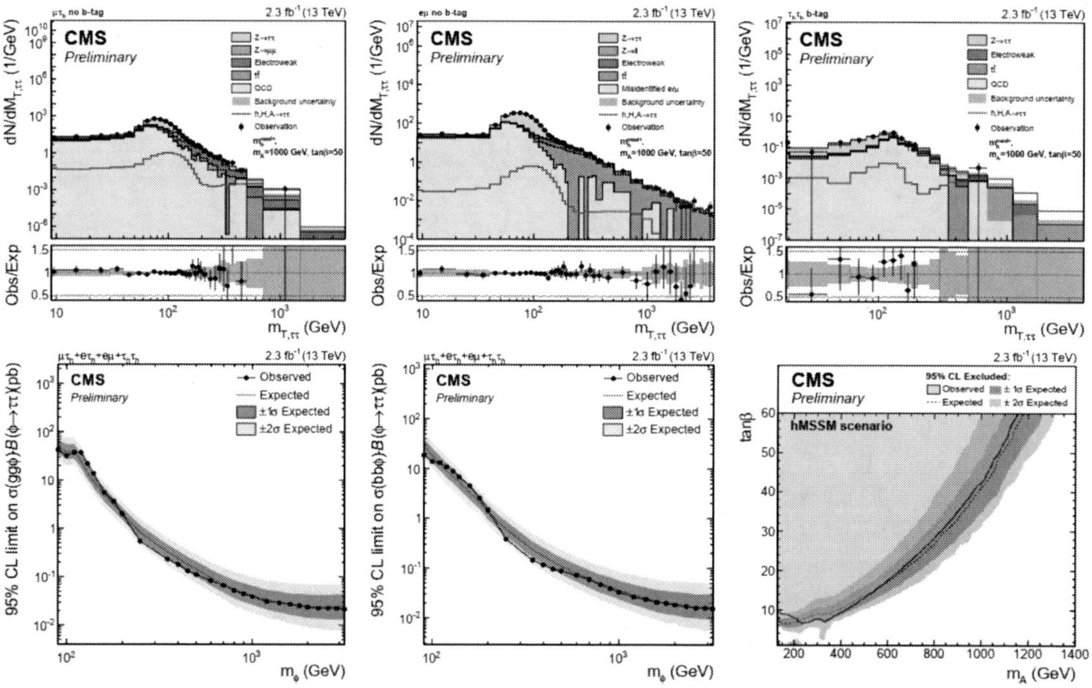

Figure 8. Selected results from the CMS $H/A \to \tau\tau$ search [18]. Top: Distributions of $m_{T,\tau\tau}$ for data and expected background for the (left) untagged $\tau_\mu\tau_{\rm had}$, (middle) untagged τ_e, τ_μ, and (right) tagged $\tau_{\rm had}\tau_{\rm had}$ channels. Bottom left, middle: 95% $\sigma \times$ BR limits as a function of m_ϕ (where ϕ denotes either H or A) for (left) production via gluon-gluon fusion and (middle) associated production with b quarks. Bottom right: 95% exclusions in $\tan\beta/m_A$ for the hMSSM scenario.

in which a control sample, not selected for tau leptons, is convoluted with the measured probability for a jet to be misidentified as a tau lepton. Selected results are shown in Figure 9.

DOI: 10.1051/epjconf/201716405001

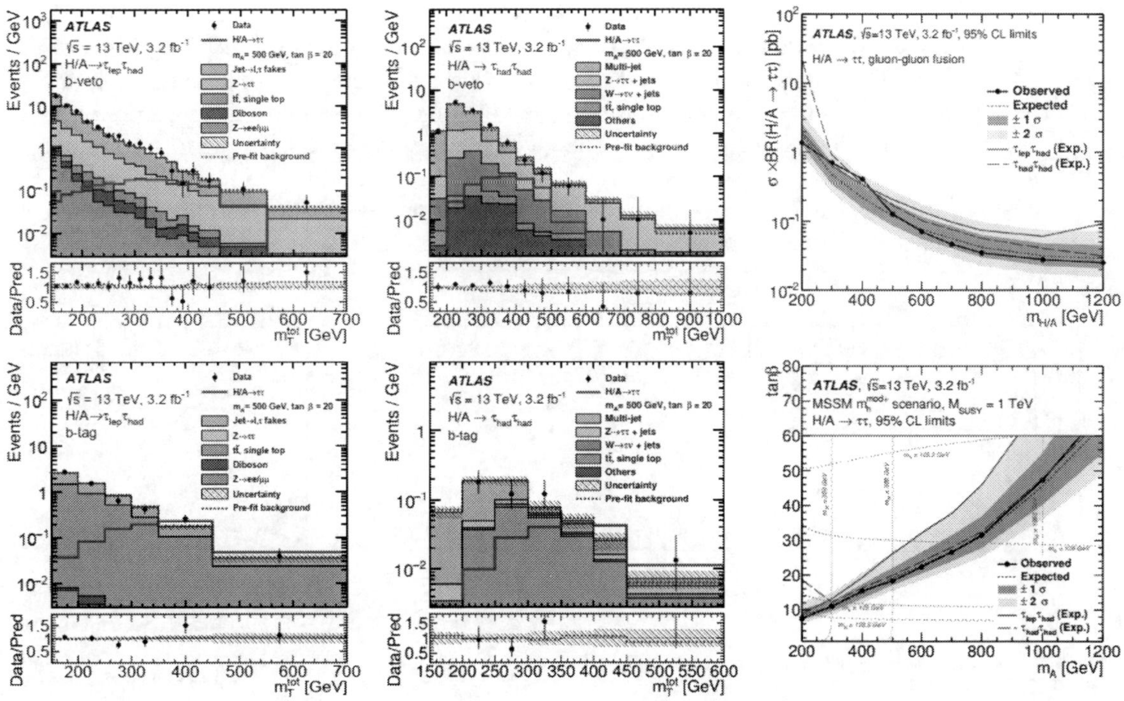

Figure 9. Selected results from the ATLAS $H/A \rightarrow \tau\tau$ search [20]. Left, middle: Distributions of m_T^{tot} for the (left) $\tau_\ell\tau_{\text{had}}$ and (right) $\tau_{\text{had}}\tau_{\text{had}}$ channels for the (top) b-veto and (bottom) b-tag categories. Top right: 95% $\sigma\times$BR limit for $H/A \rightarrow \tau\tau$ as a function of the scalar mass, assuming production via gluon-gluon fusion. Bottom right: 95% limits in $\tan\beta/m_A$ for the MSSM $m_h^{\text{mod+}}$ scenario.

5 Higgs boson decays to invisible particles

There is good evidence from astrophysics that so-called "dark matter" makes up a substantial fraction of the universe. If some portion of this is composed of a new, massive particle, then it should couple with the Higgs sector and, depending on its mass, may have a significant decay branching ratio. In fact, some models such as supersymmetry [21] and large extra dimensions [22, 23], naturally include heavy scalars coupling to invisible particles.

CMS has searched for a heavy scalar decaying to undetected particles, in both cases assuming that the heavy scalar decays exclusively to such particles. Two strategies are used to tag such decays. In the first, CMS searches for an invisibly-decaying scalar produced in association with a Z boson that decays into two leptons [24]. This search requires a $Z \rightarrow \ell\ell$ decay with $m_{\ell\ell}$ within 15 GeV of the nominal Z boson mass, $E_{\text{T}}^{\text{miss}} > 100$ GeV, no more than one jet with $p_T > 30$ GeV, and no b-tagged jets or additional leptons. Additional requirements are $\Delta\phi(\ell\ell, E_{\text{T}}^{\text{miss}}) > 2.8$ and $|E_{\text{T}}^{\text{miss}} - p_T^{\ell\ell}|/p_T^{\ell\ell} < 0.4$. Events are split into zero- and one-jet categories. The final discriminating variable is $m_T(\ell\ell, E_{\text{T}}^{\text{miss}})$, which is required to be above 200 GeV. The dominant background is Z + jets, which is estimated using a control sample of γ + jets events. Other backgrounds are estimated from a control sample in which the two leptons are different flavors. Results are shown in Figure 10; $\sigma \times$ BR limits are given for the heavy scalar mass range 125–600 GeV.

The second CMS search is for an invisibly-decaying heavy scalar produced via vector boson fusion (VBF) [25]. This search requires the presence of a VBF jet pair consisting of two forward jets with $\Delta\eta_{jj} > 3.6$ and $m_{jj} > 1100$ GeV. In addition, there must be no leptons, $E_{\mathrm{T}}^{\mathrm{miss}} > 200$ GeV, and $\Delta\phi(j, E_{\mathrm{T}}^{\mathrm{miss}}) > 2.3$. The dominant W/Z + jets background is evaluated using four control regions. One of these requires the presence of a $Z \to \mu\mu$ decay, and the other three require the presence of a single electron, muon, or tau lepton, respectively. The QCD multijet background is addressed with a fifth control region defined by inverting the $\Delta\phi$ requirement. The data are fit to simulated backgrounds in all five control regions simultaneously along with the signal region and the result found by counting the number of data and fit background events in the signal region. Results are shown in Figure 10.

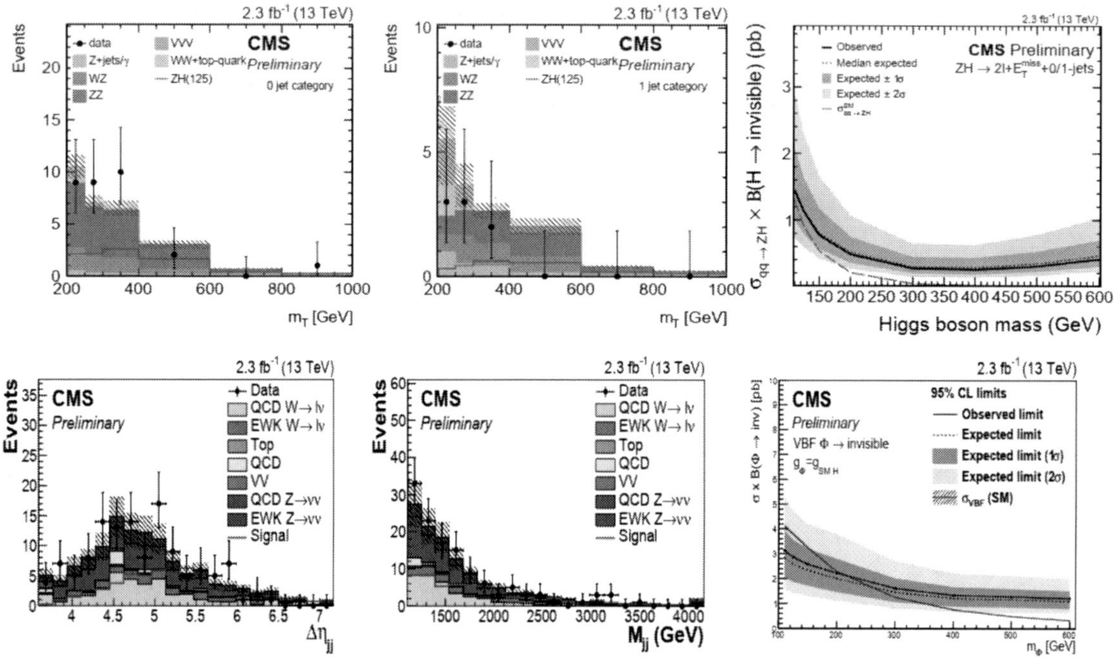

Figure 10. Selected results from the CMS searches for a heavy scalar decaying to invisible particles. Top row: Associated production with $Z \to \ell\ell$ [24]. Left/middle: $m_T(\ell\ell, E_{\mathrm{T}}^{\mathrm{miss}})$ for the 0- and 1-jet categories. Right: 95% $\sigma \times$ BR limits. Bottom row: Production via vector boson fusion [25]. Left/middle: Distributions of $\Delta\eta_{jj}$ and m_{jj} for the VBF pair. Right: 95% $\sigma \times$ BR limits.

6 Summary

ATLAS and CMS have completed an extensive set of searches for new heavy scalars using the ~ 3 fb^{-1} of $\sqrt{s} = 13$ TeV data recorded in 2015. No significant excesses above SM expectations have been found so far. However, the LHC continues to perform exceptionally well, and the searches will continue with future data.

This work is supported in part by the U.S. Department of Energy under contract DE-AC02-98CH10886 with Brookhaven National Laboratory.

References

[1] ATLAS Collaboration, Phys. Lett. B **716**, 1 (2012), 1207.7214
[2] CMS Collaboration, Phys. Lett. B **716**, 30 (2012), 1207.7235
[3] G. Branco et al., Phys. Rep. **516**, 1 (2012), 1106.0034
[4] M. Carena et al., Eur. Phys. J. **C73**, 2552 (2013), 1302.7033
[5] A. Djouadi et al., Eur. Phys. J. **C73**, 2650 (2013), 1307.5205
[6] G. Halladjian, *Search for new resonances with masses between* 500 GeV *and* 1 TeV *with final states other than two photons (ATLAS+CMS)*, in these Proceedings
[7] CMS Collaboration, *Studies of Higgs boson production in the four-lepton final state at* $\sqrt{s} = 13$ *TeV*, CMS-PAS-HIG-15-004 (2015), http://cdsweb.cern.ch/record/2139978
[8] ATLAS Collaboration, *Search for high-mass resonances decaying into a Z boson pair in the* $\ell\ell\nu\nu$ *final state in pp collisions at* $\sqrt{s} = 13$ *TeV with the ATLAS detector*, ATLAS-CONF-2016-012 (2016), http://cdsweb.cern.ch/record/2140833
[9] CMS Collaboration, *Search for a heavy scalar boson decaying into a pair of Z bosons in the* $2\ell 2\nu$ *final state*, CMS-PAS-HIG-16-001 (2016), http://cdsweb.cern.ch/record/2140099
[10] ATLAS Collaboration, *Search for ZZ resonances in the* $\ell\ell qq$ *final state in pp collisions at* $\sqrt{s} = 13$ *TeV with the ATLAS detector*, ATLAS-CONF-2016-016 (2016), http://cdsweb.cern.ch/record/2141005
[11] X. Sun, *Search for HH production by the ATLAS collaboration*, in these Proceedings
[12] A. Pozdnyakov, *Search for HH production by the CMS collaboration*, in these Proceedings
[13] CMS Collaboration, *Search for* $pp \to H \to l^+l^-b\bar{b}$ *with 2015 data*, CMS-PAS-HIG-16-010 (2016), http://cdsweb.cern.ch/record/2140613
[14] ATLAS Collaboration, *Search for a CP-odd Higgs boson decaying to Zh in pp collisions at* $\sqrt{s} = 13$ *TeV with the ATLAS detector*, ATLAS-CONF-2016-015 (2016), http://cdsweb.cern.ch/record/2141003
[15] CMS Collaboration, JHEP **11**, 018 (2015), 1508.07774
[16] ATLAS Collaboration, Phys. Lett. B **759**, 555 (2016), 1603.09203
[17] ATLAS Collaboration, JHEP **03**, 127 (2016), 1512.03704
[18] CMS Collaboration, *Search for a neutral MSSM Higgs boson decaying into* $\tau\tau$ *at* 13 TeV, CMS-PAS-HIG-16-006 (2016), http://cdsweb.cern.ch/record/2160252
[19] CMS Collaboration, JHEP **10**, 160 (2014), 1408.3316
[20] ATLAS Collaboration (2016), 1608.000890
[21] G. Belanger et al., Phys. Lett. **B519**, 93 (2001), hep-ph/0106275
[22] G.F. Giudice et al., Nucl. Phys. **B595**, 250 (2001), hep-ph/0002178
[23] M. Battaglia et al., *The Invisible Higgs decay width in the add model at the LHC*, in *Physics at TeV colliders. Proceedings, Workshop, Les Houches, France, May 26-June 3, 2003* (2004), hep-ph/0402062
[24] CMS Collaboration, *Search for invisible Higgs bosons in* $pp \to ZH \to 2\ell + E_T^{miss}$ *channels at* $\sqrt{s} = 13$ TeV, CMS-PAS-HIG-16-008 (2016), http://cdsweb.cern.ch/record/2138371
[25] CMS Collaboration, *Search for invisible decays of a Higgs boson produced via vector boson fusion at* $\sqrt{s} =$ 13 TeV, CMS-PAS-HIG-16-009 (2016), http://cdsweb.cern.ch/record/2142460

EPJ Web of Conferences **164**, 05002 (2017) DOI: 10.1051/epjconf/201716405002
ICNFP 2016

Pair Production of Beyond the Standard Model Higgs Bosons

Ramona Gröber[1,2,a]

[1] *Institute of Particle Physics Phenomenology, Department of Physics, Durham University, DH1 3LE, UK*
[2] *INFN, Sezione di Roma Tre, Via della Vasca Navale 84, 00146 Roma, Italy*

Abstract. Higgs pair production is not only sensitive to the trilinear Higgs self-coupling, but it can give access to other anomolous couplings, as e.g. a novel $hht\bar{t}$ coupling. In Composite Higgs Models, this coupling usually leads to a large increase of the cross section. In such a framework an interesting question is, whether it might be possible to observe new physics for the first time in Higgs pair production. This question will be addressed by taking into account projected sensitivities for Higgs coupling measurements and for direct searches of new vector-like quarks.
Higher order corrections to Higgs pair production via gluon fusion are sizeable. It is hence not only important to compute them in the Standard Model (SM) but also in its extensions. Here, the computation of the QCD corrections in the SM with dimension 6 operators as well as the SUSY-QCD corrections to Higgs pair production via gluon fusion in the MSSM are presented.

1 Introduction

An experimental verification of the SM electroweak symmetry breaking mechanism requires the determination of the Higgs self-couplings which can be measured in multi-Higgs production processes.[1] While the the quartic Higgs self-coupling is out of reach of the LHC due to the small signal cross section of the triple Higgs production process [2], the trilinear Higgs self-coupling is accessible in Higgs pair production processes. The Higgs pair production cross section is typically three orders of magnitude smaller than the one of single Higgs production. This makes a measurement very challenging and requires high luminosities. Phenomenological studies point out that the $b\bar{b}\gamma\gamma$ channel is the most promising final state [3, 4]. First experimental efforts have started, see e.g. refs. [5], but are not yet sensitive to the SM cross section. Nevertheless, they can already restrict certain beyond the SM (BSM) scenarios with largely increased cross section.

The dominant Higgs pair production process is gluon fusion. The process is mediated by triangle and box diagrams of heavy fermions. Apart from the trilinear Higgs self-coupling that can be measured in this process, new physics can manifest itself in several different ways as e.g. in terms of coupling modifications to the top quark Yukawa coupling, a new $hht\bar{t}$ coupling [6, 7], new particles in the loop [8] or by the exchange of a new resonance [9]. In particular in models with an exchange of a new resonance or a $hht\bar{t}$ coupling the cross section can be strongly increased. Focusing on the

[a] e-mail: ramona.groeber@durham.ac.uk

[1] An indirect way to determine the trilinear Higgs self-coupling is via higher order corrections to single Higgs production, see refs. [1].

latter case, an interesting question to ask is whether the modification can be so large that new physics might be observed for the first time in Higgs pair production. In this contribution, I will address this question in terms of Composite Higgs Models by taking into account projected sensitivities on Higgs couplings and direct searches for vector-like quarks.

As for single Higgs production, the higher order corrections to Higgs pair production are rather large and nearly double the cross section. It is hence desirable to include higher order corrections also for beyond the SM (BSM) studies. Here, I will discuss the impact of the higher order corrections in the SM extended with dimension (dim) 6 operators [10] and in the minimal supersymmetric extension of the SM (MSSM) [11].

After reviewing the theoretical status of the SM calculation of the gluon fusion process in section 2, pair production of BSM Higgs bosons will be discussed with a special focus on the question whether new physics might be for the first time observed in Higgs pair production (section 3) and on higher order corrections to BSM Higgs pair production (section 4).

2 Higgs pair production via gluon fusion in the Standard Model

The leading order (LO) SM cross section in full top quark mass dependence is known since the late 80's [12]. Since the process is already at the LO mediated by one-loop triangle and box diagrams, the computation of the next-to-leading order (NLO) corrections involves two-loop diagrams with multiple scales. This makes a full computation of the NLO corrections technically very difficult. A useful approximation to compute higher order corrections is the infinite top mass limit. In ref. [13] this limit has been employed to compute the NLO cross section. Whereas in single Higgs production the infinite top mass limit works rather well, for Higgs pair production it is only valid if the invariant mass of the Higgs boson pair m_{hh} is much smaller than twice the top quark mass m_t, $m_{hh} \ll 2\,m_t$. The results of ref. [13] have been improved by factoring out the exact Born cross section. In ref. [14, 15] higher terms in the expansion in the large top mass have been computed; in ref. [16] analytic results for this expansion have been presented. The full real radiation corrections have been computed in ref. [17]. Recently, the NLO results in full top mass dependence were given in refs. [18]. It turned out that compared to the Born improved NLO cross section in the $m_t \to \infty$ limit, the NLO cross section in the full top mass dependence is reduced by approximately 14%.

The next-to-next-to leading order (NNLO) corrections in the infinite top mass limit have been presented in ref. [19], and in ref. [15] higher order terms in an expansion in small external momenta have been computed. The differential cross section at NNLO has been given in [20]. Threshold resummation at next-to-next-to-leading logarithmic accuracy computed in the heavy quark mass limit further increases the cross section [21]. Resummation of the Higgs transverse momentum spectrum at next-to-leading logarithmic accuracy in full top quark mass dependence has been performed in ref. [22]. The QCD corrections are typically doubling the cross section. The size of the cross section and the associated uncertainties are listed in detail in ref. [23].

3 Can new physics be seen for the first time in Higgs pair production?

In this section I will address the question whether the new physics can be seen for the first time in Higgs pair production, meaning whether it could be that we miss new physics in direct searches or indirectly, by e.g. Higgs coupling measurements, and nevertheless see a sizeable deviation in Higgs pair production. This discussion is based on ref. [24].

The question has to be addressed in concrete models. In case of a new resonance that decays predominately into Higgs bosons a large increase of the cross section can be expected and the likely

DOI: 10.1051/epjconf/201716405002

answer to this question is yes. I will instead consider another case without a new resonance, but with a new $hht\bar{t}$ coupling. Types of models where such a coupling emerges are Composite Higgs Models [7].

The idea of Composite Higgs Models is, that the Higgs boson arises as a pseudo-Goldstone boson of a breaking of a global symmetry at a scale f. Effectively the model can be described by a non-linear σ model. The minimal (custodial symmetric) model is based on a global SO(5) symmetry that is broken to a SO(4). The two-derivative Lagrangian then reads

$$\mathcal{L} = \frac{f^2}{2}(D_\mu\Sigma)(D^\mu\Sigma)^T \quad \text{with} \quad \Sigma = exp(-i\sqrt{2}T^{\hat{a}}h^{\hat{a}}(x)), \tag{1}$$

where $T^{\hat{a}}$ with $\hat{a} = 1...4$ are the generators of the SO(5)/SO(4) coset and $h^{\hat{a}}(x)$ are the four Goldstone fields (the three Goldstone bosons associated with the longitudinal degrees of freedom of the SM gauge bosons and the Higgs boson). From the covariant derivatives $D_\mu\Sigma$ the coupling of the Higgs boson to the gauge bosons can be obtained. They are modified compared to the SM by a parameter $\xi = v^2/f^2 = \sin\langle H\rangle/f$, with v denoting the vacuum expectation value of the spontanous breaking of the electroweak SU(2)×U(1) symmetry. The Higgs self-couplings and the Higgs fermion couplings depend on the fermion embedding into the symmetry group. In the following, I will consider three models that differ by their fermion embedding, the MCHM4 [25], the MCHM5 and the MCHM10 [26]. For the first two models only pure Higgs non-linearities will be considered. A table showing the modifications of the couplings with respect to the SM can be found in ref. [24]. All couplings are completely determined once a value for ξ is chosen. Both models have a $hht\bar{t}$ coupling.

Whenever referred to the MCHM10, an explicit set of new vector-like fermions is introduced in a $\mathbf{10}_{2/3}$ of SO(5). Details on the model can be found in ref. [27]. The model has three additional parameters (mass of the $\mathbf{10}_{2/3}$, coupling of the $\mathbf{10}_{2/3}$ with the Goldstone field and a mixing angle) with respect to the MCHM4 and the MCHM5. The $\mathbf{10}_{2/3}$ contains also bottom partners. This will lead to a dependence of the $hb\bar{b}$ coupling and the $gg \to h$ and $h \to \gamma\gamma$ rates on all model parameters.[2]

In order to answer the initial question, we will have to assume that new physics will not be discovered at any point before the LHC will be sensitive to the SM Higgs pair production cross section. This means for the models that are simply modified by Higgs non-linearities that Higgs coupling modifications need to be below the experimental sensitivity for an integrated luminostiy of $L = 300$ fb^{-1} (or $L = 3000$ fb^{-1} respectively). Projected sensitivities for the coupling measurements can be found in ref. [29]. Two final states of the Higgs boson pair will be considered, $b\bar{b}\tau^+\tau^-$ and $b\bar{b}\gamma\gamma$ with the acceptances A estimated from ref. [4]. Then, the model is said to be distinguishable from the SM if

$$S_{SM} + 3\sqrt{S_{SM}} \leq S \qquad \text{or} \qquad S_{SM} - 3\sqrt{S_{SM}} \geq S \tag{2}$$

where

$$S = \sigma \cdot BR \cdot A \cdot L\,, \tag{3}$$

and BR is the respective branching ratio to $b\bar{b}\gamma\gamma$ or $b\bar{b}\tau^+\tau^-$ obtained with a modified version of `HDECAY` [30]. In table 1 the NLO cross section is given for the two final states $b\bar{b}\gamma\gamma$ and $b\bar{b}\tau^+\tau^-$ for the maximal possible values of ξ that lead to Higgs coupling modifications below the expected sensitivity at $L = 300$ fb^{-1} or $L = 3000$ fb^{-1}, respectively. As can be inferred from the table, in all cases the cross section does not deviate by more than 3σ from the SM signal rate.

This indicates that for the considered simple models where the deviations are driven by just one parameter we cannot hope to see any deviations from the SM for the first time in Higgs pair production.

[2]This has to be seen in contrast to the MCHM5 with a $\mathbf{5}_{2/3}$ of vector-like fermions where $gg \to h$ and $h \to \gamma\gamma$ rates basically only depend on ξ [28].

DOI: 10.1051/epjconf/201716405002

		$\sigma_{b\bar{b}\gamma\gamma}$ [fb]	$\Delta_{3\sigma}$	$\sigma_{b\bar{b}\tau^+\tau^-}$ [fb]	$\Delta_{3\sigma}$
MCHM4	$\xi = 0.076$ (LHC300)	0.114	no	3.13	no
	$\xi = 0.051$ (LHC3000)	0.112	no	3.07	no
MCHM5	$\xi = 0.068$ (LHC300)	0.175	no	3.96	no
	$\xi = 0.015$ (LHC3000)	0.119	no	3.14	no

Table 1. Values of the cross section times branching ratio in the MCHM4 and MCHM5 for the projected values on the sensitivity of Higgs coupling measurements at $L = 300$ fb^{-1} and $L = 3000$ fb^{-1} of ref. [29]. In the 4th or 6th row, respectively, a "no" indicates that the deviation from the SM rate is below 3σ.

This analysis can be performed now also for the MCHM10 with fermionic top and bottom partners where the larger parameter space allows for more freedom. In addition to the projected sensitivities on the Higgs couplings, one also needs to consider now that the vector-like fermions can be discovered in direct searches. The lightest new fermion in this model is typically a 5/3 charged fermion, denoted by χ. The reach of the indirect searches can be estimated from the excluded cross section limit at 8 TeV [31] with

$$r = \sqrt{\frac{\sigma_{BKG}(14\ \text{TeV})}{\sigma_{BKG}(8\ \text{TeV})}\frac{L_{LHC8}}{L_{LHC14}}}\,, \tag{4}$$

where σ_{BKG} denotes the cross section of the dominant background $t\bar{t}W^{\pm}$. The reach on the lightest 5/3 charged fermion can then be estimated to $m_\chi = 1.37$ TeV at $L = 300$ fb^{-1} and $m_\chi = 1.55$ TeV at $L = 3000$ fb^{-1}.

In a scan over the parameter space only points that are compatible with the estimated sensitivities on the Higgs coupling measurements and the indirect searches are kept. In addition we only generate points that are compatible with electroweak precision tests (EWPTs) at 99% C.L. [27] and $|V_{tb}| > 0.92$ [32]. The Higgs pair production cross section was computed using the NLO QCD corrections as given in [24]. The results can be found in fig. 1. As can be inferred from the plots, for $L = 300$ fb^{-1} only in the $b\bar{b}\tau^+\tau^-$ certain parameter points can be distinguished from the SM at 3σ, whereas for $L = 3000$ fb^{-1} in both of the final states there are parameter points that can be distinguished at 3σ from the SM cross section.

4 Higher order corrections to beyond the Standard Model Higgs pair production

In this section NLO QCD corrections to BSM Higgs pair production will be discussed: for the SM with effective dim 6 operators [10] in subsection 4.1 and the SUSY-QCD corrections in the MSSM [11] in subsection 4.2. Reference [11] also gives results for the NMSSM, which will however not be discussed here. Furthermore, NLO QCD corrections to Higgs pair production via gluon fusion to the singlet extension of the SM are given in ref. [33], for the two-Higgs doublet model in ref. [34] and for Composite Higgs Models with and without fermion partners in ref. [24].

DOI: 10.1051/epjconf/201716405002

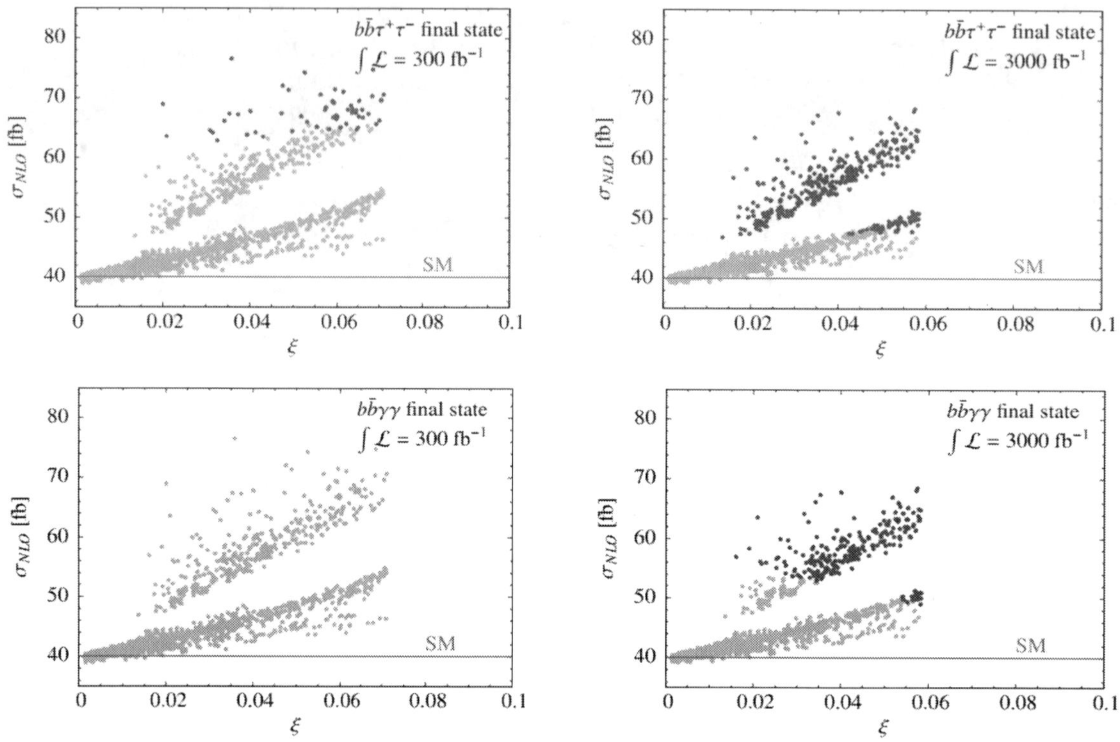

Figure 1. The NLO gluon fusion cross section to a Higgs boson pair in the MCHM10 for a scan over the parameter space. All points of the scan pass the EWPTs, fulfill the projected direct search limits for new fermionic resonances at the LHC and allow only for deviations in the Higgs boson couplings that are smaller than the expected sensitivity at the LHC for the respective luminosity. The blue points indicate that the MCHM10 cross section can be distinguished from the SM one at 3σ in the $b\bar{b}\tau^+\tau^-$ final state (upper) and the $b\bar{b}\gamma\gamma$ final state (lower) at an integrated luminosity $\int\mathcal{L} = 300$ fb^{-1} (left) and $\int\mathcal{L} = 3000$ fb^{-1} (right), whereas the grey points cannot be distinguished from the SM at 3σ. The pink line is the SM prediction for the gluon fusion cross section at NLO.

4.1 The SM with dim 6 operators

The non-linear effective Lagrangian with the relevant dim 6 operators that affect the Higgs pair production cross section are

$$\Delta\mathcal{L}_{\text{non-lin}} \supset -m_t\bar{t}t\left(c_t\frac{h}{v}+c_{tt}\frac{h^2}{2v^2}\right)-c_3\frac{1}{6}\left(\frac{3M_h^2}{v}\right)h^3+\frac{\alpha_s}{\pi}G^{a\mu\nu}G^a_{\mu\nu}\left(c_g\frac{h}{v}+c_{gg}\frac{h^2}{2v^2}\right), \qquad (5)$$

with $G^a_{\mu\nu}$ denoting the gluon field strength tensor, M_h the Higgs boson mass and α_s the strong coupling constant.

In ref. [10] the NLO QCD corrections for Higgs pair production via gluon fusion allowing for all of the operators of eq. (5) were computed in the infinite top mass limit following ref. [13]. The effective operators enter in the Born cross section and in the reducible two-loop contributions that are given by double triangle integrals. In addition, it has to be taken into account that the operators with Wilson coefficients c_g and c_{gg} do not contribute to the effective gluon coupling at $O(\alpha_s^2)$.

DOI: 10.1051/epjconf/201716405002

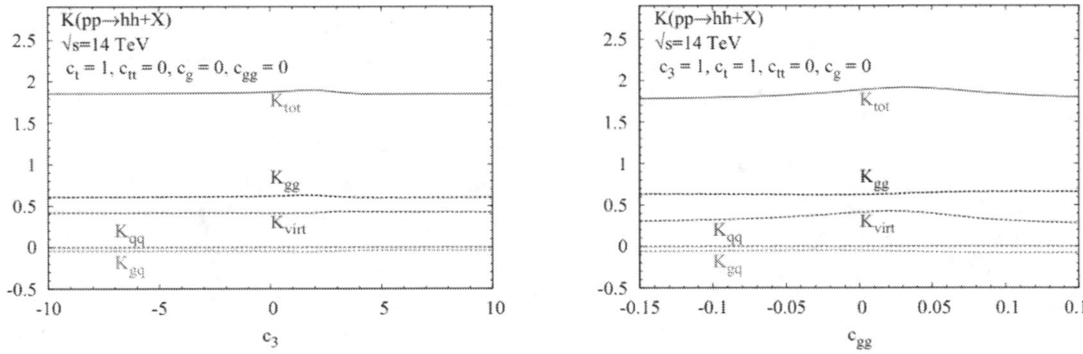

Figure 2. K-factor as a function of c_3 (left) and c_{gg} (right). The individual K-factors of the different contributions are defined as $K_i = \sigma_i/\sigma_{LO}$ with $i = virt, gg, gq, q\bar{q}$.

The results for $K = \sigma_{NLO}/\sigma_{LO}$ for a variation of c_3 and c_{gg} are given in fig. 2. From the figures it can be interferred that the SM K-factor approximates well the K-factor when dim 6 operators are included. The maximal difference to the SM K-factor is around 5%. Note, however, that on the total cross section effective operators can have a large effect. The results of this subsection are publicly available in the code `HPAIR` [35].

4.2 The MSSM

This discussion summerizes the computation of the top and bottom squark contributions to the gluon fusion cross section to a pair of CP-even Higgs bosons in the MSSM of ref. [11].

The QCD corrections for Higgs pair production via gluon fusion in the MSSM in the infinite top mass limit are known since long [13]. This computation includes, however, only the top and bottom contributions. At LO, the contributions of top squarks and bottom squarks are given in refs. [36]. At two-loop order, one can make use of the triangle form factors computed in refs. [37] for single Higgs production. The box form for the top/stop contributions can be computed via the low energy theorem (LET) for Higgs interactions [38]. The LET connects the form factor $\mathcal{H}^t_{ij}$ (with i and j the indices of the Higgs field in the interaction basis) for the interaction of two Higgs bosons with two gluons at vanishing external momentum with the second derivative of the top/stop contributions to the gluon self-energy Π^t with respect to the Higgs fields

$$\mathcal{H}_{ij} = \frac{2\pi v^2}{\alpha_s T_F} \frac{\partial \Pi^t(0)}{\partial h_i \partial h_j}, \qquad (6)$$

where $T_F = 1/2$ is a color factor. The sbottom contributions cannot be computed via the LET since the gluino contribution contains the bottom quark, whose mass is much smaller than the external momenta. The diagrams have hence to be computed explicitly via an asymptotic expansion in large sparticle masses. The limit of small external momenta used both for the top and bottom squark contributions is expected to work well for the pair production of a pair of SM-like Higgs bosons for diagrams with quartic squark couplings or gluons and squarks in realistic MSSM scenarios. Only

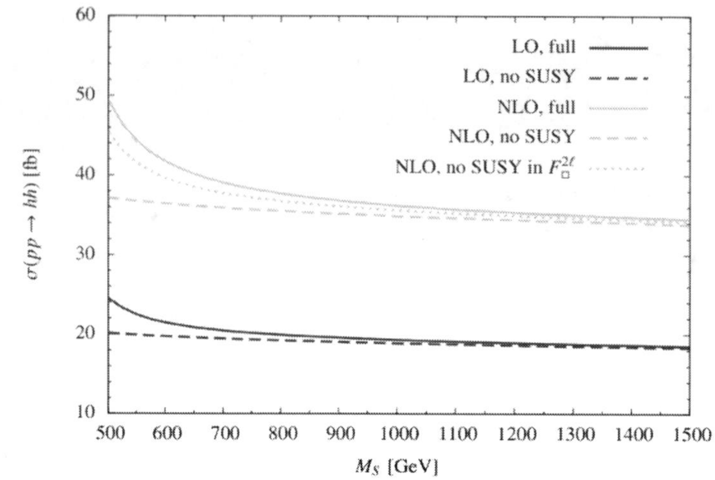

Figure 3. Cross section for Higgs pair production via gluon fusion as a function of the squark mass scale. In dark-blue the LO cross section is shown, in light blue the NLO cross section. Dashed lines show the cross section without the stop/sbottom contributions, full lines include also SUSY contributions. The dotted dark-blue line does not include the two-loop SUSY contributions to the box form factor.

contributions containing gluinos, top quarks and squarks will have, similar to the SM contribution, as discussed in section 2, thresholds at $m_{hh} = 2\,m_t$.

The results of the computation have been implemented into a private version of HPAIR [35]. In order to exemplify the impact of the SUSY-QCD corrections, the parameter point

$$\begin{aligned} &\tan\beta = 10, \quad m_A = 500\text{ GeV}, \quad \mu = -400\text{ GeV}, \quad M_3 = 1500\text{ GeV}, \\ &X_t = 2\,M_S\ , \quad m_{\tilde{t}_L} = m_{\tilde{t}_R} = m_{\tilde{b}_R} = M_S\ , \end{aligned} \tag{7}$$

is chosen. In eq. (7), $\tan\beta$ is the ratio of the vevs of the two Higgs doublets, m_A the mass of the pseudoscalar Higgs boson, $m_{\tilde{t}_L}$, $m_{\tilde{t}_R}$ and $m_{\tilde{b}_R}$ are the soft-SUSY breaking masses of the squarks, M_3 the soft-SUSY breaking parameter of the gluino, μ the Higgs/higgsino mass term in the superpotential and $X_t = A_t + \mu\cot\beta$ with A_t denoting the soft SUSY breaking trilinear Higgs-top squark coupling. The squark mass scale M_S is allowed to vary. The results for the Higgs pair production cross section of the SM-like Higgs boson can be found in fig. 3. From the figure it can be inferred that the SUSY contributions for light squark mass scale increase the cross section by up to 30%, whereas for large squark mass scale, the contributions of the stop/sbottoms become small.

5 Conclusion

Higgs pair production is not only interesting as its measurement allows to determine the trilinear coupling but it can constrain new physics in many different ways. In this contribution I have discussed whether it might be possible that new physics is for the first time observed in Higgs pair production in the context of Composite Higgs models. Taking into account projected sensitivities on Higgs couplings' determination, projected sensitivity on direct searches for vector-like quarks and constraints from electroweak precision data and the measurement of V_{tb}, it turns out that in a Composite Higgs model with both sizeable admixtures of bottom partners and top partners with the bottom or top quark, respectively, this can indeed be the case. The reason that the model can be distinguished from the SM can mainly be traced back to a large increase of the cross section due to a new $hht\bar{t}$ coupling.

Higher order corrections to Higgs pair production via gluon fusion are numerically rather large. Hence it is not only important to know the cross section as precisely as possible in the SM, but also for SM extensions higher order corrections should be included. Here, I have discussed two examples: the SM extended with dim 6 operators and the MSSM. In the former case it turns out that the K-factor of the SM is a good approximation. For the MSSM instead, the SUSY contributions can be important for light squark masses.

6 Acknowledgements

I would like to thank the organizers of the Mini-Workshop on BSM Higgs for inviting me to such a nice and stimulating workshop in a very beautiful setting. I am grateful to A. Agostini, G. Degrassi, P. Giardino, M. Mühlleitner, P. Slavich, M. Spira and J. Streicher for discussion and collaboration on various projects on (BSM) Higgs pair production. In addition, I would like to thank M. Spira for his comments on the manuscript.

References

[1] M. Gorbahn and U. Haisch, JHEP **1610** (2016) 094 [arXiv:1607.03773 [hep-ph]]; G. Degrassi, P. P. Giardino, F. Maltoni and D. Pagani, [arXiv:1607.04251 [hep-ph]]; W. Bizon, M. Gorbahn, U. Haisch and G. Zanderighi, [arXiv:1610.05771 [hep-ph]].

[2] T. Plehn and M. Rauch, Phys. Rev. D **72** (2005) 053008 [hep-ph/0507321]; T. Binoth, S. Karg, N. Kauer and R. Ruckl, Phys. Rev. D **74** (2006) 113008 [hep-ph/0608057].

[3] U. Baur, T. Plehn and D. L. Rainwater, Phys. Rev. D **69** (2004) 053004 [hep-ph/0310056]; W. Yao, arXiv:1308.6302 [hep-ph]; V. Barger, L. L. Everett, C. B. Jackson and G. Shaughnessy, Phys. Lett. B **728** (2014) 433 [arXiv:1311.2931 [hep-ph]]; A. Azatov, R. Contino, G. Panico and M. Son, Phys. Rev. D **92** (2015) no.3, 035001 [arXiv:1502.00539 [hep-ph]]; C. T. Lu, J. Chang, K. Cheung and J. S. Lee, JHEP **1508** (2015) 133 [arXiv:1505.00957 [hep-ph]]; F. Kling, T. Plehn and P. Schichtel, [arXiv:1607.07441 [hep-ph]].

[4] J. Baglio, A. Djouadi, R. Gröber, M. M. Mühlleitner, J. Quevillon and M. Spira, JHEP **1304** (2013) 151 [arXiv:1212.5581 [hep-ph]].

[5] *Some recent searches are:* CMS Collaboration, CMS-PAS-HIG-16-032; ATLAS collaboration, ATLAS-CONF-2016-071; ATLAS collaboration [ATLAS Collaboration], ATLAS-CONF-2016-049.

[6] C. O. Dib, R. Rosenfeld and A. Zerwekh, JHEP **0605** (2006) 074 [hep-ph/0509179].

[7] R. Gröber and M. Mühlleitner, JHEP **1106** (2011) 020 [arXiv:1012.1562 [hep-ph]]; R. Contino, M. Ghezzi, M. Moretti, G. Panico, F. Piccinini and A. Wulzer, JHEP **1208** (2012) 154 [arXiv:1205.5444 [hep-ph]].

[8] S. Dawson, A. Ismail and I. Low, Phys. Rev. D **91** (2015) no.11, 115008 [arXiv:1504.05596 [hep-ph]].

[9] M. J. Dolan, C. Englert and M. Spannowsky, Phys. Rev. D **87** (2013) no.5, 055002 [arXiv:1210.8166 [hep-ph]]; C. Y. Chen, S. Dawson and I. M. Lewis, Phys. Rev. D **91** (2015) no.3, 035015 [arXiv:1410.5488 [hep-ph]].

[10] R. Gröber, M. Mühlleitner, M. Spira and J. Streicher, JHEP **1509** (2015) 092 [arXiv:1504.06577 [hep-ph]].

[11] A. Agostini, G. Degrassi, R. Gröber and P. Slavich, JHEP **1604** (2016) 106 [arXiv:1601.03671 [hep-ph]].

[12] E. W. N. Glover and J. J. van der Bij, Nucl. Phys. B **309** (1988) 282.

[13] S. Dawson, S. Dittmaier and M. Spira, Phys. Rev. D **58** (1998) 115012 [hep-ph/9805244].

[14] J. Grigo, J. Hoff, K. Melnikov and M. Steinhauser, Nucl. Phys. B **875** (2013) 1 [arXiv:1305.7340 [hep-ph]].

[15] J. Grigo, J. Hoff and M. Steinhauser, Nucl. Phys. B **900** (2015) 412 [arXiv:1508.00909 [hep-ph]].

[16] G. Degrassi, P. P. Giardino and R. Gröber, Eur. Phys. J. C **76** (2016) no.7, 411 [arXiv:1603.00385 [hep-ph]].

[17] R. Frederix, S. Frixione, V. Hirschi, F. Maltoni, O. Mattelaer, P. Torrielli, E. Vryonidou and M. Zaro, Phys. Lett. B **732** (2014) 142 [arXiv:1401.7340 [hep-ph]]; F. Maltoni, E. Vryonidou and M. Zaro, JHEP **1411** (2014) 079 [arXiv:1408.6542 [hep-ph]].

[18] S. Borowka, N. Greiner, G. Heinrich, S. P. Jones, M. Kerner, J. Schlenk, U. Schubert and T. Zirke, Phys. Rev. Lett. **117** (2016) no.1, 012001 Erratum: [Phys. Rev. Lett. **117** (2016) no.7, 079901] [arXiv:1604.06447 [hep-ph]]; S. Borowka, N. Greiner, G. Heinrich, S. P. Jones, M. Kerner, J. Schlenk and T. Zirke, JHEP **1610** (2016) 107 [arXiv:1608.04798 [hep-ph]].

[19] D. de Florian and J. Mazzitelli, Phys. Rev. Lett. **111** (2013) 201801 [arXiv:1309.6594 [hep-ph]]; J. Grigo, K. Melnikov and M. Steinhauser, Nucl. Phys. B **888** (2014) 17 [arXiv:1408.2422 [hep-ph]].

[20] D. de Florian, M. Grazzini, C. Hanga, S. Kallweit, J. M. Lindert, P. Maierhöfer, J. Mazzitelli and D. Rathlev, JHEP **1609** (2016) 151 [arXiv:1606.09519 [hep-ph]].

[21] D. de Florian and J. Mazzitelli, JHEP **1509** (2015) 053 [arXiv:1505.07122 [hep-ph]].

[22] G. Ferrera and J. Pires, arXiv:1609.01691 [hep-ph].

[23] D. de Florian *et al.* [LHC Higgs Cross Section Working Group Collaboration], arXiv:1610.07922 [hep-ph].

[24] R. Gröber, M. Mühlleitner and M. Spira, JHEP **1606** (2016) 080 [arXiv:1602.05851 [hep-ph]].

[25] K. Agashe, R. Contino and A. Pomarol, Nucl. Phys. B **719** (2005) 165 [hep-ph/0412089].

[26] R. Contino, L. Da Rold and A. Pomarol, Phys. Rev. D **75** (2007) 055014 [hep-ph/0612048].

[27] M. Gillioz, R. Gröber, A. Kapuvari and M. Mühlleitner, JHEP **1403** (2014) 037 [arXiv:1311.4453 [hep-ph]].

[28] A. Falkowski, Phys. Rev. D **77** (2008) 055018 [arXiv:0711.0828 [hep-ph]]; I. Low and A. Vichi, Phys. Rev. D **84** (2011) 045019 [arXiv:1010.2753 [hep-ph]]; A. Azatov and J. Galloway, Phys. Rev. D **85** (2012) 055013 [arXiv:1110.5646 [hep-ph]]; M. Gillioz, R. Gröber, C. Grojean, M. Mühlleitner and E. Salvioni, JHEP **1210** (2012) 004 [arXiv:1206.7120 [hep-ph]].

[29] C. Englert, A. Freitas, M. M. Mühlleitner, T. Plehn, M. Rauch, M. Spira and K. Walz, J. Phys. G **41** (2014) 113001 [arXiv:1403.7191 [hep-ph]].

[30] A. Djouadi, J. Kalinowski and M. Spira, Comput. Phys. Commun. **108** (1998) 56 [hep-ph/9704448]; J. M. Butterworth *et al.*, [arXiv:1003.1643 [hep-ph]];
for an implementation of the MCHM4/MCHM5 in HDECAY see also: R. Contino, M. Ghezzi, C. Grojean, M. Mühlleitner and M. Spira, Comput. Phys. Commun. **185** (2014) 3412 [arXiv:1403.3381 [hep-ph]].

[31] G. Aad *et al.* [ATLAS Collaboration], Phys. Rev. D **91** (2015) no.11, 112011 [arXiv:1503.05425 [hep-ex]].

DOI: 10.1051/epjconf/201716405002

[32] S. Chatrchyan *et al.* [CMS Collaboration], JHEP **1212** (2012) 035 [arXiv:1209.4533 [hep-ex]].

[33] S. Dawson and I. M. Lewis, Phys. Rev. D **92** (2015) no.9, 094023 [arXiv:1508.05397 [hep-ph]].

[34] B. Hespel, D. Lopez-Val and E. Vryonidou, JHEP **1409** (2014) 124 [arXiv:1407.0281 [hep-ph]].

[35] M. Spira, HPAIR, available on `http://tiger.web.psi.ch/hpair/`.

[36] A. Belyaev, M. Drees, O. J. P. Eboli, J. K. Mizukoshi and S. F. Novaes, Phys. Rev. D **60** (1999) 075008 [hep-ph/9905266]; A. A. Barrientos Bendezu and B. A. Kniehl, Phys. Rev. D **64** (2001) 035006 [hep-ph/0103018].

[37] C. Anastasiou, S. Beerli, S. Bucherer, A. Daleo and Z. Kunszt, JHEP **0701** (2007) 082 [hep-ph/0611236]; U. Aglietti, R. Bonciani, G. Degrassi and A. Vicini, JHEP **0701** (2007) 021 [hep-ph/0611266]; M. Mühlleitner and M. Spira, Nucl. Phys. B **790** (2008) 1 [hep-ph/0612254]; R. Bonciani, G. Degrassi and A. Vicini, JHEP **0711** (2007) 095 [arXiv:0709.4227 [hep-ph]]; C. Anastasiou, S. Beerli and A. Daleo, Phys. Rev. Lett. **100** (2008) 241806 [arXiv:0803.3065 [hep-ph]]; M. Mühlleitner, H. Rzehak and M. Spira, PoS RADCOR **2009** (2010) 043 [arXiv:1001.3214 [hep-ph]]; R. V. Harlander and M. Steinhauser, Phys. Lett. B **574** (2003) 258 [hep-ph/0307346]; R. V. Harlander and M. Steinhauser, JHEP **0409** (2004) 066 [hep-ph/0409010]; G. Degrassi and P. Slavich, Nucl. Phys. B **805** (2008) 267 [arXiv:0806.1495 [hep-ph]]; G. Degrassi and P. Slavich, JHEP **1011** (2010) 044 [arXiv:1007.3465 [hep-ph]]; R. V. Harlander, F. Hofmann and H. Mantler, JHEP **1102** (2011) 055 [arXiv:1012.3361 [hep-ph]]; G. Degrassi, S. Di Vita and P. Slavich, Eur. Phys. J. C **72** (2012) 2032 [arXiv:1204.1016 [hep-ph]].

[38] J. R. Ellis, M. K. Gaillard and D. V. Nanopoulos, Nucl. Phys. B **106** (1976) 292; M. A. Shifman, A. I. Vainshtein, M. B. Voloshin and V. I. Zakharov, Sov. J. Nucl. Phys. **30** (1979) 711 [Yad. Fiz. **30** (1979) 1368]; B. A. Kniehl and M. Spira, Z. Phys. C **69** (1995) 77 [hep-ph/9505225].

DOI: 10.1051/epjconf/201716406001

Highlights from the heavy-ion program in STAR

Petr Chaloupka[1,a] for the STAR collaboration

[1] *Czech Technical University in Prague, Faculty of Nuclear Sciences and Physical Engineering*

Abstract. The experiments at RHIC have produced convincing evidence that strongly interacting partonic matter is created in the collisions of heavy ions. The unique flexibility of RHIC to collide different nuclear species over a wide range of collision energies together with STAR's wide acceptance and particle identification are ideally suited for systematic exploration of the properties of this QCD matter.
STAR collaboration has successfully completed the Beam Energy Scan, program focused on searching for the onset of the QGP signatures and studying the nature of the phase transition, indicating that the region of interests for critical point and the first-order phase transition is within the reach of RHIC experiments. Moreover, with its two newly installed detector upgrades, STAR has launched a comprehensive heavy-flavor program which allows high precision measurements of the properties of the partonic matter.

1 Introduction

Collisions of heavy ions at ultra-relativistic energies have been used to create a new state of matter called the "Quark Gluon Plasma (QGP)", in which deconfined quarks and gluons are the relevant degrees of freedom [1, 2]. The detailed studies of this strongly interacting matter, as it is currently performed by the STAR experiment at RHIC, provide important tests of Quantum Chromodynamics (QCD) and its emergent properties.

The RHIC's unique capability to collide different nuclear species over a broad range of collision energies allows to systematically explore the QCD phase diagram. At the highest RHIC collision energies the system at mid-rapidity attains the highest temperature and vanishingly small net baryon density. In these collisions the high statistics data, which were collected thanks to luminosity upgrades of RHIC, are used to study the quantitative properties of the QGP such as viscosity, transport parameters, heavy quark diffusion coefficients, and others. At lower collision energies, where baryon density is finite, STAR experiment seeks to answer questions about the actual phase transition from the hadronic matter into the QGP, most importantly: where is the onset of the QGP formation? What is the nature of the phase transition and is there a critical point in the QCD phase diagram?

In order to reach these goals STAR collaboration has successfully finished two major detector upgrades. In 2014 the Muon Telescope Detector (MTD) [3] was fully installed in STAR. Its purpose is detection and identification of muons with $p_T > 1.2$ GeV/c and $|y| < 0.5$. In the same year was also installed a new vertex detector called Heavy Flavor Tracker (HFT) [4]. This high-precision inner

[a]e-mail: petr.chaloupka@fjfi.cvut.cz

tracker can determine low momentum open heavy quark decay vertices in the environment of high-multiplicity heavy-ion collisions. These upgrades have significantly enhanced STAR's capabilities for heavy flavor measurements including open heavy quark hadrons and quarkonia. Moreover from 2010 to 2014 RHIC has conducted a Beam Energy Scan (BES) program aimed at exploring the nuclear matter phase diagram. In this review we highlight and discuss some of the results from the BES program as well as results obtained using the two new STAR upgrades.

2 Results from Beam Energy Scan

The Beam Energy Scan (BES) program has been carried out at RHIC in order to study the structure of the QCD phase diagram. The program has covered beam energies of $\sqrt{s_{NN}}$=7.7, 11.5, 14.5, 19.6, 27, 39 GeV between the years 2010 to 2014, together with 62.4, 130, 200 GeV from previous years. This collision energy range means that the STAR experiment can effectively map the QCD phase diagram in the region of baryon chemical potential of $20 < \mu_B < 420$ MeV. This detailed exploration is expected to reveal several key features of the QCD phase diagram[5, 6].

It is well established that at the highest collision energy the strongly coupled QGP has been created [1, 2]. The created medium shows a strong opacity to colored probes and very small ratio of shear viscosity over entropy density. Lowering the collision energy allows to study the emergent properties of the partonic matter and search for the onset of its formation.

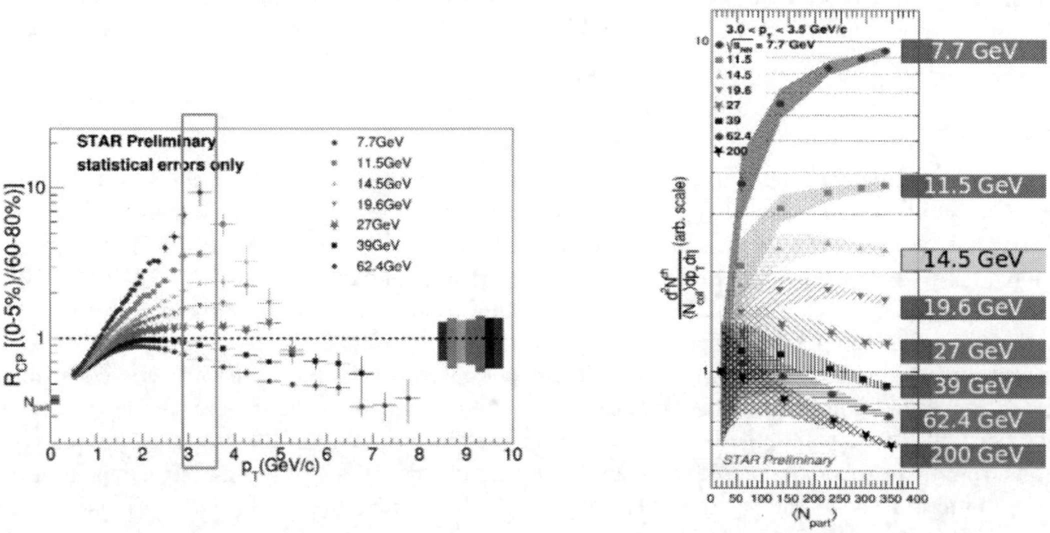

Figure 1. (Left) Charged hadron R_{CP} in Au+Au collisions at different energies. (Right) Integrated yields ($3 < p_T < 3.5$ GeV) per number of binary collisions as a function of number of participants at different collision energies. The yields are normalized so that the yield in most peripheral bin is unity.

The suppression of high p_T hadrons is one of the most pronounced signatures of energy loss of partons traversing the QGP. Figure 1 shows on the left a nuclear modification factor R_{CP} (the binary-collision-scaled p_T yields) of charged hadron at different BES energies [7]. As can be seen the R_{CP} transitions from a strong suppression at high energies to an enhancement at lower beam energies. However, the measured value of R_{CP} is also affected by competing enhancement from Cronin effect which gets stronger at lower collision energies. This means that QGP can be formed even at collision energies where R_{CP} is not below one. More differential study can help to disentangle these contributions. In right part of Figure 1 are shown STAR preliminary results on p_T integrated yields ($3<p_T<3.5$ GeV) per number of binary collisions [7] as a function of number of participants. For every collision energy the yields are scaled by the corresponding most peripheral bin. As expected the suppression effects quickly dominate at high collision energies. On the other hand at lowest energies the enhancement is clearly present in the most central collisions. The competing effects approximately cancel each other at the at $\sqrt{s_{NN}}$=14.5 GeV.

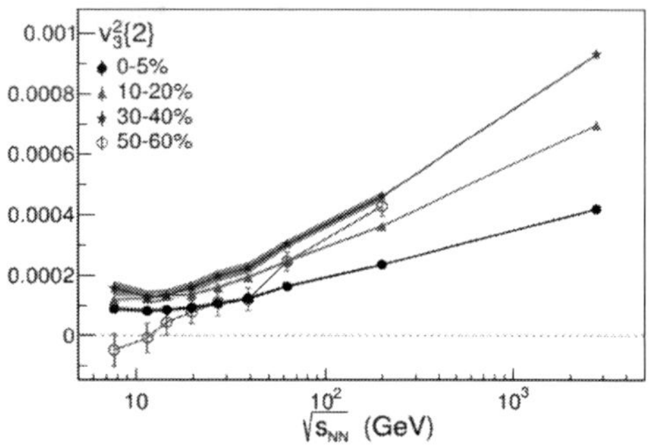

Figure 2. Collision energy dependence of $v_3^2\{2\}$ in Au+Au collisions from STAR and in Pb+Pb collisions from ALICE [8].

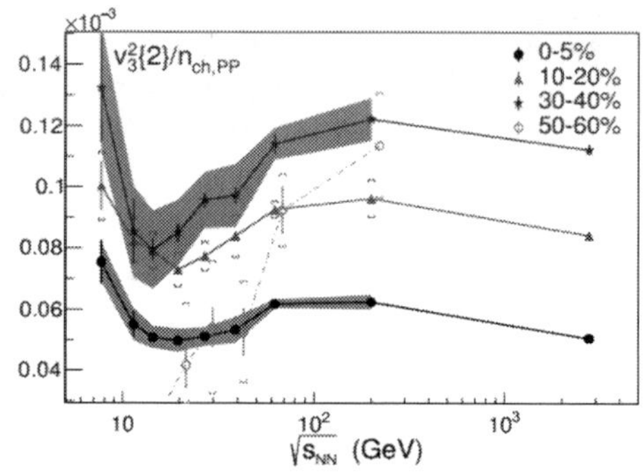

Figure 3. Collision energy dependence of $v_3^2\{2\}$ normalized by the multiplicity per participant pair ($n_{ch,PP} = dN_{ch}/dy/(N_{part}/2)$.

Another signature related to the existence of the QGP phase is the observation of triangular flow v_3 [9] since its non-zero value requires the presence of a low viscosity phase early in the collision. STAR has recently published results on measurements of $v_3^2\{2\}$ from the BES [10]. In Figure 2 are shown STAR measurements of charged hadron $v_3^2\{2\}$ in Au+Au together with ALICE measurement in Pb+Pb collisions at $\sqrt{s_{NN}}$=2.76 TeV [8]. In central and mid-central collisions we observe non-zero v_3 at collision energies which may suggest that the QGP phase exists even at the low collision energies of the BES. On the other hand, in peripheral collisions the observed v_3 is consistent with zero for energies below 14.5 GeV. It is expected that the ability of the system to convert initial state geometry fluctuations into the final state v_3 scales with multiplicity of the produced particles. Deviations from that expectation could indicate interesting physics like a softening of the equation-of-state [11]. Figure 3 shows energy dependence of $v_3^2\{2\}$ divided by the multiplicity per participant pair ($n_{ch,PP} = dN_{ch}/dy/(N_{part}/2)$. The $n_{ch,PP}$ is proportional to the system energy density. As can be seen the data show a local minimum in the region of $\sqrt{s_{NN}}$=15-20 GeV which could indicate an interesting trend in the pressure developed inside the system.

One of the main motivations for the BES program at RHIC is to search for signatures of the first-order phase transition and the possible QCD critical point. It has been argued that observation of minimum in the slope of directed flow (dv_1/dy) of protons and its double sign-change for net protons in the STAR data [12] can be linked to the softening of the equation of state [13]. So far models had hard time reproducing this structure in dv_1/dy which is expected to result from an interplay between hydro, baryon transport dynamics and baryon/anti-baryon annihilation. STAR has recently measured

 DOI: 10.1051/epjconf/201716406001

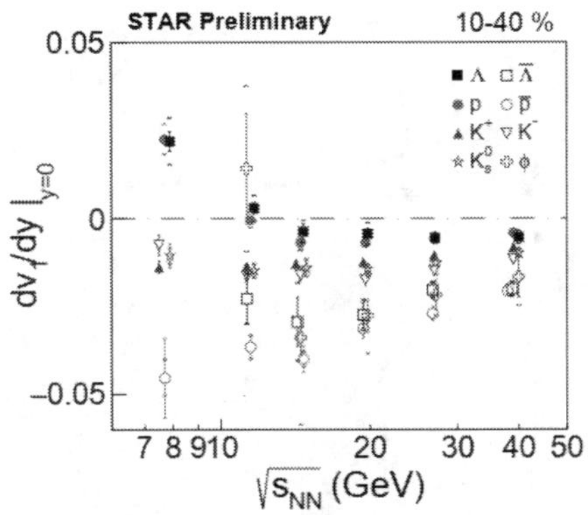

Figure 4. Beam energy dependence of dv_1/dy for p, $\bar{p}$, $K^{\pm}$, K_S^0, ϕ, Λ and $\bar{\Lambda}$ in 10-40% Au+Au collisions.

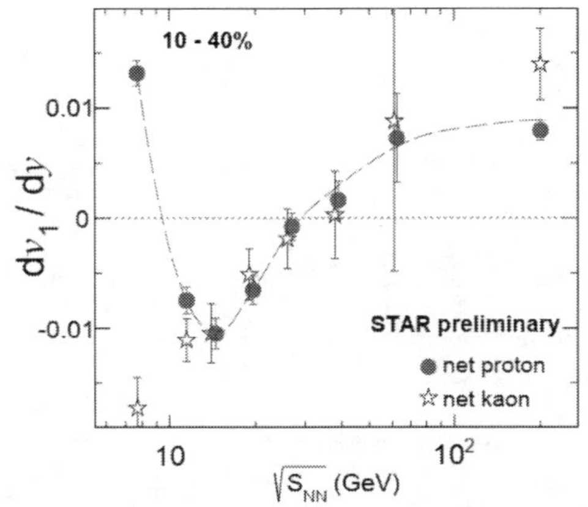

Figure 5. Beam Energy dependence of net-particle dv_1/dy in 10-40% Au+Au collisions.

directed flow of Λ, $\bar{\Lambda}$, $K^{\pm}$, K_S^0 and ϕ in Au+Au collisions at the BES energies [14]. Since these particles contain different constituent quarks the measurements will help to understand the QCD phase transition and it will allow to disentangle the role of produced and transported quarks in heavy-ion collisions.

Figure 4 shows the dv_1/dy for different particle species as measured by the STAR experiment. As can be seen, the dv_1/dy of Λ is consistent with that of the proton and similarly shows a change in sign near $\sqrt{s_{NN}}$=11.5 GeV. For the higher collision energies above 14.5 GeV, where antibaryon production is significant, the $\bar{\Lambda}$ and ϕ follow the trend of antiprotons above 14.5 GeV. However, below 14.5 GeV the dv_1/dy of ϕ is, within large uncertainties, consistent with zero. Figure 5 shows collision energy dependence of net-proton and net-kaon dv_1/dy. Such a study is expected to be sensitive to the difference between the contributions from produced quarks and quarks that are transported from the initial state. As can be seen the net-p and net-K dv_1/dy follow a similar trend for $\sqrt{s_{NN}}$=14.5-200 GeV. At lower collision energies they start to strongly deviate from each other.

The type of the transition between the hadronic and partonic phase depends on where we are in the QCD phase diagram with regard to the temperature and the baryonic chemical potential of the system. While smooth cross-over is expected at the vanishing baryonic chemical potential, at higher net-baryon region model calculations predict the existence of a first-order phase transition. This implies that there should be a critical point between these two regions [15].

One of the main goals of BES at the RHIC is to study the QCD phase diagram over a wide range of chemical potential (μ_B) and search for the QCD critical point. The critical point is characteristic by large correlation length of the dynamical system which manifests in large event-by-event fluctuations of conserved quantities, such as net-baryon number (B), net-charge (Q), and net-strangeness(S) [16–18]. The experimentally accessible order moments of these quantities, variance (σ^2), skewness (S), and kurtosis (κ), are directly connected to the theoretically calculated thermodynamic susceptibilities. However, since it is very hard to measure the net-baryon (ΔN_B) and the net-strangeness (ΔN_S) distributions it is common to use net-proton (ΔN_P) and net-kaon (ΔN_K) as proxies. STAR has performed analysis of cumulants and cumulant ratios of the net-proton, net-kaon and net-charge multiplicity distributions in Au+Au collisions from BES [14]. Figure 6 shows the energy dependence of the vol-

 DOI: 10.1051/epjconf/201716406001

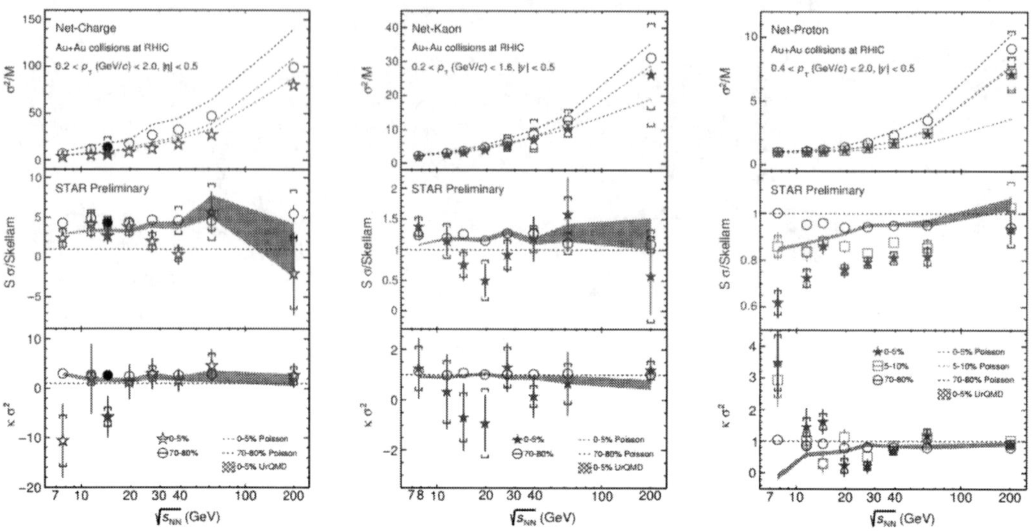

Figure 6. Energy dependence of cumulant ratios ($\sigma^2/M, S\sigma/Skellam, \kappa\sigma^2$) of net-charge, net-kaon, and net-proton multiplicity distributions for top 0-5% central (red stars), 5-10% central (green squares), and 70-80% peripheral (black circles) collisions. The Poisson expectations are denoted as dotted lines and UrQMD calculations are shown as blue bands.

ume independent cumulant ratios ($\sigma^2/M, S\sigma/Skellam, \kappa\sigma^2$) for net-proton, net-kaon, and net-charge multiplicity distributions in Au+Au collisions from BES as measured by the STAR experiment. The experimental data are compared to predictions from UrQMD calculations (blue bands) which do not include critical physics processes. The net-charge and net-kaon cumulants are observed to be within uncertainties consistent with Poisson expectations (dashed lines in the figure). Both the data and the UrQMD simulations show no energy dependence for $S\sigma^2$/Skellam and $\kappa\sigma^2$ in net-charge and net-kaon measurement. However, the values of net-proton $S\sigma^2$/Skellam and $\kappa\sigma^2$ in central collisions clearly exhibit a deviation from the Poisson expectation of unity as well as from the UrQMD calculations. More precise systematic studies will be possible in future with the data from the planned second phase of the Beam Energy Scan (BES-II).

3 Measurements with Heavy Flavor Tracker

Heavy quarks which are predominantly produced in the early stages of the collision experience the full evolution of the system while being mostly unaffected by the QCD medium due to their large masses. Study of heavy quark propagation through the QGP gives an information about transport properties of the medium.

Heavy Flavor Tracker (HFT) micro-vertexing detector has been built and installed into STAR successfully before the RHIC year 2014 running. The HFT consists of four layers of silicon detectors. While outer layers are silicon strip detectors the two innermost layers are made from state-of-the-art ultra-thin CMOS Monolithic Active Pixel Sensors (MAPS). The HFT is a first application of the CMOS MAPS detector in a collider experiment. The proximity to the collision location and the low material budget allow HFT to achieve a track pointing resolution of less than 50 μm with p_T=750 MeV/c.

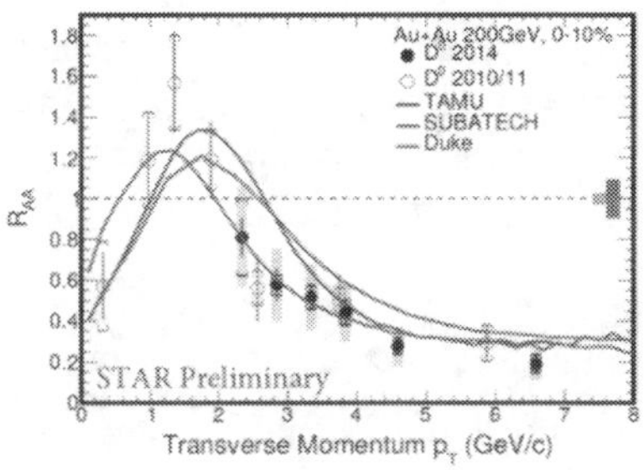

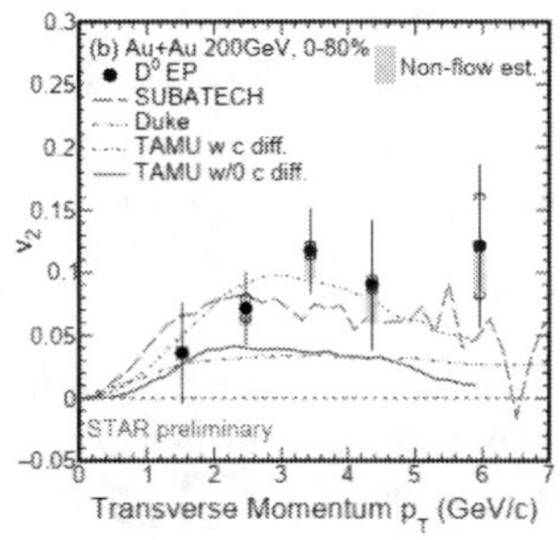

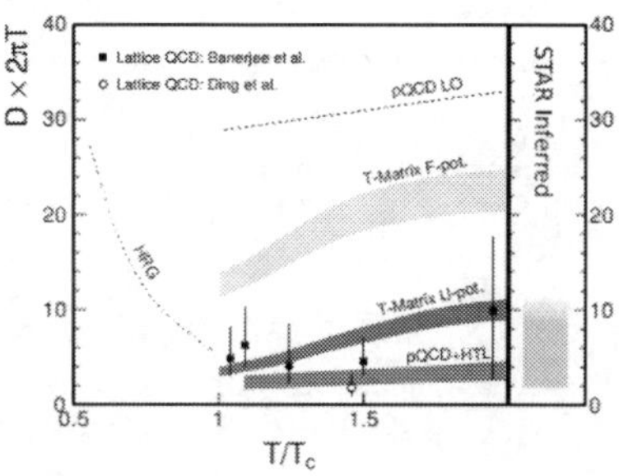

Figure 7. STAR results on nuclear modification factor R_{AA} of D^0 with and without the HFT, compared to pions.

Figure 8. Comparison of D^0 v_2 in 0-80% most central events with model calculations.

Figure 9. Diffusion coefficient from model calculations and the inferred range from STAR measurements.

STAR has reported preliminary results on open charm production in Au+Au collisions at $\sqrt{s_{NN}}$=200 GeV from analysis where charmed mesons were topologically reconstructed using the HFT [19, 20] . In the Figure 7 are shown results on D^0 R_{AA} in 0-10% most central Au+Au collisions at $\sqrt{s_{NN}}$=200 GeV. Open red circles denote already published STAR results from 1.1 billion events recorded in 2010 and 2011 without the HFT [21]. Solid black circles show the recent measurement with the HFT from approximately 780 million minimum bias events. The use of HFT significantly improves the precision and results clearly confirm the strong suppression of D^0 production at high p_T. The enhancement in the low-p_T region can be explained by coalescence of charmed quarks with a flowing medium.

STAR has also performed first measurements of elliptic flow of open charm hadrons at RHIC [19, 20, 22]. In Figure 8 is shown D^0 v_2 in 0-80% Au+Au collisions extracted with the use of HFT. Significant non-zero v_2 is observed for p_T>2 GeV/c. However, its magnitude is systematically lower than v_2 of lighter hadrons, as discussed in detail in [20].

These measurements of heavy-quark energy loss and elliptic flow allow to extract transport coefficients as a further characterization of the medium. Both Figures 7, and 8 show comparison to model calculations which include charm diffusion [23–25]. A detailed description of the models and comparison to data can be found in [20]. Figure 9 shows the comparison of extracted diffusion coefficient from the different model calculations compared to the range of inferred values that are compatible with the presented STAR measurements (yellow band in the figure). The compatible values of the charm diffusion coefficient $2\pi T D_S$ are between 2 and 12 which is also compatible with results from Lattice QCD.

4 Quarkonia measurements with MTD

Suppression of heavy quarkonia due to Debye-like screening of the quark-antiquark potential, has been predicted to be an indicator sensitive to the presence of the QGP [26]. The Debye screening length depends on the temperature attained by the QGP medium. Since different quarkonium states have different binding energies (hence sizes), they are expected to melt at different QGP temperatures. The observed quarkonium suppression pattern can then be linked to the thermodynamical properties of the created QGP.

However, there are other effects that can alter the quarkonium production with respect to the baseline of p+p collisions. These effects include secondary production of quarkonia in the QGP via

DOI: 10.1051/epjconf/201716406001

heavy quark recombination together with cold nuclear matter effects (CNM) such as shadowing/anti-shadowing of parton distribution functions, initial-state parton energy loss and final state nuclear absorption. Since these mechanisms have different dependencies on variables such as p_T or energy density, differential measurements of quarkonium production at different collision energies, collision systems, and centralities can help to disentangle the interplay of these effects and to study properties of the created medium.

It should be noted that the experimentally measured yields of J/ψ and Υ contain contributions from feed-down. Inclusive J/ψ production is a combination of prompt and non-prompt J/ψ. The prompt J/ψ are either direct ($\sim$ 60%) or from feed-down from higher excites states ($\psi(2S)$and χ_C). The non-prompt J/ψ originate from B-hadron decays. This contribution has been estimated by the STAR experiment to be 10 ~ 25% in the range of $4<p_T<12$ GeV/c [27]. It is hence important to study the production of higher charmonium states in both p+p and nucleus-nucleus collisions.

The Muon Telescope Detector (MTD) [3] which was fully installed in 2014 is designed to identify muons and also to provide triggering on di-muons from heavy flavor decays. This allows STAR to perform high-precision quarkonia measurement at mid-rapidity in the di-muon channel which is of an advantage since the di-muon decay channel is less affected by bremsstrahlung.

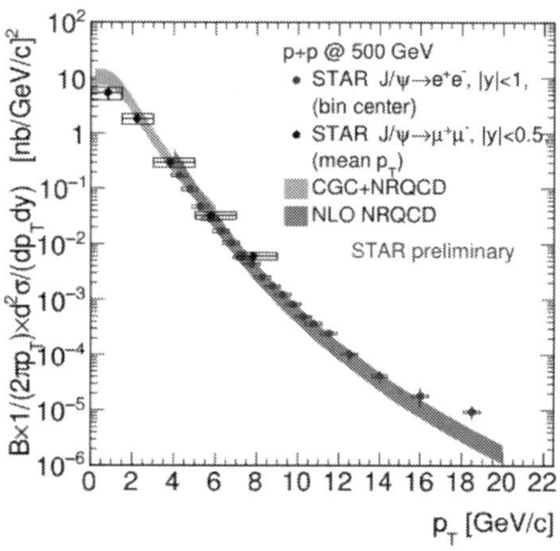

Figure 10. J/ψ cross section at $\sqrt{s_{NN}}$ = 500 GeV as a function of p_T in the di-muon decay channel (black circle) and in the di-electron decay channel (red circle).

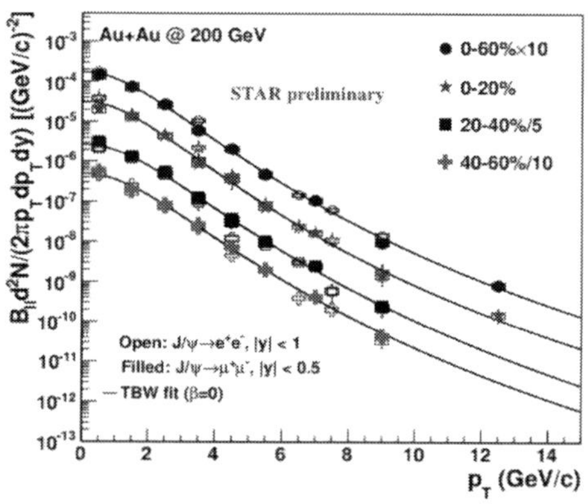

Figure 11. The J/ψ invariant yields in Au+Au collisions at $\sqrt{s_{NN}}$=200 GeV scaled by branching ratio B_{ll} as a function of p_T from di-muon decay channel (solid points) and from di-electron decay channel (open points).

Figure 10 shows the results on J/ψ production cross section in p+p collision at $\sqrt{s}$=500 GeV as a function of p_T up to p_T=20 GeV/c. The MTD measurements from the di-muon decay channel extend the p_T reach down to 0 GeV/c. They are compatible with previous measurement from di-electron decay channel in the overlapping p_T region. The experimental data are in a good agreement with CGC+NRQCD calculations at low p_T [28] and with NLO NRQCD calculations at high p_T [29].

STAR experiment has also measured J/ψ invariant yields in Au+Au collisions at $\sqrt{s_{NN}}$=200 GeV previously in di-electron and recently in di-muon channel. In Figure 11 are shown invariant yields of J/ψ in di-muon channel as a function of p_T together with results from di-electron decay channel for different collision centralities. These new results from the di-muon decay channel are consistent with the STAR published results from the di-electron decay channel [27, 30] within uncertainties.

DOI: 10.1051/epjconf/201716406001

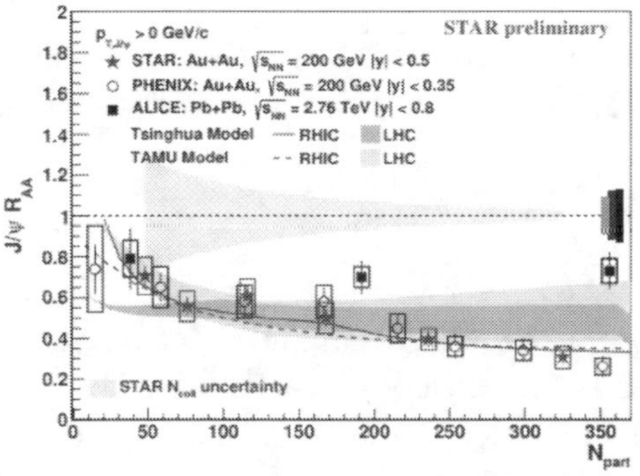

Figure 12. J/ψ nuclear modification factor for integrated p_T in $\sqrt{s_{NN}}$=200 GeV Au+Au collisions compared to LHC results as function of N_{part}.

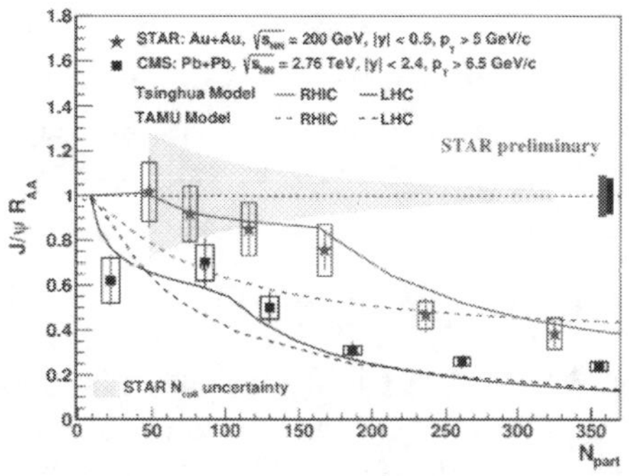

Figure 13. The J/ψ nuclear modification factor for integrated p_T>5 GeV/c in $\sqrt{s_{NN}}$=200 GeV Au+Au collisions compared to LHC results as function of N_{part}.

In Figure 12 is shown centrality dependence of J/ψ R_{AA} = $\frac{\sigma_{inel}\ d^2N_{AA}/dydp_T}{\langle N_{coll}\rangle\ d^2\sigma_{pp}/dydp_T}$ from the MTD data (red stars) for p_T>0 GeV/c and in Figure 13 for p_T>5 GeV/c. The STAR results are compared to PHENIX results at the same collision energy (black open circles) [31], as well as results from ALICE [32] and CMS [33] in Pb+Pb collisions at $\sqrt{s_{NN}}$=2.76 TeV (blue solid circles). For the integrated p_T the STAR results are consistent with the PHENIX results (black open circles) showing a decrease with increasing centrality. This differs from ALICE results which are almost centrality independent and less suppressed. Such difference can be possibly explained by larger regeneration contributions at low p_T due to higher $c\bar{c}$ yields at LHC collision energies. On the other hand, for p_T>5 GeV/c in Figure 13 RHIC data show less suppression compared to LHC for all centralities. This can be an effect of higher energy density and temperature of the system created at the LHC which in turn leads to stronger J/ψ suppression.

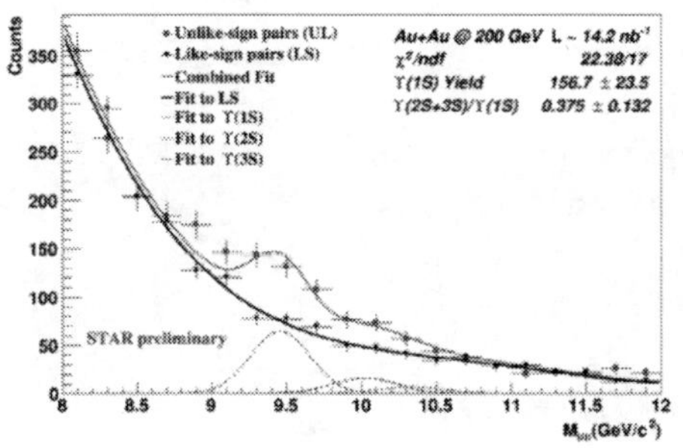

Figure 14. Υ signal from di-muon decay channel.

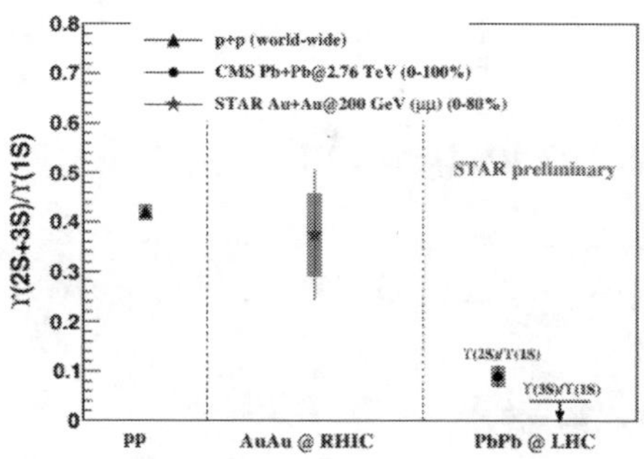

Figure 15. Comparison of Υ(2S+3S)/Υ(1S) in p+p and heavy-ion collisions at RHIC and LHC.

Compared to J/ψ, measurements with Υ states have the advantage that they do not suffer from the regeneration contribution since $b\bar{b}$ pairs have much smaller cross section at RHIC energies when compared to $c\bar{c}$ pairs. In Figure 14 is shown reconstructed Υ signal from di-muon decay channel in Au+Au collisions at $\sqrt{s_{NN}}$=200 GeV. A signal of Υ(2S+3S) in the di-muon decay channel can be observed. Figure 15 shows the $\Upsilon(2S + 3S)/\Upsilon(1S)$ ratio which is obtained by simultaneous fit to the like-sign and unlike-sign distributions. The obtained ratio is compared with the world-wide p+p data [34] and results from CMS [35, 36]. The preliminary results indicate that a there less melting of $\Upsilon(2S + 3S)$ relative to $\Upsilon(1S)$ at RHIC.

5 Summary and outlook

STAR experiment at RHIC has carried out the Beam Energy Scan program during which data were collected at seven different collision energies in order to systematically study the phase diagram by changing baryon chemical potential (μ_B) and temperature of the created matter. This extremely successful period has brought new insights and understanding of the QCD phase diagram. The BES provided first indications that the transition from hadronic to quark-gluon degrees of freedom occurs within the collision energy range of $10 \leq \sqrt{s_{NN}} \leq 20$ GeV which is well within a reach of the RHIC collider. This offers a unique opportunity to probe QCD matter properties from below to above the transition.

Moreover, STAR has successfully completed two major detector upgrades which greatly boosted its heavy flavor program. High resolution Heavy Flavor Tracker allowed to perform first analyses with topologically reconstructed open-charm mesons showing that there is a significant flow of charmed quarks in the medium. The presented results of quarkonium suppression from Muon Telescope Detector in di-muon decay channel indicate melting of J/ψ and Υ states at RHIC top energy.

It is clear that more systematic studies are needed in the region of collision energies of interest where the current results indicate the onset of the phase transition and possibly location of the critical point. With this motivation RHIC will be in 2019-2020 conducting a second phase of the Beam Energy Scan (BES-II). This program will take advantage of RHIC's increased luminosity thanks to electron cooling and STAR is hence expected to collect high statistics data for 5-20 GeV Au+Au collisions. Meanwhile, STAR will also carry out an inner-TPC (iTPC) upgrade, which will extend the pseudo-rapidity coverage, increase low-p_T reach in tracking of particles for QGP bulk property study, and improve the dE/dx resolution of proton/kaon separations at high p_T for jet fragmentation measurements. The refined high-precision measurements will enable BES-II to further improve our understanding of the nuclear matter phase diagram.

6 Acknowledgement

This work was supported by the grant No. 13-20841S of Grant Agency of the Czech Republic and project No. LM2015054 of Czech Ministry of Education, Youth and Sports.

References

[1] J. Adams et al. (STAR), Nucl. Phys. **A757**, 102 (2005), nucl-ex/0501009
[2] K. Adcox et al. (PHENIX), Nucl. Phys. **A757**, 184 (2005), nucl-ex/0410003
[3] L. Ruan et al., J. Phys. **G36**, 095001 (2009), 0904.3774
[4] G. Contin (STAR), Nucl. Instrum. Meth. **A831**, 7 (2016)
[5] P. Braun-Munzinger, J. Wambach, Rev. Mod. Phys. **81**, 1031 (2009), 0801.4256
[6] B. Mohanty, New J. Phys. **13**, 065031 (2011), 1102.2495
[7] S. Horvat (STAR), Nucl. Phys. **A956**, 838 (2016), 1601.01644

[8] K. Aamodt et al. (ALICE), Phys. Rev. Lett. **107**, 032301 (2011), 1105.3865
[9] J. Auvinen, H. Petersen, Phys. Rev. **C88**, 064908 (2013), 1310.1764
[10] L. Adamczyk et al. (STAR), Phys. Rev. Lett. **116**, 112302 (2016), 1601.01999
[11] H. Sorge, Phys. Rev. Lett. **82**, 2048 (1999), nucl-th/9812057
[12] L. Adamczyk et al. (STAR), Phys. Rev. Lett. **112**, 162301 (2014), 1401.3043
[13] Y. Nara, H. Niemi, A. Ohnishi, H. Stoecker, Phys. Rev. **C94**, 034906 (2016), 1601.07692
[14] S. Singha (STAR) (2016), 1610.07423
[15] Y. Aoki, G. Endrodi, Z. Fodor, S.D. Katz, K.K. Szabo, Nature **443**, 675 (2006), hep-lat/0611014
[16] M.A. Stephanov, Phys. Rev. Lett. **102**, 032301 (2009), 0809.3450
[17] M. Asakawa, S. Ejiri, M. Kitazawa, Phys. Rev. Lett. **103**, 262301 (2009), 0904.2089
[18] M.A. Stephanov, Phys. Rev. Lett. **107**, 052301 (2011), 1104.1627
[19] M.R. Lomnitz (STAR), Nucl. Phys. **A956**, 256 (2016), 1601.00743
[20] M.R. Lomnitz, these proceedings
[21] L. Adamczyk et al. (STAR), Phys. Rev. Lett. **113**, 142301 (2014), 1404.6185
[22] L. He (STAR), J. Phys. Conf. Ser. **736**, 012007 (2016)
[23] S. Cao, G.Y. Qin, S.A. Bass, Phys. Rev. **C88**, 044907 (2013), 1308.0617
[24] M. He, R.J. Fries, R. Rapp, Phys. Rev. **C86**, 014903 (2012), 1106.6006
[25] M. Nahrgang, J. Aichelin, S. Bass, P.B. Gossiaux, K. Werner, Phys. Rev. **C91**, 014904 (2015), 1410.5396
[26] T. Matsui, H. Satz, Phys.Lett. **B178**, 416 (1986)
[27] L. Adamczyk et al. (STAR), Phys.Lett. **B722**, 55 (2013), 1208.2736
[28] Y.Q. Ma, R. Venugopalan, Phys. Rev. Lett. **113**, 192301 (2014), 1408.4075
[29] H.S. Shao, H. Han, Y.Q. Ma, C. Meng, Y.J. Zhang, K.T. Chao, JHEP **05**, 103 (2015), 1411.3300
[30] L. Adamczyk et al. (STAR), Phys.Rev. **C90**, 024906 (2014), 1310.3563
[31] A. Adare et al. (PHENIX), Phys. Rev. Lett. **98**, 232301 (2007), nucl-ex/0611020
[32] B.B. Abelev et al. (ALICE), Phys. Lett. **B734**, 314 (2014), 1311.0214
[33] S. Chatrchyan et al. (CMS), JHEP **05**, 063 (2012), 1201.5069
[34] W. Zha, C. Yang, B. Huang, L. Ruan, S. Yang, Z. Tang, Z. Xu, Phys. Rev. **C88**, 067901 (2013), 1308.4720
[35] S. Chatrchyan et al. (CMS), Phys. Rev. Lett. **109**, 222301 (2012), 1208.2826
[36] S. Chatrchyan et al. (CMS), JHEP **04**, 103 (2014), 1312.6300

EPJ Web of Conferences **164**, 06002 (2017) DOI: 10.1051/epjconf/201716406002
ICNFP 2016

Nucleon structure observables with $\overline{\text{P}}$ANDA

María Carmen Mora Espí[1,2,a],
on behalf of the $\overline{\text{P}}$ANDA collaboration

[1] *Institut für Kernphysik, Johannes Gutenberg-Universität Mainz, Johann-Joachim-Becher-Weg 45, D-55128 Mainz, Germany.*
[2] *Helmholtz-Institut Mainz, Staudingerweg 18, 55128 Mainz, Germany.*

Abstract. The $\overline{\text{P}}$ANDA detector will be built as a part of the future FAIR facility in Darmstadt. The availability of an antiproton beam with beam momenta up to 15 GeV/c will make possible a broad nuclear physics program. Topics like hadron spectroscopy in the charmonium mass region, the property of hadrons inside nuclear matter, hypernuclear physics, or nucleon properties using electromagnetic processes are part of the physics program of $\overline{\text{P}}$ANDA. The main part of this contribution concentrates on the feasibility of measurement of nucleon structure observables, such as electromagnetic form factors or transition distribution amplitudes, via experiments using electromagnetic processes in $\overline{\text{P}}$ANDA.

1 Introduction

Understanding Quantum Chromodynamics (QCD), describing the strong force, is one of the most challenging fields of physics. The interaction mechanisms between fundamental particles can be described using perturbative QCD (p-QCD) at short distances with very high precision and predictive power. However, at long distances a series of spectacular phenomena, *e.g.* generation of hadron masses, formation of hadronic matter and color confinement, appear, and these cannot be described by basic QCD principles. The ultimate goal for the $\overline{\text{P}}$ANDA (anti$\overline{\text{P}}$roton ANnihilations at DArmstadt) experiment will be to explore the strong interaction at precisely the energy interval corresponding to the transition between the perturbative and the non-perturbative regimes of QCD. The high-intensity and high-resolution antiproton beam of the High Energy Storage Ring (HESR), ranging from $p_{\bar{p}} = 1.5$ GeV/c to $p_{\bar{p}} = 15$ GeV/c, of the future Facility for Antiproton and Ion Research (FAIR) in Darmstadt will be an excellent instrument to fulfill the ambitious physics program foreseen by the $\overline{\text{P}}$ANDA collaboration [1]. This includes several fields of research:

- **Hadron spectroscopy**: The study of QCD bound states (charmonium, *D*-meson and baryon spectroscopy, exotic states...) is of fundamental importance for a better, quantitative understanding of QCD. Unique precision measurements will help to distinguish between the different theory approaches used for the computation of particle spectra. The study of exotic bound states such as glueballs or hybrids will be also addressed in the $\overline{\text{P}}$ANDA collaboration.

[a] e-mail: mamoraes@uni-mainz.de

DOI: 10.1051/epjconf/201716406002

- **Hyperon production dynamics**: The measurement of reactions involving hyperons can help studying the creation mechanism of quark-antiquark pairs and their arrangement to hadrons. The self-analysing weak decays of most hyperons give access to spin degrees of freedom for these reactions.
- **Study of Hadrons in Nuclear Matter**: The study of medium modifications of hadrons embedded in hadronic matter is aimed at understanding the origin of hadron masses in the context of spontaneous chiral symmetry breaking in QCD and its partial restoration in a hadronic environment.
- **Hypernuclear Physics**: The availability of antiproton beams at FAIR will allow efficient production of hypernuclei with more than one strange hadron. In particular, the production of Ξ can be used to form double Λ hypernuclei, which will offer a unique opportunity to study basic properties of the potential between nucleons and hyperons in a bound state.
- **Electroweak Physics**: With the high intensity antiproton beam available at HESR, a large number of D-mesons can be produced. This gives the possibility to observe rare weak decays of these mesons allowing studying electroweak physics by probing predictions of the Standard Model and searching for enhancements introduced by processes beyond the Standard Model.
- **Nucleon structure**: $\overline{\text{P}}$ANDA will enable the investigation of the structure of the nucleon using electromagnetic processes. Nucleon structure observables, *e.g.* the Electromagnetic Form Factors (EMFF) or the Transition Distribution Amplitudes (TDA) will be measured with $\overline{\text{P}}$ANDA. Detailed information about these and other processes are presented later in this paper.

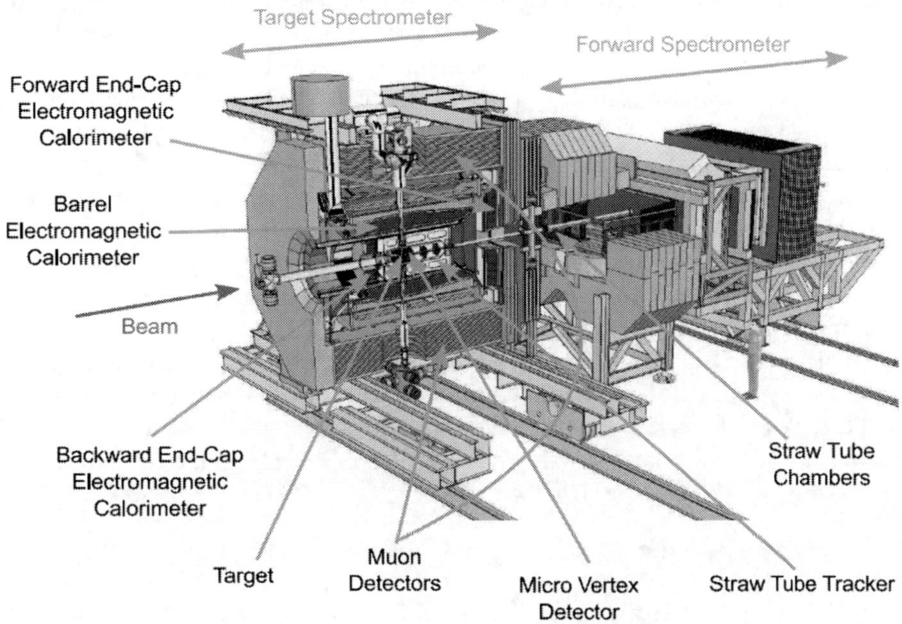

Figure 1. Sketch of the $\overline{\text{P}}$ANDA detector, divided into the target spectrometer and the forward spectrometer. Both spectrometers contain modular detectors for tracking, particle identification, electromagnetic calorimetry and muon identification. The main detectors in the target spectrometer are signaled with arrows.

In order to successfully achieve the previously mentioned goals, $\overline{\text{P}}$ANDA needs to be an almost 4π detector with very good energy resolution, at a few percent level, and excellent particle identification (PID) capabilities, with a fast data acquisition system (DAQ) and high radiation hardness. The basic concept of the setup is a fixed-target experiment divided into a target spectrometer, with a 2 T solenoid around the target area to detect particles with low momenta, and a forward spectrometer based on a 2 T dipole in order to ensure particle detection below 2° for faster particles. Modular detectors for tracking, particle identification, electromagnetic calorimetry and muon identification are designed for both spectrometers. The modularity will enable an easy replacement of parts of the tracking system and the backward Electromagnetic Calorimeter (EMC) with the hypernuclear setup. The latter has dedicated targets and gamma-ray detectors that will be used in a more advanced stage of the measurement program. Figure 1 shows a sketch of the $\overline{\text{P}}$ANDA detector depicting various subsystems foreseen to achieve the desired detection capabilities. A complete description of the different setups can be found in various technical design reports [2–7].

2 Nucleon structure observables

2.1 Proton Electromagnetic Form Factors

The annihilation of $\bar{p}p$ into an e^+e^- pair (or more generally, a $\ell^+\ell^-$ pair, with $\ell = e, \mu$) via a virtual photon of mass $q^2 = s$ (with s being the squared total energy available in the annihilation) gives access to the hadron-electromagnetic vertex $\gamma^\star \bar{p}p$. This vertex gives information about the proton EMFF, *i.e.* analytical functions of q^2 parameterizing the hadronic current at the vertex. The kinematic region accessible by $\overline{\text{P}}$ANDA is the time-like (TL) region of momentum transfer squared, $q^2 > 0$, also reachable by the time-reversed reaction $e^+e^- \to \bar{p}p$ (in general, $\ell^+\ell^- \to \bar{p}p$). This reaction is related by crossing symmetry to elastic electron-proton scattering (see Fig. 2), which gives access to the space-like (SL) region of momentum transfer squared, $q^2 < 0$. In the SL region, the EMFF can be interpreted as the Fourier transform of the distribution of charge and magnetization in the Breit-frame. Both regions are connected due to the analyticity and unitarity of the EMFF via the dispersion relations. A full review on EMFF can be found for example in Ref. [8].

The proton EMFF in the SL region have been studied since many years and abundant and precise data exist in this region. However, recent analyses using polarization methods show results in contradiction with older results from Rosenbluth separation. This has given rise to a lot of discussion about the knowledge of the SL EMFF. In addition, the data collected in the TL region are still very limited. Any measurement here would contribute to expand the knowledge on these functions, not only in the TL region but also in the SL region. The $\overline{\text{P}}$ANDA experiment will offer an excellent opportunity to extend the, until now, scarce data bank on TL EMFF, and therefore will help disentangling the questions arisen about the EMFF in the last years (see also Refs. [9, 10]).

The current world-wide data available on the EMFF in the TL region are shown in Fig. 3. Due

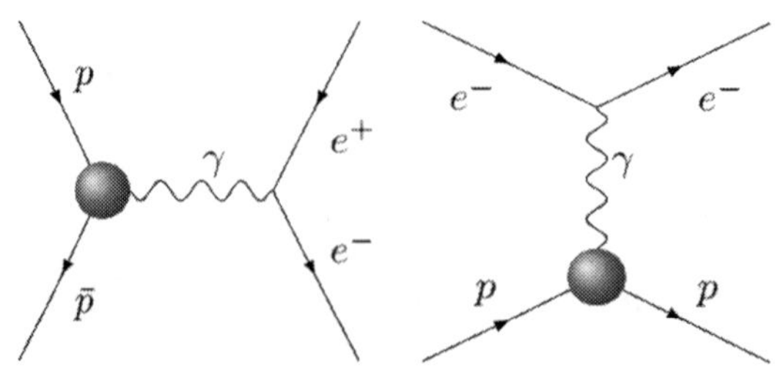

Figure 2. Feynman diagrams of the reactions $\bar{p}p \to e^+e^-$ and $e^-p \to e^-p$. The balls represent the hadron electromagnetic vertex. Both diagrams are connected by crossing symmetry.

to the small data samples, only the measurement of the ratio R or the measurement of an effective form factor under the assumption $|G_E| = |G_M|$ was possible. The results extracted by BaBar [11], from measurements of initial state radiation (ISR) processes ($e^+e^- \to \bar{p}p\gamma$) are not in agreement with the previous results from LEAR ($\bar{p}p \to e^+e^-$) [12]. More recent measurements by BESIII [13] at different energies using $e^+e^- \to \bar{p}p$ have large uncertainties. Other measurements do not add more relevant information [14, 15]. It is clear that the puzzle of the EMFF in the TL region is far from being understood. The high intensity and high resolution beam of $\overline{\text{P}}$ANDA will enable the measurement of the differential cross section of the process $\bar{p}p \to e^+e^-$, which will grant not only the ratio between $|G_E|$ and $|G_M|$, but also the chance to separate them. Feasibility studies for the measurement of the TL EMFF with $\overline{\text{P}}$ANDA using the reactions $\bar{p}p \to e^+e^-$ and $\bar{p}p \to \mu^+\mu^-$ have been performed and are summarised in the following.

Concerning the studies for the reaction $\bar{p}p \to e^+e^-$ [16], two independent simulations were performed. Method-I makes use of a dedicated event generator based on the theoretical cross section [17],

$$\frac{d\sigma}{d\cos\theta} = \frac{\pi\alpha^2}{8EP}\left[|G_M|^2(1+\cos^2\theta) + \frac{|G_E|^2}{\tau}(1-cos^2\theta)\right], \tag{1}$$

with E and P are the energy and momentum of the antiproton in the center of mass (CM) frame, α is the fine structure constant, $\tau = \frac{q^2}{4M_p^2}$, and θ is the lepton emission polar angle in the proton-antiproton CM system. In these simulations, $|G_E|/|G_M| = 1$ is assumed, and the expected amount of events is generated at each energy. In Method-II, 10^6 events are simulated using a flat angular distribution and the results are later scaled to the expected size of the data sample using a given model (*i.e.* Eq. 1 with $|G_E| = |G_M|$). In both methods an independent simulated sample with 10^6 events is used for the study of the efficiency. In Method-I, a fit using the cross section formula in Eq. 1 to the reconstructed, efficiency corrected distribution, as a function of $\cos^2\theta$ is used to extract the values of the EMFF. In Method-II, a linear fit to

$$y = \sigma_0 + \sigma_0\mathcal{A}\cos^2\theta, \tag{2}$$

with

$$\sigma_0 = \frac{\pi\alpha^2}{2\beta s}\left(|G_M|^2 + \frac{1}{\tau}|G_E|^2\right) \text{ and } \mathcal{A} = \frac{\tau - R^2}{\tau + R^2}, \tag{3}$$

is used. The main challenge of this analysis is the elimination of the huge hadronic background, due to the 10^6 times larger cross section of the channel, $\bar{p}p \to \pi^+\pi^-$. For the studies of the suppression of the main background channel, an event generator based on the calculations in Ref. [18] was developed in Mainz [19]. Other channels (*e.g.* $\bar{p}p \to K^+K^-$) are easily suppressed using kinematic constraints.

Both simulations arrive to the same conclusion: $\overline{\text{P}}$ANDA will be able to measure the ratio between the electric and magnetic Form Factors (FF), $R = |G_E|/|G_M|$ in a broad range of q^2 values up to $q^2 = 14\,(\text{GeV/c})^2$, improving the precision of the existing results in at least one order of magnitude. In addition, the separation between the electric and magnetic FF will be possible. The measurement of the proton effective FF can be extended to higher q^2 values according to the experimental efficiency. The results of these simulations are shown in Tab. 1 and Tab. 2, and also in Fig. 3 together with the results from previous experiments.

The difference between the error bars extracted with each method at $q^2 = 13.9\,\text{GeV/c}^2$ is explained by the different size of the data samples considered in each simulation. Given an integrated luminosity of $2\,\text{fb}^{-1}$, Method-I takes into account the expected number of events for the simulation at each energy, considering thus the statistical fluctuations within the data sample, whereas Method-II smears this effect by simulating higher statistics and then rescaling to the same amount of events used in the

 DOI: 10.1051/epjconf/201716406002

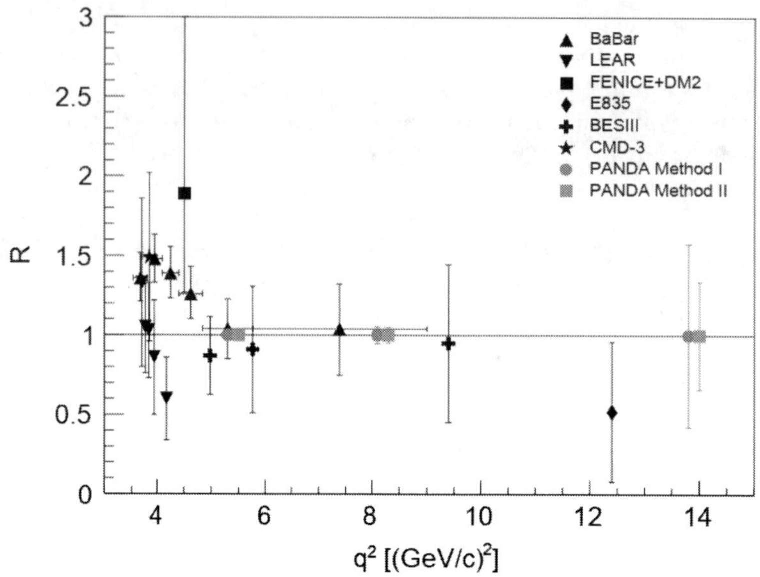

Figure 3. Expected statistical precision on the determination of the proton FF ratio from the simulations as reported in Ref. [16] (blue squares and red circles) for $R = 1$ as a function of q^2, compared with the existing data. Data are from references [11–15].

Table 1. Expected uncertainties for the measurement of the $|G_E|$, $|G_M|$ and the ratio R (Method-I). For more information, we refer to Ref. [16].

q^2 [(GeV/c)^2]	$\Delta\|G_E\|/\|G_E\|$ [%]	$\Delta\|G_M\|/\|G_M\|$ [%]	$\Delta R/R$ [%]
5.4	2.2	3.5	3.3
8.2	5.4	2.6	6.6
13.9	48	9.7	57

Table 2. Expected uncertainties for the measurement of the ratio R and the angular asymmetry on the proton FF (Method-II). For more information, we refer to Ref. [16].

q^2 [(GeV/c)^2]	R	ΔR	$\mathcal{A}$	$\Delta\mathcal{A}$
5.40	1	0.014	0.210	0.014
8.21	1	0.050	0.400	0.042
13.9	1	0.407	0.590	0.264

first simulation. The error bars of Method-II represent the most probable uncertainty if the measurement is done only once. That means, repeating the experiment multiple times and plotting the error distribution, we should get a Gaussian distribution with a sigma equal to the uncertainty extracted by Mehtod-II. This was confirmed by repeating the simulation using Method-I several times and fitting the distribution of errors from the single simulations. Taking this into account, in Method-II the un-

DOI: 10.1051/epjconf/201716406002

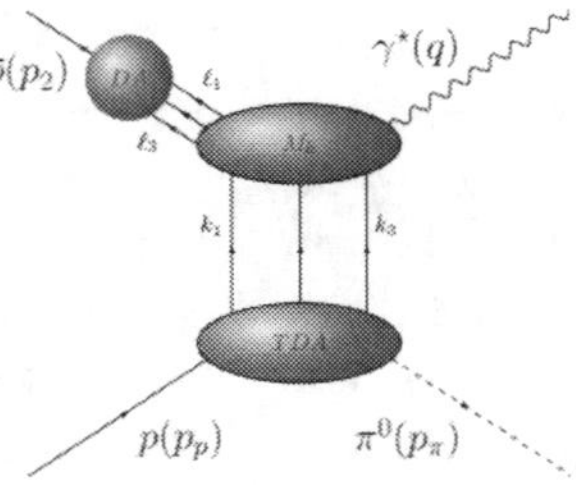

Figure 4. Feynman diagram of the process $\bar{p}p \to \gamma^\star\pi^0 \to e^+e^-\pi^0$. This process can be measured in $\overline{\text{P}}$ANDA to access the TDA.

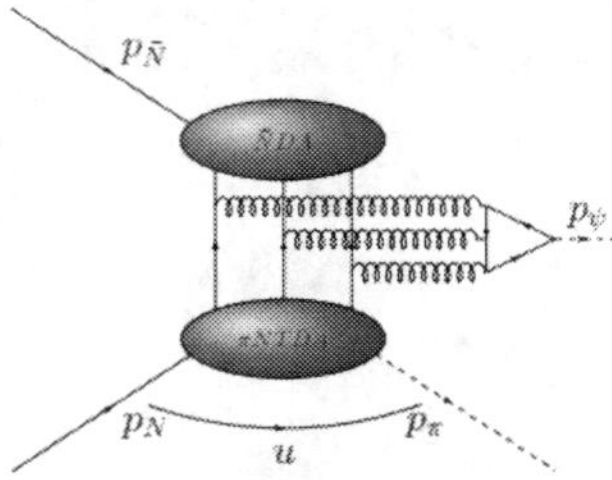

Figure 5. Feynman diagram of the process $\bar{p}p \to \pi^0 J/\psi$. This process can be measured in $\overline{\text{P}}$ANDA to access the TDA.

certainty expected in $\overline{\text{P}}$ANDA is extracted, while the results of Method-I show just one possible value for the error bars.

With $\overline{\text{P}}$ANDA it will be also possible to determine for the first time the EMFF through the reaction $\bar{p}p \to \mu^+\mu^-$ [20]. For this channel the cross section and the background to signal ratio considerations are the same as for the electron channel. The measurement with muons will be useful to verify the results obtained with the measurement of $\bar{p}p \to e^+e^-$ and to study the systematic effect in the electron channel due to higher order radiative corrections from the final state. The suppression of the main background channel, $\bar{p}p \to \pi^+\pi^-$, is more difficult in this case due to the similar masses of pions and muons. Simulations of the signal and background channels have been performed with the same dedicated event generators described for the simulations discussed earlier. Following a similar strategy, by applying a set of cuts on kinematical variables and detector observables combined with a vertex fitting, the desired background suppression for this measurement cannot be reached. A method based on a multivariate data classification delivers a huge improvement with respect to the simple analysis based on hard cuts. A signal to background ratio of 1:4 is achieved, in comparison to the previously reached 1:30. At this level of contamination, a background subtraction based on the previously measured background channel can be applied. An independently generated background sample was used to study how the background substraction has to be performed. For a beam momentum of $p_{\bar{p}} = 1.7$ GeV/c, a statistical precision of $\Delta|G_E|/|G_E| \sim 8.6\,\%$, $\Delta|G_M|/|G_M| \sim 4.1\,\%$ and $\Delta R/R \sim 5.1\,\%$ can be achieved.

2.2 Transition Distribution Amplitudes

TDA are parameterizations of the hadronic matrix elements that occur in perturbative QCD calculations of a certain family of reactions within the framework of collinear factorization. The cross section of this processes can be divided into a perturbative part, calculable thanks to the high momentum transferred in the process, and a non-perturbative, but universal and measurable part, which is described by TDA. In particular, with $\overline{\text{P}}$ANDA, the study of π-NTDA will be possible via the proposed mechanism reported in Ref. [21], by measuring the cross section of $\bar{p}p \to \gamma^\star\pi^0 \to e^+e^-\pi^0$ with the emission of the π^0 in the very forward or backward limit (see Fig. 4). This constraint is necessary in order to ensure the hard scale in which the perturbative expansion of the hard subprocess is applicable. Feasibility studies to access the π-NTDA following the cross section calculations in Ref. [21] have been described in Ref. [22]. The study of the channel $\bar{p}p \to \pi^0 J/\psi$, shown in Fig. 5, is a natural complement to the first analysis due to its higher cross section and because it can also be used to access the TDA. Therefore, it has been analysed in Ref. [23]. The measurement of these channels with $\overline{\text{P}}$ANDA will represent a test of QCD factorization, and the validity and universality of the TDA approach. For the first channel ($\pi^0\gamma^\star$), simulations were performed at two values of the CM energy squared, s. The cross section can be determined with an accuracy $\Delta\sigma/\sigma$ between 12 % to 24 %. The studies for the second channel $\bar{p}p \to \pi^0 J/\psi$, have been performed at three different values of s. The results show a background contamination lower than 1 %, coming from the channels $\bar{p}p \to \pi^0\pi^+\pi^-$, $\bar{p}p \to \pi^0\pi^+\pi^+\pi^-\pi^-$, $\bar{p}p \to \pi^0\pi^0\pi^+\pi^-$, $\bar{p}p \to \pi^0\pi^0 J/\psi$, and $\bar{p}p \to \pi^0\gamma^\star \to e^+e^-$, with a signal efficiency reconstruction of about 5 % to 10 %.

2.3 Other structure functions

The physics program of $\overline{\text{P}}$ANDA also comprises the measurement of other structure functions as *e.g.* the Generalized Distribution Amplitudes (GDA) and the Transverse Momentum Dependent Parton Distribution Functions (TMD-PDF). The GDA are the TL counterparts of the Generalized Parton Distributions (GPD) which are measured in the SL region via Wide Angle Compton Scattering (WACS) in reactions like $\gamma p \to \gamma p$. With $\overline{\text{P}}$ANDA, the reactions $\bar{p}p \to \gamma\gamma$ and $\bar{p}p \to \gamma\pi^0$ can be used for probing the GDA. Preliminary studies [1] show that it will be possible to achieve a signal to background ratio between 1 and 2 and efficiencies ranging between 25 % and 50 %.

TMD-PDF are experimentally very well known functions in the SL region. $\overline{\text{P}}$ANDA will shed light on of these functions in the TL region. In particular they can be measured via Drell-Yan (DY) processes. Preliminary studies [1, 24] show a production of DY processes of the order of $1.3 \cdot 10^5$ DY/month with a reconstruction efficiency of ca. 33 %. More studies on the measurement of these functions are nevertheless needed to determine the real potential of $\overline{\text{P}}$ANDA in this area.

2.4 Summary

A series of feasibility studies for the measurement of different structure functions with the $\overline{\text{P}}$ANDA detector have been presented. The measurement of the EMFF in the TL region using the reaction $\bar{p}p \to e^+e^-$ will represent the most precise measurement so far done of these observables. The precision will be improved in at least one order of magnitude with respect to *e.g.* BaBar. In addition $|G_E|$ and $|G_M|$, will be measured separately. $\overline{\text{P}}$ANDA will also perform pioneering measurement of the EMFF using the reaction $\bar{p}p \to \mu^+\mu^-$, for the verification of the measurements made using electrons and to study the asymmetries introduced by radiative corrections.

The measurement of the TDA has been studied for two different reactions $\bar{p}p \to \gamma^\star\pi^0$ and $\bar{p}p \to J/\psi\pi^0$, and the results are very promising.

Other nucleon structure observables, such as GDA and TMD-PDF, are also considered to be measured with $\overline{\text{P}}$ANDA and the preliminary results are encouraging.

Among these measurements, $\overline{\text{P}}$ANDA will produce results on charmonium spectroscopy, hypernuclear physics, electroweak interactions, and it will be able to study hadrons in a nuclear medium.

DOI: 10.1051/epjconf/201716406002

In conclusion, $\overline{\mathsf{P}}$ANDA will be an excellent experiment for the expansion of the knowledge on nuclear physics and QCD.

3 Acknowledgments

This work was supported by the Bundesministerium für Bildung und Forschung (BMBF) through grant 05P12UMFP9.

References

[1] W. Erni et al. (PANDA) (2009), arXiv:0903.3905
[2] W. Erni et al. (PANDA) (2008), FAIR TDR 4-01, arXiv:0810.1216
[3] W. Erni et al. (PANDA) (2009), FAIR TDR 4-02, arXiv:0907.0169
[4] W. Erni et al. (PANDA) (2012), FAIR TDR 4-03, arXiv:1207.6581
[5] W. Erni et al. (PANDA), Eur. Phys. J. **A49**, 25 (2013), FAIR TDR 4-04
[6] W. Erni et al. (PANDA) (2013), FAIR TDR 4-05
[7] W. Erni et al. (PANDA) (2014), FAIR TDR 4-06
[8] S. Pacetti, R. Baldini Ferroli, E. Tomasi-Gustafsson, Phys. Rept. **550-551**, 1 (2015)
[9] E. Tomasi-Gustafsson, M.P. Rekalo (2008), arXiv:0810.4245
[10] E. Tomasi-Gustafsson et al., Eur. Phys. J. **A24**, 419 (2005)
[11] J.P. Lees et al. (BaBar), Phys. Rev. **D88**, 072009 (2013)
[12] G. Bardin et al., Nucl. Phys. **B411**, 3 (1994)
[13] M. Ablikim et al. (BESIII), Phys. Rev. **D91**, 112004 (2015)
[14] R.R. Akhmetshin et al. (CMD-3), Phys. Lett. **B759**, 634 (2016)
[15] R. Baldini et al., Eur. Phys. J. **C46**, 421 (2006)
[16] B. Singh et al. (PANDA), Eur. Phys. J. **A52**, 325 (2016)
[17] A. Zichichi et al., Nuovo Cim. **24**, 170 (1962)
[18] J. Van de Wiele, S. Ong, Eur. Phys. J. **A46**, 291 (2010)
[19] M. Zambrana et al. (2012), PANDA Internal Note
[20] I. Zimmermann, A. Dbeyssi, D. Khaneft (PANDA), AIP Conf. Proc. **1735**, 080004 (2016)
[21] J.P. Lansberg, B. Pire, L. Szymanowski, Phys. Rev. **D76**, 111502 (2007)
[22] B.P. Singh et al. (PANDA), Eur. Phys. J. **A51**, 107 (2015)
[23] B. Singh et al. (PANDA) (2016), arXiv:1610.02149
[24] M. Destefanis (PANDA), EPJ Web Conf. **73**, 02012 (2014)

Open Heavy Flavor and Quarkonia Results at RHIC

Rachid Nouicer[1,2,a]

[1] *Physics Department, Brookhaven National Laboratory, Upton, New York 11973, United States*
[2] *RIKEN-BNL Research Center, Brookhaven National Laboratory, Upton, New York 11973, United States*

Abstract. RHIC experiments carry out a comprehensive physics program which studies open heavy flavor and quarkonium production in relativistic heavy-ion collisions. The discovery at RHIC of large high-p_T suppression and flow of electrons from heavy quarks flavors have altered our view of the hot and dense matter formed in central Au + Au collisions at $\sqrt{s_{NN}}$ = 200 GeV. These results suggest a large energy loss and flow of heavy quarks in the hot, dense matter. In recent years, the RHIC experiments upgraded the detectors; (1) PHENIX Collaboration installed silicon vertex tracker (VTX) at mid-rapidity region and forward silicon vertex tracker (FVTX) at the forward rapidity region, and (2) STAR Collaboration installed the heavy flavor tracker (HFT) and the muon telescope detector (MTD) both at the mid-rapidity region. With these new upgrades, both experiments have collected large data samples. These new detectors enhance the capability of heavy flavor measurements via precision tracking. The PHENIX experiments established measurements of $\psi(1S)$ and $\psi(2S)$ production as a function of system size, $p + p$, p + Al, p + Au, and ^{3}He + Au collisions at $\sqrt{s_{NN}}$ = 200 GeV. In p/^{3}He + A collisions at forward rapidity, we observe no difference in the $\psi(2S)/\psi(1S)$ ratio relative to $p + p$ collisions. At backward rapidity, where the comoving particle density is higher, we find that the $\psi(2S)$ is preferentially suppressed by a factor of two. STAR Collaboration presents the first J/ψ and Υ measurements in the di-muon decay channel in Au + Au collisions at $\sqrt{s_{NN}}$ = 200 GeV at mid-rapidity at RHIC. We observe clear J/ψ R_{AA} suppression and qualitatively well described by transport models simultaneously accounting for dissociation and regeneration processes.

1 Physics Motivation

At the heart of each atom, there is a nucleus consisting of distinct nucleons (protons and neutrons). In turn, these nucleons consist of quarks bound together by the strong interaction, mediated by the exchange of gluons. Quantum Chromodynamics (QCD) is considered the fundamental theory for such strong interactions. According to QCD, at ordinary temperatures or densities this force simply confines the quarks into composite hadrons. However, when the temperature reaches the QCD energy scale (T of the order of 10^{12} Kelvin) or its density rises to the point where the average inter-quark separation is less than 1 fm (quark chemical potential μ_q around 400 MeV), hadronic matter under extremely dense and hot conditions undergoes a phase transition to form a Quark Gluon Plasma (QGP) in which quarks and gluons no longer are confined to the size of a hadron [1, 2]. Knowing the exact

[a] e-mail: rachid.nouicer@bnl.gov

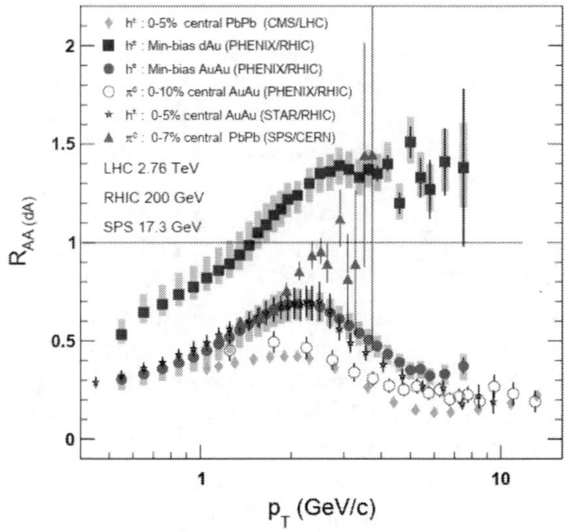

Figure 1. The evolution of the nuclear modification factor with center-of-mass energy, from the SPS to RHIC and then to the LHC. The error bars correspond to the statistical errors. For clarity, the systematic errors are shown as vertical bands. For more details see Ref. [4].

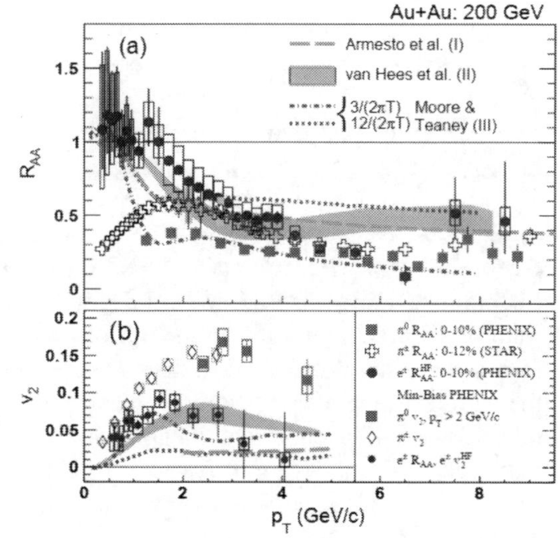

Figure 2. The nuclear modification factor, R_{AA}^{HF}, for HF electrons compared with the R_{AA} of π^0 in central Au+Au collisions at $\sqrt{s_{NN}} = 200\,\text{GeV}$, see panel (a). Panel (b) considers the anisotropic flow of HF electrons v_2^{HF} with that of v_2 of π^0 and $\pi^\pm$ in minimum-bias Au+Au collisions. For more details see Ref. [4]

nature of quark confinement in hadrons is crucial, and yet this remains a poorly understood aspect of the quark-gluon description of matter. Collision experiments of heavy nuclei at high energies provide the unique opportunity to create and study the properties of the QGP in the laboratory.

Hadrons carrying heavy quarks, i.e. charm (c) or bottom (b), are consider important probes of the hot and dense medium, QGP, created in relativistic heavy-ion collisions. Heavy quark-antiquark pairs are mainly produced in initial hard scattering processes of partons. While some of the produced pairs form bound quarkonia, the vast majority hadronize into open heavy flavor particles. They interact through the medium and are expected to be sensitive to its energy density through the mechanism of parton energy loss. Due to the large mass of heavy quarks the suppression of small-angle gluon radiation should reduce their energy loss, and consequently any suppression of heavy-quark mesons like D and B mesons at high-p_T is expected to be smaller than that observed for hadrons consisting of light quarks [3]. We quantify the medium effects on high-p_T production in nucleus-nucleus collisions, A + A, with the nuclear modification factor which is defined as following:

$$R_{AA}(p_T) = \frac{1}{\langle N_{coll} \rangle} \times \frac{\text{yield per A + A collision}}{\text{yield per } p + p \text{ collision}} = \frac{1}{\langle N_{coll} \rangle} \times \frac{d^2N^{A+A}/dydp_T}{d^2N^{p+p}/dydp_T} \quad (1)$$

This factor reflects the deviation of measured distributions of nucleus-nucleus, A + A, transverse momentum, at given impact parameter b, from measured distributions of an incoherent superposition of nucleon-nucleon ($p + p$) transverse momentum, scaled by the average number of expected binary collisions' $\langle N_{coll} \rangle$. This normalization often is known as "binary collisions scaling".

The rise of the nuclear modification factor with center-of-mass energy, from the SPS to RHIC and then to the LHC is presented in figure 1. We observe that in the presented p_T region, charged hadron production at the LHC is found to be about 50% more suppressed than at RHIC, and has a similar suppression value as for neutral pions (π^0) measured by PHENIX. These measurements of $R_{AA}(p_T)$ at RHIC and LHC is consistent with a large energy loss in the medium causing it to become opaque to the propagation of high momentum partons [4, 5]. In contrast, $R_{AA}(p_T)$ for π^0 at SPS energy exhibit enhancement, as shown in figure 1. This effect of $R_{AA}(p_T)$ suppression of the high-p_T hadron spectra at the mid-rapidity region in central A + A collisions, proposed by Bjorken, Gyulassy, and others [6, 7] rests on the expectation of a large energy loss of high momentum partons scattered in the initial stages of collisions in a medium with a high density of free color charges. According to the QCD theory, colored objects will lose energy via bremsstrahlung radiation of gluons. Such a mechanism strongly would degrade the energy of leading partons, as reflected in the reduced transverse momentum of leading particles in the jets emerging after their fragmentation into hadrons. This energy loss is expected to depend strongly on the color charge density of the created system, and the path length traversed by the propagating parton. It has been recognized that the elastic energy loss is too small to engender significant attenuation of the jet in the created medium [8]. Adding to this discovery of suppression of particles at high-p_T, two very striking results were seen for open heavy flavor from the PHENIX experiment via the measurement of electrons from the semi-leptonic decays of hadrons carrying charm or bottom quarks. First, heavy mesons, despite their large mass, exhibit a suppression at high transverse momentum compared to that expected from $p + p$ collisions [4, 9–11]. This suppression is similar to that of light mesons, which implies a substantial energy loss of fast heavy quarks while traversing the medium, see figure 2(a). Second, an elliptic flow is observed for heavy mesons, electrons from semi-leptonic decays of hadrons carrying charm or bottom quarks, that is similar to that of light mesons like pions, as shown in figure 2(b).

2 PHENIX Recent Heavy Flavor Results

Heavy quark production has been studied by the PHENIX experiment via measurements of single leptons from semi-leptonic decays in both the electron channel at mid-rapidity and in the muon channel at forward-rapidity. Recently, the PHENIX Collaboration has installed a silicon vertex tracker both in central rapidity (VTX) and in forward rapidity (FVTX) regions, and has collected large data samples [12, 13]. These two silicon trackers enhance the capability of heavy flavor measurements via precision tracking [14].

2.1 Open Heavy Flavor - Single Electron/Muon Measurements

The PHENIX experiment quantifies open heavy flavor production through the measurement of inclusive electrons or muons [15–18]. These analyses use a cocktail method to remove fake and real electrons/muons from the data sample. i.e. for the electron measurement, the electrons that come from either meson decay or photon conversions are measured and subtracted from the inclusive spectrum and the remainder is attributed to electrons ($e^{\pm}_{HF}$) coming from the semi-leptonic decay of D and B mesons. The effects of CNM are expected to be present in the initial state of A + A collisions; however, this CNM enhancement is convolved with the suppressing effects of hot nuclear matter. Figure 3(a) shows $R_{dAu}(p_T)$ and $R_{AA}(p_T)$ for $e^{\pm}_{HF}$ and the neutral pion (π^0), for which only small CNM effects are observed between $e^{\pm}_{HF}$ and π^0 for $p_T < 5$ GeV/c. Above $p_T \approx 5$ GeV/c, where the CNM effects on both species, $e^{\pm}_{HF}$ and π^0, are small, their $R_{AA}(p_T)$ values are consistent within uncertainties. However, in the range where CNM enhancement is large for $e^{\pm}_{HF}$ and small on π^0, the corresponding $e^{\pm}_{HF}$ $R_{AA}(p_T)$ values are consistently above the π^0 values. This could suggest that the difference in the initial state cold nuclear matter effects due to the mass-dependent Cronin enhancement is reflected in the final

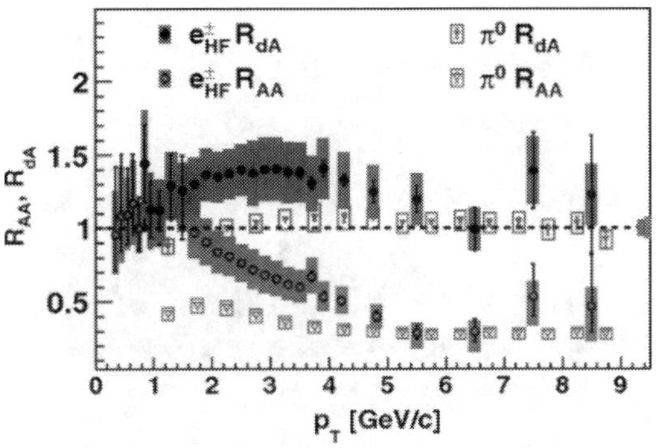

(a) $e^{\pm}_{HF}$ and π^0: Au + Au vs. d + Au at $\sqrt{s_{NN}}$ = 200 GeV

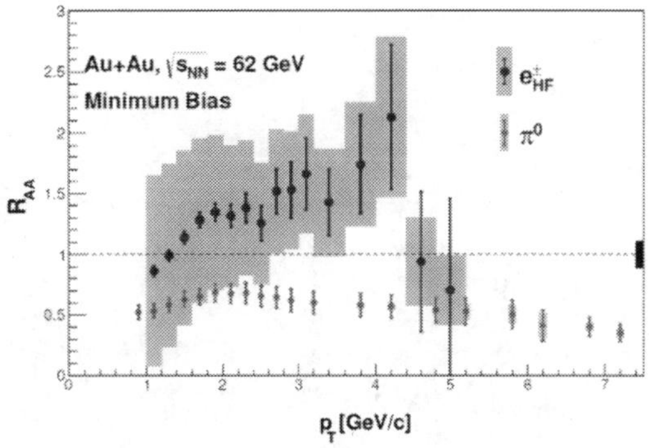

(b) $e^{\pm}_{HF}$ and π^0: Cu + Cu at $\sqrt{s_{NN}}$ = 62.4 GeV

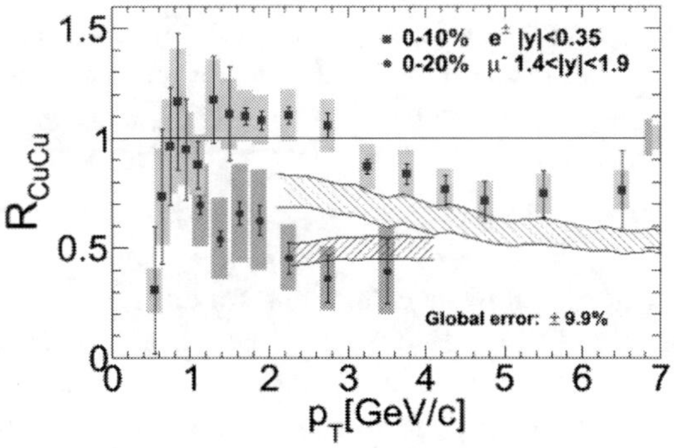

(c) $e^{\pm}_{HF}$ and μ^{-}_{HF}: Cu + Cu at $\sqrt{s_{NN}}$ = 62.4 GeV

Figure 3. The nuclear modification factors $R_{dAu}(p_T)$ and $R_{AA}(p_T)$ for the π^0, $e^{\pm}_{HF}$ and μ^{-}_{HF}: panels (a) and (b) for minimum-bias d + Au, Au + Au and panel (c) for central collisions Cu + Cu [15, 16].

state spectra of these particles in Au + Au collisions, although alternate explanations involving mass-dependent partonic energy loss in the hot medium are not ruled out. To study the interplay between initial-state effects and final-states effects for heavy flavor ($e^{\pm}_{HF}$) productions, PHENIX measured the $R_{AA}(p_T)$ in Au + Au collisions at low beam energy, $\sqrt{s_{NN}}$ = 62.4 GeV. These $R_{AA}(p_T)$ values for $e^{\pm}_{HF}$ in Au + Au collisions at 62.4 GeV are compared to π^0 $R_{AA}(p_T)$ at the same energy as shown in figure 3(b). At 62.4 GeV, the competition, if present, favors heavy flavor enhancement over suppression. This is consistent with previous results with hadrons where the Cronin enhancement increases as the collision energy decreases [19]. PHENIX measurements of heavy flavor muons (μ^{-}_{HF}) at forward rapidity (1.4 < y < 1.9) in central Cu+Cu collisions at $\sqrt{s_{NN}}$ = 62.4 GeV show a significant suppression [19]. The magnitude of this suppression at forward-rapidity in Cu + Cu (shown in figure 3(c)) is compared to the suppression of $e^{\pm}_{HF}$ in central Au+Au collisions at mid-rapidity at the same energy. As open heavy flavor is significantly more suppressed at forward-rapidity than at mid-rapidity in Cu + Cu, additional nuclear effects, such as gluon shadowing at low-Björken-x or partonic energy loss in the nucleus, may be significant. The heavy flavor $e^{\pm}_{HF}$ and μ^{-}_{HF} are compared in figure 3(c) to a theoretical prediction that combines the effects of partonic energy loss, energy loss from fragmentation and dissociation, and includes nuclear matter effects such as shadowing and Cronin enhancement due to parton scattering in the nucleus [20]. While consistent within uncertainties, the model predicts more suppression for heavy flavor electrons than seen in the data.

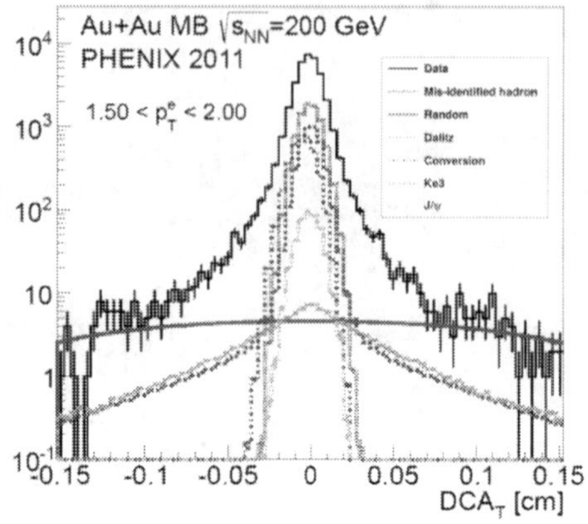

Figure 4. The DCA$_T$ distribution for electrons with $1.5 < p_T^e < 2.0$ GeV/c in Au + Au minimum-bias collisions at $\sqrt{s_{NN}} = 200$ GeV. Also shown are the normalized contributions for the various background components detailed in Ref. [22].

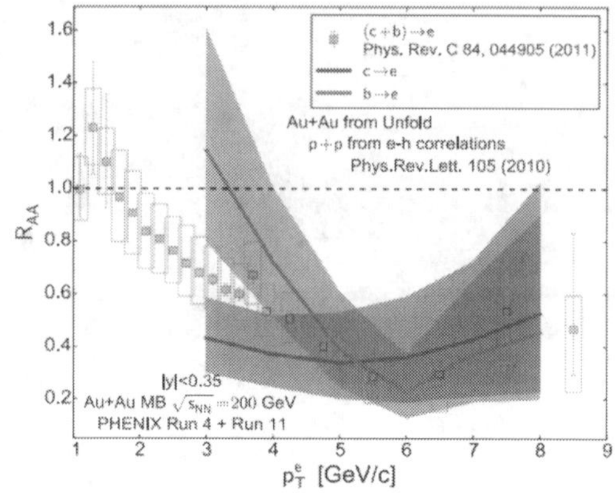

Figure 5. The nuclear modification factors R_{AA} of electrons from charm ($c{\to}e$), bottom ($b{\to}e$) decays and combined heavy flavor as a function of p_T^e in Au + Au minimum-bias collisions at $\sqrt{s_{NN}} = 200$ GeV. For more details see Ref. [22].

The large suppression R_{AA} of $e^{\pm}_{HF}$ at high-p_T in Au + Au collisions at $\sqrt{s_{NN}} = 200$ GeV, shown in figures 2 and 3, and similar to that of light mesons π^0, is the more striking results because perturbative QCD calculations indicate a substantial contribution from bottom quark decays for $p_T > 5$ GeV/c [21]. For the specific purpose of separating the contributions of charm and bottom quarks at mid-rapidity, the PHENIX collaboration has added micro-vertexing capabilities in the form of a silicon vertex tracker (VTX) [12, 13]. The different lifetimes and kinematics for charm and bottom hadrons decaying to electrons enable separation of their contributions with measurements of displaced tracks [14, 22]. Figure 4 shows the DCA$_T$ distribution (in transverse plane) for electrons ($e^{\pm}$) in $1.5 < p_T^e < 2.0$ GeV/c range obtained in Au + Au minimum-bias collisions at $\sqrt{s_{NN}} = 200$ GeV using data set of RHIC Run 2011. In order to extract the heavy flavor contributions, all background components, as shown in figure 4, were fully accounted for and their DCA$_T$ shapes as a function of p_T incorporated. The nuclear modifications, $R_{AA}^{c\to e}$, and $R_{AA}^{b\to e}$, for charm and bottom hadron decays respectively are presented in figure 5 and detailed in Ref [22].

2.2 $\psi(1S)$ and $\psi(2S)$ Mesons Production Versus System Size

Figure 6 presents the measured di-muon mass spectra from $p + p$, p + Al, p + Au, and ^{3}He + Au collisions at $\sqrt{s_{NN}} = 200$ GeV. These distributions comprise peaks at the $\psi(1S)$ and $\psi(2S)$ masses on top of correlated background from charm and bottom hadron decays and Drell-Yan pairs, plus combinatorial background from light meson decays and hadrons, which are not stopped in the absorbers [24]. From figure 6 it is apparent that the $\psi(2S)$ peaks are suppressed relative to the $\psi(1S)$ peak in the columns on the left (in the A-going direction). Quantitative comparisons are accomplished by calculating the double ratio of $\psi(2S)/\psi(1S)$ production in p/He+A collisions to the ratio found in $p + p$ collisions, as shown in figure 7. At forward rapidity, the double ratio is consistent with unity in all three collision systems, indicating that any possible nuclear effects on the two charmonium states are comparable. Because these states are not fully formed until after they exit the nucleus, the fact that any nuclear effects have an equal magnitude on both states suggests that there are no significant final

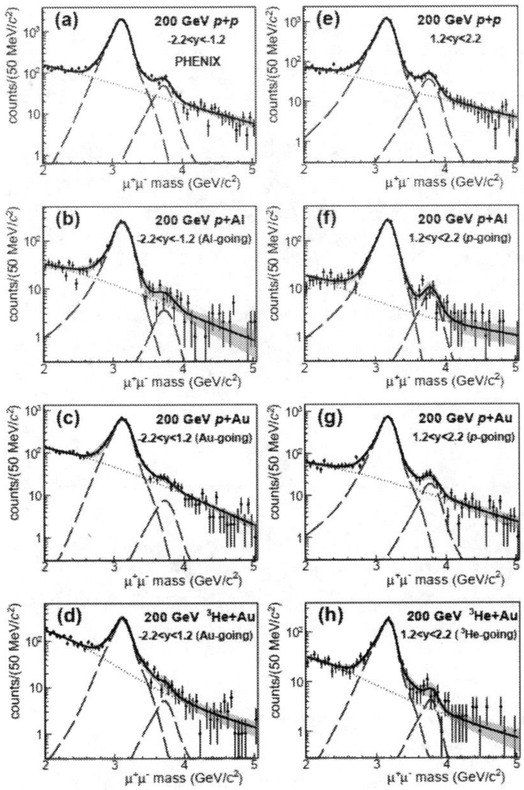

Figure 6. The measured di-muon mass spectra with fits from the (a–d) South and (e–h) North PHENIX muon arms, for $p + p$, p + Al, p + Au, and ^{3}He + Au collisions at $\sqrt{s_{NN}}$ = 200 GeV. The total fit is the solid [black] line with a shaded [gray] band representing the 90% confidence level. The dashed [blue] and dotted [red] lines represent the contributions from the resonances and background, respectively. For more details see Ref. [24].

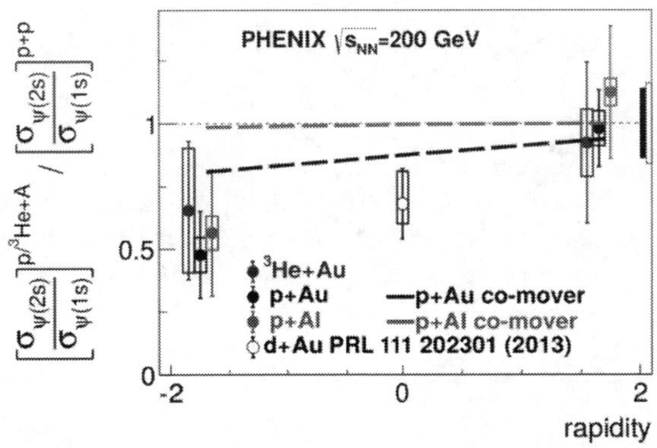

Figure 7. The double ratio of $\psi(2S)/\psi(1S)$ mesons measured in in p + Al, p + Au, and ^{3}He + Au collisions to that same ratio in $p + p$ collisions at $\sqrt{s_{NN}}$ = 200 GeV, with a calculation based on breakup by comoving particles. The bars (boxes) on the data points represent the statistical (systematic) uncertainties, and the shaded (open) box around unity represents the global uncertainty on the forward/backward (mid) rapidity data. For more details see Ref. [24].

state effects on the pair which occur in this rapidity region. At backward rapidity, the ratios in all collision systems are suppressed by a factor of two. The mechanism for this preferential suppression of the $\psi(2S)$ relative to the $\psi(1S)$ is expected to occur after the $\psi(1S)$ formation time. A significant difference in the late stages of the collision between this region and forward rapidity suggests the presence of a larger number of comoving hadrons [24].

3 STAR Recent Heavy Flavor Results

The study of heavy flavor in relativistic nuclear collisions follows two different approaches: (1) The direct reconstruction of the heavy flavor meson, and (2) the identification of electrons and muons from semi-leptonic decays of such mesons. However, the rare production of heavy quarks and large combinatorial background in heavy-ion collisions makes the measurements of heavy flavor particles

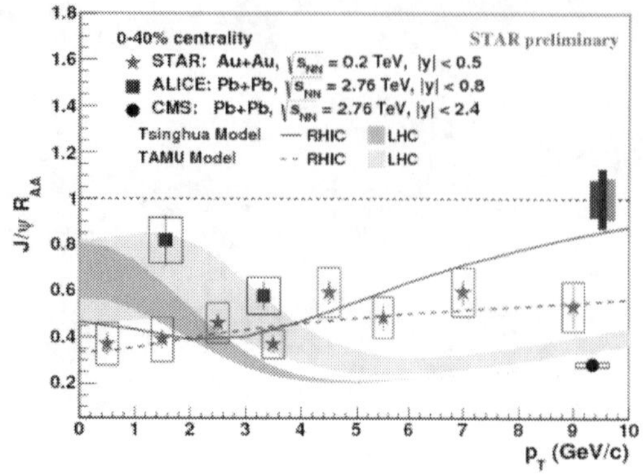

Figure 8. The nuclear modification factors R_{AA} as a function of p_{T} at RHIC (red star) and at LHC (blue square, black circle). The lines and bands indicate transport model calculations for RHIC and LHC energies [25].

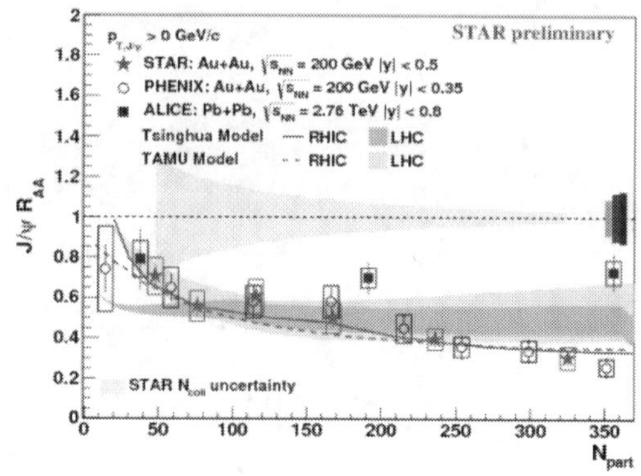

Figure 9. The nuclear modification factors R_{AA} for integrated p_{T} as a function of $\mathrm{N_{part}}$ in Au + Au and Pb + Pb collisions at RHIC and LHC energies respectively [25].

very difficult with the STAR detectors. To overcome the challenge of the combinatorial background, the STAR Collaboration upgraded the detector by adding the heavy flavor tracker (HFT) and the muon telescope detector (MTD) in 2014.

3.1 J/ψ and Υ Measurements in Au + Au collisions at Mid-rapidity

Figure 8 presents the nuclear modification factor of J/ψ R_{AA} measurements in the di-muon decay channel in 0-40% central Au + Au collisions at $\sqrt{s_{NN}}$ = 200 GeV at mid-rapidity region, compared with LHC results [25–27]. We observe a strong suppression at RHIC at low p_{T} indicating that dissociation plays a significant role in this p_{T} range. The increasing trend of R_{AA} at RHIC at high-p_{T} region could be interpreted by formation-time effects and feed-down of B hadrons. The least suppression of J/ψ at LHC at low-p_{T} indicates larger regeneration contribution due to higher charm cross section, while more suppression of J/ψ at LHC at high-p_{T} indicates larger dissociation rate due to higher temperature of the medium. Transport Models [28–31], including dissociation and regeneration effects, describe qualitatively the R_{AA} of J/ψ as a function of p_{T} at RHIC and LHC given energies [25]. Figure 9 presents the J/ψ R_{AA} centrality dependence for integrated p_{T}. We observe that both models reproduce centrality dependence at RHIC, but aim to overestimate suppression (or underestimate R_{AA} of J/ψ) at LHC. The di-muon mass spectrum in Υ state mass range in Au + Au collisions at $\sqrt{s_{NN}}$ = 200 GeV is shown in figure 10. We notice signs of an indication of $\Upsilon\,(2S + 3S)$ signals in the di-muon decay channel at mid-rapidity region. Using simultaneous fit to the like-sign and unlike-sign distributions, the raw yields of Υ states are obtained [25]. The comparison of the fitted $\Upsilon\,(2S + 3S)/\Upsilon\,(1S)$ ratio with the world-wide $p + p$ data [32] and CMS data [33, 34] is shown in figure 11.

4 Conclusions

Understanding the properties of strongly interacting matter at high temperatures has been a central goal of the STAR and PHENIX experiments at RHIC. The large suppression R_{AA} of electrons from

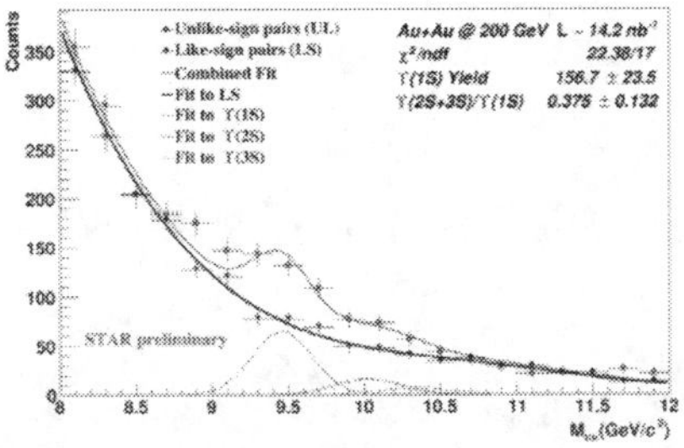

Figure 10. The di-muon mass spectrum in the Υ state mass range from Au + Au collisions at $\sqrt{s_{NN}} = 200$ GeV in mid-rapidity region [25].

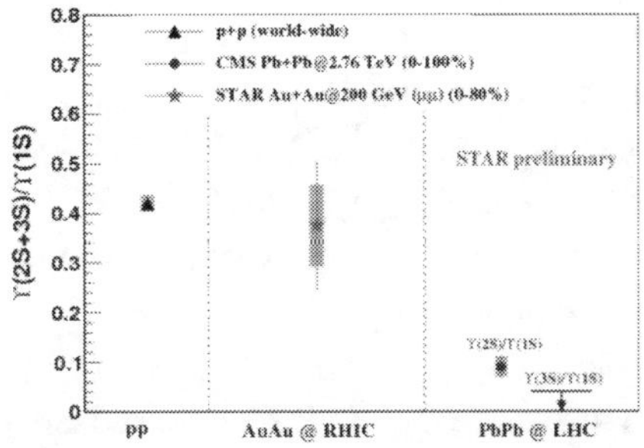

Figure 11. Υ $(2S + 3S)/\Upsilon$ $(1S)$ ratio for world-wide $p + p$ data, and for heavy-ion collisions at RHIC, and LHC energies [25].

the semi-leptonic decays of hadrons carrying charm or bottom quarks at high-p_T in Au + Au collisions at $\sqrt{s_{NN}} = 200$ GeV, and similar to that of light mesons π^0, is the more striking results because perturbative QCD calculations indicate a substantial contribution from bottom quark decays for $p_T > 5$ GeV/c. The PHENIX experiments established measurements of $\psi(1S)$ and $\psi(2S)$ production as a function of system size, $p + p$, p + Al, p + Au, and ^{3}He + Au collisions at $\sqrt{s_{NN}} = 200$ GeV. In p/^{3}He + A collisions at forward rapidity, we observe no difference in the $\psi(2S)/\psi(1S)$ ratio relative to $p + p$ collisions. At backward rapidity, where the comoving particle density is higher, we find that the $\psi(2S)$ is preferentially suppressed by a factor of two. STAR Collaboration presents the first J/ψ and Υ measurements in Au + Au collisions at $\sqrt{s_{NN}} = 200$ GeV in the di-muon decay channel at mid-rapidity region. We observe a clear J/ψ R_{AA} suppression and qualitatively well described by transport models, including dissociation and regeneration simultaneously.

References

[1] E. Shuryak, Phys. Rep. **61**, 71 (1980)
[2] L. McLerran, Rev. Mod. Phys. **58**, 1021 (1986)
[3] Y. L. Dokshitzer and D. E. Kharzeev, Phys. Lett. B **519**, 199 (2001)
[4] R. Nouicer, European Physical Journal Plus **131**, 70 (2016)
[5] S. M. Pari *et al.*, International Journal of Modern Physics E, **25**, No. 6, 1650040 (2016)
[6] J. D. Bjorken, Phys. Rev. D **27**, 140 (1983)
[7] M. Gyulassy *et al.* Phys. Lett. B **243**, 432 (1990)
[8] M. H. Thoma *et al.*, Nucl. Phys. B **351**, 491 (1991)
[9] S.S. Adler *et al.* (PHENIX Collaboration), Phys. Rev. Lett. **98**, 172302 (2007)
[10] S.S. Alder *et al.* (PHENIX Collaboration), Phys. Rev. Lett. **96**, 032301 (2006)
[11] J. Adams *et al.* (STAR Collaboration), Phys. Rev. Lett. **94**, 062301 (2005)
[12] R. Nouicer *et al.*, Journal of Instrumentation (JINST), **4**, P04011 (2009)
[13] R. Nouicer *et al.*, Nuclear Instruments and Methods in Physics Research B **261**, 10671071 (2007)
[14] R. Nouicer *et al.* (PHENIX Collaboration), Nuclear Physics A **904-905C**, 647 (2013)
[15] R. Nouicer *et al.* (PHENIX Collaboration), Nuclear Physics A **862**, 64 (2011)
[16] R. Nouicer *et al.* (PHENIX Collaboration), Journal of Physics : Conf. Series **420**, 012021 (2013)

[17] A. Adare *et al.* (PHENIX Collaboration), Phys. Rev. Lett. **109**, 242301 (2012)
[18] A. Adare *et al.* (PHENIX Collaboration), Phys. Rev. Lett. **112**, 252301 (2014)
[19] A. Adare *et al.* (PHENIX Collaboration), Phys. Rev. C **86**, 024909 (2012)
[20] R. Sharma *et al.*, Phys. Rev. C **80**, 054902 (2009)
[21] M. Cacciari *et al.*, Phys. Rev. Lett. **95**, 122001 (2005)
[22] A. Adare *et al.* (PHENIX Collaboration), Phys. Rev. Lett. **93**, 034904 (2016)
[23] H. Satz, Journal of Physics G **32**, R25 (2006)
[24] A. Adare *et al.* (PHENIX Collaboration), arXiv:1609.06550 [nucl-ex]
[25] T. Todoroki *et al.* (STAR Collaboration), arXiv:1612.02499
[26] B. Abelev *et al.* (ALICE Collaboration), Phys. Let. B **734**, 314 (2014)
[27] S. Chatrchyan *et al.* (CMS Collaboration), JHEP **05**, 063 (2012)
[28] Y. Liu *et al.*, Phys. Let. B **678**, 72 (2009)
[29] K. Zhou *et al.*, Phys. Rev. C**89**, 054911 (2014)
[30] X. Zhao *et al.*, Phys. Rev. C **82**, 064905 (2010)
[31] X. Zhao *et al.*, Nucl. Phys. A **859**, 114 (2011)
[32] W. Zha *et al.*, Phys. Rev. C **88**, 067901 (2013)
[33] S. Chatrchyan *et al.* (CMS Collaboration), Phys. Rev. Lett. **109**, 222301 (2012)
[34] S. Chatrchyan *et al.* (CMS Collaboration), JHEP **04**, 103 (2014)

 DOI: 10.1051/epjconf/201716406004

Strangeness from SPS to FAIR

Searching for the onset of deconfinement

Volker Friese[1,a]

[1] *GSI Helmholtzzentrum für Schwerionenforschung mbH, Planckstraße 1, 64291 Darmstadt, Germany*

Abstract. Since the early days of heavy-ion physics, strangeness has been considered a sensitive probe of the state of matter created in nuclear collisions. This assessment still holds today, where we are witnessing renewed interest in collisions at moderate energies, manifested in the running or projected experimental programmes at RHIC, SPS, FAIR, and NICA. In this article, we will review the current understanding of strangeness production at lower energies and discuss how far future measurement of strange particles can contribute to understanding the properties of dense QCD matter and to the search for the onset of deconfinement.

1 Introduction

Nuclear matter, when sufficiently heated or compressed, is expected to undergo a transition to a deconfined state, where the degrees-of-freedom are not hadronic, but partonic. In this "Quark-Gluon Plasm", the quarks as basic constituents of matter are not bound into hadrons, but can move quasi-free over a volume much larger than the typical hadron size. Numerical simulations of Quantum-Chromodynamics on a discrete space-time lattice predict the deconfinement transition to take place at a critical temperature of about 160 MeV at vanishing net-baryon density.

In the terrestrial laboratory, heated or compressed QCD matter can be produced by the collision of heavy nuclei, although only for small time intervals and volumes. In the course of such collisions, a hot and dense zone is formed in the overlap area of the two colliding nuclei ("fireball"), which, depending on the collision energy, may be in the deconfined phase. After expansion and cooling, the system decomposes into a multitude of hadrons. It is the task of heavy-ion physics to connect the measured properties of this hadronic final state with the state of the system in the early stage of the collision.

Experimentally, heavy-ion physics has a history of several decades by now, from first experiments in the early 1970's at the Bevalac to the currently ongoing nuclear collision programme at the LHC. Progress was achieved by the availability of new accelerator facilities, giving access to higher beam energies and, consequently, larger energy densities in the reaction zone. The experimental results strongly suggest that in heavy-ion reactions at RHIC energies ($\sqrt{s_{\mathrm{NN}}} = 200\,\mathrm{GeV}$), and probably already at top SPS energy ($\sqrt{s_{\mathrm{NN}}} = 17.3\,\mathrm{GeV}$), a deconfined state of matter is indeed created. The properties of this state are currently being investigated in detail at the highest available collisions energies at the LHC.

[a] e-mail: v.friese@gsi.de

In parallel to these investigations of QCD matter at highest energies, a number of experimental programmes at lower, moderate collisions energies are proposed or already under way: the beam energy scan programme at RHIC, the NA61 experiment at the CERN-SPS, the CBM experiment at GSI-FAIR, and the MPD at JINR-NICA (the latter being facilities under construction). Because of the nature of baryon stopping, collisions at lower energies produce a medium with moderate temperature but high net-baryon density. Varying the beam energy thus allows to a certain extent to study strongly interacting matter at different conditions. This is illustrated in Fig. 1, showing a speculative phase diagram of QCD matter. This diagram may contain a rich structure in particular at high net-baryon densities: a first-order chiral transition, possible distinguished from the deconfinement transition by a third, "quarkyonic" phase, and a critical point separating the region of first-order chiral transition from the crossover at low density. Only a small part of the phase diagram is theoretically accessible from first-principle calculations (lattice QCD). For the larger rest, effective models are employed to study the properties of QCD matter. The experimental confirmation of the conjectured properties and structures would provide a breakthrough in the understanding of QCD.

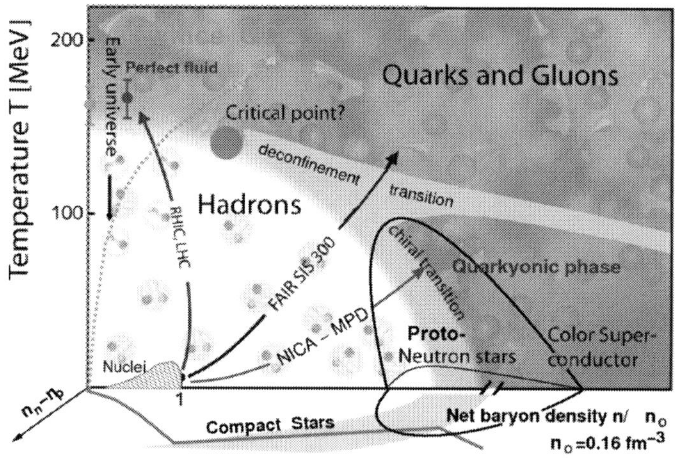

Figure 1. Conjectured phase diagram of strongly interacting matter in terms of net-baryon density and temperature. (Courtesy A. Sorin)

In the following, we discuss the role of the measurement of strange particles in the experimental study of the phase diagram, in particular in the study of the transition from confined to deconfined matter. We begin with a brief review of the current understanding of strangeness production at moderate energies, outline the open questions, and finally propose some observables which are most promising to address the issue of deconfinement.

2 The interest in strangeness

Since the beginning of the search for the Quark-Gluon Plasma (QGP), particles containing strange quarks were considered a key probe. A straightforward and qualitative reason for this is of course that there is no strangeness in the entrance channel of the nucleus-nucleus collision; all strange quarks seen in the final state hadrons are thus produced in the reaction and are not remnants of the incoming nucleons.

Quantitatively, it was shown that the different production mechanisms of strangeness in a partonic scenario, compared to "conventional" hadronic processes, lead to quite different expectations for the amount of strangeness produced in the collision. The reason for this can be traced to the mass of the lightest strangeness carrier, which is the Kaon ($m \sim 500$ MeV) in the hadronic case, and the s quark ($m \sim 100$ MeV) on the partonic side. Producing strange hadrons in hadronic processes like, e.g., $p + p \to pK^+\Lambda$, is thus energetically much more expensive than the partonic production of $s\bar{s}$ pairs, e.g., $q\bar{q} \to s\bar{s}$. In fact, partonic processes were shown to saturate the amount of strange quarks in a (sufficiently hot) QGP on a time scale of a few fm, well within the lifetime of the fireball [1, 2]. The most effective process was found to be the gluon fusion ($g + g \to s\bar{s}$). Hadronisation is expected to preserve the amount of strange quarks, which show up in strange hadrons in the final state. The prediction thus was that there is more strangeness production in nucleus-nucleus reactions relative to p+p collisions if a deconfined phase is formed. This so-called "strangeness enhancement" was the earliest predicted signature for the Quark-Gluon Plasma.

An enhanced production of strange hadrons in central Pb+Pb reactions relative to p+p collisions was in fact observed by experiments at the CERN-SPS in the late 1990s at a bombarding energy of 158 GeV ($\sqrt{s_{NN}} = 17.3$ GeV) [3, 4]. The enhancement observed a strict hierarchy determined by the number of strange valence quarks: while Kaons and Λ with one strange valence quark are found to be enhanced by a factor of about two, an enhancement factor of 15 was observed for Ω (sss) (see Fig. 2).

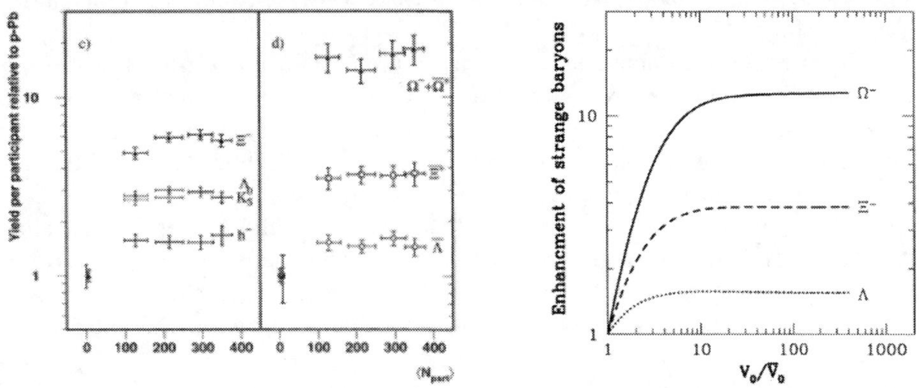

Figure 2. Left: yields of strange hadrons (normalised to the number of participant nucleons) as a function of centrality, expressed by the number of participants [4]. The abscissa is normalised to the value measured in p+Pb collision ($N_{part} = 2$). Right: Strangeness enhancement for hyperons as a function of volume, calculated with the canonical formulation of the hadron gas model [6].

The original perception of strangeness enhancement underwent a shift in interpretation when it was shown that all observed hadron abundances at top SPS energy, including those of strange hadrons, can be reproduced by a hadron resonance gas in chemical equilibrium [5]. This formally purely hadronic description uses only three parameters – volume, temperature, and baryo-chemical potential – to fit particle yields or yield ratios. The temperature obtained from these fits is about 160 MeV, stunningly close to both the Hagedorn temperature and the critical temperature as obtained from lattice QCD calculations.

In the context of thermal models, the difference in the relative strangeness content in nuclear and elementary collisions is understood as an effect of exact strangeness conservation, which suppresses strangeness production in small systems like resulting from p + p reactions [6]. Strangeness conservation is accounted for by the canonical formulation of the thermal model w.r.t. strangeness produc-

 DOI: 10.1051/epjconf/201716406004

tion, which shows a strong volume effect. For large systems like those produced in central Pb + Pb collisions, the grand-canonical formulation, which is technically easier to solve, gives a sufficiently precise approximation. The thermal model in its canonical formulation also reproduces the hierarchy of strangeness enhancement as observed experimentally (see Fig. 2). So, "strangeness enhancement" has become "canonical suppression" in small systems.

The question remains how the apparent chemical equilibrium is obtained. A-priori calculations [2] as well as microscopic hadronic transport simulations like UrQMD show that inelastic hadronic two-body collisions are not efficient enough to equilibrate strange hadrons – in particular not heavy hyperons like Ξ or Ω – within the finite lifetime of the fireball. Two explanations were brought forward. The first argues that the transition from the deconfined to the confined state, i.e., from the QGP to the hadron gas, taking place when the system expands and cools below the critical temperature, must populate the hadronic phase space statistically, whatever the detailed mechanisms of hadronisation may be [7]. The phase transition thus creates a hadronic state of maximum entropy, i.e., an equilibrium state, which in this view is not the result of a hadronic relaxation process, but a consequence of the nature of hadronisation.

An alternative explanation of the apparent chemical equilibrium took into account the production of strange particles in collisions with more than two hadrons in the entrance channel, e.g., $2\pi + 3K \rightarrow \bar{N}\Omega$ [8]. Such processes require three or more particles to be close in space-time and, hence, show a very strong dependence on temperature and density. They are thus only effective very close to the critical temperature, i.e., near the phase transition. There, however, they are sufficient to equilibrate even Ω baryons within a few fm.

The two pictures outlined above agree in the opinion that a phase transition from a QGP to a hadron gas is prerequisite for a hadro-chemical equilibrium state, and that the hadron abundances do not change significantly after hadronisation, i.e., hadronisation and chemical freeze-out coincide. Consequently, the observed chemical equilibrium is taken as an indirect proof of the transition to a deconfined state in the course of the collision.

Strange hadron yields measured in nuclear collisions at much higher energies as available at BNL-RHIC and CERN-LHC are also in very good agreement with a thermal description. Strong evidence for the creation of a QGP here comes also from very different observables like jet quenching and flow [9–12]. Strangeness at top SPS and higher energies thus seems to be understood.

3 The onset of deconfinement

If a deconfined state is formed in heavy-ion collisions at top SPS energy, the question arises at which collision energy the deconfinement transition is first achieved. Following the argument that strangeness equilibration is indicative for the formation of a QGP, it seems natural to search for the onset of deconfinement by testing the applicability of the hadron gas model at lower collision energies. The thermal model is able to describe data obtained at lower SPS and at AGS energies; here, however, only a small number of hadron species was experimentally accessible up to now. In particular, data on multi-strange hyperons are quite scarce, owing to their low multiplicity at moderate energies.

At very low energy, the HADES collaboration at GSI-SIS18 reported a strong deviation – by a factor of about 15 – of the Ξ^- yield from the thermal fit to the hadron yields measured in Ar + Cl collisions at 1.76*A* GeV [13]. This is the first indication that the statistical description of strange hadrons yields does not hold at low energies. It should be noted, however, that at this energy, Ξ^- is produced deep sub-threshold, presumably by secondary collisions involving previously produced K and Λ, e.g., $\Lambda + K^- \rightarrow \pi^0 + \Xi^-$ or $\Lambda + \Lambda \rightarrow p + \Xi^-$.

A systematic investigation of strangeness production at lower SPS energies was performed by the NA49 collaboration in the years 1999 - 2002, with apparently intriguing results: both the K^+/π^+ and the Λ/π^- ratios show a narrow maximum at a beam energy of 30*A* GeV [14, 15]. Such a maximum was predicted by a the so-called Statistical Model of the Early Stage (SMES) [16], assuming the deconfinement to set in at this energy. The association of the observed maximum in relative strangeness production to the onset of deconfinement, however, was heavily debated subsequently. In particular, it was shown that the hadron gas model, with the parameters T and μ_B smoothly parametrised as function of collision energy, also results in a strangeness maximum at 30*A* GeV, which is attributed to the transition from a baryon-dominated to a meson-dominated hadronic final state [17]. Still, the original description of the energy dependence of the K^+/π^+ ratio by the hadron gas model was rather poor [18], but it was improved by including high-mass resonances into the model [19]. The situation remains unclear (see Figure). The deviation of the K^+/π^+ and Λ/π^- ratios from the statistical description does not appear sufficient for a strong claim for "new physics"; the Ξ^- measurements lacks precision, and for the Ω^-, only one data point at 40*A* GeV exists.

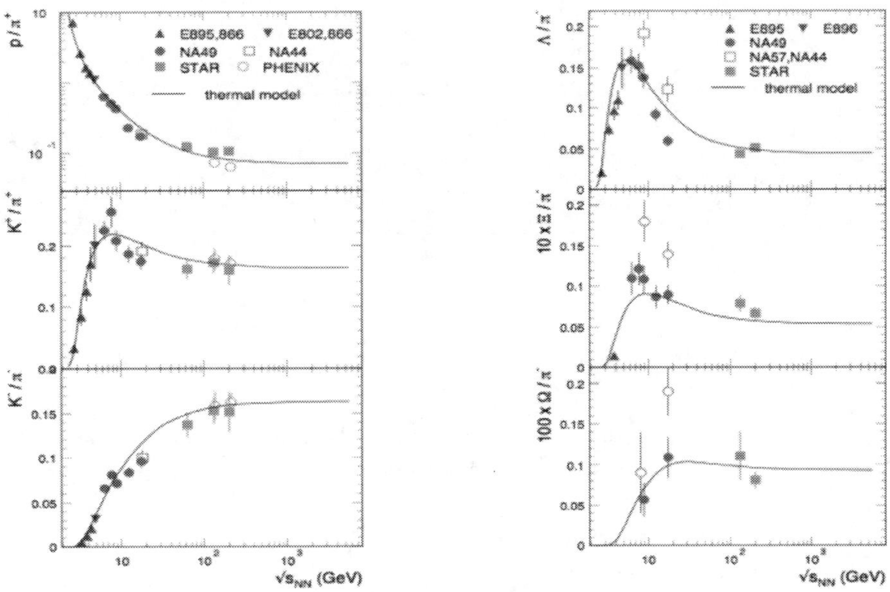

Figure 3. Energy dependence of the relative yield of strange hadrons. The full line shows the parametrisation by the the hadron gas model. (Courtesy NA49 collaboration)

The predictions of the hadron gas model and the Statistical Model of the Early Stage are quite different: in the hadron gas model, the strangeness maximum is due to the maximum in net-baryon density. The effect thus diminishes with increasing strangeness content: the maximum for Ω/π is shallower than that for Ξ/π, which is shallower than Λ/π. Moreover, the position of the maximum shifts to higher energies with increasing strangeness content [20]. In the SMES, the situation is the opposite: the effect is due to the overall strangeness production and thus more pronounced for Ω than for Λ or K; and the maxima are predicted to be at the same energy. So, yields of multi-strange hyperons are discriminative, and precision data on the Ξ and Ω yields as function of collision energy can be expected to lead to a definite conclusion on the issue. Such data can be expected from future experiments at NICA and FAIR-SIS300.

 DOI: 10.1051/epjconf/201716406004

4 The ϕ meson

In the general context of strangeness production, the ϕ meson plays a particular role owing to its valence quark content $s\bar{s}$, which makes it strangeness-neutral as a hadron, but sensitive to partonic strangeness production scenarios. In particular, strongly enhanced ϕ production (w.r.t. proton-proton reactions) was suggested as indicator for the QGP [22]. The respective measurements at SIS-18, AGS, and RHIC energies, however, do not reveal a spectacular behaviour; they are again well described by the statistical hadronisation scenario. In the SPS energy range, in contrast, the description of the hadron gas model fails considerably, and so do the results of microscopic transport simulations (Fig. 4). ϕ production in this energy range, hosting the maximum in net-baryon density and, possibly, the onset of deconfinement, still lacks a sound interpretation. In contrast to multi-strange hyperons, the data on ϕ yields are rather precise, such that future measurements will not help elucidate the situation. The ball here is surely on the theory side.

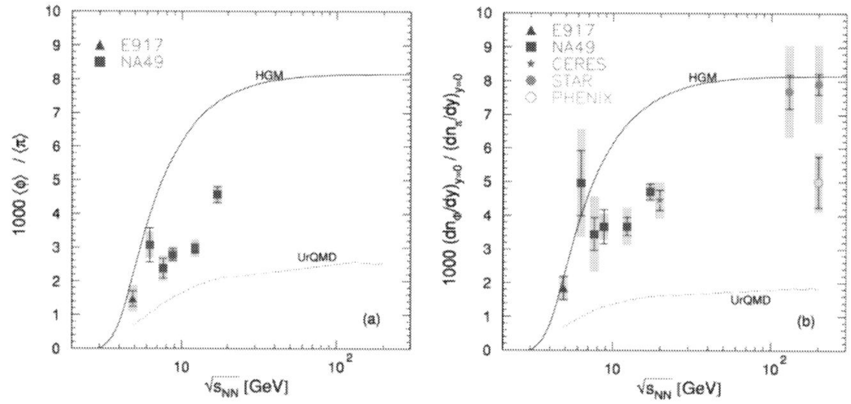

Figure 4. ϕ/π ratio in full phase space (left) and at mid-rapidity (right) measured in central Pb+Pb (Au+Au) collisions, as function of collision energy. The full line shows the prediction of the hadron gas model, the dotted line that of the UrQMD model. (Taken from [21].)

5 Strange anti-baryons

In the energy range under discussion here, the measurement of anti-baryons is notoriously difficult, since their yield is very low – the production of anti-baryons is strongly suppressed in a baryon-dense environment. Consequently, the current data situation is scarce. In particular, data on mutli-strange anti-baryons do not exist below top SPS energy. Still, strange anti-baryons might be a very sensitive probe of the matter produced in heavy-ion collisions. The reason is that the suppression of their production is released in a deconfined medium. This was demonstrated by the microscopic transport model pHSD, which switches to partonic production mechanisms in regions of the fireball where a critical energy density is exceeded [23]. The volume fraction of deconfined to confined matter is a function of collision energy; it is very small at e.g., 5*A* GeV, but close to unity at top SPS energy (158*A* GeV). The effect of the deconfined medium on the production of Ω^- and $\bar{\Omega}^+$ is shown in Fig. 5 by comparison to the results obtained with HSD, without any partonic phase, in the FAIR energy range. While the yield of Ω^- is only slightly affected, its anti-particle appears very sensitive, and the

 DOI: 10.1051/epjconf/201716406004

effect is most pronounced at the lower energies, where it amounts to more than an order of magnitude. The reason for the Ω not being affected appreciably is that the fraction of QGP is very low and does not significantly add to the hadronic production. This changes at higher energies, where partonic production eventually becomes dominant. The hadronic anti-Ω production, in contrast, is so low that event a tiny amount of deconfined medium suffices to enhance it by factors. This microscopic picture is quite different from the concept of an onset of deconfinement as discussed in section 3; it rather predicts a gradual increase of the fraction of deconfined medium until the entire fireball consists of QGP.

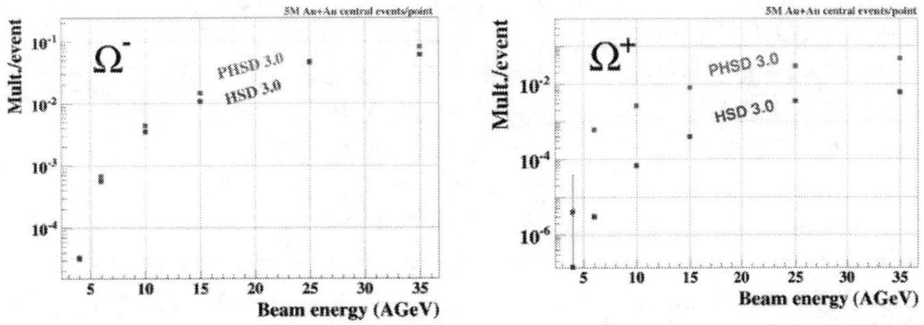

Figure 5. Multiplicity of Ω^- (left) and $\bar{\Omega}^+$ in central Au+Au collisions as predicted by the HSD (blue) and the pHSD (red) models. The latter includes a partonic production scenario in fireball regions exceeding a critical energy density. (Courtesy CBM collaboration)

Following this model prediction, the measurement of the yield of multi-strange anti-hyperons is most promising for the study of deconfinement even (or in particular) at low energies as will become available with the FAIR facility. Experimentally, the very low expected anti-hyperon yields necessitate extreme event statistics, which seem only feasible with the CBM experiment [24].

6 Summary

Strangeness production below top SPS energy still poses a number of open questions: Does thermalisation hold even at low energies? Can the "horn" in the K/π ratio be understood in terms of the hadron gas model? What happens at maximal net-baryon density (30*A* GeV)? What are the production mechanisms near or below the threshold in elementary collisions? Why can the ϕ yield at SPS energies neither be described by the hadron gas model nor by microscopic transport? What is the energy dependence of the production of multi-strange (anti-)hyperons?

Strangeness continues to be one of the most important probes for the study of the QCD phase diagram at high net-baryon densities and in particular for the search for the first-order phase transition. Many new data will become available in the near future by ongoing activities at RHIC and SPS, and in the next decade by the new experimental programmes CBM at FAIR and MPD at NICA. The latter two promise to decisively improve the data situation on multi-strange hyperons and, in particular, anti-hyperons. Such data are likely one of the prime keys to the understanding of deconfinement in dense QCD matter.

References

[1] J. Rafelski and B. Müller, Phys. Rev. Lett. **48**, 1066 (1982)

[2] P. Koch, B. Müller and J. Rafelski, Phys. Rep. **142**, 167 (1986)
[3] F. Sikler *et al.* (NA49 collaboration), Nucl. Phys. **A 661**, 45c (1999)
[4] E. Andersen *et al.* (WA97 collaboration), Phys. Lett. **B 449**, 401 (1999)
[5] J. Stachel, Nucl. Phys. **A 654** 199c (1999)
[6] S. Hamieh, K. Redlich and A. Tounsi, Phys. Lett. **B 486**, 61 (2000)
[7] R. Stock, Phys. Lett. **B 456**, 277 (1999)
[8] P. Braun-Munzinger, J. Stachel and C. Wetterich, Phys. Lett. **B 596**, 61 (2004)
[9] I. Arsene *et al.* (BRAHMS collaboration), Nucl. Phys. **A 757**, 1 (2005)
[10] B. B. Back *et al.* (PHOBOS collaboration), Nucl. Phys. **A 757**, 28 (2005)
[11] J. Adams *et al.* (STAR collaboration), Nucl. Phys. **A 757**, 102 (2005)
[12] K. Adcox *et al.* (PHENIX collaboration), Nucl. Phys. **A 757**, 184 (2005)
[13] G. Agakishiev *et al.* (HADES collaboration), arXiv:1512.07070 [nucl-ex]
[14] C. Alt *et al.* (NA49 collaboration), Phys. Rev. **C 77**, 024903 (2008)
[15] C. Alt *et al.* (NA49 collaboration), Phys. Rev. **C 78**, 034918 (2008)
[16] M. Gazdzicki and M. Gorenstein, Act. Phys. Pol. **B 30**, 2705 (1999)
[17] P. Braun-Munzinger *et al.*, Nucl. Phys. **A 697**, 902 (2002)
[18] A. Andronic, P. Braun-Munzinger and J. Stachel, Nucl. Phys. **A 772**, 167 (2006)
[19] A. Andronic, P. Braun-Munzinger and J. Stachel, Phys. Lett. **B 673**, 142 (2009)
[20] J. Cleymans *et al.*, arXiv:1603.09553 [hep-ph]
[21] C. Alt *e*t al., Phys. Rev. **C 78**, 044907 (2008)
[22] A. Shor, Phys. Rev. Lett. **54**, 1122 (1985)
[23] W. Cassing and E. Bratkovskaya, Nucl. Phys. **A 831**, 215 (2009)
[24] V. Friese (for the CBM collaboration), J. Phys. Conf. Ser. **668**, 012014 (2016)

AUTHOR INDEX

AUTHOR INDEX

AUTHOR INDEX

AUTHOR INDEX

EDP Sciences – Web of Conferences
17, Avenue du Hoggar
Parc d'Activité de Courtabouf
BP 112
F-91944 Les Ulis Cedex A
France

ISBN 978-1-5108-5347-8
9 781510 853478

5th International Conference on New Frontiers in Physics 2016

EPJ Web of Conferences Volume 164 (2017)

Crete, Greece
6 - 14 July 2016

Part 2 of 2

edp sciences

5th International Conference on New Frontiers in Physics 2016

EPJ Web of Conferences Volume 164 (2017)

Crete, Greece
6 - 14 July 2016

Part 2 of 2

Editors:

Larissa Bravina
Yiota Foka
Sonja Kabana

ISBN: 978-1-5108-5347-8

Printed from e-media with permission by:

Curran Associates, Inc.
57 Morehouse Lane
Red Hook, NY 12571

Printed by Curran Associates, Inc. (2018)

For additional information, please contact EDP Sciences – Web of Conferences at the address below.

EDP Sciences – Web of Conferences
17, Avenue du Hoggar
Parc d'Activité de Courtabœuf
BP 112
F-91944 Les Ulis Cedex A
France

Phone: +33 (0) 1 69 18 75 75
Fax: +33 (0) 1 69 28 84 91

contact-edps@webofconferences.org

Additional copies of this publication are available from:

Curran Associates, Inc.
57 Morehouse Lane
Red Hook, NY 12571 USA
Phone: 845-758-0400
Fax: 845-758-2633
Email: curran@proceedings.com
Web: www.proceedings.com

TABLE OF CONTENTS

PART 1

PART 2

EPJ Web of Conferences **164**, 06005 (2017) DOI: 10.1051/epjconf/201716406005
ICNFP 2016

Results from the RHIC Beam Energy Scan

A. Taranenko[1,a]

[1] *National Research Nuclear University MEPhI (Moscow Engineering Physics Institute), Kashirskoe highway 31, Moscow, 115409, Russia*

Abstract. The first phase of the Beam Energy Scan (BES-I) program at the Relativistic Heavy Ion Collider (RHIC) is based on Au+Au collision data collected between 2010 and 2014 at center-of-mass energies $\sqrt{s_{\rm NN}}$=7.7, 11.5, 14.5, 19.6, 27, and 39 GeV. The BES-I program has four physics goals: search for the turning off of the signatures of the Quark Gluon Plasma (QGP), the search for the possible first-order phase transition between hadronic gas and QGP, the search for the possible critical end point and the study of the transport properties of the strongly interacting matter as a function of the temperature T and baryon chemical potential μ_B. In this article, we briefly review and discuss the main results and their understanding from the BES-I program and future plans for the BES-II program (2019-2020).

1 Introduction

The first decade of RHIC running (2000-2010) at top energy of $\sqrt{s_{\rm NN}}$ =200 GeV has established the existence of a strongly coupled Quark Gluon Plasma (sQGP), a new state of nuclear matter with partonic degrees of freedom [1, 2] and with low specific shear viscosity η/s, i.e. the ratio of shear viscosity η to entropy density s [3, 4]. The left panel of Figure 1 shows a cartoon of the phase diagram of nuclear matter. Lattice QCD calculations [5] indicate that the quark-hadron transition is a smooth crossover at top RHIC energies and above (small μ_B). The possible region of a first order phase transition is expected at larger values of μ_B (lower beam energies), suggesting the existence of a critical end point (CEP) [6]. Thus, a current strategy for experimental mapping of the phase diagram is centered on beam energy scans, which sample reaction trajectories with the broadest possible range of μ_B and T values. The RHIC BES-I program (2010-2014) has four physics goals [7–9]: search for the turning off of the signatures of the QGP well established at top RHIC energies, the search for the possible first-order phase transition between hadronic gas and QGP and for critical end point, and the study of the transport properties of the strongly interacting matter: e.g $\eta/s(T,\mu_B)$. The $\sqrt{s_{\rm NN}}$ energies from BES-I program (200-7.7 GeV) correspond to baryon chemical potentials (μ_B) in the range 25 to about 400 MeV [8]. The midrapidity transverse-energy distributions, $dE_T/d\eta$, for Au+Au collisions at $\sqrt{s_{\rm NN}}$ = 7.7-200 GeV were measured by PHENIX collaboration [10]. The right panel of Fig. 1 shows estimates of the initial energy density times the formation time of a QGP τ_0 for various Au-Au collision centralities. Lattice calculations show that the QGP should form when the energy density gets much higher than 0.6 GeV/fm^3 [11]. For τ_0 = 1 fm, for the Au+Au collisions at $\sqrt{s_{\rm NN}}$=7.7 GeV

[a] e-mail: AVTaranenko@mephi.ru

can be expected to create a QGP according to this estimate. Even for $\tau_0 = 2$ fm, the central 7.7 GeV collisions would reach energy densities high enough for QGP formation.

2 Search for the turning off of the signatures of the QGP

Key evidence for the formation of a strongly coupled Quark Gluon Plasma (sQGP) in Au+Au collisions at top RHIC energies was the observation of jet quenching and large azimuthal anisotropic flow signals for all produced particles [1, 2]. The results of STAR measurements show that the strong suppression of high p_T charged particles, as indicated by comparisons of binary scaled yields, is no longer evident in Au+Au collisions at $\sqrt{s_{NN}} < 14.5\ GeV$ for any centrality [12]. This may indicate that hard scattering effects become rare at lower beam energies, as the high-p_T regions are much less populated [8].

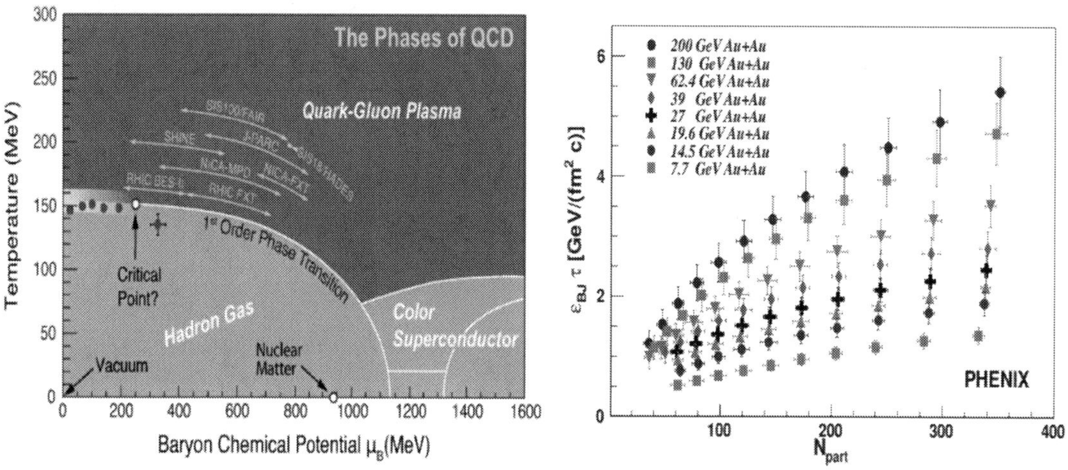

Figure 1. Left: Schematic view of QCD phase diagram plotted as temperature T versus baryon chemical potential μ_B. Various details, as the position of the critical end point and order of the phase transition curve are highly uncertain. The figure is taken from [9]. Right: Estimates of the initial energy density for a variety of energies time the formation time. Figure from [10].

The anisotropic flow, as manifested by the anisotropic emission of particles in the plane transverse to the beam direction, is one of the important observables sensitive to the transport properties of sQGP: the equation of state, the speed of sound, and the value of specific shear viscosity η/s. The azimuthal anisotropy of produced particles can be quantified by the Fourier coefficients v_n in the expansion of the particles' azimuthal distribution as: $dN/d\phi \propto 1 + \sum_{n=1} 2v_n \cos(n(\phi - \Psi_n))$ [13–15], where n is the order of the harmonic, ϕ is the azimuthal angle of particles of a given type, and Ψ_n is the azimuthal angle of the nth-order event plane. The nth-order flow coefficients v_n can be calculated as $v_n = \langle\cos[n(\varphi - \Psi_n)]\rangle$, where the brackets denote an average over particles and events. Relativistic viscous hydrodynamics has been successful in describing the observed flow anisotropy v_n for the produced particles in the collision of heavy-ion systems like Au+Au at 200 GeV (RHIC) and Pb+Pb at 2.7 TeV (LHC) [3, 4, 16]. In this model framework, the values of the coefficients v_n (for $p_T < 3 - 4$ GeV/c) have been attributed to an eccentricity-driven hydrodynamic expansion of the plasma produced in the collision zone. That is, a finite eccentricity moment ε_n drives uneven pressure

 DOI: 10.1051/epjconf/201716406005

gradients in- and out of the event plane ψ_n, and the resulting expansion leads to the anisotropic flow of particles about this plane. The event-by-event geometric fluctuations in its initial density distribution are found to be responsible for finite elliptic flow signal v_2 in the collisions with almost zero impact parameter, and the presence of odd harmonic moments in the initial geometry ε_n and final momentum anisotropy v_n [14, 15]. The proportionality constant between v_n and ε_n is found to be sensitive to the transport properties of the matter such as as the equation of state and the specific shear viscosity η/s. The shear viscosity suppresses higher order harmonic flow coefficients $v_{n>2}$ more strongly than the elliptic flow signal v_2. The v_n data at top RHIC energies and LHC seem to follow the "acoustic scaling" for anisotropic flow, which suggests that viscous corrections to v_n/ε_n grow exponentially as n^2 and $1/(\bar{R}T)$ [17–19]:

$$\frac{v_n(p_T,\text{cent})}{\varepsilon_n(\text{cent})} \propto \exp\left[-n^2\beta'\right],\ \beta' \propto \frac{\eta}{s}\frac{1}{\bar{R}T} \propto \frac{\eta}{s}\frac{1}{(dN_{ch}/d\eta)^{1/3}} \tag{1}$$

where T is the temperature, $\bar{R}$ is the transverse size of the collision zone and ε_n is the n-th order eccentricity moment. For a given harmonic number n, this equation indicates a characteristic linear dependence of $\ln(v_n/\varepsilon_n)$ on $1/(\bar{R}T)$, with a slope proportional to η/s [17–19]. The dimensionless size $\bar{R}T$ is approximately proportional to $(dN_{ch}/d\eta)^{1/3}$, where $dN_{ch}/d\eta$ is the charge particle multiplicity density [20].

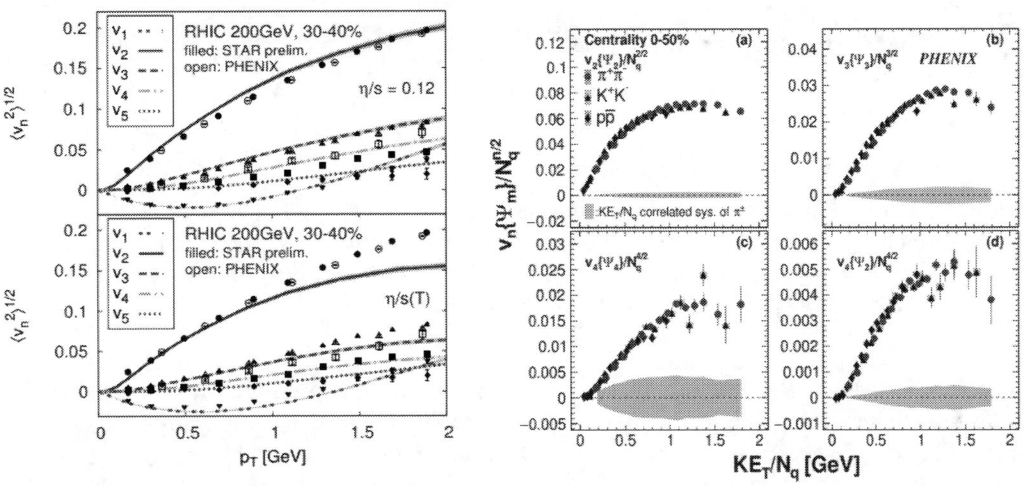

Figure 2. Left: Comparison of viscous hydro calculations and $v_n(p_T)$ data for charged hadrons at top RHIC energy that can be used to better constrain the temperature dependence of η/s. The figure is taken from [3]. Right: Quark-number (n_q) scaling for 0-50% central Au+Au collisions at $\sqrt{s_{NN}}$ = 200 GeV, where n_q is the constituent valence quark number of each hadron and KE_T is the transverse kinetic energy of the hadron [22].

The data for top RHIC energy show that, for a given centrality, $v_{n>1}$ for all observed hadrons scale to a single curve when plotted as $v_n/n_q^{n/2}$ versus KE_T/n_q, where n_q is the number of constituent quarks in a given hadron species and $KE_T = m_T - m_0$ is the transverse kinetic energy for these hadrons [21, 22]. The observed scaling may indicate that the bulk of the anisotropic flow at top RHIC energies is partonic, rather than hadronic. Lowering the collision energy and studying the energy dependence of anisotropic flow allows a search for the onset of the transition to a phase with partonic degrees of freedom at an early stage of the collision.

Figure 3(left) shows the p_T dependence of v_2 of charged hadrons calculated from the four-particle cumulant method measured for several centralities and for energies ranging from 7.7 GeV up to 2.76 TeV[23]. It is striking how little the elliptic flow signal for inclusive charged hadron changes over such a wide range of energies and initial energy densities: the energy changes by a factor 400 and initial energy density changes by nearly a factor of 10. Figure 3(right) shows quark-number (n_q) scaled elliptic flow, v_2/n_q versus $(m_T - m_0)/n_q$, for identified particles produced in 0-80% central Au+Au collisions at $\sqrt{s_{NN}}$ = 7.7-62.4 GeV as published by STAR collaboration [24]. This work shows that the scaling holds separately for most of the particles and anti-particles withing 10-15%. The ϕ meson may not follow the trend of the other identified hadrons at 7.7 and 11.5 GeV, but more data are needed before any conclusions can be drawn. Hybrid model calculations show that the weak dependence of $v_2(p_T)$ on the beam energy may result from the interplay of the hydrodynamic and hadronic transport phase [26]. The trianglular flow v_3 is more sensitive to the viscous damping and might be an ideal observable to probe the formation of a QGP and the pressure gradients in the early plasma phase.

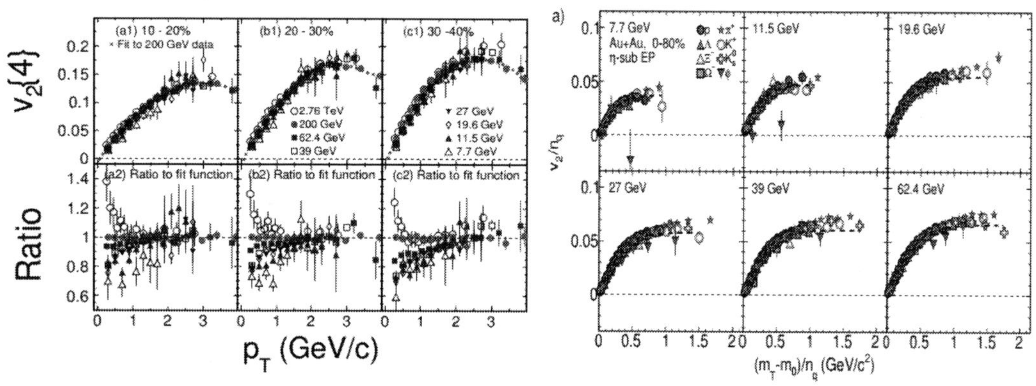

Figure 3. Left: The variation of $v_2(p_T)$ for charged hadrons produced at mid-rapidity from 7.7 GeV up to 2.76 TeV [23]. Right: Quark-number (n_q) scaled elliptic flow, v_2/n_q versus $(m_T - m_0)/n_q$, for identified particles produced in 0-80% central Au+Au collisions at $\sqrt{s_{NN}}$ = 7.7-62.4 GeV, where n_q is the constituent valence quark number of each hadron and $(m_T - m_0)$ is the transverse kinetic energy of the particle [24].

Figure 4 left shows the variation of $v_3^2\{2\}$ for charged hadrons produced at mid-rapidity from 7.7 GeV up to 2.76 TeV for different bins in collision centrality [25]. At low energies $\sqrt{s_{NN}} < 14.5\ GeV$, the $v_3^2\{2\}$ become consistent with zero in 50-60% peripheral collisions. This result is consistent with the idea of absence of a low viscosity QGP phase in low energy peripheral collisions [26]. For more central collisions, the values of $v_3^2\{2\}$ are positive and change little from 19.6 GeV to 7.7 GeV. For the energies above 19.6 GeV, the values of $v_3^2\{2\}$ linearly increase with the $log(\sqrt{s_{NN}})$ for all bins in collision centrality [25]. Figure 4 right shows the preliminary data from STAR collaboration for v_3 of identified charged hadrons from Au+Au collisions at $\sqrt{s_{NN}}$ = 200 GeV (upper panels) and 39 GeV (lower panels) [27]. The left panels represent the results for particles and right panels for anti-particles. For 200 GeV the measured v_3 values follow the $v_3/n_q^{3/2}$ versus $(m_T - m_0)/n_q$ scaling. However, for 39 GeV data one can see that the scaling is not so perfect.

The left panel of Fig. 5 shows the preliminary STAR data for the $\sqrt{s_{NN}}$ dependence of the p_T integrated v_n of charged hadrons from 0-40% central Au+Au collisions [19]. It shows an essentially monotonic trend for v_2, v_3 and v_4 with $\sqrt{s_{NN}}$ as might be expected for a temperature increase as $\sqrt{s_{NN}}$ increases [19]. The same authors applied the "acoustic scaling" to v_n data in order to estimate the $\sqrt{s_{NN}}$ dependence of η/s. Using the observation that ε_n changes very slowly with beam energy and

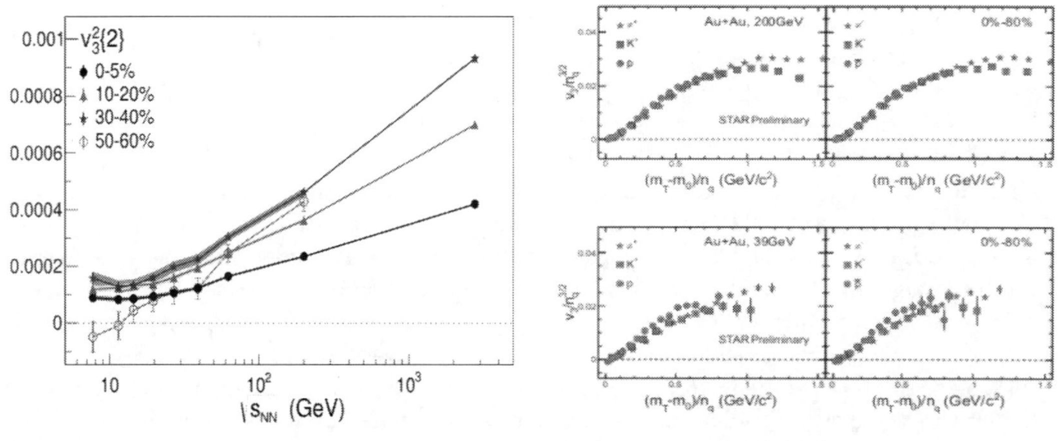

Figure 4. Left: The variation of $v_3^2\{2\}$ for charged hadrons produced at mid-rapidity from 7.7 GeV up to 2.76 TeV [23] for different bins in collision centrality. The figure is taken from [25] Right: Quark-number (n_q) scaled triangular flow, $v_3/n_q^{3/2}$ versus $(m_T - m_0)/n_q$, for charged pions, kaons and (anti)protons emerged from 0-80% central Au+Au collisions at $\sqrt{s_{NN}}$ = 200 GeV (top panels) and at $\sqrt{s_{NN}}$ = 39 GeV (bottom panels). The left panels represent the results for particles and right panels for anti-particles. The figure is taken from [27].

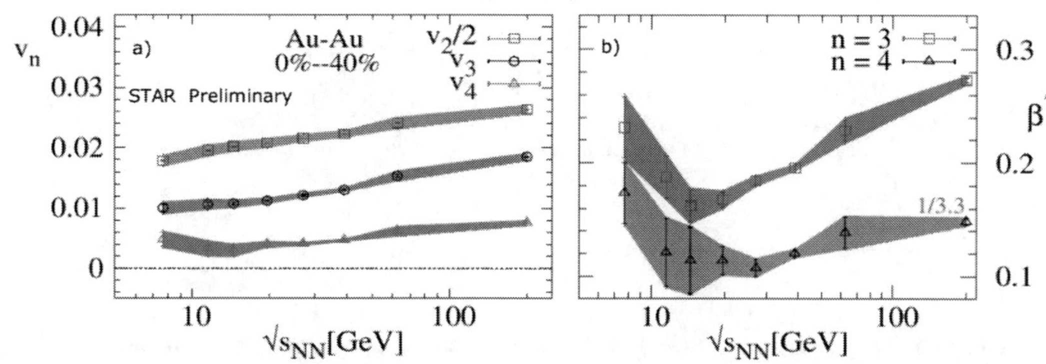

Figure 5. $\sqrt{s_{NN}}$ dependence of the p_T -integrated v_n (left panel) and the estimated viscous coefficient $\beta'' \propto \eta/s$ (right panel). Results are shown for 0-40% central Au+Au collisions; the shaded lines are the systematic uncertainty. The figure is taken from [19].

the measurements for two different harmonics$n$ and n' ($n \neq n'$) the Eq.1 can be simplified [19] to the viscous coefficient $\beta'' \propto (\eta/s) \propto (dN_{ch}/d\eta)^{1/3} ln(v_n^{1/n}/v_{n'}^{1/n'})$. The right panel of Fig. 5 shows the $\sqrt{s_{NN}}$ dependence of the viscous parameter β'' extracted from the $ln(v_3^{1/3}/v_2^{1/2})$ and $ln(v_4^{1/4}/v_2^{1/2})$ results from the left panel of Fig. 5. In contrast to v_n, the excitation function of the viscous parameter β'' shows a non-monotonic behaviour over the same beam energy range. A similar non-monotonic trend for η/s has been observed in the hybrid viscous hydrodynamical calculations [28], tuned to describe the STAR BES-I v_2 data.

DOI: 10.1051/epjconf/201716406005

3 Search for the first-order phase transition

The next BES-I goal is to search for the predicted first-order phase transition [12, 13] between hadronic and QGP phases. Such a transition can also be characterized by a dramatic drop in the pressure, or a softening of the Equation of State (EOS) [14, 15]. The signals like anisotropic flow are very promising due to their sensitivity to EOS. The directed flow (v_1) can probe the very early stages of the collision as it is generated during the passage time of the two colliding nuclei [29]. Both hydrodynamic [21, 22] and transport model [23, 24] calculations indicate that the directed ow of charged particles, especially baryons at midrapidity, is very sensitive to the equation of state. The slope of the rapidity dependence dv_1/dy close to mid-rapidity is a convenient way to characterize the overall magnitude of the directed flow signal. The left panel of Fig. 6 shows slope of net-proton and net-kaon directed flow (v1) as a function of $\sqrt{s_{\rm NN}}$ measured by the STAR. The non-monotonic behavior with a double sign-change in dv_1/dy and with the minimum around $\sqrt{s_{\rm NN}}$=14.5 GeV is observed for net-protons. The slope of net-kaon directed flow shows a monotonic decrease trend with decreasing energy. The observed double sign-change in dv_1/dy for net-protons could be related to the softening of equation-of-state due to the first order phase transition. However, the final-state interaction effects may also play an important role in the observed v_1 signals and overall agreement between theory and data is very poor.

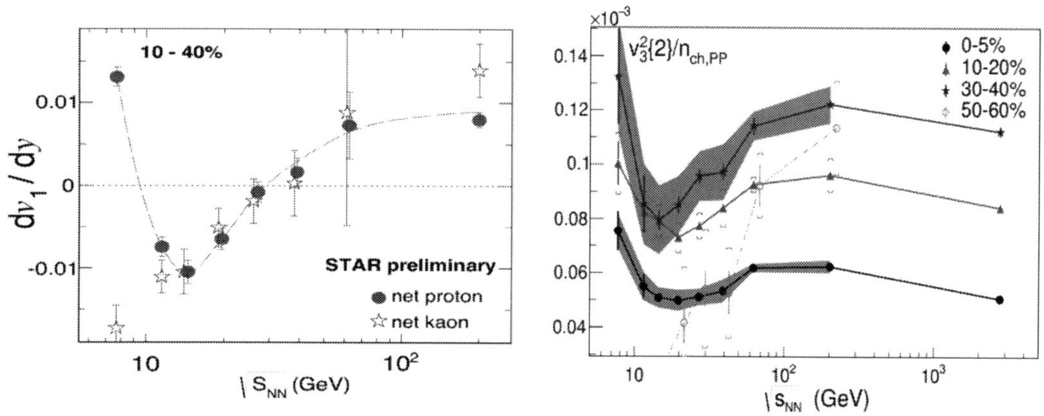

Figure 6. Left: Beam Energy dependence of directed flow slope dv_1/dy near mid-rapidity for net-protons and net-kaons from 10-40% central Au+Au collisions measured by STAR. The figure is taken from [9] Right: Beam energy dependence of $v_3^2\{2\}$ of charged hadrons divided by the mid-rapidity, charged particle multiplicity, pseudo-rapidity density per participant pair in Au+Au and Pb+Pb (2.76 TeV) collisions. The figure is taken from [25].

The next promising probe which may be sensitive to the pressure produced at early stage is the $v_3^2\{2\}$ signal, please see the previous section. The right panel of Fig. 6 shows $\sqrt{s_{\rm NN}}$ dependence of the $v_3^2\{2\}$ for charged hadrons from Au+Au collisions (data from left panel of Fig. 4) scaled by the charged particle multiplicity per participant pair $n_{ch,\rm PP} = \frac{2}{N_{\rm part}} dN_{ch}/d\eta$ for four bins in collision centrality [25]. The central and semi-central collisions data exhibit a local minimum in the $\sqrt{s_{\rm NN}}$ range around 15-20 GeV which is absent for peripheral collisions. This is the consequence of a relatively flat trend for $v_3^2\{2\}$ and monotonically increasing trend for the $n_{ch,\rm PP}$ in the energy range 7.7 < $\sqrt{s_{\rm NN}}$ < 20 GeV. If the general increase of $v_3^2\{2\}$ is driven by ever increasing pressure gradients in ever denser systems at higher energies, then the local minimum in $v_3^2\{2\}/n_{\rm ch,PP}$ could be an indication of an anomalously low pressure inside the matter created in collisions with energies near 15-20 GeV, where a minimum is also observed for the slope of net-proton directed flow [25]. The interpretation of data in this energy regime is complicated by changes in meson to baryon ratio, baryon stopping, non-flow effects and fluctuations, and longer passing time at lower energies [25].

4 Search for the QCD critical end point

A common strategy for locating the critical end point (CEP) is to scan the phase diagram by varying the $\sqrt{s_{\rm NN}}$, and look for an increase/divergence or a non-monotonic behavior of experimental observables sensitive to the correlation length (ξ) [31]. Large fluctuations in the event-by-event distributions of conserved quantities, such as net-charge, net-baryon number, and net-strangeness have been proposed as possible experimental signatures of the QCD CEP [36]. The moments of these distributions can be related to the correlation length (ξ) and can be directly connected to the susceptibility of the system computed in Lattice QCD [32, 33]. The higher-order moments are predicted to be even more sensitive to the CEP since they are proportional to higher powers of the correlation length. The STAR experiment has measured various order fluctuations of net-proton ($N_p - N_{\bar{p}}$, proxy for net-baryon), net-charge and net-kaon (proxy for net-strangeness) numbers in the Au+Au collisions at $\sqrt{s_{\rm NN}}$=7.7-200 GeV [7].

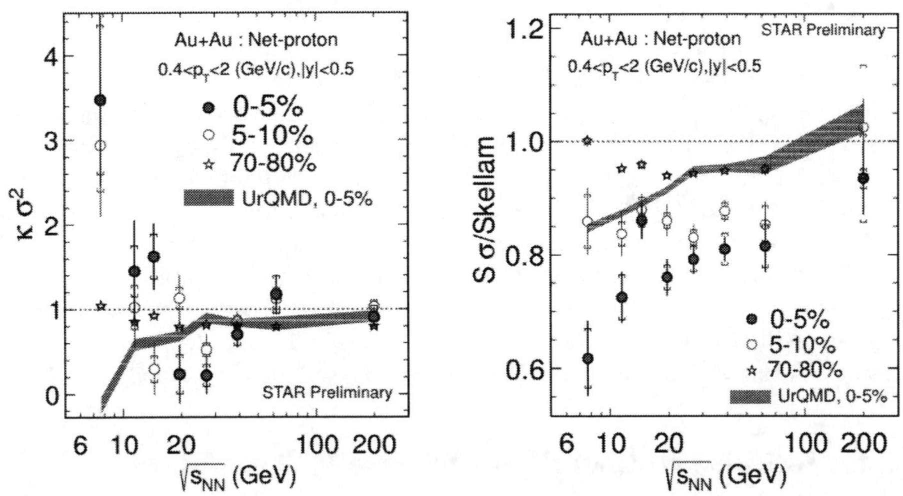

Figure 7. Left: $\sqrt{s_{\rm NN}}$ dependence of $\kappa\sigma^2$ of net-proton distributions and Right: $S\sigma$ divided by Skellam (Poisson) expectations for 0-5%, 5-10% and 70-80% centralities of Au+Au collisions measured by STAR. The experimental data is compared with Poisson expectations (dashed lines) and the UrQMD transport model calculations (shade bands). The protons and anti-protons numbers are measured with transverse momentum $0.4 < p_{\rm T} < 2$ GeV/c and at mid-rapidity $|y| < 0.5$. The figure is taken from [7].

The moment products ($S\sigma$ and $\kappa\sigma^2$) are related to the ratio of baryon number susceptibilities, $S\sigma \sim \chi_B^{(3)}/\chi_B^{(2)}$ and $\kappa\sigma^2 \sim \chi_B^{(4)}/\chi_B^{(2)}$ [36, 37]. Here, σ, S and κ are the variance, skewness and kurtosis (respectively) of the event-by-event net-proton multiplicity distributions. Figure 7 left shows the efficiency corrected $\kappa\sigma^2$ of net-proton distributions as a function of $\sqrt{s_{NN}}$ for 0-5%, 5-10% and 70-80% centralities of Au+Au collisions measured by STAR [7]. The $\kappa\sigma^2$ shows a clear non-monotonic variation with $\sqrt{s_{NN}}$ for 0-5% centrality with a minimum around 20 GeV, with a steep rise as the beam energy is scanned further down, ending with a value about 2 standard errors above the baseline at 7.7 GeV. The $\kappa\sigma^2$ results are qualitatively consistent with a QCD based model including a critical point [21]. The new data from RHIC BES-II should help to check the prediction that $\kappa\sigma^2$ moments will return to unity at $\sqrt{s_{NN}} < 7.7$ GeV as they do at high $\sqrt{s_{NN}}$. The right panel of Fig. 7 shows the $S\sigma$ of net-proton distributions divided by the skellam baselines [7] as a function of $\sqrt{s_{NN}}$ for three centralities measured by STAR. For 0-5% centrality, the $S\sigma$/Skellam shows significantly deviate below unity and monotonically decrease with decreasing energies except for showing a peak at 14.5 GeV. The results from UrQMD model without critical point, show a monotonic decreasing trend as a function of $\sqrt{s_{NN}}$ and can not reproduce the structures observed in $\kappa\sigma^2$ and $S\sigma$/Skellam of net-proton distributions measured by STAR.

An emitting system produced in the vicinity of the critical end point would also be subject to the influence of a divergence in the compressibility of the medium, resulting in a collateral increase in the emission duration [39]. One can expect to see the non-monotonic dependence in the excitation functions ($\sqrt{s_{NN}}$) for the difference between the Gaussian emission source radii ($R^2_{\rm out} - R^2_{\rm side}$), related to the time duration of emission [40]. The left panel of Fig. 8 shows the $\sqrt{s_{NN}}$ dependence of the charged pion femtoscopy parameters ($R^2_{\rm out} - R^2_{\rm side}$) for 0-5%, 5-10%, 10-20%, 30-40%, 40-50% and 50-60% central Au+Au and Pb+Pb collisions from 7.7 GeV to 2.76 TeV. The observed non-monotonic behavior of ($R^2_{\rm out} - R^2_{\rm side}$) may indicate an increase in the emission duration in the vicinity of the peak at $\sqrt{s_{NN}}$ = 20-40 GeV. The magnitude of the peaks decrease with increasing centrality and positions of the peaks shift to lower values of $\sqrt{s_{NN}}$ with an increase in centrality, the width of the distributions increases with centrality. R.A. Lacey has attributed these qualitative patterns for ($R^2_{\rm out} - R^2_{\rm side}$) distributions to the Finite-Size Scaling (FSS) effects expected for the deconfinement phase transition [40]. A finite-size scaling (FSS) analysis of these data suggests a second order phase transition with the estimates $T^{cep} \sim 165MeV$ and $\mu_B^{cep} \sim 95MeV$ for the location of the critical end point. The first interesting observation is that all ($R^2_{\rm out} - R^2_{\rm side}$) data points for different values of $\sqrt{s_{NN}}$ and centralities (from the left panel of Fig. 8) will collapse onto a single curve in the FSS scaled variables, see the right panel of Fig. 8. The second observation is that the same parameters: $T^{cep} \sim 165MeV$ and $\mu_B^{cep} \sim 95MeV$ will work for recent STAR measurements of the excitation function for the moment products $S\sigma$ and $\kappa\sigma^2$ of the event-by-event net-proton multiplicity distributions [41]. The data for 7 centrality bins will also collapse onto a single curve in the FSS scaled variables. The change in parameter of μ_B^{cep} from 95 MeV to 205 MeV will break the scaling for $S\sigma$ and $\kappa\sigma^2$ [41].

5 Plans for RHIC Beam Energy Scan Phase-II

The STAR Collaboration has planned a second phase of the Beam Energy Scan program at RHIC (BES-II), scheduled for 2019-2020, to explore the high baryon density region of the QCD phase diagram [42, 43]. The upgrades for STAR detector and RHIC accelerator have been started. The stochastic electron cooling technique and long beam bunches will be applied to accelerate gold beam, which will improve the luminosity by a factor about 10 at $\sqrt{s_{NN}}$ =19.6 GeV to 25 at $\sqrt{s_{NN}}$ =7.7 GeV compared to BES-I. The proposed Fixed-Target program (FXT)[44], which uses only one beam on an internal target will allows STAR to reach higher μ_B (μ_B=720 MeV for $\sqrt{s_{NN}} = 3GeV$) with high luminosity. There are three detector upgrades are ongoing at STAR: the iTPC upgrade [45], the Event Plane Detector (EPD) upgrade [46] and the eTOF upgrade [47], see left panel of Fig. 9. The inner TPC (iTPC) of STAR[45] is being upgraded to improve the energy loss dE/dx resolution, momentum

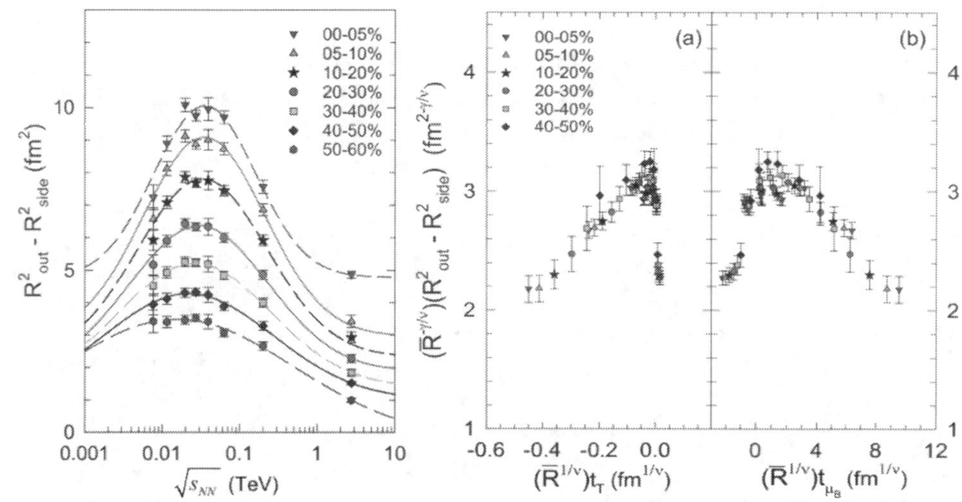

Figure 8. Left: Pion femtoscopy parameters $(R^2_{\text{out}} - R^2_{\text{side}})$ vs. $\sqrt{s_{NN}}$ for 0-5%, 5-10%, 10-20%, 30-40%, 40-50% and 50-60% central Au+Au and Pb+Pb collisions for $m_T = 0.26$ GeV and 0.29 GeV. Right: (a) $\bar{R}^{-\gamma/\nu}\times(R^2_{\text{out}} - R^2_{\text{side}})$ vs. $\bar{R}^{1/\nu} \times t_T$. (b) $\bar{R}^{-\gamma/\nu} \times (R^2_{\text{out}} - R^2_{\text{side}})$ vs. $\bar{R}^{1/\nu} \times t_{\mu_B}$. The $(R^2_{\text{out}} - R^2_{\text{side}})$ values are the same as those in Left panel. The figures are taken from [40].

resolution, the pseudo-rapidity acceptance from $|\eta| < 1$ to $|\eta| < 1.5$ and acceptance for low p_T particles from 125 MeV/c to 60 MeV/c. The EPD is a new dedicated event-plane and centrality detector placed in the forward rapidity region $2 < |\eta| < 4$. With 16 radial segments and 24 azimuthal segments, the detector will provide precise measurements of both the collision centrality and the event plane for all possible harmonics. The Endcap Time-Of-Flight (eTOF) is the result of a CBM and STAR collaboration [47]; 10% of the full CBM ToF system will be installed on one side of STAR, covering $-1.1 < y < -1.6$, as part the FAIR Phase-0 program. In combination with the new inner TPC sectors, particle identification will now be possible at forward rapidities, see right panel of Fig. 9.

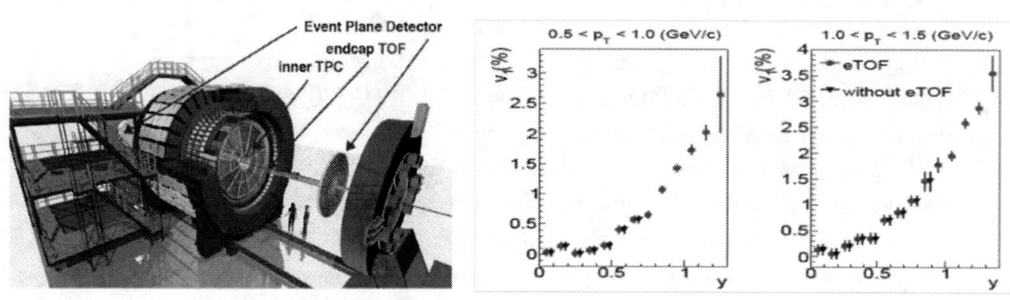

Figure 9. (Color online) Left: Upgrades of the STAR detector for the second phase of beam energy scan at RHIC. Right: The expected improvement in the rapidity dependence of the proton v_1 due to the inclusion of the iTPC and eTOF upgrades for the BES-II [47].

6 Acknowledgment

This work was partially supported by the Ministry of Science and Education of the Russian Federation, grant N 3.3380.2017/4.6, and by the National Research Nuclear University MEPhI in the framework of the Russian Academic Excellence Project (contract No. 02.a03.21.0005, 27.08.2013).

References

[1] K. Adcox *et al.* [PHENIX Collaboration], Nucl. Phys. A **757** (2005) 184
[2] J. Adams *et al.* [STAR Collaboration], Nucl. Phys. A **757** (2005) 102
[3] C. Gale, S. Jeon, B. Schenke, P. Tribedy and R. Venugopalan, Phys. Rev. Lett. **110** (2013) no.1, 012302
[4] U. Heinz and R. Snellings, Ann. Rev. Nucl. Part. Sci. **63** (2013) 123
[5] Y. Aoki, G. Endrodi, Z. Fodor, S. D. Katz and K. K. Szabo, Nature **443** (2006) 675
[6] S. Ejiri, Phys. Rev. D **78** (2008) 074507 ; E. S. Bowman, J. I. Kapusta, Phys. Rev. C **79** (2009) 015202.
[7] X. Luo, Nucl. Phys. A **956** (2016) 75
[8] D. Keane, J. Phys. Conf. Ser. **878** (2017) no.1, 012015.
[9] H. Caines, Nucl. Phys. A **967** (2017) 121.
[10] A. Adare *et al.* [PHENIX Collaboration], Phys. Rev. C **93** (2016) no.2, 024901
[11] A. Bazavov *et al.* [HotQCD Collaboration], Phys. Rev. D **90** (2014) 094503
[12] L. Adamczyk *et al.* [STAR Collaboration], arXiv:1707.01988 [nucl-ex].
[13] S. Voloshin and Y. Zhang, Z. Phys. C **70** (1996) 665
[14] S. A. Voloshin, A. M. Poskanzer and R. Snellings, arXiv:0809.2949 [nucl-ex].
[15] R. Snellings, J. Phys. G **41** (2014) no.12, 124007
[16] R. Derradi de Souza, T. Koide and T. Kodama, Prog. Part. Nucl. Phys. **86** (2016) 35
[17] R. A. Lacey, A. Taranenko, J. Jia, D. Reynolds, N. N. Ajitanand, J. M. Alexander, Y. Gu and A. Mwai, Phys. Rev. Lett. **112** (2014) no.8, 082302
[18] R. A. Lacey, D. Reynolds, A. Taranenko, N. N. Ajitanand, J. M. Alexander, F. H. Liu, Y. Gu and A. Mwai, J. Phys. G **43** (2016) no.10, 10LT01
[19] N. Magdy [STAR Collaboration], J. Phys. Conf. Ser. **779** (2017) no.1, 012060.
[20] R. A. Lacey *et al.* , arXiv:1601.06001 [nucl-ex].
[21] R. A. Lacey and A. Taranenko, PoS CFRNC **2006** (2006) 021
[22] A. Adare *et al.* [PHENIX Collaboration], Phys. Rev. C **93** (2016) no.5, 051902
[23] L. Adamczyk *et al.* [STAR Collaboration], Phys. Rev. C **86** (2012) 054908
[24] L. Adamczyk *et al.* [STAR Collaboration], Phys. Rev. C **88** (2013) 014902
[25] L. Adamczyk *et al.* [STAR Collaboration], Phys. Rev. Lett. **116** (2016) no.11, 112302
[26] J. Auvinen and H. Petersen, Phys. Rev. C **88** (2013) no.6, 064908
[27] X. Sun [STAR Collaboration], J. Phys. Conf. Ser. **535** (2014) 012005.
[28] I. A. Karpenko, P. Huovinen, H. Petersen and M. Bleicher, Phys. Rev. C **91** (2015) no.6, 064901
[29] S. Singha, P. Shanmuganathan and D. Keane, Adv. High Energy Phys. **2016** (2016) 2836989
[30] M. Asakawa, S. Ejiri and M. Kitazawa, Phys. Rev. Lett. **103** (2009) 262301
[31] M. Asakawa *et al.*, Phys. Rev. Lett. **103**, 262301 (2009); M. A. Stephanov, Phys. Rev. Lett. **102**, 032301 (2009).

[32] A. Bazavov *et al.*, Phys. Rev. Lett. **109**, 192302 (2012); S. Borsanyi *et al.*, Phys. Rev. Lett. **111**, 062005 (2013); P. Alba *et al.*, Phys. Lett. B **738**, 305 (2014). F. Karsch *et al.*, Nucl. Phys. A, these proceedings. [arXiv:1512.06987].
[33] M. Cheng *et al.*, Phys. Rev. D **79**, 074505 (2009). A. Bazavov, *et al.*, Phys. Rev. D **86**, 034509 (2012).
[34] L. Adamczyk *et al.* [STAR Collaboration], Phys. Rev. Lett. **112** (2014) 032302
[35] M. A. Stephanov, K. Rajagopal and E. V. Shuryak, Phys. Rev. D **60** (1999) 114028
[36] M. A. Stephanov, Phys. Rev. Lett. **102** (2009) 032301
[37] C. Athanasiou, K. Rajagopal and M. Stephanov, Phys. Rev. D **82** (2010) 074008
[38] M. A. Stephanov, Phys. Rev. Lett. **107** (2011) 052301
[39] M. A. Lisa, S. Pratt, R. Soltz and U. Wiedemann, Ann. Rev. Nucl. Part. Sci. **55** (2005) 357
[40] R. A. Lacey, Phys. Rev. Lett. **114** (2015) no.14, 142301
[41] R. A. Lacey, P. Liu, N. Magdy, B. Schweid and N. N. Ajitanand, arXiv:1606.08071 [nucl-ex].
[42] STAR Note 598. https://drupal.star.bnl.gov/STAR/starnotes/public/sn0598
[43] C. Yang [STAR Collaboration], Nucl. Phys. A **967** (2017) 800.
[44] K. Meehan [STAR Collaboration], Nucl. Phys. A **967** (2017) 808
[45] STAR Note 619. https://drupal.star.bnl.gov/STAR/starnotes/public/sn0619
[46] STAR Note 666. https://drupal.star.bnl.gov/STAR/starnotes/public/sn0666
[47] [STAR Collaboration and CBM eTOF Group], arXiv:1609.05102 [nucl-ex].

EPJ Web of Conferences **164**, 06006 (2017) DOI: 10.1051/epjconf/201716406006
ICNFP 2016

The First Asymmetry Measurements in High-Energy Polarized Proton-Nucleus Collision at PHENIX-RHIC

Itaru Nakagawa[1,2,a] for the PHENIX Collaboration

[1] *RIKEN, 2-1 Hirosawa, Wako, Saitama, 351-0198, Japan*
[2] *Riken-BNL Research Center, 20 Pennsylvania Avenue Bldg.510A, Upton, NY 11973, U.S.A.*

Abstract. The single spin asymmetries in very forward neutron production had been first observed about a decade ago at RHIC in transversely polarized proton + proton collision at $\sqrt{s}$ = 200 GeV. Although neutron production near zero degrees is well described by the one-pion exchange (OPE) framework, the OPE appeared to be not satisfactory to describe the observed analyzing power A_N. The absorptive correction to the OPE generates the asymmetry as a consequence of a phase shift between the spin flip and non-spin flip amplitudes. However the amplitude predicted by the OPE is too small to explain the large observed asymmetries. Only the model which introduces interference between major pion and small a_1-Reggeon exchange amplitudes has been successful in reproducing the experimental data. During RHIC Run-15, RHIC delivered polarized proton collisions with Au and Al for the first time, enabling the exploration of the mechanism of transverse single-spin asymmetries with nuclear collisions. A very striking A-dependence was discovered in very forward neutron production at PHENIX in transversely polarized proton + nucleus collision at $\sqrt{s}$ = 200 GeV. Such a dependence has not been predicted from the existing framework which has been succesful in proton + proton collision. In this report, experimental and theoretical efforts are discussed to disentangle the mysterious A-dependence in the very forward neutron asymmetry.

1 Introduction

The single spin asymmetry of very forward (almost zero degree) was first discovered[1] at RHIC in transversely polarized proton+proton collision at $\sqrt{s}$ = 200 GeV. Neutrons were measured by a zero-degree calorimeter (ZDC) [2] with a position-sensitive shower-maximum detector (SMD).

The magnitude of observed asymmetry of the neutron which has the energy fraction of incident proton $x_F = E_n/E_p \geq 0.4$ was several percent. This asymmetry was unexpectedly large. The production mechanism of the such a neutron is driven by soft process and of which cross section is well described by one pion exchange (OPE) model. However the predicted asymmetry within the OPE framework appeared to be very tiny and far underestimated the data[3]. The forward neutron asymmetry is formulated as

$$A_N \propto \phi_{\rm flip}\phi_{\rm non-flip} \sin\delta \quad (1)$$

[a] e-mail: itaru@riken.jp

where ϕ_{flip} ($\phi_{\text{non-flip}}$) is spin flip (non-flip) amplitude between incident proton and out-going neutron, and δ is the relative phase between these two amplitudes. Although the OPE can contribute to both spin flip and non-flip amplitudes, resulting A_N is small due to the small relative phase. The several percent amplitude was generated only by introducing the interference between spin flip π exchange and spin non-flip a_1-Reggeon exchange which has large phase shift in between[3]. Shown in Fig. 1 are observed forward neutron asymmetries in three different collision energies 64, 200, and 500GeV and plotted as a function of transverse momentum p_T of detected neutron. The data demonstrate the absolute amplitude of the asymmetries grow as a function of p_T. These data are well reproduced by the model calculations[3].

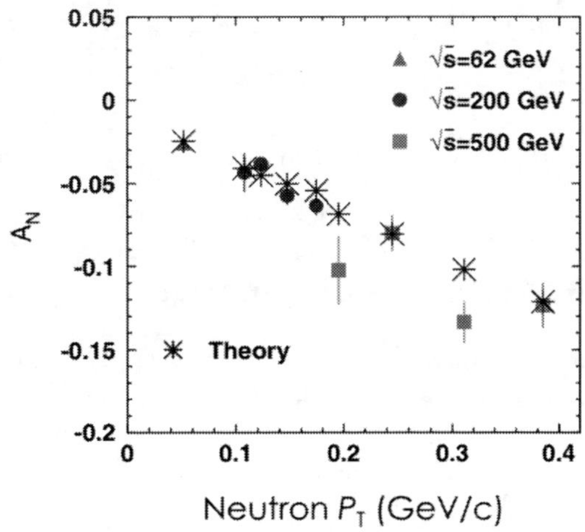

Figure 1. The forward neutron asymmetries observed p+p collision system at three different collision energies 64, 200, and 500GeV plotted as a function of transverse momentum p_T. The data points are well reproduced by the model calculations assumes the interference between π and a_1-Reggeon[3].

2 Experimental Setup

The ZDC is composed of Cu-W alloy absorbers with PMMA-based communication grade optical fibers and it corresponds to 1.7 nuclear interaction length for one module. A photomultiplier collects Cherenkov lights via the optical fibers in one module. As shown in Fig. 2, three ZDCs are located in series at ± 18 m away from the collision point within the small acceptance, covering 10 cm in the transverse plane. It has 5.1 nuclear interaction length and 149 radiation length by combining three ZDC modules, and has energy resolution about 20% for 100 GeV neutron.

The SMD consists of x-y scintillator strip hodoscopes and they are inserted between the first and second ZDC modules at the position of maximum hadronic shower approximately. The x-coordinate (horizontal) is sampled by 7 scintillator strips of 15 mm width, while the y-coordinate (vertical) is sampled by 8 strips of 20 mm width, tilted by 45 degree. It gives a position by calculating the center of gravity of shower generated in the first ZDC module. Its position resolution is estimated to be about 1 cm for 50 GeV neutron in the simulation study.

The data was collected by triggers employing the ZDCs and Beam Beam Counters (BBCs). One BBC consists of 64 sets of PMTs and 3 cm thick quartz Cherenkov radiator. BBCs are mounted

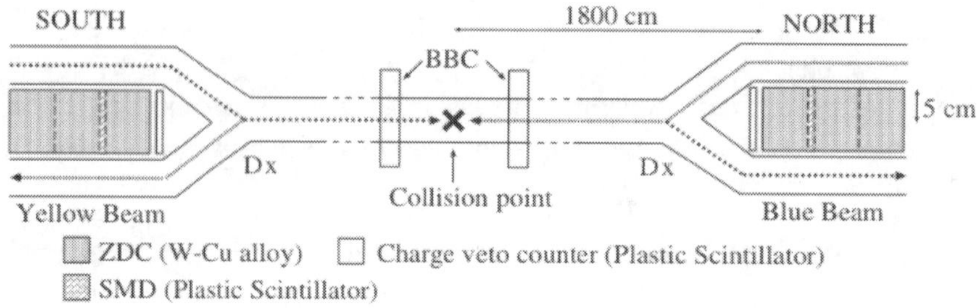

Figure 2. The experimental setup for ZDC and BBC detectors. The ZDCs were located ±18 m downstream from the collision point. The BBCs were installed around the beam pipe at ±18 m downstream of the collision point.

around beam pipe located ±144 cm away from the collision point. They cover ±(3.0–3.9) and 2π in η and ϕ spaces, respectively.

3 New Forward Neutron Asymmetries in p+A

In year 2015, the first attempt was made to collide the transversely polarized proton and nucleus in RHIC at $\sqrt{s} = 200$ GeV. In order to explore the atomic mass number (A) evolution of the asymmetry, the forward (p-going side) neutron asymmetries were measured for p+Au and p+Al collisions. Since the proton beam is unpolarized at LHC, the measurement is the high energy frontier of the polarized proton and nucleus collision. The existing polarized proton+nucleus experiments are carried out in one order of magnitude lower $\sqrt{s}$, for instance, at RHIC[5] and Fermilab [11] using polarized proton beam and fixed nuclear targets.

The asymmetry analysis basically followed the same procedures and conditions as previous p+p measurements. The measurement in p+p collision system was repeated in year 2015 and resulting asymmetry was consistent with previous measurements. The ZDC acceptance was segmented into eight azimuthal bins and evaluated left-right raw asymmetries ϵ_N^{raw} defined by the Eqn. (2) in the each bin independently.

$$\epsilon_N^{raw}(\phi) = \frac{N^+(\phi) - N^-(\phi)}{N^+(\phi) + N^-(\phi)} \tag{2}$$

where $N^+(\phi)$ and $N^-(\phi)$ are number of neutrons observed in the given azimuthal ϕ bin when protons are polarized up and down, respectively.

The observed ϕ dependent asymmetries normalized by the proton polarization P are plotted in Fig. 4 as a function of ϕ. The $\phi = 0$ is set by the polarization direction of the proton. The analyzing power A_{N} is evaluated by fitting the data with a sine function:

$$\frac{\epsilon_N^{raw}(\phi)}{P} = A_{\mathrm{N}} \sin(\phi - \phi_0) \tag{3}$$

where ϕ_0 allows a deviation of the maximum asymmetry axis from vertical.

Fig.4 shows preliminary results of forward neutron A_{N} measurements in 2015. The red points are the results of forward neutron A_{N} inclusive measurements. Surprisingly, they show unexpectedly strong mass number (A) dependence. The asymmetry even flips the sign from p+p, p+Al to p+Au.

 DOI: 10.1051/epjconf/201716406006

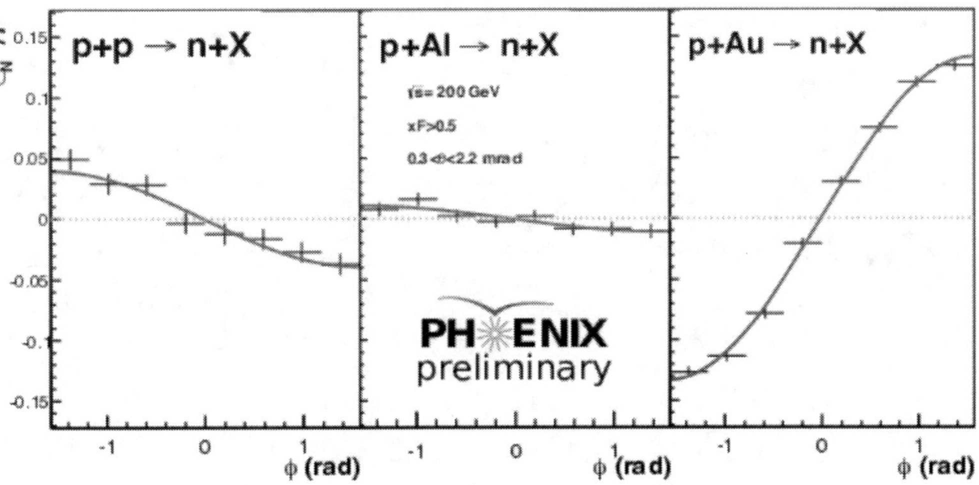

Figure 3. (Observed azimuthal angular modulations of the inclusive forward neutron production in p+p (left), p+Al (middle) and p+Au (right) collisions in year 2015.

The p+Au data point shows magnificently large A_N of about 0.18 which is three times larger than that of p+p in absolute amplitude. The existing π and a_1-Reggeon interference framework predicts only moderate evolution as growing A and does not have any mechanism to flip the sign of A_N in any p+A collision systems.

More interestingly, another drastic dependence of A_N was observed in correlation measurements in addition to the inclusive neutron. In these measurements, another out-going charged particle was either tagged or vetoed within the acceptance of the BBC in both North and South arms. The BBCs cover such a limited acceptance as described in Section 2, but the resulting asymmetries behaved remarkably contradicts. Once BBC hits (BBC tagging) are required in both arms (green data points), the drastic behavior of inclusive A_N is vanished and no flipping sign was observed between p+p and p+Au. On the contrary, the asymmetries are pushed even more positive for p+Al and p+Au data points once no hits in BBC are required (BBC vetoed) as represented by blue data points.

4 Ultra-Peripheral Collision (Primakoff) Effects

The electro-magnetic (EM) interaction was ignored in the forward neutron production in the case of p+p collision system. However EM interaction may not be ignorable for p+Au case. In fact the importance of EM interaction was pointed out in high energy forward hadron production [11] and elastic p+A scattering [5] experiments. Due to the smallness of the four momentum transfers of the present kinematics, i.e. $-t \leq 0.5$ $(\mathrm{GeV}/c)^2$, the EM interaction may play a role which becomes increasingly important in large atomic number nucleus. The EM field of the nucleus becomes rich source of exchanging photons between the polarized proton. This is known as Primakoff effect or the ultra-peripheral collision (UPC) in heavy ion collider experiments. In the UPC process, there is no charge exchange at the collision vertex unlike π or a_1 meson exchange. The forward neutron in the

 DOI: 10.1051/epjconf/201716406006

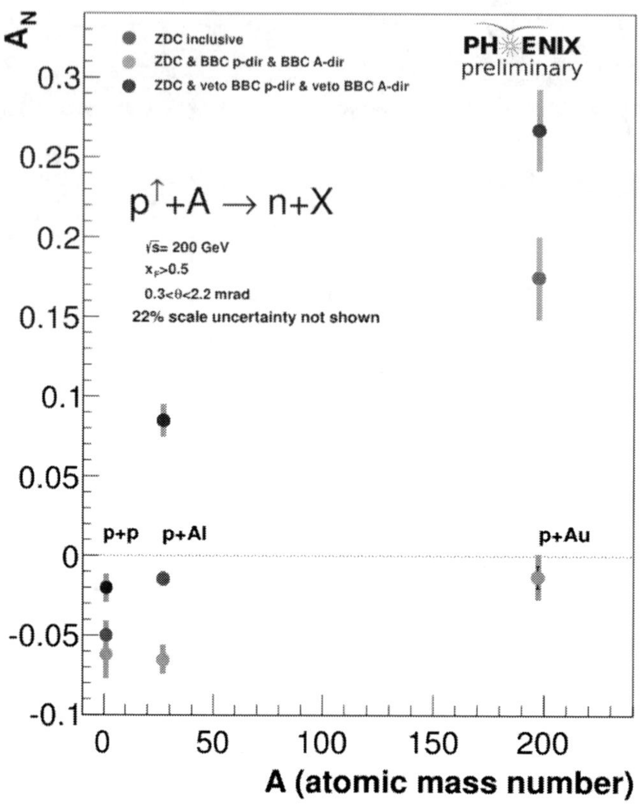

Figure 4. (Observed forward neutron A_N in transversely polarized proton-nucleus collisions. Data points are A=1, A=27, and A=197 are results of p+p, p+Al, and p+Au, respectively. Red, Blue and Green data points are neutron inclusive, neutron + BBC veto, and BBC tagged events, respectively

final state then can be produced through the neutron decay channel from the nucleon excitation states as shown in diagrams in Fig. 5.

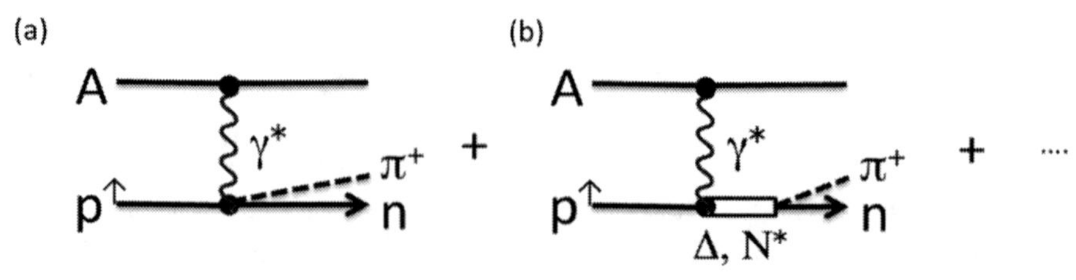

Figure 5. Diagrams of Primakoff process to leave the high-z neutron in the final state to the proton going directions. (a) threshold photo pion production (b) Δ or N^* excitation and its decay into n+π^+ channel.

The description of A_N is thus extended from Eqn. (1) to Eqn. 5, which includes not only hadronic but also EM amplitudes:

$$\begin{aligned} A_N \propto & \phi^{had}_{flip}\phi^{had}_{non-flip}\sin\delta_1 + \phi^{EM}_{flip}\phi^{had}_{non-flip}\sin\delta_2 \\ + & \phi^{had}_{flip}\phi^{EM}_{non-flip}\sin\delta_3 + \phi^{EM}_{flip}\phi^{EM}_{non-flip}\sin\delta_4 \end{aligned} \quad (4)$$

where 'EM' and 'had' stand for electromagnetic and hadronic interactions, and $\delta_1 \sim \delta_4$ are relative phases, respectively. The second and the third terms are known as Coulomb nuclear interference (CNI) and to be discussed in details in Section 5.

A Monte-Carlo (MC) study was made to predict the kenematic behavior of the process through the Primakoff Effect. The nucleon excitation through Primakoff effect can be quantified into two steps; 1) equivalent photon flux from nuclear Coulomb field, and 2) photo-nucleon excitation. We employed STARLIGHT[6] for modeling nuclear Coulomb field and used Weizsacker-Williams formula[7] to calculate the equivalent photon flux. Among many models are available in the market, we employed SOPHIA model[8] which covers higher excitation states above Δ region. While the MC is modeling higher excitation states, the first excitation states such as Δ is rather important in this case because the equivalent photon flux rapidly drops as a function of the photon energy. Threshold pion production even below resonance may play non-negligible role. According to the MC study [9], the neutron and its counter part π^+ via Primakoff process is substantially boosted towards the proton beam direction and therefore the fragmenting π^+ are mostly emitted in even higher rapidity region than BBC as shown in Fig. 6. Only small fraction (6%) of π^+ are detected by BBC. Thus EM processes are suppressed in the BBC tagging events while enhanced in the BBC vetoed events. As a consequence, one may draw a hypothesis that the moderate A_N evolution of the BBC tagging data are consistent with the prediction based on π and a_1-Reggeon interference scenario because the data are dominated by the hadronic amplitude. Although it is not trivial to predict the sign of the A_N caused by EM amplitude, the observed growing feature of A_N for BBC vetoed events makes sense if EM amplitude guides to opposite sign of hadronic interactions at least qualitatively. Nevertheless it is not known at all at this moment why EM should have opposite sign nor should have such a large amplitude or large relative phases between hadronic/EM spin-flip and spin-nonflip amplitudes to produce such the large asymmetries. It remains mystery.

5 Coulomb-Nuclear Interference

It is known to have finite A_N in the elastic proton + proton (Carbon) scattering process using fixed target with transversely polarized high energy proton beam. In fact the polarization of the proton beams at RHIC are determined by measuring the asymmetry of the elastic scattering on fixed targets. The magnitude of the $|A_N|$ is $< 5\%$ asymmetry in the momentum transfer region of $0.002 < -t < 0.012$ in p + p, and p + A [5] as shown in Fig. 7 . Because of the smallness of the momentum transfer, the EM interaction plays important role in these processes as shown in the Fig.8 (b), while the hadron interaction amplitude is mainly driven by Pomeron exchange as shown in Fig.8 (a). As expected from the nature of EM interaction, stronger EM field in the large A nucleus results in the distinctive A-dependence in the observables. Since this is the elastic scattering, the final states are the same in both EM and hadronic interactions by definition. Thus these two amplitudes interfere and their non-zero relative phase creates finite asymmetry as observed.

The CNI interference in the very forward neutron asymmetry is not trivial as the elastic case since the final states of both EM and hadronic interactions have to exactly match among infinite number

DOI: 10.1051/epjconf/201716406006

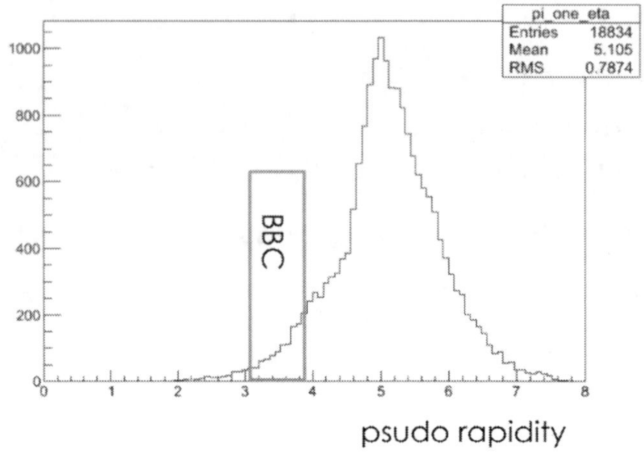

Figure 6. Psudorapidity distribution of π^+ production through Primakoff process predicted by MC[9]. The π^+ are mostly emitted to forward psudo rapidity region than BBC covers 3.1<η<3.9.

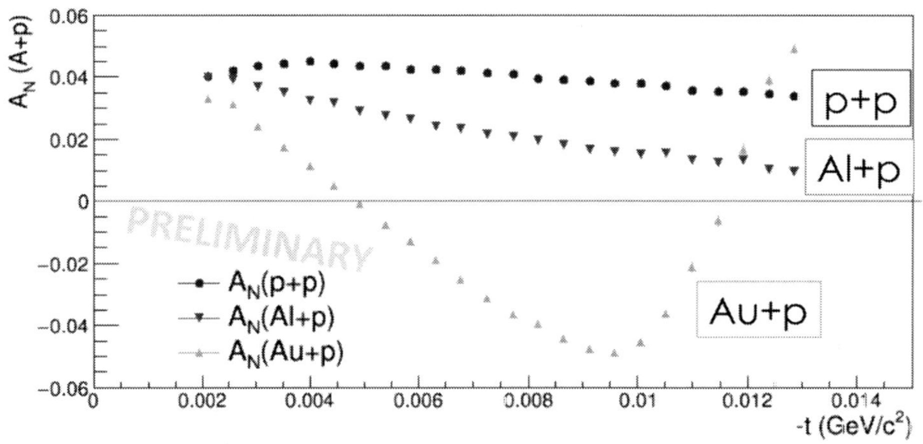

Figure 7. Elastic A_N for p + p, Al + p, and Au + p plotted as a function of 4-momentum transfer $-t$.

of possible multi-particle states. One of the possible leading order diagrams are drawn in Fig.8 (c) for the hadronic and (d) for the EM interactions, respectively. (c) is the inelastic neutron production via Pomeron-π scattering, and (d) is $n + \pi^+$ decay channel of photo-Δ or N^* excitation. Due to the requirement on the same final state, it is natural to consider the lower the multiplicity in the final state, the higher the chance to contribute to CNI. Thus the neutron production via diffractive process is expected to play rather important role CNI terms. On the other hand, the magnitude of $|A_N|$ was only < 5% level as a consequence of CNI in the even in the Au + p elastic case and it is not clear yet if CNI can induce the $|A_N|$ in the order of 25 % for the neutron production.

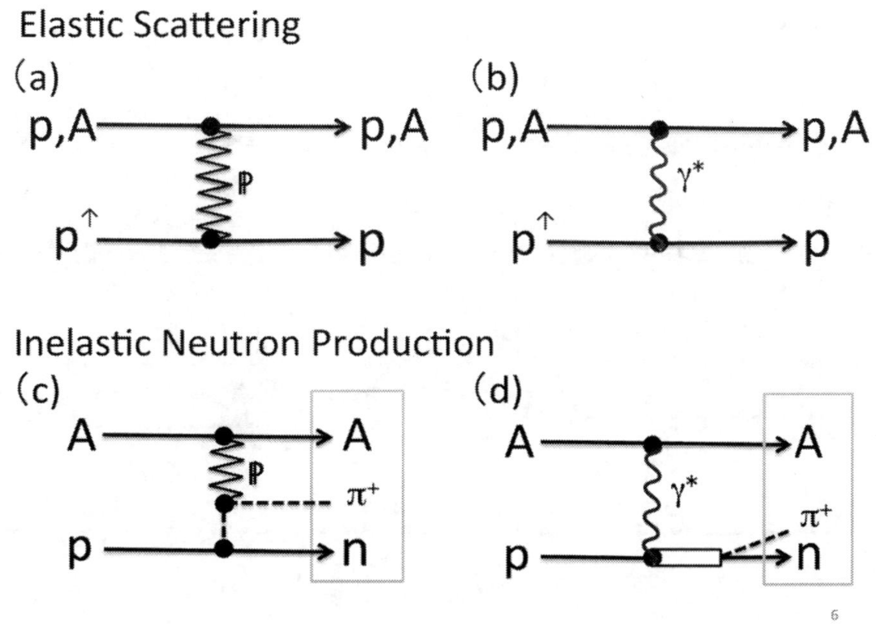

Figure 8. Feynman diagrams of the elastic proton scattering via Pomeron exchange (a) and one-photon exchange (b). The inelastic neutron production via Pomeron-π scattering (c), and $\pi + n$ decay channel of photo- Δ or N^* excitation (d).

6 Fermilab Fixed Target Experiment

There is one possibly related experimental results were reported by E704 collaboration at Fermilab using a fixed target [11]. They measured A_N in inclusive π_0 production using 185 GeV transversely polarized proton beam on a lead target. The center of mass energy was $\sqrt{s}$ = 19.5 GeV, which is one order of magnitude lower than the present experiment. They reported a large $A_N = -0.57 \pm 0.12$(statistical) $\pm^{+0.21}_{-0.18}$(scale uncertainty due to the dilution factor) was observed in π^0 very small $|t'| < 0.001$ (GeV/c)^2 region[1] in $p + Pb \rightarrow p + \pi^0 + X$ reaction. Their claim was the cause of the asymmetry comes from the interference between Δ and N^*(Roper) via Primakoff process. Unlike the present experiment, they measured out-going proton in coincidence with pion, they can determine the final state kinematics assuming two body decay of the proton and the lead nucleus remains grand state. The observed A_N neither in the larger $|t'| > 0.001$(GeV/c)^2 nor in the invariant mass M region of Δ, i.e. $M < 1.36$ GeV were zero consistent within their precision. The only invariant mass region $1.36 < M < 1.52$(GeV/c^2) above Δ and small $|t'| < 0.001$ (GeV/c)^2 region showed finite and large asymmetry as described above. They didn't claim the CNI contribution to the finite asymmetry because their $|t'|$ was believed to be sufficiently small. However, the importance of CNI effect was pointed out later by other author [12].

Although the origin of A_N in the Fermilab experiment is still remain unresolved as of today, it may provide some hints to reveal the origin of the present neutron A_N. To some extent, the Fermilab and the present experiments are similar if we consider as follows. The Fermilab experiment measures

[1]See reference [11] for the definition of t'.

 DOI: 10.1051/epjconf/201716406006

$$A_N \propto \phi^{had}_{non-flip}\phi^{had}_{flip}\sin\delta_1 + \phi^{EM}_{non-flip}\phi^{had}_{flip}\sin\delta_2 + \phi^{had}_{non-flip}\phi^{EM}_{flip}\sin\delta_3 + \phi^{EM}_{non-flip}\phi^{EM}_{flip}\sin\delta_4$$

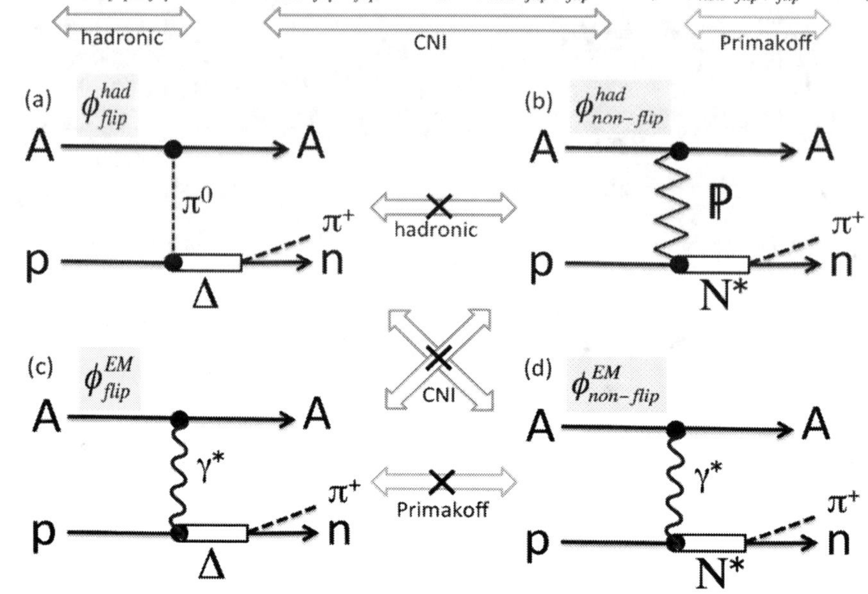

Figure 9. Leading diagrams for the hadronic interaction (top) and EM (bottom) for (n+π^+) decay channel assuming similar interference effect between Δ and N^* as the Fermilab experiment. (a) spin-flipping Δ excitation via π^0-exchange (b) spin non-flipping N^* excitation via Pomeron exchange, (c) spin-flipping Δ excitation and (d) spin non-flipping N^* excitation via one photon exchange diagrams. The hadronic, CNI, and Primakoff terms of $A_{\rm N}$ can be given by the interference between spin-flip and non spin-flip amplitudes of hadronic interaction, hadronic and EM interactions, and EM interactions, respectively.

the (p+π^0) decay channel whereas the present experiment measure (n+π^+) one. Since their claim was the origin of the asymmetry was the interference between Δ and N^* via Primakoff, the corresponding leading diagrams in the case of hadronic interaction can be one π^0 exchange as shown in Fig. 9(a) for spin flip, and Pomeron exchange as (b) for non-spin flip amplitudes. The diagram (c) and (d) shows the (n+π^+) decay channel of the spin flipping photo-Δ excitation and non spin-flipping photo-N^* excitations, respectively. The hadronic, CNI, and Primakoff terms of $A_{\rm N}$ can be given by the interference between spin-flip and non spin-flip amplitudes of interaction, hadronic and EM interactions, and EM interactions, respectively. These diagrams are not consistent to the leading order diagram raised in Fig. 8(c). In order to be conclusive to judge which diagram is important, quantitative calculation is necessary.

7 Summary

We have observed surprisingly strong A-dependence in the forward neutron asymmetry in transversely polarized proton-nucleus collision at $\sqrt{s}$=200GeV in PHENIX using ZDC detector. Such a drastic A-dependence in the forward neutron asymmetry $A_{\rm N}$ is absolutely unexpected within the current framework although it was successful in reproducing the $A_{\rm N}$ data in p+p. Another drastic dependence has been observed in the semi-inclusive measurements by requiring both North and South BBCs

$3.1 < |\eta| < 3.9$ to be fired or not-fired in addition to forward neutron. The EM effect may play key role to disentangle these drastic behavior because 1) its amplitude glows as a function of atomic number Z, and 2) the fraction of EM process events are suppressed in BBC fired events while enhanced in BBC vetoed events. 3) Possibly related experiments [5] and [11] points out the importance of EM effects. Further experimental dependency study is certainly be a help to distinguish the hadronic and EM interaction due to their different dependence in impact radius. The corresponding efforts are now underway by segmenting p_T into multiple bins.

References

[1] Y. Fukao et al., Phys. Lett. B 650, 325 (2007), A. Adare et al.: Phys. Rev. D 88, 032006 (2013).
[2] C. Adler, A. Denisov, E. Garcia, M. J. Murray, H. Strobele, and S. N. White, Nucl. Instrum. Methods Phys. Res., A 470, 488 (2001).
[3] B. Z. Kopeliovich, I. K. Potashnikova, and Ivan Schmidt: Phys. Rev. Lett. 64, 357 (1990).
[4] D. C. Carey et al.: Phys. Rev. D 79, 094014 (2009).
[5] I. G. Alekseev et al.: Phys. Rev. D 79, 094014 (2009).
[6] S. R. Klein, J. Nystrand, Phys. Rev. C 60, 014903 (1999), STARLIGHT webpage, http://starlight.hepforge.org/
[7] J. D. Jackson," *Classical Electrodynamics* " , John Wiley & Sons, Inc. (1999).
[8] A. Mucke, R. Engel, J.P. Rachen, R.J. Protheroe, and T. Stanev, Comput. Phys. Commun. 123, 290-314 (2000); SOPHIA webpage, http://homepage.uibk.ac.at/ c705282/SOPHIA.html
[9] G. Mitsuka, Eur. Phys. J. C 75, 614 (2015).
[10] O. Eyser, PSTP workshop, Bochum, Sept. (2015).
[11] D. C. Carey et al.(FNAL-E704 Collaboration), Phys. Rev. Lett. 64, 357 (1990).
[12] Elliot Leader, "*Spin in Particle Physics*", Cambridge University Press (2001).

EPJ Web of Conferences **164**, 07001 (2017) DOI: 10.1051/epjconf/201716407001
ICNFP 2016

4D Sommerfeld quantization of the complex extended charge

Igor E. Bulyzhenkov[1,2,a]

[1] *Moscow Institute of Physics and Technology, 9 Institutskiy per., Dolgoprudny, Moscow reg., 141700, Russia*
[2] *Lebedev Physics Institute RAS, 53 Lininsky pros., Moscow, 119991, Russia*

Abstract. Gravitational fields and accelerations cannot change quantized magnetic flux in closed line contours due to flat 3D section of curved 4D space-time-matter. The relativistic Bohr-Sommerfeld quantization of the imaginary charge reveals an electric analog of the Compton length, which can introduce quantitatively the fine structure constant and the Plank length.

1 London superfluid motion

In 1939 Einstein first inferred the logical failure of the Schwarzschild curved metric with singularities for physical reality from a *gedanken* experiment [1]. Gravitation or acceleration analogs of the Aharonov-Bohm Effect were never found in SQUID experiments that indirectly confirms flatness of 3D space. Strict quantization of magnetic flux and steady SQUID measurements in time-varying gravitational and acceleration fields have refuted non-Euclidean metric approaches to 3D geometry in advanced metric theories of gravitational fields.

The superfluid motion of Cooper pairs in electromagnetic fields $\mathbf{A}$ can be formally described by the macroscopic Schrodinger equation with a complex continuous function $\Psi = n^{1/2}exp(i\chi/\hbar c)$. This approach of condensed matter theory results in the well-tested London supercurrent,

$$\mathbf{j} \equiv \frac{i\hbar q}{2m}(\Psi^*\nabla\Psi - \Psi\nabla\Psi^*) + \frac{q^2}{2mc}|\Psi|^2\mathbf{A} = \frac{qn}{mc}(\nabla\chi - q\mathbf{A}). \tag{1}$$

Reffered by London's 3-currents, we start 4-vector operations with the canonical four-momentum density, $n(x)P_\mu(x) \equiv -n(x)\nabla_\mu\chi/c = -n(x)\partial_\mu\chi/c$, of superelectrons using the phase function gradient $\nabla_\mu\chi$ and the covariant derivative ∇_μ in metric space-time, $\mu \to \{0, 1, 2, 3\}$. The symmetrical Christoffel coefficients in Einstein's General Relativity (GR) provide the universal covariant equality $\nabla_\mu P_\nu - \nabla_\nu P_\mu \equiv \partial_\mu P_\nu - \partial_\nu P_\mu \equiv -(\partial_\mu\partial_\nu - \partial_\nu\partial_\mu)\chi \equiv 0$ for charged elementary matter independently from its density function $n(x)$. This material density, as was assumed by London in 1934, is to be finite in all space points where the single-valued phase χ of superfluid matter is defined.

[a] e-mail: bulyzhenkov.ie@mipt.ru

2 4D path quantization of extended charged particles

One may conventionally associate canonical energy-momentum densities of a distributed Cooper pair with relativistic gravitational and electromagnetic fields,

$$-\frac{n}{c}\partial_\mu\chi = ng_{\mu\nu}\left(mc\frac{dx^\nu}{ds} + \frac{q}{c}A^\nu\right) \equiv n\{P_o, P_i\} \equiv n\left\{\frac{mc\sqrt{g_{oo}}}{\sqrt{1-v^2c^{-2}}} + \frac{g_{o\nu}qA^\nu}{c}, \frac{-m(\gamma_{ij}v^j + \sqrt{g_{oo}}g_i c)}{\sqrt{1-v^2c^{-2}}} + \frac{g_{i\nu}qA^\nu}{c}\right\}, \quad (2)$$

where $v^i \equiv cdx^i(g_{oo})^{-1/2}(dx^o - g_i dx^i)^{-1} \equiv dx^i/d\tau$ is the relativistic three-velocity of the local analytical density $n = n(x)$ of the continuous charge $q = \int qn\sqrt{|det\gamma_{ij}|}dV$ and the continuous mass $m = \int mn\sqrt{|det\gamma_{ij}|}dV$, $v^2 \equiv \gamma_{ij}v^i v^j$, $\gamma_{ij} \equiv g_i g_j g_{oo} - g_{ij}$, $g_i \equiv -g_{oi}/g_{oo}$, $i \to \{1,2,3\}$.

Now one can generalize the Bohr-Sommerfeld quantization rule on the self-coherent four-momentum densities (2) of a distributed elementary particle with continuous mass, $mn(x)\sqrt{|det\gamma_{ij}|}$, and charge, $qn(x)\sqrt{|det\gamma_{ij}|}$, densities by taking into account the GR time synchronization, $d\tau \equiv \sqrt{g_{oo}}(dx^o - g_i dx^i)/c = 0$ or $dx^o = g_i dx^i$, for neighboring space points of closed line contours,

$$\pm 2\pi\hbar N = \frac{1}{c}\oint dx^\mu \nabla_\mu \chi_s \equiv \oint\left[-P_i dx^i - P_o dx^o\right] \equiv \oint_{d\tau=o}\left[\frac{m(\gamma_{ij}v^j + c\sqrt{g_{oo}}g_i)}{\sqrt{1-v^2c^{-2}}} - \frac{q(g_{io}A^o + g_{ij}A^j)}{c}\right]dx^i$$
$$-\oint_{d\tau=o}\left(\frac{mc\sqrt{g_{oo}}}{\sqrt{1-v^2c^{-2}}} + \frac{q(g_{oo}A^o + g_{oj}A^j)}{c}\right)g_i dx^i \equiv \oint_{d\tau=o}\left[\frac{mv^j}{\sqrt{1-v^2c^{-2}}} + \frac{qA^j}{c}\right]\gamma_{ij}dx^i. \quad (3)$$

We arrived for the curved space-time to the very Bohr-Sommerfeld quantisation rule from the Special Theory of Relativity. The electric potential $A^o = A^o(x)$ is not relevant to the mechano-magnetic fluxoid quantization (3) for the charged elementary mass in electromagnetic and gravitational fields. Indeed, Coulomb electric fields $A^o(x)$ of external charges have never destroyed quantization of any elementary particle, including the Cooper boson of two superelectrons. Similarly, Newton gravitational field, or $g_{oo}(x)$, should also be irrelevant in practice to quantization of elementary matter. It is not surprising that this metric component has vanished in (3) for the relativistic contour quantization in 4D space-time.

On the other side, the 4D quantization rule (3) still corresponds to gravity-dependent 3D space with the inhomogeneous metric tensor $\gamma_{ij}(x)$. But gravity-dependent phase shifts along different path lines between two material points of the same continuous Cooper boson, for example, contradict to a single valued phase requirement for always self-coherent states of free elementary matter. Consequently, such path-varying shifts would destroy steady Bohr-Sommerfeld quantization in closed line contours. The observed magnetic flux quantization in superconducting loops has confirmed, in fact, strict spatial flatness of quantum reality, where $\gamma_{ij}(x) = \delta_{ij}$.

3 Complex charge reveals meaning of $\alpha = 1/137$ and $l_p = 1.62 \times 10^{-35}m$

The curved 3D metric solution was adopted in 1916 due to the use of the unrealistic empty space paradigm and the Newton point particle references rather than the continuous particle references from the 1925 nondual wave mechanics. Nowadays Einstein's metric formalism may replace the operator material density $\delta(\mathbf{x} - \mathbf{x}_e)$ of the postulated point particle with an analytical density of the continuous particle from the non-empty space paradigm. Author can propose, for instance, the elementary density solution $n_e(\mathbf{x}, \mathbf{x}_e, t) = r_o/4\pi(\mathbf{x} - \mathbf{x}_e)^2(r_o + |\mathbf{x} - \mathbf{x}_e|)^2$ to both Maxwell and Einstein equations with extended radial electron. The theory parameter $r_o = Q\sqrt{G}/c^2$ for electron's radial density in the classical (probability-free) is defined by the elementary charge $Q = (ie + \sqrt{G}m_o)$ having the center of spherical symmetry at $\mathbf{x}_e$ [2].

Under such a non-empty-space approach, all metric components $g_{\mu\nu}$ of the strongly warped 4D space-time interval for distributed radial matter can depend on gravitational fields, but keep six inherent metric bounds or symmetries $g_{oi}g_{oj}g_{oo}^{-1}(x) = g_{ij}(x) + \delta_{ij}$. Pseudo-Riemannian 4-geometry with 6 inherent metric symmetries $\gamma_{ij}(x) = \delta_{ij}$ has the universal, Euclidean 3D space sub-interval in nonlocal GR equations and provides new physical interpretation of the known gravitational tests [3]. Strict spatial flatness of the warped space-time manifold reinforces the surface independent magnetic flux (3) over a closed line, the gauss electric flux over a closed surface, and gravity/inertia independent quantization of elementary matter in covariant relativistic relations.

Despite the nonlocal nature of radial sources in non-empty material space, the continuous electron behaves in collisions with energy exchanges like a material point, because half of electron's mass and charge is concentrated within the ultra small radius $l_{1/2} = |r_o| = 1.38 \times 10^{-36} m$ (which is far behind the top resolution $10^{-19} m$ of space measurements in CERN).

Electron's relativistic 4-current $Qn_e(\mathbf{x}, \mathbf{x}_e, t)u_\mu$ is proportional to the complex energy 4-flow, $\varphi_o Qn_e(\mathbf{x}, t)u_\mu$, where $Q = (ie + \sqrt{G}m_o)$ is the complex charge of the elementary energy carrier and $\varphi_o = c^2 / \sqrt{G} = 1.04 \times 10^{27} V$ is the universal self-potential for the mass and the electric charge. Complex self- energy $E(\mathbf{x}_e)$ of such a distributed charge centered around $\mathbf{x}_e$ can be found from the volume integral

$$E(\mathbf{x}_e) = \int d^3x \varphi_o Qn_e(\mathbf{x}, \mathbf{x}_e, t)u_\mu u^\mu = \varphi_o Q = m_o c^2 + iec^2 G^{-1/2} = 511 KeV + i1.04 \times 10^{27} KeV. \quad (4)$$

The action of this complex elementary energy is given by the line integral over the 4-path x_e^μ

$$S = -\int \frac{\varphi_o Q u_{e\mu} dx_e^\mu}{c} \Rightarrow (m_o c + \frac{iec}{\sqrt{G}}) \oint_{\tau=0} \sqrt{d\tau^2 - dl^2} = \pm nh \pm ikh. \quad (5)$$

From here one can introduce as the well known Compton length, $\bar{l}_C = \hbar / m_o c = 3.86 \times 10^{-13} m$, as well as its electrical analog for the electron, $\bar{l}_e = \hbar \sqrt{G} / ec = 1.89 \times 10^{-34} m$. By compare this fundamental length with the half-charge length $l_{1/2}$ of the radial distribution,

$$\begin{cases} \bar{l}_e / l_{1/2} = \hbar c / e^2 \equiv \alpha^{-1} = 137, \\ \bar{l}_e \cdot l_{1/2} = \hbar c / \varphi_o^2 = \hbar G / c^3 \equiv l_p^2 = (1,62 \times 10^{-35} m)^2. \end{cases} \quad (6)$$

one can knowledgeable introduce into the extended electron theory the fine structure constant α and the Plank length l_P.

References

[1] A. Einstein, *Annals of Mathematics*, **40**, 922 (1939).

[2] I. E. Bulyzhenkov, *Complex Charge Densities Unify Particles with Fields and Gravitation with Electricity*, Bullet. Lebedev Phys. Inst, **43**, 140 (2016).

[3] I. E. Bulyzhenkov, *Geometrization of radial particles in non-empty space complies with tests of General Relativity,* Journal of Modern Physics, **3**, N10, 1465 (2012).

 DOI: 10.1051/epjconf/201716407002

A Measurement of the Absolute Reactor Antineutrino Flux and Spectrum at Daya Bay

Fengpeng An[1,a] on behalf of the Daya Bay Collaboration

[1] Meilong Road 130, East China University of Science and Technology, Shanghai, 200237, China

Abstract. The Daya Bay Reactor Neutrino Experiment uses an array of eight underground detectors to study antineutrinos from six reactor cores with different baselines. Since the start of data-taking from late 2011, Daya Bay has collected the largest sample of reactor antineutrino events to date, and has made the most precise measurement of the neutrino oscillation parameters $sin^2 2\theta_{13}$ and Δm^2_{ee}. Using the data from the four detectors in the near experimental halls, Daya Bay has made a high statistics measurement of the absolute reactor antineutrino flux and spectrum. In this paper we will present this measurement and its comparison to predictions based on different flux models.

1 Introduction

Reactor neutrino experiments played an important role in neutrino physics since the first experimental observation of antineutrinos by Reines and Cowan et al.[1]. In 2012 Daya Bay experiment observed non-zero neutrino mixing angle θ_{13} with a significance more than 5 σ [2], consistent with recent measurements from other experiments[3–6]. The near future reactor neutrino experiments are aiming to determine the neutrino mass hierarchy, improve the precision of neutrino mixing parameters measurement, and search for short baseline neutrino oscillation. In reactors, electron antineutrinos are emitted primarily from the fissions of four isotopes: ^{235}U, ^{238}U, ^{239}Pu and ^{241}Pu. The modeling of the reactor antineutrino spectrum of the four isotopes are based on the measurements of the beta spectra at ILL Grenoble for ^{235}U, ^{239}Pu and ^{241}Pu [7–9], and the theoretical calculation from Vogel for ^{238}U [10], before 2011, and the predictions of the reactor antineutrino flux and spectrum based on the ILL+Vogel model agreed well with experiment data. In 2011, Huber and Muller carried out re-evaluation of the antineutrino spectrum model with improved theoretical treatments [11, 12] and the predictions of the reactor antineutrino flux based on the Huber+Muller model was shown to be higher than the experimental data, implying a discrepancy which is commonly referred as the "Reactor Antineutrino Anomaly". Recent reactor neutrino experiments also observed an excess of detected events comparing to predictions in the antineutrino energy region of 5-7 MeV, raising the suspicion of the accuracy of existing reactor antineutrino prediction models. It is thus important to precisely measure the reactor antineutrino flux and spectrum and compare with model predictions. This paper will report the recent absolute reactor antineutrino flux and spectrum measurement of Daya Bay experiment based on 621 days of data taking, with more than 1.2 million antineutrino candidates collected from eight antineutrino detectors.

[a] e-mail: anfengpeng@gmail.com

 DOI: 10.1051/epjconf/201716407002

2 The Daya Bay Reactor Neutrino Experiment

The Daya Bay experiment is designed to precisely measure the neutrino mixing angle θ_{13} employing relative measurements between multiple detectors. The Daya Bay nuclear power complex is built at Daya Bay in southern China, about 55 kilometers to Hong Kong. As shown in Fig. 1 the power complex consists of three nuclear power plants (NPPs) and each NPP has a pair of reactor cores generating 2.9 GW thermal power each. The reactor power plants are designed based on the French Framatome Advanced Nuclear Power three cooling loop design. The antineutrino detectors (ADs) were built in three underground experimental halls (EHs): EH1 and EH2 each consists of two identical ADs and EH3 has four ADs.

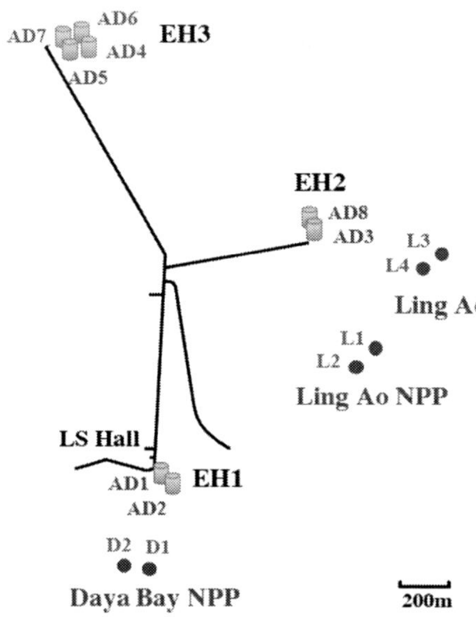

Figure 1: Layout of the full configuration of the Daya Bay experiment with eight antineutrino detectors (ADs) installed in three underground experimental halls (EHs). The dots represent reactor cores, labeled as D1, D2, L1, L2, L3 and L4.

The reactors are pure antineutrino sources. Almost all antineutrinos are produced from the fission of ^{235}U, ^{238}U, ^{239}Pu and ^{241}Pu, fissions of other isotopes contribute less than 0.3%. The reactor antineutrino spectrum with energy E at time t can be predicted by:

$$\frac{d^2\phi(E,t)}{dEdt} = \frac{W_{th}(t)}{\sum_i f_i(t)e_i}\sum_i f_i(t)S_i(E)c_i^{ne}(E,t) + S_{\mathrm{SNF}}(E,t) \qquad (1)$$

where i is the index of the four isotopes, W_{th} is the thermal power of the reactor core measured by the systems of the NPP with 0.5% uncertainty, f_i is the fission fraction of the isotope i provided by the NPP using commercial reactor fuel evolution simulation package SCIENCE based on APOLLO2 code [13] as shown in Fig. 2, e_i is the energy release per fission of isotope i [14], $S_i(E)$ is the antineutrino spectrum of fission of isotope i from prediction models, c_i^{ne} is the correction factor for the non-equilibrium effects of long-lived fission fragments, and S_{SNF} is the antineutrino contribution from the spent nuclear fuel.

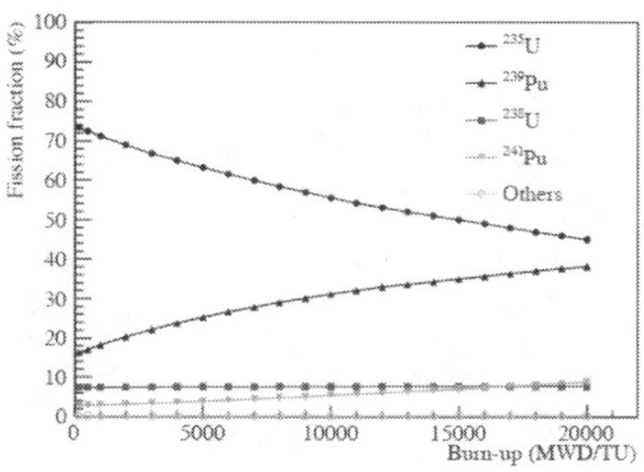

Figure 2: Fission fractions of isotopes in reactor core D1 as a function of cycle burn-up from a simulation of a complete refueling cycle. Other isotopes contribute less than 0.3%.

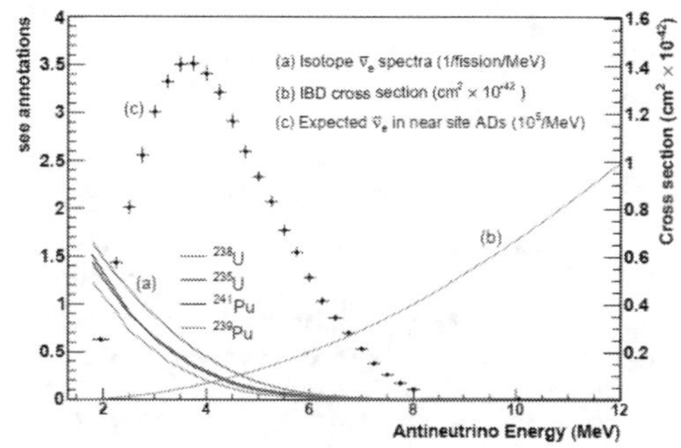

Figure 3: (a) The antineutrino spectra for four isotopes in Huber+Mueller model.(b) The inverse beta decay (IBD) cross section. (c) The expected antineutrino spectrum weighted by the IBD cross section without oscillation in the near site ADs. The error bars are systematic only.

The antineutrinos are detected via inverse beta decay (IBD) interactions: $\bar{\nu}_e + p \to e^+ + n$ in the gadolinium-doped liquid scintillator (GdLS) of the Daya Bay ADs. The expected antineutrino spectrum in detector d is calculated by:

$$S_d(E_\nu, t) = \sum_r \frac{1}{4\pi L_{dr}^2} \frac{d^2\phi(E,t)}{dEdt} \epsilon^d N_p^d \sigma(E_\nu), \qquad (2)$$

where L_{dr} is the baseline from reactor r to detector d, ϵ^d is the detection efficiency of detector d, N_p^d is the target proton number of detector, and $\sigma(E_\nu)$ is the IBD cross section. Fig. 3 shows the expected antineutrino spectrum without oscillation in the near site ADs.

The inverse beta decay candidates were selected by applying a series cuts, requiring a time coincidence between a prompt signal from the positron and a following delayed signal from the neutron captured on Gd 30 μs later on average. The selection and cuts are described in [15]. The IBD selection efficiency was evaluated by carrying out detailed Mote Carlo simulation using GEANT4 based simulation package. By considering the IBD selection criteria and effects of target proton number, an absolute detection efficiency common to all detectors was defined and was evaluated to be 80.6%. The total systematic uncertainty of detection efficiency is 1.93%.

The data taking has two periods, during the first period from December 2011 to July 2012, six ADs were deployed for data taking, two ADs were at Daya Bay site(EH1), one at the Ling Ao site(EH2), and three were at the far site (EH3). During the second period from October 2012 to November 2013, the full eight-AD detector configuration was deployed with two ADs at each near sites and four ADs at the far site. With 621 days data taking time, more than 1.2 million inverse beta decay (IBD) candidates were collected, forming the largest reactor antineutrino sample.

3 Measurement of Reactor Antineutrino Flux

The reactor flux can be measured by the near site ADs. A normalization factor R was defined to scale the measured reactor antineutrino rate to the predicted rate. Two methods were used to extract the reactor antineutrino flux. In the first method, the value of R together with the $sin^2 2\theta_{13}$ were determined by a χ^2 fitting using all data of the eight ADs. Fig. 4 shows the daily averaged rate of IBD events per AD in each experimental halls during the data taking time.

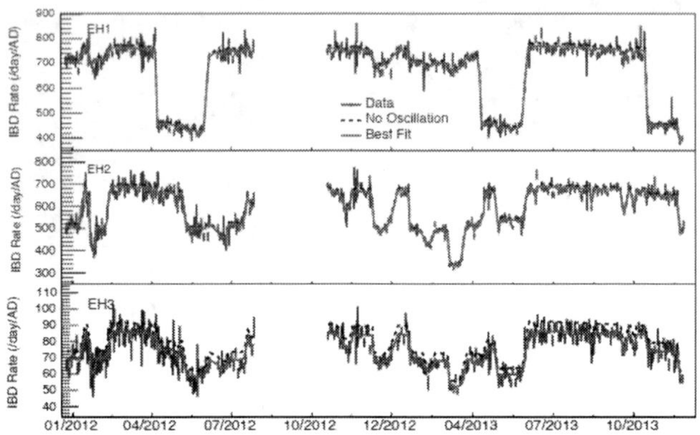

Figure 4: Daily averaged rates of IBD candidate events per AD in the three experimental halls as a function of time. The dotted curves represent no-oscillation predictions based on reactor antineutrino flux analyses and detector simulation. The predictions incorporated the best-fit normalization parameter (R). The red solid curves represent the rates predicted with the best-fit $\sin^2 2\theta_{13}$.

The χ^2 function using the integrated rate information was constructed as below:

$$\chi^2 = \sum_{d=1}^{8} \frac{[M_d - R \cdot T_d(1 + \epsilon_D + \sum_r \omega_r^d \alpha_r + \epsilon_d) + \eta_d]^2}{M_d + B_d} + \sum_{r=1}^{6} \frac{\alpha_r^2}{\sigma_r^2} + \sum_{d=1}^{8} \left(\frac{\epsilon_d^2}{\sigma_d^2} + \frac{\eta_d^2}{\sigma_{B,d}^2} \right) + \frac{\epsilon_D^2}{\sigma_D^2} \tag{3}$$

where M_d is the measured IBD event number in detector d (backgrounds subtracted), B_d is the number of background events, T_d is the predicted IBD event number by a reactor model with neutrino oscillations, and ω_r^d is the fractional IBD contribution from the r-th reactor to the d-th detector. σ_r (0.9%) is the uncorrelated reactor uncertainty, σ_d (0.2%) is the uncorrelated detection uncertainty, $\sigma_{B,d}$ is the background uncertainty listed in Ref. [16], and σ_D (1.93%) is the correlated detection efficiency uncertainty. α_r, ϵ_d, η_d, and ϵ_D are the corresponding nuisance parameters respectively.

The minimization of the χ^2 function gives the best-fit value of R and $\sin^2 2\theta_{13}$. The best-fit value of R is $0.946 \pm 0.020(0.992 \pm 0.021)$ with respect to the Huber+Mueller (ILL+Vogel) model. The best-fit value of $\sin^2 2\theta_{13} = 0.085 \pm 0.006$.

In the second method, we use the near site data and correct the oscillation using the value of measured $\sin^2 2\theta_{13}$. The measured reactor antineutrino flux is expressed in two model-independent ways: the measured IBD events in each detector are normalized to σ_f with unit: $cm^2 fission^{-1}$ and Y with unit: $cm^2 GW^{-1} day^{-1}$.

σ_f for each AD can be determined by solving the following equation:

$$M_d = \sum_{r=1}^{6} \frac{N_r^f}{4\pi L_{dr}^2} \sigma_f^d N_d^{\mathrm{T}} P_{sur}^{dr} \epsilon_d^D, \tag{4}$$

 DOI: 10.1051/epjconf/201716407002

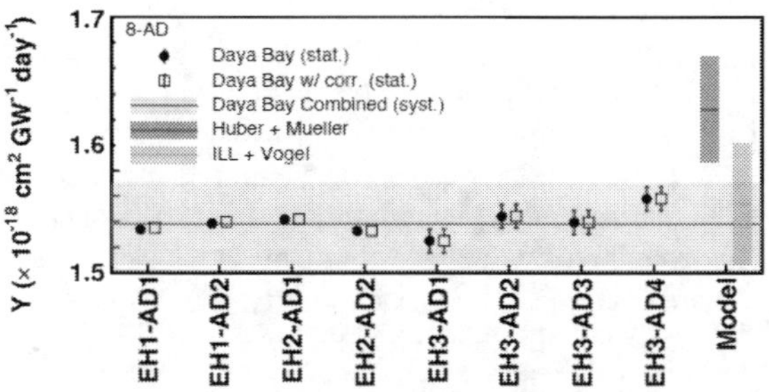

Figure 5. Yield Y for the IBD events with corrections of 3-flavor oscillations (closed circles), and additional corrections to the variations of flux-weighted fission fractions at different sites (open squares). The horizontal line is the average yield of the near detectors, and the gray band is its 1 σ systematic uncertainty. The rate predicted by the Huber+Mueller (ILL+Vogel) model and its uncertainty are shown in blue (orange) region.

where N_r^f is the predicted number of fission number of reactor core r using average thermal power and average fission fraction, N_d^{T} is the target proton number, P_{sur}^{dr} is the survival probability of the antineutrinos from reactor r to d, and ϵ_d^D is the detection efficiency.

The IBD yield per GW thermal power per day is defined as $Y_d \equiv \sigma_f^d N_r^f / W_r$. Fig. 5 shows the fission yield of the ADs. Table. 1 summaries the IBD fission Yield of the near site ADs, the flux normalization R with respect to Huber+Muller (ILL+Vogel) Model, and the average fission fraction.

Table 1: The average IBD yields (Y and σ_f) of the near halls, the flux normalization with respect to different reactor model predictions, and the flux-weighted average fission fractions of the near halls.

IBD Yield	
Y (cm^2/GW/day)	$(1.53 \pm 0.03) \times 10^{-18}$
σ_f (cm^2/fission)	$(5.91 \pm 0.12) \times 10^{-43}$
Data / Prediction	
R (Huber+Mueller)	0.946 ± 0.020 (exp.)
R (ILL+Vogel)	0.992 ± 0.021 (exp.)
^{235}U : ^{238}U : ^{239}Pu : ^{241}Pu	0.561 : 0.076 : 0.307 : 0.056

We also performed a global fit of the past reactor neutrino experiments, comparing the measurements with prediction using Huber+Muller model as the reference, including 21 short baseline measurements using the data from [17] and CHOOZ[18] and Palo Verde [19] . The final result is:

$$R_{past} = 0.942 \pm 0.009(exp.) \pm 0.023(model), \tag{5}$$

The Daya Bay result $R_{DYB} = 0.946 \pm 0.020(exp.)$ is consistent with this global average. The results of the global fit and the Daya Bay measurement are shown in Fig. 6.

4 Measurement of Reactor Antineutrino Spectrum

We extend our study from the reactor antineutrino flux to the energy spectrum. The measured positron spectrum of IBD events from the four ADs of the near sites were combined and compared with predictions using Huber+Muller model. The predicted antineutrino energy spectrum was transformed into prompt energy spectrum utilizing the detector response matrix. The response matrix was built considering the IBD prompt energy shift, the effects of energy loss in the inactive acrylic vessels, the non-linearity effect, and the energy resolution. The response matrix was evaluated with two methods:

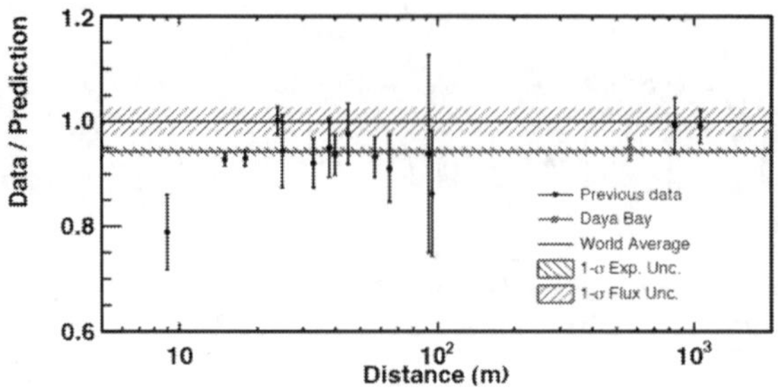

Figure 6. The measured reactor $\bar{\nu}_e$ rate as a function of the distance from the reactor, normalized to the prediction using Huber+Mueller model. The rate is corrected with neutrino oscillations at the distance of each experiment. The purple shaded region represents the global average and its 1 σ uncertainty. The 2.4% model uncertainty is shown as a band around unity. The measurements at the same baseline are combined together for clarity. The Daya Bay measurement is shown at the flux-weighted baseline (573 m) of the two near halls.

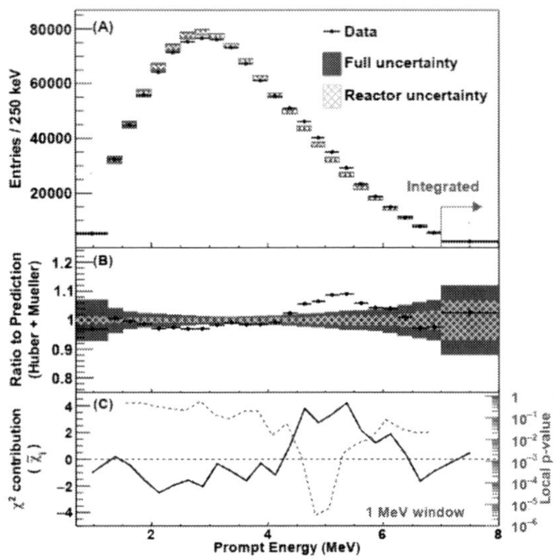

Figure 7. (A) Comparison of predicted and measured prompt energy spectra. The error bars on the data points represent the statistical uncertainty. The hatched and red filled bands represent the square-root of diagonal elements of the covariance matrix ($\sqrt{(V_{ii})}$) for the reactor related and the full systematic uncertainties, respectively. (B) Ratio of the measured prompt energy spectrum to the predicted spectrum. (C) The defined χ^2 distribution ($\widetilde{\chi}_i$) of each bin (black solid curve) and local P-values for 1 MeV energy windows (magenta dashed curve).

a step-by-step analytical method and a full detector simulation method base on GEANT4. Both methods produced consistent results. Fig. 7 shows the measured near site combined prompt spectrum and its comparison with the prediction. The prediction based on the Huber+Mueller model is normalized to the number of measured events.

From Fig. 7 , a "bump" like structure is clearly visble. A χ^2 function was constructed to qualify the discrepancy of the measurement and the prediction:

$$\chi^2 = \sum_{i,j}(N_i^{\text{obs}} - N_i^{\text{pred}})(V^{-1})_{ij}(N_j^{\text{obs}} - N_j^{\text{pred}}), \tag{6}$$

where $N_i^{\text{obs(pred)}}$ is the observed (predicted) number of events at the i-th prompt energy bin and V is the covariance matrix that including all uncertainties. The energy ranges from 0.7 to 12 MeV. Fitting the χ^2 function yields $\chi^2/dof = 46.6/24$, corresponding to a 2.9 σ discrepancy over all the energy range.

To qualify the local discrepancy of the bump structure, another two methods were deployed to evaluate the significance of the deviation. The first method calculates the χ^2 contribution of each energy bin:

$$\widetilde{\chi}_i = \frac{N_i^{\text{obs}} - N_i^{\text{pred}}}{|N_i^{\text{obs}} - N_i^{\text{pred}}|}\sqrt{\sum_j \chi_{ij}^2},$$
$$\chi_{ij}^2 = (N_i^{\text{obs}} - N_i^{\text{pred}})(V^{-1})_{ij}(N_j^{\text{obs}} - N_j^{\text{pred}}). \quad (7)$$

From Fig. 7 panel (C), the enhanced contribution around 5 MeV can be seen clearly. The second method introduces N (number of bins) nuisance parameters with no pull terms to the oscillation fitter. The χ^2_{min} difference before and after introducing the N nuisance parameters is expected to follow a χ^2 distribution with N-1 d.o.f., therefore a P-value can be calculated. Fig. 7 panel (C) shows the P-values with 1 MeV sliding energy window, and around 5 MeV the discrepancy is $> 4\ \sigma$. Further more, the local significance for the 2 MeV window between 4 and 6 MeV was evaluated, and we obtained $\Delta\chi^2/dof = 37.4/8$, corresponding to the P-value of $9.7 \times 10^{-6}(4.4\ \sigma)$. The events around 5 MeV were carefully examined, the events match all IBD event characteristics, more checks ruled out detector effects, simple background contribution and other interaction channels as explanations. This study suggests that the deviation is due to the imperfection of the modeling of the reactor antineutrino spectrum.

A generic reactor antineutrino spectrum for the IBD reaction was extracted from the Daya Bay measured spectrum. This spectrum is independent with the Daya Bay detector response and could be used as an input to predict the reactor antineutrino flux and spectrum for other experiments. The near site measured prompt spectrum were combined together, and then the detector response effects were removed by utilizing unfolding to the prompt spectrum to obtain the antineutrino spectrum. Two unfolding algorithms, the Singular Value Decomposition(SVD) and Bayes iteration algorithms were applied and got almost same results. In addition, the independence of unfolded spectrum to Monte Carlo input of response matrix, the minimization of variance plus bias were all tested.

Oscillation effects was then removed from the unfolded antineutrino spectrum and further more each energy bin of the antineutrino spectrum was normalized to $cm^2/fission/MeV$ using the reactor information, making the antineutrino spectrum to be a model-independent spectrum:

$$S_{\text{generic}}(E) = \frac{S_{\text{combined}}(E)}{P_{\text{sur}}(E, L) \cdot N_P \cdot F_{\text{total}}}, \quad (8)$$

where S_{combined} is the combined IBD prompt spectrum of the near site ADs, $P_{\text{sur}}(E, L)$ is the average survival probability of the $\bar{\nu}_e$ calculated with the fluxes from the six reactors to the four detectors, N_P is the number of target protons of the average target mass, and F_{total} is a normalization factor based on the baseline-weighted total number of fissions. Fig. 8 shows the obtained generic antineutrino spectrum. As a comparison, the predicted spectrum of the near site antineutrino spectrum was processed with the same steps as in Equation (8) and was plotted in Fig. 8 Panel (B), which has the same rate deficit as the flux measurement and similar spectral deviations as in the comparison of measured and predicted IBD prompt spectra.

A possible application of the generic spectrum is to predict antineutrino spectrum for other reactor experiments. A simple example is shown below. Suppose a reactor antineutrino experiment A, the predicted antineutrino spectrum of experiment A can be expressed as:

$$S_A = S_{dyb} + \sum_i (f_{A_i} - f_{dyb_i}) S_{mod_i}, \quad (9)$$

where i is the index of the four fissile isotopes, S_{dyb} is the Daya Bay generic spectrum $S_{generic}(E)$, f_{dyb} and f_A are the average fission fractions of the Daya Bay experiment and the reactor antineutrino

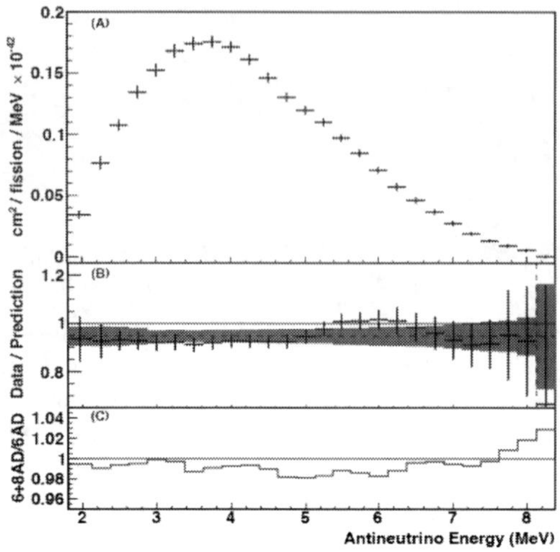

Figure 8. (A) The antineutrino spectrum weighted by the IBD cross section. The last bin is integrated up to 12 MeV. (B) Ratio of the extracted reactor antineutrino spectrum to the Huber+Mueller prediction. The error bars of the data points are the square-roots of the diagonal elements of the antineutrino spectrum covariance matrix. The solid red band represents the square-roots of the diagonal elements of the prediction covariance matrix, including both reactor and Huber+Mueller model uncertainties. (C) the ratio of the spectra from the 6+8 AD periods used in this analysis and the 6 AD period used in the previous analysis [20].

experiment A, and S_{mod} are the isotope antineutrino spectra from models, such as ILL+Vogel, Huber+Mueller, etc. This method works when the difference of the average fission fraction between experiment A and Daya Bay is relative small, which is usually true for PWR reactor experiments.

5 Summary

With 621 days of data taken by eight ADs, more than 1.2 million IBD events were collected by the Daya Bay experiment. The measured absolute reactor antineutrino flux is consistent with previous measurements. The measured IBD yield is $(1.53 \pm 0.03) \times 10^{-18}$ cm^2/GW/day or $(5.91 \pm 0.12) \times 10^{-43}$ cm^2/fission. The ratio of measured flux to the predictions is 0.946 ± 0.020 (0.992 ± 0.021) for the Huber+Mueller (ILL+Vogel) model. The predicted and measured spectra were compared and found a deviation of 2.9 σ. In the region of 4-6 MeV the event excess has a local significance of 4.4 σ. A generic reactor antineutrino spectrum was extracted from the measured IBD prompt spectrum, providing a model-independent prediction option for future reactor antineutrino experiments.

Acknowledgement

Part of work in this paper is supported by the Fundamental Research Funds for the Central Universities (Program No.222201414032), and by the National Natural Science Foundation of China (11405057, 11390383).
The Daya Bay Experiment is supported in part by the Ministry of Science and Technology of China, the United States Department of Energy, the Chinese Academy of Sciences, the CAS Center for Excellence in Particle Physics, the National Natural Science Foundation of China, the Guangdong provincial government, the Shenzhen municipal government, the China General Nuclear Power Group, the Research Grants Council of the Hong Kong Special Administrative Region of China, the Ministry of Science and Technology and the Ministry of Education in Taiwan, the U.S. National Science Foundation, the Ministry of Education, Youth and Sports of the Czech Republic, the Joint Institute of Nuclear Research in Dubna, Russia, the NSFC-RFBR joint research program, the National Commission for Scientific and Technological Research of Chile. We acknowledge Yellow River Engineering Consulting Co., Ltd. and China Railway 15th Bureau Group Co., Ltd. for building the underground laboratory.

We are grateful for the ongoing cooperation from the China Guangdong Nuclear Power Group and China Light & Power Company.

References

[1] F. Reines, et al., Phys. Rev. 117 (1960) 159.
[2] F. P. An et al. (Daya Bay Collaboration), Phys. Rev. Lett. 108 (2012) 171803
[3] K. Abe et al. (T2K Collaboration), Phys. Rev. D 88, 032002 (2013).
[4] P. Adamson et al. (MINOS Collaboration), Phys. Rev. Lett. 110, 171801 (2013).
[5] Y. Abe et al. (Double Chooz), Phys. Rev. Lett. 108, 131801 (2012).
[6] J. K. Ahn et al. (RENO), Phys. Rev. Lett. 108, 191802 (2012).
[7] K. Schreckenbach, G. Colvin, W. Gelletly, and F. Von Feilitzsch, Phys.Lett. B160, 325 (1985).
[8] F. Von Feilitzsch, A. A. Hahn, and K. Schreckenbach, Phys.Lett. B118, 162 (1982).
[9] A. A. Hahn et al., Phys.Lett. B218, 365 (1989).
[10] P. Vogel, G. K. Schenter, F. M. Mann, and R. E. Schenter, Phys.Rev. C24, 1543 (1981).
[11] T. A. Mueller et al., Phys.Rev. C83, 054615 (2011).
[12] P. Huber, Phys.Rev. C84, 024617 (2011).
[13] R. Sanchez et al., Nucl. Eng. Tech. 42, 474 (2010).
[14] X. B. Ma et al., Phys.Rev. C88, 014605 (2013).
[15] F. P. An et al. (Daya Bay Collaboration), Chin.Phys. C37, 011001 (2013).
[16] F. P. An et al. (Daya Bay Collaboration), Phys. Rev. Lett. 115, 111802 (2015).
[17] G. Mention et al., Phys.Rev. D83, 073006 (2011).
[18] M. Apollonio et al., Phys.Lett. B466, 415 (1999).
[19] F. Boehm et al., Phys.Rev. D64, 112001 (2001).
[20] F. P. An et al. (Daya Bay Collaboration), Phys. Rev. Lett. 116, 061801 (2016).

EPJ Web of Conferences **164**, 07003 (2017) DOI: 10.1051/epjconf/201716407003
ICNFP 2016

ATLAS Trigger and DAQ Upgrades for High-Luminosity LHC

M. ISHINO on behalf of the ATLAS collaboration[1,a]

[1] *ICEPP, The University of Tokyo, Bunkyo-ku Hongo 7-3-1, Tokyo, Japan*

Abstract. The ATLAS Trigger-DAQ system will be significantly upgraded to explore new physics beyond the standard model in the High-Luminosity LHC environment with instantaneous luminosity of up to $7.5 \times 10^{34} cm^{-2} s^{-1}$. Two general trigger schemes are under study and their prospects and concerns are discussed. The muon trigger and readout electronics will be totally replaced to cope with a longer latency and a higher trigger rate. At the same time, new trigger algorithms using the precision tracker will be introduced to sharpen efficiency turn-ons and suppress fake triggers. The algorithm and the expected performance of the muon trigger are described. The hardware based track trigger is one of the major items introduced in the upgrade program. It will improve the trigger performance of various trigger items with a combination of the calorimeter or muon signature. The procedure for processing track information and an expected performance on the electron trigger are presented.

1 Introduction

The ATLAS Collaboration [1] is planning make significant upgrades to the trigger-DAQ system to achieve the best reach to new physics beyond the standard model in the High-Luminosity LHC program. The system is designed to maintain the trigger thresholds at the peak luminosity of $7.5 \times 10^{34} cm^{-2} s^{-1}$ at similar levels to the LHC Run-1 period when the instantaneous luminosity was $0.7 \times 10^{34} cm^{-2} s^{-1}$ and keep the physics acceptance as large as possible.

A simplified example of the trigger menu is shown in Table 1. The list of trigger items with corresponding thresholds for offline analyses, geometrical acceptance in pseudo-rapidity as well as trigger rates for each trigger level at a luminosity of $7.5 \times 10^{34} cm^{-2} s^{-1}$ and $\sqrt{s} = 14$ TeV are presented. As an example the single lepton "*isolated single electron*" and "*single* μ" objects have p_T thresholds of 22 GeV and 20 GeV respectively, compared to the Run-1 threshold of 25 GeV.

Two options, a single-level and a two-level scheme, are under consideration, described in Section 2. The latency of the hardware trigger and the maximum output rate of the hardware and software triggers will be increased, with latencies of $10\,\mu s$ for the Level-0 trigger and $60\,\mu s$ for the Level-1 trigger allowing the hardware trigger to use similar algorithms to the current high-level trigger.

Information from the precision Monitored Drift-Tube (MDT) tracker will be used in the Level-0 muon trigger. The sharp turn-on of efficiency in terms of muon p_T of the MDT trigger will allow effective rejection of candidates below threshold, controlling the rate, whilst maximising the trigger efficiency. The general scheme of the Level-0 muon trigger is discussed in Section 3.

[a] e-mail: Masaya.Ishino@cern.ch

Table 1. Simplified and illustrative trigger menu at an instantaneous luminosity of $7.5 \times 10^{34} cm^{-2} s^{-1}$. There are considerable overlaps in the rates, especially for the hadronic triggers. The totals also allow for pre-scaled supporting triggers which are not listed for the tables. For hadronic items, the Event Filter rates assume additional rejection from b-tagging and other event-level analysis algorithms. Forward jets also include multi jet triggers with lower thresholds and requirements on other quantities such as invariant mass.

Item	Offline p_T Threshold [GeV]	Offline $\|\eta\|$	L0 Rate [kHz]	L1 Rate [kHz]	EF Rate [kHz]
isolated single e	22	< 2.5	200	40	2.20
forward e	35	2.4 – 4.0	40	8	0.23
single γ	120	< 2.4	66	33	0.27
single μ	20	< 2.4	40	40	2.20
di-γ	25	< 2.4	8	4	0.18
di-e	15	< 2.5	90	10	0.08
di-μ	11	< 2.4	20	20	0.25
$e - \mu$	15	< 2.4	65	10	0.08
single τ	150	< 2.5	20	10	0.13
di-τ	40,30	< 2.5	200	30	0.08
single jet	180	< 3.2	60	30	0.60*
large-R jet	375	< 3.2	35	20	0.35*
four-jet	75	< 3.2	50	25	0.50*
H_T	500	< 3.2	60	30	0.60*
E_T^{miss}	200	< 4.9	50	25	0.50*
jet + E_T^{miss}	140,125	< 4.9	60	30	0.30*
forward jet**	180	3.2 - 4.9	30	15	0.30*
Total			~1000	~400	~10

In the Level-1 trigger, track information from the ATLAS inner tracker will be available on the hardware trigger. Combining this with the output of the Level-0 calorimeter and muon signature will significantly reduce the trigger rate of a number of trigger items. As an example the expected performance of the single electron trigger is described in Section 4, and a summary is given in the last section.

2 The general schemes of the ATLAS Trigger and DAQ

A new trigger and DAQ scheme will be introduced for the high-luminosity LHC program (HL-LHC) to maintain trigger thresholds similar to those in the Run-1 period. Two options on the trigger and DAQ scheme are under study and they are presented in Figure 1. The first option in Figure 1-(a) consists of two hardware trigger levels and the Event Filter (EF) with the associated data readout path (DAQ). The second option has a single hardware trigger level and all the accepted events are sent to the EF and DAQ as shown in Figure 1-(b).

2.1 Two-level hardware trigger scheme

In the two-level hardware scheme, a Level-0 trigger rate of 1 MHz and a Level-1 trigger rate of 400 kHz are assumed. The Level-0 trigger latency is set to 10 μs and the Level-1 trigger latency is set to 60 μs. Following the diagram in the Figure 1-(a) from top to bottom, the calorimeter and muon detector information is provided to the Level-0 calorimeter (L0Calo) and muon trigger (L0Muon).

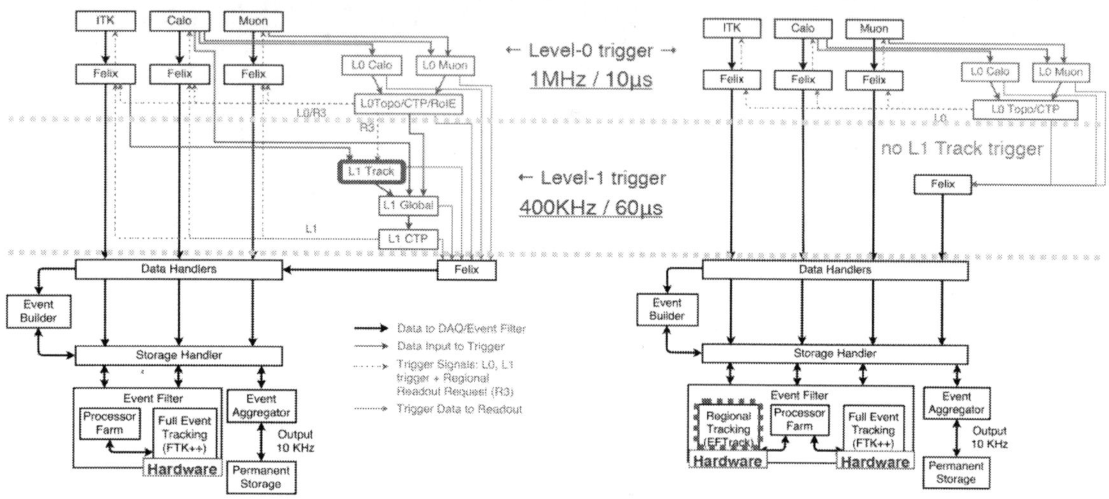

(a) (b)

Figure 1. High-level design diagram of the ATLAS trigger and DAQ system is shown. The two hardware level scheme in the Figure (a), while single level hardware scheme which does not include Level-1 trigger is shown in Figure (b). The maximum trigger rate and the maximum possible latency are indicated.

The L0Calo trigger identifies and measures the energy of electron, tau, jet candidates and calculates missing transverse-energy (E_T^{miss}). The L0Muon trigger finds muon tracks in the spectrometer and processes and determines their transverse-momentum (p_T). A new feature of the L0Muon trigger for HL-LHC is to include the precision tracker, MDT, information to determine the $p_T$of muon candidates. The Level-0 Topological trigger (L0Topo) receives a list of objects from L0Calo and L0Muon together with the information of energy and location (Region-of-Interest : RoI) in the detector, and then combines them to more complex signatures, for example invariant mass, opening angle and so on. The Level-0 Central Trigger Processor (L0CTP) receives the output of L0Topo and makes a final Level-0 decision. A pre-scale factor assigned to each trigger item and dead-time handling are taken care by the L0CTP system.

The Region-of-Interest Engine (RoIE) processes the R3 (Regional-Readout-Request) signal according to the information given by L0CTP and L0Topo and provides R3 to the data readout system (FELIX). A part of the inner-tracker (ITK) data and the calorimeter information specified in the R3 signal are sent to the Level-1 track trigger (L1Track) and the Level-1 global trigger (L1Global). The L1Track trigger finds tracks and estimates momenta on the special hardware and the output is passed to the L1Global trigger. The L1Global trigger combines the track information and the finer-grained calorimeter information, and then processes refined signatures of electrons, photons, taus, jets, $E_T^{miss}$and other hadronic signatures. The Level-1 Central Trigger Processor (L1CTP) receives the output of L1Global and makes a final Level-1 decision according to the trigger menu configuration and pre-scales.

Depending on the detector system the full event data are readout by FELIX on either the Level-0 or Level-1 trigger accept signal and transmitted to the Data Handler, where raw data is formatted. The Storage Handler receives the full set of event data and buffers until the Event Filter (EF) makes a final decision to record the event or not.

The Event Filter (EF) is placed in the last step of the trigger chain and selects events to be finally recorded or not. The EF system consists of a large scale computing farm and a dedicated hardware tracking system (FTK++) which provides a full track list using hit information of the ITK detectors. Because of the high track density in a single event, the help of special hardware is required. The event accept rate of the EF is 10 kHz.

The event data selected by the EF system is transferred from the Storage Handler to the Permanent Storage system by the Event Aggregator. The size of the raw data is estimated as 5 MB per event. The DAQ system needs to handle the event rate of 400 kHz up to the Storage Handler, so the data flow rate is 2 TB per second. The amount of recording data is 50 GB per second.

2.2 Single-level hardware trigger scheme

An alternative scheme is also under consideration. As all the detectors are able to accommodate a data readout rate of 1 MHz, a single-level hardware trigger scheme is considered and shown in Figure 1-(b).

The L0CTP system distributes the Level-0 accept signal at 1 MHz and all data is readout from the detector front-end electronics and transferred to the EF via the DAQ system. The Level-0 trigger system is the same as the two-level scheme described in the previous sub-section. The Level-1 trigger system disappears in this single-level option but the same hardware system as L1Track for regional tracking is moved to the initial step of the EF system to help reduce the rate from 1 MHz to 400 kHz, with the data flow into EF increasing to 5 TB per second.

3 Level-0 Muon Trigger

The Level-0 muon trigger (L0Muon) is a part of the ATLAS Level-0 trigger system and a major upgrade on both on-detector and off-detector electronics is planned to profit from the longer Level-0 trigger latency of 10 μs and maximum trigger rate of 1 MHz. A high-level diagram of L0Muon and DAQ is shown in Figure 2.

The ATLAS muon spectrometer consists of the trigger chambers and the precision trackers. Regarding the trigger chambers, there are Resistive-Plate-Chambers (RPC) in the barrel, Thin-Gap-Chambers (TGC) in the end-cap and the New-Small-Wheel trigger (NSW-trigger) in the end-cap inner station. Regarding the precision trackers, MDTs have been installed in the entire spectrometer as well as NSW-trackers in the end-cap inner station. In addition to the muon chambers, the outermost layer of the sampling calorimeter made of tile-scintillators and iron absorbers (TileCal) has a good signal-to-noise separation for muon signatures and the cell energy is used as input to the muon trigger.

In the endcap of the spectrometer, hit information of TGCs, vector information processed by the NSW-trigger, a 2 out of 3 layers coincidence of the RPCs (BIS-78) and the cell energies of the TileCal are sent to the TGC Sector-Logic board. The TGC Sector-Logic processes trigger signals using a

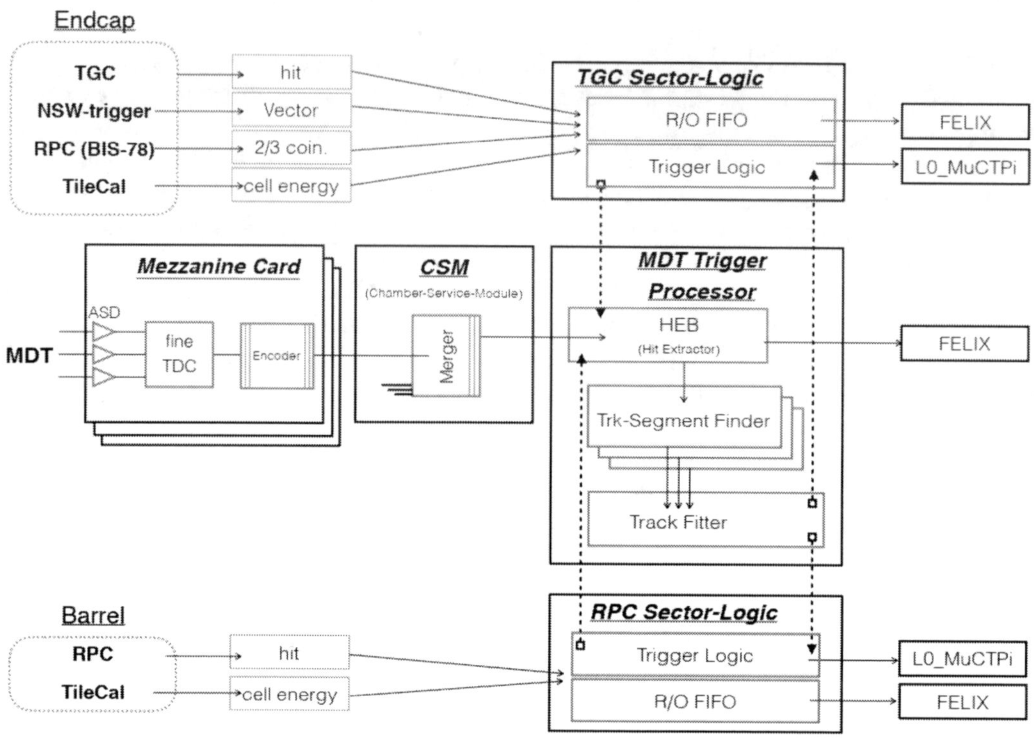

Figure 2. High-level diagram of the Level-0 muon trigger

combination of the coincidence between TGCs and the other detector systems depending on the eta range. The bending angle of the muon track due to the magnetic-field is measured and translated to p_T, and then compared with thresholds. The estimated $p_T$of the muon track, Region-of-Interest (RoI) and the triggered Bunch-Crossing Identification number (BCID) are passed to the MDT trigger processor.

The barrel trigger system works similarly as the endcap. Hit information of the RPCs as well as cell energies of TileCal are sent to the RPC Sector-Logic board and the $p_T$of the muon track is estimated using the sagitta. The trigger signal is processed and passed to the MDT trigger processor.

To date the MDT has only been used for precision tracking in the ATLAS muon spectrometer because the maximum drift-time of about 800 ns does not meet the latency allowed for the ATLAS hardware trigger. The use of MDT drift-time information in the Level-0 hardware trigger becomes feasible in the ATLAS trigger and DAQ upgrade because of the longer trigger latency of $10\,\mu$s. The precise spatial information of the MDTs has a significant impact on the momentum resolution, leading to a sharper efficiency turn-on and a consequent reduction of the muon trigger rate.

All the on-detector electronics (Mezzanine cards and Chamber Service Modules (CSM)) will be replaced with new hardware. All the TDC information is sent to the off-detector electronics board, the MDT data processor, via high-speed optical links. In the first step of the MDT data processor, the Hit-Extractor (HEB) selects MDT hits matched to the TGC or RPC RoI. At the same time, TDC information is converted to drift-time according to the BCID triggered by the TGC or RPC. In the second step, track-segments are processed in each MDT station. Several approaches, using Microprocessors on FPGA, introducing Associative-Memory chips and employing a combination of simple coincidence logic and Look-Up-Tables, are proposed and under development. Finally the track-segments from three MDT stations are combined and the transverse momentum of the muon candidate is evaluated. A trigger threshold is applied and the output is sent to the MuCTPi system (L0MuCTPi) via the TGC or RPC Sector-Logic board. According to current studies, the trigger rate is reduced by 70% with respect to the trigger condition before the HL-LHC program starts, when the trigger threshold is set at 20 GeV, as shown in the Figure 3.

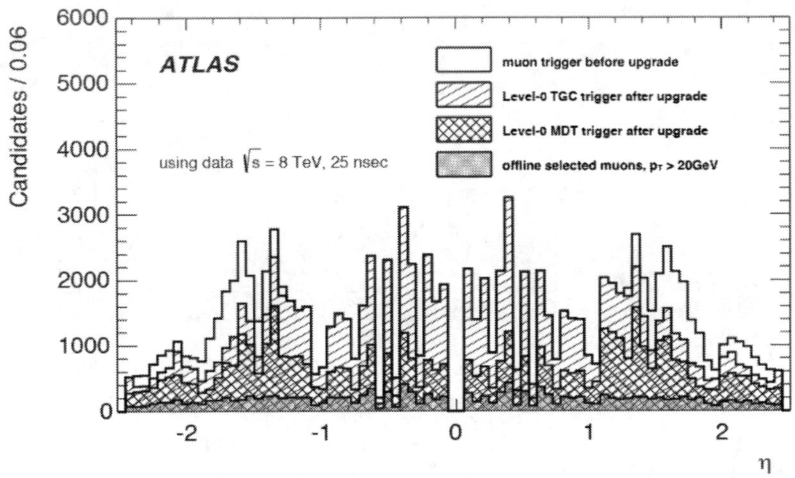

Figure 3. A plot showing η distributions of single muon trigger candidates with a p_T threshold of 20 GeV/c [2]. The histogram in white is the situation before the muon trigger is upgraded. Applying a new TGC trigger algorithm in the endcap, the number of entries is reduced by 25% and the distribution is changed (Red). Further reduction will be expected with the MDT trigger and further trigger rate reduction as shown in the plot (Blue). As a reference, the offline reconstructed tracks having a p_T greater than 20 GeV/c is shown in green.

4 Level-1 Track-Trigger

Association of track information with trigger signatures from the calorimeter and muon spectrometers plays a crucial role in improving trigger performance and controlling the trigger rate in the HL-LHC environment. One of the major items in the ATLAS trigger upgrade for the HL-LHC program is a hardware based track trigger aiming to use track information and make a global decision in the ATLAS Level-1 trigger. In order to meet the Level-1 trigger latency of 60 μs, the track trigger only receives hit information from the pixel and silicon-strip sensors in the inner tracker from the regions around where the Level-0 calorimeter or muon trigger fires (Region-of-Interest : RoI). At the first step of track-finding, multiple pixels or multiple strips are grouped into a single channel, super-cluster (SC). Using an Associative-Memory chip, the hit pattern of SCs of multiple layers is compared with a set of pre-defined track patterns whose transverse momentum is greater than 4 GeV/c. If one of the pre-defined SC pattern matches the input, the hit information with full detector granularity is restored and linear fitting is performed to evaluate track information such as transverse momentum, track direction(ϕ, η) and impact parameter.

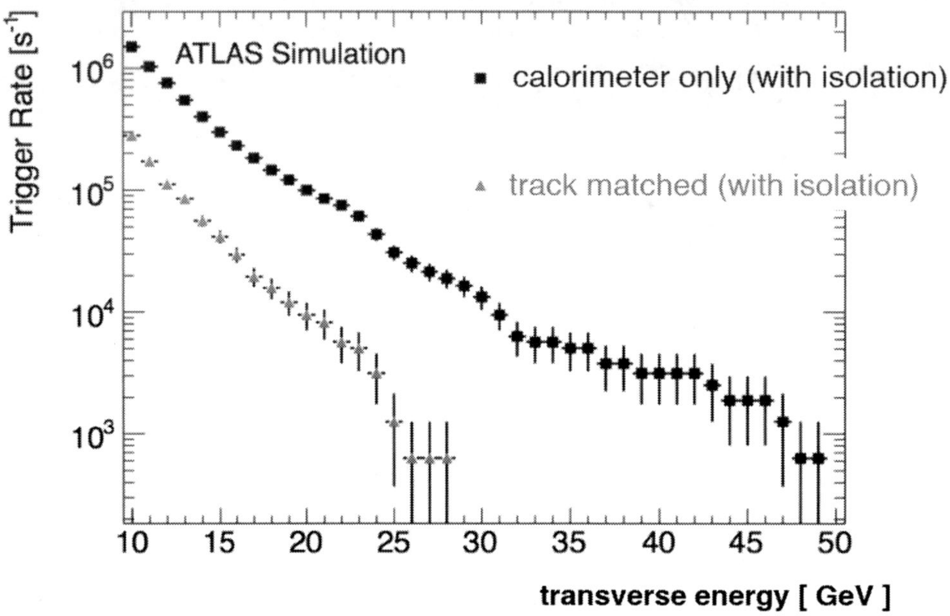

Figure 4. Plot showing expected rates of single electron trigger as a function of the transverse energy observed in the calorimeter [3]. Two different trigger algorithms with and without track information are overlaid in the same plot. A significant trigger rate reduction of a factor of five is found.

Taking the electron trigger as example, the advantage using track information in addition to the calorimeter signature is seen in the first row of Table 1. The ATLAS electro-magnetic calorimeter has fine granularity and electrons are selected based on the lateral shower-shape and a calorimeter-based isolation cut. The expected trigger rate of the Level-0 isolated single-electron is 200 kHz at the threshold of 22 GeV. Requiring a position matching between the calorimeter cluster and the track position as well as the ratio of measured cluster energy over transverse momentum of the track (E/p) to be greater than threshold gives a large suppression of backgrounds from charged hadrons and neutral pions and the Level-1 electron trigger rate is reduced by a factor five to some 40 kHz. The single electron trigger rate as a function of E_T threshold is shown in Figure 4 for both with and without a requirement of track matching. The study uses simulated minimum bias events with $\mu = 70$ and shows a rejection factor of five to ten which may be achieved by requiring a track matching [3]. Further realistic studies under HL-LHC environment, $\mu = 200$ need to be performed.

5 Summary

The ATLAS trigger and DAQ upgrade for the High-Luminosity LHC program is presented. The target is to keep the physics acceptance in terms of kinematic threshold and geometrical acceptance similar or better than the current trigger conditions. The increased maximum trigger rate as well as the longer latency allows to introduce sophisticated algorithms in hardware which have previously been employed in the software based trigger. The Level-0 muon trigger will be totally renewed with information from the precision tracker, MDTs, being introduced in the first step of the hardware trigger. Another new trigger scheme being investigated for the Level-1 trigger is a track trigger using information from the pixel and silicon-strip detectors near the interaction point. The track information is combined with the calorimeter or muon signatures and a trigger decision is taken with an offline-like sophisticated algorithm which leads to a significant trigger rate reduction.

References

[1] The ATLAS Experiment at the CERN Large Hadron Collider, ATLAS Collaboration, JINST 3 (2008) S08003
[2] ATLAS Phase-II Upgrade Scoping Document, ATLAS Collaboration, CERN-LHCC-2015-020
[3] Letter of Intent for the Phase-II Upgrade of the ATLAS Experiment, ATLAS Collaboration, CERN-LHCC-2012-022 (2013)

EPJ Web of Conferences **164**, 07004 (2017) DOI: 10.1051/epjconf/201716407004
ICNFP 2016

Azimuthal angular correlations of D mesons with charged particles in pp collisions at $\sqrt{s}$ = 7 TeV and p-Pb collisions at $\sqrt{s_{NN}}$ = 5.02 TeV with the ALICE experiment at the LHC

Fatiha Lehas[1,a] (for the ALICE collaboration)

[1] *Institute for Subatomic Physics, Utrecht University, Utrecht, the Netherlands.*
and SUBATECH, Ecole des Mines de Nantes, Université de Nantes, CNRS-IN2P3, Nantes, France

Abstract. Heavy flavour studies played an important role in the understanding of energy loss in the formed medium in heavy-ion collisions through different observables (the nuclear modification factor R_{AA}, the elliptic flow v_2). In this paper, results on heavy-flavour correlations will be presented. This observable is sensitive to the fragmentation processes of heavy flavours as well as the system of the collisions. First, the reason why studies on correlations are interesting will be discussed. The ALICE experimental step will be presented. Then the analysis strategy is explained. Finally, the results are presented.

1 Introduction

Several heavy-flavour observables were studied in pp and p-Pb collisions at the LHC, in particular the nuclear modification factor R_{AA}, the elliptic flow v_2 and two-particle correlations. In this paper, we are interested in correlations between D mesons with charged particles. Heavy flavours (charm and beauty quarks) have played an important role to understand and probe the medium that is created in heavy-ion collisions. They are also exploited to investigate the medium created in p-Pb collisions that could shed light on cold nuclear matter effects. In pp collisions, where measurements are used as a reference to those obtained in p-Pb and Pb-Pb collisions, heavy flavours allow us to verify Quantum Chromodynamics calculations in the perturbative regime by comparing the measured p_T-differential cross sections to the ones obtained from theoretical predictions. This was done in ALICE (see figure. 1) where the cross section of D mesons (D^{*+}, D^0 and D^+) obtained from FONLL and GM-VFNS was compared to the one from measurements in ALICE [1]. Measured cross sections show a good compatibility with theoretical predictions. Due to the large mass of heavy flavours, the strong coupling constant is smaller than 1 which makes perturbative QCD calculations applicable. Particles containing heavy flavours are characterised by a short life time, even shorter than the formation time of the Quark Gluon Plasma (QGP) and may help to reveal the properties of this dense and hot matter created in heavy ion collisions [2]. The purpose of correlations in the ALICE experiment is to provide differential observables to study heavy-quark production

[a] e-mail: fatiha.lehas@cern.ch

mechanisms and fragmentation properties in pp collisions and to understand possible medium effects in p-Pb and Pb-Pb collisions.

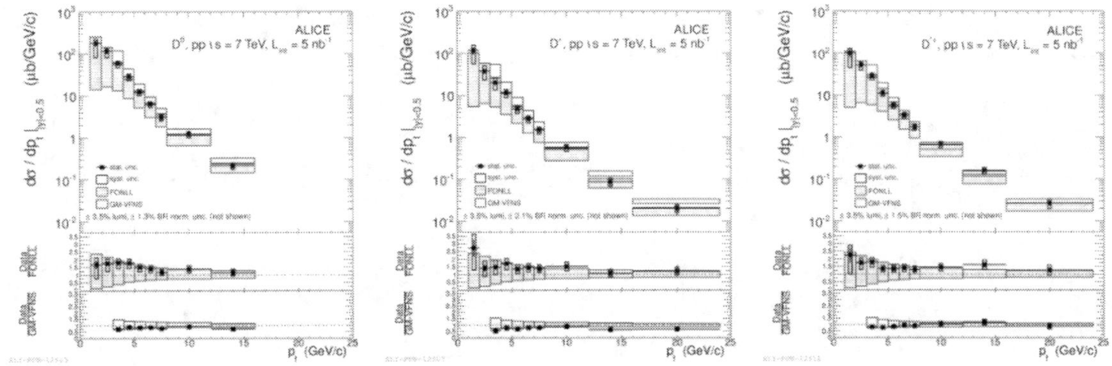

Figure 1. Measured differential production cross section as a function of the transverse momentum p_T of D^0(left panel), D^+(middle panel) and D^{*+}(right panel) in pp collisions at 7 TeV compared to predictions FONLL and GM-VFNS [1].

2 ALICE detector

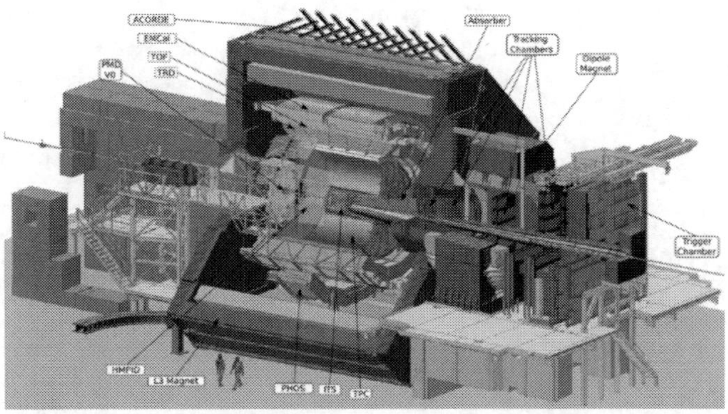

Figure 2. Layout of the ALICE detector [3].

Data from pp collisions at $\sqrt{s} = 7$ TeV and p-Pb collisions at $\sqrt{s_{NN}} = 5.02$ TeV were collected with the ALICE experiment. ALICE detector was designed to study the observables useful to characterize the hot and dense medium formed in heavy-ion collisions to study the QGP, the state of matter that existed in the primordial universe in the first micro-seconds after the Big Bang. ALICE consists of three main parts: the central barrel, the muon spectrometer and some forward detectors used for triggering and event characterization (figure. 2) [3]. The data used for correlation analysis were collected in the central barrel, precisely three detectors were used to reconstruct D mesons: the Inner Tracking System (ITS), the Time Projection Chamber (TPC) and the Time-Of-Flight (TOF). The ITS is the closest detector

DOI: 10.1051/epjconf/201716407004

to the interaction point and it consists of six layers: two layers of Silicon Pixel Detectors (SPD), two layers of Silicon Drift Detectors (SDD) and two layers of Silicon Strip Detectors (SSD). The principal role of the ITS is the determination of primary and secondary vertices. The ITS is used also for tracking and particle identification. The TPC is used for tracking by matching the reconstructed tracks in the ITS with those reconstructed in the TPC. The TPC is the main detector used for particle identification based on the determination of the energy loss which depends on the mass of the particle passing through the gas that fills the TPC drift volume. The TOF is mainly used for particle identification based on the measurement of the time of flight of particles passing through the detector and which is also mass dependent. More details about particles identification in ALICE can be found in [4].

3 Analysis strategy

The final results are obtained by applying the following steps:

- **Reconstruction of D mesons:** D^0, D^+ and D^{*+} mesons are reconstructed in their hadronic decay channels $D^0 \to K^- \pi^+$, $D^+ \to K^- \pi^+ \pi^+$ and $D^{*+} \to D^0 \pi^+$ with the branching ratios: 3.88%, 9.13% and 67.7%, respectively. The decay products are reconstructed applying some selection criteria. D meson candidates are reconstructed by optimizing a number of parameters (topological cuts) in order to maximise the significance of the signal which is defined by $\frac{S}{\sqrt{S+B}}$ where S and B are the signal and the background, respectively. Among the main parameters to be used in the candidate selection, we can mention the transverse momentum of the daughters (π, K), the track impact parameter d_0 (K, π) which is the perpendicular closest distance of the trajectory of K or π to the primary vertex, the product of the impact parameters $d_0(K) \times d_0(\pi)$, the distance of the closest approach (DCA) between the pion and the kaon, the pointing angle (θ_{point}) which is the angle between the reconstructed D^0 momentum and the vector pointing from the primary to the secondary vertex. The obtained mass spectra of the three D mesons are shown in figure. 3. For more details see [5]. The range of the signal in the invariant mass spectra is $[\mu - 2\sigma, \mu + 2\sigma]$ where μ and σ are the peak position and width of the peak, respectively and they are obtained by doing a fit with a Gaussian function for the signal and an exponential term for the background. The sideband range is $[\mu + 4\sigma,\ \mu + 8\sigma]$ for D^{*+} mesons. For D^0 and D^+, the range of the sideband is $[\mu + 4\sigma,\ \mu + 8\sigma]$ and $[\mu - 4\sigma,\ \mu - 8\sigma]$.

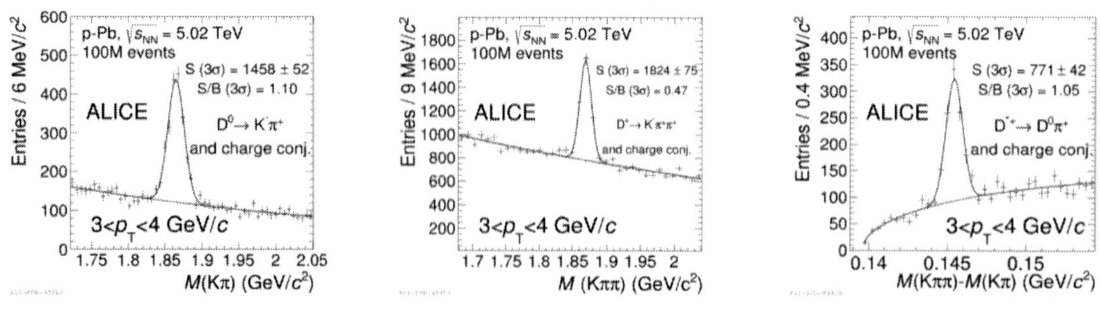

Figure 3. Invariant mass spectra of D^0 (left panel), D^+ (middle panel) and D^{*+}(right panel) mesons measured in p-Pb collisions at $\sqrt{s_{NN}} = 5.02$ TeV.

- **Reconstruction of the associated tracks:** The reconstructed tracks that fulfill the selection criteria are associated with the D-meson candidates.
- **Single and mixed events:** In this step, each D meson (trigger particle) is correlated with all the charged particles (associated particles). For each correlated pair D meson - charged particle, the difference of the azimuthal angle and the pseudo-rapidity is calculated,

$$\Delta\varphi = \varphi^{assoc} - \varphi^{trigg}, \Delta\eta = \eta^{assoc} - \eta^{trigg} \tag{1}$$

If the D meson and the associated track are from the same event, the obtained distribution is a single event distribution. If they are from two different events, we obtain a mixed event distribution. Both single and mixed events distributions are obtained in the signal region and sideband region if D meson candidates are from the signal and the sideband region, respectively. The mixed event technique is used in correlation analysis to correct for inhomegeneities that could exist in the detector as well as for the limited acceptance of the detector.

An example of the single event and normalised mixed event distributions is presented in the left and middle panels of figure. 4. Every single event distribution is corrected by the corresponding normalised mixed event distribution as it is stated in the equation,

$$\frac{dN^{corr}(\Delta\varphi, \Delta\eta)}{d\Delta\varphi\Delta\eta} = \frac{\frac{dN^{SE}(\Delta\varphi,\Delta\eta)}{d\Delta\varphi\Delta\eta}}{\frac{dN^{ME}(\Delta\varphi,\Delta\eta)}{d\Delta\varphi\Delta\eta} / \frac{dN^{ME}(0,0)}{d\Delta\varphi\Delta\eta}} \tag{2}$$

Where $\frac{dN^{SE}(\Delta\varphi,\Delta\eta)}{d\Delta\varphi\Delta\eta}$ is the single event distribution and $\frac{dN^{ME}(\Delta\varphi,\Delta\eta)}{d\Delta\varphi\Delta\eta} / \frac{dN^{ME}(0,0)}{d\Delta\varphi\Delta\eta}$ is the normalised mixed event distribution to the central bin $(\Delta\varphi,\Delta\eta)=(0,0)$. The obtained distribution using this correction is presented in the right panel of figure. 4.

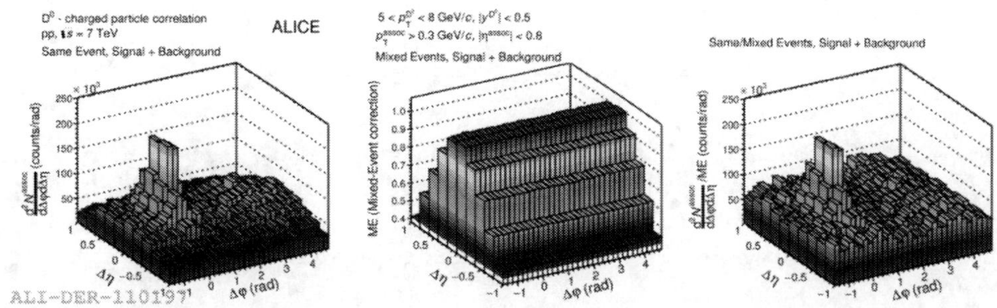

Figure 4. Example of single event distribution (left panel), mixed event distribution (middle panel) and corrected single event distribution (right panel) obtained in signal region for D^0-charged particles correlations in the range $5 < p_T(D^0) < 8$ GeV/c in pp collisions at 7 TeV.

- **Correction for the selection efficiency for D mesons and associated tracks:** The single and mixed event distributions and the invariant mass spectra of D mesons should be corrected for the efficiency of finding a candidate D meson and the associated tracks. The efficiency is evaluated as a function of p_T and multiplicity for D mesons and as a function of p_T, Z vertex position and rapidity for the associated tracks. The previous distributions (single and mixed events and invariant mass spectra) are weighted by the inverse of the efficiency, $1/\epsilon_D$ for D mesons and $1/(\epsilon_D \times \epsilon_{tracks})$ for single and mixed event distributions. The efficiency × acceptance as a function of p_T for D^0 is shown as an example in figure. 5 [6] where the acceptance factor also includes the detector acceptance for D meson reconstruction.

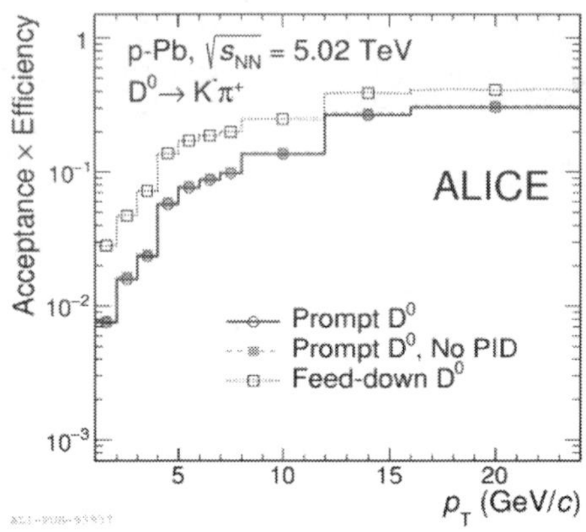

Figure 5. Acceptance × Efficiency of D^0 mesons as a function of the transverse momentum obtained in p-Pb collisions at $\sqrt{s_{NN}} = 5.02$ TeV.

- **Projection on $\Delta\varphi$:** Each single event is corrected using the mixed event technique. Two distributions are obtained: the corrected single event distribution in the signal region and in the sideband region. Due to the limited available statistics, the correlations are projected onto the $\Delta\varphi$ axis. The obtained $\Delta\varphi$ distribution in the sideband region is scaled to the one in the signal region using a scale factor as a weight. The scale factor is defined by the equation,

$$Scale_{factor} = \frac{\int_{Peak} Background}{\int_{Sideband} Background} \tag{3}$$

  Where $\int_{Peak} Background$ is the integral of the invariant mass spectra in the signal region and $\int_{Sideband} Background$ is the integral in the sideband region. The obtained correlation in the sideband region is then subtracted from the one obtained in the signal region and the final correlation is normalised to the number of triggers (D mesons) in the considered p_T range.

- **Correction for the B feed-down:** In this step, since we are interested in D mesons coming from the fragmentation of charm quarks only, those coming from B mesons decay are removed. This correction is based on Monte Carlo simulations using PYTHIA event generator, used to obtain a template of D-hadron correlations coming from B-meson decay. The obtained distribution is then subtracted from the inclusive D-hadron correlation distribution from data to obtain the correlations of prompt D-meson with charged particles. This is achieved using the following equation,

$$\tilde{C}_{promptD}(\Delta\varphi) = \frac{1}{f_{prompt}} \left(\tilde{C}_{inclusive}(\Delta\varphi) - (1 - f_{prompt}) \tilde{C}^{MCtempl}_{feed-down}(\Delta\varphi) \right) \tag{4}$$

where f_{prompt} is the fraction of prompt D mesons and is evaluated from FONLL calculations of charm and beauty quarks. More details can be found in [7].

- **Correction for the purity:** Only the primary charged particles are taken into account. To remove the contribution of secondary tracks we run on Monte Carlo to obtain the purity defined as:

$$Purity = \frac{\int_0^{DCA_{XY}^{cut}} f_{prim}(DCA_{XY})dDCA_{XY}}{\int_0^{DCA_{XY}^{cut}} f_{prim}(DCA_{XY}) + f_{sec}(DCA_{XY})dDCA_{XY}} \quad (5)$$

Where DCA_{XY} is the distance of the closest approach of the tracks trajectory to the primary vertex in the plane XY. The purity is then used as a weight for the correlation distribution obtained in the previous step as it is stated in the following equation,

$$C_{promptD}(\Delta\varphi) = purity \times \widetilde{C}_{promptD}(\Delta\varphi) \quad (6)$$

More details can be found in [7].

- **Average of D mesons (D^{*+}, D^0, D^+):** after the application of all corrections on each D meson separately, we do the average of the three D mesons.

4 Results

The analysis strategy was applied on a data sample of 3×10^8 minimum bias events with integrated luminosity of 5 nb^{-1} in pp collisions at $\sqrt{s} = 7$ TeV and 10^8 minimum bias events with integrated luminosity of 50 nb^{-1} in p-Pb collisions at $\sqrt{s_{NN}} = 5.02$ TeV.
We obtain the correlations of prompt D mesons with primary charged particles in the two systems: pp collisions and p-Pb collisions. D mesons are reconstructed in a wide transverse momentum range [3,16] GeV/c that allowed to define 3 p_T bins [3,5], [5,8] and [8,16] GeV/c. The associated tracks were reconstructed in a transverse momentum range $p_T > 0.3$ GeV/c and that was divided into [0.3,1] GeV/c and $p_T > 1$ GeV/c.

Correlation distributions are expected to have two peaks around $\Delta\varphi = 0, \pi$ which are interpreted as the presence of particles in the same/opposite direction relative to the trigger particles (D mesons). A baseline is expected as well and it is interpreted as coming from the underlying event. In figure. 6 the subtracted baseline azimuthal D-hadron correlations are shown for both collision systems. The results for D-meson p_T ranges [5,8] and [8,16] GeV/c are compatible within present uncertainties once the baseline is subtracted. This suggests that cold nuclear matter effects are smaller compared to the current level of uncertainties.

Other interesting results are shown in figure. 7. The left panel presents correlations in pp collisions at $\sqrt{s} = 7$ TeV for the three p_T ranges of D mesons (3 columns) and the three ranges of the associated tracks (3 rows) where correlations from data were compared to those obtained using different event generators: PYTHIA6 with different tunes, PYTHIA8 and POWHEG + PYTHIA6. There is no significant difference between different event generators and the measurements and the Monte Carlo simulations are in agreement within uncertainties.

The baseline, the width of the near side peak as well as the associated yields were extracted from the correlations for pp collisions at $\sqrt{s} = 7$ TeV using a fit function consisting

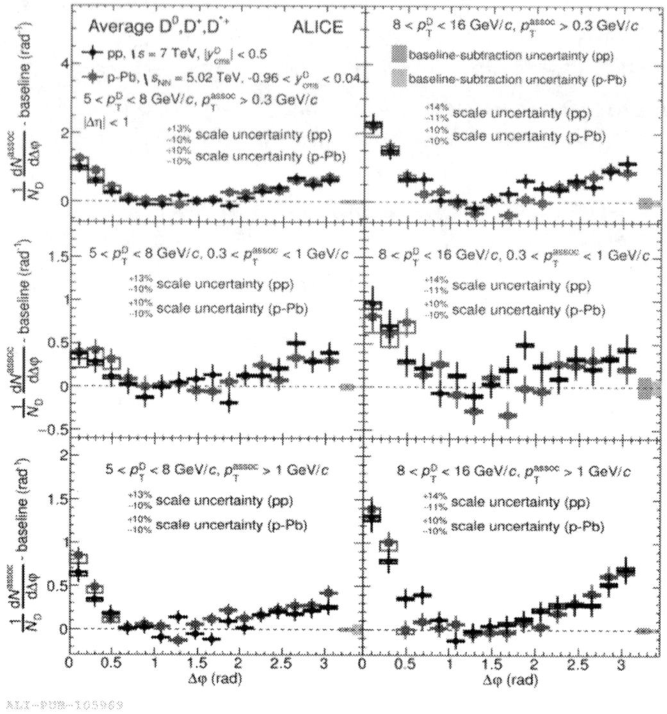

Figure 6. Average azimuthal correlations of D^0, D^+, D^{*+} mesons with charged particles measured in pp collisions at $\sqrt{s} = 7$ TeV (black) compared to those obtained in p-Pb collisions at $\sqrt{s_{NN}} = 5.02$ TeV (red) (with subtracted baseline) in the p_T ranges [5,8] GeV/c and [8,16] GeV/c.

of a constant term (baseline) and two Gaussians (Near side and away side peaks). They are all shown in figure. 7 where they are also compared to those obtained with different event generators. The measured baseline is well reproduced by Monte Carlo simulations. The width and the associated yield are in agreement within uncertainties with simulations. More results and details can be found in [7].

5 Conclusions and outlook

The azimuthal correlations between D mesons with charged particles were measured in pp collisions at $\sqrt{s} = 7$ TeV and in p-Pb collisions at $\sqrt{s_{NN}} = 5.02$ TeV. The similarity between correlations in pp collisions and p-Pb collisions indicates that with the current level of uncertainties no modifications due to cold nuclear matter effects are observed. The results from measurements and Monte Carlo simulations (using PYTHIA6, PYTHIA8 and POWHEG + PYTHIA6 event generators) are in agreement within uncertainties. Better results are expected in Run 2 with higher statistics from pp and p-Pb collisions and in Run 3 with the higher statistics and the upgrade of the ALICE ITS [8] .

DOI: 10.1051/epjconf/201716407004

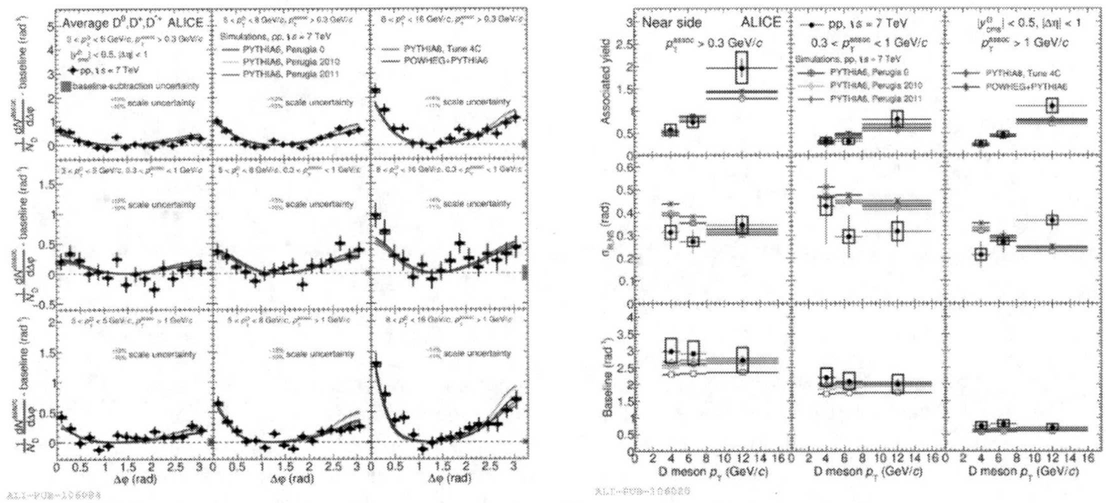

Figure 7. Left panel: Azimuthal correlations of D mesons (average D^0, D^+, D^{*+}) with charged particles measured in pp collisions at $\sqrt{s} = 7$ TeV compared to azimuthal correlations obtained using Monte Carlo simulations with different event generators (different tunes of PYTHIA6, PYTHIA8, PYTHIA6+POWHEG) where the baseline is subtracted in both measurements and simulations. Right panel: Comparison between the associated yield (upper row), the width of the near side peak (middle row) and the baseline (lower row) obtained from measurements (black) and Monte Carlo simulations (colour lines) in pp collisions at $\sqrt{s} = 7$ TeV.

References

[1] Abelev, B. et al. [ALICE Collaboration], JHEP 1201 (2012) 128.

[2] R. Rapp, H. Van Hess, Heavy quarks in Quark-Gluon Plasma, arXiv:0903.1096v2 [hep-ph], World Scientific, p.111(2010).

[3] K. Aamodt et al. [ALICE Collaboration], The ALICE experiment at the CERN LHC, JINST 3 (2008) S08002.

[4] J. Adam et al. [ALICE Collaboration], Eur. Phys. J. Plus (2016) 131: 168.

[5] B. Abelev et al. [ALICE Collaboration], PRL 113 (2014) 232301.

[6] J. Adam et al. [ALICE Collaboration], arXiv:1605.07569v1[nucl-ex].

[7] J. Adam et al. [ALICE Collaboration], arXiv:1605.06963v1.

[8] Abelev, B. et al. [ALICE Collaboration], J. Phys. G: Nucl. Part. Phys. 41 (2014) 087002, Upgrade of the ALICE Inner Tracking System, CERN-LHCC-2013-024.

EPJ Web of Conferences **164**, 07005 (2017)
ICNFP 2016
DOI: 10.1051/epjconf/201716407005

B physics and Quarkonia in CMS

Sara Fiorendi[1,a] on behalf of the CMS Collaboration

[1] *Università degli Studi di Milano - Bicocca and INFN*

Abstract. The heavy-flavor sector offers the opportunity to perform indirect tests of beyond the Standard Model physics through precision measurements and of quantum chromodynamics (QCD) through particle production studies. A review of recent measurements on heavy flavors from the CMS experiment is presented. Results are based on data collected during LHC Run I and Run II and include measurements of heavy flavor production and properties, rare decays, CP violation, exotic and standard quarkonia.

1 Introduction

The heavy flavor program of the CMS experiment [1] at the CERN LHC is very competitive and differentiated. The main objectives include testing the Standard Model (SM) predictions via high precision measurements of heavy flavor properties (decay rates, lifetimes, CP violation) and production spectra, searching for new exotic quarkonia states, new hadrons and indications of new physics through rare decay studies. Selected recent results, based on data collected at 8 and 13 TeV, are reviewed in this Letter.

2 Measurement of B^+ production cross section at 13 TeV

The CMS measurement of the B^+ production cross section at 13 TeV [6] provides a test of theoretical calculations, complementing the measurements carried out at 7 and 8 TeV by the CMS [3], LHCb [4, 5] and ATLAS [2] Collaborations.

The inclusive B^+ differential production cross sections as a function of the B^+ meson transverse momentum and rapidity are measured using 50.8 pb^{-1} of data collected in 2015 with 50 ns bunch spacing. The measurement is based on the inclusive channel $pp \rightarrow B^+X \rightarrow J/\psi K^+X$, where $J/\psi \rightarrow \mu^+\mu^-$, and covers the phase space region where $10 < p_T^B < 100$ GeV and $|y^B| < 2.4$. Data have been collected using a displaced dimuon trigger path, which requires two muons coming from the same vertex and with invariant mass compatible with the J/ψ one. Candidate B^+ mesons are reconstructed by combining a J/ψ with an additional track, assumed to be a kaon.

The detection efficiency and geometrical acceptance are evaluated through a Monte Carlo simulation, while trigger and reconstruction efficiencies are measured directly from data.

The signal yield is extracted with an extended unbinned maximum likelihood fit to the invariant mass distribution of the B^+ candidates, in each p_T^B or $|y^B|$ bins. The signal component is modeled by

[a] e-mail: sara.fiorendi@cern.ch

 DOI: 10.1051/epjconf/201716407005

the sum of two Gaussians functions, while the combinatorial background component is described by an exponential function. The background from mis-reconstructed $B \to J/\psi K^+ X$ decays is modeled by an error function.

Systematic uncertainties have been evaluated on signal yield extraction, p_T and y resolution, p_T^B and $|y^B|$ simulated distributions, B^+ lifetime, trigger and muon/track reconstruction, size of the simulation, and luminosity measurement.

Results for the differential cross sections as a function of p_T^B, integrated within $|y^B| < 2.4$, and as a function of $|y^B|$, integrated within $10 < p_T^B < 100$ GeV, are shown in Fig. 1, together with the corresponding 7 TeV measurements [3]. FONLL (shaded boxes) and PYTHIA (dashed lines) calculations are also reported on the plots, showing reasonable agreement both in terms of shape and normalization.

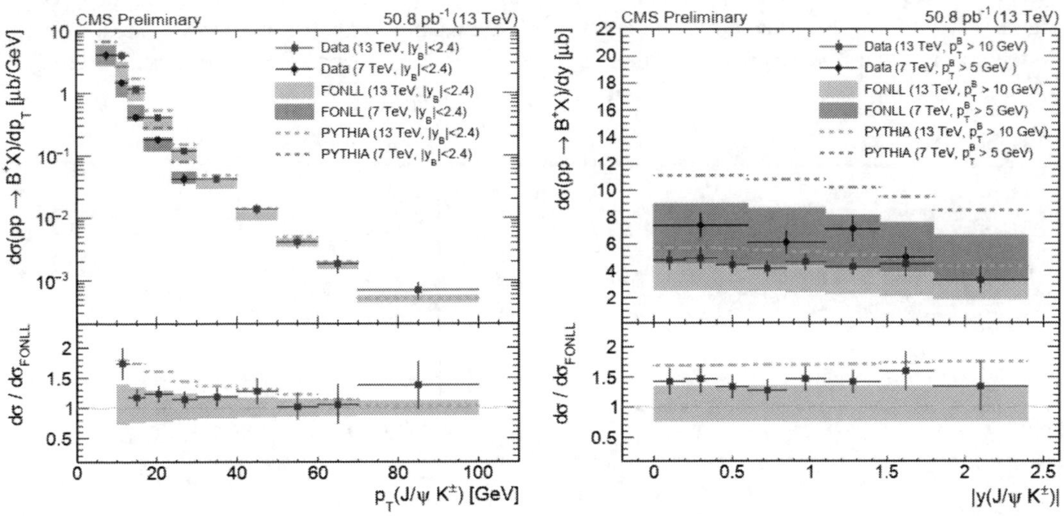

Figure 1. B^+ differential production cross sections at 13 TeV as a function of p_T^B for $|y^B| < 2.4$, and rapidity for $10 < p_T^B < 100$ GeV. The previous CMS measurements from 7 TeV data [3] (circles) are also shown. The vertical bars show the total uncertainty in the measured cross sections, and the horizontal bars represent the bin width. The calculations from FONLL and PYTHIA are shown as shaded boxes and dashed lines, respectively. The bottom panels display the ratio of the data at 13 TeV to the FONLL predictions (points) and the ratios of the PYTHIA to the FONLL calculations (dashed lines), with the shaded region displaying the uncertainties in the FONLL predictions.

3 Quarkonium production cross section at 13 TeV

Precise measurements of cross sections and polarizations for five quarkonium states (J/ψ, $\psi(2S)$, $\Upsilon(nS)$) were provided by the LHC experiments already during Run I [7–16]. Measurements at 13 TeV offer an interesting opportunity to test Non-Relativistic QCD at higher energies and extend the comparison to theoretical calculations to higher p_T ranges.

The analysis [17] is based on data collected in 2015, corresponding to an integrated luminosity of 2.4 fb^{-1} for J/ψ, and 2.7 fb^{-1} for the other mesons. Quarkonium states are reconstructed through their decay into two opposite-sign muons. The measurement is restricted to the central region, where $|\eta^{\mu\mu}| < 1.25$.

The acceptance is evaluated through simulated samples, assuming unpolarized dimuons. Muon reconstruction, identification and trigger efficiencies and dimuon efficiency are measured on data as a function of muon (dimuon) p_T and pseudorapidity.

Yields are extracted through an unbinned maximum likelihood fit to the dimuon invariant mass distribution. The three Y(nS) signal peaks are modeled by Crystal Ball functions, and the background is described by an exponential distribution. For Charmonium states, the additional non-prompt component, originating from b-hadron decays, is taken into account. A simultaneous fit to mass and pseudo proper-decay-length distributions is used to disentangle the amount of prompt and non-prompt mesons.

Systematic uncertainties on dimuon acceptance and efficiencies, determination of the signal yields and luminosity measurement are evaluated.

Cross section results for the J/ψ and $\psi(2S)$ states are shown in the left plot of Fig. 2 as a function of the dimuon p_T, together with the corresponding 7 TeV measurements, for the rapidity-integrated range $|y| < 1.2$. The right plot shows the cross section measurements for the Y(nS) states. The 13 TeV cross sections of all five quarkonium states are factors of 2 to 3 larger than the corresponding 7 TeV cross sections, with slow changes as a function of the dimuon p_T.

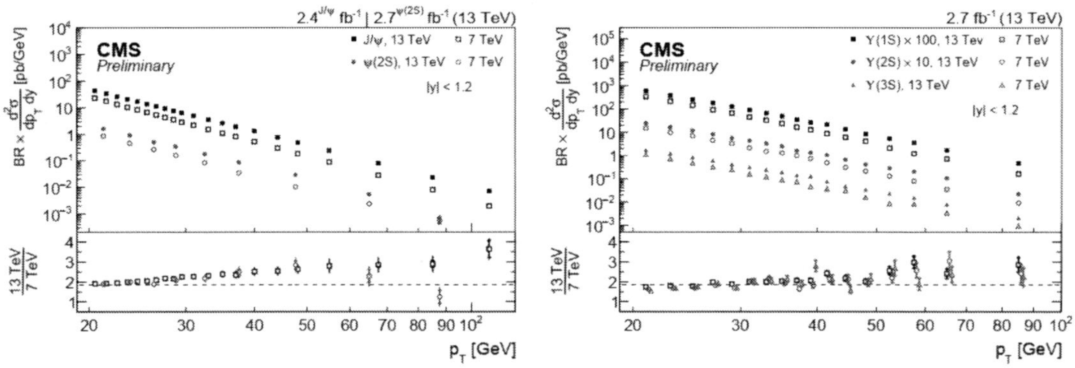

Figure 2. Differential cross sections times branching ratios for 7 TeV and 13 TeV CMS data [17], for J/ψ and $\psi(2S)$ (left) and Y(nS) (right), for $|y| < 1.2$. The inner error bars represent the statistical uncertainty while the total errors show the statistical and systematic uncertainties. The uncertainty on the luminosity measurement is not included in the plots.

4 Observation of Y(1S) pair production

The first observation of the simultaneous production of Y(1S) pairs has been reported by CMS in [18]. Measurements of the production of quarkonium pairs are considered the first step in the search for tetra-quark states. Though the event yield of this observation is limited, future measurements with higher statistics can provide information on the production mechanism (Single or Double Parton Scattering).

The analysis is based on 20.7 fb^{-1} of data collected in 2012 at $\sqrt{s} = 8$ TeV. Events are selected by requiring four muons with zero total charge, each muon having $p_T > 3.5$ GeV. Each Y candidate is reconstructed by performing a kinematic fit to the oppositely charged muon pairs and requiring a minimum vertex fit χ^2 probability. Two kinematic variables are defined: the dimuon invariant mass of the higher mass Y candidate, $M^1_{\mu\mu}$ and that of the lower mass Y candidate, $M^2_{\mu\mu}$.

The signal yields of Y(1S)Y(1S) and Y(1S)Y(2S) are extracted through a two-dimensional unbinned maximum likelihood fit to $M^1_{\mu\mu}$ and $M^2_{\mu\mu}$ (Fig. 3). The invariant mass distribution of each muon pair is modeled by a signal and background component, the signal model consisting in the sum of two Crystal Ball functions, and the background in a first order Chebychev Polynomial. The signal yield of Y(1S)Y(1S) is measured to be 38 ± 7 events and the statistical significance for pp → Y(1S)Y(1S) production is determined to be 9.6 standard deviations.

The inclusive cross section is measured in the region |y(Y)| < 2.0 and p_T(Y) < 50 GeV. Efficiency and acceptance corrections are evaluated on an event-by-event basis using data-embedding methods to minimize the possible model dependence. Several sources of systematic uncertainties have been considered: shape of the signal model, uncertainties on trigger and reconstruction efficiency and acceptance, luminosity measurement.

Assuming isotropic Y(1S) decays into muons, the total cross section of Y(1S) pair production is measured to be σ_T = 68.8 ± 12.7 (stat) ± 7.4 (syst) ± 2.8 (BR) pb. It is also measured that extreme polarization scenarios could vary the results by +36% (100% longitudinal polarization) or -38% (100% transverse polarization).

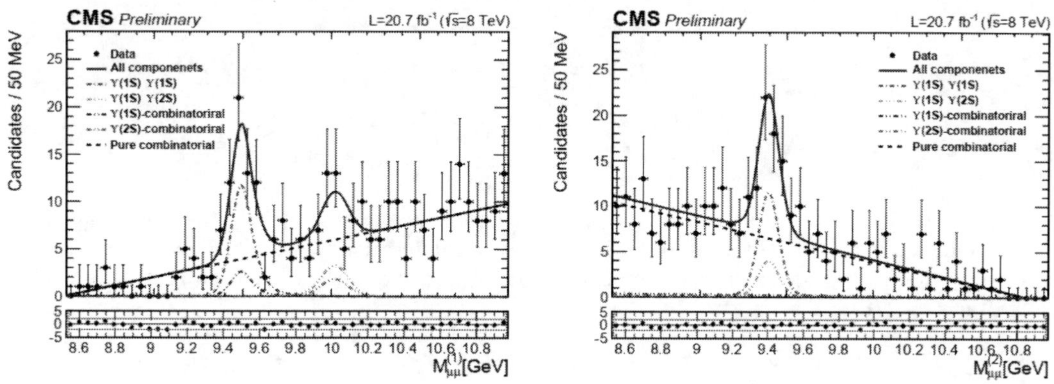

Figure 3. Projections of the two dimensional fit to $M^1_{\mu\mu}$ (left) and $M^2_{\mu\mu}$ (right). The lower pads show the difference between the total fit curve and data, with red lines corresponding to a difference of ±2 standard deviations.

5 Angular analysis of $B^0 \to K^{*0}\mu^+\mu^-$

The flavor-changing neutral-current decay $B^0 \to K^{*0}\mu^+\mu^-$, highly suppressed in the SM, is particularly sensitive to new physics contributions. They could reveal themselves through modifications of the decay properties, such as the branching fraction, muon forward-backward asymmetry (A_{FB}) and longitudinal polarization fraction of the K^{*0} (F_L). These physical parameters are measured by the CMS Collaboration on a 20.5 fb^{-1} data sample collected at 8 TeV [19]. The decays $B^0 \to J/\psi K^{*0}$ and $B^0 \to \psi(2S)K^{*0}$ are used as control channels.

The $B^0 \to K^{*0}\mu^+\mu^-$ decay is fully described by three angles (the helicity angle for the dimuon system, θ_L, the helicity angle of the K^{*0} candidate, θ_K, and the angle between the K^{*0} and dimuon planes, ϕ) and the dimuon invariant mass squared (q^2). The results of the analysis are extracted from unbinned extended maximum likelihood fits, in bins of q^2, to the invariant mass of the B^0 candidate and the two angles θ_K and θ_L. In Fig. 4 the results are compared with two sets of SM predictions [20–22], which differ in the calculation of the form factors. Good agreement is observed between data and theoretical calculations.

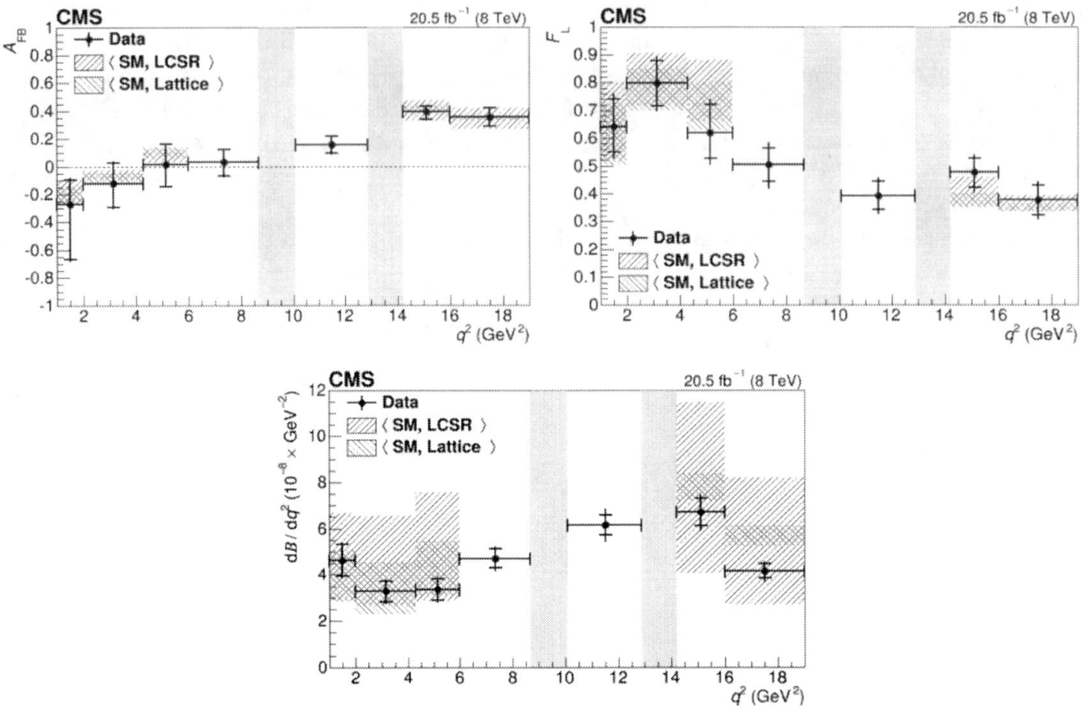

Figure 4. Results of the measurement of A_{FB}, F_L and the differential branching fraction versus the dimuon q^2. The inner error bars on data points represent the statistical uncertainty, while the total errors show the total uncertainty. The gray shaded regions correspond to the J/ψ and $\psi(2S)$ resonances. Two sets of Standard Model predictions [20–22] are also shown.

6 Measurement of ϕ_s and $\Delta\Gamma_s$

The interference between direct B_s decays to $J/\psi\phi$ and decays through $B^0_s - \bar{B}^0_s$ mixing gives rise to the CP violating weak phase ϕ_s. The SM predictions for ϕ_s are very precise, so any measured deviation would be an indication of new physics contributions.

The decay $B_s \to J/\psi\Phi(1020) \to \mu^+\mu^-K^+K^-$ is exploited to measure ϕ_s and the decay-width difference $\Delta\Gamma_s$ between light and heavy B_s mass eigenstates [23]. A time-dependent and flavor-tagged angular analysis is used to disentangle the CP-odd and CP-even components of the final state. Three angles are defined to describe the decay: θ_T (ϕ_T) is the polar (azimuthal) angle defined by the μ^+ in the J/ψ rest frame, while the helicity angle ψ_T is the angle of the K^+ in the $\Phi(1020)$ rest frame with respect to the negative J/ψ momentum direction. The flavor of each B_s candidate at production time is determined with an opposite-side lepton tagging algorithm.

An unbinned maximum likelihood fit to the data is performed by including the information on the B_s invariant mass, the three decay angles, the flavor tag decision, the proper decay time ct and its uncertainty.

The analysis is performed on a 20 fb^{-1} dataset collected at 8 TeV. The measured values for ϕ_s and $\Delta\Gamma_s$ are ϕ_s = -0.075 ± 0.097 (stat) ± 0.031 (syst) rad and $\Delta\Gamma_s$ = 0.095 ± 0.013 (stat) ± 0.007 (syst) ps^{-1}. The left plot of Fig. 5 shows the measured values together with SM predictions. The right plot shows the combination of the results from different experiments, as provided by the HFAG group [24].

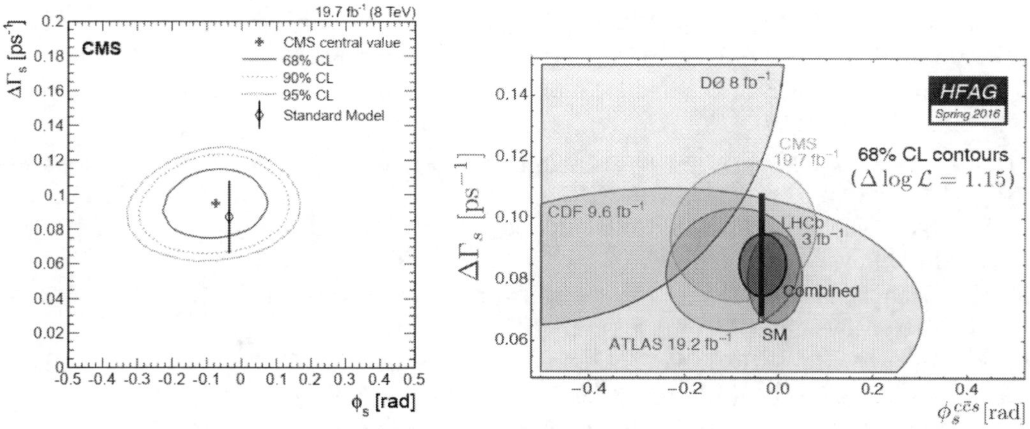

Figure 5. Left: ϕ_s and $\Delta\Gamma_s$ as measured by CMS [23], together with 68%, 90%, and 95% CL contours. SM predictions are also shown. Right: individual 68% confidence-level contours of ATLAS, CMS, CDF, D0 and LHCb measurements, in the ($\phi_s^{c\bar{c}s}$, $\Delta\Gamma_s$) plane [24]. Their combined contour (solid line and shaded area), and the Standard Model predictions (thin black rectangle) are also shown.

7 Conclusions

The Run II of the LHC will provide a significant statistics of heavy flavor events, improving the sensitivity reach of many analyses. The first results from CMS on 13 TeV data have been summarized in this Letter, together with selected measurements carried out on Run I data.

References

[1] S. Chatrchyan *et al.* [CMS Collaboration], JINST **3** (2008) S08004. doi:10.1088/1748-0221/3/08/S08004

[2] G. Aad *et al.* [ATLAS Collaboration], JHEP **1310**, 042 (2013) doi:10.1007/JHEP10(2013)042 [arXiv:1307.0126 [hep-ex]].

[3] V. Khachatryan *et al.* [CMS Collaboration], Phys. Rev. Lett. **106**, 112001 (2011) doi:10.1103/PhysRevLett.106.112001 [arXiv:1101.0131 [hep-ex]].

[4] R. Aaij *et al.* [LHCb Collaboration], JHEP **1204**, 093 (2012) doi:10.1007/JHEP04(2012)093 [arXiv:1202.4812 [hep-ex]].

[5] R. Aaij *et al.* [LHCb Collaboration], JHEP **1308**, 117 (2013) doi:10.1007/JHEP08(2013)117 [arXiv:1306.3663 [hep-ex]].

[6] CMS Collaboration, CMS-PAS-BPH-15-004, https://cds.cern.ch/record/2114806

[7] V. Khachatryan *et al.* [CMS Collaboration], Phys. Rev. D **83**, 112004 (2011) doi:10.1103/PhysRevD.83.112004 [arXiv:1012.5545 [hep-ex]].

[8] S. Chatrchyan *et al.* [CMS Collaboration], Phys. Lett. B **727**, 101 (2013) doi:10.1016/j.physletb.2013.10.033 [arXiv:1303.5900 [hep-ex]].

[9] V. Khachatryan *et al.* [CMS Collaboration], Phys. Rev. Lett. **114**, no. 19, 191802 (2015) doi:10.1103/PhysRevLett.114.191802 [arXiv:1502.04155 [hep-ex]].

[10] V. Khachatryan *et al.* [CMS Collaboration], Phys. Lett. B **749**, 14 (2015) doi:10.1016/j.physletb.2015.07.037 [arXiv:1501.07750 [hep-ex]].

[11] G. Aad *et al.* [ATLAS Collaboration], Phys. Rev. D **87**, no. 5, 052004 (2013) doi:10.1103/PhysRevD.87.052004 [arXiv:1211.7255 [hep-ex]].

[12] R. Aaij *et al.* [LHCb Collaboration], Eur. Phys. J. C **72**, 2100 (2012) doi:10.1140/epjc/s10052-012-2100-4 [arXiv:1204.1258 [hep-ex]].

[13] R. Aaij *et al.* [LHCb Collaboration], Eur. Phys. J. C **72**, 2025 (2012) doi:10.1140/epjc/s10052-012-2025-y [arXiv:1202.6579 [hep-ex]].

[14] S. Chatrchyan *et al.* [CMS Collaboration], Phys. Rev. Lett. **110**, no. 8, 081802 (2013) doi:10.1103/PhysRevLett.110.081802 [arXiv:1209.2922 [hep-ex]].

[15] S. Chatrchyan *et al.* [CMS Collaboration], Phys. Lett. B **727**, 381 (2013) doi:10.1016/j.physletb.2013.10.055 [arXiv:1307.6070 [hep-ex]].

[16] R. Aaij *et al.* [LHCb Collaboration], Eur. Phys. J. C **73**, no. 11, 2631 (2013) doi:10.1140/epjc/s10052-013-2631-3 [arXiv:1307.6379 [hep-ex]].

[17] CMS Collaboration, CMS-PAS-BPH-15-005, http://cds.cern.ch/record/2145411

[18] CMS Collaboration, CMS-PAS-BPH-14-008, http://cds.cern.ch/record/2154307

[19] V. Khachatryan *et al.* [CMS Collaboration], Phys. Lett. B **753** (2016) 424 doi:10.1016/j.physletb.2015.12.020 [arXiv:1507.08126 [hep-ex]].

[20] C. Bobeth, G. Hiller and D. van Dyk, JHEP **1007**, 098 (2010) doi:10.1007/JHEP07(2010)098 [arXiv:1006.5013 [hep-ph]].

[21] C. Bobeth, G. Hiller and D. van Dyk, Phys. Rev. D **87**, no. 3, 034016 (2013) [Phys. Rev. D **87**, 034016 (2013)] doi:10.1103/PhysRevD.87.034016 [arXiv:1212.2321 [hep-ph]].

[22] C. Bobeth, G. Hiller, D. van Dyk and C. Wacker, JHEP **1201** (2012) 107 doi:10.1007/JHEP01(2012)107 [arXiv:1111.2558 [hep-ph]].

[23] V. Khachatryan *et al.* [CMS Collaboration], Phys. Lett. B **757**, 97 (2016) doi:10.1016/j.physletb.2016.03.046 [arXiv:1507.07527 [hep-ex]].

[24] Y. Amhis *et al.* [Heavy Flavor Averaging Group (HFAG) Collaboration], arXiv:1412.7515 [hep-ex] and online update at http://www.slac.stanford.edu/xorg/hfag

Beauty production measurements in pp, p–Pb and Pb–Pb collisions with the ALICE detector

Minjung Kim for the ALICE collaboration[1,a]

[1] *Inha University, Incheon, Republic of Korea*

Abstract. Beauty production has been measured in the ALICE experiment via its semi-electronic decays and non-prompt J/ψ at mid-rapidity. A review of results on beauty production at mid-rapidity in pp collisions at $\sqrt{s}$ = 7 TeV and at $\sqrt{s}$ = 2.76 TeV, in p-Pb collisions at $\sqrt{s_{\rm NN}}$ = 5.02 TeV and in Pb-Pb collisions at $\sqrt{s_{\rm NN}}$ = 2.76 TeV are reported, along with the current status of b-jet tagging studies in ALICE. Prospects of beauty production measurements with RUN2 and RUN3-4 are outlined, focusing on the upgraded Inner Tracking System (ITS) and the new Muon Forward Tracker (MFT).

1 Introduction

In ultra-relativistic heavy-ion collisions, the hot and dense medium of colour-deconfined quarks and gluons, called Quark-Gluon plasma (QGP), is expected to be created with a formation of $\tau_{\rm QGP} \sim 0.3\,{\rm fm}/c$ and a of lifetime $\sim 10\,{\rm fm}/c$ at LHC energies [1, 2]. Heavy quarks (charm and beauty) are predominantly produced in the early stages of the collisions via initial hard scattering processes due to their large masses ($m_{\rm Q} \gg \Lambda_{\rm QCD}$) [3]. Since the time-scale of their formation is expected to be smaller than $\tau_{\rm QGP}$, heavy quarks can propagate through the QGP medium and experience the full system evolution, interacting with its constituents and loosing part of their energy within the medium. According to Quantum Chromodynamics (QCD), due to their smaller colour coupling to the medium with respect to gluons, the medium-induced parton energy loss of quarks is smaller than that of gluons [4–6]. Additionally, theoretical model calculations predict the mass dependence of medium-induced parton energy loss in radiative energy loss process [7–9] and collisional energy loss process [10]. In many models, the mass-dependence of parton energy loss is given as a function of m/E and it becomes negligible when the ratio m/E approaches to zero. Consequently, beauty quarks can be sensitive probes for testing the mass dependence of medium-induced parton energy loss in broader parton energy range with respect to charm quarks [11].
The sensitive observable for investigating the medium-induced parton energy loss is the nuclear modification factor ($R_{\rm AA}$), which is defined as the ratio of the yield in Pb-Pb collisions to that observed in the pp collisions scaled by the average number of binary collisions. $R_{\rm AA}$ of heavy-flavour hadrons in comparison with $R_{\rm AA}$ of light-flavour hadrons indicate possible colour-charge and mass dependence of medium-induced energy loss. For that reason, the measurements of open heavy-flavour hadron yields in pp collisions are essential, as the references for corresponding measurements in Pb-Pb collisions. In addition, the measurements in pp collisions provide the tests of perturbative Quantum

[a] e-mail: minjung.kim@cern.ch

 DOI: 10.1051/epjconf/201716407006

Chromodynamics (pQCD) calculations. The R_{AA} measurements possess not only the final-state effects from hot and dense medium, but also the initial-state effects due to the presence of nuclei, so called cold nuclear matter (CNM) effect. Similar to R_{AA} in Pb-Pb collisions, the nuclear modification factor R_{pA}, defined as the ratio of the yield in p-Pb collisions to that observed in the pp collisions scaled by the number of nucleons in the nucleus, can quantify the possible CNM effects, such as the modification of parton distribution functions in the nucleus [12] and the transverse momentum broadening from multiple scattering [13].
ALICE (A Large Ion Collider Experiment) at the LHC is dedicated to studying the QGP properties [14]. The central barrel detectors of ALICE provide electron identification capability down to very low p_T and excellent momemtum and track impact parameter resolution [15, 16]. Accordingly, beauty-hadron production at mid-rapidity have been measured via electrons from semi-leptonic decays of beauty hadrons and J/ψ from weak decays of beauty hadrons (non-prompt J/ψ) through di-electron decay channel, exploiting displaced secondary vertices of beauty hadrons due to their long lifetime ($c\tau \sim 500\,\mu$m). As a complementary approach in different kinematic range, the measurements of jets coming from beauty-quark fragmentation (b-jet) are ongoing.
In the following, the review of results on beauty production at mid-rapidity in pp collisions at $\sqrt{s} = 7$ TeV and $\sqrt{s} = 2.76$ TeV, in p-Pb collisions at $\sqrt{s_{NN}} = 5.02$ TeV and in Pb-Pb collisions at $\sqrt{s_{NN}} = 2.76$ TeV from data collected during the LHC-RUN1 are presented, together with the status of b-jet analysis. Prospects of beauty production measurements with RUN2 and RUN3-4 are outlined, focusing on the upgraded Inner Tracking System (ITS) and the new Muon Forward Tracker (MFT), which will be installed during the second long shutdown (LS2) between RUN2 and RUN3 [17].

2 Measurements of beauty production in ALICE

For charged particles reconstruction, the Inner Tracking System (ITS), which consists of 6-layered silicon detector and Time Projection Chamber (TPC) are used, together with a magnetic field of 0.5 T [14]. They provide a transverse momentum (p_T) resolution better than 2.5% for $p_T \leq 30$ GeV/c and transverse impact parameter (d_0) resolution better than 75 μm for $p_T > 1$ GeV/c, where the d_0 is defined as the track distance of closest approach to the primary vertex in the plane transverse to the beam axis [18]. For measurements of electrons from beauty-hadron decays and of non-prompt J/ψ, electrons were identified using specific energy loss (dE/dx) measured in the TPC. The ITS, Time-Of-Flight (TOF) detector or Electromagnetic Calorimeter (EMCal) were additionally used for measurements of electrons from beauty-hadron decays, depending on the collision systems and the p_T ranges. V0 scintillator arrays located at $-3.7 < \eta < -1.7$ and at $2.8 < \eta < 5.1$ were used for centrality determination in Pb-Pb measurements.

2.1 Electrons from beauty-hadron decays

The long lifetime of beauty hadrons leads to larger d_0 of electrons from beauty-hadron decays than those of background electrons, which are mainly from Dalitz decays of light-neutral mesons, from conversions of photons in the detector materials and from semi-electronic decays of charm hadrons. After electron identification two different analysis methods, exploiting the distinguishable d_0 distributions of electrons from different origins, were applied to extract electrons from beauty-hadron decays. In pp collisions and p-Pb collisions, the signal were extracted by requiring minimum d_0 for electron candidates to increase signal to backgroud ratio and subtracting remaining background electrons, which were estimated based on ALICE measured hadron spectra [19–21]. In Pb-Pb collisions, the signal was extracted by fitting d_0 distribution of electron candidates with signal and background templates, obtained from the Monte Carlo simulations. The correction on d_0 distributions in Monte Carlo simulations was done based on data. For instance, p_T spectra of charm hadrons in Monte Carlo simulations were re-weighted using the ALICE measured D meson spectrum [22]. The remaining uncertainty from imperfection of Monte Carlo simulations on describing data was propagated to the systematic uncertainties [21].

Apart from the methods based on impact parameter, the relative contribution of beauty to the electrons from heavy-flavour decays could be measured using azimuthal angular correlation of electrons from heavy-flavour decays to charged hadrons. The width of the near-side peak in the correlation distribution is broader for beauty-hadron decays than for charm-hadron decays due to the their different decay kinematics. The fraction of electrons from beauty-hadron decays in inclusive heavy-flavour decays electrons was extracted by fitting the correlation distribution with Monte Carlo templates [20].

In pp collisions, the p_T-differential invariant cross section of electrons from beauty-hadron decays in $|y| < 0.8$ was measured at $\sqrt{s} = 7$ TeV and at $\sqrt{s} = 2.76$ TeV as shown in Figure 1 are well described by pQCD-based calculations within the experimental and theoretical uncertainties.

In p-Pb collisions and in the 20% most central Pb-Pb collisions, the nuclear modification factor (R_{pPb} and R_{AA}, respectively) was measured as a function of p_T as shown in Figure 2. Within uncertainties,

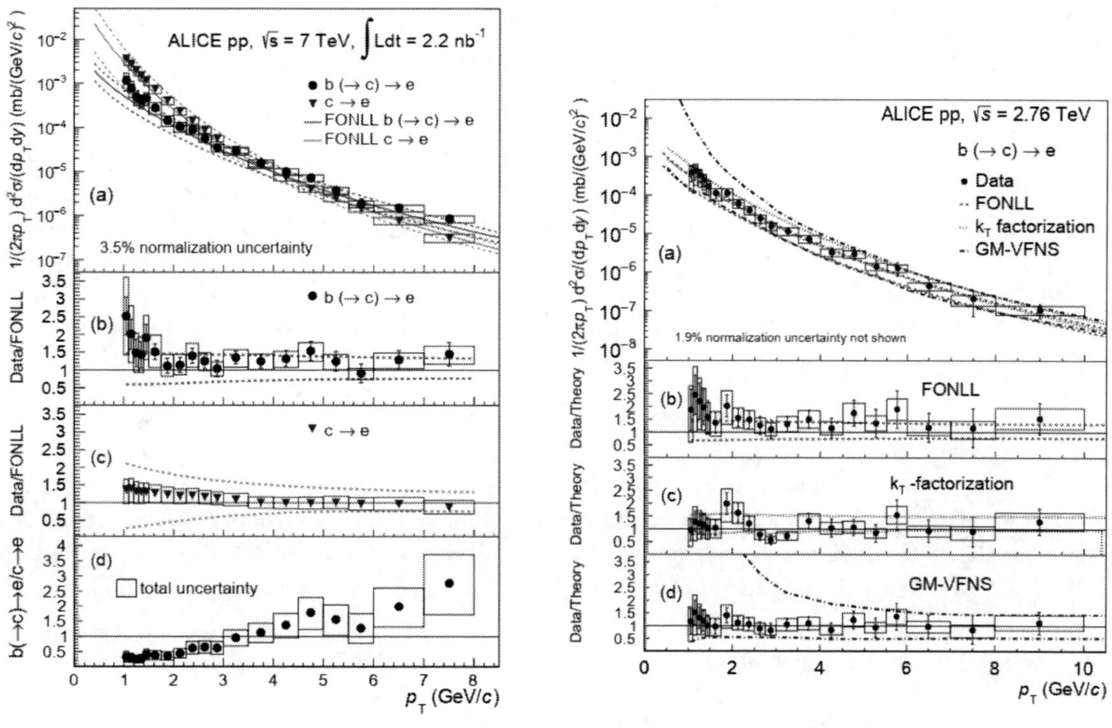

Figure 1: Production cross section of electrons from beauty- and charm-hadron decays and their ratio in pp collisions at $\sqrt{s} = 7$ TeV [19], compared to Fixed-Order plus Next-to-Leading Log (FONLL) pQCD calculations [23] (left), and production cross section of electrons from beauty-hadron decays in pp collisions at $\sqrt{s} = 2.76$ TeV [20], compared to FONLL [23], k_T-factorization [24] and GM-VFNS [25] predictions (right).

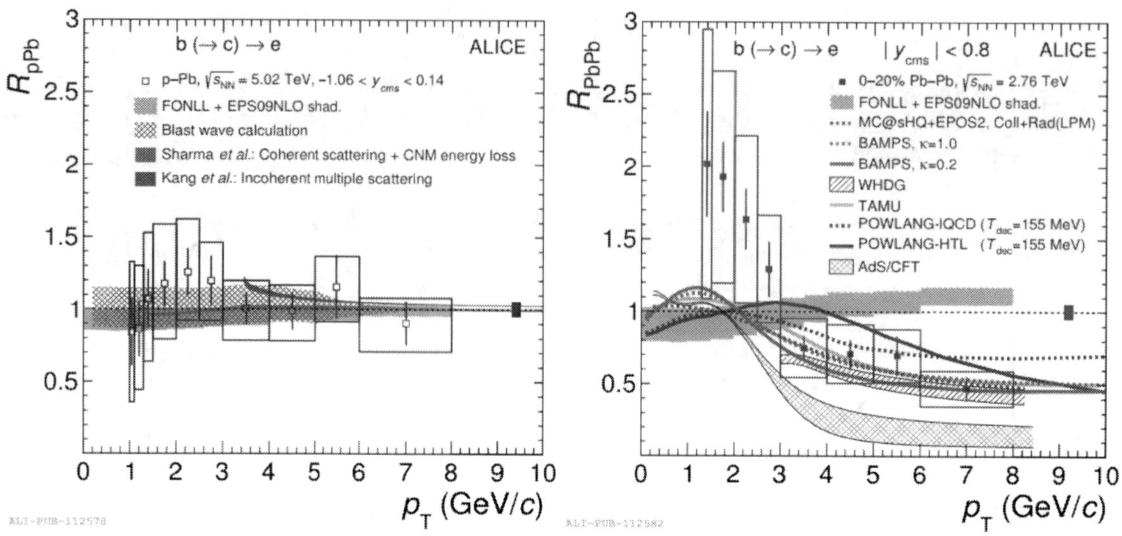

Figure 2: Nuclear modification factor of electrons from beauty-hadron decays at mid-rapidity in p-Pb collisions at $\sqrt{s_{\rm NN}}$ = 5.02 TeV (left) and in Pb-Pb collisions at $\sqrt{s_{\rm NN}}$ = 2.76 TeV in the 20% most central collisions (right) with different theoretical predictions [21]. The normalization uncertainty is shown as a filled box for both plots.

the $R_{\rm pPb}$ is compatible with unity and is well described by theoretical predictions, such as FONLL pQCD calculations [23, 26, 27] with the EPS09NLO nPDF parameterisation [12], model calculations including CNM energy loss, nuclear shadowing and mutiple scattering [28, 29] and blast-wave model calculations considering possible hydrodynamic effects [30]. The $R_{\rm AA}$ shows the suppression of electrons from beauty-hadron decays in the 20% most central Pb-Pb collisions with respect to pp collisions for $p_{\rm T} > 3\,{\rm GeV}/c$. Therefore, the suppression of electrons from beauty-hadron decays for $p_{\rm T} > 3\,{\rm GeV}/c$ in the 20% most central Pb-Pb collisions can be interpreted as an effect of the hot and dense medium on beauty quarks.

2.2 Non-prompt J/ψ

The long lifetime of beauty hadrons leads to larger flight distance of J/ψ from the decay of beauty hadrons than prompt J/ψ, which are direct produced in collisions and J/ψ from decays of heavier charmonium states. Consequently, non-prompt J/ψ is more likely to have large and positive pseudo-proper decay length (x), which is defined as $x = \frac{c \cdot L_{xy} \cdot m_{J/\psi}}{p_{\rm T}^{J/\psi}}$, where $L_{\rm xy}$ is the projection of the flight distance onto its transverse momentum vector and $m_{J/\psi}$ is the mass of the J/ψ. After electron identification, an unbinned 2-dimensional simultaneous fit for invariant mass distributions of J/ψ and x was used to determine the ratio of non-prompt to inclusive J/ψ [31].
The fraction of non-prompt J/ψ measured in pp collisions at $\sqrt{s}$ = 7 TeV and in Pb-Pb collisions at $\sqrt{s_{\rm NN}}$ = 2.76 TeV is presented in left panel of Figure 3 [31, 32]. The fraction in Pb-Pb collisions has a similar $p_{\rm T}$ dependence as in pp collisions. It is an open question whether the similarity comes from compensation of the medium effects on the prompt J/ψ, such as J/ψ dissociation and recombination or from medium-induced effects on the non-prompt J/ψ, related to beauty-quark energy loss, or from an interplay between those effects. Based on the measurement of inclusive J/ψ $R_{\rm AA}$ and the fraction of

DOI: 10.1051/epjconf/201716407006

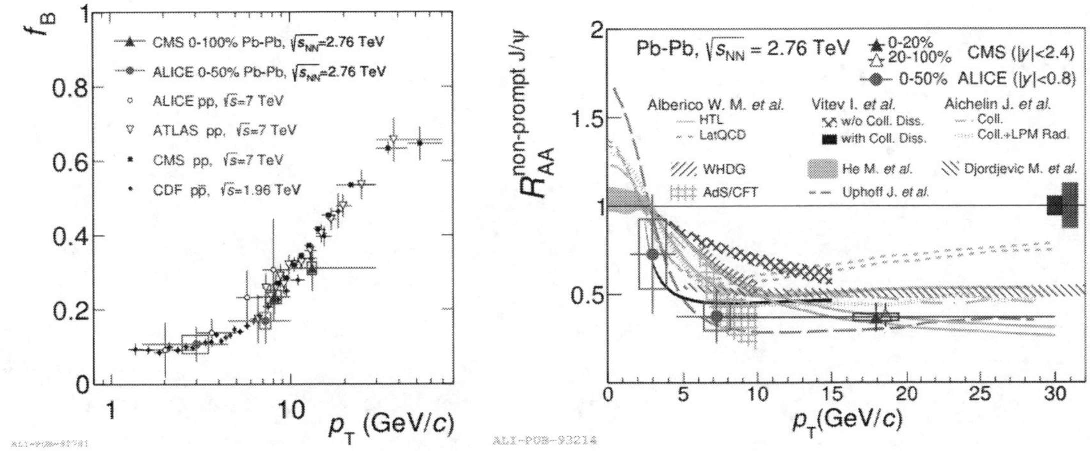

Figure 3: The fraction of J/ψ from the decay of beauty-hadrons measured at mid-rapidity in pp collisions and Pb-Pb collisions as a function of p_{T} compared with measurements by CMS [33, 34], ATLAS [35] and CDF [36] results (left) and the nuclear modification factor of J/ψ from the decay of beauty-hadrons at mid-rapidity in the 50% most central Pb-Pb collisions at $\sqrt{s_{\mathrm{NN}}} = 2.76\,\mathrm{TeV}$ together with CMS results [33] and theoretical predictions (right) [32].

non-prompt J/ψ, the R_{AA} of non-prompt J/ψ was obtained as shown in the right panel of Figure 3 [32]. The R_{AA} extends the kinematic coverage of CMS [33] down to low p_{T}. In $4.5 < p_{\mathrm{T}} < 10\,\mathrm{GeV}/c$, the suppression of non-prompt J/ψ tends to be stronger than what is predicted by most of the models.

2.3 b-jet

Tagging the jets coming from the fragmentation of beauty quarks (b-jet) represents an unbiased selection on the kinematics of the hard scattering, even in heavy-ion collisions. In addition, b-jet tagging with different jet radii allow the study of beauty-quark energy loss redistribution in Pb-Pb collisions. After jet reconstruction with charged tracks, b-jet can be identified exploiting long lifetime and large mass of beauty hadrons. In the following, two methods for b-jet tagging used in ALICE are presented. One way to tag b-jets among reconstucted jets is based on the signed impact parameter of the tracks in the reconstructed jet, where the sign of impact parameter is determined with the sign of the scalar product of the jet direction and the vector from primary vertex to the point of DCA. The tagging procedure considered using the signed impact parameter is the following. First, sort tracks in a jet by decreasing values of the signed impact parameter. Based on the fact that the probability to have several tracks with high positive signed impact parameter is high for b-jet due to the displaced secondary vertex of beauty hadron, assign the impact parameter of the Nth most displaced track as discriminator for a given jet and select jets which have large discriminator exceeding certain threshold (d_0). Here, N and d_0 are selected depending on the datasets since they should be optimised for S/B and statistical significance. In the left panel of Figure 4, the signed impact parameter distribution of the 3rd most displaced track in b-jet features distinct shape from that of c-jet.
The other way to tag b-jet is relied on the reconstruction of secondary vertex. After reconstructing secondary vertices (SV) with 2 or 3 tracks within a jet, use the signed flight distance significance ($L_{\mathrm{xy}}/\sigma_{L_{\mathrm{xy}}}$) and SV dispersion ($\sigma_{\mathrm{vtx}}$) as discriminator of a given jet. As shown in middle panel of

DOI: 10.1051/epjconf/201716407006

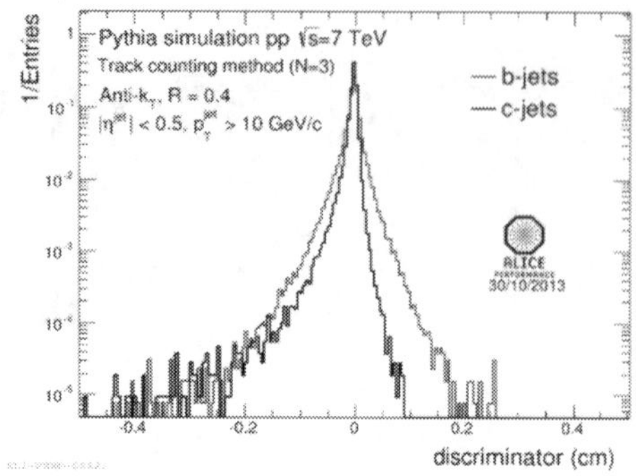

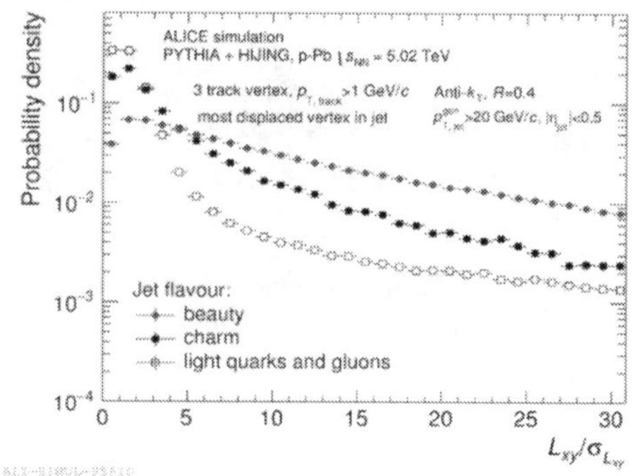

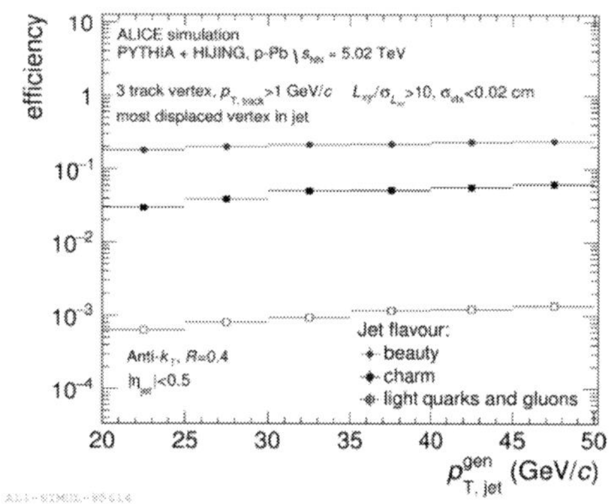

Figure 4: Signed impact parameter distributions of the 3rd most displaced track in given flavour jets in pp collisions at $\sqrt{s} = 7\,\text{TeV}$ (left, top), flight distance significance distributions of the most displaced secondary vertex found in given flavour jets in p-Pb collisions at $\sqrt{s_{\text{NN}}} = 5.02\,\text{TeV}$ (right, top) and b-jet tagging efficiency and mis-tagging rates based on reconstruction of secondary vertex in p-Pb collisions at $\sqrt{s_{\text{NN}}} = 5.02\,\text{TeV}$ (left, bottom). All plots are obtained from the Monte Carlo simulations.

Figure 4, $L_{\text{xy}}/\sigma_{L_{\text{xy}}}$ distribution of the most displaced SV in b-jet is broader compared to those in c-jet and jet from light quarks and gluons. σ_{vtx}, the dispersion of the tracks in the vertex, qualify the reconstructed secondary vertex. The right panel in Figure 4 shows the b-jet tagging efficiency and mis-tagging rates of c-jet and jet from light quarks and gluons calculated for requiring $L_{\text{xy}}/\sigma_{L_{\text{xy}}} > 10$ and $\sigma_{\text{vtx}} < 0.02\,\text{cm}$. They do not strongly vary with p_{T} and the b-tagging efficiency is higher about factor 100 compared to the jet from light quarks and gluons and about 3 to 5 compared to c-jet.

3 Prospects with RUN2 & RUN3

RUN2 of the LHC, from 2015 to 2018, offers the new higher collision energy for pp collisions up to $\sqrt{s} = 13\,\text{TeV}$ and for Pb-Pb collisions up to $\sqrt{s_{\text{NN}}} = 5\,\text{TeV}$. The increase of statistics compared to RUN1 allows to reduce not only statistical uncertainty but also systematic uncertainties. More precise beauty production measurements in an extended p_{T} coverage are expected.

Upgrading major detectors and computing framework is planned in LS2 between RUN2 and RUN3 [17]. Increase the readout capabilities of the detectors, especially TPC readout upgrade together with introducing new combined Online-Offline system for calibration and data compression, will increase the readout capabilities up to a rate of 50 kHz for Pb-Pb collisions and 200 kHz for pp and p-Pb collisions [37]. Improvement of tracking precision is expected via upgraded ITS at mid-rapidity and newly installed Muon Forward Tracker (MFT) at forward rapidity. As a result, better precision with more statistics and extended kinematic coverage measurements of beauty production

 DOI: 10.1051/epjconf/201716407006

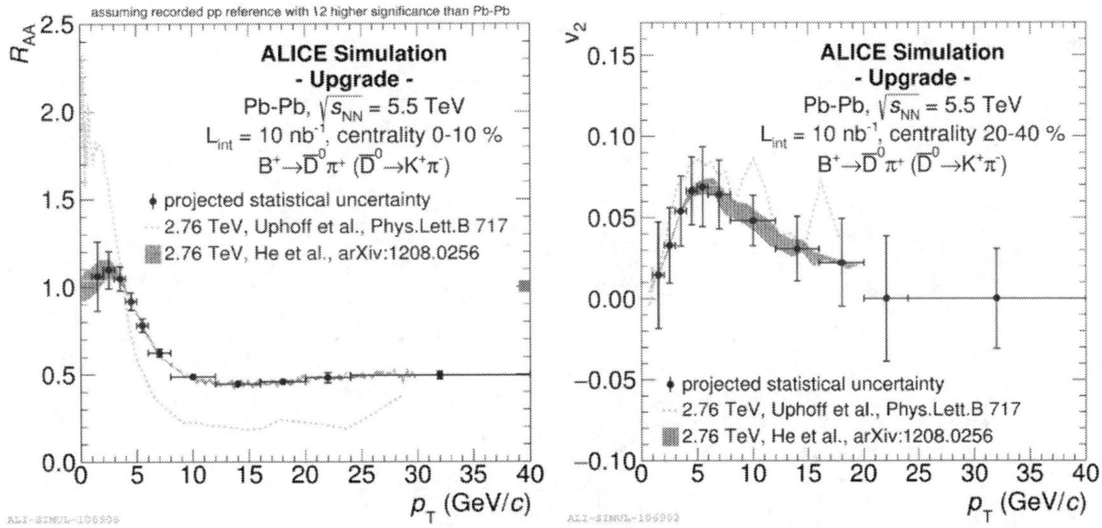

Figure 5: Nuclear modification factor of B^+ as a function of p_T compared to theoretical predictions (left) and azimuthal anisotropy, v_2 of B^+ as a function of p_T compared to theoretical predictions.(right)

with new observables will be possible. Brief description of upgraded ITS and MFT and their expected performances for beauty measurements are presented below.

3.1 Upgraded ITS

The upgraded ITS consists of 7 cylindrical layers of Monolithic Active Pixel Sensors (MAPS) [38]. The impact parameter resolution will be improved by a factor of 3 in $r\phi$-direction and 6 in z-direction with closer position of the first layer to interaction point (from r = 39 mm to r = 23 mm), reduced material budget (from $x/X_0 = 1.14$ % to $x/X_0 = 0.3$ %) and smaller pixel size (from 50 $\mu m \times$ 425 μm to 28 $\mu m \times$ 28 μm). The tracking efficiency and p_T resolution will be further improved by increasing layers from 6 to 7. Together with increased statistics in RUN3, more precise measurement of current observables such as R_{AA} of non-prompt J/ψ with extended p_T will be possible. Moreover, distinct decay topologies and kinematics allow for efficient full reconstruction of beauty-hadrons, for example, $B^+ \rightarrow D^0 + \pi^-$ ($D^0 \rightarrow K^+ + \pi^-$), $B^+ \rightarrow J/\psi + K^+$ ($J/\psi \rightarrow e^+ + e^-$) and $\Lambda_b^0 \rightarrow \Lambda_c^+\pi^-$ ($\Lambda_c^+ \rightarrow pK^-\pi^+$). Figure 5 shows the expected R_{AA} and azimuthal anisotropy (v_2) of B^+ under the assumption of an integrated luminosity of $L_{int} = 10$ nb^{-1}.

3.2 MFT

In current ALICE apparatus, the muon spectrometer is blind to the details of the vertex region due to the hadron absorber. Muon Forward Tracker (MFT) composed with 5 layers (each layer consists of two half-disks) of silicon pixel sensors using same technology as the upgraded ITS will be installed during LS2 [39]. Muon tracks reconstructed in the muon spectrometer will be matching with the track segments reconstructed in MFT planes. High pointing accuracy of muons at forward rapidity region, which would be enough to separate muons from semi-leptonic decays of beauty-hadrons exploiting their displaced secondary vertex is expected. It also allows to measure non-prompt J/ψ down to $p_T = 0$ at forward rapidity as shown in Figure 6.

 DOI: 10.1051/epjconf/201716407006

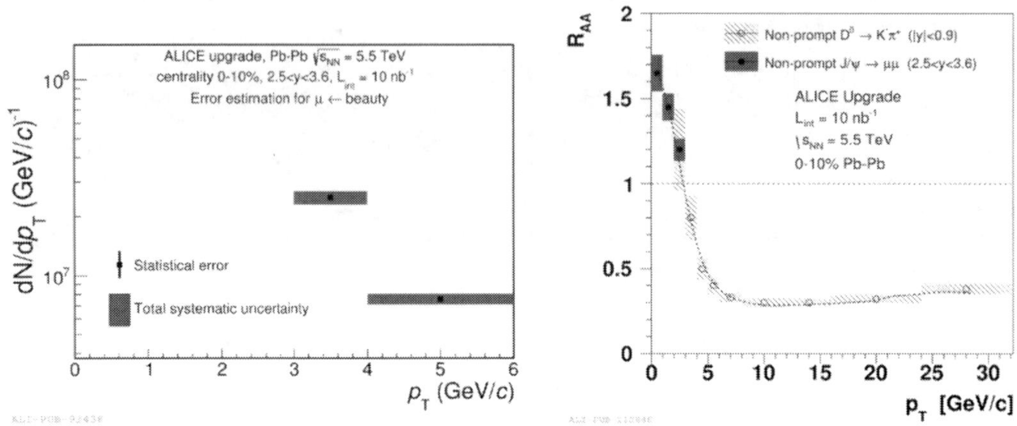

Figure 6: Transverse momentum distributions of muons from beauty-hadron decays (left) and nuclear modification factor of non-prompt J/ψ at forward rapidity compared with nuclear modification factor of non-prompt D^0 at mid-rapidity region (right) [39].

4 Conclusions

Thanks to the excellent tracking, vertexing and particle-identification capabilities provided by ALICE, beauty production is studied via semi-eletronic decay channels in pp, p-Pb and Pb-Pb collisions and non-prompt J/ψ in the e^+e^- decay channel in pp and Pb-Pb collisions at mid-rapidity.

The p_T-differential cross sections of electrons from beauty-hadron decays measured in pp collisions at $\sqrt{s} = 7$ TeV and $\sqrt{s} = 2.76$TeV are reproduced by several pQCD-based calculations. Since the R_{pA} of electrons from beauty-hadron decays is close to unity for $1 < p_T < 8\,\mathrm{GeV}/c$, the suppression of R_{AA} shown in $p_T > 3$ GeV/c provides a hint of the energy loss of beauty quarks in the hot and dense medium.

In the fraction of non-prompt J/ψ measured in Pb-Pb collisions, a similar p_T dependence as in pp collisions is observed. The suppression shown in R_{AA} of non-prompt J/ψ at $4.5 < p_T < 10$ GeV/c tends to be stronger than what predicted by most of the models.

In order to study of in-medium energy loss of beauty without bias on the kinematics of the hard scattering and explore the energy loss redistribution of beauty quarks, b-jet tagging is ongoing in ALICE.

Better precision, more statistics and extended kinematic coverage measurements will be possible with the upcoming Run2 and RUN3-4 data. Especially after installation of the upgraded ITS and the MFT, beauty production measurements will be significantly improved with new observables, such as R_{AA} and v_2 of exclusively reconstructed beauty hadrons.

Acknowledgements

This work was supported by National Research Foundation of Korea (NRF), Basic Science Research Program through the National Research Foundation of Korea funded by the Ministry of Education, Science and Technology (Grant number: 2013S1A2A2035612) and by Inha University Research Grant (INHA-47297).

References

[1] F.M. Liu, S.X. Liu, Phys. Rev. C **89**, 034906 (2014)
[2] K. Aamodt et al., (ALICE Collaboration), Phys. Lett. **B696**, 328 (2013), arXiv:1012.4035
[3] A. Andronic et al., Eur. Phys. J. **C76**, 107 (2016), arXiv:1506.03981
[4] M. Gyulassy, X.N. Wang, Nucl. Phys. **B420**, 583 (1994), nucl-th/9306003
[5] R. Baier, Y.L. Dokshitzer, A.H. Mueller, S. Peigne, D. Schiff, Nucl. Phys. **B484**, 265 (1997), hep-ph/9608322
[6] U.A. Wiedemann, Nucl. Phys. **B588**, 303 (2000), hep-ph/0005129
[7] Y.L. Dokshitzer, D.E. Kharzeev, Phys. Lett. **B519**, 199 (2001), hep-ph/0106202
[8] M. Djordjevic, M. Gyulassy, Nucl. Phys. **A733**, 265 (2004), nucl-th/0310076
[9] N. Armesto, C.A. Salgado, U.A. Wiedemann, Phys. Rev. **D69**, 114003 (2004), hep-ph/0312106
[10] H. van Hees, V. Greco, R. Rapp, Phys. Rev. **C73**, 034913 (2006), 0508055
[11] N. Armesto, A. Dainese, C.A. Salgado, U.A. Wiedemann, Phys. Rev. D **71**, 054027 (2005)
[12] K.J. Eskola, H. Paukkunen, C.A. Salgado, JHEP **04**, 065 (2009), arXiv:0902.4154
[13] I. Vitev, Phys. Rev. C **75**, 064906 (2007)
[14] K. Aamodt et al., (ALICE Collaboration), JINST **3**, S08002 (2008)
[15] F. Carminati et al., (ALICE Collaboration), J. Phys. G: Nucl. Part. Phys. **30**, 1517 (2004)
[16] B. Alessandro et al., (ALICE Collaboration), J. Phys. G: Nucl. Part. Phys. **32**, 1295 (2004)
[17] B. Abelev et al., (ALICE Collaboration), J. Phys. G: Nucl. Part. Phys. **41**, 087001 (2014)
[18] B. Abelev et al., (ALICE Collaboration), Int. J. Mod. Phys. **A29**, 1430044 (2014), arXiv:1402.4476
[19] B. Abelev et al., (ALICE Collaboration), Phys. Lett. **B721**, 13 (2013), arXiv:1208.1902
[20] B. Abelev et al., (ALICE Collaboration), Phys. Lett. **B738**, 97 (2014), arXiv:1405.4144
[21] J. Adam et al., (ALICE Collaboration) (2016), arXiv:1609.03898
[22] B. Abelev et al., (ALICE Collaboration), JHEP **09**, 112 (2012), arXiv:1203.2160
[23] M. Cacciari, S. Frixione, N. Houdeau, M.L. Mangano, P. Nason, G. Ridolfi, Journal of High Energy Physics **2012**, 1 (2012)
[24] R. Maciuła, A. Szczurek, Phys. Rev. D **87**, 094022 (2013)
[25] P. Bolzoni, G. Kramer, Nucl. Phys. B **872**, 253 (2013)
[26] M. Cacciari, M. Greco, P. Nason, JHEP **05**, 007 (1998), hep-ph/9803400
[27] M. Cacciari, S. Frixione, P. Nason, JHEP **03**, 006 (2001), hep-ph/0102134
[28] R. Sharma, I. Vitev, B.W. Zhang, Phys. Rev. **C80**, 054902 (2009), arXiv:0904.0032
[29] Z.B. Kang, I. Vitev, E. Wang, H. Xing, C. Zhang, Phys. Lett. **B740**, 23 (2015), arXiv:1409.2494
[30] A.M. Sickles, Phys. Lett. **B731**, 51 (2014), arXiv:1309.6924
[31] B. Abelev, etal, Journal of High Energy Physics **2012**, 1 (2012)
[32] J. Adam et al., (ALICE Collaboration), JHEP **07**, 051 (2015), arXiv:1504.07151
[33] S. Chatrchyan et al., (CMS Collaboration), JHEP **05**, 063 (2012), arXiv:1201.5069
[34] V. Khachatryan et al., (CMS Collaboration), Eur. Phys. J. **C71**, 1575 (2011), arXiv:1011.4193
[35] G. Aad et al., (ATLAS Collaboration), Nucl. Phys. **B850**, 387 (2011), arXiv:1104.3038
[36] D. Acosta et al., (CDF Collaboration), Phys. Rev. **D71**, 032001 (2005), hep-ex/0412071
[37] P. Antonioli, A. Kluge, W. Riegler (ALICE Collaboration), Tech. Rep. CERN-LHCC-2013-019. ALICE-TDR-015 (2013), https://cds.cern.ch/record/1603472
[38] B. Abelev et al., (ALICE Collaboration), Tech. Rep. CERN-LHCC-2013-024. ALICE-TDR-017 (2013), https://cds.cern.ch/record/1625842

DOI: 10.1051/epjconf/201716407006

[39] B. Abelev et al., (ALICE Collaboration), Tech. Rep. CERN-LHCC-2015-001. ALICE-TDR-018 (2015), https://cds.cern.ch/record/1981898

 DOI: 10.1051/epjconf/201716407007

Constraints of beyond Standard Model parameters from the study of neutrinoless double beta decay

Sabin Stoica[1,2,a]

[1] *Horia Hulubei Foundation, P.O. Box MG12, and*
[2] *Horia Hulubei Institute of Physics and Nuclear Engineering, P.O. Box MG6, 077125 Bucharest-Magurele, Romania*

Abstract. Neutrinoless double beta ($0\nu\beta\beta$) decay is a beyond Standard Model (BSM) process whose discovery would clarify if the lepton number is conserved, decide on the neutrinos character (are they Dirac or Majorana particles?) and give a hint on the scale of their absolute masses. Also, from the study of $0\nu\beta\beta$ one can constrain other BSM parameters related to different scenarios by which this process can occur. In this paper I make first a short review on the actual challenges to calculate precisely the phase space factors and nuclear matrix elements entering the $0\nu\beta\beta$ decay lifetimes, and I report results of our group for these quantities. Then, taking advance of the most recent experimental limits for $0\nu\beta\beta$ lifetimes, I present new constraints of the neutrino mass parameters associated with different mechanisms of occurrence of the $0\nu\beta\beta$ decay mode.

1 Introduction

Despite the fact that neutrinos are among the most abundant particles in the universe, their fundamental properties as the absolute mass and mass hierarchy, their character (are they Dirac or Majorana particles?), the number of neutrino flavors other than the three active ones, etc. are still unknown. There are different processes/measurements from which one can get information about these issues, one of them being the neutrinoless double beta ($0\nu\beta\beta$) decay [1] -[6]. Double beta decay (DBD) is the rarest spontaneous nuclear decay measured until now, by which an e-e nucleus transforms into another e-e nucleus with its nuclear charge changed by two units. It occurs whatever single beta decays can not occur due to energetic reasons or it is highly forbidden by angular momentum selection rules. Usually, the DBD process occurs with the emission of two neutrinos or two anti-neutrinos besides two positrons or two electrons, and hence with the lepton number conservation (LNC). However, since the LNC is not a consequence of a known gauge symmetry, BSM theories allow DBD to occur without emission of anti-neutrinos/neutrinos (the so called $0\nu\beta\beta$ decay modes), as well. In this contribution I refer to the $0\nu\beta\beta$ with the emission of two electrons in the final states. This process can provide us with decisive information about the neutrino character, absolute mass of neutrino and mass hierarchy, and it still is at present the most sensitive process to check the lepton number conservation. The lifetime expression of this process can be factorized in a good approximation as a product of three factors: the phase space factor (PSF) that embeds the distortion of the electron wave functions (w.f.) by the nuclear

[a] e-mail: stoica@nipne.ro

Coulomb potential, the nuclear matrix elements (NME) that depend on the nuclear structure of the nuclei involved in decay and a BSM factor related to the mechanism that contributes to $0\nu\beta\beta$. At present, the NME are currently computed with several nuclear structure methods, such as the proton-neutron Quasi Random Phase Approximation (pnQRPA) [7]-[10], Shell Model(ShM) [11]-[12], Interacting Boson Approximation (IBA) [13]-[14], Projected Hartree Fock Bogoliubov (PHFB) [15] and Energy Density Functional (EDS) method [16]. The computation of the NME brings the largest uncertainties in the theoretical study of double beta decay (DBD) process. Each method has its advantages and disadvantages and there has been and still is a long debate in literature on their calculations. Unlike the NME, the PSF have been considered to be computed with enough precision [17]-[19]. However, recently, they were recalculated within an improved approach by using exact electron w.f., solution of the Dirac equations, and taking into account finite nuclear size and electron screening effects [20]. The authors found differences between their results and those calculated previously with approximate electron w.f., especially for heavier nuclei. We also recalculated the PSF by developing new accurate routines for computing the relativistic (Dirac) electron w.f. and for integrating the PSF expressions. In addition to the previous similar calculations, we use a Coulomb potential derived from a realistic proton density distribution in the daughter nuclei [21]-[22]. Hence, the NME and PSF are theoretical key quantities for predicting $0\nu\beta\beta$ decay half-lives. On the other hand, because the $0\nu\beta\beta$ decay has not been discovered so far, there are only experimental lifetime limits for this process. Using these limits and precise calculations of the PSF and NME, one can derive upper limits of the BSM parameters associated with specific decay mechanisms.

2 Theoretical formalism

We shortly present the theoretical framework for the derivation of neutrino mass parameters from the $0\nu\beta\beta$ decay analysis. In the hypothesis that this DBD mode occurs through two mechanisms, namely i) light left- handed (LH) neutrinos and ii) heavy right-handed (RH) neutrinos, between two nucleons inside the nucleus, the contribution of the two mechanisms do not interfere with one another and the lifetime formula can be expressed as a sum of two terms:

$$\left(T^{0\nu}_{1/2}\right)^{-1} = G^{0\nu}(Q_{\beta\beta}, Z)\left(\mid M^{0\nu}_{\nu} \mid^2 (\eta_\nu)^2 + \mid M^{0\nu}_{N} \mid^2 (\eta_N)^2\right) \quad (1)$$

where $G^{0\nu}$ is the PSF for this decay mode and depends on the energy decay $Q_{\beta\beta}$ and nuclear charge Z. $M^{0\nu}_{\nu,N}$ are the NME for the light LH (ν) and heavy RH (N) neutrino exchange mechanisms, respectively, and $< \eta >$ are the neutrino mass parameters which are related to the neutrino masses:

$$< \eta_\nu >= \left(\frac{\langle m_\nu \rangle}{m_e}\right)^2 \quad (2)$$

with

$$< m_\nu >= \sum_i^{light} |U_{ei}|^2 m_i e^{i\phi_i} (3)$$

for the "light" mechanism and

$$< \eta_N >= \sum_k^{heavy} |U_{ek}|^2 \frac{m_p}{m_N} e^{i\phi_k} (4)$$

for the "heavy" mechanism. In the above formulas the m_i and m_N are the neutrino (light, heavy) masses, $\phi(i,k)$ Majorana phases, m_p the proton mass and the $U_{e(i,k)}$ are the mixing matrix elements associated to LH weak interactions. We adopted the hypothesis that only one flavor of "'heavy"' neutrino exist.

As seen, for deriving the $< \eta >$ parameters we need accurate calculations of both PSF and NME, for each mechanism and each isotope for which the experimental lifetime limits exist.

For the $0\nu\beta\beta$ decay, the expression of the PSF is [20]-[21]:

$$G_{0\nu} = \frac{2}{4g_A^4 R^2 \ln 2} \int_{m_e c^2}^{T_0 - m_e c^2} f_{11}^{(0)} w_{0\nu} d\epsilon_1 \quad (5)$$

where

$$w_{0\nu} = \frac{g_A^4 (G \cos\theta_C)^4}{16\pi^5} (m_e c^2)^2 (\hbar c^2)(p_1 c)(p_2 c)\epsilon_1 \epsilon_2 \quad (6)$$

In the above formulas g_A is the axial-vector coupling constant, R is the nuclear radius, $G = 1.16637 \times 10^{-5}$ GeV^{-2} is the Fermi constant, θ_C is the Cabbibo angle, m_e is the electron rest mass, while p_1, p_2 and ϵ_1, ϵ_2 are the electron momenta and electron energies, respectively. It is worth to note that in Eq.(5) the PSF expression was redefined, such that this quantity be independent on the g_A constant and correlates (by dividing by $4R^2$) the dimension of $G_{0\nu}$ with the NME which are dimensionless, such that the PSF be expressed in $[yr^{-1}]$.

The electron phase factors $f_{jk}^{(0)}$ have the expressions:

$$f_{11}^{(0)} = | f^{-1-1} |^2 + | f_{11} |^2 + | f^{-1}{}_1 |^2 + | f_1{}^{-1} |^2 \quad (7)$$

with

$$f^{-1-1} = g_{-1}(\epsilon_1) g_{-1}(\epsilon_2) \; ; \; f_{11} = f_1(\epsilon_1) f_1(\epsilon_2), f^{-1}{}_1 = g_{-1}(\epsilon_1) f_1(\epsilon_2) \; ; \; f_1{}^{-1} = f_1(\epsilon_1) g_1(\epsilon_2) \quad (8)$$

The radial functions f and g are solutions of the Dirac equations [20]), [22]:

$$\frac{dg_\kappa(\epsilon, r)}{dr} = -\frac{\kappa}{r} g_\kappa(\epsilon, r) + \frac{\epsilon - V + m_e c^2}{c\hbar} f_\kappa(\epsilon, r) \quad (9)$$

$$\frac{df_\kappa(\epsilon, r)}{dr} = -\frac{\epsilon - V - m_e c^2}{c\hbar} g_\kappa(\epsilon, r) + \frac{\kappa}{r} f_\kappa(\epsilon, r) \quad (10)$$

The expression of the NME can be written as a sum of three components [10]-[24]:

$$M^{0\nu} = M_{GT}^{0\nu} - \left(\frac{g_V}{g_A}\right)^2 \cdot M_F^{0\nu} - M_T^{0\nu} \quad (11)$$

where $M_{GT}^{0\nu}$, $M_F^{0\nu}$ and $M_T^{0\nu}$ are components associated with the matrix elements of the Gamow-Teller (GT), Fermi(F) and Tensor(T) DBD operators, respectively. These are defined as follows:

$$M_\alpha^{0\nu} = \sum_{m,n} \left\langle 0_f^+ || \tau_{-m} \tau_{-n} O_{mn}^\alpha || 0_i^+ \right\rangle \quad (12)$$

where O_{mn}^α are DBD transition operators ($\alpha = GT, F, T$) and the summation is over all the nucleon states. The NME calculations can be divided in the computation of two quantities, the two-body transitions densities and the reduced matrix elements of the two-body transition operators O^α. The most difficult is the computation of the radial part of the two-body transition operators, which contains the neutrino potentials. These potentials are defined by integrals of momentum carried by the virtual neutrino exchanged between the two nucleons inside the nucleus [10], [23]-[24]:

$$H_\alpha(r) = \frac{2R}{\pi}\int_0^\infty j_i(qr)\frac{h_\alpha(q)}{\omega}\frac{1}{\omega+\langle E\rangle}q^2 dq$$
$$\equiv \int_0^\infty j_i(qr)V_\alpha(q)q^2 dq \qquad (13)$$

,
where $R = r_0 A^{1/3}$ fm ($r_0 = 1.2fm$), $\omega = \sqrt{q^2+m_\nu^2}$ is the neutrino energy and $j_n(qr)$ is the spherical Bessel function (i = 0, 0 and 2 for GT, F, and T, respectively). Usually, in $0\nu\beta\beta$ calculations one uses the closure approximation which consists in replacing the energies of the excited states of the intermediate odd-odd nucleus contributing to the decay, by an average value $\langle E\rangle$. This approximation works good in the case of $0\nu\beta\beta$ decay modes (but not in the case of $2\nu\beta\beta$) and simplifies much the calculations. The expressions of $h_\alpha(\alpha = GT, F, T)$ are [10]

$$h_F = G_V^2(q^2) \qquad (14)$$

$$h_{GT}(q^2) = \frac{G_A^2(q^2)}{g_A^2}\left[1 - \frac{2}{3}\frac{q^2}{q^+ m_\pi^2} + \frac{1}{3}\left(\frac{q^2}{q^2+m_\pi^2}\right)^2\right] +$$
$$+\frac{2}{3}\frac{G_M^2(q^2)}{g_A^2}\frac{q^2}{4m_p^2} \qquad (15)$$

and

$$h_T(q^2) = \frac{G_A^2(q^2)}{g_A^2}\left[\frac{2}{3}\frac{q^2}{q^2+m_\pi^2} - \frac{1}{3}\left(\frac{q^2}{q^2+m_\pi^2}\right)^2\right] +$$
$$+\frac{1}{3}\frac{G_M^2(q^2)}{g_A^2}\frac{q^2}{4m_p^2} \qquad (16)$$

where m_π is the pion mass, m_p is the proton mass and

$$G_M(q^2) = (\mu_p - \mu_n)G_V(q^2) \qquad (17)$$

with $(\mu_p - \mu_n) = 4.71$.

The expressions (14)-(17) include nuclear ingredients that have to be taken into account for a precise computation of the NME, such as the inclusion of higher order currents in the nuclear interaction (HOC) and finite nucleon size effect (FNS). Inclusion of HOC effects brings additional terms in the H_{GT} and H_F components and apparition of the H_T component in the expressions of the neutrino potentials. FNS effect is taken into account through G_V and G_A form factors [10], [12]:

$$G_A\left(q^2\right) = g_A\left(\frac{\Lambda_A^2}{\Lambda_A^2+q^2}\right)^2, \; G_V\left(q^2\right) = g_V\left(\frac{\Lambda_V^2}{\Lambda_V^2+q^2}\right)^2 \qquad (18)$$

For the vector and axial coupling constants many calculations take $g_V = 1$ and the unquenched value $g_A = 1.25$, while the values of the vector and axial vectors form factors are $\Lambda_V = 850MeV$ and $\Lambda_A = 1086MeV$ [1], respectively.

To compute the radial matrix elements $\langle nl|H_\alpha|n'l'\rangle$ one uses harmonic oscillators w. f. $\psi_{nl}(lr)$ and $\psi_{n'l'}(r)$ corrected by a factor $[1+f(r)]$, in order to take into account the short range correlations (SRC) induced by the nuclear interaction:

$$\psi_{nl}(r) \rightarrow [1 + f(r)]\, \psi_{nl}(r) \tag{19}$$

For the correlation function we take the functional form

$$f(r) = -c \cdot e^{-ar^2}\left(1 - br^2\right) \tag{20}$$

where a, b and c are constants that have particular values for different parameterizations [10], [25]-[26]. For $c = 1$, we get the Jastrow prescription of inclusion of the SRC effects, and in this case the a and b constants are given by the Miller-Spencer (MS) parametrization [18].

3 Numerical results and discussions

The NME calculations are performed with a new ShM code [23]. In a previous paper [24] we shortly presented a new improved (fast, efficient) ShM code that reduces substantially the computing time of calculation of the two-body matrix elements (TBME) of the transition operators for the $0\nu\beta\beta$ decay. It incorporates all the relevant nuclear effects. The radial NME are double integrals, over coordinates and momentum, that require long CPU time to be performed. The main improvement of the code consists in performing analytically the radial integrals over the coordinates, by rearranging the terms depending on "r" in the TBME expression and using recurrence properties of the Laguerre polynomials and Bessel functions. Therefore, only the integration over the momentum remains to be performed numerically. Testing this code we found it about 30 times faster than our previous code [12], for the same calculations, and it can be of much help for investigating the quenching of the Gammow-Teller (GT) operator in ShM calculations of the NME, in large model spaces. The computation of the NME is performed with nuclear ingredients chosen according to a general consensus, as follows. The short range correlations between nucleons are introduced according to the CD-Bonn parametrization [10], the tensor contribution, due to the inclusion of HOC effects, is taken into account, the FNS effect is involved in calculation through the form factors g_V and g_A with the cut-off parameters $\Lambda_V = 0.85$ GeV and $\Lambda_A = 1.09$ GeV. For g_A we adopt the unquenched value 1.25 and the nuclear radius $R = r_0 A^{1/3}$ is calculated with the constant $r_0 = 1.2$ fm.

To compute PSF for DBD modes we obtained the w.f. of the electrons emitted in the decay, which are distorted by the Coulomb potential of the nucleus. We use exact Dirac electron w.f. with the inclusion of the FNS and screening effects. This last effect was taken into account by multiplying the expression of the Coulomb potential with a function $\phi(r)$, which is solution of the Thomas Fermi-type equation [20], [22]. In addition to other previous calculations, we use a Coulomb-type potential derived from a realistic proton density distribution in nucleus. Also, we develop own routines for solving the Dirac equations and integrate the PSF expressions with improved numerical accuracy as compared with older codes, and we use new Q-values reported in ref. [27].

Our results are presented in In Tables I and II. In Table I upper limits for the electron-neutrino mass parameter m_ν, associated with the light LH neutrino exchange mechanism, are derived in the case of three isotopes, ^{48}Ca, ^{76}Ge and ^{82}Se, together with corresponding NME and Q-values and experimental lifetime limits. The effective NN interactions used are: GXPF1A [28] for ^{48}Ca and JUN45 [29] for ^{76}Ge and ^{82}Se. As seen from the Table I, the most stringent upper limit for the neutrino mass parameter, which is related to the electron-neutrino absolute mass, is obtained from the study of DBD of ^{76}Ge and is of 0.27 eV. This is due both to the accurate calculation of the NME (with advanced ShM methods) and to the experimental lower limit of the half-life (~ 2 x 10^{25} yr), in the case of this isotope. Indeed, the DBD experiments on ^{76}Ge, besides those on ^{136}Xe are at present the most sensitives ones and provide us with the largest lower limits for $0\nu\beta\beta$ decay half-lives. These limits, of ~ 0.2 eV, for the "light" neutrino parameters can be checked by the next generation of DBD experiments. In Table II upper limits of the neutrino parameter (coupling strength) η_N, associated with the exchange of a heavy RH neutrino mechanism are presented. Again, due to the same reason, the most stringent upper limits for the "heavy" neutrino mass parameter is extracted from the ^{76}Ge isotope.

Table 1. Upper limits of the Majorana light neutrino mass parameter together with the other components of the DBD lifetime: Q values, experimental lifetime limits, PSF and NME.

	$Q[MeV]$	$T^{0\nu}_{exp}[yr]$	$G^{0\nu}[yr^{-1}]$	$M^{0\nu}_{\nu}$	$\langle m_\nu \rangle\,[eV]$
^{48}Ca	4.272	> 5.8 10^{22}[30]	2.46E-14	0.81	< 9.62
^{76}Ge	2.039	> 2.1 10^{25}[31]	2.37E-15	3.13	< 0.27
^{82}Se	2.995	> 3.6 10^{23}[32]	1.01E-14	2.88	< 1.19

Table 2. Upper limits of the Majorana heavy neutrino mass parameter together with experimental lifetime limits, PSF and NME.

	$T^{0\nu}_{exp}[yr]$	$G^{0\nu}[yr^{-1}]$	$M^{0\nu}_N$	$\langle \eta_N \rangle\,[eV]$
^{48}Ca	> 5.8 10^{22}[30]	2.46E-14	70.2	$< 3.77 \cdot 10^{-7}$
^{76}Ge	> 2.1 10^{25}[31]	2.37E-15	199.2	$< 0.22 \cdot 10^{-7}$
^{82}Se	> 3.6 10^{23}[32]	1.01E-14	184.5	$< 0.90 \cdot 10^{-7}$

4 Conclusions

In this work I make first a short review on the challenges encountered in the computation of the PSF and NME, the two key ingredients in the study of the $0\nu\beta\beta$ decay mode. Then, I present new constraints of the neutrino mass parameters associated with the "light" and "heavy" neutrino exchange mechanisms of occurrence of the $0\nu\beta\beta$ decay mode. The NME and PSF for three isotopes, ^{48}Ca, ^{76}Ge and ^{82}Se, are calculated with advanced codes developed by our group, while the most recent experimental limits for $0\nu\beta\beta$ lifetimes were taken from literature. The most strigent upper limits for the neutrino parameters were obtained for the ^{76}Ge isotope, for which both accurate calculations for NME and PSF, and the largest experimental lifetime value, exist.

Acknowledgment

This work was done with the support of the MENCS and UEFISCDI through the project IDEI-PCE-3-0318, contract Nr. 58/28.10/2011.

References

[1] F. T. Avignon, S.R. Elliott and J. Engel, Rev. Mod. Phys. **80**, 481 2008.
[2] H. Ejiri. Prog. Part. Nucl. Phys., **4**, 249 (2010).

[3] J. Vergados, H. Ejiri and F. Simkovic, Rep. Prog. Phys. **75**, 106301 (2012).
[4] W. Rodejohann, J.Phys. G: Nucl. Part. Phys. **39**, 124008 (2012).
[5] A. Faessler, V. Rodin, F. Simkovic, J. Phys. G **39**, 124006 (2012).
[6] P. Vogel, J.Phys. G: Nucl. Part. Phys. **39**, 124002 (2012).
[7] S. Stoica and H.V. Klapdor-Kleingrothaus, Nucl. Phys.**A 694** (2001) 269.
[8] M. Kortelainen and J. Suhonen, Phys. Rev. C **75**, 051303 (2007); Phys. Rev. C **76**, 024315 (2007).
[9] F. Simkovic, A. Faessler, V.A. Rodin, P. Vogel and J. Engel, Phys. Rev. C **77**, 045503 (2008)
[10] F. Simkovic, A. Faessler, H. Muther, V. Rodin, and M. Stauf, Phys. Rev. C **79**, 055501 (2009).
[11] J. Menendez, A. Poves, E. Caurier, F. Nowacki, and A. Poves, Nuclear Physics **A 818** 139 (2009).
[12] M. Horoi and S. Stoica, Phys. Rev. **81**, 024321 (2010).
[13] J. Barea and F. Iachello, Phys. Rev. C **79**, 044301 (2009).
[14] J. Barea, J. Kotila, F. Iachello, Phys. Rev. C **87**, 014315 (2013).
[15] P.K. Rath, R. Chandra, K. Chaturvedi, P.K. Raina, J.G. Hirsch, Phys. Rev. C **82**, 064310 (2010).
[16] T.R. Rodriguez and G. Martinez-Pinedo, Phys. Rev. Lett **105**, 252503 (2010).
[17] M. Doi, T. Kotani and E. Takasugi, Prog. Theor. Phys. Suppl. **83**, 1 (1985).
[18] T. Tomoda, Rep. Prog. Phys. **54**, 53 (1991).
[19] J. Suhonen and O. Civitarese, Phys. Rep. **300**, 123 (1998).
[20] J. Kotila and F. Iachello, Phys. Rev. C **85**, 034316 (2012).
[21] S. Stoica and M. Mirea, Phys. Rev. C **88**, 037303 (2013).
[22] M. Mirea, T. E. Pahomi and S. Stoica, Rom. Rep. Phys. 67, 872 (2015).
[23] A. Neacsu, S. Stoica, J. Phys. G:Nucl. Part. Phys. **41**, 015201 (2014).
[24] A. Neacsu, S. Stoica and M. Horoi, Phys. Rev. C **86**, 067304 (2012).
[25] C. Giusti, H. Muther, F. D. Pacati, and M. Stauf, Phys. Rev. C **60**, 054608 (1999).
[26] H. Muther and A Polls, Phys. Rev. C 61, 014304 (1999) Part. Nucl. Phys. **45**, 243 (2000).
[27] M. Wang, G. Audi, A.H. Wapstra, F.G. Kondev, M. MacCornick, X. Xu, B. Pfeiffer, Chin. Phys. C 36, 1603 (2012).
[28] M. Honma, T. Otsuka, B. A. Brown, and T. Mizusaki, Eur. Phys. J. A **25**, Suppl. 1, 499 (2005).
[29] M. Honma, T. Otsuka, T. Mizusaki and M. Hjorth-Jensen, Phys. Rev. C **80**, 064323 (2009).
[30] S. Umehara et al., Phys Rev C **78**, 058501 (2008).
[31] C. Macolino et al. (GERDA Collaboration), Mod. Phys. Lett. A **29**, 1430001 (2014).
[32] A. S. Barabash, and Brudanin V. B., Phys. At. Nucl. **74**, 312 (2011).

CP violation in D meson decays at Belle

K. Prasanth[1,a]
(*On behalf of Belle collaboration*)

[1] *Indian Institute of Technology Madras, Chennai – 600036*

Abstract. CP violation and branching fraction measurements in D decays are interesting topics as any difference with respect to the Standard Model prediction would be an indication of new physics. With the large data sample collected by the Belle detector, which sits at the interaction point of KEKB asymmetric e^+e^- collider in Japan, we present the results of searches for CP violation in $D^0 \to V\gamma$ ($V = \phi, \overline{K}^{*0}, \rho^0$) decay and the rare D decay $D^0 \to \gamma\gamma$.

1 Introduction

CP violation in the charm sector is an interesting topic as it is very sensitive to the new physics. This is because Standard Model CP violation in the D meson system is very small $O(10^{-3})$. Therefore even a small enhancement in CP asymmetry will be due to new physics. Any enhancement comes from loop-level contributions from non-Standard Model particles or interactions. Here we present the preliminary branching fraction ($\mathcal{B}$) and CP asymmetry (A_{CP}) results for $D^0 \to V\gamma$ ($V = \phi, \overline{K}^{*0}, \rho^0$) decay and the measurements of branching fraction ($\mathcal{B}$) of $D^0 \to \gamma\gamma$ decay with the Belle data set corresponding to an integrated luminosity of 943 fb^{-1} and 832 fb^{-1}, respectively.

2 $D^0 \to V\gamma$ decay

The decay $D^0 \to V\gamma$ ($V = \phi, \overline{K}^{*0}, \rho^0$) is sensitive to new physics via A_{CP} measurements. This is because of the contributions from the chromomagnetic dipole operators [1], [2] that can enhance A_{CP} from the expected value of zero. So far no A_{CP} measurements have been reported for this mode. The previous $\mathcal{B}$ measurements [3], [4] of $D^0 \to V\gamma$ mode are shown in Table 1. With the 943 fb^{-1} Belle data set, which is very much larger than the one used in the previous measurements, we expect to get a more precise measurement of $\mathcal{B}$ for this decay channel. This is the first analysis to measure the $\mathcal{B}$ for $D^0 \to \rho^0\gamma$ mode.

2.1 Selection criteria, fitting and signal extraction

Selection criteria are defined using a Monte Carlo (MC) simulation study performed on data produced by the `EVTGEN` [5] and `GEANT3` [6] packages; the former includes the effects of final state

[a]e-mail: prasanthkp@physics.iitm.ac.in

 DOI: 10.1051/epjconf/201716407008

Table 1. Previous $\mathcal{B}$ measurements in $D^0 \to V\gamma$ decays.

Collaboration	Luminosity	Decay mode	Branching fraction ($\mathcal{B}$)
Belle[3]	78 fb^{-1}	$D^0 \to \phi\gamma$	$\left(2.60^{+0.70}_{-0.61}(\text{stat})^{+0.15}_{-0.17}(\text{syst})\right) \times 10^{-5}$
BABAR[4]	387 fb^{-1}	$D^0 \to \phi\gamma$	$(2.73 \pm 0.30(\text{stat}) \pm 0.26(\text{syst})) \times 10^{-5}$
BABAR[4]	387 fb^{-1}	$D^0 \to \overline{K}^{*0}\gamma$	$(3.22 \pm 0.20(\text{stat}) \pm 0.27(\text{syst})) \times 10^{-4}$

radiations (FSR). The D mesons are required to come from the $D^{*\pm}$ meson by $D^{*\pm} \to D^0\pi^{\pm}_{\text{slow}}$ decay. As the final state is the same for both D^0 and its conjugate $\overline{D^0}$ decay, we use the charge of $\pi^{\pm}_{\text{slow}}$ to distinguish the flavour of the D meson. The π_{slow} is so called because it carries very little momentum compared to the D meson. The vector mesons are reconstructed from the following decay channels: $\phi \to K^+K^-$, $\overline{K}^{*0} \to K^-\pi^+$ and $\rho^0 \to \pi^+\pi^-$. The selection criteria for the variables are chosen to maximize the significance, defined as $\frac{N_{\text{sig}}}{\sqrt{N_{\text{sig}}+N_{\text{bkg}}}}$, where N_{sig} and N_{bkg} are the number of signal and background events in the defined signal region. A tight mass window of 11 MeV/c^2 is applied for the ϕ candidates around its nominal mass [7] due to the narrow resonance. The mass window for the $\overline{K}^{*0}$ and ρ^0 candidates are 60 MeV/c^2 and 150 MeV/c^2 respectively. The photon candidates are selected with energy greater than 540 MeV. In order to suppress the merged photon cluster, we made a cut on the ratio of 3×3 array (E_9) to 5×5 array (E_{25}) of Electromagnetic Calorimeter (ECL) crystals to be greater than 0.94. A vertex fit is performed for D^* and D mesons where the candidates with confidence level less than 10^{-3} are rejected. The D^* meson fit also has an Interaction-Point (IP) constraint, which ensures that the D^* daughter particles originate from the e^+e^- interaction region. We defined a variable $q \equiv M(D^{*+}) - M(D^0) - M(\pi^+)$, where $M(D^{*+})$, $M(D^0)$, $M(\pi^+)$ are the masses of D^* meson, D meson and π meson, respectively, which is the total energy released in a D^* decay. The D^* candidates with a q value within 0.6 MeV/c^2 of the nominal value [7] are selected. A cut on the momentum of D^* meson in the center-of-mass (CMS) system is also applied for the modes, which are 2.42 GeV/c for the ϕ mode, 2.17 GeV/c for the $\overline{K}^{*0}$ mode and 2.72 GeV/c for the ρ^0 mode.

The measurements of $\mathcal{B}$ and A_{CP} are performed by using control samples for all the three modes. This procedure ease the systematic studies as many common effects get cancelled. These control samples are $D^0 \to K^+K^-$ for the ϕ mode, $D^0 \to K^-\pi^+$ for the $\overline{K}^{*0}$ mode and $D^0 \to \pi^+\pi^-$ for the ρ^0 mode. The branching fraction is calculated as

$$\mathcal{B}_{\text{sig}} = \mathcal{B}_{\text{norm}} \times \frac{N_{\text{sig}}}{N_{\text{norm}}} \times \frac{\epsilon_{\text{norm}}}{\epsilon_{\text{sig}}}, \tag{1}$$

where N_{sig} (N_{norm}), $\mathcal{B}_{\text{sig}}$ ($\mathcal{B}_{\text{norm}}$) and ϵ_{sig} (ϵ_{norm}) are the yields from the fit, branching fractions and reconstruction efficiencies for the signal (control sample) modes.

The CP asymmetry can be derived from the following equation

$$A_{\text{raw}} = A_{CP} + A_{FB} + A^{\pm}_{\epsilon}, \tag{2}$$

where $A_{\text{raw}} = \frac{N(D^0)-N(\overline{D^0})}{N(D^0)+N(\overline{D^0})}$ is the raw asymmetry ($N(D^0)$, $N(\overline{D^0})$ are the yields of D^0 and $\overline{D^0}$ from the fit to the data), A_{FB} [7] is the asymmetry in the production of D^{*+} and D^{*-} due to the interference between γ^* and Z^0 bosons in the $e^+e^- \to c\overline{c}$ process and $A^{\pm}_{\epsilon}$ is the asymmetry in the efficiency of the reconstruction of positive and negative charged particles, respectively. Thus A_{CP} can be extracted by using the information from the control samples in the respective signal modes. Since A_{FB} and $A^{\pm}_{\epsilon}$ is common to both signal and control sample modes, they are cancelled out. Finally,

$$A_{CP}^{\rm sig} = A_{\rm raw}^{\rm sig} - A_{\rm raw}^{\rm norm} + A_{CP}^{\rm norm}, \tag{3}$$

where subscript sig (norm) corresponds to the signal (control) sample.

The dominant background for the signal mode comes from the decay of a π^0 meson to a pair of photons. Here one of the the photons from the π^0 meson is misreconstructed as the signal candidate. To reduce these events, a π^0 mass veto is applied with a neural network variable [8], [9] obtained from the two mass veto variables. The signal photon is combined with all other photons in an event with an energy 30 (75) MeV and when the diphoton invariant mass is close to the nominal mass of π^0 meson, the pair is fed to the neural network. A selection on the output of the neural network results in the retention of 85% of the signal while rejecting 60% of the background.

To extract the signal yield and then A_{CP}, a simultaneous two-dimensional unbinned extended maximum likelihood fit is performed with the variables M_{D^0} and $\cos\theta_H$, where θ_H is the the angle between one of the V daughter particles with D meson in the V rest frame. The signal $\cos\theta_H$ distribution is expected to have the form $1-\cos^2\theta_H$ due to the conservation of angular momentum while the background distributions do not. The range for M_{D^0} is $1.67 < M_{D^0} < 2.06$ GeV/c^2 for all the three modes. A tighter cut of $-0.8 < \cos\theta_H < 0.4$ is also applied in $\overline{K}^{*0}$ and ρ^0 modes to suppress certain peaking backgrounds. The signal reconstruction efficiency is estimated from signal MC to be 9.7%, 7.8% and 6.8% for ϕ, $\overline{K}^*$ and ρ^0 modes, respectively.

The invariant mass distribution of signal events in ϕ and ρ^0 modes are modelled with Crystal-Ball function [10] whereas Crystal Ball with two Gaussians is used for the $\overline{K}^{*0}$ mode. To account for the data-MC difference, a free offset and scale factor are implemented for the mean and width of the $\overline{K}^{*0}$ PDF, and the obtained values are used for the other two signal modes. The π^0 and η M_{D^0} background distributions are modelled in one of two ways: (1) a single Crystal Ball, or (2) combinations of either Crystal Ball or logarithmic Gaussians [11] with upto two Gaussians. There is only one π^0 type background in the ϕ mode, which is $D^0 \to \phi\pi^0$. The π^0 and η type backgrounds in the $\overline{K}^{*0}$ modes are: $D^0 \to \overline{K}^{*0}\pi^0$, $K^-\rho^+$, $K_0^*(1430)^-\pi^+$, $K^{*-}\pi^+$, non-resonant $K^-\pi^+\pi^0$, $\overline{K}^{*0}\eta$ and non-resonant $K^-\pi^+\eta$. The backgrounds in the ρ^0 mode are $\rho^0\pi^0$, $\rho\pi$ and $K^-\rho^+$ due to the misidentification of a kaon as a pion. In all of the three modes, apart from the above mentioned backgrounds, another category is defined for all other decays with correctly reconstructed D^0. Two more backgrounds are present in the $\overline{K}^{*0}$ mode: non-resonant decay to $K^-\pi^+(\pi^+\pi^-)$ with the photon being emitted as FSR, and the decay to $K^-\rho^+$ with the photon being emitted from a radiative decay of the charged ρ meson. The M_{D^0} distributions are similar to that of signal in these modes as there are no missing particles, which are fixed from MC and known branching fractions. The remaining combinatoric backgrounds in M_{D^0} is modelled with an exponential function in the ϕ mode and a second order Chebyshev polynomial in the $\overline{K}^{*0}$ and ρ^0 modes.

The M_{D^0} distribution for the dominant background is calibrated by reconstructing $D^0 \to K_S^0\gamma$ in both simulation and data. Since the decay is forbidden, this yields mostly the backgrounds of the type $D^0 \to K_S^0\pi^0$ and $D^0 \to K_S^0\eta$. The same selection cuts as in the ϕ mode are applied here. K_S^0 candidates have a mass cut of ± 9 MeV/c^2 around the nominal K_S^0 mass.

The $\cos\theta_H$ distribution is modelled with $1-\cos^2\theta_H$, where all the parameters are fixed to MC values for all the three signal modes. The $V\pi^0$ and $V\eta$ categories are parametrized by $\cos^2\theta_H$ for the ϕ mode, a third order Chebyshev polynomial for the $\overline{K}^{*0}$ mode and a second order polynomial for the ρ^0 mode. The other background categories are parametrized with respect to their shapes. The fit results are shown in Fig. 1, Fig. 2 and Fig. 3 for the ϕ, $\overline{K}^{*0}$ and ρ^0, respectively. The signal component is

 DOI: 10.1051/epjconf/201716407008

shown by dotted red line. The signal yield obtained from the fits are: 524 ± 35 for ϕ mode, 9104 ± 396 for $\overline{K}^{*0}$ mode and 500 ± 85 for ρ^0 mode, respectively. Thus the raw asymmetries are −0.091 ± 0.066 (stat), −0.002 ± 0.020 (stat), 0.056 ± 0.151 (stat) for ϕ, $\overline{K}^{*0}$ and ρ^0 modes, respectively.

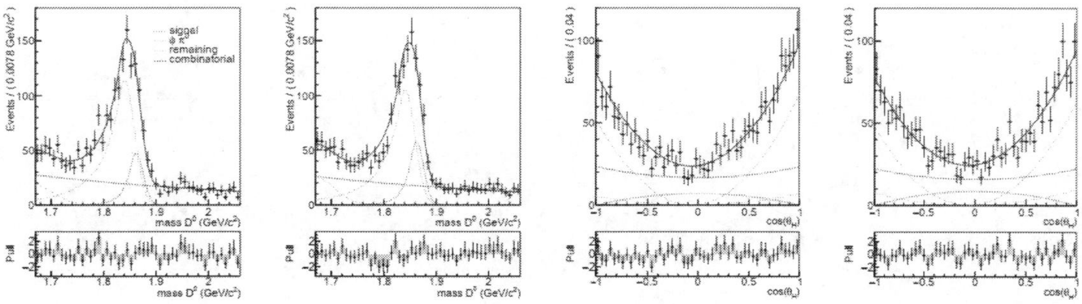

Figure 1. M_{D^0} distributions for D^0 (first) and $\overline{D^0}$ (second) and $\cos\theta_H$ distributions for D^0 (third) and $\overline{D^0}$ (fourth) in the ϕ mode. The dotted red, cyan, blue and yellow lines are signal, $\phi\pi^0$, combinatoric and remaining components, respectively.

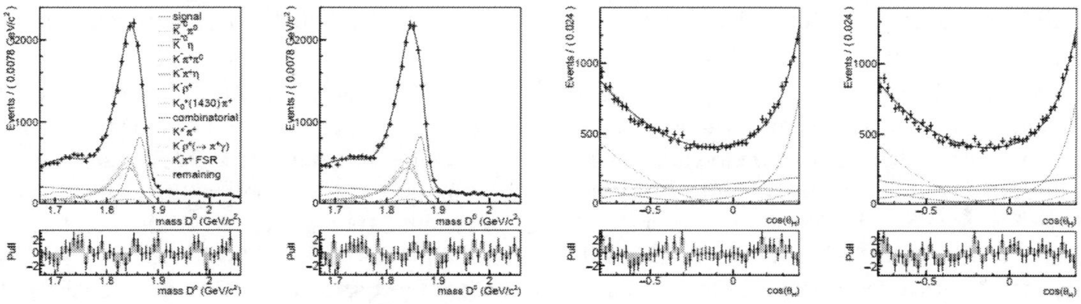

Figure 2. M_{D^0} distributions for D^0 (first) and $\overline{D^0}$ (second) and $\cos\theta_H$ distributions for D^0 (third) and $\overline{D^0}$ (fourth) in the $\overline{K}^{*0}$ mode. The dotted red, blue lines are the signal and combinatoric components, respectively. The other components are also shown in different colours.

The control samples are analysed in the similar way as done earlier by the Belle collaboration [12]. The selection criteria are the same as in the signal modes and the signal extraction is performed via sideband subtraction method, where the signal (SW), lower (LW) and upper (UW) windows are defined in M_{D^0}. The efficiencies from the simulation is obtained as 22.7%, 27.0% and 21.4% for K^+K^-, $K^-\pi^+$ and $\pi^+\pi^-$ samples respectively. The extracted signal yields are 362,274 for K^+K^- mode, 4.02×10^6 for $K^-\pi^+$ mode and 127,683 for $\pi^+\pi^-$ mode. The raw asymmetries are $(2.2 \pm 1.7) \times 10^{-3}$ for K^+K^- mode, $(1.3 \pm 0.5) \times 10^{-3}$ for $K^-\pi^+$ mode and $(8.1 \pm 3.0) \times 10^{-3}$ for $\pi^+\pi^-$ mode.

2.2 Systematics

The list of various sources of systematics is shown in Table 2. There are two main sources: one is due to the applied selection criteria and the other due to the signal extraction method for both the

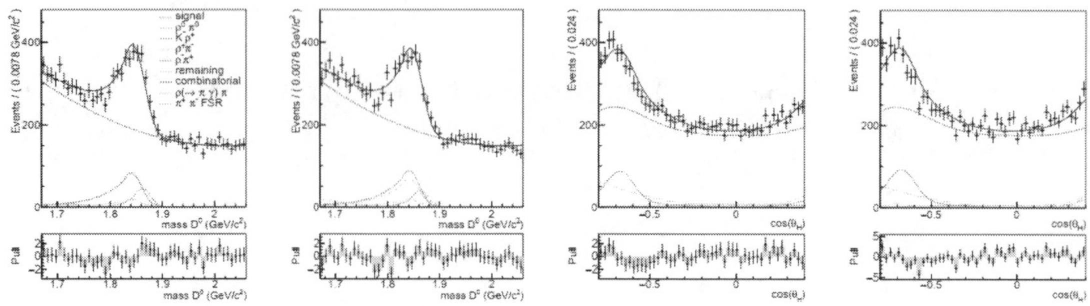

Figure 3. M_{D^0} distributions for D^0 (first) and $\overline{D^0}$ (second) and $\cos\theta_H$ distributions for D^0 (third) and $\overline{D^0}$ (fourth) in the ρ^0 mode. The dotted red, blue lines are the signal and combinatoric components, respectively. The other components are also shown in different colours.

signal and the control sample modes. The uncertainties common to both signal and normalization modes cancel. A 2% uncertainty is given to the photon reconstruction efficiency [13]. The systematic effect from the variable q is due to its low resolution in the signal mode compared to the control sample mode. This is evaluated with the help of the control sample $D^0 \to \overline{K}^{*0}\pi^0$. The systematic uncertainty for the q cut is found to be 1.16%. The systematic for the π^0 veto is calculated with the mode $D^0 \to K^0_S\pi^0$. The veto is performed with the first daughter of π^0 with all other photons except the second π^0 daughter. The ratio R between the yields for the signal and the control sample is calculated for both simulation and the data and by taking the double ratio, $\frac{R_{\rm MC}}{R_{\rm DATA}}$ we assigned the systematic for the π^0 veto. Similarly we extracted the systematics for the $\frac{E_9}{E_{25}}$ variable with $D^0 \to K^0_S\gamma$ as the control sample. A systematic uncertainty is assigned to the fit procedure where the fixed parameters are varied with respect to their errors. The biggest difference between the so obtained $\mathcal{B}$ and A_{CP} and the mean value is taken as the systematic error. An uncertainty is assigned to the dominant backgrounds of π^0 type for the chosen width of the smearing, varying the width by ± 1 MeV/c^2. For the normalization mode systematics, the procedure is repeated with different sidebands, ± 25 MeV/c^2 for M_{D^0}. The statistical error on the fraction f, which is the fraction of background events in the signal window compared to all events in the sideband, is also taken into account. The difference between the data and simulation could also affect f, so a similar procedure for the calibration of the π^0 background is also performed. A systematic uncertainty is assigned for the case when f is obtained from simulation smeared by a Gaussian of width 1.6 MeV.

2.3 Results

The preliminary measurements of the branching fraction in $D^0 \to V\gamma$ mode are

$$\mathcal{B}(D^0 \to \phi\gamma) = (2.76 \pm 0.20 \pm 0.08) \times 10^{-5},$$
$$\mathcal{B}(D^0 \to \overline{K}^{*0}\gamma) = (4.66 \pm 0.21 \pm 0.18) \times 10^{-4},$$
$$\mathcal{B}(D^0 \to \rho^0\gamma) = (1.77 \pm 0.30 \pm 0.08) \times 10^{-5},$$

where the uncertainties are statistical and systematic, respectively. The branching fraction for the ϕ mode is consistent with the previous measurements. There is a 3.3σ difference with the BABAR

Table 2. List of various sources of systematics and their contributions for $\mathcal{B}$ and A_{CP} in $D^0 \to V\gamma$ study.

Source	$D^0 \to \phi\gamma$		$D^0 \to \overline{K}^{*0}\gamma$		$D^0 \to \rho\gamma$	
	$\mathcal{B}$ (%)	$A_{CP} \times 10^{-3}$	$\mathcal{B}$ (%)	$A_{CP} \times 10^{-3}$	$\mathcal{B}$ (%)	$A_{CP} \times 10^{-3}$
γ rec. eff	2	–	2	–	2	–
ΔM	1.16	–	1.16	–	1.16	–
π^0 veto	0.5	–	0.5	–	0.5	–
E_9/E_{25}	0.96	–	0.96	–	0.96	–
Signal shape	1.39	0.32	–	–	2.33	4.29
Background shape	0.95	0.30	2.81	0.41	3.00	3.78
Norm modes systematics	0.05	0.46	0.00	0.01	0.14	0.54
Total	3.06	0.64	3.80	0.41	4.58	5.74

result for the branching fraction of $D^0 \to \overline{K}^{*0}\gamma$ mode. We also report the first branching fraction measurement for the $D^0 \to \rho^0\gamma$ mode with $> 5\sigma$ significance. The observed value of $\mathcal{B}$ for the ρ^0 mode is very close to that of ϕ mode, agreeing with the theory predictions. We also measure the first-ever A_{CP} measurements for $D^0 \to V\gamma$ modes. The preliminary results are

$$A_{CP}(D^0 \to \phi\gamma) = -0.094 \pm 0.066 \pm 0.001,$$
$$A_{CP}(D^0 \to \overline{K}^{*0}\gamma) = -0.003 \pm 0.020 \pm 0.000,$$
$$A_{CP}(D^0 \to \rho^0\gamma) = \;\; 0.056 \pm 0.151 \pm 0.006,$$

where the uncertainties are statistical and systematic, respectively. No CP asymmetry is observed.

3 $D^0 \to \gamma\gamma$ decay

Flavour changing neutral currents (FCNC) in the Standard Model are forbidden at the tree level of an interaction, while it can happen through higher orders. There are many new physics models that allow FCNC even at the tree level by a $c \to u\gamma\gamma$ transition. The branching fraction is expected to be very small ($O(10^{-8})$) [14], [15], [16]. But with the minimal supersymmetric Standard Model (MSSM), it has been predicted that the branching fraction can be enhanced due to the exchange of a gluino to $O(10^{-6})$ [17]. Thus by measuring the mode $D^0 \to \gamma\gamma$, we could identify any new physics contributions [18].

Previous measurements were carried out by BABAR [19], CLEO [20] and BESIII [21] collaborations. The most stringent limit is set by the BABAR collaboration: 2.2×10^{-6} with a confidence level of 90%. Our analysis is with the 832 fb^{-1} data collected at the $\Upsilon(4S)$ and $\Upsilon(5S)$ resonances. As in the $D^0 \to V\gamma$ analysis, we also performed a Monte Carlo simulation for the selection criteria and background studies.

3.1 Selection criteria, fitting and signal extraction

The analysis is similar to that of $D^0 \to V\gamma$ decay. We use the mode $D^* \to D^0\pi_{\rm slow}$ to suppress the large combinatoric background. But there are peaking backgrounds due to the decays of a π^0 and/or η meson, which decays to a pair of photons. These are $D^0 \to \pi^0\pi^0$, $D^0 \to \eta\pi^0$, $D^0 \to \eta\eta$, $D^0 \to K^0_S\pi^0$ and $D^0 \to K^0_L\pi^0$. To suppress these peaking backgrounds, a dedicated $\pi^0(\eta)$ veto is applied and the

DOI: 10.1051/epjconf/201716407008

suppression of merged clusters in ECL by E_9/E_{25}. The chosen control sample mode is $D^0 \to K_S^0\pi^0$ to measure the branching fraction.

The signal extraction is performed with a two-dimensional fit between M_{D^0} and ΔM variables. The efficiency is found to be 7.3%. We obtained a signal yield 4 ± 15 from the fit. The fit is shown in Fig. 4. The similar procedure is done for the control sample mode and we obtained a signal yield 343050 ± 673.

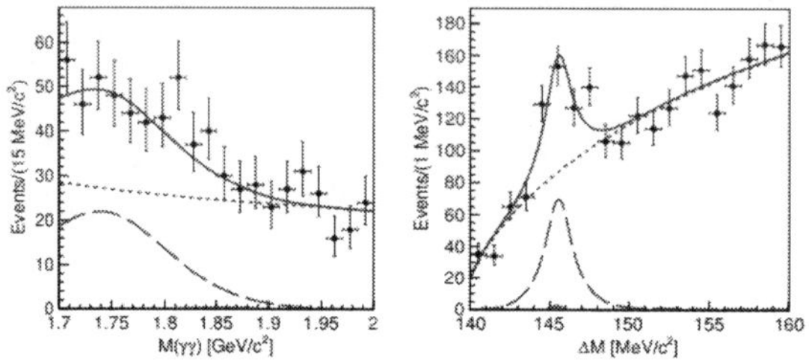

Figure 4. 2D fit between ΔM (left) and M_{D^0} (right) where the dashed blue, purple lines are combinatoric, peaking backgrounds respectively and the red histogram is the signal.

3.2 Systematics

The dominant contribution is coming from the cut variation of $E_{\gamma 2}$, A_E and $\mathcal{P}(\pi^0)$, which are the energy of lower energy photon, energy asymmetry between the two photons and the probability of π^0, respectively. For $E_{\gamma 2}$, we calculated $\frac{N}{\epsilon}$, where N is signal yield and ϵ is the detection efficiency, with and without any photon requirement on the photon energy in the $D^0 \to \phi\gamma$ control sample. The change with respect to the nominal value is taken as the systematic error. $\mathcal{P}(\pi^0)$ systematics is estimated with the same control sample where we relaxed the selection to $\mathcal{P}(\pi^0) < 0.7$ from $\mathcal{P}(\pi^0) < 0.15$. We double the above two systematic uncertainties as we have two photons in the final state. For A_E, as we do not have any proper control sample, we fit to the data without any requirement on A_E and take the resulting change in the upper limit as the systematic error. The list of all the sources of systematics are shown in Table 3.

Source	Contribution
cut variation	± 6.8 %
signal shape	$^{+4.0}_{-2.4}$ events
γ rec. eff	± 4.4 %
K_S^0 reconstruction	± 0.7 %
π^0 identification	± 4.0 %
$\mathcal{B}(D^0 \to K_S^0\pi^0)$	± 3.3 %

Table 3. Systematic uncertainties for $D^0 \to \gamma\gamma$ study.

3.3 Results

As the signal is absent in this analysis, a frequentist method is used to set the upper limit on the branching fraction of $D^0 \to \gamma\gamma$ decay with 90% confidence level. The result is

$$\mathcal{B}(D^0 \to \gamma\gamma) < 8.4 \times 10^{-7}. \tag{4}$$

This is the most stringent limit of this mode to date, as illustrated in Fig. 5.

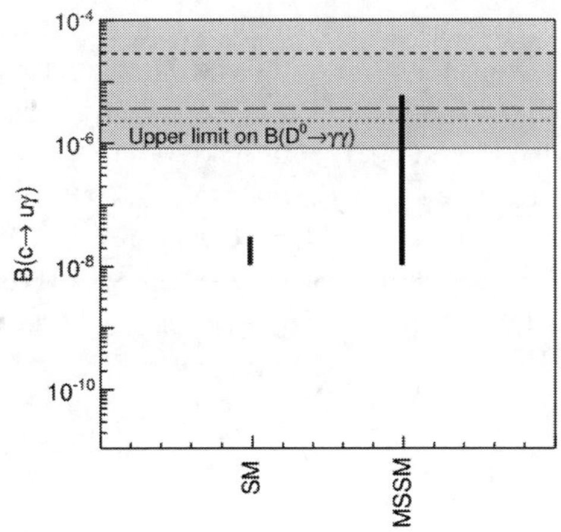

Figure 5. $\mathcal{B}(c \to u\gamma)$ prediction in SM and MSSM. Our result is shown in purple line. Dotted green, red and black lines are the results from BABAR, BESIII and CLEO collaborations, respectively.

4 Acknowledgements

We thank the KEKB group and all institutes and agencies that support the work of the members of the Belle collaboration. We also extend our gratitude to Indian Institute of Technology Madras (IIT Madras) and International and Alumni relations, IIT Madras for their financial support to attend the conference.

References

[1] G. Isidori and J. F. Kamenik, Phys. Rev. Lett. **109**, 171801 (2012), arXiv:1205.3164 [hep-ph].

[2] J. Lyon and R. Zwicky, (2012), arXiv:12210.6546 [hep-ph].

[3] K. Abe *et al.* (Belle collaboration), Phys. Rev. Lett. **92**, 101803 (2004), arXiv:hep-ex/0308037 [hep-ex].

[4] B. Aubert *et al.* (BABAR collaboration), Phys. Rev. D **78**, 071101 (2008), arXiv:0808.1838 [hep-ex].

[5] D. J. Lange, Nucl. Instrum. Meth. A **462**, 152 (2001).

[6] R. Brun *et al.*, CERN-DD-EE-84-1.

[7] K. A. Olive *et al.* (Particle Data Group), Chin. Phys. C **38**, 090001 (2014).
[8] M. Feindt, arXiv:physics/0402093 [physics].
[9] M. Feindt, U. Kerzel, Nuclear Instruments and Methods in Physics Research Section A: Accelerators, Spectrometers, Detectors and Associated Equipment. **559**, 190 (2006).
[10] T. Skwarnicki, *A study of the radiative CASCADE transitions between the Upsilon-Prime and Upsilon resonances*, Ph.D. thesis, Cracow, INP (1986).
[11] H. Ikeda *et al.* (Belle collaboration), Nucl. Instrum. Meth. A **441**, 401 (2000).
[12] M. Staric *et al.* (Belle collaboration), Phys. Lett. B **670**, 190 (2008), arXiv:0807.0148 [hep-ex].
[13] N. K. Nisar *et al.* (Belle collaboration), Phys. Rev. D **93**, 051102 (2016), arXiv:1512.02992 [hep-ex].
[14] C. Greub, T. Hurth, M. Misiak and D. Wyler, Phys. Lett. B **382**, 415 (1996).
[15] S. Fajfer, P. Singer and J. Zupan, Phys. Rev. D **64**, 074008 (2001).
[16] G.Burdman, E. Golowich, J. A. Hewett and S. Pakvasa, Phys. Rev. D **66**, 014009 (2002).
[17] S. Prelovsek and D. Wyler, Phys. Lett. B **500**, 304 (2001).
[18] A. Paul, I. I. Bigi and S. Recksiegel, Phys. Rev. D **82**, 094006 (2010).
[19] J. P. Lees *et al.* (BABAR collaboration), Phys. Rev. D **85**, 091107 (2012).
[20] T. E. Coan *et al.* (CLEO collaboration), Phys. Rev. Lett. **90** 101801 (2003).
[21] M. Ablikim *et al.* (BESIII collaboration), Phys. Rev. D **91**, 112015 (2015).

EPJ Web of Conferences **164**, 07009 (2017) DOI: 10.1051/epjconf/201716407009
ICNFP 2016

Deformations of spacetime and internal symmetries

Niels G. Gresnigt[1,a] and Adam B. Gillard[1]

[1] *Department of Mathematical Sciences, Xi'an Jiaotong-Liverpool University, 111 Ren'ai Road, Suzhou SIP, Jiangsu, China 215123*

Abstract. Algebraic deformations provide a systematic approach to generalizing the symmetries of a physical theory through the introduction of new fundamental constants. The applications of deformations of Lie algebras and Hopf algebras to both spacetime and internal symmetries are discussed. As a specific example we demonstrate how deforming the classical flavor group $SU(3)$ to the quantum group $SU_q(3) \equiv U_q(su(3))$ (a Hopf algebra) and taking into account electromagnetic mass splitting within isospin multiplets leads to new and exceptionally accurate baryon mass sum rules that agree perfectly with experimental data.

1 Introduction

In this short article we discuss some of the applications of algebraic deformations to spacetime and internal symmetries. Deformation theory provides a systematic method of generalizing a physical theory through imposing additional fundamental constants. The passage from Galilean relativity to special relativity, with fundamental scale c, and from classical mechanics to quantum mechanics, with fundamental scale $\hbar$, can both be understood using deformations within the category of Lie algebras.

The combined Poincaré-Heisenberg Lie algebra structure that underlies the standard model (SM) may be deformed to a semi-simple algebra through the introduction of two additional fundamental length scales, at which point there no longer exist any non-trivial deformation within the category of Lie algebras and so no further fundamental scales can be introduced.

The resulting algebra provides a natural candidate for the symmetry algebra in the low energy limit of the quantum-gravitational realm and an adaptation of physics to a framework that respects this algebraic structure provides a logical step toward an extension of the SM. Its energy momentum sector corresponds to the (anti) de Sitter algebra (a deformations of the Poincaré algebra). Among the possible new physics are the singleton representations, irreducible unitary representations on the anti de Sitter asymptotic boundary.

A second class of deformations are q-deformations (q for quantization). These are deformations of the universal enveloping algebra of a semi-simple Lie algebra into quantum groups. That is, these are deformations in the category of Hopf algebras. Both spacetime and internal symmetries may be q-deformed and quantum groups have found manifold applications in physics, including quantum gravity and gauge theories.

[a] e-mail: niels.gresnigt@xjtlu.edu.cn

DOI: 10.1051/epjconf/201716407009

As a specific example of the application of q-deformations to internal symmetries, we consider a recent result by one of the authors that shows that replacing the flavor symmetry group $SU(3)$ by its quantum group counterpart $SU_q(3)$ and accounting the for mass splittings within baryon isospin multiplets leads to octet and decuplet baryon mass formulas of exceptional accuracy. We consider this as strong evidence for the conjecture that quantum groups play an important role in high energy physics.

2 Lie-type deformations and the importance of stability

Much of the formal theory relating to deformations of Lie algebras dates back to the works of Gerstenhaber [1], Nijenhuis and Richardson [2] in the 1960s. We provide here a brief overview only.

Given a Lie algebra L_0 with basis $\{X_i\}$ and bracket $\left[X_i, X_j\right]_0 = if_{ij}^k X_k$, a one parameter deformation of L is defined through the deformed commutator

$$\left[X_i, X_j\right]_t = \left[X_i, X_j\right]_0 + \sum_{m=1}^{\infty} \phi_m(X_i, X_j)t^m, \tag{1}$$

where t is the deformation parameter. Such a deformation is said to be trivial if L_0 is isomorphic to the deformed Lie algebra L_t.

A trivial deformation is equivalent to saying there is an invertible linear transformation $T_t : V \longrightarrow V$ such that

$$T_t\left(\left[X_i, X_j\right]_t\right) = \left[T_t X_i, T_t X_j\right]_0 . \tag{2}$$

Under the linear transformations $X_i \longrightarrow T_t X_i$, the structure constants transform as:

$$f_{ij}^{(t)\,k} = T_i^u T_j^v (T^{-1})_w^k f_{uv}{}^w, \tag{3}$$

and so the deformed commutators under a trivial deformation may be rewritten in terms of the t-dependent structure constants as

$$\left[X_i, X_j\right]_t = if_{ij}^{(t)\,k} X_k. \tag{4}$$

A Lie algebra is said to be stable (or rigid) if all infinitesimal deformations result in isomorphic algebras. In particular, semi-simple Lie algebras are stable.

This concept of stability provides insight into the robustness of a physical theory (the spacetime symmetries of which are encoded in a Lie algebra) or the need to generalize it. Given a Lie algebra describing symmetries of a physical theory, some of the structure constants depend on certain physical constants (for example c for the Poincaré algebra and $\hbar$ for the Heisenberg algebra). The numerical values of these physical constants must be determined experimentally and are therefore not known without some error. A stable Lie algebra guarantees that the physics is not dependent on the precise value of these physical constants as any small deformation of the structure constants returns an isomorphic algebra. The same is not true for an algebra that is not stable. The concept of stability therefore provides insight into the validity of a physical theory or the need to generalize the theory. A theory with symmetries encoded within an unstable Lie algebra should be deformed until a stable Lie algebra is reached which encodes symmetries of wider validity and represents robust physics free of fine tuning issues. The importance of Lie-algebraic stability was first promoted by Mendes [3]. Since then, several others have similarly argued that the stability of a physically relevant Lie algebra should be considered a physical principle [4–8].

2.1 Lie-type deformations of spacetime symmetries

Algebraically, much of the success of modern physics can be attributed to the Poincaré and Heisenberg algebras. It was noted by Faddeev (albeit in hindsight) that both the quantum and relativistic revolutions of the 20th century can be considered as Lie-algebraic stabilizations of the unstable algebras of classical mechanics and Galilean relativity respectively [9]. The stabilization of these algebras give the Heisenberg and Poincaré algebras, which are both individually stable[1]. The deformation parameters introduced in the process of deforming an algebra correspond to new invariant scales, c and $\hbar$ in the examples above. Lie algebra deformations therefore provide a systematic way of introducing new invariant scales.

It has been argued in [7] that in quantum cosmology/gravity, an operationally defined view of physical space will inevitably ask for both a minimal (Planck order) length scale as well as a cosmological length scale, and furthermore, that algebraically there should be a mechanism that describes how the Poincaré -algebraic description of present day spacetime relates to the conformal algebraic description of spacetime in an early universe where quarks and leptons had yet to acquire mass.

Mathematically, a Lie algebra that incorporates these requirements already exists. Although the Poincaré and Heisenberg algebras are separately stable, the combined Poincaré-Heisenberg algebra, consisting of the Poincaré algebra extended by the position generators and their commutation relations with momenta, lacks the desired stability. Its stabilized form was found by [3]. The resulting stabilized Poincaré-Heisenberg algebra (SPHA) turned out to be (up to various signs) the same algebra arrived at by Yang [10] in 1947 based on the work of Snyder who demonstrated that the assumption that space be a continuum is not required for Lorentz invariance earlier the same year [11]. Uniqueness, an issue that was not addressed by Mendes, was later demonstrated by Chryssomalakos and Okon [4]. SPHA introduces in addition to c and $\hbar$, two invariant length scales ℓ_P and ℓ_C, the former on the order of the Planck length, the latter a cosmological length scale. Furthermore, it was shown in [7] that in a certain limit, this algebra satisfies the commutation relations of the conformal algebra of spacetime. In the contraction $\ell_P \to 0$, the algebra reduces to the (anti) de Sitter algebra; the deformation of the Poincaré algebra.

2.2 Singleton physics

The Poincaré symmetries of Minkowski spacetime are deformed to those of (anti) de Sitter spacetime through the introduction of the length scale $\ell_C = \Lambda^{-\frac{1}{2}}$. This introduces interesting new features. Elementary particles in Minkowski spacetime are associated with unitary irreducible representations of the Poincaré group (the isometry group of Minkowski spacetime). In the present case, elementary particles should likewise be associated with the unitary irreducible representations not of the Poincaré group, but of the (anti) de Sitter group.

The representations of the anti de Sitter group, like the Poincaré group, admit a positive minimum energy. This means that representations of this group naturally lends themselves to a particle interpretation. This is not the case for the de Sitter group. The most fundamental irreducible representations of the anti de Sitter group were first discovered by Dirac [12], and are called the singleton representations. In anti de Sitter space, massless particles are composed of two singletons [13]. These singletons are naturally confined (in a kinematic sense) which has also led to the question of whether perhaps singletons take the role of quarks [14]. Compatibility between the singleton representations and quantum electrodynamics (QED) was demonstrated by in [15].

[1] Technically it is the semi-simple Lorentz algebra that is stable. Adding the abelian translations renders the Poincaré algebra unstable again. Its stabilized version is the (anti) de Sitter algebra. However, the instability of the Poincaré algebra may be ignored as long as one restricts oneself to the tangent space.

The two singleton representations, named *Di* and *Rac* are given by

$$Rac = D(\frac{1}{2}, 0), \qquad Di = D(1, \frac{1}{2}). \tag{5}$$

These singleton representations have the interesting property that a direct product of two positive energy singletons reduces to a sum of massless representations of $SO(2,3)$ as follows (s is the spin) [13]

$$Rac \otimes Rac = \oplus_{s=0,1,...} D(s+1, s), \tag{6}$$
$$Rac \otimes Di = \oplus_{2s=1,3,...} D(s+1, s), \tag{7}$$
$$Di \otimes Di = \oplus_{s=1,2,...} D(s+1, s) \oplus D(2,0). \tag{8}$$

The *Di* and *Rac* themselves do not have contractions to representations of the Poincaré group. In the flat space limit the singletons reduce to vacua. It may then be possible to consider all leptons as *Rac-Di* and the vector mesons of the electroweak model as *Rac-Rac* composites [16].

3 q-deformations

A stable Lie algebra cannot be deformed any further within the category of Lie algebras. It is however possible to deform the completed universal enveloping algebra of a Lie algebra. Such a q-deformation is a deformation in the category of Hopf algebras and deforms the universal enveloping algebra into a quantum group.

Quantum groups (which are algebras rather than groups) provide a generalization of familiar symmetry concepts through extending the domain of classical group theory. Quantum groups depend on a deformation parameter q with the value $q = 1$ returning the undeformed universal enveloping algebra but for general q giving the structure of a Hopf algebra. First formalized by Jimbo [17] and Drinfeld [18] as a class of Hopf algebras, quantum groups have found many applications in theoretical physics, see [19–22] and references therein.

3.1 The quantum groups $SU_q(n) \equiv U_q(su(n))$

The literature on quantum groups and algebras is extensive. For an excellent introduction the reader is directed to [23].

The quantum (enveloping) algebra $SU_q(n) \equiv U_q(su(n))$ corresponding to a one-parameter deformation of the universal enveloping algebra of $su(n)$, is a Hopf algebra with unit **1** and generators H_i, $X_i^{\pm}$, $i = 1, 2, ..., n-1$, defined through the commutation relations in the Cartan-Chevalley basis as

$$\left[H_i, H_j\right] = 0 \tag{9}$$
$$\left[H_i, X_j^{\pm}\right] = a_{ij} X_j^{\pm} \tag{10}$$
$$\left[X_i^{+}, X_j^{-}\right] = \delta_{ij}[H_i]_q \equiv \delta_{ij} \frac{q^{H_i} - q^{-H_i}}{q - q^{-1}}, \tag{11}$$

together with the quadratic and cubic deformed q-Serre relations

$$\left[X_i^{\pm}, X_j^{\pm}\right] = 0, \quad j \neq i \pm 1, \quad 1 \leq i, j \leq n-1 \tag{12}$$

and

$$(X_i^{\pm})^2 X_j^{\pm} - [2]_q X_i^{\pm} X_j^{\pm} X_i^{\pm} + X_j^{\pm} (X_i^{\pm})^2 = 0, \quad j = i \pm 1, \quad 1 \leq i, j \leq n-1 \tag{13}$$

respectively [17, 24]. Here a_{ij} is an element of the Cartan matrix

$$a_{ij} = \begin{cases} 2 & j = i \\ -1 & j = i \pm 1 \\ 0 & \text{otherwise.} \end{cases}$$

The q-number

$$[N]_q = \frac{q^N - q^{-N}}{q - q^{-1}} \tag{14}$$

is defined for both operators (as in equation (11)) and real numbers[2] (as in equation (13)). The definition of the algebra is completed by the Hermiticity properties

$$(H_i)^\dagger = H_i, \qquad (X_i^\pm)^\dagger = X_i^\mp. \tag{15}$$

The quantum algebra $SU_q(n)$ has the structure of a Hopf algebra admitting a coproduct, counit and antipode. These are not used here and so we do not define them (see [24]). In the limit $q = 1$ the above relations approach the relations for the universal enveloping algebra $U(su(n))$ but for general q they represent a deformation of the universal enveloping algebra of $su(n)$.

3.2 q-deformations of spacetime symmetries

Quantum groups as well as non-commutative spaces have been utilized in many approaches to quantum gravity. There is strong evidence that in both 2+1 and 3+1 quantum gravity, the symmetry algebra must be q-deformed. Quantum gravity often requires a nonzero cosmological constant to be well defined. In loop quantum gravity (LQG) for example, the only known exact state in the connection basis that has semi-classical properties, the Kodama state, exists only when Λ is nonzero [25].

In the quantization of general relativity, the local gauge symmetry of the spacetime connection must be q-deformed from $SU(2)$ to $SU_q(2)$ [26]. As explained in [27], for a universe characterized by a maximal $\ell_C = \Lambda^{\frac{1}{2}}$ and minimal length ℓ_P the local rotational symmetry is better described by the quantum groups $SU_q(2)$ than by $SU(2)$. In this case the deformation parameter q given by

$$q = \exp(i\ell_P^2\Lambda). \tag{16}$$

It is interesting that it is precisely these two scales that are obtained in the deformation of the Poincaré -Heisenberg algebra. It seems therefore that both Lie deformations and q-deformations play an important role in quantum spacetime.

Quantum spacetime is generally accepted to be non-commutative. The simplest example of a non-commutative geometry is provided by the quantum plane. One may quantize the classical vector space of commuting coordinates x and y to obtain a quantum vector space by assuming that the coordinates no longer commute with each other. This noncommutativity of the quantized coordinates X and Y of the quantum plane can be modelled by

$$XY = qYX, \tag{17}$$

where q is the deformation parameter. In the classical limit $q \to 1$, one recovers the classical plane. As in the classical case, it turns out to be possible to define a differential calculus on the quantum plane that is invariant under a generalization of the classical group $SL(2)$. This generalization of the classical group is the quantum group $SL_q(2)$. Thus, the theory of quantum groups is part of the program of noncommutative geometry.

[2] In this paper we will only have to deal with integer values of N. The definition however holds for real numbers.

3.3 q-deformed electroweak theory

Finkelstein in a series of papers, see [28] and references therein, developed a solitonic model of elementary particles in terms of quantized knots. The model is constructed by replacing the electroweak gauge group $SU(2) \times U(1)$ by the quantum group $SUq(2)$. Quantum groups are closely related to the Jones polynomial of knots and $SU_q(2)$ is the algebra of oriented knots. It is this close relation between quantum groups and knots that allows for a deformed gauge theory based on $SU_q(2)$ in which the quantum numbers of solitonic elementary particles are topologically defined as certain knot invariants.

Assuming that the most elementary particles are also the most elementary knots, Finkelstein identified the four classes of fermions in the SM are identified with the four quantum trefoils. The electroweak gauge bosons are represented as compositions (knot sums) of trefoil knots. The quantized knots are labelled by the irreducible representations of $SU_q(2)$ with the elementary fermions described by the trefoil representations of $SU_q(2)$. The three femions of each family (for example electron, muon, and tau) correspond to the three lowest states in the excitation spectrum of the knots. The model successfully matches knots with the electroweak field quanta.

The model was later expanded into a preon model in which the knotted field has a composite structure of three or more preons that are described by the fundamental representation of $SU_q(2)$. Interestingly, the preons in this model correspond to twisted loops rather than non-trivial knots. Given the close relation between knots and braids (knots being the closure of a braid) these ideas also seem to be closely related to another preon inspired model, called the Helon model, in which the simplest non-trivial braids consisting of three ribbons and two crossings map precisely to the first generation of SM leptons and quarks [29]. It was later demonstrated that these braids may be embedded within spin networks, which makes it compatible with loop quantum gravity in the spin network basis [30, 31].

3.4 $SU_q(3)$ flavor symmetry

As a more detailed application of q-deformations to particle physics, we review the recent result by one of us, see [32], where it is shown that replacing the flavor symmetry group $SU_q(3)$ by the quantum group $SU_q(3)$ and accounting for the electromagnetic mass splitting of baryons within isospin multiplets leads to exceptionally accurate baryon mass formulas.

The standard Gell-Mann-Okubo mass formula [33, 34] is

$$M = \alpha_0 + \alpha_1 S + \alpha_2 \left[I(I+1) - \frac{1}{4} S^2 \right], \tag{18}$$

where M is the mass of a hadron within a specific multiplet, S and I are the strangeness and isospin respectively, and $\alpha_0, \alpha_1, \alpha_2$ are free parameters. By eliminating the free parameters $\alpha_0, \alpha_1, \alpha_2$, one obtains mass relations between the different baryons within a given (isospin) multiplet. For the case of octet baryons one obtains the standard relation

$$N + \Xi = \frac{3}{2}\Lambda + \frac{1}{2}\Sigma, \tag{19}$$

whereas for the decuplet baryons $+\frac{3}{2}$ one obtains the equal spacing rule

$$\Delta - \Sigma^* = \Sigma^* - \Xi^* = \Xi^* - \Omega \tag{20}$$

These formulas hold to first order in flavor symmetry breaking only. Using the most recent data formula (19) is accurate to about 0.6%. The equal spacing rule (20) is less accurate however a modified relation $\Omega - \Delta = 3(\Xi^* - \Sigma^*)$ due to Okubo [35] is accurate to about 1.4%.

The idea of replacing the classical flavor symmetry group by its quantum group counterpart was first considered in [36–38], and its authors found meson and baryon mass sum rules of improved accuracy. By fixing a definite value for the deformation parameter q (through fitting the data), the modified baryon mass sum rules are accurate to an impressive 0.06% for the octet baryons and 0.32% for the decuplet baryons.

The basic approach of the construction is the representation theory of $U_q(su(n))$ [38, 39]. q-Deformed mass sum rules are computed from the expectation value of the mass operator which is defined in terms of the generators of the dynamical algebras (quantum groups) $U_q(u(n+1))$ or $U_q(u(n,1))$. The expectation values are computed from the matrix elements of these generators. Utilizing the q-algebras $U_q(u(n+1))$ or $U_q(u(n,1))$ of dynamical symmetry, breaking of n-flavor symmetries up to exact isospin symmetry $SU_q(2)$ are realized and the q-analogues of mass sum rules for baryon multiplets are derived.

The masses used in both the standard as well as the q-deformed mass formulas are the averages of the isospin multiplets (isoplets). At the level of accuracy of the q-deformed mass relations, the mass splittings within isoplets become significant and can no longer be ignored. For example $\Xi^{*-} - \Xi^{*0} =$ 3.2MeV and $\Sigma^{*-} - \Sigma^{*+} = 4.4$MeV, representing about $\sim 0.2 - 0.3\%$ of the average isoplet mass. The impressive accuracy of the deformed mass formulas lose their significance when these mass splittings (due to electromagnetic contributions to the masses of baryons) are not considered.

The electromagnetic contributions to baryon masses are determined within the QCD general parametrization scheme in the spin-flavour space considered by Morpurgo [40] and to zeroth order symmetry breaking the electromagnetic contributions to the octet baryon masses are given in terms of four parameters [41].

By replacing the classical flavor group with its associated quantum group as well as accounting for the electromagnetic mass splittings within isospin multiplets, charge specific baryon mass formulas with $SU_q(3)$ flavor symmetry were found by one of us [32]. For the octet baryons the new mass formula is

$$p + \frac{\Xi^0}{[2]_q - 1} = \frac{\Lambda^0}{[2]_q - 1} + (2\Sigma^+ - \Sigma^0), \tag{21}$$

where $q = e^{i\pi/n}$ for integer n, is the deformation parameter. Substituting experimental data [42] we find that the closest fit to the data is when $n = 7$ so that $q_7 = e^{i\pi/7}$, in agreements with [36]. The formula is

$$p + \frac{\Xi^0}{[2]_{q_7} - 1} = \frac{\Lambda^0}{[2]_{q_7} - 1} + (2\Sigma^+ - \Sigma^0). \tag{22}$$

The formula gives LHS=2577.87MeV and RHS=2577.33MeV, which has an error of only 0.02%. This is a significant improvement on the standard octet relations (19) which holds to 0.57% as well as the q-deformed relation [37] which (for $q = e^{i\pi/7}$) holds to 0.06%.

For the decuplet baryons, the mass formula is

$$\Omega^- - \Delta^- = ([2]_q + 1)(\Xi^{*-} - \Sigma^{*-}), \tag{23}$$

where q is again of the form $q = e^{i\pi/n}$ for integer n. In [36] is was suggested that choosing $q = e^{i\pi/14}$ provides a good fit to data. Although this remains true in the present case, there are other values of q for which the error is smaller. In particular, we consider the case where $q = e^{i\pi/21}$ [3].

[3] Although this is not the absolute best choice for q, because 21 is a multiple of 7, it allows for an elegant new relation between octet and decuplet baryons [32] (this is also the reason why $n = 14$ is chosen in [36, 37]).

The new baryon mass formulas based on an $SU_q(3)$ flavor symmetry are accurate to well within the experimental uncertainty. It may be that nature's true flavor symmetry is not $SU(3)$ but rather $SU_q(3)$.

4 Conclusion

The deformations of spacetime and internal symmetries were considered. For deformation within the Lie algebra framework, the concept of stability provides a important role in generalizing the symmetries of a physical theory in a systematic way. An unstable symmetry algebra should be deformed, via the introduction of additional invariant scales, into a stable one. The resulting theory will describe robust physics and be free of fine tuning issues.

Deformations within the category of Lie algebras are not the only type of deformations that have been considered in the literature. For example, in double special relativity (DSR) the deformations considered introduce non-linearity into the theory (i.e. the Lie-algebraic framework is abandoned) [43–45]. The advantage of Lie-type deformations is that they can be handled systematically without introducing non-linearity. Indeed, it has been shown that through a suitable redefinition of the generators, triply special relativity [46] can be brought into linear form [47].

Deforming the combined Poincaré plus Heisenberg algebra introduces two new invariant length scales. The energy-momentum sector of the stable algebra is the (anti) de Sitter algebra. For the case of the anti de Sitter spacetime this leads to interesting new physics, where particles correspond to the unitary irreducible representations of not the Poincaré group but rather the anti de Sitter group $SO(2,3)$. The singleton representations of this group suggest that massless particles may be considered as singleton composites.

Once a stable algebra is obtained, no further non-trivial deformations are possible within the framework of Lie algebras. One can still however deform in the category of Hopf algebras. This gives rise to quantum groups. There is evidence from quantum gravity that spacetime symmetries should be q-deformed and in this case the deformation parameter can be expressed in terms of the two length scales introduced in the deformation of the Poincaré -Heisenberg algebra.

Quantum groups also likely play a role in generalizing the symmetries of SM particles. The use of quantum groups in gauge theories allows for a solitonic description of elementary particles where a particle's quantum numbers are described in terms of knot invariants. Taken as a flavor symmetry, the quantum group $SU_q(3)$ gives rise to exceptionally accurate baryon mass formulas that agree perfectly with the experimental data.

Acknowledgments

This work is supported in part by XJTLU research grant RDF-14-03-13 and Natural Science Foundation of China grant RR0116.

References

[1] M. Gerstenhaber, The Annals of Mathematics **79**, 59 (1964)
[2] A. Nijenhuis, R.W. Richardson, J. Math. Mech **17**, 89 (1967)
[3] R.V. Mendes, Journal of Physics A: Mathematical and General **27**, 8091 (1994)
[4] C. Chryssomalakos, E. Okon, [Arxiv preprint hep-th/0410212] (2004)
[5] D.V. Ahluwalia-Khalilova, Classical and Quantum Gravity **22**, 1433 (2005)

DOI: 10.1051/epjconf/201716407009

[6] N.G. Gresnigt, P.F. Renaud, P.H. Butler, Int. J. Mod. Phys. D [arXiv:hep-th/0611034] **16**, 1519 (2007)
[7] D.V. Ahluwalia, N.G. Gresnigt, A.B. Nielsen, D. Schritt, T.F. Watson, International Journal of Modern Physics D **17**, 495 (2008)
[8] N.G. Gresnigt, A.B. Gillard, arXiv:1512.04339 (2015)
[9] L.D. Faddeev, *Mathematician's view on the development of physics, Frontiers in Physics: High technology and mathematics* (Singapore: Word Scientific, 1989)
[10] C.N. Yang, Physical Review **72**, 874 (1947)
[11] H.S. Snyder, Physical Review **71**, 38 (1947)
[12] P.A.M. Dirac, Journal of Mathematical Physics **4**, 901 (1963)
[13] M. Flato, C. Fronsdal, Letters in Mathematical Physics **2**, 421 (1978)
[14] M. Flato, C. Fronsdal, Physics Letters B **172**, 412 (1986)
[15] M. Flato, C. Fronsdal, Journal of Mathematical Physics **22**, 1100 (1981)
[16] D. Sternheimer, in *From geometry to quantum mechanics* (Springer, 2007), pp. 287–301
[17] M. Jimbo, Letters in Mathematical Physics **10**, 63 (1985)
[18] V. Drinfeld, Lett. Math. Phys **10**, 63 (1985)
[19] R.J. Finkelstein, arXiv:1205.1026 (2012)
[20] S. Majid, H. Ruegg, Physics Letters B **334**, 348 (1994)
[21] J. Lukierski, A. Nowicki, International Journal of Modern Physics A **18**, 7 (2003)
[22] L. Castellani, J. Wess, *Quantum groups and their applications in physics*, Vol. 127 (IOS Press, 1996)
[23] R. Jaganathan, arXiv preprint math-ph/0105002 (2001)
[24] C. Quesne, Journal of Physics A: Mathematical and General **25**, 5977 (1992)
[25] L. Smolin, arXiv preprint hep-th/0209079 (2002)
[26] J. Kowalski-Glikman, L. Smolin, Phys. Rev. D **70**, 065020 (2004)
[27] E. Bianchi, C. Rovelli, Physical Review D **84**, 027502 (2011)
[28] R.J. Finkelstein, arXiv preprint arXiv:1011.2545 (2010)
[29] S.O. Bilson-Thompson, arXiv preprint hep-ph/0503213 (2005)
[30] S.O. Bilson-Thompson, F. Markopoulou, L. Smolin, Classical and Quantum Gravity **24**, 3975 (2007)
[31] S. Bilson-Thompson, *Braided Topology and the Emergence of Matter*, in *Journal of Physics: Conference Series* (IOP Publishing, 2012), Vol. 360, p. 012056
[32] N.G. Gresnigt, arXiv preprint arXiv:1607.01632 (2016)
[33] M. Gell-Mann, Tech. rep., California Inst. of Tech., Pasadena. Synchrotron Lab. (1961)
[34] S. Okubo, Progress of Theoretical Physics **27**, 949 (1962)
[35] S. Okubo, Physics Letters **4**, 14 (1963)
[36] A. Gavrilik, arXiv preprint hep-ph/9712411 (1997)
[37] A. Gavrilik, Nuclear Physics B-Proceedings Supplements **102**, 298 (2001)
[38] A. Gavrilik, arXiv preprint hep-ph/0402082 (2004)
[39] A. Gavrilik, I. Kachurik, A. Tertychnyj, arXiv preprint hep-ph/9504233 (1995)
[40] G. Morpurgo, Physical Review D **40**, 2997 (1989)
[41] G. Morpurgo, Physical review letters **68**, 139 (1992)
[42] K.A. Olive, P.D. Group et al., Chinese Physics C **38**, 090001 (2014)
[43] G. Amelino-Camelia, International Journal of Modern Physics D **11**, 35 (2002)

[44] J. Kowalski-Glikman, S. Nowak, Physics Letters B **539**, 126 (2002)
[45] F. Girelli, E. Livine, D. Oriti, Nuclear Physics B **708**, 411 (2005)
[46] J. Kowalski-Glikman, L. Smolin, Physical Review D **70**, 65020 (2004)
[47] C. Chryssomalakos, E. Okon, [Arxiv preprint hep-th/0407080] (2004)

EPJ Web of Conferences **164**, 07010 (2017) DOI: 10.1051/epjconf/201716407010
ICNFP 2016

Detection of mirror-meson decays at CERN

George Triantaphyllou
National Technical University of Athens
58, SINA Street, 10672 ATHENS, GREECE

Abstract. The existence of mirror partners (katoptrons) of Standard-Model fermions offers a viable alternative to a fundamental BEH mechanism, with the coupling corresponding to the mirror generation gauge symmetry becoming strong at around 1 TeV. The resulting non-perturbative processes produce dynamical katoptron masses and create mirror mesons with masses ranging approximately from 0.1 to 2.8 TeV. Since the corresponding phenomenology expected at the LHC is particularly rich, some relevant detection methods of mirror mesons are explored in order to obtain a deeper understanding of strongly-interacting mirror fermions.

1 Introduction

Several decades after the discovery of massive gauge bosons corresponding to the electroweak symmetry, the origin of their mass remains a mystery. The recent discovery of a 125-GeV particle at the LHC, considered to be the central actor of the BEH mechanism, is theoretically complicated by the well-known fact that the mass of an elementary scalar particle is unstable under quantum corrections, unless a new mechanism, like a natural low-energy cut-off, is introduced in the theory. Efforts to have a new strong interaction supply such a cut-off as a mass-stabilization mechanism [1], in which case the agents of the BEH mechanism would be several new composite bosons which should be observable at CERN, are confronted by experimental data since the early 1990s, which indicate via the smallness of the S parameter that the size of new fermion representations under possible new gauge symmetries, while in parallel also coupled to the electroweak symmetry, should be quite limited [2]. These data constrain severely QCD-like strongly-coupled gauge theories providing a dynamical BEH mechanism.

However, the introduction of new fermions in the theory, usually called mirror fermions [3] or katoptrons, having quantum-number assignments "mirroring" the ones of known Standard-Model (SM) fermions, i.e. the same after interchanging left- and right-handed components, in addition to transforming under an initially unbroken $SU(3)_K$ gauged generation symmetry, might offer a solution to this problem. At very high energies, the role of $SU(3)_K$ is to prevent the mixing of mirror fermions with SM ones, which would otherwise give them unacceptably heavy masses. Assuming that the strength α_K of the $SU(3)_K$ gauge coupling is equal to the strengths of the gauge couplings of all the SM gauge groups near the Planck energy scale [4], something that would suggest unification of all the gauge symmetries at that scale and was first attempted in [5] under a different symmetry group, the renormalization group equations governing the running of α_K suggest that $SU(3)_K$ becomes naturally strongly coupled near 1 TeV, i.e. near the electroweak-symmetry breaking scale [6]. Although this fact renders it a serious candidate for an agent of dynamical electroweak-symmetry breaking, in which way might this new mirror sector couple to the SM fermions?

Attempting to answer this question leads one to the next crucial step involving the study of mirror-fermion-condensates of the form $\mathbf{3} \times \mathbf{3} \rightarrow \mathbf{3}$ due to non-perturbative $SU(3)_K$ dynamics around 1 TeV, which, apart from breaking dynamically the electroweak symmetry, lead to the self-breaking of $SU(3)_K$ down to $SU(2)_K$, which in its turn self-breaks at energies around six times lower via similar condensates. Their formation has several interesting consequences. On one hand, the breaking of the katoptron generation symmetry allows the mixing of mirror fermions with their SM counterparts. The operators mixing these two kinds of fermions are isospin singlets, which allows for the lifting of the mass degeneracy of SM-fermion isospin doublets without introducing too large contributions to the $\Delta\rho$ parameter. On the other hand, mirror fermions acquire dynamical masses due to $SU(3)_K$. Therefore, the diagonalization of mass matrices including both SM and mirror fermions not only provides mass to the SM fermions, which is consistent with their appearance in three generations, but also supplies a mechanism for the natural emergence of the CKM and neutrino-mixing matrices. In parallel, the smallness of neutrino masses can be explained by the standard see-saw mechanism [7]. The resulting mass matrices allow for the introduction of CP-violating phases possibly being at the root of the baryon asymmetry of the Universe, while their form has the potential to solve the strong CP problem. Moreover, their diagonalization after the self-breaking of the mirror generation symmetry $SU(3)_K$ leads to an explicit breaking of mirror-fermion chiral symmetry, providing thus masses to the relevant pseudo-Nambu-Goldstone bosons, i.e. mirror pions, as well as to other mirror mesons, on the order of at least 100 GeV.

Furthermore, the sequential breaking of $SU(3)_K$ offers additionally three considerable advantages. First, it leads to a correct order-of-magnitude mass hierarchy between the SM-fermion generations. In fact, an inductive analysis within the framework of dynamical symmetry breaking could lead one to argue in favour of a broken gauge symmetry of a new strongly-coupled theory based exactly on the large mass hierarchy between SM fermion generations. Second, it may produce contributions to the S parameter which are compatible with present experimental limits. The reason is that mirror-pion decay-constant constraints stemming from the first Weinberg sum rule can be largely fulfilled by the heaviest mirror fermion generation, leaving the decay constants pertaining to the two lightest mirror generations relatively unconstrained, possibly leading to near-zero or even negative contributions to the S parameter [8]. Third, it renders the existence of a low-lying 125 GeV composite Higgs boson, considered here to be a mirror meson, compatible with a dynamically-generated electroweak energy scale roughly estimated via the Pagels-Stokar formalism, since this boson is accompanied by several additional heavier mirror mesons. Last, note that problems related to flavor-changing neutral currents within this framework are expected to be minimal since mirror fermions are not contained within the same representation of a gauge group at low energies. Having listed the main advantages of strongly-interacting mirror fermions in the context briefly described above, we proceed by presenting the corresponding Lagrangian $\mathcal{L}_K$ which is expected to reproduce the SM at lower energies.

2 The Lagrangian

As usual, $\mathcal{L}_K$ is the sum of gauge kinetic and self-interaction terms $\mathcal{L}_{YM}$ and of gauge-fermion interaction terms $\mathcal{L}_{int}$ expressed respectively as [8]

$$\begin{aligned}
\mathcal{L}_{YM} &= -\frac{1}{4}B_{\mu\nu}B^{\mu\nu} - \frac{1}{4}W^a_{\mu\nu}W^{a\,\mu\nu} - \frac{1}{4}G^e_{\mu\nu}G^{e\,\mu\nu} - \frac{1}{4}G^{K\,e}_{\mu\nu}G^{K\,e\,\mu\nu} \\
\mathcal{L}_{int} &= i\sum_{j,k}\left[(\bar{\psi}^{j|k}_u, \bar{\psi}^{j|k}_d)\gamma_\mu \mathcal{D}^\mu_k \begin{pmatrix} \psi^{j|k}_u \\ \psi^{j|k}_d \end{pmatrix} + (\bar{\hat{\psi}}^{j|k}_u, \bar{\hat{\psi}}^{j|k}_d)\gamma_\mu \hat{\mathcal{D}}^\mu_k \begin{pmatrix} \hat{\psi}^{j|k}_u \\ \hat{\psi}^{j|k}_d \end{pmatrix}\right]
\end{aligned} \qquad (1)$$

in flat space-time and at energies above electroweak-symmetry breaking, where γ_μ are Dirac matrices,

$$\begin{aligned} B_{\mu\nu} &= \partial_\mu B_\nu - \partial_\nu B_\mu, & W^a_{\mu\nu} &= \partial_\mu W^a_\nu - \partial_\nu W^a_\mu - g_2 f_2^{abc} W^b_\mu W^c_\nu, \\ G^e_{\mu\nu} &= \partial_\mu G^e_\nu - \partial_\nu G^e_\mu - g_3 f_3^{efg} G^f_\mu G^g_\nu, & G^{K\,e}_{\mu\nu} &= \partial_\mu G^{K\,e}_\nu - \partial_\nu G^{K\,e}_\mu - g_{3K} f_3^{efg} G^{K\,f}_\mu G^{K\,g}_\nu \end{aligned} \tag{2}$$

are the curvatures of the gauge symmetries $U(1)_Y$, $SU(2)_L$, $SU(3)_C$ and $SU(3)_K$ with coupling strengths $g_{1,2,3,3K}$ respectively, $\mu, \nu = 0, ...3$ are space-time indices, f_2^{abc} are the structure functions of $SU(2)_L$ with $a, b, c = 1, 2, 3$, f_3^{efg} are the structure functions of $SU(3)$ and $SU(3)_K$ with $e, f, g = 1, ...8$, the SM-fermion generations are denoted by $j = 1, 2, 3$ and the space-time dependence of the various fields is omitted for simplicity.

The index k introduced above, $k = 1, ..., 4$ ($k = 1, 2$ for SM fermions and $k = 3, 4$ for mirror fermions), defines fermion fields consisting of two sets of Weyl fermions of opposite chirality:

$$\begin{aligned} \psi_u^{j|k} &= (N^j_L, U^j_L, N^K_R \delta^{3j}, U^K_R \delta^{3j}), & \psi_d^{j|k} &= (E^j_L, D^j_L, E^K_R \delta^{3j}, D^K_R \delta^{3j}), \\ \hat{\psi}_u^{j|k} &= (N^j_R, U^j_R, N^K_L \delta^{3j}, U^K_L \delta^{3j}), & \hat{\psi}_d^{j|k} &= (E^j_R, D^j_R, E^K_L \delta^{3j}, D^K_L \delta^{3j}) \end{aligned} \tag{3}$$

where neutrinos are denoted by N^j, charged leptons by E^j, up-type quarks by U^j and down-type quarks by D^j. The superscript "K" denotes mirror fermions, the subscripts "L" and "R" denote fermion chirality, Kronecker's δ^{3j} prevents multiple counting of mirror generations under summation, while the Weyl-spinor, color and mirror-generation indices carried by fermions are omitted for notational simplicity.

The correct assignment of quantum numbers to the fermions of the theory leads to the following definitions for the covariant derivatives:

$$\begin{aligned} \mathcal{D}^\mu_k &= \hat{\mathcal{D}}^\mu_k + \frac{ig_1 Y_k}{2} B^\mu + \frac{ig_2 \tau_a}{2} W^\mu_a \\ \hat{\mathcal{D}}^\mu_1 &= \partial^\mu + \frac{ig_1 \hat{Y}_1}{2} B^\mu \\ \hat{\mathcal{D}}^\mu_2 &= \partial^\mu + \frac{ig_1 \hat{Y}_2}{2} B^\mu + \frac{ig_3 \lambda_e}{2} G^\mu_e \\ \hat{\mathcal{D}}^\mu_3 &= \hat{\mathcal{D}}^\mu_1 + \frac{ig_{3K} \lambda_e}{2} G^{K\,\mu}_e \\ \hat{\mathcal{D}}^\mu_4 &= \hat{\mathcal{D}}^\mu_2 + \frac{ig_{3K} \lambda_e}{2} G^{K\,\mu}_e \\ \text{with } Y_k &= \begin{pmatrix} -1/2 & 0 \\ 0 & 1/2 \end{pmatrix}, \hat{Y}_1 = \begin{pmatrix} 0 & 0 \\ 0 & -1 \end{pmatrix}, \hat{Y}_2 = \begin{pmatrix} 2/3 & 0 \\ 0 & -1/3 \end{pmatrix}, \end{aligned} \tag{4}$$

where the 2×2 unit matrix in isospin space multiplying ∂^μ and the $SU(3)$, $SU(3)_K$ gauge fields is omitted. The $SU(2)$ and $SU(3)$, $SU(3)_K$ generators are denoted by τ_a and λ_e respectively. In the above, we omit for simplicity not only an extra $U(1)'$ interaction possibly felt only by katoptrons [6] but also neutrino Majorana mass terms responsible for a neutrino see-saw mechanism [7], as well as possible additional sterile-neutrino terms implied by the embedding of this model within larger gauge symmetries [6].

As already mentioned, the $SU(3)_K$-coupling renormalization assuming unification renders mirror fermions strongly coupled around 1 TeV. Apart from breaking the fermion chiral and electroweak gauge symmetries, this is expected to lead to the formation of composite spin-0 and spin-1 bosons similarly to QCD. Leaving aside most parity-even and excited states of these bosons, estimates of the masses of the lowest-lying parity-odd bosons along with the analogue of the QCD σ resonance follow next. With regards to states similar to the ρ meson in QCD, one expects two groups of spin-1 mesons, the heavier one corresponding to the third mirror generation (denoted by B) and the lighter corresponding to the two lighter mirror generations (denoted by A) with the following mass ranges [8]:

$$M_{\rho_A^K} \approx 0.25 - 0.5 \text{ TeV}, \qquad M_{\rho_B^K} \approx 1.4 - 2.8 \text{ TeV}. \tag{5}$$

With regards to QCD singlet spin-zero mirror mesons, one has the following rough order of magnitude estimates (σ^K denoting scalar and π^K pseudoscalar particles)[8]:

$$M_{\sigma_A^K, \pi_A^K} \approx 0.1 - 0.2 \text{ TeV}, \qquad M_{\sigma_B^K, \pi_B^K} \approx 0.57 - 1.15 \text{ TeV}. \tag{6}$$

Last, with regards to QCD colored spin-zero mirror mesons, one has the following rough order of magnitude estimates (π_3^K denoting QCD triplets and π_8^K QCD octets)[8]:

$$\begin{aligned} M_{\pi_{3\,A}^K} &\sim 0.11 - 0.23 \text{ TeV}, \qquad M_{\pi_{3\,B}^K} \sim 0.64 - 1.3 \text{ TeV}, \\ M_{\pi_{8\,A}^K} &\sim 0.13 - 0.26 \text{ TeV}, \qquad M_{\pi_{8\,B}^K} \sim 0.72 - 1.45 \text{ TeV}. \end{aligned} \tag{7}$$

The order-of-magnitude mass ranges presented above may provide a rough guide for experiments in order to detect excesses of events in comparison to the SM at specific center-of-mass ranges. Note that one cannot exclude *a priori* the possibility of a doubling of the group-A mirror-meson spectrum [8]. We proceed below by presenting some approximate cross-sections involving production and decay of these mirror mesons at the LHC.

3 The Cross-sections

The processes presenting particular interest at the LHC most frequently involve mirror-meson production via gluon-gluon fusion. Taking this into account in the narrow-width approximation and neglecting proton quark distribution functions, an order-of-magnitude estimate of the cross-section of a mirror meson R of mass M and total width Γ_{tot} produced via proton collisions and decaying into a final state X is given by

$$\begin{aligned} \sigma(\bar{p}p \longrightarrow R \longrightarrow X) &= \mathcal{L}(M)\frac{c\Gamma^{gg}\Gamma^X}{M\Gamma_{tot}}, \\ \text{with } \mathcal{L}(M) &\approx 4(M/\text{TeV})^{-(6+1.6\log_{10}(M/\text{TeV}))} \text{ nb} \end{aligned} \tag{8}$$

a luminosity function fitting in a very rough approximation the proton gluon distribution functions [9], c an appropriate QCD-color factor and $\Gamma^{gg,X}$ the relevant mirror-meson production and decay widths.

Given that mirror-meson decay widths to SM fermion pairs are expected to be proportional to the square of the final fermion masses, just the most interesting kinematically allowed processes are considered here, which include on one hand group-A mirror meson decays to $\bar{\tau}\tau$ and $\bar{b}b$ pairs and on the other hand group-B mirror meson decays to $\bar{\tau}\tau$ and $\bar{t}t$ pairs. In addition, mirror-meson decays to photon pairs are also studied since they are expected to produce clear signals free of QCD background. The relevant expressions using a specific assumption for the total mirror meson decay widths are given by [8]:

$$\begin{aligned} \sigma(\bar{p}p \longrightarrow \sigma_A^K \longrightarrow \bar{\tau}\tau) &= \mathcal{L}(M_{\sigma_A^K})\frac{c_{\sigma g\,A}}{\pi}\left(\frac{\alpha_{s\,A}(M_{\sigma_A^K})M_{\sigma_A^K}m_\tau}{36\pi v m_b}\right)^2 \\ \sigma(\bar{p}p \longrightarrow \sigma_A^K \longrightarrow \bar{b}b) &= \mathcal{L}(M_{\sigma_A^K})\frac{3c_{\sigma g\,A}}{\pi}\left(\frac{\alpha_{s\,A}(M_{\sigma_A^K})M_{\sigma_A^K}}{36\pi v}\right)^2 \\ \sigma(\bar{p}p \longrightarrow \sigma_A^K \longrightarrow \gamma\gamma) &= \mathcal{L}(M_{\sigma_A^K})\frac{3c_{\sigma g\,A}c_{\sigma\gamma\,A}}{2\pi}\left(\frac{\alpha_{s\,A}(M_{\sigma_A^K})\alpha(M_{\sigma_A^K})M^2_{\sigma_A^K}}{486\pi^2 v m_b}\right)^2 \end{aligned}$$

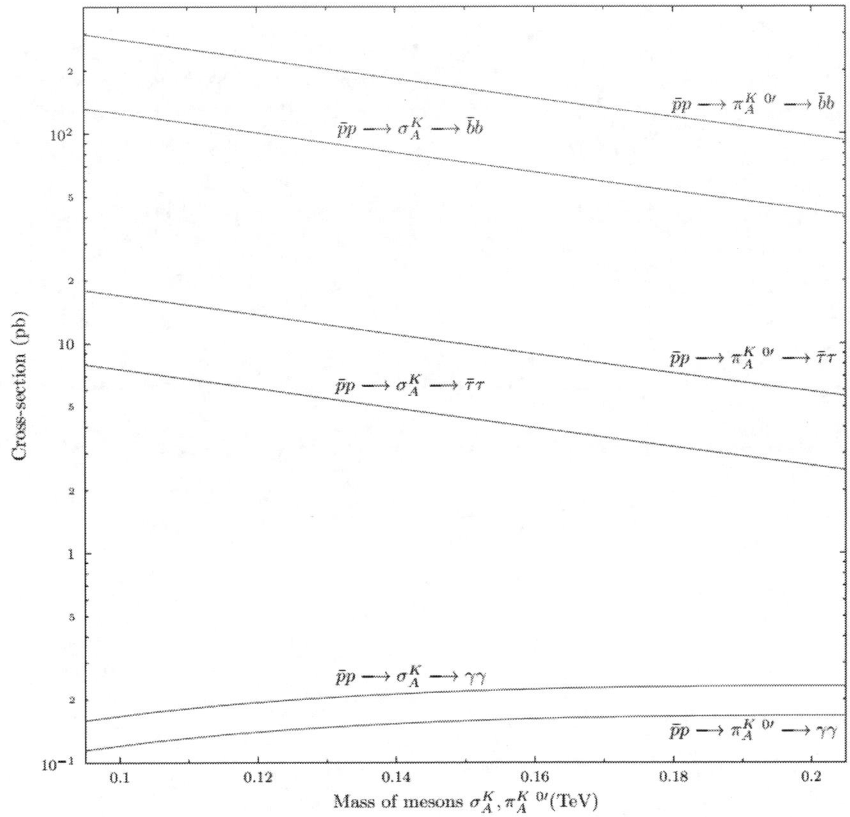

Figure 1. Cross sections of group-A spin-0 color-singlet mirror mesons versus their invariant mass. The Higgs particle discovered in 2012 should correspond to σ_A^K [8].

$$
\begin{aligned}
\sigma(\bar{p}p \longrightarrow \pi_A^{K\,0\prime} \longrightarrow \bar{\tau}\tau) &= \mathcal{L}(M_{\pi_A^{K\,0\prime}})\frac{c_{\pi g\;A}}{\pi}\left(\frac{\alpha_{s\;A}(M_{\pi_A^{K\,0\prime}})M_{\pi_A^{K\,0\prime}}m_\tau}{24\pi v m_b}\right)^2 \\
\sigma(\bar{p}p \longrightarrow \pi_A^{K\,0\prime} \longrightarrow \bar{b}b) &= \mathcal{L}(M_{\pi_A^{K\,0\prime}})\frac{3c_{\pi g\;A}}{\pi}\left(\frac{\alpha_{s\;A}(M_{\pi_A^{K\,0\prime}})M_{\pi_A^{K\,0\prime}}}{24\pi v}\right)^2 \\
\sigma(\bar{p}p \longrightarrow \pi_A^{K\,0\prime} \longrightarrow \gamma\gamma) &= \mathcal{L}(M_{\pi_A^{K\,0\prime}})\frac{3c_{\pi g\;A}c_{\pi\gamma\;A}}{2\pi}\left(\frac{\alpha_{s\;A}(M_{\pi_A^{K\,0\prime}})\alpha(M_{\pi_A^{K\,0\prime}})M^2_{\pi_A^{K\,0\prime}}}{216\pi^2 v m_b}\right)^2
\end{aligned}
$$

DOI: 10.1051/epjconf/201716407010

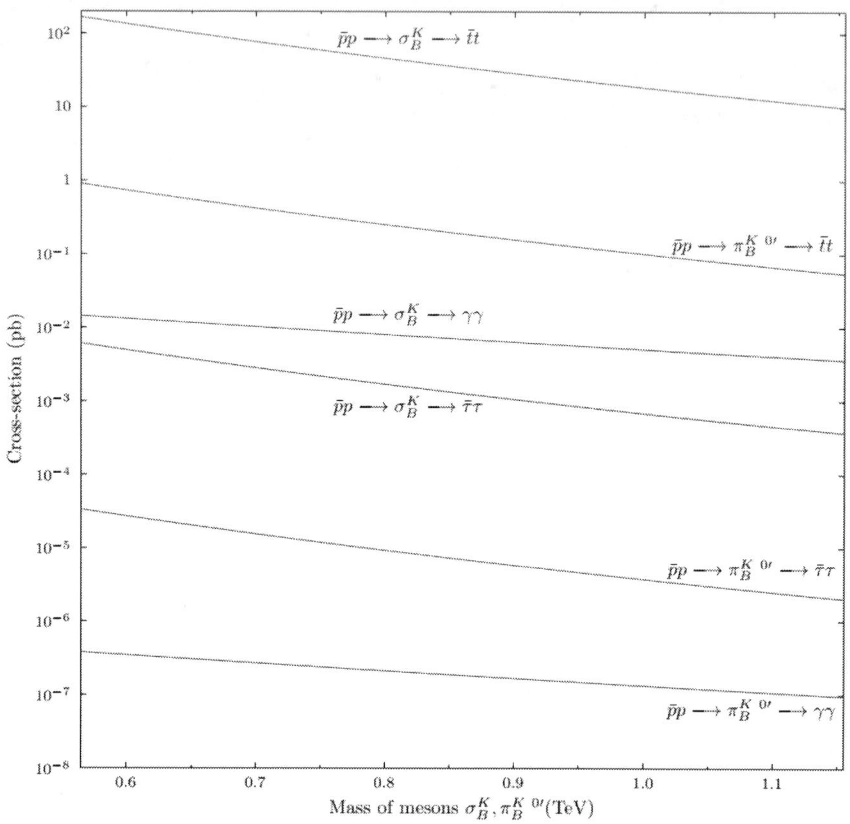

Figure 2. Cross sections of group-B spin-0 color-singlet mirror mesons versus their invariant mass [8]. The excess of diphoton events around 750 GeV reported in 2016 by LHC might be related to σ_B^K, but it should be followed by an excess of $\bar{t}t$ events over the QCD background at the same mass.

$$\sigma(\bar{p}p \longrightarrow \sigma_B^K \longrightarrow \bar{\tau}\tau) = \mathcal{L}(M_{\sigma_B^K})\frac{c_{\sigma g\ B}}{\pi}\left(\frac{\alpha_{s\ B}(M_{\sigma_B^K})M_{\sigma_B^K}m_\tau}{36\pi v m_t}\right)^2$$

$$\sigma(\bar{p}p \longrightarrow \sigma_B^K \longrightarrow \bar{t}t) = \mathcal{L}(M_{\sigma_B^K})\frac{3c_{\sigma g\ B}}{\pi}\left(\frac{\alpha_{s\ B}(M_{\sigma_B^K})M_{\sigma_B^K}}{36\pi v}\right)^2$$

$$\sigma(\bar{p}p \longrightarrow \sigma_B^K \longrightarrow \gamma\gamma) = \mathcal{L}(M_{\sigma_B^K})\frac{3c_{\sigma g\ B}c_{\sigma\gamma\ B}}{2\pi}\left(\frac{\alpha_{s\ B}(M_{\sigma_B^K})\alpha(M_{\sigma_B^K})M^2_{\sigma_B^K}}{486\pi^2 v m_t}\right)^2$$

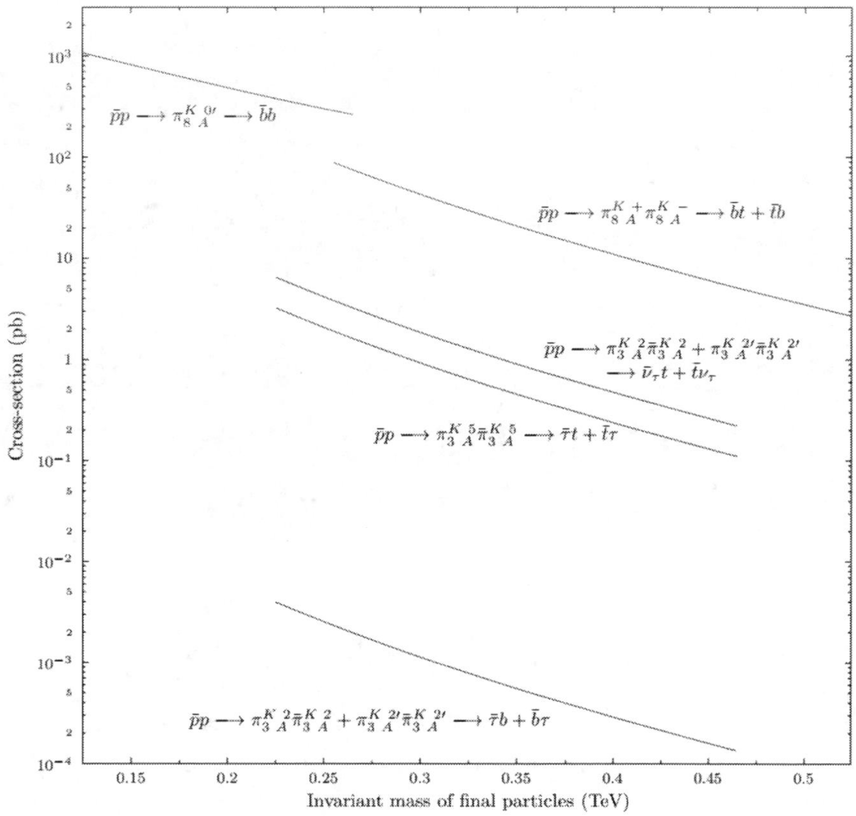

Figure 3. Cross sections involving single or pair production of group-A pseudoscalar spin-0 color-octet and color-triplet mirror mesons versus their total invariant mass [8].

$$\sigma(\bar{p}p \longrightarrow \pi_B^{K\ 0\prime} \longrightarrow \bar{\tau}\tau) = \mathcal{L}(M_{\pi_B^{K\ 0\prime}})\frac{c_{\pi g\ B}}{\pi}\left(\frac{\alpha_{s\ B}(M_{\pi_B^{K\ 0\prime}})M_{\pi_B^{K\ 0\prime}}m_\tau}{24\pi v m_t}\right)^2$$
$$\sigma(\bar{p}p \longrightarrow \pi_B^{K\ 0\prime} \longrightarrow \bar{t}t) = \mathcal{L}(M_{\pi_B^{K\ 0\prime}})\frac{3c_{\pi g\ B}}{\pi}\left(\frac{\alpha_{s\ B}(M_{\pi_B^{K\ 0\prime}})M_{\pi_B^{0\prime}}}{24\pi v}\right)^2$$
$$\sigma(\bar{p}p \longrightarrow \pi_B^{K\ 0\prime} \longrightarrow \gamma\gamma) = \mathcal{L}(M_{\pi_B^{K\ 0\prime}})\frac{3c_{\pi g\ B}c_{\pi\gamma\ B}}{2\pi}\left(\frac{\alpha_{s\ B}(M_{\pi_B^{K\ 0\prime}})\alpha(M_{\pi_B^{K\ 0\prime}})M^2_{\pi_B^{K\ 0\prime}}}{216\pi^2 v m_t}\right)^2 \quad (9)$$

 DOI: 10.1051/epjconf/201716407010

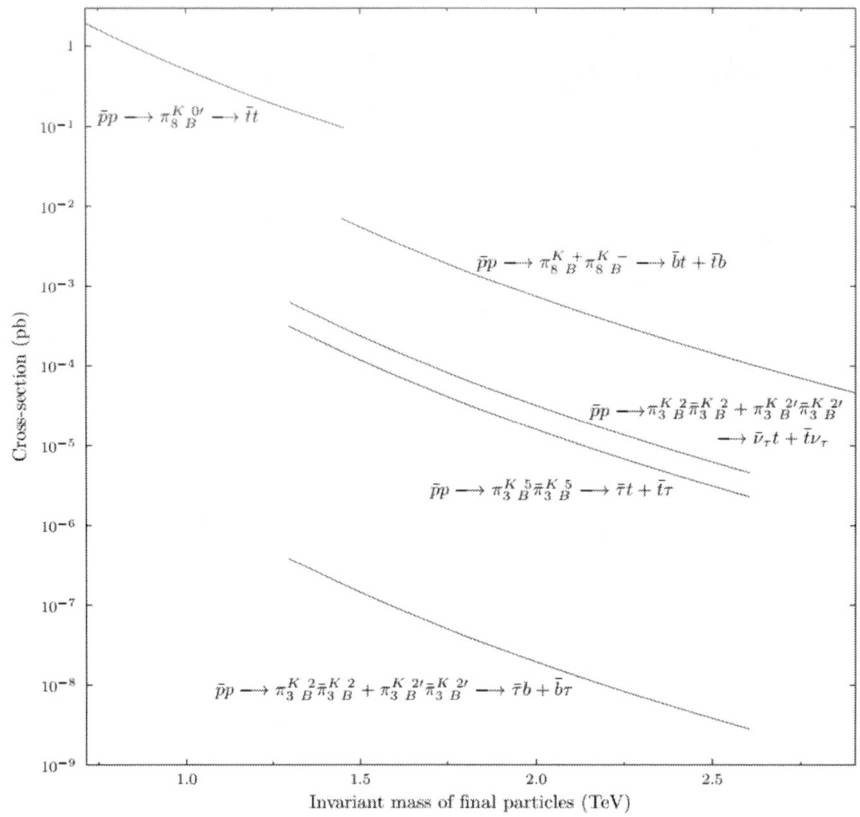

Figure 4. Cross sections involving single or pair production of group-B pseudoscalar spin-0 color-octet and color-triplet mirror mesons versus their total invariant mass [8].

$$
\begin{aligned}
\sigma(\bar{p}p \longrightarrow \pi^{K\ 0\prime}_{8\ A} \longrightarrow \bar{b}b) &= \mathcal{L}(M_{\pi^{K\ 0\prime}_{8\ A}})\frac{15c_{\pi g\ A}}{\pi}\left(\frac{\alpha_{s\ A}(M_{\pi^{K\ 0\prime}_{8\ A}})M_{\pi^{K\ 0\prime}_{8\ A}}}{24\pi v}\right)^2 \\
\sigma(\bar{p}p \longrightarrow \pi^{K\ 0\prime}_{8\ B} \longrightarrow \bar{t}t) &= \mathcal{L}(M_{\pi^{K\ 0\prime}_{8\ B}})\frac{15c_{\pi g\ B}}{\pi}\left(\frac{\alpha_{s\ B}(M_{\pi^{K\ 0\prime}_{8\ B}})M_{\pi^{K\ 0\prime}_{8\ B}}}{24\pi v}\right)^2 .
\end{aligned}
\tag{10}
$$

In the above, v denotes the weak scale, $\alpha(M)$ corresponds to the electromagnetic coupling at energy M and $\alpha_{s\ A,B}(M)$ corresponds to the QCD coupling at energy M, with the A, B subscripts reminding the different gauge coupling renormalization according to the number of particles participating in the renormalization loops.

Similarly, expressions involving production of pairs of mirror color-octet mesons and mirror lepto-quarks (differentiated from each-other by numerical superscripts) are given below, neglecting production enhancements stemming from color-octet vector mirror mesons and using a specific assumption for the total mirror meson decay widths [8]:

$$\begin{aligned}
\sigma(\bar{p}p \longrightarrow \pi^{K\,+}_{8\,A,B}\,\pi^{K\,-}_{8\,A,B} \longrightarrow \bar{b}t + \bar{t}b) &\approx \mathcal{L}(2M_{\pi^{K\,\pm}_{8\,A,B}})27c_w\pi\left(\frac{\alpha_s(2M_{\pi^{K\,\pm}_{8\,A,B}})}{32}\right)^2 \\
\sigma(\bar{p}p \longrightarrow \pi^{K\,5}_{3\,A,B}\,\bar{\pi}^{K\,5}_{3\,A,B} \longrightarrow \bar{\tau}t + \bar{t}\tau) &= \mathcal{L}(2M_{\pi^{K\,5}_{3\,A,B}})\frac{7c_w\pi}{12}\left(\frac{\alpha_s(2M_{\pi^{K\,5}_{3\,A,B}})}{32}\right)^2 \\
\sigma(\bar{p}p \longrightarrow \pi^{K\,2}_{3\,A,B}\,\bar{\pi}^{K\,2}_{3\,A,B} + \pi^{K\,2\prime}_{3\,A,B}\,\bar{\pi}^{K\,2\prime}_{3\,A,B} &\longrightarrow \bar{\nu}_\tau t + \bar{t}\nu_\tau) = \\
&= \mathcal{L}(2M_{\pi^{K\,2}_{3\,A,B}})\frac{7c_w\pi}{6}\left(\frac{\alpha_s(2M_{\pi^{K\,2}_{3\,A,B}})}{32}\right)^2 \\
\sigma(\bar{p}p \longrightarrow \pi^{K\,2}_{3\,A,B}\,\bar{\pi}^{K\,2}_{3\,A,B} + \pi^{K\,2\prime}_{3\,A,B}\,\bar{\pi}^{K\,2\prime}_{3\,A,B} &\longrightarrow \bar{\tau}b + \bar{b}\tau) = \\
&= \mathcal{L}(2M_{\pi^{K\,2}_{3\,A,B}})\frac{7c_w\pi}{6}\left(\frac{\alpha_s(2M_{\pi^{K\,2}_{3\,A,B}})m_b}{32m_t}\right)^2 . \qquad (11)
\end{aligned}$$

The prefactors $c_{\sigma \text{ or } \pi, g \text{ or } \gamma, A \text{ or } B}$ above contain information on the content of quantum loops contributing to the corresponding processes and c_w quantifies the effect of a certain cut of the final-particle total invariant mass [8]. Using certain assumptions in order to approximate these prefactors, figures 1-4 present the relevant order-of-magnitude results corresponding to the mirror-meson mass ranges reported in Section 2. The results in figures 1 and 2 indicate that one should mostly expect excess production of $\bar{b}b$ and $\bar{t}t$ pairs over the relevant QCD background at specific invariant mass bins. This is also true for the results pertaining to single spin-zero color-octet mirror meson production, as can be clearly seen in figures 3 and 4. Moreover, the results of figures 3 and 4 regarding pair production of spin-zero color-octet and color-triplet mirror mesons underline the importance of detection of acollinear $\bar{b}b$ and $\bar{t}t$ jets exceeding relevant estimates from QCD.

Since most of the processes above are obscured by significant QCD backgrounds, these results may be mostly used in order to have just a rough order-of-magnitude indication of the strength of the relevant signals and an estimate of their relative importance under specific assumptions. Crucial for their phenomenological importance would be their comparison with the corresponding QCD backgrounds, obviously using the same gluon luminosity function $\mathcal{L}$. However, the relevant signal-to-background ratios can be largely enhanced by judicious rapidity and transverse momentum cuts that may be best studied by computer simulations and which are therefore left for future work. Regarding mirror-meson decays to photon pairs which are free from QCD background, their strength varies considerably according to the aforementioned prefactors depending on effective mirror-meson couplings of non-perturbative origin, which are therefore *a priori* unknown. Further experimental input is obviously crucial for determining their relative importance.

4 Conclusions

In general, efforts to interpret physical phenomena of persistently unclear origin lead inexorably to the introduction of new degrees of freedom based on novel assumptions. In the present case, the desire to obtain a natural explanation for the appearance of three massive chiral fermion generations interacting according to gauge symmetries including a broken one in the SM leads often to radical proposals which are ultimately connected to the fundamental structure of spacetime and are usually based on the introduction of an extra-dimensional space of non-trivial topology. The theory presented in this study might be based on a discrete version of such a multi-dimensional space, where mirror fermions carrying the required quantum numbers are expected to emerge naturally [6]. Partially restoring the left-right symmetry of fermion representations in the way briefly described here, apart from solving several important theoretical puzzles, entails numerous phenomenological consequences, some of which are timely studied in the present work since they are in principle accessible at the LHC.

In such a framework, quantities possibly exhibiting non-negligible deviations from their SM expected values include the CKM-matrix element $|V_{tb}|$, third-generation SM-fermion weak couplings, the muon magnetic moment and B-meson branching ratios, not to mention the proton-decay prediction due to unification considerations [10]. In the present work however, our interest is focused on the production and decay of some of the lowest-lying bound states of mirror fermions formed due to their interacting with a strongly-coupled gauged mirror-generation symmetry. The results obtained indicate that experiments should place particular emphasis on detecting spin-0 and spin-1 mirror mesons which are either QCD-color singlets or triplets or octets, belonging in two groups separated from each-other by a mass hierarchy of around six. Their existence is expected to be inferred mainly by excesses of $\bar{b}b$ and $\bar{t}t$ production over the QCD background within specific invariant-mass ranges, noting that in cases of mirror-meson pair production these jets are generally acollinear. Possible excesses of Z^0, $W^\pm$ production over the SM background within specific invariant-mass bins, in case they can be traced to decays of QCD-color singlet mirror mesons, depend on quark distribution functions and are left for future work. Moreover, lest large QCD backgrounds at the LHC do not permit a solid confirmation of the processes described here, the case for a leptonic - electron or muon - collider with center-of-mass energies of around 3-4 TeV would appear more compelling, since around such energies the $SU(3)_K$ interactions between mirror fermions are expected to weaken to an extent possibly allowing for a measurement of their inverted left-right asymmetry [10].

Acknowledgements

The author would like to cordially thank the organizers of the 5th International Conference on New Frontiers in Physics 2016, Sonja Kabana, Yiota Foka and Larisa Bravina for their invitation and warm hospitality in a particularly interesting Conference and for making this presentation possible.

References

[1] S. Weinberg, *Phys. Rev.* **D13** 974 (1976); L. Susskind, *Phys. Rev.* **D20** 2619 (1979).

[2] M. E. Peskin, T. Takeuchi, *Phys. Rev. Lett.* **65** 964 (1990).

[3] T. D. Lee, C. N. Yang, *Phys. Rev.* **104** 254 (1956).

[4] J. C. Pati, A. Salam, *Phys. Rev.* **D8** 1240 (1973); H. Georgi, S. L. Glashow, *Phys. Rev. Lett.* **32** 438 (1974).

[5] F. Wilczek, A. Zee, *Phys. Rev.* **D25** 553 (1982).

[6] G. Triantaphyllou, *El. J. Th. Phys.* **10** 135 (2013); *ibid., Eur. Phys. J. Web Conf.* **70** 00051 (2014).

[7] G. Triantaphyllou, *Eur. Phys. J.* **C10** 703 (1999).

[8] G. Triantaphyllou, *El. J. Th. Phys.* **13** 115 (2016).

[9] C. Quigg, Fermilab Preprint FERMILAB-FN-0839-T, August 2009.

[10] G. Triantaphyllou, *Int. J. Mod. Phys.* **A15** 265 (2000).

Do we live in the best of all possible worlds? The fine-tuning of the constants of nature

Thomas Naumann[1,a]

[1]Deutsches Elektronen-Synchrotron DESY, Platanenallee 6, 15738 Zeuthen, Germany

Abstract. Our existence depends on a variety of constants which appear to be extremely fine-tuned to allow for the existence of life. These include the number of spatial dimensions, the strengths of the forces, the masses of the particles, the composition of the Universe and others. On the occasion of the 300th anniversary of the death of G.W. Leibniz we discuss the question of whether we live in the "Best of all Worlds". The hypothesis of a multiverse could explain the mysterious fine tuning of so many fundamental quantities. Anthropic arguments are critically reviewed.

1 The best of all worlds - From Leibniz to Einstein

The Electress of Brandenburg and later Queen of Prussia Sophie Charlotte was a lifelong friend to Gottfried Wilhelm Leibniz who died 300 years ago. They inspired the king to create the Kurfürstlich Brandenburgische Societät der Wissenschaften zu Berlin in 1700. Leibniz became the first lifetime President of this academy renamed in 1701 to Royal Prussian Society of Sciences and known today as Berlin-Brandenburgische Akademie der Wissenschaften.

Since Leibniz tended to never finish his many activities and to spread his ideas over thousands of loose notes the Queen asked him to summarize his world view in a popular pamphlet. As he was very much indebted to her he completed in 1710 the only coherent text published during his lifetime, his "Essays on the Goodness of God, the Freedom of Man and the Origin of Evil" [1, 2].

In this treatise he tried what one might call the impossible: to unify and harmonise science, metaphysics and theology. In § 8 of its first part he writes: „this supreme wisdom, united to a goodness that is no less infinite, cannot but have chosen the best. … if there were not the best among all possible worlds, God would not have produced any... there is an infinitude of possible worlds among which God must have chosen the best, since he does nothing without acting in accordance with supreme reason..." [2]. The important point for modern science is that Leibniz assumes the existence of a multitude of possible worlds although by this he meant imaginary or intended but not real worlds.

In modern physics it was Einstein who in a discussion with his assistant Banesh Hoffmann in the late thirties asked a similar question: "When I am judging a theory, I ask myself whether, if I were God, I would have arranged the world in such a way." [3] And to Ernst Gabor Straus, his assistant from 1940-48, Einstein made the famous remark: "What

[a] Corresponding author: Thomas.Naumann@desy.de

DOI: 10.1051/epjconf/201716407011

really interests me is whether God could have created the world any differently." [4] This reiterates Leibniz's question whether we live in the best of all worlds in the context of modern physics.

2 Had God a choice?

An early observation of the fine-tuning of the fundamental cosmological parameters (Hubble and gravitational constants H_0 and G_N and mean density Ω of the Universe) was made by Dicke in 1961 [5]. If any of several fundamental physical constants were only slightly different, the Universe would be unlikely to lead to the formation of nuclear, atomic or cosmic structures which allow for chemical elements, stellar burning and finally life.

The Universe indeed appears extremely fine-tuned for life. We will illustrate this phenomenon by asking the question: "What if we change some of the constants of nature?" proceeding from the most general to more specific constants:

- the dimensionality of the Universe
- symmetries: the matter-antimatter asymmetry
- the cosmic inventory: mean density, Hubble and cosmological constant, curvature
- What if no Higgs? The masses of weak bosons, quarks and the electron.

The amazing fine-tuning of nuclear processes visible in the deuterium bottleneck, the triple-alpha process of stellar burning and nucleon-nucleon binding has to be left for a separate discussion.

2.1 The dimensionality of the Universe

A world with less than three dimensions was discussed by Barrow and Tipler [6] and Hawking [7]. In such a world, the digestive tract of any higher animal would separate it into disjoint upper and lower parts. Also nerves and blood vessels would cross in two dimensions, and food absorption would have to proceed by breaking a one-dimensional boundary.

The case of a world with more than three spatial dimensions was discussed by Ehrenfest in 1917 [8, 9]. He considered the scattering of a small on a large mass. The probe particle is either trapped by the large mass - or it escapes to infinity. In more than three dimensions no stable orbits of planets are possible! The same is true for electrons in classical atoms. In 1963 Tangherlini [10] showed that also in quantum mechanics in more than three spatial dimensions there are no stable atoms.

2.2 The matter-antimatter asymmetry

We have to assume that the Universe is neutral and that all charges are conserved. Hence the Big Bang should have started from equal amounts of matter and antimatter. A matter-antimatter symmetric universe would however have rapidly annihilated to an empty light bubble while we live today in a matter dominated Universe. Related to the density of relic photons the number of matter particles amounts only to about one in several billions. The matter content of our Universe is hence characterized by a tiny but decisive misbalance. A significantly larger matter density would have caused a rapid recollapse of the Universe without any chance for structure formation and life.

2.3 The cosmic inventory

DOI: 10.1051/epjconf/201716407011

As already observed by Dicke in 1961 [5] the evolution of the Universe is characterized by a delicate balance of its inventory, a balance between repulsion and attraction, between expansion and contraction. Since gravity is purely attractive its action sums up throughout the whole Universe. The strength of this attraction is defined by the gravitational constant G_N and by an environmental parameter, the density Ω_M of (dark and baryonic) matter. The strength of the repulsion is defined by two parameters: the initial impetus of the Big Bang parameterized by the Hubble constant H_0 and the cosmological constant Λ.

As determined from data of the PLANCK satellite combined with other cosmological observations the total density of the Universe is $\Omega_{tot} = \rho / \rho_{crit} = 1.0002 \pm 0.0026$ and thus extremely close to its critical density $\rho_{crit} = 3H_0^2/8\pi G_N$ with a contribution from a curvature of the Universe of $\Omega_k = 0.000 \pm 0.005$ [11]. However, the balance between the ingredients of the Universe is time-dependent. Any deviation from zero curvature after inflation is amplified by many orders of magnitude. Hence, a cosmic density fine-tuned to flatness today to less than a per mille must have been initially fine-tuned to tens of orders of magnitude.

Another mystery is the smallness of the cosmological constant. First of all, the Planck scale is the only natural energy scale of gravitation: $m_{Pl} = (\hbar c/G_N)^{1/2} = 1.2 \cdot 10^{19}$ GeV/c^2. Compared to this natural scale the cosmological constant or dark energy density Λ is tiny:

$$\Lambda \sim (10\ meV)^4 \sim (10^{-30}\ m_{Pl})^4 = 10^{-120}\ m_{Pl}^4 . \tag{1}$$

The cosmological constant is also much smaller than expected from the vacuum expectation value of the Higgs field which like the inflaton field or dark energy is an omnipresent scalar field:

$$\langle \Phi^4 \rangle \sim m_H^4 \sim (100\ GeV)^4 \sim 10^{52}\ \Lambda \quad . \tag{2}$$

As observed by Weinberg in 1987 there is an 'anthropic' upper bound on the cosmological constant Λ. He argued "that in universes that do not recollapse, the only such bound on Λ is that it should not be so large as to prevent the formation of gravitationally bound states." [12]

2.4 What if no Higgs? The fine-tuning of particle masses

2.4.1 Boson masses

The carriers of the strong and electromagnetic interaction, the gluon and the photon, are massless. After spontaneous breaking of the electroweak symmetry the weak bosons acquire a mass which at low energies suppresses the weak interaction. Without this symmetry breaking, i.e. without a Higgs field, the weak bosons would be massless, and the weak interaction would be about as strong as the electromagnetic one. Star burning would not last billions of years but rather resemble a nuclear explosion with no time left for the evolution of planetary systems and life.

On a more quantitative level the mass $m_W = 80.4$ GeV of the W bosons governs the fusion process $pp \rightarrow d\ e^+\ \nu_e$ in stars. The cross section of this fusion process which corresponds to inverse beta decay depends on the inverse of the fourth power of the weak boson mass: $\sigma \sim (m_W)^{-4}$. An increase of the weak boson mass by a factor of two would let the Sun and similar stars burn out within roughly a billion years. Such a rapid star burning would make the evolution of higher life almost impossible.

2.4.2 Fermion masses

DOI: 10.1051/epjconf/201716407011

The neutron is heavier than the proton and hence unstable: m_n - m_p = 1.3 MeV ≈ 1‰ $m_{nucleon}$. This is due to the difference of 3-4 MeV between the masses of the down and up quark (this difference is partially compensated by the stronger electric repulsion of the additional up quark in the proton). In the second and third quark families, however, the up quarks are significantly heavier than the down quarks. So one might ask: What if also in the first family the up quark would be heavier than the down quark?

Then the proton would be heavier than the neutron and decay: $p \rightarrow n\ e^+\ \nu_e$. The positron would subsequently annihilate with an electron $e^+ e^- \rightarrow \gamma\ \gamma$. The result would be a neutral universe consisting of photons, neutrons and neutrinos without a chance for the formation of nuclei and atoms and hence life. Since $m_p - (m_n + m_e) > 2.2$ MeV = E_b^d the free energy of this decay is larger than the binding energy E_b^d of the neutron. This also makes the deuteron unstable $d \rightarrow 2n\ e^+\ \nu_e$ so that the decaying protons cannot be trapped in deuterium and form helium like in our world.

Since the Rydberg energy is $R = 0.5\ \boldsymbol{\alpha}^2\ m_e$ and the Bohr radius $r_B = 1\ /\ (\boldsymbol{\alpha}\ m_e)$ the mass of the electron governs energy levels and distances of the atomic world. An increase of the electron mass to 5 MeV would have the same effect as making the up quark heavier than the down quark. Protons would decay, and we would end up in a neutral world with only photons, neutrons and neutrinos left over.

3 The anthropic principle

The observation of the fine-tuning of the Universe is closely related to the anthropic principle. An early expression of this principle was done by Dicke in 1961: "The existence of physicists… [is] sufficient to demand that… relations between the three numbers [G_N, H_0, Ω] be satisfied." Lord Martin Rees, cosmologist and Astronomer Royal in his popular book "Just Six Numbers" [13] added the number of dimensions and the relative strength of the electromagnetic to the gravitational interaction to his discussion of the fine tuning of the constants of nature. The subtitle of the German translation nicely summarizes this book: "Hatte Gott eine Wahl?" - "Had God a choice?"

In 1986 in their ground-breaking book "The Anthropic Cosmological Principle" [6] Barrow and Tipler explain: this principle holds that the fundamental structure of the Universe is determined by the existence of intelligent observers. The Universe is as it is because if it were otherwise, observers could not exist. In short: Man not only fits to the Universe. The Universe also fits to man. Hawking in his "Brief History of Time" pointed out in 1988: "The remarkable fact is that the values of these numbers seem to have been very finely adjusted to make possible the development of life." [14]

The anthropic principle has been under intense discussions and criticism over the last decades. Steven Weinberg who repeatedly referred to it reportedly warned: "A physicist talking about the anthropic principle runs the same risk as a cleric talking about pornography: No matter how much you say you are against it, some people will think you are a little too interested." [15]

The big epistemological problem behind the anthropic principle is that for the moment we only observe one Universe. Hence, we cannot apply the scientific method to make repeated experiments with the object of our investigation in order to extract regularities and laws. The anthropic principle also runs the danger to end up in a tautology or a causal circle: We only observe universes that allow an observer. Another critical point is that the anthropic principle makes no predictions but only postdictions.

Another open question should be carefully considered. What exactly is fine-tuned in our Universe: underlying laws or parameters, and are the parameters fundamental or environmental ones?

DOI: 10.1051/epjconf/201716407011

4 From Universe to Multiverse

The problem of the uniqueness of our Universe can be avoided going from a single and unique universe to a multiverse consisting of a huge number of different universes. At the extreme energy densities at Planck time ($\sim 10^{-43}$ s) immediately after the Big Bang due to quantum gravity the Universe is thought to be dominated by vacuum fluctuations of space-time. Soon ($\sim 10^{-36}$ s) after, in the cosmological model of inflation vacuum fluctuations of a scalar (Higgs-like) inflaton field of the size of a tiny fraction of a proton radius could have inflated by many orders of magnitude to a region larger than the event horizon which later expanded to today's observable Universe.

At these extreme energies all non-gravitational interactions are still thought to be unified to one primordial force. At decreasing energies these grand unified symmetries can break down to the symmetries of separate forces in many different ways. In our world we observe $SU(3)_C$, $SU(2)_W$ and $U(1)$. On top of that the superstring theory developed in the 1980ies by Green, Schwarz, Witten and others [16] assumes six (or seven) extra spatial dimensions which are curled up at the small distance scales relevant in the early Universe. These theories give rise to a huge ($\sim 10^{500}$) [17] number of possible compactifications to the three dimensions which we observe at the low energies in our Universe today.

Each of the inflationary bubbles may have evolved into a universe with a different realization of the physics potential hidden in grand unified and superstring theories. So the physics in these universes can be different from the physics in our Universe. This vast field of possibilities Susskind called "The anthropic landscape of string theory" [18].

In his essay "Particle Physics and Inflationary Cosmology" [19] Linde who in the 1980ies together with Guth [20], Steinhardt and others had developed the inflationary scenario explained: "Instead of a Universe with a single law of physics operating everywhere we are discussing an eternally existing self-reproducing Multiverse which consists of many different parts where all possibilities can be realized." Applying the anthropic principle to the string landscape Weinberg in 2007 in an article "Living in the Multiverse" [21] stated that this principle "may explain how the constants of nature that we observe can take values suitable for life without being fine-tuned by a benevolent creator."

Steinhardt however criticized that no experiment can rule out the multiverse hypothesis since it allows for all possible physics scenarios: "...the inflationary paradigm is so flexible that it is immune to experimental and observational tests. First, inflation is driven by a hypothetical scalar field, the inflaton, which has properties that can be adjusted to produce effectively any outcome. Second, inflation does not end with a universe with uniform properties, but almost inevitably leads to a multiverse with an infinite number of bubbles, in which the cosmic and physical properties vary from bubble to bubble. The part of the multiverse that we observe corresponds to a piece of just one such bubble. Scanning over all possible bubbles in the multiverse, everything that can physically happen does happen an infinite number of times. No experiment can rule out a theory that allows for all possible outcomes. Hence, the paradigm of inflation is unfalsifiable." [22]

In his discussion „Universe Or universe? It All Depends On The Multiverse" of whether to capitalize the word 'Universe' Gleiser [23] mentions that "we may obtain information of neighbouring universes, if they have collided with ours in the past" [24] but also insists that we have only our Universe.

4.1 Learning from history: Kepler's *Harmonices Mundi*

In a similar spirit as Kepler at his time asked the question: "Why are there just six planets?" we ask today: "Why are there just three fermion families?" Without deeper knowledge it is impossible to decide whether such a question is a fundamental or an environmental one. In his main works *Mysterium Cosmographicum* (1596) and *Harmonices Mundi* (1619) Kepler was looking for God's plan of the geometry of the Universe and tried to explain the ratios of the planetary orbits from first (and divine) principles. Nesting and encasing the spheres inscribed and circumscribed to the five Platonic solids within one another he produced six shells the radii of which miraculously corresponded to the sizes of the orbits of the six known planets.

Since Kepler did not know that there are more planets in our solar system, millions of planetary systems in our galaxy and billions of galaxies in our Universe he considered an accidental fine-tuning to be fundamental. He thus tried to answer an irrelevant question. Also, Kepler's model relied on the orbits being circles which clearly contradicted his most important discovery that the orbits are ellipses. To save God's plan Kepler had to compromise his Platonic model.

So a sufficient number of alternatives can easily solve an apparent fine-tuning problem. The same may be true for our present anthropic problem: We are fascinated by our fine-tuned Universe which may be completely accidental in view of a sufficiently large number of alternative universes.

Kepler's dilemma, however, besides the statistical solution also had a fundamental solution: based on his laws Newton was able to create his fundamental theory of gravity. So in Kepler's case two solutions of the dilemma existed which were both viable and even not contradicted each other: the statistical and the fundamental one.

We may be in a similar situation today. The existence within a multiverse of a large number of universes with different dimensionalities, symmetry breaking schemes, strengths of the fundamental forces, particle masses, inflation scenarios and asymmetries (like matter-antimatter) may allow for exactly the fundamental parameters needed for the formation of stable structures, chemical elements and finally life in our world. Fine-tuning would then be a result of a natural cosmic selection of our world out of many worlds as the best habitat for the development of life (see e.g. Smolin [25]). Or as Gell-Mann once put it: "If we really live in a multiverse, physics will have been reduced to an environmental science like botany." [26]

So the indefiniteness and openness which Steinhardt and others deplore become very productive: they open the rich field of possibilities needed for the development of life.

4.2 From physics to metaphysics: the ultimate dethronement of Man

The physics in our world could thus be one of many solutions of more general physical laws. Based on his treatise *Physics* Aristotle subsequently composed his *Metaphysics*. In a similar spirit we might consider the physics of a multiverse as a meta-physics in its original positive sense: a meta-physics encompassing many possible physics scenarios realized in a huge number of universes of a multiverse.

Copernicus assumed that the Earth is not at the center of the Universe but only one of many planets orbiting around the Sun. Later Darwin proposed that Man is not the crown and goal of creation but an animal like others and a product of biological evolution. Hubble showed that our Milky Way is only one out of many galaxies in the Universe. Modern science repeatedly pushed Man off his throne into some remote corner of the Universe. Today, we may get rid of the anthropic problem assuming that our Universe is only one out of a multitude of universes in a multiverse.

This would be the ultimate dethronement of Man in the history of science. Monod described this aspect of human existence as follows: "The ancient alliance has been broken. Man at last knows he is alone in the unfeeling immensity of the Universe, out of which he has emerged only by chance. His destiny is nowhere spelled out, nor is his duty. The kingdom above or the darkness below; it is for him to choose." [27]

Finally, a word of caution is needed. The multiverse is not an established theory but a hypothesis. Asking the question "Does the Multiverse Really Exist?" the astrophysicist George Ellis said [28]: "As skeptical as I am, I think the contemplation of the multiverse is an excellent opportunity to reflect on the nature of science and on the ultimate nature of existence: why we are here... In looking at this concept, we need an open mind, though not too open. It is a delicate path to tread. Parallel universes may or may not exist; the case is unproved. We are going to have to live with that uncertainty. Nothing is wrong with scientifically based philosophical speculation, which is what multiverse proposals are. But we should name it for what it is."

Hypotheses have been very fruitful in the history of science, and it may take a lot of time and effort to verify or falsify them. The history of the atomic hypothesis is a good example. Based on his observation of the law of multiple proportions in 1808 John Dalton in his essay „A New System of Chemical Philosophy" formulated the hypothesis that all matter consists of indivisible atoms. The final confirmation of the atomic hypothesis took almost one century, and the theories of Maxwell, Boltzmann and Einstein had to stand against the criticism of positivists like Mach.

In order to keep science alive we should therefore aim to keep an open mind for hypotheses even if we cannot today give prescriptions how to verify or falsify them.

References

1. G. W. Leibniz, *Essais de théodicée sur la bonté de Dieu, la liberté de l'homme, et l'origine du mal* (Mortier, Amsterdam, 1710)
2. G. W. Leibniz, *Essays on the Goodness of God, the Freedom of Man and the Origin of Evil.* In: The Philosophical Works of G.W. Leibniz, Translated by E.M. Huggard from C.J. Gerhardt, Die Philosophischen Schriften von G.W. Leibniz, Berlin 1875-90; (Tuttle, Morehouse & Taylor, New Haven, 1890)
3. B. Hoffmann, quoted in: A. Calaprice, ed., *The Ultimate Quotable Einstein*, p. 407 (Princeton University Press, Princeton, 2010)
4. E. G. Straus, quoted in: A. Calaprice, ed., *The Ultimate Quotable Einstein*, p. 344 (Princeton University Press, Princeton, 2010)
5. R. H. Dicke, *Dirac's Cosmology and Mach's Principle*, Nature 192 (1961) 4801, 440
6. J. D. Barrow and F. Tipler, *The Anthropic Cosmological Principle* (Oxford University Press, New York 1986). See also: J. D. Barrow, *The Constants of Nature* (Pantheon Books, New York, 2003)
7. S. Hawking, *A Brief History of Time* (Bantam Press, London, 1988)
8. P. S. Ehrenfest, *In what way does it become manifest in fundamental laws of physics that space has three dimensions?*, Proc. Royal Acad. Amsterdam **20**, 200 (1917)
9. P. S. Ehrenfest, *How do the fundamental laws of physics make manifest that space has three dimensions?*, Annalen der Physik Vol. **61** Nr. 5, 440 (1920)
10. F. R. Tangherlini, *Schwarzschild field in n dimensions and the dimensionality of space problem*, Nuovo Cimento **27**, 636 (1963)
11. P.A.R. Ade et al., Planck Collab. *2015 Results XIII*, Astronomy and Astrophysics **594**, A13 (2016). arXiv:1502.01589
12. S. Weinberg, *Anthropic Bound on the Cosmological Constant*, Phys. Rev. Lett. Vol. **59** Nr. 22, 2607 (1987)
13. M. Rees, *Just Six Numbers. The Deep Forces That Shape The Universe* (Basic Books, New York, 2000)
14. S. Hawking, *A Brief History of Time* (Bantam Press, London, 1988)
15. S. Weinberg, quoted from: G. F. Giudice in: *Perspectives on LHC Physics*, G. L. Kane, A. Pierce (Eds.), p. 172 (World Scientific, Singapore, 2008)

16. M. Green, J. Schwarz and E. Witten, *Superstring theory*, two volumes (Cambridge University Press, 2012).
17. M. Douglas, *The statistics of string/M theory vacua*, JHEP 0305, **046** (2003). arXiv:hep-th/0303194;
S. Ashok and M. Douglas, JHEP 0401, **060** (2004). arXiv:hep-th/0307049
18. L. Susskind, *The anthropic landscape of string theory*, arXiv:hep-th/0302219.
19. A. D. Linde, *Particle Physics and Inflationary Cosmology* (Harwood, Chur, Switzerland, 1990). See also: arXiv:hep-th/0503203
20. For a popular introduction see e.g.: A. Guth, *The Inflationary Universe: The Quest for a New Theory of Cosmic Origins* (Perseus Basic Books, New York, 1997)
21. S. Weinberg, Living in the Multiverse, in: *Universe or Multiverse?,* ed. B. Carr (Cambridge University Press 2007). See also: arXiv:hep-th/0511037
22. P. Steinhardt, *Big Bang blunder bursts the Multiverse bubble*, Nature 510, 9 (June 3, 2014)
23. M. Gleiser, *Universe Or universe? It All Depends On The Multiverse.* npr commentary, February 20, 2013, www.npr.org/sections/13.7/2013/02/19/172391249/universe-or-universe-it-all-depends-on-the-multiverse
24. M. Gleiser, *The Universe Next Door*, npr commentary, January 9, 2013, www.npr.org/blogs/13.7/2013/01/09/168863713/good-news-bad-news-the-universe-next-door
25. L. Smolin, *The Life of the Cosmos* (Oxford University Press, New York, 1997)
26. M. Gell-Mann, quoted by M. Duff in: New Scientist, 30. June 1988, **Vol. 118**, No. 1619, p. 70.
27. J. Monod, *Chance and Necessity*, p. 180 (Alfred Knopf, New York, 1971)
28. G. Ellis, Scientific American, 1 August 2011, **305** 38-43 (2011)

EPJ Web of Conferences **164**, 07012 (2017) DOI: 10.1051/epjconf/201716407012
ICNFP 2016

Electromagnetic effects as a new source of information on the space-time evolution of heavy ion collisions

Nikolaos Davis[1,a], Andrzej Rybicki[1], and Antoni Szczurek[1,2]

[1] *H. Niewodniczański Institute of Nuclear Physics, Polish Academy of Sciences, Radzikowskiego 152, 31-342 Kraków, Poland*
[2] *University of Rzeszów, Rejtana 16, 35-959 Rzeszów, Poland*

Abstract. We review our studies of spectator-induced electromagnetic (EM) effects on charged pion emission in ultrarelativistic heavy ion collisions. These effects are found to consist in the electromagnetic charge splitting of pion directed flow as well as very large distortions in spectra and ratios of produced charged particles. As it emerges from our analysis, they offer sensitivity to the actual distance, d_E, between the pion formation zone at freeze-out and the spectator matter. As a result, this offers a new possibility of studying the space-time evolution of dense and hot matter created in the course of the collision. Having established that d_E traces the longitudinal evolution of the system and therefore rapidly decreases as a function of pion rapidity, we investigate the latter finding in view of pion feed-over from intermediate resonance production. As a result, we obtain a first estimate of the pion decoupling time from EM effects which we compare to existing HBT data. We conclude that spectator-induced EM interactions can serve as a new tool for studying the space-time characteristics and longitudinal evolution of the system. We discuss the future perspectives for this activity on the basis of existing and future data from the NA61/SHINE experiment.

1 Introduction

In the present work, we review various aspects of the final state electromagnetic (EM) interactions between the charged pions produced during an ultrarelativistic heavy-ion collision and the positively charged nuclear remnant ("spectator system") of the colliding nuclei, which does not participate directly in the collision. We argue that the study of the induced EM effects provides a useful new source of information on the space-time evolution of the system, valid in a broad range of collision centrality.

2 EM effects in heavy ion collisions

It is well established that charged spectator systems in non-central heavy ion collisions are the source of EM fields of considerable strength, which have a significant effect on the final state distributions of positively and negatively charged pions produced during freeze-out, as well as influence a variety of other observable quantities, such as the π^+/π^- ratios and directed flow as a function of the longitudinal

[a] e-mail: nikolaos.davis@ifj.edu.pl

momentum of pions [1, 2]. Below, we will review specific aspects of these phenomena, focusing mainly on the applicability of EM effects as a new source of information on the space-time evolution of the system.

The effect of spectator EM fields on the phase-space distributions of produced pions is evident in Fig. 1. The measurements were performed by the NA49 experiment (CERN SPS) in peripheral and intermediate centrality Pb+Pb collisions at $\sqrt{s_{NN}} = 17.3$ GeV. In panel 1a (1b), we see that the invariant double differential spectra of positive (negative) pions reveal a suppression (enhancement) of produced pions for low values of transverse momentum, p_T, in the vicinity of longitudinal momentum corresponding to pions comoving with the spectator system, $x_F = 0.15 = m_\pi/m_p$. This isospin symmetry-breaking distortion of pion spectra becomes even more apparent if we examine the π^+/π^- ratios as a function of transverse momentum for $x_F = 0.15$, Fig. 1(c). We notice that the distortion of ratios is very similar for peripheral and intermediate centrality collisions. The weak centrality dependence of this effect may at first seem surprising, however it can be explained as a consequence of the reduced charge of spectators partially compensated by the decrease of impact parameter b with increasing collision centrality. Another noteworthy effect observed in NA49 data is the significant enhancement, for low p_T values, of the ratio R of invariant particle densities of positive and negative pions produced in Pb+Pb collisions, compared to the same quantity in p+p collisions, as a function of longitudinal pion momenta, Fig. 2. As can be seen, the effect actually consists of a strong enhancement for π^- alone and a suppression for π^+, indicating a strong isospin symmetry-breaking effect, possibly of largely EM origin.

The influence of EM effects can also be detected in the behavior of collective azimuthal anisotropies (directed flow) of emitted charged pions with regards to the reaction plane, as detailed in [4]. In Fig. 3, directed flow v_1 is plotted as a function of the ratio of pion to beam rapidity in the c.m.s., for peripheral collisions recorded by the STAR and WA98 experiments.

By combining the results of the above measurements, we can obtain information on the dependence of EM effects of spectators on the rapidity of pions produced during freeze-out. Results were interpreted by means of a simplified one-point pion emission model, where the propagation of charged pions through the spectator EM field was numerically computed using classical relativistic equations of motion [1], while the spectators were represented as Lorentz-contracted homogeneously charged spheres, Fig. 4(a). The crucial free parameter in this model was the "pion emission distance" d_E between the pion emission point and the spectator system. As an example, Fig. 4(c) shows Monte Carlo results for the π^+/π^- ratio for a specific choice of d_E. It is evident that by fine-tuning d_E, we can obtain a fairly good agreement with the measured NA49 ratios, Fig. 4(b). Similarly, Fig. 3(b) shows the values of d_E furnished by the MC for different pion rapidities, y/y_{beam}.

3 Sensitivity of EM effects to the space-time evolution of the system

The final state distributions of pions measured in the detector are a direct result of the distortions induced by the spectator EM field on the trajectories of pions produced during the freeze-out. It is therefore not surprising that final state spectra appear sensitive to the space-time evolution of the particle production process. As was demonstrated in [1, 2], the general features of the distortions of pion spectra, as well as charge-dependent effects on directed flow, are sensitive to the distance d_E between the longitudinal position of the pion formation zone at freeze-out and the longitudinal position of the spectator system at the same moment in time (in the c.m.s. reference frame), as schematically illustrated in Fig. 5 (a). Results furnished by our one-point pion emission model simulations indicate that the π^+/π^- ratio distortions shown in Fig. 4 correspond to a pion formation zone of pions comoving with spectators ($y \approx y_{beam}$) being about $d_E \approx 0.75$ fm behind the spectator system in the longitudinal

 DOI: 10.1051/epjconf/201716407012

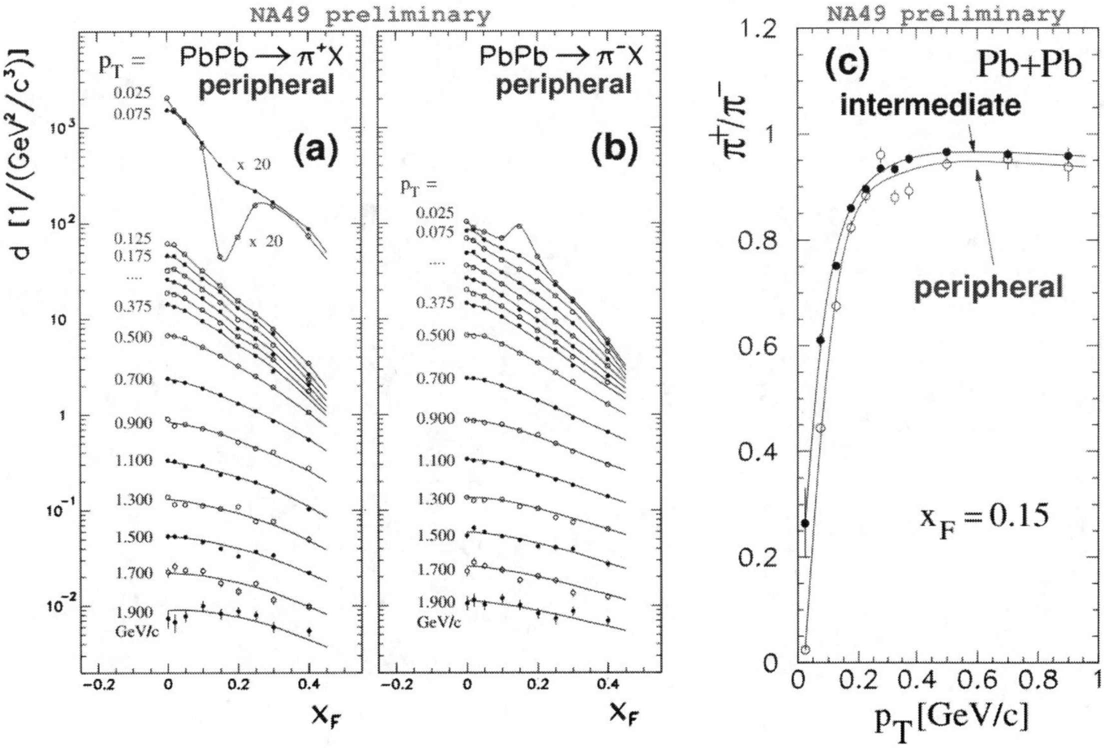

Figure 1. Spectator-induced EM effects on final state charged π meson spectra and π^+/π^- ratios in Pb+Pb collisions at $\sqrt{s_{NN}} = 17.3$ GeV. Double differential $d = E\frac{d^3N}{dp^3}$ distributions of (a) positive and (b) negative pions drawn as a function of $x_F = \frac{2p_L}{\sqrt{s_{NN}}}$ and transverse momentum p_T (both quantities in the collision c.m.s.), and (c) ratio of π^+ over π^- distributions at $x_F = 0.15$. Plot from [3].

direction. Similarly, the charge-splitting of directed flow observed in STAR Collaboration Au+Au collisions at $\sqrt{s_{NN}} = 7.7$ GeV [5], Fig. 3(a), corresponds to approximately $d_E \approx 3$ fm for pions produced at low c.m.s. rapidities, whereas WA98 experiment [7] measurements of very large π^+ directed flow seen in target vicinity in Pb+Pb collisions at the top SPS energy, Fig. 3(b), would suggest $d_E \approx 0.5$ fm at still higher rapidities. In Fig. 5 (d), these results are summarized, and we clearly notice a trend of decreasing d_E with increasing pion rapidity.

In order to further investigate this result, a Monte-Carlo model was constructed incorporating existing knowledge about the production and decay properties of baryonic (Δ) and mesonic (ρ) resonances, as described in [9]. In Fig. 5(b), a sketch of the model is given where the simplifying assumption is made that resonances are produced at the interaction point *instantly*, that is at $\tau = 0$. As is evident from Fig. 5(d), this fails to reproduce the observed values of d_E obtained from EM effects. In Fig. 5(c), a different model is sketched, for which the intermediate system is given a proper lifetime τ, which can be adjusted to fit the observed d_E curve. This suggests that quantitative information can be obtained on the spacetime properties and evolution of the hot and dense matter produced in a collision by studying spectator-induced EM effects. In Fig. 5(e), we explore the range of uncertainties introduced by incomplete knowledge of resonance production, by significantly varying the parameters of our simulation. Results indicate that the uncertainties are well within the tentative error bars

DOI: 10.1051/epjconf/201716407012

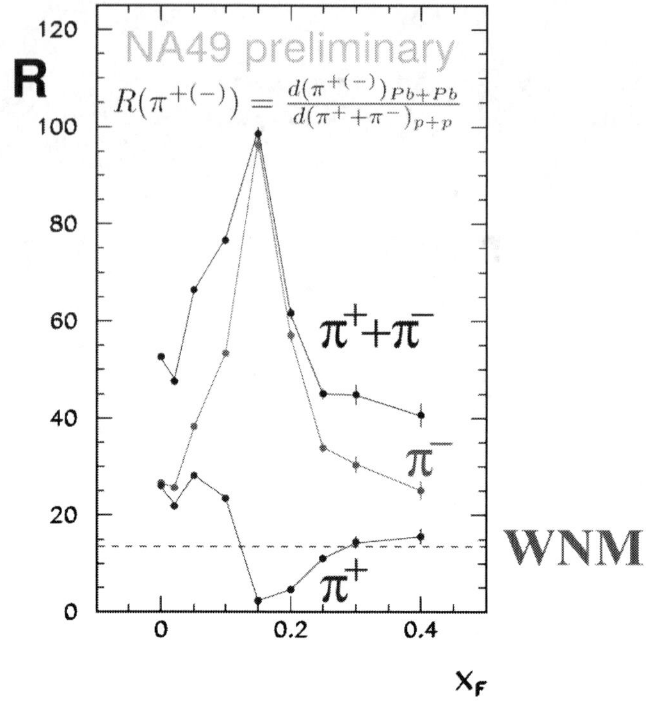

Figure 2. The ratios R(π^+) of positive , R(π^-) of negative and R($\pi^+ + \pi^-$) of summed invariant particle densities for charged pions produced in Pb+Pb peripheral versus p+p collisions in the NA49 experiment, as a function of x_F. The dashed line indicates the Wounded Nucleon Model prediction for $R(\pi^+) = R(\pi^-)$, neglecting isospin effects.

associated with our present d_E estimations, and can further be reduced by future experimental and theoretical investigation. We therefore expect that the study of EM effects can improve our knowledge of the space-time evolution of the system up to very high rapidities (where the latter is apparently least known).

At the present moment, the space-time characteristics of this system, tentatively established from these effects in the collision energy range $\sqrt{s_{NN}} = 7.7 - 17.3$ GeV discussed here and shown in Figs 5(d) and 5(c), can give an independent estimate for the time of pion emission as a function of its rapidity. At midrapidity this time being equal to $d_E(y/y_{\text{beam}} = 0)/\beta$, where β is the spectator velocity, our preliminary estimate gives 5.3 ± 2.2 fm/c for the time of pion emission in this region. For comparison, the pion decoupling times compiled in the published HBT analysis by the ALICE Collaboration [10] would indicate a value of approximately $\tau_f \approx 6$ fm/c for the same range of collision energies.

4 Conclusions and outlook

On the basis of the above analysis, we can conclude that spectator-induced EM interactions can serve as a useful tool for studying the space-time evolution of heavy-ion collision systems, and in particular the longitudinal evolution from "hot" QGP matter at $y \approx 0$ to "cold" hadronic/nuclear matter at $y \approx y_{beam}$. Our MC simulations of the production and decay of baryonic and mesonic resonances indicate

DOI: 10.1051/epjconf/201716407012

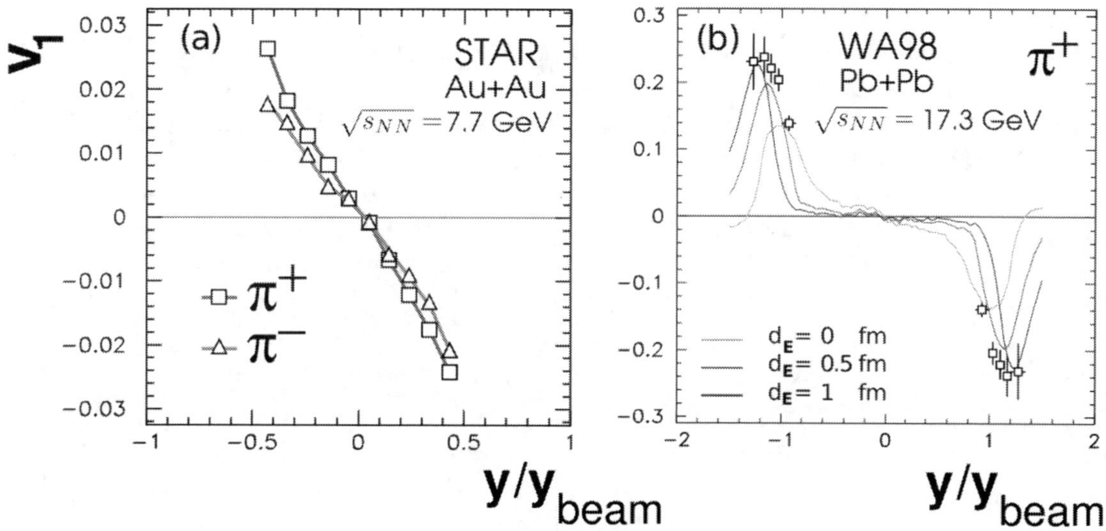

Figure 3. Directed flow v_1 of π^+, π^- as a function of the ratio of pion to beam rapidity in the c.m.s., y/y_{beam}, for (a) Au+Au collisions at $\sqrt{s_{NN}}$ = 7.7 GeV (STAR Collaboration [5]), plot from [6] (see also [4]), and (b) Pb+Pb collisions at $\sqrt{s_{NN}}$ = 17.3 GeV (WA98 Collaboration [7]), plot from [2].

that the space-time evolution of pion production at $y = y_{beam}$, i.e. non-zero d_E, can be understood as resulting from the space-time evolution of resonance decay. Further analysis of new experimental data on EM effects, on flow as well as π^+/π^-ratios, augmented by MC simulations, would improve our phenomenological knowledge and reduce the estimated uncertainties of the dependence of d_E on pion rapidity y/y_{beam}, and possibly p_T. What is needed is precise data on the dependence of EM effects on such variables as collision geometry, reaction type, centrality, energy, particle type, rapidity and p_T.

To this end, we are actively seeking to further our analysis and MC simulation on EM spectator effects as members of the NA61/SHINE Collaboration [11], a heavy ion fixed target collision experiment and the successor to the NA49 experiment. The NA61/SHINE scan program includes collisions of heavy and intermediate size nuclei in the $\sqrt{s_{NN}}$ = 6.3 – 16.8 GeV energy range; a systematic analysis of EM effects on non-central collision data would cast new light on the behavior of d_E with pion rapidity and transverse momentum, and hence on the space-time evolution of heavy ion collisions.

Acknowledgments: This work was supported by the National Science Centre, Poland (grant no. 2014/14/E/ST2/00018).

References

[1] A. Rybicki, A. Szczurek, *Phys. Rev. C* **75**, 054903 (2007)
[2] A. Rybicki and A. Szczurek, *Phys. Rev. C* **87**, 054909 (2013)
[3] A. Rybicki, A. Szczurek, M. Klusek-Gawenda, N. Davis, V. Ozvenchuk and M. Kielbowicz, to appear in EPJ Web of Conf.
[4] A. Rybicki, A. Szczurek and M. Kłusek-Gawenda, *Acta Phys. Polon. B* **46**, no. 3, 737 (2015), and references therein
[5] STAR Collab., L. Adamczyk *et al.*, *Phys. Rev. Lett.* **112**, 162301 (2014)

[6] A. Rybicki, A. Szczurek and M. Klusek-Gawenda, EPJ Web Conf. **81** (2014) 05024. doi:10.1051/epjconf/20148105024
[7] WA98 Collab., H. Schlagheck, *Nucl. Phys. A* **663**, 725 (2000)
[8] A. Rybicki, *Acta Phys. Polon.* **B42**, 867 (2011).
[9] A. Rybicki, A. Szczurek, M. Kiełbowicz, N. Davis and V. Ozvenchuk, *Acta Phys. Polon. Supp.* **9**, 303 (2016), and references therein
[10] ALICE Collab., K. Aamodt *et al.*, *Phys. Lett. B* **696**, 328 (2011)
[11] NA61 Collab., N. Abgrall *et al.*, CERN-SPSC-2015-038; SPSC-P-330-ADD-8

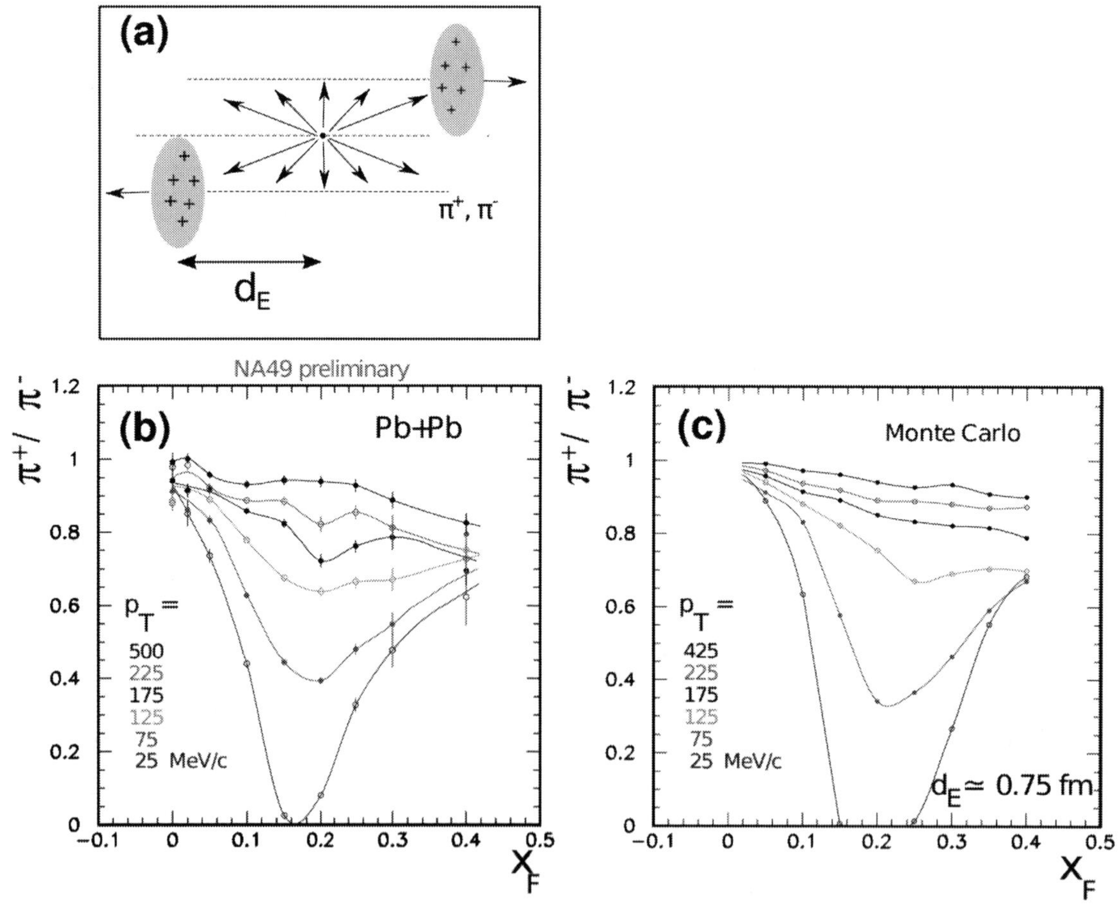

Figure 4. (a) Schematic representation of one-point emission model for spectator EM effects. (b) ratio of π^+ over π^- distributions as a function of x_F for various p_T values, for Pb+Pb collisions recorded in the NA49 experiment at $\sqrt{s_{NN}}$ = 17.3 GeV. (c) ratio of π^+ over π^- distributions for pions simulated by one-point emission model MC, for $d_E \approx 0.75$ fm. Panels (b),(c) originally from [8].

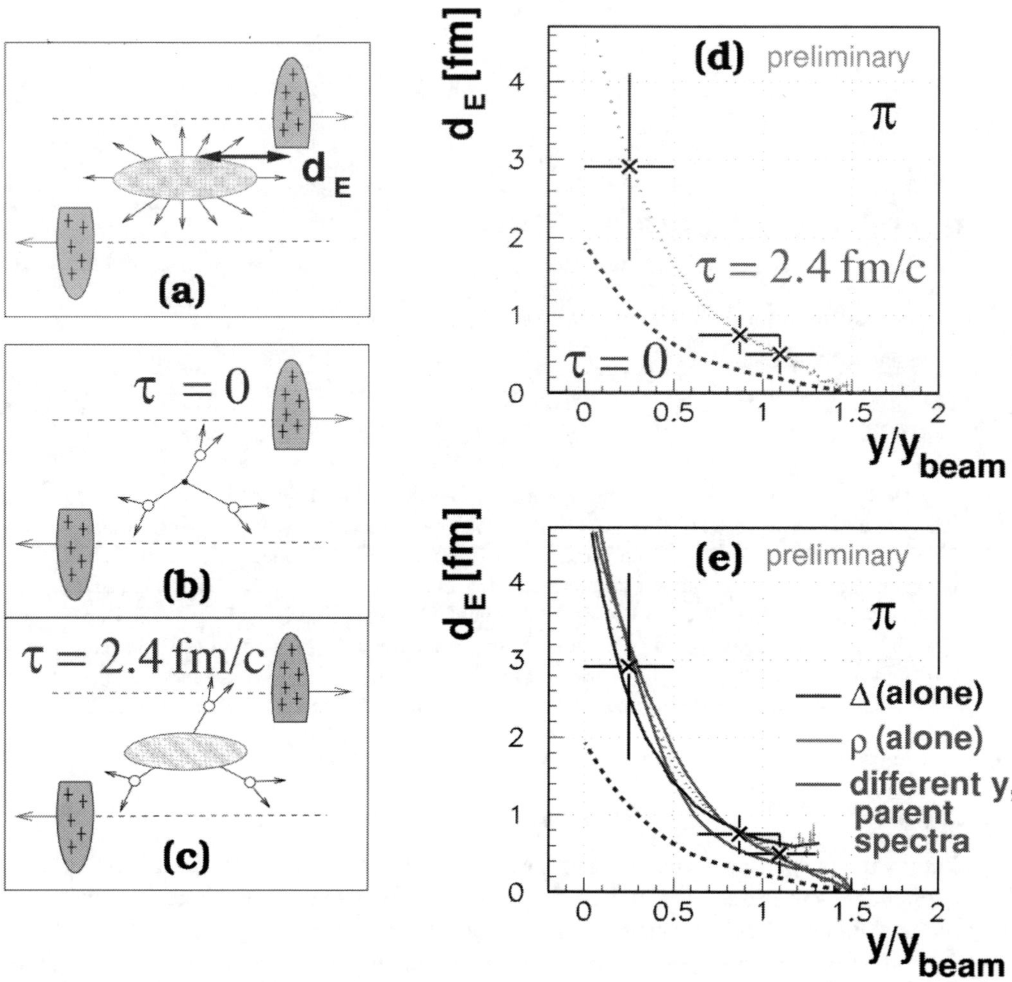

Figure 5. Information obtained on the basis of EM effects described in section 2: (a) schematic definition of d_E, (b) simulation of resonances produced instantly at the moment of the collision ($\tau = 0$) and (c) produced from an intermediate system characterized by a given proper lifetime τ, (d) dependence of d_E on pion rapidity, deduced from EM effects (data points), compared to the results of our Monte Carlo simulation illustrated in panels (b) and (c), respectively. (e) uncertainty of the present analysis induced by our incomplete knowledge of resonance production, estimated by assuming different characteristics for the input mix of parent Δ, ρ resonances. All the panels originate from [9].

EPJ Web of Conferences **164**, 07013 (2017) DOI: 10.1051/epjconf/201716407013
ICNFP 2016

Examination of the heavy-ion collisions using EPOS model in the frame of BES program at RHIC

Maria Stefaniak[1,a] and Hanna Zbroszczyk[1]

[1] *Warsaw University of Technology, Poland*

Abstract. EPOS generator allows one to simulate various types of collisions (nuclei-nuclei or even more complex ion-ion) with different initial parameters. It is based on Gribov-Regge theory. So far the EPOS model has been used to describe higher collision energies obtained with RHIC[b] or LHC[c] data. However, there is another program under investigation: Beam Energy Scan conducted at Brookhaven National Laboratory. The beams of gold ions are collided at energies such as $\sqrt{s_{NN}} - 7.7, 11.5, 19.6, 27, 39,$ and 62.4 GeV in order to collect and analyze data needed to examine QGP Phase Diagram. The EPOS model could become a useful tool in such studies. Simulated with EPOS data will be verified using elliptic flow analyses and two-particles femtoscopic correlations, which allow one to measure the size of sources determined by newly created particle.

1 Introduction

One way to study the properties of the nuclear matter are collisions heavy-ions in major experiments. Scientists explain such processes in detailed theories. One of such descriptions is included in the phenomenological model EPOS. It considers the pardon-based Gribov-Regge theory [1]. EPOS gives possibility to study different observables what helps to understand even complex reactions with heavy-ions. So far the EPOS model has been used to describe higher collision energies obtained with RHIC or LHC. However, there is another interesting program currently under investigation at RHIC: Beam Energy Scan (BES), conducted at Brookhaven National Laboratory. Main goals of this project are to examine the Phase Diagram, to study the characteristics of the first-order phase transition between Hadron Gas and Quark-Gluon Plasma phases of nuclear matter, to search signatures of turn-off the QGP. The search of Critical Point between first-order phase transition and transition of "cross-over" is another interesting topic. RHIC collides beams of Au nuclei at selected energies as: $\sqrt{s_{NN}} - 7.7, 11.5, 19.6, 27, 39,$ and 62.4 GeV. The variety of initial conditions provides covering as widest part of Phase Diagram of nuclear matter as possible.

2 EPOS

Energy conserving quantum mechanical multiple scattering approach, based on **P**artons (parton ladders), **O**ff-shell remnants, and **S**plitting of parton ladders [1].

[a] e-mail: mzstefaniak@gmail.com
[b] Relativistic Heavy-Ion Collider
[c] Large Hadron Collider

DOI: 10.1051/epjconf/201716407013

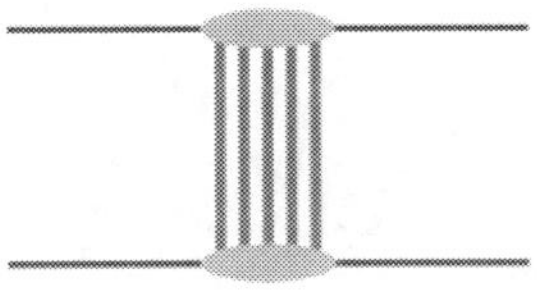

Figure 1: Hadron - hadron multiple scattering. Navy blue lines are hadrons and each of pink line illustrate Pomeron.[1]

EPOS is a generator based on Parton-Based Gribov-Regge Theory which is the mixture of two various theories. First one, Gribov-Regge theory uses effective field principies and it focuses on soft aspects of particle collision. The interactions are described with theoretical objects called *Pomerons*, which are crucial in calculation the cross-section in multiple scattering (Figure 1). The eikonalized parton model introduces quarks and gluons, what results with a more precise description of the interactions such as a gluon exchange. The QCD[1] and QED[2] are included in the description. [1] [2]

The combination of Gribov - Regge theory and parton model ensures that the energy is shared properly between both the particle production and the cross section. With the parton model there are hard processes added as well. The production of partons is represented by Lund model strings [3]. There are introduced objects called *parton ladders*. The open ones contribute to inelastic scattering which supports particles production. Meanwhile the closed ladders illustrates elastic scattering (Figure 2). In third version of EPOS there is applied the procedure consists hydrodynamic effects described in [4].

3 Beam Energy Scan program

Beam Energy Scan is a project run on the RHIC complex. The main goal, besides investigation of properties of QGP, is to understand the QCD phase diagram structure (Fig. [3]). Only a few pieces of information have been discovered yet. The QCD phase diagram is being mapped out in order to look for the evidences of:

- critical point fluctuations
- signals of 1st order phase transition
- turn-off the signatures of QGP [5]

The gold ions are collided with such energies as: $\sqrt{s_{NN}} = 7.7, 11.5, 19.6, 27, 39, 62.4 GeV$.

One of the tools used with a view to studying occurring transitions of matter is femtoscopic correlations.

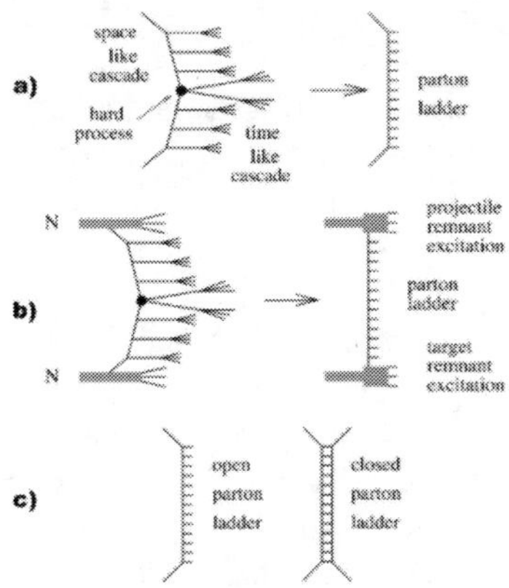

Figure 2: a) Elementary parton - parton scattering and representation as parton ladder. b) Full presentation of scattering, completed with remnants. c) Open and closed parton ladder [2]

4 Femtoscopy correlations

4.1 Two-particle correlations

The method inspired by HBT[3] interferometry [7] focuses on the smallest sizes in nature, about 10^{-15}m (femtometre), the sizes of nuclei. Studies of two-particle correlations allows one to examine space-time extension of the emitting source created during the collisions of nuclei. The size of area where the given kind of particles are produced is called *homogeneity length* [8].

[1]Quantum Chromodynamics
[2]Quantum Electrodynamic
[3]R. Hanbury-Brown and R. Q. Twiss

DOI: 10.1051/epjconf/201716407013

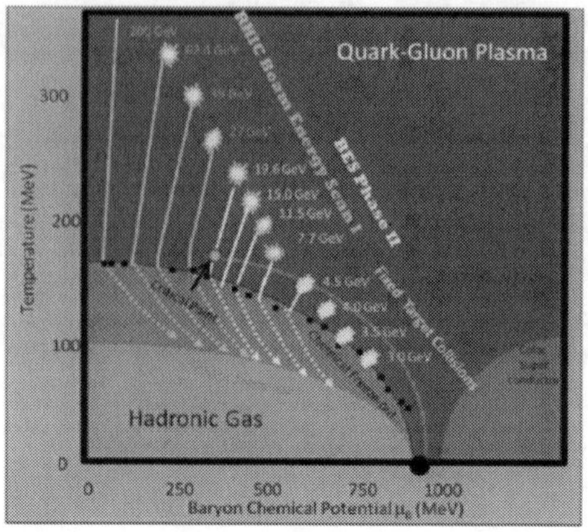

Figure 3: *Studies of phase diagram peformed in a frame of BES program [6]*

4.1.1 Correlation function

A two-particle correlation function defines ratio between probability of detecting two particles with given momenta and position simultaneously to product of detecting them separately [8].

$$C(p_1, p2) = \frac{P_2(p_1, p_2)}{P_1(p_1)P_1(p_2)} \quad (1)$$

The source is often treated as a three-dimensional sphere and described with parametrization proposed by G. Bertsch and S. Pratt. The correlation function is represented in the -out -side -long (o-s-l) coordinate system [9]. Kopylov and Podgoretsky proposed the three-vector $\vec{q} = \vec{p_1} - \vec{p_2}$ so that the correlation function is represented in the following form:

$$C(q_{out}, q_{side}, q_{long}, \lambda) = 1 + \lambda exp(-q_{out}^2 r_{out}^2 - q_{side}^2 r_{side}^2 - q_{long}^2 r_{long}^2) \quad (2)$$

where $r_{out,side,long}$ are lengths of the source in *out*, *side* and *long* directions (three-dimensional femtoscopy radii) and λ is a chaosity parameter [Fig. 4].
The out-direction width of the correlation function is inversely proportional to the lifetime of source (duration of particle emission). The side-direction measurements are inverse to the transverse size of source. The R_{side} and R_{out} are only proportional to the system average size or lifetime. The measurements of radii are affected by hydrodynamical flow or it depends on the chosen reference frame. In order to make such effects more irrelevant or even cancel them out, one can study the ratio R_{out}/R_{side}.The system size is approximately constant, while the lifetime of source varies depending on presense of a phase transition. Hence, the R_{out}/R_{side} is considered to be optimal in studies of the

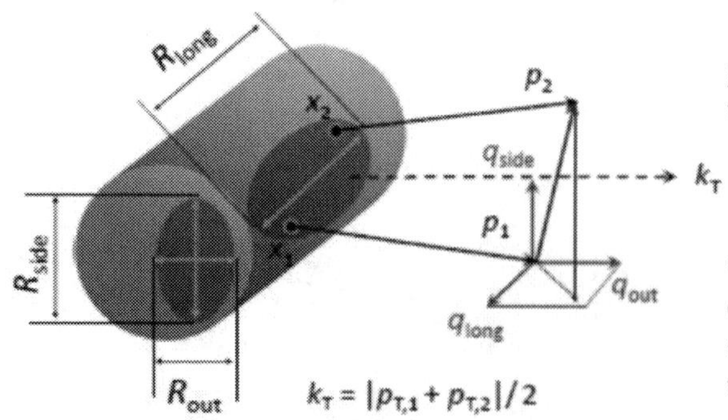

Figure 4: Definition the three-dimensional femtoscopy radii.

Dealing with experimental data one has to parametrize the two-particle correlation function with equation:

$$C_2(\vec{q}) = \frac{A(\vec{q})}{B(\vec{q})} \quad (3)$$

where A (signal) is a momentum distribution of particles built in the same event, while B (background) is a reference momentum distribution obtained from different events.

5 Azimuthal anisotropy

The azimuthal anisotropy characterizes expansion of highly-compressed matter created in heavy-ion collisions [Fig. 5]. It is one of the most significant tool used in studying hot and dense matter, described more detailed in [11].
One of the standard methods used in order to calculate flow is *event plane method*. The crucial part is to estimate the reaction plane from the measured particles. It is defined as a *event plane* in range $\Psi_n(-\pi/n \geq \Psi_n < \pi/n)$. For elliptic flow $n = 2$ and in order to calculate *event plane* one can use equation:

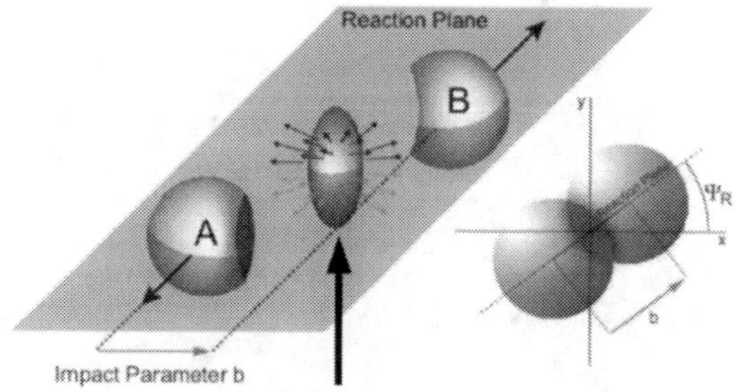

Figure 5: *Nucleus after collision.*

$$\Phi_2 = \tan^{-1}\left(\frac{\sum_i w_i sin(2\phi_i)}{\sum_i w_i cos(2\phi_i)}\right)/2 \tag{4}$$

where: i - selected particle, ϕ_i - azimuthal angle of particles, w_i -weight of i particle [12]. In proposed $\eta - sub$ method, from all measured particles there are selected two groups with "forward" and "backward" pseudorapidity with a gap between them. This method is introduced with a view to reducing the effect of "non flow" (e.g. decay of resonances to several charged daughter particles, jets or HBT correlations). [13]
To express the observed v_2 of particles with respect to already investigated event plane one uses:

$$v_2^{obs}(p_T, y) = \langle \cos[2(\phi_i - \Phi_2)] \rangle \tag{5}$$

This is average over all particles in all events. As a consequence of final multiplicity limitation in the investigation of the angle of the reaction plane, the correction of v_2 with *event plane* resolution have to be done.

$$R_2 = \sqrt{\langle \cos[2(\Phi_2^A - \Phi_2^B)] \rangle} \tag{6}$$

where: Φ_n^A - *event plane* calculated only using "forward-pseudorapidity" particles while Φ_n^B - with "backward-pseudorapidity" ones. The final result of v_2 calculation is given by:

$$v_2 = \frac{v_2^{obs}}{R_2} \tag{7}$$

6 Results

There were three different kinds of analyze performed: one-particle analyze - p_T spectra, azimuthal anisotropy - v_2, two-particles analyze - femtoscopy correlations, in order to compare simulated and experimental data.

6.1 Transverse momentum spectra

Transverse momentum distribution is one of the elementary tools in High-Energy analyze. In this examination it was used to make basic comparison between simulated and experimental data published in [14]. Such measurements were done for Au + Au collisions at two centrality ranges: $0 - 5\%$ (central) and $60\% - 80\%$ (peripheral events). Particles with rapidity $|y| < 0.5$ were taken into consideration. Below there are placed the obtained p_T distributions for $\frac{h_+ + h_-}{2}$, the sum of positive-charged hadrons (p, π^+, K^+) and negative once $(\bar{p}, \pi^-, K^-)$.

 DOI: 10.1051/epjconf/201716407013

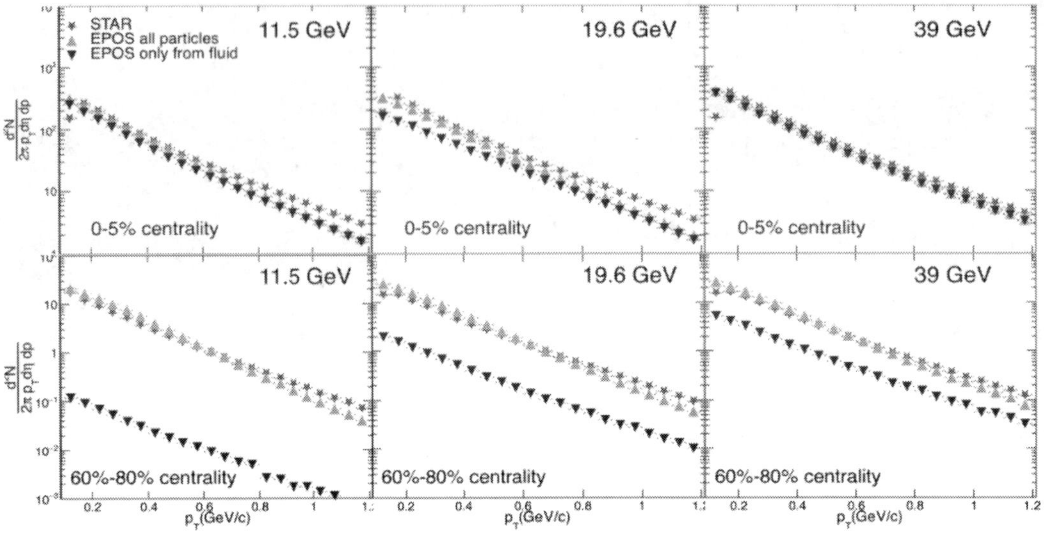

Figure 6: p_T spectra for energies $\sqrt{s_{NN}} = 11.5, 19.6, 39 GeV$ at $0-5\%$ and $60\%-80\%$ centrality ranges.

There is high agreement between experimental and theoretical data registered for the most central collisions. However, for lower energies for higher values of p_T the slopes of plotted data are slightly different. In case of peripheral collisions EPOS data based on fluid equations values are considerably lower.

6.2 Correlation function

In this research there was performed positively charged pions femtoscopic analysis in Au+Au collisions. In order to study beam energy in femtoscopy the were calculated and parametrized three-dimensional correlation functions. The π^+ pairs were taken into consideration at $\sqrt{s_{NN}}$ = 11.5GeV 19.6 GeV at different centralities or k_T ranges. With a view to inferring size of emitting source the fitting procedure was used (equation 2). Obtained radii and λ parameter from calculation of simulated data were compared with published experimental results and presented in different combinations.

With the decrease of energy of collision $\sqrt{s_{NN}}$ the differences becomes more relevant. Calculated values of R_{out} are in the range of expectations. The duration of particle emission is similar for compared systems. In case of measurements of R_{side} the experimental results are higher then the simulated once, especially for lower k_T bins. The sizes of sources created in STAR are bigger

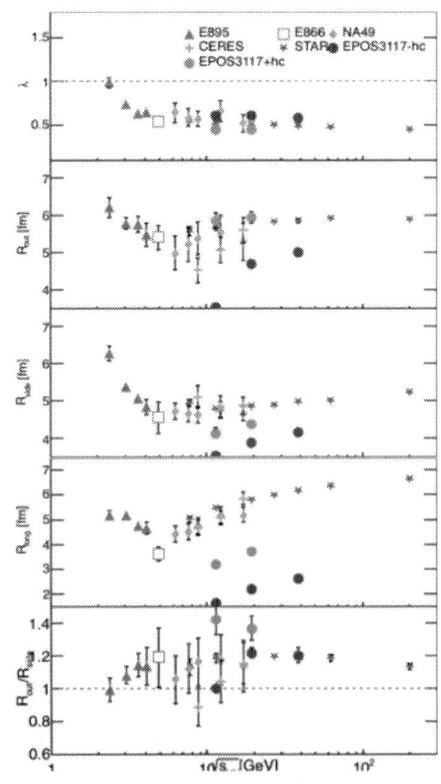

Figure 7: λ and radii dependence on energy $\sqrt{s_{NN}}$ for different experiments and EPOS data ($k_T \approx 0.225 GeV/c$). [15]

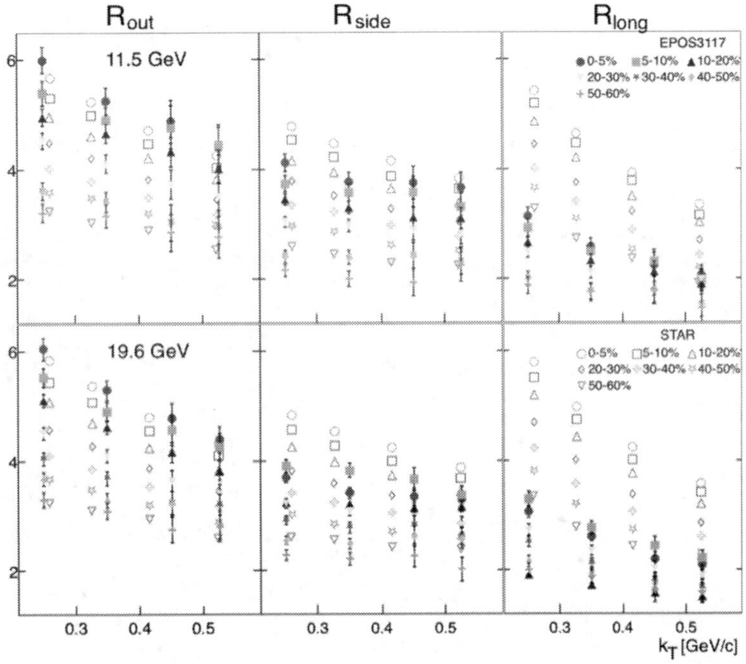

Figure 8: Radii calculated for various centralities and k_T bins[15]

then simulated in EPOS. However, the most visible discrepancy is in studies of R_{long}. In order to explain these differences the elliptic flow analyze was performed.

It is possible in EPOS generator to turn off some physics processes occur in heavy-ion collisions. In order to study the impact of hadron cascades on size of emitting source, they were turn off and turn on in simulation. The results are presented in Figure 7. Hadron cascades have relevant influence on values of calculated radii and λ parameter. When they are present the size of the source is bigger and more similar to experimental studies. The λ parameter and R_{out} are in the range of expectations, while rest of parameters are considerably lower then in experimental results. Too high value of ratio R_{out}/R_{side} means that the liftime of the source is shorter for simulated collisions than those obtained in experiments.

6.3 Elliptic flow

The studies of the elliptic flow were preformed in order to examine the description of matter expansion included in EPOS model. The data simulated with the EPOS were compared with STAR results published in [13]. The studies of beam energy were performed at mid-rapidity region for p, $\bar{p}$, π^+, π^-, K^+, K^- at $\sqrt{s_{NN}} = 11.5, 19.6, 39GeV$ in minimum bias $(0-80\%)$ using η-*method*. Particles with $|\eta| < 1$, momentum between 0.15 and 5 GeV/c were taken into consideration. The calculated values for pions and kaons coming from simulation are slightly higher than experimental data. The same tendency is visible for antiprotons. The elliptic flow of protons in all p_T bins are comparable for both data sets.

 DOI: 10.1051/epjconf/201716407013

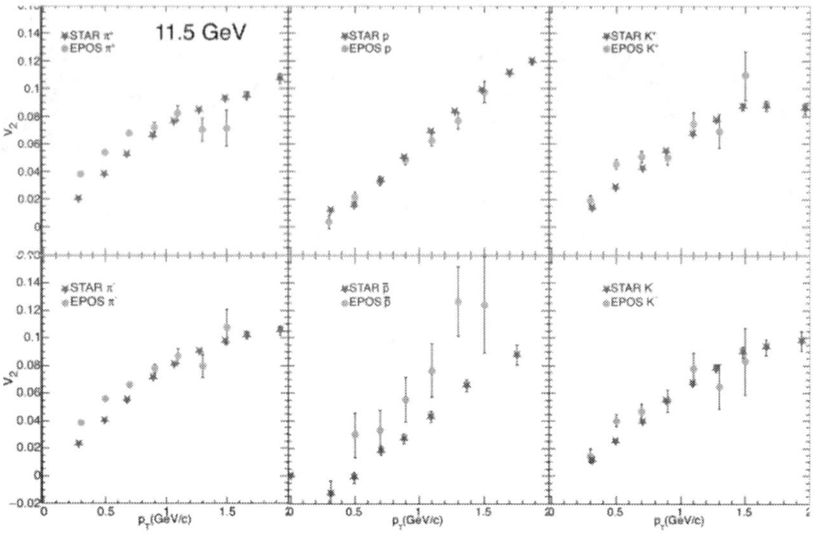

Figure 9: Elliptic flow dependence on p_T [13].

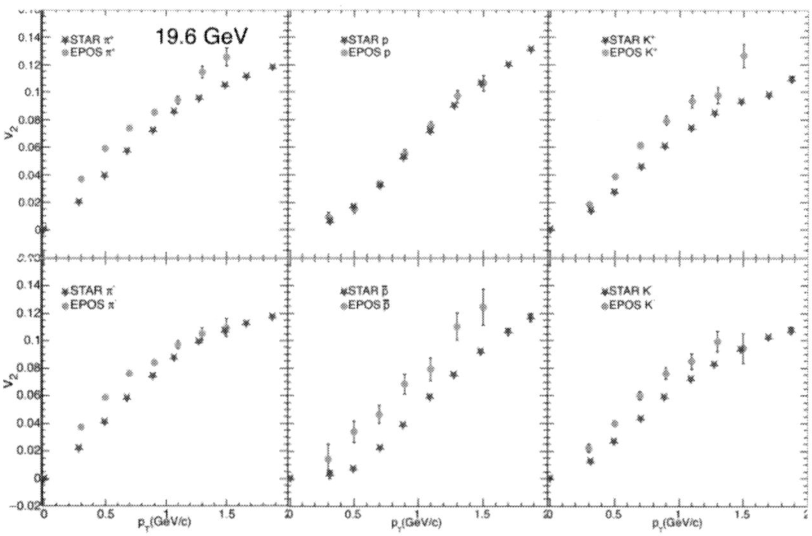

Figure 10: Elliptic flow dependence on p_T [13].

7 Summary

The EPOS model predictions for Au+Au collisions at BES energy spectrum are verified with results obtained with experimental data. The p_T spectra, correlation functions for identical pions and elliptic flow at 11.5 and 19.6 GeV for both simulated and STAR data were compared and presented. Observed discrepancies can be caused by too small impact of fluid created in the collision. The transparency of the collision should be investigated as well. The work is still in progress.

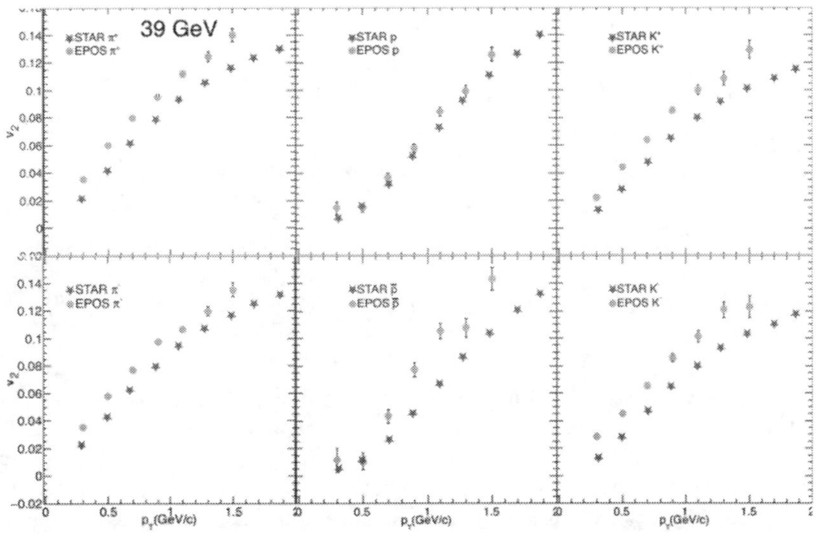

Figure 11: Elliptic flow dependence on p_T [13].

References

[1] H.J. Drescher, M. Hladik, S. Ostapchenko, T. Pierog, K. Werner, Phys. Rept. **350**, 93 (2001), hep-ph/0007198

[2] K. Werner, F.M. Liu, T. Pierog, Phys. Rev. **C74**, 044902 (2006), hep-ph/0506232

[3] M. Anderson, B. Pennington (1985)

[4] K. Werner, I. Karpenko, M. Bleicher, T. Pierog, S. Porteboeuf-Houssais, Phys. Rev. **C85**, 064907 (2012), 1203.5704

[5] G. Odyniec, *"results and future of beam energy scan program at rhic"*

[6] D. Cebra, *"exploring the qcd phase diagram: Rhic beam energy scan ii"*, http://www.c-ad.bnl.gov/ardd/LEeC/Presentations/Cebra_LEReC_BESII_ver2.pdf, accessed: 2016-08-16

[7] U. Gordon Baym, University of Illinois, *Hanbury brown – twiss intensity interferometry: from stars to nuclei to atoms and electrons*, http://ithes.science-server.com/content/images/colloquium/pdf/Baym.pdf, accessed: 2016-10-18

[8] M.A. Lisa, S. Pratt, R. Soltz, U. Wiedemann, Ann. Rev. Nucl. Part. Sci. **55**, 357 (2005), nucl-ex/0505014

[9] H. Tilsner, Ph.D. thesis, Heidelberg U. (2002), http://www.ub.uni-heidelberg.de/archiv/2474

[10] D.H. Rischke, M. Gyulassy (1996), 9606039v1

[11] H. Masui, A. Schmah, A.M. Poskanzer, Nucl. Instrum. Meth. **A833**, 181 (2016), 1212.3650

[12] S.A. Voloshin, A.M. Poskanzer, R. Snellings (2008), 0809.2949

[13] L. Adamczyk et al. (STAR), Phys. Rev. **C88**, 014902 (2013), 1301.2348

[14] S.P. Horvat (STAR), J. Phys. Conf. Ser. **446**, 012017 (2013), 1303.7260

[15] L. Adamczyk et al. (STAR), Phys. Rev. **C92**, 014904 (2015), 1403.4972

EPJ Web of Conferences **164**, 07014 (2017) DOI: 10.1051/epjconf/201716407014
ICNFP 2016

Forward Physics with the CMS Experiment at LHC

D. Sunar Cerci[1,a,b]

[1] *Adiyaman University, Faculty of Arts and Sciences, Department of Physics, 02040 Adiyaman, Turkey.*

Abstract. Forward physics measurements with the Compact Muon Solenoid (CMS) experiment, one of the two large multi-purpose experiments at the Large Hadron Collider (LHC) at CERN, cover a wide range of physics subjects. The forward calorimeters of CMS, HF and CASTOR, are used to collect data up to a pseudo-rapidity of 6.6. These detectors provide sensitivity to a large part of the total inelastic cross section, including diffractive events that produce particles only at forward rapidity, with the exception of very low mass diffraction. The results obtained with a centre-of-mass energy of 13 TeV are presented. The measurements are compared to model predictions and provide valuable input for tuning of Monte Carlo models used to describe high-energy hadronic interactions.

1 Introduction

The LHC at CERN has started to deliver proton-proton collisions at the highest energy of $\sqrt{s} = 13$ TeV ever reached by any other accelerator since 2015. Therefore, the CMS detector [1] at the LHC is a suitable machine which provides the opportunity to study forward physics. The forward physics programme at CMS mainly focuses on the physics that can be reached with the forward detectors HF and CASTOR at $3.0 < |\eta| < 5.2$ and $-6.6 < \eta < -5.2$, respectively. Measurements in the forward region can be used to improve the knowledge of the content of the proton at small values of the momentum fraction x carried by the partons and to provide inputs for parton density functions (PDFs). The hard scattering events are produced by the strong interactions of partons in a region where non-perturbative QCD and multiple partonic interactions (MPI) play an important role. Measurements performed in the forward region can shed light on MPI and the underlying events (UE) as well as provide useful information on the tuning of Monte Carlo (MC) event generators available.

2 Measurements

In the following the measurements performed with the data taken at a centre-of-mass energy $\sqrt{s} = 13$ TeV in 2015 within the forward physics program of CMS are presented.

[a] e-mail: deniz.sunar.cerci@cern.ch
[b] On behalf of the CMS Collaboration

2.1 Inelastic pp cross section

One of the fundamental observables in high energy particle physics is the hadronic total cross section which has been measured in many experiments, covering several orders of magnitude in centre-of-mass energy $\sqrt{s}$. The hadronic cross section can be divided into two parts as elastic and inelastic interactions. In inelastic interactions, which can be described by non-perturbative and phenomenological models, either one or both protons dissociate as a result of colorless (diffractive) or colored (non-diffractive) exchange. Precise measurements of the hadronic cross sections provide valuable input for the phenomenological models as well as for the tuning of MC event generators.

The inelastic proton-proton cross section measurement at a centre-of-mass energy of $\sqrt{s}$ = 13 TeV is performed with the CMS detector, particularly with information from forward calorimeters, HF and CASTOR, at pseudorapidities of $3.0 < \eta < 5.2$ and $-6.6 < \eta < -3.0$ [2]. Fig. 1 shows the comparison of corrected cross sections in various phase space regions to different models (left) and the same cross sections normalized to the cross section obtained for $\xi > 10^{-6}$ (right). A visible cross section of $\sigma(\xi_X > 10^{-7}\text{or } \xi_Y > 10^{-6}) = 66.9 \pm 0.4$ (exp.) $\pm$ 2.0 (lum.) mb is measured for $M_X > 4.1$ GeV and $M_Y > 13$ GeV where M_X and M_Y are the invariant masses of the two hadronic systems separated by the largest rapidity gap in the event. The visible cross section is extrapolated to the full inelastic phase space domain and is obtained as 71.3 ± 0.5 (exp.) $\pm$ 2.1 (lum.) $\pm$ 2.7 (ext.) mb. The increase from $\xi > 10^{-6}$ to $\xi_X > 10^{-7}$ or $\xi_Y > 10^{-6}$ observed in data is reasonably described by most of the models used in the analysis. However, the hadronic scattering models predict higher cross section than the measured one.

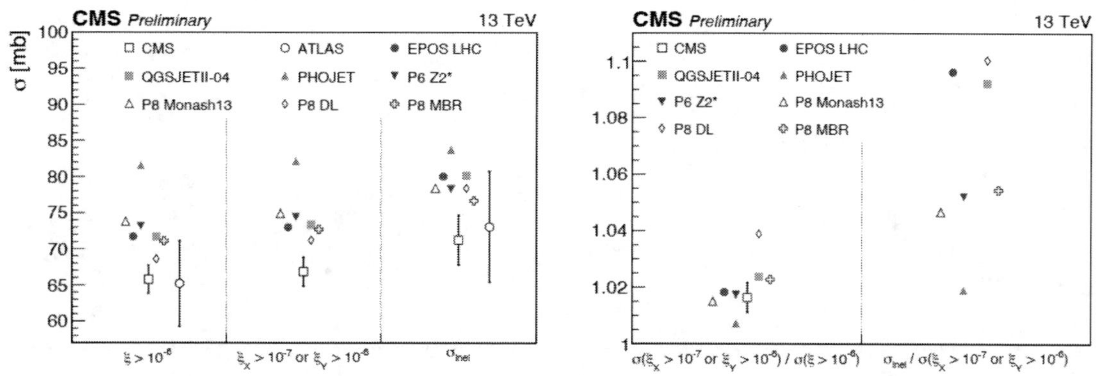

Figure 1. Left: Comparison of corrected cross sections in various phase space regions to different models. Right: the same cross sections scaled to the cross section obtained for $\xi > 10^{-6}$ [2].

2.2 Forward energy flow and transverse energy density

The forward energy flow measurement is sensitive to the Underlying Event activity (UE) and parton shower dynamics. Therefore the measurement provides an opportunity to study multiple parton interactions (MPI), which are at the origin of UE activity. On the other hand, such a measurement can provide useful input for the tuning of existing hadronic interaction models.

The energy flow and transverse energy measurement [4] is performed with low luminosity pp data at $\sqrt{s}$ = 13 TeV collected in 2015 in the pseudorapidity range $3.15 < |\eta| < 6.6$. The forward energy flow is defined as the total energy measured by the HF and CASTOR calorimeters by using the sum of all energy deposits of calorimeter towers above noise threshold (5 GeV). The forward energy flow measurements presented here are obtained for two different event selections: soft inclusive inelastic events (INEL) allowing significant fraction of diffractive events and non-single diffractive enhanced events (NSD-enhanced) where diffractive contributions are effectively suppressed.

In order to have a comparison with model predictions the energy flow measured in calorimeters is corrected to the particle level by using GEANT4 [3] based MC detector simulation. The correction factors are defined as ratios of MC predictions at particle level and detector level where the detector level MC is defined as a fully simulated and reconstructed sample which goes through the same analysis chain as data. The average value of these correction factors is taken for the final correction. For INEL events, the event selection at the particle level is defined using the kinematic variable $\xi = M^2/s > 10^{-6}$, which is the scaled mass of the largest (diffractively dissociated) system separated from the rest of the observed final state by the largest rapidity gap. NSD-enhanced event selection at particle level is done with a requirement of at least one particle except muon or neutrino on both sides with respect to the interaction point. The comparison of energy flow measurement as a function of η for soft-inclusive-inelastic events to various MC models is shown in Fig. 2 (left). The ratio of data to MC models is shown in Fig. 2 (right). The gray band represents the total systematic uncertainties. Similar plots are also shown for non-single diffractive enhanced events in Fig. 3. In Fig. 4, the measured transverse energy density as a function of shifted pseudorapidity variable, $\eta' = \eta - y_{beam}$, is compared to published CMS data at lower centre-of-mass energies and MC models. The results at different $\sqrt{s}$ are consistent in terms of $\eta - y_{beam}$, and the trends observed both in data and MC models support the limiting fragmentation hypothesis.

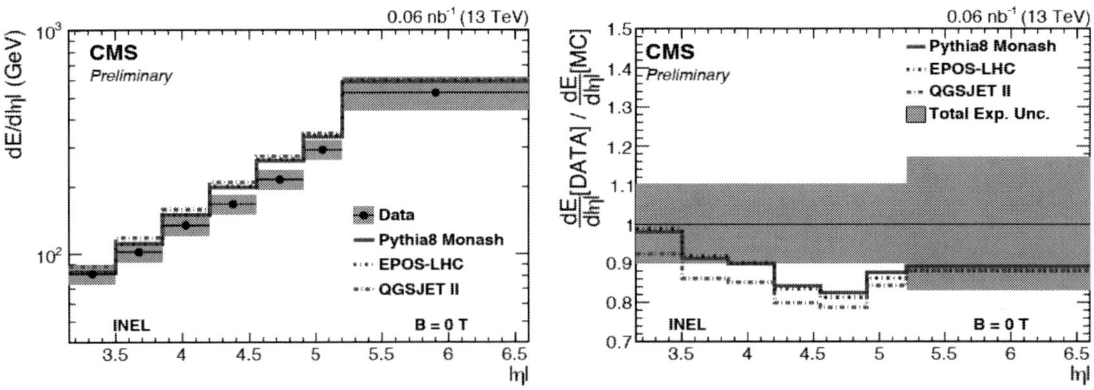

Figure 2. Left: Energy flow for soft-inclusive-inelastic events compared to various MC models: PYTHIA8, EPOS-LHC and QGSJET II. Right: Ratio of data to MC models [4].

2.3 Pseudorapidity distributions of charged-particles

A measurement of charged-particle pseudorapidity distributions, $dN_{ch}/d\eta$, produced in pp collisions at $\sqrt{s} = 13$ TeV is performed by CMS [6]. The charged-particles with a transverse momentum $p_T > 0.5$ GeV in the pseudorapidity range $|\eta| < 2.4$ are required. Four different event classes corresponding to different topologies of the final state particles and the activity measured in the forward region of the detector are selected. Pseudorapidity distributions of charged-particles for the inelastic-enhanced, single diffractive enhanced and non-single diffractive enhanced event samples are shown

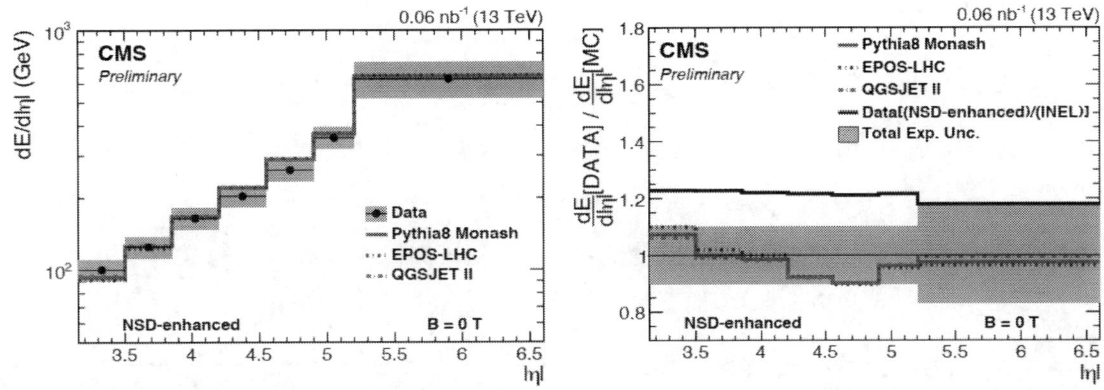

Figure 3. Left: Energy flow for non-single diffractive enhanced events compared to various MC models. Right: Ratio of data to MC models [4].

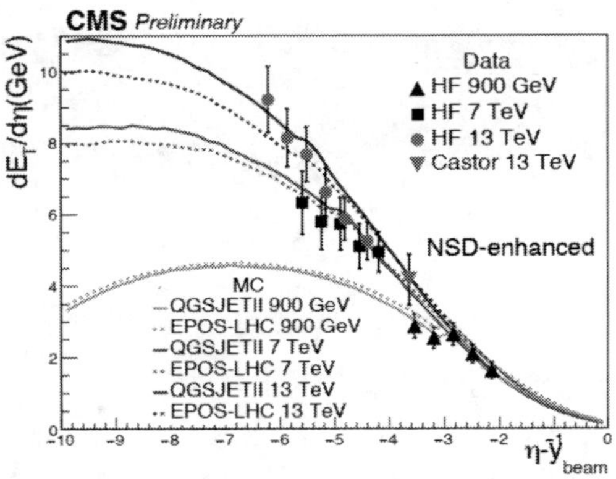

Figure 4. Comparison of transverse energy density measured by CMS at different centre-of-mass energies [5] and MC models [4].

in Fig. 5. The shaded bands around the data points represent the total systematic uncertainties. The measurements are compared to various MC event generators predictions and a reasonable agreement is observed between data and the MC predictions.

2.4 Energy distribution in the very forward direction

The energy spectrum, $\mathrm{d}N/\mathrm{d}E$, in the very forward direction $-6.6 < \eta < -5.2$ is measured with the CMS detector in pp collisions at $\sqrt{s} = 13$ TeV [7]. The hadronic and electromagnetic energy spectra are also presented. The data corrected to the particle level is compared to models. A sensitivity is found on data to the production of neutral and charged mesons, which is an important input for the modeling of cosmic ray induced extensive air showers at ultra-high energies. The comparison of normalized total energy spectra in the pseudorapidity range $-6.6 < \eta < -5.2$ to different MC and cosmic ray models is shown in Fig. 6. Despite the large systematic uncertainties where the energy

DOI: 10.1051/epjconf/201716407014

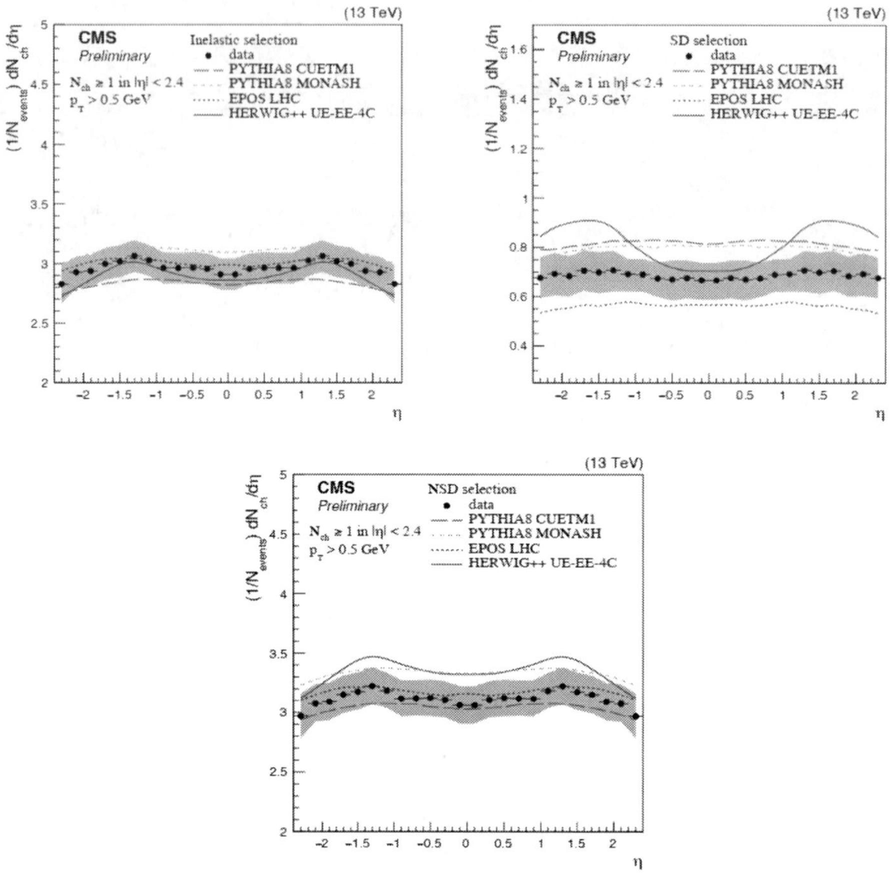

Figure 5. Pseudorapidity distributions of charged-particles for the inelastic-enhanced (top-left), single diffractive enhanced (top-right) and non-single diffractive enhanced (bottom) event samples [6].

scale is the dominant one, all models show significant differences with the data within the uncertainty band. The best agreement is observed between data and SIBYLL 2.1.

2.5 Very forward inclusive jet cross section

The very forward inclusive jet cross section is sensitive to radiation of partons, underlying partonic QCD processes and to the parton density functions (PDF) of the incoming hadrons. As shown by the previous measurements, jet cross sections can be described by perturbative QCD over several orders of magnitude in Q^2 and x, where Q is the scale of scattering and x is the longitudinal momentum fraction of the parton.

The very forward inclusive jet cross section measurement presented here is based on the very forward detector of CMS, CASTOR, and pp collisions recorded in 2015 at $\sqrt{s} = 13$ TeV. Jets are reconstructed using the anti-k_t algorithm [9] with a radius parameter of $R = \sqrt{\Delta\phi^2 + \Delta\eta^2} = 0.5$ and are required to have their axis between $-6.6 < \eta < -5.2$. The transverse momentum of the jets must be in the range $3 < p_T < 13$ GeV and various jet quality criteria are applied to remove fake jets. The

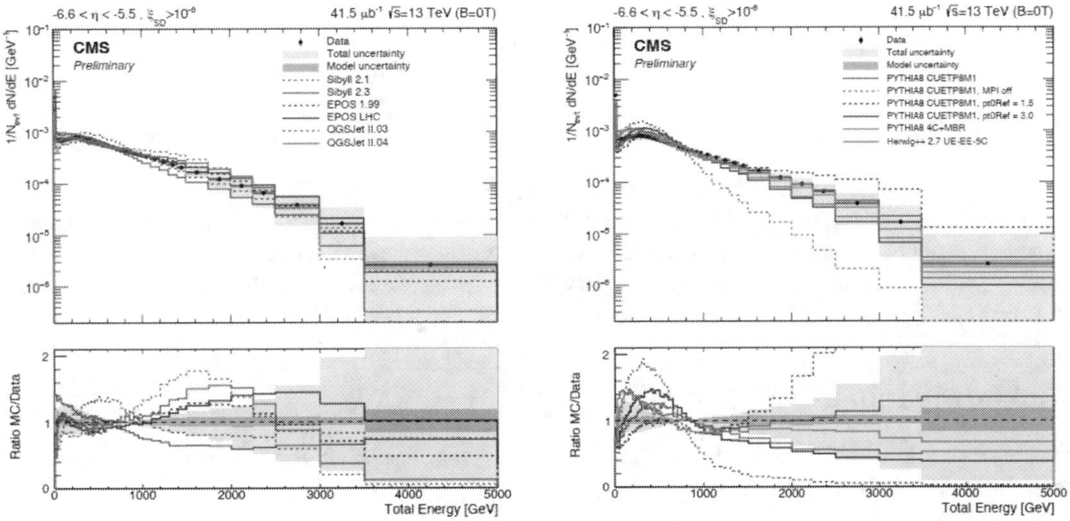

Figure 6. Comparison of normalized total energy spectrum to various MC models (left) and to PYTHIA8 tunes (right) [7].

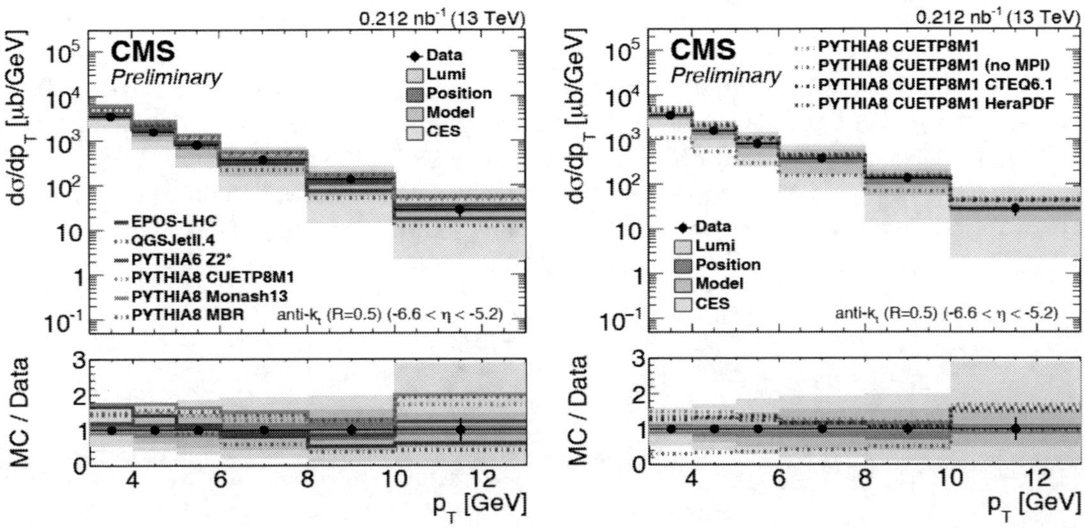

Figure 7. The measured very forward inclusive jet cross-section at $\sqrt{s}$ = 13 TeV compared to various MC models (left) and to PYTHIA8 tune CUETP8M1 with different PDFs (right) [8].

corrected cross section measurement of the very forward inclusive jet production is normalized by the luminosity and is shown in Fig. 7. The yellow band represents the total systematic uncertainty. The dominating systematic uncertainties on the very forward inclusive jet cross section measurement are the jet energy scale of CASTOR (CES), model dependence and the alignment of CASTOR. Within the current experimental uncertainty all models are consistent with the data. However, an overestimation is observed in all the PYTHIA tunes whereas the cosmic ray MC models EPOS-LHC and QGSJETII.4 underestimate the data. A high sensitivity to MPI is observed, but the measurement is found to have a moderate sensitivity to the underlying PDF set of the model.

3 Summary

An extensive list of forward physics measurements performed with data taken in 2015 at $\sqrt{s} = 13$ TeV is presented. The measurements presented here can be used as input for tuning of existing hadronic interaction models. The collection of measurements in the forward region is an important baseline for more exclusive analyses and shows that forward physics is well understood within the systematic uncertainties at a $\sqrt{s} = 13$ TeV scale collected by CMS.

4 Acknowledgments

The author would like to acknowledge the Adiyaman University Scientific Activity Foundation and the organizers of ICNFP2016 for the funding.

References

[1] S. Chatrchyan *et al.* [CMS Collaboration], JINST **3** S08004 (2008)
[2] CMS Collaboration, CMS Physics Analysis Summary **CMS PAS FSQ-15-005**, (2016)
[3] S. Agostinelli *et al.*, Nuclear Instruments and Methods in Physics Research Section A: Accelerators, Spectrometers, Detectors and Associated Equipment **506**, no. 3, 250 – 303, (2003)
[4] CMS Collaboration, CMS Physics Analysis Summary **CMS PAS FSQ-15-006**, (2016)
[5] S. Chatrchyan *et al.* [CMS Collaboration], JHEP **11** 148, Erratum JHEP 02 (2012) 055, (2011)
[6] CMS Collaboration, CMS Physics Analysis Summary **CMS PAS FSQ-15-008**, (2016)
[7] CMS Collaboration, CMS Physics Analysis Summary **CMS PAS FSQ-16-002**, (2016)
[8] CMS Collaboration, CMS Physics Analysis Summary **CMS PAS FSQ-16-003**, (2016)
[9] M. Cacciari, G. P. Salam, and G. Soyez, JHEP **04** 063, (2008)

EPJ Web of Conferences **164**, 07015 (2017) DOI: 10.1051/epjconf/201716407015
ICNFP 2016

Forward-backward multiplicity correlations in proton-proton collisions from several GeV to LHC energies

L. Bravina[1,a], J. Bleibel[2,3], R. Emaus[1], and E. Zabrodin[1,4,5]

[1]*Department of Physics, University of Oslo, PB 1048 Blindern, N-0316 Oslo, Norway*
[2]*Max-Planck-Institut für Intelligente Systeme, Heisenbergstr. 3, D-70569 Stuttgart, Germany*
[3]*Institut für Augewandte Physik, Universität Tübingen, Auf der Morgenstelle 10, D-72076 Tübingen, Germany*
[4]*Skobeltsyn Institute of Nuclear Physics, Moscow State University, RU-119991 Moscow, Russia*
[5]*National Research Nuclear University "MEPhI" (Moscow Engineering Physics Institute), Kashirskoe highway 31, Moscow, RU-115409, Russia*

Abstract. Forward-backward multiplicity correlations in pp collisions at LHC energies are studied with the quark-gluon string model. Comparison with experimental data and with model calculations for lower energies is performed. The model correctly reproduces the linear slope of the correlations, $\langle n_B(n_F)\rangle = a + b_{corr} n_F$, in the whole energy interval. Positive correlations arise because of mixing of sub-processes with different mean multiplicities. The increase of b_{corr} with rising collision energy is linked to the increase of the variety of sub-processes going via the soft and hard multi-Pomeron exchanges. For the events with fixed amount of Pomerons the correlation slope b_{corr} is shown to be essentially zero.

1 Introduction

The first observation of significant positive correlations between the multiplicity of charged particles emitted in forward and backward hemispheres in $\bar{p}p$ collisions at ISR energies was considered as evidence of long range correlations between the "clusters" of fragmenting system [1]. This phenomenon has attracted a lot of attention, see e.g. [2–8] and references therein. Positive forward-backward (FB) correlations in multiplicity were found in hadronic interactions at energies from several GeV to $\sqrt{s} = 1.8$ TeV [9–12]. In contrast, no significant FB correlations were observed in e^+e^- annihilation at energies up to $\sqrt{s} = 93$ GeV [13] and 133 GeV [14].

It appeared soon that the FB multiplicity correlations possessed the following properties:

- the linear dependence of the averaged multiplicity of charged particles emitted in forward (or backward) hemisphere on the multiplicity of charged particles emitted in the opposite hemisphere, i.e.,

$$\langle n_B(n_F)\rangle = a + b_{corr} n_F \ . \qquad (1)$$

[a]e-mail: larissa.bravina@fys.uio.no

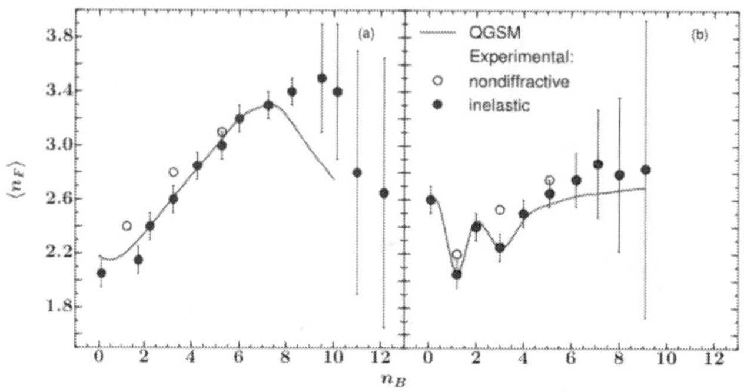

Figure 1. **(a)**: Dependence $\langle n_F \rangle (n_B)$ of charged particles in inelastic (solid circles) and nondiffractive (open circles) $\bar{p}p$ collisions at 32 GeV/c. Solid curve denotes the QGSM calculations. **(b)**: The same as **(a)** but for pp interactions at 32 GeV/c.

The slope parameter b_{corr} is defined as

$$b_{corr} = \frac{\langle (n_F - \langle n_F \rangle)(n_B - \langle n_B \rangle)\rangle}{\left[\langle (n_F - \langle n_F \rangle)^2 \rangle \langle (n_B - \langle n_B \rangle)^2 \rangle\right]^{1/2}}, \qquad (2)$$

where n_F and n_B represent multiplicities of charged particles in forward and backward hemispheres, respectively.

- the correlations reveal positive slope b_{corr} for particles from the central region $|x_F| < 0.1$, whereas for particles from the fragmentation regions $|x_F| > 0.1$ the slope is consistent with zero
- the strength of the correlations increases with rising collision energy
- for events with very high particle multiplicity the correlation strength is weakened.

On a phenomenological level these features were explained by mixture of emitting clusters with different mean multiplicity [4]. To provide a linear dependence given by Eq. (1) the clusters should obey the Poisson distribution [8].

We would like to present here our study of forward-backward multiplicity correlations within the quark-gluon string model (QGSM) [15] and its Monte Carlo version [16, 17]. The details of the model can be found in these Proceedings in [18]. QGSM was successfully applied for the description of FB correlations in pp and $\bar{p}p$ collisions at $p_{lab} = 32$ GeV/c [10], and recently in pp collisions at LHC energies 900 GeV $\leq \sqrt{s} \leq$ 13 TeV [19]. Results of the both studies are presented below.

2 FB correlations in pp and $\bar{p}p$ interactions at intermediate energies

In [10] the FB multiplicity correlations were studied in both pp and $\bar{p}p$ collisions at $p_{lab} = 32$ GeV/c. The experiments were carried out on the big bubble chamber "Mirabelle" (ITEP, Serpukhov). Figure 1 shows the $\langle n_F \rangle (n_B)$ dependence of charged particles measured in inelastic and in nondiffractive pp and $\bar{p}p$ interactions. Calculations of QGSM are plotted onto the experimental results as well. The model correctly reproduces the data. One can see that the slope b_{corr} of the distributions in $\bar{p}p$ collisions is steeper than that in pp collisions. For both reactions, the FB multiplicity correlations seem to be stronger in inelastic collisions compared to non-single diffractive (NSD) ones.

DOI: 10.1051/epjconf/201716407015

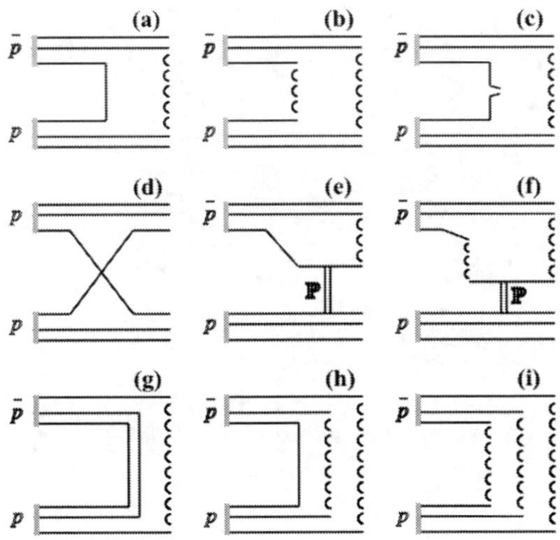

Figure 2. Diagrams taken into account in QGSM in the modeling of $\bar{p}p$ and pp interactions at intermediate energies: (a) planar, (b) cylinder, (c) undeveloped cylinder, (d) binary, (e)-(f) single diffraction with low-mass and high-mass excitation, (g)-(j) annihilation diagrams.

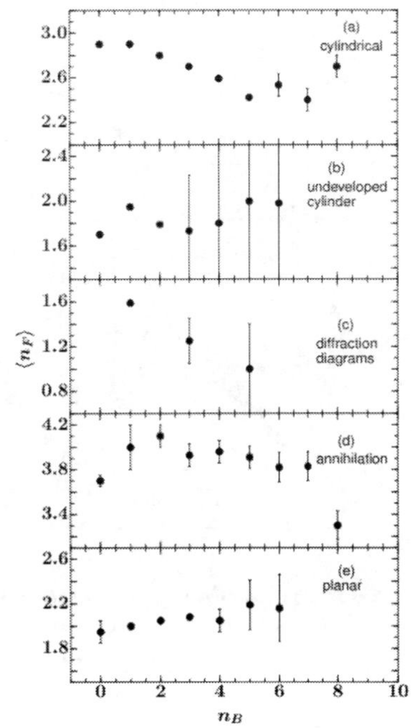

Figure 3. Dependencies $\langle n_F \rangle (n_B)$ for individual mechanisms of $\bar{p}p$ interactions at 32 GeV/c in the QGSM: cylinder diagrams (a), undeveloped cylinder diagrams (b), diffraction diagrams (c), annihilation diagrams (d), and planar diagrams (e).

Table 1. Slope parameters b_{corr} of the forward-backward correlations in inelastic and NSD pp and $\bar{p}p$ collisions at 32 GeV/c.

Reaction	pp	$\bar{p}p$
non-single diffraction	0.070±0.009	0.177±0.010
inelastic	0.098±0.006	0.192±0.007
inelastic, $\lvert x_F \rvert < 0.1$	0.237±0.007	0.175±0.006
inelastic, $\lvert x_F \rvert > 0.1$	0.013±0.006	-0.073±0.006

To present this more distinctly, we list the slope parameters in Table 1 for inelastic and NSD events. The data were also extracted for central ($|x_F| < 0.1$) and for non-central ($|x_F| > 0.1$) regions in inelastic collisions. The positive FB correlations are observed for central region $|x_F| < 0.1$ only, whereas for the regions with $|x_F| > 0.1$ there are practically no correlations.

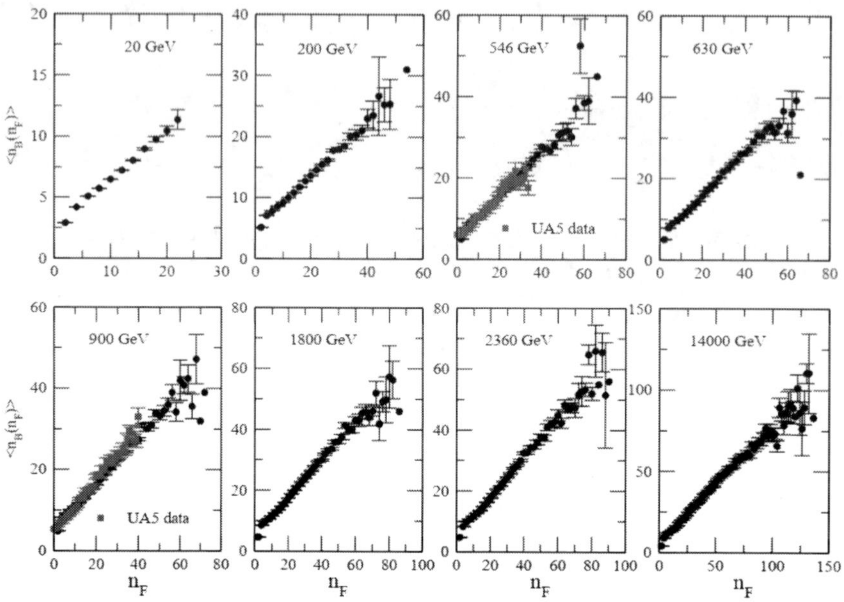

Figure 4. Dependencies $\langle n_B\rangle(n_F)$ for rapidity interval $0 \leq \eta \leq 4$ in NSD pp collisions at 20 GeV$\leq \sqrt{s} \leq$ 14 TeV. Data for 546 GeV and 900 GeV (from [3]) are shown by full squares.

For better understanding of the reason why the FB correlations in $\bar{p}p$ collisions are stronger at this energy than the FB correlations in pp collisions, we plot in Fig. 2 the set of diagrams employed in QGSM for treatment of both types of collisions at intermediate energies. Obviously, the variety of sub-processes in $\bar{p}p$ interactions is richer compared to that in pp ones, because, e.g., the planar diagrams (a) and annihilation diagrams (g)-(i) are absent in pp processes. Since more sub-processes with different mean multiplicities contribute to particle production in $\bar{p}p$ collisions, the correlation slope is steeper, i.e. $b_{corr}^{\bar{p}p} > b_{corr}^{pp}$. Similarly, because of the lack of single diffraction diagrams the correlation strength in non-single diffractive (NSD) processes is smaller compared to that in inelastic ones. Figure 3 displays the FB correlations for different sub-processes shown in Fig. 2, namely, cylinder diagram, undeveloped cylinder, single diffraction, $\bar{p}p$ annihilation and planar diagram. Only the planar sub-processes demonstrate weak positive FB correlations, whereas the correlation slopes for cylinder and diffractive diagrams are slightly negative.

The cross section of annihilation process drops rapidly with rising energy of the collisions. Therefore, the slopes of FB multiplicity correlations in pp and $\bar{p}p$ interactions become similar after some collision energy threshold. The strength of the correlations continues to increase, as shown in Fig. 4. Here QGSM calculations of FB correlations in pp interactions are displayed for energies ranging from $\sqrt{s}$ = 20 GeV to $\sqrt{s}$ = 14 TeV. Comparison with the UA5 data on $\bar{p}p$ collisions at $\sqrt{s}$ = 546 GeV and 900 GeV [3, 11] demonstrates a good agreement between the model calculations and the experimental data. For all reactions the slope of the FB correlations remains almost linear. We have mentioned already the vanishing of annihilation diagrams, as well as other so-called pre-asymptotic diagrams, with the increase of collision energy. However, the only reason of enlargement of the correlation strength b_{corr} in the model is the rise of the variety of different sub-processes with different mean multiplic-

 DOI: 10.1051/epjconf/201716407015

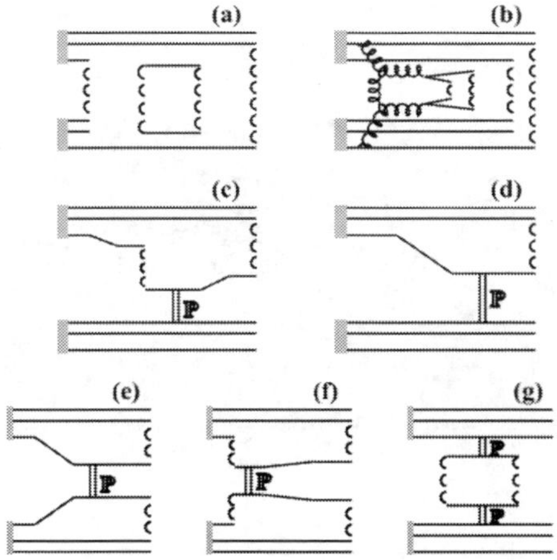

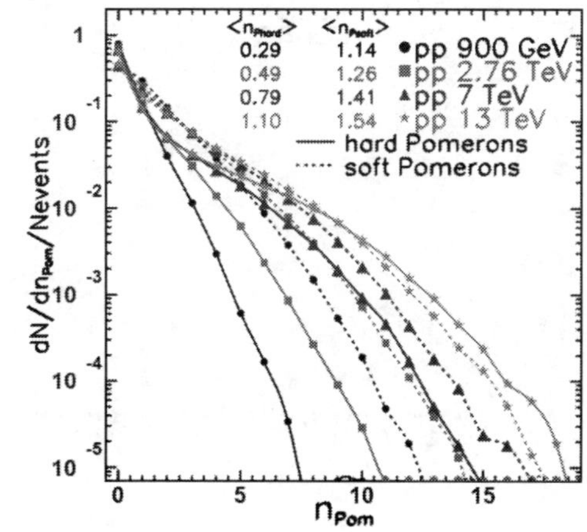

Figure 5. Diagrams taken into account in QGSM in the modeling of pp interactions at ultrarelativistic energies: (a) multi-Pomeron exchange, (b) (semi)hard gluon-gluon interaction and soft Pomeron exchange, (c)-(d) single diffraction with high-mass and low-mass excitation, (e)-(f) double diffraction with low-mass and high-mass excitation, (g) central diffraction.

Figure 6. Average numbers of soft (dotted curves) and hard (solid curves) Pomerons per event in pp collisions at $\sqrt{s}$ = 0.9, 2.76, 7, and 13 TeV, respectively.

ity. These sub-processes and their role in formation of FB multiplicity correlations at ultra-relativistic energies are discussed in 3.

3 FB correlations at LHC energies

The set of diagrams describing the pp interactions at ultra-relativistic energies is shown in Fig.5. The first two diagrams represent (a) the soft (multi)Pomeron exchanges and (b) processes going via the formation of hard Pomerons. Other diagrams deal with the processes of single diffraction, (c) and (d), and double diffraction, (e)-(g). The main contribution to particle multiplicity comes from the processes with soft and hard Pomeron exchanges. Their numbers increase with rising collision energy, as it can be seen in Fig. 6. This figure displays the relative amounts of soft and hard Pomerons in a single pp-event at energies $\sqrt{s}$ = 900 GeV, 2.76 TeV, 7 TeV and 13 TeV, respectively. At $\sqrt{s}$ = 900 GeV the number of soft Pomerons exceeds the number of hard Pomerons by factor 4, whereas at $\sqrt{s}$ = 13 TeV their ratio drops to 1.4. As it follows from Fig. 6, the maximum number of both soft and hard Pomerons in a single event gradually increases with collision energy. The variety of sub-processes containing all possible combinations of soft and hard Pomerons becomes more extensive. Since these processes have different mean multiplicities of produced hadrons, the strength of the FB correlations measured for the whole sample of events increases, as shown in Fig. 7. One can see that the FB correlations are not very strong at low multiplicity area, $n^{ch} \leq 15$. The slopes of the distributions are also becoming less steep after a certain multiplicity threshold.

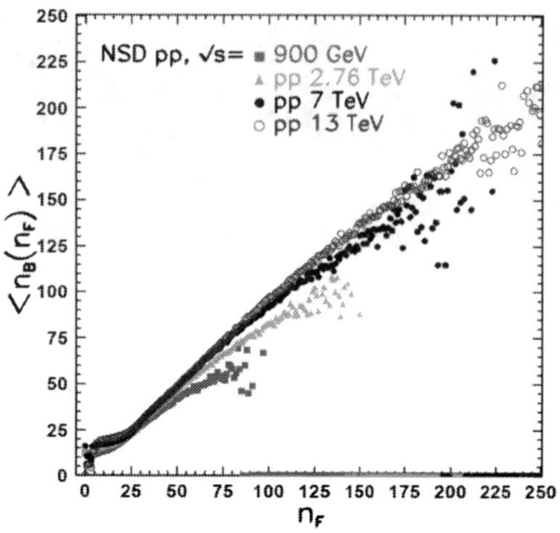

Figure 7. Forward-backward multiplicity correlations in NSD pp collisions at $\sqrt{s} = 0.9, 2.76, 7$, and 13 TeV.

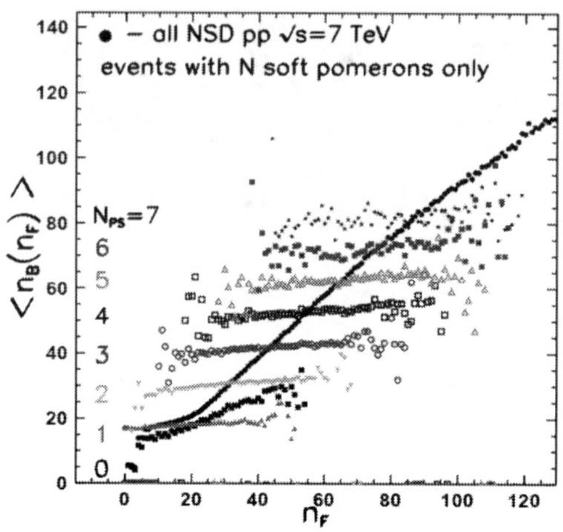

Figure 8. Dependence of average multiplicity $\langle n_B \rangle$ on n_F (solid circles) in NSD pp collisions at $\sqrt{s} = 7$ TeV in QGSM. Curves indicate $\langle n_B(n_F) \rangle$ distributions for sub-events with only soft Pomerons, $N_{s\_Pom} = 0, 1, \ldots, 7$.

Figure 8 displays the $\langle n_B \rangle (n_F)$ dependencies calculated in QGSM for pp events at $\sqrt{s} = 7$ TeV which proceed only via the soft Pomeron exchanges. The FB correlations for the sub-processes with fixed amount of soft Pomerons, varying from 1 to 7, are also plotted in Fig. 8. It is worth noting that FB correlations within each of the selected sample of events have zero slope, although the charged particle multiplicity measured on event-by-event basis changes from few hadrons up to one hundred. Not all topologies contribute to event with very small and very big multiplicity. This circumstance leads to reduction of the correlation strength in both multiplicity intervals.

At LHC energies the FB correlations in pp interactions were studied by the ALICE [21] and ATLAS [22] Collaborations. The analysis is performed in terms of gaps in pseudorapidity, η_{gap}, between the hadrons in forward and backward hemispheres, and widths of pseudorapidity bins, $\delta\eta$. Figure 9 presents the comparison of model calculations of the b_{corr} as a function of $\delta\eta$ at zero rapidity gap with the ALICE data for $\sqrt{s}$ = 900 GeV, 2.76 TeV and 7 TeV. The slope b_{corr} increases with broadening of $\delta\eta$ for all three energies. Note, that the strength of FB correlations drops with rising midrapidity gap [19].

ALICE Collaboration has also studied the FB correlations between different azimuthal sectors. Parameters of the study are as follows. The azimuthal angle of the sectors is $\varphi = \pi/4$ and the width of the bin is $\delta\eta = 0.2$. The data obtained for pp interactions at 900 GeV and 7 TeV are shown in Fig. 10 in comparison with the QGSM calculations. The correlations at $\sqrt{s} = 7$ TeV are twice stronger than that at 900 GeV; other characteristics are pretty similar.

4 Conclusions

We apply the quark-gluon string model, based on Reggeon Field Theory, for the description of forward-backward multiplicity correlations in proton-proton collisions at ultrarelativistic energies. It

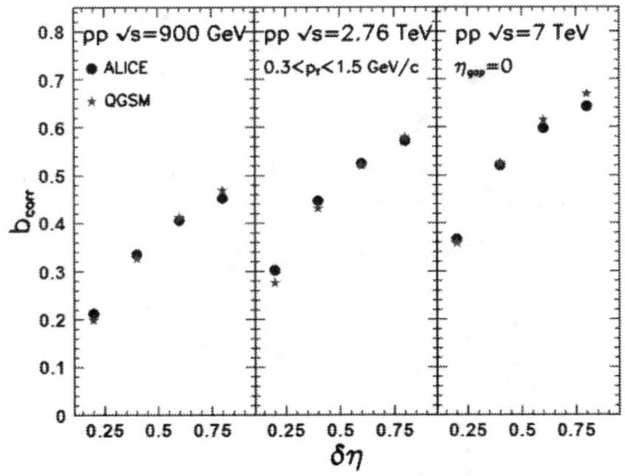

Figure 9. FB correlation parameter b_{corr} at $\eta_{gap} = 0$ in four rapidity bins: $\delta\eta = 0.2, 0.4, 0.6, 0.8$ in pp collisions at $\sqrt{s} = 900$ GeV, 2.76 TeV, 7 TeV and 13 TeV. Stars and circles denote the model calculations and the ALICE data from [21], respectively.

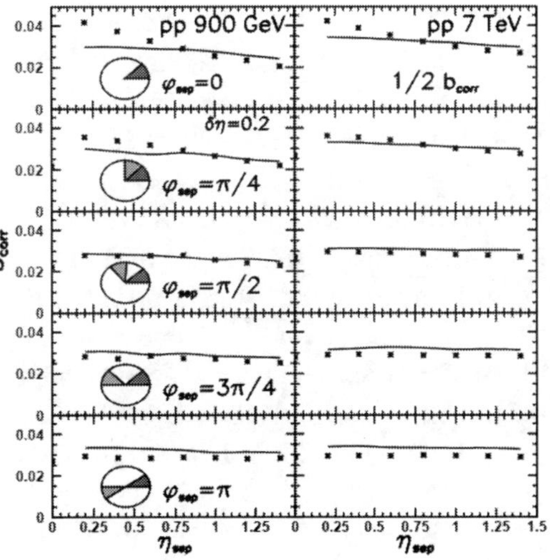

Figure 10. FB correlation parameter b_{corr} for azimuthally separated sectors of $\varphi = \pi/4$ as a function of midrapidity gap at fixed $\delta\eta = 0.2$ in pp collisions at $\sqrt{s} = 900$ GeV (left) and 7 TeV (right). Results for 7 TeV are reduced by factor 2. Lines and asterisks denote the model calculations and the data from [21], respectively.

is shown that positive FB correlations arise in QGSM because of addition of different sub-processes with different mean multiplicities. For the individual sub-processes, their $\langle n_B\rangle(n_F)$ distributions are remarkably flat, although the event-by-event multiplicity can vary from few particles up to more than one hundred.

At c.m. energies about 10 GeV the number of diagrams describing $\bar{p}p$ collisions is larger than that for pp interactions. Therefore, FB correlations are stronger in $\bar{p}p$ case. At c.m. energies higher than 100 GeV the sets of diagrams for $\bar{p}p$ and pp collisions are similar. The further rise of the correlation strength occurs due to increasing number of soft and hard Pomerons allowed for a single event.

The correlation dependence is linear, $\langle n_B\rangle(n_F) = a + b_{corr} n_F$, for $\bar{p}p$ and pp collisions at all energies in question. However, for low and for very high multiplicities n_F the slopes b_{corr} are not so steep. Finally, the FB correlations take place mainly in the region $|x_F| < 0.1$, whereas for $|x_F| > 0.1$ the correlations are almost absent.

Acknowledgements

Fruitful discussions with M. Bleicher, L. Csernai and V. Vechernin are gratefully acknowledged. L.B. acknowledges financial support of the Alexander von Humboldt Foundation. J.B. thanks the German Research Foundation (DFG) for the financial support through the Project BL 1286/2-1.

References

[1] S. Uhlig, I. Derado, R. Meinke, H. Preissner, Nucl. Phys. B **132**, 15 (1978)

[2] K. Alpgaard *et al.*, UA5 Collab., Phys. Lett. B **123**, 361(1983)
[3] G.G. Alner *et al.*, UA5 Collab., Phys. Rep. **154**, 247 (1987)
[4] T.T. Chou, C.N. Yang, Phys. Lett. B **135**, 175 (1984)
[5] A. Capella, A. Krzywicki, Phys. Rev. D **18**, 4120 (1978)
[6] A. Capella, J.Tran Thanh Van, Z. Phys. C **18**, 85 (1983)
[7] J. Dias de Deus, Phys. Lett. B **100**, 177 (1981)
[8] M.A. Braun, C. Pajares, V.V. Vechernin, Phys. Lett. B **493**, 54 (2000)
[9] E.G. Boos *et al.*, Phys. Scripta **15**, 305 (1977)
[10] L.V. Bravina *et al.*, Sov. J. Nucl. Phys. **50**, 245 (1989)
[11] R.E. Ansorge *et al.*, UA5 Collab., Z. Phys. C **37**, 191 (1988)
[12] T. Alexopoulos *et al.*, E735 Collab., Phys. Lett. B **353**, 155 (1995)
[13] W. Braunschweig *et al.*, TASSO Collab., Z. Phys. C **45**, 193 (1989)
[14] R. Akers *et al.*, OPAL Collab., Phys. Lett. B **320**, 417 (1994)
[15] A.B. Kaidalov, Phys. Lett. B **116**, 459 (1982);
A.B. Kaidalov and K.A. Ter-Martirosyan, Phys. Lett. B **117**, 247 (1982)
[16] N.S. Amelin, L.V. Bravina, Sov. J. Nucl. Phys. **51**, 133 (1990)
[17] J. Bleibel, L.V. Bravina, E.E. Zabrodin, Phys. Rev. D **93**, 114012 (2016)
[18] E.E. Zabrodin *et al.*, these Proceedings.
[19] L.V. Bravina, J. Bleibel, E.E. Zabrodin, Phys. Lett. B (submitted).
[20] G. J. Alner *et al.*, UA5 Collab., Phys. Rep. **154**, 247 (1987)
[21] J. Adam *et al.*, ALICE Collab., J. High Energy Phys. **JHEP05**, 097 (2015)
[22] G. Aad *et al.*, ATLAS Collab., J. High Energy Phys. **JHEP07**, 019 (2012)

EPJ Web of Conferences **164**, 07016 (2017) DOI: 10.1051/epjconf/201716407016
ICNFP 2016

Higgs decay to fermions (CMS)

Pooja Saxena[1,a] on behalf of the CMS colloboration

[1] *Deutsches Elektronen-Synchrotron, Notkestrasse 85, 22607 Hamburg, Germany*

Abstract. A search for high mass Higgs boson of the Minimal Supersymmetric Standard Model decaying into two fermions using the first 2015 data at 13 TeV is presented. The four final decay channels of $\mu\tau_h$, $e\tau_h$, $\tau_h\tau_h$ and $e\mu$ is used. The limits on production cross section times branching ratio has been set. Other results from Run1 and different searches and measurements involving Higgs decays fermions will also be reviewed.

1 Introduction

On 4th of July 2012, after the discovery of a particle with a mass near 125 GeV by CMS and ATLAS collaboration, [1] [2] [3], many precise measurements of these properties have been performed: the signal strength, the latest mass measurement and the couplings with bosons ($\gamma\gamma$, ZZ and WW) [4] [5]. All these measurements are compatible with a standard model (SM) Higgs boson [4] [5]. However, to claim that this newly observed particle is the SM Higgs boson, the couplings to the fermions has to be verified as well.

At the LHC, the Higgs boson can be produced in several ways, mainly by gluon-gluon fusion (ggH), also by Vector Boson Fusion (VBF), in association with a top quark pair (ttH), or in association with a W/Z boson (VH). It mostly decays into a pair of b-quarks (57%), in a pair of two τ leptons (6.3%), and more rare into a pair of muons (0.02%). In this document, results from Run-I on measurements involving Higgs decays to fermions ($\tau\tau$, bb and combination of $\tau\tau$ and bb) will be reviewed in section 2. Measurement of fermionic coupling is discussed in section 3. Further, measurements on universality of Higgs coupling flavour is address in section 4. Section 5 will report on the results on lepton flavour violation. A search for high mass Higgs boson of the MSSM decaying into two fermions using the first 2015 data at 13 TeV is presented in section 6. Finally, conclusions are given in section 7.

2 Higgs (125 GeV) Decay to fermions

In this section, Higgs decay in fermionic decay mode ($\tau\tau$, bb and combination) will be presented. The results of bb measurement in vector boson fusion (VBF) mode will also be shown for first 2015 data at 13 TeV.

[a] e-mail: pooja.saxena@desy.de

2.1 Higgs boson decay to $\tau\tau$

Among the Higgs decay to a di-lepton pair, the $\tau\tau$ is the one with the largest BR, because of the large event rate expected in the SM compared to the other leptonic decay modes. This section reports on the results of a search for a SM Higgs boson using final states with a pair of τ leptons in proton-proton collisions at $\sqrt{s}$ = 7 and $\sqrt{s}$ = 8 TeV at the LHC. Entire dataset collected in 2011 and 2012 by the CMS experiment corresponding to an integrated luminosity of 4.9 fb^{-1} at a centre of mass energy of 7 TeV and 19.7 fb^{-1} at 8 TeV is being used [7]. All the possible decay modes of the di-tau system are used: $e\mu$, ee, $\mu\mu$, $e\tau_h$ (τ_h represent the hadronic decay mode of τ), $\mu\tau_h$ and $\tau_h\tau_h$. The main backgrounds are represented by $Z \to \tau\tau$, $Z \to ll$, W+jets events and QCD processes. For most of the channels, the signal extraction is done with a likelihood fit on the di-tau mass. Events are categorized exploiting the number of jets, boost of the Higgs system and the sensitivity to ggH production according to the expected S/B. The "0-jet" categories are mostly used to constrain the main backgrounds. The combined observed and predicted $m_{\tau\tau}$ distributions for the $\mu\tau_h$, $e\tau_h$, $\tau_h\tau_h$, and $e\mu$ channels is shown in Fig 1 (left). An excess of events over the background-only hypothesis is observed with a local significance of 3.2 standard deviations at m_H=125 GeV, while the expected significance is 3.7 standard deviations and the best-fit value for the signal strength, $\mu = \sigma/\sigma_{SM}$, is 0.78 ± 0.27 times the SM prediction, see Fig 1 (right), from which it can be conclude that there is clear evidence of the Higgs boson decays to pair of τ lepton.

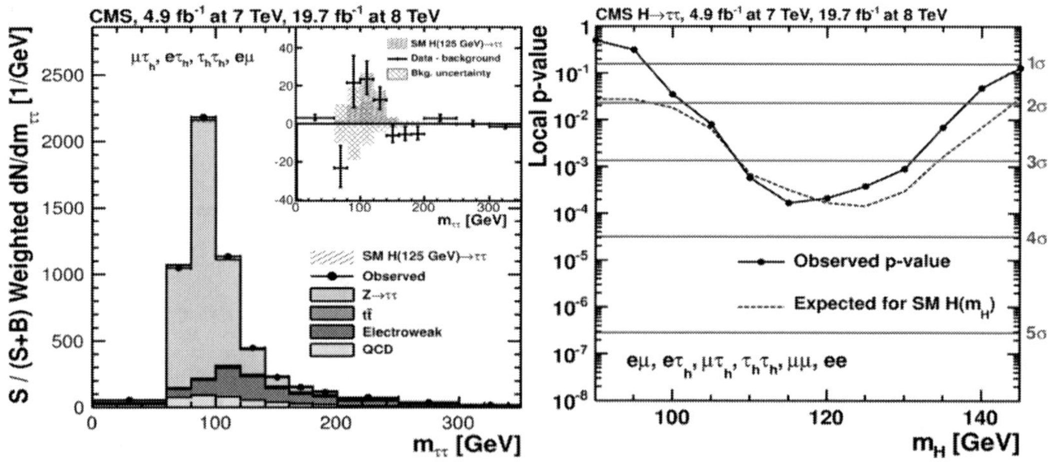

Figure 1. Combined observed and predicted $m_{\tau\tau}$ distributions for the $\mu\tau_h$, $e\tau_h$, $\tau_h\tau_h$, and $e\mu$ channels (left). Local p-value and significance in number of standard deviations as a function of the SM Higgs boson mass hypothesis (right)

2.2 Higgs boson decay to bb

At the LHC, in the mass range considered in this study, the vector boson fusion (VBF) process has the second largest production cross section following gluon fusion (GF) [8]. Furthermore, for a SM Higgs boson with a mass m_H < 135 GeV, the expected dominant decay mode is bb. Because of the huge QCD background, the ggH production is not considered. The signal topology of the production mechanism is exploited for VH and VBF production mode.

2.2.1 VH mode

All the six channels for the leptonic decay of V are considered: Z(ll)H(bb) with l=e,μ, Z($\nu\nu$)H(bb) and W(lν)H(bb) with l=e,mu,τ. Data control regions are used to normalize the main contributions (V+jets, tt and VV) To increase the background rejection, the resolution on the b-jet pair mass (jets arising from the hadronization of a b-quark) is improve by 15-25% by applying a Multi Variate Analysis (MVA) regression. The event categorization is based on the high boost (100 GeV) of the V bosons, with bins in pT (V), and multivariate discriminators are used for signal extraction in 14 regions defined. An excess of events is observed with a local significance of 2.1 standard deviations at mH=125 GeV (the expected significance is 2.1 standard deviations), and the best-fit value for the signal strength is 1.0 ± 0.5 times the SM prediction (see Fig. 2), and therefore shows a very good agreement with SM expectations.

Fig. 2 shows the combination of all channels into a single distribution. Events are sorted in bins of similar expected signal to background ratio, as given by the value of the output of their corresponding BDT discriminant (trained with a Higgs boson mass hypothesis of 125 GeV).

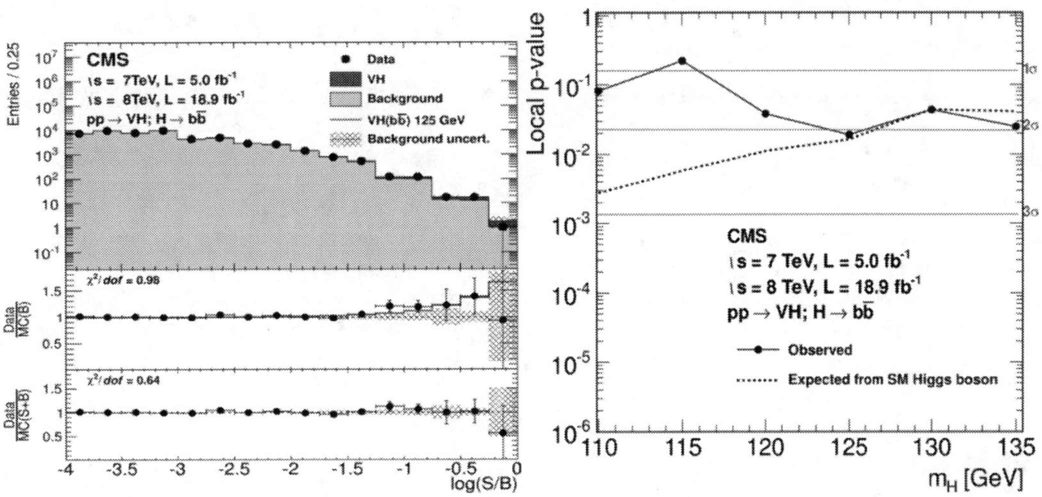

Figure 2. Combination of all channels into a single distribution (left). Local p values and corresponding significance (measured in standard deviations) for the background-only hypothesis to account for the observed excess of events in the data (right).

2.2.2 VBF mode

The data sample for this analysis corresponds to an integrated luminosity of 2.3 fb^{-1} recorded in 2015 with the CMS experiment [9]. The prominent feature of the VBF process is the presence of four energetic jets in the final state. Further, the dominant background arise from QCD production of multijet events. Other backgrounds arise from hadronic decays of Z or W bosons produced in association with additional jets, hadronic decays of singly produced top quarks and hadronic decays of top quark pairs. The search is performed on selected four-jet events that are characterized by the response of a multivariate discriminant trained to separate signal events from background without making use of kinematic information on the two b-jet candidates. Distributions of the invariant mass of the two VBF-jet candidates for events in the DoubleB samples is shown in Fig. 3 (left). Observed and SM-expected likelihood profile of the signal strength $\mu = \sigma/\sigma_{SM}$ with m_H = 125 GeV, using 8 TeV data of Run-I and 13 TeV data of Run 2 and for the combination of 8 TeV and 13 TeV data (right). The expected upper limits in the absence of a signal is 5.0 times the SM prediction, while the observed upper limit is 3.0, and the fitted signal strength is $\mu = \sigma/\sigma_{SM} = -3.7^{+2.4}_{-2.5}$. The combination of the results obtained in this search with the similar CMS search with Run 1 data yields an observed (expected) upper limit of 3.4 (2.3) times the SM prediction. The combined fitted signal strength is $\mu = 1.3^{+1.2}_{-1.1}$, with a signal significance of 1.2 standard deviations for m_H = 125 GeV, see Fig. 3 (right).

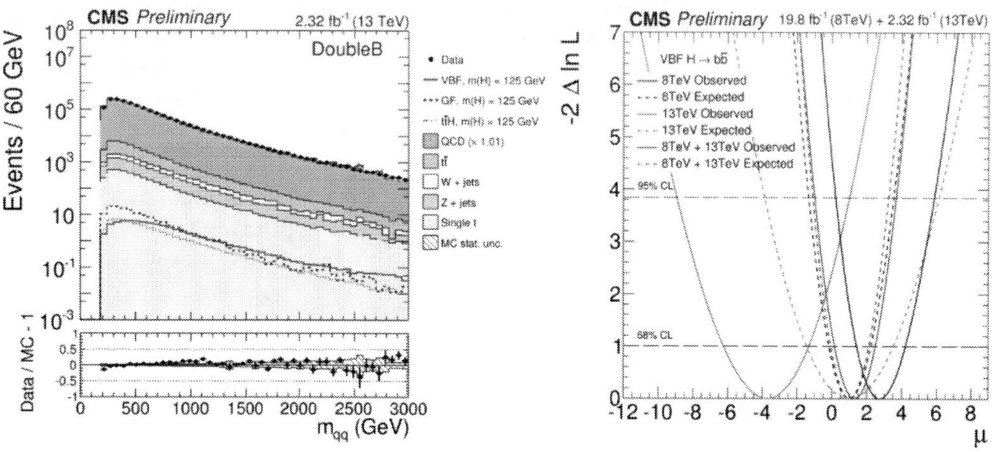

Figure 3. Distributions of the invariant mass of the two VBF-jet candidates (mqq) for events in the DoubleB samples (left). Observed and SM-expected likelihood profile of the signal strength $\mu = \sigma/\sigma_{SM}$ with m_H = 125 GeV, using 8 TeV data of Run-I and 13 TeV data of Run 2 and for the combination of 8 TeV and 13 TeV data (right).

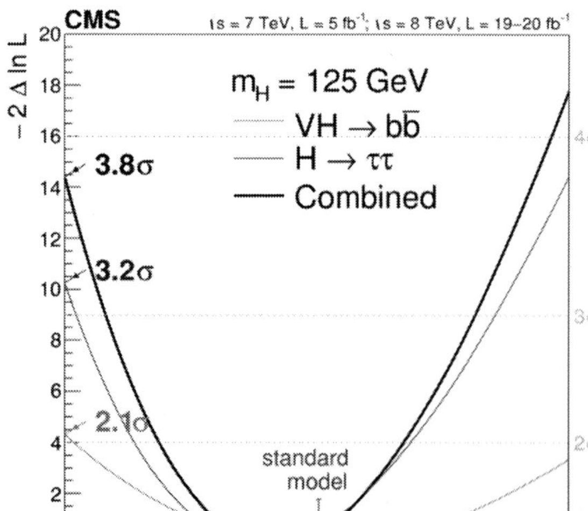

Figure 4. Scan of the profile likelihood as a function of the signal strength relative to the expectation for the production and decay of a standard model Higgs boson, μ , for m_H=125 GeV.

2.3 Fermionic combination

A combination of the Higgs boson decays to $\tau\tau$ channel and Higgs boson decays to bb (VH only) is performed [10]. This combination results in strong evidence for the direct coupling of the 125 GeV Higgs boson to fermions, with an observed (expected) significance of 3.8 (4.4) standard deviations (see Fig. 4).

3 Measurement of fermionic coupling

Fig 5 shows fits for deviations in the coupling for a generic five-parameter model [14] assuming effective loop couplings according to the SM. For the fermions, the values of the fitted Yukawa couplings are shown, while for vector bosons the square-root of the coupling for the hVV vertex divided by twice the vacuum expectation value of the Higgs boson field is shown. The linear dependence on the mass parameter nicely illustrates the validity of the Higgs mechanism providing the masses of the fundamental particles in the SM.

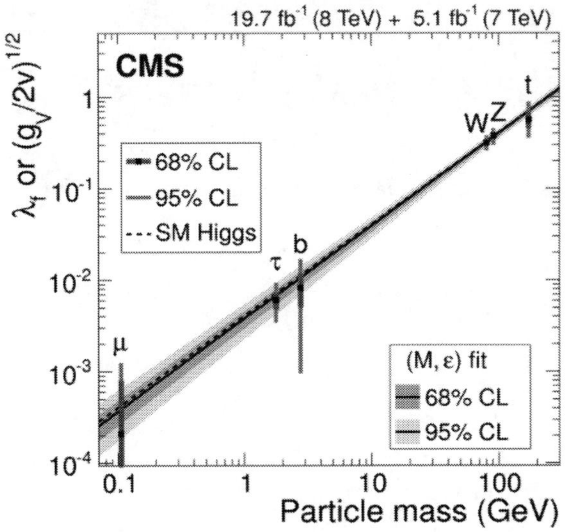

Figure 5. Summary of the fits for deviations in the coupling using the SM loop induced couplings, expressed as function of the particle mass.

4 Universality of Higgs coupling flavour

In this section, measurement of Higgs boson coupling to first and second generation leptons will be presented [12]. The analysis is performed by directly combining the full $\sqrt{s}$ = 7 TeV and $\sqrt{s}$ = 8 TeV datasets (5.0 + 19.7 fb^{-1}), using the ggH and VBF production modes (for the VH and ttH production mode the branching ratio is too small). The SM predicts the coupling of the new boson to leptons is flavour-universal or proportional to the lepton mass. However, the branching ratio for $H \to \mu\mu$ is very low (2.2×10^{-4}), and even smaller for $H \to ee$. One of the important feature of this analysis is the clean signature in the detector.

Events are categorized based on the number or reconstructed jets attempting to separate the gluon-gluon fusion and VBF production components. The signal is extracted by means of a fit to the $m_{\mu\mu}$ distribution using signal and background shapes. As no significant $H \to \mu\mu$ is observed, upper limits on the cross section times the $H \to \mu\mu$ branching ratio are set. The combined observed (expected) limit on the signal strength, for a Higgs boson with a mass of 125 GeV, is 7.4 ($6.5^{+2.8}_{-1.9}$), Fig. 6 (left).

Assuming the SM production cross section, this corresponds to an upper limit of 0.0016 on $\beta(H \to \mu\mu)$. For $H \to ee$, SM Higgs boson decays are far too rare to detect, and no signal is observed. For a Higgs boson mass of 125 GeV, a 95% CL upper limit of 0.041 pb is set on $\sigma\beta(H \to ee)$ at 8 TeV, see Fig. 6 (right).

These results, together with the above quoted evidence for the 125 GeV boson coupling to τ leptons confirm the SM prediction that the leptonic couplings of the new boson are not flavour universal.

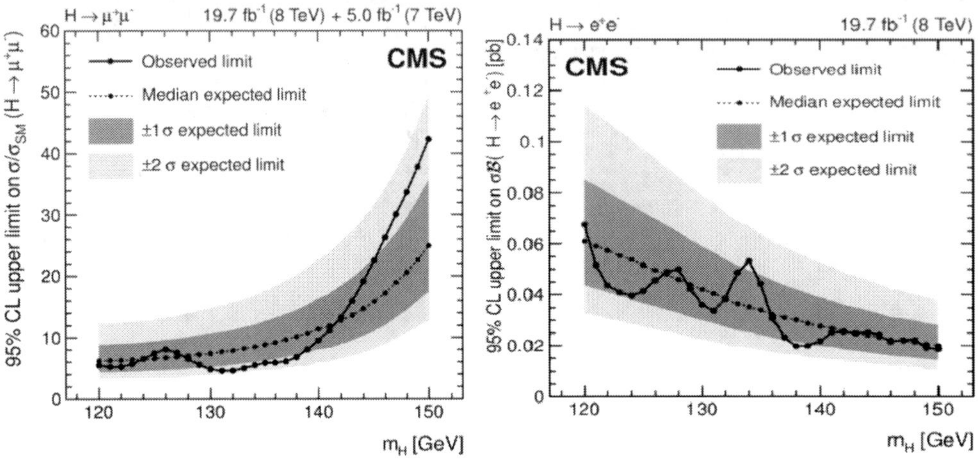

Figure 6. Mass scan for the background-only expected and observed combined exclusion limits for $H \to \mu\mu$ (left). Exclusion limits on $\sigma\beta$ shown for $H \to ee$ (right).

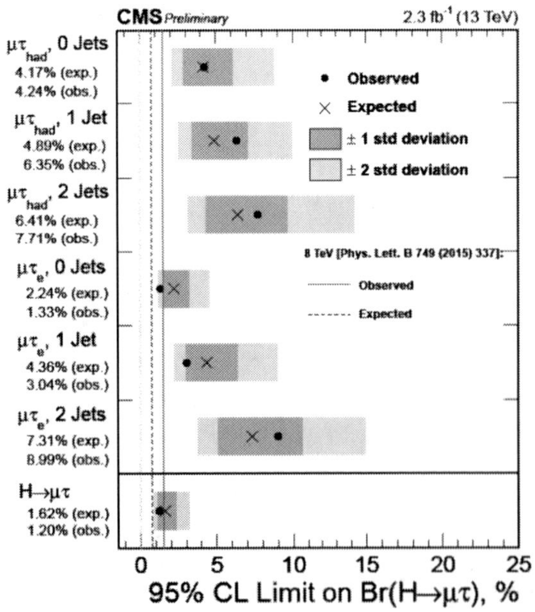

Figure 7. Observed and expected 95% CL upper limits on the $B(H \to \mu\tau)$ for each individual category and combined.

5 Lepton flavour violation (LFV)

In this section, a direct search for lepton flavour violating decays of the Higgs boson in the $H \to \mu\tau$ channel using 2.3 fb^{-1} data at $\sqrt{s}$= 13 TeV with the CMS experiment [11]. In the SM, LFV decays of the Higgs boson are forbidden, which make it an interesting observation. The search is performed in two channels, $H \to \mu\tau_e$ and $H \to \mu\tau_h$, where τ_e and τ_h are tau leptons reconstructed in the electronic and hadronic decay channels, respectively. This analysis follow the procedure of the $\sqrt{s}$ = 8 TeV analysis, where a small excess was observed, corresponding to 2.4 σ, with an analysis based on an integrated luminosity of 19.7 fb^{-1} that yielded an expected 95% CL limit on the branching fraction of 0.75%. No excess is observed in $\sqrt{s}$= 13 TeV analysis. The best-fit branching fraction is $B(H \to \mu\tau) = -0.76^{+0.81}_{-0.84}\%$ and an upper limit of $B(H \to \mu\tau) < 1.20\%$ (1.62% expected) is set at 95% CL, see Fig. 7

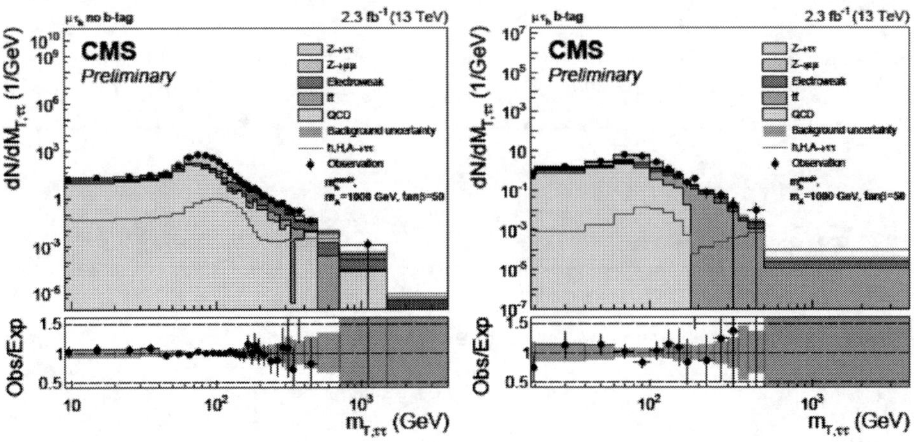

Figure 8. Post–fit plot of the transverse mass distribution in the no b–tag category (left) and the b–tag category of the $e\tau_h$ channel (right).

6 Minimal Supersymmetric Standard Model (MSSM)

This search is performed on a dataset corresponding to an integrated luminosity of 2.3 fb^{-1} of pp collision data at a centre-of-mass energy of 13 TeV, collected by CMS in 2015 [13]. As number of arguments favour the existence of theories beyond the SM, many versions of which include an extended Higgs sector. One of the simplest extensions of the SM is the MSSM.

Four final states of the tau pair are considered: $e\mu$, $e\tau_h$, $\mu\tau_h$ and $\tau\tau$. The methedology of this analysis is to stay close to analysis performed at 8 TeV, though the physics objects, selection and analysis tools are re-optimized. The dominant production mode is gluon fusion (ggf) and b-associated production (bbf).

A binned maximum likelihood fit is performed simultaneously to the transverse mass distributions in the b–tag and no b–tag categories and all four channels Fig. 8. As no significant excess of data over the background predictions is found, we set 95% CL upper limits on cross-section times branching fraction for the two dominant production modes, ggf and bbf. Fig. 9 shows these limits as a function of m_ϕ. Finally, Fig. 10 shows the model dependent limits for two different benchmark models, the m_h^{mod+} and hMSSM scenarios.

DOI: 10.1051/epjconf/201716407016

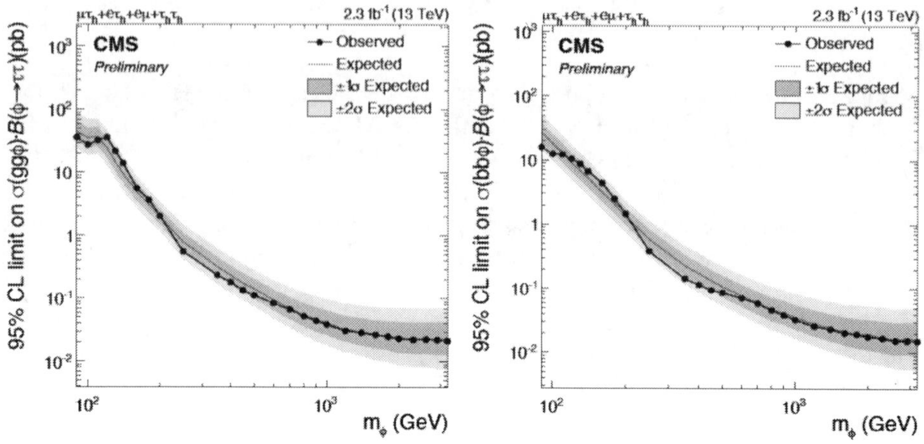

Figure 9. Expected and observed limits on cross-section times branching fraction for the gluon fusion process (left) and the b–associated production process (right), resulting from the combination of all four channels.

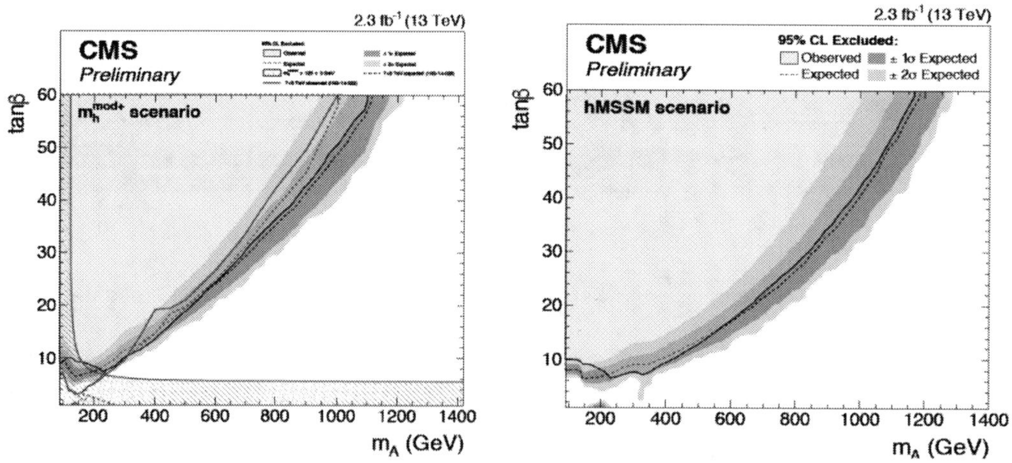

Figure 10. Model dependent exclusion limits in the $m_A - tan\beta$ plane, combining all channels, for the m_h^{mod+} (left) and hMSSM (right) scenarios.

7 Summary

Searches for a SM Higgs boson decaying to fermions have been performed in CMS for different channels, $H \rightarrow \tau\tau$, $H \rightarrow bb$ and $H \rightarrow \mu\mu$, using different production modes, and there is a strong evidence of the Higgs decays to fermions. The excess seen for $H \rightarrow \tau\tau$ is of 3.2 σ and constitutes by itself an evidence, the excess in the $H \rightarrow bb$ (VH channel) is of 2.1 σ, and combining these two channels, it leads to an evidence of 3.8 σ for the Higgs decaying to fermions.

References

[1] CMS Collaboration, Phys. Lett. B716 **30** (2012).
[2] CMS Collaboration, JHEP06 **081** (2013) .
[3] ATLAS Collaboration, Phys. Lett. B716 **1** (2012).
[4] CMS Collaboration, JHEP01 **096** (2014).
[5] CMS Collaboration, Phys. Rev. D89 **092007** (2014).
[6] CMS Collaboration, JINST3 **S08004** (2008)
[7] CMS collaboration, J. High Energ. Phys. **104** (2014), doi:10.1007/JHEP05(2014)104
[8] LHC Higgs Cross Section Working Group, CERN Report CERN-2012-002, 2012. 534 doi:10.5170/CERN-2012-002, arXiv:1201.3084.
[9] CMS Collaboration, CMS-PAS-HIG-16-003.
[10] CMS collaboration, Nature Physics 10, 557–560 (2014) doi:10.1038/nphys3005
[11] CMS Collaboration, CMS-PAS-HIG-16-005.
[12] CMS Collaboration, Phys. Lett. B744 (2015) 184–207
[13] CMS Collaboration, CMS-PAS-HIG-16-006.

ICARUS detector: present and future

F. Varanini (on behalf of the ICARUS Collaboration)
[1] *INFN Sezione di Padova, via Marzolo 8, Padova, Italy*

Abstract. The ICARUS-T600 detector is the first large-scale liquid Argon TPC ever operated; its successful data-taking campaign at Gran Sasso laboratory, with CNGS beam and atmospheric neutrinos, proved the maturity of this innovative detection technology. In particular, it performed a search for sterile neutrinos through ν_μ to ν_e oscillations, observing no excess w.r.t. backgrounds and allowing to strongly constrain the parameter space, indicating a narrow region in agreement with all experimental data.
A definitive answer to the sterile neutrino puzzle will be given by the SBN experiment at the Booster beam at FNAL, where ICARUS-T600 will act as far detector, by performing both appearance and disappearance measurements.

1 The sterile neutrino puzzle

Neutrino oscillations represent one of the greatest discoveries in recent particle physics; the vast majority of experimental results in the last 20 years appear consistent with a 3-neutrino scenario, characterized by three flavor eigenstates (ν_e, ν_μ, ν_τ) and three corresponding mass eigenstates, with small mass differences ($\Delta m^2_{12} \sim 8 \cdot 10^{-5} eV^2, \Delta m^2_{23} \sim 2.5 \cdot 10^{-3} eV^2$). However, a few anomalous results pointed out to possible oscillations mediated by a much larger Δm^2, suggesting the existence of (at least) a fourth mass eigenstate:

- The observation of a ν_e excess in ν_μ beams from accelerators, at experiments such as LSND ($\sim 3.8\sigma$ effect) and MiniBoone;
- A deficit w.r.t. expectations (also $\sim 3\sigma$) in the $\overline{\nu_e}$ flux from nuclear reactors;
- A similarly lower signal from high-intensity ν_e sources used in the calibration of solar neutrino experiments.

While these results are not statistically compelling, the experimental scenario concerning the possible sterile neutrinos is far from clear, and requires a full clarification.

2 The ICARUS-T600 experiment at LNGS

The ICARUS-T600 data-taking run at the underground LNGS laboratory, lasting from 2010 to 2013, represents the first large-scale physics experiment with a liquid Argon Time Projection Chamber (*LAr-TPC*), with a detector active mass of $\sim$ 470 tons.

The LAr-TPC technique exploits the ionization of LAr, generating electrons ($\sim$ 5000 per mm, for a m.i.p.) that drift towards an anodic wire-plane array; the stereoscopic, non-destructive read-out of wire signals allows a 3D reconstruction of any ionizing track with a millimeter-scale resolution. Moreover, the VUV scintillation light produced in the Argon, detected by PMTs immersed in the detector, is used for signal timing and trigger generation.

The ICARUS-T600 detector [1, 2] was exposed to the CNGS neutrino beam from CERN, collecting a total statistics of $8.6 \cdot 10^{19}$ protons on target, with a detector live-time higher than 93%. It also recorded atmospheric neutrinos with a total exposure of 0.73 kton × year.

The concentration of electronegative impurities (essentially O_2) in LAr is a crucial parameter for the operation of a LAr-TPC; a large contamination would absorb a significant fraction of drift electrons, resulting in a degradation of signals. Significant experimental effort must be devoted to the Argon purification and recirculation; in the ICARUS-T600 case, the drift electron lifetime was kept higher than 7 ms during the whole data-taking period, as shown in figure 1; this results, obtained through continuous Argon recirculation both in the liquid and gas state, ensures a small signal attenuation during the drift (12% for a drift length of 1.5 meters) and represents an important milestone for future LAr-TPC with much larger drift lengths [3].

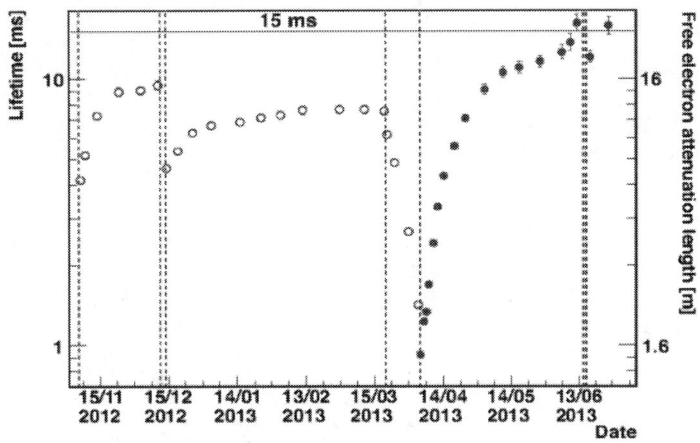

Figure 1. Electron lifetime as a function of time, during the last 7 months of the ICARUS-T600 run at LNGS. Dashed lines indicate periods were recirculation was not active, while the increase in the final weeks is due to the installation of a new, more powerful recirculation pump.

2.1 Search for atmospheric neutrinos

The analysis of the sample of atmospheric neutrinos recorded by ICARUS-T600 at LNGS (~ 0.73 kt) is ongoing; about 25% of the collected statistics has been analyzed so far. The search concentrated on multi-prong interaction vertex topologies with at least two charged particles, corresponding to ~ 30% of ν_μCC events.

An automatic filter algorithm allows identifying this kind of topologies, reducing the amount of events to be visually scanned by a factor ~ 200; in the analyzed statistics, 12 atmospheric neutrino interactions (2 electron-like, 3 muon-like) have been identified, in agreement with the expectation of 10 ± 2 events.

 DOI: 10.1051/epjconf/201716407017

2.2 Measurement of muon momentum via multiple Coulomb scattering

Multiple Coulomb scattering (*MCS*) is the only way to compute the momentum of non-contained muons, such as the ones from ν_μ charged current interactions. A dedicated algorithm developed for ICARUS-T600 was tested on a sample of ~ 1000 *stopping muons*, i.e. muons from CNGS neutrino interactions in the upstream rock, entering the active TPC volume and stopping or decaying there. This sample is an ideal benchmark for the MCS measurement, providing an independent calorimetric estimate of momentum, in an energy range (1÷ 5 GeV/c) relevant for future neutrino experiments.

The MCS and calorimetric estimates result in good agreement, as shown in Figure 2; the average resolution of the MCS estimates on the stopping muon sample is about 15%. A small (~ 15%) underestimation of the MCS measurement w.r.t. calorimetry, appearing for $p \gtrsim 3$ GeV/c, is related to the non-perfect planarity of the TPC cathode. This distortion was directly and independently measured after the end of the LNGS run, showing deviations that extend to ~ 3 cm (see Figure 3). They generate a non-uniformity of the electric field, resulting in spurious track deflections that mimic the effect of MCS. A full 3D field simulation showed that this effect can result in a momentum underestimation of ~ 10%, largely accounting for the observed underestimation.

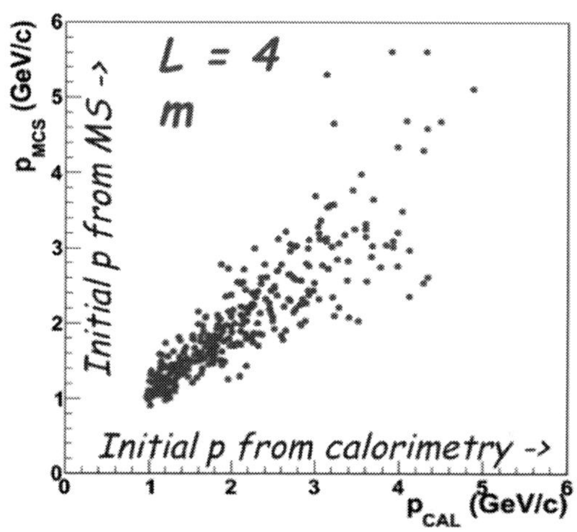

Figure 2. Momentum measured from MCS compared with momentum estimated by calorimetry, for the CNGS stopping muon sample. Muons with track lengths larger than 5 meters were considered, and only the first 4 meter of the muon tracks were used for MCS measurement.

2.3 Electron/photon separation

Since electron neutrino interactions usually represent the main signal in many present and future neutrino oscillation experiments, an efficient identification of electrons is a crucial feature of neutrino detectors; in particular, it is very important to efficiently reject the main background given by neutral current interactions with π^0 production.

A unique feature of LAr-TPCs is the very efficient electron-photon separation capability, due to the good spatial and calorimetric resolution; in particular, the accurate measurement of the ionization density dE/dx allows to discriminate showers initiated by electrons (characterizing genuine ν_e interactions) from ones initiated by photons, as illustrated in Figures 4 and 5.

DOI: 10.1051/epjconf/201716407017

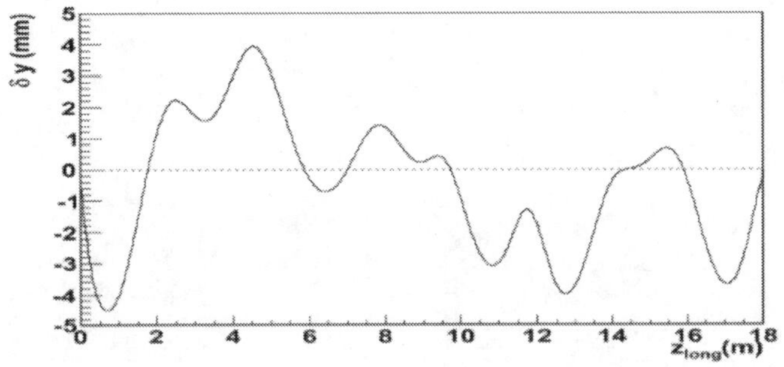

Figure 3. Deviations δy from cathode planarity as a function of the longitudinal coordinate z, at a height of 75 cm and a distance of 30 cm from the cathode plane. The δy values have been interpolated from direct measurements with a laser-meter, performed during the T600 refurbishing at CERN.

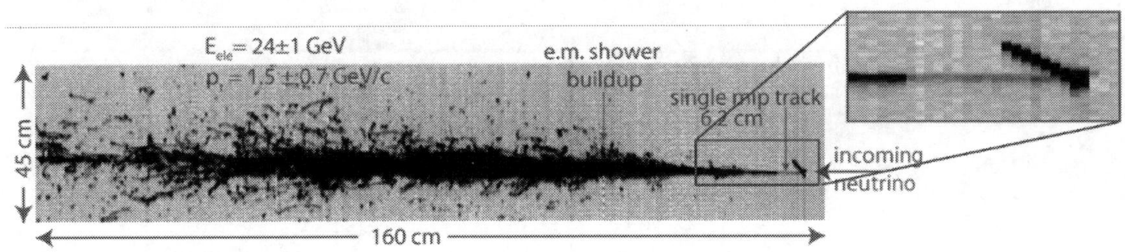

Figure 4. Example of an identified CNGS ν_e charged current interaction in ICARUS-T600 at LNGS.

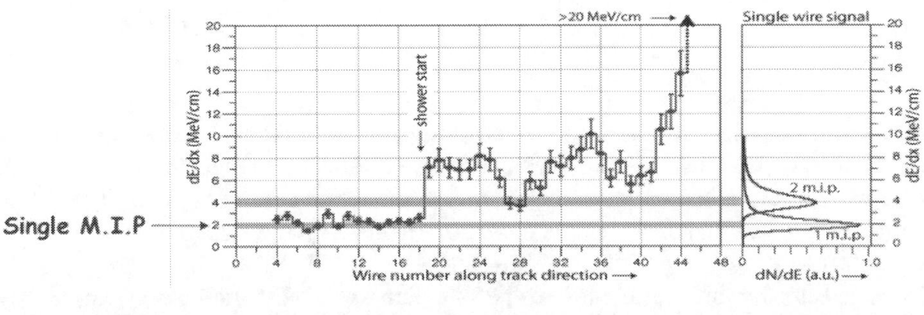

Figure 5. Ionization density as a function of wire number (along the track direction) for the first wires of the event in Figure 4. The grey bands represents dE/dx values corresponding to a single m.i.p. (~ 2.1 MeV/cm) and a double m.i.p.

3 Search for sterile neutrinos at the CNGS beam

ICARUS-T600 searched for possible oscillations mediated by a sterile neutrino in the range identified by LSND ($\sim 1eV^2$), by looking for a ν_e eccess in the CNGS ν_μ beam. A sample of $\sim 7.9 \cdot 10^{19}$ pot, corresponding to ~2650 neutrinos, was considered; the analysis was limited to the energy range between 10 and 30 GeV. In this sample, 7 electron neutrino events were found, compared to an expected background of 8.5 ± 1.1, dominated by the intrinsical ν_e beam contamination; the ν_e identification

 DOI: 10.1051/epjconf/201716407017

efficiency was estimated from a detailed MonteCarlo study, resulting ~ 74% with a negligible contamination from neutral current backgrounds. No excess was therefore observed, allowing to strongly constrain the allowed region for sterile neutrino oscillation in the $(\sin^2 2\theta, \Delta m^2)$ parameter space, identifying a small area $(\sin^2 2\theta, \Delta m^2) \sim (0.005, 0.5eV^2)$ where all available experimental data can be accommodated at 90% confidence level [4], as summarized in figure 6.

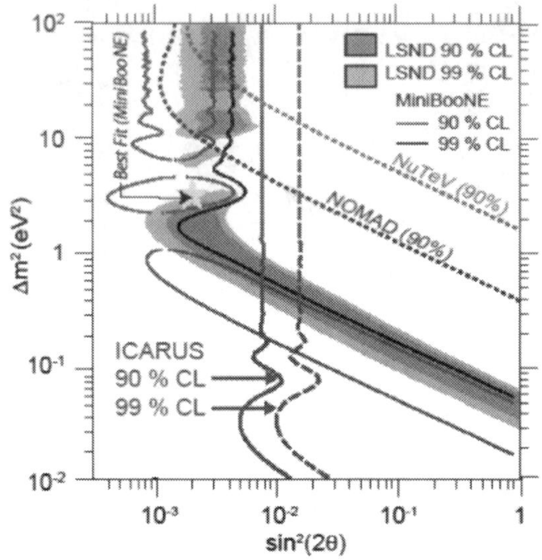

Figure 6. Limits in the sterile neutrino parameter space obtained by the ICARUS-T600 $\nu_\mu \rightarrow \nu_e$ appearance search, compared with LSND and MiniBoone allowed regions and limits from previous experiments.

4 The SBN experiment

The phenomenological scenario of sterile neutrino results still shows tension between different experimental results, and requires a solid verification of the flavour change "hints" provided by several experiments. In this complex scenario, the SBN experiment has the goal to clarify this issue by performing both appearance and disappearance measurements with the same beam and detectors.

In this experiment, neutrinos from the Booster beam at FNAL, with an average energy of ~ 800 MeV, will be detected by three different LAr-TPC detectors (SBND, MicroBoone and ICARUS-T600) at different distances from the target (about 100, 470 and 600 m respectively). This approach will allow a greater oscillation sensitivity, based on comparison of spectra at the near and far detector; in absence of anomalous signals due to sterile neutrinos, signals in these detectors should be identical, except for an obvious solid-angle factor, as illustrated in Figure 7. Moreover, the use of the same technology for all detectors will allow a strong reduction of systematics related to ν-Ar cross sections and to event identification and selection.

The appearance measurement will allow to fully cover the currently allowed region in the parameter space with 99% confidence level in 3 years of data taking, corresponding to $6.6 \cdot 10^{20}$ pot, as shown in Figure 8. On the other hand, no positive sterile neutrino signal has been recorded so far in the ν_μ disappearance channel; the SBN experiment will be able to extend the current limits by about one order of magnitude. The appearance and disappearance measurements will be dependent on different

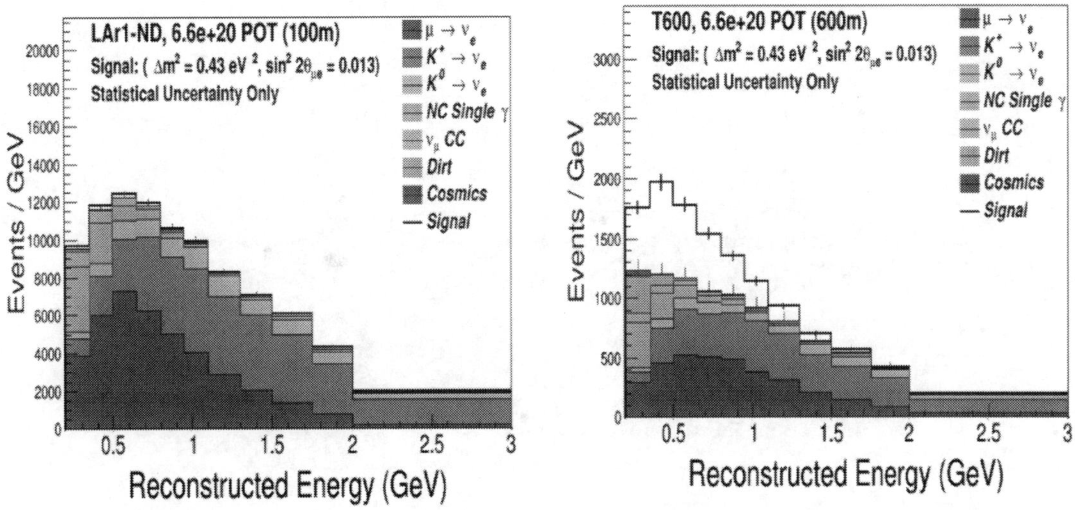

Figure 7. Energy spectrum of all ν_e events expected at SBND near detector (left) and ICARUS-T600 far detector (right) for 3 years of data taking at the Booster beam. The expected signal for a reference sterile neutrino case is shown in white, while the various background components are in colors.

mixing angles, parametrizing the mixing of the active flavors to the hypothetical sterile neutrino; this will allow to test both these mixings with the same experiment.

5 ICARUS-T600 overhauling at CERN

The ICARUS-T600 detector is currently undergoing a massive overhauling at CERN, in view of its use for the SBN experiment. This program (WA104) will introduce substantial improvements while keeping the present detector performance. In particular, the refurbishing work will involve several detector subsystems:

- New cold vessels, with a purely passive insulation;
- Renovated cryogenics and purification equipment;
- Reduction of the cathode non-planarity discussed in section 2.2 to a few millimeters;
- Upgrade of the light collection system, in order to achieve a high timing resolution (few ns), needed for the precise localization of events within the drift window; this is required for an efficient rejection of cosmic background. The number of PMTs will also be increased to 360, corresponding to $\sim 5\%$ coverage.
- New faster, higher-performance read-out electronics. In particular, the synchronous sampling on the whole detector will allow to improve MCS momentum resolution to $\sim 12\%$.

The T600 detector is scheduled to be transported to FNAL during 2017, to start installation and commissioning.

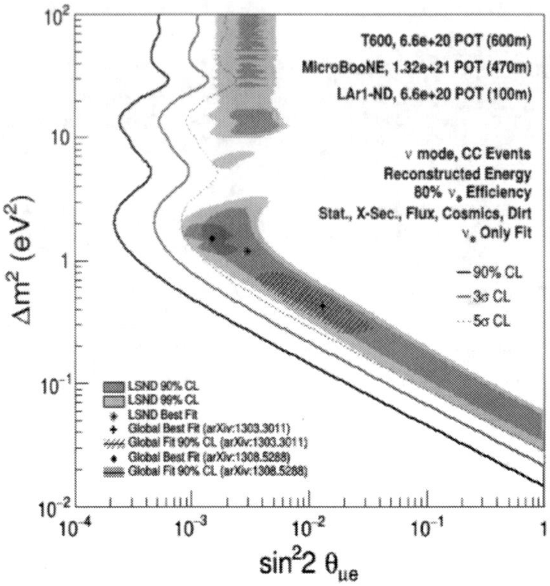

Figure 8. Expected sensitivity of the SBN program $\nu_\mu \rightarrow \nu_e$ appearance search for 3 years of data taking, compared with the LSND allowed region and two different global fits to all experimental data (see figure for references).

6 Conclusions

The successful ICARUS-T600 run at LNGS proved the maturity of the LAr-TPC technology, and represents a milestone towards future larger-scale detectors. It also provided interesting physics results, including strong constraints on sterile neutrinos.

Nevertheless, the sterile neutrino puzzle remains open and further experimental tests of this hypothesis are needed. The SBN experiment will definitely clarify the situation, by performing high-statistics, accurate searches in both the appearance ($\nu_\mu \rightarrow \nu_e$) and disappearance ($\nu_\mu \rightarrow \nu_\mu$) channels.

References

[1] S. Amerio et al. (ICARUS Coll.), NIM **A527**, 329 (2004)
[2] C. Rubbia et al. (ICARUS Coll.), JINST **6**, P07011 (2011)
[3] M. Antonello et al. (ICARUS Coll.), JINST **9**, P12006 (2014)
[4] M. Antonello et al. (ICARUS Coll.), Eur.Phys.J **C73**, 2599 (2013)
[5] R. Acciarri et al., *A Proposal for a Three Detector Short-Baseline Neutrino Oscillation Program in the Fermilab Booster Neutrino Beam*, arXiv:1503.01520

DOI: 10.1051/epjconf/201716407018

Introducing HEP to high-school and university students through ATLAS event analysis tools

Dimitris Fassouliotis[1], Christine Kourkoumelis[1], Stylianos Vourakis[1]

[1] Physics Faculty and Institute of Accelerating Systems and Applications, University of Athens Panepistimioupoli, Ilissia 15771, Greece

Abstract. Several EU outreach projects have been running for a few years now and have created a large number of inquiry based educational resources for high-school teachers and students. Their goal is the promotion of science education in schools though new methods built on the inquiry based education techniques, involving large consortia of European partners and implementation of large-scale pilots in a very large number of European schools. Until recently there has been a shortage of educational scenaria addressed to university students to be implemented in the framework of the laboratory courses. Two such scenaria were introduced recently at the National and Kapodistrian University undergraduate labs and are described below.

1 Introduction

The European Union (EU) has long recognized the decreasing interest of young people in science subjects and has been very supportive towards promoting outreach projects which would bridge the gap between school science education and active research.

The authors of this contribution have been involved in several EU outreach projects for more than a decade and have been developing several resources and good practices for use in high schools.

All resources were developed according to the Inquiry Based pedagogical framework, involving the phases of hypothesis, experimentation and conclusions. The authors have been promoting the scenaria in a large number of Greek high schools during half day mini-masterclasses. Up to very recently, there was a lack of more advanced scenaria for the university undergraduate labs.

As a result, the university students were not getting educated in an effective and efficient way in line with the most advanced HEP research and the state-of-the-art experimental and technological achievements. Consequently, the authors launched an effort to develop new laboratory exercises which would permit the undergraduate physics students to interact with the data taken at the world's most powerful accelerator the LHC. The intervention for university students has been designed in the context of the HYPATIA tool created by the authors.

2 Description of the "HYPATIA" data analysis tool

HY.P.A.T.I.A (Hybrid Pupil's Analysis Tool for Interactions in ATLAS) [1], is an event visualization tool that uses real data from the ATLAS experiment, namely offers a graphical representation of the products of proton collisions. The user can interactively analyse the LHC data collected almost in real-time. In addition, the tool has the flexibility which permits the user to perform different analysis tasks: from simple event browsing, to searches for the Higgs boson. The analysis paths are mainly based on virtual inspection of the events and the primary target audience is senior high school students. Two versions of HYPATIA exist and are both widely used. The off-line version [2] which has been used by IPPOG's International Masterclasses since 2008 in the so called "Z-path" [3] by thousands of high-school students in all continents each year. It uses java and has to be downloaded and run locally. The on-line version of HYPATIA [4] was first launched in 2010 by the University of Athens, in an effort to address all devices, including mobile phones and tablets; it is simpler to use, yet all the different searches can be performed. It includes four different parts each with an increasing level of complexity and difficulty. This version has been continuously developed in the context of the EU outreach projects [5-10] and has been used in about 100 Greek schools across the country during the last five years. In 2016 the HYPATIA online version was awarded the GOLC award [11] for the "best visualized experiment".

3 The HYPATIA tool for a university laboratory

There is a general lack of resources for university laboratory exercises which introduce students to state-of-the art technological achievements as well as to the "Big Science" experimental effort pursuit at CERN. The Higgs boson was discovered more than four years ago after a fifty year-long effort and although it attracted huge media coverage and generated great public interest, in most universities worldwide there is not a single experimental exercise to allow the students to experience the discovery themselves.

For the above reasons, the authors decided to introduce activities where the university students would directly connect themselves with the most advanced research done at CERN. These activities are designed in the context of HYPATIA but the university students are asked to perform more complex tasks. Since 2010 the HYPATIA tool has been used in a lab course at the National and Kapodistrian University of Athens (NKUA) for third year undergraduate physics students, majoring in HEP [12]. Each year about 50 students take part in the two three-hour long courses performing all the steps of event analysis mimicking the work of the ATLAS researchers (Figure 1). The first of the three hour lab guides the students to become familiar with the HYPATIA tool, visually identify the signatures which the different leptons leave in the detector and combine decay products to reconstruct the parent particles which in this case are the Z and Higgs bosons. In addition, after the end of the first part of the lab course, they have to hand-in a homework paper where they try to answer advanced questions which are related to the decays modes of the bosons, their widths, masses and experimental resolution and use statistical tools to investigate the significance of discoveries.

The second part of the lab course which has been recently implemented involves the students' analysis of large samples of events both simulated and real ones collected by ATLAS in the 2011-2012 period. ATLAS has released for public use $1fb^{-1}$ [13] of the Run I data and these are used by the students in the so called "batch" mode [14] of the HYPATIA tool. The main philosophy behind the "batch" analysis is to allow the students to realistically repeat the exact analysis which the ATLAS researches do by optimizing a number of event selection criteria. The goal of this process is to improve the signal to background rejection and eventually make discoveries.

DOI: 10.1051/epjconf/201716407018

Figure 1. Third year physics students at the NKUA working on the HYPATIA exercises and analyzing ATLAS events.

There are essentially three different analysis paths possible:

a) The optimization of criteria for the Z to two lepton decay in order to enhance the signal over background ratio.
b) The optimization of criteria for the Higgs to four lepton decay in order to make a "discovery" by suppressing the reducible background.
c) Simple analysis and studies of kinematical distributions of real events.

For the first two analysis paths three kinds of samples are required: simulated signal, simulated background and real events. For the third path -which is the simpler- mainly real events should be used.

 DOI: 10.1051/epjconf/201716407018

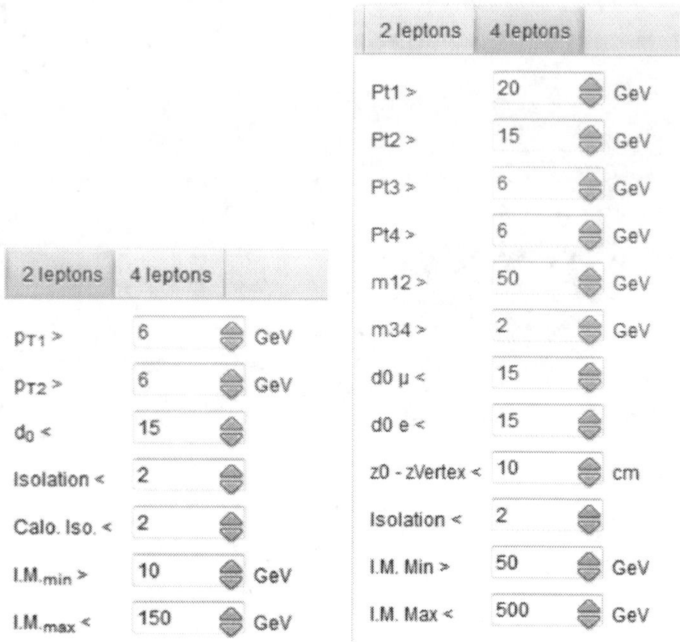

Figure 2. The sets of available optimization criteria. Left for the Z→two lepton decay and right for the Higgs→four lepton decay.

Below only the first two paths will be described in detail. In both of these a group of criteria is given using a simple GUI (Figure 2) and the students can optimize each one separately by inspecting three plots provided for each criterion: the simulated signal distribution as a function of the value of the criterion, the simulated background distribution and the value of the significance as a function of the criterion value. The student is guided to choose as an optimization value, the value where the significance $\sqrt{2*\left(\left(S+B\right)*\ln\left(1+\frac{S}{B}\right)-S\right)}$ is maximized and then proceed to the next criterion. The procedure should be repeated a few times for each criterion. The starting point of each iteration should be the most recent values of all criteria while optimizing one at a time. Figure 3 shows the optimization of the isolation criterion for the four lepton sample. The "isolation" provides an indication of whether or not there are tracks (or energy for "calo isolation") accompanying the leptons of the event. Leptons originating from the Z or Higgs events tend to be isolated whereas leptons from $b\acute{b}$ semileptonic decays -which are part of the reducible background- are as they are typically parts of a particle jet.

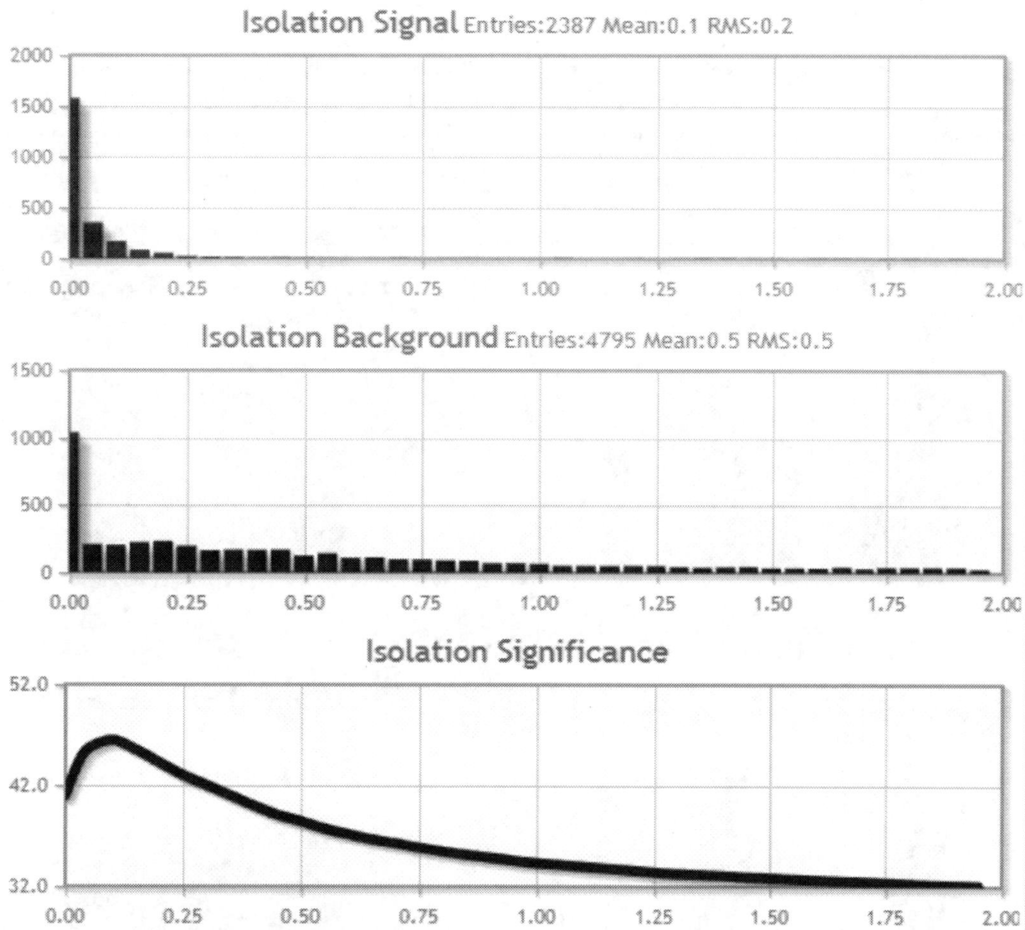

Figure 3. Histogram of the isolation criterion (for definition see above) .The top plot is the simulated signal, in the middle the simulated background and at the bottom the significance (for definition see above) as a function of the criterion value.

The criteria which are given to the students to optimize were chosen as close as possible to the ones which the ATLAS researches use in the analysis. These are (as shown on Figure 2)

a) For the two lepton analysis: the momenta of the pre-selected two leptons $P_{T1,2}$, the largest impact parameter of the lepton in the event (divided by its error) and the largest isolation parameter of the leptons in the event (both for accompanying tracks and energy). The impact parameter gives an indication of how far a secondary vertex is (if there is one) from the primary vertex of the event. Background events -especially in the case of jets from b quarks- tend to have well separated secondary vertexes.
b) For the four lepton analysis: The momenta of all pre-selected four leptons $P_{T1,2,3,4}$, the mass of the lepton pair (opposite sign, same flavor) closest in value to the Z mass (m_{12}), the mass of the next lepton pair (opposite sign, same flavor) closest in value to the Z mass (m_{34}), the largest impact parameter of the leptons in the event divided by its error (separately for muons and electrons) and the largest isolation parameter of the leptons in the event.

During the optimization iterations, the students can compare histograms of the invariant mass of the signal they are after -two leptons or four leptons- with the background before and after the application of the criteria. They can also compare the combined simulated signal plus background with the real data. Their ultimate goal -after they define the optimum set of cuts- is to get the simulated combined data after the application of the optimized criteria to look as close as possible to the histograms of real data after the application of the criteria which ATLAS researchers defined after their in depth analysis. For the Z analysis, the data released by the ATLAS experiment are enough for this comparison. In the Higgs case, since the Higgs published signal for Run I contains only a few tens of events [15] and the data released correspond to 1/25 of the whole sample, this is not possible (Figure 4). Instead the students are asked to compare their combined simulated four lepton histogram of Figure 4 with the ATLAS publication [15].

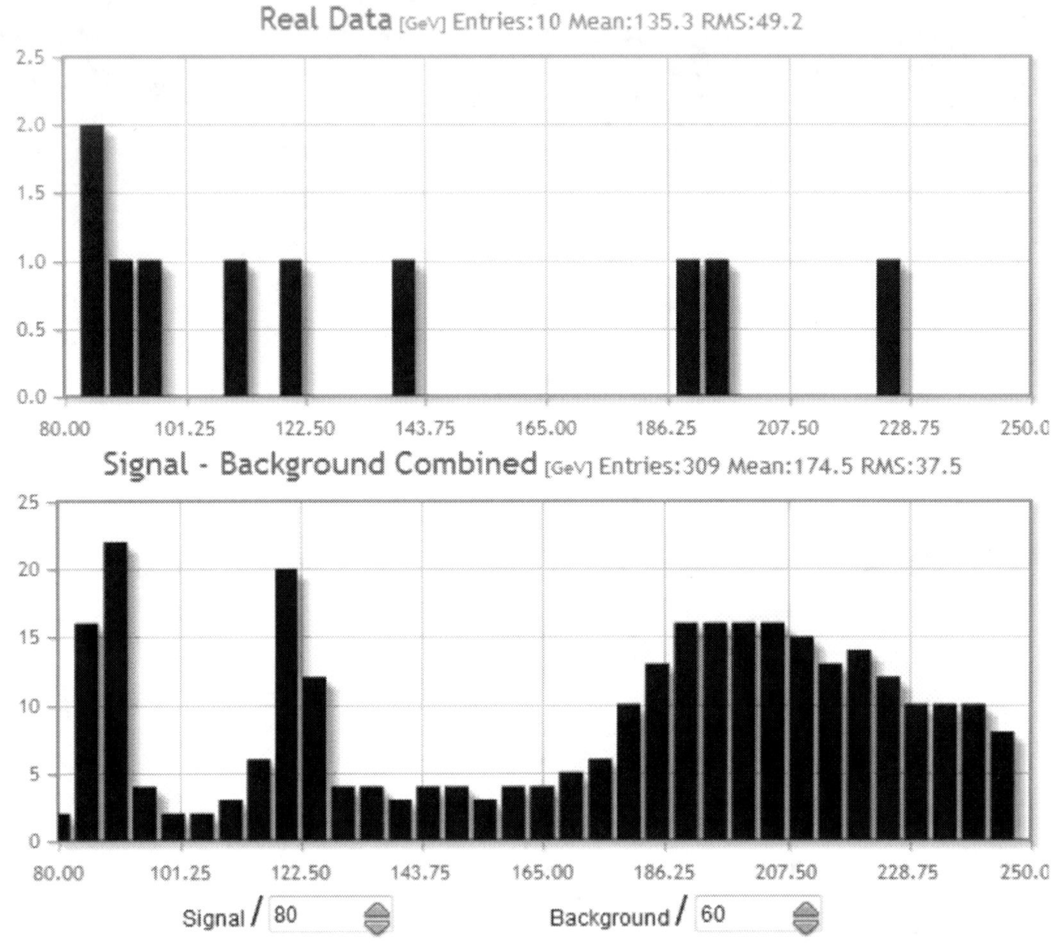

Figure 4. Histogram of four lepton invariant mass after application of the optimum criteria. Top: real data corresponding to 1 fb^{-1}. Bottom: signal plus background both simulated and combined according to their cross sections.

During the academic years 2015-2016 and 2016-2017 about 100 students in total have followed the recently designed lab course with great success. The part which includes the real ATLAS data was added only a few months ago and has been tried out by both the NKUA as well as very recently by the University of Birmingham students as the hands-on exercise of their 4th year Experimental Particle Physics Techniques course.

4 Conclusions

A two part university laboratory course has been developed which involves the analysis and selection criteria optimization of ATLAS events in a way which mimics the experiment's analysis. This way the undergraduate students are exposed to top-level research and grow a real enthusiasm for it.

The exercise is based on the HYPATIA event visualization tool and detailed instructions are available online [16]. It is also listed as an external resource [17] of the ATLAS opendata portal. The exercises have been presented in multiple outreach meetings and have attracted the interest of several universities.

5 Acknowledgements

This work was partially funded by the European Union in the context of the CREATIONS project (Grant Agreement no. 665917) under the Horizon 2010 program. This document does not represent the opinion of the European Union, and the European Union is not responsible for any use that might be made of its content. The authors would also like to thank Prof. K. Nikolopoulos for his efforts to create a mirror source of HYPATIA at Birmingham and include the exercise as part of his course.

6 References

1. Kourkoumelis, C., & Vourakis, S. (2014). Hypatia-An online tool for ATLAS event visualization. Physics Education, 49, 21-32
2. [Online] http://hypatia.phys.uoa.gr/
3. [Online] http://atlas.physicsmasterclasses.org/en/zpath.htm
4. [Online] http://hypatia.iasa.gr/
5. [Online] http://www.learningwithATLAS-portal.eu.
6. [Online] http://www.pathway-project.eu/
7. [Online] http://www.discoverthecosmos.eu/
8. [Online] http://www.go-lab-project.eu/
9. [Online] http://inspiringscience.eu/
10. [Online] http://creations-project.eu/
11. Global Online Laboratory Consortium: [Online] http://online-engineering.org/GOLC_about.php
12. D. Fassouliotis, C. Kourkoumelis, S.Vourakis, The Inspiring Science Education project and the resources for HEP analysis by university students, 4th ICNFP Kolympari 2015,EPJ Web of Conferences 126, 02017 (2016),http://dx.doi.org/10.1051/epjconf/201612602017
13. [Online] http://opendata.atlas.cern/
14. [Online] http://hypatia.iasa.gr/en/batch.html
15. G.Aad et al (ATLAS collaboration) http://dx.doi.org/10.1016/j.physletb.2013.08.010
16. [Online] http://hypatia.iasa.gr/en/Exercise_6_eng.pdf
17. [Online] http://opendata.atlas.cern/externals/

EPJ Web of Conferences **164**, 07019 (2017)
ICNFP 2016
DOI: 10.1051/epjconf/201716407019

Luminescence of water or ice as a new detection method for magnetic monopoles

Anna Obertacke Pollmann[1,a] for the IceCube Collaboration[b]

[1] *Dept. of Physics, University of Wuppertal, D-42119 Wuppertal, Germany*

Abstract. Cosmic ray detectors use air as a radiator for luminescence. In water and ice, Cherenkov light is the dominant light producing mechanism when the particle's velocity exceeds the Cherenkov threshold, approximately three quarters of the speed of light in vacuum.

Luminescence is produced by highly ionizing particles passing through matter due to the electronic excitation of the surrounding molecules. The observables of luminescence, such as the wavelength spectrum and decay times, are highly dependent on the properties of the medium, in particular, temperature and purity. The results for the light yield of luminescence of previous measurements vary by two orders of magnitude.

It will be shown that even for the lowest measured light yield, luminescence is an important signature of highly ionizing particles below the Cherenkov threshold. These could be magnetic monopoles or other massive and highly ionizing exotic particles. With the highest observed efficiencies, luminescence may even contribute significantly to the light output of standard model particles such as the PeV IceCube neutrinos.

We present analysis techniques to use luminescence in neutrino telescopes and discuss experimental setups to measure the light yield of luminescence for the particular conditions in neutrino detectors.

1 Magnetic monopoles

The existence of magnetic monopoles, particles carrying a single magnetic charge, is motivated by various theories which extend the Standard Model of particles, such as Grand Unified Theories (GUTs), String theory, Kaluza-Klein, and M-Theory [1]. The elementary magnetic charge g_D can be found by basic considerations to be dependent on the elementary electric charge $g_D = e/2\alpha \approx 68.5\,e$ where e is the elementary electric charge and α is the fine structure constant [2]. Other parameters of magnetic monopoles, such as their mass and predicted flux, highly depend on the details of the models regarding particle creation. In most theories the mass of magnetic monopoles is in the order of $10^7\,\mathrm{GeV}/c^2$ to $10^{13}\,\mathrm{GeV}/c^2$ (intermediate mass monopoles), or $10^{13}\,\mathrm{GeV}/c^2$ to $10^{19}\,\mathrm{GeV}/c^2$ (GUT /some SUSY monopoles). The mass is so high that monopoles cannot be created in any foreseeable particle accelerator but rather shortly after Big Bang through (intermediate steps of) gauge symmetry breaking [3]. The proposed inflationary phase of the early universe leads to a dilution of the monopole density.

[a] e-mail: anna.pollmann@uni-wuppertal.de
[b] http://icecube.wisc.edu

DOI: 10.1051/epjconf/201716407019

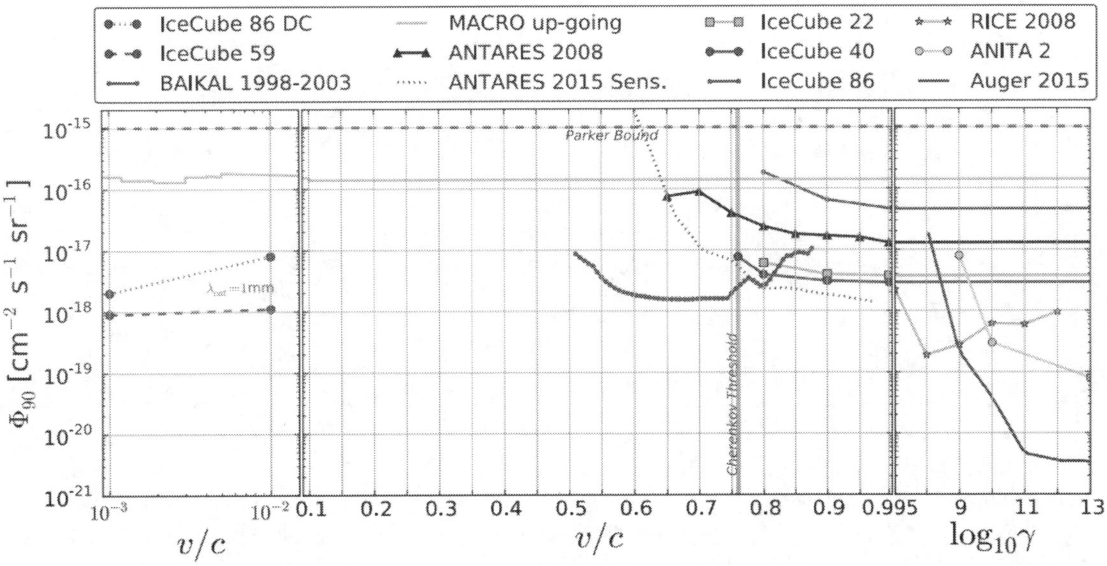

Figure 1. Current upper limits on the flux of magnetic monopoles in dependence on the speed of monopoles at the detector [7–15]. Best limits for cosmic monopoles are achieved by general purpose detectors only. However, low relativistic speeds, from $0.1\,c$ to $0.5\,c$, are not covered by them yet so that the latest limit in this region is almost two orders of magnitude higher than for all other speed ranges

Current experimental limits on the monopole flux are stronger than theoretical bounds. An overview of the most recent experimental limits as a function of the monopole speed at the detectors is given in Fig. 1. Monopoles at relativistic speeds are expected to have intermediate masses since only in this mass range they can be sufficiently accelerated in cosmic magnetic fields [4].

Special purpose detectors for monopoles use induction, visible damage of plastic targets, and time of flight measurement such as MACRO or MoEDAL [5, 6]. General purpose particle detectors are used for monopole searches if they have larger effective areas, leading to an increased sensitivity compared with special purpose detectors. This condition is fulfilled by all large scale neutrino telescopes, Baikal, ANTARES, and IceCube as shown in Fig. 1.

2 Monopole detection with IceCube

IceCube is a cubic-kilometer neutrino detector installed in the ice at the geographic South Pole [16] between depths of 1450 m and 2450 m. It comprises 86 strings each with 60 digital optical modules (DOMs). Neutrino reconstruction relies on the optical detection of Cherenkov radiation emitted by secondary particles produced in neutrino interactions in the surrounding ice or the nearby bedrock. The DeepCore sub-array, as defined in this work, includes 8 more densely instrumented strings optimized for low energies plus 12 adjacent standard strings.

The detector is triggered when a DOM and its nearest or next-to-nearest DOMs record a hit within a $1\,\mu$s window. Then all hits in the detector within a window of $10\,\mu$s will be readout and combined into one event [17]. A series of data filters are run on-site in order to select potentially interesting events for further analysis, reducing the amount of data to be transferred via satellite.

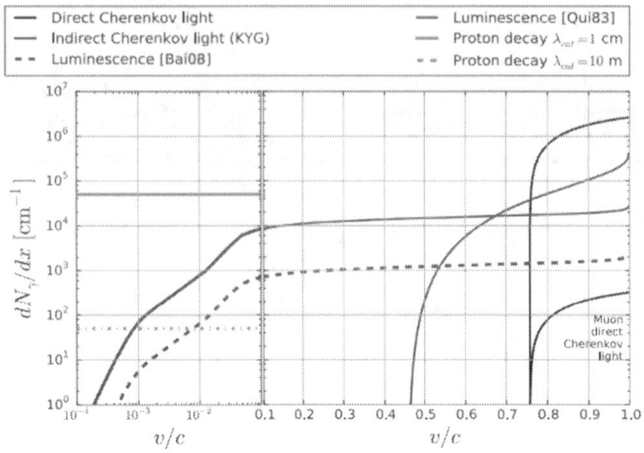

Figure 2. Light yield for the different detection mechanisms of magnetic monopoles in neutrino telescopes (here assuming the refraction index of ice). For comparison the direct Cherenkov light emitted by a bare muon is shown. The luminescence light yield is calculated by multiplying the luminescence efficiency with the energy loss of magnetic monopoles taken from Ref. [18]. Quenching effects are not taken into account because they are not measured for water yet

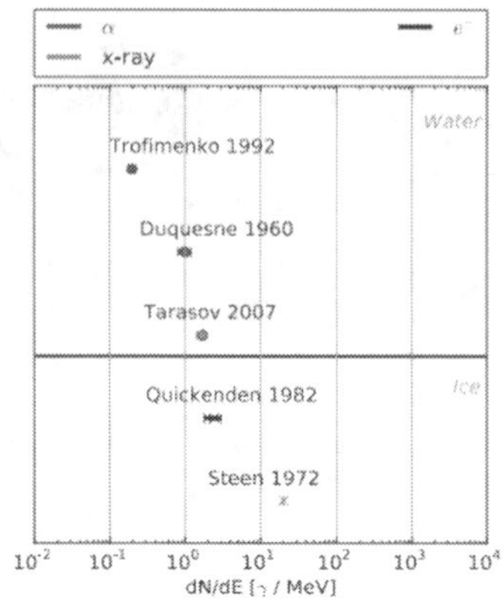

Figure 3. Overview of luminescence measurements at different phases of matter and using different irradiations, taken from Refs. [19–23]. Further measurements are summarized in Ref. [22]

Magnetic monopoles faster than the Cherenkov threshold in ice, $v_C \approx 0.76\,c$, emit Cherenkov light analogous to electrically charged particles. In addition monopoles ionize the surrounding matter, and most of the knocked-off δ-electrons are ejected with energies sufficient to produce Cherenkov light. This mechanism is called indirect Cherenkov light emission and produces light down to a monopole velocity of $\gtrsim 0.45\,c$. At ultra-relativisitc speeds radiative losses start to dominate the light production. In water these are dominantly photonuclear reactions.

Some theories predict the catalysis of proton decay by magnetic monopoles [24]. The decay products start particle cascades which are visible through Cherenkov radiation of the particles. If realized in nature, the Rubakov-Callan mechanism is the dominating light production mechanism at low monopole speeds.

An overview over the light yield of magnetic monopoles from different production mechanisms in dependence on the particle speed is given in Fig. 2. Low relativistic speeds, from $0.1\,c$ to $0.45\,c$, are not covered by the described mechanisms, but may be covered by luminescence.

3 Luminescence of water

Luminescence is defined in this text as the excitation of transparent media by ionizing radiation resulting in visible light. It was first concluded in 1953 that the luminescence, measured in water, originates from the water molecules itself and not from impurities [25]. Subsequently the effect was investigated in regard of the luminescence efficiency, lifetime of excited states, and wavelength spectrum to find the underlying mechanisms. The early results are summarized in Ref. [22]. Later measurements of the luminescence efficiency, or measurements not covered in this reference, are summarized in Fig. 3.

 DOI: 10.1051/epjconf/201716407019

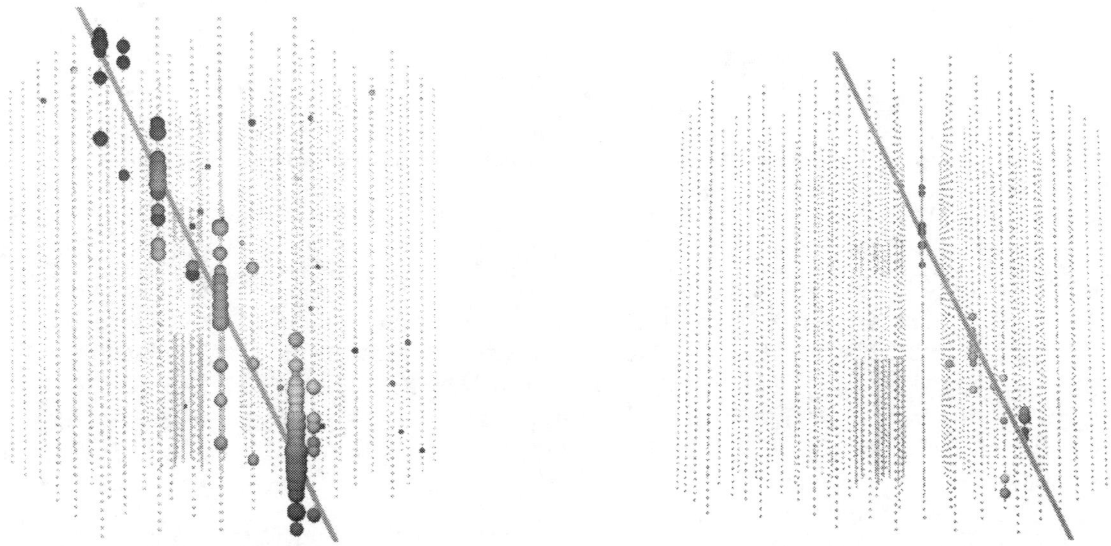

Figure 4. Event view of simulated magnetic monopoles. The gray dots represent the 5160 digital optical modules (DOMs) of IceCube. When a DOM has seen light it is colored from red (early) to blue (late) regarding hit time. The more light was recorded, the larger the radius of a DOM is enhanced. Both events develop up-going from the bottom right to the top left. **Left:** The simulated speed is $v = 0.52\,c$. The light originates from indirect Cherenkov radiation only. When luminescence light is added according to the light yield given in Fig. 2, the signature would be even brighter. **Right:** The simulated speed is $v = 0.3\,c$. The light originates from luminescence assuming a conservative efficiency of $dN_\gamma/dE = 0.2\,\gamma/\mathrm{MeV}$, as given in Ref. [8], integrated over wavelengths from 300 nm to 600 nm and with an assumed lifetime of 500 ns

The spectrum of luminescence peaks at visible wavelengths around 400 nm [22]. The measurements of the lifetimes of excited states are inconclusive. It was debated that the luminescence originates from transitions of excited OH^- or H_3O^+ molecules [26].

In Ref. [22] it is concluded that impurities in water enhance the luminescence efficiency. Also it is shown that the luminescence efficiency is highly dependent on the temperature [27].

4 Luminescence induced by magnetic monopoles

For Fig. 2 the light yield of monopoles through luminescence is estimated using the measurement results $dN_\gamma/dE = 0.2\,\gamma/\mathrm{MeV}$, given in Ref. [8], and $dN_\gamma/dE = 2.4\,\gamma/\mathrm{MeV}$, given in Ref. [22]. This values were chosen to include both a conservative and an optimistic value, for the actual conditions at the neutrino detectors. As shown, luminescence can even complement the Cherenkov light from proton decays at non relativistic speeds, with the additional advantage that it is independent of the monopole model and catalysis cross section.

Event signatures of magnetic monopoles with low relativistic speeds between $0.1\,c$ and $0.5\,c$ inducing luminescence light are simulated for the current data acquisition and trigger of IceCube. Event views are shown in Fig. 4 and the trigger efficiency is shown in Tab. 1. Adjustments of the triggers for higher sensitivity to slow and faint events requires an upgrade in computing resources which is feasible in short term. However, the IceCube detector in its current configuration is already capable of detecting luminescence signatures induced by magnetic monopoles at the considered speeds.

To use luminescence of water as a particle detection method, the exact properties have to be measured. In addition, the IceCube filters have already been extended to address this additional signature.

5 New IceCube filter for monopoles

As shown in Fig. 4, the signature of low relativistic magnetic monopoles inducing luminescence is a track which develops slowly through the whole detector. The brightness of the track decreases with lower speeds.

To achieve best filter efficiencies for both, faint and bright tracks, two selection criteria were developed for the new monopole filter. The *IceCube selection* (IC) is applied on hits recorded with all IceCube strings but the denser instrumented strings belonging to the sub-array DeepCore. The *DeepCore selection* (DC) is applied on the DeepCore strings and the adjacent IceCube strings. Both selections apply cuts on the minimum number of hit DOMs (IC/DC: 6), the maximum reconstructed speed (IC: $0.8\,c$, DC: $0.6\,c$) [28] to reduce background from muon signatures, and the minimal event time length[1] (IC: 4000 ns, DC: 2750 ns). In addition, the IceCube selection requires a minimal hit separation[2] of 250 m and a maximum gap[3] of 200 m. The DeepCore selection additionally requires an FWHM time[4] of minimum 2500 ns.

Table 1. Efficiency of the standard IceCube triggers and the new monopole filter in keeping simulated monopoles inducing luminescence. The filter efficiency is given with respect to the rate at trigger level. Efficiencies are given for the lowest (left in each column) and highest (right in each column) expected luminescence efficiencies. Below the speed $0.2\,c$ most faint events are selected by the DeepCore selection

		Standard trigger efficiency		Monopole filter efficiency	
luminescence efficiency		$0.2\,\gamma/\text{MeV}$	$2.4\,\gamma/\text{MeV}$	$0.2\,\gamma/\text{MeV}$	$2.4\,\gamma/\text{MeV}$
monopole speed	$0.1\,c$	9%	60%	25%	45%
	$0.3\,c$	36%	68%	58%	60%
	$0.5\,c$	46%	70%	64%	71%

The filter efficiency, shown in Tab. 1, is up to twice the efficiency compared to the usage of the previously installed filters not targeting monopoles. The filter rate is 30.8 Hz adding 1.4 GB to the bandwidth of about 100 GB which are transfered every day via satellite from the South Pole to the northern computer farms. The filter was installed in May 2016. The pass rate of low relativistic monopoles could be enhanced by accounting for this signature in the detector design and data acquisition systems closer to the hardware than the filters when a detector upgrade is due.

6 Measuring the properties of luminescence

To use luminescence as a detection method, one needs to measure the luminescence efficiency, the lifetimes of excited states, and the wavelength spectrum at the given conditions from the detectors. Two types of measurements are planned, one in situ and one in the laboratory, which complement each other. Since there are significant discrepancies between the results of previous measurements, it is mandatory to have reliable results of this work for further use of luminescence.

[1] Time difference between the last and the first hit of an event. Only the first pulse per DOM is taken into account.

[2] Separation length of the center-of-gravity of the first and last quantile of hits which are within a cylinder with radius of 100 m around the reconstructed track.

[3] Longest gap between hits within a cylinder with radius of 100 m projected on the reconstructed track.

[4] All first pulses per DOM are ordered in time. The charge distribution is then fit with a Gaussian and the full width of half maximum of this fit is taken as the FWHM time.

Luminescence depends on the radiation type and charge due to quenching effects. Since the luminescence of water itself is hardly studied, the dependency of quenching on the energy loss of the incident particle charge Z is unknown. The planned procedure of the laboratory measurement is to mimic a single magnetic charge g_D with a corresponding electric charge Ze with Z close to $g_D/e \approx 68.5$. If a heavy ion of this charge penetrates a water probe from the detector site with speed smaller than the Cherenkov threshold and comparable energy loss, the emitted light is comparable to the light emitted by a magnetic monopole. For preparation, measurements are planned using radioactive sources and an electron gun as well as a target of ultra pure water. The latter is done to study the luminescence of pure water itself and the influence of different added substances. In addition, the temperature dependence will be measured in detail.

The standard model particles measured in neutrino telescopes also induce the emission of luminescence light. However, the contribution of luminescence to the whole light output is at most 10% assuming the optimistic luminescence efficiency of $dN_\gamma/dE = 2.4\,\gamma/\mathrm{MeV}$. The contribution of luminescence can be identified in the time distribution of recorded light due to the long lifetimes. There are two kind of events which will be used for that. The first type are the astrophysical neutrino signatures which have the largest deposited energy, thus the largest luminescence contribution. The second type are minimal ionizing vertical muons moving close and along a string of DOMs. This event type is abundant in IceCube and the Cherenkov light can be used as a trigger.

7 Summary and outlook

The feasibility, to use luminescence as a new detection method for magnetic monopoles in neutrino detectors, is studied with IceCube. The signature of monopoles inducing luminescence is very likely detectable, thus a search for monopoles with low relativistic speeds would be possible. For this purpose, a new filter was developed and installed at IceCube, to identify the corresponding event signatures and to collect the events for future analysis. In addition, measurements are planned to improve knowledge of the properties of water and ice luminescence which allow a proper interpretation of the corresponding IceCube results.

References

[1] Y.M. Shnir, *Magnetic Monopoles* (Springer, 2005)
[2] P. Dirac, Proc. Roy. Soc. **A 133**, 60 (1931)
[3] A. Rajantie, Contemp. Phys. **53**, 195 (2012)
[4] S.D. Wick, T.W. Kephart, T.J. Weiler, P.L. Biermann, Astropart. Phys. **18**, 663 (2003)
[5] M. Ambrosio et al. (MACRO Collaboration), Eur. Phys. J. C **25**, 511 (2002)
[6] B. Acharya et al. (MoEDAL Collaboration) (2016)
[7] D.P. Hogan et al., Phys. Rev. D **78**, 075031 (2008)
[8] V. Aynutdinov et al. (BAIKAL Collaboration), Astrophys. J. **29**, 366 (2008)
[9] M. Detrixhe et al. (ANITA Collaboration), Phys. Rev. D **83**, 023513 (2011)
[10] S. Adrián-Martínez et al. (ANTARES Collaboration), Astropart. Phys. **35**, 634 (2012)
[11] R. Abbasi et al. (IceCube Collaboration), Phys. Rev. D **87**, 022001 1 (2013)
[12] M.G. Aartsen et al. (IceCube Collaboration), Eur. Phys. J. C **74**, 2938 (2014)
[13] I.E. Bojaddaini, G.E. Pavalas (ANTARES Collaboration), in *ICRC* (2015), POS1097
[14] T. Fujii (Pierre Auger Collaboration), in *ICRC* (2015), PoS(ICRC2015)319
[15] M.G. Aartsen et al. (IceCube Collaboration), Eur. Phys. J. C **76**, 1 (2016)

[16] A. Achterberg et al. (IceCube Collaboration), Astropart. Phys. **26**, 155 (2006)
[17] R. Abbasi et al. (IceCube Collaboration), Nucl. Instrum. Meth. A **601**, 2994 (2009)
[18] J. Derkaoui et al., Astropart. Phys. **9**, 173 (1998)
[19] I.I. Trofimenko (1992), preprint INR 765/92 (in Russian), later published in [8]
[20] M. Duquesne, I. Kaplan, Phys. Radium **21**, 708 (1960), (French)
[21] M.D. Tarasov et al., Instrum. Exp. Tech. **50**, 761 (2007)
[22] T.I. Quickenden, S.M. Trotman, D.F. Sangster, J. Chem. Phys. **77**, 3790 (1982)
[23] H.B. Steen, O.I. Sorensen, J.A. Holteng, Int. J. Radiat. Phys. Chem. **4**, 75 (1972)
[24] V.A. Rubakov, Rep. Prog. Phys. **51**, 189 (1988)
[25] E.H. Belcher, Proc. R. Soc. Lond. A **216**, 90 (1953)
[26] S.M. Trotman, T.I. Quickenden, D.F. Sangster, J. Chem. Phys. **85** (1986)
[27] T.I. Quickenden et al., J. Chem. Phys. **95**, 8843 (1991)
[28] M.G. Aartsen et al. (IceCube Collaboration), Nucl. Instrum. Meth. A **736**, 143 (2014)

NEWSdm: Nuclear Emulsions for WIMP Search with directional measurement

A. Di Crescenzo[1,2,a] on behalf of the NEWSdm Collaboration

[1] *Dipartimento di Fisica dell'Università Federico II di Napoli*
[2] *INFN Sezione di Napoli*

Abstract. Direct Dark Matter searches are nowadays one of the most exciting research topics. Several experimental efforts are concentrated on the development, construction, and operation of detectors looking for the scattering of target nuclei with Weakly Interactive Massive Particles (WIMPs). The measurement of the direction of WIMP-induced nuclear recoils is a challenging strategy to extend dark matter searches beyond the neutrino floor and provide an unambiguous signature of the detection of Galactic dark matter. Current directional experiments are based on the use of gas TPC whose sensitivity is strongly limited by the small achievable detector mass. We present an innovative directional experiment based on the use of a solid target made by newly developed nuclear emulsions and read-out systems reaching a position resolution of the order of 10 nm.

1 Introduction

Current experimental efforts in the field of direct Dark Matter (DM) searches are devoted to the search for rare interactions between galactic halo WIMPs and nuclei in a detector on Earth. As dark matter detectors are rapidly improving in sensitivity, they will encounter the neutrino background, at which point Solar, atmospheric, and diffuse supernova neutrinos can mimic the dark matter signal. Neutrinos are therefore the ultimate background for WIMP direct detection searches as they cannot be shielded against and produce recoils with similar rates and energy spectra. Moreover, the recent controversy in the low-mass WIMP region O(10 GeV/c^2), where some dark matter hints are inconsistent with null results, highlights the need for additional discrimination power between WIMP events and backgrounds in order to clearly authenticate a genuine WIMP signal.
New generation detectors capable of measuring the direction of a nuclear recoil track resulting from the elastic scattering of a target nucleus by an incoming WIMP (Weakly Interacting Massive Particle), would provides a route towards the discrimination of neutrino background and the unambiguous identification of WIMPs as being responsible for the galactic dark matter.
From the experimental point of view several approaches have been proposed. The more diffuse strategy is based on the use of low pressure gaseous Time Projection Chambers (TPCs) [1]. Experiments based on this technology have provided the first directional limits in the Spin-Dependent (SD) case and the collaborations aim at developing larger detectors to be operated in a medium term time scale. Nevertheless this technology is hardly scalable to very large detectors masses needed to reach a good

[a] e-mail: dicrescenzo@na.infn.it

sensitivity to the Spin-Independent (SI) case.
The use of a solid target for directional searches would overcome the mass limitation of gaseous TPC approach thus allowing to reach an high sensitivity in the low cross section sectors of the SI case. Nevertheless, in a solid medium, the track of the WIMP-scattered nuclear recoil will have a path length of the order of a few hundred nanometers, much shorter than in the case of a gaseous target where the recoil length is expected to be of the order of a few millimeters. A detector with high tracking resolution is therefore needed.

2 The NEWSdm experiment

The approach proposed by the NEWSdm Collaboration [2] consists in the use of a nuclear emulsion-based detector acting both as target and as tracking device. The project foresees the employment of a novel emulsion technology called Nano Imaging Trackers (NIT) [3, 4] featuring a position resolution of an order of magnitude higher than that of the emulsion currently used in the OPERA experiment [5]. The detector is conceived as a bulk of NIT surrounded by a shield to reduce the external background. The detector is then placed on an equatorial telescope in order to absorb the Earth rotation, thus keeping fixed the detector orientation with respect to the incoming apparent WIMP flux. The angular distribution of the WIMP-scattered nuclei is therefore expected to be strongly anisotropic with a peak centered in the forward direction.

2.1 NIT: Nano Imaging Trackers

Nuclear emulsions are made of silver halide crystals embedded in a gelatine matrix. When ionizing particles pass through it, some of the halide crystals are modified in such a way that they are turned into grains of silver when a *developing* process is performed. The three-dimensional trajectory of passing through particles can be reconstructed with an optical microscope by connecting all the silver grains.
NIT are a novel type of emulsion films with grain diameters down to a few tens of nm, one order of magnitude smaller than conventional ones. They have a linear density of crystals of about 11 crystals/μm [3] and they are made by the following nuclei (mass fraction in %): H (1.6), C (10.1), N (2.7), O (7.4), S (0.3), Br (32.0), Ag (44.0), I (1.9). The total density is 3.2 g/cm^3. They make the reconstruction of trajectories with path lengths shorter than 100 nm possible, if analyzed by means of microscopes with enough resolution.
The presence in the emulsion gel of lighter nuclei such as carbon, oxygen and nitrogen, in addition to the heavier nuclei of silver and bromine, is a key feature of the NEWS project, resulting in a good sensitivity to WIMPs with both light and heavy masses.

2.2 Read-out system

In the NEWS experiment the expected WIMP signal will consist of short-path, anisotropically distributed, nuclear recoils over an isotropically distributed background. The search for signal candidates requires the scanning of the whole emulsion volume. The read-out system has therefore to fulfill two main requirements: a fast, completely automated, scanning system is needed to analyse the target volume over a time scale comparable with the exposure; the spatial resolution has to be improved by more than one order of magnitude compared to that achieved with standard emulsion films, reaching the challenging value of a few tens of nanometers, in order to ensure high efficiency and purity in the selection of signal candidates.

The analysis of NIT emulsions is performed with a two-step approach: a fast scanning with a state-of-the-art resolution for the signal preselection followed by a pin-point check of preselected candidates with unprecedented nanometric resolution to further enhance the signal to noise ratio and perform very accurate measurements of the range and the recoil direction. In the first phase a fast scanning is performed by means of an improved version of the optical microscope used for the scanning of the OPERA films [7, 9]. In the last years an R&D program aimed at improving the ESS performances was carried by INFN groups, leading to prototypes with resolution improved by one order of magnitude, achieving a speed of almost 200 cm^2/h [8]. A new system is being developed in Japan (the Super-Ultra Track Selector), aiming at increasing the scanning speed up to 5000 cm^2/h.
The starting point of the emulsion scanning is the image analysis to collect clusters made of dark grains at several depths across the emulsion plate thickness. Given the intrinsic resolution of the optical microscope ($\sim$ 200 nm), the sequence of several grains making a track of a few hundred nanometers, appears as a single cluster. Therefore, the key element to distinguish clusters made of several grains from clusters made of a single grain produced by thermal excitation (*fog*) is the analysis of their shape. A cluster made of several grains indeed tends to have an elliptical shape with the major axis along the direction of the trajectory, while a cluster produced by a single grain tends to have a spherical shape.
In order to simulate the effect of a WIMP-induced nuclear recoil and to measure the efficiency and the resolution of the new optical prototype, a test beam with low velocity ions was performed. Kr ion beam with energies of 200 and 400 keV [10] and a C ion beam with energies of 60, 80 and 100 keV were used.
When analysed with the optical microscope, submicrometric tracks produced by Kr and C ions appear as shown in Fig. 1(a). Although silver grains belonging to the tracks are not distinguishable and appear as a single cluster, the elongated form of the cluster is clearly visible [11]. An elliptical fit of the cluster shape allows a clear separation between fog grains and signal tracks: the latter ones are expected to have ellipticity larger than a given threshold, typically 1.25 or higher.
In order to evaluate the intrinsic angular resolution of the scanning system we analysed an emulsion sample exposed to a 2.8 MeV neutron beam at the Fusion Neutronics Source (FNS) of the Japan Atomic Energy Agency (JAEA). An intrinsic angular resolution of 230 mrad was achieved with fully automated scanning systems, by far the best resolution achieved with direction sensitive detectors in this energy range. The simulation shows that this result is compatible with the measurement reported above when the scattering contribution is included.
The second phase is performed exploiting the resonance effect occurring when nanometric metal grains are dispersed in a dielectric medium [12]. The polarization dependence of the resonance frequencies strongly reflects the shape anisotropy and can be used to infer the presence of non-spherical nanometric silver grains. NEWS will use this technology to retrieve track information in NIT emulsions beyond the optical resolution. Images of the same cluster taken with different polarization angles will show a displacement of the position of its barycenter. The analysis of the displacements allows to distinguish clusters made of a single grain from those made of two (or more) grains.
The evaluation of the position accuracy was performed by analysing images of single grains. The unprecedented accuracy of about 10 nm can be achieved in both coordinates (Fig. 1(b)).

3 Background

Background sources for dark matter searches are α and β particles, γ-rays and neutron induced recoils, while NIT are essentially not sensitive to minimum ionizing particles (MIP).
The main sources of α-particles are U and Th radioactive chains and Radon. The α-particles produced in those processes have energies of the order of MeV and their range in emulsion is of the order of tens

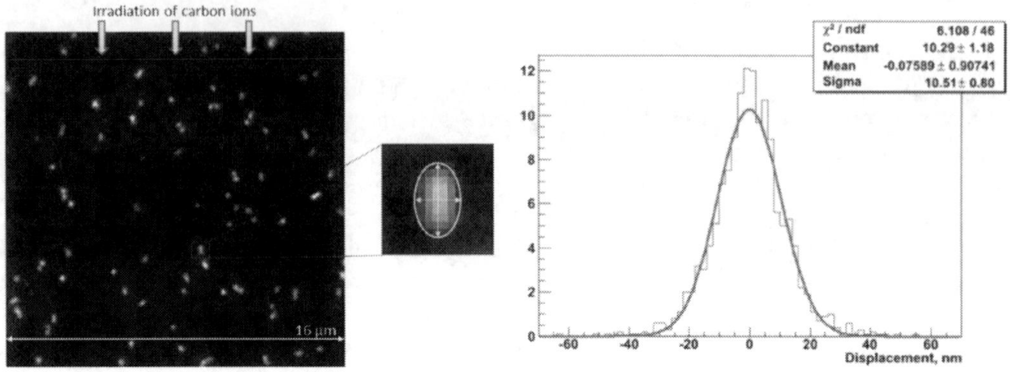

(a) (b)

Figure 1. 1(a) Kr ions implanted on NIT films. The image is taken with an optical microscope. The selection of candidate tracks is based on the elliptic fit of the clusters. 1(b) Position accuracy of about 10 nm with the resonant light scattering.

of microns, by far longer than WIMP-induced nuclear recoils. α-particles can therefore be identified and discarded in the emulsions by an upper cut on the track length.
The β-rays produced in ^{14}C decay constitute a non-negligible contribution to the total background budget. This kind of background is anyway less critical for NIT emulsion with respect to other sources: electrons can be rejected by properly regulating the emulsion response, in terms of number of sensitized crystals per unit path length (i.e. the sensitivity), through a chemical treatment of the emulsion itself. Moreover possible improvements in the rejection power can be achieved exploiting the response of β-rays to the polarized light scattering or performing a cryogenic exposure and by exploiting the phonon effect.
Neutron induced recoils are the main background source because they are not distinguishable from the expected WIMP signal, except for the isotropic angular distribution and for the typical track length, largely exceeding the range expected for WIMP-induced recoils. Three types of neutron sources affect underground experiments: radiogenic neutrons in the MeV range produced in (α, n) and spontaneous fission reactions in the detector due to its intrinsic radioactive contaminants, cosmogenic neutrons with energy spectrum extending to GeV energies induced by muons penetrating underground through the rock, neutrons induced by environmental radioactivity.
While the external neutron flux can be reduced to a reasonable level with an appropriate shielding, the intrinsic emulsion radioactivity would be responsible of an irreducible neutron yield through (α, n) and ^{238} U spontaneous fission reaction. In order to estimate this contribution, the activities of U and Th in the emulsion components has been measured with the Inductively Coupled Plasma Mass Spectrometry (ICP-MS) and with the γ-spectrometry. By using a dedicated MC simulation based on the SOURCES code [13], the corresponding neutron yield has been estimated to be of the order of 1.2± 0.4 n/yr/kg. The neutron energy spectrum, as calculated with SOURCES, was then used as the input for a GEANT4-based simulation in order to estimate the fraction of neutrons interacting in the emulsion and laying in the signal region (100 nm < L < 1 μm). The detectable neutron-induced background would be 0.02 ÷ 0.03 per year per kilogram [14].

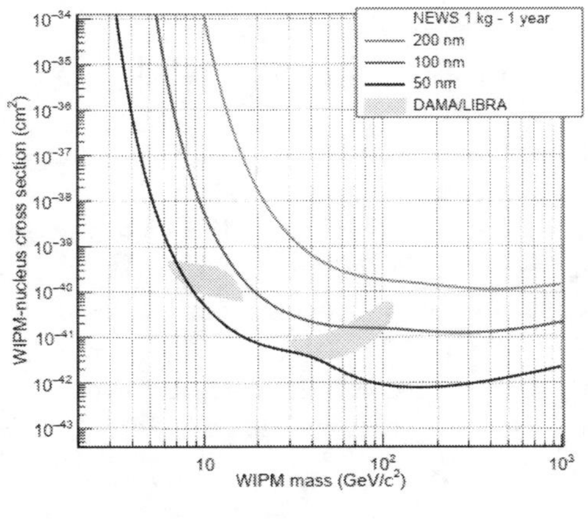

Figure 2. The 90% C.L. upper limits for a NIT detector with an exposure of 1 kg × year, a threshold ranging from 200 nm down to 50 nm, in the zero background hypothesis. The directionality information is not included.

4 Experiment time-scale and sensitivity

We plan to perform the first exposure with a target mass of 1 kg and the corresponding analysis of the data on a time scale of five years. The first year of the project will be devoted to the realization of a pilot exposure of 10 gr in order to confirm the estimations of the overall background budget of the experiment. The second and third year of the project will be devoted to the construction of the infrastructures, the production of the emulsion target and of the detector shield. The exposure of 1 Kg detector is foreseen by the beginning of 2020 and will last one year. We plan to get first analysis results by the end of 2021.
In Fig. 2 the 90% C.L. upper limit, in case of null observation for an exposure of 1 kg×year of NIT emulsions, with a minimum detectable track length ranging from 200 nm down to 50 nm and in the hypothesis of zero background, is shown. Even not including the directionality discrimination of the signal and assuming to reach a negligible background level, such an experiment would cover a large part of the parameter space indicated by the DAMA/LIBRA results with a small (1 kg) detector mass, using a powerful and complementary approach.

5 Conclusions

NEWS is meant to be the first experiment with a solid target for directional dark matter searches: the use of a nuclear emulsion based detector, acting both as target and tracking device, would allow to extend dark matter searches beyond the neutrino floor and provide an unambiguous signature of the detection of Galactic dark matter.
The novel emulsion technology, based on the use of nuclear emulsion with nanometric AgBr crystals (NIT), makes it possible to record the sub-micrometric tracks produced by the WIMP scattering off a target nucleus. The presence, in the emulsion components, of light and heavy nuclei results in an enhanced sensitivity to both light and heavy WIMP masses.
The final signal confirmation is obtained with powerful optical microscope equipped with a light polarizer: exploiting the different response of non spherical grain clusters to different polarization

angles, the unprecedented spatial resolution of 10 nm is obtained. This resolution allows to resolve grains belonging to a few hundred of nanometer long tracks thus providing the
final signal confirmation with a very high signal to noise ratio.
We plan to perform a pilot experiment with a 1 kg mass target on a time scale of five years: even using a rather small detector mass we would be able to explore the region indicated by the DAMA experiment with a powerful and complementary approach.
The actual intrinsic radioactive level allows to scale the target mass and exposure time up to one order of magnitude. A careful selection of the emulsion components and a better control of their production could further increase the radiopurity, thus allowing larger detector mass. The reduction of the fog density and further developments of the optical microscopy with polarized light would allow to reduce the detection threshold down to 50 nm.
Improvements both in the mechanics (use of piezoelectric-driven objective) and in the image acquisition (use of multiple image sensors) envisage the possibility to analyse with such a resolution a volume of 100 kg or larger.

References

[1] J.B.R. Battat et al., Phys. Rep **662** (2016) 1
[2] A. Alexandrov et al. (NEWS Collaboration), arXiv:1604.04199, LNGS-LOI 48/15
[3] M. Natsume et al., Nucl. Instr. Meth. **A575** (2007) 439
[4] T. Naka et al., Nucl. Instrum. Meth. **A718** (2013) 519
[5] A. Agafonova et al. (OPERA Collaboration), JINST **4** (2009) P04018.
[6] B. Morgan, A.M. Green and N.J.C. Spooner, Phys. Rev. D **71** (2005) 103507
[7] N. Armenise et al., Nucl. Instr. Meth. **A551** (2005) 261
L. Arrabito et al., Nucl. Instr. Meth. **A568** (2006) 578
[8] A. Alexandrov, V. Tioukov, M. Vladymyrov, JINST **9** (2014) C02034
[9] K. Morishima and T. Nakano, JINST **5** (2010) P04011
S. Aoki et al., Nucl. Instrum. Meth. B **51** (1990) 466
[10] T. Naka et al., EAS Publ. Ser. **53** (2012) 51
[11] M. Kimura and T. Naka, Nucl. Instrum. Meth. **A680** (2012) 12
[12] H. Tamaru et al., Applied Phys. Lett. **80** (2002) 1826
[13] W.B. Wilson et al., LA-13639-MS, Los Alamos (1999)
[14] A. Alexandrov et al., Astropart. Phys. **80** (2016) 16

Nonlinearity of the forward-backward correlation function in the model with string fusion

Vladimir Vechernin[1,⋆]

[1]*Saint Petersburg State University, 7/9 Universitetskaya nab., St.Petersburg, 199034 Russia*

Abstract. The behavior of the forward-backward correlation functions and the corresponding correlation coefficients between multiplicities and transverse momenta of particles produced in high energy hadronic interactions is analyzed by analytical and MC calculations in the models with and without string fusion. The string fusion is taking into account in simplified form by introducing the lattice in the transverse plane. The results obtained with two alternative definitions of the forward-backward correlation coefficient are compared. It is shown that the nonlinearity of correlation functions increases with the width of observation windows, leading at small string density to a strong dependence of correlation coefficient value on the definition. The results of the modeling enable qualitatively to explain the experimentally observed features in the behavior of the correlation functions between multiplicities and mean transverse momenta at small and large multiplicities.

1 Introduction

The investigations of long range rapidity correlations give the information about the initial stage of high energy hadronic interactions [1]. In paper [2] the study of the correlations between multiplicities in two separated rapidity windows (the long range forward-backward correlations) to find a signature of the string fusion and percolation phenomenon [3–6] in ultrarelativistic heavy ion collisions has been proposed. Later it was realized [7–12] that the investigations of the forward-backward correlations involving intensive observables, such e.g. as the event-mean transverse momentum, enable to obtain more clear signal about the initial stage of hadronic interaction, including the process of string fusion, compared to usual forward-backward multiplicity correlations.

In the present work, correlation functions and correlation coefficients between the observables in two separated rapidity intervals in ultrarelativistic heavy ion collisions are analyzed by analytical and MC calculations in a simple model which enables to take into account the string fusion effects. It is shown that the nonlinearity of correlation function increases with the width of observation windows, leading at small string density to the strong dependence of correlation coefficient value on the definition. Whereas the forward-backward correlation function itself is uniquely defined and carries more physical information than the correlation coefficients deduced from it.

⋆e-mail: v.vechernin@spbu.ru

2 Definitions

The forward-backward (FB) correlation function, for the correlation between two observables, F and B, measured event-by-event in rapidity intervals, $\delta\eta_F$ and $\delta\eta_B$, separated by some rapidity gap, η_{gap}, gives the dependence of the mean value of the variable B on the variable F:

$$\langle B\rangle_F = f(F)\,, \tag{1}$$

In present paper we restrict ourself to the study of long range FB correlations between the multiplicities of charged particles (n-n) and between the event-mean transverse momentum and the multiplicity (p_t-n). So as F we use the multiplicity of charged particles, n_F, measured in forward rapidity window and as B we use the multiplicity of charged particles, n_B, measured in backward rapidity window or the event-mean transverse momentum of these particles, p_B:

$$p_B = \frac{1}{n_B}\sum_{i=1}^{n_B} |\mathbf{p}^i_{tB}|\,. \tag{2}$$

In the last case of the p_t-n correlation we expect smaller influence on results of the trivial fluctuations in the number of sources (strings) - the so called "volume" fluctuations. Being an intensive variable, the p_B does not depends directly on the number of strings. It is sensitive to changes in the characteristics of the sources, e.g. due to the string fusion processes.

Usually the strength of the FB correlation is characterized by the correlation coefficient b_{FB}. Unfortunately the definition of the correlation coefficient is not unique (see e.g. [13] for a review). One of the most popular definition [14, 15] is

$$b_{FB} = \frac{\langle FB\rangle - \langle F\rangle\langle B\rangle}{\langle F^2\rangle - \langle F\rangle^2} = \frac{\mathrm{cov}(F,B)}{D_F}\,. \tag{3}$$

Another possible option [8, 9]:

$$b'_{FB} = \left.\frac{d\langle B\rangle_F}{dF}\right|_{F=\langle F\rangle}\,. \tag{4}$$

In the case of an integer variable F, as n_F, one has to interpret this definition as follows

$$b'_{FB} = \langle B\rangle_{[\langle F\rangle]} - \langle B\rangle_{[\langle F\rangle]+1}\,, \tag{5}$$

where the $[\langle F\rangle]$ is an entire part of the $\langle F\rangle$. It is clear that the definitions of b_{FB} and b'_{FB} are the same for the linear correlation function

$$\langle B\rangle_F = a + b_{FB}F\,, \tag{6}$$

the so-called linear regression, differing from each other in the general case.

Due to the locality of strong interaction in rapidity the FB correlation naturally divides into two parts: the short- and the long-range ones. The short-range correlation takes place only between particles produced from a same source (string). It goes to zero for large enough rapidity gaps, $\eta_{gap} \gtrsim 2$. The origin of the long-range correlations at large η_{gap} is event-by-event variance in the number and properties of particle sources (strings [13], cut pomerons [16] and so on), formed in high energy hadronic interaction. In present paper we will consider only these long-range part of the correlations, as we are interested in the study of the variations in number and properties of the strings, produced at the initial stage of the hadronic interaction.

3 The model

We will take into account the string fusion effects in simplified form by introducing a lattice in the transverse plane. This discrete version of string fusion model [5, 6] was introduced in [17] and then was exploited for a description of various phenomena (correlations, anisotropic azimuthal flows, the ridge) in high energy hadronic collisions [8–12, 19–21]. In this approach one splits the impact parameter plane into M cells with the area equal to the transverse area of single string, σ_{str}, and supposes the fusion of all strings with the centers in a given cell. Then each string configuration is characterized by the set of integers:

$$C_\eta = \{\eta_1, ..., \eta_M\} \ , \tag{7}$$

where η_i is a number of initial strings fused in a given i-th cell . This leads to the splitting of the transverse area into domains with different, fluctuating values of color field within them. What is similar to the attempts to take into account the density variation in transverse plane in models based on the BFKL evolution [22] and on the CGC approach [23].

In this model one supposes that the η_i in each cell fluctuates around the same mean value η with a scaled variance ω_η. Then in accordance with the string fusion prescription [5, 6] the mean number of charged particles in forward and backward observation windows, produced from the fragmentation of fused strings in the i-th cell, is given by the expression

$$\overline{n}_i^F = \mu_F \sqrt{\eta_i} \ , \qquad \overline{n}_i^B = \mu_B \sqrt{\eta_i} \ , \tag{8}$$

where μ_F is a mean number of particles produced from decay of a single string in the forward window $\delta\eta_F$. Clear that μ_F is proportional to the window width: $\mu_F = \mu_0\, \delta\eta_F$ and the same for the backward window. Note that in the framework of the present model we can simultaneously calculate the results without taking into account the string fusion effects simply using instead of (8) the dependence $\overline{n}_i^F = \mu_F \eta_i$.

Because we study only LRC, we also suppose the independent fragmentation of the i-th fused string in forward and backward windows. So the corresponding numbers of charge particles n_i^F and n_i^B fluctuates independently event-by-event around the above mean values with some scaled variance ω_μ.

Finally, in accordance with the string fusion ideas [5, 6] we suppose that the transverse momentum distribution of particles, produced from the η_i fused strings, has enlarged mean transverse momentum:

$$\overline{p}(\eta_i) = p_0 \sqrt[4]{\eta_i} \ , \tag{9}$$

where p_0 is a mean transverse momentum for particles produced from a decay of a single string.

Basically in this model one can calculate the correlation functions and the correlation coefficients by direct MC simulations using the definitions (1)-(5). To calculate n-n correlation one needs to generate the following configuration

$$C = \left\{C_\eta, C_n^F, C_n^B\right\} \ , \tag{10}$$

where

$$C_n^F = \left\{n_1^F, ..., n_M^F\right\} \ , \quad C_n^B = \left\{n_1^B, ..., n_M^B\right\} \ . \tag{11}$$

Here n_i^F and n_i^B are the numbers of charged particles, produced in forward and backward windows from the fragmentation of fused string in i-th cell. For p_t-n correlation one needs also to generate the transverse momenta p_{ti}^{jB} of all n_i^B particles produced in backward window for each cell:

$$C_p^B = \left\{p_{t1}^{1B}, ..., p_{t1}^{n_1^B B}; ...; p_{tM}^{1B}, ..., p_{tM}^{n_M^B B}\right\} \ , \tag{12}$$

supposing some transverse momentum distribution in accordance with (9). The physical observables, defined in previous Sect. 2, are expressed through these generated values:

$$n_F = \sum_{i=1}^{M} n_i^F \,, \qquad n_B = \sum_{i=1}^{M} n_i^B \,, \qquad p_B = \frac{1}{n_B}\sum_{i=1}^{M}\sum_{j=1}^{n_i^B} p_{ti}^{jB} \,. \tag{13}$$

As we will see below such direct MC approach to the calculation of the correlation functions and coefficients in not optimal. The approach, in which part of the computations is performed analytically, proves to be more practical. It allows considerably reduce the amount of MC simulations, enabling more precise results for the correlation function in the regions of very small or large multiplicity n_F. Whereas the direct MC approach needs a huge number of simulations to generate configurations with n_F, which is considerably different from $\langle n_F \rangle$.

These simplifying analytical calculations we will perform using the representation for the long-range FB correlation function, exploited in [15]:

$$\langle B \rangle_F = \frac{\sum_C \langle B \rangle_C \, P(C)\, P_C(F)}{\sum_C P(C)\, P_C(F)} \,, \tag{14}$$

where $P(C)$ is the probability of realization of the configuration C in a given event. The $\langle B \rangle_C$ and $P_C(F)$ are correspondingly, the mean value of the observable B in the backward window and the probability to have the value F for the observable in the forward window for given configuration C.

4 Correlation between multiplicities

In paper [17] the non-monotonic dependence of the correlation coefficients b'_{nn} and b'_{p_tn} on string density η at small string densities was found. What is interesting the effect took place only for the extended rapidity windows. We start the analysis of this effect from the consideration of the n-n correlation. In this case at small string density as a first step we can neglect string fusion and check whether there is this effect without a string fusion.

4.1 The correlation coefficient b_{nn}

In the version of the model without string fusion the allocation of strings on cells does not play any role and the configuration C_η is characterized simply by the total number of strings:

$$N = \sum_{i=1}^{M} \eta_i \,. \tag{15}$$

In this case, acting as in [24], we can obtain the explicit analytical formula for b_{nn}, (3):

$$b_{nn} = \frac{\omega_N - \lambda\overline{N}}{\omega_{\mu_F}/\mu_F + \omega_N - \lambda\overline{N}} \,, \tag{16}$$

where ω_{μ_F} and ω_N are the corresponding scaled variances:

$$\omega_{\mu_F} = D_{\mu_F}/\mu_F \,, \qquad \omega_N = D_N/\overline{N} = \omega_\eta \,, \qquad D_N = \overline{N^2} - \overline{N}^2 \,, \qquad \overline{N} = \eta M \tag{17}$$

and

$$\lambda = P(0)/[1 - P(0)] \,, \qquad P(0) = p^M(0) \,. \tag{18}$$

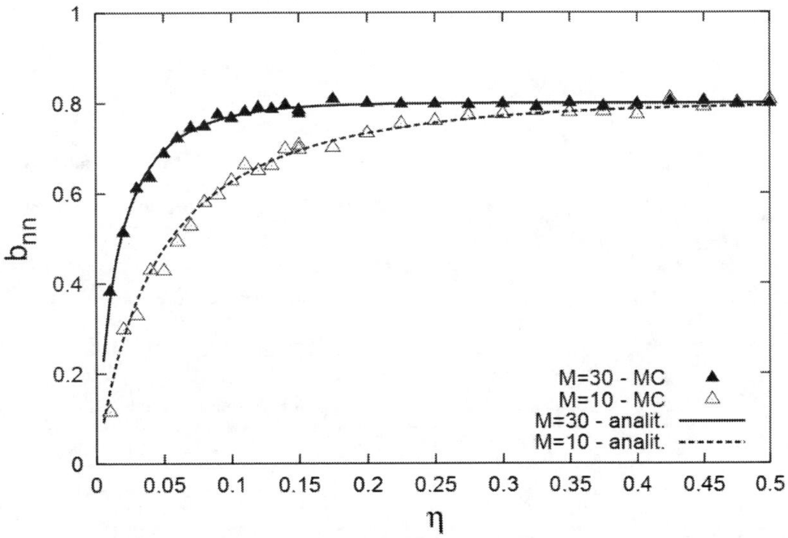

Figure 1. The FB multiplicity correlation coefficient, b_{nn}, (3), calculated by direct MC simulations (points) and analytically by formula (21) (curves) for lattices with $M = 10$ and 30 cells as a function of η (the η/σ_{str} is a string density), using the wide observation windows, corresponding to $\mu_F = 4$. The poissonian distribution for the number of particles produced from a string ($\omega_{\mu_F} = 1$) and the modified (19) poissonian distribution for the fluctuation in the number of strings were supposed. The string fusion processes are not taken into account.

Here $p(0)$ is the probability to have zero number of string in a given cell. In addition to the consideration in [24] we take into account that modeling inelastic collisions we have to discard the events with zero total number of strings. This corresponds to the following simple modification ($P(N) \to P'(N)$) of event string distribution:

$$P'(0) = 0\,; \qquad P'(N) = P(N)/1 - P(0)\,, \qquad N \geq 1\,. \tag{19}$$

We denote as $\overline{N}$ and $\langle N \rangle$ the mean values calculated correspondingly with $P(N)$ and $P'(N)$.

In paper [17] the non-monotonic dependence on string density was obtained for the modified poissonian distribution of the total number of strings. In this case

$$P(0) = p^M(0) = e^{-\eta M} = e^{-\overline{N}}\,, \qquad \langle N^k \rangle = \overline{N^k}/(1 - e^{-\overline{N}}) \tag{20}$$

and the formula (16) converts into

$$b_{nn} = \frac{1 - \lambda\eta M}{\omega_{\mu_F}/\mu_F + 1 - \lambda\eta M}\,, \qquad \lambda = \frac{e^{-\eta M}}{1 - e^{-\eta M}}\,. \tag{21}$$

We see that the dependence of the correlation coefficient on M takes place only in the region $\eta \lesssim 1/M$ (see figure 1). At other values of η we have the well known result $b_{nn} = \mu_F/(\mu_F + \omega_{\mu_F})$ [25].

It follows from (21) that $b_{nn} \to 0$ at $\eta \to 0$. The physical reason for this decrease is clear, at small string densities the events with single string start to dominate (recall that we have discarded the configurations with zero number of strings). The decays of the string in forward and backward windows are independent leading to zero value of the correlation coefficient b_{nn}, because we study only the long-range part of the correlation and suppose that the observation windows are separated by a sufficient rapidity gap.

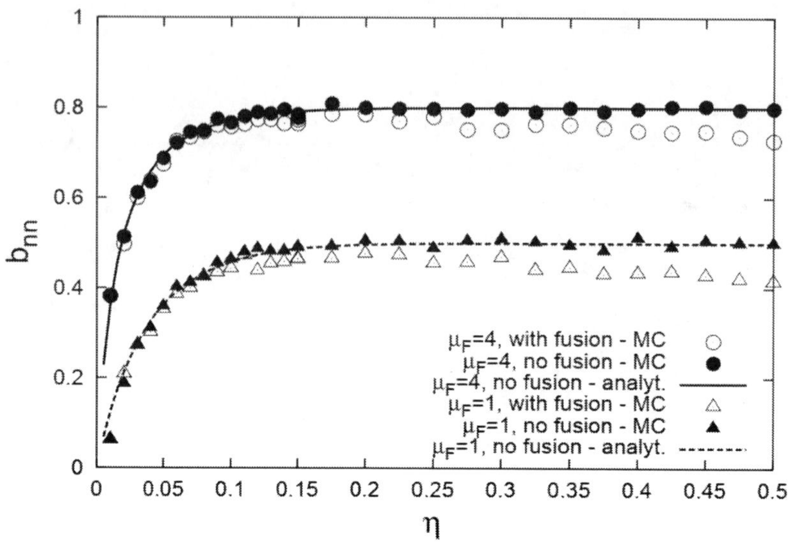

Figure 2. The same as in figure 1, but for the two widths of observation windows, $\mu_F = 1$ and 4, for the lattice with $M = 30$ cells. The open points demonstrate the influence of string fusion processes.

The analysis shows also that this decrease with η is monotonic at any value of the parameter μ_F, i.e. for the observation windows of arbitrary width in rapidity (see figure 2). So it can not explain the non-monotonic behavior of b'_{nn} observed in [17] for wide rapidity windows, corresponding $\mu_F = 4$. The taking into account the string fusion processes does not change the situation for b_{nn} at small values of η (see figure 2).

We have to suppose now that this non-monotonic behavior of b'_{nn} arises due to nonlinearity of the forward-backward correlation function for wide observation windows, which leads to the difference between the correlation coefficients b_{nn} and b'_{nn}, defined by (3) and (4). To calculate the correlation coefficient b'_{nn}, we have at first to calculate the correlation function $\langle n_B \rangle_{n_F}$.

4.2 The correlation function $\langle n_B \rangle_{n_F}$

For the case without string fusion taking into account considerations of the previous subsection 4.1 we can present the formula (14) as

$$\langle B \rangle_F = \frac{\sum_{N=1} \langle B \rangle_N \, P'(N) \, P_N(F)}{\sum_{N=1} \, P'(N) \, P_N(F)} \,, \tag{22}$$

with

$$\langle B \rangle_N = \mu_B N \,, \qquad \langle F \rangle_N = \mu_F N \,. \tag{23}$$

In this subsection to be short we imply that $F = n_F$ and $B = n_B$, but do not fulfill this substitution explicitly. As in paper [17] we will suppose the poissonian distribution for charge particles from a string decay. Then the distribution from decay of N string also will be poissonian $P_N(F)$:

$$P_N(F) = e^{-\mu_F N} \, (\mu_F N)^F / F! \,, \tag{24}$$

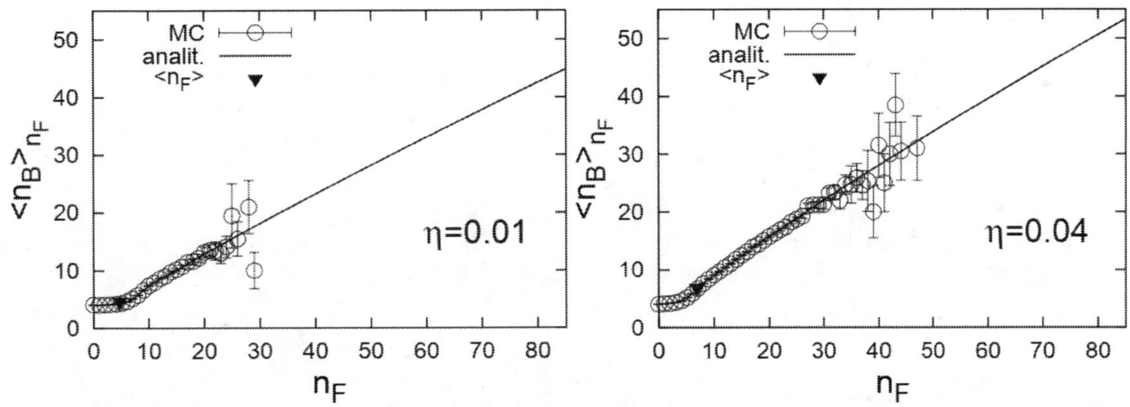

Figure 3. The correlation function between the multiplicities of charged particles produced in the backward, n_B, and forward, n_F, rapidity intervals at string densities, corresponding $\eta = 0.01$ (left panel) and $\eta = 0.04$ (right panel), for large observation windows with $\mu_F = 4$. The points are the results obtained in the model on the base of 800 000 direct MC simulations, the curves are obtained by the analytical calculations using formulas (25) and (29). The black triangular marks the mean value $\langle n_F \rangle$.

Substituting in (22) we get:

$$\langle B \rangle_F = \mu_B \frac{\Phi(F+1)}{\Phi(F)} , \qquad \Phi(F) \equiv \sum_{N=1} P(N) N^F e^{-\mu_F N} . \tag{25}$$

This result is in agreement with the formula $\langle B \rangle_F = (F+1)P(F+1)/P(F)$, obtained using generating function technique for the case $\mu_F = \mu_B$ in [15]. Note the nonlinear dependence of the correlation function on μ_F.

If we will use as in the previous subsection 4.1 and in paper [17] the modified poissonian distribution $P'(N)$ for the total number of strings, then omitting common constant factor, we can use as $\Phi(F)$

$$\Phi(F) = \sum_{N=1} N^F \frac{a^N}{N!} = \left(a \frac{d}{da} \right)^F e^a , \qquad F \geq 1 , \tag{26}$$

$$\Phi(0) = e^a - 1, \qquad a \equiv \overline{N} e^{-\mu_F} = \eta M e^{-\mu_F} \tag{27}$$

for the calculation of $\langle B \rangle_F$ in (25). By (26) we can get explicit expression:

$$\Phi(F) = e^a \sum_{k=1}^{F} \frac{(-1)^k a^k}{k!} \left[\sum_{m=1}^{k} (-1)^m C_k^m m^F \right] . \tag{28}$$

But practical calculations by the formula (28) is not convenient, as the sum is alternating. More suitable to use the following recurrence relation:

$$\Phi(F) = e^a S_F , \qquad S_F = \sum_{k=1}^{F} b_k^{(F)} a^k , \qquad F \geq 1 , \tag{29}$$

$$b_k^{(F+1)} = k\, b_k^{(F)} + b_{k-1}^{(F)} , \quad k = 2, ..., F ; \qquad b_1^{(F)} = 1 , \quad b_F^{(F)} = 1 .$$

 DOI: 10.1051/epjconf/201716407021

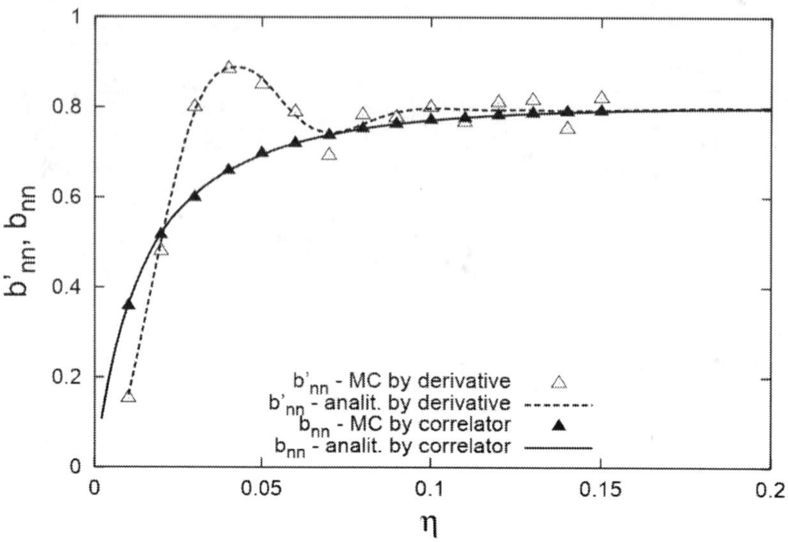

Figure 4. The comparison of the FB correlation coefficients b_{nn} (filled points, solid curve) and b'_{nn} (open points, dash curve), defined by (3) and (4), as a function of string density. The points are the results of direct 800 000 MC simulations, the curves are analytically calculated using the formulas (21), (25) and (29).

The results of the calculation of the correlation function $\langle n_B\rangle_{n_F}$ by these analytical formulae and by direct MC simulations are presented in figure 3. We see the good agreement and that the analytical approach enables to calculate the FB correlation function at large multiplicities n_F, when direct MC simulations can not produce a result even after 800 000 simulations.

4.3 The correlation coefficient b'_{nn}

Now by (5) we can calculate the correlation coefficient b'_{nn}. The results obtained by analytical calculations and by direct MC simulations are presented in figure 4. We see that even without string fusion for wide rapidity windows ($\mu_F = 4$) we have the non-monotonic behavior of b'_{nn} at small string density, observed in [17].

We can understand the origin of this non-monotonic behavior of b'_{nn} comparing the left and right panel plots in figure 3. In these plots we have marked the point $\langle n_F\rangle$, in which the correlation coefficient b'_{nn} is calculated. We see that with the increase of η the form of the correlation function $\langle n_B\rangle_{n_F}$ remains almost unchanged, whereas the point $\langle n_F\rangle$ moves to higher values. Due to nonlinearity of the correlation function $\langle n_B\rangle_{n_F}$ at small values of n_F this leads to the non-monotonic behavior of b'_{nn} with the increase of η.

The detail analysis shows that this nonlinearity at small values of n_F arises only for the observation windows which are wide in rapidity e.g. it takes place at $\mu_F = 4$ and does not take place at $\mu_F = 1$. The reason is that for wide rapidity windows the contributions to the correlation function $\langle n_B\rangle_{n_F}$ from configurations with one and two strings are not almost overlapping.

5 Correlation between transverse momentum and multiplicity

In the present model the correlations between event-mean transverse momentum p_B, (13), in the backward window and the multiplicity n_F in the forward window arises only when we take into

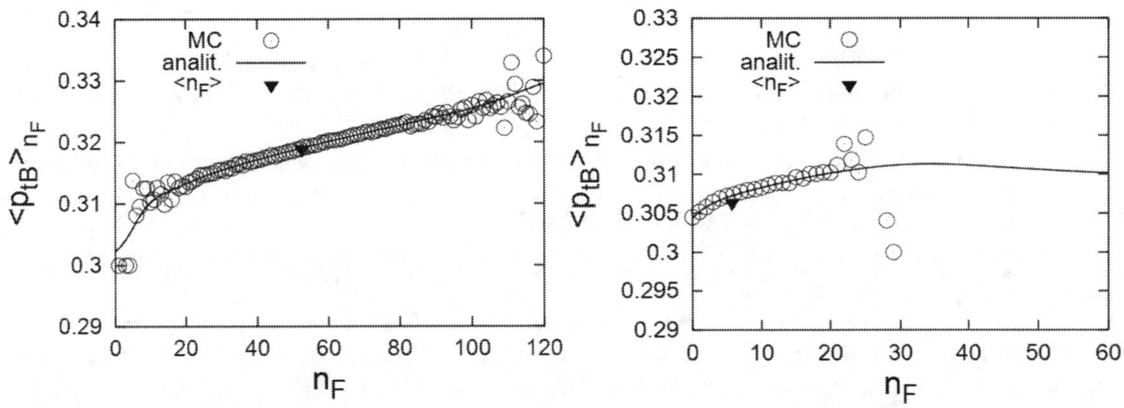

Figure 5. The function of long-range FB correlation between the event-mean transverse momentum, p_B, of particles in the backward rapidity window, (2), and the multiplicity, n_F, in the forward window, arising due to string fusion processes, for two sets of parameters: $\eta = 0.5$, $\mu_F = 4$ (left panel) and $\eta = 0.2$, $\mu_F = 1$ (right panel). The parameter $p_0 = 0.3$ GeV/c, the value of the mean transverse momentum for a single string in formula (9), was chosen. The circles are the results obtained in the model on the base of 800 000 direct MC simulations, the curves are obtained by the semi-analytical calculations using formulas (32), (33) and (34). The black triangular marks the mean value $\langle n_F \rangle$.

account the string fusion process as it was explained above. We again can simplify the MC simulations performing part of the work analytically. As in the case of *n*-*n* correlation, considered in Sect. 4, we start from the general formula (14):

$$\langle p_B \rangle_{n_F} = \frac{\sum_C \langle p_B \rangle_C \, P(C) \, P_C(n_F)}{\sum_C P(C) \, P_C(n_F)} \tag{30}$$

Recall that now $C = \left\{C_\eta, C_n^F, C_n^B, C_p^B\right\}$ by the remark before formula (12).

Averaging of the p_B, given by the formula (13), over configurations C_p^B we have

$$\langle p_B \rangle_{C_\eta C_n^B} = \left\langle \frac{\sum_{i=1}^M n_i^B \overline{p}(\eta_i)}{\sum_{i=1}^M n_i^B} \right\rangle_{C_\eta C_n^B} , \tag{31}$$

where $\overline{p}(\eta_i)$ is given by (9). In paper [8] it is shown that for the poissonian distribution in n_i^B averaging over configurations C_n^B leads to:

$$\langle p_{tB} \rangle_{C_\eta} = \left\langle \frac{\sum_{i=1}^M \overline{n}_i^B(\eta_i) \overline{p}(\eta_i)}{\sum_{i=1}^M \overline{n}_i^B(\eta_i)} \right\rangle_{C_\eta} , \tag{32}$$

with $\overline{n}_i^B(\eta_i)$ given by (8) (see also [26]). After that the general formula (30) is reduced to

$$\langle p_B \rangle_{n_F} = \frac{\sum_{C_\eta} \langle p_B \rangle_{C_\eta} \, P(C_\eta) \, P_{C_\eta}(n_F)}{\sum_{C_\eta} P(C_\eta) \, P_{C_\eta}(n_F)} , \tag{33}$$

where for poissonian distribution in n_i^F:

$$P_{C_\eta}(n_F) = P_{\langle n_F \rangle}(n_F) = e^{-\langle n_F \rangle} \frac{\langle n_F \rangle^{n_F}}{n_F!} , \qquad \langle n_F \rangle = \mu_F \sum_{i=1}^M \sqrt{\eta_i} . \tag{34}$$

The residual averaging over configurations C_η in (33) can be easily done by MC simulations.

An example of the calculation of the $\langle p_B \rangle_{n_F}$ correlation function by this semi-analytical approach and by direct MC simulations is presented in figure 5. We see again the good agreement and that the semi-analytical approach enables to calculate the FB correlation function in wider region of n_F, than direct MC simulations.

We would like to draw attention to an interesting detail. In right panel of figure 5 we observe the slight decrease of the p_t-n correlation function at extremely large multiplicities. One can find the signature of this effect in figure 7 in paper [27] for the correlation between transverse momentum and multiplicity of charged particles produced in the same window (in the acceptance of ALICE TPC $|\eta| < 0.8$) in 900 GeV pp collisions.

In the framework of the present model this effect has the following explanation. In events with the highest number of initial strings, the fluctuations in their distribution on transverse plane, compared to the uniform distribution, increase the event-mean transverse momentum and decrease the event multiplicity, due to string fusion effects, what leads to the negative correlation between them in this region of large n_F.

6 Concluding remarks

We have analyzed the behavior of the forward-backward correlation functions and the corresponding correlation coefficients between multiplicities and transverse momenta by analytical and MC calculations in the model with a string fusion on a transverse lattice. We see that in many cases the FB correlation function, $\langle B \rangle_F$, occurs nonlinear, containing the physical information, which can not be reduced to the one number - the correlation coefficient. As consequence the value of the correlation coefficient in these cases can depend on its definition (e.g. $b_{nn} \neq b'_{nn}$), whereas the corresponding correlation function, $\langle n_B \rangle_{n_F}$, is uniquely defined.

This analysis of the nonlinearity of FB correlation functions is urgent, because, for example, due to the huge arrays of events registered at the LHC we can analyze the experimental distributions in very wide region of n_F - far away from $\langle n_F \rangle$. (Clear that for n_F in a vicinity of $\langle n_F \rangle$ we can always limit ourselves to the linear regression, (6).)

As a result we found that the nonlinearity is greater for large observation rapidity intervals, for which the correlation between particles, produced from the same string, leads to the deviation of the distribution in n_F from the poissonian one [13]. The results of the modeling enable qualitatively to explain the experimentally observed features in the behavior of the correlation functions between multiplicities and mean transverse momenta at small and large multiplicities.

In particular we found that the string fusion processes lead to the conversion of the increase of the $\langle p_B \rangle_{n_F}$ correlation function to the decrease at very large values of the multiplicity n_F. The developed methods can be applied to the analysis of the long-range rapidity correlations involving heavy flavor particles. In this case we expect stronger nonlinear effects in the correlation function, assuming the sensitivity of the heavy flavor particle yields to the string fusion processes.

The research was funded by the grant of the Russian Science Foundation (project 16-12-10176).

References

[1] A. Dumitru, F. Gelis, L. McLerran, R. Venugopalan, Nucl. Phys. A **810**, 91 (2008)
[2] N.S. Amelin et al., Phys. Rev. Lett. **73**, 2813 (1994)
[3] T.S. Biro, H.B. Nielsen, J. Knoll, Nucl. Phys. B **245**, 449 (1984)
[4] A. Bialas, W. Czyz, Nucl. Phys. B **267**, 242 (1986)

[5] M.A. Braun, C. Pajares, Phys. Lett. B **287**, 154 (1992)
[6] M.A. Braun, C. Pajares, Nucl. Phys. B **390**, 542 (1993)
[7] M.A. Braun, C. Pajares, Phys. Rev. Lett. **85**, 4864 (2000)
[8] M.A. Braun, R.S. Kolevatov, C. Pajares, V.V. Vechernin, Eur. Phys. J. C **32**, 535 (2004)
[9] ALICE collaboration et al., J. Phys. G **32**, 1295 (2006), [Sect. 6.5.15]
[10] V.V. Vechernin, R.S. Kolevatov, Phys. Atom. Nucl. **70**, 1797 (2007)
[11] V.V. Vechernin, R.S. Kolevatov, Phys. Atom. Nucl. **70**, 1809 (2007)
[12] V.V. Vechernin, Theor. Math. Phys., **184**, 1271 (2015)
[13] V. Vechernin, Nucl. Phys. A **939**, 21 (2015)
[14] J. Adam et al. [ALICE Collab.], JHEP **05**, 097 (2015)
[15] M.A. Braun, C. Pajares, V.V. Vechernin, Phys. Lett. B **493**, **54** (2000).
[16] A. Capella, A. Krzywicki, Phys. Rev. D **18**, 4120 (1978)
[17] V.V. Vechernin, R.S. Kolevatov, Vestn. Peterb. Univ., Ser.4: Fiz. Khim., No.4, 11 (2004), arXiv: hep-ph/0305136
[18] M.A. Braun, C. Pajares, Eur. Phys. J. C **16**, 349 (2000)
[19] M.A. Braun, C. Pajares, Eur. Phys. J. C **71**, 1558 (2011)
[20] M.A. Braun, C. Pajares, V.V. Vechernin, Nucl. Phys. A **906**, 14 (2013)
[21] M.A. Braun, C. Pajares, V.V. Vechernin, Eur. Phys. J. A **51**, 44 (2015)
[22] E. Levin, A.H. Rezaeian, Phys.Rev. D **84**, 034031 (2011)
[23] A.Kovner., M. Lublinsky, Phys.Rev. D **83**, 034017 (2011)
[24] V.V. Vechernin, *in Relativistic Nuclear Physics and Quantum Chromodynamics, Proceedings of the Baldin ISHEPP XX vol.2*, JINR, Dubna (2011) 10-14, arXiv:1012.0214.
[25] V.V. Vechernin, R.S. Kolevatov, Vestn. Peterb. Univ., Ser.4: Fiz. Khim., No.2, 12 (2004), arXiv:hep-ph/0304295
[26] V.V. Vechernin, Theor. Math. Phys. (2016) [in press]
[27] K. Aamodt et al. [ALICE Collab.], Phys. Lett. B **693**, 53 (2010)

EPJ Web of Conferences **164**, 07022 (2017) DOI: 10.1051/epjconf/201716407022
ICNFP 2016

Nuclear and hadron matter equation of state within the induced surface tension approach

V.V. Sagun[1,a], K.A. Bugaev[1], A.I. Ivanytskyi[1], D.R. Oliinychenko[1,2], and I.N. Mishustin[2,3]

[1] *Bogolyubov Institute for Theoretical Physics, Metrologichna str. 14^B, Kiev 03680, Ukraine*
[2] *FIAS, Goethe-University, Ruth-Moufang Str. 1, 60438 Frankfurt upon Main, Germany*
[3] *Kurchatov Institute, Russian Research Center, Akademika Kurchatova Sqr., Moscow, 123182, Russia*

Abstract. We present a novel equation of state which is based on the virial expansion for the multicomponent mixtures with hard core repulsion. The suggested equation of state explicitly contains the surface tension which is induced by particle interaction. At high densities such a surface tension vanishes and in this way it switches the excluded volume treatment of hard core repulsion to its eigen volume treatment. The great advantage of the developed model is that the number of equations to be solved is two and it does not depend on the number of independent hard-core radii. Using the suggested equation of state we obtained a high quality fit of the hadron multiplicities measured at AGS, SPS, RHIC and ALICE energies and studied the properties of the nuclear matter phase diagram. It is shown the developed equation of state is softer than the gas of hard spheres and remains causal up to the several normal nuclear densities. Therefore, it could be applied to the neutron star interior modeling.

1 Introduction

Investigation of the strongly interacting matter equation of state (EoS) is in focus of several physical communities. It is necessary to reliably model the process of nucleus-nucleus (A+A) collisions at intermediate and high energies, and to study the properties of neutron star interiors. The hard-core repulsion is an important component of the hadronic and nuclear matter EoS because without it one cannot describe their properties at moderate and high baryonic densities. However, in the vast majority of models such a repulsion is treated using the Van der Waals approach which is inapplicable at the particle densities that are close to the transition region to quark gluon plasma (QGP). The problem is rooted in the wrong values of the third, the fourth and higher virial coefficients generated by the Van der Waals EoS. As it was shown in [1] a possible solution of this problem requires to account for the fact that at low densities an interparticle hard-core repulsion is well described by the excluded volume approximation, whereas the high density regime is controlled by the eigen volume of particles. The Van der Waals prescription is unable to switch between these two regimes and, therefore, it requires for an improvement. Another problem which is typical for all EoS with the hard-core repulsion is their non-causal behavior at high particle densities.

[a] e-mail: sagun@bitp.kiev.ua

These facts motivated us to work out a novel EoS based on the analysis of the virial expansion for the multicomponent mixtures, i.e. for any number of hard-core radii of particle species. The main requirement for such an EoS was that it should be able to reproduce at least third and fourth virial coefficients of the gas of hard spheres in order to go beyond the Van der Waals approximation. Below we show that the latter requirement allows us to formulate a thermodynamically consistent EoS which obeys the causality condition up to seven or eight normal nuclear densities. It is necessary to mention that the developed EoS has a remarkable advantage which is important for practical applications: compared to other EoS with multicomponent hard-core repulsion the new one has a form of two coupled nonlinear equations for any, even infinite, number of particle species. The contribution of surface tension induced by the particle interaction is a principally new element of suggested approach.

In addition, using the novel EoS we would like to reinvestigate the question raised in [2] whether or not the deeper minimum of χ^2/dof exist at very high temperature. Since, this question is very important for the whole heavy ion phenomenology, here we reanalyze the ALICE data on hadron multiplicities [3–9] with the novel model, which is able to go beyond the usual Van der Waals approximation.

The work is organized as follows. In the next section the theoretical basis of the present model is briefly outlined. In Sections 3 and 4 we present the application of the proposed model to the hadron and nuclear matter EoS, respectively. Section 5 is devoted to the conclusions.

2 Model formulation

The present model is formulated using a consistent treatment of the second virial coefficients for an ensemble of any, even infinite, number N of hard-core radii either nuclear or hadron fragments of all sizes. Such an approach allows one to explicitly account for the many-body effects and to deduce that the hard-core interaction between the constituents generate an additional contribution into the (induced) surface tension free energy. Thermodynamically consistent EoS developed in [1] is a system of coupled equations between the pressure p of considered system and the induced surface tension coefficient Σ which has the form

$$p = T\sum_{k=1}^{N}\phi_k \exp\left[\frac{\mu_k}{T} - V_k\frac{p}{T} - S_k\frac{\Sigma}{T}\right], \quad (1)$$

$$\Sigma = T\sum_{k=1}^{N}R_k\phi_k \exp\left[\frac{\mu_k}{T} - V_k\frac{p}{T} - S_k\alpha\frac{\Sigma}{T}\right]. \quad (2)$$

Here μ_k, m_k and R_k are, respectively, the chemical potential, the mass and the hard-core radius of the k-sort of particles. Evidently, $V_k = \frac{4}{3}\pi R_k^3$ and $S_k = 4\pi R_k^2$ denote, respectively, the hard-core eigen volume and hard-core eigen surface of particle of sort k. The actual parameterization of the one-particle thermal density $\phi_k(T, m, g)$ corresponding to the particle of the mass m_k and the degeneration factor g_k depends on the nature of constituents and, hence, it is discussed below in details. The summations in Eqs. (1) and (2) are made over all sorts of particles and antiparticles which are considered as independent species.

The dimensionless parameter α is introduced in (2) due to the non-uniquness of the Van der Waals extrapolation to high densities [1]. This parameter accounts for the high density terms which modify the Van der Waals EoS to a more realistic one. As it was shown in [1] to reproduce the physically correct phase diagram properties of nuclear matter such a parameter should obey the inequality $\alpha > 1$. The physical meaning of α is a switcher between the excluded volume and the eigen volume regimes.

DOI: 10.1051/epjconf/201716407022

Indeed, using the following relation

$$\Sigma \;=\; pR\exp\left[-4\pi R^2\cdot(\alpha-1)\frac{\Sigma}{T}\right], \tag{3}$$

between the total pressure p and the induced surface tension coefficient Σ for the one component mixture, i.e. in case when all particles have the same hard-core radius R, one obtains Eq. (3) from the system (1-2). With the help of (3) one can rewrite the system pressure for the one component case as

$$p \;=\; T\sum_{k=1}^{N}\phi_k\exp\left[\frac{\mu_k}{T}-v^{eff}\frac{p}{T}\right], \quad v^{eff} \;=\; v\left[1+3\cdot\exp\left(-3v\cdot(\alpha-1)\frac{\Sigma}{TR}\right)\right], \tag{4}$$

where we introduced an effective excluded volume of hadrons v^{eff} which is defined by their eigen volume $v = \frac{4}{3}\pi R^3$. It is easy to see that in the low density limit $\mu_k \to -\infty$ and, hence, one finds $\frac{\Sigma v}{TR} \to 0$ and $v^{eff} \simeq 4v$, i.e. Eq. (4) for v^{eff} correctly reproduces the excluded volume of the one component case. On the other hand, in the high density limit $\mu_k/T \gg 1$ and, hence, $\frac{\Sigma v}{TR} \gg 1$, i.e. for $\alpha > 1$ the exponential function on the right hand side of Eq. (4) vanishes and the effective excluded volume becomes equal to the eigen volume, i.e. $v^{eff} \simeq v$.

The value of α was determined by comparing the system (1-2) with the induced surface tension (IST EoS hereafter) for the point-like pions and for baryons having the same hard-core radius 0.4 fm with the famous Carnahan-Starling (CS) EoS [10]. As one can see from left panel on Fig. 1 up to the baryonic density $\rho \sim 0.65$ fm^{-3} the IST EoS with α =1.25 reproduces the speed of sound c_S of the CS EoS. From the right panel of Fig. 1 one can see that for temperature 150 MeV the proposed EoS obeys causality up to baryonic density $\rho \sim 1.2$ fm^{-3}, i.e up to seven and a half values of the normal nuclear density $\rho_0 \simeq 0.16$ fm^{-3}.

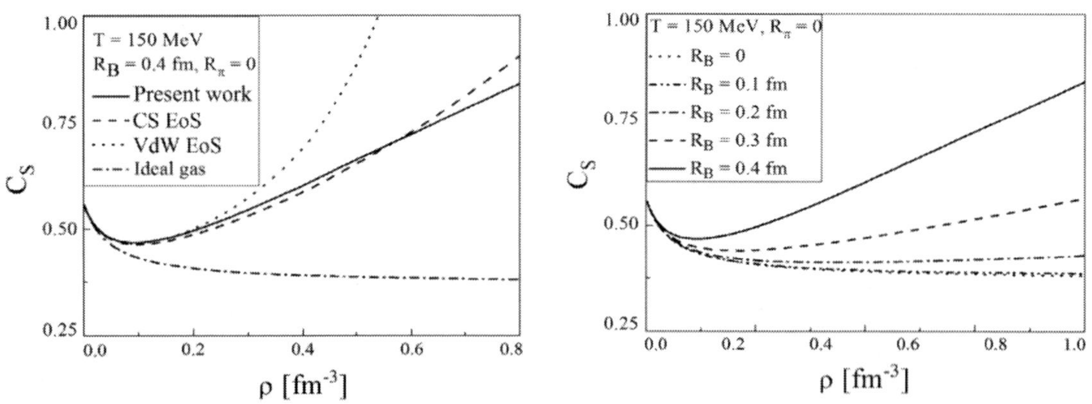

Figure 1. The speed of sound as a function of baryonic density for T = 150 MeV is shown for a mixture of point-like pions and baryons having the hard-core radius of 0.4 fm for several EoS (left panel) and different radii of baryons (right panel) for IST EoS. In the left panel the ideal gas EoS (dash dotted curve), the Van der Waals EoS (dotted curve), the CS EoS (long dashed curve) and the IST EoS (solid curve) curves are shown.

After fixing the parameter α one can apply the IST EoS to study the properties of the hadron resonance gas and the ones of nuclear matter. The new EoS is essentially more effective compared to the traditional multicomponent Hadron Resonance Gas Model (MHRGM) [11–14] and traditional

 DOI: 10.1051/epjconf/201716407022

Statistical Multifragmentation Model (SMM) [15–18], since it can be easily used for any number of independent hard-core radii.

3 HRGM with the induced surface tension

Now we apply the IST EoS to the description of the hadron multiplicities measured in the central collisions of heavy ions. The total chemical potential of hadron sort k is

$$\mu_k = \mu_B B_k + \mu_{I3} I_{3k} + \mu_S S_k \,, \tag{5}$$

where $B_k, \mu_B, S_k, \mu_S, I_{3k}, \mu_{I3}$ are, respectively, the baryonic, the strange and the isospin third projection charges and chemical potentials. All these parameters describe the chemical freeze-out (CFO) curve on which all multiplicities form. The one-particle thermal density ϕ_k in Eqs. (1) and (2) accounts for the Breit-Wigner mass distribution and is written in the Boltzmann approximation

$$\phi_k = g_k \gamma_S^{|s_k|} \int\limits_{M_k^{Th}}^{\infty} \frac{dm}{N_k(M_k^{Th})} \frac{1}{(m-m_k)^2 + \Gamma_k^2/4} \int \frac{d^3 p}{(2\pi)^3} \exp\left[-\frac{\sqrt{p^2+m^2}}{T} \right] . \tag{6}$$

Here g_k and m_k are, respectively, the degeneracy factor and the mass of the k-sort of hadrons, γ_S is the strangeness suppression factor [19], $|s_k|$ is the number of valence strange quarks and antiquarks in this kind of hadrons, $N_k(M_k^{Th}) \equiv \int\limits_{M_k^{Th}}^{\infty} \frac{dm}{(m-m_k)^2+\Gamma_k^2/4}$ denotes a corresponding normalization, while M_k^{Th} the decay threshold mass of the hadrons of k-sort.

Experimentally detected hadron multiplicity of each hadron is the sum of a thermal and the decay contributions, i.e. $n_X^{tot} = n_X^{th} + n^{decay} = n_X^{th} + \sum_Y n_Y^{th} Br(Y \rightarrow X)$, where $Br(Y \rightarrow X)$ is the decay branching ratio of the Y-th hadron into the hadron X (for more details see [14]). In addition, it is supposed for convenience that $BR(X \rightarrow X) = 1$. All the parameters used in the fitting of data (the masses m_i, the widths Γ_i, the degeneration factors g_i and the probabilities of decays for all strong decay channels) were taken from the particle tables of the thermodynamic code THERMUS [20].

The best fit criterion is a minimum of $\chi^2 = \sum_k \frac{(r_k^{theor} - r_k^{exp})^2}{\sigma_k^2}$, where r_k^{exp} is an experimental value of k-th particle ratio, r_k^{theor} is our prediction and σ_k is a total error of experimental value. Using the value α =1.25 we employed the IST EoS to fit the 111 independent hadronic multiplicity ratios measured in the central nuclear collisions for the center of mass collision energies $\sqrt{s_{NN}}$ = 2.7, 3.3, 3.8, 4.3, 4.9, 6.3, 7.6, 8.8, 9.2, 12, 17, 62.4, 130 and 200 GeV were taken from Ref. [14]. For the AGS, SPS, RHIC data the γ_S parameter was included in the fit.

The lowest χ^2/dof value corresponds to following hard core radii of baryons R_b=0.365 fm, mesons R_m=0.42 fm, pions R_π=0.15 fm, kaons R_K=0.395 fm and Λ-hyperons R_Λ=0.085 fm within $\chi^2/dof = 57.099/55 \simeq 1.038$ which is about 9% larger than the χ^2/dof found earlier in [14] for the MHRGM. Compared to the values found by the MHRGM, one sees that only pionic hard-core radius increased by 50 %, while the hard-core radius of Λ hyperons diminished by 20 %. The most important thing is that these radii remain essentially smaller compared to R_b, R_m and R_K. The latter hard-core radii are practically unchanged.

Exactly the same set of hard-core radii was used to the description of ALICE data. The IST EoS was applied to the description of the hadron multiplicities at midrapidity $\frac{dN}{dy}\|_{|y|<0.5}$ measured by the ALICE detector at $\sqrt{s_{NN}}$ = 2.76 TeV in Pb + Pb collisions. The experimental data were taken from Refs. [3–9, 21]. All hadron multiplicities were p_T-integrated and those ones at low p_T were

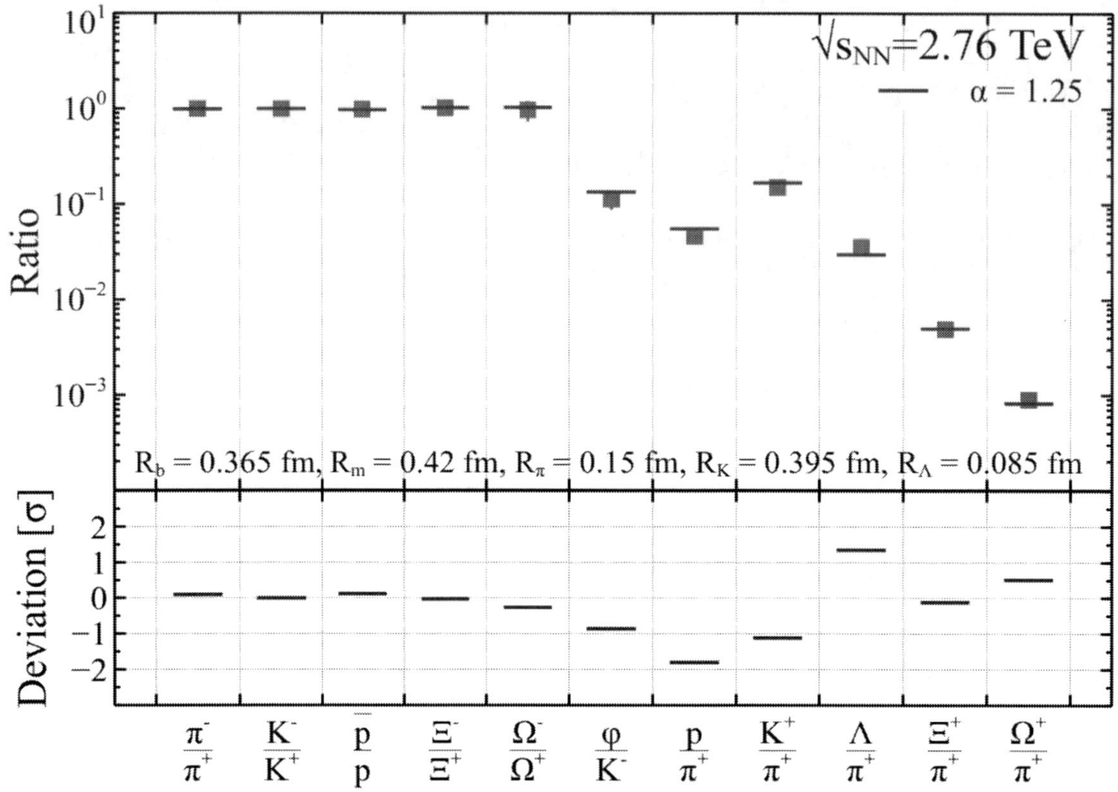

Figure 2. ALICE data fit by the IST EOS with the new set of hard-core radii. The found CFO temperature is $T_{CFO} \simeq 154 \pm 7$ MeV. The (anti)nuclei rations are not included in the fit and its quality is $\chi^2/dof \simeq 7.7/5 \simeq 1.54$. The upper panel shows the fit of the ratios, while the lower panel shows the deviation between data and theory in the units of estimated error.

extrapolated by the blast wave fit in each centrality bin while extrapolation error was included in the systematic errors.

The ratios of hadron multiplicities $r = \frac{A}{B}$ were constructed from species $A + \Delta A$ and $B + \Delta B$, while the estimation for a relative error $\epsilon_r \equiv \frac{\Delta r}{r}$ as $\sqrt{\epsilon_A^2 + \epsilon_B^2}$ [22], where $\epsilon_A = \frac{\Delta A}{A}$ and $\epsilon_B = \frac{\Delta B}{B}$. It is, certainly, a slight overestimation, because a part of systematic errors, such as errors related to detector acceptance, usually is cancelled in the experimentally measured ratios. If the statistical and systematic errors ΔA_{stat} and ΔA_{sys} are given, we add them as $\Delta A = \sqrt{\Delta A_{stat}^2 + \Delta A_{sys}^2}$ [22].

For the ALICE data we fixed $\gamma_S = 1$, while all chemical potentials are set to zero. The best global fit of all hadronic multiplicity ratios is found for the CFO temperature $T_{CFO} \simeq 154 \pm 7$ MeV with $\chi^2/dof \simeq 7.7/5 \simeq 1.54$ (see Fig. 2).

Using the IST EOS we found no traces of the high temperature minimum in comparison to Vovchenko-Stoecker results [2]. Our analysis shows that the high temperature minimum of χ^2/dof found in [2] is a consequence of extrapolating the Van der Waals EOS far beyond the bounds of its applicability.

4 SMM with the induced surface tension

Originally, the IST EoS was first applied to the description of the nuclear matter properties on the basis of an exactly solvable version of the SMM [17, 18]. Such a model is dealing with the nucleons with the mass $m \simeq 940$ MeV and the eigen volume $V_1 = \rho_0^{-1}$ (where ρ_0 denotes the normal nuclear density at $T = 0$ and zero pressure) and the composite nuclear fragments of any number of nucleons $k \geq 2$. Their eigen volume is $V_k = kV_1$ and the corresponding surface area is $S_k \sim k^{\frac{2}{3}}$.

To connect the above system of equations (1-2) for pressure and induced surface tension coefficient with the gaseous phase pressure of the SMM, we used the parameterization of the one-particle thermal densities of all k-nucleon fragments defined as

$$\phi_1 = z_1 \left[\frac{mT}{2\pi}\right]^{\frac{3}{2}} \exp\left[-\frac{\sigma(T)}{T}\right], \tag{7}$$

$$\phi_{k\geq 2} = g \left[\frac{mT}{2\pi}\right]^{\frac{3}{2}} \frac{1}{k^{\tau}} \exp\left[\frac{(k\,p_L V_1 - \mu_k)}{T} - \frac{\sigma(T)}{T} k^{\frac{2}{3}}\right]. \tag{8}$$

Here $z_1 = 4$ is the degeneracy factor of nucleons, while the degeneracy factor for other fragments g is, for simplicity, chosen to be 1 (see a discussion in [16]). In (8) $\mu_k = k\mu$ is the baryonic chemical potential of k-nucleon fragment, $\tau \simeq 1.9$ is the Fisher topological exponent and $\sigma(T)$ is the T-dependent eigen surface tension coefficient with the following parameterization

$$\sigma(T) = \sigma_0 \left[\frac{T_{cep} - T}{T_{cep}}\right] \mathrm{sign}(T_{cep} - T)\,, \tag{9}$$

with critical temperature $T_{cep} = 18$ MeV and eigen surface tension at zero temperature $\sigma_0 = 18$ MeV. In contrast to the Fisher droplet model [23] and the usual SMM [15], in the IST SMM the value of the eigen surface tension (9) is negative above the critical temperature T_{cep}. An extended discussion on the validity of such a parameterization can be found in [24]. In order to consider compressible nuclear liquid the following parameterization of its pressure

$$p_L = \frac{W(T) + \mu + a_2(\mu - \mu_0)^2 + a_4(\mu - \mu_0)^4}{V_1}\,, \tag{10}$$

was suggested in [24]. Here $W(T) = W_0 + \frac{T^2}{W_0}$ denotes the usual temperature dependent binding energy per nucleon with $W_0 = 16$ MeV [17] and the constants $\mu_0 = -W_0$, $a_2 \simeq 1.233 \cdot 10^{-2}$ MeV^{-1} and $a_4 \simeq 4.099 \cdot 10^{-7}$ MeV^{-3}. These constants are fixed in order to reproduce the properties of normal nuclear matter, i.e. at vanishing temperature $T = 0$ and normal nuclear density $\rho = \rho_0$ the liquid pressure must be zero. It is worth to note that such a parametrization of the nuclear liquid pressure describes a compressible nuclear liquid and, in contrast to the original SMM formulation [15], it leads to a nonzero isothermal compressibility $K_T \equiv \frac{1}{\rho}\frac{d\rho}{dp}\,|_T$.

The IST SMM was solved analytically and the first order phase transition of the liquid-gas type was found in [24]. It was proven that such a model has a tricritical endpoint with the temperature $T = 18$ MeV and the baryonic density $\rho_{cep} = \rho_0/3$. The resulting phase diagram of the IST SMM for two sets of variables is shown in Fig. 3. As one can see from this figure the developed model with the surface tension induced by the repulsive interaction between the nuclear fragments in combination with a finite incompressibility of liquid phase has rather rich phase structure of the nuclear matter phase diagram.

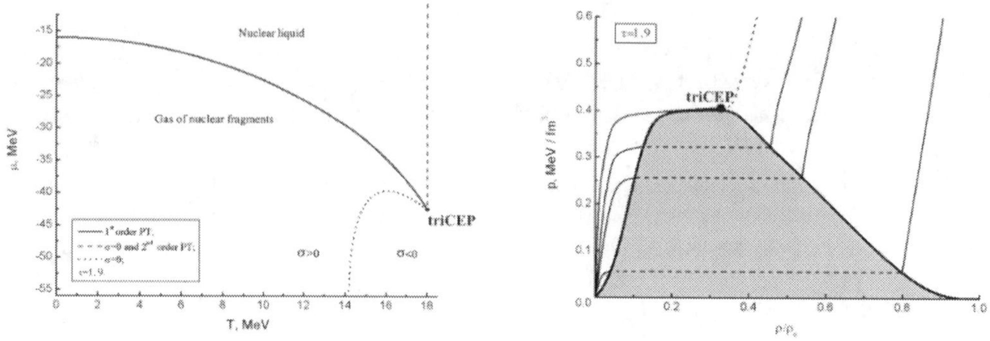

Figure 3. Phase diagram in $T - \mu$ plane (left panel) and $\rho - p$ plane (right panel) for $\nu = 2$ and $\tau = 1.9$. The 1-st order PT corresponds to a solid curve (left panel) and grey area of a mixed phase (right panel). The long dashed line on the left panel shows a 2-nd order PT, while the short dashed curve indicates the nil line of the surface tension coefficient. The isotherms on the right panel are shown for T=11,16,17,18 MeV from bottom to top. While at the critical temperature T_{cep}=18 MeV and density $\rho/\rho_0 = 1/3$ there is a triCEP.

5 Conclusions

In this work we present a thermodynamically consistent approach which is able to account for the effects of hard-core repulsion in the ensemble of constituents (clusters) of different sizes. It is necessary to stress that the developed EoS allows one to go far beyond the usual Van der Waals approximation. We showed that the hard-core repulsion between the clusters unavoidably leads to an additional equation for the induced surface tension coefficient. Due to the non-uniqueness of the Van der Waals extrapolation to high particle densities a novel parameter α was introduced. We found that this parameter is the "switcher" between the excluded and eigen volume regimes. A detailed comparison with the famous CS EoS clearly demonstrates the validity of the IST EoS for the baryonic densities up to 0.65 fm^{-3} and, while at higher densities it is softer than the CS EoS. In contrast to the conventional MHRGM the present one is mathematically much simpler because it consists only of two coupled equations independently of the number of different hard-core radii.

In order to demonstrate these advantages in practical applications, the IST EoS was used to study the properties of hadronic and nuclear matter. Using the developed approach we performed fit of the ALICE data measured at the center of mass collision energy $\sqrt{s_{NN}}$= 2.76 TeV with overall fit quality $\chi^2/dof \simeq 1.54$. The found value of the CFO temperature is $T_{CFO} \simeq 154 \pm 6$ MeV. In comparison to the [2] we found no traces of the high temperature minimum for the ALICE data. The quality of the AGS, SPS and RHIC data description achieved in the present work is $\chi^2/dof \simeq 64.8/60 \simeq 1.038$.

The IST EoS allowed us to formulate a more realistic version of the SMM in which for the first time the nuclear liquid is compressible. This property enabled us to generates the tricritical endpoint at the one third of the normal nuclear density which is in line with the liquid-gas phase transition in the ordinary liquids. It is necessary to stress that in such SMM the non-monotonic isotherms in the mixed phase region which are typical for the mean-field models are simply absent. Due to these new features the present model is more realistic than the standard SMM.

The discussed features of the proposed EoS make it applicable to the description of properties of hadronic, nuclear and neutron matter.

DOI: 10.1051/epjconf/201716407022

5.1 Acknowledgments

V.V.S., K.A.B. and A.I.I. are thankful for the partial support of the program "On perspective fundamental research in high-energy and nuclear physics" launched by the Section of Nuclear Physics of National Academy of Sciences of Ukraine.

References

[1] V. V. Sagun, A. I. Ivanytskyi, K. A. Bugaev and I. N. Mishustin, Nucl. Phys. A **924**, 24 (2014).
[2] V. Vovchenko and H. Stöcker, arXive:1512.08046v2 [hep-ph].
[3] B. Abelev *et al.* [ALICE Collaboration], Phys. Rev. C **88**, 044910 (2013).
[4] B. B. Abelev *et al.* [ALICE Collaboration], Phys. Lett. B **728**, 216 (2014); Erratum: [Phys. Lett. B **734**, 409 (2014)]
[5] B. B. Abelev *et al.* [ALICE Collaboration], Phys. Rev. Lett. **111**, 222301 (2013).
[6] A. G. Knospe [ALICE Collaboration], J. Phys. Conf. Ser. **509**, 012087 (2014).
[7] J. Adam *et al.* [ALICE Collaboration], Phys. Rev. C **93**, 024917 (2016).
[8] B. Dönigus [ALICE Collaboration], EPJ Web Conf. **97**, 00013 (2015).
[9] J. Adam *et al.* [ALICE Collaboration], Phys. Lett. B **754**, 360 (2016).
[10] N. F. Carnahan and K. E. Starling, J. Chem. Phys. **51**, 635 (1969).
[11] A. Andronic, P. Braun-Munzinger and J. Stachel, Nucl. Phys. A **772**, 167 (2006).
[12] K. A. Bugaev, D. R. Oliinychenko, A. S. Sorin and G. M. Zinovjev, Eur. Phys. J. A **49**, 30 (2013).
[13] K. A. Bugaev et al., Europhys. Lett. **104**, 22002 (2013).
[14] V. V. Sagun, Ukr. J. Phys. **59**, 755 (2014).
[15] J. P. Bondorf et al., Phys. Rep. **257**, 131 (1995) and references therein.
[16] K. A. Bugaev, Phys. Part. Nucl. **38** 447, (2007).
[17] K. A. Bugaev, M. I. Gorenstein, I. N. Mishustin and W. Greiner, Phys. Rev. C **62**, 044320 (2000).
[18] P. T. Reuter and K. A. Bugaev, Phys. Lett. B **517**, 233 (2001).
[19] J. Rafelski, Phys. Lett. B **62**, 333 (1991).
[20] S. Wheaton, J. Cleymans and M. Hauer, Comput. Phys. Commun. **180**, 84 (2009).
[21] J. Stachel, A. Andronic, P. Braun-Munzinger and K. Redlich, J. Phys. Conf. Ser. **509**, 012019 (2014).
[22] J. R. Taylor, *"An introduction to error analysis"*, University Science Book Mill Valley, California (1982).
[23] M. E. Fisher, Physics **3**, 255 (1967).
[24] K. A. Bugaev, A. I. Ivanytskyi, V. V. Sagun and D. R. Oliinychenko, Phys. Part. Nucl. Lett. **10**, 832 (2013).

EPJ Web of Conferences 164, 07023 (2017) DOI: 10.1051/epjconf/201716407023
ICNFP 2016

Overview of the CMS detector performance in LHC run II

L. Borrello[1,a] on behalf of the CMS collaboration

[1] *University and INFN Pisa, Italy*

Abstract. The CMS experiment at the LHC successfully collected data in 2015 during proton-proton collision at 13 TeV center of mass energy and lead-lead collision at 5.02 TeV/nucleon. An intense activity has been performed to further improve the detector and the trigger before the LHC restart in 2016. Report on the performance of the trigger, on the commissioning of the main detector components and on the physics objects will be presented based on the latest collected data.

1 Introduction

The CMS (Compact Muon Solenoid) experiment [1] successfully recorded data during the Large Hadron Collider (LHC) operation in Run 1 which covered 2010-2013 period.
Important physics results have been achieved including the discovery of the Higgs boson announced in July 2012 [2]. Despite the experiment reached this important milestone in the physics program, there are fundamental questions which are not yet answered by the Standard Model. The LHC operation at 13 TeV center of mass energy started in 2015 and opened a new energy domain which should allow to address some of these fundamental questions.
In order to fully exploit the physics potential, major improvements of the CMS experiment has been implemented to cope with the challenging condition of the LHC machine in Run 2. This paper describes the commissioning of the main detector components, the new trigger system and the performance of the physics objects.

2 Upgrade of the CMS experiment

During the long shutdown, lasted from March 2013 up to February 2015, the CMS experiment has undergone major changes involving the detector, the data acquisition and the trigger system.
The data acquisition (DAQ) [3] system has been completely redesigned to take the advantage of the latest developments in the computing industry and to cope with the upgraded electronics used for some sub-detectors. In addition, the High Level Trigger (HLT) processing is now file based to allow the DAQ and the HLT systems to be independent. Following the upgrade of the data acquisition system, the online monitoring system used to check the data quality has been completely re-organized.
The Level-1 (L1) trigger system was upgraded to increase the calorimeter tower-level precision reducing the out-of-time pile-up contribution and to combine the candidates from the three muon subdetectors providing a higher efficiency and robustness.

[a] e-mail: laura.borrello@cern.ch

Many activities were done on the detector affecting both the hardware and software. A non exhaustive list include the upgrade of the tracker detector cooling to operate at low temperature, the addition of two new planes of muon chambers to improve the trigger acceptance, the replacement of the readout electronics of the hadron calorimeter and the first upgrade of the trigger system.
Further activities were done during the end of the year technical stop (YETS) mainly to consolidate the cryogenic system of the CMS magnet and to complete the upgrade of the trigger system.
In 2015, the cryogenic system of the CMS solenoid showed some issues and the solenoid could not be operated stably. Hence a detailed plan was performed to refurbish the cryogenic system for stable magnet operation. The cold box of the cryogenic system was cleaned to remove traces of the oil contaminant and the primary oil removal system was fully replaced. After completing the scheduled interventions, the nominal magnetic field of 3.8 T was reached at the beginning of LHC physics operation and the cryogenic system was stable during 2016.
The CMS L1 trigger architecture upgrade was completed in order to maintain and improve the trigger performance under the extreme LHC operating conditions.

3 Status of the CMS experiment at the start of 2016

Before resuming the LHC operation, cosmic data and first collision events were used to perform the commissioning of the trigger system, detectors and to produce the first alignment constants of the Tracker subsystem.
The commissioning period was successful as proved considering the excellent status of the CMS experiment. The detector availability is shown in figure 1 where the fraction of active channel per subsystem in 2016 is compared with the number observed before the technical stop. Status is compa-

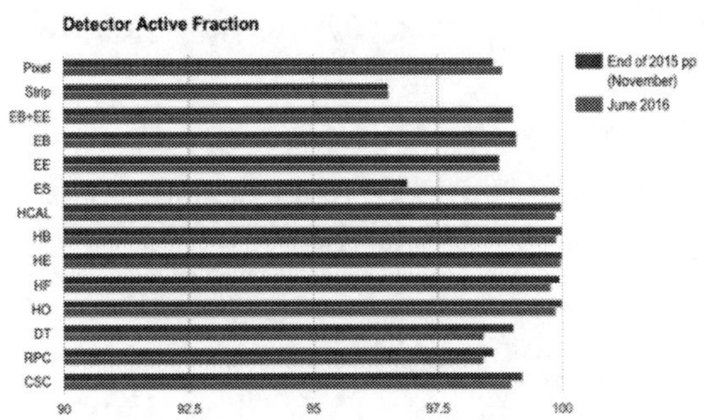

Figure 1: Fraction of active channel operational for each detector

rable for all subsystems while a clear improvement is shown for the preshower detector where 3% of the channels have been recovered before the intervention on the hardware.
The tracker detector, made of Pixel and Silicon Strip Tracker, is running stably at -15°C since the beginning of Run 2. About 1.5 million cosmic data have been collected to update the alignment constants while first collision data have been used for timing optimization purpose.
The readout system of the electromagnetic calorimeter and the algorithm used for its local reconstruction have been tuned to cope with higher pile-up operation while trigger and readout timing were fully validated. Calibration stream were deployed and first data analysed to update the alignment and intercalibration constants.
The Hadron Calorimeter (HCAL) was improved with new fast readout (10Gb/s links) for barrel and

 DOI: 10.1051/epjconf/201716407023

endcap system. A new data quality monitoring framework was deployed to improve the detection of data integrity problems. First collision events confirmed the timing synchronization and the good matching between data and trigger primitives.
The muon detectors (DT, CSC, RPC) performed well during first collisions showing for all subsystems good efficiency and a time resolution which is comparable or even better than in 2015.

The operation of the CMS experiment is summarized in figure 2 where the integrated luminosity

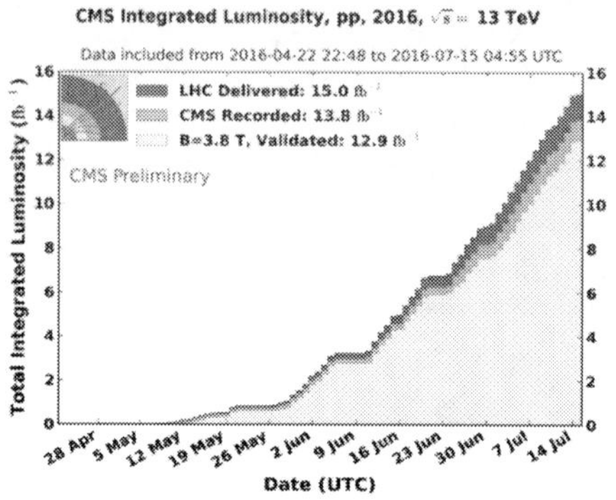

Figure 2: Cumulative curves for the luminosity delivered by LHC (azure), recorded by CMS (orange) and certified as good for physics analysis during stable beams (light orange) for proton-proton collisions at 13 TeV center of mass energy.

delivered by the LHC machine is reported together with the luminosity recorded by the CMS detector and the data validated for physics analysis. The efficiency of data taking and data considered good for physics analysis is higher than 90% confirming the excellent work done by the CMS collaboration.

4 Performance of physics object

The L1 trigger architecture was fully upgraded to maintain and improve the trigger performance in Run 2. First collision data were used to commission the L1 trigger which proved to be effective. Fig-

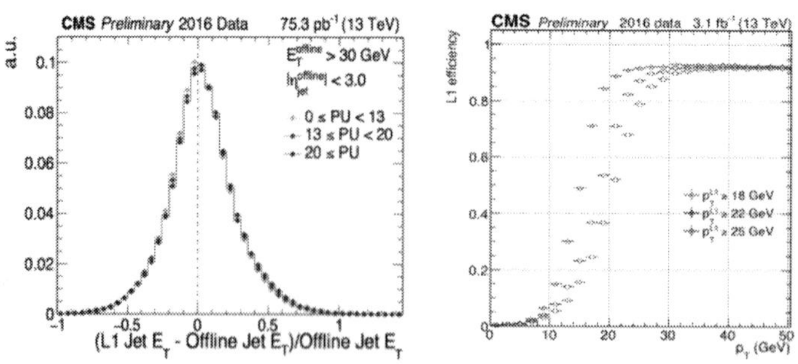

Figure 3: Left: Jet transverse energy resolution for central jets as a function of different pile-up events. Rigth: Trigger efficiency for the single muon path as a function of the offline reconstructed muon transverse momentum

ure 3 (left) shows the difference between Level-1 and offline jet transverse energy for central jets. The

distribution is split into different pile-up (PU) ranges and the dependency on the number of PU events is negligible. For jets in the forward calorimeter, there is a small dependency on the number of pile up collisions which does not impact the performance significantly.
The L1 muon trigger was redesigned to work with three regional track finders based on the pseudorapidity to exploit the detector redundancy at a lower level. Trigger objects are then merged and cleaned to remove duplicate object by a dedicated global muon trigger. Figure 3 (right) shows the trigger efficiency for the single muon path as a function of the offline reconstructed muon transverse momentum emphasising the turn on part of efficiency. The efficiency is measured with a tag and probe method on a dataset of events recorded using a single muon trigger and the measured value is above 90%.

The electron reconstruction is good as shown in figure 4 where the di-electron invariant mass for

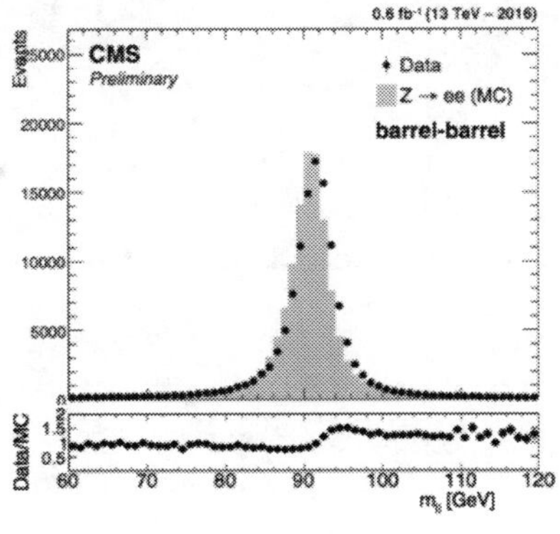

Figure 4: Di-electron invariant mass for electron pairs reconstructed in the barrel region before scale calibration is applied.

electron pairs in the barrel region is reconstructed. An energy shift is observed in data since energy scale and smearing are not applied and the ECAL calibration is not yet updated.
The jet energy response is measured in simulated proton collisions and in data using Z/γ + jet events.

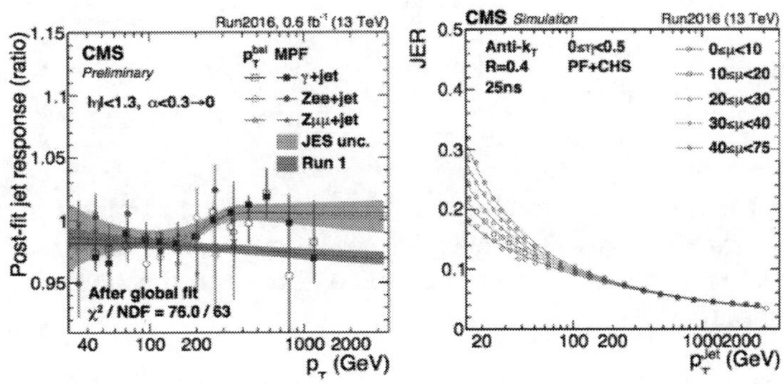

Figure 5: Left: post-fit jet response as a function of the jet transverse momentum. Right: jet energy resolution as a function of the jet transverse momentum for different pile-up events.

The ratio of the data to simulated jet energy response is shown in the left panel of figure 5. The kink observed around 150 GeV is attributed to changes in hadronic calorimeter local reconstruction.

Performance are similar to Run 1 while the jet energy resolution proved to be robust against pile-up events as shown in the right panel of figure 5.

The performance of the missing transverse energy (E_T^{miss}) were studied in events with an identified

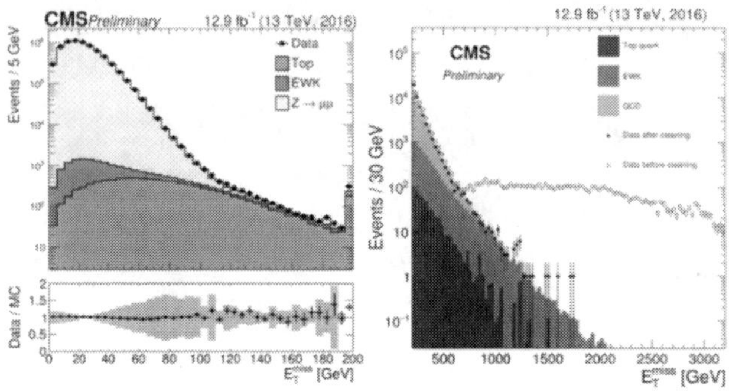

Figure 6: Left: Distributions of E_T^{miss} in $Z \to \mu^+\mu^-$ events. The points in the lower panel show the data/MC ratio, including the statistical uncertainties. The last bin contains overflow content. Right: The E_T^{miss} distributions for events passing the dijet selection after applying the cleaning algorithms.

Z boson or an isolated photon. The distribution of E_T^{miss} for events passing a $Z \to \mu\mu$ selection is shown in figure 6 where is evident the good agreement between data and simulation. The right plot in figure 6 shows the good result of the cleaning procedure applied to remove the long MET tail which is produced by different phenomena in the calorimeters.

Also the high-level object identification algorithms have been reoptimized for Run 2. The hadronic τ decays is well identified as shown in the left panel of figure 7. The improved MVA-based identifica-

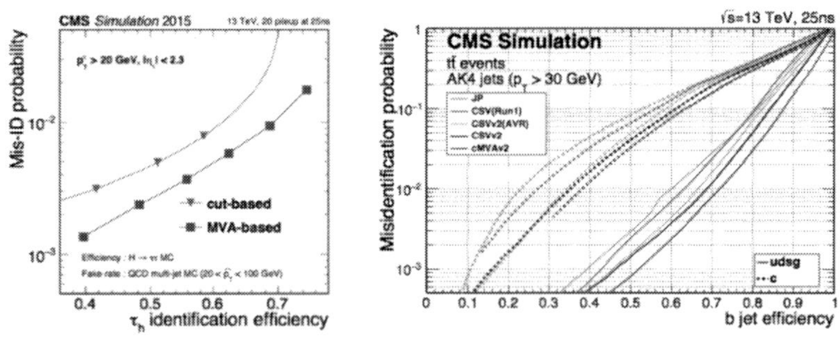

Figure 7: Left: misidentification probability as a function of the hadronic τ identification efficiency in multijet events (for the background) and Higgs decays to two τ leptons (for the signal) at 13 TeV. Rigth: Misidentification probability for c and d, u, s, g jets as a function of the b jet identification efficiency in $t\bar{t}$ events at 13 TeV.

tion efficiency is performing significantly better than the cut-based strategy, with a 20% difference in relative efficiency.

Right panel of figure 7 shows the performance of different algorithms compared with the Combined Secondary Vertex (CSV) used in Run 1. The comparison highligths the better performance of the algorithms which have been modified for Run 2 operation. As an example, the CSVv2 algorithm, which combines the secondary vertex and track-based lifetime information with a neural network instead of a likelihood ratio, provides an identification probability which is about 10% higher than the value obtained with the previous algorithm.

5 Conclusions

We discussed the activity done in the CMS experiment in preparation of LHC Run 2 operation and presented the performance of the detector and the physics object at the start of 2016.
The CMS detector is running stably with excellent performance and the collaboration is ready to exploit the full physics potential collecting and analyzing the large datasets provided by the LHC.

References

[1] CMS Collaboration, JINST 3 3 S08004 (2008).
[2] S. Chatrchyan et al. (CMS collaboration), Phys.Lett. B716, 30-61 (2012)
[3] T. Bawej et al., IEEE Transactions on Nuclear Science, VOL. 62, NO. 3, 1099, (2015)

Parton distribution functions (PDFs) from dipole models and HERA data - predictions for the LHC

Agnieszka Luszczak[1,a]

[1] T.Kosciuszko Cracow University of Technology

Abstract. We analyse, within a dipole model, the final, inclusive HERA DIS cross section data in the low x region, using fully correlated errors. We show, that these highest precision data are very well described within the dipole model framework starting from Q^2 values of 3.5 GeV^2 to the highest values of $Q^2 = 250$ GeV^2. We made predictions for the LHC data using parton distributions functions (PDFs) in particular - gluon distributions from BGK dipole model fits to HERA data

1 Introduction

Many investigations have shown that HERA inclusive and diffractive DIS cross sections are very well described by the dipole models [1–3], which provide a natural description of QCD reaction in the low-x and low Q^2 region. They allow a simultaneous description of many different physics reactions, like inclusive DIS processes, inclusive diffractive processes, exclusive J/ψ, ρ, ϕ production, diffractive jet production, or diffractive and non-diffractive charm production. Due to the optical theorem, all these processes are determined by the same, universal, gluon density [4–6]. The understanding of the properties of the gluon density and its precise knowledge is very important because the QCD-evolved gluon density determines the cross sections of most relevant physics processes, e.g. Higgs production at LHC. Any significant deviation of the predicted cross section from their Standard Model value could be a sign of new physics.

The precise determination of the gluon density relies on the analysis of high quality, inclusive, DIS data taken over the full, experimentally accessible x and Q^2 region. Therefore, the H1 and ZEUS experiments have combined their inclusive DIS cross sections which, due to a substantial reduction of systematic measurements errors, led to an increase of precision by about a factor two [7]. More recently, also the final set of HERA data was released which provides a full set of correlated errors. These errors contain the most complete information about the data and lead to very restrictive fits [8]. The full information about the data is made now accessible within the xFitter facility [9].

2 Dipole models

The dipole picture was first derived, in the low x limit of QCD, by Nikolaev and Zaharov [10]. They have shown that the deep inelastic scattering can be viewed as a two stage process; first the

[a] e-mail: agnieszka.luszczak@ifj.edu.pl

virtual photon fluctuates into a dipole, which consists of a quark-antiquark pair (or a $q\bar{q}g$ or $q\bar{q}gg$ ... system) and in the second stage the dipole interacts with the proton. Dipole denotes a quasi-stable quantum mechanical state, which has a very long life time ($\approx 1/m_p x$) and a size r, which remains unchanged during scattering. The wave function Ψ determines the probability to find a dipole of size r within a photon. This probability depends on the value of external Q^2 and the fraction of the photon momentum carried by the quarks forming the dipole, z. Neglecting the z dependence, in a very rough approximathion, $Q^2 \sim 1/r^2$.

The scattering amplitude is a product of the virtual photon wave function, Ψ, with the dipole cross section, $\sigma_{\rm dip}$, which determines a probability of the dipole-proton scattering. Thus, within the dipole formulation of the $\gamma^* p$ scattering

$$\sigma_{T,L}^{\gamma^* p}(x, Q^2) = \int dr^2 \int dz \Psi^*_{T,L}(Q, r, z) \sigma_{\rm dip}(x, r) \Psi_{T,L}(Q, r, z), \tag{1}$$

where T, L denotes the virtual photon polarization and $\sigma_{T,L}^{\gamma^* p}$ the total inclusive DIS cross section.

This simple and intuitive approach became then a basis of many dipole many models, [11–17]. which have been developed to test various aspects of data. They vary due to different assumption made about the physical behavior of dipole cross sections. In the following we will shortly review some them to motivate the choice of the model used for present investigation.

2.1 GBW model

The dipole model became an important tool in investigations of deep-inelastic scattering due to the initial observation of Golec-Biernat and Wüesthoff (GBW) [1], that a simple ansatz for the dipole cross section was able to describe simultaneously the total inclusive and diffractive cross sections.

In the GBW model the dipole-proton cross section $\sigma_{\rm dip}$ is given by

$$\sigma_{\rm dip}(x, r^2) = \sigma_0 \left(1 - \exp\left[-\frac{r^2}{4R_0^2(x)} \right] \right), \tag{2}$$

where r corresponds to the transverse separation between the quark and the antiquark, and R_0^2 is an x dependent scale parameter which has a meaning of saturation radius, $R_0^2(x) = (x/x_0)^{\lambda_{GBW}} / GeV^{-2}$. The free fitted parameters are: the cross-section normalisation, σ_0, as well as x_0 and λ_{GBW}. In this model saturation is taken into account in the eikonal approximation and the saturation radius is intimately related to the gluon density, see below. The exponent λ_{GBW} determines the growth of the total and diffractive cross section with decreasing x. For dipole sizes which are large in comparison to the saturation radius, R_0, the dipole cross section saturates by approaching a constant value σ_0, i.e. saturation damps the growth of the gluon density at low x.

The GBW model provided a good description of data from medium Q^2 values ($\approx$ 30 GeV^2) down to low Q^2 ($\approx$ 0.1) GeV^2). Despite its success and its appealing simplicity the model has some shortcomings; in particular it describes the QCD evolution by a simple x dependence, $\sim (1/x)^{\lambda}_{BGW}$, i.e the Q^2 dependence of the cross section evolution is solely induced by the saturation effects. Therefore, it does not match with DGLAP QCD evolution, which is known to describe data very well from $Q^2 \approx 4$ GeV^2 to very large $Q^2 \approx 10000$ GeV^2.

2.2 BGK model

The evolution ansatz of the GBW model was improved in the model proposed by Bartels, Golec-Biernat and Kowalski, (BGK) [2], by taking into account the DGLAP evolution of the gluon density

 DOI: 10.1051/epjconf/201716407024

in an explicit way. The model preserves the GBW eikonal approximation to saturation and thus the dipole cross section is given by

$$\sigma_{\text{dip}}(x, r^2) = \sigma_0 \left(1 - \exp\left[-\frac{\pi^2 r^2 \alpha_s(\mu^2) xg(x, \mu^2)}{3\sigma_0}\right]\right). \tag{3}$$

The evolution scale μ^2 is connected to the size of the dipole by $\mu^2 = C/r^2 + \mu_0^2$. This assumption allows to treat consistently the contributions of large dipoles without making the strong coupling constant, $\alpha_s(\mu^2)$, un-physically large. This means also that we can extend the model, keeping its perturbative character, to the data at low Q^2, because the external Q^2 and the internal μ^2 scales are connected only by the wave function.

The gluon density, which is parametrized at the starting scale μ_0^2, is evolved to larger scales, μ^2, using LO or NLO DGLAP evolution. We consider here two forms of the gluon density:

- the *soft* ansatz, as used in the original BGK model

$$xg(x, \mu_0^2) = A_g x^{-\lambda_g}(1 - x)^{C_g}, \tag{4}$$

- the *soft* + *hard* ansatz

$$xg(x, \mu_0^2) = A_g x^{-\lambda_g}(1 - x)^{C_g}(1 + D_g x + E_g x^2), \tag{5}$$

The free parameters for this model are σ_0 and the parameters for gluon A_g, λ_g, C_g or additionally D_g, E_g, Their values are obtained by a fit to the data. The fit results were found to be independent on the parameter C, which was therefore fixed as $C = 4$ GeV^2, in agreement with the original BGK fits. It is also possible to vary the parameter μ_0^2. However, to assure that the evolution is performed in the perturbative region and to be compatible with the standard pdf fits we took as a starting scale $\mu_0^2 = 1.9$ or 1.1 GeV^2. In the BGK model, the μ_0^2 scale is the same as the Q_0^2 scale of the standard QCD pdf fits.

3 Fits to the dipole model

The results of the BGK fit with valence quarks and the *soft* gluon density, are shown in Table 1, 2 and 3. The best fit is obtained using valence quarks and gluon density of the *soft* + *hard* type, see Table 4 and 6. In Table 1, 2 and 4 the starting QCD scale is $Q_0^2 = 1.9$ GeV^2, in Table 6 it is $Q_0^2 = 1.1$ GeV^2,

In Table 1, 4 and 6 the valence quark contribution was taken from the standard pdf fit, shown below, in Table 2 the valence quarks were fitted. In these tables, N_{df} denotes the number of degrees of freedom, which is equal to the number of measured data points minus the number of free parameters used in the fit. The parameters σ_0 of the dipole model and the starting parameters for gluon A_g, λ_g, C_g are obtained from the fit. The value of the parameter C was fixed, as explained above. To limit the fit to the perturbative region only we took $Q^2 \geq Q^2_{min}$ with $Q^2_{min} = 3.5$ (or 8.5) GeV^2. For the x region, we took $x \leq 0.01$. There are 538 (or 452) measured points in this region.

Q^2_{min} [GeV^2]	$\sigma_0[mb]$	A_g	λ_g	C_g	N_{df}	χ^2	χ^2/N_{df}
3.5	87.0±	2.32±	-0.056±	8.21±	534	551.1	1.03
	8.9	0.009	0.11	0.80			
8.5	72.4±	2.77±	-0.042±	6.54±	448	452.5	1.01
	7.4	0.009	0.123	0.632			

Table 1. BGK fit with fixed valence quarks for σ_r for H1ZEUS-NC data in the range $Q^2 \geq 3.5$ or 8.5 GeV^2 and $x \leq 0.01$. NLO fit. *Soft gluon.* $m_{uds} = 0.14, m_c = 1.3$ GeV. $Q_0^2 = 1.9$ GeV^2.

Q^2_{min} [GeV^2]	$\sigma_0[mb]$	A_g	λ_g	C_g	N_{df}	χ^2	χ^2/N_{df}
3.5	89.99±	2.44±	-0.079±	7.24±	530	540.35	1.02
	9.2	0.145	0.099	0.61			

Table 2. BGK fit with fitted valence quarks for σ_r for H1ZEUS-NC data in the range $Q^2 \geq 3.5$ GeV^2 and $x \leq 0.01$. NLO fit. *Soft gluon.* $m_{uds} = 0.14, m_c = 1.3$ GeV. $Q_0^2 = 1.9$ GeV^2.

No	$Auv(fix)$	Buv	Cuv	Euv	$Adv(fix)$	Bdv	Cdv
1	4.073(sum rule)	0.892±	5.832±	17.997±	3.151(sum rule)	0.840±	3.480±
		0.019	0.341	0.876		0.012	0.056

Table 3. Parameters of the valence quark contribution fitted in the BGK fit of Table 2.

Q^2_{min} [GeV^2]	$\sigma_0[mb]$	A_g	λ_g	C_g	D_g	E_g	N_{df}	χ^2	χ^2/N_{df}
3.5	77,6±	2.62±	-0.064±	37.1±	3.06 ±	1406.4±	532	534.2	1.00
	18,6	0.16	0.0087	5.06	6.51	552.7			
8.5	63.5 ±	2.11±	-0.054±	21.3±	1.10 ±	867.2±	448	439.0	0.98
	18.5	0.10	0.0065	4.062	5.76	423.7			

Table 4. BGK fit with valence quarks for σ_r for H1ZEUS-NC data in the range $Q^2 \geq 3.5$ or 8.5 GeV^2 and $x \leq 0.01$. NLO fit. *Soft + hard gluon.* $m_{uds} = 0.14, m_c = 1.3$ GeV. $Q_0^2 = 1.9$ GeV^2.

4 Gluon density from BGK dipole model

In Figure 1 we compare the gluon densities obtained from fits with and without saturation ansatz of Tables 4 and 5. Note that for $x < 0.001$, the gluon densities obtained in the dipole approach are higher than that of the standard pdf fit, and that of the non-saturated one, see Fig 1. This is expected and is due to damping of gluon density by saturation effects.

Q^2_{min} [GeV^2]	Q^2_0 [GeV^2]	A_g	λ_g	C_g	D_g	E_g	Ndf	χ^2	χ^2/Np
3.5	1.9	2.33± 0.10	-0.094± 0.006	14.8± 11.5	9.80 ± 14.7	-99.5± 74.830	533	556.17	1.04
3.5	1.1	3.80± 0.22	0.10± 0.01	32.5± 1.6	-25.2 ± 3.49	1868 ± 252	533	539.2	1.01

Table 5. BGK fit with valence quarks for σ_r for H1ZEUS-NC data in the range $Q^2 \geq 3.5$ GeV^2 and $x \leq 0.01$. NLO fit. *Soft + hard gluon.* $m_{uds} = 0.14, m_c = 1.3$ GeV, non-saturation ansatz. $Q_0^2 = 1.9$ or 1.1 GeV^2.

Q^2_{min} [GeV^2]	Q^2_0 [GeV^2]	$\sigma_0[mb]$	A_g	λ_g	C_g	D_g	E_g	N_{df}	χ^2	χ^2/N_{df}
3.5	1.1	220 ± 122	3.57± 0.38	0.082± 0.017	31.3± 5.4	11.0 ± 9.8	1360± 690	532	532	1.00

Table 6. BGK fit with valence quarks for σ_r for H1ZEUS-NC data in the range $Q^2 \geq 3.5$ GeV^2 and $x \leq 0.01$. NLO fit. *Soft + hard gluon.* $m_{uds} = 0.14, m_c = 1.3$ GeV. $Q_0^2 = 1.1$ GeV^2.

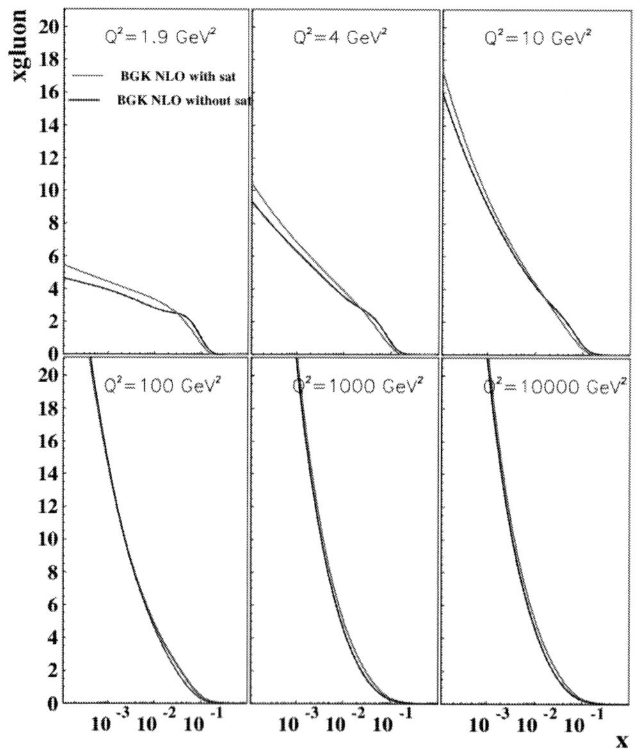

Figure 1. Comparison between the gluon densities obtained from fits in the $Q^2 > 3.5$ GeV^2 region, with and without the saturation ansatz, Table 4 and 5.

5 Predictions with gluon from BGK dipole fit for the LHC

In Figure 2 we show predictions for the LHC using the gluon density obtained from fits with saturation ansatz of Tables 4. We see that there is a large discrepancy between ATLAS data for jets and predictions from BGK dipole model.

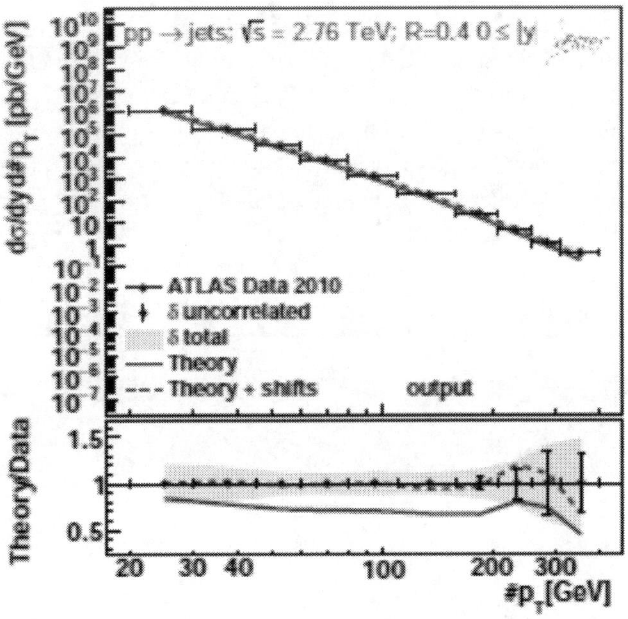

Figure 2. Predictions for jets in ATLAS using gluon density from BGK dipole model

6 Acknowledgement

I would like to thank Henri Kowalski (DESY) for various help and careful reading of the manuscript. This work is supported by the Polish Ministry under program Mobility Plus, no. 1320/MOB/IV/2015/0.

References

[1] K. Golec-Biernat, M. Wuesthoff, Phys. Rev. **D59**, 014017 (1999); Phys. Rev. **D60**, 114023 (1999).
[2] J. Bartels, K.Golec-Biernat and H. Kowalski, Phys.Rev. **D66,** (2002) 19
[3] E. Iancu, K. Itakura, and S. Munier. *Phys. Lett.*, B590:199–208, 2004.
[4] S. Munier, A. M. Staśto and A. H. Mueller, Nucl. Phys. B **603** (2001) 427.
[5] H. Kowalski and D. Teaney, Phys. Rev. **D68** (2003) 114005.

[6] H. Kowalski, L. Motyka and G. Watt, Phys. Rev. **D74** (2006) 074016.
[7] F. D. Aaron *et al.* [H1 and ZEUS Collaborations], JHEP 1001 (2010) 109.
[8] H. Abramowicz *et al.* [H1 and ZEUS Collaborations], Eur. Phys. J. **C75** (2015) nb.12.
[9] "HERAFitter, Open Source QCD Fit Project" By S. Alekhin at al., DESY Report 14-188, 7 Nov 2014, Published in EPJC (2015), 75: 304, e-Print: arXiv:1410.4412 [hep-ex]
[10] N. N. Nikolaev and B. G. Zakharov, Z. Phys. **C49** (1991) 607; Z. Phys. **C53** (1992) 331.
[11] J. Nemchik, N. N. Nikolaev, E. Predazzi and B. G. Zakharov, Z. Phys. C **75** (1997) 71.
[12] E. Gotsman, E. Levin and U. Maor, Nucl. Phys. B **464** (1996) 251.
[13] H. G. Dosch, T. Gousset, G. Kulzinger and H. J. Pirner, Phys. Rev. D **55** (1997) 2602.
[14] A. C. Caldwell and M. S. Soares, Nucl. Phys. A **696**, 125 (2001).
[15] J. R. Forshaw, R. Sandapen and G. Shaw, Phys. Rev. D **69** (2004) 094013.
[16] L. Frankfurt, M. Strikman and C. Weiss, Ann. Rev. Nucl. Part. Sci. **55** (2005) 403.
[17] H. Kowalski, T. Lappi, C. Marquet and R. Venugopalan, Phys. Rev. **C78** (2008) 045201.

Parton distribution functions (PDFs) from dipole models and HERA data - predictions for the LHC

Agnieszka Luszczak[1,a]

[1] T.Kosciuszko Cracow University of Technology

Abstract. We analyse, within a dipole model, the final, inclusive HERA DIS cross section data in the low x region, using fully correlated errors. We show, that these highest precision data are very well described within the dipole model framework starting from Q^2 values of 3.5 GeV^2 to the highest values of $Q^2 = 250$ GeV^2. We made predictions for the LHC data using parton distributions functions (PDFs) in particular - gluon distributions from BGK dipole model fits to HERA data

1 Introduction

Many investigations have shown that HERA inclusive and diffractive DIS cross sections are very well described by the dipole models [1–3], which provide a natural description of QCD reaction in the low-x and low Q^2 region. They allow a simultaneous description of many different physics reactions, like inclusive DIS processes, inclusive diffractive processes, exclusive J/ψ, ρ, ϕ production, diffractive jet production, or diffractive and non-diffractive charm production. Due to the optical theorem, all these processes are determined by the same, universal, gluon density [4–6]. The understanding of the properties of the gluon density and its precise knowledge is very important because the QCD-evolved gluon density determines the cross sections of most relevant physics processes, e.g. Higgs production at LHC. Any significant deviation of the predicted cross section from their Standard Model value could be a sign of new physics.

The precise determination of the gluon density relies on the analysis of high quality, inclusive, DIS data taken over the full, experimentally accessible x and Q^2 region. Therefore, the H1 and ZEUS experiments have combined their inclusive DIS cross sections which, due to a substantial reduction of systematic measurements errors, led to an increase of precision by about a factor two [7]. More recently, also the final set of HERA data was released which provides a full set of correlated errors. These errors contain the most complete information about the data and lead to very restrictive fits [8]. The full information about the data is made now accessible within the xFitter facility [9].

2 Dipole models

The dipole picture was first derived, in the low x limit of QCD, by Nikolaev and Zaharov [10]. They have shown that the deep inelastic scattering can be viewed as a two stage process; first the

[a] e-mail: agnieszka.luszczak@ifj.edu.pl

virtual photon fluctuates into a dipole, which consists of a quark-antiquark pair (or a $q\bar{q}g$ or $q\bar{q}gg$ ... system) and in the second stage the dipole interacts with the proton. Dipole denotes a quasi-stable quantum mechanical state, which has a very long life time ($\approx 1/m_p x$) and a size r, which remains unchanged during scattering. The wave function Ψ determines the probability to find a dipole of size r within a photon. This probability depends on the value of external Q^2 and the fraction of the photon momentum carried by the quarks forming the dipole, z. Neglecting the z dependence, in a very rough approximathion, $Q^2 \sim 1/r^2$.

The scattering amplitude is a product of the virtual photon wave function, Ψ, with the dipole cross section, $\sigma_{\rm dip}$, which determines a probability of the dipole-proton scattering. Thus, within the dipole formulation of the $\gamma^* p$ scattering

$$\sigma^{\gamma^* p}_{T,L}(x, Q^2) = \int dr^2 \int dz \Psi^*_{T,L}(Q, r, z)\sigma_{\rm dip}(x, r)\Psi_{T,L}(Q, r, z), \tag{1}$$

where T, L denotes the virtual photon polarization and $\sigma^{\gamma^* p}_{T,L}$ the total inclusive DIS cross section.

This simple and intuitive approach became then a basis of many dipole many models, [11–17]. which have been developed to test various aspects of data. They vary due to different assumption made about the physical behavior of dipole cross sections. In the following we will shortly review some them to motivate the choice of the model used for present investigation.

2.1 GBW model

The dipole model became an important tool in investigations of deep-inelastic scattering due to the initial observation of Golec-Biernat and Wüesthoff (GBW) [1], that a simple ansatz for the dipole cross section was able to describe simultaneously the total inclusive and diffractive cross sections.

In the GBW model the dipole-proton cross section $\sigma_{\rm dip}$ is given by

$$\sigma_{\rm dip}(x, r^2) = \sigma_0 \left(1 - \exp\left[-\frac{r^2}{4R_0^2(x)}\right]\right), \tag{2}$$

where r corresponds to the transverse separation between the quark and the antiquark, and R_0^2 is an x dependent scale parameter which has a meaning of saturation radius, $R_0^2(x) = (x/x_0)^{\lambda_{GBW}} /GeV^{-2}$. The free fitted parameters are: the cross-section normalisation, σ_0, as well as x_0 and λ_{GBW}. In this model saturation is taken into account in the eikonal approximation and the saturation radius is intimately related to the gluon density, see below. The exponent λ_{GBW} determines the growth of the total and diffractive cross section with decreasing x. For dipole sizes which are large in comparison to the saturation radius, R_0, the dipole cross section saturates by approaching a constant value σ_0, i.e. saturation damps the growth of the gluon density at low x.

The GBW model provided a good description of data from medium Q^2 values ($\approx$ 30 GeV^2) down to low Q^2 ($\approx$ 0.1) GeV^2). Despite its success and its appealing simplicity the model has some shortcomings; in particular it describes the QCD evolution by a simple x dependence, $\sim (1/x)^{\lambda}_{BGW}$, i.e the Q^2 dependence of the cross section evolution is solely induced by the saturation effects. Therefore, it does not match with DGLAP QCD evolution, which is known to describe data very well from $Q^2 \approx 4$ GeV^2 to very large $Q^2 \approx 10000$ GeV^2.

2.2 BGK model

The evolution ansatz of the GBW model was improved in the model proposed by Bartels, Golec-Biernat and Kowalski, (BGK) [2], by taking into account the DGLAP evolution of the gluon density

DOI: 10.1051/epjconf/201716407024

in an explicit way. The model preserves the GBW eikonal approximation to saturation and thus the dipole cross section is given by

$$\sigma_{\mathrm{dip}}(x, r^2) = \sigma_0 \left(1 - \exp\left[-\frac{\pi^2 r^2 \alpha_s(\mu^2) xg(x, \mu^2)}{3\sigma_0}\right]\right). \tag{3}$$

The evolution scale μ^2 is connected to the size of the dipole by $\mu^2 = C/r^2 + \mu_0^2$. This assumption allows to treat consistently the contributions of large dipoles without making the strong coupling constant, $\alpha_s(\mu^2)$, un-physically large. This means also that we can extend the model, keeping its perturbative character, to the data at low Q^2, because the external Q^2 and the internal μ^2 scales are connected only by the wave function.

The gluon density, which is parametrized at the starting scale μ_0^2, is evolved to larger scales, μ^2, using LO or NLO DGLAP evolution. We consider here two forms of the gluon density:

- the *soft* ansatz, as used in the original BGK model

$$xg(x, \mu_0^2) = A_g x^{-\lambda_g} (1 - x)^{C_g}, \tag{4}$$

- the *soft* + *hard* ansatz

$$xg(x, \mu_0^2) = A_g x^{-\lambda_g} (1 - x)^{C_g} (1 + D_g x + E_g x^2), \tag{5}$$

The free parameters for this model are σ_0 and the parameters for gluon A_g, λ_g, C_g or additionally D_g, E_g, Their values are obtained by a fit to the data. The fit results were found to be independent on the parameter C, which was therefore fixed as $C = 4$ GeV^2, in agreement with the original BGK fits. It is also possible to vary the parameter μ_0^2. However, to assure that the evolution is performed in the perturbative region and to be compatible with the standard pdf fits we took as a starting scale $\mu_0^2 = 1.9$ or 1.1 GeV^2. In the BGK model, the μ_0^2 scale is the same as the Q_0^2 scale of the standard QCD pdf fits.

3 Fits to the dipole model

The results of the BGK fit with valence quarks and the *soft* gluon density, are shown in Table 1, 2 and 3. The best fit is obtained using valence quarks and gluon density of the *soft* + *hard* type, see Table 4 and 6. In Table 1, 2 and 4 the starting QCD scale is $Q_0^2 = 1.9$ GeV^2, in Table 6 it is $Q_0^2 = 1.1$ GeV^2,

In Table 1, 4 and 6 the valence quark contribution was taken from the standard pdf fit, shown below, in Table 2 the valence quarks were fitted. In these tables, N_{df} denotes the number of degrees of freedom, which is equal to the number of measured data points minus the number of free parameters used in the fit. The parameters σ_0 of the dipole model and the starting parameters for gluon A_g, λ_g, C_g are obtained from the fit. The value of the parameter C was fixed, as explained above. To limit the fit to the perturbative region only we took $Q^2 \geq Q^2_{min}$ with $Q^2_{min} = 3.5$ (or 8.5) GeV^2. For the x region, we took $x \leq 0.01$. There are 538 (or 452) measured points in this region.

Q^2_{min} [GeV^2]	$\sigma_0[mb]$	A_g	λ_g	C_g	N_{df}	χ^2	χ^2/N_{df}
3.5	87.0± 8.9	2.32± 0.009	-0.056± 0.11	8.21± 0.80	534	551.1	1.03
8.5	72.4± 7.4	2.77± 0.009	-0.042± 0.123	6.54± 0.632	448	452.5	1.01

Table 1. BGK fit with fixed valence quarks for σ_r for H1ZEUS-NC data in the range $Q^2 \geq 3.5$ or 8.5 GeV^2 and $x \leq 0.01$. NLO fit. *Soft gluon.* $m_{uds} = 0.14, m_c = 1.3$ GeV. $Q_0^2 = 1.9$ GeV^2.

Q^2_{min} [GeV^2]	$\sigma_0[mb]$	A_g	λ_g	C_g	N_{df}	χ^2	χ^2/N_{df}
3.5	89.99± 9.2	2.44± 0.145	-0.079± 0.099	7.24± 0.61	530	540.35	1.02

Table 2. BGK fit with fitted valence quarks for σ_r for H1ZEUS-NC data in the range $Q^2 \geq 3.5$ GeV^2 and $x \leq 0.01$. NLO fit. *Soft gluon.* $m_{uds} = 0.14, m_c = 1.3$ GeV. $Q_0^2 = 1.9$ GeV^2.

No	$Auv(fix)$	Buv	Cuv	Euv	$Adv(fix)$	Bdv	Cdv
1	4.073(sum rule)	0.892± 0.019	5.832± 0.341	17.997± 0.876	3.151(sum rule)	0.840± 0.012	3.480± 0.056

Table 3. Parameters of the valence quark contribution fitted in the BGK fit of Table 2.

Q^2_{min} [GeV^2]	$\sigma_0[mb]$	A_g	λ_g	C_g	D_g	E_g	N_{df}	χ^2	χ^2/N_{df}
3.5	77,6± 18,6	2.62± 0.16	-0.064± 0.0087	37.1± 5.06	3.06 ± 6.51	1406.4± 552.7	532	534.2	1.00
8.5	63.5 ± 18.5	2.11± 0.10	-0.054± 0.0065	21.3± 4.062	1.10 ± 5.76	867.2± 423.7	448	439.0	0.98

Table 4. BGK fit with valence quarks for σ_r for H1ZEUS-NC data in the range $Q^2 \geq 3.5$ or 8.5 GeV^2 and $x \leq 0.01$. NLO fit. *Soft + hard gluon.* $m_{uds} = 0.14, m_c = 1.3$ GeV. $Q_0^2 = 1.9$ GeV^2.

4 Gluon density from BGK dipole model

In Figure 1 we compare the gluon densities obtained from fits with and without saturation ansatz of Tables 4 and 5. Note that for $x < 0.001$, the gluon densities obtained in the dipole approach are higher than that of the standard pdf fit, and that of the non-saturated one, see Fig 1. This is expected and is due to damping of gluon density by saturation effects.

Q^2_{min} [GeV^2]	Q^2_0 [GeV^2]	A_g	λ_g	C_g	D_g	E_g	Ndf	χ^2	χ^2/Np
3.5	1.9	2.33± 0.10	-0.094± 0.006	14.8± 11.5	9.80 ± 14.7	-99.5± 74.830	533	556.17	1.04
3.5	1.1	3.80± 0.22	0.10± 0.01	32.5± 1.6	-25.2 ± 3.49	1868 ± 252	533	539.2	1.01

Table 5. BGK fit with valence quarks for σ_r for H1ZEUS-NC data in the range $Q^2 \geq 3.5$ GeV^2 and $x \leq 0.01$. NLO fit. *Soft + hard gluon.* $m_{uds} = 0.14, m_c = 1.3$ GeV, non-saturation ansatz. $Q^2_0 = 1.9$ or 1.1 GeV^2.

Q^2_{min} [GeV^2]	Q^2_0 [GeV^2]	$\sigma_0[mb]$	A_g	λ_g	C_g	D_g	E_g	N_{df}	χ^2	χ^2/N_{df}
3.5	1.1	220 ± 122	3.57± 0.38	0.082± 0.017	31.3± 5.4	11.0 ± 9.8	1360± 690	532	532	1.00

Table 6. BGK fit with valence quarks for σ_r for H1ZEUS-NC data in the range $Q^2 \geq 3.5$ GeV^2 and $x \leq 0.01$. NLO fit. *Soft + hard gluon.* $m_{uds} = 0.14, m_c = 1.3$ GeV. $Q^2_0 = 1.1$ GeV^2.

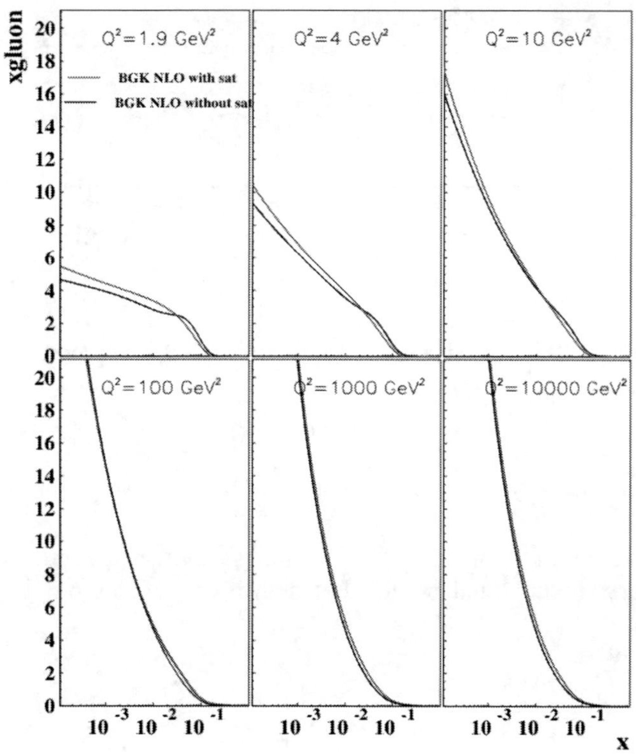

Figure 1. Comparison between the gluon densities obtained from fits in the $Q^2 > 3.5$ GeV^2 region, with and without the saturation ansatz, Table 4 and 5.

5 Predictions with gluon from BGK dipole fit for the LHC

In Figure 2 we show predictions for the LHC using the gluon density obtained from fits with saturation ansatz of Tables 4. We see that there is a large discrepancy between ATLAS data for jets and predictions from BGK dipole model.

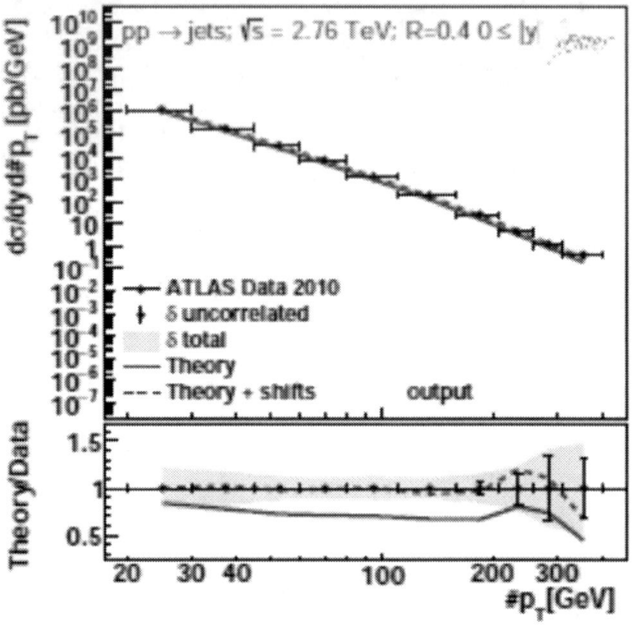

Figure 2. Predictions for jets in ATLAS using gluon density from BGK dipole model

6 Acknowledgement

I would like to thank Henri Kowalski (DESY) for various help and careful reading of the manuscript. This work is supported by the Polish Ministry under program Mobility Plus, no. 1320/MOB/IV/2015/0.

References

[1] K. Golec-Biernat, M. Wuesthoff, Phys. Rev. **D59**, 014017 (1999); Phys. Rev. **D60**, 114023 (1999).
[2] J. Bartels, K.Golec-Biernat and H. Kowalski, Phys.Rev. **D66,** (2002) 19
[3] E. Iancu, K. Itakura, and S. Munier. *Phys. Lett.*, B590:199–208, 2004.
[4] S. Munier, A. M. Staśto and A. H. Mueller, Nucl. Phys. B **603** (2001) 427.
[5] H. Kowalski and D. Teaney, Phys. Rev. **D68** (2003) 114005.

[6] H. Kowalski, L. Motyka and G. Watt, Phys. Rev. **D74** (2006) 074016.
[7] F. D. Aaron *et al.* [H1 and ZEUS Collaborations], JHEP 1001 (2010) 109.
[8] H. Abramowicz *et al.* [H1 and ZEUS Collaborations], Eur. Phys. J. **C75** (2015) nb.12.
[9] "HERAFitter, Open Source QCD Fit Project" By S. Alekhin at al., DESY Report 14-188, 7 Nov 2014, Published in EPJC (2015), 75: 304, e-Print: arXiv:1410.4412 [hep-ex]
[10] N. N. Nikolaev and B. G. Zakharov, Z. Phys. **C49** (1991) 607; Z. Phys. **C53** (1992) 331.
[11] J. Nemchik, N. N. Nikolaev, E. Predazzi and B. G. Zakharov, Z. Phys. C **75** (1997) 71.
[12] E. Gotsman, E. Levin and U. Maor, Nucl. Phys. B **464** (1996) 251.
[13] H. G. Dosch, T. Gousset, G. Kulzinger and H. J. Pirner, Phys. Rev. D **55** (1997) 2602.
[14] A. C. Caldwell and M. S. Soares, Nucl. Phys. A **696**, 125 (2001).
[15] J. R. Forshaw, R. Sandapen and G. Shaw, Phys. Rev. D **69** (2004) 094013.
[16] L. Frankfurt, M. Strikman and C. Weiss, Ann. Rev. Nucl. Part. Sci. **55** (2005) 403.
[17] H. Kowalski, T. Lappi, C. Marquet and R. Venugopalan, Phys. Rev. **C78** (2008) 045201.

EPJ Web of Conferences **164**, 07025 (2017) DOI: 10.1051/epjconf/201716407025
ICNFP 2016

Performance and calibration of b-tagging with the ATLAS experiment at LHC Run-2

Ian Connelly[1,a], on behalf of the ATLAS Collaboration

[1] *Royal Holloway, University of London*

Abstract. The identification of b-flavoured jets is key to many physics analyses at the LHC, including measurements involving Higgs bosons or top quarks, and searches for physics beyond the Standard Model. The capacity of ATLAS to efficiently tag b-jets has been enhanced for Run-2 with the addition of the Insertable B Layer (IBL), and improvements in the tracking and b-tagging algorithms. In the algorithm optimisation special emphasis has been placed in improving the performance for reconstructing high-p_{T} b-jets, addressing the challenges posed by track and vertex reconstruction in such an environment. The efficiency and rejection power of these algorithms have been calibrated on data taken in 2015, in particular by exploiting the copious production of b-jets in top quark decays, complemented by studies in multi-jet events.

1 Introduction

The Large Hadron Collider (LHC) has provided proton-proton collisions with centre-of-mass energies of 7 and 8 TeV during Run-1 and at 13 TeV during Run-2. Using these datasets, the ATLAS experiment has recorded and released over 500 publications which analyse the collisions with immense precision. Some of these analyses require the ability to identify whether a collimated group of particles, collectively known as a hadronic jet, originated from the decay of a b-hadron.

A hadronic jet which originated from the production of a b-hadron is commonly referred to as a b-jet. Their identification is a crucial factor in a range of physics analyses; measurements of top-quark properties (where a top-quark will decay to a b-quark with a branching fraction close to unity), searches for supersymmetric particles (where the production of the lightest supersymmetric particle in a model will often result in a b-jet) and searches and measurements of the Higgs boson (where the favoured decay is to a pair of b-quarks).

2 The ATLAS Detector

The ATLAS detector[1] is a general purpose detector providing an inclusive picture of proton collisions at the LHC. The ATLAS detector has been recording data at $\sqrt{s} = 13$ TeV since June 2015.

[a]e-mail: ian.connelly@cern.ch

[1]ATLAS uses a right-handed coordinate system with its origin at the nominal interaction point (IP) in the centre of the detector and the z-axis along the beam pipe. The x-axis points from the IP to the centre of the LHC ring, and the y-axis points upward. Cylindrical coordinates (r, ϕ) are used in the transverse plane, ϕ being the azimuthal angle around the z-axis. The pseudorapidity is defined in terms of the polar angle θ as $\eta = -\ln\tan(\theta/2)$.

DOI: 10.1051/epjconf/201716407025

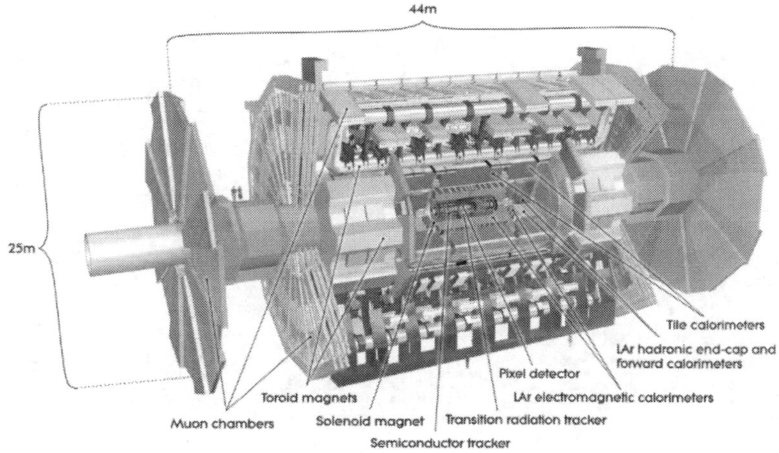

Figure 1. Cross-section of the ATLAS detector, highlighting the different detector components [1].

The ATLAS detector can be broadly categorised into three detector components: inner tracker, calorimeters and muon spectrometers. For the purposes of b-tagging, the performance of the inner tracker and the charged track reconstruction efficiency are the key considerations.

The inner detector consists of four silicon pixel layers, four silicon strip layers and a transition radiation tracker. The inner detector has an acceptance of $|\eta| < 2.5$ and sits in a 2 T axial magnetic field. The closest pixel layer is known as the insertable b-layer (IBL) which was installed during the long shutdown between Run-1 and Run-2. The IBL was installed 33.25 mm radially away from the beam line and has a smaller pixel size compared to the other pixel layers ($50 \times 250\,\mu\text{m}^2$ compared to $50 \times 400\,\mu\text{m}^2$ in the R-z plane). The addition of a new pixel layer added increased redundancy to the existing detector, providing an additional space-time point to reconstruct charged particle tracks close to the beam line. This closer proximity also improves the robustness of the tracking algorithms as the impact of multiple scattering is much reduced when extrapolating tracks back to the interaction region.

When reconstructing a track and extrapolating back towards to the interaction point, the point of closest approach is known as the impact parameter. This value is measured in the transverse plane (d_0) and in the longitudinal direction (z_0). The resolution of these variables feeds directly into the ability to identify b-jets as typically tracks originating from a b-hadron decay will have larger impact parameters as they originate from a different production vertex. The transverse and longitudinal impact parameter resolution is shown in Figure 2a and 2b respectively. The comparison is shown between Run-1 data collected in 2012 at 8 TeV and Run-2 data collected in 2015 at 13 TeV. The improvement at low transverse momentum comes from the inclusion of the IBL which reduced the impact of multiple scattering. The improvement in the longitudinal impact parameter across the entire transverse momentum range is due to the improved granularity in the z-direction. One additional change from Run-1 to Run-2 has been an improvement in the track reconstruction algorithms. Neural network pixel hit clustering has improved the reconstruction of tracks that share the same hit in the pixel layers. This has primarily improved the reconstruction of tracks in the core of high-p_{T} jets [2].

 DOI: 10.1051/epjconf/201716407025

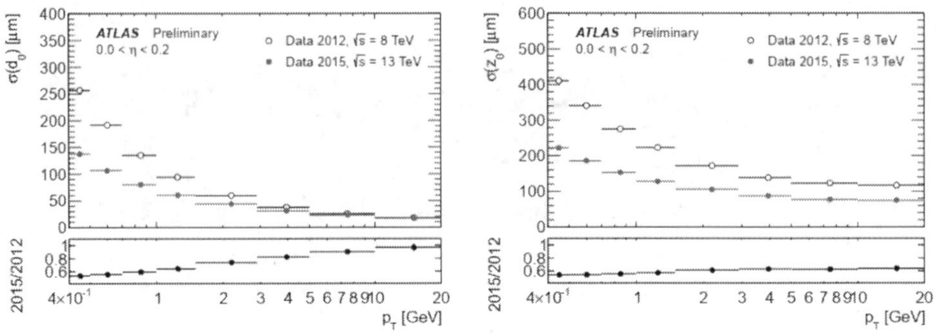

(a) (b)

Figure 2. Comparison of impact parameter resolutions ($\sigma(d_0)$, $\sigma(z_0)$) between Run-1 and Run-2 [3].

3 b-Tagging in ATLAS

b-tagging is defined as the identification of a hadronic jet which originated from the fragmentation and hadronisation of a b-hadron, with the resulting jet being known as a b-jet. b-tagging allows for precise Standard Model measurements, plays a key role in a number of Higgs searches, and opens the door to exploring New Physics scenarios with heavy flavours.

The rate at which a true b-jet is identified defines the b-tagging efficiency for a particular b-tagging algorithm. The true flavour of a jet is defined in simulated data using a spatial ΔR matching between stable hadrons and reconstructed jets. A hierarchal matching is performed, first checking whether a b-hadron can be matched, followed by a charm hadron and followed by a τ-lepton. This matching procedure results in a jet being classified respectively as either a b-jet, c-jet, τ-jet or a light-jet (if no match is found).

b-hadrons have a collection of unique properties which can be exploited in order to be identified in data. An illustration of the production of a b-jet is shown in Figure 3. The decay of a b-hadron is suppressed by a CKM factor which results in a longer flight path before decaying in the inner detector compared to charm and light hadrons. In the lab frame, this means that a b-hadron will travel on the order of a few millimetres before subsequently decaying[2]. This gives rise to a displaced secondary vertex which is a key signature of a b-hadron decay. In addition, the b-quark is the heaviest quark which can form hadrons which results in a heavy hadronic decay signature. Finally, the decay of a b-hadron will on average have a higher charged track multiplicity in the decay than other hadron decays.

3.1 b-Tagging Algorithms

There are three main categories of b-tagging algorithms commonly used in ATLAS. There are impact parameter based algorithms (IP2D, IP3D); inclusive secondary vertex reconstruction algorithms (SV); and decay chain reconstruction algorithms (JetFitter). These algorithms contribute complementary information and can be combined using a multivariate function to create a single b-tagging discriminant (MV). Whilst b-tagging is a status used to describe a hadronic jet, the b-tagging algorithms use information relating to tracks which have been associated to a jet. This association is performed using a p_T

[2] A B^+ hadron has $c\tau$ of approximately 0.5 mm. If this hadron was produced with p_T of 100 GeV, the resultant distance travelled radially into the detector is given by $\langle l \rangle = \beta\gamma c\tau \approx 10$ mm.

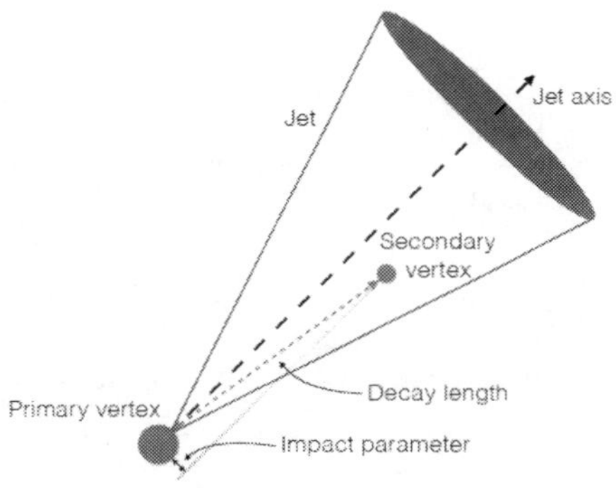

Figure 3. Illustration of the production of a b-jet.

dependent spatial ΔR matching between tracks and jets. The tracks are then subjected to an algorithm-dependent selection in order to determine whether they are used by the b-tagging algorithms.

The impact parameter algorithms use the signed impact parameter significance,

$$\mathrm{sig}(x) = \frac{x}{\sigma_x} \tag{1}$$

of the tracks which are matched to a jet. The sign is defined as positive (negative) if the point of closest approach of the track to the primary vertex is in front (behind) the primary vertex with respect to the jet direction. The IP2D algorithm uses templates of the signed impact parameter significance in the transverse direction for different jet flavours. The IP3D algorithm uses templates of the signed impact parameter significance in both the transverse and longitudinal directions. The distributions for the d_0 and z_0 signed impact parameter significances are shown in Figure 4a and Figure 4b respectively. These distributions provide probability distribution functions (PDFs) for tracks coming from different jet flavour hypotheses. A ratio can be defined between the b-jet hypothesis and light jet hypothesis to combine this information into a single likelihood-ratio discriminant. The final IP2D and IP3D discriminants can be seen in Figure 5a and Figure 5b for three different flavour hypotheses. It can be seen that one could choose to cut on this variable to select a sample in data with a high b-jet purity.

The secondary vertex algorithms attempt to reconstruct an inclusive displaced secondary vertec which is formed by selected tracks within a jet. All possible two-track vertices are identified from the track candidates. Vertices which are likely from long-lived particles (K_s, Λ), photon conversions or hadronic interactions with the detector material are removed. A single inclusive vertex is reconstructed by iteratively removing outlier tracks until a good vertex candidate is identified. Once reconstructed, the kinematic properties of the tracks can be used to offer discrimination between the b-jet and light-jet hypotheses. The invariant mass of the reconstructed secondary vertex is shown in Figure 6 where a b-jet typically has a higher value compared to non-b-jets.

The multi-vertex decay chain algorithm attempts to reconstruct the weak decay chain of a b-hadron produced at the primary vertex which subsequently decays to a charm hadron (producing a tertiary vertex). A Kalman filter is used to identify the b-hadron flight path using the assumption that the tertiary vertex lies along the same flight path. The tracks which are associated to the reconstructed vertices can also provide discriminating information, in a similar way to the SV algorithm. The invariant mass of tracks fitted to one or more of the displaced vertices is shown in Figure 7a and the fraction of energy carried by the fitted tracks compared to the total energy of the jet is shown in Figure 7b.

 DOI: 10.1051/epjconf/201716407025

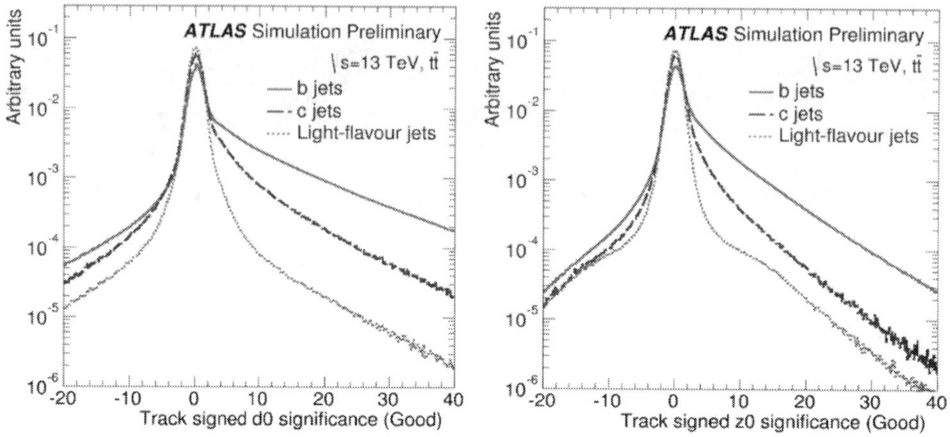

(a) (b)

Figure 4. Impact parameter significances (d_0, z_0) for b, c and light-jets [4].

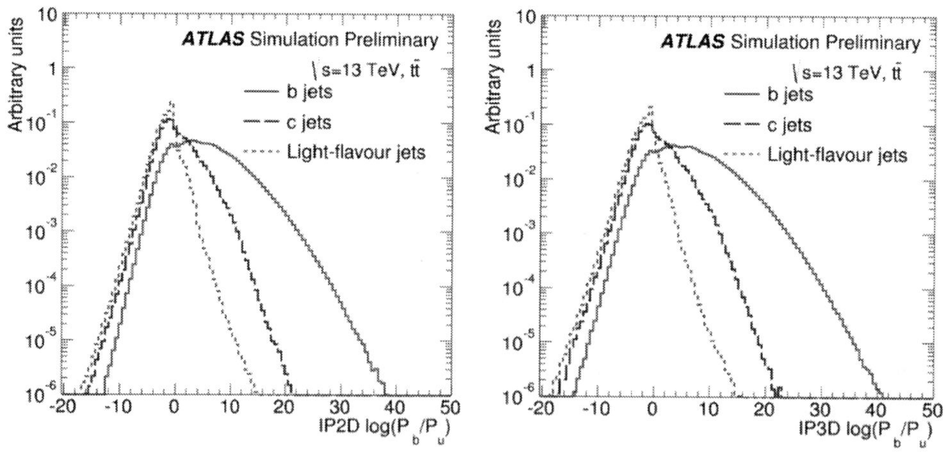

(a) (b)

Figure 5. Impact parameter b-tagging algorithms (IP2D, IP3D) for b, c and light-jets [4].

3.2 Performance in 2016

The three types of algorithms provide input to a multivariate classifier (MV). In Run-2 this classifier is a boosted decision tree (BDT) trained to discriminate b-jets from light-jets. The background samples of light-jets can contain an admixture of charm and light-jets in order to improve the charm-jet rejection. The discrimination performance is shown in Figure 8a for the MV2 algorithm with a 10% charm-jet admixture (MV2c10). The performance between the Run-1 MV algorithm (MV1) and the Run-2 MV algorithm (MV2) has seen an improvement of 10%, primarily driven by the inclusion of the IBL. Between 2015 and 2016, retraining of the b-tagging classifier has further improved the

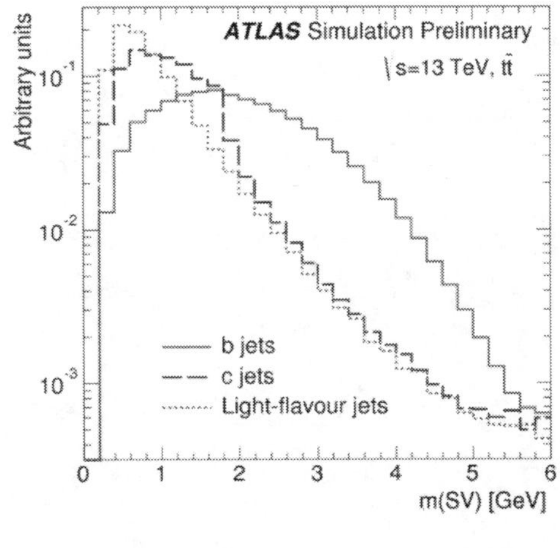

Figure 6. Invariant mass of the displaced vertex identified with the SV algorithm [5].

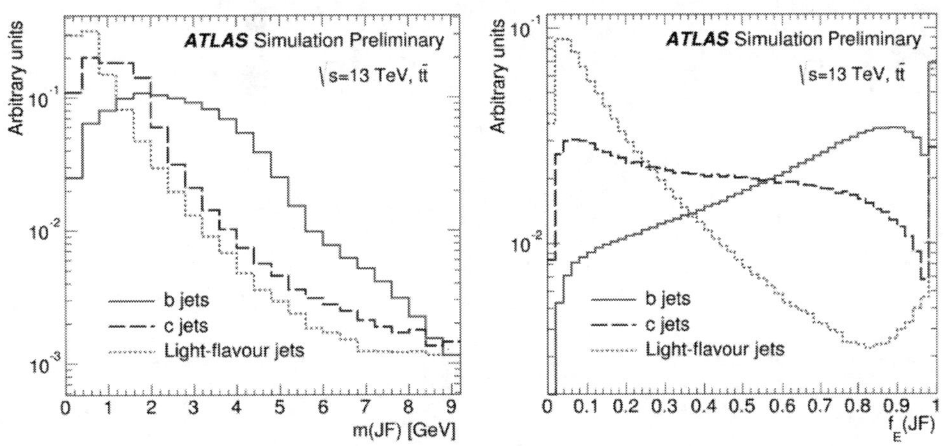

(a) (b)

Figure 7. Distributions showing the invariant mass of all tracks associated with the identified displaced vertices using the JetFitter algorithm and the energy fraction of tracks identified with the JetFitter algorithm compared to the jet [4].

charm-jet rejection by around 40% at the 77% working point. The charm-jet rejection rate as a function of b-tagging efficiency is shown in Figure 8b. Jets which have been selected using a top-pair event selection are shown in Figure 8c comparing the performance of the b-tagging weight (MV2c10) in simulation and data, where a good agreement can be observed.

4 Calibration Techniques

Calibrations of b-tagging algorithms derive a correction factor applied to simulated data to match the b-tagging efficiency measured in data. This correction factor is dependent on the true flavour

DOI: 10.1051/epjconf/201716407025

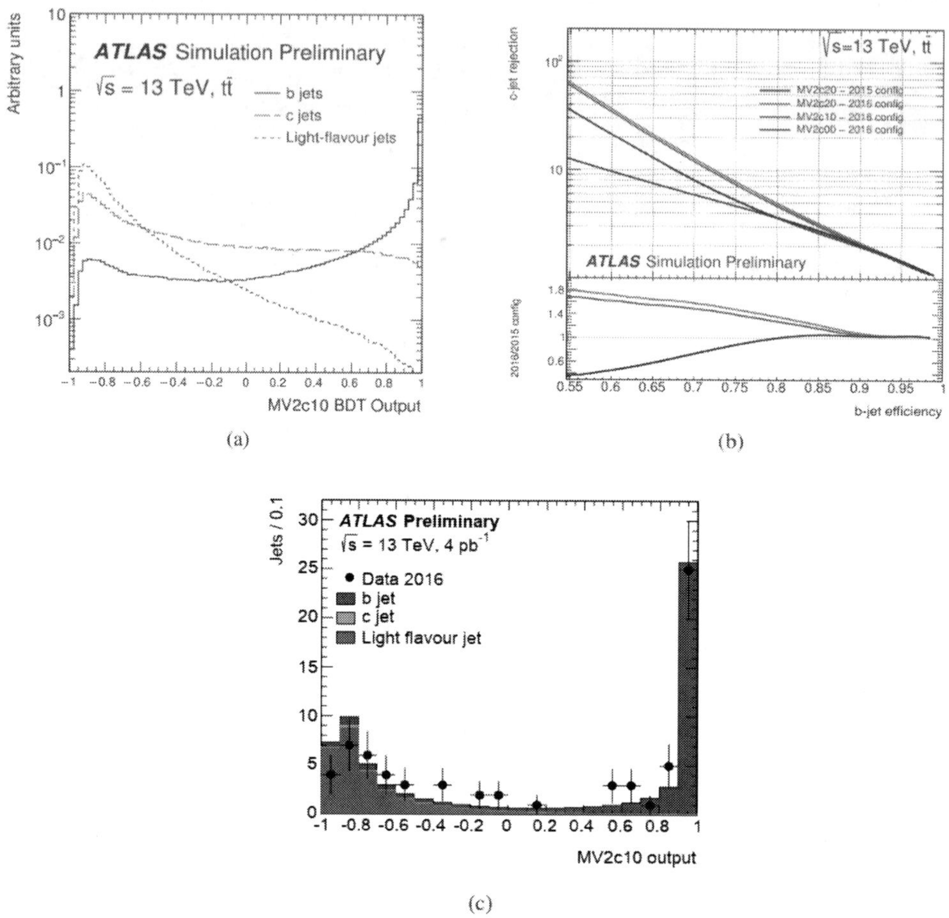

Figure 8. The performance of the MV2c10 algorithm is shown for different jet flavours [5], a comparison between the MV2c algorithms trained in 2015 and trained in 2016 [5] and the performance of the MV2c10 algorithm compared to 2016 data [7].

of the simulated jet and the jet kinematics. A correction factor is derived both for true b-jet which are correctly identified as well as for light and charm-jets which are incorrectly b-tagged. In order to derive these scale factors, the performance of the b-tagging algorithm needs to be measured using events in data which are enriched in a particular jet flavour whilst correctly modelling the backgrounds.

To measure the b-tagging efficiency, events are measured using dileptonically decaying top-pair events as these events are pure in b-jets using the Run-1 combinatorial likelihood method [6]. Events with one electron, one muon and two or three jets are used to measure the efficiency. Events with two same-flavour leptons are used to derive data-driven corrections to the Z+jets and top-pair samples. A combinatorial likelihood method is applied to extract the b-tagging efficiency in data by applying a two-dimensional maximum likelihood fit as a function of leading and sub-leading jet p_T, which allows the method to include correlations between the two jets. The current b-tagging efficiency scale factors calculated with this method as a function of jet p_T is shown in Figure 9 where the scale factors agree

DOI: 10.1051/epjconf/201716407025

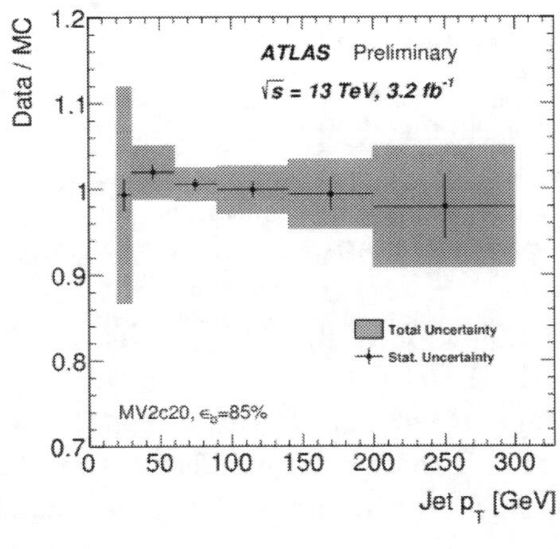

Figure 9. Scale factors derived from the preliminary Run-2 b-tagging efficiency measurement on dilepton top-pair events in 2015 data [9].

with unity within errors. The intermediate jet range has a total uncertainty of 2-4% and is dominated by the effect of top-pair modelling.

Charm and light-jet calibrations are underway using the Run-2 dataset. The techniques used in Run-1 are being applied to the Run-2 dataset [8]. The charm-jet calibration uses the D^* method, where a sample of events enriched in charm-jets are selected by reconstructing the decay chain of $D^* \rightarrow D_0(\kappa\pi)\pi_s$. The yields before and after applying a b-tagging requirement comes from some fraction of true b-jets correctly selected with an efficiency and some fraction of charm-jets incorrectly selected with an efficiency. By fitting the proper lifetime of the events, the contamination from b-hadron decays can be estimated and used to extract the charm-tagging efficiency. The light tagging efficiency has become harder to estimate since the performance of the b-tagging algorithms has improved in Run-2 compared to Run-1. The negative-tag method was used in Run-1 where parameters such as the signed impact parameter significance and the signed decay length significance have their sign requirements inverted. This enhances the fraction of light-jets selected by the IP and SV algorithms. The inclusive distribution of the negatively tagged light-jets has a similar shape to the mis-tagged light-jets, and once the contamination from heavy flavour and long-lived particles is accounted for, can be used to extract the light-jet tagging efficiency.

5 Concluding Remarks

The b-tagging perfomance in Run-2 shows a marked improvement compared to Run-1. The inclusion of the IBL has improved the track reconstruction efficiency and accounts for the majority of the 10% improvement in b-tagging efficiency. The optimisation of tracking algorithms and b-tagging algorithms has improved the performance at high p_T and the latest 2016 optimisation has improved the charm-jet rejection factor by around 40%. Efforts are currently underway to calibrate the b-tagging algorithms with the 13 TeV data collected in 2015 and 2016. Preliminary b-tagging efficiencies have been successfully extracted from 13 TeV data and charm-tagging efficiency measurements are being finalised.

References

[1] ATLAS Collaboration, The ATLAS experiment at the CERN Large Hadron Collider, JINST 08003 (2008)

[2] ATLAS Collaboration, The Optimization of ATLAS Track Reconstruction in Dense Environments, ATL-PHYS-PUB-2015-006 (2015)

[3] ATLAS Collaboration, Impact parameter resolution Run-1 vs. Run-2, ITDR-2015-007 (2015)

[4] ATLAS Collaboration, Expected performance of the ATLAS b-tagging algorithms in Run-2, ATL-PHYS-PUB-2015-022 (2015)

[5] ATLAS Collaboration, Optimisation of the ATLAS b-tagging performance for the 2016 LHC Run, ATL-PHYS-PUB-2016-012 (2016)

[6] ATLAS Collaboration, Calibration of b-tagging using dileptonic top pair events in a combinatorial likelihood approach with the ATLAS experiment, ATLAS-CONF-2014-004 (2014)

[7] ATLAS Collaboration, b-tagging performance plots in a $t\bar{t}$-dominated sample from early 2016 ATLAS data, FTAG-2016-001 (2016)

[8] ATLAS Collaboration, Calibration of the performance of b-tagging for c and light-flavour jets in the 2012 ATLAS data, ATLAS-CONF-2014-046 (2014)

[9] ATLAS Collaboration, b-tagging calibration plots using dileptonic $t\bar{t}$ events produced in pp collisions at $\sqrt{s} = 13$ TeV and a combinatorial likelihood approach, FTAG-2016-002 (2016)

DOI: 10.1051/epjconf/201716407026

Perturbative unitarity bounds on di-boson scalar resonances

Luca Di Luzio[1,a], Jernej F. Kamenik[2,3,b], and Marco Nardecchia[4,c]

[1] *Institute for Particle Physics Phenomenology, Department of Physics, Durham University, DH1 3LE, UK*
[2] *Jožef Stefan Institute, Jamova 39, 1000 Ljubljana, Slovenia*
[3] *Faculty of Mathematics and Physics, University of Ljubljana, Jadranska 19, 1000 Ljubljana, Slovenia*
[4] *CERN, Theoretical Physics Department, Geneva, Switzerland*

Abstract. We consider the constraints implied by partial wave unitarity on new physics in the form of di-boson resonances at LHC. We derive the scale where the effective description in terms of the SM supplemented by a single scalar resonance is expected to break down depending on the resonance width and signal cross-section.

1 Introduction

Perturbative unitarity is a powerful theoretical tool for inferring the range of validity of a given effective field theory (EFT), with notable examples of applications both in the physics of strong and electroweak interactions. Perhaps most famously, constraints imposed by perturbative unitarity in WW scattering have been used in the past to infer an upper bound on the Higgs boson mass or, alternatively, on the scale where the standard model (SM) description of weak interactions would need to be completed in the ultraviolet (UV) in terms of some new strongly coupled dynamics [1].

The recently rekindled interest in new physics (NP) in the form of (possibly broad) di-photon resonances [2, 3] at the LHC motivated us to reconsider the implications of perturbative unitarity for EFT interpretations of resonances decaying to di-boson final states. In particular, focusing on promptly produced scalar SM singlets decaying to two SM gauge bosons we aim to address the following question: at which maximal energies do we expect the effective description in terms of the SM supplemented by a single scalar to break down? Assuming that a scalar resonance is within the reach of the present LHC run, can a next generation hadron collider potentially probe the on-shell effects of new degrees of freedom responsible for the restoration of unitarity?

After a brief recap of partial wave unitarity arguments in Sect. 2, we apply them in Sect. 3 to the EFT case where a di-boson resonance is the only new degree of freedom beyond the SM. The scale of unitarity violation is hence interpreted as an upper bound on the mass scale of new degrees of freedom UV completing the effective low-energy description and regularizing (unitarizing) the amplitudes' growth. The present contribution is largely based on Ref. [4], to which the reader is referred for further details.

[a] e-mail: luca.di-luzio@durham.ac.uk
[b] e-mail: jernej.kamenik@cern.ch
[c] e-mail: marco.nardecchia@cern.ch

2 Partial wave unitarity

Let us denote by $\mathcal{T}_{fi}(\sqrt{s}, \cos\theta)$ the matrix element of a $2 \to 2$ scattering amplitude in momentum space, defined via

$$(2\pi)^4\delta^{(4)}(P_i - P_f)\mathcal{T}_{fi}(\sqrt{s}, \cos\theta) = \langle f|T|i\rangle\,, \tag{1}$$

where T is the interacting part of the S-matrix, $S = 1 + iT$. The dependence of the scattering amplitude on $\cos\theta$ is eliminated by projecting it onto partial waves of total angular momentum J

$$a^J_{fi} = \frac{\beta_f^{1/4}(s, m^2_{f1}, m^2_{f2})\beta_i^{1/4}(s, m^2_{i1}, m^2_{i2})}{32\pi s}\int_{-1}^{1} d(\cos\theta)\, d^J_{\mu_i\mu_f}(\theta)\,\mathcal{T}_{fi}(\sqrt{s}, \cos\theta)\,, \tag{2}$$

where $d^J_{\mu_i\mu_f}$ is the J-th Wigner d-function appearing in the Jacob-Wick expansion [5], while $\mu_i = \lambda_{i1} - \lambda_{i2}$ and $\mu_f = \lambda_{f1} - \lambda_{f2}$ are defined in terms of the helicities of the initial $(\lambda_{i1}, \lambda_{i2})$ and final $(\lambda_{f1}, \lambda_{f2})$ states. The function $\beta(x, y, z) = x^2 + y^2 + z^2 - 2xy - 2yz - 2zx$ is a kinematical factor related to the momentum (to the fourth power) of a given particle in the center of mass frame. The right hand side of Eq. (2) must be further multiplied by a $\frac{1}{\sqrt{2}}$ factor for any identical pair of particles either in the initial or final state.

When restricted to a same-helicity state (zero total spin), the Wigner d-functions reduce to the Legendre polynomials, i.e. $d^J_{00} = P_J$. In practice, we will only focus on $J = 0$ ($d^0_{00} = P_0 = 1$), since higher partial waves typically give smaller amplitudes, unless $J = 0$ amplitudes are suppressed or vanish for symmetry reasons. Hence, the quantity we are interested in is

$$a^0_{fi} = \frac{\beta_f^{1/4}(s, m^2_{f1}, m^2_{f2})\beta_i^{1/4}(s, m^2_{i1}, m^2_{i2})}{32\pi s}\int_{-1}^{1} d(\cos\theta)\,\mathcal{T}_{fi}(\sqrt{s}, \cos\theta)\,. \tag{3}$$

In the high-energy limit, $\sqrt{s} \to \infty$, one has $\beta_f^{1/4}\beta_i^{1/4}/s \to 1$. The unitarity condition on the S-matrix, $SS^\dagger = 1$, gives

$$\frac{1}{2i}\left(a^J_{fi} - a^{J*}_{if}\right) \geq \sum_h a^{J*}_{hf} a^J_{hi}\,, \tag{4}$$

where the sum over h is restricted to 2-particle states, which slightly underestimates the left hand side. For $i = f$ Eq. (4) reduces to

$$\mathrm{Im}\, a^J_{ii} \geq |a^J_{ii}|^2\,. \tag{5}$$

Hence, a^J_{ii} must lie inside the circle in the Argand plane defined by

$$\left(\mathrm{Re}\, a^J_{ii}\right)^2 + \left(\mathrm{Im}\, a^J_{ii} - \frac{1}{2}\right)^2 \leq \frac{1}{4}\,, \tag{6}$$

which implies

$$|\mathrm{Im}\, a^J_{ii}| \leq 1 \qquad \text{and} \qquad |\mathrm{Re}\, a^J_{ii}| \leq \frac{1}{2}\,. \tag{7}$$

Under the assumption that the tree-level amplitude is real, Eq. (7) suggests the following perturbativity criterium

$$|\mathrm{Re}\,(a^J_{ii})^{\mathrm{Born}}| \leq \frac{1}{2}\,. \tag{8}$$

In fact, a Born value of $\mathrm{Re}\, a^J_{ii} = \frac{1}{2}$ and $\mathrm{Im}\, a^J_{ii} = 0$ needs at least a correction of 40% in order to restore unitarity, thus signalling the breakdown of the (E/Λ) power expansion of the EFT, where E is the typical energy involved in a process and Λ is the EFT cut-off scale.

3 Effective field theory of a scalar resonance

We consider the EFT of a gauge singlet spin-0 resonance, S with mass M_S, coupled to the SM fields. Assuming CP invariance, we choose S to transform as a scalar.[1] Apart from renormalizable terms coupling S to the Higgs in the scalar potential, the $d = 5$ Lagrangian reads

$$\mathcal{L}^{(5)}_{\text{int.}} = -\frac{g_3^2}{2\Lambda_g} S G^2_{\mu\nu} - \frac{g_2^2}{2\Lambda_W} S W^2_{\mu\nu} - \frac{g_1^2}{2\Lambda_B} S B^2_{\mu\nu} - \frac{1}{\Lambda_H} S \left(D_\mu H\right)^\dagger D^\mu H - \frac{1}{\Lambda'_H} S \left(H^\dagger H\right)^2$$
$$- \frac{1}{\Lambda_d} S \overline{Q}_L d_R H - \frac{1}{\Lambda_u} S \overline{Q}_L u_R H^c - \frac{1}{\Lambda_e} S \overline{L}_L e_R H + \text{h.c.}\,, \tag{9}$$

where we have suppressed flavor indices. This parametrization makes it clear that the leading interactions of a scalar singlet with the SM fields, directly relevant for di-boson resonances at the LHC, are all due to non-renormalizable $d = 5$ operators. Their effects are thus expected to be enhanced at high energies eventually leading to the breakdown of perturbative unitarity. In order to quantify this simple observation in the following subsections we evaluate the relevant scattering amplitudes involving SM gauge bosons, Higgs and quarks at the respective leading orders in perturbation theory. Moreover, since we are interested in studying $2 \to 2$ scattering processes at energies $\sqrt{s} \gg M_S \gg v \simeq 246$ GeV, we can safely set all the massive parameters (including M_S) to zero and work within the unbroken SM theory. This also implies that we can neglect any h-S mixing effects and set the masses of the final state SM particles to zero. We distinguish between two classes of tree-level processes characterized by a different energy scaling of the amplitude: scalar mediated scatterings and $d = 5$ contact interactions.

3.1 Scalar mediated boson scattering

Let us start, as an example, by considering the $\gamma\gamma \to \gamma\gamma$ scattering amplitude due to the effective operator

$$-\frac{e^2}{2\Lambda_\gamma} S F^2_{\mu\nu}\,. \tag{10}$$

In the $(++, --)$ helicity basis we find [4]

$$\mathcal{T} = -\frac{e^4}{\Lambda_\gamma^2} \begin{pmatrix} \frac{\frac{s^2}{s-M_S^2}}{\frac{s^2}{s-M_S^2} + \frac{t^2}{t-M_S^2} + \frac{u^2}{u-M_S^2}} & \frac{\frac{s^2}{s-M_S^2} + \frac{t^2}{t-M_S^2} + \frac{u^2}{u-M_S^2}}{\frac{s^2}{s-M_S^2}} \end{pmatrix} \overset{\sqrt{s} \gg M_S}{\simeq} -\frac{e^4 s}{\Lambda_\gamma^2} \begin{pmatrix} 1 & 0 \\ 0 & 1 \end{pmatrix}, \tag{11}$$

where in the last step we took the high-energy limit, and only the s-channel survives at high energies.

The projection on the $J = 0$ partial waves is obtained by applying Eq. (3) and by multiplying by a $1/2$ factor which takes into account the presence of identical particles both in the initial and final states. In the high-energy limit we get

$$a^0 \simeq -\frac{e^4 s}{32\pi\Lambda_\gamma^2} \begin{pmatrix} 1 & 0 \\ 0 & 1 \end{pmatrix}, \tag{12}$$

which, confronted with Eq. (8), leads to the tree-level unitarity bound

$$\sqrt{s} \lesssim \sqrt{16\pi}\frac{\Lambda_\gamma}{e^2}\,. \tag{13}$$

[1] The pseudo-scalar case leads to analogous conclusions as far as unitarity bounds are concerned, hence in the following we will not consider it separately.

DOI: 10.1051/epjconf/201716407026

As a matter of fact, the bound above can be made stronger if one considers the full $VV \to V'V'$ scattering matrix, where V and V' are any of the $8+3+1$ (transversely polarized) SM gauge bosons of the effective Lagrangian in Eq. (9). In such a case, the previous calculation is readily generalized in the high-energy limit where only the s-channel survives. To this end, we note that a scattering amplitude in the s-channel can be written as

$$m_{ij} = \frac{a_i a_j}{s - M_S^2}\,, \tag{14}$$

where a_i and a_j are obtained by cutting any $i \to j$ diagram in two parts along the s-channel propagator. The matrix in Eq. (14) has rank 1 and its non-zero eigenvalue is given by the trace. Hence, denoting by $\tilde{a}^0$ the eigenvalue of the $VV \to V'V'$ scattering matrix, in the high-energy limit we get

$$\tilde{a}^0 \simeq -\frac{s}{32\pi}\left(\frac{8g_3^4}{\Lambda_g^2} + \frac{3g_2^4}{\Lambda_W^2} + \frac{g_1^4}{\Lambda_B^2}\right). \tag{15}$$

Correspondingly, the tree-level unitarity bound is given by

$$\frac{s}{32\pi}\left(8\frac{g_s^4}{\Lambda_g^2} + 3\frac{g_2^4}{\Lambda_W^2} + \frac{g_1^4}{\Lambda_B^2}\right) \lesssim \frac{1}{2}\,. \tag{16}$$

We remark that in deriving these bounds we consider only the transverse polarizations of the W and Z gauge bosons. Generally, scattering amplitudes involving longitudinally polarized massive vector bosons can grow as positive powers of $E/m_{W,Z}$ implying apparently stronger dependence on s. However, as it can be easily verified (through an explicit calculation of the processes at hand or more generally via a clever gauge choice [6]), the scattering amplitudes involving longitudinally polarized states sourced by the gauge field strengths in Eq. (9) are *suppressed* by powers of $m_{W,Z}/E$ and thus do not lead to relevant unitarity constraints at high s.

3.2 Fermion-scalar contact interactions

Next we consider the contact interaction

$$-\frac{1}{\Lambda_d} S\,\overline{Q}_L d_R H = \left[-\frac{1}{\Lambda_d}\delta_a^b \delta_i^j\right] S\,(\overline{Q}_L)^{ai}(d_R)_b H_j\,, \tag{17}$$

where we have explicitly factored out the color and $SU(2)_L$ group structure. In this case the leading scattering process is $\overline{Q}d \to SH$. By explicitly writing the polarization and gauge indices in the amplitude, one finds

$$\mathcal{T} = -\frac{\delta_a^b \delta_i^j}{2\Lambda_d}\bar{v}^s(k)\,(1+\gamma_5)\,u^r(p)\,. \tag{18}$$

Only the ++ and −− polarizations survive. By explicit evaluation we get [4]

$$\mathcal{T}_{++} = \frac{\delta_a^b \delta_i^j}{\Lambda_d}(E + p^3) \overset{\sqrt{s} \gg M_S}{\simeq} \delta_a^b \delta_i^j \frac{\sqrt{s}}{\Lambda_d}\,, \tag{19}$$

$$\mathcal{T}_{--} = \frac{\delta_a^b \delta_i^j}{\Lambda_d}(E - p^3) \overset{\sqrt{s} \gg M_S}{\simeq} 0\,. \tag{20}$$

At high energies the $J = 0$ partial wave is obtained by considering the color singlet channel for a state in the linear combination $\frac{1}{\sqrt{2}}\left(|\overline{Q}d\rangle + |SH\rangle\right)$, which gives

$$a^0 \simeq \frac{1}{16\pi}\frac{\sqrt{s}}{\Lambda_d}. \tag{21}$$

Correspondingly, the tree-level unitarity bound reads

$$\sqrt{s} \lesssim 8\pi\Lambda_d. \tag{22}$$

Similarly, from the other two contact interactions in the last row of Eq. (9) we get $\sqrt{s} \lesssim 8\pi\Lambda_{u,e}$.

3.3 Unitarity bounds

As an exemplification we consider a scalar resonance S with mass M_S and total width Γ_S appearing in a di-photon final state at the LHC.[2] Expanding the effective Lagrangian in Eq. (9) around the broken electroweak (EW) vacuum, the part relevant for S production at the LHC is

$$\mathcal{L}^{(5)}_{\text{int.}} \supset -\frac{g_3^2}{2\Lambda_g} S G^2_{\mu\nu} - \frac{e^2}{2\Lambda_\gamma} S F^2_{\mu\nu} - \sum_q y_{qS} S \overline{q} q, \tag{23}$$

whose operators give rise to the decay widths

$$\Gamma_{\gamma\gamma} \equiv \Gamma(S \to \gamma\gamma) = \pi\alpha^2_{\text{EM}}\frac{M_S^3}{\Lambda_\gamma^2}, \tag{24}$$

$$\Gamma_{gg} \equiv \Gamma(S \to gg) = 8\pi\alpha_s^2\frac{M_S^3}{\Lambda_g^2}, \tag{25}$$

$$\Gamma_{q\overline{q}} \equiv \Gamma(S \to q\overline{q}) = \frac{3}{8\pi} y^2_{qS} M_S \left(1 - \frac{4m_q^2}{M_S^2}\right)^{3/2}. \tag{26}$$

The matching between the operators in Eq. (23) and Eq. (9) then yields

$$\frac{1}{\Lambda_\gamma} = \frac{1}{\Lambda_B} + \frac{1}{\Lambda_W}, \qquad y_{qS} = \frac{v}{\sqrt{2}\Lambda_q}. \tag{27}$$

In the narrow width approximation the prompt S production at the LHC can also be fully parametrized in terms of the relevant decay widths

$$\sigma(pp \to S) = \frac{1}{M_S s}\left[\sum_{\mathcal{P}} C_{\mathcal{P}\overline{\mathcal{P}}}\Gamma_{\mathcal{P}\overline{\mathcal{P}}}\right], \tag{28}$$

where $\sqrt{s}$ is the LHC pp collision energy and $C_{\mathcal{P}\overline{\mathcal{P}}}$ parametrize the relevant parton luminosities.

For illustration purposes in the following we consider in turn either gg and $\gamma\gamma$ induced processes or alternatively $b\overline{b}$ and $\gamma\gamma$ rates at a benchmark mass of $M_S = 750$ GeV. The remaining possibilities

[2] Analogous analysis can be performed also for other EW gauge boson final states with the slight complication of disentangling the transverse and longitudinal gauge boson polarizations, as they are sourced by different terms in the EFT Lagrangian ($\Lambda_{B,W}$ and Λ_H, respectively).

lie in between these two limiting cases considering the values of relevant parton luminosities (their values at $\sqrt{s} = 8$ TeV and 13 TeV LHC can be found e.g. in [7]). In the former case given a 13 TeV cross-section $\sigma_{\gamma\gamma} \equiv \sigma(pp \to S)\mathcal{B}_{\gamma\gamma}$ one obtains the relation

$$\frac{\Gamma_{\gamma\gamma}}{M_S}\frac{\Gamma_{gg}}{M_S} \simeq 1.4 \times 10^{-7}\frac{\sigma_{\gamma\gamma}}{\text{fb}}\frac{\Gamma_S}{M_S}, \tag{29}$$

while for the latter we obtain

$$\frac{\Gamma_{\gamma\gamma}}{M_S}\frac{\Gamma_{b\bar{b}}}{M_S} \simeq 2.3 \times 10^{-5}\frac{\sigma_{\gamma\gamma}}{\text{fb}}\frac{\Gamma_S}{M_S}. \tag{30}$$

These relations define the phenomenological benchmarks for the resonance partial widths into gauge boson and quark final states, to be subjected to constraints from perturbative unitarity.

To make contact with the EFT unitarity discussion of the preceding subsections we use Eqs. (24)–(25) and trade Λ_g, Λ_W and Λ_B for Γ_{gg}, $\Gamma_{\gamma\gamma}$ and the ratio $r \equiv \Lambda_B/\Lambda_W$. In particular, we get

$$\frac{1}{\Lambda_g^2} = \frac{\Gamma_{gg}}{8\pi\alpha_s^2 M_S^3}, \qquad \frac{1}{\Lambda_W^2} = \frac{\Gamma_{\gamma\gamma}}{\pi\alpha_{\text{EM}}^2 M_S^3}\left(\frac{r}{1+r}\right)^2, \qquad \frac{1}{\Lambda_B^2} = \frac{\Gamma_{\gamma\gamma}}{\pi\alpha_{\text{EM}}^2 M_S^3}\left(\frac{1}{1+r}\right)^2, \tag{31}$$

which inserted back into Eq. (16) yield

$$\sqrt{s} \lesssim M_S\left(\frac{\Gamma_{gg}}{M_S} + f(r)\frac{\Gamma_{\gamma\gamma}}{M_S}\right)^{-1/2}, \tag{32}$$

with

$$f(r) = \frac{3r^2 s_W^{-4} + c_W^{-4}}{(1+r)^2}. \tag{33}$$

In the phenomenologically interesting range $r \in [-0.3, 15]$ [7, 8], the function $f(r)$ has the minimal value 1.6 for $r = 0.030$ and the maximal one 50 at $r = 15$. Hence, we can set the following unitarity bounds

$$\sqrt{s} \lesssim 32\, M_S\left(\frac{\Gamma_{gg}/M_S}{10^{-3}}\right)^{-1/2}, \tag{34}$$

$$\sqrt{s} \lesssim (15 \div 80)\, M_S\left(\frac{\Gamma_{\gamma\gamma}/M_S}{10^{-4}}\right)^{-1/2}, \tag{35}$$

where the values 15 and 80 in the last equation correspond respectively to $r = 15$ and 0.030.

In the case of quark scattering, we use Eq. (26) and Eq. (27). Thus the bound in Eq. (22) translates into

$$\sqrt{s} \lesssim 2\sqrt{3\pi}v\left(\frac{\Gamma_{q\bar{q}}}{M_S}\right)^{-1/2} \simeq 4.8\ \text{TeV}\left(\frac{\Gamma_{q\bar{q}}/M_S}{0.1}\right)^{-1/2}, \tag{36}$$

where on the r.h.s. we have normalized the partial width in $q\bar{q}$ to a broad resonance scenario.

In Fig. 1 we display the scale of unitarity violation Λ_U [TeV] in the $\mathcal{B}_{\gamma\gamma}$ vs. $\sigma_{\gamma\gamma}$ plane, for either gg or $b\bar{b}$ production and assuming either a broad or narrow resonance. In particular, for gg production we have

$$\Lambda_U = M_S\left[\frac{\Gamma_{gg}}{M_S} + f(r)\frac{\Gamma_{\gamma\gamma}}{M_S}\right]^{-1/2}, \tag{37}$$

while for $b\bar{b}$ production

$$\Lambda_U = \min\left\{2\sqrt{3\pi}v\left(\frac{\Gamma_{b\bar{b}}}{M_S}\right)^{-1/2}, M_S\left[f(r)\frac{\Gamma_{\gamma\gamma}}{M_S}\right]^{-1/2}\right\}. \tag{38}$$

As reference values we take $M_S = 750$ GeV and $f(r) = 30$ The horizontal lines from top to bottom indicate a cross-section signal of 6, 0.6 and 0.2 fb, assuming the same significance of the signal over the three integrated luminosities $\int\mathcal{L} = 3.2$, 300 and 3000 fb^{-1}. The red curve denotes instead the reference value $\Lambda = 20$ TeV, corresponding to the typical squark-gluino reach of a futuristic 100 TeV collider [9], which applies in the case of coloured new physics generating the effective operators. Hence, if a signal is observed above the red curve it basically means that a 100 TeV collider could potentially probe the physics responsible for the restoration of unitarity. We observe that such low-scale violation of unitarity are more readily obtained in the large width scenario and that for any given $\sigma_{\gamma\gamma}$ and $\mathcal{B}_{\gamma\gamma}$, unitarity violation sets in earlier for $b\bar{b}$ induced production, compared to gluon fusion processes, due to much smaller PDFs.

4 Conclusions

In the case of a TeV-scale scalar di-boson resonance observable at the LHC we have, under some very basic and natural assumptions on the structure of the EFT (mainly that S is a spin-0 SM gauge singlet and that the $d = 5$ operators in Eq. (9) are the most relevant ones for the decay of S), demonstrated a potential violation of tree-level unitarity in the scattering of SM fields at energy scales of few tens of TeV. This scale can be interpreted as an upper bound on the mass of new degrees of freedom responsible for the unitarization of the scattering amplitudes.

We conclude that in the event of an experimental observation of a scalar di-boson resonance at the LHC, while our estimates cannot provide a guarantee to see on-shell effects of additional new degrees of freedom at the LHC, they would immediately imply the existence of additional phenomena within the energy reach of the next generation 50-100 TeV hadron colliders, thus making a strong physics case for their construction.

Acknowledgments

L.D.L. wishes to thank the organizers of the ICNFP 2016 for providing a wonderful atmosphere on the Crete island. J.F.K. acknowledges the financial support from the Slovenian Research Agency (research core funding No. P1-0035).

References

[1] B.W. Lee, C. Quigg, H.B. Thacker, Phys. Rev. Lett. **38**, 883 (1977)
[2] M. Aaboud et al. (ATLAS), JHEP **09**, 001 (2016), 1606.03833
[3] V. Khachatryan et al. (CMS), Phys. Rev. Lett. **117**, 051802 (2016), 1606.04093
[4] L. Di Luzio, J.F. Kamenik, M. Nardecchia (2016), 1604.05746
[5] M. Jacob, G.C. Wick, Annals Phys. **7**, 404 (1959), [Annals Phys.281,774(2000)]
[6] A. Wulzer, Nucl. Phys. **B885**, 97 (2014), 1309.6055
[7] R. Franceschini, G.F. Giudice, J.F. Kamenik, M. McCullough, A. Pomarol, R. Rattazzi, M. Redi, F. Riva, A. Strumia, R. Torre, JHEP **03**, 144 (2016), 1512.04933
[8] J.F. Kamenik, B.R. Safdi, Y. Soreq, J. Zupan, JHEP **07**, 042 (2016), 1603.06566
[9] T. Golling et al. (2016), 1606.00947

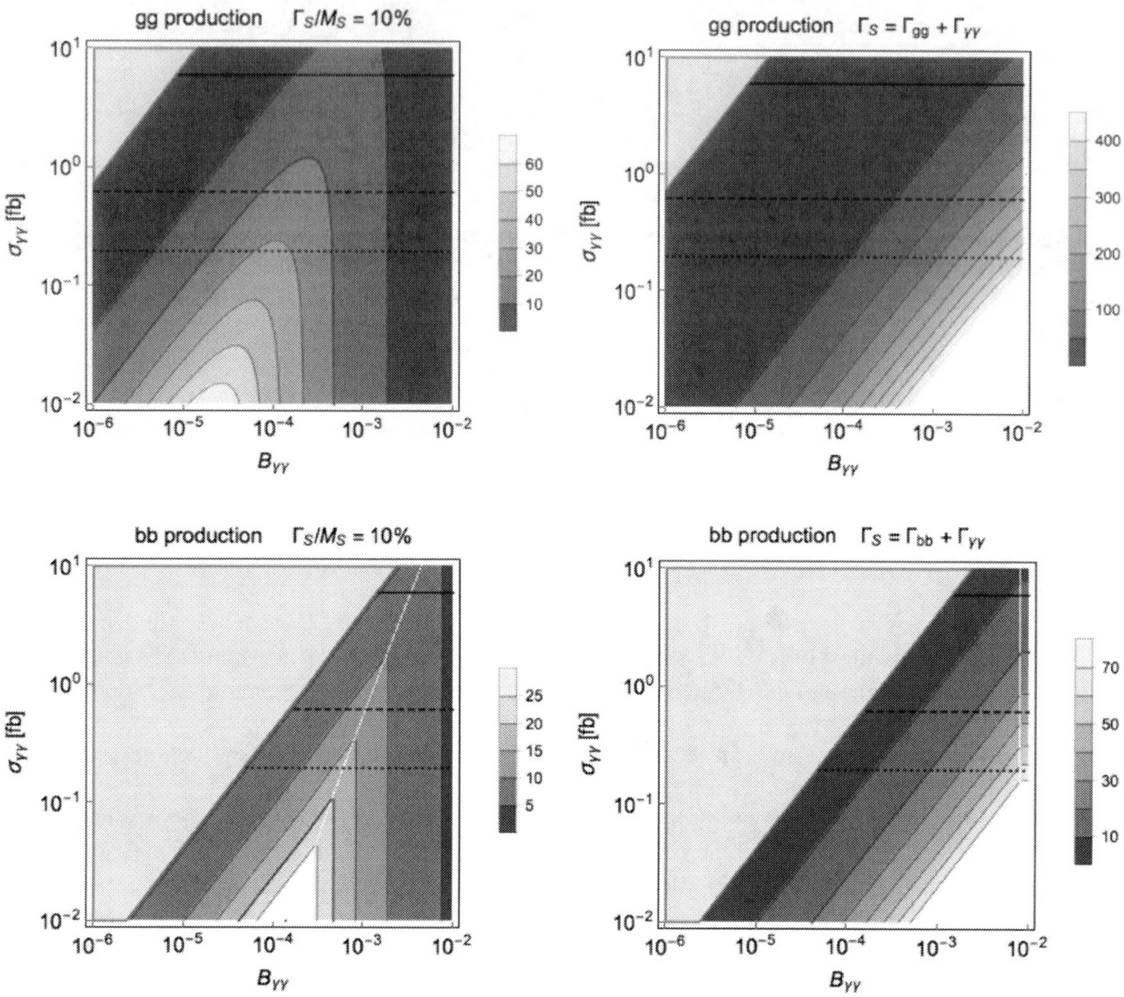

Figure 1: Scale of unitarity violation Λ_U in TeV in the $(\mathcal{B}_{\gamma\gamma}, \sigma_{\gamma\gamma})$ plane (cf. Eqs. (37)–(38)). Upper/lower plots corresponding to $gg/b\bar{b}$ production, while left/right plots to the large/small width scenario. As reference values we assume $M_S = 750$ GeV and $f(r) = 30$. The red curve denotes the new physics scale accessible at a futuristic 100 TeV collider, $\Lambda = 20$ TeV, while the three horizontal lines from top to bottom are three reference cross-sections, namely 6, 0.6 and 0.2 fb. The yellow triangle on the top-left of each figure is the region in parameter space where $\Gamma_S/M_S > 10$ %.

Probing the BFKL dynamics in inclusive three jet production at the LHC

F. Caporale[1,a], F. G. Celiberto[1,2,b], G. Chachamis[1,c], D. Gordo Gómez[1,d], and A. Sabio Vera[1,e]

[1] *Instituto de Física Teórica UAM/CSIC, Nicolás Cabrera 15 & Universidad Autónoma de Madrid, E-28049 Madrid, Spain.*
[2] *Dipartimento di Fisica, Università della Calabria & Istituto Nazionale di Fisica Nucleare, Gruppo Collegato di Cosenza, I-87036 Arcavacata di Rende, Cosenza, Italy.*

Abstract. We propose the study of new observables in LHC inclusive events with three tagged jets, one in the forward direction, one in the backward direction and both well-separated in rapidity from the each other (Mueller-Navelet jets), together with a third jet tagged in central regions of rapidity. Since non-tagged associated mini-jet multiplicity is allowed, we argue that projecting the cross sections on azimuthal-angle components can provide several distinct tests of the BFKL dynamics. Realistic LHC kinematical cuts are introduced.

1 Introduction

The Balitsky-Fadin-Kuraev-Lipatov (BFKL) resummation program in the leading logarithmic (LL) [1–6] and next-to-leading logarithmic (NLL) approximation [7, 8] may be applied for phenomenological studies at hadronic colliders when the final-state is characterised by jets that are produced at large relative rapidities. Mueller-Navelet jets [9] is an key example, specifically, for observables that are based on the azimuthal angle formed by the two outermost in rapidity tagged jets, ϕ. The precise form of the observables is built by considering ratios of projections on the azimuthal angle $\mathcal{R}_n^m = \langle\cos(m\,\phi)\rangle/\langle\cos(n\,\phi)\rangle$. Comparison of different NLL predictions against LHC experimental data for these observables has been quite successful [10–30].

New LHC observables, that may be seen as a generalisation of the Mueller-Navelet jets, were proposed recently for inclusive three-jet [31, 32] and four-jet production [33, 34]. In this work we discuss only the observables for inclusive three-jet production. These are defined by the generalised ratios [31]

$$\mathcal{R}_{PQ}^{MN} = \frac{\langle\cos(M\,\phi_1)\cos(N\,\phi_2)\rangle}{\langle\cos(P\,\phi_1)\cos(Q\,\phi_2)\rangle}\,, \tag{1}$$

[a] e-mail: francesco.caporale@uam.es
[b] e-mail: francescogiovanni.celiberto@fis.unical.it
[c] e-mail: chachamis@gmail.com
[d] e-mail: david.gordo@csic.es
[e] e-mail: a.sabio.vera@gmail.com

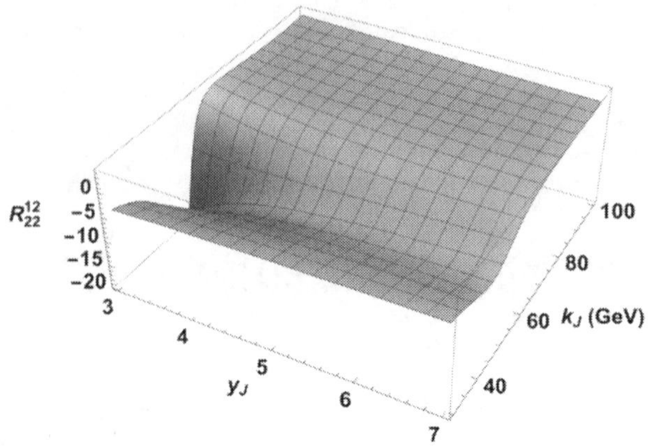

Figure 1. 3D plot of the partonic $\mathcal{R}_{22}^{12}$ as a function of the rapidity y_J and the momentum k_J of the central jet for $k_A = 40$ GeV, $k_B = 50$ GeV and $\Delta Y_{A,B} = 10$.

where ϕ_1 is the azimuthal angle difference between the forward and the central jet and ϕ_2 the azimuthal angle difference between the central jet and the backward in rapidity jet. The ratios $\mathcal{R}_{PQ}^{MN}$ in Eq. (1) are actually partonic level quantities and therefore, cannot be readily compared to experimental data. Therefore, we define the hadronic level observables R_{PQ}^{MN} [32] and study their stability once we introduce corrections beyond the LL accuracy. For that, we produce the two outermost in rapidity jets within the collinear factorization scheme, each of them associated with a forward "jet vertex" [35]. Then we link these vertices and the central jet using two BFKL gluon Green's functions. At the end, the partonic differential cross-section is convoluted with collinear parton distribution functions and is integrated over the momenta of all produced jets in order to calculate the ratios R_{PQ}^{MN}. For the integration over the momenta of the jets we use standard LHC experimental cuts. The rapidity of the central jet takes values close to the middle of the rapidity distance between the two outermost tagged jets.

2 Hadronic inclusive three-jet production in multi-Regge kinematics

Let us first remember some of the notation defined in [31, 32]. Assuming that the transverse momenta of the outermost jets are $\vec{k}_{A,B}$, their rapidity difference, Y, is large and the central jet has transverse momentum $\vec{k}_J$. We allow for mini-jet activity between the three tagged jets so that the process[1] we need to study is

$$\text{proton}(p_1) + \text{proton}(p_2) \to \text{jet}(k_A) + \text{jet}(k_J) + \text{jet}(k_B) + \text{minijets}\,. \tag{2}$$

Firstly, we define the two relative azimuthal angles between the outermost jets and the central jet as $\Delta\theta_{\widehat{AJ}} = \theta_A - \theta_J - \pi$ and $\Delta\theta_{\widehat{JB}} = \theta_J - \theta_B - \pi$ respectively. Then the projection on azimuthal-angle components gives the mean value

$$\begin{aligned} C_{MN} &= \langle \cos(M(\theta_A - \theta_J - \pi)) \cos(N(\theta_J - \theta_B - \pi)) \rangle \qquad (3) \\ &= \frac{\int_0^{2\pi} d\theta_A d\theta_B d\theta_J \cos(M(\theta_A - \theta_J - \pi)) \cos(N(\theta_J - \theta_B - \pi)) d\sigma^{3\text{-jet}}}{\int_0^{2\pi} d\theta_A d\theta_B d\theta_J d\sigma^{3\text{-jet}}}, \end{aligned}$$

[1] Another interesting idea, suggested in [36] and investigated in [37, 38], is the study of the production of two charged light hadrons, $\pi^{\pm}$, $K^{\pm}$, p, $\bar{p}$, with large transverse momenta and well separated in rapidity.

where M, N are positive integers) and $d\sigma^{3-\text{jet}}$ the partonic deferential cross-section for three-jet production defined in [31].

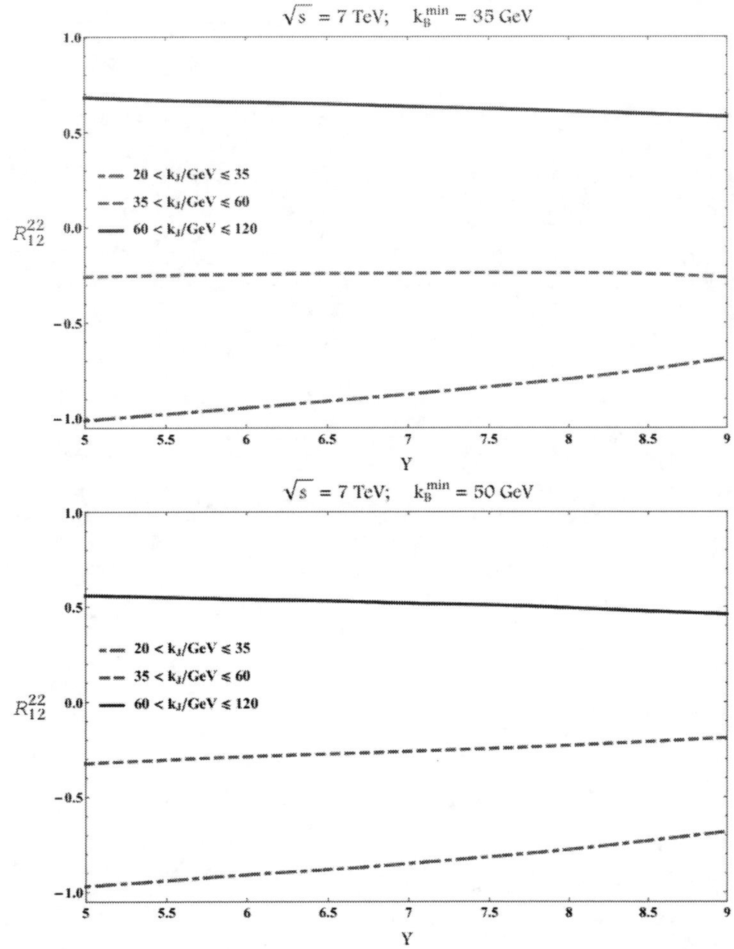

Figure 2. Y-dependence of the LL R_{12}^{22} for $\sqrt{s} = 7$ TeV. Symmetric cut $k_B^{\min} = 35$ GeV (top) and asymmetric cut $k_B^{\min} = 50$ GeV (bottom).

In order to compute theoretical estimates that may be compared against current and future experimental data, we integrate $C_{M,N}$ over the momenta of the tagged jets in the form

$$C_{MN} = \int_{Y_A^{\min}}^{Y_A^{\max}} dY_A \int_{Y_B^{\min}}^{Y_B^{\max}} dY_B \int_{k_A^{\min}}^{k_A^{\max}} dk_A \int_{k_B^{\min}}^{k_B^{\max}} dk_B \int_{k_J^{\min}}^{k_J^{\max}} dk_J \delta\left(Y_A - Y_B - Y\right) C_{MN}, \tag{4}$$

where the rapidity of the forward jet takes values in the range $Y_A^{\min} = 0$ and $Y_A^{\max} = 4.7$ and that of the backward jet in the range $Y_B^{\min} = -4.7$ and $Y_B^{\max} = 0$ while their difference $Y \equiv Y_A - Y_B$ is kept fixed at definite values in the range $5 < Y < 9$. We calculate C_{MN} for two different center-of-mass energies, $\sqrt{s} = 7$ and $\sqrt{s} = 13$ TeV and we introduce two typical kinematical cuts previously used in the study of Mueller-Navelet jets at the LHC. Specifically, we use both a symmetric and an asymmetric cut [19, 29]:

1. $k_A^{\min} = 35$ GeV, $k_B^{\min} = 35$ GeV, $k_A^{\max} = k_B^{\max} = 60$ GeV (symmetric);
2. $k_A^{\min} = 35$ GeV, $k_B^{\min} = 50$ GeV, $k_A^{\max} = k_B^{\max} = 60$ GeV (asymmetric).

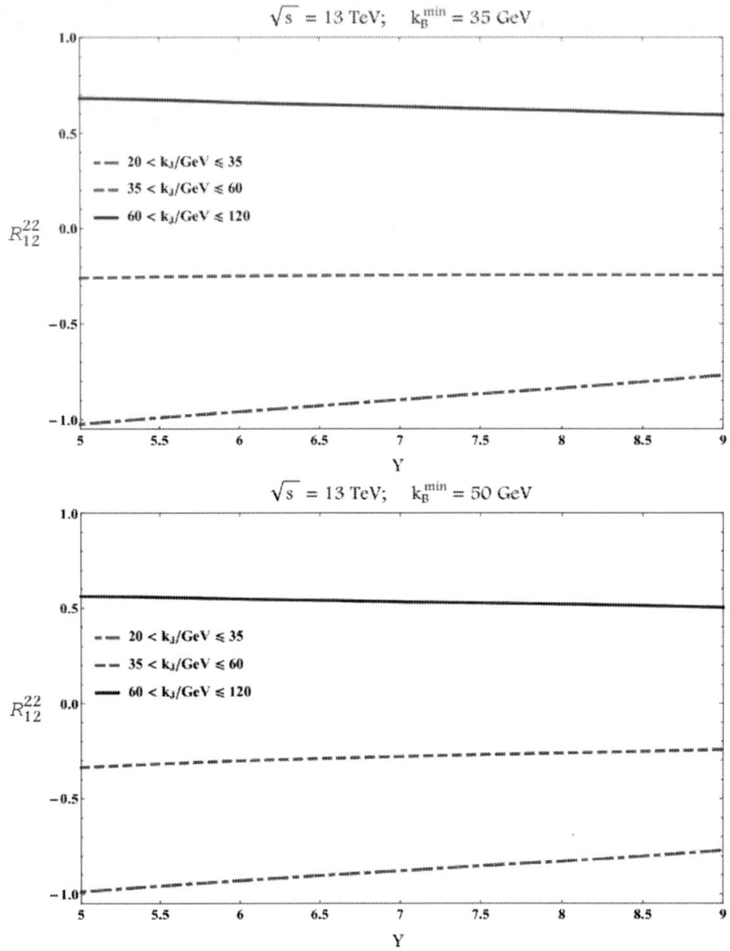

Figure 3. Y-dependence of the LL R_{12}^{22} for $\sqrt{s} = 13$ TeV. Symmetric cut $k_B^{\min} = 35$ GeV (top) and asymmetric cut $k_B^{\min} = 50$ GeV (bottom).

We are interested in maximising the stability with respect to higher order effects (beyond LL) in our results (see [15]), therefore, we remove the zeroth conformal spin contribution of the BFKL kernel by considering the ratios

$$R_{PQ}^{MN} = \frac{C_{MN}}{C_{PQ}}, \qquad M, N, P, Q > 0, \tag{5}$$

which have no $n = 0$ dependence. Thus, we can study the ratios $R_{PQ}^{MN}(Y)$ in Eq. (5) as functions of the rapidity difference Y between the outermost jets for some typical values of M, N, P, Q. We define three different p_T ranges (bins) for the allowed momentum of the central jet:
bin-1 = [20 GeV $< k_J <$ 35 GeV] (k_J 'smaller' that k_A, k_B),

DOI: 10.1051/epjconf/201716407027

bin-2 = [35 GeV < k_J < 60 GeV] (k_J 'similar' to k_A, k_B),
bin-3 = [60 GeV < k_J < 120 GeV] (k_J 'larger' than k_A, k_B).
This permits the discrimination of different behaviours of the $R_{PQ}^{MN}(Y)$ by using as a criterion the relative size of the central jet. In Fig. 1 we show the behaviour of $\mathcal{R}_{22}^{12}$ as we change the size of the central jet and its position in rapidity. We notice that while a small variation in y_J around the central rapidity value $\Delta Y_{A,B}/2 = 5$ does not result in significant changes for a fixed k_J, a change in the value of k_J may have a big impact for a fixed y_J. A number of different ratios was presented in [32], here

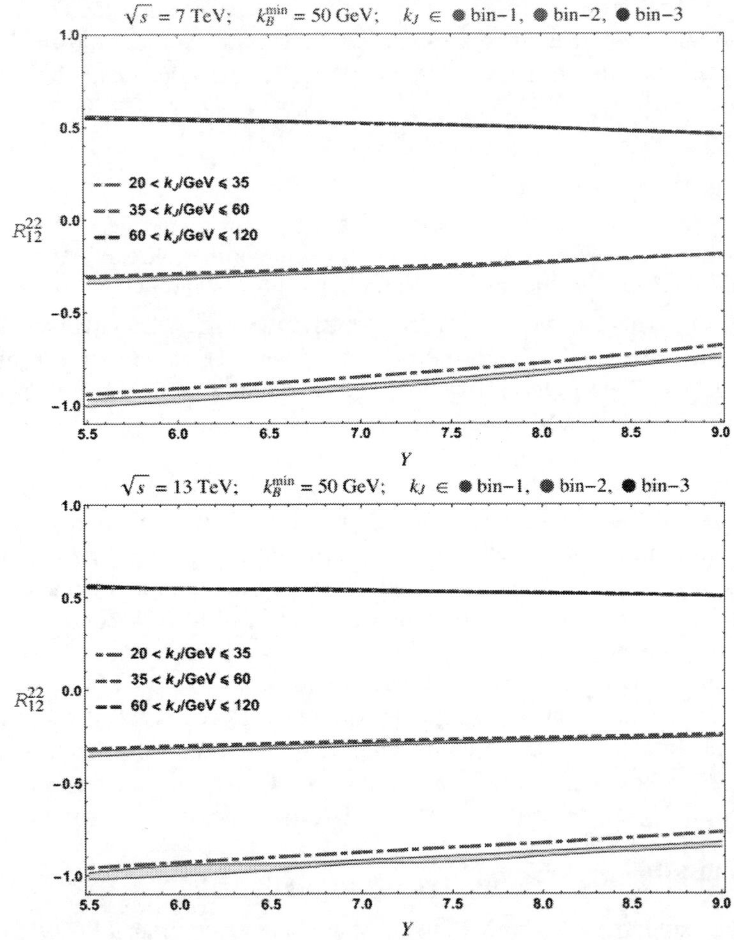

Figure 4. Y-dependence of the LL (dashed lines) and NLL (BLM) predictions (continuous bands) for R_{12}^{22} in the asymmetric cut at $\sqrt{s}$ = 7 TeV (top) and $\sqrt{s}$ = 13 TeV (bottom). The blue NLL band is very narrow and lies on top of the LL line.

we are focussing on ratios that involve the coefficients C_{12} and C_{22}. In Figs. 2 and 3 we see the LL accuracy results for R_{12}^{22}. Generally, the dependence of the different observables on the rapidity difference between k_A and k_B is rather smooth whereas the slope of the three curves depends on the particular observable. For R_{12}^{22} we see that shifting from a symmetric to an asymmetric cut makes no noticeable difference. Moreover, there are no important changes when we change the colliding energy from $\sqrt{s} = 7$ TeV to $\sqrt{s} = 13$ TeV. The latter is crucial as it suggests that R_{12}^{22} is already within some sort of asymptotic regime for the specific kinematical configurations.

Apart from the stability of the observable with regard to an increase of the colliding energy, another important question is the stability with respect to effects that go beyond the LL approximation [39]. A first important step towards a full NLL computation is to take into account the NLL contributions to the two gluon Green's functions that connect the three jets. In Fig. 4 we present exactly these corrections obtained by using the Brodsky-Lepage-Mackenzie (BLM) prescription [40] for the R_{12}^{22} coefficient in the asymmetric cut. In particular, we have used the MOM scheme and chosen the renormalisation scale such that the β_0-dependence of the given observable vanishes, following the BLM prescription. The dashed lines represent the LL predictions and the coloured bands represent the NLL BLM predictions. It is impressive that the NLL values are almost on top of the LL ones which gives us great confidence that the observables R_{PQ}^{MN} are indeed excellent BFKL probes at the LHC.

3 Summary & Outlook

We have presented a first BFKL driven hadronic-level phenomenological work on the recently proposed observables R_{PQ}^{MN} that depend on the azimuthal-angles of the jets in inclusive three-jet production at the LHC. In particular, we concentrated on the ratio R_{12}^{22}. Our major task was to study the variation of R_{12}^{22} when we set the rapidity difference Y between the outermost jets to different fixed values in the range $5 < Y < 9$. Generally, we notice a smooth functional dependence of the ratio R_{12}^{22} on Y. A key observation is that R_{12}^{22} and other similar ratio observables do not exhibit a significantly different behaviour when one changes the energy configuration from 7 to 13 TeV. This gives us confidence that these ratios pinpoint the most important characteristics of the azimuthal behaviour of the tagged jets within the BFKL framework. Moreover, one of the two big parts of radiative corrections beyond the LL approximation, namely, the NLL contributions to the two gluon Green's functions that link the three tagged jets, do not significantly change the functional behaviour of R_{12}^{22} on Y. The other big part of the beyond the LL corrections would be the NLO corrections to the jet vertices and these need also to be taken into account. In addition, it is crucial to investigate whether fixed order calculations and studies with the BFKL inspired Monte Carlo **BFKLex** [41–48] give similar results. Predictions from the usual all-purpose collinear Monte Carlo codes are also needed to complete the picture from the theoretical side. To conclude with, a dedicated experimental analysis on the proposed ratio observables R_{PQ}^{MN} based on existing (and future) LHC data will answer the question of whether these observables qualify as new probes for the BFKL dynamics.

Acknowledgements

GC acknowledges support from the MICINN, Spain, under contract FPA2013-44773-P. ASV acknowledges support from Spanish Government (MICINN (FPA2010-17747,FPA2012-32828)) and, together with FC and FGC, to the Spanish MINECO Centro de Excelencia Severo Ochoa Programme (SEV-2012-0249). FGC thanks the Instituto de Física Teórica (IFT UAM-CSIC) in Madrid for warm hospitality.

References

[1] L. N. Lipatov, Sov. Phys. JETP **63** (1986) 904 [Zh. Eksp. Teor. Fiz. **90** (1986) 1536].
[2] I. I. Balitsky and L. N. Lipatov, Sov. J. Nucl. Phys. **28** (1978) 822 [Yad. Fiz. **28** (1978) 1597].

[3] E. A. Kuraev, L. N. Lipatov and V. S. Fadin, Sov. Phys. JETP **45** (1977) 199 [Zh. Eksp. Teor. Fiz. **72** (1977) 377].

[4] E. A. Kuraev, L. N. Lipatov and V. S. Fadin, Sov. Phys. JETP **44** (1976) 443 [Zh. Eksp. Teor. Fiz. **71** (1976) 840] [Erratum-ibid. **45** (1977) 199].

[5] L. N. Lipatov, Sov. J. Nucl. Phys. **23** (1976) 338 [Yad. Fiz. **23** (1976) 642].

[6] V. S. Fadin, E. A. Kuraev and L. N. Lipatov, Phys. Lett. B **60** (1975) 50.

[7] V. S. Fadin and L. N. Lipatov, Phys. Lett. B **429** (1998) 127 [hep-ph/9802290].

[8] M. Ciafaloni and G. Camici, Phys. Lett. B **430** (1998) 349 [hep-ph/9803389].

[9] A. H. Mueller and H. Navelet, Nucl. Phys. B **282** (1987) 727.

[10] V. Del Duca and C. R. Schmidt, Phys. Rev. D **49** (1994) 4510 [hep-ph/9311290].

[11] W. J. Stirling, Nucl. Phys. B **423** (1994) 56 [hep-ph/9401266].

[12] L. H. Orr and W. J. Stirling, Phys. Rev. D **56** (1997) 5875 [hep-ph/9706529].

[13] J. Kwiecinski, A. D. Martin, L. Motyka and J. Outhwaite, Phys. Lett. B **514** (2001) 355 [hep-ph/0105039].

[14] M. Angioni, G. Chachamis, J. D. Madrigal and A. Sabio Vera, Phys. Rev. Lett. **107**, 191601 (2011) [arXiv:1106.6172 [hep-th]].

[15] F. Caporale, B. Murdaca, A. Sabio Vera and C. Salas, Nucl. Phys. B **875** (2013) 134 [arXiv:1305.4620 [hep-ph]].

[16] F. Caporale, D. Y. Ivanov, B. Murdaca and A. Papa, Nucl. Phys. B **877** (2013) 73 [arXiv:1211.7225 [hep-ph]].

[17] C. Marquet and C. Royon, Phys. Rev. D **79**, 034028 (2009) [arXiv:0704.3409 [hep-ph]].

[18] D. Colferai, F. Schwennsen, L. Szymanowski and S. Wallon, JHEP **1012**, 026 (2010) [arXiv:1002.1365 [hep-ph]].

[19] B. Ducloue, L. Szymanowski and S. Wallon, JHEP **1305**, 096 (2013) [arXiv:1302.7012 [hep-ph]].

[20] B. Ducloue, L. Szymanowski and S. Wallon, Phys. Lett. B **738**, 311 (2014) [arXiv:1407.6593 [hep-ph]].

[21] A. H. Mueller, L. Szymanowski, S. Wallon, B. W. Xiao and F. Yuan, JHEP **1603**, 096 (2016) [arXiv:1512.07127 [hep-ph]].

[22] N. Cartiglia *et al.* [LHC Forward Physics Working Group Collaboration], CERN-PH-LPCC-2015-001, SLAC-PUB-16364, DESY-15-167.

[23] G. Chachamis, arXiv:1512.04430 [hep-ph].

[24] A. Sabio Vera, Nucl. Phys. B **746** (2006) 1 [hep-ph/0602250].

[25] A. Sabio Vera and F. Schwennsen, Nucl. Phys. B **776** (2007) 170 [hep-ph/0702158 [HEP-PH]].

[26] B. Ducloue, L. Szymanowski and S. Wallon, Phys. Rev. Lett. **112** (2014) 082003 [arXiv:1309.3229 [hep-ph]].

[27] F. Caporale, D. Y. Ivanov, B. Murdaca and A. Papa, Eur. Phys. J. C **74** (2014) 3084 [arXiv:1407.8431 [hep-ph]].

[28] F. Caporale, D. Y. Ivanov, B. Murdaca and A. Papa, Phys. Rev. D**91** (2015) 11, 114009 [arXiv:1504.06471 [hep-ph]].

[29] F. G. Celiberto, D. Yu. Ivanov, B. Murdaca and A. Papa, Eur. Phys. J. C **75** (2015) 292 [arXiv:1504.08233 [hep-ph]].

[30] F. G. Celiberto, D. Y. Ivanov, B. Murdaca and A. Papa, Eur. Phys. J. C **76**, no. 4, 224 (2016) [arXiv:1601.07847 [hep-ph]].

 DOI: 10.1051/epjconf/201716407027

[31] F. Caporale, G. Chachamis, B. Murdaca and A. Sabio Vera, Phys. Rev. Lett. **116**, no. 1, 012001 (2016) [arXiv:1508.07711 [hep-ph]].

[32] F. Caporale, F. G. Celiberto, G. Chachamis, D. G. Gomez and A. Sabio Vera, arXiv:1603.07785 [hep-ph].

[33] F. Caporale, F. G. Celiberto, G. Chachamis and A. Sabio Vera, Eur. Phys. J. C **76**, no. 3, 165 (2016) [arXiv:1512.03364 [hep-ph]].

[34] F. Caporale, F. G. Celiberto, G. Chachamis, D. G. Gomez and A. Sabio Vera, arXiv:1606.00574 [hep-ph].

[35] F. Caporale, D. Yu. Ivanov, B. Murdaca, A.Papa, A.Perri, JHEP **1202** (2012) 101; [arXiv:1212.0487 [hep-ph]].

[36] D. Y. Ivanov and A. Papa, JHEP **1207**, 045 (2012) [arXiv:1205.6068 [hep-ph]].

[37] F. G. Celiberto, D. Y. Ivanov, B. Murdaca and A. Papa, arXiv:1604.08013 [hep-ph].

[38] F. G. Celiberto, D. Y. Ivanov, B. Murdaca and A. Papa, work to be released soon.

[39] F. Caporale, F. G. Celiberto, G. Chachamis, D. G. Gomez and A. Sabio Vera, work to be released soon.

[40] S. J. Brodsky, G. P. Lepage and P. B. Mackenzie, Phys. Rev. D **28**, 228 (1983). doi:10.1103/PhysRevD.28.228

[41] G. Chachamis, M. Deak, A. Sabio Vera and P. Stephens, Nucl. Phys. B **849** (2011) 28 [arXiv:1102.1890 [hep-ph]].

[42] G. Chachamis and A. Sabio Vera, Phys. Lett. B **709** (2012) 301 [arXiv:1112.4162 [hep-th]].

[43] G. Chachamis and A. Sabio Vera, Phys. Lett. B **717** (2012) 458 [arXiv:1206.3140 [hep-th]].

[44] G. Chachamis, A. Sabio Vera and C. Salas, Phys. Rev. D **87** (2013) 1, 016007 [arXiv:1211.6332 [hep-ph]].

[45] F. Caporale, G. Chachamis, J. D. Madrigal, B. Murdaca and A. Sabio Vera, Phys. Lett. B **724** (2013) 127 [arXiv:1305.1474 [hep-th]].

[46] G. Chachamis and A. Sabio Vera, arXiv:1511.03548 [hep-ph].

[47] G. Chachamis and A. Sabio Vera, JHEP **1602** (2016) 064 [arXiv:1512.03603 [hep-ph]].

[48] G. Chachamis and A. Sabio Vera, Phys. Rev. D **94**, no. 3, 034019 (2016) [arXiv:1606.07349 [hep-ph]].

EPJ Web of Conferences **164**, 07028 (2017) DOI: 10.1051/epjconf/201716407028
ICNFP 2016

Production spectra of zero-degree neutral particles measured by the LHCf experiment

A. Tiberio[1,2,a], on behalf of the LHCf Collaboration

[1] *University of Florence, Italy*
[2] *INFN Section of Florence, Italy*

Abstract. The LHC-forward (LHCf) experiment, situated at the LHC accelerator, has measured neutral particles production in a very forward region (pseudo-rapidity $\eta > 8.4$) in proton-proton and proton-lead collisions. The main purpose of the LHCf experiment is to test hadronic interaction models used in ground based cosmic rays experiments to simulate cosmic rays induced air-showers in the Earth's atmosphere.
The experiment is composed of two independent detectors located at 140m from the ATLAS interaction point (IP1) on opposite sides; each detector is composed of two sampling and position sensitive calorimeters.
In this paper, latest published physics results from p-p and p-Pb collisions (at $\sqrt{s} = 7$, 2.76 TeV and $\sqrt{s_{NN}} = 5.02$ TeV, respectively) compared with Monte Carlo predictions of DPMJET, EPOS, PYTHIA, QGSJET and SIBYLL event generators will be presented. In particular, the inclusive energy spectra of neutrons in p-p collisions and the transverse and longitudinal momentum spectra of neutral pions for different pseudo-rapidity ranges in p-p and p-Pb collisions will be shown; then, test of Feynman scaling hypothesis using neutral pion spectra will be discussed. Preliminary results of photon inclusive energy spectra in p-p collisions at $\sqrt{s} = 13$ TeV will be also presented.

1 Introduction

The LHC-forward experiment (LHCf) has measured neutral particles production in a very forward region in proton-proton and proton-lead collisions at the Large Hadron Collider. The main purpose of LHCf is to improve hadronic interaction models of Monte Carlo (MC) simulations used in cosmic rays indirect measurements. Highest energy cosmic rays can only be detected from secondary particles which are produced in the interaction of the primary particle with nuclei of the atmosphere, the so-called air showers. Studying the development of air showers, it is possible to reconstruct the type and kinematic parameters of primary particles. To reproduce the development of air showers, MC simulations with accurate hadronic interaction models are needed. Since the energy flow of secondary particles is concentrated in the forward direction, measurements of particle production in the high pseudo-rapidity region (i.e. small angles) are very important. Furthermore, in the forward region soft QCD interactions (non perturbative regime) dominates and MC simulations of air showers are based on phenomenological models, so inputs from experimental data are crucial for the tuning of that

[a]e-mail: alessio.tiberio@fi.infn.it

models. LHC accelerator gives the possibility to study a wide range of collision energies, from 0.9 TeV to 13 TeV in the center of mass frame, which corresponds to an energy range in the laboratory frame from 10^{14} eV to 0.9×10^{17} eV. This energy range covers the "knee" region of cosmic rays spectrum, which occurs around 10^{15} eV.

2 The detector

LHCf is composed of two independent detectors, called *Arm1* and *Arm2*. Arm1 is located 140 meters away from ATLAS interaction point (IP1) in the IP8 direction, while Arm2 is placed 140 meters away from IP1 in the opposite direction (toward IP2). Detectors are placed inside Target Neutral Absorber (TAN), where the beam pipe turns into two separates tubes. Since charged particles are deviated by the D1 dipole magnet (which bends colliding beams into the two separate beam pipes), only neutral particles, mainly photons and neutrons, reach the detector.

Each detector was made of two sampling and imaging calorimeters (called *towers* hereafter). Each tower is composed of 16 tungsten layers and 16 scintillator layers to measure the energy deposit and it also contains 4 position sensitive layers. During 0.9 TeV, 2.76 TeV and 7 TeV operations at the LHC, plastic scintillators (EJ-260) were used. Arm1 detector used scintillating fiber (SciFi) to measure position, while Arm2 used silicon microstrip detectors. For 13 TeV operation both detectors were upgraded: all the plastic scintillators were replaced by Gd_2SiO_5 (GSO) scintillators because of their radiation hardness; also the Arm1 SciFi were replaced by GSO bars. In Arm2 the signal of silicon detectors was reduced using a new bonding scheme of the microstrips to avoid saturation of readout electronics due to the higher energy deposit expected at $\sqrt{s} = 13$ TeV.

Transverse cross sections of towers are 20×20 mm^2 and 40×40 mm^2 for Arm1 and 25×25 mm^2 and 32×32 mm^2 for Arm2. Longitudinal dimension of towers is of 44 radiation lengths, which correspond to 1.6 nuclear interaction lengths. Energy resolution is better than 5 % for photons and of about 40 % for neutrons. Position resolution for photons is 200 μm and 40 μm for Arm1 and Arm2 respectively, while position resolution for neutrons is of about 1 mm. Smaller tower of each detector is placed on the beam center and covers the pseudo-rapidity range $\eta > 9.6$, while larger tower covers the pseudo-rapidity range $8.4 < \eta < 9.4$. More detailed descriptions of detector performance are reported elsewhere [1–3].

3 Experimental results

LHCf collected data in three different low-luminosity dedicated runs in 2010, 2013 and 2015. Proton-proton collisions at $\sqrt{s} = 900$ GeV (2010), $\sqrt{s} = 2.76$ TeV (2013), $\sqrt{s} = 7$ TeV (2010) and $\sqrt{s} = 13$ TeV (2015) were produced; proton-lead collisions at $\sqrt{s_{NN}} = 5.02$ TeV were produced in 2013.

Results for single photon energy spectra in p-p collisions at $\sqrt{s} = 900$ GeV, 7 TeV and for π^0 transverse momentum spectra in p-p and p-Pb collisions at $\sqrt{s} = 7$ TeV and $\sqrt{s_{NN}} = 5.02$ TeV, respectively, have already been published [4–7]. Latest published results are presented in this paper: neutron energy spectra in p-p collisions at $\sqrt{s} = 7$ TeV [8] are shown in Sect. 3.1 while results from π^0 transverse (P_T) and longitudinal (P_Z) momentum analysis in p-p at $\sqrt{s} = 2.76$, 7 TeV and p-Pb collisions at $\sqrt{s_{NN}} = 5.02$ TeV [9] are shown in Sect. 3.2. Finally, also preliminary result from photon analysis at $\sqrt{s} = 13$ TeV will be presented in Sect. 3.3.

3.1 Neutron energy spectra in p-p collisions

In this section results from neutron analysis in p-p collisions at $\sqrt{s} = 7$ TeV are presented [8]. Data were taken during LHC Fill #1104 on May 15, 2010. Integrated luminosities were 0.68 nb^{-1} and 0.53

DOI: 10.1051/epjconf/201716407028

nb^{-1} for Arm1 and Arm2 respectively, with a typical luminosity of $6.4 \times 10^{28} cm^{-2} s^{-1}$, derived from LHCf front counters [10].

Due to the limited dimensions of calorimeter, hadronic showers are not fully contained in the tower and the study of hadronic neutral particles is less precise and more complicated than photons. Hadronic particles (mostly neutrons) are selected using information of longitudinal distribution of energy deposit in the calorimeter. Since the energy resolution for neutrons is $\sim 40\%$, the true energy spectra are smeared by the detector response. Smeared energy spectra of neutrons measured by Arm1 and compared with MC simulation predictions of several models are shown in figure 1 in three pseudo-rapidity ranges.

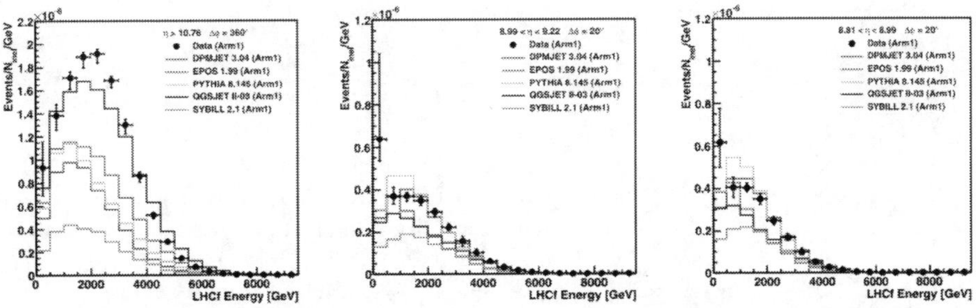

Figure 1. Energy distribution of neutrons in p-p collisions at 7 TeV before unfolding in three pseudo-rapidity regions: coloured histograms show results from Monte Carlo simulations with different hadronic interaction models, black points show experimental data of Arm1. Particles above maximum available energy (3500 GeV) are clearly due to the smearing effect of the detector.

Unfolding technique is therefore needed to estimate the true energy spectra: a multidimensional-spectra unfolding Bayesian method [11] relying on energy and P_T measured variables was used. Results after unfolding are shown in figure 2. It is evident from the results that in the very forward pseudo-rapidity region ($\eta > 10.76$) there is a large amount of high energy neutrons, only predicted by the QGSJETII-03 [12] model. In other regions data lie between models predictions, but there is not an unique model that represents data.

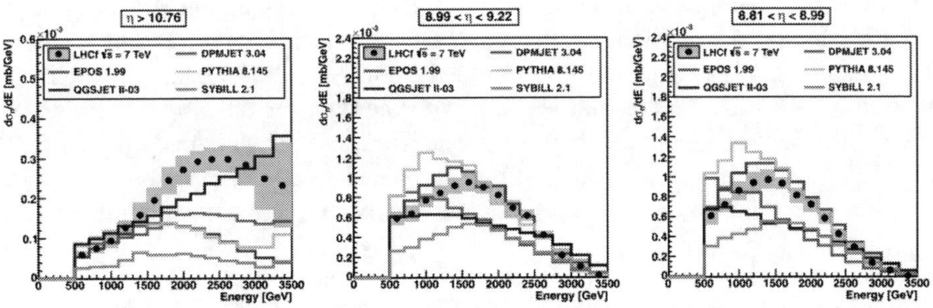

Figure 2. Energy distribution of neutrons in p-p collisions at 7 TeV after unfolding in three pseudo-rapidity regions: coloured histograms show results from Monte Carlo simulations with different hadronic interaction models, black points show combined experimental data of Arm1 and Arm2. A large amount of very-forward ($\eta > 10.76$) high energy neutrons is predicted by QGSJETII-03 and confirmed by data.

3.2 Neutral pion P_T and P_Z spectra in p-p and p-Pb collisions

Published results [9] for π^0 transverse and longitudinal momentum spectra in proton-proton collisions at $\sqrt{s} = 7$ TeV will be shown in Sect. 3.2.1. Results from comparison with proton-lead collisions data collected at $\sqrt{s_{NN}} = 5.02$ TeV will be presented in Sect. 3.2.2. A test for the Feynman scaling rule will be shown in Sect. 3.2.3.

In previous published results [6, 7], π^0s were reconstructed from energy and incident position of their decay photons in events with one incident photon in each tower ("Type I" event), while events with two incident photons in the same tower ("Type II" event) were rejected. In this analysis Type II events were included for the first time. The inclusion of Type II events allowed to extend the coverage in the $P_T - P_Z$ phase space as shown in figure 3.

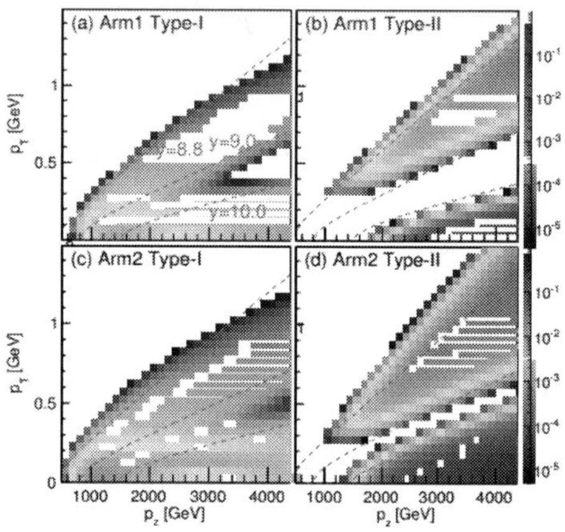

Figure 3. Acceptance map of π^0s detected by LHCf in $P_T - P_Z$ phase space for Arm1 (top figures) and Arm2 (bottom figures). Type II π^0s (right figures) extended the phase space coverage up to $P_T \sim 1$ GeV, while with only Type I events the maximum detectable P_T was ~ 0.5 GeV.

3.2.1 P_T and P_Z distributions at $\sqrt{s} = 7$ TeV

The data in p-p collision at $\sqrt{s} = 7$ TeV were obtained during LHC Fills #1104, #1107, #1112 and #1117 from May 15 to 22, 2010. The integrated luminosity was 2.67 nb^{-1} for Arm1 and 2.10 nb^{-1} for Arm2.

Measured P_T spectra are compared with predictions from different MC interaction models in several regions of the rapidity y, as shown in figure 4. QGSJET II-04 model [12] shows a good overall agreement with LHCf data, while EPOS-LHC [13] has a good agreement at low $P_T(< 0.5$ GeV) and shows a slightly harder distribution for $P_T > 0.5$ GeV. Both models are based on the parton-based Gribov-Regge approach [14, 15] and they are tuned using ALICE, ATLAS, CMS and TOTEM data. The prediction of SIBYLL 2.1 [16] agrees well with the data for $8.8 < y < 9.2$ and $P_T < 0.4$ GeV while its absolute yield is about half that of the data for $y > 9.2$. The predictions of DPMJET 3.06 [17] and PYTHIA 8.185 [18, 19] are compatible with data for $9.0 < y < 9.8$ and $P_T < 0.2$ GeV, while for $P_T > 0.2$ GeV they become significantly harder than both LHCf data and the predictions of other models.

P_Z spectra in several P_T intervals are presented and compared with models predictions in figure 5. There is again an overall agreement between LHCf data and the QGSJET prediction, especially for 0.0 $< P_T < 0.2$ GeV. The EPOS prediction is compatible with data for $P_Z < 2$ TeV, while showing a hard slope for $P_Z > 2$ TeV in all P_T regions. The predictions by DPMJET and PYTHIA agree with data for $P_T < 0.2$ GeV and $P_Z < 1.6$ TeV, while showing a harder distribution for the higher P_Z regions. SIBYLL predicts a smaller production of π^0 s for $P_T < 0.2$ GeV and shows a similar behaviour with DPMJET and PYTHIA with increasing P_T.

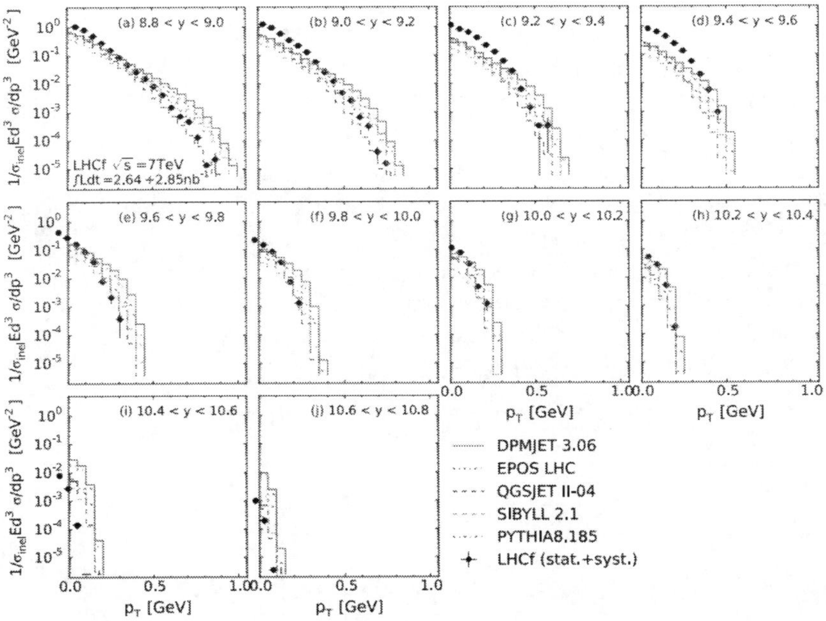

Figure 4. Transverse momentum distribution of neutral pions in several rapidity regions at $\sqrt{s} = 7$ TeV: coloured histograms show results from Monte Carlo simulations with different hadronic interaction models, black points show experimental data.

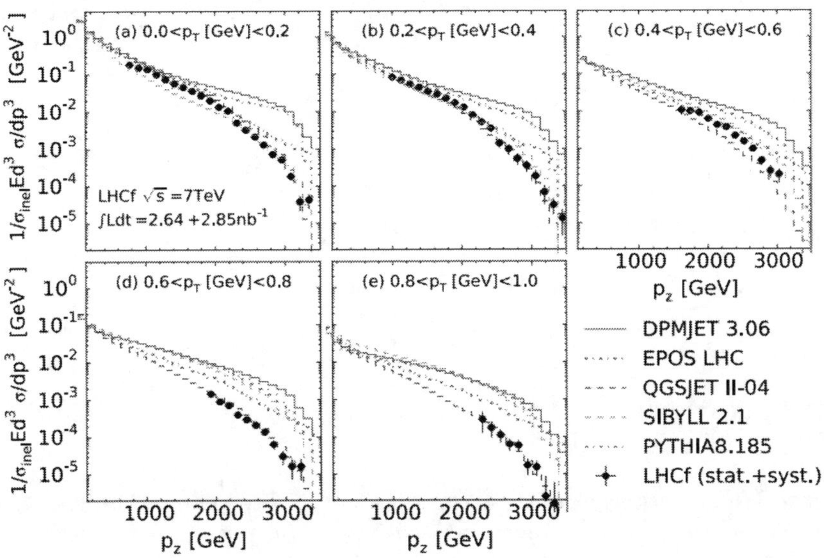

Figure 5. Longitudinal momentum distribution of neutral pions in several P_T regions at $\sqrt{s} = 7$ TeV: coloured histograms show results from Monte Carlo simulations with different hadronic interaction models, black points show experimental data.

3.2.2 Nuclear modification factor

During the p-Pb run in 2013, only Arm2 detector was installed because of two different reasons: to minimize interference with ATLAS ZDC operation and because MC simulations predicted too high multiplicity of neutrons in Pb-remnant side. As a consequence, Arm2 were installed on p-remnant side for almost all the time. Data recorded on p-remnant side in LHC Fills #3478 and #3481 on January 21-22, 2013 were used in the analysis, with an integrated luminosities of 0.63 nb^{-1}.

Proton-lead collisions are very important to understand nuclear effects on forward particles production, because the first collision of an air-shower happens between the primary cosmic ray (a proton

DOI: 10.1051/epjconf/201716407028

or a nucleus) and a nucleus of the atmosphere. Comparing spectra in p-p and p-Pb collisions it is possible to estimate nuclear effects; in particular it is possible to measure the nuclear modification factor (R_{pPb}) which is defined as

$$R_{pPb} \equiv \frac{\sigma^{pp}_{inel}}{\langle N_{coll}\rangle \sigma^{pPb}_{inel}} \frac{Ed^3\sigma^{pPb}/dp^3}{Ed^3\sigma^{pp}/dp^3}$$

where $Ed^3\sigma^{pp}/dp^3$ and $Ed^3\sigma^{pPb}/dp^3$ are the inclusive cross sections of π^0 production in p-p and p-Pb collisions at 5.02 TeV respectively, and $\langle N_{coll}\rangle$ is the mean number of binary nucleon-nucleon collisions. $\langle N_{coll}\rangle$ is estimated as 6.9 from MC simulations. Since there are no data in p-p collisions at $\sqrt{s}$ = 5.02 TeV, the distribution $Ed^3\sigma^{pp}/dp^3$ is obtained interpolating results at 7 TeV and 2.76 TeV in p-p collisions. Nuclear modification factor as a function of P_T in different rapidity regions is shown in figure 6. MC models predict a strong suppression ($R_{pPb} \sim 0.1$) over all range of P_T and it is consistent with the LHCf data within the uncertainties.

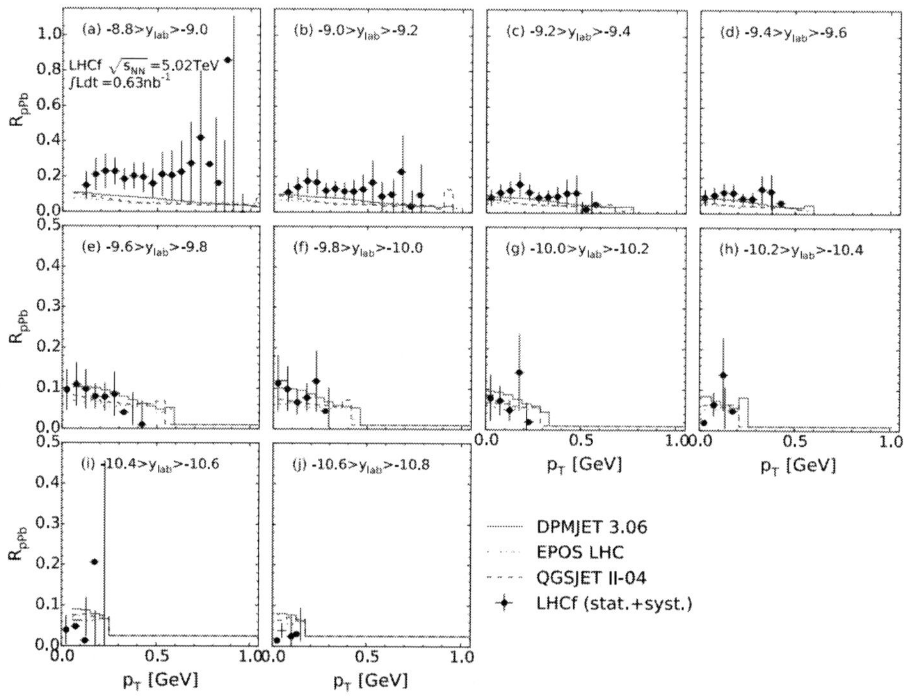

Figure 6. Nuclear modification factor for neutral pions in p-Pb collisions at 5.02 TeV: coloured histograms show results from Monte Carlo simulations with different hadronic interaction models, experimental data are shown as black points.

3.2.3 Test of Feynman scaling

The Feynman scaling rule [20] states that the production cross sections of secondary particles as a function of the Feynman-x variable ($X_F \equiv 2P_Z/\sqrt{s}$) were independent of the incident energy in the forward region. If the Feynman scaling holds, the differential cross section $(X_F/\sigma_{inel})(d\sigma/dX_F)$ should be independent of the center of mass energy for $X_F \gtrsim 0.2$.

DOI: 10.1051/epjconf/201716407028

This scaling rule was tested with the LHCf data at $\sqrt{s}$ = 2.76 TeV and $\sqrt{s}$ = 7 TeV. The test was possible only in the P_T range 0.0 GeV < P_T < 0.4 GeV because there are no data at 2.76 TeV outside this range. The X_F distributions of data at 7 TeV and 2.76 TeV are shown in figure 7 for two different P_T ranges. The comparison of the two data sample indicates that the Feynman scaling holds at the ~ 20% level.

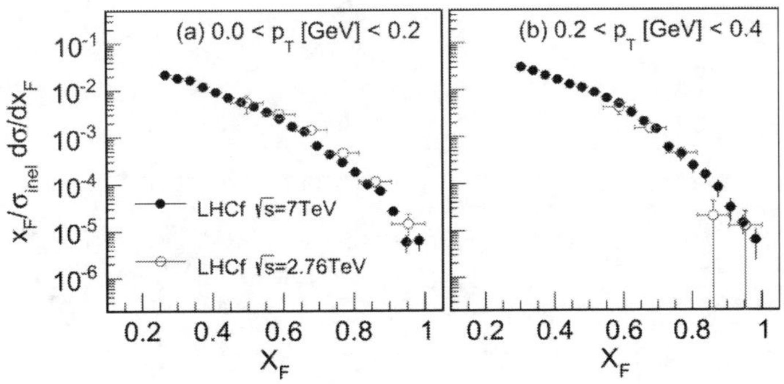

Figure 7. X_F distributions of LHCf data at $\sqrt{s}$ = 7 TeV (black points) and $\sqrt{s}$ = 2.76 TeV (red points) in two P_T ranges. The distributions are consistent at the ~ 20% level.

3.3 Photon inclusive energy spectra in p-p collisions at $\sqrt{s}$ = 13 TeV

Proton-proton collisions at $\sqrt{s}$ = 13 TeV were produced for the first time in 2015 at LHC. LHCf had a dedicated low-luminosity run from 9th to 13th of June 2015, with an instantaneously luminosity of $0.5 \div 1.6 \times 10^{29}\text{cm}^{-2}\text{s}^{-1}$. The preliminary analysis of photon spectrum was done on the data obtained during the LHC Fill #3855. The integrated luminosity was 0.23 nb^{-1} for both Arm1 and Arm2. The preliminary inclusive energy spectra of photon produced in p-p collisions at $\sqrt{s}$ = 13 TeV are presented in figure 8 for two pseudo-rapidity ranges together with the predictions of QGSJET, EPOS, SIBYLL, DPMJET and PYTHIA hadronic interaction models. The LHCf data lie between MC predictions but there is not an unique model with a good agreement in the whole energy range.

4 Conclusions

LHCf experiment performed measurements on forward production of neutral particles in proton-proton and proton-lead collisions at LHC accelerator. These measurements are necessary to calibrate hadronic interaction models used in cosmic rays physics to understand the development of atmospheric showers. LHCf performed measurements at $\sqrt{s}$ = 0.9, 2.76, 7 and 13 TeV in p-p collisions and at $\sqrt{s_{NN}}$ = 5.02 TeV in p-Pb collisions.

In this paper latest published results about π^0 analysis (p-p at 2.76 and 7 TeV, p-Pb at 5.02 TeV) and neutron analysis (p-p at 7 TeV) were presented. Preliminary results from neutron analysis indicate a large amount of high-energy hadrons in the very forward region, predicted only by QGSJETII-03 model. π^0 P_T and P_Z measured spectra show a better agreement with QGSJET II-04 (in the whole range) and EPOS-LHC (for P_T < 0.5 GeV and P_Z < 2 TeV). π^0 analysis in p-Pb indicates a strong suppression of π^0 production compared to p-p case; nuclear modification factor from data is consistent with MC models predictions within errors. Feynman scaling hypotesis was tested comparing π^0 spectra at 7 and 2.76 TeV; the results indicate that the rule holds at the ~ 20% level. Preliminary photon energy spectra indicate that there is not an unique model representing the data, but the measured data lie between MC predictions.

 DOI: 10.1051/epjconf/201716407028

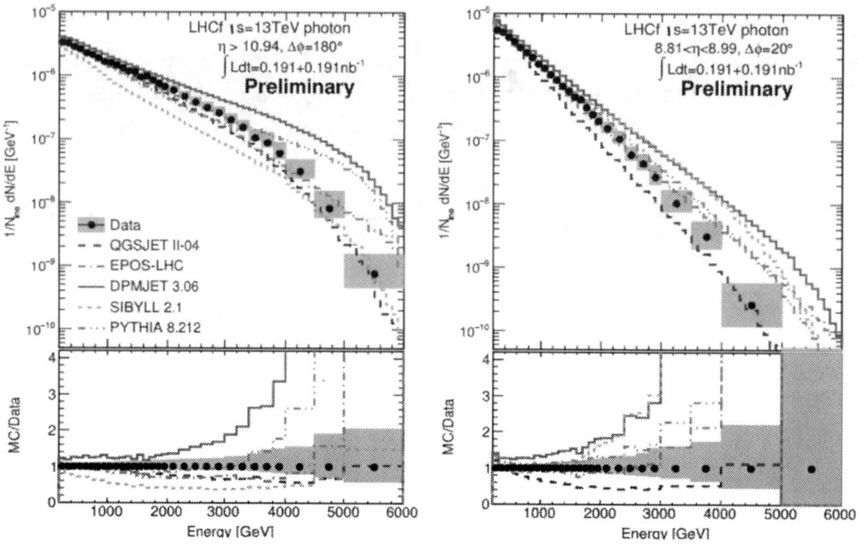

Figure 8. Preliminary inclusive photon energy spectrum for pseudorapidity $\eta > 10.94$ (left figure) and $8.81 < \eta < 8.99$ (right picture). Data are represented by black point while MC prediction from several models are represented by coloured histograms. Green shaded area represents statistical+systematic errors of data.

References

[1] O. Adriani et al., JINST **3**, S08006 (2008).
[2] O. Adriani et al., JINST **5**, P01012 (2010).
[3] K. Kawade et al., JINST **9**, P03016 (2014).
[4] O. Adriani et al., Physics Letters B **715**, 298-303 (2012).
[5] O. Adriani et al., Physics Letters B **703**, 128-134 (2011).
[6] O. Adriani et al., Physical Review D **86**, 092001 (2012).
[7] O. Adriani et al., Physical Review C **89**, 065209 (2014).
[8] O. Adriani et al., Physics Letters B **750**, 360-366 (2015).
[9] O. Adriani et al., Physical Review D **94**, 032007 (2016).
[10] K. Taki et al., JINST **7**, T01003 (2012).
[11] G. D' Agostini, Nucl. Instrum. Meth. A **362**, 487 (1995).
[12] S. Ostapchenko, Nucl. Phys. B, Proc. Suppl. **151**, 143 (2006).
[13] K. Werner, F.-M. Liu, and T. Pierog, Phys. Rev. C **74**, 044902 (2006).
[14] V. Gribov, Sov. Phys. J. Exp. Theor. Phys. **26**, 414 (1968).
[15] T. Regge, Nuovo Cimento **14**, 951 (1959
[16] E.-J. Ahn, R. Engel, T. K. Gaisser, P. Lipari, and T. Stanev, Phys. Rev. D **80**, 094003 (2009).
[17] F. W. Bopp, J. Ranft, R. Engel, and S. Roesler, Phys. Rev. C **77**, 014904 (2008); R. Engel, J. Ranft, and S. Roesler, Phys. Rev. D **55**, 6957 (1997).
[18] T. Sjöstand, S. Mrenna, and P. Skands, J. High Energy Phys. **05** (2006) 026.
[19] T. Sjöstand, S. Mrenna, and P. Skands, Comput. Phys. Commun. **178**, 852 (2008).).
[20] R. P. Feynman, Physical Review Letters **23**, 1415 (1969).

EPJ Web of Conferences **164**, 07029 (2017) DOI: 10.1051/epjconf/201716407029
ICNFP 2016

QCD Physics with the CMS Experiment

S. Cerci[1,a,b]

[1]*Adiyaman University, Faculty of Arts and Sciences, Department of Physics, 02040 Adiyaman, Turkey.*

Abstract. Jets which are the signatures of quarks and gluons in the detector can be described by Quantum Chromodynamics (QCD) in terms of parton-parton scattering. Jets are abundantly produced at the LHC's high energy scales. Measurements of inclusive jets, dijets and multijets can be used to test perturbative QCD predictions and to constrain parton distribution functions (PDF), as well as to measure the strong coupling constant α_S. The measurements use the samples of proton-proton collisions collected with the CMS detector at the LHC at various center-of-mass energies of 7, 8 and 13 TeV.

1 Introduction

Quantum chromodynamics (QCD) is the theory of strong interaction describing the interactions between quarks and gluons, so called partons. The QCD physics can be classified into two as "soft QCD" and "hard QCD" depending on the scale of the parton-parton interaction. The soft QCD takes into account the multiple partonic interaction (MPI), underlying event (UE) and fragmentation whereas the hard QCD takes into account the perturbative QCD theory, parton density functions (PDFs), parton shower modeling and initial and final state radiation (ISR and FSR). Many measurements in recent years have been performed for understanding the QCD both in experimental and theoretical aspects. In the theory side, particularly perturbative calculations at next-to-leading order (NLO) are available for most of the observables and Monte Carlo event generators are tuned more precisely with the measurements. In terms of the experimental side, many measurements have been performed with the data taken with the hadron colliders, which can be used to probe the theory in a large phase space.

2 Soft QCD measurements

In a high-energy hadron-hadron collision the underlying event (UE) is the activity that is not part of the parton-parton scattering in the collision. Understanding of the UE at high energies accurately is important to have precise measurements of standard model processes and searches for new physics. The underlying event activity is measured with the pp collisions at the centre-of-mass energy of $\sqrt{s}$ = 13 TeV by the CMS Collaboration [1]. The leading charged-particles as well as leading charged-particle jets are used as reference objects in the pseudorapidity region ($|\eta| < 2$). The direction of the leading charged-particle / charged-particle jets is taken into account for defining two transverse regions in $\eta - \phi$ space. In order to construct four observables (transMAX, transMIN, transDIF and

[a]e-mail: Salim.Cerci@cern.ch

[b]On behalf of the CMS Collaboration

transAVE) that are useful for describing UE activity the two transverse regions are utilized. The four observables are descibed as following: transMAX (transMIN) is the density in the transverse side with higher (lower) particle density. The difference between transMAX and transMIN densitites is called as transDIF whereas the average of transMAX and transMIN regions is named as transAVE. Figure 1 shows the comparisons of leading jet transverse momentum p_T data taken at different centre-of mass energies $\sqrt{s}$ = 0.9, 2.76, 7, and 13 TeV with PYTHIA8 (version 8.153) CUETP8M1 [2], CUETP8S1 [2] and MONASH [3] tunes and HERWIG++ (version 2.7.0) [4] CUETHS1 tune. A strong rise in the UE activity with the increasing centre-of-mass energy is observed as predicted by the MC model predictions.

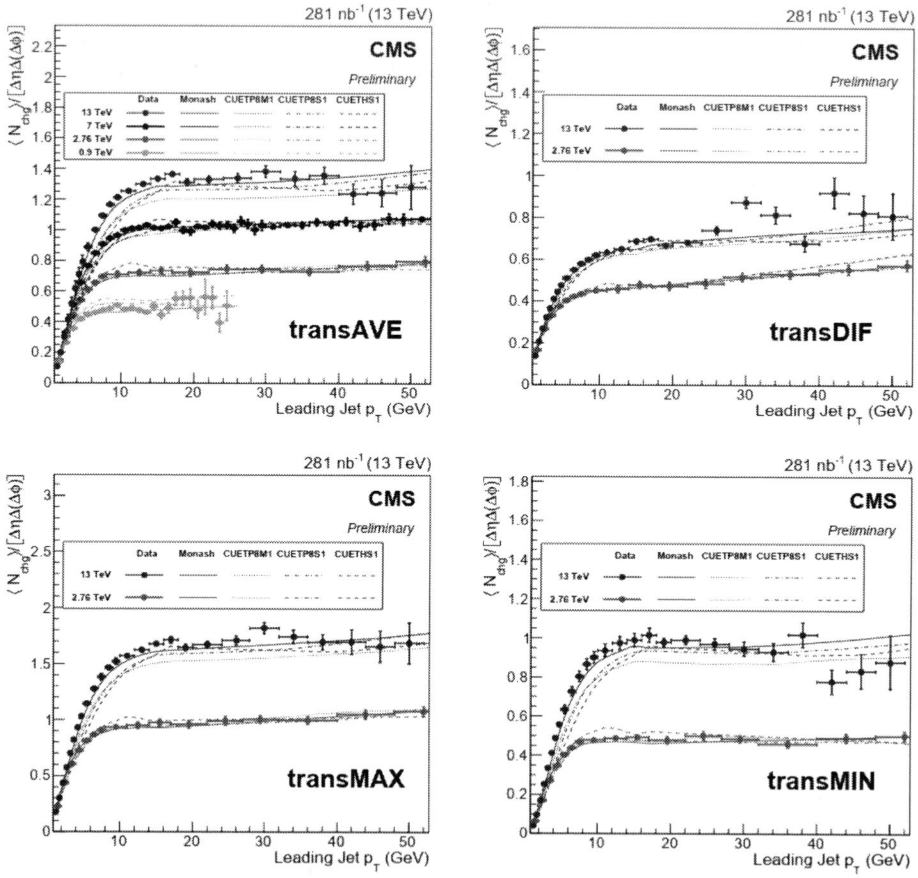

Figure 1. Comparisons between data at different centre-of-mass energies and various MC simulated samples for transAVE, transDIF, transMAX, and transMIN densities.

MPI, the parton saturation scale and the transition between soft and hard QCD may affect the particle and jet transverse momentum spectra at low p_T where the fixed-order partonic cross section in theory diverges. The CMS experiment has measured the leading charged particles and leading charged-particle jets with the low transverse momenta in pp collisions at $\sqrt{s}$ = 8 TeV [5].

Azimuthal angle decorrelation between the hadronic jets, which are the visible manifestations of the energetic partons emerging from the underlying processes, is a sensitive probe for better understanding of the QCD radiation in hard processes in high energy particle collisions. Such a measure-

DOI: 10.1051/epjconf/201716407029

ment may allow to test various models of parton showers or analytic pQCD resummation methods. According to the QCD prediction the production of two partons has a back-to-back topology in azimuthal plane. Hence, the two jets show a strong correlation in their azimuthal angle. When the higher-order processes are taken into account, the decorrelation, deviation from the back-to-back topology, occurs and more partons manifest in the final state. In the BFKL [6–8] framework such a final state is described with hard parton radiation ordered in rapidity and with no ordering on p_T. In contrast to DGLAP [9–13], BFKL predicts a strong growth of the inclusive dijet cross section with increasing rapidity interval between the low-p_T jets. The measurement of decorrelation in the azimuthal angle between the most forward and the most backward jets (Mueller-Navelet jets) is perfomed by CMS in pp collisions at $\sqrt{s} = 7$ TeV [14]. The jets are required to have transverse momenta $p_T > 35$ GeV and rapidity $|y| < 4.7$, allowing a rapidity separation betwen the MN jets of up to $\Delta y = 9.4$. The anti-k_t [15] algorithm with a distance parameter $R = 0.5$ is used for the reconstruction of jets. Decorrelations are measured as normalised differential cross-sections in $\Delta\phi$ for three rapidity separation in $\Delta y < 9.4$, the average cosines of $(\pi - \Delta\phi)$, $2(\pi - \Delta\phi)$, and $3(\pi - \Delta\phi)$, and ratios of these cosines (C_2/C_1 and C_3/C_2). Here $\Delta\phi = \Delta\phi_1 - \Delta\phi_2$ represents the difference between the azimuthal angles ϕ_1 and ϕ_2 of the MN jets. Figure 2 shows comparison of the average cosines as a function of Δy with the LL DGLAP-based MC generator SHERPA, NLL BFKL-motivated MC generator HEJ with hadronisation performed with ARIADNE and analytical NLL BFKL calculation.

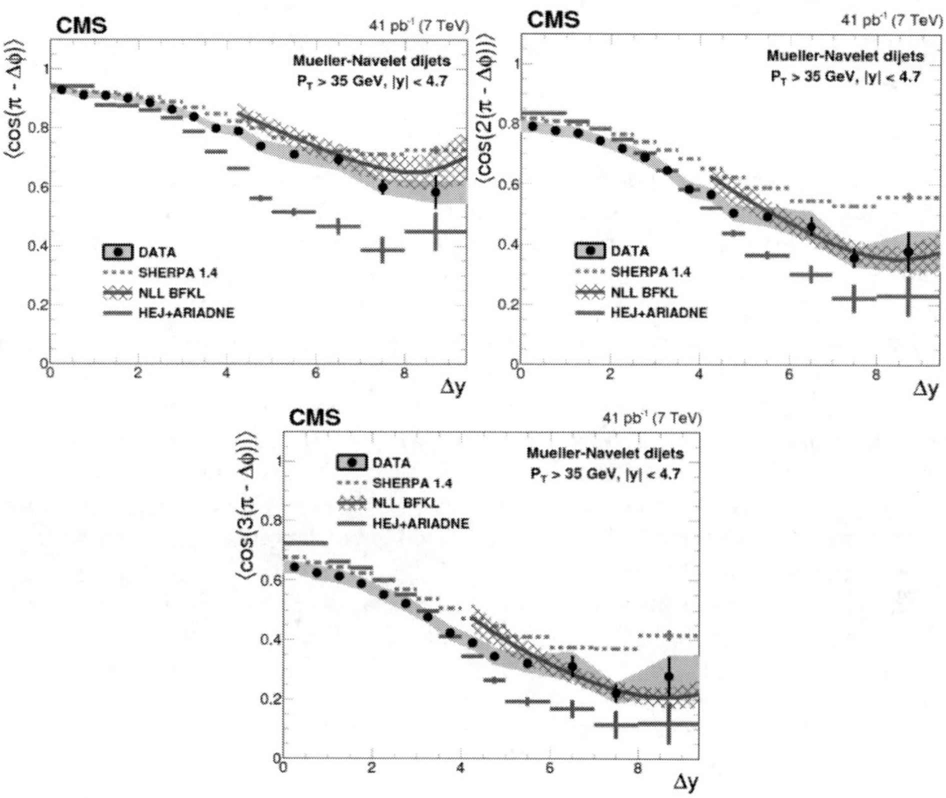

Figure 2. Average cosines as a function of Δy compared with the LL DGLAP-based MC generator SHERPA, NLL BFKL-motivated MC generator HEJ with hadronisation performed with ARIADNE and analytical NLL BFKL calculations

DOI: 10.1051/epjconf/201716407029

The measured ratios C_2/C_1 and C_3/C_2 as a function of Δy are shown in Figure 3. The analytical NLL BFKL calculations gives the best description of data within the experimental and theoretical uncertainties.

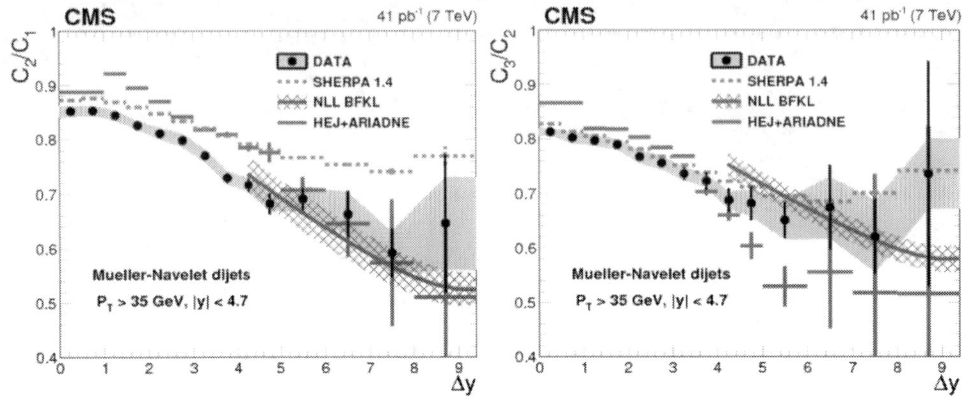

Figure 3. Comparison of measured ratios C_2/C_1 (left) and C_3/C_2 (right) as a function of Δy with the LL DGLAP-based MC generator SHERPA, NLL BFKL-motivated MC generator HEJ with hadronisation performed with ARIADNE and analytical NLL BFKL calculations.

3 Hard QCD measurements

Most of the measurements performed in LHC extend well into the TeV range, with potential sensitivity to new physics. At the LHC, the principal colliding objects are quarks and gluons and hence QCD becomes the most dominant interaction process with very large cross-section. QCD processes take role as a major background to other standard model processes as well as the new physics searches. Jets are defined as the streams of particles clustered together and observed signatures of quarks and gluons in the detector. Measurements of jet cross section as a function of the jet rapidity y and the jet transverse momentum (p_T) can provide information about all the aspects of the theory and can optimize the hadronization and parton shower models. Such measurements also can be used to constrain the parameters of PDF's at a fixed order of perturbation theory, which for most calculations today is next-to-leading order (NLO) and provide a way of measuring α_S at various momentum-transfer scales. The inclusive jet cross-section gives an opportunity to access a wide range of proton momentum fraction, x. The large x jet production is dominated by quark jets whereas the lower x jet production is dominated by gluons.

The double-differential inclusive jet cross section as a function of the jet transverse momentum p_T and the absolute jet rapidity $|y|$ is measured with the data collected by the CMS detector in pp collisions at $\sqrt{s}$ = 8 TeV [16] and $\sqrt{s}$ = 13 TeV [17]. The $\sqrt{s}$ = 8 TeV measurement is carried out using the anti-k_t jet clustering algorithm with distance parameter $R = 0.7$ in the range of $|y| < 4.7$ for jet transverse momenta $21 < p_T < 2000$ GeV. The corrected inclusive jet cross section is compared with the predictions from pQCD at NLO using various sets of PDFs.

Figure 4 shows the measured double-differential cross sections for inclusive jet production as a function of jet p_T in the various $|y|$ ranges. Open points represent the data of low-p_T analysis, filled points are for the high-p_T one. CT10 PDF set is corrected with the non-perturbative (NP) factor for the low-p_T data (solid line) and is corrected with the NP and electroweak correction factors for the

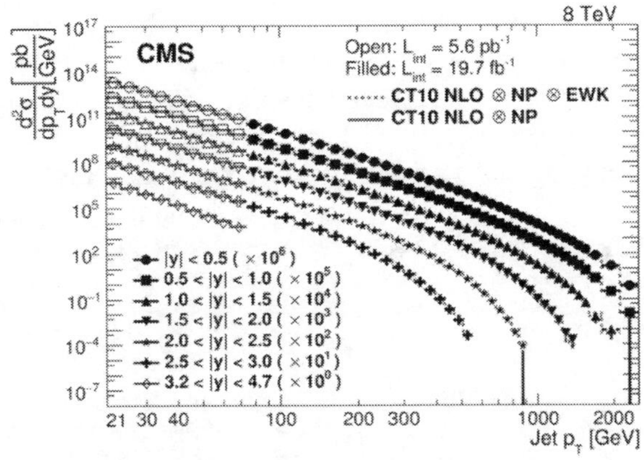

Figure 4. Double-differential inclusive jet cross section as a function of jet p_T at $\sqrt{s}$ = 8 TeV.

high-p_T data (dashed line). Electroweak correction factors are found to be negligible for low-p_T. Figure 5 shows the ratios of data and alternative predictions to the theory prediction that uses the CT10 PDF set. For comparison, predictions employing five other PDF sets are shown in addition to the total theoretical and total experimental systematic uncertainties. The error bars represent the statistical uncertainty of the data.

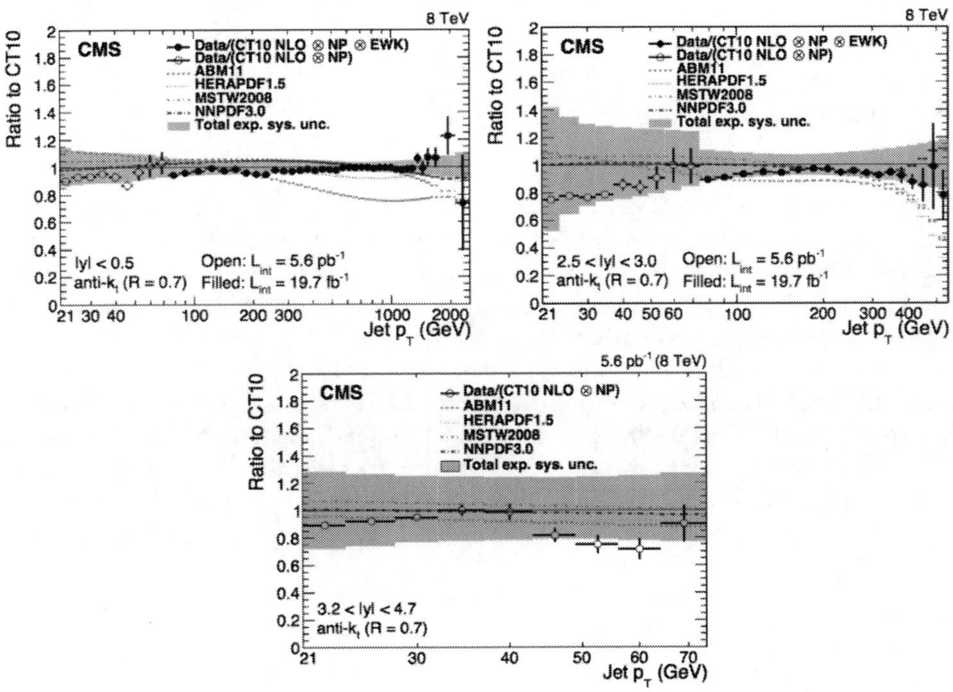

Figure 5. Comparisons of the ratios of data and various predictions to the theory prediction using the CT10 PDF set: for $|y| < 0.5$ (top-left), $2.5 < |y| < 3.0$ (top-right) and $3.2 < |y| < 4.7$ (bottom).

DOI: 10.1051/epjconf/201716407029

The CMS experiment has recently measured the double-differential cross sections at $\sqrt{s} = 13$ TeV [17]. The anti-k_t jet clustering algorithm is used for jet reconstruction for two distance parameters, $R = 0.4$ and 0.7, in a phase space region covering 114 GeV $< p_T <$ 2 TeV and $|y| < 4.7$. Double-differential inclusive jet cross section as function of jet p_T is shown in Figure 6. The results are compared to the NLO++ predictions using CT14 PDF set, and this theory prediction is corrected with the NP and electroweak effects. Ratio of measured values to theoretical prediction from NLO-Jet++ using the CT14 PDF set is shown in Figure 7. Predictions employing three other PDF sets are also shown for comparison. The data are found to be consistent with the predictions within the range of $114 < p_T < 2000$ GeV.

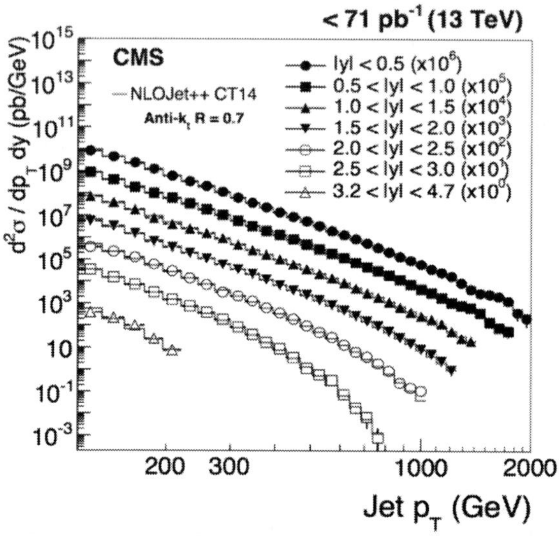

Figure 6. Double-differential inclusive jet cross section as a function of jet p_T at $\sqrt{=}13$ TeV.

The strong coupling constant α_S and PDFs are a key ingredient for precision measurements at hadron colliders. Determination of $\alpha_s(M_Z)$ is based on the measurement of the inclusive jet production cross section as function of jet p_T and rapidity. Such a measurement is performed by CMS as it is shown in Figure 8. The present measurement is in very good agreement with results obtained by previous experiments. The present analysis constrains the $\alpha_S(Q)$ running for Q between 86 GeV and 1.5 TeV, which is the highest scale at which α_S has been measured, to date.

HERA DIS cross section data with the CMS inclusive jet cross section data at $\sqrt{s} = 8$ TeV, provide important constraints on the gluon and valence-quark distributions in the kinematic range studied. These constraints are shown in Figure 9, where the distributions of the gluon and valence quarks are shown at the scales of $Q^2 = 1.9$ GeV^2. Gluon PDF uncertainty is reduced in the wide range of x, while for quark PDF uncertainty improvement is demonstrated for large x.

4 Summary

CMS has been delivering a wealth of precise standard model results since the start of the LHC Run 2 era. The soft and hard QCD results presented here provide a deep insight to understand both perturbative and non-perturbative QCD physics in LHC as it reaches higher energy. Results can also be utilized in improving MC models, determining strong coupling constant and constraining PDFs.

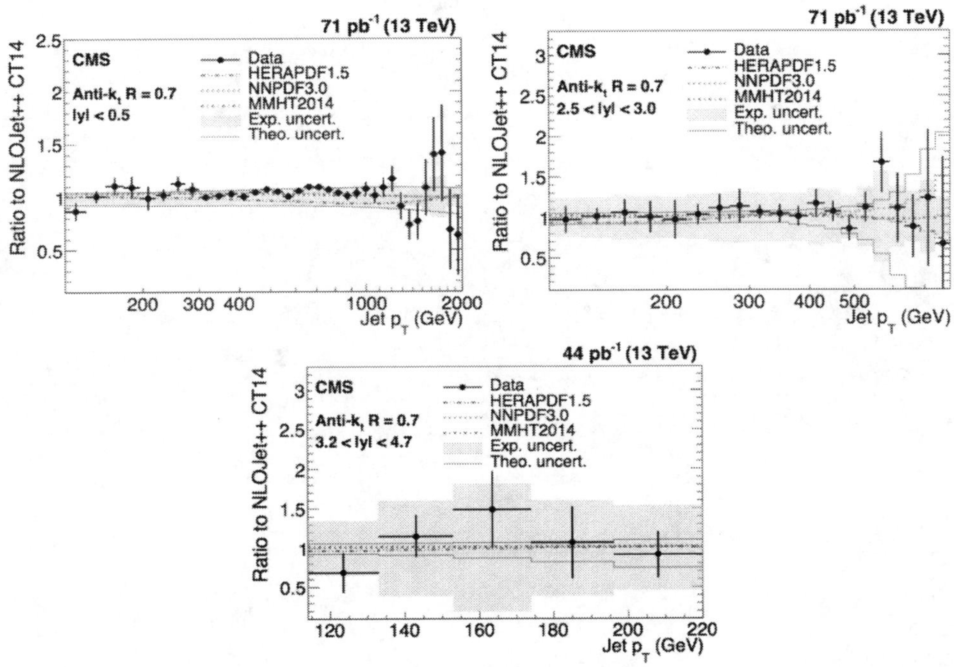

Figure 7. Comparisons of the ratios of data and various predictions to the theory prediction using the CT14 PDF set: for $|y| < 0.5$ (top-left), $2.5 < |y| < 3.0$ (top-right) and $3.2 < |y| < 4.7$ (bottom).

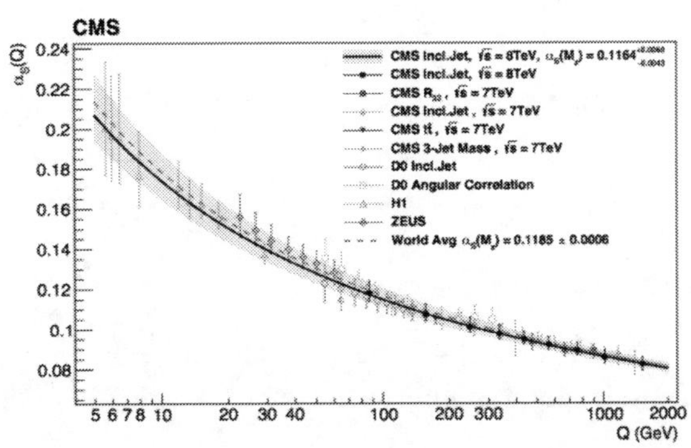

Figure 8. Running of the strong coupling constant α_S as measured in hadron colliders: Tevatron (D0), HERA (H1 and ZEUS) and LHC (CMS).

The CMS collaboration has already provided extensive list of measurements on the jets as well as the strong coupling constant based on different observables. All measurements are found to have good agreement with the world average value and the running of α_S is confirmed at the TeV scale. The inclusive jet spectrum is used to provide additional constraints to PDFs, thus improving the gluon and the valence-quark distributions. A significant reduction of PDF uncertainties has been observed.

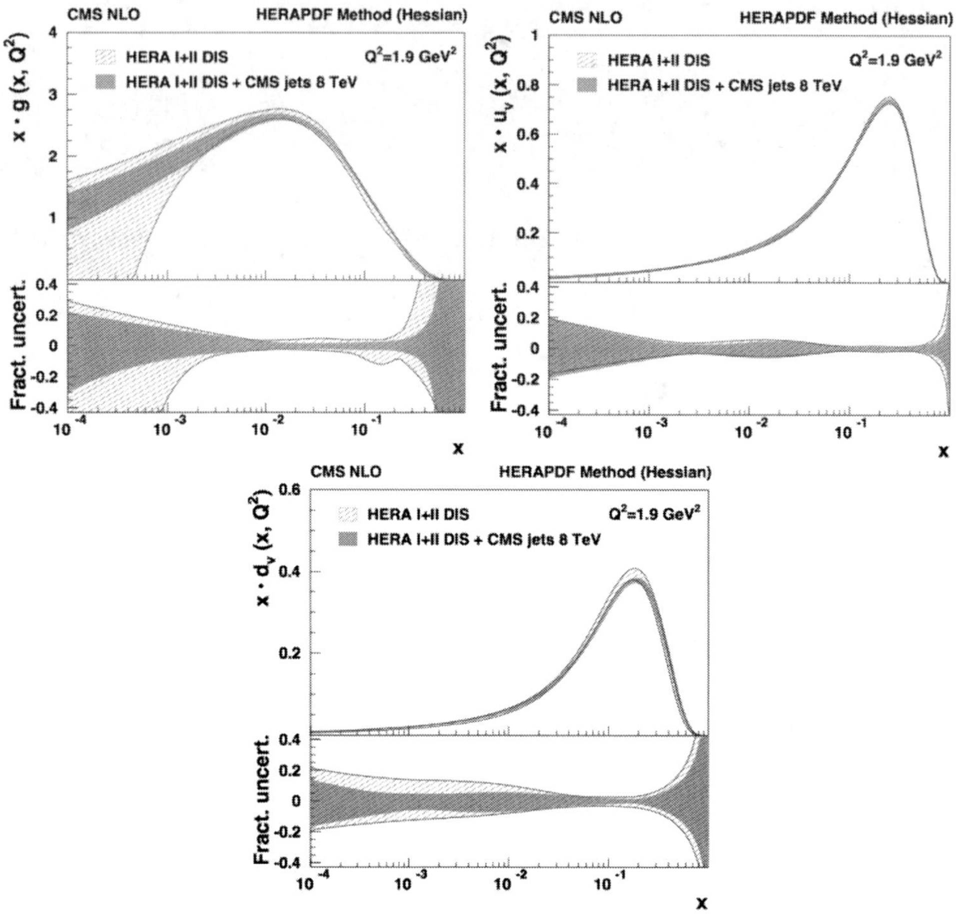

Figure 9. Distributions of gluon (top-left), u-valence quark (top-right), and d-valence quark (bottom) as functions of x at $Q^2 = 1.9$ GeV^2.

5 Acknowledgments

The author would like to acknowledge the Adiyaman University Scientific Activity Foundation and the organizers of ICNFP2016 for the funding.

References

[1] CMS Collaboration, CMS Physics Analysis Summary **CMS PAS FSQ-15-007**, (2016)
[2] V. Khachatryan *et al.* [CMS Collaboration], Eur. Phys. J. C **76**, no. 3, 155, (2016)
[3] P. Skands, S. Carrazza, and J. Rojo, EPJ C **74**, no. 8 (2014)
[4] M. Bahr *et al.*, Eur. Phys. J. C **58**, 639 (2008)
[5] V. Khachatryan *et al.* [CMS Collaboration], Phys. Rev. D **92**, 112001 (2015)
[6] E.A. Kuraev, L.N. Lipatov and V.S. Fadin, Sov. Phys. JETP **44** 443, (1976)
[7] E.A. Kuraev, L.N. Lipatov and V.S. Fadin, Sov. Phys. JETP **45** 199, (1977)
[8] I.I. Balitsky and L.N. Lipatov, Sov. J. Nucl. Phys. **28** 822, (1978)
[9] V.N. Gribov and L.N. Lipatov, Sov. J. Nucl. Phys. **15** 438, (1972)
[10] V.N. Gribov and L.N. Lipatov, Sov. J. Nucl. Phys. **15** 675, (1972)
[11] L.N. Lipatov, Sov. J. Nucl. Phys. **20** 94, (1975)
[12] G. Altarelli and G. Parisi, Nucl. Phys. B **126** 298, (1977)
[13] Y.L. Dokshitzer, Sov. Phys. JETP **46** 641, (1977)
[14] V. Khachatryan et al. [CMS Collaboration], JHEP **08** 139, (2016)
[15] M. Cacciari, G.P. Salam and G. Soyez, JHEP **04** 063, (2008)
[16] V. Khachatryan *et al.* [CMS Collaboration], Submitted to JHEP, arXiv:1609.05331 [hep-ex], (2016)
[17] V. Khachatryan *et al.* [CMS Collaboration], Eur. Phys. J. C **76** 451, (2016)

Quantum Optics Properties of QCD Vacuum

V. Kuvshinov[1,a], V. Shaparau[1,b], E. Bagashov[1,c]

[1] *Joint Institute for Power and Nuclear Research - Sosny*
National Academy of Science of Belarus
PO box 119, 220109 Minsk, BELARUS

Abstract. Theoretical justification of the occurrence of multimode squeezed and entangled colour states in QCD is given. We show that gluon entangled states which are closely related with corresponding squeezed states can appear by the four-gluon self-interaction. Correlations for the collinear gluons are revealed two groups of the colour correlations which is significant at consider of the quark-antiquark pair productions.
It is shown that the interaction of colour quark with the stochastic vacuum of QCD leads to the loss of information on the initial colour state of the particle, which gives a new perspective regarding the confinement of quarks phenomenon. The effect is demonstrated for a single particle and in the multiparticle case is proposed.
Quantum characteristics (purity and von Neumann entropy) are used to analyse the process of interaction.

1 Introduction

Many experiments at $e^+e^-, p\bar{p}, ep$ colliders are devoted to hadronic jet physics, since detailed studies of jets are important for better understanding and testing both perturbative and non-perturbative QCD and also for finding manifestations of new physics. Although the nature of jets is of a universal character, e^+e^-- annihilation stands out among hard processes, since jet events admit a straightforward and clear-cut separation in this process.

In the reaction $e^+e^- \rightarrow hadron$ four evolution phases are recognized at various time and space scales (Fig.1). These are (I) the production of a quark-antiquark pair: $e^+e^- \rightarrow q\bar{q}$; (II) the emissions of gluons from primary partons — perturbative evolution of the quark-gluon cascade; (III) the non-perturbative evolution and the hadronization of quarks and gluons; (IV) the decays of unstable particles.

The second phase of e^+e^-- annihilation has been well understood and sufficiently accurate predictions for it have been obtained within the perturbative QCD (PQCD) [1]. But predictions of the PQCD are limited by small effective coupling $\alpha(Q^2) < 1$ and third phase is usually taken into account either through a constant factor which relates partonic features with hadronic ones (within local parton-hadron duality) or through the application of various phenomenological models of hadronization. As a consequence, theoretical predictions for jet characteristics remain unsatisfactory. For example, the

[a] e-mail: v.kuvshinov@sosny.bas-net.by
[b] e-mail: shaparau@sosny.bas-net.by
[c] e-mail: bagashov@sosny.bas-net.by

width of the multiplicity distribution (MD) according to the predictions of PQCD is larger than the experimental one. The discrepancies between theoretical calculations and experimental data suggest that the non-perturbative evolution of the quark-gluon cascade plays important role. New gluon states, generated at the non-perturbative stage, contribute to various features of jets. In particular, such a contribution to MD can be made in the form of the sub-Poissonian distribution [2, 3]. Therefore we must take into account both perturbative and non-perturbative stage of the jet evolution.

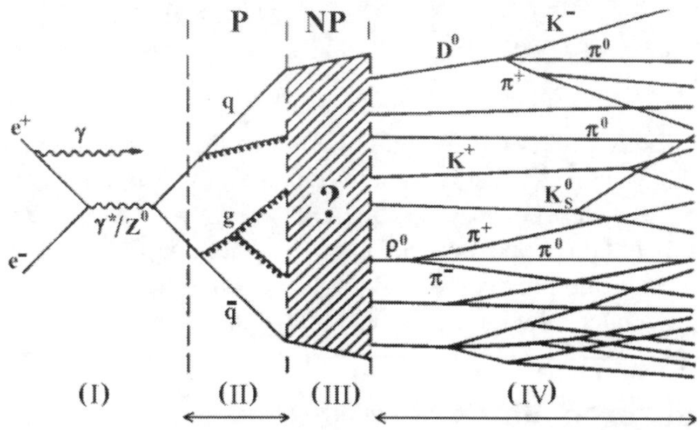

Figure 1. Jet evolution

It is known that such property is inherent for the squeezed states (SS), which are well studied in quantum optics (QO) [4]-[6]. Squeezed states have uncommon properties: they display a specific behaviour of the factorial and cumulant moments [7] and can have both sub-Poissonian and super-Poissonian statistics corresponding to antibunching and bunching of photons. Moreover oscillatory behaviour MD of photon SS is differentiated from Poissonian and Negative binomial distributions (NBD).

Therefore we believe that the non-perturbative stage of gluon evolution can be one of sources of the gluon SS in QCD by analogy with nonlinear medium for photon SS in QO. Gluon MD in the range of the small transverse momenta (thin ring of jet) is Poissonian [8]. Quark-gluon MD in the whole jet at the end of the perturbative cascade can be represented as a combination of Poissonian distributions each of which corresponds to a coherent state. Studying a further evolution of gluon states at the non-perturbative stage of jet evolution we obtain new gluon states. These states are formed as a result of non-perturbative self-interaction of the gluons expressed by nonlinearities of Hamiltonian. Using the Local parton hadron duality it is easy to show that in this case behaviour of hadron multiplicity distribution in jet events is differentiated from the negative binomial one that is confirmed by experiments for $pp, p\bar{p}$-collisions [9]-[11].

At finite squeezed r (continuous variable) entangled state is known from quantum optics as a two-mode squeezed state [4, 18]

$$|f\rangle = \hat{S}_{12}(r)|0\rangle_1|0\rangle_2 = \frac{1}{\cosh r}\sum_{n=0}^{\infty}(\tanh r)^n\,|n\rangle_1|n\rangle_2, \tag{1}$$

where $\hat{S}_{12}(r) = \exp\{r(\hat{a}_1^+\hat{a}_2^+ - \hat{a}_1\hat{a}_2)\}$ is operator of two-mode squeezing. It is not difficult to demonstrate that the state vector $|f\rangle$ describes the entangled state. Each of these entangled states has a uncommon property: a measurement over one particle have an instantaneous effect on the other, possibly located at a large distance.

The dimensionless coefficient

$$y = \left[\frac{|\overline{\hat{a}_1 \hat{a}_2^+}|^2 + |\overline{\hat{a}_1 \hat{a}_2}|^2}{2(\overline{\hat{a}_1^+ \hat{a}_1} + 1/2)(\overline{\hat{a}_2^+ \hat{a}_2} + 1/2)} \right]^{1/2} \tag{2}$$

is the measure of entanglement for two-mode states [19], $0 \le y < 1$ (entanglement is not observed when $y = 0$). Here $\overline{\hat{a}_i \hat{a}_j^+} = \langle \hat{a}_i \hat{a}_j^+ \rangle - \langle \hat{a}_i \rangle \langle \hat{a}_j^+ \rangle$, $\hat{a}_i, \hat{a}_j^+$ are the annihilation and creation operators correspondingly. Averaging the annihilation and creation operators in the expression (2) over the vector $|\mathrm{f}\rangle$ (1) at small squeeze factor we have

$$y \simeq \sqrt{2} r. \tag{3}$$

Two-mode gluon states with two different colours can lead to $q\bar{q}$-entangled states. Interaction of the quark entangled states with stochastic vacuum (quantum measurement) has a remarkable property, namely, as soon as some measurement projects one quark onto a state with definite colour, the other quark also immediately obtains opposite colour that leads to coupling of quark-antiquark pair, string tension inside $q\bar{q}$-pair and free propagation of colourless hadrons. Therefore the investigation of the gluon entangled states connected with the corresponding squeezed ones is the issue of the day.

2 Multimode squeezed states of the gluons

By analogy with QO [6] a multimode squeezing condition for gluons with different colors $i_1, \ldots, i_p$ is written as

$$\hat{N} \left\langle \left(\Delta (\hat{X}_\lambda^{i_1,\ldots,i_p})_{\frac{1}{2}} \right)^2 \right\rangle < 0, \tag{4}$$

where $\Delta(\hat{X}_\lambda^{i_1,\ldots,i_p}) = \hat{X}_\lambda^{i_1,\ldots,i_p} - \left\langle \hat{X}_\lambda^{i_1,\ldots,i_p} \right\rangle$, $\hat{N}$ is a normal ordering operator, the phase-sensitive Hermitian operators $(\hat{X}_\lambda^{i_1,\ldots,i_p})_1 = \frac{1}{2\sqrt{p}} \sum_{j=1}^{p} [\hat{b}_\lambda^{i_j} + (\hat{b}_\lambda^{i_j})^+]$ and $(\hat{X}_\lambda^{i_1,\ldots,i_p})_2 = \frac{1}{2i\sqrt{p}} \sum_{j=1}^{p} [\hat{b}_\lambda^{i_j} - (\hat{b}_\lambda^{i_j})^+]$ are linear combination of the annihilating (creating) operators $\hat{b}_\lambda^{i_j}$ $(\hat{b}_\lambda^{i_j+})$, $i_1, \ldots, i_p = \overline{1,8}$ are gluon color charges, λ is a polarization index. Averaging in (4) is performed over final state vector

$$|\mathrm{f}\rangle \simeq |\mathrm{in}\rangle - it \left(\hat{H}_\mathrm{I}^{(3)}(0) + \hat{H}_\mathrm{I}^{(4)}(0) \right) |\mathrm{in}\rangle \tag{5}$$

which describes gluon system later small time t. Operators $\hat{H}_\mathrm{I}^{(3)}(t)$ and $\hat{H}_\mathrm{I}^{(4)}(t)$ describing the three- and four-gluon selfinteractions include combinations of three and four annihilating and creating operators [20].

$$\begin{aligned}
\hat{H}_\mathrm{I}^{(3)}(t) = {} & i \frac{g}{(2\pi)^{3/2}} f_{abc} \sum_{\lambda_1,\lambda_2,\lambda_3} \int d\tilde{k}_1 d\tilde{k}_2 d\tilde{k}_3 \times \Big\{ \vec{k}_1 \vec{\varepsilon}_{\lambda_2}(k_2) \varepsilon^{\nu}_{\lambda_1}(k_1) \varepsilon^{\lambda_3}_{\nu}(k_3) \delta(\vec{k}_1 + \vec{k}_2 + \vec{k}_3) \\
& \times [\hat{b}^a_{\lambda_1}(k_1) \hat{b}^b_{\lambda_2}(k_2) \hat{b}^c_{\lambda_3}(k_3)\, e^{-2i(k_{01}+k_{02}+k_{03})t} - \hat{b}^{a+}_{\lambda_1}(k_1) \hat{b}^{b+}_{\lambda_2}(k_2) \hat{b}^{c+}_{\lambda_3}(k_3)\, e^{2i(k_{01}+k_{02}+k_{03})t}] \\
& + \hat{b}^{a+}_{\lambda_1}(k_1) \hat{b}^b_{\lambda_2}(k_2) \hat{b}^c_{\lambda_3}(k_3)\, e^{2i(k_{01}-k_{02}-k_{03})t}\ \delta(\vec{k}_1 - \vec{k}_2 - \vec{k}_3) \\
& \times \Big[\vec{k}_3 \vec{\varepsilon}_{\lambda_1}(k_1) \varepsilon^{\nu}_{\lambda_2}(k_2) \varepsilon^{\lambda_3}_{\nu}(k_3) + \vec{k}_2 \vec{\varepsilon}_{\lambda_3}(k_3) \varepsilon^{\nu}_{\lambda_1}(k_1) \varepsilon^{\lambda_2}_{\nu}(k_2) - \vec{k}_1 \vec{\varepsilon}_{\lambda_2}(k_2) \varepsilon^{\nu}_{\lambda_1}(k_1) \varepsilon^{\lambda_3}_{\nu}(k_3) \Big] \\
& + \hat{b}^{a+}_{\lambda_1}(k_1) \hat{b}^{b+}_{\lambda_2}(k_2) \hat{b}^c_{\lambda_3}(k_3)\, e^{2i(k_{01}+k_{02}-k_{03})t}\ \delta(\vec{k}_1 + \vec{k}_2 - \vec{k}_3) \\
& \times \Big[\vec{k}_1 \vec{\varepsilon}_{\lambda_3}(k_3) \varepsilon^{\nu}_{\lambda_1}(k_1) \varepsilon^{\lambda_2}_{\nu}(k_2) + \vec{k}_3 \vec{\varepsilon}_{\lambda_1}(k_1) \varepsilon^{\nu}_{\lambda_2}(k_2) \varepsilon^{\lambda_3}_{\nu}(k_3) - \vec{k}_1 \vec{\varepsilon}_{\lambda_2}(k_2) \varepsilon^{\nu}_{\lambda_1}(k_1) \varepsilon^{\lambda_3}_{\nu}(k_3) \Big] \Big\},
\end{aligned} \tag{6}$$

$$\hat{H}_{\mathrm{I}}^{(4)}(t) = \frac{g^2}{4(2\pi)^3} \sum_{\lambda_1,\lambda_2,\lambda_3,\lambda_4} \int d\tilde{k}_1 d\tilde{k}_2 d\tilde{k}_3 d\tilde{k}_4 \Big\{ \varepsilon^{\mu}_{\lambda_1}(k_1)\varepsilon^{\lambda_3}_{\mu}(k_3)\varepsilon^{\nu}_{\lambda_2}(k_2)\varepsilon^{\lambda_4}_{\nu}(k_4) f_{abc} f_{ade}$$
$$\times\Big[\delta(\vec{k}_1+\vec{k}_2+\vec{k}_3+\vec{k}_4)\Big(\hat{b}^{b}_{\lambda_1}(k_1)\hat{b}^{c}_{\lambda_2}(k_2)\hat{b}^{d}_{\lambda_3}(k_3)\hat{b}^{e}_{\lambda_4}(k_4)\, e^{-2i(k_{01}+k_{02}+k_{03}+k_{04})t}$$
$$+\hat{b}^{b+}_{\lambda_1}(k_1)\hat{b}^{c+}_{\lambda_2}(k_2)\hat{b}^{d+}_{\lambda_3}(k_3)\hat{b}^{e+}_{\lambda_4}(k_4)\, e^{2i(k_{01}+k_{02}+k_{03}+k_{04})t}\Big)$$
$$+4\hat{b}^{b+}_{\lambda_1}(k_1)\hat{b}^{c}_{\lambda_2}(k_2)\hat{b}^{d}_{\lambda_3}(k_3)\hat{b}^{e}_{\lambda_4}(k_4)\, e^{2i(k_{01}-k_{02}-k_{03}-k_{04})t}\,\delta(\vec{k}_1-\vec{k}_2-\vec{k}_3-\vec{k}_4) \tag{7}$$
$$+4\hat{b}^{b+}_{\lambda_1}(k_1)\hat{b}^{c+}_{\lambda_2}(k_2)\hat{b}^{d+}_{\lambda_3}(k_3)\hat{b}^{e}_{\lambda_4}(k_4)\, e^{2i(k_{01}+k_{02}+k_{03}-k_{04})t}\,\delta(\vec{k}_1+\vec{k}_2+\vec{k}_3-\vec{k}_4)\Big]$$
$$+2\hat{b}^{b+}_{\lambda_1}(k_1)\hat{b}^{c+}_{\lambda_2}(k_2)\hat{b}^{d}_{\lambda_3}(k_3)\hat{b}^{e}_{\lambda_4}(k_4)\, e^{2i(k_{01}+k_{02}-k_{03}-k_{04})t}\,\delta(\vec{k}_1+\vec{k}_2-\vec{k}_3-\vec{k}_4)$$
$$\times\Big[\varepsilon^{\mu}_{\lambda_1}(k_1)\varepsilon^{\lambda_3}_{\mu}(k_3)\varepsilon^{\nu}_{\lambda_2}(k_2)\varepsilon^{\lambda_4}_{\nu}(k_4)(f_{abc}f_{ade}+f_{abe}f_{adc})$$
$$+\varepsilon^{\mu}_{\lambda_1}(k_1)\varepsilon^{\lambda_2}_{\mu}(k_2)\varepsilon^{\nu}_{\lambda_3}(k_3)\varepsilon^{\lambda_4}_{\nu}(k_4) f_{abd} f_{ace}\Big]\Big\}.$$

Here g is a self-interaction constant, $d\tilde{k} = \dfrac{d^3k}{\sqrt{2k_0}}$, k_0 is a gluon energy, $\varepsilon^{\mu}_{\lambda}$ is a polarization vector, f_{ahb} are structure constants of $SU_c(3)$ group.

Initial state vector $|\mathrm{in}\rangle$ describes gluon system at end of perturbative stage [8] and is product of the coherent states of the gluons with different colours and polarization indexes

$$|\mathrm{in}\rangle \equiv |\alpha\rangle = \prod_{\lambda=1}^{3}\prod_{b=1}^{8} |\alpha^{b}_{\lambda}\rangle. \tag{8}$$

By analogy with QO gluon coherent state vector $|\alpha^{b}_{\lambda}\rangle$ is the eigenvector of the corresponding annihilation operator $\hat{b}^{b}_{\lambda}$ with the eigenvalue $\alpha^{b}_{\lambda_1}$ which can be written in terms of the gluon coherent field amplitude $|\alpha^{b}_{\lambda_1}|$ and phase γ^{b}_{λ} of the given gluon field $\alpha^{b}_{\lambda_1} = |\alpha^{b}_{\lambda_1}|e^{i\gamma^{b}_{\lambda_1}}$. In each gluon coherent state $|\alpha^{b}_{\lambda}\rangle$ the gluon number with fixed colour b and polarization λ is arbitrary (the average multiplicity of given gluon is equal to square of the gluon coherent field amplitude $\langle n^{b}_{\lambda}\rangle = |\alpha^{b}_{\lambda}|^2$) and phase of considering state γ^{b}_{λ} is fixed.

Averaging the annihilation and creation operators $\hat{b}^{i_j}_{\lambda}, \hat{b}^{i_j+}_{\lambda}$ in (4) over the evolved vector $|\mathrm{f}\rangle$ which is defined according to (5) and taking into account of chosen initial state vector we write the multimode squeezing condition in the form

$$\left\langle \hat{N}\left(\Delta(\hat{X}^{i_1,\ldots,i_p}_{\lambda})_{\frac{1}{2}}\right)^2 \right\rangle = \mp\frac{it}{4p}\sum_{j,k=1}^{p}\Big\{\langle\alpha|[\hat{b}^{i_k}_{\lambda},[\hat{b}^{i_j}_{\lambda},\hat{H}_{\mathrm{I}}(0)]]|\alpha\rangle - \langle\alpha|[[\hat{H}_{\mathrm{I}}(0),\hat{b}^{i_j+}_{\lambda}],\hat{b}^{i_k+}_{\lambda}]|\alpha\rangle\Big\} < 0, \tag{9}$$

where $\hat{H}_{\mathrm{I}}(0) = \hat{H}^{(3)}_{\mathrm{I}}(0) + \hat{H}^{(4)}_{\mathrm{I}}(0)$.

It can be shown that only the four-gluon self-interaction can yield the multimode squeezing effect since

$$[[\hat{H}^{(3)}_{\mathrm{I}}(0),\hat{b}^{i_j+}_{\lambda}],\hat{b}^{i_k+}_{\lambda}] = 0, \quad [\hat{b}^{i_k}_{\lambda},[\hat{b}^{i_j}_{\lambda},\hat{H}^{(3)}_{\mathrm{I}}(0)]] = 0. \tag{10}$$

Indeed, the multimode squeezing condition can be written in the explicit form as

$$\left\langle \hat{N}\left(\Delta(\hat{X}_{\lambda}^{i_1,\dots,i_p})_{\frac{1}{2}}\right)^2\right\rangle = \pm\frac{i\,t}{8\,p\,k_0}\frac{g^2}{(2\pi)^3}\sum_{j,k=1}^{p}\sum_{\lambda_1,\lambda_2=1}^{3}\int\prod_{i=1}^{2}\frac{d^3k_i}{\sqrt{2k_{0i}}}$$
$$\times\left[\left(\varepsilon_{\mu}^{\lambda_1}(k_1)\,\varepsilon_{\lambda_2}^{\mu}(k_2)\right)\left(\varepsilon_{\nu}^{\lambda}(k)\,\varepsilon_{\lambda}^{\nu}(k)\right)-\left(\varepsilon_{\mu}^{\lambda_1}(k_1)\varepsilon_{\lambda}^{\mu}(k)\right)\left(\varepsilon_{\nu}^{\lambda_2}(k_2)\varepsilon_{\lambda}^{\nu}(k)\right)\right]$$
$$\times\left\{f_{ai_j b}f_{ai_k c}\left[\langle\alpha|\hat{b}_{\lambda_1}^{b+}(k_1)\hat{b}_{\lambda_2}^{c+}(k_2)|\alpha\rangle-\langle\alpha|\hat{b}_{\lambda_1}^{b}(k_1)\hat{b}_{\lambda_2}^{c}(k_2)|\alpha\rangle\right]\right.$$
$$\times\left[\delta(2\vec{k}-\vec{k}_1-\vec{k}_2)-\delta(2\vec{k}+\vec{k}_1+\vec{k}_2)\right]+\left(f_{ai_j b}f_{ai_k c}+f_{ai_k b}f_{ai_j c}\right)$$
$$\left.\times\langle\alpha|\hat{b}_{\lambda_1}^{b+}(k_1)\hat{b}_{\lambda_2}^{c}(k_2)|\alpha\rangle\left[\delta(2\vec{k}-\vec{k}_1+\vec{k}_2)-\delta(2\vec{k}+\vec{k}_1-\vec{k}_2)\right]\right\}<0. \quad (11)$$

In particular for the collinear gluons we have the condition

$$\left\langle\hat{N}\left(\Delta(X_{\lambda}^{i_1,\dots,i_p})_{\frac{1}{2}}\right)^2\right\rangle=\pm\frac{g^2\,t}{8\,p\,(2\pi)^3\,k_0^2}\sum_{j,k=1}^{p}\sum_{\lambda_1\neq\lambda}f_{ai_k b}f_{ai_j c}|\alpha_{\lambda_1}^{b}||\alpha_{\lambda_1}^{c}|\sin(\gamma_{\lambda_1}^{b}+\gamma_{\lambda_1}^{c})<0, \quad (12)$$

Here $|\alpha_{\lambda_1}^{b}|$ and $\gamma_{\lambda_1}^{b}$ are an amplitude and a phase of the initial gluon coherent field, f_{ahb} is a structure constant of the color group $SU_c(3)$. The multimode squeezing condition (12) is fulfilled for any cases apart from if all initial gluon coherent fields are real or imaginary. Obviously, the larger are both the amplitudes of the initial gluon coherent fields with different colour and polarization indexes and coupling constant, the larger is multimode squeezing effect.

Thus non-perturbative four-gluon selfinteraction is source of the multimode squeezing effect.

3 Gluon entangled states

One of entangled condition is

$$0<y<1, \quad (13)$$

where entangled measure is defined by dimentionless value (2)

$$y=\left[\frac{|\overline{\hat{b}_{\lambda}^{h}\hat{b}_{\lambda}^{g+}}|^2+|\overline{\hat{b}_{\lambda}^{h}\hat{b}_{\lambda}^{g}}|^2}{2(\overline{\hat{b}_{\lambda}^{h+}\hat{b}_{\lambda}^{h}}+1/2)(\overline{\hat{b}_{\lambda}^{g+}\hat{b}_{\lambda}^{g}}+1/2)}\right]^{1/2} \quad (14)$$

Entangled condition for collinear gluons is

$$0<\left|t\,\frac{\alpha_s\pi}{\sqrt{2}k_0}(f_{ahb}f_{ahc}+f_{agb}f_{agc}+f_{ahb}f_{agc}+f_{agb}f_{ahc})\sum_{\lambda_1\neq\lambda}|\alpha_{\lambda_1}^{b}||\alpha_{\lambda_1}^{c}|\sin(\gamma_{\lambda_1}^{b}+\gamma_{\lambda_1}^{c})\right|<1. \quad (15)$$

Obviously squeezed gluon states are simultaneously entangled if the amplitudes of the initial gluon coherent fields are small enough.

Thus by analogy with quantum optics as a result of four-gluon self-interaction we obtain two-mode squeezed gluon states which are also entangled.

4 Interaction of quark colour states with QCD vacuum

We might investigate the dynamics of individual quarks and multi-quark systems, using the similar information-centered approach. The model of stochastic vacuum of QCD is known to be a possible phenomenological solution for the confinement of quarks problem [12–14], has been proven to be applicable by lattice calculations [15] and thus could serve as an effective basis for such an attempt.

While using this model, we consider that only the second field correlators are giving a non-negligible contribution (Gauss domination) and also use the fact that the interactions of some quantum system with the environment can be effectively represented by additional stochastic terms in the Hamiltonian of the system. Then the density matrix of the system is obtained by averaging with respect to degrees of freedom of the quantum environment - the stochastic vacuum of QCD. Interactions with the environment and the corresponding non-unitary evolution result in decoherence and loss of quantum superpositions. So the information on the initial state of the quantum system (in this case - the colour state of quarks) is lost after sufficiently large time.

Let us now consider the interaction of colour particle (quark) with QCD stochastic vacuum. Assume this quark propagates along some fixed path γ. The amplitude of such process is obtained by [21]

$$|\phi(\gamma)\rangle = \mathcal{P}\exp\left(i\int\limits_{\gamma} dx^{\mu}\hat{A}_{\mu}\right)|\phi_{in}\rangle, \tag{16}$$

where $\mathcal{P}$ is the path ordering operator, $\hat{A}_{\mu}$ is the gauge field vector and $|\phi_{in}\rangle$ is the initial state vector. In order to consider mixed states we introduce the colour density matrix, taking into account both colour particle and QCD stochastic vacuum (environment):

$$\hat{\rho}(\gamma) =< |\phi(\gamma)\rangle\langle\phi(\gamma)| > . \tag{17}$$

Here we average over all implementations of stochastic gauge field (environment degrees of freedom). In order to keep the gauge invariance, the path γ should represent a closed loop. The averaging yields [22]:

$$\hat{\rho}(\gamma) = N_c^{-1} + (|\phi_{in}\rangle\langle\phi_{in}| - N_c^{-1})W_{adj}(\gamma). \tag{18}$$

Here N_c is the number of colours, W_{adj} is the Wilson loop in the adjoint representation.

In confinement region Wilson loop decays exponentially with the area spanned by loop, so for the rectangular loop spanning the time interval T and distance R we get:

$$\hat{\rho}(\gamma) = N_c^{-1} + (\hat{\rho}_{in} - N_c^{-1})\exp(-\sigma_{adj}RT), \tag{19}$$

$$\hat{\rho}(\gamma : RT \to \infty) = N_c^{-1}, \tag{20}$$

where σ_{adj} is the string tension in the adjoint representation.

As follows from (20), all colour states are mixed with equal probabilities and all information on initial color state is lost. The same result might be acquired for the interaction of colour superposition with stochastic QCD vacuum [23].

5 Quantum optical characteristics

The rate of transition from pure colour states to white mixture state can be estimated on the base of purity [24]:

$$P = \mathrm{Tr}\,\hat{\rho}^2. \tag{21}$$

 DOI: 10.1051/epjconf/201716407030

This characteristic represents the closeness of a quantum state to a pure one. In our case,

$$P = N_c^{-1} + (1 - N_c^{-1})\exp\left(-\sigma_{adj}RT\right). \tag{22}$$

When RT tends to 0, $P \to 1$, which corresponds to a pure state. When the composition RT tends to infinity, the purity tends to N_c^{-1}, which corresponds to the white mixture.

It can be inferred from (18) and (22) that the larger the loop is, the quicker the initial state tends to white mixture as a result of interaction with the QCD stochastic vacuum. Thus the information of quark colour states is lost in hadrons due to interactions between quarks and confining non-Abelian gauge fields.

Von Neumann entropy can be used as a measure of the loss of information:

$$S = -\mathrm{Tr}\,(\hat{\rho}\ln\hat{\rho}). \tag{23}$$

Initial density matrix gives us $S = 0$ and we will have $S = \ln N_c$ for large RT.

In order to obtain a more suitable characteristic to work with, we would define the information on the basis of von Neumann entropy. As it can be seen, the entropy ranges from 0 (for a pure, unperturbed quantum state, which in our approach corresponds to the maximum amount of quantum information) to $\ln N_c$. So it would be quite natural to define the information measure as

$$I = \frac{\ln N_c - S}{\ln N_c} = 1 - \frac{S}{\ln N_c}. \tag{24}$$

So the overall range of this measure is [0, 1]: in case of zero entropy it is equal to 1 and in case of maximum entropy it is equal to 0. The latter case corresponds to the asymptotically big values of RT (the confinement region).

Thus, as we might see, during the interaction with QCD vacuum the entropy increases and the information is being lost.

6 Multiparticle States and QCD Vacuum

Density matrix of a system is dependent on the quantity of its degrees of freedom, which is dependent on the number of particles in the system. Accordingly, the density matrix dimesions for multiparticle states would be increased from (N_c)x(N_c) to $(N_c^{N_p})$x$(N_c^{N_p})$, where N_p is the number of particles. So the expression (19) in the multiparticle case would look like

$$\hat{\rho} = N_c^{-N_p}\hat{I} + (\hat{\rho}_{in} - N_c^{-N_p}\hat{I})\exp(-\sigma_{adj}RT). \tag{25}$$

And the expression (20) can be written as

$$\hat{\rho}(\gamma : RT \to \infty) = N_c^{-N_p}\hat{I}. \tag{26}$$

The purity (21) would change to:

$$\mathrm{P} = N_c^{-N_p} + (1 - N_c^{-N_p})W_{adj}^2(\gamma), \tag{27}$$

with maximum von Neumann entropy (23) being:

$$S = N_p \ln N_c. \tag{28}$$

The proposed information measure (24) might then be generalized as

Table 1. Purity and entropy of different states of multiparticle system before the interaction with vacuum

State:	pure separable	mixed separable	pure entangled	mixed entangled
P (purity)	1	$\frac{1}{N_c^{N_p}} \leq P < 1$	1	$\frac{1}{N_c^{N_p}} < P < 1$
S (entropy)	0	$0 < S \leq N_p \ln N_c$	0	$0 < S < N_p \ln N_c$

Table 2. Purity and entropy of different states of multiparticle system after the interaction with vacuum

State:	pure separable	mixed separable	pure entangled	mixed entangled
P (purity)	$\frac{1}{N_c^{N_p}}$	$\frac{1}{N_c^{N_p}}$	$\frac{1}{N_c^{N_p}}$	$\frac{1}{N_c^{N_p}}$
S (entropy)	$N_p \ln N_c$	$N_p \ln N_c$	$N_p \ln N_c$	$N_p \ln N_c$

$$I = 1 - \frac{S}{N_p \ln N_c}. \tag{29}$$

Here is the list of possible types of states, in which the system may be found (excluding some specific types e.g. as the squeezed states):

- By purity: 1) pure; 2) mixed.
- By separability: 1) separable; 2) non-separable (entangled).

Table 1 represents the quantum characteristics for the four types of states (before the interaction with vacuum). The strictness of the inequalities in the last column is due to the fact that a fully mixed state is always separable.

The same quantities after the interaction are shown in table 2.

Thus we can see that the final state is defined by a number of particles (and the number of possible colours for any individual particle) in the system. But in any case the entropy during the interaction with QCD vacuum rises to its maximum value and the information is being lost in this process. In the limit $W(\gamma) \to 0$ we get fully mixed state of a form (26).

7 Conclusion

Investigating of the gluon fluctuations we have proved theoretically the possibility of existence of the multimode gluon squeezed states. The emergence of such remarkable states becomes possible owing to the four-gluon self-interaction. The three-gluon self-interaction does not lead to squeezing effect.

We have shown that QCD jet non-perturbative evolution leads both to squeezing and entanglement of gluons. It should be noted that the greater are both the amplitudes of the initial gluon coherent fields with different colour and polarization indexes and coupling constant, the greater is multimode squeezing effect of the colour gluons. We have demonstrated that entanglement condition of the gluon states with fixed two colours and polarization is defined by the corresponding squeeze factor which is dependent on the amplitudes and phases of initial coherent gluon fields.

Because two-mode gluon states with two different colours can lead to $q\bar{q}$-entangled states role of colour correlations could be very significant for explanation of the hadronization phenomenon.

Vacuum of quantum chromodynamics can be considered as environment for colour particles. Density matrix, purity and von Neumann entropy for colour particles depend on Wilson loop averaged through QCD vacuum degrees of freedom. The interaction of colour particle with QCD vacuum leads to the decoherence of pure colour states into a mixed white states.

For multiparticle states (pure separable, mixed separable and entangled) in the asymptotic of large time and space intervals we obtain diagonalization of density matrix, decreasing of purity and increasing of von Neumann entropy.

References

[1] Yu.L. Dokshitzer, V.A. Khoze, A.H. Mueller, S. I. Troyan, *Basics of Perturbative QCD* (Frontières, France, 1991)
[2] V.I. Kuvshinov, Acta Phys. Pol. B **10**, 19 (1979)
[3] E.S. Kokoulina, V.I. Kuvshinov, Acta Phys. Pol. B **13**, 553 (1982)
[4] D.F. Walls, G.J. Milburn, *Quantum Optics* (Springer-Verlag, NY., USA, 1995)
[5] M.O. Scully, M.S. Zubairy, *Quantum Optics* (Cambridge University Press, 1997)
[6] S.Ya. Kilin, *Quantum Optics* (Minsk, 1990). (in Russian)
[7] I.M. Dremin, Phys. Lett. A **193**, 209 (1994)
[8] S. Lupia, W. Ochs and J. Wosiek, Nucl. Phys. B **540**, 405 (1999)
[9] G.J. Alner et al. (UA5 coll.), Phys. Rep. **154**, 247 (1987)
[10] P. Abreu et al. (DELPHI coll.), Z. Phys. C **50**, 185 (1991)
[11] P.D. Acton et al. (OPAL coll.), Z. Phys. C **53**, 539 (1992)
[12] Yu. A. Simonov. The Confinement. Phys. Usp. **39**, 313 (1996).
[13] J. Ambjørn, P. Olesen. On the Formation of a Color Magnetic Quantum Liquid in QCD. Nuclear Physics B. **170**, no. 1, 60-78 (1980).
[14] D. S. Kuzmenko, Y. A. Simonov, et al. Vacuum, Confinement and QCD Strings in the Vacuum Correlator Method. Uspekhi Fizicheskih Nauk. **174**, no. 1 (2004) (in Russian).
[15] A. D. Giacomo, H. Dosch, V. I. Shevchenko, and Y. A. Simonov, Physics Reports **372**, no. 4, 319-368 (2002).
[16] J.S. Bell, Physics **1**, 195 (1964)
[17] A. Einstein, B. Podolsky and N. Rosen , Phys. Rev. **45**, 777 (1935)
[18] E.A. De Wolf , Progress in Optics. **42**, 1 (2001)
[19] V.V. Dodonov, A.S.M. de Castro, S.S. Mizrahi, Phys. Lett. A **296**, 73 (2002)
[20] V.I. Kuvshinov, V.A. Shaparau, Ukrainian Journal of Physics **48**, 1255 (2003)
[21] V. Kuvshinov, P. Buividovich. White Mixed States in QCD Stochastic Vacuum. Nonlinear Phenomena in Complex Systems. **8**, no.3, 313-316 (2005).
[22] V. Kuvshinov, A. Kuzmin. *Gauge Fields and Theory of Deterministic Chaos*. (Belorussian Science, Minsk, 2006, in Russian).
[23] V. I. Kuvshinov, E. G. Bagashov. Evolution of colour superposition in the stochastic QCD vacuum. Nonlinear Phenom. Complex Syst., **16**, No. 3, 242-246 (2013).
[24] F. Haake. *Quantum Signatures of Chaos*. (Springer-Verlag, Berlin, 1991).

DOI: 10.1051/epjconf/201716407031

Quantum spin correlations in relativistic Møller scattering

Paweł Caban[1], Jakub Rembieliński[1], Marta Włodarczyk[2,a], Jacek Ciborowski[2], Michał Drągowski[2], and Artem Poliszczuk[2]

[1] *University of Lodz, Department of Theoretical Physics, Pomorska 149/153, 90-236 Łódź*
[2] *University of Warsaw, Department of Physics, Pasteura 5, 02-093 Warszawa*

Abstract. We present the relativistic spin correlation function (and the corresponding probabilities) for a pair of polarized electrons originating from the Moller scattering. This particular state is easy to prepare experimentally; therefore, the results are discussed in view of a possible measurement. We also discuss the state after the Moller scattering in terms of entanglement and polarization transfer.

1 Introduction

1.1 The EPR paradox

In 1935 Albert Einstein, Boris Podolsky and Nathan Rosen (EPR) formulated their famous paradox [1] which, according to their reasoning, indicated that quantum mechanical description of physical reality was incomplete. It considered two entangled particles in such a state, that the distance between them was precisely known and equal to L, and their total momentum was equal to 0. Now if the pair was investigated by two arbitrairly distant ovservers, Alice and Bob, each one having access to only one particle, we could imagine that Alice performed the position measurement on her particle. But knowing the distance between the particles, in that very moment she would have gained information about the position of the Bobs particle, too. On the other hand she could have peformed the momentum measurement, simultaneously determining the momentum of Bob's particle. EPR claimed that nothing that happens to Alices particle can have impact on the state of the Bobs one (which can be arbitrarily distant) because no information or interaction can propagate with infinite speed. It would however mean that the state (in this case – both position and momentum) of the Bobs particle has been determined all the time, which is not reflected in the quantum mechanical description of physical reality, so (i) either qunatum mechanical description of physical reality is incomplete and the complete state of the system (results of all possible measurements) is precisely determined at all times or (ii) EPR were wrong and measurements performed on one particle of the entangled pair can alter the state of the other particle instantaneously. Einstein strongly disagreed with the second possibility, which he referred to as "spooky action at the distance".

[a]e-mail: marta.wlodarczyk@gmail.com

1.2 Bell type inequalities and the correlation function

For nearly thirty years the discussion whether spooky action at a distance exists was an academic debate, until in 1964 John Stuart Bell formulated his famous theorem [2] stating that every theory including the principle of local realism (i.e. not containing spooky action at the distance) must fulfill a certain class of inequalities, which are now called the Bell-type inequalities. In other words, outcomes of measurements performed on two space-like separated systems cannot be correlated arbitrarily strong. The strength of this correlation can be "measured" by means of so called *correlation function*. In a special case of a dychotomic observable, whose eigenvalues can take values ± 1, the correlation function can be defined as the difference between the probability that Alice and Bob get the same outcome and the probability that their outcomes are opposite

$$C(\vec{a}, \vec{b}) = P_{++} + P_{--} - P_{+-} - P_{-+}, \tag{1}$$

where $P_{\pm\pm}$ is the joint probability that both Alice and Bob measure ± 1.

Bell noticed that if theory includes three assumptions of (i) locality (nothing that happens with one particle can influence the other one instantanously), (ii) realism (outcomes of unperformed measurements do exist despite we don't know them), (iii) free will (Alice and Bob can freely choose what kind of experiment they want to perform), the following inequality must be fullfilled[1]

$$|C(\vec{a}, \vec{b}) + C(\vec{c}, \vec{b} + C(\vec{c}, \vec{d}) - C(\vec{a}, \vec{d}))| \leq 2. \tag{2}$$

It turns out that quantum mechanics can violate the Bell-type inequalities, which means that it is a non–local theory so it is either incorrect (and not only incomplete, as stated by EPR) or the principle of local realism is wrong. In particular the CHSH inequality is violated for a singlet state of two fermions, for which the correlation function has a simple form $C(\vec{a}, \vec{b}) = -\vec{a} \cdot \vec{b}$, and which has bin described by David Bohm in 1951 in his spin version of the EPR paradox [4]. It can be seen in Fig. 1

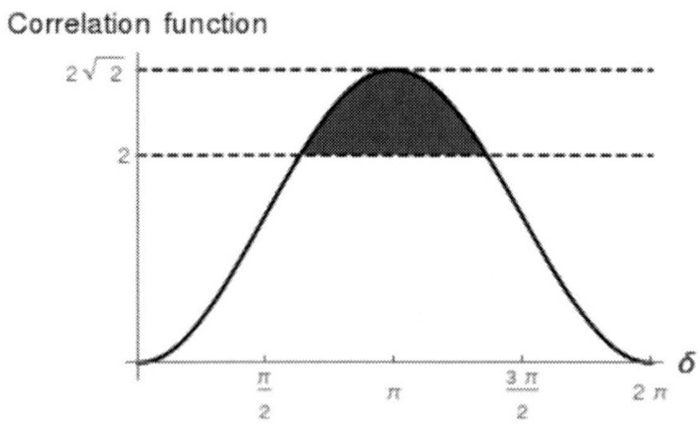

Figure 1: The left-hand-side of the inequality (2) as a function of the angle δ between the X axis and $\vec{d}$, where $\alpha = \pi/4, \beta = \pi/2, \gamma = 3\pi/4$, for a singlet state of two fermions. It reaches its maximum equal to $2\sqrt{2}$ for $\delta = \pi$.

that when the directions $\vec{a}$, $\vec{b}$. $\vec{c}$ and $\vec{d}$ are separated with $\pi/4$ angle, respectively, the value of the left-hand-side of the inequality (2) reaches value $2\sqrt{2} > 2$.

[1]It is not the original Bell inequality, which is valid only for maximally entangled state, but more general CHSH inequality [3].

This outcome raised a question, whether real quantum systems violate the Bell-type inequalities or not. The first experiment to verify this was performed with use of fully entangled photons by Alain Aspect [5, 6] in 1982 and its outcome was that the Bell-type inequalities are violated by 5 standard deviations. It was a beautiful confirmation of correctness and completeness of quantum mechanics – its non–locality is not just a feature of the formalism, but a fundamental property of Nature.

2 Relativistic correlation functions

If one wanted to perform a Bell-type experiment using massive fermions (as in Bohm version of the paradox) instead of photons, they would have to take into account that they should move apart with great velocities to assure their space–like separation during the measurement and calculate the theoretical correlation function in relativistic regime.

It has been shown that the relativistic correlation function exhibits a significantly different behavior than in the non—relativistic case. It depends on particles momenta [7–9], moreover in some energy ranges this dependence may be non—monotonic (in systems containing at least one massive particle) [8–11]. It can be seen in Fig 2.

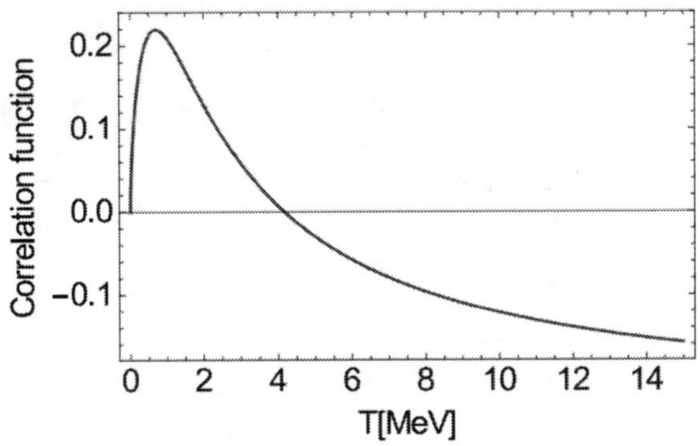

Figure 2: The relativistic correlation function for a pure singlet state as a function of kinetic energy of one particle.The kinetic energy of the second particle equals 5MeV, $\vec{a} = (0,0,1)$, $\vec{b} = (1,0,0)$, $\frac{\vec{p}}{|\vec{p}|} = \frac{1}{\sqrt{2}}(0,1,1)$ and $\frac{\vec{k}}{|\vec{k}|} = \left(-\frac{1}{4}, \frac{\sqrt{3}}{4}, \frac{\sqrt{3}}{2}\right)$. The non-monotonic behaviour of the function can be clearly seen.

None of the experiments performed so far could observe relativistic effects in correlation function. Up to our best knowledge, only three correlation experiments using massive fermions have been performed so far [12–14]. They were all focused on measuring the violation of the Bell-type inequalities for maximally entangled protons in a singlet state. In this case the correlation function reads:

$$C^{k,p}(\vec{a},\vec{b}) = -\vec{a}\cdot\vec{b} + \frac{(\vec{k}\times\vec{p})}{m^2+kp}\cdot\left((\vec{a}\times\vec{b}) + \frac{(\vec{a}\cdot\vec{k})(\vec{b}\times\vec{p}) - (\vec{b}\cdot\vec{p})(\vec{a}\times\vec{k})}{(k^0+m)(p^0+m)}\right), \tag{3}$$

where k and p are the four-momenta of Alices and Bobs particle, respectively. To measure the relativistic correction to the non-relativistic correlation function $-\vec{a}\cdot\vec{b}$, the kinetic energies of the particles used should be at least of order of their rest mass. In all the aforementioned experiments the energies were to low to distinguish between the relativistic and the non-relativistic case.

3 2POL experiment

In this situation the groups from University of Lodz, University of Warsaw and Technische Universität Darmstadt decided to perform a pioneering experiment that could allow for distinguishing between the relativistic and non-relativistic correlation function for the first time.

3.1 Initial state preparation

The first problem was the choice of the initial state in which the correlation function is to be measured. It would be optimal to use a pure singlet state, as in the aforementioned experiments with protons. It can be achieved (for protons) by using nuclear reactions, but in such a case the energy of the pair cannot be high enough. This is why we decided to use a state originating from Møller scattering of a polarized electron beam on a target. It is neither maximally entangled nor pure, but it is well defined and it has been shown [15] that the difference between relativistic and non-relativistic correlation function is measurable in this particular case.

Although the state before the scattering is fully separable, after the scattering the particles become entangled. If we take negativity, N, as an entanglement measure

$$N(\hat{\rho}^{in}) = \Sigma_i \frac{|\lambda_i| - \lambda_i}{2}, \tag{4}$$

where λ_i are the eigenvalues of the density matrix $\hat{\rho}^{in}$, and analyze its dependence on the scattering angle and the beam kinetic energy (in the simplest case of an unpolarized target), we can see that for a given beam energy negativity reaches its maximum for the symmetric scattering angle. On the other hand if we focus on the symmetric scattering only, the degree of the entanglement increases with the decrease of the beam kinetic energy (see Fig. 3). Although due to their entanglement, Møller electrons do not have a well determined polarization, it is possible to calculate their mean polarization vectors using the spin density matrix of the final state. It can be shown that although before the scattering one of the electrons was unpolarized, after the Møller scattering both electrons have nonzero mean polarization vectors (see Fig 4). In case of the symmetric scattering (due to their indistinguishability) they are equally polarized. Moreover, in this specific case the length of the mutual polarization vector takes its smallest value (see Fig. 5), which corresponds to maximum of negativity. In section 4 it will be shown that this configuration maximizes also the correlation function.

3.2 Spin-projection measurement

To be able to measure and compare the relativistic correlation function to the theoretical predictions, one must have (i) a proper relativistic spin-projection observable, $\hat{\vec{S}}$, and (ii) a corresponding method of measuring spin projection on a given direction. Problem of defining a proper relativistic spin operator no satisfactionary solution so far and has been widely discussed elsewhere [16, 17]. It can however been shown that only one of the operators proposed in the literature (i) reduces to non-relativistic form in rest frame: $\hat{\vec{S}} = \frac{\hat{\vec{W}}}{m}$, where $\hat{W}$ is the Pauli-Lubanski four-vector, (ii) is a three-vector: $\left[\hat{J}_i, \hat{S}_j\right] = i\epsilon_{ijk}\hat{S}_k$, (iii) fulfills standard commutation relations: $\left[\hat{S}_i, \hat{S}_j\right] = i\epsilon_{ijk}\hat{S}_k$. and (iv) is a linear combination of Pauli-Lubanski four-vector components. It is the Newton-Wigner spin operator of form:

$$\hat{\vec{S}} = \frac{1}{m}\left(\hat{\vec{W}} - \hat{W}^0 \frac{\hat{\vec{P}}}{\hat{P}^0 + m}\right), \tag{5}$$

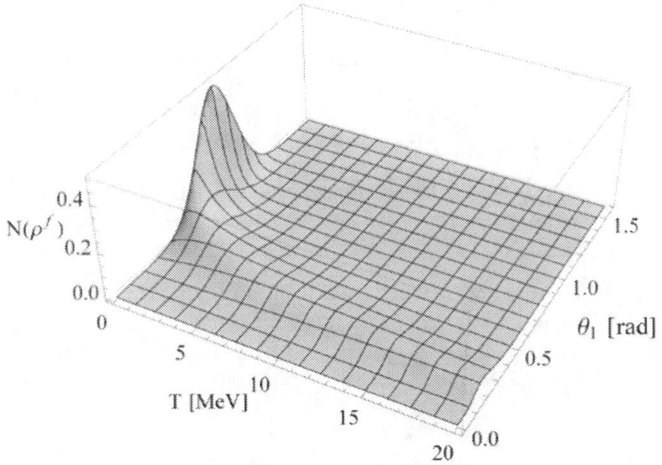

Figure 3: Negativity of a state of two electrons originating from Møller scattering of a polarized beam on an unpolarized target as the function of the scattering angle and the beam kinetic energy. For a given energy, the maximum of entanglement corresponds to the symmetric scattering and for symmetric scattering the less energetic beam, the more entangled the state after the scattering is. Note that before the scattering the beam and the target electrons were in a fully separable state.

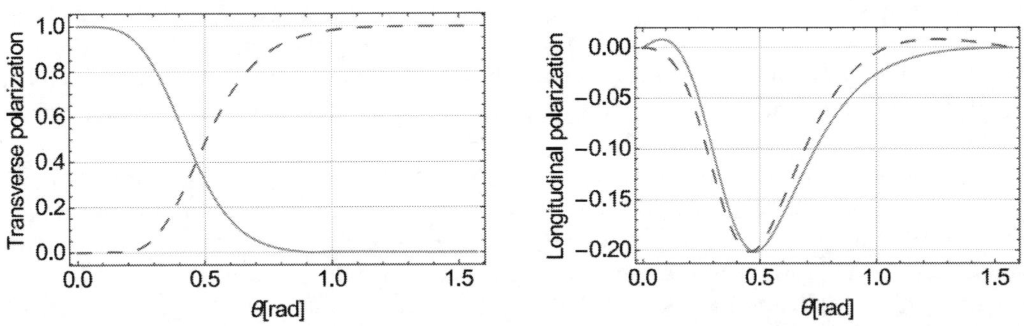

(a) transverse polarization in Møller scattering (b) longitudinal polarization in Møller scattering

Figure 4: The dependence of transverse and longitudinal polarization on the scattering angle in the case of Møller scattering of a 100% transversly polarized beam scattering off an unpolarized target. The scattering angle θ is the angle between the initial beam direction and the direction of the electron with subscript A.

where $\hat{P}$ is the four-momentum operator.

On the other hand a spin-projection measurement procedure had to be chosen. Basically there are two main methods allowing for determining polarization of an electron beam – Møller and Mott [18] polarimetry. For the case of electrons at energies up to about 10 MeV, Mott polarimetry has been established as a standard tool.

DOI: 10.1051/epjconf/201716407031

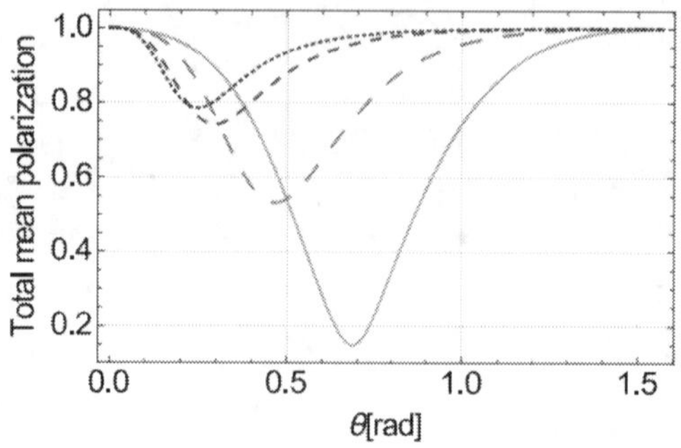

(a) Mutual polarization vector length as a function of scattering angle for beam kinetic energies equal to: 0.5MeV (solid line), 3MeV (large dashing), 10MeV (medium dashing), 15MeV (tiny dashing)

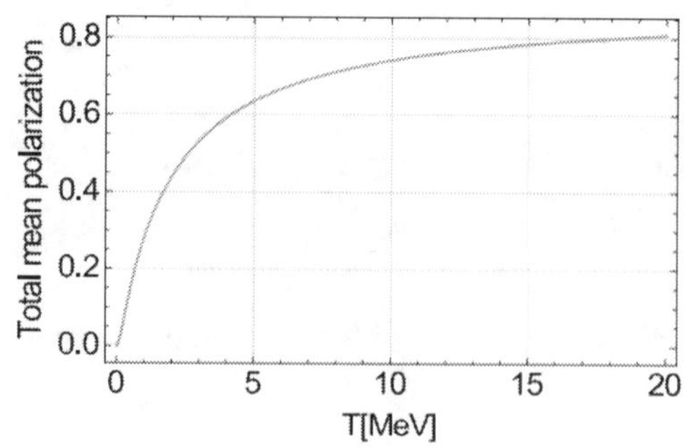

(b) mutual polarization vector lenght as a function of the beam kinetic energy, symmetric scattering

Figure 5: Mutual polarization vector length.

Mott scattering is an electron - nucleus scattering and Mott polarimetry is realized by scattering an electron beam off targets made of heavy elements. For a given scattering angle, Mott cross section depends on the beam polarization due to the spin-orbit coupling. It results in azimuthal asymmetry of scattered electrons. If one places two detectors (conventionally called left and right) symmetrically about the beam axis, the polarization of the beam can be determined. In the correlation experiment a double Mott polarimeter shall be used. The four probabilities P_{++}, P_{+-}, P_{-+} and P_{--}, constituing the correlation function, can be determined by measuring four other probabilities: P_{LL}, P_{LR}, P_{RL} and P_{RR}, where P_{IJ} are the probabilities that the first electron will be scattered towords I-th and the second one towards J-th detector. The latter probabilities can be found by counting corresponding coincidences (LL, LR etc.).

There are two main problems that we need to face in the context of Mott polarimetry. The first one is that the method is most sensitive in case of backscattering, which makes coincidences at a few MeV energies extremely rare. The other one is that it is sensitive only to spin projection on a direction perpendicular to the Mott scattering plane, and in that case the relativistic correction to the correlation function is rather small, as will be shown in the next section.

4 Correlation function in the case of scattering of a polarized beam on a target

In order to design an optimal experiment, the relativistic correlation function for a pair of electrons originatin from scattering of a polarized electron beam off both unpolarized [15] and polarized target has been calculated. In the case of an unpolarized target, the absolute value of the correlation function is small, yet measurable, and reaches its maximum in the case of the symmetric scattering (compare section 3.1). It does not depend on the beam polarization, on the contrary to the joint probabilities P_{++}, P_{+-}, P_{-+} and P_{--} that constitute it. The dependence of the correlation function and the probabilities on the beam kinetic energy for symmetric Møller scattering and vectors $\vec{a}$ and $\vec{b}$ in the Møller scattering plane can be seen in Figs 6 and 7. This configuration has been chosen for experimental reasons for the planned measurement and all the plots in the following sections will be presented in this configuration, unless stated otherwise.

DOI: 10.1051/epjconf/201716407031

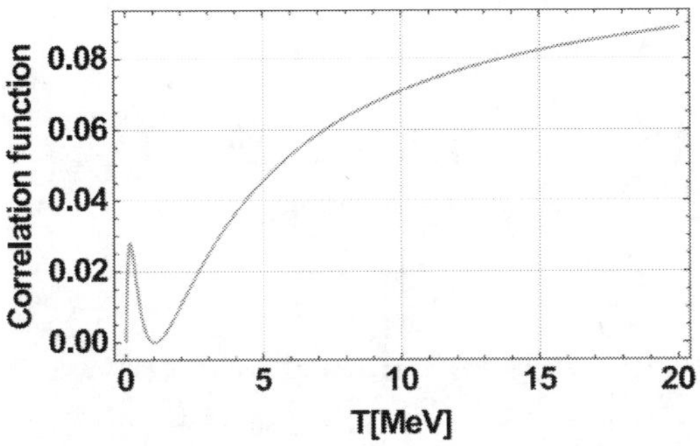

Figure 6: The correlation function vs kinetic energy of the beam, polarized electron beam, unpolarized target. Symmetric Møller scattering, $\vec{a}$ and $\vec{b}$ in the Møller scattering plane. The correlation function does not depend on the polarization.

The dependence of the probabilities on the beam polarization is the reason why we chose to use polarized electron beam in the experiment, despite the correlation function is not sensitive to it.

4.1 Polarized target

Although it is possible to observe the relativistic effects by measuring joint probabilities for a few MeV polarized electron beam, one could ask whether polarizing the target would not cause increase of the absolute value of the correlation function, letting perhaps for measuring also the violation of Bell-type inequalities. We analysed the dependence of the correlation function on the beam kinetic energy in the case of targets polarized in (i) 8% and (ii) 85%. The first degree of polarization could be arranged experimentally, the other one is rather unrealistic and we investigated it only for scientific reasons. The results are the following (i) the polarization of the target electron influences the behaviour of the correlation function only if spins of both scattering electrons are paralell to each other, (ii) in case of realisticly polarized target, the absolute value of the correlation function does not change much, so violating theBell-type inequalities is also impossible in this case. It can be seen in Fig 8. The same conclusions are valid for the probabilities, see Fig 9 If the target could be polarized to the same amount as the beam itself, the increase of the absolute value of the correlation function would be much more profaund.

5 Conclusions

Our aim is to design and perform a first experiment allowing to distinguish between relativistic and non-relativistic quantum correlations. It is not going to be a Bell-test experiment, rather a verification of the existing theory. We decided that the correlation function will be measured for a pair of electrons originating from Møller scattering of a polarized electron beam off a stationary target, which is in a well-determined and partilally entangled state. For the spin-projection measurement Mott polarimetry will be used.

We have shown that in the simple case of an unpolarized target the correlation function is small, yet measurable and although it does not depend on the beam polarization, the probabilities that constitute it, do. For that reason the experiment will be performed at Technische Universität Darmstadt, where up to 85% polarized beam can be provided. We have also shown that polarizing the target would not

DOI: 10.1051/epjconf/201716407031

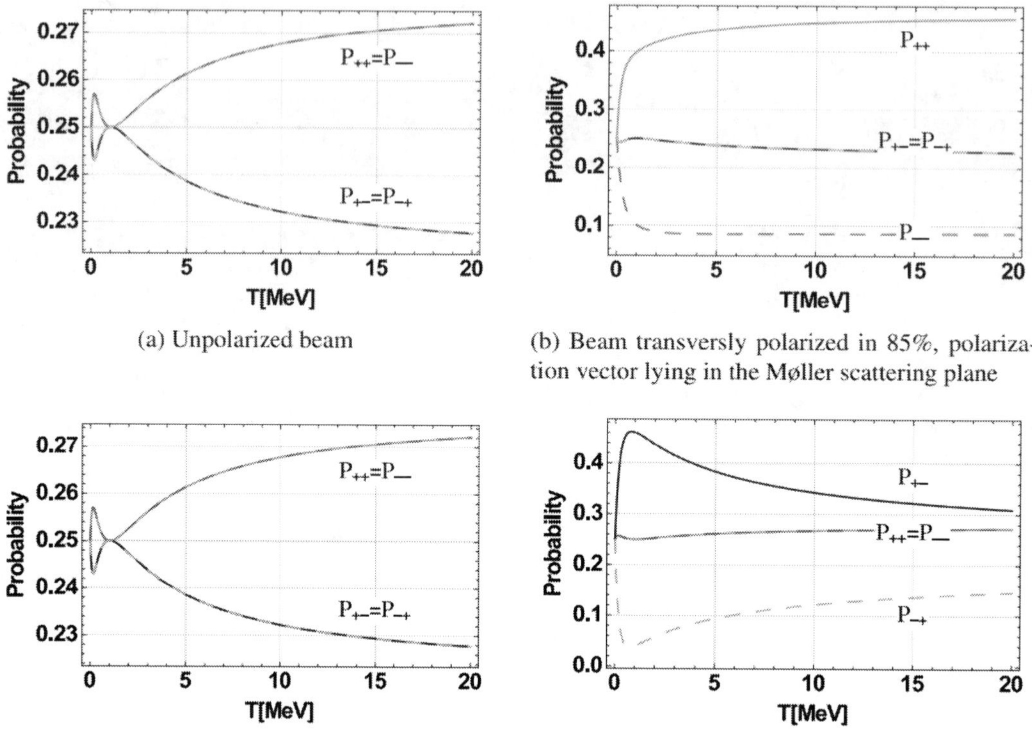

(a) Unpolarized beam

(b) Beam transversly polarized in 85%, polarization vector lying in the Møller scattering plane

(c) Beam transversly polarized in 85%, polarization vector perpendicular to the Møller scattering plane

(d) Beam longitudinally polarized in 85%

Figure 7: Probabilities as functions of kinetic energy for different beam polarizations. Note that the case of beam polarized in direction perpendicular to the Møller scattering plane is equivalent to the case of an unpolarized beam.

bring much improvement to the measurement, this is why the beryllium target we shall use as the Møller target, will be unpolarized.

6 Acknowledgments

This work was supported from the funds of the National Science Centre: (1) as a part of the research project, contract/decision DEC-2012/06/M/ST2/00430; (2) as a postdoc project (M.W.) contract/decision DEC-2013/08/S/ST2/00551; (3) as a part of the research project (P.C., J.R) contract/decision 2014/15/B/ST2/00117.

References

[1] A. Einstein and B. Podolsky and N. Rosen, Phys. Rev. **47**, 777-780 (1935)
[2] J. S. Bell, Physics **1**, 195-200 (1965)

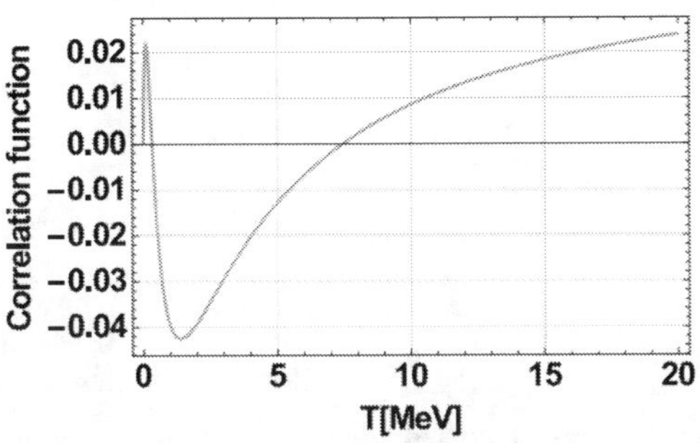

(a) Beam and target electrons transversly polarized in 85% and 8%, respectively, polarization vector lying in the Møller scattering plane.

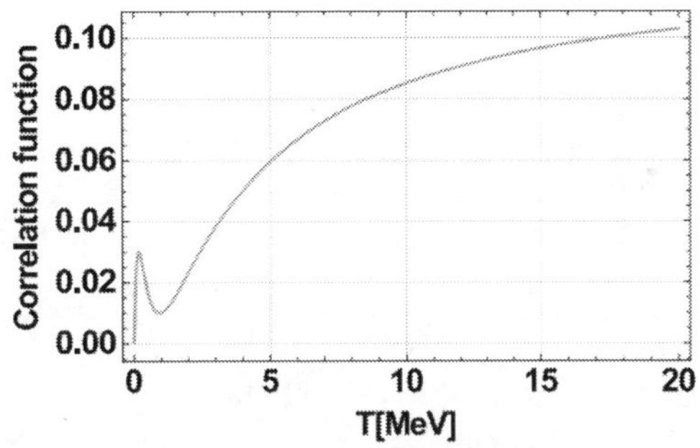

(b) Beam and target electrons longitudinally polarized in 85% and 8%, respectively.

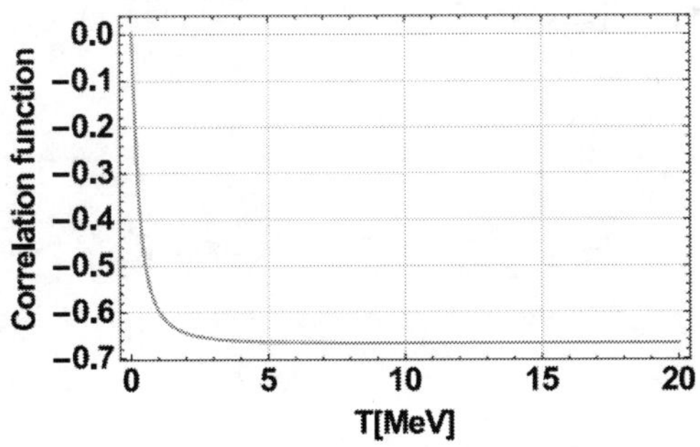

(c) Beam and target electrons transversly polarized in 85%, polarization vector lying in the Møller scattering plane.

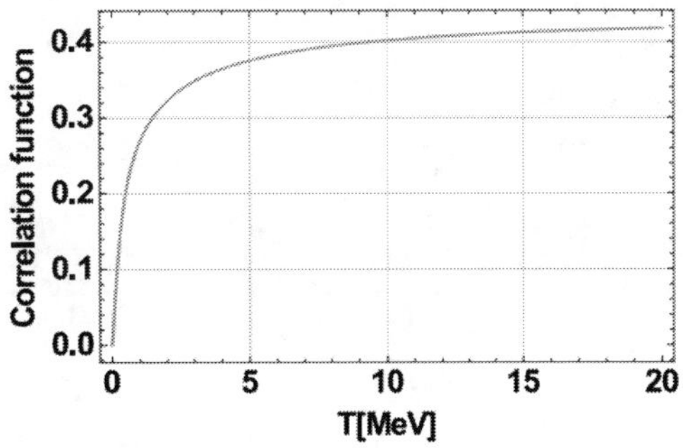

(d) Beam and target electrons longitudinally polarized in 85%

Figure 8: Correlation function in terms of kinetic energy for different beam and target polarizations.

[3] J. F. Clauser and M. A. Horne and A. Shimony and R. A. Holt, Phys. Rev. Lett. **23**, 880-884 (1969)

[4] D. Bohm, *Quantum Theory*, (Prentice-Hall,Englewood Cliffs ,1951)

[5] A. Aspect and P. Grangier and G. Roger, Phys. Rev. Lett. **47**, 460-463 (1981)

[6] A. Aspect and P. Grangier and G. Roger, Phys. Rev. Lett. **49**, 91-94 (1982)

[7] M. Czachor, Phys. Rev. A **55**, 72-77 (1997)

[8] P. Caban and J. Rembieliński, Phys. Rev. A **72**,012103 (2005)

[9] P. Caban and J. Rembieliński, Phys. Rev. A **74**, 042103 (2006)

[10] P. Caban and J. Rembieliński and M. Włodarczyk, Phys. Rev. A **77**, 012103 (2008)

[11] P. Caban and J. Rembieliński and M. Włodarczyk, Phys. Rev. A **79**, 014102 (2009)

[12] M. Lamehi-Rachti and W. Mittig, Phys. Rev. D **14**, 2543 (1976)

[13] S. Hamieh et. al., J. Phys. G: Nucl. Part. Phys. **30**, 481 (2004)

[14] H. Sakai et. al., Phys. Rev. Lett. **97**, 150405 (2006)

[15] P. Caban and J. Rembieliński and M. Włodarczyk, Phys. Rev. A **88**, 032116 (2013)

[16] P. Caban and J. Rembieliński and M. Włodarczyk, OSID **19**, 1250027 (2012)

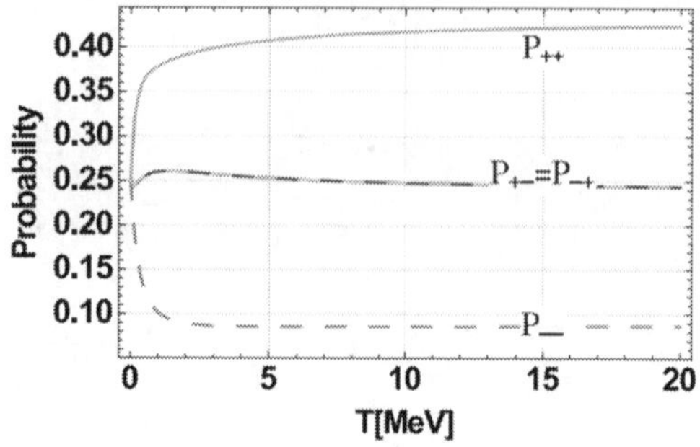

(a) Target electrons transversly polarized in 8%, polarization vector lying in the Møller scattering plane

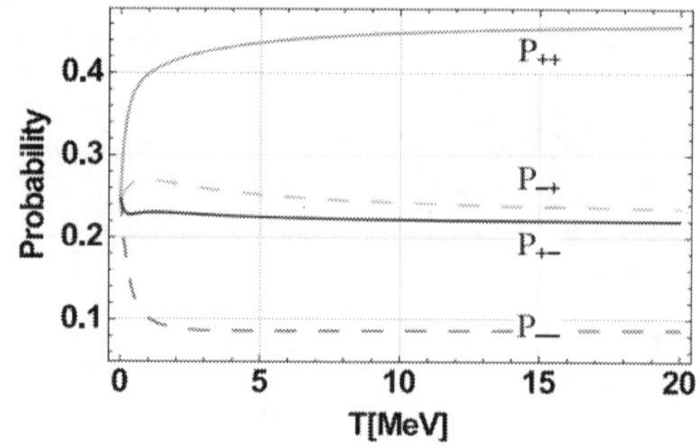

(b) Target electrons longitudinally polarized in 8%

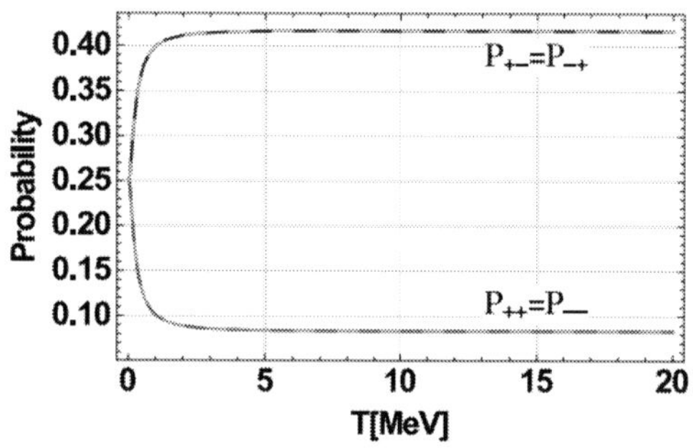

(c) Target electrons transversly polarized in 85%, polarization vector lying in the Møller scattering plane

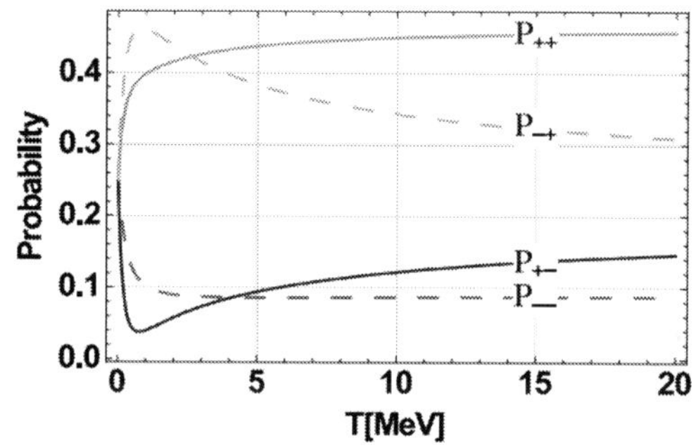

(d) Target electrons longitudinally polarized in 85%

Figure 9: Probabilities as functions of the beam kinetic energy for beam transversly polarized in 85%, polarization vector lying in the Møller scattering plane and for different target polarizations

[17] P. Caban and J. Rembieliński and M. Włodarczyk, Phys. Rev. A **88**, 022199 (2013)
[18] T. J. Gay et al., Rev. Sci. Instr. B **279**, 114-130 (1992)

EPJ Web of Conferences **164**, 07032 (2017) DOI: 10.1051/epjconf/201716407032
ICNFP 2016

Recent NA61/SHINE measurements performed for the T2K experiment

Magdalena Posiadała-Zezula for the NA61/SHINE Collaboration[1,a]

[1] *University of Warsaw, Faculty of Physics, Pasteura 5, 02-093 Warsaw, Poland*

Abstract. The neutrino programme of the NA61/ SHINE experiment at the CERN SPS is aiming to deliver precise hadron production measurements for improving calculations of the initial neutrino beam flux in the long-baseline neutrino oscillation experiments. The first receiver of such measurements is the T2K neutrino oscillation project in Japan. New results on $\pi^{\pm}$, $K^{\pm}$, p, K_S^0 and Λ production from the NA61/SHINE 2009 thin target data analyses with smaller statistical and systematic errors are discussed. They enable us to reduce further the flux uncertainties in T2K for neutrino and antineutrino beam mode. We also report on the first corrected $\pi^{\pm}$ results obtained for T2K replica target (a 90 cm long cylinder of 2.6 cm diameter, about $1.9\lambda_I$). Up to 90% of the neutrino flux can be constrained by such measurements as compared to 60% for the thin target measurements that are sensitive only to primary hadron interactions.

1 Introduction

The NA61/SHINE (SHINE = SPS Heavy Ion and Neutrino Experiment) combines a rich physics program in three different fields: auxiliary measurements for neutrino experiments, cosmic-ray simulations, and the behaviour of strongly interacting matter at high density. In its first stage of data taking (years 2007, 2009 and 2010) NA61/SHINE performed measurements of p+C interactions at 31 GeV/c needed for neutrino experiment Tokai-2-Kamioka (T2K).

2 The T2K experiment

The T2K is a long baseline neutrino oscillation experiment which uses an intense proton beam produced by the J-PARC (Japan Proton Accelerator Research Complex) in Tokai, Japan. It is composed of neutrino beamline, near detector complex, both of which were newly constructed at J-PARC and far detector Super-Kamiokande , located 295 km away in Mozumi mine. Comprehensive review of the instrumentation aspects of the T2K experiment may be found in Ref. [1]. The T2K can measure neutrino oscillation parameters in two channels. The first is muon neutrino disappearance, which provides measurements of the parameters: θ_{23} and the largest mass neutrino splitting. The second is electron neutrino appearance, which provides measurements of the parameters: θ_{13} and, also, the CP-violating phase δ_{CP}.

[a] e-mail: Magdalena.Posiadala@fuw.edu.pl

2.1 The T2K beam production

The J-PARC neutrino beamline has two complementary subsystems. These are: the primary and secondary beamlines. The main aim of the primary beamline is to tune proton beam coming from J-PARC accelerator, bent it towards the direction of Kamioka, and focus the beam onto the target. In the secondary beamline we can distinguish three sections: the target station, decay volume and beam dump. The target station is equipped with three magnetic horns and the graphite target, 90 cm in length, which is inserted inside the first one. When proton beam is impinging on the graphite target many hadrons are produced, amongst of which charged pions and kaons are crucial as neutrino parents. At the beginning of the T2K data taking horns were working at 250 kA current pulse in order to focus positively charged hadrons to produce a muon neutrino enhanced beam ν_μ. An anti-neutrino enhanced beam $\bar{\nu}_\mu$ is produced by T2K starting from 2013. Experimentally it is achieved by reversing the current direction in the focusing horns in order to select negatively charged particles. Precision of the T2K beam simulation relies heavily on the knowledge of the hadron production of neutrino parents produced in p+C interactions at 31 GeV/c. This is the reason why neutrino programme of the NA61/SHINE experiment is essential for T2K.

3 The NA61/SHINE experiment

The NA61/SHINE apparatus is a wide-acceptance hadron spectrometer at the CERN SPS. The detector is built around five Time Projection Chambers (TPCs). Two Vertex TPCs are placed in the magnetic field produced by two superconducting dipole magnets and two Main-TPCs are located downstream symmetrically with respect to the beamline. An additional small TPC is placed between Vertex TPCs, covering the very forward region, and is referred to as the GAP TPC (GTPC). The GTPC is used for tracking in the analysis described here allowing to extend the kinematic coverage compared to our previously published results. Since 2007 pilot run the experiment has been updated with a new forward time-of-flight detector (ToF-F) in order to extend the acceptance for pion and kaon identification as required for T2K measurements [2, 3]. The NA61/SHINE experimental facility is described in detail in Ref. [5]

Data needed for the T2K experiment were collected using two graphite targets:

- the 2 cm long (about 0.04 of interaction length λ_I) so called thin target,
- the T2K replica target (a 90 cm long cylinder of 2.6 cm diameter, about $1.9\lambda_I$).

A summary of NA61/SHINE data collected for T2K is presented in Table 1. Results of $\pi^\pm$ and K^+ yields from proton-carbon interactions at 31 GeV/c taken in 2007 pilot run with a thin graphite target were already used by the T2K experiment to tune neutrino beam simulations and reduce uncertainties([6–11]). In this article we want to emphasise recently published results from the thin target data set obtained from 2009 run [12]. Moreover, T2K replica target results, from 2009 dataset, which were also recently accepted for publication [14] are discussed.

3.1 The NA61/SHINE 2009 thin target results

New thin target results from the 2009 data set on $\pi^\pm$, $K^\pm$, p, K^0_S and Λ were obtained in bins of momentum p and production angle θ. In Fig. 1 we present the kinematic region of interest of neutrino parents for the T2K. Four panels represent the momentum p and θ distributions of $\pi^\pm$ and $K^\pm$ whose daughter neutrinos pass through the Super-Kamiokande detector. These studies apply to neutrino beam mode configuration in T2K. On all figures the NA61/SHINE coverage from 2007 pilot run (dashed line) and

target	year	triggers $\times 10^6$	status of the NA61/SHINE analysis	usage in the T2K beam MC
thin target	2007	0.7	published: $\pi^\pm$ [2], K^+ [3], K^0_S, Λ [4]	has already been used [11]
2 cm ($0.04\lambda_I$)	**2009**	5.4	$\pi^\pm$, $K^\pm$, p, K^0_S, Λ [12]	currently being used
T2K replica	2007	0.2	published: $\pi^\pm$ [13]	method developed
target	2009	2.8	$\pi^\pm$ analysis finalised [14]	to be integrated
90 cm ($1.9\lambda_I$)	2010	7.2	analysis currently on-going	—

Table 1: A summary of NA61/SHINE data collected for T2K physics goals.

2009 run (solid line) is also included. One may conclude that the improved statistics of the 2009 data (by a factor of 8 larger with respect to 2007 data set) allows for the first measurements of negatively charged kaons within NA61/SHINE. Overall precision of the measurements is also improved. Moreover, the acceptance of the pions is also extended in comparison to 2007 data set. Similar results for T2K anti-neutrino beam mode may be found in [12].

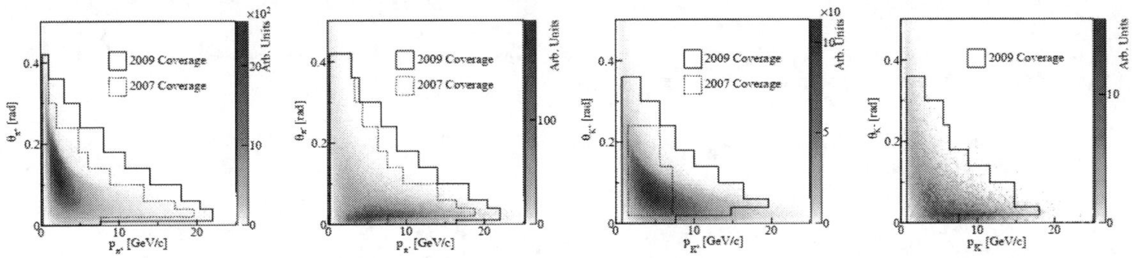

(a) π^+ (b) π^- (c) K^+ (d) K^-

Figure 1: (Colour online) The $\{p, \theta\}$ phase space of $\pi^\pm$, $K^\pm$, contributing to the predicted neutrino flux at Super-Kamiokande in the neutrino beam mode, and the regions covered by the previously published NA61/SHINE measurements [2, 3] and by the new 2009 results [12].

Particle identification in NA61/SHINE relies on measurements of the specific energy loss (dE/dx) in the TPCs and the time-of-flight measurements (tof) using the ToF-F detector. Details on spectra analysis techniques, particle identification methods as well as correction factors and uncertainties are described in great detail in Ref. [12]. Below in Figs. 2 and 3, as an example, we present the momentum distributions of π^+ (K^+) and π^- (K^-) identified in p+C interactions at 31 GeV/c for several polar angle intervals (θ).

Impact of the NA61/SHINE thin target measurements on the T2K flux predictions

Details of the T2K beam simulation and the way of the tunning of flux with NA61/SHINE results may be found in [11]. Below, we present the impact of the thin target NA61/SHINE measurements on the total T2K flux uncertainties, which are shown in Fig. 4 for ν_μ (left) and $\bar{\nu}_\mu$ (right) enhanced beam mode. All the sources which have contributed to the flux predictions are also shown. Clearly one can see that for ν_μ beam we can reduce the total error on the flux, in the region of energy peak around 0.6 GeV, from 11% (obtained with 2007 NA61/SHINE data - dashed black line) below 10% (obtained with

DOI: 10.1051/epjconf/201716407032

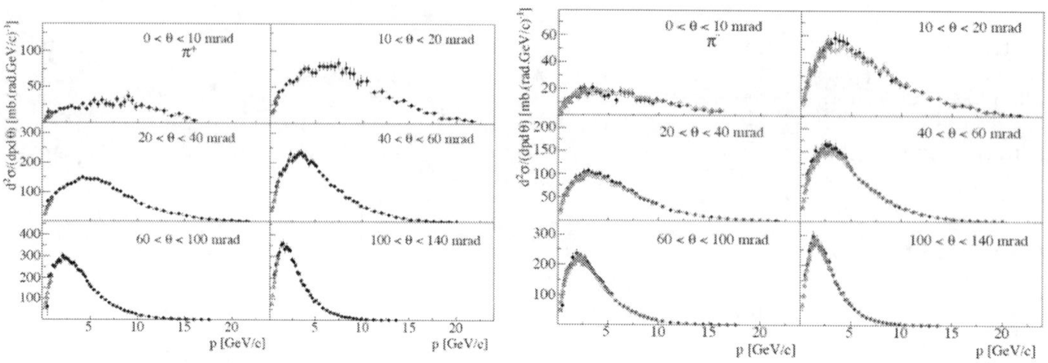

(a) π^+ (b) π^-

Figure 2: (Colour online) Laboratory momentum distributions of π^+ and π^- produced in p+C interactions at 31 GeV/c production processes in selected polar angle intervals (θ). Results obtained with two (left) and three (right) different analysis techniques are presented: open blue triangles are the *dE/dx* analysis, open red circles are the h^- analysis and full black circles are the *tof-dE/dx* analysis [12].

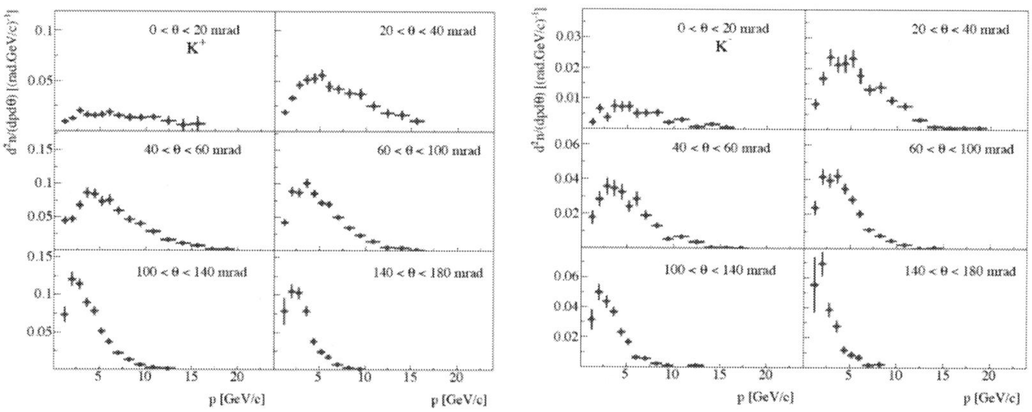

(a) K^+ (b) K^-

Figure 3: (Colour online) Laboratory momentum distributions of K^+ and K^- produced in p+C interactions at 31 GeV/c production processes in selected polar angle intervals (θ). Results obtained with *tof-dE/dx* analysis [12].

2009 NA61/SHINE data - solid back line). In addition the 2009 NA61/SHINE data set was for the first time used to tune $\bar{\nu}_\mu$ enhanced beam, and current status presents again below 10% total uncertainty on the flux at the energy peak. However, hadron interactions uncertainties, marked with the red lines, are still the most dominant sources in the flux predictions for both neutrino and anti-neutrino modes. We are limited with reduction of the flux uncertainties even with the dedicated measurements done on the thin target, because these measurements do not account for the re-interactions in the 90 cm long T2K target. The flux uncertainties can only be further improved using the NA61/SHINE measurements with the T2K replica target.

DOI: 10.1051/epjconf/201716407032

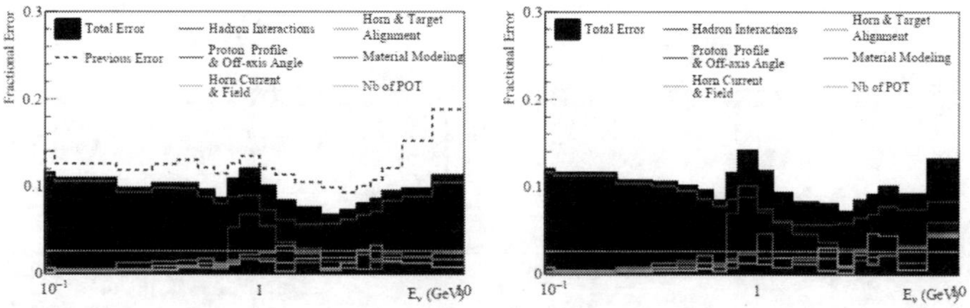

Figure 4: (Colour online) The total T2K flux uncertainties are shown for ν_μ (left) and $\bar{\nu}_\mu$ (right) enhanced beam modes together with all the sources which contribute to the T2K flux predictions.

3.2 The NA61/SHINE 2009 replica target results

In addition to the published measurements on a thin target the NA61/ SHINE collaboration also performed studies of hadron emission in interactions of a 31 GeV/c proton beam with a full-size replica of the T2K target using data taken in 2007 (see Tab.1). Uncorrected differential yields of π^+ at the surface of the replica target and their ratios with respect to the predictions of the model used to simulate hadronic interactions in the T2K target were published [13]. Moreover, there are available new measurements of fully-corrected differential yields of $\pi^\pm$ mesons from the surface of the T2K replica target [14]. These results were obtained using data collected during a high-statistics run performed also in 2009 (see Tab.1). Compared to the previous NA61/SHINE thin-target analyses discussed above, the 90 cm long T2K replica target was divided into 6 longitudinal sections, which contribute differently to the final neutrino flux (see Fig. 5). Hence, the analysis of the T2K replica target is conducted in $\{p, \theta, z\}$ bins. The longitudinal z binning was determined by a study performed together with the T2K beam group. It was found that five longitudinal bins and one additional downstream bin are sufficient to obtain required neutrino flux prediction.

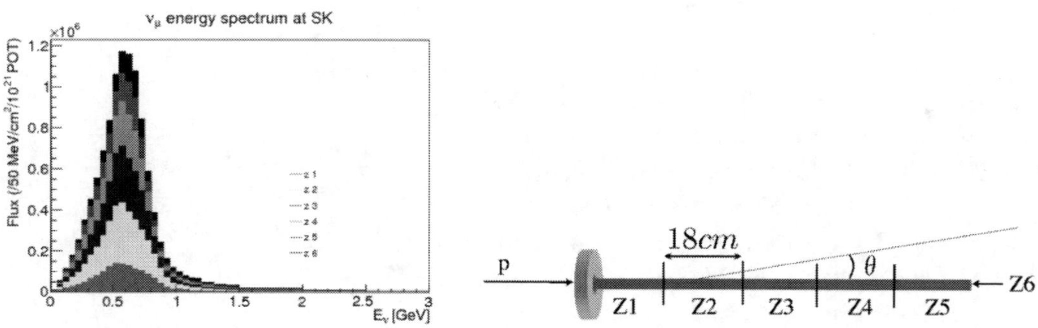

Figure 5: (Colour online) Right figure: histograms showing the contribution of each of the 6 longitudinal target bins to the muon neutrino flux at Super-Kamiokande . Left figure: a sketch of the longitudinal binning of the T2K replica target [14].

In Fig. 6 we present spectra of π^+ at the surface of the T2K replica target, for selected polar angle intervals and for the six longitudinal bins z. All results which were obtained on T2K replica

DOI: 10.1051/epjconf/201716407032

target from 2009 data set with details of the analysis, systematic error studies and more may be found in Ref. [14]. The work to use these measurements for the T2K (anti-)neutrino flux predictions in on-going.

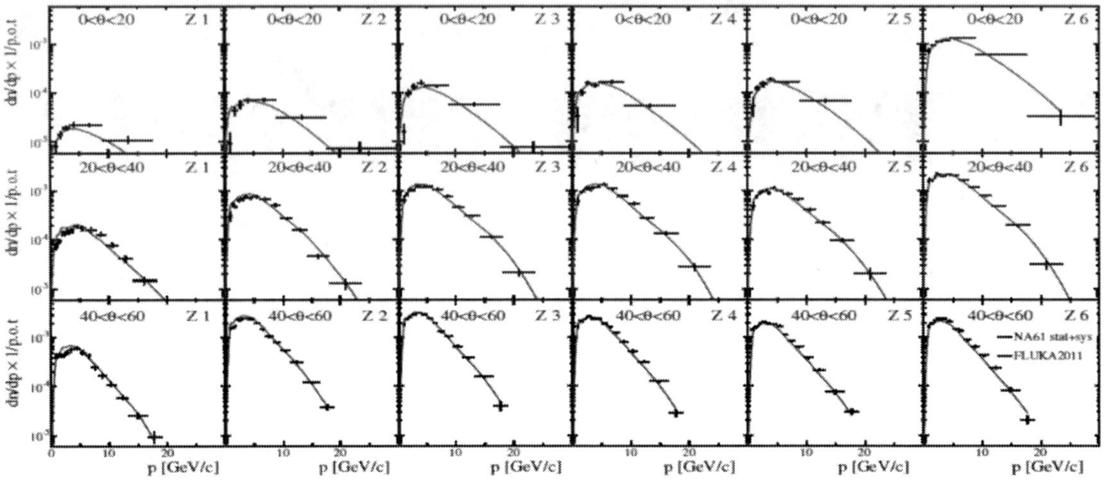

Figure 6: (Colour online) Spectra of positively charged pions at the surface of the T2K replica target, in the polar angle range from 0 to 60 mrad, and for the six longitudinal bins z as a function of momentum. The FLUKA2011 prediction is overlaid [14].

4 Acknowledgments

Author wishes to acknowledge financial support from the National Science Centre (NCN) of Poland grant number 2014/14/M/ST2/00850.

References

[1] K. Abe *et al.*, **Nucl. Instrum. Meth.** A659 (2011), 106-135
[2] N. Abgrall *et al.*, **Phys. Rev. C84** (2011) 034604
[3] N. Abgrall *et al.*, **Phys. Rev. C85** (2012) 035210
[4] N. Abgrall *et al.*, **Phys. Rev. C89** (2014), 025205
[5] N. Abgrall *et al.*, **JINST** 9 (2014) P06005
[6] K. Abe *et al.*, **Phys. Rev. Lett** 112 (2014), 061802
[7] K. Abe *et al.*, **Phys. Rev. D88** (2013), no. 3, 032002
[8] K. Abe *et al.*, **Phys. Rev. Lett** 107 (2011), 041801
[9] K. Abe *et al.*, **Phys. Rev. Lett** 112 (2014) no. 18, 181801
[10] K. Abe *et al.*, **Phys. Rev. Lett** 111 (2013) no. 21, 211803
[11] K. Abe *et al.*, **Phys. Rev. D87** 112 (2013) no. 1, 012001
[12] N. Abgrall *et al.*, **Eur. Phys. J.** C76 (2016) no.2, 84
[13] N. Abgrall *et al.*, **Nucl. Instrum. Meth.** A701 (2013) 99-114
[14] N. Abgrall *et al.*, **arXiv:1603.06774** to appear in **Eur. Phys. J. C** (2016)

EPJ Web of Conferences **164**, 07033 (2017) DOI: 10.1051/epjconf/201716407033
ICNFP 2016

Recent results from the strong interactions program of NA61/SHINE

Szymon Pulawski[1,a]

[1] *Institute of Physics, University of Silesia, Katowice, Poland*

Abstract. The NA61/SHINE experiment studies hadron production in hadron+hadron, hadron+nucleus and nucleus+nucleus collisions. The strong interactions program has two main purposes: study the properties of the onset of deconfinement and search for the signatures of the critical point of strongly interacting matter. This aim is pursued by performing a two-dimensional scan of the phase diagram by varying the energy/momentum (13A-158A GeV/c) and the system size (p+p, Be+Be, Ar+Sc, Xe+La) of the collisions. This publication reviews recent results from p+p, Be+Be and Ar+Sc interactions. Measured particle spectra are discussed and compared to NA49 results from Pb+Pb collisions. The results illustrate the progress towards scanning the phase diagram of strongly interacting matter.

1 The NA61/SHINE facility

The layout of the NA61/SHINE detector [1] is presented in Fig. 1. It consists of a large acceptance hadron spectrometer with excellent capabilities in charged particle momentum measurements and identification by a set of five Time Projection Chambers as well as Time-of-Flight detectors. The high resolution forward calorimeter, the Projectile Spectator Detector, measures energy flow around the beam direction, which in nucleus-nucleus reactions is primarily a measure of the number of spectator (non-interacted) nucleons and is thus related to the centrality of the collision. An array of beam detectors identifies beam particles and measures precisely their trajectories. Primary and secondary hadron as well as ion beams are used by the experiment.

2 Particle ratios in inelastic p+p collisions and inverse slope parameter T

The excellent particle identification based on the Time-off-Flight (ToF) and energy loss (dE/dx) measurements allows to calculate the ratio of yields K^+/π^+. The energy dependence of the K^+/π^+ ratio at midrapidity for inelastic p+p interactions and central Pb+Pb/Au+Au collisions is presented in Fig. 2 (left). The NA61/SHINE data suggest that even in inelastic p+p interactions the energy dependence of the K^+/π^+ ratio exhibits rapid changes in the SPS energy range. However, the horn structure [2] observed in central Pb+Pb collisions is reduced to a step. Data obtained beyond the SPS energy range, namely at RHIC and LHC [3–7], continue the trend seen at the SPS.

[a]e-mail: s.pulawski@cern.ch On behalf of the NA61/Shine Collaboration

 DOI: 10.1051/epjconf/201716407033

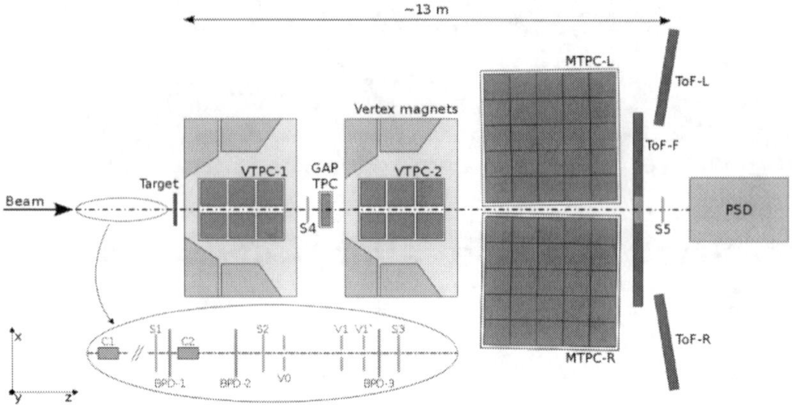

Figure 1. Schematic view of the NA61/SHINE detector system.

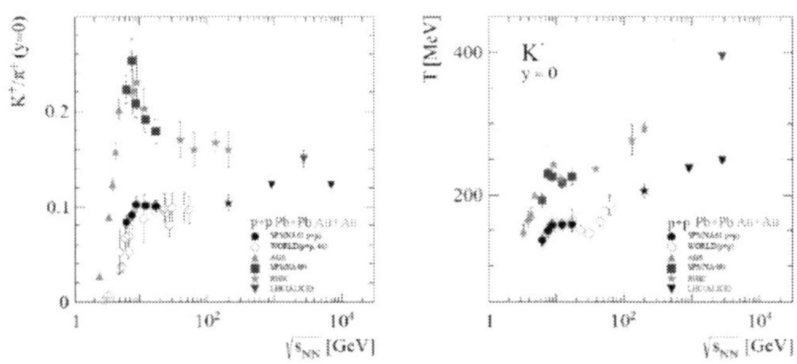

Figure 2. Energy dependence of the K^+/π^+ ratio (left) and of the inverse slope parameter T of K^- transverse mass spectra (right) in p+p interactions compared to Pb+Pb and Au+Au collisions.

The inverse slope parameters T for π, K and p of the spectra of transverse mass $m_T = \sqrt{p_T^2 + m^2}$ were fitted in inelastic p+p reactions. Fig. 2 (right) presents the energy dependence of T for K⁻ in p+p collision. The NA61/SHINE results from inelastic p+p collisions exhibit rapid changes like those observed in central Pb+Pb interactions. World data for p+p and Pb+Pb/Au+Au reactions are plotted for a comparison and were taken from Refs. [6, 8–10].

The dependence of the mean pion multiplicity divided by the mean number of wounded nucleons (participants) as function of collision energy, the so called kink plot, is shown for inelastic p+p, Ar+Sc and A+A interactions in Fig. 3. The pion multiplicity $\langle \pi \rangle$ in the SPS energy range increases faster in central Pb+Pb than in p+p collisions. The two dependencies cross at about 40*A* GeV/c. For high SPS energies results from Ar+Sc reactions follow the Pb+Pb trend while for low SPS energies Ar+Sc follows the p+p tendency. The situation is opposite for Be+Be collisions.

Spectra of transverse mass $m_T - m_0$ of π^- mesons in Ar+Sc are shown and compared to measurements for p+p, Be+Be, and Pb+Pb collisions in Fig. 4. Spectra in p+p reactions are exponential. A concave shape is observed in Pb+Pb, Ar+Sc and Be+Be collisions. The shape difference between p+p and A+A reactions is attributed to transverse collective flow in A+A.

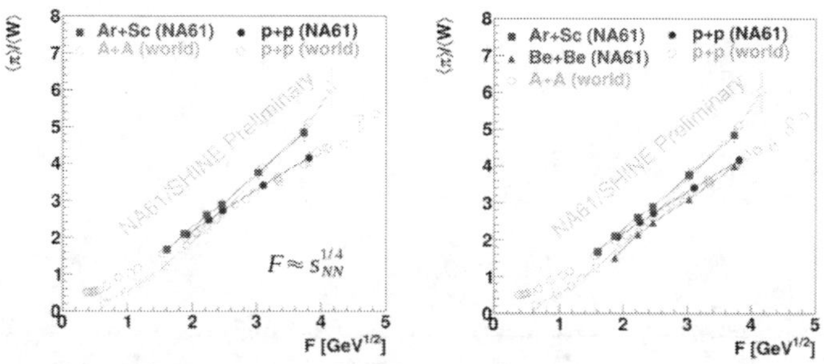

Figure 3. Mean pion multiplicity divided by the mean number of wounded nucleons as function of collision energy, the so called kink plot, for inelastic p+p, Ar+Sc and A+A interactions. The Fermi variable $F \approx s_{NN}^{0.25}$ was used as energy measure.

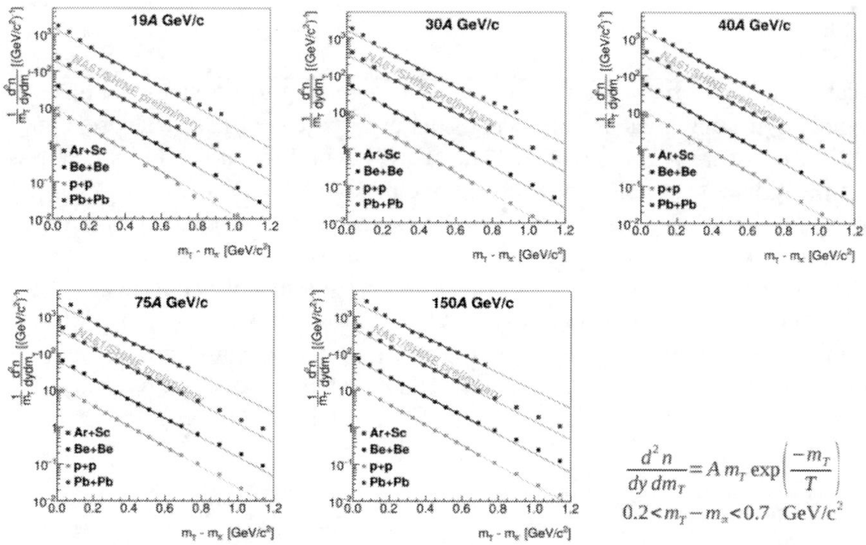

Figure 4. Spectra of transverse mass $m_T - m_0$ of π^- mesons in Ar+Sc compared to p+p, Be+Be, and Pb+Pb collisions.

3 Λ spectra in p+p interactions

The transverse momentum distributions of Λ hyperons shown in Fig. 5 (left) are for inelastic p+p collisions at a beam momentum of 40 GeV/c. The yields are corrected for acceptance and for losses due to the topological and track selection cuts. Integration and extrapolation of the transverse momentum distributions provide the data points for the distributions of rapidity dn/dy and the Feynman variable dn/dx_F. The Feynman variable distribution of Λ hyperons produced in inelastic p+p reactions at 40 GeV/c is shown in Fig.5 (right).

The Λ cross-section produced in inelastic p+p interactions at 40 GeV is presented in Fig. 5 (left). Results are shown as function of transverse momentum in rapidity intervals. Fig. 5 (right) depicts the dependence of the Λ production cross-section as function of x_F together with the extrapolation line for determining the total yield.

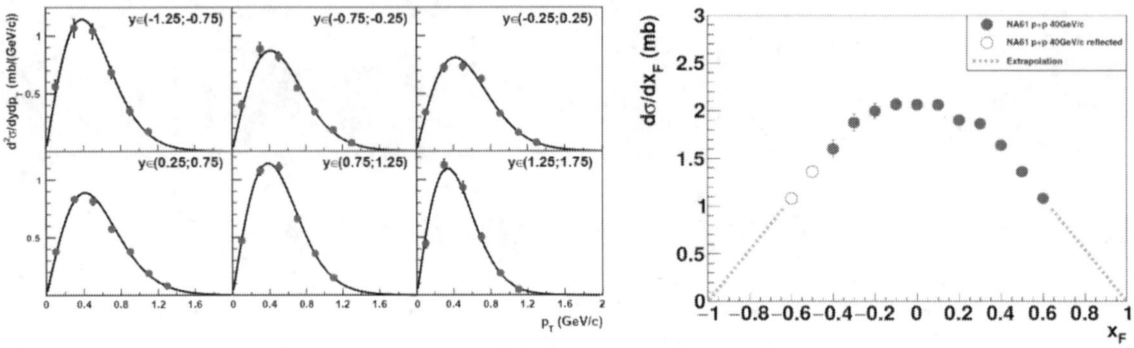

Figure 5. Λ production cross section in p+p interactions at 40 GeV/c beam momentum. Left: p_T distributions in rapidity intervals. The lines are the result of fits to the experimental points. Right: p_T integrated cross-section as function of x_F. The dotted line was used to extrapolate to the total Λ yield.

Extrapolation of the x_F distributions into the unmeasured region provides estimates of the Λ yields. The corresponding multiplicities (average yields per event) are compared to world data in Fig. 6 (left) as function of center-of-mass energy $\sqrt{s_{NN}}$. The NA61 data follow the trend set by the world data and at the same time reduce the uncertainties of the yields significantly at the two energies analysed by NA61. Above $\sqrt{s_{NN}}$ = 19 GeV the experimental data have large errors and lie consistently below the EPOS model calculation. It is desirable to obtain new experimental data in this energy range. Below 10 GeV EPOS seems to overpredict Λ yields. However, this model is not meant to provide reliable results at this energy and below. The ratio $\langle\Lambda\rangle$ / $\langle\pi\rangle$ is shown in Fig. 6 (right) as function of $\sqrt{s_{NN}}$. The presented NA61/SHINE results agree with the world data. Measurements in A+A collisions at the AGS and by NA49 show a different behaviour than that observed in p+p reactions.

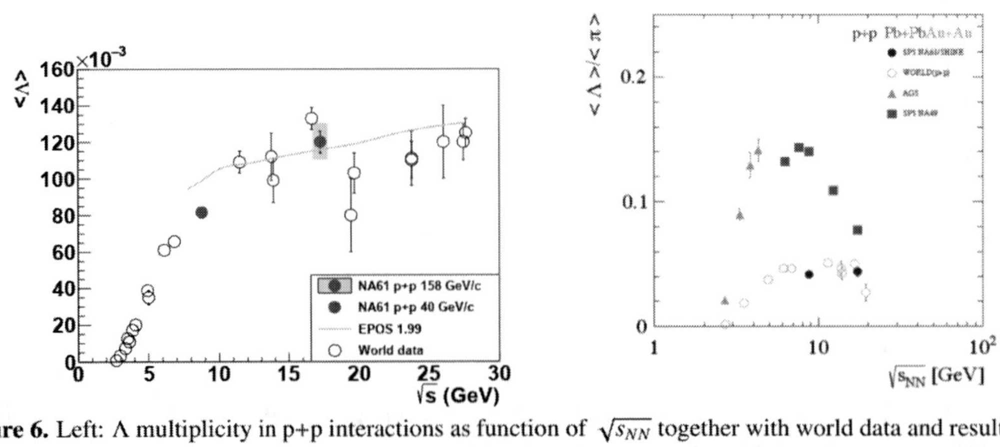

Figure 6. Left: Λ multiplicity in p+p interactions as function of $\sqrt{s_{NN}}$ together with world data and results from calculations with the EPOS model. Right: mean multiplicity of Λ hyperons divided by the mean π multiplicity.

Acknowledgments

This work was supported by the Polish Ministry of Science and Higher Education (grants 667/N-CERN/2010/0, NN 202 48 4339 and NN 202 23 1837), the Polish National Center for Science (grants 2013/11/N/ST2/03879, 2014/12/T/ST2/00692, and 2015/18/M/ST2/00125),

References

[1] N. Abgrall *et al.*, *JINST 9*, P06005, (2014)
[2] C. Alt *et al.*, *Phys. Rev. C 77*, 024903, 2008
[3] M. Gazdzicki and D. Roehrich, *Z. Phys. C 65*, 215, 1995
[4] M. Gazdzicki and D. Roehrich, *Z. Phys. C 71*, 55???64, 1996
[5] I. Arsene *et al.*, *Phys. Rev. C 72*, 014908, 2005
[6] K. Aamodt *et al.*, *Eur. Phys. J. C 71*, 1655, 2011
[7] B. Abelev *et al.*, *Phys. Rev. Lett. 109*, 252301, 2012
[8] M. Kliemant, B. Lungwitz and M. Gazdzicki, *Phys. Rev. C 69*, 044903, 2004
[9] B.I. Abelev *et al.*, *Phys. Rev. C 79*, 034909, 2009
[10] B.B. Abelev *et al.*, *Phys. Lett. B 736*, 196, 2014

EPJ Web of Conferences **164**, 07034 (2017) DOI: 10.1051/epjconf/201716407034
ICNFP 2016

Recent results on doubly-polarised pion photoproduction and the GDH sum rule on the nucleon at MAMI

Susanna Costanza[1,2,a] for the A2 Collaboration

[1] *Dipartimento di Fisica, Università degli Studi di Pavia, I-27100 Pavia, Italy*
[2] *INFN, Sezione di Pavia, I-27100 Pavia, Italy*

Abstract. New measurements of the helicity dependence of the total inclusive photoabsorption cross section and of the partial cross sections for several reaction channels on the proton and on the neutron were carried out at MAMI (Mainz) in the energy region $200 < E_\gamma < 1500$ MeV. The experiments were performed at the tagged photon beam facility of the MAMI accelerator in Mainz, using circularly and longitudinally polarised photons on longitudinally polarised proton, deuteron and ^{3}He targets.
These new doubly-polarised pion-photoproduction data sets give new input to the partial wave analyses and allow to constrain the multipole solution of the different analyses; they also allow to study the double-polarisation observables G and E. Furthermore, the results obtained give information on the well-known Gerasimov-Drell-Hearn sum rule.

1 Introduction

In the past 50 years, the structure of the nucleus has been a central issue of nuclear and particle physics and the subject of intense research, both from the theoretical and experimental point of view. Of particular interest is the study of the excited states of the nucleon: as the excitation spectra of atoms reflect the properties of the electromagnetic interaction, the excited states of the nucleon are related to the fundamental properties of the strong interaction.
Since Quantum Chromodynamics (QCD) is non-perturbative at this energy scale, we have to rely on phenomenological quark models to interpret the experimental results. This approach has limited success, if one thinks about the "missing resonances" issue: at present, the experimental findings are quite in agreement with what models predict in the low mass region of the spectrum, but there is still a pletora of higher lying states predicted by theories and not yet observed.
From the experimental point of view, the reason for this mismatch can be due to data analysis relying mostly on pion scattering, that will therefore miss states that couple only weakly to πN: it may be that the higher lying states couple preferentially to decay channels involving heavier mesons or decay via intermediate excited states [1].
For a better understanding of the nucleon excitation spectra, the alternative is to use electromagnetic-induced reactions, like meson photoproduction: they allow to explore multiple meson production states, like $\pi\pi$, $\pi\eta$, ... and access resonances that decay mainly via intermediate excited states (Fig. 1). In particular, photoproduction reactions of single pseudoscalar mesons allow to measure a set of 16

[a]e-mail: susanna.costanza@pv.infn.it

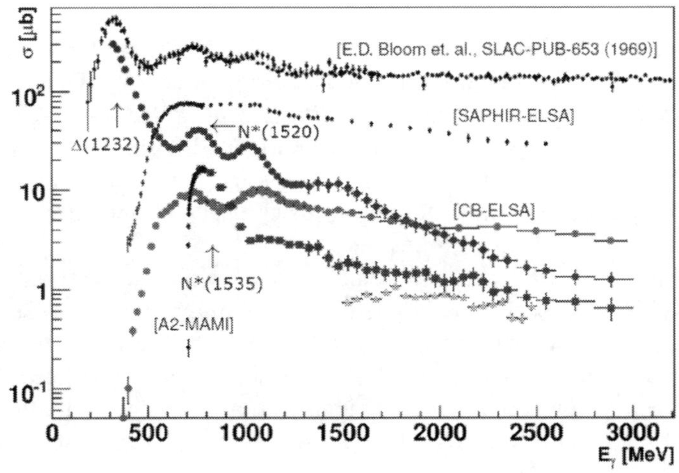

Figure 1. Cross section distributions for different reaction channels, showing the contributions of the different resonances: $\gamma p \to X$ and $\gamma p \to p\pi^-\pi^+$ (black), $\gamma p \to p\pi^0\pi^0$ (red), $\gamma p \to p\pi^0$ (blue circles), $\gamma p \to p\eta$ (blue squares), $\gamma p \to p\eta'$ (green triangles).

polarisation observables, for every fixed value of energy and angle (W, θ). In addition to the unpolarised cross section σ, there are 3 single polarisation observables and 12 double polarisation observables, grouped in beam-target, beam-recoil nucleon and target-recoil polarisation classes (Table 1). The measurement of 7 (8) properly chosen observables is necessary to yield a model-independent complete analysis [2, 3].

Table 1. Polarisation observables.

Photon polarisation		Target polarisation			Recoil nucleon polarisation			Target and recoil polarisation			
		X	Y	$Z_{(beam)}$	X′	Y′	Z′	X′ X	X′ Z	Z′ X	Z′ Z
unpolarised	σ	-	T	-	-	P	-	$T_{X'}$	$L_{X'}$	$T_{Z'}$	$L_{Z'}$
linear	-Σ	H	(-P)	-G	$O_{X'}$	(-T)	$O_{Z'}$	($-L_{Z'}$)	($T_{Z'}$)	($L_{X'}$)	($-T_{X'}$)
circular	-	F	-	-E	$C_{X'}$	-	$C_{Z'}$	-	-	-	-

2 Double polarisation observables G and E

As described in Table 1, the polarisation observables G and E require the polarisation of both the photon beam and the target. In the following, the results obtained by the A2 Collaboration will be shown.

2.1 The experimental setup

The first experiment for the measurement of the double-polarisation observables E and G simultaneously using a longitudinally polarised electron beam in combination with a diamond radiation, thus resulting in an elliptically polarised photon beam, has been performed by the A2 Collaboration at the tagged photon facility of MAMI in Mainz (Germany).
The photon beam was produced via bremsstrahlung of an electron beam of 1557 MeV, provided by the

DOI: 10.1051/epjconf/201716407034

MAMI accelerator and polarised up to ~75-78%. The photons were then tagged using the Glasgow-Mainz magnetic spectrometer with an energy resolution of ~2 MeV.
For this experiment, a frozen-spin butanol (C_4H_9OH) [4] target was used: the 2 cm long target cell was located in a $^3He/^4He$ dilution refrigerator and the target nucleons were polarised via Dynamic Nuclear Polarisation (DNP) [5] to an initial polarisation degree up to 90%. During the data taking, a magnetic field of 0.68 T together with the 25 mK temperature provided by the dilution refrigerator ensured long relaxation times (up to 2000 h) before repolarisation was required. Since only the hydrogen nuclei of the butanol can be polarised, a carbon foam target was used in a dedicated beamtime to study background contributions from unpolarised carbon nuclei.
The target cell was placed inside the central detector system, devoted to the detection of the reaction products. It was composed by the Crystall Ball (CB) NaI spectrometer, a large solid angle, highly segmented photon and hadron spectrometer, and it was complemented by the Multi-Wire Proportional Chambers (MWPCs), used to discriminate between the charged and the neutral particles detected in CB. Precise energy information, as well as angle and particle identification in the azimuthal (ϕ) and polar (θ) angular regions from 0° to 360° and from 21° to 159°, respectively, are obtained by combining the information from these three detectors.

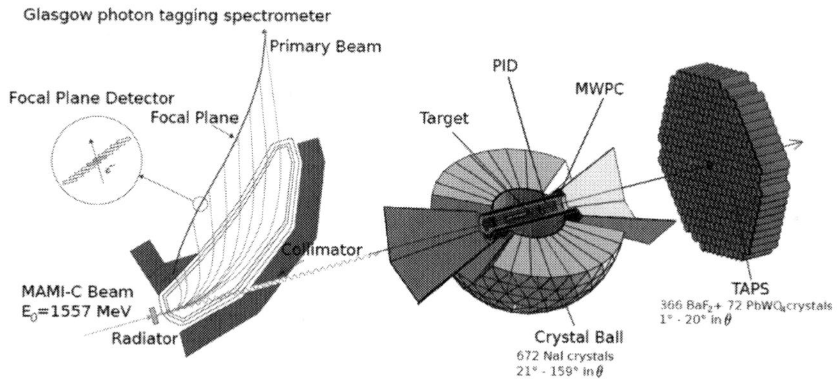

Figure 2. Schematic view of the A2 experimental setup.

2.2 The experimental results

The double polarisation observables E and G were calculated by using the differential cross section for pseudo-scalar meson photoproduction; in the case of elliptically polarised photons in combination with a longitudinally polarised target, it is given by:

$$\frac{d\sigma}{d\Omega}(\theta,\phi) = \frac{d\sigma}{d\Omega_0}(\theta)\left[1 - P_{lin}\Sigma\cos(2(\alpha-\phi)) - P_z(-P_{lin}G\sin(2(\alpha-\phi)) + P_{circ}E)\right]. \quad (1)$$

By integrating Eq. 1 over ϕ, it is possible to extract the E observable:

$$N_B\Big|_{\pm\alpha}^{\pm P_z}(\theta) = N_B(\theta)\cdot[1 - P_{circ}P_zE] \quad \rightarrow \quad E = \frac{\sigma^{1/2}-\sigma^{3/2}}{\sigma^{1/2}+\sigma^{3/2}} = \frac{N_B^{1/2}-N_B^{3/2}}{N_B^{1/2}+N_B^{3/2}}\cdot\frac{1}{d}\cdot\frac{1}{P_{circ}P_z}, \quad (2)$$

where $N_B^{1/2}$ and $N_B^{3/2}$ are the helicity-dependent count rates, d is the dilution factor and $P_{circ}P_z$ the target and beam polarisation degrees. As the data were obtained with a butanol target, both the polarisable free protons in hydrongen nuclei and the unpolarised bound nucleons in carbon and oxygen

nuclei contribute to the count rates. This is taken into account by the dilution factor d, which specifies the amount of polarisable free protons in the data ($d = N_{free}/(N_{free} + N_{bound})$). The factor d was determined during a dedicated data taking using a carbon foam target, by scaling the coplanarity distribution of the carbon data (N_C) to the one of the butanol data (N_B): $d = 1 - s \cdot N_C/N_B$, where s takes into account possible differences in acceptance, photon flux and target density during the carbon and butanol beamtimes.
Similarly, by integrating Eq. 1 over all possible helicity states, one can get the observable G:

$$N_B\Big|_{\pm\alpha}^{\pm P_z}(\theta, \phi) = N_B(\theta) \cdot [1 - P_{lin}\Sigma_B \cos(2(\alpha - \phi)) + dP_{lin}P_z G \sin(2(\alpha - \phi))]\,. \tag{3}$$

The double polarisation observables G and E were extracted from the data according to Eqs. 2 and 3 as a function of the beam energy E_γ for the entire angular range $-1 \leq \cos\theta_\pi \leq 1$. Preliminary results for the photoproduction reaction $\vec{\gamma}\vec{p} \rightarrow p\pi^0$ are shown in Figs. 3 and 4. The results are compared to data from the CBELSA/TAPS collaboration ([6] for the G observable and [7] for E) and to different partial wave analyses: for both the double polarisation observables there is a good agreement with the existing data and mostly with the different PWA models, especially with the BnGa2014-02 PWA solution.

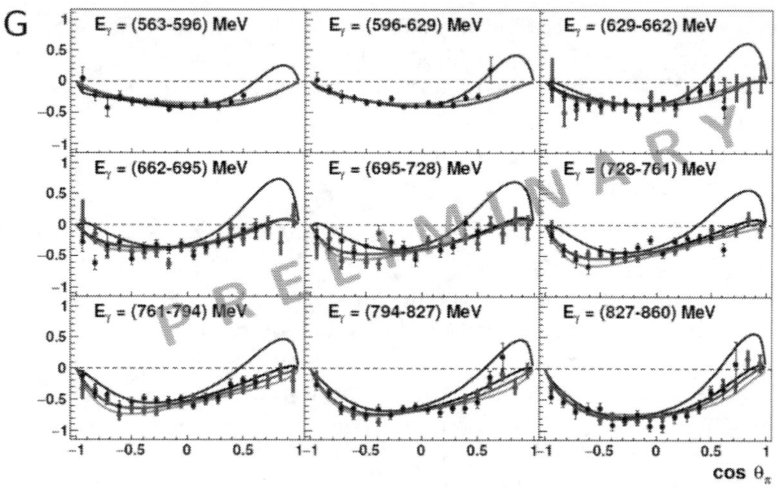

Figure 3. Results for G in π^0-photoproduction for selected energy bins, from 560 MeV to 860 MeV. The experimental results from A2 (blue circles) are compared to the CBELSA/TAPS data (green circles) [6], to PWA fits (BnGa_2014_02 [8], red line and BnGa_2014_01, black line) and PWA predictions from MAID2007 [9] (green line) and SAID-CM12 [10] (blue line).

3 The GDH sum rule

Sum rules are important tools to study the excited states of the nucleons, since they connect information from all energies to fundamental parameters of the interaction models of interest.
The Gerasimov-Drell-Hearn (GDH) sum rule [11, 12] relates the nucleon anomalous magnetic moment (AMM) κ, the spin S and the mass M of a nucleon to the integral over the weighted helicity difference of the total absorption cross section for circularly polarised photons on a longitudinally polarised nucleon target:

$$I_{GDH} = \int_{\nu_{th}}^{\infty} \frac{\sigma_p - \sigma_a}{\nu} d\nu = 4\pi^2\kappa^2 \frac{e^2}{M^2} S, \tag{4}$$

where ν is the photon energy, and σ_p and σ_a denote the total absorption cross section for parallel and antiparallel orientation of photon and particle spins, respectively. The lower limit of the integral, ν_{th},

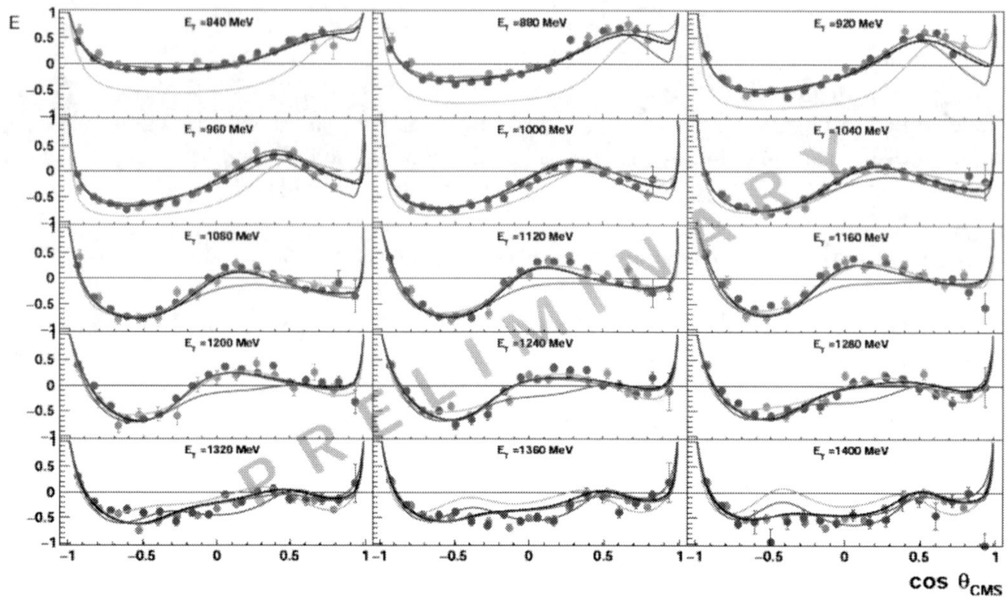

Figure 4. Results for E in π^0-photoproduction for selected energy bins from 840 MeV to 1400 MeV. The experimental results from A2 (blue circles) are compared to the CBELSA/TAPS data (green circles) [7], to PWA fits (BnGa_2014_02 [8], red line and BnGa_2014_01, black line) and PWA predictions from MAID2007 [9] (green line) and SAID-CM12 [10] (blue line).

corresponds to pion production and photodisintegration threshold for a nucleonic and nuclear target, respectively.

Since the GDH sum rule is derived from very general fundamental physical principles, like Lorentz invariance, the optical and the low energy theorems, the test of this relation represents a fundamental check of our knowledge of the γ−nucleon interaction, as well as of the physics of strongly interacting systems and of the existing photo-reaction models. In addition, through the helicity dependent $N\pi$ channels, it is possibile to access new observables and study the baryonic resonances, as well as test multipole models.

A comprehensive experimental program is carried out in Mainz since a few years, in order to give a better insight into the GDH sum rule on the proton and on the neutron, and to perform an accurate investigation of the properties of baryonic resonances.

3.1 The experimental setup

The experiments were carried out by the A2 Collaboration at the tagged photon facility of the MAMI accelerator in Mainz, using circularly polarised photons obtained by bremsstrahlung of longitudinally polarised electrons with an average polarisation of 75%.

For the studies on the proton, the frozen-spin butanol target data collected for the E and G double polarisation observables were used.

Polarised deuteron or high pressure ^{3}He gas targets were used as polarised neutron sources. Longitudinally polarised deuterons were provided by a frozen spin target using deuterated butanol (C_4OD_{10}): thanks to the use of a new doping material [13] it was possible to get a starting value of target polari-

 DOI: 10.1051/epjconf/201716407034

sation up to 70%, with a relaxation time of ~200 hours. Gaseous ^{3}He was polarised via the method of metastability exchange optical pumping (MEOP [14]) and it was possible to obtain initial polarisation values up to 70%, with total relaxation time of ~20 hours.
During the data taking, the photoemitted hadrons were registered by the large acceptance detector described in 2.1.

3.2 Results on the proton and deuteron

The preliminary results of the helicity dependence of the total inclusive cross section $\Delta\sigma_{tot} = (\sigma_a - \sigma_p)$ on the proton and on the neutron are depicted in Fig. 5 (blue circles). They are compared to the results obtained by the GDH Collaboration [15, 16] (red circles) and are in good agreement within the errors. Fig. 6 shows the results for the partial channels $\gamma p \rightarrow \pi^0 X$ and $\gamma d \rightarrow \pi^0 X$: also in these plots, the

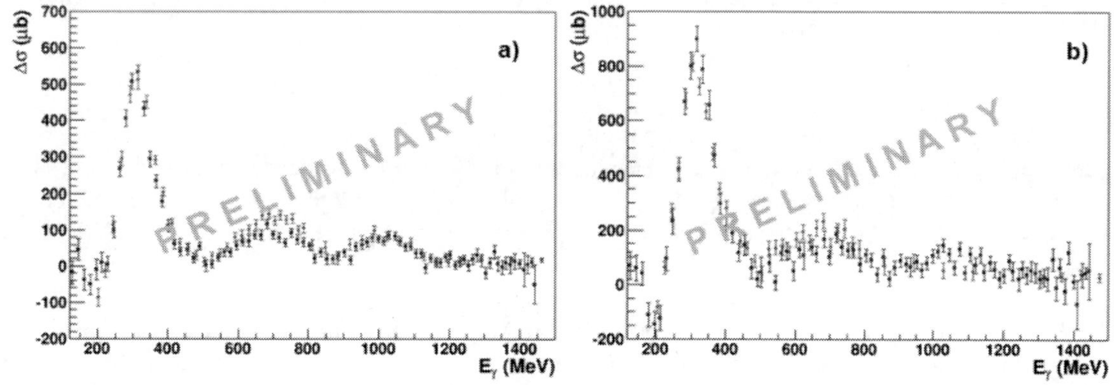

Figure 5. Helicity dependent total inclusive cross section for the $\gamma p \rightarrow X$ (a) and $\gamma d \rightarrow X$ channels (b). The A2 results (blue circles) are compared to the results from the GDH Collaboration (red circles) [15, 16].

A2 data (blue circles) are compared to the GDH Collaboration results, which are available only in the Δ-resonance region, where the agreement is good. It is therefore clear from the plots that our recent results can provide a useful contribution in an energy region where no other data are available.
The A2 results are also compared to the predictions of the MAID multipole analysis: since the model is a free nucleon estimation, it does not take into account the nuclear effects in the deuteron case, where the discrepancy with the data is evident.

3.3 Results on ^{3}He

In Figure 7, the unpolarised and helicity-dependent total cross sections for the partial channels $\gamma^3He \rightarrow \pi^0 X$ and $\gamma^3He \rightarrow \pi^\pm X$ are shown in the energy region $200 < E_\gamma < 500$ MeV [17].
The data are compared to the prediction of the Fix-Arenhövel (FA) model (red line) and of a simple plane-wave impulse approximation (PWIA) model (blue line): the first model is an extention of the model on the deuteron [18]; in the second model, the cross sections are evaluated as an incoherent sum of quasi-free single nucleon contributions, determined using the MAID multipole analysis and the momentum distribution of the nucleons inside ^{3}He as parametrised in [19]. The difference between the two models, i.e. the role of nuclear effects, results in damping and broadening the peak corresponding to the Δ resonance excitation.

 DOI: 10.1051/epjconf/201716407034

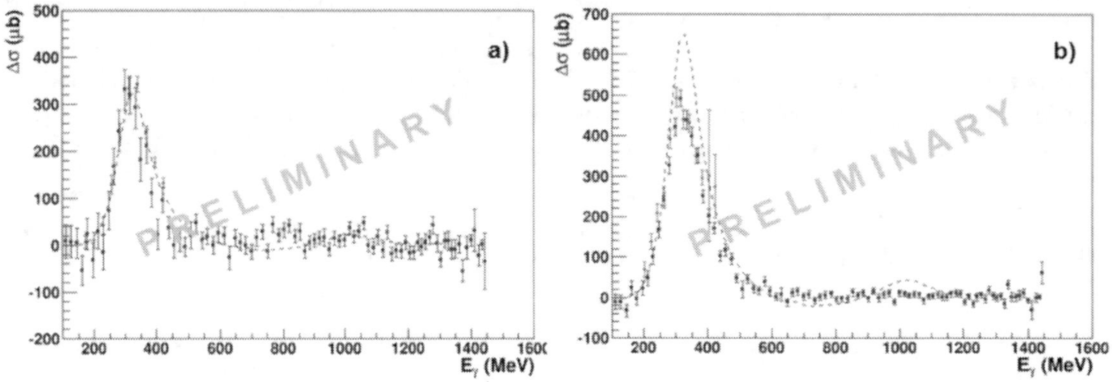

Figure 6. Helicity dependent total cross section for the semi-exclusive channels $\gamma p \rightarrow \pi^0 X$ (a) and $\gamma d \rightarrow \pi^0 X$ channels (b). The A2 results (blue circles) are compared to the results from the GDH Collaboration (red circles) [15, 16] and to the MAID model (dashed line).

In the unpolarised case, for the neutral partial channel the FA model is in good agreement with our data for $E_\gamma \geq 250$ MeV, while it underestimated them at lower photon energies, where the coherent $\gamma^3He \rightarrow \pi^0\ {}^3He$ reactions is expected to play a dominant role. On the contrary, in the $\pi^\pm X$ channel the FA model describes the measured cross section less well and for $E_\gamma \geq 350$ MeV the PWIA model does better.
In the helicity dependent cross section case, as in the unpolarised case, the FA model describes the $\gamma^3He \rightarrow \pi^0 X$ data at higher photon energies while it does not reproduce the shape of the experimental data of the charged partial channel. The PWIA model, instead, reproduces the data at higher E_γ values for both reaction reasonably well.
In order to better understand the origin of the observed discrepancies between the experimental models and the data, an analysis of both the unpolarised and polarised differential cross sections for the $\gamma^3He \rightarrow \pi^0 X$ and $\gamma^3He \rightarrow \pi^\pm X$ has been performed: the results can be found in [20].

4 Conclusions

Double polarised pion photoproduction experiments are a useful tool for the understanding of the nucleon excitation spectra.
The A2 Collaboration at MAMI has performed the first simultaneous measurement of the double polarisation observables E and G, by using a longitudinally polarised electron beam incident on a diamond radiator and a longitudinally polarised frozen-spin target. The preliminary results available up to $E_\gamma = 1500$ MeV are in good agreement with existing data from the CBELSA/TAPS Collaboration and provide an original contribution in the $\Delta(1232)$ resonance region.
The experimental data obtained from these experiments can be used also to study the Gerasimov-Drell-Hearn sum rule: the A2 Collaboration has carried out a comprehensive experimental program to study it on the proton, on the neutron and also on ^{3}He. Preliminary results of the helicity dependent total inclusive $\gamma p \rightarrow X$ and $\gamma d \rightarrow X$ cross sections, and of the helicity dependent total cross sections for the semi-exclusive channels $\gamma p \rightarrow \pi^0 X$ and $\gamma d \rightarrow \pi^0 X$ are in reasonable agreement with the existing data from the GDH Collaboration. Furthermore, results on the helicity dependent $\gamma^3He \rightarrow \pi X$ reactions have been shown up to $E_\gamma = 500$ MeV. These unprecedented data of this kind have proved the feasibility of the use of polarised ^{3}He gas target to check the GDH sum rule on the neutron.

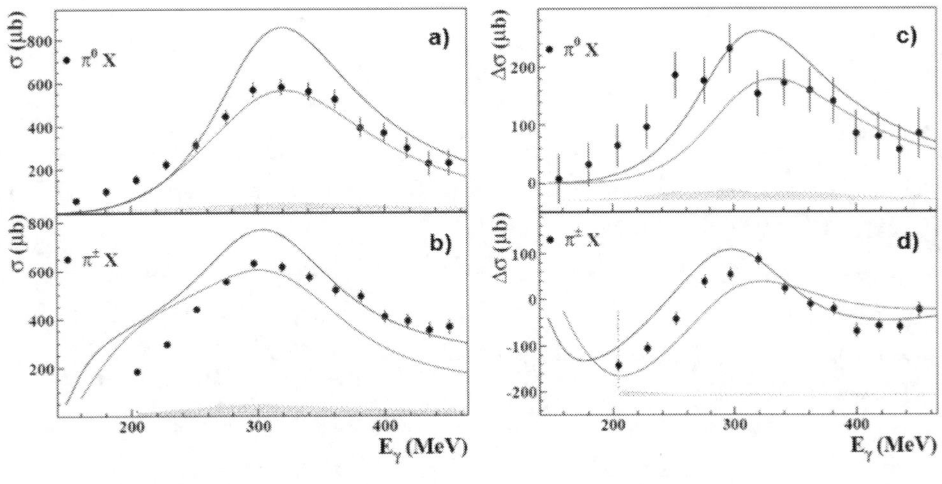

Figure 7. Unpolarised (left) and polarised (right) total cross section for $\gamma^3He \to \pi^0 X$ (a and c) and $\gamma^3He \to \pi^\pm X$ (b and d) channels. The error bars are statistical and the hatched bands show the systematic uncertainties. The measured cross sections are compared to the FA model (red line) and to our PWIA model (blue line).

References

[1] B. Krusche and S. Schadmand, arXiv:nucl-ex/0306023v1
[2] I. S. Barker, A. Donnachie, J. K. Storrow, Nucl. Phys. **B95**. 347 (1975)
[3] W.-T. Chiang and F. Tabakin, Phys. Rev. C **55**, 2054 (1997)
[4] D. Goertz et al., Proceedings of PSTP205, PoS(PSTP2015)009 (2016)
[5] D.G. Crabb and W. Meyer, Annu. Rev. Nucl. Part. Sci. 47, 67 (1997)
[6] A. Thiel et al., Phys. Rev. Lett. **109**, 102001 (2012)
[7] M. Gottschall et a., Phys. Rev. Lett. **112**, 012003 (2014)
[8] E. Gutz et al., Eur. Phys. J. A 50, 74 (2014)
[9] D. Drechsel, S.S. Kamalov and L. Tiator, Eur. Phys. J. A **34**, 69 (2007)
[10] R.L. Workman, M.W. Paris, W.J. Briscoe, I.I. Strakovsky, Phys. Rev. C **86**, 015202 (2012)
[11] S. B. Gerasimov, Sov. J. Nucl. Phys. **2**, 430 (1966)
[12] S. D. Drell and A. C. Hearn, Phys. Rev. **96**, 1428 (1966)
[13] S. Goertx et al., Nucl. Instr. Meth. A **526**, 43 (2004)
[14] F. Colegrove et al., Phys. Rev. Lett. **132**, 2561 (1963)
[15] J. Ahrens et al.,Phys. Lett. B **672**, 328 (2009)
[16] H. Dutz et al., Phys. Rev. Lett. **94**, 162001 (2005)
[17] P. Aguar Bartolomè et al., Phys. Lett. B **723** (2013) 71-77
[18] H. Arenhövel, A. Fix, Phys. Rev. C **72** (2005) 064004
[19] R. Schiavilla, C. Pandharipande, Nucl. Phys. A **449** (1986) 219
[20] S. Costanza et al., Eur. Phys. J. A (2014) 50: 173

EPJ Web of Conferences **164**, 07035 (2017) DOI: 10.1051/epjconf/201716407035
ICNFP 2016

Recent results on soft QCD topics, and jet, V+jet and photon production from ATLAS

Ewa Stanecka[1,2,a]

[1] *Institute of Nuclear Physics Polish Academy of Sciences*
[2] *on behalf of the ATLAS Collaboration*

Abstract. This paper presents selection of recent QCD related measurements from the ATLAS Experiment at the Large Hadron Collider at CERN. The results on the charged particle production, underlying event properties, prompt photon production, charged particle multiplicity inside jets and Z boson production in association with up to four jets are presented. The measurements allowed us to cross-check several Monte Carlo generators with data at centre-of-mass energies of 8 and 13 TeV.

1 Introduction

The ATLAS experiment [1] performs studies of proton-proton (pp) collisions at the Large Hadron Collider (LHC) [2] at CERN. The LHC collides bunches of protons at the unprecedented centre of mass energy ($\sqrt{s}$) of 7, 8 and 13 TeV. The peak instantaneous luminosity recorded by the time of the conference was about $12 \times 10^{33}\text{cm}^{-1}\text{s}^{-2}$ [3].

The ATLAS detector is composed of several sub-detectors organised in layers. The closest to the proton beams collision point is the high precision tracking system, next come calorimeters and then muon chambers. The ATLAS experiment is a general purpose detector designed to investigate a wide range of particle physics, which can be generally divided into searches for new phenomena and the precision measurements. The precision measurements are very important as they improve our understanding of the Standard Model (SM) of particle physics, give better knowledge of background processes for new physics signals and also provide an important benchmark for the detector performance. For these reasons the precision measurements are repeated at each newly accessible pp collision energy. In this report a selection of recently published ATLAS SM measurements related to the production of primary charged particles, hadronic jets, photons, and jets associated with a Z boson is reviewed.

2 Measurements of the underlying-event at 13 TeV.

The underlying event (UE) is the activity accompanying any hard scattering in a collision event, including multiple parton interactions, final state radiation, initial state radiation as well as the partons not participating in hard scattering process. The UE is an irreducible background at the LHC as it is not possible to separate UE processes from the hard process on an event by event basis.

[a] e-mail: ewa.stanecka@ifj.edu.pl

 DOI: 10.1051/epjconf/201716407035

The underlying event activity was measured by the ATLAS experiment with 13 TeV data with a leading particle method [4]. The leading particle is defined as a charged particle with the highest p_T. The direction of the leading particle is used to define regions in the azimuthal plane that have different sensitivity to the UE. The azimuthal angular difference ($\Delta\phi$) between charged particles and the leading particle is used to define three azimuthal regions: toward ($|\Delta\phi| < 60°$), transverse ($60° < |\Delta\phi| < 120°$) and away regions ($|\Delta\phi| > 120°$). The transverse region is sensitive to the UE activity since it is perpendicular to the direction of the hard scatter and hence it is expected to have less activity from the hard scattering process compared to the away region.

A measurement of charged-particle distributions sensitive to the properties of the underlying event was performed on the data sample corresponding to an integrated luminosity of 170 μb^{-1}. The data was collected during a special LHC fill with a low expected mean number of interactions per bunch crossing, same as for the analysis described in Section 3. The event and track selections used in the analysis follow those described in [5], with the only additional requirement which is the presence of the leading particle with p_T of at least 1 GeV. The measured distributions are compared with predictions from Monte Carlo event generators. In Figure 1 the average scalar sum of charged-particle transverse momenta, Σp_T, for transverse and toward regions is presented as a function of the leading particle p_T. In the transverse region sum p_T shows a gradual increase, rising to an approximately constant plateau for p_T of the leading particle greater than 6 GeV. For higher values of the leading particle p_T, the toward region includes contributions from the jet-like activity, which results in a gradual rise of the average sum p_T density [4].

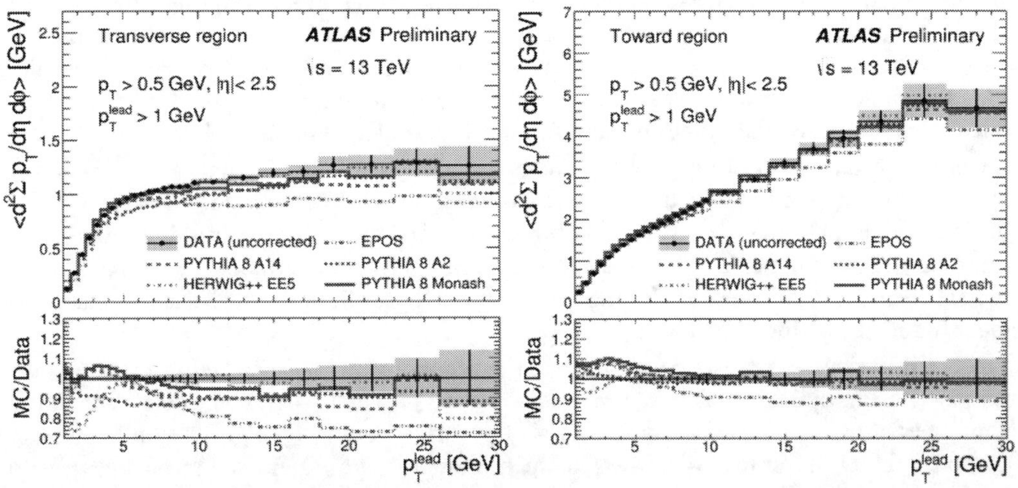

Figure 1: Comparison of detector level data and MC predictions for average charged-particle multiplicity density values (left column) and average scalar p_T sum density of tracks (right column) as a function of leading particle transverse momentum, p_T^{lead}, in the transverse (left) and toward (right) regions. The bottom panels in each plot show the ratio of MC predictions to data. The shaded bands represent the combined statistical and systematic uncertainties, while the error bars show the statistical uncertainties [4].

Several Monte Carlo (MC) predictions were compared with the data. Pythia 8 A14, Monash and Herwig++ UEEE5 models are closest to data in the plateau part of the transverse region, but none of them describe the initial part well. The Epos generator predicts significantly less activity at

 DOI: 10.1051/epjconf/201716407035

higher transverse momenta of the leading particle, which is a consequence of the absence of semi-hard minimum bias processes in the model. In general, none of the models is very discrepant from the data, building a confidence in the multiple parton interactions (MPI) description used in those generators [4].

3 Measurement of charged-particle spectra at 13 TeV

Soft interactions are characterised by low transverse momentum transfer and cannot be calculated by perturbative Quantum Chromodynamics (QCD). The predictions for low momentum transfer processes are made with phenomenological models inspired by QCD, implemented in MC event generators. The observables connected to the production of charged-particles provide experimental probes of strong interactions at low momentum transfers and good benchmark for phenomenological models.

Measurement of the charged particles production in proton–proton collisions at $\sqrt{s}$=13 TeV was performed by the ATLAS experiment [6]. The data used for the analysis correspond to an integrated luminosity of 151 μb^{-1}. The data were recorded during special runs with low beam currents and reduced focusing to give a mean number of interactions per bunch crossing of 0.005. Particles selected for the analysis are required to have a transverse momentum (p_T) greater than 100 MeV and an absolute pseudorapidity ($|\eta|$) less than 2.5 . Events containing at least two charged particles satisfying the above kinematic criteria were used to measure the charged-particle multiplicity (N_{ch}), its dependence on transverse momentum and pseudorapidity, and the dependence of the mean transverse momentum on multiplicity. Figure 2 presents these measurements (black dots) corrected for detector effects and compared to the predictions from several MC event generators (coloured curves). The vertical bars represent the statistical uncertainties, while the shaded areas show statistical and systematic uncertainties added in quadrature. The lower panel in each figure shows the ratio of the MC simulation to data. The MC generators and their tunes used in the analysis are listed in Table 1.

The measured mean number of charged particles per unit pseudorapidity in the region $|\eta| < 0.2$ is 6.422±0.096 with a negligible statistical uncertainty. Significant differences are observed between the measured distributions and tested MC predictions. Amongst the models considered, Epos has the best overall description of the data. Pythia 8 A2 and Pythia 8 Monash provide a reasonable overall description, whereas Qgsjet-ii does not describe $< p_T >$ vs. N_{ch} well, but provides a reasonable level of agreement for other distributions [6].

The presented measurement was combined with previous results to shed light on the evolution of charged-particle multiplicities with centre-of-mass energy, which so far is poorly constrained. The mean number of primary charged-particles after the correction for the contribution from strange baryons (6.500 ± 0.099) is compared to previous measurements at different centre-of-mass energies in Figure 3.

Table 1: Summary of MC generators used for comparison with the data. The generator, its version, the corresponding tune and the parton distribution function are given [6].

Generator	Version	Tune	PDF
Pythia 8	8.185	A2	MSTW2008LO
Pythia 8	8.186	Monash	NNPDF2.3LO
Epos	LHCv3400	LHC	–
Qgsjet-ii	II-04	default	–

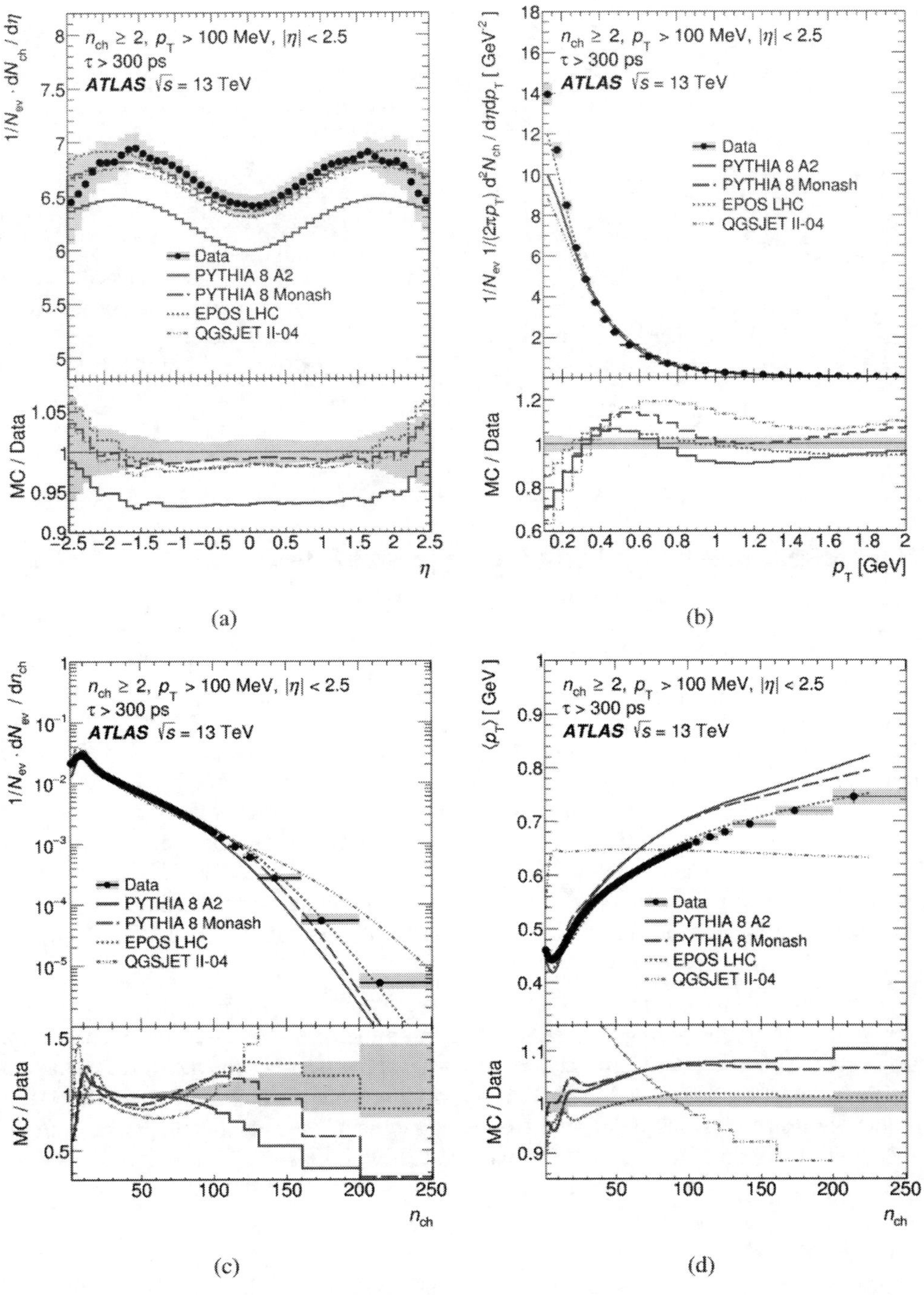

Figure 2: Primary charged-particle multiplicities as a function of (a) pseudorapidity η and (b) transverse momentum p_T. The distribution of the primary charged-particle multiplicity n_{ch} (c) and the mean transverse momentum $< p_T >$ versus n_{ch} (d) for events with at least two primary charged-particles with $p_T > 100$ MeV and $|\eta| < 2.5$ [6].

DOI: 10.1051/epjconf/201716407035

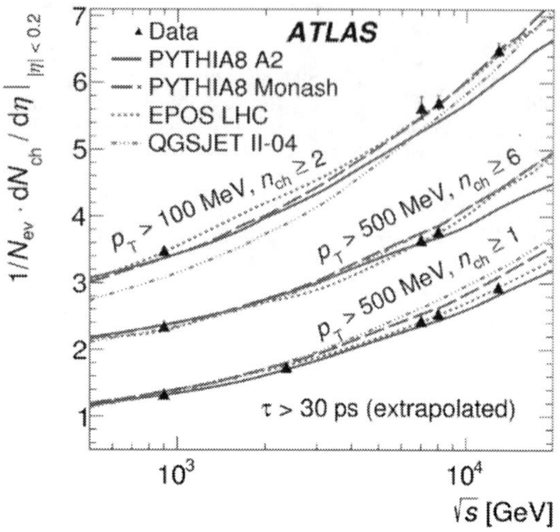

Figure 3. The average primary charged-particle multiplicity in pp interactions per unit of pseudorapidity η for $|\eta|< 0.2$ as a function of the centre-of-mass energy $\sqrt{s}$. The values for the other pp centre-of-mass energies are taken from previous ATLAS analyses. The data are shown as black triangles with vertical errors bars representing the total uncertainty. They are compared to various MC predictions which are shown as coloured lines [6].

4 Charged-particle multiplicity inside jets at 8 TeV

The number of charged particles inside jets is often used as a discriminant for identifying the quark or gluon nature of the initiating parton and it is sensitive to both the perturbative and non-perturbative components of the jet formation.A measurement of the average number of charged particles with $p_T > 500$ MeV inside high-momentum jets was performed on a data sample of 20.3 fb^{-1} recorded in pp collisions at $\sqrt{s} = 8$ TeV [7]. The events selected for analysis are required to have at least two central jets ($|\eta| < 2.1$), so that their charged particles are within the acceptance of the tracking system. The jets considered are clustered using the anti-k_t jet algorithm with the radius parameter R = 0.4 and have transverse momenta from 50 GeV up to and beyond 1.5 TeV. The jet with smaller absolute pseudorapidity is classified as more central, and the jet with larger $|\eta|$ is classified as a more forward jet. The more forward jet tends to be correlated with the parton with higher longitudinal momentum fraction x, and is less likely to be a gluon-initiated jet.

The reconstructed charged-particle multiplicity distribution is unfolded to remove distortions from detector effects and the resulting charged-particle multiplicity is compared to several jet formation models, as shown in Figure 4a. Comparisons at a particle level between the measured average charged-particle multiplicity and various MC models show significant differences between the simulations using tunes for Run 1 (first season of the LHC data taking) and the data, but the Run 2 (current season of the LHC data taking) tunes for both Pythia 8 and Herwig++ significantly improve the modelling of the average $n_{charged}$.

The analysis shows the difference in the average charged-particle multiplicity between the more forward and the more central jets. This effect is sensitive to the difference between the number of quark and gluon constituents. The average difference, combined with the gluon fraction, was used to extract the average charged-particle multiplicity for quark- and gluon-initiated jets separately [7]. The charged-particle multiplicity in the extracted gluon-initiated jets is higher than the corresponding quantity for the quark-initiated jets and the MC generators accurately model the jet p_T dependence observed in the data, see Figure 4b.

DOI: 10.1051/epjconf/201716407035

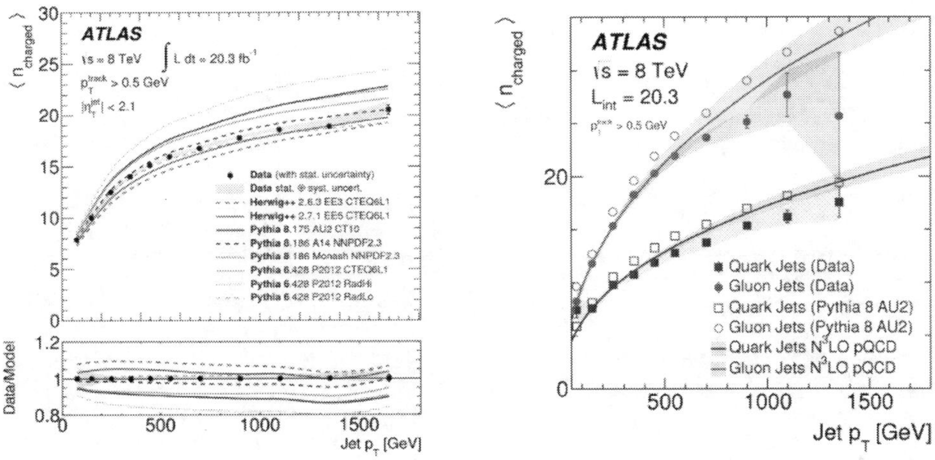

(a) (b)

Figure 4: (a) The measured average charged-particle multiplicity (left) as a function of the jet p_T, combining the more forward and the more central jets for $p_T{}^{track}$>0.5 GeV. (b)The jet p_T dependence of the average charged-particle multiplicity ($p_T{}^{track}$> 0.5 GeV) for quark- and gluon-initiated jets, extracted with the gluon fractions from Pythia 8.175 with the CT10PDF [7].

5 Inclusive isolated-photon production

Inclusive isolated-photon production ($pp \rightarrow \gamma + X$) in pp collisions provides a testing environment for perturbative QCD. It was studied by the ATLAS experiment at $\sqrt{s}$ = 13 TeV [8]. The data sample used for analysis corresponds to an integrated luminosity of 6.4 pb^{-1}. A sample of photons with high transverse energies has been selected using the characteristics of the shower shapes in the calorimeter. Both converted and unconverted photons were included in the sample. The photons were required to be isolated by rejecting those candidates in which the transverse energy in a cone of ΔR=0.4[1] around their direction was larger than 4.8 GeV+4.2 $\cdot$ $10^{-3} E_T{}^{\gamma}$. A data-driven subtraction method was applied to remove the background from multi-jet processes, in which a jet is misidentified as a photon. As a result of this analysis, a clear signal of inclusive isolated photons at $\sqrt{s}$ = 13 TeV has been observed for the first time in the phase-space region given by for $E_T{}^{\gamma} > 125$ GeV and $|\eta^{\gamma}| < 2.37$, excluding the region of the ATLAS calorimeter crack between the barrel and the end-caps ($1.37 < |\eta_{\gamma}| < 1.56$). The kinematic distributions of the signal yield as functions of $E^{\gamma}{}_T$ and $|\eta^{\gamma}|$ are shown in Figure 5. The measured distributions are compared to the simulated Sherpa events normalised to the total yield observed in the data. The simulation provides a good description of the shape of the measured kinematic distributions.

[1] $\Delta R = \sqrt{(\Delta\eta)^2 + (\Delta\phi)^2}$ defines the distance between objects in (η-ϕ) space.

 DOI: 10.1051/epjconf/201716407035

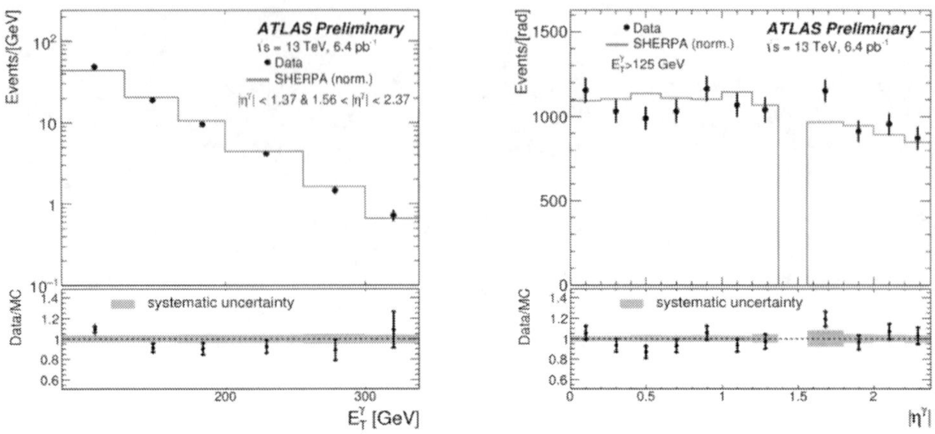

Figure 5: Measured isolated-photon yield (dots) as a function of $E_T{}^{\gamma}$ (left) and as a function of $|\eta^{\gamma}|$ for $E_T{}^{\gamma} > 125$ GeV (right); the error bars represent the statistical uncertainty of the data. The Sherpa MC simulations, normalised to the integrated measured signal yield, (histograms) are also included. The bottom panels of the plots show the ratio of the data and the MC simulation (dots). The inner error bars represent the statistical uncertainty of the data and the outer error bars display the uncertainty due to the background subtraction added in quadrature to the statistical uncertainty. The shaded band represents the uncertainty due to the photon energy scale, energy resolution and identification efficiency as well as the trigger efficiency and the modelling of the isolation transverse energy in the MC added in quadrature [8].

6 Production cross-sections of a Z boson in association with jets

The measurement of the weak boson (Z) production cross-section gives another good benchmark for current understanding of the perturbative QCD and electroweak processes. The weak boson production at LHC pp collisions has a relatively large cross-section and boson decays to leptonic final states are a clean experimental signature. Therefore it can be measured with a high precision. The processes with weak bosons production constitute an important background for Higgs boson and new physics searches.

The production cross-section of Z boson in association with up to four jets was measured using pp collision data collected at $\sqrt{s} = 13$ TeV and corresponding to a total integrated luminosity of approximately 85 pb^{-1} [9]. The events with Z bosons decaying to electron or muon pairs, produced in association with one or more jets in the kinematic range of $p^{jet}{}_T > 30$ GeV and $|y^{jet}| < 2.5$, were analysed. The kinematic cuts used in the analysis are described in detail in [9]. The fiducial production cross-sections for Z bosons and inclusive cross-section to up to four jet multiplicities were measured with a precision ranging from approximately 10% to approximately 20% as presented in Figure 6. Ratios of cross-sections for successive jet multiplicities were derived and are also shown in Figure 6. The measured fiducial cross-sections for the various jet multiplicities and the ratios of successive jet multiplicities were compared to predictions from MC models - Sherpa and MadGraph. The predictions show a reasonable agreement with the measured cross-sections and cross-section ratios within the uncertainties.

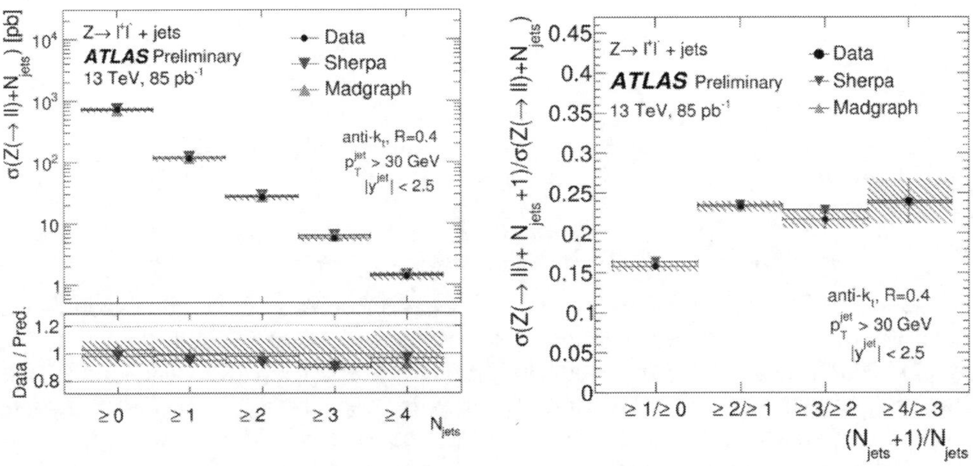

Figure 6: Combined Z to *ll* plus N jets fiducial cross-sections (left), and cross-section ratios for successive jet multiplicities (right). The results are compared to predictions from SHERPA and MADGRAPH. The hatched error band corresponds to the total uncertainty of the results: systematic, statistical, and luminosity uncertainties for the fiducial cross-section results (left), systematic and statistical uncertainties for the ratio results (right) [9].

7 Summary

Several recent measurements related to the soft SM processes from the ATLAS experiment are presented for *pp* collisions at 8 and 13 TeV at the LHC at CERN. Monte Carlo generators, tuned in lower energy regimes, describe the collected data reasonably well. The latest ATLAS public results can be found at https://twiki.cern.ch/twiki/bin/view/AtlasPublic/StandardModelPublicResults.

8 Acknowledgements

This work was in part supported in Poland by the National Science Centre grant 2015/19/B/ST2/00906 and by PL-Grid Infrastructure.

References

[1] ATLAS Collaboration, 2008 JINST 3 S08003.

[2] Evans, L. and Bryant, P., 2008 JINST 3 S08001.

[3] ATLAS Collaboration, https://twiki.cern.ch/twiki/bin/view/AtlasPublic/LuminosityPublicResultsRun2.

[4] ATLAS Collaboration, ATL-PHYS-PUB-2015-019, http://cds.cern.ch/record/2037684.

[5] ATLAS Collaboration, Phys. Lett. B **758** (2016) 67 [arXiv:1602.01633 [hep-ex]].

[6] ATLAS Collaboration, Eur. Phys. J. C **76** (2016) no.9, 502 [arXiv:1606.01133 [hep-ex]].

[7] ATLAS Collaboration, Eur. Phys. J. C **76**, no. 6, 322 (2016) [arXiv:1602.00988 [hep-ex]].

[8] ATLAS collaboration, ATLAS-CONF-2015-034, http://cdsweb.cern.ch/record/2037667.

[9] ATLAS collaboration, ATLAS-CONF-2015-041, http://cdsweb.cern.ch/record/2048104.

DOI: 10.1051/epjconf/201716407036

Recoil Directionality Studies in Two-Phase Liquid Argon TPC Detectors

Matteo Cadeddu[1,2,a], Giovanni Batignani[3,4], Walter Marcello Bonivento[2], Bianca Bottino[5,6], Luigi Campajola[7,8], Mauro Caravati[1,2], Sergio Catalanotti[7,8], Corrado Cicalò[2], Alfredo Cocco[8], Giovanni Covone[7,8], Gianfranca De Rosa[7,8], Alberto Devoto[1,2], Carlo Dionisi[9,10], Giuliana Fiorillo[7,8], Stefano Giagu[9,10], Marisa Gulino[11,12], Michael Kuss[4], Marcello Lissia[2], Luca Lista[8], Giuseppe Longo[7,8], Marco Pallavicini[5,6], Luciano Pandola[12], Marco Razeti[2], Marco Rescigno[9], Biagio Rossi[8,14], Nicola Rossi[15], Gemma Testera[6], Pasquale Trinchese[7,8], Susan Walker[7,8], and Maurizio Zullo[9]

[1] *Physics Department, Università degli Studi, Cagliari 09042, Italy*
[2] *Istituto Nazionale di Fisica Nucleare, Sezione di Cagliari, Cagliari 09042, Italy*
[3] *Physics Department, Università degli Studi di Pisa, Pisa 56127, Italy*
[4] *Istituto Nazionale Fisica Nucleare, Sezione di Pisa, Pisa 56127, Italy*
[5] *Physics Department, Università degli Studi di Genova, Genova 16146, Italy*
[6] *Istituto Nazionale di Fisica Nucleare, Sezione di Genova, Genova 16146, Italy*
[7] *Physics Department, Università degli Studi Federico II, Napoli 80126, Italy*
[8] *Istituto Nazionale di Fisica Nucleare, Sezione di Napoli, Napoli 80126, Italy*
[9] *Istituto Nazionale di Fisica Nucleare, Sezione di Roma 1, Roma 00185, Italy*
[10] *Physics Department, Sapienza Università di Roma, Roma 00185, Italy*
[11] *Università di Enna KORE, Enna 94100, Italy*
[12] *Istituto Nazionale Fisica Nucleare, Laboratori Nazionali del Sud, 95123 Catania, Italy*
[13] *Università di Enna KORE, Enna 94100, Italy*
[14] *Physics Department, Princeton University, Princeton, NJ 08544, USA*
[15] *Istituto Nazionale Fisica Nucleare, Laboratori Nazionali del Gran Sasso, Assergi (AQ) 67100, Italy*

Abstract. Projects attempting the direct detection of WIMP dark matter share the common problem of eliminating sources of background or using techniques to distinguish background events from true signals. Although experiments such as DarkSide have achieved essentially background free exposures through careful choice of materials and application of efficient veto techniques, there will still be a high burden of proof to convince the greater scientific community when a discovery is claimed. A directional signature in the data would provide extremely strong evidence to distinguish a true WIMP signal from that of an isotropic background. Two-phase argon time projection chambers (TPCs) provide an experimental apparatus which can both be scaled to the ton-scale size required to accommodate the low cross-section expected for WIMP interactions and have an anisotropy that could be exploited to evaluate the polar angles of the resulting nuclear recoils from WIMP collisions with target nuclei. Our studies show that even a modest resolution in the polar angle reconstruction would offer a powerful tool to detect a directional signature. In this contribution, the status of the ReD experiment, which is under construction at Naples University, will be also shown. The aim of the project

[a] e-mail: matteo.cadeddu@ca.infn.it

is to assess and enhance the directionality of two-phase argon TPCs. ReD will use a small TPC exposed to a beam of mono-energetic neutrons to study the so called "columnar recombination" in liquid argon. This development could have high impact on the future experiments in the field, opening up the potential to find conclusive evidence for dark matter or disprove the WIMP hypothesis at and above the mass range explored by planned accelerator experiments.

1 Introduction

At present, a plausible explanation of the nature of Dark Matter (DM) is that it is constituted by yet undetected Weakly Interacting Massive Particles (WIMPs). Nowadays, it is believed that every visible galaxy is surrounded by a roughly spherical halo of DM, extending far beyond the visible disk. The standard non-directional direct detection, which exploits the elastic scattering of WIMPs on nuclear targets in the detector, is experimentally very challenging in particular in what concerns background rejection. An important additional information that could strengthen the DM signature hypothesis is a possible annual modulation of the interaction rate due to the motion of the Earth around the Sun. Indeed, the solar system orbits around the center of the Galaxy, with a velocity $V_{SG} \simeq 220$ km/s. Earth moves around the Sun with a velocity $V_{ES} \simeq 30$ km/s in an orbit inclined about 60° relative to the disk of the Galaxy. The combination of these motions results in a cosine dependence of the count rate with time. Unfortunately, this modulation is quite small and it is very hard to disentangle from spurious effects, leaving doubts on the authenticity of the few DM observation claims [1]. In addition, numerous radioactive backgrounds are known to exhibit a modulation over the course of the year.

If detectors were able to distinguish the recoil direction of the scattered nucleus after a WIMP interaction, then the signal would have a large and characteristic angular dependence within a sidereal day[1]. Indeed, due to the Earth's rotation around its axis, the detector would face a change in the direction of the WIMP wind and accordingly the angular distribution of the recoil nuclei will vary with time. This modulation in arrival direction produces a much more sizeable and unambiguous signature with respect to a non-directional annual modulation. In principle, assuming the most advantageous experimental conditions, the effect of this modulation can be as large as 100%. The most attractive feature of this kind of modulation is that no known background is correlated with the sidereal period. Actually, a sidereal rate modulation goes out of phase with the solar time, avoiding possible accidental day/night effects which could take place in the laboratory.

Nowadays, a detector capable of measuring the direction of the scattered nucleus has not yet been developed. There are few prototypes [2] of detectors capable of reconstructing the tracks of the recoil nucleus but, since they exploit the usage of rarefied gases, this technology clashes with the necessity of building multi-tonne detectors. For this reason a promising approach is to explore a detector based on the so-called Columnar Recombination (CR) effect on a Noble Liquid [3], that despite a far from perfect angular resolution can reach large target masses.

2 A directional LAr TPC

By filling a gap in the literature, the sensitivity of an argon detector to the galactic WIMP directional signal has been studied. In particular, a two-phase Liquid Argon Time Projection Chamber (LAr TPC) has been considered, where recoiling nuclei cause argon excitation and the formation of Ar dimers.

[1]The sidereal day is about four minutes shorter than the solar day. It correspond to $23^{h}\ 56^{m}\ 04^{s}.090$.

Each event is then detected in two ways: the prompt scintillation light, S1, from argon dimers de-excitation, and the signal from free ionization electrons. They drift drift towards the top of the TPC, thanks to an electric field, and are extracted into the argon gas phase, confined in the so-called "gas pocket", close to the top array of photosensors. Accelerated electrons produce a secondary light pulse, S2 (through electroluminescence). A fraction of free electrons, however, recombine with argon ions in the ion-electron cloud produced by the ionizing track itself.

CR models suggest that the magnitude of the recombination effect should vary with the angle between the field and the track direction. Namely, when the ionizing track is parallel to the electric field the electrons have to drift in a "column" of ion-electron pairs produced by the ionizing track, maximizing the recombination probability. When it is perpendicular, the "column" needs to be crossed and the recombination probability is minimized. The net effect of the increased electron-ion recombination is the reduction of the ionization signal (S2), and the consequent enhancing of the primary scintillation pulse (S1). An accurate measurement of these two signals and their combination (the sum S1+S2 and the ratio S1/S2) may provide information on the nuclear track direction and thus on the WIMP directionality. The results shown below were obtained considering an Earth-bound detector, located at Laboratori Nazionali del Gran Sasso (LNGS), with a 20 tonne target mass of argon.

3 Distribution of observables

In a given reference frame, $\mathbf{v_i}$ is the the velocity of the incoming WIMP of mass m_W, $\mathbf{u}$ the velocity of the recoiling nucleus of mass M_N, $\mathbf{q} = M_N\mathbf{u}$ and $E_r = q^2/(2M_N)$ the corresponding momentum and energy, respectively. The azimuthal and zenith angles of the recoiling nucleus are ϕ_r and θ_r, while $\cos\vartheta = \hat{\mathbf{v}}_i \cdot \hat{\mathbf{q}}$ is the cosine of the angle between the incoming WIMP direction and the recoiling nucleus.

For a spin-independent interaction with equal couplings for neutrons and protons the WIMP-nucleus cross section, σ_{W-N}, can be expressed in terms of WIMP-nucleon cross section, σ_n as $\sigma_{W-N}/\mu_N^2 = A^2\sigma_n/\mu_n^2$, where A is the atomic mass and μ_N and μ_n are the WIMP-nucleus and the WIMP-nucleon reduced masses, respectively. The finite size of the nucleus is taken into account thanks to the standard Helm form-factor, $F(q)$.

In the frame where the target nucleus is at rest, the double differential cross section becomes dependent on $\cos\vartheta$; the azimuthal symmetry of the scattering around the WIMP arrival direction gives $d\Omega = 2\pi\, d\cos\vartheta$, thus

$$\frac{d^2\sigma(q, \hat{\mathbf{v}}_i \cdot \hat{\mathbf{q}})}{dq^2 d\Omega} = \frac{d^2\sigma(q, \cos\vartheta)}{2M_N dE_r 2\pi d\cos\vartheta} = \frac{\sigma_{W-N}}{8\pi\mu_N^2 v_i} F^2(q)\,\delta\left(\mathbf{v}_i \cdot \hat{\mathbf{q}} - \frac{q}{2\mu_N}\right) . \tag{1}$$

Given a normalized velocity distribution for the incoming WIMP, $f(\mathbf{v}_i)$, and a WIMP mass density, ρ, the double differential recoil rate per unit mass as a function of the nuclear recoil energy, E_r, and of the recoil direction $\hat{\mathbf{q}}$ is

$$\frac{d^2R(E_r, \hat{\mathbf{q}})}{dE_r d\Omega_r} = \frac{\rho}{m_W}\frac{\sigma_{W-N}F^2(q)}{4\pi\mu_N^2}\hat{f}(v_{min}, \hat{\mathbf{q}}) , \tag{2}$$

where $v_{min} = q/(2\mu_N) = \sqrt{2M_N E_r}/(2\mu_N)$ is the minimal WIMP velocity that can give momentum q or energy E_r to the recoiling nucleus and $\hat{f}(v_{min}, \hat{\mathbf{q}})$ is the 3-dimensional Radon transform of the velocity distribution $f(\mathbf{v})$.

In this work, the Standard Halo Model, i.e. an Isotropic Maxwell-Boltzmann (IMB) WIMP velocity distribution of width σ_v in a inertial reference frame at rest with respect to the Galactic center,

 DOI: 10.1051/epjconf/201716407036

is assumed. The corresponding Radon transform [4] is

$$\hat{f}(v_{min}, \hat{\mathbf{q}}) = \frac{1}{\sqrt{2\pi\sigma_v^2}} \exp\left[-\frac{1}{2}\left(\frac{v_{min} + \hat{\mathbf{q}} \cdot \mathbf{V}}{\sigma_v}\right)^2\right] . \tag{3}$$

Therefore, if recoils are measured in a Galactic rest frame, $\mathbf{V} = 0$ and the rate is isotropic. On the other hand, when measured in a frame at rest with respect to the Sun, $\mathbf{V}$ is the Sun velocity relative to the galactic center, $\mathbf{V}_{SG}$, which points towards the galactic coordinates (ℓ_c=90°, b_c=0°), roughly in the direction of the Cygnus constellation. It has magnitude $V_{SG} \approx v_0$, where v_0 is the Galactic orbital speed at the Sun position. For an Earth-bound laboratory the velocity $\mathbf{V}$ can be decomposed as $\mathbf{V} = \mathbf{V}_{SG} + \mathbf{V}_{ES}$, where $\mathbf{V}_{ES}$ is the Earth velocity relative to the Sun.

Given E_{th} and E_{max} the threshold and maximum detector energy, respectively, such that the detector energy range is $E_{th} < E_r < E_{max}$, the direction dependent recoil rate per unit mass is obtained by substituting the Radon transform introduced in Eq. (3) inside Eq. (2) and integrating over the energy range

$$\frac{dR(E_{th}, E_{max}, \hat{\mathbf{q}})}{d\Omega_r} = \int_{E_{th}}^{E_{max}} dE_r \frac{d^2R(E_r, \hat{\mathbf{q}})}{dE_r d\Omega_r} . \tag{4}$$

In general, a CR directional detector should be able to discriminate only between tracks with different inclination relative to the vertical (electric field) direction. For this reason, the only measurable angle is the polar angle θ_r, and the azimuthal angle should be integrated out. The relevant recoil rate is

$$\frac{dR(E_{th}, E_{max}, \cos\theta_r)}{d\cos\theta_r} = \int_0^{2\pi} d\phi_r \int_{E_{th}}^{E_{max}} dE_r \frac{d^2R(E_r, \hat{\mathbf{q}})}{dE_r d\Omega_r} . \tag{5}$$

In addition, since a detector based on CR cannot distinguish signals from recoil tracks differing by 180° , i.e. events that differ by 180° are summed together, the relevant rate is the so-called "folded" angular recoil rate

$$\frac{dR_F}{d\cos\theta}(|\cos\theta_r|) \equiv \frac{dR}{d\cos\theta_r}(\cos\theta_r) + \frac{dR}{d\cos\theta_r}(-\cos\theta_r) , \tag{6}$$

which depends only on $|\cos\theta_r|$. In this work the results are shown using the reference values m_W = 200 GeV, $M_N = 0.923\,A$, where $A = 39.9$ is the argon atomic mass, $\rho = 0.3$ GeV cm^{-3}, and the IMB $\sigma_v = v_0/\sqrt{2}$. Rates are shown for a reference cross section $\sigma_n = 10^{-46}$ cm^2, which is of the same magnitude of the most stringent limit set by the LUX collaboration [5], and for a recoil energies range from E_{th} = 50 keV to E_{max} = 200 keV.

Note that the anisotropy of all rates in Eqs. (4), (5), and (6) depends only on the velocity $\mathbf{V}$. In a given frame, which fixes $\mathbf{V}$, one can choose different coordinate systems, yielding different components for $\mathbf{V}$. The coordinates can be also time dependent. In a frame at rest with respect to the Earth and using Galactic coordinates, $\mathbf{V}_{SG}$ is constant and only $\mathbf{V}_{ES}$ rotates with the annual periodicity of the Earth revolution. Since $\mathbf{V}_{ES}$ is an order of magnitude smaller than $\mathbf{V}_{SG}$, the recoil rate points back to a direction that rotates with an opening angle about one tenth of a radiant and annual periodicity around the $\mathbf{V}_{SG}$ direction. In this frame the peaked angular distribution is the main signature of the WIMP signal and allows background reduction. In a coordinate system fixed with a laboratory, the coordinates and, therefore, the apparent direction of $\mathbf{V}$ makes an additional rotation with the periodicity of a sidereal day and an amplitude that depends on the latitude. This specific periodicity is a characteristic signature and provides more background suppression.

DOI: 10.1051/epjconf/201716407036

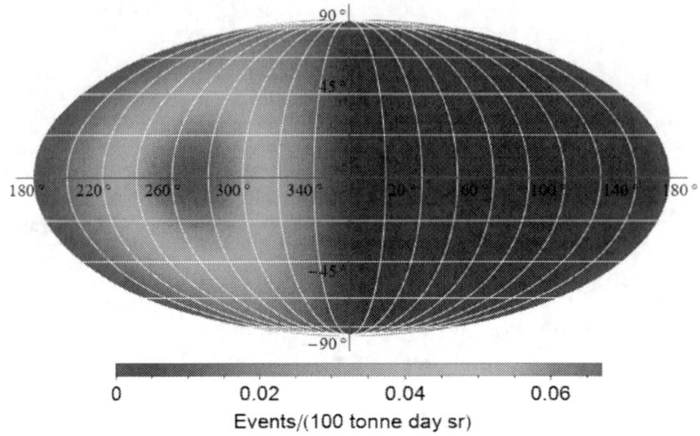

Figure 1. Directional differential recoil rate in argon integrated over the energy range 50÷200 keV, plotted in Mollweide equal area projection map of the celestial sphere in Galactic coordinates (ℓ, b). A WIMP mass of 200 GeV and a WIMP-nucleon cross section of 10^{-46} cm^2 is considered.

Most of our results will be shown for a detector located at the latitude of LNGS in the local coordinate system where the polar axis points in the vertical direction. A reference frame at rest with respect to the Sun is used only to introduce the potentialities of a directional detector, as it has been extensively done in literature [6].

4 Recoil spectrum in galactic coordinates

The general potentialities of a directional detector and, more specifically, the signature in the angular recoil rate of the detector motion through the WIMP wind are best illustrated in Galactic coordinates. In this coordinate system $\hat{x}$ points from the Sun towards the Galactic center, $\hat{y}$ in the direction of the Solar motion and $\hat{z}$ towards the Galactic north pole; therefore, $\mathbf{V} = v_0\hat{y}$. In Fig. 1 it is shown the angular recoil rate of Eq. (4) on a Mollweide equal area projection map in Galactic coordinates. The horizontal axis is the Galactic longitude, $0° < \ell < 360°$, and the vertical axis the Galactic latitude, $-90° < b < 90°$. The recoil rate is clearly anisotropic and points towards coordinates (ℓ=270°, b=0°), opposite to the direction of the Sun motion throughout the Galaxy.

5 Recoil directional signals at LNGS

In this section, the expected rates of WIMP scattering, Eqs. (4-6), is illustrated as a function of $\cos\theta_r$ for different times of an arbitrary chosen day, corresponding to the Summer Solstice (SS). Obviously, there is a strict correlation between the expected signal and the position of the Cygnus in the sky as seen at the LNGS latitude since it corresponds to the main direction of the WIMP wind. During most of the day, the Cygnus constellation is over the horizon, reaching the zenith at ~4:00 a.m. and setting around 12 hours later, remaining near the horizon just for few hours. Fig. 2 shows on the left the recoil rate, Eq. (5), as a function of $\cos\theta_r$ for different times of the day and the daily average. There is a strong dependence on the time of the day, since the asymmetry is larger when the Cygnus is high in the sky (00:00 $\leq t_{day}$[hr] $\leq$ 08:00), smaller when it is close to the horizon (12:00 $\leq t_{day}$[hr] $\leq$ 20:00). This

DOI: 10.1051/epjconf/201716407036

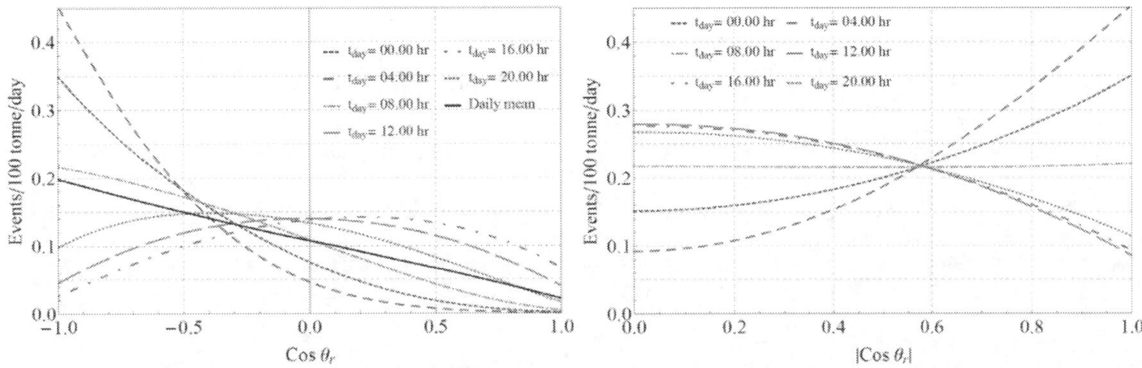

Figure 2. Differential (left) and folded differential (right) recoil rate as a function of the polar angle $\cos\theta_r$ at the latitude of LNGS. The y axis represents events per day for an exposure of 100 tonne year Argon detector with a 50 keV energy threshold for a cross section $\sigma_n = 10^{-46}$ cm^2. The solid curve is the daily average, while the other six curves show the rate at times of the day four-hour apart.

implies that it is possible to discriminate the WIMP nature of the events against possible backgrounds. Indeed, even the average rate varies considerably since it changes from ~0.2 events/(100 tonne day) for $\cos\theta_r = -1$, to ~0.02 events/(100 tonne day) for $\cos\theta_r = 1$. In the right panel of Fig. 2 it is shown the "folded" differential recoil rate introduced in Eq. (6), which should be the rate relevant for a CR detector. The angular and time dependences of the rate remain quite strong even without the information on where the track is pointing. At times when the Cygnus is close to the zenith (horizon) the rate is peaked at $|\cos\theta_r| \sim 1$ ($|\cos\theta_r| \sim 0$).

5.1 A realistic CR detector

A realistic CR detector have a finite angular resolution. Here, it is studied how much the angular signature of the signal is reduced assuming to have only a minimal amount of angular information. Events are divided in only two classes. Horizontal events (HOR), defined by $|\cos\theta_r| < 0.5$ or $60° < \theta_r < 120°$, and vertical events (VER=UP+DOWN), defined by $|\cos\theta_r| > 0.5$, as shown schematically in the left panel of Fig. 3. From an experimental point of view, ratios of event rates are useful quantities in order to keep the systematic uncertainties budget under control. Thus, the ratio, R, of HOR and VER events is defined, which should be constant and equal to one for an isotropic signal. On the right panel of Fig. 3 the time dependence of this ratio and of the single components is shown. The dashed lines represent the daily variation of the HOR and VER event rates, and their ratio, R, for four equidistant sidereal days. Use of the sidereal day (s.d.) as a unit is convenient because after a s.d. the Cygnus constellation returns exactly in the same position in the sky and the angle between the WIMP wind and the detector axis has the same value. Apart from a small variation in the velocity of the WIMP wind caused by the Earth revolution around the Sun, a similar behaviour of the HOR and VER components during different sidereal days is expected. Therefore it is possible to make the annual averages of the daily rates in order to accumulate statistics. The results are shown in the right panel of Fig. 3 in which the blue (orange) band represents the overall variation registered in the HOR (VER) component among different sidereal days of the year, while the solid curves represent the averages over different sidereal days. Considering the annual mean, the variation of the single component is of the order of 35% (29%), with respect to the average daily value. Moreover, the

DOI: 10.1051/epjconf/201716407036

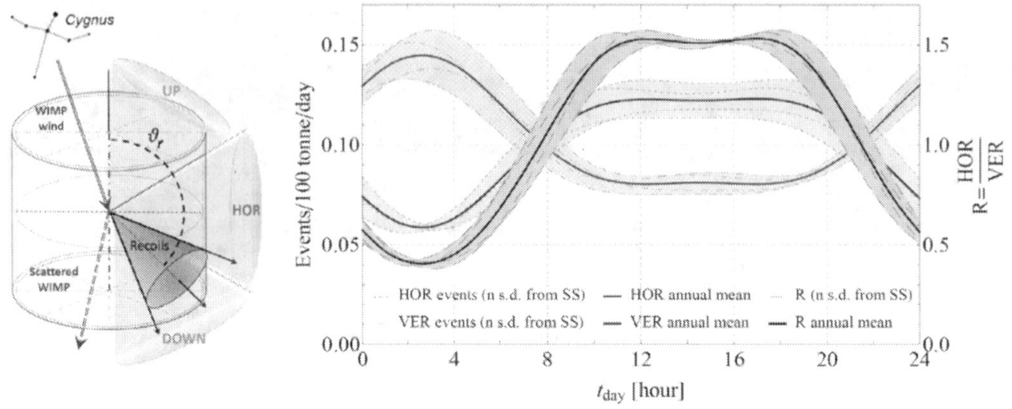

Figure 3. Left: Schematic representation of the recoils distribution between horizontal (HOR $\equiv |\cos\theta_r| < 0.5$) and vertical (VER=UP+DOWN $\equiv |\cos\theta_r| > 0.5$) events in a cylindrical LAr TPC for a particular positions of the Cygnus constellation in the sky. Right: Daily variation of HOR (VER) event rate in blue dashed (orange dotted) lines for different sidereal days of the year. The solid blue (orange) line represents the annual mean of the HOR (VER) events. They refer to the left scale. Ratio of HOR and VER events for different sidereal days (dashed black lines) and the corresponding annual mean (solid black line) refer to the right scale.

annual mean ratio, R, shows a remarkable deviation from one since it varies during the day by a factor $f \approx 4$. For comparison, the seasonal modulation of the event rate that can be observed under the same assumptions, for the case of a non-directional WIMP detection is only about 8%. Thus, even for this minimal angular resolution, it is confirmed that adding the directional information increases significantly the discovery potential of a WIMP signature.

6 ReD experiment

In this section, the ReD experiment will be briefly described. The aim of this project is to assess and enhance the directionality of two-phase argon TPC in order to provide an experimental proof for the CR effect. This represents a first fundamental step into the developing of a Dark Matter detector with directional sensitivity.

Initial ReD experiments will be carried out at Università degli Studi di Napoli "Federico II", using a $d - d$ neutron generator delivering monoenergetic neutrons of 2.45 MeV. Fig. 4 shows a schematic drawing of the experiment. LAr-TPC signals from neutron elastic scattering nuclear recoils of energy 10 keV to 200 keV in LAr will be recorded in the LAr-TPC, in coincidence with signals from the scattered neutron interacting in neutron detectors placed at various scattering angles relative to the beam. The coincidence kinematics allow the energy and direction of the nuclear recoils occurring in the LAr target to be precisely known event-by-event, given the beam energy and the scattered neutron angle. As demonstrated in SCENE [7], the coincidence technique combined with time-of-flight and the Pulse Shape Discrimination capabilities of both the LAr-TPC and the neutron detectors can give a very clean selection of single-scatter neutron elastic events. A directional dependence would manifest itself as a difference in the scintillation and/or ionization response for nuclear recoils of the same energy but different track orientations, selected from data of a single run according to which neutron detector fired.

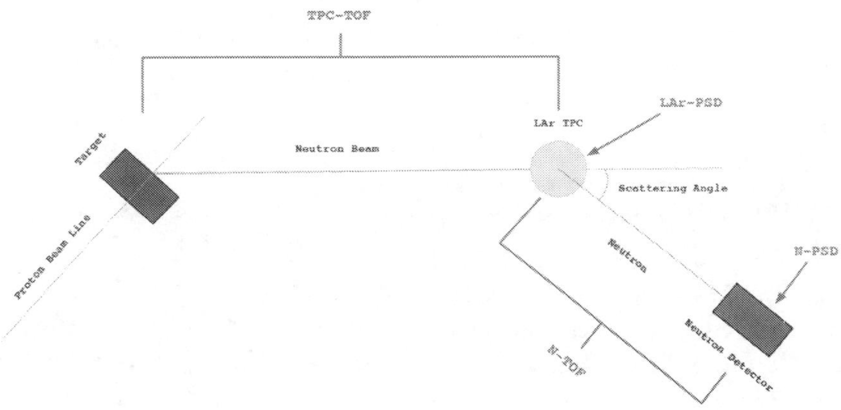

Figure 4. Schematic drawing of the ReD experimental setup

The ReD experimental setup is designed to minimize all neutron interactions other than single scatters in the liquid argon. This is achieved in two ways: by reducing as much as possible the material between the neutron beam and the LAr, and with an innovative structure for the LAr detector which minimizes the presence of inactive volumes.

Preliminary measurements will be performed using a simple neutron detector, in order to chose the most favorable reaction for the ReD purpose and to study the room background, to minimize parasitic interactions which would spoil the global performance. To measure the scattered neutrons the ReD experiment will use an array of liquid scintillator neutron detectors that can measure the neutron scattering angles and the neutron time-of-flight (TOF). In this way, for monochromatic neutrons single-scattered from LAr, the kinematics is fully determined and the ^{40}Ar recoil energy and direction can be reconstructed, and background from direct neutrons coming from the production target or random coincidences of neutrons with signals in the TPC can be suppressed. The proposed neutron spectrometer is based on 3" liquid scintillator counters from Scionix, using EJ309 liquid scintillator coupled to ET 982B PMTs. These tubes are fast enough to provide accurate TOF with a precision of around 1 ns. These detectors can provide separation between neutron-induced and other signals, by using of pulse shape discrimination, from 1.5 σ to 2 σ. The plan is to use 8 to 10 of such counters in the array, allowing us to record data for different recoil kinematics simultaneously.

7 Conclusions

In these proceedings, it is discussed how a directional detector located at LNGS could provide a very clear signature for WIMP signal. This signature remains sizeable even for a minimal angular resolution which should be easily obtained by a detector based on columnar recombination. In particular, the daily modulation shown in Fig. 3 is an order of magnitude larger than the annual modulation obtained ignoring the angular information. Thus, a realistic detector able to use the angular information to reject the background would provide a strong and unambiguous evidence of DM, the latter due to the fact that no known background is correlated with the sidereal period of the DM modulation.

The status of the ReD experiment, which is under construction at Naples University, has also been reported. The success of the ReD experiment could have a crucial impact on the broader program of next generation direct Dark Matter detection experiment. The discovery of sensitivity to the direction of recoils in liquid argon would revolutionize the field of Dark Matter direct searches.

References

[1] R. Bernabei, et al., *Eur. Phys. J.* C **56**, 333-355 (2008)
[2] S. Ahlen et al, *Int. J. Mod. Phys.* A **25**, 1–51 (2010)
[3] G. Jaffé, *Ann. Phys. (Leipzig)* **42**, p. 303 (1913)
[4] P. Gondolo, *Phys. Rev.* D **66**, 103513 (2002)
[5] D. S. Akerib et al., (the LUX Collaboration) *ArXiv:astro-ph.CO*, **1608.07648** (2016)
[6] F. Billard, J. Mayet J, F. Macias-Perez and D. Santos, *Phys. Lett.* B **691**, 156-162 (2010)
[7] H. Cao et al., (The SCENE Collaboration), *Phys. Rev.* D **91**, 092007 (2015)

 DOI: 10.1051/epjconf/201716407037

Review of recent heavy flavor measurements in STAR

Michael R. Lomnitz[1,2,a] (for the STAR collaboration)

[1] *Nuclear Science Division, Lawrence Berkeley National Lab, MS70R0319, One Cyclotron Road, Berkeley, CA 94720, USA*
[2] *Department of Physics, Kent State University, 800 E. Summit St., Kent, OH 44240, USA*

Abstract. Heavy-ion collisions at RHIC provide a unique environment to study the behavior of nuclear matter under extreme conditions. In particular, heavy quarks, which are produced during the early stages of a collision, provide an exceptional probe in understanding the hot and dense medium created in such collisions. The Heavy Flavor Tracker and Muon Telescope Detector at the STAR experiment at RHIC have been successfully installed since early 2014 and have significantly improved the experimental capabilities in measuring both open and hidden heavy flavor hadrons in heavy-ion collisions. We present an overview of recent heavy flavor results obtained at STAR using these two dedicated detectors.

1 Introduction

Heavy flavor quarks are suggested to be an excellent probe to study the strongly coupled quark-gluon plasma (sQGP) as they are predominantly produced early in heavy-ion collisions through hard scattering processes. They experience the full evolution of the system, while their large masses are mostly unaffected by the QCD medium. Measurements of open heavy flavor and quarkonia at the Relativistic Heavy Ion Collider (RHIC) have revealed several interesting features such as: the enhancement (suppression) in D^0 production at intermediate (high) transverse momenta (p_T) in Au+Au collisions at $\sqrt{s_{\mathrm{NN}}}$ = 200 GeV relative to p+p collisions [1]; the weak collision energy dependence of the J/ψ suppression in Au+Au collisions from $\sqrt{s_{\mathrm{NN}}}$ = 39 to 200 GeV [2]. While the former can be described by theoretical frameworks including strong charm-medium interactions and charm-light quark coalescence, the latter is attributed to the interplay of color screening in the QGP and J/ψ regeneration from deconfined $c\bar{c}$ pairs.
Despite significant progress in the last several decades, precise measurements are still needed to fully understand the properties of the sQGP and the parton-QGP interactions. The STAR experiment has recently completed collecting data with two new detectors designed for heavy flavor studies: the Heavy Flavor Tracker (HFT) [3] and the Muon Telescope Detector (MTD) [4]. In these proceedings we will present the first set of physics results using the HFT and the MTD.

[a] e-mail: mrlomnitz@lbl.gov

2 The STAR experiment

STAR is a large acceptance, multi-purpose detector at RHIC designed to discover and study the QGP. It possesses full azimuthal coverage at mid-rapidity ($|\eta| < 1$) using the Time Projection Chamber (TPC) to reconstruct tracks inside a uniform 0.5 T magnetic field. By measuring the energy loss (dE/dx), the TPC can also provide excellent PID for charged hadrons over a broad range in p_T. Combining with information from the Time Of Flight (TOF) detector, STAR can provide accurate identification over ~ 98% of the charged hadrons produced in heavy-ion collisions.
Since the beginning of 2014, two new detectors, the HFT and MTD, have been fully installed at STAR. The HFT consists of four layers of silicon detectors located inside the TPC. With thinned Monolithic Active Pixel Sensors (MAPS), the HFT greatly improves the track pointing resolution to be less than 50 μm for kaons with $p_T = 750$ MeV/c, and thus permits topological reconstruction of open heavy flavor decays through hadronic channels.
The MTD is installed outside of the STAR magnet, utilizing Multi-gap Resistive Plate Chamber (MRPC) technology. By performing precise time-of-flight measurements with resolution of ~ 100 ps, the MTD can trigger on and identify muons with $p_T > 1.2$ GeV/c and $|\eta| < 0.5$. The HFT and MTD have collected large-statistics data sets in p+p, p+Au, d+Au, and Au+Au collisions at $\sqrt{s_{\rm NN}} = 200$ GeV between 2014 and 2016. Results presented below are obtained using around 1/4 of the Minimum-Bias (MB) data with the HFT and 1/2 of the dimuon triggered data with the MTD in Au+Au collisions.

3 Open heavy flavor production

The charm production cross-sections at mid-rapidity ($|y| < 1$) have been measured by STAR in p+p and Au+Au collisions at $\sqrt{s_{\rm NN}} = 200$ GeV, and were found to scale with the number of binary nucleon collisions [1]. This observation supports the picture in which heavy flavor quarks are predominantly produced in initial hard parton scatterings at RHIC. Moreover, because of their relatively long lifetimes, heavy flavor quarks can experience the entire evolution of the system, serving as a calibrated probe to study the QGP. Below we present results on D^0 meson production reconstructed through the hadronic decay channels.
Using the HFT, the combinatorial background is greatly suppressed by selecting on the following topological variables: decay length (distance between primary and decay vertices), distance of closest approach (DCA) between daughter tracks, DCA between reconstructed D^0 candidate and the primary vertex (PV), and DCA between daughter tracks and the PV. Shown in figure 1a are the nuclear modification factor (R_{AA}) of D^0 in 0-10% Au+Au collisions at $\sqrt{s_{\rm NN}} = 200$ GeV extracted from about 780 million MB events with the HFT (black solid circles). Also shown are the STAR published results from 1.1 billion MB events recorded in 2010 and 2011 without the HFT (red open circles) [1], as well as the R_{AA} for charged pions. The vertical bars (brackets) represent statistical (systematic) uncertainties, while the grey shaded areas depict the systematic uncertainties from the p+p reference uncertainties for the R_{AA} obtained from the HFT data. The latter are not shown for the published results due to the cancellation of efficiency corrections between the p+p and 2010/2011 Au+Au data, both of which had the same experimental set-up. The colored boxes centered at $R_{AA} = 1$ on the far right of the panel correspond to global normalization uncertainties for 0-10% Au+Au (left) and p+p (right) collisions. Despite using a smaller data sample and finer $p_T$bins, the HFT results have significantly better precision than the published ones. The new results confirm the strong suppression of D^0 yields (relative to p+p) at high transverse momenta, which is comparable to the suppression observed for light hadrons, an indication of strong charm-medium interactions.
Leveraging the HFT also enables the first measurement of the second-order azimuthal anisotropy (v_2)

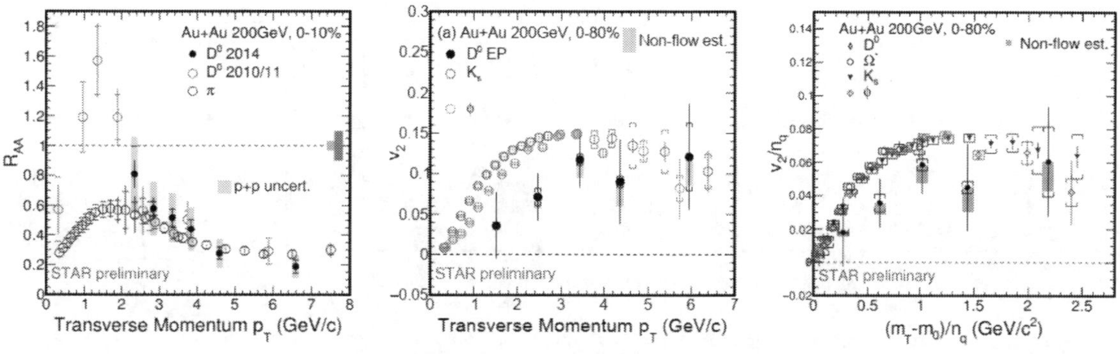

(a) (b) (c)

Figure 1: a) Measured D^0 nuclear modification factor, R_{AA}, with and without the HFT, compared to that of pions. b) Measured D^0 v_2 compared to that of light hadrons. c) v_2/n_q vs. $(m_T - m_0)/n_q$ for D^0 and other particle species.

for D^0 at RHIC. Figure 1b shows the D^0 v_2 in 0-80% Au+Au collisions which are significantly above zero for $p_T > 2$ GeV/c suggesting charm quarks flow with the medium. The vertical bars (brackets) represent statistical (systematic) uncertainties, while the grey shaded areas correspond to the estimate of contributions from non-flow effects. The rightmost panel, figure 1c, shows v_2/n_q as a function of $(\sqrt{m_0^2 + p_T^2} - m_0)/n_q$, where m_0 is the hadron mass and n_q the number of constituent quarks. The D^0 v_2/n_q is systematically below those of light hadrons [5], but still consistent within uncertainties.

Figures 2a and 2b show the comparison between the D^0 R_{AA} and v_2 to three theoretical models. The TAMU model [6] uses a non-perturbative T-matrix approach assuming that the two-body interactions can be described by a potential as a function of the transferred 4-momentum. In the case of v_2, two curves from TAMU are shown: the scenario including charm diffusion (blue) agrees with the data while the predictions without charm diffusion (magenta) are systematically lower. The calculation by the SUBATECH group [7] employs pQCD with Hard Thermal Loop approximation for soft collisions and can also describe both measurements over the whole p_T range. In the model developed by the Duke group [12] the diffusion coefficient $2\pi T D_s$ is a free parameter which, in the case of the red curve shown, has been constrained using the R_{AA} measured at LHC to be roughly 7. While both the SUBATECH and TAMU models show good agreement with the data in both observables, the calculation from the DUKE group under-predicts the v_2 observed in our data. The third panel, figure 2c, shows the extracted diffusion coefficient from different model calculations compared to the yellow band on the far right, which indicates the range of inferred values that are compatible with the measurements presented here. It is worth noting that lattice QCD predictions [13], within large uncertainties, are consistent with the inferred range from data.

4 Quarkonium production

While the heavy quark production can be calculated by pQCD, how they evolve into quarkonia, a bound quark-antiquark pair, is intrinsically non-perturbative. For this reason, precise measurements of quarkonium production in p+p collisions can help understand quarkonium production mechanisms.

DOI: 10.1051/epjconf/201716407037

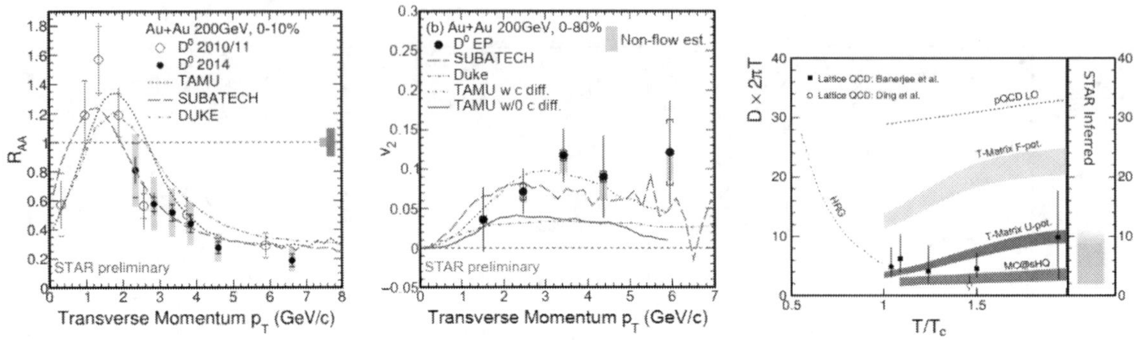

(a) (b) (c)

Figure 2: a) D^0 R_{AA} in 0-10% centrality compared to model calculations. b) D^0 v_2 in 0-80% centrality compared to model calculations. c) Diffusion coefficient from model calculations and the inferred range from STAR measurements.

On the other hand, in heavy-ion collisions quarkonia may dissociate due to the color screening of the heavy quark-antiquark potential by the QGP. This dissociation should show dependence on the temperature of the QGP, and the binding energies of the different quarkonium states. Therefore the comparison between quarkonium production in heavy-ion collisions relative to p+p collisions can be used to experimentally deduce the temperature of the QGP medium once the color screening, regeneration, and cold nuclear matter effects are fully understood. Below, we present J/ψ and Υ measurements at STAR using the newly installed MTD described in previous sections.

4.1 J/ψ production

J/ψ invariant yields in Au+Au collisions at $\sqrt{s_{NN}}$ = 200 GeV have been determined via the dimuon channel using the data recorded in 2014. The results (solid points) are shown as a function of the transverse momenta for different centrality classes in figure 3a, which are seen to be consistent with the STAR published results in the dielectron channel (open points). Figure 3b (3c) shows the R_{AA} for $p_T > 0$ (5) GeV/c determined from the above mentioned MTD data (red stars) compared to the results from PHENIX [9] for Au+Au collisions at $\sqrt{s_{NN}}$ = 200 GeV (black open circles) as well as the measurements from ALICE and CMS in Pb+Pb collisions at $\sqrt{s_{NN}}$ = 2.76 TeV (blue squares). For $p_T > 0$ GeV/c, the STAR results are consistent with the PHENIX results showing a decreasing trend towards more central collisions. On the other hand the ALICE results are almost independent of the centrality class and are significantly higher than those at RHIC in the central collisions. This difference could be due to a larger contribution from regeneration at low p_T due to higher $c\bar{c}$ yields, increasing J/ψ production at the LHC. At high p_T, the R_{AA} at both RHIC and LHC have a decreasing trend towards more central collisions, and the LHC results are lower than those at RHIC. This is likely due to higher temperature reached at the LHC which leads to stronger J/ψ dissociation. Two different transport models [10, 11], both including dissociation and regeneration, can qualitatively describe the measured R_{AA}. There is, however, tension between the model calculations and experimental data at high transverse momenta.

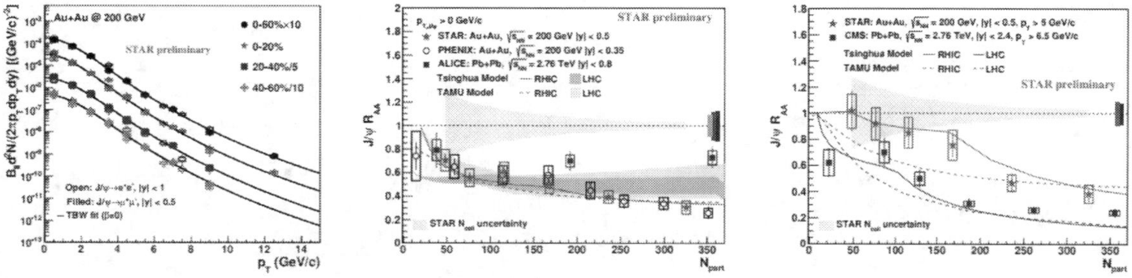

(a) (b) (c)

Figure 3: a) J/ψ invariant yields in Au+Au collisions at $\sqrt{s_{NN}}$ = 200 GeV in four centrality classes. b) J/ψ R_{AA} vs. N_{part} for $p_T > 0$ GeV/c in Au+Au (Pb+Pb) at $\sqrt{s_{NN}}$ = 200 GeV (2.76 TeV). c) J/ψ R_{AA} vs. N_{part} for $p_T > 5$ GeV/c in Au+Au (Pb+Pb) at $\sqrt{s_{NN}}$ = 200 GeV(2.76 TeV).

4.2 Υ production

Using the MTD data described above, STAR has studied Υ production in Au+Au collisions at $\sqrt{s_{NN}}$ = 200 GeV. Compared to the dielectron channel, the dimuon channel has the advantage of better invariant mass resolution due to the significantly suppressed bremsstrahlung radiation. Figure 4a shows the invariant mass spectrum of dimuon pairs. The unlike-sign (red points) distribution is fitted to the sum of Υ(1S), Υ(2S) and Υ(3S) signal distributions, $B\overline{B}$, Drell-Yan and the combinatorial background distributions. The shapes of the signal distributions are Gaussian functions with means fixed to the PDG values, and widths determined from detector simulations, while the shape of $B\overline{B}$ and Drell-Yan background is determined from PYTHIA. Both the shape and normalization of the combinatorial background are obtained by fitting the like-sign (black points) spectrum. Shown in figure 4b are the ratio of the excited state yield to the ground state yield, obtained from the previous fit, together with the world-wide p+p data and CMS measurements [12, 14] in Pb+Pb collisions at $\sqrt{s_{NN}}$ = 2.76 TeV. There is an indication of larger $\Upsilon(2S+3S)/\Upsilon(1S)$ ratio at RHIC than that at the LHC.

5 Summary

STAR has carried out the first heavy flavor measurements in heavy-ion collisions using the newly installed HFT and MTD detectors. The measured charmed meson v_2 in Au+Au collisions is found to be non-zero, but systematically below the v_2 of light hadrons in 0-80% centrality, while the measurement of the R_{AA} using the HFT shows a similar level of suppression for the charmed meson as the light hadrons at high transverse momenta. Comparison to a series of models shows that they are able to describe both the v_2 and R_{AA} simultaneously, favouring a scenario where charm quarks flow with the medium. These comparisons can be used to infer a range, between 2 and ~12, for the value of the charm diffusion coefficient $2\pi T D_s$.
The measurements of J/ψ with the MTD via the dimuon channel yield consistent results as obtained through the dielectron channel and show a distinct rising trend in the R_{AA} with transverse momenta in 20-60% centrality. The suppression observed at high $p_T$in 0-20% central collisions is consistent with dissociation due to color screening.

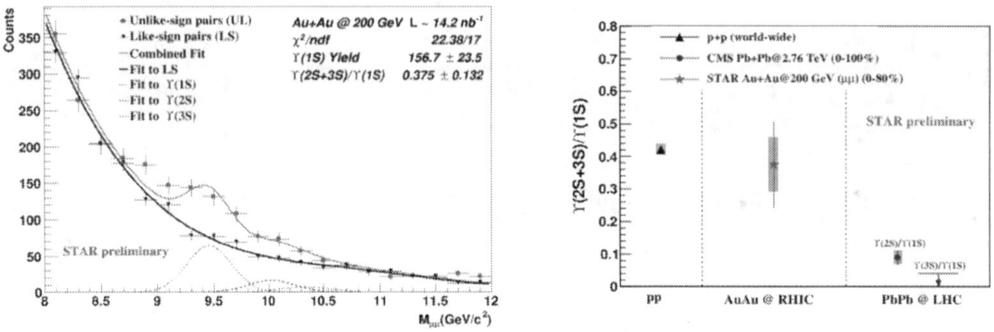

(a) (b)

Figure 4: a) Invariant mass distribution of dimuon pairs in Au+Au collisions at $\sqrt{s_{NN}} = 200$ GeV. b) Ratio of excited to ground ϒ states from world data in p+p collisions, Au+Au collisions at $\sqrt{s_{NN}} = 200$ GeV and Pb+Pb collisions at $\sqrt{s_{NN}} = 2.76$ TeV.

Acknowledgements

This material is based upon work supported by the U.S. Department of Energy, Office of Science, Office of Workforce Development for Teachers and Scientists, Office of Science Graduate Student Research (SCGSR) program. The SCGSR program is administered by the Oak Ridge Institute for Science and Education for the DOE under contract number DE-AC05-06OR23100.

References

[1] L. Adamczyk et al. (STAR Collaboration), Phys. Rev. Lett. **113** (2014) 142301.
[2] L. Adamczyk et al. (STAR Collaboration), arXiv 1607.07517 (2016).
[3] Technical Design Report: The STAR Heavy Flavor Tracker (2011).
[4] L. Ruan et al., J. Phys. G: Nucl. Part. Phys. **36** (2009) 095001.
[5] B. I. Abelev et al. (STAR Collaboration), Phys. Rev. C **77**(5) (2008) 054901
[6] M. He, R. J. Fries and R. Rapp, Phys. Rev. C **86**(1) (2012) 014903
[7] M. Nahrgang et. al., Phys. Rev. C **91**(1) (2015) 014904
[8] Cao S, Qin G Y and Bass S A, Phys. Rev. C **88**(4) (2013) 044907
[9] A. Adare et al. (PHENIX collaboration), Phys. Rev. D **82** (2010) 012001.
[10] Y. Liu et. al. , Phys. Let. B **678** (2009) 72; K. Zhou et. al. , Phys. Rev. C **89** (2014) 054911.
[11] X. Zhao and R. Rapp, Phys. Rev. C **82** (2010) 064905.
[12] W. Zha et. al., Phys. Rev. C **88** (2013) 067901
[13] A. Andronic et. al., The Eur. Phys. J. C **76** (2016) 3 107
[14] S. Chatrchyan et al. (CMS Collaboration), PRL **109** (2012) 222301

EPJ Web of Conferences **164**, 07038 (2017) DOI: 10.1051/epjconf/201716407038
ICNFP 2016

Search for $K^+ \rightarrow \pi^+ \nu\bar{\nu}$ at NA62

Dario Soldi[1,a]
On behalf of the NA62 Collaboration

[1] *University of Turin*

Abstract. $K^+ \rightarrow \pi^+ \nu\bar{\nu}$ is one of the theoretically cleanest meson decay where to look for indirect effects of new physics complementary to LHC searches. The NA62 experiment at the CERN SPS is designed to measure the branching ratio of this decay with 10% precision. NA62 took data in pilot runs in 2014 and 2015 reaching the final designed beam intensity. The quality of data acquired in view of the final measurement will be presented.

1 Introduction

The NA62 experiment is located in the North Area High Intensity Facility at CERN. It is a high intensity kaon beam fixed target experiment [1]. The 3-year data-taking program at the CERN Super Proton Synchrotron (SPS) has started in 2015. The main goal of the experiment is to measure the branching ratio of the ultra-rare decay: $K^+ \rightarrow \pi^+ \nu\bar{\nu}$ with 10% accuracy. This decay is driven by Flavor Changing Neutral Current (FCNC), and the branching ratio, highly suppressed by GIM mechanism, is predicted in Standard Model with high precision (see figure 1). This makes this process very sensitive to new physics. The actual Standard Model prediction of the branching ratio is $(8.4 \pm 1.0) \times 10^{-11}$ [2]. Should the experimental value be different from the prediction, new physics contributions have to be present.

The only experimental value has been obtained combining datasets collected by Brookhaven National Laboratory (BNL) E787 and E949 experiments [3, 4]. The measured branching ratio, based on 7 events, is:

$$BR_{SM}(K^+ \rightarrow \pi^+ \nu\bar{\nu}) = (17.3^{+11.5}_{-10.5} \times 10^{-11})$$

2 NA62 Experiment

The NA62 experiment [5], sketched in figure 2, exploits the 400 GeV/c proton beam delivered by the CERN SPS. The protons impinging on a beryllium target produce a secondary hadron beam with a momentum 75 GeV/c composed of: 71% pions, 23% protons and 6% kaons. Kaons are recognized with a differential Cherenkov counter detector: the KTAG. The KTAG combines a time resolution of ~100ps and identification efficiency higher than 95%. The kaon direction momentum and time are

[a]e-mail: dario.soldi@cern.ch

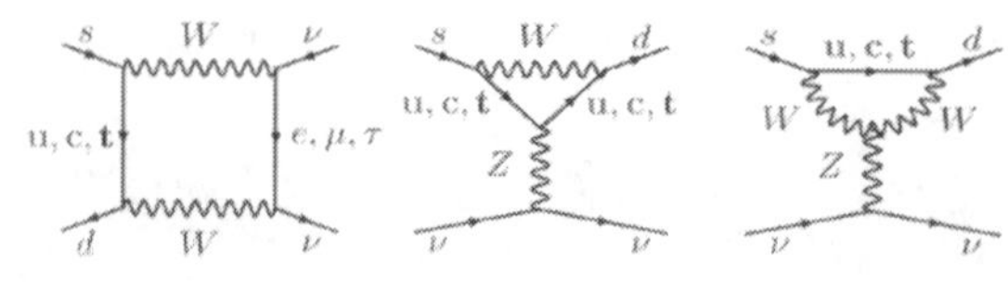

Figure 1. Flavor Changing Neutral Current (FCNC) loop processes, highly suppressed by GIM mechanism, but theoretically very clean [2].

measured by the GigaTracKer (GTK) detector. The GTK operates at ~750 MHz rate. It is composed by three stations interspersed by two pairs of dipole magnets. Each station is made of 200 μm thick silicon pixel matrix (300×300μm^2) coupled with 100 μm thick readout chip for a total of 0.5 X_0 per station with time resolution better than 200 ps and momentum resolution 0.2%. The decay products momentum is measured by the Straw spectrometer (STRAW). This detector is composed of 4 stations made of 1792 straw tubes positioned in 4 views (X, Y, U, V). They are located in vacuum before and after a dipole magnet which produces a vertical field of 0.36 T. The momentum resolution is better than 1%. A first pion/muon separation is obtained with the RICH detector, wich is a Cherenkov detector filled with Neon at atmospheric pressure. It is designed to work in the momentum range of $15 \div 35$ GeV/c. Muon are recognised with 99% of efficiency. Time matching between primary and secondary tracks and reference time for the trigger system require RICH time resolution less than 100psec. The Charged particle HODoscope (CHOD) consisting of 128 scintillator slabs arranged in 2 planes (horizontal and vertical) measures the track crossing time with ~200 ps resolution. The photon veto system allows to reject decays with π^0 in the final state with an inefficiency better than 10^{-8}. The system covers the solid angle up to 50 mrad and it includes 4 different detectors:

- Large Angle Veto (LAV) covering the angle region $8.5 \div 50$ mrad and composed of 12 station made of lead glass blocks;
- Liquid Kripton electromagnetic calorimeter (LKr) reused from the NA48 experiment, vetoing the region $1 \div 8.5$ mrad;
- Small Angle Calorimeter (SAC) and Inner Radius Calorimeter (IRC) covering the angle region < 1 mrad.

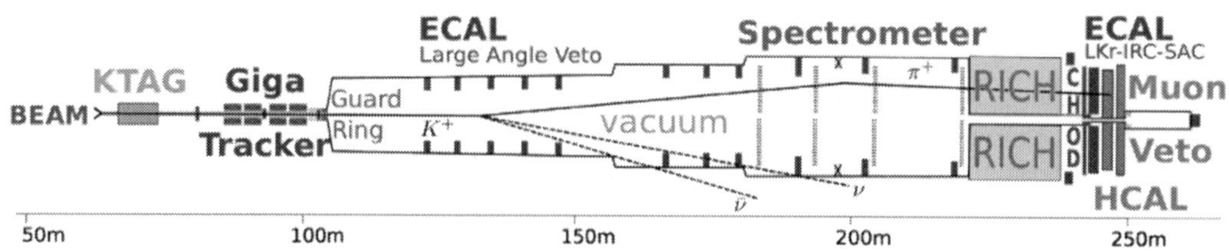

Figure 2. Scheme of the NA62 experimental setup.

Upstream, inelastic interactions and muon halo are vetoed by the CHANTI detector placed after the third station of GTK (referred as Guard Ring in figure 2).

The iron/scintillator hadronic calorimeters (HCAL) provide further pion/muon separation. Together with RICH, they guarantee a separation inefficiency of 10^{-7}. Finally the fast Muon Veto

 DOI: 10.1051/epjconf/201716407038

detector composed of an iron wall followed by 5 cm thick scintillators, coupled to PMTs, identifies muons. Muon Veto provides also signals for the trigger system.

In order to reduce the raw data rate, NA62 implements a multi-level trigger system consisting of a hardware L0 trigger processor, followed by a multi stage software based High-Level Trigger (HLT, L1 and L2). The purpose of the trigger system is to reduce the data rate to a manageable level, rejecting as much as possible all the main decay modes of the K^+ with a large efficiency for the $K^+ \to \pi^+ \nu\bar{\nu}$ signal. L0 logic has been implemented in hardware (FPGA), with the goal of reducing the trigger rate from 10MHz to ~1MHz. On a positive L0 decision data are transferred from the detector readout electronics to a dedicated farm of computers, which implements the HLT algorithms and other data acquisition functionality, such as the event building. L1 and L2 are software triggers expected to reduce the L0 rate from ~1MHz to ~10 kHz. The L1 algorithm aims at rejecting multi-track events, presently relying only on CHOD and LAV data fragments. A positive L1 decision issues a request to the LKr calorimeter in order to gather the full event data in the farm.

Assuming 10% signal acceptance, at least 10^{13} K^+ decays are required to reach the designed precision level. In this framework the fully functional experimental apparatus is expected to detect ~ 45 events per year.

3 Analysis method

The signature of the $K^+ \to \pi^+ \nu\bar{\nu}$ signal is one track in the GTK making a vertex with a charged track identified as π^+ in the detector downstream. A key variable used to distinguish the signal from the background is the squared missing mass defined as:

$$m^2_{miss} = (p_{K^+} - p_{\pi^+})^2$$

where p_{K^+} and p_{π^+} are the four momenta of incoming kaon and outgoing pion respectively. It is possible to identify two regions in the m^2_{miss} distribution where the signal to background ratio is most favorable (see figure 3).

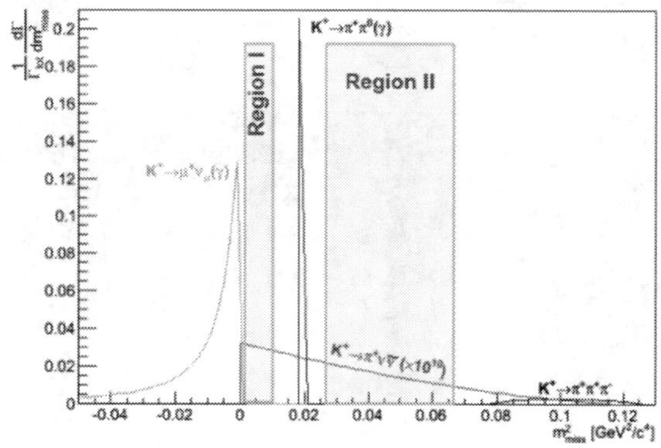

Figure 3. Theoretical m^2_{miss} distribution for signal and backgrounds from the main K^+ decay modes. In the figure the two regions where the signal/ratio is larger are highlighted.

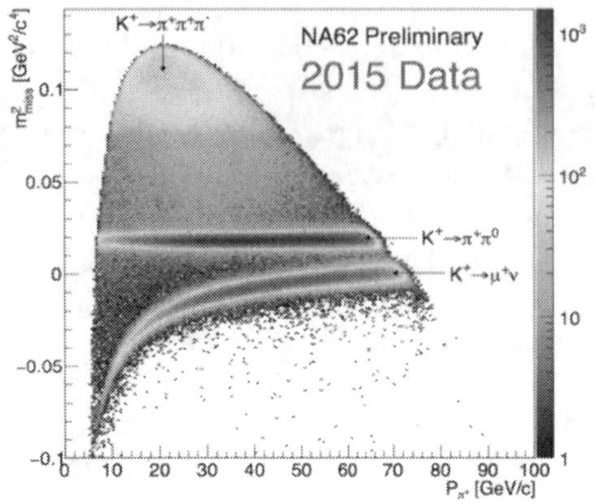

Figure 4. m^2_{miss} versus the momentum of the downstream track in single-track events confirmed by GTK and KTAG signals.

3.1 2015 data quality analysis

In this section we present some results from a study on data quality based on data collected in 2015 and relevant for the measurement of the $K^+ \to \pi^+ \nu\bar{\nu}$ decay. A single outgoing track has to form a vertex with the incoming kaon, required to be in the 65m long region between the last GTK and the first straw station. A quality check on the vertex is based on the distance of closest approach (CDA) between GTK and Straw tracks. CDA has to be less than 1.5 cm. Further selection requirements to have a valid signal are:

- the track projection needs to spatially intersect an energy deposits in the calorimeters and the corresponding hit in the CHOD; and
- the associated CHOD hits have to be in time with the KTAG candidate.

The squared missing mass of tracks that satisfy all those constraints are reported in figure 4 as a function of the track momentum as measured with the spectrometer. The data shown were recorded in 2015 at around 3% of the nominal intensity.

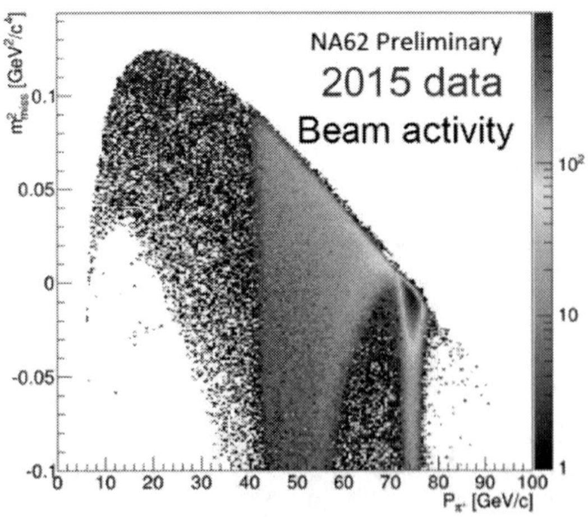

Figure 5. Plot of m^2_{miss} vs P_{π^+} momentum under π^+ mass hypothesis. The downstream track was required to be in anti-coincidence with the kaon track in KTAG.

The KTAG can also be used in anti-coincidence with a Gigatracker track in order to select single track events not related to kaons (Figure 5). This technique allows to study the background generated by the non-kaon beam. The figure 5 shows the main sources of non-kaon tracks:

 DOI: 10.1051/epjconf/201716407038

- decay from beam π^+;
- elastic scattering of beam particles in the material along the beam line (KTAG and Gigatracker stations); and
- inelastic scatterings in the last GTK station are the main sources of non-kaon tracks.

3.2 Kinematic results

The resolution of the $K^+ \to \pi^+\pi^0$ squared missing mass peak is essential to reduce the tails that can span across the two signal regions. In figure 6 is shown the squared missing mass resolution as a function of momentum. The resolution measured is 1.2×10^{-3} GeV$^2/c^4$, close to the design value represented by the solid black line.

In figure 6 can also be seen that the resolution is three times larger without the GTK detector.

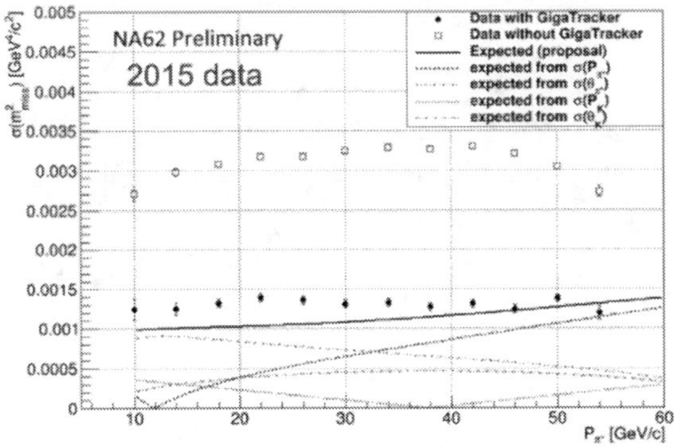

Figure 6. $\sigma(m^2_{miss})$ distribution obtained with individual kaon track measured with GTK (black points) and assuming kaon track aligned on the central beam axis (empty squares).

3.3 Particle identification

The particle identification of NA62 is designed to separate π^+ from μ^+ and e^+ in order to guarantee at least 7 orders of magnitude suppression of $K^+ \to \mu^+\nu_\mu$ in addition to the kinematic rejection.

The RICH is designed to contribute with a factor 100 to that rejection. To study the performance, $K^+ \to \mu^+\nu_\mu$ and $K^+ \to \pi^+\pi^0$ decays were selected applying the kinematic constraints, Muon Veto, and calorimetric information. The positive tracks were selected to be between 15 and 35 Gev/c. In fig 7, it is shown that remaining muon contamination of 0.01 corresponds to a pion efficiency of 80%, close to the design target. This figures are expected to be improved in 2016 run.

3.4 Photon veto

The photon veto system has to provide a suppression factor of 10^{-8} to reject π^0 from the decay: $K^+ \to \pi^+\pi^0$ ($\pi^0 \to \gamma\gamma$).

The rejection of such events is implemented by requesting at least a photon in one of the electromagnetic calorimeters, LAV, LKr, IRC or SAC. The target of a 10^8 suppression factor, is fulfilled by requiring that the photon detection has an inefficiency lower than 10^{-5} for energy greater than 10 GeV . This is true for pions that have at least 35 GeV/c momentum. Figure 8 shows the inefficiency for the detection of π^0 when only the LKr calorimeter is used, when LKr and LAV are used and when the full photon veto system is employed. The measurement of the efficiency with 2015 data results is less than 10^{-6} at 90% C.L. but it is statistically limited. New measurements are ongoing with 2016 data.

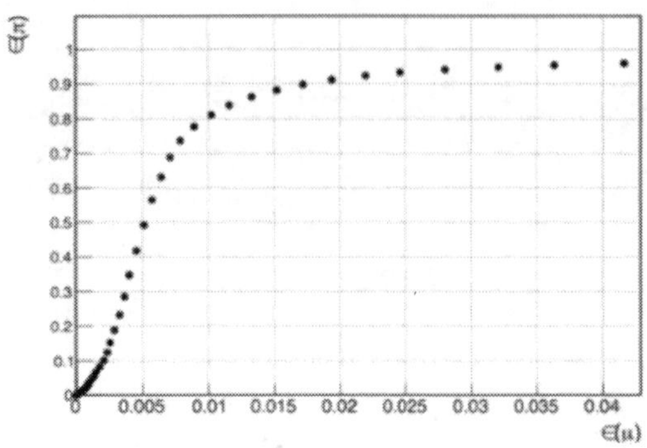

Figure 7. π^+ efficiency as a function of μ^+ remaining contamination measured by the RICH.

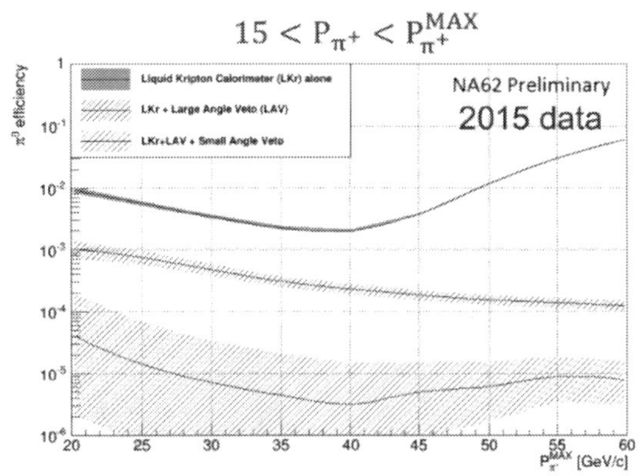

Figure 8. π^0 veto efficiency for different combinations of the calorimeters information.

4 Conclusions and Prospects

The $K^+ \to \pi^+ \nu\bar{\nu}$ branching ratio is one of the key observables to test the SM. The NA62 experiment is designed to measure its branching ratio with a 10% precision. To conclude, the data taken in 2015 shows performances close to the design sensitivity. Refined analyses are ongoing and data taking has restarted in April 2016.

References

[1] Proposal to Measure the Rare Decay $K^+ \to \pi^+ \nu\bar{\nu}$ at the CERN SPS, CERN-SPSC-2005-013 (2005)
[2] Andrzej J. Buras, Journal of High Energy Physics **1511** 033 (2015)
[3] S. Adler et al. (E949 and E787 Collaborations), Phys. Rev. D **77**, 052003 (2008).
[4] S. Adler et al. (E949 and E787 Collaborations), Phys. Rev. D **79**, 092004 (2009).
[5] NA62-Collaboration, Technical Design Document , 092004 (2010).

EPJ Web of Conferences **164**, 07039 (2017) DOI: 10.1051/epjconf/201716407039
ICNFP 2016

Search for new physics beyond the Standard Model in final states with jets and leptons+jets at CMS

Francesco Santanastasio[1,a] on behalf of the CMS collaboration

[1] *Sapienza, University of Rome and INFN Rome*

Abstract. Searches for new physics beyond the Standard Model performed by the CMS experiment in final states with jets and leptons+jets are presented. The document focuses on a selection of recent results from searches for heavy dijet resonances, leptoquarks, and heavy majorana neutrinos using proton-proton collision data at both $\sqrt{s}$ =8 and 13 TeV. The novel CMS datascouting technique, used to probe the low-mass region in hadronic final states, is introduced and results from a Run1 analysis are shown. Two searches in the electrons+jets final states showing an excess of events in Run1 data are also discussed.

1 Introduction

Searches for new physics beyond the Standard Model performed by the CMS experiment [1] at the CERN LHC are presented in these conference proceedings. A selection of searches in final states with jets and leptons+jets is shown[1]. Results are obtained using proton-proton collision data collected in Run1 (2012) at $\sqrt{s}$ = 8 TeV and Run2 (2015) at $\sqrt{s}$ = 13 TeV. Section 2 presents results of the search for dijet resonances at high-mass using 2015 Run2 data and results of the low-mass analysis with the Run1 *data scouting* technique. Section 3 reports the results from searches for leptoquarks at $\sqrt{s}$ = 13 TeV. In Section 4 a few interesting excesses of events in data observed in the Run1 searches for leptoquarks and heavy majorana neutrinos are discussed. Finally the future prospects for these searches are presented in Section 5.

2 Dijet resonances

The search for new resonances decaying to pairs of jets is among the most important ones at LHC because any hypothetical new particle that might be produced originates from the colliding protons and therefore it must couple to quarks and/or gluons.

The analysis strategy consists in reconstructing the invariant mass of the dijet system (m_{jj}) and searching for a resonant peak in its spectrum. Wide-jets (corresponding to a jet cone size of $\Delta R = \sqrt{\Delta\eta^2 + \Delta\phi^2} = 1.1$) are used to reconstruct the energy of the initial partons from the resonance decay. This algorithm improves the energy resolution by collecting final state radiation from

[a] e-mail: francesco.santanastasio@cern.ch

[1] The complete list of publications and preliminary results on these subject are available in the CMS public web pages [2].

the outgoing partons. Events used for the search are required to have $\Delta\eta_{jj} = |\eta_{j1} - \eta_{j2}| < 1.3$ reducing significantly the QCD multijet background in t-channel with respect to the s-channel process of resonance production. The search is performed in two different dijet mass regions depending on the trigger strategy used: high-mass ($m_{jj} > 1$ TeV) and low-mass (500 GeV $< m_{jj} <$ 1 TeV).

2.1 High-mass dijet search

The high-mass analysis uses events collected by standard hadronic triggers employed in many CMS analyses. The trigger requires that events have $H_T > 800$ GeV, where H_T is the scalar sum of all jets with p_T >40 GeV reconstructed at high level trigger (HLT). Offline events are required to pass a cut on the dijet mass at approximately $m_{jj} > 1$ TeV for which the trigger is fully efficient on events in data passing the final dijet selection.

Figure 1 (left) shows a display of the highest dijet-mass event passing the full selection, recorded by CMS during 2015 data taking. The reconstructed invariant mass of the dijet system is about 6 TeV. Figure 1 (right) shows the dijet mass spectrum in 2.4 fb^{-1} of data collected in 2015 at $\sqrt{s} = 13$ TeV [3]. The background is estimated by fitting the data with a smoothly falling function. The parametrization used is

$$\frac{d\sigma}{dm_{jj}} = \frac{P_0(1-x)^{P_1}}{x^{P_2+P_3 ln(x)}} \tag{1}$$

where $x = m_{jj}/\sqrt{s}$. This function models well the QCD multijet simulation and the data and it has been successfully employed in previous searches at hadron colliders at different $\sqrt{s}$ values. No sign of new resonances is observed in the dijet mass spectrum, and 95% confidence level (CL) upper limits are set on the resonance production cross section (σ) time branching ratio to dijet (BR) times acceptance (A) as a function of the hypothesized resonance mass. The limits are reported in Figure 2 (left) for different final states (quark-quark, quark-gluon, gluon-gluon depending on the resonance decay products). The upper limits on $\sigma \times BR \times A$ are compared to theoretical predictions for several new physics models including strongly coupled resonances (such as string resonances, excited quarks, axigluons, etc.) and weakly coupled resonances (such as Z' and W' bosons and Randall-Sundrum Gravitons), covering a wide range of cross sections to which this analysis is sensitive. Lower limits are set on the mass of these resonances as summarized in Figure 2 (right). The limits of this analysis significantly exceed the ones from the previous Run1 search thanks to the increased center-of-mass energy in Run2.

2.2 Low-mass dijet search with data scouting

It is important to extend the dijet search in the mass region below 1 TeV in order to probe hypothetical hadronic resonances with small couplings to quarks and gluons that similar searches performed at previous colliders could not find yet. The main experimental difficulties at LHC originate from the large cross section of multijet events at low dijet mass and the finite computing resources for processing and storing these data. To solve this issue, a new technique known as *data scouting* was first proposed by CMS in 2011 [4]. This technique, by significantly reducing the event size compared to the standard CMS data stream (about a factor 100), enabled CMS to relax the trigger thresholds and record fully hadronic events at a rate of few KHz to extend the search in the sub-TeV mass region. With this approach, the analysis is performed using jets reconstructed online in the CMS trigger computing farm.

This novel research strategy was fully integrated in the CMS physics program in 2012 [5] allowing to collect data corresponding to almost 19 fb^{-1} of integrated luminosity at $\sqrt{s} = 8$ TeV in the sub-TeV

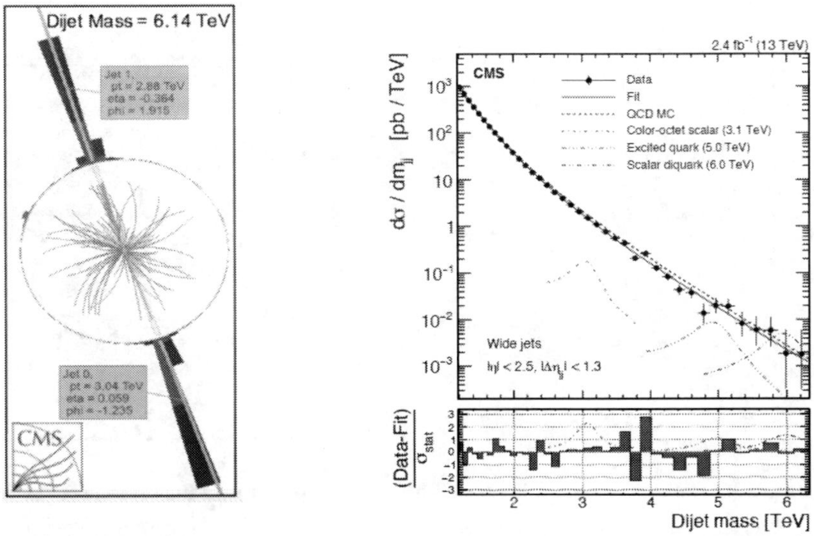

Figure 1. Left: Display of the event with the highest dijet mass (6.14 TeV) after the full selection. The transverse view with respect to the beam axis is shown. The kinematic quantities of the two wide jets are reported. Right: Dijet mass spectrum (points) compared to a fitted parameterization (solid curve) and to the prediction of the PYTHIA8 QCD MC event generator including simulation of the detector (dashed curve). The lower panel shows the difference between the data and the fitted parametrization, divided by the statistical uncertainties. The predicted distributions of narrow resonance signals for three models, with resonance mass values corresponding to the respective 95% confidence level exclusion limit, are shown in both panels (dash-dotted curves).

dijet mass region. The new trigger allows to go down to 400 GeV in dijet mass. The offline analysis strategy used for the high-mass analysis is also employed in this low-mass search. Figure 3 (left) shows the dijet mass spectrum in scouting data in the region between 400 and 1900 GeV. No sign of new resonances in observed and upper limits on $\sigma \times BR \times A$ are set as a function of the resonance mass. The analysis results has been published on the PRL journal in 2016 [6].

It is interesting to compare the limits of this search with those obtained by previous experiments at hadronic colliders with different center-of-mass energies. This can be done assuming a common theoretical model of a new resonance Z' that decays to a pair of quarks as described in Ref. [7]. This simplified model is characterized by only two free parameters: the resonance mass $M_{Z'}$ and the coupling g_B of the resonance to the quarks. The upper limits on signal cross section can be translated into limits in the coupling vs mass plane as shown in Figure 3 (right). The CMS dijet analysis with data scouting extends the search below 1 TeV and provides, at the time of the publication, the most stringent limits to date in the region between 500 and 800 GeV. In the region below 500 GeV the best limits still come from CDF and UA2 experiments. ATLAS and CMS have initiated recently a physics program targeting the mass region below 500 GeV using various techniques that exploit the emission of initial state radiation (quarks/gluons or photons) to trigger these events.

DOI: 10.1051/epjconf/201716407039

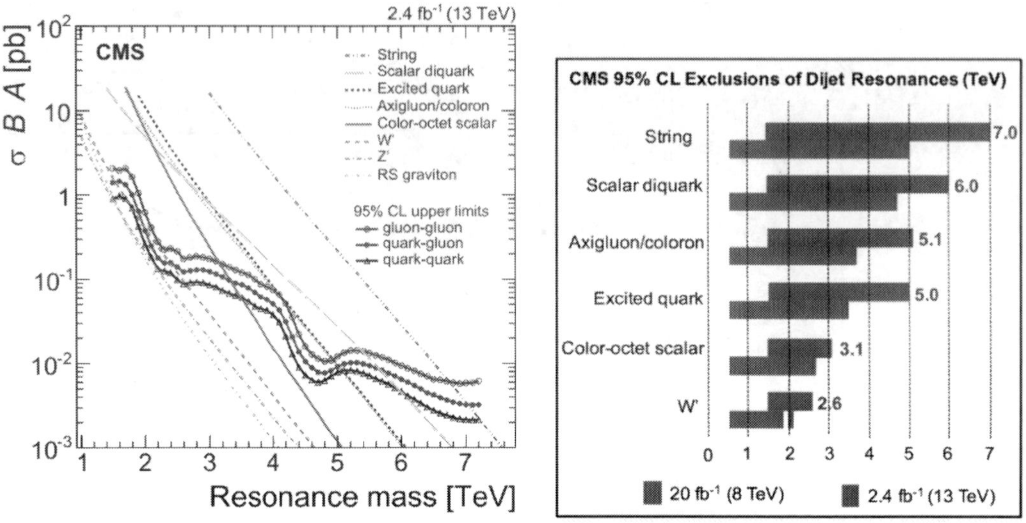

Figure 2. Left: The observed 95% CL upper limits on the product of the cross section, branching fraction, and acceptance for quark-quark, quark-gluon, gluon-gluon type dijet resonances. The limits are compared to the predicted cross sections of several resonance models. Right: Observed 95%CL lower limits on resonance mass for the listed models obtained by the CMS experiment in Run 1 (green) and Run 2 (red) analyses.

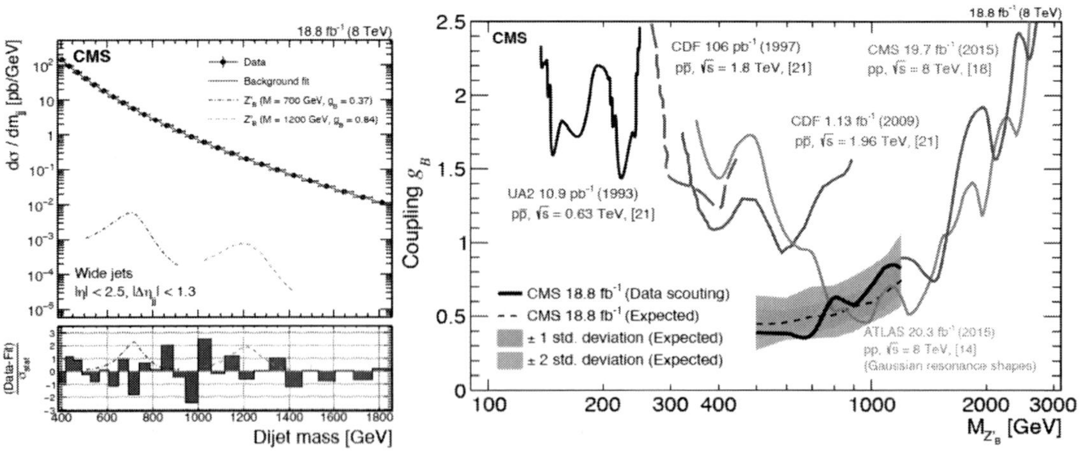

Figure 3. Left: Dijet mass spectrum using data scouting (points) compared to a fitted parameterization (solid curve). The bottom panel compares the data and the fit result, normalized by the statistical uncertainty in the data, for each bin. The predicted distributions of narrow resonance signals for a hypothetical leptophobic resonance Z' with two different mass and coupling values are shown in both panels (dash-dotted curves). This dijet mass distribution complements that observed in the high-mass analysis. Right: Observed 95% CL upper limits on the coupling g_B of the Z' resonance as a function of its mass. The results are compared to those obtained with similar searches at different collider energies.

3 Leptoquarks

The Standard Model contains hints that the lepton and quark sectors are related by a fundamental symmetry. Many beyond the Standard Model theories (such as grand unified theories, composite models, technicolor, SUSY RPV, and others) include such a symmetry, and predict the existence of new massive bosons called leptoquarks (LQ). Leptoquarks can have spin 0 or 1, are coloured particles, have fractional electric charge, and carry both baryon and lepton number. In order to avoid proton decay, baryon and lepton numbers must be conserved separately. In addition constrains from rare decays involving flavour-changing-neutral-currents suggests that LQs should only couple with leptons and quarks within the same generation.

In the simplest case, the model parameters are three: the mass of the LQ (M_{LQ}), the branching ratio of the LQ decay into charged lepton and quark (β), and the LQ-lepton-quark coupling (λ). The branching ratio of the LQ decay into neutrino and quark is $1-\beta$. At LHC, LQs can be produced singly or in pairs. The pair production, which happens mainly via gluon-gluon fusion, is particularly interesting because it does not depend on the unknown coupling λ. The pair production cross section depends on the strong coupling and it is calculated at next-to-leading order.

Depending on the value of the parameter β different decay channels are possible from the pair production of LQs. They differ for the branching ratio (BR) and the observed final state, thus producing a wide range of experimental signatures:

- $\ell\ell jj$ channel, $BR = \beta^2$, two charged leptons plus two jets;
- $\ell\nu jj$ channel, $BR = 2\beta(1-\beta)$, one charged leptons, missing transverse energy (MET) plus two jets;
- $\nu\nu jj$ channel, $BR = (1-\beta)^2$, two jets plus MET.

Feynman diagrams for the different final states are shown in Figure 4.

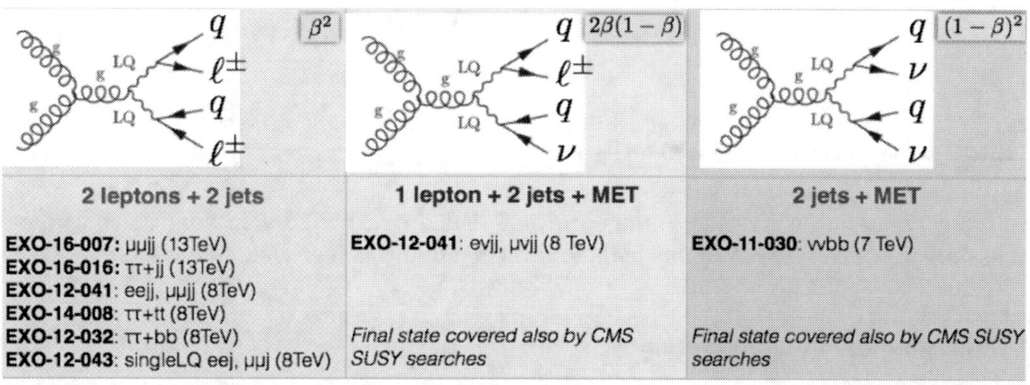

Figure 4. Feynman diagrams for different final states in LQ pair production processes. The labels EXO-XX-YYY indicates the identifier of the CMS publications or preliminary results describing the most recent searches in the corresponding channel [2].

The most recent searches for LQ at CMS have been performed using about 2 fb^{-1} of $\sqrt{s} = 13$ TeV data collected in 2015.

A search for second generation LQs is performed in final states with two muons and two jets [8]. The event selection, optimized for each LQ mass hypothesis, is based on three main kinematic observables: the di-muon invariant mass, the $S_T = p_T(\mu_1) + p_T(\mu_2) + p_T(jet_1) + p_T(jet_2)$ variable, and the lepton-jet mass $M_{\mu j}^{min}$ defined as the smaller of the two reconstructed LQ masses which minimizes

DOI: 10.1051/epjconf/201716407039

the LQ-LQ mass difference. In presence of a signal, the $M_{\mu j}^{min}$ distribution would show a narrow peak in correspondence of the LQ mass, as shown in Figure 5 (left).

A search for third generation LQs is performed in final states with two taus (with reconstructed hadronic decays) and two jets [9]. The main physics observable to discriminate between signal and background is the variable $S_T = p_T(\tau_1) + p_T(\tau_2) + p_T(jet_1) + p_T(jet_2)$ shown in Figure 5 (right).

No excess of events is observed in data in these searches and lower limits are set on the mass of second and third generation LQs at 1150 GeV and 740 GeV, respectively, assuming $\beta = 1$.

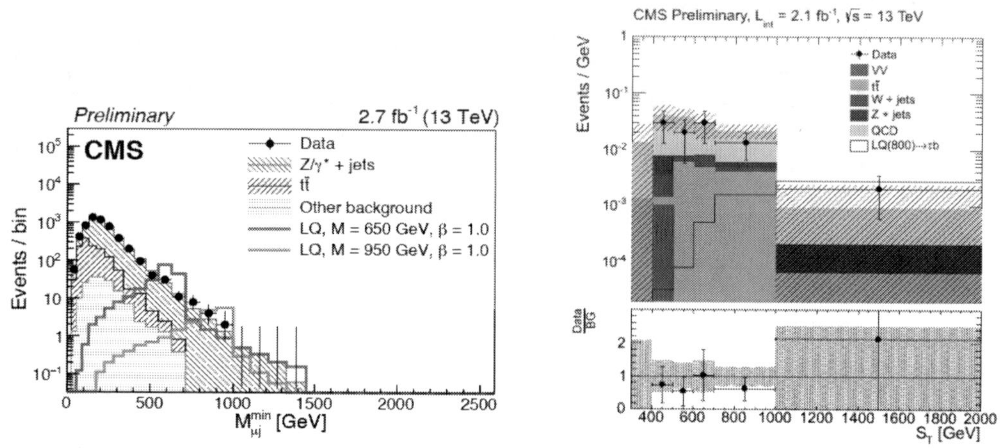

Figure 5. The $M_{\mu j}^{min}$ distribution in events with two muons and two jets (left). The S_T distribution in events with two taus decaying hadronically and two jets (right). Distributions for the data, the background predictions, and the expected LQ signals are shown.

4 Excess of events in Run1 data

Two searches for new physics performed in Run1 data at $\sqrt{s} = 8$ TeV in final states with electrons and jets reported an excess of events in data compared to the background prediction.

4.1 First generation leptoquarks

A search for first generation leptoquarks in the $e\nu jj$ channel have been performed with about 20 fb^{-1} of data [10]. The analysis is designed as a counting experiment in a signal region defined by an event selection optimized for each LQ mass hypothesis. The reconstructed observables used in the selection to separate signal from background are the electron-neutrino transverse mass, the variable $S_T = p_T(e_1) + MET + p_T(jet_1) + p_T(jet_2)$, and the electron-jet mass M_{ej}. Figure 6 (left) shows the electron-jet mass distribution after a selection optimized for the search of a LQ with 650 GeV mass. A broad excess of data events is visible above 600 GeV compared to the background expectation. This reflects also in the upper limits on signal cross section times branching ratio shows in Figure 6 (right) in the same LQ mass region. The local significance of the excess is 2.6 standard deviations. No excess is observed in the $\mu\nu jj$ channel.

 DOI: 10.1051/epjconf/201716407039

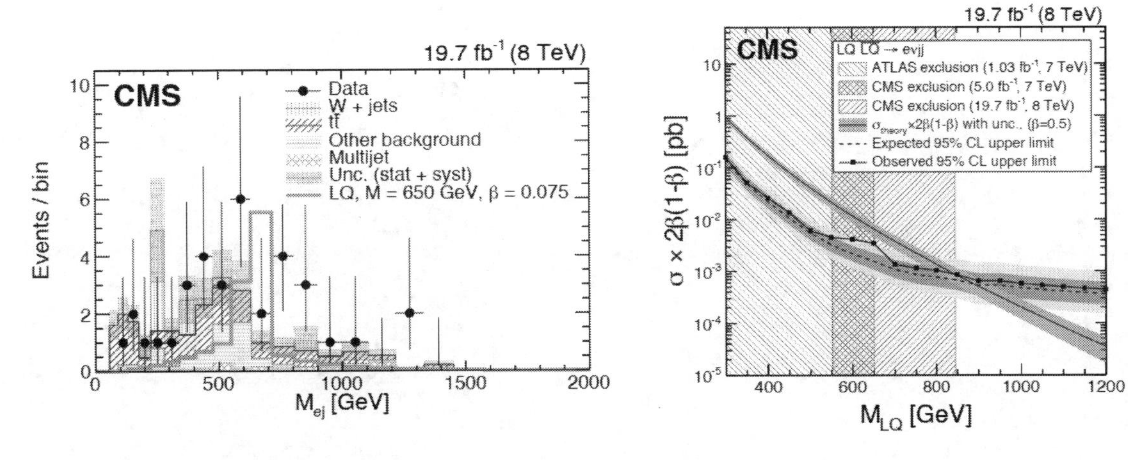

Figure 6. Left: The M_{ej} distribution in the $e\nu jj$ channel after the selection criteria optimized for a LQ mass of 650 GeV have been applied. Right: the expected and observed upper limits at 95% CL on the LQ pair production cross section times branching ratio as a function of the first generation LQ mass obtained in the $e\nu jj$ channel.

4.2 W_R and heavy Majorana neutrinos

The second analysis is a search for W_R and heavy Majorana neutrinos [11]. These new heavy particles are predicted in left-right symmetric extensions of the Standard Model. The model introduces an additional right-handed $SU_R(2)$ symmetry group to the SM, which includes heavy charged ($W_R^{\pm}$) and neutral (Z_R) gauge bosons in addition to heavy right-handed Majorana neutrinos (N_e, N_μ, N_τ). The latter ones provide a possible explanation for the mass of SM neutrinos through the so called *see-saw* mechanism. Figure 7 (left) shows a diagram of the production and decay of a massive W_R gauge boson decaying to an heavy Majorana neutrino. The final state contains two charged, same-flavour leptons and two jets. In presence of a signal the invariant mass of the four-body system $M_{\ell\ell jj}$ peaks at the mass of the W_R boson. Figure 7 (right) shows the M_{eejj} distribution in data events with two electrons and two jets. An excess is visible at around 2 TeV corresponding to a local significance of 2.8 standard deviations. No excess is observed in the two muons plus two jets channel.

5 Future prospects

Searches in jets and leptons+jets final states at CMS are ongoing using the recent 2016 data. A datasample of about 40 fb^{-1} should be recorded at $\sqrt{s}$ = 13 TeV for both ATLAS and CMS experiments. The sensitivity to new physics of the searches presented in this document will improve significantly compared to the current results thanks to a dataset 20 times larger than 2015.

At the date of this conference, the two searches discussed in Section 4 have not being repeated yet at $\sqrt{s}$ = 13 TeV neither by ATLAS or CMS experiments. Although the observed excesses are still compatible with statistical fluctuations of the background, it is very interesting to see if the new analyses will confirm the results observed in the 8 TeV data. New results from the CMS collaboration are expected in 2017.

 DOI: 10.1051/epjconf/201716407039

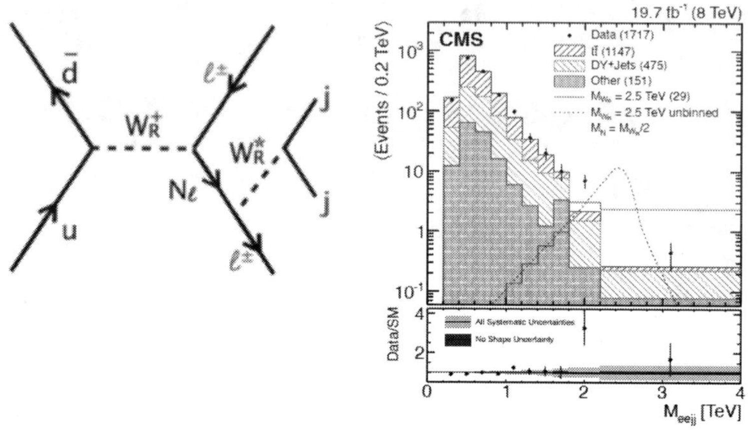

Figure 7. Left: Diagram of production of W_R and decay to heavy Majorana Neutrino. Right: M_{eejj} distribution in events with two electrons and two jets after full selection.

References

[1] CMS Collaboration, "The CMS experiment at the CERN LHC," JINST **3**, S08004 (2008).

[2] https://cms-results.web.cern.ch/cms-results/public-results/publications/EXO/index.html

[3] CMS Collaboration, "Search for narrow resonances decaying to dijets in proton-proton collisions at $\sqrt(s) = 13$ TeV," Phys. Rev. Lett. **116**, no. 7, 071801 (2016), arXiv:1512.01224.

[4] CMS Collaboration, "Search for Narrow Resonances using the Dijet Mass Spectrum in pp Collisions at sqrt s of 7 TeV," CMS-PAS-EXO-11-094.

[5] CMS Collaboration, "Data Parking and Data Scouting at the CMS Experiment," CMS-DP-2012-022.

[6] CMS Collaboration, "Search for narrow resonances in dijet final states at $\sqrt(s) = 8$ TeV with the novel CMS technique of data scouting," Phys. Rev. Lett. **117**, no. 3, 031802 (2016), arXiv:1604.08907.

[7] B. A. Dobrescu and F. Yu, "Coupling-mass mapping of dijet peak searches," Phys. Rev. D **88**, no. 3, 035021 (2013), Erratum: [Phys. Rev. D **90**, no. 7, 079901 (2014)], arXiv:1306.2629.

[8] CMS Collaboration, "Search for pair-production of second-generation scalar leptoquarks in pp collisions at $\sqrt{s} = 13$ TeV with the CMS detector," CMS-PAS-EXO-16-007.

[9] CMS Collaboration, "Search for heavy neutrinos and third-generation leptoquarks in final states with two hadronically decaying τ leptons and two jets in proton-proton collisions at $\sqrt{s} = 13$ TeV," CMS-PAS-EXO-16-016.

[10] CMS Collaboration, "Search for pair production of first and second generation leptoquarks in proton-proton collisions at sqrt(s) = 8 TeV," Phys. Rev. D **93**, no. 3, 032004 (2016) arXiv:1509.03744.

[11] CMS Collaboration, "Search for heavy neutrinos and W bosons with right-handed couplings in proton-proton collisions at $\sqrt{s} = 8$ TeV," Eur. Phys. J. C **74**, no. 11, 3149 (2014) arXiv:1407.3683.

EPJ Web of Conferences **164**, 07040 (2017) DOI: 10.1051/epjconf/201716407040
ICNFP 2016

Searches for BSM physics in dilepton, multilepton and lepton+MET final states at CMS

L. Benato[1,2,a] on behalf of CMS Collaboration

[1] *Università degli Studi di Padova*
[2] *INFN sezione di Padova*

Abstract. Numerous new physics models, e.g., theories with extra dimensions and various gauge-group extensions of the standard model, predict the existence of new particles decaying to dilepton, multilepton, and lepton+MET final states. This talk presents the recent results from searches for new physics in the leptonic final states at CMS.

1 Introduction

Many Beyond Standard Model (BSM) physics models foresee the appearance of particles at high mass range, decaying into dilepton, multilepton and lepton + neutrino final states. Leptons considered in these searches are electrons and muons; neutrinos are revealed as missing transverse energy in the detector (E_T^{miss}). The leptonic final states are historically channels of discovery and precision measurements, because they show very clean signatures; dealing with high energetic leptons coming from heavy resonances, however, implies careful studies of the objects and their properties. Generally these models predict very small branching ratio for leptonic decays: as a consequence, the statistical uncertainties dominate, so it is necessary to accurately evaluate the backgrounds. Results obtained from the LHC Run 2 data collected in 2015 by the CMS experiment [1] are presented: going up to $\sqrt{s}$ = 13 TeV makes the parton luminosities of processes $q\bar{q} \rightarrow X$ increase substantially, especially at high mass region, improving the previous limits set on these models by LHC Run 1 data. [2]-[5]

2 CMS detector

CMS is a multi-purpose experiment composed of many sub-systems enclosed in a superconducting magnet. Informations coming from different sub-detectors can be used together to improve the energy and momentum determination of the particles involved.

3 High energy objects

As a general approach, isolation requirements are applied to charged leptons, in order to reduce backgrounds coming from objects (mainly jets) misidentified as leptons. A lepton is isolated if the ratio of the energy deposits around its track divided by its transverse momentum (p_T) is below a certain threshold.

[a] e-mail: lisa.benato@cern.ch

3.1 Muons

Muons are challenging objects to reconstruct at high energy values: in these conditions, their trajectory in the magnetic field is almost straight, complicating the determination of their p_T and charge. The precision of their p_T measurement, furthermore, is strongly influenced by the alignment of the external muon gas detectors and the inner silicon tracker. In addition, high energetic muons radiate in the detector materials, causing a loss of resolution at increasing p_T. Many studies in this sense have been performed on the cosmic muons collected by CMS. Different high-p_T muon reconstruction algorithms have been exploited: they generally combine informations from muon chambers and tracker, in order to improve the identification efficiency as much as possible and to discard the segments in the muon chambers where radiactive phenomena occurred.

3.2 Electrons

Electrons are reconstructed using informations coming from all the sub-detectors. Dedicated selections for high-p_T electrons are applied in these analyses, in order to achieve the highest detection efficiency. Unlike muons, electron energy and p_T resolutions are approximatively constant at increasing energy.

3.3 E_T^{miss}

The experimental signature of the presence of a non-interacting particle, the neutrino, is represented by the missing transverse energy, E_T^{miss}. The quantity $\vec{p}_T^{miss}$ is defined as the negative sum $-\Sigma\vec{p_T}$ of all reconstructed particles; E_T^{miss} is the modulus of $\vec{p}_T^{miss}$. Its magnitude is influenced by all the other objects, including jets and their corrections.

4 Heavy spin-1 BSM bosons searches in dilepton and lepton plus E_T^{miss} final states

A common feature of models admitting new physics consists into enlarging the Standard Model gauge symmetry group, $SU(3) \times SU(2) \times U(1)$, into a larger one that, once broken into the SM group, will result into new residual groups: new spin-1 heavy bosons are expected to appear. The most naïf extension is the Sequential Standard Model [6], in which two W'_{SSM} and a Z'_{SSM} are expected to be the heavy replicas of W and Z bosons, with the same couplings to leptons as their standard model partners. More exotic spin-1 heavy bosons predicted by super string inspired models, like a Z'_ψ coming from $E(6)$ breaking [7]-[8], are searched for in the following analysis.

4.1 $Z' \rightarrow \ell^+\ell^-$ [9]

The analysis strategy consists into reconstructing the invariant mass of the heaviest $\ell^+\ell^-$ pair, after applying adequate selections to all the leptonic candidates, in order to suppress the background contributions in favour of the expected signal. The main source of irreducible background events is represented by the Drell-Yan process, $Z/\gamma^* \rightarrow \ell^+\ell^-$, that is well estimated by Monte Carlo simulated events. Monte Carlo simulations also work for the second category of backgrounds, which contains real prompt leptons coming from processes like $t\bar{t}$, tW, WW, WZ, ZZ; these predictions are validated on data in $e\mu$ control region, where no signal is expected. Non-prompt backgrounds, that are objects (mainly jets) misidentified as leptons, are predicted via data-driven methods. Cosmic muons are suppressed applying a cut on the impact parameter with respect to the primary vertex of the hard interaction and on the 3D angle between the muons.

The observed and the predicted SM backgrounds invariant mass spectra are shown in Fig. 1 for muons (left) and electrons (right). No discrepancy is seen among the two distributions. The highest mass event is observed in the electron channel, and it is located at 2.9 TeV. The expected number of events above 2.8 TeV is 0.036 ± 0.009 ee, 1 is seen. The local p-value for the null-hypothesis calculated as the Poisson probability to observe at least one event in the electron channel above 2.8 TeV is 0.036.

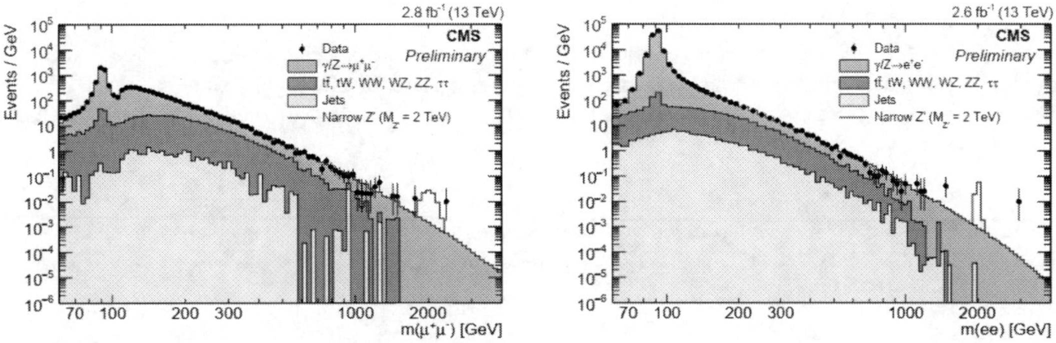

Figure 1. $Z' \rightarrow \ell^+\ell^-$ *search.* The invariant mass spectrum of $\mu^+\mu^-$ (left) and e^+e^- (right) events. The points with error bars represent the data. The histograms represent the expectations from standard model processes: Drell-Yan, other sources of prompt leptons ($t\bar{t}$, tW, WW, WZ, ZZ, $Z \rightarrow \tau\tau$), and non-prompt backgrounds. The green curve represents an hypothetical Z' of mass 2 TeV.

Limits at 95% confidence level on the ratio between the cross-section times branching ratio for the Z' divided by the same quantity for the Z are extracted:

$$\frac{\sigma(pp \rightarrow Z' + X \rightarrow \ell\ell + X)}{\sigma(pp \rightarrow Z/\gamma^* + X \rightarrow \ell\ell + X)}.$$

The ratio cancels the luminosity uncertainties. A Bayesian unbinned likelihood approach is applied; the likelihood is integrated using the Metropolis-Hastings algorithm. The main uncertainties related to signal efficiency and cross-section are due to lepton identification (6% for electrons, 7% for muons). The background uncertainty comes mainly from PDF uncertainties (up to 20% at 3 TeV). A mass scale uncertainty (1% for e, 3% for μ) is applied while combining the channels. The signal is modelled as the convolution of a Breit-Wigner and a Gaussian distribution, under the narrow width approximation: this means that only the on-shell region of the resonance is considered, because the interference is strongly model-dependent. In Fig. 2 (left), expected and observed limits for different signal widths are shown, ranging from 0%, to 0.6% of the expected resonance mass (as predicted by Z'_Ψ), to 3% of the mass (as predicted by Z'_{SSM}), combining the e and μ channel. The final expected and observed limits for the combination of electron and muon channels under the hypothesis of a width of 0.6% are displayed in the right panel of Fig. 2. In Tab. 1, a more detailed summary of the mass limits for each channel and model is reported.

 DOI: 10.1051/epjconf/201716407040

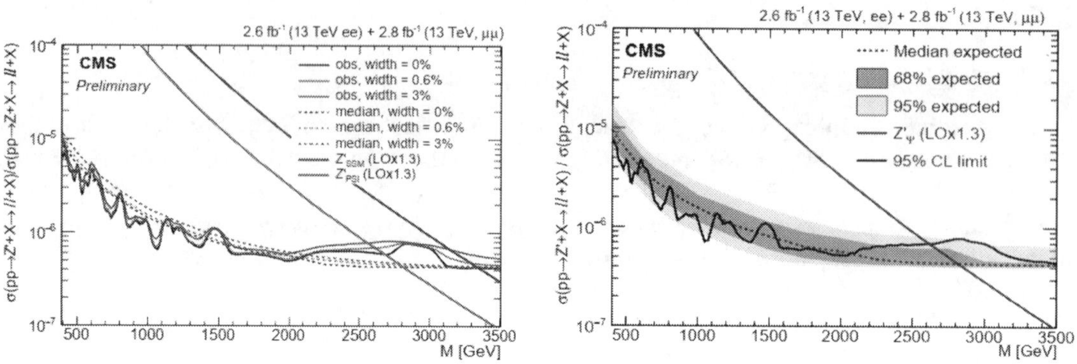

Figure 2. $Z' \rightarrow \ell^+\ell^-$ *search.* Left: 95% confidence level observed and expected limits on the cross-section ratio for different resonance widths; they are obtained combining the electron and muon channels. Right: limits set at 95% confidence level for a Z' whose natural width is the 0.6% of its mass. The theoretical curves of the discussed models are included in the plots.

Table 1. $Z' \rightarrow \ell^+\ell^-$ *search.* Observed limits on the mass of Z'_Ψ and Z'_{SSM} for each channel and for the combination.

Channel	Z'_Ψ	Z'_{SSM}
ee	2.40 TeV	2.75 TeV
$\mu\mu$	2.40 TeV	3.00 TeV
combination	2.60 TeV	3.15 TeV

4.2 $W' \rightarrow \ell\nu$ [10]

In this analysis, since a neutrino is expected in the final state, the discriminant variable is the transverse invariant mass, which takes into account only the transverse quantities:

$$M_T = \sqrt{2p_T^\ell E_T^{miss}\left(1 - \cos\left[\Delta\phi(\vec{p}_T^{\,\ell}, \vec{p}_T^{\,miss})\right]\right)}.$$

In order to achieve a back-to-back topology between the lepton and the neutrino, a cut is applied on their azimuthal angles, $|\Delta\phi(\vec{p}_T^{\,\ell}, \vec{p}_T^{\,miss})| > 2.5$. A further selection, $0.4 < p_T/E_T^{miss} < 1.5$, allows to reject background contamination.

The dominant source of background is represented by the standard model $W \rightarrow \ell\nu$ process, and it is fully predicted by Monte Carlo simulations. Secondary backgrounds like Drell-Yan processes, $t\bar{t}$, single t, WW, WZ, ZZ are also taken from simulations. QCD multijet backgrounds are estimated from data. Cosmic muons contamination is suppressed with a cut on impact parameter with regards to the primary vertex, like in the previous analysis.

In Fig. 3, the transverse mass distributions are shown for muon (left) and electron (right) channel. Data and Monte Carlo spectra are in agreement.

Limits at 95% confidence level on cross-section $\sigma(pp \rightarrow W' + X \rightarrow \ell\nu + X)$ are extracted via the modified frequentist approach (CL_s method [11]-[12]). Limits are provided in a region over 1 TeV, where Run 2 data are more sensitive due to the increased center of mass energy and, consequently, increased cross-section for the considered process. The main uncertainties in the muon channel arise

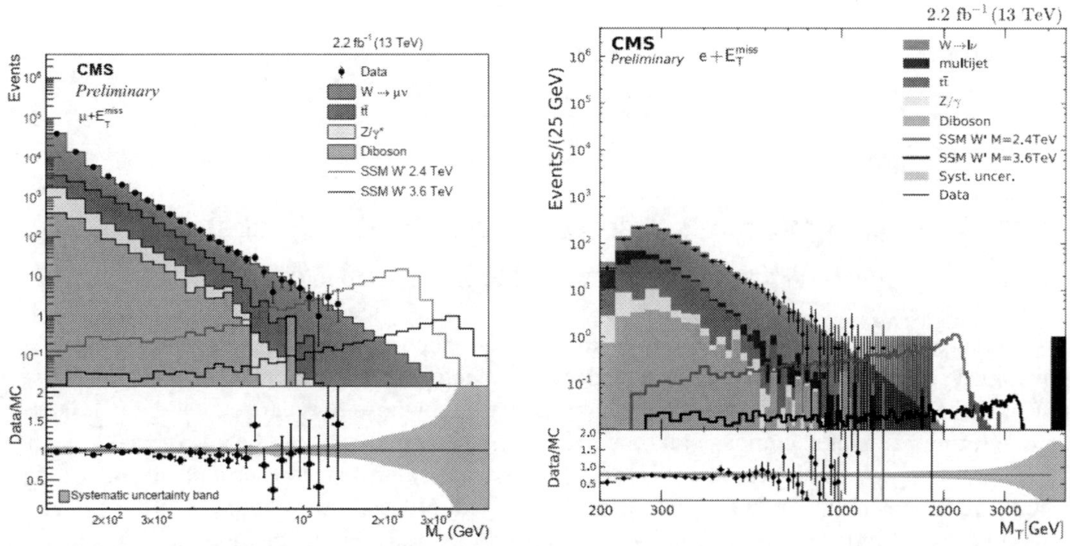

Figure 3. *$W' \to \ell\nu$ search.* The transverse mass spectrum of $\mu\nu$ (left) and $e\nu$ (right) events. The points with error bars represent the data. The histograms represent the expectations from standard model processes: W decays, other sources of prompt leptons (Z/γ^*, $t\bar{t}$, tW, WW, WZ, ZZ, $Z \to \tau\tau$), and non-prompt backgrounds. The dark orange and black curves represent two hypothetical W' of mass 2.4 and 3.6 TeV.

from the muon momentum scale and p_T resolution, that both propagate to E_T^{miss}, as pointed out in Sec. 3.1 and 3.3. The electron channel is mainly affected by PDF uncertainties, especially at high M_T.

The expected and observed limits for the cross-section obtained combining the two channels are shown in Fig. 4; the final limits on the W'_{SSM} mass for each channel are summarized in Tab. 2.

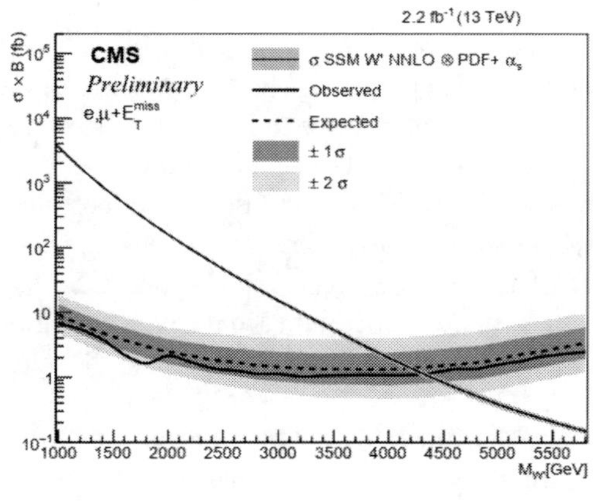

Figure 4. *$W' \to \ell\nu$ search.* 95% confidence level observed and expected limits on the cross-section compared to the theoretical curve predicted for a W'_{SSM}. Limits are obtained by combining the electron and muon channels.

Table 2. *$W' \to \ell\nu$ search.* Observed limits on the mass of a W'_{SSM} for each channel and for the combination.

Channel	W'_{SSM}
e	3.8 TeV
μ	4.0 TeV
combination	4.4 TeV

5 Seesaw type-III mechanism search in multilepton final states [13]

A remarkable puzzle left unsolved by the Standard Model is related to neutrino masses. An appealing possibility is represented by the seesaw mechanism: it introduces new heavy particles coupling to leptons and to Higgs doublets, accounting for both the neutrino masses and their smallness (six or more orders of magnitude smaller than that of the electron). In particular, in the type-III seesaw model [14], the neutrino is considered a Majorana particle whose mass arises via the mediation of a massive fermionic triplet, composed by the heavy Dirac charged $\Sigma^{\pm}$, and the heavy Majorana neutral Σ^0. In high energy proton-proton collisions, the heavy fermions may be produced in pairs through electroweak interactions. In this search, a final state with at least three charged leptons (electrons or muons) is considered: the heavy triplet can decay into leptons via electroweak or Higgs bosons. In Fig. 5, one of the 27 possible decay modes is pictured. The decay rate of a Σ to a particular flavour ℓ is proportional to $\frac{V_\ell}{\sqrt{|V_e|^2+|V_\mu|^2+|V_\tau|^2}}$, where each V_ℓ represents the coupling of the heavy fermion to the considered leptonic flavour. In the flavour democratic scenario, V_ℓ is the same for e, μ and τ.

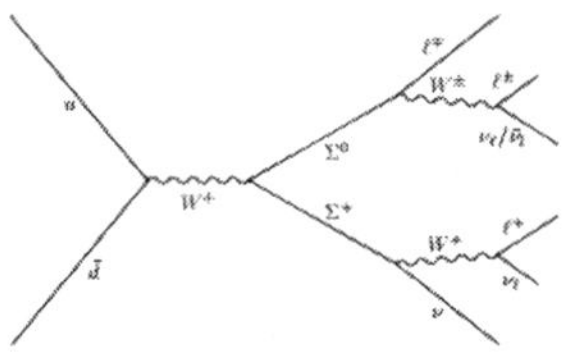

Figure 5. *Seesaw type-III search.* Feynman diagram example of the fermion Σ production and decay in leptonic final states in the context of the type-III seesaw model.

The analysis strategy is based on grouping the final states in categories, depending on the number of reconstructed leptons, the number of opposite sign same flavour pairs (OSSF), zero, one or two, and on the invariant mass of the pair, on or under-above the nominal mass of the Z boson. A promising cut that has been studied to reject backgrounds is the scalar sum of the p_T of all the considered leptons, L_T, added to E_T^{miss}, requiring $L_T + E_T^{miss} > 350$ GeV: this guarantees high efficiency for both channels involving decays where the momentum is carried by leptons, like $\Sigma^{\pm} \to Z\ell^{\pm} \to \ell^{\pm}\ell^{+'}\ell^{-'}$, and decays where there is a large amount of missing transverse energy, like $\Sigma^0 \to H\nu \to WW\nu$. Considering the most significant channels in terms of expected signal given the luminosity, the analysis consists in keeping 4 distributions, depending on the lepton properties: 3 leptons without an OSSF pair, 3 leptons with one OSSF pair on-Z, 3 leptons with one OSSF pair above-Z, and 4 leptons with at least one OSSF pair. In left Fig.6, the spectrum of $L_T + E_T^{miss}$ for the category 1 OSSF pair, 3 leptons, $m_{\ell^+\ell^-}$ above-Z is displayed, whilst in right Fig.6 there is the spectrum for the category where there is at least 1 OSSF, 4 leptons, no requirements on $m_{\ell^+\ell^-}$: it is noticeable that the statistics is very low. Data and Monte Carlo spectra are in agreement.

Backgrounds strongly depend on the categories considered. 51% of the total expected background is represented by the standard model decay $WZ \to \ell\ell\ell$, that is predicted by Monte Carlo simulations

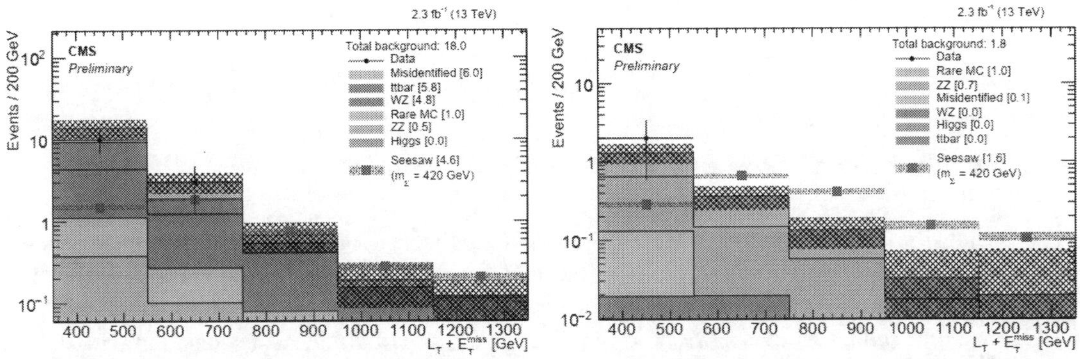

Figure 6. *Seesaw type-III search.* Left: spectrum of $L_T + E_T^{miss}$ for the category 1 OSSF pair, 3 leptons, $m_{\ell^+\ell^-}$ above-Z. Right: spectrum of $L_T + E_T^{miss}$ for the category at least 1 OSSF pair, 4 leptons, no requirements on $m_{\ell^+\ell^-}$. The points with error bars represent the data. The histograms represent the expectations from standard model processes: WZ, $t\bar{t}$, ZZ, Higgs decays, rare processes and misidentified leptons. The red points represent an hypothetical Σ of mass 420 GeV.

and validated on data. The kinematics of the leptonic $t\bar{t}$ decays, with a misidentified lepton from a b-jet, is predicted by Monte Carlo samples; the misidentification rate, instead, is measured in data control region. $Z \rightarrow \ell\ell$ plus a misidentified lepton in jet is predicted with a data-driven method. Other processes, like $ZZ \rightarrow 4\ell$, $t\bar{t}Z$, $t\bar{t}W$, $H \rightarrow 4\ell$, are evaluated via Monte Carlo simulations.

Limits at 95% confidence level on the cross-section sum for the production of seesaw heavy fermionic pairs $\Sigma^0\Sigma^+, \Sigma^0\Sigma^-, \Sigma^+\Sigma^-$ and on their mass are extracted with the asymptotic CL_s method, under the assumptions of a degenerate m_Σ, and of the flavour democratic scenario, where $V_e = V_\mu = V_\tau = 10^{-6}$. They are reported in Fig. 7 and in Tab. 3. The p-value for the observation in all the considered channels, assuming there is no new physics, is 0.93. The uncertainties are strongly dominated by statistics; systematic uncertainties are small and mainly due to PDF and to renormalization and factorization uncertainties related to E_T^{miss} range in the dominant WZ background.

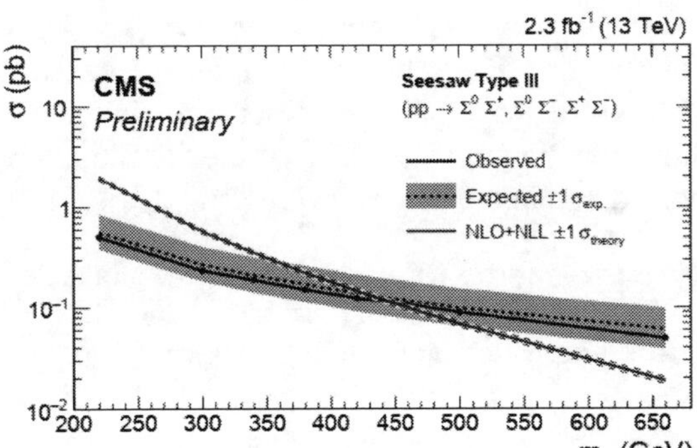

Figure 7. *Seesaw type-III search.* Expected and observed limits at 95% confidence level on the cross-section sum for the production of seesaw heavy fermion pairs $\Sigma^0\Sigma^+, \Sigma^0\Sigma^-, \Sigma^+\Sigma^-$. The theoretical curve of the discussed model is included in the plot.

DOI: 10.1051/epjconf/201716407040

Table 3. *Seesaw type-III search.* Observed limit on the mass of the seesaw type-III mediators.

Channel	m_Σ
combination	440 GeV

6 Quantum black holes search in $e\mu$ final state [15]

Many extensions of the Standard Model predict the existence of heavy resonances decaying into lepton flavour violating final states, like $e\mu$. If a fundamental Planck scale could be observable at TeV, the production of microscopic black holes decaying into couples of particles should be possible at the LHC, once the production mass threshold M_{th} is reached. In particular, in the considered models [16]-[17], quantum black holes (QBH) are spin-0, colourless, neutral particles, and lepton flavour violation is allowed. A search in this sense has been performed; the interpretation of the results depends on the number of extra-dimensions n involved: for $n = 1$, a Randall-Sundrum scenario is considered, while for $n \geq 2$, we lay in the ADD scenarios.

The analysis strategy consists in reconstructing the invariant mass distribution of the $e\mu$ pairs, without requiring to the leading leptons to be opposite sign, in order to avoid an efficiency loss due to the wrong assignment of the electron charge in high energy range. The misidentification of the objects involved is reduced by rejecting all the electrons close (in a cone $\Delta R < 0.1$) to muons that produce bremsstrahlung. In left Fig.8, the invariant mass spectrum of the $e\mu$ pairs is presented; the right plot is the cumulative distribution of the previous one. Data and Monte Carlo spectra are in agreement.

The backgrounds contributing to the final state can be divided into two categories: the first comprises events with at least two prompt, isolated leptons; the second includes jets and photons misidentified as isolated leptons or jets with non-prompt leptons coming from heavy-flavour decays. In the former category, processes like $t\bar{t}$, WW, WZ, ZZ, single t, Drell-Yan are included, and they are estimated from Monte Carlo samples; in the latter category, the dominant contribution is represented by W + jets processes, and all these backgrounds are predicted with data-driven methods.

An upper limit at 95% confidence level on the cross-section times branching ratio is determined using a binned likelihood Bayesian approach, and it is shown in Fig. 9. The non-resonant signal is modelled as a broad shape smeared by the detector resolution, with a sharper edge at M_{th}. The uncertainties, entering the limit extraction as nuisance parameters, are modelled with log-normal distributions. The uncertainty on the muon momentum scale (10% per TeV) propagates to an uncertainty on the background yield of 30% at 3 TeV. The dominant uncertainty in the background yield prediction is due to the uncertainty on the shape of the $e\mu$ distribution in the $t\bar{t}$ background, that ranges up to 31% at 2 TeV. The final limits on the threshold mass of the QBH are reported in Tab. 4, as a function of the extra-dimensions considered in the theoretical model.

Table 4. *Quantum black holes in $e\mu$ search.* Observed limits on the production mass threshold of a quantum black hole, considering a different number of extra-dimensions involved, from $n = 1$ (Randall-Sundrum scenario), up to $n = 6$ (ADD scenarios).

n	M_{th}
1 (RS)	2.5 TeV
4 (ADD)	4.2 TeV
5 (ADD)	4.3 TeV
6 (ADD)	4.5 TeV

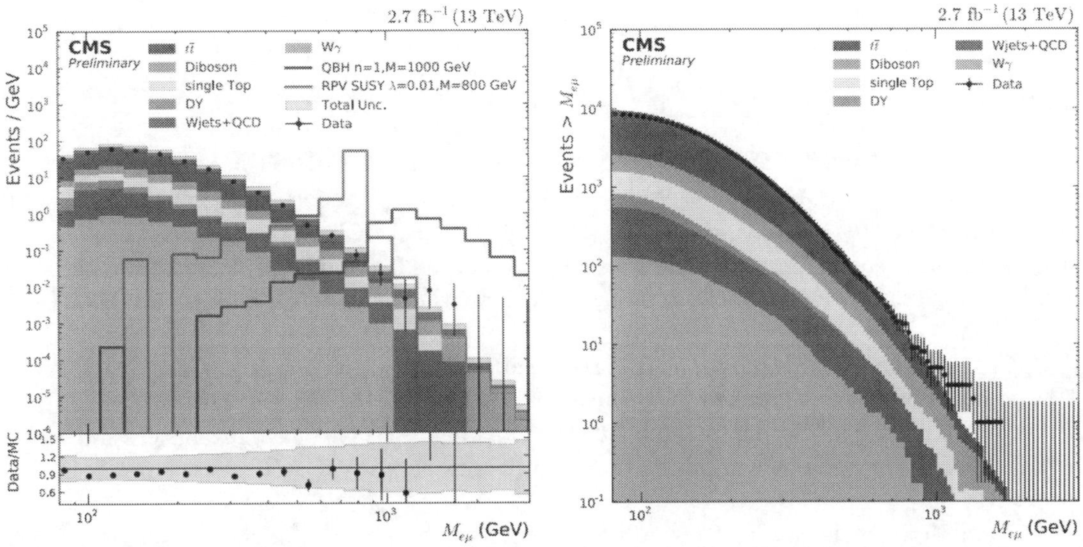

Figure 8. *Quantum black holes in eμ search.* Left: invariant mass spectrum of the $e\mu$ pairs. Right: cumulative distribution of the invariant mass spectrum of the $e\mu$ pairs. The points with error bars represent the data. The histograms represent the expectations from standard model processes: W+jets, $W\gamma$, $t\bar{t}$, WW, WZ, ZZ, single t, Drell-Yan. The red curve in the left plot represents an hypothetical QBH of mass 1 TeV in a Randall-Sundrum scenario ($n = 1$).

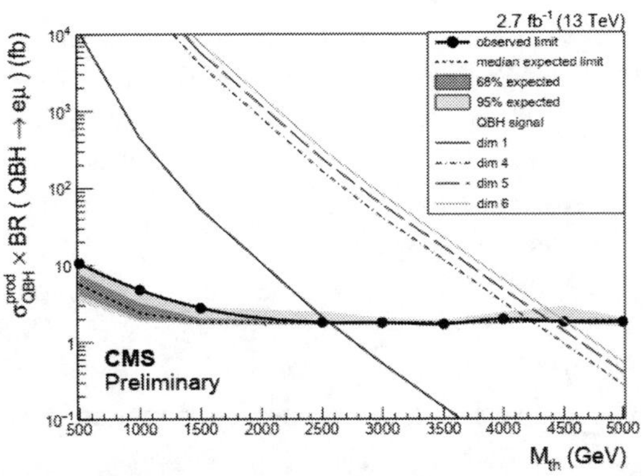

Figure 9. *Quantum black holes in eμ search.* Expected and observed limits at 95% confidence level on the cross-section times branching ratio for the quantum black hole production. The theoretical curves of the discussed models are included in the plot.

7 Conclusions

The LHC collider is facing its "search era": searching for heavy resonances predicted by BSM models that decay into leptons and neutrinos is considered a high-priority task nowadays. Many searches have been performed during Run 2 in 2015 by the CMS collaboration: the increased $\sqrt{s}$ allowed improvements with regards to Run 1 analyses, pushing the exclusion limits to higher masses of the considered states. These searches are being repeated in 2016 data: more luminosity is expected and more interesting results are coming soon.

References

[1] CMS Collaboration, "The CMS experiment at the CERN LHC", JINST **3**, S08004 (2008)

[2] CMS Collaboration, "Search for physics beyond the standard model in dilepton mass spectra in proton-proton collisions at $\sqrt{s} = 8$ TeV", JHEP **04**, 025 (2015)

[3] CMS Collaboration, "Search for physics beyond the standard model in final states with a lepton and missing transverse energy in proton-proton collisions at $\sqrt{s} = 8$ TeV", Phys. Rev. **D91**, 092005 (2015)

[4] CMS Collaboration, "Search for heavy lepton partners of neutrinos in proton-proton collisions in the context of the type III seesaw mechanism", Phys. Lett. **D718**, 348 (2012)

[5] CMS Collaboration, "Search for lepton flavour violating decays of heavy resonances and quantum black holes to an e-mu pair in proton-proton collisions at $\sqrt{s} = 8$ TeV", Eur. Phys. J. **C76**, 317 (2016)

[6] G. Altarelli, B. Mele, and M. Ruiz-Altaba, "Searching for new heavy vector bosons in $p\bar{p}$ colliders", Z. Phys. **C45**, 109 (1989)

[7] A. Leike, "The phenomenology of extra neutral gauge bosons", Phys. Rept. **317**, 143 (1999)

[8] J. L. Hewett and T. G. Rizzo, "Low-energy phenomenology of superstring-inspired E6 models", Physics Reports **183**, 193 (1989)

[9] CMS Collaboration, "Search for a Narrow Resonance Produced in 13 TeV pp collisions decaying to electron pair or muon pair final states", CMS Physics Analysis Summary CMS-PAS-EXO-15-005, 2015

[10] CMS Collaboration, "Search for SSM W' production, in the lepton+MET final state at a center-of-mass energy of 13 TeV", CMS Physics Analysis Summary CMS-PAS-EXO-15-006, 2015

[11] A. L. Read, "Presentation of search results: the CL s technique", J. Phys. G: Nucl. Part. Phys. **28**, 2693 (2002)

[12] T. Junk, "Confidence level computation for combining searches with small statistics", Nucl. Instrum. Meth. **A434**, 435 - 443 (1999)

[13] CMS Collaboration, "Search for type-III seesaw heavy fermions with multilepton final states using 2.3 $\text{fb}^{-1}$of $\sqrt{s} = 13$ TeV proton–proton collision data", CMS Physics Analysis Summary CMS-PAS-EXO-16-002, 2016

[14] R. Foot, H. Lew, X. He, and G. Joshi, "See-saw neutrino masses induced by a triplet of leptons", Z. Phys. **C44**, (1989)

[15] CMS Collaboration, "Search for high-mass resonances and quantum black holes in the $e\mu$ final state in proton-proton collisions at $\sqrt{s} = 13$ TeV", CMS Physics Analysis Summary CMS-PAS-EXO-16-001, 2016

[16] P. Meade and L. Randall, "Black Holes and Quantum Gravity at the LHC", JHEP **05**, 003 (2008)

[17] X. Calmet, W. Gong, and S. D. H. Hsu, "Colorful quantum black holes at the LHC", Phys. Lett. **B668**, 20 - 23 (2008)

EPJ Web of Conferences **164**, 07041 (2017) DOI: 10.1051/epjconf/201716407041
ICNFP 2016

Signals for invisible matter from solar - terrestrial observations

Sergio Bertolucci[1,2,a], Konstantin Zioutas[1,3,b], Sebastian Hofmann[4], and Marios Maroudas[3]

[1] *CERN, Geneva, Switzerland,*
[2] *INFN, LNF, Italy,*
[3] *University of Patras, Patras, Greece,*
[4] *Munich, Germany*

Abstract. Gravitational lensing of invisible streaming matter towards the Sun or the Earth could be the explanation of puzzling solar/terrestrial phenomena. We have analyzed solar flares, EUV emission and also the global ionization content of the Earth atmosphere. Assuming that this invisible matter has some form of interaction with normal matter and that there exist preferred directions in its flow, then one would expect an enhanced activity at certain planetary longitudes, which is also observed. The broad velocity spectrum of the assumed constituents makes it difficult at this stage to identify the origin of the stream(s) or the nature of its constituents.

1 Introduction

The detection of the constituents of dark matter (DM) is one of the central challenges in modern physics. While large scale DM distribution in space is relatively uniform, also local streams have been discussed [1]. The existence of such streams of invisible constituents could be at the origin of the puzzling active Sun or the annual anomaly of the ionization of the Earth atmosphere. We assume here that their triggering is due to planetary lensing of invisible matter stream(s). This scenario is totally different from the models based on tidal forces, which have been attempted with very little success since the discovery of the first large flare some 155 years ago. If this seminal idea holds, there will be ways to explore it further in the future, due to its implications in other ongoing DM searches. A detailed description of the reasoning behind this idea is given in ref. [2]. Here we present also new results.

2 The Observations

2.1 Solar Flares

In this work we provide a statistical analysis of time (=position in the ecliptic) distributions of the inner 3 planets and the Moon, in association with a) the daily occurrence of 6091 M-class and the 491 X-class flares (1976-2015) [3], b) the continuous full disk solar emission in the EUV (1999-2015),

[a] e-mail: Sergio.Bertolucci@cern.ch
[b] e-mail: Konstantin.Zioutas@cern.ch

 DOI: 10.1051/epjconf/201716407041

and c) the electron content of the Earth ionosphere (1995-2012) [4]. Interestingly, for comparison, this period includes the extreme solar minimum of 2008-2009.

The heliocentric longitudinal distribution for Earth, Venus and Mercury of M- and X-class Flares is shown in Figure 1. For the Earth, the peak around 45° (33°) shows the strongest clustering of such solar energetic bursts, while the peaks at 255° and in particular that at 291° are also of statistically significant. In all the planetary spectra no correction was performed for their orbital eccentricity, which is 6%, 2% and 223% for Earth, Venus and Mercury, respectively. Venus and Mercury show a strong and wide peak around 260°-270° in longitude as well as a second one at ~ 150°. Interestingly, a strong excess around 250°-290° is happening in all 3 planetary longitudinal distributions, despite the fact that Mercury, Venus and Earth have gone through the same longitudes mostly at different times. In the case of Mercury, a smaller peak appears around 33°. Due to the large orbital eccentricity of Mercury, the expected modulation has been subtracted. The derived residual spectrum makes the aforementioned peaks better visible (Figure 1).

As next, we introduced in the analysis code planetary spatial correlations. A possible combined lensing efficiency might enhance the observed effects on the Sun or the ionosphere. In fact, this has been also observed with statistically significant differences both in intensity and in spectral shape (see Figures 2 and 4). Remarkably, the excess observed with M-Flares and EUV (Figures 2) peaks at exactly at the same values around 241° and 290°. A similar behavior is also shown by the X-Flares, despite the poorer statistics (see [2]).

Figure 3 shows additional results obtained with the flaring Sun and its full disk EUV emission as well as a correlation between both. Surprisingly, the EUV irradiance preceding the onset of 7 huge X-class Flares is suppressed for a period lasting 1-2 months. It is intriguing that this suppression depends on the planetary configuration (not shown), which might be a relevant signature for the trigger of the unpredictable flares.

2.2 The Ionosphere

The total electron content (TEC) of the ionosphere depends primarily on the solar EUV irradiance, but its variations show anomalies, which have not been understood since several decades (see [2]). For example, the measured TEC during December exceeds that around June, which cannot be explained by the annual solar irradiance modulation due to the varying Sun-Earth distance. We investigated the possibility that these anomalies could be connected to planetary lensing, using the same reasoning described above.

Figures 4 shows the results obtained for combined planetary constraints of the global electron content of the ionosphere, derived from an uninterrupted sequence of 6573 daily averaged TEC measurements (1995-2013) [4]. Also this data taken period includes the extremely deep solar minimum between 2008 and 2009, which induced a more quiet behaviour of the electron content of the ionosphere. As before, the time binning of the TEC data is equal to 1 day. Surprisingly, we observed correlations with the position in the ecliptic of Mercury and Venus, including also the orbital positions of the Moon around the Earth. The first two columns in Figure 4 show the longitudinal distribution of Mercury and Venus, requiring a second planet (among the inner 3 planets) to be in a wide heliocentric longitudinal range 20°-140° vs. 200°-320°. Three out of four plots refer to the solar maximum period (1997-2006), while a similar comparison is shown for the extreme solar minimum period. Since the derived rates are normalized to 1 day, longitudinal modulations due to the planetary eccentricity are corrected. Considering long observation time periods of 10 or even 18 years, the derived distributions should be rather isotropic. This is, however, in contrast with observation.

We then addressed the annual TEC anomaly at the solstices, selecting the TEC data in 30° segments around December and June. Figure 4 (*right*, UP) shows the TEC distributions of these data

DOI: 10.1051/epjconf/201716407041

as a function of the Moon phase. Note that during the Earth's propagation by 30° in longitude, the Moon has completed one geocentric orbit. Taking into account the 223 orbits the Moon has performed around the Earth during the 18 years described by the dataset, randomly occurring TEC excursions should average out, or at least both distributions in Figure 4 (*right*, UP) should have similar shape. Figure 4 (*right*, DOWN) shows the calculated difference between the two distributions, exhibiting a modulation amplitude by a factor of 6. Remarkably, the position of the maximum coincides with the *New Moon*. This correlation fits well the scenario of this work, assuming a stream of invisible matter, with speed of about 50 km/s, coming from the direction of the Sun gravitationally focused by the interposed Moon onto the Earth.

3 Discussion

By analyzing the occurrence of X- and M-flares during the last 4 solar cycles and the full disk EUV irradiance of the Sun we find a strong evidence that the occurrence of these phenomena is strongly modulated by the position of Earth, Venus and Mercury around their heliocentric longitude. A preferred direction around 270° is common to all three planets, when their lensing effect is studied independently. The effect is further enhanced, as we show for the case of Mercury and Venus, when one takes into account the relative position of the two planets. This observation supports our working hypothesis that the activity of the Sun is triggered by influx of invisible matter and that this matter has some preferred direction or stream, which gets gravitationally lensed by the planets.

In addition, averaged daily GPS TEC measurements of the Earth ionosphere show also a marked planetary correlation, supporting our lensing scenario, which is further reinforced by the observation of the effect of the position of the Moon in modulating the TEC of the ionosphere. We bring to the attention that the anomalously high electron content of the ionosphere in December coincides with the alignment Galactic Center - Sun - Earth, i.e. the same direction around 270° previously described and that this effect becomes even more enhanced when the Moon is aligned around the same direction and interposed between the Sun and the Earth.

The identification of the nature of the predicted slow moving stream is left for future work: of course, one has to suppose that this invisible matter has some form of interaction with normal matter, but at this point we cannot infer any useful indication of its nature. Also, the (re)-analysis of the experiments searching for DM following our approach might give unexpected results.

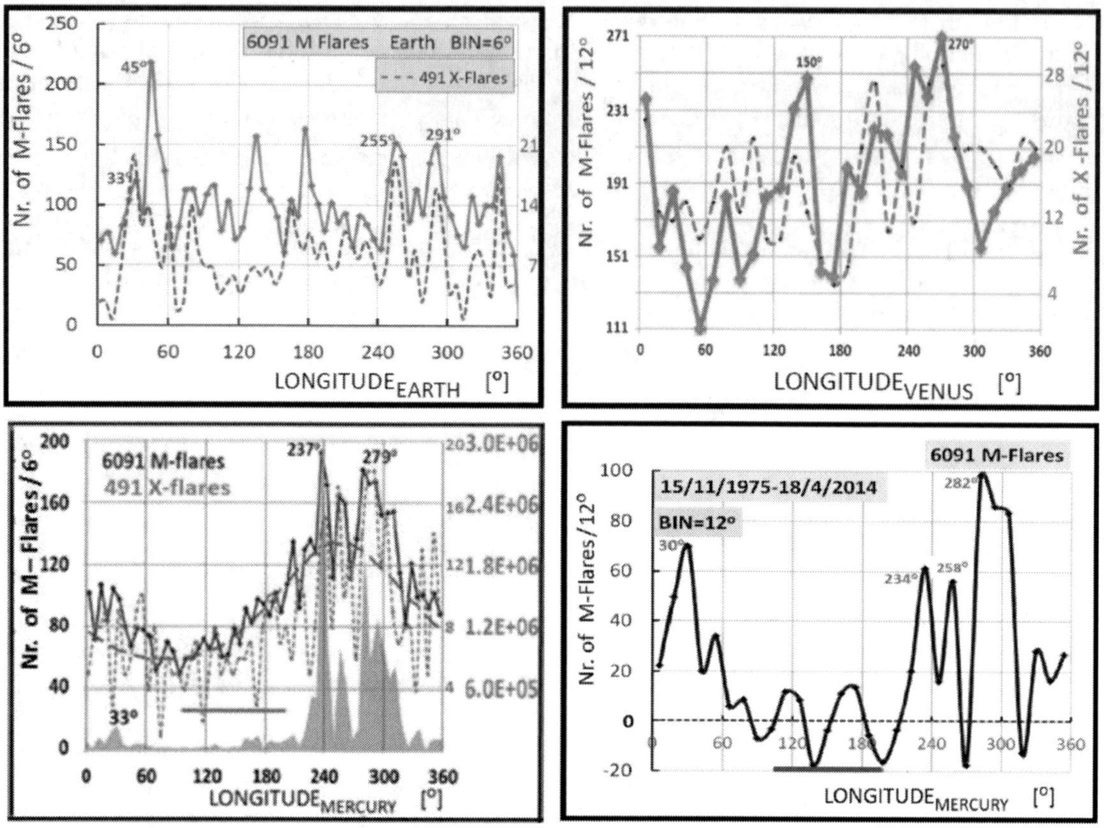

Figure 1. UP, Earth (*left*), Venus (*right*): Number of M-class Flares and X-class flares (dashed lines) as a function of the heliocentric longitude of the **Earth** (01/01/1976 – 05/05/2015) and the **Venus** (15/11/1975 – 18/04/2014). DOWN, *left*: Number of M-class flares (blue) and X-class flares (dashed) as a function of **Mercury** heliocentric longitude of the 6091 M-class Flares. The dashed red line is the normalised stay time due to Mercury orbital eccentricity. The horizontal red bar shows the region used for normalization. In orange is shown the "multiplication spectrum" of the 4 solar cycles (M-Flares). On the *right* is shown the residual spectrum (blue - red dashed on the *left*), with the 3 peaks around 30°, 240° and 290° becoming better visible. Note, that we use $BIN = 6°$ (*left column*) and $BIN = 12°$ (*right column*).

DOI: 10.1051/epjconf/201716407041

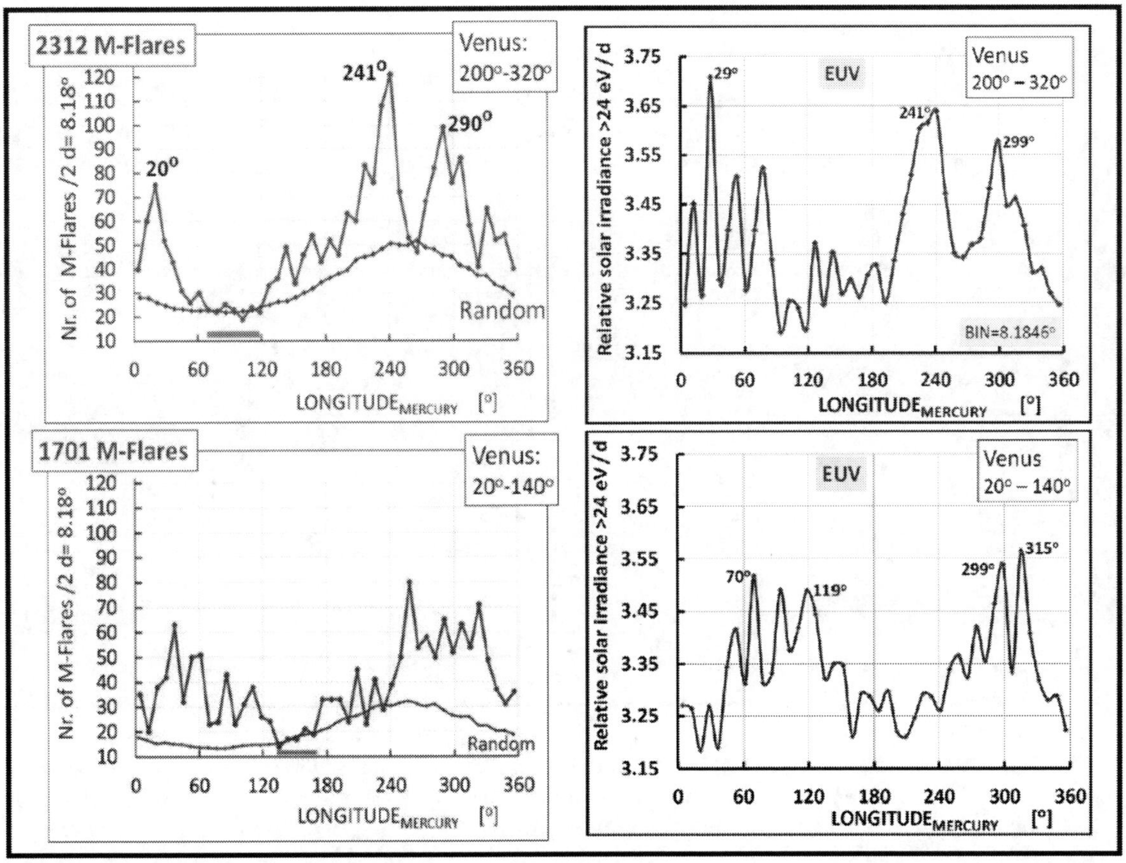

Figure 2. UP: Distributions of M-class Flares (*left*) and full disk EUV irradiance (*right*) as a function of **Mercury** heliocentric longitude with the constraint of Venus being at longitude between 200°-320°. DOWN: The same plots for Venus being between 20° and 140°. The smooth blue lines represent the expected normalized number of flares if equally distributed in time due to the large eccentricity of Mercury. The green bar shows the region used for normalisation. The EUV spectra are time normalised and therefore the eccentricity effect is out.

 DOI: 10.1051/epjconf/201716407041

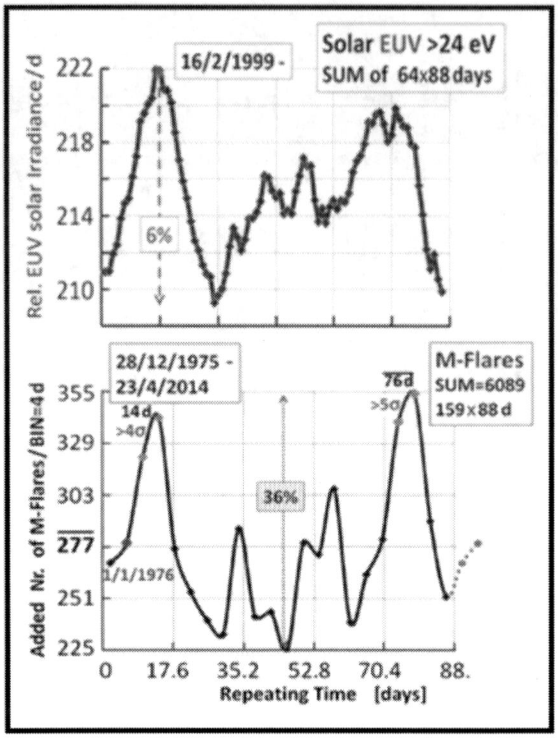

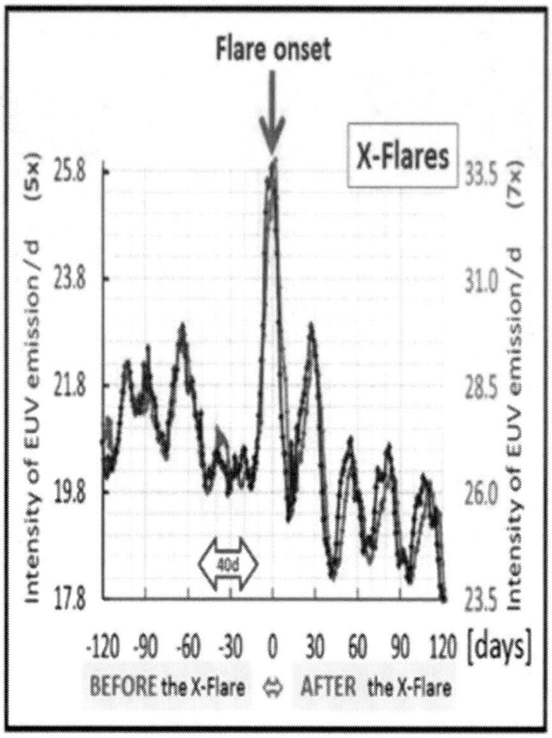

Figure 3. LEFT, UP: The sum of the daily averaged full disk solar EUV emission in 64 consecutive time intervals of 88 days (1999-2015). LEFT DOWN: The sum of M-class Flares averaged over 4 days in 159 consecutive intervals of 88 days (1976-2014). It is apparent the similar spectral shape of both solar activities. RIGHT: The start time of the biggest 7 X-class Flares during the previous solar cycle (1996-2007) is used as the reference point (*time* = 0) for the emission of full disk solar EUV during 4 months before and after the flare onset, The flares occurred far from each other, excluding double counting effects. The gap of about 40 days before the flare "explosion" is intriguing. Note, for 2 X-flare pairs, being too close to each other, an averaged onset time was used; the initial correlation remains unaffected (blue line)

DOI: 10.1051/epjconf/201716407041

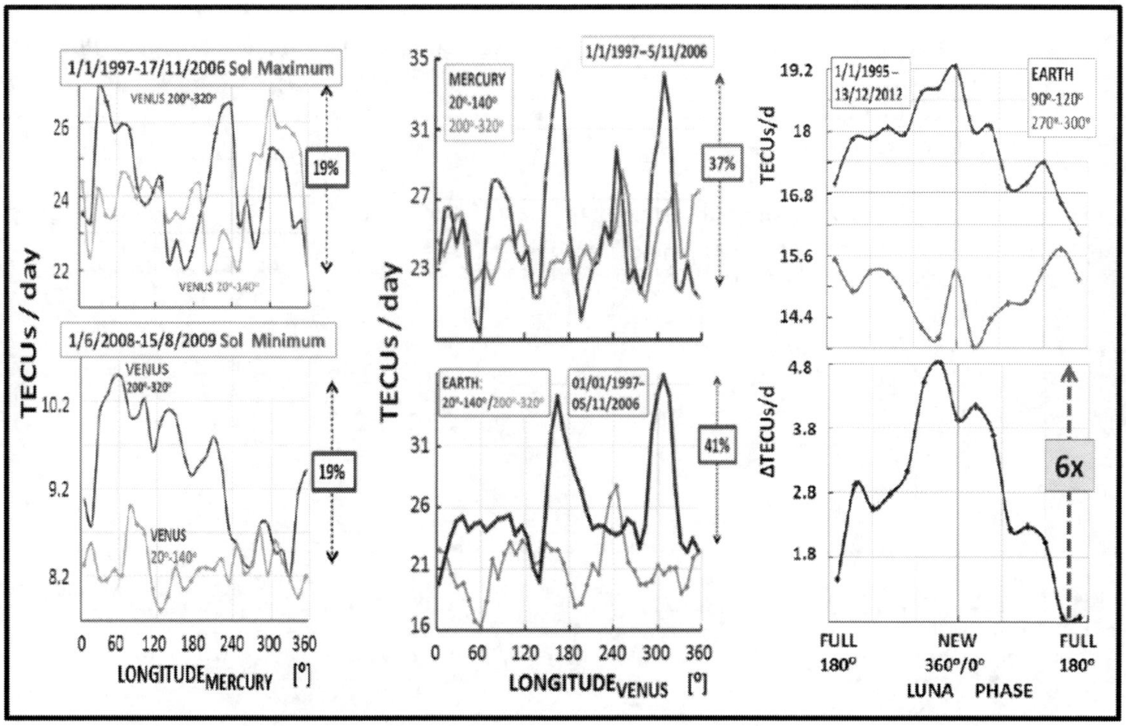

Figure 4. The daily measured longitudinal distributions of Earth's atmospheric total electron content [1] for different planetary configurations and time periods of the solar cycle. LEFT: The total electron content [TECUs] as a function of **Mercury** heliocentric longitude during the solar maximum period 1997-2006 (UP) and during the recent extremely deep solar minimum 2008-2009 (DOWN). The thick blue and the thin grey lines are associated with Venus being in one of the two opposite 120° wide orbital arcs. MIDDLE: TECUs as a function of the **Venus** longitude during the same solar maximum requiring Mercury (UP) and Earth (DOWN) to be in one of the 120° segments. RIGHT: TECUs as a function of the **Moon Phase**, while the Earth is in one of the two 30° orbital segments around the solstices (UP) and the difference between the winter-summer solstices (DOWN). Note, the Moon is between the Earth and the Sun at *New Moon*.

References

[1] C. A. J. O'Hare and A. M. Green, "Directional detection of dark matter streams," Phys. Rev. D 90 (**2014**) no.12, 123511, doi:10.1103/PhysRevD.90.123511 [arXiv:1410.2749 [astro-ph.CO]], and references therein.

[2] S. Bertolucci, K. Zioutas, S. Hofmann, M. Maroudas, "The Sun and its Planets as detectors for invisible matter," *Submitted for publication* (**2016**), [arXiv:1602.03666v4], and references therein.

[3] Astrid Veronig, private communication **2014**.

[4] Leonid Didkovsky, private communication **2014**.

Space-time development of in-medium hadronization : scenario for leading hadrons

B. Guiot[1,a] and B. Kopeliovich[1]

[1] *University Técnica Frederico Santa Maria (UTFSM), Valparaiso, Chile*

Abstract. Lepton deep inelastic scattering off nuclei at medium energies gives the opportunity to study the space-time development of hadronization. Indeed, for these kinematics, the production length is comparable to the nuclear size. Based on the Berger model [1] and dipole phenomenology, we built a model for vacuum and in-medium hadronization. The model, which includes vacuum energy loss, induced energy loss and nuclear absorption in a parameter-free way successfully describes Hermes data [2, 3]. In a future publication, Fermi motion will be taken into account and the model will be applied to CLAS Eg2 data, at Jefferson laboratory.

1 Introduction

The study of semi-inclusive production of hadrons in deep-inelastic scattering (DIS) off nuclei is interesting for at least two reasons. It gives a better understanding of cold nuclear matter and provides information on the spacetime development of hadronization. The main observables are the hadron p_t-broadening (see section 7) and the multiplicity ratio, defined as :

$$R_A(z, Q^2, E) \simeq \frac{1}{A}\frac{D^A(z, Q^2, E)}{D(z, Q^2, E)} \tag{1}$$

which shows nuclear suppression (enhancement) of hadrons. Here, D^A and D are the in-medium and vacuum fragmentation functions, respectively. z, Q^2 and E are the fraction of photon energy carried by the final hadron, the photon virtuality and the photon energy. Models based on induced energy loss [4, 5] will for instance provide information on the transport coefficient for nuclear matter ($\hat{q}$) and on the production time of the pre-hadron. Other models [7, 8] try to explain the nuclear suppression in terms of absorption of the (pre)hadron.[1] They provide information on production and formation times as well as on the spacetime dependence of the pre-hadron cross section. Both can describe reasonably well data on DIS off nuclei which have been published by the European Muon collaboration (EMC) [6] and by the HERMES collaboration at DESY.

As underlined in [4], these models use quantities which are poorly constrained : $\hat{q}$ and the pre-hadron cross section. Consequently, it is hard to make some conclusions on the respective

[a] e-mail: benjamin.guiot@usm.cl
[1] Here and in the following, (pre)hadron stands for either the hadron or the pre-hadron.

contributions of induced energy loss and nuclear absorption of the (pre)hadron to the multiplicity ratio. One of our aims is to disentangle these effects.

In this article, we present a simple model based on perturbative QCD (pQCD) for the propagation and hadronization of the leading quark in cold matter. It includes vacuum energy loss, induced energy loss and nuclear absorption. Our results are then compared to Hermes data [2, 3] (and to CLAS Eg2 data in a future publication).

2 pQCD based hadronization

Hadronization of the leading hadron is based on the Berger model [1], modified by higher order considerations [9]. In the Born approximation, the leading quark emits a gluon which splits into a $q\bar{q}$ pair. Then the $\bar{q}$ and the leading quark form the leading hadrons, as depicted in figure 1.

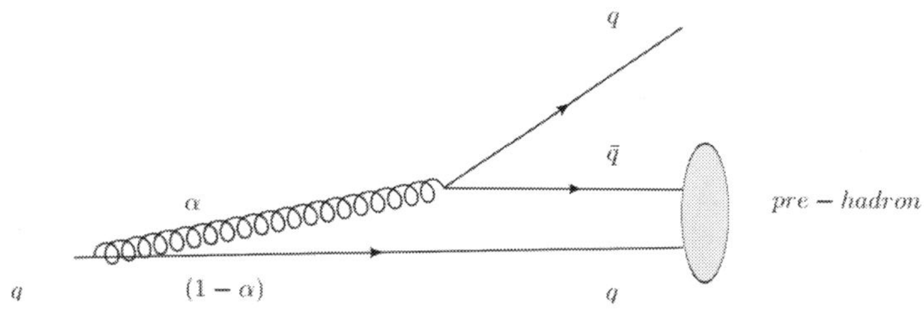

Figure 1. Berger mechanism of leading pion production in Born approximation. there, α is the fraction of quark energy taken by the gluon.

In this description of hadronization and in the limit $(1-z) \ll 1$, $k_t \ll Q^2$, one obtains the following fragmentation function for pion :

$$\frac{\partial D^{Born}}{\partial k_t}(z, k_t) \propto \frac{(1-z)^2}{k_t^4} \tag{2}$$

with k_t the gluon transverse momentum relative to the leading quark direction. As explained in [9], taking into account energy loss (see figure 2), we obtain the following expression for the pion fragmentation function at large z :

$$\frac{\partial D}{\partial k_t}(z, k_t) \propto \frac{(1-\tilde{z})^2}{k_t^4} \tag{3}$$

The energy loss results in a shift of the fragmentation variable z :

$$\tilde{z} = \frac{z}{1-\Delta E/E} \tag{4}$$

with ΔE being the total energy loss. In the assumption that the $q\bar{q}$ pair shares equally the gluon energy, we can write :

$$z = \frac{E_\pi}{E} = \frac{(\tilde{\alpha}/2 + 1 - \tilde{\alpha})(E-\Delta E)}{E} = (-\tilde{\alpha}/2 + 1)(1 - \Delta E/E) \tag{5}$$

where $\tilde{\alpha}$ is the fraction of $(E - \Delta E)$ taken by the gluon. Reversing this equation we have :

$$\tilde{\alpha} = 2\left(1 - \frac{z}{1-\Delta E/E}\right) = 2(1-\tilde{z}) \tag{6}$$

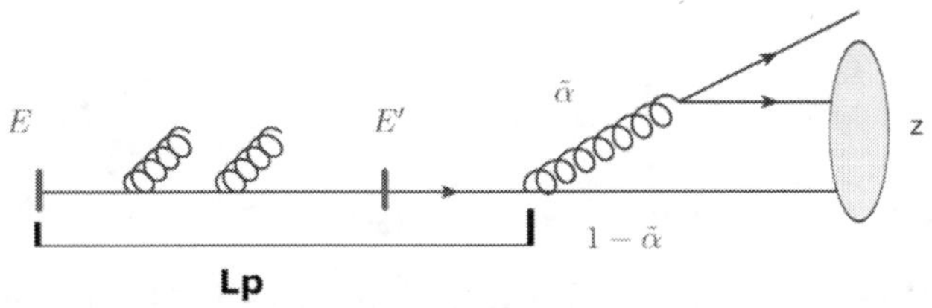

Figure 2. Berger mechanism with energy loss $\Delta E = E - E'$. $\tilde{\alpha}$ is the quark energy fraction taken by gluon after energy loss ΔE.

One can see that for $\tilde{\alpha} \in [0,1]$, we have $z \in [0.5, 1]$. It comes from our assumption of equal sharing for the $q\bar{q}$ pair, but this is not an issue since we want to focus on high z values. Another assumption of our model is to identify the production length of the pre-hadron with the radiation time of the gluon :

$$Lp = \frac{2\tilde{\alpha}E'}{k_t^2} = \frac{4E'(1-\tilde{z})}{k_t^2} \tag{7}$$

Energy loss gives a shorter production length ($\tilde{z} > z$) and the effect is larger for smaller value of z. For $z = 1$, there is no energy loss allowed and we find back the result given by the Born approximation. In terms of production length, the fragmentation function, eq. (3), is given by :

$$\frac{\partial D}{\partial Lp}(z, E, Q^2, Lp) \propto (1-\tilde{z}) \tag{8}$$

3 Vacuum energy loss

There is a perturbative and non-perturbative contribution. The perturbative one is given by :

$$\Delta E_{pert}(L, z, Q^2) = E\int_{\lambda^2}^{Q^2} dq_t^2 \int_0^1 d\beta\beta \frac{dn_g}{dq_t^2 d\beta}\Theta(L - l_c^g)\Theta(1 - z - \beta) \tag{9}$$

$\lambda = 0.7$ GeV is a cut-off (see [9]), q_t the gluon transverse momentum and β is the energy fraction taken by the radiated gluon. The last step function maintains energy conservation; none of the emitted gluons can have energy larger than $(1-z)E$. The first step function takes into account gluons radiation time :

$$l_c^g = \frac{2\beta E}{q_t^2} \tag{10}$$

and $\frac{dn_g}{dq_t^2 d\beta}$ is the usual gluon number distribution (see [9] for more details). In terms of gluon radiation time, the perturbative energy loss is :

$$\Delta E_{pert}(L, z, Q^2) = E\int_{\lambda/E}^{1-z} d\beta \int_{l_{min}}^{l_{max}} \beta \frac{dn_g}{dl d\beta} dl \tag{11}$$

with boundaries for integration given by :

$$l_{min} = \frac{2E\beta}{Q^2}\ ,\ l_{max} = min[\frac{2E\beta}{\lambda^2}, L] \tag{12}$$

DOI: 10.1051/epjconf/201716407042

Due to energy conservation, energy loss goes to zero when $z \to 1$. For $L > L_{max}$, the maximal length for perturbative energy loss given by :

$$L_{max} = \frac{2E(1-z)}{\lambda^2} \tag{13}$$

the perturbative energy loss is a constant of L.

Non-perturbative energy loss due to color flux tube formation is based on the second model in [9]. That is :

$$\Delta E_{np}(L,z) = \kappa \frac{\lambda^2 L^2}{E} \Theta(L_{max} - L) + \left[\kappa \frac{\lambda^2 L^2_{max}}{E} + \kappa(L - L_{max})\right] \Theta(L - L_{max}) \tag{14}$$

where $\kappa = 1$ GeV/fm is the string tension.

4 Fragmentation function

In order to compare with data, one has to integrate the differential fragmentation function (8) over the production length Lp :

$$D(z,Q^2,E) \propto \int_{Lp_{min}}^{Lp_{max}} dLp \frac{\partial D}{\partial Lp}(z,Q^2,E,Lp) \tag{15}$$

with Lp_{min} and Lp_{max} given by the equations :

$$Lp_{min} = \frac{2E'(1 - \tilde{z}(Lp_{min}))}{Q^2} \; ; \; Lp_{max} = \frac{2E'(1 - \tilde{z}(Lp_{max}))}{\lambda^2} \tag{16}$$

Since $\tilde{z} < z$, we see that the production length is shorted by energy loss. These equations are solved numerically.

5 In medium hadronization

At first approximation, we are taking into account only nuclear absorption. Then the nuclear fragmentation function is simply the convolution of the vacuum fragmentation function, equation (8), with a suppression factor :

$$D^A(z,Q^2,E) \propto \int d^2b \int dz_l \, \rho(b,z_l) \int_{Lp_{min}}^{Lp_{max}} dLp \frac{\partial D}{\partial Lp}(z,Q^2,E,Lp) Tr(z,Q^2,E,b,z_l + Lp, \infty) \tag{17}$$

Here b is the two dimensional impact parameter; ρ the nuclear density, taken from [11]; z_l the longitudinal coordinate of the DIS process and $Tr()$ the suppression factor due to dipole absorption by the nuclear medium.

At low energies, the dipole size can fluctuate and the eikonal approximation can't be used. The $q\bar{q}$ propagation throughout the medium is achieved with the light cone Green function $G(z_2, \vec{r_2}, z_1, \vec{r_1})$, where z_1, z_2 correspond to initial and final times, respectively, and $\vec{r_1}$, $\vec{r_2}$ represent the initial and final dipole sizes. This Green function obeys the two dimensional light cone Schrödinger equation, described in [12] together with its solution :

$$i \frac{d}{dz_2} G(z_2, \vec{r_2}, z_1, \vec{r_1}) = \left[\frac{\epsilon^2 - \Delta_{r_2}}{2p\beta(1-\beta)} + V_{q\bar{q}}(z2, \vec{r_2}, \beta)\right] G(z_2, \vec{r_2}, z_1, \vec{r_1}) \tag{18}$$

Using the Green function, The expression for the transparency factor is :

$$Tr(z, Q^2, E, b, z_1, z_2) = \left| \frac{\int d^2r_1 d^2r_2\, \psi_h^*(r_2) G(z_2, r_2, z_1, r_1) \psi_{q\bar{q}}(r_1)}{\int d^2r\, \psi_h^*(r) \psi_{q\bar{q}}(r)} \right|^2 \tag{19}$$

For the hadronic wave function, we use a parametrization in the form of the asymptotic light-cone meson wave function [13]

$$\psi_h(r) = f(\beta) \exp(-a^2(\beta) r^2/2) \tag{20}$$

$$a^2(\beta) = \frac{4(\beta(1-\beta) + a_0)}{\langle r_\pi^2 \rangle} \tag{21}$$

In [8] it is found that for the value $a_0 = 1/12$, the previous wave function reproduces correctly the pion mean radius squared. In our model, the light cone fraction β of the quark inside the pion wave function is related to z by

$$\beta = \frac{2\tilde{z} - 1}{\tilde{z}} \tag{22}$$

This wave function is solution of the Schrödinger equation (18) if the real part of the potential is given by :

$$\Re\, V_{q\bar{q}}(z_2, \vec{r}, \beta) = \frac{a^4(\beta) r^2}{2p\beta(1-\beta)} \tag{23}$$

The imaginary part of the potential responsible for absorption is given by :

$$\Im\, V_{q\bar{q}}(z_2, \vec{r}, \beta) = -\frac{\sigma_{q\bar{q}}(r)}{2} \rho_A(z_2) \tag{24}$$

where $\sigma_{q\bar{q}}(r)$ is the dipole cross section with nucleon. For simplicity, we use the small r expression :

$$\sigma_{q\bar{q}}(r) = C(s) r^2 \tag{25}$$

with $\sqrt{s}$ the pre-hadron nucleon center of mass energy :

$$s = 2zM_pE + M_p^2 + m_\pi^2 \tag{26}$$

The energy dependent factor $C(s)$ is constrained since the averaged dipole cross section has to reproduce the total meson (π^+ in our case) cross section with nucleon

$$\left\langle \sigma_{q\bar{q}} \right\rangle = \int d^2r\, C(s) r^2 |\psi_{q\bar{q}}(r)|^2 = C(s) \left\langle r_\pi^2 \right\rangle = \sigma_{tot}^{\pi p}(s) \tag{27}$$

Here we use $\sigma_{tot}^{\pi p}(s) = 23.6\,(s/s_0)^{0.079} + 1.432\,(s/s_0)^{-0.45}$ mb [14] with $s_0 = 1000$ GeV^2 for $\sqrt{s} > 2.5$ GeV^2 and a table from Igor Strakovsky otherwise [15]. Additionally, we have the relation $\left\langle r_\pi^2 \right\rangle = \frac{8}{3} \left\langle r_\pi^2 \right\rangle_{em}$, $\left\langle r_\pi^2 \right\rangle_{em} = 0.44$ fm^2 being the pion mean charge radius squared [16].

The last ingredient for the color transparency factor, equation (19), is the dipole wave function. Since we want a continuous transition between the dipole and pion wave functions, we will use :

$$\psi_{\bar{q}q}(z, Q^2, E, Lp, r) = f(\beta) \exp(-b^2(\beta) r^2/2) \tag{28}$$

with

$$b^2(\beta) = \frac{4(\beta(1-\beta) + a_0)}{\left\langle r_{q\bar{q}}^2 \right\rangle} \tag{29}$$

Then for $\left\langle r_{q\bar{q}}^2 \right\rangle = \left\langle r_\pi^2 \right\rangle$, dipole and pion wave functions are equal.

6 Multiplicity ratio

We now have all the ingredients in order to compute the suppression factor $Tr()$ and the multiplicity ratio defined in equation (1). Even though the dipole transverse size evolves with time, it is interesting to start with the approximation $\langle r^2_{q\bar{q}} \rangle = \langle r^2_\pi \rangle$ since it gives the maximal contribution of nuclear absorption. Our result for the HERMES experiment [2] is presented in figure 3.

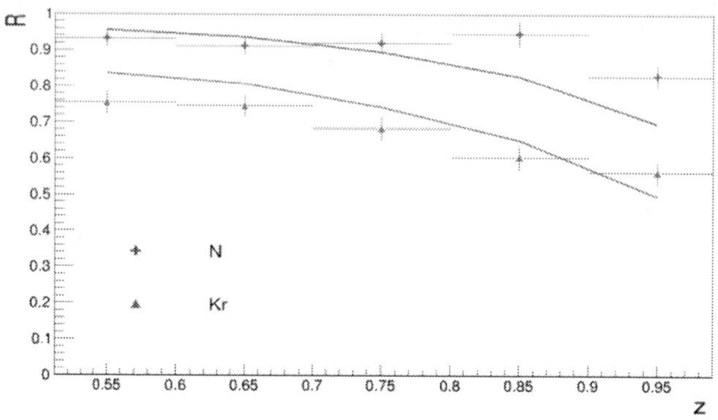

Figure 3. Multiplicity ratio for nitrogen and krypton, compared to HERMES data [2].

At $z = 1$, due to energy conservation, there is no energy loss. Therefore, the suppression comes only from the nuclear absorption. Unsurprisingly, our prediction is below the data, which is the consequence of our approximation on the dipole size. This approximation was also expected to work better for big nuclei. At $z = 0.5$, the fact that we are still above the data could be an indication that the contribution of induced energy loss is missing.

7 Broadening and induced energy loss

As it propagates through the nucleus, the leading quark undergoes multiple scattering which generates induced energy loss and p_t-broadening. Following [17, 18], the two main formulae are :

$$\delta \langle p_t^2 \rangle (s, z_q, b, L) = 2C(s)\rho(z_q, b)L \tag{30}$$

$$-\frac{dE_{ind}}{dz_q} = \frac{3}{4}\alpha_s \delta \langle p_t^2 \rangle \tag{31}$$

where z_q is the longitudinal coordinate of the quark. Since ρ depends on z_q, the correct expression for p_t-broadening on a finite length is given by :

$$\Delta \langle p_t^2 \rangle (s, z_l, b, L_p) = 2C(s) \int_{z_l}^{z_l+L_p} \rho(l, b) dl \tag{32}$$

where z_l is the longitudinal coordinate of the DIS point. Total induced energy loss is given by :

$$\Delta E_{ind}(s, z_l, b, L_p) = \frac{3}{4}\alpha_s \int_0^{L_p} \Delta \langle p_t^2 \rangle (s, z_l, b, l) dl \tag{33}$$

The DIS point is distributed over the whole nucleus, so we have to average equation (32) with $\frac{1}{A}\int d^2b\, dz_1\, \rho(z_1, b)$. We also have a distribution for the length L_p, therefore, we average equation (32) with $\frac{1}{N_L}\int dL_p \frac{\partial D}{\partial L_p}$. Then, for the quark, the experimental p_t-broadening is given by :

$$\Delta\left\langle p_t^2\right\rangle_{exp}(z, Q^2, E) = \frac{2C}{A}\int d^2b \int dz_1 \rho(z_1, b)\int_{L_{min}}^{L_{max}} dL_p \frac{1}{N_L}\frac{\partial D}{\partial L_p} T_A(b, z_1, z_1 + L_p) \quad (34)$$

with

$$T_A(b, z_1, z_1 + L_p) = \int_{z_1}^{z_1+L_p} dl\rho(l, b) \quad (35)$$

and the normalization

$$N_L(z, Q^2, E) = \int_{L_{min}}^{L_{max}} dL_p \frac{\partial D}{\partial L_p}(z, Q^2, E, L_p) \quad (36)$$

Finally, the pion p_t-broadening we are interested in is given by

$$\Delta\left\langle k_t^2\right\rangle_\pi (z, Q^2, E) = \tilde{z}^2 \Delta\left\langle p_t^2\right\rangle_{exp}(z, Q^2, E) \quad (37)$$

To take into account the induced energy loss, we need to replace $\tilde{z} \rightarrow \tilde{z} = \frac{z}{1-(\Delta E_{vac}+\Delta E_{ind})/E}$, when computing in-medium fragmentation function D^A. Results for p_t-broadening and multiplicity ratio modified by induced energy loss are shown in Figure 4.

For p_t-broadening, our result does a reasonably good description of data. It should be noticed that HERMES do a correction (based on Monte-Carlo) for decays of the ρ^0 meson into pions. For the higher bin in z, more than 50% of pions are supposed to come from ρ^0 decays. This is likely an overestimation of this number which results in a negative p_t-broadening for high z. Notice also that one argument for the negative p_t-broadening is that the p_t distribution of quarks inside a nucleus is larger for smaller nucleus. Then at $z = 1$ (no broadening from quark multiple interaction) we could have $p_{t,A} - p_{t,Deterium} < 0$. But in this case we should have $\Delta p_{t,Xe} < \Delta p_{t,Kr} < \Delta p_{t,Ne} < 0$ which is not the case.

Our result for the multiplicity ratio has been improved by induced energy loss, particularly for smaller value of z, as it was expected. In a future publication, we will show that the implementation of induced energy loss based on the model presented in [4] gives a similar result (using $\hat{q} = 0.25$). At $z = 0.5$, the correction of induced energy loss is of the order of 15%. One of the reason why in our case, the contribution of induced energy loss appears smaller than in models based solely on this effect is probably because they use $\hat{q} \sim 0.70$ while the value for cold matter is estimated to be $\hat{q} \sim 0.25$ [18]. Another reason is that the DIS point is taken to be at the front edge of the nucleus, giving a bigger path for the quark inside nuclear matter.

8 Conclusion

Based on Berger's model, we can build vacuum and in-medium fragmentation functions. The suppression factor, $Tr()$, which takes into account nuclear absorption of the dipole, is computed based on quantum mechanics, using an assumption for the dipole and pion waves functions. The induced energy loss has been implemented, using its relation with the p_t-broadening. With our no free parameter model, we are able to describe HERMES data for p_t-broadening and multiplicity ratio. The contribution of induced energy loss for the multiplicity ratio is found to be 15% at $z = 0.5$. One obvious improvement of the model, is the implementation of the dipole size evolution based on the Green function. The consequence will be a smaller suppression, particularly at $z = 1$.

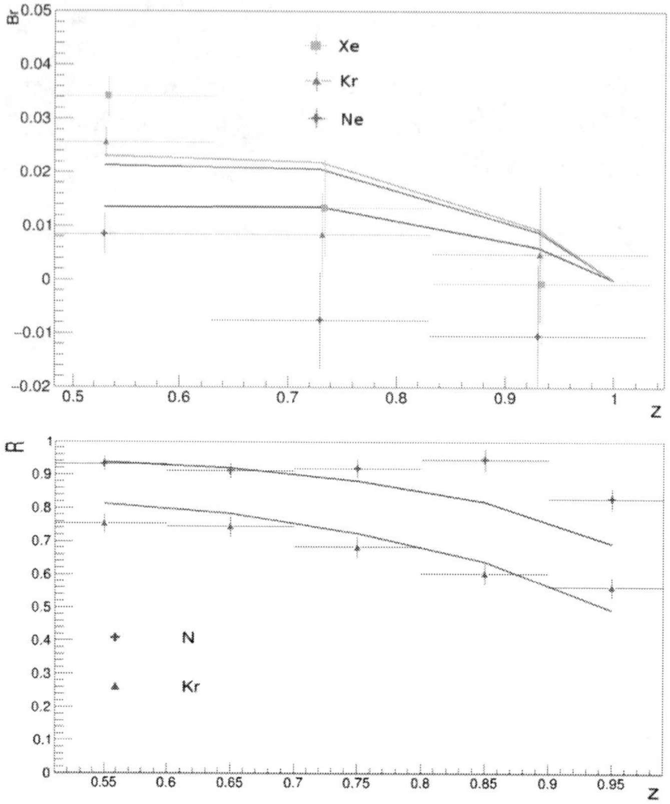

Figure 4. Top : p_t-broadening compared to HERMES data [3]. bottom : multiplicity ratio taking into account induced energy loss.

9 Acknowledgment

B. G. gratefully acknowledges generous support from Chilean FONDECYT grants 3160493. B. G. acknowledges support by the Basal project FB0821.

References

[1] E.L. Berger, Phys. Lett. B **89**, 241 (1980)

[2] HERMES Collaboration, A. Airapetian, et al., Phys. Lett. B **577** 37 (2003)

[3] HERMES Collaboration, A. Airapetian, et al., Phys. Lett. B **684** 114 (2010)

[4] F. Arleo, Eur. Phys. J. C **30** 213 (2003)

[5] E. Wang, X. N. Wang, Phys. Rev. Lett. **89** 162301 (2002)

[6] EMC Collaboration, J. Ashman et al., Z. Phys. C **52** 1 (1991);
EMC Collaboration, M. Arneodo et al., Nucl. Phys. B **321** 541 (1989)

[7] K. Gallmeister and U. Mosel, Nucl.Phys. A **801** 68 (2008)

[8] B. Z. Kopeliovich et al., Nuclear Physics A **740** 211 (2004)

[9] B. Z. Kopeliovich, H. J. Pirner, I.K. Potashnikova, I. Schmidt, Physics Letters B **662** 117 (2008)

[10] Y. Dokshitzer, Phil.Trans.Roy.Soc.Lond.A **359** 309 (2001)
[11] H. DE VRIES, C. W. DE JAGER and C. DE VRIES, ATOMIC DATA AND NUCLEAR DATA TABLES **36** 495-536 (1987)
[12] B.Z. Kopeliovich, A. Schäfer, and A.V. Tarasov, Phys. Rev. D **62** 054022 (2000)
[13] B.Z. Kopeliovich, J. Nemchik, A. Schäfer, A.V. Tarasov, Phys. Rev. C **65** (2002) 035201
[14] R.M. Barnett, et al., Phys. Rev. D **54**, 191 (1996)
[15] I. Strakovsky, total and total elastic cross sections for $\pi^+ p$ collision, private communication
[16] S.R. Amendolia, et al., Phys. Lett. B **178** 435 (1986)
[17] M.B. Johnson, B.Z. Kopeliovich and A.V. Tarasov, Phys. Rev. C **63** 35203 (2001)
[18] R. Baier, Yu.L. Dokshitzer, A.H. Mueller, S. Peigne and D. Schiff, Nucl. Phys. B**484** 265 (1997)

Spin Gauge Interactions as a Topological Mechanism of Superconductivity

Ishita Dutta Choudhury[1,a]

[1] *S N Bose National Centre for Basic Sciences, Block- JD, Sector-III, Salt Lake, Kolkata-700106, India*

Abstract. We talk about a low energy, effective, topological theory of superconductivity in which a topological mass term is radiatively induced in one loop effective action. In this field theoretic model, an antisymmetric tensor field couples with the vorticity current of charged Dirac fermions in the Lagrangian. The fermion loop generates a coupling between the gauge field and the antisymmetric tensor field below an ultraviolet cut-off. The spin interactions mediated by the antisymmetric tensor field induces a mass for the photon field indicating Meissner effect. The dual antisymmetric tensor field produces a current which satisfies the relativistic version of the London equations of superconductivity. In the non-relativistic limit, the static effective potential shows a linear, always attractive term between two electrons. Thus, the theory can be considered as an alternative, low energy, effective field theory of superconductivity without spontaneous symmetry breaking.

1 Introduction

The discovery of fractional Quantum Hall liquids (FQHL) led us to the developement of a new sector in the theory of solid state physics[1–4]. The usual superconductivity mechanism has a local, spin zero order parameter and the theory involves spontaneous symmetry breaking. On the contrary, description of the mechanism of incompressible FQHL demands the presence of a new kind of order parameter[5, 6] related to the topology of the system and the mechanism cannot be explained by ordinary theory of spontaneous symmetry breaking[7].The standard theory of superconductivity was due to Bardeen, Cooper and Schrieffer. According to BCS theory, Meissner effect can be explained by the bosonization of the electrons in the form of Cooper pairs[8–10]. The interaction between the electrons and the phonons effectively generates a mutual attraction between two electrons in a limited shell in the momentum space near the Fermi surface. When this mutual attraction dominates over Coulomb repulsion, bound pairs form. The theory has a local, spin zero order parameter and the system goes through spontaneous symmetry breaking. However, in topological superconductors, the system may not have any local order parameter and there should be an associated topological field theory without involving spontaneous symmetry breaking[11, 12].

In (3 + 1)-dimensions, the topological theory of superconductivity involves a derivative coupling between the photon field and an antisymmetric tensor field $B_{\mu\nu}$[13, 14]. The coefficient of this interaction plays the role of a topological mass of the photon field. The field equations of the antisymmetric tensor represent the relativistic London equations. The mechanism is similar to the famous

[a] e-mail: ishitadutta.choudhury@bose.res.in, iduttachoudhury2@gmail.com

 DOI: 10.1051/epjconf/201716407043

Chern-Simons theory in (2 + 1)-dimensions[15–17]. However, the Chern-Simons term breaks P and T invariance while BF term preserves these symmetries.

In (2 + 1)-dimensions, the radiative induction of Chern-Simons term made it possible to connect the theory with the physical systems[18, 19]. Likewise, in (3 + 1)-dimensions also, it is important to know the actual form of the interaction which effectively leads to the topological mass generation of the photon[20, 21]. In this talk, we deal with a long range interaction between the fermions which is mediated by the antisymmetric tensor field. This non-local interaction effectively generates topological BF term in the one loop effective action. This is a low energy theory and therefore is valid for the energy scales well below an ultraviolet cut-off. In the non-relativistic limit, the static effective potential has a linear attractive term between the interacting fermions[22].

2 Spin Current

The effective induction of topological term in the Lagrangian requires a fermion coupling with the antisymmetric tensor field $B_{\mu\nu}$. In other words, an interaction between the fermions via the antisymmetric tensor field $B_{\mu\nu}$ effectively produces topological mass for the photon. We choose the interaction as

$$\mathcal{L}^B_{\text{int}} = gB_{\mu\nu}J^{\mu\nu}\,, \tag{1}$$

where

$$J^{\mu\nu} = m\,\epsilon^{\alpha\mu\nu\sigma}\frac{\partial_\alpha}{\Box}(\bar{\psi}\gamma_\sigma\psi)\,, \tag{2}$$

with m, the mass of the fermion. Here, the coupling of $B_{\mu\nu}$ with the topologically conserved current $J^{\mu\nu}$ has some other properties which act in favour of choosing it to use in our theory. The coupling $B_{\mu\nu}J^{\mu\nu}$, along with P and T symmetry has an additional vector gauge invariance under the transformation

$$B_{\mu\nu} \longrightarrow B_{\mu\nu} + \partial_\mu\kappa_\nu - \partial_\nu\kappa_\mu\,. \tag{3}$$

Moreover, in non-relativistic limit, the spin contribution to the $0i$-th components J^{0i} are the spin magnetic moment densities of the fermions if the fermions are localised.

Thus, a non-local spin interaction between the fermions, mediated by antisymmetric tensor field can be used to generate a topological photon mass.

3 Calculations

The original form of the complete Lagrangian density used to construct the fully relativistic model is given by

$$\mathcal{L} = \bar{\psi}\gamma^\mu\left(i\partial_\mu + eA_\mu\right)\psi - m\bar{\psi}\psi + gB_{\mu\nu}J^{\mu\nu} - \frac{1}{4}F_{\mu\nu}F^{\mu\nu} + \frac{1}{12}H_{\mu\nu\alpha}H^{\mu\nu\alpha}\,, \tag{4}$$

where $H_{\mu\nu\alpha} \equiv \partial_{[\mu}B_{\nu\alpha]}$ is the Kalb-Ramond [24] gauge invariant field strength for the antisymmetric tensor and g is the dimensionless coupling constant for the spin interactions. As in [21], this model is a low-energy effective theory, valid for energy scales below some ultraviolet cutoff Λ. By partial integration, the antisymmetric tensor interaction can be written as

$$\mathcal{L}^B_{\text{int}} = gB_{\mu\nu}J^{\mu\nu} = \frac{2mg}{\Box}F_\mu J^\mu\,, \tag{5}$$

DOI: 10.1051/epjconf/201716407043

where $F_\mu = (1/2)\epsilon_{\mu\nu\alpha\beta}\partial^\nu B^{\alpha\beta}$ is the dual Kalb-Ramond field strength. One can combine the two terms for the interaction of the fermions, one with the dual Kalb -Ramond field strength and the other with ordinary gauge field, to write the total interaction term of the fermions in terms of an effective photon field $A^\mu_{\rm eff}$ which is a combination of the ordinary photon field and the inverse d'Alembertian of the dual Kalb-Ramond field strength.

Integrating out the fermions, the one-loop induced action can be obtained as

$$\Gamma^{1-\rm loop} = \frac{e^2}{4}\frac{1}{12\pi^2}\ln\frac{\Lambda^2}{m^2}\int d^4x(F_{\rm eff})_{\mu\nu}(F_{\rm eff})^{\mu\nu}\,, \tag{6}$$

where $F^{\mu\nu}_{\rm eff}$ is the field strength of the effective photon field.

Expansion of the one-loop induced action in terms of the original gauge fields present in the action produces three terms. Combining those three terms with the original action the 1-loop effective Lagrangian can be written as

$$\mathcal{L} = -\frac{1}{4e^2_{\rm ph}}F_{\mu\nu}F^{\mu\nu} + \frac{g_{\rm ph}m}{2\pi}B_{\mu\nu}\epsilon^{\mu\nu\rho\sigma}\partial_\rho A_\sigma + \frac{1}{12}H_{\mu\nu\alpha}H^{\mu\nu\alpha} + \frac{6g^2_{\rm ph}m^2}{\ln\frac{\Lambda^2}{m^2}}F_\mu\frac{1}{\Box}F^\mu\,, \tag{7}$$

where

$$e^2_{\rm ph} = e^2\left(1 + \frac{e^2}{12\pi^2}\ln\frac{\Lambda^2}{m^2}\right)\,, \tag{8}$$

is the QED renormalized charge and

$$g_{\rm ph} = \frac{g}{6\pi}\ln\frac{\Lambda^2}{m^2}\,. \tag{9}$$

Here, the second term in Eq. (7) can be identified as the topological BF term. The induced non-local term is represented as a Gaussian integral over a new massless vector gauge field. Using a proper set of combinations of the auxilliary gauge field and the photon field the Lagrangian is finally diagonalized.

Thus, the resulting one-loop Lagrangian decouples into two parts,

$$\mathcal{L} = \mathcal{L}_1 + \mathcal{L}_2\,,$$

$$\mathcal{L}_1 = -\frac{1}{4e^2_1}F_{\mu\nu}F^{\mu\nu} + \frac{g_{\rm ph}m}{2\pi}B_{\mu\nu}\epsilon^{\mu\nu\rho\sigma}\partial_\rho A_\sigma + \frac{1}{12}H_{\mu\nu\alpha}H^{\mu\nu\alpha}\,, \tag{10}$$

$$\mathcal{L}_2 = -\frac{1}{4e^2_2}G_{\mu\nu}G^{\mu\nu}\,, \tag{11}$$

where

$$e^2_1 = e^2_{\rm ph}\frac{1 + \frac{4e^2_{\rm ph}}{12\pi^2}\ln\frac{\Lambda^2}{m^2}}{\frac{4e^2_{\rm ph}}{12\pi^2}\ln\frac{\Lambda^2}{m^2}}\,,$$

$$e^2_2 = e^2_{\rm ph}\left(1 + \frac{4e^2_{\rm ph}}{12\pi^2}\ln\frac{\Lambda^2}{m^2}\right)\,. \tag{12}$$

Eq. (10) corresponds to the topological mass generation model with an effective charge e_1 that reduces to the usual renormalized electron charge, $e_1 \rightarrow e_{\text{ph}}$, for $m \ll \Lambda$. The second part written in Eq. (11)has a free, massless gauge field with field strength $G_{\mu\nu}$.

The equations of motion from the Eq. (10) can be written as

$$\partial_\mu F^{\mu\nu} = -\frac{e_1^2 g_{\text{ph}} m}{6\pi} \epsilon^{\nu\mu\alpha\beta} H_{\mu\alpha\beta} , \tag{13}$$

$$\partial_\mu H^{\mu\alpha\beta} = \frac{g_{\text{ph}} m}{2\pi} \epsilon^{\alpha\beta\mu\nu} F_{\mu\nu} , \tag{14}$$

from which one can derive [7, 13, 14, 25]:

$$\left[\Box + \left(\frac{e_1 g_{\text{ph}} m}{\pi}\right)^2\right] F_{\mu\nu} = 0 . \tag{15}$$

Eq.(15) shows that the field strength $F_{\mu\nu}$ for the photon satisfies the Klein-Gordon equation with mass given by $\frac{e_1 g_{\text{ph}} M}{\pi}$. By a similar analysis we find that $H^{\mu\nu\alpha}$ also satisfies the equation of motion with the same mass.

From eq.(13) we also recognize that the dual of the Kalb-Ramond field strength acts as charged current for the photon field:

$$\begin{aligned} \partial_\mu F^{\mu\nu} &= J^\nu , \\ J^\nu &\equiv \frac{-e_1^2 g_{\text{ph}} m}{\pi} H^\nu , \\ &= \frac{-e_1^2 g_{\text{ph}} m}{6\pi} \epsilon^{\nu\mu\alpha\beta} H_{\mu\alpha\beta} . \end{aligned} \tag{16}$$

Substituing eq.(16) in the equation eq.(14) we obtain:

$$\begin{aligned} \epsilon^{\mu\nu\alpha\beta} \partial_\alpha J_\beta &= -\left(\frac{e_1 g_{\text{ph}} m}{\pi}\right)^2 \tilde{F}_{\mu\nu} , \\ \tilde{F}_{\mu\nu} &= \frac{1}{2} \epsilon^{\mu\nu\alpha\beta} F_{\alpha\beta} \end{aligned} \tag{17}$$

which is nothing but a relativistic version of the London equations for superconductivity.

4 Static effective potential

The interaction between the fermions mediated by the antisymmetric tensor field $B_{\mu\nu}$ in our low energy, effective theory for Meissner effect, has a non-local nature. This non-local nature of the current provokes an interesting question regarding the semi-classical picture of the theory, because of another well-known fact about the antisymmetric tensor field. Just as the ordinary gauge potential couples to worldlines of charged particles, the antisymmetric tensor field $B_{\mu\nu}$ couples with worldsheets which are traced out by one dimensional strings. Actually, the antisymmetric tensor field $B_{\mu\nu}$ was first postulated to describe interstring interactions. Our fermion coupling with the B field is not localized at a point and therefore, the question arises such as whether the system contains string-like objects and what would the strings be made of.

For calculating the static effective potential between the interacting fermions, we integrate out the gauge fields to obtain the effective action as

$$S[\psi,\bar{\psi}] = S_D[\psi,m] + S_A[\psi,\bar{\psi}] + S_B[\psi,\bar{\psi}], \tag{18}$$

where $S_D[\psi,m]$ corresponds to the kinetic term and the mass term for the fermion, and

$$S_A[\psi,\bar{\psi}] = \frac{1}{2}\int \frac{d^4k}{(2\pi)^4}\, J^\sigma(-k)\,\frac{e^2}{k^2}\, J_\sigma(k),$$
$$S_B[\psi,\bar{\psi}] = \int \frac{d^4k}{(2\pi)^4}\, J^\sigma(-k)\,\frac{g^2m^2}{k^4}\, J_\sigma(k). \tag{19}$$

In the non-relativistic limit, we expand the fermion fields in terms of three dimensional creation amd annihilation operators and note that the leading contribution to the effective potential is from the J^0 component. The reason is that the components J^i are made of the lower components of the Dirac spinor and therefore can be neglected in the non-relativistic limit as their energies are much lower than their mass. In three dimensional momentum space, the resulting static potential has the form

$$V(\mathbf{k}) = \frac{e^2}{|\mathbf{k}|^2} - \frac{g^2m^2}{|\mathbf{k}|^4}, \tag{20}$$

which leads to a linear term in the expression for the static potential in three dimensional coordinate space as

$$V(r) = \frac{e^2}{4\pi r} + \frac{g^2m^2r}{4\pi}. \tag{21}$$

The sign of the linear potential remains same for interactions between fermions with like charges and for those with opposite charges. A linear potential is commonly associated with string-like objects.

5 Conclusion

We have found that a long range interaction between fermions, mediated by an antisymmetric tensor field effectively induces a derivative coupling between the photon field and the antisymmetric tensor field. The equations of motion of the photon field can be used to write the Klein- Gordon equation as a massive wave equation which explicitly shows the presence of a topological mass of the photon field. The dual Kalb-Ramond field strength acts as a current to the photon field and that current satisfies the relativistic version of the London equations of superconductivity. In the non-relativistic limit, the long range interaction mediated by the antisymmetric tensor field gives rise to linear term in the static effective potential between two electrons. This linear term is independent of the spin orientations of the interacting electrons and is always attractive in nature irrespective of the electric charges of the interaction particles. Whenever this attractive force dominates over Coulomb repulsion, bound paires may be found. Thus, the theory can be considered as an alternative, low energy, effective field theory for the superconductivity which does not involve spontaneous symmetry breaking.

References

[1] S. M. Girvin, A. H. MacDonald, Phys. Rev. Lett. **58**, 1252 (1987).
[2] S. C. Zhang, T. H. Hansson, S. Kivelson, Phys. Rev. Lett. **62**, 82 (1989).

DOI: 10.1051/epjconf/201716407043

[3] S. C. Zhang, Int. J. Mod. Phys. B **06**, 25 (1992).
[4] X. L. Qi, E. Witten and S. C. Zhang, Phys. Rev. B **87**, 134519 (2013).
[5] X. G. Wen, Advances in Physics **44**, 405 (1995).
[6] X.-G. Wen, Int. Jour. Mod. Phys. B **6**, 1711 (1992).
[7] M. C. Diamantini, P. Sodano and C. A. Trugenberger, Eur. Phys. J. B **53**, 19 (2006).
[8] L. N. Cooper, Phys. Rev. **104**, 1189 (1956).
[9] J. Bardeen, L. N. Cooper and J. R. Schrieffer, Phys. Rev. **108**, 1175 (1957).
[10] N. N. Bogolyubov, Sov. Phys. JETP **7**, 41 (1958) [Zh. Eksp. Teor. Fiz. **34**, 58 (1958)] [Front. Phys. **6**, 399 (1961)].
[11] T. H. Hansson, V. Oganesyan and S. L. Sondhi, Annals Phys. **313**, 497 (2004).
[12] D. Birmingham, M. Blau, M. Rakowski and G. Thompson, Phsy. Rep. **209** 129(1991).
[13] T. J. Allen, M. J. Bowick and A. Lahiri, Mod. Phys. Lett. A **6**, 559 (1991).
[14] M. Bergeron, G. W. Semenoff and R. J. Szabo, Nucl. Phys. B **437**, 695 (1995).
[15] R. Jackiw and S. Templeton, Phys. Rev. D **23**, 2291 (1981).
[16] S. Deser, R. Jackiw and S. Templeton, Phys. Rev. Lett. **48**, 975 (1982).
[17] S. Deser, R. Jackiw and S. Templeton, Annals Phys. **140**, 372 (1982) Erratum: [Annals Phys. **185**, 406 (1988)].
[18] A. N. Redlich, Phys. Rev. Lett. **52**, 18 (1984).
[19] A. N. Redlich, Phys. Rev. D **29**, 2366 (1984).
[20] M. C. Diamantini, G. Guarnaccia and C. A. Trugenberger, J. Phys. A **47**, 092001 (2014).
[21] I. D. Choudhury, M. C. Diamantini, G. Guarnaccia, A. Lahiri and C. A. Trugenberger, JHEP **1506**, 081 (2015).
[22] Chandrasekhar Chatterjee, I. D. Choudhury and Amitabha Lahiri, [In preparation].
[23] M. Leblanc, R. MacKenzie, P. K. Panigrahi and R. Ray, Int. J. Mod. Phys. A **9**, 4717 (1994).
[24] M. Kalb and P. Ramond, Phys. Rev. D **9**, 2273 (1974).
[25] M. C. Diamantini, P. Sodano and C. A. Trugenberger, New J. Phys. **14**, 063013 (2012).

EPJ Web of Conferences **164**, 07044 (2017)   DOI: 10.1051/epjconf/201716407044
ICNFP 2016

Stochastic Navier-Stokes equation and advection of a tracer field: One-loop renormalization near $d = 4$

N. V. Antonov[1,a], N. M. Gulitskiy[1,b], M. M. Kostenko[1,c], and T. Lučivjanský[2,d]

[1]*Department of Theoretical Physics, Faculty of Physics, Saint-Petersburg State University, 7/9 Universitetskaya nab., St. Petersburg, 199034 Russia*
[2]*Faculty of Sciences, P.J. Šafárik University, Moyzesova 16, 040 01 Košice, Slovakia*

Abstract. The renormalization group approach and the operator product expansion technique are applied to the model of a tracer field advected by the Navier-Stokes velocity ensemble for a compressible fluid. The model is considered in the vicinity of the specific space dimension $d = 4$. The properties of the equal-time structure functions are investigated. The multifractal behaviour of various correlation functions is established. All calculations are performed in the leading one-loop approximation.

1 Introduction

The investigation of such behavior as anomalous scaling requires a lot of thorough, even meticulous, analysis to be carried out. The phenomenon manifests itself in a singular (arguably, power-like) behaviour of some statistical quantities (correlation functions, structure functions, etc.) in the inertial-convective range in the fully developed turbulence regime [1].

Very useful and computationally efective approach to the problems with many interacting degrees of freedom on different scales is the field-theoretic renormalization group (RG) approach which can be subsequently accompanied by the operator product expansion (OPE); see the monographs [2–5]. One of the greatest challenges is the investigation of the Navier-Stokes equation for a compressible fluid, and, in particular, passive scalar field advection by this velocity ensemble. The first relevant discussion and analyses of the problem of passive advection emerged a few decades ago for the Kraichnan's velocity ensemble [6–8]. Further studies are connected with its more realistic generalizations [9–18]; see also review paper [19]. The RG+OPE technique were also applied to more complicated models, in particular, to the compressible case [20–33].

The present study is devoted to the investigation of the passive scalar tracer field advected by the Navier-Stokes velocity ensemble with the compressibility taken into account near the specific space dimension $d = 4$. Usually d plays a passive role – see [20] for Navie-Stokes equation itself and [21, 22] for advection of scalar fields – but if $d = 4$, the situation is similar to the analysis of the incompressible Navier-Stokes equation near space dimension $d = 2$. In this case an additional

[a]e-mail: n.antonov@spbu.ru
[b]e-mail: n.gulitskiy@spbu.ru
[c]e-mail: kontramot@mail.ru
[d]e-mail: tomas.lucivjansky@uni-due.de

divergence appears in the 1-irreducible Green's function $\langle v'v'\rangle_{1\text{-ir}}$, see [34–36]. This feature allows us to employ a double expansion scheme, in which the formal expansion parameters are y, which describes the scaling behaviour of a random force, and $\varepsilon = 4 - d$, i.e., a deviation from the space dimension $d = 4$.

The paper is a logical continuation of our previous studies [37, 38] and is organized as follows. In the introductory Section 2 a brief discussion of the fixed points' structure and possible scaling regimes is presented. In the following section the tracer field is introduced and the expressions for its critical dimensions (for each of two scaling regimes) are presented. In Section 4 the renormalization of a certain composite field is considered. In Section 5 OPE is applied to the equal-time structure functions constructed of the tracer fields; the anomalous exponents are calculated. The concluding section is devoted to the brief discussion.

2 Navier-Stokes equation, scaling regimes

Consider first an advection of a scalar field by a compressible fluid. The fluid within the model used is assumed to be compressible, so the Navier-Stokes equation and the continuity equation have the forms

$$\nabla_t v_i = \nu_0[\delta_{ik}\partial^2 - \partial_i\partial_k]v_k + \mu_0\partial_i\partial_k v_k - \partial_i\phi + f_i, \quad \nabla_t \equiv \partial_t + v_i\partial_i, \tag{1}$$

$$\nabla_t\phi = -c_0^2\partial_i v_i. \tag{2}$$

Here $\partial_t \equiv \partial/\partial t$, $\partial_i \equiv \partial/\partial x_i$, $v_i = v_i(t, \boldsymbol{x})$ is a velocity field, $\phi = \phi(t, \boldsymbol{x})$ is a pressure field and $\partial^2 = \partial_i\partial_i$ is the Laplace operator. The constants ν_0 and μ_0 are two independent molecular viscosity coefficients, c_0 is an adiabatic speed of sound. The statistics of the random force f isconveniently given in the Fourier representation by the relation

$$\langle f_i(t,\boldsymbol{x})f_j(t',\boldsymbol{x}')\rangle = \frac{\delta(t-t')}{(2\pi)^d}\int_{k>m} \mathrm{d}^d\boldsymbol{k}\, D_{ij}(\boldsymbol{k})\mathrm{e}^{i\boldsymbol{k}\cdot(\boldsymbol{x}-\boldsymbol{x}')}, \tag{3}$$

where the delta function ensures Galilean invariance of the model and m provides us with infrared (IR) cutoff. Further, the kernel function $D_{ij}(\boldsymbol{k})$ reads

$$D_{ij}(\boldsymbol{k}) = g_{10}\nu_0^3 k^{\varepsilon-y}\Big\{P_{ij}(\boldsymbol{k}) + \alpha Q_{ij}(\boldsymbol{k})\Big\} + g_{20}\nu_0^3\delta_{ij}. \tag{4}$$

A term proportional to the charge g_{10} ensures a steady input of energy into the system, which is needed in order to counterbalance dissipation processes due to a viscosity; the definition $\varepsilon = 4 - d$ is introduced here. The term proportional to g_{20} is not dictated by the physics of the fluid, but rather by a proper renormalization treatment [39].

In [37] three IR attractive fixed points, which defines possible scaling regimes of the system, were discussed. The character of the IR behaviour depends on the mutual relation between y and ε – two formally small quantities which were introduced in the correlator of the random force in the Navier-Stokes equation. In practical calculations they constitute parameters into which universal quantities are expanded. This is done in a similar fashion as calculation of critical exponents in ϕ^4 theory, see [2–5].

The fixed point FPI (the trivial or Gaussian point) is stable if $y, \varepsilon < 0$. The coordinates are

$$g_1^* = 0, \quad g_2^* = 0. \tag{5}$$

The fixed point FPII, which is stable if $\varepsilon > 0$ and $y < 3\varepsilon/2$, has the following coordinates:

$$g_1^* = 0, \quad g_2^* = \frac{8\varepsilon}{3}. \tag{6}$$

The fixed point FPIII (stable if $y > 0$ and $\varepsilon < 2y/3$) is

$$g_1^* = \frac{16y(2y - 3\varepsilon)}{9[y(2+\alpha) - 3\varepsilon]}, \quad g_2^* = \frac{16\alpha y^2}{9[y(2+\alpha) - 3\varepsilon]}. \tag{7}$$

The crossover between the two nontrivial points (6) and (7) takes place across the line $y = 3\varepsilon/2$, which is in accordance with results of [24].

Depending on the values of y and ε, the different values of the critical dimension for various quantities F are obtained. They can be calculated via the expression

$$\Delta[F] = d_F^k + \Delta_\omega d_F^\omega + \gamma_F^*, \tag{8}$$

where d_F^ω is the canonical frequency dimension, d_F^k is the momentum dimension, γ_F^* is the anomalous dimension at the critical point (FPII or FPIII), and $\Delta_\omega = 2 - \gamma_\nu^*$ is the critical dimension of frequency.

In [37] the critical dimension of the passive scalar density field θ and the field θ' were obtained for the fixed points FPII and FPIII:

$$\Delta_\theta = -1 + \varepsilon/4, \quad \Delta_{\theta'} = d + 1 - \varepsilon/4 \quad \text{for the fixed point FPII;} \tag{9}$$

$$\Delta_\theta = -1 + y/6, \quad \Delta_{\theta'} = d + 1 - y/6 \quad \text{for the fixed point FPIII.} \tag{10}$$

3 Tracer field

There are two permissible kinds of passive scalar fields in nature: the density field (density of some pollutant) and the tracer field which describes the temperature or entropy [40]. The advection of the density field is described by the stochastic equation

$$\partial_t\theta + \partial_i(v_i\theta) = \kappa_0\partial^2\theta + f, \tag{11}$$

while the advection of the tracer field is governed by

$$\partial_t\theta + (v_i\partial_i)\theta = \kappa_0\partial^2\theta + f. \tag{12}$$

Here κ_0 is the molecular diffusivity coefficient and $f \equiv f(x)$ is a Gaussian noise with zero mean and given covariance,

$$\langle f(x)f(x')\rangle = \delta(t - t')\, C(\mathbf{r}/L), \quad \mathbf{r} = \mathbf{x} - \mathbf{x}', \tag{13}$$

where $C(\mathbf{r}/L)$ is some function finite at $(\mathbf{r}/L) \to 0$ and rapidly decaying for $(\mathbf{r}/L) \to \infty$. For the case of the incompressible fluid the difference between the density and tracer fields is indistinguishable since an additional condition for the velocity field arises (namely, transversality condition $\partial_i v_i = 0$), which makes the terms $\partial_i(v_i\theta)$ and $(v_i\partial_i)\theta$ identical.

The aim of this study is to investigate the critical behaviour of various statistical quantities which are constructed in terms of the tracer field advected by the compressible fluid near the special space dimension $d = 4$. The IR (large-scale) behavior of the structure functions is the major focus of this paper. Within the method used the composite field theory and OPE are employed. This approach requires a strong RG analysis aimed at finding (or discovering) the critical dimension of the tracer field.

Fortunately, despite the difference between the formalism for density and tracer fields in the case of a compressible fluid, the critical dimensions for these two fields have the same form; for detailed analysis see [21]. Thus, expressions (9) and (10) remain true for tracer field as well. In what follows the composite fields and OPE will be discussed.

4 Composite fields

The measurable quantities are some correlation functions or structure functions of composite operators. A local composite operator is a monomial or polynomial constructed from the primary fields $\theta(x)$ and their finite-order derivatives at a single space-time point x. In the Green's functions with such objects, new UV divergences arise due to the coincidence of the field arguments. They can be removed by the additional renormalization procedure.

The simplest case of a composite operator is the scalar operator $F(x) = \theta^n(x)$. But in contrast to the density model, if the field $\theta(x)$ obeys stochastic equation (12), the real index of divergency for the diagrams with such operators is positive. This means, that in case of tracer such operators do not diverge and renormalization constants for them are trivial.

More interesting objects are composite operators with derivatives, namely

$$F^{(n,l)}_{i_1\ldots i_l} = \partial_{i_1}\theta \cdots \partial_{i_l}\theta\,(\partial_i\theta\partial_i\theta)^s + \ldots \tag{14}$$

Here l is the number of the free vector indices (the rank of the tensor) and $n = l+2s$ is the total number of the fields θ entering the operator. The ellipsis stands for the subtractions with the Kronecker's delta symbols that make the operator irreducible (so that a contraction with respect to any pair of the free tensor indices vanish). For instance,

$$F^{(2,2)}_{ij} = \partial_i\theta\partial_j\theta - \frac{\delta_{ij}}{d}\,(\partial_k\theta\partial_k\theta). \tag{15}$$

For practical calculations, it is convenient to contract the tensors (14) with an arbitrary constant vector $\lambda = \{\lambda_i\}$. The resulting scalar operator takes the following form:

$$F^{(n,l)} = (\lambda_i w_i)^l (w_i w_i)^s + \ldots, \qquad w_i \equiv \partial_i\theta, \tag{16}$$

where the subtractions, denoted by the ellipsis, necessarily include the factors of $\lambda^2 = \lambda_i\lambda_i$.

In order to calculate the critical dimension of the operator, one has to renormalize it. The operators (14) can be treated as multiplicatively renormalizable, $F^{(n,l)} = Z_{(n,l)}F^{(n,l)}_R$, with certain renormalization constants $Z_{(n,l)}$ (see [21]). The counterterm to $F^{(n,l)}$ must have the same rank as the operator itself. It means that the terms containing λ^2 should be excluded since the contracted fields $w_i w_i$, standing near them, reduce the number of free indices. It is sufficient to retain only the principal monomial, explicitly shown in (16), and to discard in the result all the terms with factors of λ^2. The renormalization constants $Z_{(n,l)}$ are determined by the finiteness of the 1-irreducible Green's function $\Gamma_{nl}(x;\theta)$, which in the one-loop approximation is diagrammatically represented as follows:

$$\Gamma_{nl}(x;\theta) = F^{(n,l)} + \frac{1}{2}\ \ \ , \tag{17}$$

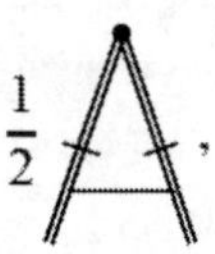

where numerical factor 1/2 is a symmetry factor of the graph and the thick dot with two lines attached denotes the operator vertex

$$V(x;x_1,x_2) = \frac{\delta^2 F^{(n,l)}}{\delta(\partial_i\theta)\,\delta(\partial_j\theta)}. \tag{18}$$

The expressions for the propagators and vertices at the bottom of the diagram can be found in [37, 39]. Then using the chain rule and up to irrelevant terms the vertex (18) for the operator $F^{(n,l)}$ can be presented in the form

$$V(x;x_1,x_2) = \frac{\partial^2 F^{(n,l)}}{\partial w_i \partial w_j}\,\partial_i\delta(x-x_1)\,\partial_j\delta(x-x_2). \tag{19}$$

The differentiation yields

$$\partial^2 F^{(n,l)}/\partial w_i \partial w_j = 2s(w^2)^{s-2}(\lambda w)^l \left[\delta_{ij} w^2 + 2(s-1)w_i w_j\right] + l(l-1)(w^2)^s(\lambda w)^{l-2}\lambda_i\lambda_j + + 2ls(w^2)^{s-1}(\lambda w)^{l-1}(w_i\lambda_j + w_j\lambda_i), \quad (20)$$

where $w^2 = w_k w_k$ and $(\lambda w) = \lambda_k w_k$. Two more factors $w_p w_r$ are attached to the bottom of the diagram due to the derivatives coming from the vertices $\theta'(v\partial)\theta$. The ultraviolet divergence is logarithmic and one can set all the external frequencies and momenta equal to zero; then the core of the diagram takes the form

$$\int \frac{d\omega}{2\pi} \int_{k>m} \frac{d^d\mathbf{k}}{(2\pi)^d} k_i k_j D_{pr}(\omega, \mathbf{k}) \frac{1}{\omega^2 + w^2 \nu^2 k^4}. \quad (21)$$

Here the first factor comes from the derivatives in (18), $w = \kappa/\nu$, D_{pr} is the velocity correlation function [see (4)], and the last factor comes from the two propagators $\langle\theta'\theta\rangle_0$.

After the integration, combining all the factors, contracting the tensor indices and expressing the result in terms of $n = l + 2s$ and l, one obtains:

$$\Gamma_n(x;\theta) = F^{(n,l)}(x)\Big\{1 - \frac{1}{4wd(d+2)}\Big[\frac{Q_1}{(1+w)}\Big(g_{10}\frac{m^{-y}}{y} + g_{20}\frac{m^{-\varepsilon}}{\varepsilon}\Big) + \frac{Q_2}{u(u+w)}\Big(\alpha g_{10}\frac{m^{-y}}{y} + g_{20}\frac{m^{-\varepsilon}}{\varepsilon}\Big)\Big]\Big\}. \quad (22)$$

Here the following abbreviations have been used:

$$Q_1 = -n(n+d)(d-1) + (d+1)l(l+d-2), \qquad Q_2 = -n(3n+d-4) + l(l+d-2). \quad (23)$$

Then the renormalization constants $Z_{(n,l)}$ calculated in the MS scheme read

$$Z_{(n,l)} = 1 - \frac{1}{2d(d+2)}\left[\frac{Q_1}{2w(1+w)}\left(\frac{g_{10}}{y} + \frac{g_{20}}{\varepsilon}\right) + \frac{Q_2}{2wu(u+w)}\left(\alpha\frac{g_{10}}{y} + \frac{g_{20}}{\varepsilon}\right)\right]. \quad (24)$$

For the corresponding anomalous dimension one obtains

$$\gamma_{(n,l)} = \frac{1}{2d(d+2)}\left\{\frac{Q_1}{2w(1+w)}(g_{10} + g_{20}) + \frac{Q_2}{2wu(u+w)}(\alpha g_{10} + g_{20})\right\}. \quad (25)$$

In order to evaluate the critical dimension, one needs to substitute the coordinates of the fixed points (5) – (7) into the expression (25) and then use the relation (8). For the fixed point FPII the critical dimension is

$$\Delta_{(n,l)} = \frac{n}{4}\varepsilon + \frac{Q_1 + Q_2}{3d(d+2)}\varepsilon. \quad (26)$$

For the fixed point FPIII it is

$$\Delta_{(n,l)} = \frac{n}{6}y + \frac{2y}{d(d+2)} \frac{Q_1(\alpha y + 2y - 3\varepsilon) + 3\alpha Q_2(y-\varepsilon)}{9[y(2+\alpha) - 3\varepsilon]}. \quad (27)$$

Both expressions (26) and (27) suppose higher order corrections in y and ε.

Therefore, the infinite set of operators with negative critical dimensions, whose spectra is unbounded from below, is observed.

5 Operator Product Expansion

Consider the equal-time structure functions

$$S_n(r) = \langle [\theta(t,\mathbf{x}) - \theta(t,\mathbf{x}']^{2n} \rangle = (\nu\mu^2)^{-n}\eta(\mu r, mr), \quad \text{where} \quad r = |\mathbf{x}' = \mathbf{x}|. \tag{28}$$

The second equality involving the dimensionless functions η follows from dimensionality considerations. Solving the RG equations gives the asymptotic expressions in the limit $\mu r \gg 1$:

$$S_n(r) = (\nu\mu^2)^{-n}(\mu r)^{-2n\Delta_\theta}\zeta(mr), \tag{29}$$

where ζ is some unknown scaling function. The inertial-convective range corresponds to the additional condition $mr \ll 1$. The behavior of the functions ζ at $mr \to 0$ can be studied by means of the OPE.

The tracer field is determined by the stochastic equation (12), therefore, the model is invariant with respect to the constant shift $\theta(x) \to \theta(x) + \text{const}$. This means, that the operators entering the corresponding OPE,

$$[\theta(t,\boldsymbol{x}) - \theta(t,\boldsymbol{x}']^{2n} \simeq \sum_F C_F(mr)\,F\Big(t, \frac{\boldsymbol{x}+\boldsymbol{x}'}{2}\Big), \tag{30}$$

must be also invariant. Therefore, that they can involve the field θ only in the form of derivatives. Clearly, the leading term of the behaviour at small mr should be determined by the scalar operator with maximal possible number of the fields θ (namely, $2n$ for a given S_n) and the minimal possible number of spatial derivatives (namely, $2n$ – one derivative for each θ). Consequently, this is the operator $F^{(2n,0)} = (\partial_i\theta\partial_i\theta)^n$; see the definition (14). Thus, the leading-order expression for S_n in the inertial range is

$$S_n(r) \sim (\nu\mu^2)^{-n}(\mu r)^{-2n\Delta_\theta}(mr)^{\Delta_{(2n,0)}}, \tag{31}$$

with the dimensions $\Delta_{(2n,0)}$ given by the expressions (26) and (27) with the substitution $l = 0$. Other types of fields or other numbers of derivatives generate the more "distant" corrections. Expression (31) together with expressions (9), (10) and (26), (27) means that existence of anomalous scaling, i.e., singular power-like dependence on the integral scale L, was established.

Considering OPE for the correlation functions $\langle F^{(p,0)}F^{(k,0)}\rangle$ with $n = p + k$, where $F^{(n,l)}$ is the operator of the type (14), one can observe that the leading contribution to the expansion is determined by the operator $F^{(n,0)}$ from the same family. Therefore, in the inertial range these correlation functions acquire the form

$$\langle F^{(p,0)}(t,\mathbf{x})F^{(k,0)}(t,\mathbf{x}')\rangle \sim r^{-\Delta_{(p,0)}-\Delta_{(k,0)}+\Delta_{(n,0)}}. \tag{32}$$

The inequality $\Delta_{(n,0)} < \Delta_{(p,0)} + \Delta_{(k,0)}$, which follows from both explicit one-loop expressions (26) and (27), indicates, that the operators $F^{(n,0)}$ demonstrate a "multifractal" behavior; see [41].

6 Conclusion

In the present paper the advection of the passive scalar tracer field by the Navier-Stokes velocity ensemble has been examined. The fluid was assumed to be compressible and the space dimension was close to $d = 4$. The problem has been investigated by means of renormalization group and operator product expansion; the double expansion in y [see (4)] and $\varepsilon = 4 - d$ was constructed. The present study has been aimed at the investigation of the anomalous scaling in the equal-time structure functions for the tracer field.

There are two IR stable fixed points in this model and, therefore, the critical behaviour in the inertial range demonstrates two different regimes depending on the relation between the exponents y

and ε. The expressions for the critical exponents of the tracer field θ were obtained in the one-loop approximation.

In order to find the anomalous exponents of the structure functions, the composite fields (14) were renormalized. The critical dimensions of them were evaluated. It turned out that there is an infinite number of the dangerous operators, i.e., the operators with negative critical dimensions. Besides, OPE allowed us to derive the explicit expressions for the critical dimensions of the structure functions. The existence of the anomalous scaling in the inertial-convective range was established for both possible scaling regimes. Another very interesting result is that some kinds of operators exhibit the "multifractal" behavior.

With regard to future research, it would be interesting to go beyond the one-loop approximation and to analyze the behaviour more precisely on the higher level of accuracy. To have a closer look at the vector fields,both passive and active, is also a challenge to be further investigated.

Acknowledgements

The authors are indebted to M. Yu. Nalimov, L. Ts. Adzhemyan, M. Hnatič, J. Honkonen, and V. Škultéty for discussions.

The work was supported by VEGA grant No. 1/0345/17 of the Ministry of Education, Science, Research and Sport of the Slovak Republic and by the Russian Foundation for Basic Research within the Project 16-32-00086. N. M. G. acknowledges the support from the Saint Petersburg Committee of Science and High School. N. M. G. and M. M. K. were also supported by the Dmitry Zimin's "Dynasty" foundation.

References

[1] U. Frisch, *Turbulence: The Legacy of A. N. Kolmogorov* (Cambridge University Press, Cambridge, 1995)

[2] A. N. Vasil'ev, *The Field Theoretic Renormalization Group in Critical Behavior Theory and Stochastic Dynamics* (Boca Raton, Chapman Hall/CRC, 2004)

[3] J. Zinn-Justin, *Quantum Field Theory and Critical Phenomena* (4th edition, Oxford University Press, Oxford, 2002)

[4] L. Ts. Adzhemyan, N. V. Antonov, A. N. Vasil'ev: *The Field Theoretic Renormalization Group in Fully Developed Turbulence* (Gordon & Breach, London, 1999)

[5] U. Täuber, *Critical Dynamics: A Field Theory Approach to Equilibrium and Non-Equilibrium Scaling Behavior* (Cambridge University Press, New York, 2014)

[6] R.H. Kraichnan, Phys. Fluids **11**, 945 (1968)

[7] K. Gawędzki and A. Kupiainen, Phys. Rev. Lett. **75**, 3834 (1995);
D. Bernard, K. Gawędzki, and A. Kupiainen, Phys. Rev. E **54**, 2564 (1996);
M. Chertkov and G. Falkovich, Phys. Rev. Lett. **76**, 2706 (1996)

[8] L. Ts. Adzhemyan, N. V. Antonov, and A. N. Vasil'ev, Phys. Rev. E **58**, 1823 (1998)

[9] N.V. Antonov, J. Phys. A: Math. Gen. **39**, 7825 (2006)

[10] E. Jurčišinova and M. Jurčišin, Phys. Rev. E **91**, 063009 (2015)

[11] N. V. Antonov, A. Lanotte, and A. Mazzino, Phys. Rev. E **61**, 6586 (2000);
N. V. Antonov and N. M. Gulitskiy, Theor. Math. Phys., **176**(1), 851 (2013)

[12] H. Arponen, Phys. Rev. E, **79**, 056303 (2009)

[13] N. V. Antonov and N. M. Gulitskiy, Lecture Notes in Comp. Science, **7125/2012**, 128 (2012);
N. V. Antonov and N. M. Gulitskiy, Phys. Rev. E **85**, 065301(R) (2012);
N. V. Antonov and N. M. Gulitskiy, Phys. Rev. E **87**, 039902(E) (2013)

[14] E. Jurčišinova and M. Jurčišin, J. Phys. A: Math. Theor., **45**, 485501 (2012);
E. Jurčišinova and M. Jurčišin, Phys. Rev. E **88**, 011004 (2013)
[15] E. Jurčišinova and M. Jurčišin, Phys. Rev. E **77**, 016306 (2008);
E. Jurčišinova, M. Jurčišin and R. Remecky, Phys. Rev. E **80**, 046302 (2009)
[16] N. V. Antonov and N. M. Gulitskiy, Phys. Rev. E **91**, 013002 (2015);
N. V. Antonov and N. M. Gulitskiy, Phys. Rev. E **92**, 043018 (2015);
N. V. Antonov and N. M. Gulitskiy, AIP Conf. Proc. **1701**, 100006 (2016);
N. V. Antonov and N. M. Gulitskiy, EPJ Web of Conf. **108**, 02008 (2016)
[17] E. Jurčišinova and M. Jurčišin, Phys. Rev. E **94**, 043102 (2016)
[18] M. Hnatič and P. Zalom, Phys. Rev. E **94**, 053113 (2016)
[19] M. Hnatič, J. Honkonen, T. Lučivjanský, Acta Physica Slovaca **66**, 69 (2016)
[20] N. V. Antonov, M. Yu. Nalimov and A. A. Udalov, Theor. Math. Phys. **110**, 305 (1997)
[21] N. V. Antonov and M. M. Kostenko, Phys. Rev. E **90**, 063016 (2014)
[22] N. V. Antonov and M. M. Kostenko, Phys. Rev. E **92**, 053013 (2015)
[23] L. Ts. Adzhemyan, N. V. Antonov, J. Honkonen, and T. L. Kim, Phys. Rev. E **71**, 016303 (2005)
[24] N. V. Antonov, Phys. Rev. Lett. **92**, 161101 (2004)
[25] N. V. Antonov, N. M. Gulitskiy, and A. V. Malyshev, EPJ Web of Conf. **126**, 04019 (2016)
[26] E. Jurčišinova, M. Jurčišin, R. Remecky, Phys. Rev. E **93**, 033106 (2016)
[27] M. Vergassola and A. Mazzino, Phys. Rev. Lett. **79**, 1849 (1997)
[28] A. Celani, A. Lanotte, and A. Mazzino, Phys. Rev. E **60** R1138 (1999)
[29] M. Chertkov, I. Kolokolov, and M. Vergassola, Phys. Rev. E. **56**, 5483 (1997)
[30] M. Hnatich, E. Jurčišinova, M. Jurčišin, and M. Repašan, J. Phys. A: Math. Gen. **39**, 8007 (2006)
[31] V. S. L'vov and A. V. Mikhailov, Preprint No. 54, Inst. Avtomat. Electron., Novosibirsk (1977)
[32] I. Staroselsky, V. Yakhot, S. Kida, and S. A. Orszag, Phys. Rev. Lett., **65**, 171 (1990)
[33] S. S. Moiseev, A. V. Tur, and V. V. Yanovskii, Sov. Phys. JETP **44**, 556 (1976)
[34] J. Honkonen and M. Yu. Nalimov, Z. Phys. B **99**, 297 (1996)
[35] L. Ts. Adzhemyan, J. Honkonen, M. V. Kompaniets et al., Phys. Rev. E **71**, 036305 (2005)
[36] L. Ts. Adzhemyan, M. Hnatich and J. Honkonen, Eur. Phys. J B **73**, 275 (2010)
[37] N.V. Antonov, N. M. Gulitskiy, M. M. Kostenko, T. Lučivjanský EPJ Web of Conf. **125**, 05006 (2016)
[38] N.V. Antonov, N. M. Gulitskiy, M. M. Kostenko, T. Lučivjanský EPJ Web of Conf. **137**, 10003 (2017)
[39] N. V. Antonov, N. M. Gulitskiy, M. M. Kostenko, and T. Lučivjanský, Phys. Rev. E **95**, 033120 (2017)
[40] L. D. Landau and E. M. Lifshitz, *Fluid Mechanics* (Pergamon Press, Oxford)
[41] B. Duplantier and A. Ludwig, Phys. Rev. Lett. **66**, 247 (1991);
G. L. Eyink, Phys. Lett. A **172**, 355 (1993)

Studies of light mesons at COMPASS

Sebastian Uhl[1,a,b], on behalf of the COMPASS Collaboration

[1] *Technische Universität München, James-Franck-Straße, 85748 Garching, Germany*

Abstract. To investigate the spectrum of light mesons, the COMPASS experiment has recorded data with a 190 GeV/c negative pion beam impinging on a liquid hydrogen target. Diffractive dissociation reactions at squared four-momentum transfers to the target between 0.1 and 1.0 (GeV/c)^2 are studied. The flagship channel is the $\pi^-\pi^+\pi^-$ final state, for which COMPASS has recorded the at present world's largest data sample. A new axial-vector signal, the $a_1(1420)$, has been found with unusual properties, pointing to an exotic nature. The findings are confirmed by the analysis of the $\pi^-\pi^0\pi^0$ final state.

1 Introduction

The **Co**mmon **M**uon and **P**roton **A**pparatus for **S**tructure and **S**pectroscopy (COMPASS) experiment is a fixed-target experiment located at CERN's Super Proton Synchrotron (SPS). One of its goals is to study the spectrum of light mesons. A 190 GeV/c negative hadron beam, consisting mostly of pions with a small admixture of kaons and anti-protons, impinges on a liquid hydrogen target, and diffractively produces the mesons under study. The decay products of these mesons are detected in a two-stage magnetic spectrometer (see figure 1). Each of the two spectrometer stages is equipped with an electromagnetic calorimeter. This setup provides nearly full phase-space coverage for charged and neutral particles, resulting in a homogenous acceptance over a wide kinematic range [1].

The reaction under study here is the diffractive production of mesons decaying either into the $\pi^-\pi^-\pi^+$ or the $\pi^-\pi^0\pi^0$ final state. To this end, exclusive events with a squared four-momentum transfer to the target t' between 0.1 (GeV/c)^2 and 1.0 (GeV/c)^2 have been selected in a three-pion mass range between 0.5 GeV/c^2 and 2.5 GeV/c^2. The selected dataset consists of 50 million $\pi^-\pi^-\pi^+$ and 3.5 million $\pi^-\pi^0\pi^0$ events. The invariant mass distributions of the two final states indicate a rich spectrum of contributing states (see figure 2).

2 Partial-wave analysis

To decompose the three-pion mass spectra into individual spin-parity components, a partial-wave analysis employing the isobar model is used. Figure 3 sketches the process under study. The beam

[a] e-mail: sebastian.uhl@mytum.de

[b] The author acknowledges support by the German Bundesministerium für Bildung und Forschung (BMBF), by the Maier-Leibnitz-Laboratorium der Universität und der Technischen Universität München, by the DFG Cluster of Excellence "Origin and Structure of the Universe" (Exc153), and by the computing facilities of the Computational Center for Particle and Astrophysics (C2PAP).

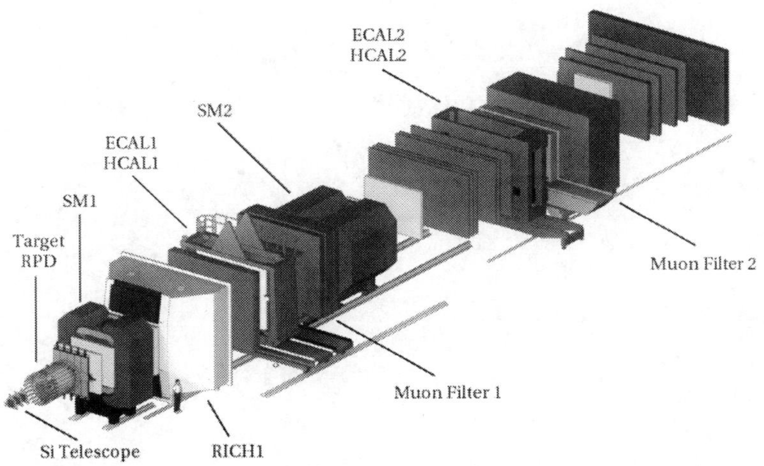

Figure 1. Sketch of the experimental setup. The beam enters the experiment from the lower left corner of the figure. (Figure from [1].)

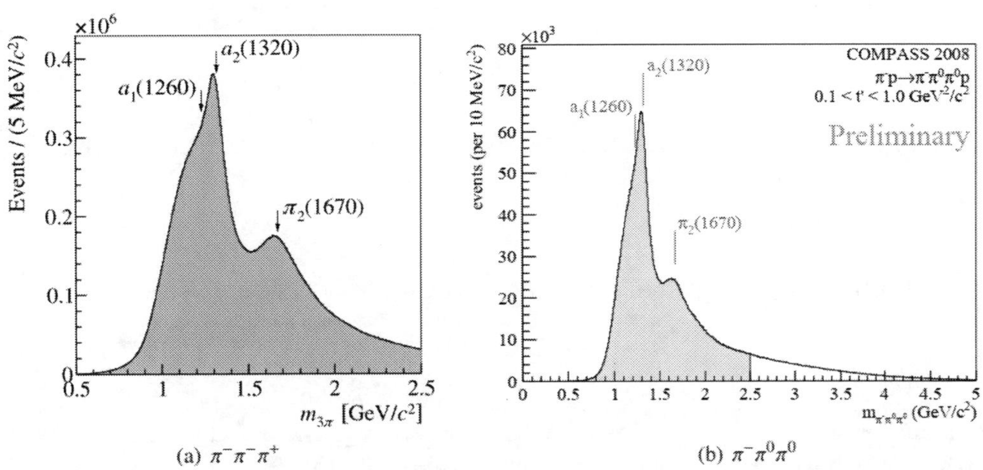

Figure 2. Invariant mass spectra of the two studied final states. (Figure (a) from [2].)

π^- diffractively scatters off the target proton producing an intermediate state X^-, which subsequently decays into a two-pion isobar $R_{\pi\pi}$ and a bachelor pion with a relative orbital angular momentum L. The isobar then decays into two pions. By exploiting the distribution in the five phase-space variables that define the kinematics of the decay products at a given invariant mass of the three-pion state, information on the spin J, the parity P, and the spin-projection M of X^-, and on the naturality ε of the exchange particle can be extracted [2].

The mass spectra of the two-pion subsystems of the $\pi^-\pi^0\pi^0$ final state are shown in figure 4. The isobars decaying to $\pi^-\pi^0$ have isospin $I = 1$. The $\rho(770)$ is clearly visible in figure 4(a). A smaller

DOI: 10.1051/epjconf/201716407045

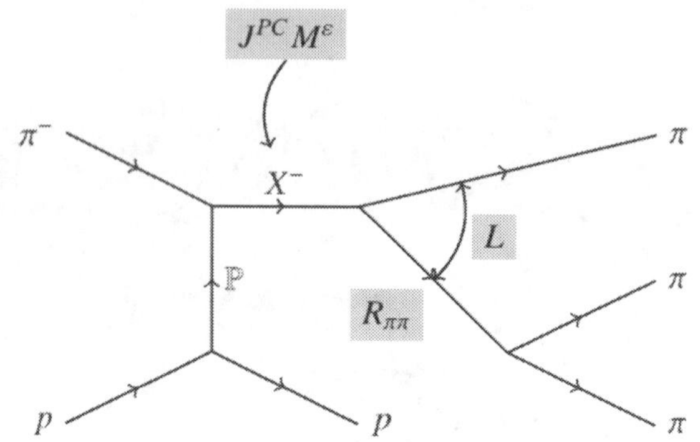

Figure 3. Sketch of the process under study.

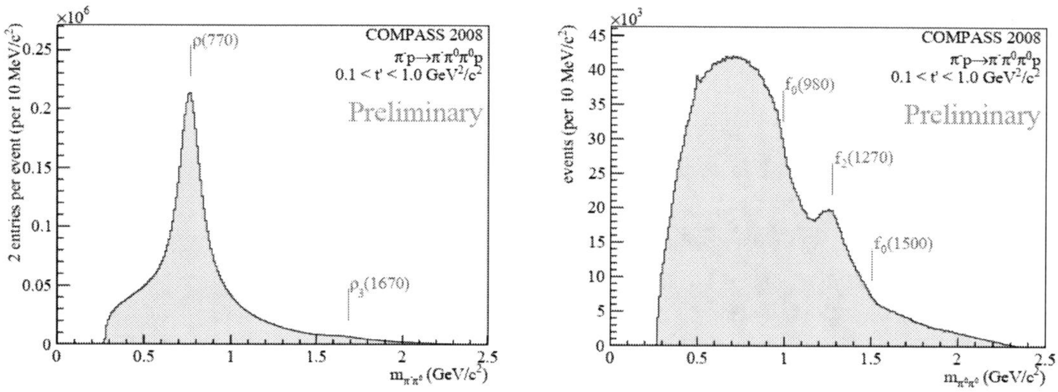

(a) $\pi^-\pi^0$ mass spectrum

(b) $\pi^0\pi^0$ mass spectrum

Figure 4. Invariant mass spectra of the two possible two-pion subsystems of the $\pi^-\pi^0\pi^0$ final state.

shoulder at higher masses suggests to also use the $\rho_3\,(1690)$ as an isobar in the partial-wave analysis model. The $\pi^0\pi^0$ isobars have isospin $I = 0$. Here, the $f_0\,(980)$, the $f_2\,(1270)$, and the $f_0\,(1500)$ are identified by bumps and kinks on top of a broad $\pi\pi$ S-wave component $(\pi\pi)_S$ in figure 4(b). For the $\pi^-\pi^-\pi^+$ final state, the two different isospin contributions cannot be separated and appear both in the $\pi^-\pi^+$ subsystem. The same isobars as for the $\pi^-\pi^0\pi^0$ final state are used in the partial-wave analysis.

Each partial wave in the partial-wave model corresponds to a state X^- with quantum numbers $J^{PC}M^{\varepsilon}$ decaying via a particular decay chain $R_{\pi\pi}\ \pi\ L$. For the analyses presented in the following, waves with spin J up to 6 are included. Also the orbital angular momentum L between the isobar $R_{\pi\pi}$ and the bachelor pion can go up to 6. Out of the possible combinations, 87 waves with non-negligible intensity are kept for the final model: 80 waves with a positive naturality $\varepsilon = +1$ of the exchange particle and 7 with $\varepsilon = -1$. In addition, one incoherent wave with an isotropic angular distribution is included [2]. The same set of 88 partial waves is used in the two analyses of the different three-pion final states.

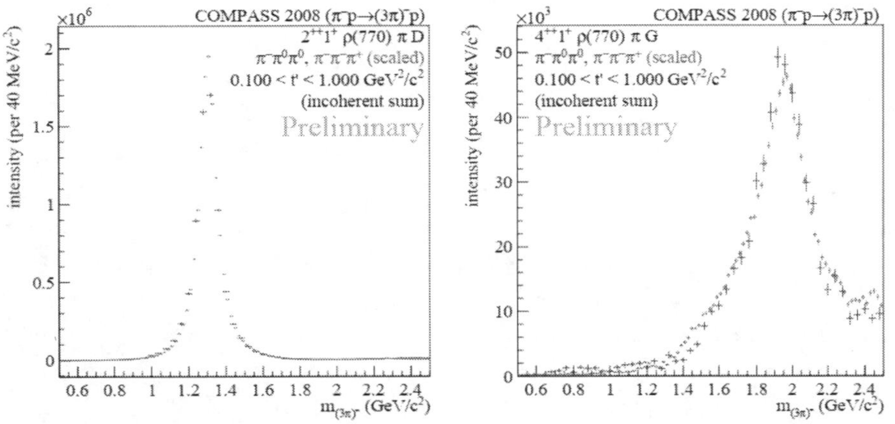

(a) $2^{++}1^{+}\rho\,(770)\,\pi D$ (b) $4^{++}1^{+}\rho\,(770)\,\pi G$

Figure 5. Sum of partial-wave intensities over all t' bins.

3 Spin-parity Decomposition

As a first analysis step, the partial-wave amplitudes are extracted in bins of the three-pion mass and the squared four-momentum transfer t' to the target assuming that all waves are fully coherent (rank-1 fit). Figure 5 shows the sum of the intensities over the individual t' bins for the $2^{++}1^{+}\rho\,(770)\,\pi D$ wave, clearly showing the $a_2\,(1320)$, and for the $4^{++}1^{+}\rho\,(770)\,\pi G$ wave, clearly showing the $a_4\,(2040)$. The results from the $\pi^-\pi^-\pi^+$ channel (red markers) and the $\pi^-\pi^0\pi^0$ channel (blue markers) are in agreement. The two data sets are normalized using the intensity integrals in each individual plot so that the peak shapes can directly be compared. As expected, no dependence of the peak shapes on t' is observed for these two partial waves. Good agreement of the partial-wave intensities between the two final states is also found for other waves with large intensities, like e.g. the $1^{++}0^{+}\rho\,(770)\,\pi S$ and $2^{-+}0^{+}f_2\,(1270)\,\pi S$ waves [3].

Also for waves with small intensities good agreement between the two final states is observed. One particularly interesting signal is observed in the $1^{++}0^{+}f_0\,(980)\,\pi P$ wave around $1.4\,\mathrm{GeV}/c^2$ (see figure 6). This narrow structure, never observed before, could correspond to a possible new a_1 state. It is seen in the decay via the $f_0\,(980)$ isobar for both channels and for all t' bins. Different parameterizations for the isobars have been tested to exclude possible artifacts from the employed model.

For a resonant interpretation of this signal, in addition to the intensities, also the phase difference between partial waves needs to be considered. Figure 7 shows the phase difference between the $1^{++}0^{+}f_0\,(980)\,\pi P$ and the $4^{++}1^{+}\rho\,(770)\,\pi G$ wave. A clear rapidly rising phase motion is visible around $1.4\,\mathrm{GeV}/c^2$ at the position of the peak in the intensity spectrum.

In order to exclude possible artifacts caused by the particular choice for the parametrization of the isoscalar $J^{PC} = 0^{++}$ isobars, a partial-wave analysis was performed, in which the waves with broad $(\pi\pi)_S$, $f_0\,(980)$, and $f_0\,(1500)$ isobars are replaced by a set of amplitudes that are piecewise constant in the two-pion mass [2]. This novel approach allows for a rather model-independent extraction of the 0^{++} isobar amplitudes as function of the invariant mass of the three-pion system and its quantum numbers. The analysis shows a clear correlation between the new $a_1\,(1420)$ signal and the $f_0\,(980)$ as a decay product.

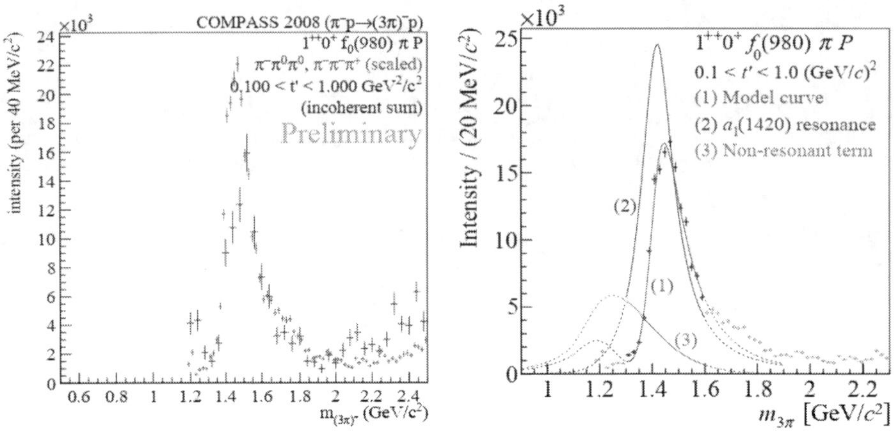

(a) comparison between $\pi^-\pi^-\pi^+$ and $\pi^-\pi^0\pi^0$ (b) fit to spin-density matrix of $\pi^-\pi^-\pi^+$

Figure 6. Sum of partial-wave intensities over all t' bins of the $1^{++}0^+ f_0(980)\pi P$ wave. (Figure (b) from [4].)

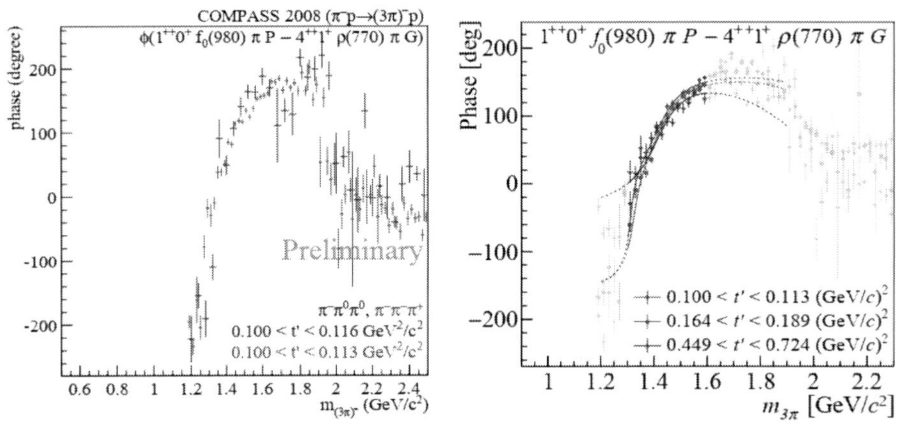

(a) comparison between $\pi^-\pi^-\pi^+$ and $\pi^-\pi^0\pi^0$ (b) fit to spin-density matrix of $\pi^-\pi^-\pi^+$

Figure 7. Phase difference between the $1^{++}0^+ f_0(980)\pi P$ and the $4^{++}1^+\rho(770)\pi G$ wave (a) for the lowest t' bins of the $\pi^-\pi^-\pi^+$ and $\pi^-\pi^0\pi^0$ final states, and (b) for different t' bins of the $\pi^-\pi^-\pi^+$ final state overlaid with the result of a fit to the spin-density matrix. (Figure (b) from [4].)

4 Resonance-model fit to spin-density matrix

To further study this new signal, and to extract Breit-Wigner parameters, the mass dependence of a sub-set of the spin-density matrix was extracted for the $\pi^-\pi^-\pi^+$ channel [4]. In addition to the intensities, also the interferences between the waves were taken into account. The three waves shown in figures 5 and 6 were used. The resonance model includes one Breit-Wigner resonance in each wave: the $a_2(1320)$ in the $2^{++}1^+\rho(770)\pi D$ wave, the $a_4(2040)$ in the $4^{++}1^+\rho(770)\pi G$ wave, and a new $a_1(1420)$ in the $1^{++}0^+ f_0(980)\pi P$ wave. The Breit-Wigner parameters were extracted by simultaneously fitting the spin-density submatrices of the three partial waves in all eleven t' bins. In addition to the Breit-Wigner components used to describe the resonances, also a coherent t'-dependent non-resonant contribution was allowed in each wave. The parameters of the Breit-Wigner functions and of the coherent backgrounds were the same in all t' bins, only the complex-valued couplings were allowed to change with t'.

The partial-wave intensity of the $1^{++}0^{+} f_0(980)\,\pi P$ wave (blue markers in figure 6(b)) is described by a resonant contribution for the new $a_1(1420)$ (blue curve) interfering with a non-resonant contribution (green curve). The total intensity of the model is represented by the red curve. The fit range is indicated by the continuous curve, the dashed curve shows the extrapolation of the model. Figure 7(b) shows the interference between the $1^{++}0^{+} f_0(980)\,\pi P$ and the $4^{++}1^{+}\rho(770)\,\pi G$ wave. The three different colors represent three of the in total eleven t' bins. Again, the markers show the result from the partial-wave decomposition, while the curves show the behaviour of the resonance model. The parameters of the Breit-Wigner amplitude describing the new $a_1(1420)$ are $M = 1414\,\mathrm{MeV}/c^2$ with a systematic uncertainty of $^{+15}_{-13}\,\mathrm{MeV}/c^2$ for the mass, and $\Gamma = 153\,\mathrm{MeV}/c^2$ with a systematic uncertainty of $^{+8}_{-23}\,\mathrm{MeV}/c^2$ for the width. The statistical uncertainties are negligible.

5 Conclusions

The COMPASS experiment has recorded a large dataset to study excited light-quark resonances produced in diffractive dissociation into three pions. The large number of events allowed us to increase the number of waves used in the partial-wave analysis. In addition, the analysis is performed in narrow bins of t'. Breit-Wigner parameters have been extracted by performing a resonance-model fit to the result of the spin-parity decomposition in mass bins. Apart from the well-known resonances, COMPASS found a new signal in the $1^{++}0^{+} f_0(980)\,\pi P$ wave around $1.4\,\mathrm{GeV}/c^2$. It is a rather narrow object with a well-defined mass and width, and appears only in the $f_0(980)$ decay mode. The nature of this signal is still unclear, it could be the isospin partner of the $f_1(1420)$, but also other explanations have been put forward. The $a_1(1420)$ could be a genuine resonance in the form of a two-quark-tetraquark mixed state [5] or a tetraquark with mixed flavor symmetry [6]. Also two explanations connecting the new $a_1(1420)$ to the known $a_1(1260)$ have been proposed. The first explains the observed intensity spectra and phases in the $1^{++}0^{+} f_0(980)\,\pi P$ wave by a two-channel unitarized Deck amplitude interfering with the direct production of the $a_1(1260)$ [7]. The second explains the signal by a singularity in a triangle diagram describing rescattering in the $K^- \bar{K}^*$ decay of the $a_1(1260)$ [8, 9].

References

[1] COMPASS collaboration, P. Abbon *et al.*, NIM A**779**, 69 (2015), arXiv:1410.1797.

[2] COMPASS collaboration, C. Adolph *et al.*, submitted to Phys. Rev. D, arXiv:1509.00992.

[3] S. Uhl on behalf of the COMPASS Collaboration, PoS(Hadron 2013)087, arXiv:1401.4943.

[4] COMPASS collaboration, C. Adolph *et al.*, Phys. Rev. Lett. **115**, 082001 (2015), arXiv:1501.05732.

[5] Z.G. Wang, arXiv:1401.1134.

[6] H.X. Chen *et al.*, Phys. Rev. D**91**, 094022 (2015).

[7] J.L. Basdevant, E. Berger, Phys. Rev. Lett. **19**, 192001 (2015).

[8] M. Mikhasenko, B. Ketzer, A. Sarantsev, Phys. Rev. D**91**, 094015 (2015).

[9] F. Aceti, L.R. Dai, E. Oset, arXiv:1606.06893.

EPJ Web of Conferences **164**, 07046 (2017)   DOI: 10.1051/epjconf/201716407046
ICNFP 2016

Temperature dependence of shear viscosity of $SU(3)$–gluodynamics within lattice simulation

N.Yu. Astrakhantsev[1,2,a], V.V. Braguta[1,2,3,4,b], and A.Yu. Kotov[1,5,c]

[1] *Institute for Theoretical and Experimental Physics, Moscow, 117218 Russia*
[2] *Moscow Institute of Physics and Technology, Dolgoprudny, 141700 Russia*
[3] *Institute for High Energy Physics NRC "Kurchatov Institute", Protvino, 142281 Russian Federation*
[4] *Far Eastern Federal University, School of Biomedicine, 690950 Vladivostok, Russia*
[5] *National Research Nuclear University MEPhI (Moscow Engineering Physics Institute), 115409, Russia, Moscow, Kashirskoe highway, 31*

Abstract. This report is devoted to the study of temperature dependence of shear viscosity in $SU(3)$-gluodynamics. To calculate shear viscosity we measured the correlation function of the energy-momentum tensor $T_{12}T_{12}$ for a set of temperatures in the region $T/T_c \in (0.9, 1.5)$. The measurements were carried out using multilevel algorithm which considerably improves the accuracy of the data. For determination of the shear viscosity two methods were used: physically motivated ansatz and non-parametrical Backus-Gilbert method. Both methods give consistent results. The calculation allows to determine temperature dependence of the ratio of shear viscosity to the entropy density η/s.

1 Introduction

One of the main purposes of heavy ion collision experiments at RHIC and LHC is the experimental study of Quark-gluon plasma (QGP). It was shown that experimental results can be described within hydrodynamical approach[1, 2]. In hydrodynamics transport coefficients such as shear viscosity, bulk viscosity, conductivity, etc. determine the dynamics of conserved charges. These transport coefficients can be determined from the experiment or calculated theoretically. For instance, the elliptic flow measurement [3, 4] allows to determine the shear viscosity of QGP. It turns out that the ratio of shear viscosity η to entropy density s lies in range $\eta/s = (1-3)\times 1/4\pi$[5], which is close to $N = 4$ Super Yang Mills (SYM) theory prediction at strong coupling $\eta/s = 1/4\pi$ [6] and differs from a weak coupling result $\eta/s \sim c/(g^4 \log(1/g)) \sim 1$ [7, 8]. Thus a ratio η/s is very small and cannot be described within perturbation theory. It would be interesting to study the value of the shear viscosity within such nonperturbative approach as lattice simulations of QCD.

There were only few attempts to calculate shear viscosity of SU(3)–gluodynamics[9–13] and SU(2)–gluodynamics[14, 15]. In this paper we are going to study temperature dependence of shear viscosity of SU(3)–gluodynamics in the vicinity of confinement/deconfinement phase transition $T/T_c \in [0.9, 1.5]$.

[a] e-mail: nikita.astrakhantsev@itep.ru
[b] e-mail: braguta@itep.ru
[c] e-mail: kotov@itep.ru

2 Details of the calculation

Shear viscosity is related to the Euclidean correlation function of the energy-momentum tensor $T_{\mu\nu}$ (here we omitted for simplicity trace anomaly):

$$T_{\mu\nu} = \frac{1}{4}\delta_{\mu\nu}F^a_{\alpha\beta}F^a_{\alpha\beta} - F^a_{\mu\alpha}F^a_{\nu\alpha} \quad (1)$$

$$C(x_0) = T^{-5}\int d^3\mathbf{x}\langle T_{12}(0)T_{12}(x_0,\mathbf{x})\rangle, \quad (2)$$

where T is the temperature of the system. The correlation function (2) can be written in terms of the spectral function $\rho(\omega)$ as follows

$$C(x_0) = T^{-5}\int_0^\infty \rho(\omega)\frac{\cosh\omega(\frac{1}{2T}-x_0)}{\sinh\frac{\omega}{2T}}d\omega. \quad (3)$$

In order to find shear viscosity from the spectral function one uses the Kubo formula [16]:

$$\eta = \pi\lim_{\omega\to 0}\frac{\rho(\omega)}{\omega}, \quad (4)$$

Lattice calculation of shear viscosity can be divided into two parts. The first part is the measurement of the correlation function $C(x_0)$ with sufficient accuracy. This part of the calculation requires large computational resources but for the gluodynamics the accuracy of the correlator can be dramatically improved with the help of the two-level algorithm [17]. The second part is the determination of the spectral function $\rho(\omega)$ from the correlation function $C(x_0)$. The last part of the calculation is probably the most complicated, since one should determine continuous spectral function $\rho(\omega)$ from integral equation (3) for the set of $O(10)$ values of the function $C(x_0)$ measured in lattice simulation.

First we recall very general properties: the positivity of the spectral function $\rho(\omega)/\omega \geq 0$ and oddness: $\rho(-\omega) = -\rho(\omega)$. At large frequencies one expects asymptotic freedom behaviour. For this reason it is also important to write the expression for the spectral function in the leading-order approximation in strong coupling constant expansion [18]:

$$\rho^{LO}(\omega) = \frac{1}{10}\frac{d_A}{(4\pi)^2}\frac{\omega^4}{\tanh(\frac{\omega}{4T})} + \left(\frac{2\pi}{15}\right)^2 d_A T^4\ \omega\delta(\omega), \quad (5)$$

where $d_A = N_c^2 - 1 = 8$ for the SU(3)-gluodynamics.

One also knows next-to-leading order expression for the spectral function at large ω [19]:

$$\lim_{\omega\to\infty}\rho^{NLO}(\omega) = \frac{1}{10}\frac{d_A}{(4\pi)^2}\omega^4\left(1 - \frac{5\alpha_s N_c}{9\pi}\right) \quad (6)$$

It should be noted here that at large ω the spectral function scales as $\rho(\omega) \sim \omega^4$, what leads to a large perturbative contribution to the correlation function for all values of Euclidean time x_0. Calculation shows that even at the $x_0 = 1/(2T)$ the tree level contribution is $\sim 80-90\%$ of the total value of the correlation function. Note also that large ω behaviour of the spectral function leads to a fast decrease of the correlation function $C(x_0) \sim 1/x_0^5$ for small x_0. For this reason the signal/noise ratio for the $C(x_0)$ is small at $x_0 \gg a$ and lattice measurement of the correlation function at $x_0 \sim 1/(2T)$ becomes computationally very expensive.

In numerical simulation we use Wilson gauge action for the SU(3)–gluodynamics

$$S_g = \beta\sum_{x,\mu<\nu}\left(1 - \frac{1}{3}Re\,\mathrm{Tr}\,U_{\mu,\nu}(x)\right), \quad (7)$$

where $U_{\mu,\nu}(x)$ is the product of the link variables along elementary rectangular (μ,ν), which starts at x.

For the tensor $F_{\mu\nu}$ we use the clover discretization scheme:

$$F^{(clov)}_{\mu\nu}(x) = \frac{1}{4iga^2}\Big(V_{\mu,\nu}(x) + V_{\nu,-\mu}(x) + V_{-\mu,-\nu}(x) + V_{-\nu,\mu}(x)\Big),\ V_{\mu,\nu}(x) = \frac{1}{2}(U_{\mu,\nu}(x) - U_{\nu,\mu}(x)). \quad (8)$$

It causes no difficulties to build energy-momentum tensor from Eq. (1) having expression for the tensor $F_{\mu\nu}$.

To calculate shear viscosity one should measure the correlation function (2) using two-level algorithm [17] to decrease computation time. Note also that instead of the correlation function $\langle T_{12}(x)T_{12}(y)\rangle$ we measure the correlation function $\frac{1}{2}(\langle T_{11}(x)T_{11}(y)\rangle - \langle T_{11}(x)T_{22}(y)\rangle)$. Both correlation functions are equal in the continuum limit[9].

It is conventional to present the value of shear viscosity as the ratio viscosity-to-entropy density η/s. For homogeneous systems the entropy density s can be expressed as $s = \frac{\epsilon+p}{T}$, where ϵ is the energy density and p is the pressure. These thermodynamic quantities were measured with the method described in [20].

Energy-momentum tensor in the continuum theory is a set of Noether currents which are related to the rotational invariance of the action. In lattice formulation of field theory rotational invariance does not exist and therfore energy-momentum tensor should be renormalized. For the components of energy-momentum tensor considered in this paper the renormalization is multiplicative [21]. Renormalization factors depend on the discretization scheme. For instance, for the diagonal component of $T_{\mu\nu}$ (when $\mu = \nu$) and the plaquette-based discretization of $T_{\mu\nu}$: $T_{\mu\mu} = \frac{2}{a^4g^2}\left(-\sum_{\nu\neq\mu}\mathrm{Tr}\,U_{\mu,\nu}(x) + \sum_{\nu,\sigma\neq\mu,\sigma>\nu}\mathrm{Tr}\,U_{\sigma,\nu}(x)\right)$ the renormalization factors are related to the anisotropy coefficients [22, 23]: $T^{(ren)}_{\mu\nu} = Z^{(plaq)}T^{(plaq)}_{\mu\nu}$, $Z^{(plaq)} = 1 - \frac{1}{2}g_0^2(c_\sigma - c_\tau)$, where c_σ and c_τ are defined in [20] and for $SU(2)$ are computed on the lattice in [24].

Using the renormalization factors for the plaquette-based discretization of T_{00}, we can find the renormalization factors for the clover discretization simply by fitting the vacuum expectation values of the renormalized T_{00}: $Z^{(plaq)}\langle T^{(plaq)}_{00}\rangle = Z^{(clov)}\langle T^{(clov)}_{00}\rangle$.

The spectral function from fitting procedure

We measured the correlation functions $C(x_0)$ for the temperatures $T/T_c = 0.9, 0.925, 0.95, 1.0, 1.1, 1.2, 1.35, 1.425, 1.5$. Application of two-level algorithm allowed us to get the uncertainty $\sim 2-3\%$ at the distance $Tx_0 = 1/2$ for all lattices. For the other points the accuracy is better. In Fig. 1 we plot the correlation functions for various temperatures.

The next part of the calculation is the extraction of the spectral function from the integral equation (3). In this section we are going to use physically motivated ansatz for the spectral function with unknown parameters that can be determined through the fitting procedure.

In the calculation of viscosity we use ansatz motivated by QCD sum rules [25]. We join hydrodynamic behaviour at small frequencies with asymptotic freedom at large frequencies [1]

$$\rho_1(\omega) = BT^3\,\omega\,\theta(\omega_0 - \omega) + A\rho_{lat}(\omega)\,\theta(\omega - \omega_0). \quad (9)$$

In the last formula $\rho_{lat}(\omega)$ is a tree level lattice expression for the spectral function calculated for the correlation function $\sim \frac{1}{2}(\langle T_{11}(x)T_{11}(y)\rangle - \langle T_{11}(x)T_{22}(y)\rangle)$ with clover discretization of the tensor $F_{\mu\nu}$ at lattice with fixed L_t and $L_s \to \infty$. The function $\rho_{lat}(\omega)$ was calculated in paper [15].

[1]Note that the frequency ω is measured in physical units.

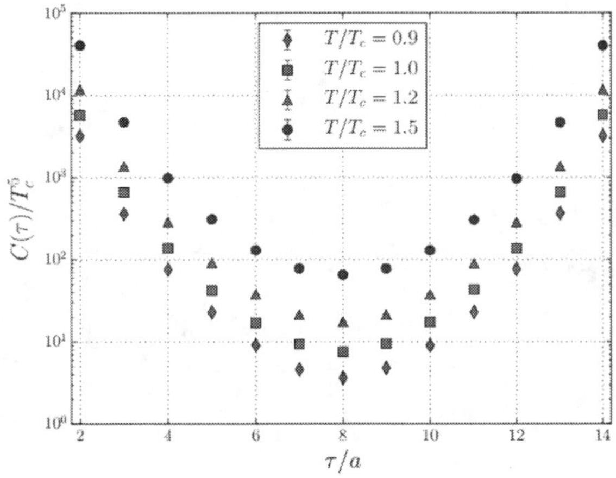

Figure 1. The correlation functions $C(x_0)$ for different temperatures.

The correlation function $C(x_0)$ is very sensitive to the ultraviolet properties of the spectral function, so we use $\rho_{lat}(\omega)$ instead of the continuum expression (5). Application of the function $\rho_{lat}(\omega)$ considerably improves the quality of the fit.

Evidently the inclusion of interactions modifies tree level formula, so we multiplied the $\rho_{lat}(\omega)$ by some factor A which effectively takes into account radiative corrections at large frequencies. Introduction of the factor A is very important for successful description of lattice data.

The obtained values $\omega_0 \sim 3$ GeV are physically well motivated. The value of the strong coupling constant is $\alpha_s(\omega_0) \sim 0.2 - 0.3$. This allows us to expect that perturbative expression for the spectral function is applicable for $\omega > \omega_0$. The values of the factor A for all temperatures, which takes into account the effect of interactions, are close to unity what confirms applicability of the asymptotic freedom at high frequencies.

Low frequency part of the spectral function (9) is given by the first-order hydrodynamic expression $\sim \omega$. Comparison of the spectral functions of the energy-momentum tensor correlation functions obtained in $N = 4$ SYM [26] and the first-order hydrodynamic expressions allows us to expect that this approximation works well up to $\omega \leq \pi T \simeq 1$ GeV[27].

On the other hand high frequency perturbative expression for the spectral function is rigidly fixed and works well up to $\omega \geq \omega_0 \sim 3$ GeV. The form of the spectral function in the region 1 GeV $\leq \omega \leq$ 3 GeV is not clear. This is the main source of the uncertainty of the calculation based on the fitting procedure. In order to get rid of these uncertainties, one should apply non-parametrical estimation techniques.

3 The Backus-Gilbert method for the spectral function

In this section we are going to determine the ratio η/s using the Backus-Gilbert(BG) method[28, 29] [2]. This method does not require assumptions about the form of the spectral function.

[2]In QCD this approach was recently applied in papers [30, 31]. The Backus-Gilbert was also recently applied in the lattice calculation of graphene conductivity [32].

The method can be formulated as follows[3]. One needs to solve equation (3). To do this we rewrite it in the following form

$$C(x_i) = \frac{1}{T^5}\int_0^\infty d\omega \frac{\rho(\omega)}{f(\omega)} K(x_i, \omega), \tag{10}$$

where $f(\omega)$ is some function with the property $f(\omega)|_{\omega\to 0} \sim \omega$ and $K(x, \omega)$ is a rescaled kernel of the integral equation

$$K(x, \omega) = f(\omega)\frac{\cosh \omega\left(\frac{1}{2T} - x\right)}{\sinh \frac{\omega}{2T}}, \tag{11}$$

Our aim is to determine the $\rho(\omega)$. In the BG method instead of the true function $\rho(\omega)$ we consider the estimator of this function $\bar{\rho}(\bar{\omega})$ which can be written as

$$\bar{\rho}(\bar{\omega}) = f(\bar{\omega})\int_0^\infty d\omega \delta(\bar{\omega}, \omega)\frac{\rho(\omega)}{f(\omega)}, \tag{12}$$

where the function $\delta(\bar{\omega}, \omega)$ is called the resolution function. This function has a peak around the point $\bar{\omega}$ and it is normalized as $\int_0^\infty d\omega \delta(\bar{\omega}, \omega) = 1$. The sharper the peak, the closer $\bar{\rho}$ is to ρ. The function is expanded over the $K(x_i, \omega)$ as

$$\delta(\bar{\omega}, \omega) = \sum_i q_i(\bar{\omega}) K(x_i, \omega). \tag{13}$$

If a resolution functon is chosen this way, the estimator is a linear combination of the values of the correlation function

$$\bar{\rho}(\bar{\omega}) = f(\bar{\omega}) \sum_i q_i(\bar{\omega}) C(x_i). \tag{14}$$

The above formulae are typical for all linear non-parametrical estimation procedures. The specifics of the BG–method are the following: one minimizes the sum of functionals $\mathcal{A} + \mathcal{B}$, where $\mathcal{A}$ ensures that the estimator reproduces correlation function, the $\mathcal{B}$ functional punishes estimator for being too dependent on the data (prevents overfittng).

To get a better approximation for the $\rho(\omega)$ by the estimator $\bar{\rho}(\bar{\omega})$ one should sharpen the $\delta(\bar{\omega}, \omega)$ function. This minimizes the $\mathcal{A}$ functional, and the BG–method uses the second moment of the resolution function as the measure of its width: $\int_0^\infty d\omega \delta(\bar{\omega}, \omega)(\omega - \bar{\omega})^2$.

Minimization requirement gives the following values of the coefficients

$$q_i(\omega) = \frac{\sum_j W_{ij}^{-1}(\bar{\omega}) R(x_j)}{\sum_{kj} R(x_k) W_{kj}^{-1}(\bar{\omega}) R(x_j)}, \tag{15}$$

$$W_{ij}(\bar{\omega}) = \int_0^\infty d\omega K(x_i, \omega)(\omega - \bar{\omega})^2 K(x_j, \omega), \tag{16}$$

$$R(x_i) = \int_0^\infty d\omega K(x_i, \omega). \tag{17}$$

The matrix $W_{ij}(\bar{\omega})$ is poorly conditioned and there are large cancellations between different terms in equation (14). To make estimator less dependent on the data, one should take the $\mathcal{B}$ functional int account. This regularizes the matrix $W_{ij}(\bar{\omega})$ by the covariance matrix of our data S_{ij}:

$$W_{ij} \to \lambda W_{ij} + (1 - \lambda) S_{ij}. \tag{18}$$

[3]Here we follow the designations of [30, 31]

 DOI: 10.1051/epjconf/201716407046

The λ parameter represents the trade-off between stability and data fitting. The exact λ value should be chosen depending on the matrix W condition number.

To determine the spectral density at small frequencies $\omega \ll \Lambda_{QCD}$ it is reasonable to choose the function $f(\omega) = \omega$. In this case the ratio $\rho(\omega)/\omega|_{\omega\to 0} = \frac{\eta}{\pi}$.

To calculate shear viscosity, we choose $\lambda = 0.002$. For this value of the λ the uncertainty in the restored spectral density at zero frequency is less than 1% for all temperatures.

As already noted, we expect that hydrodynamic approximation works well up to $\omega \sim \pi T$[27]. At the same time from the width of the resolution function at $\lambda = 0.002$ is $\Delta\omega \sim 4T$. We conclude that the convolution of the spectral density with the resolution function at $\lambda = 0.002$ averages over reasonable region and gives good approximation for the shear viscosity.

According to the results of previous section asymptotic freedom starts to work for frequencies $\omega/T \geq 7-8$. The resolution function is considerably suppressed in this region. However, the spectral function at large frequencies rises very quickly $\rho(\omega) \sim \omega^4$, so this convolution contributes to the answer. So, to get reliable estimate one should subtract the asymptotic freedom.

The simplest way to do this is to subtract the convolution

$$\rho_{uv}(0) = \int \delta(0,\bar{\omega})\rho_{uv}(\bar{\omega})d\bar{\omega},$$

where the $\rho_{uv}(\bar{\omega}) = A\theta(\bar{\omega} - \omega_0)\rho_{lat}(\bar{\omega})$.

The determination of A and ω_0 parameters can be done by many procedures, the simplest way is to fit correlation function $C(\tau)$ for low τ, where it is completely governed by the UV-contribution. Despite the fact that this method of subtraction makes assumptions about the UV spectral function form, it still leaves the intermediate and hydrodynamic regions untouched and the usage of BG method remains sensible.

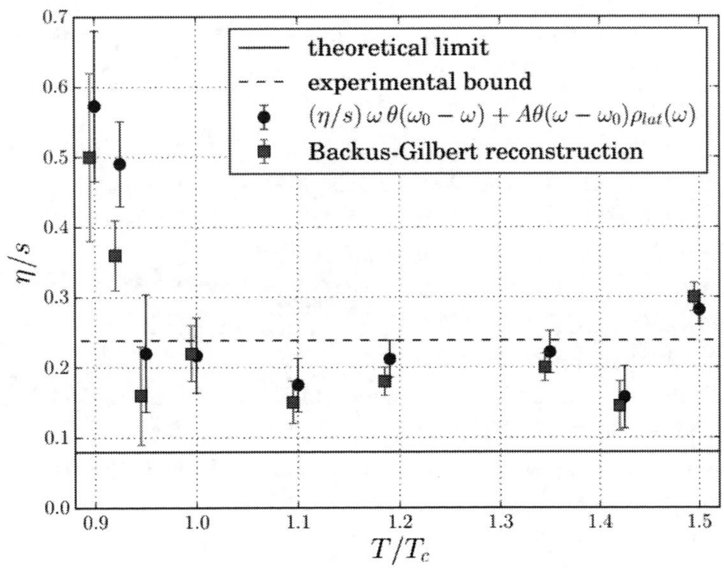

Figure 2. The ratio η/s temperature dependence for parametrical and BG procedures.

On Fig. [2] we plot η/s temperature dependence for parametrical and non-parametrical (Backus-Gilbert) methods of estimation. We also show theoretical lower bound and experimental bound. Both results coincide within errors and agree with the experimental data and theory, which means that the ratio η/s is robust to the choice of middle-range spectral function form ansatz.

4 Conclusion

In this paper we studied the energy-momentum tensor correlation function $\langle T_{12}(0)T_{12}(x)\rangle$ for SU(3)-gluodynamics using lattice simulation of QCD. We carried out the measurements of this correlation function at a set of temperatures $T/T_c \in (0.9, 1.5)$ using two-level algorithm, which gave accuracy $\sim 2-3\%$ at $x_0 = 1/2T$. For the other points the accuracy is much better. Using lattice data for the correlation function we studied the spectral function. It was shown that physically motivated ansatz which joins the first-order hydrodynamical behaviour at small frequencies with asymptotic freedom at large frequencies fits our data very well for all temperatures. In addition, non-parametrical Backus-Gilbert method was applied in order to make model-independent estimation that would not make assumptions about low-frequencies spectral function shape. The results obtained be both methods coincide within errors and agree with the experiment and $N = 4$ SYM.

Acknowledgements

The simulations were performed at ITEP supercomputer and Supercomputing Center of Lomonosov Moscow State University. This work was completed due to RSF grant (project 15-12-20008).

References

[1] P. F. Kolb and U. W. Heinz, In *Hwa, R.C. (ed.) et al.: Quark gluon plasma* 634-714.
[2] J. Y. Ollitrault, Eur. J. Phys. **29**, 275 (2008).
[3] K. H. Ackermann *et al.* [STAR Collaboration], Phys. Rev. Lett. **86**, 402 (2001).
[4] C. Adler *et al.* [STAR Collaboration], Phys. Rev. C **66**, 034904 (2002).
[5] D. A. Teaney (2009).
[6] G. Policastro, D. T. Son, and A. O. Starinets, Phys. Rev. Lett. **87**, 081601 (2001).
[7] P. B. Arnold, G. D. Moore and L. G. Yaffe, JHEP **0011**, 001 (2000).
[8] P. B. Arnold, G. D. Moore and L. G. Yaffe, JHEP **0305**, 051 (2003).
[9] F. Karsch and H. W. Wyld, Phys. Rev. **D35**, 2518 (1987).
[10] A. Nakamura and S. Sakai, Phys. Rev. Lett. **94**, 072305 (2005).
[11] H. B. Meyer, Phys. Rev. **D76**, 101701 (2007).
[12] H. B. Meyer, Nucl. Phys. **A830**, 641C (2009).
[13] S. W. Mages, S. Borsányi, Z. Fodor, A. Schäfer and K. Szabó, PoS LATTICE **2014**, 232 (2015).
[14] V. V. Braguta and A. Yu. Kotov, JETP Lett. **98**, 127 (2013).
[15] N. Y. Astrakhantsev, V. V. Braguta and A. Y. Kotov, JHEP **1509**, 082 (2015).
[16] R. Kubo, J. Phys. Soc. Jap. **12**, 570 (1957).
[17] H. B. Meyer, JHEP **01**, 048 (2003).
[18] H. B. Meyer, JHEP **08**, 031 (2008).
[19] A. L. Kataev, N. V. Krasnikov, and A. A. Pivovarov, Nucl. Phys. **B198**, 508 (1982).
[20] J. Engels, F. Karsch, and T. Scheideler, Nucl. Phys. **B564**, 303 (2000).

[21] H. B. Meyer, Eur. Phys. J. **A47**, 86 (2011).
[22] J. Engels, F. Karsch, H. Satz, and I. Montvay, Nucl. Phys. **B205**, 545 (1982).
[23] H. B. Meyer, Nucl. Phys. **B795**, 230 (2008).
[24] J. Engels, F. Karsch, and K. Redlich, Nucl. Phys. **B435**, 295 (1995).
[25] M. A. Shifman, A. I. Vainshtein, and V. I. Zakharov, Nucl. Phys. **B147**, 385 (1979).
[26] P. Kovtun and A. Starinets, Phys. Rev. Lett. **96**, 131601 (2006).
[27] H. B. Meyer, PoS **LATTICE2008**, 017 (2008).
[28] G. Backus and F. Gilbert, Geophysical Journal of the Royal Astronomical Society 16, 169205.
[29] G. Backus and F. Gilbert, Philosophical Transactions of the Royal Society of London A : Mathematical, Physical and Engineering Sciences 266, 123-192.
[30] B. B. Brandt, A. Francis, H. B. Meyer and D. Robaina, Phys. Rev. D **92**, no. 9, 094510 (2015).
[31] B. B. Brandt, A. Francis, B. Jäger and H. B. Meyer, Phys. Rev. D **93**, no. 5, 054510 (2016).
[32] D. L. Boyda, V. V. Braguta, M. I. Katsnelson and M. V. Ulybyshev, Phys. Rev. B **94**, 085421 (2016).

The CUORE and CUORE-0 experiments at LNGS

A. D'Addabbo[1,a], C. Alduino[2], K. Alfonso[3], D. R. Artusa[1,3], F. T. Avignone III[2], O. Azzolini[4], T. I. Banks[5,6], G. Bari[7], J.W. Beeman[8], F. Bellini[9,10], A. Bersani[11], M. Biassoni[12,13], A. Branca[14], C. Brofferio[12,13], C. Bucci[1], A. Camacho[4], A. Caminata[11], L. Canonica[1], X. G. Cao[15], S. Capelli[12,13], L. Cappelli[1,11,16], L. Carbone[13], L. Cardani[9,10], P. Carniti[12,13], N. Casali[9,10], L. Cassina[12,13], D. Chiesa[12,13], N. Chott[2], M. Clemenza[12,13], S. Copello[11,17], C. Cosmelli[9,10], O. Cremonesi[13], R. J. Creswick[2], J. S. Cushman[18], I. Dafinei[10], C. J. Davis[18], S. Dell'Oro[1,19], M. M. Deninno[7], S. Di Domizio[11,17], M. L. Di Vacri[1,20], A. Drobizhev[5,6], D. Q. Fang[14], M. Faverzani[12,13], G. Fernandes[11,17], E. Ferri[12,13], F. Ferroni[9,10], E. Fiorini[12,13], M. A. Franceschi[21], S. J. Freedman[6,5], B. K. Fujikawa[6], A. Giachero[13], L. Gironi[12,13], A. Giuliani[22], L. Gladstone[23], P. Gorla[1], C. Gotti[12,13], T. D. Gutierrez[24], E. E. Haller[8,25], K. Han[17,26], E. Hansen[3,23], K. M. Heeger[18], R. Hennings-Yeomans[5,6], K. P. Hickerson[3], H. Z. Huang[3], R. Kadel[27], G. Keppel[4], Yu. G. Kolomensky[5,6,27], A. Leder[23], C. Ligi[21], K. E. Lim[18], X. Liu[3], Y. G. Ma[15], M. Maino[12,13], L. Marini[11,17], M. Martinez[9,10,28], R. H. Maruyama[18], Y. Mei[6], N. Moggi[7,29], S. Morganti[10], P. J. Mosteiro[10], T. Napolitano[21], C. Nones[30], E. B. Norman[31,32], A. Nucciotti[12,13], T. O'Donnell[5,6], F. Orio[10], J. L. Ouellet[5,6,23], C. E. Pagliarone[1,16], M. Pallavicini[11,17], V. Palmieri[4], L. Pattavina[1], M. Pavan[12,13], G. Pessina[13], V. Pettinacci[10], G. Piperno[9,10], C. Pira[4], S. Pirro[1], S. Pozzi[12,13], E. Previtali[13], C. Rosenfeld[2], C. Rusconi[13], S. Sangiorgio[31], D. Santone[1,20], N. D. Scielzo[31], V. Singh[5], M. Sisti[12,13], A. R. Smith[6], L. Taffarello[14], M. Tenconi[22], F. Terranova[12,13], C. Tomei[10], S. Trentalange[3], M. Vignati[10], S. L. Wagaarachchi[5,6], B. S. Wang[31,32], H. W. Wang[15], J. Wilson[2], L. A. Winslow[23], T. Wise[17,33], A. Woodcraft[34], L. Zanotti[12,13], G. Q. Zhang[15], B. X. Zhu[3], S. Zimmermann[35], and S. Zucchelli[7,36]

[1] *INFN - Laboratori Nazionali del Gran Sasso, Assergi (L'Aquila) I-67010 - Italy*
[2] *Department of Physics and Astronomy, University of California, Los Angeles, CA 90095 - USA*
[3] *Department of Physics and Astronomy, University of South Carolina, Columbia, SC 29208 - USA*
[4] *INFN - Laboratori Nazionali di Legnaro, Legnaro (Padova) I-35020 - Italy*
[5] *Department of Physics, University of California, Berkeley, CA 94720 - USA*
[6] *Nuclear Science Division, Lawrence Berkeley National Laboratory, Berkeley, CA 94720 - USA*
[7] *INFN - Sezione di Bologna, Bologna I-40127 - Italy*
[8] *Materials Science Division, Lawrence Berkeley National Laboratory, Berkeley, CA 94720 - USA*
[9] *Dipartimento di Fisica, Sapienza Università di Roma, Roma I-00185 - Italy*
[10] *INFN - Sezione di Roma, Roma I-00185 - Italy*
[11] *INFN - Sezione di Genova, Genova I-16146 - Italy*
[12] *Dipartimento di Fisica, Università di Milano-Bicocca, Milano I-20126 - Italy*
[13] *INFN - Sezione di Milano Bicocca, Milano I-20126 - Italy*
[14] *INFN - Sezione di Padova, Padova I-35131 - Italy*
[15] *Shanghai Institute of Applied Physics, Chinese Academy of Sciences, Shanghai 201800 - China*
[16] *Dip. di Ingegneria Civile e Meccanica, Università degli Studi di Cassino e del Lazio Meridionale, Cassino I-03043 - Italy*

[a] e-mail: antonio.daddabbo@lngs.infn.it

[17] *Dipartimento di Fisica, Università di Genova, Genova I-16146 - Italy*
[18] *Department of Physics, Yale University, New Haven, CT 06520 - USA*
[19] *INFN - Gran Sasso Science Institute, L'Aquila I-67100 - Italy*
[20] *Dipartimento di Scienze Fisiche e Chimiche, Università dell'Aquila, L'Aquila I-67100 - Italy*
[21] *INFN - Laboratori Nazionali di Frascati, Frascati (Roma) I-00044 - Italy*
[22] *CSNSM, Univ. Paris-Sud, CNRS/IN2P3, Université Paris-Saclay, 91405 Orsay, France*
[23] *Massachusetts Institute of Technology, Cambridge, MA 02139 - USA*
[24] *Physics Department, California Polytechnic State University, San Luis Obispo, CA 93407 - USA*
[25] *Dep. of Materials Science and Engineering, University of California, Berkeley, CA 94720 - USA*
[26] *Department of Physics and Astronomy, Shanghai Jiao Tong University, Shanghai 200240 - China*
[27] *Physics Division, Lawrence Berkeley National Laboratory, Berkeley, CA 94720 - USA*
[28] *Laboratorio de Fisica Nuclear y Astroparticulas, Universidad de Zaragoza, Zaragoza 50009 - Spain*
[29] *Dip. di Scienze per la Qualità della Vita, Alma Mater Studiorum - Università di Bologna, Bologna I-47921 - Italy*
[30] *Service de Physique des Particules, CEA / Saclay, 91191 Gif-sur-Yvette - France*
[31] *Lawrence Livermore National Laboratory, Livermore, CA 94550 - USA*
[32] *Department of Nuclear Engineering, University of California, Berkeley, CA 94720 - USA*
[33] *Department of Physics, University of Wisconsin, Madison, WI 53706 - USA*
[34] *SUPA, Institute for Astronomy, University of Edinburgh, Blackford Hill, Edinburgh EH9 3HJ - UK*
[35] *Engineering Division, Lawrence Berkeley National Laboratory, Berkeley, CA 94720 - USA*
[36] *Dip. di Fisica e Astronomia, Alma Mater Studiorum - Università di Bologna, Bologna I-40127 - Italy*

Abstract. The Cryogenic Underground Observatory for Rare Events (CUORE) is a 1-ton scale bolometric experiment devoted to the search of the neutrinoless double-beta decay ($0\nu\beta\beta$) in ^{130}Te. The CUORE detector consists of an array of 988 TeO_2 crystals operated at 10 mK. CUORE-0 is the CUORE demonstrator: it has been built to test the performance of the upcoming CUORE experiment and represents the largest ^{130}Te bolometric setup ever operated. CUORE-0 has been running at Laboratori Nazionali del Gran Sasso (Italy) from 2013 to 2015. The final CUORE-0 analysis on $0\nu\beta\beta$ and the corresponding detector performance are presented. The present status of the CUORE experiment, now in its final construction and commissioning phase, are discussed. The results from assembly of the detector and the commissioning of the cryostat are reported.

1 Introduction

Neutrinoless double-beta decay ($0\nu\beta\beta$) is a second order nuclear decay in which two beta decays occur simultaneously in a nucleus (even A, even Z), with the emission of two electrons and no neutrinos $[(A,Z) \to (A,Z+2) + 2e^-]$. This process was first hypothesized in [1] and has never been observed so far. It would be possible only if the anti-neutrino ($\bar{\nu}_e$) emitted in one of the two beta decays, is re-absorber in the other one as a neutrino (ν_e). The $0\nu\beta\beta$ therefore represents a window on some fundamental parameters in Neutrino Physics. Its observation would demonstrate that neutrinos are Majorana particles, that is neutrino and anti-neutrino are the same particle ($\nu_e = \bar{\nu}_e$), a feature that has never been so long osserved in any of the known matter constituent. Moreover, it would represent the first evidence of lepton number violation, supporting the leptogenesis as the matter-antimatter asymmetry generation mechanism. Finally, it would constrain the absolute neutrino mass scale, an information not accessible by oscillation experiments which can measure only the difference between the squared neutrino masses. By measuring the $0\nu\beta\beta$ decay rate $\Gamma_{0\nu}$, and so its half-life $T^{0\nu}_{1/2} = (\Gamma_{0\nu})^{-1}$, it is possible to measure the so-called "effective Majorana mass" $\langle m_{\beta\beta}\rangle = |\sum_{i=1,2,3} U^2_{ei} m_i|$, where m_i

are the values of the three neutrino masses, and U_{ei} is the first row of the leptonic mixing matrix between flavour and mass eigenstates. These two quantities are linked by the relation $\Gamma_{0\nu} \propto \frac{\langle m_{\beta\beta}\rangle^2}{m_e^2}$, where m_e is the electron mass and the proportionality factor mainly depends on the nuclear matrix element adopted. Even without any $0\nu\beta\beta$ signal, it is possible to estimate a lower and an upper limit for $T_{1/2}^{0\nu}$ and $\langle m_{\beta\beta}\rangle$, respectively.

Given its importance, an intense experimental effort is ongoing to search for this decay in several nuclei [2]. The Cryogenic Underground Observatory for Rare Events (CUORE) [3], presently in the final stages of construction at the Laboratori Nazionali del Gran Sasso(LNGS), will be one of the most sensitive upcoming $0\nu\beta\beta$ experiments. CUORE aims to reach a sensitivity on the half-life of $0\nu\beta\beta$ of ^{130}Te $T_{1/2}^{0\nu} = 9.5 \cdot 10^{25}$ yr in 5 years (at 90 % C.L.). Section 2 describes the CUORE experimental setup, including the cryostat and the final detector.

The first tower built following the CUORE procedures was operated at the LNGS as a stand alone experiment, named CUORE-0, from March 2013 to March 2015. CUORE-0 has been built on the experience gained from its demonstrator, CUORICINO, operated at the LNGS from 2003 to 2008. The CUORE-0 experiment was custom designed to validate the CUORE detector performances and to set a world best limit on the half life of $0\nu\beta\beta$ decay of ^{130}Te. The CUORICINO and CUORE-0 techniques and results will be discussed in section 3.

Thanks to the experience and lessons learnt by CUORE-0, CUORE is currently approaching the final stage of its assembly and commissioning. The latest achievements and upgrades towards the completion of the CUORE setup will be commented in section 4.

2 The CUORE experiment

CUORE is a bolometric experiment devoted to the search of $0\nu\beta\beta$ in ^{130}Te. It aims to monitor a total ^{130}Te mass $M = 206kg$, cooled down to 10 mK in an ultra low background environment.

2.1 The detector

The CUORE detector is an array of 988 TeO_2 bolometers arranged in a compact structure of 19 towers, each one containing 52 TeO_2 crystals, disposed on 13 floors. Each bolometer is a TeO_2 crystal with dimensions $5 \times 5 \times 5\ cm^3$ and weight 750 g.

Each bolometer consists of three main parts: an energy absorber, a temperature sensor, a weak thermal link ($G = pW/mK$) to the heat sink (see fig. 1). The energy provided by the decay (ΔE) is released into the crystal, so that it acts either as the source and the absorber. This energy is then read by a Neutron-Transmutation Doped (NTD) Ge thermistor as a temperature increase $\Delta T \sim \Delta E/C$, where C is the heat capacitance of the crystal. For $T = 10mK \rightarrow C = 2nJ/K$ and in turn the responsivity is $\sim 100\mu K/MeV$. The thermistor temperature variation is converted into a variation of the electric resistance read as a voltage drop. This is amplified and readout at $125Hz$ acquisition rate by a warm electronics. The energy release generates a pulse with one or more decay constants, with typical values of $C/G \sim 100ms$. The pulse amplitude is calibrated using several emission peaks of a ^{232} Th source that can face the detector if necessary. The energy spectrum of the events is reconstructed using the different calibration peaks, whose principal is the 2615 keV line of ^{208}Tl. The energy spectrum of the events we measure is actually the sum energy spectrum of the emitted electrons. As a consequence, the expected $0\nu\beta\beta$ signal is a peak centred at the Q-value of the $0\nu\beta\beta$ in ^{130}Te, that corresponds to 2527.515 ± 0.013 keV [4]. We usually refer to this portion of the spetrum as Region Of Interest (ROI).

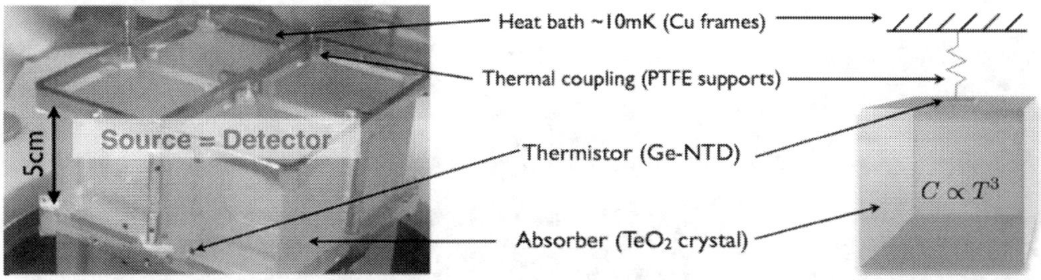

Figure 1. A picture of a CUORE tower floor (4 TeO_2 crystals) with the corresponding bolometer scheme.

2.2 The cryostat

The detector array is cooled down to 10 mK by a multistage cryostat unique of its kind: the first stage at $\sim 4K$ is obtained by means of 5 two-stage Cryomech PT415 Pulse Tubes (PTs), with nominal cooling power of 1.5 W at 4.2 K and 40 W at 45 K each.The base temperature is then provided by a DRS-CF3000 continuous-cycle $^3He/^4He$ Dilution Refrigerator (DR), customized by Leiden Cryogenics for CUORE purpouses. Its nominal cooling power is 5 μW at 12 mK (3 mW at 120 mK), and the minimum reachable temperature is around 5 mK (for more details on the CUORE cryostat see [5]). The dimensions ($> 10\ m^3$), the mass ($\sim 25tons$) and the cooling power ($3\mu W$ at $10mK$) make it the today's biggest and most powerful dilution cryostat in the world.

The cryostat (see Fig. 2) is composed by 6 coaxial vessels, kept at 300 K, 40 K, 4 K, 600 mK, 50 mK, 10 mK, respectively. 300 K and 4 K shields are vacuum tight. The detector is placed inside the 10 mK vessel. The cryostat is suspended from above to a structure called Main Support Plate (MSP). A Y-Beam, isolated from the MSP through Minus-K special suspensions, supports the detector. In this way, vibrations transmission from the cryostat to the crystals are minimized. To suppress external γ background, two different lead shields are placed inside the cryostat (see sec. 2.4).

2.3 The CUORE challenge

The aim of future neutrinoless double-beta decay experiments, such as CUORE, is to approach the inverted hierarchy region of the neutrino masses. A very common way to evaluate the sensitivity of the measure is given by $T^{0\nu}_{1/2} \propto \sqrt{\frac{M \cdot t}{b \cdot \delta E}}$. Therefore, to achieve the necessary sensitivity, the key experimental parameters are: i) large mass of candidate nuclei M, ii) long live time t, iii) good detector energy resolution δE, and iv) very low radioactive background b. The CUORE goals for them are: M = 206 kg of ^{130}Te, t = 5 years, $\delta E \leq 5keV$ and $b \leq 10^{-2} c/keV/kg/y$.

The requirement i) is matched by the CUORE design. The status of detector assembling is reported in section 4.

To match challenges ii), iii) and iv) in CUORE, it is crucial the capability of cool down to 10 mK roughly 1 ton of detector, delivering a uniform and stable base temperature in a low vibration and low background environment. The CUORE cryostat has already proved its base temperature capability and stability over a long period. More details on the status of the cryostat commissioning can be found in section 4. Thanks to the very low heat capacitance at low temperatures, the TeO_2 crystals has demonstrated a very good energy resolution ($\delta E/E(ROI) \sim 0.002 \rightarrow \delta E \sim 5keV$) at 10 mK (see section 3.2). Here we briefly comment on the main background sources.

 DOI: 10.1051/epjconf/201716407047

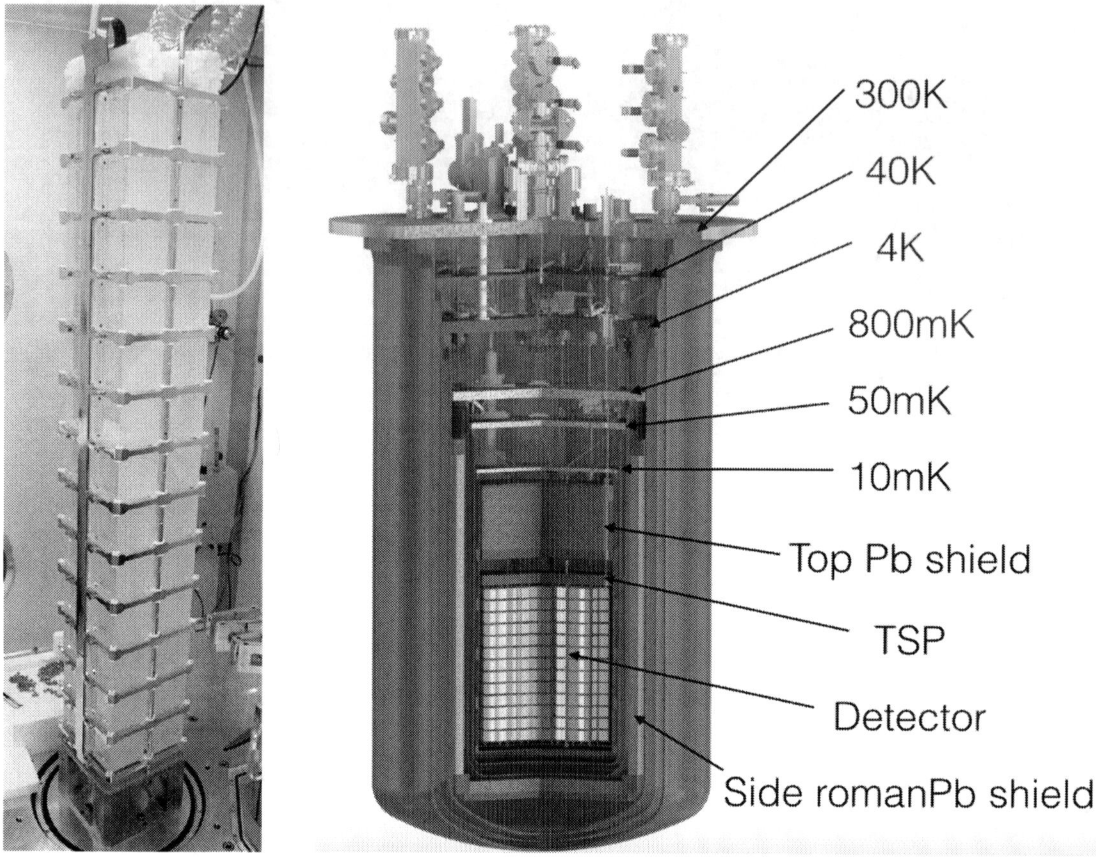

Figure 2. Left: the CUORE-0 detector during the assembling: the 52 TeO_2 crystals are arranged in 13 floors of 4 crystals each. Right: sketch of the CUORE cryostat with the CUORE detector surrounded by the vessels of increasing temperature and the two cryogenic lead shields.

2.4 Background

The excellent energy resolution of TeO_2 crystals makes the double beta decay with neutrinos emission ($2\nu\beta\beta$), that is the unavoidable intrinsic background source, easily distinguishable from $0\nu\beta\beta$ at the endpoint of the spectrum.

Besides that, one of the most common sources of background are the cosmic rays and the cosmogenic activation that they induce in the materials of the cryostat. For this reason the CUORE experiment is host in the Laboratori Nazionali del Gran Sasso (LNGS), an underground facility naturally shielded by roughly 1600 m of rocks ($\sim$ 3800 $m.w.e.$) which provide a reduction of the atmospheric muons flux from $1.5 \times 10^{-2}\ cm^{-2}s^{-1}$ to about $3 \times 10^{-8}\ cm^{-2}s^{-1}$.

Another relevant background source is given by the γs coming from the natural radioactive decay chain of U and Th, that are present as contaminants in the cryostat. To suppress this contribution a set of lead shields has been installed inside the cryostat. Suspended between the 10mK plate and the TSP, it is placed a 30-cm-thick cylindrical shielding, made of lead sheets. This is the only shielding wall from the top side for a total weight of about 2 tons. It is hang from the 10 mK by the mean of three kevlar strips and it is thermalized at 50 mK. Another cold lead shielding, kept at 4 K, is a 6-cm-thick wall that surrounds the 600 mK vessel. The lateral shield is made of about 5 tons of ^{210}Pb-free ancient roman lead. The aim of the cold shields is to prevent radioactivity contamination from the external vessels (300, 40, and 4 K) and from all the cryostat plates. This contribution is suppressed by accurate radio-purity controls on the materials, with a background evaluated in $\sim 6 \times 10^{-3} c/keV/kg/y$ by the CUORE-0 experiment (see sec. 3.2). To shield the detector for the γs sources external to the cryostat, an external lead shield ($\sim$ 80 tons, thickness = 25 cm) can be raised in operation to dress the outer cryostat vessel at room temperature.

The main and more critical contribution in the ROI is due to the so-called "degraded alphas". These are α-particles emitted by surface contaminations of the crystals and of the copper of which the tower structure is made. These αs release part of their energy in the surface that is emitting them (source), and the rest in the one that absorbs them (target). We refer to such an process as to a "source-target" event. Through anti-coincidence analysis is it possible to reject the "crystal-crystal" events, where both the source and the target are active surfaces, but it is not possible to tag "crystal-copper" nor "copper-crystal" events, which give rise to a continuos background in the ROI.

The reduction of this contribution was a crucial part of the intense CUORE R&D program, which is discussed in section 3. The main improvements have been: i) the increased radio-purity of the copper and crystal surfaces as well as in the assembly environment, ii) thicker shields using low-activity ancient Roman lead, and iii) the granularity and self shielding of the 19-towers array affords providing a better anti-coincidence coverage (a discussion of the different contribution can be found in [6]). Thanks to these improvement, the background level in CUORE is expected to be as good as 0.01 counts/keV/kg/y [7].

3 The CUORE precursors

CUORE has been designed on the experience of the predecessor experiments CUORICINO [8] and CUORE-0 [9], operated at the LNGS from 2003 to 2008, and from 2013 to 2015, respectively.

3.1 The CUORICINO experience

The demonstrator experiment CUORICINO [8] was composed by 62 TeO_2 bolometers, for a total mass of 40.7 kg and an acquired statistics of 19.75 kg(^{130}Te) $\cdot$ y. The average energy resolution of the detector at 2615 keV (^{208}Tl line) was 6.3 ± 2.5 keV [10] and the background level in the $0\nu\beta\beta$ energy region was 0.169 ± 0.006 c/keV/kg/y. No $0\nu\beta\beta$ signal was found. The corresponding lower limit on the $0\nu\beta\beta$ half-life of ^{130}Te is 2.8×10^{24} y (90% C.L.). This limit translates into an upper limit on the neutrino effective Majorana mass ranging from 300 to 710 meV, depending on the nuclear matrix elements considered in the computation.

The scaling from CUORICINO to CUORE aims at improving the sensitivity to the $0\nu\beta\beta$ half life of ^{130}Te. This goal can be achieved by increasing the exposure (increasing the active mass) and by reducing the background in the ROI. After CUORICINO, the achievement of these goals were part of the intense CUORE R&D program.

A challenge related to the detector performances is the uniformity of their behaviour. A crucial requirement is the capability of completing 988 TeO_2 bolometers and assembly them into 19 towers in clean and reproducible conditions. In CUORICINO, we observed a large spread in the pulse shape among the bolometers. This leads to a complication in the data analysis that could be more challenging in case of an experiment, like CUORE, with about 1000 single detectors.

For these reasons, before starting the construction of the 19 CUORE towers, an additional tower, named CUORE-0, was produced according to the CUORE requirements.

3.2 The CUORE-0 experiment

CUORE-0 is a single CUORE-like tower, the first one built using the low-background assembly techniques developed for CUORE [11]. It is made of 52 TeO_2 bolometers, for a total mass of 39 kg. To improve the detector response uniformity special care has been devoted in the realisation of a new detector assembly system.

Each TeO_2 detector is instrumented with a NTD-Ge thermistor glued on its surface, to measure the temperature change of the absorber and convert it into an electric signal. The thermal response of the detector depends on its temperature. For this reason each crystal is instrumented also with a silicon resistor, called "heater", to generate reference pulses of ~ 3 MeV. The tower was operated in Hall A of LNGS, in the same dilution refrigerator that previously hosted the CUORICINO experiment, and it ran between March 2013 and March 2015. Technical details are reported in [12] and [9], while the CUORE-0 physics results can be found in [13].

3.2.1 Detector performance

The detector performances of the CUORE-0 bolometers have been improved with respect to CUORICINO (see table 1). One of the main improvement is a new layout of the detector towers, where the 52 TeO_2 crystals are held in an ultra-pure copper frame by PTFE supports and arranged in 13 floors, with 4 crystals per floor (see fig. 2). In order to reduce to the background events in the ROI, more stringent material selection criteria and strictly cleaning and handling procedures were implemented for the detectors and for the cryostat components, as well as for the assembly environment. Moreover, dedicated processes have been developed to suppress the degraded-αs induced background, increasing the radio-purity of the copper and crystal surfaces. Finally, thicker shields using low-activity ancient Roman lead have been used for the first time in the CUORE-0 cryostat.

CUORE-0 acquired data for the $0\nu\beta\beta$search collecting a total exposure of 9.8 kg·y of ^{130}Te. Data are grouped in month-long blocks called datasets. At the beginning and end of each dataset we calibrate the detector by placing a ^{232}Th source next to the outer vessel of the cryogenic system.

We use the calibration line with the highest intensity and next to the ROI, 2615 keV from ^{208}Tl, in order to study the detector response function to a mono energetic energy deposit for each bolometer and dataset. We estimate the shape parameters of the 2615 keV line with a simultaneous, Unbinned Extended Maximum Likelihood (UEML) fit to calibration data. The physics exposure-weighted effective mean of the FWHM values for each bolometer and dataset is 4.9 keV, with a corresponding RMS of 2.9 keV. We evaluate the background level in the alpha-dominated region (2700-3900) keV to be 0.016 ± 0.001 counts/keV/kg/y), 6 times smaller with respect to the CUORICINO background in the same region (see fig. 3 left).

Table 1. A comparison between the CUORICINO and CUORE-0 resolution and background performances.

	Resolution [keV] at ^{208}Tl peak (2615 keV)		Background [c/keV/kg/y] 2.7-3.9 MeV	ROI
CUORICINO	6.3	FWHM rms = 2.5	0.110 ± 0.001	0.169 ± 0.006
CUORE-0	4.9	FWHM rms = 2.9	0.016 ± 0.001	0.058 ± 0.004

3.2.2 $0\nu\beta\beta$ decay result

We search for $0\nu\beta\beta$ signal in ^{130}Te in the final CUORE-0 energy spectrum performing a simultaneous UEML fit in the energy region 2470-2570 keV (Fig. 3 right). The fit function is composed by three parameters: a posited signal peak at the Q-value of the transition, a peak at ~ 2507 keV from ^{60}Co double-gammas, and a smooth continuum background attributed to multi-scatter Compton events from ^{208}Tl and surface decays. The best-fit values are $\Gamma_{0\nu} = 0.01 \pm 0.12(\text{stat}) \pm 0.01(\text{syst}) \times 10^{-24}\text{yr}^{-1}$ for the $0\nu\beta\beta$ decay rate and $0.058 \pm 0.004(\text{stat}) \pm 0.002(\text{syst})$ counts/keV/kg/y for the background index in the ROI. This result is 3 times lower than the Cuoricino background, 0.169 ± 0.006 counts/keV/kg/y, in the same ROI. Using a Bayesian approach, we set a 90% C.L. lower bound on the decay half-life of 2.7×10^{24}yr [13]. When combined with the 19.75 kg·y exposure of ^{130}Te from the Cuoricino experiment, we find a Bayesian 90% C.L. limit of $T_{0\nu} > 4.0 \times 10^{24}$yr, which is the most stringent limit to date on the ^{130}Te $0\nu\beta\beta$ half-life. Additional details on the analysis techniques can be found in [14].

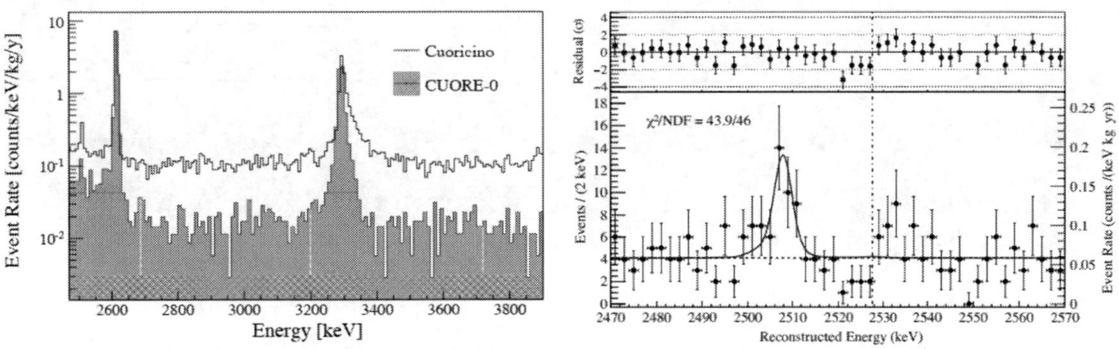

Figure 3. Left: Comparison of CUORE-0 and CUORICINO spectra in the flat alpha region (2700-3900 keV). Right bottom: The best-fit model from the UEML fit (solid blue line) overlaid on the spectrum of $\beta\beta0\nu$ decay candidates in CUORE-0 (data points); the data are shown with Gaussian error bars. Dotted black line shows the continuum background component of the best-fit model. Right top: normalized residuals of the best-fit model and the binned data. The vertical dot-dashed black line indicates the position of $Q_{\beta\beta}$

4 CUORE commissioning

The commissioning of the CUORE experiment has two main deliverables: i) the cryostat capability to cool the detector down to the expected base temperature (<10 mK), in stable conditions and for a long period, and ii) the assembly and installation of the 988 TeO_2 bolometric detectors at the coldest stage of the cryostat.

The commissioning of the CUORE cryostat was concluded in March 2016. It consisted in a sequence of could run of increasing complexity in the cryostat. The last cold run was performed with the full cryostat with all the components installed but the detector. An 8 TeO_2 bolometer array, named Mini-Tower, was built, mounted, and connected to the electronics to check bolometer functionality. The CUORE cryogenic system demonstrated to be able to cool down an experimental volume of ~ $1m^3$ at a base temperature of $6.3mK$ for more then 70 days, with temperature variations of the order of $0.2mK$ rms, a really important and unprecedented experimental achievement. The performance of the detectors measured on the Mini-Tower were encouraging and a full debugging of the electronics, DAQ, temperature stabilization, and detector calibration systems was performed.

The assembly of the CUORE detector towers started in late February 2013 and has been completed in July 2014. Following the experience of CUORE-0, the assembling was performed under nitrogen atmosfere in a dedicated assembling line. Each tower has been stored inside an individual cage and constantly flushed with nitrogen until their installation. The installation of the detectors in the CUORE cryostat started at the beginning of August 2016 and has been successfully completed on 26th August. Strictly cleaning procedures and a radon abatement system were adopted during all the installation period.

We are currently closing the cryostat vessels and finalizing all the subsystems. The final CUORE cool down will start by the end of November 2016.

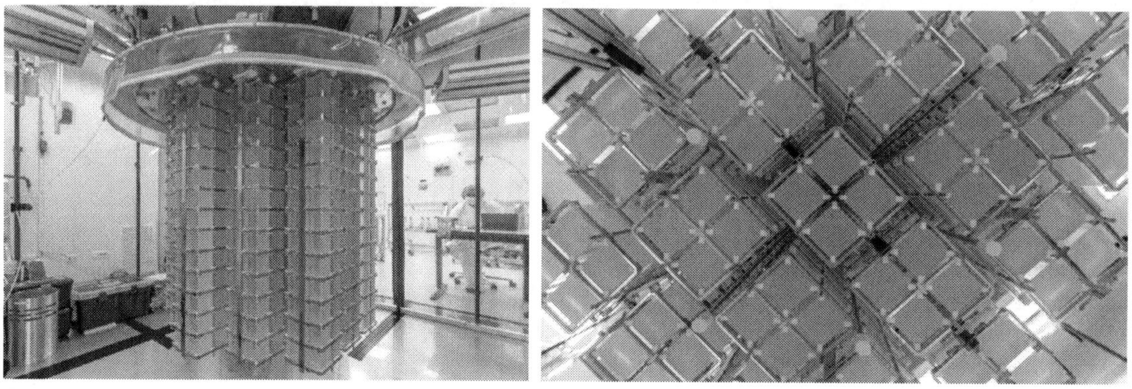

Figure 4. Two pictures of the CUORE detector after its installation in August 2016. Left: a picture of the detector attached to the MSP inside the CUORE cryostat cleanroom. Right: a bottom view of the 19 CUORE towers.

Acknowledgments

The CUORE Collaboration thanks the directors and staff of the Laboratori Nazionali del Gran Sasso and the technical staff of our laboratories. This work was supported by the Istituto Nazionale di Fisica Nucleare (INFN), the National Science Foundation, the Alfred P. Sloan Foundation, the University of Wisconsin Foundation, and Yale University. This material is also based upon work supported by the US Department of Energy (DOE) Office of Science and by the DOE Office of Nuclear Physics. This research used resources of the National Energy Research Scientific Computing Center (NERSC). More details can be found at: http: http://cuore.lngs.infn.it/support

References

[1] M.Goeppert-Mayer, *Phys. Rev.* **48**, 6 (1935)
[2] M.Agostini [GERDA Collaboration], *et al Phys. Rev. Lett.* **12**, 111 (2013)
[3] C. Arnaboldi, *et al.* [CUORE Collaboration], *Nucl. Instrum. Methods A* **518**, 775 (2004)
[4] M. Redshaw, *et al*, *Phys. Rev. Lett.* **102**, 212502 (2009)
[5] C.Ligi *et al.*, *Jour. Low Temp. Phys.* **184**, no.3-4, 590-596 (2016)

[6] F.Bellini *et al.*, *Astropart. Phys.*, **33**, 169 (2010)
[7] F. Alessandria, *et al.*, *Astropart. Phys.*, **35**, 839 (2012)
[8] C.Arnaboldi *et al.* [CUORE Collaboration], *Phys. Rev. Lett.* **78**, no.3, 035502 (2008)
[9] C.Alduino *et al.* [CUORE Collaboration], *JINST* **11** no.07, P07009 (2016)
[10] E.Andreotti *et al.*, *Astropart. Phys.* **34**, 822 (2011)
[11] E.Buccheri *et al.*, *Nucl. Instrum. Meth. A* **768** 13 (2014)
[12] D.R.Artusa *et al.* [CUORE Collaboration], *Eur. Phys. J. C* **74** 8 (2014)
[13] K.Alfonso *et al.* [CUORE Collaboration], *Phys.Rev. Lett.* **115** no.10, 102502 (2015)
[14] C.Alduino *et al.* [CUORE Collaboration], *Phys.Rev. Lett.* **93** no.4, 045503 (2016)

 DOI: 10.1051/epjconf/201716407048

The proton radius puzzle

Maurizio Bonesini[1,a]

[1] *Sezione INFN Milano Bicocca, Dipartimento di Fisica G. Occhialini, Milano, Italy*
(on behalf of the **FAMU Collaboration***)*

Abstract. The FAMU (**F**isica degli **A**tomi **Mu**onici) experiment has the goal to measure precisely the proton Zemach radius, thus contributing to the solution of the so-called proton radius "puzzle". To this aim, it makes use of a high-intensity pulsed muon beam at RIKEN-RAL impinging on a cryogenic hydrogen target with an high-Z gas admixture and a tunable mid-IR high power laser, to measure the hyperfine (HFS) splitting of the 1S state of the muonic hydrogen. From the value of the exciting laser frequency, the energy of the HFS transition may be derived with high precision ($\sim 10^{-5}$) and thus, via QED calculations, the Zemach radius of the proton. The experimental apparatus includes a precise fiber-SiPMT beam hodoscope and a crown of eight $LaBr_3$ crystals and a few HPGe detectors for detection of the emitted characteristic X-rays. Preliminary runs to optimize the gas target filling and its operating conditions have been taken in 2014 and 2015-2016. The final run, with the pump laser to drive the HFS transition, is expected in 2018.

1 Introduction: the proton radius puzzle

Many properties of the proton - the primary visible universe building block- such as its radius and anomalous magnetic moment, are not completely understood. The so-called proton radius "puzzle" [1] refers to the 7σ discrepancy between the electron and muon determination of the proton charge radius. This discrepancy may be due to a violation of the electron-muon universality or simply to not well understood experimental problems.

The precise measurement of the Zemach radius of the proton with muons may shed new light on the problem. Table 1 resumes the current experimental situation for both the proton charge (r_{ch}) and Zemach radius (r_Z) determination with muons or electrons.

2 The FAMU experimental method

The experimental method to be used by the FAMU Collaboration [7] is shown schematically in figure 1. It makes use of a high intensity pulsed low-energy muon beam, stopping in a hydrogen target, to produce muonic hydrogen (in a mixture of singlet F=0 and triplet F=1 states) and a tunable mid-IR high power laser to excite the hyperfine splitting (HFS) transition of the 1S muonic hydrogen (from F=0 to F=1 states). Exploiting the muon transfer from muonic hydrogen to another higher-Z gas in the target (such as O_2 or Ar), the $\mu^- p_{1S}$ HFS transition will be recognized by an increase of the number of X-rays from the (μZ^*) cascade, while tuning the laser frequency ν_0 ($\Delta E_{HFS} = h\nu_0^{res}$).

[a] e-mail: maurizio.bonesini@mib.infn.it

Table 1. The present situation of the proton charge and Zemach radius measurements.

	charge radius r_{ch}(fm)	Zemach radius r_Z (fm)
e-p scattering and hydrogen spectroscopy	$< r_{ch} >= 0.8751(61)$ [6]	$r_Z = 1.037(16)$ [2] (2003) $r_Z = 1.086(12)$ [3] (2004) $r_Z = 1.047(16)$ [4] (2005) $r_Z = 1.045(4)$ [5] (2011)
μ^--p scattering and and Lamb Shift spectroscopy	$< r_{ch} >= 0.84087(39)$ [1]	$r_Z = 1.082(37)$ [1]

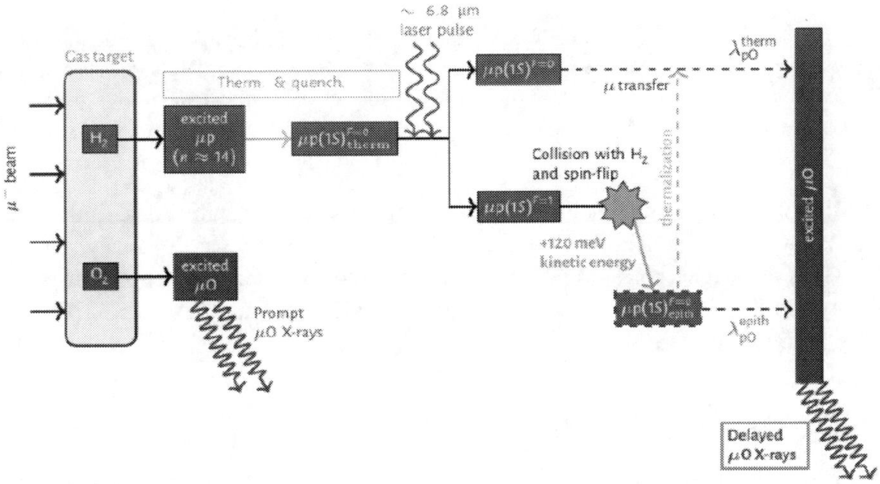

Figure 1. Schematic illustration of the principle of the FAMU experiment. After the initial phase of thermalization, a laser pulse will induce the singlet to triplet transitions in the muonic hidrogen atoms. Further collisions with H_2 molecules will convert back F=1 states to F=0 states with ~ 0.2 keV kinetic energy (~ ΔE^{HFS}). Finally, the muon is transferred from $(\mu^- p)$ to the higher Z admixture gas and the characteristic X-ray K-lines from (μZ^*) cascade are observed.

One can thus measure ΔE_{HFS} with a relative accuracy ~ 10^{-5} and determine, via QED calculations, the proton Zemach radius with high precision.

The FAMU experiment will be performed in steps, starting from the study of the transfer rate from muonic hydrogen to another higher-Z gas (2014-2016) and ending with a full working setup (2018 $\mapsto$ ...) with the pump laser and a cavity to determine the proton Zemach radius. The preliminary steps will allow determining the best mixture to be used inside the cryogenic target and optimize the operating conditions.

3 The experimental setup

The RIKEN-RAL muon facility is located at the Rutherford Appleton laboratory at Didcot (UK) and provides high intensity pulsed muon beams at four experimental ports. Its schematic layout is shown in the left panel of figure 2. The ISIS primary proton beam at 800 MeV/c, with a 50 Hz repetition rate, impinges from the left on a secondary carbon target producing a high-intensity pulsed muon beam,

DOI: 10.1051/epjconf/201716407048

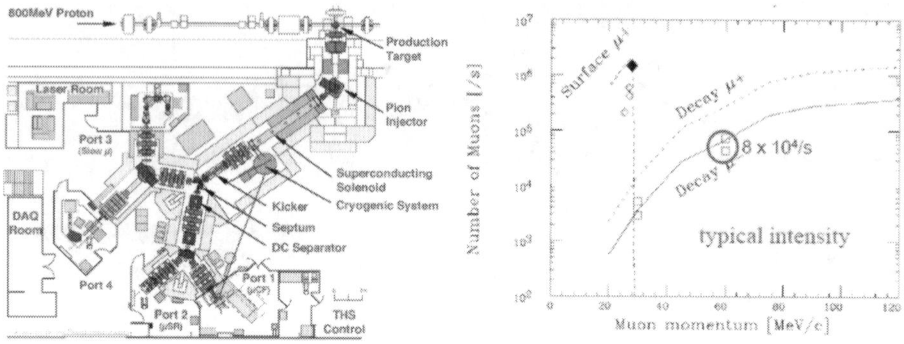

Figure 2. Left panel: layout of the RIKEN-RAL facility at RAL, with its 4 experimental ports. The FAMU esperiment presently uses PORT 4 and will move to PORT 1 for the final run. Right panel: estimated intensity of the RIKEN-RAL muon beam for surface muons and decay muons in a typical square 4×4 cm^2 area.

that reflects the primary beam structure: a double pulse structure with 70 ns pulse width (FWHM) and 320 ns peak to peak distance. The FAMU experiment makes use of the decay muon beam (muons produced outside the target from pion decay and collected by a decay solenoid) at ~ 60 MeV/c. The intensity is around $8 \times 10^4 \mu^-/s$, as shown in figure 2, with an energy spread ~ 10% and an angular divergence ~ 60 mrad.

3.1 The 2014 and 2015-2016 preliminary runs

The setup for the 2014 (R484) and 2015-2016 (R582) FAMU data taking are schematically shown in figures 3 and 4. A fiber-SiPMT beam hodoscope is the most upstream detector of the setup and had a

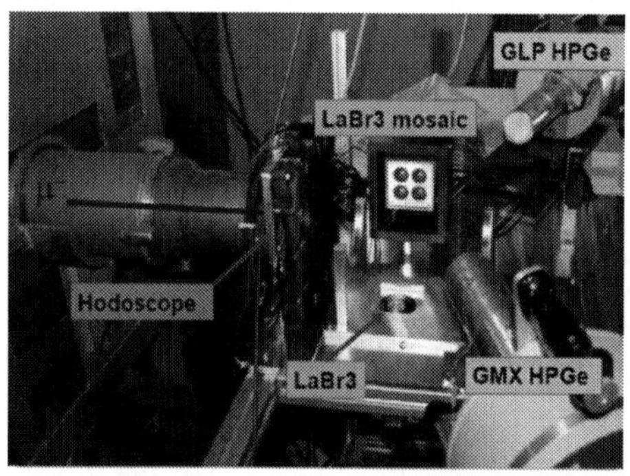

Figure 3. Preliminary setup for the 2014 FAMU run at RiKEN-RAL. From left to right, along the beamline, the 3mm pitch beam hodoscope, the target surrounded by two LaBr3 detectors and two HpGe detectors

3 mm pitch in the first 2014 run and a 1 mm pitch in the following ones. It was used to optimize beam steering inside the target. Both beam hodoscopes (with 1 mm and 3 mm pitch) have a similar structure based on BCF12 Bicron square single-clad scintillating fibers along X/Y directions (32+32 channels) read at one end by Advansid SiPMT. Difficulties encountered in developing the mechanics of the two beam hodoscopes (especially the one with 1 mm pitch) were solved via 3D printing. Details of the

 DOI: 10.1051/epjconf/201716407048

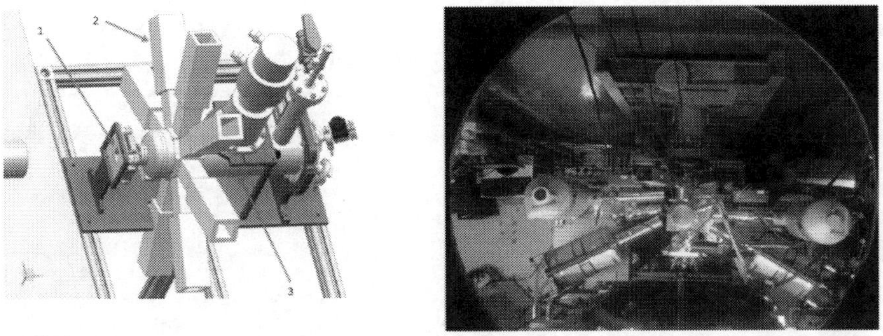

Figure 4. Left panel: layout of the FAMU experimental setup for the 2015-2016 run at RIKEN-RAL: (1) is the 1 mm pitch beam hodoscope; (2) is the crown of eight LaBr3 crystals with PMT readout, for fast X-rays detection; (3) is the cryogenic target. Right panel: view from the top of the experimental setup: the four HpGe detectors, for precise X-rays detection, are also shown.

construction and preliminary performances of the first beam hodoscope were reported in references [8] and [9].

The 1 mm pitch beam hodoscope (active area $\sim 3.2 \times 3.2$ cm^2) was developed to be put in front of the thin Beryllium window of the cryogenic target, used in the 2015-2016 run. An effort was done to minimize materials present in the beamline. Thus Bicron BCF12 square 1 mm^2 scintillating fibers, with white EMA coating to avoid light cross-talk, were used. Scintillating fibers were cut at Cern with a Fiberfin4 machine, that produced ready-to-use fibers of the right length with polished ends. The main mechanics components and a picture of the detector after the beam collimator are shown in figure 5.

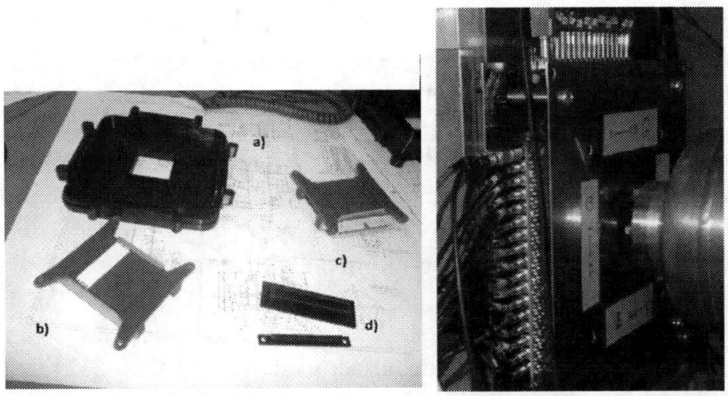

Figure 5. Left panel: main mechanical components of the 1mm pitch beam hodoscope: (a) one of the two light tight beam hodoscope cover boxes, (b),(c) X/Y plane fiber holders: in (b) some fibers are mounted, while in (c) the flat cable connector is shown , (d) cable feed-through. Right panel: picture of the beam hodoscope after the beam collimator

Scintillating fibers were read from one edge only, alternating up/down or left/right sides, by RGB 1×1 mm^2 Advansid SiPMT with 40 μm cells. As the breakdown voltages ($\sim$ 29 V) were very similar for all SiPMTs: within 100 mV, it was possible to use a simplified front-end electronics with respect

 DOI: 10.1051/epjconf/201716407048

to the one of the previous 3 mm pitch beam hodoscope, derived from the INFN TPS project [1]. All channels of the same plane (x,y) of the 1mm pitch beam hodoscope were instead powered by a single HV channel and the output signals were directly fed into a fast FADC [2]. Signals were routed from the PCB's, where the SiPMTs were mounted, to the external MCX cables, going to the FADC, via a short flat cable. The impedance mismatch of this cable was tested and found negligible. Figure 6 shows the FADC waveform for a typical channel, where the two pulses beam structure is evident (left panel) and the reconstructed x,y beam profiles from the integrated charge information. As expected, the beam

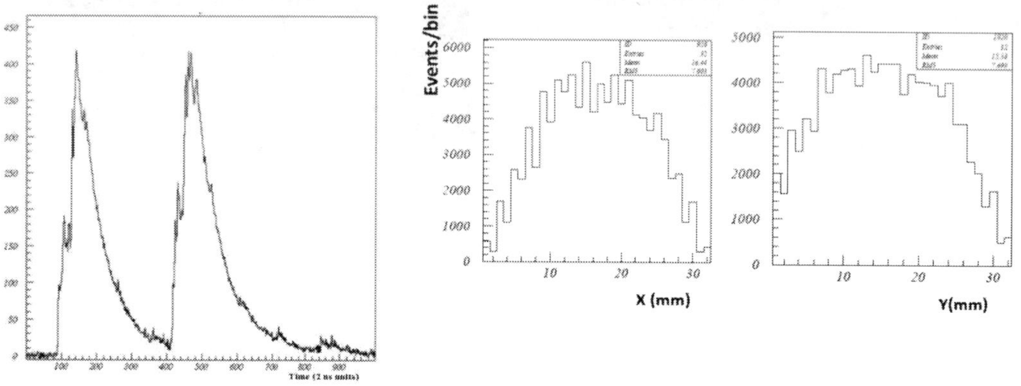

Figure 6. Left panel: waveform, as recorded by V1742 FADCs, for a typical channel of the 1 mm pitch beam hodoscope. Right panel: x/y beam profile, with standard optics, as recorded by integrating the waveform signals during the 2016 data taking.

profile is dominated by the size of the collimator and has an RMS $\sim 7-8$ mm.

The target, used in the 2014 run, was an Al cylindrical vessel [3] filled with a high-pressure gas. In the following runs, it was replaced by a cryogenic one. The layout of the cryogenic target is shown in the left panel of figure 7, where the percentage of stopped muons as a function of the beam momentum is shown nearby. Emitted X-rays are detected with high precision Germanium detectors (HpGe) and fast LaBr3:Ce detectors. Two HPGe detectors (one small area ORTEC GLP detector and one ORTEC GMX detector) were used in the 2014 run, while in the following runs two additional HPGe detectors from RIKEN-RAL were also used. In the 2014 run a mosaic of four 0.5-inch diameter, 0.5-inch length $LaBr_3(5\%Ce)$ read by Hamamatsu R11265-200 UBA photomultipliers and a single 1-inch diameter, 1-inch length Brillance 380 $LaBr_3$ crystal read by a Phillips XP2060 photomultiplier were used. In the following runs, the $LaBr_3$ mosaic was replaced by a crown of eight 1-inch $LaBr_3$ detectors read by Hamamatsu R11265-200 UBA photomultipliers, with an active divider, as shown in figure 7. The $LaBr_3$ detectors were read through 500 Mhz, 14-bit digitizer (CAEN V1730) with a $5\mu s$ window after the trigger, provided by the RF of the beamline, while for the HPGe detectors a slower 100 MHz, 14-bit digitizer (CAEN V1724) was used. In addition, during the 2016 data new X-rays detectors, developed for PET, based on PrLuAG and CeCAAG crystals and read out by Hamamatsu TSV SiPMT arrays, were introduced to instrument the volume under the cryogenic target [10].

[1]each TPS electronic board, with 8 individual channels, provided individual SiPMT voltage regulation, signal amplification and shaping and signal discrimination and trigger capability. Individual output signals were then fed into a CAEN V792 QADC, to measure the charge integrated signal

[2]CAEN V1742 FADC with 5 Gs/s, 12 bit, 1 Vpp dynamic range in VME standard

[3]with a 125 mm bare diameter and a 260 mm length, it had a 2.8 liters volume. The walls had a thickness of 7 mm, except the entrance window thinned to 4 mm.

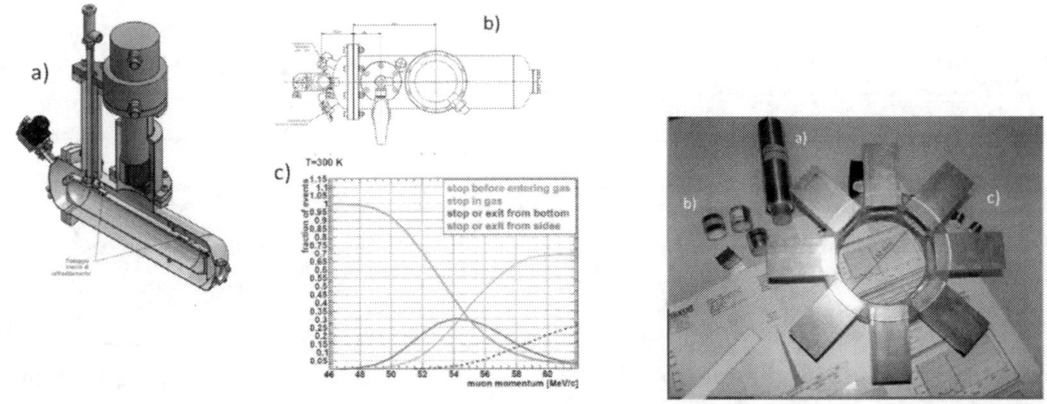

Figure 7. Right panel: (a) layout of the cryogenic target; (b) top view of the cryogenic target; (c) stopping of muons inside the target at T=300K. Left panel: a picture of the eight 1" LaBr3(Ce) detectors arranged in a star. (a) Hamamatsu R11265-200 UBA PMTs; (b) 1" LaBr3(Ce) crystals; (c) mechanics arrangement

3.2 The setup for the final step of the FAMU experiment.

To excite the HFS transition, a laser tunable at a wavelength around $\lambda_0 \sim \frac{2\pi hc}{\Delta E_{HFS}} \sim 6.785\ \mu$m will be used. In reference [11], the main requirements on the laser system, such as energy (≥ 0.25 mJ) and linewidth precision ($\lambda_0 \leq 0.07$ nm) are discussed. The layout of the proposed laser system, under development at the Electra laboratory (Ts), is schematically shown in figure 8. A scheme based on direct difference frequency generation in nonlinear crystals, using a Q-switched Nd-Yag laser and a tunable Cr:forsterite laser will be used and is presently under test [12].

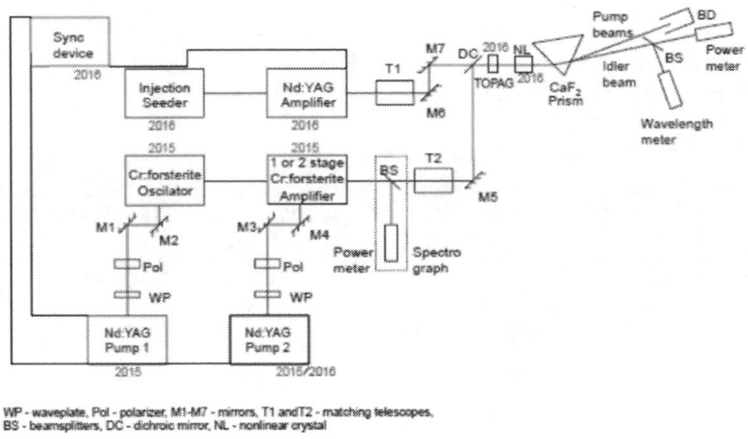

Figure 8. (Structure of the proposed mid-IR laser, under development at Electra (Ts). Wp: waveplate, Pol: polarizer; M1-M7: mirrors; T1-T2: matching telescopes; BS: beamsplitter; DC: dichroic mirror; NL: nonlinear crystal.

 DOI: 10.1051/epjconf/201716407048

To increase the probability of the spin-flip (F=0 to F=1) process either the target temperature may be lowered or a multipass cavity may be used: see reference [13] for an example. With a multipass cavity that provides $\sim 2 \times 10^3$ reflections, the spin-flip probability would reach a reasonable $\sim$ 12% value, that makes the experiment feasible. Studies of a suitable multipass cavity, that may accomodate the available cryogenic target, are under way inside the FAMU collaboration.

4 Detector performances and preliminary results

The aim of the 2014 run (R482) was to test the suitability of the proposed setup for the foreseen measurement: in particular the detection of the characteristic X-rays in a large background environment. The detection of characteristic X-rays from $LaBr_3$ detectors was cross-checked with two high precision HpGe detectors. As an example, for an H_2 + $2\%Ar$ filled target, figure 9 shows the spectra detected by the HPGe detectors and the $LaBr_3$ crystals. All characteristic X-ray lines from the Ar admixture are detected, showing the correctness of the method.

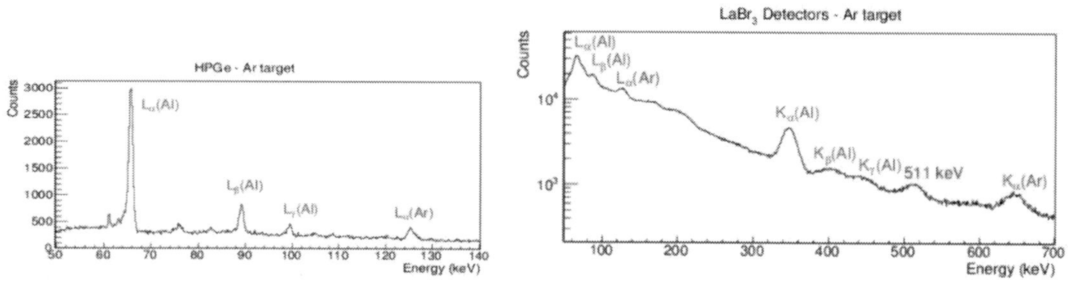

Figure 9. Left panel: characteristic X-rays lines, as detected by the HPGe detectors. Right panel: the same as detected by the $LaBr_3$ detectors.

Regarding the time distribution of the events, figure 10 shows an example of the time evolution of the X-ray spectrum for the $H_2 - CO_2$ target. The prompt peaks, corresponding to the two beam spills, are followed by a tail populated by the products of muon decay. From the events in the tails, the lifetime of muonic atoms may be estimated. Results are reported in table 2 and are compatible with previous ones, as reported in reference [14].

Table 2. Lifetimes of some muonic atoms in ns, as determined in the FAMU 2014 run.

	FAMU [8] (ns)	T. Suzuki *et al.* [14] and references therein (ns)
μC	2011 ± 16	2026 ± 1.5
μAl	879 ± 28	864 ± 2
μO	1824 ± 46	1795 ± 2
μAr	564 ± 14	537 ± 32
μp	2141 ± 98	2194.53 ± 0.11

Studies are under way to determine the transfer rate from hydrogen to the gas mixture contaminants (Argon or Oxygen), using the time evolution of the X-rays events inside the Argon (Oxygen) peak.

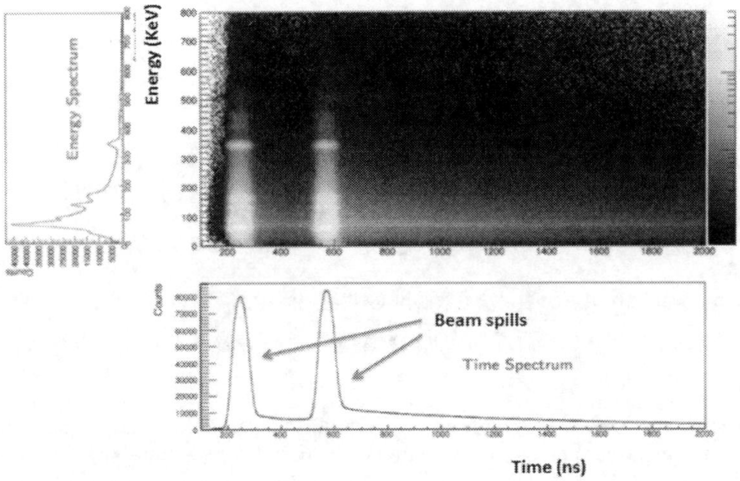

Figure 10. A snapshot of the detected X-ray spectrum (energy vs time). The two peaks are due to the two pulses structure in the incoming muon beam.

5 Conclusions

The feasibility of the method proposed by the FAMU Collaboration to determine the proton Zemach radius has been demonstrated with preliminary measurements at RIKEN-RAL: in particular the possibility to use a high-rate X-rays detection system based on $LaBr_3$ crystals.

The mid-IR pulsed laser and the multipass cavity are under development, showing that a final data taking in 2018 is feasible.

References

[1] R.Pohl *et al.*, Nature **466** 213-216 (2010); Antognini *et al.* Science **339** 417-420 (2013).

[2] Dupays *et al.* Physical Review **A68**, 052503 (2003).

[3] J L Friar and I Sick, Phys. Lett. **B579** 285-289 (2004).

[4] A.V. Volotka *at al.* Eur. Phys. J. **D33** 23-27 (2005).

[5] Distler *et al.* Phys. Lett. **B696** 343-347 (2011).

[6] P.J.Mohr *et al.*, Rev. Mod. Phys. **84**, 1527-1605 (2012) [CODATA].

[7] D.Bakalov *et al.*, Phys. Lett. **A172** 277-280 (1993); A. Vacchi *et al.*, SPIE Newsroom (2012) DOI: 10.1117/2.1201207.004274.

[8] A. Adamczak *et al.*, JINST **11/05** P05007 (2016).

[9] R. Carbone *et al.*, JINST **10/3** C08007 (2015).

[10] M.Bonesini *et al.*, PoS EPS-HEP2015 **244** (2015).

[11] A. Adamczak *et al.*, Nucl. Instr. Meth. **B281** 72-76 (2012).

[12] L.I.Stoichev *et al.*, Proc. SPIE **9135**, 91350J (2014).

[13] J.Vogelsang *et al.*, Optics Express **22**, 13050 (2014).

[14] T. Suzuki *et al.*, Phys. ReV. **C35** 2212-2224 (1987).

The XENON project for dark matter direct detection at LNGS

Andrea Molinario[1,a] on behalf of the XENON Collaboration

[1] *INFN-Laboratori Nazionali del Gran Sasso, via G. Acitelli 22, 67100 Assergi (AQ), Italy*

Abstract. The XENON project at INFN Laboratori Nazionali del Gran Sasso, Italy, aims at dark matter direct detection with liquid xenon dual-phase time projection chambers. Latest results of XENON100 detector exclude various models of leptophilic dark matter. A search for low mass weakly interacting massive particles was also performed, lowering the energy threshold for detection to 0.7 keV for nuclear recoils. The multi-ton XENON1T detector is fully installed and operating. It is expected to reach a sensitivity a factor 100 better than XENON100 with a 2 ton·year exposure.

1 Introduction

The XENON project at INFN Laboratori Nazionali del Gran Sasso (LNGS), Italy, aims at the direct detection of dark matter in the form of weakly interacting massive particles (WIMPs). In the XENON time projection chambers (TPCs), liquid xenon (LXe) is used both as a target and self-shielding medium for background reduction. The WIMP elastic scattering off Xe nuclei would produce low energy nuclear recoils at extremely low interaction rates.

The working principle of the XENON detectors is presented in figure 1. Photomultipler tubes (PMTs) arrays are placed on top and bottom of the TPC. Cathode, gate, and anode grids define electric fields in the TPC. When a particle interacts with Xe in the LXe volume, the interaction creates the excitation and ionization of Xe. Direct scintillation (S1) from the excitation and electron-ion recombination is detected by the PMTs immediately after the interaction. The electrons, escaping the recombination, drift upward due to the electric field. Those electrons are extracted into the gas volume by a high electric field between the gate and the anode, and then accelerated in gaseous xenon (GXe). This creates Xe excitation in the gas volume. The scintillation signal in the gas phase (S2) is delayed relative to S1 by the drift time of the electrons through the liquide. The three-dimensional position of the event is reconstructed using the drift time and the S2 hit pattern on the top PMT array. This enables background suppression by selection of a fiducial volume. The ratio S2/S1 is different for electronic recoils (ERs), induced by electrons and γ, and nuclear recoils (NRs), induced by neutrons and dark matter particles. This feature allows to reject the most abundant electromagnetic backround and provides powerful background discrimination.

The XENON detectors have been operating at LNGS at an average depth of 3600 m water equivalent since 2006. The first operating detector has been XENON10 [1], with a total Xe mass of 25 kg.

[a] e-mail: andrea.molinario@lngs.infn.it

DOI: 10.1051/epjconf/201716407049

It was followed, since 2007, by XENON100 [2] with a total Xe mass of 161 kg. This detector set a limit to the WIMP-nucleon spin-independent cross section of $\sigma_{SI} < 1.1 \times 10^{-45}\mathrm{cm}^2$ for a WIMP mass of 50 GeV/c^2 [3].

The multi-ton scale phase of the project, XENON1T, with a total Xe mass of 3.2 tons, is fully installed and operating.

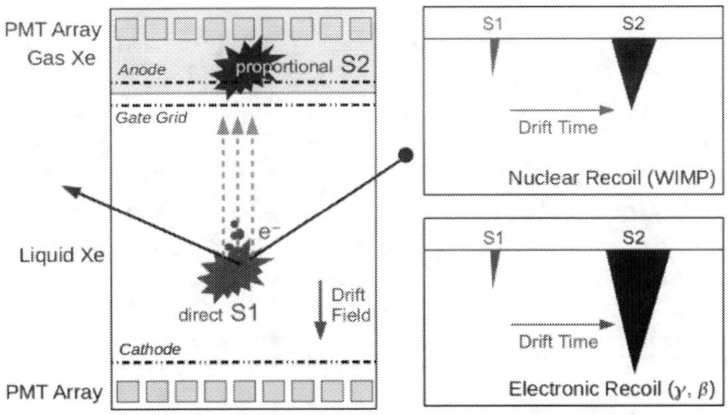

Figure 1. Left: working principle of the XENON liquid-gas dual-phase TPC. Right: sketch of the waveforms of nuclear recoils (WIMPs or neutrons) and electronic recoils (γ or β background), showing the different ratio of the charge (S2) and light (S1) signals for the two types of events [2].

2 Latest results from XENON100

Latest results from XENON100 data analysis are presented in this section. All these analysis were performed on the 225 live-days data taking period with which the collaboration was able to set stringent limits on the spin-independent (SI) and spin-dependent (SD) WIMP-nucleon cross section [3] [4] [5], as well as for axion-induced electronic recoils [6].

2.1 Exclusion of leptophilic dark matter models

We tested the possibility that DAMA/LIBRA annual modulation [7] in the 2-6 keV energy range can be due to dark matter scattering on electrons, instead of scattering on nuclei [8]. In particular we contrast the DAMA/LIBRA signal, interpreted as WIMPs coupling to electrons through axial-vector interactions [9] [10], with XENON100 data. For this analysis, the events belonging to the electronic recoil band in XENON100 and occurring in the 34 kg inner fiducial volume have been considered. Moreover, the study is performed using events from 70 summer live-days only, where the modulation effect is stronger. Figure 2 shows the comparison among the converted DAMA/LIBRA and the measured XENON100 energy spectra. A simple comparison of the integral counts in the energy interval from 2 to 14 PE, which corresponds to 2.0 to 5.9 keV electronic recoil energy, excludes the DAMA/LIBRA signal as axial-vector coupling between WIMPs and electrons at 4.4σ significance level, even considering all events from XENON100 background as signal candidates. XENON100 is also able to exclude DAMA/LIBRA signal as due to other two models. Luminous dark matter [11] is excluded at 4.6σ and mirror dark matter [12] is excluded at 3.6 σ.

DOI: 10.1051/epjconf/201716407049

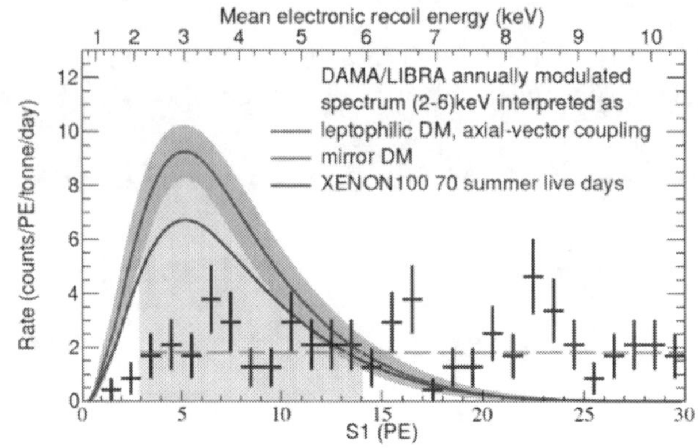

Figure 2. The DAMA/LIBRA modulated spectrum (red), interpreted as WIMPs scattering through axial-vector interactions, as it would be seen in the XENON100 detector. The 1σ band includes statistical and systematic uncertainties. The DAMA/LIBRA modulated spectrum interpreted as luminous dark matter is very similar, whereas the interpretation as mirror dark matter is indicated separately (dark red). Blue points are XENON100 data from the 70 summer live days with their statistical uncertainty. The expected average XENON100 rate is also shown (dashed cyan). The shaded region from (3 to 14) PE was used to quantify the confidence level of exclusion [8].

2.2 Search for event rate modulation

XENON100 is the first dual-phase TPC operated stably for long-enough period to allow performing a study of annual modulation [13]. The study is performed again on the events belonging to the electronic recoil band in XENON100 and occurring in the 34 kg inner fiducial volume. The time modulation is tested at first with no constraints on the period for the single scatter events in the 2.0-5.8 keV energy range. A rise in significance for a modulation at long periods is observed, but it is still below 3σ local and 1σ global (figure 3 left). This rise in significance is observed both in the single scatter and in the multiple scatters sample, in contrast with what expected if the modulation was due to dark matter scattering. If the period is fixed to 1 year, there is a 2.8 σ local significance. The phase of this modulation disfavors the dark matter interpretation at 2.5σ, and the DAMA/LIBRA signal is disfavored at 4.8 σ, see figure 3 right.

2.3 Search for low mass dark matter

A dedicated analysis has been performed to search for low mass dark matter [14]. In order to lower the energy threshold, this analysis has been performed by using only the ionization signal S2. This choice prevents on one hand to discriminate the ER from the NR (through the S2/S1 ratio), on the other hand to get the vertical position of one event (using the time difference between S2 and S1). A NR equivalent energy scale has been reconstructed, with a charge yield based on the model in [15], with a cutoff below 0.7 keV. Without a background model, we interprete all events surviving quality cuts as potential dark matter candidates. At a WIMP mass of 6 GeV/c^2, SI WIMP-nucleon interaction cross sections of $1.4 \cdot 10^{-41}$cm^2 are excluded, figure 4.

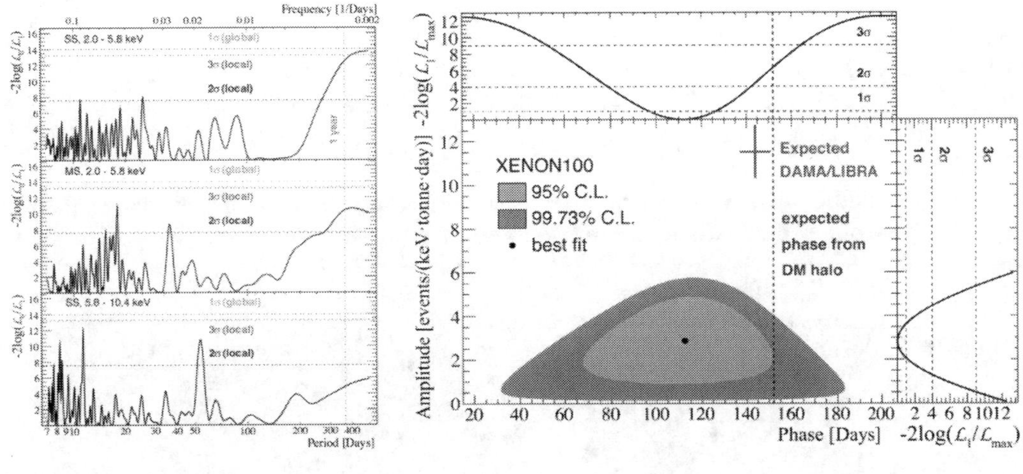

Figure 3. Left: The significance as a function of modulation period for single scatters (SS) in the low-E region (top panel), multiple scatters (MS) in the low-E region (middle panel) and single scatters (SS) in the higher energy region (bottom panel). The phase is unconstrained. Right: The XENON100 best-fit and 95% and 99.73% confidence level contours as a function of amplitude and phase relative to January 1, 2011, for period P = 1 year. The expected DAMA/LIBRA signal with statistical uncertainties only and the phase expected from a standard dark matter halo are overlaid for comparison. Top and side panels show significance as a function of phase and amplitude, respectively [13].

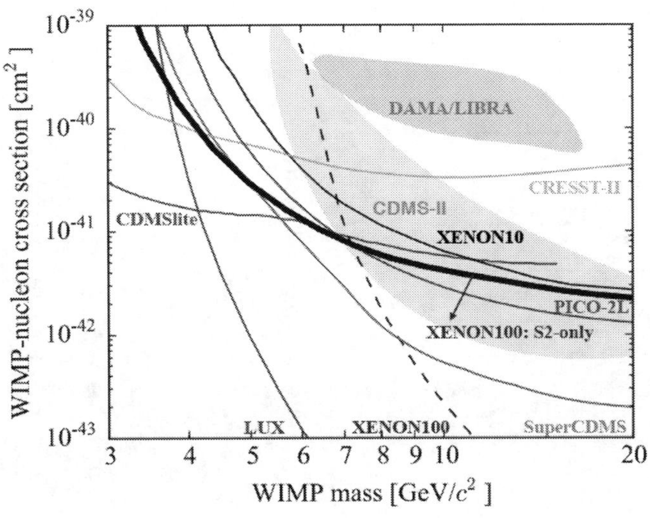

Figure 4. WIMP exclusion limit on the SI WIMP-nucleon scattering cross section at 90% C.L. The limit from the search for low mass dark matter is shown with the thick blue line [14].

DOI: 10.1051/epjconf/201716407049

3 XENON1T detector and expected sensitivity

XENON1T is the current phase of the XENON program. With a total Xe mass of 3.2 tons, XENON1T is expected to reach a sensitivity about a factor 100 higher than XENON100 in 2 ton·year exposure.

The TPC of XENON1T is 96 cm height and 96 cm diameter, filled with ~2 tons of liquid xenon (figure 5, middle). A total of 248 PMTs are distributed at the top and bottom of the TPC. The detector is hosted in a 10 m height and 10 m diameter water tank acting as a passive shield against natural radioactivity. Additionally, the water tank is operating as an active muon veto, being instrumented with 84 PMTs to detect both passing cosmic muons and muon-induced cascades by Cherenkov effect [16].

Figure 5. Left: a global view of the XENON1T infrastructure at LNGS. Middle: the TPC of XENON1T. Right: the cryostat hanging in the center of the water tank.

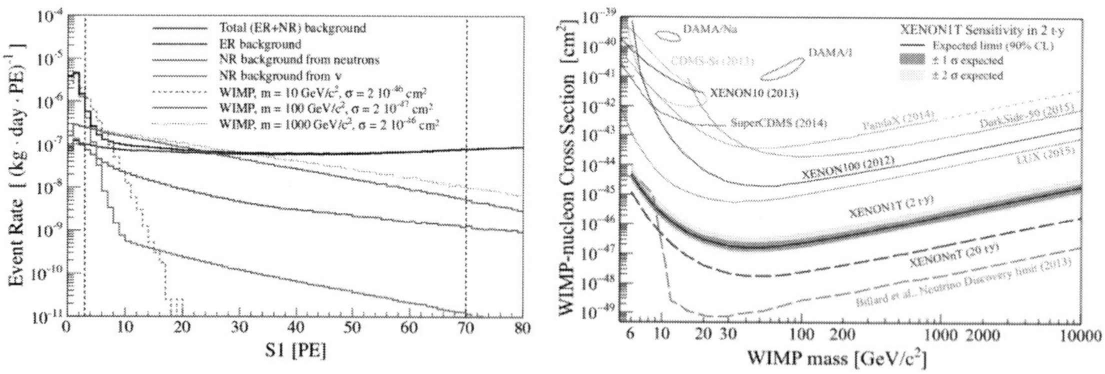

Figure 6. Left: Simulated S1 spectra of background compared with the expectations from WIMP of different masses. These are obtained considering single scatter in 1 t fiducial volume, energy range [1-12] keV_{ee} [4-50] keV_{nr}, 99.75% ER discrimination, 40% NR acceptance. Right: XENON1T sensitivity to SI WIMP-nucleon interaction with a 2 ton·year exposure [17].

A dedicated Geant4-based XENON1T Monte Carlo simulation has been developed, including radioactivities of all detector materials, and energy deposition into S1 and S2 [17]. Figure 6 shows the expected S1 spectra of backgrounds and WIMP signals as a function of S1 (left), and the expected sensitivity to SI WIMP-nucleon interaction (right). The expected XENON1T 90% C.L. SI WIMP-nucleon cross section upper limit reaches better than 2×10^{-47} cm^2 at WIMP mass of 50 GeV/c^2 with 2 ton·year exposure.

4 Summary

The XENON project aims at dark matter direct detection with liquid xenon dual-phase TPCs. Latest analysis of XENON100 data excluded various models of leptophilic dark matter. A search for low mass WIMPs was also performed, looking at ionization signals (S2) in order to lower the energy threshold to 0.7 keV$_{nr}$. The XENON1T detector is fully installed and operating at LNGS. It is expected to reach a sensitivity a factor 100 better than XENON100 with 2 ton·year exposure.

References

[1] J. Angle et al. (XENON Collaboration), Phys. Rev. Lett. **100** 021303 (2008)
[2] E. Aprile et al. (XENON Collaboration), Astropart. Phys. **35**, 573-590 (2012)
[3] E. Aprile et al. (XENON Collaboration), Submitted to PRD. arXiv:1609.06154
[4] E. Aprile et al. (XENON Collaboration), Phys. Rev. Lett. **109**, 181301 (2012)
[5] E. Aprile et al. (XENON Collaboration), Phys. Rev. Lett. **111**, 021301 (2013)
[6] E. Aprile et al. (XENON Collaboration), Phys. Rev. D Part. Fields Gravit. Cosmol. **90**, 062009 (2014)
[7] R. Bernabei et al., Eur. Phys. J. C **73**, 2648 (2013)
[8] E. Aprile et al. (XENON Collaboration), Science 2015 vol. **349** no. 6250 pp. 851-854
[9] J. Kopp, V. Niro, T. Schwetz, J. Zupan, Phys. Rev. D Part. Fields Gravit. Cosmol. **80**, 083502 (2009)
[10] N. F. Bell, Y. Cai, R. K. Leane, A. D. Medina, Phys. Rev. D Part. Fields Gravit. Cosmol. **90**, 035027 (2014)
[11] B. Feldstein, P. W. Graham, S. Rajendran, Phys. Rev. D Part. Fields Gravit. Cosmol. **82**, 075019 (2010)
[12] R. Foot, Int. J. Mod. Phys. A **29**, 1430013 (2014)
[13] E. Aprile et al. (XENON Collaboration), Phys. Rev. Lett. **115**, 091302 (2015)
[14] E. Aprile et al. (XENON Collaboration), Phys. Rev. D **94**, 092001 (2016)
[15] F. Bezrukov, F. Kahlhoefer, M. Lindner, Astropart. Phys. **35**, 119-127 (2011)
[16] E. Aprile et al. (XENON Collaboration), JINST **9**, P11006 (2014)
[17] E. Aprile et al. (XENON Collaboration), JCAP **04** (2016) 027

EPJ Web of Conferences **164**, 07050 (2017)
ICNFP 2016
DOI: 10.1051/epjconf/201716407050

Threshold $e^- p \to n\nu_e$ scattering and the electron neutrino mass

Jacek Ciborowski[1,a] and Jakub Rembieliński[2]

[1] *University of Warsaw, Faculty of Physics, L. Pasteura 5, 02-093 Warsaw, Poland*
[2] *University of Łódź, Faculty of Physics and Applied Informatics, Pomorska 149/153, 90-236 Łódź, Poland*

Abstract. The most precise measurement of the electron neutrino mass has been obtained from the shape of the electron energy spectrum near the endpoint in tritium decay. The Mainz and Troitsk experiments indicated an excess instead of expected depletion of counts in that region. Results derived from such measurements are subject to numerous atomic corrections which are absent in the scattering $e^- p \to n\nu_e$. This new idea is presented in the article, with its advantages and difficulties, and is compared to the method of tritium decay.

1 Introduction

In this article a new idea, regarding the issue of measuring the electron neutrino mass, is presented and its main advantages as well as difficulties discussed in short. Since it is practically unachievable to measure the velocity of the neutrino by a time-of-flight method, and given the energy – its mass, the only avialable methodology must rely on studying a process in which a finite neutrino mass would lead to a modification of kinematics in a particular region of the phase space. To date, the most fruitful method, which allowed to determine an upper limit on the electron neutrino mass with eV-precision, was the study of the endpoint region of the electron kinetic energy spectrum in a β-decay of tritium. In this decay, ${}^3H \to {}^3 He\ e^-\ \bar{\nu}_e$, the maximum electron kinetic energy is rather low (18.6 keV); moreover, the decay is characterised by an acceptable half-life time of 12.3 years. A finite neutrino mass would manifest as a depletion of counts, compared to the case of the massless neutrino, in the electron kinetic energy distribution near the endpoint. The latest results for the limit on the electron neutrino mass have been derived over a decade ago from data collected in two eV-precision experiments: at Troitsk (Russia) [1–4] (m_ν < 2.05 eV at 95% c.l.) and Mainz (Germany) [5, 6], [7] ($m_\mu^2 = -0.6 \pm 2.2(\text{stat}) \pm 2.1(\text{syst})\ \text{eV}^2$ or $m_\nu < 2.3$ eV at 95% c.l.). A new generation experiment, KATRIN, is currently at the commissioning stage, aiming at a sub-eV precision constraint or measurement of the electron neutrino mass [8]. A number of other experiments, for the time being less precise, are also currently in progress, including bolometric experiments MIBETA [9] ($m_\mu^2 = -141 \pm 211(\text{stat}) \pm 90(\text{syst})\ \text{eV}^2$), MANU [10] and MARE experiment [11] to measure the endpoint in the β decay of ^{187}Re and the ECHo experiment [12] making use of the electron capture in ^{163}Ho. The accuracy of determining the neutrino mass in a β decay is limited by the precision of measuring the electron energy. A new related method has recently been demonstrated using a radio-frequency spectrometer for single electrons with the operation principle based on the relativistic shift in the cyclotron frequency [13], opening a new

[a] e-mail: cib@fuw.edu.pl

perspective for the electron neutrino mass determination. It is a common feature of the results for m_ν^2 that they come out systematically negative in most measurements. This feature is a sign either of an effect commonly unaccounted for in the data analysis or, if physical, a need of a new concept of the neutrino [14]. Therefore a measurement using a yet different technique might be of importance in this respect.

2 Threshold $e^- p \to n\nu_e$ scattering

We have considered another process, in which the fact of a finite neutrino mass would be reflected in its kinematics: $e^- p \to n\nu_e$ (a charged current process, CC, compared to a neutral current elastic $e^- p$ scattering). Again, a depletion of counts, compared to the massless neutrino case, in the energy dependence of the cross section for this channel close above the threshold is expected. The above elementary process has been chosen chosen by us for simplicity but it is understood that, in general, one can consider a process: $e^- A \to B\nu_e$, where A and B denote nuclei (although it has been known from the tritium decay experience that dealing with nuclei involves numerous nuclear and atomic corrections).

The differential cross section for the process $e^- p \to n\nu_e$, as a function of the electron-proton c.m.s. energy, W, takes the following form:

$$\frac{d\sigma}{d\Omega} = \frac{|\mathcal{M}|^2}{256\pi^2} \frac{1}{W^2} \sqrt{\frac{(W^2 - m_n^2 + m_\nu^2)^2 - 4m_\nu^2 W^2}{(W^2 - m_p^2 - m_e^2)^2 - 4m_p^2 m_e^2}}, \tag{1}$$

where $\mathcal{M}$ denotes the matrix element and m_n, m_p, m_e and m_ν – the neutron, proton, electron and neutrino masses, respectively. The leading kinematical dependence of the cross section is contained in the W^2-dependent phase space term since the energy dependence of the matrix element is not strong near the threshold. Therefore we focus on the kinematical (W-dependent) term in (1) in further considerations which is sufficient for illustrating of the underlying idea presented in this paper. The dependence of the cross section in terms of the c.m.s. energy, W, near the threshold for the process $e^- p \to n\nu_e$ is shown in Fig. 1 for the massive and massless neutrino. In this particular calculation

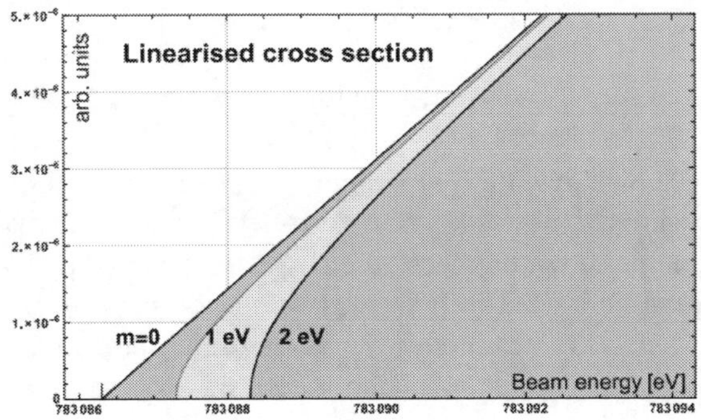

Figure 1. Energy dependence of the cross section for the process $ep \to n\nu$ near the threshold for a massive (m_ν = 1, 2 eV) and massless electron neutrino.

we used the following values for the masses of the electron, the proton and the neutron [15]: m_e = 0.510998928 MeV, m_p = 938.272046) MeV and m_n = 939.565378 MeV, respectively. The threshold value of the centre-of-mass energy for the massive and massless neutrino reads: $W_{th}^m = m_n + m_\nu$ and

DOI: 10.1051/epjconf/201716407050

$W^0_{th} = m_n$, respectively. If the neutrino were massive, the threshold energy would be shifted towards higher values by m_ν and one would observe a depletion of counts compared to the massless case, as shown in the figure. The measurement would have to involve a beam energy scan, covering a region from below to above the threshold and the possible neutrino mass could be determined by fitting the functional dependence (1).

There are several difficulties associated with the method presented in this article.

- A low energy dispersion beam (order of eV), within an appropriate energy range below and above the energy threshold, is required since the accuracy of measuring the neutrino mass depends strongly on this parameter; the threshold kinetic energy of an electron beam on a stationary proton target amounts to approx. 0.783 MeV. The dispersion should be of order 10^{-6} of the beam energy which would be a challenging task to achieve in the case of the discussed process;
- A dominating elastic background is present, $e^-p \rightarrow e^-p$, requiring an efficient procedure of rejection; $\sigma_{CC}/\sigma_{el} \approx O(10^{-12})$, depending on acceptance, where σ_{CC} denotes the cross section for the studied process;
- Low CC cross section near threshold, $\sigma_{CC} \approx G_F^2 s \approx O(10^{-4})$ nb, and in consequence low event rates;
- Detection of a low energy neutron in the final state via a secondary process, like ^2D formation with an accompanying γ (2.22 MeV).

A comparison of the two methods: tritium decay and the threshold scattering, is summarised in table 1.

Table 1. Comparison of key features characterising the two methods, tritium decay and threshold e^-p scattering, of measuring the electron neutrino mass.

Process	tritium decay	$e^-p \rightarrow n\nu_e$
Region of study	below endpoint	above threshold
Effect searched in count rates	depletion	depletion
Method	final state e^- energy measurement	• e^- beam energy setting (scan) • final state neutron detection
Precision related device	high precision e^- spectrometer	low dispersion e^- beam line
Device dimension, scale	3D, very large	1D, moderate
Background processes	none	elastic e^-p scattering
Experience status	decades of practice	intention stage

In summary, we have shown that the energy dependence of the cross section for the process $e^-p \rightarrow n\nu_e$ above threshold is sensitive to a possibly non-zero electron neutrino mass – an effect quantitatively similar to that near the endpoint of the electron energy spectrum in tritium decay. The new method requires further elaboration and a feasibility study while the main difficulties, which can be listed a priori, are: order of eV electron beam energy setting in the threshold range, abundant elastic e^-p scattering background, low counting rates of the measured process and necessity of detecting slow neutrons.

This work was partially supported from the funds of the National Science Centre, contract DEC-2012/06/M/ST2/00430.

References

[1] A. Belesev *et al.*, JETP Lett. **97**, 67 (2013)

DOI: 10.1051/epjconf/201716407050

[2] A. Belesev *et al.*, J. Phys. **G41**, 015001 (2014)

[3] V. M. Lobashev *et al.*, Phys. Lett. **B460**, 227 (1999)

[4] V. N. Aseev *et al.*, Phys. Rev. **D84**, 112003 (2011)

[5] Ch. Kraus *et al.*, Eur. Phys. J. **C73**, 2323 (2013)

[6] Ch. Kraus *et al.*, Eur. Phys. J. **C40**, 447 (2005)

[7] E. W. Otten and Ch. Weinheimer, Rep. Prog. Phys. **71** 086201 (2008)

[8] Ch. Weinheimer for the KATRIN collaboration, Progress in Particle and Nuclear Physics, **48**, 141 (2002); http://www.katrin.kit.edu/ and references therein

[9] M. Sisti *et al.*, NIM A **520** (2004) 125

[10] F. Gatti, Nucl. Phys. **B91** 293 (2001)

[11] S. Sangiorgio *et al.* Prog. Part. Nuc. Phys. **57**, 68-70 (2006); A. Nucciotti *et al.*, Nuclear Physics B (Proc. Suppl.) **229–232** 155–159 (2012)

[12] L. Gastaldo *et al.*, J. Low Temp. Phys. **176**, 876 (2014); S. Eliseev *et al.*, Phys. Rev. Lett. **115**, 062501 (2015)

[13] D. M. Asner *et al.*, Phys. Rev. Lett. **114**, 162501 (2015)

[14] J. Ciborowski and J. Rembieliński, Eur. Phys. J. **C8** 157 (1999)

[15] Review of Particle Physics, K. Olive *et al.*, Particle Data Group, Chin. Phys. **C38** 0900001 (2014); pdg.lbl.gov

 DOI: 10.1051/epjconf/201716407051

Top Physics at CMS/LHC

Georgios Daskalakis[1,a], On behalf of the CMS Collaboration

[1] *Institute of Nuclear and Particle Physics (INPP), NCSR Demokritos, Aghia Paraskevi, Greece*

Abstract. Recent results on the inclusive and differential production cross sections of top-quark pair and single top-quark processes are presented, obtained using data from proton-proton collisions collected with the CMS detector at the LHC. The large centre-of-mass energies available at LHC allow for the copious production of top-quark pairs in association with other final state particles at high transverse momentum. Measurements of such processes as well as of the top-quark mass and other properties will be discussed. The results are compared with the most up-to-date standard model theory predictions.

1 Introduction

Today, more than 20 years after the discovery of the top quark, particle physicists still investigate its properties by performing precision measurements. The top quark is the heaviest known elementary particle with a very short lifetime (τ_t=1/Γ_t ~3.3×10^{-25}s) [1]. It decays semi-weakly into a real W boson and a b quark before hadronization ($\tau_{\text{hadronization}}$ ~10^{-24}s) can occur. Due to its short lifetime, there is no sufficient time to depolarize its spin ($\tau_{\text{spin decorrelation}}$ ~10^{-21}s) [2], so its spin information is transferred on to its decay products. Therefore, it is possible to measure observables that depend on the top quark spin, providing a unique environment for tests of the Standard Model (SM) and for searches for physics beyond the Standard Model (BSM). In addition, it is the only quark whose Yukawa coupling to the Higgs boson is of the order of unity so it is often speculated that it might play a special role in the mechanism of electroweak symmetry breaking. Its phenomenology provides a unique laboratory where our understanding of the strong interactions can be tested mainly through the top quark pair production. At the same time, weak interactions can be tested through the top quark decay and single top quark production. Top quarks are a major source of background for almost all searches for BSM physics at the LHC. Precise understanding of the top quark cross sections and kinematics is crucial before claiming new physics.

The final states for the leading pair-production process can be divided into three classes depending on the decay mode of the W bosons. The $t\bar{t}$ events in which both, one or none W bosons decay hadronically are referred to as the all-jets, lepton+jets (l+jets) or dilepton channels, respectively. Their relative contributions, including hadronic corrections, are 45.7%, 43.8% and 10.5%. Most of the analyses focus on final states with muons and electrons since the τ leptons are more difficult to reconstruct. In addition to the quarks resulting from the top-quark decays, extra radiation from quarks and gluons in the event can lead to extra jets in the final state.

[a] e-mail: Georgios.Daskalakis@cern.ch

2 Inclusive Top-quark pair and Single-top production cross sections

In pp collisions, top quarks are produced dominantly in pairs through the processes gg → $t\bar{t}$ and $q\bar{q}$ → $t\bar{t}$, at leading order (LO) in quantum chromodynamics (QCD). At $\sqrt{s}$=13 TeV, approximately 85.7% of the production cross section is from gluon-gluon fusion, with the remainder from $q\bar{q}$ annihilation.

The most precise CMS measurements of the $t\bar{t}$ production cross section at $\sqrt{s}$=13 TeV comes from the dilepton channel [3]. The measurement used data that correspond to an integrated luminosity of 2.2 fb^{-1} analyzing events with one electron and one muon (eμ channel) and at least two jets, one of which is required to be identified as a b-jet. The measured cross section is $\sigma_{t\bar{t}}$ = 793±8(stat)±38(syst)±21(lumi) pb, corresponding to a total relative uncertainty of 5.6%. The corresponding SM prediction for the $t\bar{t}$ production cross section is $\sigma_{t\bar{t}}^{\mathrm{NNLO+NNLL}}$ = 832±29(scale)±35(PDF+α_s) pb, as calculated with the TOP++ program [4] at next-to-next-to-leading-order (NNLO) in perturbative QCD, including soft-gluon resummation at next-to-next-to-leading-log (NNLL) order [5], assuming m_t = 172.5 GeV. The first uncertainty is an estimate of the effect of missing higher-order corrections and reflects uncertainties in the factorization and renormalization scales, while the second one is associated with possible choices of PDFs and the strong coupling constant (α_s). The measured cross section is therefore in agreement with the expectations from the SM. Further analyses in the l+jets channel and the all-jets channel, yield consistent results.

In Fig. 1 (left) the summary of the cross section measurements of top pair production in CMS is presented in comparison with the theory calculation at NNLO with NNLL soft gluon resummation. The Tevatron measurements are also shown.

The associate production of $t\bar{t}$ along with an electroweak gauge boson (W or Z) have been also studied. The cross sections of these processes are typically much smaller than the $t\bar{t}$, yet can provide key information on the properties of the top quark. In these final states, according to the SM, the W boson is expected to be produced via initial state radiation, while the Z-boson can also be radiated from a final-state top-quark and hence provides sensitivity to the top-quark neutral current weak gauge coupling. Both $\sigma_{t\bar{t}W}$ and $\sigma_{t\bar{t}Z}$ would be altered in a variety of new physics models that can be parameterized by dimension-six operators added to the SM Lagrangian. CMS measured the $t\bar{t}W$ and $t\bar{t}Z$ production cross sections using pp collision data at $\sqrt{s}$=8 TeV corresponding to an integrated luminosity of 19.5 fb^{-1} [6]. Signatures from different decay modes of the top quark pair resulting in final states with two, three, and four leptons have been analyzed. The measured cross section in same-sign dilepton and three-lepton final states is 382^{+117}_{-102} fb, with an observed (expected) significance of 4.8 (3.5) standard deviations from the background-only hypothesis. Combining opposite-sign dilepton, three-lepton, and four-lepton channels, the $t\bar{t}Z$ cross section is measured to be 242^{+65}_{-55} fb, an observation with a significance of 6.4 standard deviations from the background-only hypothesis, and in agreement with the SM expectation. The measured cross sections are used to constrain the vector and axial couplings of the Z boson to the top quark and to set bounds on the Wilson coefficients of five dimension-six operators that would affect the $t\bar{t}W$ and $t\bar{t}Z$ cross sections by parameterizing new physics. The measurements are compatible with the SM predictions. Recently, the cross section of $t\bar{t}$ production in association with a Z boson was measured using pp collisions at $\sqrt{s}$ = 13 TeV [7]. The data sample used corresponds to an integrated luminosity of 2.7 fb^{-1}. The measurement is performed in three- and four-lepton final states where the jet and b-jet multiplicities were exploited to enhance the signal over background ratio. The $t\bar{t}Z$ production cross section was measured to be 1065^{+352}_{-313}(stat. $)^{+168}_{-142}$(sys.) fb with a significance of 3.6 standard deviations from the background-only hypothesis.

The electroweak single top-quark production mechanisms, namely from $q\bar{q}'$ → $t\bar{b}$, qb → q′t, mediated by virtual s-channel and t-channel W bosons, and tW-associated production, through bg → tW^-, lead to smaller cross sections with respect to $t\bar{t}$ production. The t-channel, while suppressed by the weak coupling with respect to the strong pair production, results in a sizable cross section at the

DOI: 10.1051/epjconf/201716407051

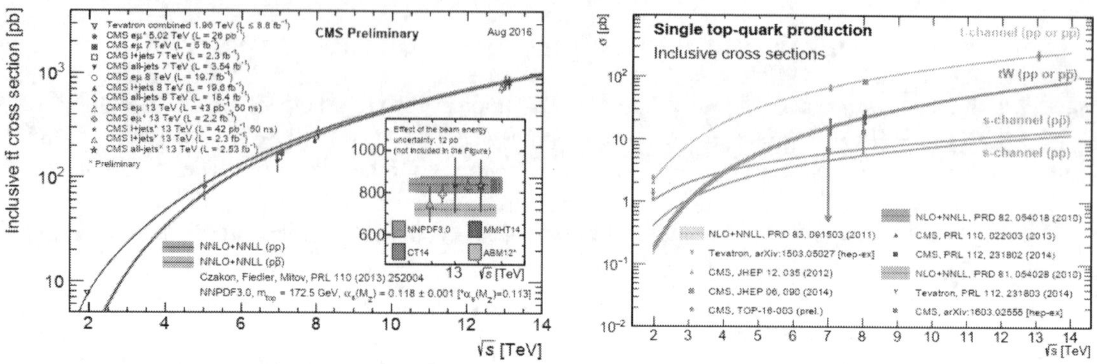

Figure 1. Left: Summary of the top quark pair cross section measurements from CMS, versus the centre-of-mass energy, in comparison with the theory calculation at NNLO+NNLL accuracy. **Right**: Summary of the single top cross section measurements from CMS, versus the centre-of-mass energy, in comparison with the theory calculation at NLO+NNLL accuracy. The Tevatron measurements are also shown.

LHC energies. The t- and s-channel cross sections of top and anti-top are not identical, due to the charge-asymmetric initial state. Measurements of the s- and t-channel production cross sections as well as the tW mechanism though the t-channel allow an extraction of the CKM matrix element V_{tb}.

A recent measurement of the t-channel single top quark production cross section was performed by CMS at $\sqrt{s}$=13 TeV [8], analyzing data that correspond to an integrated luminosity of 2.3 fb^{-1}. The muon+jets signature was explored and several kinematic variables were combined into one multivariate discriminator optimized to separate signal from background events. A fit to the distribution of the discriminating variable yields cross sections of :

$$\sigma_{t-ch.,t} = 141.5\pm6.7(\text{stat.})\pm9.4(\text{exp.})^{+19.3}_{-19.6}(\text{theo.})\pm3.8(\text{lumi.})\ \text{pb and}$$
$$\sigma_{t-ch.,\bar{t}} = 81.0\pm6.2(\text{stat.})\pm8.1(\text{exp.})^{+10.9}_{-10.9}(\text{theo.})\pm2.2(\text{lumi.})\ \text{pb}$$

for the production of single top quarks and single top anti-quarks, respectively. The cross section ratio between top quark and anti-quark production is measured to be: $R_{t-ch.}$ = 1.75±0.16(stat.)±0.21 (syst.). A comparison between the measured ratio and the prediction of different PDF sets is shown on the left plot of Fig. 2. In case of MSTW2008NLO [9] and of NNPDF [10] the fixed four-flavour scheme (4FS) PDFs are used together with the POWHEG 4FS calculation. The POWHEG calculation in the five-flavour scheme (5FS) is used for all other PDFs, as they are variable flavor scheme PDFs.

Assuming $|V_{tb}|$ much greater than $|V_{td}|$, $|V_{ts}|$ the cross sections for single top production are proportional to $|V_{tb}|^2$. From the measured inclusive cross section, $\sigma_{t-ch.}$ = $227.8\pm9.1(\text{stat.})\pm14.0(\text{exp.})^{+28.7}_{-27.7}(\text{theo.})\pm6.2(\text{lumi.})$ pb and its predicted value, the CKM matrix element $|V_{tb}|$ is calculated to be $|V_{tb}|$ = 1.02±0.07(exp.)±0.02(theo.). All results are in agreement with the SM predictions.

CMS performed a search for single top quark production in the s-channel in decay modes of the top quark containing a muon or an electron in the final state [11]. The signal is extracted through a maximum-likelihood fit to the distribution of a multivariate discriminant defined to separate the expected signal contribution from background processes. The analysis uses data corresponding to integrated luminosities of 5.1 fb^{-1} at 7 TeV and 19.7 fb^{-1} at 8 TeV, respectively. The measured cross sections of 7.1±8.1 pb at 7 TeV and 13.4±7.3 pb at 8 TeV result in a best fit value of 2.0±0.9 for the combined ratio of the measured and expected values. The signal significance is 2.5 standard

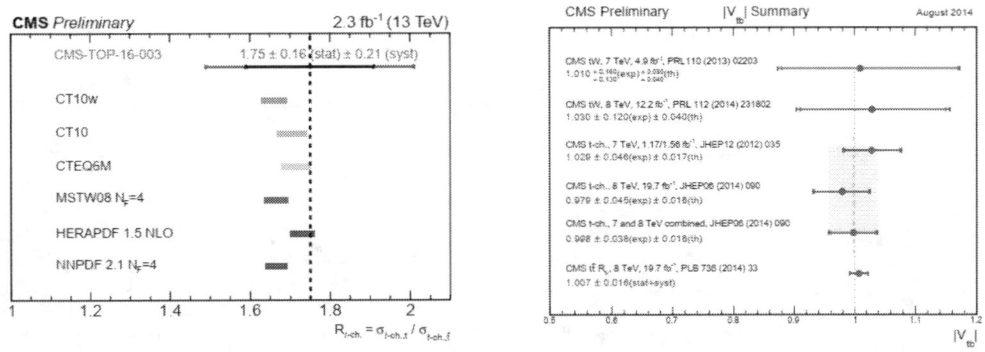

Figure 2. **Left**: Comparison of the measured $R_{t-ch.} = \sigma_{t-ch.,t}/\sigma_{t-ch.,\bar{t}}$ with the prediction obtained using different PDF sets. The nominal value for the top quark mass is 173.0 GeV. Error bars for the different PDF sets include the statistical uncertainty, the uncertainty on the factorization and renormalization scales, and the uncertainty on the top quark mass. **Right**: Summary of $|V_{tb}|$ measurements in CMS.

deviations, and the upper limit on the rate relative to the SM expectation is 4.7 at 95% confidence level.

The tW process is of interest because it probes the tWb vertex in a different kinematic region than s- and t-channel production. The definition of tW production in perturbative QCD at NLO mixes with top-quark pair production so special treatment is necessary in order to disentangle the two processes [12, 13]. At measurement level the selection cuts are defined such that the interference effects are small. The associated production of a single top quark and a W boson was measured in CMS by using a data set of 12.2 fb^{-1} of pp collisions at $\sqrt{s}$ = 8 TeV [14]. Events with two leptons and a jet originating from a b quark are selected. A multivariate analysis based on kinematic and topological properties is used to separate the signal from the dominant $t\bar{t}$ background. An excess consistent with the signal hypothesis is observed, with a significance which corresponds to 6.1 standard deviations above a background-only hypothesis. The measured production cross section is 23.4±5.4 pb, in agreement with the SM prediction.

In Fig. 1 (right) a summary of single top cross-section measurements at CMS are presented as a function of the center-of-mass energy in comparison with the theory calculation at NLO+NNLL accuracy. The Tevatron measurements are also shown. All cross-section measurements are very well described by the theory calculations within their uncertainty. In Fig. 2 (right) a summary of V_{tb} determinations by CMS is presented.

3 Differential cross sections

Thanks to the large available luminosity, CMS performed differential cross-section measurements in $t\bar{t}$ production. Such measurements are crucial, as they allow even more stringent tests of perturbative QCD as description of the production mechanism and enhance the sensitivity to possible new physics contributions, especially now that NNLO predictions for the main differential observables in $t\bar{t}$ production have become available [15]. Furthermore, such measurements reduce the uncertainty in the description of $t\bar{t}$ final state as background in Higgs physics, searches for rare processes, or BSM physics.

Differential cross-sections are typically measured by applying a kinematic reconstruction on the selected candidate events and subsequent unfolding the obtained event counts in bins of kinematic

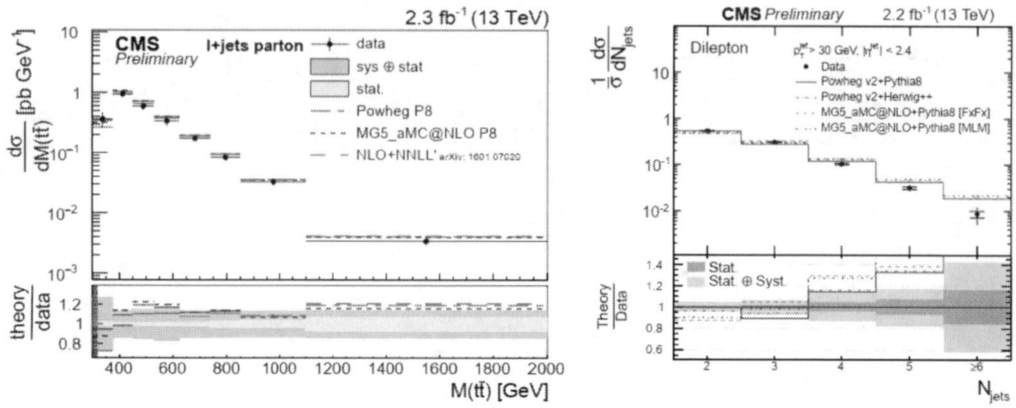

Figure 3. Left: Differential cross sections at parton level as a function of $M_{t\bar{t}}$ compared to the SM predictions. The measurements are performed in the lepton+jets decay channel with an electron or a muon in the final state. **Right**: Normalized differential $t\bar{t}$ production cross section as a function of the jet multiplicity. The inner (outer) error bars indicate the statistical (combined statistical and systematic) uncertainty. The measurements have been performed in dilepton decay channels.

distributions in order to correct for detector resolution effects, acceptance and migration effects. By using pp collisions at 13 TeV with 2.3 fb^{-1} of integrated luminosity CMS measured the differential $t\bar{t}$ cross section as a function of various kinematic variables of the $t\bar{t}$ system as well as jet multiplicities. The measurements are performed in the lepton+jets decay channel with an electron or a muon in the final state [16]. The differential cross sections are presented at particle level, within a phase space close to the experimental acceptance, and at parton level in the full phase space. The measurement at parton level is dominated by uncertainties in the parton shower and hadronization model. The dependency on these theoretical models is reduced for the particle level measurement, for which the experimental uncertainties of jet energy calibration and b tagging efficiency are the most dominant. The results are compared to several SM predictions of approximate NNNLO [17, 18] and NLO+NNLL′ [19] calculations as shown in the left plot of Fig. 3. All calculations are compatible with the measured results. Similar measurements have been performed in dilepton decay channels at 13 TeV [20] using data that correspond to an integrated luminosity of 2.2 fb^{-1}. The results are compared with several perturbative QCD calculations as shown in the right plot of Fig. 3 and are found to be broadly in agreement with the predictions.

Differential cross-section measurements were also performed by CMS in single-top t-channel production. Such measurements are extremely useful as they test our understanding of both QCD and EW top-quark interactions. Using pp collisions at a centre-of-mass energy of 13 TeV and analyzing data that corresponded to an integrated luminosity of 2.3 fb^{-1} the differential cross sections were measured as functions of the transverse momentum and the absolute value of the rapidity of the top quark [21]. A maximum-likelihood fit to a multivariate discriminator is used to infer the signal and background fractions from the data. Unfolding to parton level is performed in this case. The measured cross sections are compared with theoretical predictions to next-to-leading order matched with parton showering as implemented in Monte Carlo generators. General agreement is found within uncertainties as shown in Fig. 4.

DOI: 10.1051/epjconf/201716407051

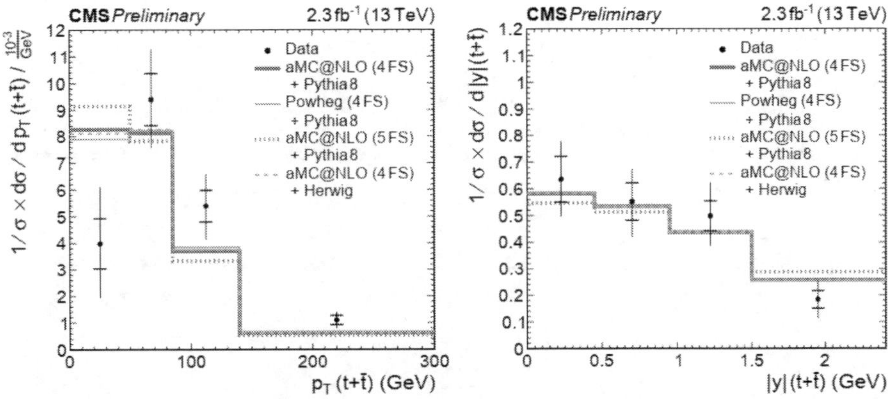

Figure 4. Left: Measured differential cross section of t-channel single top quark production as a function of the top quark transverse momentum. **Right**: Measured differential cross section of t-channel single top quark production as a function of the top quark rapidity. Horizontal ticks on the error bars indicate the statistical uncertainty and vertical bars indicate the total uncertainty per bin.

4 Top-quark mass measurements

The mass of the top quark (m_t) is one of the fundamental parameters of the SM. With the discovery of a Higgs boson at the LHC with a mass of about 126 GeV [22, 23], the precision measurement of the top-quark mass takes a central role in the question of the stability of the electroweak vacuum. Recent NNLO calculations [24, 25] together with the central values of the Higgs and top-quark masses put the electroweak vacuum in the metastable region. The uncertainty is dominated by the precision of the top-quark mass measurement and its interpretation as the pole mass.

CMS has measured the top quark mass using pp data at $\sqrt{s}$ = 8 TeV corresponding to a luminosity of 19.7 fb^{-1} [26]. In the measurement all decay channels were used. The measurements use analysis techniques in which either m_t alone is determined or m_t and the overall jet energy scale factor are determined simultaneously. When the results from the $\sqrt{s}$ = 8 TeV analysis are combined with the published CMS results at $\sqrt{s}$ = 7 TeV, a top quark mass measurement of 172.44±0.13(stat.)±0.47(syst.) GeV is achieved. Given the precision of the experimental results, a detailed understanding of the relationship between the measurements and the value of m_t in different theoretical schemes is needed. Current indications are that the present measurements based on the kinematic reconstruction of the top quark mass correspond approximately to the pole ("on-shell") mass to within a precision of about 1 GeV [27]. A summary of the CMS m_t measurements and their combination is shown in the left plot of Fig. 5. The top-quark mass can be extracted by using alternative techniques or observables that might have sensitivity on the m_t like the normalized invariant mass distribution of the $t\bar{t}$+jet system, the peak position of the energy distribution of b jets produced from top-quark decays and several others. A long list of alternative measurements is shown in the right hand-side plot of Fig. 5 together with the references to the corresponding documentation.

5 Top polarization, $t\bar{t}$ spin correlations and charge asymmetry

One of the unique features of the top quark is that it decays before its spin can be flipped by the strong interaction. Thus the top-quark polarization is directly observable via the angular distribution of its

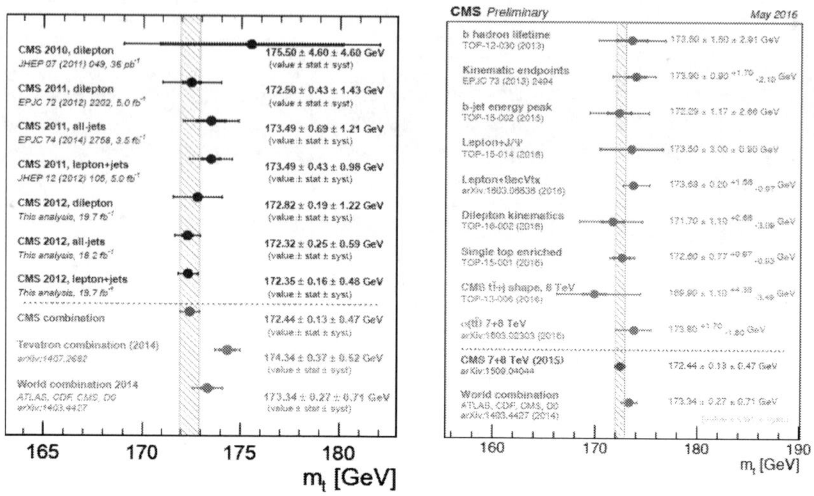

Figure 5. **Left**: Summary of the CMS m_t measurements and their combination. The thick error bars show the statistical uncertainty and the thin error bars show the total uncertainty. Also shown are the current Tevatron [28] and world average [29] combinations. **Right**: Summary of Run-I CMS alternative m_t measurements.

decay products. In the SM, top quarks from pair production have only a small net polarization arising from electroweak corrections to the QCD-dominated production process, but the pairs have significant spin correlations [30]. Hence, it is possible to define and measure observables sensitive to the top-quark spin and its production mechanism. For low $t\bar{t}$ invariant masses, the production is dominated by the fusion of pairs of gluons with the same helicities, resulting in the creation of top quark pairs with antiparallel spins in the $t\bar{t}$ center-of-mass frame. For larger $t\bar{t}$ invariant masses, the dominant production is via the fusion of gluons with opposite helicities, resulting in $t\bar{t}$ pairs with parallel spins. For BSM models, couplings of the top quark to new particles can alter both the top quark polarization and the strength of the spin correlations in the $t\bar{t}$ system [30–33].

CMS has measured the top quark-antiquark spin correlations and the top quark polarization for $t\bar{t}$ pairs produced in pp collisions at $\sqrt{s} = 8$ TeV [34]. The data correspond to an integrated luminosity of 19.5 fb^{-1}. The measurements are performed using events with two oppositely charged leptons (electrons or muons) and two or more jets, where at least one of the jets is identified as originating from a bottom quark. The spin correlations and polarization are measured from the angular distributions of the two selected leptons, both inclusively and differentially, with respect to the invariant mass, rapidity, and transverse momentum of the $t\bar{t}$ system. The measurements are unfolded to the parton level and found to be in agreement with predictions of the SM as can be seen in Fig. 6.

The $t\bar{t}$ charge asymmetry is sensitive to BSM contributions that might affect the $t\bar{t}$ production process. The dominant $t\bar{t}$ production mechanism is the charge-symmetric gluon-gluon fusion. For the sub-dominant qq production mechanism, the symmetric pp collision does not define a forward and backward direction. Instead, the larger average momentum fraction of the valence quarks leads to an excess of top quarks produced in the forward and backward directions, while the top antiquarks are produced more centrally. This makes the difference in the absolute values of the rapidities of the top quark and anti-quark, $\Delta|y| = |y_t| - |y_{\bar{t}}|$, a suitable observable to measure the $t\bar{t}$ charge asymmetry at the LHC. The charge asymmetry, A_C, is defined as $A_C = (N^+ - N^-)/(N^+ + N^-)$, where N^+ and N^- represent the number of events with positive and negative values of $\Delta|y|$, respectively. To shed light

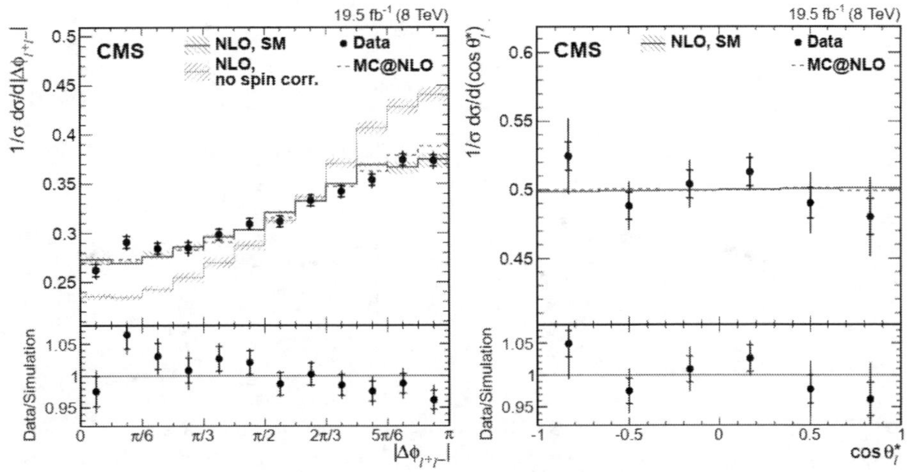

Figure 6. Left: Normalized differential cross section as a function of $|\Delta\phi_{l^+,l^-}|$. **Right**: Normalized differential cross section as a function of $\cos\theta_l^*$. The plots are shown data (points); parton-level predictions from MC@NLO [35, 36] (dashed histograms); and theoretical predictions at NLO [30, 37] with (SM) and without (no spin corr.) spin correlations (solid and dotted histograms, respectively). For the $\cos\theta_l^*$ distribution, CP conservation is assumed in the combination of the $\cos\theta_l^*$ measurements from positively and negatively charged leptons. The ratio of the data to the MC@NLO prediction is shown in the lower panels. The inner and outer vertical bars on the data points represent the statistical and total uncertainties, respectively.

on the possible existence and the nature of new physics contributions, it is crucial to measure not only the inclusive asymmetry but also A_C as a function of variables magnifying the $t\bar{t}$ charge asymmetry.

CMS has measured the $t\bar{t}$ charge asymmetry at 8 TeV with an integrated luminosity of 19.7 fb^{-1} [38]. Events are selected if they contain one electron or muon and four or more jets, at least one of which is identified as originating from the hadronization of a bottom quark. The measurement is based on the reconstructed four-momenta of the t and $\bar{t}$ quarks in each event. The observed distributions are then corrected for background contributions, resolution, and selection efficiency. Acceptance corrections are applied, correcting either to the fiducial phase space or to the full phase space. The inclusive charge asymmetry is found to be 0.0010±0.0068(stat)±0.0037(syst). Measurements were also performed in a reduced fiducial phase space of top quark pair production, with an integrated result of -0.0035±0.0072(stat)±0.0031(syst). All measurements are consistent within two standard deviations with zero asymmetry as well as with the predictions of the SM.

A summary of the charge asymmetry measurements on ATLAS and CMS at 8 TeV is shown on the left plot of figure 7. The plot shows both the inclusive measurements and the measurement using boosted events, compared to the respective theory predictions. The right hand-side plot of figure 7 shows the corrected asymmetry as a function of $m_{t\bar{t}}$. The results correspond to the full phase space. The measured values are compared to NLO predictions for the SM based on calculations by Kühn and Rodrigo (K&R) [39] and Bernreuther and Si (B&S) [37, 40], as well as to the predictions of a model featuring an effective axial-vector coupling of the gluon (EAG) [41, 42]. The inner bars indicate the statistical uncertainties, while the outer bars represent the statistical and systematic uncertainties added in quadrature. The differential measurements show a good agreement with the SM predictions. For the benchmark model involving an effective axial-vector coupling of the gluon, the measurement at high $m_{t\bar{t}}$ excludes new physics scales below 1.5 TeV at the 95% confidence level.

DOI: 10.1051/epjconf/201716407051

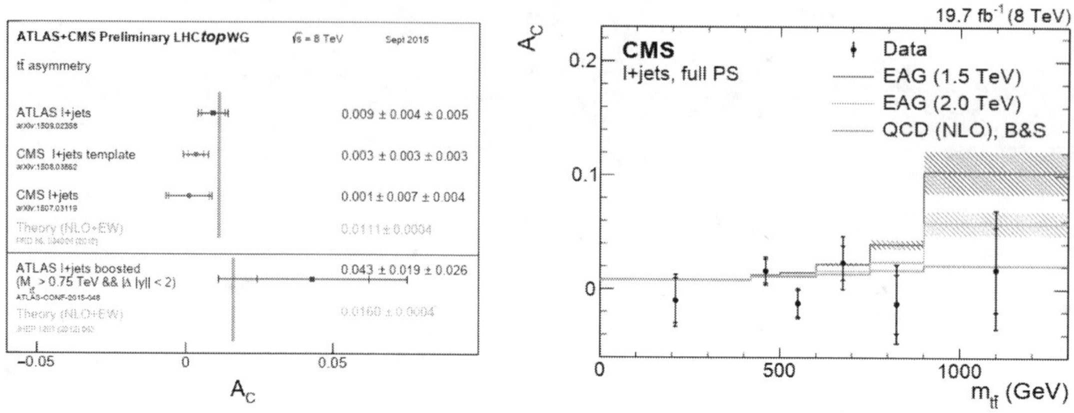

Figure 7. Left: Summary of the charge asymmetry measurements on ATLAS and CMS at 8 TeV showing both the inclusive measurements and the measurement using boosted events, compared to the respective theory predictions. The uncertainty on the theory predictions is shown but is very small. **Right**: Corrected asymmetry as a function of $m_{t\bar{t}}$. The results correspond to the full phase space. The measured values are compared to NLO predictions for the SM, as well as to the predictions of a model featuring an effective axial-vector coupling of the gluon (EAG). The inner bars indicate the statistical uncertainties, while the outer bars represent the statistical and systematic uncertainties added in quadrature.

6 Summary

Top-quark physics at hadron colliders has developed into precision physics. Various properties of the top quark have been measured with high precision. Up to now, all measurements are consistent with the SM predictions and allow stringent tests of the underlying production mechanisms by strong and weak interactions. Given the large event samples available at the CMS experiment, top-quark properties has been determined both in $t\bar{t}$ as well as in electroweak single top-quark production. Furthermore, the $t\bar{t}$ associated production with heavy bosons provides further information on the top-quark electroweak couplings. At the same time various models of physics beyond the SM involving top-quark production are being constrained. As the first results from LHC Run-II at a higher center-of-mass energy and much higher luminosity starting to be released, top-quark physics has the potential to shed light on open questions and new aspects of physics at the TeV scale.

References

[1] D0 Collaboration, Phys. Rev. D **85**, 091104 (2012)
[2] G. Mahlon and S. J. Park, Phys. Rev. D **81**, 074024 (2010)
[3] CMS Collaboration, CMS-PAS-TOP-16-005, CERN, Geneva (2016)
[4] M. Czakon and A. Mitov,Computer Physics Communications, **185**, 2930 (2014)
[5] M. Czakon, P. Fiedler and A. Mitov, Phys. Lett. B **110**, 252004 (2013)
[6] CMS Collaboration, J. High Energy Phys. **01** (2016) 096
[7] CMS Collaboration, CMS-PAS-TOP-16-009, CERN, Geneva (2016)
[8] CMS Collaboration, CMS-PAS-TOP-16-003, CERN, Geneva (2016)
[9] A. Martin, W. Stirling, and G. Watt, Eur. Phys. J. C **63**, 189 (2009)

[10] R. Ball, V. Bertone, F. Cerutti et al., Nucl. Phys. B **855**, (2012) 608
[11] CMS Collaboration, J. High Energy Phys. **09** (2016) 027
[12] S. Frixione, E. Laenen, P. Motylinski et al., J. High Energy Phys. **07**, (2008) 029
[13] T. M. P. Tait, Phys. Rev. D **61**, (1999) 034001
[14] CMS Collaboration, Phys. Rev. Lett. **112**, 231802 (2014)
[15] M. Czakon, D. Heymes, and A. Mitov, Phys. Rev. Lett. **116**, 082003 (2016)
[16] CMS Collaboration, CMS-PAS-TOP-16-008, CERN, Geneva (2016)
[17] N. Kidonakis, Phys. Rev. D **90**, 014006 (2014)
[18] N. Kidonakis, Phys. Rev. D **91**, 031501 (2015)
[19] B. Pecjak, D. Scott, X. Wang, and L. L. Yang, Phys. Rev. Lett. **116**, 202001 (2016)
[20] CMS Collaboration, CMS-PAS-TOP-16-011, CERN, Geneva (2016)
[21] CMS Collaboration, CMS-PAS-TOP-16-004, CERN, Geneva (2016)
[22] ATLAS Collaboration, Phys. Lett. B **716**, 1 (2012)
[23] CMS Collaboration, Phys. Lett. B **716**, 30 (2012)
[24] G. Degrassi, et al., J. High Energy Phys. **08**, 98 (2012)
[25] A. V. Bednyakov, et al., Phys. Rev. Lett. **115**, 201802 (2015)
[26] CMS Collaboration, Phys. Rev. D **93**, 072004 (2016)
[27] A. Buckley, et al., Phys. Rep. **504**, 145 (2011)
[28] The Tevatron EWK Working Group, CDF and D0 Collaborations, arXiv:1407.2682
[29] ATLAS, CDF, CMS and D0 Collaborations, arXiv:1403.4427
[30] W. Bernreuther and Z.-G. Si, Phys. Lett. B **725**, 115 (2013)
[31] D. Krohn, T. Liu, J. Shelton, and L.-T. Wang, Phys. Rev. D **84**, 074034 (2011)
[32] S. Fajfer, J. F. Kamenik, and B. Melic, J. High Energy Phys. **08** (2012) 114
[33] J. A. Aguilar-Saavedra and M. Perez-Victoria, J. Phys. Conf. Ser. **447**, 012015 (2013)
[34] CMS Collaboration, Phys. Rev. D **93**, 052007 (2016)
[35] S. Frixione and B. R. Webber, J. High Energy Phys. **06**, 029 (2002)
[36] S. Frixione, P. Nason, and B. R. Webber, J. High Energy Phys. **08**, 007 (2003)
[37] W. Bernreuther and Z.-G. Si, Phys. Lett. B **837**, 90 (2010)
[38] CMS Collaboration, Phys. Lett. B **757**, 154 (2016)
[39] J.H. Kühn, G. Rodrigo, J. High Energy Phys. **01**, 063 (2012)
[40] W. Bernreuther, Z.-G. Si, Phys. Rev. D **86**, 034026 (2012)
[41] E. Gabrielli, M. Raidal, A. Racioppi, Phys.Rev. D **85**, 074021 (2012)
[42] E. Gabrielli, M. Raidal, Phys. Rev. D **84**, 054017 (2011)

EPJ Web of Conferences **164**, 07052 (2017) DOI: 10.1051/epjconf/201716407052
ICNFP 2016

Top quark production measurements using the ATLAS detector at the LHC

Naim Bora Atlay on behalf of the ATLAS Collaboration[1,a]

[1] *University of Siegen*
Walter-Flex Str. 3, 57072, Siegen, Germany

Abstract. A selection from the inclusive top-quark pair production and single top-quark production cross-section measurements in proton-proton collisions with the ATLAS detector at the Large Hadron Collider is presented, with data collected in 2012 and early 2015 at centre-of-mass energies of 8 TeV and 13 TeV, respectively.

1 Introduction

The top-quark, discovered over twenty years ago [1, 2], is the heaviest known fundamental particle and close to the scale of electroweak symmetry breaking. Having a lifetime at the order of 10^{-25} s, the top-quark decays before it hadronises which enables the decay products to carry its "bare quark" properties. The study of its production and decay properties forms a core part of the LHC physics programme. Due to the high amount of top-quarks the LHC can produce, the ATLAS experiment [3] is advantageous in comparison to former experiments. The LHC, being a proton-proton collider, the top-quark production is dominated by gluon-gluon fusion process.

2 Top-quark pair production

At the LHC, top quarks can be produced in either quark-antiquark pairs ($t\bar{t}$) or as single quarks, where the former is the primary production. In this section, three different measurements of the top-quark pair production are presented.

2.1 Measurement of the $t\bar{t}$ production cross-section using $e\mu$ events with b-tagged jets

This measurement [4] uses a data sample of 3.2 $\text{fb}^{-1}$of proton-proton collisions at a centre-of-mass energy of $\sqrt{s}$= 13 TeV and is performed using dileptonic events with an opposite-charge electron-muon pair in the final state. The algorithm used to tag jets containing b-quarks exploits track impact parameters and reconstructed secondary vertices. The $t\bar{t}$ cross-section is extracted by counting the $e\mu$ events with exactly one and exactly two b-tagged jets. The cross-section is extracted simultaneously with the combined probability to reconstruct and tag a b-jet from the top-quark decay, which minimises the associated systematic uncertainties. The comparison between data and simulations for

[a]e-mail: naim.bora.atlay@cern.ch

 DOI: 10.1051/epjconf/201716407052

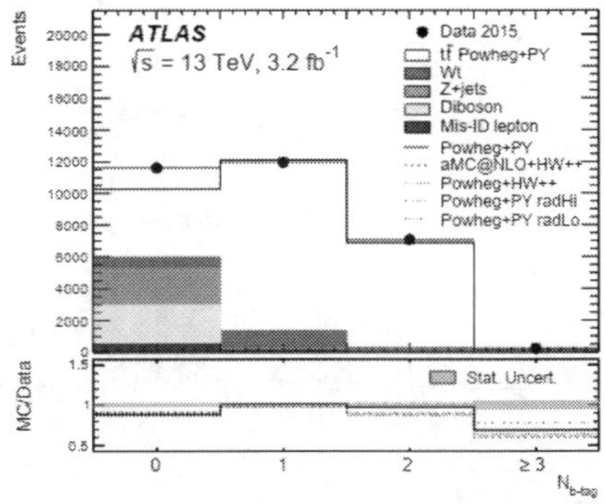

Figure 1: Distribution of the number of b-tagged jets. The data is compared to the expectation from signal and background simulations. The lower part of the figure shows the ratio of simulation to data, using various tt signal samples, and the cyan band indicates the statistical uncertainty [4].

signal and backgrounds in terms of the number of b-tagged jets is shown in figure 1. The extracted cross-section [4] is

$$\sigma_{t\bar{t}} = 818 \pm 8(\text{stat.}) \pm 27(\text{syst.}) \pm 19(\text{lumi}) \pm 12(\text{beam})\text{pb}, \quad (1)$$

where the theoretical prediction to be compared with is $\sigma_{t\bar{t}} = 832^{40}_{46}$ pb [4].

2.2 Measurements of top-quark pair differential cross-sections in the lepton+jets channel

Differential cross-section measurements of top-quark pair production with respect to different kinematic variables are important to study the characteristics of $t\bar{t}$ production. In addition, effects beyond the Standard Model can be observed as deviations of differential distributions with respect to the Standard Model predictions.

With this motivation, a set of measurements [5] of the $t\bar{t}$ production cross-section as a function of different properties of the reconstructed top quark and of the $t\bar{t}$ system has been performed. The results are unfolded both to a fiducial particle-level phase space and to the full phase space. Unfolding to a fiducial particle-level phase space and using variables directly related to detector observables allows precision tests of QCD by avoiding large model-dependent extrapolation corrections to the parton-level top-quark and to a phase space region outside the detector sensitivity.

The measurement is done using a data sample of 20.3 fb^{-1}collected at a centre-of-mass energy of $\sqrt{s} = 8$ TeV. The baseline observables to perform the differential measurements are the transverse momentum ($p_T^{t,had}$), the absolute value of the rapidity($|y^{t,had}|$) of the hadronically decaying top-quark, and the transverse momentum ($p_T^{t\bar{t}}$), the absolute value of the rapidity $|y^{t\bar{t}}|$ and the invariant mass $m^{t\bar{t}}$ of the $t\bar{t}$ system. The observables have been chosen to emphasise the $t\bar{t}$ production process. The observables are also sensitive to effects of initial and final-state radiation, to the different parton distribution functions, and to non-resonant processes and higher-order corrections. The measurement uses events in the lepton+jets channel, requiring exactly one charged lepton and at least four jets with at least two jets tagged as originating from a b-quark. Distributions of two of the baseline

DOI: 10.1051/epjconf/201716407052

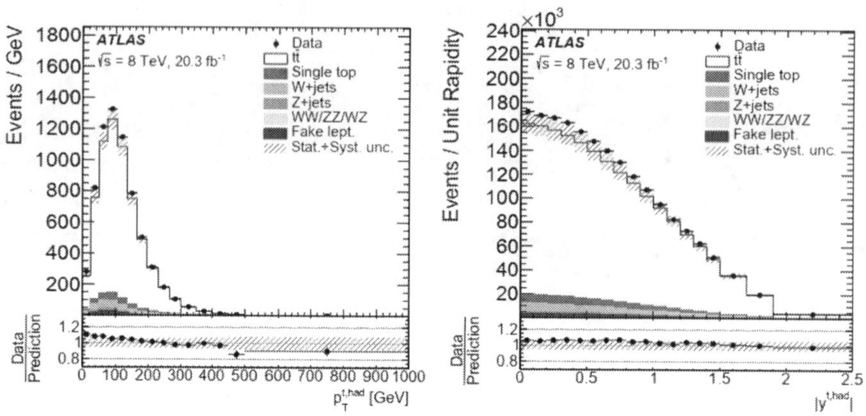

Figure 2: Distributions of observables of the combined electron and muon selections at detector level: hadronic top-quark transverse momentum, p_T^t, (left) and absolute value of the rapidity, $|y^{t,had}|$, (right). Data distributions are compared to predictions, using Powheg+PYTHIA as the $t\bar{t}$ signal model. The lower part of the figure shows the ratio data to simulation. The hashed area indicates the combined statistical and systematic uncertainties on the total prediction, excluding systematic uncertainties related to the modelling of the $t\bar{t}$ system [5].

observables for the combined electron and muon selections at detector level are shown in figure 2. A fair agreement between the measurement and the predictions over a wide kinematic range is to be found. However, most generators predict a harder top-quark transverse momentum distribution at high values than what is observed in the data. The agreement of the data with the prediction for this observable improves, when the NNLO corrections are taken into account. Distributions for the full phase-space normalised differential cross-sections as a function of the transverse momentum, $p_T^{t\bar{t}}$ and for absolute value of the rapidity, $|y^{t,had}|$ are shown in figure 3.

2.3 Measurement of the top-quark production cross-sections in association with a massive vector-boson in multi-lepton final states

The rate of top-quark pair production in association with a massive vector boson(W or Z boson), $t\bar{t}Z$ and $t\bar{t}W$, could be altered in the presence of physics beyond the Standard Model, such as vector-like quarks, strongly coupled Higgs bosons or technicolor. The cross-section measurements [6] of $t\bar{t}Z$ and $t\bar{t}W$ play therefore an important role for the validity check of the Standard Model at this new energy regime. In addition, a measurement of $\sigma_{t\bar{t}Z}$ allows to extract information on the neutral current coupling of the top-quark. The measurements of $\sigma_{t\bar{t}Z}$ and $\sigma_{t\bar{t}W}$ are performed using 3.2 $\text{fb}^{-1}$of proton-proton collision data at $\sqrt{s}$= 13 TeV. Depending on the decay modes of the top-quarks, W, and Z bosons, the final state can include between zero and four isolated, prompt leptons. The measurement considers final states with two same-sign charged muons, or three and four leptons. The analysis channels and the targeted decay modes of the $t\bar{t}Z$ and $t\bar{t}W$ processes are presented in table 1.

Same-sign dimuon analysis

This region targets the $t\bar{t}W$ process and since electrons have a much larger charge misidentification probability, inducing a significant background from top-quark pairs, this region has the highest sensitivity among all same sign dilepton regions [7].

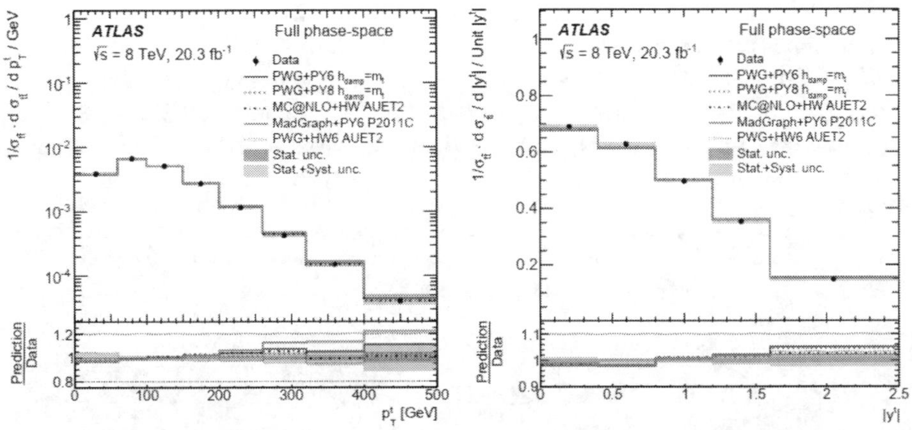

Figure 3: Full phase-space normalised differential cross-sections as a function of the transverse momentum, p_T^t, (left) and of the absolute value of the rapidity, $|y^{t,had}|$, (right). The lower part of the figure shows the ratio data to simulation. The grey bands indicate the total uncertainty on the data in each bin. The Powheg+PYTHIA generator with $h_{damp} = m_t$, which effectively regulates the high-p_T radiation, and the CT10nlo PDF is used as the nominal prediction to correct for detector effects [5].

Table 1: List of $t\bar{t}W$ and $t\bar{t}Z$ decay modes and analysis channels targeting them [6].

Process	$t\bar{t}$decay	Boson decay	Channel
$t\bar{t}W^{\pm}$	$(\mu^{\pm}\nu b)\ (q\bar{q}b)$	$\mu^{\pm}\nu$	SS dimuon
	$(l^{\pm}\nu b)\ (l^{\mp}\nu b)$	$l^{\pm}\nu$	Trilepton
$t\bar{t}Z$	$(\mu^{\pm}\nu b)\ (q\bar{q}b)$	$l^{+}l-$	Trilepton
	$(l^{\pm}\nu b)\ (l^{\mp}\nu b)$	$l^{+}l^{-}$	Tetralepton

Trilepton analysis

This region is divided into four signal regions with exactly three leptons. Three of these, out of which each requires an opposite-sign same flavour (OSSF) pair of leptons, whose mass is within 10 GeV of the Z boson mass, are sensitive to $t\bar{t}Z$. The signal regions are categorised by their jet and b-jet multiplicities. The 3l-Z-1b4j region requires at least four jets, exactly one of which is b-tagged. The 3l-Z-2b3j region requires exactly three jets with at least two b-tagged jets. The 3l-Z-2b4j region requires at least four jets, from which at least two are b-tagged and the 3l-noZ-2b region requires at least two and at most four jets, of which at least two jets are b-tagged. No OSSF lepton pair is allowed in the Z mass window. The sum of the lepton charges must be one. This region primarily targets the $t\bar{t}W$ process but has $t\bar{t}Z$ contribution comparable with that of the 3l-Z-2b3j region. In addition to the signal regions, a control region is used to constrain the normalisation of the WZ background in data with exactly three leptons: 3l-WZ-CR.

Tetralepton analysis

The target process of this region is $t\bar{t}Z$, where both W bosons resulting from top-quark decays and the Z boson decay leptonically. Existence of two pairs of opposite-sign leptons is required, of which at least one pair having the same flavour. The lepton pair with opposite-sign same flavour and

DOI: 10.1051/epjconf/201716407052

reconstructed invariant mass closest to the Z boson mass is attributed to the Z boson decay. According to the relative flavour of the other remaining two leptons, different flavour (DF) or same flavour (SF), and the number of b-tagged jets: one, or at least two (1b, 2b), four signal regions are defined: 4*l*-DF-1b, 4*l*-DF-2b, 4*l*-SF-1b and 4*l*-SF-2b. As in the trilepton analysis, a control region used to constrain the ZZ normalisation is defined to have exactly four reconstructed leptons: 4*l*-ZZ-CR.

The cross-sections for the two processes are determined by performing a simultaneous fit to the eight signal and two control regions defined above. The post-fit distributions for the expected yields

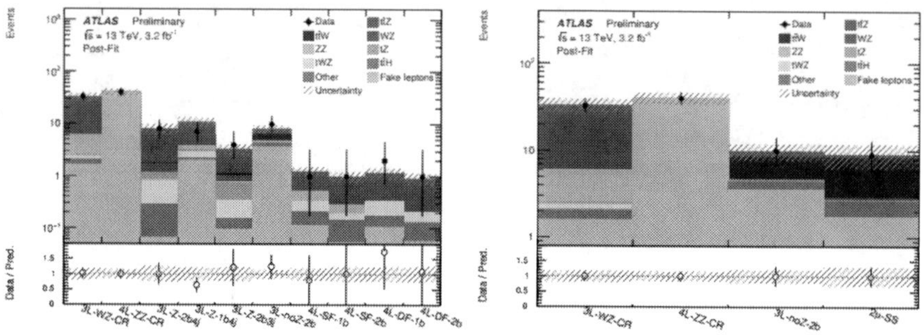

Figure 4: Expected yields after the fits. Data compared to signal and MC simulations for the $t\bar{t}Z$ (left) and $t\bar{t}W$(right) fit in the relevant signal regions and the two control regions used to constrain the WZ and ZZ backgrounds. The lower part of the figure shows the ratio data to simulation. The hatched area corresponds to the total uncertainty on the predicted yields [6].

are presented in figure 4. The determined cross-sections [6] are:

$$\sigma_{t\bar{t}Z} = 0.9 \pm 0.3 \text{ pb and } \sigma_{t\bar{t}W} = 1.4 \pm 0.8 \text{ pb.} \tag{2}$$

The theoretical predictions [6] to be compared with are:

$$\sigma^{theo}_{t\bar{t}Z} = 0.76 \pm 0.08 \text{ pb and } \sigma^{theo}_{t\bar{t}W} = 0.57 \pm 0.06 \text{ pb.} \tag{3}$$

3 Single top-quark production

The production measurement of the single top-quark is important to understand the properties of the *Wtb* vertex. In the leading-order perturbation theory, the single production of the top-quark is described by three subprocesses: the *t*-channel, the *Wt*-channel and the *s*-channel. The subprocesses are distinguished by the virtuality of the exchanged W boson. This section presents an inclusive cross-section measurement in the *t*-channel at $\sqrt{s}$= 13 TeV and in the *Wt* channel at $\sqrt{s}$= 8 TeV as well as the first evidence of the single top-quark production in the *s*-channel at 8 TeV.

3.1 Measurement of the inclusive cross-section of single top-quark *t*-channel production at $\sqrt{s}$ = 13 TeV

The single top-quark production in the *t*-channel is the dominant process of the channels mentioned above. It occurs by exchange of a virtual W boson between a light quark from one of the colliding protons and a *b*-quark from the other colliding proton. The cross-section of the single top-quark production induced by an up-quark is expected to be about twice the cross-section of the single-antiquark

 DOI: 10.1051/epjconf/201716407052

production induced by the down-quark since the density of the up-quark is twice the density of down-quark.

The measurement [8] is performed to determine the *t*-channel single top-quark production cross-section in the lepton+jets channel using 3.2 $\text{fb}^{-1}$of pp collision data at a centre-of-mass energy of 13 TeV. For the measurement, events with an isolated muon and two jets with high transverse momenta are selected where one jet is required to pass the threshold of dedicated b-tagging algorithms. The discrimination between the signal and the background is achieved using a neural network (NN). The most discriminating variables are found to be the reconstructed top-quark mass and the invariant mass of the two jets. Distributions of the two discriminating variables are shown in

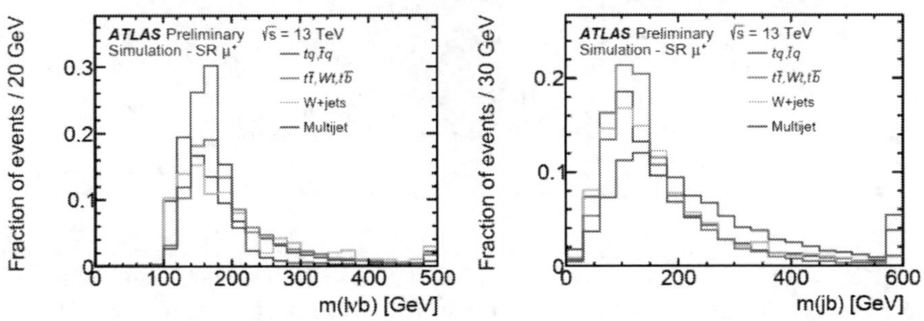

Figure 5: Distributions of the two most discriminating variables, the reconstructed top-quark mass (left) and the invariant mass of the jet pair (right) [8].

figure 5. The cross-section is measured by performing a binned maximum-likelihood (ML) fit to the NN output distribution. The cross-sections [8] determined are:

$$\sigma_{tq} = 133 \pm 6(\text{stat.}) \pm 24(\text{syst.}) \pm 7(\text{lumi})\ \text{pb}, \quad (4)$$

$$\sigma_{\bar{t}q} = 96 \pm 5(\text{stat.}) \pm 23(\text{syst.}) \pm 5(\text{lumi})\ \text{pb}. \quad (5)$$

3.2 Measurement of the production cross-section of a single top-quark in association with a W boson at $\sqrt{s}$ = 8 TeV

While the other two channels of the single top-quark production have been observed at the Tevatron, the single top-quark production in association with a W boson has been observed for the first time at the LHC. Production of single top-quark in association with a W boson occurs via *b*-quark-induced partonic channels. This measurement [9] is performed with a data of 20.3 fb^{-1}collected at 8 TeV using dilepton events. The analysis requires two opposite-sign high p_T leptons, missing transverse energy, one high p_T jet which is required contain a *b*-hadron (*b*-tagged). The separation of the signal from the background is performed through the use of boosted decision tree (BDT) algorithm. Each of the BDT combines a number of discriminating variables into one classifier. The training of the BDTs is done separately in three regions, 1jet-1tag, 2jets-1tag and 2jets-2tag using Wt (signal), and $t\bar{t}$ (main background) MC samples. The cross-section is then extracted by performing a profile likelihood fit to the classifier output distributions. Distributions of the boosted decision tree response for the three regions mentioned above are shown in figure 6. The measured cross-section [9] is:

$$\sigma_{Wt} = 23 \pm 1.3(\text{stat.})^{+3.2}_{-3.5}(\text{syst.}) \pm 1.1(\text{lumi})\ \text{pb}, \quad (6)$$

DOI: 10.1051/epjconf/201716407052

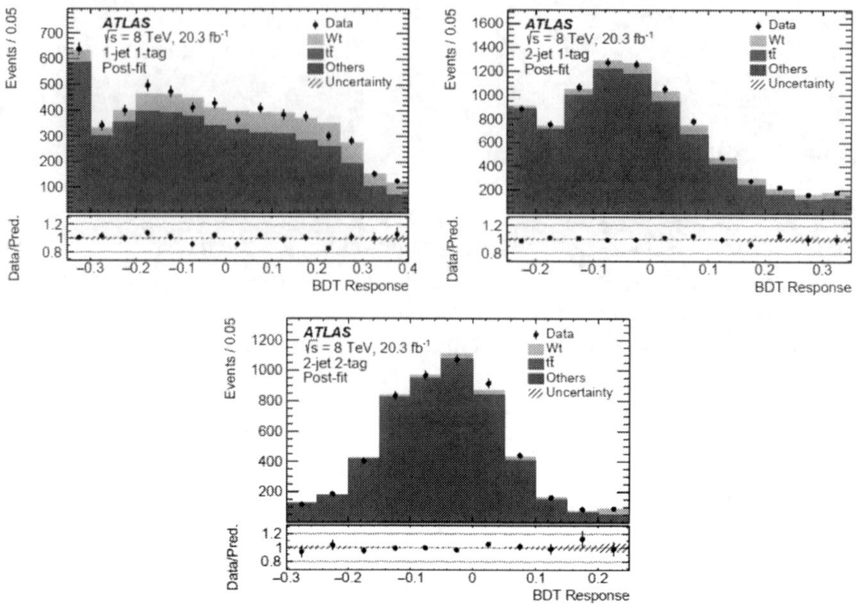

Figure 6: Boosted decision tree response for 1-jet 1-tag(upper left), 2-jet 1-tag (upper right) and 2-jet 2-tag (lower) events. Each contribution is normalised to its expectation. The lower part of the figure shows the ratio data to simulation. The hatched area represents the sum in quadrature of the statistical and systematic uncertainties [9].

where the theoretical prediction for the Wt production at NLO order with NNLL soft gluon corrections is $\sigma_{Wt}^{theo} = 22.37 \pm 1.52(\text{stat.})$ pb [10].

3.3 Evidence for single top-quark production in the s-channel in proton-proton collisions at $\sqrt{s}$ = 8 TeV using the Matrix Element Method

The single top-quark production via s-channel is sensitive to new particles proposed by models of physics beyond the Standard Model such as charged Higgs boson or W' boson.

The measurement [11] is performed using 20.3 fb^{-1} of data collected at the centre-of-mass energy of 8 TeV. Events with one isolated electron or muon, large missing transverse momentum and exactly two b-tagged jets are used. In addition to the signal region, two control regions are defined for the validation of $t\bar{t}$ and $W + jets$ backgrounds. Extraction of the signal is achieved by a maximum-likelihood fit to a discriminant. The discriminant is based on the matrix element method. The distribution of the matrix element discriminant for the $t\bar{t}$ validation region is shown in figure 7. The result of the maximum-likelihood [11] fit is:

$$\sigma_{Wt} = 4.8 \pm 0.8(\text{stat.})^{+1.6}_{-1.3}(\text{syst.})\ \text{pb}. \tag{7}$$

The result is in agreement with the Standard Model prediction. The observed significance of the result is 3.2 standard deviations, where the expected significance is 3.9 standard deviations. The post-fit distribution of the ME discriminant in the signal region is presented in figure 8.

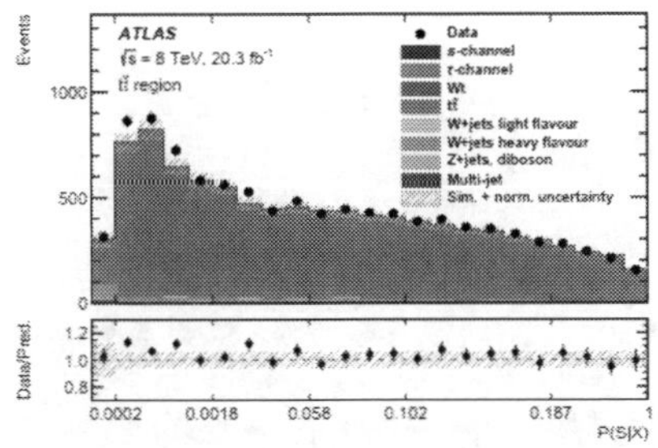

Figure 7: Matrix element discriminant for the $t\bar{t}$ validation region. Except for the data-driven multi-jet background all samples are scaled to their SM expectation. The lower part of the figure shows the ratio of data to simulation. The hatched bands represent the predicted normalisation uncertainty of all processes added in quadrature with the statistical uncertainty of the simulation [11].

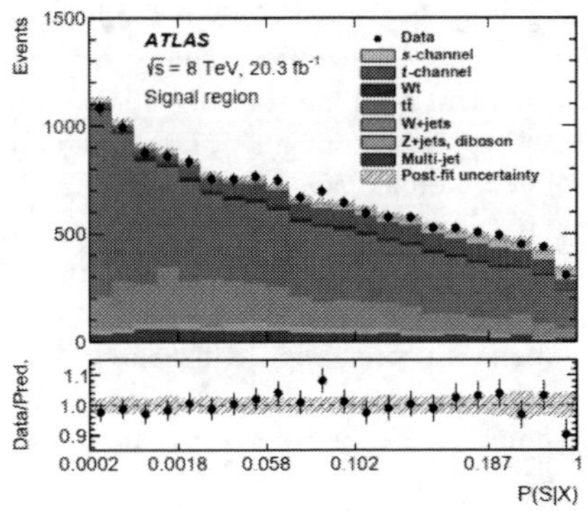

Figure 8: Post-fit distribution of the matrix element discriminant in the signal region. All samples are scaled by the fit result utilising all fit parameters. The lower part of the figure shows the ratio data to simulation. The hatched bands indicate the total uncertainty of the post-fit result including all correlations [11].

References

[1] CDF Collaboration, Phys.Rev.Lett., 74:2626-2631, 1995.
[2] D0 Collaboration, Phys.Rev.Lett., 74:2632-2637, 1995.
[3] ATLAS collaboration, JINST 3, S08003, 2008.
[4] ATLAS Collaboration, Phys. Lett. B761 (2016) 136.
[5] ATLAS Collaboration, arXiv:1511.04716v2 [hep-ex], 2016.
[6] ATLAS Collaboration, ATLAS-CONF-2016-003, 2015.
[7] ATLAS Collaboration, JHEP 1511 (2015) 172.
[8] ATLAS Collaboration, ATLAS-CONF-2015-079, 2015.
[9] ATLAS Collaboration, JHEP01 (2016) 064.
[10] N. Kidonakis Phys. Rev. D 82, 054018, 2010
[11] ATLAS Collaboration, Physics Letters B 756 (2016) 228-246.

EPJ Web of Conferences **164**, 07053 (2017)
ICNFP 2016

DOI: 10.1051/epjconf/201716407053

Top quark properties using the ATLAS detector at the LHC

M.J. Costa on behalf of the ATLAS Collaboration[1,a]

[1] *Instituto de Física Corpuscular - Centro Mixto Consejo Superior de Investigaciones Científicas - Universidad de Valencia, Edificio Institutos de Investigación, C/Catedrático Beltrán 2, E-46980 Paterna, Spain*

Abstract.
The latest measurements of the properties of the top quark using the ATLAS experiment are presented. The top quark mass is one of the fundamental parameters of the Standard Model. A new measurement using the template method in dilepton events in which the event selection was optimised to achieve the smallest total uncertainty is presented. It is combined with a measurement on a multi-dimensional template fit that can constrain the uncertainties on the energy measurements of jets. In addition, measurements aiming to measure the mass in a well-defined scheme are presented. The top quark pair charge asymmetry is an asymmetry predicted to occur beyond leading-order QCD in the Standard Model, and may be significantly enhanced by the presence of new physics. The $t\bar{t}$ production charge asymmetry is measured inclusively and differentially using the 8 TeV ATLAS datasets in the dilepton channel and with lepton+jets events including boosted topology. Making use of the large number of top quark pairs collected, we also present measurements of the spin correlation between top and anti-top quarks and discuss their sensitivities to new physics.

1 Introduction

The top quark is the heaviest known fundamental particle, leading to important theoretical and experimental implications. The large top quark mass implies a large coupling to the Higgs boson and raises the possibility that it plays a special role in the mechanism of electroweak symmetry breaking. It also implies that the top quark decays rapidly through electroweak interactions before it can form a hadronic bound state, which provides the unique opportunity to study the properties from distributions of its decay products. It is therefore essential to measure with high precision the top quark properties using data from the ATLAS experiment [1] and confront them with the Standard Model (SM) predictions.

2 The top quark mass

The top quark mass is a free parameter of the SM which together with the W boson mass constrains the Higgs boson mass through electroweak quantum corrections, allowing for stringent tests of the SM. It is also an important input for the studies of electroweak vacuum stability.

[a] e-mail: Maria.Jose.Costa@ific.uv.es

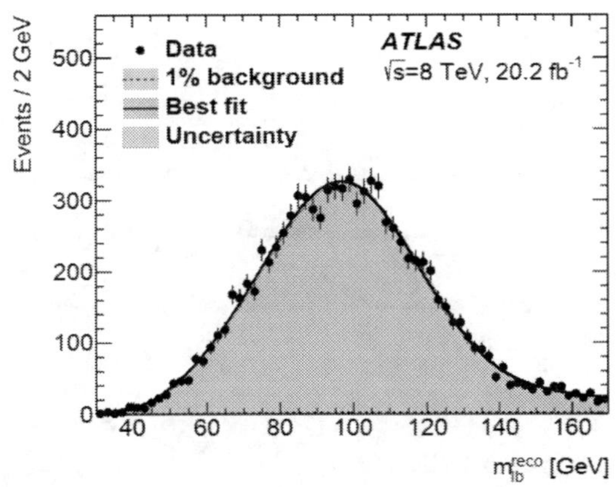

Figure 1. The reconstructed $m_{\ell b}$ observable for data together with the fitted probability density functions for background alone and for the sum of signal and background [6].

The most precise results are obtained using the template method, in particular when applied in the $t\bar{t}$ lepton+jets and dilepton channels. For a chosen observable sensitive to the top quark mass parameter (m_{top}), simulated distributions (templates) are produced using a number of discrete values of that parameter. These templates are fitted to distributions that interpolate between different input values of the top quark mass parameter. A fit to the observed data distribution is then performed to obtain the value of m_{top} that best describes data. Therefore, the mass determined with this procedure is the mass as defined in the Monte Carlo generator. It is expected that the difference between this mass definition and the pole mass is of order 1 GeV [2–4].

The measurement in the lepton+jets channel is performed using data from proton-proton collisions at $\sqrt{s}$ = 7 TeV [5]. The mass is extracted from a three-dimensional template fit to the reconstructed invariant mass of the hadronic top quark decay products, the invariant mass of the two jets coming from the W boson to constrain the jet energy scale uncertainty, and an observable sensitive to the relative b-to-light jet energy scale to reduce the otherwise dominant b-jet energy scale uncertainty. The top quark mass is simultaneously determined with a global light-jet energy scale factor (JSF) and a relative b-to-light-jet energy scale factor (bJSF), accounting for residual differences of data and simulation in the light-jet and b-to-light-jet energy scale. The result obtained is $m_{\text{top}} = 172.33 \pm 0.75\,(\text{stat} + \text{JSF} + \text{bJSF}) \pm 1.02\,(\text{syst})$ GeV, dominated by statistical uncertainties.

In the dilepton channel, measurements have been performed with data collected at both 7 TeV and 8 TeV center-of-mass energies [5, 6]. In this channel, the top quark mass is extracted from a one-dimensional template method, where the templates are constructed from the $m_{\ell b}$ observable, defined as per-event average invariant mass of the two lepton-b-jet systems from the decay of the top quarks (see Figure 1). Due to the underconstrained kinematics associated with the dilepton final state, no in situ constraint of the jet energy scales is performed.

The results obtained are $m_{\text{top}} = 173.79 \pm 0.54\,(\text{stat}) \pm 1.30\,(\text{syst})$ GeV and $m_{\text{top}} = 172.99 \pm 0.41\,(\text{stat}) \pm 0.74\,(\text{syst})$ GeV for the 7 and 8 TeV datasets respectively. The result at $\sqrt{s}$ = 8 TeV is about 40% more precise than the one obtained from the $\sqrt{s}$ = 7 TeV data and the most precise single result from ATLAS. The increased precision is partially driven by a better knowledge of the jet energy scale. In addition, the larger dataset at $\sqrt{s}$ = 8 TeV allowed for an optimised event selection

DOI: 10.1051/epjconf/201716407053

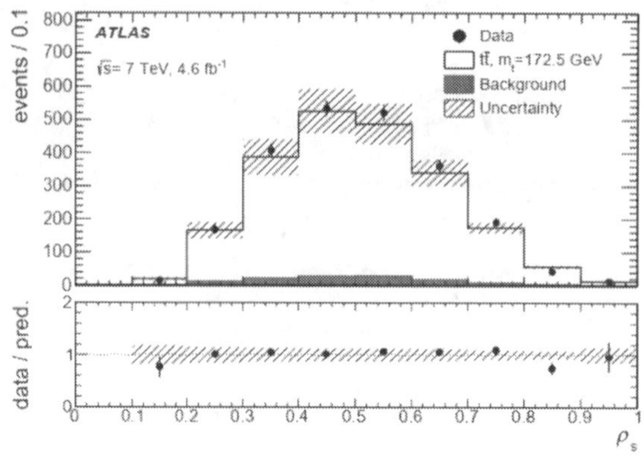

Figure 2. Number of reconstructed events as a function of $\rho_s = \frac{2m_0}{\sqrt{s_{t\bar{t}+1\text{-jet}}}}$ (m_0 = 170 GeV) related to the inverse of the invariant mass of the $t\bar{t}$+1-jet system. The data are compared to the signal plus background expectations [10].

significantly reducing the total systematic uncertainty, mostly due to a lower impact of the jet energy scale and theory modelling uncertainties.

These three most precise measurements obtained by ATLAS have been combined using the best linear unbiased estimate (BLUE) method [7, 8] in which both statistical and systematic uncertainties, and their correlations are taken into account. The correlations of the measurements are evaluated for each source of systematic uncertainty. Using a dedicated mapping of uncertainty categories, the combination results in $m_{\text{top}} = 172.84 \pm 0.34\,(\text{stat}) \pm 0.61\,(\text{syst})$ GeV, with a total relative precision of 0.4%, representing a 17% improvement with respect to the most precise single input measurement.

As mentioned above, the measurements presented so far correspond to the mass definition used in the Monte Carlo simulation. An alternative approach is to extract the mass from measurements of observables whose mass dependence can be calculated beyond leading-order in a perturbative expansion. With this method the top quark mass scheme is unambiguously defined in the theoretical calculations. The results obtained so far have however a lower precision.

Using this approach, the top quark pole mass has been determined from measurements of two different observables: The inclusive cross section for top quark pair production and the normalised differential cross section for $t\bar{t}$ production with at least one additional jet, $t\bar{t}$+1-jet, as a function of the inverse of the invariant mass of the $t\bar{t}$+1-jet system (see Figure 2). The top quark mass dependence for these observables is calculated at NNLO+NNLL accuracy for the former and at NLO for the latter. The differential observable presents a larger sensitivity to the top quark pole mass. Its sensitivity is due to the mass dependence of the amount of gluon radiation, with large effects in the phase space region relatively close to the $t\bar{t}$+1-jet production threshold.

The value extracted from the measurements of the inclusive $t\bar{t}$ cross section at $\sqrt{s}$ = 7 TeV and $\sqrt{s}$ = 8 TeV is $m_{\text{top}}^{\text{pole}} = 172.9^{+2.5}_{-2.6}$ GeV [9]. From the differential measurement, the mass is measured to be $m_{\text{top}}^{\text{pole}} = 173.7 \pm 1.5\,(\text{stat}) \pm 1.4\,(\text{syst})^{+1.0}_{-0.5}\,(\text{theo})$ GeV, reaching a relative uncertainty of 1.3% [10].

3 Spin correlations in top quark pair production

Since the top quark decays before it can form hadrons, the orientations of the top and anti-top quark spins are transferred to the decay products and can be measured directly via their angular distributions.

DOI: 10.1051/epjconf/201716407053

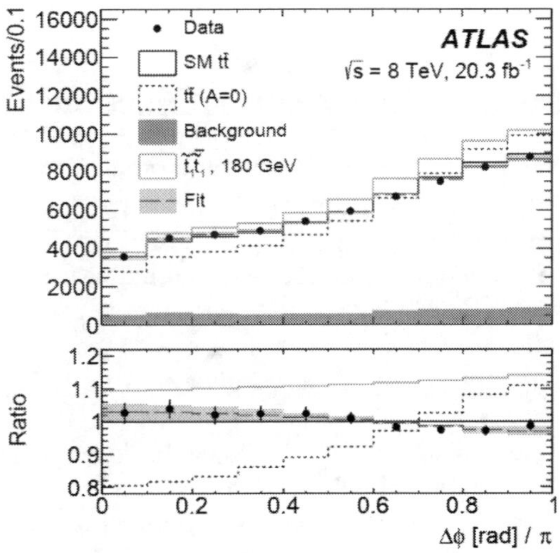

Figure 3. Reconstructed $\Delta\phi$ distribution. The prediction for background (blue histogram) plus SM $t\bar{t}$ production (solid black histogram) and background plus $t\bar{t}$ prediction with no spin correlation (dashed back histogram) is compared to data and to the result of the fit to the data (red dashed histogram) with orange band representing the total systematic uncertainty on $f_{\rm SM}$. The prediction for top squark pair production plus SM $t\bar{t}$ production plus background is also shown (solid green histogram). The lower plot shows those distributions (except for background only) divided by the SM $t\bar{t}$ plus background prediction [11].

Angular distributions of charged leptons are particularly interesting because the polarisation-analysing power of the charged leptons is effectively 100%.

Within the SM QCD top quark pair production, the top quarks are expected to have a negligible polarisation, but the spins of the top and anti-top are expected to be correlated. Many scenarious of physics beyond the SM predict different spin correlations.

The latest and most precise measurements provided by ATLAS in the dilepton channel use two different observables to extract the spin correlation strength: $A_{\rm helicity} = \frac{N_{\rm like}-N_{\rm unlike}}{N_{\rm like}+N_{\rm unlike}}$, where $N_{\rm like}$ ($N_{\rm unlike}$) is the number of events where the top quark and top anti-quark spins are parallel (anti-parallel). One of the observables is the difference in azimuthal angle $\Delta\phi$ between the two charged leptons, defined in the laboratory frame. The second measured observable is the double differential distribution of $\cos\theta_1 \cdot \theta_2$ where θ_1 and θ_2 are for each lepton the polar angle between the direction of the lepton in the top rest frame and the top quark direction in the $t\bar{t}$ center-of-mass frame.

Figure 3 shows the reconstructed $\Delta\phi$ distribution obtained using $\sqrt{s} = 8$ TeV data, compared with Monte Carlo predictions with and without spin correlations. These predictions are then used in a template fit to the data to obtain a coefficient $f_{\rm SM}$ that measures the degree of spin correlations with respect to the SM prediction. The extracted value is $f_{\rm SM} = 1.20 \pm 0.05\,(\text{stat}) \pm 0.13\,(\text{syst})$, that can be translated into $A_{\rm helicity} = 0.38 \pm 0.04$ assuming that the $t\bar{t}$ sample is composed of top quark pairs as predicted by the SM but with varying spin correlation [11]. The measurement agrees with the NLO QCD prediction to within two standard deviations.

The distribution of $\cos\theta_1 \cdot \theta_2$ is measured using $\sqrt{s} = 7$ TeV data. After background subtraction, the distribution is unfolded to parton level using an iterative Bayesian method. The unfolded distri-

 DOI: 10.1051/epjconf/201716407053

bution is in good agreement with the SM prediction and yields to $A_{\text{helicity}} = 0.315 \pm 0.061\,(\text{stat}) \pm 0.049\,(\text{syst})$ [12].

4 Charge asymmetry in top quark pair production

At leading-order, $t\bar{t}$ production is expected to be symmetric under the exchange of top and anti-top quarks. Beyond leading-order and for $q\bar{q}$ initiated processes, due to an interference between the Born and NLO diagrams and between processes with initial and final state gluon emissions, there is a connection between the direction of the top and the initial quark and between the anti-top and the initial anti-quark. At the LHC, this leads to top quarks being produced more forward and anti-top quarks more central. This effect is very small within the SM, at the 1% level, because the dominant top quark pair production is via gluon fusion, which is always symmetric. However, it can be enhanced in several BSM scenarious.

The observable used to study this effect is the forward-backward asymmetry $A_{\text{C}}^{t\bar{t}}$ on the distribution of the difference between the absolute rapidities of the top and anti-top quarks. For the case in which both top quarks decay leptonically, the asymmetry $A_{\text{C}}^{\ell\ell}$ on the distribution of the difference between absolute rapidities of the positive and negative charged leptons is also measured. This observable has a slightly diluted sensitivity but has the advantage that no reconstruction of the $t\bar{t}$ is needed.

Measurements of these two observables have been performed inclusively and differentially as a function of several kinematic variables (as the invariant mass, transverse momentum and longitudinal boost of the $t\bar{t}$ system). They have been done in the full phase space, so that they can be directly compared with the NLO QCD predictions [13, 14], and also in a fiducial region in order not to rely on extrapolations to regions of the phase space that are not within the detector acceptance.

The latest measurements are performed using the $\sqrt{s} = 8$ TeV data in the lepton+jets and dilepton channels [15–17]. For the lepton+jets channel there is a dedicated measurement in events with highly boosted top quarks with an invariant mass of the $t\bar{t}$ system in the TeV range. This kinematic region has a higher sensitivity to the SM asymmetry as well as for BSM models that introduce massive new states. The measured distributions are corrected for acceptance and detector resolution effects using a full Bayesian unfolding method.

All results are found to be compatible with the SM predictions, being the precision limited by statistics. These measurements have allowed to set limits on the masses and couplings of new particles within BSM models. The results obtained for the inclusive top quark based measurements, including also the fiducial measurement in the boosted regime, are summarised in Figure 4 and compared to the CMS measurements. Figure 5 shows the values of $A_{\text{C}}^{t\bar{t}}$ and $A_{\text{C}}^{\ell\ell}$ from the inclusive measurements in the full phase space in the dilepton channel, compared to the SM predictions and a BSM model compatible with the Tevatron results.

5 Conclusion

The latest and most precise measurements performed using Run-1 data from the ATLAS experiment of the top quark mass and angular distributions in top quark pair events sensitive to spin correlations and charge asymmetry effects have been presented. These measurements allow to probe with high precision the SM, being also sensitive to new physics effects.

The most precise measurements of the top quark mass are achieved using a template method, reaching a relative precision of 0.4% on the Monte Carlo top quark mass by combining the measurements performed in the lepton+jets and dilepton channels. Direct determinations of the pole top quark mass have also been performed, reaching a precision of 1.3%.

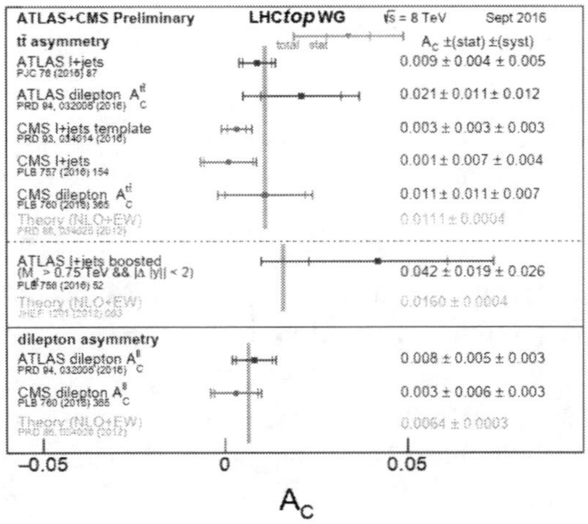

Figure 4. Summary of the charge asymmetry $A_C^{t\bar{t}}$ measurements provided by the ATLAS and CMS experiments, including both inclusive measurements and the measurement in the boosted regime in a restricted phase space region [18].

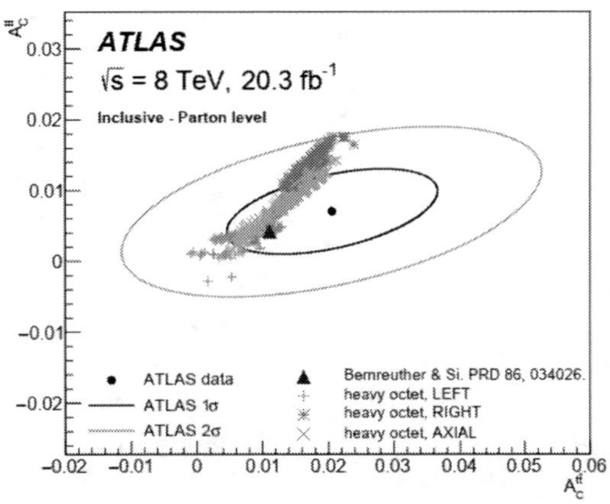

Figure 5. Comparison of the inclusive $A_C^{t\bar{t}}$ and $A_C^{\ell\ell}$ measurement values in the full phase space in the dilepton channel, compared to the SM predictions and to a benchmark BSM model with a heavy octed with mass beyond the reach of the LHC, for various couplings. Ellipses correspond to the 1σ and 2σ uncertainty of the measurement [17].

The measurements of top quark spin correlations and charge asymmetry in $t\bar{t}$ production are all compatible with the SM predictions, allowing to set limits on BSM scenarious.

References

[1] ATLAS Collaboration, JINST 3 (2018) S08003.

[2] A. Buckley et al., Phys. Rept. **504** (2011) 145.
[3] A.H. Hoang, arXiv:1412.3649 [hep-ph].
[4] S. Moch et al., arXiv:1405.4781 [hep-ph].
[5] ATLAS Collaboration, Eur. Phys. J. C **75** (2015) 330.
[6] ATLAS Collaboration, Phys. Lett. B **761** (2016) 350.
[7] L. Lyons et al., Nucl. Instr. Meth. A **270** (1988) 110.
[8] R. Nisius, Eur. Phys. J. C **74** (2014) 3004.
[9] ATLAS Collaboration, Eur. Phys. J. C **74** (2014) 3109.
[10] ATLAS Collaboration, JHEP **10** (2015) 121.
[11] ATLAS Collaboration, Phys. Rev. Lett. **114** (2015) 142001.
[12] ATLAS Collaboration, Phys. Rev. D **93** (2016) 012002.
[13] J.H. Kuhn and G. Rodrigo, JHEP **1201** (2012) 063.
[14] W. Bernreuther and Z.-G.Si, Phys. Rev. D **86** (2012) 034026.
[15] ATLAS Collaboration, Eur. Phys. J. C **76** (2016) 87.
[16] ATLAS Collaboration, Phys. Lett. B **756** (2016), 52.
[17] ATLAS Collaboration, arXiv:1604.05538 [hep-ex].
[18] https://twiki.cern.ch/twiki/bin/view/LHCPhysics/LHCTopWGSummaryPlots.

EPJ Web of Conferences **164**, 07054 (2017)
ICNFP 2016
DOI: 10.1051/epjconf/201716407054

Triangular flow of negative pions emitted in PbAu collisions at top SPS energy

J. Milosevic[1,a]

[1] *University of Belgrade and Vinca Institute of Nuclear Sciences, M. Petrovica Alasa 12-14, 11001 Belgrade, Serbia*

Abstract. The CERES/NA45 experiment measured for the first time triangular flow (v_3) as a function of transverse momentum (p_T) of negative pions emitted around midrapidity in central PbAu collisions at $\sqrt{s_{NN}}$= 17.3 GeV. The v_3, determined within 0.05 < p_T < 2 GeV/c, is corrected for the HBT effects. It is shown that the triangular flow magnitude is smaller by a factor of about 2 than the one measured by the PHENIX experiment at RHIC and the ALICE experiment. Within the analyzed centrality bins, the v_3 does not shows significant centrality dependence. The obtained $v_3(p_T)$ dependence is well described by a viscous hydrodynamic calculation combined with an UrQMD cascade model in the late stages of the collision.

1 Introduction

One of main observables used to study properties of hot and dense systems created in heavy-ion collisions is the single-particle azimuthal anisotropy. In an ideal circle-like geometry of the collisions of two nuclei, the overlapping region has an almond shape which manifests itself in the appearance of the elliptic flow anisotropy [1] driven by strong interactions among constituents of the expanding medium. If the initial-state energy density would be smooth in the transverse plane then only the elliptic flow, v_2, would appears. In a real collision positions of the colliding nucleons fluctuate. As a consequence, beside the event plane derived from the elliptic anisotropy appear event planes of higher-order symmetries and thus also higher-order anisotropies may appear [2]. These higher-order azimuthal anisotropies have been measured at the Relativistic Heavy Ion Collider (RHIC) [3–5] and at the Large Hadron Collider (LHC) [6–8]. The most famous higher-order harmonic, v_3, is called the triangular flow.

Contrary to the elliptic flow which strongly depends on the collision centrality, the triangular flow is nearly independent of centrality. Except for the most central collisions [9], the differential $v_2(p_T)$ at SPS [10–12], although similar in shape to those obtained at the RHIC and LHC experiments, stay below ideal hydrodynamics calculations [13]. This can be explained by insufficient number densities at very early collision stages [14] and strong dissipative effects at the late hadronic stages [15–18].

The negative pions' differential $v_3(p_T)$ measurement from central PbAu collisions at the top SPS energy is presented in this contribution. The results are compared with those from the PHENIX and STAR experiment at RHIC and the ALICE experiment at LHC and also with a hydrodynamics

[a]e-mail: Jovan.Milosevic@cern.ch

calculation coupled with a UrQMD cascade model [19] to describe the late stages. These findings might shed some light on the late stage of collective expansion.

2 Experment and data used

About 30 millions of rather central PbAu collisions at the top SPS energy were collected. CERES detector worked in the magnetic filed of 0.5 T which enable precise measurement of particle momenta with the relative momentum resolution between 2% and 8%. The pseudorapidity[1], η, coverage is close to mid-rapidity ($y_{mid} = 2.91$). This, together with a full azimuthal, ϕ, coverage well suits for the flow study. In this analysis only negative pions are used. They are identified using the differential energy loss dE/dx along their tracks in the radial-drift Time Projection Chamber (TPC). A mixture of three centrality triggers has been used for data collection in the range 0 – 30% of σ/σ_{geo} with an average centrality of 5.5%. A detailed description of the CERES detector is given in [20].

3 Analysis and results

The third-order harmonic coefficient, v_3, of the azimuthal particle distribution is measured with respect to Ψ_3, the azimuthal angle of the 3rd-order participant event-plane. The angle Ψ_3 is determined as:

$$\Psi_3 = \frac{1}{3} \arctan \frac{\sum_{i=0}^{n} w_i(p_{Ti}) \sin(3\phi_i)}{\sum_{i=0}^{n} w_i(p_{Ti}) \cos(3\phi_i)} \tag{1}$$

The ϕ_i in Eq. (1) denotes the azimuthal angle of the i-th particle out of n used for event-plane reconstruction, and $w_i(p_{Ti}) = p_{Ti}$ are weights used to increase the event-plane resolution. As in [9], the ϕ coordinates are divided into 100 adjacent equal slices spanning the full azimuth. In this way the trivial autocorrelation effect is avoided, and up to some extent the contribution from short-range correlations are removed. The local detector inefficiency are corrected by applying a shifting and flattening procedure to ensure an azimuthally isotropic event-plane distribution in the laboratory system. The azimuthal anisotropy of particle is then measured with respect to the 3rd-order event plane constructed in the above described way. The correction factor, calculated as $(2\langle\cos[3(\Psi_3^{(a)} - \Psi_3^{(b)})]\rangle)^{-1/2}$ from the two nonadjacent sliced subevents, a and b, is used to compensate the observed raw v_3 for finite event-plane resolution. Depending on multiplicity, the value of the correction factor goes from 8 till 11.

As only negative pions are used in the analysis, a spurious triangular flow appears as a consequence of the Hanbury Brown & Twiss effect (HBT) of identical bosons. The HBT effect produces a space-momentum correlation between two identical bosons when $|\vec{p}_2 - \vec{p}_1| \leq \hbar/R$. In the central PbAu collisions R is about 7 fm and thus $\hbar/R \approx 30$ MeV/c. This is much smaller than the mean pion transverse momentum $\langle p_\mathrm{T}\rangle \approx 400$ MeV/c. The HBT correlation is short range also in azimuth, and it becomes significant only if $|\phi_1 - \phi_2| \leq \hbar/Rp_\mathrm{T} \simeq 0.1$. As identical bosons are used in this analysis, the correlation is positive like flow itself, and therefore the HBT correlations produce a spurious flow.

In order to remove the non-flow HBT contribution, the standard Bertsch-Pratt parametrization and procedure described in [21] has been used. The size of the corrections applied to the raw $v_3(p_T)$ shown in Fig. 1 is large at low p_T, but decreases quickly with an increase of p_T. The systematic uncertainty

[1]Pseudorapidity η is defined as $-ln\tan(\theta/2)$ where θ is the polar angle.

in the determination of the HBT contribution is derived by varying all source parameters by $\pm 1\ \sigma$ together and independently, and then taking the error of the mean of the resulting distribution as the systematic uncertainty for each p_T bin. It has the biggest size (0.4%) just in the p_T region where the HBT effect is largest, and becomes negligible for $p_T > 0.8$ GeV/c. The final, HBT-corrected v_3 increases roughly linearly, starting from zero at transverse momenta close to zero up to 0.04 at p_T around 2 GeV/c.

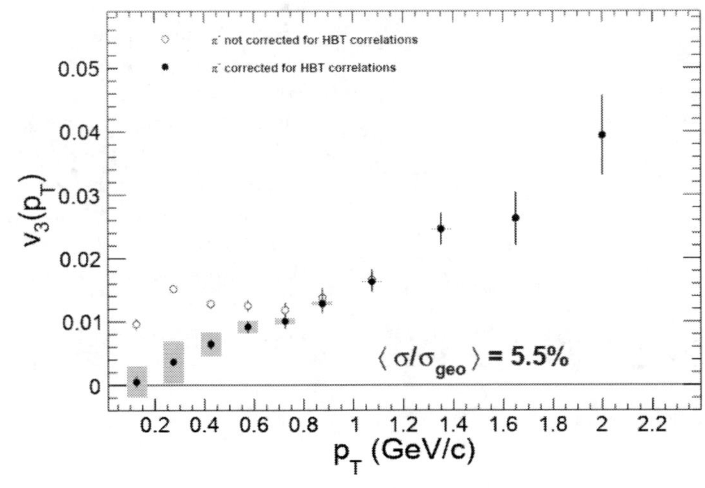

Figure 1. The v_3 as a function of negative pion p_T in PbAu collisions at $\sqrt{s_{NN}} = 17.3$ GeV. Averaged centrality is 5.5%. Open green circles denote v_3 results before, while closed circles those after correcting for the HBT effect. Statistical uncertainties are represented with the error bars, while systematic ones are indicated by gray boxes.

In Fig. 2 are compared CERES v_3 results with those from the PHENIX and ALICE Collaborations at $\sqrt{s_{\rm NN}} = 200$ GeV and $\sqrt{s_{\rm NN}} = 2.76$ TeV [3, 22], respectively. The PHENIX and ALICE results are limited within the p_T range accessible to CERES and at comparable centrality. The Fig. 2 shows that the $v_3(p_T)$ at RHIC and at LHC energy are nearly equal [23]. In contrast, the $v_3(p_T)$ at the top SPS energy reach only about one half of the corresponding values at LHC energy. The p_T range of the analyzed SPS data is small with respect to that covered by ALICE data [22], and within this restricted p_T range the data suggest a linear $v_3(p_T)$ dependence starting from zero.

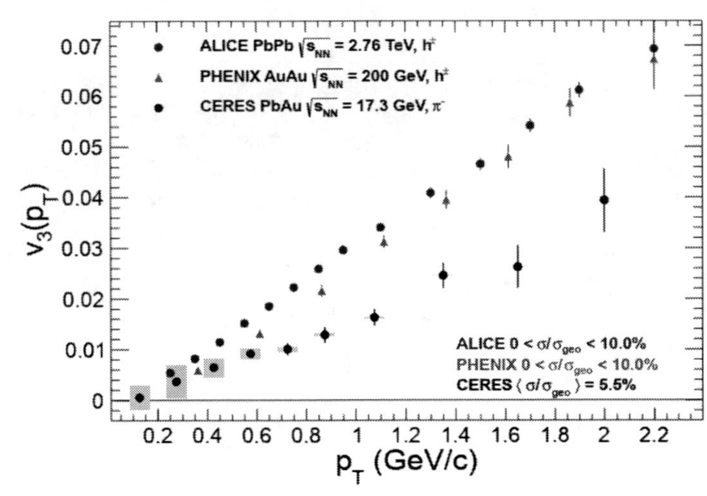

Figure 2. Comparison of negative pions' $v_3(p_T)$ measured at $\sqrt{s_{NN}} = 17.3$ GeV PbAu collisions (CERES, solid black circles) with those from AuAu collisions at $\sqrt{s_{NN}} = 200$ GeV (PHENIX, red triangles) and from PbPb collisions at $\sqrt{s_{NN}} = 2.76$ TeV (ALICE, solid blue circles) at comparable centrality. Statistical uncertainties are represented with the error bars, while systematic uncertainties of the CERES results are indicated by gray boxes.

The ALICE uses large gaps in η between tracks used to measure v_3. In this way non-flow contributions from jet fragmentation have been effectively suppressed. It is shown in [24] that in very central PbAu collisions at SPS energies the total jet yield is about 0.02 per event which is more than an order of magnitude smaller with respect to the corresponding yield at the LHC energy, while the charged particle $dN/d\eta$ is only 4 times smaller [25]. This is quite fortunate, since, due to the limited η acceptance, was impossible to employ a pseudo-rapidity gap in CERES data.

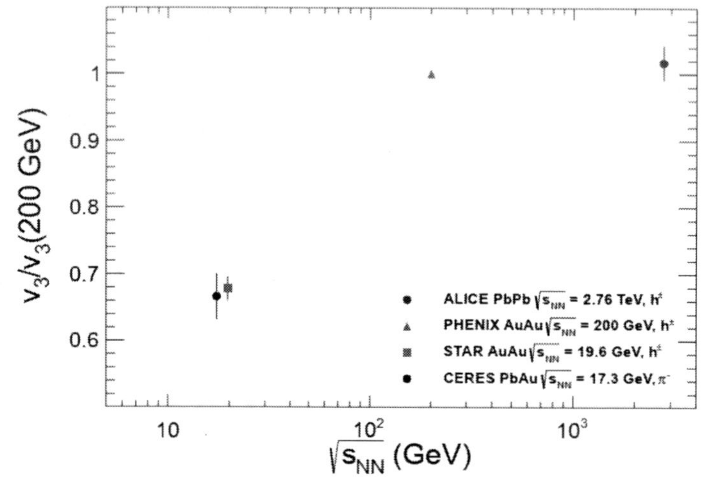

Figure 3. Ratios of the v_3 taken at different collision energies with respect to the v_3 measured in PHENIX AuAu collisions at $\sqrt{s_{NN}}$ = 200 GeV. The v_3 are obtained by integrating the corresponding differential $v_3(p_T)$ over $0.3 < p_T < 2.1$ GeV/c. The data from ALICE, PHENIX and STAR were taken from [22], [3] and [5] respectively.

The p_T-integrated ($p_T > 0.2$ GeV/c) two-particle Fourier coefficients, V_n i.e. the squared v_3 magnitude, as a function of $\sqrt{s_{NN}}$ energy have a shallow minimum between 10 and 20 GeV [5]. The ratio between the STAR $v3$ at 19.6 GeV, which is quite close to the top SPS energy of 17.3 GeV/c, with respect to the v_3 measured at 200 GeV is about 0.63. In Fig. 3 are shown corresponding ratios for 17.3, 19.6, 200 and 2760 GeV/c where in all cases the p_T integration has been done within the range $0.3 < p_T < 2.1$ GeV/c. The v_3 ratio between the top SPS and the top RHIC energy is about 0.66 which is quite close to the one found in [5].

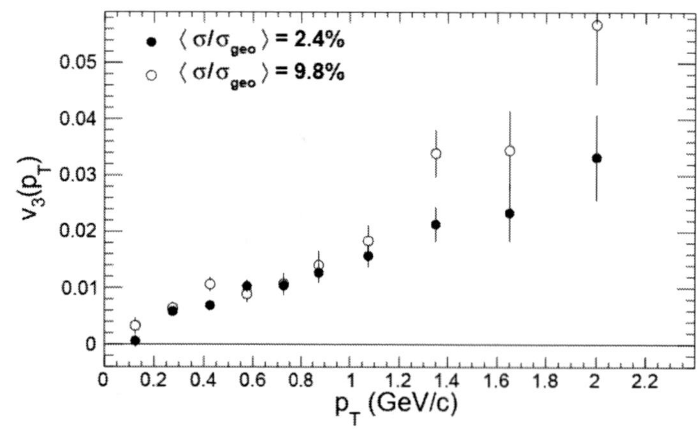

Figure 4. The p_T dependence of the v_3 measured in top-central (closed circles) and mid-central (open circles) PbAu collisions at $\sqrt{s_{NN}}$ = 17.3 GeV.

In difference to the elliptic flow, which reflects the initial anisotropy of the fireball and thus shows a strong centrality dependence (see Fig. 24 in [9]), triangular flow appears entirely from the initial-state

DOI: 10.1051/epjconf/201716407054

fluctuations. In Fig. 4 is shown that the v_3 magnitude is nearly equal for mid-central and top-central collisions, with mean averaged centralities of 2.4% and 9.8%, respectively. A rather weak centrality dependence has also been reported by ALICE [22, 23] where a very slight increase of v_3 with centrality has been observed.

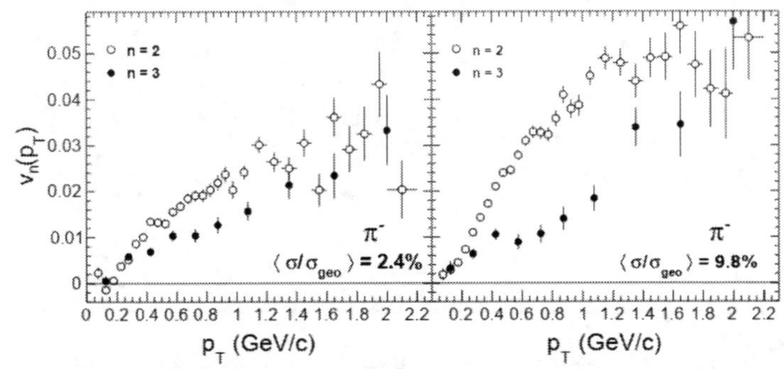

Figure 5. The comparison between the $v_2(p_T)$ (open circles) and $v_3(p_T)$ (closed circles) for top-central (left panel) and mid-central (right panel) PbAu collisions at $\sqrt{s_{NN}} = 17.3$ GeV.

The difference between the centrality dependence between the elliptic and the triangular flow can be seen also from the corresponding p_T-dependencies which are shown in Fig. 5.

The hybrid models, which combines the relativistic hydrodynamics with transport models, have been used for description of heavy-ion collisions at ultra-relativistic energies. The ideal fluid dynamics is used to describe the evolution of the hot and dense quark-gluon plasma, while hadron transport to describe the late sparse hadron gas. The model calculations [19] shown in this analysis are performed using the vHLLE viscous hydrosolver [26] combined with UrQMD hadron cascade [27]. Within it, kinetic and chemical freeze-outs are described dynamically by the UrQMD hadron cascade without a unique freeze-out temperature. The transition from fluid to cascade part, so called the 'particlization' [19], is set to take place on a constant energy density surface with $\epsilon = 0.5$ GeV/fm^3. But, as the net baryon density is not uniform on such a surface, this density does not correspond to a single temperature. Within the chiral model of the Equation of State (EoS) used in this hydrodynamics description, the value of the switching density ϵ_{sw} corresponds to $T \approx 175$ MeV at $\mu_B = 0$. The rest of parameters used in the model are the two Gaussian radii for the initial distribution of energy, and the starting time for the hydrodynamic phase. Their values, together with the value of the switching density, $\epsilon_{sw} = 0.5$ GeV/fm^3, are based on reproduction of the data in collisions at RHIC energies, and are kept unchanged at the SPS for simplicity. In Ref. [19], the authors of the hybrid model investigated parameter dependence of the model results in the case of 0-5% centrality AuAu collisions at $\sqrt{s_{NN}} =$ 19.6 GeV. It is shown that the model results change less than 10% when the parameters of the model are varied by 10%. Within the model [19] is studied the dependence of the elliptic and triangular flow magnitude on the collision energy.

In Fig. 6, the comparison between the predictions from this hydrosolver+UrQMD model and our $v_3(p_T)$ measurement of negative pions in PbAu collisions at $\sqrt{s_{NN}} = 17.3$ GeV is shown. The model predictions are calculated for hadrons within $0.2 < p_T < 2.0$ GeV/c and $-1 < \eta < 1$, which is very close to the experimental acceptance. The model calculation is performed within the centrality samples which roughly correspond to the experimental ones. The hybrid model predictions are in a rather good agreement with the experimental results, except in the p_T region between 0.3 and 0.6 GeV/c where the model slightly underpredicts the experimental data.

DOI: 10.1051/epjconf/201716407054

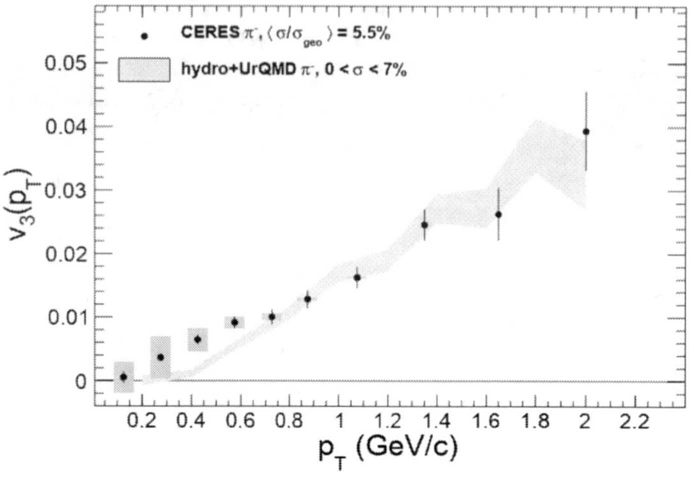

Figure 6. Comparison between the p_T dependence of negative pion v_3 measured in PbAu collisions at $\sqrt{s_{NN}}$ = 17.3 GeV with hydrosolver+UrQMD model predictions. The statistical errors of the model predictions are shown as yellow band. Statistical uncertainties of the experimental results are represented with the error bars, while systematic ones are indicated by gray rectangles.

4 Summary

The triangular flow appears as a hydrodynamic response of the system created in heavy-ion collision to the fluctuation of the positions of the overlapping nucleons at the moment of impact. This is the first differential triangular flow $v_3(p_T)$ measurement at the top SPS energy. The magnitudes of v_3 achieve about one half of the ones measured at the top RHIC and LHC energies. The v_3 measured by CERES at SPS energy of $\sqrt{s_{\rm NN}}$ = 17.3 GeV is similar to the one measured by STAR at RHIC energy of $\sqrt{s_{\rm NN}}$ = 19.6 GeV. The hydrosolver+UrQMD model is able to reproduce the experimental data rather well except in the lowest-p_T region. This comparison could shed some light on the dynamics of the system created in heavy-ion collisions at top SPS energy.

5 Acknowledgments

We would like to thank to Pasi Huovinen for his help concerning the comparison to hydrodynamics. We acknowledge the support by the Ministry of Education, Science and Technological Development of the Republic of Serbia throughout the project 171019.

References

[1] J-Y. Ollitrault, Phys. Rev. **D 46**, 229 (1992)
[2] B. Alver, and G. Roland, Phys. Rev. **C 81**, 054905 (2010)
[3] PHENIX Collaboration, Phys. Rev. Lett. **107** 252301 (2011)
[4] STAR Collaboration, Nucl. Phys. **A 931** 1194 (2014)
[5] STAR Collaboration, Phys. Rev. Lett. **116** 112302 (2016)
[6] ALICE Collaboration, Phys. Rev. Lett. **107** 032301 (2011)
[7] ATLAS Collaboration, Phys. Rev. **C 86** 014907 (2012)
[8] CMS Collaboration, Phys. Rev. **C 89**, 044906 (2014)
[9] NA45/CERES Collaboration, Nucl. Phys. **A 894** 41 (2012)
[10] NA49 Collaboration, Phys. Rev. **C 68** 034903 (2003)
[11] NA45/CERES Collaboration, Phys. Rev. Lett. **92** 032301 (2004)

[12] WA98 Collaboration, Nucl. Phys. **A 762** 129 (2005)
[13] P. Huovinen,P.F. Kolb,U. Heinz,P.V. Ruuskanen,S.A. Voloshin, Phys. Lett. **B 503** 58 (2001)
[14] U. Heinz, and P. Kolb, J. Phys. **G 30** S1229 (2004)
[15] M. Gyulassy, and L. McLerran, Nucl. Phys. **A 750** 30 (2005)
[16] T. Hirano, U. Heinz, D. Kharzeev, R. Lacey, and Y. Nara, Phys. Lett. **B 636** 299 (2006)
[17] D. Teaney, Phys. Rev. **C 68** 034913 (2003)
[18] H. Niemi, G.S. Denicol, P. Huovinen, E. Molnar, and D.H. Rischke, Phys. Rev. Lett. **106** 212302 (2011)
[19] I.A. Karpenko, P. Huovinen, H. Petersen, and M. Bleicher, Phys. Rev. **C 91** 064901 (2015)
[20] A. Marin *et al.*, NA45/CERES Collaboration, J. Phys. **G 30** S709 (2004
[21] P.M. Dinh, N. Borghini, and J.-Y. Ollitrault, Phys. Lett. **B 477** 51 (2000)
[22] ALICE Collaboration, Phys. Lett. **B 719** 18 (2013)
[23] C. Gale *et al.*, Int. J. Mod. Phys. **A 28** 1340011 (2013)
[24] NA45/CERES Collaboration, Phys. Rev. Lett. **92** 032301 (2004)
[25] ALICE Collaboration, Phys. Rev. Lett. **105** 252301 (2010)
[26] I. Karpenko, P. Huovinen, and M. Bleicher, Comput. Phys. Commun. **185** 3016 (2014)
[27] S.A. Bass *et al.*, Prog. Part. Nucl. Phys. **41** 255 (1998); M. Bleicher *et al.*, J. Phys. **G 25** 1859 (1999)

ttH production at 13 TeV (CMS)

Milos Dordevic[1] on behalf of the CMS Collaboration

[1] *Vinca Institute of Nuclear Sciences (University of Belgrade), PO Box 522, 11001 Belgrade, Serbia*

Abstract. First results of a search for the Standard Model Higgs boson produced in association with a top quark-antiquark pair (ttH) in proton-proton collisions at a center of mass energy of 13 TeV is presented. The decays of Higgs boson into two photons, bottom quark-antiquark pair and multileptons via WW, ZZ and tautau pairs are analysed separately and then combined. The results are presented in terms of the best-fit signal strength relative to the Standard Model prediction and the observed and expected upper limits at the 95% confidence level are extracted.

1 Introduction

The discovery of the Higgs boson at CERN [1, 2] is one of the major achievements in the history of particle physics and now the focus is on the precise measurement of its properties, in particular its couplings to other particles. The top quark and Higgs boson are the heaviest particles yet discovered and the leading experiments of today, such as ATLAS and CMS at CERN, are able to provide a direct measurement of their coupling through the study of the associated production in the ttH process. The indirect measurement of the top-Higgs coupling is also possible from the leading Higgs production mechanism, gluon-gluon fusion, with the assumption of no additional particles (beyond SM) contributing to the interaction loop. Thus, a direct measurement of top-Higgs coupling in the associated production can set constraints on the new physics with respect to the $gg \rightarrow$ Higgs process. The top quark is the most strongly coupled to the Higgs boson of all the SM particles and the top-Higgs Yukawa coupling is predicted to be close to unity. It is the top quark that is expected to be the main responsible for the instability of the Higgs mass against the radiative corrections and it could assume a key role in the electroweak symmetry breaking (EWSB). The top-Higgs Yukawa coupling is also a probe for new physics beyond the SM, tightly related to the EWSB. The higher dimension operators of the top and Higgs field are little tested and particularly sensitive to the new physics associated to the EWSB.

The combined result of the search for $t\bar{t}H$ process, using the 7 and 8 TeV data from LHC Run I, provided the observed significance of 4.4σ and the expected significance of 2.0σ, that is obtained from a combined fit to all the analysed channels, the $t\bar{t}H(\gamma\gamma)$, $t\bar{t}H(b\bar{b})$ and $t\bar{t}H(multilepton)$.

This report presents the first study of ttH production at 13 TeV from the CMS Collaboration, using the dataset recorded in 2015 by the CMS experiment [3]. Three different final states are considered, corresponding to the decays of Higgs boson to pair of photons, pair of bottom quarks or the multi-lepton final state from the decays to WW, ZZ and $\tau\tau$. These processes, although very rare, represent some of the cleanest signatures for studying the associated top-Higgs production. In addition to its

DOI: 10.1051/epjconf/201716407055

topology, the ttH cross section increases by a factor of 4 when moving from 8 TeV to 13 TeV of center of mass energy, while the main backgrounds increase only by a factor of 3, thus enhancing the sensitivity at the new collision energy. After being analysed separately, the combination of the results of these three signatures is performed.

2 Analysis and Results

The three ttH analysis at 13 TeV are performed separately, but they have a common dominant background process being the $t\bar{t}$ production in association with other particles (vector bosons and jets). For each of the analysis the irreducible and reducible background yields are dominating the signal after the final selection. The events are further categorised to achieve a better signal to background ratio.

2.1 $t\bar{t}H(\gamma\gamma)$

The $H \rightarrow \gamma\gamma$ channel has a small branching ratio of 0.2%, but offers a very clean final state [5]. The excellent photon identification and energy resolution enables a precise reconstruction of the diphoton invariant mass peak. In the inclusive $H \rightarrow \gamma\gamma$ analysis [5] the events are categorised according to the production mode and the Vector Boson Fusion (VBF) and ttH production are considered. In addition to the inclusive gluon fusion process, in focus here are the diphoton events accompanied by two bottom quarks and two W bosons originating from the decay of top quark pair. Aside of the two photons, the final state may contain leptons and additional jets. The analysis used the $2.7 fb^{-1}$ of the CMS data, preselected with diphoton triggers.

The main backgrounds are the irreducible $t\bar{t} + \gamma\gamma$ and $t\bar{t} + \gamma + jet$ which is the reducible component and where the jets are mis-identified as photons. One of the crucial aspects of the analysis is to suppress the mis-identified (fake) photons. A dedicated photon reconstruction and energy calibration is performed, but also the association to the primary vertex is made in the high pileup environment. The events are categorised according to the presence of at least one lepton ("leptonic tag") or no leptons are reconstructed at all ("hadronic tag"). At least two jets of which at least one b-tagged is required in the "leptonic tag" category, while at least five jets with at least one b-tagged are selected for the "hadronic tag" category. The photon identification is performed combining the different information into the Boosted Decision Tree (BDT). Another BDT is used to further select the events with good diphoton mass resolution in each of the two categories.

The signal extraction strategy is the same as in the inclusive $H \rightarrow \gamma\gamma$ analysis, where a diphoton mass resonance is being searched for on top of a non-resonant smoothly falling background. The analytic function is used to fit the signal events in each category and for each simulated Higgs mass point. Parametric fit functions with several functional forms, where a large set of function families are treated as a discrete parameter in a likelihood fit, are used to model the background.

The signal model for the Higgs boson mass of 125 GeV is shown in Fig. 1 for the "hadronic tag" category, while the data with signal plus background model fit is shown in Fig. 2 for the same category. The 1σ (green) and 2σ (yellow) bands are shown for the background component of the fit and include the uncertainty of the fit parameters. A small excess of events around the Higgs mass is observed in the "hadronic tag" category, while for the "leptonic tag" the analysis is still statistically very limited and only three events are passing the final selection, while none of them correspond to the mass of the Higgs boson. The observed signal strength obtained by combining the two categories is $3.8^{+4.5}_{-3.6}$ for the $m_H = 125.09 GeV$.

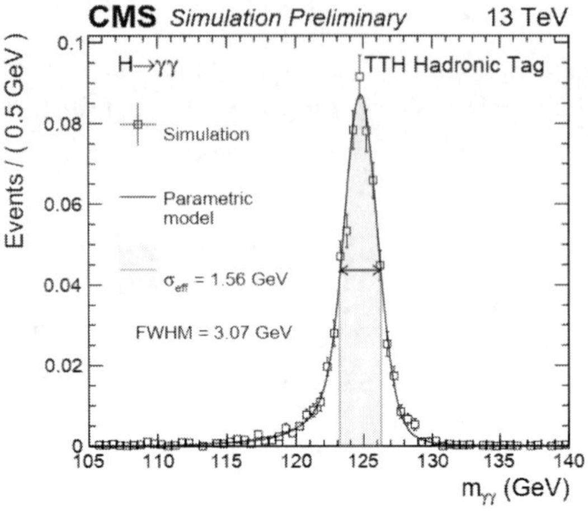

Figure 1. The signal shape parametrization for the "hadronic tag" and Higgs boson mass of 125 GeV. The black points with the uncertainties represent weighted simulation events, while the blue line is the corresponding parametric model.

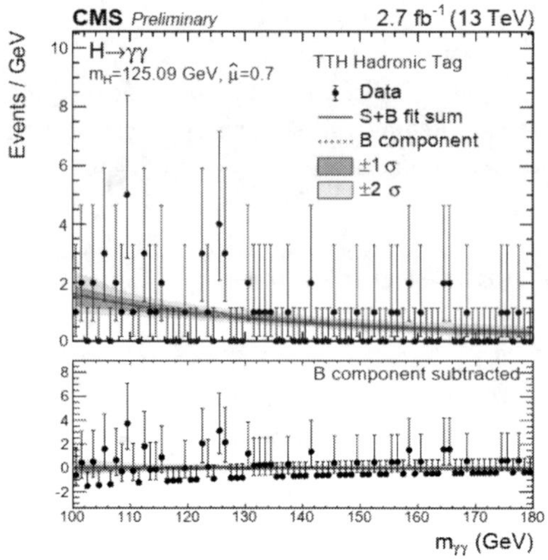

Figure 2. Diphoton mass distribution for the "hadronic tag" category, with the data (black points) and signal plus background fit results (red line), for $m_H = 125.09 GeV$. The pull distribution is obtained after the background is subtracted.

2.2 $t\bar{t}H(b\bar{b})$

This final state profits from the largest branching ratio of 0.56±0.02 for the Higgs boson of 125 GeV [6]. The most dominant background process is the $t\bar{t}+jets$, of which the $t\bar{t}+bb$ (with two b-tagged jets) is an irreducible component and it is experimentally very challenging. The limited mass resolution for $H \to b\bar{b}$ and the presence of jets with similar kinematical properties makes it difficult to assign them properly to the $t\bar{t}$ system or to the Higgs boson itself. The jets that are associated with the top quark pairs are split according to their flavour and are treated independently in the analysis, as different subsamples stem from different physics processes thus having different systematic uncertainties.

The events are categorised in order to enhance the presence of the signal over the dominating background contributions. Two main categories are defined: "lepton+jets" having exactly one lepton in the final state and "dilepton" with exactly two leptons of the opposite sign. The "lepton+jets" category benefits from higher statistics, while still being able to trigger on the presence of one lepton and suppress the QCD background. In the "dilepton" category the contribution of the backgrounds other than $t\bar{t}$ is minimal and the jet combinatorics is reduced as well.

The signal ttH events usually have more (b-tagged) jets than the background and thus the events are further categorised based on the number of jets and number of b-tagged jets. As a new feature with respect ot the 8 TeV analysis [6], the category with boosted jets from hadronic decays of top quark or $H \to b\bar{b}$ with large transverse momenta is introduced. The "lepton+jets" category is split into eight subcategories, while the "dilepton" candidate events are further separated into five subcategories. In each of these categories, the machine learning (BDT) and physics motivated (matrix element - MEM) methods, or their combination, are used to achieve the best separation of the signal from the background processes. For example, in the category with 1 lepton, at least 6 jets of which 3 are b-tagged, the BDT is trained using the MEM discriminator as one of its input variables and the output distribution is shown in Fig. 3. The category with 1 lepton, al least 4 jets of which all 4 are b-tagged is the one for which the MEM is the most suitable and has the best performance, as shown in Fig. 4.

The simultaneous binned maximum-likelihood fit to data in all categories is performed. In the "lepton+jets" category the signal strength is $-4.7^{+3.7}_{-3.8}$, while in the "dileptons" category it is -0.4 ± 2.1. The combination of the two categories yield with the signal strength of -2.0 ± 1.8 and this result is 1.7σ below the SM expectation, which is compatible with the assumption of the absence of signal.

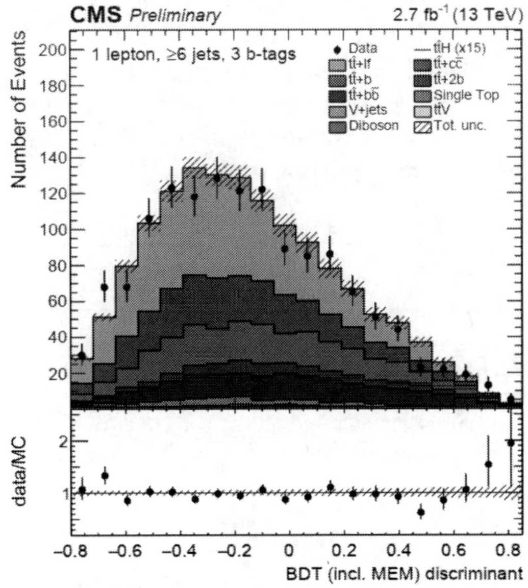

Figure 3. The distribution of the BDT output that includes also the MEM discriminator as one of its input variables, for the 1 leptons, at least 6 jets of which at least 3 are b-tagged. The expected background contributions are nomalised to the integrated luminosity and stacked.

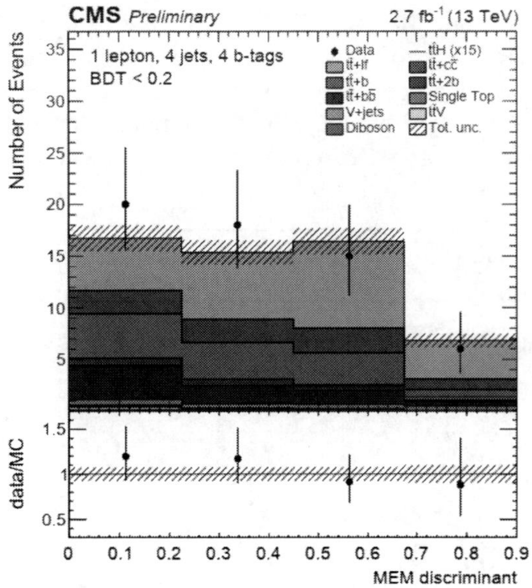

Figure 4. The distribution of the MEM discriminator used exclusively in 1 lepton, at least four jets (all b-tagged) category. The background histograms are stacked and normalised to the luminosity of $2.7 fb^{-1}$.

2.3 $t\bar{t}H(multilepton)$

This analysis considers finals states from Higgs decays to WW, ZZ or $\tau\tau$ where at least one of the vector bosons or taus decays to leptons [7]. This channel has a small branching ratio but the presence of the additional one or two leptons from top quark decays leads to clean experimental signatures: two same-sign leptons (electrons or muons) or at least three leptons plus b-tagged jets. There are two main background components: $t\bar{t}W$ and $t\bar{t}Z$ that is irreducible and is estimated from the MC and the reducible $t\bar{t} + jets$ estimated using a dedicated fake rate method from data. The fake rate is estimated from QCD multijet and Z+jets events. Control regions are defined to model the non-prompt lepton background by relaxing the lepton identification requirement and then the events are weighted by the function of the mis-identification probability of the leptons. Charge mis-identification for electrons is evaluated in a sample of electrons from Z decays, that is split in the opposite- and same-sign pairs, where the charge mis-reconstruction probability is measured as a function of $|p_T|$ and $|\eta|$. This probability ranges from 0.03% in barrel and 0.4% in the endcap region of the CMS detector.

DOI: 10.1051/epjconf/201716407055

The events are preselected with single, double or triple lepton (electron or muon) triggers. At least four jets are required in the same-sign dilepton category; for the three or more leptons event category at least two jets are required. The events are further categorised according to the lepton flavour, by the presence of hadronically decaying tau leptons and by the b-tagging identification criterion. Two BDTs are trained in order to maximise the presence of signal with respect to the $t\bar{t}$ and $t\bar{t}W/Z$ processes. The two-dimensional plane spanning the output of the two BDT trainings is divided into several bins and the signal and background content of each bin is folded to the one-dimensional histogram. The signal extraction is performed by a fit to the distribution of events among these bins. The Fig. 5 and Fig. 6 show the distribution of the output of two-dimensional BDT in the two same-sign and at least three leptons categories, respectively, after the final event selection.

The best fit of the signal strength in the two same-sign lepton and three lepton category is $-0.5^{+1.0}_{-0.7}$ and $5.8^{+3.3}_{-2.7}$, respectively. The combination of the two categories gives a value of $0.6^{+1.4}_{-1.1}$, in agreement with the SM prediction. Slight deficit of events and small excess is observed in the two same-sign and trilepton category, yet not significant for any indication of possible deviation from the SM expectation.

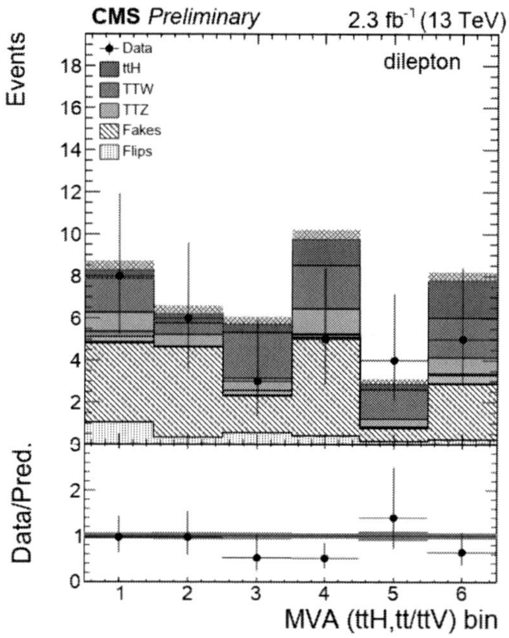

Figure 5. Distribution of the final BDT discriminator, obtained after combining the two BDT trainings against $t\bar{t}$ and $t\bar{t}W/Z$ processes, shown for the two same-sign leptons category.

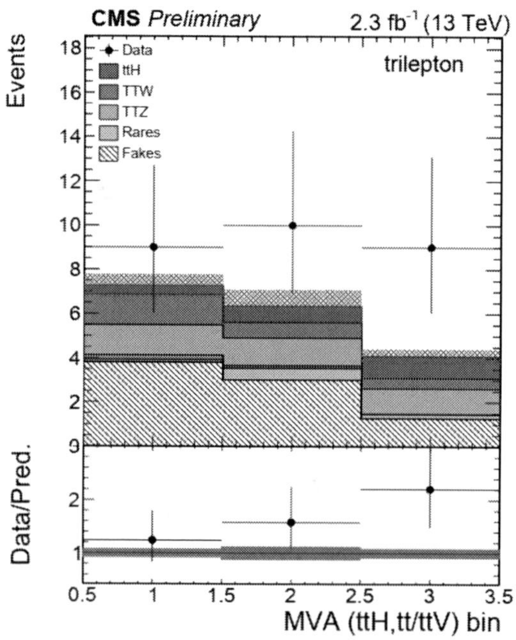

Figure 6. Distribution of the final BDT discriminator, obtained after combining the two BDT trainings against $t\bar{t}$ and $t\bar{t}W/Z$ processes, shown for the at least three leptons category.

2.4 Combination

The three analysis are combined assuming the mass of Higgs boson at 125 GeV and taking into account the correlation of the common systematic uncertainties. The best fit of the signal strength of $0.15^{+0.95}_{-0.81}$ is obtained, in agreement with the SM prediction of $1.00^{+0.96}_{-0.81}$, as shown in Fig. 7. The 95% CL limits on the signal strength in each of the three analysis and also their combination, that has an observed (expected) value of 2.1 (1.9), are shown in Fig. 8.

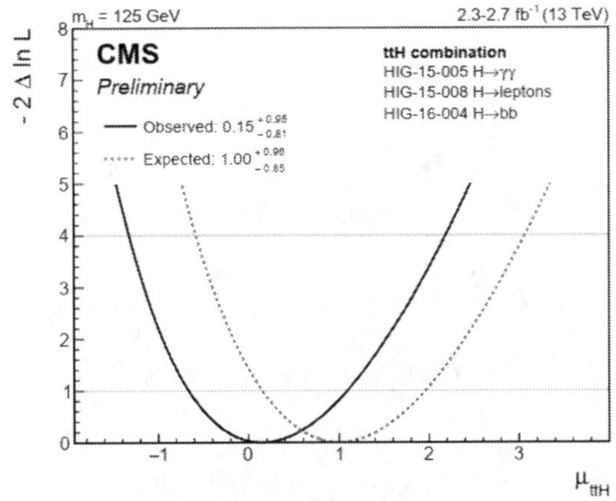

Figure 7. The measurement of the combined signal strength of the $t\bar{t}H$, where the Higgs boson decays to WW, ZZ and $\tau\tau$ are considered.

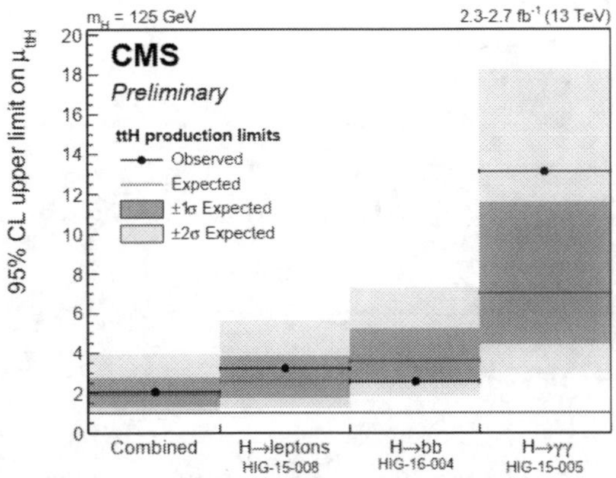

Figure 8. The 95% CL upper limits on the rate of $t\bar{t}H$ production, relative to SM expectation of 1, for each decay mode and the combination.

2.5 Conclusion

The first results of studying the ttH production at 13 TeV using the 2015 data recorded by the CMS experiment at the center of mass energy of 13 TeV have been presented in this report. Three independent channels are analysed separately: $t\bar{t}H(\gamma\gamma)$, $t\bar{t}H(b\bar{b})$ and $t\bar{t}H(multilepton)$ and then combined. The combination of the three analyses results with a best fit to the signal strength of $0.15^{+0.95}_{-0.81}$, in agreement with the SM. Results of the individual channels, as well as the combination, are comparable to the Run I in terms of the signal sensitivity, although almost an order of magnitude less data is used, resulting mainly from the increase of the ttH production cross section by a factor of 4 at the higher center of mass energy, but also from the various improvements and optimization in each of the analyses.

References

[1] The ATLAS Collaboration, Phys. Lett. B 716, 1 (2012).
[2] The CMS Collaboration, Phys. Lett. B 716, 30 (2012).
[3] The CMS Collaboration, JINST 3 S08004 (2008).
[4] The ATLAS and CMS Collaborations, ATLAS-CONF-2015-044, CMS-PAS-HIG-15-002.
[5] The CMS Collaboration, CMS-PAS-HIG-15-005 (2016).
[6] The CMS Collaboration, CMS-PAS-HIG-16-004 (2016).
[7] The CMS Collaboration, CMS-PAS-HIG-15-008 (2016).
[8] CMS public webpage: https://twiki.cern.ch/twiki/bin/view/CMSPublic/TTHCombMoriond2016.

Vacuum alignment and radiatively induced Fermi scale

Tommi Alanne[1,a]

[1] *CP³-Origins, University of Southern Denmark, Campusvej 55, DK-5230 Odense M, Denmark*

Abstract. We extend the discussion about vacuum misalignment by quantum corrections in models with composite pseudo-Goldstone Higgs boson to renormalisable models with elementary scalars. As a concrete example, we propose a framework, where the hierarchy between the unification and the Fermi scale emerges radiatively. This scenario provides an interesting link between the unification and Fermi scale physics.

1 Introduction

In models with enhanced global symmetries, radiative effects have to be taken into account to determine the actual alignment between the electroweak (EW) gauge group and the stability group related to the spontaneous global symmetry breaking. This was first observed in the context of technicolour by Peskin [1] and Preskill [2]: They discovered that the EW gauge group prefers to be unbroken, and thus tends to destabilise the technicolour vacuum. Later, it was noticed that the corrections from the Standard-Model (SM) fermion sector, most notably the top quark, prefer the fully broken EW gauge group, see e.g. [3, 4], and therefore the actual vacuum alignment depends strongly on the underlying model. Kaplan and Georgi [5, 6] realised that the vacuum misalignment problem in technicolour models can be used to realise the Higgs doublet of the SM as doublet of dynamically generated Goldstone bosons (GBs). Furthermore, since in such scenarios the Fermi scale, $v_{\mathrm{w}} = 246$ GeV, originates due to the vacuum misalignment, the actual compositeness scale can be much higher than in traditional technicolour models.

This phenomenon is not only relevant for underlying composite dynamics, but is present also in models with elementary scalars whenever the scalar potential carries an enhaced global symmetry. It turns out, however, that assuming renormalisability changes the vacuum alignment phenomenology with respect to composite-Higgs case [7]. Analysing the vacuum alignment issue in the framework of renormalisable models, thus, opens up interesting prospects for models with an elementary pseudo-GB (pGB) Higgs boson and radiative Fermi scale.

The discussion is organised as follows: In Sec. 2, we outline the vacuum alignment analysis in the framework of renormalisable models with elementary scalars and make a comparison with the composite case. In Sec. 3, we provide in the Pati–Salam-unification framework a concrete model, where the Fermi scale is radiatively generated and has an origin near the unification scale. Finally, we conclude in Sec. 4.

[a] e-mail: alanne@cp3.sdu.dk

 DOI: 10.1051/epjconf/201716407056

2 Vacuum alignment with elementary scalars

Whenever the scalar sector features a larger global symmetry than the full Lagrangian, radiative effects must be taken into account to find out the actual alignment of the vacuum with respect to the gauge group. It is well known that in technicolour models, the EW gauge group prefers to be unbroken [1, 2], but the actual vacuum alignment crucially depends on the embedding of the SM-fermion masses.

In [7] it was shown that the different UV structures of the composite-Higgs models and renormalisable scenarios with elementary scalars lead to different vacuum alignment phenomenology. This can be understood in terms of different effective-potential structures: In the composite case, the low-energy effective theory below the compositeness scale depends on the underlying strongly-interacting model and the SM-fermion-mass mechanism, and the leading contributions to the effective potential are quadratic in the compositeness scale, i.e.

$$V^{\text{eff}}_{\text{comp}} \sim M^2\Lambda^2 + \dots \tag{1}$$

On the other hand, in a renormalisable model with elementary scalars, the radiative corrections are calculable, and the one-loop scalar potential is of the form of the renormalised Coleman–Weinberg potential, i.e.

$$V^{1-\text{loop}}_{\text{elem}} \sim M^4(\log\frac{M^2}{\mu_0^2} - C). \tag{2}$$

In either case, the true vacuum depends on the underlying model, and should be determined when the radiative corrections are correctly evaluated. In general, the actual vacuum is a linear combination of the vacuum preserving the EW gauge group, E_0, and the one fully breaking it to electromagnetism, E_{B}. In the former case, the stability group related to the global symmetry breaking fully contains the EW gauge group, whereas in the latter case the two subgroups are maximally misaligned. It is convenient to parameterise this misalignment by an angle, θ, and thus write the true vacuum as

$$E_\theta = \cos\theta E_0 + \sin\theta E_{\text{B}}. \tag{3}$$

The value of the angle, θ, is undetermined at the tree-level, but a preferred value is picked once radiative corrections are taken into account. The amount of the breaking of the EW subgroup in the spontaneous symmetry breaking thus depends on the vacuum alignment, and the Fermi scale is given by $v_{\text{w}} = v\sin\theta$. In particular, if a non-zero but small value of the angle, $\theta \ll 1$, is preferred, the Fermi scale is much smaller than the actual symmetry breaking scale, v.

2.1 SO(N) → SO($N-1$) template

To illustrate the main features of the vacuum alignment problem with elementary scalars, let us consider the SO(N) → SO($N-1$) breaking pattern. The SO(N)-symmetric scalar sector can be written as

$$V_0 = \frac{m^2}{2}\Phi^\dagger\Phi + \frac{\lambda}{4!}(\Phi^\dagger\Phi)^2. \tag{4}$$

When the scalar Φ acquires a vev, the symmetry is broken spontaneously to SO($N-1$) leaving behind $N-1$ GBs. It is convenient to parameterise Φ in terms of these GBs and an SO($N-1$) singlet, σ, as

$$\Phi = (\sigma + \mathrm{i}\Pi^a X^a)\, E, \tag{5}$$

where X^a are the broken generators corresponding to the vacuum alingment, E.

Whenever $N > 4$, we can embed $\mathrm{SO}(4) \cong \mathrm{SU}(2)_\mathrm{L} \times \mathrm{SU}(2)_\mathrm{R}$ as a subgroup of $\mathrm{SO}(N-1)$ so that EW is unbroken at the tree level. We gauge the EW subgroup by introducing the covariant derivative

$$D_\mu \Phi = \partial_\mu \Phi - ig A_\mu^a \tau^a \cdot \Phi, \tag{6}$$

and parameterise the true vacuum as in Eq. (3), $E \equiv E_\theta = \cos\theta E_0 + \sin\theta E_\mathrm{B}$, where E_0 preserves EW and E_B breaks it to $\mathrm{U}(1)_Q$. Thus, when Φ acquires a vev, $\langle\Phi\rangle = vE$, the gauge bosons obtain masses

$$\mu_W^2 = \frac{1}{4} g^2 v^2 \sin^2\theta, \quad \mu_Z^2 = \frac{1}{4}(g^2 + g'^2) v^2 \sin^2\theta. \tag{7}$$

Furthermore, we add the SM-like Yukawa interaction for the top quark, i.e. couple it to the left doublet of Φ, such that as $\langle\Phi\rangle = vE$, the top quark gets a mass

$$m_t = \frac{1}{\sqrt{2}} y_t v \sin\theta. \tag{8}$$

To study the vacuum structure, we calculate the one-loop Coleman–Weinberg potential in the $\overline{\mathrm{MS}}$ scheme,

$$V_{1-\mathrm{loop}} = V_1^{\mathrm{scalar}} + V_1^{\mathrm{gauge}} + V_1^{\mathrm{fermion}}, \tag{9}$$

where

$$\begin{aligned} V_1^{\mathrm{scalar}} &= \frac{1}{64\pi^2} \mathrm{Tr}\left[M^4(\varphi) \left(\log \frac{M^2(\varphi)}{\mu_0^2} - \frac{3}{2} \right) \right], \\ V_1^{\mathrm{gauge}} &= \frac{3}{64\pi^2} \mathrm{Tr}\left[\mu^4(\varphi) \left(\log \frac{\mu^2(\varphi)}{\mu_0^2} - \frac{5}{6} \right) \right], \\ V_1^{\mathrm{fermion}} &= -\frac{4}{64\pi^2} \mathrm{Tr}\left[\left(m^\dagger(\varphi) m(\varphi) \right)^2 \left(\log \frac{m^\dagger(\varphi) m(\varphi)}{\mu_0^2} - \frac{3}{2} \right) \right], \end{aligned} \tag{10}$$

and $M(\varphi)$, $\mu(\varphi)$, and $m(\varphi)$ are the background-dependent scalar, gauge boson, and fermion mass matrices, respectively.

To find out the preferred value for the vacuum angle, θ, at the vacuum, we minimize the the full effective potential, $V_\mathrm{eff} = V_0 + V_{1-\mathrm{loop}}$ both with respect to the fields and the angle. As described in detail in [7], the minimum of the potential is at $\theta = 0$. Therefore, the EW symmetry remains intact, and there are no non-trivial solutions that would allow for a pGB Higgs.

2.2 Vacuum misalignment

In order to break the EW symmetry and obtain a pGB Higgs, we need to extend the model in some way. In the composite framework, one invokes some further explicit breaking of the global symmetry induced by e.g. the underlying fermion-mass mechanism. With elementary scalars, there is also another possibility, as described in [7]: We can obtain a non-trivial vacuum alignment by adding a singlet scalar acquiring some of its mass from the vev of Φ. The most minimal extension of this kind is to add a real Z_2-symmetric singlet scalar and respectively extend the scalar potential to

$$V_0 = \frac{1}{2} m^2 \Phi^\dagger \Phi + \frac{1}{2} m_S^2 S^2 + \frac{\lambda}{4!} (\Phi^\dagger \Phi)^2 + \frac{\lambda_{\sigma S}}{4} (\Phi^\dagger \Phi) S^2 + \frac{\lambda_S}{4!} S^4. \tag{11}$$

In the limit of very heavy singlet, $M_S \ll v$, we find

$$\sin^2\theta = \lambda_{\sigma S}\frac{v_w^2\left(3A+2B+2A\log\dfrac{g^2v_w^2}{M_S^2}\right)}{4M_S^2\left(2A+B+A\log\dfrac{g^2v_w^2}{M_S^2}\right)} + O\left((v/M_S)^4\right), \tag{12}$$

where

$$\begin{aligned} A &= \frac{1}{16g^4}\left(3g^4+2g^2g'^2+g'^4-16y_t^4\right), \\ B &= \frac{1}{16g^4}\left[(g^2+g'^2)^2\log\left(\frac{1}{4}\left(\frac{g'^2}{g^2}+1\right)\right)-4g^4\log 2-16y_t^4\log\frac{y_t^2}{2g^2}\right. \\ &\quad\left.-\frac{5}{6}\left(3g^4+2g^2g'^2+g'^4\right)+24y_t^4\right]. \end{aligned} \tag{13}$$

It is worth noting that this reproduces the EW symmetry in the excactly decoupled limit. For finite values of M_S, the portal coupling is directly responsible for a non-zero value of θ.

Another possibility is to add an additional global-symmetry-breaking term in the similar manner as in the composite framework. The minimal of this kind is given by

$$V_B = C_B v^3 E_0^\dagger \Phi, \tag{14}$$

where C_B is a dimensionless constant.

2.3 Comparison with the composite case

The leading EW-gauge and fermion-Yukawa contributions to the effective potential in the composite framework are given by:

$$\widetilde{V}_1^{\rm gauge} = \frac{3}{64\pi^2}\sum_K C_K {\rm Tr}\left[\mu^2\right]\Lambda^2, \quad \text{and} \quad \widetilde{V}_1^{\rm ferm} = -\frac{4}{64\pi^2}\sum_F C_F {\rm Tr}\left[m_F^\dagger m_F\right]\Lambda^2, \tag{15}$$

where Λ is the physical cut-off, and C_K, C_F are the form factors related to gauge group K and fermion F, resp. Identifying Λ with the compositeness scale $\Lambda \sim 4\pi f$, where f is the "pion decay constant" for the underlying composite model, we obtain for the SM gauge group and top-quark embedding:

$$\widetilde{V}_1^{\rm gauge}\Big|_{\rm vac} = \frac{3}{16}\left(3g^2C_g + g'^2C_{g'}\right)\sin^2\theta f^4, \tag{16}$$

$$\widetilde{V}_1^{\rm top}\Big|_{\rm vac} = -\frac{3}{2}y_t^2 C_t \sin^2\theta f^4. \tag{17}$$

The gauge part has a minimum at $\theta = 0$, and thereby the gauge sector prefers to be unbroken as already noticed by Peskin [1] and Preskill [2]. The top sector, on the other hand, prefers the minimum to be at $\theta = \pi/2$. This contribution was considered only later in recent technicolour and composite-Higgs models, see e.g. [3, 4]. Since the top contribution dominates over the gauge part, it aligns the vacuum in the direction where the electroweak symmetry is fully broken. Therefore, to achieve a pGB Higgs scenario, new sources of vacuum misalignment are needed.

A minimal solution to achieve the desired vacuum alignment is to add an explicit symmetry breaking operator taking the form:

$$V_B = -C_B f^4 E_0^\dagger \Sigma = -C_B f^4 \cos\theta + \ldots, \tag{18}$$

where C_B is a positive dimensionless constant.

3 Unification with radiative Fermi scale

Unification of (some of) the SM interactions is one of the long-standing paradigms beyond the SM. These scenarios, however, imply a large hierarchy between the EW symmetry breaking at the Fermi scale, $v_w = 246$ GeV, and the breaking of the unified symmetry at some much higher scale. These symmetry breakings are typically modelled via ad-hoc scalar sectors, and there is no symmetry reason to prohibit the portal interactions between these two sectors. However, unless the portal coupling is highly suppressed compared to the SM Higgs self-coupling, the symmetry breaking at the unification scale would induce a large mass for the SM Higgs already at the tree level.

In [8] we proposed a framework to explain this large hierarchy by vacuum misalignment. The minimal scenario utilises global-symmetry-breaking pattern SU(4) ≅ SO(6) → Sp(4) ≅ SO(5) in the Pati–Salam-unification [9] framework. This scheme unifies quarks and leptons by promoting the lepton number to the fourth colour, and therefore the full symmetry group is then $\mathrm{SU}(4)_{\mathrm{PS}} \times \mathrm{SU}(4)_{\mathrm{global}}$. In this scenario, the proton does not decay via gauge interactions, as is the case in the Georgi–Glashow [10] framework, where one unifies strong and electroweak interactions. Instead, the spin-one leptoquarks mediate $K_L \to \mu^\pm e^\mp$. Strong experimental limits on these decays lead to a lower bound $M > 1.5 \cdot 10^6$ GeV on the leptoquark masses [11] translating into the lower bound on the Pati–Salam-unification scale $\Lambda_{PS} \gtrsim 1.9 \cdot 10^6$ GeV.

The SU(4)/Sp(4) model with elementary scalars has been studied in both non-supersymmetric [12, 13] and supersymmetric frameworks [14]. The breaking pattern is obtained by a scalar, M, transforming in the six-dimensional antisymmetric representation of SU(4), and can be conveniently parameterised in terms of the GBs, Π^a, and the Sp(4) singlet, σ, as

$$M = (\frac{\sigma}{2} + \mathrm{i}\sqrt{2}\Pi^a X^a)E, \tag{19}$$

where X^a are the broken generators with respect to vacuum E.

We embed $\mathrm{SU}(2)_L \times \mathrm{SU}(2)_R \cong \mathrm{SO}(4)$ into SU(4) by identifying the left and right generators

$$T_L^i = \frac{1}{2}\begin{pmatrix} \sigma_i & 0 \\ 0 & 0 \end{pmatrix}, \quad \text{and} \quad T_R^i = \frac{1}{2}\begin{pmatrix} 0 & 0 \\ 0 & -\sigma_i^T \end{pmatrix}, \tag{20}$$

where σ_i are the Pauli matrices. The generator of the hypercharge is then identified with the third generator of the $\mathrm{SU}(2)_R$ group, $T_Y = T_R^3$. After embedding the EW group, we can identify the EW preserving and breaking vacua, E_0 and E_B, resp.:

$$E_0 = \begin{pmatrix} \mathrm{i}\sigma_2 & 0 \\ 0 & -\mathrm{i}\sigma_2 \end{pmatrix}, \quad E_B = \begin{pmatrix} 0 & 1 \\ -1 & 0 \end{pmatrix}. \tag{21}$$

As discussed in Sec. 2.1 , the misalignment between the EW group and the stability group, Sp(4), can be parameterised by and angle θ, and the actual vacuum can again be written as $E_\theta = \cos\theta E_0 + \sin\theta E_B$.

To break the $\mathrm{SU}(4)_{\mathrm{PS}}$-leptocolour group to $\mathrm{SU}(3)_c \times U(1)_{B-L}$, we need to introduce an addtional scalar multiplet. The minimal extension of the scalar sector contains a scalar multiplet, $P = p_a T^a$, transforming in the adjoint representation under the leptocolour group but is singlet under the global $\mathrm{SU}(4)_{\mathrm{glo}}$. This additional scalar multiplet thus, in addition to breaking the leptocolour, serves as the minimal EW-singlet extension allowing for a non-tirivial vacuum alignment as discussed in Sec. 2.2. The most general renormalisable scalar potential including these scalar multiplets then reads

$$\begin{aligned} V =& \frac{1}{2}m_M^2 \mathrm{Tr}[M^\dagger M] + m_P^2 \mathrm{Tr}[P^2] + \frac{\lambda_M}{4}\mathrm{Tr}[M^\dagger M]^2 + \lambda_{P1}\mathrm{Tr}[P^2]^2 \\ &+ \lambda_{P2}\mathrm{Tr}[P^4] + \frac{\lambda_{MP}}{2}\mathrm{Tr}[M^\dagger M]\mathrm{Tr}[P^2], \end{aligned} \tag{22}$$

and the desired breaking pattern occurs as the scalars acquire vevs $\langle M\rangle = \frac{v_0}{2}E$, and $\langle P\rangle = b_0 T^{15}$.

To find the actual vacuum alignment angle, we calculate the one-loop Coleman–Weinberg potential and minimize the full effective potential. We fix the renormalization scale such that the tadpole contributions in the σ direction vanish, i.e. the vev $v = \langle\sigma\rangle$ is given by the tree-level value v_0, while the other vev, $\langle p_{15}\rangle = b$, is determined by minimising the full one-loop potential along with the dynamical value of θ.

Fixing the mass of the Higgs boson further constrains the parameter space. Three states, Π_4, σ and p_{15}, have the same quantum numbers as Higgs, and the lightest eigenstate, turning out to be dominantly the pGB Π_4, is identified with the Higgs boson. The Higgs phenomenology constraints were investigated previously [12, 13], and it was shown that the elementary-Goldstone-Higgs paradigm reproduces the phenomenological success of the SM. Similar analysis applies here.

The numerical analysis shows that a small value of θ is preferred, implying a mostly pGB Higgs boson. Furthermore, we find that the preferred values of v are roughly of the order of b, which was fixed just above the experimental bound for the unification scale, $b = 2.5 \cdot 10^6$ GeV. We checked that this conclusion is independent of the specific value of b, and the same conclusion holds for larger values of the unification scale as well. Furthermore, the values of the quartic couplings are overall very small, and in particular there is no large hierarchy between them. This feature of tiny quartic couplings originates from the relation between the couplings of the scalar potential and the vacuum angle θ given by the minimisation conditions. In the limit of equal self-couplings, and $v = b$, this relation is given by $\lambda \propto \sin^2\theta$. Note that this coincides with Eq. (12).

4 Conclusions

We argued that the different UV structures of a composite-Higgs scenario and a renormalisable model with elementary scalars lead to different vacuum alignment phenomenology in the presence of enhanced global symmetries. Furthermore, in the elementary-scalar case, it is possible to achieve a pGB-like Higgs by extending the scalar sector with an EW-singlet without introducing further explicit global-symmetry-breaking operators.

This provides an interesting connection between the Fermi scale physics and unification, since the scalar multiplet that breaks the unification symmetry generally couples to the EW-breaking scalar multiplet via portal interactions. Therefore, the heavier scalar multiplet can additionally play an important role in aligning the vacuum towards the pGB-like Higgs. We have provided a concrete model example in the Pati–Salam-unification framework and found out that once the quantum effects are taken into account, the enhanced global symmetry of the Higgs sector indeed tends to break near the unification scale, and the actual Fermi scale is radiatively generated via the vacuum-misalignment phenomenon. Another interesting feature of this kind of scenario is that all the quartic scalar couplings are generally tiny, but there is no large hierarchy between them. This opens interesting paths to be explored in the future ranging from neutrino masses to cosmic inflation.

Acknowledgments

I thank the organizers of the 5th International Conference on New Frontiers in Physics for the opportunity to present this work. The CP^3-Origins centre is partially funded by the Danish National Research Foundation, grant number DNRF90. The author acknowledges partial funding from a Villum foundation grant.

 DOI: 10.1051/epjconf/201716407056

References

[1] M. E. Peskin, Nucl. Phys. B **175** (1980) 197. doi:10.1016/0550-3213(80)90051-6

[2] J. Preskill, Nucl. Phys. B **177** (1981) 21. doi:10.1016/0550-3213(81)90265-0

[3] J. Galloway, J. A. Evans, M. A. Luty and R. A. Tacchi, JHEP **1010** (2010) 086 doi:10.1007/JHEP10(2010)086 [arXiv:1001.1361 [hep-ph]].

[4] G. Cacciapaglia and F. Sannino, JHEP **1404** (2014) 111 doi:10.1007/JHEP04(2014)111 [arXiv:1402.0233 [hep-ph]].

[5] D. B. Kaplan and H. Georgi, Phys. Lett. B **136** (1984) 183. doi:10.1016/0370-2693(84)91177-8

[6] D. B. Kaplan, H. Georgi and S. Dimopoulos, Phys. Lett. B **136** (1984) 187. doi:10.1016/0370-2693(84)91178-X

[7] T. Alanne, H. Gertov, A. Meroni and F. Sannino, arXiv:1608.07442 [hep-ph].

[8] T. Alanne, A. Meroni, F. Sannino and K. Tuominen, Phys. Rev. D **93** (2016) no.9, 091701 doi:10.1103/PhysRevD.93.091701 [arXiv:1511.01910 [hep-ph]].

[9] J. C. Pati and A. Salam, Phys. Rev. D **10** (1974) 275 Erratum: [Phys. Rev. D **11** (1975) 703]. doi:10.1103/PhysRevD.10.275, 10.1103/PhysRevD.11.703.2

[10] H. Georgi and S. L. Glashow, Phys. Rev. Lett. **32** (1974) 438. doi:10.1103/PhysRevLett.32.438

[11] M. K. Parida, R. L. Awasthi and P. K. Sahu, JHEP **1501** (2015) 045 doi:10.1007/JHEP01(2015)045 [arXiv:1401.1412 [hep-ph]].

[12] T. Alanne, H. Gertov, F. Sannino and K. Tuominen, Phys. Rev. D **91** (2015) no.9, 095021 doi:10.1103/PhysRevD.91.095021 [arXiv:1411.6132 [hep-ph]].

[13] H. Gertov, A. Meroni, E. Molinaro and F. Sannino, Phys. Rev. D **92** (2015) no.9, 095003 doi:10.1103/PhysRevD.92.095003 [arXiv:1507.06666 [hep-ph]].

[14] T. Alanne, H. Rzehak, F. Sannino and A. E. Thomsen, arXiv:1606.03411 [hep-ph].

EPJ Web of Conferences **164**, 07057 (2017) DOI: 10.1051/epjconf/201716407057
ICNFP 2016

ϕ Meson Production at Forward Rapidity with the PHENIX Detector at RHIC

Murad Sarsour[1,a] on behalf of the PHENIX Collaboration
[1] *Georgia State University, Atlanta, Georgia 30303, USA*

Abstract. The ϕ meson production in p+p collisions is an important tool to study QCD, providing data to tune phenomenological QCD models, while in high-energy heavy-ion collisions it provides key information on the hot and dense state of the strongly interacting matter produced in such collisions. It is sensitive to the medium-induced effects such as strangeness enhancement, a phenomenon associated with soft particles in bulk matter. Measurements in the dilepton channels are especially interesting since leptons interact only electromagnetically, thus carrying the information from their production phase directly to the detector. Measurements in different nucleus-nucleus collisions allow us to perform a systematic study of the nuclear medium effects on ϕ meson production. The PHENIX detector provides the capabilities to measure the ϕ meson production in a wide range of transverse momentum and rapidity to study various cold nuclear effects such as soft multiple parton rescattering and modification of the parton distribution functions in nuclei.
In this proceeding, we report the most recent PHENIX results on ϕ meson production in p+p, d+Au and Cu+Au collisions.

1 Introduction

Ever since RHIC announced the discovery of the hot and dense state of strongly interacting matter called Quark-Gluon Plasma (QGP) [1, 2], the main objective has been to quantify its properties. This is accomplished by looking at as many observables as possible, such as the nuclear modification of ϕ meson production in the QGP environment. ϕ meson is an excellent probe for studying QGP (in Au+Au collisions) because it is sensitive to several aspects of the collision, including modifications of strangeness production in bulk matter. Owing to its small inelastic cross section for interaction with nonstrange hadrons, the ϕ meson is less affected by late hadronic rescattering and may reflect the initial evolution of the system. Being composed of a nearly pure strange antistrange ($s\bar{s}$) state, the ϕ meson puts additional constraints on models of quark recombination in the QGP. However, to gauge the QGP related modifications, we need to know the ϕ meson production in p+p collisions as a baseline. There are additional effects from the nuclear medium itself and they are accessed by studying ϕ production in d+Au collisions.

The lepton decay channel is of particular interest because of the absence of strong interactions between muons and the surrounding hot hadronic matter. At forward rapidity, it allows us to study the

[a] e-mail: msar@gsu.edu

DOI: 10.1051/epjconf/201716407057

rapidity dependence of ϕ production and especially in asymmetric heavy-ion collisions provides the means for accessing different mixtures of initial and final state effects.

In p+p collisions, the ϕ meson production is also important to study because it may have similar production mechanisms to other onia like J/ψ and Υ. In addition, it provides a tool to study effects that scale with mass (e.g. collective effects) since it is the heaviest easily accessible meson made of light quarks.

2 Experimental Setup

The PHENIX detector [3] has a high rate capability utilizing a fast DAQ and specialized triggers, high granularity detectors, and good mass resolution and particle ID. The $\mu^{\pm}$ are detected through the forward spectrometers consisting of Muon ID and Muon Tracker. The very forward beam-beam counters (BBC) are used to determine the collision vertex position and time, the beam luminosity and form a minimum bias trigger.

The PHENIX collaboration collected data from a variety of collisions provided by the Relativistic Heavy Ion Collider (RHIC) at Brookhaven National Laboratory (BNL). Results from p+p, d+Au and Cu+Au data sets are presented in this proceeding.

3 Results

The differential production cross sections of ϕ meson have been extracted as a function of transverse momentum ($1 < p_T < 7$ GeV/c) and rapidity ($1.2 < |y| < 2.2$), in p+p collisions as shown in Fig. 1 [4]. The results are compared to different PYTHIA tunes and PHOJET [5] simulations. ATLAS-CSC [6] and PHENIX [7] tunes of PYTHIA reproduce the ϕ cross section reasonably well while the other simulations under-predict it by a factor 2.

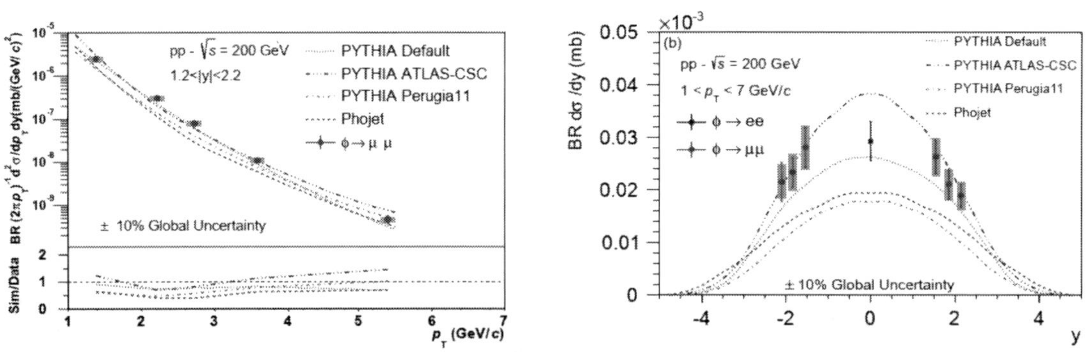

Figure 1. Left: (Top) p_T-dependent differential cross section vs p_T of ϕ at rapidity, $1.2 < |y| < 2.2$. The data are compared with the PYTHIA ATLAS-CSC [6], default [7] and PERUGIA-11 [8] tunes and PHOJET [5]. (Bottom) Ratio between data and models. Right: Rapidity dependent differential cross section of ϕ along with previous PHENIX results [9] summed over the p_T range, $1 < p_T < 7$ GeV. The data are compared with the PYTHIA ATLAS-CSC and PERUGIA-11 tunes and PHOJET. In both panels, the error bars represent the quadratic sum of the statistical uncertainties and point-to-point fluctuating uncertainties, and the gray shaded band represents the quadratic sum of p_T correlated systematic uncertainties.

The left panel of Fig. 2 shows the ϕ meson invariant yield in Cu+Au collisions as a function of the number of participating nucleons, N_{part}, while the right panel shows the dependence of the invariant

DOI: 10.1051/epjconf/201716407057

yield on p_T [10]. More ϕ mesons are produced in the Au-going direction ($-2.2 < y < -1.2$) than in the Cu-going direction ($1.2 < y < 2.2$). This may be explained by the larger multiplicity in the Au-going direction coupled with a mixture of both hot nuclear matter (HNM) and cold nuclear matter (CNM) effects.

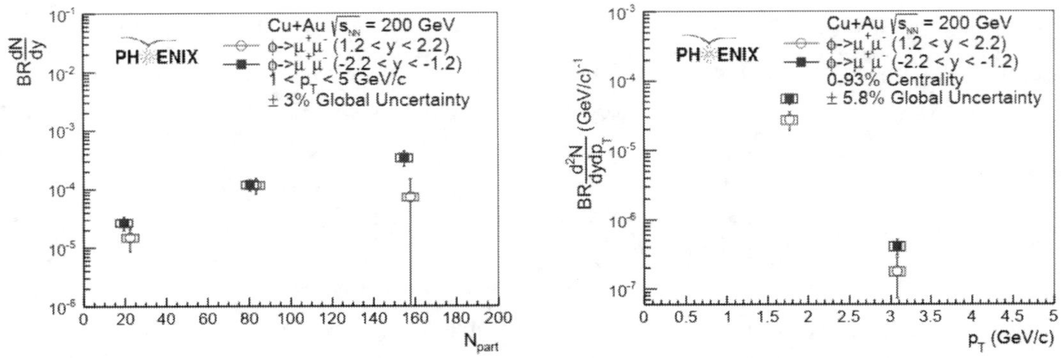

Figure 2. Left: The ϕ meson invariant yield as a function of the number of participating nucleons for $1.2 < |y| < 2.2$ and $1 < p_T < 5$ GeV/c. The data points at $1.2 < y < 2.2$ are shifted along the x-axis to $N_{\text{part}} + 3$ for clarity. Right: The ϕ meson invariant yield as a function of transverse momentum for $1.2 < |y| < 2.2$ and 0%–93% centrality. The Cu-going direction corresponds to the forward rapidity, $1.2 < y < 2.2$, while the Au-going direction corresponds to the backward rapidity, $-2.2 < y < -1.2$.

To gain insight into nuclear medium effects and particle production mechanisms in A+B collisions, the ratio of the ϕ meson yields in A+B collisions to p+p collisions scaled by the number of nucleon-nucleon collisions in the A+B system, N_{coll} [11], is calculated as:

$$R_{AB} = \frac{d^2N_{AB}/dydp_T}{N_{\text{coll}} \times d^2N_{pp}/dydp_T}, \tag{1}$$

where $d^2N_{AB}/dydp_T$ is the per-event yield of particle production in heavy ion collisions and $d^2N_{pp}/dydp_T$ is the per-event yield of the same process in p+p collisions. The p+p invariant yield used in the A+B calculation for the $\mu^+\mu^-$ decay channel is the p+p differential cross section divided by the p+p total cross section, 42.2 mb.

Figure 3 shows the ϕ meson $R_{d\text{Au}}$ as a function of p_T for different centralities [12]. The observed nuclear modification is very similar to that of the heavy flavor muons [13]. In most central collisions, a significant enhancement in the Au-going direction is observed that is more pronounced at intermediate p_T, a characteristic of the Cronin effect [14], while in the d-going direction a suppression at low-p_T is observed. However, in the peripheral collisions no modification is observed. At midrapidity, the $R_{d\text{Au}}$ remains consistent with unity for p_T above 1 GeV/c. This is consistent between the measurements done in the e^+e^- and in the K^+K^- decay channels. The ϕ meson enhancement in the Au-going direction and the suppression in the d-going direction are consistent with what is observed by ALICE in p+Pb collisions at $\sqrt{s_{NN}}$=5.02 TeV in $-4.46 < y < -2.96$ and $2.03 < y < 3.53$ [15].

The ϕ meson $R_{d\text{Au}}$, measured as a function of rapidity, is compared with that of heavy flavor leptons [13, 16] and J/ψ meson [11, 17], as shown in Fig. 4 [12]. It is very interesting to observe the similarity with the nuclear modification of heavy flavor quark production given that the heavy flavor quark production is expected to be dominated by hard processes over the accessed p_T range. In

DOI: 10.1051/epjconf/201716407057

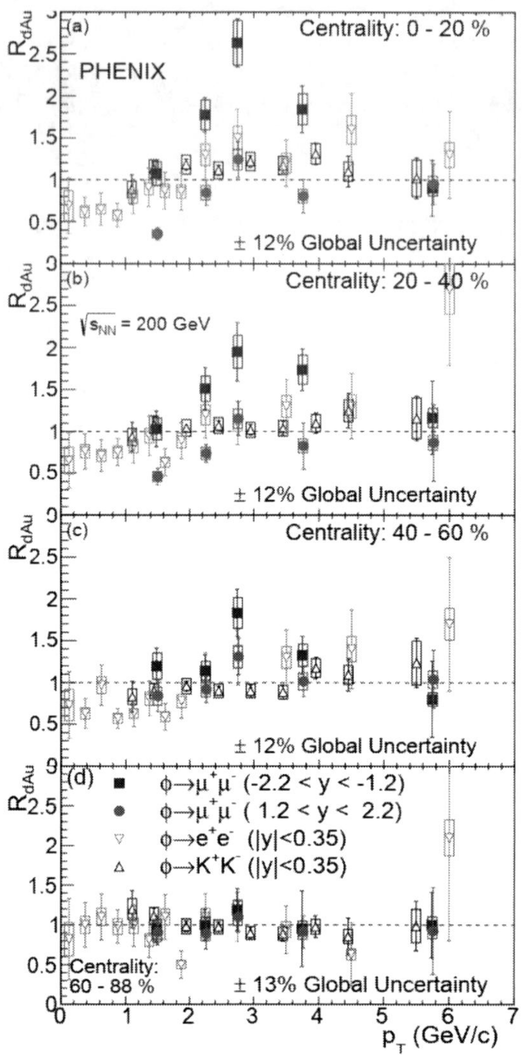

Figure 3. The ϕ meson $R_{d\mathrm{Au}}$ as a function of p_T for the shown centrality classes in the Au-going direction (solid blue squares) and the d-going direction (solid red circles). At midrapidity, the upright black triangles are from $\phi \rightarrow K^+K^-$ while the upside down triangles are from $\phi \rightarrow e^+e^-$. The ±12%–13% uncertainty is the associated global scaling uncertainty.

contrast, one expects a significant contribution to ϕ meson production from soft processes, particularly at low p_T where the yield is dominant. This enhancement (suppression) in the Au-going (d-going) direction is also observed in d+Au charged hadron density results measured by PHOBOS [18]. The pattern observed in the charged hadron measurement is often considered as a result of a rapidity shift in the Au-going direction via soft processes [19]. Figure 4 also shows that the J/ψ meson suffers from additional suppression at backward and midrapidity relative to the ϕ meson and heavy-flavor decay leptons. These differences could be attributed to a larger J/ψ break up cross section, effects

DOI: 10.1051/epjconf/201716407057

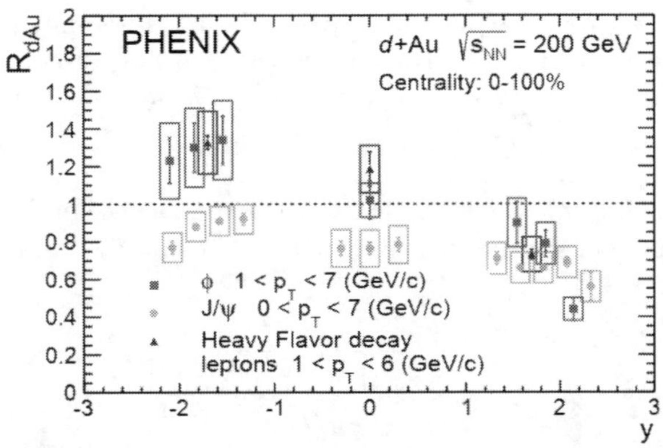

Figure 4. J/ψ meson (solid green circles), heavy flavor decay leptons (solid blue triangles) and ϕ meson (solid red squares) nuclear modification factors, R_{dAu}, as a function of rapidity. The global scaling systematic uncertainties associated with heavy flavor and J/ψ meson measurements are 10% and 8%, respectively.

in the higher energy- density backward-rapidity region, or changes between soft and hard production mechanisms between the two mesons.

The nuclear-modification factor is also studied to evaluate the effects of hot and cold nuclear matter on ϕ meson production in Cu+Au collisions at $\sqrt{s_{NN}}$= 200 GeV. The nuclear-modification factor as a function of N_{part} is shown in the left panel of Fig. 5 [10]. There is a dependence of R_{CuAu} on both centrality and rapidity. In the Au-going direction, the R_{CuAu} is greater than unity for all centralities. The rapidity dependence has a similar trend to that observed in d+Au collisions, as well as measurements made by ALICE in p+Pb collisions at 5.02 TeV [15]. To further understand the

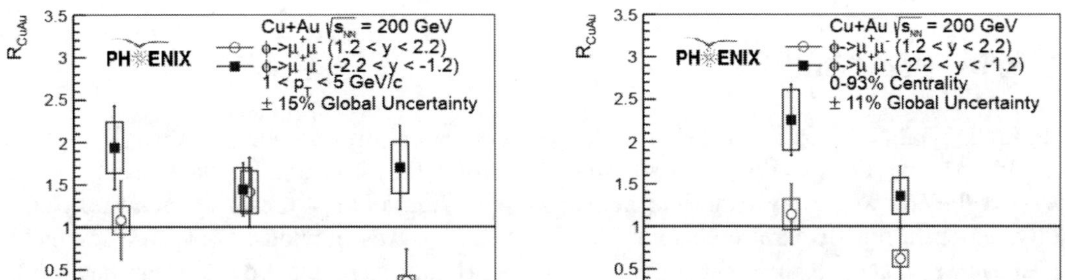

Figure 5. Left: The nuclear-modification factor R_{CuAu} as a function of the number of participating nucleons for $1.2 < |y| < 2.2$ and $1 < p_T < 5$ GeV/c. The data points at $1.2 < y < 2.2$ are shifted along the x-axis to N_{part} + 3 for clarity. Right: The nuclear-modification factor R_{CuAu} as a function of transverse momentum for $1.2 < |y| < 2.2$ and 0%–93% centrality.

relative roles of different nuclear matter effects in this collision system, the transverse momentum dependence of the nuclear-modification factor is shown in the right panel of Fig. 5. Here the nuclear modification is calculated over integrated centrality, but it should be noted that the data are dominated by central collisions. There is an enhancement at low p_T in the Au-going direction. In the Cu-going direction, R_{CuAu} is consistent with unity. The enhancement in the Au-going direction is similar in scale to that observed in the Au-going direction in d+Au collisions [12], indicating similar nuclear modification between the two collision systems.

Figure 6 shows the nuclear modification factor R_{CuAu} as a function of rapidity [10]. The rapidity dependence of R_{CuAu} shows a similar trend to that observed in the case of $p(d)$+Au collisions, shown in Fig. 4. The J/ψ meson yield is strongly suppressed in the Au-going direction compared to the ϕ

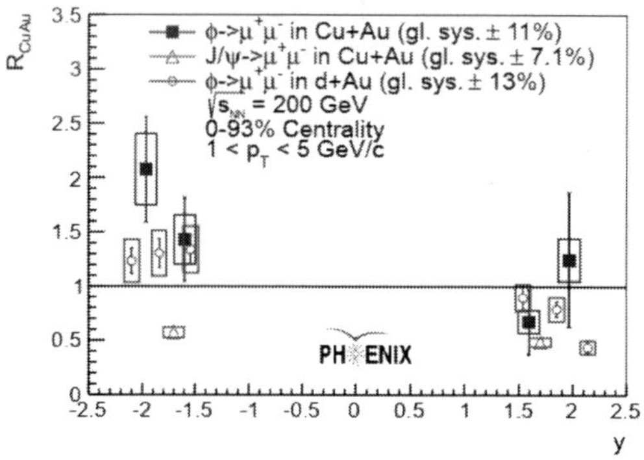

Figure 6. The nuclear modification factor R_{CuAu} as a function of rapidity for $1 < p_T < 5$ GeV/c and 0%–93% centrality. Also included are previous PHENIX results for ϕ mesons in d+Au collisions [12] represented by open circles and J/ψ mesons in Cu+Au collisions [20] represented by open triangles.

meson yield at the same rapidity. This is similar to the differences previously observed between J/ψ and ϕ meson nuclear modification in d+Au collisions [12].

4 Summary and Outlook

The PHENIX collaboration measured the ϕ meson production, over a wide p_T range in the forward and backward rapidities in d+Au and Cu+Au collisions to study cold and hot nuclear matter effects. In d+Au collisions, we observed an enhancement (suppression) of the ϕ meson at backward (forward) rapidity region in most central collisions. Similar behavior was previously observed for inclusive charged hadrons and open heavy flavor muons which may suggest similar cold nuclear matter effects. In Cu+Au collisions, there is an enhancement over all centralities in the Au-going direction, while a suppression is observed for the most central collisions in the Cu-going direction. A comparison with the J/ψ meson in both d+Au and Cu+Au collisions shows different modification in Au-going direction which suggests that the ϕ meson follows a different production mechanism.

New data sets from p+Au and p+Al collisions collected in 2015 will allow ϕ measurement at backward and forward rapidities in less complicated p+Au and p+Al systems. The wealth of small system data sets (d+Au, p+Au and p+Al) along with the introduction of the forward vertex detector (FVTX) in 2015 data sets will allow studying the different CNM effects using models like AMPT and EPOS.

References

[1] K. Adcox *et al.* (PHENIX Collaboration), Nucl. Phys. A **757**, 184 (2005).
[2] J. Adams *et al.* (STAR Collaboration), Nucl. Phys. A **757**, 102 (2005).
[3] K. Adcox *et al.* (PHENIX Collaboration), Nucl. Instrum. Methods Phys. Res., Sec. A **499**, 469 (2003).
[4] A. Adare *et al.* (PHENIX Collaboration), Phys. Rev. D **90**, 052002 (2014).
[5] R. Engel and J. Ranft, Phys. Rev. D **54**, 4244 (1996).
[6] C. Buttar *et al.*, Acta Phys. Pol. B **35**, 433 (2004).
[7] T. Sjostrand *et al.*, Comput. Phys. Commun. **135**, 238 (2001).
[8] P. Z. Skands, Phys. Rev. D **82**, 074018 (2010).
[9] A. Adare *et al.* (PHENIX Collaboration), Phys. Rev. D **83**, 052004 (2011).
[10] A. Adare *et al.* (PHENIX Collaboration), Phys. Rev. C **93**, 024904 (2016).
[11] A. Adare *et al.* (PHENIX Collaboration), Phys. Rev. C **87**, 034904 (2013).
[12] A. Adare *et al.* (PHENIX Collaboration), Phys. Rev. C **92**, 044909 (2015).
[13] A. Adare *et al.* (PHENIX Collaboration), Phys. Rev. Lett. **112**, 252301 (2014).
[14] J. W. Cronin, H. J. Frisch, M. J. Shochet, J. P. Boymond and R. Mermod, *et al.*, Phys. Rev. C **11**, 3105 (1975).
[15] J. Adam *et al.* (ALICE Collaboration), arXiv:1506.09206.
[16] A. Adare *et al.* (PHENIX Collaboration), Phys. Rev. Lett. **109**, 242301 (2012).
[17] A. Adare *et al.* (PHENIX Collaboration), Phys. Rev. Lett. **107**, 142301 (2011).
[18] B. B. Back *et al.* (PHENIX Collaboration), Phys. Rev. C **72**, 031901 (2013).
[19] B. B. Back *et al.*, Phys. Rev. Lett. **93**, 082301 (2004).
[20] A. Adare *et al.* (PHENIX Collaboration), Phys. Rev. C **90**, 064908 (2014).

EPJ Web of Conferences **164**, 08001 (2017) DOI: 10.1051/epjconf/201716408001
ICNFP 2016

Anomaly induced transport in non-anomalous currents *

Eugenio Megías[1,2,a]

[1] *Max-Planck-Institut für Physik (Werner-Heisenberg-Institut), Föhringer Ring 6, D-80805, Munich, Germany*
[2] *Departamento de Física Teórica, Universidad del País Vasco UPV/EHU, Apartado 644, 48080 Bilbao, Spain*

Abstract. Quantum anomalies are one of the subtlest properties of relativistic field theories. They give rise to non-dissipative transport coefficients in the hydrodynamic expansion. In particular a magnetic field can induce an anomalous current via the chiral magnetic effect. In this work we explore the possibility that anomalies can induce a chiral magnetic effect in non-anomalous currents as well. This effect is implemented through an explicit breaking of the symmetries.

1 Introduction

The basic ingredients in the hydrodynamic approach are the constitutive relations, which are expressions of the energy-momentum tensor $T^{\mu\nu}$, and the charge currents J^μ, in terms of fluid quantities [1]. These relations are supplemented with the hydrodynamic equations, which correspond to the conservation laws of the currents. However, in presence of chiral anomalies the currents are no longer conserved, i.e. $\partial_\mu T^{\mu\nu} \neq 0$ and $\partial_\mu J^\mu \neq 0$. This leads to very interesting non-dissipative phenomena that already appear at first order in the hydrodynamic expansion: the *chiral magnetic effect*, responsible for the generation of an electric current parallel to a magnetic field [2], and the *chiral vortical effect*, in which the current is induced by a vortex [3]. The constitutive relation for the charge currents then read

$$\langle J^\mu \rangle = nu^\mu + \sigma^{\mathcal{B}} B^\mu + \sigma^{\mathcal{V}} \omega^\mu + \cdots , \tag{1}$$

where n is the charge density, u^μ is the local fluid velocity, $B^\mu = \frac{1}{2}\epsilon^{\mu\nu\rho\lambda}u_\nu F_{\rho\lambda}$ is the magnetic field, with the field strength of the gauge field defined as $F_{\mu\nu} = D_\mu A_\nu - D_\nu A_\mu$, and $\omega^\mu = \epsilon^{\mu\nu\rho\lambda}u_\nu D_\rho u_\lambda$ is the vorticity vector. The transport coefficients responsible for the chiral magnetic and vortical effects, $\sigma^{\mathcal{B}}$ and $\sigma^{\mathcal{V}}$, have been studied in a wide variety of methods: these include Kubo formulae [4–6], diagrammatic methods [7], fluid/gravity correspondence [8–11], and the partition function formalism [12–15]. It is already clear from these studies that the axial anomaly [16] and the mixed gauge-gravitational anomaly [17] are responsible for the previously mentioned non-dissipative effects.

In this work we are going to focus on the Kubo formalism. The most significant result of anomalies is that they produce equilibrium currents, and the corresponding conductivities are defined via Kubo

*Presented by E. Megías at the 5th International Conference on New Frontiers in Physics (ICNFP 2016), 6-14 July 2016, Kolymbari, Crete, Greece.

[a] e-mail: emegias@mppmu.mpg.de

formulae that involve retarded correlators at zero frequency. The Kubo formula for the chiral magnetic conductivity was derived in [16], and it reads

$$\sigma^{\mathcal{B}} = \lim_{p_c \to 0} \frac{i}{2p_c} \sum_{a,b} \epsilon_{abc} \langle J^a J^b \rangle |_{\omega=0} \,. \tag{2}$$

A similar formula involving the energy-momentum tensor was derived in [4] for the chiral vortical conductivity. In this work we use this formalism to study the chiral magnetic effect of anomalous conductivities. We will study the possibility that, under certain circumstances, this effect might be present in non-anomalous currents as well.

2 The chiral magnetic and separation effects

The chiral magnetic (CME) and chiral separation (CSE) effects are examples of anomalous transport. In the context of heavy ion collisions, a very strong magnetic field produced during a non-central collision induces a parity-odd charge separation which can be modelled by an axial chemical potential, and as a consequence an electric current parallel to the magnetic field is generated, leading to the CME [2, 16, 18]. On the other hand, chirally restored quark matter might give rise to an axial current parallel to a magnetic field, known as CSE [19]. These effects have been predicted as well in condensed matter systems, see e.g. [20, 21].

Let us consider a theory of N chiral fermions transforming under a global symmetry group G generated by matrices $(T_A)^f{}_g$. The chemical potential for the fermion Ψ^f is given by $\mu^f = \sum_A q_A^f \mu_A$, while the Cartan generator is $H_A = q_A^f \delta^f{}_g$ where q_A^f are the charges. The general form of the anomalous induced currents by a magnetic field is

$$\vec{J}^a = (\sigma^{\mathcal{B}})_{ab} \vec{B}_b \,, \tag{3}$$

where $\vec{B}_b$ is the magnetic field corresponding to symmetry b. The 1-loop computation of the chiral magnetic conductivity by using the Kubo formula of Eq. (2) leads to [6, 16, 17]

$$(\sigma^{\mathcal{B}})_{ab} = d_{abc} \frac{\mu^c}{4\pi^2} \,, \qquad d_{abc} = \frac{1}{2} [\mathrm{tr}(T_a \{T_b, T_c\})_R - \mathrm{tr}(T_a \{T_b, T_c\})_L] \,, \tag{4}$$

where d_{abc} is the group theoretic factor related to the axial anomaly, which typically appears in the computation of the anomalous triangle diagram corresponding to three non-abelian gauge fields coupled to a chiral fermion. The subscripts R, L stand for the contributions of right-handed and left-handed fermions. Anomalies are responsible for a non-vanishing value of the divergence of the current, that reads in this case [22]

$$D_\mu J^\mu_a = \epsilon^{\mu\nu\rho\lambda} \frac{d_{abc}}{32\pi^2} F^b_{\mu\nu} F^c_{\rho\lambda} \,. \tag{5}$$

Let us particularize Eq. (3) to the symmetry group $U_V(1) \times U_A(1)$. Then there are vector and axial currents induced by the magnetic field of the vector fields, i.e.

$$\vec{J}_V = \frac{\mu_A}{2\pi^2} \vec{B}_V \,, \qquad \vec{J}_A = \frac{\mu_V}{2\pi^2} \vec{B}_V \,, \tag{6}$$

which correspond to the CME and CSE respectively.

The question then arises: is it possible to get a *chiral magnetic effect* for a non-anomalous symmetry w? This means to have an induced current in symmetry w, i.e.

$$\vec{J}_w \neq 0 \qquad \text{with} \qquad d_{wab} = d_{awb} = d_{abw} = 0 \qquad \forall a, b. \tag{7}$$

In the rest of the manuscript we will study the possibility that anomalies can induce transport also in non-anomalous currents.

3 Holographic model

The Kubo formula Eq. (2) has been computed in [23–25] at strong coupling within a Einstein-Maxwell model in 5 dim. In order to account for the anomalous effects, the model is supplemented with Chern-Simons (CS) terms. In this work we will restrict to a pure gauge CS term, which mimics the axial anomaly. The action reads

$$S = \frac{1}{16\pi G}\int d^5x\sqrt{-g}\Big[R + 12 - \frac{1}{4}F_V^2 - \frac{1}{4}F_A^2 - \frac{1}{4}F_W^2 + \frac{\kappa}{3}A\wedge F_V\wedge F_V\Big] + S_{GH}\,, \tag{8}$$

where S_{GH} is the usual Gibbons-Hawking boundary term. V_M and A_M are vector and axial gauge fields, respectively, and W_M is an extra gauge field associated to the non-anomalous symmetry w. The anomalous term mixes the V_M and A_M fields. [1] A computation of the currents with this model leads to

$$\tilde{J}^\mu_V = \frac{\partial S}{\partial V_\mu} = -\lim_{r\to\infty}\frac{\sqrt{-\gamma}}{16\pi G}\left[F_V^{r\mu} + 6\kappa\epsilon^{\mu\nu\rho\lambda}A_\nu F^V_{\rho\lambda}\right] = J^\mu_V + K^\mu\,, \tag{9}$$

$$\tilde{J}^\mu_A = \frac{\partial S}{\partial A_\mu} = -\lim_{r\to\infty}\frac{\sqrt{-\gamma}}{16\pi G}F_A^{r\mu} = J^\mu_A\,, \tag{10}$$

$$\tilde{J}^\mu_W = \frac{\partial S}{\partial W_\mu} = -\lim_{r\to\infty}\frac{\sqrt{-\gamma}}{16\pi G}F_W^{r\mu} = J^\mu_W\,. \tag{11}$$

$\tilde{J}^\mu$ stands for the holographic *consistent* currents, which are not gauge covariant in general. The *covariant* version of the currents, denoted by J^μ, is the usual one appearing in the constitutive relations, and it corresponds to the covariant part, dropping the CS current K^μ. An analysis of the divergence of the covariant currents leads to the *"covariant" anomalies*:

$$D_\mu J^\mu_V = -3\kappa\epsilon^{\mu\nu\rho\lambda}F^A_{\mu\nu}F^V_{\rho\lambda}\,,\qquad D_\mu J^\mu_A = -\frac{3}{2}\kappa\epsilon^{\mu\nu\rho\lambda}F^V_{\mu\nu}F^V_{\rho\lambda}\,,\qquad D_\mu J^\mu_W = 0\,. \tag{12}$$

From this result, it is clear that while one expects the existence of anomalous transport effects in J^μ_V and J^μ_A, this is not the case for J^μ_W. In a holographic computation of the conductivities with this model, one gets the result of Eq. (6) for J^μ_V and J^μ_A, but a vanishing value for J^μ_W.

3.1 Holographic model with symmetry breaking

In the following we are going to study the possibility that the constitutive relation for $\langle J^\mu_W\rangle$ receives anomalous contributions. Let us extend the model of Eq. (8) with the contribution ($S_{\rm tot} = S + S_\phi$)

$$S_\phi = \frac{1}{16\pi G}\int d^5x\sqrt{-g}\Big(-|D_M\phi|^2 - m^2\phi^2\Big)\,,\qquad D_M\phi = [\partial_M - i(A_M - W_M)]\,\phi\,, \tag{13}$$

where ϕ is a scalar field with a tachyonic bulk mass $m^2 = \Delta(\Delta-4)$, and $0 \le \Delta \le 4$. S_ϕ produces an explicit breaking of A and W symmetries via the scalar field ϕ. From the *AdS/CFT dictionary*, the model is the holographic dual of a Conformal Field Theory (CFT) with a deformation

$$\mathcal{L} = \mathcal{L}^{CFT} + \lambda O\,, \tag{14}$$

where O is an operator dual of the scalar field with $\dim O = \Delta$, and λ is the source of the operator with $\dim\lambda = 4-\Delta$. The near boundary expansion of ϕ reads

$$\phi(r) = \phi_{4-\Delta}r^{\Delta-4} + \phi_\Delta r^{-\Delta} + \cdots\,,\qquad r\to\infty\,, \tag{15}$$

[1] We use the notation for capital indices $M\in\{r,t,x,y,z\}$, and Greek indices $\mu\in\{t,x,y,z\}$.

where $\phi_{4-\Delta}$ is interpreted as the source λ, and ϕ_Δ as the condensate $\langle O \rangle$. In the following we will choose $\Delta = 3$, so that the bulk mass is $m^2 = -3$. The explicit breaking of symmetries is realized via the boundary term $\lim_{r\to\infty} r \cdot \phi(r) = M$, where we have identified the mass parameter M with the source in Eq. (15). Our goal is to study the induced currents

$$\vec{J}_V = \sigma_V(M)\vec{B}\,, \qquad \vec{J}_A = \sigma_A(M)\vec{B}\,, \qquad \vec{J}_W = \sigma_W(M)\vec{B}\,, \tag{16}$$

where one expects a dependence of the conductivities in M. A non zero value for $\sigma_W(M)$ would signal the existence of a non-anomalous current induced by anomalies.

3.2 Background equations of motion

We will work in the probe limit, so that the metric fluctuations are neglected. The equations of motion of the background $(\delta g^{MN}, \delta V^M, \delta A^M, \delta W^M, \delta\phi)$ can be solved by considering the AdS Schwarzschild solution

$$ds^2 = -r^2 f(r)dt^2 + \frac{dr^2}{r^2 f(r)} + r^2\left(dx^2 + dy^2 + dz^2\right)\,, \qquad f(r) = 1 - \frac{r_h^4}{r^4}\,, \tag{17}$$

and the background gauge fields

$$V = V_t(r)dt\,, \qquad A = A_t(r)dt\,, \qquad W = W_t(r)dt\,. \tag{18}$$

The chemical potentials are computed as $\mu_Y \equiv Y_t(r\to\infty) - Y_t(r_h)$ with $Y = V, A, W$. Then, the fields have the following near boundary expansion

$$\lim_{r\to\infty} r\cdot\phi(r) = M\,, \tag{19}$$

$$\lim_{r\to\infty} V_t(r) = \mu_V + V_t(r_h)\,, \quad \lim_{r\to\infty} A_t(r) = \mu_A + A_t(r_h)\,, \quad \lim_{r\to\infty} W_t(r) = \mu_W + W_t(r_h)\,. \tag{20}$$

It is convenient to define in the following new fields $A_1 = A - W$ and $A_2 = A + W$, so that the covariant derivative writes $D_M\phi = [\partial_M - i(A_1)_M]\,\phi$. The boundary expansions of $A_{1,2}$ write as in Eq. (20), with chemical potentials $\mu_{1,2} \equiv \mu_A \mp \mu_W$. Then, the equations of motion of the background read

$$0 = V_t'' + \frac{3}{r}V_t'\,, \qquad 0 = (A_2)_t'' + \frac{3}{r}(A_2)_t'\,, \tag{21}$$

$$0 = (A_1)_t'' + \frac{3}{r}(A_1)_t' - \frac{4\phi^2}{r^2 f(r)}(A_1)_t\,, \tag{22}$$

$$0 = \phi'' + \left(\frac{5}{r} + \frac{f'}{f}\right)\phi' + \left(\frac{(A_1)_t^2}{r^4 f^2} - \frac{m^2}{r^2 f}\right)\phi\,. \tag{23}$$

The solutions of V_t and $(A_2)_t$ can be obtained analytically, and the result is

$$V_t(r) = c_V - \mu_V\frac{r_h^2}{r^2}\,, \qquad (A_2)_t(r) = c_2 - \mu_2\frac{r_h^2}{r^2}\,, \tag{24}$$

where c_V and c_2 are integration constants. The solutions of A_1 and ϕ can be obtained numerically by solving the coupled system of differential equations, Eqs. (22)-(23). From these equations one can easily see that regularity of the solution near the horizon demands $(A_1)_t(r_h) = 0$.

4 Conductivities in presence of symmetry breaking

The Kubo formulae for the conductivities appearing in Eq. (16) are:

$$\sigma_Y = \lim_{p_z \to \infty} \frac{i}{p_z} \langle (J_Y)_x (J_V)_y \rangle|_{\omega=0}, \qquad \text{with} \qquad Y = V, A, W. \tag{25}$$

These involve retarded correlators which can be computed by using the AdS/CFT dictionary, see e.g. [5, 26–28]. We will explain in this section the computation in some details.

4.1 Fluctuations

Without loss of generality we consider perturbations of momentum p in the z-direction at zero frequency. To study the effect of anomalies it is enough to consider the shear sector, i.e. transverse momentum fluctuations,

$$\mathcal{K}_\mu = K_\mu(r) + \epsilon k_\mu(r) e^{ip_z z}, \qquad \mu = x, y, \tag{26}$$

where $K_\mu(r)$ refers to the background solutions computed in Sec. 3.2 for any of the fields V, $A_{1,2}$ (or V, A and W), and $k_\mu(r)$ are the corresponding fluctuations v, $a_{1,2}$ (or v, a and w).

In studying the fluctuations it is useful to organize the equations of motion according to their helicity under the transverse SO(2) rotation, which is a left-over symmetry. The equations for the fluctuations of the gauge fields are then classified into helicity ± 1, and at order $O(\epsilon)$ they read [2]

$$0 = v''_\pm + \left(\frac{3}{r} + \frac{f'}{f}\right) v'_\pm - \frac{p_z}{3r^4 f}(3p_z \pm 4\kappa r\,((A_1)'_t + (A_2)'_t))\,v_\pm \mp \kappa \frac{4p_z}{3r^3 f}\,((a_1)_\pm + (a_2)_\pm)\,V'_t\,, \tag{27}$$

$$0 = (a_1)''_\pm + \left(\frac{3}{r} + \frac{f'}{f}\right)(a_1)'_\pm - \frac{1}{r^4 f}(p_z^2 + 4r^2\phi^2)(a_1)_\pm \mp \kappa \frac{8p_z}{3r^3 f} V'_t v_\pm\,, \tag{28}$$

$$0 = (a_2)''_\pm + \left(\frac{3}{r} + \frac{f'}{f}\right)(a_2)'_\pm - \frac{p_z^2}{r^4 f}(a_2)_\pm \mp \kappa \frac{8p_z}{3r^3 f} V'_t v_\pm\,, \tag{29}$$

where the fields are defined as $v_\pm = v_x \pm i v_y$ and $(a_{1,2})_\pm = (a_{1,2})_x \pm i (a_{1,2})_y$. In order to obtain the retarded correlation functions we should perform a low momentum expansion of the fluctuation solutions, i.e.

$$v_\pm = v_\pm^{(0)} + p_z v_\pm^{(1)} + \cdots, \qquad (a_{1,2})_\pm = (a_{1,2})_\pm^{(0)} + p_z (a_{1,2})_\pm^{(1)} + \cdots. \tag{30}$$

From Eqs. (27)-(29) one gets the corresponding equations at order $O(p_z^0)$ and $O(p_z)$, which can be solved by imposing the appropriate boundary conditions for the fluctuation fields. In general we should consider regularity of the fields up to the horizon, and the sourceless condition, which means that the fluctuations cannot modify the background fields at the boundary. Then

$$k_\mu(r_h) = \text{finite}, \qquad k_\mu(r \to \infty) = 0, \tag{31}$$

for any of the fluctuation fields $k_\mu = v_\pm$ or $(a_{1,2})_\pm$. The only fluctuation that can be computed analytically is $(a_2)_\pm$, and the solution at order $O(p_z)$ reads

$$(a_2)_\pm^{(1)}(r) = \mp c\kappa \frac{4}{3}\frac{\mu_V}{r_h^2} \log\left(1 + \frac{r_h^2}{r^2}\right), \tag{32}$$

[2] The equation for the scalar field is only affected by these perturbations at order $O(\epsilon^2)$.

where c is an integration constant. In a near boundary expansion, the fields behave as

$$v_{\pm}^{(1)} = C_{0,v} + \frac{\widetilde{C}_{2,v}}{r^2}\log r + \frac{C_{2,v}}{r^2} + \cdots , \quad (33)$$

$$(a_1)_{\pm}^{(1)} = C_{0,a_1} + \frac{\widetilde{C}_{2,a_1}}{r^2}\log r + \frac{C_{2,a_1}}{r^2} + \cdots , \quad (34)$$

$$(a_2)_{\pm}^{(1)} = C_{0,a_2} + \frac{\widetilde{C}_{2,a_2}}{r^2}\log r + \frac{C_{2,a_2}}{r^2} + \cdots . \quad (35)$$

The sourceless condition for the fluctuations imply $C_{0,v} = C_{0,a_1} = C_{0,a_2} = 0$. The other coefficients $\widetilde{C}_2$, C_2 can be obtained from a numerical solution of the equations of motion of the fluctuations.

4.2 Conductivities

The correlators of Eq. (25) are contained in the coefficients $C_{2,v}$, C_{2,a_1} and C_{2,a_2}. In particular, we have the following contributions

$$C_{2,v} = \lim_{p_z\to 0}\frac{i}{p_z}\sum_{Y=V,1,2}\langle J_V J_Y\rangle|_{\omega=0}\,,\quad C_{2,a_1} = \lim_{p_z\to 0}\frac{i}{p_z}\sum_{Y=V,1,2}\langle J_1 J_Y\rangle|_{\omega=0}\,,\quad C_{2,a_2} = \lim_{p_z\to 0}\frac{i}{p_z}\sum_{Y=V,1,2}\langle J_2 J_Y\rangle|_{\omega=0}\,. \quad (36)$$

Using the relation between the conductivities in the (A, W) and (A_1, A_2) basis

$$\sigma_{1,2} \equiv \lim_{p_z\to 0}\frac{i}{p_z}\langle J_{1,2}J_V\rangle = \sigma_A \mp \sigma_W\,, \quad (37)$$

one can express them as

$$\sigma_V = \lim_{p_z\to 0}\frac{i}{p_z}\langle J_V J_V\rangle|_{\omega=0}\,, \qquad \sigma_{A,W} = \lim_{p_z\to 0}\frac{i}{2p_z}\left(\langle J_2 J_V\rangle \pm \langle J_1 J_V\rangle\right)|_{\omega=0}\,, \quad (38)$$

where the sign $\pm$ corresponds to the case σ_A and σ_W respectively. We show in Fig. 1 the result for the chiral conductivities of Eq. (38) after solving numerically the equations of motion of the fluctuations. We find a non zero value for $\sigma_W(M)$ in presence of explicit symmetry breaking, i.e. $M \neq 0$. In particular, we observe the following properties:

- $\sigma_W(0) = 0$.
- $\sigma_W(\infty) = \sigma_A(\infty) = \frac{1}{2}\sigma_A(0)$.

The first property is just the expected behavior of the conductivity of the non-anomalous symmetry W in absence of symmetry breaking. The second one can be understood intuitively in the following way: In the basis (A_1, A_2) the CS term in the action is

$$S_{CS} = \frac{1}{16\pi G}\int d^5x\sqrt{-g}\left[\frac{\kappa}{6}\left(A_1\wedge F_V\wedge F_V + A_2\wedge F_V\wedge F_V\right)\right], \quad (39)$$

so that both symmetries A_1 and A_2 have a CS interaction. The scalar field however breaks only A_1. At $M = 0$ there will be CSE for both A_1 and A_2, but for large M, A_1 will be badly broken and the CSE in the J_1 current goes to zero. [3] Since σ_1 vanishes at $M \to \infty$, we find from Eq. (37) the second property.

[3] See [29] for a similar effect.

DOI: 10.1051/epjconf/201716408001

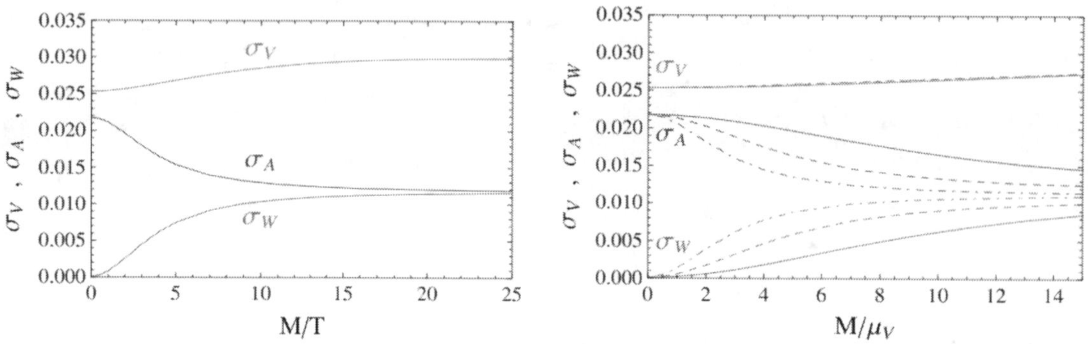

Figure 1. *Vector σ_V, axial σ_A and induced σ_W conductivities as a function of M/T (left) and M/μ_V (right). Left panel: we have considered $\mu_V = 0.4$, $\mu_A = 0.5$, $\mu_W = 0.6$ and $T = 1$. Right panel: We have considered the same values of the chemical potentials, for temperatures $T = 1$ (solid), $T = 0.5$ (dashed) and $T = 0.3$ (dot dashed).*

5 Discussion and outlook

In this work we have studied the role played by the axial anomaly in the hydrodynamics of relativistic fluids in presence of *external magnetic fields*. The anomalous conductivities have been computed by using Kubo formulae. The most interesting result is the characterization of a novel phenomenon related to the possibility that, when symmetries are explicitly broken, anomalies can induce transport effects not only in anomalous currents, but also in non-anomalous ones, i.e. those with a vanishing divergence. We have studied this phenomenon at strong coupling in a holographic Einstein-Maxwell model in 5 dim supplemented with a pure gauge Chern-Simons term. The symmetry breaking is introduced through a scalar field which is dual to an operator O with $\dim O = 3$.

We plan to extend this study to other anomalous coefficients, like the chiral vortical conductivity. In addition, it would be interesting to check whether the mixed gauge-gravitational anomaly could induce any effect as well in non-anomalous currents. These and other issues will be addressed in detail in a forthcoming publication [30].

Acknowledgements

I would like to thank S.D. Chowdhury, J.R. David, K. Jensen, and especially K. Landsteiner for valuable discussions. This work has been supported by Plan Nacional de Altas Energías Spanish MINECO grant FPA2015-64041-C2-1-P, and by the Spanish Consolider Ingenio 2010 Programme CPAN (CSD2007-00042). I thank the Instituto de Física Teórica UAM/CSIC, Madrid, Spain, for their hospitality during the completion of the final stages of this work. The research of E.M. is supported by the European Union under a Marie Curie Intra-European fellowship (FP7-PEOPLE-2013-IEF) with project number PIEF-GA-2013-623006, and by the Universidad del País Vasco UPV/EHU, Bilbao, Spain, as a Visiting Professor.

References

[1] P. Kovtun, J. Phys. **A45**, 473001 (2012)
[2] K. Fukushima, D.E. Kharzeev, H.J. Warringa, Phys. Rev. **D78**, 074033 (2008)
[3] D.T. Son, P. Surowka, Phys. Rev. Lett. **103**, 191601 (2009)
[4] I. Amado, K. Landsteiner, F. Pena-Benitez, JHEP **1105**, 081 (2011)

[5] K. Landsteiner, E. Megias, F. Pena-Benitez, Lect. Notes Phys. **871**, 433 (2013)
[6] S.D. Chowdhury, J.R. David, JHEP **11**, 048 (2015)
[7] J.L. Manes, M. Valle, JHEP **01**, 008 (2013)
[8] S. Bhattacharyya, V.E. Hubeny, S. Minwalla, M. Rangamani, JHEP **0802**, 045 (2008)
[9] J. Erdmenger, M. Haack, M. Kaminski, A. Yarom, JHEP **01**, 055 (2009)
[10] N. Banerjee et al., JHEP **01**, 094 (2011)
[11] E. Megias, F. Pena-Benitez, JHEP **05**, 115 (2013)
[12] N. Banerjee, J. Bhattacharya, S. Bhattacharyya, S. Jain, S. Minwalla, T. Sharma, JHEP **09**, 046 (2012)
[13] K. Jensen, Phys. Rev. **D85**, 125017 (2012)
[14] K. Jensen, M. Kaminski, P. Kovtun, R. Meyer, A. Ritz, A. Yarom, Phys. Rev. Lett. **109**, 101601 (2012)
[15] E. Megias, M. Valle, JHEP **11**, 005 (2014)
[16] D.E. Kharzeev, H.J. Warringa, Phys. Rev. **D80**, 034028 (2009)
[17] K. Landsteiner, E. Megias, F. Pena-Benitez, Phys. Rev. Lett. **107**, 021601 (2011)
[18] B. Keren-Zur, Y. Oz, JHEP **06**, 006 (2010)
[19] G.M. Newman, D.T. Son, Phys. Rev. **D73**, 045006 (2006)
[20] G. Basar, D.E. Kharzeev, H.U. Yee, Phys. Rev. **B89**, 035142 (2014)
[21] K. Landsteiner, Phys. Rev. **B89**, 075124 (2014)
[22] T. Kumura, Prog. Theor. Phys. **42**, 1191 (1969)
[23] K. Landsteiner, E. Megias, L. Melgar, F. Pena-Benitez, JHEP **09**, 121 (2011)
[24] K. Landsteiner, E. Megias, L. Melgar, F. Pena-Benitez, J. Phys. Conf. Ser. **343**, 012073 (2012)
[25] K. Landsteiner, E. Megias, F. Pena-Benitez, Phys. Rev. **D90**, 065026 (2014)
[26] D.T. Son, A.O. Starinets, JHEP **09**, 042 (2002)
[27] C.P. Herzog, D.T. Son, JHEP **03**, 046 (2003)
[28] M. Kaminski, K. Landsteiner, J. Mas, J.P. Shock, J. Tarrio, JHEP **02**, 021 (2010)
[29] A. Jimenez-Alba, K. Landsteiner, Y. Liu, Y.W. Sun, JHEP **07**, 117 (2015)
[30] E. Megias, work in progress (2016)

EPJ Web of Conferences **164**, 08002 (2017)
ICNFP 2016 DOI: 10.1051/epjconf/201716408002

Correlation between heavy flavour production and multiplicity in string fusion approach

Vladimir Kovalenko[1] and Vladimir Vechernin[1,a]

[1] *Saint Petersburg State University, 7/9 Universitetskaya nab., St. Petersburg, 199034 Russia.*

Abstract. The multiplicity dependence of heavy flavour production in pp-collisions at LHC energies is studied in the framework of string fusion approach. We applied a Monte Carlo model [1, 2], with the string configurations simulating event-by-event. We assumed that the heavy flavour yield is proportional to the number of initial strings, whereas a total charged particle multiplicity is influenced by the string fusion process and can be obtained according to the standard string fusion prescriptions. We show that under this assumption the faster-than-linear growth of the open charm production, observed in experiment [3], can be related to the reduction of the total multiplicity due to string overlapping and fusion. The influence of this effect on forward-backward correlations involving heavy flavours is also discussed.

1 Introduction

The recent measurements on multiplicity dependence of the heavy quark production, performed by the ALICE Collaboration [3], demonstrated unexpected results. It was obtained that the D meson yield N_{D}, plotted as a function of charged multiplicity N_{ch} in relative variables, shows faster-than-linear behaviour.

This data have been interpreted in EPOS 3 model [4] with the hydrodynamical stage of the evolution, and a simple string percolation model [5]. In both models, the growth of charmed yield faster than N_{ch} is related to some physical process, which dumps multiplicity of bulk particles, but has almost no effect on c-$\bar{\mathrm{c}}$ pair production.

We should note, that in the simple string percolation model [5] it is assumed that the string density is the same for every p-p collision and doesn't fluctuate. In the present report, we performed the calculations in more realistic Monte Carlo model with string fusion [1, 2], where the string configurations are simulated event-by-event.

2 The model

The present model [1, 2] is based on the partonic picture of pp-scattering. The partons are distributed in transverse plane with Gauss distribution relative to the center of nucleon and arranged into dipoles.

[a] e-mail: v.vechernin@spbu.ru

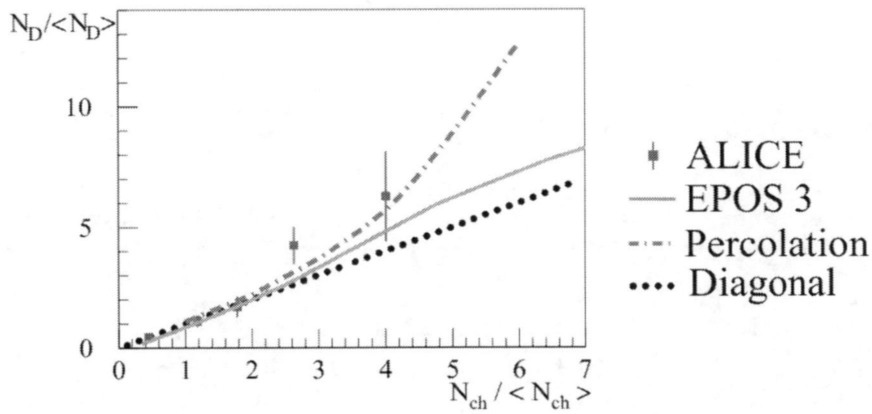

Figure 1. Average D-meson relative yield for $1 < p_t < 2$ GeV as a function of the relative charged multiplicity in pp collisions at $\sqrt{s}$=7 TeV. Experimental data [3] compared to EPOS 3 [4] and string percolation model [5].

Interaction probability amplitude of two dipoles with transverse coordinates $(\mathbf{r}_1\mathbf{r}'_1)$ and $(\mathbf{r}_2\mathbf{r}'_2)$ is given by [6, 7]:

$$f = \frac{\alpha_S^2}{8} \ln^2 \frac{(\mathbf{r}_1 - \mathbf{r}'_1)^2(\mathbf{r}_2 - \mathbf{r}'_2)^2}{(\mathbf{r}_1 - \mathbf{r}'_2)^2(\mathbf{r}_2 - \mathbf{r}'_1)^2},$$

Note, that two dipoles interact more probably, if their ends are close to each other, and (others equal) if they are wide. The model also incorporates confinement effects.

The charged multiplicity is calculated in the approach of colour strings, taking into account their finite rapidity length and interactions due to non-zero transverse radius of string r_{str} – string fusion [8, 9]. Mean multiplicity of charged particles originated from k overlapping strings is modified as $\langle\mu\rangle_k = \mu_0 \sqrt{k}$, where μ_0 is a mean charged multiplicity from one single string.

Parameters of the model are fixed from the pp, p-Pb and Pb-Pb data on the total inelastic cross section and charged multiplicity in wide energy range [10].

We note, that contrary to the total charged particle multiplicity, which is influenced by string fusion process, the heavy flavour yield is assumed to be proportional to the number of initial strings.

3 Results and discussion

The results of the calculation are shown in Figure 2.

The results demonstrate that reduction of total charged multiplicity due to string fusion gives faster-than-linear behaviour of relative charmed meson yield, observed in the experiment.

At largest multiplicities, however, another mechanism of charm enhancement may play an important role, which is related to Schwinger (or thermal-like) production of c-$\bar{c}$ pairs [11]. We suppose that the effect may be more pronounced also in p-Pb collisions at LHC energy, where the string density is larger, and allowing to produce charm quarks by strings of high string tension.

These two mechanisms of the charm growth could be discriminated in the experiment by studying the long-range correlations of the relative D meson yield in one rapidity interval with the average p_t or relative strangeness in another window [12, 13]. The hard mechanism of the charm production

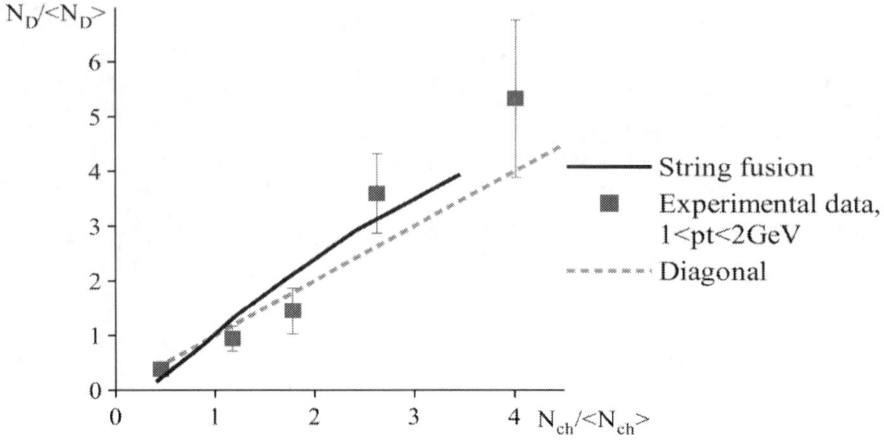

Figure 2. Relative average D-meson yield as a function of charged multiplicity, calculated in MC string fusion model, compared to experimental data [3] for pp collisions at $\sqrt{s}$=7 TeV.

will contribute to the negative correlation, and the one related to the string tension increase will give positive correlation.

Acknowledgments

The research was funded by the grant of the Russian Science Foundation (project 16-12-10176).

References

[1] V. N. Kovalenko, Phys. Atom. Nucl. **76**, 1189 (2013), arXiv:1211.6209 [hep-ph].
[2] V. Kovalenko, V. Vechernin, PoS (Baldin ISHEPP XXI) 077, arXiv:1212.2590 [nucl-th], 2012.
[3] J. Adam, et al. (ALICE Collaboration), JHEP **09**, 148 (2015), arXiv:1505.00664 [nucl-ex].
[4] K. Werner, B. Guiot, Iu. Karpenko, T. Pierog, G. Sophys, DESY Conf. Proc. 2016-01, p. 66-70, arXiv:1602.03414 [nucl-th] (2016)
[5] E. G. Ferreiro and C. Pajares, arXiv:1501.03381 [hep-ph] (2015).
[6] C. Flensburg, G. Gustafson, L. Lonnblad, Eur. Phys. J. C **60**, 233 (2009).
[7] G. Gustafson, Acta Phys. Polon. B **40**, 1981 (2009), arXiv:0905.2492 [hep-ph].
[8] M. A. Braun, C. Pajares, V. V. Vehernin, Phys. Lett. B **493**, 54 (2000).
[9] M. Braun, R. Kolevatov, C. Pajares, V. Vehernin, Eur. Phys. J. C **32**, 535 (2004).
[10] V. Kovalenko, PoS (QFTHEP 2013) 052 (2013).
[11] J. Schwinger, Phys. Rev. **82**, 664 (1951); T. S. Biro, H. B. Nielsen, and J. Knoll, Nucl. Phys. B. **245**, 449 (1984).
[12] V. Kovalenko, V. Vechernin, J. Phys. Conf. Ser. **668**, 012065 (2016), arXiv:1509.06696 [hep-ph].
[13] V. N. Kovalenko, A. M. Puchkov, G. A. Feofilov, Bull. Russ. Ac. Sc. Phys. **80**, 966 (2016) preprint http://hdl.handle.net/11701/1743 .

EPJ Web of Conferences **164**, 08003 (2017) DOI: 10.1051/epjconf/201716408003
ICNFP 2016

Counting of oscillatory modes of valence quarks forming qqq baryons for 3 quark flavors u,d,s

Sonia Kabana[1,2,a] and Peter Minkowski[2,b]

[1] *University of Nantes, Nantes, France*
[2] *Albert Einstein Center for Fundamental Physics - ITP, University of Bern, Bern, Switzerland*

Abstract. We present a selection the unique properties of oscillatory modes of $N_{fl} = 3$ light quarks $-u, d, s-$, using the SU $\left(2\, N_{fl} = 6\right) \times$ SO3 $\left(\vec{L}\right)$ broken symmetry classification. $\vec{L} = \sum_{n=1}^{N_{fl}} \vec{L}_n$ stands for the space rotation group generated by the sum of the three individual angular momenta of quarks in their c.m. system. The baryonic multiplets are shown to emerge from the picture of oscillating quarks in 3 space dimensions in the center of mass system of the baryons. All oscillatory modes are fully relativistic with a finite number of oscillators and ths is forming the unique harmonic oscillators with these properties. The density of states as a function of mass-square is calculated. This estimate is of relevance for the accounting of the missing states of unobserved hadrons, as the here estimated baryonic multiplets include both the observed and the unobserved (or "missing") hadrons. The estimate is conceptually different from Hagedorn's model and is based on field theory of QCD.

1. Introduction

The underlying field theory of strong interaction, QCD, assigns a local spin 1/2 field to every valence quark and antiquark with the quantum numbers color and flavor, interacting locally with spin 1 color octet gauge fields, called gluons. The gauge interactions of the gluons are exactly conserved gauge transformations based on the local gauge group SU(3) of color. The field theoretic concepts were developed as core topics of an in depth investigation together with the oscillatory modes of quarks suggested by linear Regge trajectories [1].
The linear Regge trajectories (e.g. ref. [1] and [2]) with exception of the pomeron, suggest oscillatory behaviour, whereas the difficulty to describe them is due to the fact that the field theory variables fall into the perturbatively non accessible domain. Lattice QCD in principle is set to solve the non-perturbative problems ref. [3] .
Within the theory of Quantum Chromo Dynamics the mesons and baryons are arranged in multiplets while it was already found since the mid-1960'ies that many states are missing refs. [4], [5], [6] in the sense that they have not been detected by experiments even though they are predicted by the quark model. The non observation of some states can be due for example to resonance states overlapping or being very wide. The missing states remain missing up to the present.

[a]e-mail: kabana@mail.cern.ch
[b]e-mail: mink@itp.unibe.ch

In this paper we are constructing the full theoretically expected multiplets of baryons including those measured and those that are missing, and count their states, based on the oscillatory modes for quarks and antiquarks.
The oscillatory modes of valence quarks and antiquarks in mesons and baryons were developed by one of us (P.M.) in CALTECH from 1975 onwards and published in ref. [1], ref. [7], ref. [8] . In reference [8] only non strange hadrons have been considered, which have been compared to a good compilation of missing states of non strange baryons, as can be found in PDG 1980. However there are missing states also among the strange hadrons ref. [9] or ref. [10] .
The main idea of this approach is to use relative coordinates and associated canonically conjugate momenta in the center of mass of $q\overline{q}^{\,\prime}$ and $q\,q^{\prime}\,q^{\prime\prime}$ systems to describe harmonic oscillations of the valence quarks and antiquarks, with full special relativistic invariance using the zero width approximation for resonances.
We are separating quark antiquark spins combining them to baryonic sub-multiplets.
Oscillator modes mean by definition gaussian wavefunction in the ground state, either in configuration space or in momentum space.
With respect to light quark masses u,d,s we take the chiral limit. Further the limit considered is the long range approximation between quarks and antiquarks which involves large main oscillatory quantum number N.
The model is valid strictly for large N. For small and finite N the multiplets are not degenerate for different I, J, P, C quantum numbers of particles, deviating from the large N asymptoytic behaviour.
Many modes are unstable against strong decays and are approximated in the zero width limit.
A perturbatively renormalizable field theory does not allow an exponential growth of mass- and/or masssquare-density of such states because it would lead to an infinite number of free constants, which render the theory non-renormalizable. In string theory there is an infinite number of independent oscillators (like many springs) one each for the ground frequency ω, 2ω, 3ω etc. This cannot be described by a field theory which contains only a finite number of degrees of freedom. The string theory allows a spectrum of density growing exponentially with mass(square).
The prediction of Hagedorn's bootstrap model [11], was an exponential spectrum of hadron resonances , which later was shown to be realized in string theory, refs. [12], [13], [14], [15]. In addition the limiting temperature resulting from the exponential hadron resonance spectrum in Hagedorn's model was associated with the critical temperature of the QCD phase transition between hadrons and partons [16] . However if a limiting temperature exists from the hadron spectrum this is not nessecarily associated with a phase transition.
In string theory there is a limiting temperature à la Hagedorn, which is determined from the string tension i.e. of the order of the Planck mass and it is not nessecarily associated with a phase transition.
An exponentially growing spectrum of hadron resonances would not agree with the expectation for an underlying field theory of strong interactions.
The approximate use of a broken extended symmetry combining $SU6 = SU\left(2\left(\sum_q \vec{S}_q\right) \times 3_{fl}\right)$ with $SO3\left(\sum_q \vec{L}_q\right)$ was vigorously pursued up to the end of the 60-ies, whence the three valence quarks wave function is conceived in their c.m. frame. Other orbital $q\overline{q}^{\,\prime}$ and 3-quark wave functions were also considered, as e.g. the $p_z \to \infty$ states, where a boosted $SU6_w$ symmetry is introduced instead of $SU6$. In this respect the main quantum number N was not related to genuine oscillator degrees of freedom but instead inferred from recurrences along Regge trajectories from the relation

$$\alpha^{\prime} M^2(J) = J + J_0 \text{ with } \alpha^{\prime} \Delta M^2 = \Delta J = \Delta N \tag{1}$$

In eq. 1 $\vec{J} = \vec{L} + \vec{S}$ denotes total angular momentum.

There is some literature quoted in ref. [8], which is however very limited, and among the many papers on that matter we mention here few representative ones by J. Schwinger [4], R. Dashen and M. Gell-Mann [5] and by G. Zweig [6], illustrating the search for local field variables underlying the strong interactions at that time.
Historically, one result from the investigations and efforts towards identifying the known baryon resonances, was the finding that several states were missing. The counting of oscillatory modes of baryons with light flavored u, d, s quarks, that we report here, is a new investigation.
The estimate and counting of oscillatory modes of mesons with light flavored u, d, s quarks and corresponding antiquarks has been published previously in ref. [17]. In the following we will elaborate on the oscillating modes of baryons and compare with the measured ones.

Oscillatory modes of quarks in baryons reveal many properties going beyond the quark model, e.g. the numbers of modes with fixed main quantum number N.
We do not discuss here the techniques derived from Young tableaux, applied to $SU\left(2N_{fl}=6\right) \times SO3\left(\vec{L}\right)$. Instead let us quote the book by Hermann Weyl [18] .

2. Oscillatory modes for u,d,s baryons

4. Dynamics - Poincaré-invariant harmonic oscillator modes for baryons in the space of barycentric coordinates for 3 valence quarks u , d , s

Seen from the time line of the first years of the 70-th it took about 9 years until an explicit construction of oscillator modes of baryons exhibiting a density function

$$\varrho_n\left(m^2\right) = \frac{\partial N\left(m^2\right)}{\partial m^2} \quad ; \; N = m^2 \cdot \alpha' \tag{2}$$

where α' is the slope of Regge trajectories, and

$$\varrho_n\left(m^2\right) \sim \left(\frac{m^2}{m_0^2}\right)^u \quad \text{for } m \to \infty \tag{3}$$
$$m_0 ,\, u \; : \; \text{characteristic parameters and } u = \text{positive integer}$$

for valence quark (antiquark) configurations in mesons and baryons
– ref. [8] – compatible with a finite number of degrees of freedom .
Figure 1 shows the logarithm of the density of states of baryons defined in formula 3, together with the density of states of mesons from ref. [17], [19] as a function of the main quantum number N plus one. This calculation (shown here for N+1 until 10) includes only oscillatory modes for u,d,s quarks and antiquarks for qqq baryons and $q\bar{q}$ for mesons, i.e. not including exotics. The shown density of states is valid in the limit of large N. Furthermore the here derived formulae do not include the intercept and assume a universal slope of Regge trajectories. This does not affect the counting of the number of these states. We observe that the baryon density of states grows faster than the meson density and is always larger than the latter. The data used in Fig. 1 come from Table 1 below.

DOI: 10.1051/epjconf/201716408003

N	z(Meson)	z(Baryon)
0	36	56
1	108	210
2	216	1044
3	360	2780
4	540	5766
5	756	12510
6	1008	22366
7	1296	38052
8	1620	62826
9	1980	97540

Table 1: Density of states for each main quantum number N until N=9, for mesons and baryons with u,d,s quarks and antiquarks.

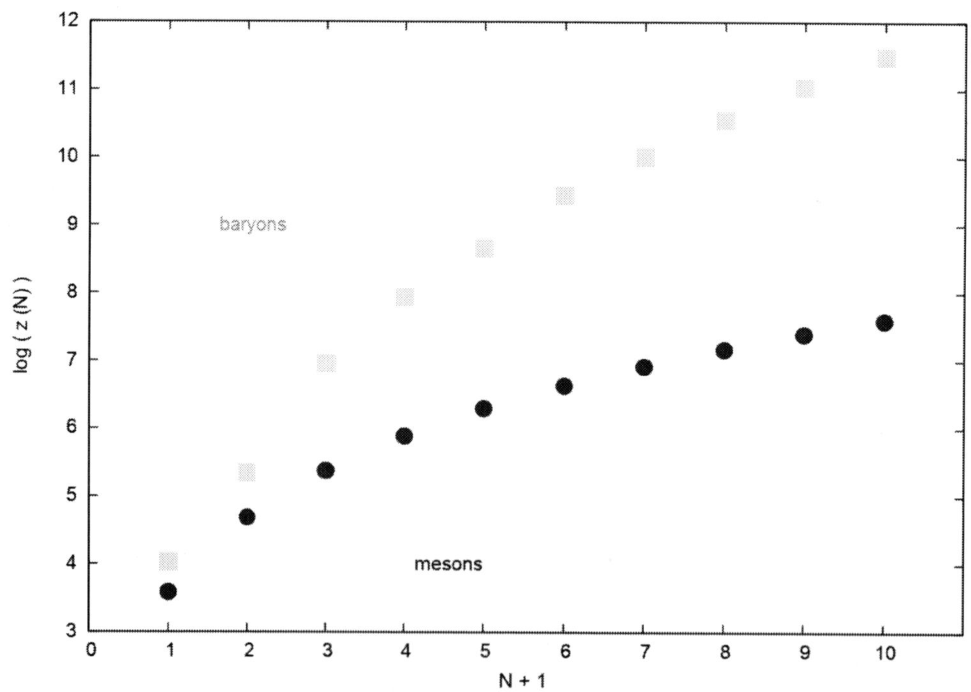

Figure 1. Fig. 1 : Logarithm of the density of states of baryons defined in eq. 3 together with the density of states of mesons from [17] as a function of the main quantum number N plus one.

We envisage baryons within QCD , allowing *temporarily* the local color gauge group to be generalized from 3 colors to $1 < N_c \rightarrow N$ to simplify notation , whenever no confusion arises . Baryons are : fermions for N odd , bosons for N even . A concise sketch of barycentric canonically conjugate momentum and spatial coordinates – pairs of three vectors

$$(\vec{\pi}_\nu , \vec{z}_\nu) \; ; \; \nu = 1 , 2 , \cdots , N \tag{4}$$

follows .

Barycentric coordinates in the overall c.m. system

$$\begin{array}{ll}
\vec{\pi}_1 = \frac{1}{\sqrt{2}} (\vec{p}_1 - \vec{p}_2) \quad , & \vec{z}_1 = \frac{1}{\sqrt{2}} (\vec{x}_1 - \vec{x}_2) \\
\vec{\pi}_2 = \frac{1}{\sqrt{6}} (\vec{p}_1 + \vec{p}_2 - 2 \vec{p}_3) \; , & \vec{z}_2 = \frac{1}{\sqrt{6}} (\vec{x}_1 + \vec{x}_2 - 2 \vec{x}_3) \\
\cdot & \cdot \\
\cdot & \cdot \\
\vec{\pi}_\nu = (\nu(\nu+1))^{-1/2} \begin{pmatrix} \sum_{\alpha=1}^{\nu} \vec{p}_\alpha \\ -\nu \, \vec{p}_{\nu+1} \end{pmatrix} , & \vec{z}_\nu = (\nu(\nu+1))^{-1/2} \begin{pmatrix} \sum_{\alpha=1}^{\nu} \vec{x}_\alpha \\ -\nu \, \vec{x}_{\nu+1} \end{pmatrix} \\
\cdot & \cdot \\
\cdot & \cdot \\
\vec{\pi}_{N-1} = \cdots \quad , & \vec{z}_{N-1} = \cdots \\
\hline
\vec{\pi}_N = N^{-1/2} \sum_{\alpha=1}^{N} \vec{p}_\alpha \rightarrow 0 \quad , & \vec{z}_N = N^{-1/2} \sum_{\alpha=1}^{N} \vec{x}_\alpha \rightarrow 0
\end{array} \tag{5}$$

The last line in eq. 5 refers to c.m. momenta and positions , both fixed to vanish, they do not represent genuine degrees of freedom .

4.1. Completing canonically conjugate variables in the pairs in $\mathcal{L}_N$, t

In the local Lagrangean density setting of QCD the external quark masses, denoted M_q , appropriate multipliers of the scalar densities $\overline{q}q$ composing the mass term, enter in the form, also discussed recently in refs. 20 – [20] – and 21 – [21] –

$$- \mathcal{L}_{q-mass} = \sum_{flavors} \frac{z_q}{z_M} M_q \overline{q}^{\,c} q^{\,c} \tag{6}$$

The quark masses M_q are independent of global c.m. frame time t and of configuration space variables $\vec{z}_\nu$; $\nu = 1, 2, \cdots , N - 1$; the *mass functions* m_q do depend on the entire spacial configurarion of the (resonance-) state considered – always reduced to the overall c.m. system – and *therein* of time t .

For the dynamical meaning of c.m. time the implication of the last relation in eq. 5 repeated below

$$\vec{\pi}_N = N^{-1/2} \sum_{\alpha=1}^{N} \vec{p}_\alpha \rightarrow 0 \quad , \quad \vec{z}_N = N^{-1/2} \sum_{\alpha=1}^{N} \vec{x}_\alpha \rightarrow 0 \tag{7}$$

is central : the 'prima facie' canonical pair $\vec{\pi}_N$, $\vec{z}_N$ in eq. 7 is *removed* from the set of dynamical canonical (3-vector-)variable pairs

$$(\vec{\pi}_1 , \vec{z}_1) \, , \cdots , (\vec{\pi}_{N-1} , \vec{z}_{N-1}) \; ; \; \vec{\pi}_N = 0 \, , \vec{z}_N = 0 \tag{8}$$

 DOI: 10.1051/epjconf/201716408003

maintaining *kinematic* simultaneous vanishing displayed in eq. 8 .
Configuration space variables $\vec{x}_\nu$; $\nu = 1, \cdots N$ inherit from the relations displayed in eqs. 7 and 8 the following c.m.-equivalence holds

$$\vec{x}_\nu \underset{c.m.}{\simeq} \vec{x}_\nu - \vec{X} \; ; \; \nu = 1, \cdots, N \; ; \; \vec{X} = \frac{1}{N} \left(\Sigma_{\varrho=1}^{N} \vec{x}_\varrho \right) \tag{9}$$

which implies that each $\vec{x}_\nu$ is equivalent to a homogeneous linear combination of the barycentric coordinates $\vec{z}_\kappa$; $1 \leq \kappa \leq N-1$. c.m. time t is defined as *common* time coordinate of four vectors

$$\begin{array}{l} \vec{x}_\nu \rightarrow \{ x^\mu \}_\nu = (t , \vec{x}_\nu) \; ; \quad \begin{array}{l} \mu = 0,1,2,3 \\ \nu = 1, \cdots, N \end{array} \\ \text{consistent with the constraint : } \Sigma_{\nu=1}^{N} \vec{x}_\nu = 0 \; ; \\ \text{remembering } N = N_c \end{array} \tag{10}$$

4.2. Orbits in configuration space and velocities

The configuration space *variables* obeying c.m.-constraints and c.m. equivalence are together with c.m. time, as displayed in eq. 10

$$(t , \vec{x}_1) , (t , \vec{x}_2) , \cdots , (t , \vec{x}_N) \; ; \; \Sigma_{\varrho=1}^{N} \vec{x}_\varrho = 0 \tag{11}$$

with $\vec{x}_\nu$; $\nu = 1, \cdots, N$ conceived *independent* of t – whereas an *orbit* is formed by binding the space-variables $\vec{x}_\nu = f_\nu (t) \equiv \vec{x}_\nu (t)$, as expressed in the next equation

$$\begin{array}{l} (t, \vec{x}_1) \quad , (t, \vec{x}_2) \quad , \cdots , (t, \vec{x}_N) \quad ; \Sigma_{\varrho=1}^{N} \vec{x}_\varrho \quad = 0 \\ (t, \vec{x}_1 (t)), (t, \vec{x}_2 (t)), \cdots , (t, \vec{x}_N (t)) ; \Sigma_{\varrho=1}^{N} \vec{x}_\varrho (t) = 0 \end{array} \tag{12}$$

The definition of *orbit* in the second relation in eq. 12 leads to the definition of *velocities* maintaining c.m.-constraints and c.m.-equivalence

$$\begin{array}{l} \vec{v}_\nu (t) = \dfrac{d}{d t} \vec{x}_\nu (t) \equiv \dot{\vec{x}}_\nu \; ; \; \nu = 1 , \cdots , N \; ; \\ \text{with : } \Sigma_{\varrho=1}^{N} \vec{v}_\varrho (t) = 0 \end{array} \tag{13}$$

4.3. The main quantities $\mathcal{L}_N$ depends on

These quantities are compiled in the list of 3 items below

1) a universal scale parameter of dimension mass-square $\Lambda = \left(2 \alpha' \right)^{-1}$ related to the inverse slope of Regge trajectories , as well as quark masses, minimally M_q ; $q = u , d , s$, composing the mass term in the local QCD Lagrangean given in eq. 6 , where the finite renormalization constants z_q , z_m are to be chosen to implement exact Ward identities , which in turn specify M_q – refs. [20], [21] .

2) quark mass functions $m_\varrho \left(M_q ; \{ \vec{z}_\nu \} \right)$; $\nu = 1 , \cdots , N-1$; $\varrho = 1, \cdots, N$

3) velocities $\vec{v}_{\nu}(t) \equiv \dot{\vec{x}}_{\nu}$; $\nu = 1, \cdots, N$; with : $\sum_{\varrho=1}^{N} \vec{v}_{\varrho}(t) = 0$

This leads to the Ansatz – ref. [8] – for light and heavy quark flavors remembering that we use rational units with $\hbar = c = 1$.

$$\begin{aligned} \mathcal{L}_N &= -\sum_{\alpha=1}^{N} \left[m_{\alpha}^{2} - \sum_{\beta=1}^{N} \sum_{\gamma=1}^{N} Q_{\beta\gamma}^{\alpha} \vec{v}_{\beta} \vec{v}_{\gamma} \right]^{1/2} \\ m_{\alpha} &= m_{\alpha} \left(M_{\delta}, \Lambda ; \underline{z} \right) \; ; \; Q_{\beta\gamma}^{\alpha} = Q_{\beta\gamma}^{\alpha} \left(M_{\delta}, \Lambda ; \underline{z} \right) \\ \underline{z} &= \{ \vec{z}_1, \cdots, \vec{z}_{N-1} \} \end{aligned} \tag{14}$$

The limiting quantities in the *chiral* limit obtain as follows

$$\begin{aligned} & m_{\alpha} \left(M_{\delta}, \Lambda ; \underline{z} \right) \sim \overset{\circ}{m}_{\alpha} \left(\Lambda ; \underline{z} \right) + M_{\alpha} \qquad + o(M_{\delta}) \\ & Q_{\beta\gamma}^{\alpha} \left(M_{\delta}, \Lambda ; \underline{z} \right) \sim \overset{\circ}{Q}{}_{\beta\gamma}^{\alpha} \left(\Lambda ; \underline{z} \right) + M_{\varrho} A_{\beta\gamma}^{\varrho \mid \alpha} \Lambda^{\frac{1}{2}} + o\left(M_{\delta} \Lambda^{\frac{1}{2}} \right) \\ & \text{for } M_{\delta} \to 0 \end{aligned} \tag{15}$$

In ref. [8] the quantities'

$$\begin{aligned} m_{\alpha} \left(M_{\delta}, \Lambda ; \underline{z} \right) \quad &\underset{\text{chiral limit}}{\longrightarrow} \quad \overset{\circ}{m}_{\alpha} \left(\Lambda ; \underline{z} \right) \\ Q_{\beta\gamma}^{\alpha} \left(M_{\delta}, \Lambda ; \underline{z} \right) \quad &\longrightarrow \quad \overset{\circ}{Q}{}_{\beta\gamma}^{\alpha} \left(\Lambda ; \underline{z} \right) \end{aligned} \tag{16}$$

defined in eq. 15 were identified but shall be kept distinct here , for clarity of logic.

4.4. Remarks with respect to the chiral limit

$\chi 1$ The discussion of oscillatory modes of baryons in the chiral limit is simplified but the phenomenological application is limited to the light flavors u,d,s.

$\chi 2$ In the zero width approximation the appearance of 8 massless pseudoscalar goldstone bosons

$$\begin{aligned} \overset{\circ}{\pi}_{\sigma} &= \left(\pi^{+}, \pi^{0}, \pi^{-} ; \overline{K}^{0}, \overline{K}^{-} ; K^{+}, K^{0} ; \eta \right) ; \\ \sigma &= 1, \cdots, 8 ; \; m_{\overset{\circ}{\pi}_{\sigma}} = 0 \end{aligned} \tag{17}$$

does not enter directly , but the modes with large masses will decay to lower ones and one ore more $\overset{\circ}{\pi}$ - s – also in cascades . This is after removing infrared divergencies not significantly different for the realistic case of finite π_{σ} - masses . As a consequence such cascades of excited baryon resonance decays generate a pronounced feeding problem , aggravated if heavy ion collisions are performed at todays energy frontiers at RHIC and LHC , less so at FAIR .

$\chi 3$ In addition at high energy also hevay flavors c , b , t are or will be copiously produced aggravating the study of light flavored
(anti-)baryons .

On the phenomenological side the Δ Regge trajectory can be found in ref. 22 – ref. [22] – . The additional trajectories were erased in our subsequent abalysis.

DOI: 10.1051/epjconf/201716408003

4.5. $\mathcal{L}_N \rightarrow \overset{\circ}{\mathcal{L}}_N$ in the chiral limit

For logical consistency we can consider the chiral limit of arbitrary $N = N_c$ and N_{fl} light flavors of quark , as long as $N > 1$ and $N > \frac{2}{11} N_{fl}$, ensuring asymptotic (ultraviolet) freedom .
The Ansatz for $\overset{\circ}{\mathcal{L}}_N$ becomes

$$\overset{\circ}{Q}{}^{\alpha}_{\beta\gamma} = \overset{\circ}{m}{}^2_{\alpha} \frac{1}{K_N} \delta_{\beta\gamma} \; ; \; K_N \text{ : dimensionless , positive constant} \tag{18}$$

Eq. 14 reduces to

$$\begin{array}{c} \overset{\circ}{\mathcal{L}}_N = - \left(\sum_{\alpha=1}^{N} \overset{\circ}{m}_\alpha \right) \left[1 - \frac{1}{K_N} \, \upsilon^2 \right]^{\frac{1}{2}} \; ; \; \upsilon^2 = \sum_{\beta=1}^{N} \vec{\upsilon}_\beta \vec{\upsilon}_\beta \\ \hline \longrightarrow \overset{\circ}{m} = \sum_{\alpha=1}^{N} \overset{\circ}{m}_\alpha \; ; \; \overset{\circ}{m}_\alpha = \overset{\circ}{m}_\beta = \overset{\circ}{m} \, / N \quad \forall \; \alpha , \beta = 1, \cdots , N \end{array} \tag{19}$$

In eq. 19 $\upsilon = \sqrt{\upsilon^2}$ defines a Euclidean distance in a space over real numbers $R^{\widehat{d}}$ of dimension $\widehat{d} = d_{space} \times N$, with $d_{space} \rightarrow 3$, yielding the associated scalar product

$$\underline{w} = \{ \vec{w}_1 , \cdots , \vec{w}_N \} \longrightarrow \left(\underline{\upsilon}_2 \middle| \underline{\upsilon}_1 \right) = \sum_{\beta=1}^{N} \vec{\upsilon}_{2\beta} \vec{\upsilon}_{1\beta} \tag{20}$$

In $R^{\widehat{d}}$ the transformation to barycentric coordinates defined in eq. 5

$$\begin{array}{l} \underline{x} = \{ \vec{x}_1 , \cdots , \vec{x}_N \} \longrightarrow \underline{z} = \underline{z} \left(\underline{x} \right) = \{ \vec{z}_1 , \cdots , \vec{z}_N \} \\ \hline \vec{z}_1 = \frac{1}{\sqrt{2}} (\vec{x}_1 - \vec{x}_2) \\ \vec{z}_2 = \frac{1}{\sqrt{6}} (\vec{x}_1 + \vec{x}_2 - 2 \vec{x}_3) \\ \cdots \\ \vec{z}_\nu = (\nu(\nu+1))^{-1/2} \left(\sum_{\alpha=1}^{\nu} (\vec{x}_\alpha - \vec{x}_{\nu+1}) \right) \\ \cdots \\ \vec{z}_{N-1} = \cdots \; ; \; (\nu = N - 1) \\ \vec{z}_N = N^{-1/2} \sum_{\alpha=1}^{N} \vec{x}_\alpha = N^{1/2} \vec{X}_{c.m.} \end{array} \tag{21}$$

is an orthogonal $O\widehat{d}$ mapping, i.e. preserves the scalar product defined in eq. 20. [1]
The reduction from the $O\widehat{d}$ transformation , to be proven orthogonal , to its form in $R^{\widehat{d}} = R^N \otimes R^{d_{space}}$

$$\begin{array}{l} O^{(\widehat{d})} = O^{(N)} \otimes \P_{d_{space} \times d_{space}} \; ; \; O^{(N)} \longleftrightarrow O_{\mu\nu} \; ; \\ \mu , \nu = 1, \cdots , N \text{ with } \vec{z}_\mu = O_{\mu\nu} \vec{w}_\nu \end{array} \tag{22}$$

is not presented here , additional details can be found in ref. [23] and references cited therein.

The proof of orthogonality of $O^{(\widehat{d})}$ reduces to establish orthogonality of the matrix $O^{(N)}$ displayed in eq. 22 .

[1] in the *approximately* realistic case we have $N \equiv N_c = d_{space} = N_{fl} = 3$ *exactly* .

Two relations follow independently from the c.m. constraint in configuration space $\vec{X}_{c.m.} \sim 0$ from eqs. 18 - 22

$$\begin{array}{rcl} \sum_{\alpha=1}^{N} \vec{x}_{\alpha}^{\,2} & = & \sum_{\beta=1}^{N-1} \vec{z}_{\beta}^{\,2} + N \vec{X}_{c.m.}^{\,2} \\ & = & \sum_{\alpha=1}^{N} \left(\vec{x}_{\alpha} - \vec{X}_{c.m.} \right)^2 + N \vec{X}_{c.m.}^{\,2} \\ \hline \longrightarrow \sum_{\beta=1}^{N-1} \vec{z}_{\beta}^{\,2} & = & \sum_{\alpha=1}^{N} \left(\vec{x}_{\alpha} - \vec{X}_{c.m.} \right)^2 \end{array} \tag{23}$$

4.6. $\overset{\circ}{\mathcal{L}}_N \longrightarrow$ constructing canonical variables of Hamiltonian quantum mechanics

$$\begin{array}{l} \vec{p}_{\beta} = \overset{\circ}{\mathcal{L}}_{N\,,\,\vec{v}_{\beta}} = \dfrac{\overset{\circ}{m}}{K_N} \left[1 - \omega^2 \right]^{-1/2} \frac{1}{2} v^2_{\,,\,\vec{v}_{\beta}} \\ \qquad = \dfrac{\overset{\circ}{m}}{K_N} \left[1 - \omega^2 \right]^{-1/2} \vec{v}_{\beta} \qquad ; \\ \beta = 1, \cdots .N \\ \omega^2 = \dfrac{1}{K_N} \sum_{\gamma=1}^{N} v_{\gamma}^2 \; ; \; K_N = \dfrac{N}{N-1} \; ; \\ \overset{\circ}{m} = \sum_{\alpha=1}^{N} \overset{\circ}{m}_{\alpha} \\ \hline \end{array} \tag{24}$$

Eq. 24 establishes the N canonically conjugate pairs of $d_{space}-$vectors

$$\left\{ \vec{p}_{\beta} , \vec{x}_{\beta} \right\} \; ; \; \beta = 1, \cdots , N \tag{25}$$

generated through $\overset{\circ}{\mathcal{L}}_N$ with general properties outlined in eq. 19. These variables do *not* satisfy the c.m. equivalence $\longrightarrow$ constraints defined in eqs. 7 - 9 collected in eq. 26 below

$$\begin{array}{l} \vec{\pi}_N = N^{-1/2} \sum_{\alpha=1}^{N} \vec{p}_{\alpha} \to 0 \quad , \quad \vec{z}_N = N^{-1/2} \sum_{\alpha=1}^{N} \vec{x}_{\alpha} \to 0 \\ (\vec{\pi}_1 , \vec{z}_1) , \cdots , (\vec{\pi}_{N-1} , \vec{z}_{N-1}) \; ; \; \vec{\pi}_N = 0 , \vec{z}_N = 0 \\ \vec{x}_{\nu} \underset{c.m.}{\sim} \vec{x}_{\nu} - \vec{X} \; ; \; \nu = 1 , \cdots , N \; ; \; \vec{X} = \dfrac{1}{N} \left(\sum_{\varrho=1}^{N} \vec{x}_{\varrho} \right) \end{array} \tag{26}$$

Omitting many details , we proceed to construct *the* main conserved quantity, following the original ref. 8 – ref. [8]

$$\begin{array}{l} \mathcal{H}_N = \sum_{\alpha=1}^{N} \vec{v}_{\alpha} \vec{p}_{\alpha} - \overset{\circ}{\mathcal{L}}_N \; ; \; \sum_{\alpha=1}^{N} \vec{v}_{\alpha} \vec{p}_{\alpha} = \omega^2 \left[1 - \omega^2 \right]^{-1/2} \\ \overset{\circ}{\mathcal{L}}_N = - \overset{\circ}{m} \left[1 - \omega^2 \right]^{1/2} \longrightarrow \mathcal{H}_N = \overset{\circ}{m} \left[1 - \omega^2 \right]^{-1/2} \end{array} \tag{27}$$

From eqs. 24 and 27 we infer

$$K_N \vec{p}_{\beta}^{\,2} = \mathcal{H}_N^2 \frac{1}{K_N} \vec{v}_{\beta}^{\,2} \longrightarrow \begin{array}{rcl} K_N \sum_{\gamma=1}^{N} \vec{p}_{\gamma}^{\,2} & = & \mathcal{H}_N^2 \omega^2 \\ & = & \mathcal{H}_N^2 - \overset{\circ}{m}^2 \end{array} \tag{28}$$

The last relation on the lower right side of eq. 28 follows from the decomposition $\omega^2 = 1 - \left(1 - \omega^2 \right)$ and eqs. 24 and 27 . The structure of a *genuine* harmonic oscillator is brought a step nearer rearranging eq. 28

$$K_N \sum_{\gamma=1}^{N} \vec{p}_{\gamma}^{\,2} + \overset{\circ}{m}^2 = \mathcal{H}_N^2 \tag{29}$$

Since eq. 29 derives from a variational principle and yields consistent equations of motion both in the classical as well as the quantized interpretation the Ansatz for $\overset{\circ}{m}{}^2$ realizing a system of $s = (N-1)\, d_{space}$ linear harmonic oscillators [2]
implies the choice

$$\begin{aligned} &\overset{\circ}{m}{}^2 \propto \Sigma_{\gamma=1}^{N} \left(\vec{x}_\beta - \vec{X}_{c.m.} \right)^2 + \delta_{m^2} ; \\ &\delta_{m^2} : \text{ configuration space independent} \end{aligned} \tag{30}$$

The constant transforming the proportionality in eq. 30 into an equality has to be determined in agreement with canonically conjugate momenta and positions .
The result is, using also the relations in eq. 23

$$\begin{aligned} \mathcal{H}_N^2 &= \left[K_N \, \Sigma_{\alpha=1}^{N} \left(\vec{p}_\alpha \right)^2 \Big|_{\Sigma_{\beta=1}^{N} \vec{p}_\beta = 0} + \overset{\circ}{m}{}^2 \left(x_\gamma - X \right) \right] \\ &= \left[\begin{array}{l} K_N \, \Sigma_{\alpha=1}^{N} \left(\vec{p}_\alpha \right)^2 \Big|_{\Sigma_{\beta=1}^{N} \vec{p}_\beta = 0} + \\ + \dfrac{\Lambda^2}{K_N} \, \Sigma_{\alpha=1}^{N} \left(\vec{x}_\alpha \right)^2 \Big|_{\Sigma_{\beta=1}^{N} \vec{x}_\beta = 0} + \delta_{m^2} \end{array} \right] \end{aligned} \tag{31}$$

Eq. 31 reveals that $\mathcal{H}_N^2$ so defined *does* satisfy the c.m.-constraints , and thus by Poincaré invariance is the mass-square operator , henceforth denoted $\mathcal{M}^2 \equiv \overset{\circ}{\mathcal{M}}{}^2$. For simplicity of notation the superfix $\circ$, which stands for the chiral limit , in $\mathcal{M}^2$, is suppressed .
Next we transform momenta and positions to barycentric coordinates and cast the relations in eqs. 29 - 31 into the form

$$\overset{\circ}{m}{}^2 \left(x_\gamma - X \right) = \frac{\Lambda^2}{K_N} \, \Sigma_{\alpha=1}^{N-1} \vec{z}_\alpha^{\,2} + \delta_{m^2} \tag{32}$$

In eq. 32 Λ , of dimension mass-square , is the spring-tension [3].
Λ^2 is related to the gauge boson condensate of QCD ; a detailed discussion can be found in refs. [19, 23].
We arrive at the relation for $\mathcal{M}^2$ expressed through
$\widehat{d}_{c.m} = (N-1)\, d_{space}$ unconstrained , canonically conjugate pairs
$(\vec{\pi}_1 , \vec{z}_1) , \cdots , (\vec{\pi}_{N-1} , \vec{z}_{N-1})$ defined in eq. 5

$$\begin{aligned} &\mathcal{M}^2 = \left[K_N \, \Sigma_{\alpha=1}^{N-1} \left(\vec{\pi}_\alpha \right)^2 + \frac{\Lambda^2}{K_N} \, \Sigma_{\alpha=1}^{N-1} \left(\vec{z}_\alpha \right)^2 + \delta_{m^2} \right] \\ &\vec{\pi}_\alpha = \tfrac{1}{i} \partial_{\vec{z}_\alpha} \; ; \; \alpha = 1, \cdots , N-1 \; ; \; \left[\vec{\pi}_{\beta k} , \vec{z}_{\gamma l} \right] = \tfrac{1}{i} \delta_{\beta\gamma} \delta_{km} \P \\ &k , l = 1, \cdots , d_{space} \end{aligned} \tag{33}$$

The structure of the canonical variables in eq. 33 is intrinsically related to the trace anomaly in QCD, as derived in ref. [23] , [24].

---

[2] For the realistic case $N = d_{space} = 3$, s is = 6 .
[3] Spring is not string.

6. Conclusions and outlook

In the present paper we calculated the density of states for baryons and compared it to our previous calculation for mesons [17]. The counting of baryons is done here for the first time in a complete fashion and for the mesons it has been published in a previous paper [17]. This calculation is valid in the limit of large N. It includes only oscillatory modes for u,d,s quarks and antiquarks for qqq baryons and $q\overline{q}$, i.e. is not including exotics and heavy quarks. Furthermore the here derived formulae do not include the intercept and assume a universal slope of the Regge trajectory. Within this approximation all Regge trajectory intercepts are neglected. This however does not affect the counting of the number of these states. For N=0 we find 36 states of mesons and 56 states of baryons as expected from more general Quark Model considerations for u,d,s quarks and antiquarks, and the 56 states of the baryon octet with spin=1/2 and the baryon decuplet with spin=3/2.
The density of states of baryons and mesons rise as a power law following the construction of oscillatory modes of quarks. We observe that the baryon density of states grows faster than the meson density and is always larger than the latter. The mass square density for qqq baryons and $q\overline{q}$ mesons within the oscillatory modes approach goes like a power determined by the number of independent oscillators and the leading powers are 5 for baryons and 2 for mesons.
The calculation predicts the density of baryonic states including the missing states and does not grow exponentially as in Hagedorn's model, with respect to N (or mass-square). The fact that there is no limiting temperature in the oscillatory modes approach does not mean that there is no critical temperature for the QCD phase transition .
The estimate of missing hadron states is important to take into account in thermal models that fit the hadron yields and ratios in heavy ion collisions to extract the temperature and chemical potentials of the particle source. Such fits have been performed by many groups for example in references [25], [26], [27], [28], [29].
Furthermore, not only there are missing states of *qqbar* mesons and qqq baryons and their antiparticles, but probably also missing states of glueballs, tetraquarks, pentaquarks etc., which also influence the feeding and the thermal fits, while there are candidates for non excited glueball states, and strong candidates for tetraquarks and pentaquarks.
An aim for the future would be therefore to obtain main consequences from including the missing hadronic states for the description of thermal equilibrium as applying to the hadronic phase of QCD . A development of the current model would be to proceed along the path of counting oscillatory modes of heavy flavored qqq and $q\overline{q}$ hadrons as well as for multiplets for exotic states like tetraquarks, pentaquarks and glueballsi .

Acknowledgments

We would like to warmly thank the members of the Theory Division of CERN, and especially Dr. Gian Giudice, Dr Wolfgang Lerche, Prof. Ignatios Antoniades and Dr Luis Alvarez-Gaume, as Theory group leaders, and Ms. Elena Gianolio, Ms. Anne-Marie Perrin, Ms. Michelle Connor and Ms. Jeanne Rostant for their support. We also thank Prof. Moshe Moshe from the Technion University in Haifa, Israel for the information on the paper [30] .

References

[1] P. Minkowski and F. Halzen, 'Regge Parametrization for πN Scattering', Il Nuovo Cimento, Vol. 1A, N. 1, 1. January 1971, received 1.July 1970 .
[2] R.P. Feynman, M. Kislinger, F. Ravndal, Phys. Rev. D, vol. 3, number 11, 1 june 1971, p. 2706.
[3] K. G. Wilson, 'Confinement of quarks', Phys. Rev. D 10, 2445, 15. Oct. 1974 .

[4] J. Schwinger, 'A Ninth Baryon ?', Phys. Rev. Lett. 12 (1964) 237 .

[5] Roger F. Dashen, Murray Gell-Mann (Caltech), 1965, 'Approximate symmetry and the algebra of current components', Phys.Lett. 17 (1965) 142-145, DOI: 10.1016/0031-9163(65)90277-5 .

[6] G. Zweig , "Meson classification in the quark model", in "Meson Spectroscopy", edited by C. Baltay and A. H. Rosenfeld, 1968 , 485-496 .

[7] P. Minkowski, Asymptotic freedom-infrared instability, Bern University preprint BE-78-0360, September 1978, 24. pp., in New phenomena in lepton-hadron physics, ed. D. Fries and J. Wess (Plenum Press, New York, 1979) p. 315.

[8] P. Minkowski, 'On The Oscillatory Modes Of Quarks In Baryons', Nucl.Phys. B174 (1980) 258-268, DOI: 10.1016/0550-3213(80)90202-3.

[9] J. Beringer et al. (Particle Data Group), Phys. Rev. D 86 (2012) 010001 .

[10] C. Amsler, T. De Grand and B. Krusche, Quark Model, pdg.lbl.gov/2012/reviews/rpp2012-rev-quark-model.pdf .

[11] R. Hagedorn , Statistical thermodynamics of strong interactions at high energy, Nuovo Cimento Suppl. 3 (1965) 147-186 .

[12] G. Veneziano , 'Construction of a crossing - symmetric, Regge behaved amplitude for linearly rising trajectories', Nuovo Cim. A57 (1968) 190-197 .

[13] J. Schwarz , 'Superstring Theory', Physics Reports 89 (1982) 223 .

[14] Enrique Alvarez (CERN), M.A.R. Osorio (Madrid, Autonoma U.), 'Superstrings at Finite Temperature', Oct 1986, 22 pp., CERN-TH-4571 , // Phys.Rev. D36 (1987) 1175 .

[15] Steven S. Gubser (Princeton U.), Sergei Gukov (Princeton U. & Caltech & CIT-USC), Igor R. Klebanov (Princeton U.), Mukund Rangamani (Princeton U. & Caltech & CIT-USC), Edward Witten (Caltech & CIT-USC & Princeton, Inst. Advanced Study), 'The Hagedorn transition in non-commutative open string theory', Sep 2000, PUPT-1949, CALT-68-2296, CITUSC-00-050, J.Math.Phys. 42 (2001) 2749-2764.

[16] U. Heinz et al., 'The Hot QCD White Paper: Exploring the Phases of QCD at RHIC and the LHC', A White Paper on Future Opportunities in Relativistic Heavy Ion Physics, Yasuyuki Akiba (Nishina Ctr., RIKEN), Aaron Angerami (Columbia U.), Helen Caines (Yale U.) , Anthony Frawley (Florida State U.) , Ulrich Heinz (Ohio State U.) , Barbara Jacak (UC, Berkeley) , Jiangyong Jia (SUNY, Stony Brook, Chem. Dept.), Tuomas Lappi (Jyvaskyla U.) , Wei Li (Rice U.) , Abhijit Majumder (Wayne State U.), David Morrison (Brookhaven) , Mateusz Ploskon (LBNL, NSD) , Joern Putschke (Wayne State U.) , Krishna Rajagopal (MIT) , Ralf Rapp (Texas A-M) , Gunther Roland (MIT) , Paul Sorensen (Brookhaven) , Urs Wiedemann (CERN), Nu Xu (LBL, Berkeley) , W.A. Zajc (Columbia U.), arXiv:1502.02730 [nucl-ex] .

[17] 'Counting of oscillatory modes of valence quarks forming q-qbar mesons', Sonia Kabana, Peter Minkowski, Int.J.Mod.Phys. A31 (2016) no.07, 1650023.

[18] H. Weyl, 'The Theory of Groups and Quantum Mechanics' , Dover Books on Mathematics , June 1 , 1950 amazon.com

[19] P. Minkowski and S. Kabana, Oscillatory modes of quarks in baryons for 3 quark flavors u,d,s, EPJ Web of Conferences, Volume 71, 2014, 2nd International Conference on New Frontiers in Physics, ICNFP2013.

[20] P. Minkowski, 'Light quark mass ratios (mu:md:ms) from meson and bayyon mass splittings', Modern Physics Letters A Vol. 28, No. 25 (2013) 1360015 and references cited therein.

[21] J. H. Kühn, 'Precise heavy quark masses', Modern Physics Letters A 28, No. 25 (2013) 1360019.

[22] Shuchi Bisht, Navjot Hothi† and Gaurav Bhakuni, 'Phenomenological Analysis of Hadronic Regge Trajectories', EJTP 7, No. 24 (2010) 299–318.

[23] P. Minkowski, Embedding oscillatory modes of quarks in baryons in QCD, ccsem.infn.it/issp2013/index.html,and references cited therein.

[24] P. Minkowski, On the anomalous divergence of the dilatation current in gauge theories, Bern preprint 1976, URL: mink.itp.unibe.ch .

[25] Mapping out the QCD phase transition in multiparticle production, Sonia Kabana, Peter Minkowski (Bern U.), Oct 2000, 53 pp., New J.Phys. 3 (2001) 4, BUHE-00-09, BUTP-2000-26, DOI: 10.1088/1367-2630/3/1/ 304, e-Print: hep-ph/0010247 .

[26] The Strange border of the QCD phases, Sonia Kabana (Bern U.). Mar 2001. 36 pp. Published in Eur.Phys.J. C21 (2001) 545-555 BUHE-01-01, DOI: 10.1007/s100520100728, e-Print: hep-ph/0104001 .

[27] J. Stachel (Heidelberg U.), A. Andronic (Darmstadt, EMMI), P. Braun-Munzinger (Darmstadt, EMMI & Darmstadt, Tech. U. & Frankfurt U., FIAS), K. Redlich (Wroclaw U.), 'Confronting LHC data with the statistical hadronization model', Nov 19, 2013. 2 pp., J.Phys.Conf.Ser. 509 (2014) 012019, DOI: 10.1088/1742-6596/509/1/012019, Conference: C13-07-22.1 Proceedings, arXiv:1311.4662 .

[28] Andronic A, Braun-Munzinger P and Stachel J, Dec 2008, 9 pp., Thermal hadron production in relativistic nuclear collisions: The Hadron mass spectrum, the horn, and the QCD phase transition, Phys. Lett. B 673 (2009) 142.

[29] Status of the Thermal Model and Chemical Freeze-Out, J. Cleymans (UCT-CERN Res. Ctr. and Cape Town U.), Dec 22 2014, 5 pp. Published in EPJ Web Conf. 95 (2015) 03004, DOI: 10.1051/epjconf/20159503004, arXiv:1412.7045 [hep-ph].

[30] A Petermann 1965 Nuclear Physics 63 349.

Magnetic Field Effects on In-Medium Υ Dissociation

Johannes Hoelck[1,a], Felix Nendzig[1], and Georg Wolschin[1]
[1] *Institut für Theoretische Physik, Universität Heidelberg, Germany, EU*

Abstract. The electromagnetic fields during relativistic heavy ion collisions are calculated using a simple model which characterises the emerging quark-gluon medium by its conductivity only. An estimate of the average magnetic field strength experienced by the $b\bar{b}$ mesons produced in the collision is made. In a sufficiently strong magnetic field, the individual spins of $b\bar{b}$ mesons can align with the field leading to quantum mixing of the singlet and triplet spin configurations. The extent of this intermixture, however, is found to be negligible at field strengths occurring in heavy ion collisions at LHC energies.

1 Introduction

The suppression of Υ meson production in heavy ion collisions can be used as a probe for the properties of the quark-gluon medium expected to be formed during these events. Being part of the more massive $b\bar{b}$ meson family, the Υ is less prone to recombination than the relatively lightweight J/ψ meson which might offer the opportunity for a cleaner examination of the underlying suppression mechanisms.

We have interpreted recent LHC data [1] for the $\Upsilon(1S)$ state using a model with collisional damping, screening of the $b\bar{b}$ binding potential, gluon-induced dissociation, and hydrodynamic expansion of the fireball [2]. Concerning the $\Upsilon(2S)$ state, however, the model underestimates suppression in peripheral collisions which indicates the need for further suppression mechanisms to be included. Promising candidates are magnetic field effects which are expected to possess the required centrality dependence.

Following the calculation of the electromagnetic fields generated in heavy ion collisions, we will make a first attempt to investigate their effect on the $b\bar{b}$ mesons produced in the collision by studying the magnetically induced mixing of the spin singlet and triplet states.

2 Electromagnetic fields in heavy ion collisions

We start by examining a single point charge e moving with constant velocity $\boldsymbol{v}$ through a conducting quark-gluon medium. For a resting observer at position $\boldsymbol{r} = x\hat{\boldsymbol{e}}_x + y\hat{\boldsymbol{e}}_y + z\hat{\boldsymbol{e}}_z$ and time t, the charge's worldline is given by

$$\boldsymbol{L}(t) = t\hat{\boldsymbol{e}}_t + vt\hat{\boldsymbol{e}}_z\,, \tag{1}$$

[a] e-mail: hoelck@thphys.uni-heidelberg.de

 DOI: 10.1051/epjconf/201716408004

where $\boldsymbol{v} = v\hat{\boldsymbol{e}}_z$ and $\boldsymbol{L}(0) = \boldsymbol{0}$ by our choice of coordinates. For this observer, Maxwell's equations for the charge's electromagnetic fields $\boldsymbol{B}_1, \boldsymbol{E}_1, \boldsymbol{D}_1, \boldsymbol{H}_1$ take the form

$$\nabla \cdot \boldsymbol{B}_1 = 0\,, \tag{2}$$

$$\nabla \times \boldsymbol{E}_1 = -\frac{\partial \boldsymbol{B}_1}{\partial t}\,, \tag{3}$$

$$\nabla \cdot \boldsymbol{D}_1 = e\delta(x)\delta(y)\delta(z - vt)\,, \tag{4}$$

$$\nabla \times \boldsymbol{H}_1 = \frac{\partial \boldsymbol{D}_1}{\partial t} + \sigma \boldsymbol{E}_1 + ev\delta(x)\delta(y)\delta(z - vt)\hat{\boldsymbol{e}}_z\,. \tag{5}$$

These equations can easily be solved in Fourier space but a re-transform to real space is infeasible since the dispersion relations of the quark-gluon medium's permittivity ε, permeability μ, and conductivity σ are unknown to date. To circumvent this problem, we adopt a very simple model for the medium's electromagnetic properties [3]: We neglect the medium's permittivity and permeability, i. e. set $\varepsilon \equiv \mu \equiv 1$, and characterise it by its finite conductivity only. The conductivity, in turn, is set to the constant value $\sigma = 5.8\,\mathrm{MeV}$ [3, 4].

Using this simple model, exploiting cylindrical symmetry, and considering only high-relativistic settings ($\gamma_v = 1/\sqrt{1 - v^2} \gg 1$), we arrive at the approximate solution

$$E_1^\rho = \frac{e}{4\pi}\frac{\rho\sigma}{2t_{\mathrm{eff}}^2}\mathrm{e}^{-\frac{\rho^2\sigma}{4t_{\mathrm{eff}}}}\,, \qquad B_1^\rho = 0\,, \tag{6}$$

$$E_1^\varphi = 0\,, \qquad B_1^\varphi = \mathrm{sgn}(v)E_\rho\,, \tag{7}$$

$$E_1^z = \mathrm{sgn}(v)\frac{e}{4\pi}\frac{\rho^2\sigma/4 - t_{\mathrm{eff}}}{\gamma_v^2 t_{\mathrm{eff}}^3}\mathrm{e}^{-\frac{\rho^2\sigma}{4t_{\mathrm{eff}}}}\,, \qquad B_1^z = 0\,, \tag{8}$$

$$\boldsymbol{D}_1 \equiv \boldsymbol{E}_1\,, \qquad \boldsymbol{H}_1 \equiv \boldsymbol{B}_1\,, \tag{9}$$

where we use the shorthand $t_{\mathrm{eff}} = t - z/v$ and cylindrical coordinates $\rho = \sqrt{x^2 + y^2}$, $\varphi = \mathrm{atan}(y/x)$.

One can think of a heavy ion nucleus as a nearly continuous distribution n_Z of point charges e normalised to the number of protons Z. At LHC beam rapidities, nuclei are Lorentz-contracted to almost 2-dimensional slabs. Thus, we model n_Z as being localised in z-direction while choosing a Woods-Saxon shape in the (x, y)-plane,

$$n_Z(\boldsymbol{r}) \propto \frac{\delta(z)}{1 + \exp\left(\frac{\rho - R}{a}\right)}\,, \tag{10}$$

where R and a denote the radius and diffuseness of the nucleus, respectively. Due to the linearity of Maxwell's equations, the nucleus' electromagnetic fields $\boldsymbol{E}_Z, \boldsymbol{B}_Z$ are given by the convolution of n_Z with the point charge fields $\boldsymbol{E}_1, \boldsymbol{B}_1$,

$$\begin{bmatrix}\boldsymbol{E}_Z\\ \boldsymbol{B}_Z\end{bmatrix}(t, \boldsymbol{r}) = \int \mathrm{d}^3r'\, n(\boldsymbol{r} - \boldsymbol{r}')\begin{bmatrix}\boldsymbol{E}_1\\ \boldsymbol{B}_1\end{bmatrix}(t, \boldsymbol{r}')\,. \tag{11}$$

Since n_Z is radially symmetric, $\boldsymbol{E}_Z, \boldsymbol{B}_Z$ and $\boldsymbol{E}_1, \boldsymbol{B}_1$ share the same cylindrical symmetries and vanishing components. As can be seen in Fig. 1, the general shape of the field components as functions of ρ and t_{eff} is not substantially altered by the convolution. The kink which can be seen in the plots of $|E_1^z|$ and $|E_Z^z|$ is the result of a change in sign occurring in the longitudinal field components when the point charge (or nucleus) passes by the observer. Due to time dilation, the observer perceives the sign

DOI: 10.1051/epjconf/201716408004

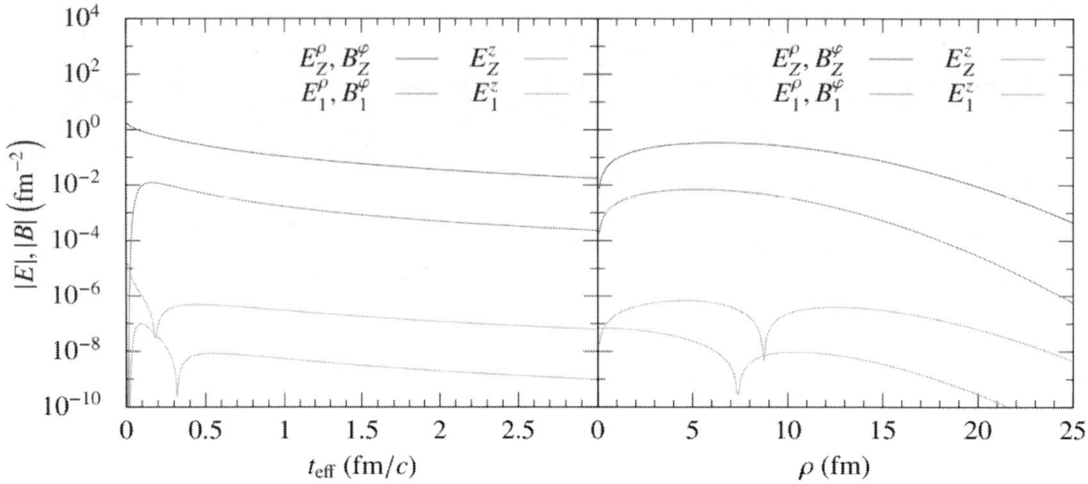

Figure 1. Electromagnetic field components for a single point charge and a Pb nucleus with artanh(v) = 7.987 as a function of t_{eff} with $\rho = 6.62$ fm (left) and as a function of ρ with $t_{\text{eff}} = 0.4$ fm/c (right).

change at $t_{\text{eff}} > 0$ for $\rho > 0$. The exact point in time (and space) is slightly shifted for the convoluted fields because of the now finite-sized charge distribution.

By summing the fields of two antiparallel moving nuclei with velocities $\pm v_{\text{beam}}$, shifted in x-direction by the impact parameter $\boldsymbol{b} = b\hat{\boldsymbol{e}}_x$, we finally obtain the total electromagnetic field in a heavy ion collision,

$$\begin{bmatrix}\mathcal{E}\\ \mathcal{B}\end{bmatrix}(t,\boldsymbol{r}) = \begin{bmatrix}\boldsymbol{E}_Z\\ \boldsymbol{B}_Z\end{bmatrix}_{v=+v_{\text{beam}}}(t,\boldsymbol{r}+\boldsymbol{b}/2) + \begin{bmatrix}\boldsymbol{E}_Z\\ \boldsymbol{B}_Z\end{bmatrix}_{v=-v_{\text{beam}}}(t,\boldsymbol{r}-\boldsymbol{b}/2)\,. \tag{12}$$

3 Effective average field strength for b$\bar{\text{b}}$ mesons

The electromagnetic field representations derived so far are only valid for observers resting in the coordinate system (t, x, y, z) which can be identified with the laboratory frame in heavy ion collisions. To examine field effects on moving observers, we need to change to their rest frames and transform the fields accordingly.

If we neglect any transverse momentum, all b$\bar{\text{b}}$ mesons produced in the initial collision share the following properties:

- A constant velocity $\boldsymbol{u}$,
- movement parallel to the heavy ion nuclei, $\boldsymbol{u} = u\hat{\boldsymbol{e}}_z$, and
- a starting position in the $(z = 0)$-plane at $t = 0$, the time of the initial collision.

The restframe of this kind of observers coincides with the *longitudinally co-moving frame* (LCF) [5], which is reached through the coordinate transform $(t, x, y, z) \mapsto (\tau, x, y, \vartheta)$ with

$$\tau = \sqrt{t^2 - z^2}\,, \qquad \vartheta = \operatorname{artanh}(z/t)\,. \tag{13}$$

For the observers, the coordinate τ is always equal to their proper time $\sqrt{1-u^2}t = t/\gamma_u$ and the coordinate ϑ corresponds to their rapidity artanh(u).

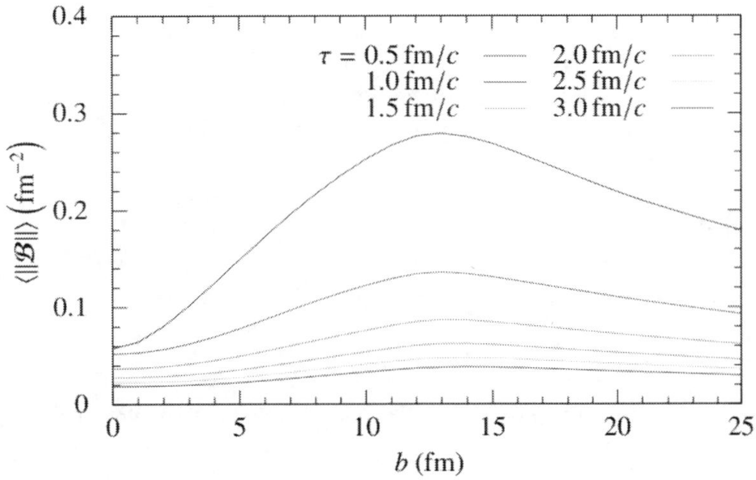

Figure 2. Average magnetic field strength $\langle\|\mathcal{B}\|\rangle$ in a PbPb collision with $\mathrm{artanh}(v_{\mathrm{beam}}) = 7.987$ (corresponds to $\sqrt{s_{\mathrm{NN}}} = 2.76\,\mathrm{TeV}$) as a function of the impact parameter b at different proper times τ in the shared $\mathrm{b\bar{b}}$ restframe. Converting natural to SI units, $1\,\mathrm{fm}^{-2}$ corresponds to approximately 2×10^{14} Tesla.

An electromagnetic field $\boldsymbol{E}_{\mathrm{lab}}, \boldsymbol{B}_{\mathrm{lab}}$ in the laboratory frame translates into the new coordinates as

$$E^{x,y}_{\mathrm{LCF}} = \left[\cosh(\vartheta)\boldsymbol{E}_{\mathrm{lab}} + \sinh(\vartheta)\hat{\boldsymbol{e}}_z \times \boldsymbol{B}_{\mathrm{lab}}\right] \cdot \hat{\boldsymbol{e}}_{x,y}\,, \qquad E^{\vartheta}_{\mathrm{LCF}} = E^{z}_{\mathrm{lab}}\,, \tag{14}$$

$$B^{x,y}_{\mathrm{LCF}} = \left[\cosh(\vartheta)\boldsymbol{B}_{\mathrm{lab}} - \sinh(\vartheta)\hat{\boldsymbol{e}}_z \times \boldsymbol{E}_{\mathrm{lab}}\right] \cdot \hat{\boldsymbol{e}}_{x,y}\,, \qquad B^{\vartheta}_{\mathrm{LCF}} = B^{z}_{\mathrm{lab}}\,. \tag{15}$$

From our moving observers' perspective, this is equivalent to a Lorentz boost, since

$$\cosh[\mathrm{artanh}(u)] = \gamma_u\,, \qquad \sinh[\mathrm{artanh}(u)] = u\gamma_u\,. \tag{16}$$

In the following, we will need an estimate of the magnetic field strength experienced by $\mathrm{b\bar{b}}$ mesons produced in heavy ion collisions. For this purpose, we average the magnetic field magnitude with respect to the $\mathrm{b\bar{b}}$ meson distribution $n_{\mathrm{b\bar{b}}}$,

$$\langle\|\mathcal{B}\|\rangle\,(\tau, b) = \frac{\int \mathrm{d}x\,\mathrm{d}y\,\mathrm{d}\vartheta\; n_{\mathrm{b\bar{b}}}(x,y,\vartheta,b)\,\|\mathcal{B}_{\mathrm{LCF}}(\tau,x,y,\vartheta)\|}{\int \mathrm{d}x\,\mathrm{d}y\,\mathrm{d}\vartheta\; n_{\mathrm{b\bar{b}}}(x,y,\vartheta,b)}\,, \tag{17}$$

where we have transformed $\mathcal{B}$ from Eq. (12) according to (14). We assume $n_{\mathrm{b\bar{b}}}$ to be proportional to the number of nucleon-nucleon collisions T_{AA} in real space and Gaussian-shaped in rapidity-space,

$$n_{\mathrm{b\bar{b}}}(x,y,\vartheta,b) \propto T_{AA}(x,y,b)\exp\!\left(-\frac{\vartheta^2}{2\sigma_\vartheta^2}\right), \tag{18}$$

where $\sigma_\vartheta \approx 2.7$ based on a fit of recent CMS data on LHC $\mathrm{b\bar{b}}$ production [6].

In Fig. 2, the result is shown as function of the impact parameter b and selected values of τ. The portrayed time frame $0.5\,\mathrm{fm}/c \le \tau \le 3.0\,\mathrm{fm}/c$ is approximately equal to the lifetime of the quark-gluon medium, starting from the moment of $\mathrm{b\bar{b}}$ meson formation. For fixed τ, the magnetic field strength reaches its maximum at $b \approx 2R$ when the two nuclei barely touch. For smaller impact parameters, the individual magnetic fields interfere destructively, whereas for large b the nuclei field strengths decrease due to the large distances (cf. Fig. 1). With increasing proper time, $\langle\|\mathcal{B}\|\rangle$ decreases monotonically.

DOI: 10.1051/epjconf/201716408004

4 Magnetically induced Υ/η_b mixing

Since bottom quarks are spin-½ particles, $b\bar{b}$ mesons exist in singlet (η_b) and triplet (Υ) spin configurations.[1] We denote the latter by $\Upsilon^0, \Upsilon^+, \Upsilon^-$ corresponding to the associated spin projection quantum number $m_S \in \{0, \pm 1\}$. In analogy to positronium, we assume a $b\bar{b}$ interacts with magnetic fields primarily by its quark spins coupling to the field [7] with a combined magnetic moment

$$\hat{\mu}_{b\bar{b}} = \frac{Q_b}{m_b}\hat{s}_b + \frac{Q_{\bar{b}}}{m_{\bar{b}}}\hat{s}_{\bar{b}} = \frac{Q_b}{m_b}(\hat{s}_b - \hat{s}_{\bar{b}}) \tag{19}$$

subject to charge $Q_b = -Q_{\bar{b}}$, mass $m_b = m_{\bar{b}}$, and spin $s_b, s_{\bar{b}}$ of quark and antiquark. Hence, the Hamiltonian for a $b\bar{b}$ meson resting in a magnetic field $\boldsymbol{B}$ is given by

$$\hat{H} = \hat{H}\big|_{\boldsymbol{B}=\boldsymbol{0}} - \hat{\mu}_{b\bar{b}} \cdot \boldsymbol{B}\,. \tag{20}$$

The eigenstates of $\hat{H}\big|_{\boldsymbol{B}=\boldsymbol{0}}$ in spin space are well-known from its positronium analogue,

$$\left|\eta_b\right\rangle = \frac{|\uparrow\downarrow\rangle - |\downarrow\uparrow\rangle}{\sqrt{2}}, \qquad \left|\Upsilon^0\right\rangle = \frac{|\uparrow\downarrow\rangle + |\downarrow\uparrow\rangle}{\sqrt{2}}, \qquad \left|\Upsilon^+\right\rangle = |\uparrow\uparrow\rangle\,, \qquad \left|\Upsilon^-\right\rangle = |\downarrow\downarrow\rangle\,. \tag{21}$$

The symbols $|\uparrow\downarrow\rangle, |\downarrow\uparrow\rangle, |\uparrow\uparrow\rangle, |\downarrow\downarrow\rangle$ denote the spin alignments of the constituent quarks with the first (second) entry representing the (anti-)bottom quark spin. In the absence of magnetic fields, the spin part of $\hat{H}$ is isotropic and the actual orientation of $\uparrow, \downarrow$ irrelevant. If a field is present, we choose $\uparrow$ ($\downarrow$) to be equal (antiparallel) to the orientation of $\boldsymbol{B}$.

Treating $\hat{\boldsymbol{\mu}}_{b\bar{b}} \cdot \boldsymbol{B}$ as a perturbation, we can compute the spin eigenstates of $\hat{H}$ for finite $\boldsymbol{B}$ [8],

$$\begin{bmatrix} \left|\Psi_s\right\rangle \\ \left|\Psi_t^0\right\rangle \\ \left|\Psi_t^+\right\rangle \\ \left|\Psi_t^-\right\rangle \end{bmatrix} = \begin{bmatrix} \cos(\alpha) & -\sin(\alpha) & 0 & 0 \\ \sin(\alpha) & \cos(\alpha) & 0 & 0 \\ 0 & 0 & 1 & 0 \\ 0 & 0 & 0 & 1 \end{bmatrix} \cdot \begin{bmatrix} \left|\eta_b\right\rangle \\ \left|\Upsilon^0\right\rangle \\ \left|\Upsilon^+\right\rangle \\ \left|\Upsilon^-\right\rangle \end{bmatrix}. \tag{22}$$

As can be seen, the perturbed and unperturbed eigenstate basis are related by a rotation matrix mixing the singlet and $m_S = 0$ triplet states. The mixing angle α depends on the energy difference and magnetic momentum of η_b and Υ^0,

$$\tan(\alpha) = \frac{x}{1 + \sqrt{1 + x^2}}, \qquad x = \frac{2\boldsymbol{\mu} \cdot \boldsymbol{B}}{\left|E_{\eta_b} - E_{\Upsilon^0}\right|}, \qquad \boldsymbol{\mu} = \left\langle \eta_b \middle| \hat{\boldsymbol{\mu}}_{b\bar{b}} \middle| \eta_b \right\rangle = \left\langle \Upsilon^0 \middle| \hat{\boldsymbol{\mu}}_{b\bar{b}} \middle| \Upsilon^0 \right\rangle. \tag{23}$$

The $m_S = \pm 1$ spin states remain unaffected, $|\Psi_t^+\rangle = |\Upsilon^+\rangle$, $|\Psi_t^-\rangle = |\Upsilon^-\rangle$, since

$$\left\langle \Upsilon^+ \middle| \hat{\boldsymbol{\mu}}_{b\bar{b}} \middle| \Upsilon^+ \right\rangle = \boldsymbol{0} = \left\langle \Upsilon^- \middle| \hat{\boldsymbol{\mu}}_{b\bar{b}} \middle| \Upsilon^- \right\rangle. \tag{24}$$

To get an idea of the order of magnitude of the mixing, we roughly approximate

$$\boldsymbol{\mu} \cdot \boldsymbol{B} \approx \|\boldsymbol{\mu}\| \left\langle \|\mathcal{B}\| \right\rangle \tag{25}$$

using our estimate of the average magnetic field strength derived in Eq. (17). In Fig. 3, the overlap

[1] Strictly speaking, the names η_b and Υ only refer to states with angular momentum $L = 0$, while states with $L = 1$ are termed h_b (singlet) and χ_b (triplet), respectively. However, in the following, we will use the labels η_b and Υ for all states regardless of L to keep notation simple.

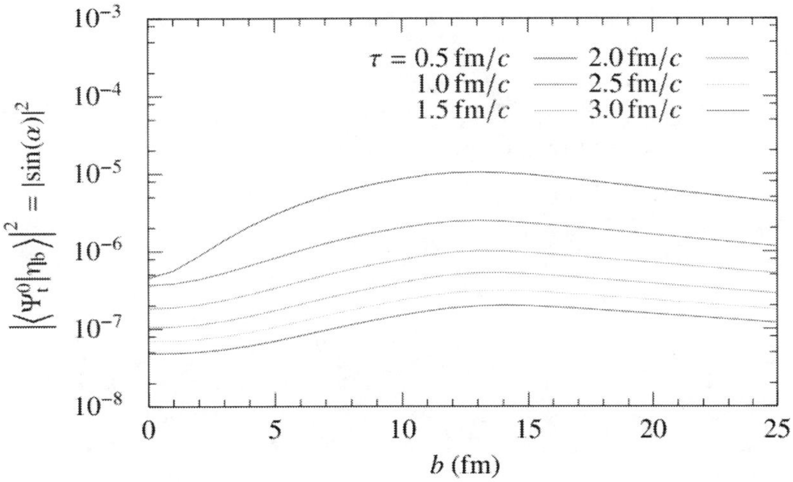

Figure 3. Overlap between the spin states $|\Psi_t^0\rangle$ and $|\eta_b\rangle$ with $N = 1, L = 0$ as function of the impact parameter b at different proper times τ using $\langle \|\mathcal{B}\| \rangle$ as magnetic field strength.

of the Ψ_t^0 and η_b spin states $|\langle \Psi_t^0 | \eta_b \rangle|^2 = |\sin(\alpha)|^2$ is plotted as a function of the impact parameter b and proper time τ. Being a monotonous function of $\langle \|\mathcal{B}\| \rangle$, the general shape of the overlap is similar to Fig. 2. However, even for the (b, τ) configuration yielding the highest magnetic field strength, the overlap is $\lesssim 10^{-5}$ and therefore negligible. In consequence, the magnetically perturbed states Ψ_s, Ψ_t^0 in relativistic heavy ion collisions are expected to behave virtually identically to their unperturbed counterparts η_b, Υ^0.

5 Outlook

Due to exposure to strong magnetic fields, the $b\bar{b}$ mesons in relativistic heavy ion collisions might behave differently from the ones produced in pp events where almost no electromagnetic fields are present. The magnetically perturbed spin singlet state $|\Psi_s\rangle$ and $m_S = 0$ triplet state $|\Psi_t^0\rangle$ are each composed of a superposition of the two unperturbed spin states $|\eta_b\rangle$ and $|\Upsilon^0\rangle$. Hence, properties of the perturbed states related to spin, such as the decay widths between excited states and the final decay into dimuon pairs, can differ compared to their unperturbed counterparts. However, the extent of this effect is found to be negligible at LHC energies.

Therefore, future investigation will focus on other possible interactions between Υ mesons and electromagnetic fields. Especially the fields' impact on $b\bar{b}$ formation and thereby the initial meson populations could prove important since the highest field strengths occur immediately after the collision where meson formation takes place. In addition, photon-gluon induced dissociation of $b\bar{b}$ pairs and dissociation due to increased potential energy may significantly affect in-medium decay widths for the excited $b\bar{b}$ states.

References

[1] S. Chatrchyan, et al., Phys. Rev. Lett. **109**, 222301 (2012)
[2] F. Nendzig, and G. Wolschin, J. Phys. G Nucl. Part. Phys. **41**, 095003 (2014)

[3] K. Tuchin, Phys. Rev. C **88**, 024911 (2013)
[4] H.-T. Ding, A. Francis, O. Kaczmarek, F. Karsch, E. Laermann, and W. Soeldner, Phys. Rev. D **83**, 034504 (2011)
[5] J. D. Bjorken, Phys. Rev. D **27**, 140 (1983)
[6] M. Jo, et al., Proc. Quark Matter Conf. Kobe 2015, in preparation; CMS-PAS-HIN-15-001
[7] P. Filip, Proc. Sci. **CPOD 2013**, 035 (2013)
[8] A. Rich, Rev. Mod. Phys. **53**, 127 (1981)

 DOI: 10.1051/epjconf/201716408005

Polarized parton densities and spin dependent structure function of the nucleon

M. Salimi-Amiri[1,a], A. Khorramian[1,b], H. Abdolmaleki[1,c], and A. Ghaffari-Turan[1,d]

[1] *Faculty of Physics, Semnan University, 35131-19111, Semnan, Iran*

Abstract. We study the polarized parton distribution functions (PPDFs) from recent experimental data at the next-to-leading order (NLO) approximation in the fixed-flavor number scheme. In this analysis, we can compare our results with experimental data such as very recent COMPASS data. It would be very worth to study the PPDFs parametrization form when we want to use this new COMPASS data in QCD analysis on polarized proton structure and parton distribution functions.

1 Introduction

Determination of fundamental spin structure in terms of their polarized quarks and gluons is one of the main challenges in polarized hadron physics. Despite the spin physics theory of polarized hadrons has been compatible with the experimental data of polarized nucleon, some fundamental aspects of the polarized parton distribution and structure function are still weakly determined. One of the important points in polarized hadron structure is understanding of nucleon's total angular momentum in terms of the single contributions of polarized parton. The total angular momentum of the polarized nucleon in terms of its parton spin, i.e. quark (Σ) and gluon spin(G), and also quark and gluon orbital momentum (L_q, L_g)is as following

$$\frac{1}{2} = \frac{1}{2}\Sigma(Q^2) + G(Q^2) + L_q(Q^2) + L_g(Q^2)\,, \tag{1}$$

In this equation, the first two terms present the fraction of the proton spin which carried by quarks and gluons. In fact, it would be worth to be able to measure each contribution in above equation. In this regard, we need to have the polarized parton distribution functions (PPDFs) as a function of x and Q^2. Also the dependence of the PPDFs on the Q^2 can be obtained using the DGLAP evolution equations [DGLAP] in perturbative QCD. Very recent analysis on polarized splitting functions in Mellin space has been reported at the next-to-next-to-leading order (NNLO) accuracy. Very recent theoretical and experimental developments in spin physics are addressed in Ref. [1].

We have different analysis to determine PPDFs parametrization such as simple polynomials approach, neural networks for PPDFs parametrization and etc such as DSSV [2–4], NNPDF [5, 6] and

[a]e-mail: Maryam.salimi.amiri@semnan.ac.ir
[b]e-mail: Khorramiana@semnan.ac.ir,Khorramiana@theory.ipm.ac.ir
[c]e-mail: Abdolmalki@semnan.ac.ir
[d]e-mail: A.Ghafary@semnan.ac.ir

JAM collaboration. On the other hand, we have a different kind of experimental data such as inclusive and semi-inclusive DIS data. COMPASS and JLAB have been reported very recent DIS measurements [7].

In Refs.[8–16] we have extracted PPDFs from different approaches, such as Jacobi polynomials for PPDFs parameterization up to NLO and different available experimental data.

This paper is organized as follows. In Sec. 2 we present the polarized proton structure in Mellin-Moment space. The functional form for PPDFs is presented in Sec. 3 at the NLO. Finally our results and conclusions are discussed in Sec. 4.

2 Formalism

According to Mellin N-moment space, the Mellin moment of function f which is depending on momentum fraction x is defined as:

$$\mathbf{M}[f,N] \equiv \int_0^1 dx\, x^{N-1}\, f(x)\,. \tag{2}$$

In this regard, the polarized nucleon structure function is defined in terms of PPDFs at the NLO as following [17]

$$\begin{aligned}\mathbf{M}[g_1,N] &= \frac{1}{2}\sum_q e_q^2\left\{\left[1+\frac{\alpha_s}{2\pi}\Delta C_q^N\right][\delta q(N,Q^2)+\delta\bar{q}(N,Q^2)]\right.\\ &\left.+\frac{\alpha_s}{2\pi}\,2\Delta C_g^N\delta g(N,Q^2)\right\}\,.\end{aligned} \tag{3}$$

In above equation, $\delta q(N,Q^2)$,$\delta\bar{q}(N,Q^2)$ and $\delta g(N,Q^2)$ are moment of polarized parton distribution functions. Also α_s is strong coupling constant which can be determined by the following equation at NLO [18]

$$\begin{aligned}\frac{1}{a_s(Q^2)} &= \frac{1}{a_s(Q_0^2)}+\beta_0\ln\left(\frac{Q^2}{Q_0^2}\right)\\ &-\beta_1\ln\left\{\frac{a_s(Q^2)[1+\beta_1 a_s(Q_0^2)]}{a_s(Q_0^2)[1+\beta_1 a_s(Q^2)]}\right\}\,,\end{aligned} \tag{4}$$

where $a_s=\frac{\alpha_s}{4\pi}$ and $\alpha_s(Q_0^2)$ is strong coupling constant at input scale. In this analysis, we fixed $N_f=3$ in the perturbative QCD evolution at Fixed-flavor number scheme at NLO with massless partonic flavors $\{u,d,s\}$. Also we fixed $\alpha_s(Q_0^2)=0.580$ which is related to $\alpha_s(M_Z^2)=0.118$.

Now by having the moments of polarized structure function, we have the polarized structure function $xg_1(x,Q^2)$ in x-space using Jacobi polynomials approach:

$$xg_1(x,Q^2) = x^\beta(1-x)^\alpha\sum_{n=0}^{N_{max}}\Theta_n^{\alpha\beta}(x)\times\sum_{j=0}^{n}c_j^{(n)}(\alpha,\beta)\,\mathbf{M}[xg_1,j+2]\,, \tag{5}$$

where N_{max} is the number of polynomials which we have considered $N_{max}=9$ and we used $\alpha=3.0$ and $\beta=0.5$ [9–12]. Also, in above equation $\Theta_n^{\alpha\beta}(x)$ is as following

 DOI: 10.1051/epjconf/201716408005

$$\Theta_n^{\alpha,\beta}(x) = \sum_{j=0}^{n} c_j^{(n)}(\alpha,\beta)\, x^j, \tag{6}$$

So by having $\mathbf{M}[xg_1, N]$ at any values of Q^2 in Eq. 3, one can determine the polarized structure $xg_1(x, Q^2)$ using Jacobi polynomials approach according to Eq. 4. The comparison of different values of N_{max} in the NLO approximation at $Q^2 = 2.5$ GeV^2 is presented in Fig. 1.

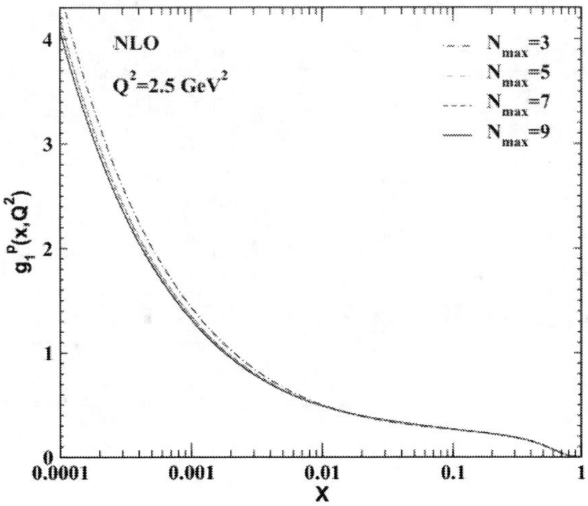

Figure 1. Our results for polarized structure function at $Q^2 = 2.5$ GeV^2 and for different values of N_{max}= 3, 5,7,9 in Jacobi polynomials approach.

3 Polarized Parton Distribution

In this analysis we used the AKS[13] parametrization at $Q_0^2 = 1$ GeV^2. According to this parametrization the polarized parton distribution functions ($\delta q = \delta u + \delta\bar{u},\ \delta d + \delta\bar{d},\ \delta\bar{u},\ \delta\bar{d},\ \delta\bar{s}$ and δg) at the NLO and the initial scale are as following:

$$x\,\delta q = \mathcal{N}_q \eta_q x^{a_q}(1-x)^{b_q}(1 + c_q x^{0.5} + d_q x)\,, \tag{7}$$

The normalization constant is $\mathcal{N}_q$

$$\frac{1}{\mathcal{N}_q} = \left(1 + d_q \frac{a_q}{a_q + b_q + 1}\right) B\left(a_q, b_q + 1\right) + c_q B\left(a_q + \frac{1}{2}, b_q + 1\right), \tag{8}$$

and $B(a, b)$ is the Euler beta function.

DOI: 10.1051/epjconf/201716408005

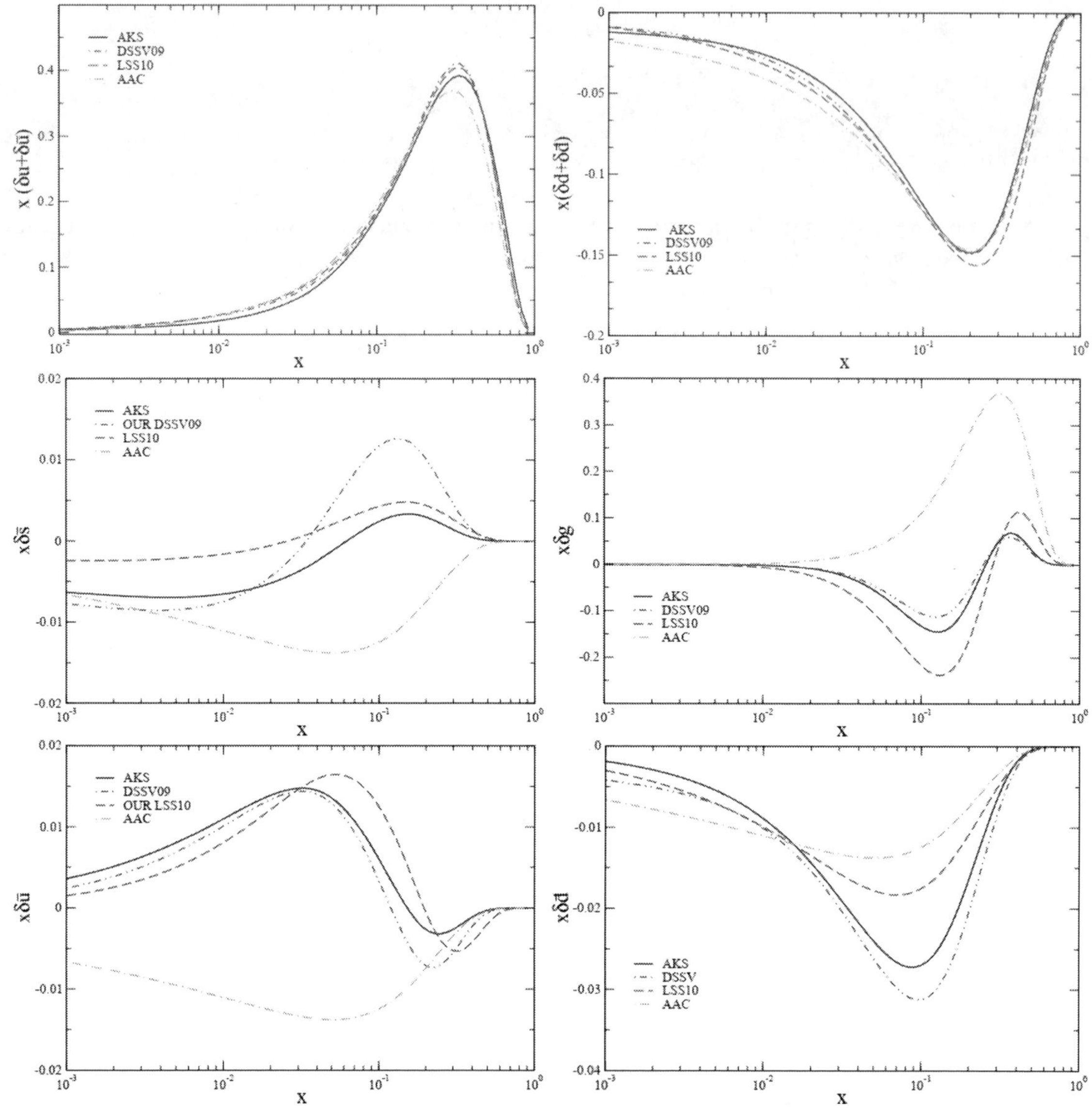

Figure 2. The comparison of parton polarized distribution functions for different parametrization in the NLO approximation, such as DSSV ,LSS, AAC, AKS with an initial scale.

In Fig. 2, we present the evolution of the distribution of polarized PDFs using AKS[13], DSSV[3], LSS[19], AAC[20] parametrization.

In Fig. 3 we show our results for polarized structure function xg_1 as a function of x at $< Q^2 >=$ 50 GeV^2 at the NLO. In this figure, we used the polarized parton distribution functions from Ref. [13] and compare our results with very recent COMPASS experimental data.

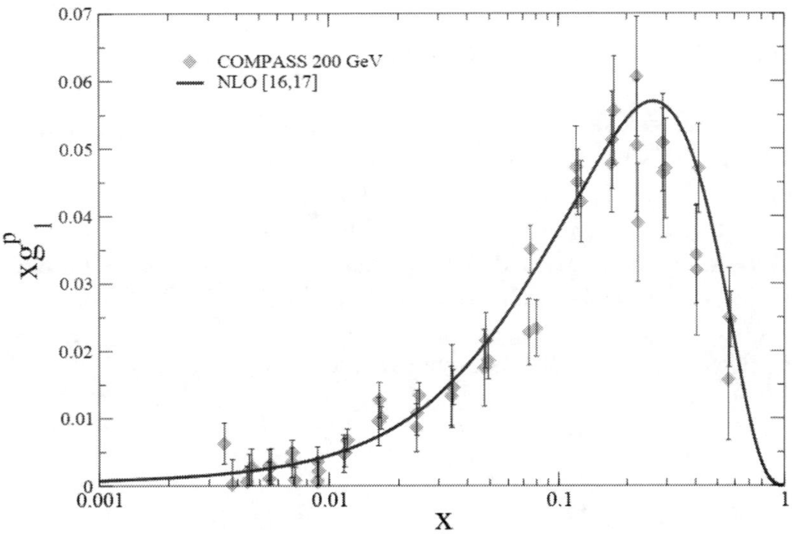

Figure 3. Comparison of polarized structure functionxg_1 with recent experimental COMPASS data at NLO approximation using the polarized parton distribution functions from Ref. [13].

4 Conclusion

In this analysis, we investigate the polarized parton distributions at the NLO to see how much compatible our previous parametrization with very recent experimental COMPASS data. Also, we found that the behavior of our parametrization is good agreement with very recent COMPASS data. Our analysis show us the Jacobi polynomial approach is very good approach for extracting the PPDFs in QCD analysis as well.

References

[1] E. Nocera and S. Pisano, arXiv:1608.08575 [hep-ph].
[2] D. de Florian, R. Sassot, M. Stratmann and W. Vogelsang, Phys. Rev. Lett. **101**, 072001 (2008)
[3] D. de Florian, R. Sassot, M. Stratmann and W. Vogelsang, Phys. Rev. D **80**, 034030 (2009)
[4] A. Bazavov *et al.*, Phys. Rev. Lett. **113**, no. 7, 072001 (2014)
[5] R. D. Ball *et al.* [NNPDF Collaboration], Nucl. Phys. B **874**, 36 (2013)
[6] E. R. Nocera *et al.* [NNPDF Collaboration], Nucl. Phys. B **887**, 276 (2014)
[7] C. Adolph *et al.* [COMPASS Collaboration], Phys. Lett. B **753**, 18 (2016)
[8] S. Taheri Monfared, Z. Haddadi and A. N. Khorramian, Phys. Rev. D **89**, no. 7, 074052 (2014) Erratum: [Phys. Rev. D **89**, no. 11, 119901 (2014)]
[9] A. N. Khorramian, S. Atashbar Tehrani and A. Mirjalili, Nucl. Phys. Proc. Suppl. **152**, 83 (2006).
[10] S. Atashbar Tehrani and A. N. Khorramian, JHEP **0707**, 048 (2007)
[11] A. N. Khorramian, S. Atashbar Tehrani, S. Taheri Monfared, F. Arbabifar and F. I. Olness, Phys. Rev. D **83**, 054017 (2011)
[12] A. N. Khorramian and S. Atashbar Tehrani, AIP Conf. Proc. **915**, 420 (2007).
[13] F. Arbabifar, A. N. Khorramian and M. Soleymaninia, Phys. Rev. D **89**, no. 3, 034006 (2014)

[14] F. Arbabifar, A. N. Khorramian and M. Soleymaninia, Acta Phys. Polon. Supp. **7**, no. 3, 585 (2014).

[15] F. Arbabifar, A. N. Khorramian, H. Khanpour and S. A. Tehrani, Nucl. Phys. Proc. Suppl. **234**, 66 (2013)

[16] A. N. Khorramian, S. Atashbar Tehrani, F. Olness, S. Taheri Monfared and F. Arbabifar, Nucl. Phys. Proc. Suppl. **207-208**, 65 (2010).

[17] B. Lampe and E. Reya, Phys. Rept. **332**, 1 (2000)

[18] A. Vogt, Comput. Phys. Commun. **170**, 65 (2005)

[19] E. Leader, A. V. Sidorov and D. B. Stamenov, Phys. Rev. D **82**, 114018 (2010) doi:10.1103/PhysRevD.82.114018 [arXiv:1010.0574 [hep-ph]].

[20] Y. Goto *et al.* [Asymmetry Analysis Collaboration], Phys. Rev. D **62**, 034017 (2000) doi:10.1103/PhysRevD.62.034017 [hep-ph/0001046].

EPJ Web of Conferences **164**, 08006 (2017) DOI: 10.1051/epjconf/201716408006
ICNFP 2016

Precise determination of the muon reconstruction efficiency in ATLAS at Run-II

Despoina Sampsonidou[1,a] on behalf of the ATLAS Collaboration

[1] *Aristotle University of Thessaloniki*

Abstract. In Run-2 of the LHC, the ATLAS experiment reconstruction algorithm has been improved and extended compared to the one used in Run-1. In this presentation, we will discuss the precise measurement of the muon reconstruction efficiency measured in pp collisions at $\sqrt(s)$ = 13 TeV in 2015 and 2016 using samples of $J/\psi \to \mu\mu$ and $Z \to \mu\mu$ decays. The reconstruction efficiency is measured using different methods in the various regions of the detector and for muon momenta between 6 and hundreds of GeV.

1 Introduction

These proceedings report on the precise measurement of the muon reconstruction efficiency measured in pp collisions at sqrt(s)= 13 TeV in 2015 using samples of $J/\psi \to \mu\mu$ and $Z \to \mu\mu$ decays. The reconstruction efficiency is measured using different methods in the various regions of the detector and for muon momenta between 6 and hundreds of GeV. Precise determination of the reconstruction efficiency of muons is essential in particle physics to accurately test Standard Model [2–4]processes or the search for processes beyond the Standard Model[5, 6].

2 The ATLAS Detector

A full description of the ATLAS detector can be found in Ref. [1]. Muons are identified and reconstructed by exploiting information from the Inner Detector (ID) and the Muon Spectrometer (MS) and to a lesser extent the calorimeters. The ID is the detector closest to the interaction point. It is designed to provide high precision tracking and high resolution momentum measurements for charged particles, covering pseudorapidity of $|\eta| < 2.5$. It uses magnetic field of 2T and consists of three subdetectors: 4 layers of pixel sensors and 4 layers of silicon strips with a pseudorapidity coverage up to 2.5, and 72 straw layers of Transition Radiation Tracker (TRT) with a pseudorapidity range up to 2.0.

The MS forms the outer part of the ATLAS detector. It provides precise measurement of muon momenta up to several TeV, covering pseudorapidity range of $|\eta| < 2.7$, and trigger coverage in the pseudo rapidity range of $|\eta| < 2.4$. The MS includes also magnetic field provided by three toroid magnet systems. An air-core toroid in the barrel region $0 < |\eta| < 1.6$ and two toroids in the end

[a] e-mail: despoina.sampsonidou@cern.ch

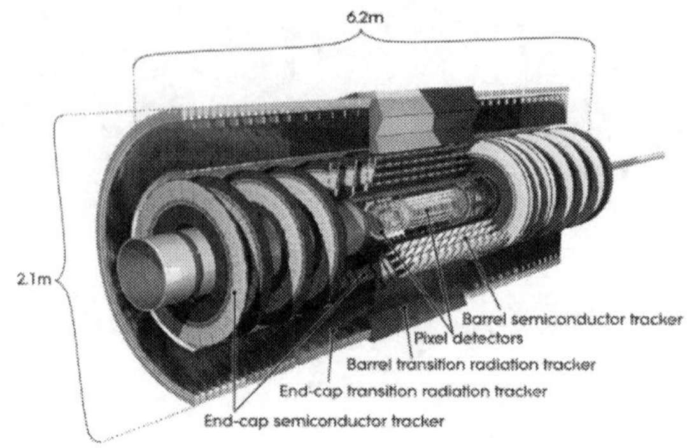

Figure 1. The Inner Detector

caps region of $1.4 < |\eta| < 2.7$ provide a mean value of 0.5T, allowing for stand-alone momentum measurement in principle without the need of combining the MS and the ID track reconstruction [9]. The MS consists of a barrel and two endcap sections, with three stations of trigger chambers, Resistive Plate Chambers (RPC) in the barrel and Thin Gap Chambers (TGC) in the end-caps and three stations of precision chambers, Monitored Drift Tube chambers (MDT). The innermost station in the endcaps was replaced by Cathode Strip Chambers (CSC).

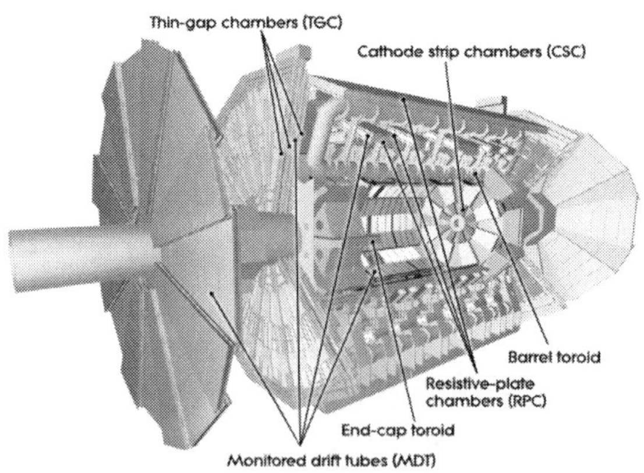

Figure 2. The Muon Spectrometer

3 Muon Identification

Muon identification is performed by combining information from the MS, the ID and the calorimeter. Reconstructed muons can be grouped into four different complementary categories depending on the sub detector systems used in the reconstruction:

 DOI: 10.1051/epjconf/201716408006

Stand-alone muons (MS): muon tracks that are reconstructed only in the MS and are extrapolated to the beam line accounting for energy losses. The MS magnetic field allows for stand-alone momentum measurements up to $|\eta| < 2.7$.

Combined muons (CB): They are muon tracks that are reconstructed in the MS and are associated to ID tracks. The best purity and momentum resolution is provided by the combination of the information from both detectors.

Segment-tagged muons: They are ID tracks which are extrapolated to the MS. These tracks are usually associated to track segments in the first layer of the MS that were not used in the reconstruction of a full track in the MS. The segment-tagged muons improve the total reconstruction efficiency since they correspond to low-pT muons which do not reach the middle and outer stations due to bending effects of the magnetic field.

Calorimeter-tagged muons: They are ID tracks which are reconstructed in a region with no coverage from the MS. This is a small region around $\eta = 0.0$ ($|\eta| < 0.1$). ID tracks associated with an energy deposit in the calorimeter which is consistent with a minimum ionizing particle, are considered as muon candidates. These muons are used to recover the acceptance loss in the non-instrumented region of the MS, even though they have the lowest purity of all types of muons.

4 Muon Reconstruction Efficiency

Depending on the kinematics and desired purity, the muons form four identification selections:

Loose muons: The Loose identification criteria are designed to maximise the reconstruction efficiency while providing good-quality muon tracks. They are specifically optimised for reconstructing Higgs boson candidates in the four-lepton final state. All muon types are used.

Medium muons: This selection minimises the systematic uncertainties associated with muon reconstruction and calibration. Only CB and MS muons are used.

Tight muons: Tight muons are selected to maximise the purity of muons at the cost of some loss in efficiency. Only CB muons with hits in at least two stations of the MS and satisfying the Medium selection criteria are considered.

High-p_T muons: The High-pT selection aims to maximise the momentum resolution for muons with transverse momentum above 100 GeV. The selection is optimised for searches for high-mass Z' and W' resonances [8, 9]. CB muons passing the medium selection and having at least three hits in three MS stations are selected.

In the region $|\eta| < 2.5$, the muon reconstruction in the ID and the MS is performed independently, a precise determination of the muon reconstruction efficiency is obtained by using the tag-and-probe method. In the region $2.5 < |\eta| < 2.7$, the reconstruction is performed solely in the MS, so a different methodology is used to estimate the efficiency in this eta region.

4.1 Reconstruction Efficiency for $|\eta| < 2.5$

The reconstruction efficiency is measured using a tag-and-probe method on $Z \to \mu\mu$ and $J/\psi \to \mu\mu$ events [7]. A selection of an almost pure muon sample from $Z \to \mu\mu$ and $J/\psi \to \mu\mu$ events is made. The Z decays provide a sample of probes with $p_T > 10$ GeV and the J/ψ decays provide a sample of 5 GeV< $p_T < 15$ GeV. A leg of the decay is required to be a medium muons that triggers the event (tag). The second leg of the decay is required to be an ID track or an Id track with calorimeter tagging (if the reconstruction efficiency of the MS is measured) or an MS track (if the reconstruction efficiency of the ID is measured). The fraction of probe tracks that were successfully reconstructed as muons measures the efficiency. In order for a track to be considered as successfully reconstructed muon, a reconstructed muon is required to be found within a cone in the $\eta - \phi$ plane of size $\Delta R = \sqrt{\Delta\eta^2 + \Delta\phi^2} < 0.05$ around the probe track. The efficiency measurement is carried out in data and simulation and a scale factor is computed as the ratio of the two. The scale factors are applied to the MC in order to correct for a possible mismodelling of the muon reconstruction efficiency. In Fig. 3,4 the measured muon reconstruction efficiencies for varying η for medium and high-p_T muons from $Z \to \mu\mu$ samples are presented whereas in Fig. 5 the muon reconstruction efficiencies with varying η for loose muons from $J/\psi \to \mu\mu$ is presented. In Fig. 6, the muon reconstruction efficiency as a function of transverse momentum is presented, with measurements from both $Z \to \mu\mu$ and $J/\psi \to \mu\mu$ events.

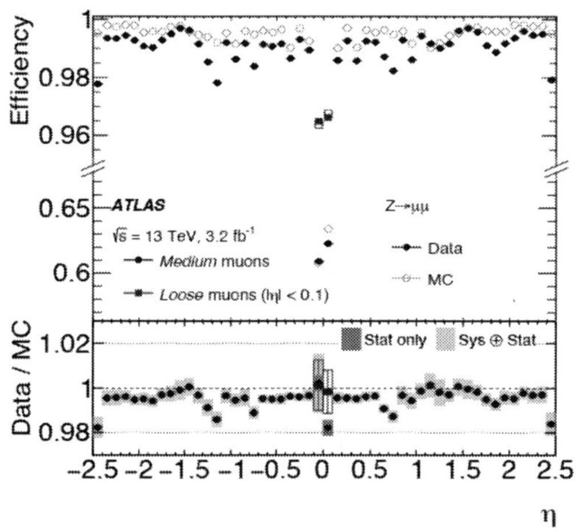

Figure 3. Measured reconstruction efficiency for varying eta of the ATLAS muon spectrometer for medium muons, from $Z \to \mu\mu$. The panel at the bottom shows the ratio of the measured to predicted efficiencies, with statistical and systematic uncertainties. Taken from Ref. [11] .

4.2 Reconstruction Efficiency for $|\eta| > 2.5$

The ID acceptance restricts the reconstruction of the combined muons to pseudorapidity region $|\eta| < 2.5$. For $2.5 < |\eta| < 2.7$, a measurement of the efficiency SF is performed by using the double ratio method. The number of muons observed in $Z \to \mu\mu$ decays in the high eta region is normalized to the number of muons observed in the region $2.2 < |\eta| < 2.5$, both for data and simulation [10].The high eta scale factor is derived as the ratio of the two results and it is presented in Fig. 7, as a function of the detector phi.

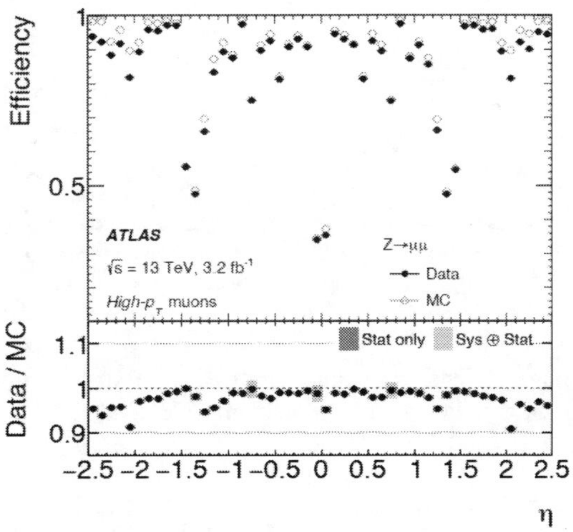

Figure 4. Measured reconstruction efficiency for varying eta of the ATLAS muon spectrometer for High-p_T muons, from $Z \rightarrow \mu\mu$ sample. The panel at the bottom shows the ratio of the measured to predicted efficiencies, with statistical and systematic uncertainties. Taken from Ref. [11] .

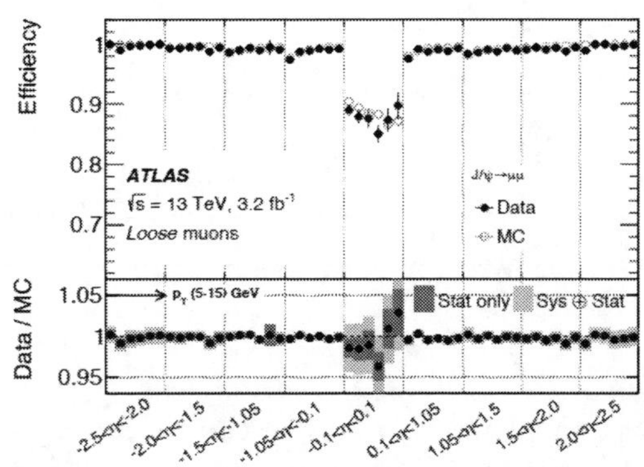

Figure 5. Measured reconstruction efficiency for varying eta of the ATLAS muon spectrometer for Loose muons with $5GeV < p_T < 15GeV$, from $J/\psi \rightarrow \mu\mu$ sample.The panel at the bottom shows the ratio of the measured to predicted efficiencies, with statistical and systematic uncertainties. Taken from Ref. [11] .

5 Conclusion

These proceedings summarized the determination of the muon reconstruction efficiency in the various regions of the detector, and for muon momenta between 6 and hundreds of GeV. The reconstruction efficiency is found to be in good agreement with the simulation.

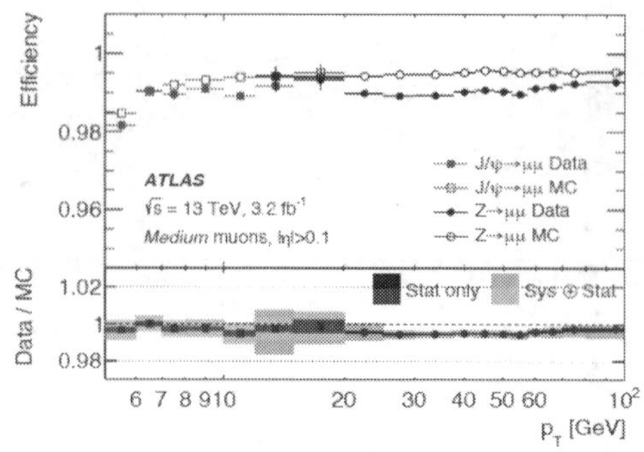

Figure 6. Muon reconstruction efficiency as a function of the transverse momentum. The panel at the bottom shows the ratio of the measured to predicted efficiencies, with statistical and systematic uncertainties. Taken from Ref. [11] .

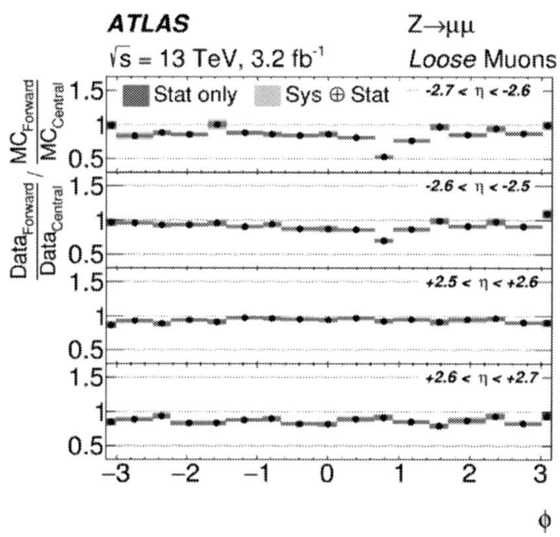

Figure 7. Double ratio for the forward muons for varying phi. Taking into account all corrections to the simulation for the central pseudorapidity region, this quantity measures the reconstruction efficiency scale factor for the forward region, where the standard tag-and-probe technique cannot be used due to missing coverage by the Inner Detector. Taken from Ref. [11] .

References

[1] ATLAS Collaboration, JINST 3, S08003 (2008).
[2] ATLAS Collaboration, Phys. Rev. D 85, 072004 (2012). arXiv:1109.5141 [hep-ex]
[3] ATLAS Collaboration, JHEP 1406, 112 (2014). arXiv:1404.1212 [hep-ex]
[4] ATLAS Collaboration, JHEP 1405, 068 (2014). arXiv:1402.6263 [hep-ex]
[5] ATLAS Collaboration, Phys. Rev. D 87, 015010 (2013). arXiv:1211.1150 [hep-ex]
[6] ATLAS Collaboration, Phys. Rev. D 88, 072001 (2013). arXiv:1308.4075 [hep-ex]

 DOI: 10.1051/epjconf/201716408006

[7] ATLAS Collaboration, Eur. Phys. J. C (2016). arXiv:1603.05598 [hep-ex]

[8] ATLAS Collaboration, ATL-PHYS-PUB-2015-037 (2015),http://atlas.web.cern.ch/Atlas/GROUPS/PHYSI(2015-10/

[9] ATLAS Collaboration, The ATLAS Experiment at the CERN Large Hadron Collider, JINST 3 (2008) S08003

[10] ATLAS Collaboration,Eur.Phys.J. C74 (2014)

[11] Eur. Phys. J. C (2016) 76:292

EPJ Web of Conferences **164**, 08007 (2017) DOI: 10.1051/epjconf/201716408007
ICNFP 2016

QCD analysis of non-singlet structure functions

A. Ghaffari-Turan[1,a], A. Khorramian[1,b], and H. Abdolmaleki[1,c]

[1] *Faculty of Physics, Semnan University, 35131-19111, Semnan, Iran*

Abstract. We present the results of the leading order (LO) and next-to-leading order (NLO) QCD analysis of experimental data in deep-inelastic scattering which relate to non-singlet structure functions. We use HERAFitter package to study the non-singlet xF_3 structure function. In this regards, we use the non-singlet structure function in x space and we combine CCFR experimental data with NuTeV, CDHSW and CHORUS experimental data as well. Our results at LO and NLO are compatible with the available theoretical models.

1 Introduction

To obtain the nucleon structure functions in lepton-nucleon scattering or a physical cross section in nucleon-nucleon scattering, we need to have parton distribution functions (PDFs). The PDFs can not be extracted from the theory and in fact, they must be determined from the fits of the experimental data. These distributions are functions of x and Q^2, and they are not only correlated by measurements from a wide variety of different processes and experiments, but also are correlated by theoretical constraints.

One of the most important structure functions in the neutrino-nucleon deep inelastic scattering, is the non-singlet structure function. However, the non-singlet structure function xF_3 is the main source to study the valence quark distributions $xu_v(x, Q^2)$ and $xd_v(x, Q^2)$ in QCD. The Q^2 dependence of these non-singlet distributions can obtain using the DGLAP evolution equations [1]. In this regard, the experimental data in the special range of deep inelastic scattering (DIS), is very important to determine the flavor of non-singlet parton distribution functions.

In recent years, several results of our QCD analysis about non-singlet structure functions in different cases are reported in Refs. [2-11]. In Ref. [2], we extracted valence quark distributions using QCD analysis of non-singlet parton distributions based on polynomials approach, such as Bernstein polynomials, and only CCFR data [12]. In the present paper, several data sets such as NuTeV [13] and CDHSW [14] and CHORUS [15] are included in the new analysis. So we can calculate the non-singlet structure function to determine valence quark distributions, using fits of the mentioned experimental data which are sensitive for determination of valence quark distributions.

Generally, the QCD analysis of parton distribution functions of DIS are accomplished at next-to-leading order (NLO). Although the NLO approximation, as a standard approximation, is important

[a]e-mail: A.Ghafary@semnan.ac.ir
[b]e-mail: Khorramiana@semnan.ac.ir, Khorramiana@theory.ipm.ac.ir
[c]e-mail: Abdolmaleki@semnan.ac.ir

DOI: 10.1051/epjconf/201716408007

for the most processes in QCD, but next-to-next-to-leading order (NNLO) approximation is useful to have a higher precise [4]. In Ref. [16] non-singlet coefficient functions for charged-current DIS is reported in the third order as well.

The plan of the paper is as following. Formalism of the non-singlet structure function in QCD is given in Sec. 2. Our parametrization of parton densities are also presented in this section. Sec. 3 we summarize our results and discussions.

2 Formalism

In this section, we present the non-singlet structure function in x-space. The xF_3 non-singlet of structure function, represents the momentum distribution function of valence quarks.

In the LO approximation, the average of neutrino and anti-neutrino structure functions is related to non-singlet $xF_3(x)$ structure function and can be defined as [2]

$$xF_3(x) = \frac{xF_3^{\nu N} + xF_3^{\bar{\nu}N}}{2} , \tag{1}$$

here $xF_3^{\nu N}$ and $xF_3^{\bar{\nu}N}$ is combination of PDFs as

$$xF_3^{\nu N} = xu_v(x) + xd_v(x) + 2xs(x) - 2xc(x) , \tag{2}$$
$$xF_3^{\bar{\nu}N} = xu_v(x) + xd_v(x) - 2xs(x) + 2xc(x) . \tag{3}$$

In above equations u_v and d_v are the valence quark distribution functions. So the average of the neutrino and anti-neutrino structure functions in above equations, is as following

$$xF_3(x) = \frac{xF_3^{\nu N} + xF_3^{\bar{\nu}N}}{2} = xu_v(x) + xd_v(x) , \tag{4}$$

so it is clear that the $xF_3(x)$ structure function is only depend on the combination of valence quark distributions. In this analysis, and at the input scale of $Q_0^2 = 1.4$ GeV^2 we choose the following valence quark distributions:

$$xu_v(x, Q_0^2) = N_{u_v} x^{a_{uv}} (1 - x)^{b_{uv}} (1 + c_{u_v} \sqrt{x} + d_{u_v} x) , \tag{5}$$

$$xd_v(x, Q_0^2) = \frac{N_{d_v}}{N_{u_v}} (1 - x)^{b_{dv}} xu_v(x, Q_0^2) . \tag{6}$$

In the above equations, the term of $x^{a_{uv}}$ controls the low-x behavior parton distribution, and the term of $(1 - x)^{b_{uv}}$ is important in the large x values. Other polynomial factors control the parton distribution for additional medium-x values. Also the parameter b_{d_v} is important to control the behavior of xd_v for large x value.

The above parametrization is similar to parametrization which is reported in Ref. [2]. The difference between QCD analysis in Ref. [2] and our new analysis is the different kinds of experimental data. In Ref. [2] we chose the CCFR experimental data and we used the QCD evolution of valance quarks distribution functions in N-Mellin space. In this analysis, we use the non-singlet structure function in x space and also we combine CCFR with NuTeV, CDHSW and CHORUS experimental data as well. According to Ref. [17], the benefit of x-space approach is that we can choose any functional form at the input parton distribution in QCDnum package.

In above parametrization, the normalization constants N_{u_v} and N_{d_v} are fixed by :

$$\int_0^1 u_v dx = 2\,, \tag{7}$$

$$\int_0^1 d_v dx = 1\,, \tag{8}$$

so the normalization constants N_{u_v} and N_{d_v} are as following

$$N_{u_v} = \frac{2}{B(a_{u_v}, 1 + b_{u_v}) + c_{u_v} B(1/2 + a_{u_v}, 1 + b_{u_v}) + d_{u_v} B(1 + a_{u_v}, 1 + b_{u_v})}\,, \tag{9}$$

$$N_{d_v} = \frac{1}{B(a_{u_v}, 1 + b_{u_v} + b_{d_v}) + c_{u_v} B(1/2 + a_{u_v}, 1 + b_{u_v} + b_{d_v}) + d_{u_v} B(1 + a_{u_v}, 1 + b_{u_v} + b_{d_v})}\,. \tag{10}$$

In above, $B(a, b)$ is the Euler Beta function.

The normalization constants N_{u_v} and N_{d_v} are very effective to determine unknown parameters in Eqs. (5,6) via the QCD fitting procedure. We use also the evolution of strong coupling up to NLO which is given [17]:

$$\frac{da_s(\mu^2)}{d\ln\mu^2} = -\sum_{i=0}^{1} \beta_i\, a_s^{i+2}(\mu)\,, \tag{11}$$

here $a_s = \alpha_s/2\pi$ and β_i denote the expansion coefficients of the QCD β-function which are depend on the number of active quark flavor n_f, and in the $\overline{\text{MS}}$ scheme they are given by [18, 19]

$$\beta_0 = \frac{11}{2} - \frac{1}{3} n_f\,, \tag{12}$$

$$\beta_1 = \frac{51}{2} - \frac{19}{6} n_f\,. \tag{13}$$

Note that, an analytical solution of Eq. (11) can be written at the leading order,

$$\frac{1}{\alpha_s(\mu^2)} = \frac{1}{\alpha_s(\mu_0^2)} + \beta_0 \ln(\frac{\mu^2}{\mu_0^2})\,. \tag{14}$$

By definition of Λ_{QCD}, we can ignore the first term at the right-hand side of above equation, where at this scale the coupling constant could be infinite, i.e.,

$$\frac{1}{\alpha_s(\mu^2)} = \beta_0 \ln(\frac{\mu^2}{\Lambda^2_{\text{QCD}}})\,. \tag{15}$$

Since our QCD analysis is performed up to NLO, it is convenient to take $\alpha_s(m_Z^2)$ as a reference. The values of α_s at any other scales are obtained using a numerical integration of Eq. (11) instead of approximate analytical solutions. The NNLO QCD Beta function is also reported in Ref. [20].

So the five unknown parameters can be extracted by the fitting procedure of the experimental data, with using HERAFitter package [21]. We add the mentioned experimental data which we need to study non-singlet structure function in HERAFitter package and we put our parametrization in this package as well.

DOI: 10.1051/epjconf/201716408007

3 Conclusion

In this section, we describe our QCD fit results using the experimental data which we choose to study of non-singlet structure function. As we mentioned before, the experimental data which we used in this analysis at LO and NLO approximation are CCFR, NuTeV, CDHSW and CHORUS. In fact we include the experimental data which collected around 283 data points. During the fitting process, it was noted that functional form is unchanged from which used in our previous analysis [2]. In Fig. 1, the quality of our QCD fit to the data of non-singlet structure function are illustrated at LO and NLO. According to our fits, non-singlet xF_3 structure functions from experimental data are compared with the LO and NLO our fit results as a function of x and at constant values of Q^2. In this case, we can survey the impact of NuTeV, CDHSW and CHORUS data when we combine them to CCFR experimental data as well.

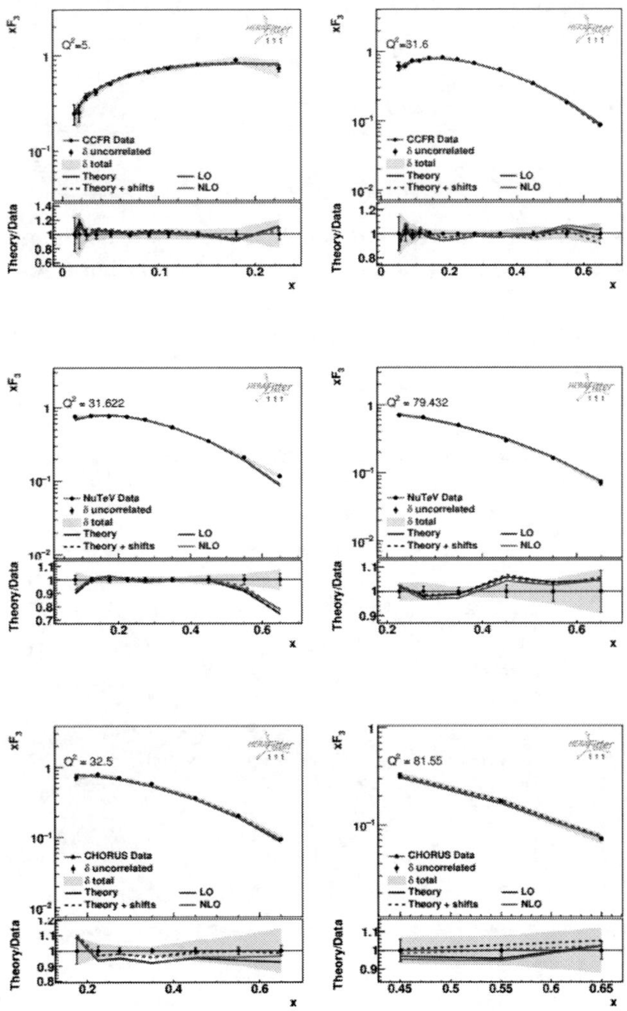

Figure 1. Comparison of xF_3 non-singlet structure function data with our QCD fit at LO and NLO approximation, as a function of x and for fixed Q^2 values.

The parameter values in Eqs. (5,6) and their uncertainties for LO and NLO at the input scale of Q_0^2 are given in Table 1, with the parameters that are listed without uncertainty and fixed by constrains of Eqs. (5,6). We fixed the strong coupling at the scale of M_Z as a default value $\alpha_s(M_Z) = 0.1185$, using the world average values which reported by the Particle Data Group [22].

Table 1. The fit results for parameter values at LO and NLO approximation and the input scale of Q_0^2 as defined in Eqs. (5,6) using the experimental CCFR, NuTeV, CDHSW and CHORUS data.

Parameter	LO	NLO
N_{u_v}	0.3236	2.5355
a_{u_v}	0.355 ± 0.074	0.635 ± 0.096
b_{u_v}	3.227 ± 0.058	3.180 ± 0.060
c_{u_v}	12.9 ± 5.2	−0.84 ± 0.78
d_{u_v}	3.4 ± 1.9	4.6 ± 3.3
N_{d_v}	0.2081	1.5163
b_{d_v}	1.65 ± 0.43	0.87 ± 0.21

According to Eqs. (5,6) and using the extracted parameters from the fits of experimental data at the input scale of Q_0^2, which are shown in Table 1, the Q^2 evolution for xu_v and xd_v valence quark distributions are possible. In this regards, the LO and NLO results for xu_v and xd_v valence quark distributions as a function of x and for fixed different values of $Q^2 = 1, 4, 100, 8317$ GeV^2 are compared in Fig. (2) and Fig. (3) respectively.

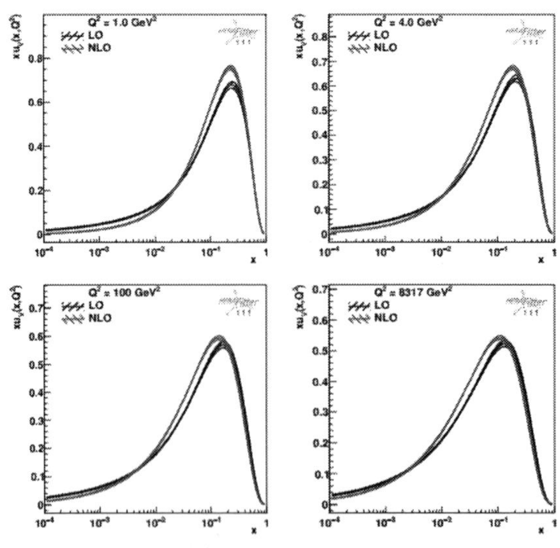

Figure 2. xu_v valence quark distribution as a function of x and for different values of $Q^2 = 1, 4, 100, 8317$ GeV^2.

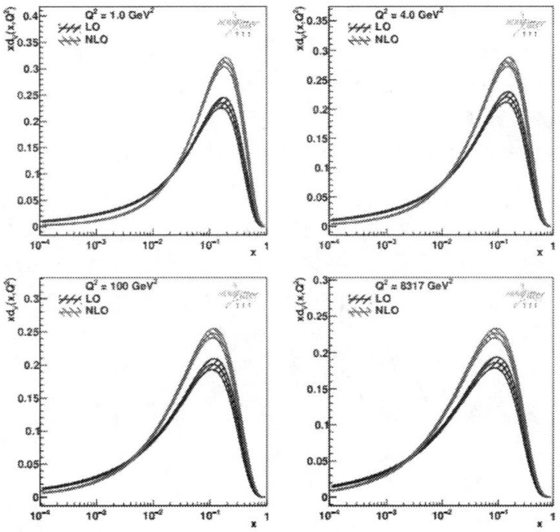

Figure 3. xd_v valence quark distribution as a function of x and for different values of $Q^2 = 1, 4, 100, 8317$ GeV^2.

4 Acknowledgment

We thank R. Placakyte for useful discussion on the xFitter framework. H. A. is grateful to the JINR (Joint Institute For Nuclear Research) and xFitter group for their hospitality and scholarship in xFitter meeting in Dubna.

References

[1] G. Altarelli and G. Parisi, Nucl. Phys. B **126**, 298 (1977).
[2] A. N. Khorramian and S. Atashbar Tehrani, JHEP **0703**, 051 (2007) [hep-ph/0610136].
[3] A. N. Khorramian and S. A. Tehrani, Phys. Rev. D **78**, 074019 (2008) [arXiv:0805.3063 [hep-ph]].
[4] A. N. Khorramian, H. Khanpour and S. A. Tehrani, Phys. Rev. D **81**, 014013 (2010) [arXiv:0909.2665 [hep-ph]].
[5] S. Atashbar Tehrani and A. N. Khorramian, Nucl. Phys. Proc. Suppl. **186**, 58 (2009).
[6] A. N. Khorramian and S. A. Tehrani, J. Phys. Conf. Ser. **110**, 022022 (2008).
[7] A. N. Khorramian, S. A. Tehrani and M. Ghominejad, Acta Phys. Polon. B **38**, 3551 (2007).
[8] A. N. Khorramian, S. A. Tehrani and A. Mirjalili, Nucl. Phys. Proc. Suppl. **164**, 34 (2007).
[9] A. N. Khorramian and S. Atashbar Tehrani, eConf C **0605151**, 0014 (2006) [hep-ph/0610172].
[10] A. N. Khorramian, H. Khanpour and S. Atashbar Tehrani, PoS EPS **-HEP2009**, 393 (2009).
[11] M. Soleymaninia and A. N. Khorramian, Int. J. Mod. Phys. A **26**, 686 (2011).
[12] W. G. Seligman *et al.*, Phys. Rev. Lett. **79**, 1213 (1997) [hep-ex/9701017].
[13] D. Naples *et al.* [NuTeV Collaboration], [hep-ex/0307005].
[14] J. P. Berge *et al.*, Z. Phys. C **49**, 187 (1991).
[15] G. Onengut *et al.* [CHORUS Collaboration], Phys. Lett. B **632**, 65 (2006).
[16] J. Davies, A. Vogt, S. Moch and J. A. M. Vermaseren, arXiv:1606.08907 [hep-ph].

[17] M. Botje, Comput. Phys. Commun. **182**, 490 (2011) [arXiv:1005.1481 [hep-ph]].
[18] W. Furmanski and R. Petronzio, Z. Phys. C **11**, 293 (1982).
[19] O. V. Tarasov, A. A. Vladimirov and A. Y. Zharkov, Phys. Lett. **93B**, 429 (1980).
[20] S. A. Larin and J. A. M. Vermaseren, Phys. Lett. B **303**, 334 (1993).
[21] S. Alekhin *et al.*, Eur. Phys. J. C **75**, no. 7, 304 (2015) [arXiv:1410.4412 [hep-ph]].
[22] K. A. Olive *et al.* [Particle Data Group Collaboration], Chin. Phys. C **38**, 090001 (2014).

EPJ Web of Conferences **164**, 08008 (2017) DOI: 10.1051/epjconf/201716408008
ICNFP 2016

Search for WIMP dark matter produced in association with a Z boson with the ATLAS detector

Artem Basalaev[1,a], on behalf of the ATLAS Collaboration

[1] *National Research Centre "Kurchatov Institute" B.P.Konstantinov Petersburg Nuclear Physics Institute, 188300, Gatchina, Leningrad district, Russia.*

Abstract. The search for weakly interacting dark matter particle (WIMP) candidates produced in association with a Z boson with the ATLAS detector at the LHC is presented. Events with large missing transverse momentum and consistent with the decay of a Z boson into oppositely charged electron or muon pairs were selected in analysis. Background estimates and corresponding systematic uncertainties are shown. The limits on the mass scale of the contact interaction as a function of the dark matter particle mass and the limits on the coupling and scalar particle mediator mass for 8 TeV proton-proton collisions data are presented. Prospects for analysis using 13 TeV proton-proton collisions data are discussed.

1 Introduction

Astrophysical measurements provide compelling evidence for the existence of nonbaryonic dark matter (DM) [1]. However, no experiments to date have revealed its particle nature nor proved its non-gravitational interactions, if they exist [2]. At hadron colliders, weakly interacting massive particles (WIMPs), the promising candidates for dark matter particles, are predicted to be produced in pairs in some Beyond the Standard Model (BSM) theories. A WIMP would escape as undetected particle, but could be identified by a signature involving missing transverse momentum $E_{\mathrm{T}}^{\mathrm{miss}}$. Selecting events with a clear Z boson signature results in a well-studied $E_{\mathrm{T}}^{\mathrm{miss}}$ distribution. A search was performed with the ATLAS detector [3] at the LHC for deviations from $E_{\mathrm{T}}^{\mathrm{miss}}$ distribution expected from Standard Model processes [4].

The results were examined in terms of effective field theory (EFT) models [5] and models in which the intermediate state is specified [6]. The EFTs considered in this analysis (shown on Figure 1) are expressed in terms of two parameters: mass of the DM particle m_χ and a mass scale M_*, which parameterizes the coupling between the WIMP and SM particles[1]. EFT operators D1 (scalar, spin independent), D5 (vector, spin independent), and D9 (tensor, spin dependent), following the definitions in [5], are representative of the full set of operators in which the Z boson is emitted as ISR. Two examples of the dimension-7 $ZZ_{\chi\chi}$ operator mixtures are considered: one in which the $Z_{\gamma*}$ contribution is negligible and one in which it is maximal.

[a] e-mail: artem.basalaev@cern.ch

[1] $M_* = \frac{M}{\sqrt{g_\chi g_q}}$, where M is mediator mass, g_q is SM-mediator coupling and g_χ is mediator-DM coupling.

While the use of EFTs is appealing because the intermediate state mediating the parton-WIMP interaction is unknown, there are strong restrictions on EFTs validity. In particular, momentum transfer should generally be much lower than the mass scale and EFT is not a UV-complete theory. For a detailed analysis regarding validity of the EFTs for searches at the LHC please refer to [7]. To complement the EFT analysis, results were also examined in terms of a model with a scalar mediator η, with mass m_η, and a scalar-WIMP coupling strength f, shown on Figure 1(c). The production cross section is proportional to f^4. The same final state signature as predicted by the EFT, $Z_{\chi\bar{\chi}}$, is produced.

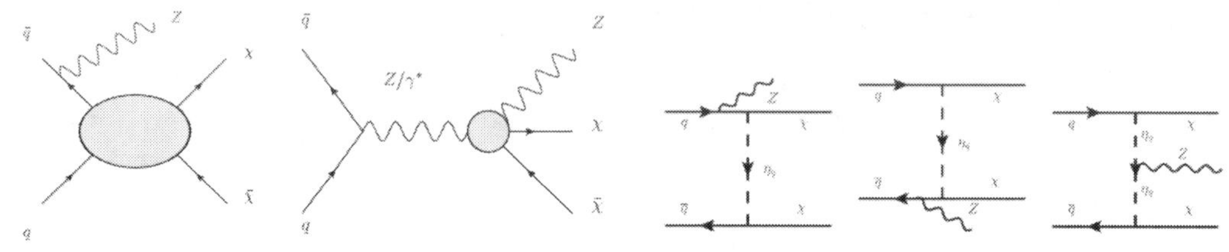

(a) (b) (c)

Figure 1. Diagrams are shown for EFTs with an ISR operator (a), with a $ZZ_{\chi\bar{\chi}}$ operator (b), and for models with specified mediator (c).

2 Event Selection and Background Estimation

Electron candidates are reconstructed from isolated energy deposits in the electromagnetic calorimeter with a shower shape consistent with electrons or photons, matched to inner detector tracks [8]. Muon candidates are reconstructed combining tracks independently found in the muon spectrometer and inner tracking detector [9]. Jets are reconstructed using the anti-k_t algorithm [10] with a radius parameter $R = 0.4$. Basic kinematic cuts are also applied, along with some additional selection to ensure good separation between reconstructed objects [4].

The invariant mass of a lepton pair is required to be consistent with that of a Z boson, 76 GeV $\leq$ $M_{\ell\ell} \leq$ 106 GeV. Events are required to pass event quality checks for errors in sub-detectors during recording, and at least one of the trigger requirements: single-lepton with higher p_T, or lepton-pair with lower p_T. The E_T^{miss} is expected to be back-to-back with the Z boson. This is accounted for in the $\Delta\phi(Z, E_T^{miss})$ variable, which is required to be greater than 2.5. The absolute value of the pseudorapidity of the dilepton system, $|\eta_{ll}|$, must be less than 2.5. Any possible imbalance, implying the presence of fake E_T^{miss}, is accounted for by an upper threshold on the fractional p_T difference variable, defined as $|E_T^{miss} - p_T^{\ell\ell}|/p_T^{\ell\ell} < 0.5$. Events are removed if they contain jets with $p_T > 25$ GeV to suppress top-quark pair background. Events containing a third lepton with $p_T > 7$ GeV, satisfying looser identification requirements are removed to suppress diboson background. The dark matter models considered here have different E_T^{miss} spectra, leading to a variety of optimal lower thresholds of E_T^{miss}. Four inclusive signal regions are defined with lower thresholds in E_T^{miss} of 150, 250, 350, and 450 GeV.

The dominant background process is $ZZ \rightarrow \ell^+\ell^-\nu\bar{\nu}$, an irreducible background. Reducible backgrounds may have jets produced in association with two leptons, where the jets are misidentified or unreconstructed, such as $t\bar{t} \rightarrow \ell^+\nu b\ell^-\bar{\nu}\bar{b}$, $ZZ \rightarrow \ell^+\ell^-q\bar{q}$, $WZ \rightarrow \bar{q}q'\ell^+\ell^-$ or Z + jets (including decays to τ leptons). The other important background is $WW \rightarrow \ell^+\nu\ell^-\bar{\nu}$, which may only be reduced through the mass window requirement. Additional reducible sources may produce events with more than two leptons where the additional leptons are misidentified or not reconstructed, or less than two leptons where jets are misidentified as leptons, such as $WZ \rightarrow \ell\nu\ell^{+}\ell^{-}$ and W + jets, respectively.

Data-driven techniques for background estimation were used where possible. Otherwise, MC simulation samples were used, obtained with MC generators such as SHERPA, POWHEG. Detailed information on background estimations is provided in [4].

The ZZ and WZ backgrounds are estimated from MC simulations. The WW, $t\bar{t}$, Wt, and $Z \rightarrow t\bar{t}$ backgrounds are estimated from data using the absence of signal in the $e\mu$ channel and the relative production rate of 1 : 1 : 2 for the ee, $\mu\mu$, and $e\mu$ channels. An $e\mu$ control region similar to the signal region is defined, and the background estimate for the ee and $\mu\mu$ signal regions is obtained from the number of $e\mu$ events in the control region after correcting for different acceptances and efficiencies. The Z + jets background is estimated using two data-driven techniques: *ABCD* method [11] using uncorrelated variables $E_{\mathrm{T}}^{\mathrm{miss}}$ and η_{ll}, and by fitting the distributions of $\Delta\phi(Z, E_{\mathrm{T}}^{\mathrm{miss}})$ and $E_{\mathrm{T}}^{\mathrm{miss}}$ at small values and extrapolating them to the signal regions. The W + jets background is estimated by reversing the electron isolation condition and loosening identification requirements for one electron in order to obtain a data sample enriched in jets reconstructed as electrons.

3 Results and Conclusions

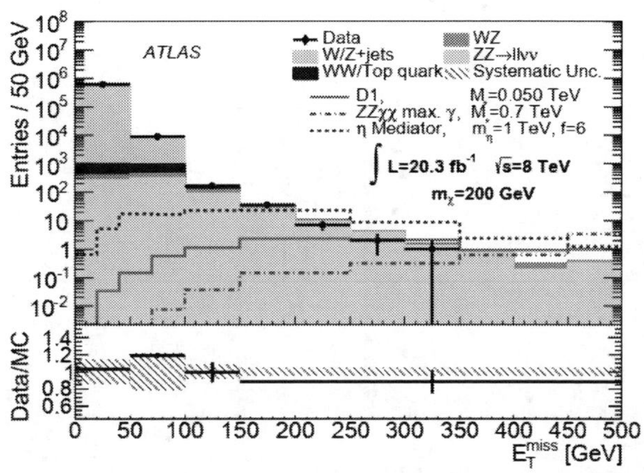

Figure 2. $E_{\mathrm{T}}^{\mathrm{miss}}$ distribution is presented after all event selections other than the $E_{\mathrm{T}}^{\mathrm{miss}}$ thresholds for the observed data [4].

The search was performed on 20.3 fb^{-1} of ATLAS dataset collected in 2012 at $\sqrt{s}$ =8 TeV. Theoretical and experimental systematic uncertainties are estimated for all studied backgrounds, including uncertainties associated with pileup reweighting, luminosity calculation, object and $E_{\mathrm{T}}^{\mathrm{miss}}$ reconstruction algoritms and methods of background estimation. Data yields with associated statistical and systematic uncertainties are shown on Table 1. The $E_{\mathrm{T}}^{\mathrm{miss}}$ distribution after applying all selection requirements other than the $E_{\mathrm{T}}^{\mathrm{miss}}$ thresholds for the observed data, the expected SM backgrounds, and the hypothetical $pp \rightarrow Z_{\chi\bar{\chi}}$ signals for various values of the mass scale are shown on Figure 2. No significant deviations from the SM predictions have been observed.

DOI: 10.1051/epjconf/201716408008

Table 1. Observed yields and expected SM backgrounds in each signal region. Statistical, systematic, and luminosity uncertainties are combined [4].

	E_T^{miss} Threshold, GeV			
Process	150	250	350	450
ZZ	41 ± 15	6.4 ± 2.4	1.3 ± 0.5	0.3 ± 0.1
WZ	8.0 ± 3.1	0.8 ± 0.4	0.2 ± 0.1	0.1 ± 0.1
WW, $t\bar{t}$, $Z \to \tau^+\tau^-$	1.9 ± 1.4	$0^{+0.7}_{-0.0}$	$0^{+0.7}_{-0.0}$	$0^{+0.7}_{-0.0}$
Z + jets	0.1 ± 0.1	...	...	...
W + jets	0.5 ± 0.3	...	...	...
Total	52 ± 18	7.2 ± 2.8	1.4 ± 0.9	$0.4^{+0.7}_{-0.4}$
Data	**45**	**3**	**0**	**0**

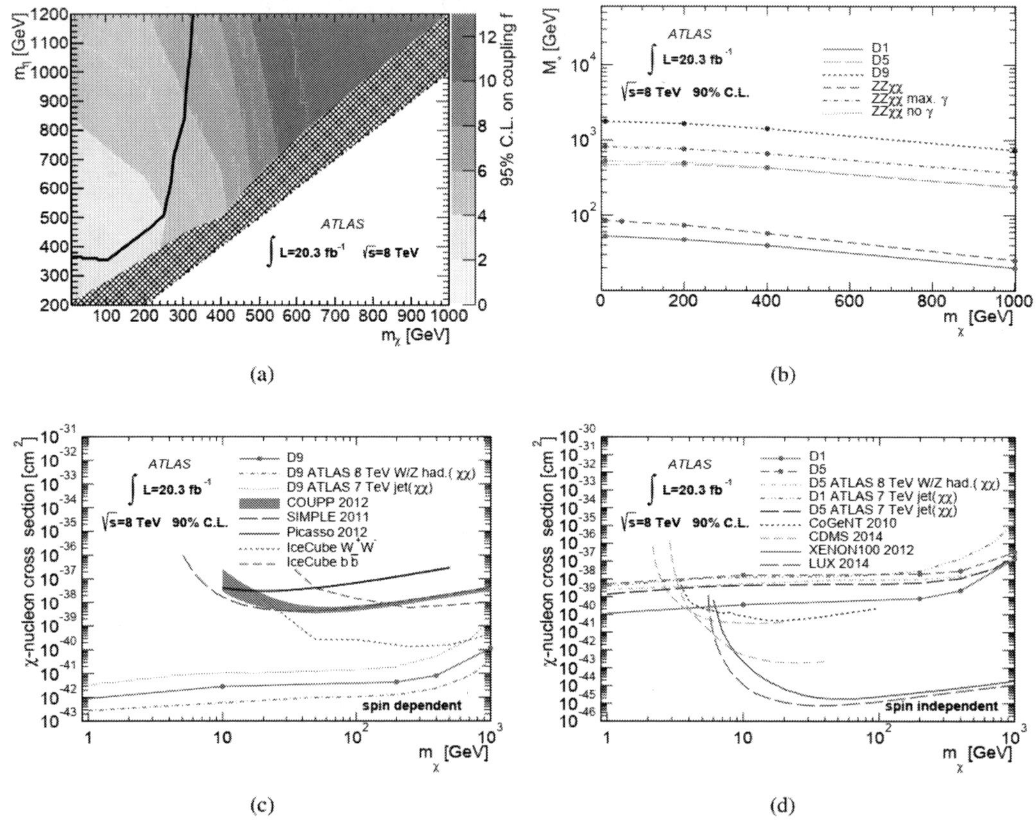

Figure 3. Observed upper limits at 95% CL are shown on the coupling constant f of the scalar-mediator theory as a function of mediator mass m_η and dark matter particle mass m_χ [4] (a). Observed 90% C.L. lower limits are shown on the mass scale M_* of considered effective field theories as a function of m_χ [4] (b) and on the χ-nucleon scattering cross section as a function of m_χ for the spin-dependent (c) and spin-independent (d) effective operators [4].

The 95% C.L. upper limits on the on the coupling constant f of the scalar-mediator theory and the 90% C.L. lower limits on the mass scale M_* of the effective field theory, which parameterizes the coupling between the WIMP and SM particles, are presented on Figure 3(a) and Figure 3(b) correspondingly.

Additionally the limits are translated into χ-nucleon scattering cross section and compared with several other ATLAS searches and direct searches for spin-depentent (Figure 3(c)) and spin-independent (Figure 3(d)) cases.

3.1 Prospects in Run 2

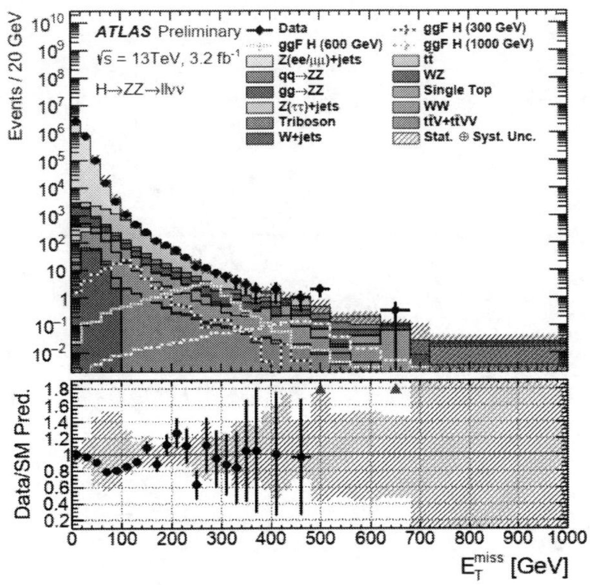

Figure 4. $E_{\mathrm{T}}^{\mathrm{miss}}$ distribution is shown after the dilepton mass requirement for the search for additional heavy Higgs boson in $\ell\ell\nu\nu$ final state, performed in Run 2 on 2015 data [12].

Run 2 of the LHC is steadily ongoing, with proton-proton collision energy reaching unprecedented $\sqrt{s}$ = 13 TeV. The integrated luminosity collected in Run 1 has already been surpassed, promising increase in sensitivity for many analyses, including the search discussed here. Presented on Figure 4 are the background estimates with 2015 data for the search for additional heavy Higgs boson, which is utilizing the same $\ell\ell\nu\nu$ final state and the same processes are considered as background.

References

[1] E. Komatsu et al., Astrophys. J. Suppl. Ser. **192**, 18 (2011)
[2] K.A. Olive et al. (Particle Data Group), Chin. Phys. **C38**, 090001 (2014)
[3] ATLAS Collaboration, JINST **3**, S08003 (2008)
[4] ATLAS Collaboration, Phys. Rev. D **90**, 012004 (2014)
[5] J. Goodman, M. Ibe, A. Rajaraman, W. Shepherd, T.M.P. Tait, H.B. Yu, Phys. Rev. D **82**, 116010 (2010)
[6] N. Bell, J. Dent, A. Galea, T. Jacques, L. Krauss, T. Weiler, Phys. Rev. D **86**, 096011 (2012)

[7] D. Abercrombie et al., ArXiv e-prints (2015), 1507.00966
[8] ATLAS Collaboration, Eur. Phys. J. C **72**, 1909 (2012)
[9] ATLAS Collaboration, J. High Energy Phys. **12**, 060 (2010)
[10] M. Cacciari, G.P. Salam, G. Soyez, J. High Energy Phys. **04**, 063 (2008)
[11] ATLAS Collaboration, Phys. Lett. B **719**, 299 (2013)
[12] ATLAS Collaboration, Tech. Rep. ATLAS-CONF-2016-012 (2016), https://cds.cern.ch/record/2140833

EPJ Web of Conferences **164**, 08009 (2017) DOI: 10.1051/epjconf/201716408009
ICNFP 2016

Towards an emergent model of solitonic particles from non-trivial vacuum structure

Adam B. Gillard[1,a] and Niels G. Gresnigt[1]

[1] *Department of Mathematical Sciences, Xi'an Jiaotong-Liverpool University, 111 Ren'ai Road, Suzhou HET, Jiangsu, China 215123*

Abstract. We motivate and introduce what we refer to as the principles of Lie-stability and Hopf-stability and see what the physical theories must look like. Lie-stability is needed on the classical side and Hopf-stability is needed on the quantum side. We implement these two principles together with Lie-deformations consistent with basic constraints on the classical kinematical variables to arrive at the form of a theory that identifies standard model fermions with quantum solitonic trefoil knotted flux tubes which emerge from a flux tube vacuum network. Moreover, twisted unknot fluxtubes form natural dark matter candidates.

1 Introduction

The goals of this paper are to motivate the need for both Lie and Hopf-type deformations in physics, define and explain the need for Lie and Hopf-stability, implement these broad but powerful physical principles, and see what the class of physical theories look like that result from satisfying these important principles.

We present our ideas in two stages. Firstly we look at physics from a classical viewpoint and discuss why we are forced to consider Lie-type deformations of underlying Lie-algebraic symmetries. Secondly we look at physics from a quantum viewpoint, in two stages. In the first stage we consider what the classical limit of a background independent fully emergent quantum theory should look like. In the second stage we use these considerations, enlightened by our previous classical level considerations, to arrive at a fully quantum theory of spacetime and matter. We sketch the motivational outline of the theory in this brief conceptual essay and leave the more lengthy details to be presented elsewhere.

Our model is built on two significant underlying assumptions.

1. There exist fundamental quantum states which define everything, including space and time (and hence gravity), and that all physical phenomena can be thought of as emergent

[a] e-mail: Adam.Gillard@xjtlu.edu.cn

from the fundamental self-interacting quantum states.

This is essentially identical to the operational viewpoint of loop quantum gravity except in that case matter fields are thought of as being both additional and fundamentally different from the quantum gravitational fields. In our view, the matter field quantum states *become* additional when closed knotted flux tubes become free (i.e. emerge) from the spacetime and vacuum defining flux tube network.

2. The universally observed cosmological scale features of spacetime, i.e. the cosmological constant and the cosmic microwave background radiation, are intimately linked with the underlying structure of the physical background independent vacuum.

Loop quantum gravity operationally takes this viewpoint when it comes to the cosmological constant.[1] We believe that perhaps we should be taking, *additionally*, the cosmic microwave background radiation into similar analogous consideration.

These assumptions together with our Lie and Hopf-algebraic stability considerations (which we discuss next and throughout the rest of this paper) leads to a physical model which is physically significantly different to the usual loop quantum gravity with cosmological constant and matter fields, which are purely thought of as being both physically distinct and additional to the quantum gravitational fields.

We need to define Lie-stability. If a Lie algebra g_0 with basis $\{X_i\}$ and bracket $[X_i, X_j] = if_{ij}^k X_k$ and a one parameter deformation of g_0 to g_t defined by $[X_i, X_j]_t \equiv [X_i, X_j]_0 + \sum_{m=1}^{\infty} \phi_m(X_i, X_j)t^m$ with g_t isomorphic to g_0, the Lie algebra is Lie-algebraically stable, i.e. Lie-stable [2, 3].[2] Physically the Lie-stability of a classical symmetry algebra is saying that the theory is robust and free of fine tuning issues.

This is so because the structure constants in a Lie algebra of a physical theory often depend on physical constants. The numerical values of these physical constants must be determined experimentally and are therefore not known without some error. Lie-stability therefore provides insight into the validity of a physical theory or the need to generalize the theory. An unstable theory might be deformed until a stable theory is reached which is likely to be of wider validity than that of the original unstable theory.

The importance of Lie-stability in physical theories was first promoted by Mendes [3]. Since then, several others, most notably Chryssomalakos and Okon [4], and Ahluwalia [5] have similarly argued that the stability of a physically relevant Lie algebra should be considered a physical principle, on an equal footing with the principle of local gauge invariance, where, since each fiber of a vector bundle is isomorphic to $\mathbb{R}^n$ and a canonical isomorphism is absent, the laws of physics should not depend on any particular choice of local trivialization of any vector bundles.

The deformation parameters that arise in the stabilization of an algebra correspond to new invariant scales. These invariant scales constrain the permissible values that certain kinematic variables can take and therefore restricts the theory. We note however that the numerical values of these invariant scales are not obtained via algebraic considerations but must be determined in the laboratory. Lie-type deformations therefore provide a systematic method of introducing additional invariant scales. Lie-stability is both a physically sensible

[1]The Kodama state [1] is predominantly interpreted as a quantum gravitational ground state, which in the classical limit is identified with de Sitter spacetime for positive cosmological constant.

[2]The Lie algebra needs to be isomorphic under every Lie-deformation.

requirement in the sense that it provides an indication of the robustness of a theory, as well as mathematically unambiguous in the sense that the stable algebra is unique up to isomorphism.

If a physical theory that is Lie-unstable has a domain of validity where the theory works quite well, this suggests that the theory is not necessarily wrong as such, but rather that it merely requires deformation to a hopefully true, Lie-stable physical theory. Deforming a Lie algebra is done in part by introducing a deformation parameter corresponding to some physically invariant scale. For example, if we constrain the speed kinematical variables v^i with the physical constraint $0 \leq v^i \leq c < \infty$, we get a new type of length, $x_0 \equiv ct$ which belongs to a four-dimensional manifold, the (non-gravitational) result being the usual special relativity with Minkowski spacetime isometry group $\mathbb{R}^{1,3} \rtimes SO(1,3)$, the associated Lie algebra of which, is still Lie-unstable. It is remarkable that imposing a physically realistic constraint on one of the kinematical variables produced a new theory with a much greater domain of validity, especially given that the quantum field theory of the standard model is constructed to respect the Poincaré symmetries. It is natural then to ask whether there exist additional physically necessary constraints on the kinematical variables, and then to ask if the resulting Lie algebra is Lie-stable, which would then make it at least a possible viable candidate for the true classical limit of some true quantum theory of universal validity.

Finkelstein [6–13], showed that the standard model particles (as well as natural dark matter candidates), with their characteristic defining quantum electroweak and quantum chromodynamic quantum numbers, can be thought of as knotted quantum solitons corresponding to representations of the quantum group $SU_q(2)$.[3] Finkelstein's model is important for us because it presents important physical interpretations of the quantum groups $SL_q(2)$ and $SU_q(2)$, both quantum groups of which are important for our model. For this reason we include a brief but important appendix summarizing many key ideas in Finkelstein's work. We refer to a Hopf algebra with $q \neq 1$ as Hopf-stable, for reasons which will become clearer later.

2 Classical physics and Lie-stability

In classical Galilean relativity we have a configuration space $\mathcal{C}$ with position elements $\mathbf{x}$ on which there is a Euclidean metric which is preserved by the isometry group (ignoring the discrete symmetry transformations) $ISO(3) \equiv \mathbb{R}^3 \rtimes SO(3)$. Moreover, phase space is $T^*\mathcal{C}$, speed $|\mathbf{v}| \equiv \frac{d|\mathbf{x}|}{dt}$ is an element of the coset space $ISO(3)/SO(3)$ and time t is a free parameter. In this setting we have unconstrained kinematical variables, $0 \leq |\mathbf{x}| < \infty$, $0 \leq |\mathbf{v}| < \infty$, $0 \leq t < \infty$. Similar statements hold for energy, momentum and mass. The theory is very successful within a relatively narrow domain of validity where the distances, times and speeds are not too "extreme." Moreover, if one or more of these variables starts tending in an extreme direction, the theory is violated. The interesting point is that the degree of violation smoothly increases. This suggests that the theory is simply incomplete in a certain sense rather than outright wrong. Special relativity can be interpreted as simply the result of imposing a physically realistic constraint on the kinematic variables, specifically,

[3] A quantum group is a quantum Hopf algebra, i.e. a bialgebra with the additional notion of antipode [14, 15]. This is not a mathematical group. It can be thought of loosely as taking the symmetries which are present in a corresponding Lie-theory, incorporating them into the category of algebras, and then deforming with deformation parameter q within the category of algebras. The category of Lie groups is rigid, i.e. not deformable, hence a need to consider the Hopf algebras. The deformation parameter q provides the additional degree of freedom necessary to get away from considering pointlike particles to considering knotted quantum solitons. In the limit $q \to 1$, the degenerate case of pointlike particles is restored.

a finite upper bound c on the speeds such that now $0 \leq |\mathbf{v}| \leq c < \infty$. Here the space manifold has increased from three-dimensional to four-dimensional with a new distance $x_0 \equiv ct$ made possible by the dimensionless deformation parameter c. Nature has such a constraint deformation parameter so the imposition of this constraint on the theory increased the now constrained theory's domain of validity. Moreover the quantization of this improved theory leads directly to the standard model of particle physics. A Lie-deformation has thus turned out to be quite revolutionary to physics. This theory, although improved, is still imperfect. In our Lie-stability paradigm it can immediately be recognized that the theory cannot be perfect because the Minkowski metric has an isometry group $\mathbb{R}^{1,3} \rtimes SO(1,3)$ which is Lie-unstable. The problems of ultraviolet and infrared divergences hint at what the problem might be. The natural question to ask is whether there should exist additional constraints, $0 < l_p \leq |\mathbf{x}| \leq l_c < \infty$ (the minimal length scale l_p together with c implying a minimum time t_p such that $0 < t_p \equiv \frac{l_p}{c} \leq t < \infty$). We consider physical motivation for l_p and then consider the need for l_c.

Small distance scales require high energy to probe. There is a natural point at which the energy put into the system is sufficiently high to form a black hole so there should be a natural notion of a minimum length scale below which, no meaningful concept of external observers can be assigned. Moreover, Bronstein showed that localization to arbitrary precision is impossible in the presence of a gravitational field [16]. There exist physical grounds for taking the addition of a minimum nonzero length constraint seriously. Following the deformation procedure that took Galilean relativity to special relativity, deforming the Minkowski spacetime by imposing the additional kinematical constraint $0 < l_p \leq |\mathbf{x}|$ gives a new natural length, $x_4 \equiv l_p^{-1}A$ where A is some parameter with dimensions of area. The resulting spacetime has metric $ds^2 = c^2dt^2 + l_p^{-2}dA^2 - d\mathbf{x}^2$ which is preserved by the isometry group $\mathbb{R}^{2,3} \rtimes SO(2,3)$ which again has a Lie-unstable algebra. The introduction of both c and l_p naturally also implicitly introduces a minimal time $t_p \leq t$. We could develop this theory on a now five-dimensional manifold further but in our view it still would not be ultimately correct because the symmetry Lie algebra is still Lie-unstable.

We are not aware of any need for a nonzero minimum speed or a maximum time so the only obvious further physical constraint on the kinematical variables is to impose a maximum length scale l_c. It is natural to do this because such a scale is related to a positive cosmological constant $\Lambda \approx l_c^{-1/2}$.[4] Such a constant has been observed so we have a natural additional length $x_5 \equiv l_c\phi$ where ϕ is some dimensionless parameter, the interpretation of which we consider later. The resulting metric is then six-dimensional. This kinimatical space (which contains what we will call kinematical timelike, spacelike and lightlike intervals) is not physical however. Physical space can only be defined on the five-dimensional hypersurface[5]

$$-\left(\frac{ds}{d\phi}\right)^2 + c^2\left(\frac{dt}{d\phi}\right)^2 + l_p^{-2}\left(\frac{dA}{d\phi}\right)^2 - \left(\frac{d\mathbf{x}}{d\phi}\right)^2 = l_c^2. \tag{1}$$

We see immediately that the induced metric for timelike intervals has isometry group $SO(2,4)$ corresponding to a type of five-dimensional anti-de Sitter space AdS_5 and isomorphic to the

[4] moreover, in loop quantum gravity, classically large scale infrared divergences in the transition amplitudes are tamed by q-deforming the theory, which, in the classical limit yields Einsteins equations with positive cosmological constant [17]. This in turn then implies a related cosmological length scale l_c. This demonstrates a need for Hopf-stability.

[5] if we "divide through" by the dimensionless quantity $d\phi^2$ we get a five-dimensional hypersurface defined by quantities which are still lengths. Moreover, this space is hyperbolic which, being curved, cuts out the abelian translation group.

conformal group. This space has an asymptotic boundary on which one has a natural notion of a 1+3-dimensional conformal field theory CFT_4 by the AdS/CFT correspondence. Also, for spacelike intervals we have a five-dimensional space with induced metric having isometry group $SO(3,3)$. For lightlike intervals we have a space with induced metric having isometry group $SO(2,3)$ corresponding to a type of four-dimensional anti-de Sitter space AdS_4. This has an asymptotic boundary on which there exists a natural notion of a 1+2-dimensional conformal field theory CFT_3. The Lie algebras of these isometry groups are all Lie-stable so we have arrived at something which feasibly could be the classical limit of some true quantum gravitational ground state in some universally valid quantum theory. A natural question to ask now is: is the physical space five-dimensional, or do there exist any additional constraints? By observation we appear to live in a 1+3-dimensional universe so one would expect there to exist one additional constraint. The existence of a minimal length scale and a maximum length scale introduces a natural notion of a radius of curvature and a natural notion of a minimum angular resolution $\phi_p = l_p l_c^{-1}$. The existence of a universal radius of curvature, say R, would suggest that $\frac{d|\mathbf{x}|}{d\phi} = R = constant$. This defines a two-dimensional space S^2 with induced metric having isometry group $SO(3)$. But from Eqn. (1) this implies that

$$-\left(\frac{ds}{d\phi}\right)^2 + c^2\left(\frac{dt}{d\phi}\right)^2 + l_p^{-2}\left(\frac{dA}{d\phi}\right)^2 = \text{constant} \qquad (= l_c^2 + R^2). \tag{2}$$

For timelike intervals this defines a two-dimensional constraint surface with induced metric having isometry group $SO(2,1)$. For spacelike intervals this defines a two-dimensional constraint surface with induced metric having isometry group $SO(3,0)$. For lightlike intervals this defines a one-dimensional constraint surface with induced metric with isometry group $SO(2,0)$. Putting the two two-dimensional spaces together, for timelike intervals, we have a direct product space $AdS_2 \times S^2$ which is a four-dimensional spacetime composed of two two-dimensional bits; a 1+1-dimensional space and a two-dimensional space. We recognize this from general relativity as the Bertotti-Robinson spacetime [18, 19]. The space corresponds to the spacetime solution to the Einstein-Maxwell equations with positive cosmological constant where the energy-momentum tensor $T_{\mu\nu}$ is composed of a cosmological constant term and two terms corresponding to electromagnetism. If this were to be physically realistic we would expect the universe to be full of electromagnetic radiation. At this point we note that universal cosmological background radiation has been observed and its existence is universal just as the cosmological constant is universal. For the spacelike case the physical hypersurface we have from Eqn. (2) a two-dimensional space with induced metric having isometry group $SO(3)$, so the total four-dimensional space is $S^2 \times S^2$ with induced metric having isometry group $SO(3,0) \times SO(0,3)$. Finally, for the lightlike case we have $S^1 \times S^2$ with induced metric having isometry group $SO(2,0) \times SO(0,3)$.

We conclude that $AdS_2 \times S^2$, $S^2 \times S^2$ and $S^1 \times S^2$ constitute complementary viable candidates for a target classical limit that a true quantum theory would reproduce as a quantum gravitational ground state (i.e. in the absence of matter).

In the standard model of particle physics the underlying spacetime symmetry group is the Poincaré group $\mathcal{P} = \mathbb{R}^{1,3} \rtimes SO(1,3)$ of special relativity. The transition to quantum field theory from classical special relativity can be obtained by taking the infinite-dimensional unitary representations $U(\Lambda, a)$ of this isometry group of the fixed background structure with fixed Minkowski metric, act on a Hilbert space $\mathcal{P} : H \to H$ and invoke the quantum mechanical axioms to interpret the states in Hilbert space. Then adjoin the internal symmetry groups to get $\mathcal{P} \times SU(3)_c \times SU(2)_L \times U(1)_Y : H \to H$ where the details of the internal

DOI: 10.1051/epjconf/201716408009

symmetry group are apparently taken from experiment with no obvious theoretical origins. This theory is non-gravitational since there is no spacetime curvature dynamics in the theory. If we take the same approach here, also thought of as a non-gravitational theory for the time being, then for our timelike case we would write down $SO(2,1) \times SO(3) : H \to H$ by specifying the infinite-dimensional unitary representations (this group is also non-compact so its nontrivial unitary representations are infinite-dimensional). In practice, we would actually write $SL(2,R) \times SU(2)$ since $sl(2) \cong so(2,1)$ and $SU(2)$ is the double cover for $SO(3)$ (the former having representations while the latter only having projective representations) so we have $SL(2) \times SU(2) : H \to H$. However, thinking of our symmetry groups in this way leaves gravity out of the picture, whereas a full quantum theory however must take into account quantum gravity. For the other cases we would have respectively $SU(2) \times SU(2) : H \to H$ and $U(1) \times SU(2) : H \to H$. This lightlike case is interesting for several reasons. Firstly, it is mathematically the same as the electroweak gauge group of the standard model. Secondly, if we take the universal enveloping algebra $\mathcal{U}(u(1) \oplus su(2))$ and q-deform to the quantum group $\mathcal{U}_q(u(1) \oplus su(2))$ this is isomorphic to $\mathcal{U}_q(gl(2))$ which is also isomorphic to the braided matrices $BM_q(2)$ which in turn is isomorphic to the quantum Minkowski space $\mathbb{R}_q^{1,3}$. These isomorphisms themselves which only hold under $q \neq 1$ were discovered by Majid [20] and constitute powerful additional reasons for considering Hopf-deformations and braided geometry.

With these things in mind we now turn our attention to background and metric independent quantum theory.

3 Quantum gravity and Hopf-stability

In this section we take a more bottom-up approach to considering what any potential fully quantum theory needs, and then we harmonize the observations with those of the previous section.

Since the spacetime metric is the classical gravitational field, a quantum theory of spacetime which gives the classical structures in the classical limit must be background independent and hence also metric independent. The fundamental physical fields of nature should then be described by forms. These physical fields will have characteristic defining internal symmetries so they will be G-valued forms for some appropriate group G. Actually, they really need to be G_q-valued forms where G_q is the corresponding quantum group.[6] That is to say, the symmetries need to be Hopf-stable, i.e. there should be a principle of Hopf-stability, analogous with classical physics where we need a principle of Lie-stability. We take a few sentences to motivate this. Firstly we motivate wanting Hopf algebras instead of Lie groups and then we motivate wanting q-deformed Hopf algebras, i.e. Hopf-stable algebras, i.e. quantum Hopf algebras.

In ordinary classical physics the physical states are defined in phase space $\mathbb{R}^{2n}$ with configuration space $\mathbb{R}^n$. In ordinary quantum mechanics in contrast, information about a physical state is encoded in the wavefunction $\psi : \mathbb{R}^n \to \mathbb{C}$. Furthermore, commuting position observables $x_i : \mathbb{R}^n \to \mathbb{C}$ act on wavefunctions in a multiplicative way. Geometry in ordinary quantum mechanics is described by the algebra $\mathcal{A}(\mathbb{R}^n)$ of functions over a space $\mathbb{R}^n$ rather than being described directly by the space $\mathbb{R}^n$ itself as is the case in ordinary classical physics.

[6] We were slightly sloppy here in saying G-valued rather than g-valued for the corresponding Lie algebra g. However, in either case, under q-deformation we get a quantum group and the two cases are dual to each other.

Now suppose in classical physics there is some group G of symmetries. We would incorporate this into the classical theory by writing $G \otimes \mathbb{R}^n \to \mathbb{R}^n$. In the quantum theory, in order to be mathematically internally consistent, instead of having G directly, we should have the algebra $\mathcal{A}(G)$ of functions over G. It follows then that if we have an action ϕ of a group G on the space $\mathbb{R}^n$ in the classical theory, then in the corresponding quantum theory $\mathcal{A}(G) \otimes \mathcal{A}(\mathbb{R}^n) \xleftarrow{\phi^*} \mathcal{A}(\mathbb{R}^n)$ where ϕ^* is the pull-back and is a homomorphism of algebras and is called the coaction map.[7] The axioms of the group under the action are translated into axioms of the algebra under the coaction. The group has structure maps; multiplication, unit and inverse. These are translated respectively into comultiplication, counit and antipode. An algebra with this additional co-structure is a hopf algebra. In general there are two dual quantum Hopf algebras G_q and $\mathcal{U}_q(g)$, where g is the Lie algebra of the Lie group G and $\mathcal{U}$ denotes the universal enveloping algebra. When $q = 1$ the usual group structures are restored.

Here is a key point. In classical Newtonian physics we have a Lie-unstable algebra due to the presence of an abelian translation group conjugate to a commutative spacetime. When $q = 1$ we have Hopf algebras which are commutative and cocommutative. Our principle of emergence (i.e. everything physical is background independent and has to in a sense be "internally defined") naturally dictates that the physical fields must themselves be elements of the Hopf algebra. To get non-trivial fields the elements $T \in \mathcal{A}(G)$ must be non-cocommutative and this requires that the Hopf algebra be q-deformed. We refer to this as a Hopf-stability principle which is analogously to the quantum theory what the Lie-stability principle is to the classical theory.

We expect then, that our fully quantum theory must be of the form $G_q : H \to H$ where H denotes the physical Hilbert space of physical states. The challenge then becomes to identify the correct G_q and then elucidate all of the mathematical details, along with associated physical interpretation. If the theory is good, known observations will be explainable within the mathematical structure. The remainder of the mathematical structure would then dictate the theory's predictive power. Presumably if such a programme failed, a true theory of everything would have to be based on a different set of principles rather than quantum principles with the quantum principles emerging from the other set.

To help us determine what the correct quantum group G_q should be, we make some observations from standard loop quantum gravity. An underlying assumption in the usual theory is that the classical limit of the quantum gravitational ground state should be flat Minkowski spacetime. The fundamental variables in the spin network representation are then Lorentzian, and on a 4-simplex, given by $U_e = Pe^{\int_e \omega} \in SL(2, \mathbb{C})$ which discretizes the Lorentzian connection ω and $B_f = \int_{t_f} B \in sl(2, \mathbb{C})$ ([17]) which discretizes the (complex-valued) frame field $e : \Sigma \times \mathbb{C}^4 \to \mathbb{C}TM$. A problem is that when one tries to do quantum kinematics, i.e. calculate the transition amplitudes, the integrals are infrared divergent; infrared in the sense that the divergence comes about because of problems in the large scale (i.e. quantum gravitational high spin) structure. The standard fix is to q-deform the theory [21–24]. The classical limit of this q-deformed theory does not give flat Minkowski spacetime however. It gives a spacetime differing from Minkowski spacetime by the presence of a positive cosmological constant. This is an example of a theory which has problems when Hopf-unstable which disappear if one moves to a Hopf-stable theory. In the connection representation a viable quantum gravitational state is the Kodama state [1]. This state seems to have a natural interpretation of being a ground state, and it again crucially depends on a

[7] See [14, 15] for more mathematical details.

nonzero cosmological constant. If positive, the ground state in the classical limit corresponds to de Sitter spacetime. This Kodama state has some potential problems however [25].[8] In that same work Randono introduces a generalized set of Kodama states which solves some of the problems. The work involves trying to make a properly quantum de Sitter based quantum gravity. Interestingly, on p.111 Randono states:

The true constraint algebra is likely to be a deformation of the de Sitter Lie algebra. The deformation itself contains the local degrees of freedom of general relativity. Since the action possesses an exact de Sitter symmetry on a very large portion of the phase space (anywhere $t = 0$), the constraint algebra is likely to reduce to the de Sitter algebra on a large portion of the constraint manifold.

Summarizing the last few observations: Ordinary loop quantum gravity in the spin network representation showed a need for fixing to properly regulate quantum observables. The fix was to q-deform; the classical result being the inclusion of a positive cosmological constant. If one includes such a cosmological constant and looks at loop quantum gravity in the connection representation one finds the Kodama state. However this too is not (necessarily) problem free. A suggested fix is to q-deform the de Sitter based loop quantum gravity theory.

We propose that the classical limit of a true quantum gravitational ground state should look like $AdS_2 \times S^2$ spacetime for our timelike case, $S^2 \times S^2$ spacetime for our spacelike case and $S^1 \times S^2$ spacetime for our lightlike case. This results from our Lie-stability arguments together with the corresponding twin problem of making sure the kinematical variables are constrained and constrained in the way relevant to nature. At the quantum level however, we have argued that we want a quantum Hopf algebra so we want physical fields in general based on $SL_{q_1}(2) \times SU_{q_2}(2)$ for the timelike case. To tighten up what our theory should look like, we note that in the un-q-deformed case, we have a direct product of two two-dimensional spaces. The fact that the fundamental spaces are two-dimensional gives rise to a natural notion of q-deformations and also braiding, anyonic structures and quantum flux tubes [26], so an integrated internally consistent picture is starting to emerge. One question which remains is what the space is that the q-deformed Hopf algebra $SL_{q_1}(2) \times SU_{q_2}(2)$ acts on. It turns out ([20]) that braided spaces have quantum groups as their natural symmetries. Thus we want a braided version $B(AdS_2)\underline{\times}B(S^2)$ of $AdS_2 \times S^2$ upon which $SL_{q_1}(2) \times SU_{q_2}(2)$ acts. Here $\underline{\times}$ signifies that the tensor product between algebras has been deformed rather than the algebras themselves. This allows all relevant mathematical structures to be deformed in an internally consistent way [15, 20]. Finally, for correct braid statistics normalization under braided co-addition the quantum group must be adjoined with a dilatation operator [20].

4 Conclusion and a possible physical interpretation

We have explored the consequences of both imposing the principles of Lie-algebraic and Hopf-algebraic stability, and imposing a more complete physically realistic set of constraints on the kinematical variables. On the classical Lie-stability side we arrived at three unphysical "kinematical" five-dimensional hypersurfaces distinguished by whether the metric on the parent six-dimensional space was describing a timelike interval, a spacelike interval or a lightlike interval. We also explored the consequences of a further restriction, $\frac{d|\mathbf{x}|}{d\phi} = R$, which resulted

[8] e.g. non-renormalizability issues, negative energies, CPT violation, non-invariance under large gauge transformations.

in three product space cases, $AdS_2 \times S^2$, $S^2 \times S^2$ and $S^1 \times S^2$ corresponding to the timelike case, the spacelike case and the lightlike case respectively. The respective symmetry groups are $SO(2,1) \times SU(2)$, $SU(2) \times SU(2)$ and $U(1) \times SU(2)$. These Lie-stable spaces do not have an abelian translation group. The corresponding quantum theory should not have commuting coordinates either, so the corresponding commutative and cocommutative quantum Hopf algebras needed to be Hopf-stabilized (in other words, q-deformed). Hence our quantum symmetry groups are respectively $SL_{q_1} \underline{\times} SU_{q_2}(2)$, $SU_{q_1}(2) \underline{\times} SU_{q_2}(2)$ and $U_{q_1}(1) \underline{\times} SU_{q_2}(2)$ for the timelike, spacelike and lightlike cases. The quantum spaces that these quantum groups (along with the adjoined dilatation operator needed for correct braid statistics normalization being preserved under braided co-addition) act on must be braided versions $B(AdS_2 \underline{\times} S^2)$, $B(S^2 \underline{\times} S^2)$ and $B(S^1 \underline{\times} S^2)$ of the above mentioned classical spacetimes. In the standard model the matter particles are characterized by unitary irreducible representations of the (Poincaré) spacetime symmetry group (adjoined together with quantum numbers associated with the internal symmetry quantum electroweak and quantum chromodynamics gauge groups). Keeping the same general notion of compatibility with the physical vacuum, it is natural to demand that matter be characterized by representations of these above quantum groups.

We have seen from Finkelstein's work (see the appendix) that the standard model spin-$\frac{1}{2}$ fermions and spin-1 electroweak vector bosons can naturally be described by the representations of $SU_q(2)$ corresponding to quantum trefoil knots and quantum ditrefoil knots respectively.[9] Moreover, the simpler, twisted unknots (which Finkelstein refers to also as preons) are natural dark matter candidates, and by virtue of them being more fundamental than the quantum trefoil and ditrefoil knots, would naturally be expected to be in higher abundance which is consistent with the observation that there must be more dark matter in the universe than standard model matter.

In our theory based on the fairly minimal and natural physical requirements of Lie and Hopf-stability together with a more complete and realistic set of physical constraints on the classical kinematical variables, both the physical vacuum and matter (both standard model and dark) are all in some deep way quantum gravitationally connected via the quantum group $SL_q(2)$ Hopf algebra of knotted flux tubes and its quantum subgroup $SU_q(2)$ Hopf algebra of oriented knotted flux tubes. It will surely be of great interest to elucidate the mathematical details, identify the parts of the theory that are predictive in nature, and figure out ways of testing those predictions to either verify or falsify the theory. Since the expansion of this theory is largely mathematical in nature with only a small number of physical axioms, any predictions should be well-defined and hence testable in principle, making the theory concretely falsifiable.

5 Appendix: Finkelstein's knotted $SL_q(2)$-based extension of the standard model

We briefly review the representations of $SL_q(2)$ and their restriction to $SU_q(2)$. The elements L of the classical group $SL(2)$ satisfy

$$\epsilon_{ij} L_{im} L_{jn} = \epsilon_{mn} \det L = \epsilon_{mn}, \qquad \longrightarrow \qquad L^t \epsilon L = \epsilon. \tag{3}$$

Here L and ϵ take the forms

$$L = \begin{pmatrix} a & b \\ c & d \end{pmatrix}, \qquad \epsilon = \begin{pmatrix} 0 & 1 \\ -1 & 0 \end{pmatrix}. \tag{4}$$

[9]The case of the photon and Z^0 are slightly more complicated.

To q-deform to $SL_q(2)$ we wish to preserve the essential characteristic that $\det\epsilon = 1$. For later notational convenience we define $q_1 = q^{-1}$ and deform the matrix ϵ in the obvious way:

$$\epsilon_q = \begin{pmatrix} 0 & q_1^{\frac{1}{2}} \\ -q^{\frac{1}{2}} & 0 \end{pmatrix}. \tag{5}$$

We now want matrices $L_q \in SL_q(2)$ of the form

$$L_q = \begin{pmatrix} a & b \\ c & d \end{pmatrix} \tag{6}$$

such that

$$L_q^t \epsilon_q L_q = \epsilon_q. \tag{7}$$

Defining the q-commutator $[A, B]_q = AB - qBA$ we see that after an explicit calculation based on Eqn.(7) the algebra that the elements of L_q must satisfy is

$$[a, b]_q = 0, \quad [a, c]_q = 0, \quad [b, c] = 0, \quad [b, d]_q = 0, \quad [c, d]_q = 0, \tag{8}$$

$$ad - qbc = 1, \qquad da - q_1 bc \equiv \det_q L = 1.$$

This marks a significant shift from the usual representation theory of classical Lie groups where the matrix elements are commutative numbers, whereas here they are *noncommutative operators.* The q-binomial theorem is required in order to specify the representations of $SL_q(2)$. We have the following definitions:

$$(a+b)^n = \sum_s^n \left\langle \begin{matrix} n \\ s \end{matrix} \right\rangle_q b^s a^{n-s}, \quad (a+b)^n = \sum_s^n \left\langle \begin{matrix} n \\ s \end{matrix} \right\rangle_{q_1} a^s b^{n-s}, \tag{9}$$

$$\left\langle \begin{matrix} n \\ s \end{matrix} \right\rangle_q = \frac{\langle n \rangle_q!}{\langle n-s \rangle_q! \langle s \rangle_q!}, \quad \langle n \rangle_q = \frac{q^n - 1}{q - 1}.$$

We work in the monomial basis where for $[x_1, x_2]_q = 0$, $n_\pm = j \pm m$, $N^j_m = [\langle n_+ \rangle_{q_1}! \langle n_- \rangle_{q_1}!]^{\frac{-1}{2}}$ and $\psi^j_m \equiv N^j_m x_1^{n_+} x_2^{n_-}$. Taking $\mathbf{x}^t = [x_1, x_2]$, we write $\mathbf{x}' \equiv L_q \mathbf{x}$. On the monomial basis $\psi^j_m \equiv N^j_m x^{n_+} x^{n_-}$, assuming that the x_i commute with a, b, c, d, this transformation takes the form

$$\psi^{j'}_m = N^j_m (ax_1 + bx_2)^{n_+} (cx_1 + dx_2)^{n_-} \tag{10}$$

and by use of the q-binomial thorem we directly obtain

$$\psi^{j'}_m = N^j_m \sum_s^{n_+} \sum_t^{n_-} \left\langle \begin{matrix} n_+ \\ s \end{matrix} \right\rangle_{q_1} \left\langle \begin{matrix} n_- \\ t \end{matrix} \right\rangle_{q_1} a^s b^{n_+ - s} c^t d^{n_- - t} x_1^{s+t} x_2^{n_+ + n_- - s - t}. \tag{11}$$

By defining $n'_\pm = j \pm m'$ we make use of the Kroneckar delta $\delta(s + t, n'_+)$ to write

$$\begin{aligned} \psi^{j'}_m &= \sum_{s,t} \left(\frac{N^j_m}{N^j_{m'}} \right) \delta(s+t, n'_+) \left\langle \begin{matrix} n_+ \\ s \end{matrix} \right\rangle_{q_1} \left\langle \begin{matrix} n_- \\ t \end{matrix} \right\rangle_{q_1} a^s b^{n_+ - s} c^t d^{n_- - t} (N^j_{m'} x_1^{n'_+} x_2^{n'_-}) \\ &= \sum_{m'} D^j_{mm'} \psi^j_{m'} \end{aligned} \tag{12}$$

where

$$D^j_{mm'}(q|a,b,c,d) \equiv \sum_{s,t} \begin{pmatrix} N^j_m \\ N^j_{m'} \end{pmatrix} \delta(s+t,n'_+) \left\langle \begin{array}{c} n_+ \\ s \end{array} \right\rangle_{q_1} \left\langle \begin{array}{c} n_- \\ t \end{array} \right\rangle_{q_1} a^s b^{n_+-s} c^t d^{n_--t}. \quad (13)$$

The $SU_q(2)$ representations are obtained by the above $SL_q(2)$ representations by setting $d = \bar{a}$ and $c = -q_1\bar{b}$. For both quantum groups we have (no sum on the j)

$$\psi^j_m(x'_1, x'_2) = \sum_{m'} D^j_{mm'} \psi^j_{m'}(x_1, x_2). \quad (14)$$

These representations are not strictly uniquely determined. They are only determined up to a $U(1)_a \times U(1)_b$ gauge transformation,

$$a' = e^{i\varphi_a}a, \quad b' = e^{i\varphi_b}b, \quad c' = e^{-i\varphi_b}c, \quad d' = e^{-i\varphi_a}d \quad (15)$$

which leaves the a, b, c, d internal algebra unchanged. Explicitly, we have

$$D^j_{mm'}(q|a', b', c', d') = e^{i(\varphi_a+\varphi_b)m} e^{i(\varphi_a-\varphi_b)m'} D^j_{mm'}(q|a,b,c,d). \quad (16)$$

Unlike the undeformed case, there is a natural internal state space due to the fact that the a, b, c, d can be thought of as noncommutative creation operators on an internal Fock space.

Finkelstein proposed an extension of the standard model based on the quantum group $SL_q(2)$ by replacing the elementary standard model field operators $\Psi(x)$ with $\hat{\Psi}^j_{mm'} D^j_{mm'}$ [7]. $SL_q(2)$ is the algebra of knots and $SU_q(2)$ is the algebra of oriented knots. Taking N to be the crossing number knot invariant, w to be the writhe of a knot, r to be the rotation number of the two-dimensional projection of the corresponding classical three-dimensional knot and o to be an odd number additional knot invariant required to preserve the difference in parity between the writhe and rotation, Finkelstein forms the identification

$$(j, m, m') = \frac{1}{2}(N, w, r+o) \equiv \frac{1}{2}(N, w, \tilde{r}) \quad (17)$$

between the quantum state labels (j, m, m') and the topological knot invariants $\frac{1}{2}(N, w, \tilde{r})$. The combined quantity $\tilde{r}$ is called the *quantum rotation.* Finkelstein identifies the elementary fermions (e, μ, τ), $(\nu_e, \nu_\mu, \nu_\tau)$, (u, c, t) and (d, s, b) with the simplest nontrivial quantum knots, which are the $N = 3$ quantum trefoil knots, of which there are four kinds represented by $D^{N/2}_{\frac{w}{2}\frac{r+1}{2}}$ as $D^{3/2}_{\frac{3}{2}\frac{3}{2}}$, $D^{3/2}_{-\frac{3}{2}\frac{3}{2}}$, $D^{3/2}_{\frac{3}{2}-\frac{1}{2}}$ and $D^{3/2}_{-\frac{3}{2}-\frac{1}{2}}$. The first of these is identified with the (e, μ, τ), the second with $(\nu_e, \nu_\mu, \nu_\tau)$, the third with (d, s, b) and the fourth with (u, c, t) elementary standard model fermions. The electroweak numbers isotopic spin t, third component t_3 and electroweak hypercharge t_0 are consistent with the empirical identification

$$(j, m, m') = \frac{1}{2}(N, w, \tilde{r}) = 3(t, -t_3, -t_0)_L \quad (18)$$

for the left chiral fields and quantum trefoils. The trivial quantum knot case $N = 1$ corresponds to twisted unknots which are naturally electroweak dark and hence can be identified as natural dark matter candidates. These so-called preon states nicely match up with the preons in the Harari and Shupe preon models [27, 28]. Since the representations are invariant only up to $U_m \times U_{m'}$ transformations, and given that $m = \frac{w}{2}$ and $m' = \frac{\tilde{r}}{2}$, there are Noether writhe and rotation charges Q_w and Q_r which are related to the electromagnetic charge Q_e by $Q_e \equiv Q_w + Q_r$. In addition, one can write down $Q_e = e(t_3 + t_0) = -\frac{e}{6}(w + \tilde{r})$. The electric charge then naturally provides a measure of the sum of writhe and quantum rotation of the trefoil.

Finkelstein interprets the noncommuting a, b, c, d operator-valued elements of $D^j_{mm'}$ to be fermionic preon creation operators with a having charge $-\frac{e}{3}$, d being its antiparticle and b and c being a neutral preon-antipreon pair. The preons correspond to twisted loops (as seen in their two-dimensional projections) and their quantum realizations $\hat{\Psi}^{\frac{1}{2}}_{mm'} D^{\frac{1}{2}}_{mm'}$ are interpreted as flux tubes. The knots have orientation and framing. The flow direction in the flux tube defines the helicity.

The electroweak vectors W^+, W^- and W^3, taking the $N = 6$ representations, correspond to ditrefoil knots (which are knot sums of trefoil knots) and the electroweak quantum numbers work out analogously.

We have briefly reviewed the electroweak aspects of Finkelstein's knot model. We now briefly review the colour charge aspects and then a possible physical interpretation of the Hopf-type deformation parameter q. We have

$$D^{3/2}_{\frac{3}{2}\frac{3}{2}} \approx a^3, \quad D^{3/2}_{-\frac{3}{2}\frac{3}{2}} \approx c^3, \quad D^{3/2}_{\frac{3}{2}-\frac{1}{2}} \approx ab^2, \quad D^{3/2}_{-\frac{3}{2}-\frac{1}{2}} \approx cd^2 \tag{19}$$

for the four classes of elementary leptons and for the $D^3_{-3t_3-3t_0}$ electroweak vector bosons W^+, W^- and W^3 we have respectively

$$D^3_{-3,0} \approx c^3 d^3, \quad D^3_{3,0} \approx a^3 b^3, \quad D^3_{0,0} \approx A_1 b^3 c^3 + A_2 ab^2 c^2 d + A_3 a^2 bcd^2 + A_4 a^3 d^3 \tag{20}$$

where the precise details of the coefficients A_i do not concern us here. The important point here is that Finkelstein argues that in order to keep the Pauli exclusion principle we have to replace (a, c) with (a_i, c_i) where for charged leptons $a^3 \to \epsilon^{ijk} a_i a_j a_k$ and for neutrinos $c^3 \to \epsilon^{ijk} c_i c_j c_k$. The a_i and c_i yield a basis for the fundamental representation of $SU(3)$. Furthermore, if the b preons and d preons constitute colour singlets, it follows that (d, s, b)-type quarks $(a_i b^2)$ and (u, c, t)-type quarks $(c_i d^2)$ also provide a basis for the standard model fundamental representation of $SU(3)$. Finkelstein also points out that if one were to restrict to $SU_q(2)$, the algebra of oriented knots, where $\bar{a} = d$ and $\bar{c} = -q^{-1}b$, then a_i and c_i correspond to the $\bar{3}$-representation of $SU(3)$ while b^i and d^i correspond to the 3-representation of $SU(3)$. The electroweak vector bosons naturally are colour singlets in this model so these colour aspects of the model are internally consistent.

Regarding the physical interpretation of the Hopf-type deformation parameter q, Finkelstein arrives at the conjecture $q = \frac{e}{g}$ or $q = \frac{g}{e}$ with e, g and q being thought of as running coupling constants.

There are more interesting and important details in Finkelstein's knot model $SL_q(2)$ based extension to the standard model but we have now achieved our goal of describing the possible physical significance of the quantum groups $SL_q(2)$ and $SU_q(2)$ as describing knotted quantum flux tubes.

References

[1] H. Kodama, Prog. Theor. Phys. pp. 1024–1040 (1988)
[2] M. Gerstenhaber, The Annals of Mathematics **79**, 59 (1964)

[3] R.V. Mendes, Journal of Physics A: Mathematical and General **27**, 8091 (1994)
[4] C. Chryssomalakos, E. Okon, Int.l J. Mod. Phys. D **13**, 2003 (2004), [Arxiv:hep-th/0407080]
[5] D. Ahluwalia-Khalilova, Class. Quantum Grav. **22**, 1433 (2005)
[6] R. Finkelstein, Int. J. Mod. Phys. A pp. 4467–4480 (2007)
[7] R. Finkelstein, arXiv:1511.07919 (2015)
[8] R. Finkelstein, Int. J. Mod. Phys. A. pp. 4269–4302 (2006)
[9] R. Finkelstein, Int. J. Mod. Phys. A. pp. 6487–6494 (2005)
[10] R. Finkelstein, Int. J. Mod. Phys. A. pp. 627–650 (2003)
[11] R. Finkelstein, Lett. Math. Phys. pp. 157–163 (2000)
[12] R. Finkelstein, Mod. Phys. Lett. A pp. 1709–1716 (2000)
[13] R. Finkelstein, Lett. Math. Phys. pp. 105–114 (1999)
[14] S. Majid, *A quantum groups primer.* (Cambridge University Press, 2002)
[15] S. Majid, *Foundations of quantum group theory.* (Cambridge University Press, 1995)
[16] M. Bronstein, Physikalische Zeitschrift der Sowjetunion pp. 140–157 (1936)
[17] C. Rovelli, F. Vidotto, *Covariant loop quantum gravity.* (Cambridge University Press, 2002)
[18] B. Bertotti, Phys. Rev. pp. 1331–1333 (1959)
[19] I. Robinson, Bull. Acad. Polon. Sci. pp. 351–352 (1959)
[20] S. Majid, Proceedings of the International School of Physics: Quantum groups and their applications in physics pp. 267–345 (1994)
[21] E. Buffenoir, P. Roche, Commun. Math. Phys. pp. 669–698 (1995)
[22] E. Buffenoir, P. Roche, Commun. Math. Phys. (1999)
[23] K. Noui, P. Roche, Class. Quant. Grav. (2003)
[24] W. Fairbairn, C. Meusburger, J. Math. Phys. (2012)
[25] A. Randono, Ph.D. thesis, University of Texas at Austin Graduate School (2007)
[26] M. Frau, S. Sciuto, A. Lerda, Proceedings of the International School of Physics: Quantum groups and their applications in physics pp. 215–243 (1994)
[27] H. Harari, Phys. Lett. B **86**, 83 (1979)
[28] M. Shupe, Phys. Lett. B. pp. 87–89 (1979)

Value of H, space-time patterns, vacuum, matter, expansion of the Universe, alternative cosmologies

Luis Gonzalez-Mestres[1,a]

[1] *Cosmology Laboratory, John Naisbitt University, Belgrade and Paris*
Goce Delceva 8, 11070 Novi Beograd, Serbia

Abstract. To the experimental uncertainties on the present value H_0 of the Lundmark - Lemaître - Hubble constant, fundamental theoretical uncertainties of several kinds should also be added. In standard Cosmology, consistency problems are really serious. The cosmological constant is a source of well-known difficulties while the associated dark energy is assumed to be at the origin of the observed acceleration of the expansion of the Universe. But in alternative cosmologies, possible approaches without these problems exist. An example is the pattern based on the spinorial space-time (SST) we introduced in 1996-97 where the $H\,t = 1$ relation (t = cosmic time = age of the Universe) is automatically generated by a pre-existing cosmic geometry before standard matter and conventional forces, including gravitation and relativity, are introduced. We analyse present theoretical, experimental and observational uncertainties, focusing also on the possible sources of the acceleration of the expansion of the Universe as well as on the structure of the physical vacuum and its potential cosmological role. Particular attention is given to alternative approaches to both Particle Physics and Cosmology including possible preonic constituents of the physical vacuum and associated pre-Big Bang patterns. A significant example is provided by the cosmic SST geometry together with the possibility that the expanding cosmological vacuum releases energy in the form of standard matter and dark matter, thus modifying the dependence of the matter energy density with respect to the age and size of our Universe. The SST naturally generates a new leading contribution to the value of H. If the matter energy density decreases more slowly than in standard patterns, it can naturally be at the origin of the observed acceleration of the expansion of the Universe. The mathematical and dynamical structure of standard Physics at very short distances can also be modified by an underlying preonic structure. If preons are the constituents of the physical vacuum, as postulated two decades ago with the superbradyon (superluminal preon) hypothesis, the strongest implication would be the possibility that vacuum actually drives the expansion of the Universe. If an unstable (metastable) vacuum permanently expands, it can release energy in the form of conventional matter and of its associated kinetic energy. The SST can be the expression of such an expanding vacuum at cosmic level. We briefly discuss these and related issues, as well as relevant open questions including the problematics of the initial singularity and the cosmic vacuum dynamics in a pre-Big Bang era. The possibility to obtain experimental information on the preonic internal structure of vacuum is also considered.

This paper is dedicated to the memory of Henri Poincaré

[a] e-mail: luis.gonzalez-mestres@cosmology.megatrend.edu.rs;personale-mail:lgmsci@yahoo.fr

1 Introduction

What is the *"trouble with H_0"* recently considered by several authors [1]? *Nature News* wrote on April 2016 [2] *"Measurement of Universe's expansion rate creates cosmological puzzle. Discrepancy between observations could point to new physics."*, referring to the April 2016 paper by Adam G. Riess and other authors *"A 2.4 % Determination of the Local Value of the Hubble Constant"* [3].

Using the Wide Field Camera 3 (WFC3) on the Hubble Space Telescope (HST) [4], Riess et al. announce a best estimate of 73.24±1.74 km/sec/Mpc for the present value H_0 of the Lundmark - Lemaître - Hubble (LLH) constant H (usually called the Hubble constant, but actually due to the work of these three scientists). This result appears to contrast with previous estimates from other experiments, including *Planck* [5].

Riess et al. compare the new obtained value of H_0 with *"the prediction of 69.3±0.7 km/sec/Mpc with the combination of WMAP + ACT + SPT + BAO"* [6, 7] and with the value of *"66.93±0.7 km/sec/Mpc predicted by ΛCDM with 3 neutrinos with mass 0.06 eV and the Planck data"*. To explain the observed differences between the measured values, they consider the possibility of an additional source of dark radiation in the early Universe. Actually, the evolution of the number density of galaxies in our Universe remains an open subject [8].

To the present experimental and observational uncertainties, basic theoretical puzzles and unknowns should be also added. They concern in particular the mathematical structure of space-time, the origin of the Universe, its age and size, the density of matter, the nature and origin of the cosmological constant, the structure and dynamical properties of vacuum, a possible preonic structure...

In particular, if the arrow of time makes sense, what is the force driving the expansion of space in our Universe as the cosmic time increases? What are its nature and its origin? Standard cosmology deals with these questions in a purely phenomenological way starting from the Big Bang hypothesis of Georges Lemaître [9] and adding to it a the *ad hoc* inflationary scenario [10–12]. But alternative cosmologies involving pre-Big Bang scenarios and incorporating new physics can naturally avoid the conventional Big Bang and cosmic inflation [13, 14].

This contribution is devoted to alternative approaches to some basic questions in Cosmology and Particle Physics, with particular emphasis on space-time and on the possible internal structure of the physical vacuum (see also [15]). The potential implications of new physics, including superluminal preons as fundamental vacuum constituents, together with a pre-Big Bang era, are given particular attention including the question of a possible initial singularity.

2 The physical vacuum in an expanding universe

If the Universe expands, how does the physical vacuum evolve? How can vacuum adapt itself to this expansion of space? Or can it actually play a more "active" role in Cosmology?

The answer to these questions will depend crucially on the, by now totally unknown, internal vacuum structure and dynamics.

Is this problematics too complex or far from experiment and observation? In what follows and in [15], we consider some possible signatures of such a new physics and cosmology. Furthermore, standard concepts including Big Bang, inflation, strings, quantum gravity... have led to involved situations [16–18] while: i) direct particle physics tests become more and more difficult as the energy scale increases beyond those of accelerator experiments; ii) astrophysical and cosmological data are incomplete and difficult to interpret in an unambiguous way.

In such a situation, new physics can be present at energy scales far below the Planck energy. There is by now no evidence for a standard Grand Unification scale and the theoretical situation is uncertain [19–23]. Most likely, the Planck scale and standard quantum gravity do no longer make sense [13,

 DOI: 10.1051/epjconf/201716408010

25] in the presence of new physics. All standard forces, including gravity, are then expected to be composite phenomena and standard symmetries become low-energy approximations for conventional matter. The dynamical role of the physical vacuum is a crucial subject in this respect, together with the space-time structure [14, 24].

In standard Quantum Field Theory (QFT), the physical vacuum is a purely passive object where fields associated to conventional "elementary" particles can condensate. But boson fields involve harmonic oscillators with nontrivial "zero modes". Such an amboguity is by itself a source of theoretical and phenomenological problems when a possible "active" role of vacuum is considered.

Attempts to describe the cosmological role of vacuum using standard quantum field theory have led to the well-known cosmological constant problem [26, 27]. But as pointed out in previous papers [13, 28], the situation can be radically different with a preonic vacuum structure naturally avoiding a permanent static presence of the (composite) QFT standard condensates and bosonic zero modes [29]. If the conventional particles are actually excitations of the physical vacuum, the underlying vacuum dynamics can generate them in specific situations and in a way compatible with the successful experimental tests of QFT.

In a preonic vacuum where signals can propagate faster than light [15], the Higgs boson and the zero modes of bosonic harmonic oscillators do not need to be permanently materialized, as the vacuum dynamics can temporarily produce them when required by the presence of surrounding matter and thus make possible the standard particle interactions. In this case, the (ground) state of vacuum in regions of the Universe without matter would be substantially different from the (excited) one observed in the laboratory. Leaving aside possible controversies on the precise origin of the Casimir effect [30–32], one can expect the structure and dynamics of the physical vacuum in the laboratory, even with a locally void setup, to be influenced by the presence of surrounding matter. The fact that quantum entanglement is being successfully checked at rather large distances [33, 34] can be an indication on the propagation of signals in vacuum [15].

An important nontrivial information emerges from cosmological data: the physical vacuum appears to be able to naturally expand in space. But the energetic balance of such an expansion and the force driving it remain unknown and can play a crucial role in the present dynamics of our Universe. Patterns incorporating dark energy the a cosmological cosmological constant tacitly postulate a crucial role of vacuum in the expansion of the Universe. But what is the nature of this physical vacuum and what is its internal dynamics?

As discussed in [15], the definition itself of the physical vacuum is far fron trivial. Assuming a preonic vacuum, a practical choice can be to define vacuum as the set of preons, including virtual ones, trapped in its internal structure. Then, "real" preons able to travel over long distances will be dealt with as free particles.

If the expansion of the physical vacuum is energetically favored and standard particles are actually vacuum excitations, the evolution of our Universe can lead vacuum to release matter and matter energy as it expands. This can fundamentally change the dependence of the matter energy density with respect to the age and size of our Universe.

Then, equations involving matter alone can in principle formally describe the observed acceleration of the expansion of the matter Universe, and no conventional dark energy would be required to produce such an important effect. We present here a simple illustration of this fundamentally new situation, using the cosmology based on the spinorial space-time (SST) we introduced in [35, 36]. We also briefly consider other cosmological scenarios leading to the same phenomenon and discuss the possibility that vacuum actually leads the expansion of the Universe. A preliminary version of the work presented here can be found in two preprints contributed to this Conference [37, 38].

In all cases, it clearly appears that the hypothesis of a preonic vacuum is not more involved than the scenarios of standard cosmology and that, combined with a pre-Big Bang description of the origin of the Universe, it can lead to a much more natural cosmic dynamics. Phenomenology, experiments and observations should focuse on this possibility and search for relevant signatures.

3 The spinorial space-time (SST)

If the physical vacuum has a nontrivial internal structure, what can be the relevant space-time geometry? And if standard particles are just vacuum excitations, can they be sensitive in the laboratory to the actual space-time structure? A first question must be raised: why are the fermion wave functions representations of the covering group SU(2) of the standard space rotation group S0(3), and not of the rotation group itself?

It is a fact that the usual description of space-time is based on everyday's life and on macroscopic objects around us. It was not conceived to describe particles or the very large scale structure of the Universe. Precisely, the SST seems to be able to improve the situation in both domains.

The spinorial space-time (SST) introduced in [35, 36] automatically generates the $H\,t = 1$ relation (t = cosmic time = age of the Universe) in a purely geometric way, already before introducing conventional matter and the associated forces. Its properties have been briefly reminded in [15, 39] and previously dealt with in several papers including [13, 28], [40, 41] and [14, 42].

If ξ is a SU(2) spinor describing the cosmic SST coordinates (two complex variables instead of the standard four real ones) of a point of our space-time, the cosmic time to associated to ξ can be defined by the positive scalar $t = |\,\xi\,|$ with the relation $|\,\xi\,|^2 = \xi^\dagger \xi$ where the dagger stands for hermitic conjugate. Other definitions of t (a function of $|\,\xi\,|$) would lead to equivalent results provided the cosmic time is positive, increases with $|\,\xi\,|$ and is equal to zero for $|\,\xi\,| = 0$.

In such an approach where the cosmic space-time coordinates are no longer real numbers, the status of relativity necessarily changes. Rather than an intrinsic fundamental property of space and time defining a global space-time geometry, conventional relativity is expected to be a low-energy symmetry of standard matter similar to the effective Lorentz-like symmetry of the kinematics of low-momentum phonons or solitons in a condensed medium [43, 44]. In the present case, the preonic vacuum can play the role of the condensed medium and the SST can be the space-time geometry of the internal vacuum structure [15, 39]..

Indeed, a fundamental feature of the cosmic SST geometry is that it is not dominated by standard matter or dark energy, and that its structure is defined in a totally independent way suggesting a more primordial origin. In the SST without standard matter, space units are not required, as time provides an effective distance scale at cosmic level.

In practice, the possible connection between the cosmic spinorial space-time structure and the ultimate dynamics of matter and vacuum remains an open question requiring further fundamental research in both Particle Physics and Cosmology. Clearly, preonic and and pre-Big Bang scenarios should be considered in this respect [14, 45] together with possible sources of signatures (particle accelerators, high-energy cosmic rays, black holes, gravitational waves...). The automatic $H\ t = 1$ relation can be an encouraging signature in spite of the "trouble with H_0".

As reminded in [15] and already found in [35, 36], the SST naturally generates a local privileged space direction (PSD) for each comoving observer. The existence of the local PSD may be consistent with Planck data [46] and does not imply any anisotropy of the Universe at cosmic level, as the PSD is different for any point of the SST except for those lying on the PSD itself. Again, a signature of the SST can result from such an effect.

4 SST, matter and the acceleration of the expansion of the Universe

What do the equations of standard cosmology become if the usual real space-time is replaced by the SST? The equation $H\,t = 1$ in the absence of matter must be a basic startpoint.

In [47] and in other papers, we considered the following Friedmann-like relation for the standard matter universe within the cosmic SST geometry:

$$H^2 \;=\; 8\pi\, G\, \rho/3 \;-\; k\, R^{-2}\, c^2 \;+ t^{-2} \;+\; K \;+\; \Lambda\, c^2/3 \tag{1}$$

t is the above defined cosmic time (age of the Universe), ρ the energy density associated to standard matter, c the speed of light, k the standard curvature parameter, R the present curvature distance scale of the Universe (the curvature radius, and possibly the radius of the Universe for $k = 1$) and Λ a possible new version of the cosmological constant decreasing as the Universe expands.

Λ is now free of any cosmological constant problem. The new term t^{-2}, of cosmic geometric origin as suggested by the SST structure and the $H\,t = 1$ law in the absence of matter, is larger than the standard curvature term $-\,k\,R^{-2}\,c^2$ and has a positive sign independent of k. It in principle clearly dominates the large scale expansion of the Universe at large values of t. *Planck* measured [5]: $\mid \Omega_k \mid = \mid k\,c^2\,R^{-2}\,H_0^{-2} \mid < 0.005$.

K is a correction term that will be neglected in what follows. It can account in particular for:

- a possible small difference between the comoving frames of standard cosmology and those (pre-existing) obtained from the underlying SST cosmic geometry;
- similarly, a correction related to remnant effects from the pre-Big Bang era;
- a reaction of the nucleated standard matter to the pre-existing expansion of the Universe.

Crucial questions to deal with equation (1) are the dependence of ρ and Λ on the age and the size of the Universe. We consider here some specific unconventional approaches to this basic question.

Ignoring, to a first approximation, K and the standard curvature term, one can write:

$$H^2 \;=\; 8\pi\, G\, \rho/3 \;+\; \Lambda\, c^2/3 \;=\; t^{-2} \;+\; \Gamma \tag{2}$$

where $\Gamma \;=\; 8\pi\, G\, \rho/3 \;+\; \Lambda\, c^2/3$ is the sum of the contributions of matter (including dark matter) and of the new cosmological constant.

In what follows, we consider the possibility that Γ is basically dominated by matter, and that the matter energy density decreases more slowly with the Universe expansion than in conventional cosmology.

5 An unconventional way to accelerate the expansion of the Universe

Two basic question are: i) how do matter and the vacuum interact, and what can the preonic vacuum radiate? ; ii) how do matter and vacuum interact with a possible leading role of the SST in the Universe expansion?

In previous papers, we considered a scenario where in the early Universe the standard matter just generated reacts gravitationally to the pre-existing expansion of space generated by the cosmic SST geometry. This may have slowed the expansion of the Universe.

Then, as the matter density becomes smaller and the gravitational force decreases, the effect becomes weaker and the expansion of the Universe accelerates to reach an asymptotic regime with the limit $H\,t = 1$ at large t [47, 48]. This is just an example of possible alternative approaches to the current standard acceleration of the expansion of the Universe.

 DOI: 10.1051/epjconf/201716408010

Other alternative, or complementary, cosmic mechanisms involving the dynamics of the physical vacuum are presented in this and the following sections. The above mechanism can complete those considered below.

If the physical vacuum has a nontrivial internal structure, this structure is in all cases sensitive to the expansion of the Universe. In the most conservative hypothesis, vacuum expands like the available space and its internal structure must follow this nontrivial cosmic process.

It seems then reasonable to expect that "creating more vacuum" should in principle have a nontrivial cost in energy. This cost can, in the same conservative approach, be positive or negative and remains by now unknown. But in all realistic scenarios, such an energy cost does not prevent the space and the vacuum from expanding.

In this respect, a possible set of assumptions can be that:

- The physical vacuum releases a positive amount of energy as it expands with the present Universe evolution.

- This energy is mainly converted into matter (standard and dark), and there is no leading conventional dark energy.

- As a consequence, the cosmic matter energy density decreases more slowly than usually expected as the Universe expands.

Then, if Γ corresponds basically to the matter density, its new dependence on the age and size of the Universe can lead to nontrivial effects and, in particular, make indeed unnecessary the usual role of dark energy and of the cosmological constant. A simple way to illustrate the most essential mechanism can be to write for equation (2):

$$\Gamma = \gamma\, t^{-2} \tag{3}$$

where γ is a constant, and the dependence of the matter energy density with respect to cosmic time leads to:

$$H^2 = t^{-2}\,(1 + \gamma) \tag{4}$$

thus modifying H at the leading level.

It has been assumed in this specific example that the matter energy density varies proportionally to t^{-2}, similarly to the geometric SST term and contrary to standard patterns where matter plays a much less important role and dark energy replaces the SST term.

One then gets, using simultaneously the equation $H = a^{-1}\, da/dt$ where a is the usual cosmic distance scale:

$$da/a = (1 + \gamma)^{1/2}\, dt/t \tag{5}$$

so that, as a consequence of (3) and (4), the cosmic distance scale grows faster than the cosmic time for positive values of γ.

Equation (5) trivially leads to:

$$a = f\, t^{\lambda} \tag{6}$$

where f is a positive constant, not dimensionless, and $\lambda = (1 + \gamma)^{1/2}$. One therefore has, for positive γ, $\lambda > 1$ and $da/dt = (1 + \gamma)^{1/2}\, a/t = f\,\lambda\, t^{\lambda-1} > a/t$ implying a permanent acceleration of the expansion of the Universe.

More explicitly, one can write:

$$d^2a/dt^2 = \lambda\,(\lambda - 1)\, f\, t^{\lambda-2} \tag{7}$$

The acceleration d^2a/dt^2 of the growth of a is therefore positive for $\lambda > 1$. It is constant for $\lambda = 2$ and tends to zero at large t if $\lambda < 2$.

It then follows, in particular, that the expansion of the Universe is permanently accelerated for all positive values of γ without any need for a standard cosmological constant. A positive matter energy density varying like t^{-2} is enough to produce such an effect.

Writing, as usual, for the standard deceleration parameter q_0 :

$$q_0 = - a\, d^2a/dt^2\, (da/dt)^{-2} \tag{8}$$

the above equations lead to

$$q_0 = - f\, t^{\lambda}\, \lambda\, (\lambda - 1)\, f\, t^{\lambda-2}\, (\, f\, \lambda\, t^{\lambda-1})^{-2} = - (\lambda - 1)\, \lambda^{-1} \tag{9}$$

One gets the value $q_0 \simeq -0.05$ for $\gamma \simeq 0.1$ corresponding to $\Gamma \simeq 0.1\, t^{-2}$ (matter energy term $\simeq$ 0.1 times the SST term). An alternative is thus provided to approaches trying to use q_0 to determine the properties of dark energy [49, 50]. The choice made here for the values of γ and λ may look modest, but it appears sufficient when added to the SST term in (1).

5.1 Some phenomenological considerations

In the previous example, the limit $H\, t \rightarrow 1$ is replaced by $H\, t \rightarrow\ (1 + \gamma)^{1/2}$ with $1 + \gamma)^{1/2} \simeq 1.05$ for $\Gamma \simeq 0.1\, t^{-2}$.

It clearly follows, in particular, that the family of scenarios just considered can produce values of H_0 larger than usually expected.

The relation between H and t is thus modified with respect to conventional cosmology or to standard SST predictions. But there is by now no experimental or observational evidence against equations (3)-(4). On the contrary, the recent measurements of H_0 by Wide Field Camera 3 may tend to confirm the validity of this kind of pattern or, in any case, the need to study it closely.

Taking the age of the Universe t to be 13.799 billion years [5], the $H\, t = 1$ law would yield a value of H around 70.91 km/sec/Mpc . There is more than a standard deviation between this value and that of 73.24±1.74 km/sec/Mpc measured by the WFC3.

Then, between 70.91 km/sec/Mpc and 73.24 + 1.74 km/sec/Mpc = 74.94 km/sec/Mpc = 70.91 + 4.07 km/sec/Mpc, the difference of 4.07 km/sec/Mpc can be compatible with the value $\gamma \simeq 0.1$ $(1 + \gamma)^{1/2} \simeq 1.05)$ considered above.

This, however, does not invalidate the possibility that, instead, the $H\, t = 1$ law be correct at the leading level if all the components of equation (1) are taken into account. The difference between the value of H_0 measured by the WFC3 and the inverse of the age of the Universe is 2.33 km/sec/Mpc, which corresponds to 1.34 standard deviations.

It must also be reminded at this stage that the approach developed describes a matter Universe in terms of matter equations without explicitly considering the internal dynamics of the physical vacuum, even if the role of vacuum is tacitly important in equations (3) and (4).

5.2 A more conservative choice

The limit $H\, t \rightarrow 1$ would however be preserved, simultaneously to the positive sign of d^2a/dt^2 (positive acceleration of the expansion of the Universe), replacing equation (3) by a similar law with a less strong dynamical content:

$$\Gamma = \gamma'\, t^{-\alpha} \tag{10}$$

and

$$H^2 = t^{-2} + \gamma'\, t^{-\alpha} \tag{11}$$

where γ' is a constant and $2 < \alpha < 3$. The value of γ' can then be adjusted in terms of that of α and the results of the measurements of H_0 considering also other cosmological parameters.

The value $\alpha = 3$ is a limiting case, where one would get:

$$H^2 = t^{-2} + \gamma' t^{-3} \tag{12}$$

with

$$da/dt = a\,H = a\,(t^{-2} + \gamma' t^{-3})^{1/2} \tag{13}$$

and

$$d^2a/dt^2 = da/dt\,(t^{-2} + \gamma' t^{-3})^{1/2} - a\,(t^{-2} + \gamma' t^{-3})^{-1/2}\,(2\,t^{-3} + 3\,\gamma' t^{-4})/2 \tag{14}$$

leading to:

$$d^2a/dt^2 = a\,(t^{-2} + \gamma' t^{-3}) - a\,(t^{-2} + \gamma' t^{-3})^{-1/2}\,(2\,t^{-3} + 3\,\gamma' t^{-4})/2 \tag{15}$$

where

$$a\,(t^{-2} + \gamma' t^{-3})^{-1/2}\,(2\,t^{-3} + 3\,\gamma' t^{-4})/2 = a\,t^{-2}\,(1 + \gamma' t^{-1})^{-1/2}\,(1 + 3/2\,\gamma' t^{-1}) \tag{16}$$

Then, expanding (16) in a power series of t^{-1}, the coefficients of t^{-2} and t^{-3} in the similar expansion of (15) turn out to vanish simultaneously. This does not happen for $\alpha < 3$, in which case the term proportional to $t^{-\alpha}$ yields a (leading) positive contribution to the value of d^2a/dt^2 .

It follows from (15) and (16) that, for $\alpha = 3$, a vanishing acceleration is obtained at the first relevant order for small t^{-1}. For higher values of α, the acceleration becomes negative at large t.

For $2 < \alpha < 3$, one has:

$$d^2a/dt^2 = a\,(t^{-2} + \gamma' t^{-\alpha}) - a\,(t^{-2} + \gamma' t^{-\alpha})^{-1/2}\,(2\,t^{-3} + \alpha\,\gamma' t^{-\alpha-1})/2 \tag{17}$$

leading to:

$$d^2a/dt^2 \simeq (3 - \alpha)\,\gamma' t^{-\alpha}/2 \tag{18}$$

It seems then realistic to conclude that a matter energy density of the form (9) with $2 < \alpha < 3$ can: i) preserve the $H\,t \rightarrow 1$ limit at large t ; ii) generate the observed acceleration of the Universe expansion with a suitable value of γ' for the relevant value of α.

A more precise observational knowledge of the actual matter energy density in the Universe (including in particular dark matter and a deeper understanding of black hole dynamics) would allow to further constrain the parameters α and γ'.

5.3 Possible scenarios in other space-time geometries

The SST t^2 term can be removed from the formulae leading to (4) without altering the cosmological prediction if, for instance, (3) is replaced by:

$$\Gamma = (1 + \gamma)\,t^{-2} \tag{19}$$

In this specific example, the matter energy term would, alone, replace the role of the SST geometry in (1) and of any conventional dark energy.

Similarly, the cosmological implications of (10) and (11) can be preserved without using the t^2 term from SST if (9) is replaced by:

$$\Gamma = t^{-2} + \gamma' t^{-\alpha} \tag{20}$$

Thus, the above results for the acceleration of the expansion of the Universe apply to other cosmic space-time geometries provided suitable parameterizations are used for the matter energy density and other possible cosmological variables replacing in particular the contribution from the SST geometry.

Less strong parameterizations would not work similarly. A possible attempt without the SST term writing instead of (11) (σ is a constant):

$$H^2 = \sigma \, t^{-\alpha} \tag{21}$$

leads for $\alpha > 2$ to a negative acceleration of the expansion of the Universe and to a constant value of a in the large t limit.

The presence in equations like (10), together with the t^{-2} term, of a term proportional to $t^{-\alpha'}$ with $\alpha' \leq 2$ turns out to be necessary to obtain a positive acceleration of the expansion of the Universe and possibly explain cosmological data.

Values of α' smaller than 2 can be considered in this respect, but they would lead to an exponential expansion of the Universe at large t.

6 Further remarks

As shown in [47] and contrary to the curvature term standard patterns, the SST can account for both spherical and hyperbolic observed cosmic space curvatures with the same positive-definite dominant t^{-2} term in (1). The SST curvature term t^{-2} describes a primordial space-time geometry and not the cosmic space curvature directly seen in current observations.

Similarly, as discussed in previous papers and obtained from (11) and (18) in the approaches considered here, the observed acceleration of the expansion of the Universe is not necessarily a permanent leading phenomenon. Instead, the value $H \, t = 1$ predicted by the SST geometry can be the asymptotic limit at cosmic level for large values of t.

This is the case for some of the models discussed here, whereas other scenarios yield a large t limit of the form $H \, t = (1 + \gamma)^{1/2}$ with a positive value of γ and the matter energy density replacing the standard cosmological constant.

If the cosmic physical vacuum emits matter and matter energy as it expands, the matter energy density in our Universe will decrease more slowly than assumed in conventional patterns and can potentially be described by the models considered here for the Γ term. Matter would then naturally be at the origin of the observed acceleration of the expansion of the Universe. This unconventional hypothesis may turn out to be fully realistic when hopefully unveiling the ultimate structure of vacuum and matter [43, 44].

Then, as previously shown, dark energy and the cosmological constant would no longer be necessary, even if they can still exist in a more limited form as suggested in [47] and in subsequent work. Actually, if the acceleration of the expansion of the Universe can be generated by matter with a suitable vacuum dynamics, there is no obvious motivation for a cosmological constant.

The cosmology based on the spinorial space-time appears particularly well suited to describe a scenario where matter itself generates the observed acceleration of the expansion of the Universe. The t^{-2} term from the SST geometry in (1) and (2) directly contributes to this effect.

Simultaneously, potential evidences for the SST already exist. In particular, the PSD (local privileged space direction) may have been observed by Planck [46] through an asymmetry of the cosmic microwave background. To date, the Planck Collaboration has not modified this 2013 announcement, but its final results have not yet been made public.

A basic unknown remains: that of the global size and shape of the SST Universe in standard space units. As discussed in previous papers [14, 42], space units in the SST geometry are defined only at

DOI: 10.1051/epjconf/201716408010

a later stage when introducing standard matter and the associated SST space coordinates. Then, the SST Universe may turn out to be much larger than the space occupied by conventional matter around us, and other regions of the global Universe may exist with different properties.

More generally, the role of the actual space-time structure in Particle Physics and Cosmology remains a crucial issue [15], including possible superluminal constituents of the physical vacuum [43, 44] and associated pre-Big Bang scenarios [14, 42]. Alternative approaches like those defined by (1), (3), (10) and (20) can be at the origin of a new generation of tests.

Further work is required in all cases, including measurements with higher precision and detailed data analyses as well as relevant theoretical developments beyond standard patterns. In what follows, we discuss some possible implications of a preonic vacuum structure.

In connection with attempts [19, 52] to explain a possible high-energy diphoton excess at CERN experiments [53, 54] in terms of supersymmetry and other fashionable extensions of the standard model, it must be reminded [14, 24] that the physical vacuum with the SST geometry is expected to naturally generate Regge-like trajectories of excitations (particles) with spins spaced by 1/2 instead of 1. In this way, a large number of new particles can appear at all energies. Those with the relevant masses would potentially be produced at CERN experiments.

The properties, role and interactions of such newa particles in our Universe are also open questions. String-like scenarios involvng them cannot be excluded. Also, as the SST symmetry is a compact group, it naturally escapes constraints of the Coleman-Mandula type. New kinds of symmetries for particles can therefore be considered.

If Regge-like and string-like scenarios with a spin spacing of 1/2 apply involving simultaneously standard and new particles, a universal critical speed in vacuum equal to c appears as the natural hypothesis. More involved scenarios can also be imagined, in which case superluminal new particles would be expected to spontaneously decay.

7 Possible cosmological implications of a preonic vacuum

The Planck collaboration [51] has not yet released its final results, and previously reported possible evidences for new physics and alternative cosmologies [14, 39] have not by now been excluded. Precisely, a really new cosmology can emerge from the patterns considered here.

As discussed in [15], as well as in previous papers like [14, 39] and [24, 37], the crucial open interrogation concerning the deep internal structure and dynamics of the physical vacuum is a crucial one to understand the properties of matter and of the Universe. Not only this question is not really dealt with by standard quantum field theory, that is basically a phenomenological description of a family of vacuum excitations (the standard particles), but the description of matter itself may turn out to be incomplete as just pointed out about the spinorial Regge trajectories.

The original preon hypothesis [55] assumed preons to be direct constituents of standard particles. Relativity and Quantum Mechanics where not challenged in such a context. But in 1995 we introduced the superbradyon hypothesis [44, 56] assuming that the physical vacuum is made of superluminal preons (superbradyons) and that that conventional particles are excitations of the preonic vacuum. A new approach to preons and to the vacuum structure was thus sugggested. More than twenty years later, the vacuum properties remain unknown and there is no experimental or observational evidence against the superbradyon hypothesis.

Superbradyons (bradyons of a super-relativity) would be expected to obey a Lorentz-like symmetry with a new critical speed c_s much larger than the speed of light c . The superluminal character of such vacuum constituents can be considered as a natural property, just as the speed of light is much larger than that of phonons or solitons in condensed matter [43, 56]. But what are the structure and the properties of such an underlying "condensed" medium? And what is the precise relation between vacuum and the Universe?

Combining superbradyons with the cosmic SST geometry, a new cosmology can be built [14, 42] incorporating a pre-Big Bang scenario where relativity and Quantum Mechanics would be deformed at very high energy [42, 57] and no longer be ultimate principles of Physics. Superbradyons and the SST can be at the origin of Quantum Mechanics and relativity as low-energy approximations for standard particles [15, 45] in the presence of a more fundamental dynamics. But what can be this dynamics, and how to obtain experimental and observational signatures from it? The above mentioned new particles can provide possible signatures at accelerator experiments, but what about cosmological signatures?

As developed in [15, 38], black holes interacting with the vacuum structure and gravitational waves can provide possible signatures of a preonic vacuum. But a stronger effect must be expected from pre-Big Bang patterns where the early preonic vacuum can play a crucial dynamical role.

Already in [58], it was emphasized that *"if superluminal sectors exist and Lorentz invariance is only a sectorial property, the Big Bang scenario may become quite different"*. The existence of a critical high temperature T_0 for standard particles was postulated, associated to a high energy scale such that *"above T_0 the Universe may have contained only superluminal particles"*. A pre-Big Bang scenario was then clearly formulated, together with its basic implications (no standard Big Bang, no cosmic inflation...) without any need for the "string-driven" scenario proposed in [59, 60]. As strings are known to have a composite fundamental structure (fishnet diagrams...) [57, 61], a string approach to Cosmology can naturally lead to a superbradyonic pre-Big Bang.

The composite nature of hadronic strings was also confirmed by work on the structure of the Pomeron-like double-twisted graph of dual amplitudes [62, 63]. The kinematical properties of the intermediate states at high energy and low momentum transfer where shown to be perfectly compatible with a multiperipheral configuration of constituents. But strings cannot describe the fundamental large-angle processes involving small distance scales and "hard" interactions. A similar situation can be expected for cosmic strings and the role of vacuum preons in the early Universe.

Theoretical problems raised by the first BICEP2 announcement of a possible observation of B-modes of the cosmic microwave background (CMB) radiation were discussed in [64, 65] and subsequently in [42]. In alternative cosmologies such as those based on the SST, primordial gravitational waves from inflation would not be the natural explanation for such a signal. Further experimental work on the subject is in progress at the South Pole [66] and Atacama [67].

8 An unstable(metastable) expanding vacuum?

In [37, 38], we considered the possibility of a physical vacuum expanding in an unstable (metastable) state and releasing conventional matter and energy as a form of cosmic latent heat. If, as a consequence, the matter energy density *Gamma* decreases with time more slowly than in standard cosmology, important effects can be expected as just described. But the most important effect would come from the vacuum itself.

In such a scenario, the unstable (metastable) vacuum would actually drive the expansion of our Universe. If this is the case, the leading role of vacuum in such an expansion would most likely have been permanent since the beginning of the Universe evolution. Thus, the by now unknown internal structure and dynamics of the physical vacuum would clearly be the most important ingredient for modern Cosmology.

The cosmic SST geometry can actually be the expression of the expansion of a physical vacuum at very large distance scales, even beyond our standard Universe. It clearly emerges from the SST properties obtained without any explicit presence of matter and automatically incorporating an expansion of space with the $H\,t = 1$ law, that the question of a possible primordial origin of such a space-time geometry naturally points to the structure and dynamics of the physical vacuum.

The superbradyonic vacuum appears then as a well-suited scenario, and may even be at the origin of the recently reported evidence for quantum entanglement [33, 68] if superluminal signals can propagate inside such a physical vacuum [14, 15]. Indeed, a superbradyon critical speed c_s equal to a million times the speed of light c (just as the speed of light is around a million times the speed of sound) would be compatible (around 4 picoseconds time interval) with the observed quantum simultaneity at a distance of 1.3 km.

Direct experimental tests of quantum entanglement at larger distances may even allow to determine the value of c_s if they reach distance scales at which entanglement ceases to hold. Thus, experiments on entanglement can provide a unique exploration of the internal vacuum structure and of its dynamical and cosmological implications.

A major step to explore the vacuum structure and the possible value of c_s through direct tests of quantum entanglement can be provided by the Quantum Science Satellite recently launched [69–71].

In all cases, if the physical vacuum has a nontrivial internal structure, this structure is necessarily sensitive to the expansion of the Universe where vacuum itself expands like space and its internal structure and dynamics must follow this cosmic process. It seems then reasonable to expect that "creating more vacuum" should in principle have a nontrivial cost in energy. This cost remains by now unknown but, according to observation, it does not prevent the space (and consequently, the vacuum) from expanding.

A natural assumption can therefore be that the physical vacuum "likes expanding" (a negative cost in energy) and releases energy in this dynamical process. The implications of such a scenario are very strong from a cosmological point of view

8.1 An unstable (metastable) vacuum driving the expansion of the Universe?

If the expansion of the physical vacuum is energetically favored, it must actually be at the origin of the expansion of space and of our Universe. No other realistic explanation seems possible for such a phenomenon in the vacuum dynamics can drive it.

The observed expansion of the matter Universe would then actually be driven by that of the physical vacuum leading the expansion of the available space. A permanent interaction between vacuum and matter (they form actually a single dynamical and cosmological system [15]) is required in order to guarantee a common expansion.

Black holes are also an ingredient of such a cosmic dynamics, and vacuum directly interacts with them [15, 38].

Then, the formal cosmology of the matter Universe in terms of matter would not actually describe the deep Universe dynamics but the way matter adapts itself to the expansion of the internal vacuum structure. Instead of the standard dark energy, the expanding vacuum can emit conventional matter and associated energy, possibly leading to a matter energy density that decreases with time more slowly than in standard cosmology and is formally able by itself to accelerate the expansion of the matter Universe [37, 38].

Concerning the time spent by a pre-Big Bang era, it must be noticed that the current estimates of the age of the Universe involve error bars of the order of 20 million years [5]. A 10 million years pre-Big Bang period with a hyperspherical Universe expanding at a superluminal radial speed $\sim 10^6\,c$ would yield a hypersphere radius $\sim 10^{13}$ light years. This is far more than required to solve the horizon problem and to satisfy similar standard requirements usually leading to the standard inflationary pattern. Thus, pre-Big Bang with an expanding superbradyonic vacuum would have no actual influence

 DOI: 10.1051/epjconf/201716408010

on the estimated age of the Universe and can be expected to successfully replace the conventional description of the beginning of its evolution.

It therefore follows that there is in principle no conceptual or obvious phenomenological problem in assuming that an unstable or metastable preonic vacuum has been leading the expansion of the Universe since the earliest times of its evolution.

9 Why a pre-Big Bang?

At this stage, a basic question must be raised: why a pre-Big Bang? Can one imagine a specific reason for a pre-Big Bang phase preceding the present one in the history of the Universe?

If the expansion of our Universe has always been driven by the dynamics of a preonic vacuum, why did such an expansion have a previous phase different from the current one? Why not a single phase with the present $H\,t = 1$ law?

To potentially answer these questions, a complementary interrogation should be added: what is the highest temperature at which standard matter can exist and a matter Universe can be formed?

The conventional Big Bang model enforces from the beginning the presence and the dominant role of standard matter. But why should these constraints be compulsory, especially in the presence of a superbradyonic vacuum?

A simple hypothesis can be that the pre-Big Bang era corresponds to an initial phase of the history of the Universe, when the physical vacuum was so hot that standard matter could not be formed in a stable way and superbradyons were not trapped in rigid structures "inside" the vacuum.

In such a "fluid" early vacuum, it seems reasonable to assume that superbradyons traveled at speeds of the order of c_s or at least much larger than c, thus leading to a superluminal expansion of the pre-Big Bang preonic Universe in the absence of nucleated standard matter.

Then, as vacuum expanded, its temperature decreased and at a later stage the present phase was reached. This new phase can involve confined superbradyons and, in all cases, a significant population of standard matter. Conventional particles would be excitations of the new vacuum structure replacing the earlier unconfined motion of superbradyons in the pre-Big Bang vacuum.

In the current phase of the Universe evolution, possibly described by the cosmic SST, the internal strucure of vacuum would still expand and become colder following a law close to $H\,t = 1$ and releasing matter and energy in this process.

The pre-Big Bang phase and the transition from the pre-Big Bang to the present phase of the evolution of the Universe are contracted to the point $t = 0$ in the cosmic SST description developed above. A sepa rate description is therefore required for the pre-Big Bang.

A new cosmic SST can also be considered for pre-Big Bang with an expansion law for the pre-Big Bang Universe that can be of the type $H_{PBB}\,\Phi\,(t_{PBB}) = 1$ where t_{PBB} is time during pre-Big Bang, H_{PBB} a new LLH-like constat and *Phi* a well-suited fonction of which the $H_{PBB}\,t_{PBB} = 1$ equation would be a particular example. If the pre-Big Bang time is to be considered as completely decoupled from that of the present phase of the history of the Universe and independent of his scale, t_{PBB} can in principle be redefined.

In the hyperspherical configuration of space, this would be compatible with a pre-Big Bang Universe radius of $c_s\,t_{PBB}$ expanding at a velocity of c_s. In practice, the expansion velocity of the pre-Big Bang Universe may have been smaller than c_s and time-dependent but in any case much larger than c.

At "large" t_{PBB} , a transition to the present phase would occur and the $H_{PBB}\,\Phi\,(t_{PBB}) = 1$ equation would have to be adapted in order to operate the corresponding evolution where the superbradyonic structure will be "condensed" and standard matter will nucleate. In particular, the expansion of the Universe will become much slower but superluminal signals will still be allowed to propagate at a velocity of c_s or close to it.

If the transition from pre-Big Bang to the current phase of the Universe evolution is of type I, a nucleation center can expand very quickly in the superbradyonic vacuum where causality relations exist since the beginning between all points. The situation is clearly different from the conventional Big Bang, where only a small amount of standard matter exist at the beginning of the Universe and cosmic inflation is required for this reason.

Similarly, if the pre-Big Bang Universe undergoes a type II phase transition to reach the present evolution, a large amount of nucleation centers is expected to appear in a medium where all the points are causally related and can exchange signals at a velocity of c_s. Such a situation would be fundamentally different from the standard Big Bang.

Thus, a pre-Big Bang era appears as the natural alternative to the standard inflationary scenario built *ad hoc* to preserve the Big Bang theory and the description of the Universe referring only to conventional matter. The horizon and flatness problems are thus naturally solved.

10 The initial singularity

Has there been an initial singularity? If this is the case, what were its properties and its role in the origin of the Universe? How was it generated?

Stephen Hawking suggests [72, 73] that the Universe would have expanded from a single point borrowing energy from the gravitational field to create matter during its expansion. In [73], an initial gravitational singularity is considered and an updated version of this approach must possibly use quantum gravity at very low distances. But a preonic approach to vacuum structure would replace quantum gravity at much larger distance scales.

As early as 1993, Borde and Vilenkin argued [74, 75] thar *"a physically reasonable spacetime that is eternally inflating to the future must possess an initial singularity"*. Work on the initial singularity also concerns information and dynamics contained in imaginary time [76–79].

In particular, an initial imaginary time direction was considered in [76, 77] with an initial gravitational instanton and a pre-Big Bang scenario incorporating the generation of standard particles.

Attempts [80] to avoid the initial singularity pattern using an energy-dependent ("rainbow") spacetime metric [81] are just direct applications of the high-energy deformation of relativity considered in [35, 43], [82, 83] and in subsequent papers. In principle, the notion itself of energy dependence requires a privileged local rest frame (the "vacuum rest frame", VRF [43, 82] to be usefully applied. Already discussed in [15] and in the present contribution, new physics can be present well before Planck scale with an associated pre-Big Bang scenario.

In the presence of a superbradyonic vacuum, an alternative to the role of gravitation in the expansion of the Universe may come from new physics. Assuming that energy conservation still holds, if the internal "condensed" vacuum structure has a negative energy, it will naturally expand by reproducing itself. The matter energy density and temperature will then decrease with time.

The questions of a possible initial singularity and of the source of the energy required for the expansion of the pre-Big Bang Universe can also be raised for a superbradyonic vacuum. The initially very high temperature can account for such an expansion, but the origin of the initial state remains an interrogation. The presence of a strong attractive force can play an important role in helping vacuum to create superbradyons, slowering the expansion and operating the transition to the present phase.

It must also be noticed that, since the superbradyons of the pre-Big Bang era did not obey Quantum Mechanics, their properties as waves or particles are a complementary interrogation. They may have existed with a variable mass and have been able to split into similar objects with lower masses following the evolution of the temperature of the Universe. For a speed v such that $c_s \gg v \gg c$, the leading energy term would be close to $m\, c_s^2$ and lower values of m will be favoured as the superbradyonic pre-Big Bang Universe cools and expands.

Another possibility for the pre-Big Bang era would be that t_{PBB} actually varies between - ∞ and the positive transition time, with a suitably defined law for $H\ (t_{PBB})$. Then, the meaning of time itself and of the arrow of time may need to be reconsidered.

Multiverse approaches [84–87] can also be substantially transformed by a preonic vacuum approach. In particular, two kinds of scenarios can be considered:

- A possible approach would be to assume that the pre-Big Bang phase does not actually become a global universe in the present phase, but remains instead a superbradyonic medium where several separate universes with standard matter can nucleate and develop without any direct contact between them. Then, the energy asociated to the initial singularity of each Universe and to its further development will be provided by the evolving pre-Big Bang medium.

- An alternative can be to assume the existence of a fundamental superbradyonic medium where pre-Big Bang universes can nucleate and further evolve. In this case, the energy for the initial singularity and the subsequent evolution will be provided by such a medium.

In both cases, there is no obvious evidence that the laws of Physics for standard matter and the associated fundamental constants would be different for different universes. The dynamics of the common superbradyonic medium in the scenarios just described can potentially generate a single set of physical laws for the excitations of this medium, including those corresponding to conventional matter in all the universes.

A relevant question, however, is that of the possible time variation of the fundamental constants that remains by now under investigation [88].

Also, in the models just described, one may think that if each universe expands forever as a law like $H\ t = 1$ suggests, two or several different universes could possibly collide as they expand and become a single cosmological object. But such a collision will be avoided if the medium where these universes have been born expands fast enough, as a supebradyonic internal structure can in principle guarantee. Thus, the pattern appears consistent from this point of view.

Obviously, the subject of the possible initial singularity and of the associated energy sources is full of open questions and uncertainties in a context where nothing is really known about the structure of the physical vacuum. Again, experiments on quantum entanglement may offer a unique chance to improve this situation by reaching a phenomenological determination of c_s.

11 Conclusion and comments

This paper actually corresponds to two posters presented at the Conference [89].

The subjects convered are full of unknowns, but also of possible alternatives to standard cosmological prejudices. It clearly turns out that the observed acceleration of the expansion of the Universe does not require the introduction of the standard cosmological constant with the usual dark energy, and that simpler and more natural mechanisms can be at work.

A preonic vacuum can: i) generate alternatives to the standard Quantum Field Theory avoiding the pattern that leads to the conventional cosmological constant; and, simultaneously, ii) play a leading role in the expansion of our Universe if its own expansion is energetically favoured. In this case, the expansion of the Universe can be basically that of the physical vacuum.

The space-time geometry introduced by the SST generates a t^{-2} contribution (t = age of the Universe) to the basic formula for H^2 that changes fundamentally the situation in Cosmology. Then, instead of dark energy, a contribution from the matter energy density can basically complete the SST description of H. Other space-time geometries, together with well-suited choices for the matter energy density, can also provide useful alternatives.

The choice of a pre-Big Bang scenario with new physics instead of the standard cosmic inflation appears really well motivated and naturally solves the horizon and flatness problems. The question of the initial singularity can be discussed in this context, with the fundamental interrogations of how did the Universe start and what is the actual source of its energy.

A crucial question is that of the possibility to get experimental information on the internal structure and dynamics of the physical vacuum. As stressed in [15] and in the present paper, direct tests of quantum entanglement at very long distances can potentially provide a unique way to determine the critical speed c_s of the possible superbradyonic constituents of vacuum if entanglement is due to the transmission of superluminal signals.

The internal structure and dynamics of the physical vacuum are nowadays the most important interrogation for Particle Physics and Cosmology. Further theoretical work is required, together with devoted long-term experimental and observational programs including the direct search for "free" superbradyons and specific tests of SST predictions.

This paper is dedicated to the memory of Henri Poincaré, remembering his pioneering role in several fields of theoretical physics and mathematics, including the birth of relativity. Unfortunately, Poincaré died of a serious health problem in 1912. He was 58 years old, and he had worked until the end in spite of his illness [90, 91].

References

[1] See, for instance, J.L. Bernal, L. Verde and A. Riess, *The trouble with* H_0, JCAP **10**, 019 (2016), arXiv:1607.05617, and references therein.

[2] D. Castelvecchi, *Measurement of Universe's expansion rate creates cosmological puzzle*, Nature News, 11 April 2016, http://www.nature.com/news/measurement-of-universe-s-expansion-rate-creates-cosmological-puzzle-1.19715

[3] A. Riess et al., *A 2.4% Determination of the Local Value of the Hubble Constant, Astrophysical Journal* **826**, 56 (2016), arXiv:1604.01424

[4] See, for instance, http://sci.esa.int/hubble/34007-hubble-instruments/ (European State Agency, ESA).

[5] Planck Collaboration, *Planck 2015 results. XIII. Cosmological parameters, Astronomy & Astrophysics* **594**, A13 (2016), arXiv:1502.01589

[6] Wilkinson Microwave Anisotropy Probe, Atacama Cosmology Telescope, South Pole Telescope, Baryon Acoustic Oscillations.

[7] The authors refer to C.L. Bennett et al., *The 1% Concordance Hubble Constant*, APJ **794**, 135 (2014), arXiv:1406.1718

[8] See, for instance, C. Conselice et al. *The Evolution of Galaxy Number Density at* $z < 8$ *and its Implications*, APJ **830**, 83 (2016), arXiv:1607.03909

[9] G. Lemaître, *The Beginning of the World from the Point of View of Quantum Theory*, Letter to *Nature*, *Nature* **127**, 706, 9 May 1931.

[10] A. Linde, *Gravitational waves and large field inflation*, arXiv:1612.00020, and references therein.

[11] A. Linde, *On inflation, cosmological constant, and SUSY breaking*, arXiv:1608.00119, and references therein.

[12] See also A. Ijjas and P. Steinhardt, *Implications of Planck2015 for inflationary, ekpyrotic and anamorphic bouncing cosmologies*, *Class. Quantum Grav.* **33** 044001 (2016), arXiv: 1512.09010, and references therein.

[13] L. Gonzalez-Mestres *Pre-Big Bang, fundamental Physics and noncyclic cosmologies*, presented at the International Conference on New Frontiers in Physics, ICFP 2012, Kolymbari, Crete, June

10-16 2012, *EPJ Web of Conferences* **70**, 00035 (2014), and references therein. Preprint version at mp_arc 13-18.

[14] See, for instance, L. Gonzalez-Mestres, *Big Bang, inflation, standard Physics... and the potentialities of new Physics and alternative cosmologies*, talk given at the 4th International Conference on New Frontiers in Physics, Kolymbari, Crete, Greece, 23-30 August 2015, *EPJ Web of Conferences* **126**, 0212 (2016), and references therein.

[15] L. Gonzalez-Mestres, *Quantum Mechanics, vacuum, particles, Gödel-Cohen incompleteness and the Universe*, first-day lecture at this Conference.

[16] See, for instance, L. Smolin, *Four principles for quantum gravity*, arXiv:1610.01968 , and references therein.

[17] See, for instance, F. Quevedo, *Is String Phenomenology an Oxymoron?*, arXiv:1612.01569 , and references therein.

[18] See, for instance, J. Martin, *The Observational Status of Cosmic Inflation after Planck*, Lectures given at the II JPBCosmo school (Brazil), arXiv:1502.05733 , and references therein.

[19] See, for instance, A. Mütter, M. Ratz and P. Vaudrevange, *Grand Unification without Proton Decay*, arXiv:1606.02303

[20] See, for instance, H. Nilles and M. Winkler, *750 GeV Diphotons and Supersymmetric Grand Unification, J. High Energ. Phys.* **1605** (2016), 182, arXiv:1604.03598

[21] See, for instance, A. Chamseddine, A. Connes and W. van Suijlekom, *Grand Unification in the Spectral Pati-Salam Model, J. High Energ. Phys.* **1511** (2015), 011, arXiv:1507.08161

[22] See, for instance, Y. BenTov and A. Zee, *The Origin of Families and SO(18) Grand Unification, Phys. Rev. D* **93**, 065036 (2016), arXiv:1505.04312

[23] See, for instance, G. 't Hooft, *The Evolution of Quantum Field Theory, From QED to Grand Unification*, contribution to *The Standard Theory up to the Higgs discovery - 60 years of CERN*, arXiv:1503.05007

[24] L. Gonzalez-Mestres, *Spinorial Regge trajectories and Hagedorn-like temperatures*, presented at ICNFP 2015, EPJ Web of Conferences **126**, 05005 (2016).

[25] L. Gonzalez-Mestres, *Ultra-high energy physics and standard basic principles*, contribution the 2nd International Conference on New Frontiers in Physics, Kolymbari, Crete, Greece, August 28 - September 5, 2013, *EPJ Web of Conferences* **71**, 00062 (2014). See also the Post Scriptum to the preprint version, mp_arc 14-31.

[26] See, for instance, A. Padilla, *Lectures on the Cosmological Constant Problem*, arXiv: 1502.05296 , and references therein.

[27] S. Weinberg, *The Cosmological Constant Problems*, Talk given at Dark Matter 2000, Marina del Rey, CA, February 2000, arXiv:astro-ph/0005265 , and references therein.

[28] L. Gonzalez-Mestres, *Cosmic rays and tests of fundamental principles*, CRIS 2010 Proceedings, *Nucl. Phys. B, Proc. Suppl.* **212-213** (2011), 26, and references therein. The *arXiv.org* version arXiv:1011.4889 includes a relevant Post Scriptum.

[29] L. Gonzalez-Mestres, *Lorentz symmetry violation, dark matter and dark energy*, Invisible Universe International Conference, Paris June 29 - July 3, 2009. The *arXiv.org* version arXiv:0912.0725 includes a relevant Post Scriptum.

[30] G. Klimchitskaya and V. Mostepanenko, *Casimir and van der Waals forces: Advances and problems*, Proceedings of Peter the Great St.Petersburg Polytechnic University, **N1 (517)**, 41 (2015), arXiv:1605.1507.02393 , and references therein.

[31] H. Nikolic, *Proof that Casimir force does not originate from vacuum energy, Phys. Lett. B 761*, 197 (2016), arXiv:1605.04143 , and references therein.

[32] See also L. Gonzalez-Mestres, *Tests and prospects of new physics at very high energy*, contribution the 3rd International Conference on New Frontiers in Physics (ICNFP 2014), Kolymbari, Crete, Greece, 28 July - 6 August 2014, *EPJ Web of Conferences* **95**, 05007 (2015), and references therein.

[33] See, for instance, B. Hensen et al., *Loophole-free Bell test using electron spins in diamond: second experiment and additional analysis*, *Scientific Reports* **6**, 30289 (2016), arXiv:1603.05705 , and references therein.

[34] S. Bogdanovic et al., *Design and low-temperature characterization of a tunable microcavity for diamond-based quantum networks*, arXiv:1612.02164 , and references therein.

[35] L. Gonzalez-Mestres, *Physical and Cosmological Implications of a Possible Class of Particles Able to Travel Faster than Light*, contribution to the 28th International Conference on High Energy Physics, Warsaw 1996, arXiv:hep-ph/9610474, and references therein.

[36] L. Gonzalez-Mestres, *Space, Time and Superluminal Particles*, arXiv:physics/9702026

[37] L. Gonzalez-Mestres, *Can matter accelerate the expansion of the Universe? (I)*, mp_arc 16-33

[38] L. Gonzalez-Mestres, *Cosmological implications of a preonic vacuum (I)*, mp_arc 16-62,

[39] See, for instance, L. Gonzalez-Mestres, *Spinorial space-time and the origin of Quantum Mechanics*, contribution to the 4th International Conference on New Frontiers in Physics, Kolymbari, Crete, Greece, 23-30 August 2015, *EPJ Web of Conferences* **126**, 05006 (2016), and references therein.

[40] L. Gonzalez-Mestres, *Spinorial space-time and privileged space direction (I)*, mp_arc 13-75, and references therein.

[41] L. Gonzalez-Mestres, *Spinorial space-time and Friedmann-like equations (I)*, mp_arc 13-80, and references therein.

[42] L. Gonzalez-Mestres, *BICEP2, Planck, spinorial space-time, pre-Big Bang*, contribution the 3rd International Conference on New Frontiers in Physics, Kolymbari, Crete, Greece, August 23 - 30, 2014, *EPJ Web of Conferences* **95**, 03014 (2015), and references therein.

[43] L. Gonzalez-Mestres, *Vacuum Structure, Lorentz Symmetry and Superluminal Particles*, arXiv:physics/9704017

[44] L. Gonzalez-Mestres, *Properties of a possible class of particles able to travel faster than light*, Proceedings of the January 1995 Moriond Workshop, Ed. Frontières, arXiv:astro-ph/9505117

[45] See, for instance, L. Gonzalez-Mestres, *Spinorial space-time and the origin of Quantum Mechanics*, contribution to the 4th International Conference on New Frontiers in Physics, Kolymbari, Crete, Greece, 23-30 August 2015, *EPJ Web of Conferences* **126**, 05006 (2016), and references therein.

[46] See, for instance, the Planck Collaboration, *Planck 2015 results. XVI. Isotropy and statistics of the CMB*, *Astronomy & Astrophysics* **594**, A16 (2016), arXiv:1506.07135, and references therein.

[47] L. Gonzalez-Mestres, *Pre-Big Bang, space-time structure, asymptotic Universe*, 2nd International Conference on New Frontiers in Physics, Kolymbari, Crete, Greece, August 28 - September 5, 2013, *EPJ Web of Conferences* **71**, 00063 (2014), references therein and Post Scriptum to the preprint hal-00983005.

[48] L. Gonzalez-Mestres, *Planck data, spinorial space-time and asymptotic Universe*, mp_arc 13-33, and references therein.

[49] See, for instance, M. Putten, *A new estimator of the deceleration parameter from galaxy rotation curves*, arXiv:1604.01099, and references therein.

[50] See, for instance, M. Putten, *New estimates of the deceleration parameter in weak gravity*, arXiv:1611.05692, and references therein.

[51] Planck Collaboration, ESA site, http://www.cosmos.esa.int/web/planck

[52] See, for instance, S. Heinemeyer, *The 750 GeV diphoton excess and SUSY*, Talk given at the "Conference on New Physics at the Large Hadron Collider", 29.02. - 04.03.2016, Nanyang University, Singapore, arXiv:1605.08324, and references therein.

[53] See, for instance, B. Lenzi, *Search for a high mass diphoton resonance using the ATLAS detector*, Proceedings for the 38th International Conference on High Energy Physics (Chicago), arXiv:1611.03308, and references therein.

[54] See, for instance, CMS Collaboration, *Search for high-mass diphoton resonances in proton-proton collisions at 13 TeV and combination with 8 TeV search* arXiv:1609.02507, and references therein.

[55] See, for instance, the 1979 Nobel lecture by Abdus Salam, and references therein.

[56] L. Gonzalez-Mestres, *Cosmological implications of a possible class of particles able to travel faster than light*, Proceedings of the TAUP 1995 Conference, *Nucl. Phys. Proc. Suppl.* **48**, 131 (1996), arXiv:astro-ph/9601090

[57] L. Gonzalez-Mestres, *Preon models, relativity, quantum mechanics and cosmology (I)*, arXiv:0908.4070 , and references therein.

[58] L. Gonzalez-Mestres, *Cosmological Implications of a Possible Class of Particles Able to Travel Faster than Light (abridged version)*, arXiv:gr-qc/9508054

[59] M. Gasperini and G. Veneziano, *Pre-Big-Bang in String Cosmology*, *Astropart. Phys.* **1**, 317 (1993), arXiv:hep-th/9211021

[60] M. Gasperini and G. Veneziano, *The Pre-Big Bang Scenario in String Cosmology*, *Phys. Rept.* **373**, 1 (2003), arXiv:hep-th/0207130

[61] H. Nielsen and P. Olesen, *A parton view on dual amplitudes*, *Phys. Lett. B* **32**, 203 (1970). Preprint CERN TH. 1166 available on *Research Gate*.

[62] P. Aurenche and L. Gonzalez-Mestres, *Structure of the cylinder term in the topological expansion*, *Phys. Rev. D* **18**, 2995 (1978).

[63] P. Aurenche and L. Gonzalez-Mestres, *Glueball-Pomeron identity and the transverse structure of two-jet events*, *Z. Phys. C - Particles and Fields* **2**, 229 (1979). CERN preprint TH. 2643 available at the CERN Document Server.

[64] L. Gonzalez-Mestres, *CMB B-modes, spinorial space-time and Pre-Big Bang (I)*, mp_arc 14-16

[65] L. Gonzalez-Mestres, *CMB B-modes, spinorial space-time and Pre-Big Bang (II)*, mp_arc 14-60

[66] See, for instance, *The BICEP and Keck Array CMB Experiments*, http://bicepkeck.org/

[67] See, for instance, S. Simon et al., *Characterizing Atacama B-mode Search Detectors with a Half-Wave Plate*, arXiv:1511.04760

[68] B. Hensen et al., *Loophole-free Bell inequality violation using electron spins separated by 1.3 kilometres*, *Nature* **526**, 682 (2015), arXiv:1508.05949

[69] Pan Jianwei, *Quantum Science Satellite*, *Chinese Journal of Space Science* **34**, 547 (2014).

[70] Xinhua, August 16 (2016): *China launches first-ever quantum communication satellite*, http://news.xinhuanet.com/english/2016-08/16/c_135601010.htm

[71] E. Gibney, *Chinese satellite is one giant step for the quantum internet*, *Nature* **535**, 478 (2016).

[72] S. Hawking, *The Beginning of Time*, http://www.hawking.org.uk/the-beginning-of-time.html in Stephen Hawking Official Website.

[73] S. Hawking, *A Brief History of Time*, Bantam; 10th Anniversary edition (September 1, 1998).

[74] A. Borde and A. Vilenkin, *Eternal inflation and the initial singularity*, *Phys. Rev. Lett.* **72**, 3305 (1994), arXiv:gr-qc/9312022

[75] A. Borde, *Open and Closed Universes, Initial Singularities and Inflation*, *Phys.Rev. D* **50**, 3692 (1994), arXiv:gr-qc/9403049

[76] See, for instance, G. Bogdanoff, *Fluctuations quantiques de la signature de la métrique à l'échelle de Planck*, Thesis Université de Bourgogne (1999), https://tel.archives-ouvertes.fr/tel-00001502/ , and references therein.

[77] See, for instance, I. Bogdanoff, *Etat topologique de l'espace temps à l'échelle 0*, Thesis Université de Bourgogne (2002), https://tel.archives-ouvertes.fr/tel-00001503/ , and references therein.

[78] S. Hawking, *The Universe in a Nutshell*, Bantam 2001.

[79] S. Hawking and L. Mlodinow, *The Grand Design*, Bantam reprint 2012.

[80] See, for instance, G. Santos, G. Gubitosi and G. Amelino-Camelia, *On the initial singularity problem in rainbow cosmology*, JCAP **08**, 005 52015), arXiv:1502.02833 , and references therein.

[81] J. Magueijo and L. Smolin, *Gravity's Rainbow*, *Class. Quant. Grav.* **21**, 1725 (2004), arXiv:gr-qc/0305055

[82] L. Gonzalez-Mestres, *Absence of Greisen-Zatsepin-Kuzmin Cutoff and Stability of Unstable Particles at Very High Energy, as a Consequence of Lorentz Symmetry Violation*, Proceedings of the 25th International Cosmic Ray Conference, Potchefstroomse Universiteit 1997, Vol. **6**, p. 113. arXiv:physics/9705031.

[83] L. Gonzalez-Mestres, *Superluminal Particles and High-Energy Cosmic Rays* , Proceedings of the 25th International Cosmic Ray Conference, Potchefstroomse Universiteit 1997, Vol. **6**, p. 109. arXiv:physics/9705032.

[84] See, for instance, J. Donoghue, *The Multiverse and Particle Physics*, *Annual Reviews of Nuclear and Particle Science*, **66**, 1 (2016), arXiv:1601.05136 , and references therein.

[85] Y. Aharonov, E. Cohen and T. Shushi, *Is the Quilted Multiverse Consistent with a Thermodynamic Arrow of Time?*, arXiv: 1608.08798 , and references therein.

[86] See, for instance, A. Linde, *A brief history of the multiverse*, arXiv:1512.01203 , and references therein.

[87] See also J. Horgan, *Physicist Slams Cosmic Theory He Helped Conceive*, *Scientific American*, *Cross-Check* blog, December 1, 2014.

[88] See, for instance, F. Albareti et al., *Constraint on the time variation of the fine-structure constant with the SDSS-III/BOSS DR12 quasar sample*, *Mon. Not. Roy. Astron. Soc.* **452**, 4153 (2015), arXiv:1501.00560

[89] L. Gonzalez-Mestres, *The present status of Cosmology and new approaches to particles and Cosmos* and *Value of H, spinorial space-time and Universe's expansion*, ICNFP 2016 Conference.

[90] See, for instance, *Henri Poincaré Papers*, http://henripoincarepapers.univ-lorraine.fr/ (Université de Lorraine).

[91] An example of an article by Henri Poincaré published in 1912 is *Sur la théorie des quanta*, *Journal de physique théorique et appliqué* **2**, 5 (see [90]).

AUTHOR INDEX

AUTHOR INDEX

AUTHOR INDEX

AUTHOR INDEX

EDP Sciences – Web of Conferences
17, Avenue du Hoggar
Parc d'Activité de Courtabouf
BP 112
F-91944 Les Ulis Cedex A
France

ISBN 978-1-5108-5347-8

CURRAN ASSOCIATES INC.
proceedings.com